Viruses

Vernacular, nonscientific virus names are not written in italics, and thus the following names are not italicized.

adenovirus (ad″ē-no-vi′rus)
arbovirus (ar″bo-vi′rus)
baculovirus (bak″u-lo-vi′rus)
coronavirus (kor″o-nah-vi′rus)
cytomegalovirus (si″to-meg″ah-lo-vi′rus)
Epstein-Barr virus (ep′stīn-bar′)
hepatitis virus (hep″ah-ti′tis)
herpesvirus (her″pēz-vi′rus)
influenza virus (in″flu-en′zah)
measles virus (me′zelz)
mumps virus (mumps)
orthomyxovirus (or″tho-mik″so-vi′rus)
papillomavirus (pap″ĭ-lo″mah-vi′rus)
papovavirus (pap″o-vah-vi′rus)
paramyxovirus (par″ah-mik″so-vi′rus)
parvovirus (par″vo-vi′rus)
picornavirus (pi-kor″nah-vi′rus)
poliovirus (po″le-o-vi′rus)
polyomavirus (pol″e-o-mah-vi′rus)
poxvirus (poks-vi′rus)
rabies virus (ra′bēz)
reovirus (re″o-vi′rus)
retrovirus (re″tro-vi′rus)
rhabdovirus (rab″do-vi′rus)
rhinovirus (ri″no-vi′rus)
rotavirus (ro′tah-vi″rus)
rubella virus (roo-bel′ah)
togavirus (to″gah-vi′rus)
varicella-zoster virus (var″ĭ-sel′ah zos′ter)
variola virus (vah-ri′o-lah)

Fungi

Agaricus (ah-gar′ĭ-kus)
Amanita (am″ah-ni′tah)
Arthrobotrys (ar″thro-bo′tris)
Aspergillus (as″per-jil′us)
Blastomyces (blas″to-mi′sēz)
Candida (kan′dĭ-dah)
Cephalosporium (sef″ah-lo-spo′re-um)
Claviceps (klav′ĭ-seps)
Coccidioides (kok-sid″e-oi′dēz)
Cryptococcus (krip″to-kok′us)
Epidermophyton (ep″ĭ-der-mof′ĭ-ton)
Fusarium (fu-sa′re-um)
Histoplasma (his″to-plaz′mah)
Microsporum (mi-kros′po-rum)
Mucor (mu′kor)
Neurospora (nu-ros′po-rah)
Penicillium (pen″ĭ-sil′e-um)
Phytophthora (fi-tof′tho-rah)
Rhizopus (ri-zo′pus)
Saccharomyces (sak″ah-ro-mi′sēz)
Saprolegnia (sap″ro-leg′ne-ah)
Sporothrix (spo′ro-thriks)
Trichoderma (trik-o-der′mah)
Trichophyton (tri-kof′ĭ-ton)

Protozoa

Acanthamoeba (ah-kan″thah-me′bah)
Amoeba (ah-me′bah)
Balantidium (bal″an-tid′e-um)
Cryptosporidium (krip″to-spo-rid′e-um)
Entamoeba (en″tah-me′bah)
Giardia (je-ar′de-ah)
Leishmania (lēsh-ma′ne-ah)
Naegleria (na-gle′re-ah)
Paramecium (par″ah-me′she-um)
Plasmodium (plaz-mo′de-um)
Pneumocystis (noo″mo-sis′tis)
Tetrahymena (tet″rah-hi′mē-nah)
Toxoplasma (toks″o-plaz′mah)
Trichomonas (trik″o-mo′nas)
Trypanosoma (tri″pan-o-so′mah)

Algae

Acetabularia (as″ē-tab″u-la′re-ah)
Chlamydomonas (klah-mid″do-mo′nas)
Chlorella (klo-rel′ah)
Euglena (u-gle′nah)
Gonyaulax (gon″e-aw′laks)
Laminaria (lam″ĭ-na′re-ah)
Prototheca (pro″to-the′kah)
Spirogyra (spi″ro-ji′rah)
Volvox (vol′voks)

MICROBIOLOGY

A personal library is a lifelong source of enrichment and distinction. Consider this book an investment in your future and add it to your personal library.

MICROBIOLOGY

SECOND EDITION

Lansing M. Prescott
Augustana College

John P. Harley
Eastern Kentucky University

Donald A. Klein
Colorado State University

WCB Wm. C. Brown Publishers
Dubuque, Iowa•Melbourne, Australia•Oxford, England

Book Team

Editor *Kevin Kane*
Developmental Editor *Megan Johnson*
Production Editor *Diane E. Beausoleil*
Designer *Elise A. Burckhardt*
Art Editor *Jodi Wagner*
Photo Editor *Carrie Burger*
Permissions Editor *Karen L. Storlie*
Art Processor *Andréa Lopez Meyer*

Wm. C. Brown Publishers
A Division of Wm. C. Brown Communications, Inc.

Vice President and General Manager *Beverly Kolz*
National Sales Manager *Vincent R. Di Blasi*
Assistant Vice President, Editor-in-Chief *Edward G. Jaffe*
Director of Marketing *John W. Calhoun*
Marketing Manager *Carol J. Mills*
Advertising Manager *Amy Schmitz*
Director of Production *Colleen A. Yonda*
Manager of Visuals and Design *Faye M. Schilling*
Design Manager *Jac Tilton*
Art Manager *Janice Roerig*
Publishing Services Manager *Karen J. Slaght*
Permissions/Records Manager *Connie Allendorf*

Wm. C. Brown Communications, Inc.

Chairman Emeritus *Wm. C. Brown*
Chairman and Chief Executive Officer *Mark C. Falb*
President and Chief Operating Officer *G. Franklin Lewis*
Corporate Vice President, President of WCB Manufacturing *Roger Meyer*

About the main text cover photo: This computer enhanced electron micrograph shows a macrophage engulfing several *Candida albicans* cells. *Candida* spp. is one of the ten most prevalent pathogens in nursing homes and hospitals. Macrophages are a major defense against fungal infections in the human body.

Microbiology, second edition
Main: Courtesy of Roerig, A division of Pfizer Pharmaceuticals; background: © C. Raymond/Photo Researchers, Inc.

Volume One and Volume Two
Main: © Alfred Paseika/Peter Arnold, Inc.; background: © C. Raymond/Photo Researchers, Inc.

Volume Three
Main: © Biophoto Associates/Photo Researchers, Inc.; background: © C. Raymond/Photo Researchers, Inc.

Copyedited by *Beatrice Sussman*

The credits section for this book begins on page C1 and is considered an extension of the copyright page.

Library of Congress Catalog Card Number: 91–77745

ISBN 0–697–01372–3
ISBN Volume One: 0–697–16885–9
ISBN Volume Two: 0–697–16886–7
ISBN Volume Three: 0–697–16887–5
ISBN **Microbiology, second edition** Boxed set: 0–697–16888–3

In regard to the discussion of drug therapy and reactions and the use of equipment, every effort has been made to ensure the accuracy of the information in this text. However, with changing governmental regulations and with constantly changing technology, the reader is urged to review the packaged instructions and information that are provided by the manufacturer with each specific drug or piece of equipment to look for any changes in the instructions or for any added warnings.

Printed in the United States of America by Wm. C. Brown Communications, Inc., 2460 Kerper Boulevard, Dubuque, IA 52001

10 9 8 7

Other Titles of Related Interest from Wm. C. Brown Publishers

Microbiology

Laboratory Exercises in Microbiology, second edition (1993) by John P. Harley and Lansing M. Prescott

Microbiology Laboratory Exercises, second edition, short and complete versions (1990) by Margaret Barnett

Microbial Applications, fifth edition, complete and short versions (1990) by Harold Benson

Laboratory Manual and Workbook in Microbiology: Application in Patient Care (1991) by Josephine Morello, Helen Eckel Mizer and Marion Wilson

AIDS/STDs

The Biology of Sexually Transmitted Diseases (1992) by Gerald Stine

The Sexually Transmitted Diseases (1992) by George Wistreich

AIDS: The Biological Basis (1993) by I. Edward Alcamo

Virology

Introductory Experiments in Virology (1992) by Gerald Goldstein

Immunology

Fundamental Immunology, second edition (1992) by Robert Coleman, Mary Lombard, and Raymond Sicard

Immunology: A Laboratory Manual (1989) by Richard Myers

Cell and Molecular Biology

Experimental Cell and Molecular Biology (1992) by John S. Choinski

Introduction to Experimental Cell Biology (1992) by Holly Ahern

Introduction to Experimental Molecular Biology (1992) by Holly Ahern

A Manual of Laboratory Experiments in Cell Biology (1989) by C. Edward Gasque

Genetics

Genetics, second edition (1992) by Robert F. Weaver and Philip W. Hedrick

Basic Genetics (1991) by Robert F. Weaver and Philip W. Hedrick

Principles of Genetics, third edition (1991) by Robert Tamarin

Genetics: A Human Perspective, third edition (1992) by Linda R. Maxson and Charles H. Daugherty

The New Human Genetics (1989) by Gerald Stein

Genetics Problem Solving Guide (1992) by William R. Wellnitz

Student Study Guide in Genetics (1991) by Ken Zwicker

Biochemistry

Biochemistry, third edition (1993) by Geoffrey Zubay

Publisher's Note

Microbiology, second edition, is available as a full-length casebound text and as three paperbound separates.

Binding Options	***ISBN***
•*Microbiology* (Casebound): The full-length text, chapters 1-44	0-697-01372-3
•*Microbiology, Volume One* (Paperbound): Chapters 1-28	0-697-16885-9
•*Microbiology, Volume Two* (Paperbound): Chapters 29-39	0-697-16886-7
•*Microbiology, Volume Three* (Paperbound): Chapters 40-44	0-697-16887-5
•*Microbiology, Boxed Set* (Paperbound): The full-length text in an attractive boxed set of all three paperback "splits"	0-697-16888-3

Brief Contents

Volume One

Part One
Introduction to Microbiology

Part Two
Microbial Growth and Metabolism

Part Three
Microbial Genetics

Part Four
The Control of Microorganisms

Part Five
The Viruses

Part Six
The Diversity of the Microbial World

VOLUME TWO

Part Seven

The Nature of Symbiotic Associations

Part Eight

The Fundamentals of Immunology

Part Nine

Microbial Diseases

VOLUME THREE

Part Ten

Microorganisms and the Environment

Part Eleven

Food and Industrial Microbiology

Expanded Contents

Part One
Introduction to Microbiology

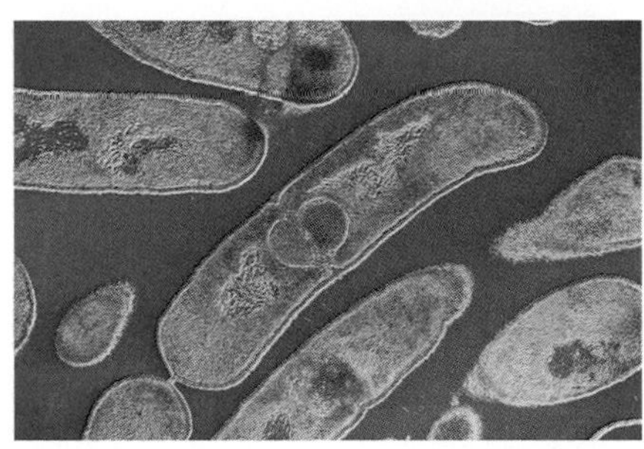

PART TWO
Microbial Growth and Metabolism

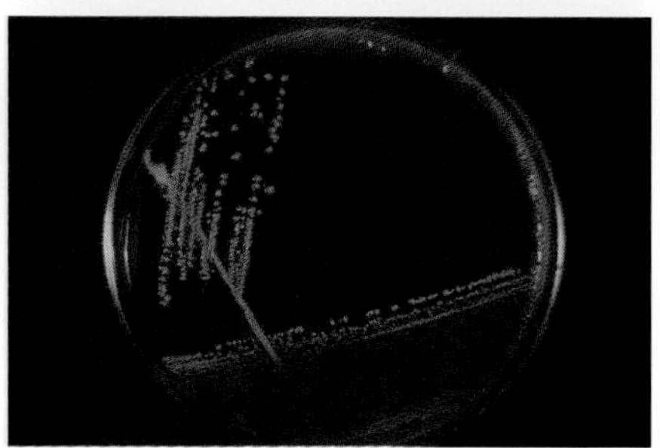

7 Metabolism: Energy and Enzymes 133

8 Metabolism: The Generation of Energy 145

9 Metabolism: The Use of Energy in Biosynthesis 171

10 Metabolism: The Synthesis of Nucleic Acids and Proteins 191

11 Metabolism: The Regulation of Enzyme Activity and Synthesis 216

PART THREE
Microbial Genetics

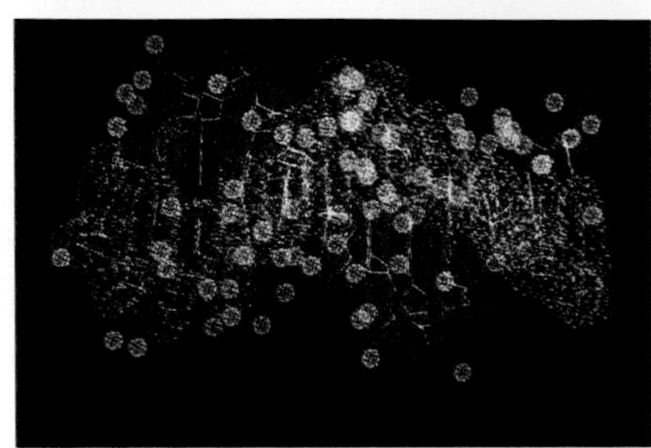

PART FOUR
The Control of Microorganisms

PART FIVE
The Viruses

PART SIX
The Diversity of the Microbial World

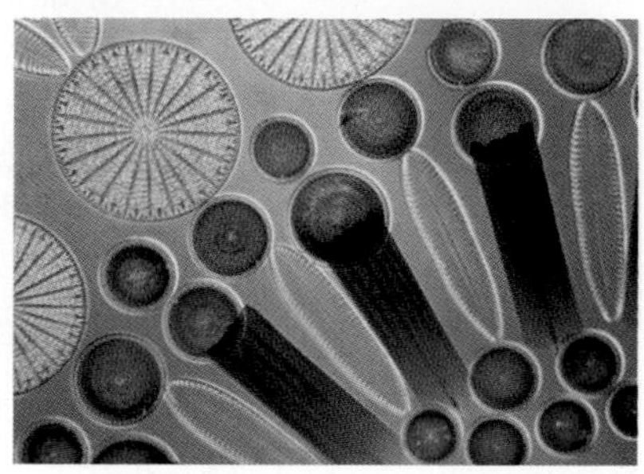

20 Microbial Taxonomy 404

21 The Bacteria: Gram-Negative Bacteria of General, Medical, or Industrial Importance 426

22 The Bacteria: Gram-Positive Bacteria Other Than Actinomycetes 450

23 The Bacteria: Remaining Gram-Negative Bacteria and Cyanobacteria 467

Part Seven
The Nature of Symbiotic Associations

29 Symbiotic Associations: Commensalism, Mutualism, and Normal Microbiota of the Human Body 564

30 Symbiotic Associations: Parasitism, Pathogenicity, and Resistance 578

Part Eight
The Fundamentals of Immunology

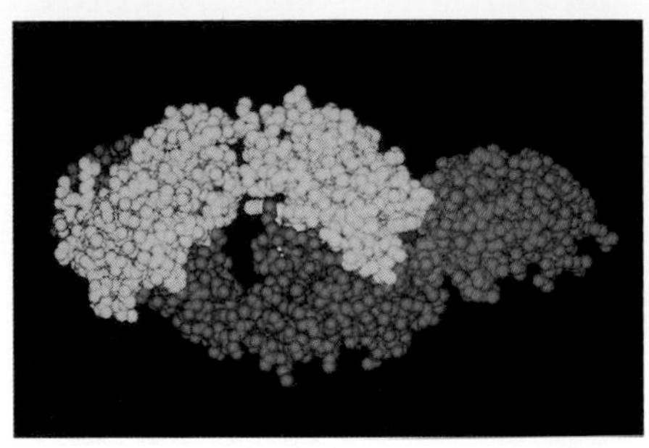

PART NINE
Microbial Diseases

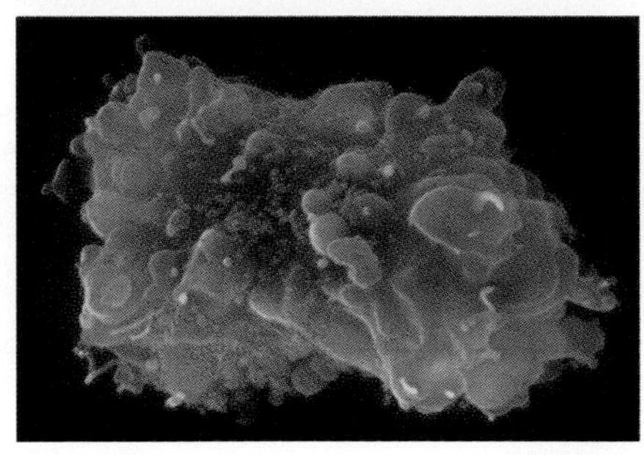

PART TEN
Microorganisms and the Environment

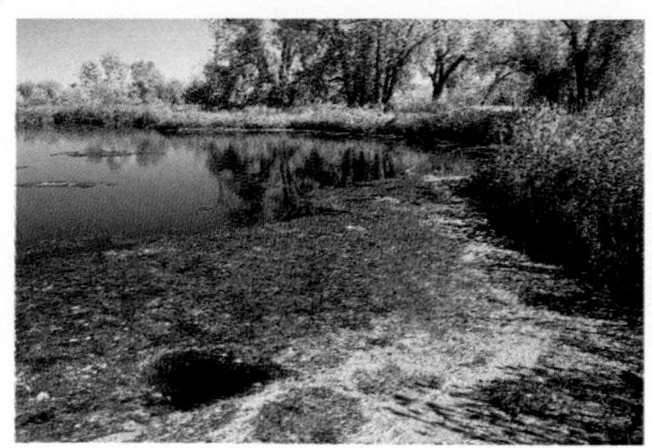

Part Eleven
Food and Industrial Microbiology

Boxed Readings

THE *MICROBIOLOGY* LEARNING SYSTEM

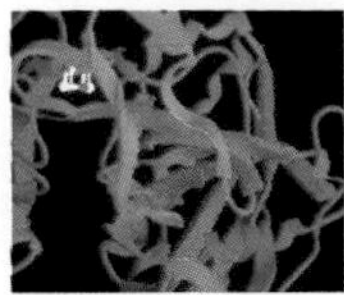

CHAPTER 9

Metabolism: The Use of Energy in Biosynthesis

Biological structures are almost always constructed in a hierarchical manner, with subassemblies acting as important intermediates en route from simple starting molecules to the end products of organelles, cells, and organisms.

—W. M. Becker and D. W. Deamer

Outline

Concepts

1. In anabolism or biosynthesis, cells use free energy to construct more complex molecules and structures from smaller, simpler precursors.
2. Biosynthetic pathways are organized to optimize efficiency by conserving genetic storage space, biosynthetic raw materials, and energy.
3. Autotrophs use ATP and NADPH from photosynthesis or from oxidation of inorganic molecules to reduce CO_2 and incorporate it into organic material.
4. Catabolic and anabolic pathways may differ in enzymes, regulation, intracellular location, and use of cofactors and nucleoside diphosphate carriers. Although many enzymes of amphibolic pathways participate in both catabolism and anabolism, some pathway enzymes are involved only in one of the two processes.
5. Phosphorus, in the form of phosphate, can be directly assimilated, whereas inorganic sulfur and nitrogen compounds must often be reduced before incorporation into organic material.
6. The tricarboxylic acid (TCA) cycle acts as an amphibolic pathway and requires anaplerotic reactions to maintain adequate levels of cycle intermediates.
7. Most glycolytic enzymes participate in both the synthesis and catabolism of glucose. In contrast, fatty acids are synthesized from acetyl-CoA and malonyl-CoA by a pathway quite different from fatty acid β-oxidation.
8. Peptidoglycan synthesis is a complex, multistep process that is begun in the cytoplasm and completed at the cell wall after the peptidoglycan repeat unit has been transported across the cell membrane.

171

Chapter Outlines and Concepts

The Chapter Outlines and Concepts at the beginning of each chapter are designed to help students identify the major topics of each chapter.

Full-Color Visuals

The dramatic, full-color visuals enhance the learning program and spark interest and discussion of important topics.

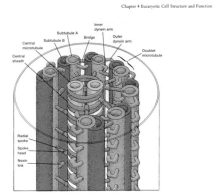

Chapter 4 Eucaryotic Cell Structure and Function 89

Inner dynein arm
Subtubule A
Subtubule B
Bridge
Outer dynein arm
Central microtubule
Doublet microtubule
Central sheath
Radial spoke
Spoke head
Nexin link

Figure 4.28 **Diagram of Cilia and Flagella Structure.** Doublets 2 and 3 removed for sake of visibility.

1. How do eucaryotic microorganisms differ from procaryotes with respect to supporting or protective structures external to the plasma membrane? Describe the pellicle and indicate which microorganisms have one.
2. Prepare and label a diagram showing the detailed structure of a cilium or flagellum. How do cilia and flagella move, and what is dynein's role in the process?

Comparison of Procaryotic and Eucaryotic Cells

A comparison of the cells in figure 4.29 demonstrates that there are many fundamental differences between eucaryotic and procaryotic cells. **Eucaryotic cells** have a membrane-enclosed nucleus. In contrast, **procaryotic cells** lack a true, membrane-

Boxed Readings

Informative boxed readings cover a wide variety of topics.

302 Part Three: Microbial Genetics

Box 14.1

Gene Expression and Kittyboo Colors

Genetic engineering techniques can yield unusual approaches to long-standing problems. To understand how the regulation of gene expression influences complex processes, the activity of more than one gene must be followed simultaneously. The use of genetic engineering techniques with luciferase genes has made this much easier.

A Jamaican beetle named the kittyboo has two light organs on its head and one on its abdomen. Recently, four luciferase genes have been isolated from the beetle. Each luciferase enzyme produces a different colored light when it acts on the substrate luciferin. Keith V. Wood has cloned the genes and inserted them into *E. coli*. When the bacteria are exposed to luciferin, they glow (Box Figure 14.1).

These luciferase genes can be used to study gene regulation. Suppose that one wishes to follow the activity of the liver gene that codes for serum albumin. The albumin gene can be replaced with a luciferase gene, and the modified cell incubated with luciferin. Whenever the albumin gene is activated, the newly inserted luciferase gene will function and the cell glow. Measurement of light intensity is easy, rapid, and sensitive. By substituting a different luciferase gene for another liver gene, one can simultaneously follow the activity and coordination of two different liver genes. It is only necessary to measure light intensity at the two wavelengths characteristic of the luciferase genes.

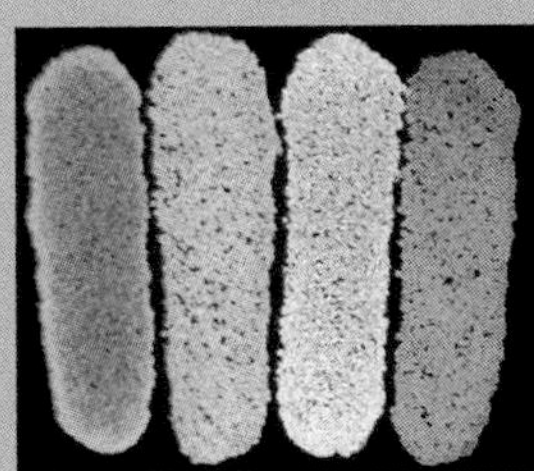

Box Figure 14.1 ***E. coli* with Luciferase Genes.** These four streaks of *E. coli* glow different colors because they contain four different luciferase genes cloned from the Jamaican click beetle or kittyboo, *Pyrophorus plagiophthalamus*.

have also been synthesized by rDNA techniques. The human growth hormone gene was too long to synthesize by chemical procedures and was prepared from mRNA as cDNA.

As mentioned earlier, introns in eucaryotic genes are not removed by bacteria and will render the final protein nonfunctional. The easiest solution is to prepare cDNA from processed mRNA that lacks introns and directly reflects the correct amino acid sequence of the protein product. In this instance, it is particularly important to fuse the gene with an expression vector since a promoter and other essential sequences will be missing in the cDNA.

Eucaryotic mRNA processing (pp. 203–4).

1. What is a transgenic animal? Describe how electroporation and gene guns are used to insert foreign genes into cells.
2. How can one prevent rDNA from undergoing recombination in a host cell?
3. List several reasons why a cloned gene might not be expressed in a host cell. What is an expression vector?
4. Briefly outline the procedure for somatostatin production.
5. Identify a way to eliminate eucaryotic introns during the synthesis of rDNA.

Applications of Genetic Engineering

Genetic engineering and biotechnology will contribute in the future to medicine, industry, and agriculture, as well as to basic research (Box 14.1). In this section, some practical applications are briefly discussed.

Medical Applications

Certainly, the production of medically useful proteins such as somatostatin, insulin, human growth hormone, and interferon is of great practical importance (table 14.4). This is particularly true of substances that previously only could be obtained from human tissues. For example, in the past, human growth hormone for treatment of pituitary dwarfism was extracted from pituitaries obtained during autopsies and was available only in limited amounts. Interleukin-2 (a protein that helps regulate the immune response) and blood-clotting factor VIII have recently been cloned, and undoubtedly other important peptides and proteins will be produced in the future. Synthetic vaccines—for instance, vaccines for malaria and rabies—are also being developed with recombinant techniques. A recombinant hepatitis B vaccine is already commercially available.

Other medical uses of genetic engineering are being investigated. Probes are now being used in the diagnosis of in-

Review Questions

Review Questions appear after each major section within a chapter.

l Genetics: Recombination and Plasmids 259

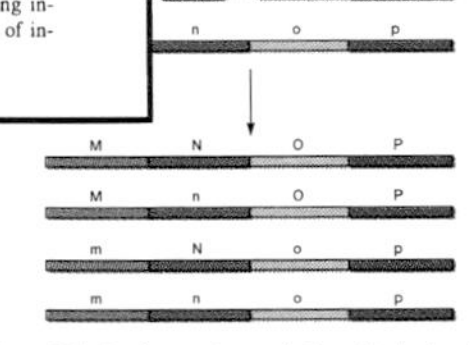

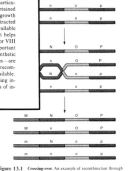

Figure 13.1 **Crossing-over.** An example of recombination through crossing-over between homologous eucaryotic chromosomes. The Nn gene pair is exchanged. This process usually occurs during meiosis.

Recombination is the process in which a new recombinant chromosome, one with a genotype different from either parent, is formed by combining genetic material from two organisms. It results in a new arrangement of genes or parts of genes and normally is accompanied by a phenotypic change. Most eucaryotes exhibit a complete sexual life cycle, including meiosis, a process of extreme importance in generating new combinations of alleles (alternate forms of a particular gene) through recombination. These chromosome exchanges during meiosis result from **crossing-over** between homologous chromosomes, chromosomes containing identical sequences of genes (figure 13.1). Until about 1945, the primary focus in genetic analysis was on the recombination of genes in plants and animals. The early work on recombination in higher eucaryotes laid the foundations of classical genetics, but it was the development of bacterial and phage genetics between about 1945 and 1965 that really triggered a rapid advance in our understanding of molecular genetics.

Meiosis (pp. 85–86).

Bacterial Recombination: General Principles

Microorganisms carry out several types of recombination. **General recombination,** the most common form, usually involves a reciprocal exchange between a pair of homologous DNA sequences. It can occur anyplace on the chromosome, and it results from DNA strand breakage and reunion leading to crossover (figure 13.2). General recombination is carried out by the products of *rec* genes such as the recA protein so important for DNA repair (*see pp. 255–56*). In bacterial transformation, a nonreciprocal form of general recombination takes place. A piece of genetic material is inserted into the chromosome through the incorporation of a single strand to form a stretch of **heteroduplex DNA** (figure 13.3). A second type of recombination, one particularly important in the in-

tegration of virus genomes into bacterial chromosomes, is **site-specific recombination.** The genetic material is not homologous with the chromosome it joins, and generally the enzymes responsible for this event are specific for the particular virus and its host. A third type, **replicative recombination,** accompanies the replication of genetic material and does not depend upon sequence homology. It is used by genetic elements that move about the chromosome.

DNA replication (pp. 195–201).

Although sexual reproduction with the formation of a zygote and subsequent meiosis is not present in bacteria, recombination can take place in several ways. In general, a piece of donor DNA, the **exogenote,** must enter the recipient cell and become a stable part of the recipient cell's genome, the **endogenote.** Two kinds of DNA can move between bacteria. If a DNA fragment is the exchange vehicle, then the exogenote must get into the recipient cell and become incorporated into the endogenote as a replacement piece (or as an "extra" piece) without being destroyed by the host. During replacement of host genetic material, the recipient cell becomes temporarily diploid for a portion of the genome and is

Cross References

Cross references refer students to major topics they may need to review to fully understand the current material.

Key Terms

Key Terms are set off in bold type when they are first used. They are also page-referenced and listed at the end of each chapter and in the glossary.

330 Part Four: The Control of Microorganisms

Table 16.3 Inhibition Zone Diameter of Selected Chemotherapeutic Drugs

Chemotherapeutic Drug	Disk Content	Zone Diameter (Nearest mm)		
		Resistant	*Intermediate*	*Susceptible*
Carbenicillin (with *Proteus* spp. and *E. coli*)	100 μg	≤17	18–22	≥23
Carbenicillin (with *Pseudomonas aeruginosa*)	100 μg	≤13	14–16	≥17
Ceftriaxone	30 μg	≤13	14–20	≥21
Chloramphenicol	30 μg	≤12	13–17	≥18
Erythromycin	15 μg	≤13	14–17	≥18
Penicillin G (with staphylococci)	10 U*	≤20	21–28	≥29
Penicillin G (with other microorganisms)	10 U	≤11	12–21	≥22
Streptomycin	10 μg	≤11	12–14	≥15
Sulfonamides	250 or 300 μg	≤12	13–16	≥17
Tetracycline	30 μg	≤14	15–18	≥19

*One milligram of penicillin G sodium = 1,600 units.

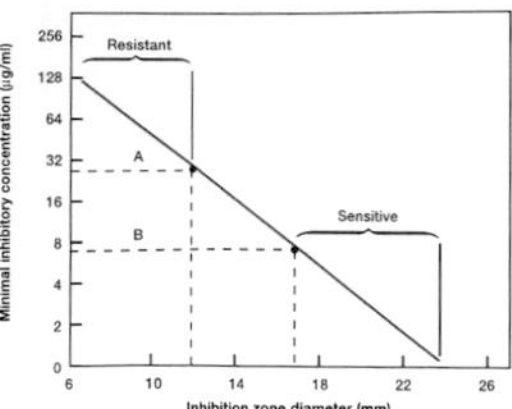

Figure 16.2 **Interpretation of Kirby-Bauer Test Results.** The relationship between the minimal inhibitory concentrations with a hypothetical drug and the size of the zone around a disk in which microbial growth is inhibited. As the sensitivity of microorganisms to the drug increases, the MIC value decreases and the inhibition zone grows larger. Suppose that this drug varies from 7–28 μg/ml in the body during treatment. Dashed line *A* shows that any pathogen with a zone of inhibition less than 12 mm in diameter will have an MIC value greater than 28 μg/ml and will be resistant to drug treatment. A pathogen with a zone diameter greater than 17 mm will have an MIC less than 7 μg/ml and will be sensitive to the drug (see line *B*). Zone diameters between 12 and 17 mm indicate intermediate sensitivity and usually signify resistance.

Mechanisms of Action of Antimicrobial Agents

The mechanisms of action of specific chemotherapeutic agents are taken up in more detail when individual drugs and groups of drugs are discussed later in this chapter. A few general observations are offered at this point. It is important to know something about the mechanisms of drug action because such knowledge helps explain the nature and degree of selective toxicity of individual drugs and sometimes aids in the design of new chemotherapeutic agents.

Antimicrobial drugs can damage pathogens in several ways, as can be seen in table 16.4, which summarizes the mechanisms of the antibacterial drugs listed in table 16.1. The most selective antibiotics are those that interfere with the synthesis of bacterial cell walls (e.g., penicillins, cephalosporins, and bacitracin). These drugs have a high therapeutic index because bacterial cell walls have a unique structure (*see chapter 3*) not found in eucaryotic cells.

Streptomycin, gentamicin, spectinomycin, clindamycin, chloramphenicol, tetracyclines, erythromycin, and many other antibiotics inhibit protein synthesis by binding with the procaryotic ribosome. Because these drugs discriminate between procaryotic and eucaryotic ribosomes, their therapeutic index is fairly high, but not as favorable as that of cell wall synthesis inhibitors. Some drugs bind to the 30S (small) subunit, while others attach to the 50S (large) ribosomal subunit. Several different steps in the protein synthesis mechanism can be affected: aminoacyl-tRNA binding, peptide bond formation, mRNA reading, and translocation. For example, fusidic acid binds to EF-G and blocks translocation, while mucopirocin inhibits isoleucyl-tRNA synthetase (*see chapter 10*).

The antibacterial drugs that inhibit nucleic acid synthesis or damage cell membranes often are not as selectively toxic as other antibiotics. This is because procaryotes and eucaryotes do not differ as greatly with respect to nucleic acid synthetic mechanisms or cell membrane structure. Good examples of drugs that affect nucleic acid synthesis or membrane structure are quinolones and polymyxins. Quinolones inhibit the DNA gyrase and thus interfere with DNA replication, repair, and transcription. Polymyxins act as detergents or surfactants and disrupt the bacterial plasma membrane.

Several valuable drugs act as **antimetabolites:** they block the functioning of metabolic pathways by competitively inhibiting the use of metabolites by key enzymes. Sulfonamides and several other drugs inhibit folic acid metabolism. Sulfonamides (e.g., sulfanilamide, sulfamethoxazole, and sulfacetamide) have a high therapeutic index because humans cannot synthesize folic acid and must obtain it in their diet. Most bacterial pathogens synthesize their own folic acid and are there-

Tables

Strategically placed tables concisely summarize important information.

Chapter 14 Recombinant DNA Technology 307

Terms

…6
…on (PCR
…nology 286
restriction enzymes 287

site-directed mutagenesis 290
Southern blotting technique 288
Ti plasmid 303
transfection 299
transgenic animal 300
vectors 288

Questions for Thought and Review

1. Could the Southern blotting technique be applied to RNA and proteins? How might this be done?
2. Why could a band on an electrophoresis gel still contain more than one kind of DNA fragment?
3. What advantage might there be in creating a genomic library first rather than directly isolating the desired DNA fragment?
4. Suppose that one inserted a simple plasmid (one containing an antibiotic resistance gene and a separate restriction site) carrying a human interferon gene into *E. coli*, but none of the transformed bacteria produced interferon. Give as many plausible reasons as possible for this result.
5. In what areas do you think genetic engineering will have the greatest positive impact in the future? Why?
6. What do you consider to be the greatest potential dangers of genetic engineering? Are there ethical problems with any of its potential applications?

Additional Reading

Anderson, W. F., and Diacumakos, E. G. 1981. Genetic engineering in mammalian cells. *Sci. Am.* 245(1):106–21.
Antebi, E., and Fishlock, D. 1985. *Biotechnology: Strategies for life.* Cambridge, Mass.: MIT Press.
Arnheim, N. and Levenson, C. H. 1990. Polymerase chain reaction. *Chem Eng. News* 68:36–47.
Atlas, R. M. 1991. Environmental applications of the polymerase chain reaction. *ASM News* 57(12):630–32.
Brown, C. M.; Campbell, I.; and Priest, F. G. 1987. *Introduction to biotechnology.* Boston: Blackwell Scientific Publications.
Chilton, M.-D. 1983. A vector for introducing new genes into plants. *Sci. Am.* 248(6):51–59.
Cohen, S. N. 1975. The manipulation of genes. *Sci. Am.* 233(1):25–33.
Emery, A. E. H. 1984. *An introduction to recombinant DNA.* New York: John Wiley and Sons.
Erlich, H. A. 1989. *PCR technology: Principles and applications of DNA amplification.* San Francisco: W. H. Freeman.
Erlich, H. A.; Gelfand, D.; and Sninsky, J. J. 1991. Recent advances in the polymerase chain reaction. *Science* 252:1643–51.
Gardner, E. J.; Simmons, M. J.; and Snustad, D. P. 1991. *Genetics*, 8th ed. New York: John Wiley and Sons.
Gasser, C. S., and Fraley, R. T. 1992. Transgenic crops. *Sci. Am.* 266(6):62–69.
Gilbert, W., and Villa-Komaroff, L. 1980. Useful proteins from recombinant bacteria. *Sci. Am.* 242(4):74–94.
Goodfield, J. 1977. *Playing God: Genetic engineering and the manipulation of life.* New York: Random House.
Grobstein, C. 1977. The recombinant-DNA debate. *Sci. Am.* 237(1):22–33.
Hansen, M.; Busch, L.; Burkhardt, J.; Lacy, W. B.; and Lacy, L. R. 1986. Plant breeding and biotechnology. *BioScience* 36(1):29–39.
Hopwood, D. A. 1981. The genetic programming of industrial microorganisms. *Sci. Am.* 245(3):91–102.
Jaenisch, R. 1988. Transgenic animals. *Science* 240:1468–74.
Kenney, M. 1986. *Biotechnology: The university-industrial complex.* New Haven, Conn.: Yale University Press.
Krimsky, S. 1982. *Genetic alchemy: The social history of the recombinant DNA controversy.* Cambridge, Mass.: MIT Press.
Lear, J. 1978. *Recombinant DNA: The untold story.* New York: Crown Publishers.
Lewin, B. 1990. *Genes*, 4th ed. New York: Oxford University Press.
Moses, P. B. 1987. Strange bedfellows. *BioScience* 37:6–10.
Mullis, K. B. 1990. The unusual origin of the polymerase chain reaction. *Sci. Am.* 262(4):56–65.
Old, R. W., and Primrose, S. B. 1989. *Principles of gene manipulation*, 4th ed. Boston: Blackwell Scientific Publications.
Pestka, S. 1983. The purification and manufacture of human interferons. *Sci. Am.* 249(2):37–43.
Peters, P. 1993. *Biotechnology: A guide to genetic engineering.* Dubuque, Iowa: Wm. C. Brown.
Primrose, S. B. 1987. *Modern Biotechnology.* Boston: Blackwell Scientific Publications.
Richards, J., ed. 1978. *Recombinant DNA: Science, ethics, and politics.* New York: Academic Press.
Rifkin, J. 1983. *Algeny.* New York: Viking Press.
Smith, J. E. 1985. *Biotechnology principles.* Washington, D.C.: American Society for Microbiology.
Teitelman, R. 1989. *Gene dreams: Wall street, academia, and the rise of biotechnology.* New York: Basic Books, Inc.
Tucker, J. B. Winter 1984–85. Gene wars. *Foreign Policy* 57:58–79.
Watson, J. D., Gilman, M.; Witkowski, J.; and Zoller, M. 1992. *Recombinant DNA*, 2d ed. San Francisco: W.H. Freeman.
Yoxen, E. 1984. *The gene business: Who should control biotechnology?* New York: Harper & Row.
Zilinskas, R. A., and Zimmerman, B. K., eds. 1986. *The gene-splicing wars: Reflections on the recombinant DNA controversy.* New York: Macmillan.

Questions for Thought and Review

Questions for Thought and Review appear at the end of each chapter and require students to review, analyze, and apply the information they have just learned.

Additional Readings

Additional Readings suggest references for further study.

Preface

Books are the carriers of civilization. Without books, history is silent, literature dumb, science crippled, thought and speculation at a standstill. They are engines of change, windows on the world, lighthouses erected in a sea of time.

Barbara Tuchman

Microbiology is an exceptionally broad discipline encompassing specialties as diverse as biochemistry, cell biology, genetics, taxonomy, pathogenic bacteriology, food and industrial microbiology, and ecology. A microbiologist must be acquainted with many biological disciplines and with all major groups of microorganisms: viruses, bacteria, fungi, algae, and protozoa. The key is balance. Students new to the subject need an introduction to the whole before concentrating on those parts of greatest concern to them. This text provides a balanced introduction to all major areas of microbiology for a variety of students. Because of this balance, the book is suitable for courses with orientations ranging from basic microbiology to medical and applied microbiology. Students preparing for careers in medicine, dentistry, nursing, and allied health professions will find the text just as useful as those aiming for careers in research, teaching, and industry. A quarter/semester or two each of biology and chemistry is assumed, and an overview of relevant chemistry is also provided in appendix I.

Organization and Approach

The book is organized flexibly so that chapters and topics may be arranged in almost any order. Each chapter has been made as self-contained as possible to promote this flexibility. Some core topics are essential to microbiology and have been given more extensive treatment. The chapters that cover these topics (chapters 2–19 and 30–33) are somewhat longer than others and may be combined with a selection of noncore chapters to achieve the desired orientation, whether basic, applied, or medical. The text is available not only in a casebound version but also as three separate paperbound volumes. Volume one covers the more general aspects of microbiology (chapters 1–28); volume two, immunology and medical microbiology (chapters 29–39); and volume three, microbial ecology and applied microbiology (chapters 40–44). Thus, the complete text is available as a boxed set, or its separate volumes can be used in courses with a specific focus.

The book is divided into 11 parts. The first 5 introduce the foundations of microbiology: the development of microbiology, the structure of microorganisms, microbial growth and metabolism, microbial genetics, the control of microorganisms, and the nature of viruses. In this second edition, the microbial genetics unit directly follows microbial growth and metabolism (Part Two) and discusses genetics before the introduction to viruses (Part Five). Part Six is a survey of the microbial world. The bacterial survey closely follows the organization of *Bergey's Manual of Systematic Bacteriology.* Because of the importance and uniqueness of archaeobacteria, the second edition contains a separate chapter on them. Although principal attention is devoted to bacteria, eucaryotic microorganisms receive more than usual coverage. Fungi, algae, and protozoa are important in their own right. The introduction to their biology in chapters 26–28 is essential to the understanding of topics as diverse as clinical microbiology and microbial ecology. Part Seven discusses symbiotic associations and parasitism in depth, providing a good foundation for the later survey of specific diseases. Three chapters in Part Eight describe in detail all major aspects of the immune response. Part Nine begins with an introduction to epidemiology and clinical microbiology, followed by a survey of the major human microbial diseases. The disease survey is primarily organized taxonomically on the chapter level; within each chapter, diseases are covered according to mode of transmission. This approach provides flexibility and allows the student easy access to information about any disease of interest. The survey is not a simple cataloguing of diseases; diseases are included because of their medical importance and their ability to illuminate the basic principles of disease and resistance. Part Ten focuses on the relationship of microorganisms to aquatic and terrestrial environments. Chapter 40 presents the general principles underlying microbial ecology and environmental microbiology so that the chapters on aquatic and terrestrial habitats can use them without excessive redundancy. Part Eleven concludes the text with an introduction to food and industrial microbiology. Five appendices aid the student with a review of some basic chemical concepts and with extra information about important topics not completely covered in the text.

Besides major organizational changes, every chapter has been thoroughly revised and updated. Examples of more significant changes are the following:

1. Sections on the polymerase chain reaction and techniques for inserting genes into eucaryotic cells have been added (chapter 14).
2. The effects of rRNA sequence studies on bacterial taxonomy are described in more detail (chapter 20).
3. New diseases, such as peptic ulcer disease (chapter 37), have been added, and the AIDS material has been combined into one section (chapter 36).
4. The Universal Precautions are described (box 34.1 and the endnotes).
5. A new unit on marine microbiology has been added (chapter 41).
6. The use of recombinant DNA techniques in industrial microbiology is described (chapter 44).

This text is designed to be an effective teaching tool. A text is only as easy for a student to use as it is easy to read. Readability has been enhanced by using a relatively simple, direct writing style, many section headings, and an organized outline format within each chapter. The level of difficulty has been carefully set with the target audience in mind. During preparation of the second edition, every sentence was carefully checked for clarity and revised when necessary. The American Society for Microbiology's *ASM Style Manual* conventions for nomenclature and abbreviations have been followed as consistently as possible.

The many new terms encountered in studying microbiology are a major stumbling block for students. This text lessens the problem by addressing and reinforcing a student's vocabulary development in three ways: (1) no new term is used without being clearly defined (often derivations are also given)—a student does not have to be familiar with the terminology of microbiology to use this text; (2) the most important terms are printed in boldface when first used; and (3) an extensive, up-to-date, page-referenced glossary is included at the end of the text.

Because illustrations are critical to a student's learning and enjoyment of microbiology, all illustrations in the second edition are full-color, and as many color photographs as possible have been used. Color not only enhances the text's attractiveness but also increases each figure's teaching effectiveness. More than 90 new illustrations have been added, and all line art has been produced under the direct supervision of the authors and designed to illustrate and reinforce specific points in the text. Consequently, every illustration is directly related to the narrative and specifically cited where appropriate. Great care has been taken to position illustrations as close as possible to the place where they are cited. Illustrations and captions have been reviewed for accuracy and clarity.

Themes in the Book

At least seven themes permeate the text, though a particular one may be more obvious at some points than are others. These themes or emphases are the following:

1. The development of microbiology as a science
2. The nature and importance of the techniques used to isolate, culture, observe, and identify microorganisms
3. The control of microorganisms and reduction of their detrimental effects
4. The importance of molecular biology for microbiology
5. The medical significance of microbiology
6. The ways in which microorganisms interact with their environments and the consequences of these interactions
7. The influences that microorganisms and microbiological applications have on everyday life

These themes help unify the text and enhance continuity. The student should get a feeling for what microbiologists do and for how this affects society.

Aids to the Student

It is hard to overemphasize the importance of pedagogical aids for the student. Accuracy is most important, but if a text is not clear, readable, and attractive, up-to-dateness and accuracy are wasted because students will not read the text. Students must be able to understand the material being presented, effectively use the text as a learning tool, and enjoy reading the book.

To be an effective teaching tool, a text must present the science of microbiology in a way that can be clearly taught and easily learned. Therefore, many aids are included to make the task of learning more efficient and enjoyable. Following the preface, a special section addressed to the student user reviews the principles of effective learning, including the SQ4R (survey, question, read, revise, record, and review) study technique. Each chapter contains the following:

1. *Opening quote:* The opening quote is designed to perk student interest and provide perspective on the chapter's contents.
2. *Chapter outline:* The chapter outline, with page numbers, includes all major headings in the chapter. This helps the reader locate particular topics of interest.
3. *Chapter concepts:* Several statements briefly summarize the most important concepts the student should master.
4. *Chapter preface:* One or two short paragraphs preview the chapter's contents and relate it to the rest of the text. The preface is not a summary but allows the student to put the chapter in perspective at the start.

5. *Boldfaced terms:* Important terms are emphasized and clearly defined when they are first used. The number of boldfaced terms has been substantially increased in this edition.
6. *Chapter summaries:* A series of brief, numbered statements is designed to serve more as a study guide than as a complete, detailed summary of the chapter.
7. *Key terms:* A list of all boldfaced terms is provided at the end of the chapter to emphasize the most significant facts and concepts. Each term is page-referenced to the page on which the term is first introduced in the chapter.
8. *Review questions and activities:* Two kinds of review questions appear in each chapter. A small box with one to five brief review questions is located at the end of most major sections. These questions help the student master the section's factual material and major concepts before continuing with the chapter. The "Questions for Thought and Review" section at the end of the chapter contains factual questions and some synthetic questions to aid the student in reviewing, integrating, and applying the knowledge gained and principles learned.
9. *Additional reading:* References are provided for further study. Most are reviews, monographs, and *Scientific American* articles rather than original research papers. Publications cited in these reviews introduce sufficiently interested students to the research literature. References through the middle of 1992 have been included.
10. *Cross-reference notes:* These notes refer the student to *major* topics that are difficult and may need review in order to understand the current material. They also point the student either forward or backward to a related item of unusual interest or importance. The second edition contains more cross-references in each chapter.
11. *Boxed readings:* Two types of boxed reading material are provided. Small boxes scattered throughout the text point out practical applications of microbiology related to the subjects being discussed. Most chapters also contain one or two larger boxes, which describe items of interest that are not essential to the primary thrust of the chapter. Topics include currently exciting research areas, the practical impact of microbial activities, items of medical significance requiring a more lengthy treatment than possible in the short boxes, historical anecdotes, and descriptions of extraordinary microorganisms.

Besides the chapter aids, the text also contains a glossary, an index, and five appendices. The *glossary* defines the most important terms from each chapter and includes page references. Where desirable, phonetic pronunciations are also given. Most of the glossary definitions have not been taken directly from the text but have been rewritten to give the student further understanding of the item. To improve ease of use, the second edition has a single *index* rather than three. It has been expanded to make text material more accessible. The *appendices* aid the student with extra review of chemical principles and metabolic pathways and provide further details about the taxonomy of bacteria and viruses. A brief introduction to helminth diseases is also included.

Supplementary Materials

Many supplementary materials are available to help instructors with their presentations and general course management.

1. An *Instructor's Manual,* written by Ralph Rascati, Kennesaw College, gives examples of lecture/reading schedules for courses with various emphases and suggests appropriate audiovisual materials. In addition, the number of test questions has been increased to about 60 for each chapter. This extensive battery of more than 2,600 test items is a powerful instructional tool.
2. The *WCB TestPak* is a computerized testing service offered free on request to adopters of this text. It provides a call-in/mail-in test preparation service. A complete test item file is also available on microcomputer diskette for use with either Apple IIe/IIc or IBM compatible computers.
3. A *Student Study Guide* by Ralph Rascati contains chapter objectives, focus questions, mastery tests, and other activities to aid student comprehension.
4. A set of 125 full-color acetate *transparencies* is available and may be used to supplement classroom lectures.
5. A set of *projection slides* provides further examples of microorganisms and diseases to supplement the illustrations in the text.
6. An *electronic study guide* by Medi-Sim is available in IBM and Macintosh formats for student purchase. This program contains approximately 2,000 page-referenced questions that review the material covered in *Microbiology.* Based on incorrect answers, this study guide provides the student a page-referenced study plan for each chapter.
7. A second edition of the *laboratory manual, Laboratory Exercises in Microbiology* by John P. Harley and Lansing M. Prescott, has been prepared to accompany the text. This manual is directly correlated with the text and designed to be used with it (although it may be used easily with other microbiology textbooks). Like the text, the laboratory manual provides a balanced introduction to laboratory techniques and principles that are important in each area of microbiology. The class-tested exercises are modular and short so that an instructor can easily choose only those exercises that fit his or her course. The second edition of the manual contains recipes for all reagents and media. Each exercise in this manual is also available as a customized, one-color separate. Specific lab exercises can be

combined with one's own materials. The local Wm. C. Brown Publishers representative should be contacted for more details on this custom publishing service.

A *lab resource guide* has been prepared by David Mardon of Eastern Kentucky University. It has complete answers to all laboratory report questions.

Acknowledgments

The authors wish to thank the reviewers, who provided detailed criticism and analysis of the first and second editions. Their suggestions greatly improved the final product.

Richard J. Alperin
Community College of Philadelphia
Susan T. Bagley
Michigan Technological University
R. A. Bender
University of Michigan
Hans P. Blaschek
University of Illinois
Dennis Bryant
University of Illinois
Arnold L. Demain
Massachusetts Institute of Technology
A. S. Dhaliwal
Loyola University of Chicago
Donald P. Durand
Iowa State University
John Hare
Linfield College
Robert B. Helling
University of Michigan–Ann Arbor
Barbara Bruff Hemmingsen
San Diego State University
R. D. Hinsdill
University of Wisconsin–Madison
John G. Holt
Michigan State University
Robert L. Jones
Colorado State University
Martha M. Kory
University of Akron
Robert I. Krasner
Providence College
Ron W. Leavitt
Brigham Young University
Glendon R. Miller
Wichita State University
Richard L. Myers
Southwest Missouri State University
G. A. O'Donovan
North Texas State University
Pattle P. T. Pun
Wheaton College
Ralph J. Rascati
Kennesaw State College
Albert D. Robinson
SUNY–Potsdam
Ronald Wayne Roncadori
University of Georgia–Athens
Ivan Roth
The University of Georgia–Athens
Thomas Santoro
SUNY–New Paltz
Ann C. Smith
University of Maryland, College Park
David W. Smith
University of Delaware
Paul Smith
University of South Dakota
James F. Steenbergen
San Diego State University
Henry O. Stone, Jr.
East Carolina University
James E. Struble
North Dakota State University
Kathleen Talaro
Pasadena City College
Thomas M. Terry
The University of Connecticut
Michael J. Timmons
Moraine Valley Community College
John Tudor
St. Joseph's University
Robert Twarog
University of North Carolina
Blake Whitaker
Bates College
Calvin Young
California State University–Fullerton

The publication of a textbook requires the effort of many people besides the authors. We would like to express special appreciation to the editorial and production staffs of Wm. C. Brown Publishers for their excellent work. In particular, we would like to thank our project editor, Megan Johnson, for her guidance and support, and our production editor, Diane Beausoleil, for her effort and attention to detail. Our copy editor, Beatrice Sussman, contributed greatly to the second edition's clarity, consistency, and readability. We would like to offer our special thanks to Raymond B. Otero, Eastern Kentucky University, for the many excellent clinical photographs he contributed to this textbook. Barbara Hemmingsen, San Diego State University, read the entire first edition and provided many helpful suggestions for revision. Our new, improved index is due to the knowledge and skill of Robert F. Boyd, Wirtz, Virginia. We also wish to thank the following for their advice, information, and photographs, which contributed significantly to the revision of Part Ten: Michael F. Allen, San Diego State University; Dwight Baker, Yale University; Keith Clay, Indiana University; A. L. Demain, MIT; Kenneth Doxtader, Colorado State University; Carol Ishimaru, Colorado State University; Derek Lovley, U.S. Geological Survey; Kirke Martin, Colorado State University; and L. K. Porter, USDA–ARS.

Finally, but most important, we wish to extend appreciation to our families for their patience and encouragement. Our wives, Linda Prescott, Jane Harley, and Sandra Klein, have been supportive on a daily basis, for which we are grateful. To them, we dedicate this book.

Lansing M. Prescott
John P. Harley
Donald A. Klein

To the Student

One of the most important factors contributing to success in college, and in microbiology courses, is the use of good study techniques. This textbook is organized to help you to study more efficiently. But even a text with many learning aids is not effective unless used properly. Thus, this section briefly outlines some practical study skills that will help ensure success in microbiology and make your use of this textbook more productive. Many of you already have the study skills mentioned here and will not need to spend time reviewing familiar material. These suggestions are made in the hope that they may be useful to those who are unaware of approaches like the SQ4R technique for studying textbooks.

Time Management and Study Environment

Many students find it difficult to study effectively because of a lack of time management and a proper place to study. Often a student will do poorly in courses because not enough time has been spent studying outside class. For best results, you should plan to spend at least an average of four to eight hours a week outside class working on each course. There is sufficient time in the week for this, but it does require time management. If you spend a few minutes early in the morning planning how the day is to be used and allow adequate time for studying, much more will be accomplished. Students who make efficient use of every moment find that they have plenty of time for recreation.

A second important factor is a proper place to study so that you can concentrate and efficiently use your study time. Try to find a quiet location with a desk and adequate lighting. If possible, always study in the same place and use it only for studying. In this way, you will be mentally prepared to study when you are at your desk. This location may be in the dorm, the library, a special study room, or somewhere else. Wherever it is, your study area should be free from distractions; including friends who drop by to socialize. Much more will be accomplished if you really study during your designated study times.

Making the Most of Lectures

Attendance at lectures is essential for success. Students who chronically miss classes usually do not do well. To gain the most from lectures, it is best to read any relevant text material beforehand. Be prepared to concentrate during lectures; do not simply sit back passively and listen to the instructor. During the lecture, record your notes in a legible way so that you can understand them later. It is most efficient to employ an outline or simple paragraph format. The use of abbreviations or some type of shorthand notation is often effective. During lecture, concentrate on what is being said, and be sure to capture all of the main ideas and concepts, and definitions of important terms. Do not take sketchy notes assuming that you will remember things because they are easy or obvious; you won't. Diagrams, lists, and terms written on the board are almost always important, as is anything the instructor clearly emphasizes by tone of voice. Feel free to ask questions during class when you don't understand something or wish the instructor to pursue a point further. Remember that if you don't understand, it is very likely that others in the class don't either, but simply aren't willing to show their confusion. As soon as possible after a lecture, carefully review your notes to be certain that they are complete and understandable. Refer to the textbook when uncertain about something in your notes; it will be invaluable in clearing up questions and amplifying major points. When studying your notes for tests, it is a good idea to emphasize the most important points with a felt-tip marker just as you would when reading the textbook.

Studying the Textbook

Your textbook is one of the most important learning tools in any course and should be very carefully and conscientiously used. Many years ago, Francis P. Robinson developed a very effective study technique called SQ3R (survey, question, read, recite, and review). More recently, L. L. Thistlethwaite and N. K. Snouffer have slightly modified it to yield the SQ4R approach (survey, question, read, revise, record, and review). This latter approach is summarized below.

1. *Survey.* Briefly scan the chapter to become familiar with its general content. Quickly read the title, introduction, summary, and main headings. Record the major ideas and points that you think the chapter will make. If there are a list of chapter concepts and a chapter outline, pay close attention to these. This survey should give you a feel for the topic and how the chapter is approaching it.
2. *Question.* As you reach each main heading or subheading, try to compose an important question or two that you believe the section will answer. This preview question will help focus your reading of the section. It is also a good idea to keep asking yourself questions as you read. This habit facilitates active reading and learning.
3. *Read.* Carefully read the section. Read to understand concepts and major points, and try to find the answer to your preview question(s). You may want to highlight very important terms or explanations of concepts, but do not indiscriminantly highlight everything. Be sure to pay close attention to any terms printed in color or boldface since the author(s) considered these to be important.
4. *Revise.* After reading the section, revise your question(s) to more accurately reflect the section's contents. These questions should be concept type questions that force you to bring together a number of details. They can be written in the margins of your text.
5. *Record.* Underline the information in the text that answers your questions, if you have not already done so. You may wish to write down the answers in note form as well. This process will give you good material to use in preparing for exams.
6. *Review.* Review the information by trying to answer your questions without looking at the text. If the text has a list of key words and a set of study questions, be sure to use these in your review. You will retain much more if you review the material several times.

Preparing for Examinations

It is extremely important to prepare for examinations properly so that you will not be rushed and tired on examination day. All textbook reading and lecture note revision should be completed well ahead of time so that the last few days can be spent in mastering the material, not in trying to understand the basic concepts. Cramming at the last moment for an exam is no substitute for daily preparation and review. By managing time carefully and keeping up with your studies, you will have plenty of time to review thoroughly and clear up any questions. This will allow you to get sufficient rest before the test and to feel confident in your preparation. Because both physical condition and general attitude are important factors in test performance, you will automatically do better. Proper reviewing techniques also aid retention of the material.

Further Reading

Grassick, P. 1983. *Making the grade: How to score high on all scholastic tests.* New York: Arco.

Shaw, H. 1976. *30 Ways to improve your grades.* New York: McGraw-Hill.

Shepherd, J. F. 1988. *RSVP: The Houghton Mifflin reading, study, and vocabulary program,* 3d ed. Boston: Houghton Mifflin.

Thistlethwaite, L. L., and Snouffer, N. K. 1976. *College reading power,* 3d ed. Dubuque, Ia.: Kendall/Hunt.

MICROBIOLOGY

Part One
Introduction to Microbiology

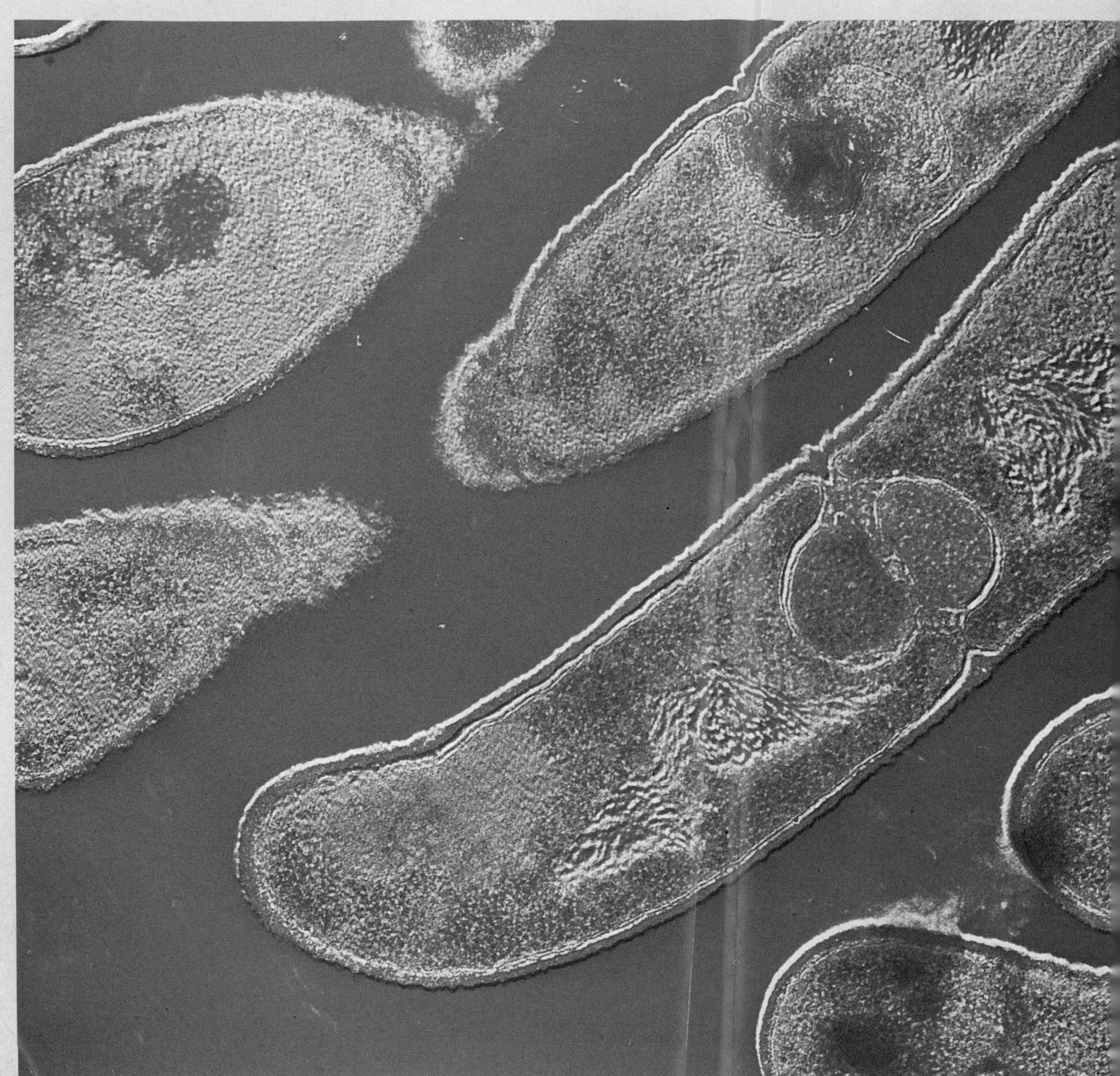

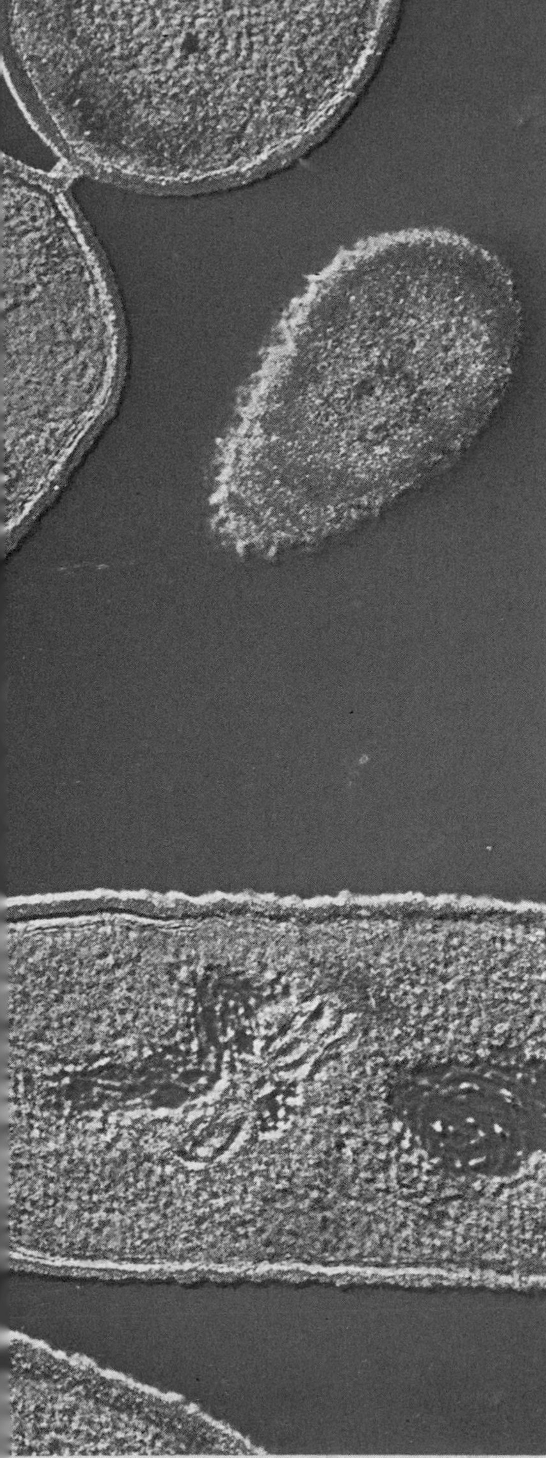

A *Listeria* colony. The genus contains aerobic and facultatively anaerobic, nonsporing, gram-positive bacteria. The bacterium in the center is dividing by binary fission. Nucleoids or nuclear regions are highlighted in blue. *L. monocytogenes* can cause meningitis and septicemia in humans. False color transmission electron micrograph (× 12,000).

CHAPTER 1
The History and Scope of Microbiology

Dans les champs de l'observation, le hasard ne favorise que les esprits prepares.

—L. Pasteur

(In the field of observation, chance favors only prepared minds.)

Outline

Concepts

1. Microbiology is the study of organisms that are usually too small to be seen by the unaided eye; it employs techniques—such as sterilization and the use of culture media—that are required to isolate and grow these microorganisms.
2. Microorganisms are not spontaneously generated from inanimate matter, but arise from other microorganisms.
3. Many diseases result from viral, bacterial, fungal, or protozoan infections. Koch's postulates may be used to establish a causal link between the suspected microorganism and a disease.
4. The development of microbiology as a scientific discipline has depended upon the availability of the microscope and the ability to isolate and grow pure cultures of microorganisms.
5. Microorganisms are responsible for many of the changes observed in organic and inorganic matter (e.g., fermentation and the carbon, nitrogen, and sulfur cycles that occur in nature).
6. Microorganisms have two fundamentally different types of cells—procaryotic and eucaryotic—and are distributed among several kingdoms.
7. Microbiology is a large discipline and has a great impact on other areas of biology and general human welfare.

One can't overemphasize the importance of microbiology. Society benefits from microorganisms in many ways. They are necessary for the production of bread, cheese, beer, antibiotics, vaccines, vitamins, enzymes, and many other important products. Indeed, modern biotechnology rests upon a microbiological foundation. Microorganisms are indispensable components of our ecosystem. They make possible the carbon, oxygen, nitrogen, and sulfur cycles that take place in terrestrial and aquatic systems, and are a source of nutrients at the base of all ecological food chains and webs.

Of course, microorganisms also have harmed humans and disrupted society since the beginning of recorded history. Microbial diseases undoubtedly played a major role in historical events such as the decline of the Roman Empire and the conquest of the New World. In the year 1347, plague or black death (*see chapter 36*) struck Europe with brutal force. By 1351, only four years later, the plague had killed 1/3 of the population (about 25 million people). Over the next 80 years, the disease struck again and again, eventually wiping out 75% of the European population. Some historians believe that this disaster changed European culture and prepared the way for the Renaissance. Today, the struggle by microbiologists and others against killers like AIDS and malaria continues.

The biology of AIDS and its impact (pp. 722–29).

In this introductory chapter, the historical development of the science of microbiology is described, and its relationship to medicine and other areas of biology is considered. The nature of the microbial world is then surveyed to provide a general idea of the organisms and agents that microbiologists study. Finally, the scope and relevance of modern microbiology are discussed.

Microbiology often has been defined as the study of organisms and agents too small to be seen clearly by the unaided eye—that is, the study of **microorganisms.** Because objects less than about one millimeter in diameter cannot be clearly seen and must be examined with a microscope, microbiology is concerned primarily with organisms this small and smaller. An extraordinary variety of organisms—viruses, bacteria, many algae and fungi, and protozoa—fits this size criterion (*see table 30.1*). Yet other members of these groups, particularly some of the algae and fungi, are larger and quite visible. For example, bread molds and filamentous algae are studied by microbiologists, yet are visible to the naked eye. The difficulty in setting the boundaries of microbiology led Roger Stanier to suggest that the field be defined not only in terms of the size of its subjects, but also in terms of its techniques. A microbiologist usually first isolates a specific microorganism from a population and then cultures it. Thus, microbiology employs techniques—such as sterilization and the use of culture media—that are necessary for successful isolation and growth of microorganisms.

The development of microbiology as a science is described in the following sections. Table 1.1 presents a summary of some of the major events in this process and their relationship to other historical landmarks.

The Discovery of Microorganisms

Even before microorganisms were seen, some investigators suspected their existence and responsibility for disease. Among others, the Roman philosopher Lucretius (about 98–55 B.C.) and the physician Girolamo Fracastoro (1478–1553) suggested that disease was caused by invisible living creatures. The earliest microscopic observations appear to have been made between 1625 and 1630 on bees and weevils by the Italian Francesco Stelluti, using a microscope probably supplied by Galileo. However, the first person to observe and describe microorganisms accurately was the amateur microscopist Antony van Leeuwenhoek (1632–1723) of Delft, Holland (figure 1.1*a*). Leeuwenhoek earned his living as a draper and haberdasher (a dealer in men's clothing and accessories), but spent much of his spare time constructing simple microscopes composed of double convex glass lenses held between two silver plates (figure 1.1*b*). His microscopes could magnify around 50 to 300 times, and he may have illuminated his liquid specimens by placing them between two pieces of glass and shining light on them at a 45° angle to the specimen plane. This would have provided a form of dark-field illumination (*see chapter 2*) and made bacteria clearly visible. Beginning in 1673, Leeuwenhoek sent detailed letters describing his discoveries to the Royal Society of London. It is clear from his descriptions that he saw both bacteria and protozoa.

TABLE 1.1 Some Important Events in the Development of Microbiology

Date	Microbiological History	Other Historical Events
1546	Fracastoro suggests that invisible organisms cause disease	Publication of Copernicus's work on the heliocentric solar system (1543)
		Shakespeare's *Hamlet* (1600–1601)
1676	Leeuwenhoek discovers "animalcules"	J. S. Bach and Handel born (1685)
1688	Redi publishes work on spontaneous generation of maggots	Isaac Newton publishes the *Principia* (1687)
		Linnaeus's *Systema Naturae* (1735)
		Mozart born (1756)
1798	Jenner introduces vaccination for smallpox	French Revolution (1789)
1799	Spallanzani attacks spontaneous generation	Beethoven's first symphony (1800)
		John Dalton publishes atomic theory (1808)
		The battle of Waterloo and the defeat of Napoleon (1815)
		Faraday demonstrates the principle of an electric motor (1821)
1838–1839	Schwann and Schleiden, the Cell Theory	England issues first postage stamp (1840)
1835–1844	Bassi discovers that a silkworm disease is caused by fungi and proposes that many diseases are microbial in origin	Marx's *Communist Manifesto* (1848)
1847–1850	Semmelweis shows that childbed fever is transmitted by physicians and introduces the use of antiseptics to prevent the disease	Velocity of light first measured by Fizeau (1849)
		Clausius states the first and second laws of thermodynamics (1850)
1849	Snow studies the epidemiology of a cholera epidemic in London	Graham distinguishes between colloids and crystalloids
		Melville's *Moby Dick* (1851)
1857	Pasteur shows that lactic acid fermentation is due to a microorganism	Otis installs first safe elevator (1854)
		Bunsen introduces the use of the gas burner (1855)
		Mendel begins his genetics experiments (1856)
1858	Virchow states that all cells come from cells	Darwin's *On the Origin of Species* (1859)
1861	Pasteur shows that microorganisms do not arise by spontaneous generation	American Civil War (1861–1865)
		Cross-Atlantic cable laid (1865)
1867	Lister publishes his work on antiseptic surgery	Dostoevski's *Crime and Punishment* (1866)
1869	Miescher discovers nucleic acids	Franco-German War (1870–1871)
1876–1877	Koch demonstrates that anthrax is caused by *Bacillus anthracis*	Bell invents telephone (1876)
		Edison's first light bulb (1879)
1881	Koch cultures bacteria on gelatin	Ives produces first color photograph (1881)
	Pasteur develops anthrax vaccine	
1882	Koch discovers tubercle bacillus	First central electric power station constructed by Edison (1882)
1884	Koch's postulates first published	Mark Twain's *The Adventures of Huckleberry Finn* (1884)
	Metchnikoff describes phagocytosis	
	Gram stain developed	
1885	Pasteur develops rabies vaccine	First motor vehicles developed by Daimler (1885–1886)
1887	Petri dish	
1887–1890	Winogradsky studies sulfur and nitrifying bacteria	Hertz discovers radio waves (1888)
1888	Beijerinck isolates root nodule bacteria	Eastman makes box camera (1888)
1890	Von Behring prepares antitoxins for diphtheria and tetanus	
1892	Ivanowsky provides evidence for virus causation of tobacco mosaic disease	First zipper patented (1895)
1895	Bordet discovers complement	Röntgen discovers X rays (1895)
1897	Buchner prepares yeast extract that ferments	Thomson discovers the electron (1897)
	Ross shows that malaria parasite is carried by the mosquito	Spanish-American War (1898)
1899	Beijerinck proves that a virus particle causes the tobacco mosaic disease	

TABLE 1.1 Continued

Date	Microbiological History	Other Historical Events
1900	Reed proves that yellow fever is transmitted by the mosquito	Planck develops the quantum theory (1900)
1902	Landsteiner discovers blood groups	First electric typewriter (1901)
1903	Wright and others discover antibodies in the blood of immunized animals	First powered aircraft (1903)
		Einstein's special theory of relativity (1905)
1906	Schaudinn and Hoffmann show *Treponema pallidum* causes syphilis	First model T Ford (1908)
	Wassermann develops complement fixation test for syphilis	Peary and Hensen reach North Pole (1909)
	Ricketts shows that Rocky Mountain spotted fever is transmitted by ticks	Rutherford presents his theory of the atom (1911)
1910	Ehrlich develops chemotherapeutic agent for syphilis	Picasso and cubism (1912)
1915–1917	D'Herelle and Twort discover bacterial viruses	World War I begins (1914)
		Einstein's general theory of relativity (1916)
1921	Fleming discovers lysozyme	Russian Revolution (1917)
1923	First edition of *Bergey's Manual*	Lindberg's transatlantic flight (1927)
1928	Griffith discovers bacterial transformation	
1929	Fleming discovers penicillin	Stock market crash (1929)
1933	Ruska develops first transmission electron microscope	Hitler becomes chancellor of Germany (1933)
1935	Stanley crystallizes the tobacco mosaic virus	Krebs discovers the citric acid cycle (1937)
	Domagk discovers sulfa drugs	
1937	Chatton divides living organisms into procaryotes and eucaryotes	
1941	Beadle and Tatum, one-gene-one-enzyme hypothesis	World War II begins (1939)
1944	Avery shows that DNA carries information during transformation	The insecticide DDT introduced (1944)
	Waksman discovers streptomycin	
		Atomic bombs dropped on Hiroshima and Nagasaki (1945)
1946	Lederberg and Tatum describe bacterial conjugation	United Nations formed (1945)
		First electronic computer (1946)
1950	Lwoff induces lysogenic bacteriophages	Korean War begins (1950)
1952	Hershey and Chase show that bacteriophages inject DNA into host cells	First hydrogen bomb exploded (1952)
		Stalin dies (1952)
	Zinder and Lederberg discover generalized transduction	First commercial transistorized product (1952)
1953	Phase-contrast microscope developed	U.S. Supreme Court rules against segregated schools (1954)
	Medawar discovers immune tolerance	
	Watson and Crick propose the double helix structure for DNA	
1955	Jacob and Wollman discover the F factor is a plasmid	Montgomery bus boycott (1955)
	Jerne and Burnet propose the clonal selection theory	Sputnik launched by Soviet Union (1957)
1959	Yalow develops the radioimmunoassay technique	Birth control pill (1960)
1961	Jacob and Monod propose the operon model of gene regulation	First humans in space (1961)
1961–1966	Nirenberg, Khorana, and others elucidate the genetic code	Cuban missile crisis (1962)
		Test ban treaty (1963)
1962	Porter proposes the basic structure for immunoglobulin G	
	First quinolone antimicrobial (nalidixic acid) synthesized	President Kennedy assassinated (1963)
		March on Washington (1964)
		Bombing of North Vietnam begun and first regular U.S. combat troops sent (1965)
		Arab-Israeli War (1967)

TABLE 1.1 Continued

Date	Microbiological History	Other Historical Events
1970	Discovery of restriction endonucleases by Arber and Smith Discovery of reverse transcriptase in retroviruses by Temin and Baltimore	M. L. King assassination (1968) Neil Armstrong walks on the moon (1969)
1973	Ames develops a bacterial assay for the detection of mutagens and carcinogens	Salt I Treaty (1972) Vietnam War ends (1973)
1975	Kohler and Milstein develop technique for the production of monoclonal antibodies Lyme disease discovered	President Nixon resigns because of Watergate cover-up (1975)
1977	Recognition of archaeobacteria as a distinct microbial group	Panama Canal Treaty (1977)
1979	Insulin synthesized using recombinant DNA techniques	Hostages seized in Iran (1978) Three Mile Island disaster (1979)
1980	Development of the scanning tunneling microscope	Home computers marketed (1980)
1982	Recombinant hepatitis B vaccine developed	AIDS first recognized (1981)
1982–1983	Discovery of catalytic RNA by Cech and Altman	First artificial heart implanted (1982)
1983–1984	The human immunodeficiency virus isolated and identified by Gallo and Montagnier The polymerase chain reaction developed by Mullis	Meter redefined in terms of distance light travels (1983)
1986	First vaccine (hepatitis B vaccine) produced by genetic engineering approved for human use	Gorbachev becomes Communist party general secretary (1985) Chernobyl reactor disaster (1986)
1990	First human gene-therapy testing begun	Berlin Wall falls (1989) Persian Gulf War with Iraq begins (1990)

The Spontaneous Generation Conflict

From earliest times, people had believed in **spontaneous generation**—that living organisms could develop from nonliving or decomposing matter. Even the great Aristotle (384–322 B.C.) thought some of the simpler invertebrates could arise by spontaneous generation. This view finally was challenged by the Italian physician Francesco Redi (1626–1697), who carried out a series of experiments on decaying meat and its ability to produce maggots spontaneously. Redi placed meat in three containers. One was uncovered, a second was covered with paper, and the third was covered with a fine gauze that would exclude flies. Flies laid their eggs on the uncovered meat, and maggots developed. The other two pieces of meat did not produce maggots spontaneously. However, flies were attracted to the gauze-covered container and laid their eggs on the gauze; these eggs did produce maggots. Thus, the generation of maggots by decaying meat resulted from the presence of fly eggs, and meat did not spontaneously generate maggots as previously believed. Similar experiments by others helped discredit the theory for larger organisms.

Leeuwenhoek's discovery of microorganisms renewed the controversy. Some proposed that microorganisms did arise by spontaneous generation even though larger organisms did not. They pointed out that boiled extracts of hay or meat would give rise to microorganisms after sitting for a while. In 1748, the English priest John Needham (1713–1781) reported the results of his experiments on spontaneous generation. Needham boiled mutton broth and then tightly stoppered the flasks. Eventually, many of the flasks became cloudy and contained microorganisms. He thought organic matter contained a vital force that could confer the properties of life on nonliving matter. A few years later, the Italian priest and naturalist Lazzaro Spallanzani (1729–1799) improved upon Needham's experimental design by first sealing glass flasks that contained water and seeds. If the sealed flasks were placed in boiling water for 3/4 of an hour, no growth took place as long as the flasks remained sealed. He proposed that air carried germs to the infusions, but also commented that the external air might be required for growth of animals already in the infusion. The supporters of spontaneous generation maintained that heating the air in sealed flasks destroyed its ability to support life.

Several investigators attempted to counter such arguments. Theodore Schwann (1810–1882) allowed air to enter a flask containing a sterile nutrient solution after the air had passed through a red-hot tube. The flask remained sterile. Subsequently, Georg Friedrich Schroder and Theodor von Dusch allowed air to enter a flask of heat-sterilized medium after it had passed through sterile cotton wool. No growth oc-

(a)

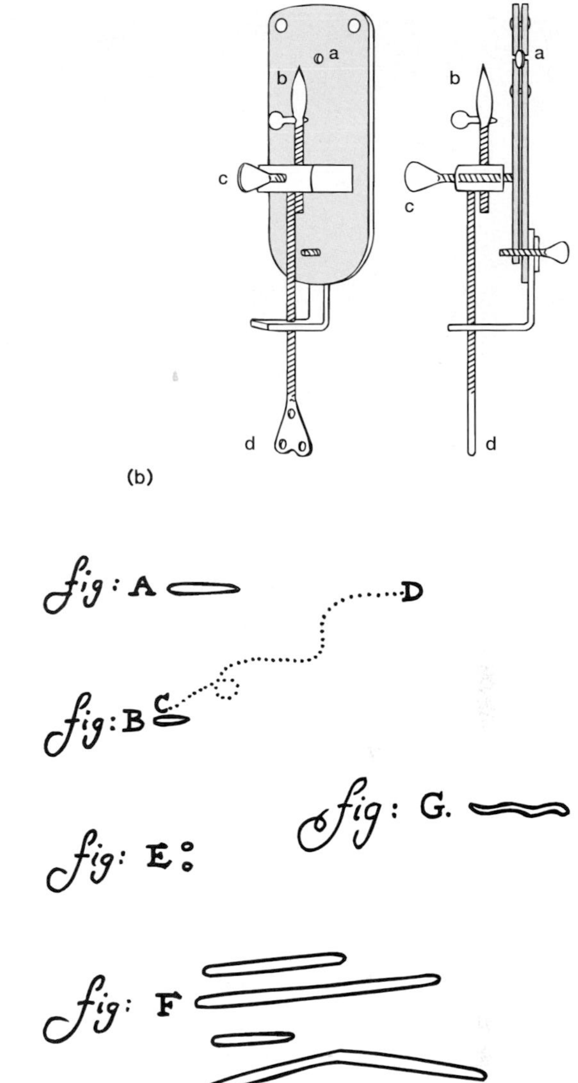

Figure 1.1 **Antony van Leeuwenhoek.** Leeuwenhoek (1632–1723) and his microscopes. (*a*) Leeuwenhoek holding a microscope. (*b*) A drawing of one of the microscopes showing the lens, *a;* mounting pin, *b;* and focusing screws, *c* and *d.* (*c*) Leeuwenhoek's drawings of bacteria from the human mouth. (*b*) *Source: C. E. Dobell,* Antony van Leeuwenhoek and His Little Animals *(1932), Russell and Russell, 1958.*

curred in the medium even though the air had not been heated. Despite these experiments, the French naturalist Felix Pouchet claimed in 1859 to have carried out experiments conclusively proving that microbial growth could occur without air contamination. This claim provoked Louis Pasteur (1822–1895) to settle the matter once and for all. Pasteur (figure 1.2) first filtered air through cotton and found that objects resembling plant spores had been trapped. If a piece of the cotton was placed in sterile medium after air had been filtered through it, microbial growth appeared. Next, he placed nutrient solutions in flasks, heated their necks in a flame, and drew them out into a variety of curves, while keeping the ends of the necks open to the atmosphere (figure 1.3). Pasteur then boiled the solutions for a few minutes and allowed them to cool. No growth took place even though the contents of the flasks were exposed to the air. Pasteur pointed out that no growth occurred because dust and germs had been trapped on the walls of the curved necks. If the necks were broken, growth commenced immediately. Pasteur had not only resolved the controversy by 1861, but also had shown how to keep solutions sterile.

The English physicist John Tyndall (1820–1893) dealt a final blow to spontaneous generation in 1877 by demonstrating that dust did indeed carry germs and that if dust were absent, broth remained sterile even if directly exposed to air. During the course of his studies, Tyndall provided evidence for the existence of exceptionally heat-resistant forms of bacteria. Working independently, the German botanist Ferdinand Cohn (1828–1898) discovered the existence of heat-resistant bacterial endospores (*see chapter 3*).

1. Describe the field of microbiology in terms of the size of its subject material and the nature of its techniques.
2. How did Pasteur and Tyndall finally settle the spontaneous generation controversy?

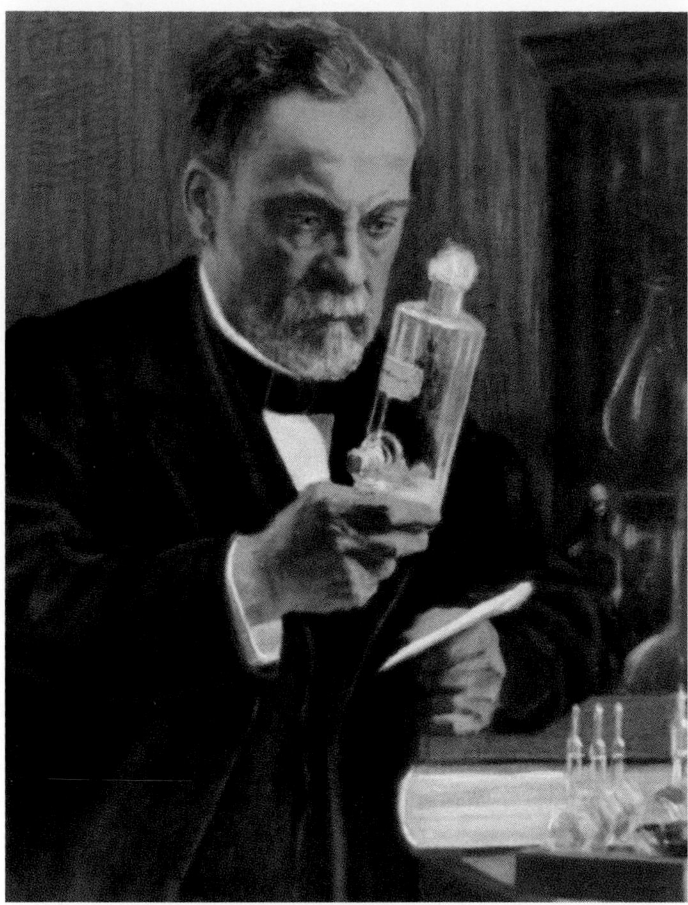

Figure 1.2 **Louis Pasteur.** Pasteur (1822–1895) working in his laboratory.

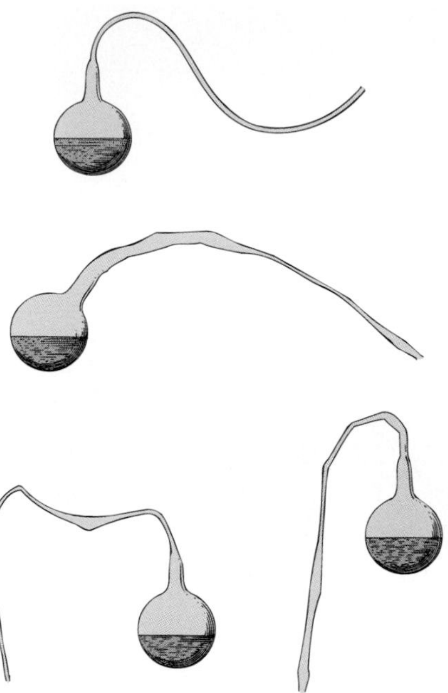

Figure 1.3 **The Spontaneous Generation Experiment.** Pasteur's swan neck flasks used in his experiments on the spontaneous generation of microorganisms. *Source:* Annales Sciences Naturelle, *4th Series, Vol. 16, pp. 1–98, Pasteur, L., 1861, "Mémoire sur les Corpuscules Organisés Qui Existent Dans L'Atmosphère: Examen de la Doctrine des Générations Spontanées."*

The Recognition of the Microbial Role in Disease

Although Fracastoro and a few others had suggested that invisible organisms produced disease, most believed that disease was due to causes such as supernatural forces, poisonous vapors called miasmas, and imbalances between the four humors thought to be present in the body. The idea that an imbalance between the four humors (blood, phlegm, choler, and melancholy) led to disease had been widely accepted since the time of the Greek physician Galen (129–199). Support for the germ theory of disease began to accumulate in the early nineteenth century. Agostino Bassi (1773–1856) first showed a microorganism could cause disease when he demonstrated in 1835 that a silkworm disease was due to a fungus infection. He also suggested that many diseases were due to microbial infections. In 1845, M. J. Berkeley proved that the great Potato Blight of Ireland was also caused by a fungus. Following his successes with the study of fermentation, Pasteur was asked by the French government to investigate the pébrine disease of silkworms that was disrupting the silk industry. After several years of work, he showed that the disease was due to a protozoan parasite. The disease was controlled by raising caterpillars from eggs produced by healthy moths.

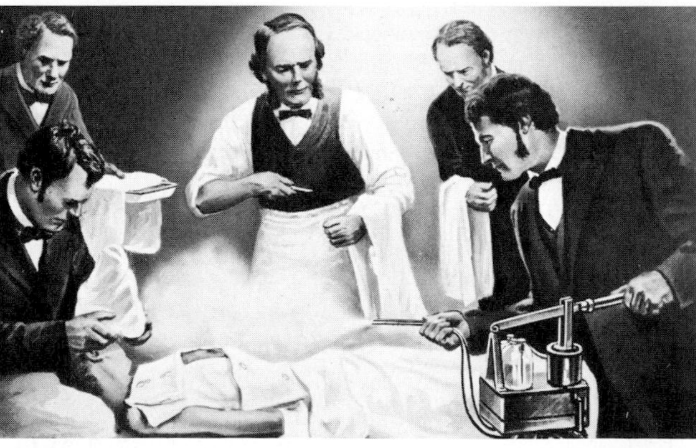

Figure 1.4 **Joseph Lister.** Lister (1827–1912) performing surgery with the use of antiseptic techniques.

Indirect evidence that microorganisms were agents of human disease came from the work of the English surgeon Joseph Lister (1827–1912) on the prevention of wound infections. Lister (figure 1.4), impressed with Pasteur's studies on the involvement of microorganisms in fermentation and putrefaction, developed a system of antiseptic surgery designed to prevent microorganisms from entering wounds. Instruments were heat sterilized, and phenol was used on surgical dressings and at times sprayed over the surgical area. The approach was remarkably successful and transformed surgery. It also pro-

Figure 1.5 **Robert Koch.** Koch (1843–1910) examining a specimen in his laboratory.

Figure 1.6 **Fannie Eilshemius (1850–1934) and Walther Hesse (1846–1911).** Fannie Hesse first proposed the use of agar in culture media.

vided strong indirect evidence for the role of microorganisms in disease because phenol, which killed bacteria, also prevented wound infections.

The first direct demonstration of the role of bacteria in causing disease came from the study of anthrax (*see chapter 37*) by the German physician Robert Koch (1843–1910). Koch (figure 1.5) used the criteria proposed by his former teacher, Jacob Henle (1809–1885), to establish the relationship between *Bacillus anthracis* and anthrax, and published his findings in 1876 (see Box 1.1 for a brief discussion of the scientific method). Koch injected healthy mice with material from diseased animals, and the mice became ill. After transferring anthrax by inoculation through a series of 20 mice, he incubated a piece of spleen containing the anthrax bacillus in beef serum. The bacilli grew, reproduced, and produced spores. When the isolated bacilli or spores were injected into mice, anthrax developed. His criteria for proving the causal relationship between a microorganism and a specific disease are known as **Koch's postulates** and can be summarized as follows:

1. The microorganism must be present in every case of the disease but absent from healthy organisms.
2. The suspected microorganism must be isolated and grown in a pure culture.
3. The same disease must result when the isolated microorganism is inoculated into a healthy host.
4. The same microorganism must be isolated again from the diseased host.

Although Koch used the general approach described in the postulates during his anthrax studies, he did not outline them fully until his 1884 publication on the cause of tuberculosis.

Koch's proof that *Bacillus anthracis* caused anthrax was independently confirmed by Pasteur and his coworkers. They discovered that after burial of dead animals, anthrax spores survived and were brought to the surface by earthworms. Healthy animals then ingested the spores and became ill.

During Koch's studies on bacterial diseases, it became necessary to isolate suspected bacterial pathogens. At first, he cultured bacteria on the sterile surfaces of cut, boiled potatoes. This was unsatisfactory because bacteria would not always grow well on potatoes. He then tried to solidify regular liquid medium by adding gelatin. Separate bacterial colonies developed after the surface had been streaked with a bacterial sample. The sample could also be mixed with liquefied gelatin medium. When the gelatin medium hardened, individual bacteria produced separate colonies. Despite its advantages, gelatin was not an ideal solidifying agent because it was digested by many bacteria and melted when the temperature rose above 28°C. A better alternative was provided by Fannie Eilshemius Hesse, the wife of Walther Hesse, one of Koch's assistants (figure 1.6). She suggested the use of agar as a solidifying agent—she had been using it successfully to make jellies for some time. Agar was not attacked by most bacteria and did not melt until reaching a temperature of 100°C. One of Koch's assistants, Richard Petri, developed the petri dish (plate), a container for solid culture media. These developments made possible the isolation of pure cultures that contained only one type of bacterium, and directly stimulated progress in all areas of bacteriology.

Isolation of bacteria and pure culture techniques (pp. 106–8).

Box 1.1

The Scientific Method

Although biologists employ a variety of approaches in conducting research, microbiologists and other experimentally oriented biologists often use the general approach known as the scientific method. They first gather observations of the process to be studied and then develop a tentative **hypothesis**—an educated guess—to explain the observations (see figure). This step often is inductive and creative because there is no detailed, automatic technique for generating hypotheses. Next, they decide what information is required to test the hypothesis and collect this information through observation or carefully designed experiments. After the information has been collected, they decide whether the hypothesis has been supported or falsified. If it has failed to pass the test, the hypothesis is rejected, and a new explanation or hypothesis is constructed. If the hypothesis passes the test, it is subjected to more severe testing. The procedure often is made more efficient by constructing and testing alternative hypotheses and then refining the hypothesis that survives testing. This general approach is often called the hypothetico-deductive method. One deduces predictions from the currently accepted hypothesis and tests them. In deduction, the conclusion about specific cases follows logically from a general premise ("if . . . , then . . . ." reasoning). Induction is the opposite. A general conclusion is reached after considering many specific examples. Both types of reasoning are used by scientists.

When carrying out an experiment, it is essential to use a control group as well as an experimental group. The control group is treated precisely the same as the experimental group except that the experimental manipulation is not performed on it. In this way, one can be sure that any changes in the experimental group are due to the experimental manipulation rather than to some other factor not taken into account.

If a hypothesis continues to survive testing, it may be accepted as a valid theory. A **theory** is a set of propositions and concepts that provides a reliable, systematic, and rigorous account of an aspect of nature. It is important to note that hypotheses and theories are never absolutely proven. Scientists simply gain more and more confidence in their accuracy as they continue to survive testing, fit with new observations and experiments, and satisfactorily explain the observed phenomena.

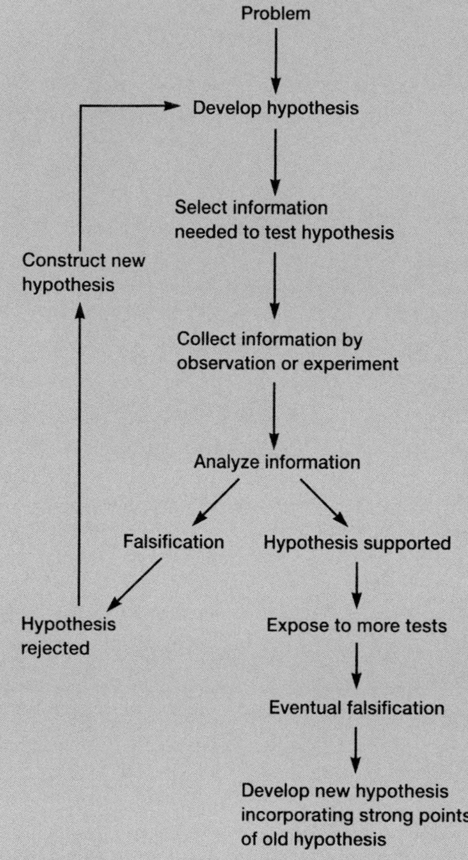

Box Figure 1.1 The Hypothetico-Deductive Method. This approach is most often used in scientific research.

Koch also developed media suitable for growing bacteria isolated from the body. Because of their similarity to body fluids, meat extracts and protein digests were used as nutrient sources. The result was the development of nutrient broth and nutrient agar, media that are still in wide use today.

By 1882, Koch had used these techniques to isolate the bacillus that caused tuberculosis. There followed a golden age of about 30 to 40 years in which most of the major bacterial pathogens were isolated (table 1.2).

The discovery of viruses and their role in disease was made possible when Charles Chamberland (1851–1908), one of Pasteur's associates, constructed a porcelain bacterial filter in 1884. The first viral pathogen to be studied was the tobacco mosaic disease virus (*see chapter 17*).

The development of virology (pp. 347–48).

During this period, progress also was made in determining how animals resisted disease and in developing techniques for protecting humans and livestock against pathogens. During his work on chicken cholera, Pasteur accidentally discovered that older cultures of the bacterium were attenuated, which meant they had lost their ability to cause the disease. If the chickens were injected with these attenuated cultures, they remained healthy but developed the ability to resist the

TABLE 1.2 The Discovery of Some of the Major Pathogens

Disease	Causative Agent[a]	Discoverer	Date
Anthrax	*Bacillus anthracis*	Koch	1876
Gonorrhea	*Neisseria gonorrhoeae*	Neisser	1879
Typhoid fever	*Salmonella typhi*	Eberth	1880
Malaria	*Plasmodium* spp.	Laveran	1880
Tuberculosis	*Mycobacterium tuberculosis*	Koch	1882
Cholera	*Vibrio cholerae*	Koch	1883
Diphtheria	*Corynebacterium diphtheriae*	Klebs and Loeffler	1883–1884
Tetanus	*Clostridium tetani*	Nicolaier	1885
Diarrhea	*Escherichia coli*	Escherich	1885
Pneumonia	*Streptococcus pneumoniae*	Fraenkel	1886
Meningitis	*Neisseria meningitidis*	Weichselbaum	1887
Undulant fever	*Brucella* spp.	Bruce	1887
Gas gangrene	*Clostridium perfringens*	Welch and Nuttal	1892
Plague	*Yersinia pestis*	Kitasato and Yersin	1894
Botulism	*Clostridium botulinum*	Van Ermengem	1896
Dysentery	*Shigella dysenteriae*	Shiga	1898
Syphilis	*Treponema pallidum*	Schaudin and Hoffmann	1905
Whooping cough	*Bordetella pertussis*	Bordet and Gengou	1906
Rocky Mountain spotted fever	*Rickettsia rickettsii*	Ricketts	1909

[a]Modern name used.

disease. He called the attenuated culture a vaccine (Latin *vacca,* cow) in honor of Edward Jenner because, many years earlier, Jenner had used vaccination with material from cowpox lesions to protect people against smallpox (*see chapter 17*). The same general approach was used to prepare an anthrax vaccine.

The development of chemotherapy (pp. 326–27).

Pasteur next prepared rabies vaccine by a different approach. The pathogen was attenuated by growing it in an abnormal host, the rabbit. After infected rabbits had died, their brains and spinal cords were removed and dried. Injection of a mixture of this material and glycerin stimulated dogs to develop resistance to rabies. Then a nine-year-old boy, Joseph Meister, who had been bitten by a rabid dog, was brought to Pasteur. Since the boy's death was certain in the absence of treatment, Pasteur agreed to try vaccination. Joseph was injected 13 times over the next 10 days with increasingly virulent preparations of the attenuated virus. He survived.

In gratitude for Pasteur's development of vaccines, people from around the world contributed to the construction of the Pasteur Institute in Paris, France. One of the initial tasks of the Institute was vaccine production.

After the discovery that the diphtheria bacillus produced a toxin, Emil von Behring (1854–1917) and Shibasaburo Kitasato (1852–1931) injected inactivated toxin into rabbits, inducing them to produce an antitoxin, a substance in the blood that would inactivate the toxin and protect against the disease. A tetanus antitoxin was then prepared and both antitoxins were used in the treatment of people.

The antitoxin work provided evidence that immunity could result from soluble substances in the blood, now known to be antibodies (humoral immunity). It became clear that blood cells were also important in immunity (cellular immunity) when Elie Metchnikoff (1845–1916) discovered that some blood leukocytes could engulf disease-causing bacteria (figure 1.7). He called these cells phagocytes and the process phagocytosis (Greek *phagein,* eating).

Figure 1.7 Elie Metchnikoff. Metchnikoff (1845–1916) shown here at work in his laboratory.

1. Discuss the contributions of Lister, Pasteur, and Koch to the germ theory of disease and to the treatment or prevention of diseases.
2. What other contributions did Koch make to microbiology?
3. Describe Koch's postulates in detail.
4. How did von Behring and Metchnikoff contribute to the development of immunology?

The Discovery of Microbial Effects on Organic and Inorganic Matter

Although Theodore Schwann and others had proposed in 1837 that yeast cells were responsible for the conversion of sugars to alcohol, a process they called alcoholic fermentation, the leading chemists of the time believed microorganisms were not involved. They were convinced that fermentation was due to a chemical instability that degraded the sugars to alcohol. In 1856, an industrialist in Lille, France, where Pasteur worked, requested his assistance. His business produced ethanol from the fermentation of beet sugars, and recently alcohol yields had declined and the product had become sour. Pasteur discovered that the fermentation was failing because the yeast normally responsible for alcohol formation had been replaced by microorganisms producing lactic acid rather than ethanol. In solving this practical problem, Pasteur demonstrated that all fermentations were due to the activities of specific yeasts and bacteria, and he published several papers on fermentation between 1857 and 1860. His success led to a study of wine diseases and the development of pasteurization (*see chapter 15*) to preserve wine during storage. Pasteur's studies on fermentation continued for almost 20 years. One of his most important discoveries was that some fermentative microorganisms were anaerobic and could live only in the absence of oxygen, whereas others were able to live either aerobically or anaerobically.

Fermentation (pp. 155–58).
The effect of oxygen on microorganisms (pp. 126–28).

A few of the early microbiologists chose to investigate the ecological role of microorganisms. In particular, they studied microbial involvement in the carbon, nitrogen, and sulfur cycles taking place in soil and aquatic habitats. Two of the pioneers in this endeavor were Sergei N. Winogradsky (1856–1953) and Martinus Beijerinck (1851–1931).

Biogeochemical cycles (pp. 809–13).

The Russian microbiologist Sergei N. Winogradsky (figure 1.8) made many contributions to soil microbiology. He discovered that soil bacteria could oxidize iron, sulfur, and ammonia to obtain energy, and that many bacteria could incorporate CO_2 into organic matter much like photosynthetic organisms do. Winogradsky also isolated anaerobic nitrogen-fixing soil bacteria and studied the decomposition of cellulose.

Martinus Beijerinck (figure 1.9) was one of the great general microbiologists who made fundamental contributions to microbial ecology and many other fields. He isolated the aerobic nitrogen-fixing bacterium *Azotobacter;* a root nodule bacterium also capable of fixing nitrogen (later named *Rhizobium*); and sulfate-reducing bacteria. Beijerinck and Winogradsky developed the enrichment-culture technique and the use of selective media (*see chapter 5*), which have been of such great importance in microbiology.

Figure 1.8 **Sergei N. Winogradsky** (1856–1953).

Figure 1.9 **Martinus W. Beijerinck** (1851–1931).

The Development of Microbiology in This Century

During the early part of the twentieth century, microbiology developed somewhat independently of other biological disciplines, at least partially because of a difference in focus. Many biologists were interested in such topics as cell structure and function, the ecology of plants and animals, the reproduction and development of organisms, the nature of heredity, and the mechanism of evolution. In contrast, microbiologists were more concerned with the agents of infectious disease, the immune response, the search for new chemotherapeutic agents, and bacterial metabolism.

Microbiology established a closer relationship with other biological disciplines during the 1940s because of its association with genetics and biochemistry. Microorganisms are extremely useful experimental subjects because they are relatively simple, grow rapidly, and can be cultured in large quantities. George W. Beadle and Edward L. Tatum studied the relationship between genes and enzymes in 1941 using mutants of the bread mold *Neurospora.* Salvadore Luria and Max Delbrück (1943) used bacterial mutants to show that gene mutations were truly spontaneous and not directed by the environment. Oswald T. Avery, Colin M. MacLeod, and Maclyn McCarty provided strong evidence in 1944 that DNA was the genetic material and carried genetic information during transformation (*see chapter 12*). The interactions between microbiology, genetics, and biochemistry soon led to the development of modern molecularly oriented genetics.

More recently, microbiology has been a major contributor to the rise of molecular biology, the branch of biology dealing with the physical and chemical aspects of living matter and its function. Microbiologists have been deeply involved in the elucidation of the genetic code and in studies on the mechanisms of DNA, RNA, and protein synthesis. Microorganisms were used in many of the early studies on the regulation of gene expression and the control of enzyme activity (*see chapter 11*). In the 1970s, new discoveries in microbiology led to the development of recombinant DNA technology and genetic engineering.

The mechanisms of DNA, RNA, and protein synthesis (chapter 10).
Recombinant DNA and genetic engineering (chapter 14).

One indication of the importance of microbiology in the twentieth century is the Nobel Prize given for work in physiology or medicine. About 1/3 of these have been awarded to scientists working on microbiological problems (table 1.3).

1. Briefly summarize the contributions of Pasteur, Winogradsky, and Beijerinck to the study of microbial effects on organic and inorganic matter.
2. What is molecular biology, and why has microbiology been so important to its development?

The Composition of the Microbial World

Although the kingdoms of organisms and the differences between procaryotic and eucaryotic cells are discussed in much more detail later, a brief introduction to the organisms a microbiologist studies is given here.

Comparison of procaryotic and eucaryotic cells (pp. 89–91).
The five-kingdom system of classification (pp. 413–15).

Two fundamentally different types of cells exist. **Procaryotic cells** (Greek *pro,* before, and *karyon,* nut or kernel; organism with a primordial nucleus) have a much simpler morphology than eucaryotic cells and lack a true membrane-delimited nucleus. All bacteria are procaryotic. In contrast, **eucaryotic cells** (Greek *eu,* true, and *karyon,* nut or kernel) have a membrane-enclosed nucleus; they are more complex morphologically and are usually larger than procaryotes. Algae, fungi, protozoa, higher plants, and animals are eucaryotic. Procaryotic and eucaryotic cells differ in many other ways as well (*see chapter 4*).

The early description of organisms as either plants or animals clearly is too simplified, and many biologists now believe that organisms should be divided into at least five kingdoms (*see chapter 20*).

1. Procaryotic organisms are found in the kingdom *Monera* or *Procaryotae.*
2. The kingdom *Protista* contains either unicellular or colonial eucaryotic organisms that lack true tissues. Protozoa, the lower fungi, and most of the smaller algae are placed in this kingdom.
3. Members of the kingdom *Fungi* are eucaryotic organisms with absorptive nutrition, and often are multinucleate.
4. The kingdom *Animalia* contains multicellular animals with ingestive nutrition.
5. Multicellular plants with walled eucaryotic cells and photosynthesis are located in the kingdom *Plantae.*

Microbiologists study primarily members of the first three kingdoms—the *Monera* or *Procaryotae, Protista,* and *Fungi.* *(See table 20.7 and figure 20.6 for a more complete summary of the five-kingdom system.)* Although they are not included in the five kingdoms, viruses are also studied by microbiologists.

Fungi (chapter 26).
Algae (chapter 27).
Protozoa (chapter 28).
Introduction to the viruses (chapters 17–19).

Recent studies on ribosomal RNA sequences and other molecular properties of procaryotes indicate there are two quite different groups of procaryotic organisms: eubacteria and archaeobacteria. The differences between eubacteria, archaeobacteria, and eucaryotic cells are so great some microbiologists have proposed that organisms should be divided among three domains or kingdoms: bacteria (the true bacteria or eubacteria), archaea (the archaeobacteria), and eucarya (all eucaryotic organisms). This proposal and the results leading to it are discussed in chapter 20.

1. Describe and contrast procaryotic and eucaryotic cells.
2. Briefly describe the five-kingdom system and give the major characteristics of each kingdom.

TABLE 1.3 Nobel Prizes Awarded for Research in Microbiology

Date	Scientist[a]	Research
1901	E. von Behring (GR)	Diphtheria antitoxin
1902	R. Ross (GB)	Cause and transmission of malaria
1905	R. Koch (GR)	Tuberculosis research
1907	C. Laveran (F)	Role of protozoa in disease
1908	P. Ehrlich (GR) E. Metchnikoff (R)	Work on immunity
1913	C. Richet (F)	Work on anaphylaxis
1919	J. Bordet (B)	Discoveries about immunity
1928	C. Nicolle (F)	Work on typhus fever
1930	K. Landsteiner (US)	Discovery of human blood groups
1939	G. Domagk (GR)	Antibacterial effect of prontosil
1945	A. Fleming (GB) E. B. Chain (GB) H. W. Florey (AU)	Discovery of penicillin and its therapeutic value
1951	M. Theiler (SA)	Development of yellow fever vaccine
1952	S. A. Waksman (US)	Discovery of streptomycin
1954	J. F. Enders (US) T. H. Weller (US) F. Robbins (US)	Cultivation of poliovirus in tissue culture
1957	D. Bovet (I)	Discovery of the first antihistamine
1958	G. W. Beadle (US) E. L. Tatum (US) J. Lederberg (US)	Microbial genetics
1959	S. Ochoa (US) A. Kornberg (US)	Discovery of enzymes catalyzing nucleic acid synthesis
1960	F. M. Burnet (AU) P. B. Medawar (GB)	Discovery of acquired immune tolerance to tissue transplants
1962	F. H. C. Crick (GB) J. D. Watson (US) M. Wilkins (GB)	Discoveries concerning the structure of DNA
1965	F. Jacob (F) A. Lwoff (F) J. Monod (F)	Discoveries about the regulation of genes
1966	F. P. Rous (US)	Discovery of cancer viruses
1968	R. W. Holley (US) H. G. Khorana (US) M. W. Nirenberg (US)	Deciphering of the genetic code
1969	M. Delbrück (US) A. D. Hershey (US) S. E. Luria (US)	Discoveries concerning viruses and viral infection of cells
1972	G. Edelman (US) R. Porter (GB)	Research on the structure of antibodies
1975	H. Temin (US) D. Baltimore (US) R. Dulbecco (US)	Discovery of RNA-dependent DNA synthesis by RNA tumor viruses; reproduction of DNA tumor viruses

Date	Scientist[a]	Research
1976	B. Blumberg (US) D. C. Gajdusek (US)	Mechanism for the origin and dissemination of hepatitis B virus; research on slow virus infections
1977	R. Yalow (US)	Development of the radioimmunoassay technique
1978	H. O. Smith (US) D. Nathans (US) W. Arber (SW)	Discovery of restriction enzymes and their application to the problems of molecular genetics
1980	B. Benacerraf (US) G. Snell (US) J. Dausset (F)	Discovery of the histocompatibility antigens
	P. Berg (US) W. Gilbert (US) & F. Sanger (GB)	Development of recombinant DNA technology (Berg); development of DNA sequencing techniques (Chemistry Prize)
1982	A. Klug (GB)	Development of crystallographic electron microscopy and the elucidation of the structure of viruses and other nucleic-acid–protein complexes (Chemistry Prize)
1984	C. Milstein (GB) G. J. F. Kohler (GR) N. K. Jerne (D)	Development of the technique for formation of monoclonal antibodies (Milstein & Kohler); theoretical work in immunology (Jerne)
1986	E. Ruska (GR)	Development of the transmission electron microscope (Physics Prize)
1987	S. Tonegawa (J)	The genetic principle for generation of antibody diversity
1988	J. Deisenhofer, R. Huber, and H. Michel (GR)	Crystallization and study of the photosynthetic reaction center from a bacterial membrane
	G. Elion (US) G. Hitchings (US)	Development of drugs for the treatment of cancer, malaria, and viral infections
1989	J. M. Bishop (US) H. E. Varmus (US)	Discovery of oncogenes
	S. Altman (US) T. R. Cech (US)	Discovery of catalytic RNA

[a]The Nobel laureates were citizens of the following countries: Australia (AU), Belgium (B), Denmark (D), France (F), Germany (GR), Great Britain (GB), Italy (I), Japan (J), Russia (R), South Africa (SA), Switzerland (SW), and the United States (US).

The Scope and Relevance of Microbiology

Microorganisms are exceptionally diverse, are found almost everywhere, and affect human society in countless ways. Thus, modern microbiology is a large discipline with many different specialties; it has a great impact on medicine, agricultural and food sciences, ecology, genetics, biochemistry, and many other fields.

Microbiology has both basic and applied aspects. Many microbiologists are interested primarily in the biology of the microorganisms themselves. They may focus on a specific group of microorganisms and be called virologists (viruses), bacteriologists (bacteria), phycologists or algologists (algae), mycologists (fungi), or protozoologists (protozoa). Others are interested in microbial morphology or particular functional processes and work in fields such as microbial cytology, microbial physiology, microbial ecology, microbial genetics and molecular biology, and microbial taxonomy. Of course, a person can be thought of in both ways (e.g., as a bacteriologist who works on taxonomic problems). Many microbiologists have a more applied orientation and work on practical problems in fields such as medical microbiology, food and dairy microbiology, and public health microbiology (basic research is also conducted in these fields). Because the various fields of microbiology are interrelated, an applied microbiologist must be familiar with basic microbiology. For example, a medical microbiologist must have a good understanding of microbial taxonomy, genetics, immunology, and physiology to identify and properly respond to the pathogen of concern.

What are some of the current occupations of professional microbiologists? One of the most active and important is medical microbiology, which deals with the diseases of humans and animals. Medical microbiologists identify the agent causing an infectious disease and plan measures to eliminate it. Frequently, they are involved in tracking down new, unidentified pathogens such as the bacterium that causes Legionnaires' disease or the virus responsible for AIDS. These microbiologists also study the ways in which microorganisms cause disease.

Public health microbiology is closely related to medical microbiology. Public health microbiologists try to control the spread of communicable diseases. They often monitor community food establishments and water supplies in an attempt to keep them safe and free from infectious disease agents.

Immunology is concerned with how the immune system protects the body from pathogens and the response of infectious agents. It is one of the fastest growing areas in science; for example, techniques for the production and use of monoclonal antibodies have developed extremely rapidly. Immunology also deals with practical health problems such as the nature and treatment of allergies and autoimmune diseases like rheumatoid arthritis.

Monoclonal antibodies and their uses (chapter 31 and Box 34.1).

Many important areas of microbiology do not deal directly with human health and disease, but certainly contribute to human welfare. Agricultural microbiology is concerned with the impact of microorganisms on agriculture. Agricultural microbiologists try to combat plant diseases that attack important food crops, work on methods to increase soil fertility and crop yields, and study the role of microorganisms living in the digestive tracts of ruminants such as cattle. Currently, there is great interest in using bacterial and viral insect pathogens as substitutes for chemical pesticides.

The field of microbial ecology is closely related to agricultural microbiology. Microbial ecologists study the relationships between microorganisms and their habitats. They are concerned with the contribution of microorganisms to the carbon, nitrogen, and sulfur cycles in soil and in fresh water. The study of pollution effects on microorganisms also is important because of the impact these organisms have on the environment.

Scientists working in food and dairy microbiology try to prevent microbial spoilage of food and the transmission of food-borne diseases such as botulism and salmonellosis (*see chapter 37*). They also use microorganisms to make foods such as cheeses, yogurts, pickles, and beer. In the future, microorganisms themselves will become a more important nutrient source for livestock and humans.

In industrial microbiology, microorganisms are used to make products such as antibiotics, vaccines, steroids, alcohols and other solvents, vitamins, amino acids, and enzymes. Microorganisms can even leach valuable minerals from low-grade ores.

Research on the biology of microorganisms occupies the time of many microbiologists and also has practical applications. Those working in microbial physiology and biochemistry study the synthesis of antibiotics and toxins, microbial energy production, the ways in which microorganisms survive harsh environmental conditions, microbial nitrogen fixation, the effects of chemical and physical agents on microbial growth and survival, and many other topics.

Microbial genetics and molecular biology focus on the nature of genetic information and how it regulates the development and function of cells and organisms. The use of microorganisms has been very helpful in understanding gene function. Microbial geneticists play an important role in applied microbiology by producing new microbial strains that are more efficient in synthesizing useful products. Genetic techniques are used to test substances for their ability to cause cancer. More recently, the field of genetic engineering (*see chapter 14*) has arisen from work in microbial genetics and molecular biology, and will contribute substantially to microbiology, biology as a whole, and medicine. Engineered microorganisms are used to make hormones, antibiotics, vaccines, and other products. New genes can be inserted into plants and animals; for example, it may be possible to give corn and wheat nitrogen-fixation genes so they will not require nitrogen fertilizers.

The future of microbiology is bright. With the advent of recombinant DNA technology and genetic engineering, microbiology will expand and change even faster in the future.

The need for microbiologists to work on problems in medicine, environmental protection, food production and preservation, and industrial development will undoubtedly grow. The microbiologist René Dubos has summarized well the excitement and promise of microbiology:

> How extraordinary that, all over the world, microbiologists are now involved in activities as different as the study of gene structure, the control of disease, and the industrial processes based on the phenomenal ability of microorganisms to decompose and synthesize complex organic molecules. Microbiology is one of the most rewarding of professions because it gives its practitioners the opportunity to be in contact with all the other natural sciences and thus to contribute in many different ways to the betterment of human life.

Summary

1. Microbiology may be defined in terms of the size of the organisms studied and the techniques employed.
2. Antony van Leeuwenhoek was the first person to observe microorganisms.
3. Experiments by Redi and others disproved the theory of spontaneous generation in regard to larger organisms.
4. The spontaneous generation of microorganisms was disproved by Spallanzani, Pasteur, Tyndall, and others.
5. Support for the germ theory of disease came from the work of Bassi, Pasteur, Koch, and others. Lister provided indirect evidence with his development of antiseptic surgery.
6. Koch's postulates are used to prove a direct relationship between a suspected pathogen and a disease.
7. Koch also developed the techniques required to grow bacteria on solid media and to isolate pure cultures of pathogens.
8. Vaccines against anthrax and rabies were made by Pasteur; von Behring and Kitasato prepared antitoxins for diphtheria and tetanus.
9. Metchnikoff discovered some blood leukocytes could phagocytize and destroy bacterial pathogens.
10. Pasteur showed that fermentations were caused by microorganisms and that some microorganisms could live in the absence of oxygen.
11. The role of microorganisms in carbon, nitrogen, and sulfur cycles was first studied by Winogradsky and Beijerinck.
12. In the twentieth century, microbiology has contributed greatly to the fields of biochemistry and genetics. It also has helped stimulate the rise of molecular biology.
13. Procaryotic cells differ from eucaryotic cells in that the former lack a membrane-delimited nucleus, and in other ways as well.
14. There are at least five different kingdoms of living organisms. Microbiologists study primarily members of the kingdoms *Monera* or *Procaryotae, Protista,* and *Fungi.* Some have proposed that the archaeobacteria constitute another kingdom of organisms.
15. There is a wide variety of fields in microbiology, and many have a great impact on society. Some of these are medical microbiology, public health microbiology, immunology, agricultural microbiology, microbial ecology, food and dairy microbiology, industrial microbiology, microbial physiology and biochemistry, and microbial genetics and molecular biology.

Key Terms

eucaryotic cell *15*
hypothesis *12*
Koch's postulates *11*
microbiology *5*
microorganism *5*
procaryotic cell *15*
spontaneous generation *8*
theory *12*

Questions for Thought and Review

1. Why was the belief in spontaneous generation an obstacle to the development of microbiology as a scientific discipline?
2. Describe the major contributions of the following people to the development of microbiology: Leeuwenhoek, Spallanzani, Fracastoro, Pasteur, Tyndall, Cohn, Bassi, Lister, Koch, Chamberland, von Behring, Metchnikoff, Winogradsky, and Beijerinck.
3. Would microbiology have developed more slowly if Fannie Hesse had not suggested the use of agar? Give your reasoning. What is a pure culture?
4. Why do you think viruses are not included in the five-kingdom system?
5. Why are microorganisms so useful to biologists as experimental models?
6. What do you think were the most important discoveries in the development of microbiology? Why?
7. List all the activities or businesses you can think of in your community that are directly dependent on microbiology.

Additional Reading

Baker, J. J. W., and Allen, G. E. 1968. *Hypothesis, prediction, and implication in biology.* Reading, Mass.: Addison-Wesley.

Brock, T. D. 1961. *Milestones in microbiology.* Englewood Cliffs, N.J.: Prentice-Hall.

Brock, T. D. 1988. *Robert Koch: A life in medicine and bacteriology.* Madison, Wis.: Science Tech Publishers.

Bulloch, W. 1979. *The history of bacteriology.* New York: Dover.

Clark, P. F. 1961. *Pioneer microbiologists of America.* Madison: Univ. of Wisconsin Press.

Collard, P. 1976. *The development of microbiology.* New York: Cambridge Univ. Press.

de Kruif, P. 1937. *Microbe hunters.* New York: Harcourt, Brace.

Dobell, C. 1960. *Antony van Leeuwenhoek and his "little animals."* New York: Dover.

Dubos, R. J. 1950. *Louis Pasteur: Free lance of science.* Boston: Little, Brown.

Gabriel, M. L., and Fogel, S., eds. 1955. *Great experiments in biology.* Englewood Cliffs, N.J.: Prentice-Hall.

Hellemans, A., and Bunch, B. 1988. *The timetables of science.* New York: Simon and Schuster.

Hill, L. 1985. Biology, philosophy, and scientific method. *J. Biol. Educ.* 19(3):227–31.

Hitchens, A. P., and Leikind, M. C. 1939. The introduction of agar-agar into bacteriology. *J. Bacteriol.* 37(5):485–93.

Lechevalier, H. A., and Solotorovsky, M. 1965. *Three centuries of microbiology.* New York: McGraw-Hill.

Lederberg, J., ed. 1992. *Encyclopedia of microbiology.* San Diego, CA: Academic Press.

McNeill, W. H. 1976. *Plagues and peoples.* Garden City, N.Y.: Anchor Press/Doubleday.

Singer, C. 1959. *A history of biology,* 3d ed. New York: Abelard-Schuman.

Singleton, P., and Sainsbury, D. 1988. *Dictionary of microbiology and molecular biology.* New York: John Wiley and Sons.

Stanier, R. Y. 1978. What is microbiology? In *Essays in microbiology,* eds. J. R. Norris and M. H. Richmond, 1/1–1/32. New York: John Wiley and Sons.

Vallery-Radot, R. 1923. *The life of Pasteur.* New York: Doubleday.

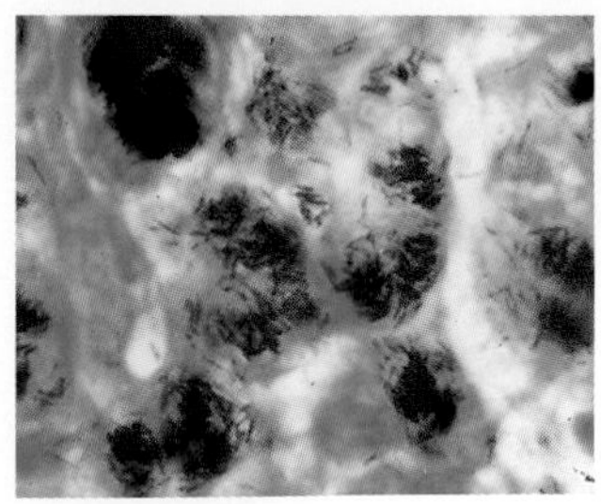

CHAPTER 2
The Study of Microbial Structure: Microscopy and Specimen Preparation

There are more animals living in the scum on the teeth in a man's mouth than there are men in a whole kingdom.

—Antony van Leeuwenhoek

Outline

Concepts

1. Light microscopes use glass lenses to bend and focus light rays and produce enlarged images of small objects. The resolution of a light microscope is determined by the numerical aperture of its lens system and by the wavelength of the light it employs; maximum resolution is about 0.2 μm.
2. The most common types of light microscopes are the bright-field, dark-field, phase-contrast, and fluorescence microscopes. Each yields a distinctive image and may be used to observe different aspects of microbial morphology.
3. Because most microorganisms are colorless and therefore not easily seen in the bright-field microscope, they are usually fixed and stained before observation. Either simple or differential staining can be used to enhance contrast. Specific bacterial structures such as capsules, endospores, and flagella also can be selectively stained.
4. The transmission electron microscope achieves great resolution (about 0.5 nm) by using electron beams of very short wavelength rather than visible light. Although one can prepare microorganisms for observation in other ways, one normally views thin sections of plastic-embedded specimens treated with heavy metals to improve contrast.
5. External features can be observed in great detail with the scanning electron microscope, which generates an image by scanning a fine electron beam over the surface of specimens rather than projecting electrons through them.

Microbiology usually is concerned with organisms so small they cannot be seen distinctly with the unaided eye. Because of the nature of this discipline, the microscope is of crucial importance; much of what is known about microorganisms has been discovered with microscopes. Thus, it is important to understand how the microscope works and the way in which specimens are prepared for examination.

The chapter begins with a detailed treatment of the standard bright-field microscope and then describes other common types of light microscopes. Next, preparation and staining of specimens for examination with the light microscope are discussed. The chapter closes with a description of transmission and scanning electron microscopes, both of which are used extensively in current microbiological research.

Lenses and the Bending of Light

In order to understand how a light microscope operates, one must know something about the way in which lenses bend and focus light to form images. When a ray of light passes from one medium to another, **refraction** occurs; that is, the ray is bent at the interface. The **refractive index** is a measure of how greatly a substance slows the velocity of light, and the direction and magnitude of bending is determined by the refractive indexes of the two media forming the interface. When light passes from air into glass, a medium with a greater refractive index, it is slowed and bent toward the normal, a line perpendicular to the surface (figure 2.1). As light leaves glass and returns to air, a medium with a lower refractive index, it accelerates and is bent away from the normal. Thus, a prism bends light because glass has a different refractive index from air, and the light strikes its surface at an angle.

Lenses act like a collection of prisms operating as a unit. When the light source is distant so that parallel rays of light strike the lens, a convex lens will focus these rays at a specific point, the **focal point** (*F* in figure 2.2). The distance between the center of the lens and the focal point is called the **focal length** (*f* in figure 2.2).

Our eyes cannot focus on objects nearer than about 25 cm or 10 inches (table 2.1). This limitation may be overcome by using a convex lens as a simple magnifier (or microscope) and holding it close to an object. A magnifying glass provides a clear image at much closer range, and the object appears larger. Lens strength is related to focal length; a lens with a short focal length will magnify an object more than a weaker lens having a longer focal length.

1. Define refraction, refractive index, focal point, and focal length.
2. Describe the path of a light ray through a prism or lens.
3. How is lens strength related to focal length?

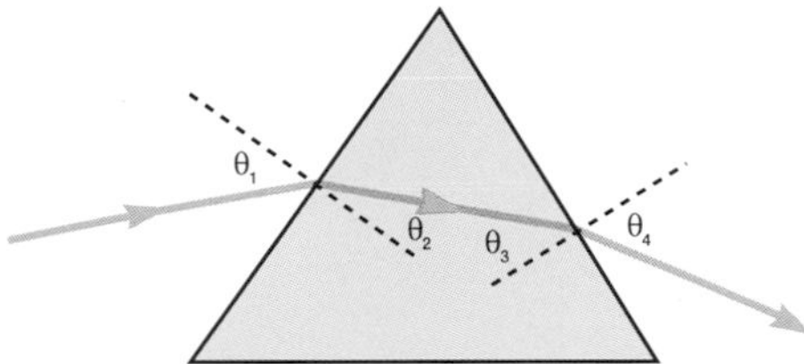

Figure 2.1 **The Bending of Light by a Prism.** Normals, lines perpendicular to the surface of the prism, are indicated by dashed lines. As light enters the glass, it is bent toward the first normal (angle θ_2 is less than θ_1). When light leaves the glass and returns to air, it is bent away from the second normal (θ_4 is greater than θ_3). As a result, the prism bends light passing through it.

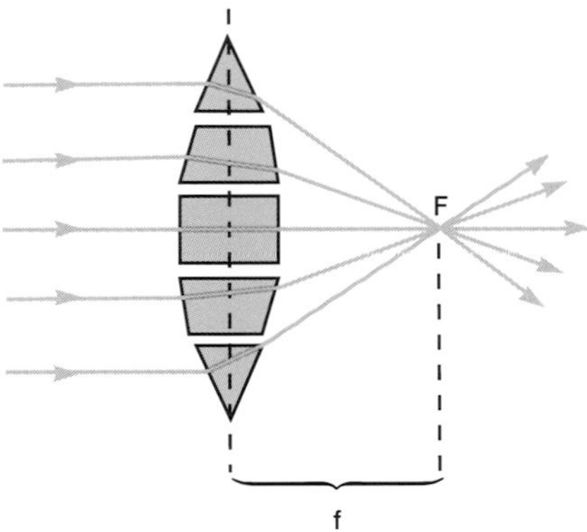

Figure 2.2 **Lens Function.** A lens functions somewhat like a collection of prisms. Light rays from a distant source are focused at the focal point *F*. The focal point lies a distance *f*, the focal length, from the lens center.

TABLE 2.1 Common Units of Measurement

Unit	Abbreviation	Value
1 centimeter	cm	10^{-2} meter or 0.394 inches
1 millimeter	mm	10^{-3} meter
1 micrometer	μm	10^{-6} meter
1 nanometer	nm	10^{-9} meter
1 Angstrom	Å	10^{-10} meter
1 picometer	pm	10^{-12} meter

The Light Microscope

Microbiologists currently employ a variety of light microscopes in their work; bright-field, dark-field, phase-contrast, and fluorescence microscopes are most commonly used. Modern microscopes are all compound microscopes. That is, the magnified image formed by the objective lens is further enlarged by one or more additional lenses.

The Bright-Field Microscope

The ordinary microscope is called a **bright-field microscope** because it forms a dark image against a brighter background. The microscope consists of a sturdy metal body or stand composed of a base and an arm to which the remaining parts are

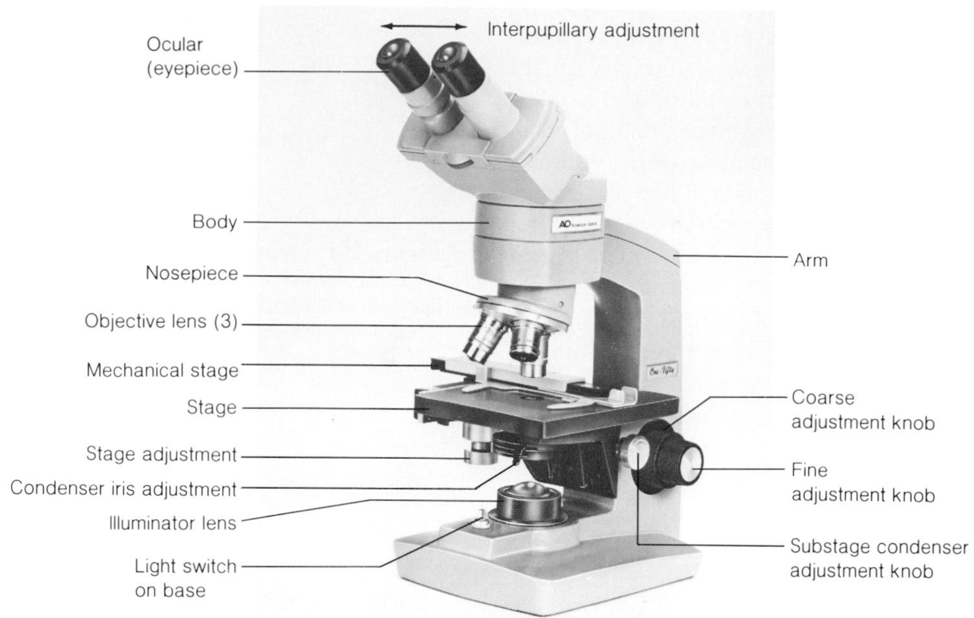

Figure 2.3 **A Bright-field Microscope.** The parts of a modern bright-field microscope. The microscope pictured is somewhat more sophisticated than those found in many student laboratories. For example, it is binocular (has two eyepieces) and has a mechanical stage, an adjustable substage condenser, and a built-in illuminator.

attached (figure 2.3). A light source, either a mirror or an electric illuminator, is located in the base. Two focusing knobs, the **fine** and **coarse adjustment knobs,** are located on the arm and can move either the stage or the nosepiece to focus the image.

The **stage** is positioned about halfway up the arm, and holds microscope slides by either simple slide clips or a mechanical stage clip. A mechanical stage allows the operator to move a slide around smoothly during viewing by use of stage control knobs. The **substage condenser** is mounted within or beneath the stage and focuses a cone of light on the slide. Its position often is fixed in simpler microscopes, but can be adjusted vertically in more advanced models.

The curved upper part of the arm holds the body assembly, to which a **nosepiece** and one or more **eyepieces** or **oculars** are attached. More advanced microscopes have eyepieces for both eyes and are called binocular microscopes. The body assembly itself contains a series of mirrors and prisms so that the barrel holding the eyepiece may be tilted for ease in viewing (figure 2.4). The nosepiece holds three to five **objectives** with lenses of differing magnifying power and can be rotated to position any objective beneath the body assembly. Ideally, a microscope should be **parfocal;** that is, the image should remain in focus when objectives are changed.

The path of light through a bright-field microscope is shown in figure 2.4. The objective lens forms an enlarged real image within the microscope, and the eyepiece lens further magnifies this primary image. When one looks into a microscope, the enlarged specimen image, called the virtual image, appears to lie just beyond the stage at about 25 cm away. The total magnification is calculated by multiplying the objective

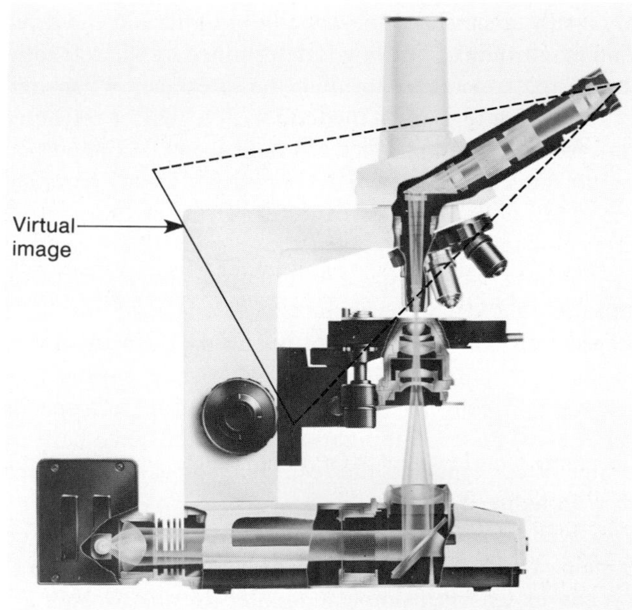

Figure 2.4 **A Microscope's Light Path.** The light path in an advanced bright-field microscope and the location of the virtual image are shown.

and eyepiece magnifications together. For example, if a 45× objective is used with a 10× eyepiece, the overall magnification of the specimen will be 450×.

Microscope Resolution

The most important part of the microscope is the objective, which must produce a clear image, not just a magnified one.

TABLE 2.2 The Properties of Microscope Objectives

Property	Scanning	Low Power	High Power	Oil Immersion
	Objective			
Magnification	4×	10×	40–45×	90–100×
Numerical aperture	0.10	0.25	0.55–0.65	1.25–1.4
Approximate focal length (*f*)	40 mm	16 mm	4 mm	1.8–2.0 mm
Working distance	17–20 mm	4–8 mm	0.5–0.7 mm	0.1 mm
Approximate resolving power with light of 450 nm (blue light)	2.3 μm	0.9 μm	0.35 μm	0.18 μm

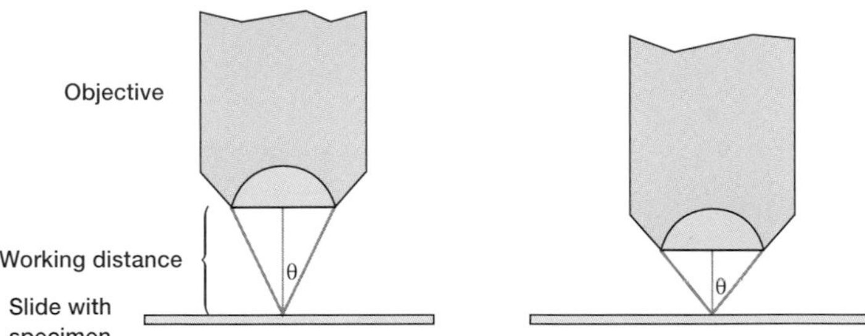

Figure 2.5 Numerical Aperture in Microscopy. The angular aperture θ is ½ the angle of the cone of light that enters a lens from a specimen, and the numerical aperture is $n \sin \theta$. In the right-hand illustration, the lens has larger angular and numerical apertures; its resolution is greater and its working distance smaller.

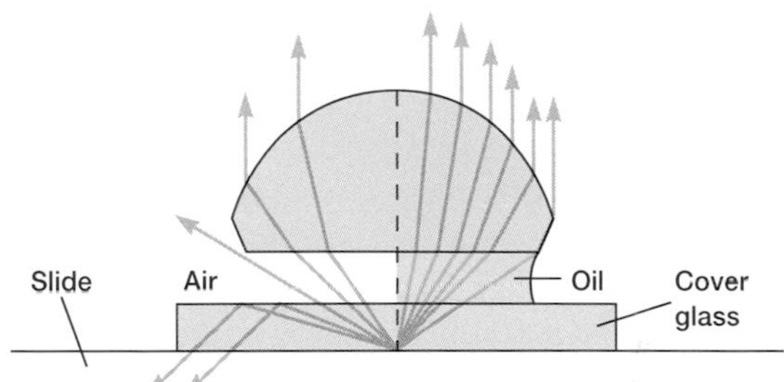

Figure 2.6 The Oil Immersion Objective. An oil immersion objective operating in air and with immersion oil.

Thus, resolution is extremely important. **Resolution** is the ability of a lens to separate or distinguish between small objects that are close together. The minimum distance (d) between two objects that reveals them as separate entities is given by the following equation, in which lambda (λ) is the wavelength of light used to illuminate the specimen and $n \sin \theta$ is the numerical aperture.

$$d = \frac{0.5\lambda}{n \sin \theta}$$

As d becomes smaller, the resolution increases, and finer detail can be discerned in a specimen.

The preceding equation indicates that a major factor in resolution is the wavelength of light used. The wavelength must be shorter than the distance between two objects or they will not be seen clearly. Thus, the greatest resolution is obtained with light of the shortest wavelength, light at the blue end of the visible spectrum (in the range of 450 to 500 nm).

The electromagnetic spectrum of radiation (p. 129).

The **numerical aperture** ($n \sin \theta$) is more difficult to understand. Theta is defined as 1/2 the angle of the cone of light entering an objective (figure 2.5). Light that strikes the microorganism after passing through a condenser is cone-shaped. When this cone has a narrow angle and tapers to a sharp point, it does not spread out much after leaving the slide and therefore does not adequately separate images of closely packed objects. The resolution is low. If the cone of light has a very wide angle and spreads out rapidly after passing through a specimen, closely packed objects appear widely separated and are resolved. The angle of the cone of light that can enter a lens depends on the refractive index (n) of the medium in which the lens works, as well as upon the objective itself. The refractive index for air is 1.00. Since $\sin \theta$ cannot be greater than one (the maximum θ is 90° and sin 90° is 1.00), no lens working in air can have a numerical aperture greater than 1.00. The only practical way to raise the numerical aperture above 1.00, and therefore achieve higher resolution, is to increase the refractive index with **immersion oil,** a colorless liquid with the same refractive index as glass (table 2.2). If air is replaced with immersion oil, many light rays that did not enter the objective due to reflection and refraction at the surfaces of the objective lens and slide will now do so (figure 2.6). An increase in numerical aperture and resolution results. For optimum performance one also must place oil between the slide and condenser, but this is not done routinely.

The limits set on the resolution of a light microscope can be calculated using the preceding equation. The maximum theoretical resolving power of a microscope with an oil immersion objective (numerical aperture of 1.25) and blue-green light is approximately 0.2 μm.

$$d = \frac{(0.5)(530\ \text{nm})}{1.25} = 212\ \text{nm or } 0.2\ \mu\text{m}$$

At best, a bright-field microscope can distinguish between two dots around 0.2 μm apart (the same size as a very small bacterium).

Normally a microscope is equipped with three or four objectives ranging in magnifying power from 4× to 100× (table 2.2). The **working distance** of an objective is the distance between the front surface of the lens and the surface of the cover glass (if one is used) or the specimen when it is in sharp focus. Objectives with large numerical apertures and great resolving power have short working distances.

The largest useful magnification increases the size of the smallest resolvable object enough to be visible. Our eye can just detect a speck 0.2 mm in diameter, and consequently, the useful limit of magnification is about 1,000 times the numerical aperture of the objective lens. Most standard microscopes come with 10× eyepieces and have an upper limit of about 1,000× with oil immersion. A 15× eyepiece may be used with good objectives to achieve a useful magnification of 1,500×. Any further magnification increase does not enable a person to see more detail. A light microscope can be built to yield a final magnification of 10,000×, but it would simply be magnifying a blur. Only the electron microscope provides sufficient resolution to make higher magnifications useful.

Proper specimen illumination also is extremely important in determining resolution. A microscope equipped with a concave mirror between the light source and the specimen illuminates the slide with a fairly narrow cone of light and has a small numerical aperture. Resolution can be improved with a substage condenser, a large light-gathering lens used to project a wide cone of light through the slide and into the objective lens, thus increasing the numerical aperture.

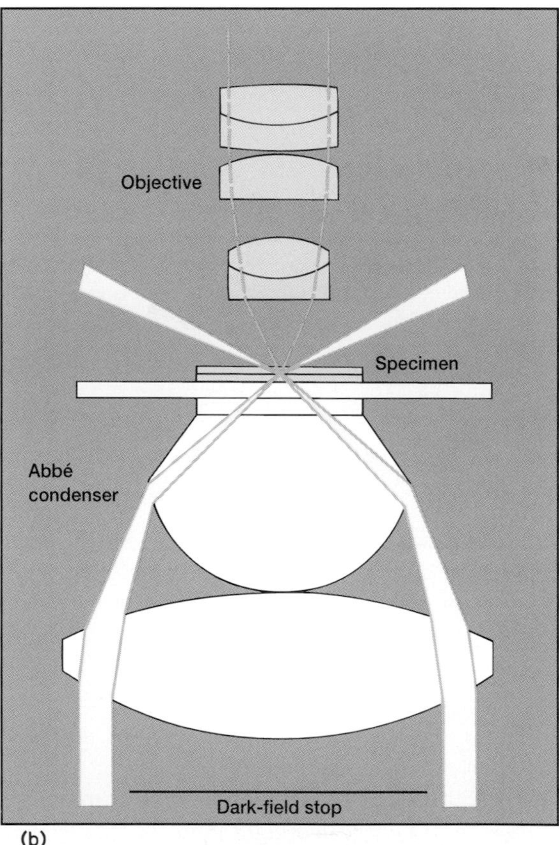

Figure 2.7 Dark-field Microscopy. The simplest way to convert a microscope to dark-field microscopy is to place (*a*) a dark-field stop underneath (*b*) the condenser lens system. The condenser then produces a hollow cone of light so that the only light entering the objective comes from the specimen.

The Dark-Field Microscope

Living, unstained cells and organisms can be observed by simply changing the way in which they are illuminated. A hollow cone of light is focused on the specimen in such a way that unreflected and unrefracted rays do not enter the objective. Only light that has been reflected or refracted by the specimen forms an image (figure 2.7). The field surrounding a specimen appears black, while the object itself is brightly illuminated (figure 2.8*a,b*); because the background is dark, this type of microscopy is called **dark-field microscopy.** Considerable internal structure is often visible in larger eucaryotic microorganisms (figure 2.8*b*). The dark-field microscope is used to identify bacteria like the thin and distinctively shaped *Treponema pallidum* (figure 2.8*a*), the causative agent of syphilis.

The Phase-Contrast Microscope

Unpigmented living cells are not clearly visible in the bright-field microscope because there is little difference in contrast between the cells and water. Thus, microorganisms must often be fixed and stained before observation to increase contrast and create variations in color between cell structures. A **phase-contrast microscope** converts slight differences in refractive index and cell density into easily detected variations in light intensity, and is an excellent way to observe living cells (figure 2.8*c–e*).

The condenser of a phase-contrast microscope has an annular stop, an opaque disk with a thin transparent ring, which produces a hollow cone of light (figure 2.9). As this cone passes

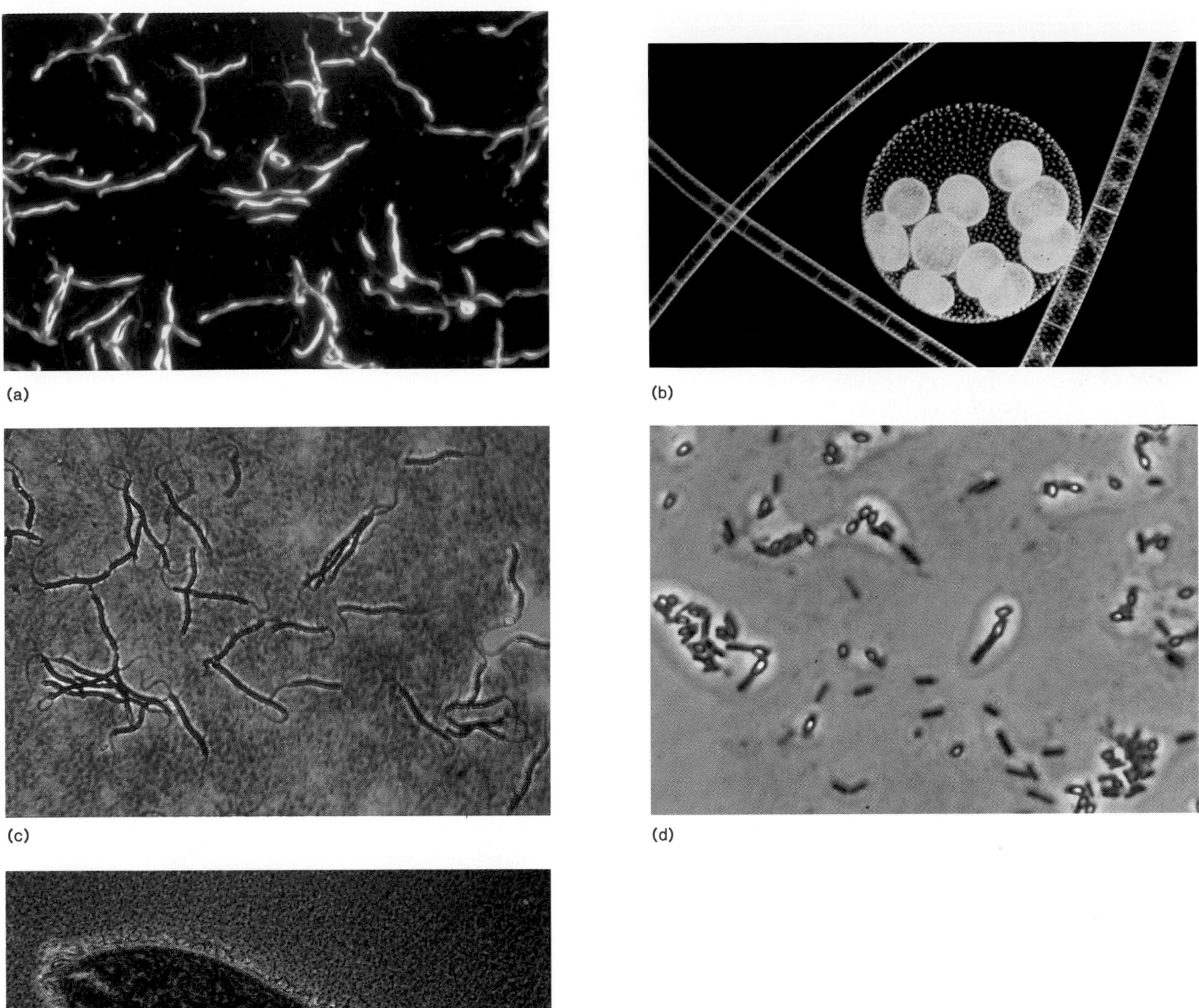

Figure 2.8 **Examples of Dark-field and Phase-contrast Microscopy.** (*a*) *Treponema pallidum,* the spirochete that causes syphilis; dark-field microscopy (×500). (*b*) *Volvox* and *Spirogyra;* dark-field microscopy (×175). Note daughter colonies within the mature *Volvox* colony (center) and the spiral chloroplasts of *Spirogyra* (left and right). (*c*) *Spirillum volutans,* a very large bacterium with flagellar bundles; phase-contrast microscopy (×210). (*d*) *Clostridium botulinum,* the bacterium responsible for botulism, with subterminal oval endospores; phase-contrast microscopy (×600). (*e*) *Paramecium* stained to show a large central macronucleus with a small spherical micronucleus at its side; phase-contrast microscopy (×100).

through a cell, some light rays are bent due to variations in density and refractive index within the specimen and are retarded by about 1/4 wavelength. The deviated light is focused to form an image of the object. Undeviated light rays strike a phase ring in the phase plate, a special optical disk located in the objective, while the deviated rays miss the ring and pass through the rest of the plate. If the phase ring is constructed in such a way that the undeviated light passing through it is advanced by 1/4 wavelength, the deviated and undeviated waves will be about 1/2 wavelength out of phase and will cancel each other when they come together to form an image (figure 2.10). The background, formed by undeviated light, is bright, while the unstained object appears dark and contrasty. This type of microscopy is called **dark-phase-contrast microscopy.** Color filters often are used to improve the image (figure 2.8*c,d*).

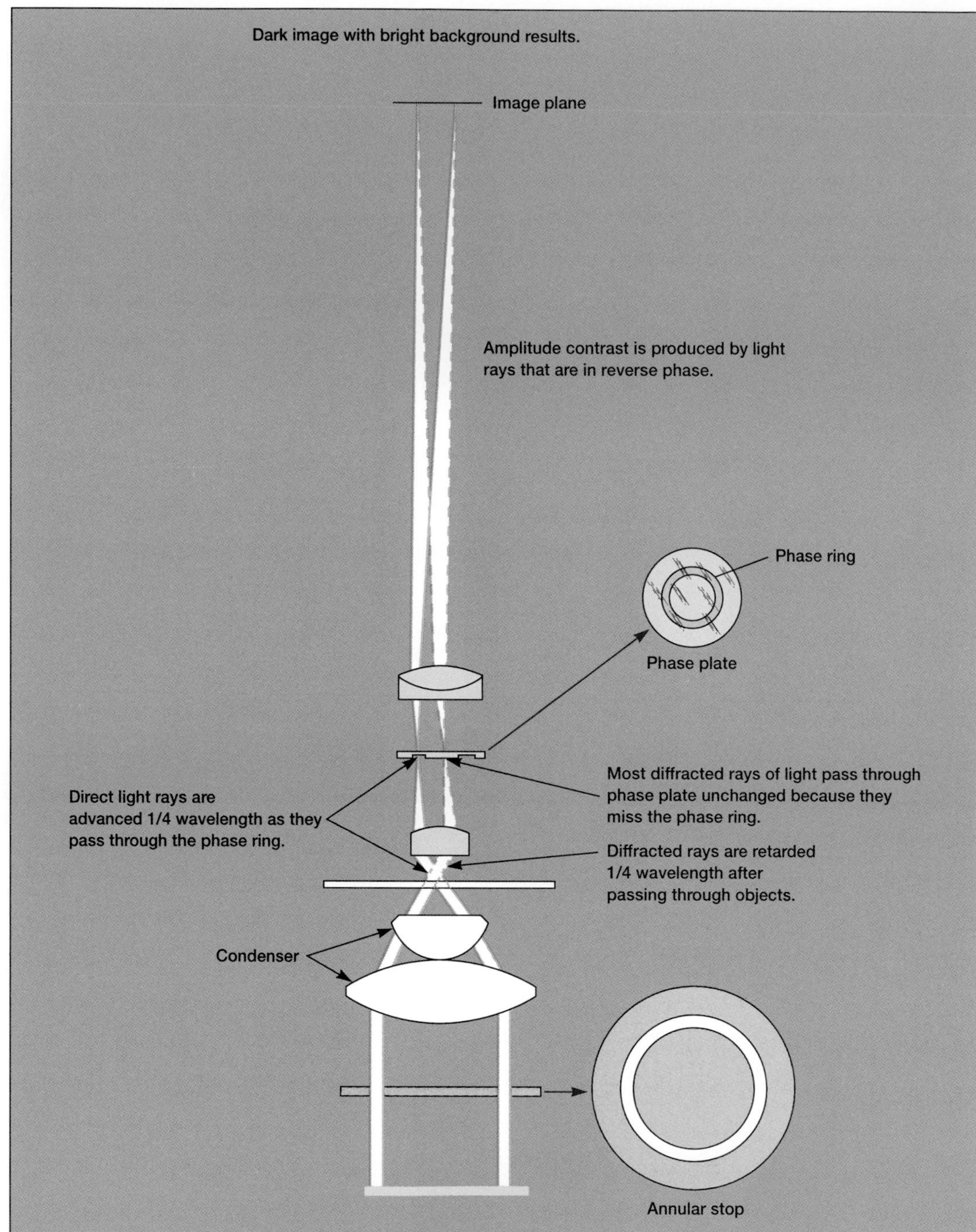

Figure 2.9 Phase-contrast Microscopy. The optics of a dark-phase-contrast microscope.

Phase-contrast microscopy is especially useful for the detection of bacterial components such as endospores and inclusion bodies containing poly-β-hydroxybutyrate, polymetaphosphate, sulfur, or other substances (*see chapter 3*). These are clearly visible (figure 2.8*d*) because they have refractive indexes markedly different from that of water. Phase-contrast microscopes also are widely used in studying eucaryotic cells.

The Fluorescence Microscope

The microscopes thus far considered produce an image from light that passes through a specimen. An object also can be seen because it actually emits light, and this is the basis of fluorescence microscopy. When some molecules absorb radiant energy, they become excited and later release much of

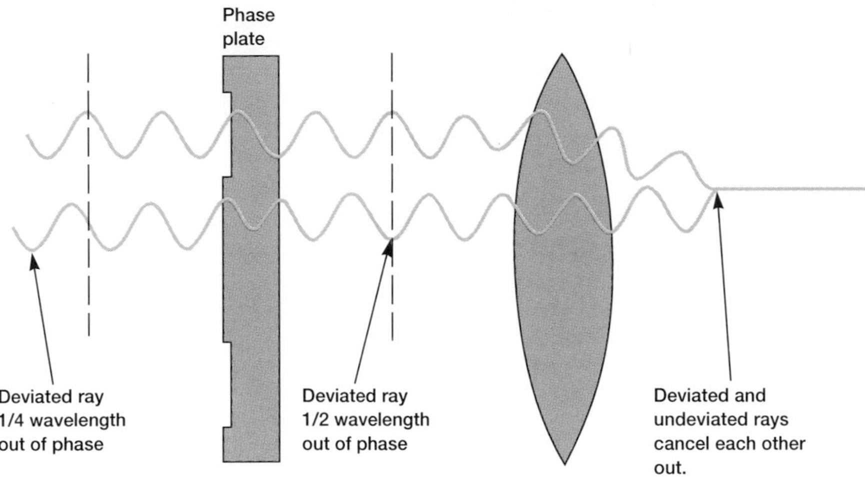

Figure 2.10 **The Production of Contrast in Phase Microscopy.** The behavior of deviated and undeviated or undiffracted light rays in the dark-phase-contrast microscope. Because the light rays tend to cancel each other out, the image of the specimen will be dark against a brighter background.

their trapped energy as light. Any light emitted by an cxcited molecule will have a longer wavelength (or be of lower energy) than the radiation originally absorbed. **Fluorescent light** is emitted very quickly by the excited molecule as it gives up its trapped energy and returns to a more stable state.

The **fluorescence microscope** (figure 2.11) exposes a specimen to ultraviolet, violet, or blue light and forms an image of the object with the resulting fluorescent light. A mercury vapor arc lamp or other source produces an intense beam, and heat transfer is limited by a special infrared filter. The light passes through an exciter filter that transmits only the desired wavelength. A darkfield condenser provides a black background against which the fluorescent objects glow. Usually, the specimens have been stained with dye molecules, called **fluorochromes,** that fluoresce brightly upon exposure to light of a specific wavelength, but some microorganisms are autofluorescing. The microscope forms an image of the fluorochrome-labeled microorganisms from the light emitted when they fluoresce (figure 2.12). A barrier filter positioned after the objective lenses removes any remaining ultraviolet light, which could damage the viewer's eyes, or blue and violet light, which would reduce the image's contrast.

The fluorescence microscope has become an essential tool in medical microbiology and microbial ecology. Bacterial pathogens (e.g., *Mycobacterium tuberculosis,* the cause of tuberculosis) can be identified after staining them with fluorochromes or specifically labeling them with fluorescent antibodies using immunofluorescence procedures. In ecological studies, the fluorescence microscope is used to observe microorganisms after tagging them with the fluorochrome acridine orange. The stained organisms will fluoresce orange or green and can be detected even in the midst of other particulate material. Thus, the microorganisms can be viewed and directly counted in a relatively undisturbed ecological niche.

Immunofluorescence and diagnostic microbiology (pp. 663–65, 678).

1. List the parts of a light microscope and their functions.
2. Define resolution, numerical aperture, working distance, and fluorochrome.
3. How does resolution depend upon the wavelength of light, refractive index, and the numerical aperture? What are the functions of immersion oil and the substage condenser?
4. Briefly describe how dark-field, phase-contrast, and fluorescence microscopes work, and the kind of image provided by each. Give a specific use for each type.

Preparation and Staining of Specimens

Although living microorganisms can be directly examined with the light microscope, they often must be fixed and stained to increase visibility, accentuate specific morphological features, and preserve them for future study.

Fixation

The stained cells seen in a microscope should resemble living cells as closely as possible. **Fixation** is the process by which the internal and external structures of cells and microorganisms are preserved and fixed in position. It inactivates enzymes that might disrupt cell morphology and toughens cell structures so that they do not change during staining and observation. A microorganism usually is killed and attached firmly to the microscope slide during fixation.

There are two fundamentally different types of fixation. (1) Bacteriologists heat-fix bacterial smears by flame heating an air-dried film of bacteria gently. This adequately preserves overall morphology, but not structures within cells. (2) Chemical fixation must be used to protect fine cellular substructure and the morphology of larger, more delicate microorganisms. Chemical fixatives penetrate cells and react with

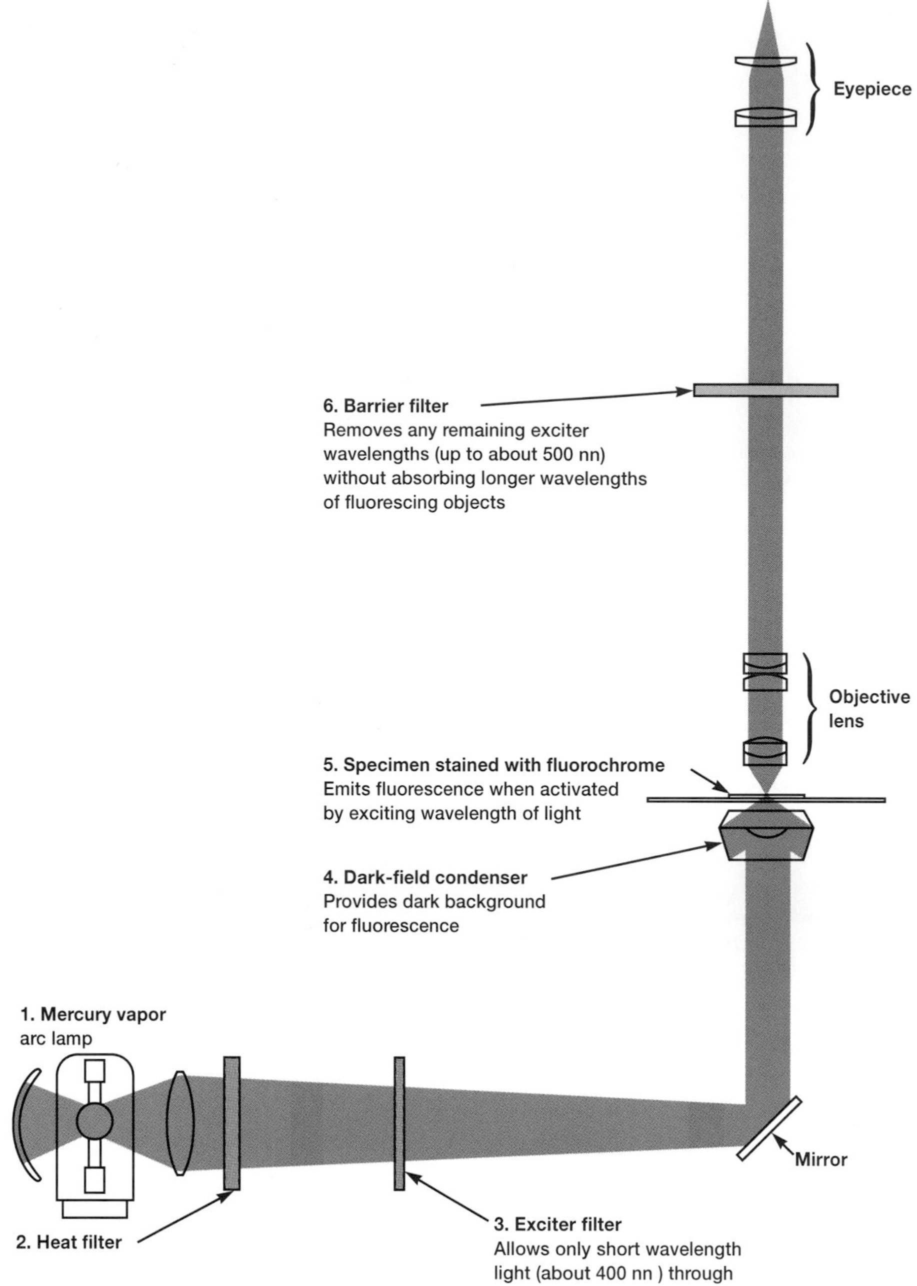

Figure 2.11 **Fluorescence Microscopy.** The principles of operation of a fluorescence microscope.

cellular components, usually proteins and lipids, to render them inactive, insoluble, and immobile. Common fixative mixtures contain such components as ethanol, acetic acid, mercuric chloride, formaldehyde, and glutaraldehyde.

Dyes and Simple Staining

The many types of dyes used to stain microorganisms have two features in common. (1) They have **chromophore groups,** groups with double bonds that give the dye its color. (2) They can bind with cells by ionic, covalent, or hydrophobic bonding. For example, a positively charged dye binds to negatively charged structures on the cell.

Ionizable dyes may be divided into two general classes based on the nature of their charged group.

1. **Basic dyes**—methylene blue, basic fuchsin, crystal violet, safranin, malachite green—are cationic or have positively charged groups (usually some form of pentavalent nitrogen) and are generally sold as chloride salts. Basic dyes bind to negatively charged molecules like nucleic acids and many proteins. Because the surfaces of bacterial cells also are negatively charged, basic dyes are most often used in bacteriology.
2. **Acid dyes**—eosin, rose bengal, and acid fuchsin—are anionic or possess negatively charged groups such as carboxyls (—COOH) and phenolic hydroxyls (—OH).

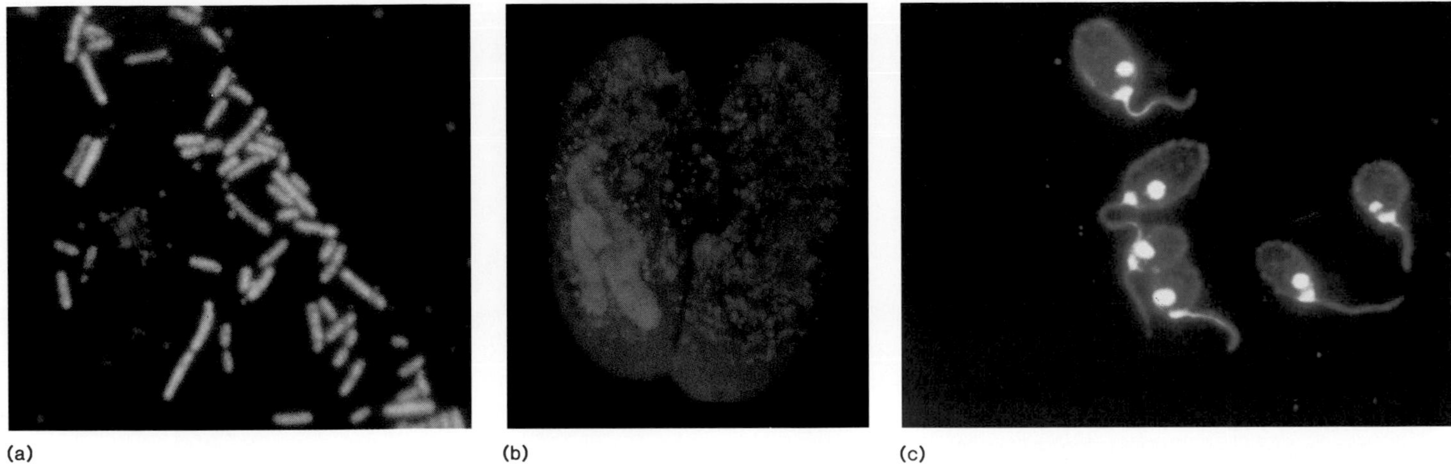

Figure 2.12 **Examples of Fluorescence Microscopy.** (*a*) *Escherichia coli* stained with fluorescent antibodies (×600). The green material is debris. (*b*) *Paramecium tetraurelia* conjugating, acridine-orange fluorescence (× 125). (*c*) The flagellate protozoan *Crithidia luciliae* stained with fluorescent antibodies to show the kinetoplast (×1,000).

Acid dyes, because of their negative charge, bind to positively charged cell structures. The pH may alter staining effectiveness since the nature and degree of the charge on cell components change with pH.

Thus, anionic dyes stain best under acidic conditions when proteins and many other molecules carry a positive charge; basic dyes are most effective at higher pHs.

Although ionic interactions are probably the most common means of attachment, dyes also bind through covalent bonds or because of their solubility characteristics. For instance, DNA can be stained by the Feulgen procedure in which Schiff's reagent is covalently attached to its deoxyribose sugars after hydrochloric acid treatment. Sudan III (Sudan Black) selectively stains lipids because it is lipid soluble, but will not dissolve in aqueous portions of the cell.

Microorganisms often can be stained very satisfactorily by **simple staining,** in which a single staining agent is used. Simple staining's value lies in its simplicity and ease of use. One covers the fixed smear with stain for the proper length of time, washes the excess stain off with water, and blots the slide dry. Basic dyes like crystal violet, methylene blue, and carbolfuchsin are frequently used to determine the size, shape, and arrangement of bacteria.

Differential Staining

Differential staining procedures divide bacteria into separate groups based on staining properties. The **Gram stain,** developed in 1884 by the Danish physician Christian Gram, is the most widely employed staining method in bacteriology. It is a differential staining procedure because it divides bacteria into two classes—gram negative and gram positive.

Gram-positive and gram-negative bacteria (pp. 421–22).

In the first step of the Gram-staining procedure (figure 2.13), the smear is stained with the basic dye crystal violet, the primary stain. It is followed by treatment with an iodine solution functioning as a **mordant.** That is, the iodine increases the interaction between the cell and the dye so that the cell is stained more strongly. The smear is next decolorized by washing with ethanol or acetone. This step generates the differential aspect of the Gram stain; gram-positive bacteria retain the crystal violet, whereas gram-negative bacteria lose their crystal violet and become colorless. Finally, the smear is counterstained with a basic dye different in color from crystal violet. Safranin, the most common counterstain, colors gram-negative bacteria pink to red and leaves gram-positive bacteria dark purple (figure 2.14).

Cell wall structure and the mechanism of the Gram stain (p. 55).

Acid-fast staining is another important differential staining procedure. A few species, particularly those in the genus *Mycobacterium* (*see chapter 22*) do not bind simple stains readily and must be stained by a harsher treatment: heating with a mixture of basic fuchsin and phenol (the Ziehl-Neelsen method). Once basic fuchsin has penetrated with the aid of heat and phenol, acid-fast cells are not easily decolorized by an acid-alcohol wash and hence remain red. This is due to the quite high lipid content of acid-fast cell walls; in particular, mycolic acid—a group of branched chain hydroxy lipids—appears responsible for acid-fastness. Non-acid-fast bacteria are decolorized by acid-alcohol and thus are stained blue by methylene blue counterstain. This method is used to identify *Mycobacterium tuberculosis* and *M. leprae* (figure 2.15), the pathogens responsible for tuberculosis and leprosy, respectively.

Staining Specific Structures

Many special staining procedures have been developed over the years to study specific bacterial structures with the light

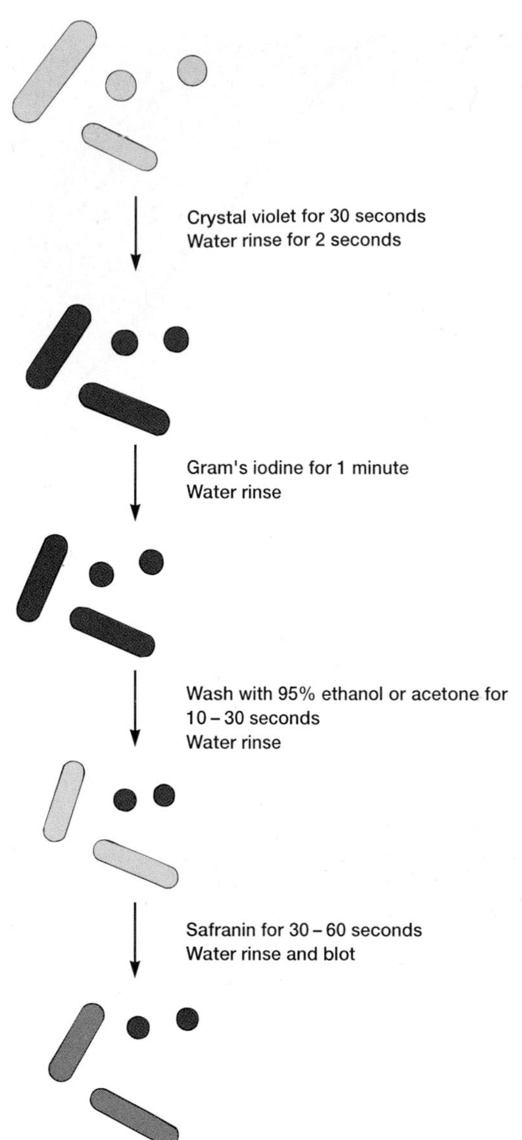
Crystal violet for 30 seconds
Water rinse for 2 seconds
Gram's iodine for 1 minute
Water rinse
Wash with 95% ethanol or acetone for
10 – 30 seconds
Water rinse
Safranin for 30 – 60 seconds
Water rinse and blot

Figure 2.13 **The Gram-staining Procedure.** Note that decolorization with ethanol or acetone removes crystal violet from gram-negative cells, but not from the gram-positive cells. The gram-negative cells then turn pink to red when counterstained with safranin.

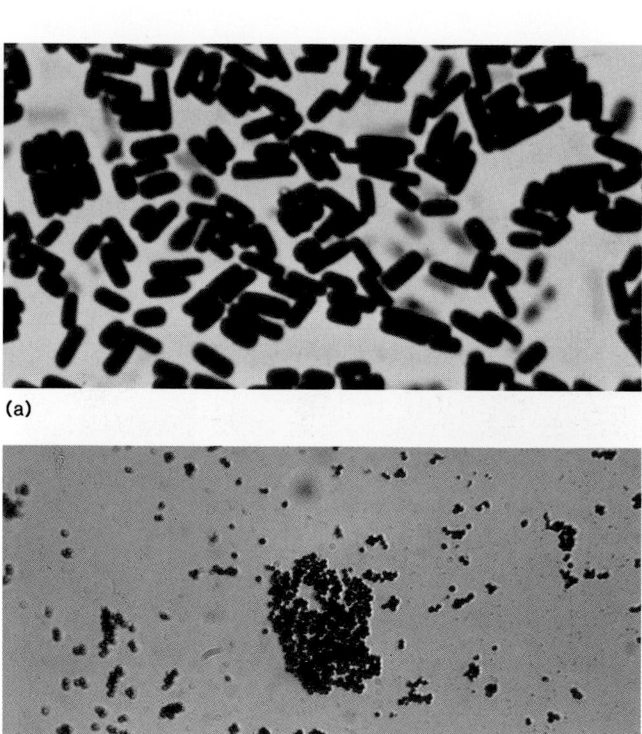
(a)

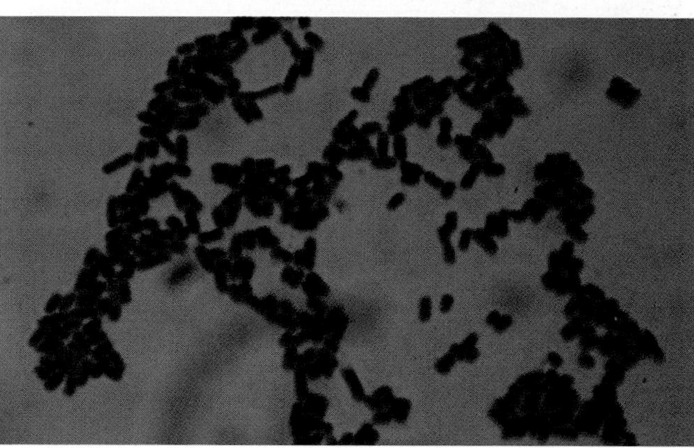
(b)

(c)

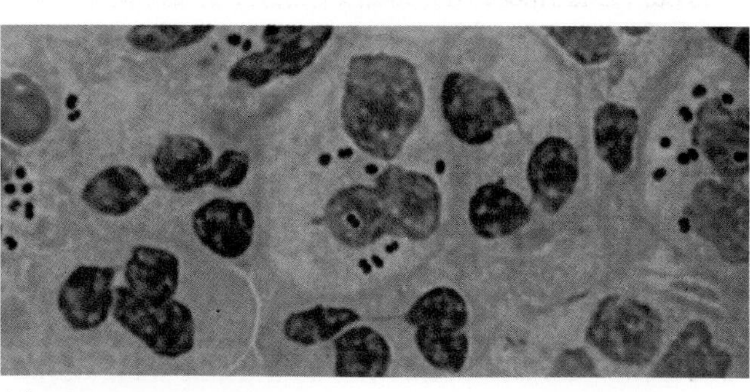
(d)

Figure 2.14 **Examples of Gram Staining.** (*a*) Gram-positive *Clostridium perfringens* (×800). Some rods have stained pink rather than purple, as often happens when gram-positive cells age. (*b*) *Staphylococcus aureus.* Gram stain, bright-field microscopy (×1,000). The gram-positive cocci associate in grapelike clusters. (*c*) *Escherichia coli,* Gram stain (×500). (*d*) *Neisseria gonorrhoeae.* The diplococci are often within white blood cells from pus (×1,000).

microscope. One of the simplest is **negative staining,** a technique that reveals the presence of the diffuse capsules surrounding many bacteria. Bacteria are mixed with India ink or Nigrosin dye and spread out in a thin film on a slide. After air-drying, bacteria appear as lighter bodies in the midst of a blue-black background because ink and dye particles cannot penetrate either the bacterial cell or its capsule. The extent of the light region is determined by the size of the capsule and of the cell itself. There is little distortion of bacterial shape, and the cell can be counterstained for even greater visibility (figure 2.16).

Capsules and slime layers (p. 56).

Bacteria in the genera *Bacillus* and *Clostridium* (*see chapter 22*) form an exceptionally resistant structure capable

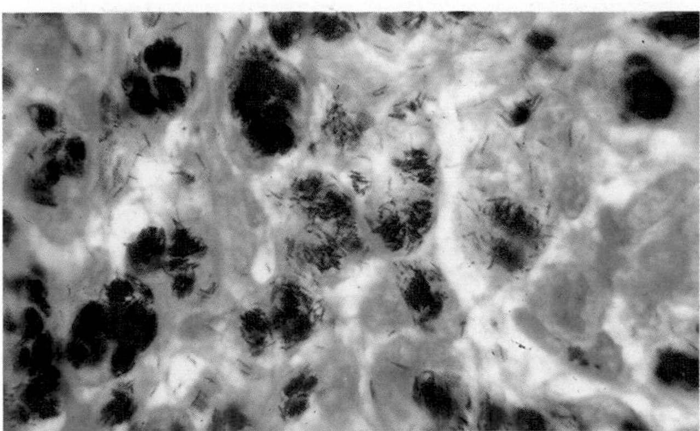

Figure 2.15 **Acid-fast Staining.** *Mycobacterium leprae.* Acid-fast stain (×380). Note the masses of red bacteria within host cells.

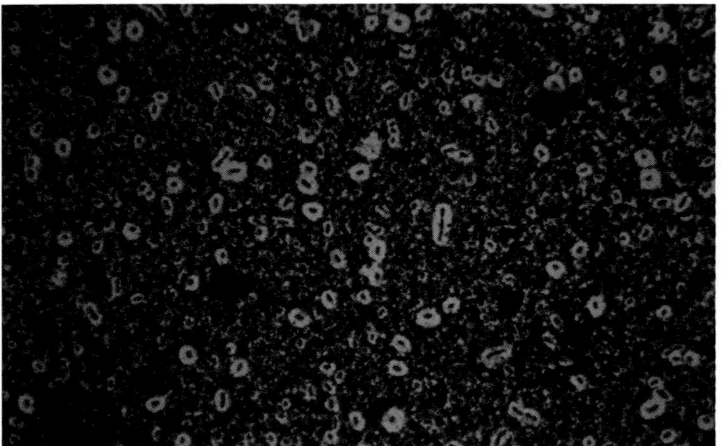

Figure 2.16 **Negative Staining.** *Klebsiella pneumoniae* negatively stained with India ink to show its capsules (×900).

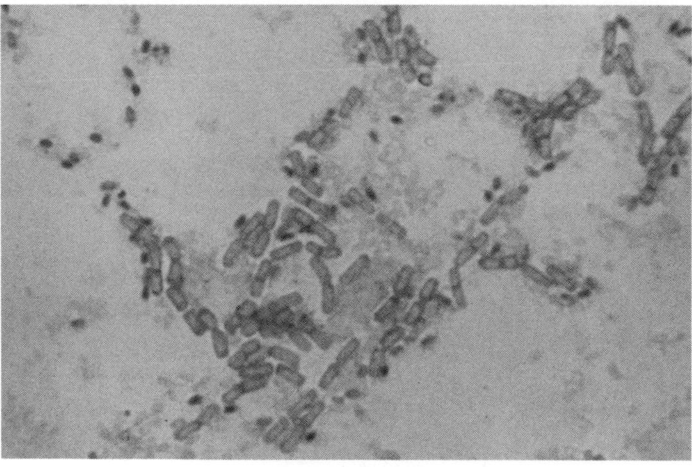

Figure 2.17 **Spore Staining.** A spore stain of *Bacillus subtilis* with elliptical, central spores stained red. Some spores have already been released (×1,000).

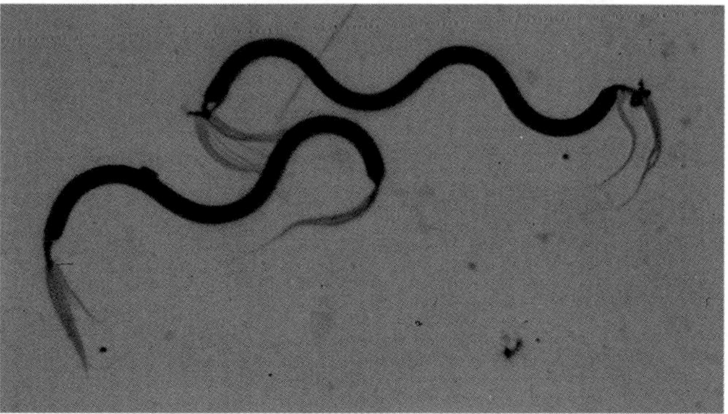

Figure 2.18 **Example of Flagella Staining.** *Spirillum volutans* with bipolar tufts of flagella (×400). *(See figure 3.30.)*

of surviving for long periods in an unfavorable environment. This dormant structure is called an endospore since it develops within the cell. Endospore morphology and location vary with species and often are valuable in identification; endospores may be spherical to elliptical and either smaller or larger than the diameter of the parent bacterium. They can be observed with the phase-contrast microscope or negative staining. Endospores are not stained well by most dyes, but once stained, they strongly resist decolorization. This property is the basis of most **spore staining** methods (figure 2.17). In the Schaeffer-Fulton procedure, endospores are first stained by heating bacteria with malachite green; then the rest of the cell is washed free of dye with water and is counterstained with safranin. This technique yields a green endospore resting in a pink to red cell.

Bacterial endospore structure (pp. 62–65).

Bacterial flagella are fine, threadlike organelles of locomotion that are so slender (about 10 to 30 nm in diameter) they can only be seen directly using the electron microscope. In order to observe them with the light microscope, the thickness of flagella is increased by coating them with mordants like tannic acid and potassium alum, and they are stained with pararosaniline (Leifson method) or basic fuchsin (Gray method). **Flagella staining** procedures provide taxonomically valuable information about the presence and location of flagella (figure 2.18; *also see figure 3.30*).

The bacterial flagellum (pp. 57–60).

---

1. Define fixation, dye, chromophore, basic dye, acid dye, simple staining, differential staining, mordant, negative staining, and acid-fast staining.
2. Describe the Gram-stain procedure and how it works. How would you visualize capsules, endospores, and flagella?

Electron Microscopy

For centuries the light microscope has been the most important instrument for studying microorganisms. More recently, the electron microscope has transformed microbiology and

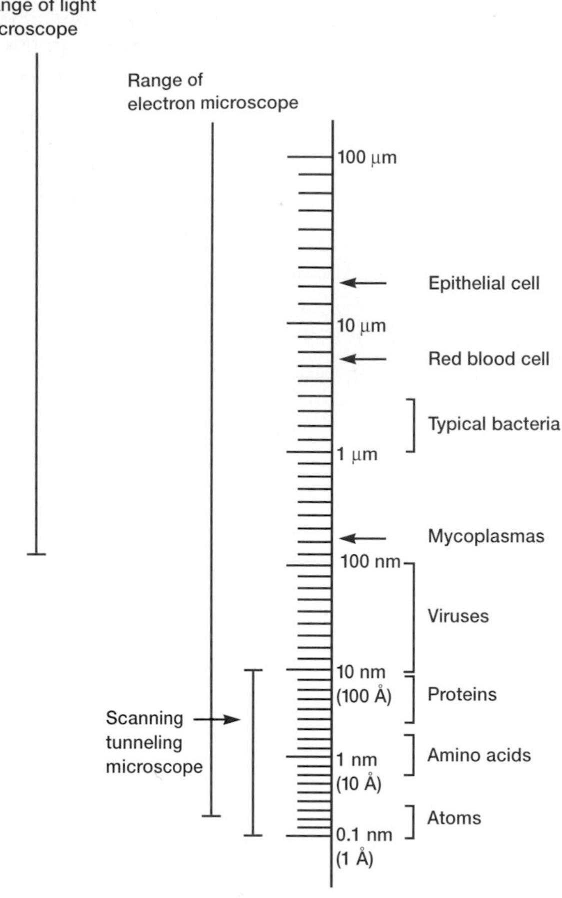

Figure 2.19 **The Limits of Microscopic Resolution.** Dimensions are indicated with a logarithmic scale (each major division representing a tenfold change in size). To the right side of the scale are the approximate sizes of cells, bacteria, viruses, molecules, and atoms.

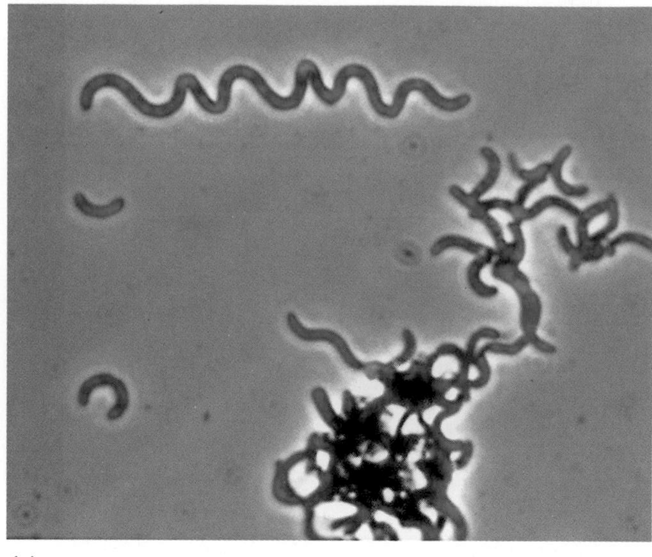

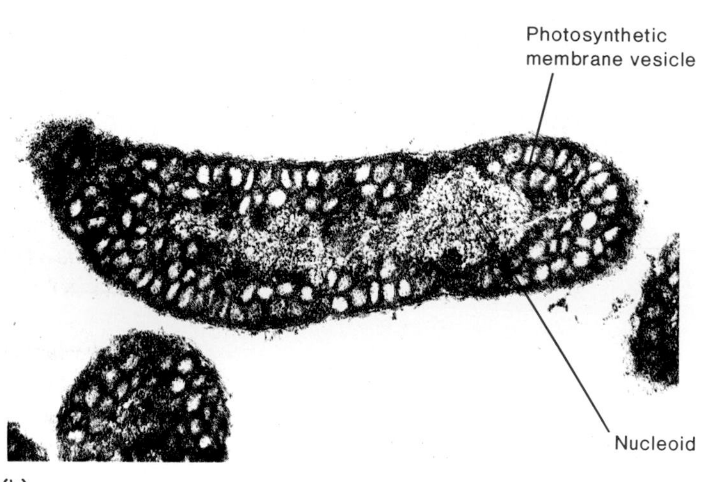

Figure 2.20 **Light and Electron Microscopy.** A comparison of light and electron microscopic resolution. (*a*) *Rhodospirillum rubrum* in phase-contrast light microscope (×600). (*b*) A thin section of *R. rubrum* in transmission electron microscope (×100,000).

added immeasurably to our knowledge. The nature of the electron microscope and the ways in which specimens are prepared for observation are reviewed briefly in this section.

The Transmission Electron Microscope

The very best light microscope has a resolution limit of about 0.2 μm. Because bacteria usually are around 1 μm in diameter, only their general shape and major morphological features are visible in the light microscope. The detailed internal structure of larger microorganisms also cannot be effectively studied by light microscopy. These limitations arise from the nature of visible light waves, not from any inadequacy of the light microscope itself.

Recall that the resolution of a light microscope increases with a decrease in the wavelength of the light it uses for illumination. Electron beams behave like radiation and can be focused much as light is in a light microscope. If electrons illuminate the specimen, the microscope's resolution is enormously increased because the wavelength of the radiation is around 0.005 nm, approximately 100,000 times shorter than that of visible light. The transmission electron microscope has a practical resolution roughly 1,000 times better than the light microscope; with many electron microscopes, points closer than 5 Å or 0.5 nm can be distinguished, and the useful magnification is well over 100,000× (figure 2.19). The value of the electron microscope is evident upon comparison of the photographs in figure 2.20; microbial morphology can now be studied in great detail.

A modern **transmission electron microscope (TEM)** is complex and sophisticated (figure 2.21), but the basic principles behind its operation can be understood readily. A heated tungsten filament in the electron gun generates a beam of electrons that is then focused on the specimen by the condenser (figure 2.22). Since electrons cannot pass through a glass lens,

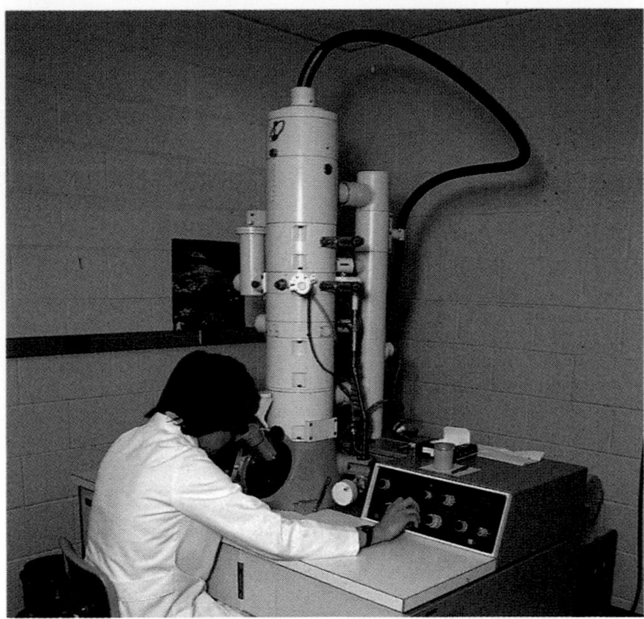

Figure 2.21 **A Modern Transmission Electron Microscope.** The electron gun is at the top of the central column, and the magnetic lenses are within the column. The image on the fluorescent screen may be viewed through a magnifier positioned over the viewing window. The camera is in a compartment below the screen.

doughnut-shaped electromagnets called magnetic lenses are used to focus the beam. The column containing the lenses and specimen must be under high vacuum to obtain a clear image because electrons are deflected by collisions with air molecules. The specimen scatters electrons passing through it, and the beam is focused by magnetic lenses to form an enlarged, visible image of the specimen on a fluorescent screen. A denser region in the specimen scatters more electrons and therefore appears darker in the image since fewer electrons strike that area of the screen. In contrast, electron-transparent regions are brighter. The screen can also be moved aside and the image captured on photographic film as a permanent record.

Specimen Preparation

Table 2.3 compares some of the important features of light and electron microscopes. The distinctive features of the TEM place harsh restrictions on the nature of samples that can be viewed and the means by which those samples must be prepared. Since electrons are quite easily absorbed and scattered by solid matter, only extremely thin slices of a microbial specimen can be viewed in the average TEM. The specimen must be around 20 to 100 nm thick, about 1/50 to 1/10 the diameter of a typical bacterium, and able to maintain its structure when bombarded with electrons under a high vacuum!

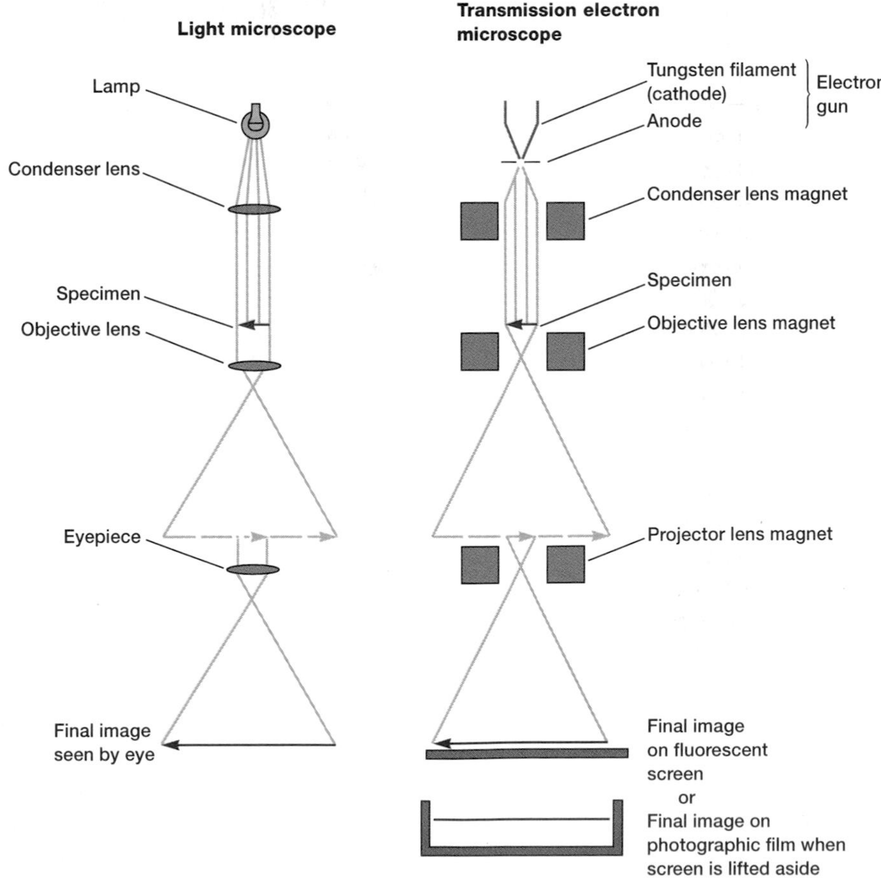

Figure 2.22 **Transmission Electron Microscope Operation.** An overview of TEM operation and a comparison of the operation of light and transmission electron microscopes.

TABLE 2.3 Characteristics of Light and Transmission Electron Microscopes

Feature	Light Microscope	Electron Microscope
Highest practical magnification	About 1,000–1,500	Over 100,000
Best resolution[a]	0.2 μm	0.5 nm
Radiation source	Visible light	Electron beam
Medium of travel	Air	High vacuum
Type of lens	Glass	Electromagnet
Source of contrast	Differential light absorption	Scattering of electrons
Focusing mechanism	Adjust lens position mechanically	Adjust current to the magnetic lens
Method of changing magnification	Switch the objective lens or eyepiece	Adjust current to the magnetic lens
Specimen mount	Glass slide	Metal grid (usually copper)

[a]The resolution limit of a human eye is about 0.2 mm.

Such a thin slice cannot be cut unless the specimen has support of some kind; the necessary support is provided by plastic. After fixation with chemicals like glutaraldehyde or osmium tetroxide to stabilize cell structure, the specimen is dehydrated with organic solvents (e.g., acetone or ethanol). Complete dehydration is essential because most plastics used for embedding are not water soluble. Next the specimen is soaked in unpolymerized, liquid epoxy plastic until it is completely permeated, and then the plastic is hardened to form a solid block. Thin sections are cut from this block with a glass or diamond knife using a special instrument called an ultramicrotome.

Cells usually must be stained before they can be seen clearly in the bright-field microscope; the same thing is true for observations with the TEM. The probability of electron scattering is determined by the density (atomic number) of the specimen atoms. Biological molecules are composed primarily of atoms with low atomic numbers (H, C, N, and O), and electron scattering is fairly constant throughout the unstained cell. Therefore, specimens are prepared for observation by soaking thin sections with solutions of heavy metal salts like lead citrate and uranyl acetate. The lead and uranium ions bind to cell structures and make them more electron opaque, thus increasing contrast in the material. Heavy osmium atoms from the osmium tetroxide fixative also "stain" cells and increase their contrast. The stained thin sections are then mounted on tiny copper grids and viewed.

Although the above procedure for preparing thin sections normally is used to reveal the internal structure of the smallest cell, there are other ways in which microorganisms and smaller objects can be readied for viewing. One very useful technique is negative staining. The specimen is spread out in a thin film with either phosphotungstic acid or uranyl acetate. Just as in negative staining for light microscopy, heavy metals do not penetrate the specimen but render the background dark, whereas the specimen appears bright in photographs. Negative staining is an excellent way to study the structure of viruses, bacterial gas vacuoles, and other similar material. A microorganism can also be viewed after **shadowing** with metal. It is coated with a thin film of platinum or other heavy metal by evaporation at an angle of about 45° from horizontal so that the metal strikes the microorganism on only one side. The

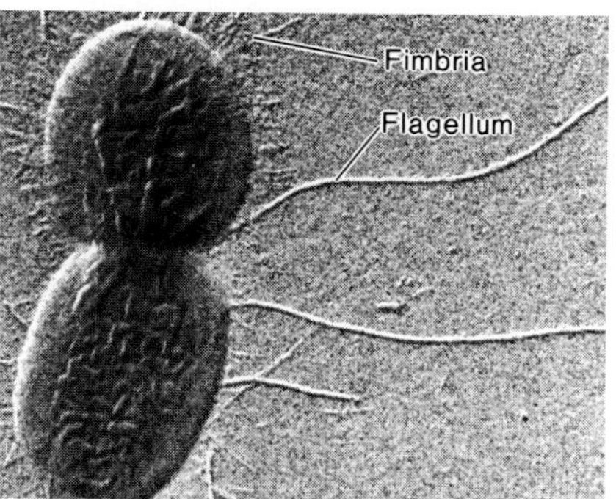

(a)

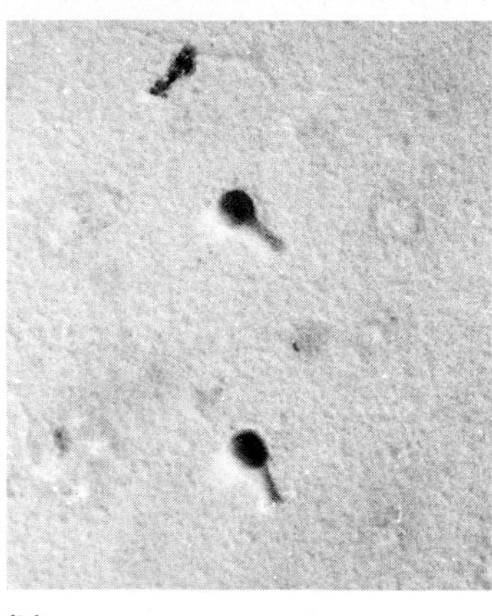

(b)

Figure 2.23 Specimen Shadowing for the TEM. Examples of specimens viewed in the TEM after shadowing with uranium metal. (*a*) *Proteus mirabilis* (×42,750); note flagella and fimbriae. (*b*) T4 coliphage (×72,000).

area coated with metal scatters electrons and appears light in photographs, whereas the uncoated side and the shadow region created by the object is dark (figure 2.23). The specimen looks much as it would if light were shining on it to cast a shadow. This technique is particularly useful in studying virus morphology, bacterial flagella, and plasmids (*see chapter 13*).

The TEM will also disclose the shape of organelles within microorganisms if specimens are prepared by the **freeze-etching**

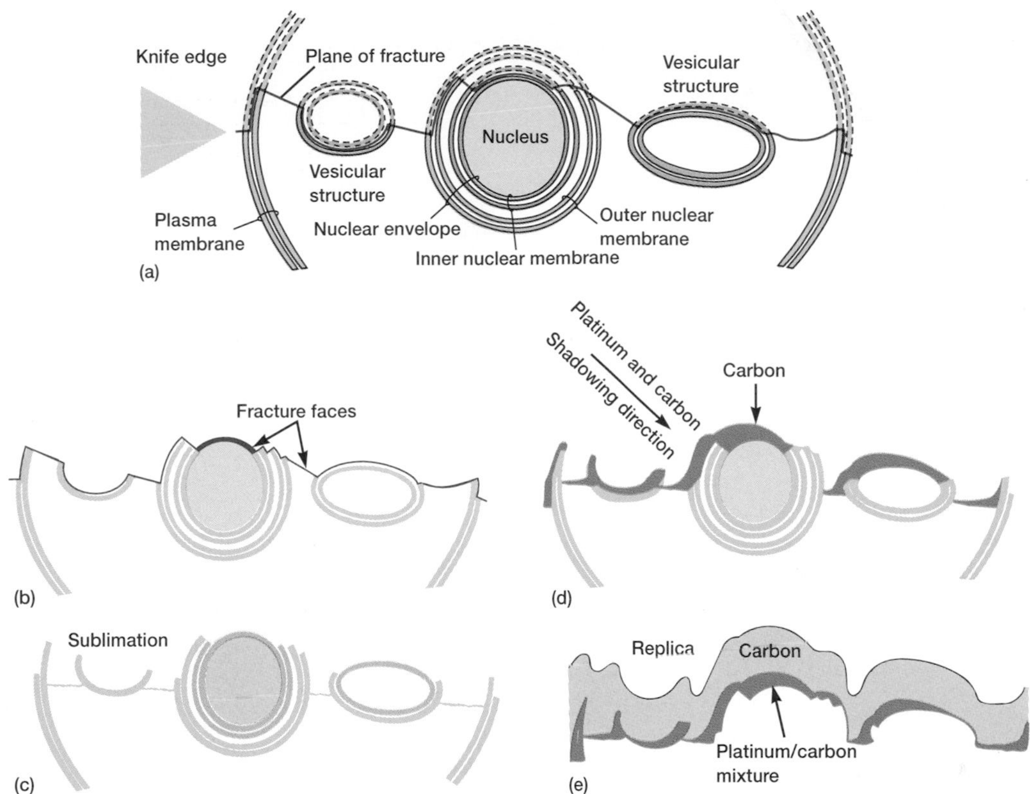

Figure 2.24 The Freeze-etching Technique. In steps *a* and *b*, a frozen eucaryotic cell is fractured with a cold knife. Etching by sublimation is depicted in *c*. Shadowing with platinum plus carbon and replica formation are shown in *d* and *e*. See text for details.

procedure. Cells are rapidly frozen in liquid nitrogen and then warmed to −100° C in a vacuum chamber. Next, a knife that has been precooled with liquid nitrogen (−196° C) fractures the frozen cells that are very brittle and break along lines of greatest weakness, usually down the middle of internal membranes (figure 2.24). The specimen is left in the high vacuum for a minute or more so that some of the ice can sublimate away and uncover more structural detail (sometimes this etching step is eliminated). Finally, the exposed surfaces are shadowed and coated with layers of platinum and carbon to form a replica of the surface. After the specimen has been removed chemically, this replica is studied in the TEM and provides a detailed, three-dimensional view of intracellular structure (figure 2.25). An advantage of freeze-etching is that it minimizes the danger of artifacts because the cells are frozen quickly rather than being subjected to chemical fixation, dehydration, and plastic embedding.

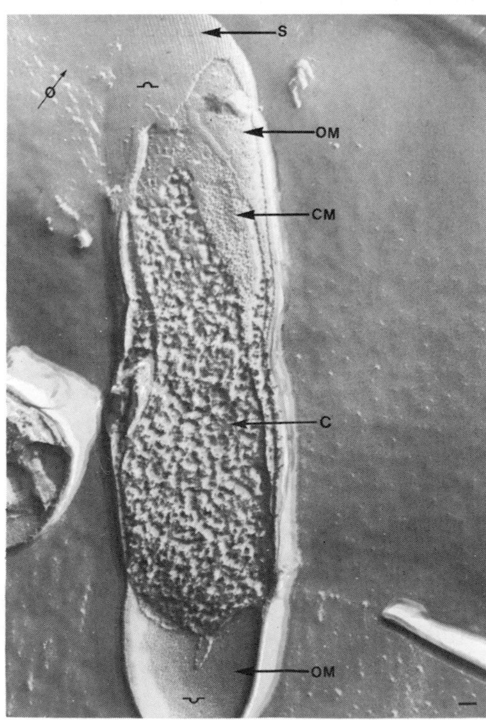

Figure 2.25 Example of Freeze-etching. A freeze-etched preparation of the bacterium *Thiobacillus kabobis*. Note the differences in structure between the outer surface, *S;* the outer membrane of the cell wall, *OM;* the cytoplasmic membrane, *CM;* and the cytoplasm *C*. Bar = 0.1 μm.

The Scanning Electron Microscope

The previously described microscopes form an image from radiation that has passed through a specimen. More recently, the **scanning electron microscope (SEM)** has been used to examine the surfaces of microorganisms in great detail; many instruments have a resolution of 7 nm or less. The SEM differs from other electron microscopes in producing an image from electrons emitted by an object's surface rather than from transmitted electrons. (See also Box 2.1.)

Box 2.1

A Microscope That Sees Atoms

Although light and electron microscopes have become quite sophisticated and reached an advanced state of development, powerful new microscopes are still being created. The **scanning tunneling microscope,** invented in 1980, is an excellent example. It can achieve magnifications of 100 million and allow scientists to view atoms on the surface of a solid. The electrons surrounding surface atoms tunnel or project out from the surface boundary a very short distance. The scanning tunneling microscope has a needlelike probe with a point so sharp that often there is only one atom at its tip. The probe is lowered toward the specimen surface until its electron cloud just touches that of the surface atoms. If a small voltage is applied between the tip and specimen, electrons flow through a narrow channel in the electron clouds. This tunneling current, as it is called, is extraordinarily sensitive to distance and will decrease about a thousandfold if the probe is moved away from the surface by a distance equivalent to the diameter of an atom.

The arrangement of atoms on the specimen surface is determined by moving the probe tip back and forth over the surface while keeping it at a constant height by adjusting the probe distance to maintain a steady tunneling current. As the tip moves up and down while following the surface contours, its motion is recorded and analyzed by a computer to create an accurate three-dimensional image of the surface atoms. The surface map can be displayed on a computer screen or plotted on paper. The resolution is so great that individual atoms are observed easily. The microscope's inventors, Gerd Binnig and Heinrich Rohrer, shared the 1986 Nobel Prize in Physics for their work. Interestingly, the other recipient of the prize was Ernst Ruska, the designer of the first transmission electron microscope.

The scanning tunneling microscope will likely have a major impact in biology. Recently, it has been used to directly view DNA (Box Figure 2.1*b*). Since the microscope can examine objects when they are immersed in water, it may be particularly useful in studying biological molecules.

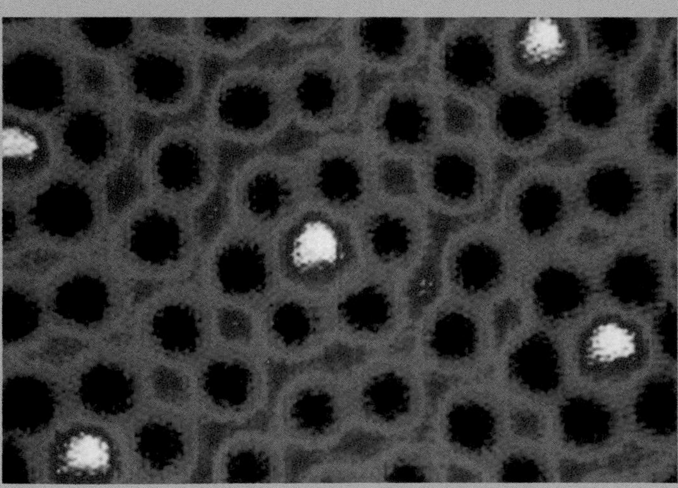

(a)

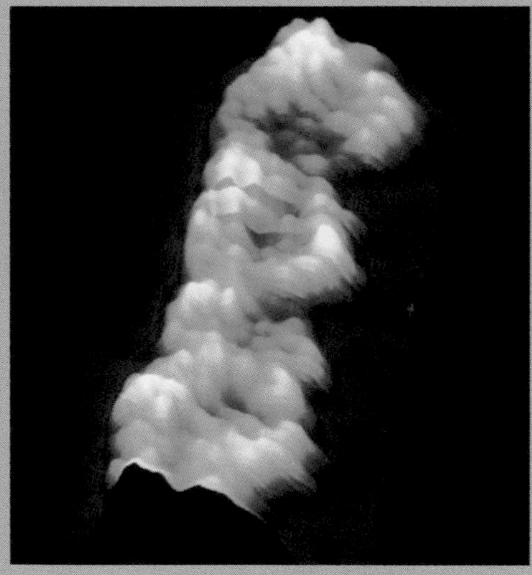

(b)

Box Figure 2.1 Examples of Scanning Tunneling Microscopy. (*a*) Silicon (Si) atoms viewed through a scanning tunneling microscope. (*b*) A scanning tunneling micrograph of the DNA double helix with approximately three turns shown (false color; ×2,000,000).

Specimen preparation is very easy, and, in some cases, air-dried material can be examined directly. Most often, however, microorganisms must first be fixed, dehydrated, and dried to preserve surface structure and prevent collapse of the cells when they are exposed to the SEM's high vacuum. Before viewing, dried samples are mounted and coated with a thin layer of metal to prevent the buildup of an electrical charge on the surface and to give a better image.

The SEM scans a narrow, tapered electron beam back and forth over the specimen (figure 2.26). When the beam strikes a particular area, surface atoms discharge a tiny shower of electrons called secondary electrons, and these are trapped by a special detector. Secondary electrons entering the detector strike a scintillator causing it to emit light flashes that a photomultiplier converts to an electrical current and amplifies. The signal is sent to a cathode-ray tube and produces an image like a television picture, which can be viewed or photographed.

The number of secondary electrons reaching the detector depends upon the nature of the specimen's surface. When the electron beam strikes a raised area, a large number of sec-

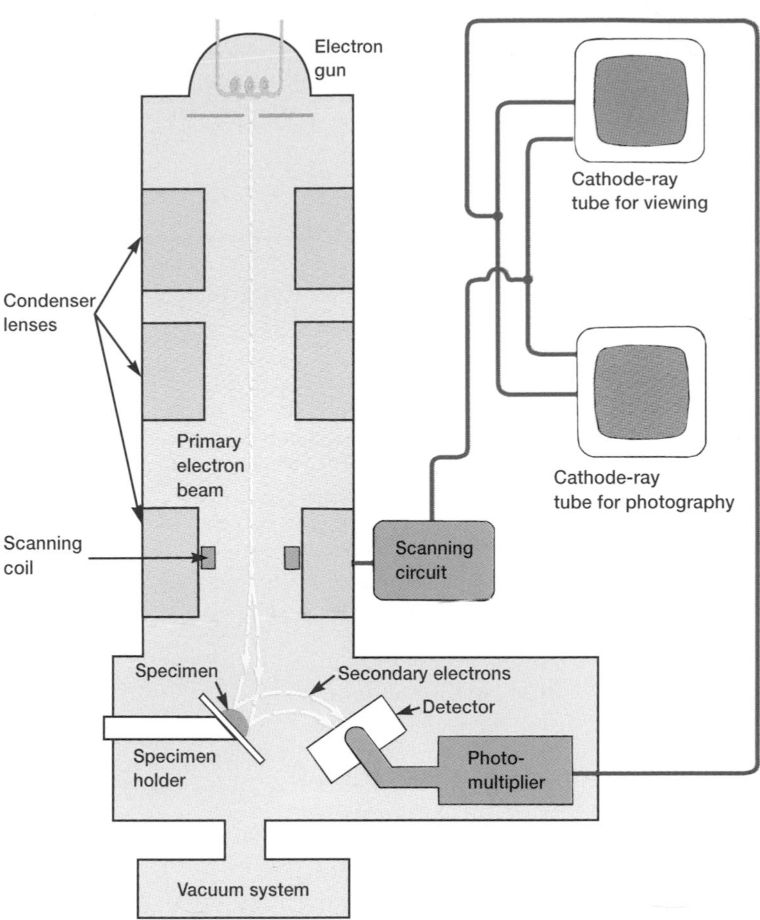

Figure 2.26 **The Scanning Electron Microscope.** *From P. Sheeler and D. E. Bianchi* Cell Biology: Structure, Biochemistry, and Function, *2d ed. Copyright © 1980 John Wiley & Sons, Inc., New York, NY.*

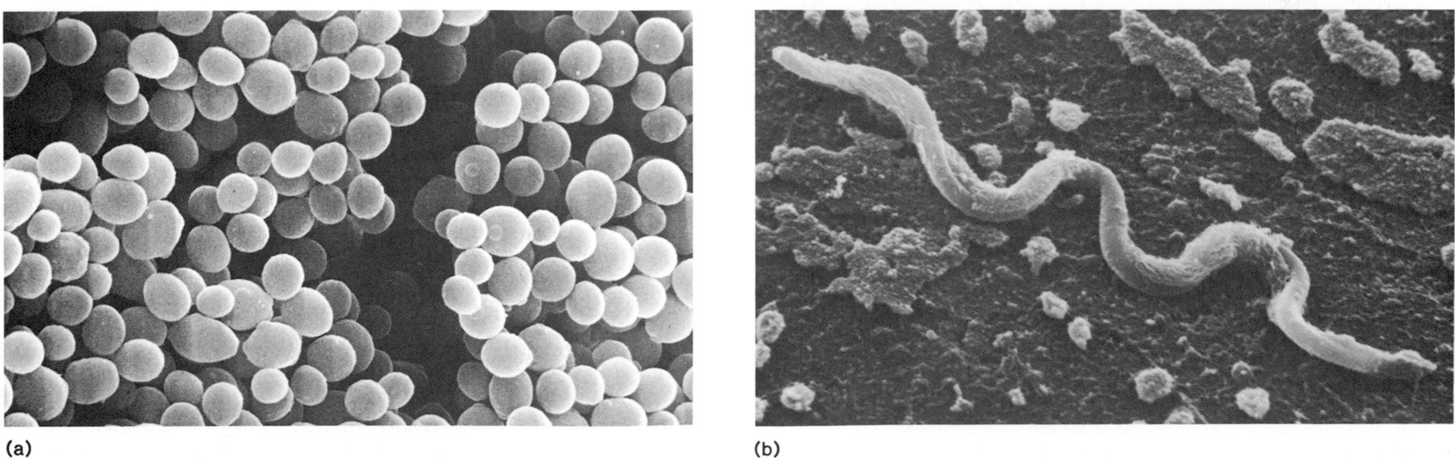

Figure 2.27 **Scanning Electron Micrographs of Bacteria.** (*a*) *Staphylococcus aureus* (×32,000). (*b*) *Cristispira,* a spirochete from the crystalline style of the oyster, *Ostrea virginica.* The axial fibrils or periplasmic flagella are visible around the protoplasmic cylinder (×6,000).

ondary electrons enter the detector; in contrast, fewer electrons escape a depression in the surface and reach the detector. Thus, raised areas appear lighter on the screen and depressions are darker. A realistic three-dimensional image of the microorganism's surface with great depth of focus results (figure 2.27). The actual *in situ* location of microorganisms in ecological niches such as the human skin and the lining of the gut also can be examined.

1. Why does the transmission electron microscope have such greater resolution than the light microscope? Describe in general terms how the TEM functions.
2. Describe how specimens are prepared for viewing in the TEM. How are sections usually stained to increase contrast? What is negative staining, shadowing, and freeze-etching?
3. How does the scanning electron microscope operate and in what way does its function differ from that of the TEM? What aspect of morphology is the SEM used to study?

Summary

1. A light ray moving from air to glass, or vice versa, is bent in a process known as refraction. Lenses focus light rays at a focal point and magnify images.
2. In a compound microscope like the bright-field microscope, the primary image is formed by an objective lens and enlarged by the eyepiece or ocular lens to yield a virtual image.
3. A substage condenser focuses a cone of light on the specimen.
4. Microscope resolution increases as the wavelength of radiation used to illuminate the specimen decreases. The maximum resolution of a light microscope is about 0.2 μm.
5. The dark-field microscope uses only refracted light to form an image, and objects glow against a black background.
6. The phase-contrast microscope converts variations in the refractive index and density of cells into changes in light intensity and thus makes colorless, unstained cells visible.
7. The fluorescence microscope illuminates a fluorochrome-labeled specimen and forms an image from its fluorescence.
8. Specimens usually must be fixed and stained before viewing them in the bright-field microscope.
9. Most dyes are either positively charged basic dyes or negative acid dyes and bind to ionized parts of cells.
10. In simple staining, a single dye mixture is used to stain microorganisms.
11. Differential staining procedures like the Gram stain and acid-fast stain distinguish between microbial groups by staining them differently.
12. Some staining techniques are specific for particular structures like bacterial capsules, flagella, and endospores.
13. The transmission electron microscope uses magnetic lenses to form an image from electrons that have passed through a very thin section of a specimen. Resolution is great because the wavelength of electrons is very short.
14. Thin section contrast can be increased by treatment with solutions of heavy metals like osmium tetroxide, uranium, and lead.
15. Specimens are also prepared for the TEM by negative staining, shadowing with metal, or freeze-etching.
16. The scanning electron microscope is used to study external surface features of microorganisms.

Key Terms

acid dyes *28*
acid-fast staining *29*
basic dyes *28*
bright-field microscope *21*
chromophore groups *28*
dark-field microscopy *24*
dark-phase-contrast microscopy *25*
differential staining procedures *29*
eyepieces *22*
fine and coarse adjustment knobs *22*
fixation *27*
flagella staining *31*
fluorescent light *27*
fluorescence microscope *27*
fluorochromes *27*
focal length *21*
focal point *21*
freeze-etching *34*
Gram stain *29*
immersion oil *23*
mordant *29*
negative staining *30*
nosepiece *22*
numerical aperture *23*
objectives *22*
oculars *22*
parfocal *22*
phase-contrast microscope *24*
refraction *21*
refractive index *21*
resolution *23*
scanning electron microscope (SEM) *35*
scanning tunneling microscope *36*
shadowing *34*
simple staining *29*
spore staining *31*
stage *22*
substage condenser *22*
transmission electron microscope (TEM) *32*
working distance *24*

Questions for Thought and Review

1. How are real and virtual images produced in a light microscope? Which one is a person actually seeing?
2. If a specimen is viewed using a 43× objective in a microscope with a 15× eyepiece, how many times has the image been magnified?
3. Why don't most light microscopes use 30× eyepieces for greater magnification?
4. Why would one expect basic dyes to be more effective under alkaline conditions?
5. What step in the Gram-stain procedure could be dropped without losing the ability to distinguish between gram-positive and gram-negative bacteria? Why?
6. Why must the TEM use a high vacuum and very thin sections?
7. Material is often embedded in paraffin before sectioning for light microscopy. Why can't this approach be used when preparing a specimen for the TEM?
8. Under what circumstances would it be desirable to prepare specimens for the TEM by use of negative staining? Shadowing? Freeze-etching?
9. Compare the microscopes described in this chapter—bright-field, dark-field, phase-contrast, fluorescence, TEM, and SEM—in terms of the images they provide and the purposes for which they are most often used.

Additional Reading

Binnig, G., and Rohrer, H. 1985. The scanning tunneling microscope. *Sci. Am.* 253(2):50–56.

Boatman, E. S.; Berns, M. W.; Walter, R. J.; and Foster, J. S. 1987. Today's microscopy. *BioScience* 37(6):384–94.

Clark, G. L. 1961. *The encyclopedia of microscopy.* New York: Van Nostrand Reinhold.

Clark, G. L., ed. 1973. *Staining procedures used by the Biological Stain Commission,* 3d ed. Baltimore: Williams & Wilkins.

Cosslett, V. E. 1966. *Modern microscopy or seeing the very small.* Ithaca, N.Y.: Cornell University Press.

Gerhard, P.; Murray, R. G. E.; Costilow, R. N.; Nester, E. W.; Wood, W. A.; Krieg, N. R.; and Phillips, G. B., eds. 1981. *Manual of methods for general bacteriology.* Washington, D.C.: American Society for Microbiology.

Gray, Peter. 1964. *Handbook of basic microtechnique,* 3d ed. New York: McGraw-Hill.

Kopp, Friedrich. 1981. Electron microscopy. *Carolina Biology Reader,* no. 105. Burlington, N.C.: Carolina Biological Supply Co.

Lillie, R. D. 1969. *H. J. Conn's biological stains,* 8th ed. Baltimore: Williams & Wilkins.

Meek, G. A. 1976. *Practical electron microscopy for biologists,* 2d ed. New York: John Wiley and Sons.

Postek, M. T.; Howard, K. S.; Johnson, A. H.; and McMichael, K. L. 1980. *Scanning electron microscopy: A student's handbook.* Burlington, Vt.: Ladd Research Industries.

Scherrer, Rene. 1984. Gram's staining reaction, Gram types and cell walls of bacteria. *Trends Biochem. Sci.* 9:242–45.

Wickramasinghe, H. K. 1989. Scanned-probe microscopes. *Sci. Am.* 261(4):98–105.

Wischnitzer, S. 1981. *Introduction to electron microscopy,* 3d ed. New York: Pergamon Press.

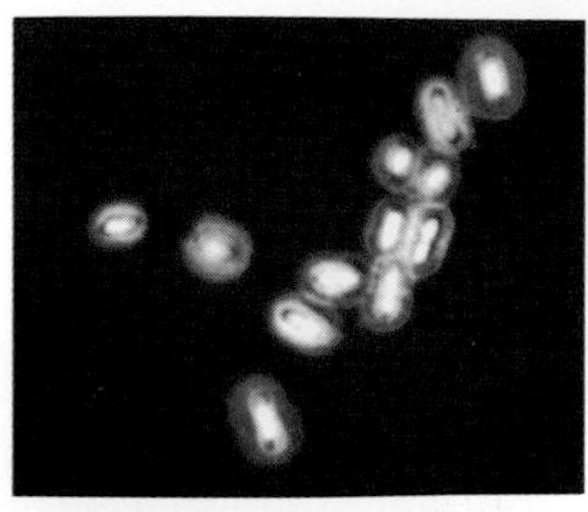

CHAPTER 3
Procaryotic Cell Structure and Function

The era in which workers tended to look at bacteria as very small bags of enzymes has long passed.

—Howard J. Rogers

Outline

Concepts

1. Bacteria are small and simple in structure when compared with eucaryotes, yet they often have characteristic shapes and sizes.
2. Although they have a plasma membrane, which is required by all living cells, bacteria generally lack extensive, complex, internal membrane systems.
3. The cytoplasmic matrix typically contains several constituents that are not membrane-enclosed: inclusion bodies, ribosomes, and the nucleoid with its genetic material.
4. The procaryotic cell wall almost always has peptidoglycan and is chemically and morphologically complex. Most bacteria can be divided into gram-positive and gram-negative forms based on their cell wall structure and response to the Gram stain.
5. Components like capsules and fimbriae are located outside the cell wall. One of these is the flagellum, which many bacteria use like a propeller to swim toward attractants and away from repellents.
6. Some bacteria form resistant endospores to survive harsh environmental conditions in a dormant state.

Even a superficial examination of the microbial world shows that bacteria are one of the most important groups by any criterion: numbers of organisms, general ecological importance, or practical importance for humans. Indeed, much of our understanding of phenomena in biochemistry and molecular biology comes from research on bacteria. Although considerable space is devoted to eucaryotic microorganisms, the major focus is on procaryotes or bacteria. Therefore, the unit on microbial morphology begins with the structure of procaryotes.

Eucaryotes, procaryotes, and the composition of the microbial world (pp. 15; 89–91).

An Overview of Procaryotic Cell Structure

Because much of this chapter is devoted to a discussion of individual cell components, a preliminary overview of the procaryotic cell as a whole is in order.

Size, Shape, and Arrangement

One might expect that small, relatively simple organisms like bacteria would be uniform in shape and size. While it is true that many bacteria are similar in morphology, there is a remarkable amount of variation (figures 3.1 and 3.2; *see also figures 2.8 and 2.14*). Major morphological patterns are described here, and interesting variants are mentioned in the bacterial survey (*see chapters 21–25*).

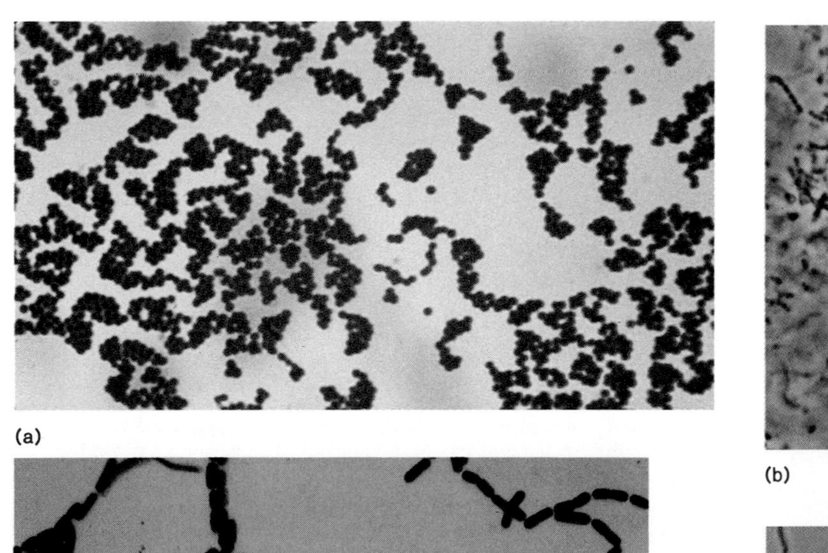

(a)

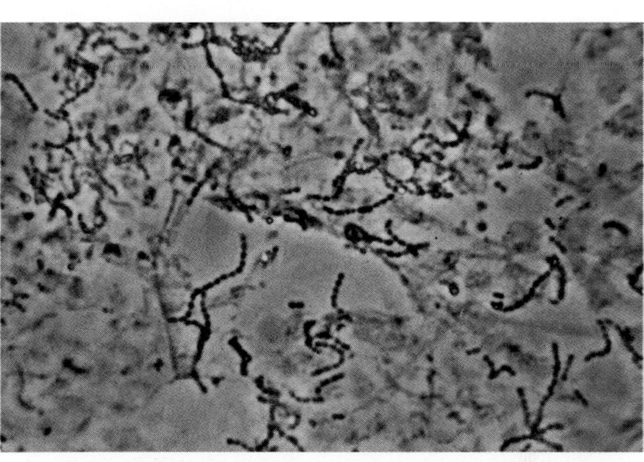

(b)

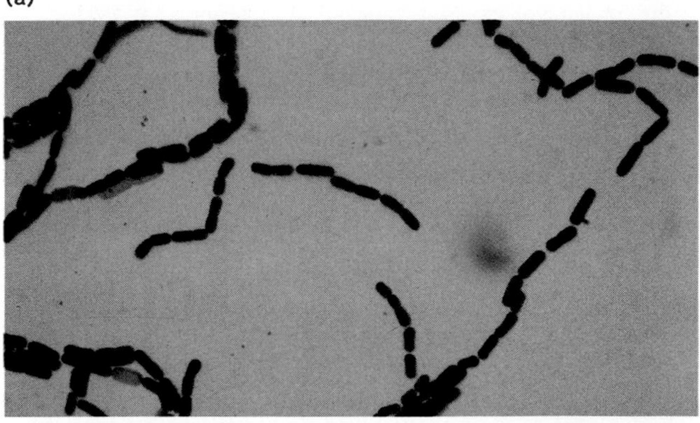

(c)

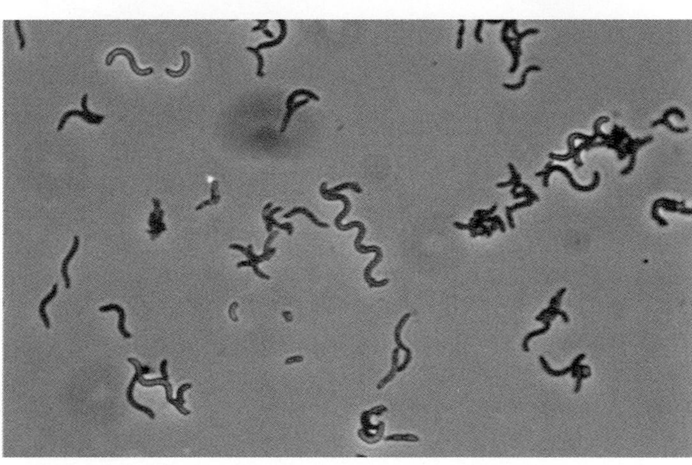

(d)

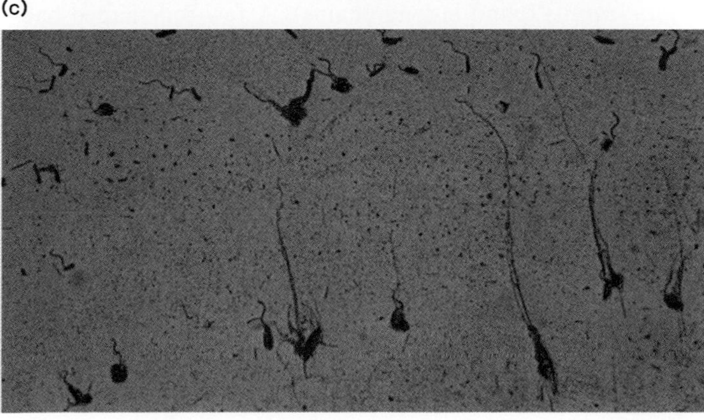

(e)

Figure 3.1 **Representative Bacteria.** Stained bacterial cultures as seen in the light microscope. (*a*) *Staphylococcus aureus*. Note the gram-positive spheres in irregular clusters. Gram stain (×1,000). (*b*) *Enterococcus faecalis*. Note the chains of cocci; phase contrast (×200). (*c*) *Bacillus megaterium*, a rod-shaped bacterium in chains. Gram stain (×600). (*d*) *Rhodospirillum rubrum*. Phase contrast (×500). (*e*) *Vibrio cholerae*. Curved rods with polar flagella (× 1,000).

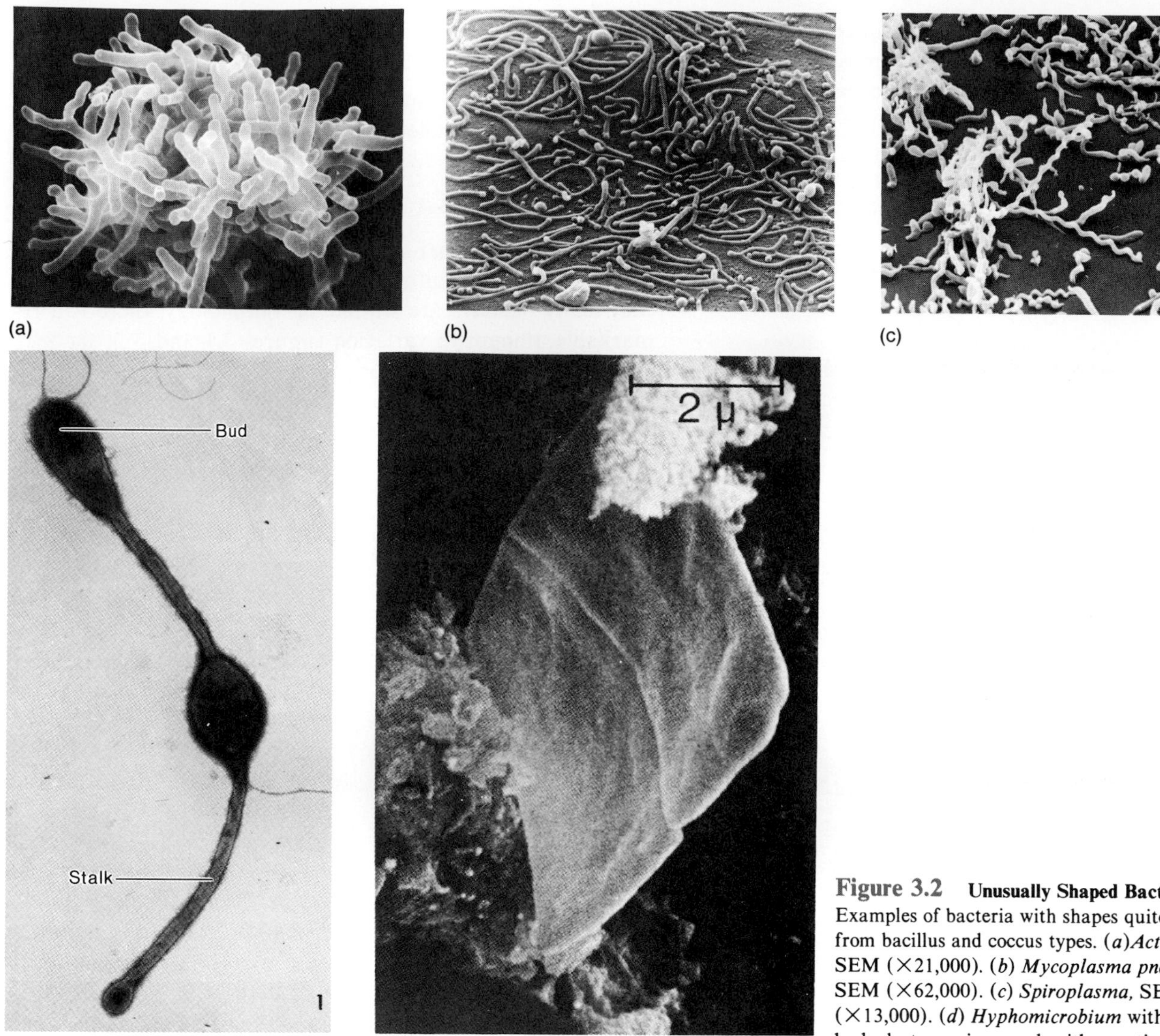

Figure 3.2 Unusually Shaped Bacteria. Examples of bacteria with shapes quite different from bacillus and coccus types. (*a*)*Actinomyces,* SEM (×21,000). (*b*) *Mycoplasma pneumoniae,* SEM (×62,000). (*c*) *Spiroplasma,* SEM (×13,000). (*d*) *Hyphomicrobium* with stalk and bud, electron micrograph with negative staining. (*e*) Walsby's square bacterium.

Most commonly encountered bacteria have one of two shapes. **Cocci** (s., **coccus**) are roughly spherical cells. They can exist as individual cells, but are also associated in characteristic arrangements that are frequently useful in bacterial identification. **Diplococci** (s., **diplococcus**) arise when cocci divide and remain together to form pairs (*Neisseria; see figure 2.14d*). Long chains of cocci result when cells adhere after repeated divisions in one plane; this pattern is seen in the genera *Streptococcus, Enterococcus,* and *Lactococcus* (figure 3.1*b*). *Staphylococcus* divides in random planes to generate irregular grapelike clumps (figure 3.1*a*). Divisions in two or three planes can produce symmetrical clusters of cocci. Members of the genus *Micrococcus* often divide in two planes to form square groups of four cells called tetrads. In the genus *Sarcina,* cocci divide in three planes producing cubical packets of eight cells.

The other common bacterial shape is that of a **rod,** often called a **bacillus** (pl., **bacilli**). *Bacillus megaterium* is a typical example of a bacterium with a rod shape (figure 3.1*c; see also figure 2.14a,c*). Bacilli differ considerably in their length-to-width ratio, the coccobacilli being so short and wide that they resemble cocci. The shape of the rod's end often varies between species and may be flat, rounded, cigar-shaped, or bifurcated. While many rods do occur singly, they may remain together after division to form pairs or chains (e.g., *Bacillus megaterium* is found in long chains). A few rod-shaped bacteria, the **vibrios,** are curved to form distinctive commas or incomplete spirals (figure 3.1*e*).

Bacteria can assume a great variety of shapes, although they are often simple spheres or rods. Actinomycetes characteristically form long multinucleate filaments or hyphae that may branch to produce a network called a **mycelium** (figure 3.2*a*). Many bacteria are shaped like long rods twisted into spirals or helices; they are called **spirilla** if rigid and **spirochetes** when flexible (figures 3.1*d*, 3.2*c; see also figure 2.8a,c*). The oval- to pear-shaped *Hyphomicrobium* (figure 3.2*d*) produces a bud at the end of a long hypha. A few bacteria actually are flat. For example, Anthony E. Walsby has discovered square bacteria living in salt ponds (figure 3.2*e*). These bac-

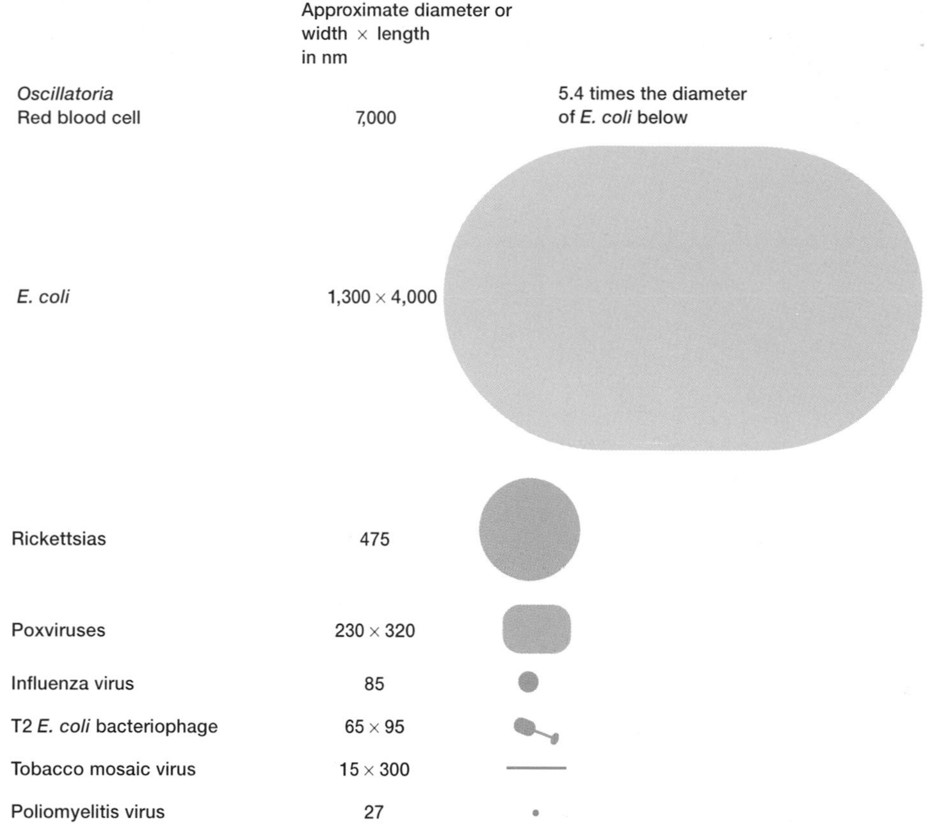

Figure 3.3 **Sizes of Bacteria and Viruses.** The sizes of selected bacteria relative to the red blood cell and viruses.

teria are shaped like flat, square-to-rectangular boxes about 2 μm by 2 to 4 μm, and only 0.25 μm thick. Finally, some bacteria are variable in shape and lack a single, characteristic form (figure 3.2*b*). These are called **pleomorphic** even though they may, like *Corynebacterium,* have a generally rodlike form.

Bacteria vary in size as much as in shape (figure 3.3). The smallest (e.g., some members of the genus *Mycoplasma*) are about 100 to 200 nm in diameter, approximately the size of the largest viruses (the poxviruses). A few become fairly large; some spirochetes occasionally reach 500 μm in length, and the cyanobacterium *Oscillatoria* is about 7 μm in diameter (the same diameter as a red blood cell). *Escherichia coli,* a bacillus of about average size, is 1.1 to 1.5 μm wide by 2.0 to 6.0 μm long.

Procaryotic Cell Organization

A variety of structures is found in procaryotic cells. Their major functions are summarized in table 3.1, and figure 3.4 illustrates many of them. Not all structures are found in every genus. Furthermore, gram-negative and gram-positive cells differ, particularly with respect to their cell walls. Despite these variations, procaryotes are consistent in their fundamental structure and most important components.

Procaryotic cells almost always are bounded by a chemically complex cell wall. Inside this wall, and separated from

TABLE 3.1 Functions of Procaryotic Structures

Structure	Function
Plasma membrane	Selectively permeable barrier, mechanical boundary of cell, nutrient and waste transport, location of many metabolic processes (respiration, photosynthesis), detection of environmental cues for chemotaxis
Gas vacuole	Buoyancy for floating in aquatic environments
Ribosomes	Protein synthesis
Inclusion bodies	Storage of carbon, phosphate, and other substances
Nucleoid	Localization of genetic material (DNA)
Periplasmic space	Contains hydrolytic enzymes and binding proteins for nutrient processing and uptake
Cell wall	Gives bacteria shape and protection from lysis in dilute solutions
Capsules and slime layers	Resistance to phagocytosis, adherence to surfaces
Fimbriae and pili	Attachment to surfaces, bacterial mating
Flagella	Movement
Endospore	Survival under harsh environmental conditions

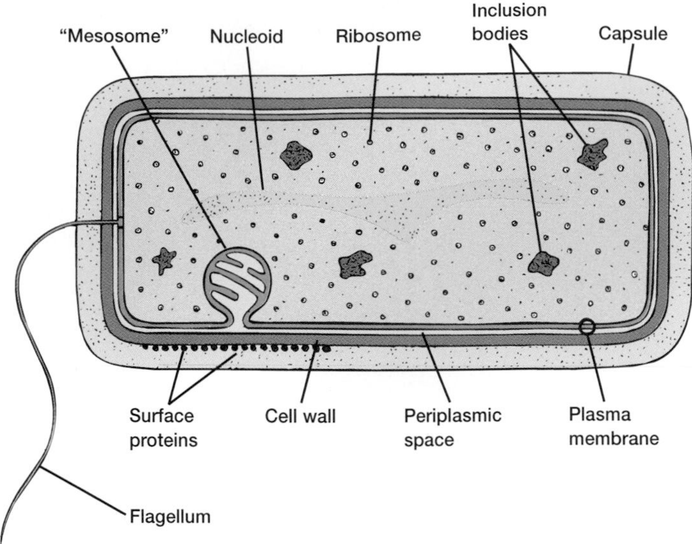

Figure 3.4 Morphology of a Gram-positive Bacterial Cell. The majority of the structures shown here are found in all gram-positive cells. Only a small stretch of surface proteins has been included to simplify the drawing; when present, these proteins cover the surface. Gram-negative bacteria are similar in morphology.

it by a periplasmic space, lies the plasma membrane. This membrane can be invaginated to form simple internal membranous structures. Since the procaryotic cell does not contain internal membrane-bound organelles, its interior appears morphologically simple. The genetic material is localized in a discrete region, the nucleoid, and is not separated from the surrounding cytoplasm by membranes. Ribosomes and larger masses called inclusion bodies are scattered about in the cytoplasmic matrix. Both gram-positive and gram-negative cells can use flagella for locomotion. In addition, many cells are surrounded by a capsule or slime layer external to the cell wall.

Procaryotic cells are morphologically much simpler than eucaryotic cells. These two cell types are compared following the review of eucaryotic cell structure (*see pp. 89–91*).

1. What characteristic shapes can bacteria assume? Describe the ways in which bacterial cells cluster together.
2. Draw a bacterial cell and label all important structures.

Procaryotic Cell Membranes

Membranes are an absolute requirement for all living organisms. Cells must interact in a selective fashion with their environment, whether it is the internal environment of a multicellular organism or a less protected and more variable external environment. Cells must not only be able to acquire nutrients and eliminate wastes, but they also have to maintain their interior in a constant, highly organized state in the face of external changes. The **plasma membrane** encompasses the cytoplasm of both procaryotic and eucaryotic cells. This membrane is the chief point of contact with the cell's environment and thus is responsible for much of its relationship with the outside world. To understand membrane function, it is necessary to become familiar with membrane structure, and particularly with plasma membrane structure.

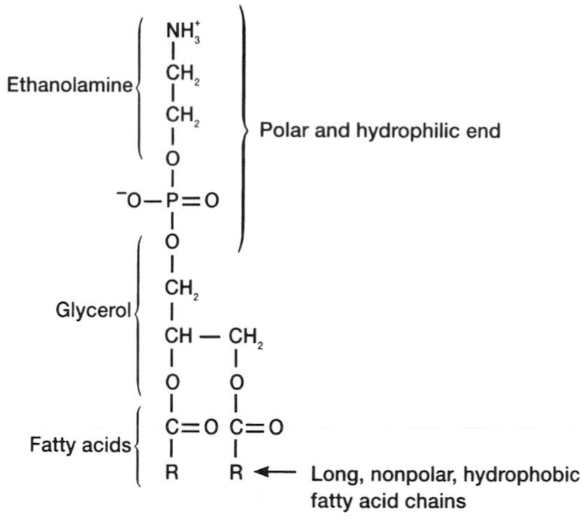

Ethanolamine: NH_3^+ – CH_2 – CH_2 – O
^-O–P=O (Polar and hydrophilic end)
Glycerol: O – CH_2 – CH – CH_2 ; O, O
Fatty acids: C=O, C=O ; R, R ← Long, nonpolar, hydrophobic fatty acid chains

Figure 3.5 The Structure of a Polar Membrane Lipid. Phosphatidylethanolamine, an amphipathic phospholipid often found in bacterial membranes. The R groups are long, nonpolar fatty acid chains.

The Plasma Membrane

Membranes contain both proteins and lipids, although the exact proportions of protein and lipid vary widely. Most membrane-associated lipids are structurally asymmetric with polar and nonpolar ends (figure 3.5) and are called amphipathic. The polar ends interact with water and are **hydrophilic;** the nonpolar **hydrophobic** ends are insoluble in water and tend to associate with one another. This property of lipids enables them to form a bilayer in membranes. The outer surfaces are hydrophilic, whereas hydrophobic ends are buried in the interior away from the surrounding water. Many of these amphipathic lipids are phospholipids (figure 3.5). Bacterial membranes usually differ from eucaryotic membranes in lacking sterols such as cholesterol (figure 3.6*a*). However, many bacterial membranes do contain pentacyclic sterol-like molecules called hopanoids (figure 3.6*b*), and huge quantities of hopanoids are present in our ecosystem (see Box 3.1). Hopanoids are synthesized from the same precursors as steroids. Like steroids in eucaryotes, they probably stabilize the bacterial membrane. The membrane lipid is organized in two layers, or sheets, of molecules arranged end-to-end (figure 3.6).

Many archaeobacterial membranes differ from other bacterial membranes in having a monolayer with lipid molecules spanning the whole membrane.

Archaeobacteria (chapter 24).

Cell membranes are very thin structures, about 5 to 10 nm thick, and can only be seen with the electron microscope. The freeze-etching technique has been used to cleave mem-

Box 3.1

Bacteria and Fossil Fuels

For many years there has been great interest in the origin of fossil fuels such as coal and petroleum. Fossil fuel formation begins when organic matter is buried before it can be oxidized to carbon dioxide by microorganisms. When organic matter is buried deeply and subjected to increasing temperature under anaerobic conditions, petroleum and coal are often formed. The quantities involved in these processes are enormous. It has been estimated that the earth contains about 10^{16} tons of carbon in its sediments.

There is increasing evidence that much of the organic material in sediments is bacterial in origin. About 90% of this material is in the form of insoluble kerogen, an organic precursor of petroleum. Recently, the hopanoid bacteriohopanetetrol (figure 3.6*b*) was isolated from kerogen, and evidence is accumulating that kerogen arises from bacterial activity. We may owe our supply of fossil fuels largely to bacteria that serve as the final degraders of the organic material in dead organisms.

It has been estimated that the total mass of hopanoids in sediments is around 10^{11-12} tons, about as much as the total mass of organic carbon in all living organisms (10^{12} tons). Hopanoids may be the most abundant biomolecules on our planet.

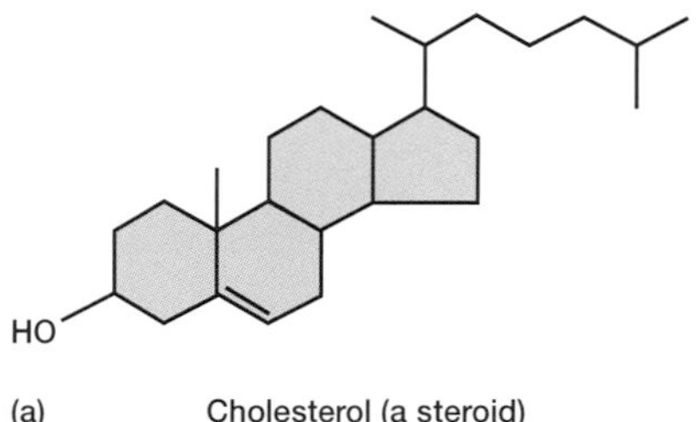

(a) Cholesterol (a steroid)

OH OH
OH OH

(b) A bacteriohopanetetrol (a hopanoid)

Figure 3.6 **Membrane Steroids and Hopanoids.** Common examples.

branes down the center of the lipid bilayer, splitting it in half and exposing the interior. In this way, it has been discovered that many membranes, including the plasma membrane, have a complex internal structure. The small globular particles seen in these membranes are thought to be membrane proteins that lie within the membrane lipid bilayer (*see figure 2.25*).

Freeze-etching technique (pp. 34–35).

The most widely accepted current model for membrane structure is the **fluid mosaic model** of S. Jonathan Singer and Garth Nicholson (figure 3.7). They distinguish between two types of membrane proteins. **Peripheral proteins** are loosely connected to the membrane and can be easily removed. They are soluble in aqueous solutions and make up about 20 to 30% of total membrane protein. About 70 to 80% of membrane proteins are **integral proteins.** These are not easily extracted from membranes, and are insoluble in aqueous solutions when freed of lipids.

Protein and lipid chemistry (Appendix I).

Integral proteins, like membrane lipids, are amphipathic; their hydrophobic regions are buried in the lipid while the hydrophilic portions project from the membrane surface (figure 3.7). Some of these proteins even extend all the way through the lipid layer. Integral proteins can diffuse laterally around the surface to new locations, but do not flip-flop or rotate through the lipid layer. Often, carbohydrates are attached to the outer surface of plasma membrane proteins and seem to have important functions.

The emerging picture of the cell membrane is one of a highly organized and asymmetric system, which is also flexible and dynamic. Although membranes apparently have a common basic design, there are wide variations in both their structure and functional capacities.

The plasma membranes of bacterial cells must fill an incredible variety of roles successfully. Many major plasma membrane functions are noted here even though they are discussed individually at later points in the text. The plasma membrane retains the cytoplasm, particularly in cells without cell walls, and separates it from the surroundings. The plasma membrane also serves as a selectively permeable barrier: it allows particular ions and molecules to pass, either into or out of the cell, while preventing the movement of others. Thus, the membrane prevents the loss of essential components through leakage while allowing the movement of other molecules. Because many substances cannot cross the plasma membrane without assistance, it must aid such movement when necessary. Transport systems can be used for such tasks as nutrient uptake, waste excretion, and protein secretion. The bacterial plasma membrane also is the location of a variety of crucial

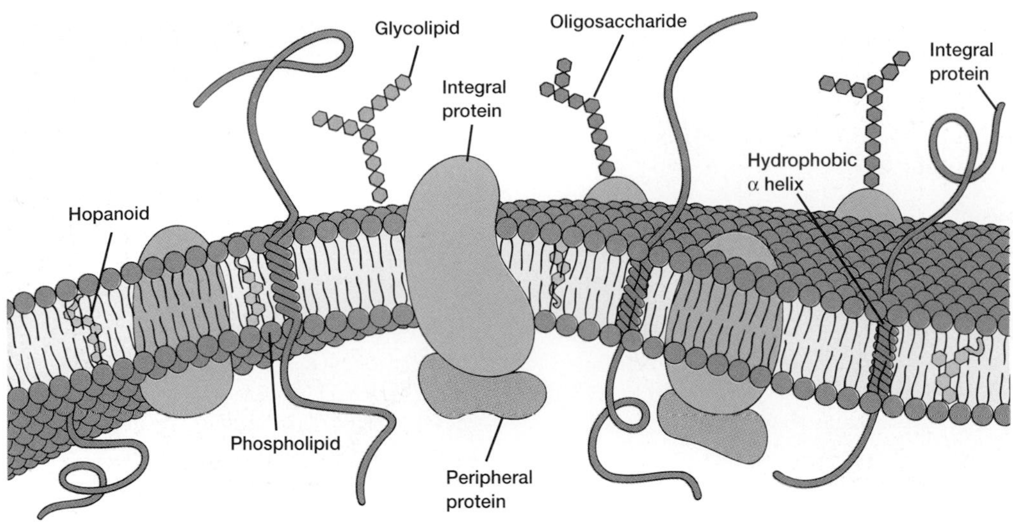

Figure 3.7 **Plasma Membrane Structure.** This diagram of the fluid mosaic model of bacterial membrane structure shows the integral proteins (blue) floating in a lipid bilayer. Peripheral proteins (purple) are associated loosely with the membrane surface. Small spheres represent the hydrophilic ends of membrane phospholipids and wiggly tails, the hydrophobic fatty acid chains. Other membrane lipids such as hopanoids (gold) may be present. For the sake of clarity, phospholipids are shown in proportionately much larger size than in real membranes.

metabolic processes: respiration, photosynthesis, and the synthesis of lipids and cell wall constituents. Finally, the membrane contains special receptor molecules that help bacteria detect and respond to chemicals in their surroundings. Clearly, the plasma membrane is essential to the survival of microorganisms.

Osmosis (pp. 55–56).
Transport of substances across membranes (pp. 100–104).

Internal Membrane Systems

Although bacterial cytoplasm does not contain complex membranous organelles like mitochondria or chloroplasts, membranous structures of several kinds can be observed. A common structure is the **mesosome.** Mesosomes are invaginations of the plasma membrane in the shape of vesicles, tubules, or lamellae (figures 3.4, 3.8, and 3.10). They are seen in both gram-positive and gram-negative bacteria, although they are generally more prominent in the former.

Despite years of research on mesosomes, their exact functions are still unknown. They are often found next to septa or cross-walls in dividing bacteria and sometimes seem attached to the bacterial chromosome. Thus, they may be involved in cell wall formation during division or play a role in chromosome replication and distribution to daughter cells. Mesosomes also may be involved in secretory processes.

Currently, many bacteriologists believe that mesosomes are artifacts generated during the chemical fixation of bacteria for electron microscopy. Possibly they represent parts of the plasma membrane that are chemically different and more disrupted by fixatives. However, mesosomes are seen occasionally in freeze-etched bacteria, and therefore sometimes may be present in living cells. Further research should resolve this controversy.

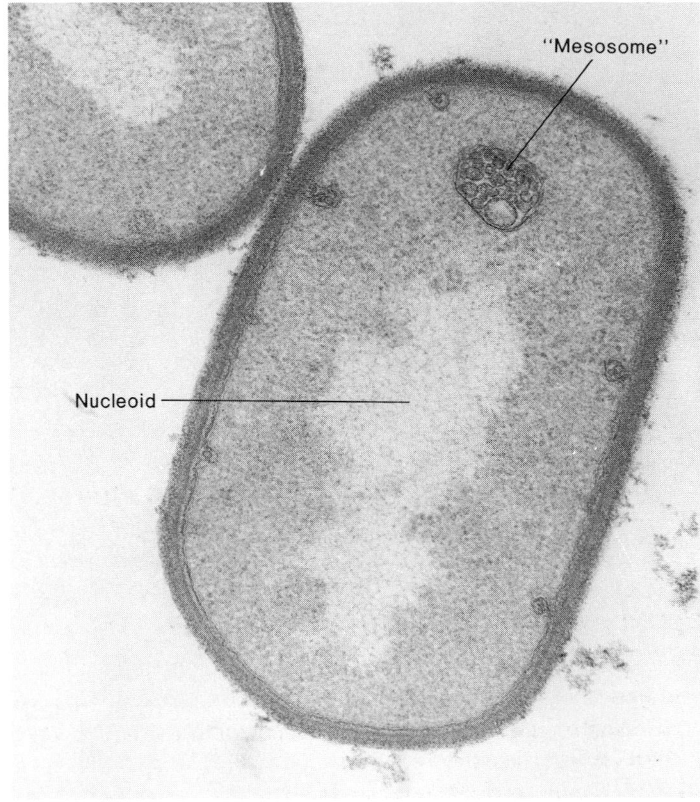

Figure 3.8 **Mesosome Structure.** *Bacillus fastidiosus* (×91,000). A large mesosome lies adjacent to the nucleoid.

Many bacteria have internal membrane systems quite different from the mesosome (figure 3.9). Plasma membrane infoldings can become extensive and complex in photosynthetic bacteria such as the cyanobacteria and purple bacteria or in bacteria with very high respiratory activity like the nitrifying

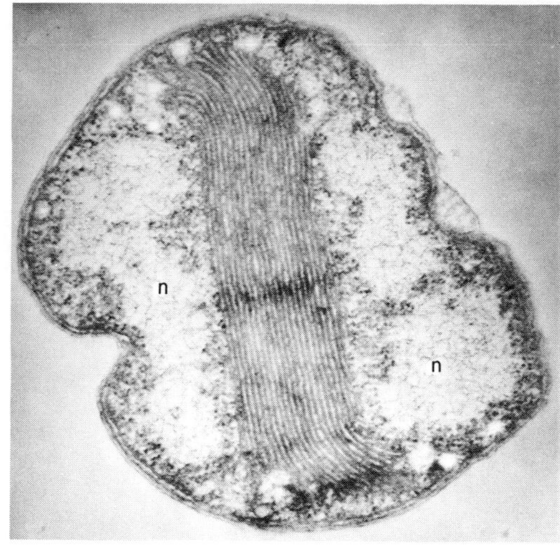

(a)

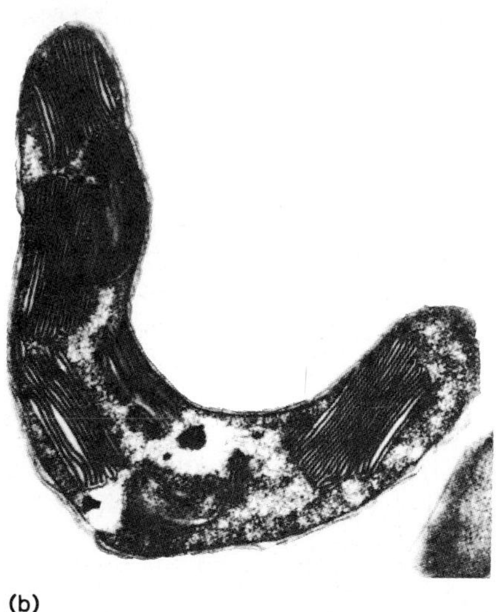

(b)

Figure 3.9 **Internal Bacterial Membranes.** Membranes of nitrifying and photosynthetic bacteria. (*a*) *Nitrocystis oceanus* with parallel membranes traversing the whole cell. Note nucleoplasm (*n*) with fibrillar structure. (*b*) *Ectothiorhodospira mobilis* with an extensive intracytoplasmic membrane system (×60,000).

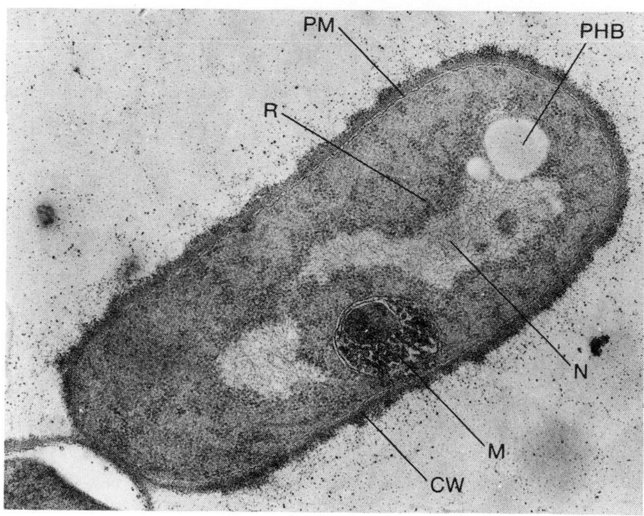

Figure 3.10 **The Structure of a Typical Gram-positive Cell.** Electron micrograph of *Bacillus megaterium* (×30,500). Note the thick cell wall, *CW;* "mesosome," *M;* nucleoid, *N;* poly-β-hydroxybutyrate inclusion body, *PHB;* plasma membrane, *PM;* and ribosomes, *R.*

bacteria (*see chapter 23*). Their function may be to provide a larger membrane surface for greater metabolic activity.

1. Describe with a labeled diagram and in words the fluid mosaic model for cell membranes.
2. List the functions of the plasma membrane.
3. Discuss the nature, structure, and possible functions of the mesosome.

The Cytoplasmic Matrix, Ribosomes, and Inclusions

Procaryotic cytoplasm, unlike that of eucaryotes, lacks unit membrane-bounded organelles. The **cytoplasmic matrix** is the substance lying between the plasma membrane and the nucleoid (*p. 50*). The matrix is largely water (about 70% of bacterial mass is water). It is featureless in electron micrographs and is often packed with ribosomes. The plasma membrane and everything within is called the **protoplast;** thus the cytoplasmic matrix is a major part of the protoplast.

A variety of **inclusion bodies,** granules of organic or inorganic material that are often clearly visible in a light microscope, is present in the cytoplasmic matrix. Some inclusion bodies are not bounded by a membrane and lie free in the cytoplasm—for example, polyphosphate granules, cyanophycin granules, and some glycogen granules. Other inclusion bodies are enclosed by a single-layered, nonunit membrane about 2.0 to 4.0 nm thick. Examples of nonunit membrane-enclosed inclusion bodies are poly-β-hydroxybutyrate granules, some glycogen and sulfur granules, carboxysomes, and gas vacuoles. Inclusion body membranes vary in composition. Some are protein in nature, whereas others contain lipid. A brief description of several important inclusion bodies follows.

Organic inclusion bodies usually contain either glycogen or poly-β-hydroxybutyrate. **Glycogen** is a polymer of glucose units composed of long chains formed by $\alpha(1\rightarrow4)$ glycosidic bonds and branching chains connected to them by $\alpha(1\rightarrow6)$ glycosidic bonds (*see appendix I*). **Poly-β-hydroxybutyrate (PHB)** contains β-hydroxybutyrate molecules joined by ester bonds between the carboxyl and hydroxyl groups of adjacent molecules. Usually only one of these polymers is found in a species, but purple photosynthetic bacteria have both. Poly-β-hydroxybutyrate accumulates in distinct bodies that are readily stained with Sudan black for light microscopy and are clearly visible in the electron microscope (figure 3.10). Glycogen is dispersed more evenly throughout the matrix as small granules and often can be seen only with the electron microscope. If cells contain a large amount of glycogen, staining with an iodine solution will turn them reddish-brown. Glycogen and

(a) (b)

Figure 3.11 **Gas Vacuoles and Buoyancy.** A hammer and stopper experiment with the cyanobacterium *Microcystis aeruginosa.* (*a*) The experimental bottle has been filled and tightly corked while the control bottle is left open to the atmosphere. (*b*) After the stopper is struck with a hammer, the cyanobacteria settle to the bottom. The undisturbed control organisms float.

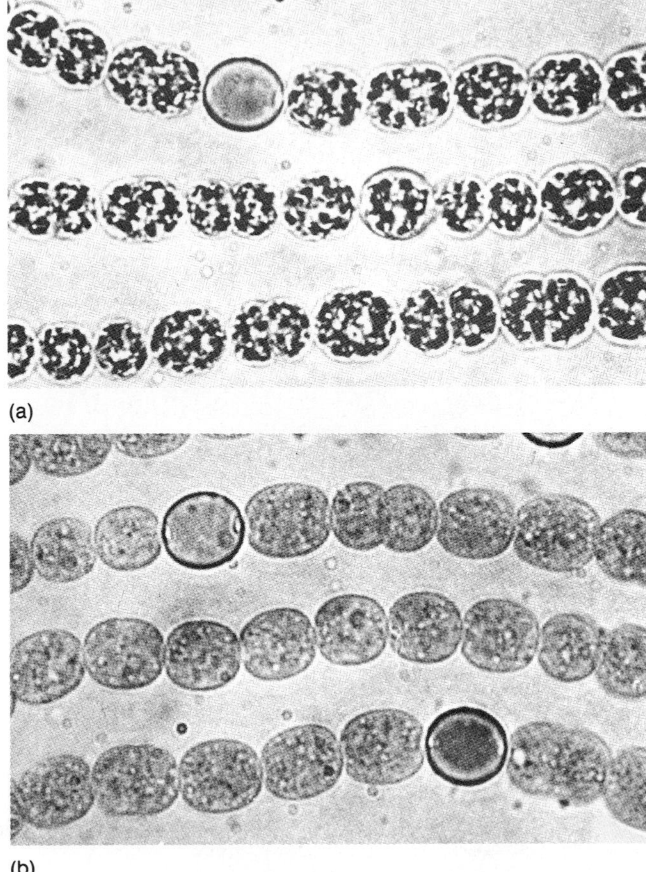

(a)

(b)

Figure 3.12 **Gas Vacuole Collapse.** Filaments of the cyanobacterium *Anabaena flos-aquae.* (*a*) Before application of pressure. (*b*) After application of pressure, the vacuoles have collapsed and are no longer visible.

PHB inclusion bodies are carbon storage reservoirs providing material for energy and biosynthesis. Many bacteria also store carbon as lipid droplets.

Cyanobacteria have two distinctive organic inclusion bodies. **Cyanophycin granules** (figure 3.14*a*) are composed of large polypeptides containing approximately equal amounts of the amino acids arginine and aspartic acid. The granules often are large enough to be visible in the light microscope and store extra nitrogen for the bacteria. **Carboxysomes** are present in many cyanobacteria, nitrifying bacteria, and thiobacilli. They are polyhedral, about 100 nm in diameter, and contain the enzyme ribulose-1,5-bisphosphate carboxylase (see p. 173) in a paracrystalline arrangement. They serve as a reserve of this enzyme and may be a site of CO_2 fixation.

A most remarkable organic inclusion body, the **gas vacuole,** is present in many cyanobacteria (*see chapter 23*), purple and green photosynthetic bacteria, and a few other aquatic forms such as *Halobacterium* and *Thiothrix.* These bacteria float at or near the surface, because gas vacuoles give them buoyancy. This is vividly demonstrated by a simple, but dramatic, experiment. Cyanobacteria held in a full, tightly stoppered bottle will float (figure 3.11*a*), but if the stopper is struck with a hammer, the bacteria sink to the bottom (figure 3.11*b*). Examination of the bacteria at the beginning and end of the experiment shows that the sudden pressure increase has collapsed the gas vacuoles and destroyed the microorganisms' buoyancy (figure 3.12).

Gas vacuoles are aggregates of enormous numbers of small, hollow, cylindrical structures called **gas vesicles** (figure 3.13). Gas vesicle walls do not contain lipid and are composed entirely of a single small protein. These protein subunits assemble to form a rigid enclosed cylinder that is hollow and impermeable to water, but freely permeable to atmospheric gases. Bacteria with gas vacuoles can regulate their buoyancy to float at the depth necessary for proper light intensity, oxygen concentration, and nutrient levels. They descend by simply collapsing vesicles and float upward when new ones are constructed.

Two major types of inorganic inclusion bodies are seen. Many bacteria store phosphate as **polyphosphate granules** or **volutin granules** (figure 3.14*a*). Polyphosphate is a linear polymer of orthophosphates joined by ester bonds. Thus, volutin granules function as storage reservoirs for phosphate, an important component of cell constituents such as nucleic acids. They are sometimes called **metachromatic granules** because they show the metachromatic effect; that is, they appear red or a different shade of blue when stained with the blue dyes methylene blue or toluidine blue. Some bacteria also store sulfur temporarily as sulfur granules, a second type of inorganic inclusion body (figure 3.14*b*). For example, purple photosynthetic bacteria can use hydrogen sulfide as a photosynthetic electron donor (*see chapter 8*) and accumulate the resulting sulfur in either the periplasmic space or in special cytoplasmic globules.

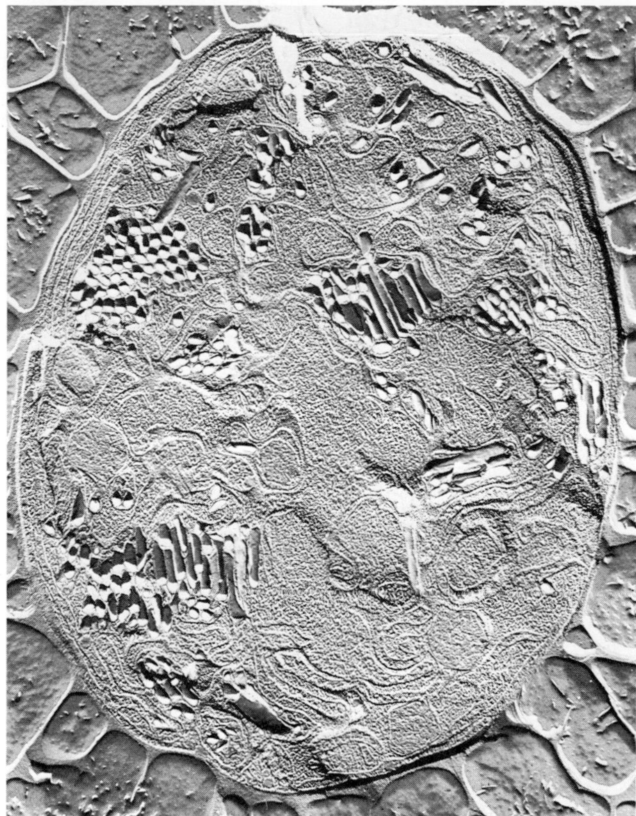

Figure 3.13 **Gas Vesicles.** A freeze-fracture preparation of the cyanobacterium *Anabaena flos-quae* (×89,000). Clusters of the cigar-shaped vesicles form gas vacuoles. Both longitudinal and cross-sectional views of gas vesicles can be seen.

As mentioned earlier, the cytoplasmic matrix often is packed with **ribosomes;** they are also loosely attached to the plasma membrane. Ribosomes look like small, featureless particles at low magnification in electron micrographs (figure 3.10), but are actually very complex objects made of both protein and ribonucleic acid (RNA). They are the site of protein synthesis; matrix ribosomes synthesize proteins destined to remain within the cell whereas the plasma membrane ribosomes make proteins for transport to the outside. Protein synthesis, including a detailed treatment of ribosomes, is discussed at considerable length in chapter 10. At present, note that procaryotic ribosomes are smaller than eucaryotic ribosomes. They commonly are called 70S ribosomes, have dimensions of about 14 to 15 by 20 nm, a molecular weight of approximately 2.7 million, and are constructed of a 50S and a 30S subunit. The S in 70S and similar values stands for **Svedberg unit.** This is the unit of the sedimentation coefficient, a measure of the sedimentation velocity in a centrifuge; the faster a particle travels when centrifuged, the greater its Svedberg value or sedimentation coefficient. The sedimentation coefficient is a function of a particle's molecular weight, volume, and shape. Heavier and more compact particles normally have larger Svedberg numbers or sediment faster. It must be emphasized that Svedberg values are not directly proportional to molecular weights: the weight of a 70S ribosome equals the sum of the 50S and 30S subunit weights even though the sum of 50 and 30 is 80, not 70. Ribosomes in the cytoplasmic matrix of eucaryotic cells are 80S ribosomes and about 22 nm in diameter. Despite their overall difference in size, both types of ribosomes are similarly composed of a large and a small subunit.

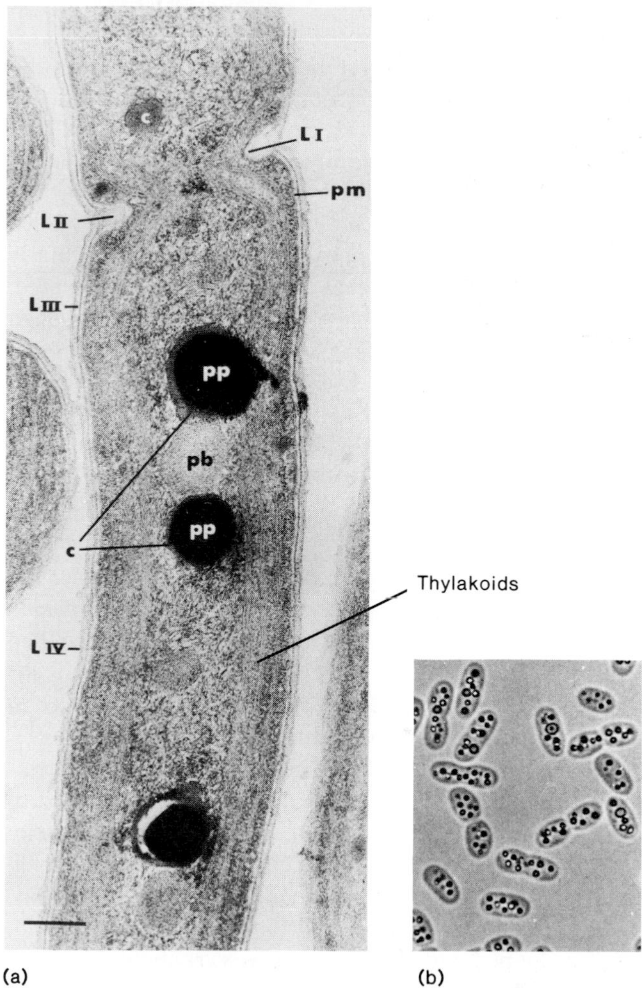

Figure 3.14 **Inclusion Bodies in Bacteria.** (*a*) Ultrastructure of the cyanobacterium *Anacystis nidulans.* The bacterium is dividing, and a septum is partially formed, *LI* and *LII.* Several structural features can be seen, including cell wall layers, *LIII* and *LIV;* the plasma membrane, *pm;* polyphosphate granules, *pp;* a polyhedral body, *pb;* and cyanophycin material, *c.* Thylakoids run along the length of the cell. The bar equals 0.1 μm. (*b*) *Chromatium vinosum,* a purple sulfur bacterium, with intracellular sulfur granules, light field (×2,000).

---

1. Briefly describe the nature and function of the cytoplasmic matrix and the ribosome. What is a protoplast?
2. What kinds of inclusion bodies do procaryotes have? What are their functions?
3. What is a gas vacuole? Relate its structure to its function.

The Nucleoid

Probably the most striking difference between procaryotes and eucaryotes is the way in which their genetic material is packaged. Eucaryotic cells have two or more chromosomes contained within a membrane-delimited organelle, the nucleus. In contrast, procaryotes lack a membrane-delimited nucleus. The procaryotic chromosome, a single circle of double-stranded **deoxyribonucleic acid (DNA)**, is located in an irregularly shaped region called the **nucleoid** (other names are also used: the nuclear body, chromatin body, nuclear region). Although nucleoid appearance varies with the method of fixation and staining, fibers often are seen in electron micrographs (figure 3.10) and are probably DNA. The nucleoid is also visible in the light microscope after staining with the Feulgen stain, which specifically reacts with DNA. A cell can have more than one nucleoid when cell division occurs after the genetic material has been duplicated.

Careful electron microscopic studies have often shown the nucleoid in contact with either the mesosome or the plasma membrane. Membranes also are found attached to isolated nucleoids. Thus, there is evidence that bacterial DNA is attached to cell membranes, and membranes may be involved in the separation of DNA into daughter cells during division.

Nucleoids have been isolated intact and free from membranes. Chemical analysis reveals that they are composed of about 60% DNA, some RNA, and a small amount of protein. In *Escherichia coli,* a rod-shaped cell about 2 to 6 μm long, the closed DNA circle measures about 1,400 μm. Obviously, it must be very efficiently packaged to fit within the nucleoid. The DNA is looped and coiled extensively, probably with the aid of nucleoid proteins (these proteins differ from the histone proteins present in eucaryotic nuclei).

Procaryotic DNA and its function (chapter 10).

Many bacteria possess **plasmids** in addition to their chromosome. These are circular, double-stranded DNA molecules that can exist and replicate independently of the chromosome or may be integrated with it; in either case, they are inherited or passed on to the progeny. Plasmids are not required for host growth and reproduction, although they may carry genes that give their bacterial host a selective advantage. Plasmid genes can render bacteria drug-resistant, give them new metabolic abilities, make them pathogenic, or endow them with a number of other properties.

Plasmids (pp. 261–64).

---

1. Characterize the nucleoid with respect to its structure and function.
2. What is a plasmid?

The Procaryotic Cell Wall

The cell wall is one of the most important parts of a procaryotic cell for several reasons. Except for the mycoplasmas (*see chapter 21*) and some archaeobacteria (*see chapter 24*), most bacteria have strong walls that give them shape and protect them from osmotic lysis (*p. 55*). The cell walls of many pathogens have components that contribute to their pathogenicity. The wall can protect a cell from toxic substances and is the site of action of several antibiotics.

After Christian Gram developed the Gram stain in 1884, it soon became evident that bacteria could be divided into two major groups based on their response to the Gram-stain procedure. Gram-positive bacteria stained purple whereas gram-negative bacteria were colored pink or red by the technique. The true structural difference between these two groups became clear with the advent of the transmission electron microscope. The gram-positive cell wall consists of a single 20 to 80 nm thick homogeneous **peptidoglycan** or **murein** layer lying outside the plasma membrane (figure 3.15). In contrast, the gram-negative cell wall is quite complex. It has a 1 to 3 nm peptidoglycan layer surrounded by a 7 to 8 nm thick **outer membrane.** Microbiologists often call all the structures outside the plasma membrane the **envelope.** This includes the wall and structures like capsules (*p. 56*) when present.

Gram-stain procedure (p. 29).

Frequently a space is seen between the plasma membrane and the outer membrane in electron micrographs of gram-negative bacteria, and sometimes a similar but smaller gap is observed between the plasma membrane and wall in gram-positive bacteria. This space is called the **periplasmic space** or **periplasm.** Recent evidence indicates that the periplasmic space may be filled with a loose network of peptidoglycan. Possibly it is more a gel than a fluid-filled space. Size estimates of the periplasmic space in gram-negative bacteria range from 1 nm to as great as 71 nm. Some recent studies indicate that it may constitute about 20 to 40% of the total cell volume (around 30 to 70 nm), but more research is required to establish an accurate value. When cell walls are disrupted carefully or removed without disturbing the underlying plasma membrane, periplasmic enzymes and other proteins are released and may be easily studied. The periplasmic space of gram-negative bacteria contains many proteins that participate in nutrient acquisition—for example, hydrolytic enzymes attacking nucleic acids and phosphorylated molecules, and binding proteins involved in transport of materials into the cell. Gram-positive bacteria do not appear to have as many periplasmic proteins; rather, they secrete several enzymes that ordinarily would be periplasmic in gram-negative bacteria. Such secreted enzymes are often called **exoenzymes.**

The recently discovered archaeobacteria differ from other bacteria in many respects (*see chapter 24*). Although they may be either gram positive or gram negative, their cell walls are

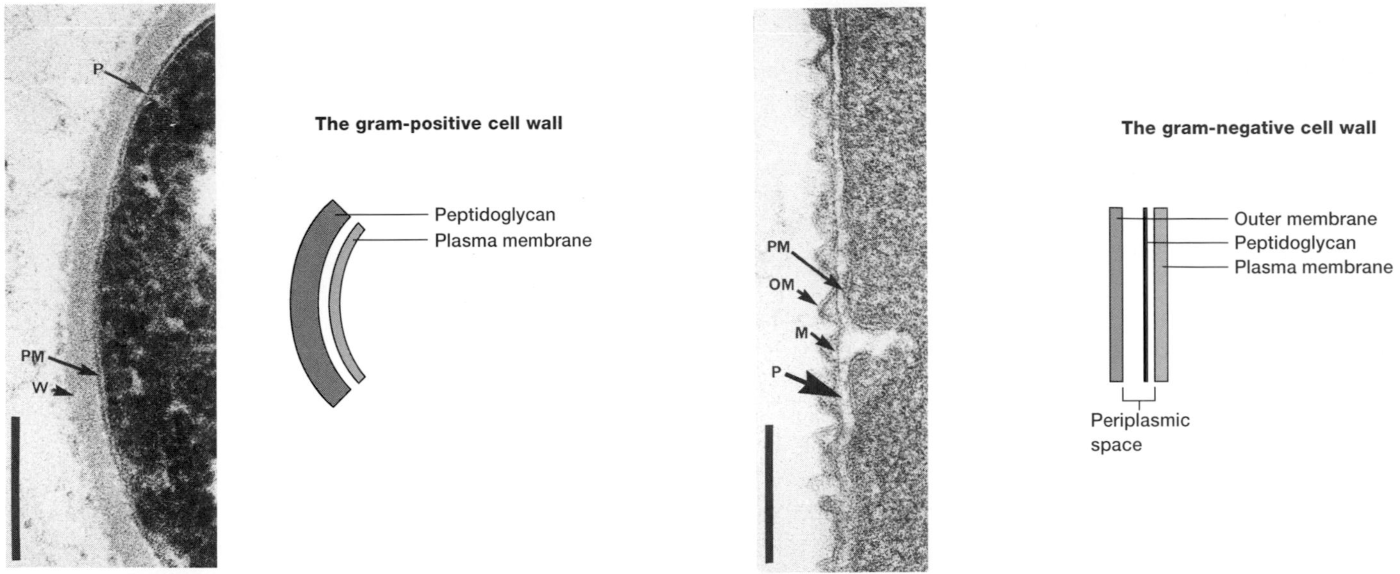

Figure 3.15 Gram-positive and Gram-negative Cell Walls. The gram-positive envelope is from *Bacillus licheniformis* (*left*), and the gram-negative micrograph is of *Aquaspirillum serpens* (*right*). *M;* peptidoglycan or murein layer; *OM,* outer membrane; *PM,* plasma membrane; *P,* periplasmic space; *W,* gram-positive peptidoglycan wall. Bar equals 100 nm.

distinctive in structure and chemical composition. The walls lack peptidoglycan and are composed of proteins, glycoproteins, or polysaccharides.

Following this overview of the envelope, peptidoglycan structure and the organization of gram-positive and gram-negative cell walls are discussed in more detail.

Peptidoglycan Structure

Peptidoglycan or murein is an enormous polymer composed of many identical subunits. The polymer contains two sugar derivatives, N-acetylglucosamine and N-acetylmuramic acid (the lactyl ether of N-acetylglucosamine), and several different amino acids, three of which—D-glutamic acid, D-alanine, and *meso*-diaminopimelic acid—are not found in proteins. The peptidoglycan subunit present in most gram-negative bacteria and many gram-positive ones is shown in figure 3.16. The backbone of this polymer is composed of alternating N-acetylglucosamine and N-acetylmuramic acid residues. A peptide chain of four alternating D- and L-amino acids is connected to the carboxyl group of N-acetylmuramic acid. Many bacteria substitute another diaminoacid, usually L-lysine, in the third position for *meso*-diaminopimelic acid (figure 3.17).

A review of the chemistry of biological molecules (appendix I).
Peptidoglycan structural variations (pp. 451–52).

Chains of linked peptidoglycan subunits are joined by cross-links between the peptides. Often the carboxyl group of the terminal D-alanine is connected directly to the amino group of diaminopimelic acid, but a **peptide interbridge** may be used instead (figure 3.18). Most gram-negative cell wall peptidoglycan lacks the peptide interbridge. This cross-linking results in an enormous peptidoglycan sac that is actually one dense, interconnected network (figure 3.19). These sacs have been isolated from gram-positive bacteria and are strong enough to retain their shape and integrity (figure 3.20), yet they are elastic and somewhat stretchable, unlike cellulose. They also must be porous, as molecules can penetrate them.

Gram-Positive Cell Walls

Normally, the thick, homogeneous cell wall of gram-positive bacteria is composed primarily of peptidoglycan, which often contains a peptide interbridge (figures 3.20 and 3.21). However gram-positive cell walls usually also contain large amounts of **teichoic acids,** polymers of glycerol or ribitol joined by phosphate groups (figures 3.21 and 3.22). Amino acids such as D-alanine or sugars like glucose are attached to the glycerol and ribitol groups. The teichoic acids are connected to either the peptidoglycan itself or to plasma membrane lipids; in the latter case, they are called lipoteichoic acids. Teichoic acids appear to extend to the surface of the peptidoglycan, and, because they are negatively charged, help give the gram-positive cell wall its negative charge. The functions of these molecules are still unclear, but they may be important in maintaining the structure of the wall. Teichoic acids are not present in gram-negative bacteria.

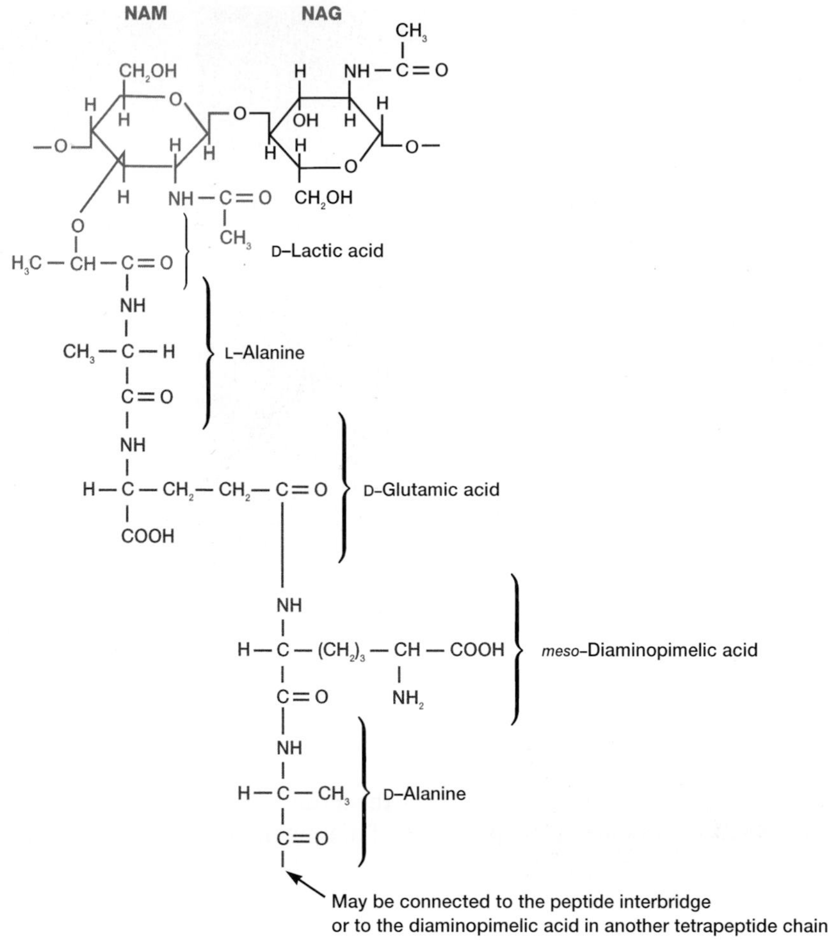

NAM NAG

D-Lactic acid

L-Alanine

D-Glutamic acid

meso-Diaminopimelic acid

D-Alanine

May be connected to the peptide interbridge or to the diaminopimelic acid in another tetrapeptide chain

Figure 3.16 Peptidoglycan Subunit Composition. The peptidoglycan subunit of *Escherichia coli*, most other gram-negative bacteria, and many gram-positive bacteria. *NAG* is N-acetylglucosamine. *NAM* is N-acetylmuramic acid (NAG with lactic acid attached by an ether linkage). The tetrapeptide side chain is composed of alternating D- and L-amino acids since *meso*-diaminopimelic acid is connected through its L-carbon. NAM and the tetrapeptide chain attached to it are shown in different shades of color for clarity.

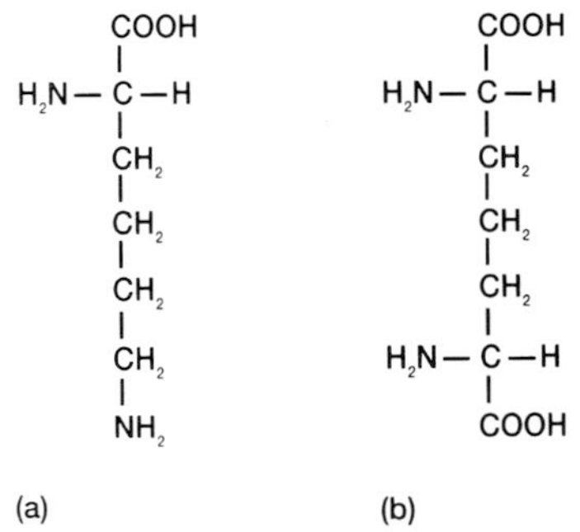

(a) COOH, H_2N-C-H, CH_2, CH_2, CH_2, CH_2, NH_2

(b) COOH, H_2N-C-H, CH_2, CH_2, CH_2, H_2N-C-H, COOH

Figure 3.17 Diaminoacids Present in Peptidoglycan. (*a*) L-lysine. (*b*) *Meso*-diaminopimelic acid.

(a) —NAM—NAG—, L-Ala, D-Glu, DAP, D-Ala — DAP, D-Glu, L-Ala, —NAM—NAG—

(b) —NAM—NAG—, L-Ala, D-GluNH$_2$, L-Lys, D-Ala — Gly — Gly — Gly — Gly — Gly — L-Lys, D-GluNH$_2$, L-Ala, —NAM—NAG—

Peptide interbridge

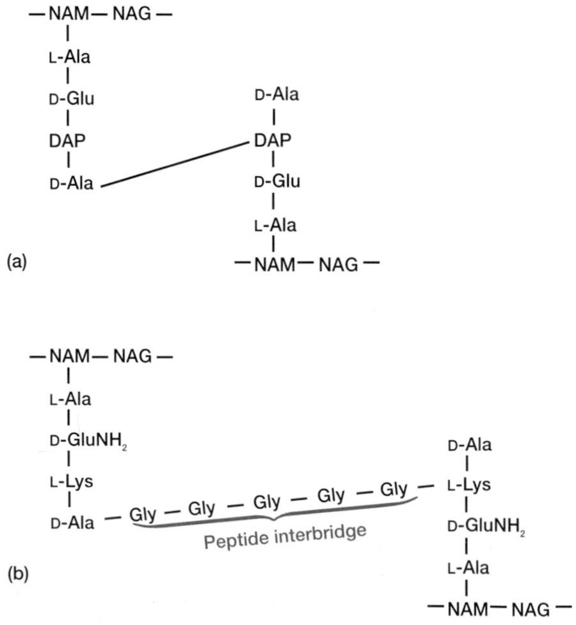

Figure 3.18 Peptidoglycan Cross-links. (*a*) *E. coli* peptidoglycan with direct cross-linking, typical of many gram-negative bacteria. (*b*) *Staphylococcus aureus* peptidoglycan. *S. aureus* is a gram-positive bacterium. *NAM* is N-acetylmuramic acid. *NAG* is N-acetylglucosamine. *Gly* is glycine.

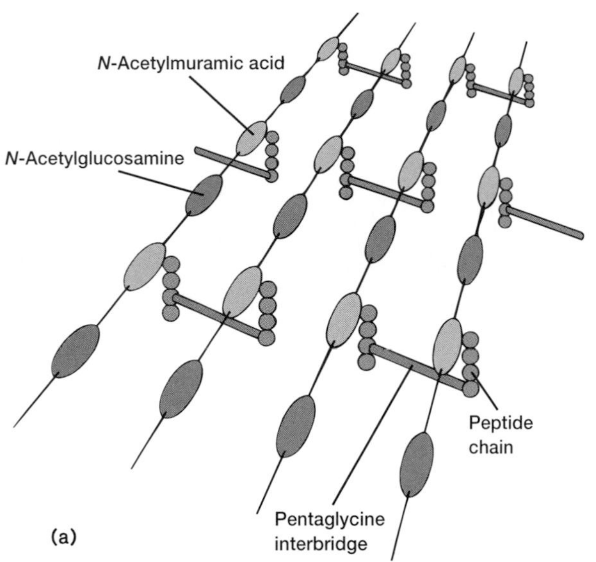

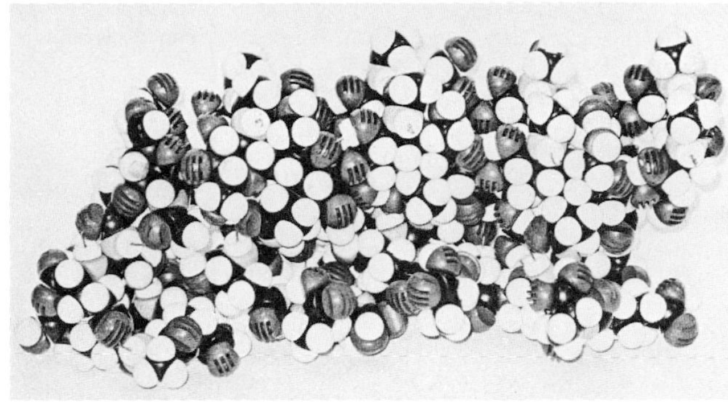

Figure 3.19 Peptidoglycan Structure. A peptidoglycan segment showing the polysaccharide chains, tetrapeptide side chains, and peptide interbridges. (*a*) A schematic diagram. (*b*) A space-filling model of murein with four repeating peptidoglycan subunits in the plane of the paper. Two chains are arranged vertical to this direction. *From Donald Voet and Judith G. Voet,* Biochemistry. *Copyright © 1990 John Wiley & Sons, Inc., New York, NY.*

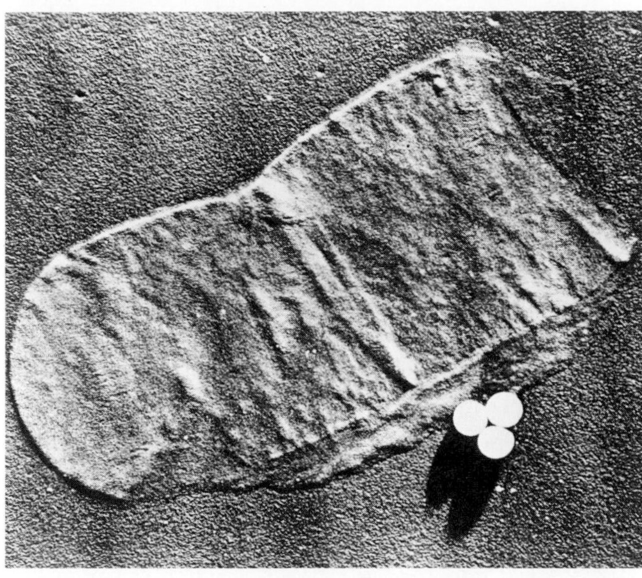

Figure 3.20 Isolated Gram-positive Cell Wall. The peptidoglycan wall from *Bacillus megaterium,* a gram-positive bacterium. The latex balls have a diameter of 0.25 μm.

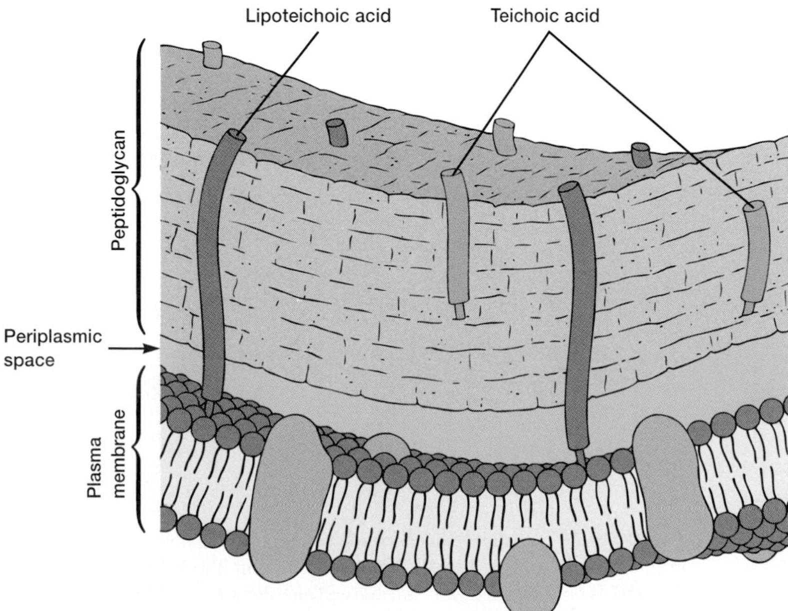

Figure 3.21 The Gram-positive Envelope.

$$
\begin{array}{c}
| \\
O \\
| \\
O{=}P{-}O^{-} \\
| \\
O \\
| \\
CH_2 \\
| \\
H{-}C{-}O{-}R \\
| \\
CH_2 \\
| \\
O \\
| \\
O{=}P{-}O^{-} \\
| \\
O \\
| \\
CH_2 \\
| \\
H{-}C{-}O{-}R \\
| \\
CH_2 \\
| \\
O \\
| \\
O{=}P{-}O^{-} \\
| \\
O \\
|
\end{array}
$$

Figure 3.22 Teichoic Acid Structure. The segment of a teichoic acid made of phosphate, glycerol, and a side chain, *R*. R may represent D-alanine, glucose, or other molecules.

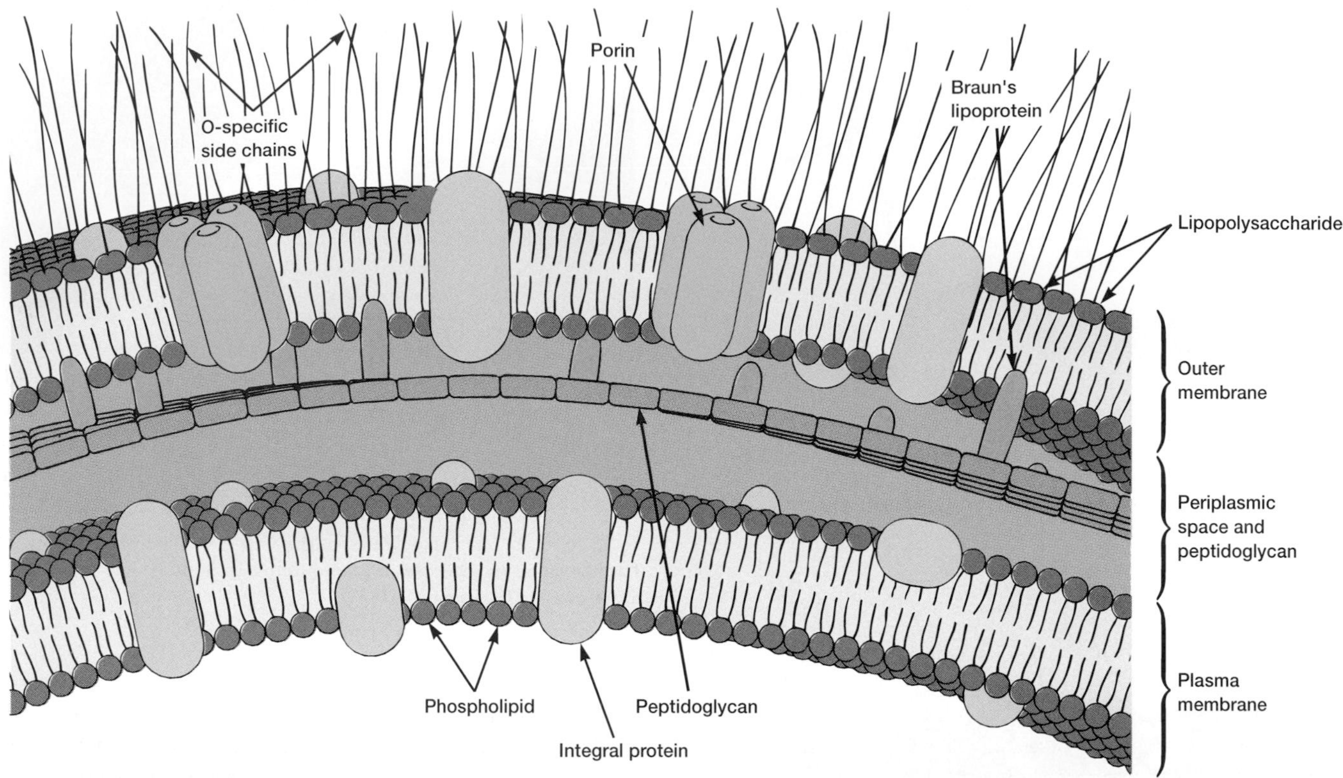

Figure 3.23 **The Gram-negative Envelope.**

Gram-Negative Cell Walls

Even a brief inspection of figure 3.15 shows that gram-negative cell walls are much more complex than gram-positive walls. The thin peptidoglycan layer next to the plasma membrane may constitute not more than 5–10% of the wall weight. In *E. coli,* it is about 1 nm thick and contains only one or two layers or sheets of peptidoglycan. As mentioned earlier, the peptidoglycan may be in the form of a gel rather than a compact layer.

The outer membrane lies outside the thin peptidoglycan layer (figure 3.23). The most abundant membrane protein is Braun's lipoprotein, a small lipoprotein covalently joined to the underlying peptidoglycan and embedded in the outer membrane by its hydrophobic end. The outer membrane and peptidoglycan are so firmly linked by this lipoprotein that they can be isolated as one unit.

Possibly the most unusual constituents of the outer membrane are its **lipopolysaccharides (LPSs)**. These large, complex molecules contain both lipid and carbohydrate, and consist of three parts: (1) lipid A, (2) the core polysaccharide, and (3) the O side chain. The LPS from *Salmonella typhimurium* has been studied most, and its general structure is described here (figure 3.24). The **lipid A** region contains two glucosamine sugar derivatives, each with three fatty acids and phosphate or pyrophosphate attached. It is buried in the outer membrane while the remainder of the LPS molecule projects from the surface. The **core polysaccharide** is joined to lipid A. In *Salmonella,* it is constructed of ten sugars, many of them unusual in structure. The **O side chain** or **O antigen** is a short

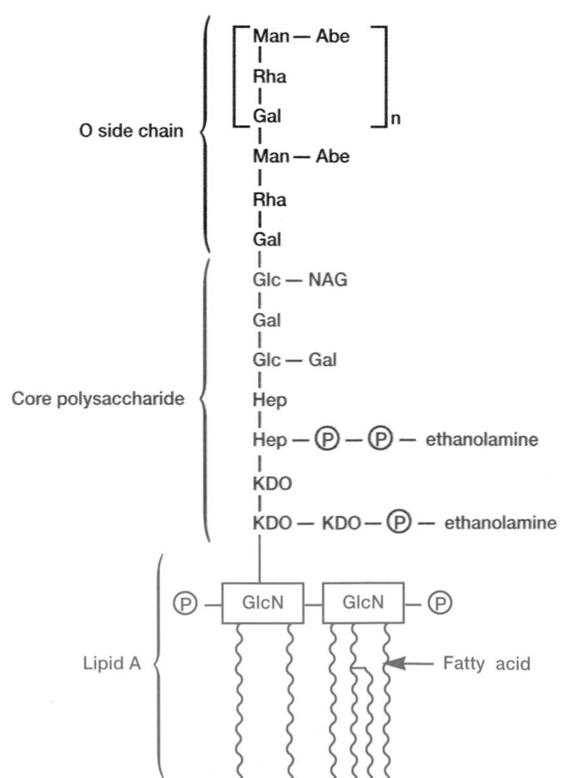

Figure 3.24 **Lipopolysaccharide Structure.** The lipopolysaccharide from *Salmonella.* This slightly simplified diagram illustrates one form of the LPS. Abbreviations: *Abe,* abequose; *Gal,* galactose; *Glc,* glucose; *GlcN,* glucosamine; *Hep,* heptulose; *KDO,* 2-keto-3-deoxyoctonate; *Man,* mannose; *NAG,* N-acetylglucosamine; *P,* phosphate; *Rha,* L-rhamnose.

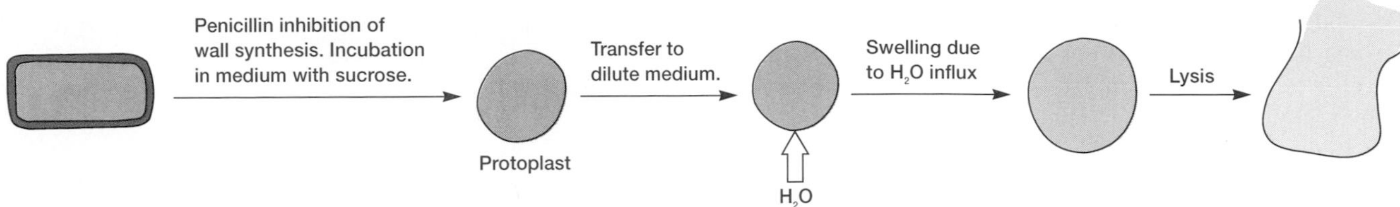

Figure 3.25 **Protoplast Formation.** Protoplast formation induced by incubation with penicillin in an isotonic medium. Transfer to dilute medium will result in lysis.

polysaccharide chain extending outward from the core. It has several peculiar sugars and varies in composition between bacterial strains. Although O side chains are readily recognized by host antibodies, gram-negative bacteria may thwart host defenses by rapidly changing the nature of their O side chains to avoid detection. Antibody interaction with the LPS before reaching the outer membrane proper may also protect the cell wall from direct attack.

Antibodies and antigens (chapters 31 and 33).

The LPS is important for several reasons other than the avoidance of host defenses. Since the core polysaccharide usually contains charged sugars and phosphate (figure 3.24), LPS contributes to the negative charge on the bacterial surface. Lipid A is a major constituent of the outer membrane, and the LPS helps stabilize membrane structure. Furthermore, lipid A often is toxic; as a result, the LPS can act as an endotoxin (*see chapter 30*) and cause some of the symptoms that arise in gram-negative bacterial infections.

A most important outer membrane function is to serve as a protective barrier. It prevents or slows the entry of bile salts, antibiotics, and other toxic substances that might kill or injure the bacterium. Even so, the outer membrane is more permeable than the plasma membrane and permits the passage of small molecules like glucose and other monosaccharides. This is due to the presence of special **porin proteins.** Three porin molecules cluster together and span the outer membrane to form a narrow channel through which molecules smaller than about 600 to 700 daltons can pass. Larger molecules such as vitamin B_{12} must be transported across the outer membrane by specific carriers. The outer membrane also prevents the loss of constituents like periplasmic enzymes.

The Mechanism of Gram Staining

Although several explanations have been given for the Gram-stain reaction results, it seems likely that the difference between gram-positive and gram-negative bacteria is due to the physical nature of their cell walls. If the cell wall is removed from gram-positive bacteria, they become gram negative. The peptidoglycan itself is not stained; instead, it seems to act as a permeability barrier preventing loss of crystal violet. During the procedure, the bacteria are first stained with crystal violet and next treated with iodine to promote dye retention. When gram-positive bacteria then are decolorized with ethanol, the alcohol is thought to shrink the pores of the thick peptidoglycan. Thus, the dye-iodine complex is retained during the short decolorization step and the bacteria remain purple. In contrast, gram-negative peptidoglycan is very thin, not as highly cross-linked, and has larger pores. Alcohol treatment also may extract enough lipid from the gram-negative wall to increase its porosity further. For these reasons, alcohol more readily removes the purple crystal violet-iodine complex from gram-negative bacteria.

The Cell Wall and Osmotic Protection

The cell wall is usually required to protect bacteria against destruction by osmotic pressure. Solutes are much more concentrated in bacterial cytoplasm than in most microbial habitats, which are hypotonic. During **osmosis,** water moves across selectively permeable membranes such as the plasma membrane from dilute solutions (higher water concentration) to more concentrated solutions (lower water concentration). Thus, water normally enters bacterial cells and the osmotic pressure may reach 20 atmospheres or 300 pounds/square inch. The plasma membrane cannot withstand such pressures and the cell will swell and **lyse,** or be physically disrupted and destroyed, without the wall that protects it by resisting cell swelling. Solutes are more concentrated in hypertonic habitats than in the cell. Thus, water flows outward, and the cytoplasm shrivels up and pulls away from the cell wall. This phenomenon is known as **plasmolysis** and is useful in food preservation because many microorganisms cannot grow in dried foods and jellies as they cannot avoid plasmolysis (*see pp. 122–23, chapter 43*).

The importance of the cell wall in protecting bacteria against osmotic lysis is demonstrated by treatment with lysozyme or penicillin. The enzyme **lysozyme** attacks peptidoglycan by hydrolyzing the bond that connects N-acetylmuramic acid with carbon four of N-acetylglucosamine. **Penicillin** inhibits peptidoglycan synthesis (*see chapter 16*). If bacteria are incubated with penicillin in an isotonic solution, gram-positive bacteria are converted to **protoplasts** that continue to grow normally when isotonicity is maintained even though they completely lack a wall. Gram-negative cells retain their outer membrane after penicillin treatment and are classified as **spheroplasts** because some of their cell wall remains. Protoplasts and spheroplasts are osmotically sensitive. If they are transferred to a dilute solution, they will lyse due to uncontrolled water influx (figure 3.25).

Although most bacteria require an intact cell wall for survival, some have none at all. For example, the mycoplasmas lack a cell wall, yet often can grow in dilute media or terrestrial environments because their plasma membranes are stronger than normal. The precise reason for this is not known, although the presence of sterols in the membranes of many species may provide added strength. Without a rigid cell wall, mycoplasmas tend to be pleomorphic or variable in shape.

The **L forms** (named after the Lister Institute in London where they were discovered) also lack cell walls. The loss may be complete or partial (some have a defective wall), and the parent organism may be either gram positive or gram negative. They are pleomorphic like mycoplasmas and continue to reproduce. These organisms can arise through spontaneous mutations or from treatments such as growth in isotonic or hypertonic media containing penicillin. If all traces of the peptidoglycan disappear, bacteria cannot resynthesize it because preexisting wall is necessary to construct new peptidoglycan. In this case, the L form may be stable; that is, it may continue to grow and reproduce after the penicillin treatment has ceased. Other L forms sometimes synthesize a wall again. L forms are not closely related to mycoplasmas and should not be confused with them.

1. Describe in some detail the composition and structure of peptidoglycan, gram-positive cell walls, and gram-negative cell walls. Include labeled diagrams in the answer.
2. Define or describe the following: outer membrane, periplasmic space, envelope, teichoic acid, lipopolysaccharide, and porin protein.
3. Explain the role of the cell wall in protecting against lysis and how this role may be experimentally demonstrated. What are protoplasts, spheroplasts, and L forms?

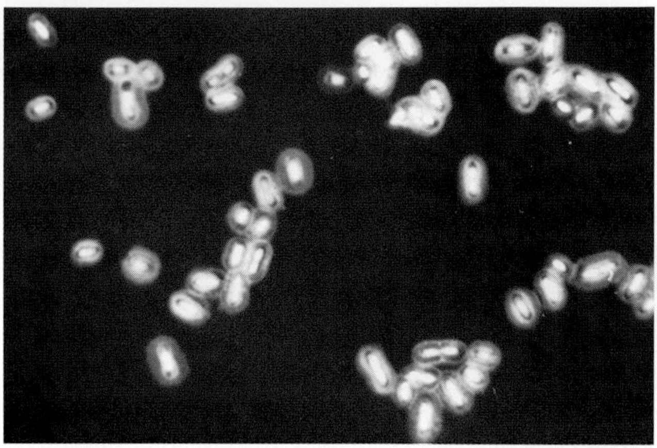

(a)

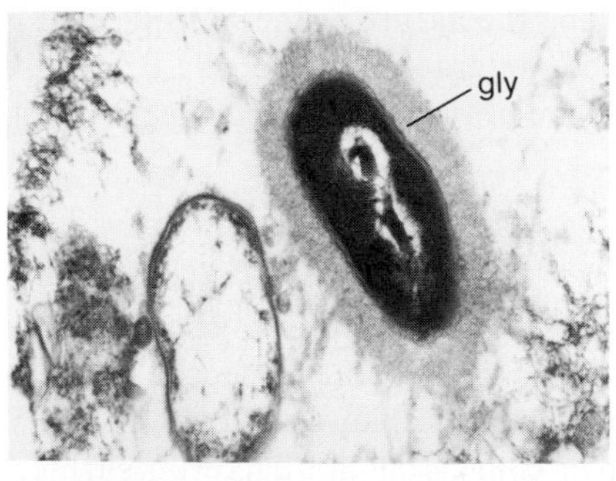

(b)

Figure 3.26 **Bacterial Capsules.** (*a*) *Klebsiella pneumoniae* with its capsule stained for observation in the light microscope (×1,500). (*b*) *Bacteroides* glycocalyx (*gly*), TEM (×71,250).

Components External to the Cell Wall

Bacteria have a variety of structures outside the cell wall that function in protection, attachment to objects, or cell movement. Several of these are discussed.

Capsules, Slime Layers, and S Layers

Some bacteria have a layer of material lying outside the cell wall. When the layer is well organized and not easily washed off, it is called a **capsule.** A **slime layer** is a zone of diffuse, unorganized material that is removed easily. A **glycocalyx** (figure 3.26*b*) is a network of polysaccharides extending from the surface of bacteria and other cells (in this sense, it could encompass both capsules and slime layers). Capsules and slime layers usually are composed of polysaccharides, but they may be constructed of other materials. For example, *Bacillus anthracis* has a capsule of poly-D-glutamic acid. Capsules are clearly visible in the light microscope when negative stains or special capsule stains are employed (figure 3.26*a*); they also can be studied with the electron microscope (figure 3.26*b*).

Although capsules are not required for bacterial growth and reproduction in laboratory cultures, they do confer several advantages when bacteria grow in their normal habitats. They help bacteria resist phagocytosis by host phagocytic cells. *Streptococcus pneumoniae* provides a classic example. When it lacks a capsule, it is destroyed easily and does not cause disease, whereas the capsulated variant quickly kills mice. Capsules contain a great deal of water and can protect bacteria against desiccation. They exclude bacterial viruses and most hydrophobic toxic materials such as detergents. The glycocalyx also aids bacterial attachment to surfaces of solid objects in aquatic environments or to tissue surfaces in plant and animal hosts (figure 3.27).

The relationship of surface polysaccharides to phagocytosis and host colonization (chapter 30).

Many gram-positive and gram-negative bacteria have a regularly structured layer called an **S layer** on their surface. S layers also are common among archaeobacteria. The S layer has a pattern something like floor tiles and is composed of protein or glycoprotein (figure 3.28). It may protect the cell against ion and pH fluctuations, osmotic stress, enzymes, or the predacious bacterium *Bdellovibrio* (*see chapter 23*). It also may contribute to the virulence of some pathogens.

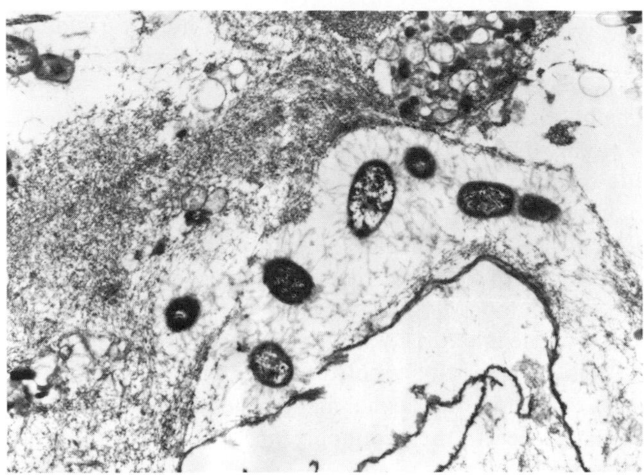

Figure 3.27 **Bacterial Glycocalyx.** Bacteria connected to each other and to the intestinal wall, by their glycocalyxes, the extensive networks of fibers extending from the cells (×17,500).

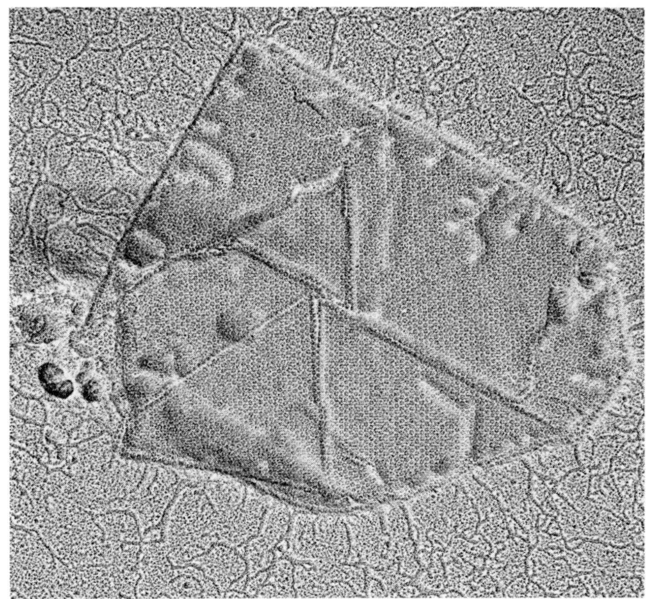

Figure 3.28 **The S Layer.** An electron micrograph of the S layer of *Deinococcus radiodurans* after shadowing.

Pili and Fimbriae

Many gram-negative bacteria have short, fine, hairlike appendages that are thinner than flagella and not involved in motility. These are usually called **fimbriae** (s., **fimbria**). They also are sometimes referred to as pili. Although a cell may be covered with up to 1,000 fimbriae, they are only visible in an electron microscope due to their small size (figure 3.29). They seem to be slender tubes composed of helically arranged protein subunits and are about 3 to 10 nm in diameter and up to several μm long. At least some types of fimbriae attach bacteria to solid surfaces such as rocks in streams and host tissues.

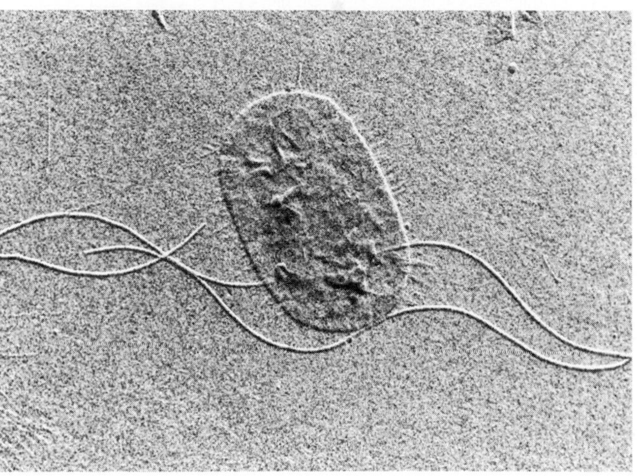

Figure 3.29 **Flagella and Fimbriae.** The long flagella and the numerous shorter fimbriae are very evident in this electron micrograph of *Proteus vulgaris* (×39,000).

Sex pili (s., **pilus**) are similar appendages, about 1 to 10 per cell, that differ from fimbriae in the following ways. Pili are often larger than fimbriae (around 9 to 10 nm in diameter). They are genetically determined by sex factors or conjugative plasmids and are required for bacterial mating (*see chapter 13*). Some bacterial viruses attach specifically to receptors on sex pili at the start of their reproductive cycle.

Flagella and Motility

Most motile bacteria move by use of **flagella** (s., **flagellum**), threadlike locomotor appendages extending outward from the plasma membrane and cell wall. They are slender, rigid structures, about 20 nm across and up to 15 or 20 μm long. Flagella are so thin they cannot be observed directly with a bright-field microscope, but must be stained with special techniques designed to increase their thickness (*see chapter 2*). The detailed structure of a flagellum can only be seen in the electron microscope (figure 3.29).

Bacterial species often differ distinctively in their patterns of flagella distribution. **Monotrichous** bacteria (*trichous* means hair) have one flagellum; if it is located at an end, it is said to be **polar** (figure 3.30*a*). **Amphitrichous** bacteria (*amphi* means "on both sides") have a single flagellum at each pole. In contrast, **lophotrichous** bacteria (*lopho* means tuft) have a cluster of flagella at one or both ends (figure 3.30*b*). Flagella are spread fairly evenly over the whole surface of **peritrichous** (*peri* means "around") bacteria (figure 3.30*c*). Flagellation patterns are very useful in identifying bacteria.

Flagellar Ultrastructure

Transmission electron microscope studies have shown that the bacterial flagellum is composed of three parts. (1) The longest and most obvious portion is the **filament,** which extends from the cell surface to the tip. (2) A **basal body** is embedded in the cell; and (3) a short, curved segment, the **hook,** links the

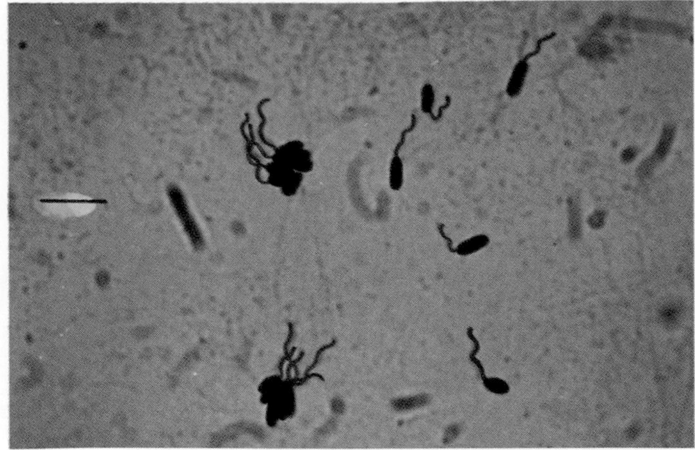

(a)

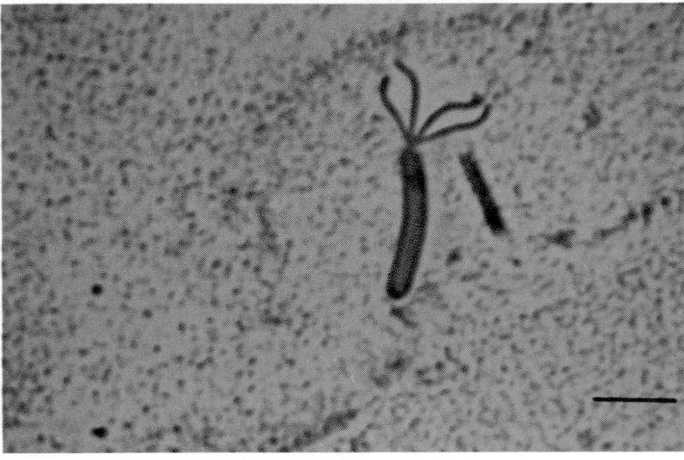

(b)

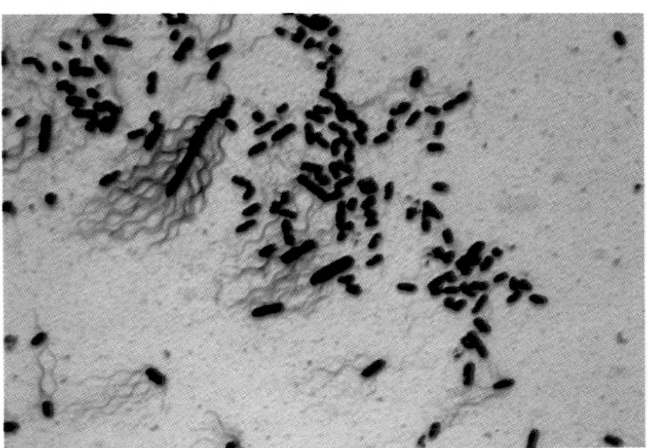

(c)

Figure 3.30 **Flagellar Distribution.** Examples of various patterns of flagellation as seen in the light microscope. (*a*) Monotrichous polar (*Pseudomonas*). (*b*) Lophotrichous (*Spirillum*). (*c*) Peritrichous (*Proteus vulgaris*, ×600). Bars equal 5 μm.

filament to its basal body. The filament is a hollow, rigid cylinder constructed of a single protein called **flagellin,** which ranges in molecular weight from 30,000 to 60,000.

The hook and basal body are quite different from the filament (figure 3.31). Slightly wider than the filament, the hook is made of different protein subunits. The basal body is the most complex part of a flagellum (figures 3.31 and 3.32). In *E. coli* and most gram-negative bacteria, the body has four rings connected to a central rod. The outer L and P rings associate with the lipopolysaccharide and peptidoglycan layers, respectively. The inner M ring contacts the plasma membrane. Gram-positive bacteria have only two basal body rings, an inner ring connected to the plasma membrane and an outer one probably attached to the peptidoglycan.

Flagellar Synthesis

The synthesis of flagella is a complex process involving at least 20 to 30 genes. Besides the gene for flagellin, 10 or more genes code for hook and basal body proteins; other genes are concerned with the control of flagellar construction or function. It is not known how the cell regulates or determines the exact location of flagella.

Bacteria can be deflagellated, and the regeneration of the flagellar filament can then be studied. It is believed that flagellin subunits are transported through the filament's hollow internal core. When they reach the tip, the subunits spontaneously aggregate so that the filament grows at its tip rather than at the base (figure 3.33). Filament synthesis is an excellent example of **self-assembly.** Many structures form spontaneously through the association of their component parts without the aid of any special enzymes or other factors. The information required for filament construction is present in the structure of the flagellin subunit itself.

The Mechanism of Flagellar Movement

Procaryotic flagella operate differently from eucaryotic flagella. The filament is in the shape of a rigid helix, and the bacterium moves when this helix rotates. Considerable evidence shows that flagella act just like propellers on a boat. Bacterial mutants with straight flagella or abnormally long hook regions (polyhook mutants) cannot swim. When bacteria are tethered to a glass slide using antibodies to filament or hook proteins, the cell body rotates rapidly about the stationary flagellum. If polystyrene-latex beads are attached to flagella, the beads spin about the flagellar axis due to flagellar rotation. Flagella can rotate at rates as fast as 40 to 60 revolutions per second.

Eucaryotic flagella and motility (pp. 87–89).

The direction of flagellar rotation determines the nature of bacterial movement. Monotrichous, polar flagella rotate counterclockwise (when viewed from outside the cell) during normal forward movement, whereas the cell itself rotates slowly clockwise. The rotating helical flagellar filament thrusts the cell forward with the flagellum trailing behind (figure 3.34).

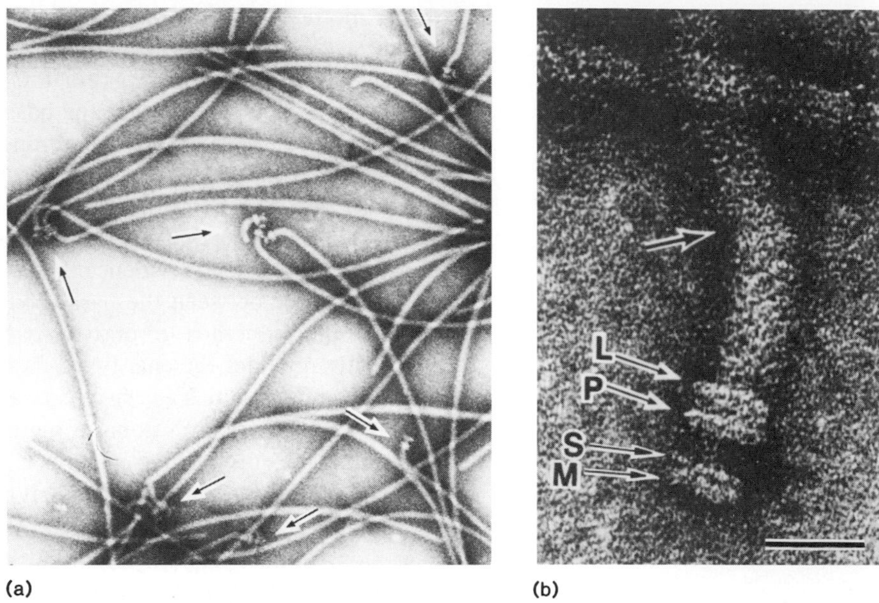

Figure 3.31 The Ultrastructure of Gram-negative Flagella. (*a*) Negatively stained flagella from *Escherichia coli* (×66,000). Arrows indicate the location of curved hooks and basal bodies. (*b*) An enlarged view of the basal body of an *E. coli* flagellum (×485,000). All four rings (*L, P, S,* and *M)* can be clearly seen. The uppermost arrow is at the junction of the hook and filament. Bar equals 30 nm.

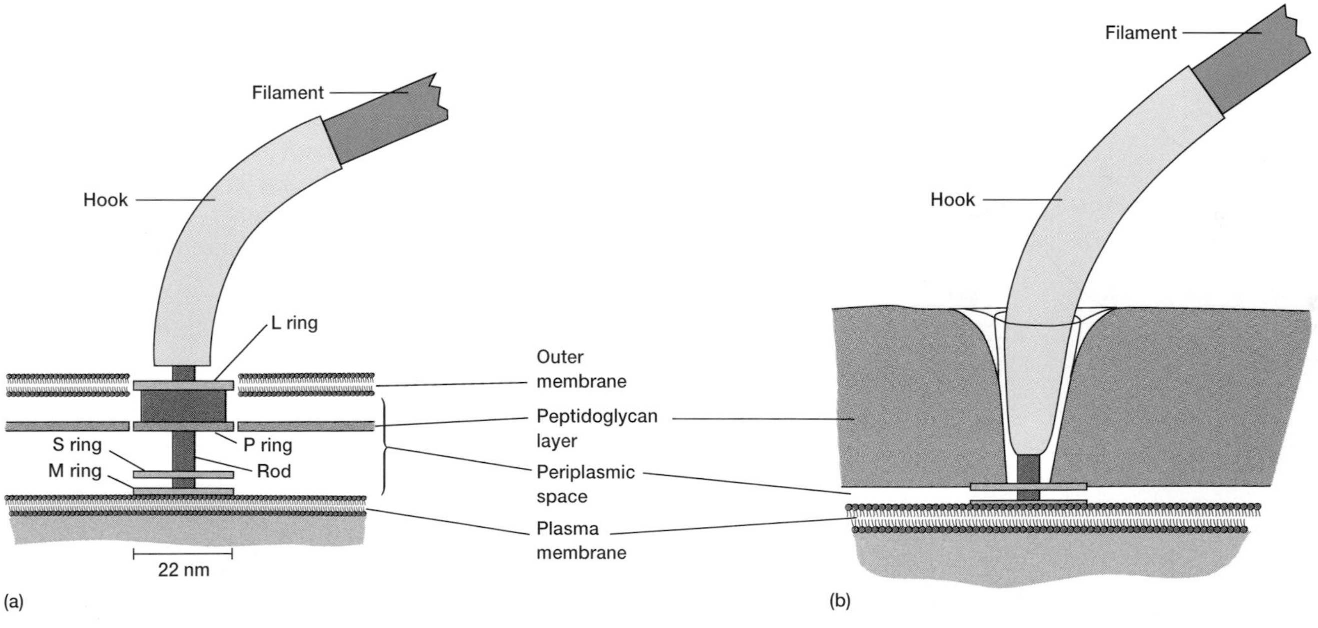

Figure 3.32 The Ultrastructure of Bacterial Flagella. Flagellar basal bodies and hooks in (*a*) gram-negative and (*b*) gram-positive bacteria.

Monotrichous bacteria stop and tumble randomly by reversing the direction of flagellar rotation. Peritrichously flagellated bacteria operate in a somewhat similar way. In order to move forward, the flagella rotate counterclockwise. As they do so, they bend at their hooks to form a rotating bundle that propels them forward. Clockwise rotation of the flagella disrupts the bundle and the cell tumbles.

Because bacteria swim through rotation of their rigid flagella, there must be some sort of motor at the base. According to one hypothesis, a flagellum rotates because of interactions between its S ring and M ring. A rod or shaft extends from the hook and ends in the M ring, which can rotate freely in the plasma membrane (figure 3.32). It is believed that the S ring is attached to the cell wall in gram-positive cells and does

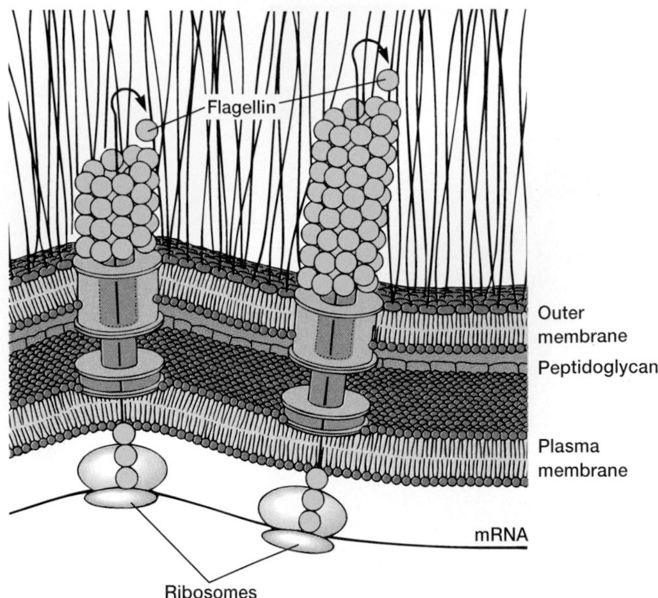

Figure 3.33 Growth of Flagellar Filaments. Flagellin subunits travel through the flagellar core and attach to the growing tip.

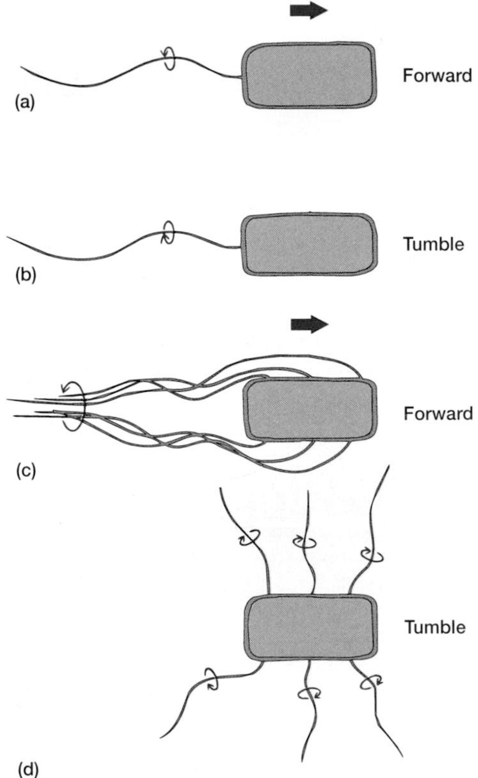

Figure 3.34 Flagellar Motility. The relationship of flagellar rotation to bacterial movement. Parts (*a*) and (*b*) describe the motion of monotrichous, polar bacteria. Parts (*c*) and (*d*) illustrate the movements of peritrichous organisms.

not rotate. Thus, a turning motion will result if the two rings interact in some way. The P and L rings of gram-negative bacteria would act as bearings for the rotating rod. In contrast, there is some evidence that the basal body is a passive structure and rotates within a membrane-embedded protein complex much like the rotor of an electrical motor turns in the center of a ring of electromagnets (the stator).

The exact mechanism that drives basal body rotation still is not clear. There is evidence that the flow of protons past the two rings or between the basal body and surrounding membrane proteins leads to rotation. It does not appear that ATP directly provides the energy for flagellar rotation in bacteria.

The flagellum is a very effective swimming device indeed. From the bacterium's point of view, swimming is quite a job because the surrounding water seems as thick and viscous as molasses. The cell must bore through the water with its helical or corkscrew-shaped flagella, and if flagellar activity ceases, it stops almost instantly. Despite such environmental resistance to movement, bacteria can swim from 20 to almost 90 μm/second. This is equivalent to traveling from 2 to over 100 cell lengths per second. In contrast, an exceptionally fast 6 ft human might be able to run around 5 body lengths per second.

Bacteria can move by mechanisms other than flagellar rotation. Spirochetes are helical bacteria that travel through viscous substances such as mucus or mud by flexing and spinning movements caused by a special **axial filament** (*see chapter 21*). A very different type of motility, **gliding motility,** is employed by many bacteria: cyanobacteria (*see chapter 23*), members of the orders *Myxobacteriales* (*see chapter 23*) and *Cytophagales* (*see chapter 23*), and some mycoplasmas (*see chapter 21*). Although there are no visible structures associated with gliding motility, these bacteria can coast along solid surfaces at rates up to 3 μm/second.

The mechanism of gliding motility (Box 23.1).

1. Briefly describe capsules, slime layers, glycocalyxes, and RS layers. What are their functions?
2. Distinguish between fimbriae and sex pili, and give the function of each.
3. Be able to discuss the following: flagella distribution patterns, flagella structure and synthesis, and the way in which flagella operate to move a bacterium.

Chemotaxis

Bacteria do not always swim aimlessly, but are attracted by such nutrients as sugars and amino acids and are repelled by many harmful substances and bacterial waste products. (Bacteria can respond to other environmental cues; see Box 3.2.) Movement toward chemical attractants and away from repellents is known as **chemotaxis.** Such behavior is of obvious advantage to bacteria.

Chemotaxis may be demonstrated by observing bacteria in the chemical gradient produced when a thin capillary tube

Box 3.2

Living Magnets

Bacteria can respond to environmental factors other than chemicals. A fascinating example is that of the aquatic **magnetotactic bacteria** that orient themselves in the earth's magnetic field. These bacteria have intracellular chains of magnetite (Fe_3O_4) particles or **magnetosomes,** around 40 to 100 nm in diameter and bounded by a membrane (see figure). Since each iron particle is a tiny magnet, the bacteria use their magnetosome chain to determine northward and downward directions, and swim down to nutrient-rich sediments or locate the optimum depth in fresh water and marine habitats. Magnetosomes also are present in the heads of birds, tuna, dolphins, green turtles, and other animals, presumably to aid navigation. Animals and bacteria share more in common behaviorally than previously imagined.

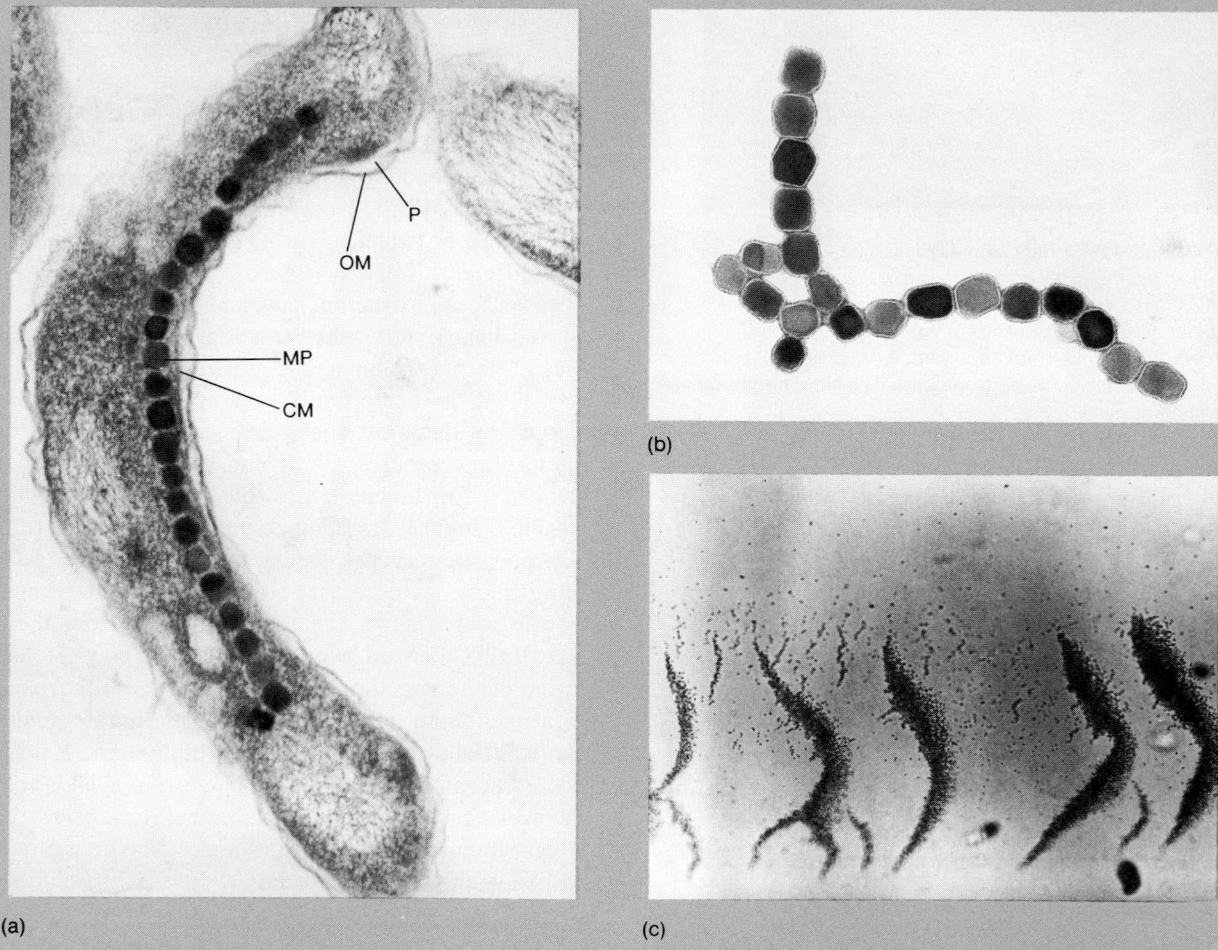

Box Figure 3.2 Magnetotactic Bacteria. (*a*) Transmission electron micrograph of the magnetotactic bacterium *Aquaspirillum magnetotacticum* (×123,000). Note the long chain of electron-dense magnetite particles, *MP.* Other structures: *OM,* outer membrane; *P,* periplasmic space; *CM,* cytoplasmic membrane. (*b*) Isolated magnetosomes (×140,000). (*c*) Bacteria migrating in waves when exposed to a magnetic field.

is filled with an attractant and lowered into a bacterial suspension. As the attractant diffuses from the end of the capillary, bacteria collect and swim up the tube. The number of bacteria within the capillary after a short length of time reflects the strength of attraction and rate of chemotaxis. Positive and negative chemotaxis also can be studied with petri dish cultures (figure 3.35). If bacteria are placed in the center of a dish of agar containing an attractant, the bacteria will exhaust the local supply and then swim outward following the attractant gradient they have created. The result is an expanding ring of bacteria. When a disk of repellent is placed in a petri dish of semisolid agar and bacteria, the bacteria will swim away from the repellent, creating a clear zone around the disk (figure 3.36).

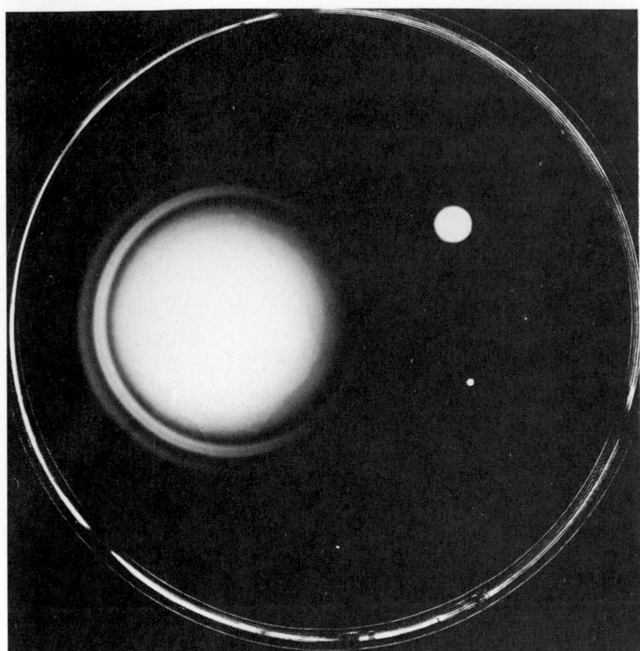

Figure 3.35 Positive Bacterial Chemotaxis. Chemotaxis can be demonstrated on an agar plate that contains various nutrients. Positive chemotaxis by *Escherichia coli* on the left. The outer ring is composed of bacteria consuming serine. The second ring was formed by *E. coli* consuming aspartate, a less powerful attractant. The upper right colony is composed of motile, but nonchemotactic mutants. The bottom right colony is formed by nonmotile bacteria.

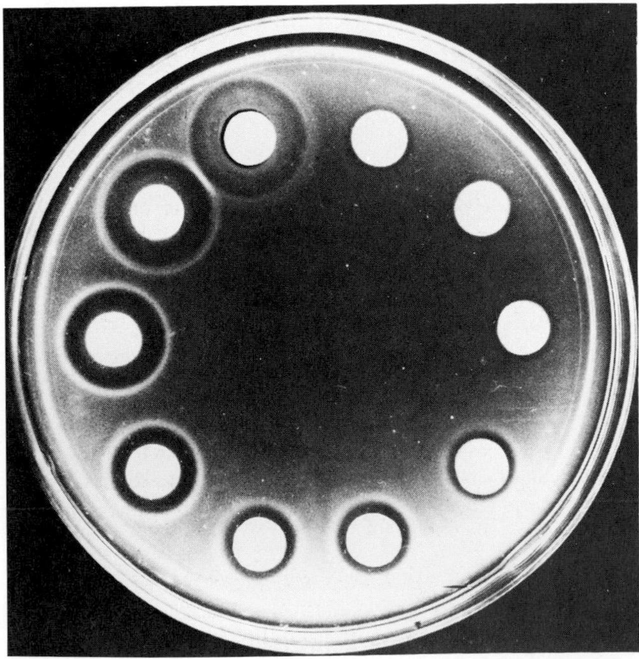

Figure 3.36 Negative Bacterial Chemotaxis. Negative chemotaxis by *E. coli* in response to the repellent acetate. The bright disks are plugs of concentrated agar containing acetate that have been placed in dilute agar inoculated with *E. coli*. Acetate concentration increases from zero at the top right to 3 M at top left. Note the increasing size of bacteria-free zones with increasing acetate. The bacteria have migrated for 30 minutes.

Bacteria can respond to very low levels of attractants (about 10^{-8} M for some sugars), the magnitude of their response increasing with attractant concentration. Usually they sense repellents only at higher concentrations. If an attractant and a repellent are present together, the bacterium will compare both signals and respond to the chemical with the most effective concentration.

Attractants and repellents are detected by **chemoreceptors,** special proteins that bind chemicals and transmit signals to the other components of the chemosensing system. About 20 attractant chemoreceptors and 10 chemoreceptors for repellents have been discovered thus far. These chemoreceptor proteins may be located in the periplasmic space or the plasma membrane. Some receptors participate in the initial stages of sugar transport into the cell.

The chemotactic behavior of bacteria has been studied using the tracking microscope, a microscope with a moving stage that automatically keeps an individual bacterium in view. In the absence of a chemical gradient, *E. coli* and other bacteria move randomly. A bacterium travels in a straight or slightly curved line, a **run,** for a few seconds; then it will stop and **tumble** or **twiddle** about. The tumble is followed by a run in a different direction (figure 3.37). When the bacterium is exposed to an attractant gradient, it tumbles less frequently (or has longer runs) when traveling up the gradient, but tumbles at normal frequency if moving down the gradient. Consequently, the bacterium moves up the gradient. Behavior is shaped by temporal changes in chemical concentration: the bacterium compares its current environment with that experienced a few moments previously; if the attractant concentration is higher, tumbling is suppressed and the run is longer. The opposite response occurs with a repellent gradient. Tumbling frequency decreases (the run lengthens) when the bacterium moves down the gradient away from the repellent. Recall that forward swimming is due to counterclockwise rotation of the flagellum while tumbling results from clockwise rotation. There may be some sort of tumble generator that controls the direction of flagellar rotation and is influenced by chemoreceptors (figure 3.38). The mechanism is complex and involves many components. Nevertheless, it allows bacteria to collect in nutrient-rich regions and at the proper oxygen level while avoiding toxic materials.

1. Define chemotaxis, run, and tumble or twiddle.
2. Explain how bacteria are attracted to substances like nutrients while being repelled by toxic materials.

The Bacterial Endospore

A number of gram-positive bacteria can form a special resistant, dormant structure called an **endospore.** Endospores develop within vegetative bacterial cells of several genera: *Bacillus* and *Clostridium* (rods), *Sporosarcina* (cocci), and others. These structures are extraordinarily resistant to environmental stresses such as heat, ultraviolet radiation, chem-

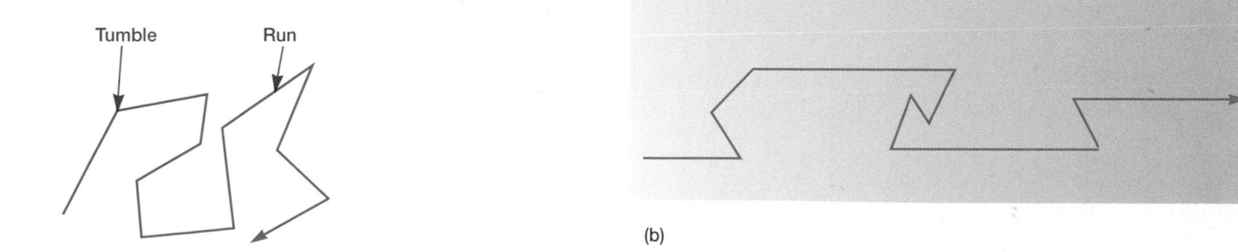

Figure 3.37 Directed Movement in Bacteria. (*a*) Random movement of a bacterium in the absence of a concentration gradient. Tumbling frequency is fairly constant. (*b*) Movement in an attractant gradient. Tumbling frequency is reduced when the bacterium is moving up the gradient. Therefore, runs in the direction of increasing attractant are longer.

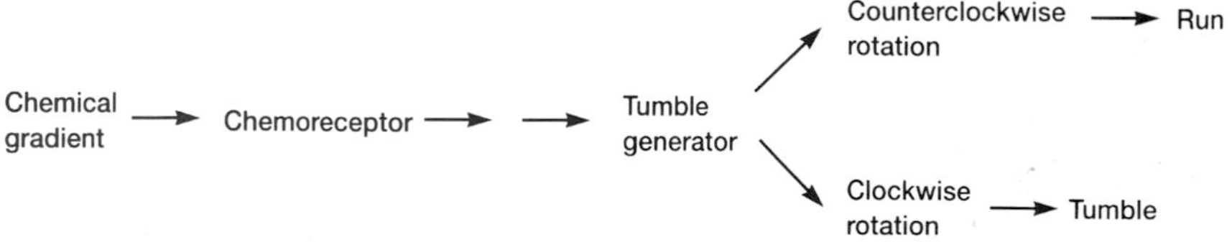

Figure 3.38 The Control of Chemotaxis. A schematic diagram of the bacterial chemotactic system.

ical disinfectants, and desiccation. In fact, some endospores have remained viable for well over 500 years, and actinomycete spores (which are not true endospores) have been recovered alive after burial in the mud for 7,500 years! Because of their resistance and the fact that several species of endospore-forming bacteria are dangerous pathogens, endospores are of great practical importance in industrial and medical microbiology. This is because it is essential to be able to sterilize solutions and solid objects. Endospores often survive boiling for an hour or more; therefore, autoclaves (*see chapter 15*) must be used to sterilize many materials. Endospores are also of considerable theoretical interest. Because bacteria manufacture these intricate entities in a very organized fashion over a period of a few hours, spore formation is well suited for research on the construction of complex biological structures. In the environment, endospores aid in survival when moisture or nutrients are scarce.

Resistance of endospores to environmental stresses (chapter 15).

Endospores can be examined with both light and electron microscopes. Because spores are impermeable to most stains, they are often seen as colorless areas in bacteria treated with methylene blue and other simple stains; special spore stains are used to make them clearly visible (*see chapter 2*). Spore position in the mother cell or **sporangium** frequently differs among species, making it of considerable value in identification. Spores may be centrally located, close to one end (subterminal), or definitely terminal (figure 3.39). Sometimes a spore is so large that it swells the sporangium.

Electron micrographs show that endospore structure is complex (figure 3.40). The spore often is surrounded by a thin, delicate covering called the **exosporium.** A **spore coat** lies beneath the exosporium, is composed of several protein layers,

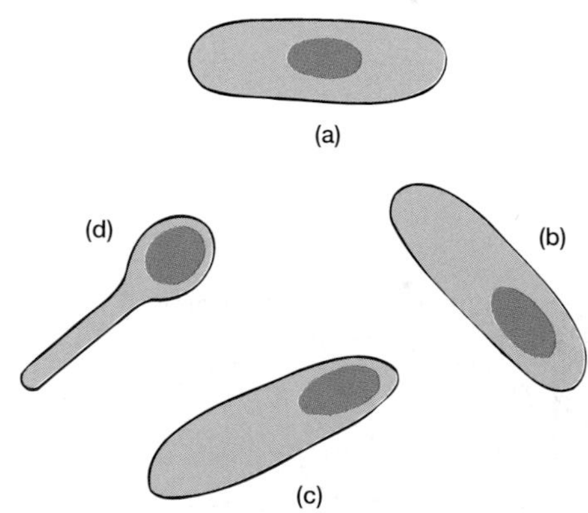

Figure 3.39 Examples of Endospore Location and Size. (*a*) Central spore. (*b*) Subterminal spore. (*c*) Terminal spore. (*d*) Terminal spore with swollen sporangium.

and may be fairly thick. It is impermeable and responsible for the spore's resistance to chemicals. The **cortex,** which may occupy as much as half the spore volume, rests beneath the spore coat. It is made of a peptidoglycan that is less cross-linked than that in vegetative cells. The **spore cell wall** (or core wall) is inside the cortex and surrounds the protoplast or **core.** The core has the normal cell structures such as ribosomes and a nucleoid.

It is still not known precisely why the endospore is so resistant to heat and other lethal agents. As much as 15% of the spore's dry weight consists of **dipicolinic acid** complexed with calcium ions (figure 3.41). It has long been thought that dipicolinic acid was directly involved in spore heat resistance, but heat-resistant mutants lacking dipicolinic acid now have been isolated. It may be that calcium-dipicolinate often stabilizes spore nucleic acids. Dehydration of the protoplast appears to be very important in heat resistance. The cortex may

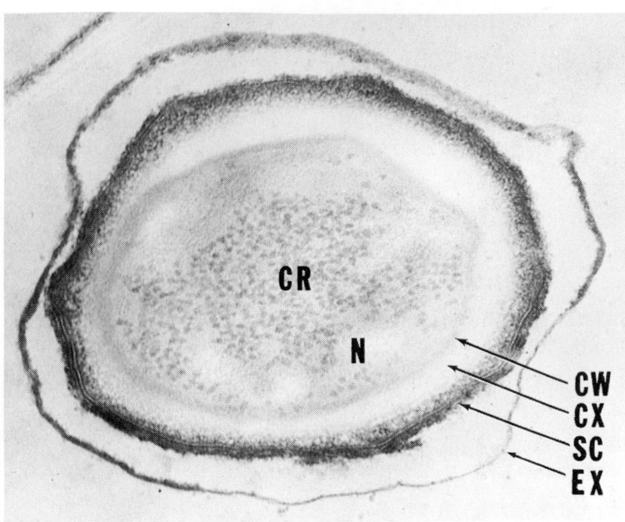

Figure 3.40 **Endospore Structure.** *Bacillus anthracis* endospore (×151,000). Note the following structures: exosporium, *EX;* spore coat, *SC;* cortex, *CX;* core wall, *CW;* and the protoplast or core with its nucleoid, *N,* and ribosomes, *CR.*

osmotically remove water from the protoplast, thereby protecting it from both heat and radiation damage. In summary, endospore heat resistance probably is due to several factors: calcium-dipicolinate stabilization of components like DNA, protoplast dehydration, the greater stability of cell proteins in bacteria adapted to growth at high temperatures, and others.

Spore formation, **sporogenesis** or **sporulation,** normally commences when growth ceases due to lack of nutrients. It is a complex process and may be divided into seven stages (figure 3.42). An axial filament of nuclear material forms (stage I),

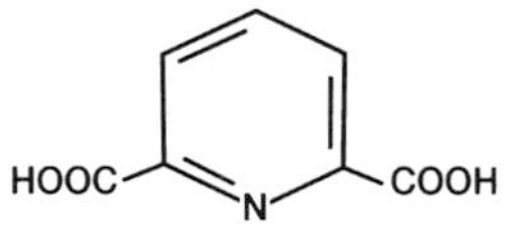

HOOC N COOH

Figure 3.41 **Dipicolinic Acid.**

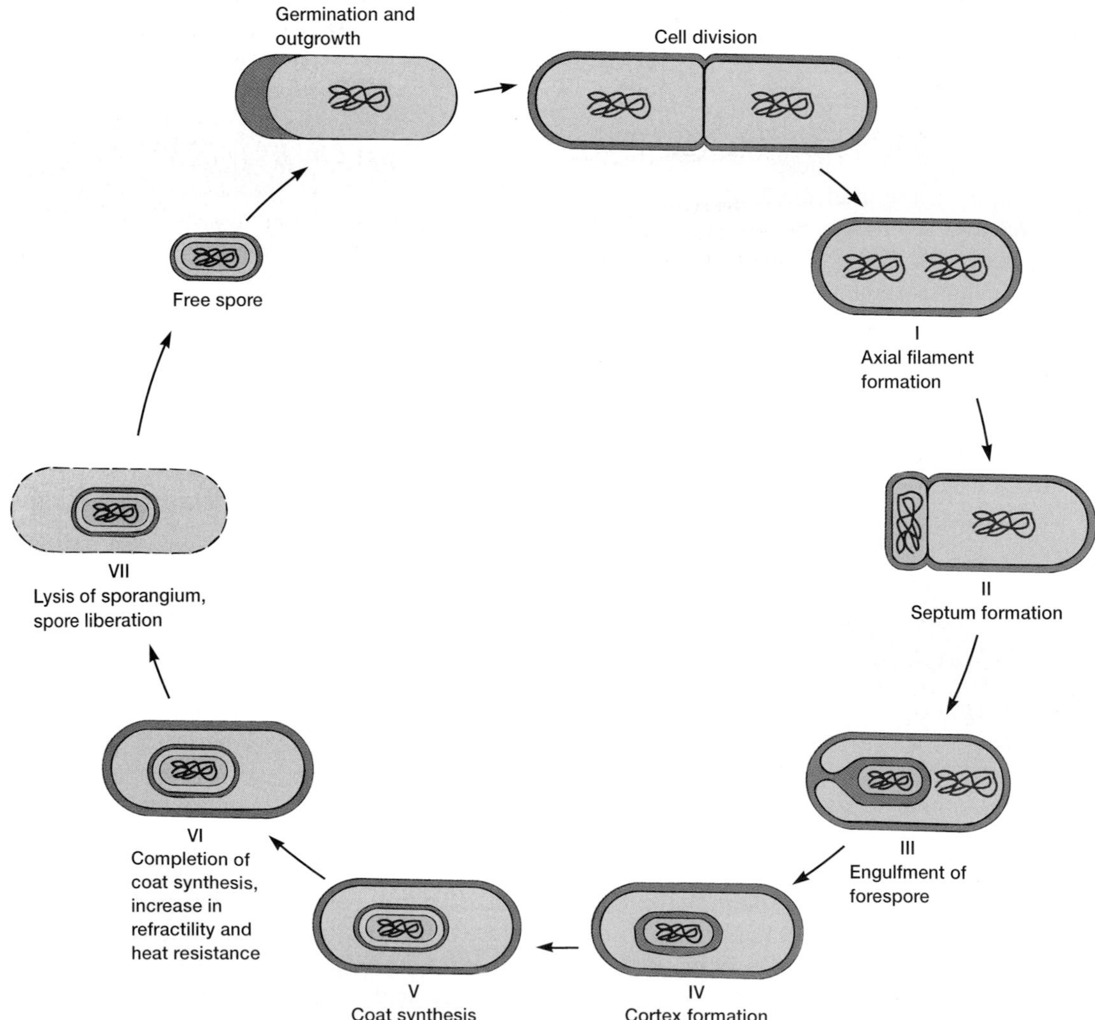

Figure 3.42 **Endospore Formation.** Life cycle of a sporulating bacterium. Stages are indicated by Roman numerals. *Reproduced, with permission, from the* Annual Review of Biochemistry, *Volume 37, © 1968 by Annual Reviews Inc.*

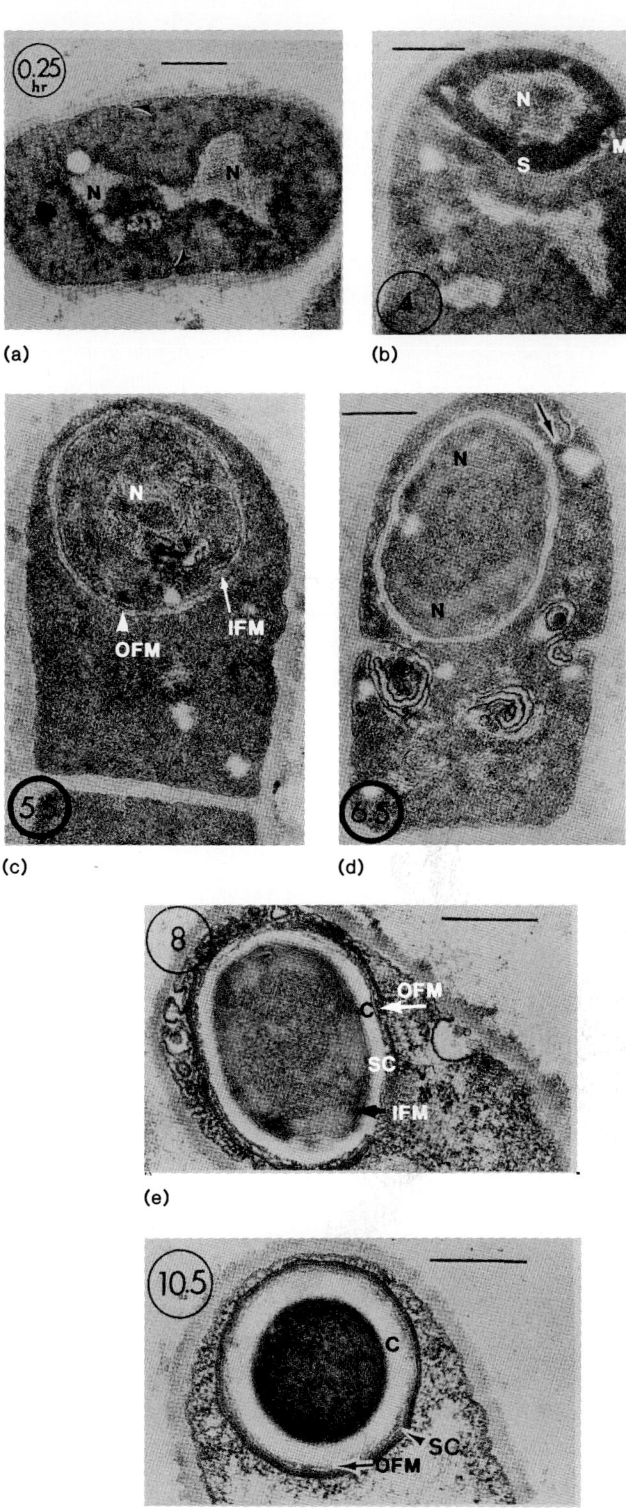

Figure 3.43 **Stages of Sporulation in *Bacillus megaterium*.** The circled numbers refer to the hours from the end of the logarithmic phase of growth. *0.25 h:* a typical vegetative cell. *4 h:* stage II cell, septation. *5.5 h:* stage III cell, engulfment. *6.5 h:* stage IV cell, cortex formation, *8 h:* stage V cell, coat formation. *10.5 h:* stage VI cell, mature spore in sporangium. *C,* cortex; *IFM* and *OFM,* inner and outer forespore membranes; *M,* mesosome; *N,* nucleoid; *S,* septum; *SC,* spore coats. The bars indicate *0.5* μm.

followed by an inward folding of the cell membrane to enclose part of the DNA and produce the forespore septum (stage II). The membrane continues to grow and engulfs the immature spore in a second membrane (stage III). Next, cortex is laid down in the space between the two membranes, and both calcium and dipicolinic acid are accumulated (stage IV). Protein coats then are formed around the cortex (stage V), and maturation of the spore occurs (stage VI). Finally, lytic enzymes destroy the sporangium releasing the spore (stage VII). Figure 3.43 shows this process taking place in *Bacillus megaterium.* Sporulation requires only about 10 hours.

The transformation of dormant spores into active vegetative cells seems almost as complex a process as sporogenesis. It occurs in three stages: (1) activation, (2) germination, and (3) outgrowth. Often an endospore will not germinate successfully, even in a rich medium packed with nutrients, unless it has been activated. Activation is a reversible process that prepares spores for germination and usually results from treatments like heating. It is followed by **germination,** the breaking of the spore's dormant state. This process is characterized by spore swelling, rupture or absorption of the spore coat, loss of resistance to heat and other stresses, loss of refractility, release of spore components, and increase in metabolic activity. Many normal metabolites or nutrients (e.g., amino acids and sugars) can trigger germination after activation. Germination is followed by the third stage, outgrowth. The spore protoplast makes new components, emerges from the remains of the spore coat, and develops again into an active bacterium (figure 3.44).

1. Describe the structure of the bacterial endospore using a labeled diagram.
2. Briefly describe endospore formation and germination. What is the importance of the endospore? What might account for its heat resistance?

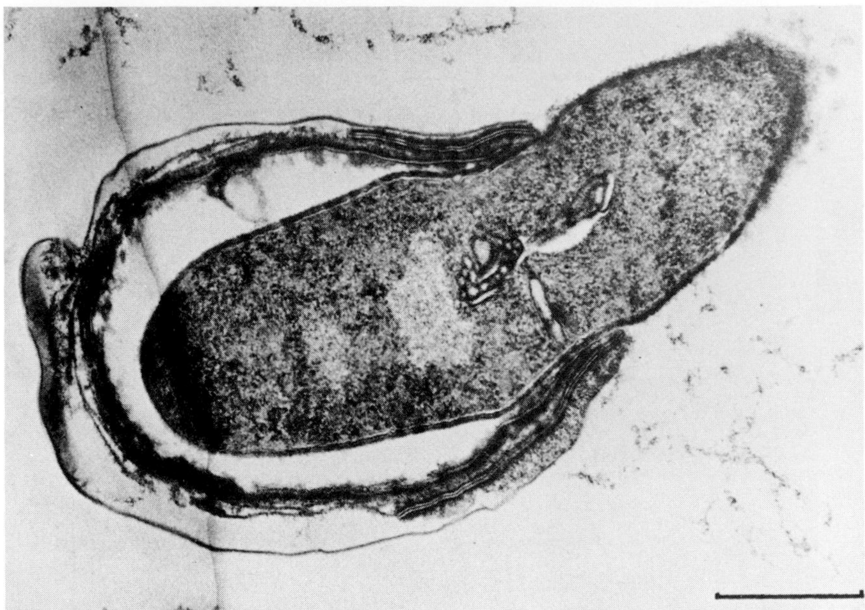

Figure 3.44 **Endospore Germination.** *Clostridium pectinovorum* emerging from the spore during germination. Bar represents 0.5 μm.

Summary

1. Bacteria may be spherical (cocci), rod-shaped (bacilli), or spiral; resemble fungi; form buds and stalks; or even have no characteristic shape at all (pleomorphic).
2. Bacterial cells can remain together after division to form pairs, chains, and clusters of various sizes and shapes.
3. All bacteria are procaryotes and much simpler structurally than eucaryotes. Table 3.1 summarizes the major functions of bacterial cell structures.
4. The plasma membrane and most other membranes are composed of a lipid bilayer in which integral proteins are buried. Peripheral proteins are more loosely attached to membranes.
5. The plasma membrane may invaginate to form some simple structures such as membrane systems containing photosynthetic and respiratory assemblies, and possibly mesosomes.
6. The cytoplasmic matrix contains inclusion bodies and ribosomes.
7. Procaryotic genetic material is located in an area called the nucleoid and is not enclosed by a membrane.
8. Most bacteria have a cell wall outside the plasma membrane to give them shape and protect them from osmotic lysis.
9. Bacterial walls are chemically complex and usually contain peptidoglycan or murein.
10. Bacteria often are classified as either gram positive or gram negative based on differences in cell wall structure and their response to Gram staining.
11. Gram-positive walls have thick, homogeneous layers of peptidoglycan and teichoic acids. Gram-negative bacteria have a thin peptidoglycan layer surrounded by a complex outer membrane containing lipopolysaccharides (LPSs) and other components.
12. Bacteria such as mycoplasmas and L forms lack a cell wall.
13. Structures such as capsules, fimbriae, and sex pili are found outside the cell wall.
14. Many bacteria are motile, usually by means of threadlike locomotory organelles called flagella.
15. Bacterial species differ in the number and distribution of their flagella.
16. The flagellar filament is a rigid helix and rotates like a propeller to push the bacterium through the water.
17. Motile bacteria can respond to gradients of attractants and repellents, a phenomenon known as chemotaxis.
18. Some bacteria survive adverse environmental conditions by forming endospores, dormant structures resistant to heat, desiccation, and many chemicals.

Key Terms

amphitrichous *57*
axial filament *60*
bacillus *42*
basal body *57*
capsule *56*
carboxysomes *48*
chemoreceptors *62*
chemotaxis *60*
coccus *42*
core *63*
core polysaccharide *54*
cortex *63*
cyanophycin granules *48*
cytoplasmic matrix *47*
deoxyribonucleic acid (DNA) *50*
dipicolinic acid *63*
diplococcus *42*
endospore *62*
envelope *50*
exoenzyme *50*
exosporium *63*
filament *57*
fimbriae *57*
flagellin *58*
flagellum *57*
fluid mosaic model *45*
gas vacuole *48*
gas vesicles *48*
germination *65*
gliding motility *60*
glycocalyx *56*
glycogen *47*
hook *57*
hydrophilic *44*
hydrophobic *44*
inclusion body *47*
integral proteins *45*
L form *56*
lipid A *54*
lipopolysaccharides (LPSs) *54*
lophotrichous *57*
lysis *55*
lysozyme *55*
magnetosomes *61*
magnetotactic bacteria *61*
mesosome *46*
metachromatic granule *48*
monotrichous *57*
murein *50*
mycelium *42*
nucleoid *50*
O antigen *54*
O side chain *54*
osmosis *55*
outer membrane *50*
penicillin *55*
peptide interbridge *51*
peptidoglycan *50*
peripheral proteins *45*
periplasm *50*
periplasmic space *50*
peritrichous *57*
plasma membrane *44*
plasmid *50*
plasmolysis *55*
pleomorphic *43*
polar flagellum *57*
poly-β-hydroxybutyrate (PHB) *47*
polyphosphate granules *48*
porin protein *55*
protoplast *47*
ribosome *49*
rod *42*
run *62*
S layer *56*
self-assembly *58*
sex pili *57*
slime layer *56*
spheroplast *55*
spirilla *42*
spirochete *42*
sporangium *63*
spore cell wall *63*
spore coat *63*
sporogenesis *64*
sporulation *64*
Svedberg unit *49*
teichoic acid *51*
tumble *62*
twiddle *62*
vibrio *42*
volutin granule *48*

Questions for Thought and Review

1. List all the major procaryotic structures discussed in this chapter and provide a brief description of the functions of each.
2. Why are procaryotic cells usually smaller than eucaryotic cells?
3. Some microbiologists believe that the plasma membrane is involved in DNA synthesis during bacterial reproduction. How might one prove that this is the case?
4. Discuss a possible mechanism of Gram staining in terms of differences in structure and chemistry between the walls of gram-positive and gram-negative bacteria.
5. What is self-assembly and why is it so important to cells?
6. How might one go about showing that a bacterium could form true endospores?

Additional Reading

Adler, J. 1976. The sensing of chemicals by bacteria. *Sci. Am.* 234(4):40–47.

Aronson, A. I., and Fitz-James, P. 1976. Structure and morphogenesis of the bacterial spore coat. *Bacteriol. Rev.* 40(2):360–402.

Balows, A.; Truper, H. G.; Dworkin, M.; Harder, W.; and Schleifer, K.-H. 1992. *The prokaryotes,* 2d ed. New York: Springer-Verlag.

Berg, H. C. 1975. How bacteria swim. *Sci. Am.* 233(2):36–44.

Beveridge, T. J. 1989. The structure of bacteria. In *Bacteria in nature,* Vol. 3, eds. J. S. Poindexter and E. R. Leadbetter, pp. 1–65. New York: Plenum.

Beveridge, T. J., and Graham, L. L. 1991. Surface layers of bacteria. *Microbiol. Rev.* 55(4):684–705.

Blakemore, R. P. 1982. Magnetotactic bacteria. *Ann. Rev. Microbiol.* 36:217–38.

Brock, T. D. 1988. The bacterial nucleus: A history. *Microbiol. Rev.* 52: 397–411.

Costerton, J. W.; Geesey, G. G.; and Cheng, K.-J. 1978. How bacteria stick. *Sci. Am.* 238(1):86–95.

Doetsch, R. N., and Sjoblad, R. D. 1980. Flagellar structure and function in eubacteria. *Ann. Rev. Microbiol.* 34:69–108.

Ferris, F. G., and Beveridge, T. J. 1985. Functions of bacterial cell surface structures. *BioScience* 35(3): 172–77.

Gest, H., and Mandelstam, J. 1987. Longevity of microorganisms in natural environments. *Microbiol. Sci.* 4(3):69–71.

Hancock, R. E. W. 1991. Bacterial outer membranes: Evolving concepts. *ASM News* 57(4):175–82.

Henning, U. 1975. Determination of cell shape in bacteria. *Ann. Rev. Microbiol.* 29:45–60.

Koch, A. L. 1990. Growth and form of the bacterial cell wall. *Am. Sci.* 78:327–41.

Koval, S. F., and Murray, R. G. E. 1986. The superficial protein arrays on bacteria. *Microbiol. Sci.* 3(12):357–61.

Lederberg, J. 1992. *Encyclopedia of microbiology.* San Diego, CA: Academic Press.

Mayer, F. 1986. *Cytology and morphogenesis of bacteria.* Berlin: Gebruder Borntraeger.

Murray, R. G. E. 1978. Form and function—I. Bacteria. In *Essays in microbiology,* eds. J. R. Norris and M. H. Richmond, 2/1–2/32. New York: John Wiley and Sons.

Neidhardt, F. C.; Ingraham, J. L.; and Schaechter, M. 1990. *Physiology of the bacterial cell: A molecular approach.* Sunderland, Mass.: Sinauer Associates.

Osborne, M. J., and Wu, H. C. P. 1980. Proteins of the outer membrane of gram-negative bacteria. *Ann. Rev. Microbiol.* 34:369–422.

Ottow, J. C. G. 1975. Ecology, physiology, and genetics of fimbriae and pili. *Ann. Rev. Microbiol.* 29:79–108.

Ourisson, G.; Albrecht, P.; and Rohmer, M. 1984. The microbial origin of fossil fuels. *Sci. Am.* 251(2):44–51.

Raetz, C. R. H. 1990. Biochemistry of endotoxins. *Ann. Rev. Biochem.* 59:129–70.

Rogers, H. J. 1983. *Bacterial cell structure.* Washington: American Society for Microbiology.

Salton, M. R. J., and Owen, P. 1976. Bacterial membrane structure. *Ann. Rev. Microbiol.* 30:451–82.

Sargent, M. G. 1985. How do bacterial nuclei divide? *Microbiol. Sci.* 2(8):235–39.

Scherrer, R. 1984. Gram's staining reaction, Gram types and cell walls of bacteria. *Trends Biochem. Sci.* 9:242–45.

Schmidt, M. B. 1988. Structure and function of the bacterial chromosome. *Trends Biochem. Sci.* 13(4):131–35.

Sharon, N. 1969. The bacterial cell wall. *Sci. Am.* 221(5):92–98.

Shively, J. M. 1974. Inclusion bodies of prokaryotes. *Ann. Rev. Microbiol.* 28:167–87.

Silverman, M., and Simon, M. I. 1977. Bacterial flagella. *Ann. Rev. Microbiol.* 31:397–419.

Slepecky, R. A. 1978. Resistant forms. In *Essays in microbiology,* eds. J. R. Norris and M. H. Richmond, 14/1–14/31. New York: John Wiley and Sons.

Stanier, R. Y.; Ingraham, J. L.; Wheelis, M. L.; and Painter, P. R. 1986. *The microbial world,* 5th ed. Englewood Cliffs, N.J.: Prentice-Hall.

Troy, F. A. 1979. The chemistry and biosynthesis of selected bacterial capsular polymers. *Ann. Rev. Microbiol.* 33:519–60.

Walsby, A. E. 1977. The gas vacuoles of blue-green algae. *Sci. Am.* 237(2):90–97.

Wittmann, H. G. 1983. Architecture of prokaryotic ribosomes. *Ann. Rev. Biochem.* 52:35–65.

CHAPTER 4
Eucaryotic Cell Structure and Function

The key to every biological problem must finally be sought in the cell.

—E. B. Wilson

Outline

Concepts

1. Eucaryotic cells differ most obviously from procaryotic cells in having a variety of complex membranous organelles in the cytoplasmic matrix and their genetic material within membrane-delimited nuclei. Each organelle has a distinctive structure directly related to specific functions.
2. A cytoskeleton composed of microtubules, microfilaments, and intermediate filaments helps give eucaryotic cells shape; microtubules and microfilaments are also involved in cell movements and intracellular transport.
3. In eucaryotes, genetic material is distributed between cells by the highly organized, complex processes called mitosis and meiosis.
4. Despite great differences between eucaryotes and procaryotes with respect to such things as morphology, they are similar on the biochemical level.

In chapter 3, considerable attention is devoted to procaryotic cell structure and function because bacteria are immensely important in microbiology and have occupied a large portion of microbiologists' attention in the past. Nevertheless, eucaryotic algae, fungi, and protozoa also are microorganisms and have been extensively studied. These organisms are often extraordinarily complex, interesting in their own right, and prominent members of the ecosystem (figure 4.1*c*). In addition, fungi (and to some extent, algae) are exceptionally useful in industrial microbiology. Many fungi and protozoa are also major human pathogens; one only need think of either malaria or African sleeping sickness (figure 4.1*c*) to appreciate the significance of eucaryotes in pathogenic microbiology. So, although this text emphasizes bacteria, eucaryotic microorganisms are discussed at many points.

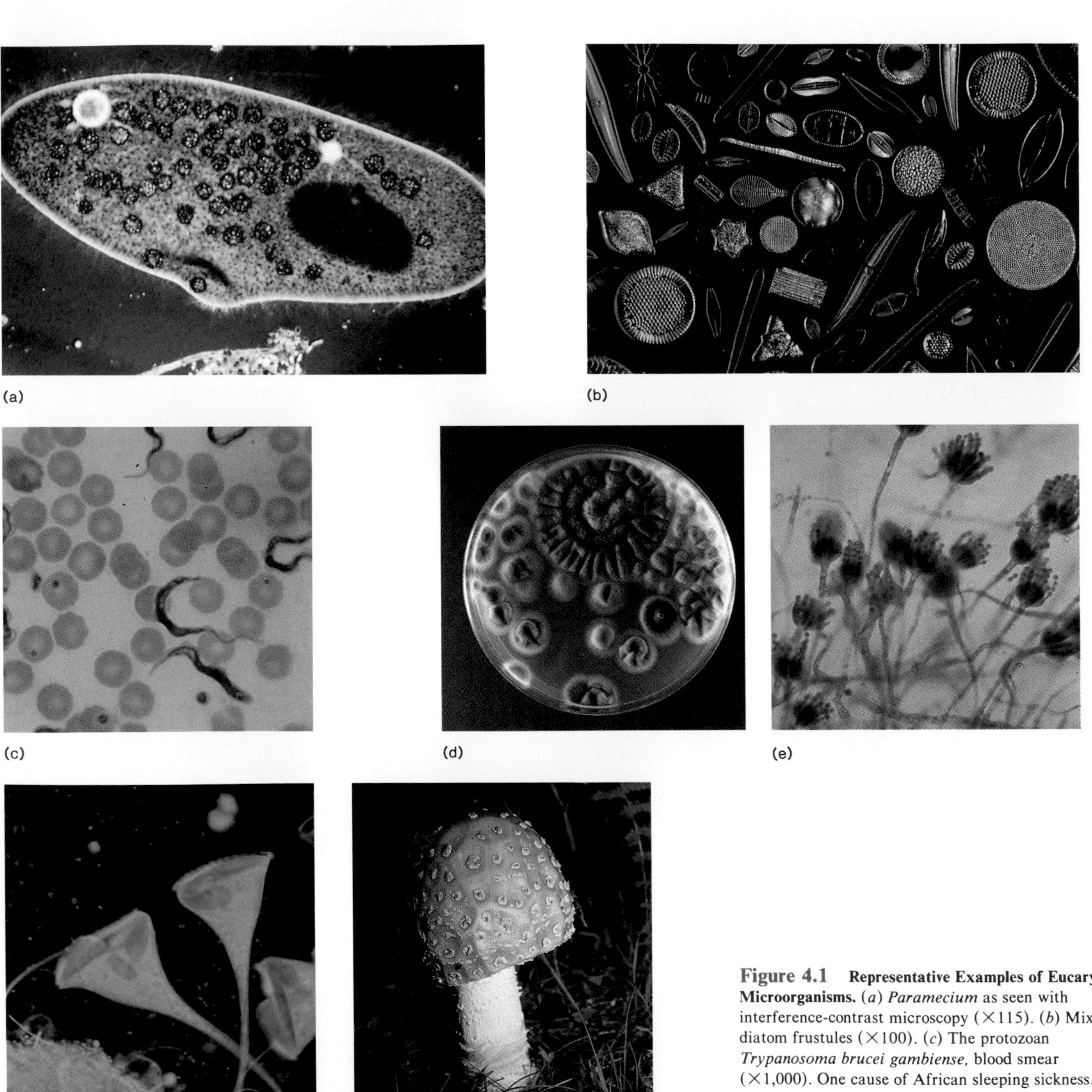

Figure 4.1 Representative Examples of Eucaryotic Microorganisms. (*a*) *Paramecium* as seen with interference-contrast microscopy (×115). (*b*) Mixed diatom frustules (×100). (*c*) The protozoan *Trypanosoma brucei gambiense,* blood smear (×1,000). One cause of African sleeping sickness. (*d*) *Penicillium* colonies, and (e) a microscopic view of the mold's hyphae and conidia (×220). (*f*) *Stentor.* The ciliated protozoa are extended and actively feeding, dark-field microscopy (×100). (*g*) *Amanita muscaria,* a large poisonous mushroom (×5).

Chapter 4 focuses on eucaryotic cell structure and its relationship to cell function. Because many valuable studies on eucaryotic cell ultrastructure have used organisms other than microorganisms, some work on nonmicrobial cells is presented. At the end of the chapter, procaryotic and eucaryotic cells are compared in some depth.

An Overview of Eucaryotic Cell Structure

The most obvious difference between eucaryotic and procaryotic cells is in their use of membranes. Eucaryotic cells have membrane-delimited nuclei, and membranes also play a prominent part in the structure of many other organelles (figures 4.2 and 4.3). **Organelles** are intracellular structures that perform specific functions in cells. The name organelle (little organ) was coined because biologists saw a parallel between the relationship of organelles to a cell and that of organs to

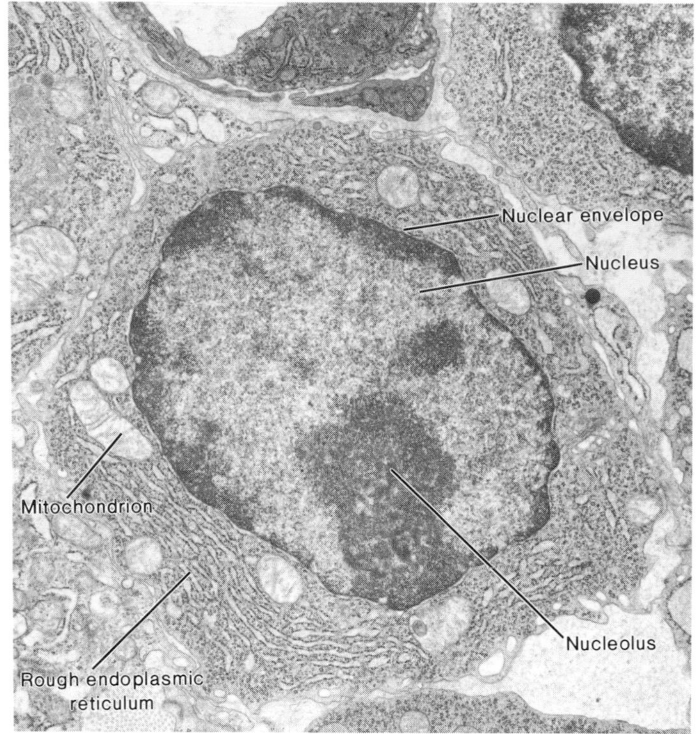

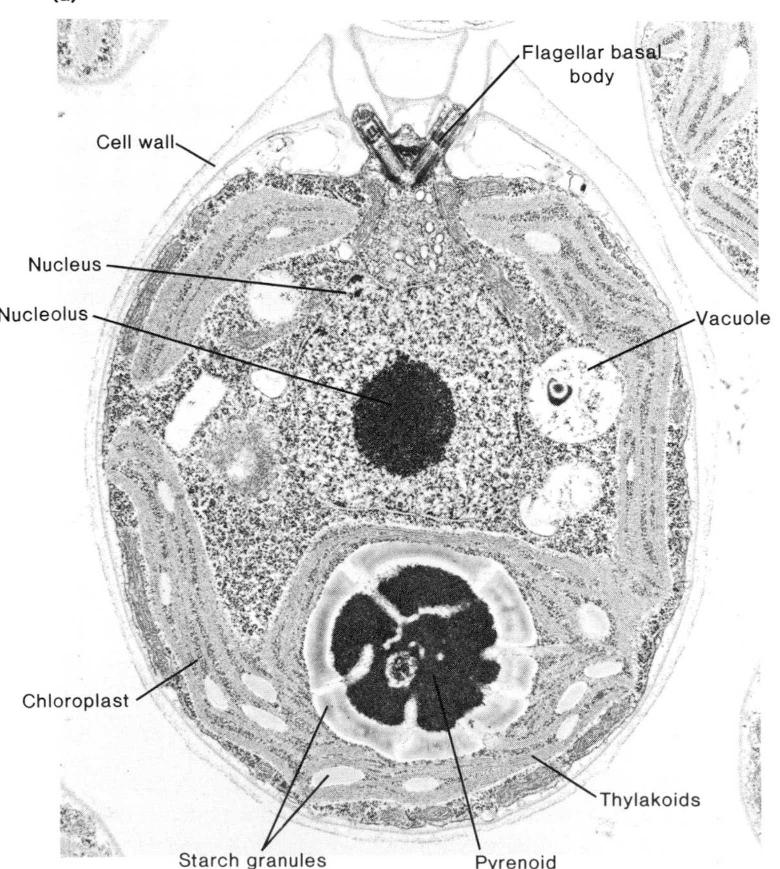

Figure 4.2 Eucaryotic Cell Ultrastructure. (*a*) A lymphoblast in the rat lymph node (×17,500). (*b*) The alga *Chlamydomonas reinhardtii,* a deflagellated cell. Note the large chloroplast with its pyrenoid body (×30,000).

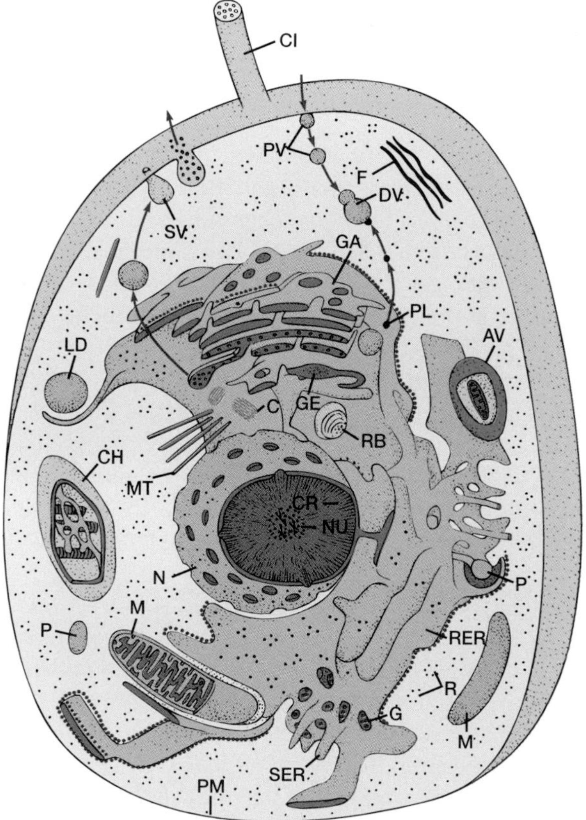

Figure 4.3 Eucaryotic Cell Ultrastructure. This is a schematic, three-dimensional diagram of a cell with the most important organelles identified in the illustration. *AV,* autophagic vacuole; *C,* centriole; *CH,* chloroplast; *CI,* cilium; *CR,* chromatin; *DV,* digestion vacuole; *F,* microfilaments; *G,* glycogen; *GA,* Golgi apparatus; *GE,* GERL; *LD,* lipid droplet; *M,* mitochondrion; *MT,* microtubules; *N,* nucleus; *NU,* nucleolus; *P,* peroxisome; *PL,* primary lysosome; *PM,* plasma membrane; *PV,* pinocytotic vesicle; *R,* ribosomes and polysomes; *RB,* residual body; *RER,* rough endoplasmic reticulum; SER, smooth endoplasmic reticulum; *SV,* secretion vacuole.

the whole body. A comparison of figures 4.2 and 4.3 with figure 3.10 (*p.* 47) shows how much more structurally complex the eucaryotic cell is. This complexity is due chiefly to the use of internal membranes for several purposes. The partitioning of the eucaryotic cell interior by membranes makes possible the placement of different biochemical and physiological functions in separate compartments so that they can more easily take place simultaneously under independent control and proper coordination. Large membrane surfaces make possible greater respiratory and photosynthetic activity because these processes are located exclusively in membranes. The intracytoplasmic membrane complex also serves as a transport system to move materials between different cell locations. Thus, abundant membrane systems probably are necessary in eucaryotic cells because of their large volume and the need for adequate regulation, metabolic activity, and transport.

Figures 4.2 and 4.3 provide a generalized view of eucaryotic cell structure and illustrate most of the organelles to be discussed. Table 4.1 briefly summarizes the functions of the major eucaryotic organelles. Those organelles lying inside the plasma membrane are first described, and then components outside the membrane are discussed.

TABLE 4.1 Functions of Eucaryotic Organelles

Organelle	Function
Cytoplasmic matrix	Environment for other organelles, location of many metabolic processes
Microfilaments, intermediate filaments, and microtubules	Cell structure and movements, form the cytoskeleton
Endoplasmic reticulum	Transport of materials, protein and lipid synthesis
Ribosomes	Protein synthesis
Golgi apparatus	Packaging and secretion of materials for various purposes, lysosome formation
Lysosomes	Intracellular digestion
Mitochondria	Energy production through use of the tricarboxylic acid cycle, electron transport, oxidative phosphorylation, and other pathways
Chloroplasts	Photosynthesis—trapping light energy and formation of carbohydrate from CO_2 and water
Nucleus	Repository for genetic information, control center for cell
Nucleolus	Ribosomal RNA synthesis, ribosome construction
Cell wall and pellicle	Strengthen cell and give it shape
Cilia and flagella	Cell movement

The Cytoplasmic Matrix, Microfilaments, Intermediate Filaments, and Microtubules

When a eucaryotic cell is examined at low power with the electron microscope, its larger organelles are seen to lie in an apparently featureless, homogeneous substance called the **cytoplasmic matrix.** The matrix, although superficially uninteresting, is actually one of the most important and complex parts of the cell. It is the "environment" of the organelles and the location of many important biochemical processes. Several physical changes seen in cells—viscosity changes, cytoplasmic streaming, and others—also are due to matrix activity.

Water constitutes about 70 to 85% by weight of a eucaryotic cell. Thus a large part of the cytoplasmic matrix is water. Cellular water can exist in two different forms. Some of it is bulk or free water; this is normal, osmotically active water.

Osmosis, water activity, and growth (pp. 55, 122–23).

Water also can exist as bound water or water of hydration. This water is bound to the surface of proteins and other macromolecules, and is osmotically inactive and more ordered than bulk water. There is some evidence that much of metabolism occurs in bound water. The protein content of cells is so high that the cytoplasmic matrix often may be semicrystalline. Usually matrix pH is around neutrality, about pH 6.8 to 7.1, but can vary widely. For example, protozoan digestive vacuoles may reach pHs as low as 3 to 4.

Probably all eucaryotic cells have **microfilaments,** minute protein filaments, 4 to 7 nm in diameter, which may be either scattered within the cytoplasmic matrix or organized into networks and parallel arrays. Microfilaments play a major role in cell motion and shape changes. Some examples of cellular movements associated with microfilament activity are the motion of pigment granules, amoeboid movement, and protoplasmic streaming in slime molds (*see chapter 26*).

The participation of microfilaments in cell movement is suggested by electron microscopic studies showing that they are frequently found at locations appropriate for such a role. For example, they are concentrated at the interface between stationary and flowing cytoplasm in plant cells and slime molds. Experiments using the drug cytochalasin B have provided additional evidence. Cytochalasin B disrupts microfilament structure, and often simultaneously inhibits cell movements. However, because the drug has additional effects in cells, a direct cause-and-effect interpretation of these experiments is sometimes difficult.

Microfilament protein has been isolated and analyzed chemically. It is an actin, very similar to the actin contractile protein of muscle tissue. This is further indirect evidence for microfilament involvement in cell movement.

A second type of small filamentous organelle in the cytoplasmic matrix is shaped like a thin cylinder about 25 nm in diameter. Because of its tubular nature this organelle is called a **microtubule.** Microtubules are complex structures constructed of two slightly different spherical protein subunits named tubulins, each of which is approximately 4 to 5 nm in diameter. These subunits are assembled in a helical arrangement to form a cylinder with an average of 13 subunits in one turn or circumference (figure 4.4).

Microtubules serve at least three purposes: (1) they help maintain cell shape, (2) are involved with microfilaments in cell movements, and (3) participate in the intracellular transport of substances. Evidence for a structural role comes from

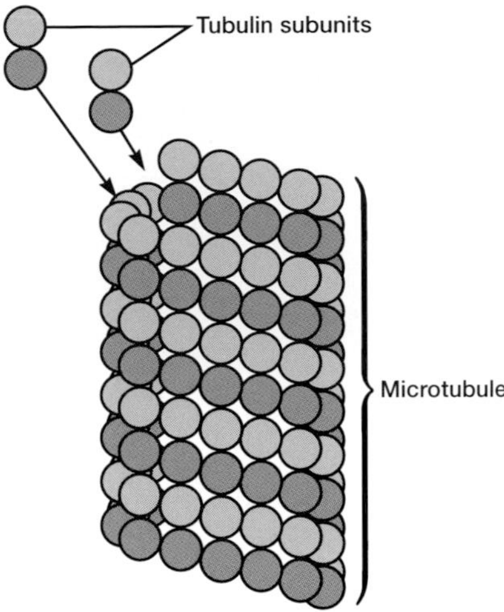

Figure 4.4 Microtubule Structure. The hollow cylinder, about 25 nm in diameter, is made of two kinds of protein subunits, α-tubulin. and β-tubulin.

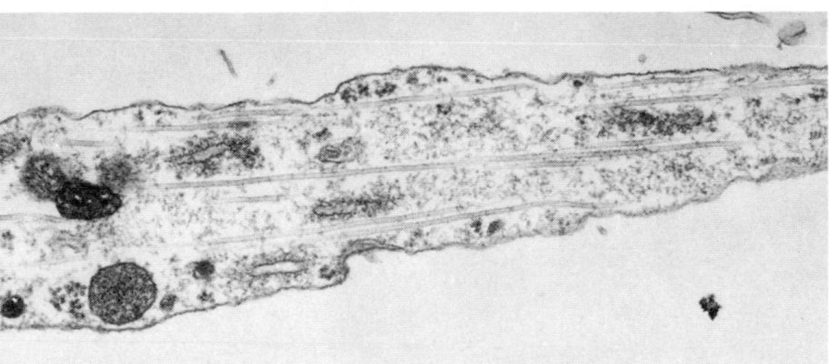
(a)

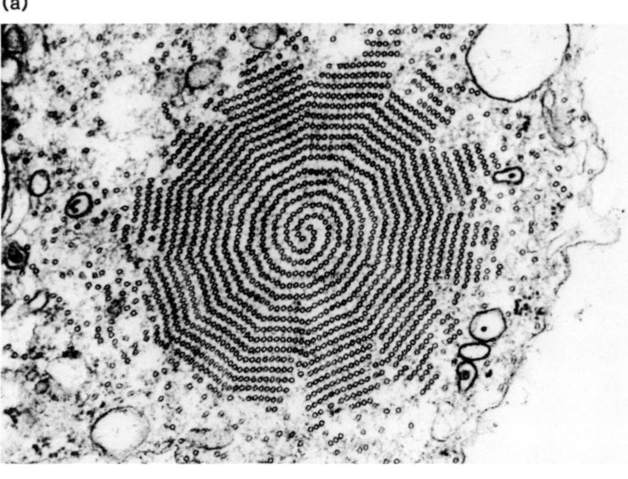
(b)

Figure 4.5 Cytoplasmic Microtubules. Electron micrographs of pseudopodia with microtubules. (*a*) Microtubules in a pseudopodium from the protozoan *Reticulomyxa* (×65,000). (*b*) A transverse section of a heliozoan axopodium (×48,000). Note the parallel array of microtubules organized in a spiral pattern.

their intracellular distribution and studies on the effects of the drug colchicine. Long, thin cell structures requiring support such as the axopodia (long, slender, rigid pseudopodia) of protozoa contain microtubules (figure 4.5). When migrating embryonic nerve and heart cells are exposed to colchicine, they simultaneously lose their microtubules and their characteristic shapes. The shapeless cells seem to wander aimlessly as if incapable of directed movement without their normal form. Their microfilaments are still intact, but due to the disruption of their microtubules by colchicine, they no longer behave normally.

Microtubules are also present in structures that participate in cell or organelle movements—the mitotic spindle, cilia, and flagella. For example, the mitotic spindle is constructed of microtubules; when a dividing cell is treated with colchicine, the spindle is disrupted and chromosome separation blocked. Microtubules are also essential to the movement of eucaryotic cilia and flagella.

Other kinds of filamentous components also are present in the matrix, the most important of which are the **intermediate filaments** (about 8 to 10 nm in diameter). The microfilaments, microtubules, and intermediate filaments are major components of a vast, intricate network of interconnected filaments called the **cytoskeleton** (figure 4.6). As mentioned previously, the cytoskeleton plays a role in both cell shape and movement. Procaryotes lack an organized cytoskeleton, and may not possess actinlike proteins.

An irregular lattice of slender, protein-rich strands also permeates the matrix and connects microtubules and other organelles (figure 4.7). This microtrabecular lattice is thought to give the matrix structure and organization, and may be involved in the movement of materials within cells. Some cell biologists believe that the lattice does not actually exist, but is an artifact generated during sample preparation. Further research should clarify its precise nature.

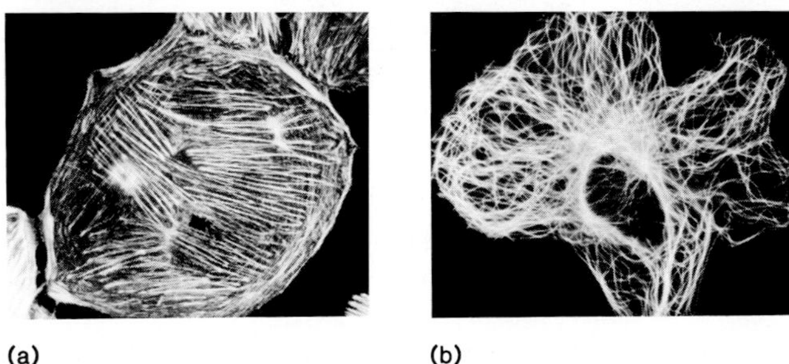
(a) (b)

Figure 4.6 The Eucaryotic Cytoskeleton. (*a*) Antibody-stained microfilament system in a mammal cell (×400). (*b*) Antibody-stained microtubule system in a mammal cell (×1,000).

1. What is an organelle?
2. Define cytoplasmic matrix, bulk or free water, bound water, microfilament, microtubule, and tubulin. Discuss the roles of microfilaments, intermediate filaments, and microtubules.
3. Describe the cytoskeleton. What are its functions?

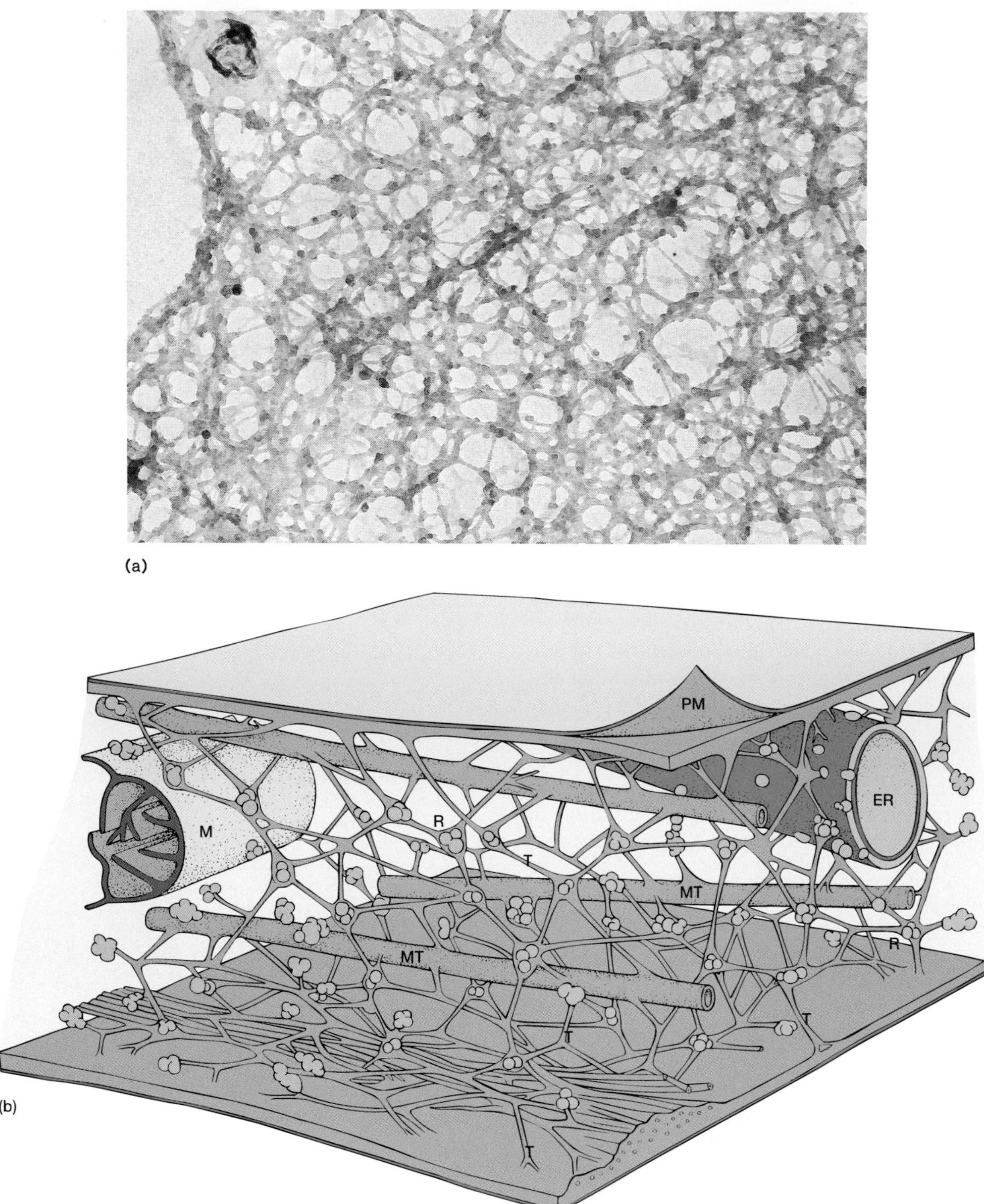

Figure 4.7 **The Microtrabecular Lattice.** (*a*) The lattice of an African green monkey kidney cell (×120,000). (*b*) A model of the microtrabecular lattice, *T,* and its relationship to other organelles: endoplasmic reticulum, *ER;* microtubules, *MT;* mitochondria, *M;* plasma membrane, *PM;* and free ribosomes or polysomes, *R* (approximately ×150,000).

The Endoplasmic Reticulum

Besides the cytoskeleton and microtrabecular lattice, the cytoplasmic matrix is permeated with an irregular network of branching and fusing membranous tubules, around 40 to 70 nm in diameter, and many flattened sacs called **cisternae** (s., **cisterna**). This network of tubules and cisternae is the **endoplasmic reticulum (ER)** (figures 4.2*a* and 4.8). The nature of the ER varies with the functional and physiological status of the cell. In cells synthesizing a great deal of protein for purposes such as secretion, a large part of the ER is studded on its outer surface with ribosomes and is called **rough** or **granular endoplasmic reticulum (RER** or **GER)**. Other cells, such as those producing large quantities of lipids, have ER that lacks ribosomes. This is **smooth** or **agranular ER (SER** or **AER)**.

The endoplasmic reticulum has many important functions. It transports proteins, lipids, and probably other materials through the cell. Lipids and proteins are synthesized by ER-associated enzymes and ribosomes. Polypeptide chains synthesized on RER-bound ribosomes may be inserted either into the ER membrane or into its lumen for transport elsewhere. The ER is also a major site of cell membrane synthesis.

New endoplasmic reticulum is produced through expansion of the old. Many biologists think the RER synthesizes new ER proteins and lipids. "Older" RER then loses its connected ribosomes and is modified to become SER. Not everyone agrees with this interpretation, and other mechanisms of growth of ER are possible.

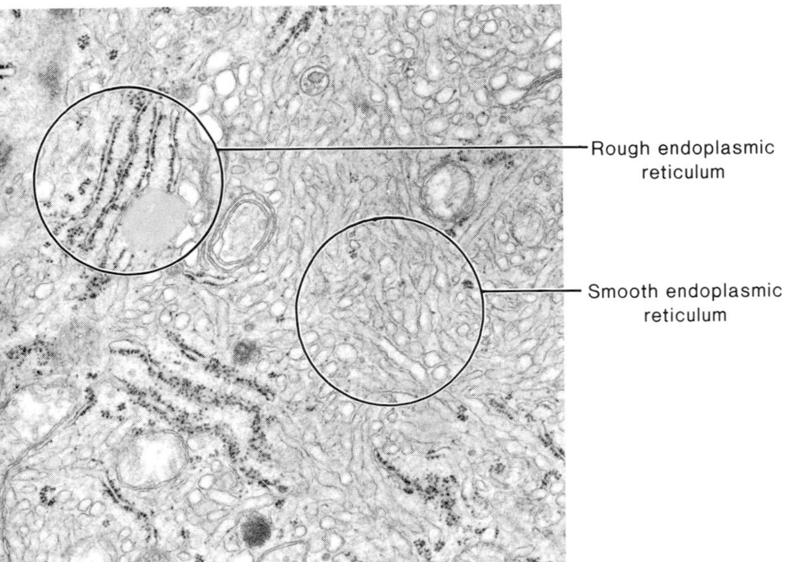

Figure 4.8 The Endoplasmic Reticulum. A transmission electron micrograph of the human corpus luteum showing structural variations in eucaryotic endoplasmic reticulum. Note the presence of both rough endoplasmic reticulum lined with ribosomes and smooth endoplasmic reticulum without ribosomes (×26,500).

The Golgi Apparatus

The **Golgi apparatus** is a membranous organelle composed of flattened, saclike cisternae stacked on each other (figures 4.2*b*, 4.9, and 4.29). These membranes, like the smooth ER, lack bound ribosomes. There are usually around 4 to 8 cisternae or sacs in a stack, although there may be many more. Each sac is 15 to 20 nm thick and separated from other cisternae by 20 to 30 nm. A complex network of tubules and vesicles (20 to 100 nm in diameter) is located at the edges of the cisternae. The stack of cisternae has a definite polarity because there are two ends or faces that are quite different from one another. The sacs on the cis or forming face are often associated with the ER and differ from the sacs on the trans or maturing face in thickness, enzyme content, and degree of vesicle formation. It appears that material is transported from cis to trans cisternae by vesicles that bud off the cisternal edges and move to the next sac.

The Golgi apparatus is present in most eucaryotic cells, but many fungi and ciliate protozoa may lack a well-formed structure. Sometimes it consists of a single stack of cisternae; however, many cells may contain up to 20, and sometimes more, separate stacks. These stacks of cisternae, often called **dictyosomes,** can be clustered in one region or scattered about the cell.

The Golgi apparatus packages materials and prepares them for secretion, the exact nature of its role varying with the organism. The surface scales of some flagellated algae and radiolarian protozoa appear to be constructed within the Golgi apparatus and then transported to the surface in vesicles. It often participates in the development of cell membranes and

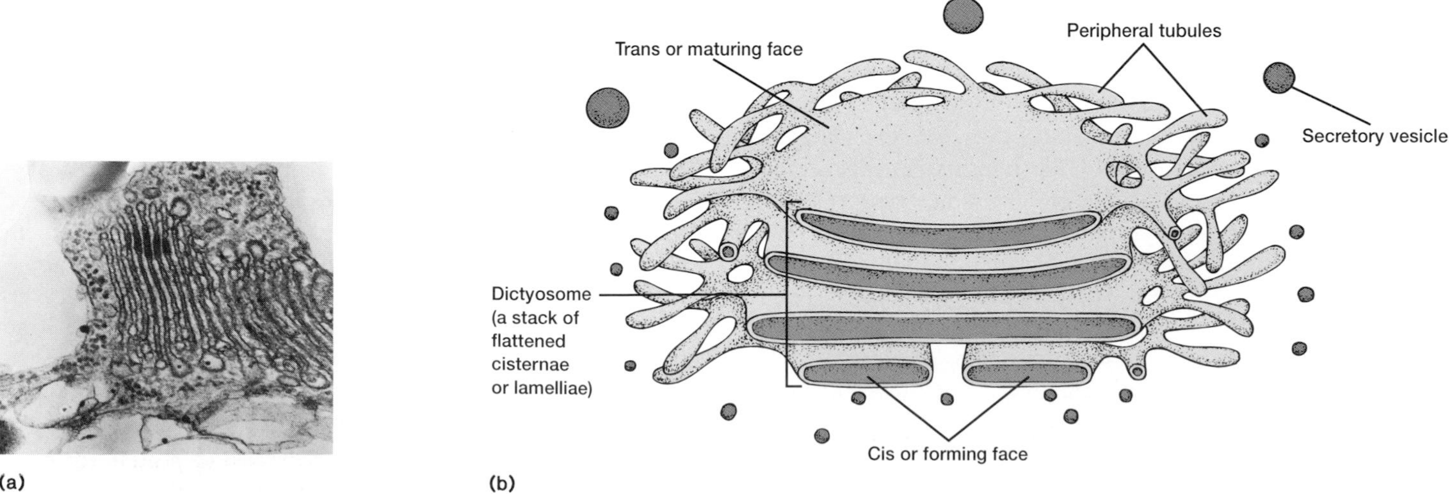

Figure 4.9 Golgi Apparatus Structure. Golgi apparatus of *Euglena gracilis*. Cisternal stacks are shown in the electron micrograph (×165,000) in (*a*) and diagrammatically in (*b*).

in the packaging of cell products. The growth of some fungal hyphae occurs when Golgi vesicles contribute their contents to the wall at the hyphal tip.

In all these processes, materials move from the ER to the Golgi apparatus. Most often, vesicles bud off the ER, travel to the Golgi apparatus, and fuse with the cis cisternae. Thus, the Golgi apparatus is closely related to the ER in both a structural and a functional sense. Most proteins entering the Golgi apparatus from the ER are glycoproteins containing short carbohydrate chains. The Golgi apparatus frequently modifies proteins destined for different fates by adding specific groups and then sends the proteins on their way to the proper location (e.g., lysosomal proteins have phosphates added to their mannose sugars).

Lysosomes and Endocytosis

A very important function of the Golgi apparatus and endoplasmic reticulum is the synthesis of another organelle, the **lysosome.** This organelle (or a structure very much like it) is found in a variety of microorganisms—protozoa, some algae, and fungi—as well as in plants and animals. Lysosomes are roughly spherical and enclosed in a single membrane; they average about 500 nm in diameter, but range from 50 nm to several μm in size. They are involved in intracellular digestion and contain the enzymes needed to digest all types of macromolecules. These enzymes, called hydrolases, catalyze the hydrolysis of molecules and function best under slightly acid conditions (usually around pH 3.5 to 5.0). Lysosomes maintain an acidic environment by pumping protons into their interior. Digestive enzymes are manufactured by the RER and packaged to form lysosomes by the Golgi apparatus. A segment of smooth ER near the Golgi apparatus also may bud off lysosomes.

Lysosomes are particularly important in those cells that obtain nutrients through **endocytosis.** In this process, a cell takes up solutes or particles by enclosing them in vacuoles and vesicles pinched off from its plasma membrane. Vacuoles and vesicles are membrane-delimited cavities that contain fluid, and often solid material. Larger cavities will be called vacuoles, and smaller cavities, vesicles. There are two major forms of endocytosis: phagocytosis and pinocytosis. During **phagocytosis,** large particles and even other microorganisms are enclosed in a phagocytic vacuole or phagosome and engulfed (figure 4.10*a*). In **pinocytosis,** small amounts of the surrounding liquid with its solute molecules are pinched off as tiny pinocytotic vesicles (also called pinocytic vesicles) or pinosomes. Often phagosomes and pinosomes are collectively called **endosomes** because they are formed by endocytosis. The type of pinocytosis, receptor-mediated endocytosis, that produces coated vesicles (*see p. 388*) is important in the entry of animal viruses into host cells.

Material in endosomes is digested with the aid of lysosomes. Newly formed lysosomes, or **primary lysosomes,** fuse with phagocytic vacuoles to yield **secondary lysosomes,** lysosomes with material being digested (figure 4.10). These phagocytic vacuoles or secondary lysosomes often are called food vacuoles. Digested nutrients then leave the secondary lysosome and enter the cytoplasm. When the lysosome has accumulated large quantities of indigestible material, it is known as a **residual body.**

Lysosomes join with phagosomes for defensive purposes as well as to acquire nutrients. Invading bacteria, ingested by a phagocytic cell, usually are destroyed when lysosomes fuse with the phagosome. This is commonly seen in leukocytes (white blood cells) of vertebrates.

Phagocytosis and resistance to pathogens (pp. 599–600).

Cells can selectively digest portions of their own cytoplasm in a type of secondary lysosome called an **autophagic vacuole** (figure 4.10*a*). It is thought that these arise by lysosomal engulfment of a piece of cytoplasm (figure 4.11), or when the ER pinches off cytoplasm to form a vesicle that subsequently fuses with lysosomes. Autophagy probably plays a role in the normal turnover or recycling of cell constituents. A cell also can survive a period of starvation by selectively digesting portions of itself to remain alive. Following cell death, lysosomes aid in digestion and removal of cell debris.

A most remarkable thing about lysosomes is that they accomplish all these tasks without releasing their digestive enzymes into the cytoplasmic matrix, a catastrophe that would destroy the cell. The lysosomal membrane retains digestive enzymes and other macromolecules while allowing small digestion products to leave.

The intricate complex of membranous organelles composed of the Golgi apparatus, lysosomes, endosomes, and associated structures seems to operate as a coordinated whole whose main function is the import and export of materials (figure 4.11). Christian de Duve has suggested that this complex be called the vacuome in recognition of its functional unity. The ER manufactures secretory proteins and membrane, and contributes these to the Golgi apparatus. The Golgi apparatus then forms secretory vesicles that fuse with the plasma membrane and release material to the outside. It also produces lysosomes that fuse with endosomes to digest material acquired through phagocytosis and pinocytosis. Membrane movement in the region of the vacuome lying between the Golgi apparatus and the plasma membrane is two-way. Empty vesicles often are recycled and returned to the Golgi apparatus and plasma membrane rather than being destroyed. These exchanges in the vacuome occur without membrane rupture so that vesicle contents never escape directly into the cytoplasmic matrix.

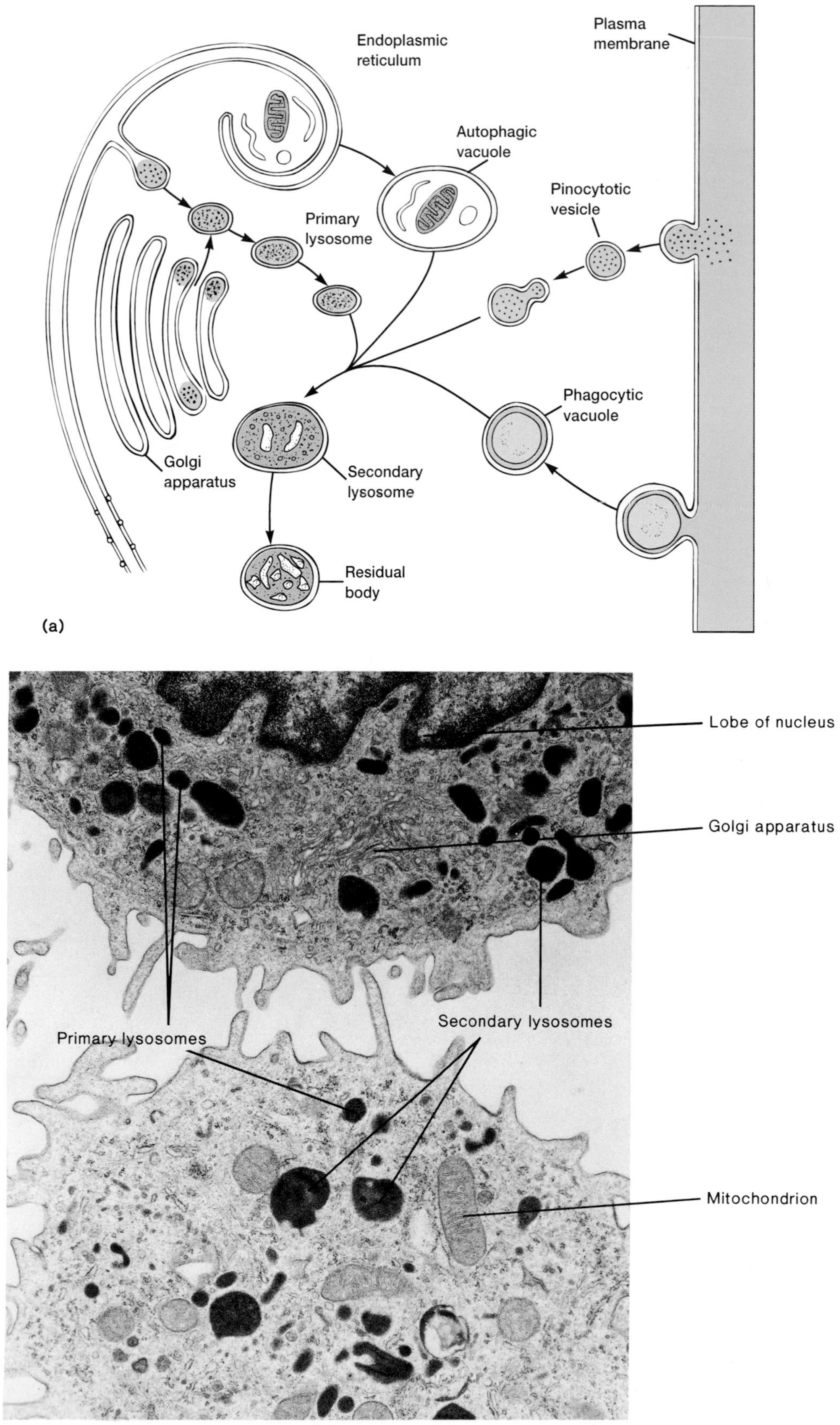

Figure 4.10 Lysosome Structure, Formation, and Function. (*a*) A diagrammatic overview of lysosome formation and function. (*b*) Lysosomes in macrophages from the lung. Secondary lysosomes contain partially digested material and are formed by fusion of primary lysosomes and phagocytic vacuoles (×14,137).

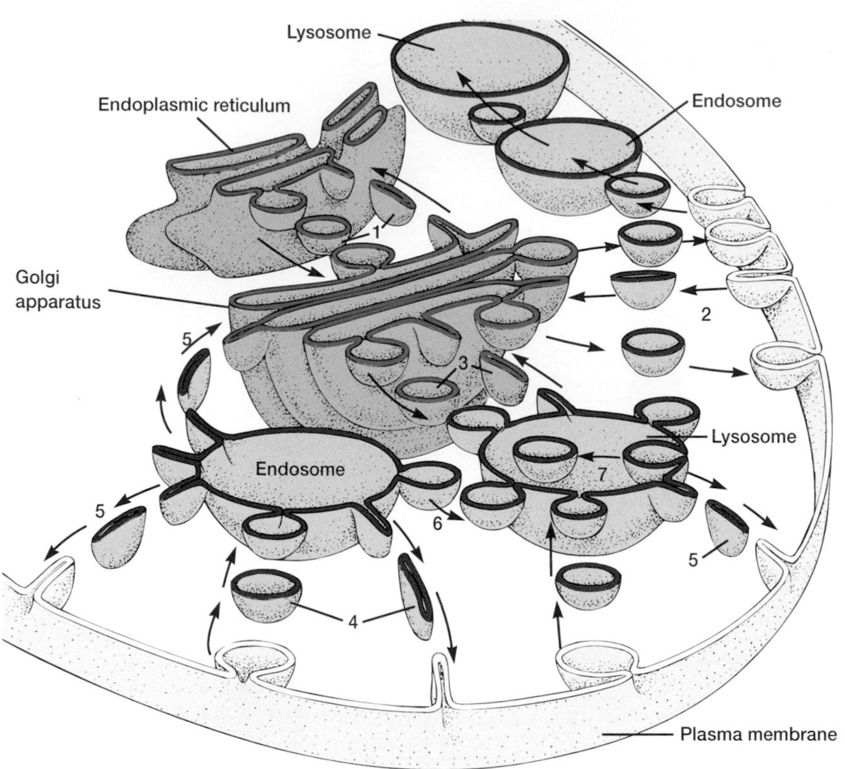

Figure 4.11 Membrane Flow in the Vacuome. The flow of material and membranes between organelles in a eucaryotic cell.
(*1*) Vesicles shuttling between the ER and Golgi apparatus.
(*2*) The Golgi-plasma membrane shuttle for secretion of materials.
(*3*) The Golgi-lysosome shuttle.
(*4*) The movement of material and membranes during endocytosis.
(*5*) Pathways of plasma membrane recovery from endosomes, lysosomes, and through the Golgi apparatus.
(*6*) Movement of vesicles from endosomes to lysosomes.
(*7*) Autophagy by a lysosome.

1. How do the rough and smooth endoplasmic reticulum differ from one another in terms of structure and function? List the processes in which the ER is involved.
2. Describe the structure of a Golgi apparatus in words and with a diagram. How do the cis and trans faces of the Golgi apparatus differ? List the major Golgi apparatus functions discussed in the text.
3. How are lysosomes formed? Describe the various forms of lysosomes and the way in which they participate in intracellular digestion. What is an autophagic vacuole? Define endocytosis, pinocytosis, and phagocytosis.

Eucaryotic Ribosomes

The eucaryotic ribosome can either be associated with the endoplasmic reticulum or be free in the cytoplasmic matrix, and is larger than the bacterial 70S ribosome. It is a dimer of a 60S and a 40S subunit, about 22 nm in diameter, and has a sedimentation coefficient of 80S and a molecular weight of 4 million. When bound to the endoplasmic reticulum to form rough ER, it is attached through its 60S subunit.

Both free and RER-bound ribosomes synthesize proteins. As mentioned earlier, proteins made on the ribosomes of the RER either enter its lumen for transport, and often for secretion, or are inserted into the ER membrane as integral membrane proteins. Free ribosomes are the sites of synthesis for nonsecretory and nonmembrane proteins. Some proteins synthesized by free ribosomes are inserted into organelles such as the nucleus, mitochondrion, and chloroplast. Several ribosomes usually attach to a single messenger RNA and simultaneously translate its message into protein. These complexes of messenger RNA and ribosomes are called **polyribosomes** or **polysomes.** Ribosomal participation in protein synthesis is dealt with later.

The role of ribosomes in protein synthesis (pp. 207–8).

1. Describe the structure of the eucaryotic 80S ribosome and contrast it with the procaryotic ribosome.
2. How do free ribosomes and those bound to the ER differ in function?

Mitochondria

Found in most eucaryotic cells, **mitochondria** (s., **mitochondrion**) frequently are called the "powerhouses" of the cell. Tricarboxylic acid cycle activity and the generation of ATP by electron transport and oxidative phosphorylation take place here. In the transmission electron microscope, mitochondria usually are cylindrical structures and measure approximately 0.3 to 1.0 μm by 5 to 10 μm. (In other words, they are about the same size as bacterial cells.) However, in living cells stained with a fluorescent dye, mitochondria often appear to be very long, threadlike objects (figure 4.12). Although cells can possess as many as 1,000 or more mitochondria, at least a few cells (some yeasts, unicellular algae, and trypanosome pro-

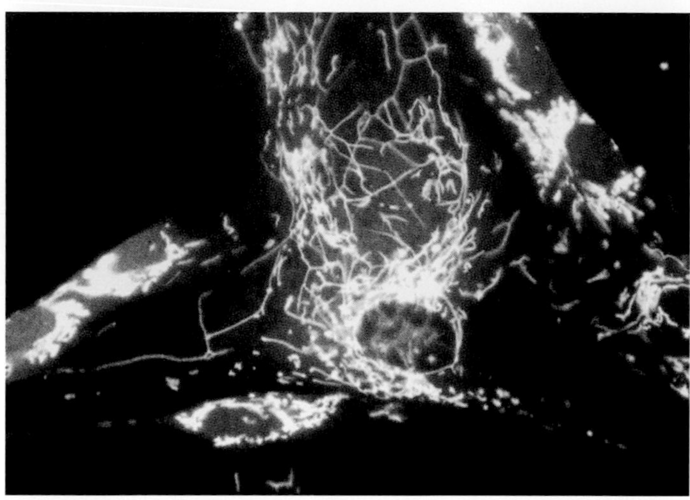

Figure 4.12 **Mitochondria in Living Cells.** A mammalian cell stained with the fluorescent dye Rhodamine 123. Note the long, threadlike mitochondria that twist and turn through the cytoplasmic matrix.

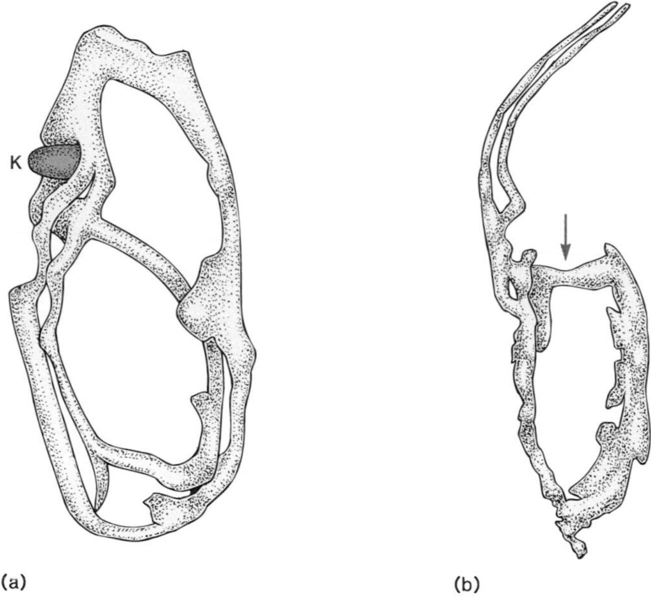

Figure 4.13 **Trypanosome Mitochondria.** The giant mitochondria from trypanosome protozoa. (*a*) *Crithidia fasciculata* mitochondrion with kinetoplast, *K*. The kinetoplast contains DNA that codes for mitochondrial RNA and protein. (*b*) *Trypanosoma cruzi* mitochondrion with arrow indicating position of kinetoplast.

tozoa) have a single giant tubular mitochondrion twisted into a continuous network permeating the cytoplasm (figure 4.13).

The tricarboxylic acid cycle, electron transport, and oxidative phosphorylation (pp. 150–54).

The mitochondrion is bounded by two membranes, an outer mitochondrial membrane separated from an inner mitochondrial membrane by a 6 to 8 nm intermembrane space (figure 4.14). Special infoldings of the inner membrane, called **cristae** (s., **crista**), greatly increase its surface area. Their shape differs in mitochondria from various species. Fungi often have platelike cristae, while euglenoid flagellates may have cristae shaped like disks. Tubular cristae are found in some fungi and most protozoa; however, amoebae can possess mitochondria with cristae in the shape of vesicles (figure 4.15). The inner membrane encloses the mitochondrial matrix, a dense matrix containing ribosomes, DNA, and often large calcium phosphate granules. Mitochondrial ribosomes are smaller than cytoplasmic ribosomes and resemble those of bacteria in several ways, including their size and subunit composition. Mitochondrial DNA is a closed circle like bacterial DNA.

Each mitochondrial compartment is different from the others in chemical and enzymatic composition. The outer and inner mitochondrial membranes, for example, possess different lipids. Enzymes and electron carriers involved in electron transport and oxidative phosphorylation (the formation of ATP as a consequence of electron transport) are located only in the inner membrane. The enzymes of the tricarboxylic acid cycle and the β-oxidation pathway for fatty acids (*see chapter 8*) are located in the matrix.

The inner membrane of the mitochondrion has another distinctive structural feature related to its function. Many small spheres, about 8.5 nm diameter, are attached by stalks to its inner surface. The spheres are called **F_1 factors** and synthesize ATP during cellular respiration.

The mitochondrion uses its DNA and ribosomes to synthesize some of its own proteins. Most mitochondrial proteins, however, are manufactured under the direction of the nucleus. Mitochondria reproduce by binary fission. Chloroplasts show similar partial independence and reproduction by binary fission. Because both organelles resemble bacteria to some extent, it has been suggested that these organelles arose from symbiotic associations between bacteria and larger cells (see Box 4.1).

Chloroplasts

Plastids are cytoplasmic organelles of algae and higher plants that often possess pigments such as chlorophylls and carotenoids, and are the sites of synthesis and storage of food reserves. The most important type of plastid is the chloroplast. **Chloroplasts** contain chlorophyll and use light energy to convert CO_2 and water to carbohydrates and O_2. That is, they are the site of photosynthesis.

Although chloroplasts are quite variable in size and shape, they share many structural features. Most often they are oval with dimensions of 2 to 4 μm by 5 to 10 μm, but some algae possess one huge chloroplast that fills much of the cell. Like mitochondria, chloroplasts are encompassed by two membranes (figure 4.16). A matrix, the **stroma,** lies within the inner membrane. It contains DNA, ribosomes, lipid droplets, starch granules, and a complex internal membrane system whose most prominent components are flattened, membrane-delimited sacs, the **thylakoids.** Clusters of two or more thylakoids are dispersed within the stroma of most algal chloroplasts

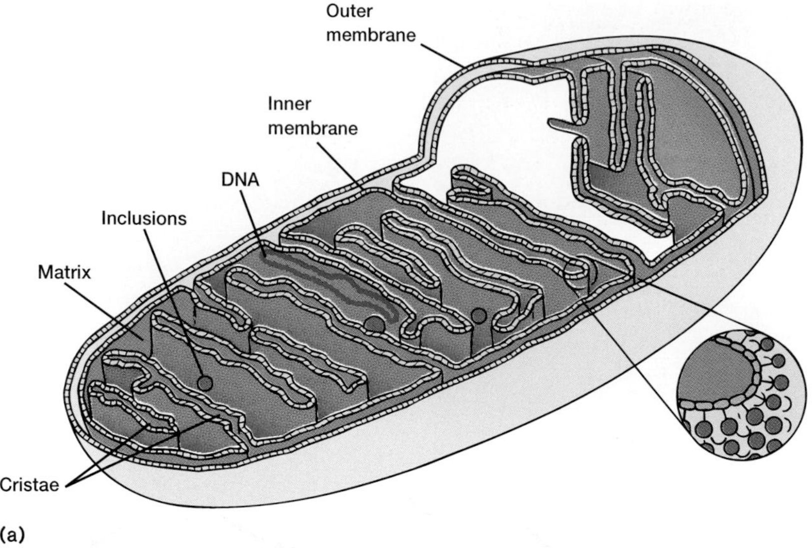

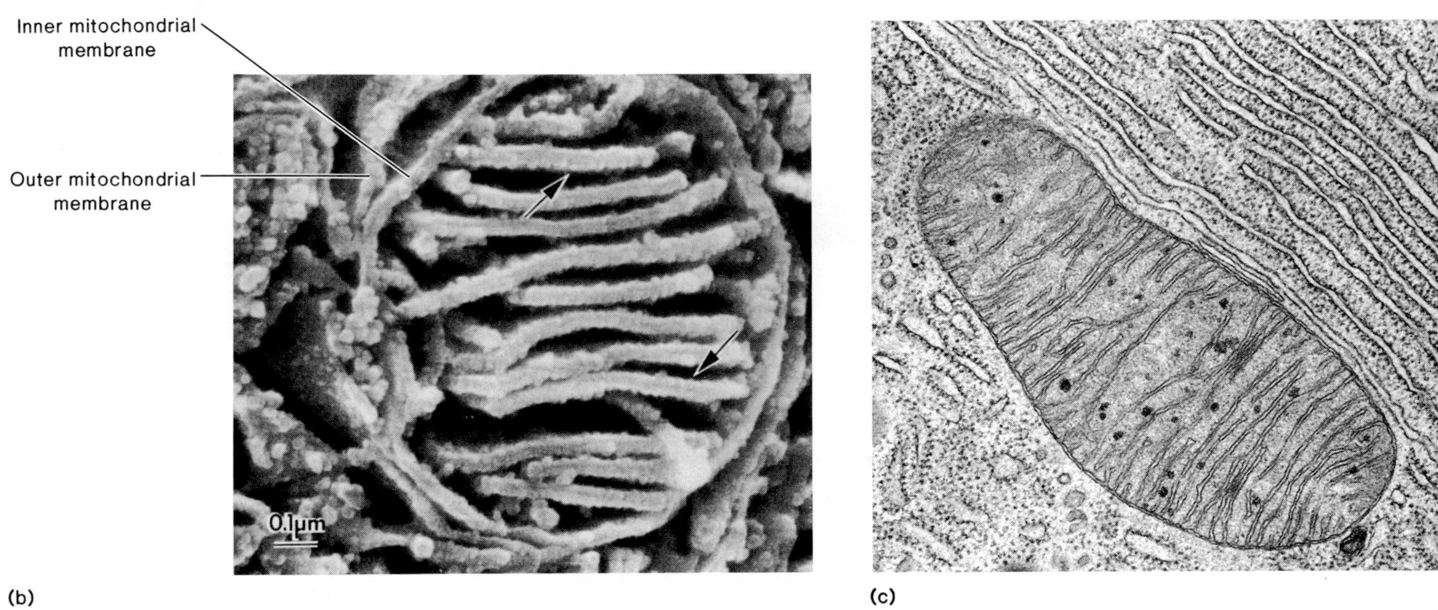

Figure 4.14 Mitochondrial Structure. (*a*) A diagram of mitochondrial structure. The insert shows F_1 particles lining the inner surface of the cristae. *From P. Sheeler and D. E. Bianchi,* Cell Biology: Structure, Biochemistry, and Function, *2d ed. Copyright © 1980 John Wiley & Sons, Inc., New York, NY.* (*b*) Scanning electron micrograph (×70,000) of a freeze-fractured mitochondrion showing the cristae (arrows). The outer and inner mitochondrial membranes are also evident. (*c*) Transmission electron micrograph of a mitochondrian from a bat pancreas (×85,000). Note outer and inner mitochondrial membranes, cristae, and inclusions in the matrix. The mitochondrian is surrounded by rough endoplasmic reticulum.

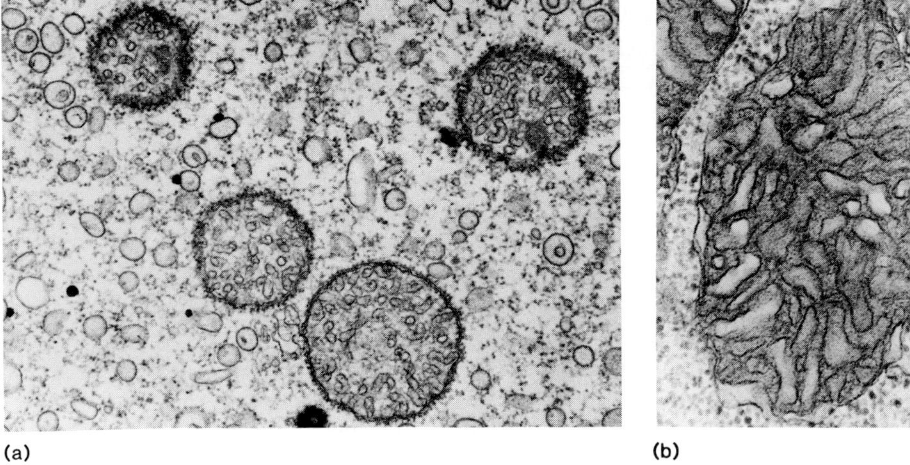

Figure 4.15 Mitochondrial Cristae. Mitochondria with a variety of cristae shapes. (*a*) Mitochondria from the protostelid *Schizoplasmodiopsis micropunctata.* Note the tubular cristae (×49,500). (*b*) The protozoan *Actinosphaerium* with vesicular cristae (×75,000).

Box 4.1

The Origin of the Eucaryotic Cell

The profound differences between eucaryotic and procaryotic cells have stimulated much discussion about how the more complex eucaryotic cell arose. Some biologists believe the original "protoeucaryote" was a large aerobic bacterium that formed mitochondria, chloroplasts, and nuclei when its plasma membrane invaginated and enclosed genetic material in a double membrane. The organelles could then evolve independently. It also is possible that a large blue-green bacterium lost its cell wall and became phagocytic. Subsequently, primitive chloroplasts, mitochondria, and nuclei would be formed by the fusion of thylakoids and endoplasmic reticulum cisternae to enclose specific areas of cytoplasm.

The most popular theory for the origin of eucaryotic cells is the **endosymbiotic theory.** In brief, it is supposed that the ancestral procaryotic cell lost its cell wall and gained the ability to obtain nutrients by phagocytosing other bacteria. When photosynthetic cyanobacteria arose, the environment slowly became aerobic. If an anaerobic, amoeboid, phagocytic procaryote—possibly already possessing a developed nucleus—engulfed an aerobic bacterial cell and established a permanent symbiotic relationship with it, the host would be better adapted to its increasingly aerobic environment. The endosymbiotic aerobic bacterium eventually would develop into the mitochondrion. Similarly, symbiotic associations with cyanobacteria could lead to the formation of chloroplasts and photosynthetic eucaryotes. Cilia and flagella might have arisen from the attachment of spirochete bacteria (*see chapter 21*) to the surface of eucaryotic cells, much as spirochetes attach themselves to the surface of the motile protozoan *Myxotricha paradoxa* that grows in the digestive tract of termites.

There is evidence to support the endosymbiotic theory. Both mitochondria and chloroplasts resemble bacteria in size and appearance, contain DNA in the form of a closed circle like that of bacteria, and reproduce semiautonomously. Mitochondrial and chloroplast ribosomes resemble procaryotic ribosomes more closely than those in the eucaryotic cytoplasmic matrix. The sequences of the chloroplast and mitochondrial genes for ribosomal RNA and transfer RNA are more similar to bacterial gene sequences than to those of eucaryotic rRNA and tRNA nuclear genes. Finally, there are symbiotic associations that appear to be bacterial endosymbioses in which distinctive procaryotic characteristics are being lost. For example, the protozoan flagellate *Cyanophora paradoxa* has photosynthetic organelles called cyanellae with a structure similar to that of cyanobacteria and the remains of peptidoglycan in their walls. Their DNA is much smaller than that of cyanobacteria and resembles chloroplast DNA. Despite such evidence, the endosymbiotic theory still is somewhat speculative and the center of much continuing research and discussion.

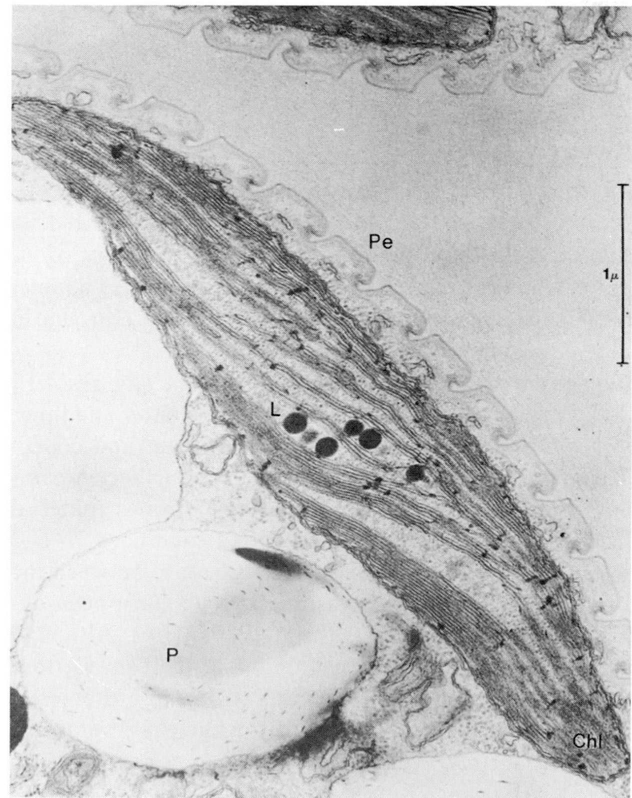

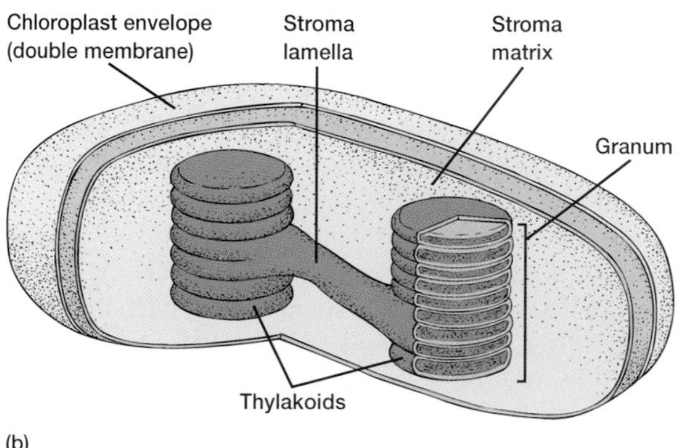

Figure 4.16 Chloroplast Structure. (*a*) The chloroplast (*Chl*), of the euglenoid flagellate *Colacium cyclopicolum.* The chloroplast is bounded by a double membrane and has its thylakoids in groups of three or more. A paramylon granule (*P*), lipid droplets (*L*), and the pellicular strips (*Pe*), can be seen (×40,000). (*b*) A diagram of chloroplast structure.

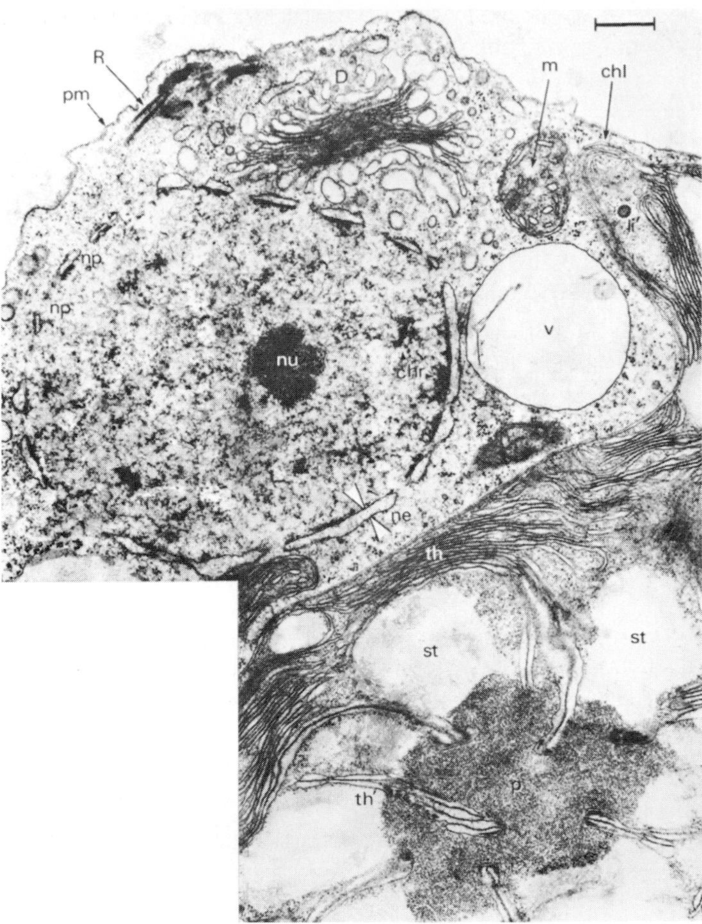

Figure 4.17 **Algal Cell Structure.** The nucleus and other organelles of the green alga *Dunaliella primolecta.* Note the following nuclear structures: heterochromatin, *chr;* nucleolus, *nu;* nuclear envelope, *ne;* and nuclear pores, *np.* Many other important organelles may be seen: dictyosome, *D;* chloroplast, *chl;* lipid inclusion, *li';* mitochondrion, *m;* pyrenoid, *p;* plasma membrane, *pm;* starch granule, *st;* thylakoids, *th;* thylakoids penetrating pyrenoid, *th';* and vacuole, *v* (×30,400). The bar represents 0.5 μm.

(figures 4.2, 4.16, and 4.17). In some groups of algae, several disklike thylakoids are stacked on each other like coins to form **grana** (s., **granum**).

Photosynthetic reactions are separated structurally in the chloroplast just as electron transport and the tricarboxylic acid cycle are in the mitochondrion. The formation of carbohydrate from CO_2 and water, the dark reaction, takes place in the stroma. The trapping of light energy to generate ATP, NADPH, and O_2, the light reaction, is located in the thylakoid membranes, where chlorophyll and electron transport components are also found.

Photosynthesis (pp. 162–68).

The chloroplasts of many algae contain a **pyrenoid** (figures 4.2 and 4.17), a dense region of protein surrounded by starch or another polysaccharide. Pyrenoids participate in polysaccharide synthesis.

1. Describe in detail the structure of mitochondria and chloroplasts. Where are the different components of these organelles' energy trapping systems located?
2. Define $\mathbf{F_1}$ factor, plastid, dark reaction, light reaction, and pyrenoid.
3. What is the role of mitochondrial DNA?

The Nucleus and Cell Division

The cell **nucleus** is by far the most visually prominent organelle. It was discovered early in the study of cell structure and was shown by Robert Brown in 1831 to be a constant feature of eucaryotic cells. The nucleus is the repository for the cell's genetic information and is its control center.

Nuclear Structure

Nuclei are membrane-delimited spherical bodies about 5 to 7 μm in diameter (figures 4.2, 4.8, and 4.17). Dense fibrous material called **chromatin** can be seen within the nucleoplasm of the nucleus of a stained cell. This is the DNA-containing part of the nucleus. In nondividing cells, chromatin exists in a dispersed condition, but condenses during mitosis to become visible as **chromosomes.** Some nuclear chromatin, the euchromatin, is loosely organized and contains those genes that are expressing themselves actively. Heterochromatin is coiled more tightly, appears darker in the electron microscope, and is not genetically active most of the time.

Organization of DNA in eucaryotic nuclei (pp. 194–95).

The nucleus is bounded by the **nuclear envelope** (figures 4.2 and 4.17), a complex structure consisting of inner and outer membranes separated by a 15 to 75 nm perinuclear space. The envelope is continuous with the ER at several points and its outer membrane is covered with ribosomes. A network of intermediate filaments, called the nuclear lamina, lies against the inner surface of the envelope and supports it. Chromatin is usually associated with the inner membrane.

Many **nuclear pores** penetrate the envelope (figures 4.17 and 4.18), each pore formed by a fusion of the outer and inner membranes. Pores are about 70 nm in diameter and collectively occupy about 10 to 25% of the nuclear surface. A complex ringlike arrangement of granular and fibrous material called the annulus is located at the edge of each pore.

The nuclear pores serve as a transport route between the nucleus and surrounding cytoplasm. Particles have been observed moving into the nucleus through the pores. Although the function of the annulus is not understood, it may either regulate or aid the movement of material through the pores. Substances also move directly through the nuclear envelope by unknown mechanisms.

The Nucleolus

Often the most noticeable structure within the nucleus is the **nucleolus** (figures 4.2*b*, 4.8, and 4.17). A nucleus may contain from one to many nucleoli. Although the nucleolus is not membrane-enclosed, it is a complex organelle with separate granular and fibrillar regions. It is present in nondividing cells, but frequently disappears during mitosis. After mitosis, the nucleolus reforms around the nucleolar organizer, a particular part of a specific chromosome.

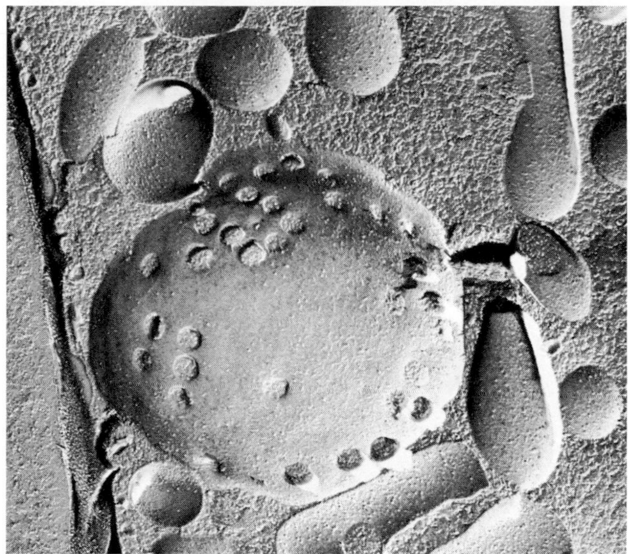

Figure 4.18 The Nucleus. A freeze-etch preparation of the conidium of the fungus *Geotrichum candidum* (×44,600). Note the large convex nuclear surface with nuclear pores scattered over it.

The nucleolus plays a major role in ribosome synthesis. The nucleolar organizer DNA directs the production of **ribosomal RNA (rRNA).** This RNA is synthesized in a single long piece that then is cut to form the final rRNA molecules. The processed rRNAs next combine with ribosomal proteins (which have been synthesized in the cytoplasmic matrix) to form partially completed ribosomal subunits. The granules seen in the nucleolus are probably these subunits. Immature ribosomal subunits then leave the nucleus, presumably by way of the nuclear envelope pores, and mature in the cytoplasm.

RNA splicing (pp. 203–4).

Mitosis and Meiosis

When a eucaryotic microorganism reproduces itself, its genetic material must be duplicated and then separated so that each new nucleus possesses a complete set of chromosomes. This process of nuclear division and chromosome distribution in eucaryotic cells is called **mitosis.** Mitosis actually occupies only a small portion of a microorganism's life as can be seen by examining the **cell cycle** (figure 4.19). The cell cycle is the

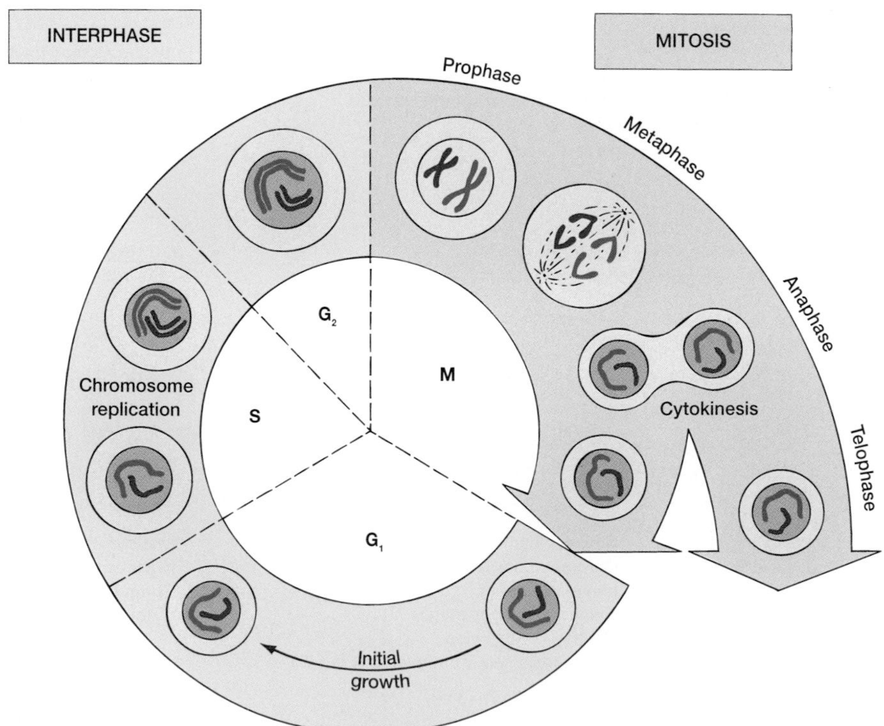

Figure 4.19 The Eucaryotic Cell Cycle. The length of the M period has been increased disproportionately in order to show the phases of mitosis. *G_1 period:* synthesis of mRNA, tRNA, ribosomes, and cytoplasmic constituents. Nucleolus grows rapidly. *S period:* rapid synthesis and doubling of nuclear DNA and histones. *G_2 period:* preparation for mitosis and cell division. *M period:* mitosis (prophase, metaphase, anaphase, telophase) and cytokinesis. *From P. Sheeler and D. E. Bianchi,* Cell Biology: Structure, Biochemistry and Function, *2d ed. Copyright © 1980 John Wiley & Sons, Inc., New York, NY.*

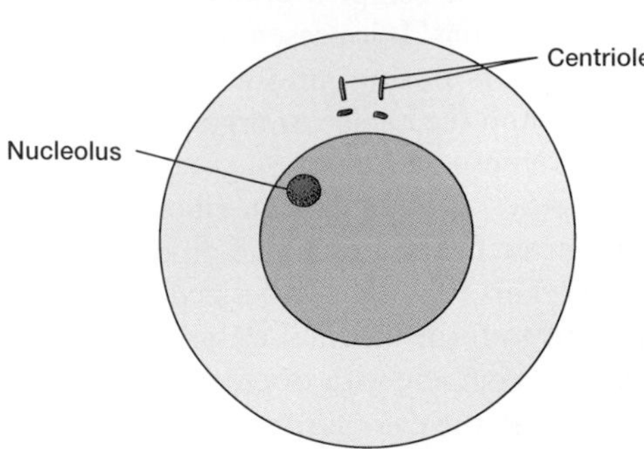

Chromosomes not seen as distinct structures.
Nucleolus visible.
The premitotic cell is 4N with two copies of each chromosome.

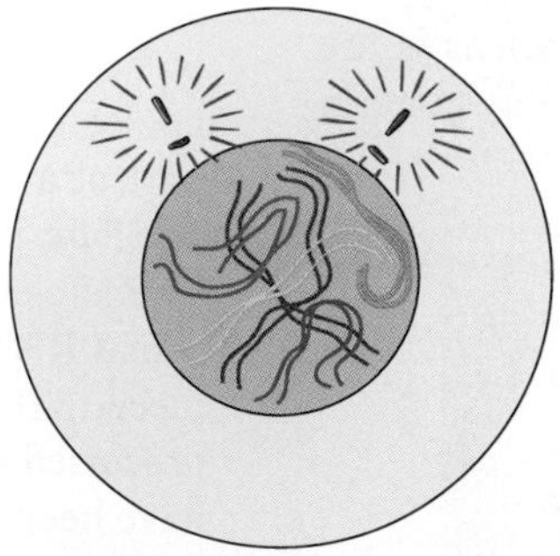

Centrioles moving apart.
Chromosomes appear as long thin threads.
Nucleolus becoming less distinct.

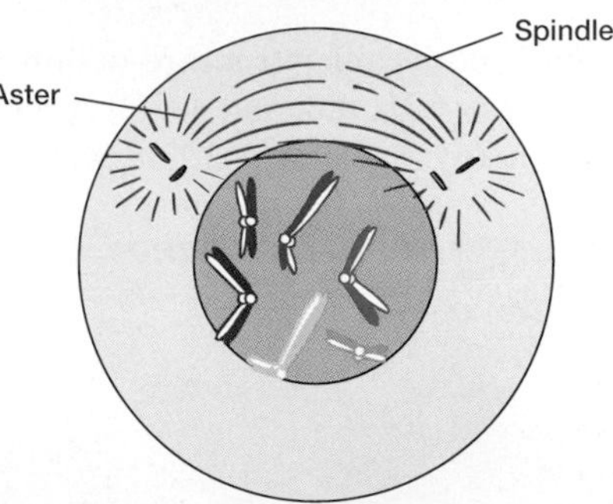

Centrioles farther apart, spindle beginning to form. Each chromosome is composed of 2 chromatids held together by their centromeres. A chromatid has 1 double-stranded DNA, 1 strand of which is newly synthesized. Thus there are 4 DNA copies in each pair of similar chromosomes.

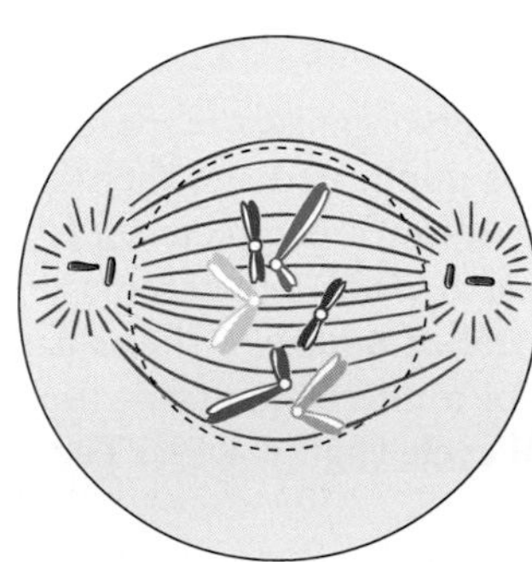

Centrioles nearly at opposite sides of nucleus.
Spindle nearly complete.
Nuclear envelope disappearing.
Chromosomes move toward equator.
Nucleolus no longer visible.

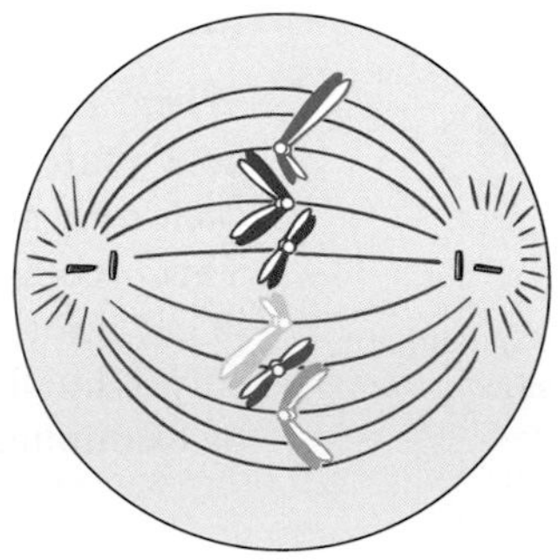

Nuclear envelope has disappeared.
Centromeres of each double-stranded chromosome attached to spindle microtubules at spindle equator.

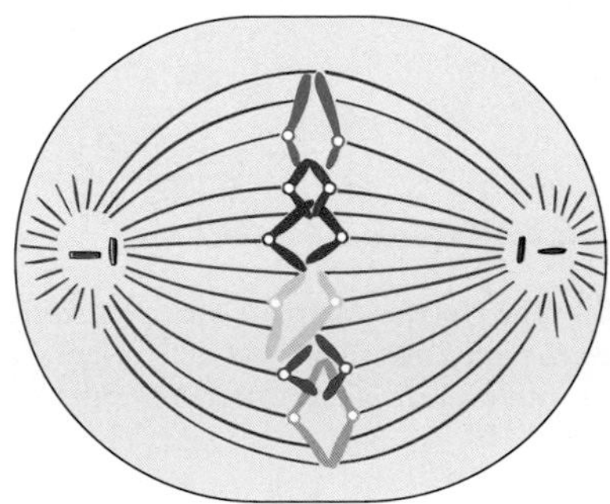

Centromeres have uncoupled and begun moving toward opposite poles of spindle.

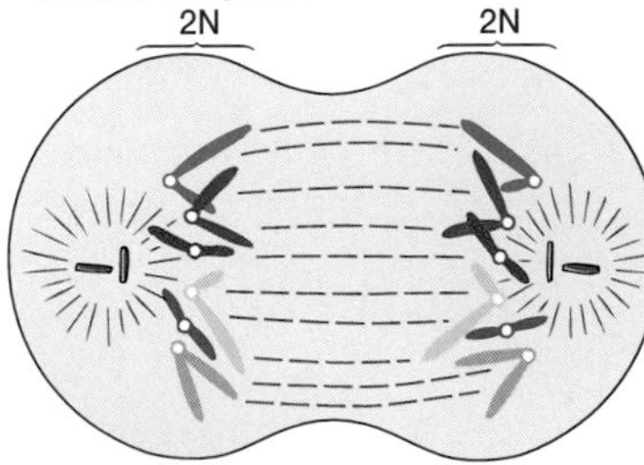

The 2 sets of new single-stranded chromosomes nearing respective poles.
Cytokinesis beginning.

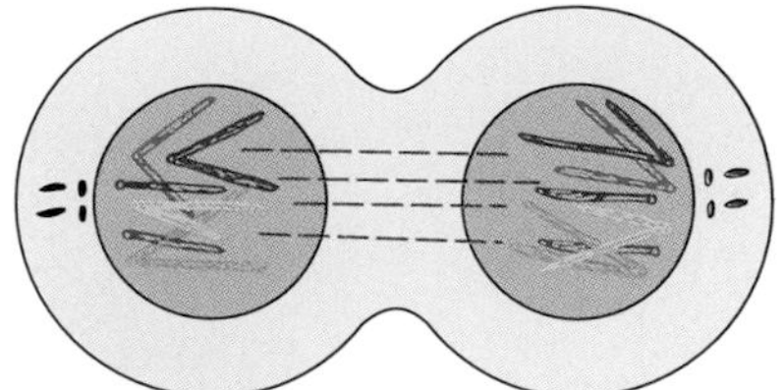

New nuclear envelope forming.
Chromosomes become longer, thinner, and less distinct.
Nucleolus reappearing.
Centrioles replicated.
Cytokinesis nearly complete.

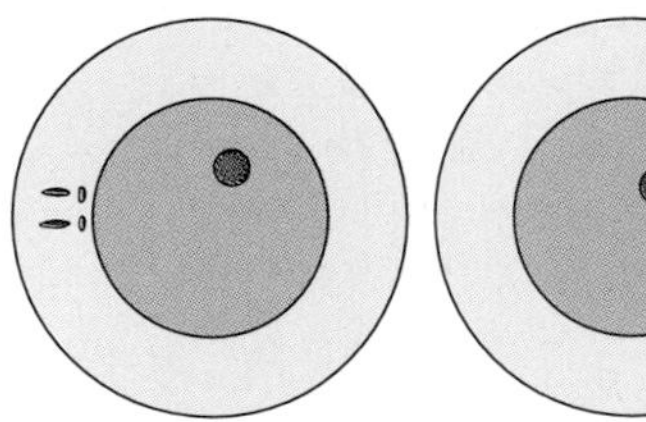

Nuclear envelope complete.
Nucleolus reformed.
Chromosomes no longer visible.
Cytokinesis complete.

Figure 4.20 **Mitosis and Cytokinesis.** The process in a typical eucaryotic cell.

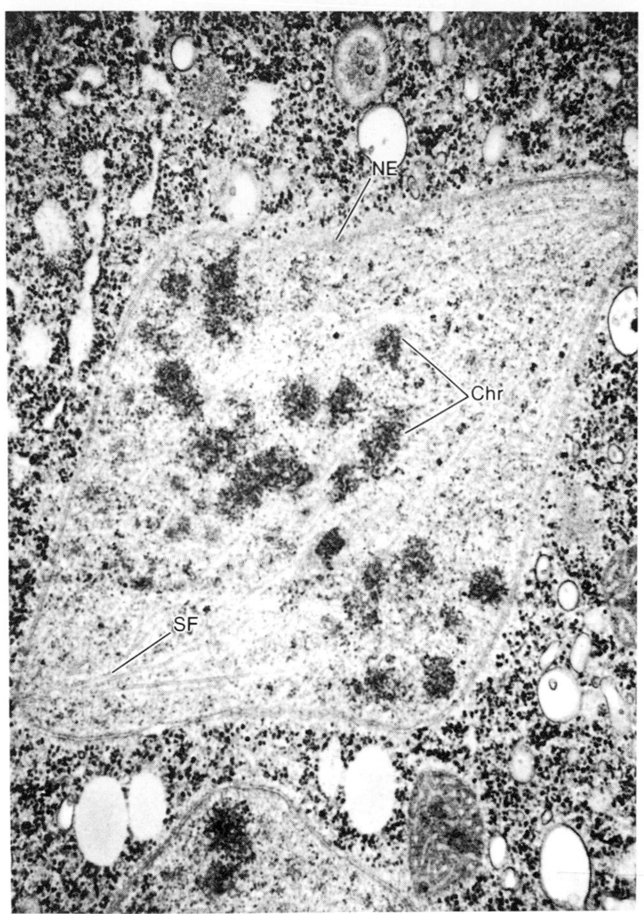

Figure 4.21 **Mitosis with an Intact Nuclear Envelope.**
Mitosis in the slime mold *Physarum flavicomum*. The nuclear envelope, *NE*, remains intact, and the spindle is intranuclear. The process is at metaphase with the chromosomes, *Chr*, aligned in the center and attached to spindle fibers, *SF* (×15,000).

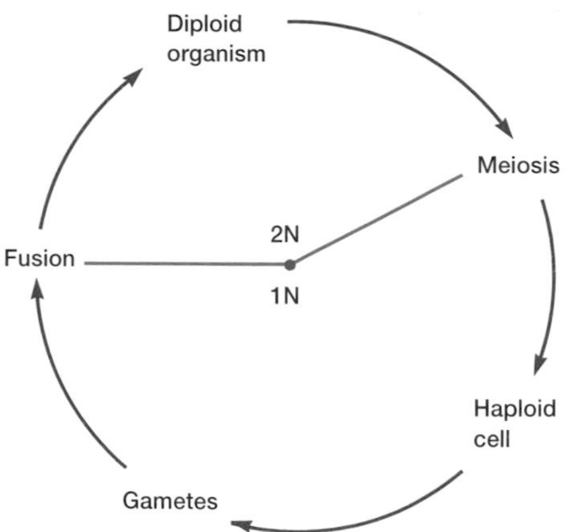

Figure 4.22 **Generalized Eucaryotic Life Cycle.**

total sequence of events in the growth-division cycle between the end of one division and the end of the next. Cell growth takes place in the **interphase,** that portion of the cycle between periods of mitosis. Interphase is composed of three parts. The G_1 period (gap 1 period) is a time of active synthesis of RNA, ribosomes, and other cytoplasmic constituents accompanied by considerable cell growth. This is followed by the S period (synthesis period) in which DNA is replicated and doubles in quantity. Finally, there is a second gap, the G_2 period, when the cell prepares for mitosis, the M period, by activities such as the synthesis of special division proteins. The total length of the cycle differs considerably between microorganisms, usually due to variations in the length of G_1.

Mitotic events are summarized in figure 4.20. During mitosis, the genetic material duplicated during the S period is distributed equally to the two new nuclei so that each has a full set of genes. Mitosis in eucaryotic microorganisms can differ from that pictured in figure 4.20. For example, the nuclear envelope does not disappear in many fungi and some protozoa and algae (figure 4.21). Frequently **cytokinesis,** the division of the parental cell's cytoplasm to form new cells, begins during anaphase and finishes by the end of telophase. However, mitosis can take place without cytokinesis to generate multinucleate or coenocytic cells.

In mitosis, the original number of chromosomes is the same after division and a diploid organism will remain **diploid** or 2N (i.e., it still has two copies of each chromosome). Frequently a microorganism reduces its chromosome number by half, from the diploid state to the **haploid** or 1N (a single copy of each chromosome). Haploid cells may immediately act as gametes and fuse to reform diploid organisms, or may form gametes only after a considerable delay (figure 4.22). The process by which the number of chromosomes is reduced in half with each daughter cell receiving one complete set of chromosomes is called **meiosis.**

Life cycles of eucaryotic microorganisms (chapters 26–28).

Meiosis is quite complex and involves two stages (figure 4.23). The first stage differs markedly from mitosis. During prophase, homologous chromosomes come together and lie side-by-side, a process known as synapsis. Then the double-stranded chromosomes from each homologous pair move to opposite poles in anaphase. In contrast, during mitotic anaphase the two strands of each chromosome separate and move to opposite poles. Consequently, the number of chromosomes is halved in meiosis, but not in mitosis. The second stage of meiosis is similar to mitosis in terms of mechanics, and single-stranded chromosomes are separated. After completion of meiosis I and meiosis II, the original diploid cell has been transformed into four haploid cells.

1. Early prophase I

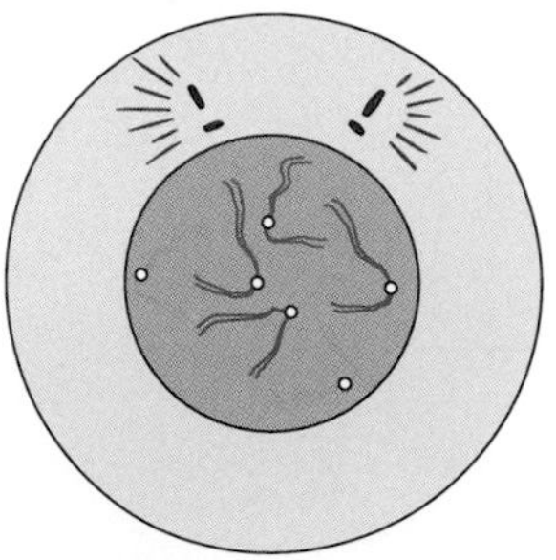

Chromosomes become visible as long, well-separated filaments; they do not appear double stranded, though other evidence indicates that replication has already occurred and the cells are 4N.

2. Middle prophase I

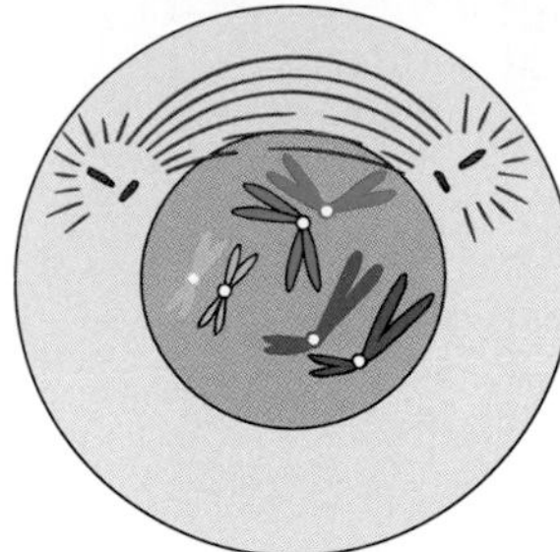

Homologous chromosomes synapse and become shorter and thicker.

3. Late prophase I

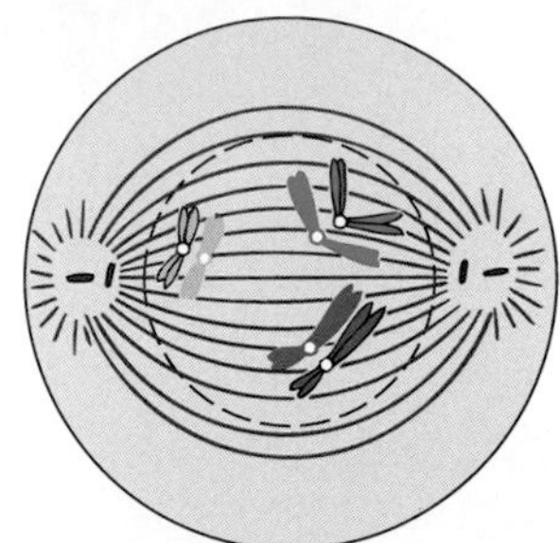

Chromosome duplication becomes visible. Nuclear envelope begins to disappear.

4. Metaphase I

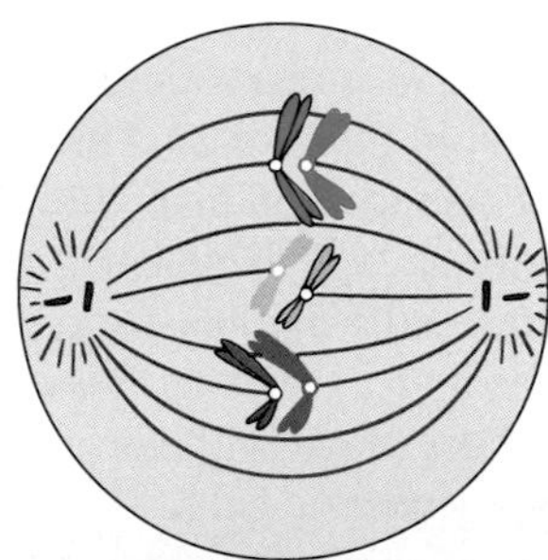

Each synaptic pair moves to the equator of the spindle as a unit.

5. Anaphase I

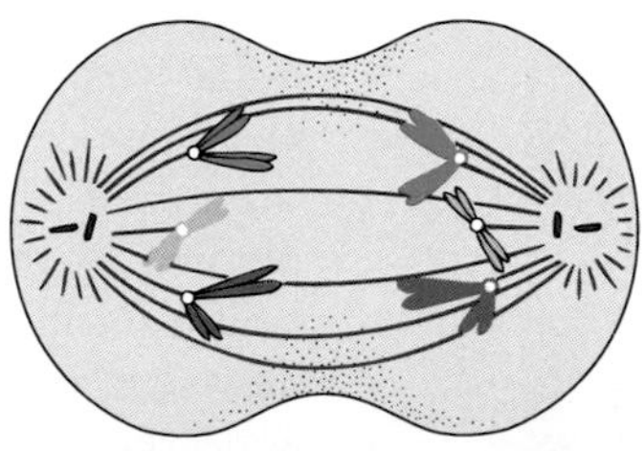

Centromeres do not uncouple. Homologous chromosomes move apart to opposite poles.

6. Telophase I

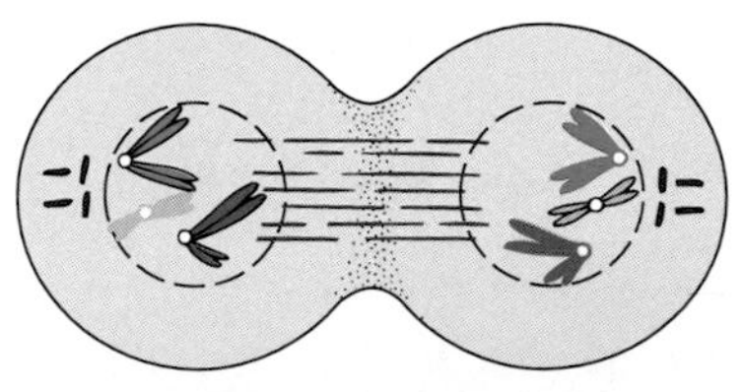

New nuclei with a haploid number of duplicated chromosomes are formed.

7. Interkinesis

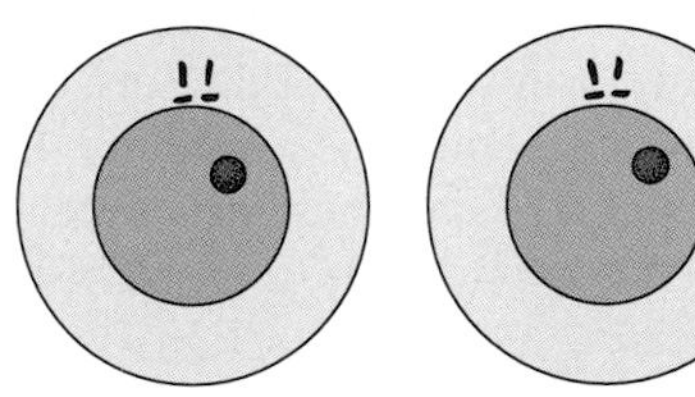

No replication of genetic material occurs.

8. Prophase II

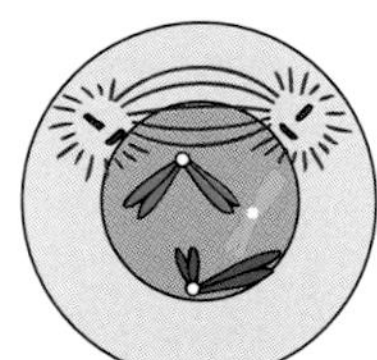

9. Metaphase II

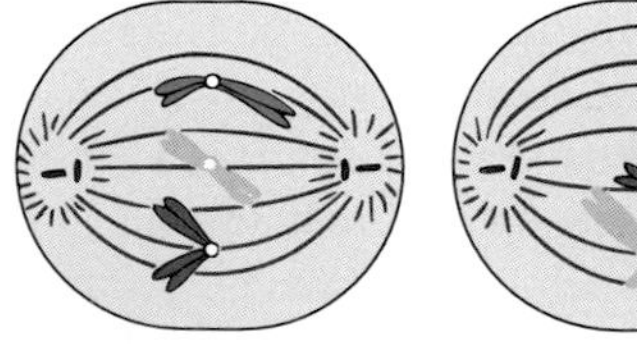

10. Anaphase II

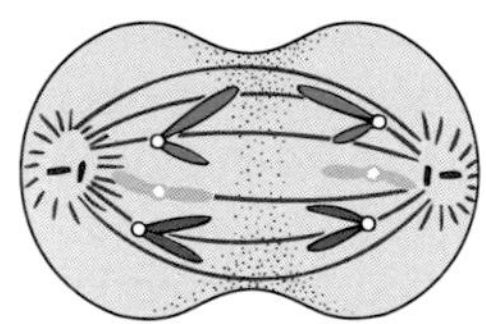
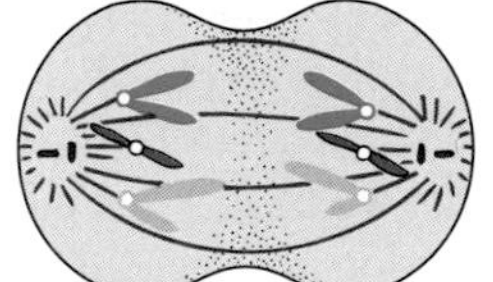

Sister chromatids separate to form a haploid number of single chromosomes.

11. Telophase II

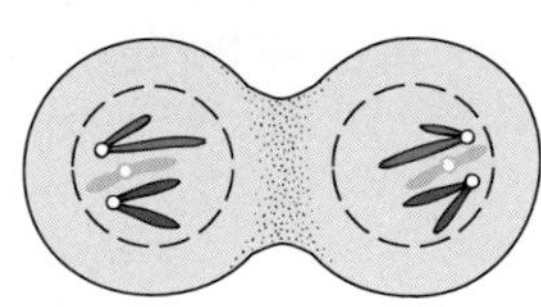
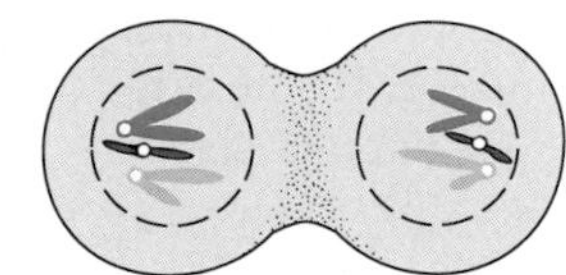

12. Interphase

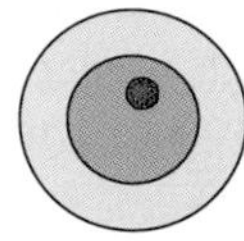
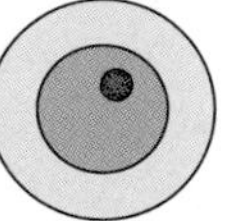
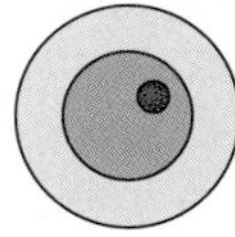
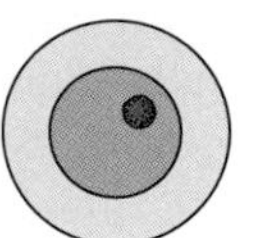

Haploid cells

Figure 4.23 **Meiosis in Eucaryotes.** There are a number of variations of this process in microorganisms.

1. Describe the structure of the nucleus. What are euchromatin and heterochromatin? What is the role of the pores in the nuclear envelope?
2. Briefly discuss the structure and function of the nucleolus. What is the nucleolar organizer?
3. Describe the eucaryotic cell cycle, its periods, and the process of mitosis. What is meiosis, how does it take place, and what is its role in the microbial life cycle?

External Cell Coverings

Eucaryotic microorganisms differ greatly from procaryotes in the supporting or protective structures they have external to the plasma membrane. In contrast with most bacteria, many eucaryotes lack an external cell wall. The amoeba is an excellent example. Eucaryotic cell membranes, unlike most procaryotic membranes, contain sterols such as cholesterol, and this may make them mechanically stronger, thus reducing the need for external support. (However, as mentioned on page 44, many procaryotic membranes are strengthened by hopanoids.) Of course, many eucaryotes do have a rigid external **cell wall.** Algal cell walls usually have a layered appearance and contain large quantities of polysaccharides such as cellulose and pectin. In addition, inorganic substances like silica (in diatoms) or calcium carbonate (some red algae) may be present. Fungal cell walls normally are rigid. Their exact composition varies with the organism; but usually, cellulose, chitin, or glucan (a glucose polymer different from cellulose) are present. Despite their nature, the rigid materials in eucaryotic walls are chemically simpler than procaryotic peptidoglycan.

Bacterial cell wall structure and chemistry (pp. 50–55).

Many protozoa and some algae have a different external structure, the **pellicle** (figure 4.16*a*). This is a relatively rigid layer of components just beneath the plasma membrane (sometimes the plasma membrane is also considered part of the pellicle). The pellicle may be fairly simple in structure. For example, *Euglena spirogyra* has a series of overlapping strips with a ridge at the edge of each strip fitting into a groove on the adjacent one. In contrast, ciliate protozoan pellicles are exceptionally complex with two membranes and a variety of associated structures. Although pellicles are not as strong and rigid as cell walls, they do give their possessors a characteristic shape.

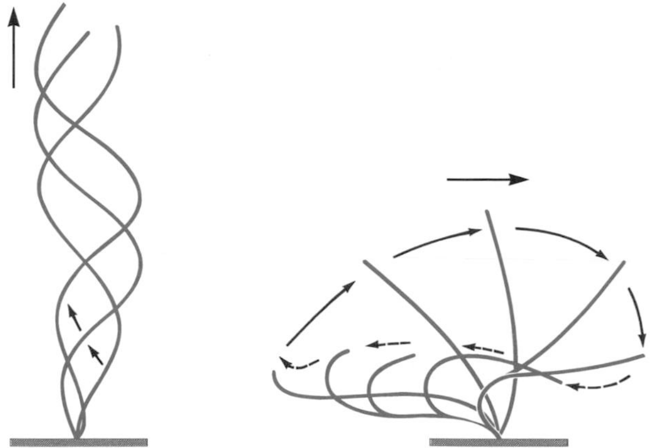

Figure 4.24 Patterns of Flagellar Movement. Flagellar movement (left illustration) often takes the form of waves that move either from the base of the flagellum to its tip or in the opposite direction. The motion of these waves propels the organism along. The beat of a cilium (right illustration) may be divided into two phases. In the effective stroke (black cilia), the cilium remains fairly stiff as it swings through the water. This is followed by a recovery stroke (colored cilia) in which the cilium bends and returns to its initial position. The large colored arrows indicate the direction of water movement in these examples.

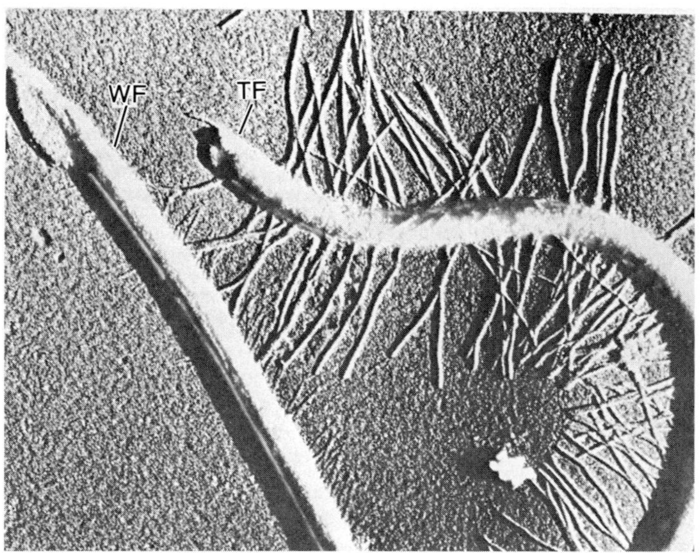

Figure 4.25 Whiplash and Tinsel Flagella. Transmission electron micrograph of a shadowed whiplash flagellum, *WF,* and a tinsel flagellum, *TF,* with mastigonemes.

Cilia and Flagella

Cilia (s., **cilium**) and **flagella** (s., **flagellum**) are the most prominent organelles associated with motility. Although both are whiplike and beat to move the microorganism along, they differ from one another in two ways. First, cilia are typically only 5 to 20 μm in length, whereas flagella are 100 to 200 μm long. Second, their patterns of movement are usually distinctive (figure 4.24). Flagella move in an undulating fashion and generate planar or helical waves originating at either the base or the tip. If the wave moves from base to tip, the cell is pushed along; a beat traveling from the tip toward the base pulls the cell through the water. Sometimes, the flagellum will have lateral hairs called flimmer filaments (thicker, stiffer hairs are called mastigonemes). These filaments change flagellar action so that a wave moving down the filament toward the tip pulls the cell along instead of pushing it. Such a flagellum often is called a tinsel flagellum, whereas the naked flagellum is referred to as a whiplash flagellum (figure 4.25). Cilia, on the other hand, normally have a beat with two distinctive phases.

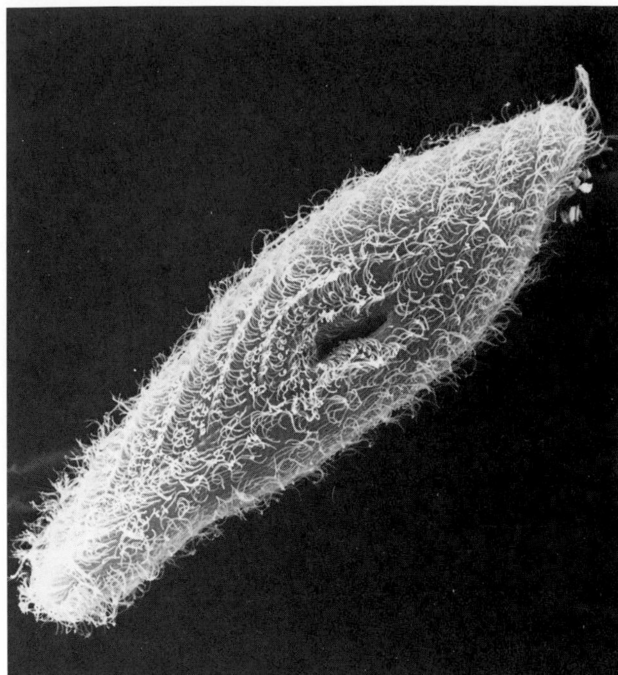

Figure 4.26 **Coordination of Ciliary Activity.** A scanning electron micrograph of *Paramecium* showing cilia (×1,500). The ciliary beat is coordinated and moves in waves across the protozoan's surface, as can be seen in the photograph.

In the effective stroke, the cilium strokes through the surrounding fluid like an oar, thereby propelling the organism along in the water. The cilium next bends along its length while it is pulled forward during the recovery stroke in preparation for another effective stroke. A ciliated microorganism actually coordinates the beats so that some of its cilia are in the recovery phase while others are carrying out their effective stroke (figure 4.26). This coordination allows the organism to move smoothly through the water.

Despite their differences, cilia and flagella are very similar in ultrastructure. They are membrane-bound cylinders about 0.2 μm in diameter. Located in the matrix of the organelle is a complex, the **axoneme,** consisting of nine pairs of microtubule doublets arranged in a circle around two central tubules (figures 4.27 and 4.28). This is called the 9 + 2 pattern of microtubules. Each doublet also has pairs of arms projecting from subtubule A (the complete microtubule) toward a neighboring doublet. A radial spoke extends from subtubule A toward the internal pair of microtubules with their central sheath. These microtubules are similar to those found in the cytoplasm. Each is constructed of two types of tubulin subunits, α- and β-tubulins, that resemble the contractile protein actin in their composition.

Bacterial flagella and motility (pp. 57–60).

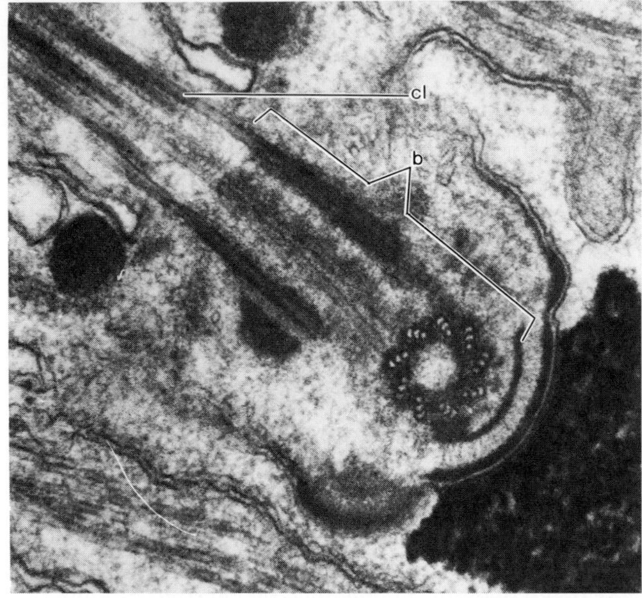

(a)

(b)

Figure 4.27 **Cilia Ultrastructure.** In micrograph (*a*) one can see the basal body, *b*, and the cilium, *cl*, itself with its central and peripheral microtubules (×100,000). The basal body is cross-sectioned at one end to show typical microtubule triplets and an absence of central microtubules. (*b*) Cross section of cilia. Note the two central microtubules surrounded by nine microbutule doublets (×160,000).

A **basal body** lies in the cytoplasm at the base of each cilium or flagellum (figure 4.27). It is a short cylinder with nine microtubule triplets around its periphery (a 9 + 0 pattern) and is separated from the rest of the organelle by a basal plate. The basal body directs the construction of these organelles. Cilia and flagella appear to grow through the addition of preformed microtubule subunits at their tips.

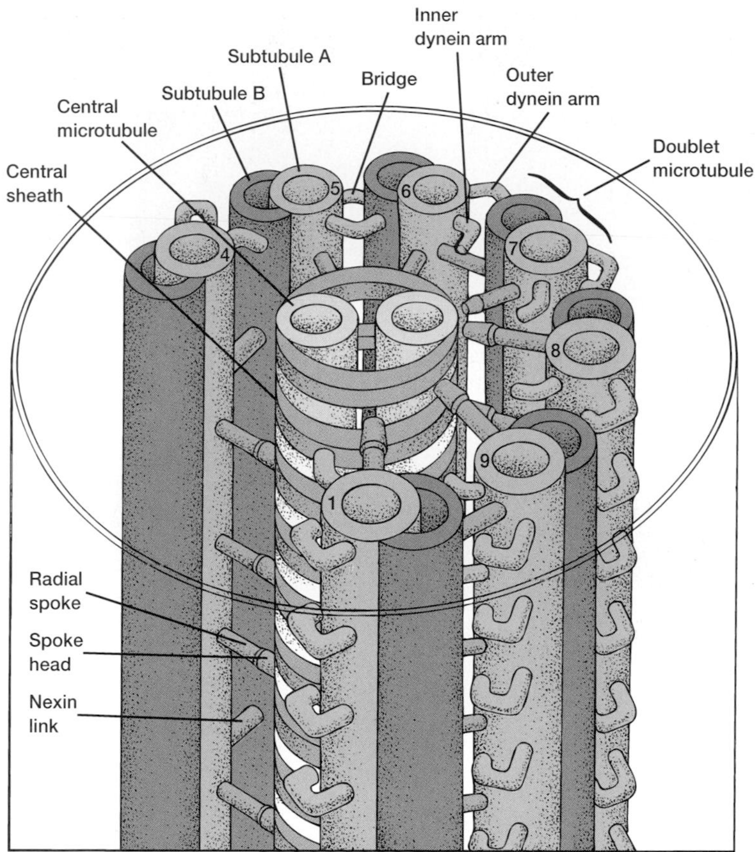

Figure 4.28 **Diagram of Cilia and Flagella Structure.** Doublets 2 and 3 removed for sake of visibility.

Cilia and flagella bend because adjacent microtubule doublets slide along one another while maintaining their individual lengths. The doublet arms (figure 4.28), about 15 nm long, are made of the protein **dynein.** ATP powers the movement of cilia and flagella, and isolated dynein hydrolyzes ATP. It appears that dynein arms interact with the B subtubules of adjacent doublets to cause the sliding. The radial spokes also participate in this sliding motion.

Cilia and flagella beat at a rate of about 10 to 40 strokes or waves per second and propel microorganisms rapidly. The record holder is the flagellate *Monas stigmatica,* which swims at a rate of 260 μm/second (approximately 40 cell lengths per second); the common euglenoid flagellate, *Euglena gracilis,* travels at around 170 μm or 3 cell lengths per second. The ciliate protozoan *Paramecium caudatum* swims at about 2,700 μm/second (12 lengths per second). Such speeds are equivalent to or much faster than those seen in higher animals.

1. How do eucaryotic microorganisms differ from procaryotes with respect to supporting or protective structures external to the plasma membrane? Describe the pellicle and indicate which microorganisms have one.
2. Prepare and label a diagram showing the detailed structure of a cilium or flagellum. How do cilia and flagella move, and what is dynein's role in the process?

Comparison of Procaryotic and Eucaryotic Cells

A comparison of the cells in figure 4.29 demonstrates that there are many fundamental differences between eucaryotic and procaryotic cells. **Eucaryotic cells** have a membrane-enclosed nucleus. In contrast, **procaryotic cells** lack a true, membrane-

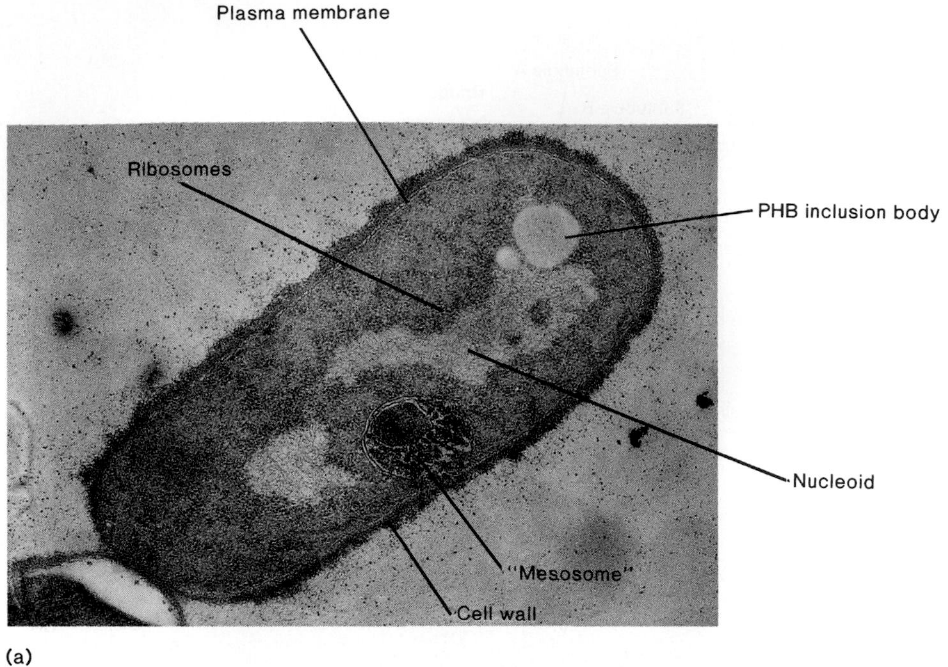

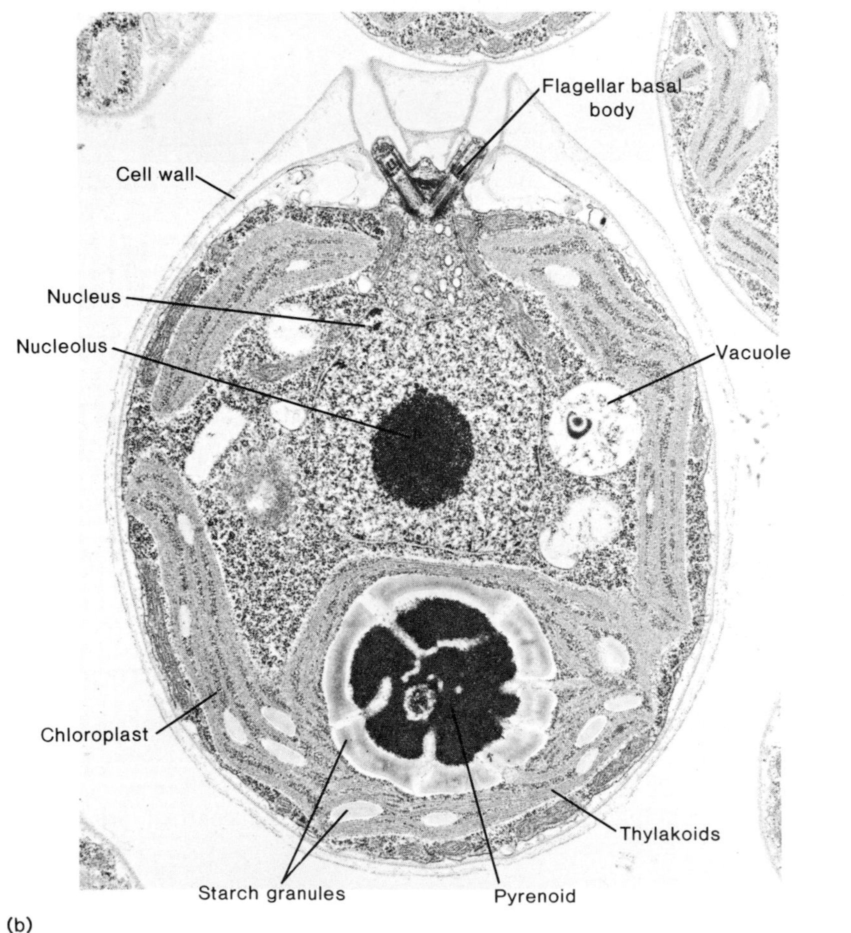

Figure 4.29 **Comparison of Procaryotic and Eucaryotic Cell Structure.** (*a*) The procaryote *Bacillus megaterium* (×30,500). (*b*) The eucaryotic alga *Chlamydomonas reinhardtii* (×30,000).

TABLE 4.2 Comparison of Procaryotic and Eucaryotic Cells

Property	Procaryotes	Eucaryotes
Organization of Genetic Material		
True membrane-bound nucleus	Absent	Present
DNA complexed with histones	No	Yes
Number of chromosomes	One[a]	More than one
Nucleolus	Absent	Present
Mitosis occurs	No	Yes
Genetic Recombination	Partial, unidirectional transfer of DNA	Meiosis and fusion of gametes
Mitochondria	Absent	Present
Chloroplasts	Absent	Present
Plasma Membrane with Sterols	Usually no[b]	Yes
Flagella	Submicroscopic in size; composed of one fiber	Microscopic in size; membrane bound; usually 20 microtubules in 9 + 2 pattern
Endoplasmic Reticulum	Absent	Present
Golgi Apparatus	Absent	Present
Cell Walls	Usually chemically complex with peptidoglycan[c]	Chemically simpler and lacking peptidoglycan
Differences in Simpler Organelles		
Ribosomes	70S	80S (except in mitochondria and chloroplasts)
Lysosomes and peroxisomes	Absent	Present
Microtubules	Absent or rare	Present
Differentiation	Rudimentary	Tissues and organs

[a]Plasmids may provide additional genetic information.

[b]Only the mycoplasmas and methanotrophs (methane utilizers) contain sterols. The mycoplasmas cannot synthesize sterols and require them preformed. Many procaryotes contain hopanoids.

[c]The mycoplasmas and archaeobacteria do not have peptidoglycan cell walls.

delimited nucleus. Bacteria are procaryotes; all other organisms—algae, fungi, protozoa, higher plants, and animals—are eucaryotic. Procaryotes normally are smaller than eucaryotic cells, often about the size of eucaryotic mitochondria and chloroplasts.

The presence of the eucaryotic nucleus is the most obvious difference between these two cell types, but several other major distinctions should be noted. It is clear from table 4.2 that procaryotic cells are much simpler structurally. In particular, an extensive and diverse collection of membrane-delimited organelles is missing. Furthermore, procaryotes are simpler functionally in several ways. They lack mitosis and meiosis, and have a simpler genetic organization. Many complex eucaryotic processes are absent in procaryotes: phagocytosis and pinocytosis, intracellular digestion, directed cytoplasmic streaming, ameboid movement, and others.

Despite the many significant differences between these two basic cell forms, they are remarkably similar on the biochemical level as discussed in succeeding chapters. Procaryotes and eucaryotes are composed of similar chemical constituents. With a few exceptions, the genetic code is the same in both, as is the way in which the genetic information in DNA is expressed. The principles underlying metabolic processes and most of the more important metabolic pathways are identical. Thus, beneath the profound structural and functional differences between procaryotes and eucaryotes, there is an even more fundamental unity: a molecular unity that is basic to life processes.

Summary

1. The eucaryotic cell has a true, membrane-delimited nucleus and many membranous organelles.
2. The cytoplasmic matrix contains microfilaments, intermediate filaments, and microtubules, small organelles partly responsible for cell structure and movement. These and other types of filaments are organized into a cytoskeleton.
3. The matrix is permeated by an irregular network of tubules and flattened sacs or cisternae known as the endoplasmic reticulum (ER). The ER may have attached ribosomes and be active in protein synthesis (rough or granular endoplasmic reticulum) or lack ribosomes (smooth or agranular ER).
4. The ER can donate materials to the Golgi apparatus, an organelle composed of one or more stacks of cisternae. This organelle prepares and packages cell products for secretion.
5. The Golgi apparatus also forms lysosomes. These vesicles contain digestive enzymes and aid in intracellular digestion of materials, including those taken up by endocytosis.
6. Eucaryotic ribosomes found free in the cytoplasmic matrix or bound to the ER are 80S ribosomes. Several may be attached to the same messenger RNA forming polyribosomes or polysomes.
7. Mitochondria are organelles bounded by two membranes, with the inner membrane folded into cristae, and are responsible for energy generation by the tricarboxylic acid cycle, electron transport, and oxidative phosphorylation.
8. Chloroplasts are the site of photosynthesis. The trapping of light energy takes place in the thylakoid membranes, whereas CO_2 incorporation is located in the stroma.
9. The nucleus is a large organelle containing the cell's chromosomes. It is bounded by a complex, double-membrane envelope perforated by pores through which materials can move.
10. The nucleolus lies within the nucleus and participates in the synthesis of ribosomal RNA and ribosomal subunits.
11. Eucaryotic chromosomes are distributed to daughter cells during regular cell division by mitosis. Meiosis is used to halve the chromosome number during sexual reproduction.
12. When a cell wall is present, it is constructed from polysaccharides, like cellulose, that are chemically simpler than procaryotic peptidoglycan. Many protozoa have a pellicle rather than a cell wall.
13. Many eucaryotic cells are motile because of cilia and flagella, membrane-delimited organelles with nine microtubule doublets surrounding two central microtubules. The doublets slide along each other to bend the cilium or flagellum.
14. Despite the fact that eucaryotes and procaryotes differ in many ways (table 4.2), they are quite similar metabolically.

Key Terms

autophagic vacuole *76*
axoneme *88*
basal body *88*
cell cycle *83*
cell wall *87*
chloroplast *79*
chromatin *82*
chromosome *82*
cilia *87*
cisternae *74*
cristae *79*
cytokinesis *85*
cytoplasmic matrix *72*
cytoskeleton *73*
dictyosome *75*
diploid *85*
dynein *89*
endocytosis *76*
endoplasmic reticulum (ER) *74*
endosome *76*
endosymbiotic theory *81*
eucaryotic cells *89*
F_1 factor *79*
flagella *87*
Golgi apparatus *75*
grana *82*
haploid *85*
intermediate filament *73*
interphase *85*
lysosome *76*
meiosis *85*
microfilament *72*
microtubule *72*
mitochondrion *78*
mitosis *83*
nuclear envelope *82*
nuclear pores *82*
nucleolus *83*
nucleus *82*
organelle *71*
pellicle *87*
phagocytosis *76*
pinocytosis *76*
plastid *79*
polyribosomes *78*
polysomes *78*
primary lysosomes *76*
procaryotic cells *89*
pyrenoid *82*
residual body *76*
ribosomal RNA (rRNA) *83*
rough or granular ER (RER or GER) *74*
secondary lysosomes *76*
smooth or agranular ER (SER or AER) *75*
stroma *79*
thylakoid *79*

Questions for Thought and Review

1. Describe the structure and function of every eucaryotic organelle discussed in the chapter.
2. Discuss the statement: "The most obvious difference between eucaryotic and procaryotic cells is in their use of membranes." What general roles do membranes play in eucaryotic cells?
3. Describe how the Golgi apparatus distributes proteins it receives from the ER to different organelles.
4. Briefly discuss how the complex of membranous organelles that de Duve calls the "vacuome" functions as a coordinated whole. What is its function?
5. Describe and contrast the ways in which flagella and cilia propel microorganisms through the water.
6. Outline the major differences between procaryotes and eucaryotes. In what way are they similar?

Additional Reading

Alberts, B.; Bray, D.; Lewis, J.; Raff, M.; Roberts, K.; and Watson, J. D. 1989. *Molecular biology of the cell,* 2d ed. New York: Garland Publishing.

Becker, W. M., and Deamer, D. W. 1991. *The world of the cell,* 2d ed. Redwood City, Calif.: Benjamin/Cummings.

Cook, G. M. W. 1980. Golgi apparatus. *Carolina Biology Reader,* no. 77, 2d ed. Burlington, N.C.: Carolina Biological Supply Co.

Darnell, J.; Lodish, H.; and Baltimore, D. 1990. *Molecular cell biology,* 2d ed. New York: Scientific American Books.

Dautry-Varsat, A., and Lodish, H. F. 1984. How receptors bring proteins and particles into cells. *Sci. Am.* 250(5):52–58.

de Duve, C. 1985. *A guided tour of the living cell.* New York: Scientific American Books.

Dingwall, C., and Laskey, R. A. 1986. Protein import into the cell nucleus. *Ann. Rev. Cell Biol.* 2:367–90.

Gray, M. W. 1983. The bacterial ancestry of plastids and mitochondria. *BioScience* 33(11):693–99.

Grivell, Leslie A. 1983. Mitochondrial DNA. *Sci. Am.* 248(3):78–89.

Helenius, A.; Mellman, I.; Wall, D.; and Hubbard, A. 1983. Endosomes. *Trends Biochem. Sci.* 8(7):245–50.

Heywood, P., and Magee, P. T. 1976. Meiosis in protists. *Bacteriol. Rev.* 40:190–240.

Holtzman, E., and Novikoff, A. B. 1984. *Cells and organelles.* 3d ed. New York: W. B. Saunders.

John, B., and Lewis, K. 1981. Somatic cell division. *Carolina Biology Reader,* no. 26, 2d ed. Burlington, N.C.: Carolina Biological Supply Co.

Karp, Gerald. 1984. *Cell biology,* 2d ed. New York: McGraw-Hill.

Margulis, L. 1971. Symbiosis and evolution. *Sci. Am.* 225(2):49–57.

McIntosh, J. R., and McDonald, K. L. 1989. The mitotic spindle. *Sci. Am.* 261(4):48–56.

Nachmias, V. T. 1984. Microfilaments. *Carolina Biology Reader,* no. 130. Burlington, N.C.: Carolina Biological Supply Co.

Newport, J. W., and Forbes, D. J. 1987. The nucleus: Structure, function, and dynamics. *Ann. Rev. Biochem.* 56:535–65.

Porter, K. R., and Tucker, J. B. 1981. The ground substance of the living cell. *Sci. Am.* 244(3):57–67.

Rothman, J. E. 1985. The compartmental organization of the Golgi apparatus. *Sci. Am.* 253(3):74–89.

Satir, P. 1983. Cilia and related organelles. *Carolina Biology Reader,* no. 123. Burlington, N.C.: Carolina Biological Supply Co.

Sheeler, P., and Bianchi, D. E. 1987. *Cell and molecular biology,* 3d ed. New York: John Wiley and Sons.

Part Two
Microbial Growth and Metabolism

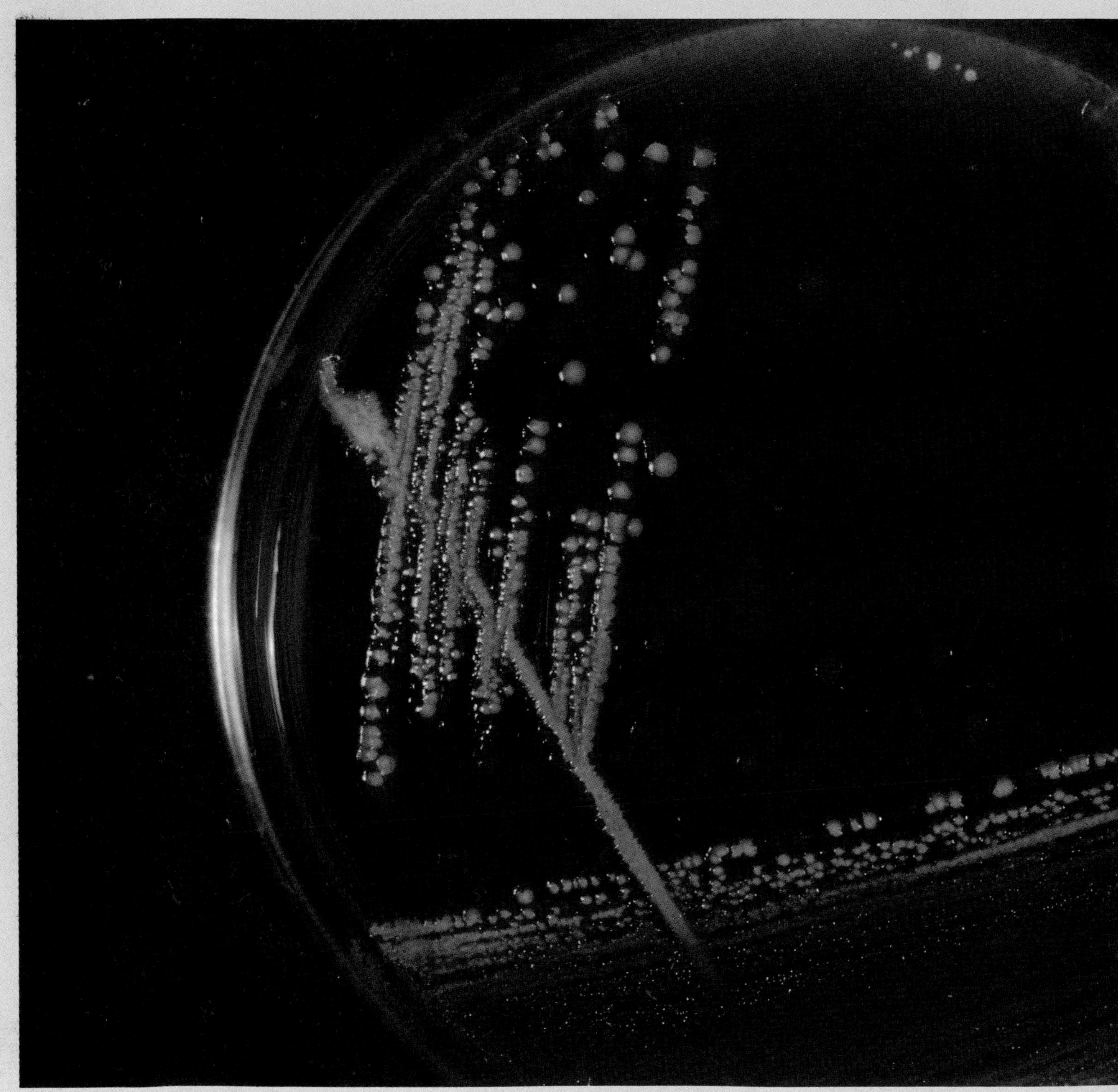

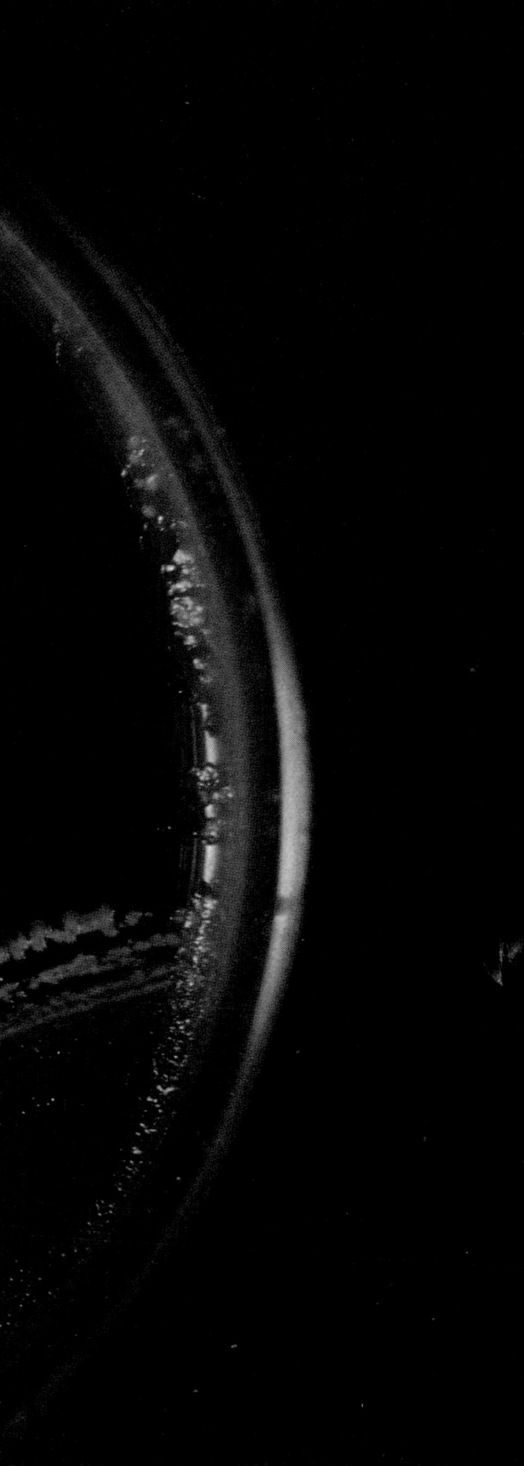

Chapter 5
Microbial Nutrition

Chapter 6
Microbial Growth

Chapter 7
Metabolism: Energy and Enzymes

Chapter 8
Metabolism: The Generation of Energy

Chapter 9
Metabolism: The Use of Energy in Biosynthesis

Chapter 10
Metabolism: The Synthesis of Nucleic Acids and Proteins

Chapter 11
The Regulation of Enzyme Activity and Synthesis

An xylose-lysine-desoxycholate agar (XLD agar) streak plate showing *Escherichia coli* and *Salmonella typhi*. XLD agar is a selective and differential agar. It is used for the selection and identification of enteric pathogens such as *Salmonella* and *Shigella*. Sodium desoxycholate inhibits the growth of gram-positive microorganisms. *E. coli* produces large yellow colonies, while *Salmonella* forms pink-red colonies with black centers.

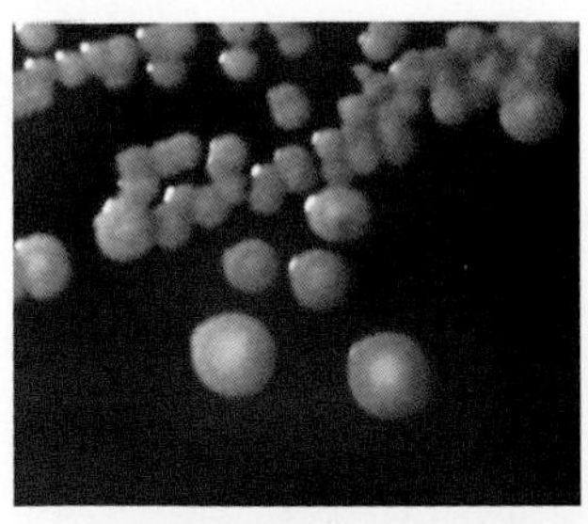

CHAPTER 5
Microbial Nutrition

The whole of nature, as has been said, is a conjugation of the verb to eat, in the active and passive.

—William Ralph Inge

Outline

Concepts

1. Microorganisms require about ten elements in large quantities, in part because they are used to construct carbohydrates, lipids, proteins, and nucleic acids. Several other elements are needed in very small amounts and are parts of enzymes and cofactors.
2. All microorganisms can be placed in one of a few nutritional categories on the basis of their requirements for carbon, energy, and hydrogen/electrons.
3. Nutrient molecules frequently cannot cross selectively permeable plasma membranes through passive diffusion and must be transported by one of three major mechanisms involving the use of membrane carrier proteins. Eucaryotic microorganisms also employ endocytosis for nutrient uptake.
4. Culture media are needed to grow microorganisms in the laboratory and to carry out specialized procedures like microbial identification, water and food analysis, and the isolation of particular microorganisms. A wide variety of media is available for these and other purposes.
5. Pure cultures can be obtained through the use of spread plates, streak plates, or pour plates and are required for the careful study of an individual microbial species.

To obtain energy and construct new cellular components, organisms must have a supply of raw materials or nutrients. **Nutrients** are substances used in biosynthesis and energy production, and therefore are required for microbial growth. This chapter describes the nutritional requirements of microorganisms, how nutrients are acquired, and the cultivation of microorganisms.

Environmental factors such as temperature, oxygen levels, and the osmotic concentration of the medium are critical in the successful cultivation of microorganisms. These topics are discussed in chapter 6 after an introduction to microbial growth.

The Common Nutrient Requirements

Analysis of microbial cell composition shows that over 95% of cell dry weight is made up of a few major elements: carbon, oxygen, hydrogen, nitrogen, sulfur, phosphorus, potassium, calcium, magnesium, and iron. These are called **macroelements** or macronutrients because they are required by microorganisms in relatively large amounts. The first six (C, O, H, N, S, and P) are components of carbohydrates, lipids, proteins, and nucleic acids. The remaining four macroelements exist in the cell as cations and play a variety of roles. For example, potassium (K^+) is required for activity by a number of enzymes, including some of those involved in protein synthesis. Calcium (Ca^{2+}), among other functions, contributes to the heat resistance of bacterial endospores. Magnesium (Mg^{2+}) serves as a cofactor for many enzymes, complexes with ATP, and stabilizes ribosomes and cell membranes. Iron (Fe^{2+} and Fe^{3+}) is a part of cytochromes and a cofactor for enzymes and electron-carrying proteins.

All microorganisms require several **trace elements** (also called microelements or micronutrients) besides macroelements. The trace elements—manganese, zinc, cobalt, molybdenum, nickel, and copper—are needed by most cells. However, cells require such small amounts that contaminants in water, glassware, and regular media components often are adequate for growth. Therefore, it is very difficult to demonstrate a trace element requirement. Trace elements are normally a part of enzymes and cofactors, and they aid in the catalysis of reactions and maintenance of protein structure. For example, zinc (Zn^{2+}) is present at the active site of some enzymes, but is also involved in the association of regulatory and catalytic subunits in *E. coli* aspartate carbamoyltransferase (*see chapter 11*). Manganese (Mn^{2+}) can aid many enzymes catalyzing the transfer of phosphate groups. Molybdenum (Mo^{2+}) is required for nitrogen fixation, and cobalt (Co^{2+}) is a component of vitamin B_{12}.

Electron carriers and enzymes (pp. 137–43).

Elements can be categorized somewhat differently with respect to nutritional requirements. The major elements (C, O, H, N, S, P) are needed in gram quantities in a liter of culture medium. The minor elements (K, Ca, Mg, Fe) often are required in milligram quantities. Trace elements (Mn, Zn, Co, Mo, Ni, Cu) must be available in microgram amounts.

Besides the common macroelements and trace elements, microorganisms may have particular requirements that reflect the special nature of their morphology or environment. Diatoms (*see figure 27.6* c,d) need silicic acid (H_4SiO_4) to construct their beautiful cell walls of silica [$(SiO_2)_n$]. Although most bacteria do not require large amounts of sodium, many bacteria growing in saline lakes and oceans (*see chapters 21 and 24*) depend upon the presence of high concentrations of sodium ion (Na^+).

Requirements for Carbon, Hydrogen, and Oxygen

The requirements for carbon, hydrogen, and oxygen are often satisfied together. Carbon is required for the skeleton or backbone of all organic molecules, and molecules serving as carbon sources usually also contribute both oxygen and hydrogen atoms. One carbon source for which this is not true is carbon dioxide (CO_2) because it is oxidized and lacks hydrogen. Probably all microorganisms can fix CO_2; that is, reduce it and incorporate it into organic molecules. However, by definition, only **autotrophs** can use CO_2 as their sole or principal source of carbon. Many microorganisms are autotrophic; many of these are photosynthetic, but some oxidize inorganic molecules to obtain energy.

Photosynthetic carbon dioxide fixation (pp. 173–75).

The reduction of CO_2 is a very energy expensive process. Thus, many microorganisms cannot use CO_2 as their sole carbon source but must rely upon the presence of more reduced, complex molecules for a supply of carbon. Organisms that use reduced, preformed organic molecules as carbon sources are **heterotrophs** (these preformed molecules normally come from other organisms). Most heterotrophs use organic nutrients as a source of both carbon and energy. For example, the glycolytic pathway traps energy as ATP and NADH, and also produces carbon skeletons for use in biosynthesis.

The glycolytic pathway (pp. 147–48).

A most remarkable nutritional characteristic of microorganisms is their extraordinary flexibility with respect to carbon sources. There is no naturally occurring organic molecule that cannot be used by some microorganism. Actinomycetes can degrade amyl alcohol, paraffin, and even rubber. Some bacteria seem able to employ almost anything as a carbon source; for example, *Pseudomonas cepacia* can use over 100 different carbon compounds. Unfortunately, many man-made substances like plastics and DDT are degraded slowly or not at all. In contrast to bacterial omnivores, some bacteria are exceedingly fastidious and catabolize few carbon compounds. Methylotrophic bacteria metabolize only methane, methanol, carbon monoxide, formic acid, and a few related one-carbon molecules. Parasitic members of the genus *Leptospira* use only long-chain fatty acids as their major source of carbon and energy.

The nutritional requirements of microorganisms vary enormously among different species. In addition, these requirements can change within a species because of gene mutations. A microorganism requiring the same nutrients as most naturally occurring members of its species is a **prototroph.** A prototrophic microorganism may mutate so that it cannot synthesize a molecule essential for growth and reproduction. It will then require that molecule, or a compound that can be converted to it, as a nutrient. A mutated prototroph that lacks the ability to synthesize an essential nutrient and therefore must obtain it or a precursor from the surroundings is an **auxotroph.** The requirement for a specific amino acid is a common form of auxotrophy. Many microorganisms can synthesize all the common amino acids needed for growth. Occasionally a mutation will block the synthesis of an essential amino acid and the microorganism becomes auxotrophic for it; that is, the amino acid must be available for growth to take place. The production of auxotrophs is useful in the study of microbial genetics.

Genetic mutations (pp. 244–52).

Nutritional Types of Microorganisms

All organisms also require sources of energy, hydrogen, and electrons for growth to take place. Microorganisms can be grouped into nutritional classes based on how they satisfy these requirements (tables 5.1 and 5.2). There are only two sources of energy available to organisms: (1) light energy trapped during photosynthesis, and (2) the energy derived from oxidizing organic or inorganic molecules. **Phototrophs** use light as their energy source; **chemotrophs** obtain energy from the oxidation of chemical compounds (either organic or inorganic). Microorganisms also have only two sources for hydrogen atoms or electrons. **Lithotrophs** (that is, "rock-eaters") use reduced inorganic substances as their electron source, whereas **organotrophs** extract electrons or hydrogen from organic compounds.

Photosynthesis light reactions (pp. 162–68).
Oxidation of organic and inorganic molecules (pp. 147–62).

Despite the great metabolic diversity seen in microorganisms, most may be placed in one of four nutritional classes based on their primary sources of energy, hydrogen and/or electrons, and carbon (table 5.2). The large majority of microorganisms thus far studied are either photolithotrophic autotrophs or chemoorganotrophic heterotrophs. **Photolithotrophic autotrophs** (often called **photoautotrophs**) use light energy and CO_2 as a carbon source. Eucaryotic algae and blue-green bacteria (cyanobacteria) employ water as the electron donor and release oxygen. Purple and green sulfur bacteria cannot oxidize water, but extract electrons from inorganic donors like hydrogen, hydrogen sulfide, and elemental sulfur. **Chemoorganotrophic heterotrophs** (often called **chemoheterotrophs,** or even heterotrophs) use organic compounds as sources of energy, hydrogen, electrons, and carbon for biosynthesis. Frequently, the same organic nutrient will satisfy all these requirements. It should be noted that essentially all pathogenic microorganisms are chemoheterotrophs. The other two classes have fewer microorganisms but often are very important ecologically. Some purple and green bacteria are photosynthetic and use organic matter as their electron donor and carbon source. These **photoorganotrophic heterotrophs** are common inhabitants of polluted lakes and streams. Some of these bacteria also can grow as photoautotrophs with molecular hydrogen as an electron donor. The fourth group, the **chemolithotrophic autotrophs,** oxidizes reduced inorganic compounds such as iron, nitrogen, or sulfur molecules to derive both energy and electrons for biosynthesis. Carbon dioxide is the carbon source. A few chemolithotrophs can derive their carbon from organic sources and thus are heterotrophic. Bacteria relying on inorganic energy sources and organic (or sometimes CO_2) carbon sources may be called **mixotrophic** (they are combining autotrophic and heterotrophic metabolic processes). Chemolithotrophs contribute greatly to the chemical transformations of elements (e.g., the conversion of ammonia to nitrate or sulfur to sulfate) that continually occur in the ecosystem.

Table 5.1 Sources of Carbon, Energy, and Hydrogen/Electrons

Carbon Sources	
Autotrophs	CO_2 sole or principal biosynthetic carbon source (*pp. 173–75*)[a]
Heterotrophs	Reduced, preformed, organic molecules from other organisms (*chapters 8 and 9*)
Energy Sources	
Phototrophs	Light (*pp. 162–68*)
Chemotrophs	Oxidation of organic or inorganic compounds (*chapter 8*)
Hydrogen or Electron Sources	
Lithotrophs	Reduced inorganic molecules (*pp. 160–62*)
Organotrophs	Organic molecules (*chapter 8*)

[a]For each category, the location of material describing the participating metabolic pathways is given within the parentheses.

Photosynthetic and chemolithotrophic bacteria (chapter 23).

Although a particular species usually belongs in only one of the four nutritional classes, some show great metabolic flexibility and alter their metabolic patterns in response to environmental changes. For example, many purple nonsulfur bacteria (*see chapter 23*) act as photoorganotrophic heterotrophs in the absence of oxygen, but oxidize organic molecules and function chemotrophically at normal oxygen levels. When oxygen is low, photosynthesis and oxidative metabolism may function simultaneously. This sort of flexibility seems complex and confusing, yet it gives its possessor a definite advantage if environmental conditions frequently change.

TABLE 5.2 Major Nutritional Types of Microorganisms

Major Nutritional Types[a]	Sources of Energy, Hydrogen/Electrons, and Carbon	Representative Microorganisms
Photolithotrophic autotrophy	Light energy Inorganic hydrogen/electron (H/e^-) donor CO_2 carbon source	Algae Purple and green sulfur bacteria Blue-green bacteria (cyanobacteria)
Photoorganotrophic heterotrophy	Light energy Organic H/e^- donor Organic carbon source (CO_2 may also be used)	Purple nonsulfur bacteria Green nonsulfur bacteria
Chemolithotrophic autotrophy	Chemical energy source (inorganic) Inorganic H/e^- donor CO_2 carbon source	Sulfur-oxidizing bacteria Hydrogen bacteria Nitrifying bacteria Iron bacteria
Chemoorganotrophic heterotrophy	Chemical energy source (organic) Organic H/e^- donor Organic carbon source	Protozoa Fungi Most nonphotosynthetic bacteria

[a]Bacteria in other nutritional categories have been found.

1. What are nutrients, and on what basis are they divided into macroelements and trace elements? Describe some ways in which macronutrients and trace elements are used by an organism.
2. Define autotroph, heterotroph, prototroph, and auxotroph.
3. Discuss the ways in which microorganisms are classified based on their requirements for energy, hydrogen, and electrons. Describe the nutritional requirements of the four major nutritional groups and give some microbial examples of each. What is a mixotroph?

Requirements for Nitrogen, Phosphorus, and Sulfur

To grow, a microorganism must be able to incorporate large quantities of nitrogen, phosphorus, and sulfur. Although these elements may be acquired from the same nutrients that supply carbon, microorganisms usually employ inorganic sources as well.

Biochemical mechanisms for the incorporation of nitrogen, phosphorus, and sulfur (pp. 176–80).

Nitrogen is needed for the synthesis of amino acids, purines, pyrimidines, some carbohydrates and lipids, enzyme cofactors, and other substances. Many microorganisms can use the nitrogen in amino acids, and ammonia often is directly incorporated through the action of such enzymes as glutamate dehydrogenase or glutamine synthetase and glutamate synthase (*see chapter 9*). Most phototrophs and many nonphotosynthetic microorganisms reduce nitrate to ammonia and incorporate the ammonia in assimilatory nitrate reduction (*p. 178*). A variety of bacteria (e.g., many cyanobacteria and the symbiotic bacterium *Rhizobium*) can reduce and assimilate atmospheric nitrogen using the nitrogenase system (*see chapter 9*).

Phosphorus is present in nucleic acids, phospholipids, nucleotides like ATP, several cofactors, some proteins, and other cell components. Almost all microorganisms use inorganic phosphate as their phosphorus source and incorporate it directly. Some microorganisms (e.g., *E. coli*) actively acquire phosphate from their surroundings. Low phosphate levels actually limit microbial growth in many aquatic environments.

Sulfur is needed for the synthesis of substances like the amino acids cysteine and methionine, some carbohydrates, biotin, and thiamine. Most microorganisms use sulfate as a source of sulfur and reduce it by assimilatory sulfate reduction (*see chapter 9*); a few require a reduced form of sulfur such as cysteine.

Growth Factors

Microorganisms, especially many photolithotrophic autotrophs, often grow and reproduce when minerals and sources of energy, carbon, nitrogen, phosphorus, and sulfur are supplied. These organisms have the enzymes and pathways necessary to synthesize all cell components required for their well-being. Many microorganisms, on the other hand, lack one or more essential enzymes. Therefore they cannot manufacture all indispensable constituents, but must obtain them or their precursors from the environment. Organic compounds required because they are essential cell components or precursors of such components and cannot be synthesized by the organism are called **growth factors.** There are three major classes of growth factors: (1) amino acids, (2) purines and pyrimidines, and (3) vitamins. Amino acids are needed for protein synthesis, purines and pyrimidines for nucleic acid synthesis. **Vitamins** are small organic molecules that usually make up all or part of enzyme cofactors (*see chapter 7*), and only very small amounts sustain growth. The functions of selected vitamins, and examples of microorganisms requiring them, are given in table 5.3. Some microorganisms require

TABLE 5.3 Functions of Some Common Vitamins in Microorganisms

Vitamin	Functions	Examples of Microorganisms Requiring Vitamin[a]
Biotin	Carboxylation (CO_2 fixation) One-carbon metabolism	*Leuconostoc mesenteroides* (B) *Saccharomyces cerevisiae* (F) *Ochromonas malhamensis* (A) *Acanthamoeba castellanii* (P)
Cyanocobalamin (B_{12})	Molecular rearrangements One-carbon metabolism—carries methyl groups	*Lactobacillus* spp. (B) *Euglena gracilis* (A) Diatoms and many other algae (A) *Acanthamoeba castellanii* (P)
Folic acid	One-carbon metabolism	*Enterococcus faecalis* (B) *Tetrahymena pyriformis* (P)
Pantothenic acid	Precursor of coenzyme A—carries acyl groups (Pyruvate oxidation, fatty acid metabolism)	*Proteus morganii* (B) *Hanseniaspora* spp. (F) *Paramecium* spp. (P)
Pyridoxine (B_6)	Amino acid metabolism (e.g., transamination)	*Lactobacillus* spp. (B) *Tetrahymena pyriformis* (P)
Niacin (nicotinic acid)	Precursor of NAD and NADP—carry electrons and hydrogen atoms	*Brucella abortus* (B) *Blastocladia pringsheimi* (F) *Crithidia fasciculata* (P)
Riboflavin (B_2)	Precursor of FAD and FMN—carry electrons or hydrogen atoms	*Caulobacter vibrioides* (B) *Dictyostelium* spp. (F) *Tetrahymena pyriformis* (P)
Thiamine (B_1)	Aldehyde group transfer (Pyruvate decarboxylation, α-keto acid oxidation)	*Bacillus anthracis* (B) *Phycomyces blakesleeanus* (F) *Ochromonas malhamensis* (A) *Colpidium campylum* (P)

[a]The representative microorganisms are members of the following groups: bacteria (*B*), fungi (*F*), algae (*A*), and protozoa (*P*).

many vitamins; for example, *Enterococcus faecalis* (a lactic acid bacterium) needs eight different vitamins for growth. Other growth factors are also seen; heme (from hemoglobin or cytochromes) is required by *Haemophilus influenzae,* and some mycoplasmas need cholesterol.

Knowledge of the specific growth factor requirements of many microorganisms makes possible quantitative growth-response assays for a variety of substances. For example, species from the bacterial genera *Lactobacillus* and *Streptococcus* can be used in microbiological assays of most vitamins and amino acids. The appropriate bacterium is grown in a series of culture vessels, each containing medium with an excess amount of all required components except the growth factor to be assayed. A different amount of growth factor is added to each vessel. The standard curve is prepared by plotting the growth factor quality or concentration against the total extent of bacterial growth. Ideally, the amount of growth resulting is directly proportional to the quantity of growth factor present; if the growth factor concentration doubles, the final extent of bacterial growth doubles. The quantity of the growth factor in a test sample is determined by comparing the extent of growth caused by the unknown sample with that resulting from the standards. Microbiological assays are specific, sensitive, and simple. They still are used in the assay of substances like vitamin B_{12} and biotin, despite advances in chemical assay techniques.

1. Briefly summarize the ways in which microorganisms obtain nitrogen, phosphorus, and sulfur from their environment.
2. What are growth factors? What are vitamins? How can microorganisms be used to determine the quantity of a specific substance in a sample?

Uptake of Nutrients by the Cell

The first step in nutrient utilization is uptake of the required nutrients by the microbial cell. Uptake mechanisms must be specific; that is, the necessary substances, and not others, must be acquired. It does a cell no good to take in a substance that it cannot use. Since microorganisms often live in nutrient-poor habitats, they must be able to transport nutrients from dilute solutions into the cell against a concentration gradient. Finally, nutrient molecules must pass through a selectively permeable plasma membrane that will not permit the free passage of most substances. In view of the enormous variety of nutrients and the complexity of the task, it is not surprising that microorganisms make use of several different transport mechanisms. The most important of these are facilitated diffusion, active transport, and group translocation. Eucaryotic microorganisms do not appear to employ group translocation, but do take up nutrients by the process of endocytosis (*see chapter 4*).

Plasma membrane structure and properties (pp. 44–46).

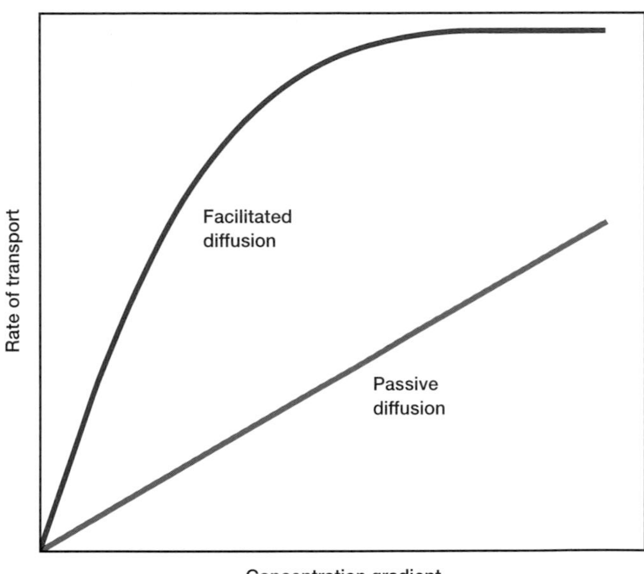

Figure 5.1 **Passive and Facilitated Diffusion.** The dependence of diffusion rate on the size of the solute's concentration gradient. Note the saturation effect or plateau at higher gradient values when a facilitated diffusion carrier is operating.

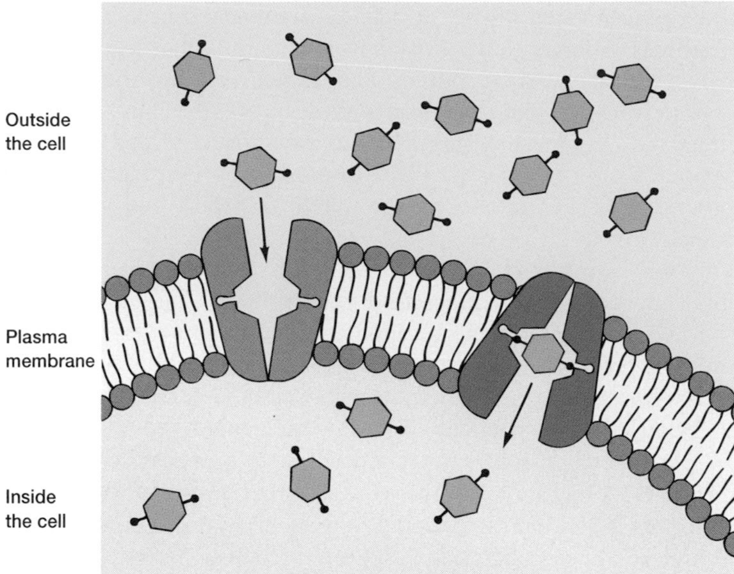

Figure 5.2 **A Model of Facilitated Diffusion.** The membrane carrier can change conformation after binding an external molecule and subsequently release the molecule on the cell interior. It then returns to the outward oriented position and is ready to bind another solute molecule. Because there is no energy input, molecules will continue to enter only as long as their concentration is greater on the outside.

Facilitated Diffusion

A few substances, such as glycerol, can cross the plasma membrane by **passive diffusion.** Passive diffusion, often simply called diffusion, is the process in which molecules move from a region of higher concentration to one of lower concentration because of random thermal agitation. The rate of passive diffusion is dependent upon the size of the concentration gradient between a cell's exterior and its interior (figure 5.1). A fairly large concentration gradient is required for adequate nutrient uptake by passive diffusion (i.e., the external nutrient concentration must be high), and the rate of uptake decreases as more nutrient is acquired unless it is used immediately. Thus, passive diffusion is an inefficient process and is not employed extensively by microorganisms. Although glycerol can enter cells by passive diffusion, another transport mechanism also operates.

The rate of diffusion across selectively permeable membranes is greatly increased by using carrier proteins, sometimes called **permeases,** which are embedded in the plasma membrane. Because a carrier aids the diffusion process, it is called **facilitated diffusion.** The rate of facilitated diffusion increases with the concentration gradient much more rapidly and at lower concentrations of the diffusing molecule than that of passive diffusion (figure 5.1). Note that the diffusion rate levels off or reaches a plateau at higher gradient levels because the carrier is saturated; that is, the carrier protein is binding and transporting as many solute molecules as possible. The resulting curve resembles an enzyme-substrate curve (*see chapter 7*) and is different from the linear response seen with passive diffusion. Carrier proteins also resemble enzymes in their specificity for the substance to be transported; each carrier is selective and will transport only closely related solutes. Although a carrier protein is involved, facilitated diffusion is truly diffusion. A concentration gradient spanning the membrane drives the movement of molecules, and no extra energy input is required. If the concentration gradient disappears, net inward movement ceases.

Although much work has been done on the mechanism of facilitated diffusion, the process is not yet understood completely. It appears that the carrier protein complex spans the membrane (figure 5.2). After the solute molecule binds to the outside, the carrier may change conformation and release the molecule on the cell interior. The carrier would subsequently change back to its original shape and be ready to pick up another molecule. The net effect is that a lipid-insoluble molecule can enter the cell in response to its concentration gradient. Remember that the mechanism is reversible; if the solute's concentration is greater inside the cell, it will move outward. Because the cell metabolizes nutrients upon entry, influx is favored.

Facilitated diffusion does not seem to be important in procaryotes. Glycerol is transported by facilitated diffusion in *E. coli, Salmonella typhimurium, Pseudomonas, Bacillus,* and many other bacteria. The process is much more prominent in eucaryotic cells where it is used to transport a variety of sugars and amino acids.

Active Transport

Although facilitated diffusion carriers can efficiently move molecules to the interior when the solute concentration is higher on the outside, they cannot take up solutes that are already more concentrated within the cell (i.e., against a concentration gradient). Microorganisms often live in habitats characterized by very dilute nutrient sources, and, to flourish, they must be able to transport and concentrate these nutrients. Thus, facilitated diffusion mechanisms are not always adequate, and other approaches must be used. The two most important transport processes in such situations are active transport and group translocation.

Active transport is the transport of solute molecules to higher concentrations, or against a concentration gradient, with the use of metabolic energy input. Because active transport involves protein carrier activity, it resembles facilitated diffusion in some ways. The carrier proteins bind particular solutes with great specificity for the molecules transported. Similar solute molecules can compete for the same carrier protein in both facilitated diffusion and active transport. Active transport is also characterized by the carrier saturation effect at high solute concentrations (figure 5.1). Nevertheless, active transport differs from facilitated diffusion in its use of metabolic energy and in its ability to concentrate substances. Metabolic inhibitors that block energy production will inhibit active transport but will not affect facilitated diffusion (at least for a short time).

Binding protein transport systems employ special substrate binding proteins located in the periplasmic space of gram-negative bacteria (*see figure 3.23*). These periplasmic proteins, which also participate in chemotaxis (*see chapter 3*), bind the molecule to be transported and then interact with the membrane transport proteins to move the solute molecule inside the cell. The energy source is ATP, although other high-energy phosphate compounds may drive uptake in some transport systems. *E. coli* transports a variety of sugars (arabinose, maltose, galactose, ribose) and amino acids (glutamate, histidine, leucine) by this mechanism. ATP-driven transport is present in gram-positive bacteria, but is not as well understood as in gram-negative organisms.

Bacteria also use protonmotive force (usually in the form of a proton gradient generated during electron transport) to drive active transport. Membrane-bound transport systems that lack special solute-binding proteins participate. The lactose permease of *E. coli* is a well-studied example. The permease is a single protein having a molecular weight of about 30,000. It transports a lactose molecule inward as a proton simultaneously enters the cell (a higher concentration of protons is maintained outside the membrane by electron transport chain activity). Such linked transport of two substances in the same direction is called **symport.** Here, energy stored as a proton gradient drives solute transport. Although the mechanism of transport is not completely understood, it is thought that binding of a proton to the transport protein changes its shape and affinity for the solute to be transported. *E. coli* also uses proton symport to take up amino acids and organic acids like succinate and malate.

The chemiosmotic hypothesis and protonmotive force (p. 153).

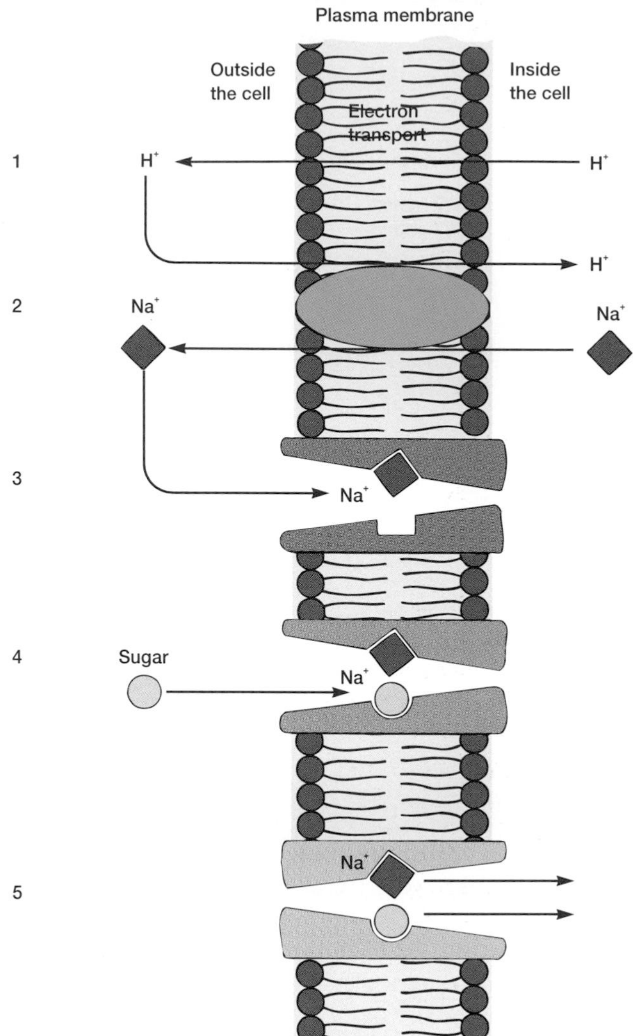

Figure 5.3 Active Transport Mechanism. The use of proton and sodium gradients in active transport. (*1*) Electron transport energy is used to pump protons to the outside of the plasma membrane. (*2*) The proton gradient drives sodium ion expulsion by an antiport mechanism. (*3*) Sodium binds to the carrier protein complex. (*4*) The shape of the solute binding site changes, and it binds the solute (e.g., a sugar or amino acid). (*5*) The carrier's conformation then alters so that sodium is released on the inside of the membrane. This is followed by solute dissociation from the carrier.

Protonmotive force can power active transport indirectly, often through the formation of a sodium ion gradient. For example, an *E. coli* sodium transport system pumps sodium outward in response to the inward movement of protons (figure 5.3). Such linked transport in which the transported substances move in opposite directions is termed **antiport.** The sodium gradient generated by this proton antiport system then drives the uptake of sugars and amino acids. A sodium ion could attach to a carrier protein, causing it to change shape. The carrier would then bind the sugar or amino acid tightly and orient its binding sites toward the cell interior. Because of the low intracellular sodium concentration, the sodium ion would dissociate from the carrier, and the other molecule would

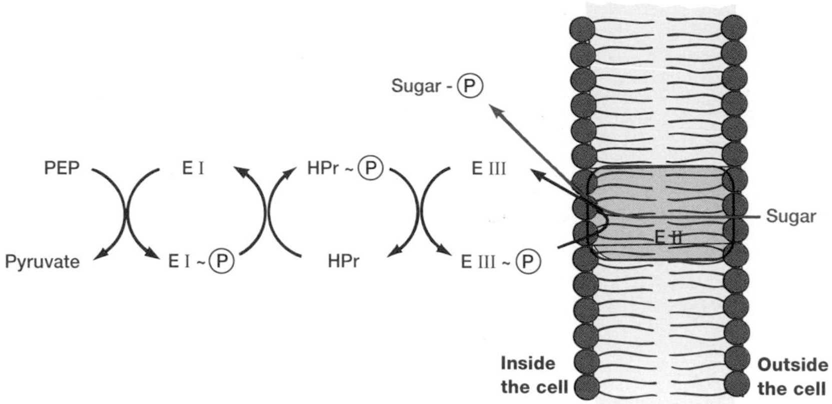

Figure 5.4 **Bacterial PTS Transport.** The phosphoenolpyruvate: sugar phosphotransferase system (PTS) of *E. coli* and *S. typhimurium*. The following components are involved in the system: phosphoenolpyruvate, *PEP;* enzyme I, *E I;* the low molecular weight heat-stable protein, *HPr;* enzyme II, *E II;* and enzyme III, *E III.*

follow. *E. coli* transport proteins carry the sugar melibiose and the amino acid glutamate when sodium simultaneously moves inward. Sodium symport or cotransport is also an important process in eucaryotic cells where it is used in sugar and amino acid uptake. ATP, rather than protonmotive force, usually drives sodium transport in eucaryotic cells.

It seems reasonable that a microorganism would have only one transport system for each nutrient, but often this is not so, as can be seen with *E. coli.* This bacterium has at least five transport systems for the sugar galactose, three systems each for the amino acids glutamate and leucine, and two potassium transport complexes. When there are several transport systems for the same substance, the systems differ in such properties as their energy source, their affinity for the solute transported, and the nature of their regulation. Presumably this diversity gives its possessor an added competitive advantage in a variable environment.

Group Translocation

In active transport, solute molecules move across a membrane without modification. Many procaryotes also can take up molecules by **group translocation,** a process in which a molecule is transported into the cell while being chemically altered. The best-known group translocation system is the **phosphoenolpyruvate: sugar phosphotransferase system (PTS).** It transports a variety of sugars into procaryotic cells while phosphorylating them using phosphoenolpyruvate (PEP) as the phosphate donor.

$$\text{PEP} + \text{sugar (outside)} \longrightarrow \text{pyruvate} + \text{sugar} - \text{P (inside)}$$

The PTS is quite complex. In *E. coli* and *Salmonella typhimurium,* it consists of three enzymes and a low molecular weight heat-stable protein (HPr). HPr and enzyme I are cytoplasmic, enzyme II is an integral membrane protein, and enzyme III is either cytoplasmic or a peripheral protein loosely bound to the membrane. A high-energy phosphate is transferred from PEP to enzyme III with the aid of enzyme I and HPr (figure 5.4). Then, as a sugar molecule is carried across the membrane by enzyme II, it is phosphorylated. Enzymes II and III transport only specific sugars and vary with PTS, while enzyme I and HPr are common to all PTSs.

PTSs are widely distributed in procaryotes. Except for some species of *Bacillus* that have both the Embden-Meyerhof pathway and phosphotransferase systems, aerobic bacteria seem to lack PTSs. Members of the genera *Escherichia, Salmonella, Staphylococcus,* and other facultatively anaerobic bacteria have phosphotransferase systems; some obligately anaerobic bacteria (e.g., *Clostridium*) also have PTSs. Many sugars are transported by these systems. *E. coli* and *S. typhimurium* take up glucose, fructose, lactose, mannitol, sorbitol, galactitol, and other sugars by group translocation. Besides their role in transport, PTS proteins can act as chemoreceptors for chemotaxis.

Iron Uptake

Almost all microorganisms require iron for use in cytochromes and many enzymes. Iron uptake is made difficult by the great insolubility of ferric iron (Fe^{3+}) and its derivatives, which leaves little free iron available for transport. Many bacteria and fungi have overcome this difficulty by secreting siderophores (Greek for iron bearers). **Siderophores** are low molecular weight molecules that are able to complex with ferric iron and supply it to the cell. These iron-transport molecules are normally either hydroxamates or phenolates-catecholates. Ferrichrome is a hydroxamate produced by many fungi; enterobactin is the catecholate formed by *E. coli* (figure 5.5*a,b*). It appears that three siderophore groups complex with iron orbitals to form a six-coordinate, octahedral complex (figure 5.5*c*).

Microorganisms secrete siderophores when little iron is available in the medium. Once the iron-siderophore complex

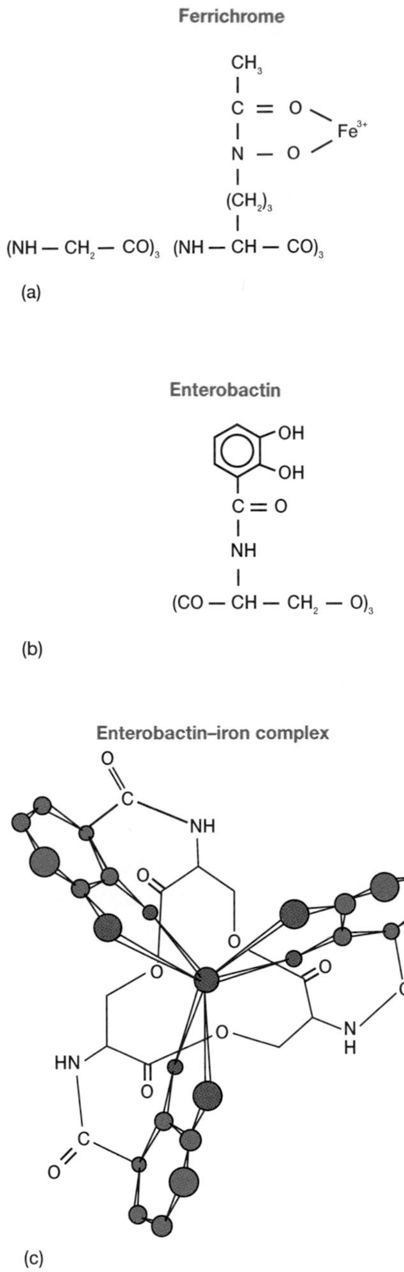

Figure 5.5 Siderophore Ferric Iron Complexes. (*a*) Ferrichrome is a cyclic hydroxamate [—CO — N(O^-) —] molecule formed by many fungi. (*b*) *E.coli* produces the cyclic catecholate derivative, enterobactin. (*c*) Ferric iron probably complexes with three siderophore groups to form a six-coordinate, octahedral complex as shown in this illustration of the enterobactin-iron complex.

has reached the cell surface, it binds to a siderophore-receptor protein. Then the iron is either released to enter the cell directly or the whole iron-siderophore complex is transported inside. In *E. coli,* the siderophore receptor is in the outer membrane of the cell envelope; when the iron reaches the periplasmic space, it moves through the plasma membrane with the aid of other proteins. After the iron has entered the cell, it is reduced to the ferrous form (Fe^{2+}). Iron is so crucial to microorganisms that they may use more than one route of iron uptake to ensure an adequate supply.

1. Describe facilitated diffusion, active transport, and group translocation in terms of their distinctive characteristics and mechanisms.
2. How do binding protein transport systems and membrane-bound transport systems differ with respect to energy sources? What are symport and antiport processes?
3. How are siderophores involved in iron transport?

Culture Media

Much of microbiology depends upon the ability to grow and maintain microorganisms in the laboratory, and this is possible only if suitable culture media are available. In addition, specialized media are essential in the isolation and identification of microorganisms, the testing of antibiotic sensitivities, water and food analysis, industrial microbiology, and other activities. Although all microorganisms need sources of energy, carbon, nitrogen, phosphorus, sulfur, and various minerals, the precise composition of a satisfactory medium will depend upon the species one is trying to cultivate because nutritional requirements vary so greatly. Knowledge of a microorganism's normal habitat is often useful in selecting an appropriate culture medium because its nutrient requirements reflect its natural surroundings. Frequently, a medium is used to select and grow specific microorganisms or to help identify a particular species. In such cases, the function of the medium also will determine its composition.

Synthetic or Defined Media

Some microorganisms, particularly photolithotrophic autotrophs such as cyanobacteria and eucaryotic algae, can be grown on relatively simple media containing CO_2 as a carbon source (often added as sodium carbonate or bicarbonate), nitrate or ammonia as a nitrogen source, sulfate, phosphate, and a variety of minerals (table 5.4). Such a medium in which all components are known is a **defined medium** or **synthetic medium.** Many chemoorganotrophic heterotrophs also can be grown in defined media with glucose as a carbon source and an ammonium salt as a nitrogen source. Not all defined media are as simple as the examples in table 5.4, but may be constructed from dozens of components. Defined media are used widely in research, as it is often desirable to know what the experimental microorganism is metabolizing.

Complex Media

Media that contain some ingredients of unknown chemical composition are **complex media.** Such media are very useful, as a single complex medium may be sufficiently rich and complete to meet the nutritional requirements of many different microorganisms. In addition, complex media are often needed because the nutritional requirements of a particular microorganism are unknown, and thus a defined medium cannot be constructed. This is the situation with many fastidious bacteria, some of which may even require a medium containing blood or serum.

TABLE 5.4 Examples of Defined Media

BG-11 Medium for Cyanobacteria	Amount (g/l)
$NaNO_3$	1.5
$K_2HPO_4 \cdot 3H_2O$	0.04
$MgSO_4 \cdot 7H_2O$	0.075
$CaCl_2 \cdot 2H_2O$	0.036
Citric acid	0.006
Ferric ammonium citrate	0.006
EDTA (Na_2Mg salt)	0.001
Na_2CO_3	0.02
Trace metal solution[a]	1.0 ml/l
Final pH 7.4	
Medium for *Escherichia coli*	**Amount (g/l)**
Glucose	1.0
Na_2HPO_4	16.4
KH_2PO_4	1.5
$(NH_4)_2SO_4$	2.0
$MgSO_4 \cdot 7H_2O$	200.0 mg
$CaCl_2$	10.0 mg
$FeSO_4 \cdot 7H_2O$	0.5 mg
Final pH 6.8–7.0	

Sources: Data from Rippka, et al. *Journal of General Microbiology*, 111:1–61, 1979; and S. S. Cohen, and R. Arbogast, *Journal of Experimental Medicine*, 91:619, 1950.

[a]The trace metal solution contains H_3BO_3, $MnCl_2 \cdot 4H_2O$, $ZnSO_4 \cdot 7H_2O$, $Na_2Mo_4 \cdot 2H_2O$, $CuSO_4 \cdot 5H_2O$, and $Co(NO_3)_2 \cdot 6H_2O$.

TABLE 5.5 Some Common Complex Media

Nutrient Broth	Amount (g/l)
Peptone (gelatin hydrolysate)	5
Beef extract	3
Tryptic Soy Broth	
Tryptone (pancreatic digest of casein)	17
Peptone (soybean digest)	3
Glucose	2.5
Sodium chloride	5
Dipotassium phosphate	2.5
MacConkey Agar	**Amount (g/l)**
Pancreatic digest of gelatin	17.0
Pancreatic digest of casein	1.5
Peptic digest of animal tissue	1.5
Lactose	10.0
Bile salts	1.5
Sodium chloride	5.0
Neutral red	0.03
Crystal violet	0.001
Agar	13.5

Complex media contain undefined components like peptones, meat extract, and yeast extract. **Peptones** are protein hydrolysates prepared by partial proteolytic digestion of meat, casein, soya meal, gelatin, and other protein sources. They serve as sources of carbon, energy, and nitrogen. Beef extract and yeast extract are aqueous extracts of lean beef and brewer's yeast, respectively. Beef extract contains amino acids, peptides, nucleotides, organic acids, vitamins, and minerals. Yeast extract is an excellent source of B vitamins as well as nitrogen and carbon compounds. Three commonly used complex media are (1) nutrient broth, (2) tryptic soy broth, and (3) MacConkey agar (table 5.5).

If a solid medium is needed for surface cultivation of microorganisms, liquid media can be solidified with the addition of 1.5% agar (Box 5.1). **Agar** is a sulfated polymer composed mainly of D-galactose, 3,6-anhydro-L-galactose, and D-glucuronic acid. It usually is extracted from red algae (*see figure 27.8*). Agar is well suited as a solidifying agent because after it has been melted in boiling water, it can be cooled to about 40 to 42°C before hardening, and will not melt again until the temperature rises to about 80 to 90°C. Agar is also an excellent hardening agent because most microorganisms cannot degrade it.

Other solidifying agents are sometimes employed. For example, silica gel is used to grow autotrophic bacteria on solid media in the absence of organic substances and to determine carbon sources for heterotrophic bacteria by supplementing the medium with various organic compounds.

Types of Media

Media like tryptic soy broth and tryptic soy agar are called general purpose media because they support the growth of many microorganisms. Blood and other special nutrients may be added to general purpose media to encourage the growth of fastidious heterotrophs. These specially fortified media (e.g., blood agar) are called enriched media.

Selective media favor the growth of particular microorganisms. Bile salts or dyes like basic fuchsin and crystal violet favor the growth of gram-negative bacteria by inhibiting the growth of gram-positive bacteria without affecting gram-negative organisms. Endo agar, eosin methylene blue agar, and MacConkey agar (Table 5.5), three media widely used for the detection of *E. coli* and related bacteria in water supplies and elsewhere, contain dyes that suppress gram-positive bacterial growth. MacConkey agar also contains bile salts. Bacteria also may be selected by incubation with nutrients that they specifically can use. A medium containing only cellulose as a carbon and energy source is quite effective in the isolation of cellulose-digesting bacteria. The possibilities for selection are endless, and there are dozens of special selective media in use.

Differential media are media that distinguish between different groups of bacteria and even permit tentative identification of microorganisms based on their biological characteristics. Blood agar is both a differential medium and an enriched one. It distinguishes between hemolytic and nonhemolytic bacteria. Hemolytic bacteria (e.g., many streptococci and staphylococci isolated from throats) produce clear zones around their colonies because of red blood cell destruction. MacConkey agar is both differential and selective. Since it contains lactose and neutral red dye, lactose-fermenting colonies appear pink to red in color and are easily distinguished from colonies of nonfermenters.

1. Describe the following kinds of media and their uses: defined or synthetic media, complex media, general purpose media, enriched media, selective media, and differential media. Give an example of each kind.
2. What are peptones, yeast extract, beef extract, and agar? Why are they found in media?

Box 5.1

The Discovery of Agar as a Solidifying Agent and the Isolation of Pure Cultures

The earliest culture media were liquid, which made the isolation of bacteria to prepare pure cultures extremely difficult. In practice, a mixture of bacteria was diluted successively until only one organism, as an average, was present in a culture vessel. If everything went well, the individual bacterium thus isolated would reproduce to give a pure culture. This approach was tedious, gave variable results, and was plagued by contamination problems. Progress in isolating pathogenic bacteria understandably was slow.

The development of techniques for growing microorganisms on solid media and efficiently obtaining pure cultures was due to the efforts of the German bacteriologist Robert Koch and his associates. In 1881, Koch published an article describing the use of boiled potatoes, sliced with a flame-sterilized knife, in culturing bacteria. The surface of a sterile slice of potato was inoculated with bacteria from a needle tip, and then the bacteria were streaked out over the surface so that a few individual cells would be separated from the remainder. The slices were incubated beneath bell jars to prevent airborne contamination, and the isolated cells developed into pure colonies. Unfortunately, many bacteria would not grow well on potato slices.

At about the same time, Frederick Loeffler, an associate of Koch's, developed a meat extract peptone medium for cultivating pathogenic bacteria. Koch decided to try solidifying this medium. Koch was an amateur photographer—he was the first to take photomicrographs of bacteria—and was experienced in preparing his own photographic plates from silver salts and gelatin. Precisely the same approach was employed for preparing solid media. He spread a mixture of Loeffler's medium and gelatin over a glass plate, allowed it to harden, and inoculated the surface in the same way he had inoculated his sliced potatoes. The new solid medium worked well, but it could not be incubated at 37°C (the best temperature for most human bacterial pathogens) because the gelatin would melt. Furthermore, some bacteria digested the gelatin.

About a year later, in 1882, agar was first used as a solidifying agent. It had long been added to jelly to provide the appropriate solid consistency, and anyone who had made jelly or jam would be well acquainted with it. Not surprisingly, Fannie Eilshemius Hesse, the New Jersey-born wife of Walter Hesse, one of Koch's assistants, suggested its use when she heard of the difficulties with gelatin. Agar-solidified medium was an instant success. Five years later, Julius Richard Petri (another Koch assistant) invented the culture dish bearing his name, and the use of agar-coated glass plates ceased.

Isolation of Pure Cultures

In natural habitats, microorganisms usually grow in complex, mixed populations containing several species. This presents a problem for the microbiologist because a single type of microorganism cannot be studied adequately in a mixed culture. One needs a **pure culture,** a population of cells arising from a single cell, to characterize an individual species. Pure cultures are so important that the development of pure culture techniques by the German bacteriologist Robert Koch transformed microbiology. Within about 20 years, most pathogens responsible for the major human bacterial diseases had been isolated. There are several ways to prepare pure cultures; a few of the more common approaches are reviewed.

A brief survey of some major milestones in microbiology (chapter 1).

The Spread Plate and Streak Plate

If a mixture of cells is spread out on an agar surface so that every cell grows into a completely separate **colony,** a macroscopically visible growth or cluster of microorganisms on a solid medium, each colony represents a pure culture. The **spread plate** is an easy, direct way of achieving this result. A small volume of dilute microbial mixture containing around 100 to 200 cells or less is transferred to the center of an agar plate and spread evenly over the surface with a sterile bent-glass rod (figure 5.6). The dispersed cells develop into isolated colonies. Because the number of colonies should equal the number of viable organisms in the sample, spread plates can be used to count the microbial population.

Pure colonies also can be obtained from **streak plates.** The microbial mixture is transferred to the edge of an agar plate with an inoculating loop or swab and then streaked out over the surface in one of several patterns (figure 5.7). At some point in the process, single cells drop from the loop as it is rubbed along the agar surface and develop into separate colonies (figure 5.8). In both spread-plate and streak-plate techniques, successful isolation depends upon spatial separation of single cells.

The Pour Plate

Extensively used with bacteria and fungi, a **pour plate** also yields isolated colonies. The original sample is diluted several times to reduce the microbial population sufficiently to obtain separate colonies upon plating (figure 5.9). Then small volumes of several diluted samples are mixed with liquid agar that has been cooled to about 45°C, and the mixtures are poured immediately into sterile culture dishes. Most bacteria and fungi are not killed by a brief exposure to the warm agar.

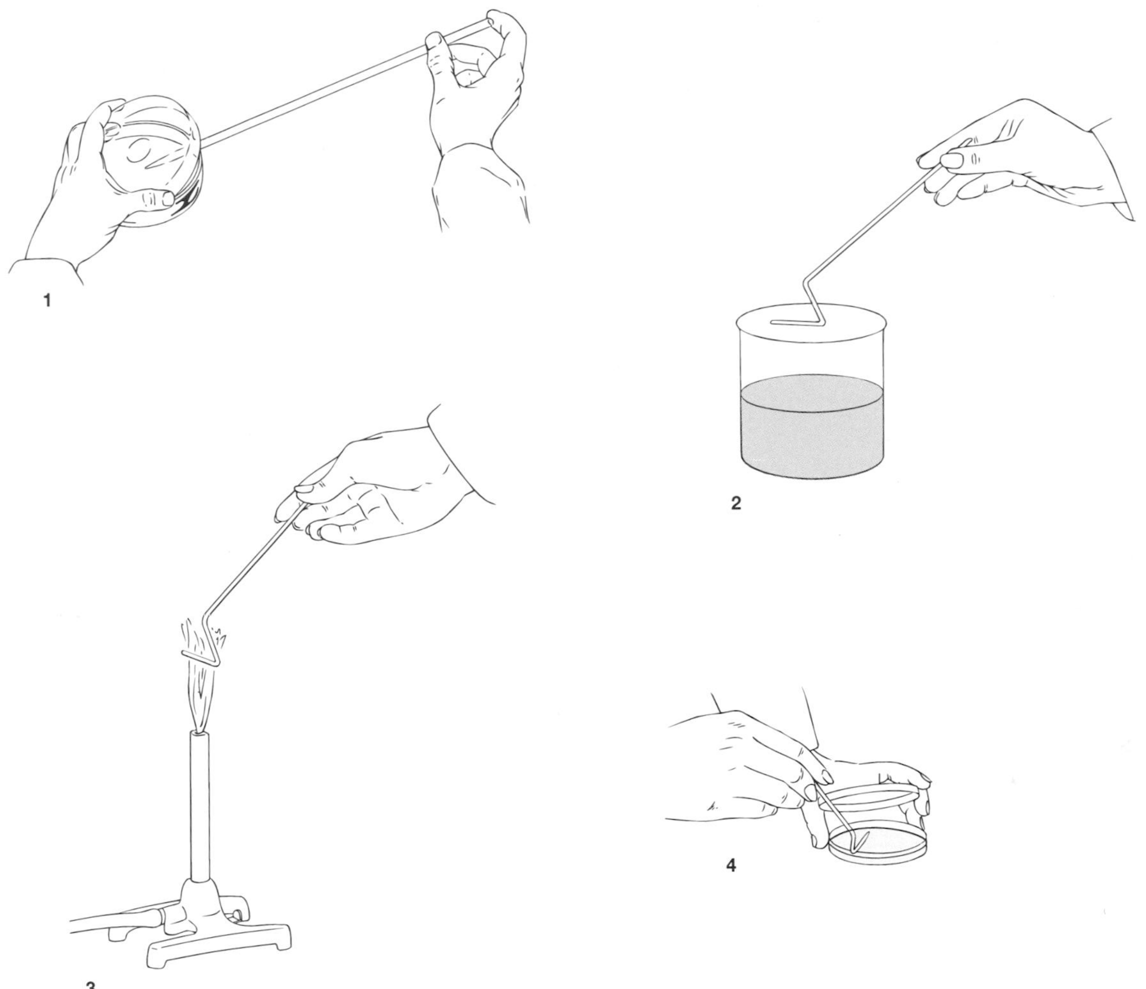

Figure 5.6 Spread-Plate Technique. The preparation of a spread plate. (*1*) Pipette a small sample onto the center of an agar medium plate. (*2*) Dip a glass spreader into a beaker of ethanol. (*3*) Briefly flame the ethanol-soaked spreader, and allow it to cool. (*4*) Spread the sample evenly over the agar surface with the sterilized spreader. Incubate.

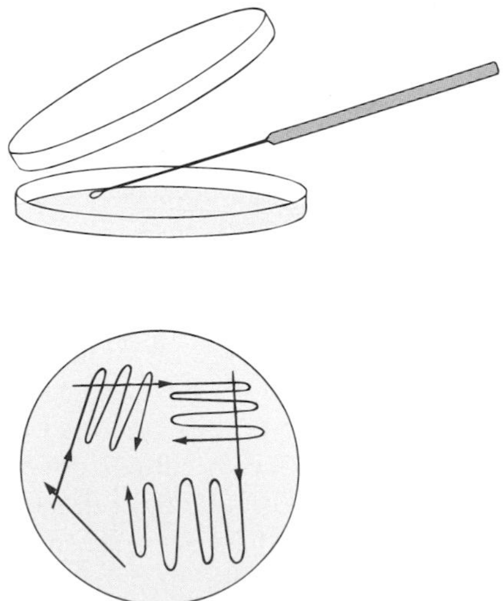

Figure 5.7 Streak-Plate Technique. Preparation of streak plates. The upper illustration shows a petri dish of agar being streaked with an inoculating loop. A commonly used streaking pattern is pictured at the bottom.

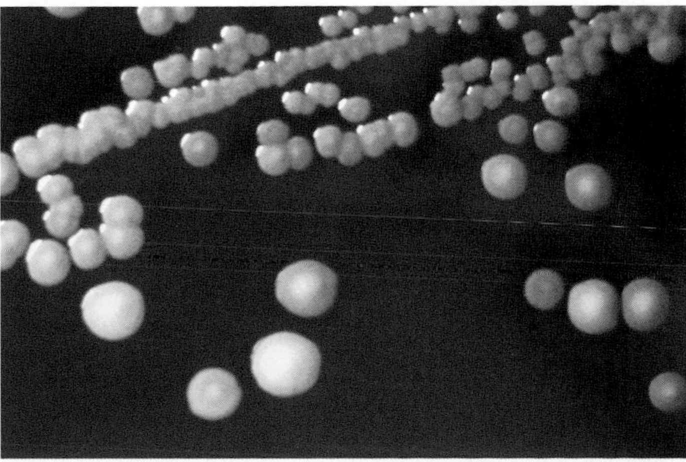

Figure 5.8 Bacterial Colonies on Agar. Colonies growing on a streak plate. A blood-agar plate has been inoculated with *Staphylococcus aureus*. After incubation, large, golden colonies have formed on the agar.

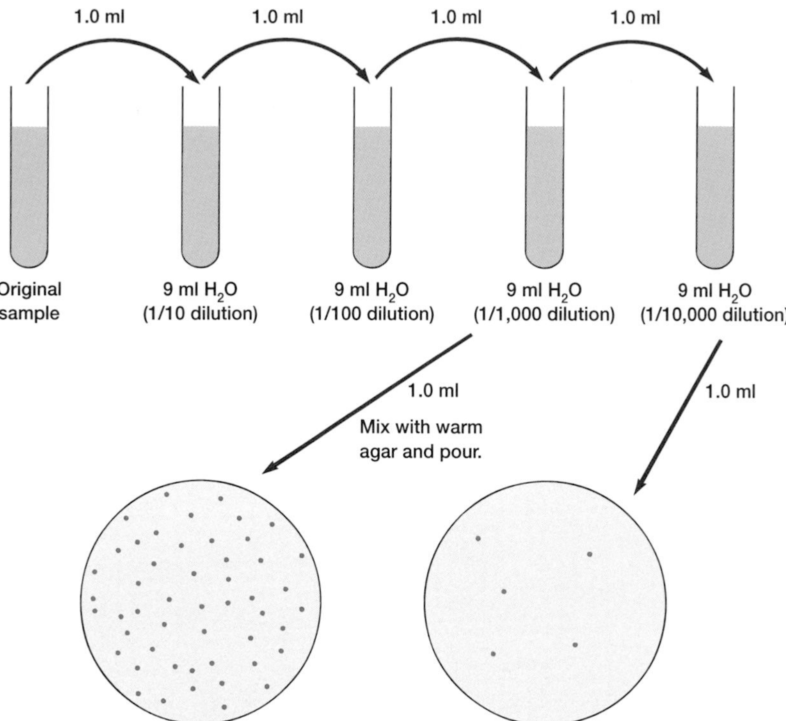

Figure 5.9 **The Pour-Plate Technique.** The original sample is diluted several times to thin out the population sufficiently. The most diluted samples are then mixed with warm agar and poured into petri dishes. Isolated cells grow into colonies and can be used to establish pure cultures. The surface colonies are circular; subsurface colonies would be lenticular or lens shaped.

After the agar has hardened, each cell is fixed in place and forms an individual colony. The total number of colonies equals the number of viable microorganisms in the diluted sample. Colonies growing on the surface also can be used to inoculate fresh medium and prepare pure cultures.

The preceding techniques require the use of special culture dishes named **petri dishes** or plates after their inventor Julius Richard Petri, a member of Robert Koch's laboratory; Petri developed these dishes around 1887. They consist of two round halves, the top half overlapping the bottom (figure 5.7). They are very easy to use, may be stacked on each other to save space, and are one of the most common items in microbiology laboratories.

The plating methods just described are even more effective in producing pure cultures when used with selective or differential media. A good example is the isolation of bacteria that degrade the herbicide 2,4-dichlorophenoxyacetic acid (2,4-D). Bacteria able to metabolize 2,4-D can be obtained with a liquid medium containing 2,4-D as its sole carbon source and the required nitrogen, phosphorus, sulfur, and mineral components. When this medium is inoculated with soil, only bacteria able to use 2,4-D will grow. After incubation, a sample of the original culture is transferred to a fresh flask of selective medium for further enrichment of 2,4-D metabolizing bacteria. A mixed population of 2,4-D degrading bacteria will arise after several such transfers. Pure cultures can be obtained by plating this mixture on agar containing 2,4-D as the sole carbon source. Only bacteria able to grow on 2,4-D form visible colonies and can be subcultured. This same general approach is used to isolate and purify a variety of bacteria by selecting for specific physiological characteristics.

Colony Morphology and Growth

Colony development on agar surfaces aids the microbiologist in identifying bacteria because individual species often form colonies of characteristic size and appearance (figure 5.10). When a mixed population has been plated properly, it sometimes is possible to identify the desired colony based on its overall appearance and use it to obtain a pure culture. The structure of bacterial colonies also has been examined with the scanning electron microscope. The microscopic structure of colonies is often as variable as their visible appearance (figure 5.11).

In nature, bacteria and many other microorganisms often grow as colonies on surfaces. Therefore, an understanding of colony growth is essential to microbial ecologists, and the growth of colonies on agar has been frequently studied. Generally, the most rapid cell growth occurs at the colony edge. Growth is much slower in the center, and cell autolysis takes place in the older central portions of some colonies (figure 5.11*b*). These differences in growth appear due to gradients of oxygen, nutrients, and toxic products within the colony. At the colony edge, oxygen and nutrients are plentiful. The colony center, of course, is much thicker than the edge. Consequently, oxygen and nutrients do not diffuse readily into the center, toxic

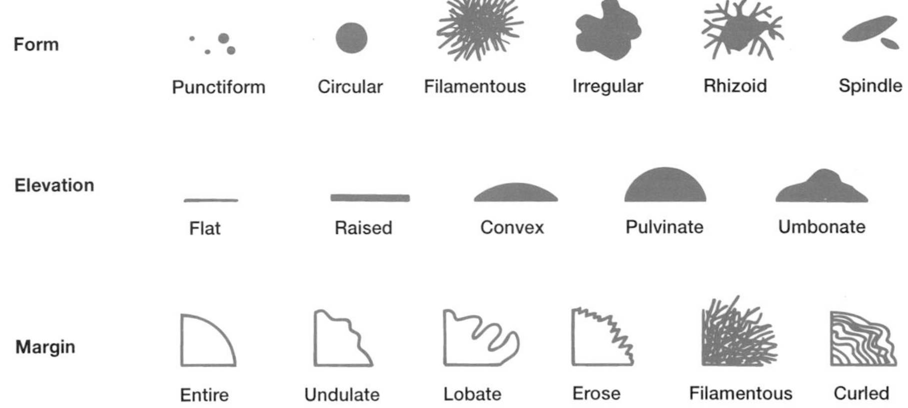

Figure 5.10 **Bacterial Colony Morphology.** Variations in bacterial colony morphology seen with the naked eye. The general form of the colony and the shape of the edge or margin can be determined by looking down at the top of the colony. The nature of colony elevation is apparent when viewed from the side as the plate is held at eye level.

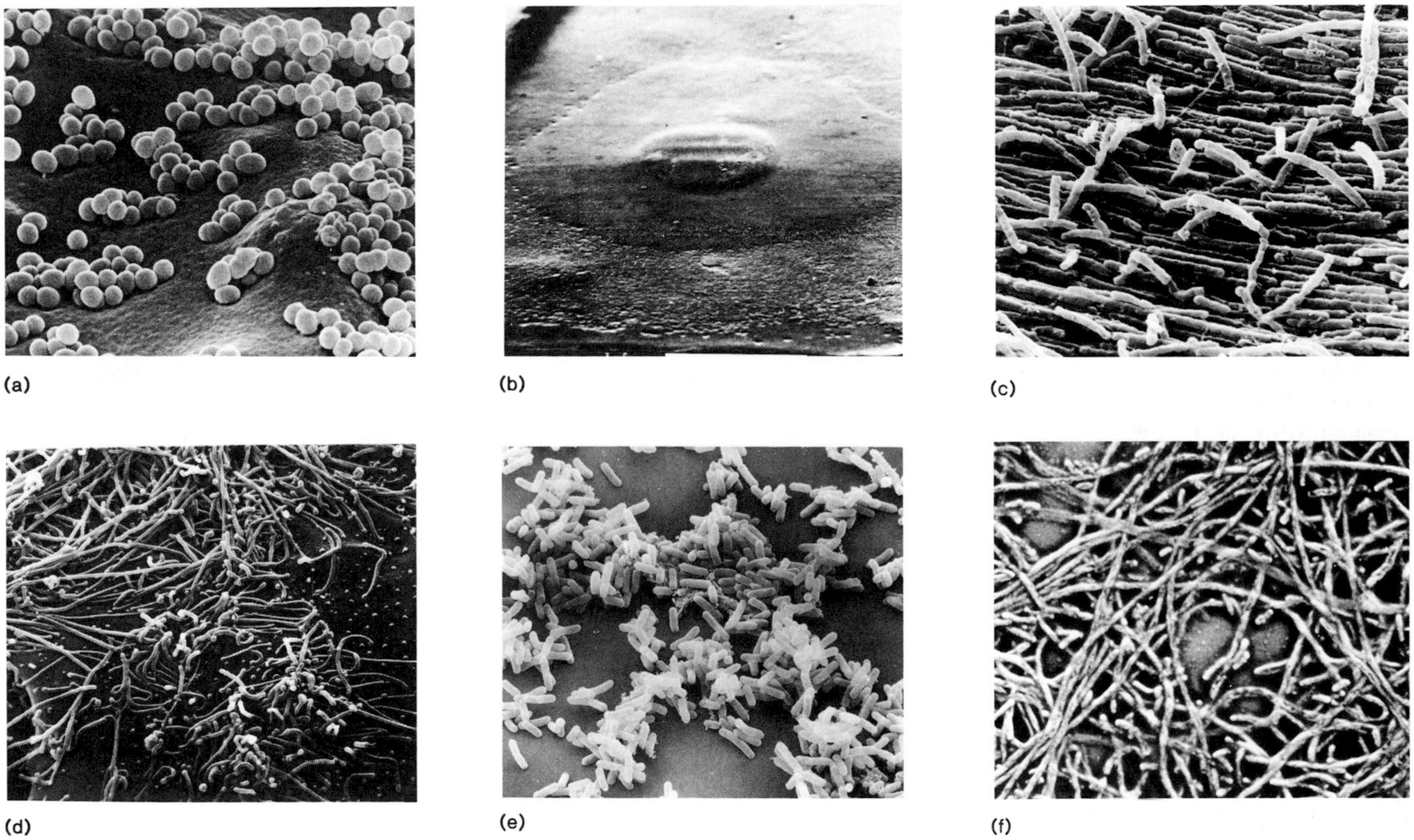

(a) (b) (c)

(d) (e) (f)

Figure 5.11 **Scanning Electron Micrographs of Bacterial Colonies.** (*a*) *Micrococcus* on agar (×31,000). (*b*) *Streptococcus pneumoniae* (×100). (*c*) *Clostridium* (×12,000). (*d*) *Mycoplasma pneumoniae* (×26,000). (*e*) *Escherichia coli* (×14,000). (*f*) *Legionella pneumophila* (×3,600).

metabolic products cannot be quickly eliminated, and growth in the colony center is slowed or stopped. Because of these environmental variations within a colony, cells on the periphery can be growing at maximum rates while cells in the center are dying.

1. What are pure cultures, and why are they important? How are spread plates, streak plates, and pour plates prepared?
2. In what way does microbial growth vary within a colony? What factors might cause these variations in growth?

Summary

1. Microorganisms require nutrients, materials that are used in biosynthesis and energy production.
2. Macronutrients or macroelements (C, O, H, N, S, P, K, Ca, Mg, and Fe) are needed in large quantities; trace elements (e.g., Mn, Zn, Co, Mo, Ni, and Cu) are used in very small amounts.
3. Autotrophs use CO_2 as their primary or sole carbon source; heterotrophs employ organic molecules.
4. Microorganisms can be classified based on their energy and electron sources. Phototrophs use light energy, and chemotrophs obtain energy from the oxidation of chemical compounds. Electrons are extracted from reduced inorganic substances by lithotrophs, and from organic compounds by organotrophs.
5. Nitrogen, phosphorus, and sulfur may be obtained from the same organic molecules that supply carbon, from the direct incorporation of ammonia and phosphate, and by the reduction and assimilation of oxidized inorganic molecules.
6. Probably most microorganisms need growth factors. Growth factor requirements make microbiological assays possible.
7. Although some nutrients can enter cells through passive diffusion, a membrane carrier protein is usually required.
8. In facilitated diffusion, the transport protein simply carries a molecule across the membrane in the direction of decreasing concentration, and no energy source is required.
9. Active transport systems use metabolic energy and membrane carrier proteins to concentrate substances actively by transporting them against a gradient. Either ATP or protonmotive force may be used as an energy source, and sodium gradients also can drive solute uptake.
10. Bacteria also transport molecules while modifying them, a process known as group translocation. For example, many sugars are transported and phosphorylated.
11. Iron is accumulated by the secretion of siderophores, small molecules able to complex with ferric iron. When the iron-siderophore complex reaches the cell surface, it is taken inside and the iron is reduced to the ferrous form.
12. Culture media can be constructed completely from chemically defined components (defined media or synthetic media) or may contain constituents like peptones and yeast extract whose precise composition is unknown (complex media).
13. Culture medium can be solidified with the addition of agar, a complex polysaccharide from red algae.
14. Culture media are classified based on function and composition as general purpose media, enriched media, selective media, and differential media.
15. The pure cultures are obtained by isolating individual cells with any of three plating techniques: the spread-plate, streak-plate, and pour-plate methods.
16. Microorganisms growing on solid surfaces tend to form colonies with distinctive morphology. Colonies usually grow most rapidly at the edge where there is an ample supply of oxygen and nutrients.

Key Terms

active transport *102*
agar *105*
antiport *102*
autotrophs *97*
auxotrophs *98*
chemoheterotrophs *98*
chemolithotrophic autotrophs *98*
chemoorganotrophic heterotrophs *98*
chemotrophs *98*
colony *106*
complex medium *104*
defined medium *104*
differential media *105*
facilitated diffusion *101*
group translocation *103*
growth factors *99*
heterotrophs *97*
lithotrophs *98*
macroelements *97*
mixotrophic *98*
nutrient *97*
organotrophs *98*
passive diffusion *101*
peptones *105*
permease *101*
petri dish *108*
phosphoenolpyruvate: sugar phosphotransferase system (PTS) *103*
photoautotrophs *98*
photolithotrophic autotrophs *98*
photoorganotrophic heterotrophs *98*
phototrophs *98*
pour plate *106*
prototrophs *98*
pure culture *106*
selective media *105*
siderophores *103*
spread plate *106*
streak plate *106*
symport *102*
synthetic medium *104*
trace elements *97*
vitamins *99*

Questions for Thought and Review

1. Why is it so difficult to demonstrate the micronutrient requirements of microorganisms?
2. Where might one expect to find microorganisms that are unable to catabolize a variety of nutrients?
3. List some of the most important uses of the nitrogen, phosphorus, and sulfur that microorganisms obtain from their surroundings.
4. Why are amino acids, purines, and pyrimidines often growth factors whereas glucose is usually not?
5. Why do microorganisms normally take up nutrients using transport proteins or permeases? What advantage does a microorganism gain by employing active transport rather than facilitated diffusion?
6. If you wished to obtain a pure culture of bacteria that could degrade benzene and use it as a carbon and energy source, how would you proceed?
7. Describe the nutritional requirements of a chemolithotrophic heterotroph. Where might you search for such a bacterium?

Additional Reading

Braun, V. 1985. The unusual features of the iron transport systems of *Escherichia coli. Trends Biochem. Sci.* 10(2):75–78.

Bridson, E. Y. 1990. Media in microbiology. *Rev. Med. Microbiol.* 1:1–9.

Conn, H. J., ed. 1957. *Manual of microbiological methods.* New York: McGraw-Hill.

Difco Laboratories. 1984. *Difco manual of dehydrated culture media and reagents for microbiology,* 10th ed. Detroit: Difco.

Dills, S. S.; Apperson, A.; Schmidt, M. R.; and Saier, M. H., Jr. 1980. Carbohydrate transport in bacteria. *Microbiol. Rev.* 44(3):385–418.

Epstein, W., and Laimins, L. 1980. Potassium transport in *Escherichia coli:* Diverse systems with common control by osmotic forces. *Trends Biochem. Sci.* 5(1):21–23.

Guirard, B. M., and Snell, E. E. 1981. Biochemical factors in growth. In *Manual of methods for general bacteriology,* ed. P. Gerhardt, 79–111. Washington, D.C.: American Society for Microbiology.

Harder, W., and Dijkhuizen, L. 1983. Physiological responses to nutrient limitation. *Ann. Rev. Microbiol.* 37:1–23.

Kaback, H. R. 1974. Transport studies in bacterial membrane vesicles. *Science* 186:882–92.

Kelly, D. P. 1992. The chemolithotrophic prokaryotes. In *The procaryotes,* 2d ed., A. Balows et al., 331–43. New York: Springer-Verlag.

Krieg, N. R. 1981. Enrichment and isolation. In *Manual of methods for general bacteriology,* ed. P. Gerhardt, 112–42. Washington, D.C.: American Society for Microbiology.

Maloney, P. C.; Ambudkar, S. V.; Anantharam, V.; Sonna, L. A.; and Varadhachary, A. 1990. Anion-exchange mechanisms in bacteria. *Microbiol. Rev.* 54(1):1–17.

Meadow, N. D.; Fox, D. K.; and Roseman, S. 1990. The bacterial phosphoenolpyruvate: glycose phosphotransferase system. *Ann. Rev. Biochem.* 59:497–542.

Neidhardt, F. C.; Ingraham, J. L.; and Schaechter, M. 1990. *Physiology of the bacterial cell: A molecular approach.* Sunderland, Mass.: Sinauer.

Neilands, J. B. 1991. Microbial iron compounds. *Ann. Rev. Biochem.* 50:715–31.

Power, D. A., ed. 1988. *Manual of BBL products and laboratory procedures,* 6th ed. Cockeysville, Md.: Becton, Dickinson and Company

Saier, M. H., Jr. 1977. Bacterial phosphoenolpyruvate: Sugar phosphotransferase systems: Structural, functional, and evolutionary interrelationships. *Bacteriol. Rev.* 41(4):856–71.

Shapiro, J. A. 1988. Bacteria in multicellular organisms. *Sci. Am.* 258(6):82–89.

Stolp, H., and Starr, M. P. 1981. Principles of isolation, cultivation, and conservation of bacteria. In *The prokaryotes,* ed. M. P. Starr et al., 135–75. New York: Springer-Verlag.

Whittenbury, R. 1978. Bacterial nutrition. In *Essays in microbiology,* ed. J. R. Norris and M. H. Richmond, 16/1–16/32. New York: John Wiley and Sons.

Whittenbury, R., and Kelly, D. P. 1977. Autotrophy: A conceptual phoenix. In *Microbial energetics,* ed. B. A. Haddock and W. A. Hamilton, 121–49. New York: Cambridge University Press.

Wilson, D. B. 1978. Cellular transport mechanisms. *Ann. Rev. Biochem.* 47:933–65.

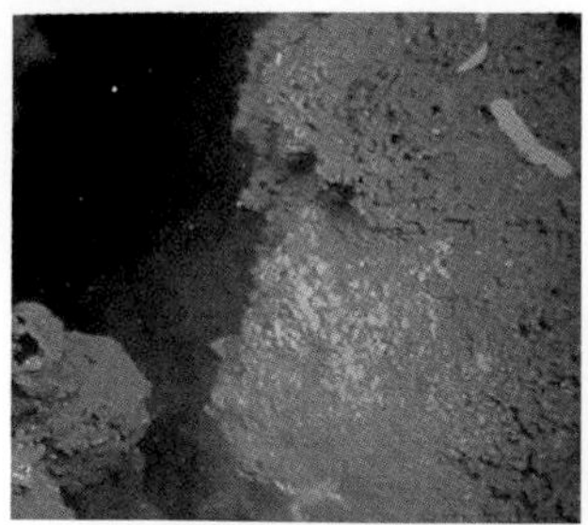

CHAPTER 6
Microbial Growth

The paramount evolutionary accomplishment of bacteria as a group is rapid, efficient cell growth in many environments.

—J. L. Ingraham, O. Maalóe, and F. C. Neidhardt

Outline

Concepts

1. Growth is defined as an increase in cellular constituents and may result in an increase in a microorganism's size, population number, or both.
2. When microorganisms are grown in a closed system, population growth remains exponential for only a few generations and then enters a stationary phase due to factors such as nutrient limitation and waste accumulation. In an open system with continual nutrient addition and waste removal, the exponential phase can be maintained for long periods.
3. A wide variety of techniques can be used to study microbial growth by following changes in the total cell number, the population of viable microorganisms, or the cell mass.
4. Water availability, pH, temperature, oxygen concentration, pressure, radiation, and a number of other environmental factors influence microbial growth. Yet many microorganisms, and particularly bacteria, have managed to adapt and flourish under environmental extremes that would destroy most organisms.

Chapter 5 emphasizes that microorganisms need access to a source of energy and the raw materials essential for the construction of cellular components. All organisms must have carbon, hydrogen, oxygen, nitrogen, sulfur, phosphorus, and a variety of minerals; many also require one or more special growth factors. The cell takes up these substances by membrane transport processes, the most important of which are facilitated diffusion, active transport, and group translocation. Eucaryotic cells also employ endocytosis.

Chapter 6 concentrates more directly on the growth that takes place in the presence of an adequate nutrient supply. The nature of growth and the ways in which it can be measured are described first, followed by consideration of some specific aspects of culture growth. An account of the influence of environmental factors on microbial growth completes the chapter.

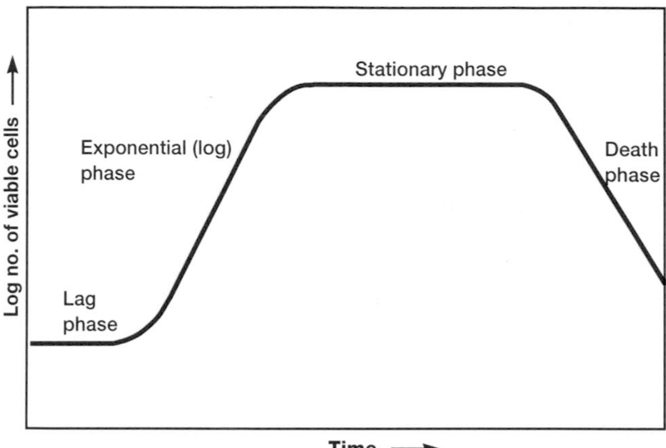

Figure 6.1 **Microbial Growth Curve in a Closed System.** The four phases of the growth curve are identified on the curve and discussed in the text.

Growth may be defined as an increase in cellular constituents. If the microorganism is **coenocytic,** that is, a multinucleate organism in which nuclear divisions are not accompanied by cell divisions, growth results in an increase in cell size but not cell number. Growth leads to a rise in cell number when microorganisms reproduce by processes like budding or binary fission. In the latter, individual cells enlarge and divide to yield two progeny of approximately equal size. It is usually not convenient to investigate the growth and reproduction of individual microorganisms because of their small size. Therefore, microbiologists normally follow changes in the total population number when studying growth.

The cell cycle (pp. 83–85; 230–31).

The Growth Curve

Population growth is studied by analyzing the growth curve of a microbial culture. When microorganisms are cultivated in liquid medium, they are usually grown in a **batch culture** or closed system; that is, they are incubated in a closed culture vessel with a single batch of medium. Because no fresh medium is provided during incubation, nutrient concentrations decline and concentrations of wastes increase. The growth of microorganisms reproducing by binary fission can be plotted as the logarithm of cell number versus the incubation time. The resulting curve has four distinct phases (figure 6.1).

Lag Phase

When microorganisms are introduced into fresh culture medium, usually no immediate increase in cell numbers or mass occurs, and therefore this period is called the **lag phase.** Although cell division does not take place right away and there is no net increase in mass, the cell is synthesizing new components. A lag phase prior to the start of cell division can be necessary for a variety of reasons. The cells may be old and depleted of ATP, essential cofactors, and ribosomes; these must be synthesized before growth can begin. The medium may be different from the one the microorganism was growing in previously. Here, new enzymes would be needed to use different nutrients. Possibly the microorganisms have been injured and require time to recover. Whatever the causes, eventually the cells retool, replicate their DNA, begin to increase in mass, and finally divide.

The lag phase varies considerably in length with the condition of the microorganisms and the nature of the medium. This phase may be quite long if the inoculum is from an old culture or one that has been refrigerated. Inoculation of a culture into a chemically different medium also results in a long lag phase. On the other hand, when a young, vigorously growing exponential phase culture is transferred to fresh medium of the same composition, the lag phase will be short or absent.

Exponential Phase

During the **exponential** or **log phase,** microorganisms are growing and dividing at the maximal rate possible given their genetic potential, the nature of the medium, and the conditions under which they are growing. Their rate of growth is constant during the exponential phase, the microorganisms dividing and doubling in number at regular intervals. Because each individual divides at a slightly different moment, the growth curve rises smoothly rather than in discrete jumps (figure 6.1). The population is most uniform in terms of chemical and physiological properties during this phase; therefore exponential phase cultures are usually used in biochemical and physiological studies.

Stationary Phase

Eventually population growth ceases and the growth curve becomes horizontal (figure 6.1). This **stationary phase** is usually attained by bacteria at a population level of around 10^9 cells per ml. Other microorganisms normally do not reach such high population densities, protozoan and algal cultures often having maximum concentrations of about 10^6 cells per ml. Of course, final population size depends on nutrient availability and other factors, as well as the type of microorganism being cultured. In the stationary phase, the total number of viable microorganisms remains constant. This may result from a balance between cell division and cell death, or the population may simply cease to divide though remaining metabolically active.

Microbial populations enter the stationary phase for several reasons. One obvious factor is nutrient limitation; if an essential nutrient is severely depleted, population growth will slow. Aerobic organisms are often limited by O_2 availability. Oxygen is not very soluble and may be depleted so quickly that only the surface of a culture will have an O_2 concentration adequate for growth. The cells beneath the surface will not be able to grow unless the culture is shaken or aerated in another way. Population growth also may cease due to the accumulation of toxic waste products. This factor seems to limit the growth of many anaerobic cultures (cultures growing in the absence of O_2). For example, streptococci can produce so much lactic acid and other organic acids from sugar fermentation that their medium becomes acidic and growth is inhibited. Streptococcal cultures also can enter the stationary phase due to depletion of their sugar supply. Thus, entrance into the stationary phase may result from several factors operating in concert.

Death Phase

Detrimental environmental changes like nutrient deprivation and the buildup of toxic wastes lead to the decline in the number of viable cells characteristic of the **death phase.** The death of a microbial population, like its growth during the exponential phase, is usually logarithmic (that is, a constant proportion of cells dies every hour). This pattern holds even when the total cell number remains constant because the cells simply fail to lyse after dying. Often, the only way of deciding whether a bacterial cell is dead is by incubating it in fresh medium; if it does not grow and reproduce, it is assumed to be dead.

Although most of a microbial population usually dies in a logarithmic fashion, the death rate may decrease after the population has been drastically reduced. This is due to the extended survival of particularly resistant cells.

The Mathematics of Growth

Knowledge of microbial growth rates during the exponential phase is indispensable to microbiologists. Growth rate studies contribute to basic physiological and ecological research and the solution of applied problems in industry. Therefore, the quantitative aspects of exponential phase growth will be discussed.

TABLE 6.1 An Example of Exponential Growth

Time[a]	Division Number	2^n	Population ($N_o \times 2^n$)	$\log_{10} N_t$
0	0	$2^0 = 1$	1	0.000
20	1	$2^1 = 2$	2	0.301
40	2	$2^2 = 4$	4	0.602
60	3	$2^3 = 8$	8	0.903
80	4	$2^4 = 16$	16	1.204
100	5	$2^5 = 32$	32	1.505
120	6	$2^6 = 64$	64	1.806

[a]The hypothetical culture begins with one cell having a 20-minute generation time.

During the exponential phase, each microorganism is dividing at constant intervals. Thus, the population will double in number during a specific length of time called the **generation time** or **doubling time.** This situation can be illustrated with a simple example. Suppose that a culture tube is inoculated with one cell that divides every 20 minutes (table 6.1). The population will be 2 cells after 20 minutes, 4 cells after 40 minutes, and so forth. Because the population is doubling every generation, the increase in population is always 2^n where n is the number of generations. The resulting population increase is exponential or logarithmic (figure 6.2).

These observations can be expressed as equations for the generation time.

Let N_o = the initial population number

N_t = the population at time t

n = the number of generations in time t

Then, inspection of the results in table 6.1 will show that

$$N_t = N_o \times 2^n.$$

Solving for n, the number of generations, where all logarithms are to the base 10,

$$\log N_t = \log N_o + n \cdot \log 2, \text{ and}$$

$$n = \frac{\log N_t - \log N_o}{\log 2} = \frac{\log N_t - \log N_o}{0.301}.$$

The rate of growth in a batch culture can be expressed in terms of the **mean growth rate constant (*k*).** This is the number of generations per unit time, often expressed as the generations per hour.

$$k = \frac{n}{t} = \frac{\log N_t - \log N_o}{0.301t}$$

The time it takes a population to double in size, that is, the **mean generation time** or mean doubling time (g), can now be calculated. If the population doubles ($t = g$), then

$$N_t = 2\,N_o.$$

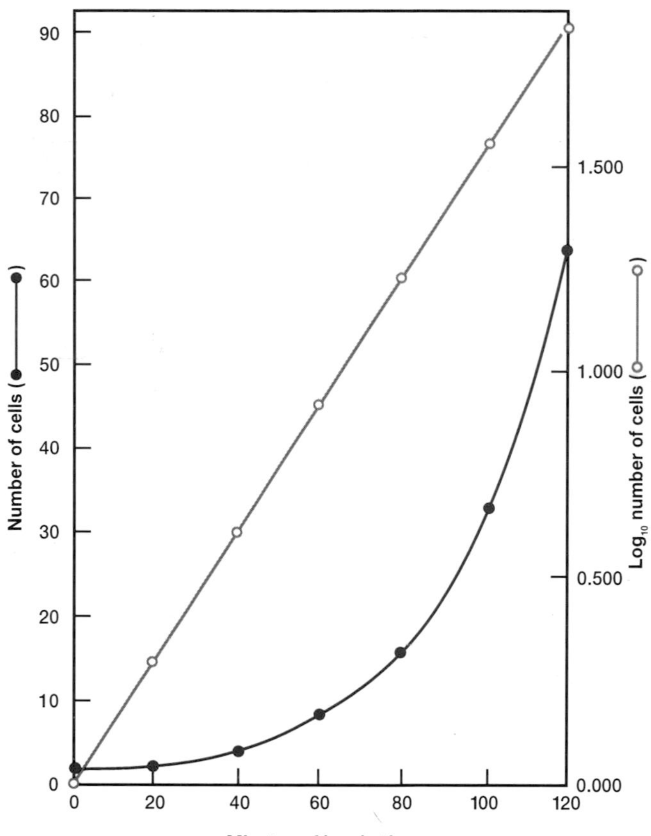

Figure 6.2 Exponential Microbial Growth. The data from table 6.1 for six generations of growth is plotted directly (●—●) and in the logarithmic form (○—○). The growth curve is exponential as shown by the linearity of the log plot.

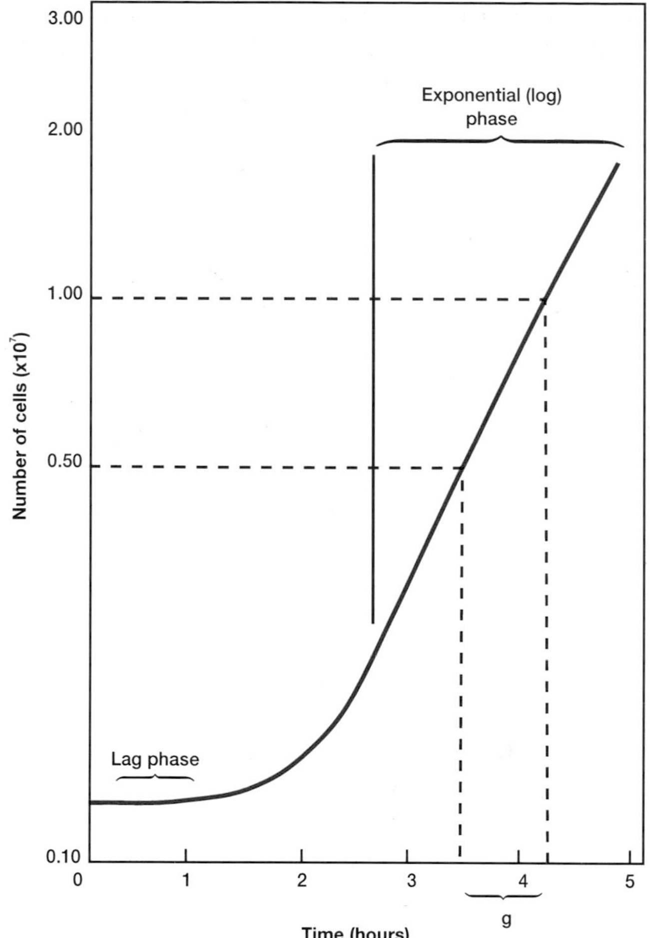

Figure 6.3 Generation Time Determination. The generation time can be determined from a microbial growth curve. The population data is plotted on semilogarithmic graph paper with the logarithmic axis used for the number of cells. The time to double the population number is then read directly from the plot. The log of the population number can also be plotted against time on regular graph paper.

Substitute $2N_0$ into the mean growth rate equation and solve for k.

$$k = \frac{\log (2N_0) - \log N_0}{0.301g} = \frac{\log 2 + \log N_0 - \log N_0}{0.301g}$$

$$k = \frac{1}{g}$$

$$g = \frac{1}{k}$$

The mean generation time can be determined directly from a semilogarithmic plot of the growth data (figure 6.3) and the growth rate constant calculated from the g value. The generation time also may be calculated directly from the previous equations. For example, suppose that a bacterial population increases from 10^3 cells to 10^9 cells in 10 hours.

$$k = \frac{\log 10^9 - \log 10^3}{(0.301)(10 \text{ hr})} = \frac{9 - 3}{3.01 \text{ hr}} = 2.0 \text{ generations/hr}$$

$$g = \frac{1}{2.0 \text{ gen./hr}} = 0.5 \text{ hr/gen. or } 30 \text{ min/gen.}$$

Generation times vary markedly with the species of microorganism and environmental conditions. They range from less than 10 minutes (0.17 hours) for a few bacteria to several days with some eucaryotic microorganisms (table 6.2). Generation times in nature are usually much longer than in culture.

1. Define growth. Describe the four phases of the growth curve in a closed system and discuss the causes of each.
2. What are the generation or doubling time and the mean growth rate constant? How can they be determined from growth data?

Table 6.2 Generation Times for Selected Microorganisms

Microorganism	Temperature (°C)	Generation Time (Hours)
Bacteria		
Beneckea natriegens	37	0.16
Escherichia coli	40	0.35
Bacillus subtilis	40	0.43
Clostridium botulinum	37	0.58
Mycobacterium tuberculosis	37	≈12
Anacystis nidulans	41	2.0
Anabaena cylindrica	25	10.6
Rhodospirillum rubrum	25	4.6–5.3
Algae		
Chlorella pyrenoidosa	25	7.75
Scenedesmus quadricauda	25	5.9
Asterionella formosa	20	9.6
Skeletonema costatum	18	13.1
Ceratium tripos	20	82.8
Euglena gracilis	25	10.9
Protozoa		
Acanthamoeba castellanii	30	11–12
Paramecium caudatum	26	10.4
Tetrahymena geleii	24	2.2–4.2
Leishmania donovani	26	10–12
Giardia lamblia	37	18
Fungi		
Saccharomyces cerevisiae	30	2
Monilinia fructicola	25	30

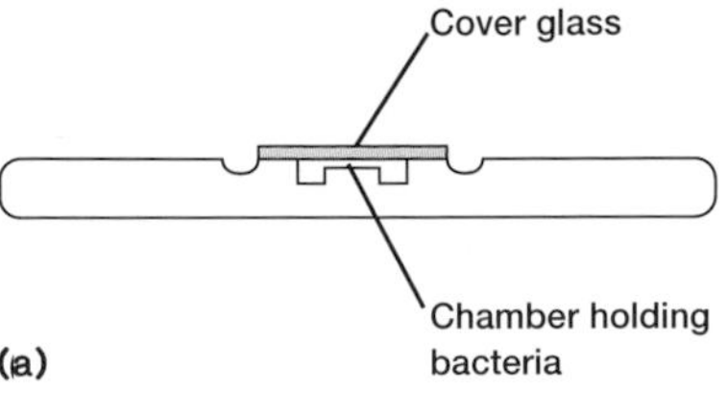

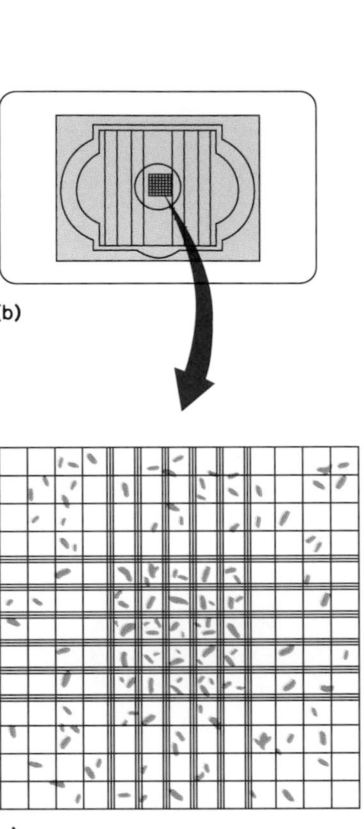

Figure 6.4 The Petroff-Hausser Counting Chamber. (*a*) Side view of the chamber showing the cover glass and the space beneath it that holds a bacterial suspension. (*b*) A top view of the chamber. The grid is located in the center of the slide. (*c*) An enlarged view of the grid. The bacteria in several of the central squares are counted, usually at 400× to 500× magnification. The average number of bacteria in these squares is used to calculate the concentration of cells in the original sample. Since there are 25 squares covering an area of 1 mm^2, the total number of bacteria in 1 mm^2 of the chamber is (number/square)(25 squares). The chamber is 0.02 mm deep and therefore,

$$\text{bacteria/mm}^3 = (\text{bacteria/square})\,(25\ \text{squares})(50)$$

The number of bacteria per cm^3 is 10^3 times this value. For example, suppose the average count per square is 28 bacteria.

$$\text{bacteria/cm}^3 = (28\ \text{bacteria})\,(25\ \text{squares})(50)(10^3) = 3.5 \times 10^7$$

Measurement of Microbial Growth

There are many ways to measure microbial growth to determine growth rates and generation times. Either population mass or number may be followed because growth leads to increases in both. The most commonly employed techniques for growth measurement are briefly examined and the advantages and disadvantages of each noted.

Measurement of Cell Numbers

The most obvious way to determine microbial numbers is through direct counting. The use of a counting chamber is easy, inexpensive, and relatively quick; it also gives information about the size and morphology of microorganisms. Petroff-Hausser counting chambers can be used for counting bacteria; hemocytometers can be used for the larger eucaryotic microorganisms. These specially designed slides have chambers of known depth with an etched grid on the chamber bottom (figure 6.4). The number of microorganisms in a sample can be calculated by taking into account the chamber's volume and any sample dilutions required. There are some disadvantages to the technique. The microbial population must be fairly large for accuracy because such a small volume is sampled. It is also difficult to distinguish between living and dead cells.

Larger microorganisms such as protozoa, algae, and nonfilamentous yeasts can be directly counted with electronic counters such as the Coulter Counter. The microbial suspension is forced through a small hole or orifice. An electrical current flows through the hole, and electrodes placed on both sides of the orifice measure its electrical resistance. Every time a microbial cell passes through the orifice, electrical resistance increases (or the conductivity drops) and the cell is counted. The Coulter Counter gives accurate results with larger cells and is extensively used in hospital laboratories to count red

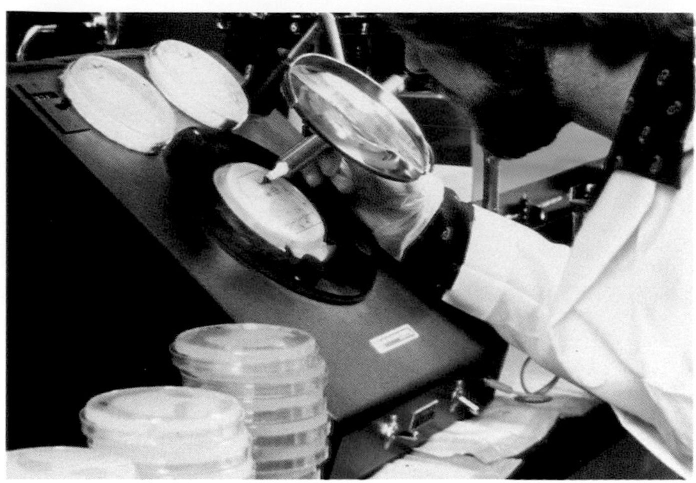

(a)

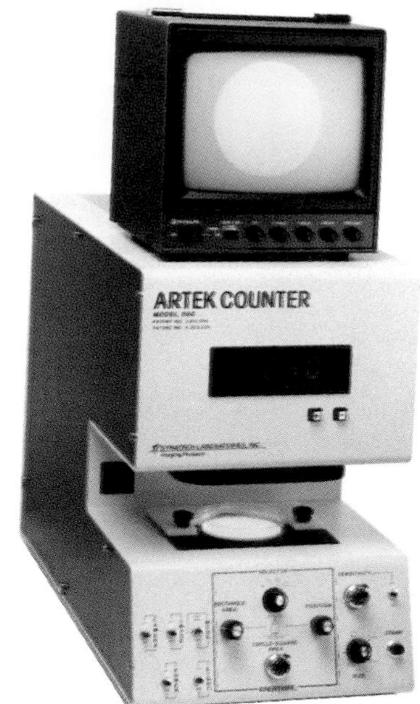

(b)

Figure 6.5 **Equipment for Counting Bacterial Colonies.** (*a*) The Quebec colony counter. The counter illuminates the petri plate uniformly from the side, and the plate is magnified for easier counting of small colonies. In this sophisticated model, an electric probe is touched to each colony to record the count. (*b*) An automated counter. The camera forms an enlarged image of the plate; all objects of the desired size range in any part of the image are counted by the instrument. The counter may be connected to a separate computer.

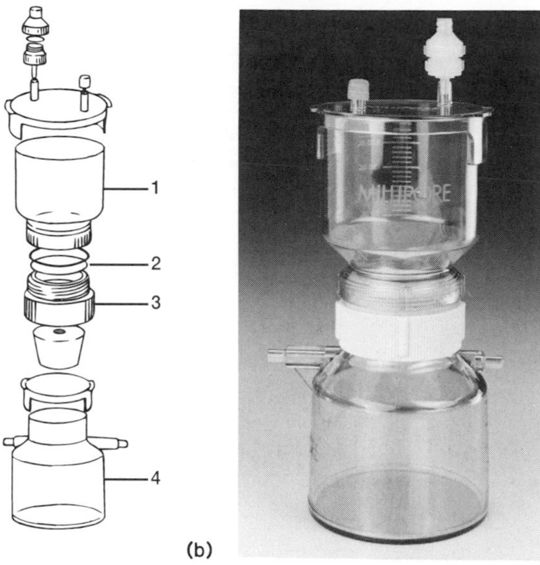

Figure 6.6 **A Millipore Membrane Filter System.** (*a*) Parts of the filter assembly. Some of the more important components are the following: (*1*) funnel, (*2*) membrane filter, (*3*) filter holder base and support, and (*4*) receiver flask. (*b*) A complete filtering system. The sample is poured into the funnel and sucked through a membrane filter with the use of vacuum.

and white blood cells. It is not as useful in counting bacteria because of interference by small debris particles, the formation of filaments, and other problems.

Counting chambers and electronic counters yield counts of all cells, whether alive or dead. There are also several viable counting techniques, procedures specific for cells able to grow and reproduce. In most viable counting procedures, a diluted sample of bacteria or other microorganisms is dispersed over a solid surface. Each microorganism or group of microorganisms develops into a distinct colony. The original number of viable microorganisms in the sample can be calculated from the number of colonies formed and the sample dilution. For example, if 1.0 ml of a 1×10^6 dilution yielded 150 colonies, the original sample contained around 1.5×10^8 cells per ml. Usually the count is made more accurate by use of a special colony counter (figure 6.5). In this way, the spread-plate and pour-plate techniques may be used to find the number of microorganisms in a sample.

Spread-plate and pour-plate techniques (pp. 106–8).

Plating techniques are simple, sensitive, and widely used to count bacteria and other microorganisms in samples of food, water, and soil. Several problems, however, can lead to inaccurate counts. Low counts will result if clumps of cells are not broken up and the microorganisms well dispersed. Because it is not possible to be absolutely certain that each colony arose from an individual cell, the results are often expressed in terms of **colony forming units** (**CFU**) rather than the number of microorganisms. The samples should yield between 25 and 250 colonies for best results. Of course, the counts will also be low if the agar medium employed cannot support growth of all the viable microorganisms present. The hot agar used in the pour-plate technique may injure or kill sensitive cells; thus, spread plates sometimes give higher counts than pour plates.

Analysis of water purity (pp. 837–40).

Microbial numbers are frequently determined from counts of colonies growing on special membrane filters having pores small enough to trap bacteria. In the membrane filter technique, a sample is drawn through a special **membrane filter** (figures 6.6 and 6.7). The filter is then placed on an agar medium or on a pad soaked with liquid media and incubated

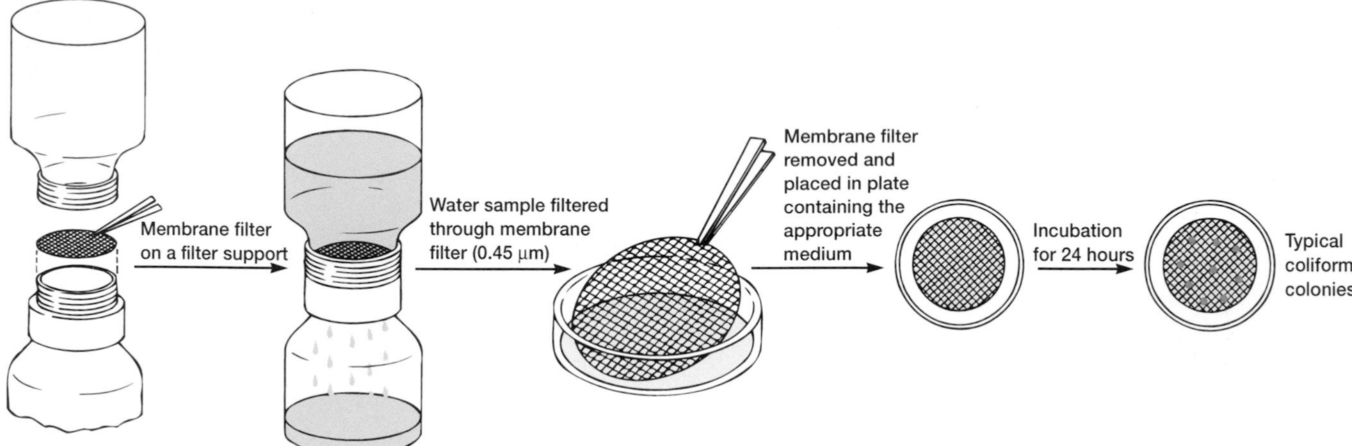

Figure 6.7 The Membrane Filtration Procedure.

until each cell forms a separate colony. A colony count gives the number of microorganisms in the filtered sample, and special media can be used to select for specific microorganisms (figure 6.8).

Membrane filters also are used to count bacteria directly. The bacterial sample is mixed with a fluorescent dye such as acridine orange and then filtered through a polycarbonate Nucleopóre filter. The filter has been stained black to provide a good background for counting the fluorescent bacteria. Acridine orange-stained microorganisms glow orange or green, and are easily counted with an epifluorescence microscope (*see chapter 2*). Usually the counts obtained with this approach are much higher than those from culture techniques.

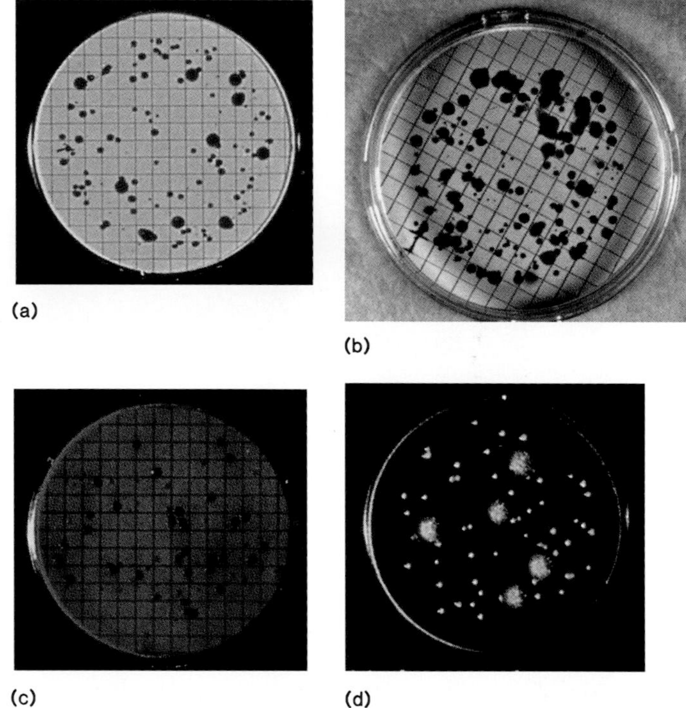

Figure 6.8 Colonies on Membrane Filters. Membrane-filtered samples grown on a variety of media. (*a*) Standard nutrient media for a total bacterial count. An indicator colors colonies red for easy counting. (*b*) Fecal coliform medium for the detection of fecal coliforms that form blue colonies. (*c*) m-Endo agar for the detection of *E. coli* and other coliforms that produce colonies with a green sheen. (*d*) Wort agar for the culture of yeasts and molds.

Measurement of Cell Mass

Increases in the total cell mass, as well as in cell numbers, accompany population growth. Therefore, techniques for measuring changes in cell mass can be used in following growth. The most direct approach is the determination of microbial dry weight. Cells growing in liquid medium are collected by centrifugation, washed, dried in an oven, and weighed. This is an especially useful technique for measuring the growth of fungi. It is time consuming, however, and not very sensitive. Because bacteria weigh so little, it may be necessary to centrifuge several hundred milliliters of culture to collect a sufficient quantity.

More rapid, sensitive techniques depend upon the fact that microbial cells scatter light striking them. Because microbial cells in a population are of roughly constant size, the amount of scattering is proportional to the concentration of cells present. When the concentration of bacteria reaches about 10 million cells (10^7) per milliliter, the medium appears slightly cloudy or turbid. Further increases in concentration result in greater turbidity and less light is transmitted through the medium. The extent of light scattering can be measured by a spectrophotometer and is almost linearly related to bacterial concentration at low absorbance levels (figures 6.9 and 6.10). Thus, population growth can be easily measured spectrophotometrically as long as the population is high enough to give detectable turbidity.

If the amount of a substance in each cell is constant, the total quantity of that cell constituent is directly related to the total microbial cell mass. For example, a sample of washed

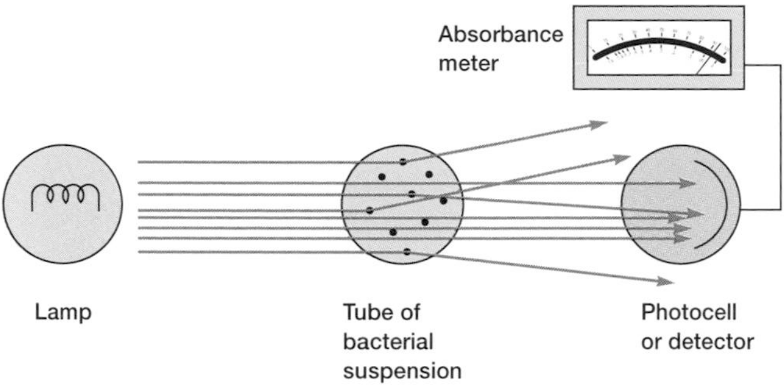

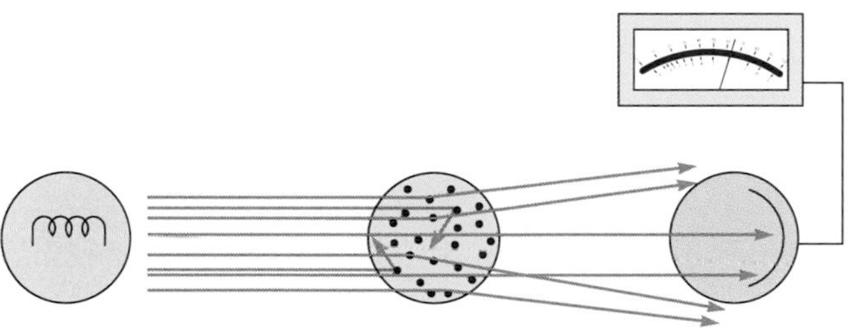

Figure 6.9 Turbidity and Microbial Mass Measurement. Determination of microbial mass by measurement of light absorption. As the population and turbidity increase, more light is scattered and the absorbance reading given by the spectrophotometer increases.

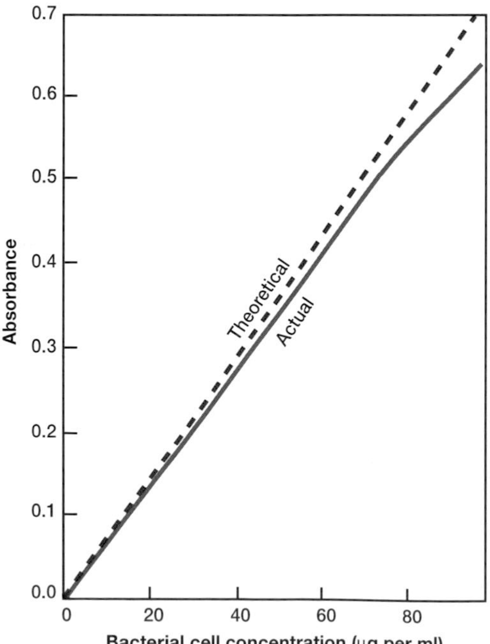

Figure 6.10 Absorbance and Cell Concentration. Absorbance of 420 nm light as a function of bacterial concentration. The concentration was determined by measurements of dry weight. Theoretically, the line should be straight as shown by the dashed line. The line is slightly curved in practice.

cells collected from a known volume of medium can be analyzed for total protein or nitrogen. An increase in the microbial population will be reflected in higher total protein levels. Similarly, chlorophyll determinations can be used to measure algal populations, and the quantity of ATP is an indication of the amount of living microbial mass present.

1. Briefly describe each technique by which microbial population numbers may be determined and give its advantages and disadvantages.
2. Why are plate count results often expressed as colony forming units?

Growth Yields and the Effects of a Limiting Nutrient

When microbial growth is limited by the low concentration of a required nutrient, the final net growth or yield of cells increases with the initial amount of the limiting nutrient present (figure 6.11*a*). This is the basis of microbiological assays for vitamins and other growth factors. The rate of growth also increases with nutrient concentration (figure 6.11*b*), but in a hyperbolic manner. The shape of the curve seems to reflect the rate of nutrient uptake by microbial transport proteins. At sufficiently high nutrient levels the transport systems are saturated, and the growth rate does not rise further with increasing nutrient concentration.

Microbiological assays (p. 100).
Nutrient transport systems (pp. 100–104).

The amount of microbial mass produced from a nutrient can be expressed quantitatively as the **growth yield (*Y*).**

$$Y = \frac{\text{mass of microorganisms formed}}{\text{mass of substrate consumed}}$$

The *Y* value is often expressed in terms of grams of cells formed per gram of substrate used or as the molar growth yield (grams of cells per mole of nutrient consumed). It is an index of the efficiency of conversion of nutrients into cell material. Aerobic microorganisms in rich media may assimilate from 20 to 50% of the sugar carbon taken up. In dilute media, some bacteria seem able to increase their efficiency and assimilate up to 80% of the sugar acquired.

Growth yields of fermenting bacteria have been studied intensively. When growing in nutritionally rich media, bacteria may use carbohydrate fermentation only to generate energy, and employ other substances like amino acids as raw material for biosynthesis. The efficiency of ATP utilization during fermentation is reflected in the Y_{ATP} value.

$$Y_{ATP} = \frac{\text{grams of cells formed}}{\text{mole ATP produced}}$$

Although Y_{ATP} values may vary at very low sugar concentrations, the Y_{ATP} for most bacteria is about 10.5 g/mole

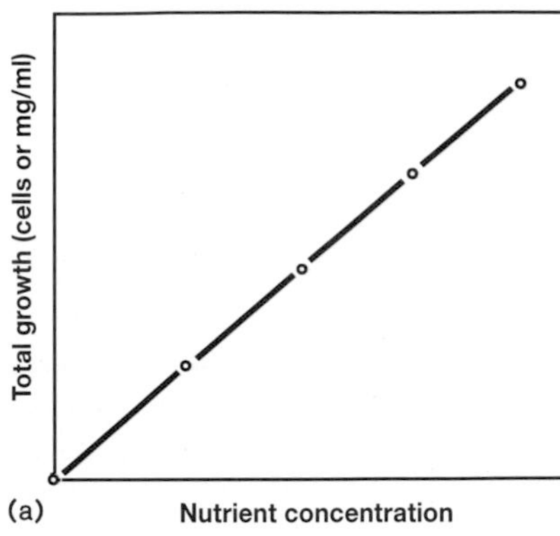

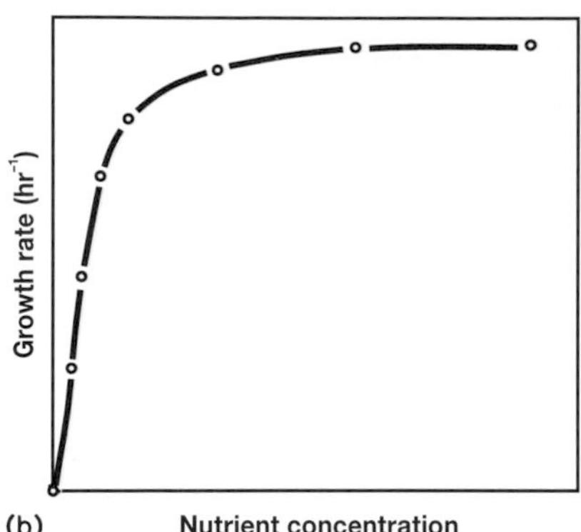

Figure 6.11 **Nutrient Concentration and Growth.** (*a*) The effect of changes in limiting nutrient concentration on total microbial yield. (*b*) The effect on growth rate.

at higher sugar levels. Thus, it appears that bacteria generally use energy in biosynthesis with a fairly constant efficiency. Furthermore, the measured Y_{ATP} value is much lower than the theoretical maximum of 32, which is the Y_{ATP} value expected if all ATP is used in biosynthesis. Much of the ATP is used in transport, mechanical work, and probably other ways.

The Continuous Culture of Microorganisms

Up to this point, the focus has been on closed systems called batch cultures in which nutrient supplies are not renewed nor wastes removed. Exponential growth lasts for only a few generations and soon the stationary phase is reached. However, it is possible to grow microorganisms in an open system, a system with constant environmental conditions maintained through continual provision of nutrients and removal of wastes. These conditions are met in the laboratory by a **continuous culture system.** A microbial population can be maintained in the exponential growth phase and at a constant biomass concentration for extended periods in a continuous culture system.

The Chemostat

There are two major types of continuous culture systems commonly in use: (1) chemostats and (2) turbidostats. A **chemostat** is constructed so that sterile medium is fed into the culture vessel at the same rate as media containing microorganisms is removed (figure 6.12). The culture medium for a chemostat possesses an essential nutrient (e.g., an amino acid) in limiting quantities. Because of the presence of a limiting nutrient, the growth rate is determined by the rate at which new medium is fed into the growth chamber, and the final cell density depends on the concentration of the limiting nutrient. The rate of nutrient exchange is expressed as the dilution rate (D), the rate at which medium flows through the culture vessel relative to the vessel volume, where f is the flow rate (ml/hr) and V is the vessel volume (ml).

$$D = f/V$$

For example, if f is 30 ml/hr and V is 100 ml, the dilution rate is 0.30 hr^{-1}.

Both the microbial population level and the generation time are related to the dilution rate (figure 6.13). The microbial population density remains unchanged over a wide range of dilution rates. The generation time shortens (i.e., the growth rate rises) as the dilution rate increases. The limiting nutrient will be almost completely depleted under these balanced conditions. If the dilution rate rises too high, the microorganisms can actually be washed out of the culture vessel before reproducing because the dilution rate is greater than the maximum growth rate. The limiting nutrient concentration rises at higher dilution rates because there are fewer microorganisms present to use it.

At very low dilution rates, an increase in D causes a rise in both cell density and the growth rate. This is because of the effect of nutrient concentration on the growth rate, sometimes called the Monod relationship (figure 6.11*b*). Only a limited supply of nutrient is available at low dilution rates. Much of the available energy must be used for cell maintenance, not for growth and reproduction. As the dilution rate increases, the amount of nutrients and the resulting cell density rise because energy is available for both maintenance and growth. The growth rate increases when the total available energy exceeds the **maintenance energy.**

The Turbidostat

The second type of continuous culture system, the **turbidostat,** has a photocell that measures the absorbance or turbidity of the culture in the growth vessel. The flow rate of media through the vessel is automatically regulated to maintain a predetermined turbidity or cell density. The turbidostat differs from the chemostat in several ways. The dilution rate in a turbidostat varies rather than remaining constant, and its culture medium lacks a limiting nutrient. The turbidostat operates best at high dilution rates; the chemostat is most stable and effective at lower dilution rates.

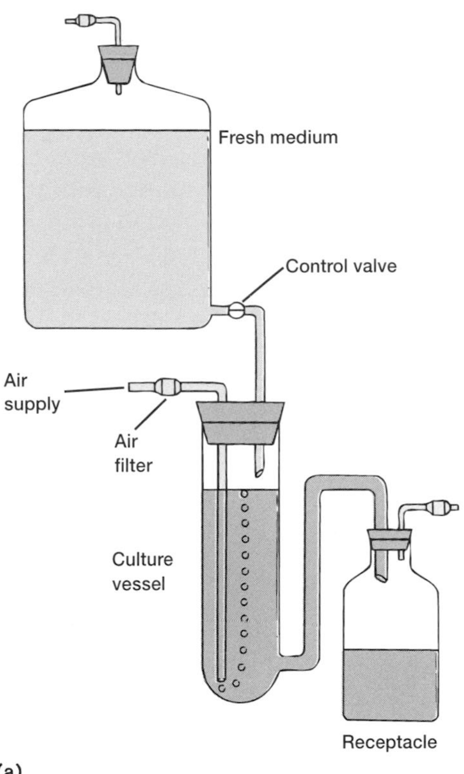

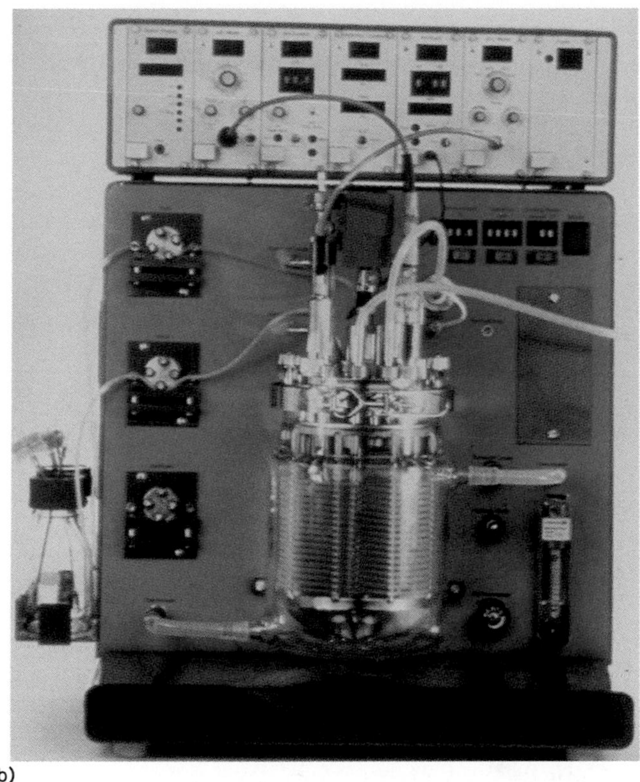

Figure 6.12 **A Continuous Culture System: The Chemostat.** (*a*) Schematic diagram of the system. The fresh medium contains a limiting amount of an essential nutrient. Growth rate is determined by the rate of flow of medium through the culture vessel. (*b*) A commercial fermenter that may be employed as a chemostat. The fermenter uses 0.75- to 1.5-liter culture vessels, and factors such as temperature, pH, and oxygen concentration can be regulated during growth.

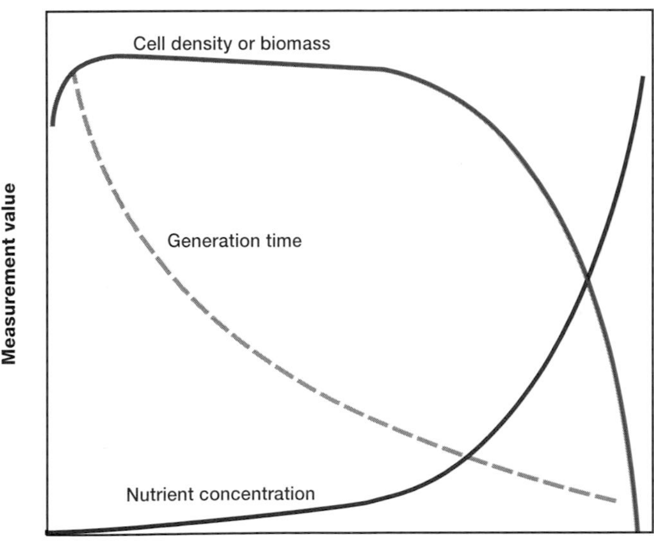

Figure 6.13 **Chemostat Dilution Rate and Microbial Growth.** The effects of changing the dilution rate in a chemostat.

Continuous culture systems are very useful because they provide a constant supply of cells in exponential phase and growing at a known rate. They make possible the study of microbial growth at very low nutrient levels, concentrations close to those present in natural environments. These systems are essential for research in many areas—for example, in studies on interactions between microbial species under environmental conditions resembling those in a freshwater lake or pond.

Balanced and Unbalanced Growth

Exponential growth, whether in batch or continuous cultures, is **balanced growth.** That is, all cellular constituents are manufactured at constant rates relative to each other. If nutrient levels or other environmental conditions change, **unbalanced growth** results because the rates of synthesis of cell components vary relative to one another until a new balanced state is reached. This response is readily observed in a shift-up experiment in which bacteria are transferred from a nutritionally poor medium to a richer one. The cells first construct new ribosomes to enhance their capacity for protein synthesis. This is followed by increases in protein and DNA synthesis. Finally, the expected rise in reproductive rate takes place.

Protein and DNA synthesis (chapter 10).

Unbalanced growth also results when a bacterial population is shifted down from a rich medium to a poor one. The organisms may previously have been able to obtain many cell components directly from the medium. When shifted to a nutritionally inadequate medium, they need time to make the enzymes required for the biosynthesis of unavailable nutrients. Consequently, cell division and DNA replication continue after the shift-down, but net protein and RNA synthesis slow. The cells become smaller and reorganize themselves metabolically until they are able to grow again. Then balanced growth is resumed.

Regulation of nucleic acid synthesis (pp. 222–30).

These shift-up and shift-down experiments demonstrate that microbial growth is under precise, coordinated control and responds quickly to changes in environmental conditions.

1. What effect does increasing a limiting nutrient have on the yield of cells and the growth rate? Explain what a growth yield value is and what information it provides.
2. How does an open system differ from a closed culture system or batch culture? Describe how the two different kinds of continuous culture systems, the chemostat and turbidostat, operate. What is the dilution rate? What is maintenance energy?
3. Define balanced growth, unbalanced growth, shift-up experiment, and shift-down experiment.

The Influence of Environmental Factors on Growth

The growth of microorganisms is greatly affected by the chemical and physical nature of their surroundings. An understanding of environmental influences aids in the control of microbial growth and the study of the ecological distribution of microorganisms.

Solutes and Water Activity

Because a selectively permeable plasma membrane separates microorganisms from their environment, they can be affected by changes in the osmotic concentration of their surroundings. If a microorganism is placed in a hypotonic solution, water will enter the cell and cause it to burst unless something is done to prevent the influx. Most bacteria, algae, and fungi have rigid cell walls that maintain the shape and integrity of the cell. Indeed, many microorganisms keep the osmotic concentration of their protoplasm above that of the habitat by the use of compatible solutes, so that the plasma membrane is always pressed firmly against their cell wall. **Compatible solutes** are solutes that are compatible with metabolism and growth when at high intracellular concentrations. Most bacteria increase their internal osmotic concentration through the synthesis or uptake of choline, betaine, proline, glutamic acid, and other amino acids; elevated levels of potassium ions are also involved to some extent. Algae and fungi employ sucrose and polyols—for example, arabitol, glycerol, and mannitol—for the same purpose. Polyols and amino acids are ideal solutes for this function because they normally do not disrupt enzyme structure and function. A few bacteria like *Halobacterium salinarium* raise their osmotic concentration with potassium ions (sodium ions are also elevated, but not as much as potassium). *Halobacterium*'s enzymes have been altered so that they actually require high salt concentrations for normal activity (*see chapter 24*). Since protozoa do not have a cell wall, they must use contractile vacuoles (*see figure 28.3*) to eliminate excess water when living in hypotonic environments.

Osmosis and the protective function of the cell wall (pp. 55–56).

When microorganisms with rigid cell walls are placed in a hypertonic environment, water leaves and the plasma membrane shrinks away from the wall, a process known as plasmolysis. This dehydrates the cell and may damage the plasma membrane; the cell usually becomes metabolically inactive and ceases to grow.

The amount of water available to microorganisms can be reduced by interaction with solute molecules (the osmotic effect) or by adsorption to the surfaces of solids (the matric effect). Because the osmotic concentration of a habitat has such profound effects on microorganisms, it is useful to be able to express quantitatively the degree of water availability. Microbiologists generally use **water activity (a_w)** for this purpose. The water activity of a solution is 1/100 the relative humidity of the solution (when expressed as a percent). It is also equivalent to the ratio of the solution's vapor pressure (P_{soln}) to that of pure water (P_{water}).

$$a_w = \frac{P_{soln}}{P_{water}}$$

The water activity of a solution or solid can be determined by sealing it in a chamber and measuring the relative humidity after the system has come to equilibrium. Suppose after a sample is treated in this way, the air above it is 95% saturated—that is, the air contains 95% of the moisture it would have when equilibrated at the same temperature with a sample of pure water. The relative humidity would be 95% and the sample's water activity, 0.95. Water activity is inversely related to osmotic pressure; if a solution has high osmotic pressure, its a_w is low.

Microorganisms differ greatly in their ability to adapt to habitats with low water activity (table 6.3). A microorganism must expend extra effort to grow in a habitat with a low a_w value because it must maintain a high internal solute concentration to retain water. Some microorganisms can do this and are **osmotolerant;** they will grow over wide ranges of water activity or osmotic concentration. For example, *Staphylococcus aureus* can be cultured in media containing any sodium chloride concentration up to about 3 M. The yeast *Saccharomyces rouxii* will grow in sugar solutions with a_w values as low as 0.6. The alga *Dunaliella viridis* tolerates sodium chloride concentrations from 1.7 M to a saturated solution.

TABLE 6.3 Approximate Lower a_w Limits for Microbial Growth

Water Activity	Environment	Bacteria	Fungi	Algae
1.00—Pure water	Blood, seawater, meat, vegetables, fruit	Most gram-negative nonhalophiles		
0.95	Bread	Most gram-positive rods	*Basidiomycetes*	Most algae
0.90	Ham	Most cocci, *Bacillus*	*Fusarium* *Mucor, Rhizopus* Ascomycetous yeasts	
0.85	Salami	*Staphylococcus*	*Saccharomyces rouxii* (in salt)	
0.80	Preserves		*Penicillium*	
0.75	Salt lakes Salted fish	*Halobacterium* *Actinospora*	*Aspergillus*	*Dunaliella*
0.70			*Aspergillus*	
	Cereals, candy, dried fruit			
0.60			*Saccharomyces rouxii* (in sugars)	
	Chocolate Honey Dried milk		*Xeromyces bisporus*	
0.55—DNA disordered				

Although a few microorganisms are truly osmotolerant, most only grow well at water activities around 0.98 (the approximate a_w for seawater) or higher. This is why drying food or adding large quantities of salt and sugar is so effective in preventing food spoilage. As table 6.3 shows, many fungi are osmotolerant and thus particularly important in the spoilage of salted or dried foods.

Food spoilage (pp. 873–76).

Halophiles have adapted so completely to saline conditions that they require high levels of sodium chloride to grow, concentrations between about 2.8 M and saturation (about 6.2 M) for extreme halophilic bacteria. The genus *Halobacterium* can be isolated from the Dead Sea (a salt lake between Israel and Jordan, and the lowest lake in the world), the Great Salt Lake in Utah, and other aquatic habitats with salt concentrations approaching saturation. *Halobacterium* and other extremely halophilic bacteria have significantly modified the structure of their proteins and membranes rather than simply increasing the intracellular concentrations of solutes, the approach used by most osmotolerant microorganisms. Halophiles are archaeobacteria, and thus very different from other bacteria. These extreme halophiles accumulate enormous quantities of potassium in order to remain hypertonic to their environment; the internal potassium concentration may reach 4 to 7 M. The enzymes, ribosomes, and transport proteins of these bacteria require high levels of potassium for stability and activity. In addition, the plasma membrane and cell wall of *Halobacterium* are stabilized by high concentrations of sodium ion. If the sodium concentration decreases too much, the wall and plasma membrane literally disintegrate. Extreme halophilic bacteria have successfully adapted to environmental conditions that would destroy most organisms. In the process, they have become so specialized that they have lost ecological flexibility and can prosper only in a few extreme habitats.

Archaeobacteria (chapter 24).

pH

pH is a measure of the hydrogen ion activity of a solution and is defined as the negative logarithm of the hydrogen ion concentration.

$$pH = -\log [H^+] = \log(1/[H^+])$$

The pH scale extends from pH 0.0 (1.0 M H^+ to pH 14.0 (1.0×10^{-14} M H^+), and each pH unit represents a tenfold change in hydrogen ion concentration. Figure 6.14 shows that the habitats in which microorganisms grow vary widely—from pH 1 to 2 at the acid end to alkaline lakes and soil that may have pH values between 9 and 10.

It is not surprising that pH dramatically affects microbial growth. Each species has a definite pH growth range and pH growth optimum (table 6.4). **Acidophiles** have their growth optimum between pH 1.0 and 5.5; **neutrophiles,** between pH 5.5 and 8.0; and **alkalophiles** prefer the pH range of 8.5 to 11.5. Extreme alkalophiles have growth optima at pH 10 or higher. In general, different microbial groups have characteristic pH preferences. Most bacteria and protozoa are neutrophiles. Most fungi prefer slightly acid surroundings, about pH 4 to 6; algae also seem to favor slight acidity. There are many exceptions to these generalizations. For example, the alga *Cyanidium caldarium* and the bacterium *Sulfolobus acidocaldarius* are common inhabitants of acidic hot springs; both grow well around pH 1 to 3 and at high temperatures.

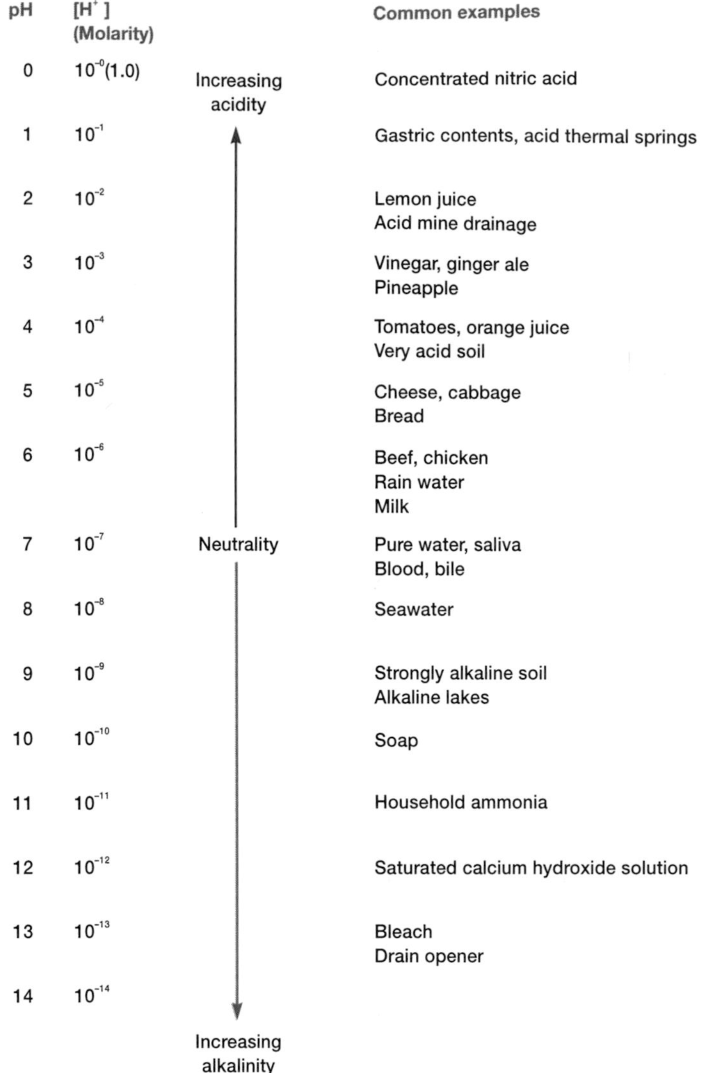

pH
[H+] (Molarity)
Common examples
0 10^-0 (1.0) Concentrated nitric acid
Increasing acidity
1 10^-1 Gastric contents, acid thermal springs
2 10^-2 Lemon juice, Acid mine drainage
3 10^-3 Vinegar, ginger ale, Pineapple
4 10^-4 Tomatoes, orange juice, Very acid soil
5 10^-5 Cheese, cabbage, Bread
6 10^-6 Beef, chicken, Rain water, Milk
7 10^-7 Neutrality Pure water, saliva, Blood, bile
8 10^-8 Seawater
9 10^-9 Strongly alkaline soil, Alkaline lakes
10 10^-10 Soap
11 10^-11 Household ammonia
12 10^-12 Saturated calcium hydroxide solution
13 10^-13 Bleach, Drain opener
14 10^-14
Increasing alkalinity

Figure 6.14 **The pH Scale.** The pH scale and examples of substances with different pHs.

Despite wide variations in habitat pH, the internal pH of most microorganisms is close to neutrality. This may result from plasma membrane impermeability to protons. Possibly, protons and hydroxyl ions are pumped out to maintain the proper intracellular pH. Extreme alkalophiles maintain their internal pH closer to neutrality by exchanging internal sodium ions for external protons.

While microorganisms will often grow over wide ranges of pH, there are limits to their tolerance. Drastic variations in pH can harm microorganisms by disrupting the plasma membrane or inhibiting the activity of enzymes and membrane transport proteins. Changes in the external pH also might alter the ionization of nutrient molecules and thus reduce their availability to the organism.

Microorganisms frequently change the pH of their own habitat by producing acidic or basic metabolic waste products. Fermentative microorganisms form organic acids from carbohydrates, while chemolithotrophs like *Thiobacillus* oxidize reduced sulfur components to sulfuric acid. Other microorganisms make their environment more alkaline by generating ammonia through amino acid degradation.

Microbial fermentations (pp. 155–58).
Sulfur-oxidizing bacteria (pp. 478–80).

Buffers are often included in media to prevent growth inhibition by large pH changes. Phosphate is a commonly used buffer and a good example of buffering by a weak acid ($H_2PO_4^-$) and its salt (HPO_4^{2-}).

$$H^+ + HPO_4^{2-} \longrightarrow HPO_4^-$$

$$OH^- + H_2PO_4^- \longrightarrow HPO_4^{2-} + HOH$$

If protons are added to the mixture they combine with the salt form to yield a weak acid. An increase in alkalinity is resisted because the weak acid will neutralize hydroxyl ions through proton donation to give water. Peptides and amino acids in complex media also have a strong buffering effect.

Temperature

Environmental temperature profoundly affects microorganisms, like all other organisms. Indeed, microorganisms are particularly susceptible because they are usually unicellular and also poikilothermic—their temperature varies with that of the external environment. For these reasons, microbial cell temperature directly reflects that of the cell's surroundings. A most important factor influencing the effect of temperature upon growth is the temperature sensitivity of enzyme-catalyzed reactions. At low temperatures, a temperature rise increases the growth rate because the velocity of an enzyme-catalyzed reaction, like that of any chemical reaction, will roughly double for every 10°C rise in temperature. Because the rate of each reaction increases, metabolism as a whole is more active at higher temperatures, and the microorganism grows faster. Beyond a certain point, further increases actually slow growth, and sufficiently high temperatures are lethal. High temperatures damage microorganisms by denaturing enzymes, transport carriers, and other proteins. Microbial membranes are also disrupted by heat; the lipid bilayer simply melts apart. Thus, although functional enzymes operate more rapidly at higher temperatures, the microorganism may be damaged to such an extent that growth is inhibited because the damage cannot be adequately repaired.

The temperature dependence of enzyme activity (p. 142–43).

Because of these opposing temperature influences, microbial growth has a fairly characteristic temperature depen-

TABLE 6.4 The Effect of pH on Microbial Growth

Microorganism	Lower Limit	pH Optimum	Upper Limit
Bacteria			
Thiobacillus thiooxidans	0.5	2.0–3.5	6.0
Lactobacillus acidophilus	4.0–4.6	5.8–6.6	6.8
Staphylococcus aureus	4.2	7.0–7.5	9.3
Escherichia coli	4.4	6.0–7.0	9.0
Pseudomonas aeruginosa	5.6	6.6–7.0	8.0
Chlorobium limicola	6.0	6.8	7.0
Nitrosomonas spp.	7.0–7.6	8.0–8.8	9.4
Bacillus pasteurii	8.5		
Nostoc muscorum		6.9–7.7	
Anabaena variabilis		6.9–9	
Microcystis aeruginosa		10	
Bacillus alcalophilus	8.5	10.6	11.5
Algae			
Cyanidium caldarium	0	2	5
Euglena gracilis	3.9	6.6	9.9
Fungi			
Acontium velatum	0		
Physarum polycephalum		5	6.8
Protozoa			
Paramecium bursaria	4.9	6.7–6.8	8.0
Tetrahymena pyriformis	4.0	7.3	9.0
Acanthamoeba castellanii	3.2	5–6	8.7

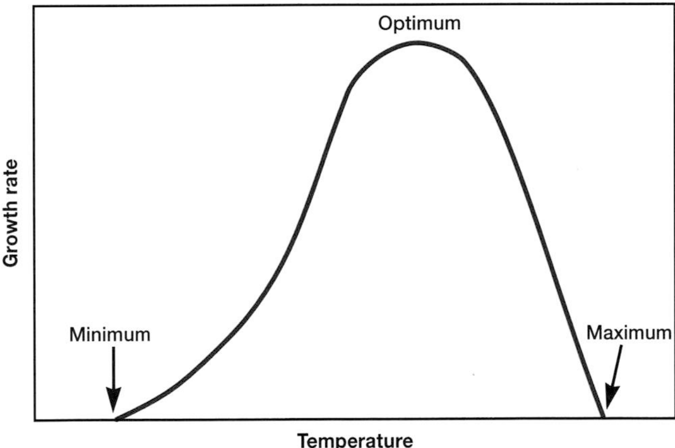

Figure 6.15 Temperature and Growth. The effect of temperature on growth rate.

dence with distinct **cardinal temperatures**—minimum, optimum, and maximum growth temperatures (figure 6.15). Although the shape of the temperature dependence curve can vary, the temperature optimum is always closer to the maximum than to the minimum. The cardinal temperatures for a particular species are not rigidly fixed, but often depend to some extent on other environmental factors such as pH and the available nutrients. For example, *Crithidia fasciculata,* a flagellated protozoan living in the gut of mosquitos, will grow in a simple medium at 22 to 27°C. However, it cannot be cultured at 33 to 34°C without the addition of extra metals, amino acids, vitamins, and lipids.

The cardinal temperatures vary greatly between microorganisms (table 6.5). Optima normally range from 0°C to as high as 75°C, while microbial growth occurs at temperatures extending from −20°C to over 100°C. The growth temperature range for a particular microorganism usually spans about 30 degrees. Some species (e.g., *Neisseria gonorrhoeae*) have a small range and are called **stenothermal;** others, like *Enterococcus faecalis,* will grow over a wide range of temperatures or are **eurythermal.** The major microbial groups differ from one another regarding their maximum growth temperature. The upper limit for protozoa is around 50°C. Some algae and fungi can grow at temperatures as high as 55 to 60°C. Bacteria have been found growing at or close to 100°C, the boiling point of water, and recently strains growing at even higher temperatures have been discovered (see Box 6.1). Clearly, procaryotic organisms can grow at much higher temperatures than eucaryotes. It has been suggested that eucaryotes are not able to manufacture organellar membranes that are stable and functional at temperatures above 60°C. The structure of water also limits the temperature range for growth. It is unlikely that microorganisms can grow in solid ice or steam—almost certainly, liquid water is necessary for microbial growth.

TABLE 6.5 Temperature Ranges for Microbial Growth

Microorganism	Cardinal Temperatures (°C) Minimum	Optimum	Maximum
Nonphotosynthetic Bacteria			
Bacillus psychrophilus	−10	23–24	28–30
Micrococcus cryophilus	−4	10	24
Pseudomonas fluorescens	4	25–30	40
Staphylococcus aureus	6.5	30–37	46
Enterococcus faecalis	0	37	44
Escherichia coli	10	37	45
Neisseria gonorrhoeae	30	35–36	38
Thermoplasma acidophilum	45	59	62
Bacillus stearothermophilus	30	60–65	75
Thermus aquaticus	40	70–72	79
Sulfolobus acidocaldarius	60	80	85
Pyrodictium occultum	82	105	110
Photosynthetic Bacteria			
Rhodospirillum rubrum		30–35	
Nostoc muscorum		32.5	
Anabaena variabilis		35	
Anacystis nidulans		41	
Oscillatoria tenuis			45–47
Synechococcus lividus			74
Synechococcus eximius	70	79	84
Eucaryotic Algae			
Chlamydomonas nivalis	−36	0	4
Fragilaria sublinearis	−2	5–6	8–9
Chlorella pyrenoidosa		25–26	29
Euglena gracilis		23	
Chilomonas paramecium	10	28	35
Skeletonema costatum	6	16–26	>28
Cyanidium caldarium	30–34	45–50	56
Fungi			
Candida scottii	0	4–15	15
Saccharomyces cerevisiae	1–3	28	40
Mucor pusillus	21–23	45–50	50–58
Protozoa			
Amoeba proteus	4–6	22	35
Naegleria fowleri	20–25	35	40
Trichomonas vaginalis	25	32–39	42
Paramecium caudatum		25	28–30
Tetrahymena pyriformis	6–7	20–25	33
Blepharisma intermedium	24–25	28	38
Cyclidium citrullus	18	43	47

Microorganisms such as those in table 6.5 can be placed in one of four classes based on their temperature ranges for growth.

1. **Psychrophiles** grow well at 0°C and have an optimum growth temperature of 15°C or lower; the maximum is around 20°C. They are readily isolated from Arctic and Antarctic habitats; because 90% of the ocean is 5°C or colder, it constitutes an enormous habitat for psychrophiles. The psychrophilic alga *Chlamydomanos nivalis* can actually turn a snowfield or glacier pink with its bright red spores. Most psychrophilic bacteria are members of the genera *Pseudomonas, Flavobacterium, Achromobacter,* and *Alcaligenes.* Psychrophilic microorganisms have adapted to their environment in several ways. Their enzymes, transport systems, and protein synthetic mechanisms function well at low temperatures. The cell membranes of psychrophilic microorganisms have high levels of unsaturated fatty acids and remain semifluid when cold. Indeed, many psychrophiles begin to leak cellular constituents at temperatures higher than 20°C because of cell membrane disruption.
2. Many species can grow at 0°C even though they have optima between 20 and 30°C, and maxima at about 35°C. These are called **psychrotrophs** or **facultative psychrophiles.** Psychrotrophic bacteria and fungi are major factors in the spoilage of refrigerated foods (*see chapter 43*).
3. **Mesophiles** are microorganisms with growth optima around 20 to 45°C and a temperature minimum of 15 to 20°C. Their maximum is about 45°C or lower. Most microorganisms probably fall within this category. Almost all human pathogens are mesophiles, as might be expected since their environment is a fairly constant 37°C.
4. Some microorganisms are **thermophiles;** they can grow at temperatures of 55°C or higher. Their growth minimum is usually around 45°C and they often have optima between 55 and 65°C. As mentioned previously, a few thermophiles have maxima above 100°C. The vast majority are bacteria although a few algae and fungi are thermophilic (table 6.5). These organisms flourish in many habitats including compost, self-heating hay stacks, hot water lines, and hot springs. Thermophiles differ from mesophiles in having much more heat-stable enzymes and protein synthesis systems able to function at high temperatures. Their membrane lipids are also more saturated than those of mesophiles and have higher melting points; therefore, thermophile membranes remain intact at higher temperatures.

---

1. How do microorganisms adapt to hypotonic and hypertonic environments? What is plasmolysis? Define water activity and briefly describe how it can be determined. Why is it difficult for microorganisms to grow at low a_w values? What are halophiles and why does *Halobacterium* require sodium and potassium ions?
2. Define pH, acidophile, neutrophile, and alkalophile. How can microorganisms change the pH of their environment, and how does the microbiologist minimize this effect?
3. What are cardinal temperatures? Why does the growth rate rise with increasing temperature and then fall again at higher temperatures? Define stenothermal, eurythermal, psychrophile, psychrotroph, mesophile, and thermophile.

Oxygen Concentration

An organism able to grow in the presence of atmospheric O_2 is an **aerobe,** whereas one that can grow in its absence is an **anaerobe.** Almost all higher organisms are completely depen-

Box 6.1

Life above 100°C

Until recently, the highest reported temperature for bacterial growth was 105°C. It seemed that the upper temperature limit for life was about 100°C, the boiling point of water. Now thermophilic bacteria have been reported growing in sulfide chimneys or "black smokers," located along rifts and ridges on the ocean floor, that spew sulfide-rich superheated vent water with temperatures above 350°C (Box Figure 6.1). Evidence has been presented that these bacteria can grow and reproduce at 115°C. It is possible that bacteria may grow at higher temperatures. The pressure present in their habitat is sufficient to keep water liquid (at 265 atm; seawater doesn't boil until 460°C).

The implications of this discovery are many. The proteins, membranes, and nucleic acids of these bacteria are remarkably temperature stable and provide ideal subjects for studying the ways in which macromolecules and membranes are stabilized. In the future, it may be possible to design enzymes that can operate at very high temperatures. Some thermostable enzymes from these bacteria may have important industrial uses.

Box Figure 6.1 **A Smoker Vent in the Galapagos Trench.** Note the worms and bacterial mats growing on the rocks.

dent upon atmospheric O_2 for growth, that is, they are **obligate aerobes.** Oxygen serves as the terminal electron acceptor for the electron transport chain in aerobic respiration. In addition, aerobic eucaryotes employ O_2 in the synthesis of sterols and unsaturated fatty acids. **Facultative anaerobes** do not require O_2 for growth, but do grow better in its presence. **Aerotolerant anaerobes** such as *Enterococcus faecalis* simply ignore O_2 and grow equally well whether it is present or not. In contrast, **strict** or **obligate anaerobes** (e.g., *Bacteroides, Fusobacterium, Clostridium pasteurianum, Methanococcus*) do not tolerate O_2 at all and die in its presence. Aerotolerant and strict anaerobes cannot generate energy through respiration and must employ fermentation or anaerobic respiration pathways for this purpose. Finally, there are a few aerobes, called **microaerophiles,** that are damaged by the normal atmospheric level of O_2 (20%) and require O_2 levels below the range of 2 to 10% for growth. The nature of bacterial O_2 responses can be readily determined by growing the bacteria in culture tubes filled with a solid culture medium or a special medium like thioglycollate broth, which contains a reducing agent to lower O_2 levels (figure 6.16).

Electron transport and aerobic respiration (pp. 152–54).
Fermentation (pp. 155–58).
Anaerobic respiration (p. 158).

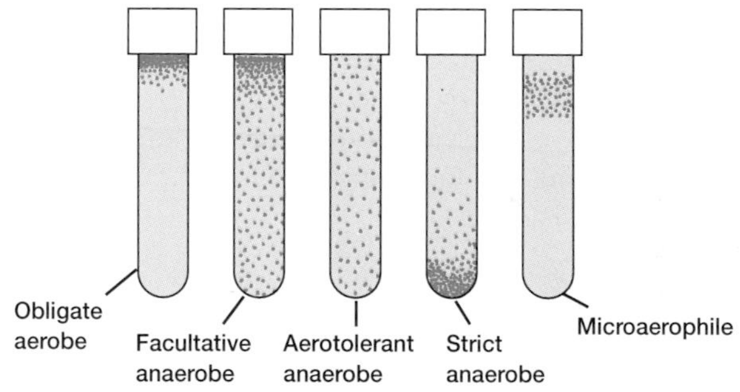

Figure 6.16 **Oxygen and Bacterial Growth.** An illustration of the growth of bacteria with varying responses to oxygen. Each dot represents an individual bacterial colony within the agar or on its surface. The surface, which is directly exposed to atmospheric oxygen, will be aerobic. The oxygen content of the medium decreases with depth until the medium becomes anaerobic toward the bottom of the tube.

A microbial group may show more than one type of relationship to O_2. All five types are found among the bacteria and protozoa. Fungi are normally aerobic, but a number of species—particularly among the yeasts—are facultative anaerobes. Algae are almost always obligate aerobes.

Although strict anaerobes are killed by O_2, they may be recovered from habitats that appear to be aerobic. In such cases, they associate with facultative anaerobes that use up the available O_2 and thus make the growth of strict anaerobes possible. For example, the strict anaerobe *Bacteroides gingivalis* lives in the mouth where it grows in the anaerobic crevices around the teeth.

These different relationships with O_2 appear due to several factors, including the inactivation of proteins and the effect of toxic O_2 derivatives. Enzymes can be inactivated when sensitive groups like sulfhydryls are oxidized. A notable example is the nitrogen-fixation enzyme nitrogenase (*see chapter 9*), which is very oxygen sensitive.

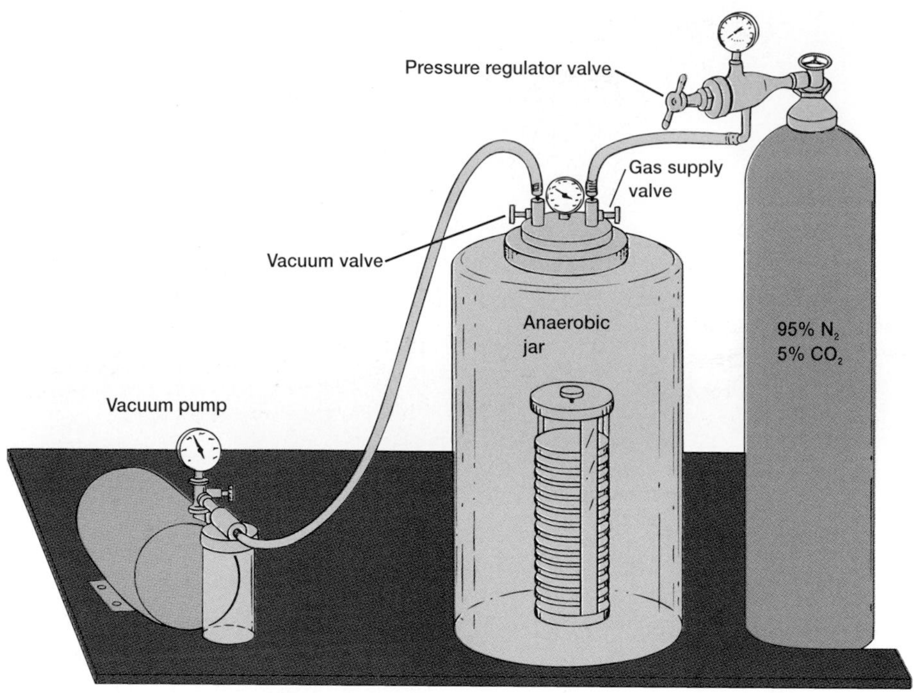

Figure 6.17 Vacuum and Gas Displacement Method of Anaerobic Culture. The anaerobic jar is evacuated at least three times and refilled each time with nitrogen or a nitrogen–carbon dioxide mixture.

Oxygen accepts electrons and is readily reduced because its two outer orbital electrons are unpaired. Flavoproteins (*see chapter 7*), several other cell constituents, and radiation (*see pp. 129–30*) promote oxygen reduction. The result is usually some combination of the reduction products **superoxide radical, hydrogen peroxide,** and **hydroxyl radical.**

$O_2 + e^- \rightarrow O_2^{\cdot -}$ (superoxide radical)
$O_2^{\cdot -} + e^- + 2H^+ \rightarrow H_2O_2$ (hydrogen peroxide)
$H_2O_2 + e^- + H^+ \rightarrow H_2O + OH\cdot$ (hydroxyl radical)

These products of oxygen reduction are extremely toxic because they are powerful oxidizing agents and destroy cellular constituents very rapidly. A microorganism must be able to protect itself against such O_2 products or it will be killed. Neutrophils and macrophages use these toxic oxygen products to destroy invading pathogens.

Oxygen-dependent killing of pathogens (p. 600).

Many microorganisms possess enzymes that afford protection against toxic O_2 products. Obligate aerobes and facultative anaerobes usually contain the enzymes **superoxide dismutase** and **catalase,** which catalyze the destruction of superoxide radical and hydrogen peroxide, respectively.

$$2O_2^{\cdot -} + 2H^+ \xrightarrow{\text{superoxide dismutase}} O_2 + H_2O_2$$

$$2\,H_2O_2 \xrightarrow{\text{catalase}} 2\,H_2O + O_2$$

Aerotolerant microorganisms may lack catalase, but almost always have superoxide dismutase. The aerotolerant *Lactobacillus plantarum* uses manganous ions instead of superoxide dismutase to destroy the superoxide radical. All strict anaerobes lack both enzymes or have them in very low concentrations and therefore cannot tolerate O_2.

Because aerobes need O_2 and anaerobes are killed by it, radically different approaches must be used when growing the two types of microorganisms. When large volumes of aerobic microorganisms are cultured, either the culture vessel is shaken to aerate the medium or sterile air must be pumped through the culture vessel. Precisely the opposite problem arises with anaerobes; all O_2 must be excluded. This can be accomplished in three ways. (1) Special anaerobic media containing reducing agents such as thioglycollate or cysteine may be used. The reducing agents will eliminate any dissolved O_2 present within the medium so that anaerobes can grow beneath its surface. (2) Oxygen also may be eliminated by removing air with a vacuum pump and flushing out residual O_2 with nitrogen gas (figure 6.17). Often CO_2 as well as nitrogen is added to the chamber since many anaerobes require a small amount of CO_2 for best growth. (3) One of the most popular ways of culturing small numbers of anaerobes is by use of a GasPak jar (figure 6.18). In this procedure, the environment is made anaerobic by using hydrogen and a palladium catalyst to remove O_2 through the formation of water. The reducing agents in anaerobic agar also remove oxygen, as mentioned previously. A laboratory may make use of all three techniques since each is best suited for different purposes.

Pressure

Most organisms spend their lives on land or on the surface of water, always subjected to a pressure of 1 atmosphere (atm),

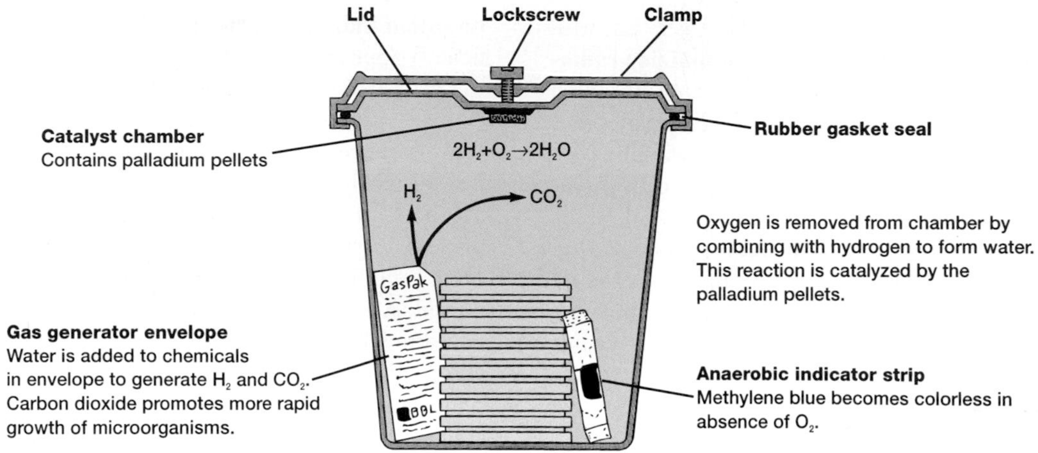

Figure 6.18 **The GasPak Anaerobic System.** Hydrogen and carbon dioxide are generated by a GasPak envelope. Palladium catalyst in the chamber lid catalyzes the formation of water from hydrogen and oxygen, thereby removing oxygen from the sealed chamber.

and are never affected significantly by pressure. Yet the deep sea (ocean of 1,000 m or more in depth) is 75% of the total ocean volume. The hydrostatic pressure can reach 600 to 1,100 atm in the deep sea, while the temperature is about 2 to 3°C. Despite these extremes, bacteria survive and adapt. Many are **barotolerant:** increased pressure does adversely affect them, but not as much as it does nontolerant bacteria. Some bacteria in the gut of deep-sea invertebrates such as amphipods and holothurians are truly **barophilic**—they grow more rapidly at high pressures. These gut bacteria may play an important role in nutrient recycling in the deep sea. One barophile has been recovered from the Mariana trench near the Philippines (depth about 10,500 m) that is actually unable to grow at pressures below about 400 to 500 atm when incubated at 2°C.

The marine environment (pp. 826–30).

Radiation

Our world is bombarded with electromagnetic radiation of various types (figure 6.19). This radiation often behaves as if it were composed of waves moving through space like waves traveling on the surface of water. The distance between two wave crests or troughs is the wavelength. As the wavelength of electromagnetic radiation decreases, the energy of the radiation increases—gamma rays and X rays are much more energetic than visible light or infrared waves. Electromagnetic radiation also acts like a stream of energy packets called photons, each photon having a quantum of energy whose value will depend upon the wavelength of the radiation.

Sunlight is the major source of radiation on the earth. It includes visible light, ultraviolet radiation (UV), infrared rays, and radio waves. Visible light is a most conspicuous and important aspect of our environment: all life is dependent upon the ability of photosynthetic organisms to trap the light energy of the sun. Almost 60% of the sun's radiation is in the infrared region rather than the visible portion of the spectrum. Infrared is the major source of the earth's heat. At the surface of the earth, one finds very little ultraviolet radiation below about 290 to 300 nm. UV radiation of wavelengths shorter than 287 nm is absorbed by O_2 in the earth's atmosphere; this process forms a layer of ozone between 25 and 30 miles above the earth's surface. The ozone layer then absorbs somewhat longer UV rays and reforms O_2. This elimination of UV is crucial because it is quite damaging to living systems (*see chapter 12*). The fairly even distribution of sunlight throughout the visible spectrum accounts for the fact that sunlight is generally "white."

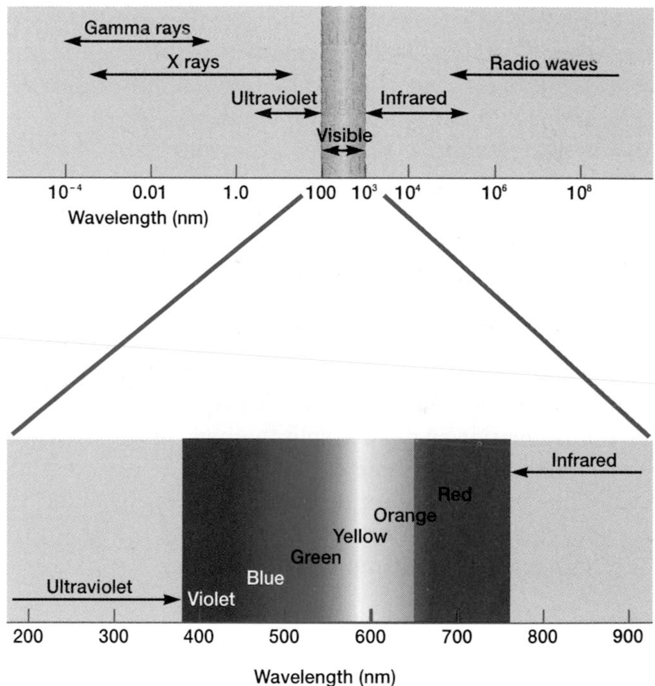

Figure 6.19 **The Electromagnetic Spectrum.** The visible portion of the spectrum is expanded at the bottom of the figure.

Microbial photosynthesis (pp. 162–68).

Many forms of electromagnetic radiation are very harmful to microorganisms. This is particularly true of **ionizing radiation,** radiation of very short wavelength or high energy, which can cause atoms to lose electrons or ionize. Two major forms of ionizing radiation are (1) X rays, which are artificially produced, and (2) gamma rays, which are emitted during radioisotope decay. Low levels of ionizing radiation will produce mutations and may indirectly result in death, while higher levels are directly lethal. Although microorganisms are more resistant to ionizing radiation than larger organisms, they will still be destroyed by a sufficiently large dose. Ionizing radiation can be used to sterilize items. Some bacteria (e.g., *Deinococcus radiodurans*) and bacterial endospores can survive large doses of ionizing radiation.

Use of radiation in destroying microorganisms (pp. 317–18).

A variety of changes in cells are due to ionizing radiation; it breaks hydrogen bonds, oxidizes double bonds, destroys ring structures, and polymerizes some molecules. Oxygen enhances these destructive effects, probably through the generation of hydroxyl radicals (OH·). Although many types of constituents can be affected, it is reasonable to suppose that destruction of DNA is the most important cause of death.

Ultraviolet radiation (UV), mentioned earlier, kills all kinds of microorganisms due to its short wavelength (approximately from 10 to 400 nm) and high energy. The most lethal UV has a wavelength of 260 nm, the wavelength most effectively absorbed by DNA. The primary mechanism of UV damage is the formation of thymine dimers in DNA. Two adjacent thymines in a DNA strand are covalently joined to inhibit DNA replication and function (*see chapter 12*). This damage is repaired in several ways. In **photoreactivation,** blue light is used by a photoreactivating enzyme to split thymine dimers. A short stretch containing the thymine dimer can also be excised and replaced. This process occurs in the absence of light and is called **dark reactivation.** When UV exposure is too heavy, the damage is so extensive that repair is impossible.

DNA repair mechanisms (pp. 253–56).

Although very little UV radiation below 290 to 300 nm reaches the earth's surface, near-UV radiation between 325 and 400 nm can harm microorganisms. Exposure to near-UV radiation induces tryptophan breakdown to toxic photoproducts. It appears that these toxic tryptophan photoproducts plus the near-UV radiation itself produce breaks in DNA strands. The precise mechanism is not known, although it is different from that seen with 260 nm UV.

Visible light is immensely beneficial because it is the source of energy for photosynthesis. Yet even visible light, when present in sufficient intensity, can damage or kill microbial cells. Usually, pigments called photosensitizers and O_2 are required. All microorganisms possess pigments like chlorophyll, bacteriochlorophyll, cytochromes, and flavins, which can absorb light energy, become excited or activated, and act as photosensitizers. The excited photosensitizer (P) transfers its energy to O_2 generating **singlet oxygen** ($^1O_2$).

$$P \xrightarrow{\text{light}} P \text{ (activated)}$$

$$P \text{ (activated)} + O_2 \longrightarrow P + \text{singlet oxygen}$$

Singlet oxygen is a very reactive, powerful oxidizing agent that will quickly destroy a cell. It is probably the major agent employed by phagocytes to destroy engulfed bacteria (*see chapter 30*).

Many microorganisms that are airborne or live on exposed surfaces use carotenoid pigments for protection against photooxidation. Carotenoids effectively quench singlet oxygen; that is, they absorb energy from singlet oxygen and convert it back into the unexcited ground state. Both photosynthetic and nonphotosynthetic microorganisms employ pigments in this way.

1. Describe the five types of O_2 relationships seen in microorganisms. For what do aerobes use O_2? Why is O_2 toxic to many microorganisms and how do they protect themselves? Describe the three ways in which anaerobes may be cultured.
2. What are barotolerant and barophilic bacteria? Where would you expect to find them?
3. List the types of electromagnetic radiation in the order of decreasing energy or increasing wavelength. How do ionizing radiation, ultraviolet radiation, and visible light harm microorganisms? How do microorganisms protect themselves against damage from UV and visible light?

Summary

1. Growth is an increase in cellular constituents and results in an increase in cell size, cell number, or both.
2. When microorganisms are grown in a closed system or batch culture, the resulting growth curve usually has four phases: the lag, exponential or log, stationary, and death phases.
3. In the exponential phase, the population number doubles at a constant interval called the doubling or generation time. The mean growth rate constant (k) is the reciprocal of the generation time.
4. Microbial populations can be counted directly with counting chambers, electronic counters, or fluorescence microscopy.
5. Viable counting techniques such as the spread plate, the pour plate, or the membrane filter can be employed.
6. Population changes also can be followed by determining variations in microbial mass through the measurement of dry weight, turbidity, or the amount of a cell component.
7. The amount of microbial mass produced from a nutrient can be expressed in terms of the growth yield (Y). Only part of the ATP generated is used in biosynthesis; the majority is used in transport, movement, and other processes.
8. Microorganisms can be grown in an open system in which nutrients are constantly provided and wastes removed.
9. A continuous culture system is an open system that can maintain a microbial population in the log phase. There are two types of these systems: chemostats and turbidostats.
10. Exponential growth is balanced growth, cell components are synthesized at constant rates relative to one another. Changes in culture conditions (e.g., in shift-up and shift-down experiments) lead to unbalanced growth. A portion of the available nutrients is used to generate maintenance energy.
11. Most bacteria, algae, and fungi have rigid cell walls and are hypertonic to the habitat because of solutes such as amino acids, polyols, and potassium ions.
12. The amount of water actually available to microorganisms is expressed in terms of the water activity (a_w).
13. Although most microorganisms will not grow well at water activities below 0.98 due to plasmolysis and associated effects, osmotolerant organisms survive and even flourish at low a_w values. Halophiles actually require high sodium chloride concentrations for growth.
14. Each species of microorganism has an optimum pH for growth and can be classified as an acidophile, neutrophile, or alkalophile.
15. Microorganisms can alter the pH of their surroundings, and most culture media must be buffered to stabilize the pH.
16. Microorganisms have distinct temperature ranges for growth with minima, maxima, and optima—the cardinal temperatures. These ranges are determined by the effects of temperature on the rates of catalysis, protein denaturation, and membrane disruption.
17. There are four major classes of microorganisms with respect to temperature preferences: (1) psychrophiles, (2) facultative psychrophiles or psychrotrophs, (3) mesophiles, and (4) thermophiles.
18. Microorganisms can be placed into at least five different categories based on their response to the presence of O_2: obligate aerobes, facultative anaerobes, aerotolerant anaerobes, strict or obligate anaerobes, and microaerophiles.
19. Oxygen is toxic because of the production of hydrogen peroxide, superoxide radical, and hydroxyl radical. These are destroyed by the enzymes superoxide dismutase and catalase.
20. Most deep-sea microorganisms are barotolerant, but some are barophilic and require high pressure for optimal growth.
21. High-energy or short-wavelength radiation harms organisms in several ways. Ionizing radiation—X rays and gamma rays—ionizes molecules and destroys DNA and other cell components.
22. Ultraviolet radiation (UV) induces the formation of thymine dimers and strand breaks in DNA. Such damage can be repaired by photoreactivation or dark reactivation mechanisms.
23. Visible light can provide energy for the formation of reactive singlet oxygen, which will destroy cells.

Key Terms

acidophile *123*
aerobe *126*
aerotolerant anaerobe *127*
alkalophile *123*
anaerobe *126*
balanced growth *121*
barophilic *129*
barotolerant *129*
batch culture *113*
cardinal temperatures *125*
catalase *128*
chemostat *120*
coenocytic *113*
colony forming units (CFU) *117*
compatible solutes *122*
continuous culture system *120*
dark reactivation *130*
death phase *114*
doubling time *114*
eurythermal *125*
exponential phase *113*
facultative anaerobe *127*
facultative psychrophiles *126*
generation time *114*
growth *113*
growth yield (Y) *119*
halophile *123*
hydrogen peroxide *128*
hydroxyl radical *128*
ionizing radiation *130*
lag phase *113*
log phase *113*
maintenance energy *120*
mean generation time *114*
mean growth rate constant (k) *114*
membrane filter *117*
mesophile *126*
microaerophile *127*
neutrophile *123*
obligate aerobe *127*
obligate anaerobe *127*
osmotolerant *122*
photoreactivation *130*
psychrophile *126*
psychrotroph *126*
singlet oxygen *130*
stationary phase *113*
stenothermal *125*
strict anaerobe *127*
superoxide dismutase *128*
superoxide radical *128*
thermophile *126*
turbidostat *120*
ultraviolet radiation (UV) *130*
unbalanced growth *121*
water activity (a_w) *122*

Questions for Thought and Review

1. Discuss the reasons why a culture might have a long lag phase after inoculation.
2. Why can't one always tell when a culture enters the death phase by the use of total cell counts?
3. Calculate the mean growth rate and generation time of a culture that increases from 5×10^2 to 1×10^8 cells in 12 hours.
4. If the generation time is 90 minutes and the initial population contains 10^3 cells, how many bacteria will there be after 8 hours of exponential growth?
5. What does the fact that the Y_{ATP} of most fermenting bacteria is about 10.5 at high sugar levels mean in terms of the efficiency of biosynthesis and the uses of ATP?
6. Why are continuous culture systems so useful to microbiologists?
7. How do bacterial populations respond in shift-up and shift-down experiments? Account for their behavior in molecular terms.
8. Does the internal pH remain constant despite changes in the external pH? How might this be achieved? Explain how extreme pH values might harm microorganisms.
9. What metabolic and structural adaptations for extreme temperatures have psychrophiles and thermophiles made?
10. Why are generation times in nature usually much longer than in culture?

Additional Reading

Atlas, R. M., and Bartha, R. 1987. *Microbial ecology: Fundamentals and applications,* 2d ed. Menlo Park, Calif.: Benjamin/Cummings.

Brock, T. D. 1978. *Thermophilic microorganisms and life at high temperatures.* New York: Springer-Verlag.

Brock, T. D. 1985. Life at high temperatures. *Science* 230:132–38.

Brock, T. D., and Darland, G. K. 1970. Limits of microbial existence: Temperature and pH. *Science* 169:1316–18.

Brown, A. D. 1976. Microbial water stress. *Bacteriol. Rev.* 40(4):803–46.

Csonka, L. N. 1989. Physiological and genetic responses of bacteria to osmotic stress. *Microbiol. Rev.* 53(1):121–47.

Fridovich, I. 1977. Oxygen is toxic! *BioScience* 27(7):462–66.

Gerhardt, P.; Murray, R. G. E.; Costilow, R. N.; Nester, E. W.; Wood, W. A.; Krieg, N. R.; and Phillips, G. B., eds. 1981. *Manual of methods for general bacteriology,* chapters 6–12. Washington, D.C.; American Society for Microbiology.

Gottschal, J. C., and Prins, R. A. 1991. Thermophiles: A life at elevated temperatures. *Trends Ecol. & Evol.* 6(5):157–62.

Harrison, D. E. F. 1978. Efficiency of microbial growth. In *Companion to microbiology,* ed. A. T. Bull and P. M. Meadow, 155–79. New York: Longman.

Inlag, J. A., and Linn, S. 1988. DNA damage and oxygen radical toxicity. *Science* 240:1302–9.

Jannasch, H. W., and Wirsen, C. O. 1977. Microbial life in the deep sea. *Sci. Am.* 236(6):42–52.

Jannasch, H. W., and Taylor, C. D. 1984. Deep-sea microbiology. *Ann. Rev. Microbiol.* 38:487–514.

Kogut, M. 1980. Are there strategies of microbial adaptation to extreme environments? *Trends Biochem. Sci.* 5(1):15–18.

Kogut, M. 1980. Microbial strategies of adaptability. *Trends Biochem. Sci.* 5(2):47–50.

Kolter, R. 1992. Life and death in stationary phase. *ASM News* 58(2):75–79.

Krieg, N. R., and Hoffman, P. S. 1986. Microaerophily and oxygen toxicity. *Ann. Rev. Microbiol.* 40:107–30.

Krulwich, T. A., and Guffanti, A. A. 1989. Alkalophilic bacteria. *Ann. Rev. Microbiol.* 43:435–63.

Kushner, D. J., ed. 1978. *Microbial life in extreme environments.* New York: Academic Press.

Le Rudulier, D.; Strom, A. R.; Dandekar, A. M.; Smith, L. T.; and Valentine, R. C. 1984. Molecular biology of osmoregulation. *Science* 224:1064–68.

Moat, A. G., and Foster, J. W. 1988. *Microbial physiology,* 2d ed. New York: John Wiley and Sons.

Morita, R. Y. 1975. Psychrophilic bacteria. *Bacteriol. Rev.* 39(2):144–67.

Neidhardt, F. C., Ingraham, J. L., and Schaechter, M. 1990. *Physiology of the bacterial cell: A molecular approach.* Sunderland, Mass.: Sinauer Associates.

Payne, W. J. 1978. Growth yield and efficiency in chemosynthetic microorganisms. *Ann. Rev. Microbiol.* 32:155–83.

Prosser, J. I., and Tough, A. J. 1991. Growth mechanisms and growth kinetics of filamentous microorganisms. Critical reviews in biotechnology 10(4):253–74.

Russell, N. J. 1984. Mechanisms of thermal adaptation in bacteria: Blueprints for survival. *Trends Biochem. Sci.* 9(3):108–12.

Schlegel, H. G., and Jannasch, H. W. 1981. Prokaryotes and their habitats. In *The prokaryotes,* ed. M. P. Starr et al., 43–82. New York: Springer-Verlag.

Stouthamer, A. H. 1977. Energetic aspects of the growth of micro-organisms. In *Microbial energetics,* ed. B. A. Haddock and W. A. Hamilton, 285–315. New York: Cambridge University Press.

Tempest, D. W. 1978. Dynamics of microbial growth. In *Essays in microbiology,* ed. J. R. Norris and M. H. Richmond, 7/1–7/32. New York: John Wiley and Sons.

Yancey, P. H.; Clark, M. E.; Hand, S. C.; Bowlus, R. D.; and Somero, G. N. 1982. Living with water stress: Evolution of osmolyte systems. *Science* 217:1214–22.

CHAPTER 7
Metabolism: Energy and Enzymes

The principles and methods of biochemistry now provide the underpinnings of all of the basic biological sciences and are the rational language for discourse in such diverse areas as ecology, clinical medicine, and agriculture.

—E. L. Smith, R. L. Hill, I. R. Lehman, and R. J. Lefkowitz

Outline

Concepts

1. Energy is the capacity to do work. Living organisms can perform three major types of work: chemical work, transport work, and mechanical work.
2. Most energy used by living organisms originally comes from sunlight trapped during photosynthesis by photoautotrophs. Chemoheterotrophs then consume autotrophic organic materials and use them as sources of energy and as building blocks.
3. An energy currency is needed to connect energy-yielding exergonic reactions with energy-requiring endergonic reactions. The most commonly used currency is ATP.
4. All living systems obey the laws of thermodynamics.
5. When electrons are transferred from a reductant with a more negative reduction potential to an oxidant with a more positive potential, energy is made available. A reversal of the direction of electron transfer—for example, during photosynthesis—requires energy input.
6. Enzymes are protein catalysts that make life possible by increasing the rate of reactions at low temperatures.
7. Enzymes do not change chemical equilibria or violate the laws of thermodynamics, but accelerate reactions by lowering their activation energy.

Chapters 3 and 4 contain many examples of an important principle: that a cell's structure is intimately related to its function. In each instance, one can readily relate an organelle's construction to its function (and vice versa). A second unifying principle in biology is that life is sustained by the trapping and use of energy, a process made possible by the action of enzymes. Because this is so crucial to our understanding of microbial function, considerable attention is given to energy and enzymes in this chapter. The metabolism of microorganisms is discussed in subsequent chapters.

This chapter begins with a brief survey of the nature of energy and the laws of thermodynamics. The participation of energy in metabolism and the role of ATP as an energy currency is considered next. The chapter closes with an introduction to the nature and function of enzymes.

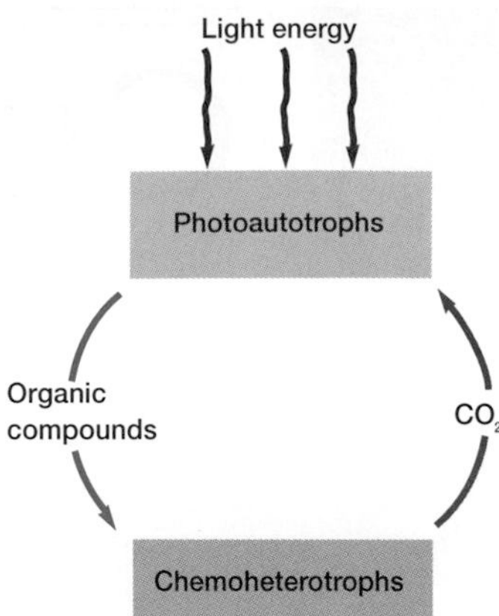

Figure 7.1 The Flow of Carbon and Energy in an Ecosystem. In this example, the flow of energy and carbon between photoautotrophs and chemoheterotrophs is shown.

Energy and Work

Energy may be most simply defined as the capacity to do work or to cause particular changes. Thus, all physical and chemical processes are the result of the application or movement of energy. Living cells carry out three major types of work, and all are essential to life processes. **Chemical work** involves the synthesis of complex biological molecules required by cells from much simpler precursors; energy is needed to increase the molecular complexity of a cell. Molecules and ions often must be transported across cell membranes against a concentration and/or electrical gradient. For example, a molecule sometimes moves into a cell even though its concentration is higher internally. Similarly, a solute may be expelled from the cell against a concentration gradient. This process is **transport work** and requires energy input in order to take up nutrients, eliminate wastes, and maintain ion balances. The third type of work is **mechanical work,** perhaps the most familiar of the three. Energy is required to change the physical location of organisms, cells, and structures within cells.

The ultimate source of all biological energy is the visible sunlight impinging on the earth's surface. Light energy is trapped during **photosynthesis,** in which it is absorbed by chlorophyll and other pigments and converted to chemical energy. Chemical energy can then be used by photoautotrophs to transform atmospheric CO_2 into glucose and other biological molecules.

Nutritional types (pp. 98–99).

The complex molecules manufactured by photosynthetic organisms (both plant and microbial producers) serve as a carbon source for chemoheterotrophs, the consumers that require complex organic molecules as a source of material and energy for building their own cellular structures. Chemoheterotrophs often use O_2 as an electron acceptor when oxidizing glucose and other organic molecules to CO_2. This process, in which O_2 acts as the final electron acceptor and is reduced to water, is called **aerobic respiration.** Much energy is released during this process. Thus, in the ecosystem, light energy is trapped by photoautotrophs; some of this energy subsequently flows to chemoheterotrophs when they use nutrients derived from photoautotrophs (figure 7.1; *see also figure 40.2*). The CO_2 produced during aerobic respiration can be incorporated again into complex organic molecules during photosynthesis. Clearly, the flow of carbon and energy in the ecosystem is intimately related.

Cells must efficiently transfer energy from their energy-generating or -trapping apparatus to the systems actually carrying out work. That is, cells must have a practical form of energy currency. In living organisms, the major currency is **adenosine 5′-triphosphate** (**ATP;** figure 7.2). When ATP breaks down to **adenosine diphosphate** (**ADP**) and orthophosphate (P_i), energy is made available for useful work. Later, energy from photosynthesis, aerobic respiration, fermentation, and other processes is used to reform ATP from ADP and P_i. An energy cycle in the cell results (figure 7.3).

Fermentation (pp. 155–58).

The Laws of Thermodynamics

To understand how energy is trapped or generated and how ATP functions as an energy currency, some knowledge of the basic principles of thermodynamics is required. The science of **thermodynamics** analyzes energy changes in a collection of matter called a system. All other matter in the universe is called the surroundings. Thermodynamics focuses on the energy differences between the initial state and the final state of a system. It is not concerned either with how the system gets from start to finish or with the rate of the process. For instance, if a pan of water is heated to boiling, only the condition of the water at the start and at boiling is important in thermodynamics, not how fast it is heated or on what kind of stove. Two important laws of thermodynamics require elucidation. The **first law** says that energy can be neither created nor destroyed. The total

(a)

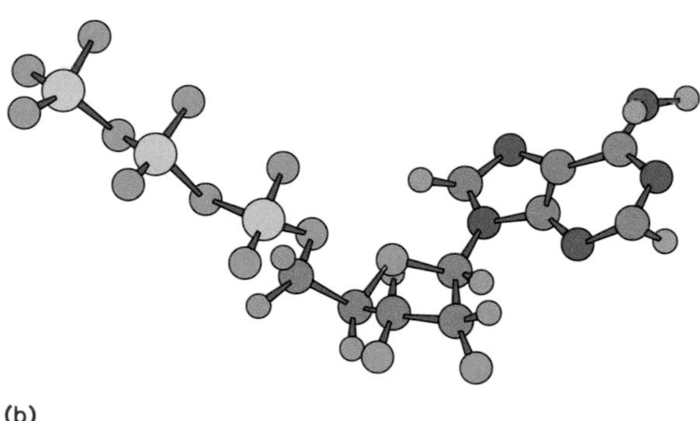

(b)

Figure 7.2 **Adenosine Triphosphate and Adenosine Diphosphate.** (*a*) Structure of ATP and ADP. The two red bonds are more easily broken or "energy rich" (see text). The pyrimidine ring atoms have been numbered. (*b*) Space-filling model of ATP. Carbon is in green; hydrogen in light blue; nitrogen in dark blue; oxygen in red; and phosphorus in orange.

energy in the universe remains constant although it can be redistributed. For example, many energy exchanges do occur during chemical reactions (e.g., heat is given off by exothermic reactions and absorbed during endothermic reactions); but these heat exchanges do not violate the first law.

It is necessary to specify quantitatively the amount of energy used in or evolving from a particular process and the two types of energy units employed. A **calorie** (cal) is the amount of heat energy needed to raise one gram of water from 14.5 to 15.5°C. The amount of energy also may be expressed in terms of **joules** (J), the units of work capable of being done. One cal of heat is equivalent to 4.1840 J of work. One thousand calories or a kilocalorie (kcal) is enough energy to boil 12 ml or about three thimbles of water. A kilojoule is enough energy to boil about 2.9 ml or 1/2 teaspoon of water, or enable a person weighing 70 kg to climb eight steps. The joule is normally used by chemists and physicists. Because biologists most often speak of energy in terms of calories, this text will employ calories when discussing energy changes.

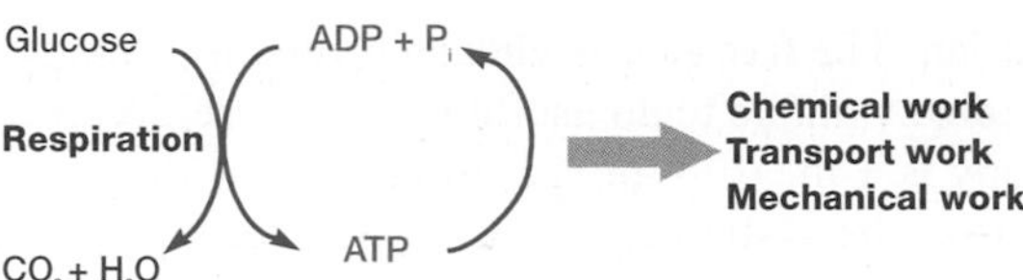

Figure 7.3 **The Cell's Energy Cycle.** *ATP* is formed from energy released during respiration. Its breakdown to *ADP* and phosphate (P_i) makes chemical, transport, and mechanical work possible.

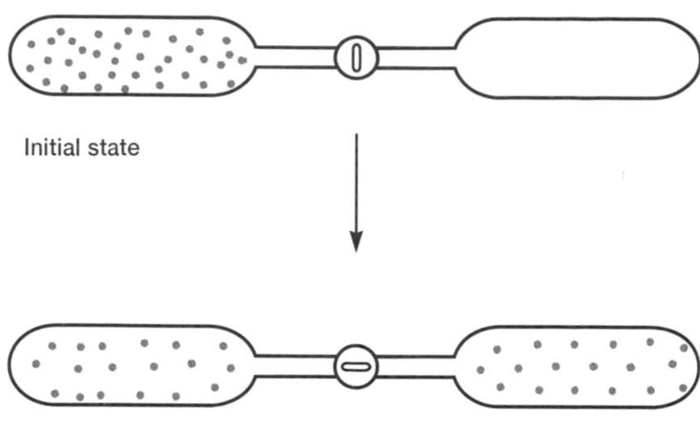

Figure 7.4 **A Second Law Process.** The expansion of gas into an empty cylinder simply redistributes the gas molecules until equilibrium is reached. The total number of molecules remains unchanged.

Although it is true that energy is conserved in the universe, the first law of thermodynamics does not account for many physical and chemical processes. A simple example may help make this clear. Suppose a full gas cylinder is connected to an empty one by a tube with a valve (figure 7.4). If the valve is opened, gas flows from the full to the empty cylinder until the gas pressure is equal on both sides. Energy has not only been redistributed but also conserved. The expansion of gas is explained by the **second law of thermodynamics** and a condition of matter called entropy. **Entropy** may be considered a measure of the randomness or disorder of a system. The greater the disorder of a system, the greater is its entropy. The second law states that physical and chemical processes proceed in such a way that the randomness or disorder of the universe (the system and its surroundings) increases to the maximum possible. Gas will always expand into an empty cylinder.

Free Energy and Reactions

The first and second laws can be combined in a useful equation, relating the changes in energy that can occur in chemical reactions and other processes.

$$\Delta G = \Delta H - T \cdot \Delta S$$

ΔG is the change in free energy, ΔH is the change in enthalpy, T is the temperature in degrees Kelvin (°C $\pm$ 273), and ΔS is the change in entropy occurring during the reaction. The change in **enthalpy** is the change in heat content, which is about the same thing as the change in total energy during the reaction. The **free energy change** is the total energy change in a system available to do useful work. Therefore, the change in entropy is a measure of the proportion of the total energy change that the system cannot use in performing work. A reaction will occur spontaneously if the free energy of the system decreases during the reaction or, in other words, if ΔG is negative. It follows from the equation that a reaction with a large positive change in entropy will normally tend to have a negative ΔG value and, therefore, occur spontaneously. A decrease in entropy will tend to make ΔG more positive and the reaction less favorable.

The change in free energy has a definite, concrete relationship to the direction of chemical reactions. Consider the following simple reaction.

$$A + B \rightleftharpoons C + D$$

If the molecules A and B are mixed, they will combine to form the products C and D. Eventually, C and D will become concentrated enough to combine and produce A and B at the same rate as they are formed from A and B. The reaction is now at **equilibrium:** the rates in both directions are equal and no further net change occurs in the concentrations of reactants and products. This situation is described by the **equilibrium constant** (**K_{eq}**), relating the equilibrium concentrations of products and substrates to one another.

$$K_{eq} = \frac{[C][D]}{[A][B]}$$

If the equilibrium constant is greater than one, the products are in greater concentration than the reactants at equilibrium; that is, the reaction tends to go to completion as written.

The equilibrium constant of a reaction is directly related to its change in free energy. When the free energy change for a process is determined at carefully defined standard conditions of concentration, pressure, pH, and temperature, it is called the **standard free energy change** ($\Delta G°$). If the pH is set at 7.0 (which is close to the pH of living cells), the standard free energy change is indicated by the symbol $\Delta G°'$. The change in standard free energy may be thought of as the maximum amount of energy available from the system for useful work under standard conditions. The use of $\Delta G°'$ values allows one to compare reactions without worrying about variations in the ΔG due to differences in environmental conditions. The relationship between $\Delta G°'$ and K_{eq} is given by the following equation.

$$\Delta G°' = -2.303RT \cdot \log K_{eq}$$

R is the gas constant (1.9872 cal/mole-degree or 8.3145 J/mole-degree), and T is the absolute temperature. Inspection of this equation shows that when $\Delta G°'$ is negative, the equilibrium constant is greater than one and the reaction goes to completion as written. It is said to be an **exergonic reaction** (figure 7.5). In an **endergonic reaction,** $\Delta G°'$ is positive and the equilibrium constant is less than one. That is, the reaction is not favorable, and little product will be formed at equilibrium under standard conditions. Keep in mind that the $\Delta G°'$ value shows only where the reaction lies at equilibrium, not how fast the reaction reaches equilibrium.

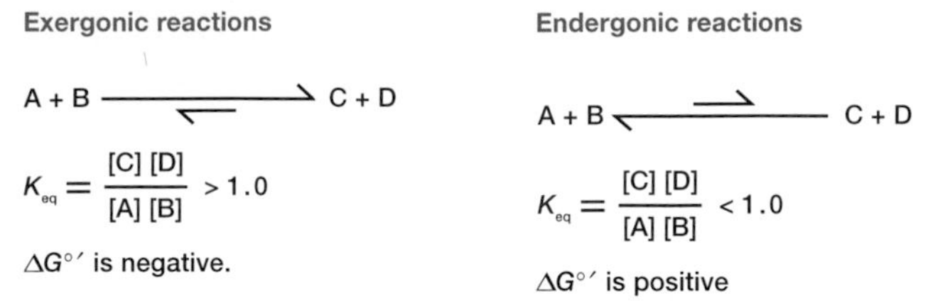

Exergonic reactions

$$A + B \rightleftharpoons C + D$$

$$K_{eq} = \frac{[C]\,[D]}{[A]\,[B]} > 1.0$$

$\Delta G°'$ is negative.

Endergonic reactions

$$A + B \rightleftharpoons C + D$$

$$K_{eq} = \frac{[C]\,[D]}{[A]\,[B]} < 1.0$$

$\Delta G°'$ is positive

Figure 7.5 **$\Delta G°'$ and Equilibrium.** The relationship of $\Delta G°'$ to the equilibrium of reactions. Note the differences between exergonic and endergonic reactions.

1. What is energy and what kinds of work are carried out in a cell? Describe the energy cycle and ATP's role in it.
2. What is thermodynamics? Summarize the first and second laws. Define free energy, entropy, and enthalpy.
3. How is the change in standard free energy related to the equilibrium constant for a reaction? What are exergonic and endergonic reactions?

The Role of ATP in Metabolism

Many reactions in the cell are endergonic and will not proceed far toward completion without outside assistance. One of ATP's major roles is to drive such endergonic reactions more to completion. ATP is a **high-energy molecule.** That is, it breaks down or hydrolyzes almost completely to the products ADP and P_i with a $\Delta G°'$ of -7.3 kcal/mole.

$$ATP + H_2O \rightleftharpoons ADP + P_i$$

In reference to ATP, the term high-energy molecule does not mean that there is a great deal of energy stored in a particular bond of ATP. It simply indicates that the removal of the terminal phosphate goes to completion with a large negative standard free energy change, or the reaction is strongly exergonic.

Thus, ATP is ideally suited for its role as an energy currency. It is formed in energy-trapping and -generating processes such as photosynthesis, fermentation, and aerobic respiration. In the cell's economy, exergonic ATP breakdown is coupled with various endergonic reactions to promote their completion (figure 7.6). In other words, ATP links energy-generating reactions, which liberate free energy, with energy-using reactions, which require free energy input to proceed toward completion. Facilitation of chemical work is the focus of the preceding example, but the same principles apply when ATP is coupled with endergonic processes involving transport work and mechanical work (figure 7.3).

Endergonic reaction alone

A + B ⇌ C + D

Endergonic reaction coupled to ATP breakdown

Figure 7.6 **ATP as a Coupling Agent.** The use of ATP to make endergonic reactions more favorable. It is formed by exergonic reactions and then used to drive endergonic reactions.

TABLE 7.1 Selected Biologically Important Redox Couples

Redox Couple	E'_o (Volts)[a]
$2H^+ + 2e^- \longrightarrow H_2$	−0.42
Ferredoxin(Fe^{3+}) + e^- ⟶ ferredoxin(Fe^{2+})	−0.42
$NAD(P)^+ + H^+ + 2e^- \longrightarrow NAD(P)H$	−0.32
$S + 2H^+ + 2e^- \longrightarrow H_2S$	−0.274
Acetaldehyde + $2H^+ + 2e^-$ ⟶ ethanol	−0.197
Pyruvate$^-$ + $2H^+ + 2e^-$ ⟶ lactate^{2-}	−0.185
$FAD + 2H^+ + 2e^- \longrightarrow FADH_2$	−0.18[b]
Oxaloacetate^{2-} + $2H^+ + 2e^-$ ⟶ malate^{2-}	−0.166
Fumarate^{2-} + $2H^+ + 2e^-$ ⟶ succinate^{2-}	0.031
Cytochrome *b* (Fe^{3+}) + e^- ⟶ cytochrome *b* (Fe^{2+})	0.075
Ubiquinone + $2H^+ + 2e^-$ ⟶ ubiquinone H_2	0.10
Cytochrome *c* (Fe^{3+}) + e^- ⟶ cytochrome *c* (Fe^{2+})	0.254
$NO_3^- + 2H^+ + 2e^- \longrightarrow NO_2^- + H_2O$	0.421
$NO_2^- + 8H^+ + 6e^- \longrightarrow NH_4^+ + 2H_2O$	0.44
$Fe^{3+} + e^- \longrightarrow Fe^{2+}$	0.771
$O_2 + 4H^+ + 4e^- \longrightarrow 2H_2O$	0.815

[a] E'_o is the standard reduction potential at pH 7.0.
[b] The value for $FAD/FADH_2$ applies to the free cofactor because it can vary considerably when bound to an apoenzyme.

Oxidation-Reduction Reactions and Electron Carriers

Free energy changes are not only related to the equilibria of "regular" chemical reactions but also to the equilibria of oxidation-reduction reactions. **Oxidation-reduction (redox) reactions** are those in which electrons move from a donor, the **reducing agent** or **reductant,** to an electron acceptor, the **oxidizing agent** or **oxidant.** By convention, such a reaction is written with the reductant to the right of the oxidant and the number (n) of electrons (e^-) transferred.

$$\text{Oxidant} + ne^- \rightleftharpoons \text{reductant}$$

The oxidant and reductant pair is referred to as a redox couple (table 7.1). When an oxidant accepts electrons, it becomes the reductant of the couple. The equilibrium constant for the reaction is called the **standard reduction potential** (E_o) and is a measure of the tendency of the reducing agent to lose electrons. The reference standard for reduction potentials is the hydrogen system with an E'_o (the reduction potential at pH 7.0) of −0.42 volts or −420 millivolts.

$$2H^+ + 2e^- \rightleftharpoons H_2$$

In this reaction, each hydrogen atom provides one proton (H^+) and one electron (e^-).

The reduction potential has a concrete meaning. Redox couples with more negative reduction potentials will donate electrons to couples with more positive potentials. Thus, electrons will tend to move from reductants at the top of the list in table 7.1 to oxidants at the bottom because they have more positive potentials. Consider the case of the electron carrier **nicotinamide adenine dinucleotide** (NAD^+). The NAD^+/NADH couple has a very negative E'_o and can therefore give electrons to many acceptors, including O_2.

$$NAD^+ + 2H^+ + 2e^- \rightleftharpoons NADH + H^+ \qquad E'_o = -0.32 \text{ volts}$$

$$1/2\, O_2 + 2H^+ + 2e^- \rightleftharpoons H_2O \qquad E'_o = +0.82 \text{ volts}$$

Because NAD^+/NADH is more negative than 1/2 O_2/H_2O, electrons will flow from NADH (the reductant) to O_2 (the oxidant).

$$NADH + H^+ + 1/2\, O_2 \longrightarrow H_2O + NAD^+$$

When electrons move from a reductant to an acceptor with a more positive redox potential, free energy is released. The $\Delta G^{\circ\prime}$ of the reaction is directly related to the magnitude of the difference between the reduction potentials of the two couples ($\Delta E'_o$). The larger $\Delta E'_o$, the greater the amount of free energy made available, as is evident from the equation

$$\Delta G^{\circ\prime} = -nF \cdot \Delta E'_o$$

in which n is the number of electrons transferred and F is the Faraday constant (23,062 cal/mole-volt or 96,494 J/mole-volt). For every 0.1 volt change in $\Delta E'_o$, there is a corresponding 4.6 kcal change in $\Delta G^{\circ\prime}$ when a two-electron transfer takes place. This is similar to the relationship of $\Delta G^{\circ\prime}$ and K_{eq} in other chemical reactions—the larger the equilibrium constant, the greater the $\Delta G^{\circ\prime}$. The difference in reduction potentials between NAD^+/NADH and 1/2 O_2/H_2O is 1.14 volts, a large $\Delta E'_o$ value. When electrons move from NADH to O_2 during aerobic respiration, a large amount of free energy is made available to synthesize ATP (figure 7.7). If energy is released when electrons flow from negative to positive reduction potentials, then an input of energy is required to move electrons in the opposite direction, from more positive to more negative potentials. This is precisely what occurs during photosynthesis (figure 7.7). Light energy is trapped and used to move electrons from water to the electron carrier **nicotinamide adenine dinucleotide phosphate** (**$NADP^+$**).

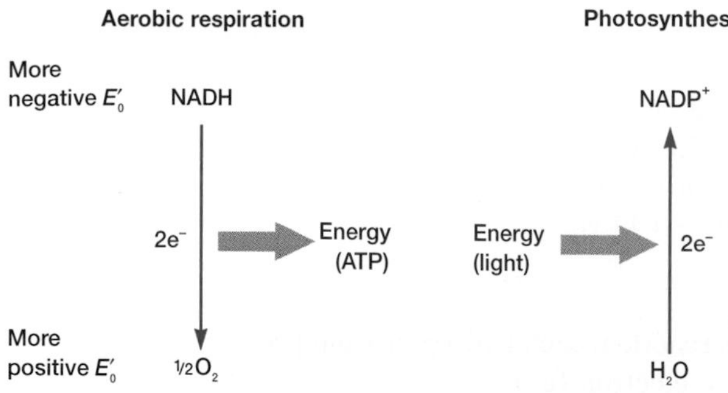

Figure 7.7 Energy Flow in Metabolism. The relationship between electron flow and energy in metabolism. Oxygen and $NADP^+$ serve as electron acceptors for NADH and water, respectively.

The cycle of energy flow discussed earlier and illustrated in figure 7.1 can be understood from a different perspective, if the preceding concept is kept in mind. Photosynthetic organisms capture light energy and use it to move electrons from water (and other electron donors) to electron acceptors, such as $NADP^+$, that have more negative reduction potentials. These electrons can then flow back to more positive acceptors and provide energy for ATP production during photosynthesis. Photoautotrophs use ATP and NADPH to synthesize complex molecules from CO_2 (*see chapter 8*). Chemoheterotrophs also make use of energy released during the movement of electrons by oxidizing complex nutrients during respiration to produce NADH. NADH subsequently donates its electrons to O_2, and the energy released during electron transfer is trapped in the form of ATP. The energy from sunlight is made available to all living organisms because of this relationship between electron flow and energy.

Photosynthesis (pp. 162–68).
Respiration and electron transport (pp. 152–54).

Electron movement in cells requires the participation of carriers such as NAD^+ and $NADP^+$, both of which can transport electrons between different locations. The nicotinamide ring of NAD^+ and $NADP^+$ (figure 7.8) accepts two electrons and one proton from a donor, while a second proton is released. There are several other electron carriers of importance in microbial metabolism (table 7.1), and they carry electrons in a variety of ways. **Flavin adenine dinucleotide** (**FAD**) and **flavin mononucleotide** (**FMN**) bear two electrons and two protons on the complex ring system shown in figure 7.9. Proteins bearing FAD and FMN are often called flavoproteins. **Coenzyme Q** (**CoQ**) or **ubiquinone** is a quinone that transports electrons and protons in many respiratory electron transport chains (figure 7.10). **Cytochromes** and several other carriers use iron atoms to transport electrons by reversible oxidation and reduction reactions.

$$Fe^{3+} \text{ (ferric iron)} + e^- \rightleftharpoons Fe^{2+} \text{ (ferrous iron)}$$

In the cytochromes, these iron atoms are part of a heme group (figure 7.11) or other similar iron-porphyrin rings. Several different cytochromes, each of which consists of a protein and an iron-porphyrin ring, are a prominent part of respiratory electron transport chains. Some iron containing electron-carrying proteins lack a heme group and are called **nonheme iron proteins. Ferredoxin** is a nonheme iron protein active in photosynthetic electron transport. Even though its iron atoms are not bound to a heme group, they still undergo reversible oxidation and reduction reactions.

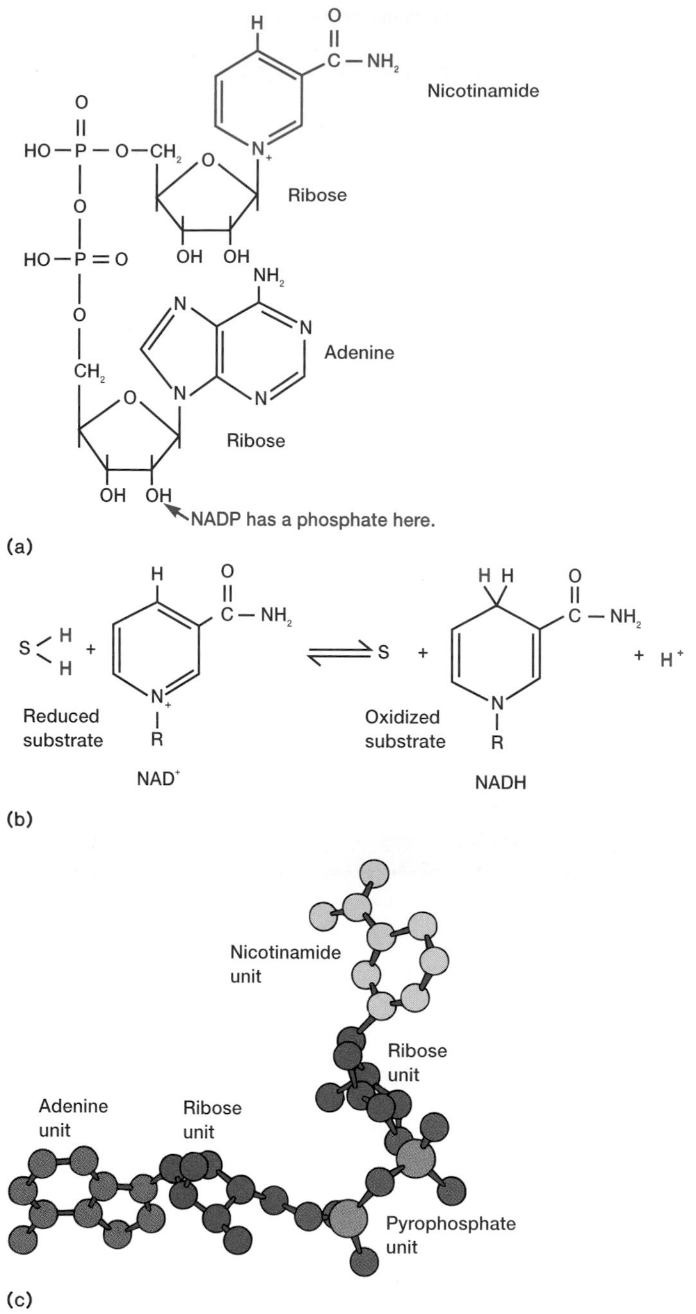

Figure 7.8 The Structure and Function of NAD. (*a*) The structure of NAD and NADP. NADP differs from NAD in having an extra phosphate on one of its ribose sugar units. (*b*) NAD can accept electrons and a hydrogen from a reduced substrate (SH_2). These are carried on the nicotinamide ring. (*c*) Model of NAD^+ when bound to the enzyme lactate dehydrogenase.

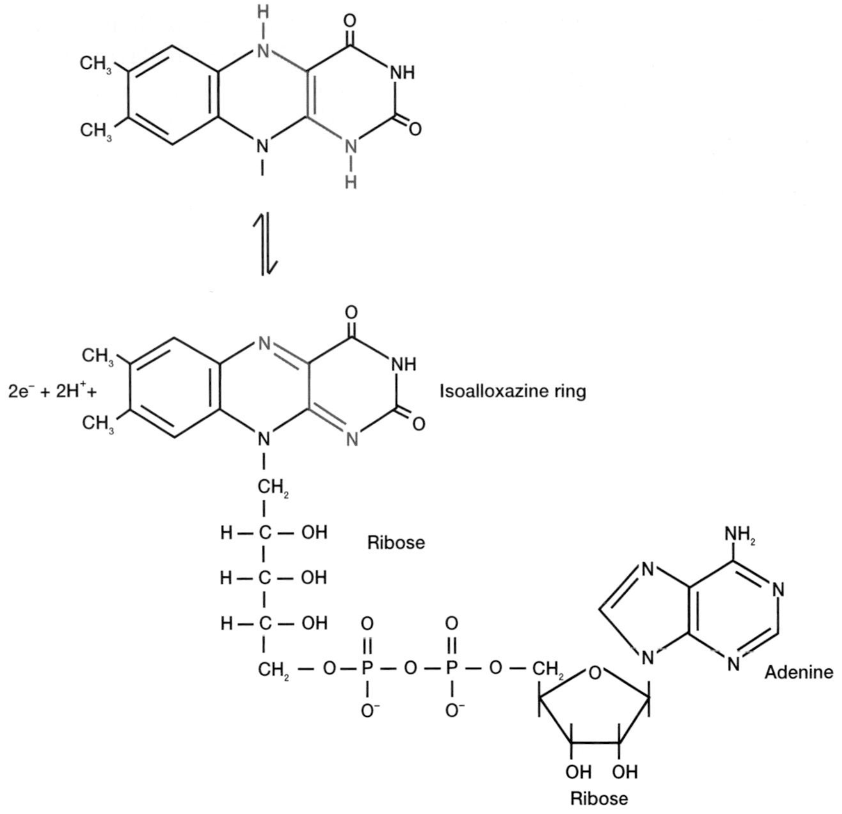

Figure 7.9 The Structure and Function of FAD. The vitamin riboflavin is composed of the isoalloxazine ring and its attached ribose sugar. FMN is riboflavin phosphate. The portion of the ring directly involved in oxidation-reduction reactions is in color.

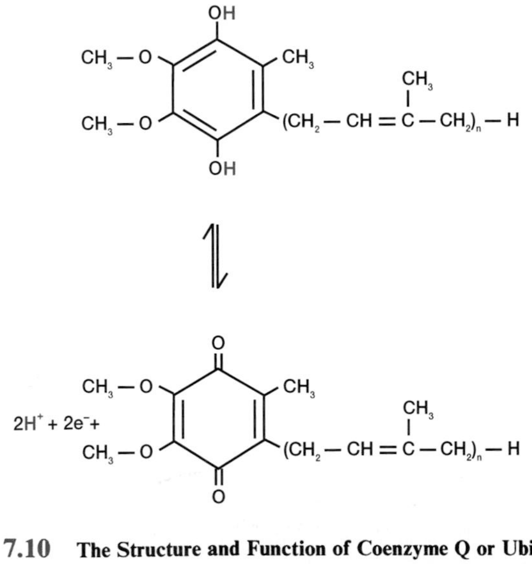

Figure 7.10 The Structure and Function of Coenzyme Q or Ubiquinone. The length of the side chain varies among organisms from n = 6 to n = 10.

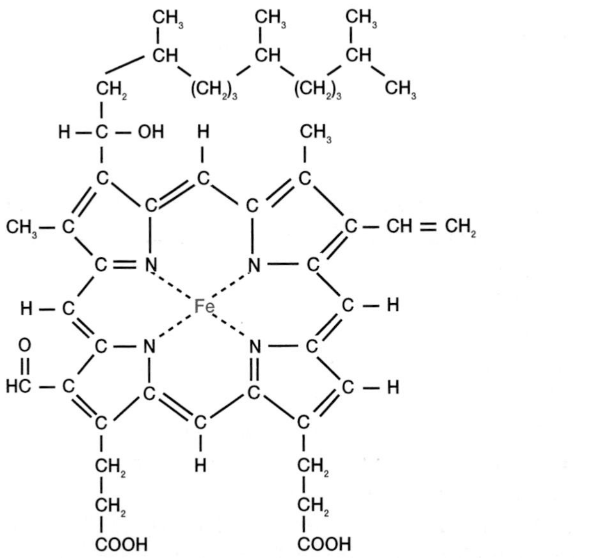

Figure 7.11 The Structure of Heme. Heme is composed of a porphyrin ring and an attached iron atom. It is the nonprotein component of many cytochromes. The iron atom alternatively accepts and releases an electron.

---

1. Why is ATP called a high-energy molecule? What is its role in the cell and how does it fulfill this role?
2. Write a generalized equation for a redox reaction. Define reductant, oxidant, and standard reduction potential.
3. How is the direction of electron flow between redox couples related to the standard reduction potential and the release of free energy? Name and briefly describe the major electron carriers found in cells.

Enzymes

Recall that an exergonic reaction is one with a negative $\Delta G^{\circ\prime}$ and an equilibrium constant greater than one. An exergonic reaction will proceed to completion in the direction written (that is, toward the right of the equation). Nevertheless, one can often combine the reactants for an exergonic reaction with no obvious result, even though products should be formed. It is precisely in these reactions that enzymes play their part.

Structure and Classification of Enzymes

Enzymes may be defined as protein catalysts that have great specificity for the reaction catalyzed and the molecules acted upon. A **catalyst** is a substance that increases the rate of a chemical reaction without being permanently altered itself. Thus, enzymes speed up cellular reactions. The reacting molecules are called **substrates,** and the substances formed are the **products.**

Protein structure and properties (appendix I).

Many enzymes are indeed pure proteins. However, many enzymes consist of a protein, the **apoenzyme,** and also a nonprotein component, a **cofactor,** required for catalytic activity. The complete enzyme consisting of the apoenzyme and its cofactor is called the **holoenzyme.** If the cofactor is firmly attached to the apoenzyme it is a **prosthetic group.** Often, the cofactor is loosely attached to the apoenzyme. It can even dissociate from the enzyme protein after products have been formed and carry one of these products to another enzyme (figure 7.12). Such a loosely bound cofactor is called a **coenzyme.** For example, NAD^+ is a coenzyme that carries electrons within the cell. Many vitamins that humans require serve as coenzymes or as their precursors. Niacin is incorporated into NAD^+ and riboflavin into FAD. Metal ions may also be bound to apoenzymes and act as cofactors.

Despite the large number and bewildering diversity of enzymes present in cells, they may be placed in one of six general classes (table 7.2). Enzymes usually are named in terms of the substrates they act on and the type of reaction catalyzed. For example, lactate dehydrogenase removes hydrogens from lactic acid.

$$\text{Lactate} + NAD^+ \rightleftharpoons \text{pyruvate} + NADH + H^+$$

Lactate dehydrogenase can also be given a more complete and detailed name, L-lactate: NAD oxidoreductase. This name describes the substrates and reaction type with even more precision.

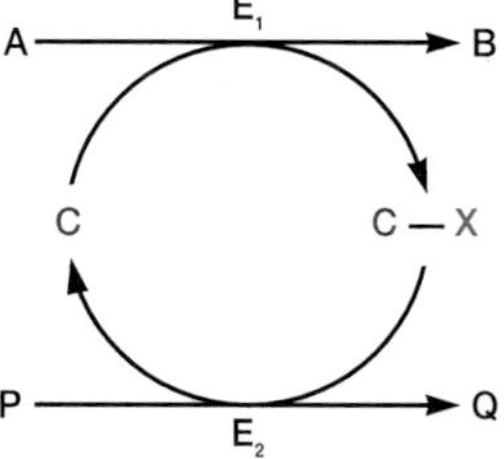

Figure 7.12 **Coenzymes as Carriers.** The function of a coenzyme in carrying substances around the cell. Coenzyme C participates with enzyme E_1 in the conversion of A to product B. During the reaction, it acquires X from the substrate A. The coenzyme can donate X to substrate P in a second reaction. This will convert it back to its original form, ready to accept another X. The coenzyme is not only participating in both reactions, but is also transporting X to various points in the cell.

TABLE 7.2 Enzyme Classification

Type of Enzyme	Reaction Catalyzed by Enzyme	Example of Reaction
Oxidoreductase	Oxidation-reduction reactions	*Lactate dehydrogenase* Pyruvate + NADH + H^+ ⇌ lactate + NAD^+
Transferase	Reactions involving the transfer of groups between molecules	*Aspartate carbamoyltransferase* Aspartate + carbamoylphosphate ⇌ carbamoylaspartate + phosphate
Hydrolase	Hydrolysis of molecules	*Glucose-6-phosphatase* Glucose 6-phosphate + H_2O → glucose + P_i
Lyase	Removal of groups to form double bonds or addition of groups to double bonds $>C{=}C< + X{-}Y \rightleftharpoons -\overset{X}{\underset{\mid}{\overset{\mid}{C}}}-\overset{Y}{\underset{\mid}{\overset{\mid}{C}}}-$	*Fumarate hydratase* L-malate ⇌ fumarate + H_2O
Isomerase	Reactions involving isomerizations	*Alanine racemase* L-alanine ⇌ D-alanine
Ligase	Joining of two molecules using ATP energy (or that of other nucleoside triphosphates)	*Glutamine synthetase* Glutamate + NH_3 + ATP → glutamine + ADP + P_i

The Mechanism of Enzyme Reactions

It is important to keep in mind that enzymes increase the rates of reactions, but do not alter their equilibrium constants. If a reaction is endergonic, the presence of an enzyme will not shift its equilibrium so that more products can be formed. Enzymes simply speed up the rate at which a reaction proceeds toward its final equilibrium.

How do enzymes catalyze reactions? Although a complete answer would be long and complex, some understanding of the mechanism can be gained by considering the course of a normal exergonic chemical reaction.

$$A + B \rightleftharpoons C + D$$

When molecules A and B approach each other to react, they form a **transition-state complex,** which resembles both the substrates and the products (figure 7.13). The **activation energy** is required to bring the reacting molecules together in the correct way to reach the transition state. The transition-state complex can then decompose to yield the products C and D. The difference in free energy level between reactants and products is $\Delta G^{\circ\prime}$. Thus, the equilibrium in our example will lie toward the products because $\Delta G^{\circ\prime}$ is negative (i.e., the products are at a lower energy level than the substrates).

Clearly, A and B will not be converted to C and D in figure 7.13 if they are not supplied with an amount of energy equivalent to the activation energy. Enzymes accelerate reactions by lowering the activation energy; therefore, more substrate molecules will have sufficient energy to come together and form products. Even though the equilibrium constant (or $\Delta G^{\circ\prime}$) is unchanged, equilibrium will be reached more rapidly in the presence of an enzyme because of this decrease in the activation energy.

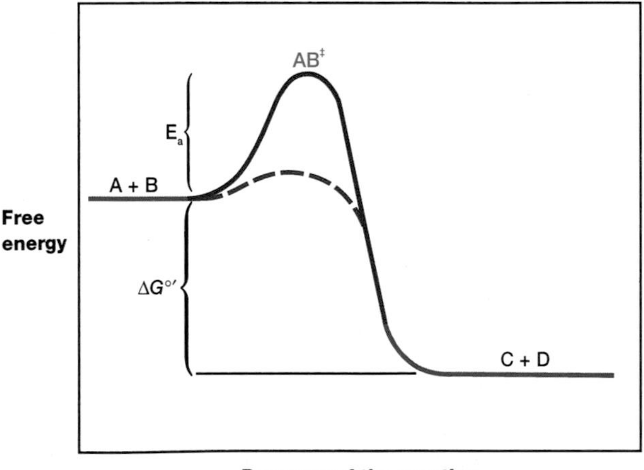

Figure 7.13 Enzymes Lower the Energy of Activation. This figure traces the course of a chemical reaction in which A and B are converted to C and D. The transition-state complex is represented by AB‡, and the activation energy required to reach it, by E_a. The colored line represents the course of the reaction in the presence of an enzyme. Note that the activation energy is much lower in the enzyme-catalyzed reaction.

Researchers have expended much effort in discovering how enzymes lower the activation energy of reactions, and the process is becoming clearer. Enzymes bring substrates together at a special place on their surface called the **active site** or **catalytic site** to form an **enzyme-substrate complex** (figures 7.14, 7.15; *see also AI.19*). The formation of an enzyme-substrate complex can lower the activation energy in many ways. For example, by bringing the substrates together at the active site, the enzyme is, in effect, concentrating them and speeding up the reaction. An enzyme does not simply concentrate its substrates, however. It also binds them so that they are correctly oriented with respect to each other in order to form a transition-state complex. Such an orientation lowers the amount of energy that the substrates require to reach the transition state. These and other catalytic site activities speed up a reaction hundreds of thousands of times, even though the reaction usually takes place at temperatures ranging from 0 to 37°C. These temperatures are not high enough to favor most organic reactions in the absence of enzyme catalysis; yet cells cannot survive at the high temperatures used by an organic chemist in routine organic syntheses. Enzymes make life possible by accelerating specific reactions at low temperatures.

The Effect of Environment on Enzyme Activity

Enzyme activity varies greatly with changes in environmental factors, one of the most important being the substrate concentration. At very low substrate concentrations, an enzyme makes product slowly because it seldom contacts a substrate

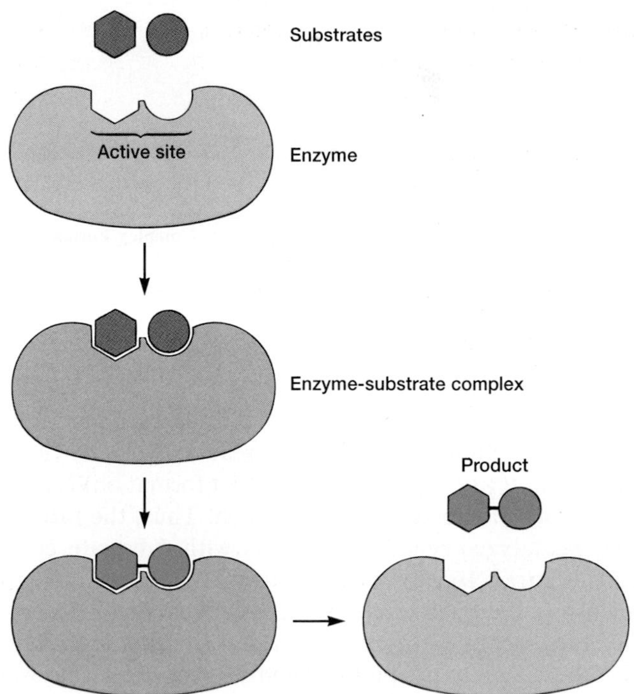

Figure 7.14 Enzyme Function. The formation of the enzyme-substrate complex and its conversion to products is shown.

(a)

(b)

Figure 7.15 **An Example of Enzyme-substrate Complex Formation.** (*a*) A space-filling model of yeast hexokinase and its substrate glucose (purple). The active site is in the cleft formed by the enzyme's small lobe (green) and large lobe (gray). (*b*) When glucose binds to form the enzyme-substrate complex, hexokinase changes shape and surrounds the substrate.

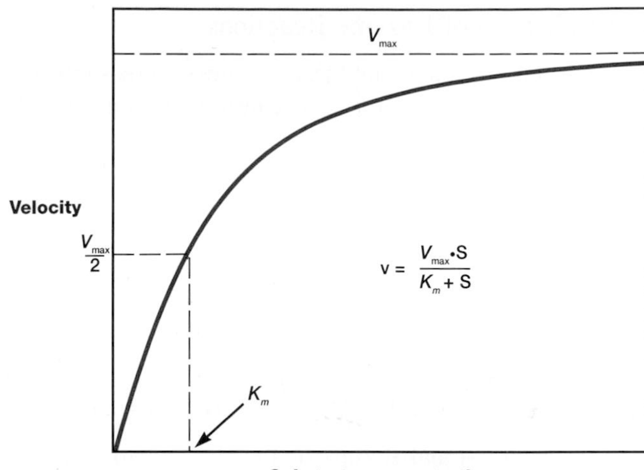

Figure 7.16 **Michaelis-Menten Kinetics.** The dependence of enzyme activity upon substrate concentration. The maximum velocity (V_{max}) is the rate of product formation when the enzyme is saturated with substrate and making product as fast as possible. The Michaelis constant (K_m) is the substrate concentration required for the enzyme to operate at half its maximal velocity. This substrate curve fits the Michaelis-Menten equation given in the figure, which relates reaction velocity (v) to the substrate concentration (S).

molecule. If more substrate molecules are present, an enzyme binds substrate more often, and the reaction velocity (usually expressed in terms of the rate of product formation) is greater than at a lower substrate concentration. Thus, the rate of an enzyme-catalyzed reaction increases with substrate concentration (figure 7.16). Eventually, further increases in substrate concentration do not result in a greater reaction velocity because the available enzyme molecules are binding substrate and converting it to product as rapidly as possible. That is, the enzyme is saturated with substrate and operating at maximal velocity (V_{max}). The resulting substrate concentration curve is a hyperbola (figure 7.16). It is useful to know the substrate concentration an enzyme needs to function adequately. Usually the **Michaelis constant (K_m)**, the substrate concentration required for the enzyme to achieve half maximal velocity, is used as a measure of the apparent affinity of an enzyme for its substrate. The lower the K_m value, the lower the substrate concentration at which an enzyme catalyzes its reaction.

Enzymes also change activity with alterations in pH and temperature (figure 7.17). Each enzyme functions most rapidly at a specific pH optimum. When the pH deviates too greatly from an enzyme's optimum, activity slows and the enzyme may be damaged. Enzymes likewise have temperature optima for maximum activity. If the temperature rises too much above the optimum, an enzyme's structure will be disrupted and its activity lost. This phenomenon, known as **denaturation,** may be caused by extremes of pH and temperature, or by other factors. The pH and temperature optima of a microorganism's enzymes often reflect the pH and temperature of its habitat. Not surprisingly, bacteria growing best at high temperatures often have enzymes with high temperature optima and great heat stability.

Enzyme Inhibition

Microorganisms can be poisoned by a variety of chemicals, and many of the most potent poisons are enzyme inhibitors. A **competitive inhibitor** directly competes with the substrate at an enzyme's catalytic site and prevents the enzyme from forming product. A classic example of this behavior is seen with the enzyme succinate dehydrogenase, which catalyzes the oxidation of succinate to fumarate in the tricarboxylic acid cycle (*see chapter 8*). Malonic acid is an effective competitive

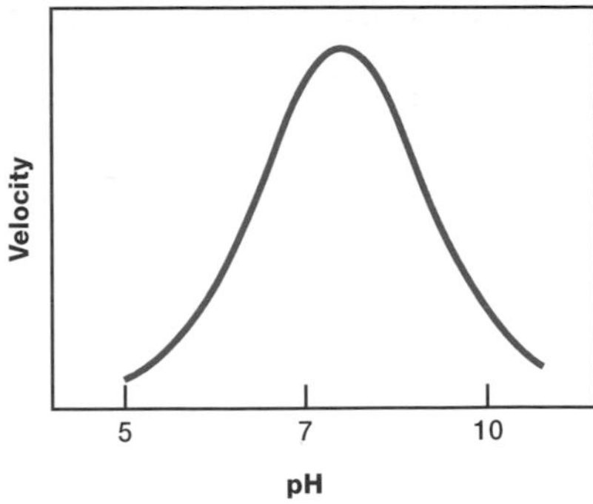

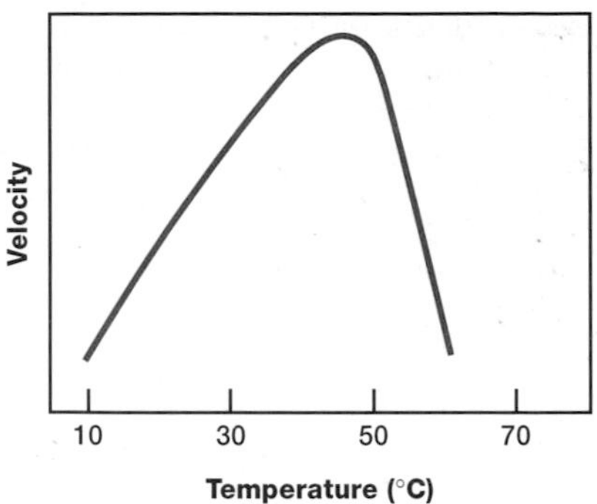

Figure 7.17 **pH, Temperature, and Enzyme Activity.** The variation of enzyme activity with changes in pH and temperature. The ranges in pH and temperature are only representative. Enzymes differ from one another with respect to the location of their optima and the shape of their pH and temperature curves.

inhibitor of succinate dehydrogenase because it so closely resembles succinate, the normal substrate (figure 7.18). After malonate binds to the enzyme, it cannot be oxidized and the formation of fumarate is blocked. Competitive inhibitors usually resemble normal substrates, but they cannot be converted to products.

COOH
|
CH_2
|
CH_2
|
COOH
Succinic acid

COOH
|
CH_2
|
COOH
Malonic acid

Figure 7.18 **Competitive Inhibition of Succinate Dehydrogenase.** A comparison of succinic acid and the competitive inhibitor, malonic acid. The colored atoms indicate the parts of the two molecules that differ.

Destruction of microorganisms by physical and chemical agents (chapter 15).

Competitive inhibitors are important in the treatment of many microbial diseases. Sulfa drugs like sulfanilamide resemble p-aminobenzoate, a molecule used in the formation of the coenzyme folic acid. The drugs compete with p-aminobenzoate for the catalytic site of an enzyme involved in folic acid synthesis. This blocks the production of folic acid and inhibits bacterial growth (*see chapter 16*). Humans are not harmed because they cannot synthesize folic acid and must obtain it in their diet.

Inhibitors also can affect enzyme activity by binding to the enzyme at some location other than at the active site. This alters the enzyme's shape, rendering it inactive or less active. These inhibitors are often called **noncompetitive** because they do not directly compete with the substrate. Heavy metal poisons like mercury frequently are noncompetitive inhibitors of enzymes.

1. What is an enzyme and how does it speed up reactions? How are enzymes named? Define apoenzyme, holoenzyme, cofactor, coenzyme, prosthetic group, active or catalytic site, and activation energy.
2. How does enzyme activity change with substrate concentration, pH, and temperature? Define Michaelis constant, maximum velocity, and denaturation.
3. What are competitive and noncompetitive inhibitors and how do they inhibit enzymes?

Summary

1. Energy is the capacity to do work. Living cells carry out three major kinds of work: chemical work of biosynthesis, transport work, and mechanical work.
2. The ultimate source of energy is sunlight trapped by photoautotrophs and used to form organic material from CO_2. Photoautotrophs are then consumed by chemoheterotrophs.
3. ATP is the major energy currency and connects energy-generating processes with energy-utilizing processes.
4. The first law of thermodynamics states that energy is neither created nor destroyed.
5. The second law of thermodynamics states that changes occur in such a way that the randomness or disorder of the universe increases to the maximum possible. That is, entropy always increases during spontaneous processes.
6. The first and second laws can be combined to determine the amount of energy made available for useful work.

$$\Delta G = \Delta H - T \cdot \Delta S$$

In this equation, the change in free energy (ΔG) is the energy made available for useful work, the change in enthalpy (ΔH) is the change in heat content, and the change in entropy is ΔS.

7. The standard free energy change ($\Delta G^{\circ\prime}$) for a chemical reaction is directly related to the equilibrium constant.
8. In exergonic reactions, $\Delta G^{\circ\prime}$ is negative and the equilibrium constant is greater than one; the reaction goes to completion as written. Endergonic reactions have a positive $\Delta G^{\circ\prime}$ and an equilibrium constant less than one.
9. In oxidation-reduction (redox) reactions, electrons move from a donor, the reducing agent or reductant, to an acceptor, the oxidizing agent or oxidant. The standard reduction potential measures the tendency of the reducing agent to give up electrons.

10. Redox couples with more negative reduction potentials donate electrons to those with more positive potentials, and energy is made available during the transfer.
11. Some most important electron carriers in cells are NAD^+, $NADP^+$, FAD, FMN, coenzyme Q, cytochromes, and the nonheme iron proteins.
12. Enzymes are protein catalysts with great specificity for the reaction catalyzed and for their substrates and products.
13. Enzymes consist of a protein component, the apoenzyme, and often a nonprotein cofactor that may be a prosthetic group, a coenzyme, or a metal activator.
14. Enzymes speed reactions by binding substrates at their active sites and lowering the activation energy.
15. The rate of an enzyme-catalyzed reaction increases with substrate concentration at low substrate levels and reaches a plateau (the maximum velocity) at saturating substrate concentrations. The Michaelis constant is the substrate concentration that the enzyme requires to achieve half maximal velocity.
16. Enzymes have pH and temperature optima for activity.
17. Enzyme activity can be slowed by competitive inhibitors and noncompetitive inhibitors.

Key Terms

activation energy *141*
active site *141*
adenosine diphosphate (ADP) *134*
adenosine 5′-triphosphate (ATP) *134*
aerobic respiration *134*
apoenzyme *140*
calorie *135*
catalyst *140*
catalytic site *141*
chemical work *134*
coenzyme *140*
coenzyme Q or CoQ (ubiquinone) *138*
cofactor *140*
competitive inhibitor *142*
cytochrome *138*
denaturation *142*
endergonic reactions *136*
energy *134*
enthalpy *135*
entropy *135*
enzyme *140*
enzyme-substrate complex *141*
equilibrium *136*
equilibrium constant (K_{eq}) *136*
exergonic reactions *136*
ferredoxin *138*
first law of thermodynamics *134*
flavin adenine dinucleotide (FAD) *138*
flavin mononucleotide (FMN) *138*
free energy change *136*
high-energy molecule *136*
holoenzyme *140*
joule *135*
mechanical work *134*
Michaelis constant (K_m) *142*
nicotinamide adenine dinucleotide (NAD^+) *137*
nicotinamide adenine dinucleotide phosphate ($NADP^+$) *137*
noncompetitive inhibitor *143*
nonheme iron protein *138*
oxidation-reduction (redox) reaction *137*
oxidizing agent (oxidant) *137*
photosynthesis *134*
product *140*
prosthetic group *140*
reducing agent (reductant) *137*
second law of thermodynamics *135*
standard free energy change *136*
standard reduction potential *137*
substrate *140*
thermodynamics *134*
transition-state complex *141*
transport work *134*

Questions for Thought and Review

1. Describe in general terms how energy from sunlight is spread throughout the biosphere.
2. Under what conditions would it be possible to create more order in a system without violating the second law of thermodynamics?
3. Suppose that a chemical reaction had a large negative $\Delta G^{\circ\prime}$ value. What would this indicate about its equilibrium constant? If displaced from equilibrium, would it proceed rapidly to completion? Would much or little free energy be made available?
4. Will electrons ordinarily move in an electron transport chain from cytochrome $c(E_0' = +210\text{mV})$ to O_2 ($E_0' = +820$ mV) or in the opposite direction?
5. If a person had a niacin deficiency, what metabolic process might well be adversely affected? Why?
6. Draw a diagram showing how enzymes catalyze reactions by altering the activation energy. What is a transition-state complex? Use the diagram to discuss why enzymes do not change the equilibria of the reactions they catalyze.
7. What special properties might an enzyme isolated from a psychrophilic bacterium have? Will enzymes need to lower the activation energy more or less in thermophiles than in psychrophiles?

Additional Reading

Becker, W. M. 1977. *Energy and the living cell: An introduction to bioenergetics.* Philadelphia: J. B. Lippincott.
Kraut, J. 1988. How do enzymes work? *Science* 242:533–39.
Mathews, C. K., and van Holde, K. E. 1990. *Biochemistry.* Menlo Park, Calif.: Benjamin/Cummings.
Neidleman, S. L. 1989. Enzymes under stress. *ASM News* 55(2): 67–70.
Rawn, J. D. 1989. *Biochemistry.* Burlington, N. C.: Carolina Biological Supply Co.
Smith, E. L.; Hill, R. L.; Lehman, I. R.; Lefkowitz, R. J.; Handler, P.; and White, A. 1983. *Principles of biochemistry: General aspects,* 7th ed. New York: McGraw-Hill.
Stryer, L. 1988. *Biochemistry,* 3d ed. New York: Freeman.
Voet, D., and Voet, J. G. 1990. *Biochemistry.* New York: John Wiley and Sons.
Zubay, G. 1993. *Biochemistry,* 3rd ed. Dubuque, IA: Wm. C. Brown Communications, Inc.

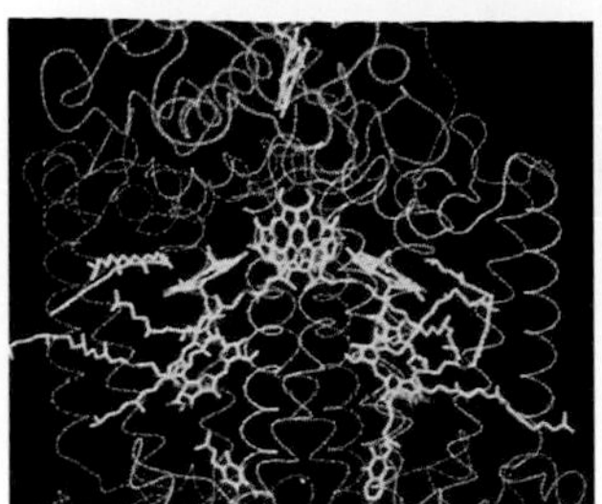

CHAPTER 8
Metabolism: The Generation of Energy

It is in the fueling reactions that bacteria display their extraordinary metabolic diversity and versatility. Bacteria have evolved to thrive in almost all natural environments, regardless of the nature of available sources of carbon, energy, and reducing power. . . . The collective metabolic capacities of bacteria allow them to metabolize virtually every organic compound on this planet. . . .

—F. C. Neidhardt, J. L. Ingraham, and M. Schaechter

Outline

Concepts

1. Metabolism, the sum total of chemical reactions occurring in the cell, can be divided into catabolism and anabolism. In catabolism, molecules are reduced in complexity and free energy is made available. Anabolism involves the use of free energy to increase the complexity of molecules.
2. During catabolism, nutrients are funneled into a few common pathways for more efficient use of enzymes (a few pathways process a wide variety of nutrients).
3. The tricarboxylic acid cycle is the final pathway for the aerobic oxidation of nutrients to CO_2.
4. The majority of energy released in catabolism is generated by the movement of electrons from electron transport carriers with more negative reduction potentials to ones with more positive reduction potentials. Thus, aerobic respiration is much more efficient than anaerobic catabolism.
5. A wide variety of electron acceptors can be used in catabolism: O_2 (aerobic respiration), organic molecules (fermentation), and oxidized inorganic molecules other than O_2 (anaerobic respiration). Furthermore, reduced inorganic molecules as well as organic molecules can serve as electron donors for electron transport and ATP synthesis. Microbial catabolism is unique in the diversity of nutrients and mechanisms employed in energy production.
6. In photosynthesis, trapped light energy boosts electrons to more negative reduction potentials or higher energy levels. These energized electrons are then used to make ATP and NAD(P)H during electron transport.

Chapter 7 introduced the basic principles of thermodynamics, the energy cycle and the use of ATP as an energy currency, and the nature and function of enzymes. With this background material in hand, microbial metabolism can be discussed. **Metabolism** is the total of all chemical reactions occurring in the cell. The flow of energy and the participation of enzymes make metabolism possible.

This chapter begins with an overview of metabolism. An introduction to carbohydrate degradation and the aerobic formation of ATP driven by electron transport follows. Next, the anaerobic synthesis of ATP through fermentation and anaerobic respiration is described. Then the breakdown of organic substances other than carbohydrates (i.e., lipids, proteins, and amino acids) is briefly surveyed. The chapter concludes with sections on the oxidation of inorganic molecules and the trapping of light energy by photosynthetic light reactions.

An Overview of Metabolism

Metabolism may be divided into two major parts. In **catabolism** (Greek *cata,* down, and *ballein,* to throw), larger and more complex molecules are broken down into smaller, simpler molecules with the release of energy. Some of this energy is trapped and made available for work; the remainder is released as heat. The trapped energy can then be used in anabolism, the second area of metabolism. **Anabolism** (Greek *ana,* up) is the synthesis of complex molecules from simpler ones with the input of energy. An anabolic process uses energy to increase the order of a system (*see chapter 7*). Although the division of metabolism into two major parts is convenient and commonly employed, note that energy-generating processes like chemolithotrophy and photosynthesis are not comfortably encompassed by this definition of catabolism unless it is expanded to include all energy-yielding metabolic processes, whether complex molecules are degraded or not.

Before learning about some of the more important catabolic pathways, it is best to look at the "lay of the land" and get our bearings. Albert Lehninger, a biochemist who worked at Johns Hopkins medical school, helped considerably in this task by pointing out that metabolism may be divided into three stages (figure 8.1). In the first stage of catabolism, larger nutrient molecules (proteins, polysaccharides, and lipids) are hydrolyzed or otherwise broken down into their constituent parts. The chemical reactions occurring during this stage do not release much energy. Amino acids, monosaccharides, fatty acids, glycerol, and other products of the first stage are degraded to a few simpler molecules in the second stage. Usually, metabolites like acetyl coenzyme A, pyruvate, and tricarboxylic acid cycle intermediates are formed. The second-stage process can operate either aerobically or anaerobically, and often produces some ATP as well as NADH and/or $FADH_2$. Finally, nutrient carbon is fed into the tricarboxylic acid cycle during the third stage of catabolism, and molecules are oxidized completely to CO_2 with the production of ATP, NADH, and $FADH_2$. The cycle operates aerobically and is responsible for the generation of much energy. Much of the ATP derived from the tricarboxylic acid cycle (and stage-two reactions) comes from the oxidation of NADH and $FADH_2$ by the electron transport chain. Oxygen, or sometimes another inorganic molecule, is the final electron acceptor.

Although this picture is somewhat oversimplified, it is useful in discerning the general pattern of catabolism. Notice that the microorganism begins with a wide variety of molecules and reduces their number and diversity at each stage. That is, nutrient molecules are funneled into ever fewer metabolic intermediates until they are finally fed into the tricarboxylic acid cycle. A common pathway often degrades many similar molecules (e.g., several different sugars). These metabolic pathways consist of enzyme-catalyzed reactions arranged so that the product of one reaction serves as a substrate for the next. The existence of a few common catabolic pathways, each degrading many nutrients, greatly increases metabolic efficiency by avoiding the need for a large number of less metabolically flexible pathways. It is in the catabolic phase that microorganisms exhibit their nutritional diversity. Most microbial biosynthetic pathways closely resemble their counterparts in higher organisms. The uniqueness of microbial metabolism lies in the diversity of sources from which it generates ATP and NADH.

Carbohydrates and other nutrients serve two functions in the metabolism of heterotrophic microorganisms: (1) they are oxidized to provide energy, and (2) they supply carbon or building blocks for the synthesis of new cell constituents. Although many anabolic pathways are separate from catabolic routes, there are **amphibolic pathways** (Greek *amphi,* on both sides) that function both catabolically and anabolically. Two of the most important are the glycolytic pathway and the tricarboxylic acid cycle. Most reactions in these two pathways are freely reversible and can be used to synthesize and degrade molecules. The few irreversible catabolic steps are bypassed in biosynthesis with special enzymes that catalyze the reverse reaction (figure 8.2). For example, the enzyme fructose bisphosphatase reverses the phosphofructokinase step when glucose is synthesized from pyruvate (pp. 175–76). The presence of two separate enzymes, one catalyzing the reversal of the other's reaction, permits independent regulation of the catabolic and anabolic functions of these amphibolic pathways.

1. Describe metabolism. How are catabolism and anabolism organized? Why are common pathways useful? What is an amphibolic pathway?

The Breakdown of Glucose to Pyruvate

Microorganisms employ several metabolic pathways to catabolize glucose and other sugars. Because of this metabolic diversity, their metabolism is often confusing. To avoid confusion as much as possible, the ways in which microorganisms degrade sugars to pyruvate and similar intermediates are intro-

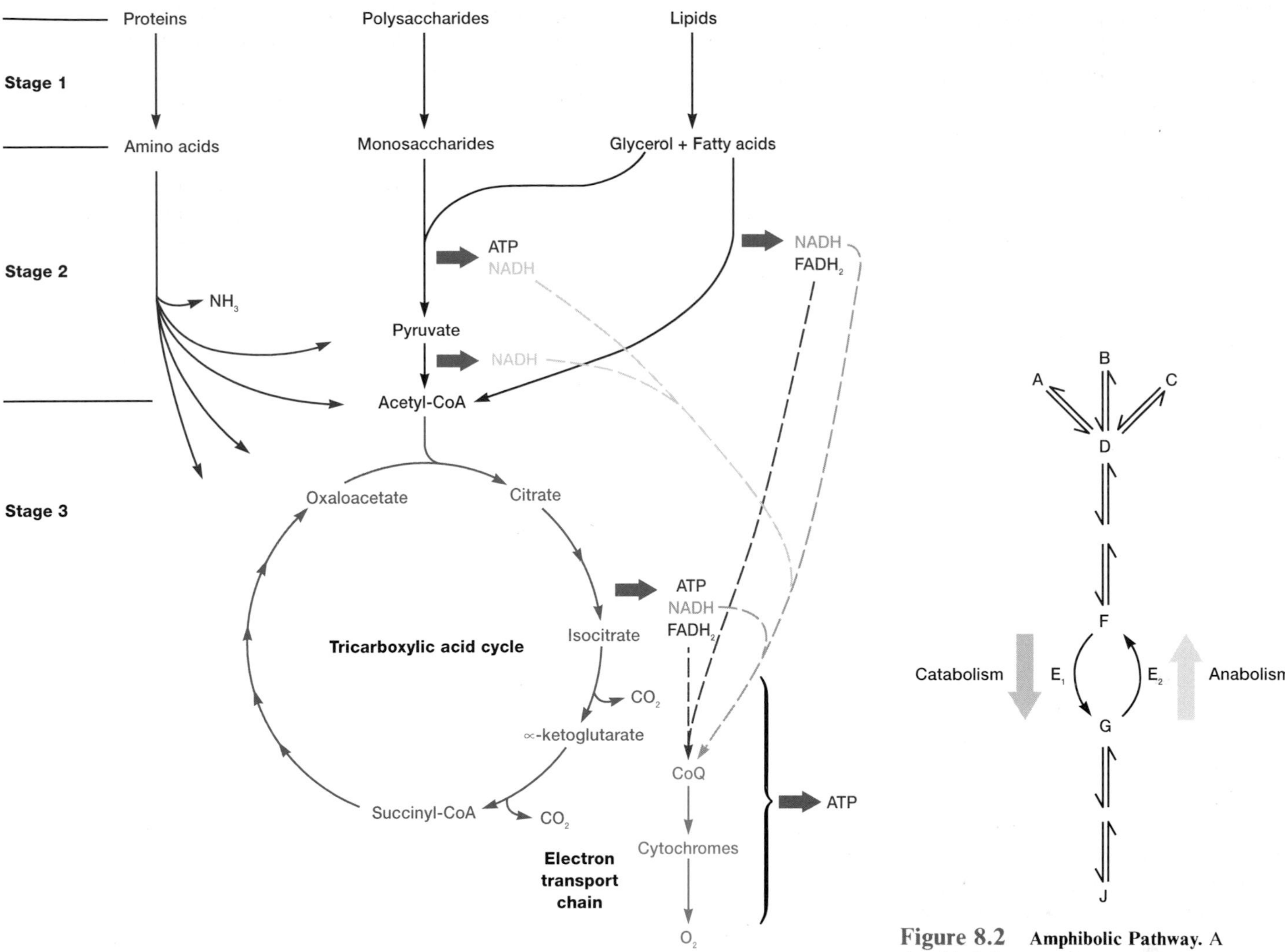

Figure 8.1 The Three Stages of Catabolism. A general diagram of catabolism showing the three stages in this process and the central position of the tricarboxylic acid cycle. Although there are many different proteins, polysaccharides, and lipids, they are degraded through the activity of a few common metabolic pathways.

Figure 8.2 Amphibolic Pathway. A simplified diagram of an amphibolic pathway such as the glycolytic pathway. Note that the interconversion of intermediates F and G is catalyzed by two separate enzymes, E_1 operating in the catabolic direction and E_2 in the anabolic.

duced by focusing on only three routes: (1) glycolysis, (2) the pentose phosphate pathway, and (3) the Entner-Doudoroff pathway. Next, the pathways of aerobic and anaerobic pyruvate metabolism are described. For the sake of simplicity, the chemical structures of metabolic intermediates are not used in pathway diagrams.

Diagrams of the glycolytic pathway and other major catabolic pathways with the structures of intermediates and enzyme names (appendix II).

The Glycolytic Pathway

The **Embden-Meyerhof** or **glycolytic pathway** is undoubtedly the most common pathway for glucose degradation to pyruvate in stage two of catabolism. It is found in all major groups of microorganisms and functions in the presence or absence of O_2. **Glycolysis** (Greek *glyco,* sweet, and *lysis,* a loosening) is located in the cytoplasmic matrix when present in eucaryotic microorganisms.

The pathway as a whole may be divided into two parts (figure 8.3 and *appendix II*). In the initial six-carbon stage, glucose is phosphorylated twice and eventually converted to fructose 1,6-bisphosphate. Other sugars are often fed into the pathway by conversion to glucose 6-phosphate or fructose 6-phosphate. This preliminary stage does not yield energy; in fact, two ATP molecules are expended for each glucose. These initial steps "prime the pump" by adding phosphates to each end of the sugar. The phosphates will soon be used to make ATP.

The three-carbon stage of glycolysis begins when the enzyme fructose 1,6-bisphosphate aldolase catalyzes the cleavage of fructose 1,6-bisphosphate into two halves, each with a phosphate group. One of the products, glyceraldehyde

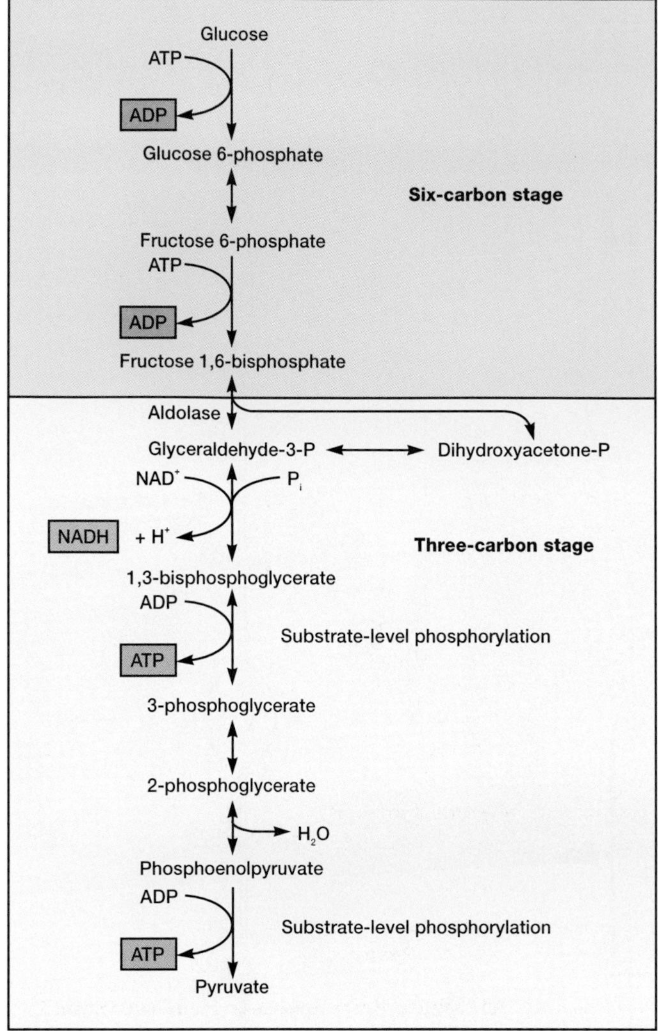

Figure 8.3 **Glycolysis.** The glycolytic pathway for the breakdown of glucose to pyruvate. The two stages of the pathway and their products are indicated.

3-phosphate, is converted directly to pyruvate in a five-step process. Because the other product, dihydroxyacetone phosphate, can be easily changed to glyceraldehyde 3-phosphate, both halves of fructose 1,6-bisphosphate are used in the three-carbon stage. Glyceraldehyde 3-phosphate is first oxidized with NAD^+ as the electron acceptor, and a phosphate is simultaneously incorporated to give a high-energy molecule called 1,3-bisphosphoglycerate. The high-energy phosphate on carbon one is subsequently donated to ADP to produce ATP. This synthesis of ATP is called **substrate-level phosphorylation** because ADP phosphorylation is coupled with the exergonic breakdown of a high-energy substrate molecule.

High-energy bonds and ATP as an energy currency (pp. 136–37).

A somewhat similar process generates a second ATP by substrate-level phosphorylation. The phosphate group on 3-phosphoglycerate shifts to carbon two, and 2-phosphoglycerate is dehydrated to form a second high-energy molecule, phosphoenolpyruvate. This molecule donates its phosphate to ADP forming a second ATP and pyruvate, the final product of the pathway.

The glycolytic pathway degrades one glucose to two pyruvates by the sequence of reactions just outlined. ATP and NADH are also produced. The yields of ATP and NADH may be calculated by considering the two stages separately. In the six-carbon stage, two ATPs are used to form fructose 1,6-bisphosphate. For each glyceraldehyde 3-phosphate transformed into pyruvate, one NADH and two ATPs are formed. Because two glyceraldehyde-3-phosphates arise from a single glucose (one by way of dihydroxyacetone phosphate), the three-carbon stage generates four ATPs and two NADHs per glucose. Subtraction of the ATP used in the six-carbon stage from that produced in the three-carbon stage gives a net yield of two ATPs per glucose. Thus, the catabolism of glucose to pyruvate in glycolysis can be represented by the following simple equation.

$$\text{Glucose} + 2\text{ADP} + 2\text{P}_i + 2\text{NAD}^+ \longrightarrow 2\text{ pyruvate} + 2\text{ATP} + 2\text{NADH} + 2\text{H}^+$$

The Pentose Phosphate Pathway

A second pathway, the **pentose phosphate** or **hexose monophosphate pathway** may be used at the same time as the glycolytic pathway or the Entner-Doudoroff sequence. It can operate either aerobically or anaerobically and is important in biosynthesis as well as in catabolism.

The pentose phosphate pathway begins with the oxidation of glucose 6-phosphate to 6-phosphogluconate followed by the oxidation of 6-phosphogluconate to the pentose ribulose 5-phosphate and CO_2 (figure 8.4 and *appendix II*). NADPH is produced during these oxidations. Ribulose 5-phosphate is then converted to a mixture of three- through seven-carbon sugar phosphates. Two enzymes unique to this pathway play a central role in these transformations: (1) transketolase catalyzes the transfer of two-carbon ketol groups, and (2) transaldolase transfers a three-carbon group from sedoheptulose 7-phosphate to glyceraldehyde 3-phosphate (figure 8.5). The overall result is that three glucose 6-phosphates are converted to two fructose 6-phosphates, glyceraldehyde 3-phosphate, and three CO_2 molecules, as shown in the following equation.

$$3\text{ glucose 6-phosphate} + 6\text{NADP}^+ + 3\text{H}_2\text{O} \longrightarrow 2\text{ fructose 6-phosphate} + \text{glyceraldehyde 3-phosphate} + 3\text{CO}_2 + 6\text{NADPH} + 6\text{H}^+$$

These intermediates are used in two ways. The fructose 6-phosphate can be changed back to glucose 6-phosphate while glyceraldehyde 3-phosphate is converted to pyruvate by glycolytic enzymes. The glyceraldehyde 3-phosphate also may be returned to the pentose phosphate pathway through glucose 6-phosphate formation. This results in the complete degra-

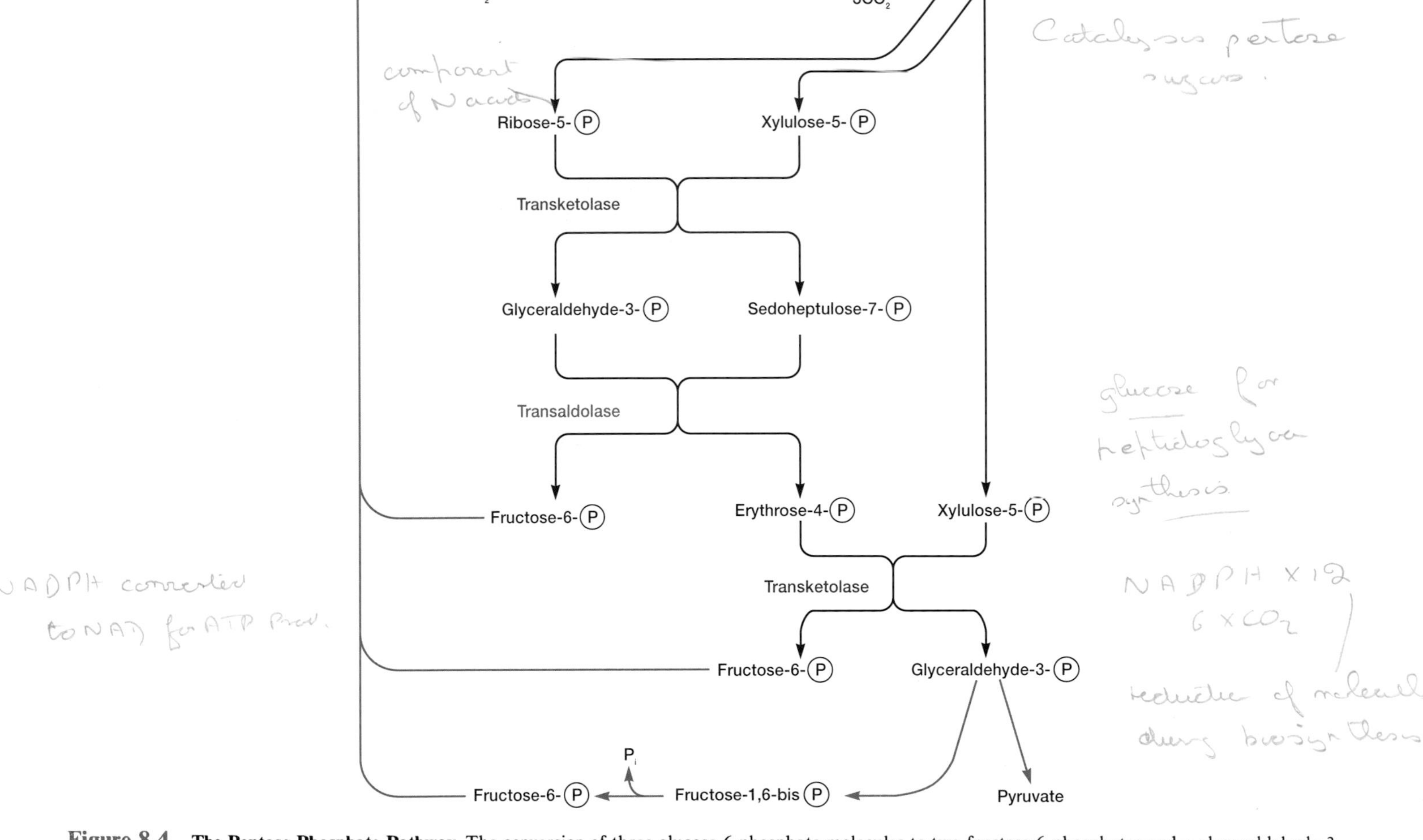

Figure 8.4 **The Pentose Phosphate Pathway.** The conversion of three glucose 6-phosphate molecules to two fructose 6-phosphates and a glyceraldehyde 3-phosphate is traced. The fructose 6-phosphates are changed back to glucose 6-phosphate. The glyceraldehyde 3-phosphate can be converted to pyruvate or combined with a molecule of dihydroxyacetone phosphate (from the glyceraldehyde 3-phosphate formed by a second turn of the pathway) to yield fructose 6-phosphate.

dation of glucose 6-phosphate to CO_2 and the production of a great deal of NADPH.

$$\text{Glucose 6-phosphate} + 12\text{NADP}^+ + 7H_2O \longrightarrow 6CO_2 + 12\text{NADPH} + 12H^+ + P_i$$

The pentose phosphate pathway has several catabolic and anabolic functions that are summarized as follows:

1. The pentose phosphate pathway may be used to produce ATP. Glyceraldehyde 3-phosphate from the pathway can enter the three-carbon stage of the glycolytic pathway and be converted to ATP and pyruvate. The latter may be oxidized in the tricarboxylic acid cycle to provide more energy. In addition, some NADPH can be converted to NADH, which yields ATP when it is oxidized by the electron transport chain. Because five-carbon sugars are intermediates in the pathway, the pentose phosphate pathway can be used to catabolize pentoses as well as hexoses.
2. NADPH from the pentose phosphate pathway serves as a source of electrons for the reduction of molecules during biosynthesis.
3. The pathway synthesizes four- and five-carbon sugars for a variety of purposes. The four-carbon sugar erythrose 4-phosphate is used to synthesize aromatic amino acids. The pentose ribose 5-phosphate is a major component of nucleic acids, and ribulose 1,5-bisphosphate is the primary CO_2 acceptor in photosynthesis. Note that when a microorganism is growing on a pentose carbon source, the pathway also can supply carbon for hexose production (e.g., glucose is needed for peptidoglycan synthesis).

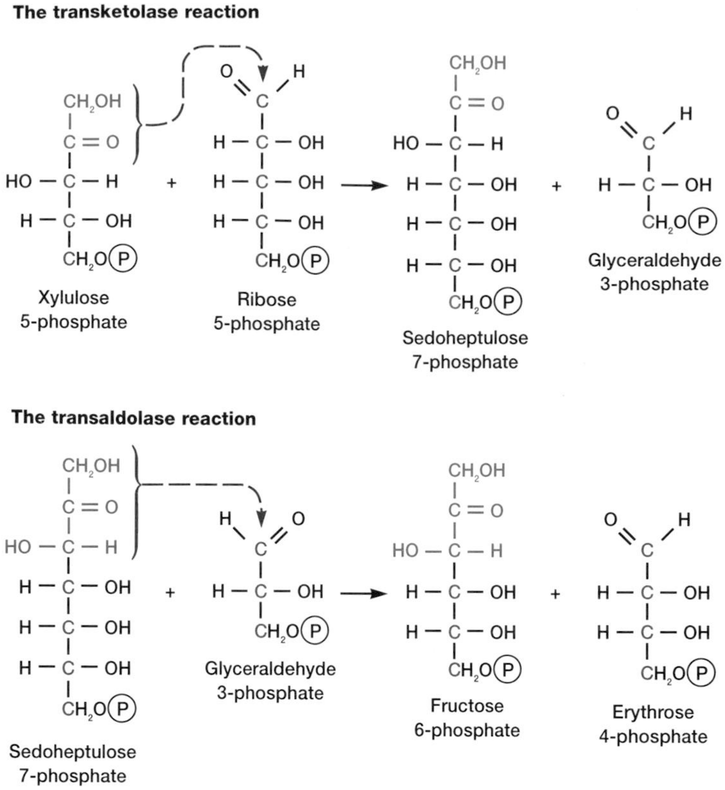

The transketolase reaction

Xylulose 5-phosphate + Ribose 5-phosphate → Sedoheptulose 7-phosphate + Glyceraldehyde 3-phosphate

The transaldolase reaction

Sedoheptulose 7-phosphate + Glyceraldehyde 3-phosphate → Fructose 6-phosphate + Erythrose 4-phosphate

Figure 8.5 Transketolase and Transaldolase. Examples of the transketolase and transaldolase reactions of the pentose phosphate pathway. The groups transferred in these reactions are in color.

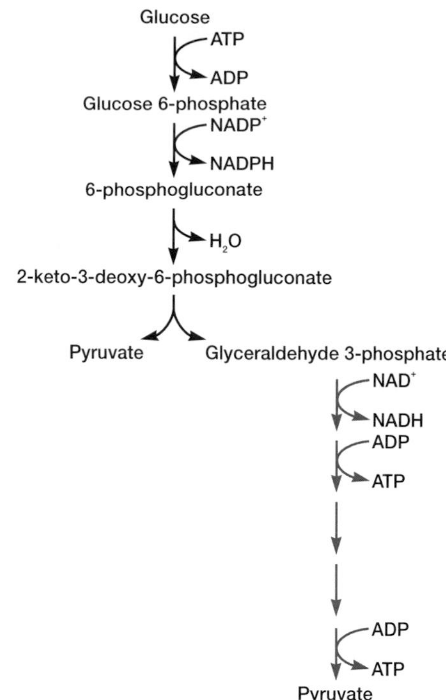

Figure 8.6 The Entner-Doudoroff Pathway. The sequence leading from glyceraldehyde 3-phosphate to pyruvate is catalyzed by enzymes common to the glycolytic pathway.

Although the pentose phosphate pathway may be a major source of energy in many microorganisms, it is more often of greater importance in biosynthesis. Several functions of the pentose phosphate pathway are mentioned again in chapter 9 when biosynthesis is considered more directly.

The Entner-Doudoroff Pathway

Although the glycolytic pathway is the most common route for the conversion of hexoses to pyruvate, another pathway with a similar role has been discovered. The **Entner-Doudoroff pathway** begins with the same reactions as the pentose phosphate pathway, the formation of glucose 6-phosphate and 6-phosphogluconate (figure 8.6 and *appendix II*). Instead of being further oxidized, 6-phosphogluconate is dehydrated to form 2-keto-3-deoxy-6-phosphogluconate or KDPG, the key intermediate in this pathway. KDPG is then cleaved by KDPG aldolase to pyruvate and glyceraldehyde 3-phosphate. The glyceraldehyde 3-phosphate is converted to pyruvate in the bottom portion of the glycolytic pathway. If the Entner-Doudoroff pathway degrades glucose to pyruvate in this way, it yields one ATP, one NADPH, and one NADH per glucose metabolized.

Most bacteria have the glycolytic and pentose phosphate pathways, but some substitute the Entner-Doudoroff pathway for glycolysis. The Entner-Doudoroff pathway is generally found in *Pseudomonas, Rhizobium, Azotobacter, Agrobacterium,* and a few other gram-negative genera. Very few gram-positive bacteria have this pathway, with *Enterococcus faecalis* being a rare exception.

1. Summarize the major features of the glycolytic pathway, the pentose phosphate pathway, and the Entner-Doudoroff sequence. Include the starting points, the products of the pathways, the critical or unique enzymes, the ATP yields, and the metabolic roles each pathway has.
2. What is substrate-level phosphorylation?

The Tricarboxylic Acid Cycle

Although some energy is obtained from the breakdown of glucose to pyruvate by the pathways previously described, much more is released when pyruvate is degraded aerobically to CO_2 in stage three of catabolism. The multienzyme system called the pyruvate dehydrogenase complex first oxidizes pyruvate to form CO_2 and **acetyl coenzyme A (acetyl-CoA)**, an energy-rich molecule composed of coenzyme A and acetic acid joined by a high energy thiol ester bond (figure 8.7). Acetyl-CoA arises from the catabolism of many carbohydrates, lipids, and amino acids (figure 8.1). It can be further degraded in the tricarboxylic acid cycle.

The substrate for the **tricarboxylic acid (TCA) cycle** or **citric acid cycle** is acetyl-CoA (figure 8.7 and *appendix II*). In the first reaction, acetyl-CoA is condensed with a four-

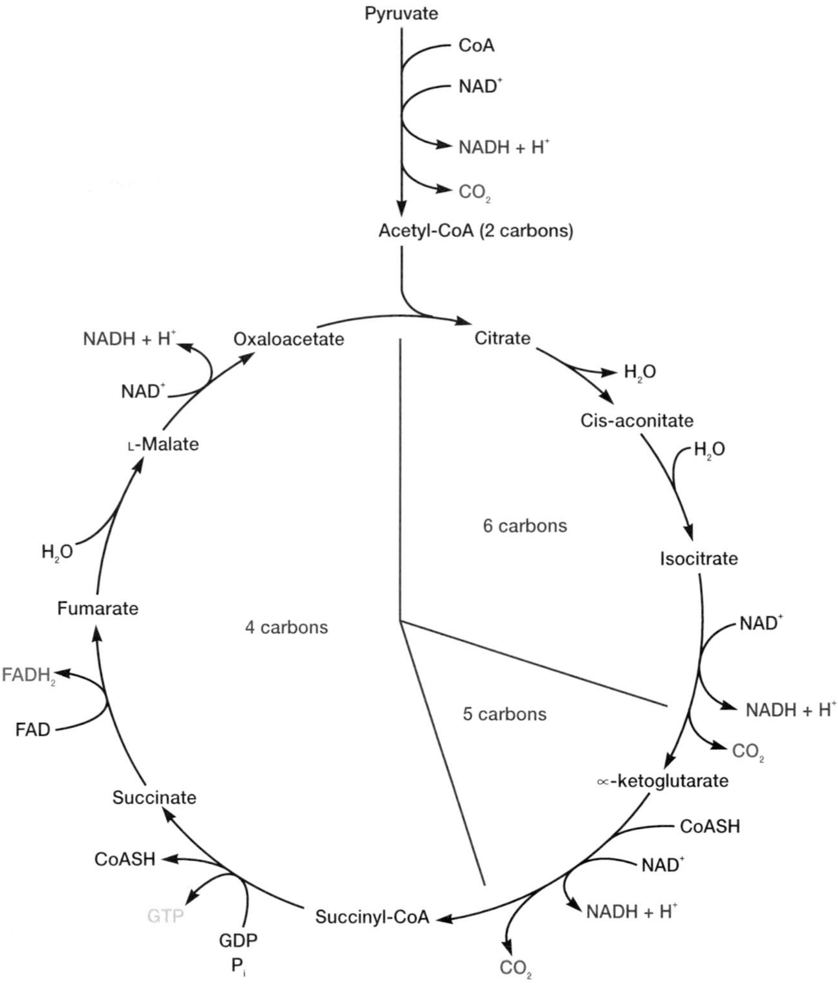

Figure 8.7 **The Tricarboxylic Acid Cycle.** The cycle may be divided into three stages based on the size of its intermediates. The three stages are separated from one another by two decarboxylation reactions (reactions in which carboxyl groups are lost as CO_2). The pyruvate dehydrogenase complex forms acetyl-CoA through pyruvate oxidation.

carbon intermediate, oxaloacetate, to form citrate and to begin the six-carbon stage. Citrate (a tertiary alcohol) is rearranged to give isocitrate, a more readily oxidized secondary alcohol. Isocitrate is subsequently oxidized and decarboxylated twice to yield α-ketoglutarate, then succinyl-CoA. It is at this point that two NADHs are formed and two carbons are lost from the cycle as CO_2. Because two carbons were added as acetyl-CoA at the start, balance is maintained and no net carbon is lost. The cycle now enters the four-carbon stage during which two oxidation steps yield one $FADH_2$ and one NADH per acetyl-CoA. In addition, GTP (a high-energy molecule equivalent to ATP) is produced from succinyl-CoA by substrate-level phosphorylation. Eventually, oxaloacetate is reformed and ready to join with another acetyl-CoA. Inspection of figure 8.7 shows that the TCA cycle generates two CO_2s, three NADHs, one $FADH_2$, and one GTP for each acetyl-CoA molecule oxidized.

TCA cycle enzymes are widely distributed among microorganisms. The cycle appears to be functional in many aerobic bacteria, free-living protozoa, and most algae and fungi. This is not surprising because the cycle is such an important source of energy. However, the facultative anaerobe *E. coli* does not use the complete TCA cycle under anaerobic conditions or when the glucose concentration is high, but does at other times. Even those microorganisms that lack the complete TCA cycle usually have most of the cycle enzymes, because one of TCA cycle's major functions is to provide carbon skeletons for use in biosynthesis.

The role of the tricarboxylic acid cycle in biosynthesis (pp. 180–81).

1. Give the substrate and products of the tricarboxylic acid cycle. Describe its organization in general terms. What are its two major functions?

Electron Transport and Oxidative Phosphorylation

Little ATP has been synthesized up to this point. Only the equivalent of four ATP molecules is directly synthesized when one glucose is oxidized to CO_2 by way of glycolysis and the TCA cycle. Most ATP generated aerobically comes from the oxidation of NADH and $FADH_2$ in the electron transport chain. The electron transport chain itself is examined first, then the mechanism of ATP synthesis is discussed.

The Electron Transport Chain

The **electron transport chain** is composed of a series of electron carriers that operate together to transfer electrons from donors, like NADH and $FADH_2$, to acceptors, such as O_2 (refer to diagram of a mitochondrial electron transport chain in figure 8.8*a*). The electrons flow from carriers with more negative reduction potentials to those with more positive potentials, and eventually combine with O_2 and H^+ to form water. The difference in reduction potentials between O_2 and NADH is large, about 1.14 volts, and makes possible the release of a great deal of energy. Despite the large overall decrease in reduction potential, there are only three places in eucaryotic transport chains where sufficient energy is made available to synthesize ATP: between (1) NADH and coenzyme Q, (2) cytochromes *b* and c_1, and (3) cytochrome *a* and O_2. The electron transport chain breaks up the large overall energy release into small steps. Some of the liberated energy is trapped in the form of ATP. As will be seen shortly, electron transport at these points may generate proton and electrical gradients. These gradients can then drive ATP synthesis.

The electron transport chain carriers reside within the inner membrane of the mitochondrion or in the bacterial plasma membrane. The mitochondrial system is arranged into four complexes of carriers, each capable of transporting electrons part of the way to O_2 (figure 8.8*b*). Coenzyme Q and cytochrome c connect the complexes with each other. The bacterial carriers are also highly organized within the membrane (figure 8.9).

The process by which energy from electron transport is used to make ATP is called **oxidative phosphorylation.** Thus, as many as three ATP molecules may be synthesized from ADP and P_i when a pair of electrons pass from NADH to an atom of O_2. This is the same thing as saying that the phosphorus to oxygen (P/O) ratio is equal to 3. Because electrons from $FADH_2$ only pass two oxidative phosphorylation points, the maximum P/O ratio for $FADH_2$ is 2. The actual P/O ratios may be less than 3.0 and 2.0 in eucaryotic mitochondria.

The preceding discussion has focused on the eucaryotic mitochondrial electron transport chain. Although some bacterial chains do resemble the mitochondrial chain, they are frequently very different. They vary in their electron carriers (e.g., in their cytochromes) and may be extensively branched. Electrons often can enter at several points and leave through several terminal oxidases. The *E. coli* chain even changes composition with alterations in the O_2 level (figure 8.9). Bacterial chains also may be shorter and have lower P/O ratios than mitochondrial transport chains. Thus, procaryotic and eucaryotic electron transport chains differ in details of construction although they operate using the same fundamental principles.

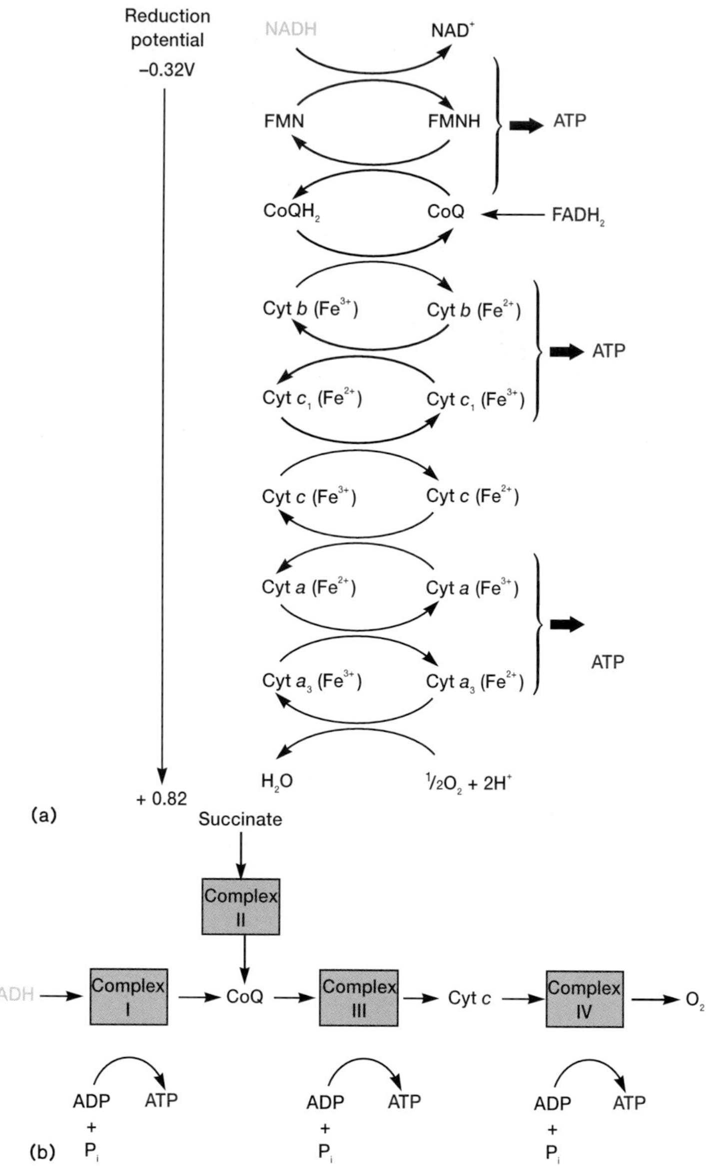

Figure 8.8 The Mitochondrial Electron Transport Chain. (*a*) This diagram is somewhat simplified for the sake of clarity, and several carriers have been excluded. Cyt is an abbreviation for cytochrome; and CoQ, for coenzyme Q. (*b*) The organization of the eucaryotic electron transport chain into four complexes.

Oxidative Phosphorylation

The mechanism by which oxidative phosphorylation takes place has been studied intensively for years. There are currently several hypotheses about how oxidative phosphorylation occurs; the two most important are the chemiosmotic hypothesis and the conformational change hypothesis.

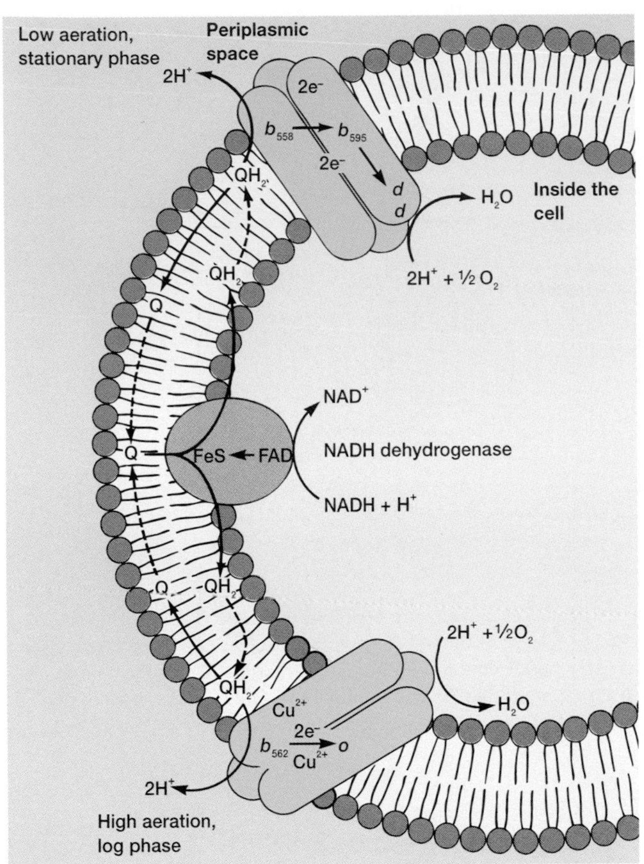

Figure 8.9 **The Aerobic Respiratory System of *E. coli*.** NADH is the electron source. Ubiquinone-8 (Q) connects the NADH dehydrogenase with two terminal oxidase systems that function at different oxygen levels. At least five cytochromes are involved: b_{558}, b_{595}, b_{562}, d, and o.

According to the **chemiosmotic hypothesis,** first formulated in 1961 by the British biochemist Peter Mitchell, the electron transport chain is organized so that, during its operation, protons move outward from the mitochondrial matrix and electrons are transported inward (figures 8.10 and 8.11). Proton movement may result either from carrier loops, as shown in figure 8.11, or from the action of special proton pumps that derive their energy from electron transport. The result is **protonmotive force (PMF),** composed of a gradient of protons and a membrane potential due to the unequal distribution of charges. When protons return to the mitochondrial matrix driven by the protonmotive force, ATP is synthesized in a reversal of the ATP hydrolysis reaction (figure 8.10). A similar process is believed to take place in bacteria, with electron flow causing the protons to move outward across the plasma membrane (figure 8.9). ATP synthesis occurs when these protons diffuse back into the cell. The protonmotive force also may drive the transport of molecules across membranes (*see chapter 5*) and the rotation of bacterial flagella (*see chapter 3*). The chemiosmotic hypothesis is accepted by most microbiologists. There is considerable evidence for the generation of proton and charge gradients across membranes. However, the evidence for proton gradients as the direct driving force for oxidative phosphorylation is not yet conclusive.

In the **conformational change hypothesis,** energy released by electron transport induces changes in the shape or conformation of the enzyme that synthesizes ATP. These shape changes drive ATP formation, possibly by changing the strength of ATP, ADP, and phosphate binding to the enzyme's catalytic site. The hypothesis is plausible because conformational changes occur in mitochondrial inner membrane proteins during electron transport.

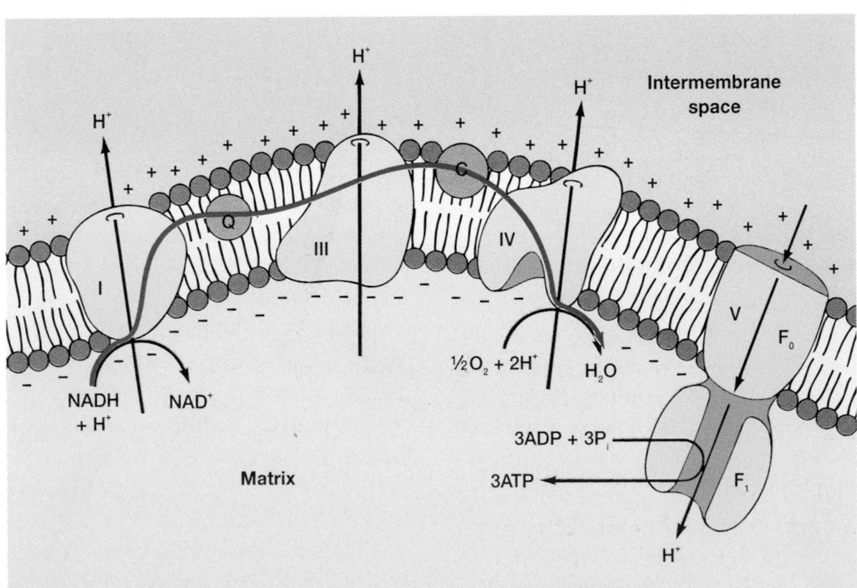

Figure 8.10 **Chemiosmosis.** The chemiosmotic hypothesis as applied to mitochondrial function. The flow of electrons from NADH to oxygen causes protons to move from the mitochondrial matrix to the intermembrane space. This generates proton and electrical gradients. When the protons move back to the matrix through the F_1F_0 complex, F_1 synthesizes ATP. *From Matthews and Van Holde,* Biochemistry. *Copyright © 1990 Benjamin/Cummings Publishing Co., Menlo Park, CA.*

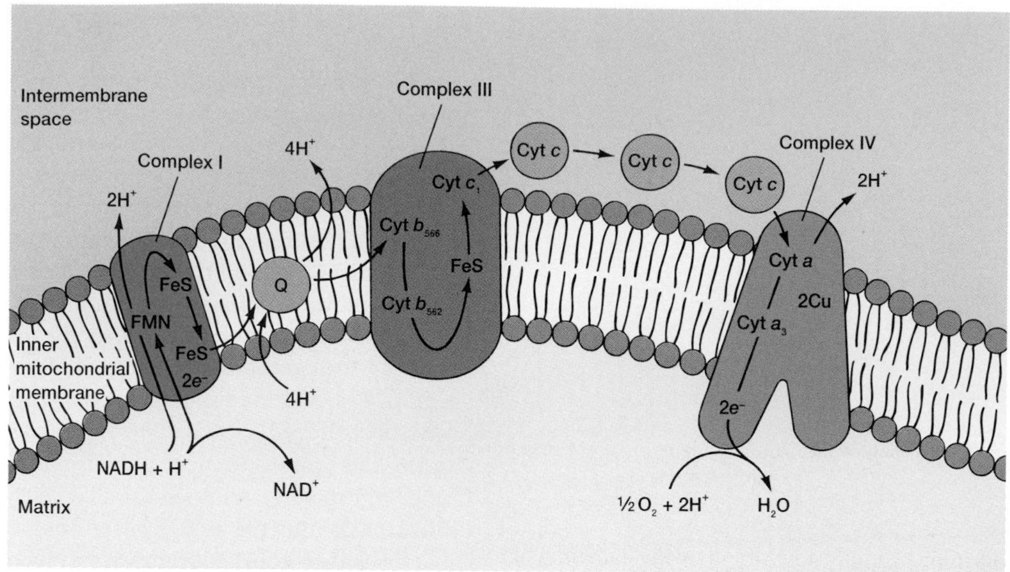

Figure 8.11 **The Chemiosmotic Hypothesis Applied to Eucaryotic Mitochondria.** In this scheme, the carriers are organized asymmetrically within the inner membrane so that protons are transported across as electrons move along the chain. Proton release in the intermembrane space occurs when electrons are transferred from carriers, such as FMN of the flavoprotein (Fp) and coenzyme Q (Q), that carry both electrons and protons to carriers like nonheme iron proteins (FeS proteins) and cytochromes (cyt) that transport only electrons. Coenzyme Q transports electrons from complex I to complex III. Cytochrome *c* moves between complexes III and IV. For the sake of simplicity, complex II is not shown. It transfers electrons from succinate to coenzyme Q. *From Donald Voet and Judith G. Voet,* Biochemistry. *Copyright © 1990 John Wiley & Sons, Inc., New York, NY.*

Research in this area is proceeding rapidly, and the precise mechanism of oxidative phosphorylation should soon become clearer. Possibly the final answer will contain features of more than one current hypothesis.

Whatever the precise mechanism, clearly ATP synthesis takes place at the F_1 ATPase or **ATP synthase** (figure 8.12). In mitochondria, this protein complex appears as a spherical structure attached to the surface of the inner membrane by a stalk and the F_0 factor, which is embedded in the inner membrane. The F_0 integral protein is believed to participate in the proton movement across the membrane associated with oxidative phosphorylation. In bacteria, the F_1 ATPase is on the inner surface of the plasma membrane.

Many chemicals inhibit the aerobic synthesis of ATP. These inhibitors generally fall into two categories. Some directly block the transport of electrons. The antibiotic piericidin competes with coenzyme Q; the antibiotic antimycin A blocks electron transport between cytochromes *b* and *c;* and both cyanide and azide stop the transfer of electrons between cytochrome *a* and O_2 because they are structural analogs of O_2. Another group of inhibitors known as **uncouplers** stops ATP synthesis without inhibiting electron transport itself. Indeed, they may even enhance the rate of electron flow. Normally, electron transport is tightly coupled with oxidative phosphorylation so that the rate of ATP synthesis controls the rate of electron transport. The more rapidly ATP is synthesized during oxidative phosphorylation, the faster the electron transport chain operates to supply the required energy. Uncouplers disconnect oxidative phosphorylation from electron transport; therefore, the energy released by the chain is given off as heat rather than as ATP. Many uncouplers like dinitrophenol and valinomycin may allow hydrogen ions, potassium ions, and other ions to cross the membrane without activating the F_1 ATPase. In this way, they destroy the pH and ion gradients. Valinomycin also may bind directly to the F_1 ATPase and inhibit its activity.

The Yield of ATP in Glycolysis and Aerobic Respiration

The maximum ATP yield in eucaryotes from glycolysis, the TCA cycle, and electron transport can be readily calculated. The conversion of glucose to two pyruvate molecules during glycolysis gives a net gain of two ATPs and two NADHs. Because each NADH can yield a maximum of three ATPs during electron transport and oxidative phosphorylation, the total aerobic yield from the glycolytic pathway is eight ATP molecules (table 8.1). Under anaerobic conditions, when the NADH is not oxidized by the electron transport chain, only two ATPs will be generated during the degradation of glucose to pyruvate. The fate of pyruvate under anaerobic conditions is the topic of the next section.

When O_2 is present and the electron transport chain is operating, pyruvate is next oxidized to acetyl-CoA, the substrate for the TCA cycle. This reaction yields 2 NADHs because 2 pyruvates arise from a glucose; therefore, 6 more ATPs are formed. Oxidation of each acetyl-CoA in the TCA cycle will yield 1 GTP (or ATP), 3 NADHs, and a single $FADH_2$ for a total of 2 GTPs (ATPs), 6 NADHs, and 2 $FADH_2$s from two acetyl-CoA molecules. As table 8.1 shows, this amounts to 24 ATPs when NADH and $FADH_2$ from the cycle are oxidized in the electron transport chain. Thus, the aerobic oxi-

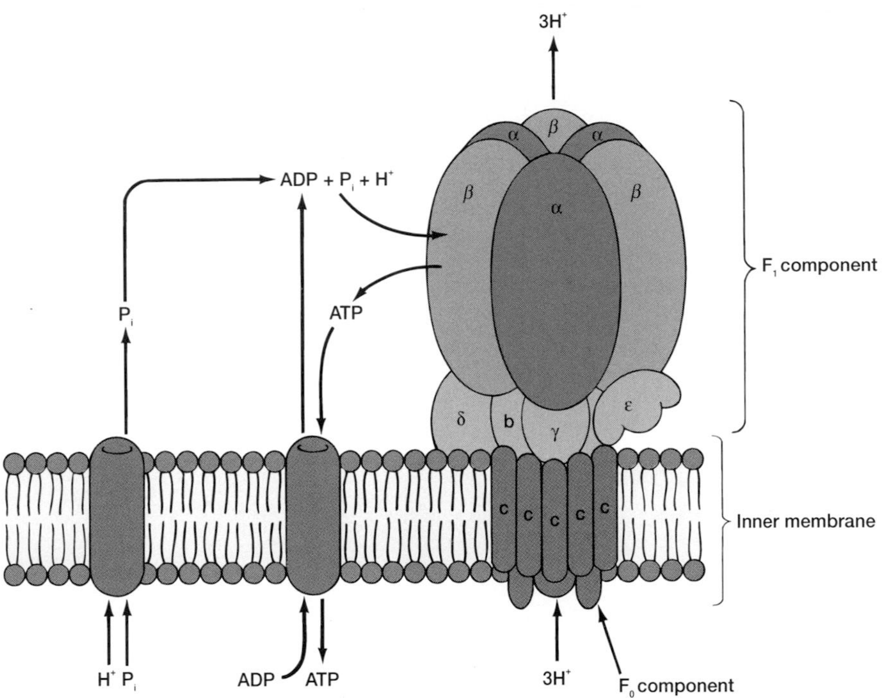

Figure 8.12 **ATP Synthesis.** The spherical structure is called the F_1 ATPase because it can catalyze the hydrolysis of ATP. This complex structure is the site of ATP synthesis. It is composed of five different subunits (α, β, ν, δ, ϵ); ATP is synthesized by the β subunit. It rests on a stalk that is connected to the F_0 complex embedded in the membrane. F_0 is made of a, b, and c subunits. ATP is synthesized when protons pass through the F_1 ATPase or ATP synthase complex under the influence of protonmotive force. ATP, ADP, and P_i are carried across the inner membrane by separate transport proteins.

TABLE 8.1 ATP Yield from the Aerobic Oxidation of Glucose by Eucaryotic Cells

Glycolytic Pathway	
Substrate-level phosphorylation (ATP)	2 ATP[a]
Oxidative phosphorylation with 2 NADH	6 ATP
2 Pyruvate to 2 Acetyl-CoA	
Oxidative phosphorylation with 2 NADH	6 ATP
Tricarboxylic Acid Cycle	
Substrate-level phosphorylation (GTP)	2 ATP
Oxidative phosphorylation with 6 NADH	18 ATP
Oxidative phosphorylation with 2 $FADH_2$	4 ATP
Total Aerobic Yield	38 ATP

[a]ATP yields are calculated with an assumed P/O ratio of 3.0 for NADH and 2.0 for $FADH_2$.

dation of glucose to 6 CO_2 molecules supplies a maximum of 38 ATPs. Only 2 ATP molecules are formed upon the anaerobic conversion of glucose to pyruvate in glycolysis. Because bacterial electron transport systems often have lower P/O ratios than the eucaryotic system being discussed, bacterial aerobic ATP yields are less.

Clearly, aerobic respiration is much more effective than anaerobic processes not involving electron transport and oxidative phosphorylation. Many microorganisms, when moved from anaerobic to aerobic conditions, will drastically reduce their rate of sugar catabolism and switch to aerobic respiration, a regulatory phenomenon known as the **Pasteur effect.** This is of obvious advantage to the microorganism as less sugar must be degraded to obtain the same amount of ATP when the more efficient aerobic process can be employed.

1. Briefly describe the structure of the electron transport chain and its role in ATP formation.
2. By what two mechanisms might ATP be synthesized during oxidative phosphorylation? What is an uncoupler?
3. Calculate the ATP yield for both glycolysis and the total aerobic oxidation of glucose. Explain your reasoning.

Fermentations

In the absence of O_2, NADH is not usually oxidized by the electron transport chain because no external electron acceptor is available. Yet NADH produced in the glycolytic pathway during the oxidation of glyceraldehyde 3-phosphate to 1,3-bisphosphoglycerate (figure 8.3) must still be oxidized back to NAD^+. If NAD^+ is not regenerated, the oxidation of glyceraldehyde 3-phosphate will cease and glycolysis will stop. Many microorganisms solve this problem by slowing or stopping pyruvate dehydrogenase activity and using pyruvate or one of its derivatives as an electron and hydrogen acceptor in the reoxidation of NADH (figure 8.13). This may lead to the production of more ATP. An energy-yielding process like this, in which organic molecules serve as both electron donors and acceptors, is called a **fermentation** (Latin *fermentare,* to cause to rise or ferment). There are many kinds of fermentations, and they

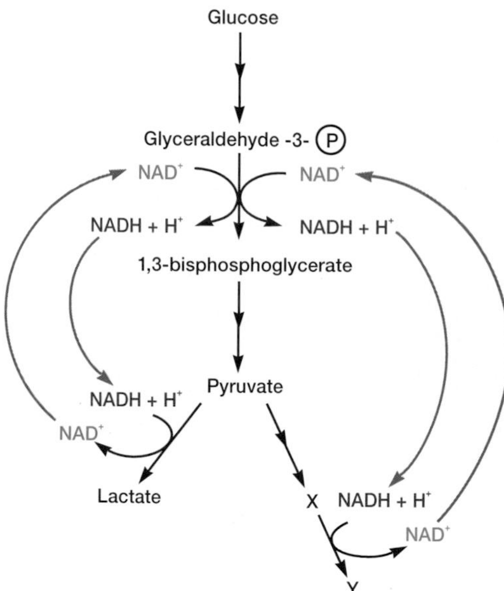

Figure 8.13 Reoxidation of NADH During Fermentation. NADH from glycolysis is reoxidized by being used to reduce pyruvate or a pyruvate derivative (X). Either lactate or reduced product Y result.

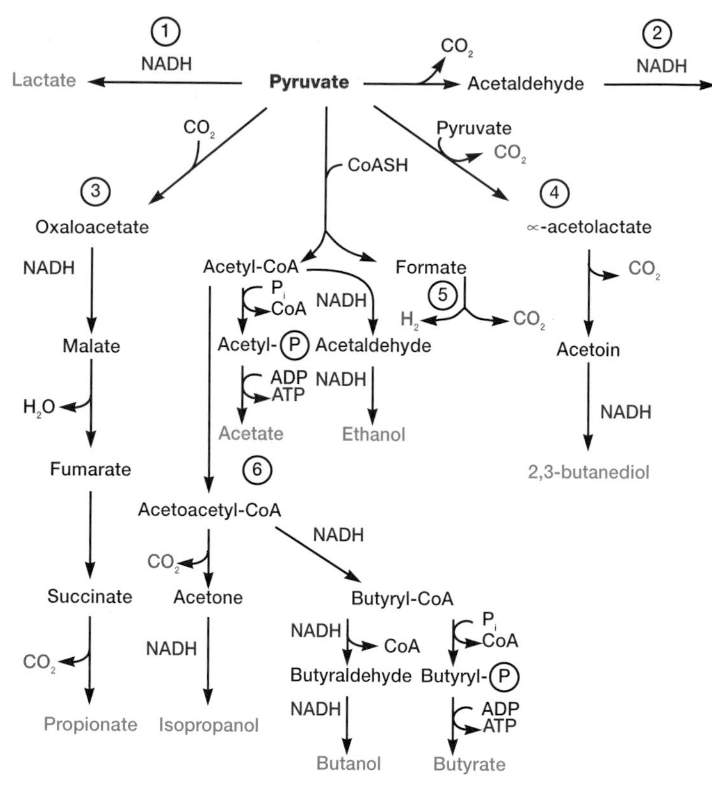

1. Lactic acid bacteria (*Bacillus, Streptococcus, Lactobacillus*)
2. Yeast, *Zymomonas*
3. Propionic acid bacteria (*Propionibacterium*)
4. *Enterobacter, Serratia, Bacillus*
5. Enteric bacteria (*Escherichia, Enterobacter, Salmonella, Proteus*)
6. *Clostridium*

Figure 8.14 Some Common Microbial Fermentations. Most of these pathways have been simplified by deletion of one or more steps and intermediates. Pyruvate and major end products are shown in color.

often are characteristic of particular microbial groups (figure 8.14). A few more common fermentations are introduced here, and several others are discussed at later points. Two unifying themes should be kept in mind when microbial fermentations are examined: (1) NADH is oxidized to NAD^+, and (2) the electron acceptor is either pyruvate or a pyruvate derivative.

Many fungi and some bacteria, algae, and protozoa ferment sugars to ethanol and CO_2 in a process called **alcoholic fermentation.** Pyruvate is decarboxylated to acetaldehyde, which is then reduced to ethanol by alcohol dehydrogenase with NADH as the electron donor (figure 8.14, number 2). **Lactic acid fermentation,** the reduction of pyruvate to lactate (figure 8.14, number 1), is even more common. It is present in bacteria (lactic acid bacteria, *Bacillus*), algae (*Chlorella*), some water molds, protozoa, and even in animal skeletal muscle. Lactic acid fermenters can be separated into two groups. **Homolactic fermenters** use the glycolytic pathway and directly reduce almost all their pyruvate to lactate with the enzyme lactate dehydrogenase. **Heterolactic fermenters** form substantial amounts of products other than lactate; many produce lactate, ethanol, and CO_2 by way of the phosphoketolase pathway (*see chapter 22*).

The Enterobacteriaceae (pp. 435–38).

Many bacteria, especially members of the family *Enterobacteriaceae,* can metabolize pyruvate to formic acid and other products in a process sometimes called the formic acid fermentation (figure 8.14, number 5). Formic acid may be converted to H_2 and CO_2 by formic hydrogenlyase (a combination of at least two enzymes).

$$HCOOH \longrightarrow CO_2 + H_2$$

There are two types of formic acid fermentation. **Mixed acid fermentation** results in the excretion of ethanol and a complex mixture of acids, particularly acetic, lactic, succinic, and formic acids (table 8.2). If formic hydrogenlyase is present, the formic acid will be degraded to H_2 and CO_2. This pattern is seen in *Escherichia, Salmonella, Proteus,* and other genera. The second type, **butanediol fermentation,** is characteristic of *Enterobacter, Serratia, Erwinia,* and some species of *Bacillus* (figure 8.14, number 4). Pyruvate is converted to acetoin, which is then reduced to 2,3-butanediol with NADH. A large amount of ethanol is also produced, together with smaller amounts of the acids found in mixed acid fermentation.

Formic acid fermentations are very useful in identification of members of the *Enterobacteriaceae.* Butanediol fermenters can be distinguished from mixed acid fermenters in three ways.

1. The Voges-Proskauer test detects the acetoin precursor of butanediol (figure 8.14) and is positive with

Box 8.1

Microbiology and World War I

The unique economic pressures of wartime sometimes provide incentive for scientific discovery. Two examples from the First World War involve the production of organic solvents by the microbial fermentation of readily available carbohydrates, such as starch or molasses.

The German side needed glycerol to make nitroglycerin. At one time, the Germans had imported their glycerol, but such imports were prevented by the British naval blockade. The German scientist Carl Neuberg knew that trace levels of glycerol were usually produced during the alcoholic fermentation of sugar by *Saccharomyces cerevisiae.* He sought to develop a modified fermentation in which the yeasts would produce glycerol instead of ethanol. Normally, acetaldehyde is reduced to ethanol by NADH and alcohol dehydrogenase (figure 8.14, pathway 2). Neuberg found that this reaction could be prevented by the addition of 3.5% sodium sulfite at pH 7.0. The bisulfite ions reacted with acetaldehyde and made it unavailable for reduction to ethanol. Because the yeast cells still had to regenerate their NAD^+ even though acetaldehyde was no longer available, Neuberg suspected that they would simply increase the rate of glycerol synthesis. Glycerol is normally produced by the reduction of dihydroxyacetone phosphate (a glycolytic intermediate) to glycerol phosphate with NADH, followed by the hydrolysis of glycerol phosphate to glycerol. Neuberg's hunch was correct, and German breweries were converted to glycerol manufacture by his procedure, eventually producing 1,000 tons of glycerol per month. Glycerol production by *S. cerevisiae* was not economically competitive under peacetime conditions and was ended. Today, glycerol is produced microbially by the halophilic alga *Dunaliella salina,* in which high concentrations of intracellular glycerol accumulate to counterbalance the osmotic pressure from the high level of extracellular salt. *Dunaliella* grows in habitats such as the Great Salt Lake of Utah and seaside rock pools.

The British side needed the organic solvents acetone and butanol. Butanol was required for the production of artificial rubber, whereas acetone was used as a solvent from nitrocellulose in the manufacture of the smokeless explosive powder cordite. Prior to 1914, acetone was made by the dry heating (pyrolysis) of wood. Between 80 and 100 tons of birch, beech, or maple wood were required to make 1 ton of acetone. When war broke out, the demand for acetone quickly exceeded the existing world supply. However, by 1915 Chaim Weizmann, a young Jewish scientist working in Manchester, England, had developed a fermentation process by which the anaerobic bacterium *Clostridium acetobutylicum* converted 100 tons of molasses or grain into 12 tons of acetone and 24 tons of butanol (most clostridial fermentations stop at butyric acid).

$$\text{2 pyruvate} \longrightarrow \text{acetoacetate} \longrightarrow \text{acetone} + CO_2$$

$$\text{Acetoacetate} \xrightarrow{\text{NADH}} \text{butyrate} \xrightarrow{\text{NADH}} \text{butanol}$$

This time, the British and Canadian breweries were converted until new fermentation facilities could be constructed. Weizmann improved the process by finding a convenient way to select high-solvent producing strains of *C. acetobutylicum.* Because the strains most efficient in these fermentations also made the most heat-resistant spores, Weizmann merely isolated the survivors from repeated 100°C heat shocks. Acetone and butanol were made commercially by this fermentation process until it was replaced by much cheaper petrochemicals in the late 1940s and 1950s. In 1948 Chaim Weizmann became the first president of the State of Israel.

butanediol fermenters, but not with mixed acid fermenters.

2. Mixed acid fermenters produce four times more acidic products than neutral ones, whereas butanediol fermenters form mainly neutral products. Thus, mixed acid fermenters acidify incubation media to a much greater extent. This is the basis of the methyl red test. The test is positive only for mixed acid fermentation because the pH drops below 4.4 and the color of the indicator changes from yellow to red.
3. CO_2 and H_2 arise in equal amounts from formic hydrogenlyase activity during mixed acid fermentation. Butanediol fermenters produce excess CO_2 and the CO_2/H_2 ratio is closer to 5:1.

TABLE 8.2 Mixed Acid Fermentation Products of *Escherichia coli*

	Fermentation Balance (μM Product/100 μM Glucose)	
	Acid Growth (pH 6.0)	Alkaline Growth (pH 8.0)
Ethanol	50	50
Formic acid	2	86
Acetic acid	36	39
Lactic acid	80	70
Succinic acid	11	15
Carbon dioxide	88	2
Hydrogen gas	75	0.5
Butanediol	0	0

Formic acid fermenters sometimes generate ATP while reoxidizing NADH. They use acetyl-CoA to synthesize acetyl phosphate, which then donates its phosphate to ADP.

$$\text{Acetyl-CoA} + P_i \longrightarrow \text{CoASH} + \text{acetyl-P}$$
$$\text{Acetyl-P} + \text{ADP} \longrightarrow \text{acetate} + \text{ATP}$$

Microorganisms carry out fermentations other than those already mentioned (Box 8.1). Protozoa and fungi often ferment sugars to lactate, ethanol, glycerol, succinate, formate, acetate, butanediol, and additional products. Substances other than sugars are also fermented. For example, many members of the genus *Clostridium* ferment amino acids, oxidizing one amino acid with a second as the electron acceptor (*see chapter 22*).

1. What are fermentations and why are they so useful to many microorganisms? Can ATP be produced during fermentation?
2. Briefly describe alcoholic, lactic acid, and formic acid fermentations. How do mixed acid fermenters and butanediol fermenters differ from each other?

Anaerobic Respiration

Electrons derived from sugars and other organic molecules are usually donated either to organic electron acceptors (fermentation) or to molecular O_2 by way of an electron transport chain (aerobic respiration). However, some bacteria have electron transport chains that can operate with inorganic electron acceptors other than O_2. The anaerobic energy-yielding process in which the electron transport chain acceptor is an oxidized inorganic molecule other than O_2 is called **anaerobic respiration.** The major electron acceptors are nitrate, sulfate, and CO_2 (table 8.3).

Some bacteria can use nitrate as the electron acceptor at the end of their electron transport chain and still produce ATP. Nitrate may be reduced to nitrite by nitrate reductase, which replaces cytochrome oxidase.

$$NO_3^- + 2e^- + 2H^+ \longrightarrow NO_2^- + H_2O$$

However, reduction of nitrate to nitrite is not a particularly effective way of making ATP, because a large amount of nitrate is required for growth (a nitrate molecule will accept only two electrons). The nitrite formed is also quite toxic. Therefore, nitrate often is further reduced all the way to nitrogen gas, a process known as **denitrification.** Each nitrate will then accept five electrons, and the product will be nontoxic.

$$2NO_3^- + 10e^- + 12H^+ \longrightarrow N_2 + 6H_2O$$

Denitrification is carried out by some members of the genera *Pseudomonas* and *Bacillus*. They use this route as an alternative to normal aerobic respiration and may be considered facultative anaerobes. If O_2 is present, these bacteria use aerobic respiration (the synthesis of nitrate reductase is repressed by O_2). Denitrification in anaerobic soil results in the loss of soil nitrogen and adversely affects soil fertility (*see chapters 40 and 42*).

TABLE 8.3 Some Electron Acceptors Used in Respiration

	Electron Acceptor	Reduced Products	Examples of Microorganisms
Aerobic	O_2	H_2O	All aerobic bacteria, fungi, protozoa, and algae
Anaerobic	NO_3^-	NO_2^-	Enteric bacteria
	NO_3^-	NO_2^-, N_2O, N_2	*Pseudomonas* and *Bacillus*
	SO_4^{2-}	H_2S	*Desulfovibrio* and *Desulfotomaculum*
	CO_2	CH_4	All methanogens
	S^0	H_2S	*Desulfuromonas* and *Thermoproteus*
	Fe^{3+}	Fe^{2+}	*Pseudomonas* and *Bacillus*

Two other major groups of bacteria employing anaerobic respiration are obligate anaerobes. Those using CO_2 or carbonate as a terminal electron acceptor are called methanogens because they reduce CO_2 to methane (*see chapter 23*). Sulfate also can act as the final acceptor in bacteria such as *Desulfovibrio*. It is reduced to sulfide (S^{2-} or H_2S), and eight electrons are accepted.

$$SO_4^{2-} + 8e^- + 8H^+ \longrightarrow S^{2-} + 4H_2O$$

Anaerobic respiration is not as efficient in ATP synthesis as aerobic respiration; that is, not as much ATP is produced by oxidative phosphorylation with nitrate, sulfate, or CO_2 as the terminal acceptors. Reduction in ATP yield arises from the fact that these alternate electron acceptors have less positive reduction potentials than O_2 (*see table 7.1*). The reduction potential difference between a donor like NADH and nitrate is smaller than the difference between NADH and O_2. Because energy yield is directly related to the magnitude of the reduction potential difference, less energy is available to make ATP in anaerobic respiration. Nevertheless, anaerobic respiration is useful because it is more efficient than fermentation and allows its possessor to make ATP by electron transport and oxidative phosphorylation in the absence of O_2.

1. Describe the process of anaerobic respiration. Is as much ATP produced in anaerobic respiration as in aerobic respiration? Why or why not? What is denitrification?

Catabolism of Carbohydrates and Intracellular Reserve Polymers

Microorganisms can catabolize many carbohydrates besides glucose. These carbohydrates may come either from outside the cell or from internal sources. Often, the initial steps in the degradation of external carbohydrate polymers differ from those employed with internal reserves.

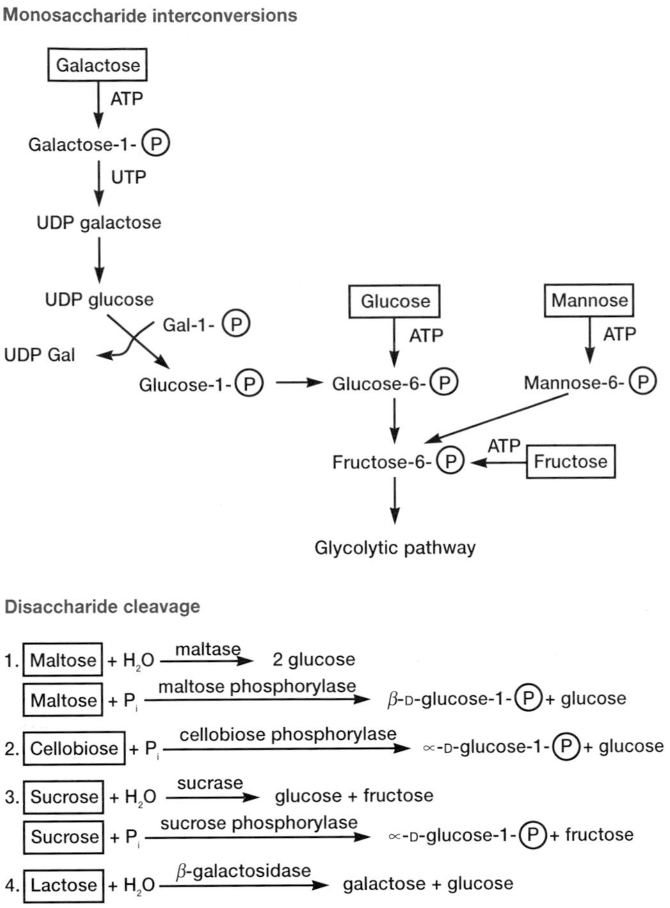

Monosaccharide interconversions

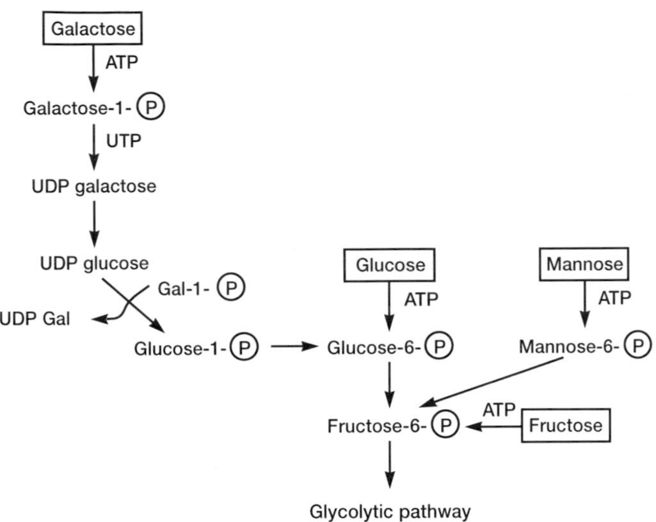

Disaccharide cleavage

1. Maltose + H_2O —maltase→ 2 glucose
   Maltose + P_i —maltose phosphorylase→ β-D-glucose-1-(P) + glucose
2. Cellobiose + P_i —cellobiose phosphorylase→ α-D-glucose-1-(P) + glucose
3. Sucrose + H_2O —sucrase→ glucose + fructose
   Sucrose + P_i —sucrose phosphorylase→ α-D-glucose-1-(P) + fructose
4. Lactose + H_2O —β-galactosidase→ galactose + glucose

Figure 8.15 **Carbohydrate Catabolism.** Examples of enzymes and pathways used in disaccharide and monosaccharide catabolism. UDP is an abbreviation for uridine diphosphate.

Carbohydrates

Figure 8.15 outlines some catabolic pathways for the monosaccharides (single sugars) glucose, fructose, mannose, and galactose. The first three are phosphorylated using ATP and easily enter the glycolytic pathway. In contrast, galactose must be converted to uridine diphosphate galactose (*see p. 175*) after initial phosphorylation, then changed into glucose 6-phosphate in a three-step process (figure 8.15).

The common disaccharides are cleaved to monosaccharides by at least two mechanisms (figure 8.15). Maltose, sucrose, and lactose can be directly hydrolyzed to their constituent sugars. Many disaccharides (e.g., maltose, cellobiose, and sucrose) are also split by a phosphate attack on the bond joining the two sugars, a process called phosphorolysis.

Polysaccharides, like disaccharides, are cleaved by both hydrolysis and phosphorolysis. Bacteria and fungi degrade external polysaccharides by secreting hydrolytic enzymes that cleave polysaccharides into smaller molecules, which can then be assimilated. Starch and glycogen are hydrolyzed by amylases to glucose, maltose, and other products. Cellulose is more difficult to digest; many fungi and a few bacteria (some gliding bacteria, clostridia, and actinomycetes) produce cellulases that hydrolyze cellulose to cellobiose and glucose. Some members of the genus *Cytophaga,* isolated from marine habitats, excrete an agarase that degrades agar. Many soil bacteria and bacterial plant pathogens degrade pectin, a polymer of galacturonic acid (a galactose derivative) that is an important constituent of plant cell walls and tissues.

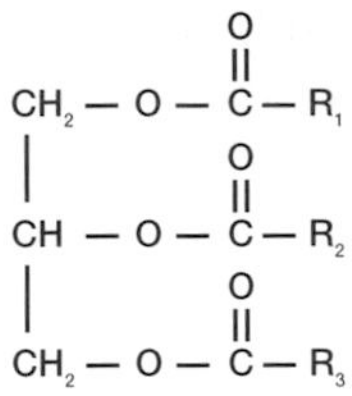

$$
\begin{array}{l}
CH_2 - O - \overset{O}{\overset{\|}{C}} - R_1 \\
\,| \\
CH - O - \overset{O}{\overset{\|}{C}} - R_2 \\
\,| \\
CH_2 - O - \overset{O}{\overset{\|}{C}} - R_3
\end{array}
$$

Figure 8.16 **A Triacylglycerol or Triglyceride.** The R groups represent the fatty acid side chains.

Reserve Polymers

Microorganisms often survive for long periods in the absence of exogenous nutrients. Under such circumstances, they catabolize intracellular stores of glycogen, starch, poly-β-hydroxybutyrate, and other energy reserves. Glycogen and starch are degraded by phosphorylases. Phosphorylases catalyze a phosphorolysis reaction that shortens the polysaccharide chain by one glucose and yields glucose 1-phosphate.

$$(\text{Glucose})_n + P_i \longrightarrow (\text{glucose})_{n-1} + \text{glucose-1-P}$$

Glucose 1-phosphate can enter the glycolytic pathway by way of glucose 6-phosphate (figure 8.15).

Poly-β-hydroxybutyrate (PHB) is an important, widespread reserve material. Its catabolism has been studied most thoroughly in *Azotobacter.* This bacterium hydrolyzes PHB to 3-hydroxybutyrate, then oxidizes the hydroxybutyrate to acetoacetate. Acetoacetate is converted to acetyl-CoA, which can be oxidized in the TCA cycle.

Lipid Catabolism

Microorganisms frequently use lipids as energy sources. Triglycerides or triacylglycerols, esters of glycerol and fatty acids (figure 8.16), are common energy sources and will serve as our example. They can be hydrolyzed to glycerol and fatty acids by microbial lipases. The glycerol is then phosphorylated, oxidized to dihydroxyacetone phosphate, and catabolized in the glycolytic pathway (figure 8.3).

Fatty acids from triacylglycerols and other lipids are often oxidized in the **β-oxidation pathway** after conversion to coenzyme A esters (figure 8.17). In this cyclic pathway, fatty acids are degraded to acetyl-CoA, which can be fed into the TCA cycle or used in biosynthesis (*see chapter 9*). One turn of the cycle produces acetyl-CoA, NADH, and $FADH_2$; NADH and $FADH_2$ can be oxidized by the electron transport chain to provide more ATP. The fatty acyl-CoA, shortened by two carbons, is ready for another turn of the cycle. Lipid fatty acids

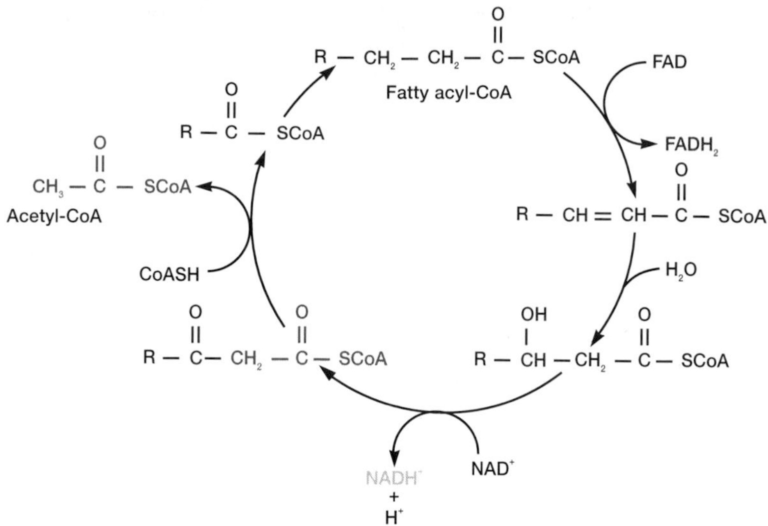

Figure 8.17 **Fatty Acid β-Oxidation.** The portions of the fatty acid being modifed are shown in color.

are a rich source of energy for microbial growth. In a similar fashion, some microorganisms grow well on petroleum hydrocarbons under aerobic conditions.

Protein and Amino Acid Catabolism

Some bacteria and fungi—particularly pathogenic, food spoilage, and soil microorganisms—can use proteins as their source of carbon and energy. They secrete **protease** enzymes that hydrolyze proteins and polypeptides to amino acids, which are transported into the cell and catabolized.

The first step in amino acid utilization is **deamination,** the removal of the amino group from an amino acid. This is often accomplished by **transamination.** The amino group is transferred from an amino acid to an α-keto acid acceptor (figure 8.18). The organic acid resulting from deamination can be converted to pyruvate, acetyl-CoA, or a TCA cycle intermediate and eventually oxidized in the TCA cycle to release energy. It also can be used as a source of carbon for the synthesis of cell constituents. Excess nitrogen from deamination may be excreted as ammonium ion, thus making the medium alkaline.

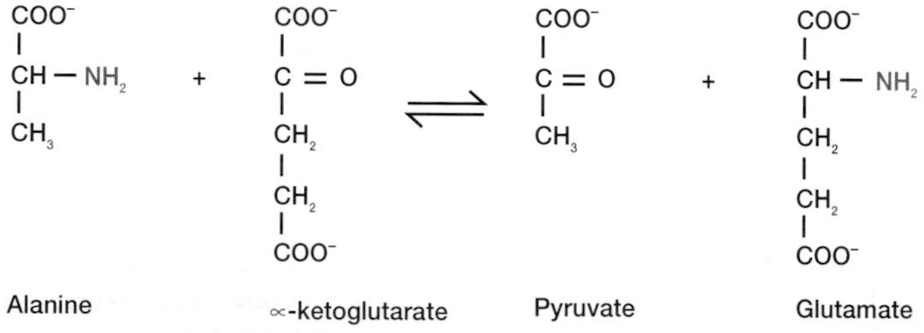

1. Briefly discuss the ways in which microorganisms degrade and use common monosaccharides, disaccharides, and polysaccharides from both external and internal sources.
2. Describe how a microorganism might derive carbon and energy from the lipids and proteins in its diet. What is β-oxidation? Transamination?

Oxidation of Inorganic Molecules

As we have seen, microorganisms can oxidize organic molecules such as carbohydrates, lipids, and proteins and synthesize ATP with the energy liberated. The electron acceptor is (1) another more oxidized organic molecule in fermentation, (2) O_2 in aerobic respiration, and (3) an oxidized inorganic molecule other than O_2 in anaerobic respiration (figure 8.19). In both aerobic and anaerobic respiration, ATP is formed as a result of electron transport chain activity. Electrons for the chain can be obtained from inorganic nutrients, and, it is possible to derive energy from the oxidation of inorganic molecules rather than from organic nutrients. This ability is confined to a small group of bacteria called **chemolithotrophs** (*see chapter 23*). Each species is rather specific in its preferences for electron donors and acceptors (table 8.4). The acceptor is usually O_2, but sulfate and nitrate are also used. The most common electron donors are hydrogen, reduced nitrogen compounds, reduced sulfur compounds, and ferrous iron (Fe^{2+}).

Chemolithotrophic bacteria are usually autotrophic and employ the Calvin cycle to fix CO_2 as their carbon source (*see chapter 9*). However, some chemolithotrophs can function as heterotrophs if reduced organic compounds are available. Considerable energy is required to reduce CO_2 to carbohydrate; incorporation of one CO_2 in the Calvin cycle requires three ATPs and two NADPHs. Moreover, much less energy is available from the oxidation of inorganic molecules (table 8.5)

Alanine ($COO^- $ / $CH—NH_2$ / CH_3) + α-ketoglutarate (COO^- / $C=O$ / CH_2 / CH_2 / COO^-) ⇌ Pyruvate (COO^- / $C=O$ / CH_3) + Glutamate (COO^- / $CH—NH_2$ / CH_2 / CH_2 / COO^-)

Figure 8.18 **Transamination.** A common example of this process. The α-amino group (blue) of alanine is transferred to the acceptor α-ketoglutarate forming pyruvate and glutamate. The pyruvate can be catabolized in the tricarboxylic acid cycle or used in biosynthesis.

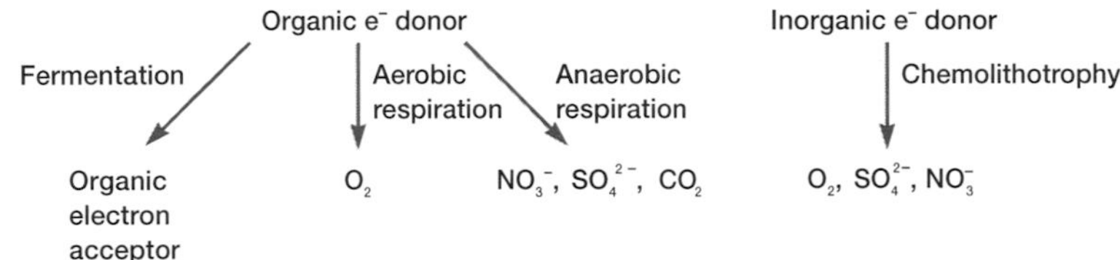

Figure 8.19 **Patterns of Energy Production.** An organic electron donor can donate electrons to an organic electron acceptor (fermentation), oxygen (aerobic respiration), or an inorganic acceptor other than oxygen (anaerobic respiration). In chemolithotrophy, a reduced inorganic molecule donates electrons for energy production.

TABLE 8.4 Representative Chemolithotrophs and Their Energy Sources

Bacteria	Electron Donor	Electron Acceptor	Products
Alcaligenes and *Pseudomonas* spp.	H_2	O_2	H_2O
Nitrobacter	NO_2^-	O_2	NO_3^-, H_2O
Nitrosomonas	NH_4^+	O_2	NO_2^-, H_2O
Thiobacillus denitrificans	S^0, H_2S	NO_3^-	SO_4^{2-}, N_2
Thiobacillus ferrooxidans	Fe^{2+}, S^0, H_2S	O_2	Fe^{3+}, H_2O, H_2SO_4

TABLE 8.5 Energy Yields from Oxidations Used by Chemolithotrophs

Reaction	$\Delta G^{o\prime}$ (kcal/mole)[a]
$H_2 + ½O_2 \longrightarrow H_2O$	− 56.6
$NO_2^- + ½O_2 \longrightarrow NO_3^-$	− 17.4
$NH_4^+ + 1½O_2 \longrightarrow NO_2^- + H_2O + 2H^+$	− 65.0
$S^0 + 1½O_2 + H_2O \longrightarrow H_2SO_4$	−118.5
$S_2O_3^{2-} + 2O_2 + H_2O \longrightarrow 2SO_4^{2-} + 2H^+$	−223.7
$2Fe^{2+} + 2H^+ + ½O_2 \longrightarrow 2Fe^{3+} + H_2O$	− 11.2

[a]The $\Delta G^{o\prime}$ for complete oxidation of glucose to CO_2 is −686 kcal/mole. A kcal is equivalent to 4.184kJ.

than from the complete oxidation of glucose to CO_2, which is accompanied by a standard free energy change of −686 kcal/mole. The P/O ratios for oxidative phosphorylation in chemolithotrophs are probably around 1.0 (in the oxidation of hydrogen it is significantly higher). Because the yield of ATP is so low, chemolithotrophs must oxidize a large quantity of inorganic material to grow and reproduce, and this magnifies their ecological impact.

Several bacterial genera (table 8.4) can oxidize hydrogen gas to produce energy because they possess a hydrogenase enzyme that catalyzes the oxidation of hydrogen.

$$H_2 \longrightarrow 2H^+ + 2e^-$$

The electrons are donated either to an electron transport chain or to NAD^+, depending upon the hydrogenase. If NADH is produced, it can be used in ATP synthesis by electron transport and oxidative phosphorylation, with O_2 as the terminal electron acceptor.

The best-studied nitrogen-oxidizing chemolithotrophs are the **nitrifying bacteria** (*see chapter 23*). These are soil and aquatic bacteria of considerable ecological significance. *Nitrosomonas* oxidizes ammonia to nitrite.

$$NH_4^+ + 1½O_2 \longrightarrow NO_2^- + H_2O + 2H^+$$

The nitrite can then be further oxidized by *Nitrobacter* to yield nitrate.

$$NO_2^- + ½O_2 \longrightarrow NO_3^-$$

When these two genera work together, ammonia in the soil is oxidized to nitrate in a process called **nitrification.**

The role of chemolithotrophs in soil and aquatic ecosystems (pp. 809–13).

Energy released upon the oxidation of both ammonia and nitrite is used to make ATP by oxidative phosphorylation. However, microorganisms need a source of electrons (reducing power) as well as a source of ATP in order to reduce CO_2 and other molecules. Since molecules like ammonia and nitrite have more positive reduction potentials than NAD^+, they cannot directly donate their electrons to form the required NADH and NADPH. This is because electrons spontaneously move only from donors with more negative reduction potentials to acceptors with more positive potentials (*see chapter 7*). Sulfur-oxidizing bacteria face the same difficulty. Both types of chemolithotrophs solve this problem by using protonmotive force or ATP to reverse the flow of electrons in their electron transport chains and reduce NAD^+ with electrons from nitrogen and sulfur donors (figure 8.20). Because energy is used to generate NADH as well as ATP, the net yield of ATP is fairly low. Chemolithotrophs can afford this inefficiency as they have no serious competitors for their unique energy sources.

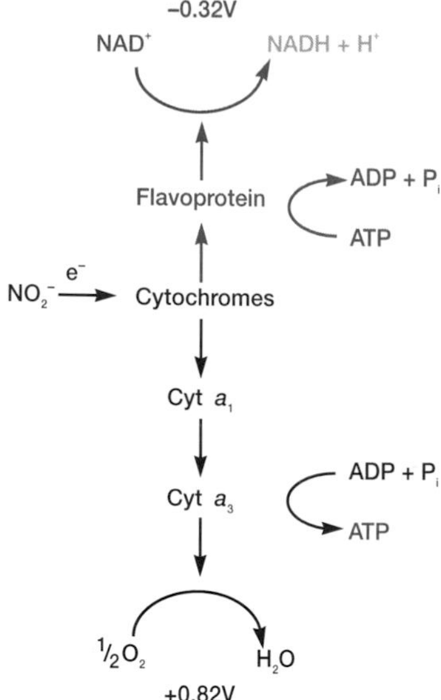

Figure 8.20 Reversed Electron Flow. The flow of electrons in the transport chain of *Nitrobacter.* Electrons flowing from nitrite to oxygen (down the reduction potential gradient) will release energy. It requires protonmotive force or ATP energy to force electrons to flow in the reverse direction from nitrite to NAD^+.

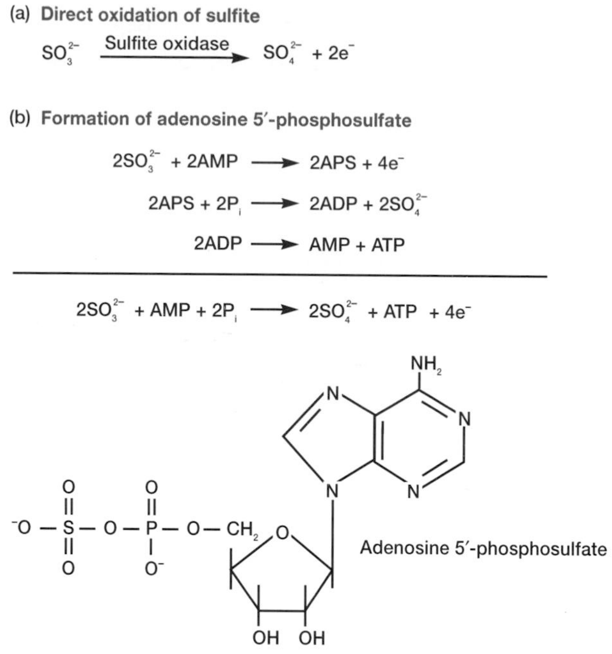

(a) Direct oxidation of sulfite

$$SO_3^{2-} \xrightarrow{\text{Sulfite oxidase}} SO_4^{2-} + 2e^-$$

(b) Formation of adenosine 5′-phosphosulfate

$$2SO_3^{2-} + 2AMP \longrightarrow 2APS + 4e^-$$

$$2APS + 2P_i \longrightarrow 2ADP + 2SO_4^{2-}$$

$$2ADP \longrightarrow AMP + ATP$$

$$2SO_3^{2-} + AMP + 2P_i \longrightarrow 2SO_4^{2-} + ATP + 4e^-$$

Adenosine 5′-phosphosulfate

Figure 8.21 Energy Generation by Sulfur Oxidation. (*a*) Sulfite can be directly oxidized to provide electrons for electron transport and oxidative phosphorylation. (*b*) Sulfite can also be oxidized and converted to APS. This route produces electrons for use in electron transport and ATP by substrate-level phosphorylation with APS.

The sulfur-oxidizing bacteria are the third major group of chemolithotrophs. The metabolism of *Thiobacillus* has been best studied. These bacteria oxidize sulfur (S^0), hydrogen sulfide (H_2S), thiosulfate ($S_2O_3^{2-}$), and other reduced sulfur compounds to sulfuric acid; therefore, they have a significant ecological impact (Box 8.2). Interestingly, they generate ATP by both oxidative phosphorylation and substrate-level phosphorylation involving **adenosine 5′-phosphosulfate (APS).** APS is a high-energy molecule formed from sulfite and adenosine monophosphate (figure 8.21).

Some of these bacteria are extraordinarily flexible metabolically. For example, *Sulfolobus brierleyi* and a few other species can grow aerobically as sulfur-oxidizing bacteria; in the absence of O_2, they carry out anaerobic respiration with molecular sulfur as an electron acceptor.

Sulfur-oxidizing bacteria, like other chemolithotrophs, can use CO_2 as their carbon source. Many will grow heterotrophically if they are supplied with reduced organic carbon sources like glucose or amino acids.

1. How do chemolithotrophs obtain their ATP and NADH? What is their source of carbon?
2. Describe energy production by hydrogen-oxidizing bacteria, nitrifying bacteria, and sulfur-oxidizing bacteria.

Photosynthesis

Microorganisms cannot only derive energy from the oxidation of inorganic and organic compounds, but many can capture light energy and use it to synthesize ATP and NADH or NADPH. This process in which light energy is trapped and converted to chemical energy is called **photosynthesis.** Usually, a photosynthetic organism reduces and incorporates CO_2, reactions that are also considered part of photosynthesis. Photosynthesis is one of the most significant metabolic processes on earth because almost all our energy is ultimately derived from solar energy (*see chapter 40*). It provides photosynthetic organisms with the ATP and NAD(P)H necessary to manufacture the organic material required for growth. In turn, these organisms serve as the base of most food chains in the biosphere. Photosynthesis is also responsible for replenishing our supply of O_2, a remarkable process carried out by a variety of organisms, both eucaryotic and procaryotic (table 8.6). Although most people associate photosynthesis with the more obvious higher plants, over half the photosynthesis on earth is carried out by microorganisms.

Photosynthesis as a whole is divided into two parts. In the **light reactions,** light energy is trapped and converted to chemical energy. This energy is then used to reduce or fix CO_2 and synthesize cell constituents in the **dark reactions.** In this section, the nature of the light reactions is discussed; the dark reactions are reviewed in the next chapter.

The photosynthetic dark reactions (pp. 173–75).

The Light Reaction in Eucaryotes and Cyanobacteria

All photosynthetic organisms have pigments for the absorption of light. The most important of these pigments are the **chlorophylls.** Chlorophylls are large planar rings composed of

Box 8.2

Acid Mine Drainage

Each year millions of tons of sulfuric acid flow down the Ohio River from the Appalachian Mountains. This sulfuric acid is of microbial origin and leaches enough metals from the mines to make the river reddish and acidic. The primary culprit is *Thiobacillus ferrooxidans,* a chemolithotrophic bacterium that derives its energy from oxidizing ferrous ion to ferric ion and sulfide ion to sulfate ion. The combination of these two energy sources is important because of the solubility properties of iron. Ferrous ion is somewhat soluble and can be formed at pH values of 3.0 or less in moderately reducing environments. However, when the pH is greater than 4.0 to 5.0, ferrous ion is oxidized to ferric ion by O_2 in the water and precipitates as a hydroxide. If the pH drops below about 2.0 to 3.0 because of sulfuric acid production by spontaneous oxidation of sulfur or sulfur oxidation by thiobacilli and other bacteria, the ferrous ion remains reduced, soluble, and available as an energy source. Remarkably, *T. ferrooxidans* grows well at such acid pHs and actively oxidizes ferrous ion to an insoluble ferric precipitate. The water is rendered toxic for most aquatic life and unfit for human consumption.

The ecological consequences of this metabolic life-style arise from the common presence of pyrite (FeS_2) in coal mines. The bacteria oxidize both elemental components of pyrite for their growth and, in the process, form sulfuric acid, which leaches the remaining minerals.

$$2FeS_2 + 7O_2 + 2H_2O \xrightarrow{\text{Autoxidation or bacterial action}} 2Fe^{2+} + 4SO_4^{2-} + 4H^+$$

$$2Fe^{2+} + \tfrac{1}{2}O_2 + 2H^+ \xrightarrow{\textit{T. ferrooxidans}} 2Fe^{3+} + H_2O$$

Pyrite oxidation is further accelerated because the ferric ion generated by bacterial activity readily oxidizes more pyrite to sulfuric acid and ferrous ion. In turn, the ferrous ion supports further bacterial growth. It is difficult to prevent *T. ferrooxidans* growth as it only requires pyrite and common inorganic salts. Because *T. ferrooxidans* gets its O_2 and CO_2 from the air, the only feasible method of preventing its damaging growth is to seal the mines to render the habitat anaerobic.

TABLE 8.6 Diversity of Photosynthetic Organisms

Eucaryotic Organisms	Procaryotic Organisms
Higher plants	Cyanobacteria (blue-green algae)
Multicellular green, brown, and red algae	Green sulfur bacteria
Unicellular algae (e.g., euglenoids, dinoflagellates, diatoms)	Green nonsulfur bacteria
	Purple sulfur bacteria
	Purple nonsulfur bacteria
	Prochloron

four substituted pyrrole rings with a magnesium atom coordinated to the four central nitrogen atoms (figure 8.22). Several chlorophylls are found in eucaryotes, the two most important of which are chlorophyll *a* and chlorophyll *b* (figure 8.22). These two molecules differ slightly in their structure and spectral properties. When dissolved in acetone, chlorophyll *a* has a light absorption peak at 665 nm; the corresponding peak for chlorophyll *b* is at 645 nm. In addition to absorbing red light, chlorophylls also absorb blue light strongly (the second absorption peak for chlorophyll *a* is at 430 nm). Because chlorophylls absorb primarily in the red and blue ranges, green light is transmitted. Consequently, many photosynthetic organisms are green in color. The long hydrophobic tail attached to the chlorophyll ring aids in its attachment to membranes, the site of the light reactions.

Other photosynthetic pigments also trap light energy. The most widespread of these are the **carotenoids,** long molecules, usually yellowish in color, that possess an extensive conjugated double bond system (figure 8.23). β-carotene is present in *Prochloron* and most divisions of algae; fucoxanthin is found in diatoms, dinoflagellates, and brown algae (*Phaeophyta*). Red algae and cyanobacteria have photosynthetic pigments called **phycobiliproteins,** consisting of a protein with a tetrapyrrole attached (figure 8.23). **Phycoerythrin** is a red pigment with a maximum absorption around 550 nm, and **phycocyanin** is blue (maximum absorption at 620 to 640 nm).

Carotenoids and phycobiliproteins are often called **accessory pigments** because of their role in photosynthesis. Although chlorophylls cannot absorb light energy effectively in the blue-green through yellow range (about 470 to 630 nm), accessory pigments do absorb light in this region and transfer the trapped energy to chlorophyll. In this way, they make photosynthesis more efficient over a broader range of wavelengths. Accessory pigments also protect microorganisms from intense sunlight, which could oxidize and damage the photosynthetic apparatus in their absence.

Chlorophylls and accessory pigments are assembled in highly organized arrays called **antennas,** whose purpose is to

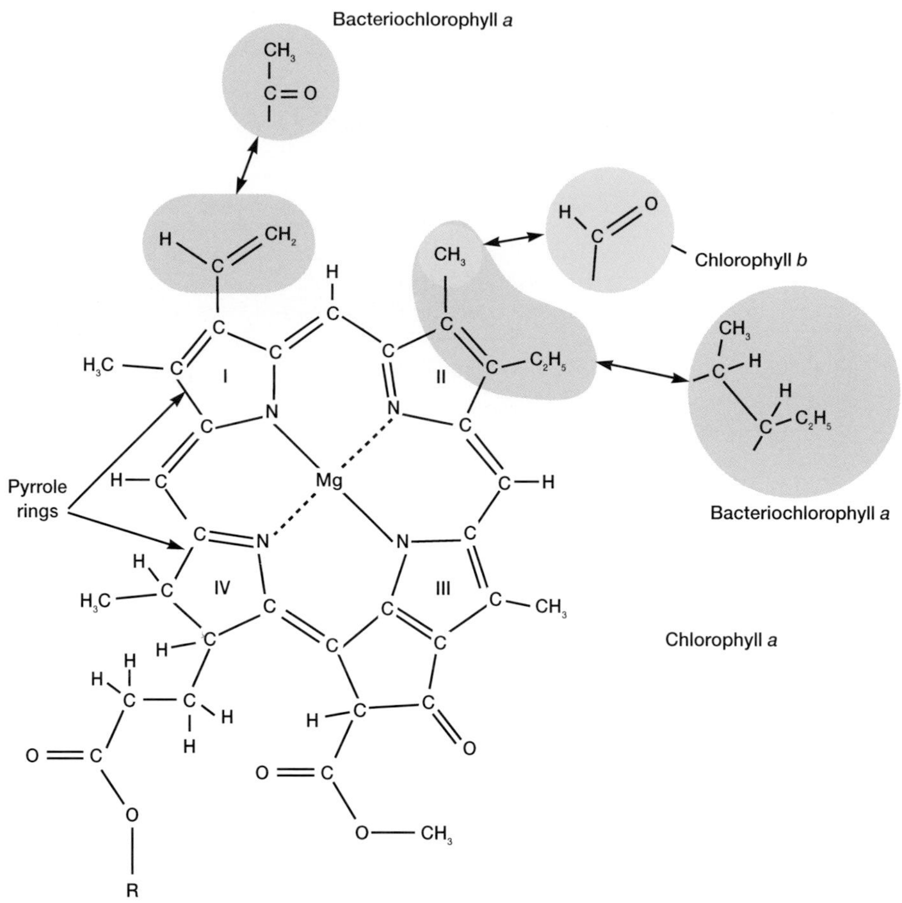

Figure 8.22 **Chlorophyll Structure.** The structures of chlorophyll *a*, chlorophyll *b*, and bacteriochlorophyll *a*. The complete structure of chlorophyll *a* is given. Only one group is altered to produce chlorophyll *b*, and two modifications in the ring system are required to change chlorophyll *a* to bacteriochlorophyll *a*. The side chain (R) of bacteriochlorophyll *a* may be either phytyl (a 20-carbon chain also found in chlorophylls *a* and *b*) or geranylgeranyl (a 20-carbon side chain similar to phytyl, but with three more double bonds).

create a large surface area in order to trap as many photons as possible. An antenna has about 300 chlorophyll molecules. Light energy is captured in an antenna and transferred from chlorophyll to chlorophyll until it reaches a special **reaction-center chlorophyll** directly involved in photosynthetic electron transport (figure 8.24). In eucaryotic cells and cyanobacteria, there are two kinds of antennas associated with two different photosystems. **Photosytem I** absorbs longer wavelength light ($\geq$ 680 nm) and funnels the energy to a special chlorophyll *a* molecule called P700. The term P700 signifies that this molecule most effectively absorbs light at a wavelength of 700 nm. **Photosystem II** traps light at shorter wavelengths ($\leq$ 680 nm) and transfers its energy to the special chlorophyll P680.

When the photosystem I antenna transfers light energy to the reaction-center P700 chlorophyll, P700 absorbs the energy and is excited; its reduction potential becomes very negative. It then donates its excited or high-energy electron to a specific acceptor, probably a special chlorophyll *a* molecule (A) or an iron-sulfur protein (figure 8.25). The electron is eventually transferred to ferredoxin and can then travel in either of two directions. In the cyclic pathway (the dashed lines in figure 8.25), the electron moves in a cyclic route through a series of electron carriers and back to the oxidized P700. The pathway is termed cyclic because the electron from P700 returns to P700 after traveling through the photosynthetic electron transport chain. ATP is formed during cyclic electron transport in the region of cytochrome b_6. This process is called **cyclic photophosphorylation** because electrons travel in a cyclic pathway and ATP is formed. Only photosystem I participates.

Electrons also can travel in a noncyclic pathway involving both photosystems. P700 is excited and donates electrons to ferredoxin as before. In the noncyclic route, however, reduced ferredoxin reduces $NADP^+$ to NADPH (figure 8.25). Because the electrons contributed to $NADP^+$ cannot be used to reduce oxidized P700, photosystem II participation is required. It donates electrons to oxidized P700 and generates

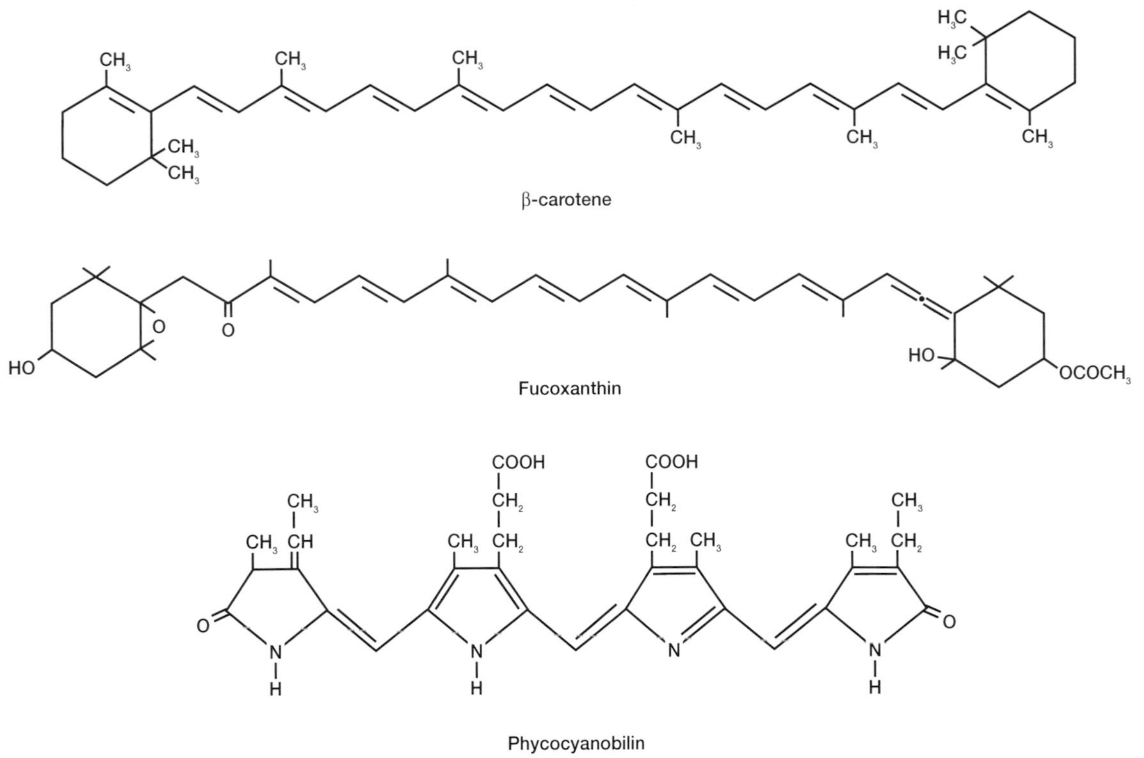

β-carotene

Fucoxanthin

Phycocyanobilin

Figure 8.23 **Representative Accessory Pigments.** Beta-carotene is a carotenoid found in algae and higher plants. Note that it has a long chain of alternating double and single bonds called conjugated double bonds. Fucoxanthin is a carotenoid accessory pigment in several divisions of algae. Phycocyanobilin is an example of a linear tetrapyrrole that is attached to a protein to form a phycobiliprotein.

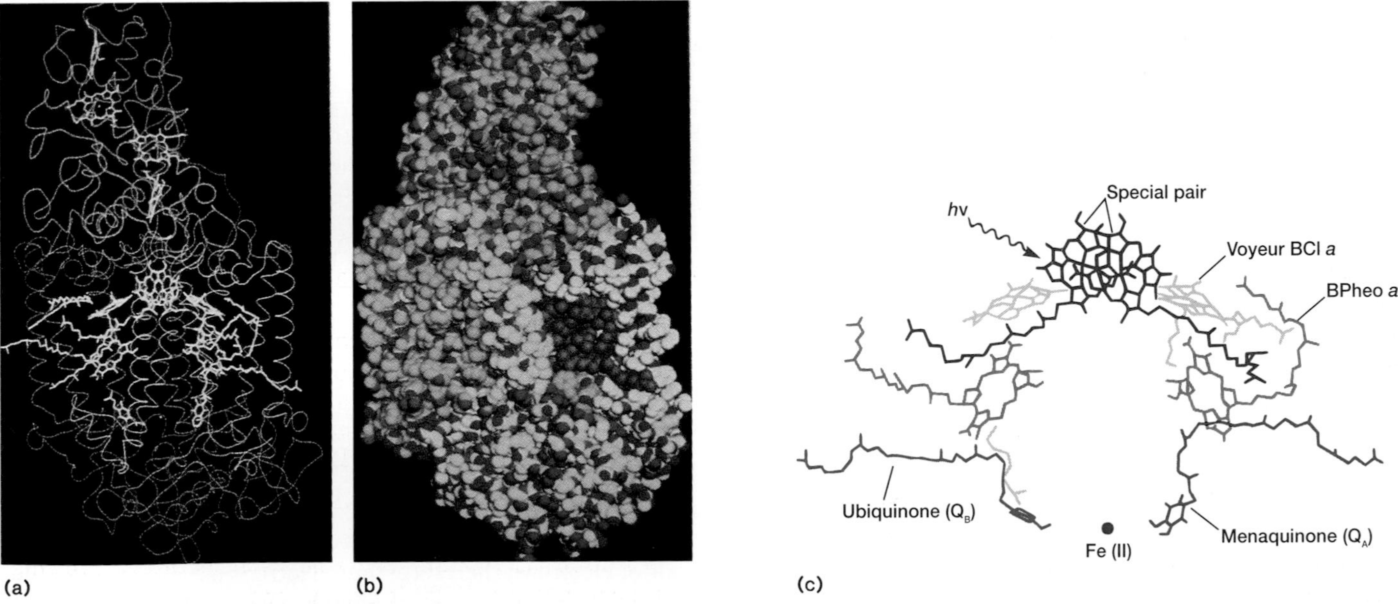

Figure 8.24 **A Photosynthetic Reaction Center.** The reaction center of the purple nonsulfur bacterium, *Rhodopseudomonas viridis.* (*a*) The structure of the C_α backbone of the center's polypeptide chains with the bacteriochlorophylls and other prosthetic groups in yellow. (*b*) A space-filling model of the reaction center. The nitrogen atoms are blue; oxygen atoms, red; and sulfur atoms, yellow. The exposed prosthetic group atoms are brown. (*c*) A close-up view of the reaction center prosthetic groups. A photon is first absorbed by the "special pair" of bacteriochlorophyll *a* molecules, thus exciting them. An excited electron then moves to the bacteriophaeophytin molecule in the right arm of the system. *(c) From Donald Voet and Judith G. Voet,* Biochemistry. *Copyright © 1990 John Wiley & Sons, Inc., New York, NY.*

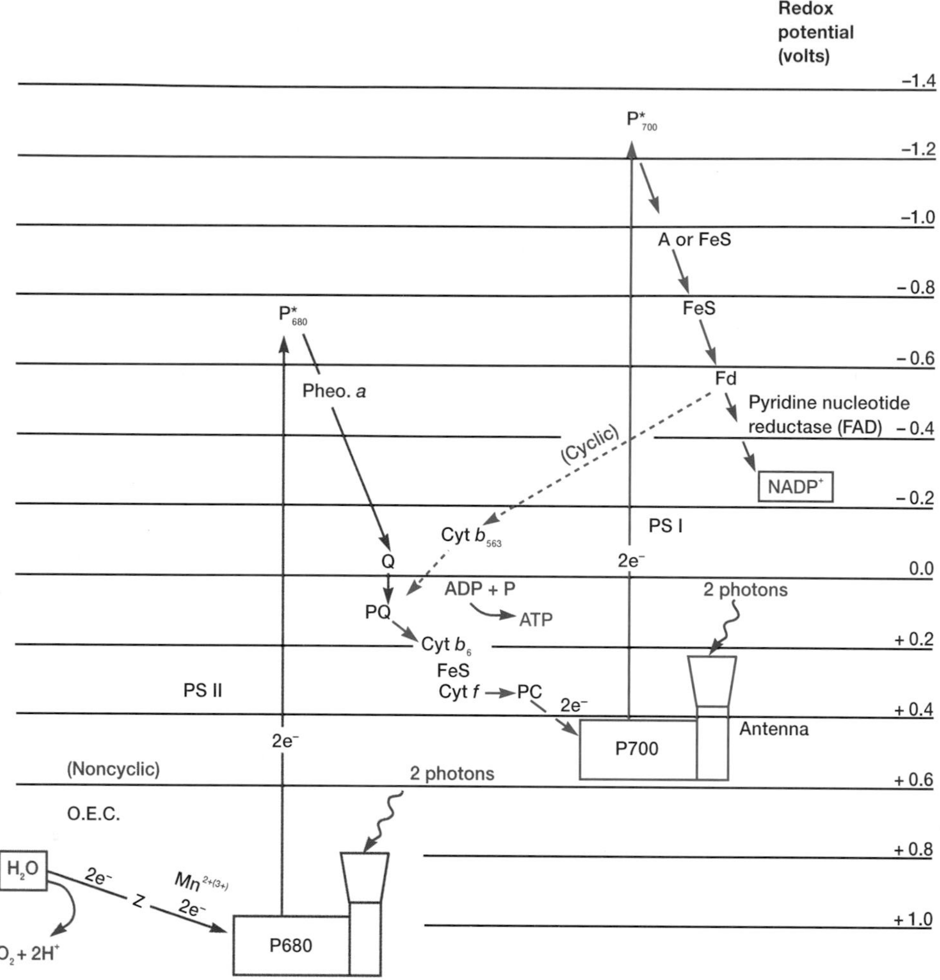

Figure 8.25 **Green Plant Photosynthesis.** Electron flow during photosynthesis in higher plants. Cyanobacteria and eucaryotic algae are similar in having two photosystems, although they may differ in some details. The carriers involved in electron transport are ferredoxin (Fd) and other FeS proteins; cytochromes b_6, b_{563}, and *f;* plastoquinone (PQ); copper containing plastocyanin (PC); pheophytin *a* (Pheo. *a*); possibly chlorophyll *a* (A); and the unknown quinone Q, which is probably a plastoquinone. Both photosystem I (PS I) and photosystem II (PS II) are involved in noncyclic photophosphorylation; only PS I participates in cyclic photophosphorylation. See the text for further details.

ATP in the process. The photosystem II antenna absorbs light energy and excites P680, which then reduces pheophytin *a*. Pheophytin *a* is chlorophyll *a* in which two hydrogen atoms have replaced the central magnesium. Electrons subsequently travel to Q (probably a plastoquinone) and down the electron transport chain to P700. Oxidized P680 then obtains an electron from the oxidation of water to O_2. Thus, electrons flow from water all the way to $NADP^+$ with the aid of energy from two photosystems, and ATP is synthesized by **noncyclic photophosphorylation.** It appears that one ATP and one NADPH are formed when two electrons travel through the noncyclic pathway.

Just as is true of mitochondrial electron transport, photosynthetic electron transport takes place within a membrane. Chloroplast granal membranes contain both photosystems and their antennas. Figure 8.26 shows a thylakoid membrane carrying out noncyclic photophosphorylation by the chemiosmotic mechanism. It is believed that stromal lamellae possess only photosystem I and are involved in cyclic photophosphorylation alone. In cyanobacteria, photosynthetic light reactions are also located in membranes.

The dark reactions require three ATPs and two NADPHs to reduce one CO_2 and use it to synthesize carbohydrate.

$$CO_2 + 3ATP + 2NADPH + 2H^+ + H_2O \longrightarrow (CH_2O) + 3ADP + 3P_i + 2NADP^+$$

The noncyclic system generates one NADPH and one ATP per pair of electrons; therefore, four electrons passing through the system will produce two NADPHs and two ATPs. A total of 8 quanta of light energy (4 quanta for each photosystem) is needed to propel the four electrons from water to $NADP^+$. Because the ratio of ATP to NADPH required for CO_2 fixation is 3:2, at least one more ATP must be supplied. Cyclic photophosphorylation probably operates independently to generate the extra ATP. This requires absorption of another 2 to 4 quanta. It follows that around 10 to 12 quanta of light energy are needed to reduce and incorporate one molecule of CO_2 during photosynthesis.

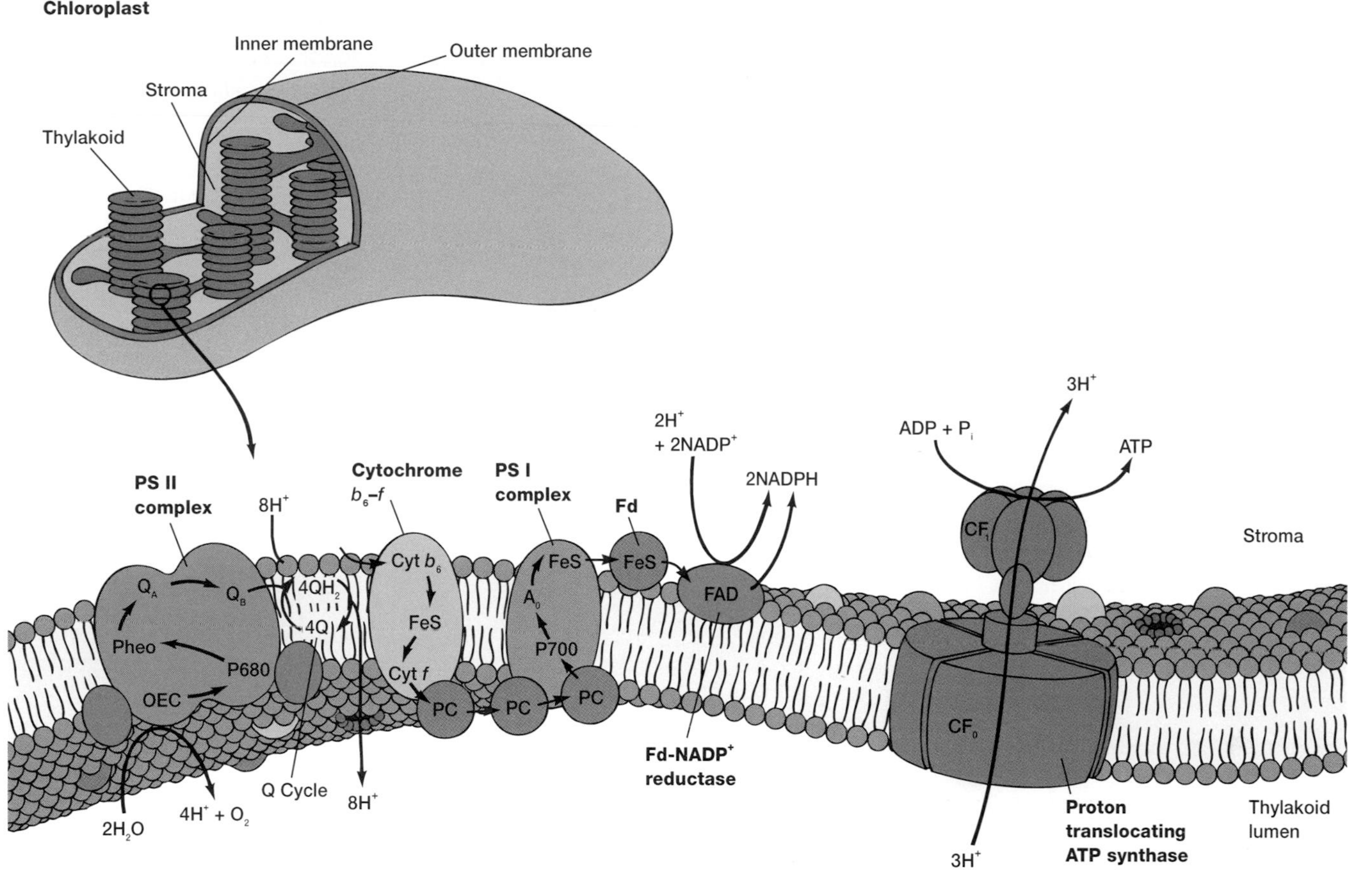

Figure 8.26 **The Mechanism of Photosynthesis.** An illustration of the chloroplast thylakoid membrane showing photosynthetic electron transport chain function and noncyclic photophosphorylation. The chain is composed of three complexes: PS I, the cytochrome b_6–f complex, and PS II. Two diffusable electron carriers connect the three complexes. Plastoquinone (Q) connects PS I with cytochrome b_6–f, and plastocyanin (PC) connects cytochrome b_6–f with PS II. The light-driven electron flow pumps protons across the thylakoid membrane and generates an electrochemical gradient, which can then be used to make ATP. Water serves as the source of electrons and the oxygen-evolving complex (OEC) produces O_2. *From Donald Voet and Judith G. Voet,* Biochemistry. *Copyright © 1990 John Wiley & Sons, Inc., New York, NY. Reprinted by permission. Originally from D. R. Ort and N. E. Good, Trends in Biochemical Sciences, 13:469, 1988. Reprinted by permission of Elsevier Science Publishers, Cambridge, UK.*

TABLE 8.7 Properties of Microbial Photosynthetic Systems

Property	Eucaryotes	Cyanobacteria	Green and Purple Bacteria
Photosynthetic pigment	Chlorophyll *a*	Chlorophyll *a*	Bacteriochlorophyll
Photosystem II	Present	Present	Absent
Photosynthetic electron donors	H_2O	H_2O	H_2, H_2S, S, organic matter
O_2 production pattern	Oxygenic	Oxygenic[a]	Anoxygenic
Primary products of energy conversion	ATP + NADPH	ATP + NADPH	ATP
Carbon source	CO_2	CO_2	Organic and/or CO_2

[a]Some cyanobacteria can function anoxygenically under certain conditions.

The Light Reaction in Green and Purple Bacteria

Green and purple photosynthetic bacteria differ from cyanobacteria and eucaryotic photosynthesizers in several fundamental ways (table 8.7). In particular, green and purple bacteria do not use water as an electron source or produce O_2 photosynthetically; that is, they are **anoxygenic.** In contrast, cyanobacteria and eucaryotic photosynthesizers are almost always **oxygenic.** NADPH is not directly produced in the photosynthetic light reaction of purple bacteria. Green bacteria can reduce NAD^+ directly during the light reaction. To synthesize NAD(P)H, green and purple bacteria must use electron donors like hydrogen, hydrogen sulfide, elemental sulfur, and organic compounds that have more negative reduction potentials than water and are therefore easier to oxidize (better electron donors). Finally, green and purple bacteria possess slightly different photosynthetic pigments, **bacteriochlorophylls** (figure 8.22), many with absorption maxima at longer

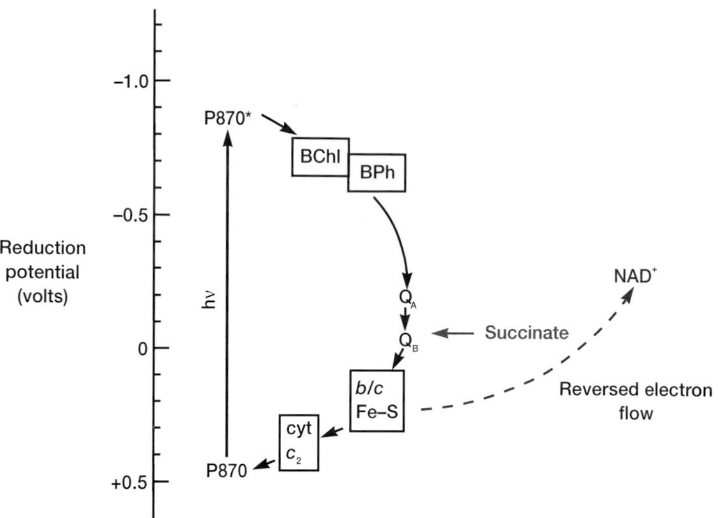

Figure 8.27 Purple Nonsulfur Bacterial Photosynthesis. The photosynthetic electron transport system in the purple nonsulfur bacterium, *Rhodobacter sphaeroides*. This scheme is very incomplete and tentative. Ubiquinone (Q_A and Q_B) is very similar to coenzyme Q. BPhe stands for bacteriopheophytin. NAD^+ and the electron source succinate are in color.

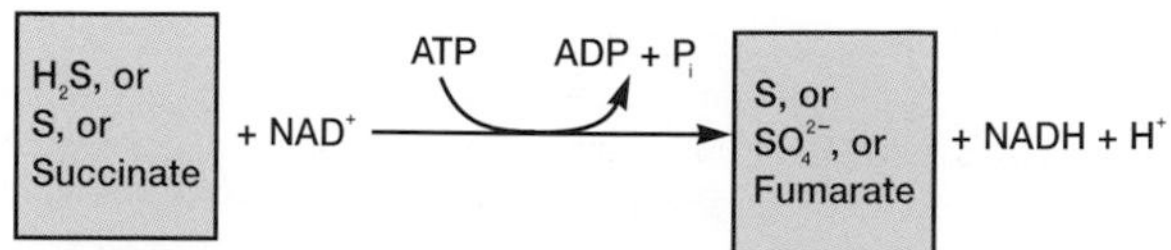

Figure 8.28 NAD Reduction in Green and Purple Bacteria. The use of reversed electron flow to reduce NAD^+. The arrow in this diagram represents an electron transport chain that is being driven in reverse by protonmotive force or ATP. That is, electrons are moving from donors with more positive reduction potentials to an acceptor (NAD^+) with a more negative potential.

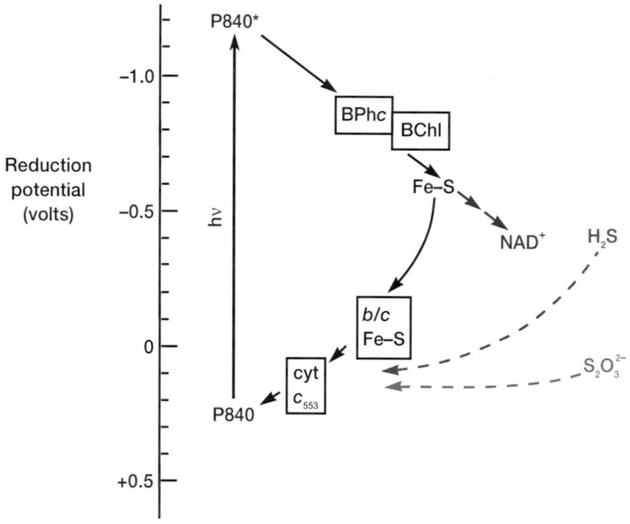

Figure 8.29 Green Sulfur Bacterial Photosynthesis. The photosynthetic electron transport system in the green sulfur bacterium, *Chlorobium limicola*. Light energy is used to make ATP by cyclic photophosphorylation and move electrons from sulfur donors (green and blue) to NAD^+ (red).

wavelengths. Bacteriochlorophylls *a* and *b* have maxima in ether at 775 and 790 nm, respectively. *In vivo* maxima are about 830 to 890 nm (bacteriochlorophyll *a*) and 1020 to 1040 nm (Bchl *b*). This shift of absorption maxima into the infrared region better adapts these bacteria to their ecological niches (*see figure 23.3*).

The biology of photosynthetic bacteria (pp. 468–76).

Many differences found in green and purple bacteria are due to their lack of photosystem II; they cannot use water as an electron donor in noncyclic electron transport. Without photosystem II, they cannot produce O_2 from H_2O photosynthetically and are restricted to cyclic photophosphorylation. Indeed, almost all purple and green sulfur bacteria are strict anaerobes. A tentative scheme for the photosynthetic electron transport chain of a purple nonsulfur bacterium is given in figure 8.27. When the special reaction-center chlorophyll P870 is excited, it donates an electron to bacteriophaeophytin. Electrons then flow to quinones and through an electron transport chain back to P870 while driving ATP synthesis.

Green and purple bacteria face a further problem because they also require NADH or NADPH for CO_2 incorporation. They may synthesize NADH in at least three ways. If they are growing in the presence of hydrogen gas, which has a reduction potential more negative than that of NAD^+, the hydrogen can be used directly to produce NADH. Like chemolithotrophs, many photosynthetic purple bacteria use ATP or protonmotive force to reverse the flow of electrons in an electron transport chain and move them from inorganic or organic donors to NAD^+ (figures 8.27 and 8.28). Green sulfur bacteria appear to carry out a simple form of noncyclic photosynthetic electron flow to reduce NAD^+ (figure 8.29).

1. Describe photosynthesis as carried out by eucaryotes and cyanobacteria. How does photosynthesis in green and purple bacteria differ?
2. Define the following terms: light reaction, chlorophyll, carotenoid, phycobiliprotein, accessory pigment, antenna, photosystems I and II, cyclic photophosphorylation, noncyclic photophosphorylation, anoxygenic and oxygenic photosynthesis, and bacteriochlorophyll.

Summary

1. Metabolism is the total of all chemical reactions in a cell and may be divided into catabolism and anabolism.
2. Amphibolic pathways like glycolysis and the tricarboxylic acid cycle have both catabolic and anabolic functions.
3. The glycolytic or Embden-Meyerhof pathway occurs by way of fructose 1,6-bisphosphate with the net production of two NADHs and two ATPs, the latter being produced by substrate-level phosphorylation.
4. In the pentose phosphate pathway, glucose 6-phosphate is oxidized twice and converted to pentoses and other sugars. It is a source of NADPH, ATP, pentoses, and tetroses.
5. In the Entner-Doudoroff pathway, glucose is oxidized to 6-phosphogluconate, which is then dehydrated and cleaved to pyruvate and glyceraldehyde 3-phosphate. The latter product can be oxidized by glycolytic enzymes to provide ATP and NADH.
6. The tricarboxylic acid cycle is the final stage of catabolism in most aerobic cells. It oxidizes acetyl-CoA to CO_2 and forms one GTP, three NADHs, and one $FADH_2$ per acetyl-CoA.
7. The NADH and $FADH_2$ produced from the oxidation of carbohydrates, fatty acids, and other nutrients can be oxidized in the electron transport chain. Electrons flow from carriers with more negative reduction potentials, to those with more positive potentials, and free energy is released for ATP synthesis by oxidative phosphorylation.
8. Bacterial electron transport chains are often different from eucaryotic chains. In eucaryotes, the P/O ratio for NADH is about 3 and that for $FADH_2$ is around 2; P/O ratios are usually much lower in bacteria.
9. Eucaryotic ATP synthesis takes place on small protein spheres, the F_1 ATPases or ATP synthases, located on the inner surface of the inner mitochondrial membrane.
10. The most widely accepted mechanism of oxidative phosphorylation is the chemiosmotic hypothesis in which protonmotive force drives ATP synthesis. However, other mechanisms have been proposed.
11. When the glycolytic pathway operates anaerobically, only 2 ATPs per glucose are formed, whereas aerobic respiration in eucaryotes can yield a maximum of 38 ATPs.
12. In the absence of O_2, a microorganism often uses an oxidized organic molecule as an electron acceptor to reoxidize the NADH formed during glycolysis. A catabolic process that employs an organic molecule as the electron acceptor is called fermentation.
13. Anaerobic respiration is the process of ATP production by electron transport in which the terminal electron acceptor is an oxidized inorganic molecule other than O_2. The most common acceptors are nitrate, sulfate, and CO_2.
14. Microorganisms catabolize many extracellular carbohydrates. Monosaccharides are taken in and phosphorylated; disaccharides may be cleaved to monosaccharides by either hydrolysis or phosphorolysis.
15. External polysaccharides are degraded by hydrolysis and the products are absorbed. Intracellular glycogen and starch are converted to glucose 1-phosphate by phosphorolysis.
16. Fatty acids from lipid catabolism are usually oxidized to acetyl-CoA in the β-oxidation pathway.
17. Proteins are hydrolyzed to amino acids that are then deaminated; their carbon skeletons feed into the TCA cycle.
18. Chemolithotrophs or chemoautotrophs synthesize ATP by oxidizing reduced inorganic compounds—usually hydrogen, reduced nitrogen and sulfur compounds, or ferrous iron—with an electron transport chain and O_2 as the electron acceptor.
19. Chemolithotrophs usually incorporate CO_2 through the Calvin cycle and produce NADH by reversing electron transport.
20. In photosynthesis, eucaryotes and cyanobacteria trap light energy with chlorophyll and accessory pigments and move electrons through photosystems I and II to make ATP and NAD(P)H (the light reactions).
21. Cyclic photophosphorylation involves the activity of photosystem I alone. In noncyclic photophosphorylation, photosystems I and II operate together to move electrons from water to $NADP^+$ producing ATP, NADPH, and O_2.
22. Photosynthetic green and purple bacteria differ from eucaryotes and cyanobacteria in possessing bacteriochlorophyll and lacking photosystem II. Thus, they cannot use the noncyclic pathway to form NADPH and O_2; they are anoxygenic.

Key Terms

accessory pigments *163*
acetyl coenzyme A (acetyl-CoA) *150*
adenosine 5′-phosphosulfate (APS) *162*
alcoholic fermentation *156*
amphibolic pathways *146*
anabolism *146*
anaerobic respiration *158*
anoxygenic photosynthesis *167*
antenna *163*
ATP synthase *154*
bacteriochlorophyll *167*
β-oxidation pathway *159*
butanediol fermentation *156*
carotenoids *163*
catabolism *146*
chemiosmotic hypothesis *153*
chemolithotroph *160*
chlorophylls *162*
citric acid cycle *150*
conformational change hypothesis *153*
cyclic photophosphorylation *164*
dark reactions *162*
deamination *160*
denitrification *158*
electron transport chain *152*
Embden-Meyerhof pathway *147*
Entner-Doudoroff pathway *150*
fermentation *155*
glycolysis *147*
glycolytic pathway *147*
heterolactic fermenters *156*
hexose monophosphate pathway *148*
homolactic fermenters *156*
lactic acid fermentation *156*
light reactions *162*
metabolism *146*
mixed acid fermentation *156*
nitrification *161*
nitrifying bacteria *161*
noncyclic photophosphorylation *166*
oxidative phosphorylation *152*
oxygenic photosynthesis *167*
Pasteur effect *155*
pentose phosphate pathway *148*
photosynthesis *162*
photosystem I *164*
photosystem II *164*
phycobiliproteins *163*
phycocyanin *163*
phycoerythrin *163*
protease *160*
protonmotive force (PMF) *153*
reaction-center chlorophyll *164*
substrate-level phosphorylation *148*
transamination *160*
tricarboxylic acid (TCA) cycle *150*
uncouplers *154*

Questions for Thought and Review

1. Why might it be desirable for a microorganism with the Embden-Meyerhof pathway and the TCA cycle also to have the pentose phosphate pathway?
2. Of what advantage would it be to a microorganism to possess an electron transport chain and oxidative phosphorylation?
3. Describe two ways in which a microorganism can continue to produce energy when O_2 is absent.
4. Why would it be wasteful for anaerobic microorganisms to operate the complete TCA cycle?
5. How do substrate-level phosphorylation and oxidative phosphorylation differ from one another?
6. Can fermentation products be used in identifying bacteria? Give some examples if the answer is yes.
7. Describe what would happen to microbial metabolism if the enzyme lactate dehydrogenase was inhibited in a homolactic fermenter growing anaerobically on a medium containing glucose as the carbon source. What effects would an inhibitor of the ATP synthase have on an aerobically respiring microorganism? An uncoupler?
8. How would you isolate a thermophilic chemolithotroph that uses sulfur compounds as a source of electrons? What changes in the incubation system would be needed to isolate bacteria using sulfur compounds in anaerobic respiration? How can one tell which process is taking place through an analysis of the sulfur molecules present in the medium?
9. Suppose that you isolated a bacterial strain that carried out oxygenic photosynthesis. What photosystems would it possess and what group of bacteria would it most likely belong to?

Additional Reading

Anraku, Y. 1988. Bacterial electron transport chains. *Ann. Rev. Biochem.* 57:101–32.

Cramer, W. A., and Knaff, D. B. 1991. *Energy transduction in biological membranes. A textbook of bioenergetics.* New York: Springer-Verlag.

Dawes, E. A. 1986. *Microbial energetics.* New York: Chapman.

Dawes, I. W., and Sutherland, I. W. 1992. *Microbial physiology,* 2d ed. London: Blackwell Scientific Publications.

Deisenhofer, J.; Michel, H.; and Huber, R. 1985. The structural basis of photosynthetic light reactions in bacteria. *Trends Biochem. Sci.* 10(6):243–48.

Ferguson, S. J. 1987. Denitrification: A question of the control and organization of electron and ion transport. *Trends Biochem. Sci.* 12(9):354–57.

Gottschalk, G. 1986. *Bacterial metabolism.* 2d ed. New York: Springer-Verlag.

Govindjee, and Coleman, W. J. 1990. How plants make oxygen. *Sci. Am.* 262(2):50–58.

Grant, W. D. 1987. The enigma of the alkaliphile. *Microbiol. Sci.* 4(8):251–55.

Hatefi, Y. 1985. The mitochondrial electron transport and oxidative phosphorylation system. *Ann. Rev. Biochem.* 54:1015–69.

Hochstein, L. I., and Tomlinson, G. A. 1988. The enzymes associated with denitrification. *Ann. Rev. Microbiol.* 42:231–61.

Jones, C. W. 1982. *Bacterial respiration and photosynthesis.* Washington, D.C.: American Society for Microbiology.

Kelly, D. P. 1985. Physiology of the thiobacilli: Elucidating the sulphur oxidation pathway. *Microbiol Sci.* 2(4):105–9.

Mandelstam, J.; McQuillen, K.; and Dawes, I. 1982. *Biochemistry of bacterial growth,* 3d ed. London: Blackwell Scientific Publications.

Mathews, C. K., and van Holde, K. E. 1990. *Biochemistry.* Redwood City, Calif.: Benjamin/Cummings.

Moat, A. G., and Foster, J. W. 1988. *Microbial physiology,* 2d ed. New York: John Wiley and Sons.

Nugent, J. H. A. 1984. Photosynthetic electron transport in plants and bacteria. *Trends Biochem. Sci.* 9(8):354–57.

Postgate, J. R. 1984. *The sulphate-reducing bacteria,* 2d ed. New York: Cambridge University Press.

Quayle, J. R., and Ferenci, T. 1978. Evolutionary aspects of autotrophy. *Microbiol. Rev.* 42(2): 251–73.

Rose, A. H. 1976. *Chemical microbiology: An introduction to microbial physiology,* 3d ed. New York: Plenum.

Schlegel, H. G., and Bowien, B., eds. 1989. *Autotrophic bacteria.* Madison, Wis.: Science Tech Publishers.

Smith, A. J., and Hoare, D. S. 1977. Specialist phototrophs, lithotrophs, and methylotrophs: A unity among a diversity of procaryotes? *Bacteriol. Rev.* 41(1):419–48.

Spencer, M. E., and Guest, J. R. 1987. Regulation of citric acid cycle genes in facultative bacteria. *Microbiol. Sci.* 4(6):164–68.

Staehelin, L. A., and Arntzen, C. J., eds. 1986. *Photosynthesis III: Photosynthetic membranes and light harvesting systems. Encyclopedia of plant physiology.* New Series. *Volume 19.* New York: Springer-Verlag.

Stewart, V. 1988. Nitrate respiration in relation to facultative metabolism in enterobacteria. *Microbiol. Rev.* 52(2):190–232.

Stryer, L. 1988. *Biochemistry,* 3d ed. New York: Freeman.

Trumpower, B. L. 1990. Cytochrome bc_1 complexes of microorganisms. *Microbiol. Rev.* 54(2):101–29.

Voet, D., and Voet, J. G. 1990. *Biochemistry.* New York: John Wiley and Sons.

Youvan, D. C., and Marrs, B. L. 1987. Molecular mechanisms of photosynthesis. *Sci. Am.* 256(6): 42–48.

Zubay, G. 1993. *Biochemistry,* 3d ed. Dubuque, Iowa: Wm. C. Brown Communications, Inc.

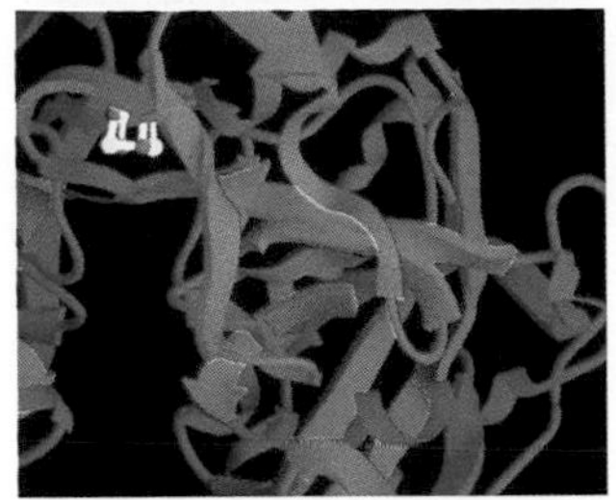

CHAPTER 9

Metabolism: The Use of Energy in Biosynthesis

Biological structures are almost always constructed in a hierarchical manner, with subassemblies acting as important intermediates en route from simple starting molecules to the end products of organelles, cells, and organisms.

—W. M. Becker and D. W. Deamer

Outline

Concepts

1. In anabolism or biosynthesis, cells use free energy to construct more complex molecules and structures from smaller, simpler precursors.
2. Biosynthetic pathways are organized to optimize efficiency by conserving genetic storage space, biosynthetic raw materials, and energy.
3. Autotrophs use ATP and NADPH from photosynthesis or from oxidation of inorganic molecules to reduce CO_2 and incorporate it into organic material.
4. Catabolic and anabolic pathways may differ in enzymes, regulation, intracellular location, and use of cofactors and nucleoside diphosphate carriers. Although many enzymes of amphibolic pathways participate in both catabolism and anabolism, some pathway enzymes are involved only in one of the two processes.
5. Phosphorus, in the form of phosphate, can be directly assimilated, whereas inorganic sulfur and nitrogen compounds must often be reduced before incorporation into organic material.
6. The tricarboxylic acid (TCA) cycle acts as an amphibolic pathway and requires anaplerotic reactions to maintain adequate levels of cycle intermediates.
7. Most glycolytic enzymes participate in both the synthesis and catabolism of glucose. In contrast, fatty acids are synthesized from acetyl-CoA and malonyl-CoA by a pathway quite different from fatty acid β-oxidation.
8. Peptidoglycan synthesis is a complex, multistep process that is begun in the cytoplasm and completed at the cell wall after the peptidoglycan repeat unit has been transported across the cell membrane.

As the last chapter makes clear, microorganisms can obtain energy in many ways. Much of this energy is used in biosynthesis or anabolism. During biosynthesis, a microorganism begins with simple precursors, such as inorganic molecules and monomers, and constructs ever more complex molecules until new organelles and cells arise (figure 9.1). A microbial cell must manufacture many different kinds of molecules; however, it is possible to discuss the synthesis of only the most important types of cell constituents.

This chapter begins with a general introduction to anabolism, then focuses on the synthesis of carbohydrates, amino acids, purines and pyrimidines, and lipids. It also describes the assimilation of CO_2, phosphorus, sulfur, and nitrogen. The chapter ends with a section on the synthesis of peptidoglycan and bacterial cell walls. Protein and nucleic acid synthesis is so significant and complex that it is described separately in chapter 10.

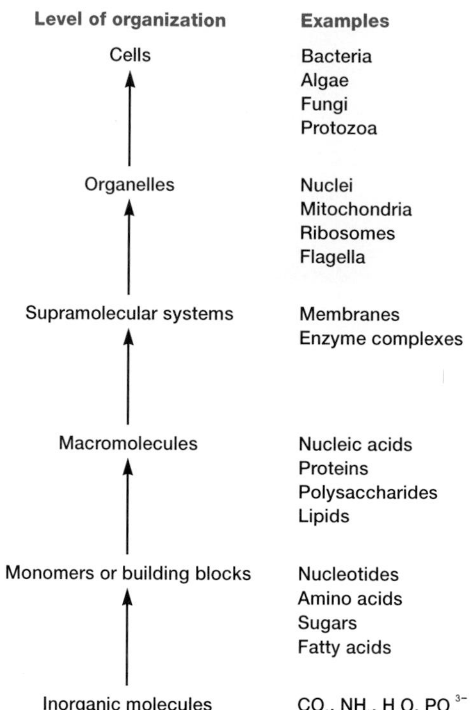

Figure 9.1 **The Construction of Cells.** The biosynthesis of procaryotic and eucaryotic cell constituents. Biosynthesis is organized in levels of ever greater complexity.

Because anabolism is the creation of order and a cell is highly ordered and immensely complex, much energy is required for biosynthesis. This is readily apparent from estimates of the biosynthetic capacity of rapidly growing *Escherichia coli* (table 9.1). Although most ATP dedicated to biosynthesis is employed in protein synthesis, ATP is also used to make other cell constituents.

Free energy is required for biosynthesis in mature cells of constant size because cellular molecules are continuously being degraded and resynthesized, a process known as **turnover.** Cells are never the same from instant to instant. Despite the continuous turnover of cell constituents, metabolism is carefully regulated so that the rate of biosynthesis is approximately balanced by that of catabolism. In addition to the energy expended in the turnover of molecules, many nongrowing cells also use energy to synthesize enzymes and other substances for release into their surroundings.

Regulation of metabolism (chapter 11).

Table 9.1 Biosynthesis in *Escherichia coli*

Cell Constituent	Number of Molecules per Cell[a]	Molecules Synthesized per Second	Molecules of ATP Required per Second for Synthesis
DNA	1	0.00083	60,000
RNA	15,000	12.5	75,000
Polysaccharides	39,000	32.5	65,000
Lipids	15,000,000	12,500.0	87,000
Proteins	1,700,000	1,400.0	2,120,000

From A. L. Lehninger, *Bioenergetics*. Copyright 1971 Benjamin/Cummings Publishing Co., Menlo Park, CA.

[a]Estimates for a cell with a volume of 2.25 μm^3, a total weight of 1×10^{-12} g, a dry weight of 2.5×10^{-13} g, and a 20-minute cell division cycle.

Principles Governing Biosynthesis

A microbial cell contains large quantities of proteins, nucleic acids, and polysaccharides, all of which are **macromolecules** or very large molecules that are polymers of smaller units joined together. The construction of large, complex molecules from a few simple structural units or **monomers** saves much genetic storage capacity, biosynthetic raw material, and energy. A consideration of protein synthesis clarifies this. Proteins—whatever size, shape, or function—are made of only 20 common amino acids joined by peptide bonds (*see appendix I*). Different proteins simply have different amino acid sequences, but not new and dissimilar amino acids. Suppose that proteins were composed of 40 different amino acids instead of 20. The cell would then need the enzymes to manufacture twice as many amino acids (or would have to obtain the extra amino acids in its diet). Genes would be required for the extra enzymes, and the cell would have to invest raw materials and energy in the synthesis of these additional genes, enzymes, and amino acids. Clearly, the use of a few monomers linked together by a single type of covalent bond makes the synthesis of macromolecules a very efficient process. Almost all cell structures are built mainly of about 30 small precursors.

The cell often saves additional materials and energy by using many of the same enzymes for both catabolism and anabolism. For example, most glycolytic enzymes are involved in the synthesis and the degradation of glucose.

Although it is true that many enzymes in amphibolic pathways (*see chapter 8*) participate in both catabolic and an-

Figure 9.2 **A Hypothetical Biosynthetic Pathway.** The routes connecting *G* with *X, Y,* and *Z* are purely anabolic because they are used only for synthesis of the end products. The pathway from *A* to *G* is amphibolic; that is, it has both catabolic and anabolic functions. Most reactions are used in both roles; however, the interconversion of *C* and *D* is catalyzed by two separate enzymes, E_1 (catabolic) and E_2 (anabolic).

abolic activities, some steps are catalyzed by two different enzymes. One enzyme catalyzes the reaction in the catabolic direction, the other reverses this conversion (figure 9.2). Thus, catabolic and anabolic pathways are never identical although many enzymes are shared. Use of separate enzymes for the two directions of a single step permits independent regulation of catabolism and anabolism. Although chapter 11 discusses this in more detail, note that the regulation of anabolism is somewhat different from that of catabolism. Both types of pathways can be regulated by their end products as well as by the concentrations of ATP, ADP, AMP, and NAD^+. Nevertheless, end product regulation generally assumes more importance in anabolic pathways.

In order to synthesize molecules efficiently, anabolic pathways must operate irreversibly in the direction of biosynthesis. Cells can achieve this by connecting some biosynthetic reactions to the breakdown of ATP and other nucleoside triphosphates. When these two processes are coupled, the free energy made available during nucleoside triphosphate breakdown drives the biosynthetic reaction to completion (Box 9.1; *see also chapter 7*).

In eucaryotic microorganisms, biosynthetic pathways are frequently located in different cellular compartments from their corresponding catabolic pathways. For example, fatty acid biosynthesis occurs in the cytoplasmic matrix while fatty acid oxidation takes place within the mitochondrion. Compartmentation makes it easier for the pathways to operate simultaneously but independently.

Finally, anabolic and catabolic pathways often use different cofactors. Usually, catabolic oxidations produce NADH, a substrate for electron transport. In contrast, when a reductant is needed during biosynthesis, NADPH rather than NADH normally serves as the donor. Fatty acid metabolism provides a second example. Fatty acyl-CoA molecules are oxidized to generate energy, whereas fatty acid synthesis involves acyl carrier protein thioesters.

After macromolecules have been constructed from simpler precursors, they are assembled into larger, more complex structures such as supramolecular systems and organelles (figure 9.1). Macromolecules normally contain the necessary information to form spontaneously in a process known as **self-assembly.** For example, ribosomes are large assemblages of many proteins and ribonucleic acid molecules; yet, they arise by the self-assembly of their components without the involvement of extra factors.

1. Define biosynthesis or anabolism and turnover.
2. List six principles by which biosynthetic pathways are organized.

The Photosynthetic Fixation of CO_2

Although most microorganisms can incorporate or fix CO_2, at least in anaplerotic reactions (pp. 181–83), only autotrophs use CO_2 as their sole or principal carbon source. The reduction and incorporation of CO_2 requires much energy. Usually autotrophs obtain energy by trapping light during photosynthesis, but some derive energy from the oxidation of reduced inorganic electron donors. Autotrophic CO_2 fixation is crucial to life on earth because it provides the organic matter upon which heterotrophs depend.

Photosynthetic light reactions and chemolithotrophy (pp. 160–68).

Almost all microbial autotrophs incorporate CO_2 by a special metabolic pathway called by several names: the **Calvin cycle,** Calvin-Benson cycle, photosynthetic carbon reduction (PCR) cycle, or reductive pentose phosphate cycle. The Calvin cycle is found in the chloroplast stroma of eucaryotic microbial autotrophs. Cyanobacteria, some nitrifying bacteria, and thiobacilli possess **carboxysomes.** These are polyhedral inclusion bodies that contain the enzyme ribulose-1,5-bisphosphate carboxylase (see following section). They may be the site of CO_2 fixation or may store the carboxylase and other proteins. It is easiest to understand the cycle if it is divided into three parts: (1) the carboxylation phase, (2) the reduction phase, and (3) the regeneration phase.

The Carboxylation Phase

Carbon dioxide fixation is accomplished by the enzyme **ribulose-1,5-bisphosphate carboxylase** (figure 9.3) that catalyzes the addition of CO_2 to ribulose 1,5-bisphosphate (RuBP), forming two molecules of 3-phosphoglycerate (PGA).

Box 9.1

The Identification of Anabolic Pathways

There are three approaches to the study of pathway organization: (1) study of the pathway *in vitro,* (2) use of nutritional mutants, and (3) incubation of cells with precursors labeled radioisotopically. In vitro (Latin, in glass) studies employ cell-free extracts to search for enzymes and metabolic intermediates that might belong to a pathway. Although this direct approach was used to work out the organization of many catabolic pathways, progress in research on biosynthesis was slow until the other two techniques were developed in the early to middle 1940s.

Techniques using nutritional mutants were developed during Beadle and Tatum's work on the genetics of the red bread mold *Neurospora.* This approach is best illustrated with a hypothetical example. Suppose that a pathway for the synthesis of end product Z is organized with E_1, E_2, and so on, representing pathway enzymes.

$$A \xrightarrow{E_1} B \xrightarrow{E_2} C \xrightarrow{E_3} Z$$

The prototroph (*see chapter 5*), which will grow in medium lacking Z, can be treated with mutagenic agents such as ultraviolet light, X rays, or chemical mutagens. Some resulting mutants will be auxotrophs that require the presence of Z for growth because one of their biosynthetic enzymes is now inactive. When E_3 is inactive, the microorganism will only grow in the presence of Z, even though it can make C from the precursor A. When grown in the presence of a small amount of Z, intermediate C (the intermediate just before the blocked step) will accumulate in the medium. In this way, a variety of mutants can be used to establish the identity of pathway intermediates. The order of intermediates can be determined by cross-feeding experiments. If E_2 has been inactivated by a mutation, the mutant will only grow when either C or Z is supplied. Because the medium in which the E_3 mutant has been cultured contains intermediate C, it will support growth of the E_2 mutant (other mutants would not produce enough C to support growth). If cross-feeding experiments are conducted with mutants of each step in the pathway, the steps can be placed in the correct order. Application of this technique quickly led to the elucidation of the pathways for the synthesis of tryptophan, folic acid, and other molecules.

Radioisotopes like ^{14}C are used in the third approach to studying pathway organization. Potential biosynthetic precursors are synthesized in the laboratory with specific atoms made radioactive. The microorganism is then incubated with culture medium containing the radioactive molecule, and the biosynthetic end product is isolated and analyzed. If the molecule truly is a precursor of the end product, the latter should be radioactive. The location of the radioactive atom will tell one what part of the product is contributed by the radioactively labeled precursor. Precisely the same approach can be employed with nonradioactive atoms like ^{15}N. This technique provided some of the first information about the nature of the purine biosynthetic pathway.

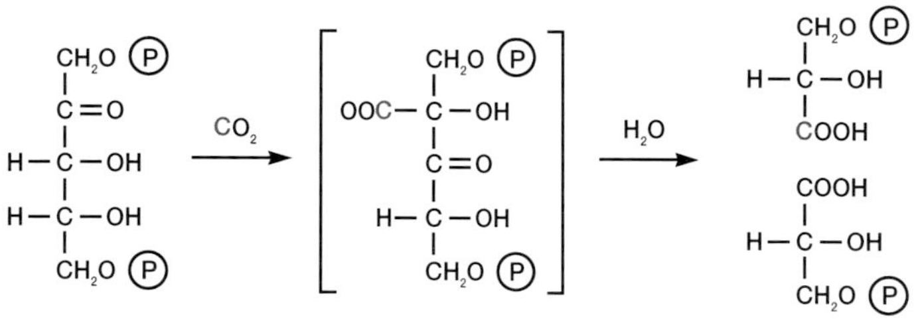

Ribulose 1,5-bisphosphate (RuBP) $\xrightarrow{CO_2}$ [unstable intermediate] $\xrightarrow{H_2O}$ 3-phosphoglycerate (PGA)

Figure 9.3 The Ribulose-1,5-Bisphosphate Carboxylase Reaction. This enzyme catalyzes the addition of carbon dioxide to ribulose 1,5-bisphosphate, forming an unstable intermediate, which then breaks down to two molecules of 3-phosphoglycerate.

The Reduction Phase

After PGA is formed by carboxylation, it is reduced to glyceraldehyde 3-phosphate. The reduction, carried out by two enzymes, is essentially a reversal of a portion of the glycolytic pathway, although the glyceraldehyde-3-phosphate dehydrogenase differs from the glycolytic enzyme in using $NADP^+$ rather than NAD^+ (figure 9.4).

The Regeneration Phase

The third phase of the PCR cycle regenerates RuBP and produces carbohydrates such as fructose and glucose (figure 9.4). This portion of the cycle is similar to the pentose phosphate pathway, and involves the transketolase and transaldolase reactions. The cycle is completed when phosphoribulokinase reforms RuBP.

To synthesize fructose 6-phosphate or glucose 6-phosphate from CO_2, the cycle must operate six times to yield the desired hexose and reform the six RuBP molecules.

$$6RuBP + 6CO_2 \longrightarrow 12PGA \longrightarrow 6RuBP + fructose\text{-}6\text{-}P$$

The incorporation of one CO_2 into organic material requires three ATPs and two NADPHs. The formation of glucose from CO_2 may be summarized by the following equation.

$$6CO_2 + 18ATP + 12NADPH + 12H^+ + 12H_2O \longrightarrow glucose + 18ADP + 18P_i + 12NADP^+$$

ATP and NADPH are provided by photosynthetic light reactions or by oxidation of inorganic molecules in chemoautotrophs. Sugars formed in the PCR cycle can then be used to synthesize other essential molecules.

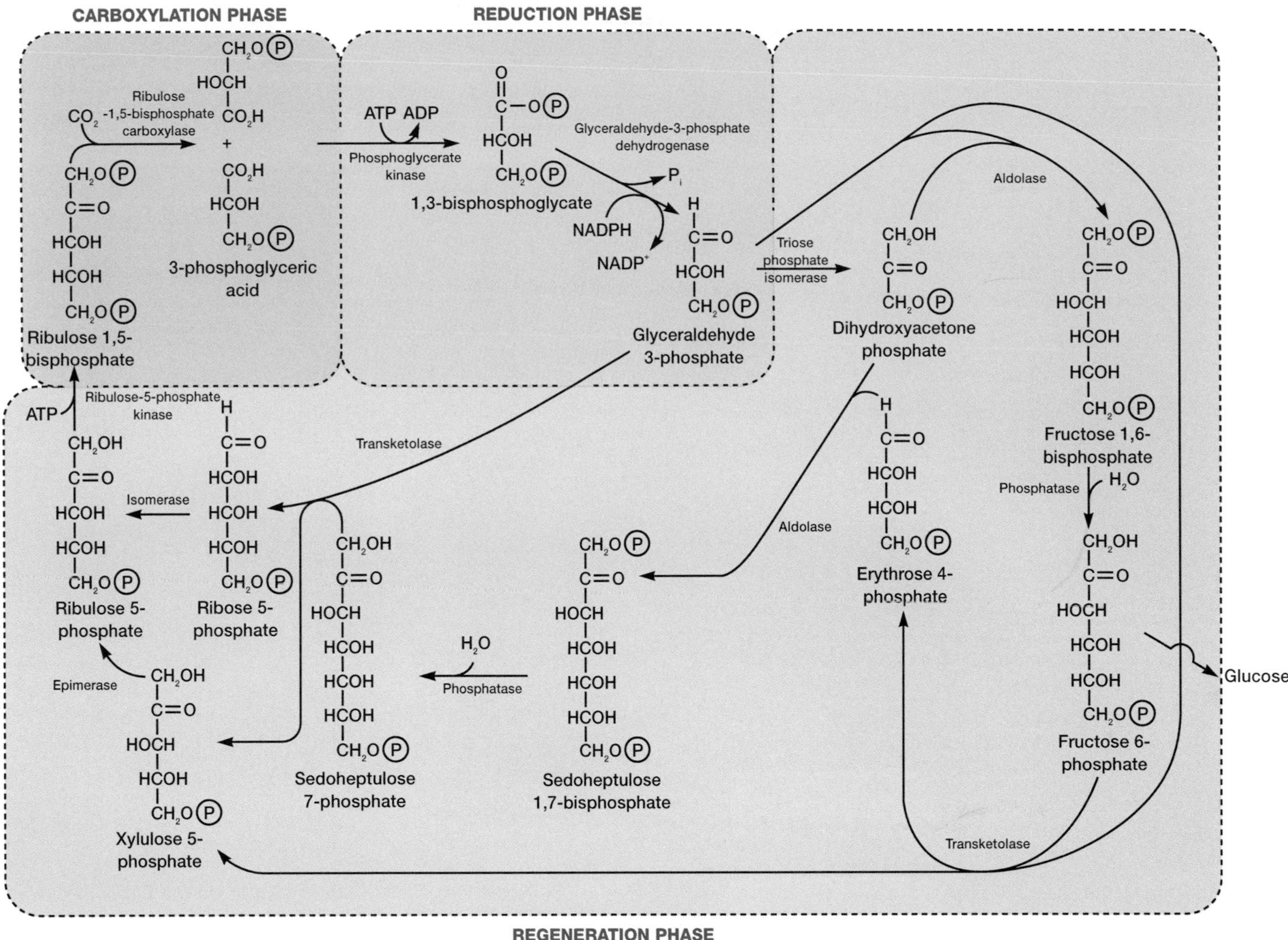

Figure 9.4 **The Calvin or Photosynthetic Carbon Reduction Cycle.**

Synthesis of Sugars and Polysaccharides

Many microorganisms cannot carry out photosynthesis and are heterotrophs that must synthesize sugars from reduced organic molecules rather than from CO_2. The synthesis of glucose from noncarbohydrate precursors is called **gluconeogenesis.** Although the gluconeogenic pathway is not identical with the glycolytic pathway, they do share seven enzymes (figure 9.5). Three glycolytic steps are irreversible in the cell: (1) the conversion of phosphoenolpyruvate to pyruvate, (2) the formation of fructose 1,6-bisphosphate from fructose 6-phosphate, and (3) the phosphorylation of glucose. These must be bypassed when the pathway is operating biosynthetically. For example, the formation of fructose 1,6-bisphosphate by phosphofructokinase is reversed by a separate enzyme, fructose bisphosphatase, which hydrolytically removes a phosphate from fructose bisphosphate. Usually at least two enzymes are involved in the conversion of pyruvate to phosphoenolpyruvate (the reversal of the pyruvate kinase step).

As can be seen in figure 9.5, the pathway synthesizes fructose as well as glucose. Once glucose and fructose have been formed, other common sugars can be manufactured. For example, mannose comes directly from fructose by a simple rearrangement.

$$\text{Fructose 6-phosphate} \rightleftharpoons \text{mannose 6-phosphate}$$

Several sugars are synthesized while attached to a nucleoside diphosphate. The most important nucleoside diphosphate sugar is **uridine diphosphate glucose (UDPG).** Glucose is activated by attachment to the pyrophosphate of uridine diphosphate through a reaction with uridine triphosphate (figure 9.6). The UDP portion of UDPG is recognized by enzymes and carries glucose around the cell for participation in enzyme reactions much like ADP bears phosphate in the form of ATP. UDP-galactose is synthesized from UDPG through a rearrangement of one hydroxyl group. A different enzyme catalyzes the synthesis of UDP-glucuronic acid through the oxidation of UDPG (figure 9.7).

Nucleoside diphosphate sugars also play a central role in the synthesis of polysaccharides such as starch and glycogen. Again, biosynthesis is not simply a direct reversal of catabolism. Glycogen and starch catabolism (*see chapter 8*) proceeds

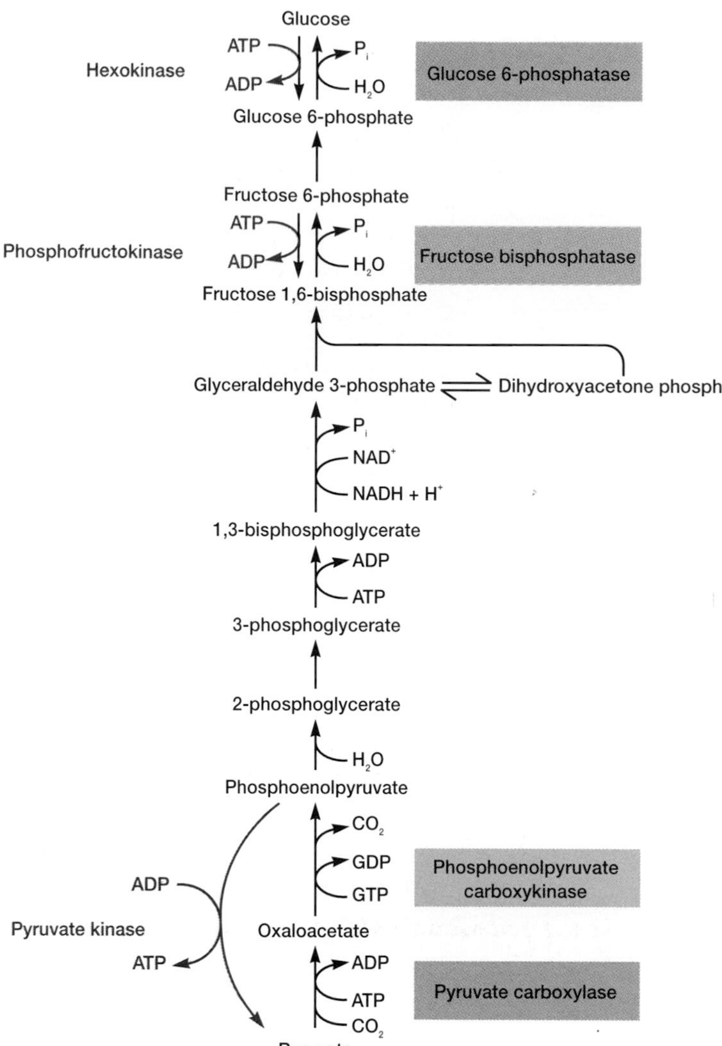

Figure 9.5 **Gluconeogenesis.** The gluconeogenic pathway used in many microorganisms. The names of the four enzymes catalyzing reactions different from those found in glycolysis are in shaded boxes. Glycolytic steps are shown in color for comparison.

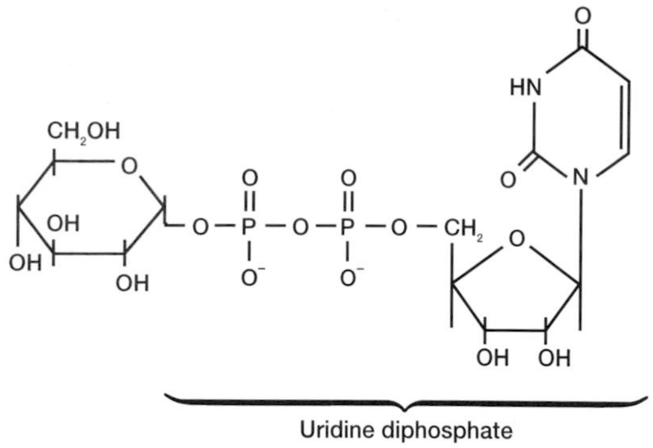

either by hydrolysis to form free sugars or by the addition of phosphate to these polymers with the production of glucose 1-phosphate. Nucleoside diphosphate sugars are not involved. In contrast, during the synthesis of glycogen and starch in bacteria and algae, adenosine diphosphate glucose is formed from glucose 1-phosphate and then donates glucose to the end of growing glycogen and starch chains.

$$\text{ATP} + \text{glucose 1-phosphate} \longrightarrow \text{ADP-glucose} + \text{PP}_i$$
$$(\text{Glucose})_n + \text{ADP-glucose} \longrightarrow (\text{glucose})_{n+1} + \text{ADP}$$

Nucleoside diphosphate sugars also participate in the synthesis of complex molecules such as bacterial cell walls (pp. 186–88).

Figure 9.6 **Uridine Diphosphate Glucose.** Glucose is in color.

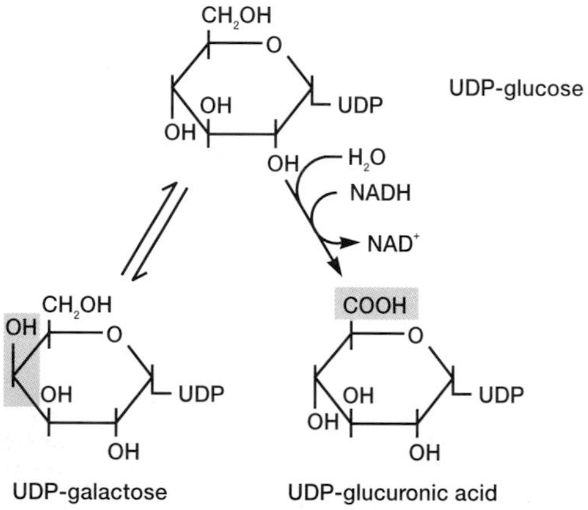

Figure 9.7 **Uridine Diphosphate Galactose and Glucuronate Synthesis.** The synthesis of UDP-galactose and UDP-glucuronic acid from UDP-glucose. Structural changes are indicated by colored boxes.

1. Briefly describe the three stages of the Calvin cycle.
2. What is gluconeogenesis and how does it usually occur? Describe the formation of mannose, galactose, starch, and glycogen. Why are nucleoside diphosphate sugars important?

The Assimilation of Inorganic Phosphorus, Sulfur, and Nitrogen

Besides carbon and oxygen, microorganisms also require large quantities of phosphorus, sulfur, and nitrogen for use in biosynthesis. Each of these is assimilated, or incorporated into organic molecules, by different routes.

Microbial nutrition (chapter 5).
Microbial participation in biogeochemical cycles (chapter 40).

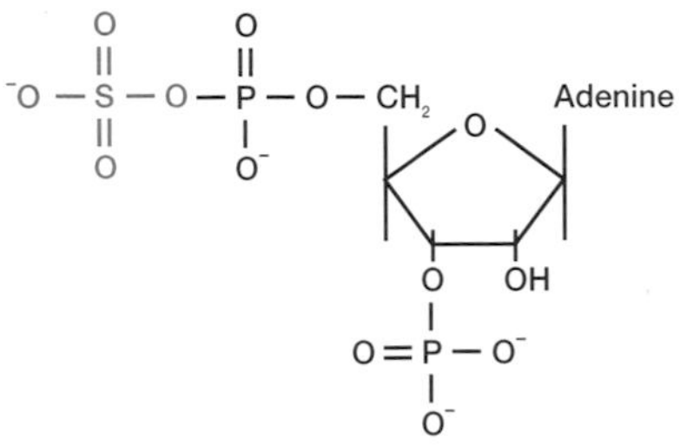

Figure 9.8 **Phosphoadenosine 5′-phosphosulfate (PAPS).** The sulfate group is in color.

Phosphorus Assimilation

Phosphorus is found in nucleic acids, proteins, phospholipids, ATP, and coenzymes like NADP. The most common sources of phosphorus are inorganic phosphate and organic phosphate esters. Inorganic phosphate is incorporated through the formation of ATP in one of three ways: by (1) photophosphorylation, (2) oxidative phosphorylation, and (3) substrate-level phosphorylation. Glycolysis provides an example of the latter process. Phosphate is joined with glyceraldehyde 3-phosphate to give 1,3-bisphosphoglycerate, which is next used in ATP synthesis.

$$\text{Glyceraldehyde-3-P} + P_i + NAD^+ \longrightarrow \text{1,3-bisphosphoglycerate} + NADH + H^+$$

$$\text{1,3-bisphosphoglycerate} + ADP \longrightarrow \text{3-phosphoglycerate} + ATP$$

Microorganisms may obtain organic phosphates from their surroundings in dissolved or particulate form. **Phosphatases** very often hydrolyze organic phosphate esters to release inorganic phosphate. Gram-negative bacteria have phosphatases in the periplasmic space between their cell wall and the plasma membrane, which allows phosphate to be taken up immediately after release. On the other hand, protozoa can directly use organic phosphates after ingestion or hydrolyze them in lysosomes and incorporate the phosphate.

Sulfur Assimilation

Sulfur is needed for the synthesis of amino acids (cysteine and methionine) and several coenzymes (e.g., coenzyme A and biotin), and may be obtained from two sources. Many microorganisms use cysteine and methionine, obtained from either external sources or intracellular amino acid reserves, as sulfur sources. In addition, sulfate can provide sulfur for biosynthesis. The sulfur atom in sulfate is more oxidized than it is in cysteine and other organic molecules; thus, sulfate must be reduced before it can be assimilated. This process is known as **assimilatory sulfate reduction** to distinguish it from the **dissimilatory sulfate reduction** that takes place when sulfate acts as an electron acceptor during anaerobic respiration (*see figure 40.10*).

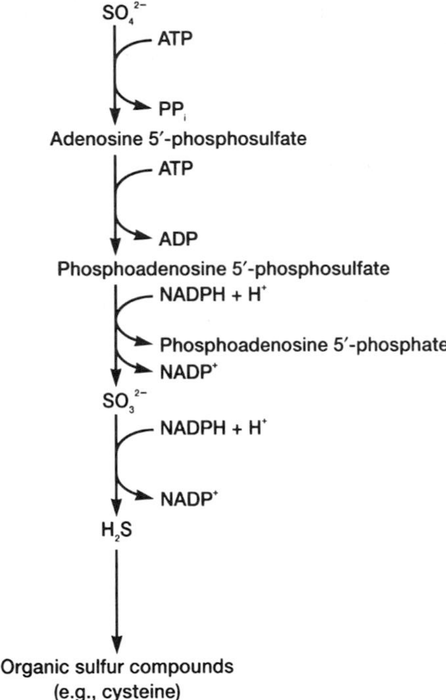

Figure 9.9 **The Sulfate Reduction Pathway.**

Anaerobic respiration (p. 158).

Assimilatory sulfate reduction involves sulfate activation through the formation of **phosphoadenosine 5′-phosphosulfate** (figure 9.8), followed by reduction of the sulfate. The process is a complex one (figure 9.9) in which sulfate is first reduced to sulfite (SO_3^{2-}), then to hydrogen sulfide. Cysteine can be synthesized from hydrogen sulfide in two ways. Fungi appear to combine hydrogen sulfide with serine to form cysteine (process 1), whereas many bacteria join hydrogen sulfide with O-acetylserine instead (2).

$$(1) \quad H_2S + \text{serine} \longrightarrow \text{cysteine} + H_2O$$

$$\text{Serine} \xrightarrow{\text{acetyl-CoA} \;\to\; \text{CoA}}$$

$$(2) \quad \text{O-acetylserine} \xrightarrow{H_2S \;\to\; \text{acetate}} \text{cysteine}$$

Once formed, cysteine can be used in the synthesis of other sulfur-containing organic compounds.

Nitrogen Assimilation

Because nitrogen is a major component of proteins, nucleic acids, coenzymes, and many other cell constituents, the cell's ability to assimilate inorganic nitrogen is exceptionally important. Although there is a great abundance of nitrogen gas in the atmosphere, few microorganisms can reduce the gas and use it as a nitrogen source. Most must incorporate either ammonia or nitrate.

Ammonia Incorporation

Ammonia nitrogen can be incorporated into organic material relatively easily and directly because it is more reduced than other forms of inorganic nitrogen. Some microorganisms form the amino acid alanine in a reductive amination reaction catalyzed by alanine dehydrogenase.

$$\text{Pyruvate} + NH_4^+ + \text{NADH (NADPH)} + H^+ \rightleftharpoons \text{L-alanine} + NAD^+ (NADP^+) + H_2O$$

The major route for ammonia incorporation often is the formation of glutamate from α-ketoglutarate (a TCA cycle intermediate). Many bacteria and fungi employ **glutamate dehydrogenase,** at least when the ammonia concentration is high.

$$\alpha\text{-ketoglutarate} + NH_4^+ + \text{NADPH (NADH)} + H^+ \rightleftharpoons \text{glutamate} + NADP^+ (NAD^+) + H_2O$$

Different species vary in their ability to use NADPH and NADH as the reducing agent in glutamate synthesis.

Once either alanine or glutamate has been synthesized, the newly formed α-amino group can be transferred to other carbon skeletons by transamination reactions (*see chapter 8*) to form different amino acids. **Transaminases** possess the coenzyme pyridoxal phosphate, which is responsible for the amino group transfer. Microorganisms have a number of transaminases, each of which catalyzes the formation of several amino acids using the same amino acid as an amino group donor. When glutamate dehydrogenase works in cooperation with transaminases, ammonia can be incorporated into a variety of amino acids (figure 9.10).

A second route of ammonia incorporation involves two enzymes acting in sequence, **glutamine synthetase** and **glutamate synthase** (figure 9.11). Ammonia is used to synthesize glutamine from glutamate, then the amide nitrogen of glutamine is transferred to α-ketoglutarate to generate a new glutamate molecule. Because glutamate acts as an amino donor in transaminase reactions, ammonia may be used to synthesize all common amino acids when suitable transaminases are present (figure 9.12). Both ATP and a source of electrons, such as NADPH or reduced ferredoxin, are required. This route is present in *Escherichia coli, Bacillus megaterium,* and other bacteria. The two enzymes acting in sequence operate very effectively at low ammonia concentrations, unlike the glutamate dehydrogenase pathway.

Assimilatory Nitrate Reduction

The nitrogen in nitrate (NO_3^-) is much more oxidized than that in ammonia. Nitrate must first be reduced to ammonia before the nitrogen can be converted to an organic form. This reduction of nitrate is called **assimilatory nitrate reduction,** which is not the same as that occurring during anaerobic respiration (*see chapters 8 and 40*). In assimilatory nitrate reduction, nitrate is incorporated into organic material and does not participate in energy generation. The process is widespread among bacteria, fungi, and algae.

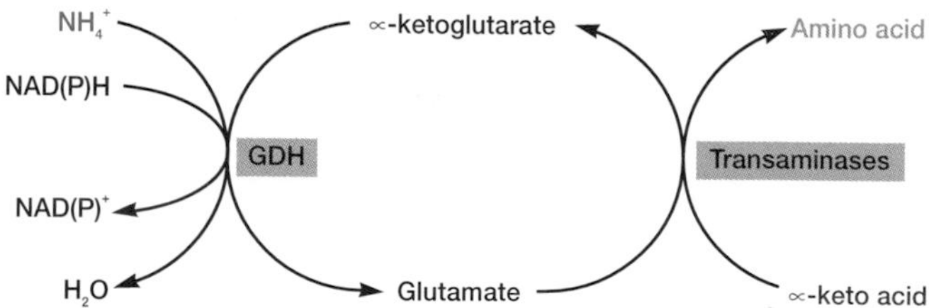

Figure 9.10 **The Ammonia Assimilation Pathway.** Ammonia assimilation by use of glutamate dehydrogenase (GDH) and transaminases. Either NADP- or NAD-dependent glutamate dehydrogenases may be involved.

The first step in nitrate assimilation is its reduction to nitrite by **nitrate reductase,** an enzyme that contains both FAD and molybdenum. NADPH is the electron source.

$$NO_3^- + NADPH + H^+ \longrightarrow NO_2^- + NADP^+ + H_2O$$

Nitrite is next reduced to ammonia with a series of two electron additions catalyzed by **nitrite reductase,** and possibly other enzymes. Hydroxylamine may be an intermediate.

$$NO_2^- \xrightarrow{2e^-} x \xrightarrow{2e^-} [NH_2OH] \xrightarrow[2H^+ \quad H_2O]{2e^-} NH_3$$

The ammonia is then incorporated into amino acids by the routes already described.

Nitrogen Fixation

The reduction of atmospheric gaseous nitrogen to ammonia is called **nitrogen fixation.** Because ammonia and nitrate levels are often low and only a few procaryotes can carry out nitrogen fixation (eucaryotic cells completely lack this ability), the rate of this process limits plant growth in many situations. Nitrogen fixation occurs in (1) free-living bacteria (e.g., *Azotobacter, Klebsiella, Clostridium,* and *Methanococcus*), (2) bacteria living in symbiotic association with plants such as legumes (*Rhizobium*), and (3) cyanobacteria (*Nostoc* and *Anabaena*). The biological aspects of nitrogen fixation are discussed in chapter 42. The biochemistry of nitrogen fixation is the focus of this section.

The biology of nitrogen-fixing microorganisms (pp. 434, 812, 851–55).

The reduction of nitrogen to ammonia is catalyzed by the enzyme **nitrogenase.** Although the enzyme-bound intermediates in this process are still unknown, it is believed that nitrogen is reduced by two-electron additions in a way similar to that illustrated in figure 9.13. The reduction of molecular nitrogen to ammonia is quite exergonic, but the reaction has a high activation energy because molecular nitrogen is an unreactive gas with very strong bonding between the two nitrogen atoms. Therefore, nitrogen reduction is expensive and

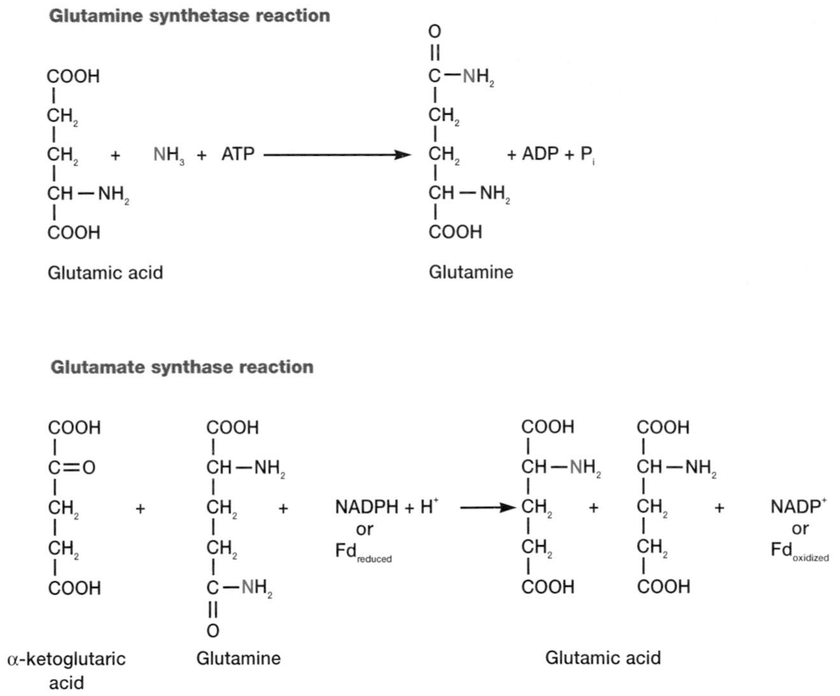

Glutamine synthetase reaction

$$\underset{\text{Glutamic acid}}{\text{COOH–CH}_2\text{–CH}_2\text{–CH(NH}_2\text{)–COOH}} + NH_3 + ATP \longrightarrow \underset{\text{Glutamine}}{\text{H}_2\text{N–C(=O)–CH}_2\text{–CH}_2\text{–CH(NH}_2\text{)–COOH}} + ADP + P_i$$

Glutamate synthase reaction

$$\underset{\alpha\text{-ketoglutaric acid}}{\text{COOH–C(=O)–CH}_2\text{–CH}_2\text{–COOH}} + \underset{\text{Glutamine}}{\text{COOH–CH(NH}_2\text{)–CH}_2\text{–CH}_2\text{–C(=O)–NH}_2} + \begin{matrix}\text{NADPH + H}^+\\ \text{or}\\ \text{Fd}_{\text{reduced}}\end{matrix} \longrightarrow \underset{\text{Glutamic acid}}{2\ \text{COOH–CH(NH}_2\text{)–CH}_2\text{–CH}_2\text{–COOH}} + \begin{matrix}\text{NADP}^+\\ \text{or}\\ \text{Fd}_{\text{oxidized}}\end{matrix}$$

Figure 9.11 Glutamine Synthetase and Glutamate Synthase. The glutamine synthetase and glutamate synthase reactions involved in ammonia assimilation. Some glutamine synthases use NADPH as an electron source; others use reduced ferredoxin (Fd). The nitrogen being incorporated and transferred is shown in color.

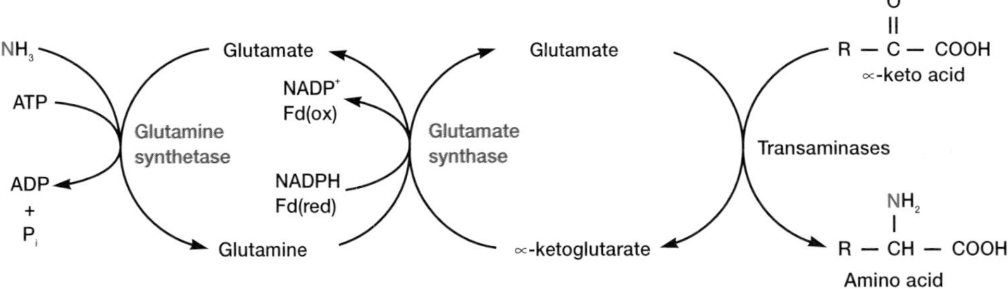

Figure 9.12 Ammonia Incorporation Using Glutamine Synthetase and Glutamate Synthase.

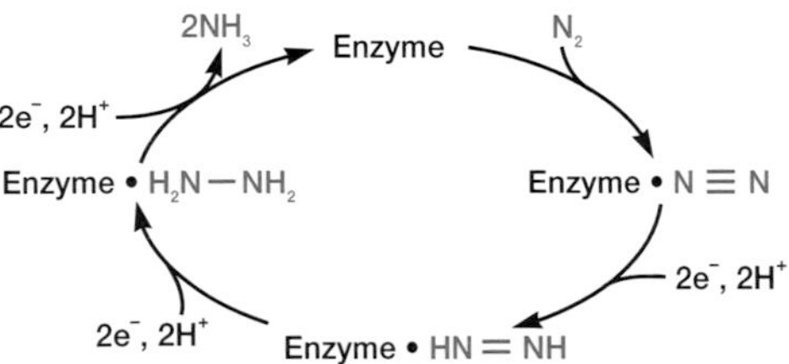

Figure 9.13 Nitrogen Reduction. A hypothetical sequence of nitrogen reduction by nitrogenase.

requires a large ATP expenditure. At least six electrons and twelve ATP molecules, four ATPs per pair of electrons, are required.

$$N_2 + 6H^+ + 6e^- + 12ATP + 12H_2O \longrightarrow 2NH_3 + 12ADP + 12P_i$$

The electrons come from ferredoxin that has been reduced in a variety of ways: by photosynthesis in cyanobacteria, respiratory processes in aerobic nitrogen fixers, or fermentations in anaerobic bacteria. For example, *Clostridium pasteurianum* (an anaerobic bacterium) reduces ferredoxin during pyruvate oxidation, whereas the aerobic *Azotobacter* uses electrons from NADPH to reduce ferredoxin.

Nitrogenase is a complex system consisting of two major protein components, a MoFe protein (MW 220,000) joined with one or two Fe proteins (MW 64,000). The MoFe protein contains 2 atoms of molybdenum and 28 to 32 atoms of iron; the Fe protein has 4 iron atoms (figure 9.14). Fe protein is first reduced by ferredoxin, then it binds ATP (figure 9.15). ATP binding lowers the reduction potential of the Fe protein, enabling it to reduce the MoFe protein. ATP is hydrolyzed when this electron transfer occurs. Finally, reduced MoFe protein

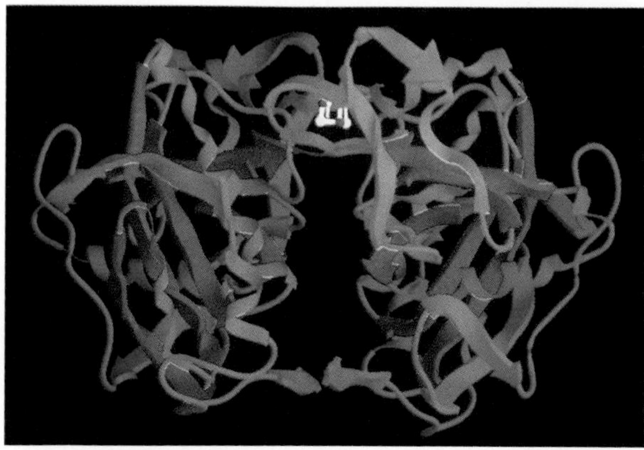

Figure 9.14 Structure of the Nitrogenase Fe Protein. The Fe protein's two subunits are arranged like a pair of butterfly wings with the iron sulfur cluster between the wings and at the "head" of the butterfly. The iron sulfur cluster is very exposed, which helps account for nitrogenase's sensitivity to oxygen. The oxygen can readily attack the exposed irons.

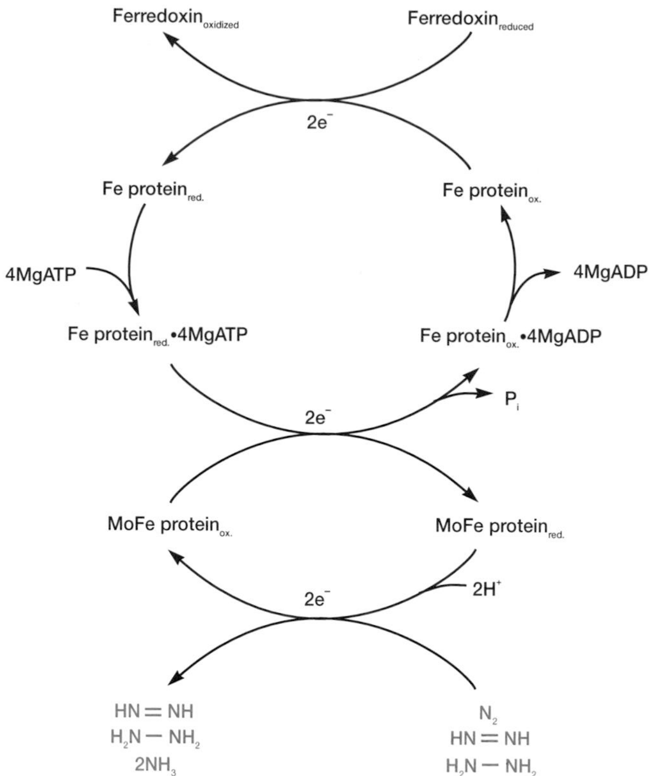

Figure 9.15 Mechanism of Nitrogenase Action. The flow of two electrons from ferredoxin to nitrogen is outlined. This process is repeated three times in order to reduce N_2 to two molecules of ammonia, and the substrate-product pairs for each step are shown at the bottom of the figure. See text for a more detailed explanation.

donates electrons to atomic nitrogen. Nitrogenase is quite sensitive to O_2 and must be protected from O_2 inactivation within the cell.

Until recently, it appeared that molybdenum was required for nitrogen fixation. It has now been shown that *Azotobacter,* when deprived of molybdenum and provided vanadium, produces a vanadium nitrogenase. This nitrogenase also fixes nitrogen, though it differs from the molybdenum nitrogenase in several ways.

Nitrogenase can reduce a variety of molecules containing triple bonds (e.g., acetylene, cyanide, and azide).

$$HC \equiv CH + 2H^+ + 2e^- \longrightarrow H_2C = CH_2$$

The rate of reduction of acetylene to ethylene is even used to measure nitrogenase activity.

Once molecular nitrogen has been reduced to ammonia, the ammonia can be incorporated into organic compounds. In the symbiotic nitrogen fixer *Rhizobium,* it appears that ammonia diffuses out of the bacterial cell and is assimilated in the surrounding legume cell. The primary route of ammonia assimilation seems to be the synthesis of glutamine via the glutamine synthetase–glutamate synthase system (figure 9.11). However, substances such as the purine derivatives allantoin and allantoic acid also are synthesized and used for the transport of nitrogen to other parts of the plant.

The Synthesis of Amino Acids

Microorganisms vary with respect to the type of nitrogen source they employ, but most can assimilate some form of inorganic nitrogen by the routes just described. Amino acid synthesis also requires construction of the proper carbon skeletons, and this is often a very complex process involving many steps. Although individual amino acid biosynthetic pathways are not described in detail, a survey of the general pattern of amino acid biosynthesis is worthwhile. Further details of amino acid biosynthesis may be found in introductory biochemistry textbooks.

The relationship of amino acid biosynthetic pathways to amphibolic routes is shown in figure 9.16. Amino acid skeletons are derived from acetyl-CoA and from intermediates of the TCA cycle, glycolysis, and the pentose phosphate pathway. To maximize efficiency and economy, the precursors for amino acid biosynthesis are provided by a few major amphibolic pathways. Sequences leading to individual amino acids branch off from these central routes. Alanine, aspartate, and glutamate are made by transamination directly from pyruvate, oxaloacetate, and α-ketoglutarate, respectively. Most biosynthetic pathways are more complex, and common intermediates are often used in the synthesis of families of related amino acids for the sake of further economy. For example, the amino acids lysine, threonine, isoleucine, and methionine are synthesized from oxalocacetate by such a branching anabolic route (figure 9.17). The biosynthetic pathways for the aromatic amino acids phenylalanine, tyrosine, and tryptophan also share many intermediates (figure 9.18).

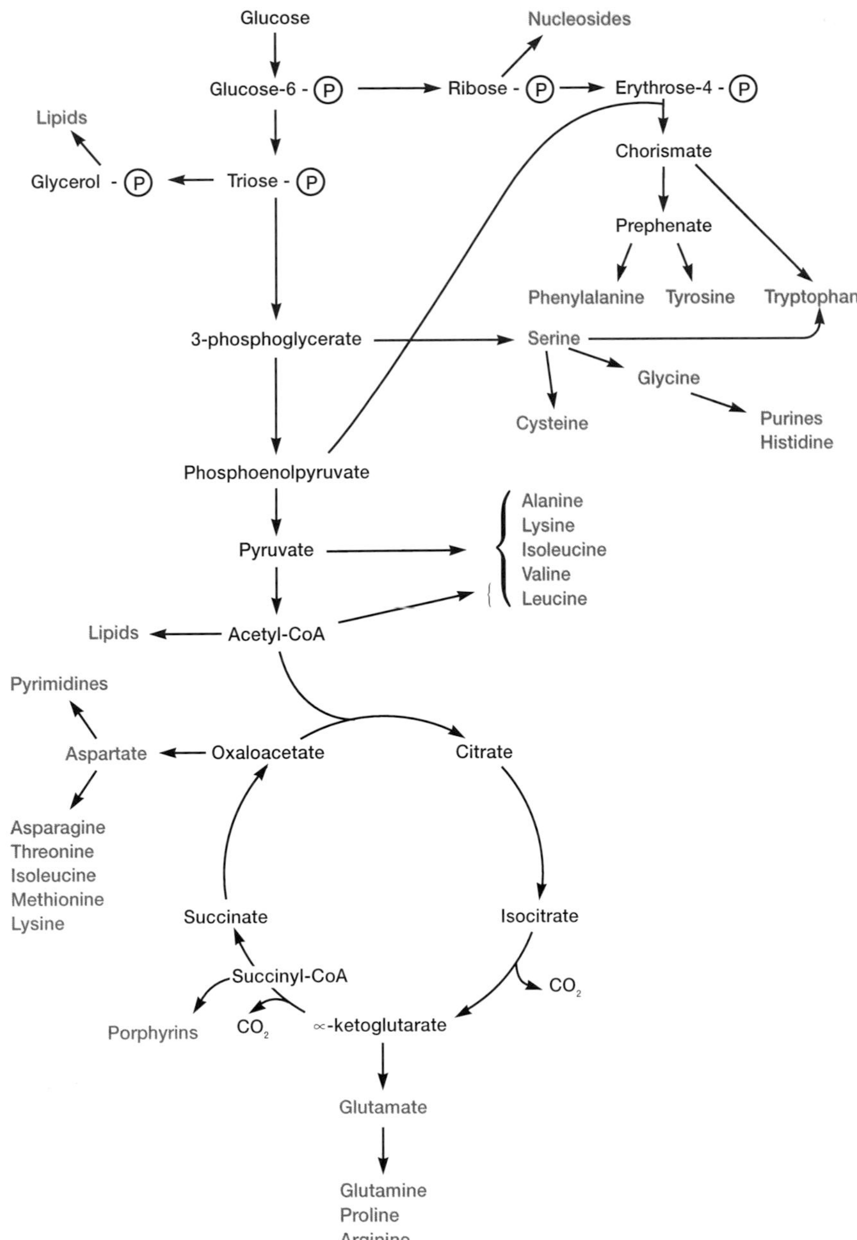

Figure 9.16 The Organization of Anabolism. Biosynthetic products (in color) are derived from intermediates of amphibolic pathways.

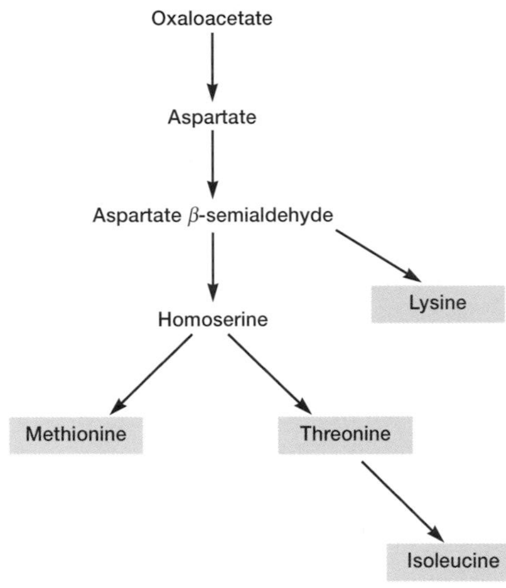

Figure 9.17 A Branching Pathway of Amino Acid Synthesis. The pathways to methionine, threonine, isoleucine, and lysine. Although some arrows represent one step, most interconversions require the participation of several enzymes.

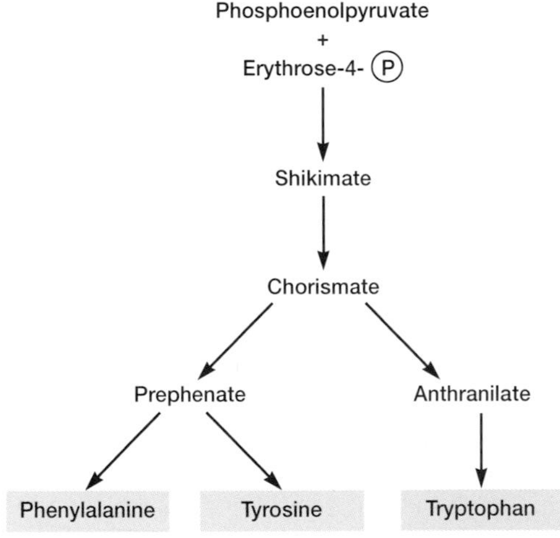

Figure 9.18 Aromatic Amino Acid Synthesis. The synthesis of the aromatic amino acids phenylalanine, tyrosine, and tryptophan. Most arrows represent more than one enzyme reaction.

1. How do microorganisms assimilate sulfur and phosphorus?
2. Describe the roles of glutamate dehydrogenase, glutamine synthetase, glutamate synthase, and transaminases in ammonia assimilation. How is nitrate incorporated by assimilatory nitrate reduction?
3. What is nitrogen fixation? Briefly describe the structure and mechanism of action of nitrogenase.
4. Summarize in general terms the organization of amino acid biosynthesis.

Anaplerotic Reactions

Inspection of figure 9.16 will show that TCA cycle intermediates are used in the synthesis of pyrimidines and a wide variety of amino acids. In fact, the biosynthetic functions of this pathway are so essential that most of it must operate anaerobically in order to supply biosynthetic precursors, even though NADH is not required for electron transport and oxidative phosphorylation in the absence of oxygen. Thus, there is a heavy demand upon the TCA cycle to supply carbon for biosynthesis, and cycle intermediates could be depleted if nothing

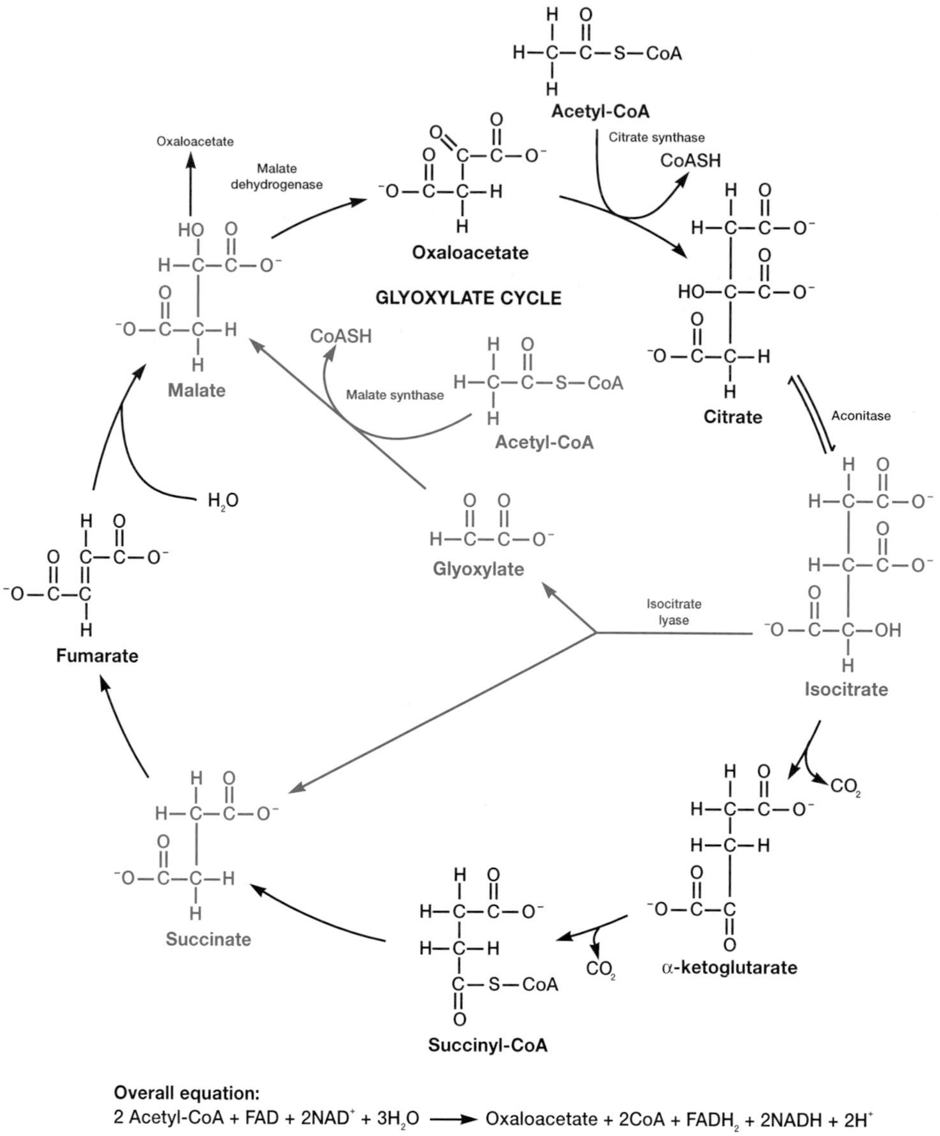

Overall equation:

2 Acetyl-CoA + FAD + $2NAD^+$ + $3H_2O$ ⟶ Oxaloacetate + 2CoA + $FADH_2$ + 2NADH + $2H^+$

Figure 9.19 **The Glyoxylate Cycle.** The reactions and enzymes unique to the cycle are shown in color. The tricarboxylic acid cycle enzymes that have been bypassed are at the bottom.

were done to maintain their levels. However, microorganisms have reactions that replenish cycle intermediates so that the TCA cycle can continue to function when active biosynthesis is taking place. Reactions that replace cycle intermediates are called **anaplerotic reactions** (Greek *anaplerotic,* filling up).

Most microorganisms can replace TCA cycle intermediates by **CO_2 fixation.** It should be emphasized that anaplerotic reactions do not serve the same function as the CO_2 fixation pathway that supplies the carbon required by autotrophs. In autotrophs, CO_2 fixation provides most or all of the carbon required for growth. Anaplerotic CO_2 fixation reactions simply replace TCA cycle intermediates and maintain metabolic balance. Usually, CO_2 is added to an acceptor molecule, either pyruvate or phosphoenolpyruvate, to form the cycle intermediate oxaloacetate. Some microorganisms (e.g., *Arthrobacter globiformis,* yeasts) use pyruvate carboxylase in this role.

$$\text{Pyruvate} + CO_2 + \text{ATP} + H_2O \xrightarrow{\text{biotin}} \text{oxaloacetate} + \text{ADP} + P_i$$

This enzyme requires the cofactor biotin and uses ATP energy to join CO_2 and pyruvate. Biotin is often the cofactor for enzymes catalyzing carboxylation reactions. Because of its importance, biotin is a vitamin for many species. Other microorganisms, such as the bacteria *Escherichia coli* and *Salmonella typhimurium,* have the enzyme phosphoenolpyruvate carboxylase, which catalyzes the following reaction.

$$\text{Phosphoenolpyruvate} + CO_2 \longrightarrow \text{oxaloacetate} + P_i$$

Some bacteria, algae, fungi, and protozoa can grow with acetate as the sole carbon source by using it to synthesize TCA cycle intermediates in the **glyoxylate cycle** (figure 9.19). This cycle is made possible by two unique enzymes, isocitrate lyase and malate synthase, that catalyze the following reactions.

$$\text{Isocitrate} \xrightarrow{\text{isocitrate lyase}} \text{succinate} + \text{glyoxylate}$$

$$\text{Glyoxylate} + \text{acetyl-CoA} \xrightarrow{\text{malate synthase}} \text{malate} + \text{CoA}$$

The TCA cycle (pp. 150–51).

The glyoxylate cycle is actually a modified TCA cycle. The two decarboxylations of the latter pathway (the isocitrate dehydrogenase and α-ketoglutarate dehydrogenase steps) are bypassed, making possible the conversion of acetyl-CoA to form oxaloacetate without loss of acetyl-CoA carbon as CO_2. In this fashion, acetate and any molecules that give rise to it can contribute carbon to the cycle.

1. Define an anaplerotic reaction and give an example.
2. How does the glyoxylate cycle convert acetyl-CoA to oxaloacetate, and what special enzymes are used?

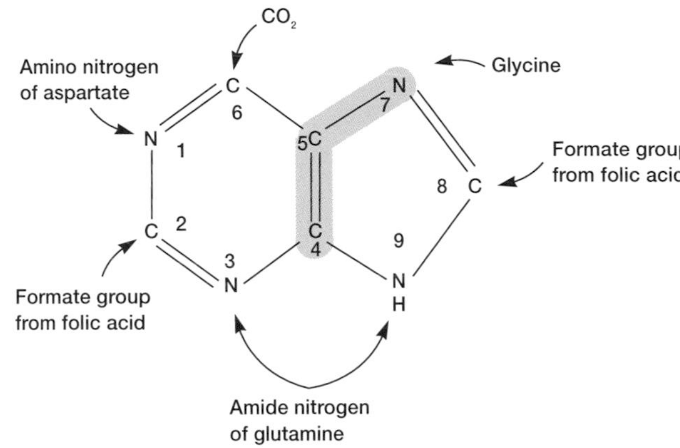

Figure 9.20 Purine Biosynthesis. The sources of purine skeleton nitrogen and carbon are indicated in color.

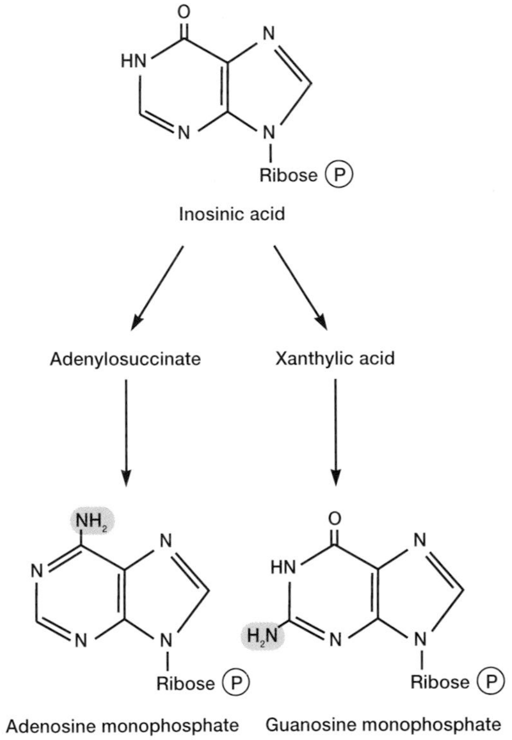

Figure 9.21 Synthesis of Adenosine Monophosphate and Guanosine Monophosphate. The shaded groups are the ones differing from those in inosinic acid.

The Synthesis of Purines, Pyrimidines, and Nucleotides

Purine and pyrimidine biosynthesis is critical for all cells because these molecules are used in the synthesis of ATP, several cofactors, ribonucleic acid (RNA), deoxyribonucleic acid (DNA), and other important cell components. Nearly all microorganisms can sythesize their own purines and pyrimidines as these are so crucial to cell function.

DNA and RNA synthesis (pp. 195–204).

Purines and **pyrimidines** are cyclic nitrogenous bases with several double bonds and pronounced aromatic properties. Purines consist of two joined rings whereas pyrimidines have only one (figures 9.20 and 9.22). The purines **adenine** and **guanine** and the pyrimidines **uracil, cytosine,** and **thymine** are commonly found in microorganisms. A purine or pyrimidine base joined with a pentose sugar, either ribose or deoxyribose, is a **nucleoside.** A **nucleotide** is a nucleoside with one or more phosphate groups attached to the sugar.

Purine Biosynthesis

The biosynthetic pathway for purines is a complex, 11-step sequence (*see appendix II*) in which seven different molecules contribute parts to the final purine skeleton (figure 9.20). Because the pathway begins with ribose 5-phosphate and the purine skeleton is constructed on this sugar, the first purine product of the pathway is the nucleotide inosinic acid, not a free purine base. The cofactor folic acid is very important in purine biosynthesis. Folic acid derivatives contribute carbons two and eight to the purine skeleton. In fact, the drug sulfonamide inhibits bacterial growth by blocking folic acid synthesis. This interferes with purine biosynthesis and other processes that require folic acid.

Once inosinic acid has been formed, relatively short pathways synthesize adenosine monophosphate and guanosine monophosphate (figure 9.21) and produce nucleoside diphosphates and triphosphates by phosphate transfers from ATP. DNA contains deoxyribonucleotides (the ribose lacks a hydroxyl group on carbon two) instead of the ribonucleotides

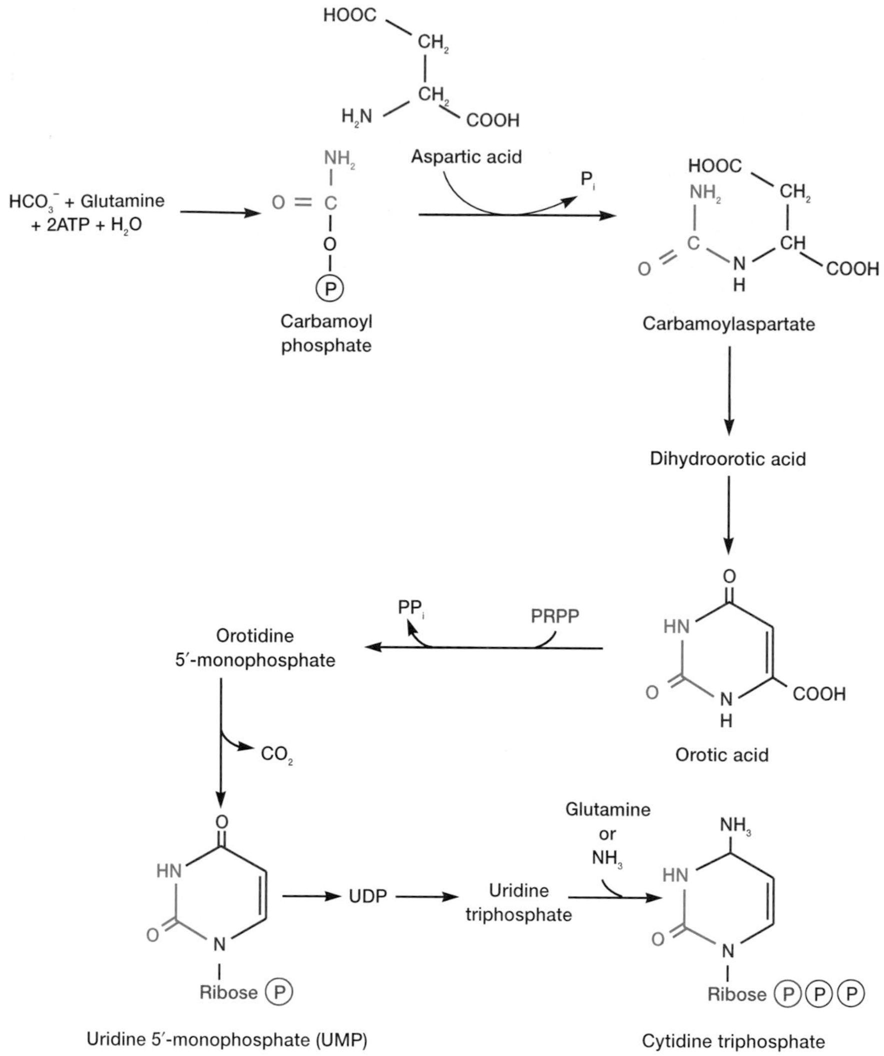

Figure 9.22 Pyrimidine Synthesis. PRPP stands for 5-phosphoribose 1-pyrophosphoric acid, which provides the ribose 5-phosphate chain.

found in RNA. Deoxyribonucleotides arise from the reduction of nucleoside diphosphates or nucleoside triphosphates by two different routes. Some microorganisms reduce the triphosphates with a system requiring vitamin B_{12} as a cofactor. Others, such as *E. coli,* reduce the ribose in nucleoside diphosphates. Both systems employ a small sulfur-containing protein called thioredoxin as their reducing agent.

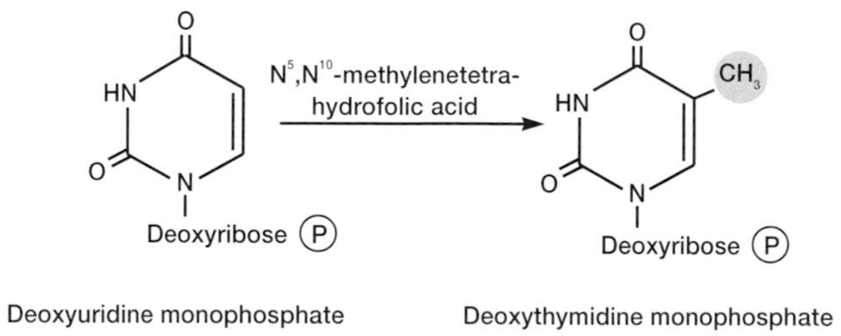

Pyrimidine Biosynthesis

Pyrimidine biosynthesis begins with aspartic acid and carbamoyl phosphate, a high-energy molecule synthesized from CO_2 and ammonia (figure 9.22). Aspartate carbamoyltransferase catalyzes the condensation of these two substrates to form carbamoylaspartate, which is then converted to the initial pyrimidine product, orotic acid.

The regulation of aspartate carbamoyltransferase activity (pp. 218–19).

After synthesis of the pyrimidine skeleton, a nucleotide is produced by the ribose 5-phosphate addition using the high-energy intermediate 5-phosphoribosyl 1-pyrophosphate. Thus, construction of the pyrimidine ring is completed before ribose is added, in contrast with purine ring synthesis, which begins with ribose 5-phosphate. Decarboxylation of orotidine monophosphate yields uridine monophosphate, and eventually uridine triphosphate and cytidine triphosphate.

Figure 9.23 Deoxythymidine Monophosphate Synthesis.

The third common pyrimidine is thymine, a constituent of DNA. The ribose in pyrimidine nucleotides is reduced in the same way as it is in purine nucleotides. Then deoxyuridine monophosphate is methylated with the use of a folic acid derivative to form deoxythymidine monophosphate (figure 9.23).

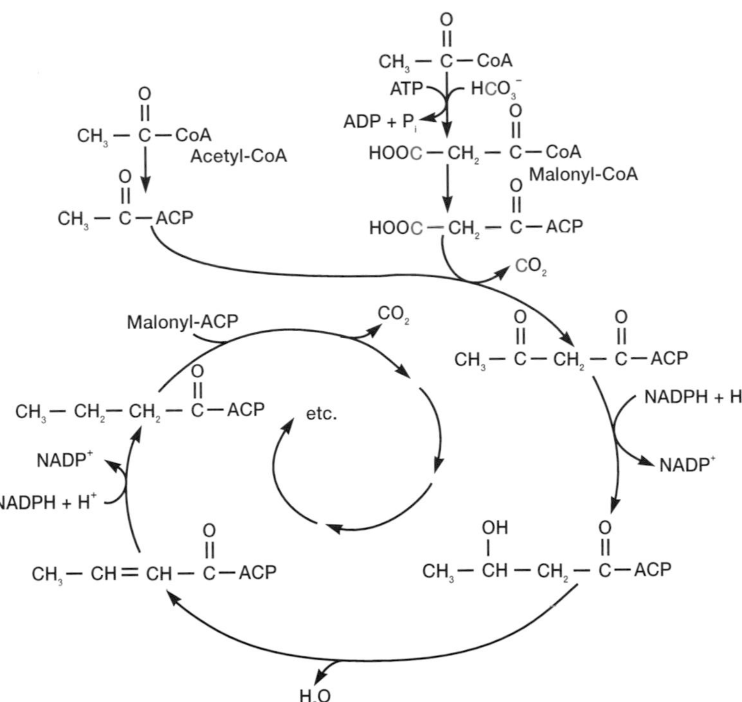

Figure 9.24 Fatty Acid Synthesis. The cycle is repeated until the proper chain length has been reached. Carbon dioxide carbon and the remainder of malonyl-CoA are shown in different colors. ACP stands for acyl carrier protein.

1. Define purine, pyrimidine, nucleoside, and nucleotide.
2. Outline the way in which purines and pyrimidines are synthesized. How is the deoxyribose component of deoxyribonucleotides made?

Lipid Synthesis

A variety of lipids are found in microorganisms, particularly in cell membranes. Most contain **fatty acids** or their derivatives. Fatty acids are monocarboxylic acids with long alkyl chains that usually have an even number of carbons (the average length is 18 carbons). Some may be unsaturated, that is, have one or more double bonds. Most microbial fatty acids are straight chained, but some are branched. Gram-negative bacteria often have cyclopropane fatty acids (fatty acids with one or more cyclopropane rings in their chains).

Lipid structure and nomenclature (appendix I).

Fatty acid synthesis is catalyzed by the **fatty acid synthetase** complex with acetyl-CoA and malonyl-CoA as the substrates, and NADPH as the reductant. Malonyl-CoA arises from the ATP-driven carboxylation of acetyl-CoA (figure 9.24). Synthesis takes place after acetate and malonate have been transferred from coenzyme A to the sulfhydryl group of the **acyl carrier protein (ACP),** a small protein that carries the growing fatty acid chain during synthesis. The synthetase adds two carbons at a time to the carboxyl end of the growing fatty acid chain in a two-stage process (figure 9.24). First, malonyl-ACP reacts with the fatty acyl-ACP to yield CO_2 and a fatty acyl-ACP two carbons longer. The loss of CO_2 drives this reaction to completion. Notice that ATP is used to add CO_2 to acetyl-CoA, forming malonyl-CoA. The same CO_2 is lost when malonyl-ACP donates carbons to the chain. Carbon dioxide is essential to fatty acid synthesis, but is not permanently incorporated. Indeed, some microorganisms require CO_2 for good growth, but they can do without it in the presence of a fatty acid like oleic acid (an 18-carbon unsaturated fatty acid). In the second stage of synthesis, the β-keto group arising from the initial condensation reaction is removed in a three-step process involving two reductions and a dehydration. The fatty acid is then ready for the addition of two more carbon atoms.

Unsaturated fatty acids are synthesized in two ways. Eucaryotes and aerobic bacteria like *Bacillus megaterium* employ an aerobic pathway using both NADPH and O_2.

$$R-(CH_2)_9-\overset{\overset{\displaystyle O}{\|}}{C}-SCoA + NADPH + H^+ + O_2 \longrightarrow$$

$$R-CH=CH-(CH_2)_7-\overset{\overset{\displaystyle O}{\|}}{C}-SCoA + NADP^+ + 2H_2O$$

A double bond is formed between carbons nine and ten, and O_2 is reduced to water with electrons supplied by both the fatty acid and NADPH. Anaerobic bacteria and some aerobes create double bonds during fatty acid synthesis by dehydrating hydroxy fatty acids. Oxygen is not required for double bond synthesis by this pathway. The anaerobic pathway is present in a number of common gram-negative bacteria (e.g., *Escherichia coli* and *Salmonella typhimurium*), gram-positive bacteria (e.g., *Lactobacillus plantarum* and *Clostridium pasteurianum*), and cyanobacteria.

Eucaryotic microorganisms frequently store carbon and energy as **triacylglycerol,** glycerol esterified to three fatty acids. Glycerol arises from the reduction of the glycolytic intermediate dihydroxyacetone phosphate to glycerol 3-phosphate, which is then esterified with two fatty acids to give **phosphatidic acid** (figure 9.25). Phosphate is hydrolyzed from phosphatidic acid giving a diacylglycerol, and the third fatty acid is attached to yield a triacylglycerol.

Phospholipids are major components of eucaryotic and most procaryotic cell membranes. Their synthesis also usually proceeds by way of phosphatidic acid. A special cytidine diphosphate (CDP) carrier plays a role similar to that of uridine and adenosine diphosphate carriers in carbohydrate biosynthesis. For example, bacteria synthesize phosphatidylethanolamine, a major cell membrane component, through the initial formation of CDP-diacylglycerol (figure 9.25). This CDP derivative then reacts with serine to form the phospholipid phosphatidylserine, and decarboxylation yields phosphatidylethanolamine. In this way, a complex membrane lipid is constructed from the products of glycolysis, fatty acid biosynthesis, and amino acid biosynthesis.

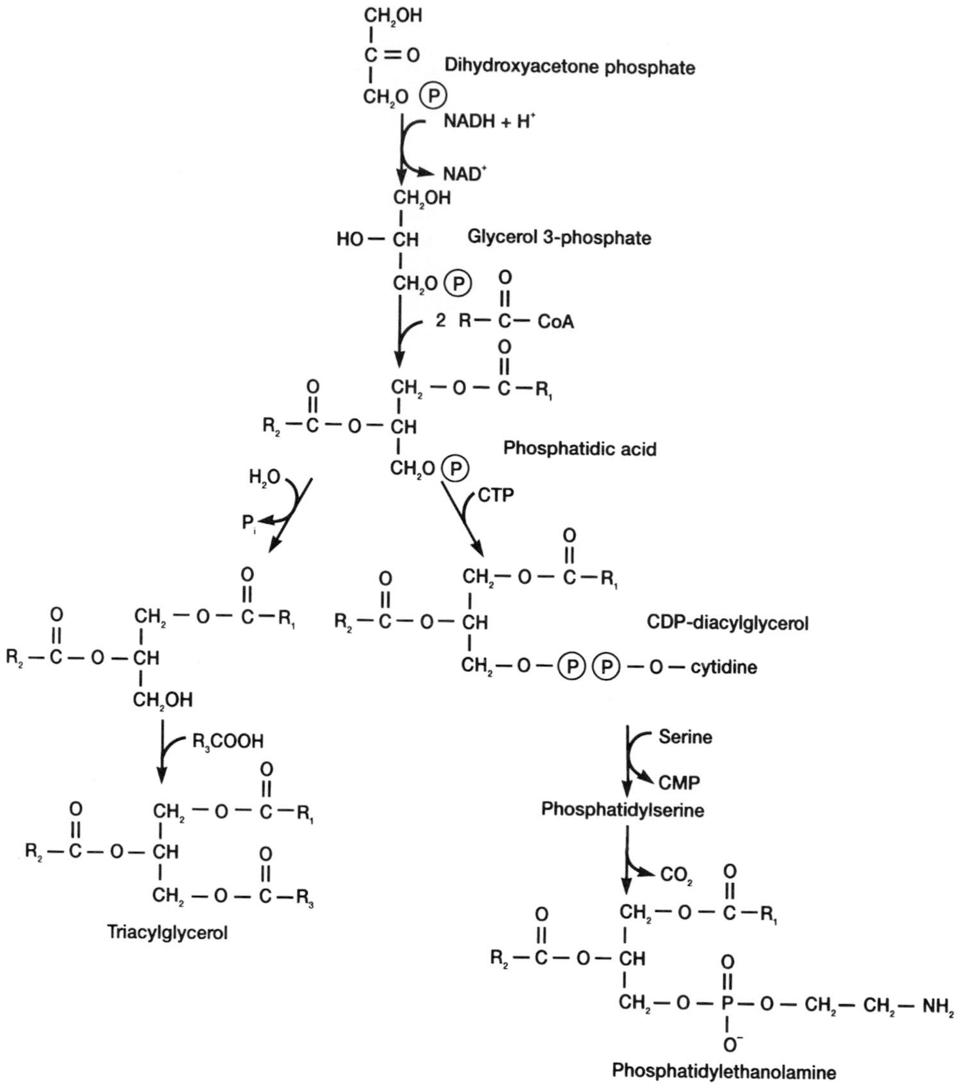

Dihydroxyacetone phosphate

NADH + H^+

NAD^+

Glycerol 3-phosphate

2 R—C(=O)—CoA

Phosphatidic acid

H_2O

P_i

CTP

CDP-diacylglycerol

R_3COOH

Triacylglycerol

Serine

CMP

Phosphatidylserine

CO_2

Phosphatidylethanolamine

Figure 9.25 **Triacylglycerol and Phospholipid Synthesis.**

---

1. What is a fatty acid? Describe in general terms how the fatty acid synthetase manufactures a fatty acid.
2. How are unsaturated fatty acids made?
3. Briefly describe the pathways for triacylglycerol and phospholipid synthesis. Of what importance are phosphatidic acid and CDP-diacylglycerol?

Peptidoglycan Synthesis

As discussed earlier, most bacterial cell walls contain a large, complex peptidoglycan molecule consisting of long polysaccharide chains made of alternating N-acetylmuramic acid (NAM) and N-acetylglucosamine (NAG) residues. Pentapeptide chains are attached to the NAM groups. The polysaccharide chains are connected through their pentapeptides or by interbridges (*see figures 3.18 and 3.19*).

Peptidoglycan structure and function (pp. 51–3).

Not surprisingly, such an intricate structure requires an equally intricate biosynthetic process, especially because the synthetic reactions occur both inside and outside the cell membrane. Peptidoglycan synthesis is a multistep process that has been best studied in the gram-positive bacterium *Staphylococcus aureus*. Two carriers participate: uridine diphosphate (UDP) and **bactoprenol** (figure 9.26). Bactoprenol is a 55-carbon alcohol that attaches to NAM by a pyrophosphate group and moves peptidoglycan components through the hydrophobic membrane.

The synthesis of peptidoglycan, outlined in figures 9.27 and 9.28, occurs in eight stages.

1. UDP derivatives of N-acetylmuramic acid and N-acetylglucosamine are synthesized in the cytoplasm.
2. Amino acids are sequentially added to UDP-NAM to form the pentapeptide chain (the two terminal D-alanines are added as a dipeptide). ATP energy is used to make the peptide bonds, but tRNA and ribosomes are not involved.

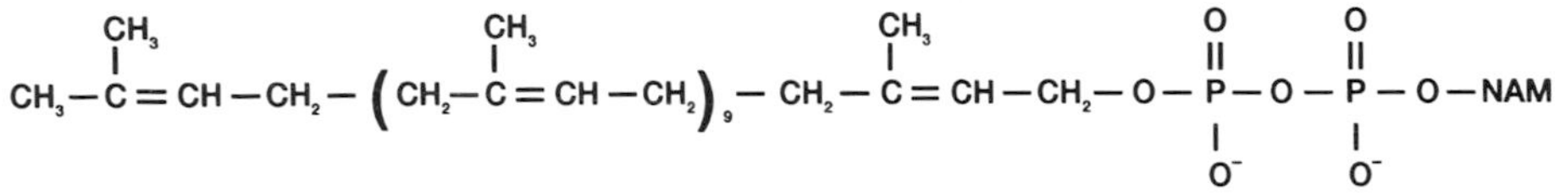

$$CH_3-\overset{CH_3}{\overset{|}{C}}=CH-CH_2-\left(CH_2-\overset{CH_3}{\overset{|}{C}}=CH-CH_2\right)_9-CH_2-\overset{CH_3}{\overset{|}{C}}=CH-CH_2-O-\underset{O^-}{\underset{|}{\overset{O}{\overset{\|}{P}}}}-O-\underset{O^-}{\underset{|}{\overset{O}{\overset{\|}{P}}}}-O-NAM$$

Figure 9.26 **Bactoprenol Pyrophosphate.** Bactoprenol pyrophosphate connected to N-acetylmuramic acid (NAM).

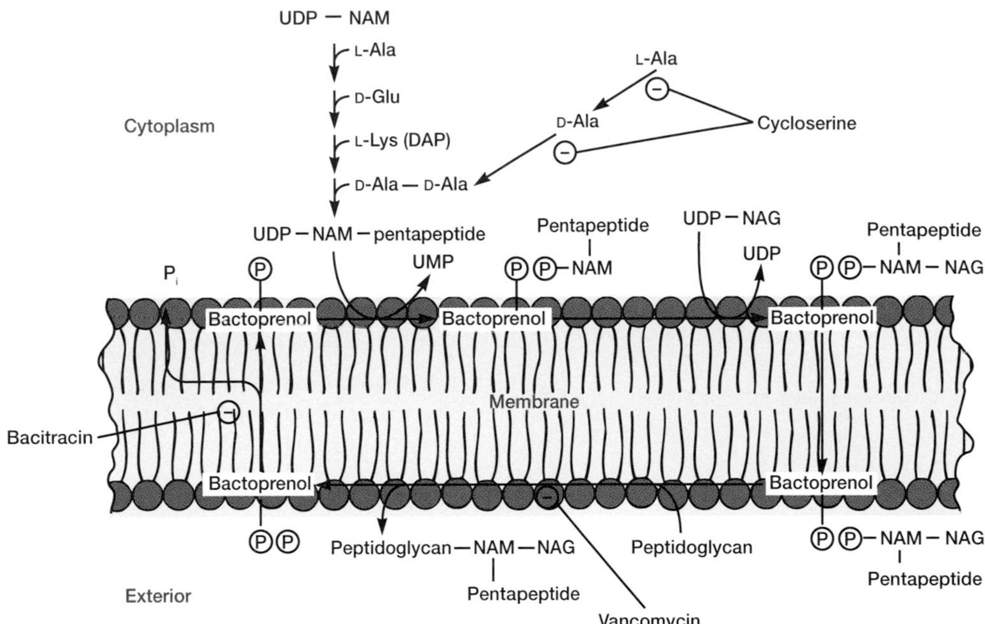

Figure 9.27 **Peptidoglycan Synthesis.** NAM is N-acetylmuramic acid and NAG is N-acetylglucosamine. The pentapeptide contains L-lysine in *S. aureus* peptidoglycan, and diaminopimelic acid (DAP) in *E. coli*.

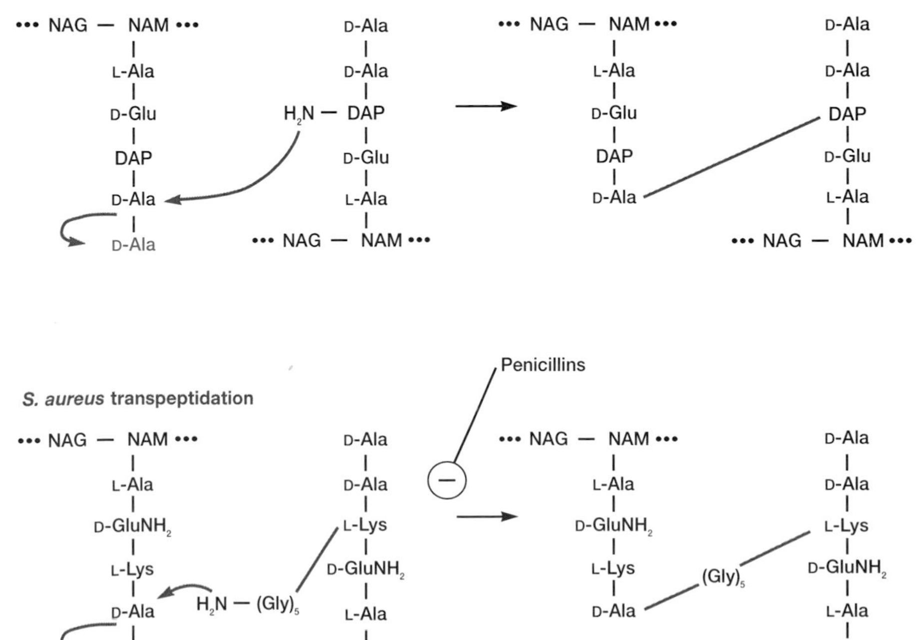

Figure 9.28 **Transpeptidation.** The transpeptidation reactions in the formation of the peptidoglycans of *Escherichia coli* and *Staphylococcus aureus*.

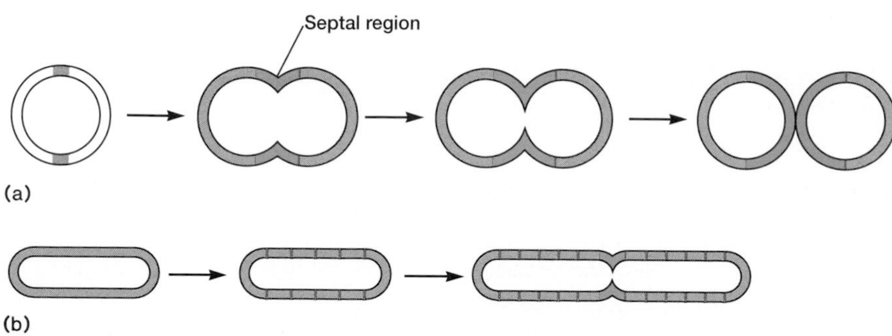

Figure 9.29 Wall Synthesis Patterns. Patterns of new cell wall synthesis in growing and dividing bacteria. (*a*) Streptococci and some other gram-positive cocci. (*b*) Synthesis in rod-shaped bacteria (*Escherichia coli, Salmonella, Bacillus*). The zones of growth are in color. The actual situation is more complex than indicated because cells can begin to divide again before the first division is completed.

3. The NAM-pentapeptide is transferred from UDP to a bactoprenol phosphate at the membrane surface.
4. UDP-NAG adds NAG to the NAM-pentapeptide to form the peptidoglycan repeat unit. If a pentaglycine interbridge is required, the glycines are added using special glycyl-tRNA molecules, not ribosomes.
5. The completed NAM-NAG peptidoglycan repeat unit is transported across the membrane to its outer surface by the bactoprenol pyrophosphate carrier.
6. The peptidoglycan unit is attached to the growing end of a peptidoglycan chain to lengthen it by one repeat unit.
7. The bactoprenol carrier returns to the inside of the membrane. A phosphate is released during this process to give bactoprenol phosphate, which can now accept another NAM-pentapeptide.
8. Finally, peptide cross-links between the peptidoglycan chains are formed by **transpeptidation** (figure 9.28). In *E. coli,* the free amino group of diaminopimelic acid attacks the subterminal D-alanine, releasing the terminal D-alanine residue. ATP is used to form the terminal peptide bond inside the membrane. No more ATP energy is required when transpeptidation takes place on the outside. The same process occurs when an interbridge is involved; only the group reacting with the subterminal D-alanine differs.

Peptidoglycan synthesis is particularly vulnerable to disruption by antimicrobial agents. Inhibition of any stage of synthesis weakens the cell wall and can lead to osmotic lysis. Many antibiotics interfere with peptidoglycan synthesis. For example, penicillin inhibits the transpeptidation reaction, and bacitracin blocks the dephosphorylation of bactoprenol pyrophosphate (figures 9.27 and 9.28).

Antibiotic effects on cell wall synthesis (pp. 334–36).

Patterns of Cell Wall Formation

To grow and divide efficiently, a bacterial cell must add new peptidoglycan to its cell wall in a precise and well-regulated way while maintaining wall shape and integrity in the presence of high osmotic pressure. Because the cell wall peptidoglycan is essentially a single enormous network, the growing bacterium must be able to degrade it just enough to provide acceptor ends for the incorporation of new peptidoglycan units. It must also reorganize peptidoglycan structure when necessary. This limited peptidoglycan digestion is accomplished by enzymes known as **autolysins,** some of which attack the polysaccharide chains, while others hydrolyze the peptide cross-links. Autolysin inhibitors keep the activity of these enzymes under tight control.

Control of cell division (pp. 230–31).

Although the location and distribution of cell wall synthetic activity varies with species, there seem to be two general patterns (figure 9.29). Many gram-positive cocci (e.g., *Enterococcus faecalis* and *Streptococcus pyogenes*) have only one to a few zones of growth. The principal growth zone is usually at the site of septum formation, and new cell halves are synthesized back-to-back. The second pattern of synthesis occurs in the rod-shaped bacteria *Escherichia coli, Salmonella,* and *Bacillus.* Active peptidoglycan synthesis occurs at the site of septum formation just as before, but growth sites also are scattered along the cylindrical portion of the rod. Thus, growth is distributed more diffusely in the rod-shaped bacteria than in the streptococci. Synthesis must lengthen rod-shaped cells as well as divide them. Presumably, this accounts for the differences in wall growth pattern.

1. Outline in a diagram the steps involved in the synthesis of peptidoglycan and show their relationship to the plasma membrane. What are the roles of bactoprenol and UDP?
2. What is the function of autolysins in cell wall peptidoglycan synthesis? Describe the patterns of peptidoglycan synthesis seen in gram-positive cocci and in rod-shaped bacteria such as *E. coli.*

Summary

1. In biosynthesis or anabolism, cells use energy to construct complex molecules from smaller, simpler precursors.
2. Many important cell constituents are macromolecules, large polymers constructed of simple monomers.
3. Although many catabolic and anabolic pathways share enzymes for the sake of efficiency, some of their enzymes are separate and independently regulated.
4. Macromolecular components often undergo self-assembly to form the final molecule or complex.
5. Photosynthetic CO_2 fixation is carried out by the Calvin cycle that may be divided into three phases: the carboxylation phase, the reduction phase, and the regeneration phase. Three ATPs and two NADPHs are used during the incorporation of one CO_2.
6. Gluconeogenesis is the synthesis of glucose and related sugars from nonglucose precursors.
7. Glucose, fructose, and mannose are gluconeogenic intermediates or made directly from them; galactose is synthesized with nucleoside diphosphate derivatives. Bacteria and algae synthesize glycogen and starch from adenosine diphosphate glucose.
8. Phosphorus is obtained from inorganic or organic phosphate.
9. Microorganisms can use cysteine, methionine, and inorganic sulfate as sulfur sources. Sulfate is reduced to sulfide during assimilatory sulfate reduction.
10. Ammonia nitrogen can be directly assimilated by the activity of transaminases and either glutamate dehydrogenase or the glutamine synthetase-glutamate synthase system.
11. Nitrate is incorporated through assimilatory nitrate reduction catalyzed by the enzymes nitrate reductase and nitrite reductase.
12. During nitrogen fixation catalyzed by the nitrogenase complex, atmospheric molecular nitrogen is reduced to ammonia.
13. Amino acid biosynthetic pathways branch off from the central amphibolic pathways.
14. Anaplerotic reactions replace TCA cycle intermediates to keep the cycle in balance while it supplies biosynthetic precursors. Many anaplerotic enzymes catalyze CO_2 fixation reactions. The glyoxylate cycle is also anaplerotic.
15. Purines and pyrimidines are nitrogenous bases found in DNA, RNA, and other molecules. The purine skeleton is synthesized beginning with ribose 5-phosphate and initially produces inosinic acid. Pyrimidine biosynthesis starts with carbamoyl phosphate and aspartate, and ribose is added after the skeleton has been constructed.
16. Fatty acids are synthesized from acetyl-CoA, malonyl-CoA, and NADPH by the fatty acid synthetase system. During synthesis, the intermediates are attached to the acyl carrier protein. Double bonds can be added in two different ways.
17. Triacylglycerols are made from fatty acids and glycerol phosphate. Phosphatidic acid is an important intermediate in this pathway.
18. Phospholipids like phosphatidylethanolamine can be synthesized from phosphatidic acid by forming CDP-diacylglycerol, then adding an amino acid.
19. Peptidoglycan synthesis is a complex process involving both UDP derivatives and the lipid carrier bactoprenol, which transports NAM-NAG-pentapeptide units acrosss the cell membrane. Cross-links are formed by transpeptidation.
20. Peptidoglycan synthesis occurs in discrete zones in the cell wall. Old peptidoglycan is selectively degraded by autolysins so new material can be added.

Key Terms

acyl carrier protein (ACP) *185*
adenine *183*
anaplerotic reactions *182*
assimilatory nitrate reduction *178*
assimilatory sulfate reduction *177*
autolysins *188*
bactoprenol *186*
Calvin cycle *173*
carboxysomes *173*
CO_2 fixation *182*
cytosine *183*
dissimilatory sulfate reduction *177*
fatty acid *185*
fatty acid synthetase *185*
gluconeogenesis *175*
glutamate dehydrogenase *178*
glutamate synthase *178*
glutamine synthetase *178*
glyoxylate cycle *182*
guanine *183*
macromolecule *172*
monomers *172*
nitrate reductase *178*
nitrite reductase *178*
nitrogenase *178*
nitrogen fixation *178*
nucleoside *183*
nucleotide *183*
phosphatase *177*
phosphatidic acid *185*
phosphoadenosine 5′-phosphosulfate *177*
purine *183*
pyrimidine *183*
ribulose-1,5-bisphosphate carboxylase *173*
self-assembly *173*
thymine *183*
transaminases *178*
transpeptidation *188*
triacylglycerol *185*
turnover *172*
uracil *183*
uridine diphosphate glucose (UDPG) *175*

Questions for Thought and Review

1. Discuss the relationship between catabolism and anabolism. How does anabolism depend upon catabolism?
2. Suppose that a microorganism was growing on a medium that contained amino acids but no sugars. In general terms, how would it synthesize the pentoses and hexoses it required?
3. Activated carriers participate in carbohydrate, lipid, and peptidoglycan synthesis. Briefly describe these carriers and their roles.
4. Which two enzymes discussed in the chapter appear to be specific to the Calvin cycle?
5. Why can phosphorus be directly incorporated into cell constituents whereas sulfur and nitrogen often cannot?
6. What is unusual about the synthesis of peptides that takes place during peptidoglycan construction?

Additional Reading

Brill, W. J. 1977. Biological nitrogen fixation. *Sci. Am.* 236(3):68–81.

Dilworth, M., and Glenn, A. R. 1984. How does a legume nodule work? *Trends Biochem. Sci.* 9(12):519–23.

Doyle, R. J.; Chaloupka, J.; and Vinter, V. 1988. Turnover of cell walls in microorganisms. *Microbiol. Rev.* 52(4):554–67.

Eady, R., Robson, R., and Postgate, J. 1987. Vanadium puts nitrogen in a fix. *New Sci.* 114(1565):59–62.

Glenn, A. R., and Dilworth, M. J. 1985. Ammonia movements in rhizobia *Microbiol. Sci.* 2(6):161–67.

Gottschalk, G. 1986. *Bacterial metabolism,* 2d ed. New York: Springer-Verlag.

Harold, F. M. 1990. To shape a cell: An inquiry into the causes of morphogenesis of microorganisms. *Microbiol. Rev.* 54(4):381–431.

Koch, A. L. 1990. Growth and form of the bacterial cell wall. *Am. Sci.* 78:327–41.

Kondorosi, E., and Kondorosi, A. 1986. Nodule induction on plant roots by *Rhizobium. Trends Biochem. Sci.* 11(7):296–99.

Mandelstam, J.; McQuillen, K.; and Dawes, I. 1982. *Biochemistry of bacterial growth,* 3d ed. London: Blackwell Scientific Publications.

Mathews, C. K., and van Holde, K. E. 1990. *Biochemistry.* Redwood City, Calif.: Benjamin/Cummings.

Moat, A. G., and Foster, J. W. 1988. *Microbial physiology,* 2d ed. New York: John Wiley and Sons.

Mora, J. 1990. Glutamine metabolism and cycling in *Neurospora crassa. Microbiol. Rev.* 54(3):293–304.

Neidhardt, F. C., Ingraham, J. L., and Schaechter, M. 1990. *Physiology of the bacterial cell: A molecular approach.* Sunderland, Mass.: Sinauer Associates.

Postgate, J. R. 1982. *The fundamentals of nitrogen fixation.* New York: Cambridge University Press.

Rose, A. H. 1976. *Chemical microbiology: An introduction to microbial physiology,* 3d ed. New York: Plenum.

Schlegel, H. G., and Bowien, B., eds. 1989. *Autotrophic bacteria.* Madison, Wis.: Science Tech Publishers.

Stryer, L. 1988. *Biochemistry,* 3d ed. New York: Freeman.

Triplett, E. W.; Roberts, G. P.; Ludden, P. W.; and Handelsman, J. 1989. What's new in nitrogen fixation. *ASM News* 55(1):15–21.

Tyler, B. 1978. Regulation of the assimilation of nitrogen compounds. *Ann Rev. Biochem.* 47:1127–62.

Voet, D., and Voet, J. G. 1990. *Biochemistry.* New York: John Wiley and Sons.

Zubay, G. 1993. *Biochemistry,* 3rd ed. Dubuque, Iowa: Wm. C. Brown Communications, Inc.

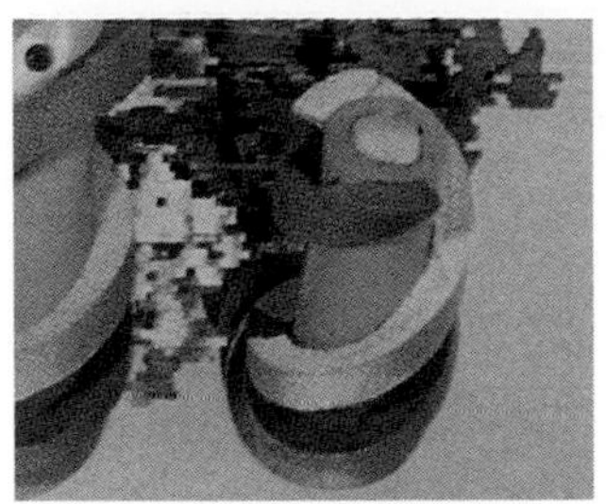

CHAPTER 10
Metabolism: The Synthesis of Nucleic Acids and Proteins

The particular field which excites my interest is the division between the living and the non-living, as typified by, say, proteins, viruses, bacteria and the structure of chromosomes. The eventual goal, which is somewhat remote, is the description of these activities in terms of their structure, i.e., the spatial distribution of their constituent atoms, in so far as this may prove possible. This might be called the chemical physics of biology.

—Francis Crick

Outline

Concepts

1. The two kinds of nucleic acid, deoxyribonucleic acid (DNA) and ribonucleic acid (RNA), differ from one another in chemical composition and structure. In procaryotic and eucaryotic cells, DNA serves as the repository for genetic information.
2. DNA is associated with basic proteins in the cell. In eucaryotes, these are special histone proteins, whereas in procaryotes, nonhistone proteins are complexed with DNA.
3. The flow of genetic information usually proceeds from DNA through RNA to protein. A protein's amino acid sequence reflects the nucleotide sequence of its mRNA. This messenger is a complementary copy of a portion of the DNA genome.
4. DNA replication is a very complex process involving a variety of proteins and a number of steps. It is designed to operate rapidly while minimizing errors and correcting those that arise when the DNA sequence is copied.
5. In transcription, the RNA polymerase copies the appropriate sequence on the DNA template sense strand to produce a complementary RNA copy of the gene. Transcription differs in a number of ways between procaryotes and eucaryotes, even though the basic mechanism of RNA polymerase action is essentially the same.
6. Translation is the process by which the nucleotide sequence of mRNA is converted into the amino acid sequence of a polypeptide through the action of ribosomes, tRNAs, aminoacyl-tRNA synthetases, ATP and GTP energy, and a variety of protein factors. As in the case of DNA replication, this complex process is designed to minimize errors.

Chapter 9 focuses on the three least complex levels of anabolism, depicted in figure 9.1. The routes for the incorporation of simple precursors, such as CO_2, ammonia, and phosphate, into organic material are introduced, and the pathways responsible for synthesis of the major biological monomers (sugars, amino acids, fatty acids, purines, and pyrimidines) are surveyed. Chapter 9 also discusses polysaccharide, lipid, and peptidoglycan synthesis.

Chapter 10 is concerned with the synthesis of the other two major classes of macromolecules: nucleic acids and proteins. These biosynthetic processes are presented in considerably more detail than polysaccharide biosynthesis, because adequate knowledge of nucleic acids and proteins is essential to the understanding of molecular biology and microbial genetics. The chapter begins with a general introduction to nucleic acid structure, in order to provide the background necessary for discussion of nucleic acid synthesis and function. This is followed by sections on DNA replication, RNA synthesis, and protein synthesis.

Because deoxyribonucleic acid carries the cell's genetic information and ribonucleic acid directs the synthesis of proteins, the structure and synthesis of nucleic acids are considered before protein synthesis.

Nucleic Acid Structure

The structure and synthesis of purine and pyrimidine nucleotides are introduced in chapter 9. These nucleotides can be combined to form nucleic acids of two kinds (figure 10.1*a*). **Deoxyribonucleic acid (DNA)** contains the 2′-deoxyribonucleosides (figure 10.1*b*) of adenine, guanine, cytosine, and thymine. **Ribonucleic acid (RNA)** is composed of the ribonucleosides of adenine, guanine, cytosine and uracil (instead of thymine). In both DNA and RNA, nucleosides are joined by phosphate groups to form long polynucleotide chains (figure 10.1*c*). The differences in chemical composition between the chains reside in their sugar and pyrimidine bases: DNA has deoxyribose and thymine; RNA has ribose and uracil.

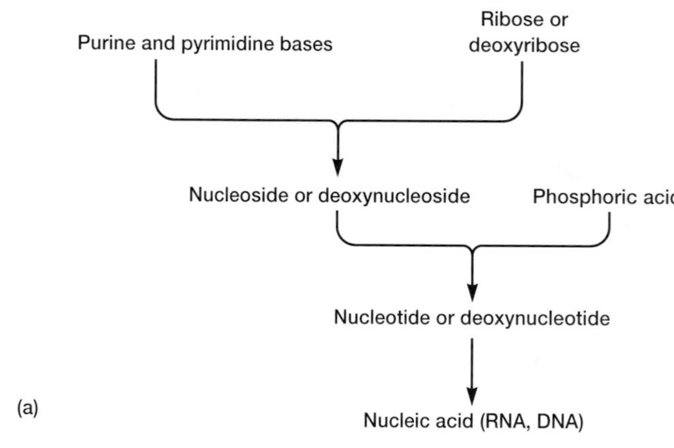

(b) Adenosine 2'-deoxycytidine

(c)

Figure 10.1 The Composition of Nucleic Acids. (*a*) A diagram showing the relationship of various nucleic acid components. Combination of a purine or pyrimidine base with ribose or deoxyribose gives a nucleoside (a ribonucleoside or deoxyribonucleoside). A nucleotide contains a nucleoside and one or more phosphoric acid molecules. Nucleic acids result when nucleotides are connected together in polynucleotide chains. (*b*) Examples of nucleosides—the purine nucleoside adenosine and the pyrimidine deoxynucleoside 2′-deoxycytidine. The carbons of nucleoside sugars are indicated by numbers with primes. (*c*) A segment of a polynucleotide chain showing two nucleosides, deoxyguanosine and thymidine, connected by a phosphodiester linkage between the 3′ and 5 ′ carbons of adjacent deoxyribose sugars.

Box 10.1

The Elucidation of DNA Structure

The basic chemical composition of nucleic acids was elucidated in the 1920s through the efforts of P.A. Levene. Despite his major contributions to nucleic acid chemistry, Levene mistakenly believed that DNA was a very small molecule, probably only four nucleotides long, composed of equal amounts of the four different nucleotides arranged in a fixed sequence. Partly because of his influence, biologists believed for many years that nucleic acids were too simple in structure to carry complex genetic information. They concluded that genetic information must be encoded in proteins because proteins were large molecules with complex amino sequences that could vary among different proteins.

As so often happens, further advances in our understanding of DNA structure awaited the development of significant new analytical techniques in chemistry. One development was the invention of paper chromatography by Archer Martin and Richard Synge between 1941 and 1944. By 1948 the chemist Erwin Chargaff had begun using paper chromatography to analyze the base composition of DNA from a number of species. He soon found that the base composition of DNA from genetic material did indeed vary among species just as he expected. Furthermore, the total amount of purines always equaled the total amount of pyrimidines; and the adenine/thymine and guanine/cytosine ratios were always one. These findings, known as Chargaff's rules, were a key to the understanding of DNA structure.

Another turning point in research on DNA structure was reached in 1951 when Rosalind Franklin arrived at King's College, London, and joined Maurice Wilkins in his efforts to prepare highly oriented DNA fibers and study them by X-ray crystallography. By the winter of 1952–1953, Franklin had obtained an excellent X-ray diffraction photograph of DNA.

The same year that Franklin began work at King's College, the American biologist James Watson went to Cambridge University and met Francis Crick. Although Crick was a physicist, he was very interested in the structure and function of DNA, and the two soon began to work on its structure. Their attempts were unsuccessful until Franklin's data provided them with the necessary clues. Her photograph of fibrous DNA contained a crossing pattern of dark spots, which showed that the molecule was helical. The dark regions at the top and bottom of the photograph showed that the purine and pyrimidine bases were stacked on top of each other and separated by 0.34 nm. Franklin had already concluded that the phosphate groups lay to the outside of the cylinder. Finally, the X-ray data and her determination of the density of DNA indicated that the helix contained two strands, not three or more as some had proposed.

Without actually doing any experiments themselves, Watson and Crick constructed their model by combining Chargaff's rules on base composition with Franklin's X-ray data and their predictions about how genetic material should behave. By building models, they found that a smooth, two-stranded helix of constant diameter could be constructed only when an adenine hydrogen bonded with thymine and when a guanine bonded with cytosine in the center of the helix. They immediately realized that the double helical structure provided a mechanism by which genetic material might be replicated. The two parental strands could unwind and direct the synthesis of complementary strands, thus forming two new identical DNA molecules (figure 10.7). Watson, Crick, and Wilkins received the Nobel Prize in 1962 for their discoveries.

DNA Structure

Deoxyribonucleic acids are very large molecules, usually composed of two polynucleotide chains coiled together to form a double helix 2.0 nm in diameter (figure 10.2). Each chain contains purine and pyrimidine deoxyribonucleosides joined by phosphodiester bridges (figure 10.1*c*). That is, two adjacent deoxyribose sugars are connected by a phosphoric acid molecule esterified to a 3′-hydroxyl of one sugar and a 5′-hydroxyl of the other. Purine and pyrimidine bases are attached to the 1′-carbon of the deoxyribose sugars, and extend toward the middle of the cylinder formed by the two chains. They are stacked on top of each other in the center, one base pair every 0.34 nm. The purine adenine (A) is always paired with the pyrimidine thymine (T) by two hydrogen bonds. The purine guanine (G) pairs with cytosine (C) by three hydrogen bonds (figure 10.3). This AT and GC base pairing means that the two strands in a DNA double helix are **complementary.** That is, the bases in one strand match up with those of the other according to the base pairing rules. Because the sequences of bases in these strands encode genetic information (*see chapter 12*), considerable effort has been devoted to determining the base sequences of DNA and RNA from many microorganisms.

Nucleic acid sequence comparison and microbial taxonomy (chapter 20).

The two polynucleotide strands fit together much like the pieces in a jigsaw puzzle because of complementary base pairing (Box 10.1). Inspection of figure 10.2*a,b*, depicting the B form of DNA (probably the most common form in cells), shows that the two strands are not positioned directly opposite one another in the helical cylinder. Therefore, when the strands twist about one another, a wide **major groove** and narrower **minor groove** are formed by the backbone. Each base pair rotates 36° around the cylinder with respect to adjacent pairs so that there are 10 base pairs per turn of the helical spiral. Each turn of the helix has a vertical length of 3.4 nm. The helix is right-handed; that is, the chains turn counterclockwise

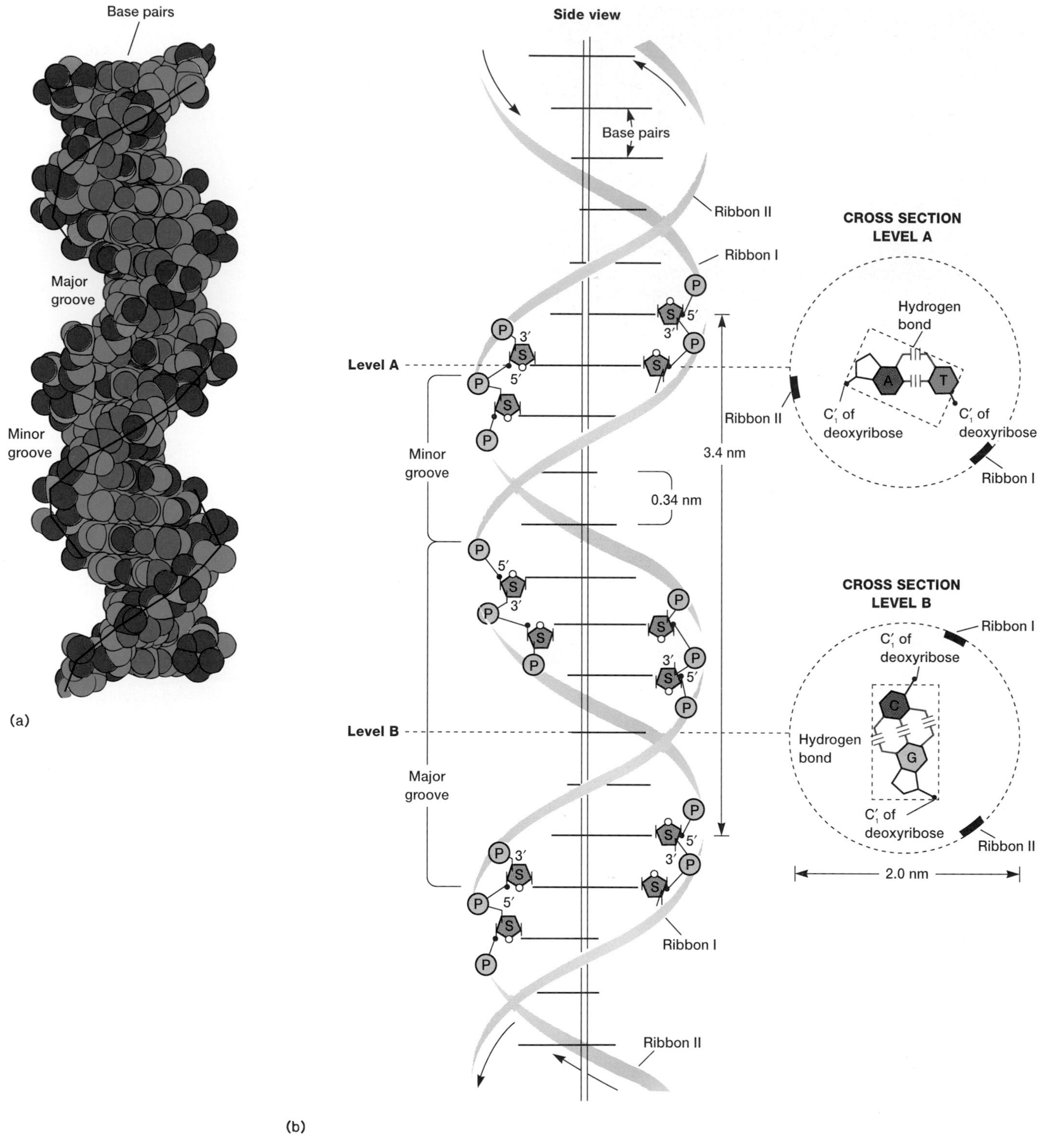

Figure 10.2 The Structure of the DNA Double Helix. (*a*) A space-filling model of the B form of DNA with the base pairs, major groove, and minor groove shown. The backbone phosphate groups, shown in color, spiral around the outside of the helix. (*b*) A diagrammatic representation of the double helix. The backbone consists of deoxyribose sugars (*S*) joined by phosphates (*P*) in phosphodiester bridges. The arrows at the top and bottom of the chains point in the 5′ to 3′ direction. The ribbons represent the sugar phosphate backbones. Base pairs are shown for two different cross sections (levels A and B). (*c*) An end view of the double helix showing the outer backbone and the bases stacked in the center of the cylinder. In the top drawing, the ribose ring oxygens are red. The nearest base pair is highlighted in white.

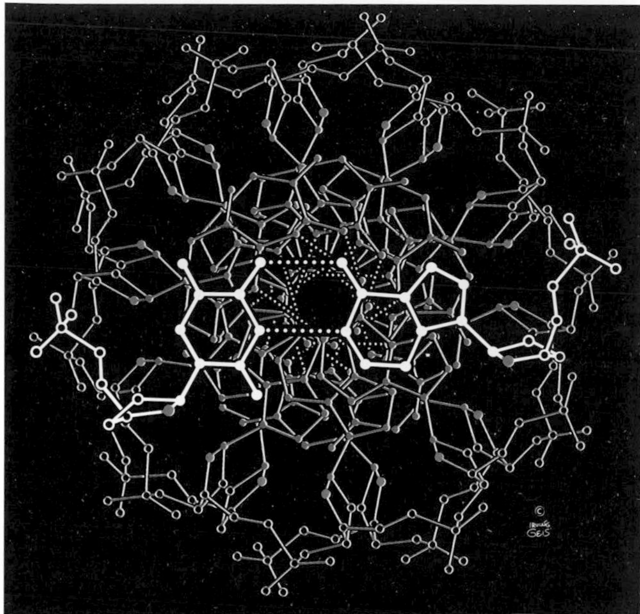

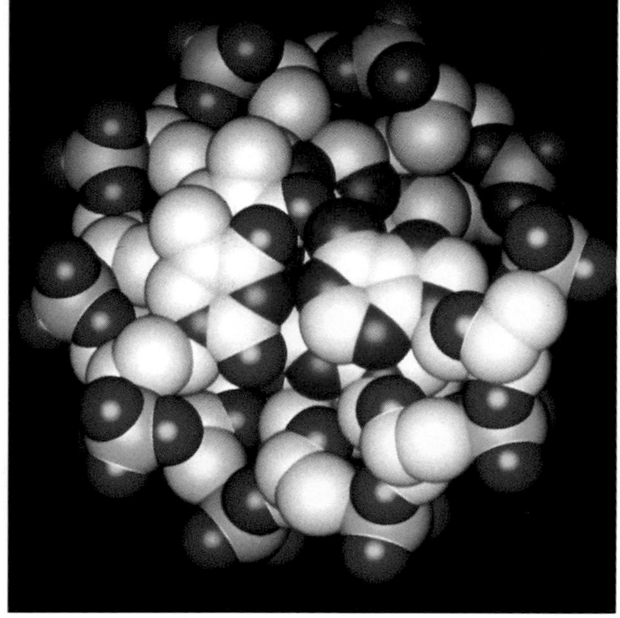

Figure 10.2 continued

Adenine Thymine

Guanine Cytosine

Figure 10.3 **DNA Base Pairs.** DNA complementary base pairing showing the hydrogen bonds (. . .).

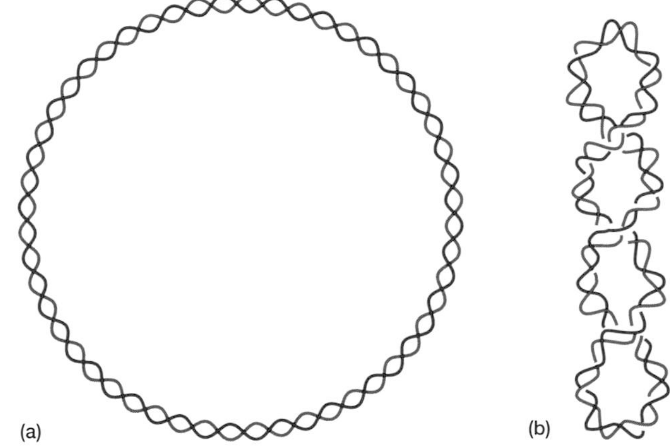

Figure 10.4 **DNA Forms.** (*a*) The DNA double helix is in the shape of a closed circle. (*b*) The circular DNA strands, already coiled in a double helix, are twisted a second time to produce supercoils.

as they approach a viewer looking down the longitudinal axis. The two backbones are antiparallel or run in opposite directions with respect to the orientation of their sugars. One end of each strand has an exposed 5′-hydroxyl, often with phosphates attached, whereas the other end has a free 3′-hydroxyl group. If the end of a double helix is examined, the 5′ end of one strand and the 3′ end of the other are visible. In a given direction, one strand is oriented 5′ to 3′ and the other, 3′ to 5′ (figure 10.2*b*).

RNA Structure

Besides differing chemically from DNA, ribonucleic acid is usually single stranded rather than double stranded like most DNA. An RNA strand can coil back upon itself to form a hairpin-shaped structure with complementary base pairing and helical organization. Cells contain three different types of RNA—messenger RNA, ribosomal RNA, and transfer RNA—that differ from one another in function, site of synthesis in eucaryotic cells, and structure.

The Organization of DNA in Cells

Although DNA exists as a double helix in both procaryotic and eucaryotic cells, its organization differs in the two cell types (*see table 4.2*). DNA is organized in the form of a closed circle in procaryotes like *E. coli.* This circular double helix is further twisted into supercoiled DNA (figure 10.4), and is associated

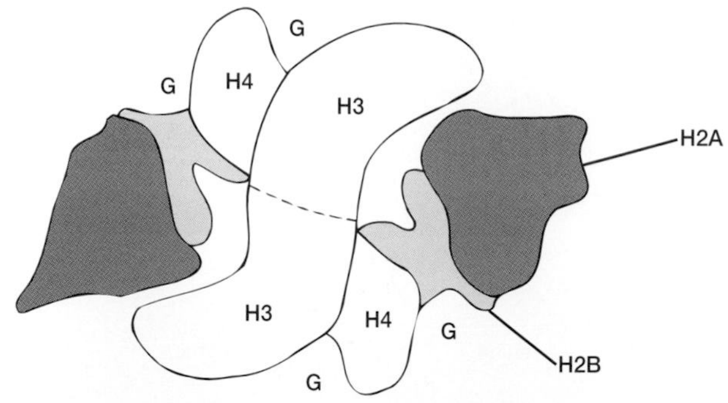

(a)

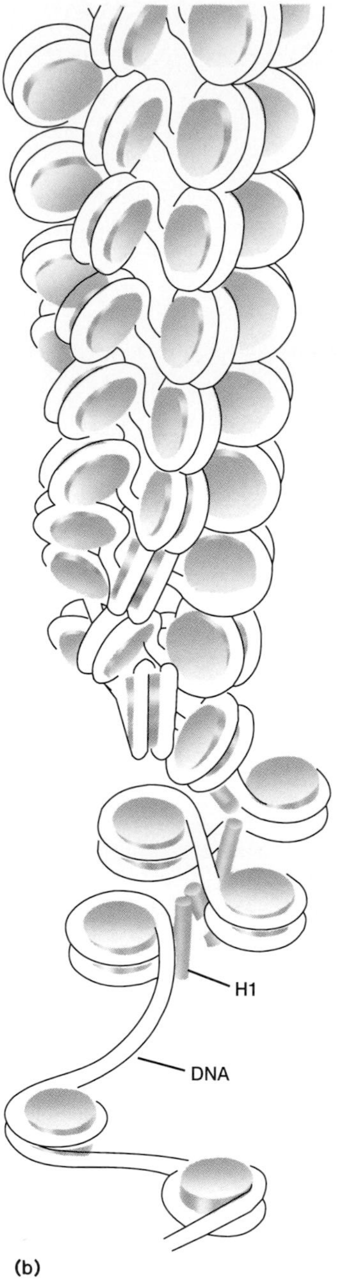

(b)

Figure 10.5 **Nucleosome Structure and Function.** (*a*) The upper diagram depicts the overall structure of nucleosomes from chicken erythrocytes. The central core is formed by the association of H3 and H4 histone dimers. H2A–H2B dimers are located at each end. The histones associate in such a way that a spiral groove (*G*) is formed, which may be the DNA binding site. The lower picture is a model of the nucleosome with DNA coiling around it in one possible arrangement. (*b*) An illustration of how nucleosomes and histone H1 might associate with DNA to form highly coiled chromatin. The nucleosomes are drawn as cylinders, though they may actually have an ellipsoidal shape.

with basic proteins but not with the histones found complexed with almost all eucaryotic DNA. These histonelike proteins do appear to help organize bacterial DNA into a coiled chromatinlike structure.

The structure of the bacterial nucleoid (p. 50).

DNA seems much more highly organized in eucaryotic chromatin (*see chapter 4*) and is associated with a variety of proteins, the most prominent of which are **histones.** These are small, basic proteins rich in the amino acids lysine and/or arginine. There are five types of histones in almost all eucaryotic cells studied: H1, H2A, H2B, H3, and H4. Eight histone molecules (two each of H2A, H2B, H3, and H4) form an ellipsoid, shaped somewhat like a rugby ball, about 11 nm long and 6.5 to 7 nm in diameter (figure 10.5*a*). DNA coils around the surface of the ellipsoid approximately 1¾ turns or 166 base pairs before proceeding on to the next. This complex of histones plus DNA is called a **nucleosome.** Thus, DNA gently isolated from chromatin looks like a string of beads. The stretch of DNA between the beads or nucleosomes, the linker region, varies in length from 14 to over 100 base pairs. Histone H1 appears to associate with the linker regions to aid the folding of DNA into more complex chromatin structures (figure 10.5*b*). When folding reaches a maximum, the chromatin takes the shape of the visible chromosomes seen in eucaryotic cells during mitosis and meiosis (*see figure 4.21*).

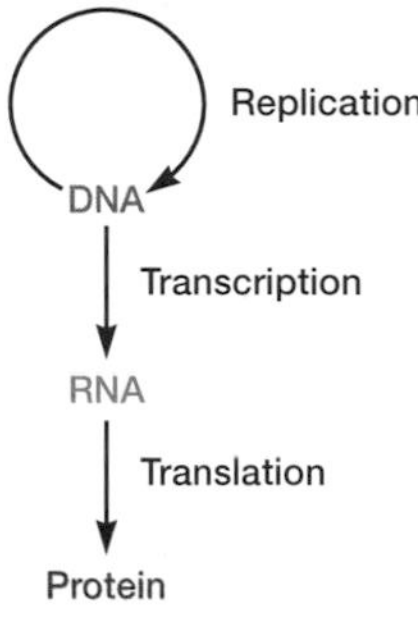

Figure 10.6 Relationships between DNA, RNA, and Protein Synthesis.

1. What are nucleic acids? How do DNA and RNA differ in structure?
2. Describe in some detail the structure of the DNA double helix. What does it mean to say that the two strands are complementary and antiparallel?
3. What are histones and nucleosomes? Describe the way in which DNA is organized in the chromosomes of procaryotes and eucaryotes.

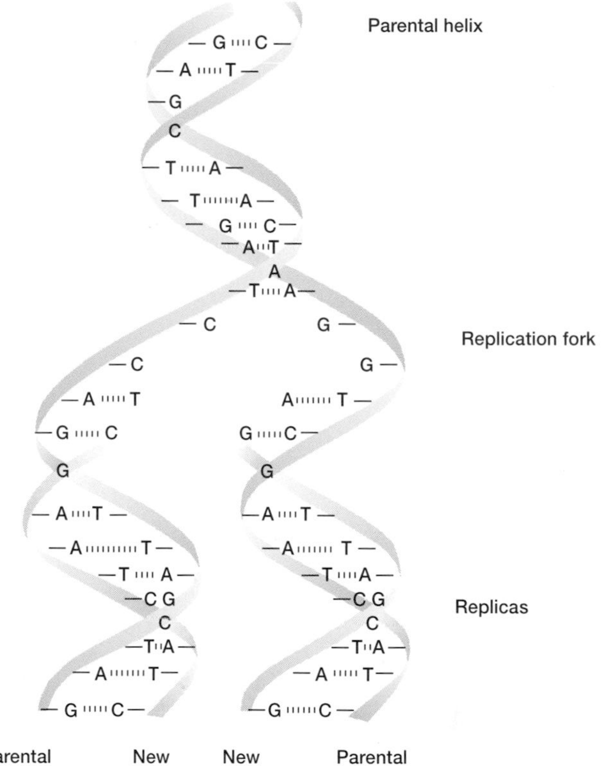

Figure 10.7 Semiconservative DNA Replication. The replication fork of DNA showing the synthesis of two progeny strands. Newly synthesized strands are in color. Each copy contains one new and one old strand. This process is called semiconservative replication.

DNA Replication

Biologists have long recognized a relationship between DNA, RNA, and protein (figure 10.6), and this recognition has guided a vast amount of research over the past decades. DNA is very precisely copied during its synthesis or **replication.** The expression of the information encoded in the base sequence of DNA begins with the synthesis of an RNA copy of the DNA sequence making up a gene (*see chapter 12*). This is called **transcription** because the DNA base sequence is being transcribed into an RNA base sequence. The RNA that carries information from DNA and directs protein synthesis is **messenger RNA (mRNA).** The last phase of gene expression is **translation** or protein synthesis. The genetic information in the form of an mRNA nucleotide sequence is translated and governs the synthesis of protein. Thus, the amino acid sequence of a protein is a direct reflection of the base sequence in mRNA. In turn, the mRNA nucleotide sequence is a complementary copy of a portion of the DNA genome.

Pattern of DNA Synthesis

Watson and Crick published their description of DNA structure in April 1953. Almost exactly one month later, a second paper appeared in which they suggested how DNA might be replicated. They hypothesized that the two strands of the double helix unwind from one another and separate (figure 10.7). Free nucleotides now line up along the two parental strands through complementary base pairing—A with T, G with C, and *vice versa* (figure 10.3). When these nucleotides are linked together by one or more enzymes, two replicas result, each containing a parental DNA strand and a newly formed strand. The replicating DNA molecule has a Y shape, just as in figure 10.7, and the actual replication process takes place at the **replication fork.** Research in subsequent years has proved Watson and Crick's hypothesis correct.

Replication patterns are somewhat different in procaryotes and eucaryotes. For example, when the circular DNA chromosome of *E. coli* is copied, replication begins at a single point, the origin. Two replication forks move outward from the origin until they have copied the whole **replicon,** that portion of the genome that contains an origin and is replicated as a unit. When the replication forks move around the circle, a structure shaped like the Greek letter theta (θ) is formed (figure 10.8). Finally, since the bacterial chromosome is a single replicon, the forks meet on the other side and two separate chromosomes are released. There is evidence that the replication fork and associated enzymes are attached to the bacterial plasma membrane. If this is the case, the DNA moves through the replication apparatus (which is fixed to the membrane) rather than the replication fork moving around the DNA.

A different pattern of DNA replication occurs during *E. coli* conjugation (*see chapter 13*) and the reproduction of bacteriophages, such as phage lambda (*see chapter 18*). In the **rolling-circle mechanism** (figure 10.9), one strand is nicked and the free 3′-hydroxyl end is extended by replication enzymes.

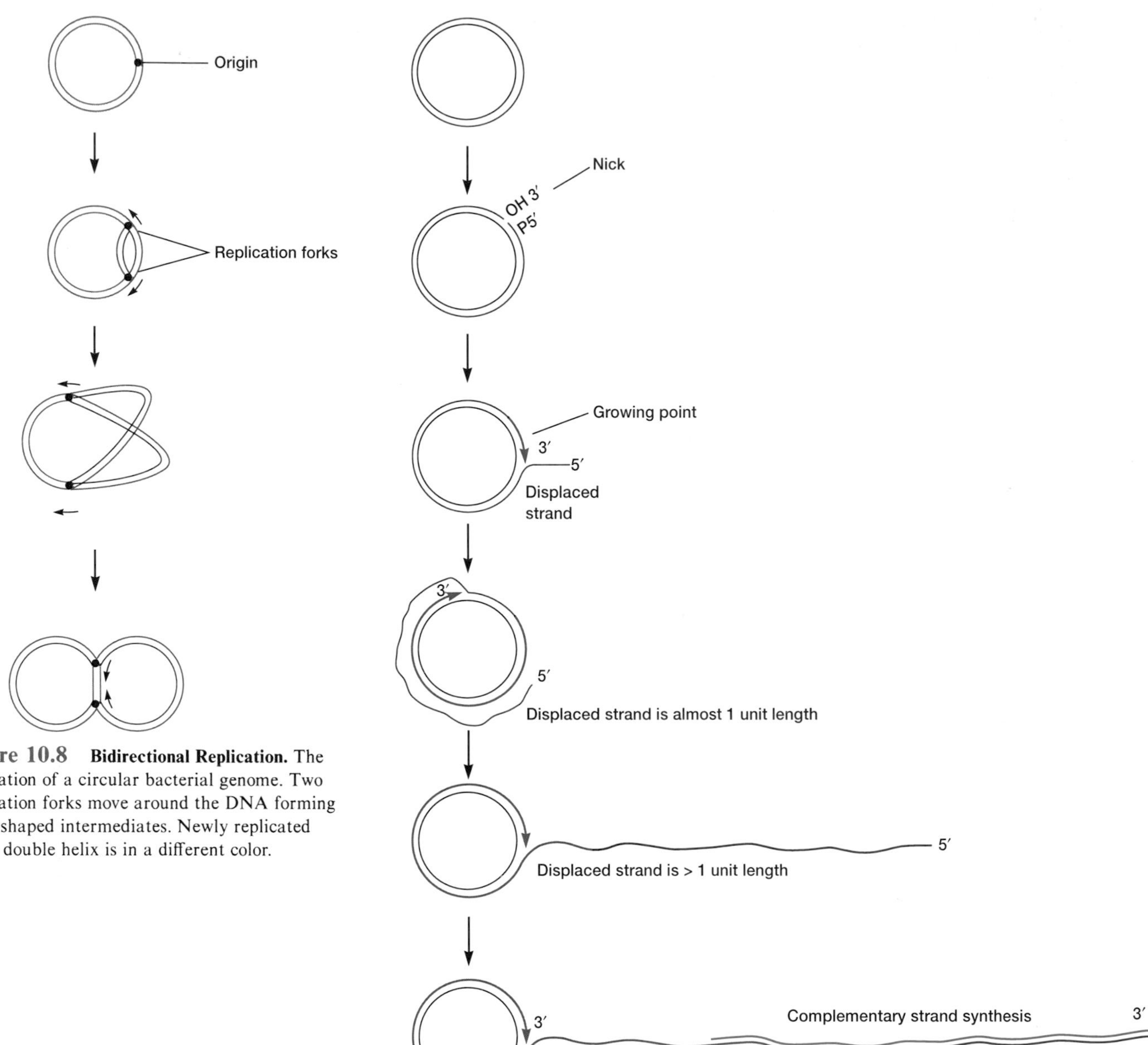

Figure 10.8 Bidirectional Replication. The replication of a circular bacterial genome. Two replication forks move around the DNA forming theta-shaped intermediates. Newly replicated DNA double helix is in a different color.

Figure 10.9 The Rolling-circle Pattern of Replication. A single-stranded tail, often composed of more than one genome copy, is generated and can be converted to the double-stranded form by synthesis of a complementary strand. The "free end" of the rolling-circle strand is probably bound to the primosome (p. 199).

As the 3′ end is lengthened while the growing point rolls around the circular template, the 5′ end of the strand is displaced and forms an ever lengthening tail. The single-stranded tail may be converted to the double-stranded form by complementary strand synthesis. This mechanism is particularly useful to viruses such as ϕX174 (p. 372) because it allows the rapid, continuous production of many genome copies from a single initiation event.

Eucaryotic DNA is linear and much longer than procaryotic DNA; *E. coli* DNA is about 1,400 μm in length, whereas the 46 chromosomes in the human nucleus have a total length of 1.8 m (almost 1,400 times longer). Clearly, many replication forks must copy eucaryotic DNA simultaneously so that the molecule can be duplicated in a relatively short period, and so many replicons are present that there is an origin about every 10 to 100 μm along the DNA. Replication forks move outward from these sites and eventually meet forks that have been copying the adjacent DNA stretch (figure 10.10). In this fashion, a large molecule is quickly copied.

Mechanism of DNA Replication

Because DNA replication is so essential to organisms, a great deal of effort has been devoted to understanding its mecha-

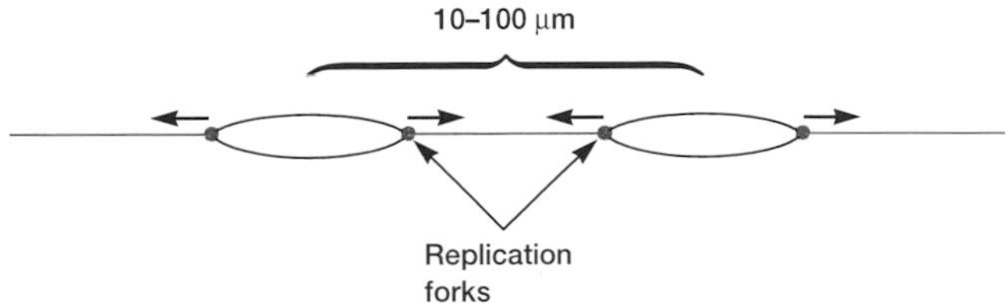

Figure 10.10 **The Replication of Eucaryotic DNA.** Replication is initiated every 10–100 μm and the replication forks travel away from the origin. Newly copied DNA is in color.

nism. The replication of *E. coli* DNA is probably best understood and is the focus of attention in this section. The process in eucaryotic cells is thought to be similar.

E. coli has three different **DNA polymerase** enzymes, each of which catalyzes the synthesis of DNA in the 5′ to 3′ direction while reading the DNA template in the 3′ to 5′ direction (figures 10.11 and 10.12). The polymerases require deoxyribonucleoside triphosphates (dATP, dGTP, dCTP, and dTTP) as substrates and a DNA template to copy. Nucleotides are added to the 3′ end of the growing chain when the free 3′-hydroxyl group on the deoxyribose attacks the first or alpha phosphate group of the substrate to release pyrophosphate (figure 10.11). DNA polymerase III plays the major role in replication, although it is probably assisted by polymerase I. It is thought that polymerases I and II participate in the repair of damaged DNA (*see chapter 17*).

During replication, the DNA double helix must be unwound to generate separate single strands. Unwinding occurs very quickly; the fork may rotate as rapidly as 75 to 100 revolutions per second. **Helicases** are responsible for DNA unwinding. These enzymes use energy from ATP to unwind short stretches of helix just ahead of the replication fork. Once the strands have separated, they are kept single through specific binding with **single-stranded DNA binding proteins (SSBs)** as shown in figure 10.12*a*. Rapid unwinding can lead to tension and formation of supercoils or supertwists in the helix, just as rapid separation of two strands of a rope can lead to knotting or coiling of the rope. The tension generated by unwinding is relieved, and the unwinding process is promoted by enzymes known as **topoisomerases.** These enzymes change the structure of DNA in such a way that it remains unaltered as its shape is changed (e.g., a topoisomerase might tie or untie a knot in a DNA strand). **DNA gyrase** is an *E. coli* topoisomerase that removes the supertwists produced during replication.

The details of DNA replication are outlined in a diagram of the replication fork (figure 10.12*b*). The replication process takes place in four stages.

1. Helicases unwind the helix with the aid of topoisomerases like the DNA gyrase. It appears that the DNAB protein is the helicase most actively involved in replication, but the n′ protein may also participate in unwinding. The single strands are kept separate by the DNA binding proteins (SSBs).

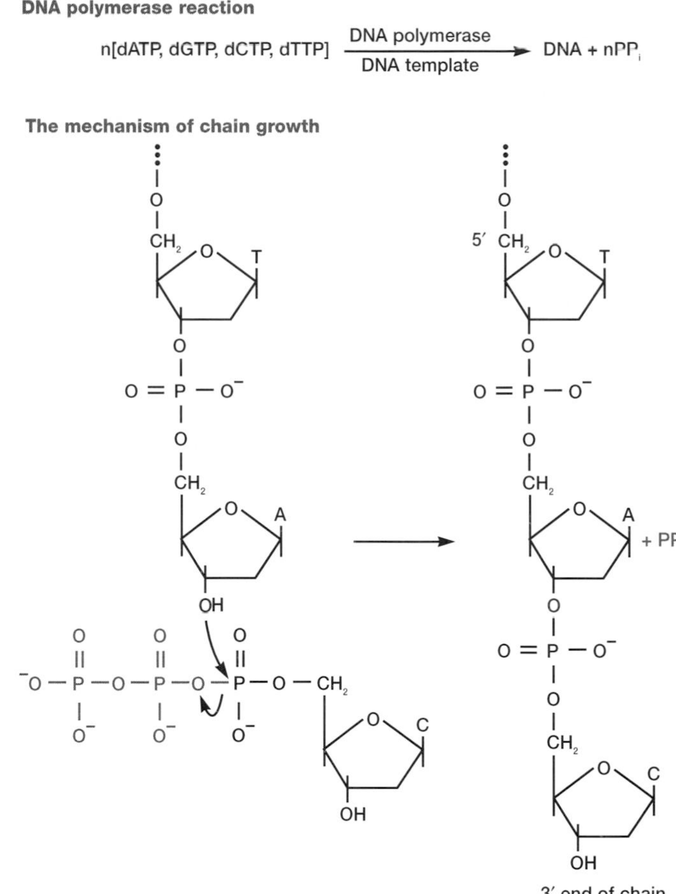

Figure 10.11 **The DNA Polymerase Reaction and its Mechanism.** The mechanism involves a nucleophilic attack by the hydroxyl of the 3′ terminal deoxyribose on the alpha phosphate group of the nucleotide substrate (in this example, cytidine triphosphate).

2. DNA is probably replicated continuously by DNA polymerase III when the leading strand is copied. Lagging strand replication is discontinuous, and the fragments are synthesized in the 5′ to 3′ direction just as in leading strand synthesis. First, a special RNA polymerase called a **primase** synthesizes a short RNA primer, usually around 1 to 5 nucleotides long, complementary to the DNA. This is necessary because DNA polymerases I and III cannot synthesize DNA without a primer. It appears that the primase requires the assistance of several other proteins, and the complex of the primase with its accessory proteins is called the **primosome.** DNA polymerase III then synthesizes complementary DNA beginning at the 3′ end of the RNA primer. Both leading and lagging strand synthesis probably occur on a single multiprotein complex, the replisome. If this is the case, the lagging strand template must be looped around the complex (figure 10.12*b*). The final fragments are around 1,000 to 2,000 nucleotides long in bacteria and approximately 100 nucleotides long in eucaryotic cells. They are called **Okazaki fragments** after their discoverer, Reiji Okazaki.

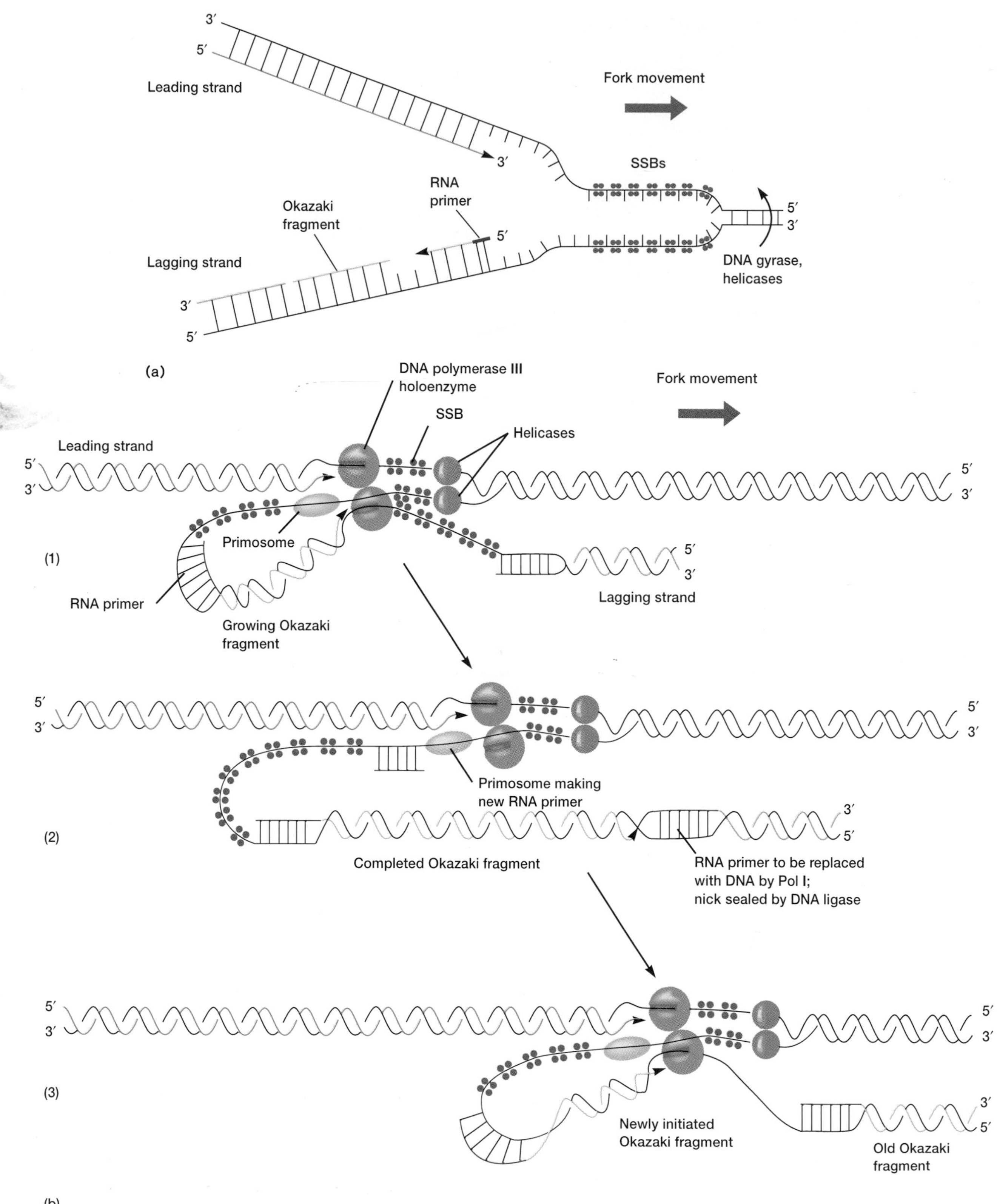

Figure 10.12 Bacterial DNA Replication. The synthesis of DNA in *E. coli* at the replication fork. (*a*) Diagram of the overall process. Bases and base pairs are represented by lines extending outward from the strands. The RNA primer is in color. See text for details. (*b*) A hypothetical model for activity at the replication fork. (*1*) The replisome has two DNA polymerase III holoenzyme complexes. One polymerase continuously copies the leading strand. The lagging strand loops around the other polymerase so that both strands can be replicated simultaneously. (*2*) When DNA polymerase III encounters a completed Okazaki fragment, it releases the lagging strand. The primosome synthesizes another RNA primer on the lagging strand. DNA polymerase I and DNA ligase fill and seal the gaps between completed fragments. (*3*) The holoenzyme binds the lagging strand at the new primer and begins to synthesize another Okazaki fragment. *Source for* (b)*: Donald Voet and Judith G. Voet,* Biochemistry. *Copyright © 1990 John Wiley & Sons, Inc., New York, NY.*

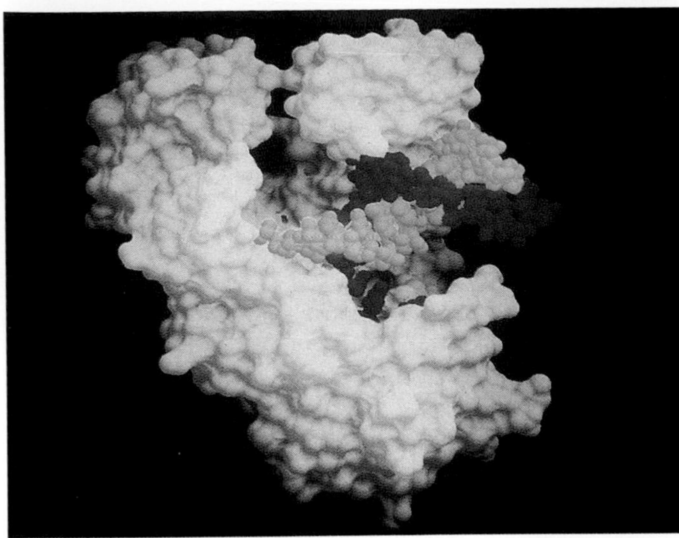

Figure 10.13 DNA Polymerase I. A model of a large fragment of *E. coli* DNA polymerase I bound to DNA. The protein is in yellow, the DNA template strand in blue, and the primer strand in red. The 3′ end of the primer strand is entering the editing exonuclease site.

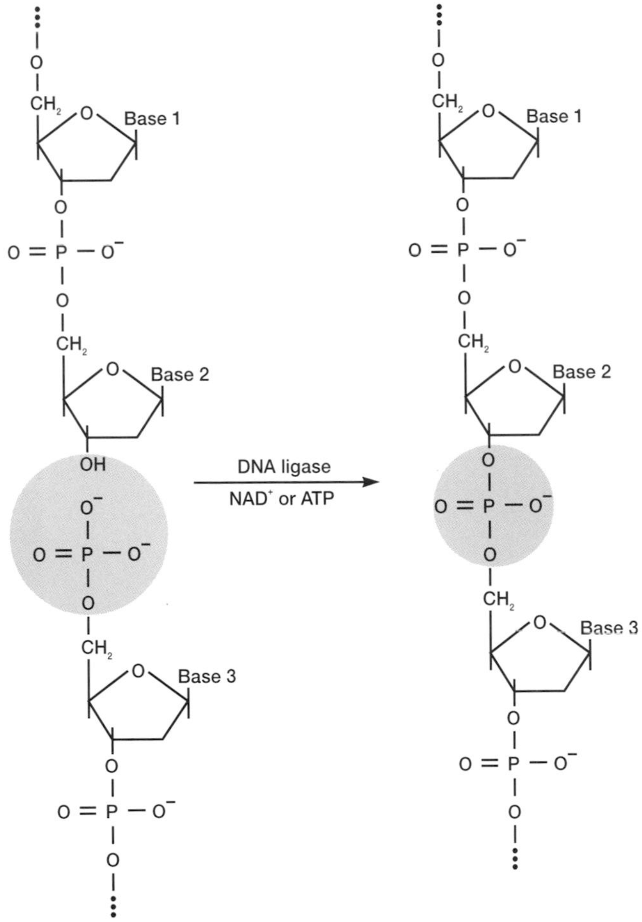

Figure 10.14 The DNA Ligase Reaction. The groups being altered are shaded in color.

3. After most of the lagging strand has been duplicated by the formation of Okazaki fragments, DNA polymerase I removes the RNA primer and synthesizes complementary DNA to fill the gap resulting from RNA deletion (figure 10.13). The polymerase appears to remove the primer one nucleotide at a time and replace it with the appropriate complementary deoxyribonucleotide.
4. Finally, the fragments are joined by the enzyme **DNA ligase,** which forms a phosphodiester bond between the 3′-hydroxyl of the growing strand and the 5′-phosphate of an Okazaki fragment (figure 10.14). Bacterial ligases use the pyrophosphate bond of NAD^+ as an energy source; many other ligases employ ATP instead.

DNA polymerase III holoenzyme, the enzyme complex that synthesizes most of the DNA copy, is a very large entity containing DNA polymerase III and several other proteins. At least one subunit, copolymerase III, is required for proper binding to the RNA primer.

DNA replication is an extraordinarily complex process. Presumably, much of the complexity is necessary for accuracy in copying DNA. It would be very dangerous for any organism to make many errors during replication because a large number of mutations would certainly be lethal. In fact, *E. coli* makes errors with a frequency of only 10^{-9} or 10^{-10} per base pair replicated (or about 10^{-6} per gene per generation). Part of this precision results from the low error rate of the copying process itself. However, DNA polymerase III (and DNA polymerase I) also can proofread the newly formed DNA. As polymerase III moves along synthesizing a new DNA strand, it recognizes any errors resulting in improper base pairing and hydrolytically removes the wrong nucleotide through a special 3′ to 5′ exonuclease activity. The enzyme then backs up and adds the proper nucleotide in its place. Polymerases delete errors by acting much like correcting typewriters.

DNA repair (pp. 253–56).

Despite its complexity and accuracy, replication occurs very rapidly indeed. In bacteria, replication rates approach 750 to 1,000 base pairs per second. Eucaryotic replication is much slower, about 50 to 100 base pairs per second. This is not surprising because eucaryotic replication also involves operations like unwinding the DNA from nucleosomes.

1. Define the following terms: replication, transcription, messenger RNA, translation, replicon, replication fork, primosome, and replisome.
2. Be familiar with the nature and functions of the following replication components and intermediates: DNA polymerases I and III, topoisomerase, DNA gyrase, helicase, single-stranded DNA binding protein, Okazaki fragment, DNA ligase, leading strand, and lagging strand.

DNA Transcription or RNA Synthesis

As mentioned earlier, synthesis of RNA under the direction of DNA is called transcription. The RNA product has a sequence complementary to the DNA template directing its synthesis (table 10.1). Thymine is not normally found in mRNA and rRNA. Although adenine directs the incorporation of thymine during DNA replication, it usually codes for uracil during RNA synthesis. Transcription generates three kinds of RNA. Messenger RNA (mRNA) bears the message for protein synthesis. **Transfer RNA (tRNA)** carries amino acids during protein synthesis, and **ribosomal RNA (rRNA)** molecules are components of ribosomes. Structure and synthesis of procaryotic mRNA is described first.

Transcription in Procaryotes

Procaryotic mRNA is a single-stranded RNA of variable length containing directions for the synthesis of one to many polypeptides. Messenger RNA molecules also have sequences that do not code for polypeptides (figure 10.15). There is a nontranslated **leader sequence** of 25 to 150 bases at the 5′ end preceding the initiation codon. In addition, polygenic mRNAs (those directing the synthesis of more than one polypeptide) have spacer regions separating the segments coding for individual polypeptides. Polygenic messenger polypeptides usually function together in some way (e.g., as part of the same metabolic pathway). At the 3′ end, following the last termination codon, is a nontranslated trailer.

Table 10.1 RNA Bases Coded for by DNA

DNA Base	Purine or Pyrimidine Incorporated into RNA
Adenine	Uracil
Guanine	Cytosine
Cytosine	Guanine
Thymine	Adenine

Messenger RNA is synthesized under the direction of DNA by the enzyme **RNA polymerase.** The reaction is quite similar to that catalyzed by DNA polymerase. ATP, GTP, CTP, and UTP are used to produce an RNA copy of the DNA sequence.

$$n[\text{ATP, GTP, CTP, UTP}] \xrightarrow[\text{DNA template}]{\text{RNA polymerase}} \text{RNA} + n\text{PP}_i$$

RNA synthesis, like DNA synthesis, proceeds in a 5′ to 3′ direction with new nucleotides being added to the 3′ end of the growing chain (figure 10.16). It should be noted that pyrophosphate is produced in both DNA and RNA polymerase reactions. Pyrophosphate is then removed by hydrolysis to orthophosphate in a reaction catalyzed by the pyrophosphatase enzyme. Removal of the pyrophosphate product makes DNA and RNA synthesis irreversible. If the pyrophosphate level were too high, DNA and RNA would be degraded by a reversal of the polymerase reactions.

A **gene** is a DNA segment or sequence that codes for a polypeptide, an rRNA, or a tRNA. Although DNA has two complementary strands, RNA polymerase copies only the **template** or **sense strand** at any particular point on DNA. If both strands of DNA were transcribed, two different mRNAs would result and cause genetic confusion. Thus, the sequence corresponding to a gene is located only on one of the two complementary DNA strands, the sense strand. Different genes may be encoded on opposite strands. The RNA polymerase opens the double helix and transcribes the sense strand to produce an RNA transcript that is complementary and antiparallel to the DNA template.

The genetic code and gene structure (pp. 239–44).

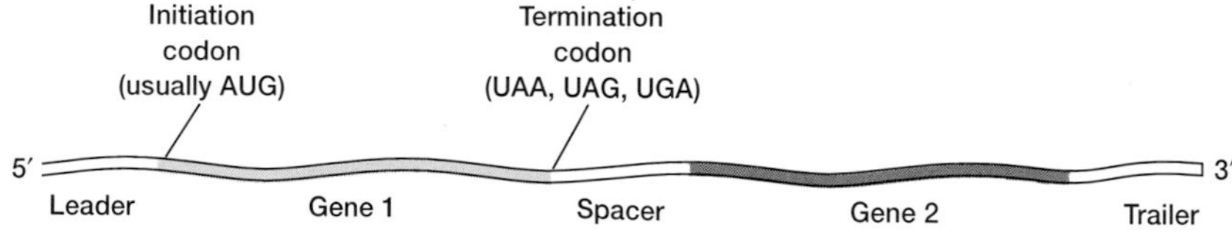

Figure 10.15 A Polygenic Bacterial Messenger RNA. See text for details.

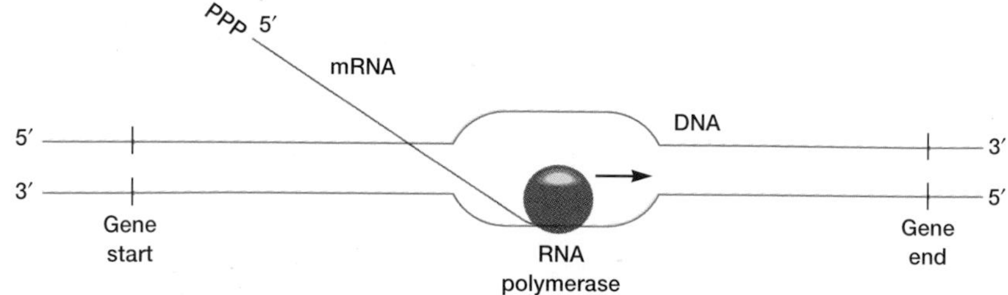

Figure 10.16 mRNA Transcription from DNA. The lower DNA strand is directing mRNA synthesis, and the RNA polymerase is moving from left to right.

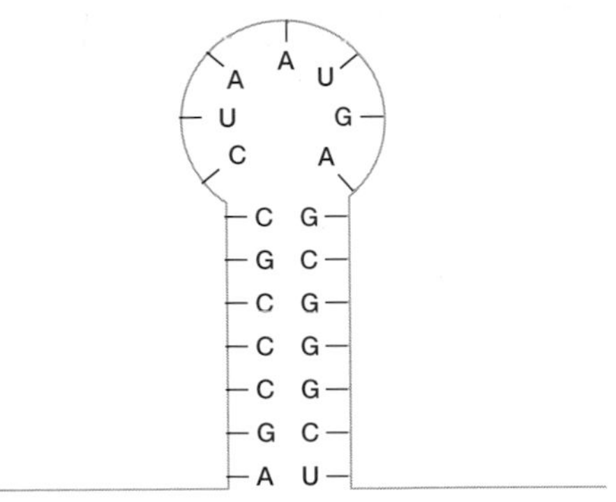

Figure 10.17 **Procaryotic Terminators.** An example of a hairpin structure formed by an mRNA terminator sequence.

The RNA polymerase of *E. coli* is a very large molecule (MW 490,000) containing four types of polypeptide chains: α, β, β', and σ. The **core enzyme** is composed of four chains (α_2, β, β') and catalyzes RNA synthesis. The **sigma factor** (σ) has no catalytic activity but helps the core enzyme recognize the start of genes. Once RNA synthesis begins, the sigma factor dissociates from the core enzyme–DNA complex and is available to aid another core enzyme. The precise functions of the α, β, and β' polypeptides are not yet clear. The binding site for DNA may be on β'. Rifampin, an RNA polymerase inhibitor, binds to the β subunit.

The region to which RNA polymerase binds with the aid of the sigma factor is called the **promoter.** The promoter sequence is not transcribed. A 6 base sequence, approximately 35 base pairs before the transcription starting point, is present in all bacterial promoters, in addition to a TATAAT sequence or **Pribnow box** about 10 base pairs before the starting point. The RNA polymerase recognizes these sequences, binds to the promoter, and unwinds a short segment of DNA beginning around the Pribnow box. Transcription starts 6 or 7 base pairs away from the 3′ end of the promoter. The first base used in RNA synthesis is usually a purine, either ATP or GTP. Since these phosphates are not removed during transcription, the 5′ end of procaryotic mRNA has a triphosphate attached to the ribose.

Promoter structure and function (pp. 241–42).

There also must be stop signals to mark the end of a gene or sequence of genes and stop transcription by the RNA polymerase. All procaryotic **terminators** contain a sequence coding for an RNA stretch that can hydrogen bond to form a hairpin-shaped loop and stem structure (figure 10.17). This structure appears to cause the RNA polymerase to pause or stop transcribing DNA. There are two kinds of stop signals or terminators. The first type contains a stretch of about six uridine residues following the mRNA hairpin, and causes the polymerase to stop transcription and release the mRNA without the aid of any accessory factors. The second kind lacks a poly-U region after the hairpin and requires the aid of a special protein, the **rho factor** (ρ). It is thought that rho binds to mRNA and moves along the molecule until it reaches the RNA polymerase that has halted at a terminator. The rho factor then causes the polymerase to dissociate from the mRNA, probably by unwinding the mRNA-DNA complex.

Table 10.2 Eucaryotic RNA Polymerases

Enzyme	Location	Product
RNA polymerase I	Nucleolus	rRNA (18S, 5.8S, 28S)
RNA polymerase II	Chromatin, nuclear matrix	mRNA
RNA polymerase III	Chromatin, nuclear matrix	tRNA, 5S rRNA

Transcription in Eucaryotes

Transcriptional processes in eucaryotic microorganisms (and in other eucaryotic cells) differ in several ways from procaryotic transcription. There are three major RNA polymerases, not one as in procaryotes. RNA polymerase II, associated with chromatin in the nuclear matrix, is responsible for mRNA synthesis. Polymerases I and III synthesize rRNA and tRNA, respectively (table 10.2). Eucaryotic promoters are also different. For example, polymerase II recognizes a TATA sequence positioned about 25 base pairs before the transcription starting point.

Eucaryotic mRNA arises from **posttranscriptional modification** of large RNA precursors, about 5,000 to 50,000 nucleotides long, called **heterogeneous nuclear RNA (hnRNA)** molecules. These are the products of RNA polymerase II activity (figure 10.18). After hnRNA synthesis, the enzyme polyadenylate polymerase catalyzes the addition of adenylic acid to the 3′ end of hnRNA to produce a poly-A sequence about 200 nucleotides long. The hnRNA is then cleaved to generate the final mRNA. Sometimes, poly-A is added after hnRNA has been cleaved to the proper length. Usually, eucaryotic mRNA also differs in having a 5′ cap consisting of 7-methylguanosine attached to the 5′-hydroxyl by a triphosphate linkage (figure 10.19). The adjacent nucleotide also may be methylated.

The 3′ poly-A and 5′ capping of eucaryotic mRNAs distinguish them from procaryotic mRNAs. In addition, eucaryotic mRNA is normally monogenic in contrast to procaryotic mRNA, which is often polygenic. The functions of poly-A and capping are still not completely clear. It is thought that poly-A protects mRNA from rapid enzymatic degradation. The 5′ cap on eucaryotic messengers may aid the initial binding of ribosomes to the messenger. The cap also may protect the messenger from enzymatic attack.

Many eucaryotic genes differ from procaryotic genes in being split or interrupted, which leads to another type of posttranscriptional processing. **Split or interrupted genes** have **exons** (*ex*pressed sequences), regions coding for RNA that end up in the final RNA product (e.g., mRNA). Exons are separated

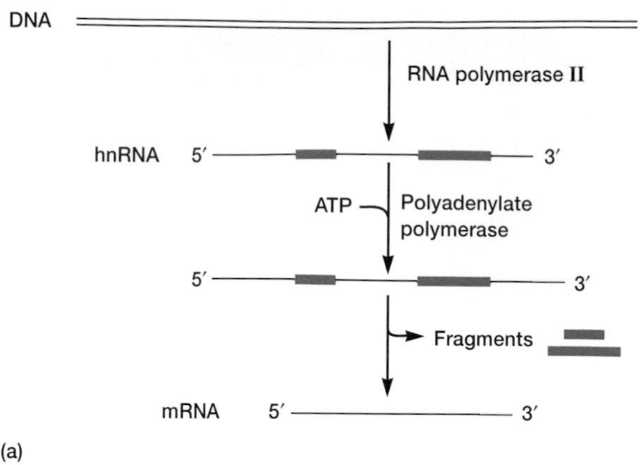

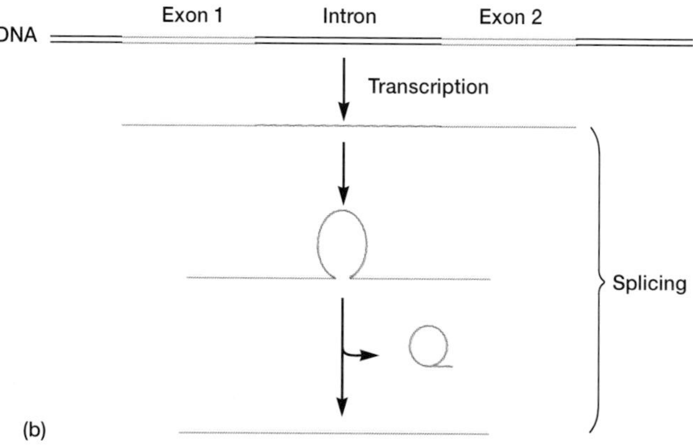

Figure 10.18 Eucaryotic mRNA Synthesis. (*a*) The production of eucaryotic messenger RNA. The addition of poly-A to the 3′ end of mRNA is included, but not the capping of the 5′ end. Poly-A sequence and introns are in color. (*b*) The splicing of interrupted genes to produce mRNA. Poly-A sequences and exons are in color. The excised intron is in the shape of a circle or lariat.

from one another by **introns** (*int*ervening sequences), sequences coding for RNA that is missing from the final product (figure 10.18*b*). The initial RNA transcript has the intron sequences present in the interrupted gene. Genes coding for rRNA and tRNA may also be interrupted. Except for cyanobacteria and archaeobacteria (*see chapters 23 and 24*), interrupted genes have not been found in procaryotes.

Introns are removed from the initial RNA transcript by a process called **RNA splicing** (figure 10.18*b*). The intron's borders must be clearly marked for accurate removal, and this is the case. Exon-intron junctions have a GU sequence at the intron's 5′ boundary and an AG sequence at its 3′ end. These two sequences define the splice junctions and are recognized by special RNA molecules. The nucleus contains several **small nuclear RNA (snRNA)** molecules, about 60 to 300 nucleotides long. These complex with proteins to form *s*mall *n*uclear *r*ibo*n*ucleo*p*rotein particles called snRNPs or snurps. Some of the snRNPs recognize splice junctions and ensure splicing accuracy. For example, U1-snRNP recognizes the 5′ splice junction, and U5-snRNP recognizes the 3′ junction. Splicing of pre-mRNA occurs in a large complex called a **spliceosome** that contains the pre-mRNA, at least five kinds of snRNPs, and non-snRNP splicing factors.

As just mentioned, a few rRNA genes also have introns. Some of these pre-rRNA molecules are self-splicing. The RNA actually catalyzes the splicing reaction and is now called a **ribozyme** (Box 10.2). Thomas Cech first discovered that pre-rRNA from the ciliate protozoan *Tetrahymena thermophila* is self-splicing. Sidney Altman then showed that ribonuclease P, which cleaves a fragment from one end of pre-tRNA, contains a piece of RNA that catalyzes the reaction. Several other self-splicing rRNA introns have since been discovered. Cech and Altman received the 1989 Nobel Prize in chemistry for these discoveries.

Although the focus has been on mRNA synthesis, it should be noted that both rRNAs and tRNAs also begin as parts of large precursors synthesized by RNA polymerases (table 10.2).

7-methylguanosine

Base 1

Base 2

Figure 10.19 The 5′ Cap of Eucaryotic mRNA.

Box 10.2

Catalytic RNA (Ribozymes)

Until recently, biologists believed that all cellular reactions were catalyzed by proteins called enzymes (*see chapter* 7). The discovery during 1981–1984 by Cech and Altman that RNA also can sometimes catalyze reactions has transformed our way of thinking about topics as diverse as catalysis and the origin of life. It is now clear that some RNA molecules, called ribozymes, catalyze reactions that alter either their own structure or that of other RNAs.

This discovery has stimulated scientists to hypothesize that the early earth was an "RNA world" in which RNA acted as both the genetic material and a reaction catalyst. Experiments showing that introns from *Tetrahymena thermophila* can catalyze the formation of polycytidylic acid under certain circumstances have further encouraged such speculations. Some have suggested that RNA viruses are "living fossils" of the original RNA world.

The best-studied ribozyme activity is the self-splicing of RNA. This process is widespread and occurs in *Tetrahymena* pre-rRNA; the mitochondrial rRNA and mRNA of yeast and other fungi; chloroplast tRNA, rRNA, and mRNA; and in mRNA from some bacteriophages (e.g., the T4 phage of *E. coli*). The 413-nucleotide rRNA intron of *T. thermophila* provides a good example of the self-splicing reaction. The reaction occurs in three steps and requires the presence of guanosine (Box Figure 10.2). First, the 3′-OH group of guanosine attacks the intron's 5′-phosphate group and cleaves the phosphodiester bond. Second, the new 3′-hydroxyl on the left exon attacks the 5′-phosphate of the right exon. This joins the two exons and releases the intron. Finally, the intron's 3′-hydroxyl attacks the phosphate bond of the nucleotide 15 residues from its end. This releases a terminal fragment and cyclizes the intron. Self-splicing of this rRNA occurs about 10 billion times faster than spontaneous RNA hydrolysis. Just as with enzyme proteins, the RNA's shape is essential to catalytic efficiency. The ribozyme even has Michaelis-Menten kinetics (p. 142).

The discovery of ribozymes has many potentially important practical consequences. Ribozymes act as "molecular scissors" and will enable researchers to manipulate RNA easily in laboratory experiments. It also might be possible to protect hosts by specifically removing RNA from pathogenic viruses, bacteria, and fungi. For example, ribozymes are already being tested against the AIDS, herpes, and tobacco mosaic viruses.

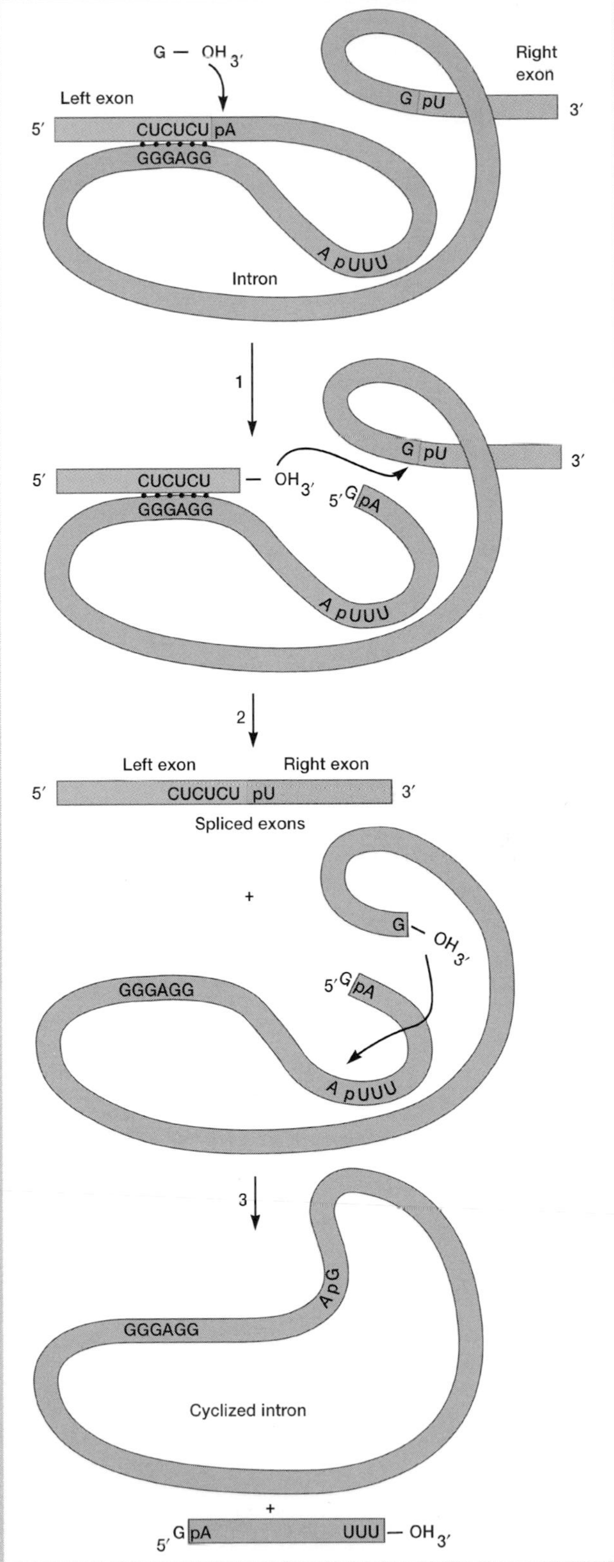

Box Figure 10.2 **Ribozyme Action.** The mechanism of *Tetrahymena* pre-rRNA self-splicing. See text for details. *From Donald Voet and Judith G. Voet,* Biochemistry. *Copyright © 1990 John Wiley & Sons, Inc., New York, NY.*

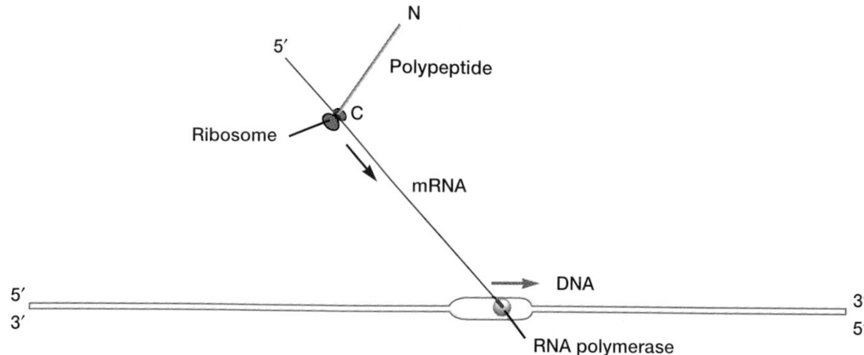

Figure 10.20 **Coupled Transcription and Translation in Bacteria.** mRNA is synthesized 5′ to 3′ by RNA polymerase while ribosomes are attaching to the newly formed 5′ end of mRNA and translating the message even before it is completed. Polypeptides are synthesized in the N-terminal to C-terminal direction.

The final rRNA and tRNA products result from posttranscriptional processing, as mentioned previously.

1. Define the following terms: leader, trailer, spacer region, polygenic mRNA, RNA polymerase core enzyme, sigma factor, promoter, sense strand, terminator, and rho factor.
2. What is a gene?
3. Define or describe posttranscriptional modification, heterogeneous nuclear RNA, 3′ poly-A sequence, 5′ capping, split or interrupted genes, exon, intron, RNA splicing, snRNA, spliceosome, and ribozyme.

Protein Synthesis

The final step in gene expression is protein synthesis or translation. The mRNA nucleotide sequence is translated into the amino acid sequence of a polypeptide chain in this step. Polypeptides are synthesized by the addition of amino acids to the end of the chain with the free α-carboxyl group (the C terminal end). That is, the synthesis of polypeptides begins with the amino acid at the end of the chain with a free amino group (the N-terminal) and moves in the C-terminal direction. Protein synthesis is not only quite accurate but also very rapid. In *E. coli,* synthesis occurs at a rate of about 300 to 450 residues per minute; eucaryotic translation is slower, about 50 residues per minute.

Polypeptide and protein structure (appendix I).

Many bacteria grow so quickly that each mRNA must be used with great efficiency in order to synthesize proteins at a sufficiently rapid rate. Frequently, bacterial mRNAs are simultaneously complexed with several ribosomes, each ribosome reading the mRNA message and synthesizing a polypeptide. At maximal rates of mRNA utilization, there may be a ribosome every 80 nucleotides along the messenger, or as many as 20 ribosomes simultaneously reading an mRNA that codes for a 50,000 dalton polypeptide. A complex of mRNA with several ribosomes is called a **polyribosome** or polysome. Polysomes are present in both procaryotes and eucaryotes. Bacteria can further increase the efficiency of gene expression through coupled transcription and translation (figure 10.20). While RNA polymerase is synthesizing an mRNA, ribosomes can already be attached to the messenger and involved in polypeptide synthesis. Coupled transcription and translation is possible in procaryotes because a nuclear envelope does not separate the translation machinery from DNA as it does in eucaryotes.

Transfer RNA and Amino Acid Activation

The first stage of protein synthesis is **amino acid activation,** a process in which amino acids are attached to transfer RNA molecules. These RNA molecules are normally between 73 and 93 nucleotides in length, and possess several characteristic structural features. The structure of tRNA becomes clearer when its chain is folded in such a way to maximize the number of normal base pairs, which results in a cloverleaf conformation of five arms or loops (figure 10.21). The acceptor or amino acid stem holds the activated amino acid on the 3′ end of the tRNA. The 3′ end of all tRNAs has the same —C—C—A sequence; the amino acid is attached to the terminal adenylic acid. At the other end of the cloverleaf is the anticodon arm, which contains the **anticodon triplet** complementary to the mRNA codon triplet. There are two other large arms: the D or DHU arm has the unusual pyrimidine nucleoside dihydrouridine; and the T or TΨC arm has ribothymidine (T) and pseudouridine (Ψ), both of which are unique to tRNA. Finally, the cloverleaf has a variable arm whose length changes with the overall length of the tRNA; the other arms are fairly constant in size.

Transfer RNA molecules are folded into an L-shaped structure (figure 10.22). The amino acid is held on one end of the L, the anticodon is positioned on the opposite end, and the corner of the L is formed by the D and T loops. Because there must be at least one tRNA for each of the 20 amino acids

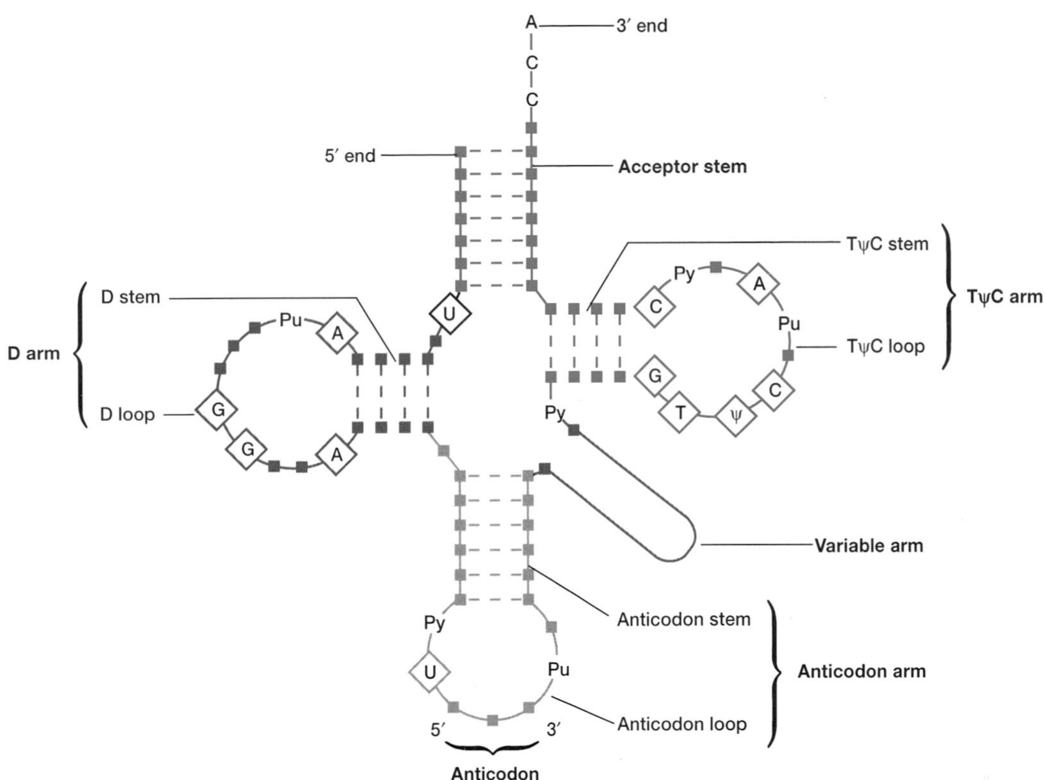

Figure 10.21 **tRNA Structure.** The cloverleaf structure for tRNA in procaryotes and eucaryotes. Bases found in all tRNAs are in diamonds; purine and pyrimidine positions in all tRNAs are labeled *Pu* and *Py,* respectively.

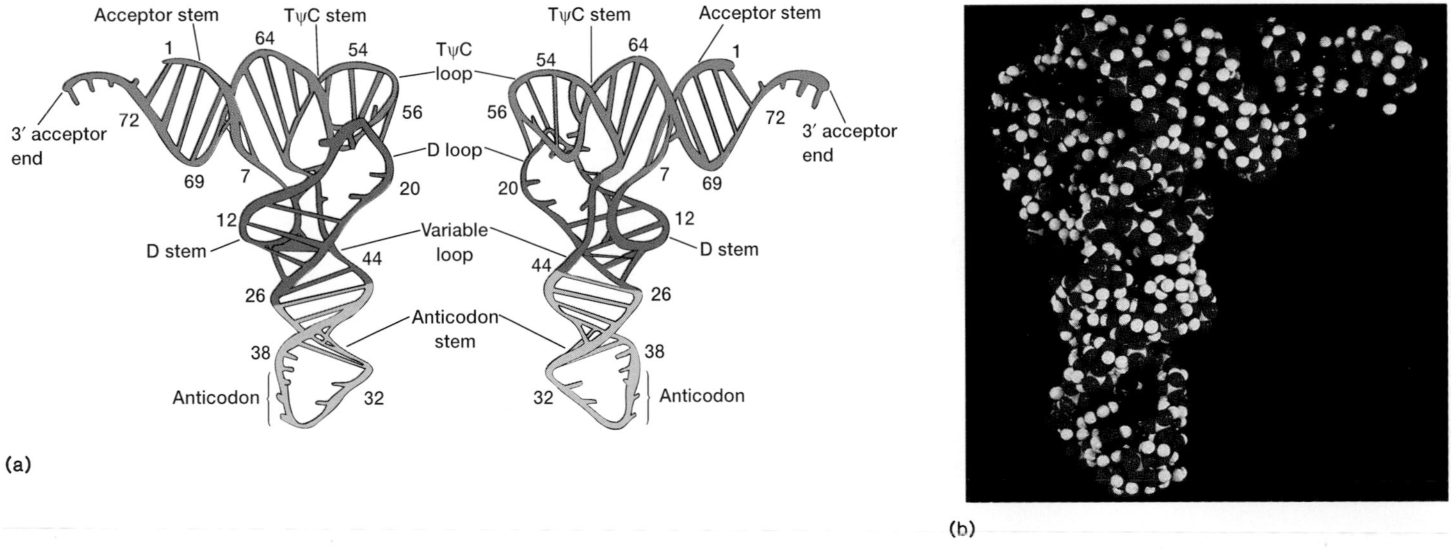

Figure 10.22 **Transfer RNA Conformation.** (*a*) The three-dimensional structure of tRNA from two side views. The different portions of the structure are distinguished by shading. (*b*) A space-filling model of yeast phenylalanine tRNA. The 3′ acceptor end is at the upper right.

incorporated into proteins, at least 20 different tRNA molecules are needed. Actually, more tRNA species than this exist (*see p. 240*).

Amino acids are activated for protein synthesis through a reaction catalyzed by **aminoacyl-tRNA synthetases** (figure 10.23).

$$\text{Amino acid} + \text{tRNA} + \text{ATP} \xrightarrow{Mg^{2+}} \text{aminoacyl-tRNA} + \text{AMP} + \text{PP}_i$$

Just as is true of DNA and RNA synthesis, the reaction is driven to completion when the pyrophosphate product is hydrolyzed to two orthophosphates. The amino acid is attached

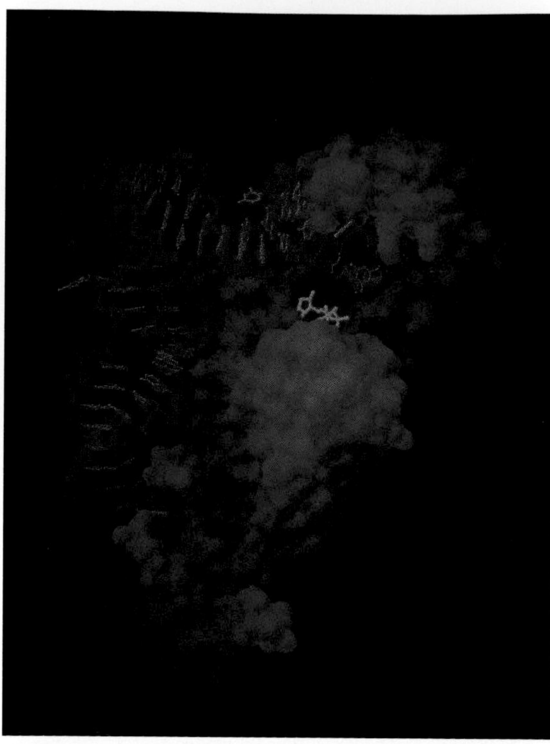

Figure 10.23 **An Aminoacyl-tRNA Synthetase.** A model of *E. coli* glutamyl tRNA synthetase complexed with its tRNA and ATP. The enzyme is in blue, the tRNA in red and yellow, and ATP in green.

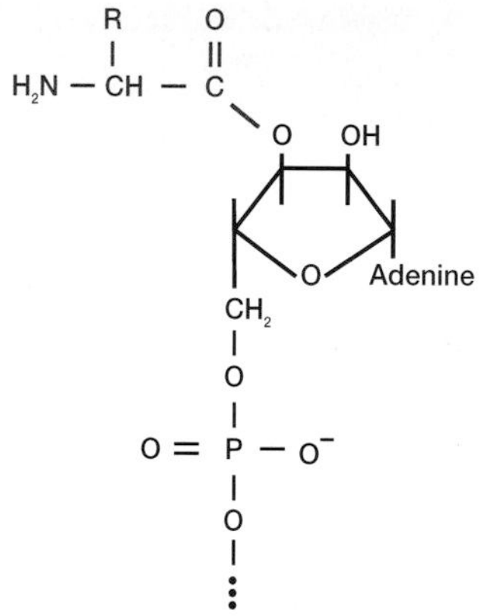

R O
H_2N — CH — C
O OH
Adenine
CH_2
O
O = P — O^-
O

Figure 10.24 **Aminoacyl-tRNA.** The 3′ end of an aminoacyl-tRNA. The activated amino acid is attached to the 3′ hydroxyl of adenylic acid by a high-energy bond.

to the 3′-hydroxyl of the terminal adenylic acid on the tRNA by a high-energy bond (figure 10.24), and is readily transferred to the end of a growing peptide chain. This is why the amino acid is called activated.

There are at least 20 aminoacyl-tRNA synthetases, each specific for a single amino acid and for all the tRNAs (cognate tRNAs) to which each may be properly attached. This specificity is critical because once an incorrect acid is attached to a tRNA, it will be incorporated into a polypeptide in place of the correct amino acid. The protein synthetic machinery recognizes only the anticodon of the aminoacyl-tRNA and cannot tell whether the correct amino acid is attached. Some aminoacyl-tRNA synthetases will even proofread like DNA polymerases. If the wrong aminoacyl-tRNA is formed, aminoacyl-tRNA synthases will hydrolyze the amino acid from the tRNA rather than release the incorrect product.

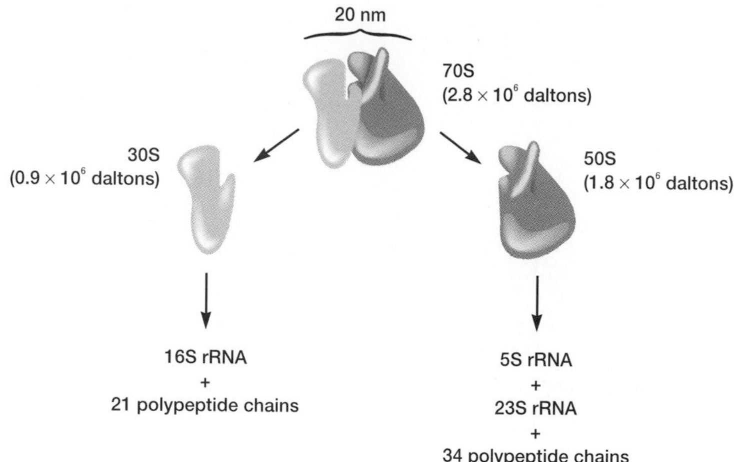

Figure 10.25 **The 70S Ribosome.** The structure of the procaryotic ribosome.

The Ribosome

The actual process of protein synthesis takes place on ribosomes that serve as workbenches, with mRNA acting as the blueprint. Procaryotic ribosomes have a sedimentation value of 70S and a mass of 2.8 million daltons. A rapidly growing *E. coli* cell may have as many as 15,000 ribosomes, about 15% of the cell mass.

Introduction to ribosomal function and the Svedberg unit (p. 49).

The procaryotic ribosome is an extraordinarily complex organelle made of a 30S and a 50S subunit (figure 10.25). Each subunit is constructed from one or two rRNA molecules and many polypeptides. The shape of ribosomal subunits and their association to form the 70S ribosome are depicted in figure 10.26. The region of the ribosome directly responsible for translation is called the translational domain (figure 10.26*d*). Both subunits contribute to this domain, located in the upper half of the small subunit and in the associated areas of the large subunit. For example, the peptidyl transferase (p. 211) is found on the central protuberance of the large subunit. The growing peptide chain emerges from the large subunit at the exit domain. This is located on the side of the subunit opposite the central protuberance in both procaryotes and eucaryotes.

Eucaryotic cytoplasmic ribosomes are 80S, with a mass of 4 million daltons, and are composed of two subunits, 40S and 60S. Many of these ribosomes are found free in the cytoplasmic matrix, whereas others are attached to membranes of the endoplasmic reticulum by their 60S subunit at a site next to the exit domain. The ribosomes of eucaryotic mito-

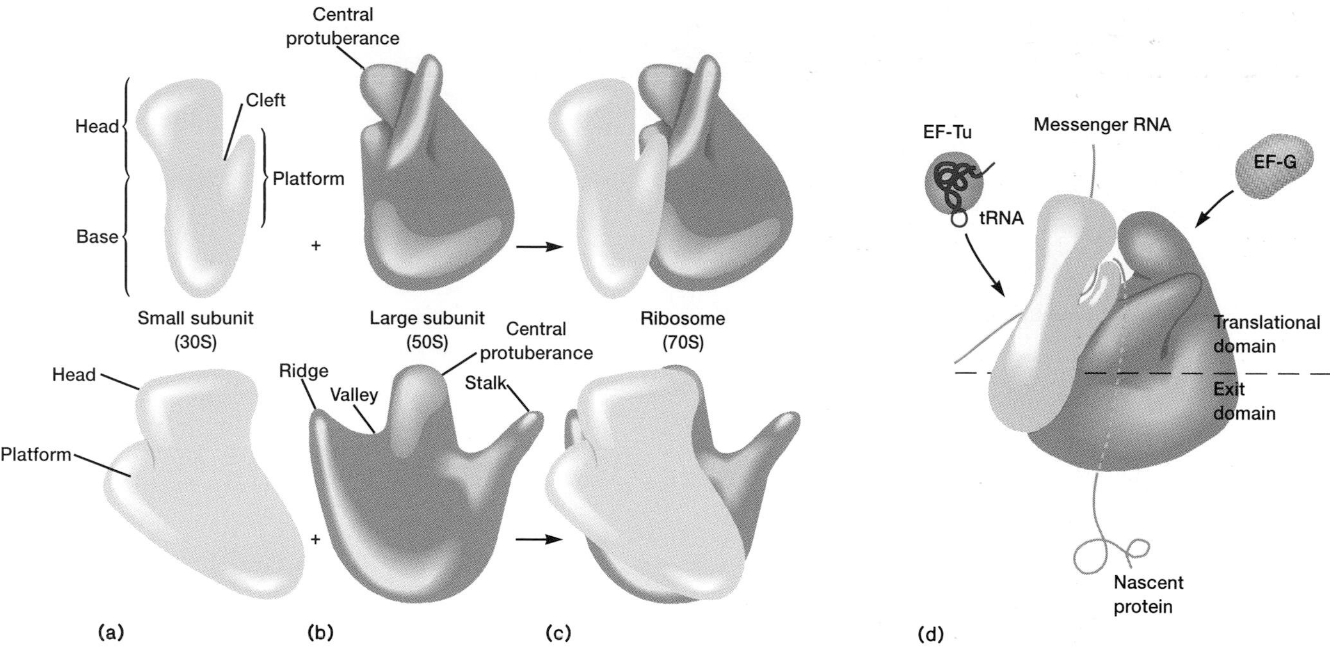

Figure 10.26 **Two Views of the *E. coli* Ribosome.** (*a*) The 30S subunit. (*b*) The 50S subunit. (*c*) The complete 70S ribosome. (*d*) Diagrammatic representation of ribosomal structure with the translational and exit domains shown. The locations of elongation factor and mRNA binding are shown. The growing peptide chain probably remains unfolded and extended until it leaves the large subunit.

chondria and chloroplasts are smaller than cytoplasmic ribosomes and resemble the procaryotic organelle.

Ribosomal RNA is thought to have two roles. It obviously contributes to ribosome structure. The 16S rRNA of the 30S subunit also may aid in the initiation of protein synthesis in procaryotes. There is evidence that the 3′ end of the 16S rRNA complexes with an initiating signal site on the mRNA and helps position the mRNA on the ribosome. Because of the discovery of catalytic RNA, some have proposed that ribosomal RNA may have a catalytic role in protein synthesis.

The use of 16S rRNA sequences in the study of bacterial phylogeny (pp. 413, 415).

Initiation of Protein Synthesis

Protein synthesis proper may be divided into three stages: initiation, elongation, and termination.

In the initiation stage, *E. coli* and most bacteria begin protein synthesis with a specially modified aminoacyl-tRNA, N-formylmethionyl-$tRNA^{fMet}$ (figure 10.27). Since the α-amino is blocked by a formyl group, this aminoacyl-tRNA can be used only for initiation. When methionine is to be added to a growing polypeptide chain, a normal methionyl-$tRNA^{Met}$ is employed. Eucaryotic protein synthesis (except in the mitochondrion and chloroplast) begins with a special initiator methionyl-$tRNA^{Met}$. Although bacteria start protein synthesis with formylmethionine, the formyl group does not remain but is hydrolytically removed. In fact, one to three amino acids may be removed from the amino terminal end of the polypeptide after synthesis.

$$CH_3-S-CH_2-CH_2-\underset{\underset{\underset{\underset{\underset{H}{|}}{C=O}}{|}}{NH}}{\overset{|}{CH}}-\overset{\overset{O}{\|}}{C}-tRNA^{fMet}$$

Figure 10.27 **Procaryotic Initiator tRNA.** The procaryotic initiator aminoacyl tRNA, N-formylmethionyl-$tRNA^{fMet}$. The formyl group is in color.

Figure 10.28 shows the initiation process in procaryotes. The initiator N-formylmethionyl-$tRNA^{fMet}$ (fMet-tRNA) binds to the free 30S subunit first. Next, mRNA attaches to the 30S subunit and is positioned properly through interactions with both the 3′ end of the 16S rRNA and the anticodon of fMet-tRNA. Messengers have a special **initiator codon** (AUG, or sometimes GUG) that specifically binds with the fMet-tRNA anticodon (*see chapter 12*). Finally, the 50S subunit binds to the 30S subunit-mRNA forming an active ribosome-mRNA complex. There is some uncertainty about the exact initiation sequence, and mRNA may bind before fMet-tRNA in procaryotes. Eucaryotic initiation appears to begin with the binding of a special initiator Met-tRNA to the small subunit, followed by attachment of the mRNA.

In procaryotes, three protein **initiation factors** are required (figure 10.28). Initiation factor 3 (IF-3) prevents 30S subunit binding to the 50S subunit and promotes the proper mRNA binding to the 30S subunit. IF-2, the second initiation

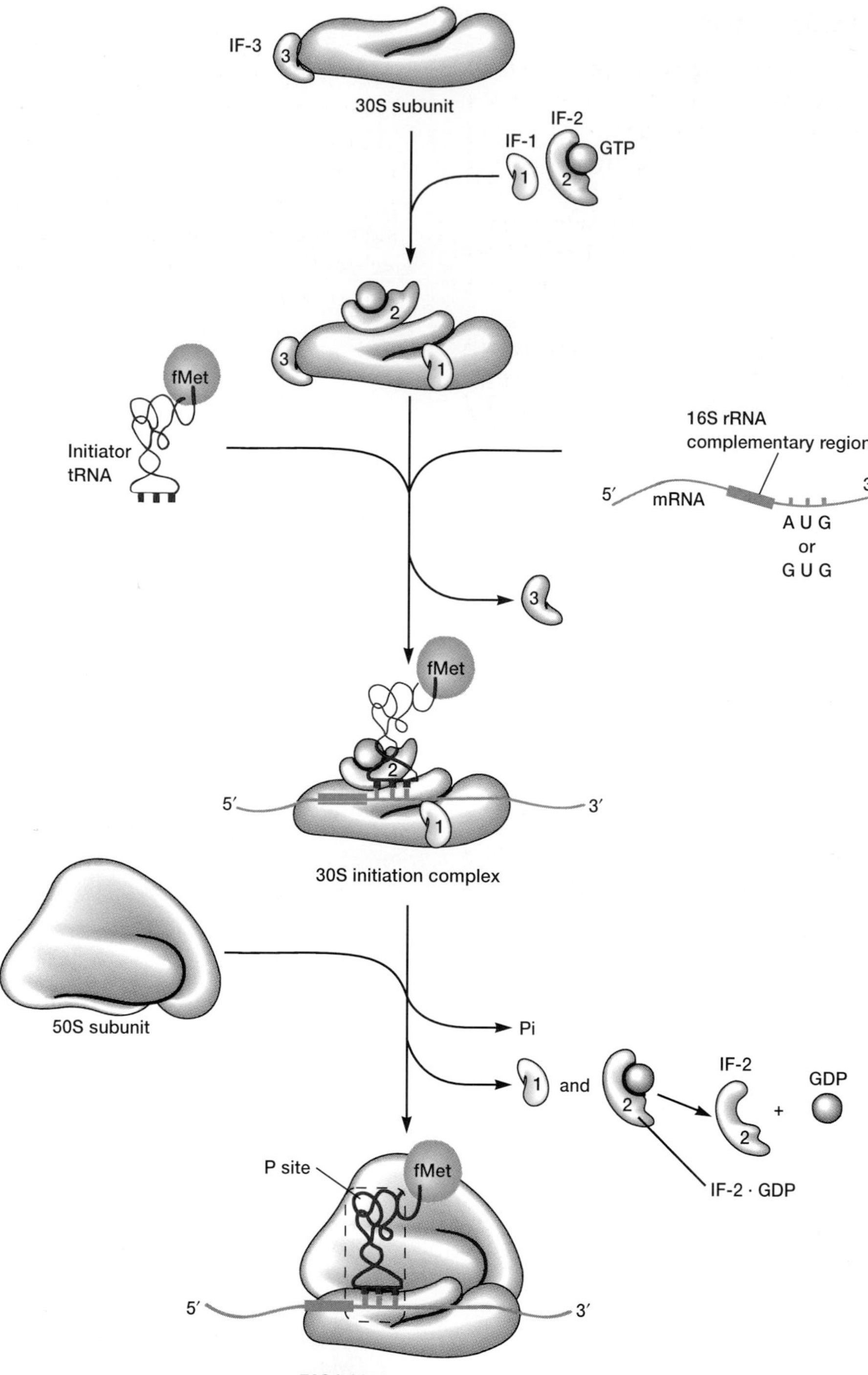

Figure 10.28 **Initiation of Protein Synthesis.** The initiation of protein synthesis in procaryotes. The following abbreviations are employed: *IF-1, IF-2,* and *IF-3* stand for initiation factors 1, 2, and 3; initiator tRNA is N-formylmethionyl-$tRNA^{fMet}$. The ribosomal locations of initiation factors are depicted for illustration purposes only. They do not represent the actual initiation factor binding sites. See text for further discussion.

factor, binds GTP and fMet-tRNA and directs the attachment of fMet-tRNA to the 30S subunit. GTP is hydrolyzed during association of the 50S and 30S subunits. The third initiation factor, IF-1 appears to be needed for release of IF-2 and GDP from the completed 70S ribosome. IF-1 also may aid in the binding of the 50S subunit to the 30S subunit. Eucaryotes require more initiation factors; otherwise, the process is quite similar to that of procaryotes.

The initiation of protein synthesis is very elaborate. Apparently, the complexity is necessary to ensure that the ribosome does not start synthesizing a polypeptide chain in the middle of a gene—a disastrous error.

Elongation of the Polypeptide Chain

Every amino acid addition to a growing polypeptide chain is the result of an **elongation cycle** composed of three phases: aminoacyl-tRNA binding, the transpeptidation reaction, and translocation. The process is aided by special protein **elongation factors** (just as with the initiation of protein synthesis). In each turn of the cycle, an amino acid corresponding to the proper mRNA codon is added to the C-terminal end of the polypeptide chain. The procaryotic elongation cycle is described next.

The ribosome has two sites for binding aminoacyl-tRNA and peptidyl-tRNA: (1) the **peptidyl** or **donor site** (the **P site**) and (2) the **aminoacyl** or **acceptor site** (the **A site**). At the beginning of an elongation cycle, the peptidyl site is filled with either N-formylmethionyl-$tRNA^{fMet}$ or peptidyl-tRNA and the aminoacyl site is empty (figure 10.29). Messenger RNA is bound to the ribosome in such a way that the proper codon interacts with the P site tRNA (e.g., an AUG codon for fMet-tRNA). The next codon (green) is located within the A site and ready to direct the binding of an aminoacyl-tRNA.

The first phase of the elongation cycle is the amino-acyl-tRNA binding phase. The aminoacyl-tRNA corresponding to the green codon is inserted into the A site. GTP hydrolysis and the elongation factor EF-Tu, which donates the aminoacyl-tRNA to the ribosome, are required for this insertion. The resulting EF-Tu·GDP complex is converted to EF-Tu·GTP with the aid of a second elongation factor, EF-Ts. Subsequently, another aminoacyl-tRNA binds to EF-Tu·GTP (figure 10.29).

Aminoacyl-tRNA binding to the A site initiates the second phase of the elongation cycle, the **transpeptidation reaction** (figures 10.29 and 10.30). This is catalyzed by the **peptidyl transferase,** located on the 50S subunit. The α-amino group of the A site amino acid nucleophilically attacks the α-carboxyl group of the C-terminal amino acid on the P site tRNA in this reaction (figure 10.30). The peptide chain grows by one amino acid and is transferred to the A site tRNA. No extra energy source is required for peptide bond formation because the bond linking an amino acid to tRNA is high in energy. Recent evidence suggests that 23S rRNA may participate in the peptidyl transferase function.

The final phase in the elongation cycle is **translocation.** Three things happen simultaneously: (1) the peptidyl-tRNA moves from the A site to the P site; (2) the ribosome moves one codon along mRNA so that a new codon is positioned in the A site; and (3) the empty tRNA leaves the P site. The intricate process requires the participation of the EF-G or translocase protein and GTP hydrolysis. The ribosome changes shape as it moves down the mRNA in the 5′ to 3′ direction.

Termination of Protein Synthesis

In the third stage, protein synthesis stops when the ribosome reaches one of three special **nonsense codons**—UAA, UAG, and UGA (figure 10.31). Three **release factors** (RF-1, RF-2, and RF-3) aid the ribosome in recognizing these codons. After the ribosome has stopped, peptidyl transferase hydrolyzes the peptide free from its tRNA, and the empty tRNA is released. GTP hydrolysis seems to be required during this sequence of events, although it may not be needed for termination in procaryotes. Next, the ribosome dissociates from its mRNA and separates into 30S and 50S subunits. IF-3 binds to the 30S subunit and prevents it from reassociating with the 50S subunit until the proper stage in initiation has been reached. Thus, ribosomal subunits come together during protein synthesis and then separate afterward. The termination of eucaryotic protein synthesis is similar to the process just described except that only one release factor appears to be active.

As a polypeptide leaves the ribosome, it is already spontaneously folding into its final shape. This folding is possible because protein conformation is a direct function of the amino acid sequence (*see appendix I*). The polypeptide is almost completely folded by the time its synthesis is completed.

Protein synthesis is a very expensive process. Two GTP molecules are used during the elongation cycle, and two ATP high-energy bonds are required for amino acid activation (ATP is converted to AMP rather than to ADP). Therefore, four high-energy bonds are required to add one amino acid to a growing polypeptide chain. GTP is also used in initiation and termination of protein synthesis (figures 10.28 and 10.31). Presumably, this large energy expenditure is required to ensure the fidelity of protein synthesis. Very few mistakes can be tolerated.

Although the mechanism of protein synthesis is similar in procaryotes and eucaryotes, procaryotic ribosomes differ substantially from those in eucaryotes. This explains the effectiveness of many important chemotherapeutic agents. Either the 30S or the 50S subunit may be affected. For example, streptomycin binding to the 30S ribosomal subunit inhibits protein synthesis and causes mRNA misreading. Erythromycin binds to the 50S subunit and inhibits peptide chain elongation.

The effect of antibiotics on protein synthesis (pp. 330, 336–37).

1. In which direction are polypeptides synthesized? What is a polyribosome and why is it useful?
2. Briefly describe the structure of transfer RNA and relate this to its function. How are amino acids activated for protein synthesis, and why is the specificity of the aminoacyl-tRNA synthetase reaction so important?
3. What are translational and exit domains? Contrast procaryotic and eucaryotic ribosomes in terms of structure. What roles does ribosomal RNA have?
4. Describe the nature and function of the following: fMet-tRNA, initiator codon, IF-3, IF-2, IF-1, elongation cycle, peptidyl and aminoacyl sites, EF-Tu, EF-Ts, transpeptidation reaction, peptidyl transferase, translocation, EF-G or translocase, nonsense codon, and release factors.

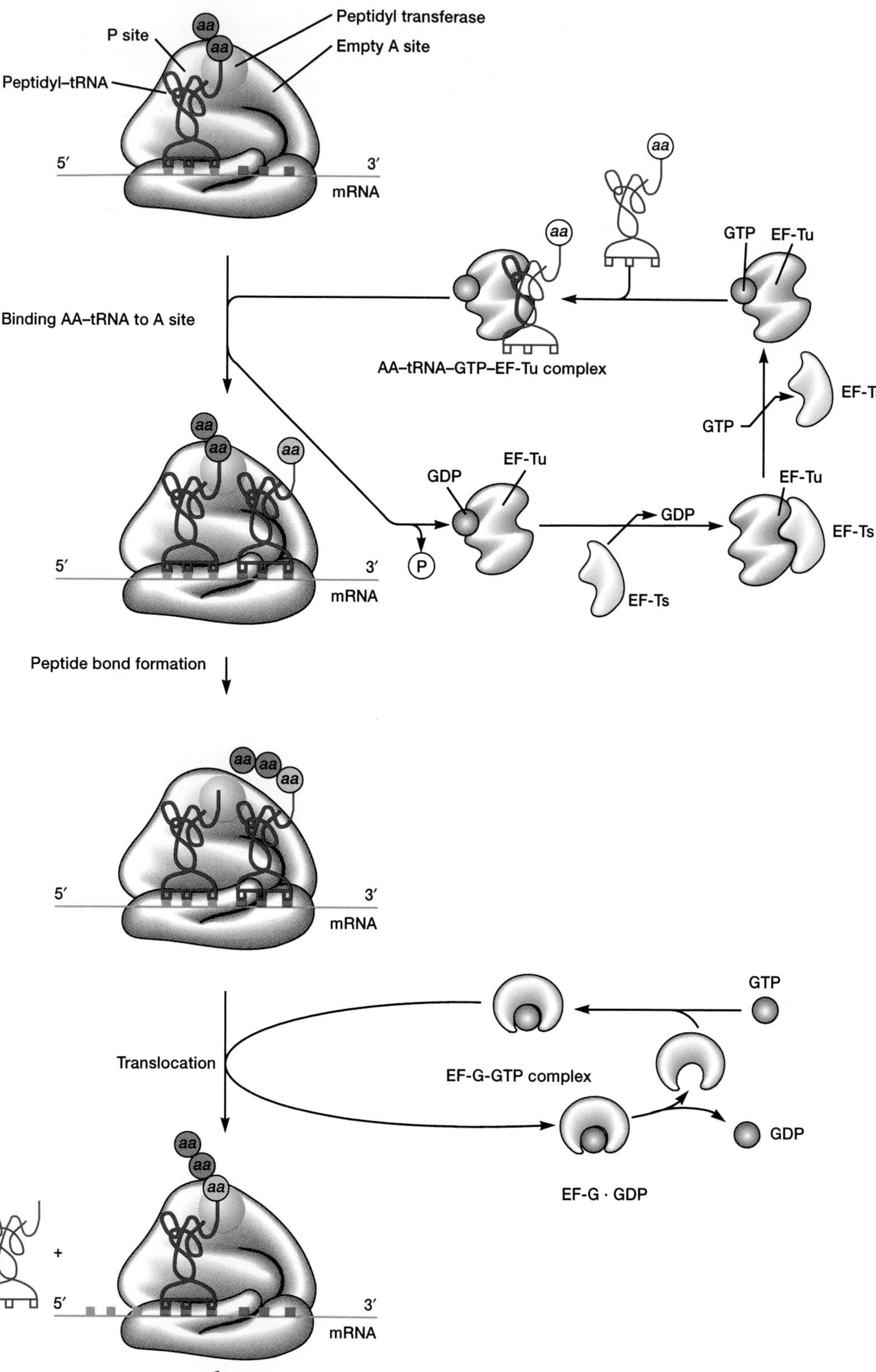

Figure 10.29 Elongation Cycle. The elongation cycle of protein synthesis. The ribosome possesses two sites, a peptidyl or donor site (P site) and an aminoacyl or acceptor site (A site). The arrow below the ribosome in the translocation step shows the direction of mRNA movement. See text for details.

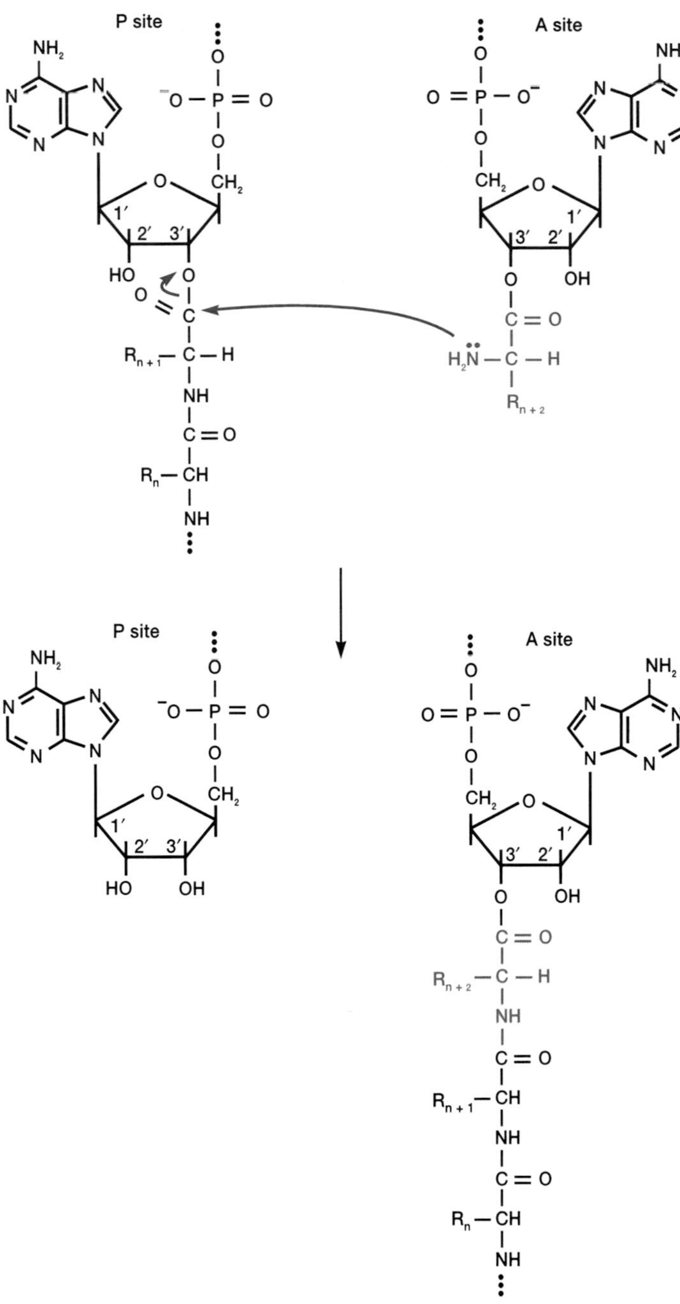

Figure 10.30 **Transpeptidation.** The transpeptidation reaction catalyzed by peptidyl transferase.

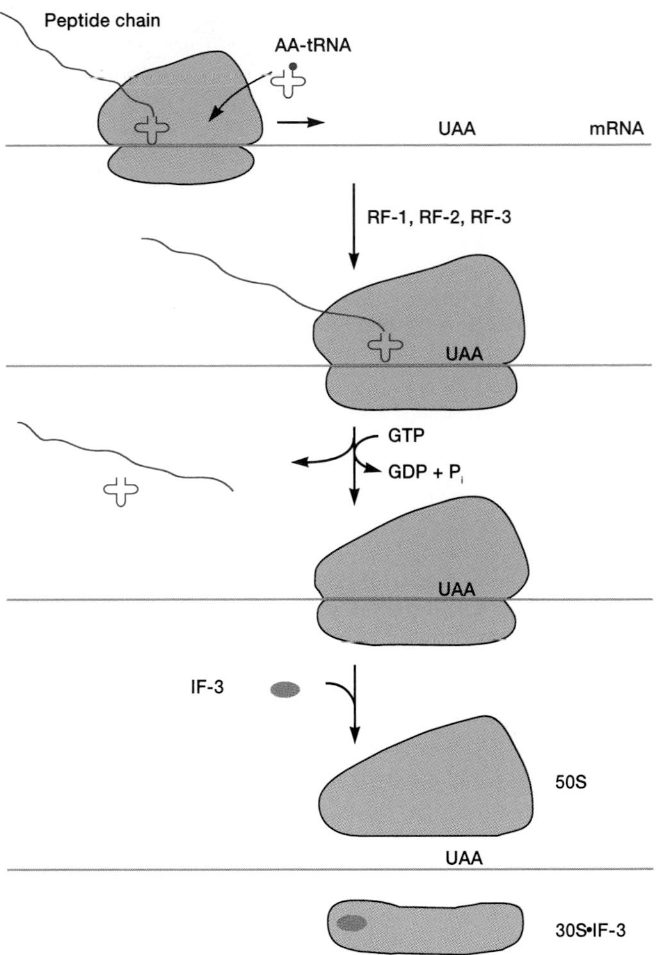

Figure 10.31 **Termination of Protein Synthesis in Procaryotes.** Although three different nonsense codons can terminate chain elongation, UAA is most often used for this purpose. Three release factors (*RF*) assist the ribosome in recognizing nonsense codons. GTP hydrolysis is probably involved in termination.

Summary

1. DNA differs in composition from RNA in having deoxyribose and thymine rather than ribose and uracil.
2. DNA is double stranded, with complementary AT and GC base pairing between the strands. The strands run antiparallel and are twisted into a right-handed double helix.
3. RNA is normally single stranded, although it can coil upon itself and base pair to form hairpin structures.
4. In procaryotes, DNA exists as a circle that is twisted into supercoils and associated with histonelike proteins.
5. Eucaryotic DNA is associated with five types of histone proteins. Eight histones associate to form ellipsoidal nucleosomes around which the DNA is coiled.
6. DNA synthesis is called replication. Transcription is the synthesis of an RNA copy of DNA and produces three types of RNA: messenger RNA (mRNA), transfer RNA (tRNA), and ribosomal RNA (rRNA).
7. The synthesis of protein under the direction of mRNA is called translation.
8. Most circular procaryotic DNAs are copied by two replication forks moving around the circle to form a theta-shaped (θ) figure. Sometimes a rolling-circle mechanism is employed instead.
9. Eucaryotic DNA has many replicons and replication origins located every 10 to 100 μm along the DNA.
10. DNA polymerase enzymes catalyze the synthesis of DNA in the 5′ to 3′ direction while reading the DNA template in the 3′ to 5′ direction.
11. The double helix is unwound by helicases with the aid of topoisomerases like the DNA gyrase. DNA binding proteins keep the strands separate.
12. DNA polymerase III synthesizes a complementary DNA copy beginning with a short RNA primer made by a primase enzyme.
13. The leading strand is probably replicated continuously, whereas DNA synthesis on the lagging strand is discontinuous and forms Okazaki fragments.
14. DNA polymerase I excises the RNA primer and fills in the resulting gap. DNA ligase then joins the fragments together.
15. Procaryotic mRNA has nontranslated leader and trailer sequences at its ends. Spacer regions exist between genes when mRNA is polygenic.
16. RNA is synthesized by RNA polymerase that copies the sequence of the DNA template sense strand.
17. The sigma factor helps the procaryotic RNA polymerase bind to the promoter region at the start of a gene.
18. A terminator marks the end of a gene. A rho factor is needed for RNA polymerase release from some terminators.
19. RNA polymerase II synthesizes heterogeneous nuclear RNA, which then undergoes posttranscriptional modification by RNA cleavage and addition of a 3′ poly-A sequence and a 5′ cap to generate eucaryotic mRNA.
20. Many eucaryotic genes are split or interrupted genes that have exons and introns. Exons are joined by RNA splicing. Splicing involves small nuclear RNA molecules, spliceosomes, and sometimes ribozymes.
21. In translation, ribosomes attach to mRNA and synthesize a polypeptide beginning at the N-terminal end. A polysome or polyribosome is a complex of mRNA with several ribosomes.
22. Amino acids are activated for protein synthesis by attachment to the 3′ end of transfer RNAs. Activation requires ATP, and the reaction is catalyzed by aminoacyl-tRNA synthetases.
23. Ribosomes are large, complex organelles composed of rRNAs and many polypeptides. Amino acids are added to a growing peptide chain at the translational domain.
24. Protein synthesis begins with the binding of fMet-tRNA (procaryotes) or an initiator methionyl-$tRNA^{Met}$ (eucaryotes) to an initiator codon on mRNA and to the two ribosomal subunits. This involves the participation of protein initiation factors.
25. In the elongation cycle, the proper aminoacyl-tRNA binds to the A site with the aid of EF-Tu and GTP. Then the transpeptidation reaction is catalyzed by peptidyl transferase. Finally, during translocation, the peptidyl-tRNA moves to the P site and the ribosome travels along the mRNA one codon. Translocation requires GTP and EF-G or translocase.
26. Protein synthesis stops when a nonsense codon is reached. Procaryotes require three release factors for codon recognition and ribosome dissociation from the mRNA.

Key Terms

Questions for Thought and Review

1. How do replication patterns differ between procaryotes and eucaryotes? Describe the operation of replication forks in the generation of theta-shaped intermediates and in the rolling-circle mechanism.
2. Outline the steps involved in DNA synthesis at the replication fork. How do DNA polymerases correct their mistakes?
3. Describe how RNA polymerase transcribes procaryotic DNA. How does the polymerase know where to begin and end transcription?
4. In what ways does eucaryotic mRNA differ from procaryotic mRNA with respect to synthesis and structure? How does eucaryotic synthesis of rRNA and tRNA resemble that of mRNA? How does it differ?
5. Draw diagrams summarizing the sequence of events in the three stages of protein synthesis (initiation, elongation, and termination) and accounting for the energy requirements of translation.

Additional Reading

Ahern, H. 1991. Self-splicing introns: Molecular fossils or selfish DNA? *ASM News* 57(5):258–61.

Bauer, W. R.; Crick, F. H. C.; and White, J. H. 1980. Supercoiled DNA. *Sci. Am.* 243(1):118–33.

Campbell, J. L. 1986. Eukaryotic DNA replication. *Ann. Rev. Biochem.* 55:733–71.

Cech, T. R. 1986. RNA as an enzyme. *Sci. Am.* 255(5):64–75.

Chase, J. W., and Williams, K. R. 1986. Single-stranded DNA binding proteins required for DNA replication. *Ann. Rev. Biochem.* 55:103–36.

Darnell, J. E., Jr. 1983. The processing of RNA. *Sci. Am.* 249(4):90–100.

Darnell, J. E., Jr. 1985. RNA. *Sci. Am.* 253(4):68–78.

Dickerson, R. E. 1983. The DNA helix and how it is read. *Sci. Am.* 249(6):94–111.

Drlica, K., and Rouviere-Yaniv, J. 1987. Histonelike proteins of bacteria. *Microbiol. Rev.* 51(3):301–19.

Felsenfeld, G. 1985. DNA. *Sci. Am.* 253(4):59–67.

Fox, J. L. 1988. Making and mistaking DNA. *ASM News* 54(3):122–26.

Freifelder, D. 1987. *Molecular biology,* 2d ed. New York: Van Nostrand Reinhold.

Guthrie, C. 1991. Messenger RNA splicing in yeast: Clues to why the spliceosome is a ribonucleoprotein. *Science* 253:157–63.

Judson, H. F. 1979. *The eighth day of creation: Makers of the revolution in biology.* London: Jonathan Cape.

Kornberg, A. 1992. *DNA replication,* 2d ed. San Francisco: W. H. Freeman.

Kornberg, R. D., and Klug, A. 1981. The nucleosome. *Sci. Am.* 244(2):52–64.

Lake, J. A. 1981. The ribosome. *Sci. Am.* 245(2):84–97.

Lake, J. A. 1985. Evolving ribosome structure: Domains in archaebacteria, eubacteria, eocytes and eukaryotes. *Ann. Rev. Biochem.* 54:507–30.

Lewin, B. 1990. *Genes,* 4th ed. New York: Oxford Univ. Press.

Mathews, C. K., and van Holde, K. E. 1990. *Biochemistry.* Redwood City, Calif.: Benjamin/Cummings.

Matson, S. W., and Kaiser-Rogers, K. A. 1990. DNA helicases. *Ann. Rev. Biochem.* 59:289–329.

McClure, W. R. 1985. Mechanism and control of transcription initiation in prokaryotes. *Ann. Rev. Biochem.* 54:171–204.

McHenry, C. S. 1988. DNA polymerase III holoenzyme of *Escherichia coli. Ann. Rev. Biochem.* 57:519–50.

Merrick, W. C. 1992. Mechanism and regulation of eukaryotic protein synthesis. *Microbiol. Rev.* 56(2):291–315.

Meyer, R. R., and Laine, P. S. 1990. The single-stranded DNA-binding protein of *Escherichia coli. Microbiol. Rev.* 54(4):342–80.

Nossal, N. G. 1983. Prokaryotic DNA replication systems. *Ann. Rev. Biochem.* 53:581–615.

Radman, M., and Walker, R. 1988. The high fidelity of DNA duplication. *Sci. Am.* 259(2):40–46.

Rich, A., and Kim, S. H. 1978. The three-dimensional structure of transfer RNA. *Sci. Am.* 238(1):52–62.

Riis, B., Rattan, S. I. S., Clark, B. F. C., and Merrick, W. C. 1990. Eukaryotic protein elongation factors. *Trends Biochem. Sci.* 15(11):420–24.

Rosbash, M., and Séraphin, B. 1991. Who's on first? The U1 snRNP-5′ splice site interaction and splicing. *Trends Biochem. Sci.* 16(5):187–90.

Steitz, J. A. 1988. "Snurps." *Sci. Am.* 258(6):56–63.

Stent, G. S., and Calendar, R. 1978. *Molecular genetics,* 2d ed. San Francisco: Freeman.

Wang, J. C. 1982. DNA topoisomerases. *Sci. Am.* 247(1):94–109.

Watson, J. D.; Hopkins, N. H.; Roberts, J. W.; Steitz, J. A.; and Weiner, A. M. 1987. *Molecular biology of the gene,* 4th ed. Menlo Park, Calif.: Benjamin/Cummings.

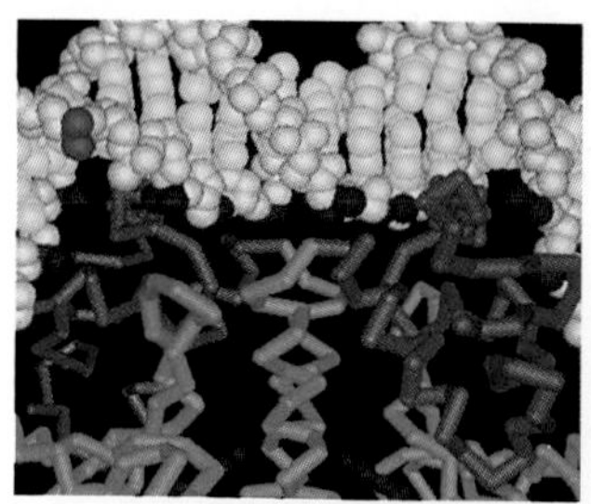

CHAPTER 11
Metabolism: The Regulation of Enzyme Activity and Synthesis

Living cells are self-regulating chemical engines, tuned to operate on the principle of maximum economy.

—A. L. Lehninger

Outline

Concepts

1. Metabolism is regulated in such a way that (1) cell components are maintained at the proper concentrations, even in the face of a changing environment, and (2) energy and material are conserved.
2. The localization of enzymes and metabolites in separate compartments of a cell regulates and coordinates metabolic activity.
3. The activity of regulatory enzymes may be changed through reversible binding of effectors to a regulatory site separate from the catalytic site or through covalent modification of the enzyme. Regulation of enzyme activity operates rapidly and serves as a fine-tuning mechanism to adjust metabolism from moment to moment.
4. A pathway's activity is often controlled by its end products through feedback inhibition of regulatory enzymes located at the start of the sequence and at branch points.
5. The long-term regulation of metabolism in bacteria is achieved through the control of transcription by repressor proteins during induction and repression, and by the attenuation of many biosynthetic operons.
6. DNA replication and cell division are coordinated in such a way that the distribution of new DNA copies to each daughter cell is ensured.

The organization of microbial metabolism is briefly described in the previous four chapters. Up to this point, a metabolic pathway has been treated simply as a sequence of enzymes functioning as a unit, with each enzyme using as its substrate a product of the preceding enzyme-catalyzed reaction. This picture is incomplete because the regulation of pathway operation has been ignored. Both regulation of the activity of individual pathways and coordination of the action of separate sequences are essential to the existence of life. Cells become disorganized and die without adequate control of metabolism. This chapter introduces the principles of metabolic regulation.

The chapter begins with a consideration of the need for metabolic regulation, and subsequently describes metabolic channeling, the regulation of the activity of critical enzymes, and the control of nucleic acid synthesis. Finally, regulation of the bacterial cell cycle is discussed.

The task of the regulatory machinery is exceptionally complex and difficult. Pathways must be regulated and coordinated so effectively that all cell components are present in precisely the correct amounts. Furthermore, a microbial cell must be able to respond effectively to environmental changes by using those nutrients present at the moment and by switching on new catabolic pathways when different nutrients become available. Because all chemical components of a cell are usually not present in the surroundings, microorganisms also must synthesize unavailable components and alter biosynthetic activity in response to changes in nutrient availability. The chemical composition of a cell's surroundings is constantly changing, and these regulatory processes are dynamic and continuously responding to altered conditions.

Regulation is essential for the cell to conserve microbial energy and material and to maintain metabolic balance. If a particular energy source is unavailable, the enzymes required for its utilization are not needed and their further synthesis is a waste of carbon, nitrogen, and energy. Similarly, it would be extremely wasteful for a microorganism to synthesize the enzymes required to manufacture a certain end product if that end product were already present in adequate amounts. Thus, both catabolism and anabolism are regulated in such a way as to maximize efficiency of operation.

The drive to maintain balance and conserve energy and material is evident in the regulatory responses of a bacterium like *E. coli*. If the bacterium is grown in a very simple medium containing only glucose as a carbon and energy source, it will synthesize the required cell components in balanced amounts. Addition of a biosynthetic end product (the amino acid tryptophan, for example) to the medium will result in the immediate inhibition of the pathway synthesizing that end product; synthesis of the pathway's enzymes also will slow or cease. If *E. coli* is transferred to medium containing only the sugar lactose, it will synthesize the enzymes required for catabolism of this nutrient. In contrast, when *E. coli* grows in a medium possessing both glucose and lactose, glucose (the sugar supporting most rapid growth) is catabolized first. The culture will use lactose only after the glucose supply has been exhausted.

The flow of carbon through a pathway may be regulated in three major ways.

1. The localization of metabolites and enzymes in different parts of a cell, a phenomenon called **metabolic channeling,** influences pathway activity.
2. Critical enzymes are often directly stimulated or inhibited to alter pathway activity rapidly.
3. The number of enzyme molecules also may be controlled. The more catalyst molecules present, the greater the pathway's activity. In bacteria, regulation is usually exerted at the level of transcription. Control of mRNA synthesis is slower than direct regulation of enzyme activity, but does result in the saving of much energy and raw material because enzymes are not synthesized when not required.

Each of these mechanisms is described in detail.

Metabolic Channeling

One of the most common channeling mechanisms is that of **compartmentation,** the differential distribution of enzymes and metabolites among separate cell structures or organelles. Compartmentation is particularly important in eucaryotic microorganisms with their many membrane-bounded organelles. For example, fatty acid oxidation is located within the mitochondrion whereas fatty acid synthesis occurs in the cytoplasmic matrix. Compartmentation makes possible the simultaneous, but separate, operation and regulation of similar pathways. Furthermore, pathway activities can be coordinated through regulation of the transport of metabolites and coenzymes between cell compartments. Suppose two pathways in different cell compartments require NAD for activity. The distribution of NAD between the two compartments will then determine the relative activity of these competing pathways, and the pathway with access to the most NAD will be favored.

Channeling also occurs within compartments such as the cytoplasmic matrix. The matrix is a structured dense material with many subcompartments. In eucaryotes, it is also subdivided by the endoplasmic reticulum and cytoskeleton (*see chapter 4*). Metabolites and coenzymes do not diffuse rapidly in such an environment, and metabolite gradients will build up near localized enzymes or enzyme systems. This occurs because enzymes at a specific site convert their substrates to products, resulting in decreases in the concentration of one or more metabolites and increases in others. For example, product concentrations will be high near an enzyme and decrease with increasing distance from it.

Channeling can generate marked variations in metabolite concentrations and, therefore, directly affect enzyme activity. Substrate levels are generally around 10^{-3} moles/liter (M) to 10^{-6} M, or even lower. Thus, they may be in the same range

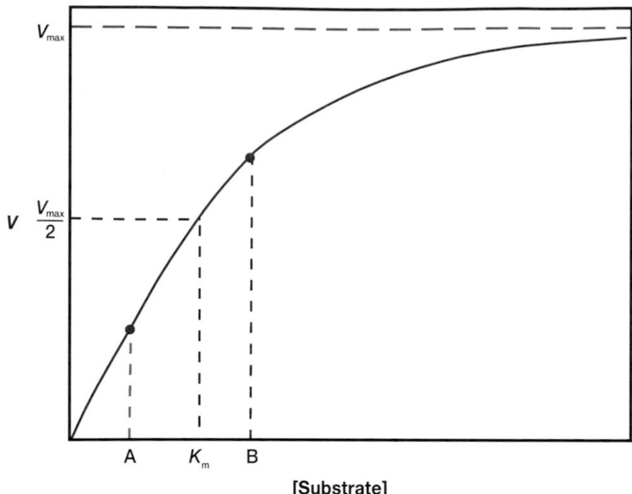

Figure 11.1 Control of Enzyme Activity by Substrate Concentration. An enzyme substrate saturation curve with the Michaelis constant (K_m) and the velocity equivalent to half the maximum velocity (V_{max}) indicated. The initial velocity of the reaction (v) is plotted against the substrate concentration [Substrate]. The maximum velocity is the greatest velocity attainable with a fixed amount of enzyme under defined conditions. When the substrate concentration is equal to or less than the K_m, the enzyme's activity will vary almost linearly with the substrate concentration. Suppose the substrate increases in concentration from level A to B. Because these concentrations are in the range of the K_m, a significant increase in enzyme activity results. A drop in concentration from B to A will lower the rate of product formation.

as enzyme concentrations and equal to or less than the Michaelis constants (K_m) of many enzymes. Under these conditions, the concentration of an enzyme's substrate may control its activity because the substrate concentration is in the rising portion of the hyperbolic substrate saturation curve (figure 11.1). As the substrate level increases, it is converted to product more rapidly; a decline in substrate concentration automatically leads to lower enzyme activity. If two enzymes in different pathways use the same metabolite, they may directly compete for it. The pathway winning this competition, the one with the enzyme having the lowest K_m value for the metabolite, will operate closer to full capacity. Thus, channeling within a cell compartment can regulate and coordinate metabolism through variations in metabolite and coenzyme levels.

Enzyme kinetics and the substrate saturation curve (pp. 141–42).

1. Give three ways in which the flow of carbon through a pathway may be regulated.
2. Define the terms metabolic channeling and compartmentation. How are they involved in the regulation of metabolism?

Control of Enzyme Activity

Adjustment of the activity of regulatory enzymes controls the functioning of many metabolic pathways. This section describes these enzymes and discusses their role in regulating pathway activity.

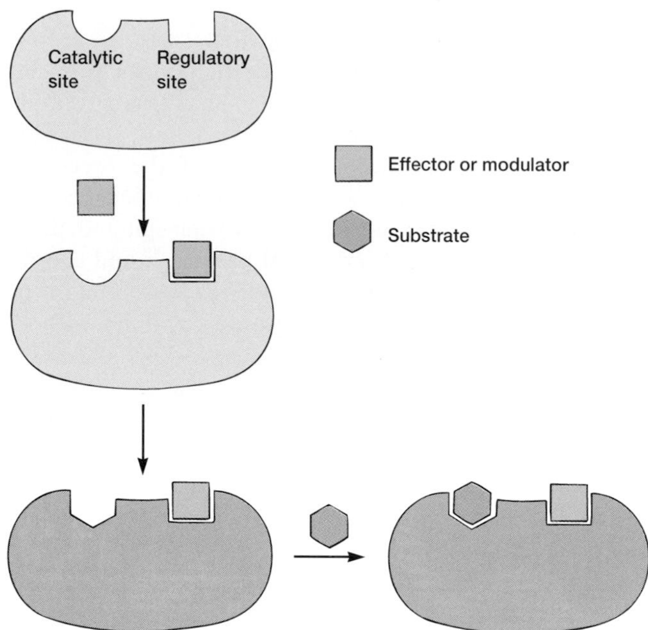

Figure 11.2 Allosteric Regulation. The structure and function of an allosteric enzyme. In this example, the effector or modulator first binds to a separate regulatory site and causes a change in enzyme conformation that results in an alteration in the shape of the active site. The active site can now more effectively bind the substrate. This effector is a positive effector because it stimulates substrate binding and catalytic activity.

Allosteric Regulation

Usually, regulatory enzymes are **allosteric enzymes.** The activity of an allosteric enzyme is altered by a small molecule known as an **effector** or **modulator.** The effector binds reversibly by noncovalent forces to a **regulatory site** separate from the catalytic site and causes a change in the shape or conformation of the enzyme (figure 11.2). The activity of the catalytic site is altered as a result. A positive effector increases enzyme activity, whereas a negative effector decreases activity or inhibits the enzyme. These changes in activity often result from alterations in the apparent affinity of the enzyme for its substrate, but changes in maximum velocity can also occur.

The kinetic characteristics of nonregulatory enzymes (*see chapter 7*) show that the Michaelis constant (K_m) is the substrate concentration required for an enzyme to operate at half its maximal velocity. This constant applies only to hyperbolic substrate saturation curves, not to the sigmoidal curves often seen with allosteric enzymes (figure 11.4). The substrate concentration required for half maximal velocity with allosteric enzymes having sigmoidal substrate curves is called the $[S]_{0.5}$ or $K_{0.5}$ value.

One of the best studied allosteric regulatory enzymes is the aspartate carbamoyltransferase (ACTase) from *E. coli.* The enzyme catalyzes the condensation of carbamoyl phosphate with aspartate to form carbamoylaspartate (figure 11.3). ACTase catalyzes the rate-determining reaction of the pyrimidine biosynthetic pathway in *E. coli.* The substrate saturation curve is sigmoidal when the concentration of either substrate is varied (figure 11.4). The enzyme has more than one active

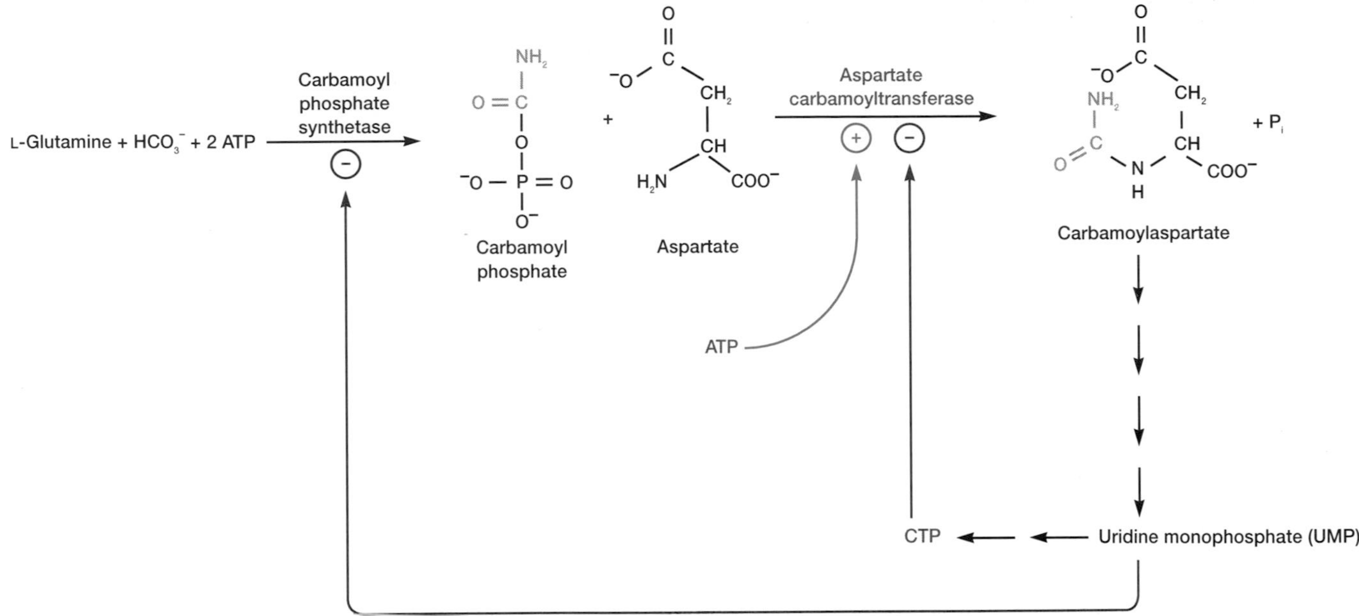

Figure 11.3 **ACTase Regulation.** The aspartate carbamoyltransferase reaction and its role in the regulation of pyrimidine biosynthesis. The end product CTP inhibits its activity (-) while ATP activates the enzyme (+). Carbamoyl phosphate synthetase is also inhibited by pathway end products such as UMP.

site, and the binding of a substrate molecule to an active site increases the binding of substrate at the other sites. In addition, cytidine triphosphate (CTP), an end product of pyrimidine biosynthesis, inhibits the enzyme and the purine ATP activates it. Both effectors alter the $K_{0.5}$ value of the enzyme, but not its maximum velocity. CTP inhibits by increasing $K_{0.5}$ or by shifting the substrate saturation curve to higher values. This allows the enzyme to operate more slowly at a particular substrate concentration when CTP is present. ATP activates by moving the curve to lower substrate concentration values so that the enzyme is maximally active over a wider substrate concentration range. Thus, when the pathway is so active that the CTP concentration rises too high, ACTase activity decreases and the rate of end product formation slows. In contrast, when the purine end product ATP increases in concentration relative to CTP, it stimulates CTP synthesis through its effects on ACTase.

Pyrimidine and purine biosynthesis (pp. 183–84).

E. coli aspartate carbamoyltransferase provides a clear example of separate regulatory and catalytic sites in allosteric enzymes. The enzyme is a large protein composed of two catalytic subunits and three regulatory subunits (figure 11.5*a*). The catalytic subunits contain only catalytic sites and are unaffected by CTP and ATP. Regulatory subunits do not catalyze the reaction, but do possess regulatory sites to which CTP and ATP bind. When these effectors bind to the complete enzyme, they cause conformational changes in the regulatory subunits and, subsequently, in the catalytic subunits and their catalytic sites. The enzyme can change reversibly between a less active T form and a more active R form (figure 11.5*b,c*). Thus, the regulatory site influences a catalytic site about 6.0 nm distant.

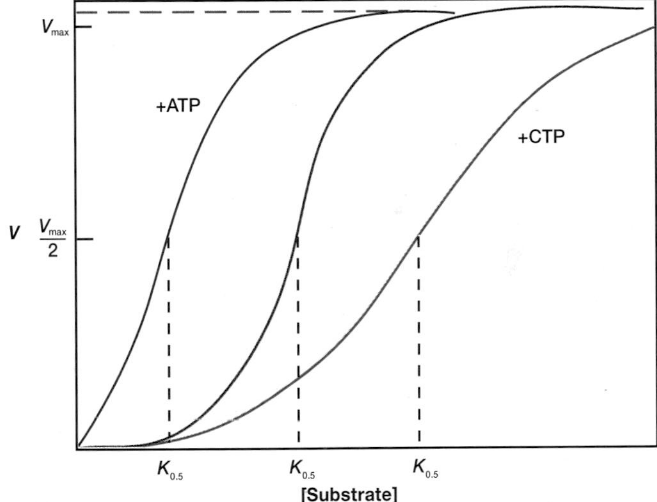

Figure 11.4 **The Kinetics of *E. coli* Aspartate Carbamoyltransferase.** *CTP,* a negative effector, increases the $K_{0.5}$ value while *ATP,* a positive effector, lowers the $K_{0.5}$. The V_{max} remains constant.

Covalent Modification of Enzymes

Regulatory enzymes also can be switched on and off by **reversible covalent modification.** Usually, this occurs through the addition and removal of a particular group, one form of the enzyme being more active than the other. For example, glycogen phosphorylase of the bread mold *Neurospora crassa* exists in phosphorylated and dephosphorylated forms called phosphorylase *a* and phosphorylase *b,* respectively (figure 11.6). Phosphorylase *b* is inactive because its required activator AMP is usually not present at sufficiently high levels. Phosphorylase *a,* the phosphorylated form, is active even

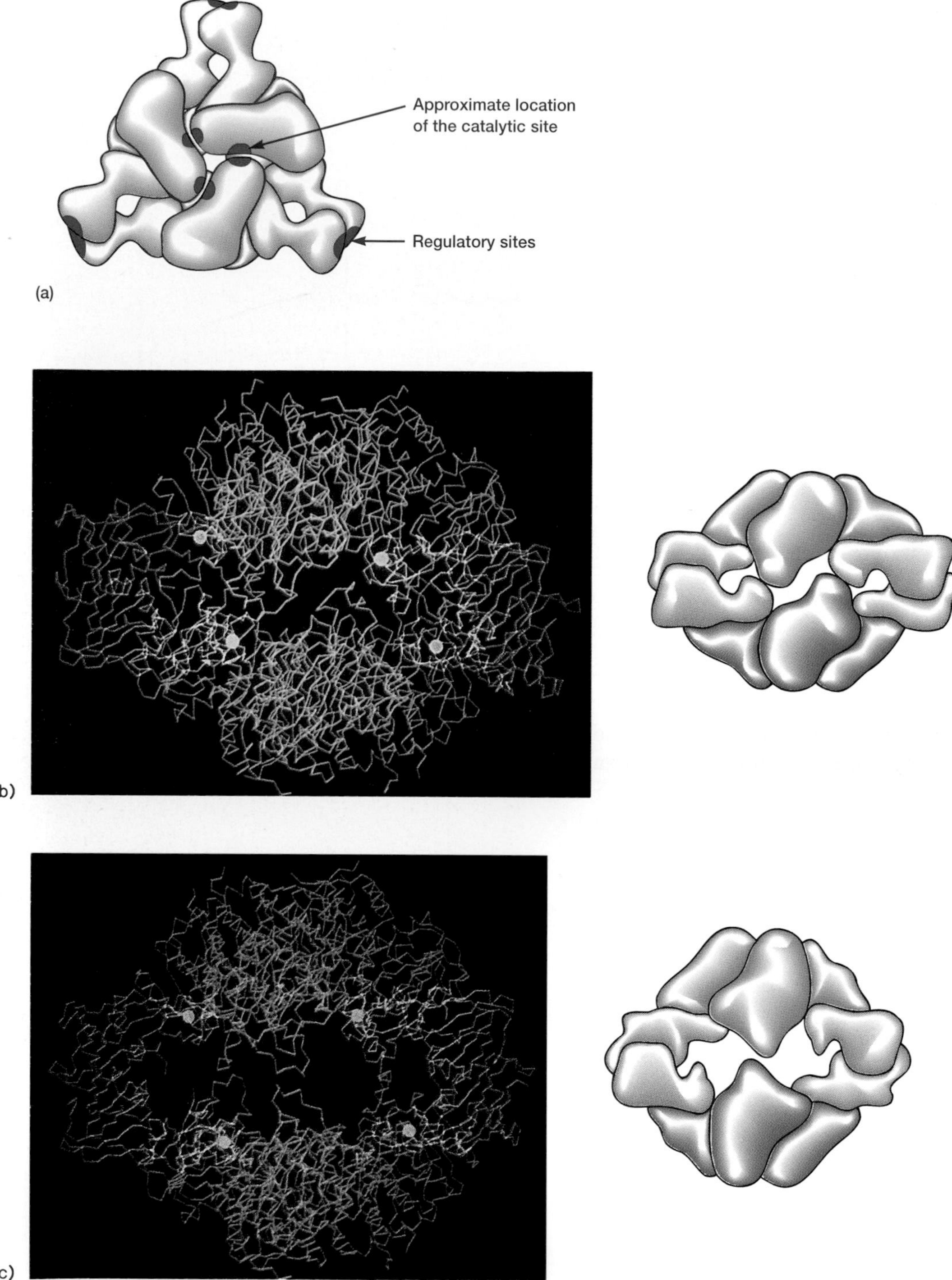

Figure 11.5 The Structure and Regulation of *E. coli* Aspartate Carbamoyltransferase. (*a*) A schematic diagram of the enzyme showing the six catalytic polypeptide chains (green and blue), the six regulatory chains (orange), and the catalytic and regulatory sites. The enzyme is viewed from the top. Each catalytic subunit contains three catalytic chains, and each regulatory subunit has two chains. (*b*) The less active T state of ACTase. The lines in the left figure represent the peptide chains. (*c*) The more active R state of ACTase. The regulatory subunits have rotated and pushed the catalytic subunits apart.

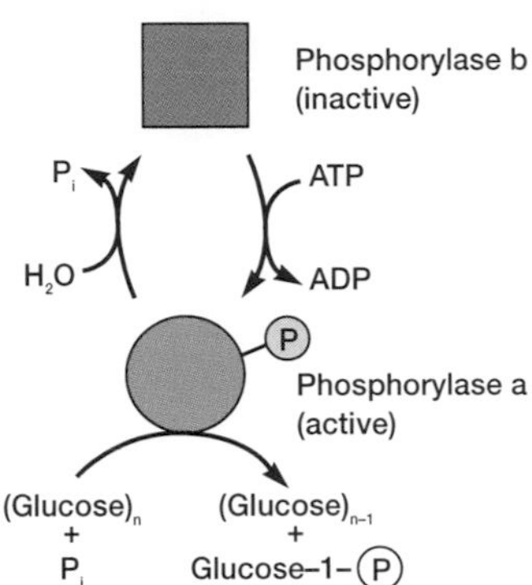

Figure 11.6 Reversible Covalent Modification of Glycogen Phosphorylase. The active form, phosphorylase *a,* is produced by phosphorylation and is inactivated when the phosphate is removed hydrolytically to produce inactive phosphorylase *b.*

without AMP. Glycogen phosphorylase is stimulated by phosphorylation of phosphorylase *b* to produce phosphorylase *a.* The attachment of phosphate changes the enzyme's conformation to an active form. Phosphorylation and dephosphorylation are catalyzed by separate enzymes, which are also regulated.

Phosphorylase and glycogen degradation (p. 159).

Enzymes can be regulated through the attachment of groups other than phosphate. One of the most intensively studied regulatory enzymes is *E. coli* glutamine synthetase, a large, complex enzyme existing in two forms (figure 11.7). When an adenylic acid residue is attached to each of its 12 subunits forming an adenylylated enzyme, glutamine synthetase is not very active. Removal of AMP groups produces the more active deadenylylated glutamine synthetase, and glutamine is formed. The glutamine synthetase system differs from the phosphorylase system in two ways: (1) AMP is used as the modifying agent, and (2) the modified form of glutamine synthetase is less active. Glutamine synthetase is also allosterically regulated.

Glutamine synthetase and its role in nitrogen metabolism (p. 178).

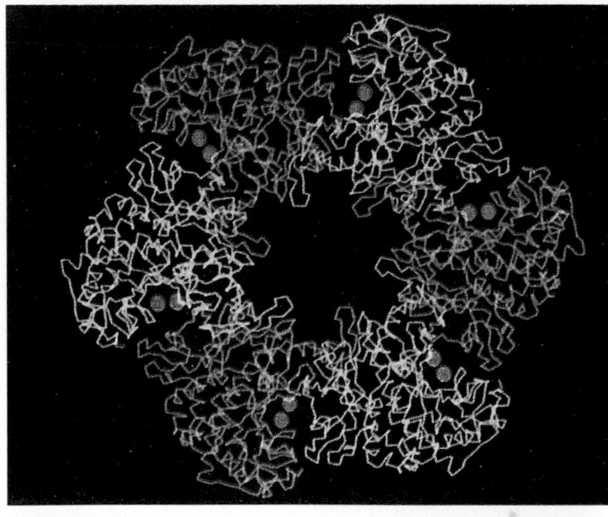

(a)

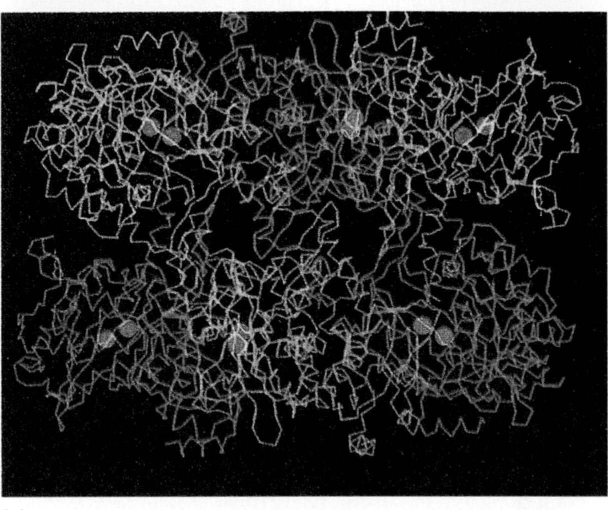

(b)

Figure 11.7 The Structure of *E. coli* Glutamine Synthetase. The enzyme contains 12 subunits in the shape of a hexagonal prism. For clarity, the subunits are colored alternating green and blue. Each of the six catalytic sites has a pair of Mn^{2+} ions (red). The tyrosine residues to which adenyl groups can be attached are colored red. (*a*) Top view of molecule. (*b*) Side view showing the 6 nearest subunits.

Feedback Inhibition

The rate of many metabolic pathways is adjusted through control of the activity of the regulatory enzymes described in the preceding section. Every pathway has at least one **pacemaker enzyme** that catalyzes the slowest or rate-limiting reaction in the pathway. Because other reactions proceed more rapidly than the pacemaker reaction, changes in the activity of this enzyme directly alter the speed with which a pathway operates. Usually, the first step in a pathway is a pacemaker reaction catalyzed by a regulatory enzyme. The end product of the pathway often inhibits this regulatory enzyme, a process known as **feedback inhibition** or **end product inhibition.** Feedback inhibition ensures balanced production of a pathway end product. If the end product becomes too concentrated, it inhibits the regulatory enzyme and slows its own synthesis. As the end product concentration decreases, pathway activity again increases and more product is formed. In this way, feedback inhibition automatically matches end product supply with the demand. The previously discussed *E. coli* aspartate carbamoyltransferase is an excellent example of end product or feedback inhibition.

Frequently, a biosynthetic pathway branches to form more than one end product. In such a situation, the synthesis of pathway end products must be precisely coordinated. It would not do to have one end product present in excess while another is lacking. Branching biosynthetic pathways usually achieve a balance between end products through the use of regulatory enzymes at branch points (figure 11.8). If an end product is present in excess, it often inhibits the branch-point enzyme on

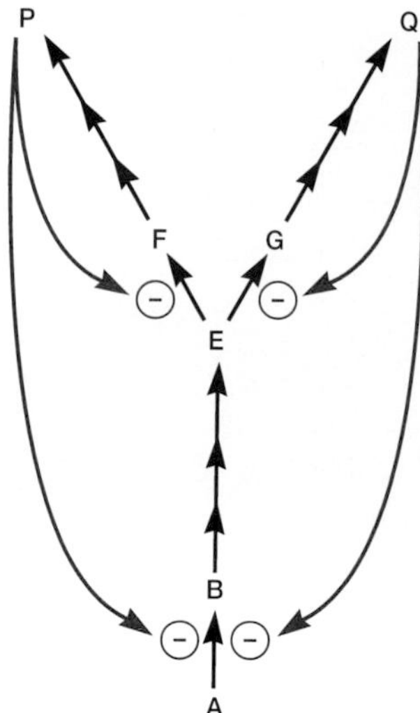

Figure 11.8 **Feedback Inhibition.** Feedback inhibition in a branching pathway with two end products. The branch-point enzymes, those catalyzing the conversion of intermediate *E* to *F* and *G*, are regulated by feedback inhibition. Products *P* and *Q* also inhibit the initial reaction in the pathway. A colored line with a minus sign at one end indicates that an end product, P or Q, is inhibiting the enzyme catalyzing the step next to the minus. See text for further explanation.

the sequence leading to its formation, in this way regulating its own formation without affecting the synthesis of other products. In figure 11.8, notice that both products also inhibit the initial enzyme in the pathway. An excess of one product slows the flow of carbon into the whole pathway while inhibiting the appropriate branch-point enzyme. Because less carbon is required when a branch is not functioning, feedback inhibition of the initial pacemaker enzyme helps match the supply with the demand in branching pathways. The regulation of multiply branched pathways is often made even more sophisticated by the presence of **isoenzymes,** different enzymes that catalyze the same reaction. The initial pacemaker step may be catalyzed by several isoenzymes, each under separate and independent control. In such a situation, an excess of a single end product reduces pathway activity, but does not completely block pathway function because some isoenzymes are still active.

1. Define the following: allosteric enzyme, effector or modulator, and $[S]_{0.5}$ or $K_{0.5}$.
2. How can regulatory enzymes be influenced by reversible covalent modification? What groups are used for this purpose with glycogen phosphorylase and glutamine synthetase, and which forms of these enzymes are active?
3. What is a pacemaker enzyme? Feedback inhibition? How does feedback inhibition automatically adjust the concentration of a pathway end product? What are isoenzymes and why are they important in pathway regulation?

Regulation of mRNA Synthesis

The control of metabolism by regulation of enzyme activity is a fine-tuning mechanism: it acts rapidly to adjust metabolic activity from moment to moment. Microorganisms also are able to control the expression of their genome, although over longer intervals. For example, the *E. coli* chromosome can code for about 2,000 to 4,000 peptide chains; yet only around 800 enzymes are present in *E. coli* growing with glucose as its energy source. Regulation of gene expression serves to conserve energy and raw material, to maintain balance between the amounts of various cell proteins, and to adapt to long-term environmental change. Thus, control of gene expression complements the regulation of enzyme activity.

Induction and Repression

The regulation of β-galactosidase synthesis has been intensively studied and serves as a primary example of how gene expression is controlled. This enzyme catalyzes the hydrolysis of the sugar lactose to glucose and galactose (figure 11.9). When *E. coli* grows with lactose as its carbon source, each cell contains about 3,000 β-galactosidase molecules, but has less than three molecules in the absence of lactose. The enzyme β-galactosidase is an **inducible enzyme;** that is, its level rises in the presence of a small molecule called an **inducer** (in this case, the lactose derivative allolactose).

The genes for enzymes involved in the biosynthesis of amino acids and other substances often respond differently from genes coding for catabolic enzymes. An amino acid present in the surroundings may inhibit the formation of the enzymes responsible for its biosynthesis. This makes good sense because the microorganism will not need the biosynthetic enzymes for a particular substance if it is already available. Enzymes whose amount is reduced by the presence of an end product are **repressible enzymes,** and metabolites causing a decrease in the concentrations of repressible enzymes are **corepressors.** Generally, repressible enzymes are necessary for synthesis and are always present unless the end product of their pathway is available. Inducible enzymes, in contrast, are required only when their substrate is available; they are missing in the absence of the inducer. It is important to emphasize that although the expression of many biosynthetic genes is regulated by repression, the expression of others is controlled by attenuation.

The Mechanism of Induction and Repression

Although variations in enzyme levels could be due to changes in the rates of enzyme degradation, most enzymes are relatively stable in growing bacteria. Induction and repression result principally from changes in the rate of transcription. When *E. coli* is growing in the absence of lactose, it often lacks mRNA molecules coding for the synthesis of β-galactosidase. In the presence of lactose, however, each cell has 35 to 50

Lactose $+ H_2O \xrightarrow{\beta\text{-galactosidase}}$ Galactose + Glucose

Figure 11.9 **The β-Galactosidase Reaction.**

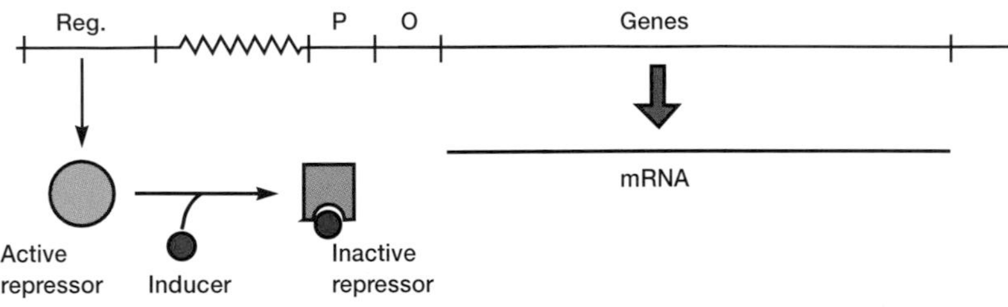

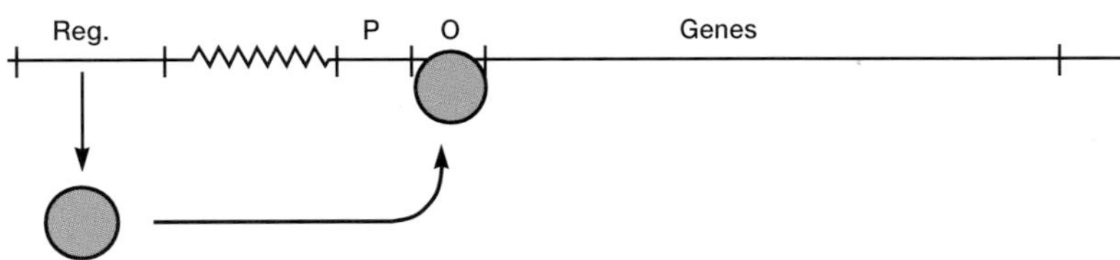

Figure 11.10 **Gene Induction.** The regulator gene, *Reg.*, synthesizes an active repressor that binds to the operator, *O*, and blocks RNA polymerase binding to the promoter, *P*, unless the inducer inactivates it. In the presence of the inducer, the repressor protein is inactive and transcription occurs.

β-galactosidase mRNA molecules. The synthesis of mRNA is dramatically influenced by the presence of lactose.

DNA transcription mechanism (pp. 201–4).

The rate of mRNA synthesis is controlled by special **repressor proteins** synthesized under the direction of regulator genes. The importance of regulator genes and repressors is demonstrated by mutationally inactivating a regulator gene to form a **constitutive mutant.** The mutant bacterium produces the enzymes in question whether or not they are needed. Thus, inactivation of repressor proteins blocks the regulation of transcription. Although the responses to the presence of metabolites are different, both induction and repression are forms of **negative control:** mRNA synthesis proceeds more rapidly in the absence of the controlling factor. The reason for this becomes clear when the mechanism of repressor action is understood.

Gene structure (pp. 240–44).

The repressor protein binds to a specific site on DNA called the **operator,** which it finds in a two-step process. First, the repressor complexes with a DNA molecule, then rapidly slides along the DNA until it reaches its operator and comes to a halt. A portion of the repressor fits into the major groove of operator-site DNA, and the N-terminal arm appears to wrap around the DNA. The shape of the repressor protein is ideally suited for specific binding to the DNA double helix.

How does a repressor inhibit transcription? The promotor to which RNA polymerase binds (*see chapter 10*) is located next to the operator; sometimes the operator even overlaps the promoter. Repressor occupation of the operator prevents the RNA polymerase from properly binding to the promoter, thus inhibiting transcription. A repressor does not affect the actual rate of transcription once it has begun.

Repressors must exist in both active and inactive forms because transcription would never occur if they were always active. In inducible systems, the regulator gene directs the synthesis of an active repressor. The inducer stimulates transcription by reversibly binding to the repressor and causing it to change to an inactive shape (figure 11.10). Just the opposite takes place in a system controlled by repression (figure 11.11). The repressor protein initially is inactive and only assumes an

Genes not transcribed

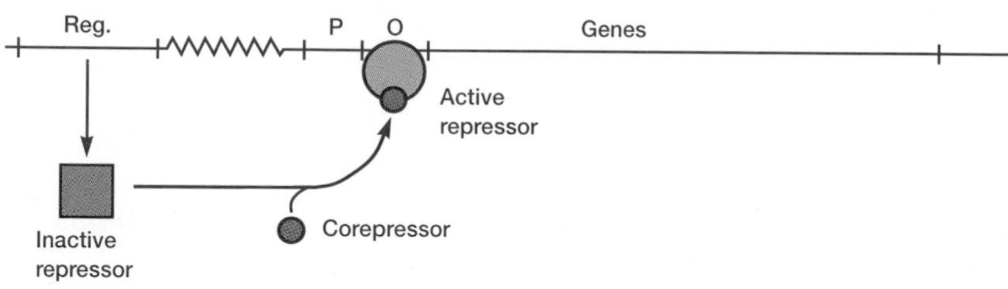

Genes transcribed

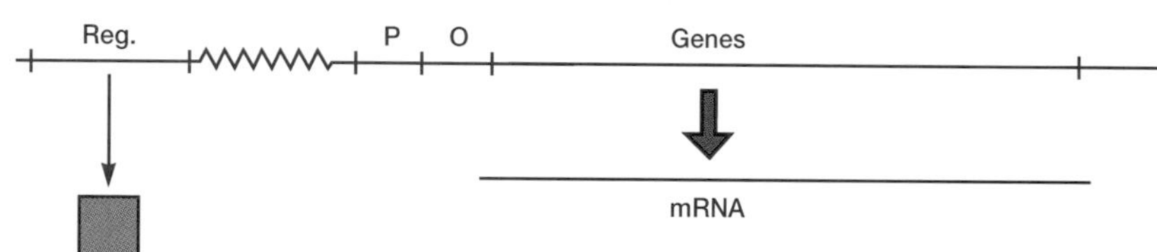

Figure 11.11 **Gene Repression.** The regulator gene, *Reg.*, synthesizes an inactive repressor protein that must be activated by corepressor binding before it can bind to the operator, *O*, and block transcription. In the absence of the corepressor, the repressor is inactive and transcription occurs.

active shape when the corepressor binds to it. The corepressor inhibits transcription by activating the repressor protein.

The synthesis of several proteins is often regulated by a single repressor. The **structural genes,** or genes coding for a polypeptide, are simply lined up together on the DNA, and a single mRNA carries all the messages. The sequence of bases coding for one or more polypeptides, together with the operator controlling its expression, is called an **operon.** This arrangement is of great advantage to the bacterium because coordinated control of the synthesis of several metabolically related enzymes (or other proteins) can be achieved. For example, the lactose or lac operon contains three structural genes. One codes for β-galactosidase; a second gene directs the synthesis of β-galactoside permease, the protein responsible for lactose uptake. The third gene codes for the enzyme β-galactoside transacetylase, whose function is still uncertain. The presence of the first two genes in the same operon ensures that the rates of lactose uptake and breakdown will vary together (Box 11.1, p. 225).

ATP

Adenyl cyclase

PP_i

cAMP

Figure 11.12 **Cyclic Adenosine Monophosphate (cAMP).** The phosphate group extends between the 3′ and 5′ hydroxyls of the ribose sugar. cAMP is formed from ATP by the enzyme adenyl cyclase.

Positive Operon Control and Catabolite Repression

The preceding section shows that operons can be under negative control, resulting in induction and repression. In contrast, some operons (e.g., the lac operon) function only in the presence of a controlling factor; that is, they are under **positive operon control.**

Lac operon function is regulated by the **catabolite activator protein** (**CAP**) or **cyclic AMP receptor protein** (**CRP**) and the small cyclic nucleotide **3′, 5′-cyclic adenosine monophosphate** (**cAMP;** figure 11.12), as well as by the lac repressor protein. The lac promoter contains a CAP site to which CAP must bind before RNA polymerase can attach to the promoter and begin transcription (figure 11.13). The catabolite acti-

Box 11.1

The Discovery of Gene Regulation

The ability of microorganisms to adapt to their environment by adjusting enzyme levels was first discovered by Emil Duclaux, a colleague of Louis Pasteur. He found that the fungus *Aspergillus niger* would only produce the enzyme that hydrolyzes sucrose (invertase) when grown in the presence of sucrose. In 1900, F. Dienert found that yeast contained the enzymes for galactose metabolism only when grown with lactose or galactose, and would lose these enzymes upon transfer to a glucose medium. Such a response made sense because the yeast cells would not need enzymes for galactose metabolism when using glucose as its carbon and energy source. Further examples of adaptation were discovered, and by the 1930s, H. Karstrom could divide enzymes into two classes: (1) adaptive enzymes that are formed only in the presence of their substrates, and (2) constitutive enzymes that are always present. It was originally thought that enzymes might be formed from inactive precursors and that the presence of the substrate simply shifted the equilibrium between precursor and enzyme toward enzyme formation.

In 1942, Jacques Monod, working at the Pasteur Institute in Paris, began a study of adaptation in the bacterium *E. coli.* It was already known that the enzyme β-galactosidase, which hydrolyzes the sugar lactose to glucose and galactose, was present only when *E. coli* was grown in the presence of lactose. Monod discovered that nonmetabolizable analogues of β-galactosides, such as thiomethylgalactoside, also could induce enzyme production. This discovery made it possible to study induction in cells growing on carbon and energy sources other than lactose so that the growth rate and inducer concentration would not depend on the lactose supply. He next demonstrated that induction involved the synthesis of new enzyme, not just the conversion of already available precursor. Monod accomplished this by making *E. coli* proteins radioactive with ^{35}S, then transferring the labeled bacteria to nonradioactive medium and adding inducer. The newly formed β-galactosidase was nonradioactive and must have been synthesized after addition of inducer.

A study of the genetics of lactose induction in *E. coli* was begun by Joshua Lederberg a few years after Monod had started his work. Lederberg isolated not only mutants lacking β-galactosidase but also a constitutive mutant in which synthesis of the enzyme proceeded in the absence of an inducer ($LacI^-$). During bacterial conjugation (*see chapter 13*), genes from the donor bacterium enter the recipient to temporarily form an organism with two copies of those genes provided by the donor. When Arthur B. Pardee, Francois Jacob, and Monod transferred the gene for inducibility to a constitutive recipient not sensitive to inducers, the newly acquired gene made the recipient bacterium sensitive to inducer again. This functional gene was not a part of the recipient's chromosome. Thus the special gene directed the synthesis of a cytoplasmic product that inhibited the formation of β-galactosidase in the absence of the inducer. In 1961, Jacob and Monod named this special product the repressor and suggested that it was a protein. They further proposed that the repressor protein exerted its effects by binding to the operator, a special site next to the structural genes. They provided genetic evidence for their hypothesis. The name operon was given to the complex of the operator and the genes it controlled. Several years later in 1967, Walter Gilbert and Benno Müller-Hill managed to isolate the lac repressor and show that it was indeed a protein and did bind to a specific site in the lac operon.

The existence of repression was discovered by Monod and G. Cohen-Bazire in 1953 when they found that the presence of the amino acid tryptophan would repress the synthesis of tryptophan synthetase, the final enzyme in the pathway for tryptophan biosynthesis. Subsequent research in many laboratories showed that induction and repression were operating by quite similar mechanisms, each involving repressor proteins that bound to operators on the genome.

vator protein is able to bind to the CAP site only when complexed with cAMP (figure 11.14). This positive control system makes lac operon activity dependent on the presence of cAMP as well as on that of lactose.

The value of this system to bacteria is shown by a simple growth experiment. If *E. coli* grows in a medium containing both glucose and lactose, it uses glucose preferentially until the sugar is exhausted. Then after a short lag, growth resumes with lactose as the carbon source (figure 11.15). This biphasic growth pattern or response is called **diauxic growth.** The cause of diauxic growth or diauxie is complex and not completely understood, but **catabolite repression** or the glucose effect probably plays a part. The enzymes for glucose catabolism are constitutive and unaffected by CAP activity. When the bacterium is given glucose, the cAMP level drops, resulting in deactivation of the catabolite activator protein and inhibition of lac operon expression. The decrease in cAMP may be due to the effect of the phosphoenolpyruvate:phosphotransferase system (PTS) on the activity of adenyl cyclase, the enzyme that synthesizes cAMP. Enzyme III of the PTS donates a phosphate to glucose during its transport; therefore, it enters the cell as glucose 6-phosphate. The phosphorylated form of enzyme III also activates adenyl cyclase. If glucose is being rapidly transported by PTS, the amount of phosphorylated enzyme III is low and the adenyl cyclase is less active, so the cAMP level drops. At least one other mechanism is involved

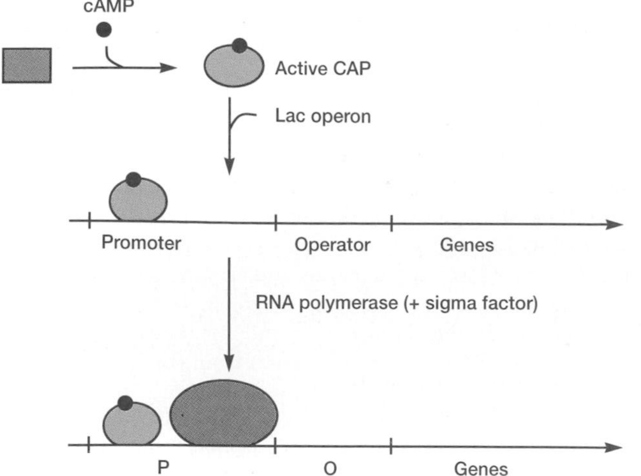

Figure 11.13 **Positive Control of the Lac Operon.** When cyclic AMP is absent or present at a low level, the CAP protein remains inactive and does not bind to the promoter. In this situation, RNA polymerase also does not bind to the promoter and transcribe the operon's genes.

in diauxic growth. When the PTS is actively transporting glucose into the cell, nonphosphorylated enzyme III is more prevalent. Nonphosphorylated enzyme III binds to the lactose permease and allosterically inhibits it, thus blocking lactose uptake.

The phosphoenolpyruvate: phosphotransferase system (p. 103).

Whatever the precise mechanism, such control is of considerable advantage to the bacterium. It will use the most easily catabolized sugar (glucose) first rather than synthesize the enzymes necessary for another carbon and energy source. These control mechanisms are present in a variety of bacteria and metabolic pathways.

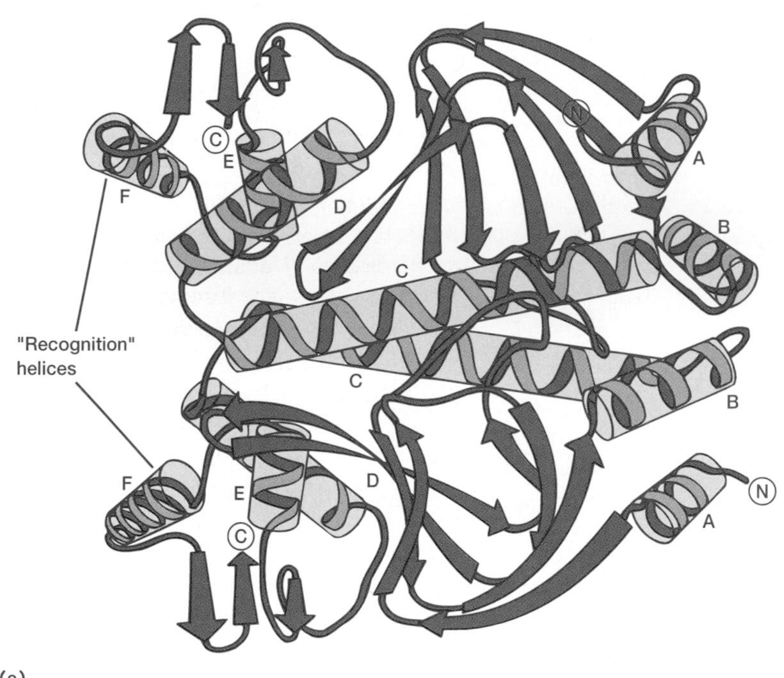

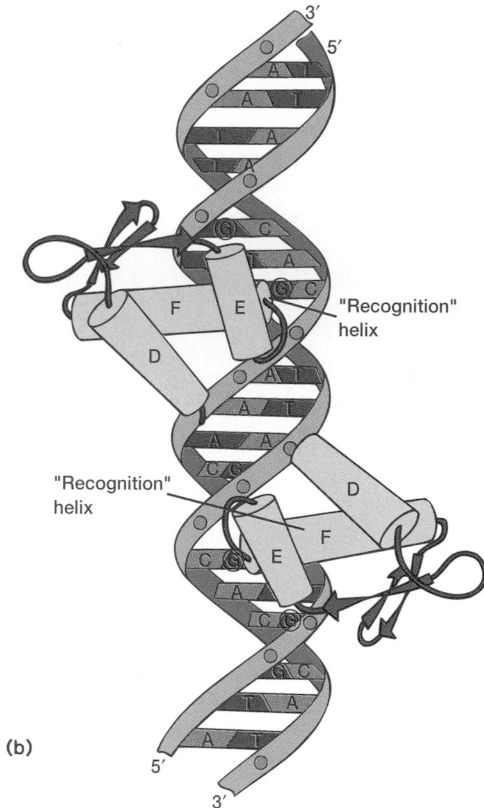

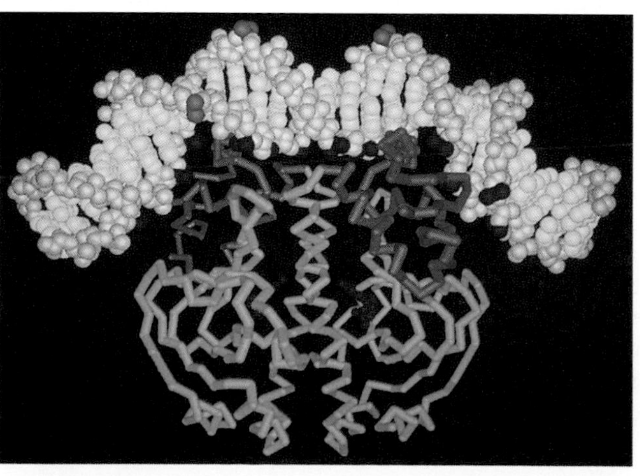

Figure 11.14 **CAP Structure and DNA Binding.** (*a*) A diagram of the CAP dimer, the active form. The C-terminal domains bind to the promoter. The cAMP binding domains are located toward the N-terminal end. (*b*) The CAP dimer binding to DNA at the lac operon promoter. The recognition helics fit into two adjacent major grooves on the double helix. (*c*) A model of the *E. coli* CAP-DNA complex derived from crystal structure studies. The cAMP binding domain is in blue and the DNA binding domain, in purple. The cAMP molecules bound to CAP are in red. Note that the DNA is bent by 90° when complexed with CAP. (a,b) *From Donald Voet and Judith G. Voet,* Biochemistry. *Copyright © 1990 John Wiley & Sons, Inc., New York, NY. Reprinted by permission. Originally from I. T. Weber and T. A. Steitz, Proceedings of the National Academy of Sciences, 81:3975, 1984. Reprinted by permission.*

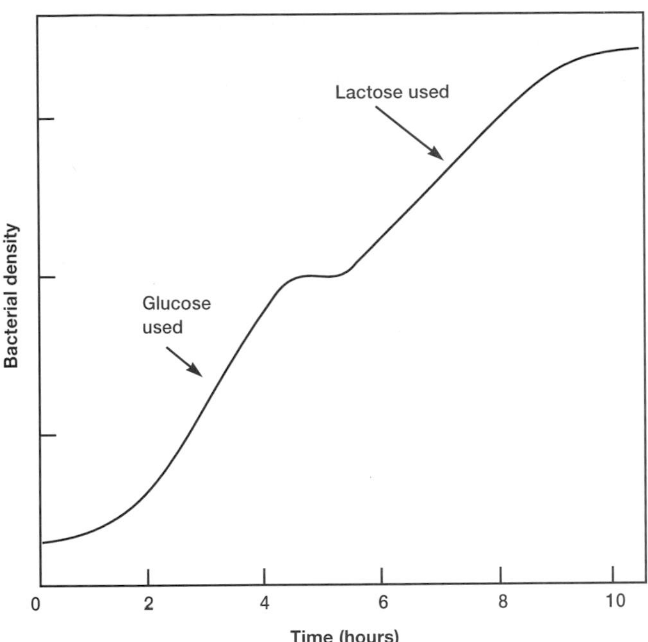

Figure 11.15 **Diauxic Growth.** The diauxic growth curve of *E. coli* grown with a mixture of glucose and lactose. Glucose is first used, then lactose. A short lag in growth is present while the bacteria synthesize the enzymes needed for lactose utilization.

Knowledge of microbial regulatory mechanisms is extremely useful in industrial microbiology. A bacterial strain that excretes large amounts of the desired metabolite (e.g., an amino acid or vitamin) can often be developed if mutations are induced that either eliminate feedback inhibition of a critical enzyme in the synthetic pathway or inactivate a regulator gene or operator to yield constitutive enzyme synthesis. Synthesis of many enzymes and end products is controlled by the presence of glucose through catabolite repression and other processes. This must be considered when bacteria are used in commercial production. For example, the enzyme glucose isomerase, which converts glucose to fructose, is important in the production of sweeteners because fructose has twice the sweetening strength of sucrose. Glucose isomerase synthesis is repressed by glucose in *Bacillus coagulans*. During log phase growth, the bacterium uses available glucose and does not synthesize the enzyme. When glucose is exhausted, other carbon sources are employed and glucose isomerase production begins, even though rapid growth has ceased.

Attenuation

Bacteria can regulate transcription in other ways, as may be seen in the tryptophan operon of *E. coli*. The tryptophan operon contains structural genes for five enzymes in this amino acid's biosynthetic pathway. As might be expected, the operon is under the control of a repressor protein coded for by the *trpR* gene (trp stands for tryptophan), and excess tryptophan inhibits transcription of operon genes by acting as a corepressor and activating the repressor protein. In addition, the continuation of transcription is also controlled.

A **leader region** lies between the operator and the first structural gene in the operon, the *trpE* gene, and is responsible for controlling the continuation of transcription after the RNA polymerase has bound to the promoter (figure 11.16*a*). The leader region contains an **attenuator** and a sequence thought to code for the synthesis of a leader peptide. The attenuator is a rho-independent termination site (*see chapter 10*) with a short GC-rich segment followed by a sequence of eight U residues. The four stretches marked off in figure 11.16*a* have complementary base sequences and can base pair with each other to form hairpin loops. In the absence of a ribosome, segments one and two pair to form a hairpin, while segments three and four generate a second loop next to the poly-U sequence (figure 11.16*b*). The hairpin formed by segments three and four plus the poly-U sequence will terminate transcription. If segment one is prevented from base pairing with segment two, segment two is free to associate with segment three. As a result, segment four remains single stranded (figure 11.16*c*) and cannot serve as a terminator for transcription. It is important to note that the sequence coding for the leader peptide contains two adjacent codons that code for the amino acid tryptophan. Thus, the complete peptide can be made only when there is an adequate supply of tryptophan. Since the leader peptide has not been detected, it must be degraded immediately after synthesis.

Ribosome behavior during translation of the mRNA regulates RNA polymerase activity as it transcribes the leader region. This is possible because translation and transcription are tightly coupled. When the active repressor is absent, RNA polymerase binds to the promoter and moves down the leader synthesizing mRNA. If there is no translation of the mRNA after the RNA polymerase has begun copying the leader region, segments three and four form a hairpin loop and transcription terminates before the polymerase reaches the *trpE* gene (figure 11.17*a*). When tryptophan is present, there is sufficient tryptophanyl-tRNA for protein synthesis. Therefore, the ribosome will synthesize the leader peptide and continue moving along the mRNA until it reaches a UGA stop codon (*see chapter 12*) lying between segments one and two. The ribosome halts at this codon and projects into segment two far enough to prevent it from pairing properly with segment three (figure 11.17*b*). Segments three and four form a hairpin loop, and the RNA polymerase terminates at the attenuator just as if no translation had taken place. If tryptophan is lacking, the ribosome will stop at the two adjacent tryptophan codons in the leader peptide sequence and prevent segment one from base pairing with segment two, because the tryptophan codons are located within segment one (figures 11.16*a* and 11.17*c*). If this happens while the RNA polymerase is still transcribing the leader region, segments two and three associate before segment four has been synthesized. Therefore, segment four will remain single stranded and the terminator hairpin will not form. Consequently, when tryptophan is absent, the RNA polymerase continues on and transcribes tryptophan operon genes. Control of the continuation of transcription by a specific aminoacyl-tRNA is called **attenuation.**

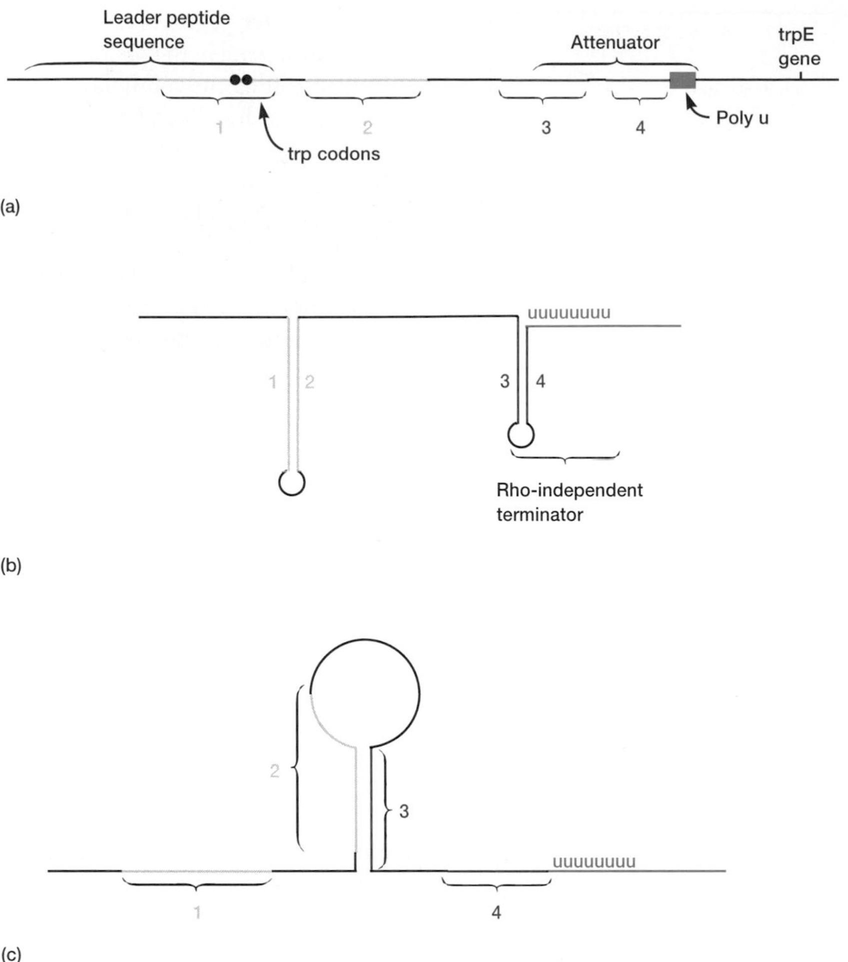

Figure 11.16 The Tryptophan Operon Leader. Organization and base pairing of the tryptophan operon leader region. The promoter and operator are to the left of the segment diagrammed, and the first structural gene (*trpE*) begins to the right of the attenuator. The stretches of DNA marked off as *1* through *4* can base pair with each other to form hairpin loops: segment *2* with *1*, and segment *3* with *2* or *4*. See text for details.

Attenuation's usefulness is apparent. If the bacterium is deficient in an amino acid other than tryptophan, protein synthesis will slow and tryptophanyl-tRNA will accumulate. Transcription of the tryptophan operon will be inhibited by attenuation. When the bacterium begins to synthesize protein rapidly, tryptophan may be scarce and the concentration of tryptophanyl-tRNA may be low. This would reduce attenuation activity and stimulate operon transcription, resulting in larger quantities of the tryptophan biosynthetic enzymes. Acting together, repression and attenuation can coordinate the rate of synthesis of amino acid biosynthetic enzymes with the availability of amino acid end products and with the overall rate of protein synthesis. When tryptophan is present at high concentrations, any RNA polymerases not blocked by the activated repressor protein probably will not get past the attenuator sequence. Repression decreases transcription about seventyfold and attenuation slows it another eight- to tenfold; when both mechanisms operate together, transcription can be slowed about 600-fold.

Attenuation seems important in the regulation of several amino acid biosynthetic pathways. At least five other operons have leader peptide sequences that resemble the tryptophan system in organization. For example, the leader peptide sequence of the histidine operon codes for seven histidines in a row and is followed by an attenuator that is a terminator sequence.

Gene Regulation by Antisense RNA

Microbiologists have known for many years that gene expression can be controlled by both regulatory proteins (e.g., repressor proteins and CAP) and aminoacyl-tRNA (attenuation). More recently, it has been discovered that the activity of some genes is controlled by a special type of small regulatory RNA molecule. The regulatory RNA, called **antisense RNA,** has a base sequence complementary to a segment of another RNA molecule and specifically binds to the target RNA. Antisense RNA binding can block DNA replication, mRNA synthesis, or translation. The genes coding for these RNAs are sometimes called antisense genes.

This mode of regulation appears to be widespread among viruses and bacteria. Examples are the regulation of plasmid

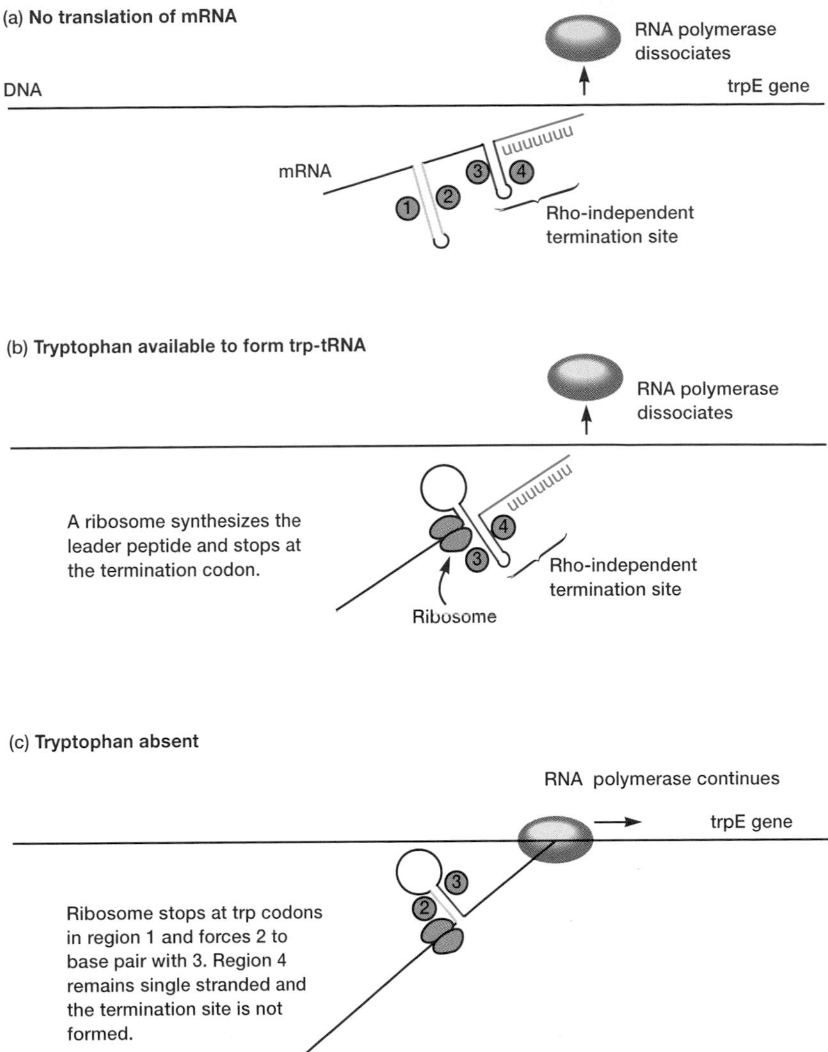

Figure 11.17 **Attenuation Control.** The control of tryptophan operon function by attenuation. See text for details.

replication and Tn10 transposition, osmoregulation of porin protein expression, regulation of λ phage reproduction, and the autoregulation of cAMP-receptor protein synthesis. Antisense RNA regulation has not yet been demonstrated in eucaryotic cells.

Plasmids and transposons (pp. 261–68).

As mentioned previously, antisense RNA can regulate gene expression in several ways. A good example of one mechanism is provided by the ColE1 plasmid of *E. coli* (see pp. 261–64). Plasmid replication is regulated so that about 10 to 30 copies of ColE1 are present in a cell. RNA I, an antisense RNA about 100 nucleotides long, binds with a precursor RNA used to form the RNA primer for plasmid DNA replication. DNA polymerase cannot copy the plasmid without the primer and replication is blocked. Thus, RNA I controls DNA replication by adjusting the number of functional RNA primers.

The regulation of *E. coli* outer membrane porin proteins provides a second example of control by antisense RNA. The outer membrane contains channels made of porin proteins (see p. 55). The two most important porins in *E. coli* are the OmpF and OmpC proteins. OmpC pores are slightly smaller and are made when the bacterium grows at high osmotic pressures. It is the dominant porin in *E. coli* from the intestinal tract. This makes sense because the smaller pores would exclude many of the toxic molecules present in the intestine. The larger OmpF pores are favored when *E. coli* grows in a dilute environment, and they allow solutes to diffuse into the cell more readily.

The *ompF* and *ompC* genes are partly regulated by a special OmpR protein that represses the *ompF* gene and activates *ompC*. In addition, the *micF* gene produces a 174-nucleotide-long antisense micF RNA that blocks *ompF* action (mic stands for *m*RNA-*i*nterfering *c*omplementary RNA). The micF RNA is complementary to *ompF* at the translation initiation site. It complexes with *ompF* mRNA and represses translation. The *micF* gene is activated by conditions such as high osmotic pressure that favor *ompC* expression. This helps ensure that OmpF protein is not produced at the same time as OmpC protein.

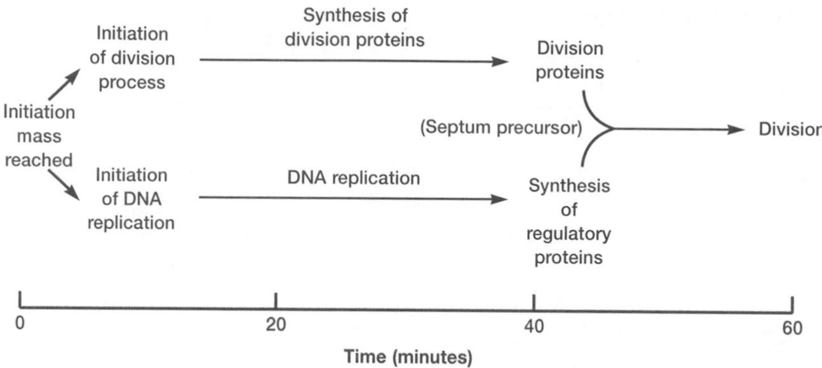

Figure 11.18 **Control of the Cell Cycle in *E. coli.*** A 60-minute interval between divisions has been assumed for purposes of simplicity (the actual time between cell divisions may be shorter). *E. coli* requires about 40 minutes to replicate its DNA and 20 minutes after termination of replication to prepare for division.

The fact that antisense RNA can bind specifically to mRNA and block its activity has great practical implications. Antisense RNA is already a valuable research tool. Suppose one desires to study the action of a particular gene. An antisense RNA that will bind to the gene's mRNA can be constructed and introduced into the cell, thus blocking gene expression. Changes in the cell are then observed. It is also possible to use the same approach with short strands of antisense DNA that bind to mRNA.

Antisense RNA and DNA may well be effective against a variety of cancers and infectious diseases. Promising preliminary results have been obtained using antisense oligonucleotides directed against *Trypanosoma brucei* (the cause of African sleeping sickness), herpesviruses, the HIV virus, tumor viruses such as the RSV and polyoma viruses, and chronic myelogenous leukemia. Although much further research is needed to determine the medical potential of these molecules, they may prove invaluable in the treatment of many diseases.

1. What are induction and repression and why are they useful? Define inducer, corepressor, repressor protein, regulator gene, constitutive mutant, operator, structural gene, and operon.
2. Define positive control, catabolite activator protein, catabolite repression, and diauxic growth.
3. Define attenuation and describe how it works in terms of a labeled diagram, such as that provided in figure 11.17. What are the functions of the leader region and the attenuator in attenuation?
4. What is antisense RNA? How does it regulate gene expression?

Control of the Cell Cycle

Although much progress has been made in understanding the control of microbial enzyme activity and pathway function, much less is known about the regulation of more complex events such as bacterial sporulation and cell division. This section briefly describes the regulation of bacterial cell division. Attention is focused primarily on *E. coli* because it has been intensively studied.

The complete sequence of events extending from the formation of a new cell through the next division is called the **cell cycle.** A young *E. coli* cell growing at a constant rate will double in length without changing in diameter, then divide into two cells of equal size by transverse fission. Because each daughter cell receives at least one copy of the genetic material, DNA replication and cell division must be tightly coordinated. In fact, if DNA synthesis is inhibited by a drug or a gene mutation, cell division is also blocked and the affected cells continue to elongate, forming long filaments. A regulatory protein may accumulate and trigger the initiation of DNA replication. The product of the *dnaA* gene is thought to be such an initiator protein. Termination of DNA replication also seems connected in some way with cell division. Although the growth rate of *E. coli* at 37°C may vary considerably, division usually takes place about 20 minutes after replication has finished. During this final interval, the genetic material must be distributed between the daughter cells. The newly formed DNA copies are attached to adjacent sites on the plasma membrane. Membrane growth first separates them, then a cross wall or septum forms between the two.

Patterns of cell wall formation (p. 188).

Current evidence suggests that two sequences of events, operating in parallel but independently, control division and end the cell cycle (figure 11.18). When a cell reaches its initiation mass or volume, both DNA replication and synthesis of division proteins are triggered. These two processes take about 40 minutes to complete. It is believed that one or more special regulatory proteins, synthesized at the end of DNA replication, interact with division proteins over the next 20 minutes to promote septum formation and division. Some kind of septum precursor, prepared at about the same time as the division and regulatory proteins are, also may be involved in the initiation of division.

The relationship of DNA synthesis to the cell cycle varies with the growth rate. If *E. coli* is growing with a doubling time of about 60 minutes, DNA replication does not take place during the last 20 minutes; that is, replication is a discontinuous process when the doubling time is 60 minutes or longer. When the culture is growing with a doubling time of less than 60 minutes, a second round of replication begins while the first round is still underway (figure 11.19). The daughter cells may actually receive DNA with two or more replication forks, and replication is continuous because the cells are always copying their DNA.

Two decades of research have provided a fairly adequate overall picture of the cell cycle in *E. coli*. Several cell division genes have been identified. Yet, it is still not known in any detail how the cycle is controlled. Future work should improve our understanding of this important process.

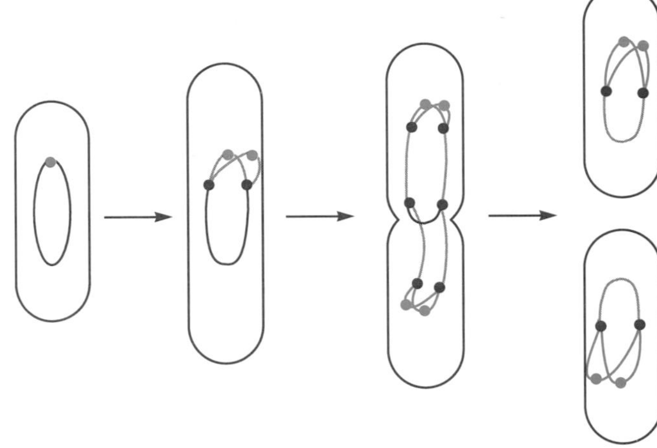

Figure 11.19 DNA Replication in Rapidly Growing Bacteria. A new round of DNA replication is initiated before the original cell divides so that the DNA in daughter cells is already partially replicated. The colored circles at the ends of the DNA loops are replication origins; the black circles along the sides are replication forks. Newly synthesized DNA is in color, with the darker shade representing the most recently synthesized DNA.

1. What is a cell cycle? Briefly describe how the cycle in *E. coli* is regulated and how cycle timing results.
2. How are the two DNA copies separated and apportioned between the two daughter cells?

Summary

1. The regulation of metabolism keeps cell components in proper balance and conserves metabolic energy and material.
2. The localization of metabolites and enzymes in different parts of the cell, metabolic channeling, influences pathway activity. A common channeling mechanism is compartmentation.
3. Regulatory enzymes are usually allosteric enzymes, enzymes in which an effector or modulator binds reversibly to a regulatory site separate from the catalytic site and causes a conformational change in the enzyme to alter its activity.
4. Aspartate carbamoyltransferase is an allosteric enzyme that is inhibited by CTP and activated by ATP.
5. Enzyme activity also can be regulated by reversible covalent modification. Two examples of such regulation are glycogen phosphorylase (phosphate addition) and glutamine synthetase (AMP addition).
6. The first enzyme in a pathway and enzymes at branch points are often subject to feedback inhibition by one or more end products. Excess end product slows its own synthesis.
7. β-galactosidase is an inducible enzyme whose concentration rises in the presence of its inducer.
8. Many biosynthetic enzymes are repressible enzymes whose levels are reduced in the presence of end products called corepressors.
9. Induction and repression result from regulation of the rate of transcription by repressor proteins coded for by regulator genes. A regulator gene mutation can lead to a constitutive mutant, which continuously produces a metabolite.
10. The repressor inhibits transcription by binding to an operator and interfering with the binding of RNA polymerase to its promoter.
11. In inducible systems, the newly synthesized repressor protein is active, and inducer binding inactivates it. In contrast, an inactive repressor is synthesized in a repressible system and is activated by the corepressor.
12. Often one repressor regulates the synthesis of several enzymes because they are part of a single operon, a DNA sequence coding for one or more polypeptides and the operator controlling its expression.
13. Positive operon control of the lac operon is due to the catabolite activator protein, which is activated by cAMP. The resulting catabolite repression probably contributes to diauxic growth when *E. coli* is cultured in the presence of both glucose and lactose.
14. In the tryptophan operon, a leader region lies between the operator and the first structural gene. It codes for the synthesis of a leader peptide and contains an attenuator, a rho-independent termination site.
15. The synthesis of the leader peptide by a ribosome while RNA polymerase is transcribing the leader region regulates transcription; therefore, the tryptophan operon is expressed only when there is insufficient tryptophan available. This mechanism of transcription control is called attenuation.
16. Small antisense RNA molecules regulate the expression of some genes. They can affect DNA replication, RNA transcription, or translation. Antisense RNA and DNA molecules are valuable research tools and may be effective against cancer and infectious disease.
17. The complete sequence of events extending from the formation of a new cell through the next division is called the cell cycle.
18. The end of DNA replication is tightly linked to cell division, so division in *E. coli* usually takes place about 20 minutes after replication is finished. Special division and regulatory proteins likely are involved.
19. The new DNA copies are attached to the plasma membrane and are separated by membrane growth before a cross wall forms.
20. In very rapidly dividing bacterial cells, a new round of DNA replication begins before the cells divide.

Key Terms

allosteric enzyme *218*
antisense RNA *228*
attenuation *227*
attenuator *227*
catabolite activator protein (CAP) *224*
catabolite repression *225*
cell cycle *230*
compartmentation *217*
constitutive mutant *223*
corepressor *222*
3′, 5′-cyclic adenosine monophosphate (cAMP) *224*
cyclic AMP receptor protein (CRP) *224*
diauxic growth *225*
effector or modulator *218*
end product inhibition *221*
feedback inhibition *221*
inducer *222*
inducible enzyme *222*
isoenzymes *222*
leader region *227*
metabolic channeling *217*
negative control *223*
operator *223*
operon *224*
pacemaker enzyme *221*
positive operon control *224*
regulatory site *218*
repressible enzyme *222*
repressor protein *223*
reversible covalent modification *219*
structural gene *224*

Questions for Thought and Review

1. How might a substrate be able to regulate the activity of the enzyme using it?
2. Describe how *E. coli* aspartate carbamoyltransferase is regulated, both in terms of the effects of modulators and the mechanism by which they exert their influence.
3. What is the significance of the fact that regulatory enzymes are often located at pathway branch points?
4. Describe in some detail the organization of the regulatory systems responsible for induction and repression, and the mechanism of their operation.
5. How is *E. coli* able to use glucose exclusively when presented with a mixture of glucose and lactose?
6. Of what practical importance is attenuation in coordinating the synthesis of amino acids and proteins? Describe how attenuation activity would vary when protein synthesis suddenly rapidly accelerated, then later suddenly slowed greatly.
7. How does the timing of DNA replication seem to differ between slow-growing and fast-growing cells? Be able to account for the fact that bacterial cells may contain more than a single copy of DNA.

Additional Reading

D'Ari, R., and Bouloc, P. 1990. Logic of the *Escherichia coli* cell cycle. *Trends Biochem. Sci.* 15:191–94.

Botsford, J. L., and Harman, J. G. 1992. Cyclic AMP in prokaryotes. *Microbiol. Rev.* 56(1):100–22.

Busby, S., and Buc, H. 1987. Positive regulation of gene expression by cyclic AMP and its receptor protein in *Escherichia coli. Microbiol. Sci.* 4(12):371–75.

Danchin, A. 1987. Membrane integration of carbohydrate transport in bacteria. *Microbiol. Sci.* 4(9):267–69.

Darnell, J.; Lodish, H.; and Baltimore, D. 1990. *Molecular cell biology,* 2d ed. New York: W. H. Freeman.

de Boer, P. A. J.; Cook, W. R.; and Rothfield, L. I. 1990. Bacterial cell division. *Ann. Rev. Genet.* 24:249–74.

Donachie, W. D. 1981. The cell cycle of *Escherichia coli.* In *The cell cycle,* ed. P. C. L. John, 63–83. New York: Cambridge University Press.

Edwards, C. 1981. *The microbial cell cycle.* Washington, D.C.: American Society for Microbiology.

Freifelder, D. 1987. *Molecular biology,* 2d ed. New York: Van Nostrand Reinhold.

Gold, L. 1988. Posttranscriptional regulatory mechanisms in *Escherichia coli. Ann. Rev. Biochem.* 57:199–233.

Gottschalk, G. 1986. *Bacterial metabolism,* 2d ed. New York: Springer-Verlag.

Green, P. J.; Pines, O.; and Inouye, M. 1986. The role of antisense RNA in gene regulation. *Ann. Rev. Biochem.* 55:569–97.

Kantrowitz, E. R., and Lipscomb, W. N. 1988. *Escherichia coli* aspartate transcarbamylase: The relation between structure and function. *Science* 241:669–74.

Koshland, D. E., Jr. 1973. Protein shape and biological control. *Sci. Am.* 229(4):52–64.

Lewin, B. 1990. *Genes,* 4th ed. New York: Oxford University Press.

Maniatis, T., and Ptashne, M. 1976. A DNA operator-repressor system. *Sci. Am.* 234(1):64–76.

Marr, A. G. 1991. Growth rate of *Escherichia coli. Microbiol. Rev.* 55(2):316–33.

Mathews, C. K., and van Holde, K. E. 1990. *Biochemistry.* Redwood City, Calif.: Benjamin/Cummings.

McClure, W. R. 1985. Mechanism and control of transcription initiation in prokaryotes. *Ann. Rev. Biochem.* 54:171–204.

McKnight, S. L. 1991. Molecular zippers in gene regulation. *Sci. Am.* 264(4):54–64.

Moat, A. G., and Foster, J. W. 1988. *Microbial physiology,* 2d ed. New York: John Wiley and Sons.

Neidhardt, F. C.; Ingraham, J. L.; and Schaechter, M. 1990. *Physiology of the bacterial cell.* Sunderland, Mass.: Sinauer Associates.

Ptashne, M. 1992. *A genetic switch,* 2d ed. Cambridge, Mass.: Blackwell Scientific Publications.

Ptashne, M., and Gilbert, W. 1970. Genetic repressors. *Sci. Am.* 222(6):36–44.

Saier, M. H., Jr.; Wu, L.-F.; and Reizer, J. 1990. Regulation of bacterial physiological processes by three types of protein phosphorylating systems. *Trends Biochem. Sci.* 15:391–95.

Saier, M. H. 1989. Protein phosphorylation and allosteric control of inducer exclusion and catabolite repression by the bacterial phosphoenolpyruvate: sugar phosphotransferase system. *Microbiol. Rev.* 53(1):109–20.

Sargent, M. G. 1985. How do bacterial nuclei divide? *Microbiol. Sci.* 2(8):235–39.

Stent, G. S., and Calendar, R. 1978. *Molecular genetics,* 2d ed. San Francisco: Freeman.

Stock, J. B.; Ninfa, A. J.; and Stock, A. M. 1989. Protein phosphorylation and regulation of adaptive responses in bacteria. *Microbiol. Rev.* 53(4):450–90.

Voet, D., and Voet, J. G. 1990. *Biochemistry.* New York: John Wiley and Sons.

Watson, J. D.; Hopkins, N. H.; Roberts, J. W.; Steitz, J. A.; and Weiner, A. M. 1987. *Molecular biology of the gene,* 4th ed. Menlo Park, Calif.: Benjamin/Cummings.

Weintraub, H. M. 1990. Antisense RNA and DNA. *Sci. Am.* 262(1):40–46.

Whittenbury, R., and Dow, C. S. 1978. Morphogenesis in bacteria. In *Companion to microbiology,* eds. A. T. Bull and P. M. Meadow, 221–63. New York: Longman.

Yanofsky, C. 1981. Attenuation in the control of expression of bacterial operons. *Nature* 289:751–58.

Part Three
Microbial Genetics

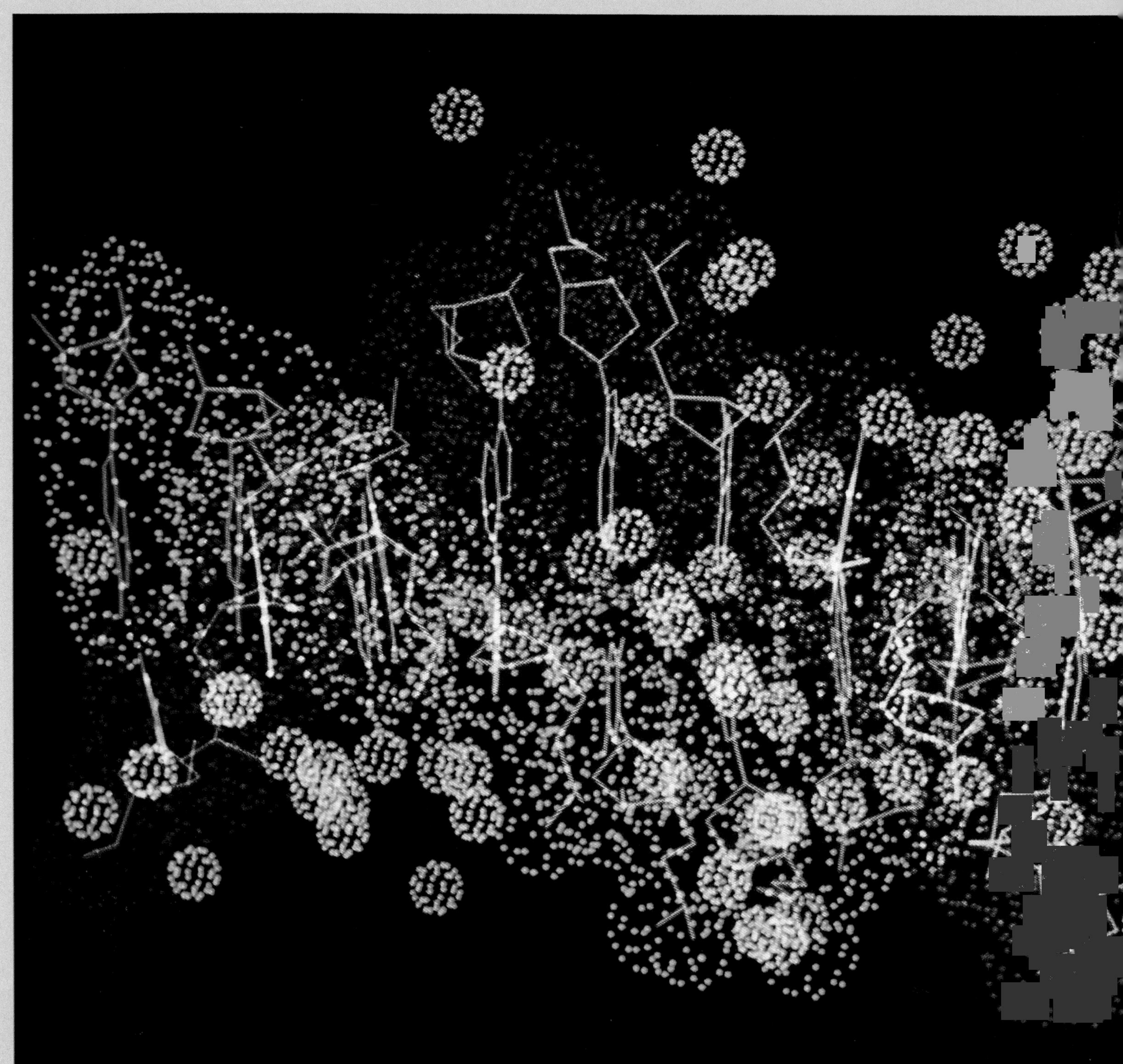

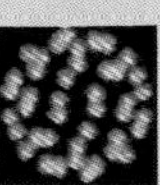

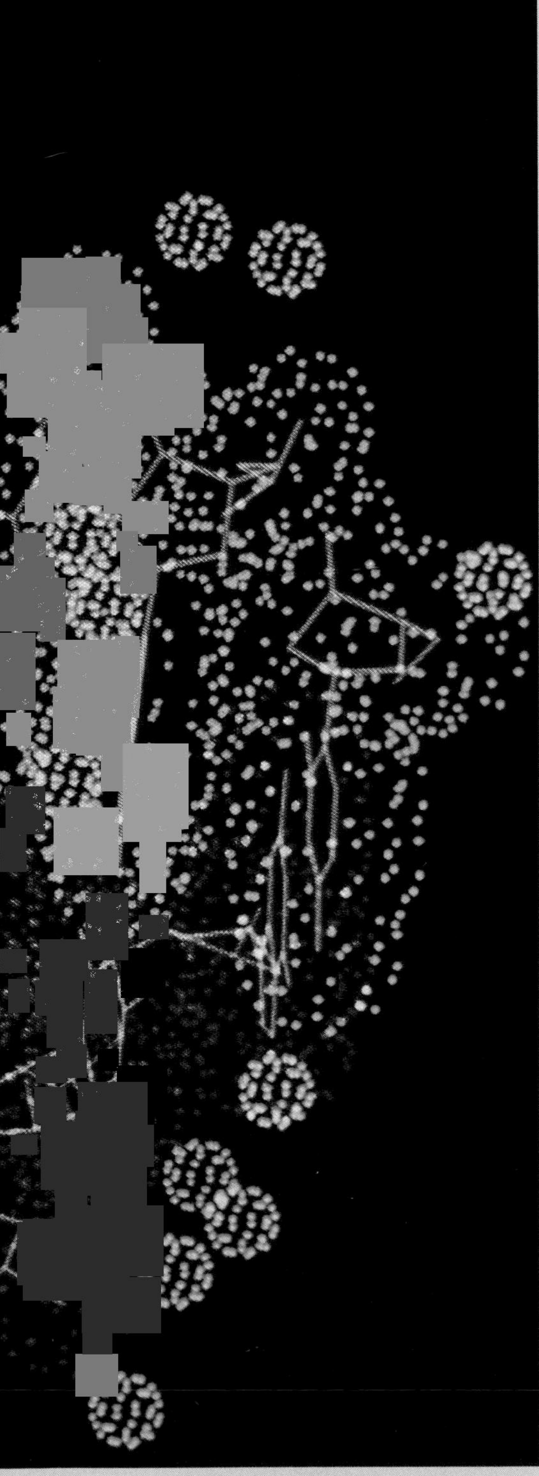

The DNA double helix. This computer graphics image shows DNA double helical structure. DNA consists of two linked polynucleotide strands (blue and red) associated through complementary base pairing and coiled into a helix. Adjacent purine and pyrimidine bases (orange and purple) hydrogen bond with each other. The green spheres show where water molecules are likely to bind to the outside of the double helix.

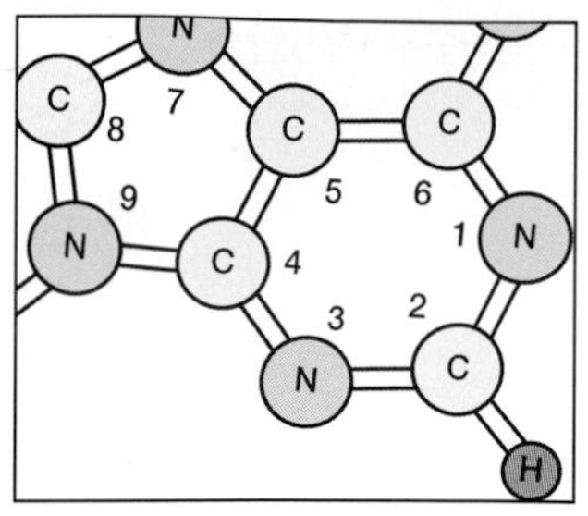

CHAPTER 12
Microbial Genetics: General Principles

But the most important qualification of bacteria for genetic studies is their extremely rapid rate of growth. . . . a single E. coli *cell will grow overnight into a visible colony containing millions of cells, even under relatively poor growth conditions. Thus, genetic experiments on* E. coli *usually last one day, whereas experiments on corn, for example, take months. It is no wonder that we know so much more about the genetics of* E. coli *than about the genetics of corn, even though we have been studying corn much longer.*

—R. F. Weaver and P. W. Hedrick

Outline

Concepts

1. Genetic information is contained in the nucleotide sequence of DNA (and sometimes RNA). When a structural gene directs the synthesis of a polypeptide, each amino acid is specified by a triplet codon.
2. A gene is a nucleotide sequence that codes for a polypeptide, tRNA, or rRNA. Controlling elements such as promoters and operators are often considered part of the gene.
3. Most bacterial genes have at least four major parts, each with different functions: promoters, leaders, coding regions, and trailers.
4. Mutations are stable, heritable alterations in the gene sequence and usually, but not always, produce phenotypic changes. Nucleic acids are altered in several different ways, and these mutations may be either spontaneous or induced by chemical mutagens or radiation.
5. It is extremely important to keep the nucleotide sequence constant, and microorganisms have several repair mechanisms designed to detect alterations in the genetic material and restore it to its original state. Often, more than one repair system can correct a particular type of mutation. Despite these efforts, some alterations remain uncorrected and provide material for evolutionary change.

The preceding chapters cover background information and concepts helpful in understanding microbial genetics: DNA, RNA, and protein synthesis; and the regulation of metabolism. This chapter reviews some basic concepts of molecular genetics: how genetic information is stored and organized in the DNA molecule, mutagenesis, and DNA repair. In addition, the use of microorganisms to identify potentially dangerous mutagenic agents in the fight against cancer is described. Much of this information will be familiar to those who have taken an introductory genetics course. Because of the importance of bacteria, primary emphasis is placed on their genetics.

Based on the foundation provided by this chapter, chapter 13 contains information on plasmids and the nature of genetic recombination in microorganisms. The section on microbial genetics concludes with an introduction to recombinant DNA technology in chapter 14.

Geneticists, including microbial geneticists, use a specialized vocabulary because of the complexities of their discipline. Some knowledge of basic terminology is necessary at the beginning of this survey of general principles. The experimental material of the microbial geneticist is the **strain** or **clone.** A clone is a population of cells that are genetically identical. Sometimes a clone is called a pure culture. The term **genome** refers to all the genes present in a cell or virus. Bacteria normally have one set of genes. That is, they are haploid. Eucaryotic microorganisms usually have two sets of genes, or are diploid (2N). The **genotype** of an organism is the specific set of genes it possesses. In contrast, the **phenotype** is the collection of characteristics that are observable by the investigator. All genes are not expressed at the same time, and the environment profoundly influences phenotypic expression. Much genetics research has focused on the relationship between an organism's genotype and phenotype.

Although genetic analysis began with the rediscovery of the work of Gregor Mendel in the early part of this century, the subsequent elegant experimentation involving both bacteria and bacteriophages actually elucidated the nature of genetic information, gene structure, the genetic code, and mutations. A few of these early experiments are reviewed before the genetic code itself is examined.

DNA as the Genetic Material

The early work of Fred Griffith in 1928 on the transfer of virulence in the pathogen *Streptococcus pneumoniae* (figure 12.1) set the stage for the research that first showed that DNA was

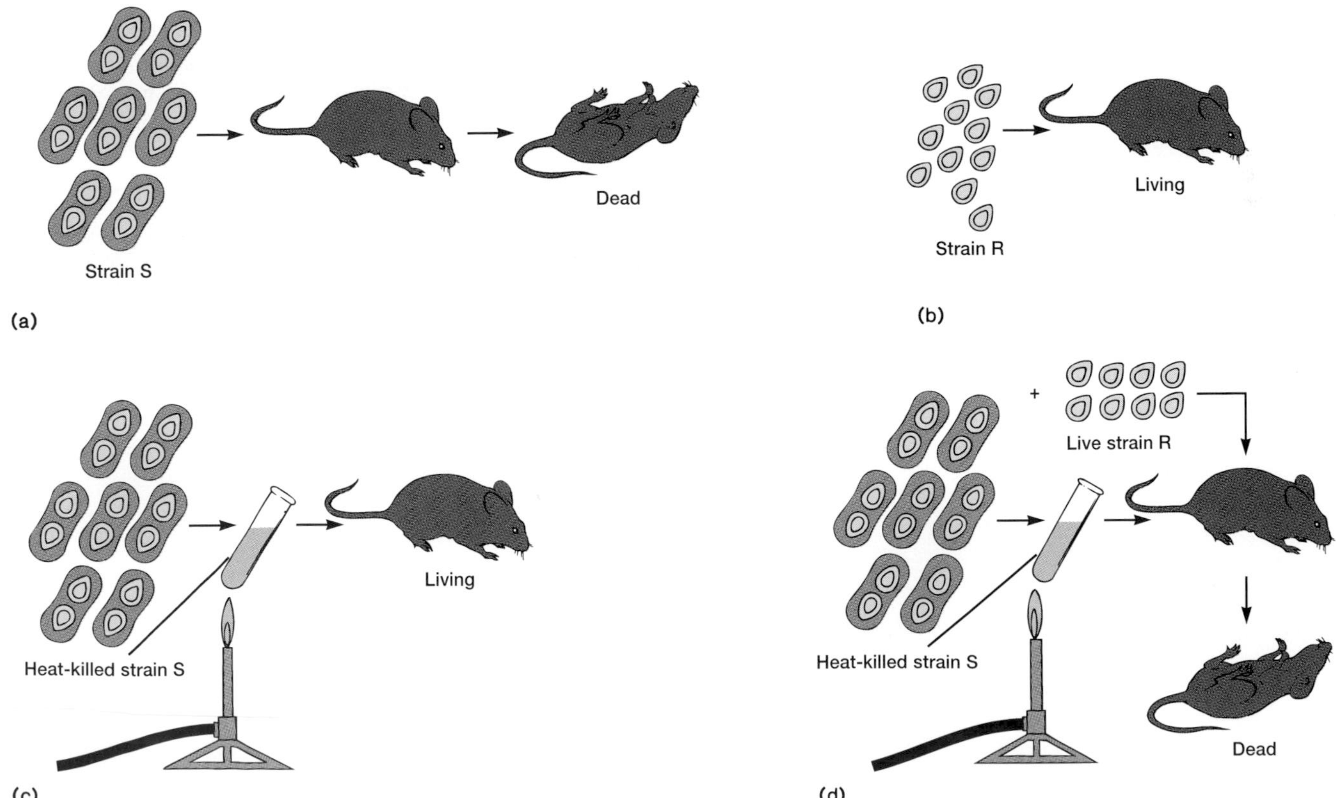

Figure 12.1 Griffith's Transformation Experiments. (*a*) Mice died of pneumonia when injected with pathogen strain S pneumococci, which have a capsule and form smooth-looking colonies. (*b*) Mice survived when injected with nonpathogenic strain R pneumococci, which lack a capsule and form rough colonies. (*c*) Injection with heat-killed S pneumococci had no effect. (*d*) Injection with live R and heat-killed S gave the mice pneumonia, and live S strain pneumococci could be isolated from the dead mice *Source: R. Sage and J. Ryan,* Cell Heredity, *1961.*

the genetic material. Griffith found that if he boiled virulent bacteria and injected them into mice, the mice were not affected and no pneumococcus could be recovered from the animals. When he injected a combination of killed virulent bacteria and a living nonvirulent strain, the mice died; moreover, he could recover living virulent bacteria from the dead mice. Griffith called this change of nonvirulent bacteria into virulent pathogens **transformation.**

Oswald T. Avery and his colleagues then set out to discover which constituent in the heat-killed virulent pneumococcus was responsible for Griffith's transformation. These investigators selectively destroyed cell constituents in extracts of virulent pneumococcus, using enzymes that would hydrolyze DNA, RNA, or protein. They then exposed nonvirulent pneumococcus strains to the treated extracts. Transformation of the nonvirulent bacteria was blocked only if the DNA was destroyed, suggesting that DNA was carrying the information required for transformation (figure 12.2). The publication of these studies by O. T. Avery, C. M. MacLeod, and M. J. McCarty in 1944 provided the first evidence that Griffith's

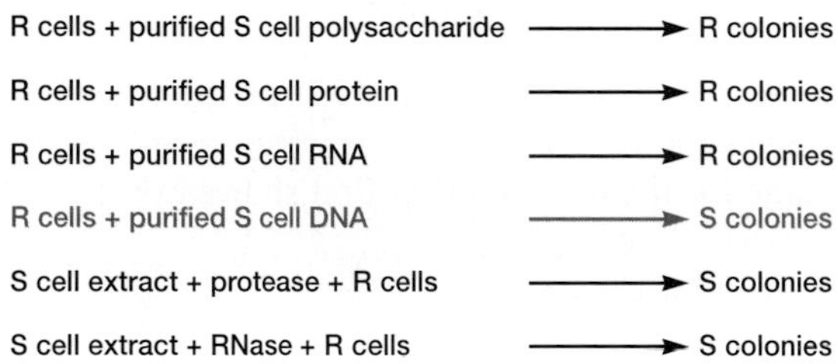

R cells + purified S cell polysaccharide → R colonies
R cells + purified S cell protein → R colonies
R cells + purified S cell RNA → R colonies
R cells + purified S cell DNA → S colonies
S cell extract + protease + R cells → S colonies
S cell extract + RNase + R cells → S colonies

Figure 12.2 Experiments on the Transforming Principle. Summary of the experiments of Avery, MacLeod, and McCarty on the transforming principle. DNA alone changed R to S cells, and this effect was lost when the extract was treated with deoxyribonuclease. Thus, DNA carried the genetic information required for the R to S conversion or transformation.

transforming principle was DNA and therefore that DNA carried genetic information.

Some years later (1952), Alfred D. Hershey and Martha Chase performed several experiments that proved that DNA was the genetic material in the T2 bacteriophage. Some luck was involved in their discovery, for the genetic material of many viruses is RNA and the researchers happened to select a DNA virus for their studies. Imagine the confusion if T2 had been an RNA virus. The controversy surrounding the nature of genetic information might have lasted considerably longer than it did. They made the virus DNA radioactive with ^{32}P or labeled the viral protein coat with ^{35}S. When the radioactive viruses were mixed with *E. coli,* the radioactive DNA was injected into the *E. coli* host cell by the virus, while most of the protein remained outside (figure 12.3). Since genetic material was injected and T2 progeny were produced, DNA must have been carrying the genetic information for T2.

The biology of bacteriophages (chapter 18).

Subsequent studies on the genetics of viruses and bacteria were largely responsible for the rapid development of molecular genetics. Furthermore, much of the new recombinant DNA technology (*see chapter 14*) has arisen from recent progress in bacterial and viral genetics. Research in microbial genetics has had a profound impact on biology as a science and on the technology that affects everyday life.

1. Define strain or clone, genome, genotype, and phenotype.
2. Briefly summarize the experiments of Griffith; Avery, MacLeod, and McCarty; and Hershey and Chase. What did each show, and why were these experiments important to the development of microbial genetics?

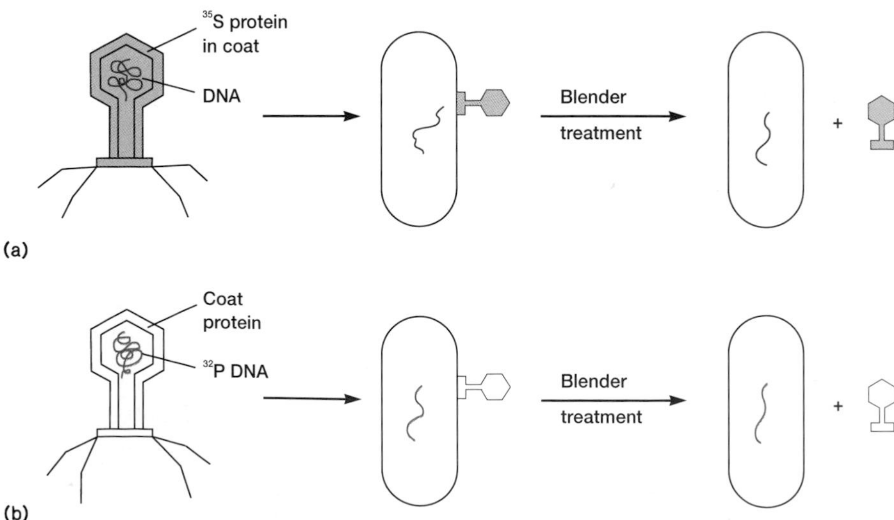

Figure 12.3 The Hershey-Chase Experiment. (*a*) When *E. coli* was infected with T2 phage containing ^{35}S protein, most of the radioactivity remained outside the host cell. (*b*) When T2 phage containing ^{32}P DNA was mixed with the host bacterium, the radioactive DNA was injected into the cell and phages were produced. Thus, DNA was carrying the virus's genetic information.

TABLE 12.1 The Genetic Code

First Position (5′ End)[a]	Second Position: U		C		A		G		Third Position (3′ End)
U	UUU	Phe	UCU	Ser	UAU	Tyr	UGU	Cys	U
	UUC		UCC		UAC		UGC		C
	UUA	Leu	UCA		UAA	STOP	UGA	STOP	A
	UUG		UCG		UAG		UGG	Trp	G
C	CUU	Leu	CCU	Pro	CAU	His	CGU	Arg	U
	CUC		CCC		CAC		CGC		C
	CUA		CCA		CAA	Gln	CGA		A
	CUG		CCG		CAG		CGG		G
A	AUU	Ile	ACU	Thr	AAU	Asn	AGU	Ser	U
	AUC		ACC		AAC		AGC		C
	AUA		ACA		AAA	Lys	AGA	Arg	A
	AUG	Met	ACG		AAG		AGG		G
G	GUU	Val	GCU	Ala	GAU	Asp	GGU	Gly	U
	GUC		GCC		GAC		GGC		C
	GUA		GCA		GAA	Glu	GGA		A
	GUG		GCG		GAG		GGG		G

From Leland G. Johnson, *Biology,* 2d ed. Copyright © 1987 Wm. C. Brown Communications, Inc., Dubuque, Iowa. All Rights Reserved. Reprinted by permission.

[a]The code is presented in the RNA form. Codons run in the 5′ to 3′ direction. See text for details.

The Genetic Code

The realization that DNA is the genetic material triggered efforts to understand how genetic instructions are stored and organized in the DNA molecule. Early studies on the nature of the genetic code showed that the DNA base sequence corresponds to the amino acid sequence of the polypeptide specified by the gene. That is, the nucleotide and amino acid sequences are colinear. It also became evident that many mutations are the result of changes of single amino acids in a polypeptide chain. However, the exact nature of the code was still unclear.

The structure of DNA and RNA (pp. 192–95).

Establishment of the Genetic Code

Since only 20 amino acids are normally present in proteins, there must be at least 20 different code words in a linear, single strand of DNA. The code must be contained in some sequence of the four nucleotides commonly found in the linear DNA sequence. There are only 16 possible combinations (4^2) of the four nucleotides if only nucleotide pairs are considered, not enough to code for all 20 amino acids. Therefore, a code word, or **codon,** must involve at least nucleotide triplets even though this would give 64 possible combinations (4^3), many more than the minimum of 20 needed to specify the common amino acids.

The actual codons were discovered in the early 1960s through the experiments carried out by Marshall Nirenberg, Heinrich Matthaei, Philip Leder, and Har Gobind Khorana. In 1968, Nirenberg and Khorana shared the Nobel Prize with Robert W. Holley, the first person to sequence a nucleic acid (phenylalanyl tRNA).

Organization of the Code

The genetic code, presented in RNA form, is summarized in table 12.1. Note that there is **code degeneracy.** That is, there are up to six different codons for a given amino acid. Only 61 codons, the **sense codons,** direct amino acid incorporation into protein. The remaining three codons (UGA, UAG, and UAA) are involved in the termination of translation and are called **stop** or **nonsense codons.** Despite the existence of 61 sense codons, there are not 61 different tRNAs, one for each codon. The 5′ nucleotide in the anticodon can vary, but generally, if the nucleotides in the second and third anticodon positions complement the first two bases of the mRNA codon, an aminoacyl-tRNA with the proper amino acid will bind to the

(a) Base pairing of one glycine tRNA with three codons due to wobble

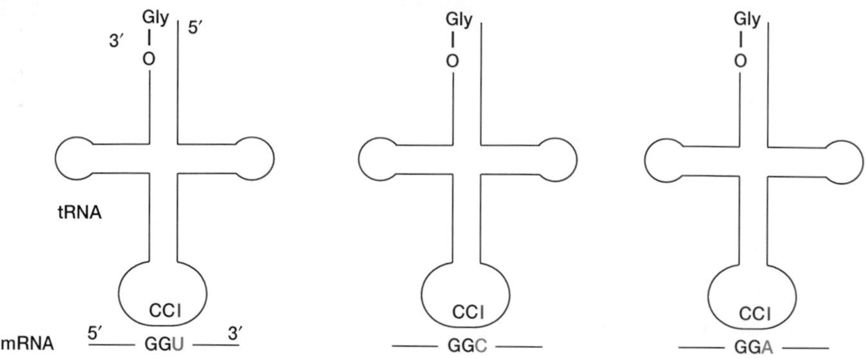

(b) Glycine codons and anticodons (written in the 5′ ⟶ 3′ direction)

Glycine mRNA codons: GGU, GGC, GGA, GGG

Glycine tRNA anticodons: ICC, CCC

Figure 12.4 **Wobble and Coding.** The use of wobble in coding for the amino acid glycine. (*a*) Inosine (*I*) is a wobble nucleoside that can base pair with uracil (*U*), cytosine (*C*), or adenine (*A*). Thus, ICC base pairs with GGU, GGC, and GGA in the mRNA. (*b*) Because of the wobble produced by inosine, two tRNA anticodons can recognize the four glycine (*Gly*) codons. ICC recognizes GGU, GGC, and GGA; CCC recognizes GGG.

mRNA-ribosome complex. This pattern is evident upon inspection of changes in the amino acid specified with variation in the third position (table 12.1). This somewhat sloppy base pairing is known as **wobble** and relieves cells of the need to synthesize so many tRNAs (figure 12.4). Wobble also decreases the ill effects of DNA mutations.

The mechanism of protein synthesis and tRNA function (pp. 204–13).

1. Why must a codon contain at least three nucleotides?
2. Define the following: code degeneracy, sense codon, stop or nonsense codon, and wobble.

Gene Structure

The **gene** has been defined in several ways. Initially, geneticists considered it to be the entity responsible for conferring traits on the organism and the entity that could undergo recombination. Recombination involves exchange of DNA from one source with that from another (*see chapter 13*) and is responsible for generating much of the genetic variability found in viruses and living organisms. Genes were typically named for some mutant or altered phenotype. With the discovery and characterization of DNA, the gene was defined more precisely as a linear sequence of nucleotides or codons (this term can be used for RNA as well as DNA) with a fixed start point and end point.

At first, it was thought that a gene contained information for the synthesis of one enzyme, the one gene-one enzyme hypothesis. This has been modified to the one gene-one polypeptide hypothesis because of the existence of enzymes and other proteins composed of two or more different polypeptide chains coded for by separate genes. The segment that codes for a single polypeptide is sometimes also called a **cistron.** More recent results show that even this description is oversimplified. Not all genes are involved in protein synthesis; some code instead for rRNA and tRNA. Thus, a gene might be defined as a polynucleotide sequence that codes for a polypeptide, tRNA, or rRNA. Some geneticists think of it as a segment of nucleic acid that is transcribed to give an RNA product. Often, controlling elements such as promoters are considered part of the gene since genes usually have noncoding regulatory sequences associated with them. Most genes consist of discrete sequences of codons that are "read" only one way to produce a single product. That is, the code is not overlapping. Chromosomes therefore usually consist of gene sequences that do not overlap one another (figure 12.5*a*). However, some viruses such as the phage ϕX174 do have overlapping genes (figure 12.5*b*), and parts of genes overlap in some bacterial genomes.

Procaryotic and viral gene structure differs greatly from that of eucaryotes. In bacterial and viral systems, the coding information within a cistron is normally continuous; however, in eucaryotic organisms, many if not most genes contain coding information (exons) interrupted periodically by noncoding sequences (introns). An interesting exception to this rule is eucaryotic histone genes, which lack introns. Because procaryotic and viral systems are the best characterized, the more detailed description of gene structure that follows will focus on *E. coli* genes.

Exons and introns in eucaryotic genes (p. 203).

Genes That Code for Proteins

Recall from the discussion of transcription that although DNA is double stranded, only one strand contains coded information

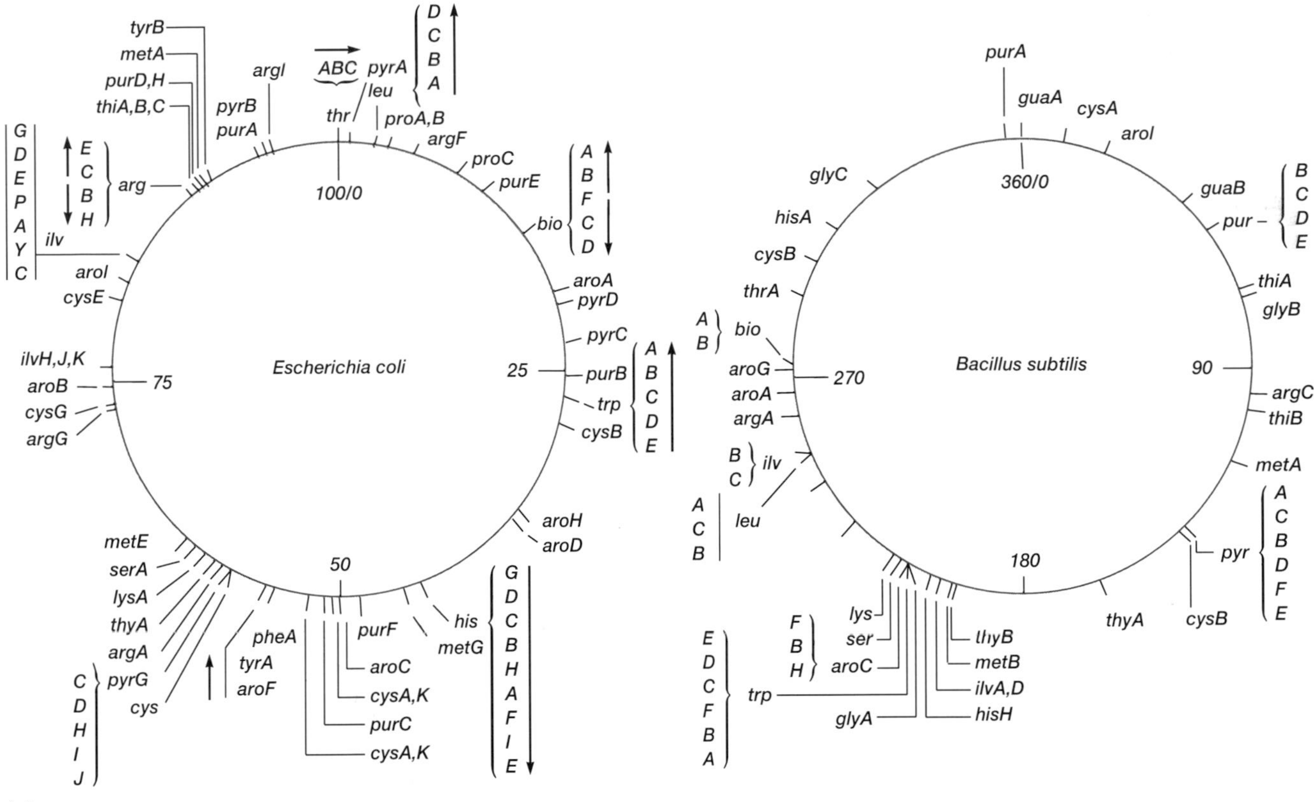

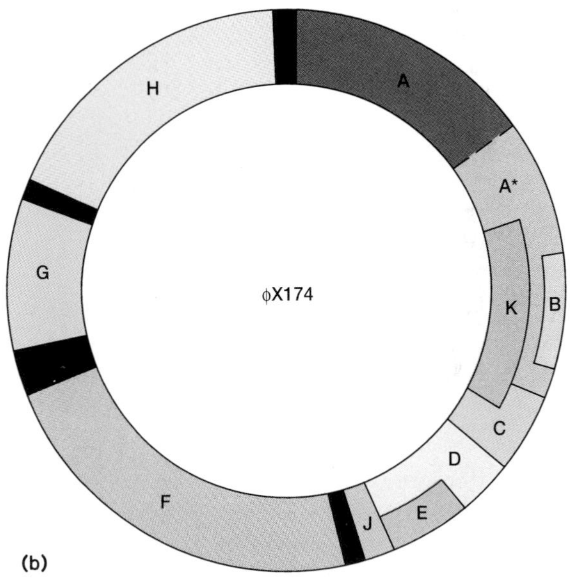

Figure 12.5 **Chromosomal Organization in Bacteria and Viruses.** (*a*) Simplified genetic maps of *E. coli* and *B. subtilis* indicating some of the genes they share. The *E. coli* map is divided into 100 minutes, the time required to transfer the chromosome from an Hfr cell to F^- at 37° C. The *B. subtilis* map is divided into 360°. (*b*) The map of phage ϕX174 showing the overlap of gene B with A, K with A and C, and E with D. The solid regions are spaces lying between genes. Protein A* consists of the last part of protein A and arises from reinitiation of transcription within gene A.

and directs RNA synthesis. This strand is called the **sense strand,** and the complementing strand is known as the antisense strand (figure 12.6). Since the mRNA is made from the 5′ to the 3′ end, the polarity of the DNA sense strand is therefore 3′ to 5′. Therefore, the beginning of the gene is at the 3′ end of the sense strand (also the 5′ end of the antisense strand). An RNA polymerase recognition/binding and regulatory site known as the **promoter** is located at the start of the gene (*see chapter 10*).

The mechanism of transcription (pp. 201–4).

Promoters are sequences of DNA on the sense strand that are usually upstream from the actual coding or transcribed region (figure 12.6); that is, the promoter is located before or upstream from the coding region in relationship to the direction of transcription (the direction of transcription is referred to as downstream). In *E. coli,* the promoter has two important functions, and these relate to two specific segments within the promoter (figures 12.6 and 12.7). Although these two segments do vary slightly in sequence between bacterial strains, they are fairly constant and may be represented by consensus sequences. These are idealized sequences composed of the bases most often found at each position when the sequences from different bacteria are compared. The RNA polymerase recognition site, with a consensus sequence of 5′TTGACA3′ on the antisense strand, is located about 35 base pairs before (the −35 region) the transcriptional start point (labeled as +1) of RNA synthesis. This sequence seems to be the site of the initial association of the RNA polymerase with the DNA. The

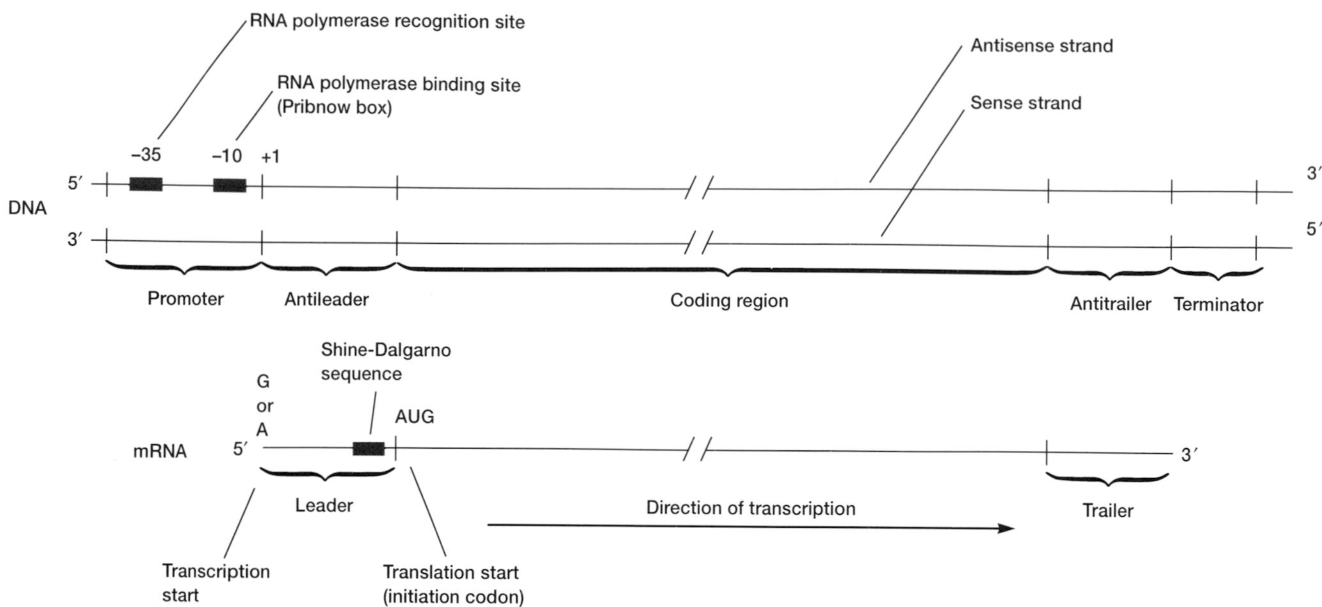

Figure 12.6 **A Bacterial Structural Gene.** The organization of a typical structural gene in bacteria. Leader and trailer sequences are included even though some genes lack one or both. Transcription begins at the +1 position in DNA, and the first nucleotide incorporated into mRNA is usually GTP or ATP. Translation of the mRNA begins with the AUG initiation codon. Regulatory sites are not shown.

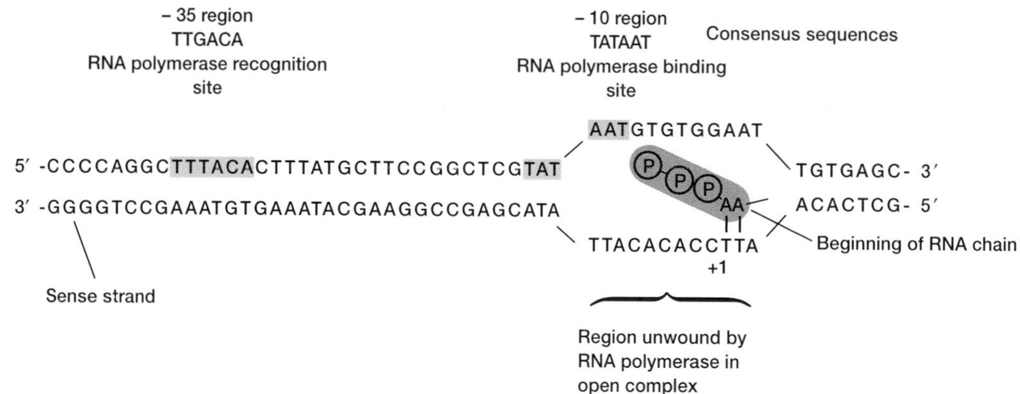

Figure 12.7 **A Bacterial Promoter.** The lactose operon promoter UV5 and its consensus sequences. The start point for RNA synthesis is labeled +1. The region around −35 is the site at which the RNA polymerase first attaches to the promoter. RNA polymerase binds and begins to unwind the DNA helix at the Pribnow box or RNA polymerase binding site, which is located in the −10 region.

RNA polymerase binding site, also known as the **Pribnow box,** is located at the −10 region and has a consensus sequence 5′TATAAT3′ (a sequence that favors the localized unwinding of DNA). This is where the RNA polymerase begins to unwind the DNA for eventual transcription. The initially transcribed portion of the gene is not necessarily sense or coding material. Rather, a **leader sequence** may be synthesized first. The leader is usually a nontranslated sequence that is important in the initiation of translation and sometimes is involved in regulation of transcription.

The leader (figure 12.6) in procaryotes generally contains a consensus sequence known as the **Shine-Dalgarno sequence,** 5′AGGA3′, the transcript of which complements a sequence on the 16S rRNA in the small subunit of the ribosome. The binding of mRNA leader with 16S rRNA properly orients the mRNA on the ribosome. The leader also sometimes regulates transcription by attenuation (*see chapter 11*). Downstream and next to the leader is the most important part of the structural gene, the coding region.

The **coding region** (figure 12.6) of genes that direct the synthesis of proteins typically begins with the sense DNA sequence 3′TAC5′. This produces the RNA translation initiation codon 5′AUG3′, which codes for N-formylmethionine. This modified form of methionine is the first amino acid incorporated in most procaryotic proteins. The remainder of the coding region of the gene consists of a sequence of codons that specifies the sequence of amino acids for that particular protein. Transcription does not stop at the translation stop codon but rather at a **terminator sequence** (*see chapter 10*). The terminator often lies after a nontranslated **trailer sequence** lo-

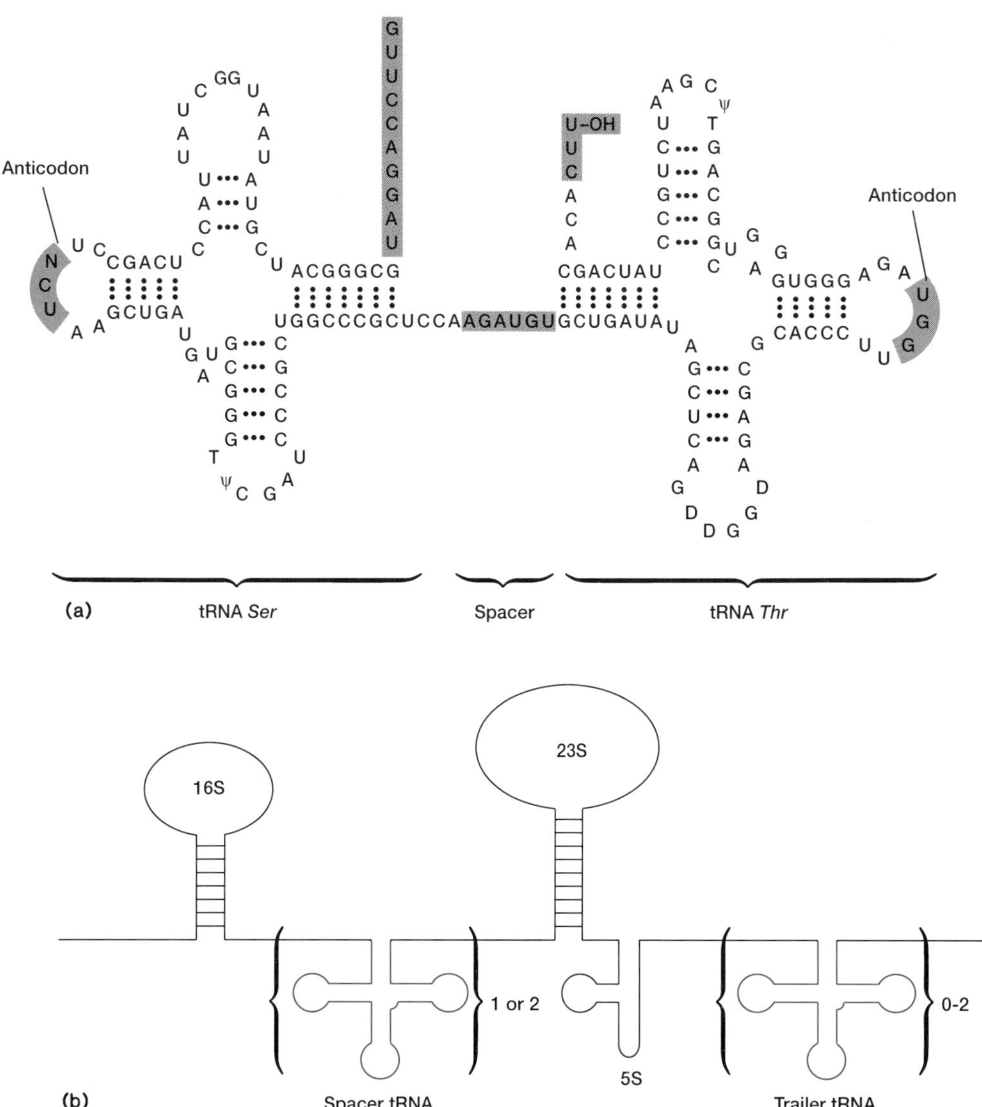

Figure 12.8 tRNA and rRNA Genes. (*a*) A tRNA precursor from *E. coli* that contains two tRNA molecules. The spacer and extra nucleotides at both ends are removed during processing. (*b*) The *E. coli* ribosomal RNA gene codes for a large transcription product that is cleaved into three rRNAs and 1–3 tRNAs. The 16S, 23S, and 5S rRNA segments are represented by red lines, and tRNA sequences are placed in brackets. The seven copies of this gene vary in the number and kind of tRNA sequences.

cated downstream from the coding region (figure 12.6). The trailer sequence, like the leader, is needed for the proper expression of the sense portion of the gene.

Besides the basic components described above—the promoter, leader, coding region, trailer, and terminator—many procaryotic genes have a variety of regulatory sites. These are locations where DNA-recognizing regulatory proteins bind to stimulate or prevent gene expression. Regulatory sites are often associated with promoter function, and some consider them to be parts of special promoters. Two such sites, the operator and the CAP binding site, are discussed in chapter 11. Certainly everything is not known about genes and their structure. With the ready availability of purified cloned genes and DNA sequencing technology, major discoveries continue to be made in this area.

The operon and transcription regulation (pp. 222–26).

Genes That Code for tRNA and rRNA

The DNA segments that code for tRNA and rRNA are also considered genes, although they give rise to structurally important RNA rather than protein. In *E. coli,* the genes for tRNA are fairly typical, consisting of a promoter and transcribed leader and trailer sequences that are removed during the maturation process (figure 12.8*a*). The precise function of the leader is not clear; however, the trailer is required for termination. Genes coding for tRNA may code for more than a single tRNA molecule or type of tRNA (figure 12.8*a*). The segments coding for tRNAs are separated by short spacer sequences that are removed after transcription by special ribonucleases, at least one of which contains catalytic RNA.

RNA splicing and ribozymes (pp. 203–5).

The genes for rRNA are also similar in organization to genes coding for proteins and have promoters, trailers, and terminators (figure 12.8*b*). Interestingly, all the rRNAs are transcribed as a single, large precursor molecule that is cut up by ribonucleases after transcription to yield the final rRNA products. *E. coli* pre-rRNA spacer and trailer regions even contain tRNA genes. Thus, the synthesis of tRNA and rRNA involve posttranscriptional modification, a relatively rare process in procaryotes.

1. Define or describe the following: gene, sense and antisense strands, promoter, consensus sequence, RNA polymerase recognition and binding sites, Pribnow box, leader, Shine-Dalgarno sequence, coding region, trailer, and terminator.
2. How do the genes of procaryotes and eucaryotes usually differ from each other?
3. Briefly discuss the general organization of tRNA and rRNA genes. How does their expression differ from that of structural genes with respect to posttranscriptional modification of the gene product?

Mutations and Their Chemical Basis

Considerable information is embedded in the precise order of nucleotides in DNA. For life to exist with stability, it is essential that the nucleotide sequence of genes is not disturbed to any great extent. However, sequence changes do occur and often result in altered phenotypes. These changes are largely detrimental but are important in generating new variability and contribute to the process of evolution. Microbial mutation rates also can be increased, and these genetic changes have been put to many important uses in the laboratory and industry.

Mutations (Latin *mutare,* to change) were initially characterized as altered phenotypes or phenotypic expressions. Long before the existence of direct proof that a mutation is the result of a stable, heritable change in the nucleotide sequence of DNA, geneticists predicted that several basic types of transmitted mutations could exist. They believed that mutations could arise from the alteration of single pairs of nucleotides and from the addition or deletion of one or two nucleotide pairs in the coding regions of a gene. In this section, the molecular basis of mutations and mutagenesis is first considered. Then the phenotypic effects of mutations, the detection of mutations, and the use of mutations in carcinogenicity testing are discussed.

Mutations and Mutagenesis

Mutations can alter the phenotype of a microorganism in several different ways. Morphological mutations change the colonial or cellular morphology. Lethal mutations, when expressed, result in the death of the microorganism. Since the microorganism must be able to grow in order to be isolated and studied, lethal mutations are recovered only if they are recessive in diploid organisms or conditional (see the following) in haploid organisms.

Conditional mutations are those that are expressed only under certain environmental conditions. For example, a conditional lethal mutation in *E. coli* might not be expressed under permissive conditions such as low temperature but would be expressed under restrictive conditions such as high temperature. Thus, the hypothetical mutant would grow normally at the permissive temperature but would die at high temperatures.

Biochemical mutations are those causing a change in the biochemistry of the cell. Since these mutations often inactivate a biosynthetic pathway, they frequently make a microorganism unable to grow on medium lacking an adequate supply of the pathway's end product (figure 12.15). That is, the mutant cannot grow on minimal medium and requires nutrient supplements. Such mutants are called **auxotrophs,** while microbial strains that can grow on minimal medium are **prototrophs.** Analysis of auxotrophy has been very important in microbial genetics due to the ease of auxotroph selection and the relative abundance of this mutational type.

Mutant detection and replica plating (pp. 250–51).

Nutrient requirements and nutritional types (pp. 97–99).

A resistant mutant is a particular type of biochemical mutant that acquires resistance to some pathogen, chemical, or antibiotic. Such mutants are also easy to select for and are very useful in microbial genetics.

Mechanisms of drug resistance (pp. 338–39).

Mutations occur in one of two ways. (1) Spontaneous mutations arise occasionally in all cells and develop in the absence of any added agent. (2) Induced mutations, on the other hand, are the result of exposure of the organism to some physical or chemical agent called a **mutagen.**

Spontaneous Mutations

Spontaneous mutations arise from several potential sources. This class of mutations may result from errors in DNA replication, damage to DNA from factors such as gamma radiation and heat, or even the action of transposons (*see chapter 13*). A few of the more prevalent mechanisms are described in the following paragraphs.

Generally, errors in replication occur when the base of a template nucleotide takes on a rare tautomeric form. Tautomerism is the relationship between two structural isomers that are in chemical equilibrium and readily change into one another. Bases typically exist in the keto form. However, they can at times take on either an imino or enol form (figure 12.9*a*). These tautomeric shifts change the hydrogen-bonding characteristics of the bases, allowing purine for purine or pyrimidine for pyrimidine substitutions that can eventually lead to a stable alteration of the nucleotide sequence (figure 12.9*b*). Such substitutions are known as **transition mutations** and are

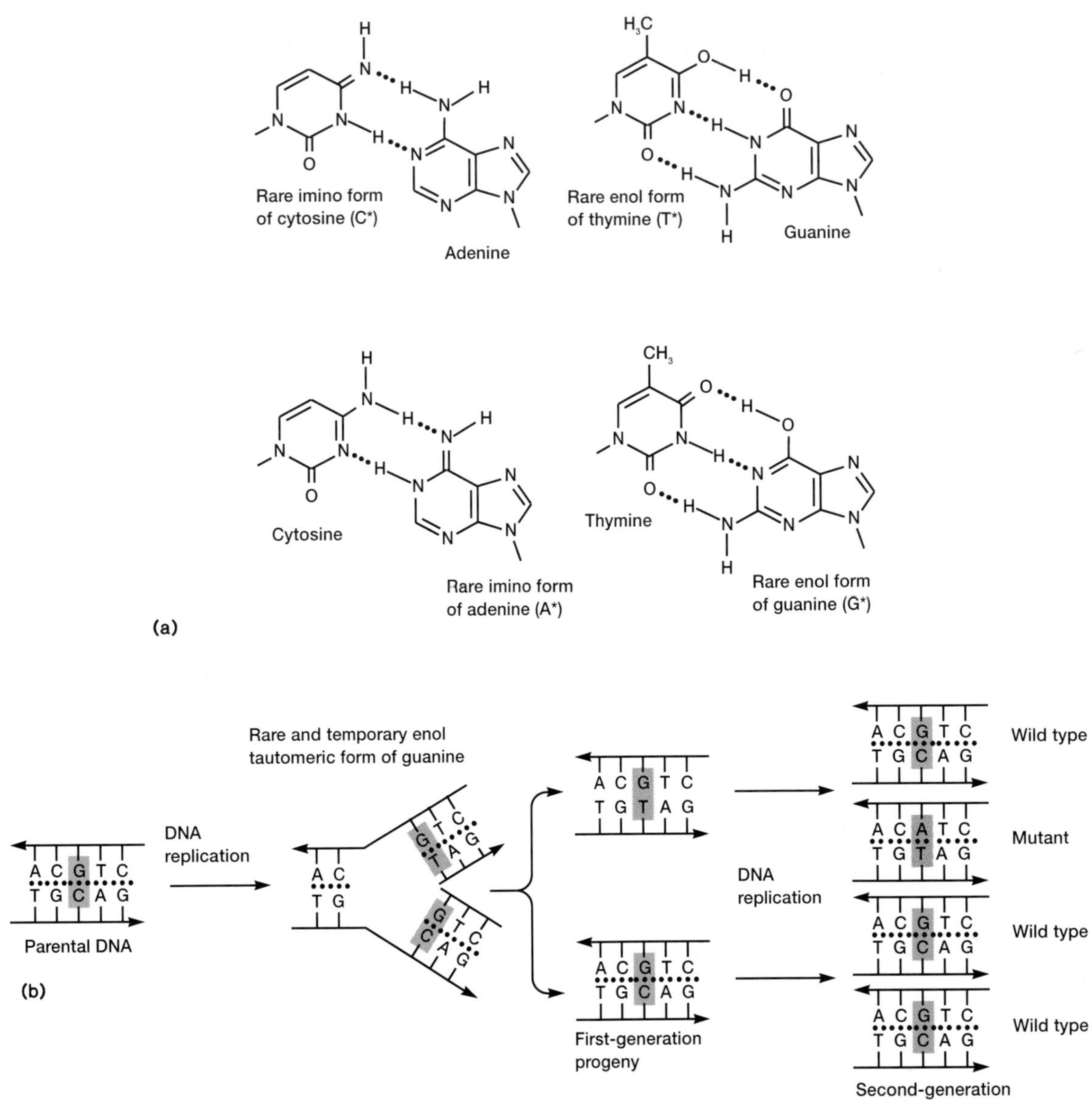

Figure 12.9 Transition and Transversion Mutations. Errors in replication due to base tautomerization. (*a*) Normally A–T and G–C pairs are formed when keto groups participate in hydrogen bonds. In contrast, enol tautomers produce A–C and G–T base pairs. (*b*) Mutation as a consequence of tautomerization during DNA replication. The temporary enolization of guanine leads to the formation of an A–T base pair in the mutant, and a G–C to A–T transition mutation occurs. The process requires two replication cycles. The mutation only occurs if the abnormal first-generation G–T base pair is missed by repair mechanisms.

relatively common, although most of them are repaired by various proofreading functions (*see pp. 199 and 253*). In **transversion mutations,** a purine is substituted for a pyrimidine, or a pyrimidine for a purine. These mutations are rarer due to the steric problems of pairing purines with purines and pyrimidines with pyrimidines.

Spontaneous mutations also arise from **frameshifts,** usually caused by the deletion of DNA segments resulting in an altered codon reading frame. These mutations generally occur where there is a short repeated nucleotide sequence. In such a location, the pairing of template and new strand can be displaced by the distance of the repeated sequence leading to a deletion of bases in the new strand (figure 12.10).

Spontaneous mutations originate from lesions in DNA as well as from replication errors. For example, it is possible for purine nucleotides to be depurinated, that is, to lose their base. This results in the formation of an **apurinic site,** which will not base pair normally and may cause a transition type mutation after the next round of replication. Cytosine can be deaminated to uracil, which is then removed to form an **apyrimidinic site.**

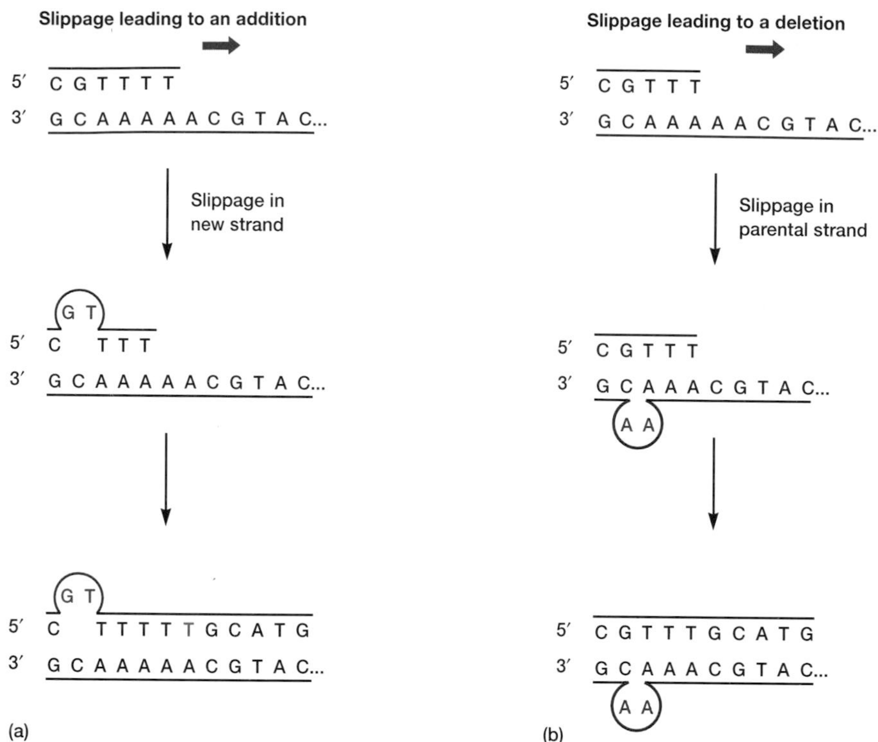

Figure 12.10 **Additions and Deletions.** A hypothetical mechanism for the generation of additions and deletions during replication. The direction of replication is indicated by the large arrow. In each case, there is strand slippage resulting in the formation of a small loop that is stabilized by the hydrogen bonding in the repetitive sequence, the AT stretch in this example. DNA synthesis proceeds to the right in this figure. (*a*) If the new strand slips, an addition of one T results. (*b*) Slippage of the parental strand yields a deletion (in this case, a loss of two Ts).

Induced Mutations

Virtually any agent that directly damages DNA, alters its chemistry, or interferes with repair mechanisms (pp. 253–56) will induce mutations. Mutagens can be conveniently classified according to their mechanism of action. Four common modes of mutagen action are incorporation of base analogs, specific mispairing, intercalation, and bypass of replication.

Base analogs are similar to normal nitrogenous bases and can be incorporated into the growing polynucleotide chain during replication. Once in place, these compounds typically exhibit base pairing properties different from the bases they replace and can eventually cause a stable mutation. A widely used base analog is 5-bromouracil (5-BU), an analog of thymine. It undergoes a tautomeric shift from the normal keto form to an enol much more frequently than does a normal base. The enol forms hydrogen bonds like cytosine and directs the incorporation of guanine rather than adenine (figure 12.11). The mechanism of action of other base analogs is similar to that of 5-bromouracil.

Specific mispairing is caused when a mutagen changes a base's structure and therefore alters its base pairing characteristics. Some mutagens in this category are fairly selective; they preferentially react with some bases and produce a specific kind of DNA damage. An example of this type of mutagen is methyl-nitrosoguanidine, an alkylating agent that adds methyl groups to guanine, causing it to mispair with thymine. A subsequent round of replication could then result in a GC-AT transition (figure 12.12). Other examples of mutagens with this mode of action are the alkylating agent ethylmethanesulfonate and hydroxylamine. Hydroxylamine hydroxylates the C-4 nitrogen of cytosine, causing it to base pair like thymine. There are many other DNA modifying agents that can cause mispairing.

Intercalating agents distort DNA to induce single nucleotide pair insertions and deletions. These mutagens are planar and insert themselves (intercalate) between the stacked bases of the helix. This results in a mutation, possibly through the formation of a loop in DNA. Intercalating agents include acridines such as proflavin and acridine orange.

Many mutagens, and indeed many carcinogens, directly damage bases so severely that hydrogen bonding between base pairs is impaired or prevented and the damaged DNA can no longer act as a template. For instance, UV radiation generates cyclobutane type dimers, usually thymine dimers, between adjacent pyrimidines (figure 12.13). Other examples are ionizing radiation and carcinogens such as aflatoxin B1 and other benzo(a)pyrene derivatives. Such damage to DNA would generally be lethal but may trigger a repair mechanism that restores much of the damaged genetic material, although with considerable error incorporation (*see pp. 253–56*).

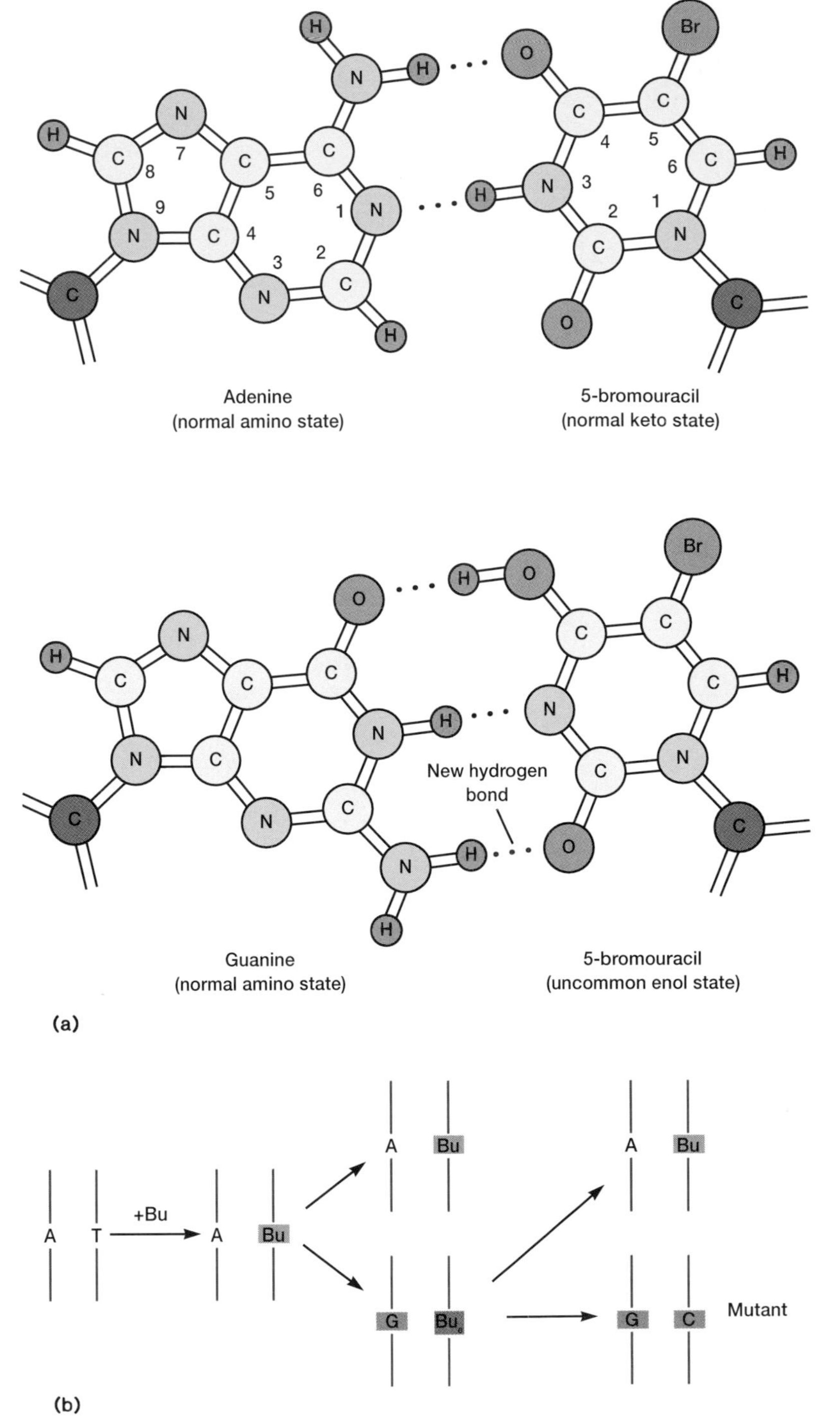

Figure 12.11 **Mutagenesis by the Base Analog 5-Bromouracil.** (*a*) The enol form of 5-BU base pairs with guanine rather than with adenine as might be expected for a thymine analog. (*b*) If the keto form of 5-BU is incorporated in place of thymine, its occasional tautomerization to the enol form (BU_e) will produce an AT to GC transition mutation.

It should be evident that retention of proper base pairing is essential in the prevention of mutations. Often the damage can be repaired before a mutation is permanently established. If a complete DNA replication cycle takes place before the initial lesion is repaired, the mutation frequently becomes stable and inheritable.

The Expression of Mutations

The expression of a mutation will only be noticed if it produces a detectable, altered phenotype. A mutation from the most prevalent gene form, the **wild type,** to a mutant form is called a **forward mutation.** Later, a second mutation may make the

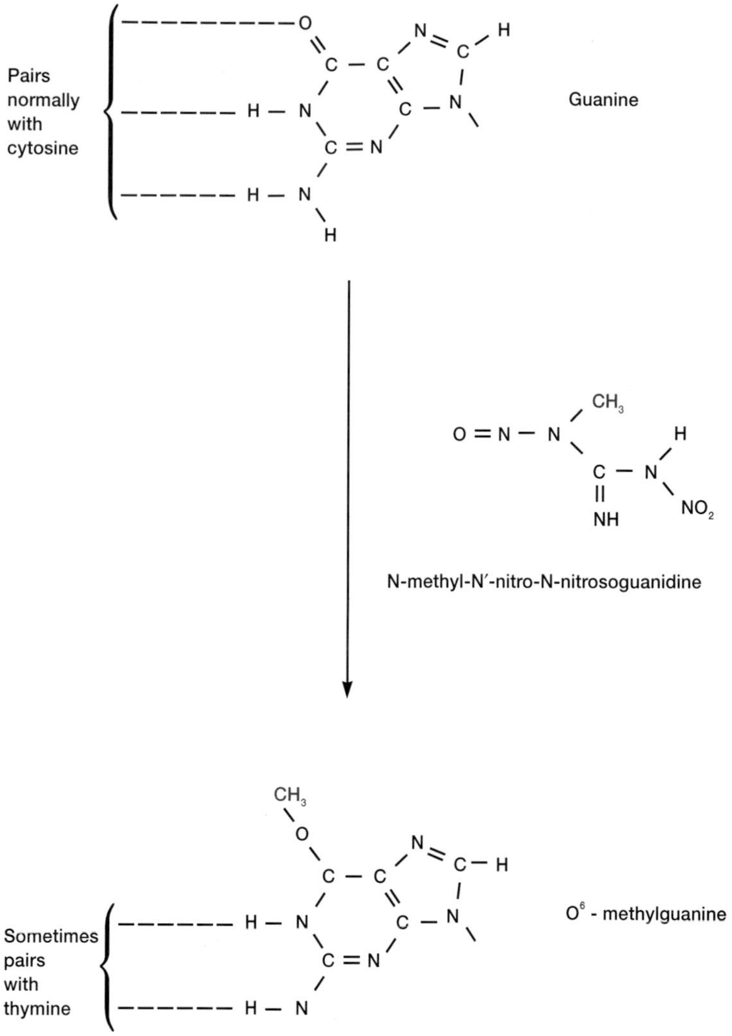

Pairs normally with cytosine

Guanine

N-methyl-N′-nitro-N-nitrosoguanidine

Sometimes pairs with thymine

O^6 - methylguanine

Figure 12.12 **Methyl-Nitrosoguanidine Mutagenesis.** Mutagenesis by methyl-nitrosoguanidine due to the methylation of guanine.

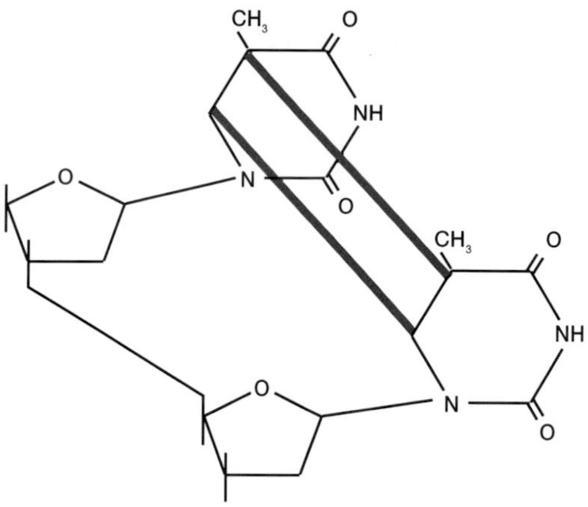

mutant appear to be a wild type organism again. Such a mutation is called a **reversion mutation** because the organism seems to have reverted back to its original phenotype. A true **back mutation** converts the mutant nucleotide sequence back to the wild type arrangement. The wild type phenotype also can be regained by a second mutation in a different gene, a **suppressor mutation,** which overcomes the effect of the first mutation. If the second mutation is within the same gene, the change may be called a second site reversion or intragenic suppression. Thus, although revertant phenotypes appear to be wild types, the original DNA sequence may not be restored (table 12.2). In practice, a mutation is visibly expressed when a protein that is in some way responsible for the phenotype is altered sufficiently to produce a new phenotype. However, mutations may occur and not alter the phenotype for a variety of reasons.

Although very large deletion and insertion mutations exist, most mutations affect only one base pair in a given location and therefore are called **point mutations.** There are several types of point mutations (table 12.2).

Figure 12.13 **Thymine Dimer.** Thymine dimers are formed by ultraviolet radiation. The enzyme photolyase cleaves the two colored bonds during photoreactivation.

One kind of point mutation that could not be detected until the advent of nucleic acid sequencing techniques is the **silent mutation.** If a mutation is an alteration of the nucleotide sequence of DNA, mutations can occur and have no visible effect because of code degeneracy. When there is more than one codon for a given amino acid, a single base substitution could result in the formation of a new codon for the same amino acid. For example, if the codon CGU were changed to CGC, it would still code for arginine even though a mutation had occurred. The expression of this mutation would not be detected except at the level of the DNA or mRNA. Since there would be no change in the protein, there would be no change in the phenotype of the organism.

A second type of point mutation is the **missense mutation.** This mutation involves a single base substitution in the DNA that changes a codon for one amino acid into a codon for another. For example, the codon GAG, which specifies glutamic acid, could be changed to GUG, which codes for valine. The expression of missense mutations can vary. Certainly, the mutation is expressed at the level of protein structure. However, at the level of protein function, the effect may range from complete loss of activity to no change at all.

Many proteins are still functional after the substitution of a single amino acid, but this depends on the type and location of the amino acid. For instance, replacement of a nonpolar amino acid in the protein's interior with a polar amino acid probably will drastically alter the protein's three-dimensional structure and therefore its function. Similarly, the replacement of a critical amino acid at the active site of an enzyme will destroy its activity. However, the replacement of one polar amino acid with another at the protein surface may have little or no effect. Missense mutations may actually play a very important role in providing new variability to drive evolution because they are often not lethal and therefore remain in the gene pool.

Protein structure (appendix I).

TABLE 12.2 Summary of Some Molecular Changes from Gene Mutations

Type of Mutation	Result and Example
Forward Mutations	
Single Nucleotide-Pair (Base-Pair) Substitutions	
At DNA Level	
Transition	Purine replaced by a different purine, or pyrimidine replaced by a different pyrimidine (e.g., AT → GC).
Transversion	Purine replaced by a pyrimidine, or pyrimidine replaced by a purine (e.g., AT → CG).
At Protein Level	
Silent mutation	Triplet codes for same amino acid: AGG → CGG both code for Arg
Neutral mutation	Triplet codes for different but functionally equivalent amino acid: AAA (Lys) → AGA (Arg)
Missense mutation	Triplet codes for a different amino acid.
Nonsense mutation	Triplet codes for chain termination: CAG (Gln) → UAG (stop)
Single Nucleotide-Pair Addition or Deletion: Frameshift Mutation	Any addition or deletion of base pairs that is not a multiple of three results in a frameshift in reading the DNA segments that code for proteins.
Intragenic Addition or Deletion of Several to Many Nucleotide Pairs	
Reverse Mutations	
True Reversion	AAA (Lys) wild type —forward→ GAA (Glu) mutant —reverse→ AAA (Lys) wild type
Equivalent Reversion	UCC (Ser) wild type —forward→ UGC (Cys) mutant —reverse→ AGC (Ser) wild type CGC (Arg, basic) wild type —forward→ CCC (Pro, not basic) mutant —reverse→ CAC (His, basic) pseudo-wild type
Suppressor Mutations	
Intragenic Suppressor Mutations Frameshift of opposite sign at site within gene. Addition of X to the base sequence shifts the reading frame from the CAT codon to XCA followed by TCA codons. The subsequent deletion of a C base shifts the reading frame back to CAT.	CATCATCATCATCATCAT (+) (−) ↓ ↓ CAT XCA TAT CAT CAT CAT y x z y y y
Extragenic Suppressor Mutations	
Nonsense suppressors	Gene (e.g., for tyrosine tRNA) undergoes mutational event in its anticodon region that enables it to recognize and align with a mutant nonsense codon (e.g., UAG) to insert an amino acid (tyrosine) and permit completion of the translation.
Physiological suppressors	A defect in one chemical pathway is circumvented by another mutation—for example, one that opens up another chemical pathway to the same product, or one that permits more efficient uptake of a compound produced in small quantities because of the original mutation.

A third type of point mutation causes the early termination of translation and therefore results in a shortened polypeptide. Such mutations are called **nonsense mutations** because they involve the conversion of a sense codon to a nonsense or stop codon. Depending upon the relative location of the mutation, the phenotypic expression may be more or less severely affected. Most proteins retain some function if they are shortened by only one or two amino acids; complete loss of normal function will almost certainly result if the mutation occurs closer to the middle of the gene.

The **frameshift mutation** is a fourth type of point mutation and was briefly mentioned earlier. Frameshift mutations arise

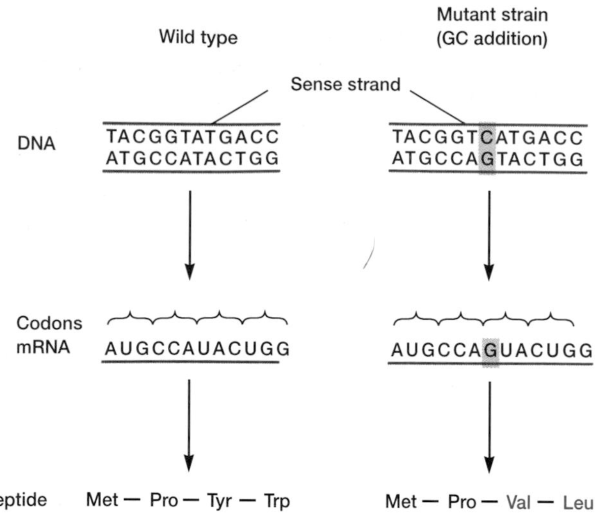

Figure 12.14 **Frameshift Mutation.** A frameshift mutation resulting from the insertion of a GC base pair. The reading frameshift produces a different peptide after the addition.

from the insertion or deletion of one or two base pairs within the coding region of the gene. Since the code consists of a precise sequence of triplet codons, the addition or deletion of fewer than three base pairs will cause the reading frame to be shifted for all codons downstream. Figure 12.14 shows the effect of a frameshift mutation on a short section of mRNA and the amino acid sequence it codes for.

Frameshift mutations are usually very deleterious and yield mutant phenotypes resulting from the synthesis of nonfunctional proteins. The reading frameshift often eventually produces a nonsense or stop codon so that the peptide product is shorter as well as different in sequence. Of course, if the frameshift occurred near the end of the gene, or if there were a second frameshift shortly downstream from the first that restored the reading frame, the phenotypic effect might not be as drastic. A second nearby frameshift that restores the proper reading frame is a good example of an intragenic suppressor mutation.

Mutations also occur in the regulatory sequences responsible for the control of gene expression and in other noncoding portions of structural genes. Constitutive lactose operon mutants in *E. coli* are excellent examples. These mutations map in the operator site and produce altered operator sequences that are not recognized by the repressor protein, and therefore the operon is continuously active in transcription. If a mutation renders the promoter sequence nonfunctional, the coding region of the structural gene will be completely normal, but a mutant phenotype will result due to the absence of a product. RNA polymerase rarely transcribes a gene correctly without a fully functional promoter.

The lac operon and gene regulation (pp. 224–27).

Mutations also occur in rRNA and tRNA genes and can alter the phenotype through disruption of protein synthesis. In fact, these mutants are often initially identified because of their slow growth. One type of suppressor mutation is a base substitution in the anticodon region of a tRNA that allows the insertion of the correct amino acid at a mutant codon (table 12.2).

1. Define or describe the following: mutation, conditional mutation, auxotroph and prototroph, spontaneous and induced mutations, mutagen, transition and transversion mutations, frameshift, apurinic site, base analog, specific mispairing, intercalating agent, thymine dimer, wild type, forward and reverse mutations, suppressor mutation, point mutation, silent mutation, missense and nonsense mutations, and frameshift mutation.
2. Give three ways in which spontaneous mutations might arise.
3. How do the mutagens 5-bromouracil, methyl-nitrosoguanidine, proflavin, and UV radiation induce mutations?
4. Give examples of intragenic and extragenic suppressor mutations.

Detection and Isolation of Mutants

In order to study microbial mutants, one must be able to detect them readily, even when there are few, and then efficiently isolate them from the parent organism and other mutants. Fortunately, this is often easy to do. This section describes some techniques used in mutant detection, selection, and isolation.

Mutant Detection

When collecting mutants of a particular organism, one must know the normal or wild type characteristics so as to recognize an altered phenotype. A suitable detection system for the mutant phenotype under study is also needed. Since mutations are generally rare, about one per 10^7 to 10^{11} cells, it is important to have a very sensitive detection system so that rare events will not be missed. Geneticists have often induced mutations to increase the probability of obtaining specific changes at high frequency (about one in 10^3 to 10^6); even so, mutations are rare.

Detection systems in bacteria and other haploid organisms are straightforward because any new allele should be seen immediately, even if it is a recessive mutation. Sometimes, detection of mutants is direct. If albino mutants of a normally pigmented bacterium are being studied, detection simply requires visual observation of colony color. Other direct detection systems are more complex. For example, the **replica plating** technique is used to detect auxotrophic mutants. It distinguishes between mutants and the wild type strain based on their ability to grow in the absence of a particular biosynthetic end product (figure 12.15). A lysine auxotroph, for instance, will grow on lysine-supplemented media but not on a medium lacking an adequate supply of lysine since it cannot synthesize this amino acid.

Once some sort of detection method is established, mutants are collected. Since a specific mutation is a rare event, it is necessary to look at perhaps thousands to millions of col-

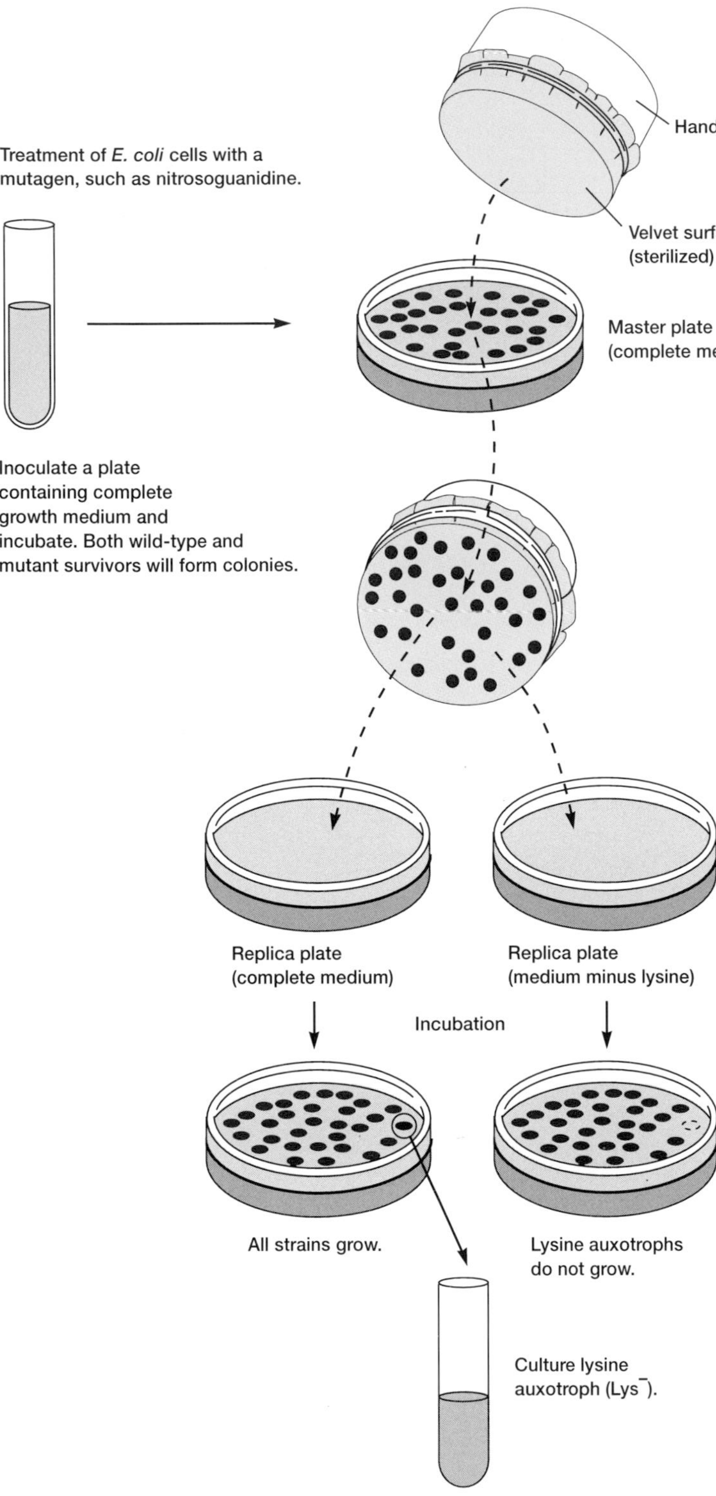
Handle
Treatment of E. coli cells with a mutagen, such as nitrosoguanidine.
Velvet surface (sterilized)
Master plate (complete medium)
Inoculate a plate containing complete growth medium and incubate. Both wild-type and mutant survivors will form colonies.
Replica plate (complete medium)
Replica plate (medium minus lysine)
Incubation
All strains grow.
Lysine auxotrophs do not grow.
Culture lysine auxotroph (Lys$^-$).

Figure 12.15 **Replica Plating.** The use of replica plating in isolating a lysine auxotroph. Mutants are generated by treating a culture with a mutagen. The culture containing wild type and auxotrophs is plated on complete medium. After the colonies have developed, a piece of sterile velveteen is pressed on the plate surface to pick up bacteria from each colony. Then the velvet is pressed to the surface of other plates and organisms are transferred to the same position as on the master plate. After determining the location of Lys$^-$ colonies growing on the replica with complete medium, the auxotrophs can be isolated and cultured.

onies or clones. Using direct detection methods, this could become quite a task, even with microorganisms. Consider a search for the albino mutants mentioned previously. If the mutation rate were around one in a million, on the average a million or more organisms would have to be tested to find one albino mutant. This probably would require several thousand plates. The task of isolating auxotrophic mutants in this way would be even more taxing with the added labor of replica plating. This difficulty can be partly overcome by using mutagens to increase the mutation rate, thus reducing the number of colonies to be examined. However, it is more efficient to use a selection system employing some environmental factor to separate mutants from wild type microorganisms.

Mutant Selection

An effective selection technique uses incubation conditions under which the mutant will grow, because of properties given it by the mutation, whereas the wild type will not. Selection methods often involve reversion mutations or the development of resistance to an environmental stress. For example, if the intent is to isolate revertants from a lysine auxotroph (Lys$^-$), the approach is quite easy. A large population of lysine auxotrophs is plated on minimal medium lacking lysine, incubated, and examined for colony formation. Only cells that have mutated to restore the ability to manufacture lysine will grow on minimal medium (figure 12.16). Thus, several million cells can be plated on a single petri dish, and many cells can be tested for mutations by scanning a few petri dishes for growth. This is because the auxotrophs will not grow on minimal medium and confuse the results; only the phenotypic revertants will form colonies. This method has proven very useful in determining the relative mutagenicity of many substances.

Resistance selection methods follow a similar approach. Often wild type cells are not resistant to phage attack or antibiotic treatment, so it is possible to grow the bacterium in the presence of the agent and look for surviving organisms. Consider the example of a phage-sensitive wild type bacterium. When the organism is cultured in medium lacking the virus and then plated out on selective medium containing phages, any colonies that form will be resistant to phage attack and very likely will be mutants in this regard. Resistance selection can be used together with virtually any environmental parameter; resistance to bacteriophages, antibiotics, or temperature are most commonly employed.

Substrate utilization mutations are also employed in bacterial selection. Many bacteria catabolize only a few primary carbon sources. With such bacteria, it is possible to select mutants by a method similar to that used in resistance selection. The culture is plated on medium containing an alternate carbon source. Any colonies that appear can use the substrate and are probably mutants.

Mutant detection and selection methods are used for purposes other than understanding more about the nature of genes or the biochemistry of a particular microorganism. One very important role of mutant selection and detection techniques is

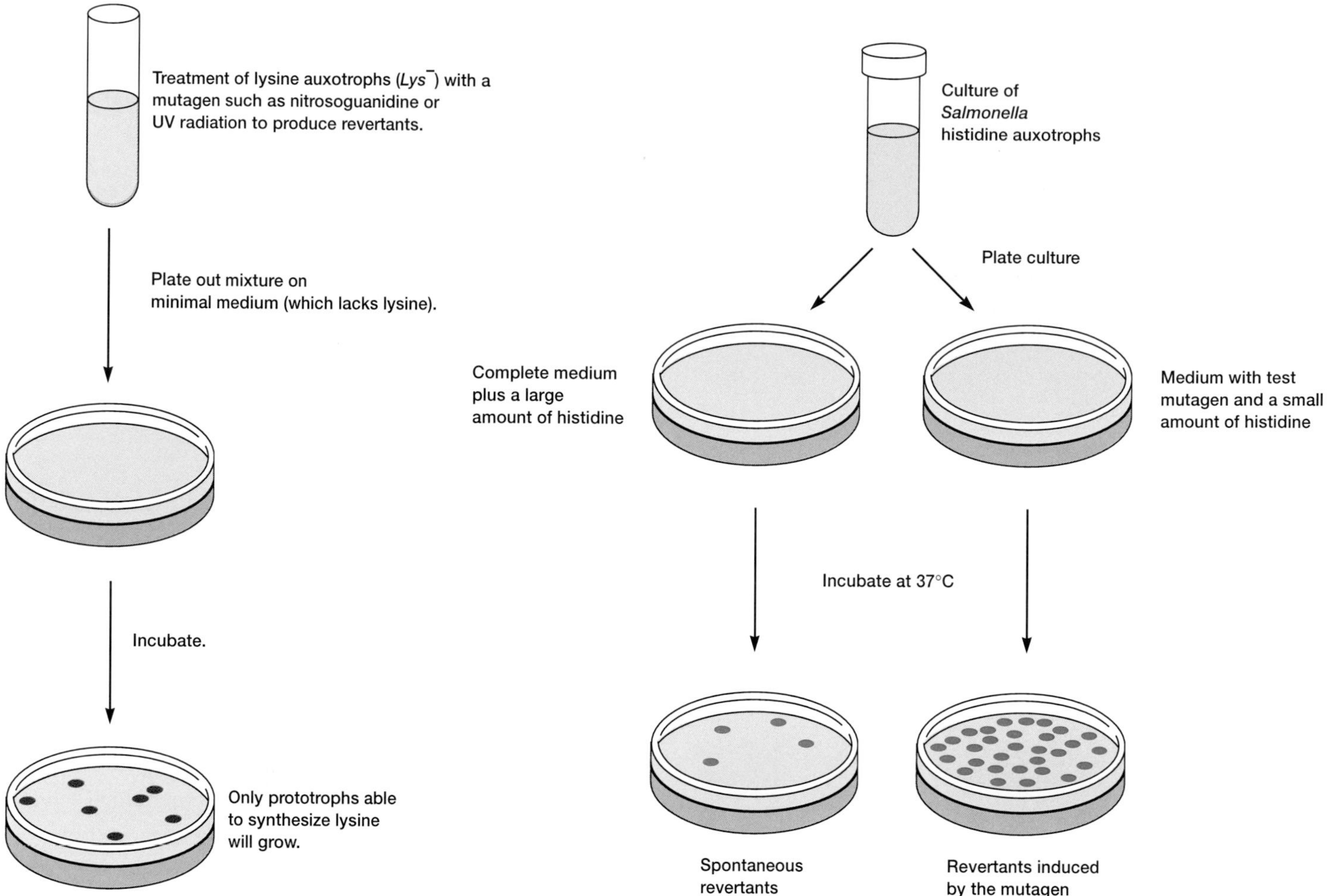

Figure 12.16 **Mutant Selection.** The production and direct selection of auxotroph revertants. In this example, lysine revertants will be selected after treatment of a lysine auxotroph culture because the agar contains minimal medium that will not support auxotroph growth.

Figure 12.17 **The Ames Test for Mutagenicity.** See text for details.

in the study of carcinogens. The next section briefly describes one of the first and perhaps best known of the carcinogen testing systems.

Carcinogenicity Testing

An increased understanding of the mechanisms of mutation and cancer induction has stimulated efforts to identify environmental carcinogens so that they can be avoided. The observation that many carcinogenic agents are also mutagenic is the basis for detecting potential carcinogens by testing for mutagenicity while taking advantage of bacterial selection techniques and short generation times. The **Ames test,** developed by Bruce Ames in the 1970s, has been widely used to test for carcinogens. The Ames test is a mutational reversion assay employing several special strains of *Salmonella typhimurium,* each of which has a different mutation in the histidine biosynthesis operon. The bacteria also have mutational alterations of their cell walls that make them more permeable to test substances. To further increase assay sensitivity, the strains are defective in the ability to carry out excision repair of DNA (p. 253) and have plasmid genes that enhance error-prone DNA repair.

In the Ames test, these special tester strains of *Salmonella* are plated with the substance being tested and the appearance of visible colonies followed (figure 12.17). To ensure that DNA replication can take place in the presence of the potential mutagen, the bacteria and test substance are mixed in dilute molten top agar to which a trace of histidine has been added. This molten mix is then poured on top of minimal agar plates and incubated for two to three days at 37°C. All of the histidine auxotrophs will grow for the first few hours in the presence of the test compound until the histidine is depleted. Once the histidine supply is exhausted, only revertants that have mutationally regained the ability to synthesize histidine will grow. The visible colonies need only be counted and compared to controls in order to estimate the relative mutagenicity of the compound: the more colonies, the greater the mutagenicity.

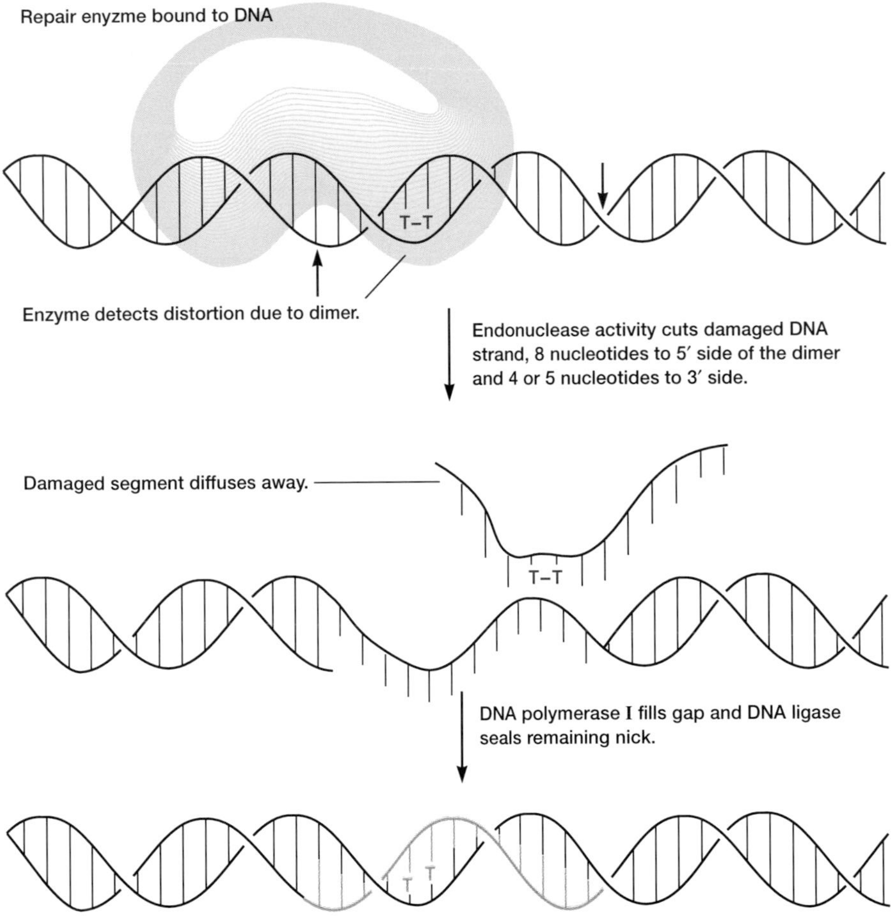

Repair enyzme bound to DNA
T–T
Enzyme detects distortion due to dimer.
Endonuclease activity cuts damaged DNA strand, 8 nucleotides to 5′ side of the dimer and 4 or 5 nucleotides to 3′ side.
Damaged segment diffuses away.
T–T
DNA polymerase I fills gap and DNA ligase seals remaining nick.
T T

Figure 12.18 Excision Repair. Excision repair of a thymine dimer that has distorted the double helix. The repair endonuclease or uvrABC endonuclease is coded for by the *uvrA, B,* and *C* genes.

A mammalian liver extract is also often added to the molten top agar prior to plating. The extract converts potential carcinogens into electrophilic derivatives that will readily react with DNA. This process occurs naturally when foreign substances are metabolized in the liver. Since bacteria do not have this activation system, liver extract is often added to the test system to promote the transformations that occur in mammals. Many potential carcinogens, such as the aflatoxins (*see pp. 873–74*), are not actually carcinogenic until they are modified in the liver. The addition of extract shows which compounds have intrinsic mutagenicity and which need activation after uptake.

1. Describe how replica plating is used to detect and isolate auxotrophic mutants.
2. Why are mutant selection techniques generally preferable to the direct detection and isolation of mutants?
3. Briefly discuss how reversion mutations, resistance to an environmental factor, and the ability to use a particular nutrient can be employed in mutant selection.
4. What is the Ames test and how is it carried out? What assumption concerning mutagenicity and carcinogenicity is it based upon?

DNA Repair

Since replication errors and a variety of mutagens can alter the nucleotide sequence, a microorganism must be able to repair changes in the sequence that might be fatal. DNA is repaired by several different mechanisms besides **proofreading** by replication enzymes (DNA polymerases can remove an incorrect nucleotide immediately after its addition to the growing end of the chain). Repair in *E. coli* is best understood and is briefly described in this section.

DNA replication and proofreading (pp. 198–201).

Excision Repair

Excision repair is a general repair system that corrects damage that causes distortions in the double helix. A repair endonuclease or uvrABC endonuclease removes the damaged bases along with some bases on either side of the lesion (figure 12.18). The resulting single-stranded gap, about 12 nucleotides long, is filled by DNA polymerase I, and DNA ligase joins the fragments (*see chapter 10*). This system can remove thymine dimers (figure 12.13) and repair almost any other injury that produces a detectable distortion in DNA.

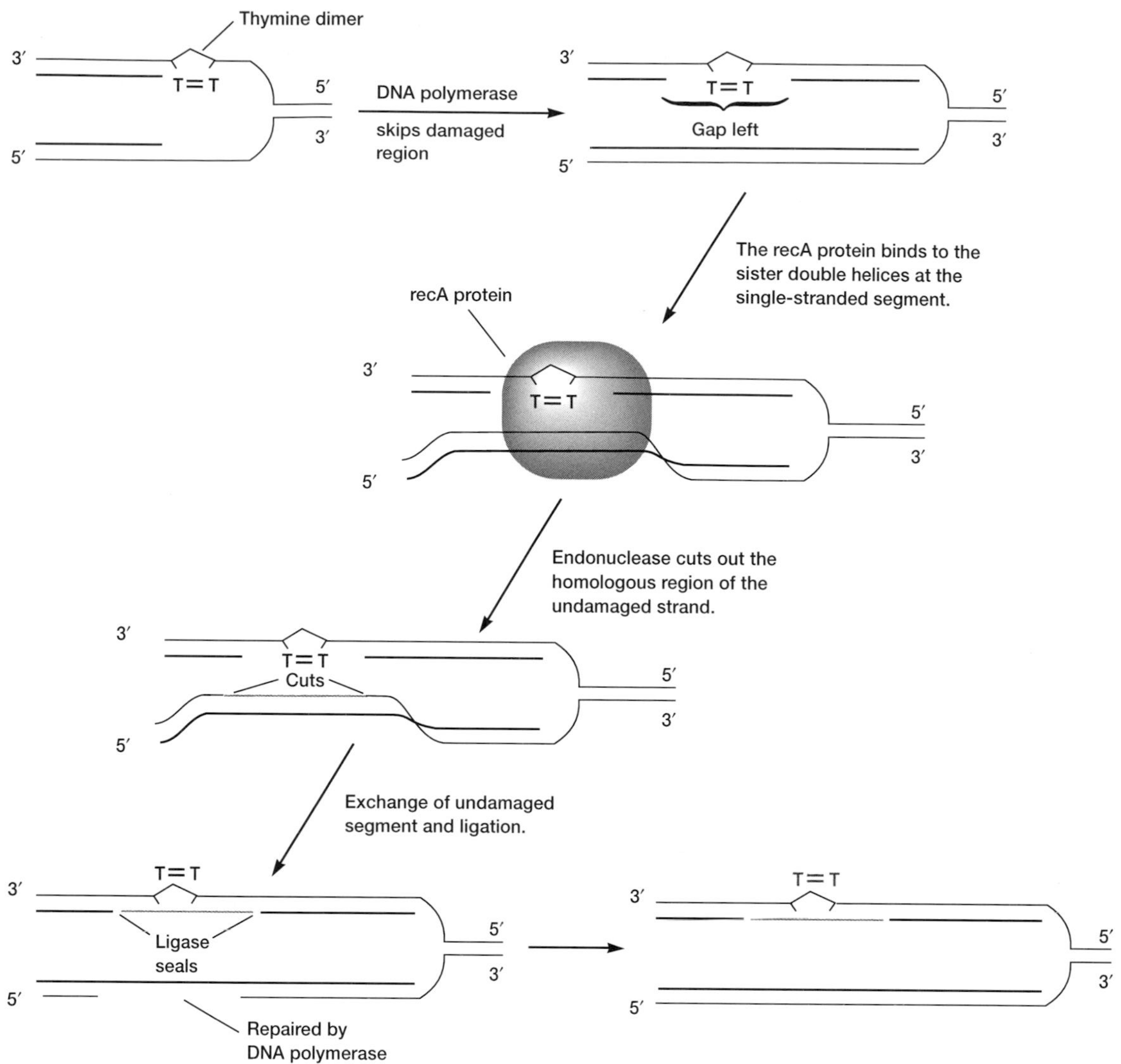

Figure 12.19 **Recombination Repair.** Recombination repair of the gap caused by a thymine dimer. The recA protein pairs the gap region with the homologous region on a sister helix. An undamaged segment is then cut out and exchanged so that the gap can be filled in.

Besides this general excision repair system, specialized versions of the system specifically excise sites on the DNA where the sugar phosphate backbone is intact but the bases have been removed to form apurinic or apyrimidinic sites (AP sites). Special endonucleases called AP endonucleases recognize these locations and nick the backbone at the site. Excision repair then commences, beginning with the excision of a short stretch of nucleotides.

Another type of excision repair employs DNA glycosylases. These enzymes remove damaged or unnatural bases yielding AP sites that are then repaired as above. Not all types of damaged bases are repaired in this way, but new glycosylases are being discovered and the process may be of more general importance than first thought.

Removal of Lesions

Thymine dimers and alkylated bases are often directly repaired. **Photoreactivation** is the repair of thymine dimers by splitting them apart into separate thymines with the help of visible light in a photochemical reaction catalyzed by the enzyme photolyase. Since this repair mechanism does not remove and replace nucleotides, it is error free.

Sometimes, damage caused by alkylation is repaired directly as well. Methyls and some other alkyl groups that have been added to the O–6 position of guanine can be removed with the help of an enzyme known as alkyltransferase or methylguanine methyltransferase. Thus, damage to guanine from mutagens such as methyl-nitrosoguanidine (figure 12.12) can be repaired directly.

Postreplication Repair

Despite the accuracy of DNA polymerase action and continual proofreading, errors still are made during DNA replication. Remaining mismatched bases and other errors are usually detected and repaired by the mismatch repair system in *E. coli.* The mismatch correction enzyme scans the newly

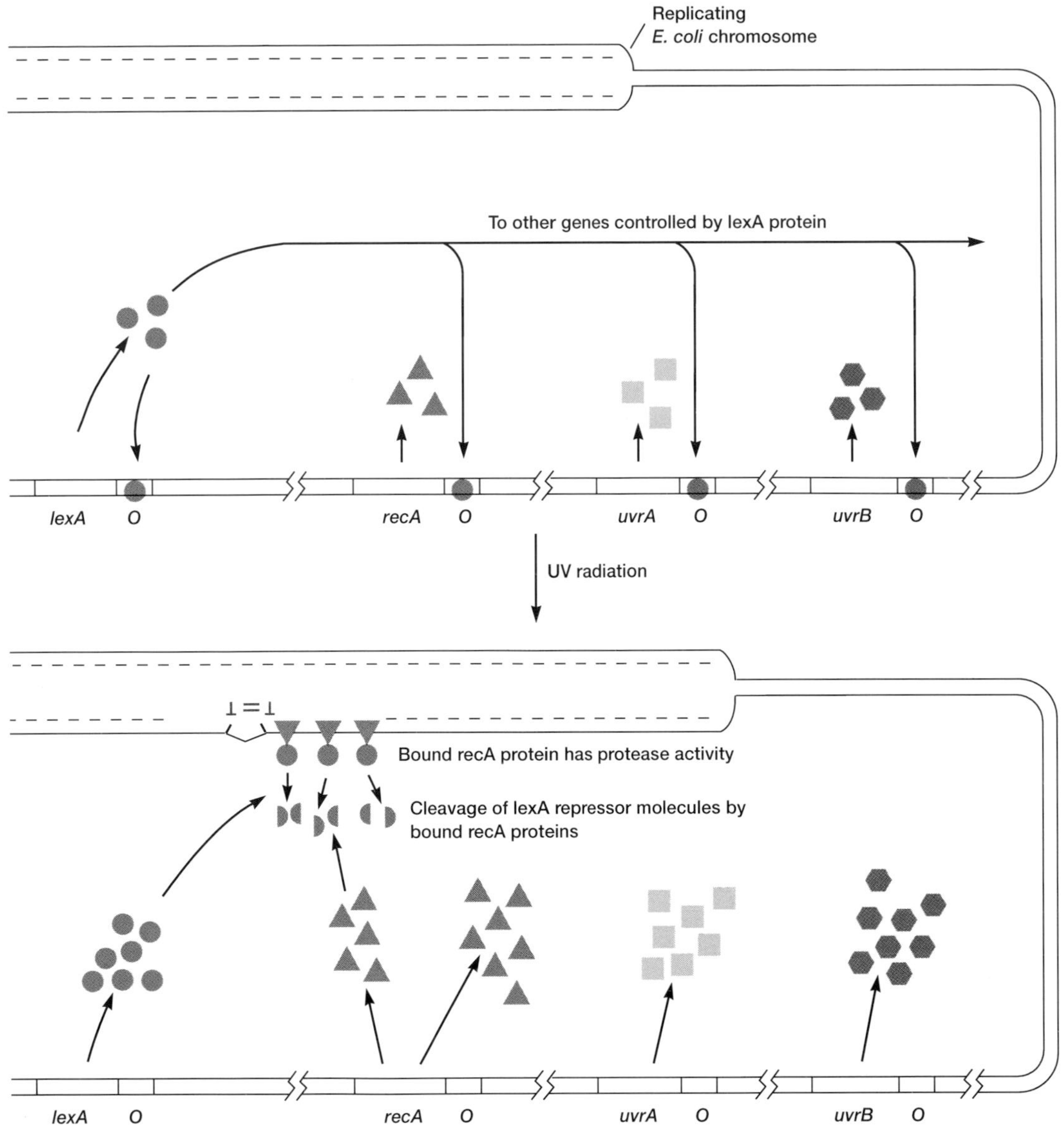

Figure 12.20 **The SOS Repair Process.** In the absence of damage, repair genes are expressed at low levels due to binding of the lexA repressor protein at their operators (*O*). When the recA protein binds to a damaged region—for example, a thymine dimer created by UV radiation—it destroys lexA and the repair genes are expressed more actively. The *uvr* genes code for the repair endonuclease or uvrABC endonuclease responsible for excision repair (see figure 12.18).

replicated DNA for mismatched pairs and removes a stretch of newly synthesized DNA around the mismatch. A DNA polymerase then replaces the excised nucleotides, and the resulting nick is sealed with a ligase. Postreplication repair is a type of excision repair (see above).

Recombination Repair

In **recombination repair,** damaged DNA for which there is no remaining template is restored. This situation arises if both bases of a pair are missing or damaged, or if there is a gap opposite a lesion. In this type of repair, the **recA protein** cuts a piece of template DNA from a sister molecule (figure 12.19) and puts it into the gap or uses it to replace a damaged strand. Although bacteria are haploid, another copy of the damaged segment is often available because either it has recently been replicated or the cell is growing rapidly and has more than one copy of its chromosome. Once the template is in place, the remaining damage can be corrected by another repair system.

The recA protein also participates in a type of inducible repair known as **SOS repair.** In this instance, the DNA damage is so great that synthesis stops completely, leaving many large gaps. RecA will bind to the gaps and initiate strand exchange. Simultaneously, it takes on a proteolytic function that destroys the lexA repressor protein, which regulates the function of many genes involved in DNA repair and synthesis (figure 12.20). As a result, many more copies of these enzymes are produced, accelerating the replication and repair processes. The system can quickly repair extensive damage caused by agents such as UV radiation, but it is error prone and does produce mutations. However, it is certainly better to have a few mutations than no DNA replication at all.

1. Define the following: proofreading, excision repair, photoreactivation, methylguanine methyltransferase, mismatch repair, recombination repair, recA protein, SOS repair, and lexA repressor.
2. Describe in general terms the mechanisms of the following repair processes: excision repair, recombination repair, and SOS repair.

Summary

1. The knowledge that DNA is the genetic material came from studies on transformation by Griffith and Avery and from experiments on T2 phage reproduction by Hershey and Chase.
2. Genetic information is carried in the form of 64 nucleotide triplets called codons; sense codons direct amino acid incorporation, and stop or nonsense codons terminate translation.
3. The code is degenerate; that is, there is more than one codon for most amino acids.
4. A gene may be defined as the nucleic acid sequence that codes for a polypeptide, tRNA, or rRNA.
5. The sense strand of DNA carries genetic information and directs the synthesis of the RNA transcript.
6. RNA polymerase binds to the promoter region, which contains RNA polymerase recognition and RNA polymerase binding sites.
7. The gene also contains a coding region and a terminator; it may have a leader and a trailer. Regulatory segments such as operators may be present.
8. The genes for tRNA and rRNA often code for a precursor that is subsequently processed to yield several products.
9. A mutation is a stable, heritable change in the nucleotide sequence of the genetic material, usually DNA.
10. Mutations can be divided into many categories based on their effects on the phenotype; some major types are morphological, lethal, conditional, biochemical, and resistance mutations.
11. Spontaneous mutations can arise from replication errors (transitions, transversions, and frameshifts) and from DNA lesions (apurinic sites).
12. Induced mutations are caused by mutagens. Mutations may result from the incorporation of base analogs, specific mispairing due to alteration of a base, the presence of intercalating agents, and a bypass of replication because of severe damage.
13. The mutant phenotype can be restored to wild type by either a true reverse mutation or a suppressor mutation.
14. There are four important types of point mutations: silent mutations, missense mutations, nonsense mutations, and frameshift mutations.
15. It is essential to have a sensitive and specific detection technique to isolate mutants; an example is replica plating for the detection of auxotrophs (a direct detection system).
16. One of the most effective isolation techniques is to adjust environmental conditions so that the mutant will grow while the wild type organism does not.
17. Because many carcinogens are also mutagenic, one can test for mutagenicity with the Ames test and use the results as an indirect indication of carcinogenicity.
18. Mutations and DNA damage are repaired in several ways; for example: proofreading by replication enzymes, excision repair, removal of lesions (e.g., photoreactivation), postreplication repair (mismatch repair), and recombination repair.

Key Terms

Ames test *252*
apurinic site *245*
apyrimidinic site *245*
auxotrophs *244*
back mutation *248*
base analog *246*
cistron *240*
code degeneracy *239*
coding region *242*
codon *239*
conditional mutation *244*
excision repair *253*
forward mutation *247*
frameshift *245*
frameshift mutation *249*
gene *240*
genome *237*
genotype *237*
intercalating agent *246*
leader sequence *242*
missense mutation *248*
mutagen *244*
mutation *244*
nonsense mutation *249*
phenotype *237*
photoreactivation *254*
point mutation *248*
promoter *241*
proofreading *253*
prototrophs *244*
recA protein *255*
recombination repair *255*
replica plating *250*
reversion mutation *248*
RNA polymerase binding site or Pribnow box *242*
sense codons *239*
sense strand *241*
Shine-Dalgarno sequence *242*
silent mutation *248*
SOS repair *255*
specific mispairing *246*
stop or nonsense codons *239*
strain or clone *237*
suppressor mutation *248*
terminator sequence *242*
trailer sequence *242*
transformation *238*
transition mutation *244*
transversion mutation *245*
wild type *247*
wobble *240*

Questions for Thought and Review

1. Suppose that you have isolated a microorganism from a sample of Martian soil. Describe how you would go about determining the nature of its genetic material.
2. Currently, a gene is described in several ways. Which definition do you prefer and why?
3. How could one use small deletion mutations to show that codons are triplet (i.e., that the nucleotide sequence is read three bases at a time rather than two or four)?
4. Sometimes a point mutation does not change the phenotype. List all the reasons you can why this is so.
5. Why might a mutation leading to an amino acid change at a protein's surface not result in a phenotypic change while the substitution of an internal amino acid will?
6. Describe how you would isolate a mutant that required histidine for growth and was resistant to penicillin.
7. How would the following mutations be repaired (there may be *more* than one way): base addition errors by DNA polymerase III during replication, thymine dimers, AP sites, methylated guanines, and gaps produced during replication?

Additional Reading

Andersson, S. G. E., and Kurland, C. G. 1990. Codon preferences in free-living microorganisms. *Microbiol. Rev.* 54(2):198–210.

Bachmann, B. J. 1990. Linkage map of *Escherichia coli* K–12, Edition 8. *Microbiol. Rev.* 54(2):130–97.

Bernstein, H. 1983. Recombinational repair may be an important function of sexual reproduction. *BioScience* 33(5):326–31.

Breathnach, R., and Chambon, P. 1981. Organization and expression of eucaryotic split genes coding for proteins. *Ann. Rev. Biochem.* 50:349–83.

Claverys, J.-P., and Lacks, S. A. 1986. Heteroduplex deoxyribonucleic acid base mismatch repair in bacteria. *Microbiol. Rev.* 50(2):133–65.

Dale, J. W. 1989. *Molecular genetics of bacteria.* New York: John Wiley and Sons.

Devoret, R. 1979. Bacterial tests for potential carcinogens. *Sci. Am.* 241(2):40–49.

Fournier, M. J., and Ozeki, H. 1985. Structure and organization of the transfer ribonucleic acid genes of *Escherichia coli* K–12. *Microbiol. Rev.* 49(4):379–97.

Freifelder, D. 1987. *Microbial genetics.* Boston: Jones and Bartlett.

Freifelder, D. 1987. *Molecular biology,* 2d ed. Boston: Jones and Bartlett.

Friedberg, E. C. 1985. *DNA repair.* San Francisco: W. H. Freeman.

Gardner, E. J.; Simmons, M. J.; and Snustad, D. P. 1991. *Principles of genetics,* 8th ed. New York: John Wiley and Sons.

Hall, B. G. 1991. Increased rates of advantageous mutations in response to environmental challenges. *ASM News* 57(2):82–86.

Hopwood, A. 1981. The genetic programming of industrial microorganisms. *Sci. Am.* 245(3):90–102.

Howard-Flanders, P. 1981. Inducible repair of DNA. *Sci. Am.* 245(5):72–80.

Lewin, B. 1990. *Genes,* 4th. New York: Oxford University Press.

Lindahl, L., and Zengel, J. M. 1986. Ribosomal genes in *Escherichia coli. Ann. Rev. Genet.* 20:297–326.

Miller, J. H. 1983. Mutational specificity in bacteria. *Ann. Rev. Genet.* 17:215–38.

Osawa, S.; Jukes, T. H.; Watanabe, K.; and Muto, A. 1992. Recent evidence for evolution of the genetic code. *Microbiol. Rev.* 56(1):229–64.

Russell, P. J. 1990. *Genetics,* 2d ed. Boston: Scott, Foresman and Company.

Sancar, A., and Sancar, G. B. 1988. DNA repair enzymes. *Ann. Rev. Biochem.* 57:29–67.

Scaife, J.; Leach, D.; and Galizzi, A., eds. 1985. *Genetics of bacteria.* New York: Academic Press.

Singer, B., and Kusmierek, J. T. 1982. Chemical mutagenesis. *Ann. Rev. Biochem.* 52:655–93.

Smith, G. R. 1988. Homologous recombination in procaryotes. *Microbiol. Rev.* 52(1):1–28.

Smith-Keary, P. 1989. *Molecular genetics of Escherichia coli.* New York: Guilford Press.

Stent, G. S., and Calender, R. 1978. *Molecular genetics: An introductory narrative,* 2d ed. San Francisco: W. H. Freeman.

Sutherland, B. M. 1981. Photoreactivation. *BioScience* 31(6):439–44.

Suzuki, D. T.; Griffiths, A. J. F.; Miller, J. H.; and Lewontin, R. C. 1989. *An introduction to genetic analysis,* 4th ed. New York: W. H. Freeman.

Van Houten, B. 1990. Nucleotide excision repair in *Escherichia coli. Microbiol. Rev.* 54(1):18–51.

Walker, G. C. 1985. Inducible DNA repair systems. *Ann. Rev. Biochem.* 54:425–57.

Watson, J. D.; Hopkins, N. H.; Roberts, J. W.; Steitz, J. A.; and Weiner, A. M. 1987. *Molecular biology of the gene,* 4th ed. Menlo Park, Calif.: Benjamin/Cummings.

Weaver, R. F., and Hedrick, P. W. 1992. *Genetics,* 2d ed. Dubuque, Iowa: Wm. C. Brown.

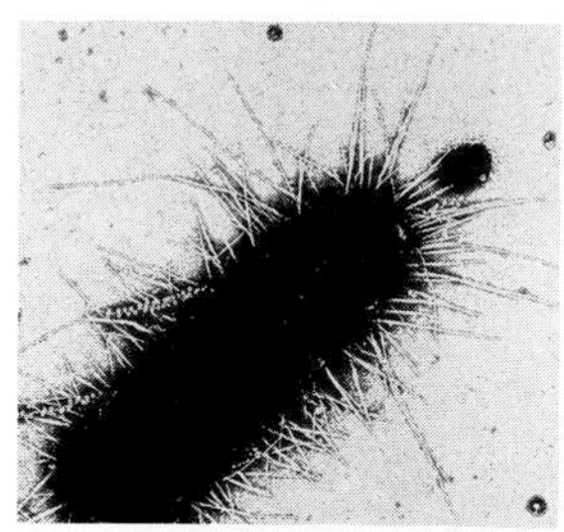

CHAPTER 13
Microbial Genetics: Recombination and Plasmids

Deep in the cavern of the

infant's breast

The father's nature lurks,

and lives anew.

—Horace, *Odes*

Outline

Concepts

1. Recombination is a one-way process in procaryotes: a piece of genetic material (the exogenote) is donated to the chromosome of a recipient cell (the endogenote) and integrated into it.
2. The actual transfer of genetic material between bacteria usually takes place in one of three ways: direct transfer between two bacteria temporarily in physical contact (conjugation), transfer of a naked DNA fragment (transformation), or transport of bacterial DNA by bacteriophages (transduction).
3. Plasmids and transposable elements can move genetic material between bacterial chromosomes and within chromosomes to cause rapid changes in genomes and drastically alter phenotypes.
4. The bacterial chromosome can be mapped with great precision, using Hfr conjugation in combination with transformational and transductional mapping techniques.
5. Recombination of virus genomes occurs when two viruses with homologous chromosomes infect a host cell at the same time.

Chapter 12 introduces the fundamentals of molecular genetics: the way genetic information is organized and stored, the nature of mutations and techniques for their isolation and study, and DNA repair. This chapter focuses on genetic recombination in microorganisms, with primary emphasis placed on recombination in bacteria and viruses.

The chapter begins with a general overview of bacterial recombination and an introduction to both bacterial plasmids and transposable elements. Next, the three types of bacterial recombination—conjugation, transformation, and transduction—are presented. Because an understanding of the techniques used to locate genes on chromosomes depends upon knowledge of recombination mechanisms, mapping the bacterial genome is discussed after the introduction to recombination. The chapter ends with a description of recombination in viruses and a brief discussion of viral chromosome mapping.

Recombination is the process in which a new recombinant chromosome, one with a genotype different from either parent, is formed by combining genetic material from two organisms. It results in a new arrangement of genes or parts of genes and normally is accompanied by a phenotypic change. Most eucaryotes exhibit a complete sexual life cycle, including meiosis, a process of extreme importance in generating new combinations of alleles (alternate forms of a particular gene) through recombination. These chromosome exchanges during meiosis result from **crossing-over** between homologous chromosomes, chromosomes containing identical sequences of genes (figure 13.1). Until about 1945, the primary focus in genetic analysis was on the recombination of genes in plants and animals. The early work on recombination in higher eucaryotes laid the foundations of classical genetics, but it was the development of bacterial and phage genetics between about 1945 and 1965 that really triggered a rapid advance in our understanding of molecular genetics.

Meiosis (pp. 85–86).

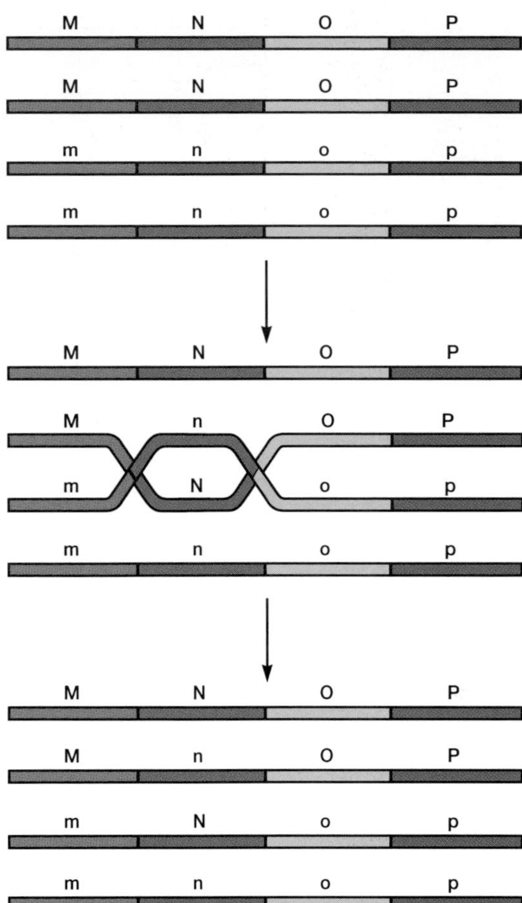

Figure 13.1 **Crossing-over.** An example of recombination through crossing-over between homologous eucaryotic chromosomes. The Nn gene pair is exchanged. This process usually occurs during meiosis.

Bacterial Recombination: General Principles

Microorganisms carry out several types of recombination. **General recombination,** the most common form, usually involves a reciprocal exchange between a pair of homologous DNA sequences. It can occur anyplace on the chromosome, and it results from DNA strand breakage and reunion leading to crossover (figure 13.2). General recombination is carried out by the products of *rec* genes such as the recA protein so important for DNA repair (*see pp. 255–56*). In bacterial transformation, a nonreciprocal form of general recombination takes place. A piece of genetic material is inserted into the chromosome through the incorporation of a single strand to form a stretch of **heteroduplex DNA** (figure 13.3). A second type of recombination, one particularly important in the integration of virus genomes into bacterial chromosomes, is **site-specific recombination.** The genetic material is not homologous with the chromosome it joins, and generally the enzymes responsible for this event are specific for the particular virus and its host. A third type, **replicative recombination,** accompanies the replication of genetic material and does not depend upon sequence homology. It is used by genetic elements that move about the chromosome.

DNA replication (pp. 195–201).

Although sexual reproduction with the formation of a zygote and subsequent meiosis is not present in bacteria, recombination can take place in several ways. In general, a piece of donor DNA, the **exogenote,** must enter the recipient cell and become a stable part of the recipient cell's genome, the **endogenote.** Two kinds of DNA can move between bacteria. If a DNA fragment is the exchange vehicle, then the exogenote must get into the recipient cell and become incorporated into the endogenote as a replacement piece (or as an "extra" piece) without being destroyed by the host. During replacement of host genetic material, the recipient cell becomes temporarily diploid for a portion of the genome and is

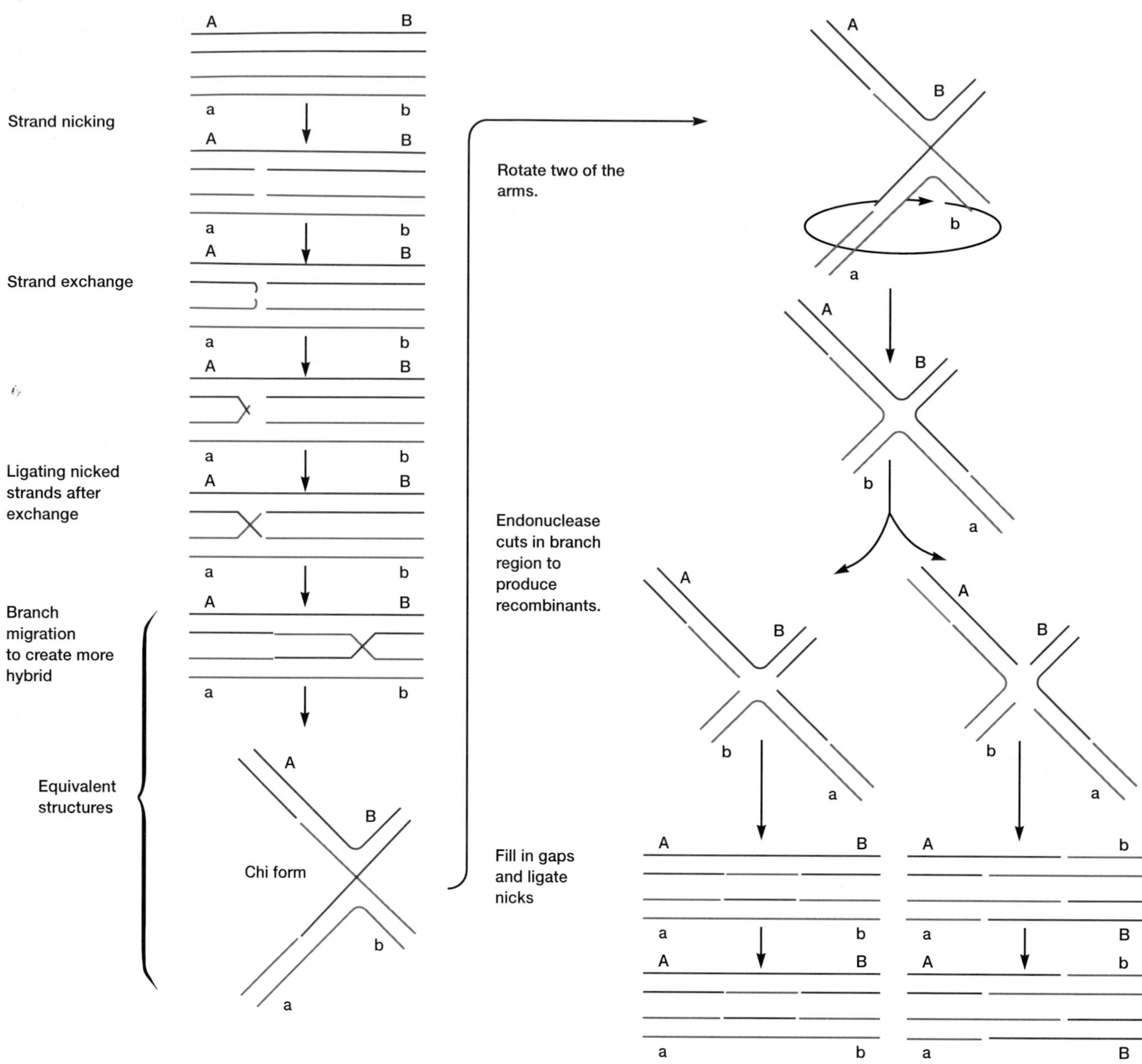

Figure 13.2 The Holliday Model for Reciprocal General Recombination. *Source: H. Potter and D. Dressler,* Proceedings of the National Academy of Sciences *73:3000, 1976, after R. Holliday,* Genetics, *78:273, 1974, and previous publication cited therein.*

called a **merozygote.** Sometimes the DNA exists in a form that cannot be degraded by the recipient cell's endonucleases. In this case, the DNA does not need to be integrated into the host genome but must only enter the recipient to confer its genetic information upon the cell. Resistant DNA, such as that in plasmids (see following), usually is circular and has sequences that allow it to maintain itself independently of the host chromosome. Thus, recombination in bacteria is a one-way gene transfer from donor to recipient. Recombination in eucaryotes tends to be reciprocal; that is, all of the DNA is conserved in the gametes that eventually arise from meiosis and recombination.

Movement of DNA from a donor bacterium to the recipient can take place in three ways: direct transfer between two bacteria temporarily in physical contact (conjugation), transfer of a naked DNA fragment (transformation), and transport of bacterial DNA by bacteriophages (transduction). Whatever the mode of transfer, the exogenote has only four possible fates in the recipient. First, when the exogenote has a sequence homologous to that of the endogenote, integration may occur; that is, it may pair with the recipient DNA and be incorporated to yield a recombinant genome. Second, the foreign DNA sometimes persists outside the endogenote and replicates to produce a clone of partially diploid cells. Third, the exogenote

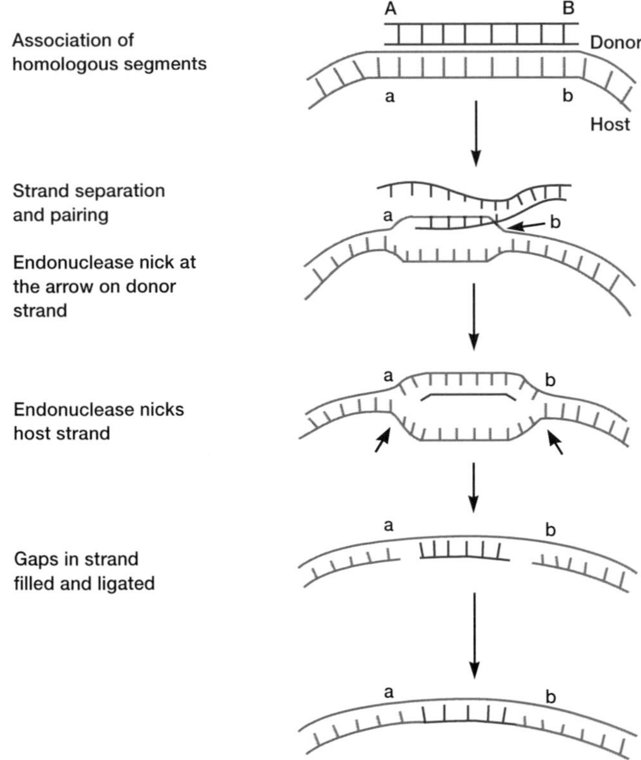
Association of homologous segments
A
B
Donor
a
b
Host
Strand separation and pairing
Endonuclease nick at the arrow on donor strand
Endonuclease nicks host strand
Gaps in strand filled and ligated

Figure 13.3 **Nonreciprocal General Recombination.** The Fox model for nonreciprocal general recombination. This mechanism has been proposed for the recombination occurring during transformation in some bacteria.

may survive, but not replicate, so that only one cell is a partial diploid. Finally, host cell nucleases may degrade the exogenote, a process called **host restriction.**

1. Define the following terms: recombination, crossing-over, general recombination, site-specific recombination, replicative recombination, exogenote, endogenote, merozygote, and host restriction.
2. Distinguish among the three forms of recombination mentioned in this section.
3. What four fates can DNA have after entering a bacterium?

Bacterial Plasmids

Conjugation, the transfer of DNA between bacteria involving direct contact, depends upon the presence of an "extra" piece of circular DNA known as a plasmid. Plasmids play many important roles in the lives of bacteria. They also have proved invaluable to microbiologists and molecular geneticists in constructing and transferring new genetic combinations and in cloning genes (*see chapter 14*). In this section, the different types of bacterial plasmids are discussed.

Plasmids are small, circular DNA molecules that can exist independently of host chromosomes and are present in many bacteria (they are also present in some yeasts and other fungi). They have their own replication origins and are autonomously replicating and stably inherited. A **replicon** is a DNA molecule or sequence that has a replication origin and is capable of being replicated. Plasmids and bacterial chromosomes are separate replicons. Plasmids have relatively few genes, generally less than 30. Their genetic information is not essential to the host, and bacteria that lack them usually function normally. Single-copy plasmids produce only one copy per host cell. Multicopy plasmids may be present at concentrations of 40 or more per cell.

Characteristically, plasmids can be eliminated from host cells in a process known as **curing.** Curing may occur spontaneously or be induced by treatments that inhibit plasmid replication while not affecting host cell reproduction. The inhibited plasmids are slowly diluted out of the growing bacterial population. Some commonly used curing treatments are acridine mutagens, UV and ionizing radiation, thymine starvation, and growth above optimal temperatures.

Plasmids may be classified in terms of their mode of existence and spread. An **episome** is a plasmid that can exist either with or without being integrated into the host's chromosome. Some plasmids, **conjugative plasmids,** have genes for pili and can transfer copies of themselves to other bacteria during conjugation. A brief summary of the types of plasmids and their properties is given in table 13.1.

Fertility Factors

A plasmid called the fertility or **F factor** plays a major role in conjugation in *E. coli* and was the first to be described. The F factor is about 94.5 kilobases long and bears genes responsible for cell attachment and plasmid transfer between specific bacterial strains during conjugation. Most of the information required for plasmid transfer is located in the *tra* operon, which contains about 21 genes. Many of these direct the formation of sex pili that attach the F^+ cell (the donor cell containing an F plasmid) to an F^- cell (figure 13.4). Other gene products aid DNA transfer.

Sex pili (p. 57).

The F factor also has several segments called insertion sequences (p. 264) that assist plasmid integration into the host cell chromosome. Thus, the F factor is an episome that can exist outside the bacterial chromosome or be integrated into it (figure 13.5).

Resistance Factors

Plasmids often confer antibiotic resistance on the bacteria that contain them. **R factors** or plasmids typically have genes that code for enzymes capable of destroying or modifying antibiotics. They are not usually integrated into the host chromosome. Genes coding for resistance to antibiotics such as ampicillin, chloramphenicol, and kanamycin have been found in plasmids. Some R plasmids have only a single resistance

TABLE 13.1 Major Types of Plasmids

Type	Representatives	Approximate Size (kb)	Copy Number (Copies/Chromosome)	Hosts	Phenotypic Features[a]
Fertility Factor[b]	F factor	95–100	1–3	*E. coli, Salmonella, Citrobacter*	Sex pilus, conjugation
R Plasmids	RP4	54	1–3	*Pseudomonas* and many other gram-negative bacteria	Sex pilus, conjugation, resistance to Ap, Km, Nm, Tc
	R1	80	1–3	Gram-negative bacteria	Resistance to Ap, Km, Su, Cm, Sm
	R6	98	1–3	*E. coli, Proteus mirabilis*	Su, Sm, Cm, Tc, Km, Nm
	R100	90	1–3	*E. coli, Shigella, Salmonella, Proteus*	Cm, Sm, Su, Tc, Hg
	pSH6	21		*Staphylococcus aureus*	Gm, Tm, Km
	pSJ23a	36		*S. aureus*	Pn, Asa, Hg, Gm, Km, Nm, Em, etc.
	pAD2	25		*Enterococcus faecalis*	Em, Km, Sm
Col Plasmids	ColE1	9	10–30	*E. coli*	Colicin E1 production
	ColE2		10–15	*Shigella*	Colicin E2
	CloDF13			*Enterobacter cloacae*	Cloacin DF13
Virulence Plasmids	Ent (P307)	83		*E. coli*	Enterotoxin production
	K88 plasmid			*E. coli*	Adherence antigens
	ColV-K30	2		*E. coli*	Siderophore for iron uptake; resistance to immune mechanisms
	pZA10	56		*S. aureus*	Enterotoxin B
Metabolic Plasmids	CAM	230		*Pseudomonas*	Camphor degradation
	SAL	56		*Pseudomonas*	Salicylate degradation
	TOL	75		*Pseudomonas putida*	Toluene degradation
	pJP4			*Pseudomonas*	2,4-dichlorophenoxyacetic acid degradation
				E. coli, Klebsiella, Salmonella	Lactose degradation
				Providencia	Urease
	sym			*Rhizobium*	Nitrogen fixation and symbiosis

[a]Abbreviations used for resistance to antibiotics and metals: Ap, ampicillin; Asa, arsenate; Cm, chloramphenicol; Em, erythromycin; Gm, gentamycin; Hg, mercury; Km, kanamycin; Nm, neomycin; Pn, penicillin; Sm, streptomycin; Su, sulfonamides; Tc, tetracycline.
[b]Many R plasmids, metabolic plasmids, and others are also conjugative.

gene, while others can have as many as eight. Often, the resistance genes are within a transposon, and thus it is possible for bacterial strains to rapidly develop multiple resistance plasmids.

R factors and antibiotic resistance (pp. 338–39).

Because many R factors are also conjugative plasmids, they can spread throughout a population, although not as rapidly as the F factor. Often, nonconjugative R factors also move between bacteria during plasmid promoted conjugation. Thus, a whole population can become resistant to antibiotics. The fact that some of these plasmids are readily transferred between species further promotes the spread of resistance. When the host consumes large quantities of antibiotics, *E. coli* and other bacteria with R factors are selected for and become more prevalent. The R factors can then be transferred to more pathogenic genera such as *Salmonella* or *Shigella,* causing even greater public health problems (*see chapter 16*).

Col Plasmids

Bacteria also harbor plasmids with genes that may give them a competitive advantage in the microbial world. **Bacteriocins** are bacterial proteins that destroy other bacteria. They usually act only against closely related strains. Bacteriocins often kill cells by forming channels in the plasma membrane, thus increasing its permeability. They may also degrade DNA and RNA, or attack peptidoglycan and weaken the cell wall. Col

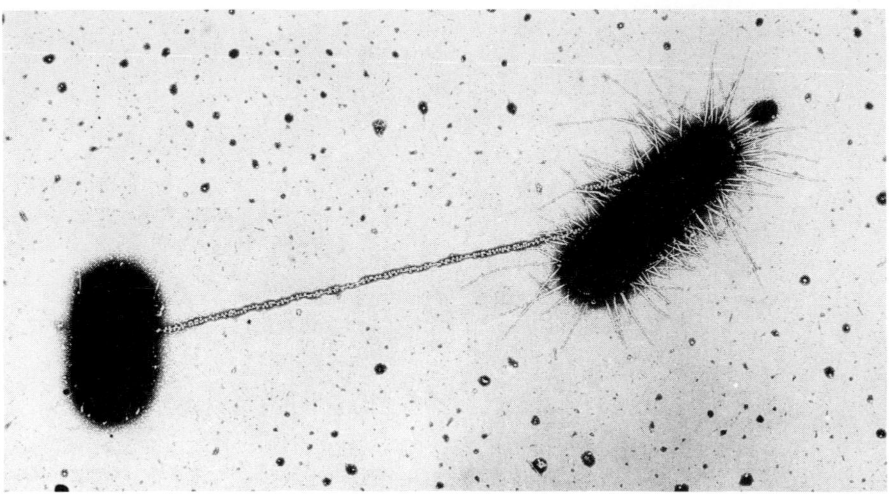

Figure 13.4 **Bacterial Conjugation.** An electron micrograph of two *E. coli* cells undergoing conjugation. The F^+ cell to the right is covered with small pili or fimbriae, and a sex pilus connects the two cells.

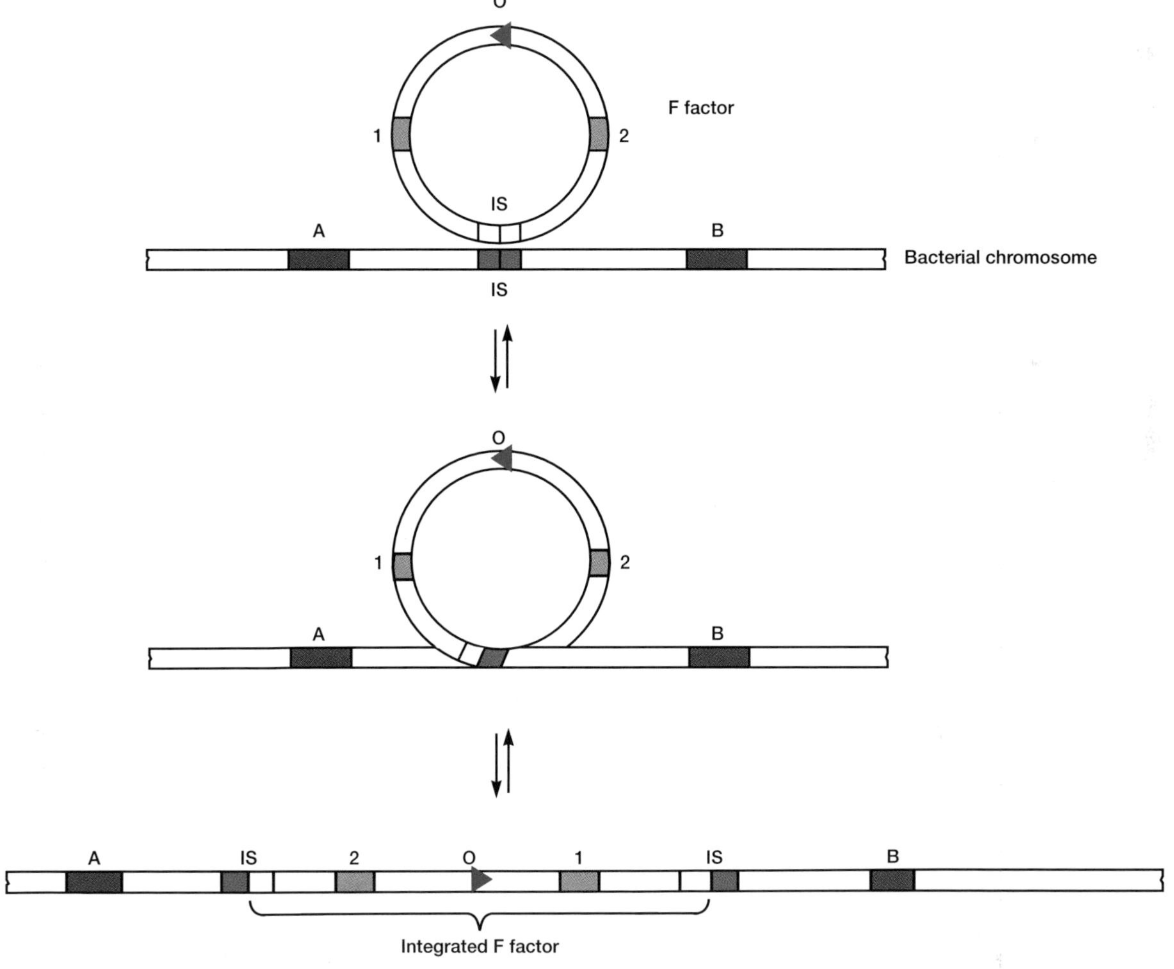

Figure 13.5 **F Plasmid Integration.** The reversible integration of an F plasmid or factor into a host bacterial chromosome. The process begins with association between plasmid and bacterial insertion sequences. The O arrow indicates the site at which the oriented transfer of chromosome to the recipient cell begins. A, B, 1, and 2 represent genetic markers.

Box 13.1

Virulence Plasmids and Disease

It is becoming increasingly evident that many bacteria are pathogenic because of their plasmids. These plasmids can carry genes for toxins, render the bacterium better able to establish itself in the host, or aid in resistance to host defenses. *E. coli* provides the best-studied example of virulence plasmids. Several strains of *E. coli* cause diarrhea. The enterotoxigenic strains responsible for traveler's diarrhea can produce two toxins: a heat-labile toxin (LT), which is a large protein very similar in structure and mechanism of action to cholera toxin (*see chapter 30*), and a heat-stable toxin (ST), a low molecular weight polypeptide. Both toxin genes are plasmid borne, and sometimes they are even carried by the same plasmid. The ST toxin gene is located on a transposon. Enterotoxigenic strains of *E. coli* also must be able to colonize the epithelium of the small intestine to cause diarrhea. This is made possible by the presence of special adhesive fimbriae encoded by genes on another plasmid. A second type of pathogenic *E. coli* invades the intestinal epithelium and causes a form of diarrhea very similar to the dysentery resulting from *Shigella* infection. This *E. coli* strain and *Shigella* contain virulence plasmids that code for special cell wall antigens and other factors enabling them to enter and destroy epithelial cells.

Some *E. coli* strains can invade the blood and organs of a host, causing a generalized infection. These pathogens often have ColV plasmids and produce colicin V. The ColV plasmid carries genes for two virulence determinants. One product increases bacterial resistance to host defense mechanisms involving complement (*see chapter 33*). The other plasmid gene directs the synthesis of a hydroxamate that enables *E. coli* to accumulate iron more efficiently from its surroundings (*see chapter 5*). Since iron is not readily available in the animal host, but is essential for bacterial growth, this is an important factor in pathogenicity.

Several other pathogens carry virulence plasmids. Some *Staphylococcus aureus* strains produce an exfoliative toxin that is plasmid borne. The toxin causes the skin to loosen and often peel off in sheets, leading to the disease staphylococcal scalded skin syndrome (*see chapter 37*). Other plasmid-borne toxins are the tetanus toxin of *Clostridium tetani* and the anthrax toxin of *Bacillus anthracis*.

plasmids contain genes for the synthesis of bacteriocins known as colicins, which are directed against *E. coli*. Similar plasmids carry genes for bacteriocins against other species. For example, the Col plasmids produce cloacins that kill *Enterobacter* species. Clearly, the host is unaffected by the bacteriocin it produces. Some Col plasmids are conjugative and also can carry resistance genes.

Bacteriocins and host defenses (p. 595).

Other Types of Plasmids

Several other important types of plasmids have been discovered. Some plasmids, called **virulence plasmids,** make their hosts more pathogenic because the bacterium is better able to resist host defense or to produce toxins. For example, enterotoxigenic strains of *E. coli* cause traveler's diarrhea because of a plasmid that codes for an enterotoxin (Box 13.1). **Metabolic plasmids** carry genes for enzymes that degrade substances such as aromatic compounds (toluene), pesticides (2,4-dichlorophenoxyacetic acid), and sugars (lactose). Metabolic plasmids even carry the genes required for some strains of *Rhizobium* to induce legume nodulation and carry out nitrogen fixation.

1. Give the major distinguishing features of a plasmid. What is an episome? A conjugative plasmid?
2. Describe each of the following plasmids and their importance: F factor, R factor, Col plasmid, virulence plasmid, and metabolic plasmid.

Transposable Elements

The chromosomes of bacteria, viruses, and eucaryotic cells contain pieces of DNA that move around the genome. Such movement is called **transposition.** DNA segments that carry the genes required for this process and consequently move about chromosomes are **transposable elements** or **transposons.** Unlike other processes that reorganize DNA, transposition does not require extensive areas of homology between the transposon and its destination site. Transposons behave somewhat like lysogenic prophages (*see pp. 375–81*) except that they originate in one chromosomal location and can move to a different location in the same chromosome. Transposable elements differ from phages in lacking a virus life cycle and from plasmids in being unable to reproduce autonomously and to exist apart from the chromosome. They were first discovered in the 1940s by Barbara McClintock during her studies on maize genetics (a discovery that won her the Nobel Prize in 1983). They have been most intensely studied in bacteria and viruses.

The simplest transposable elements are **insertion sequences** or IS elements (figure 13.6*a*). An IS element is a short sequence of DNA (around 750 to 1,600 base pairs [bp] in length) containing only the genes for those enzymes required for its transposition and bounded at both ends by identical or very similar sequences of nucleotides in reversed orientation known as inverted repeats (figure 13.6*c*). Inverted repeats are usually about 15 to 25 base pairs long and vary among IS elements so that each type of IS has its own characteristic inverted repeats. Between the inverted repeats is a gene that codes for an enzyme called **transposase** (and sometimes a gene

(a) Insertion sequence

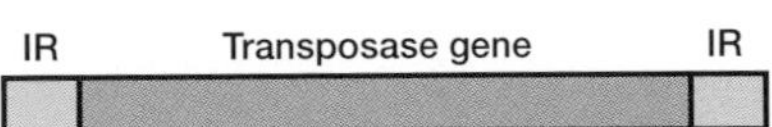

(b) A composite transposon

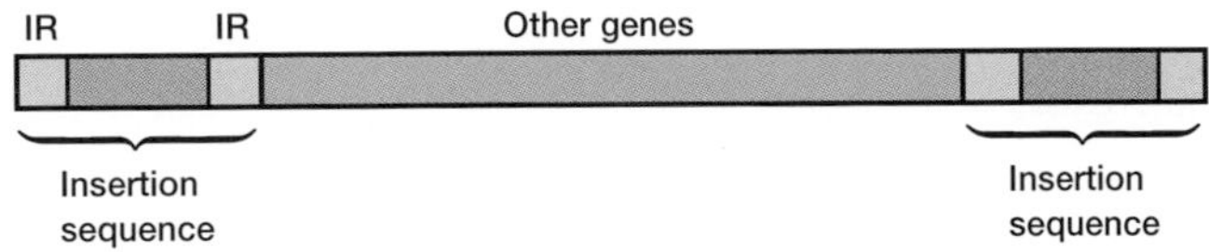

(c) A target site for the Tn3 transposon

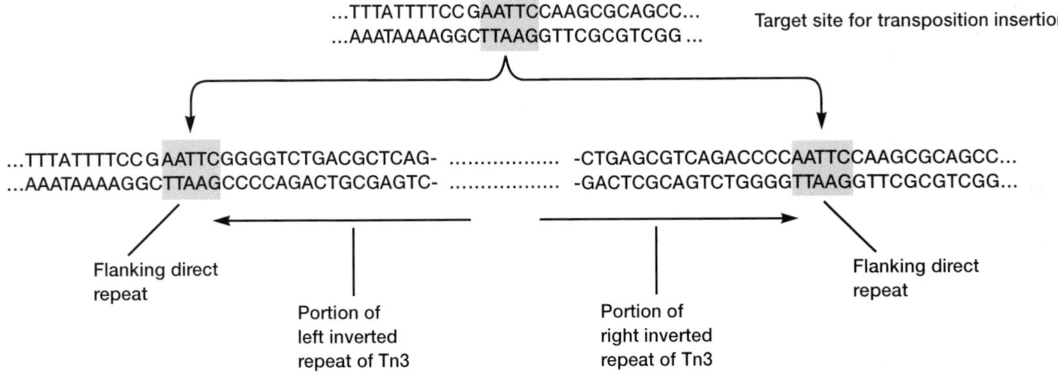

Figure 13.6 **Insertion Sequences and Transposons.** The structure of insertion sequences, composite transposons, and target sites. *IR* stands for inverted repeat. In 13.6*c*, the highlighted five-base target site is duplicated during Tn3 transposition to form flanking direct repeats. The remainder of Tn3 lies between the inverted repeats (see figure 13.9*b*).

TABLE 13.2 The Properties of Selected Insertion Sequences

Insertion Sequence	Length (bp)	Inverted Repeat (Length in bp)	Target Site (Length in bp)	Number of Copies on *E. coli* Chromosome
IS1	768	23	9 or 8	6–10
IS2	1,327	41	5	4–13(1)[a]
IS3	1,400	38	3–4	5–6(2)
IS4	1,428	18	11 or 12	1–2
IS5	1,195	16	4	10–11

[a]The value in parentheses indicates the number of IS elements on the F factor.

for another essential protein). This enzyme is required for transposition and recognizes the ends of the IS very accurately. Each type of element is named by giving it the prefix IS followed by a number. In *E. coli,* several copies of different IS elements have been observed; some of their properties are given in table 13.2.

Transposable elements also can contain genes other than those required for transposition (for example, antibiotic resistance or toxin genes). These elements are often called **composite transposons** or elements. Complete agreement about the nomenclature of transposable elements has not yet been reached. Sometimes transposable elements are called transposons when they have extra genes, and insertion sequences when they lack these. Composite transposons often consist of a central region containing the extra genes, flanked on both sides by IS elements that are identical or very similar in sequence (figure 13.6*b*). Many composite transposons are simpler in organization. They are bounded by short inverted repeats, and the coding region contains both transposition genes and the extra genes. It is believed that composite transposons are formed when two IS elements associate with a central segment containing one or more genes. This association could arise if

TABLE 13.3 The Properties of Selected Composite Transposons

Transposon	Length (bp)	Terminal Repeat Length	Terminal Module	Genetic Markers[a]
Tn3	4,957	38		Ap
Tn501	8,200	38		Hg
Tn951	16,500	Unknown		Lactose utilization
Tn5	5,700		IS50	Km
Tn9	2,500		IS1	Cm
Tn10	9,300		IS10	Tc
Tn903	3,100		IS903	Km
Tn1681	2,061		IS1	Heat-stable enterotoxin
Tn2901	11,000		IS1	Arginine biosynthesis

[a]Abbreviations for antibiotics and metals same as in table 13.1.

an IS element replicates and moves only a gene or two down the chromosome. Composite transposon names begin with the prefix Tn. Some properties of selected composites are given in table 13.3.

The process of transposition in procaryotes involves a series of events, including self-replication and recombinational processes. Typically in bacteria, the original transposon remains at the parental site on the chromosome, while a replicated copy inserts at the target DNA (figure 13.6*c*). This is called replicative transposition. Target sites are specific sequences about 5 to 9 base pairs long. When a transposon inserts at a target site, the target sequence is duplicated so that short, direct-sequence repeats flank the transposon's terminal inverted repeats (figure 13.7). This can be seen in figure 13.6*c* where the five base pair target sequence moves to both ends of the transposon and retains the same orientation.

The transposition of the Tn3 transposon (figure 13.9*b*) is a well-studied example of replicative transposition. Its mechanism is outlined in figure 13.8. In the first stage, the plasmid containing the Tn3 transposon fuses with the target plasmid to form a cointegrate molecule (figure 13.8, steps 1 to 4). This process requires the Tn3 transposase enzyme coded for by the *tnpA* gene (figure 13.9*b*). Note that the cointegrate has two copies of the Tn3 transposon. In the second stage, the cointegrate is resolved to yield two plasmids, each with a copy of the transposon (figure 13.8, steps 5 and 6). Resolution involves a crossover at the two *res* sites and is catalyzed by a resolvase enzyme coded for by the *tnpR* gene (figure 13.9*b*).

Transposable elements produce a variety of important effects. They can insert within a gene to cause a mutation or stimulate DNA rearrangement, leading to deletions of genetic material. Because some transposons carry stop codons or termination sequences, they may block translation or transcription. Other elements carry promoters and thus activate genes near the point of insertion. Eucaryotic genes as well as procaryotic genes can be turned on and off by transposon movement. Transposons are also located within plasmids and participate in such processes as plasmid fusion and the insertion of F plasmids into the *E. coli* chromosome.

In the previous discussion of plasmids, it was noted that an R plasmid can carry genes for resistance to several drugs. Transposons have antibiotic resistance genes and play a major

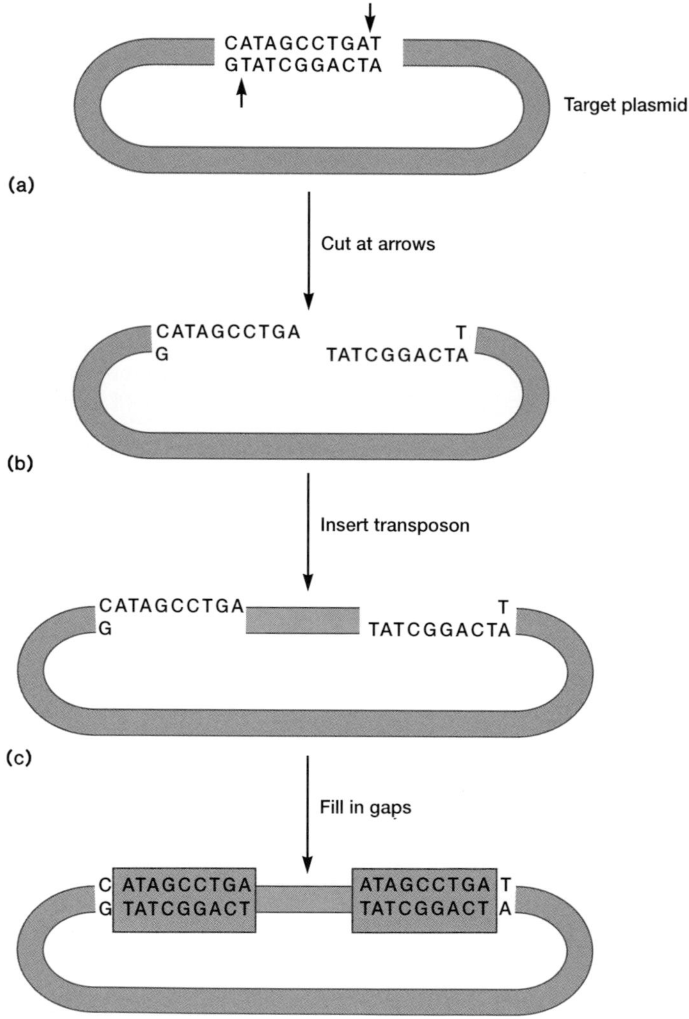

Figure 13.7 Generation of Direct Repeats in Host DNA Flanking a Transposon. (*a*) The arrows indicate where the two strands of host DNA will be cut in a staggered fashion, nine base-pairs apart. (*b*) After cutting. (*c*) The transposon (orange) has been ligated to one strand of host DNA at each end, leaving two nine-base gaps. (*d*) After the gaps are filled in, there are nine base-pair repeats of host DNA (purple boxes) at each end of the transposon.

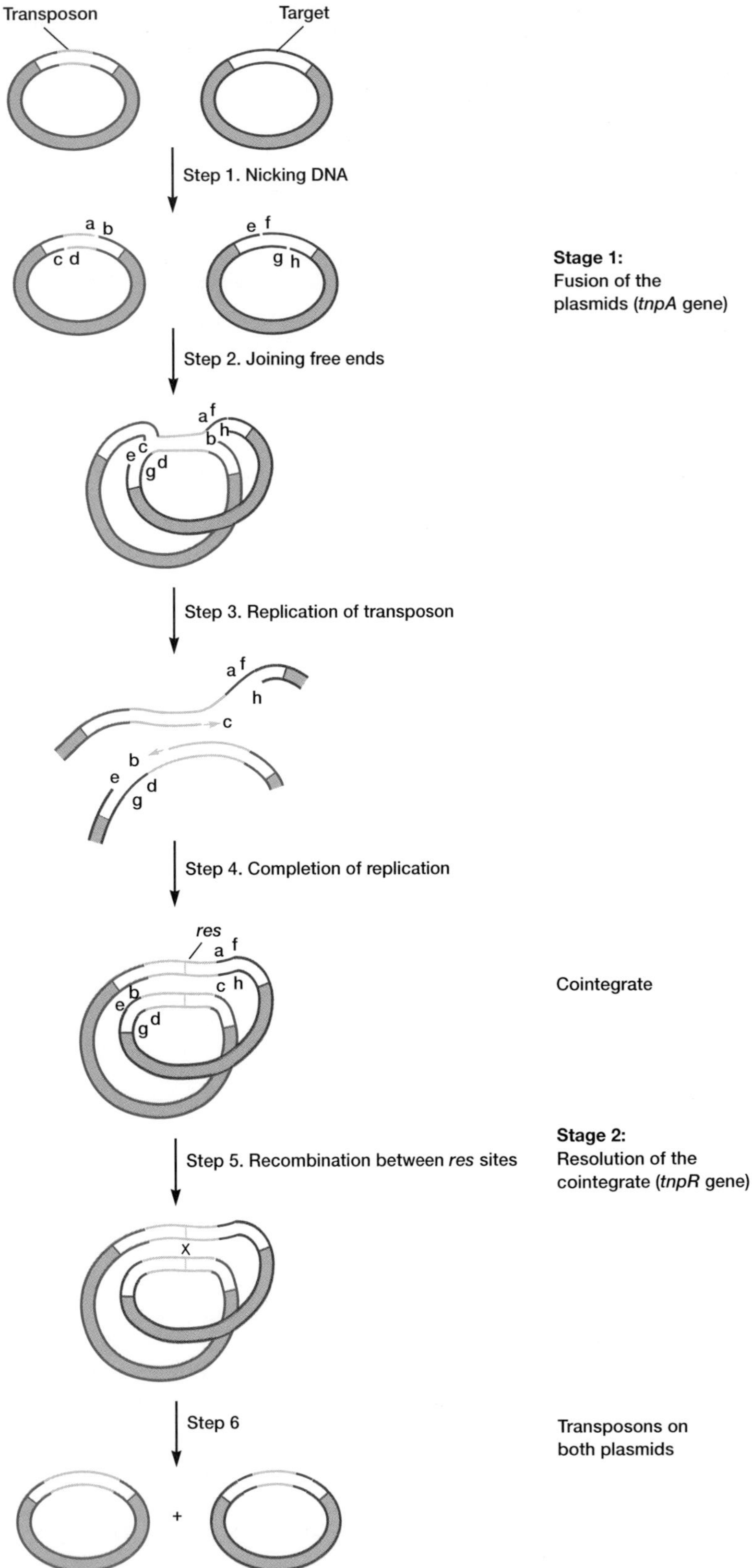

Figure 13.8 Tn3 Transposition Mechanism. Step 1: The two plasmids are nicked to form the free ends labeled a–h. Step 2: Ends a and f are joined, as are g and d. This leaves b, c, e, and h free. Step 3: Two of these remaining free ends (b and c) serve as primers for DNA replication, which is shown in a blowup of the replicating region. Step 4: Replication continues until end b reaches e and end c reaches h. These ends are ligated to complete the cointegrate. Notice that the whole transposon has been replicated. The paired *res* sites are shown for the first time here, even though one *res* site existed in the previous steps. Steps 5 and 6: A crossover occurs between the two *res* sites in the two copies of the transposon, leaving two independent plasmids, each bearing a copy of the transposon.

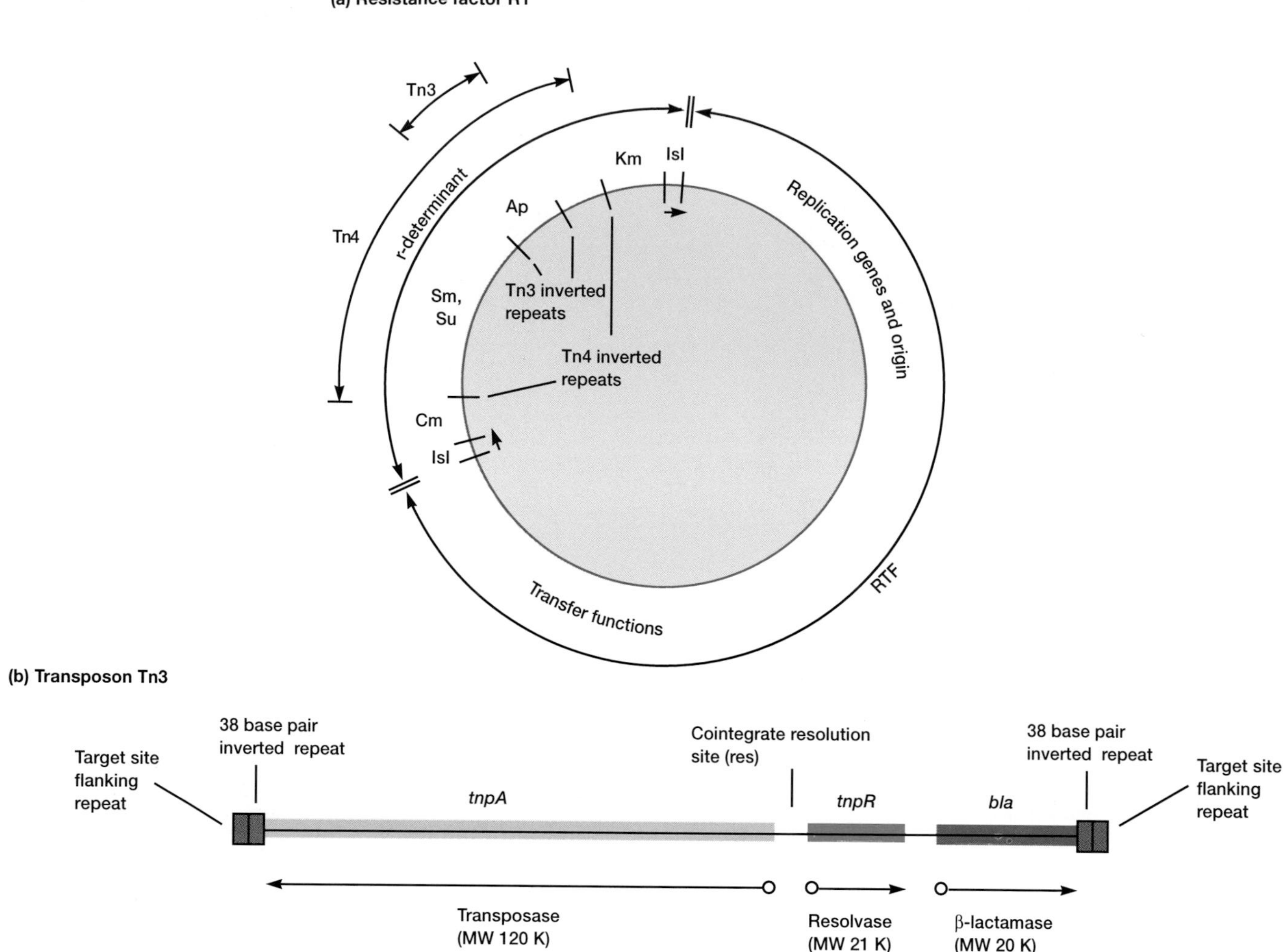

Figure 13.9 **The Structure of R Plasmids and Transposons.** (*a*) The R1 plasmid carries resistance genes for five antibiotics: chloramphenicol (*Cm*), streptomycin (*Sm*), sulfonamide (*Su*), ampicillin (*Ap*), and kanamycin (*Km*). These are contained in the *Tn3* and *Tn4* transposons. The resistance transfer factor (*RTF*) codes for the proteins necessary for plasmid replication and transfer. (*b*) The structure of Tn3. The arrows indicate the direction of gene transcription. See also figure 13.6*c*.

role in generating these plasmids. Consequently, the existence of these elements causes serious problems in the treatment of disease. Since plasmids can contain several different target sites, transposons will move between them; thus, plasmids act as both the source and the target for transposons with resistance genes. In fact, it appears that multiple drug resistance plasmids are usually produced by transposon accumulation on a single plasmid (figure 13.9). Because transposons also move between plasmids and primary chromosomes, drug resistance genes can exchange between plasmids and chromosomes, resulting in the further spread of antibiotic resistance.

Transposable elements are widespread in nature. They are present in eucaryotic as well as procaryotic systems. For example, transposable elements have been found in yeast, maize, *Drosophila,* and humans. Clearly, transposable elements play an extremely important role in the generation and transfer of new gene combinations.

1. Define the following terms: transposition, transposable element or transposon, insertion sequence, transposase, and composite transposon. Be able to distinguish between an insertion sequence and a composite transposon.
2. How does transposition usually occur in bacteria, and what happens to the target site? What is replicative transposition?
3. Give several important effects transposable elements can have on their possessors.

Bacterial Conjugation

The initial evidence for bacterial **conjugation,** the transfer of genetic information via direct cell-cell contact, came from a simple experiment performed by Joshua Lederberg and Edward L. Tatum in 1946. They mixed two auxotrophic strains, incubated the culture for several hours in nutrient

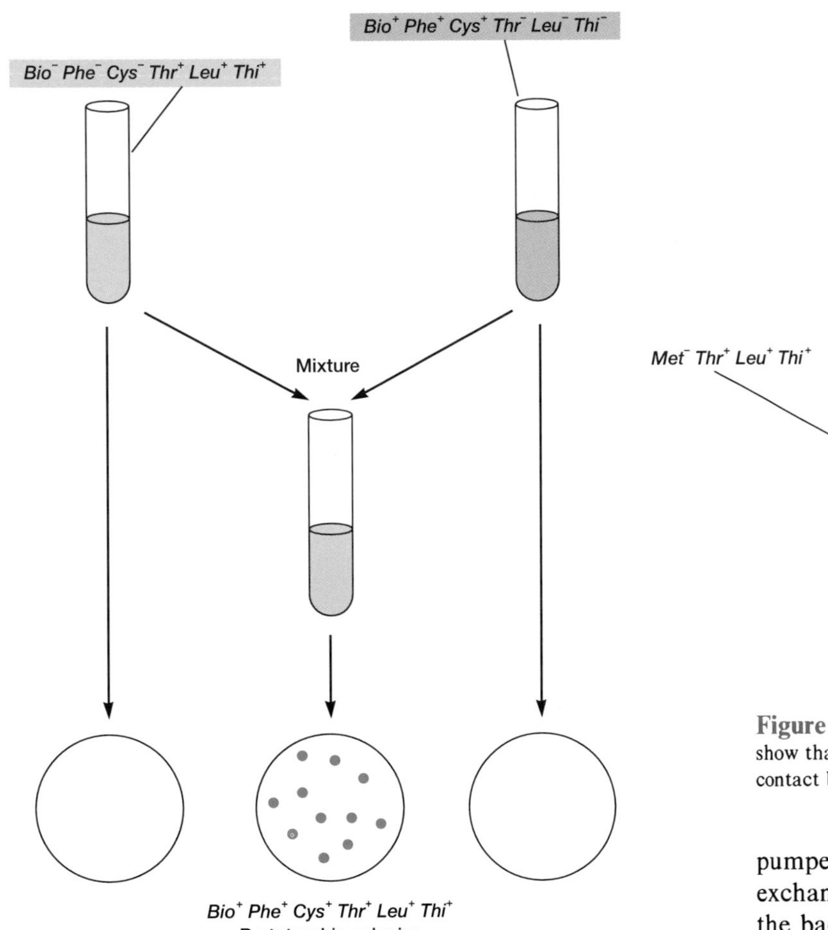

Figure 13.10 **Evidence for Bacterial Conjugation.** Lederberg and Tatum's demonstration of genetic recombination using triple auxotrophs. See text for details.

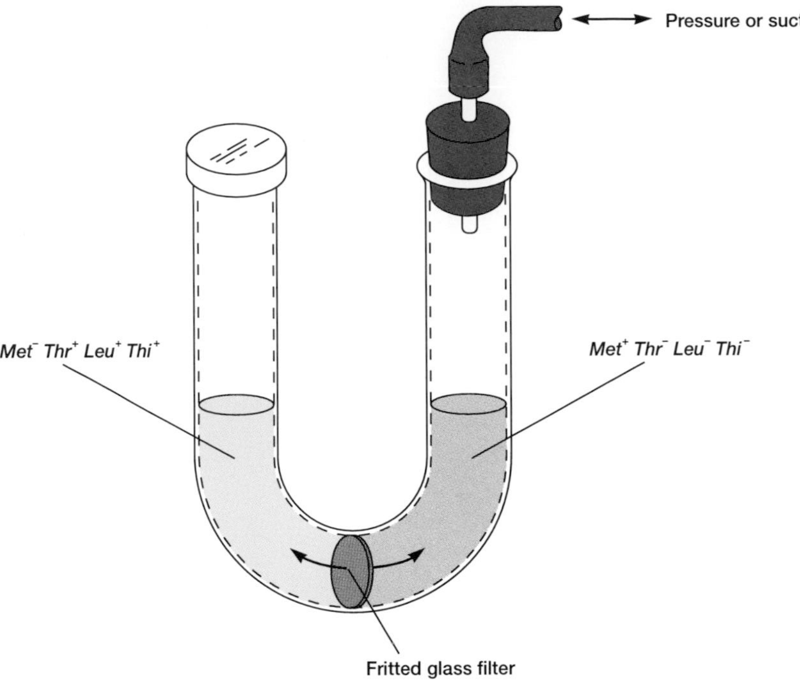

Figure 13.11 **The U-Tube Experiment.** The U-tube experiment used to show that genetic recombination by conjugation requires direct physical contact between bacteria. See text for details.

medium, and then plated it on minimal medium. To reduce the chance that their results were due to simple reversion, they used double and triple auxotrophs on the assumption that two or three reversions would not often occur simultaneously. For example, one strain required biotin (Bio^-), phenylalanine (Phe^-), and cysteine (Cys^-) for growth, and another needed threonine (Thr^-), leucine (Leu^-), and thiamine (Thi^-). Recombinant prototrophic colonies appeared on the minimal medium after incubation (figure 13.10). Thus, the chromosomes of the two auxotrophs were able to associate and undergo recombination.

Lederberg and Tatum did not directly prove that physical contact of the cells was necessary for gene transfer. This evidence was provided by Bernard Davis (1950), who constructed a U tube consisting of two pieces of curved glass tubing fused at the base to form a U shape with a fritted glass filter between the halves. The filter allows the passage of media but not bacteria. The U tube was filled with nutrient medium and each side inoculated with a different auxotrophic strain of *E. coli* (figure 13.11). During incubation, the medium was pumped back and forth through the filter to ensure medium exchange between the halves. After a four-hour incubation, the bacteria were plated on minimal medium. Davis discovered that when the two auxotrophic strains were separated from each other by the fine filter, gene transfer could not take place. Therefore, direct contact was required for the recombination that Lederberg and Tatum had observed.

$F^+ \times F^-$ Mating

In 1952, William Hayes demonstrated that the gene transfer observed by Lederberg and Tatum was polar. That is, there were definite donor (F^+) and recipient (F^-) strains, and gene transfer was nonreciprocal. He also found that in $F^+ \times F^-$ mating the progeny were only rarely changed with regard to auxotrophy (that is, bacterial genes were not often transferred), but F^- strains frequently became F^+.

These results are readily explained in terms of the F factor previously described. The F^+ strain contains an extrachromosomal F factor carrying the genes for pilus formation and plasmid transfer. During $F^+ \times F^-$ mating or conjugation, the F factor replicates by the rolling circle mechanism, and a copy of it moves to the recipient (figure 13.12*a*). The entering strand is copied to produce double-stranded DNA. Because bacterial chromosome genes are rarely transferred with the independent F factor, the recombination frequency is low. It is still not completely clear how the plasmid moves between bacteria. The **sex pilus** or F pilus joins the donor and recipient and may contract to draw them together. The channel for DNA transfer

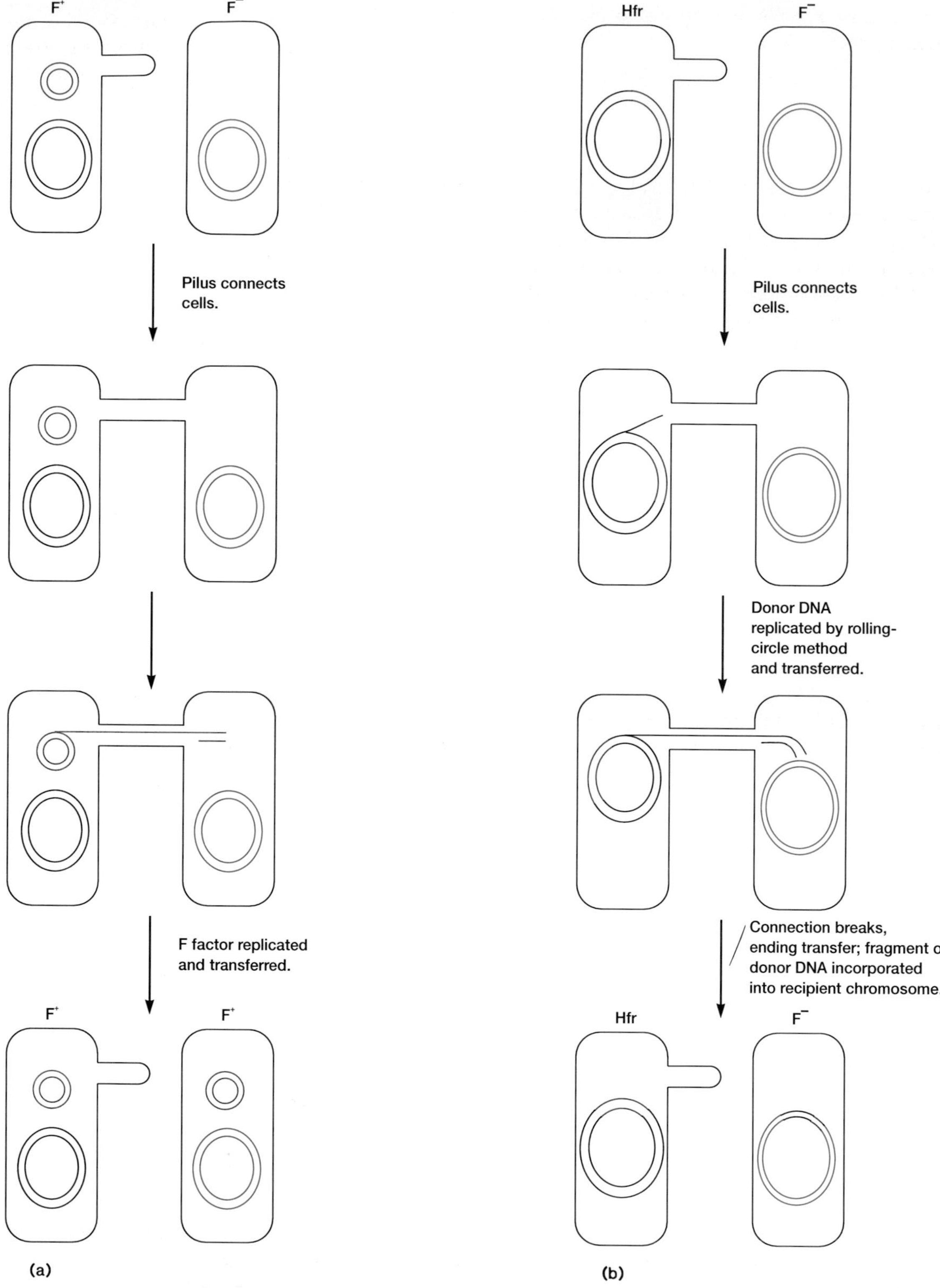

Figure 13.12 **The Mechanism of Bacterial Conjugation.** (*a*) $F^+ \times F^-$ mating. (*b*) $Hfr \times F^-$ mating (the integrated F factor is shown in color).

could be either the hollow F pilus or a special conjugation bridge formed upon contact.

The rolling-circle mechanism of DNA replication (p. 198).

Hfr Conjugation

Because certain donor strains transfer bacterial genes with great efficiency and do not usually change recipient bacteria to donors, a second type of conjugation must exist. The F factor is an episome and can integrate into the bacterial chromosome at several different locations by recombination between homologous insertion sequences present on both the plasmid and host chromosomes. When integrated, the F plasmid's *tra* operon is still functional; the plasmid can direct the synthesis of pili, carry out rolling circle replication, and transfer genetic material to an F^- recipient cell. Such a donor is called an **Hfr strain** (for high frequency of recombination) because it exhibits a very high efficiency of chromosomal gene transfer in comparison with F^+ cells. DNA transfer begins when the integrated F factor is nicked at its site of transfer origin (figure 13.12*b*). As it is replicated, the chromosome moves through the pilus or conjugation bridge connecting the donor and recipient. Because only part of the F factor is transferred at the start (the initial break is within the F plasmid), the F^- recipient does not become F^+ unless the whole chromosome is transferred. Transfer takes about 100 minutes in *E. coli,* and the connection usually breaks before this process is finished. Thus, a complete F factor is not usually transferred, and the recipient remains F^-.

As mentioned earlier, when an Hfr strain participates in conjugation, bacterial genes are frequently transferred to the recipient. Gene transfer can be in either a clockwise or counterclockwise direction around the circular chromosome, depending upon the orientation of the integrated F factor. After the replicated donor chromosome enters the recipient cell, it may be degraded or incorporated into the F^- genome by recombination. Hfr conjugation is perhaps the most efficient natural mechanism of gene transfer between bacteria.

F′ Conjugation

Because the F plasmid is an episome, it can leave the bacterial chromosome. Sometimes during this process the plasmid makes an error in excision and picks up a portion of the chromosomal material to form an **F′ plasmid** (figure 13.13*a*). It is not unusual to observe the inclusion of one or more genes in excised F plasmids. The F′ cell retains all of its genes, although some of them are on the plasmid, and still mates only with an F^- recipient. $F' \times F^-$ conjugation is virtually identical with $F^+ \times F^-$ mating. Once again, the plasmid is transferred, but usually bacterial genes on the chromosome are not (figure 13.13*b*). Bacterial genes on the F′ plasmid are transferred with it and need not be incorporated into the recipient chromosome to be expressed. The recipient becomes F′ and is a partially diploid merozygote since it has two sets of the genes carried by the plasmid. In this way, specific bacterial genes can spread rapidly throughout a bacterial population. Such transfer of bacterial genes is often called sexduction.

F′ conjugation is very important to the microbial geneticist. A partial diploid's behavior shows whether the allele carried by an F′ plasmid is dominant or recessive to the chromosomal gene. The formation of F′ plasmids is also useful in mapping the chromosome since if two genes are picked up by an F factor they must be neighbors.

---

1. What is bacterial conjugation and how was it discovered?
2. Distinguish between F^+, Hfr, and F^- strains of *E. coli* with respect to their physical nature and role in conjugation.
3. Describe in some detail how $F^+ \times F^-$ and Hfr conjugation processes proceed, and distinguish between the two in terms of mechanism and the final results.
4. What is F′ conjugation and why is it so useful to the microbial geneticist? How does the F′ plasmid differ from a regular F plasmid? What is sexduction?

DNA Transformation

The second way in which DNA can move between bacteria is through transformation, discovered by Fred Griffith. **Transformation** is the uptake by a cell of a naked DNA molecule or fragment from the medium and the incorporation of this molecule into the recipient chromosome in a heritable form. In natural transformation, the DNA comes from a donor bacterium. The process is random, and any portion of a genome may be transferred between bacteria.

The discovery of transformation (pp. 237–38).

When bacteria lyse, they release considerable amounts of DNA into the surrounding environment. These fragments may be relatively large and contain several genes. If a fragment contacts a **competent** cell, one able to take up DNA and be transformed, it can be bound to the cell and taken inside (figure 13.14*a*). The transformation frequency of very competent cells is around 10^{-3} for most genera when an excess of DNA is used. That is, about one cell in every thousand will take up and integrate the gene. Competency is a complex phenomenon and is dependent on several conditions. Bacteria need to be in a certain stage of growth; for example, *S. pneumoniae* becomes

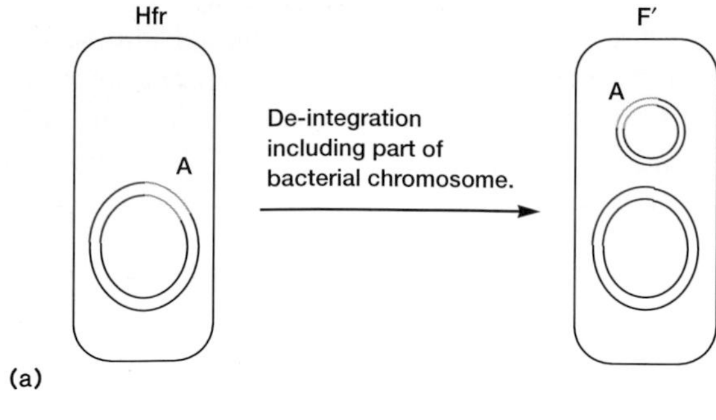

Figure 13.13 **F′ Conjugation.** (*a*) Due to an error in excision, the *A* gene of an *Hfr* cell is picked up by the *F* factor. (*b*) The *A* gene is then transferred to a recipient during conjugation.

competent during the exponential phase when the population reaches about 10^7 to 10^8 cells per ml. When a population becomes competent, bacteria such as pneumococci secrete a small protein called the competence factor that stimulates the production of 8 to 10 new proteins required for transformation. Natural transformation has been discovered so far only in certain gram-positive and gram-negative genera: *Streptococcus, Bacillus, Thermoactinomyces, Haemophilus, Neisseria, Moraxella, Acinetobacter, Azotobacter,* and *Pseudomonas.* Other genera also may be capable of transformation. Gene transfer by this process occurs in soil and marine environments, and may be an important route of genetic exchange in nature.

The mechanism of transformation has been intensively studied in *S. pneumoniae* (figure 13.15). A competent cell binds a double-stranded DNA fragment if the fragment is

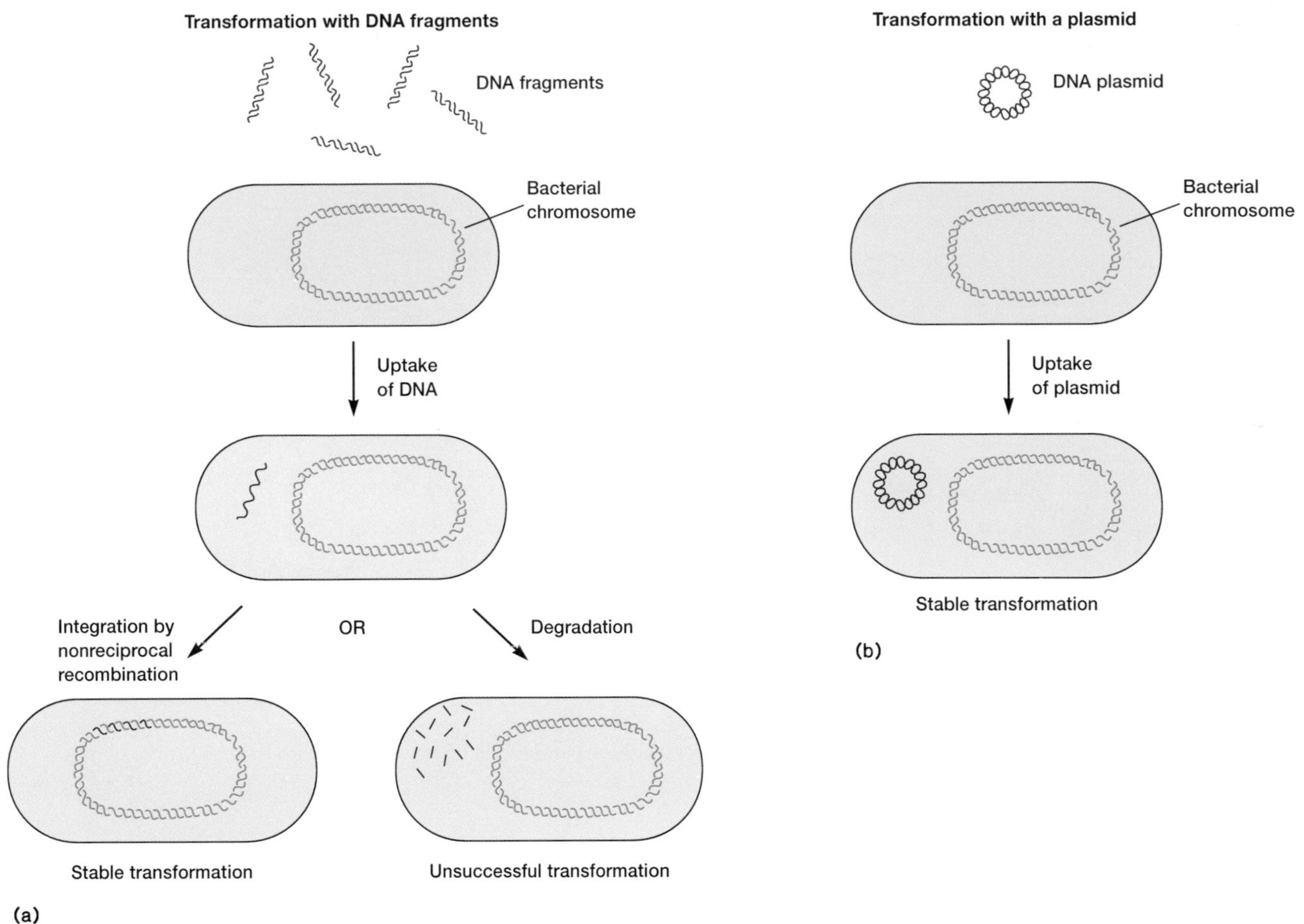

Figure 13.14 **Bacterial Transformation.** Transformation with (*a*) DNA fragments and (*b*) plasmids. Transformation with a plasmid is often induced artificially in the laboratory. See text. The transforming DNA is in red.

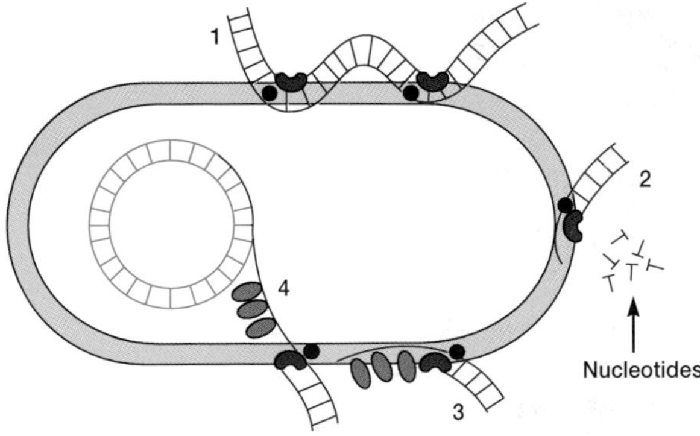

Figure 13.15 **The Mechanism of Transformation.** (*1*) A long double-stranded DNA molecule binds to the surface with the aid of a DNA-binding protein (•) and is nicked by a nuclease. (*2*) One strand is degraded by the nuclease. (*3*) The undegraded strand associates with a competence-specific protein. (*4*) The single strand enters the cell and is integrated into the host chromosome in place of the homologous region of the host DNA.

moderately large (greater than about 5×10^5 daltons); the process is random, and donor fragments compete with each other. DNA uptake requires energy expenditure. One strand is hydrolyzed by an envelope-associated exonuclease; the other strand associates with small proteins and moves through the plasma membrane. The single-stranded fragment can then align with a homologous region of the genome and be integrated, probably by a mechanism similar to that depicted in figure 13.3.

Transformation in *Haemophilus influenzae,* a gram-negative bacterium, differs from that in *S. pneumoniae* in several respects. *Haemophilus* does not produce a competence factor to stimulate the development of competence, and it takes up DNA from only closely related species (*S. pneumoniae* is less particular about the source of its DNA). Double-stranded DNA, complexed with proteins, is taken in by membrane vesicles. The specificity of *Haemophilus* transformation is due to a special 11 base pair sequence (5′-AAGTGCGGTCA-3′) that is repeated about 600 times in *H. influenzae* DNA. DNA must have this sequence to be bound by a competent cell.

Artificial transformation is carried out in the laboratory by a variety of techniques, including treatment of the cells with calcium chloride, which renders their membranes more

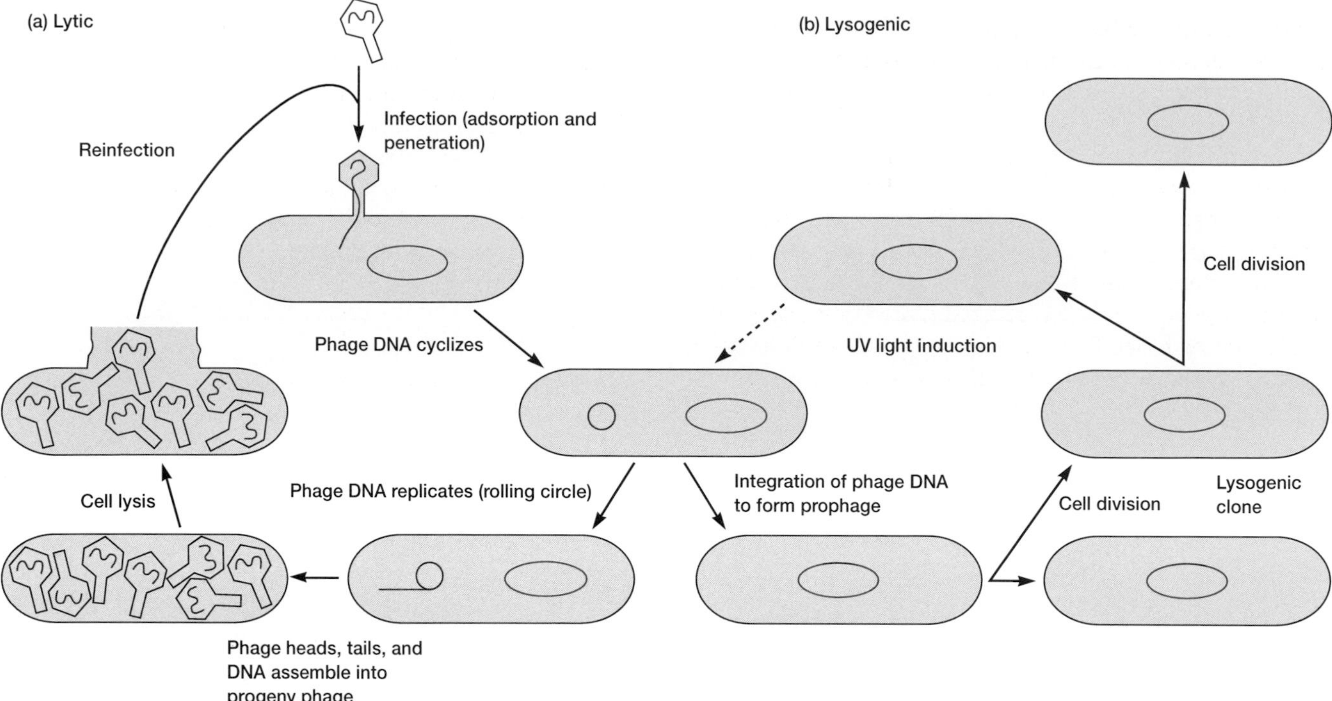

Figure 13.16 **Lytic versus Lysogenic Infection by Phage λ.** (*a*) Lytic infection. (*b*) Lysogenic infection. Viral and prophage DNA are in red.

permeable to DNA. This approach succeeds even with species that are not naturally competent, such as *E. coli.* Relatively high concentrations of DNA, higher than would normally be present in nature, are used to increase transformation frequency. When linear DNA fragments are to be used in transformation, *E. coli* is usually rendered deficient in one or more exonuclease activities to protect the transforming fragments. It is even easier to transform bacteria with plasmid DNA since plasmids are not as easily degraded as linear fragments and can replicate within the host (figure 13.14*b*). This is a common method for introducing recombinant DNA into bacterial cells (*see chapter 14*). DNA from any source can be introduced into bacteria by splicing it into a plasmid before transformation.

1. Define transformation and competence.
2. Describe how transformation occurs in *S. pneumoniae.* How does the process differ in *H. influenzae?*
3. Discuss two ways in which artificial transformation can be used to place functional genes within bacterial cells.

Transduction

Bacterial viruses or bacteriophages participate in the third mode of bacterial gene transfer. These viruses have relatively simple structures in which virus genetic material is enclosed within an outer coat, composed mainly or solely of protein. The coat protects the genome and transmits it between host cells. The morphology and life cycle of bacteriophages is not discussed in detail until chapter 18. Nevertheless, it is necessary to briefly describe the life cycle here as background for a consideration of the bacteriophage's role in gene transfer.

The lytic cycle (pp. 368–74).
Lysogeny (pp. 375–81).

After infecting the host cell, a bacteriophage (phage for short) often takes control and forces the host to make many copies of the virus. Eventually the host bacterium bursts or lyses and releases new phages. This reproductive cycle is called a lytic cycle because it ends in lysis of the host. The cycle has four phases (figure 13.16*a*). First the virus particle attaches to a specific receptor site on the bacterial surface. The genetic material, which is often double-stranded DNA, then enters the cell. After adsorption and penetration, the virus chromosome forces the bacterium to make virus nucleic acids and proteins. The third stage begins after the synthesis of virus components. Phages are assembled from these components. The assembly process may be complex, but in all cases phage nucleic acid is packed within the virus's protein coat. Finally, the mature viruses are released by cell lysis.

Bacterial viruses that reproduce using a lytic cycle are often called virulent bacteriophages because they destroy the host cell. Many DNA phages, such as the lambda phage (*p. 376*), are also capable of a different relationship with their host (figure 13.16*b*). After adsorption and penetration, the viral genome does not take control of its host and destroy it while producing new phages. Instead, the genome remains within the host cell and is reproduced along with the bacterial chromosome. A clone of infected cells arises and may grow for long periods while appearing perfectly normal. Each of these infected bacteria can produce phages and lyse under appropriate environmental conditions. This relationship between the phage and its host is called **lysogeny.** Bacteria that can produce phage particles under some conditions are said to be **lysogens** or **lysogenic,** and phages able to establish this relationship are **temperate phages.** The latent form of the virus genome that remains within the host without destroying it is called the **prophage.** The prophage is usually integrated into the bacterial genome (figure 13.16*b*). Sometimes phage reproduction is triggered in a lysogenized culture by exposure to UV radiation or other factors. The lysogens are then destroyed and new phages released. This phenomenon is called induction.

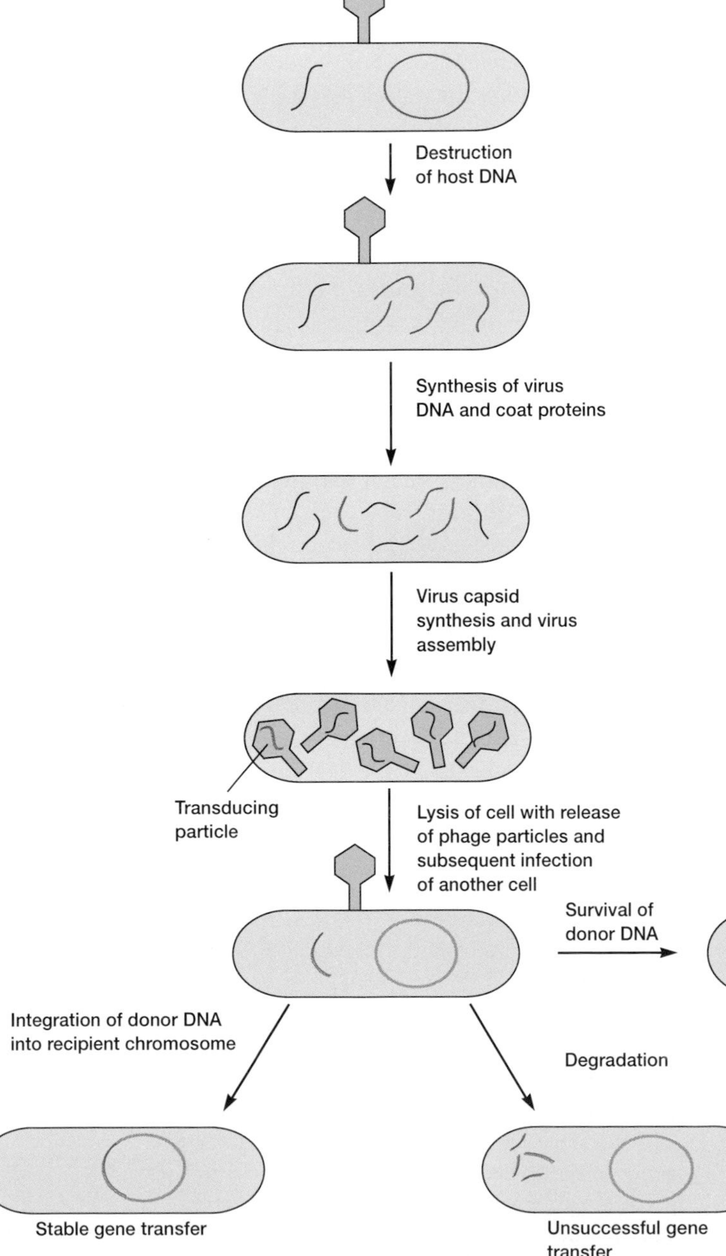

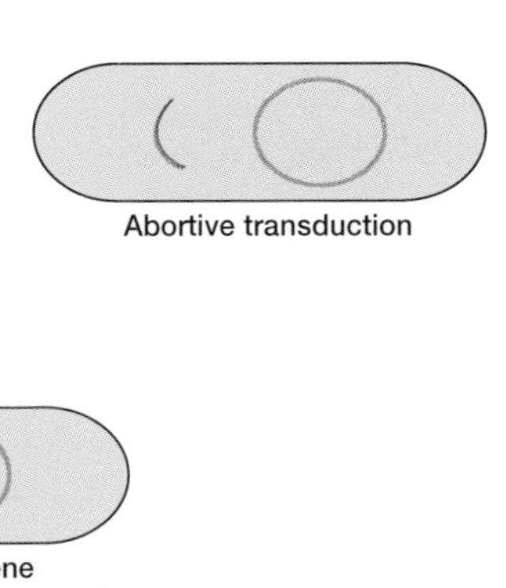

Figure 13.17 **Generalized Transduction by Bacteriophages.** See text for details.

Transduction is the transfer of bacterial genes by viruses. Bacterial genes are incorporated into a phage capsid because of errors made during the virus life cycle. The virus containing these genes then injects them into another bacterium, completing the transfer. Transduction may be the most common mechanism for gene exchange and recombination in bacteria. There are two very different kinds of transduction: generalized and specialized.

Generalized Transduction

Generalized transduction (figure 13.17) occurs during the lytic cycle of virulent and temperate phages and can transfer any part of the bacterial genome. During the assembly stage, when the viral chromosomes are packaged into protein capsids, random fragments of the partially degraded bacterial chromosome also may be packaged by mistake. Because the capsid can contain only a limited quantity of DNA, some or all of the viral DNA is left behind. The quantity of bacterial DNA carried depends primarily on the size of the capsid. The P22 phage of *Salmonella typhimurium* usually carries about 1% of the bacterial genome; the P1 phage of *E. coli* carries about 2.0 to 2.5% of the genome. The resulting virus particle often injects the DNA into another bacterial cell but does not initiate a lytic cycle. This phage is known as a **generalized transducing particle** or phage and is simply a carrier of genetic information from the original bacterium to another cell. As in transformation, once the DNA has been injected, it must be incorporated into the recipient cell's chromosome to preserve the transferred genes. The DNA remains double stranded during transfer, and both strands are integrated into the endogenote. About 70 to 90% of the transferred DNA is not integrated but is often able to survive and express itself. **Abortive transductants** are bacteria that contain this nonintegrated, transduced DNA and are partial diploids.

Generalized transduction was discovered in 1951 by Joshua Lederberg and Norton Zinder during an attempt to show that conjugation, discovered several years earlier in *E. coli,* could occur in other bacterial species. Lederberg and Zinder were repeating the earlier experiments with *Salmonella typhimurium.* They found that incubation of a mixture of two multiply auxotrophic strains yielded prototrophs at the level of about one in 10^5. This seemed like good evidence for bacterial recombination, and indeed it was, but their initial conclusion that the transfer resulted from conjugation was not borne out. When these investigators performed the U-tube experiment (figure 13.11) with *Salmonella,* they still recovered prototrophs. The filter in the U tube had small enough pores to block the movement of bacteria between the two sides but allowed the phage P22 to pass. Lederberg and Zinder had intended to confirm that conjugation was present in another bacterial species and had instead discovered a completely new mechanism of bacterial gene transfer. The seemingly routine piece of research led to surprising and important results. A scientist must always keep an open mind about results and be prepared for the unexpected.

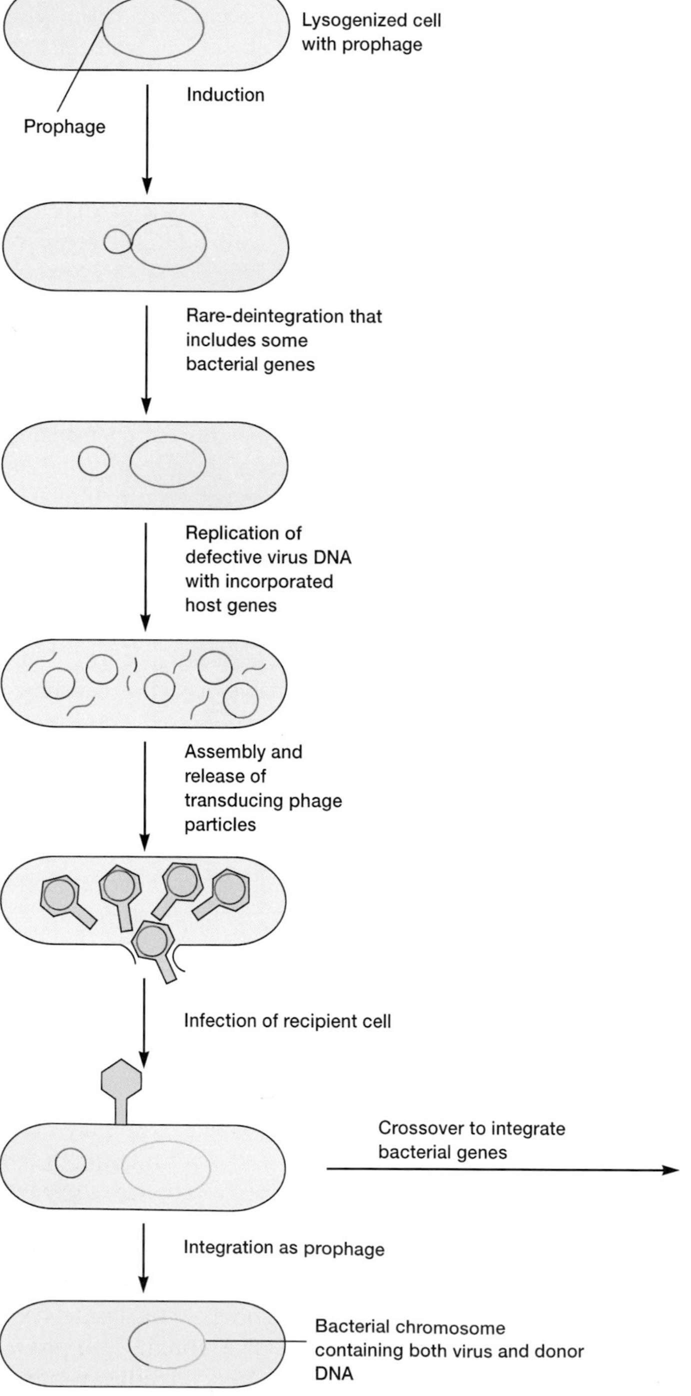

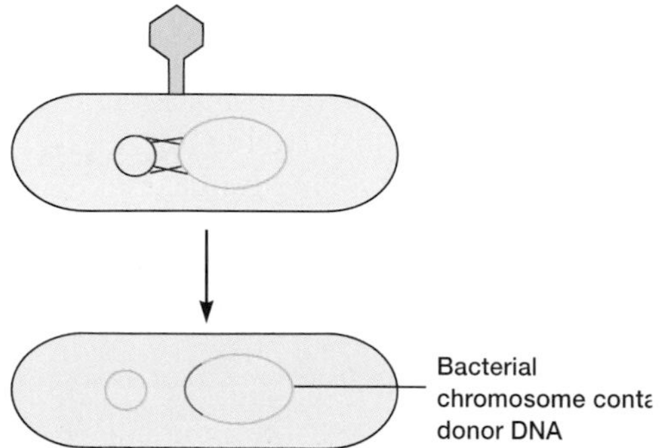

Figure 13.18

Specialized Transduction

In **specialized** or **restricted transduction,** the transducing particle carries only specific portions of the bacterial genome. Specialized transduction is made possible by an error in the lysogenic life cycle. When a prophage is induced to leave the host chromosome, excision is sometimes carried out improperly. The resulting phage genome contains portions of the bacterial chromosome (about 5 to 10% of the bacterial DNA) next to the integration site, much like the situation with F′ plasmids (figure 13.18). A transducing phage genome is usually defective and lacks some part of its attachment site. The transducing particle will inject bacterial genes into another bacterium, even though the defective phage cannot reproduce without assistance. The bacterial genes may become stably incorporated under the proper circumstances.

The best-studied example of specialized transduction is the lambda phage. The lambda genome inserts into the host chromosome at specific locations known as attachment or *att* sites (figure 13.19; *see also figure 18.16*). The phage *att* sites and bacterial *att* sites are similar and can complex with each other, although they are not identical. The *att* site for lambda is next to the *gal* and *bio* genes on the *E. coli* chromosome; consequently, specialized transducing lambda phages most often carry these bacterial genes. The lysate, or product of cell lysis, resulting from the induction of lysogenized *E. coli* contains normal phage and a few defective transducing particles. These particles are called lambda d*gal* because they carry the galactose utilization genes (figure 13.19). Because these lysates contain only a few transducing particles, they are often called **low-frequency transduction lysates (LFT lysates).** Whereas the normal phage has a complete *att* site, defective transducing particles have a nonfunctional hybrid integration

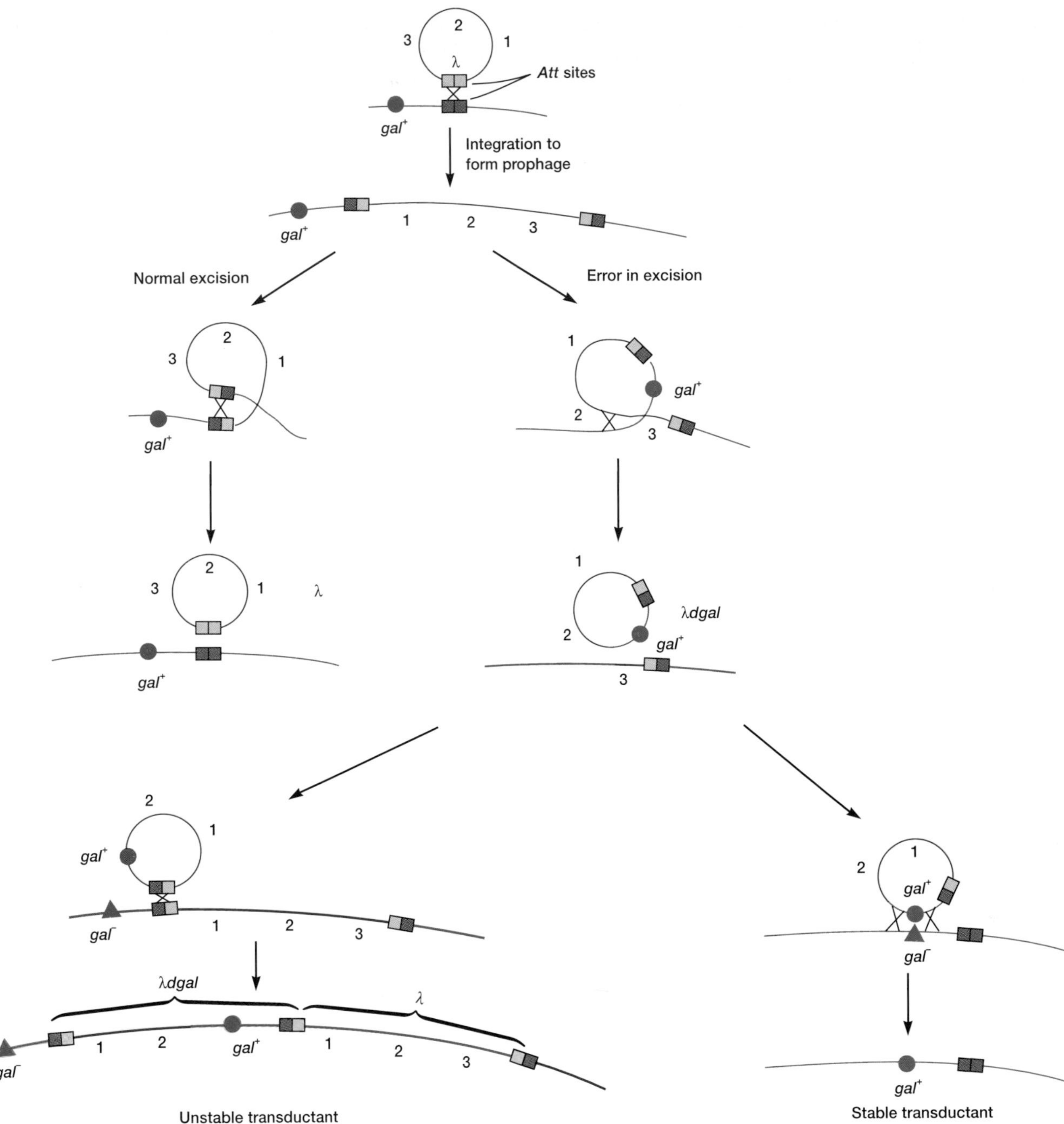

Figure 13.19 The Mechanism of Transduction for Phage Lambda and *E. coli*.

site that is part bacterial and part phage in origin. Integration of the defective phage chromosome does not readily take place. Transducing phages also may have lost some genes essential for reproduction. Stable transductants can arise from recombination between the phage and the bacterial chromosome because of crossovers on both sides of the *gal* site (figure 13.19).

The lysogenic cycle of lambda phage (pp. 376–81).

Defective lambda phages carrying the *gal* gene can integrate if there is a normal lambda phage in the same cell. The normal phage will integrate, yielding two bacterial/phage hybrid *att* sites where the defective lambda d*gal* phage can insert (figure 13.19). It also supplies the genes missing in the defective phage. The normal phage in this instance is termed the **helper phage** because it aids integration and reproduction of the defective phage. These transductants are unstable because the prophages can be induced to excise by agents such as UV radiation. Excision, however, produces a lysate containing a fairly equal mixture of defective lambda d*gal* phage and normal helper phage. Because it is very effective in transduction, the lysate is called a **high-frequency transduction lysate (HFT lysate).** Reinfection of bacteria with this mixture will result in the generation of considerably more transductants. LFT lysates and those produced by generalized transduction have one transducing particle in 10^5 or 10^6 phages; HFT lysates contain transducing particles with a frequency of about 0.1 to 0.5.

1. Briefly describe the lytic and lysogenic viral reproductive cycles. Define lysogeny, lysogen, temperate phage, prophage, and transduction.
2. Describe generalized transduction, how it occurs, and the way in which it was discovered. What is an abortive transductant?
3. What is specialized or restricted transduction and how does it come about? Be able to distinguish between LFT and HFT lysates and describe how they are formed.

Mapping the Genome

Finding the location of genes in any organism's genome is a very complex task. This section surveys approaches to mapping the bacterial genome, using *E. coli* as an example. All three modes of gene transfer and recombination have been used in mapping.

Gene structure and the nature of mutations (pp. 240–50).

Hfr conjugation is frequently used to map the relative location of bacterial genes. This technique rests on the observation that during conjugation the linear chromosome moves from donor to recipient at a constant rate. In an **interrupted mating experiment,** the conjugation bridge is broken and Hfr $\times$ F^- mating is stopped at various intervals after the start of conjugation by mixing the culture vigorously in a blender (figure 13.20*a*). The order and timing of gene transfer can be determined because they are a direct reflection of the order of genes on the bacterial chromosome (figure 13.20*b*). For example, extrapolation of the curves in figure 13.20*b* back to the x-axis will give the time at which each gene just began to enter the recipient. The result is a circular chromosome map with distances expressed in terms of the minutes elapsed until a gene is transferred. This technique can fairly precisely locate genes three minutes or more apart. The heights of the plateaus in figure 13.20*b* are lower for genes that are more distant from the F factor (the origin of transfer) because there is an ever-greater chance that the conjugation bridge will spontaneously break as transfer continues. Because of the relatively large size of the *E. coli* genome, it is not possible to generate a map from one Hfr strain. Therefore, several Hfr strains with the F plasmid integrated at different locations must be used and their maps superimposed on one another. The overall map is adjusted to 100 minutes, although complete transfer may require somewhat more than 100 minutes. Zero time is set at the threonine (*thr*) locus.

Gene linkage also can be determined from transformation by measuring the frequency with which two or more genes simultaneously transform a recipient cell. Consider the case for cotransformation by two genes. In theory, a bacterium could simultaneously receive two genes, each carried on a separate DNA fragment. However, it is much more likely that the genes reside on the same fragment. If two genes are closely linked on the chromosome, then they should be able to cotransform. The closer the genes are together, the more often they will be carried on the same fragment and the higher will be the frequency of cotransformation. If genes are spaced a great distance apart, they will be carried on separate DNA fragments and the frequency of double transformants will equal the product of the individual transformation frequencies.

Generalized transduction can be used to obtain linkage information in much the same way as transformation. Linkages are usually expressed as cotransduction frequencies, using the argument that the closer two genes are to each other, the more likely they both will reside on the DNA fragment incorporated into a single phage capsid.

Specialized transduction is used to find which phage attachment site is close to a specific gene. The relative locations of specific phage *att* sites are known from conjugational mapping, and the genes linked to each *att* site can be determined by means of specialized transduction. These data allow precise placement of genes on the chromosome.

A simplified genetic map of *E. coli* K12 is given in figure 13.21 and table 13.4. Because conjugation data are not high resolution and cannot be used to position genes that are very close together, the whole map is developed using several mapping techniques. Usually, interrupted mating data are combined with those from cotransduction and cotransformation studies. Data from recombination studies are also used. Normally, a new genetic marker in the *E. coli* genome is located within a relatively small region of the genome (10 to 15 minutes long) using a series of Hfr strains with F factor integration sites scattered throughout the genome. Once the genetic marker has been located with respect to several genes in the same region, its position relative to nearby neighbors is more accurately determined using transformation and transduction studies. Recent maps of the *E. coli* chromosome give the locations of more than a thousand genes. Remember that the genetic map only depicts physical reality in a relative sense. A map unit in one region of the genome may not be the same physical distance as a unit in another part.

Genetic maps provide useful information in addition to the order of the genes. For example, there is considerable clustering of genes in *E. coli* K12 (figure 13.21). In the regions around 2, 17, and 27 minutes, there are many genes, whereas relatively few genetic markers are found in the 33-minute region. The areas apparently lacking genes may well have undiscovered genes, but perhaps their function is not primarily that of coding genetic information. One hypothesis is that the 33-minute region is involved in attachment of the *E. coli* chro-

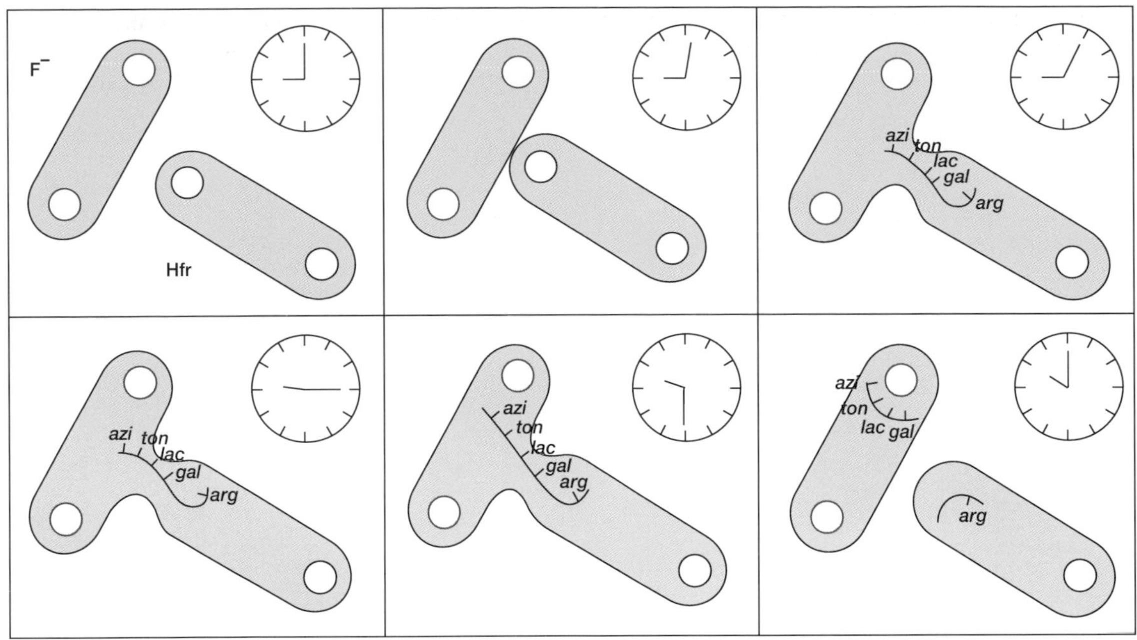

(a)

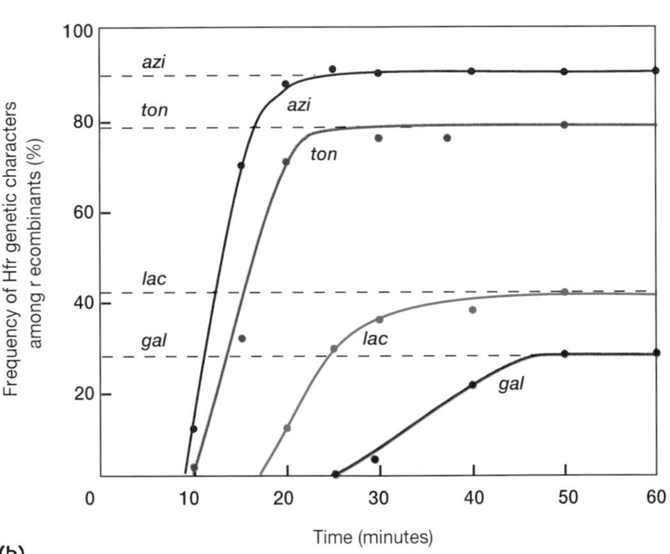

(b)

Figure 13.20 The Interrupted Mating Experiment. An interrupted mating experiment on Hfr × F⁻ conjugation. (*a*) The linear transfer of genes is stopped by breaking the conjugation bridge to study the sequence of gene entry into the recipient cell. (*b*) An example of the results obtained by an interrupted mating experiment. The gene order is *azi-ton-lac-gal*.

mosome to the membrane during replication and cell division. It is interesting that this region is almost exactly opposite the origin of replication for the chromosome (*oriC*).

1. Describe how the bacterial genome can be mapped using Hfr conjugation, transformation, generalized transduction, and specialized transduction. Include both a description of each technique and any assumptions underlying its use.
2. Why is it necessary to use several different techniques in genome mapping? How is this done in practice?

Recombination and Genome Mapping in Viruses

Bacteriophage genomes also undergo recombination, although the process is different from that in bacteria. Because phages themselves reproduce within cells and cannot recombine directly, crossing-over must occur inside a host cell. In principle, a virus recombination experiment is easy to carry out. If bacteria are mixed with enough phages, at least two virions will infect each cell on the average and genetic recombination should be observed. Phage progeny in the resulting lysate can be checked for alternate combinations of the initial parental genotypes.

Bacteriophage lytic cycle (pp. 368–74).

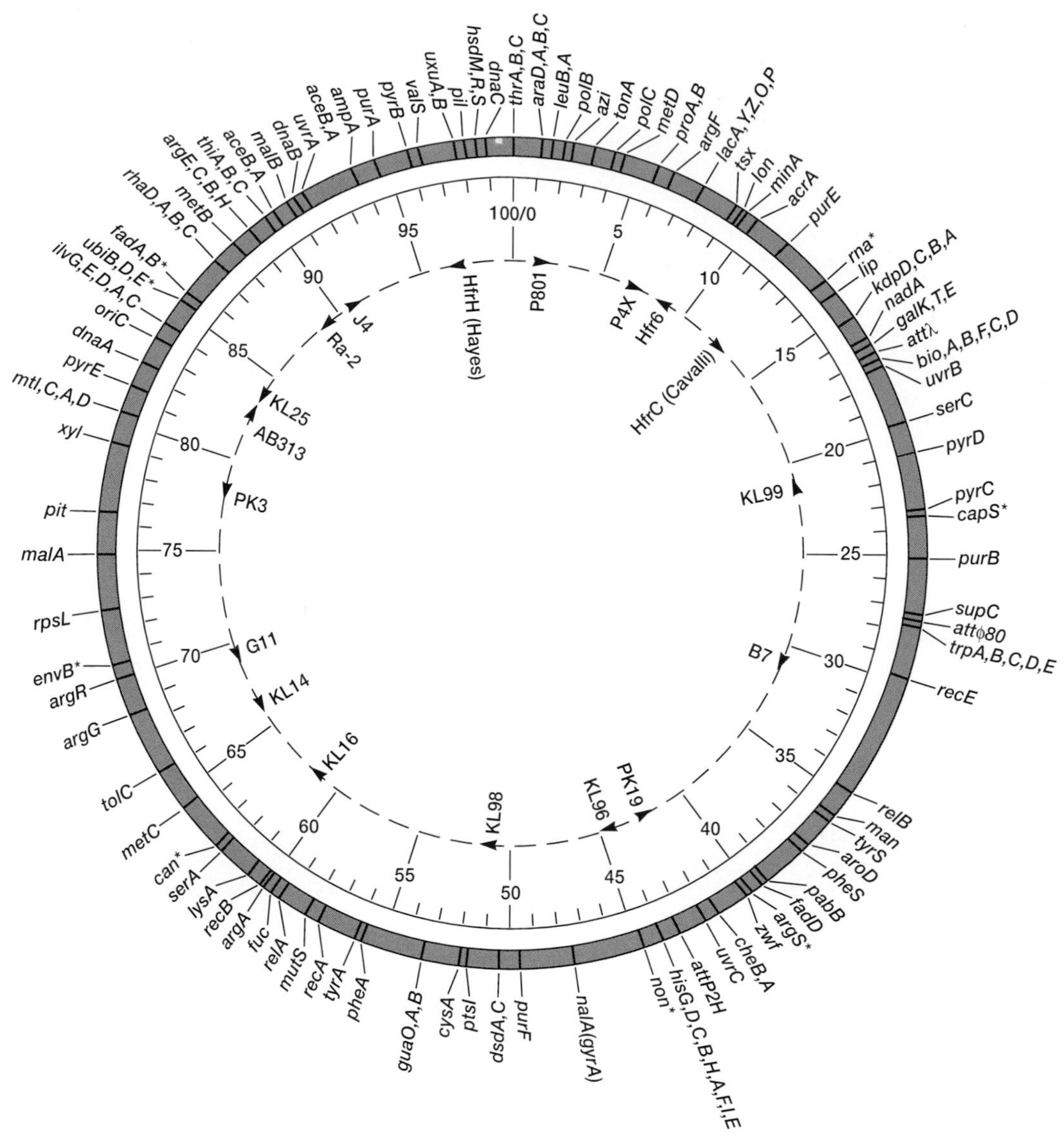

Figure 13.21 ***E. coli* Genetic Map.** A circular genetic map of *E. coli* K12 with the location of selected genes. The inner circle shows the origin and direction of transfer of several *Hfr* strains. The locations of asterisked genes are approximate. The map is divided into 100 minutes, the time required to transfer the chromosome from an Hfr cell to F^- at 37° C.

Alfred Hershey initially demonstrated recombination in the phage T2, using two strains with differing phenotypes. Two of the parental strains in Hershey's crosses were h^+r^+ and *hr*. The gene *h* influences host range; when gene *h* changes, T2 infects different strains of *E. coli*. Phages with the r^+ gene have wild type plaque morphology, while T2 with the *r* genotype has a rapid lysis phenotype and produces larger than normal plaques with sharp edges (figures 13.22*b* and 13.22*c*). In one experiment, Hershey infected *E. coli* with large quantities of the h^+r^+ and *hr* T2 strains (figure 13.22*a*). He then plated out the lysates with a mixture of two different host strains and was able to detect significant numbers of h^+r and hr^+ recombinants, as well as parental type plaques. As long as there are detectable phenotypes and methods for carrying out the crosses, it is possible to map phage genes in this way.

Plaque formation and morphology (p. 349).

Phage genomes are so small that it is often convenient to map them without determining recombination frequencies. Some techniques actually generate physical maps, which often are most useful in genetic engineering. Several of these methods require manipulation of the DNA with subsequent examination in the electron microscope. **Denaturation mapping** is an example of this approach. DNA is treated with mildly denaturing conditions so that strand separation occurs in AT-rich sequences because the base pairs are held together with only two hydrogen bonds, but not in GC-rich sequences (three hydrogen bonds per base pair). This causes the formation of bubbles that can be visualized with the electron microscope (figure 13.23). Mutants are compared to one another, and changes in the denaturation bubbles give the physical location of the mutations and their lengths.

TABLE 13.4 Selected *E. coli* Genes and Their Map Positions

Symbol	Map Position (Minutes)	Mutation (Affected Substance or Function)
aceA,B,E,F	91, 91, 3, 3	Acetate use
acrA	11	Acridine sensitivity
ade	See *pur*	Adenine
ampA	94	Penicillin resistance
araA,B,C,D,E	*1, 1, 1, 1, 61*	Arabinose use
argA,B,C,D,E,F,G, H,I,P,R,S,	61, 90, 90, 74, 90, 7, 69, 90, 97, 63, 71, 40	Arginine metabolism
aroA,B,C,D	20, 75, 51, 37	Aromatic compounds
*att*λ	17	Integration site for λ prophage
attP2H, II	44, 88	Integration sites for P2
*att*φ*80*	28	Integration site for φ80
azi	2	Azide resistance
bioA,B,C,D,F,H	17, 17, 17, 17, 17, 75	Biotin synthesis
can	63	Canavanine resistance
cheA,B	42, 41	Chemotactic motility
cysA,B,C,E,G	52, 28, 59, 81, 74	Cysteine metabolism
dnaA,B,C,E,G	83, 92, 99, 4, 67	DNA synthesis
dsdA,C	51, 51	d-serine
envA,B	2, 71	Cell envelope formation
fadA,B,D	87, 87, 40	Fatty acid degradation
fuc	60	Fucose metabolism
galE,K,P,R,T,U	17, 17, 64, 61, 17, 28	Galactose metabolism
guaA,B,C,	54, 54, 3	Guanine
hisA,B,C,D,E,F,G, H,I,R	44, 44, 44, 44, 44, 44, 44, 44, 44, 86	Histidine synthesis
hsdM,R,S	99, 99, 99	Host specificity
ilvA,B,C,D,E,F	85, 83, 85, 85, 85, 54	Isoleucine-valine
kdgK,R,T	78, 40, 88	Ketodeoxygluconate
kdpA,B,C,D	16, 16, 16, 16	Potassium dependence
lacA,I,Y,Z	8, 8, 8, 8	Lactose metabolism
leuA,B,S	2, 2, 15	Leucine synthesis
lip	15	Lipoate
lon	10	Filamentous growth
lysA,C	61, 91	Lysine synthesis
malE,F,G,K,P,Q,T	92, 92, 92, 92, 75, 75, 75	Maltose uptake and metabolism
manA	36	Mannose
metA,B,C,E,F	91, 89, 65, 86, 89	Methionine synthesis
minB	26	Minicell (no DNA)
mtlA,C,D	81, 81, 81	Mannitol
mutL,S,T	95, 59, 2	Mutator genes
nadA,B,C	17, 56, 3	NAD synthesis
nalA,B	48, 58	Nalidixic acid resistance
non	45	Blocks mucoid capsule
oriC	84	Origin of replication
pabA,B	74, 40	Para-aminobenzoate synthesis
panB,C,D	3, 3, 3	Pantothenate synthesis
pheA,S	57, 37	Phenylalanine
pil	98	Synthesis of fimbriae
pit	77	Inorganic phosphate transport
polA,B,C	87, 2, 4	DNA polymerases I, II, III
proA,B,C	6, 6, 9	Proline synthesis
ptsG,H,I	25, 52, 52	Phosphotransferase
purA,B,C,D,E,F, G,H,I	95, 25, 53, 90, 12, 50, 55, 90, 55	Purine synthesis
pyrA,B,C,D,E,F	1, 97, 24, 21, 82, 28	Pyrimidine synthesis
recA,B,C,D,E,F	58, 61, 61, 61, 30, 83	Ultraviolet sensitivity and genetic recombination
relA,B	60, 35	RNA synthesis regulation
rhaA,B,D,R	88, 88, 88, 88	Rhamnose
rna, rnb, rnc	14, 29, 55	Ribonucleases I, II, and III
rplA-Y, rpmA-J, rpsA-U	Various loci	Ribosomal proteins
rrfA-H, rrlA-H, rrnA-H, rrsA-H	Various loci	rRNA synthesis
serA,B,C,R,S	63, 100, 20, 2, 20	Serine metabolism
strC,M	7, 77	Streptomycin resistance
supB,C,D,E,F	16, 27, 43, 15, 27	Suppressors of various mutations
thiA,B,C	90, 90, 90	Thiamine (vitamin B_1)
thrA,B,C	0, 0, 0	Threonine
tolA,B,C,D	17, 17, 66, 23	Tolerance to various colicins
tonA,B	4, 28	Resistance to phages, iron uptake
trpA,B,C,D,E,R, S,T	28, 28, 28, 28, 28, 100, 74, 85	Tryptophan synthesis
tsxB	9	T6 resistance, nucleoside uptake
tyrA,R,S	57, 92, 29, 36	Tyrosine synthesis
ubiA,B,D,E,F,G	92, 87, 87, 87, 15, 48	Ubiquinone metabolism
uvrA,B,C,D	92, 18, 42, 86	Repair of ultraviolet damage
valS	97	Valine
xylA,B	80	Xylose metabolism
zwf	41	Glucose-6-phosphate dehydrogenase

A similar approach directly compares wild type and mutant viral chromosomes. In heteroduplex mapping, the two types of chromosomes are denatured, mixed, and allowed to rejoin or anneal. When joined, the homologous regions of the different DNA molecules form a regular double helix. In locations where the bases do not pair due to the presence of a mutation such as a deletion or insertion, bubbles will be visible in the electron microscope. This noncomplementary, single-stranded region is sometimes called a heteroduplex.

Several other direct techniques are used to map viral genomes or parts of them. Restriction endonucleases (*see chapter 14*) are employed together with electrophoresis to analyze

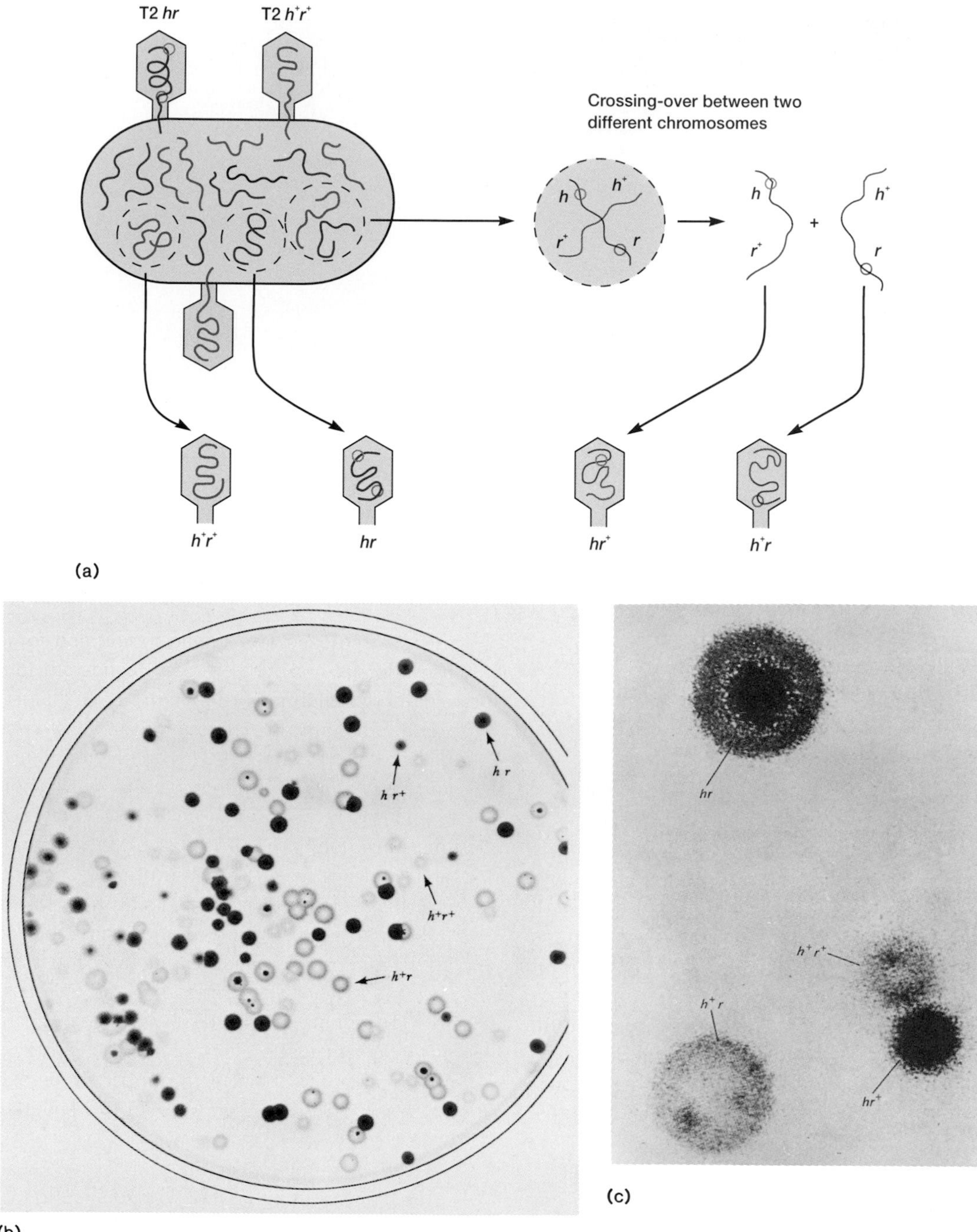

Figure 13.22 Genetic Recombination in Bacteriophages. (*a*) A summary of a genetic recombination experiment with the ***hr*** and h^+r^+ strains of the T2 phage. The *hr* chromosome is shown in color. (*b*) The types of plaques produced by this experiment on a lawn of *E. coli*. (*c*) A close-up of the four plaque types.

DNA fragments and locate deletions and other mutations that affect electrophoretic mobility. Phage genomes also can be directly sequenced to locate particular mutations and analyze the changes that have taken place.

It should be noted that many of these physical mapping techniques also have been employed in the analysis of relatively small portions of bacterial genomes. Furthermore, these methods are useful to the genetic engineer who is concerned with direct manipulation of the DNA.

1. How does recombination in viruses differ from that in bacteria? How did Hershey first demonstrate virus recombination?
2. Describe denaturation mapping and heteroduplex mapping.

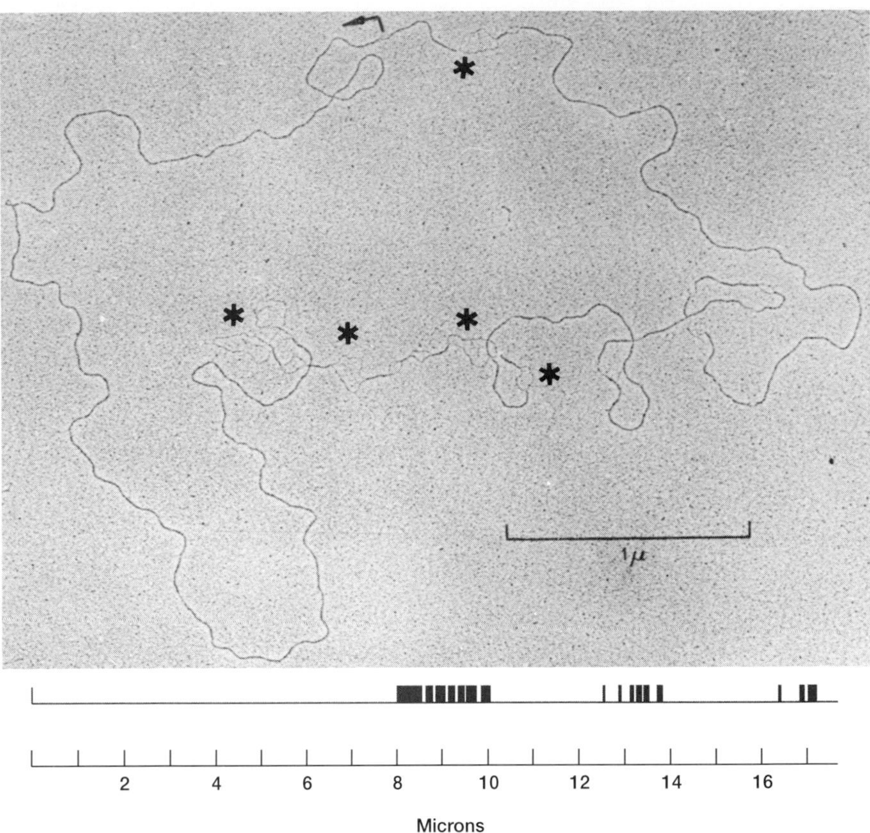

Figure 13.23 **An Example of Denaturation Mapping.** Phage lambda has been partially denatured to produce bubbles in the AT-rich regions. Some of the larger bubbles are labeled with asterisks. At the bottom, a denaturation map shows the position of these regions as vertical bars at the proper location on the ruler line. The map begins at the colored arrow and runs in a counterclockwise direction.

Summary

1. In recombination, genetic material from two different chromosomes is combined to form a new, hybrid chromosome. There are three types of recombination: general recombination, site-specific recombination, and replicative recombination.
2. Bacterial recombination is a one-way process in which the exogenote is transferred from the donor to a recipient and integrated into the endogenote.
3. Plasmids are small, circular, autonomously replicating DNA molecules that can exist independent of the host chromosome. Their genes are not required for host survival.
4. Episomes are plasmids that can be reversibly integrated with the host chromosome.
5. Many important types of plasmids have been discovered: F factors, R factors, Col plasmids, virulence plasmids, and metabolic plasmids.
6. Transposons or transposable elements are DNA segments that move about the genome in a process known as transposition.
7. There are two types of transposable elements: insertion sequences and composite transposons.
8. Transposable elements cause mutations, block translation and transcription, turn genes on and off, aid F plasmid insertion, and carry antibiotic resistance genes.
9. Conjugation is the transfer of genes between bacteria that depends upon direct cell-cell contact mediated by the F pilus.
10. In $F^+ \times F^-$ mating, the F factor remains independent of the chromosome and a copy is transferred to the F^- recipient; donor genes are not usually transferred.
11. Hfr strains transfer bacterial genes to recipients because the F factor is integrated into the host chromosome. A complete copy of the F factor is not often transferred.
12. When the F factor leaves an Hfr chromosome, it occasionally picks up some bacterial genes to become an F′ plasmid, which readily transfers these genes to other bacteria.
13. Transformation is the uptake of a naked DNA molecule by a competent cell and its incorporation into the genome.
14. Bacterial viruses or bacteriophages can reproduce and destroy the host cell (lytic cycle) or become a latent prophage that remains within the host (lysogenic cycle).
15. Transduction is the transfer of bacterial genes by viruses.
16. In generalized transduction, any host DNA fragment can be packaged in a virus capsid and transferred to a recipient.
17. Temperate phages carry out specialized transduction by incorporating bacterial genes during prophage induction and then donating those genes to another bacterium.
18. The bacterial genome can be mapped by following the order of gene transfer during Hfr conjugation; transformational and transductional mapping techniques also may be used.
19. When two viruses simultaneously enter a host cell, their chromosomes can undergo recombination.
20. Virus genomes are mapped by recombination, denaturation, and heteroduplex mapping techniques.

Key Terms

abortive transductants *275*
bacteriocin *262*
competent *271*
composite transposons *265*
conjugation *268*
conjugative plasmid *261*
crossing-over *259*
curing *261*
denaturation mapping *280*
endogenote *259*
episome *261*
exogenote *259*
F factor *261*
F′ plasmid *271*
generalized transducing particle *275*
generalized transduction *275*
general recombination *259*
helper phage *277*
heteroduplex DNA *259*
Hfr strains *271*
high-frequency transduction lysate (HFT lysate) *277*
host restriction *261*
insertion sequence *264*
interrupted mating experiment *278*
low-frequency transduction lysate (LFT lysate) *276*
lysogenic *275*
lysogens *275*
lysogeny *275*
merozygote *260*
metabolic plasmids *264*
plasmid *261*
prophage *275*
recombination *259*
restricted transduction *276*
R factors *261*
replicative recombination *259*
replicon *261*
sex pilus *269*
site-specific recombination *259*
specialized transduction *276*
temperate phage *275*
transduction *275*
transformation *271*
transposable element *264*
transposase *264*
transposition *264*
transposon *264*
virulence plasmids *264*

Questions for Thought and Review

1. How does recombination in procaryotes differ from that in most eucaryotes?
2. Distinguish between plasmids, transposons, and temperate phages.
3. How might one demonstrate the presence of a plasmid in a host cell?
4. What effect would you expect the existence of transposable elements and plasmids to have on the rate of microbial evolution? Give your reasoning.
5. How do multiple drug resistant plasmids often arise?
6. Suppose that you carried out a U-tube experiment with two auxotrophs and discovered that recombination was not blocked by the filter but was stopped by treatment with deoxyribonuclease. What gene transfer process is responsible? Why would it be best to use double or triple auxotrophs in this experiment?
7. List the similarities and differences between conjugation, transformation, and transduction.
8. How might one tell whether a recombination process was mediated by generalized or specialized transduction?
9. Why doesn't a cell lyse after successful transduction with a temperate phage?
10. Describe how you would precisely locate the recA gene and show that it was between 58 and 58.5 minutes on the *E. coli* chromosome.

Additional Reading

Chapter 12 references should also be consulted, particularly the introductory and advanced textbooks.

Bachmann, B. J. 1990. Linkage map of *Escherichia coli* K-12, Edition 8. *Microbiol. Rev.* 54(2):130–97.

Berg, C. M., and Berg, D. E. 1984. Jumping genes: The transposable DNAs of bacteria. *Am. Biol. Teach.* 46(8):431–39.

Berg, D. E., and Howe, M. M., eds. 1989. *Mobile DNA.* Washington, D.C.: American Society for Microbiology.

Brock, T. D. 1990. *The emergence of bacterial genetics.* Cold Spring Harbor, N.Y.: Cold Spring Harbor Laboratory Press.

Broda, P. 1979. *Plasmids.* San Francisco: W. H. Freeman.

Cohen, S. N., and Shapiro, J. A. 1980. Transposable genetic elements. *Sci. Am.* 242(2):40–49.

Dressler, D., and Potter, H. 1982. Molecular mechanisms in genetic recombination. *Ann. Rev. Biochem.* 51:727–61.

Elwell, L. P., and Shipley, P. L. 1980. Plasmid-mediated factors associated with virulence of bacteria to animals. *Ann. Rev. Microbiol.* 34:465–96.

Grindley, N. D. F., and Reed, R. R. 1985. Transpositional recombination in prokaryotes. *Ann. Rev. Biochem.* 54:863–96.

Hardy, K. 1986. *Bacterial plasmids,* 2d ed. Washington, D.C.: American Society for Microbiology.

Helinski, D. R. 1973. Plasmid determined resistance to antibiotics: Molecular properties of R factors. *Ann. Rev. Microbiol.* 27:437–70.

Hotchkiss, R. D. 1974. Models of genetic recombination. *Ann. Rev. Microbiol.* 28:445–68.

Ippen-Ihler, K. A., and Minkley, E. G., Jr. 1986. The conjugation system of F, the fertility factor of *Escherichia coli. Ann. Rev. Genet.* 20:593–624.

Kleckner, N. 1981. Transposable elements in prokaryotes. *Ann. Rev. Genet.* 15:341–404.

Kleckner, N. 1990. Regulation of transposition in bacteria. *Annu. Rev. Cell Biol.* 6:297–327.

Kucherlapati, R., and Smith, G. R., eds. 1988. *Genetic Recombination.* Washington, D.C.: American Society for Microbiology.

Levy, S. B., and Marshall, B. M. 1988. Genetic transfer in the natural environment. In *Release of genetically-engineered microorganisms,* eds. M. Sussman, G. H. Collins, F. A. Skinner, and D. E. Stewart-Tall, 61–76. San Diego, Calif.: Academic Press.

Lin, E. C. C.; Goldstein, R.; and Syvanen, M. 1984. *Bacteria, plasmids, and phages: An introduction to molecular biology.* Cambridge, Mass.: Harvard University Press.

Low, K. B., and Porter, R. D. 1978. Modes of gene transfer and recombination in bacteria. *Ann. Rev. Genet.* 12:249–88.

Mayer, L. W. 1988. Use of plasmid profiles in epidemiologic surveillance of disease outbreaks and in tracing the transmission of antibiotic resistance. *Clin. Microbiol. Rev.* 1(2):228–43.

McCarty, M. 1985. *The transforming principle: Discovering that genes are made of DNA.* New York: W. W. Norton.

Movable Genetic Elements. 1981. *Cold Spring Harbor symposium on quantitative biology,* vol. 45. Cold Spring Harbor, N.Y.: Cold Spring Harbor Laboratory.

Neidhardt, F. C.; Ingraham, J. L.; and Schaechter, M. 1990. *Physiology of the bacterial cell.* Sunderland, Mass.: Sinauer.

Novick, R. P. 1980. Plasmids. *Sci. Am.* 243(6):103–27.

Smith, G. R. 1989. Homologous recombination in *E. coli:* Multiple pathways for multiple reasons. *Cell* 58:807–9.

Smith, H. O.; Danner, D. B.; and Deich, R. A. 1981. Genetic transformation. *Ann. Rev. Biochem.* 50:41–68.

Stahl, F. W. 1987. Genetic recombination. *Sci. Am.* 256(2):91–101.

Stewart, G. J., and Carlson, C. A. 1986. The biology of natural transformation. *Ann. Rev. Microbiol.* 40:211–35.

Streips, U. N., and Yasbin, R. E., eds. 1991. *Modern Microbial Genetics.* New York: Wiley-Liss, Inc.

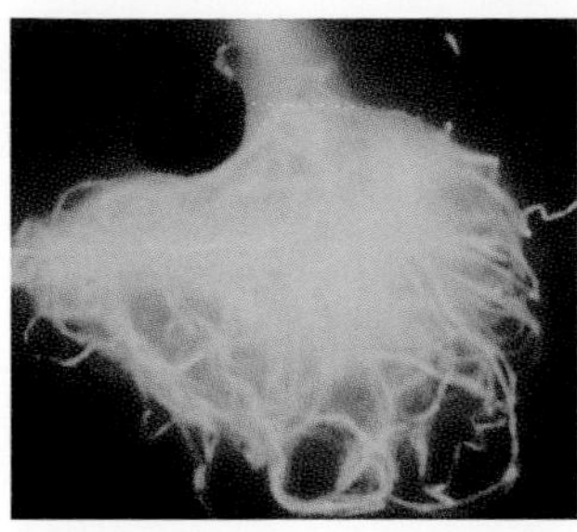

Chapter 14
Recombinant DNA Technology

The recombinant DNA breakthrough has provided us with a new and powerful approach to the questions that have intrigued and plagued man for centuries.
—Paul Berg

The concept and control of the double helix signal a new frontier of biocultural progression. A stereoscopic vision that includes both "creative pessimism" and "creative optimism" is now required.
—Clifford Grobstein

Outline

Concepts

1. Genetic engineering makes use of recombinant DNA technology to fuse genes with vectors and then clone them in host cells. In this way, large quantities of isolated genes and their products can be synthesized.
2. The production of recombinant DNA molecules depends on the ability of restriction endonucleases to cleave DNA at specific sites.
3. Plasmids, bacteriophages, and cosmids are used as vectors because they can replicate within a host cell while carrying foreign DNA and possess phenotypic traits that allow them to be detected.
4. Genetic engineering is already making substantial contributions to biological research, medicine, industry, and agriculture. Future benefits are probably much greater.
5. Genetic engineering is also accompanied by potential problems in such areas as safety, the ethics of its use with human subjects, environmental impact, and biological warfare.

Chapters 12 and 13 introduce the essentials of microbial genetics. This chapter focuses on the practical applications of microbial genetics and the technology arising from it.

Although human beings have been altering the genetic makeup of organisms for centuries by selective breeding, only recently has the direct manipulation of DNA been possible. The deliberate modification of an organism's genetic information by directly changing its nucleic acid genome is often called **genetic engineering** and is accomplished by a collection of methods known as **recombinant DNA technology.** First, the DNA responsible for a particular phenotype is identified and isolated. Once purified, the gene or genes are fused with other pieces of DNA to form recombinant DNA molecules. These are propagated (gene cloning) by insertion into an organism that need not even be in the same kingdom as the original gene donor. Recombinant DNA technology opens up totally new areas of research and applied biology. Thus, it is an essential part of **biotechnology,** which is now entering a stage of exceptionally rapid growth and development. Although the term has several definitions, in this text biotechnology refers to those processes in which living organisms are manipulated, particularly at the molecular genetic level, to form useful products. The promise for medicine, agriculture, and industry is great; yet the potential risks of this technology are not completely known and may be considerable.

Biotechnology and industrial microbiology (chapter 44).

Recombinant DNA technology is very much the result of several key discoveries in microbial genetics. The first section briefly reviews some landmarks in the development of recombinant technology (table 14.1).

Historical Perspectives

One of the first breakthroughs leading to recombinant DNA (rDNA) technology was the discovery in the late 1960s by Werner Arber and Hamilton Smith of microbial enzymes that

Table 14.1 Some Milestones in Biotechnology and Recombinant DNA Technology

1970	A complete gene synthesized *in vitro* Discovery of the first sequence-specific restriction endonuclease and the enzyme reverse transcriptase
1972	First recombinant DNA molecules generated
1973	Use of plasmid vectors for gene cloning
1975	Southern blot technique for detecting specific DNA sequences
1976	First prenatal diagnosis using a gene-specific probe
1977	Methods for rapid DNA sequencing Discovery of "split genes" and somatostatin synthesized using recombinant DNA
1978	Human genomic library constructed
1979	Insulin synthesized using recombinant DNA First human viral antigen (hepatitis B) cloned
1981	Foot-and-mouth disease viral antigen cloned
1982	Commercial production by *E. coli* of genetically engineered human insulin Isolation, cloning, and characterization of a human cancer gene Transfer of gene for rat growth hormone into fertilized mouse eggs
1985	Tobacco plants made resistant to the herbicide glyphosate through insertion of a cloned gene from *Salmonella* Development of the polymerase chain reaction technique
1987	Insertion of a functional gene into a fertilized mouse egg cures the shiverer mutation disease of mice, a normally fatal genetic disease
1988	The first successful production of a genetically engineered staple crop (soybeans) Development of the gene gun
1989	First field test of a genetically engineered virus (a baculovirus that kills cabbage looper caterpillars)
1990	Production of the first fertile corn transformed with a foreign gene (a gene for resistance to the herbicide bialaphos)
1991	Development of transgenic pigs and goats capable of manufacturing proteins such as human hemoglobin First test of gene therapy on human cancer patients

Source: Based in part on A. E. H. Emery, *An Introduction to Recombinant DNA*, John Wiley & Sons, Ltd., Sussex, England, 1984.

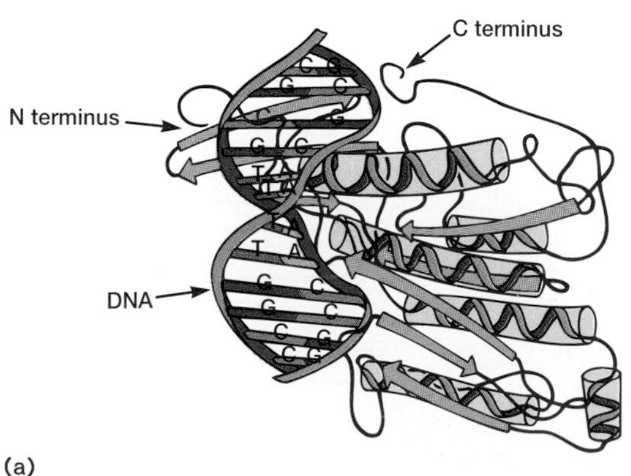

Figure 14.1 Restriction Endonuclease Binding to DNA. (*a*) A schematic diagram of one *Eco*RI endonuclease subunit bound to DNA at the recognition sequence. The enzyme binds to the major groove with the proper sequence (see figure 14.2). The N-terminal arm of each subunit wraps around the DNA helix. (*b*) A space-filling model of the complete, dimeric *Eco*RI endonuclease bound to DNA. The DNA substrate, with its green and blue strands, faces forward. The endonuclease lies behind the DNA. Its two subunits are colored orange and yellow. *From Donald Voet and Judith G. Voet,* Biochemistry. *Copyright © 1990 John Wiley & Sons, Inc., New York, NY. Reprinted by permission. Originally from J. M. Rosenberg, et al., Trends in Biochemical Sciences, 12:396, 1987.*

make cuts in double-stranded DNA. These enzymes recognize and cleave specific sequences about 4 to 6 base pairs long and are known as **restriction enzymes** or restriction endonucleases (figure 14.1). They normally protect the host cell by destroying phage DNA after its entrance. Restriction enzymes can be used to prepare DNA fragments containing specific genes or portions of genes. For example, the restriction enzyme *Eco*RI, isolated by Herbert Boyer in 1969 from *E. coli,* cleaves the DNA between G and A in the base sequence GAATTC (figure 14.2). Note that in the double-stranded condition, the base sequence GAATTC will base pair with the same sequence running in the opposite direction. *Eco*RI therefore cleaves both DNA strands between the G and the A. When the two DNA fragments separate, they contain single-stranded complementary ends, known as sticky ends or cohesive ends. There are hundreds of restriction enzymes that recognize many different specific sequences (table 14.2). Each restriction enzyme name begins with three letters, indicating the bacterium producing it. For example, *Eco*RI is obtained from *E. coli,* while *Bam*HI comes from *Bacillus amyloliquefaciens H,* and *Sal*I from *Streptomyces albus.*

In 1970, Howard Temin and David Baltimore independently discovered the enzyme reverse transcriptase that retroviruses use to produce DNA copies of their RNA genome. This

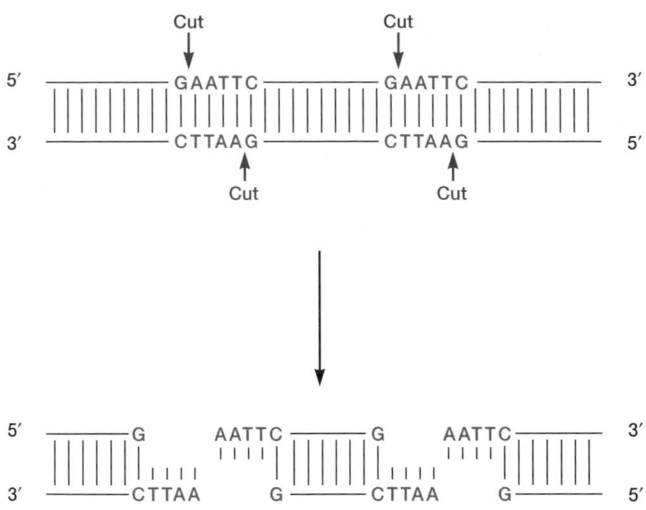

Figure 14.2 **Restriction Endonuclease Action.** The cleavage catalyzed by the restriction endonuclease *Eco*RI. The enzyme makes staggered cuts on the two DNA strands to form sticky ends.

TABLE 14.2 Some Restriction Endonucleases and Their Recognition Sequences

Enzyme	Microbial Source	Sequence[a]
*Alu*I	*Arthrobacter luteus*	5′—A—G↓C—T—3′ 3′—T—C↑G—A—5′
*Bam*HI	*Bacillus amyloliquefaciens H*	5′—G↓G—A—T—C—C—3′ 3′—C—C—T—A—G↑G—5′
*Eco*RI	*Escherichia coli*	5′—G↓A—A—T—T—C—3′ 3′—C—T—T—A—A↑G—5′
*Eco*RII	*Escherichia coli*	5′↓C—C—T—G—G—3′ 3′—G—G—A—C—C↑5′
*Hae*III	*Haemophilus aegyptius*	5′—G—G↓C—C—3′ 3′—C—C↑G—G—5′
*Hin*dIII	*Haemophilus influenzae* b	5′—A↓A—G—C—T—T—3′ 3′—T—T—C—G—A↑A—5′
*Pst*I	*Providencia stuartii*	5′—C—T—G—C—A↓G—3′ 3′—G↑A—C—G—T—C—5′
*Sal*I	*Streptomyces albus*	5′—G↓T—C—G—A—C—3′ 3′ C—A—G—C—T↑G—5′

[a]The arrows indicate the sites of cleavage on each strand.

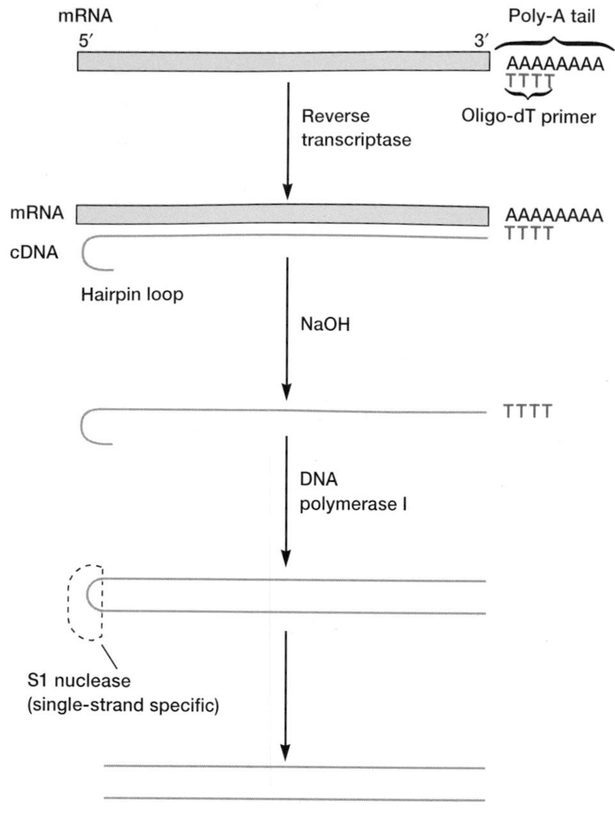

Figure 14.3 The Synthesis of Double-stranded cDNA from mRNA. *From* Recombinant DNA: A Short Course *by James D. Watson et al. Copyright © 1983 James D. Watson, John Tooze, and David T. Kurtz. Reprinted with the permission of Scientific American Books, Inc.*

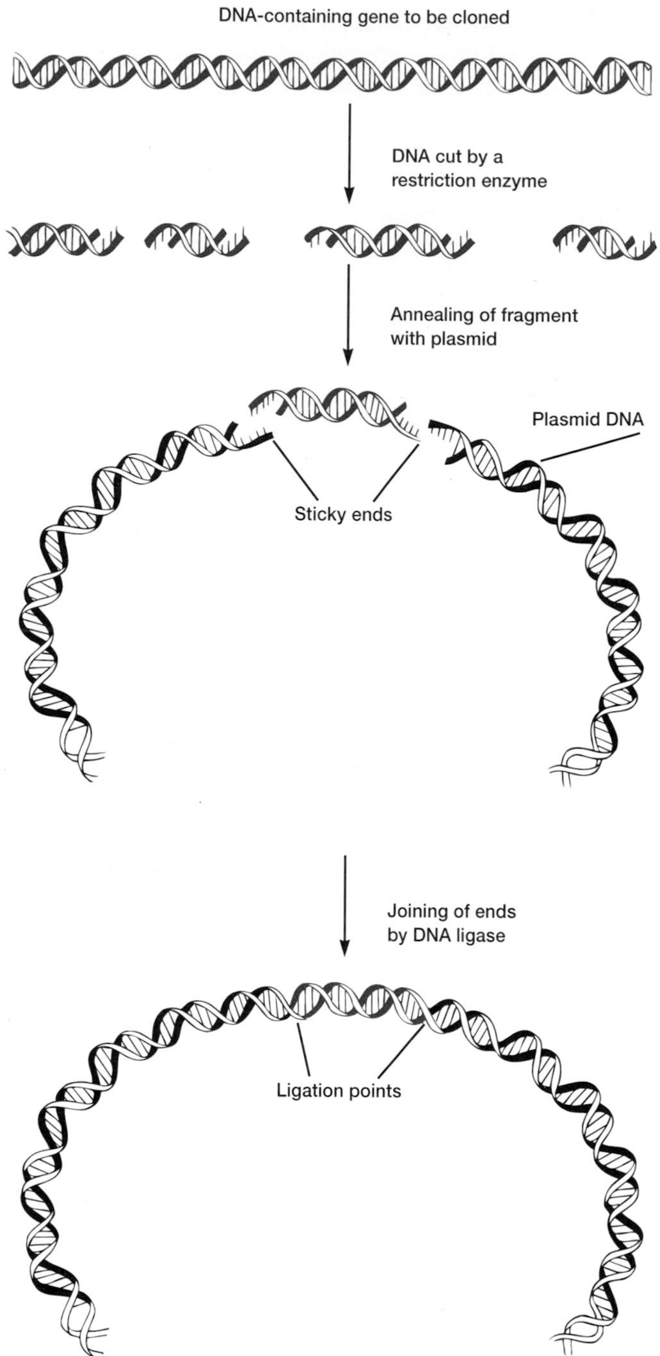

Figure 14.4 Recombinant Plasmid Construction. The general procedure used in constructing recombinant plasmid vectors is shown here.

enzyme can be used to construct a DNA copy, called **complementary DNA (cDNA),** of any RNA (figure 14.3). Thus, genes or major portions of genes can be synthesized from mRNA.

Reverse transcriptase and retroviruses (p. 389).

The next advance came in 1972, when David Jackson, Robert Symons, and Paul Berg reported that they had successfully generated recombinant DNA molecules. They allowed the sticky ends of fragments to anneal, that is, to base pair with one another, and then covalently joined the fragments with the enzyme DNA ligase. Within a year, plasmid **vectors,** or carriers of foreign DNA fragments during gene cloning, had been developed and combined with foreign DNA (figure 14.4). The first such recombinant plasmid capable of being replicated within a bacterial host was the pSC101 plasmid constructed by Stanley Cohen and Herbert Boyer in 1973 (SC in the plasmid name stands for *S*tanley *C*ohen).

In 1975, Edwin M. Southern published a procedure for detecting specific DNA fragments so that a particular gene could be isolated from a complex DNA mixture. The **Southern blotting technique** depends on the specificity of base complementarity in nucleic acids (figure 14.5). DNA fragments are first separated by size with agarose gel electrophoresis. The fragments are then denatured (rendered single stranded) and transferred to a nitrocellulose filter so that each fragment is firmly bound to the filter at the same position as on the gel. The filter is bathed with solution containing a radioactive **probe,** a piece of labeled nucleic acid that hybridizes with complementary DNA fragments and is used to locate them. Those fragments complementary to the probe become radioactive and are readily detected by means of **autoradiography.** In this technique, a sheet of photographic film is placed over the filter for several hours and then developed. The film is exposed and

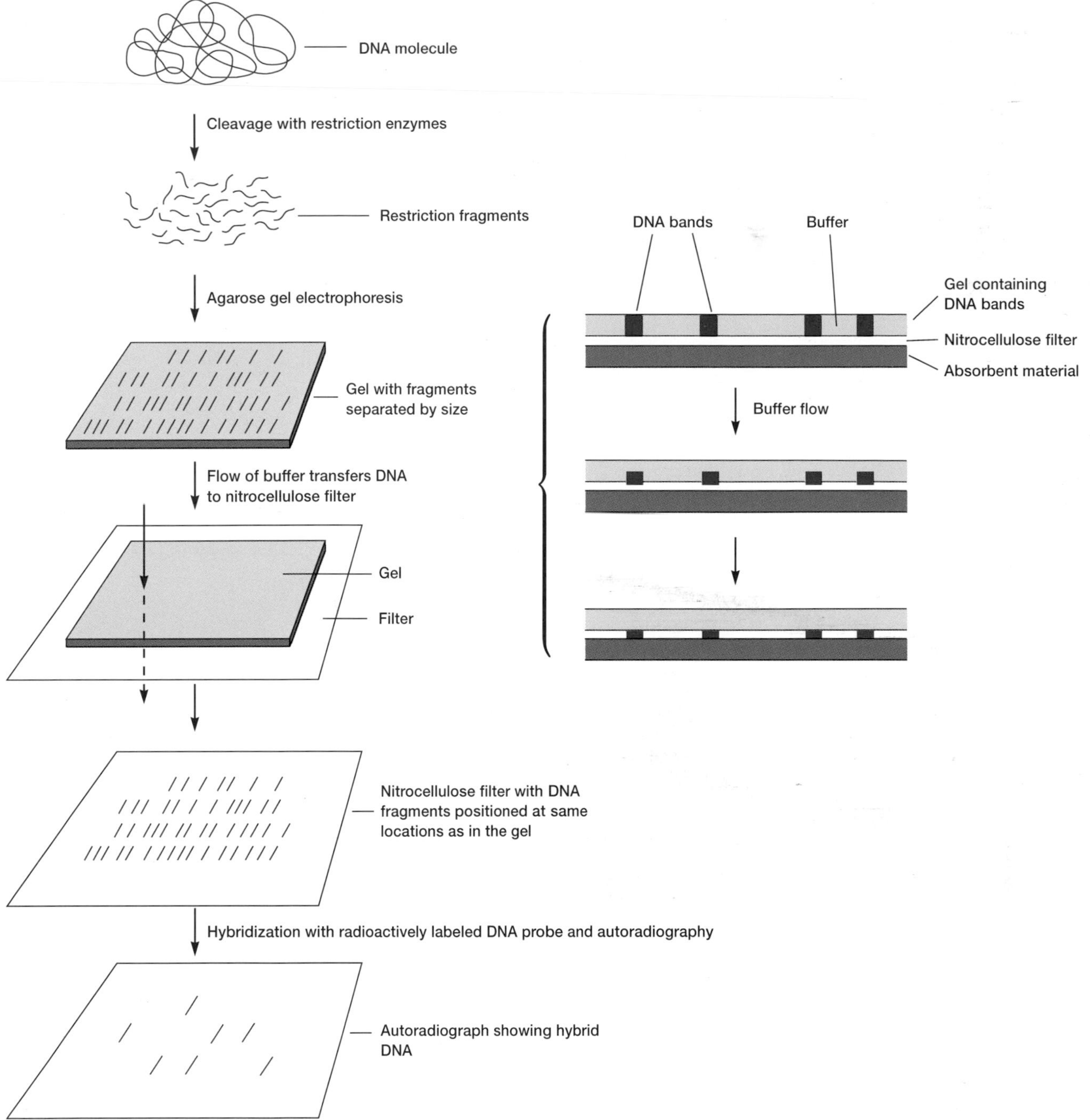

Figure 14.5 **The Southern Blotting Technique.** The insert illustrates how buffer flow transfers DNA bands to the nitrocellulose filter. See text for further details. *From Watson, Hopkins, Roberts, Steitz, and Weiner,* Molecular Biology of the Gene. *Copyright © 1987 Benjamin/Cummings Publishing Co., Menlo Park, CA.*

becomes dark everywhere a radioactive fragment is located because the energy released by the isotope causes the formation of dark-silver grains. Using Southern blotting, one can detect and isolate fragments with any desired sequence from a complex mixture.

More recently, nonradioactive probes to detect specific DNAs have been developed. In one approach, the DNA probe is linked to an enzyme such as horseradish peroxidase. After the enzyme-DNA probe has bound to a DNA fragment on the filter, a substrate that will emit light when acted on by the peroxidase is added. The probe is detected by exposing the filter to photographic film for about 20 minutes. A second technique makes use of the vitamin biotin. A biotin-DNA probe is detected by incubating the filter with either the protein avidin or a similar bacterial protein, streptavidin. The protein specifically attaches to biotin, and is visualized with a special reagent containing biotin complexed with the enzyme alkaline phosphatase (streptavidin also can be directly attached to the enzyme). The bands with the probe appear blue. These nonradioactive techniques are more rapid and safer than using radioisotopes. On the other hand, they are less sensitive than radioactively labeled probes.

By the late 1970s, techniques for easily sequencing DNA, synthesizing oligonucleotides, and expressing eucaryotic genes in procaryotes had also been developed. These techniques could then be used to solve practical problems (table 14.1). The following sections describe how the previously discussed techniques and others are used in genetic engineering.

1. Define or describe restriction enzyme, sticky end, cDNA, vector, Southern blotting, probe, and autoradiography.

Synthetic DNA

Oligonucleotides (Greek *oligo,* few or scant) are short pieces of DNA or RNA between about 2 and 20 or 30 nucleotides long. The ability to synthesize DNA oligonucleotides of known sequence is extremely useful. For example, DNA probes can be synthesized and DNA fragments can be prepared for use in a variety of molecular techniques.

DNA structure (pp. 192–94).

DNA oligonucleotides are synthesized by a stepwise process in which single nucleotides are added to the end of the growing chain (figure 14.6). The 3′ end of the chain is attached to a solid support such as a silica gel particle. A DNA synthesizer or "gene machine" carries out the solid-phase synthesis. A specially activated nucleotide derivative is added to the 5′ end of the chain in a series of steps. At the end of an additional cycle, the growing chain is separated from the reaction mixture by filtration or centrifugation. The process is then repeated to attach another nucleotide. It takes about 40 minutes to add a nucleotide to the chain, and chains as large as 50 to 100 nucleotides can be synthesized.

Advances in DNA synthetic techniques have accelerated progress in the study of protein function. One of the most effective ways of studying the relationship of protein structure to function is by altering a specific part of the protein and observing functional changes. In the past this has been accomplished either by chemically modifying individual amino acids or by inducing mutations in the gene coding for the protein under study. There are problems with these two approaches. Chemical modification of a protein is not always specific; several amino acids may be altered, not just the one desired. On the other hand, it is not always possible to produce the proper mutation in the desired gene location. Recently, these difficulties have been overcome with a technique called **site-directed mutagenesis.**

In site-directed mutagenesis, an oligonucleotide that contains the desired sequence change is synthesized. The altered oligonucleotide with its artificially mutated sequence is now allowed to bind to a single-stranded copy of the complete gene (figure 14.7). DNA polymerase is added to the gene-primer complex. The polymerase extends the primer and replicates the remainder of the target gene to produce a new gene copy with the desired mutation. If the gene is attached to a single-stranded DNA bacteriophage (*see pp. 358 and 372*) such as the M13 phage, it can be introduced into a host bacterium and cloned using the techniques to be described shortly. This will yield large quantities of the mutant protein for study of its function.

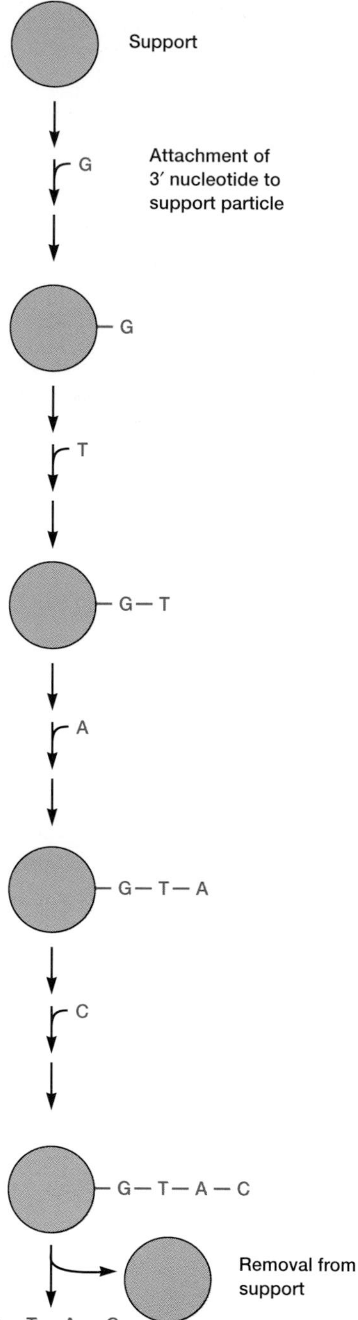

Figure 14.6 The Synthesis of a DNA Oligonucleotide. During each cycle, the DNA synthesizer adds an activated nucleotide (A, T, G, or C) to the growing end of the chain. At the end of the process, the oligonucleotide is removed from its support.

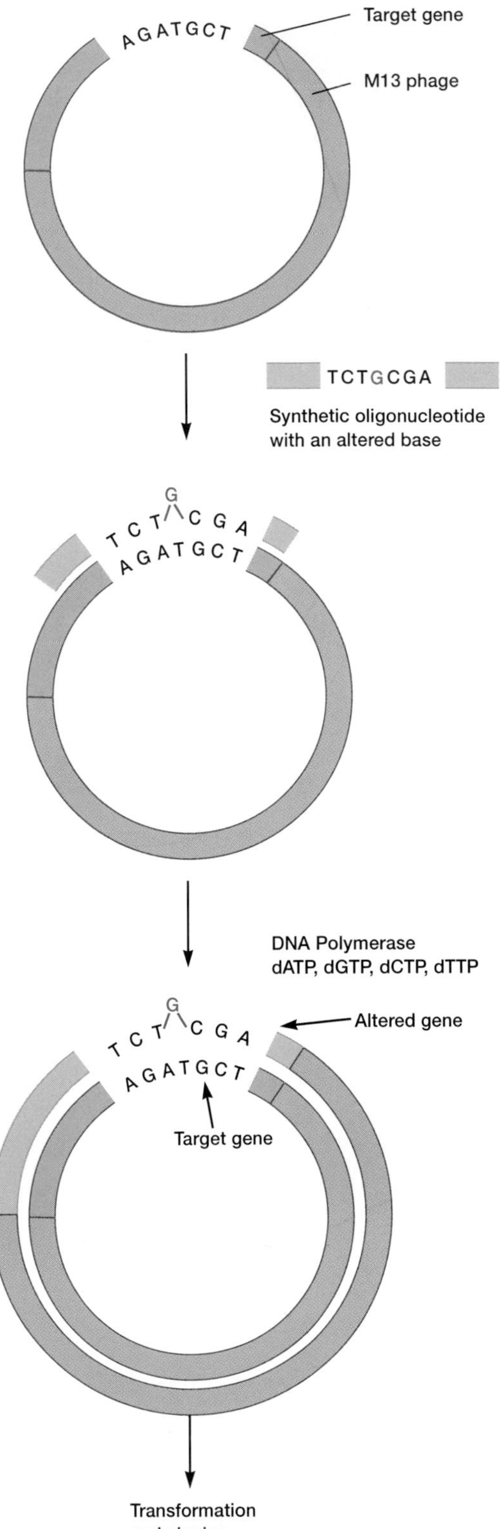

Figure 14.7 Site-directed Mutagenesis. A synthetic oligonucleotide is used to add a specific mutation to a gene. See text for details.

The Polymerase Chain Reaction

Between 1983 and 1985, Kary Mullis developed a new technique that made it possible to synthesize large quantities of a DNA fragment without cloning it. Although this chapter emphasizes recombinant DNA technology, the **polymerase chain reaction** or **PCR technique** will be introduced here because of its great practical importance and impact on biotechnology.

Figure 14.8 outlines how the PCR technique works. Suppose that one wishes to make large quantities of a particular DNA sequence. The first step is to synthesize fragments with sequences identical to those flanking the targeted sequence. This is readily accomplished with a DNA synthesizer machine as previously described. These synthetic oligonucleotides are usually about 20 nucleotides long and serve as primers for DNA synthesis. The PCR cycle itself takes place in three steps. First, the target DNA containing the sequence to be amplified is heat denatured to separate its complementary strands (step 1). Next, primers are added in excess and the temperature lowered so that they can hydrogen bond or anneal to the DNA on both sides of the target sequence (step 2). Because the primers are present in excess, the targeted DNA strands will almost always anneal to the primers rather than to each other. Finally, nucleoside triphosphates and a DNA polymerase are added to the reaction mixture. The DNA polymerase extends the primers and synthesizes copies of the target DNA sequence (step 3). Only polymerases able to function at the high temperatures employed in the PCR technique can be used. Two popular enzymes are the Taq polymerase from the thermophilic bacterium *Thermus aquaticus* and the Vent polymerase from *Thermococcus litoralis*. At the end of one cycle, the targeted sequences on both strands have been copied. When the three-step cycle is repeated (figure 14.8), the four strands from the first cycle are copied to produce eight fragments. The third cycle yields 16 products. Theoretically, 20 cycles will produce about one million copies of the target DNA sequence; 30 cycles yield around one billion copies. Pieces ranging in size from less than 100 base pairs to several thousand base pairs in length can be amplified, and only 10 to 100 picomoles of primer are required. The concentration of target DNA can be as low as 10^{-20} to 10^{-15} M (or 1 to 10^5 DNA copies per 100μl). The whole reaction mixture is often 100 μl or less in volume.

DNA replication (pp. 195–201).

The polymerase chain reaction technique has now been automated and is carried out by a specially designed machine (figure 14.9). Currently, a PCR machine can carry out 25 cycles and amplify DNA 10^5 times in as little as 57 minutes. During a typical cycle, the DNA is denatured at 94°C for 15 seconds, then the primers are annealed and extended (steps 2 and 3) at 68°C for 60 seconds.

The PCR technique has already proven exceptionally valuable in many areas of molecular biology, medicine, and biotechnology. It can be used to amplify very small quantities

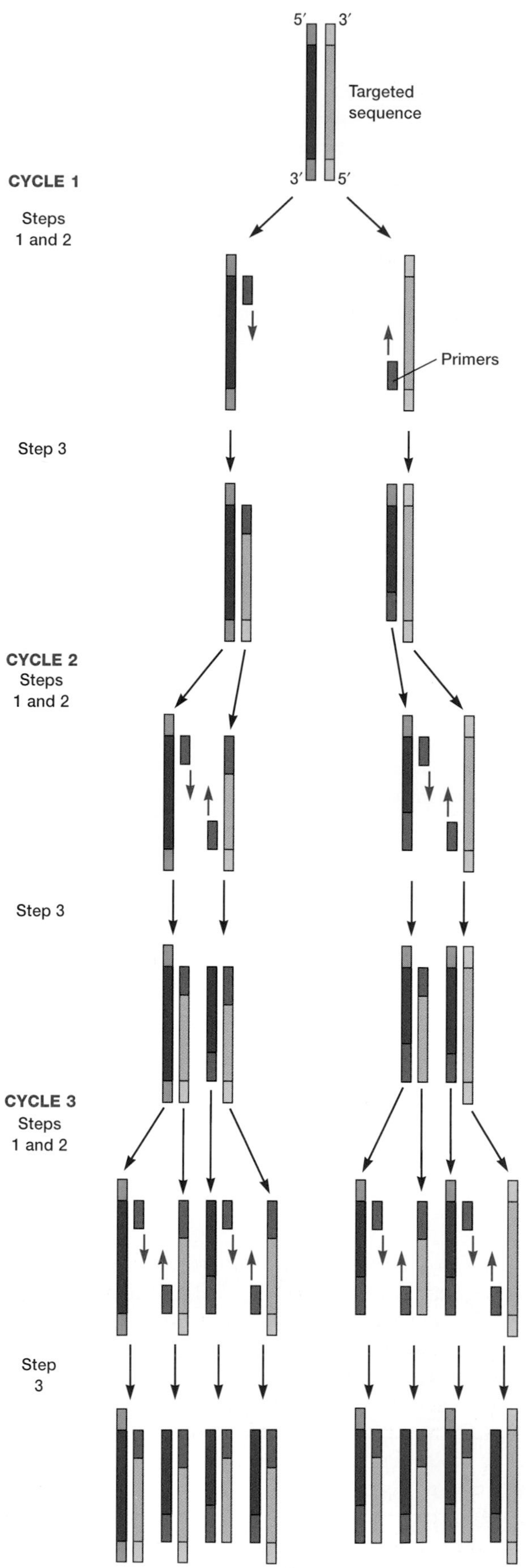

Figure 14.8 The Polymerase Chain Reaction. In three cycles, the targeted sequence has been amplified to produce eight copies. See text for details.

Figure 14.9 A Modern PCR Machine. PCR machines are now fully automated and microprocessor controlled. They can process up to 96 samples at a time.

of a specific DNA and provide sufficient material for accurately sequencing the fragment or cloning it by standard techniques. Undoubtedly PCR will be used extensively to aid genetic mapping in the human genome project. PCR-based diagnostic tests for AIDS, Lyme disease, chlamydia, the human papilloma virus, and other infectious agents and diseases are being developed. PCR is particularly valuable in the detection of genetic diseases such as sickle cell anemia, phenylketonuria, and muscular dystrophy. The technique is already having an impact on forensic science where it is being used in criminal cases. It is possible to exclude or incriminate suspects using extremely small samples of biological material discovered at the crime scene.

Preparation of Recombinant DNA

There are three ways to obtain adequate quantities of a DNA fragment. One can extract all the DNA from an organism, fragment it, isolate the fragment of interest, and finally clone it. Alternatively, all of the fragments can be cloned by means of a suitable vector, and each clone (the population of identical molecules with a single ancestral molecule) can be tested for the desired gene. One also can directly synthesize the desired DNA fragment as described earlier, and then clone it.

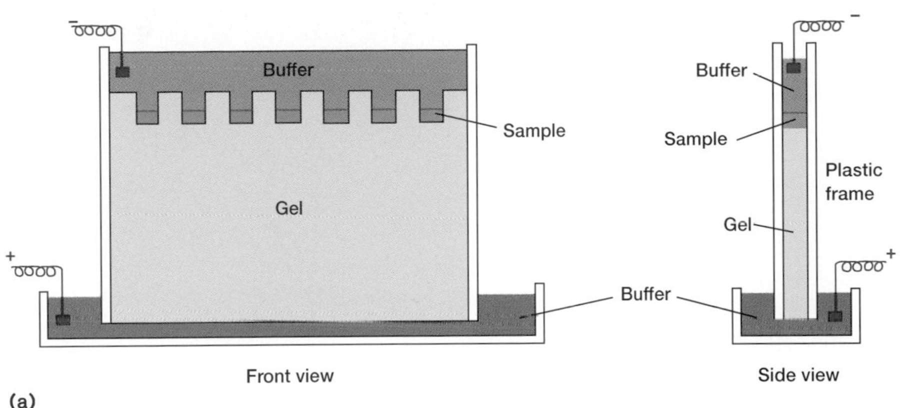

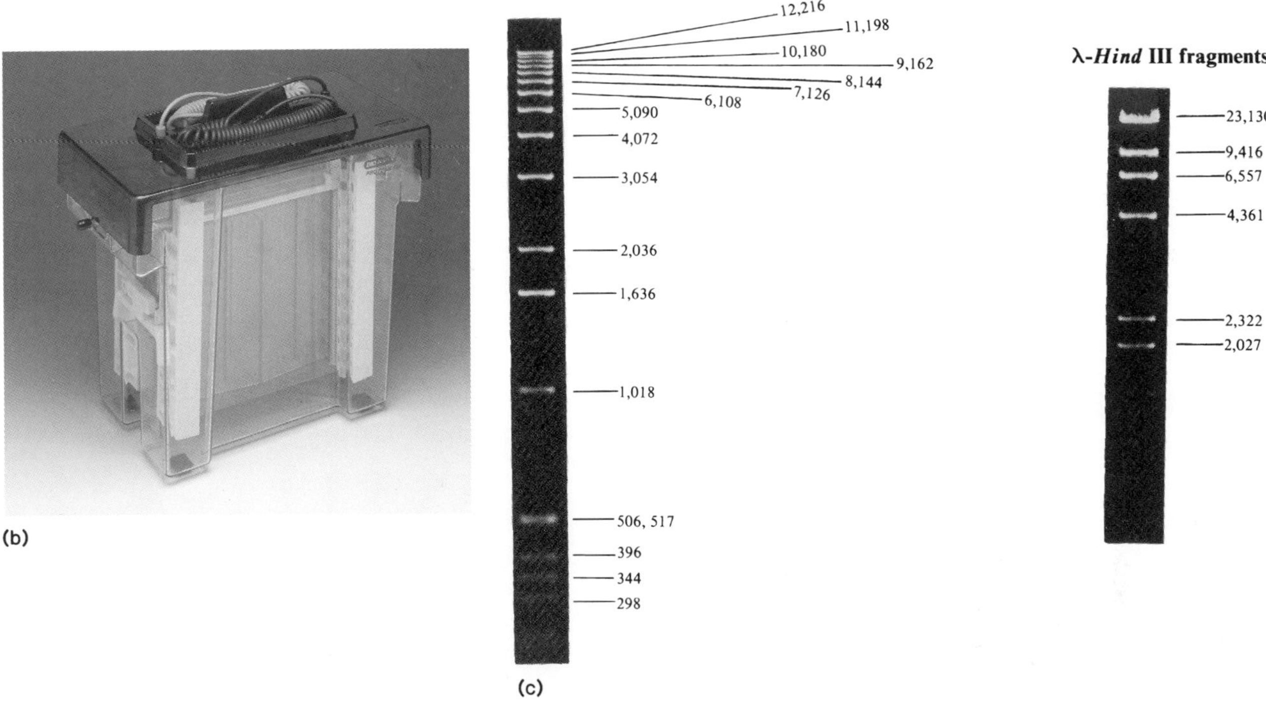

Figure 14.10 **Gel Electrophoresis of DNA.** (*a*) A diagram of a vertical gel outfit showing its parts. (*b*) A commercial gel electrophoresis outfit. (*c*) The 1 kilobase ladder is an electrophoretic gel containing a series of DNA fragments of known size. The numbers indicate number of base pairs each fragment contains. The smallest fragments have moved the farthest. The gel on the right shows the fragments that arise when lambda phage DNA is digested with the *Hind*III restriction enzyme.

Isolating and Cloning Fragments

Long, linear DNA molecules are fragile and easily sheared into fragments by passing the DNA suspension through a syringe needle several times. DNA also can be cut with restriction enzymes. The resulting fragments are either separated by electrophoresis or first inserted into vectors and cloned.

Agarose or polyacrylamide gels are usually used to separate DNA fragments electrophoretically. In **electrophoresis,** charged molecules are placed in an electrical field and allowed to migrate toward the positive and negative poles. The molecules separate because they move at different rates due to their differences in charge and size. In practice, the fragment mixture is usually placed in wells molded within a sheet of gel (figure 14.10). The gel concentration varies with the size of DNA fragments to be separated. Usually 1 to 3% agarose gels or 3 to 20% polyacrylamide gels are used. When an electrical field is generated in the gel, the fragments move through the pores of the gel toward an electrode (negatively charged DNA fragments migrate toward the positive electrode or anode). Each fragment's migration rate is inversely proportional to the log of its molecular weight, and the fragments are separated into size classes (figure 14.10*c*). A simple DNA molecule might yield only a few bands. If the original DNA was very large and complex, staining of the gel would reveal a smear representing an almost continuous gradient in fragment size. The band or section containing the desired fragment is located with

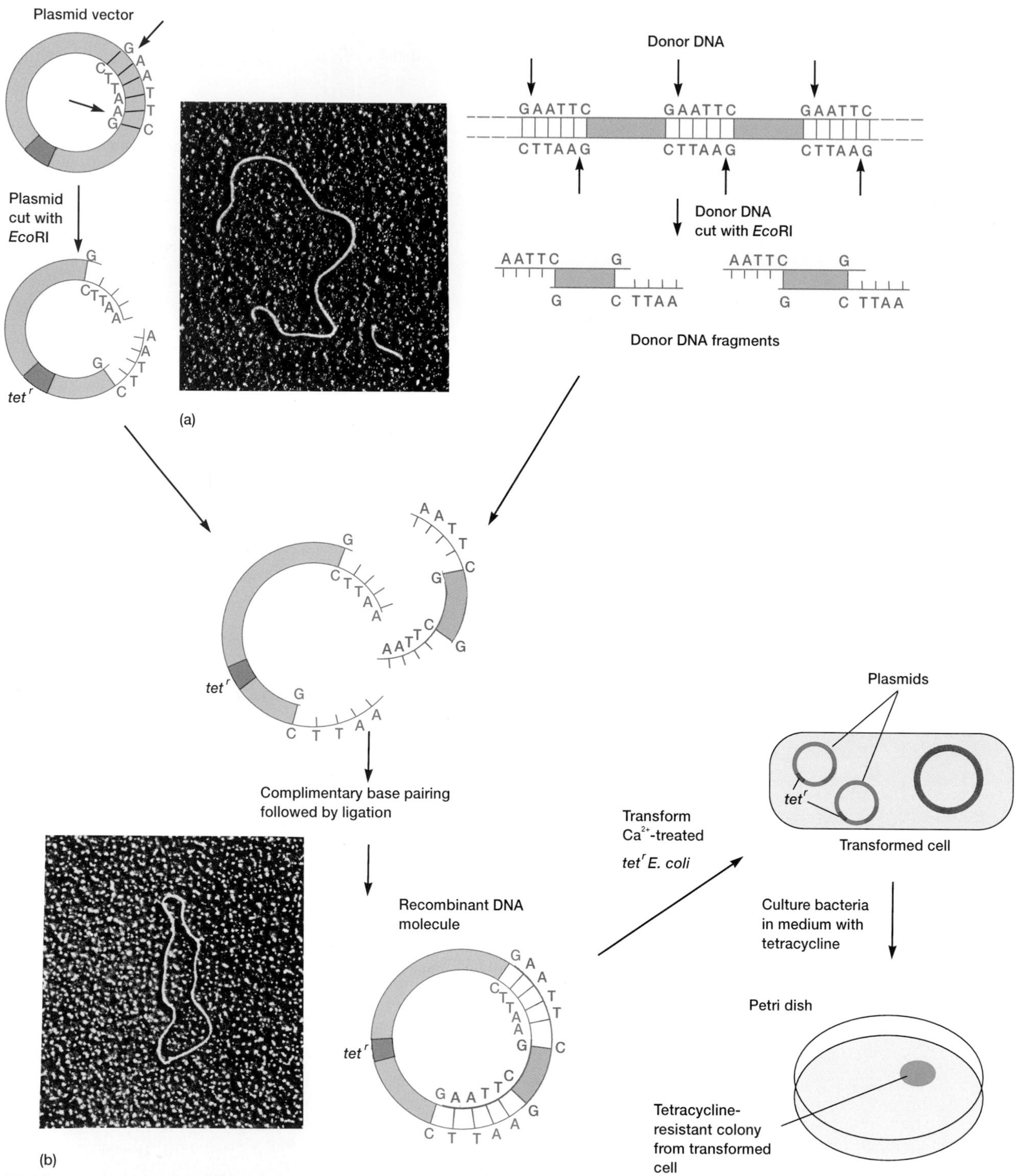

Figure 14.11 Recombinant Plasmid Construction and Cloning. The construction and cloning of a recombinant plasmid vector using an antibiotic resistance gene to select for the presence of the plasmid. The scale of the sticky ends of the fragments and plasmid has been enlarged to illustrate complementary base pairing. (*a*) The electron micrograph shows a plasmid that has been cut by a restriction enzyme and a donor DNA fragment. (*b*) The micrograph shows a recombinant plasmid. See text for details.

the Southern blotting technique (figure 14.5) and removed. Since a band may represent a mixture of several fragments, it is electrophoresed on a gel with a different concentration to separate similarly sized fragments. The location of the pure fragment is determined by Southern blotting, and it is extracted from the gel.

Once fragments have been separated, they are joined with an appropriate vector, such as a plasmid (*see chapter 13*), to form a recombinant molecule that can reproduce in a host cell. One of the easiest and most popular approaches is to cut the plasmid and donor DNA with the same restriction enzyme so that identical sticky ends are formed (figure 14.11). After a

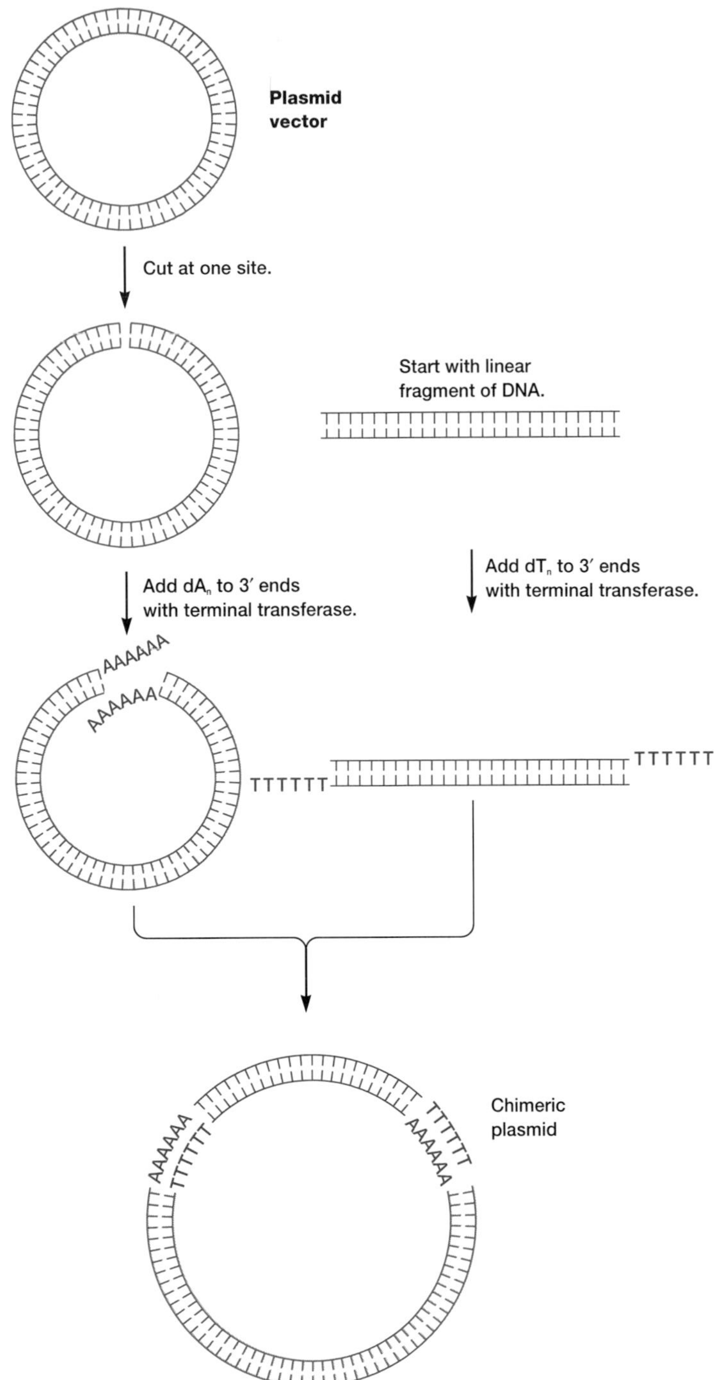

Figure 14.12 **Terminal Transferase and the Construction of Recombinant Plasmids.** The poly (dA:dT) tailing technique can be used to construct sticky ends on DNA and generate recombinant molecules.

fragment has annealed with the plasmid through complementary base pairing, the breaks are joined by DNA ligase. A second method for creating recombinant molecules can be used with fragments and vectors lacking sticky ends. After cutting the plasmid and donor DNA, one can add poly(dA) to the 3′ ends of the plasmid DNA, using the enzyme terminal transferase (figure 14.12). Similarly, poly(dT) is added to the 3′ ends of the fragments. The ends will now base pair with each

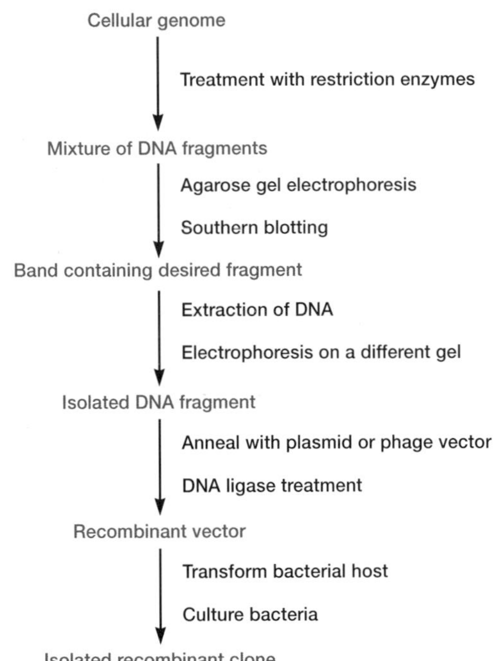

Figure 14.13 **Cloning Cellular DNA Fragments.** The preparation of a recombinant clone from previously isolated DNA fragments.

other and are joined by DNA ligase to form a recombinant plasmid. Some enzymes (e.g., the *Alu*I restriction enzyme from *Arthrobacter luteus*) cut DNA at the same position on both strands to form a blunt end. Fragments and vectors with blunt ends may be joined by T4 DNA ligase (blunt-end ligation).

The rDNA molecules are cloned by inserting them into bacteria, using transformation or phage injection (*see chapter 13*). Each strain reproduces to yield a population containing a single type of recombinant molecule. The overall process is outlined in figure 14.13. The same cloning techniques can be used with DNA fragments prepared using a DNA synthesizer machine.

Although selected fragments can be isolated and cloned as just described, it is often preferable to fragment the whole genome and clone all the fragments by using a vector. Then the desired clone can be identified. To be sure that the complete genome is represented in this collection of clones, called a genomic library, more than a thousand transformed bacterial strains must be maintained (the larger the genome, the more strains are stored). Libraries of cloned genes also can be generated using phage lambda as a vector and stored as phage lysates.

It is necessary to identify which clone in the library contains the desired gene. If the gene is expressed in the bacterium, it may be possible to assay each clone for a specific protein. However, a nucleic acid probe is normally employed in identification. The bacteria are replica plated on nitrocellulose paper and lysed in place with sodium hydroxide (figure 14.14). This yields a pattern of membrane-bound, denatured DNA corresponding to the colony pattern on the agar plate.

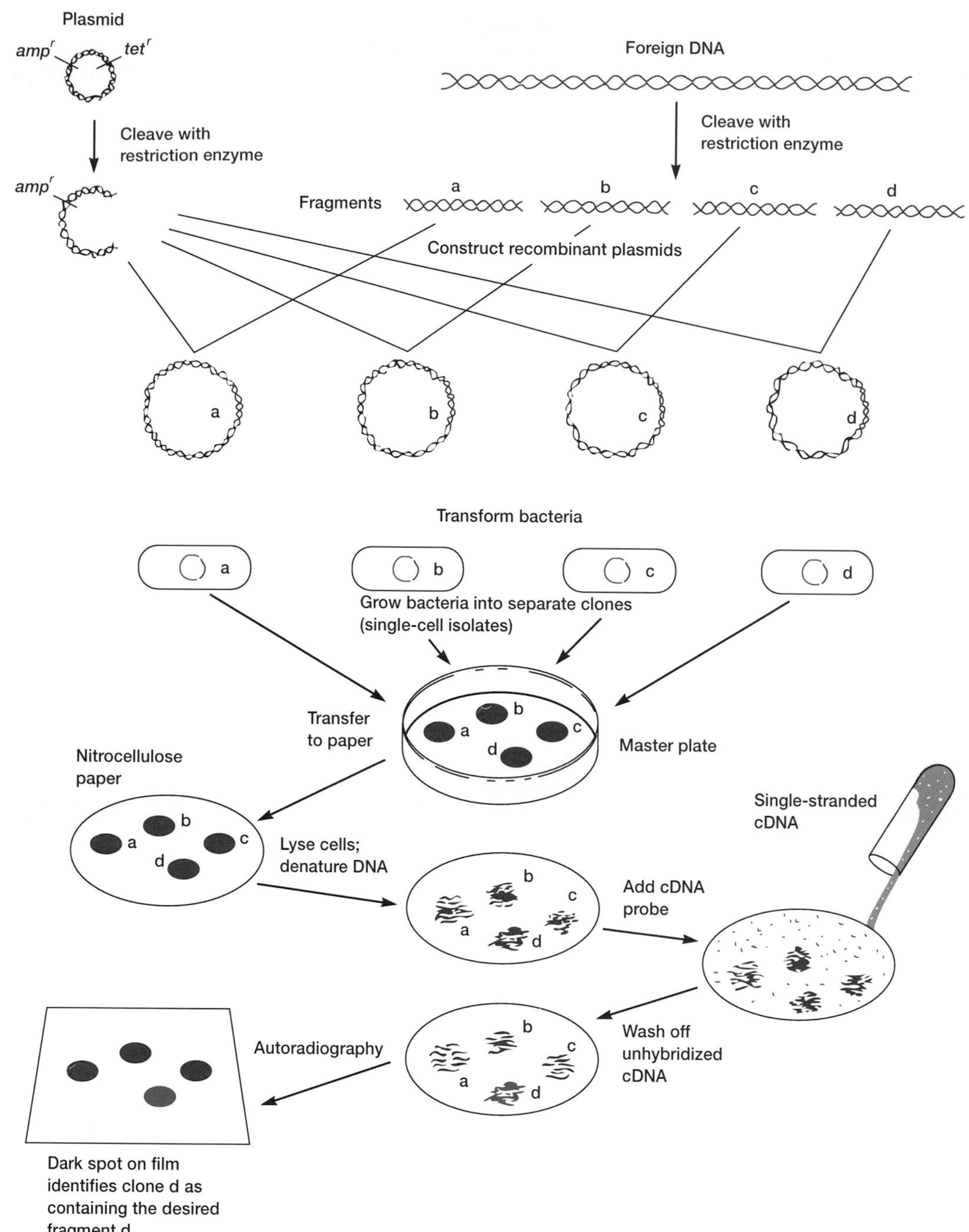

Figure 14.14 Cloning with Plasmid Vectors. The use of plasmid vectors to clone a mixture of DNA fragments. The desired fragment is identified with a cDNA probe. See text for details.

The membrane is treated with a radioactive probe as in the original Southern blotting method. The radioactive spots identify colonies on the master plate that contain the desired DNA fragment. This approach is also used to analyze a library of cloned lambda phage (figure 14.15).

Gene Probes

Success in isolating the desired recombinant clones depends on the availability of a suitable probe. Gene-specific probes are obtained in several ways. Frequently, they are constructed with cDNA clones. If the gene of interest is expressed in a specific tissue or cell type, its mRNA is often relatively abundant. For example, reticulocyte mRNA may be enriched in globin mRNA, and pancreatic cells in insulin mRNA. Although mRNA is not available in sufficient quantity to serve as a probe, the desired mRNA species can be converted into cDNA by reverse transcription (figure 14.3). The cDNA copies are purified, spliced into appropriate vectors, and cloned to provide adequate amounts of the required probe.

Probes also can be generated if the gene codes for a protein of known amino acid sequence. Oligonucleotides, about

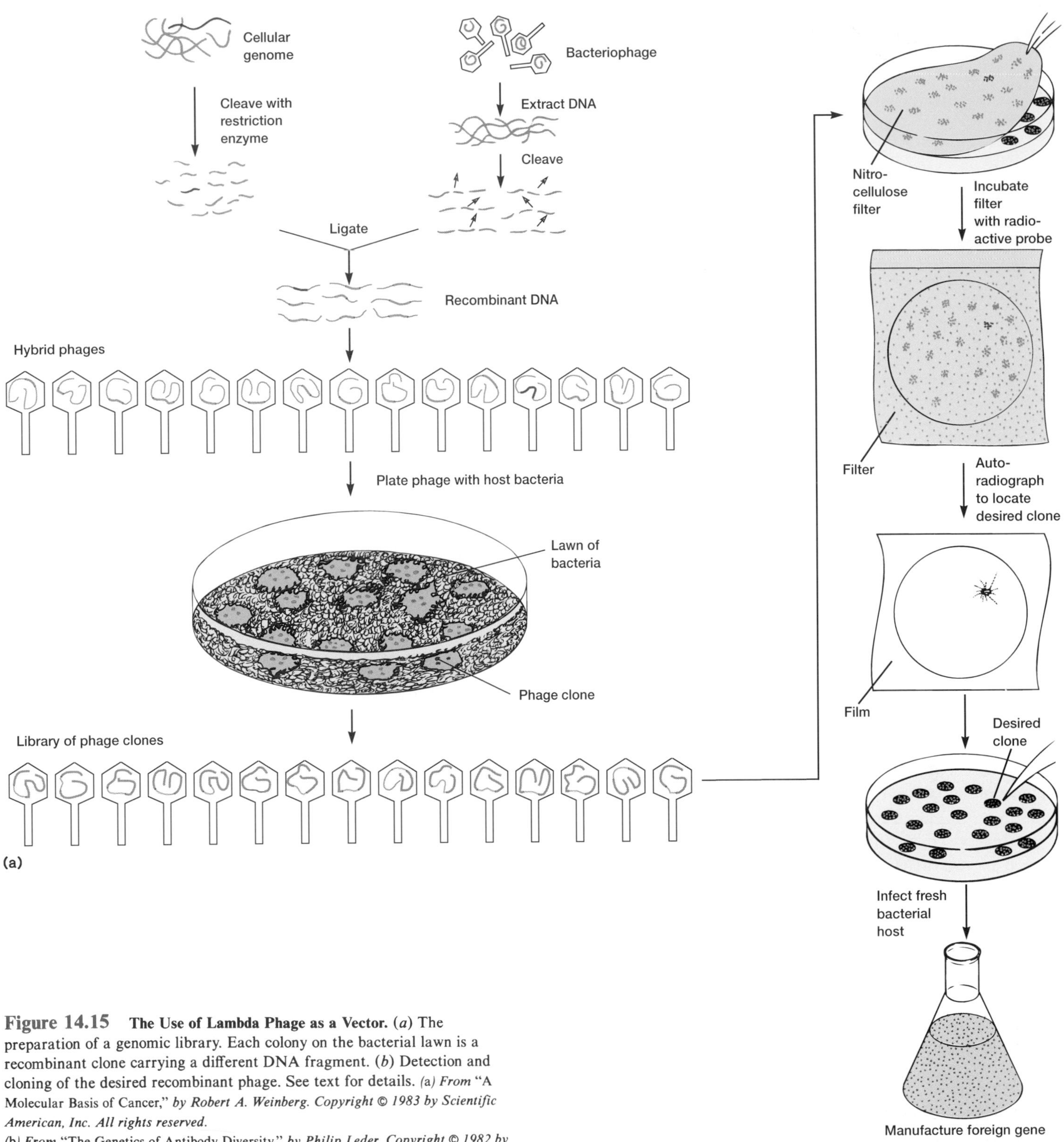

Figure 14.15 **The Use of Lambda Phage as a Vector.** (*a*) The preparation of a genomic library. Each colony on the bacterial lawn is a recombinant clone carrying a different DNA fragment. (*b*) Detection and cloning of the desired recombinant phage. See text for details. *(a) From* "A Molecular Basis of Cancer," *by Robert A. Weinberg. Copyright © 1983 by Scientific American, Inc. All rights reserved.*
(b) From "The Genetics of Antibody Diversity," *by Philip Leder. Copyright © 1982 by Scientific American, Inc. All rights reserved.*

20 nucleotides or longer, that code for a characteristic amino acid sequence are synthesized. These are often satisfactory probes since they will specifically bind to the gene segment coding for the desired protein.

Sometimes previously cloned genes or portions of genes may be used as probes. This approach is effective when there is a reasonable amount of similarity between the nucleotide sequences of the two genes.

After construction, the probe is labeled to aid detection. Often, ^{32}P is added to both DNA strands so that the radioactive strands can be located with autoradiography. Nonradioactively labeled probes may also be used.

TABLE 14.3 Some Recombinant DNA Cloning Vectors

Type	Vector	Restriction Sequences Present	Features
Plasmid (*E. coli*)	pBR322	*Bam*HI, *Eco*RI, *Hae*III, *Hind*III, *Pst*I, *Sal*I, *Xor*II	Carries genes for tetracycline and ampicillin resistance
Plasmid (yeast–*E. coli* hybrid)	pYe(CEN3)41	*Bam*HI, *Bgl*II, *Eco*RI, *Hind*III, *Pst*I, *Sal*I	Multiplies in *E. coli* or yeast cells
Cosmid (artificially constructed *E. coli* plasmid carrying lambda cos site)	pJC720	*Hind*III	Can be packaged into lambda phage particles for efficient introduction into bacteria; replicates as a plasmid; useful for cloning large DNA inserts
Virus	Charon phage	*Eco*RI, *Hind*III, *Bam*HI, *Sst*I	Constructed using restriction enzymes and a ligase, having foreign DNA as central portion, with lambda DNA at each end; carries β-galactosidase gene; packaged into lambda phage particles; useful for cloning large DNA inserts
Virus	Lambda 1059	*Bam*HI	Will carry large DNA fragments (8–21 kb); recombinant can grow on *E. coli* lysogenic for P2 phage, while vector cannot
Virus	M13	*Eco*RI	Single-stranded DNA virus; useful in studies employing single-stranded DNA insert and in producing DNA fragments for sequencing
Plasmid	Ti	*Sma*I, *Hpa*I	Maize plasmid

Isolating and Purifying Cloned DNA

After the desired clone of recombinant bacteria or phages has been located with a probe, it can be picked from the master plate and propagated. The recombinant plasmid or phage DNA is then extracted and further purified when necessary. The DNA fragment is cut out of the plasmid or phage genome by means of restriction enzymes and separated from the remaining DNA by electrophoresis.

Clearly, DNA fragments can be isolated, purified, and cloned in several ways. Regardless of the exact approach, a key to successful cloning is choosing the right vector. The next section considers types of cloning vectors and their uses.

1. How are oligonucleotides synthesized? What is site-directed mutagenesis?
2. Briefly describe the polymerase chain reaction technique. What is its importance?
3. What is electrophoresis and how does it work?
4. Describe three ways in which a fragment can be covalently attached to vector DNA.
5. Outline in detail two different ways to isolate and clone a specific gene. What is a genomic library?
6. How are gene-specific probes obtained?

Cloning Vectors

There are three major types of vectors: plasmids, bacteriophages, and cosmids (table 14.3). Each type has its own advantages. Plasmids are the easiest to work with; rDNA phages are more conveniently stored for long periods; larger pieces of DNA can be cloned with cosmids. Besides these three major types, there are also vectors designed for a specific function. For example, shuttle vectors are used in transferring genes between very different organisms and usually contain one replication origin for each host. The pYe type plasmids reproduce in both yeast and *E. coli* and can be used to clone yeast genes in *E. coli*.

All vectors share several common characteristics. They are typically small, well-characterized molecules of DNA. They contain at least one replication origin and can be replicated within the appropriate host, even when they contain "foreign" DNA. Finally, they code for a phenotypic trait that can be used to detect their presence; frequently, it is also possible to distinguish parental from recombinant vectors.

Plasmids

Plasmids were the first cloning vectors. They are easy to isolate and purify, and they can be reintroduced into a bacterium by transformation. Plasmids often bear antibiotic resistance genes, which are used to select their bacterial hosts. A recombinant plasmid containing foreign DNA is often called a **chimera,** after the Greek mythological monster that had the head of a lion, the tail of a dragon, and the body of a goat. One of the most widely used plasmids is pBR322.

The biology of plasmids (pp. 261–64).

Plasmid pBR322 has both resistance genes for ampicillin and tetracycline and many restriction sites (figure 14.16). Sev-

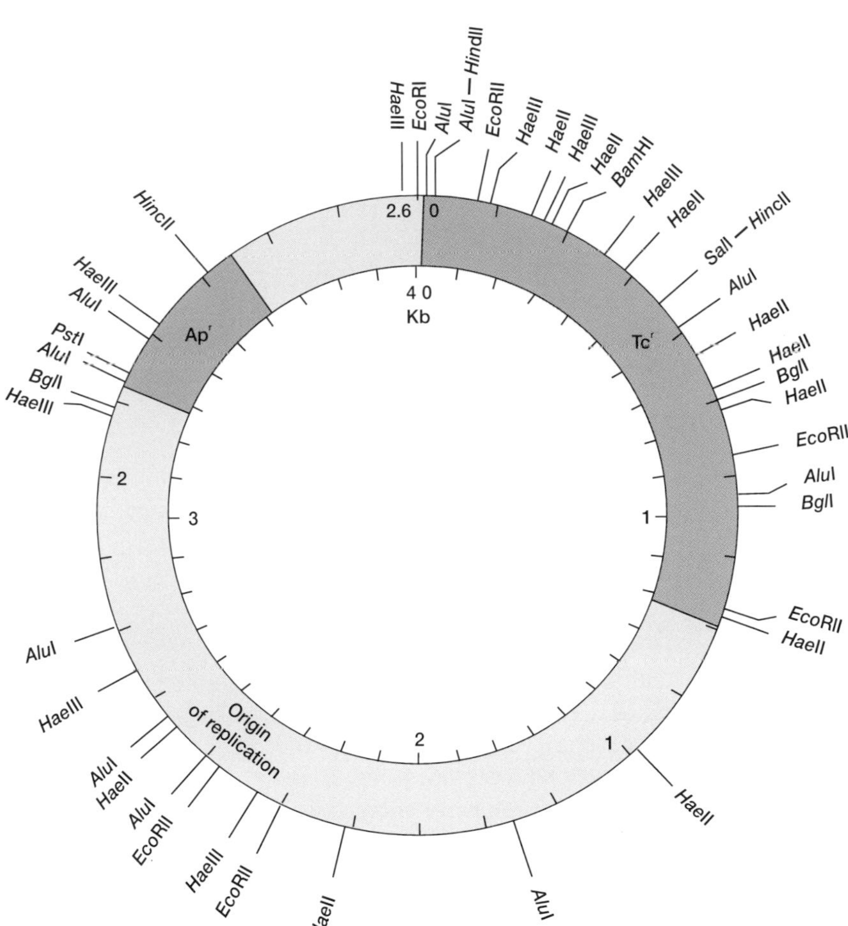

Figure 14.16 The pBR322 Plasmid. A map of the *E. coli* plasmid pBR322. The map is marked off in units of 1×10^5 daltons (*outer circle*) and 0.1 kilobases (*inner circle*). The locations of some restriction enzyme sites are indicated. The plasmid has resistance genes for ampicillin (Ap^r) and tetracycline (Tc^r).

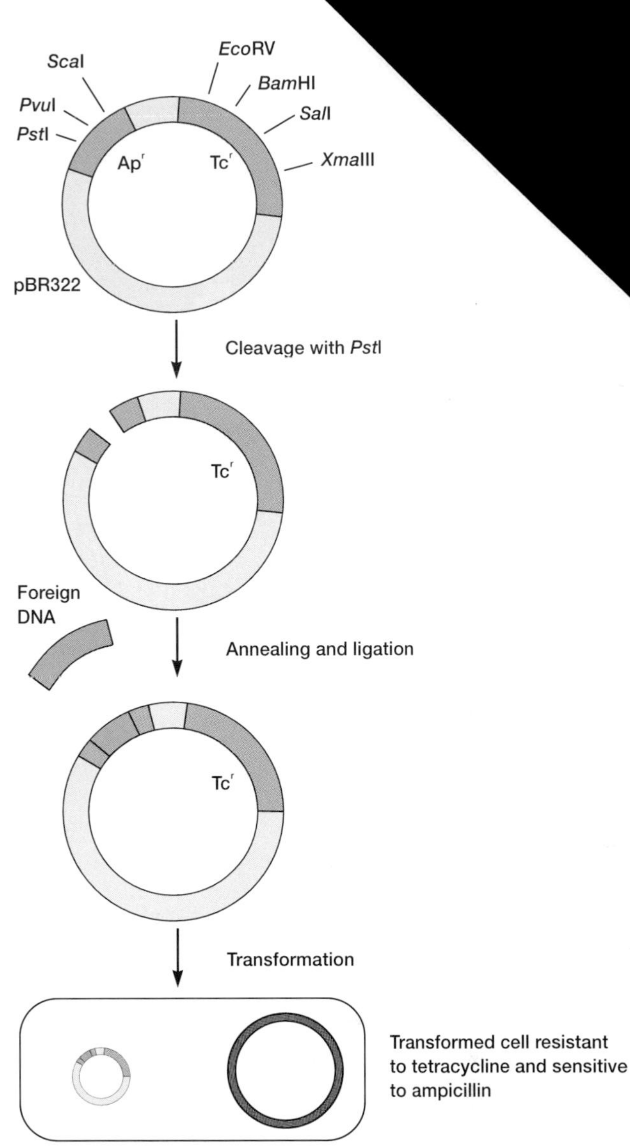

Figure 14.17 Deletion of Recombinant Plasmids. The use of antibiotic resistance genes to detect the presence of recombinant plasmids. Because foreign DNA has been inserted into the ampicillin resistance gene, the recombinant host is only resistant to tetracycline. The restriction enzymes indicated at the top of the figure cleave only one site on the plasmid.

eral of these restriction sites occur only once on the plasmid and are located within an antibiotic resistance gene. This arrangement aids detection of recombinant plasmids after transformation (figure 14.17). For example, if foreign DNA is inserted into the ampicillin resistance gene, the plasmid will no longer confer resistance to ampicillin. Thus, tetracycline-resistant transformants that lack ampicillin resistance contain a chimeric plasmid.

Phage Vectors

Both single- and double-stranded phage vectors have been employed in recombinant DNA technology. For example, lambda phage derivatives are very useful for cloning and can carry fragments up to about 45 kilobases in length. The genes for lysogeny and integration are often nonfunctional and may be deleted to make room for the foreign DNA. The modified phage genome also contains restriction sequences in areas that will not disrupt replication. After insertion of the foreign DNA into the modified lambda vector chromosome, the recombinant phage genome is packaged into viral capsids and can be used to infect host *E. coli* cells (figure 14.15). These vectors are often used to generate genomic libraries. *E. coli* also can be directly transformed with recombinant lambda DNA and produce phages. However, this approach is less efficient than the use of complete phage particles. The process is sometimes called **transfection.**

The biology of bacteriophages (chapter 18).

Cosmids

Cosmids are plasmids that contain lambda phage cos sites and can be packaged into phage capsids. The lambda genome contains a cos site at each end. When the genome is to be packaged in a capsid, it is cleaved at one cos site and the linear

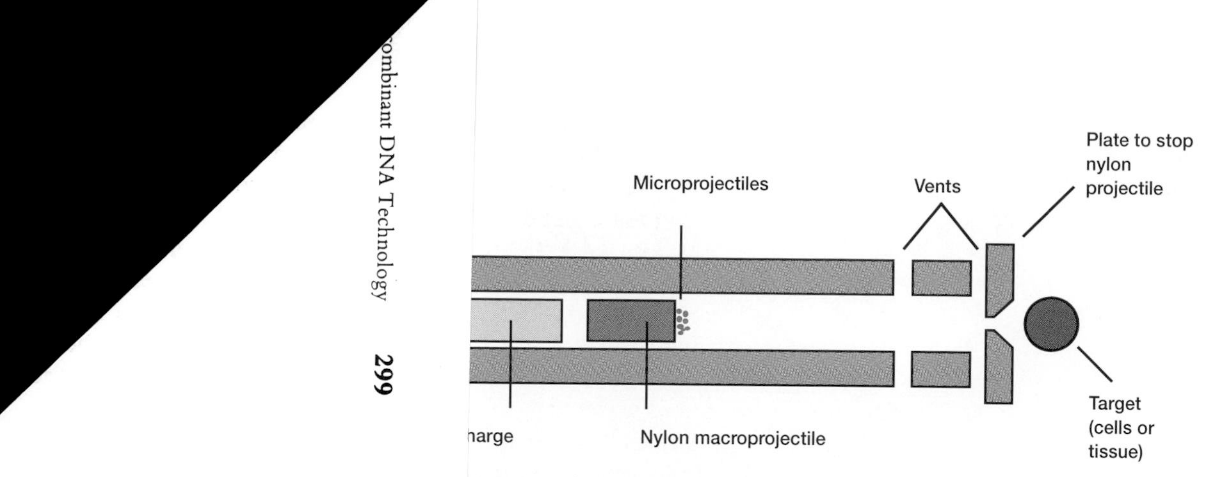

icroprojectiles are propelled into target cells with a .22-caliber blank shell. *Sources: A. M. Moffat,* Mosaic, *21(1):36, 1990; and T. P. Brock and Madigan,* Biology of Microorganisms, *6th ed., 1991.*

DNA is inserted into the capsid until the second cos site has entered. Thus, any DNA between the cos sites is packaged. Cosmids typically contain several restriction sites and antibiotic resistance genes. They are packaged in lambda capsids for efficient injection into bacteria, but they also can exist as plasmids within a bacterial host.

1. Define shuttle vector, chimera, cosmid, and transfection.
2. Describe how each of the three major types of vectors is used in genetic engineering. Give an advantage of each.
3. How can the presence of two antibiotic resistance genes be used to detect recombinant plasmids?

Inserting Genes into Eucaryotic Cells

Because of its practical importance, much effort has been devoted to the development of techniques for inserting genes into eucaryotic cells. Some of these techniques have also been used successfully to transform bacterial cells. The most direct approach is the use of microinjection. Genetic material directly injected into animal cells such as fertilized eggs is sometimes stably incorporated into the host genome to produce a **transgenic animal,** one that has gained new genetic information from the acquisition of foreign DNA.

Another effective technique that works with mammalian cells and plant cell protoplasts is **electroporation.** If cells are mixed with a DNA preparation and then briefly exposed to pulses of high voltage (from about 250 to 4,000 V/cm for mammalian cells), the cells take up DNA through temporary holes in the plasma membrane. Some of these cells will be transformed.

One of the most effective techniques is to shoot microprojectiles coated with DNA into plant and animal cells. The **gene gun,** first developed at Cornell University, operates somewhat like a shotgun. A .22-caliber blank shell shoots a spray of DNA-coated metallic microprojectiles into the cells (figure 14.18). The device has been used to transform corn and produce fertile corn plants bearing foreign genes. Other guns use either electrical discharges or high-pressure gas to propel the DNA-coated projectiles. These guns are sometimes called biolistic devices, a name derived from biological and ballistic. They have been used to transform microorganisms (yeast, the mold *Aspergillus,* and the alga *Chlamydomonas*), mammalian cells, and a variety of plant cells (corn, cotton, tobacco, onion, and poplar).

Expression of Foreign Genes in Bacteria

After a suitable cloning vector has been constructed, rDNA enters the host cell, and a population of recombinant microorganisms develops. Most often the host is an *E. coli* strain that lacks restriction enzymes and is *recA*$^-$ to reduce the chances that the rDNA will undergo recombination with the host chromosome. Plasmid vectors enter *E. coli* cells by calcium chloride-induced transformation. *Bacillus subtilis* and the yeast *Saccharomyces cerevisiae* may also serve as hosts.

A cloned gene is not always expressed in the host cell without further modification of the recombinant vector. To be transcribed, the recombinant gene must have a promoter that is recognized by the host RNA polymerase. Translation of its mRNA depends upon the presence of leader sequences and mRNA modifications that allow proper ribosome binding. These are quite different in eucaryotes and procaryotes, and a procaryotic leader must be provided to synthesize eucaryotic proteins in a bacterium. Finally, introns in eucaryotic genes must be removed because the procaryotic host will not excise them after transcription of mRNA; a eucaryotic protein is not functional without intron removal prior to translation.

The problems of expressing recombinant genes in host cells are largely overcome with the help of special cloning vectors called **expression vectors.** These vectors are often derivatives of plasmid pBR322 and contain the necessary transcription and translation start signals. They also have useful restriction sites next to these sequences so that foreign DNA can be inserted with relative ease. Some expression vectors contain portions of the lac operon and can effectively regulate the expression of the cloned genes in the same manner as the operon.

Somatostatin, the 14-residue hypothalamic polypeptide hormone that helps regulate human growth, provides an example of useful cloning and protein production. The gene for somatostatin was initially synthesized by chemical methods. Besides the 42 bases coding for somatostatin, the polynucleotide contained a codon for methionine at the 5′ end (the N-terminal end of the peptide) and two stop codons at the op-

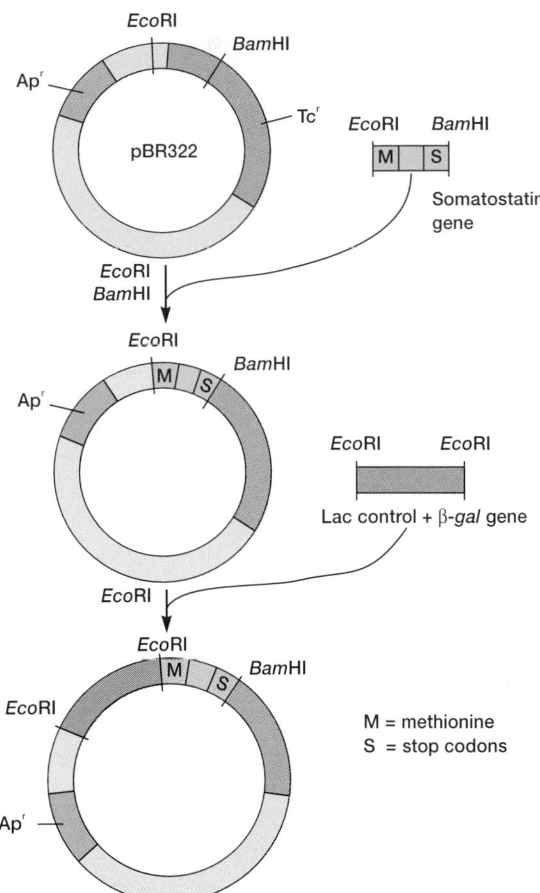

Figure 14.19 **Cloning the Somatostatin Gene.** An overview of the procedure used to synthesize a recombinant plasmid containing the somatostatin gene. See text for details. *From A. E. H. Emery,* An Introduction to Recombinant DNA. *Copyright © John Wiley & Sons, Inc., New York, NY.*

posite end. To aid insertion into the plasmid vector, the 5′ ends of the synthetic gene were extended to form single-stranded sticky ends complementary to those formed by the *Eco*RI and *Bam*HI restriction enzymes. A modified pBR322 plasmid was cut with both *Eco*RI and *Bam*HI to remove a part of the plasmid DNA. The synthetic gene was then spliced into the vector by taking advantage of its cohesive ends (figure 14.19). Finally, a fragment containing the initial part of the lac operon (including the promoter, operator, ribosome binding site, and much of the β-galactosidase gene) was inserted next to the somatostatin gene. The plasmid now contained the somatostatin gene fused in the proper orientation to the remaining portion of the β-galactosidase gene.

After introduction of this chimeric plasmid into *E. coli,* the somatostatin gene was transcribed with the β-galactosidase gene fragment to generate an mRNA having both messages. Translation formed a protein consisting of the total hormone polypeptide attached to the β-galactosidase fragment by a methionine residue. Cyanogen bromide cleaves peptide bonds at methionine residues. Treatment of the fusion protein with cyanogen bromide broke the peptide chain at the methionine and released the hormone (figure 14.20). Once free, the polypeptide was able to fold properly and become active. Since production of the fusion protein was under the control of the lac operon, it could be easily regulated.

Many proteins have been produced since the synthesis of somatostatin. A similar approach was used to manufacture human insulin. Human growth hormone and some interferons

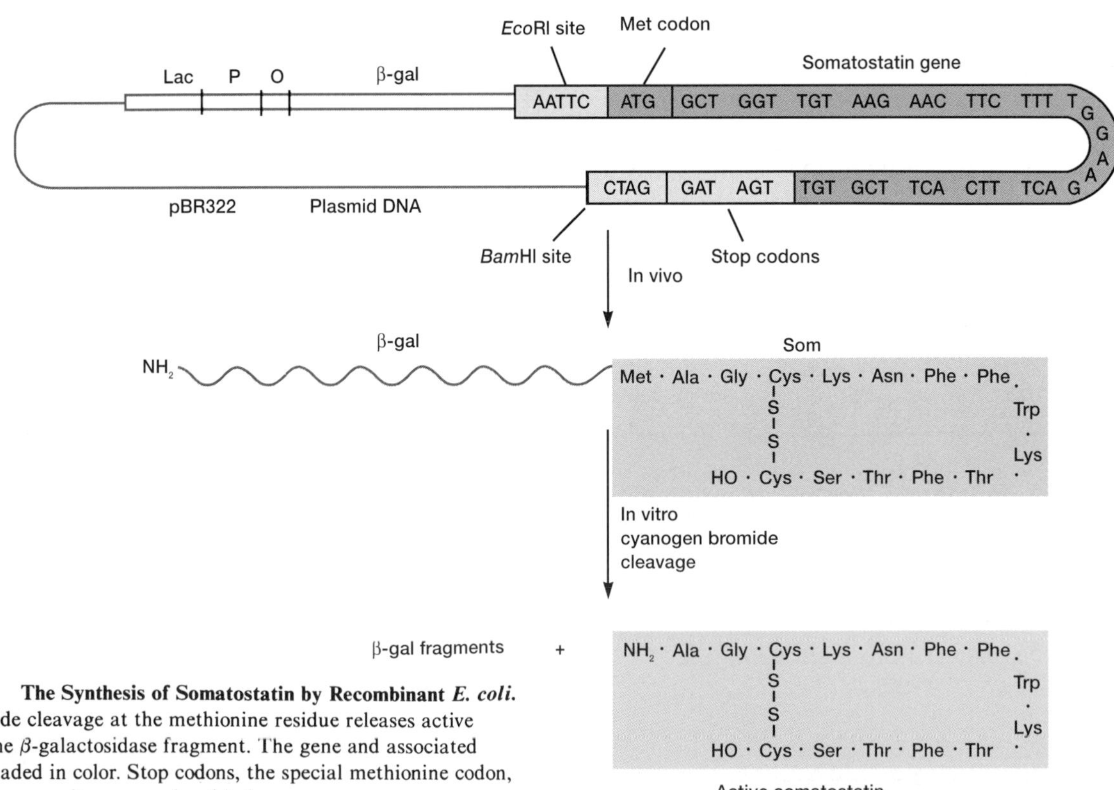

Figure 14.20 **The Synthesis of Somatostatin by Recombinant *E. coli.*** Cyanogen bromide cleavage at the methionine residue releases active hormone from the β-galactosidase fragment. The gene and associated sequences are shaded in color. Stop codons, the special methionine codon, and restriction enzyme sites are enclosed in boxes.

Box 14.1

Gene Expression and Kittyboo Colors

Genetic engineering techniques can yield unusual approaches to long-standing problems. To understand how the regulation of gene expression influences complex processes, the activity of more than one gene must be followed simultaneously. The use of genetic engineering techniques with luciferase genes has made this much easier.

A Jamaican beetle named the kittyboo has two light organs on its head and one on its abdomen. Recently, four luciferase genes have been isolated from the beetle. Each luciferase enzyme produces a different colored light when it acts on the substrate luciferin. Keith V. Wood has cloned the genes and inserted them into *E. coli.* When the bacteria are exposed to luciferin, they glow (Box Figure 14.1).

These luciferase genes can be used to study gene regulation. Suppose that one wishes to follow the activity of the liver gene that codes for serum albumin. The albumin gene can be replaced with a luciferase gene, and the modified cell incubated with luciferin. Whenever the albumin gene is activated, the newly inserted luciferase gene will function and the cell glow. Measurement of light intensity is easy, rapid, and sensitive. By substituting a different luciferase gene for another liver gene, one can simultaneously follow the activity and coordination of two different liver genes. It is only necessary to measure light intensity at the two wavelengths characteristic of the luciferase genes.

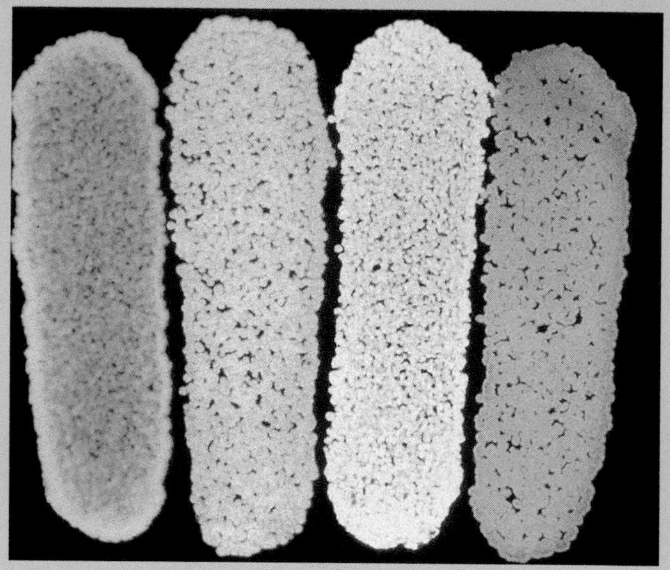

Box Figure 14.1 ***E. coli* with Luciferase Genes.** These four streaks of *E. coli* glow different colors because they contain four different luciferase genes cloned from the Jamaican click beetle or kittyboo, *Pyrophorus plagiophthalamus.*

have also been synthesized by rDNA techniques. The human growth hormone gene was too long to synthesize by chemical procedures and was prepared from mRNA as cDNA.

As mentioned earlier, introns in eucaryotic genes are not removed by bacteria and will render the final protein nonfunctional. The easiest solution is to prepare cDNA from processed mRNA that lacks introns and directly reflects the correct amino acid sequence of the protein product. In this instance, it is particularly important to fuse the gene with an expression vector since a promoter and other essential sequences will be missing in the cDNA.

Eucaryotic mRNA processing (pp. 203–4).

1. What is a transgenic animal? Describe how electroporation and gene guns are used to insert foreign genes into cells.
2. How can one prevent rDNA from undergoing recombination in a host cell?
3. List several reasons why a cloned gene might not be expressed in a host cell. What is an expression vector?
4. Briefly outline the procedure for somatostatin production.
5. Identify a way to eliminate eucaryotic introns during the synthesis of rDNA.

Applications of Genetic Engineering

Genetic engineering and biotechnology will contribute in the future to medicine, industry, and agriculture, as well as to basic research (Box 14.1). In this section, some practical applications are briefly discussed.

Medical Applications

Certainly, the production of medically useful proteins such as somatostatin, insulin, human growth hormone, and interferon is of great practical importance (table 14.4). This is particularly true of substances that previously only could be obtained from human tissues. For example, in the past, human growth hormone for treatment of pituitary dwarfism was extracted from pituitaries obtained during autopsies and was available only in limited amounts. Interleukin-2 (a protein that helps regulate the immune response) and blood-clotting factor VIII have recently been cloned, and undoubtedly other important peptides and proteins will be produced in the future. Synthetic vaccines—for instance, vaccines for malaria and rabies—are also being developed with recombinant techniques. A recombinant hepatitis B vaccine is already commercially available.

Other medical uses of genetic engineering are being investigated. Probes are now being used in the diagnosis of in-

TABLE 14.4 Some Human Peptides and Proteins Synthesized by Genetic Engineering

Peptide or Protein	Potential Use
α_1-antitrypsin	Treatment of emphysema
α-, β-, and τ-interferons	As antiviral, antitumor, and anti-inflammatory agents
Blood-clotting factor VIII	Treatment of hemophilia
Calcitonin	Treatment of osteomalacia
Erythropoetin	Treatment of anemia
Growth hormone	Growth promotion
Insulin	Treatment of diabetes
Interleukins-1, 2, and 3	Treatment of immune disorders and tumors
Macrophage colony stimulating factor	Cancer treatment
Relaxin	Aid to childbirth
Serum albumin	Plasma supplement
Somatostatin	Treatment of acromegaly
Streptokinase	Anticoagulant
Tissue plasminogen activator	Anticoagulant
Tumor necrosis factor	Cancer treatment

fectious disease. An individual could be screened for mutant genes with probes and hybridization techniques (even before birth when used together with amniocentesis). A type of genetic surgery called somatic cell gene therapy may be possible for afflicted individuals. For example, cells of an individual with a genetic disease could be removed, cultured, and transformed with cloned DNA containing a normal copy of the defective gene or genes. These cells could then be reintroduced into the individual; if they became established, the expression of the normal genes might cure the patient. Recently, an immune deficiency disease patient lacking the enzyme adenosine deaminase that destroys toxic metabolic by-products has been treated in this way. Some of the patient's lymphocytes (*pp. 609–10*) were removed, given the adenosine deaminase gene, and returned to the patient's body. It also may be possible to use a defective retrovirus (*see chapter 19*) to insert the proper genes into host cells, perhaps even specifically targeted organs or tissues.

The use of probes and plasmid fingerprinting in clinical microbiology (pp. 692–93, 694).

It now appears that livestock will also be important in medically oriented biotechnology. Pig embryos injected with human hemoglobin genes develop into transgenic pigs that synthesize human hemoglobin. Current plans are to purify the hemoglobin and use it as a blood substitute. A pig could yield 20 units of blood substitute a year. Somewhat similar techniques have produced transgenic goats whose milk contains up to 3 grams of human tissue plasminogen activator per liter. Tissue plasminogen activator (TPA) dissolves blood clots and is used to treat cardiac patients.

Recombinant DNA techniques are playing an increasingly important role in research on the molecular basis of disease. The transfer from research to practical application is often slow because it is not easy to treat a disease effectively unless its molecular mechanism is known. Thus, genetic engineering may aid in the fight against a disease by providing new information about its nature, as well as by aiding in diagnosis and therapy.

Industrial Applications

Industrial applications of recombinant DNA technology include manufacturing protein products by using bacteria, fungi, and cultured mammalian cells as factories; strain improvement for existing bioprocesses; and the development of new strains for additional bioprocesses. As mentioned earlier, the pharmaceutical industry is already producing several medically important polypeptides using this technology. In addition, there is interest in making expensive, industrially important enzymes with recombinant bacteria. Bacteria that metabolize petroleum and other toxic materials have been developed. These bacteria can be constructed by assembling the necessary catabolic genes on a single plasmid and then transforming the appropriate organism. There are also many potential applications in the chemical and food industries.

Food microbiology (chapter 43).
Industrial microbiology and biotechnology (chapter 44).

Agricultural Applications

It is also possible to bypass the traditional methods of selective breeding and directly transfer desirable traits to agriculturally important animals and plants. Potential exists for increasing growth rates and overall protein yields of farm animals. The growth hormone gene has already been transferred from rats to mice, and both *in vitro* fertilization and embryo implantation methods are fairly well developed. Recently, recombinant bovine growth hormone has been used to increase milk production by at least 10%. Perhaps farm animals' disease resistance and tolerance to environmental extremes also can be improved.

Cloned genes can be inserted into plant as well as animal cells. Presently, a popular way to insert genes into plants is with a recombinant **Ti plasmid** (tumor-inducing plasmid) obtained from the bacterium *Agrobacterium tumefaciens* (Box 14.2). It is also possible to donate genes by forming plant cell protoplasts, making them permeable to DNA, and then adding the desired rDNA. The advent of the gene gun (p. 300) will greatly aid the production of transgenic plants.

Much effort has been devoted to the transfer of the nitrogen-fixing abilities of bacteria associated with legumes to other crop plants. The genes largely responsible for the process have been cloned and transferred to the genome of plant cells; however, the recipient cells have not been able to fix nitrogen. If successful, the potential benefit for crop plants such as corn is great. Nevertheless, there is some concern that new nitrogen-fixing varieties might spread indiscriminantly like weeds or disturb the soil nitrogen cycle (*see chapters 40 and 42*).

Box 14.2

Plant Tumors and Nature's Genetic Engineer

A plasmid from the plant pathogenic bacterium *Agrobacterium tumefaciens* is responsible for much success in the genetic engineering of plants. Infection of normal cells by the bacterium transforms them into tumor cells, and the crown gall disease develops in dicotyledonous plants such as grapes and ornamental plants. Normally, the gall or tumor is located near the junction of the plant's root and stem. The tumor forms because of the insertion of genes into the plant cell genome, and only strains of *A. tumefaciens* possessing a large conjugative plasmid called the Ti plasmid are pathogenic. The Ti plasmid carries genes for virulence and the synthesis of substances involved in the regulation of plant growth. The genes that induce tumor formation reside between two 23 base pair direct-repeat sequences. This region is known as T-DNA and is very similar to a transposon. T-DNA contains genes for the synthesis of plant growth hormones (an auxin and a cytokinin) and an amino acid derivative called opine that serves as a nutrient source for the invading bacteria. In diseased plant cells, T-DNA is inserted into the chromosomes at various sites and is stably maintained in the cell nucleus.

When the molecular nature of crown gall disease was recognized, it became clear that the Ti plasmid and its T-DNA had great potential as a vector for the insertion of rDNA into plant chromosomes. In one early experiment, the yeast alcohol dehydrogenase gene was added to the T-DNA region of the Ti plasmid. Subsequent infection of cultured plant cells resulted in the transfer of the yeast gene. Since then, many modifications of the Ti plasmid have been made to improve its characteristics as a vector. Usually, one or more antibiotic resistance genes are added, and the nonessential T-DNA, including the tumor inducing genes, is deleted. Those genes required for the actual infection of the plant cell by the plasmid are left. T-DNA has also been inserted into the *E. coli* pBR322 plasmid and other plasmids to produce cloning vectors that can move between bacteria and plants (Box Figure 14.2). The gene or genes of interest are spliced into the T-DNA region between the direct repeats. Then the plasmid is returned to *A. tumefaciens,* plant culture cells are infected with the bacterium, and transformants are selected by screening for antibiotic resistance (or another trait in the T-DNA). Finally, whole plants are regenerated from the transformed cells. In this way, several plants have been made herbicide resistant (see text).

Unfortunately, *A. tumefaciens* does not produce crown gall disease in monocotyledonous plants such as corn, wheat, and other grains; and it has been used only to modify plants such as potato, tomato, celery, lettuce, and alfalfa. However, there is evidence that T-DNA is transferred to monocotyledonous plants and expressed, although it does not produce tumors. This discovery plus the creation of new procedures for inserting DNA into plant cells may well lead to the use of rDNA techniques with many important crop plants.

Attempts at rendering plants resistant to environmental stresses have been more successful. For example, the genes for detoxification of glyphosate herbicides were isolated from *Salmonella,* cloned, and introduced into tobacco cells using the Ti plasmid. Plants regenerated from the recombinant cells were resistant to the herbicide. Herbicide-resistant varieties of cotton and fertile, transgenic corn have also been developed. This is of considerable importance since many crop plants suffer stress when treated with herbicides. Resistant crop plants would not be stressed by the chemicals being used to control weeds, and yields would presumably be much greater.

Many other agricultural applications are being tested. Two examples are (1) a strain of *Pseudomonas syringae* that protects against the frost damage of plants because it cannot produce a protein that induces ice-crystal formation; and (2) a strain of *Pseudomonas fluorescens* that carries the gene for the *Bacillus thuringensis* endotoxin, which is active against many insect pests such as the cabbage looper and the European corn borer. A variety of corn with the *B. thuringensis* toxin gene is under development.

1. List several important present or future applications of genetic engineering in medicine, industry, and agriculture.
2. What is the Ti plasmid and why is it so important?

Social Impact of Recombinant DNA Technology

Despite the positive social impact of rDNA technology, dangers may be associated with rDNA work and gene cloning. The potential to alter an organism genetically raises serious scientific and philosophical questions, many of which have not yet been adequately addressed. In this section, some of the debate is briefly reviewed.

The initial concern raised by the scientific community was that recombinant *E. coli* and other microorganisms carrying

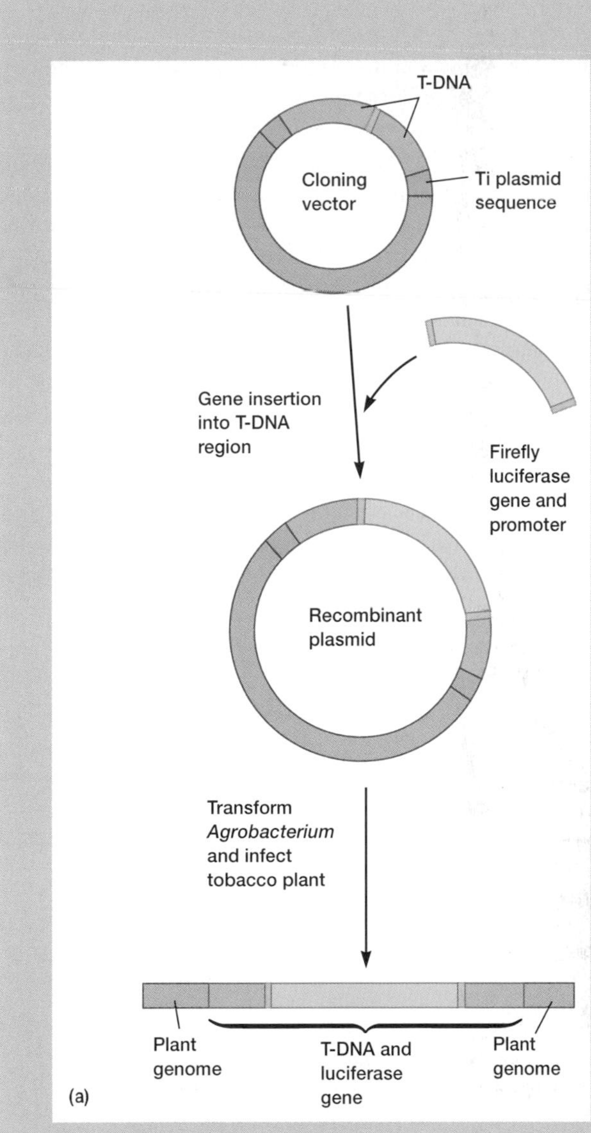

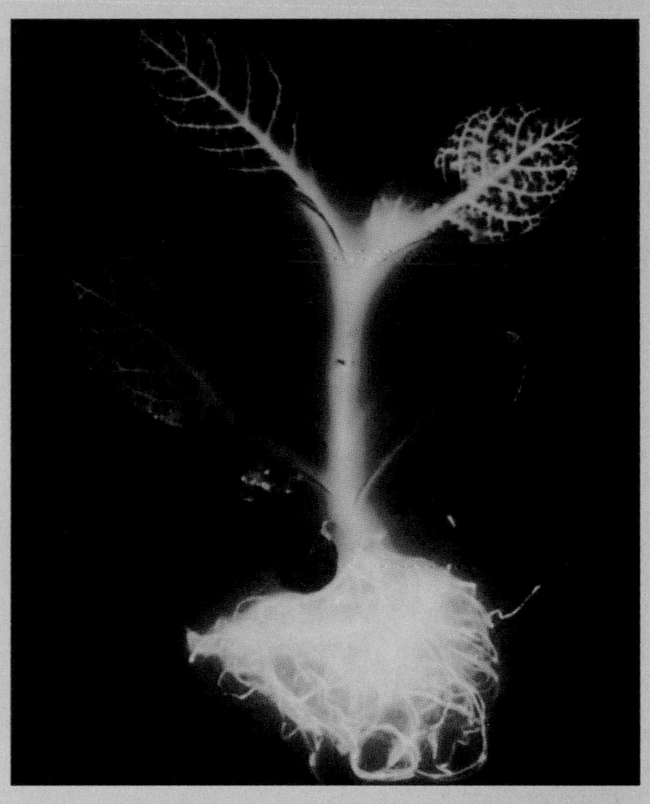

Box Figure 14.2 The Use of the Ti Plasmid Vector to Produce Transgenic Plants. (*a*) The formation of a cloning vector and its use in transformation. See text for details. (*b*) A tobacco plant, *Nicotiana tabacum,* that has been made bioluminescent by transfection with a special Ti plasmid vector containing the firefly luciferase gene. When the plant is watered with a solution of luciferin, the substrate for the luciferase enzyme, it glows. The picture was made by exposing the transgenic plant to Ektachrome film for 24 hours.

dangerous genes might escape and cause widespread infections. Because of these worries, the federal government established guidelines to limit and regulate the locations and types of potentially dangerous experiments. Physical containment was to be practiced; that is, rDNA research was to be carried out in specially designed laboratories with extra safety precautions. In addition, rDNA researchers were also to practice biological containment. Only weakened microbial hosts unable to survive in a natural environment and nonconjugating bacterial strains were to be used to avoid the spread of the vector or rDNA itself. Further experiments suggested that the dangers were not as extreme as initially conceived. Yet little is known about what would happen if recombinant organisms escaped. For example, could dangerous genes such as oncogenes (*see chapter 16*) move from a weakened strain to a hardy one that would then spread? Will there be increased risk when extremely large quantities of recombinant microorganisms are grown in industrial fermenters? Some people worry because the use of rDNA techniques has become so widespread and the guidelines on their safe use have been relaxed to a considerable extent. Biomedical rDNA research is regulated by the Recombinant DNA Advisory Committee (RAC) of the National Institutes of Health. The Environmental Protection Agency and state governments have jurisdiction over agriculturally related field experiments. Industrial rDNA research can proceed without such close regulation, and this has also caused some anxiety about potential safety hazards.

Aside from the concerns over safety, the use of recombinant technology on human beings raises ethical and moral questions. These problems are not extreme in cases of somatic cell gene therapy. There is much greater concern about attempts to correct defects or bestow traits considered to be desirable or cosmetic by introducing genes into human eggs and embryos. In 1983, two Nobel Prize winners, several other prominent scientists, and about 40 religious leaders signed a statement to Congress requesting that such alterations of human eggs or sperm not be attempted. Others argue that the good arising from such interventions more than justifies the risks of misuse and that there is nothing inherently wrong in modifying the human gene pool to reduce the incidence of genetic disease. This discussion extends to other organisms as well. Some argue that we have no right to create new life forms and further disrupt the genetic diversity that human activities have already severely reduced. United States courts have decided that researchers and companies can patent "nonnaturally occurring" living organisms, whether microbial, plant, or animal.

Another area of considerable controversy involves the use of recombinant organisms in agriculture. Many ecologists are worried that the release of unusual, recombinant organisms without careful prior risk assessment may severely disrupt the ecosystem. They point to the many examples of ecosystem disruption from the introduction of foreign organisms. Potential risks have been vigorously discussed because several recombinant organisms are either ready for field studies or are already being tested. Thus far, no obvious ecological effects have been observed.

As with any technology, the potential for abuse exists. A case in point is the use of genetic engineering in biological warfare and terrorism. Although there are international agreements that limit the research in this area to defense against other biological weapons, the knowledge obtained in such research can easily be used in offensive biological warfare. Effective vaccines constructed using rDNA technology can protect the attacker's troops and civilian population. Because it is relatively easy and inexpensive to prepare bacteria capable of producing massive quantities of toxins or to develop particularly virulent strains of viral and bacterial pathogens, even small countries and terrorist organizations might acquire biological weapons. Many scientists and nonscientists are also concerned about increases in the military rDNA research carried out by the major powers.

Recombinant DNA technology has greatly enhanced our knowledge of genes and how they function, and it promises to improve our lives in many ways. Yet, as this brief discussion shows, problems and concerns remain to be resolved. Past scientific advances have sometimes led to unanticipated and unfortunate consequences, such as environmental pollution and nuclear weapons. With prudence and forethought we may be able to avoid past mistakes in the use of this new technology.

1. Describe four major areas of concern about the application of genetic engineering. In each case, give both the arguments for and against the use of genetic engineering.

Summary

1. Genetic engineering became possible after the discovery of restriction enzymes and reverse transcriptase and the development of the Southern blotting technique and other essential methods in nucleic acid chemistry.
2. Oligonucleotides of any desired sequence can be synthesized by a DNA synthesizer machine. This has made possible site-directed mutagenesis.
3. The polymerase chain reaction allows small amounts of DNA to be increased in concentration thousands of times.
4. Agarose gel electrophoresis is used to separate DNA fragments according to size differences.
5. Fragments can be isolated and identified, then joined with plasmids or phage genomes and cloned; one also can first clone all fragments and subsequently locate the desired clone.
6. Three techniques that can be used to join DNA fragments are the creation of similar sticky ends with a single restriction enzyme, the addition of poly(dA) and poly(dT) to create sticky ends, and blunt-end ligation with T4 DNA ligase.
7. Probes for the detection of recombinant clones are made in several ways and usually labeled with ^{32}P. Nonradioactive probes are also used.
8. Several types of vectors may be used, each with different advantages: plasmids, phages, cosmids, and shuttle vectors.
9. Genes can be inserted into eucaryotic cells by techniques such as microinjection, electroporation, and the use of a gene gun.
10. The recombinant vector often must be modified by the addition of promoters, leaders, and other elements. Eucaryotic gene introns also must be removed. An expression vector has the necessary features to express any recombinant gene it carries.
11. The hormone somatostatin has been synthesized using recombinant DNA technology.
12. Recombinant DNA technology will provide many benefits in medicine, industry, and agriculture.
13. Despite the great promise of genetic engineering, it also brings with it potential problems in areas of safety, human experimentation, potential ecological disruption, and biological warfare.

Key Terms

autoradiography *288*
biotechnology *286*
chimera *298*
complementary DNA (cDNA) *288*
cosmid *299*
electrophoresis *293*
electroporation *300*
expression vector *300*
gene gun *300*
genetic engineering *286*
oligonucleotides *290*
polymerase chain reaction (PCR technique) *291*
probe *288*
recombinant DNA technology *286*
restriction enzymes *287*
site-directed mutagenesis *290*
Southern blotting technique *288*
Ti plasmid *303*
transfection *299*
transgenic animal *300*
vectors *288*

Questions for Thought and Review

1. Could the Southern blotting technique be applied to RNA and proteins? How might this be done?
2. Why could a band on an electrophoresis gel still contain more than one kind of DNA fragment?
3. What advantage might there be in creating a genomic library first rather than directly isolating the desired DNA fragment?
4. Suppose that one inserted a simple plasmid (one containing an antibiotic resistance gene and a separate restriction site) carrying a human interferon gene into *E. coli,* but none of the transformed bacteria produced interferon. Give as many plausible reasons as possible for this result.
5. In what areas do you think genetic engineering will have the greatest positive impact in the future? Why?
6. What do you consider to be the greatest potential dangers of genetic engineering? Are there ethical problems with any of its potential applications?

Additional Reading

Anderson, W. F., and Diacumakos, E. G. 1981. Genetic engineering in mammalian cells. *Sci. Am.* 245(1):106–21.

Antebi, E., and Fishlock, D. 1985. *Biotechnology: Strategies for life.* Cambridge, Mass.: MIT Press.

Arnheim, N. and Levenson, C. H. 1990. Polymerase chain reaction. *Chem Eng. News* 68:36–47.

Atlas, R. M. 1991. Environmental applications of the polymerase chain reaction. *ASM News* 57(12): 630–32.

Brown, C. M.; Campbell, I.; and Priest, F. G. 1987. *Introduction to biotechnology.* Boston: Blackwell Scientific Publications.

Chilton, M.-D. 1983. A vector for introducing new genes into plants. *Sci. Am.* 248(6):51–59.

Cohen, S. N. 1975. The manipulation of genes. *Sci. Am* 233(1):25–33.

Emery, A. E. H. 1984. *An introduction to recombinant DNA.* New York: John Wiley and Sons.

Erlich, H. A. 1989. *PCR technology: Principles and applications of DNA amplification.* San Francisco: W. H. Freeman.

Erlich, H. A.; Gelfand, D.; and Sninsky, J. J. 1991. Recent advances in the polymerase chain reaction. *Science* 252:1643–51.

Gardner, E. J.; Simmons, M. J.; and Snustad, D. P. 1991. *Genetics,* 8th ed. New York: John Wiley and Sons.

Gasser, C. S., and Fraley, R. T. 1992. Transgenic crops. *Sci. Am.* 266(6):62–69.

Gilbert, W., and Villa-Komaroff, L. 1980. Useful proteins from recombinant bacteria. *Sci. Am.* 242(4):74–94.

Goodfield, J. 1977. *Playing God: Genetic engineering and the manipulation of life.* New York: Random House.

Grobstein, C. 1977. The recombinant-DNA debate. *Sci. Am.* 237(1):22–33.

Hansen, M.; Busch, L.; Burkhardt, J.; Lacy, W. B.; and Lacy, L. R. 1986. Plant breeding and biotechnology. *BioScience* 36(1):29–39.

Hopwood, D. A. 1981. The genetic programming of industrial microorganisms. *Sci. Am.* 245(3):91–102.

Jaenisch, R. 1988. Transgenic animals. *Science* 240:1468–74.

Kenney, M. 1986. *Biotechnology: The university-industrial complex.* New Haven, Conn.: Yale University Press.

Krimsky, S. 1982. *Genetic alchemy: The social history of the recombinant DNA controversy.* Cambridge, Mass.: MIT Press.

Lear, J. 1978. *Recombinant DNA: The untold story.* New York: Crown Publishers.

Lewin, B. 1990. *Genes,* 4th ed. New York: Oxford University Press.

Moses, P. B. 1987. Strange bedfellows. *BioScience* 37: 6–10.

Mullis, K. B. 1990. The unusual origin of the polymerase chain reaction. *Sci. Am.* 262(4):56–65.

Old, R. W., and Primrose, S. B. 1989. *Principles of gene manipulation,* 4th ed. Boston: Blackwell Scientific Publications.

Pestka, S. 1983. The purification and manufacture of human interferons. *Sci. Am.* 249(2):37–43.

Peters, P. 1993. *Biotechnology: A guide to genetic engineering.* Dubuque, Iowa: Wm. C. Brown.

Primrose, S. B. 1987. *Modern Biotechnology.* Boston: Blackwell Scientific Publications.

Richards, J., ed. 1978. *Recombinant DNA: Science, ethics, and politics.* New York: Academic Press.

Rifkin, J. 1983. *Algeny.* New York: Viking Press.

Smith, J. E. 1985. *Biotechnology principles.* Washington, D.C.: American Society for Microbiology.

Teitelman, R. 1989. *Gene dreams: Wall street, academia, and the rise of biotechnology.* New York: Basic Books, Inc.

Tucker, J. B. Winter 1984–85. Gene wars. *Foreign Policy* 57:58–79.

Watson, J. D., Gilman, M.; Witkowski, J.; and Zoller, M. 1992. *Recombinant DNA,* 2d ed. San Francisco: W.H. Freeman.

Yoxen, E. 1984. *The gene business: Who should control biotechnology?* New York: Harper & Row.

Zilinskas, R. A., and Zimmerman, B. K., eds. 1986. *The gene-splicing wars: Reflections on the recombinant DNA controversy.* New York: Macmillan.

PART FOUR

The Control of Microorganisms

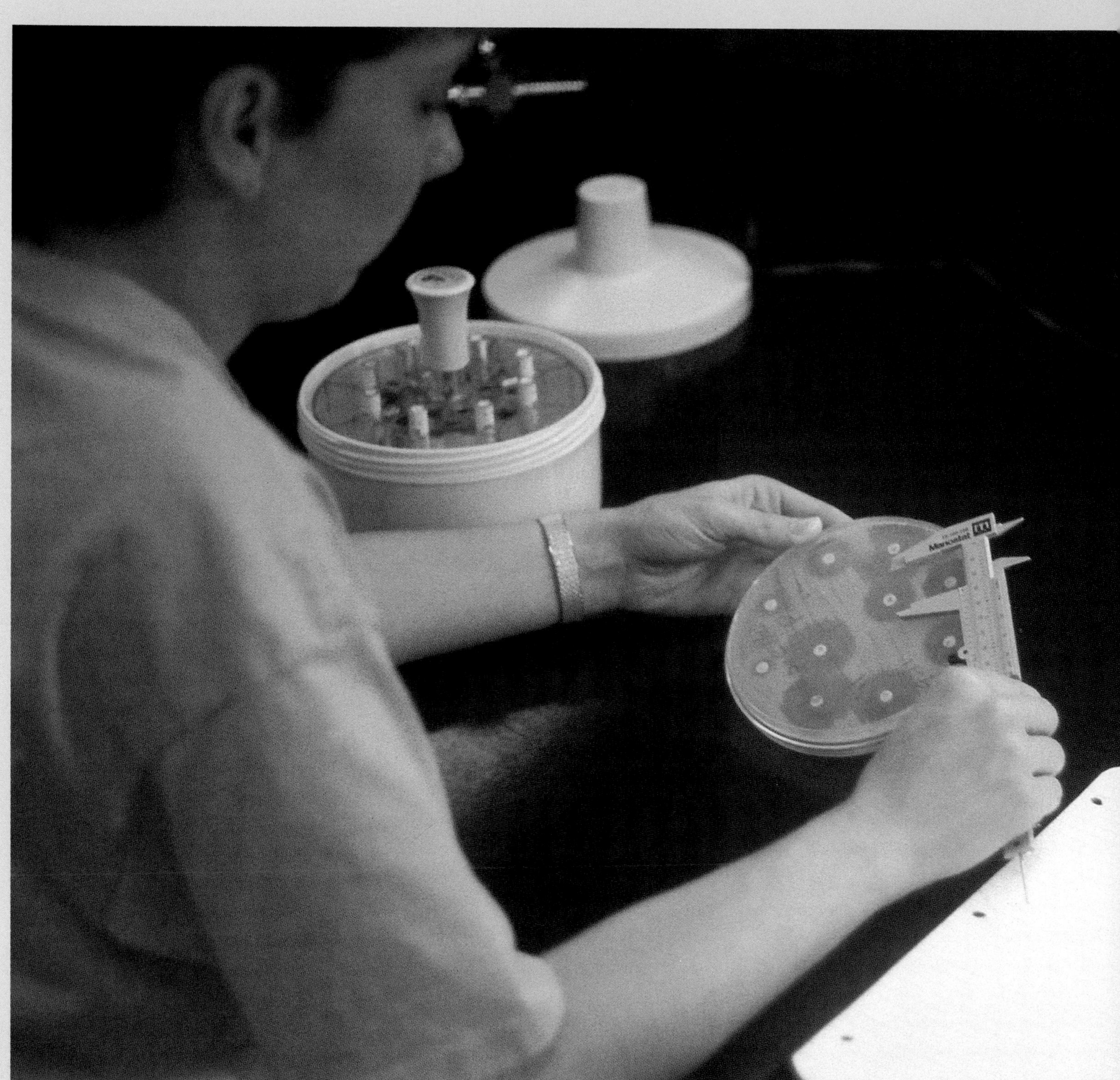

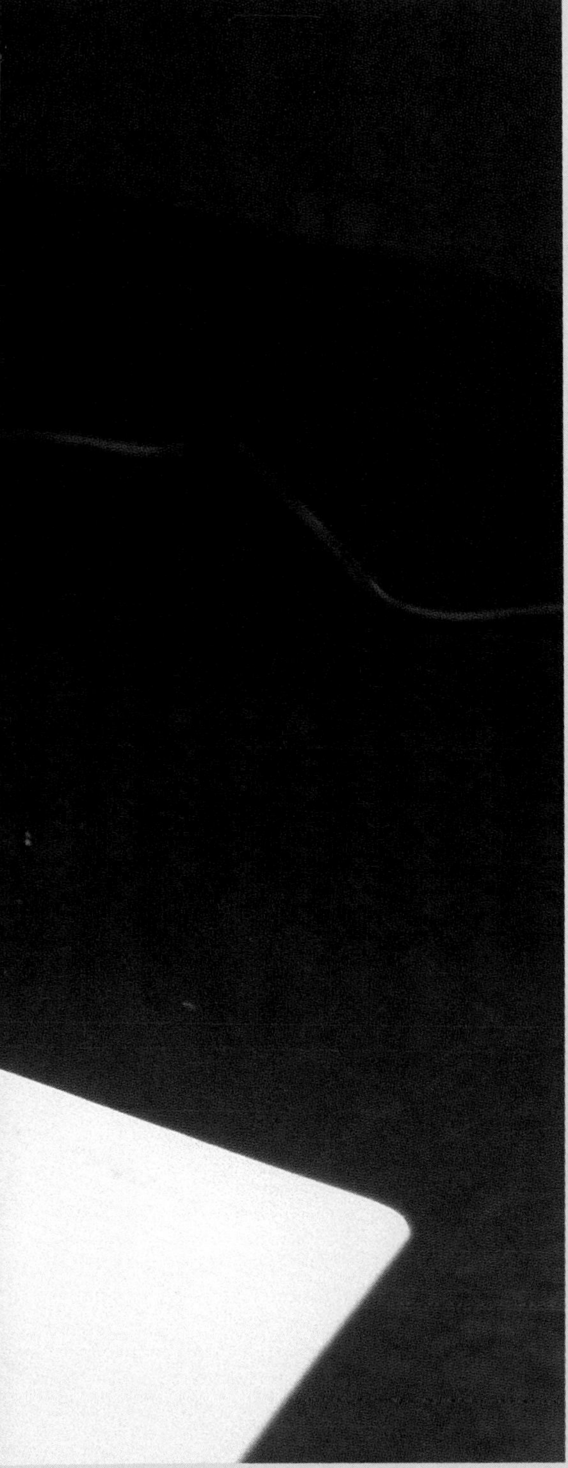

Antibiotic sensitivity testing. A microbiologist is measuring the diameter of antibiotic inhibition zones in the Kirby-Bauer test (*see p. 329*).

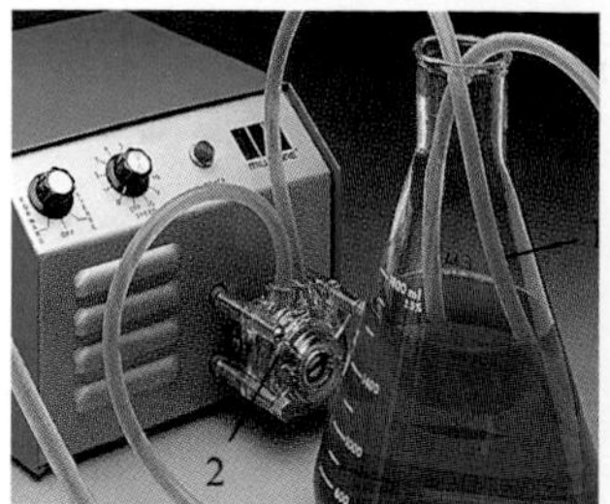

CHAPTER 15
Control of Microorganisms by Physical and Chemical Agents

Were we but able to explain
The fiefdom of the microbe—
Why one man is his serf,
Another is his lord
When all are his domain. . . .
—Anonymous

We all labour against our own cure, for death is the cure of all diseases.
—Sir Thomas Browne

Outline

Concepts

1. Microbial population death is exponential, and the effectiveness of an agent is not fixed but influenced by many environmental factors.
2. Solid objects can be sterilized by physical agents such as heat and radiation; liquids and gases are sterilized by heat, radiation, and filtration through the proper filter.
3. Most chemical agents do not readily destroy bacterial endospores and, therefore, cannot sterilize objects; they are used as disinfectants, sanitizers, and antiseptics. Objects can be sterilized by gases like ethylene oxide that destroy endospores.
4. A knowledge of methods used for microbial control is essential for personal and public safety.

The chapters in Part II are concerned with the nutrition, growth, and physiology of microorganisms. This chapter addresses the subject of their control and destruction, a topic of immense practical importance. Although many microorganisms are beneficial and necessary for human well-being, microbial activities may have undesirable consequences, such as food spoilage and disease. Therefore, it is essential to be able to kill microorganisms or inhibit their growth in order to minimize their destructive effects.

This chapter focuses on the control of microorganisms by physical and chemical agents. The following chapter introduces the use of antimicrobial chemotherapy to control disease.

From the beginning of recorded history, people have practiced disinfection and sterilization, even though the existence of microorganisms was long unsuspected. The Egyptians used fire to sterilize infectious material and disinfectants to embalm bodies, and the Greeks burned sulfur to fumigate buildings. Mosaic law commanded the Hebrews to burn any clothing suspected of being contaminated with the leprosy bacterium. Today, the ability to destroy microorganisms is no less important: it makes possible the sterile techniques used in microbiological research, the preservation of food, and the prevention of disease. The techniques described in this chapter are also essential to personal safety in both the laboratory and hospital (Box 15.1).

There are several ways to control microbial growth that have not been included in this chapter, but they should be considered for a more complete picture of how microorganisms are controlled. Chapter 6 describes the effects of osmotic activity, pH, temperature, O_2, and radiation on microbial growth and survival (*see pp. 122–30*). Chapter 43 discusses the use of physical and chemical agents in food preservation (*see pp. 868–71*).

Definition of Frequently Used Terms

Terminology is especially important when the control of microorganisms is discussed because words like disinfectant and antiseptic are often used loosely. The situation is even more confusing because a particular treatment can either inhibit growth or kill depending on the conditions.

The ability to control microbial populations on inanimate objects, like eating utensils and surgical instruments, is of considerable practical importance. Sometimes, it is necessary to eliminate all microorganisms from an object, whereas, only partial destruction of the microbial population may be required in other situations. **Sterilization** (Latin *sterilis,* unable to produce offspring or barren) is the process by which all living cells, viable spores, viruses, and viroids (*see chapter 19*) are either destroyed or removed from an object or habitat. A sterile object is totally free of viable microorganisms, spores, and other infectious agents. When sterilization is achieved by a chemical agent, the chemical is called a sterilant. In contrast, **disinfection** is the killing, inhibition, or removal of microorganisms that may cause disease. **Disinfectants** are agents, usually chemical, used to carry out disinfection, and are normally used only on inanimate objects. A disinfectant does not necessarily sterilize an object because viable spores and a few microorganisms may remain. **Sanitization** is closely related to disinfection. In sanitization, the microbial population is reduced to levels that are considered safe by public health standards. The inanimate object is usually cleaned as well as partially disinfected. For example, sanitizers are used to clean eating utensils in restaurants.

It is frequently necessary to control microorganisms on living tissue with chemical agents. Antisepsis (Greek *anti,* against, and *sepsis,* putrefaction) is the prevention of infection or sepsis and is accomplished with **antiseptics.** These are chemical agents applied to tissue to prevent infection by killing or inhibiting pathogen growth. Because they must not destroy too much host tissue, antiseptics are generally not as toxic as disinfectants.

A suffix can be employed to denote the type of antimicrobial agent. Substances that kill organisms often have the suffix **cide** (Latin *cida,* to kill): a **germicide** kills pathogens (and many nonpathogens) but not necessarily endospores. A disinfectant or antiseptic can be particularly effective against a specific group, in which case it may be called a **bactericide, fungicide, algicide,** or **viricide.** Other chemicals do not kill, but they do prevent growth. Their names end in **static** (Greek *statikos,* causing to stand or stopping)—for example, **bacteriostatic** and **fungistatic.**

Although these agents have been described in terms of their effects on pathogens, it should be noted that they also kill or inhibit the growth of nonpathogens as well. Their ability to reduce the total microbial population, not just to affect pathogen levels, is quite important in many situations.

1. Define the following terms: sterilization, sterilant, disinfection, disinfectant, sanitization, antisepsis, antiseptic, germicide, bactericide, bacteriostatic.

The Pattern of Microbial Death

A microbial population is not killed instantly when exposed to a lethal agent. Population death, like population growth, is generally exponential or logarithmic; that is, the population will be reduced by the same fraction at constant intervals (table 15.1). If the logarithm of the population number remaining is plotted against the time of exposure of the microorganism to the agent, a straight line plot will result (compare figure 15.1 with figure 6.2). When the population has been greatly reduced, the rate of killing may slow due to the survival of a more resistant strain of the microorganism.

Box 15.1

Safety in the Microbiology Laboratory

Personnel safety should be of major concern in all microbiology laboratories. It has been estimated that thousands of infections have been acquired in the laboratory, and 173 or more persons have died because of such infections. The two most common laboratory-acquired bacterial diseases are typhoid fever and brucellosis. Most deaths have come from typhoid fever (20 deaths) and Rocky Mountain spotted fever (13 deaths). Infections by fungi (histoplasmosis) and viruses (Venezuelan equine encephalitis and hepatitis B virus from monkeys) are also not uncommon. Hepatitis is the most frequently reported laboratory-acquired viral infection, especially in people working in clinical laboratories and with blood. In a survey of 426 hospital workers, 40% of those in clinical chemistry and 21% in microbiology had antibodies to hepatitis B virus, indicating their previous exposure (though only about 19% of these had disease symptoms).

Efforts have been made to determine the causes of these infections in order to enhance the development of better preventive measures. Although often it is not possible to determine the direct cause of infection, some major potential hazards are clear. One of the most frequent causes of disease is the inhalation of an infectious aerosol. An aerosol is a gaseous suspension of liquid or solid particles that may be generated by accidents or by many laboratory operations such as spills, centrifuge accidents, removal of closures from shaken culture tubes, and plunging of contaminated loops into a flame. Accidents with hypodermic syringes and needles, such as self-inoculation and spraying solutions from the needle, are also common. Hypodermics should be employed only when necessary, and then with care. Pipette accidents involving the mouth are another major source of infection; pipettes should be filled with the use of pipette aids and operated in such a way as to avoid creating aerosols.

People must exercise care and common sense when working with microorganisms. Operations that might generate infectious aerosols should be carried out in a biological safety cabinet. Bench tops and incubators should be disinfected regularly. Autoclaves must be maintained and operated properly to ensure adequate sterilization. Laboratory personnel should wash their hands thoroughly before and after finishing work.

In order to study the effectiveness of a lethal agent, one must be able to decide when microorganisms are dead, a task by no means as easy as with macroorganisms. It is hardly possible to take a bacterium's pulse. A bacterium is defined as dead if it does not grow when inoculated into culture medium that would normally support its growth (*however, see Box 40.2*). In like manner, an inactive virus cannot infect a suitable host.

1. Describe the pattern of microbial death and how one decides whether microorganisms are actually dead.

Conditions Influencing the Effectiveness of Antimicrobial Agent Activity

Destruction of microorganisms and inhibition of microbial growth are not simple matters because the efficiency of an **antimicrobial agent** (an agent that kills microorganisms or inhibits their growth) is affected by at least six factors.

1. Population size. Because an equal fraction of a microbial population is killed during each interval, a larger population requires a longer time to die than a smaller one. This can be seen in the theoretical heat-killing experiment shown in table 15.1 and figure 15.1. The same principle applies to chemical antimicrobial agents.
2. Population composition. The effectiveness of an agent varies greatly with the nature of the organisms being treated because microorganisms differ markedly in susceptibility. Bacterial endospores are much more resistant to most antimicrobial agents than are vegetative forms, and younger cells are usually more readily destroyed than mature organisms. Some species are able to withstand adverse conditions better than others. *Mycobacterium tuberculosis,* which causes tuberculosis, is much more resistant to antimicrobial agents than most other bacteria.
3. Concentration or intensity of an antimicrobial agent. Often, but not always, the more concentrated a chemical agent or intense a physical agent, the more rapidly microorganisms are destroyed. Dependence of effectiveness on concentration or intensity is generally not linear. Over a short range, a small increase in concentration leads to an exponential rise in effectiveness; beyond a certain point, further increases may not raise the killing rate much at all. Sometimes, an agent is more effective at lower concentrations. For example, 70% ethanol is more effective than 95% ethanol.
4. Duration of exposure. The longer a population is exposed to a microbicidal agent, the more organisms are killed (figure 15.1). To achieve sterilization, an exposure duration sufficient to reduce the probability of survival to 10^{-6} or less should be used.

TABLE 15.1 A Theoretical Microbial Heat-Killing Experiment

Minute	Microbial Number at Start of Minute[a]	Microorganisms Killed in 1 Minute (90 % of total)[a]	Microorganisms at End of 1 Minute	Log_{10} of Survivors
1	10^6	9×10^5	10^5	5
2	10^5	9×10^4	10^4	4
3	10^4	9×10^3	10^3	3
4	10^3	9×10^2	10^2	2
5	10^2	9×10^1	10	1
6	10^1	9	1	0
7	1	0.9	0.1	−1

[a]Assume that the initial sample contains 10^6 vegetative microorganisms per ml and that 90% of the organisms are killed during each minute of exposure. The temperature is 121°C.

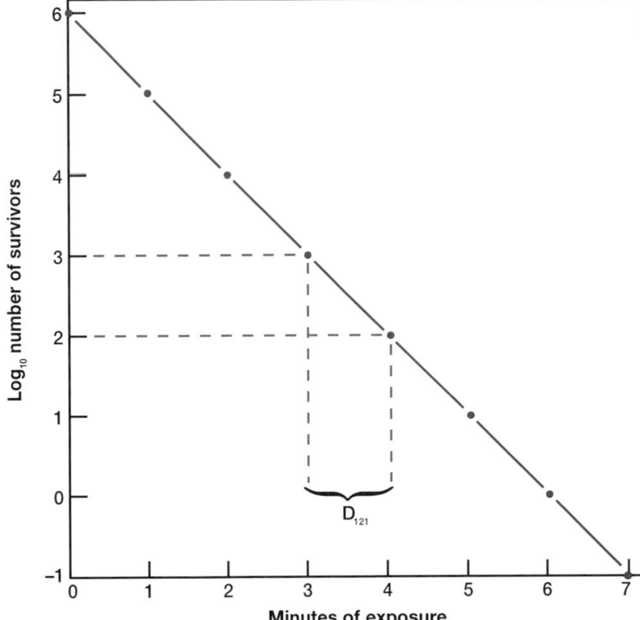

Figure 15.1 **The Pattern of Microbial Death.** An exponential plot of the survivors versus the minutes of exposure to heating at 121° C. In this example, the D_{121} value is 1 minute. The data are from table 15.1.

5. Temperature. An increase in the temperature at which a chemical acts often enhances its activity. Frequently, a lower concentration of disinfectant or sterilizing agent can be used at a higher temperature.
6. Local environment. The population to be controlled is not isolated, but surrounded by environmental factors that may either offer protection or aid in its destruction. For example, because heat kills more readily at an acid pH, acid foods and beverages such as fruits and tomatoes are much easier to pasteurize than more neutral foods like milk. A second important environmental factor is organic matter that can protect microorganisms against heating and chemical disinfectants. It may be necessary to clean an object before it is disinfected or sterilized. Syringes and medical or dental equipment should be cleaned before sterilization because the presence of too much organic matter could protect pathogens and increase the risk of infection. The same care must be taken when pathogens are destroyed during the preparation of drinking water. When a city's water supply has a high content of organic material, more chlorine must be added to disinfect it.

1. Briefly explain how the effectiveness of antimicrobial agents varies with population size, population composition, concentration or intensity of the agent, treatment duration, temperature, and local environmental conditions.

The Use of Physical Methods in Control

Heat and other physical agents are normally used to sterilize objects, as can be seen from the continual operation of the autoclave in every microbiology laboratory. The four most frequently employed physical agents are heat, filtration, ultraviolet radiation, and ionizing radiation.

Heat

Fire and boiling water have been used for sterilization and disinfection since the time of the Greeks, and heating is still one of the most popular ways to destroy microorganisms. Either moist or dry heat may be applied.

Moist heat readily kills viruses, bacteria, and fungi (table 15.2). Exposure to boiling water for 10 minutes is sufficient to destroy vegetative cells and eucaryotic spores. Unfortunately, the temperature of boiling water (100°C or 212°F) is not high enough to destroy bacterial endospores that may survive hours of boiling. Therefore, boiling can be used for disinfection of drinking water and objects not harmed by water, but boiling does not sterilize.

Because heat is so useful in controlling microorganisms, it is essential to have a precise measure of the heat-killing efficiency. Initially, effectiveness was expressed in terms of thermal death point (TDP), the lowest temperature at which a microbial suspension is killed in 10 minutes. Because TDP

Table 15.2 Approximate Conditions for Moist Heat Killing

Organism	Vegetative Cells	Spores
Yeasts	5 minutes at 50–60°C	5 minutes at 70–80°C
Molds	30 minutes at 62°C	30 minutes at 80°C
Bacteria[a]	10 minutes at 60–70°C	2 to over 800 minutes at 100°C 0.5–12 minutes at 121°C
Viruses	30 minutes at 60°C	

[a]Conditions for mesophilic bacteria.

implies that a certain temperature is immediately lethal despite the conditions, **thermal death time (TDT)** is now more commonly used. This is the shortest time needed to kill all organisms in a microbial suspension at a specific temperature and under defined conditions. However, such destruction is logarithmic, and it is theoretically not possible to "completely destroy" microorganisms in a sample, even with extended heating. Therefore, an even more precise figure, the **decimal reduction time (*D*)** or ***D* value** has gained wide acceptance. The decimal reduction time is the time required to kill 90% of the microorganisms or spores in a sample at a specified temperature. In a semilogarithmic plot of the population remaining versus the time of heating (figure 15.1), the *D* value is the time required for the line to drop by one log cycle or tenfold. The *D* value is usually written with a subscript, indicating the temperature for which it applies. *D* values are used to estimate the relative resistance of a microorganism to different temperatures through calculation of the ***z* value.** The *z* value is the increase in temperature required to reduce *D* to 1/10 its value or to reduce it by one log cycle when log *D* is plotted against temperature (figure 15.2). Another way to describe heating effectiveness is with the *F* value. The ***F* value** is the time in minutes at a specific temperature (usually 250°F or 121.1°C) needed to kill a population of cells or spores.

The food processing industry makes extensive use of *D* and *z* values. After a food has been canned, it must be heated to eliminate the risk of botulism arising from the presence of *Clostridium botulinum* spores. Heat treatment is carried out long enough to reduce a population of 10^{12} *C. botulinum* spores to 10^{0} (one spore); thus, there is a very small chance of any can having a viable spore. The *D* value for these spores at 121°C is 0.204 minutes. Therefore, it would take 12*D* or 2.5 minutes to reduce 10^{12} spores to one spore by heating at 121°C. The *z* value for *C. botulinum* spores is 10°C; that is, it takes a 10°C change in temperature to alter the *D* value tenfold. If the cans were to be processed at 111°C rather than at 121°C, the *D* value would increase by tenfold to 2.04 minutes, and the 12*D* value to 24.5 minutes. *D* values and *z* values for some common food-borne pathogens are given in table 15.3. Three

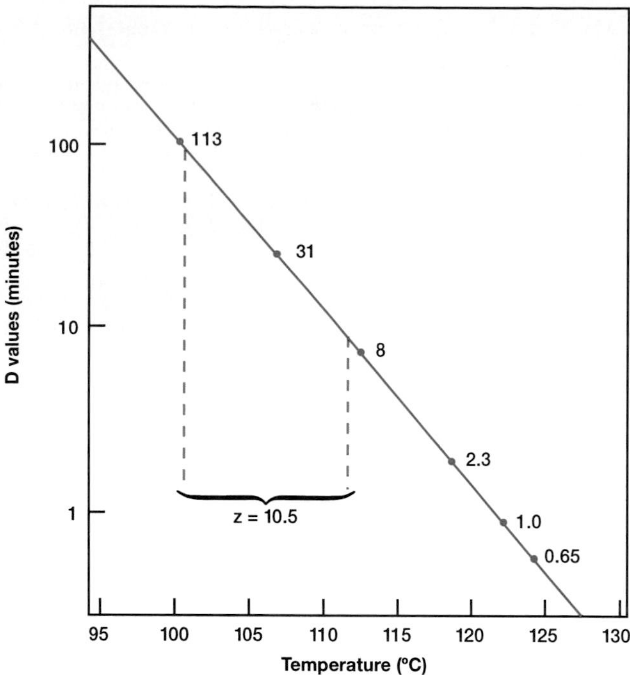

Figure 15.2 ***z* Value Calculation.** The *z* value used in calculation of time-temperature relationships for survival of a test microorganism, based on *D* value responses at various temperatures. The *z* value is the increase in temperature needed to reduce the decimal reduction time (*D*) to 10% of the original value. For this test microorganism, the *z* value is 10.5 degrees. The *D* values are plotted on a logarithmic scale.

D values are included for *Staphylococcus aureus* to illustrate the variation of killing rate with environment and the protective effect of organic material.

Food processing (pp. 868–71).
Botulism (p. 765).

Moist heat sterilization must be carried out at temperatures above 100°C in order to destroy bacterial endospores, and this requires the use of saturated steam under pressure. Steam sterilization is carried out with an **autoclave** (figure 15.3), a device somewhat like a fancy pressure cooker. Water is boiled to produce steam, which is released through the jacket and into the autoclave's chamber. The air initially present in the chamber is forced out until the chamber is filled with saturated steam and the outlets are closed. Hot, saturated steam continues to enter until the chamber reaches the desired temperature and pressure, usually 121°C and 15 pounds of pressure. At this temperature, saturated steam destroys all vegetative cells and endospores in a small volume of liquid within 10 to 12 minutes. Treatment is continued for about 15 minutes to provide a margin of safety.

Autoclaving must be carried out properly or the processed materials will not be sterile. If all air has not been flushed out of the chamber, it will not reach 121°C even though it may reach a pressure of 15 pounds. The chamber should not be

TABLE 15.3 *D* Values and *z* Values for Some Food-Borne Pathogens

Organism	Substrate	*D* Value (°C) in Minutes	*z* Value (°C)
Clostridium botulinum	Phosphate buffer	D_{121} = 0.204	10
Clostridium perfringens (heat-resistant strain)	Culture media	D_{90} = 3–5	6–8
Salmonella	Chicken à la king	D_{60} = 0.39–0.40	4.9–5.1
Staphylococcus aureus	Chicken à la king	D_{60} = 5.17–5.37	5.2–5.8
	Turkey stuffing	D_{60} = 15.4	6.8
	0.5% NaCl	D_{60} = 2.0–2.5	5.6

Values taken from F. L. Bryan, 1979, "Processes That Affect Survival and Growth of Microorganisms," *Time-Temperature Control of Foodborne Pathogens*. Atlanta: Centers for Disease Control.

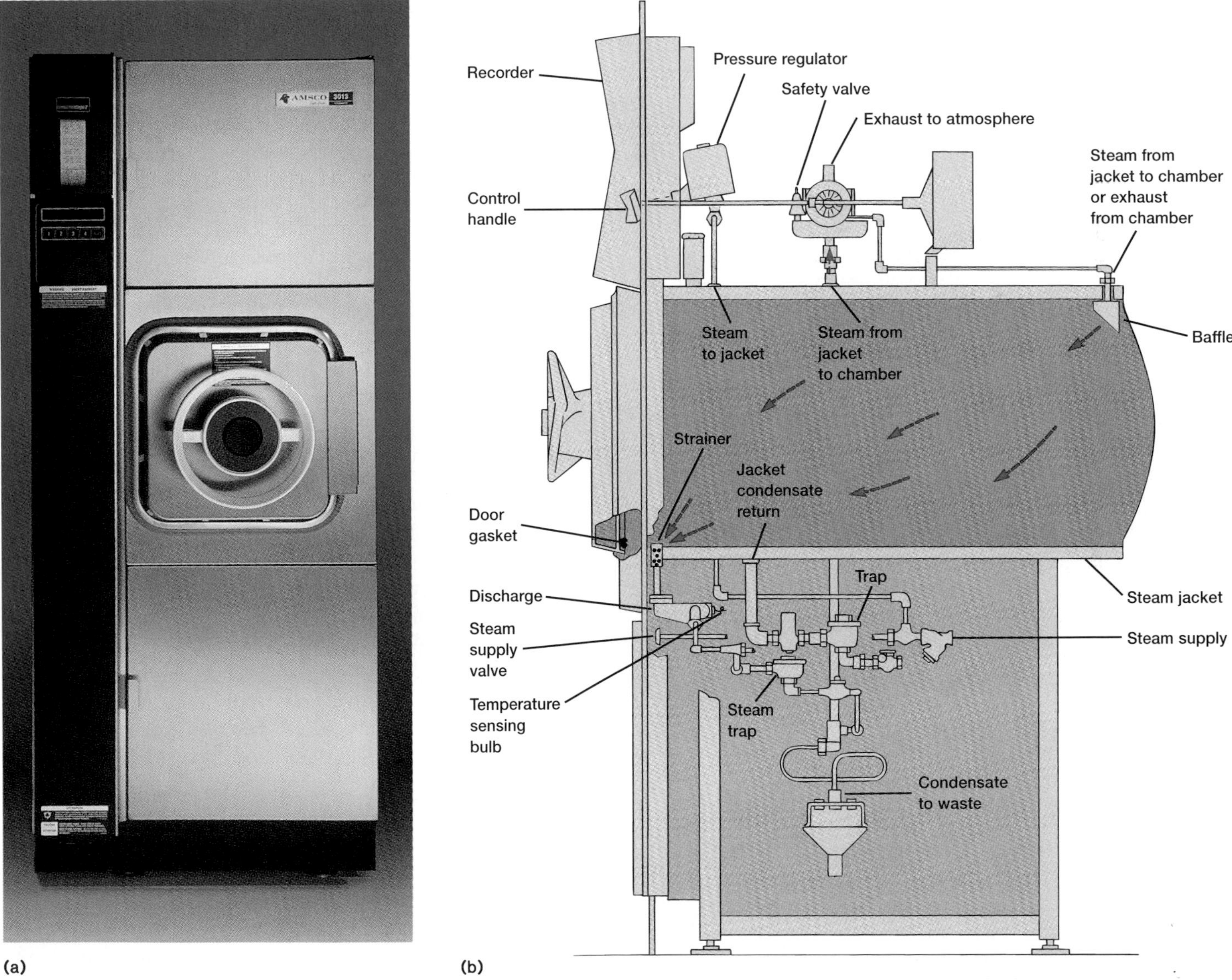

Figure 15.3 The Autoclave or Steam Sterilizer. (*a*) A modern, automatically controlled autoclave or sterilizer. (*b*) Longitudinal cross section of a typical autoclave showing some of its parts and the pathway of steam. (b) *From John J. Perkins,* Principles and Methods of Sterilization in Health Science, *2d ed., 1969. Courtesy of Charles C Thomas, Publisher, Springfield, Il.*

packed too tightly because the steam needs to circulate freely and contact everything in the autoclave. Bacterial endospores will be killed only if they are kept at 121°C for 10 to 12 minutes. When a large volume of liquid must be sterilized, an extended sterilization time will be needed because it will take longer for the center of the liquid to reach 121°C; 5 liters of liquid may require about 70 minutes. In view of these potential difficulties, a biological indicator is often autoclaved along with other material. This indicator commonly consists of a culture tube containing a sterile ampule of medium and a paper strip covered with spores of *Bacillus stearothermophilus* or *Clostridium* PA3679. After autoclaving, the ampule is aseptically broken and the culture incubated for several days. If the test bacterium does not grow in the medium, the sterilization run has been successful. Sometimes, either special tape that spells out the word *sterile* or a paper indicator strip that changes color upon sufficient heating is autoclaved with a load of material. If the word appears on the tape or if the color changes after autoclaving, the material is supposed to be sterile. These approaches are convenient and save time, but are not as reliable as the use of bacterial endospores.

Moist heat is thought to kill so effectively by degrading nucleic acids and by denaturing enzymes and other essential proteins. It also may disrupt cell membranes.

Many substances, such as milk, are treated with controlled heating at temperatures well below boiling, a process known as **pasteurization** in honor of its developer Louis Pasteur. In the 1860s, the French wine industry was plagued by the problem of wine spoilage, which made wine storage and shipping difficult. Pasteur examined spoiled wine under the microscope and detected microorganisms that looked like the bacteria responsible for lactic acid and acetic acid fermentations. He then discovered that a brief heating at 55 to 60°C would destroy these microorganisms and preserve wine for long periods. In 1886, the German chemists V. H. and F. Soxhlet adapted the technique for preserving milk and reducing milk-transmissible diseases. Milk pasteurization was introduced into the United States in 1889. Milk, beer, and many other beverages are now pasteurized. Pasteurization does not sterilize a beverage, but it does kill any pathogens present and drastically slows spoilage by reducing the level of nonpathogenic spoilage microorganisms.

Milk can be pasteurized in two ways. In the older method, the milk is held at 63°C for 30 minutes. Large quantities of milk are now usually subjected to **flash pasteurization,** which consists of quick heating to about 72°C for 15 seconds, then rapid cooling.

Pasteurization and the dairy industry (pp. 869–70).

Sometimes, a heat-sensitive material can be heat sterilized by **tyndallization** or **fractional steam sterilization.** The container with the material to be sterilized is heated at 90 to 100°C for 30 minutes on each of three consecutive days, and incubated at 37°C in between. The first heating will destroy all organisms but bacterial endospores. Most of these germinate during the subsequent 37°C incubation and are killed by the second heating period. Any remaining spores are destroyed by the second incubation and third heat treatment.

Many objects are best sterilized in the absence of water by **dry heat sterilization.** The items to be sterilized are placed in an oven at 160 to 170°C for 2 to 3 hours. Microbial death apparently results from the oxidation of cell constituents and denaturation of proteins. Although dry air heat is less effective than moist heat—*Clostridium botulinum* spores are killed in 5 minutes at 121°C by moist heat but only after 2 hours at 160°C with dry heat—it has some definite advantages. Dry heat does not corrode glassware and metal instruments as moist heat does, and it can be used to sterilize powders, oils, and similar items. Most laboratories sterilize glass petri dishes and pipettes with dry heat. Despite these advantages, dry heat sterilization is slow and not suitable for heat-sensitive materials like many plastic and rubber items.

Filtration

Filtration is an excellent way to reduce the microbial population in solutions of heat-sensitive material, and sometimes it can be used to sterilize solutions. Rather than directly destroying contaminating microorganisms, the filter simply removes them. There are two types of filters. **Depth filters** consist of fibrous or granular materials that have been bonded into a thick layer filled with twisting channels of small diameter. The solution containing microorganisms is sucked through this layer under vacuum, and microbial cells are removed by physical screening or entrapment and also by adsorption to the surface of the filter material. Depth filters are made of diatomaceous earth (Berkefield filters), unglazed porcelain (Chamberlain filters), asbestos, or other similar materials.

Recently, **membrane filters** have replaced depth filters for many purposes. These circular filters are porous membranes, a little over 0.1 mm thick, made of cellulose acetate, cellulose nitrate, polycarbonate, polyvinylidene fluoride, or other synthetic materials. Although a wide variety of pore sizes are available, membranes with pores about 0.2 μm in diameter are used to remove vegetative cells from solutions ranging in volume from 1 ml to many liters. The membranes are held in special holders (figure 15.4) and often preceded by depth filters made of glass fibers to remove larger particles that might clog the membrane filter. The solution is forced through the filter with a vacuum or with pressure from a syringe, peristaltic pump, or nitrogen gas bottle, and collected in previously sterilized containers. Membrane filters remove microorganisms by screening them out much as a sieve separates large sand particles from small ones (figure 15.5). These filters are used to sterilize pharmaceuticals, culture media, oils, antibiotics, and other heat-sensitive solutions.

The use of membrane filters in microbial counting (p. 117–18).

Air also can be sterilized by filtration. Two common examples are surgical masks and cotton plugs on culture vessels that let air in but keep microorganisms out. **Laminar flow biological safety cabinets** employing **high-efficiency particulate**

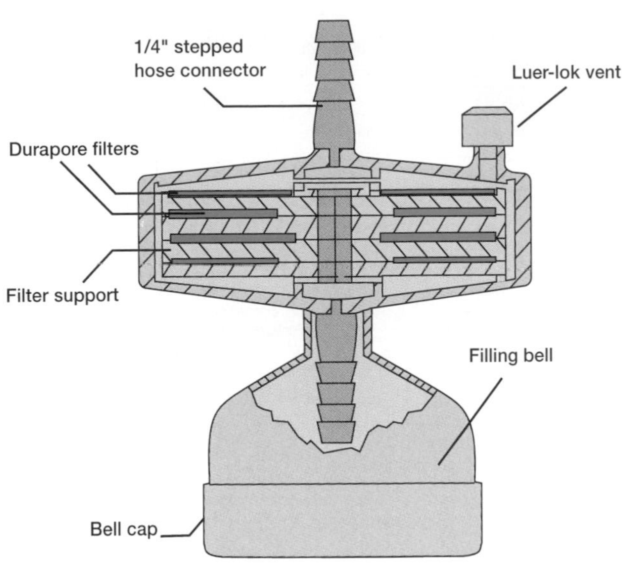

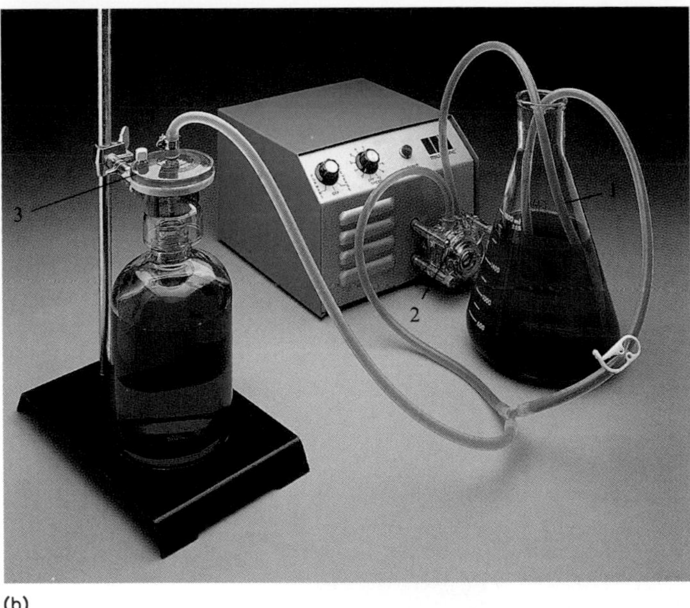

Figure 15.4 Membrane Filter Sterilization. A membrane filter outfit for sterilizing medium volumes of solution. (*a*) Cross section of the membrane filtering unit. Several membranes are used to increase capacity. (*b*) A complete filtering setup. The solution to be sterilized is kept in the Erlenmeyer flask, *1*, and forced through the filter by a peristaltic pump, *2*. The solution is sterilized by flowing through a membrane filter unit, *3*, and into a sterile container. A wide variety of other kinds of filtering outfits are also available.

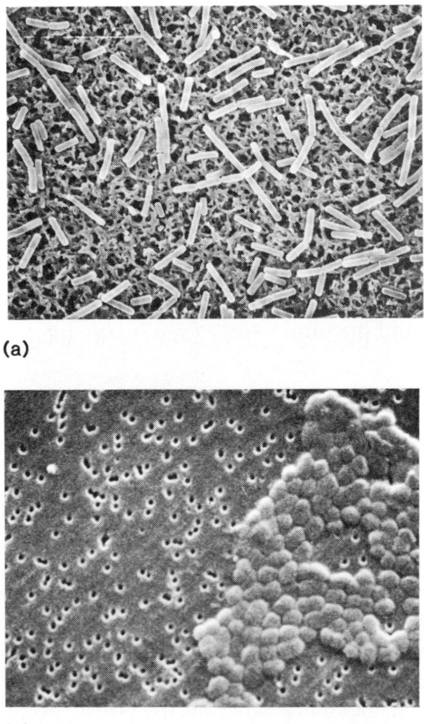

Figure 15.5 Membrane Filter Types. (*a*) *Bacillus megaterium* on a nylon membrane with a pore size of 0.2 μm (×2,000). (*b*) *Enterococcus faecalis* resting on a polycarbonate membrane filter with 0.4 μm holes (×5,900).

air (HEPA) filters, which remove 99.97% of 0.3 μm particles, are one of the most important air filtration systems. Laminar flow biological safety cabinets force air through HEPA filters, then project a vertical curtain of sterile air across the cabinet opening. This protects a worker from microorganisms being handled within the cabinet and prevents room contamination (figure 15.6). A person uses these cabinets when working with dangerous agents such as *Mycobacterium tuberculosis,* tumor viruses, and recombinant DNA. They are also employed in research labs and industries, such as the pharmaceutical industry, when a sterile working surface is needed for conducting assays, preparation of media, examination of tissue cultures, and the like.

Radiation

The types of radiation and the ways in which radiation damages or destroys microorganisms have already been discussed. The practical use of ultraviolet and ionizing radiation in sterilizing objects are briefly described next.

Radiation and its effects on microorganisms (pp. 129–30).

Ultraviolet (UV) radiation around 260 nm (*see figure 6.19*) is quite lethal but does not penetrate glass, dirt films, water, and other substances very effectively. Because of this disadvantage, UV is used as a sterilizing agent only in a few particular situations. UV lamps are sometimes placed on the ceilings of rooms or in biological safety cabinets to sterilize the air and any exposed surfaces. Because UV burns the skin and damages eyes, people working in such areas must be certain the UV lamps are off when the areas are in use. Commercial

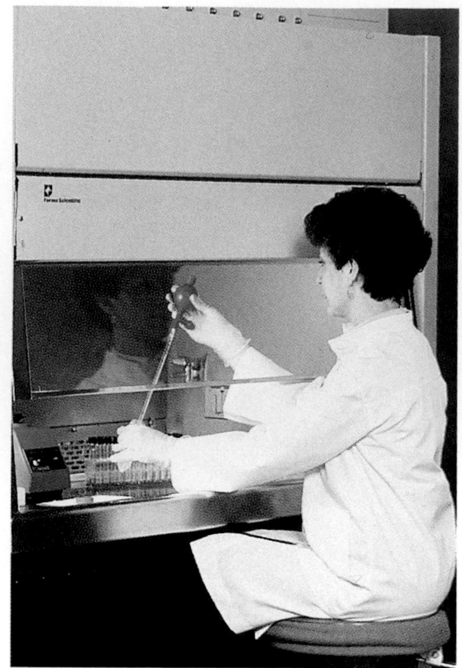

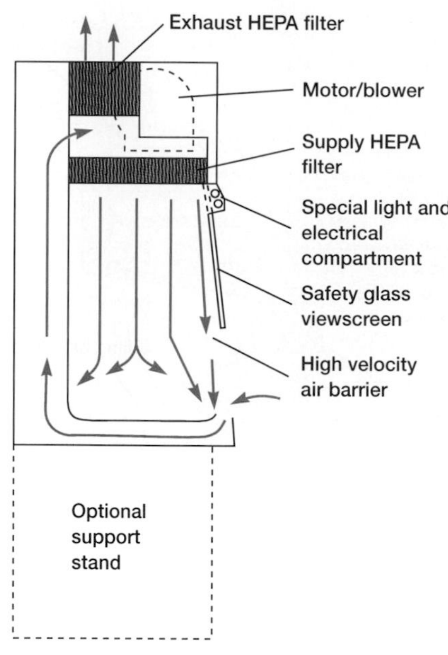

Figure 15.6 **A Laminar Flow Biological Safety Cabinet.** On the right is a schematic diagram showing the airflow pattern.

UV units are available for water treatment. Pathogens and other microorganisms are destroyed when a thin layer of water is passed under the lamps.

Ionizing radiation is an excellent sterilizing agent and penetrates deep into objects. Gamma radiation from a cobalt 60 source is used in the cold sterilization of antibiotics, hormones, sutures, and plastic disposable supplies such as syringes. Gamma radiation has also been used to "pasteurize" meat and other food. However, this process has not yet been widely employed in the United States because of the cost and concerns about the effects of gamma radiation on food. In some countries, ionizing radiation is used to treat poultry and extend the shelf life of seafoods and fruits. It will probably be more extensively employed in the future.

1. Define thermal death point (TDP), thermal death time (TDT), decimal reduction time (D) or D value, z value, and the F value.
2. Describe how an autoclave works. What conditions are required for sterilization by moist heat, and what three things must one do when operating an autoclave to help ensure success?
3. How are pasteurization, flash pasteurization, fractional steam sterilization (tyndallization), and dry heat sterilization carried out? Give some practical applications for each of these procedures.
4. What are depth filters and membrane filters, and how are they used to sterilize liquids? Describe the operation of a biological safety cabinet.
5. Give the advantages and disadvantages of ultraviolet light and ionizing radiation as sterilizing agents. Provide a few examples of how each is used for this purpose.

The Use of Chemical Agents in Control

Although objects are sometimes disinfected with physical agents, chemicals are more often employed in disinfection and antisepsis. Many factors influence the effectiveness of chemical disinfectants and antiseptics, as previously discussed. Factors such as the kinds of microorganisms potentially present, the concentration and nature of the disinfectant to be used, and the length of treatment should be considered. Dirty surfaces must be cleaned before a disinfectant or antiseptic is applied. The proper use of chemical agents is essential to laboratory and hospital safety (*see inside back cover and Box 34.1*).

The properties and uses of several groups of common disinfectants and antiseptics are surveyed next. Many of their characteristics are summarized in tables 15.4 and 15.5. Structures of some common agents are given in figure 15.7.

Besides their use as disinfectants and antiseptics, chemical agents are employed to prevent microbial growth in food. A controversial example is the nitrite used to help preserve ham, sausage, bacon, and other cured meats. Nitrite decomposes to nitric acid, which reacts with heme pigments to keep the meat red in color. More importantly, nitrite increases the safety of some meat products by inhibiting the growth of *Clostridium botulinum* and the germination of its spores, thus offering protection against botulism and reducing the rate of spoilage. Current concern about nitrite arises from the observation that it can react with amines to form carcinogenic nitrosamines. Nitrite is added in very small amounts, and eventually it may be possible to eliminate its use entirely.

TABLE 15.4 Activity Levels of Selected Germicides

Class	Use Concentration of Active Ingredient	Activity Level[a]
Gas		
Ethylene oxide	450–500 mg/l[b]	High
Liquid		
Glutaraldehyde, aqueous	2%	High to intermediate
Formaldehyde + alcohol	8 + 70%	High
Stabilized hydrogen peroxide	6–30%	High to intermediate
Formaldehyde, aqueous	6–8%	High to intermediate
Iodophors	750–5,000 mg/l[c]	High to intermediate
Iodophors	75–150 mg/l[c]	Intermediate to low
Iodine + alcohol	0.5 + 70%	Intermediate
Chlorine compounds	0.1–0.5%[d]	Intermediate
Phenolic compounds, aqueous	0.5–3%	Intermediate to low
Iodine, aqueous	1%	Intermediate
Alcohols (ethyl, isopropyl)	70%	Intermediate
Quaternary ammonium compounds	0.1–0.2% aqueous	Low
Chlorhexidine	0.75–4%	Low
Hexachlorophene	1–3%	Low
Mercurial compounds	0.1–0.2%	Low

Block, Seymour S.: *Disinfection, Sterilization and Preservation.* Lea & Febiger, Philadelphia, PA. 1983. Reprinted by permission.

[a]High-level disinfectants destroy vegetative bacterial cells including *M. tuberculosis,* bacterial endospores, fungi, and viruses. Intermediate-level disinfectants destroy all of the above except endospores. Low-level agents kill bacterial vegetative cells except for *M. tuberculosis,* fungi, and medium-sized lipid-containing viruses (but not bacterial endospores or small, nonlipid viruses).

[b]In autoclave-type equipment at 55 to 60°C.

[c]Available iodine.

[d]Free chlorine.

Phenolics

Phenol was the first widely used antiseptic and disinfectant. In 1867, Joseph Lister employed it to reduce the risk of infection during operations. Today, phenol and phenolics (phenol derivatives) such as cresols, xylenols, and orthophenylphenol are used as disinfectants in laboratories and hospitals. The commercial disinfectant Lysol is made of a mixture of phenolics. Phenolics act by denaturing proteins and disrupting cell membranes. They have some real advantages as disinfectants: phenolics are tuberculocidal, effective in the presence of organic material, and remain active on surfaces long after application. However, they do have a disagreeable odor and can cause skin irritation.

Hexachlorophene (figure 15.7) has been one of the most popular antiseptics because it persists on the skin once applied and reduces skin bacteria for long periods. However, it can cause brain damage and is now used in hospital nurseries only in response to a staphylococcus epidemic.

Alcohols

Alcohols are among the most widely used disinfectants and antiseptics. They are bactericidal and fungicidal, but not sporicidal; some lipid-containing viruses are also destroyed. The two most popular alcohol germicides are ethanol and isopropanol, usually used in about 70 to 80% concentration. They act by denaturing proteins and possibly by dissolving membrane lipids. A 10 to 15 minute soaking is sufficient to disinfect thermometers and small instruments.

Halogens

Iodine is used as a skin antiseptic and kills by oxidizing cell constituents and iodinating cell proteins. At higher concentrations, it may even kill some spores. Iodine has often been applied as tincture of iodine, 2% or more iodine in a water-ethanol solution of potassium iodide. Although it is an effective antiseptic, the skin may be damaged, a stain is left, and iodine allergies can result. More recently, iodine has been complexed with an organic carrier to form an **iodophor.** Iodophors are water soluble, stable, and nonstaining, and release iodine slowly to minimize skin burns and irritation. They are used in hospitals for preoperative skin degerming and in hospitals and laboratories for disinfecting. Some popular brands are Wescodyne for skin and laboratory disinfection and Betadine for wounds.

Chlorine is the disinfectant of choice for municipal water supplies and swimming pools, and is also employed in the dairy and food industries. It may be applied as chlorine gas, sodium hypochlorite, or calcium hypochlorite, all of which yield hypochlorous acid (HClO) and then atomic oxygen. The result is oxidation of cellular materials and destruction of vegetative bacteria and fungi, although not spores.

$$Cl_2 + H_2O \longrightarrow HCl + HClO$$

$$Ca(OCl)_2 + 2H_2O \longrightarrow Ca(OH)_2 + 2HClO$$

$$HClO \longrightarrow HCl + O$$

Death of almost all microorganisms usually occurs within 30 minutes. Since organic material interferes with chlorine action

TABLE 15.5 Relative Efficacy of Commonly Used Disinfectants and Antiseptics

Class	Disinfectant	Antiseptic	Comment
Gas			
Ethylene oxide	3–4[a]	0	Sporicidal; toxic; good penetration; requires relative humidity of 30% or more; microbicidal activity varies with apparatus used; absorbed by porous material; dry spores highly resistant; moisture must be present, and presoaking is most desirable
Liquid			
Glutaraldehyde, aqueous	3	0	Sporicidal; active solution unstable; toxic
Stabilized hydrogen peroxide	3	0	Sporicidal; use solution stable up to 6 weeks; toxic orally and to eyes; mildly skin toxic; little inactivated by organic matter
Formaldehyde + alcohol	3	0	Sporicidal; noxious fumes; toxic; volatile
Formaldehyde, aqueous	1–2	0	Sporicidal; noxious fumes; toxic
Phenolic compounds	3	0	Stable; corrosive; little inactivation by organic matter; irritates skin
Chlorine compounds	1–2	0	Fast action; inactivation by organic matter; corrosive; irritates skin
Alcohol	1	3	Rapidly microbicidal except for bacterial spores and some viruses; volatile; flammable; dries and irritates skin
Iodine + alcohol	0	4	Corrosive; very rapidly microbicidal; causes staining; irritates skin; flammable
Iodophors	1–2	3	Somewhat unstable; relatively bland; staining temporary; corrosive
Iodine, aqueous	0	2	Rapidly microbicidal; corrosive; stains fabrics; stains and irritates skin
Quaternary ammonium compounds	1	0	Bland; inactivated by soap and anionics; compounds absorbed by fabrics; old or dilute solution can support growth of gram-negative bacteria
Hexachlorophene	0	2	Bland; insoluble in water, soluble in alcohol; not inactivated by soap; weakly bactericidal
Chlorhexidine	0	3	Bland; soluble in water and alcohol; weakly bactericidal
Mercurial compounds	0	±	Bland; much inactivated by organic matter; weakly bactericidal

Block, Seymour S.: *Disinfection, Sterilization and Preservation.* Lea & Febiger, Philadelphia, PA. 1983. Reprinted by permission.

[a]Subjective ratings of practical usefulness in a hospital environment—4 is maximal usefulness; 0 is little or no usefulness; ± signifies that the substance is sometimes useful, but not always.

by reacting with chlorine and its products, an excess of chlorine is added to ensure microbial destruction.

Municipal water purification (pp. 835–37).

Chlorine is also an excellent disinfectant for individual use because it is effective, inexpensive, and easy to employ. Small quantities of drinking water can be disinfected with halazone tablets. Halazone (parasulfone dichloramidobenzoic acid) slowly releases chloride when added to water and disinfects it in about a half hour. It is frequently used by campers lacking access to uncontaminated drinking water.

Chlorine solutions make very effective laboratory and household disinfectants. An excellent disinfectant-detergent combination can be prepared if a 1/100 dilution of household bleach (e.g., 1.3 fl oz of Chlorox or Purex bleach in 1 gal or 10 ml/liter) is combined with sufficient nonionic detergent (about 1 oz/gal or 7.8 ml/liter) to give a 0.8% detergent concentration. This mixture will remove both dirt and bacteria.

Heavy Metals

For many years, the ions of heavy metals such as mercury, silver, arsenic, zinc, and copper were used as germicides. More recently, these have been superseded by other less toxic and more effective germicides (many heavy metals are more bacteriostatic than bactericidal). There are a few exceptions. A 1% solution of silver nitrate is often added to the eyes of infants to prevent ophthalmic gonorrhea (in many hospitals, erythromycin is used instead of silver nitrate because it is effective against *Chlamydia* as well as *Neisseria*). Silver sulfadiazine is used on burns. Copper sulfate is an effective algicide in lakes and swimming pools.

Heavy metals combine with proteins, often with their sulfhydryl groups, and inactivate them. They may also precipitate cell proteins.

Quaternary Ammonium Compounds

Detergents (Latin *detergere,* to wipe off or away) are organic molecules, other than soaps, that are derived from fats, that

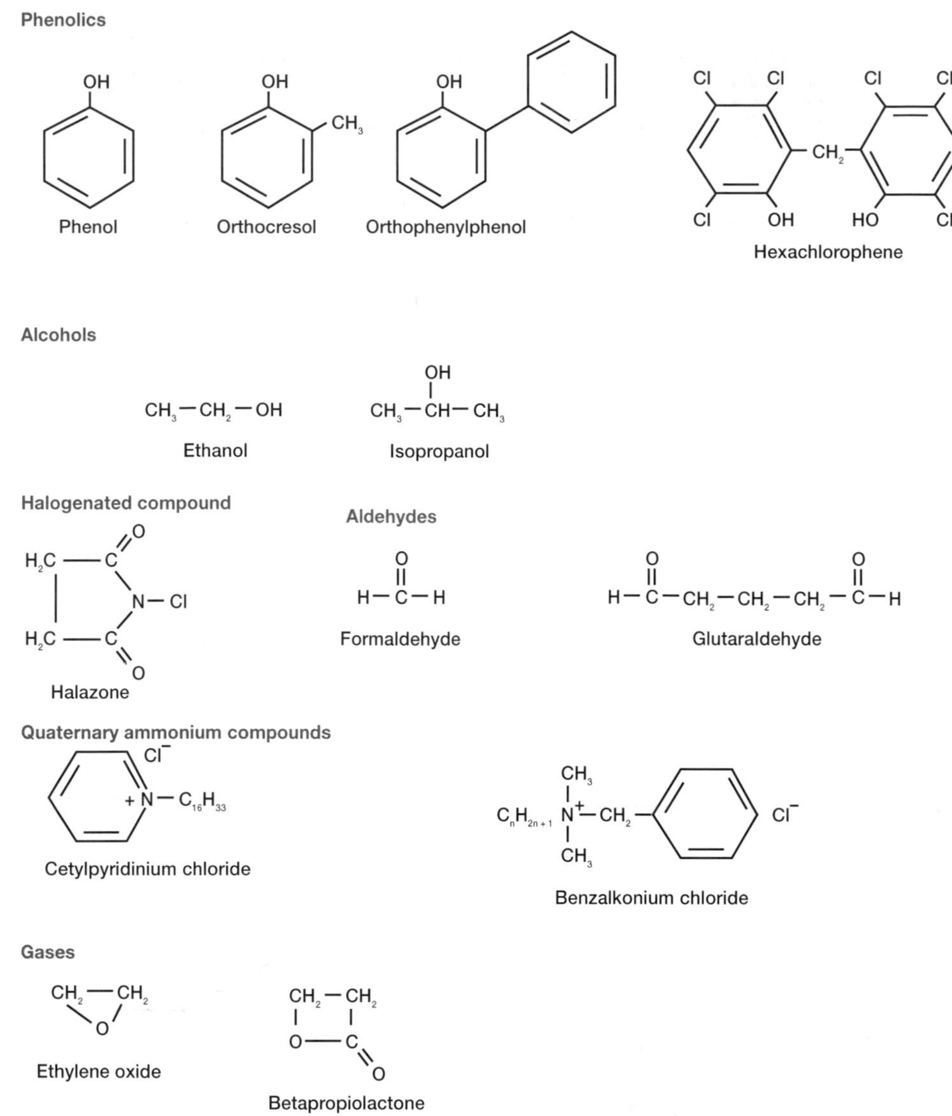

Figure 15.7 **Disinfectants and Antiseptics.** The structures of some frequently used disinfectants and antiseptics.

serve as wetting agents and emulsifiers because they have both polar hydrophilic and nonpolar hydrophobic ends. Due to their amphipathic nature (*see chapter 3*), detergents solubilize otherwise insoluble residues and are very effective cleansing agents.

Although anionic detergents have some antimicrobial properties, only cationic detergents are effective disinfectants. The most popular of these disinfectants are quaternary ammonium compounds characterized by a positively charged quaternary nitrogen and a long hydrophobic aliphatic chain (figure 15.7). They disrupt microbial membranes and may also denature proteins.

Cationic detergents like benzalkonium chloride and cetylpyridinium chloride kill most bacteria, but not *M. tuberculosis* or endospores. They do have the advantages of being stable, nontoxic, and bland; but they are inactivated by hard water and soap. Cationic detergents are often used as disinfectants for food utensils and small instruments, and as skin antiseptics. There are several brands on the market. Zephiran contains benzalkonium chloride and Ceepryn, cetylpyridinium chloride.

Aldehydes

Both of the commonly used aldehydes, formaldehyde and glutaraldehyde, are very reactive molecules that combine with proteins and inactivate them. They are sporicidal and can be used as chemical sterilants. Formaldehyde is usually dissolved in water or alcohol before use. A 2% buffered solution of glutaraldehyde is an effective disinfectant. It is less irritating than formaldehyde and is used to disinfect hospital and laboratory equipment. Glutaraldehyde usually disinfects objects within about 10 minutes, but may require as long as 12 hours to destroy all spores.

TABLE 15.6 Phenol Coefficients for Some Disinfectants

Disinfectant	Phenol Coefficients[a]	
	Salmonella typhi	*Staphylococcus aureus*
Phenol	1	1
Cetylpyridinium chloride	228	337
O-phenylphenol	5.6 (20°C)	4.0
p-cresol	2.0–2.3	2.3
Hexachlorophene	5–15	15–40
Merthiolate	600	62.5
Mercurochrome	2.7	5.3
Lysol	1.9	3.5
Isopropyl alcohol	0.6	0.5
Ethanol	0.04	0.04
2% I_2 solution in EtOH	4.1–5.2 (20°C)	4.1–5.2 (20°C)

[a]All values were determined at 37°C except where indicated.

Sterilizing Gases

Many heat-sensitive items such as disposable plastic petri dishes and syringes, heart-lung machine components, sutures, and catheters are now sterilized with ethylene oxide gas (figure 15.7). Ethylene oxide (EtO) is both microbicidal and sporicidal, and kills by combining with cell proteins. It is a particularly effective sterilizing agent because it rapidly penetrates packing materials, even plastic wraps.

Sterilization is carried out in a special ethylene oxide sterilizer, very much resembling an autoclave in appearance, that controls the EtO concentration, temperature, and humidity. Because pure EtO is explosive, it is usually supplied in a 10 to 20% concentration mixed with either CO_2 or dichlorodifluoromethane. The ethylene oxide concentration, humidity, and temperature influence the rate of sterilization. A clean object can be sterilized if treated for 5 to 8 hours at 38°C or 3 to 4 hours at 54°C when the relative humidity is maintained at 40 to 50% and the EtO concentration at 700 mg/liter. Extensive aeration of the sterilized materials is necessary to remove residual EtO because it is so toxic.

Betapropiolactone (BPL) is occasionally employed as a sterilizing gas. In the liquid form, it has been used to sterilize vaccines and sera. BPL decomposes to an inactive form after several hours and is therefore not as difficult to eliminate as EtO. It also destroys microorganisms more readily than ethylene oxide, but does not penetrate materials well and may be carcinogenic. For these reasons, BPL has not been used as extensively as EtO.

1. Why are most antimicrobial chemical agents disinfectants rather than sterilants?
2. Describe each of the following agents in terms of its chemical nature, mechanism of action, mode of application, common uses and effectiveness, and advantages and disadvantages: phenolics, alcohols, halogens (iodine and chlorine), heavy metals, quaternary ammonium compounds, aldehydes, and ethylene oxide.

Evaluation of Antimicrobial Agent Effectiveness

Testing of antimicrobial agents is a complex process under regulation by two different federal agencies. The U.S. Environmental Protection Agency regulates disinfectants, whereas agents used on humans and animals are under the control of the Food and Drug Administration. Testing of antimicrobial agents often begins with an initial screening test to see if they are effective and at what concentrations. This may be followed by more realistic in-use testing.

The best-known disinfectant screening test is the **phenol coefficient test** in which the potency of a disinfectant is compared with that of phenol. A series of dilutions of phenol and the experimental disinfectant are inoculated with the test bacteria *Salmonella typhi* and *Staphylococcus aureus,* then placed in a 20 or 37°C water bath. These inoculated disinfectant tubes are next subcultured to regular fresh medium at 5 minute intervals, and the subcultures are incubated for two or more days. The highest dilutions that kill the bacteria after a 10 minute exposure, but not after 5 minutes, are used to calculate the phenol coefficient. The reciprocal of the appropriate test disinfectant dilution is divided by that for phenol to obtain the coefficient. Suppose that the phenol dilution was 1/90 and maximum effective dilution for disinfectant X was 1/450. The phenol coefficient of X would be 5. The higher the phenol coefficient value, the more effective the disinfectant under these test conditions. A value greater than 1 means that the disinfectant is more effective than phenol. A few representative phenol coefficient values are given in table 15.6.

The phenol coefficient test is a useful initial screening procedure, but the phenol coefficient can be misleading if taken as a direct indication of disinfectant potency during normal use. This is because the phenol coefficient is determined under carefully controlled conditions with pure bacterial strains, whereas disinfectants are normally used on complex populations in the presence of organic matter and with significant variations in environmental factors like pH, temperature, and presence of salts.

To more realistically estimate disinfectant effectiveness, other tests are often used. The rates at which selected bacteria are destroyed with various chemical agents may be experimentally determined and compared. A **use dilution test** can also be carried out. Stainless steel cylinders are contaminated with specific bacterial species under carefully controlled conditions. The cylinders are dried briefly, immersed in the test disinfectants for 10 minutes, transferred to culture media, and incubated for two days. The disinfectant concentration that kills all organisms with a 95% level of confidence under these conditions is determined. Disinfectants also can be tested under conditions designed to simulate normal in-use situations. In-use testing techniques allow a more accurate determination of the proper disinfectant concentration for a particular situation.

1. Briefly describe the phenol coefficient test.
2. Why might it be necessary to employ procedures like the use dilution and in-use tests?

Summary

1. Sterilization is the process by which all living cells, viable spores, viruses, and viroids are either destroyed or removed from an object or habitat. Disinfection is the killing, inhibition, or removal of microorganisms (but not necessarily endospores) that can cause disease.
2. Many terms are used to define how microorganisms are controlled: sterilant, disinfectant, sanitization, antisepsis, and antiseptic.
3. Antimicrobial agents that kill organisms often have the suffix -cide, whereas agents that prevent growth and reproduction have the suffix -static.
4. Microbial death is usually exponential or logarithmic.
5. The effectiveness of a disinfectant or sterilizing agent is influenced by population size, population composition, concentration or intensity of the agent, exposure duration, temperature, and nature of the local environment.
6. The efficiency of heat killing is often indicated by the thermal death time or the decimal reduction time.
7. Although treatment with boiling water for 10 minutes kills vegetative forms, an autoclave must be used to destroy endospores by heating at 121°C and 15 pounds of pressure.
8. Moist heat kills by degrading nucleic acids, denaturing enzymes and other proteins, and disrupting cell membranes.
9. Heat-sensitive liquids can be preserved by pasteurization, heating at 63°C for 30 minutes or at 72°C for 15 seconds (flash pasteurization). Tyndallization or fractional steam sterilization is sometimes employed instead of pasteurization.
10. Glassware and other heat-stable items may be sterilized by dry heat at 160°C to 170°C for 2 to 3 hours.
11. Microorganisms can be efficiently removed by filtration with either depth filters or membrane filters.
12. Biological safety cabinets with high-efficiency particulate filters sterilize air by filtration.
13. Radiation of short wavelength or high energy, ultraviolet and ionizing radiation, can be used to sterilize objects.
14. Chemical agents usually act as disinfectants because they cannot readily destroy bacterial endospores. Disinfectant effectiveness depends on concentration, treatment duration, temperature, and presence of organic material.
15. Phenolics and alcohols are popular disinfectants that act by denaturing proteins and disrupting cell membranes.
16. Halogens (iodine and chlorine) kill by oxidizing cellular constituents; cell proteins may also be iodinated. Iodine is applied as a tincture or iodophor. Chlorine may be added to water as a gas, hypochlorite, or organic chlorine derivative.
17. Heavy metals tend to be bacteriostatic agents. They are employed in specialized situations such as the use of silver nitrate in the eyes of newborn infants and copper sulfate in lakes and pools.
18. Cationic detergents are often used as disinfectants and antiseptics; they disrupt membranes and denature proteins.
19. Aldehydes such as formaldehyde and glutaraldehyde can sterilize as well as disinfect because they kill spores.
20. Ethylene oxide gas penetrates plastic wrapping material and destroys all life forms by reacting with proteins. It is used to sterilize packaged, heat-sensitive materials.
21. There are a variety of ways to determine the effectiveness of disinfectants, among them the following: phenol coefficient test, measurement of killing rates with germicides, use dilution testing, and in-use testing.

Key Terms

algicide *311*
antimicrobial agent *312*
antiseptics *311*
autoclave *314*
bactericide *311*
bacteriostatic *311*
-cide *311*
D value *314*
decimal reduction time (*D*) *314*
depth filters *316*
detergent *320*
disinfectant *311*
disinfection *311*
dry heat sterilization *316*
F value *314*
flash pasteurization *316*
fractional steam sterilization *316*
fungicide *311*
fungistatic *311*
germicide *311*
high-efficiency particulate air (HEPA) filters *316*
iodophor *319*
ionizing radiation *318*
laminar flow biological safety cabinets *316*
membrane filters *316*
pasteurization *316*
phenol coefficient test *322*
sanitization *311*
-static *311*
sterilization *311*
thermal death time (TDT) *314*
tyndallization *316*
ultraviolet (UV) radiation *317*
use dilution test *323*
viricide *311*
z value *314*

Questions for Thought and Review

1. How can the D value be used to estimate the time required for sterilization? Suppose that you wanted to eliminate the risk of *Salmonella* poisoning by heating your food (D_{60}= 0.40 minutes, z value = 5.0). Calculate the $12D$ value at 60°C. How long would it take to achieve the same results by heating at 50, 55, and 65°C?
2. How can one alter treatment conditions to increase the effectiveness of a disinfectant?
3. How would the following be best sterilized: glass pipettes and petri plates, tryptic soy broth tubes, Mueller-Hinton agar medium, antibiotic solution, interior of a biological safety cabinet, wrapped package of plastic petri plates?
4. Which disinfectants or antiseptics would be used to treat the following: oral thermometer, laboratory bench top, drinking water, patch of skin before surgery, small medical instruments (probes, forceps, etc.)? List all chemicals suitable for each task.

Additional Reading

Barkley, W. E. 1981. Containment and disinfection. In *Manual of methods for general bacteriology,* ed. P. Gerhardt, 487–503. Washington, D.C.: American Society for Microbiology.

Block, S. S., ed. 1983. *Disinfection, sterilization and preservation,* 3d ed. Philadelphia: Lea and Febiger.

Borick, P. M. 1973. *Chemical sterilization.* Stroudsburg, Pa.: Dowden, Hutchinson and Ross.

Centers for Disease Control. 1987. Recommendations for prevention of HIV transmission in health-care settings. *Morbid. Mortal. Weekly Rep.* 36(Suppl. 2):3S–18S.

Centers for Disease Control and National Institutes of Health. 1988. *Biosafety in microbiological and biomedical laboratories,* 2d ed. Washington, D.C.: U.S. Government Printing Office.

Centers for Disease Control. 1988. Update: Universal precautions for prevention of transmission of human immunodeficiency virus, hepatitis B virus, and other bloodborne pathogens in health-care settings. *Morbid. Mortal. Weekly Rep.* 37(24):377–88.

Centers for Disease Control. 1989. Guidelines for prevention of transmission of human immunodeficiency virus and hepatitis B virus to health-care and public-safety workers. *Morbid. Mortal. Weekly Rep.* 38(Suppl. 6):1–37.

Collins, C. H., and Lyne, P. M. 1976. *Microbiological methods,* 4th ed. Boston: Butterworths.

Favero, M. S., and Bond, W. W. 1991. Sterilization, disinfection, and antisepsis in the hospital. In *Manual of clinical microbiology,* 5th ed., ed. A. Balows, 183–200. Washington, D.C.: American Society for Microbiology.

Groschel, D. H. M., and Strain, B. A. 1991. Laboratory safety in clinical microbiology. In *Manual of clinical microbiology,* 5th ed., ed. A. Balows, 49–58. Washington, D.C.: American Society for Microbiology.

Henderson, D. K. 1990. HIV-1 in the health-care setting. In *Principles and practice of infectious diseases,* 3d ed., eds. G. L. Mandell, R. G. Douglas, Jr., and J. E. Bennett, 2221–36. New York: Churchill Livingstone.

Korzynski, M. S. 1981. Sterilization. In *Manual of methods for general bacteriology,* ed. P. Gerhardt, 476–86. Washington, D.C.: American Society for Microbiology.

Martin, M. A., and Wenzel, R. P. 1990. Sterilization, disinfection, and disposal of infectious waste. In *Principles and practice of infectious diseases,* 3d ed., eds. G. L. Mandell, R. G. Douglas, Jr., and J. E. Bennett, 2182–89. New York: Churchill Livingstone.

Perkins, J. J. 1969. *Principles and methods of sterilization in health sciences,* 2d ed. Springfield, Ill.: Charles C. Thomas.

Pike, R. M. 1979. Laboratory-associated infections: Incidence, fatalities, causes, and prevention. *Ann. Rev. Microbiol.* 33:41–66.

Russell, A. D. 1990. Bacterial spores and chemical sporicidal agents. *Clin. Microbiol. Rev.* 3(2):99–119.

Russell, A. D.; Hugo, W. B.; and Ayliffe, G. A. J., eds. 1982. *Principles and practice of disinfection.* Oxford: Blackwell Scientific Publications.

Warren, E. 1981. Laboratory safety. In *Laboratory procedures in clinical microbiology,* ed. J. A. Washington, 729–45. New York: Springer-Verlag.

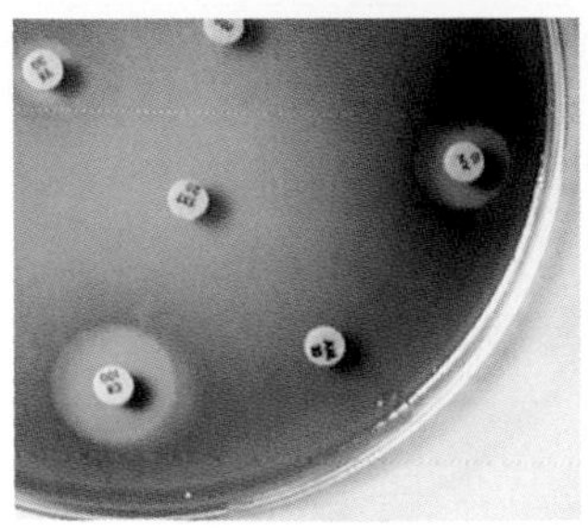

CHAPTER 16
Antimicrobial Chemotherapy

The Lord created medicines out of the earth, and the sensible will not despise them.

—*Ecclesiasticus 38:4*

It was the knowledge of the great abundance and wide distribution of actinomycetes, which dated back nearly three decades, and the recognition of the marked activity of this group of organisms against other organisms that led me in 1939 to undertake a systematic study of their ability to produce antibiotics.

—Selman A. Waksman

Outline

Concepts

1. Many diseases are treated with chemotherapeutic agents, such as antibiotics, that inhibit or kill the pathogen while harming the host as little as possible.
2. Ideally, antimicrobial agents disrupt microbial processes or structures that differ from those of the host. They may damage pathogens by hampering bacterial cell wall synthesis, inhibiting microbial protein and nucleic acid synthesis, disrupting microbial membrane structure and function, or blocking metabolic pathways through inhibition of key enzymes.
3. The effectiveness of chemotherapeutic agents depends on many factors: the route of administration and location of the infection, the presence of interfering substances, the concentration of the drug in the body, the nature of the pathogen, the presence of drug allergies, and the resistance of microorganisms to the drug.
4. The increasing number and variety of drug-resistant pathogens is a serious public health problem.
5. Although antibacterial chemotherapy is more advanced, drugs for the treatment of fungal and viral infections are also becoming increasingly available.

Chapter 15 is concerned principally with the chemical and physical agents used to treat inanimate objects in order to destroy microorganisms or inhibit their growth. In addition, the use of antiseptics is briefly reviewed. Microorganisms also grow on and within other organisms, and microbial colonization can lead to disease, disability, and death. Thus, the control or destruction of microorganisms residing within the bodies of humans and other animals is of great importance.

When disinfecting or sterilizing an inanimate object, one naturally must use procedures that do not damage the object itself. The same is true for the treatment of living hosts. The most successful drugs interfere with processes that differ between the pathogen and host, and seriously damage the target microorganism while harming its host as little as possible. This chapter introduces the principles of chemotherapy and briefly reviews the characteristics of selected antibacterial, antifungal, and antiviral drugs.

Modern medicine is dependent on **chemotherapeutic agents,** chemical agents that are used to treat disease. Chemotherapeutic agents destroy pathogenic microorganisms or inhibit their growth at concentrations low enough to avoid undesirable damage to the host. Most of these agents are **antibiotics** (Greek *anti,* against, and *bios,* life), microbial products or their derivatives that can kill susceptible microorganisms or inhibit their growth. Drugs such as the sulfonamides are sometimes called antibiotics although they are synthetic chemotherapeutic agents, not microbially synthesized.

The Development of Chemotherapy

The modern era of chemotherapy began with the work of the German physician Paul Ehrlich (1854–1915). Ehrlich reasoned that a chemical with selective toxicity that would kill pathogens and not human cells might be effective in treating disease. He hoped to find a toxic dye molecule, a "magic bullet," that would specifically bind to pathogens and destroy them; therefore, he began experimenting with dyes. By 1904, Ehrlich found that the dye Trypan Red was active against the trypanosome that causes African sleeping sickness (*see figure 28.3*a) and could be used therapeutically. Subsequently, Ehrlich and a young Japanese scientist named Sahachiro Hata tested a variety of arsenicals on syphilis-infected rabbits and found that compound number 606, arsphenamine, was active against the syphilis spirochete (*see figure 21.1*b). Arsphenamine was made available in 1910 under the trade name Salvarsan. Ehrlich's successes in the chemotherapy of sleeping sickness and syphilis established his concept of selective toxicity and led to the testing of hundreds of compounds for their therapeutic potential.

In 1927, the German chemical industry giant, I. G. Farbenindustrie, began a long-term search for chemotherapeutic agents under the direction of Gerhard Domagk. The company provided vast numbers of dyes and other chemicals that Domagk tested for activity against pathogenic bacteria and for toxicity in animals. During this screening program, Domagk discovered that Prontosil Red, a new dye for staining leather, was nontoxic for animals and completely protected mice against pathogenic streptococci and staphylococci. These results were published in 1935, and in the same year the French scientists Jacques and Therese Trefouel showed that Prontosil Red was converted in the body to sulfanilamide, the true active factor. For this discovery, Domagk received the Nobel Prize in 1939.

The story of the discovery and development of penicillin, the first antibiotic to be used therapeutically, is complex and fascinating. Although penicillin was actually discovered in 1896 by a 21-year-old French medical student named Ernest Duchesne, his work was forgotten, and penicillin was rediscovered and brought to the attention of scientists by the Scottish physician Alexander Fleming. Fleming had been interested in finding something that would kill pathogens ever since working on wound infections during the First World War. One day in September 1928, a *Penicillium notatum* spore accidentally landed on the surface of an exposed petri dish before it had been inoculated with staphylococci, and a new medical era was born. Although the precise events are still unclear, Ronald Hare has suggested that Fleming left the contaminated plate on a laboratory bench while he was on vacation. Because the first few days of the vacation were cool, the fungus grew more rapidly than the bacteria and produced penicillin. When the weather then turned warm, the bacteria began to grow and were lysed. Upon his return, Fleming noticed that a *Penicillium* colony was growing at one edge and that the staphylococci surrounding it had been destroyed. Rather than discarding the contaminated plate, he correctly deduced that his mold contaminant was producing a diffusible substance lethal to staphylococci. He began efforts to characterize what he called penicillin. He found that broth from a *Penicillium* culture contained penicillin and that the antibiotic could destroy several pathogenic bacteria. Unfortunately, Fleming's next experiments convinced him that penicillin would not remain active in the body long enough after injection to destroy pathogens. After giving a talk on penicillin and publishing several papers on the subject between 1929 and 1931, he dropped the research.

In 1939, Howard Florey, a professor of pathology at Oxford University, was in the midst of testing the bactericidal activity of many substances, including lysozyme and the sulfonamides. After reading Fleming's paper on penicillin, one of Florey's coworkers, Ernst Chain, obtained the *Penicillium* culture from Fleming and set about culturing it and purifying penicillin. Florey and Chain were greatly aided in this by the biochemist Norman Heatley. Heatley devised the original assay, culture, and purification techniques needed to produce crude penicillin for further experimentation. When the purified penicillin was injected into mice infected with streptococci or staphylococci, practically all the mice survived. Florey and Chain's success was reported in 1940, and subsequent human

TABLE 16.1 Properties of Some Common Antibacterial Drugs

Drug	Primary Effect	Spectrum	Side Effects[a]
Ampicillin	Cidal	Broad (gram +, some −)	Allergic responses (diarrhea, anemia)
Bacitracin	Cidal	Narrow (gram +)	Renal injury if injected
Carbenicillin	Cidal	Broad (gram +, many −)	Allergic responses (nausea, anemia)
Cephalosporins	Cidal	Broad (gram +, some −)	(Allergic responses, thrombophlebitis, renal injury)
Chloramphenicol	Static	Broad (gram +, −; rickettsia and chlamydia)	Depressed bone marrow function, allergic reactions
Ciprofloxacin	Cidal	Broad (gram +, −)	Gastrointestinal upset, allergic responses
Dapsone	Static	Narrow (mycobacteria)	(Anemia, allergic responses)
Erythromycin	Static	Narrow (gram +, mycoplasma)	(Gastrointestinal upset, hepatic injury)
Gentamicin	Cidal	Narrow (gram −)	(Allergic responses, nausea, loss of hearing, renal damage)
Isoniazid	Static or cidal	Narrow (mycobacteria)	(Allergic reactions, gastrointestinal upset, hepatic injury)
Methicillin	Cidal	Narrow (gram +)	Allergic responses (renal toxicity, anemia)
Penicillin	Cidal	Narrow (gram +)	Allergic responses (nausea, anemia)
Polymyxin B	Cidal	Narrow (gram −)	(Renal damage, neurotoxic reactions)
Rifampin	Static	Broad (gram +, mycobacteria)	(Hepatic injury, nausea, allergic responses)
Streptomycin	Cidal	Broad (gram +, −; mycobacteria)	(Allergic responses, nausea, loss of hearing, renal damage)
Sulfonamides	Static	Broad (gram +, −)	Allergic responses (renal and hepatic injury, anemia)
Tetracyclines	Static	Broad (gram +, −; rickettsia and chlamydia)	Gastrointestinal upset, teeth discoloration (renal and hepatic injury)
Trimethoprim	Cidal	Broad (gram +, −)	(Allergic responses, rash, nausea, leukopenia)

[a]Occasional side effects are in parentheses. Other side effects not listed may also arise.

trials were equally successful. Fleming, Florey, and Chain received the Nobel Prize in 1945 for the discovery and production of penicillin.

The discovery of penicillin stimulated the search for other antibiotics. Selman Waksman announced in 1944 that he had found a new antibiotic, streptomycin, produced by the actinomycete *Streptomyces griseus.* This discovery arose from the patient screening of about 10,000 strains of soil bacteria and fungi. Waksman received the Nobel Prize in 1952, and his success led to a worldwide search for other antibiotic-producing soil microorganisms. Microorganisms producing chloramphenicol, neomycin, terramycin, and tetracycline were isolated by 1953.

1. What are chemotherapeutic agents? Antibiotics?
2. What contributions to chemotherapy were made by Ehrlich, Domagk, Fleming, Florey and Chain, and Waksman?

General Characteristics of Antimicrobial Drugs

As Ehrlich so clearly saw, to be successful a chemotherapeutic agent must have **selective toxicity**: it must kill or inhibit the microbial pathogen while damaging the host as little as possible. The degree of selective toxicity may be expressed in terms of (1) the therapeutic dose, the drug level required for clinical treatment of a particular infection, and (2) the toxic dose, the drug level at which the agent becomes too toxic for the host. The **therapeutic index** is the ratio of the toxic dose to the therapeutic dose. The larger the therapeutic index, the better the chemotherapeutic agent (all other things being equal).

A drug that disrupts a microbial function not found in eucaryotic animal cells often has a greater selective toxicity and a higher therapeutic index. For example, penicillin inhibits bacterial cell wall peptidoglycan synthesis but has little effect on host cells because they lack cell walls; therefore, penicillin's therapeutic index is high. A drug may have a low therapeutic index because it inhibits the same process in host cells or damages the host in other ways. These undesirable effects on the host, called **side effects**, are of many kinds and may involve almost any organ system (table 16.1). Because side effects can be severe, chemotherapeutic agents should be administered with great care.

Drugs vary considerably in their range of effectiveness. Many are **narrow-spectrum drugs;** that is, they are effective only against a limited variety of pathogens (table 16.1). Others are **broad spectrum** and attack many different kinds of pathogens. Drugs may also be classified based on the general microbial group they act against: antibacterial, antifungal, antiprotozoan, and antiviral. Some agents can be used against more than one group; for example, sulfonamides are active against bacteria and some protozoa.

Chemotherapeutic agents can be synthesized by microorganisms or manufactured by chemical procedures independent of microbial activity. A number of the most commonly employed antibiotics are natural, that is, totally synthesized by one of a few bacteria or fungi (table 16.2). In contrast, sev-

TABLE 16.2 Microbial Sources of Some Antibiotics

Microorganism	Antibiotic
Bacteria	
Streptomyces spp.	Amphotericin B
	Chloramphenicol (also synthetic)
	Erythromycin
	Kanamycin
	Neomycin
	Nystatin
	Rifampin
	Streptomycin
	Tetracyclines
Micromonospora spp.	Gentamicin
Bacillus spp.	Bacitracin
	Polymyxins
Fungi	
Penicillium spp.	Griseofulvin
	Penicillin
Cephalosporium spp.	Cephalosporins

eral important chemotherapeutic agents are completely synthetic. The synthetic antibacterial drugs in table 16.1 are the sulfonamides, trimethoprim, chloramphenicol, ciprofloxacin, isoniazid, and dapsone. Many antiviral and antiprotozoan drugs are synthetic. An increasing number of antibiotics are semisynthetic. Semisynthetic antibiotics are natural antibiotics that have been chemically modified by the addition of extra chemical groups to make them less susceptible to inactivation by pathogens. Ampicillin, carbenicillin, and methicillin (table 16.1) are good examples.

Chemotherapeutic agents, like disinfectants, can be either **cidal** or **static.** As mentioned earlier (*see chapter 15*), static agents reversibly inhibit growth; if the agent is removed, the microorganisms will recover and grow again.

Although a cidal agent kills the target pathogen, its activity is concentration dependent and the agent may be static only at low levels. The effect of an agent also varies with the target species: an agent may be cidal for one species and static for another. Because static agents do not directly destroy the pathogen, elimination of the infection depends on the host's own resistance mechanisms. A static agent may not be effective if the host's resistance is too low.

Host resistance and the immune response (chapters 30–33).

Some idea of the effectiveness of a chemotherapeutic agent against a pathogen can be obtained from the **minimal inhibitory concentration** (MIC). The MIC is the lowest concentration of a drug that prevents growth of a particular pathogen. The **minimal lethal concentration** (MLC) is the lowest drug concentration that kills the pathogen. A cidal drug kills pathogens at levels only two to four times the MIC, whereas a static agent kills at much higher concentrations (if at all).

1. Define the following terms: selective toxicity, therapeutic index, side effect, narrow-spectrum drug, broad-spectrum drug, synthetic and semisynthetic antibiotics, cidal and static agents, minimal inhibitory concentration (MIC), and minimal lethal concentration (MLC).

Determining the Level of Antimicrobial Activity

Determination of antimicrobial effectiveness against specific pathogens is essential to proper therapy. Testing can show which agents are most effective against a pathogen and give an estimate of the proper therapeutic dose.

Dilution Susceptibility Tests

Dilution susceptibility tests can be used to determine MIC and MLC values. In the broth dilution test, a series of broth tubes (usually Mueller-Hinton broth) containing antibiotic concentrations in the range of 0.1 to 128 μg/ml is prepared and inoculated with standard numbers of the test organism. The lowest concentration of the antibiotic resulting in no growth after 16 to 20 hours of incubation is the MIC. The MLC can be ascertained if the tubes showing no growth are subcultured into fresh medium lacking antibiotic. The lowest antibiotic concentration from which the microorganisms do not recover and grow when transferred to fresh medium is the MLC. The agar dilution test is very similar to the broth dilution test. Plates containing Mueller-Hinton agar and various amounts of antibiotic are inoculated and examined for growth. Recently, several automated systems for susceptibility testing and MIC determination with broth or agar cultures have been developed.

Disk Diffusion Tests

If a rapidly growing aerobic or facultative pathogen like *Staphylococcus* or *Pseudomonas* is being tested, a disk diffusion technique may be used to save time and media. The principle behind the assay technique is fairly simple. When an antibiotic-impregnated disk is placed on agar previously inoculated with the test bacterium, the disk picks up moisture and the antibiotic diffuses radially outward through the agar, producing an antibiotic concentration gradient. The antibiotic is present at high concentrations near the disk and affects even minimally susceptible microorganisms (resistant organisms will grow up to the disk). As the distance from the disk increases, the antibiotic concentration drops and only more susceptible pathogens are harmed. A clear zone or ring is present around an antibiotic disk after incubation if the agent inhibits bacterial growth. The wider the zone surrounding a disk, the more susceptible the pathogen is. Zone width is also a function of the antibiotic's initial concentration, its solubility, and its diffusion rate through agar. Thus, zone width cannot be used to compare directly the effectiveness of two different antibiotics.

Currently, the disk diffusion test most often used is the **Kirby-Bauer method,** which was developed in the early 1960s at the University of Washington Medical School by William Kirby, A. W. Bauer, and their colleagues. An inoculating loop or needle is touched to four or five isolated colonies of the pathogen growing on agar and then used to inoculate a tube of culture broth. The culture is incubated for a few hours at 35° C until it becomes slightly turbid and is diluted to match a turbidity standard. A sterile cotton swab is dipped into the standardized bacterial test suspension and used to evenly inoculate the entire surface of a Mueller-Hinton agar plate. After the agar surface has dried for about 5 minutes, the appropriate antibiotic test disks are placed on it, either with sterilized forceps or with a multiple applicator device (figure 16.1). The plate is immediately placed in a 35° C incubator. After 16 to 18 hours of incubation, the diameters of the zones of inhibition are measured to the nearest mm.

Kirby-Bauer test results are interpreted using a table that relates zone diameter to the degree of microbial resistance (table 16.3). The values in table 16.3 were derived by finding the MIC values and zone diameters for many different microbial strains. A plot of MIC (on a logarithmic scale) versus zone inhibition diameter (arithmetic scale) is prepared for each antibiotic (figure 16.2). These plots are then used to find the zone diameters corresponding to the drug concentrations actually reached in the body. If the zone diameter for the lowest level reached in the body is smaller than that seen with the test pathogen, the pathogen should have an MIC value low enough to be destroyed by the drug. A pathogen with too high an MIC value (too small a zone diameter) is resistant to the agent at normal body concentrations.

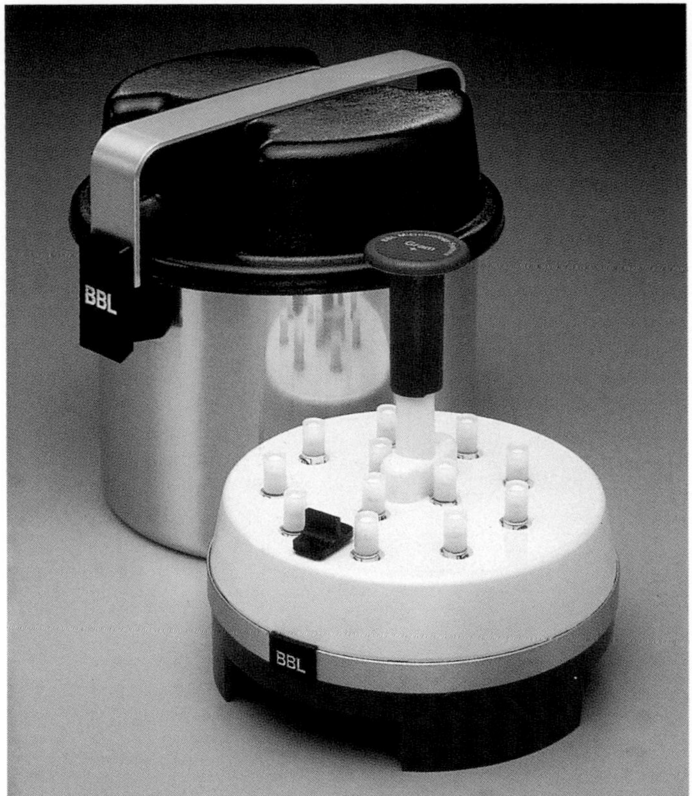

(a)

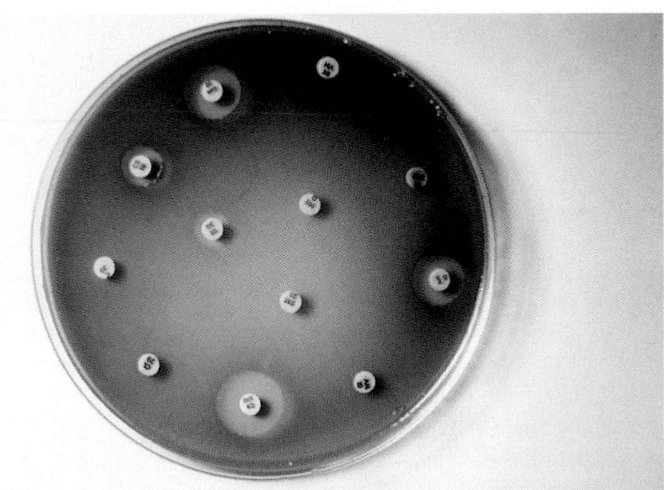
(b)

Figure 16.1 The Kirby-Bauer Method. (*a*) A multiple antibiotic disk dispenser and (*b*) disk diffusion test results.

Measurement of Drug Concentrations in the Blood

A drug must reach a concentration at the site of infection above the pathogen's MIC to be effective. In cases of severe, life-threatening disease, it is often necessary to monitor the concentration of drugs in the blood and other body fluids. This may be achieved by microbiological, chemical, immunologic, enzymatic, or chromatographic assays.

1. How can dilution susceptibility tests and disk diffusion tests be used to determine microbial drug sensitivity?
2. Briefly describe the Kirby-Bauer test and its purpose.

TABLE 16.3 Inhibition Zone Diameter of Selected Chemotherapeutic Drugs

Chemotherapeutic Drug	Disk Content	Zone Diameter (Nearest mm)		
		Resistant	*Intermediate*	*Susceptible*
Carbenicillin (with *Proteus* spp. and *E. coli*)	100 μg	≤17	18–22	≥23
Carbenicillin (with *Pseudomonas aeruginosa*)	100 μg	≤13	14–16	≥17
Ceftriaxone	30 μg	≤13	14–20	≥21
Chloramphenicol	30 μg	≤12	13–17	≥18
Erythromycin	15 μg	≤13	14–17	≥18
Penicillin G (with staphylococci)	10 U[a]	≤20	21–28	≥29
Penicillin G (with other microorganisms)	10 U	≤11	12–21	≥22
Streptomycin	10 μg	≤11	12–14	≥15
Sulfonamides	250 or 300 μg	≤12	13–16	≥17
Tetracycline	30 μg	≤14	15–18	≥19

[a]One milligram of penicillin G sodium = 1,600 units.

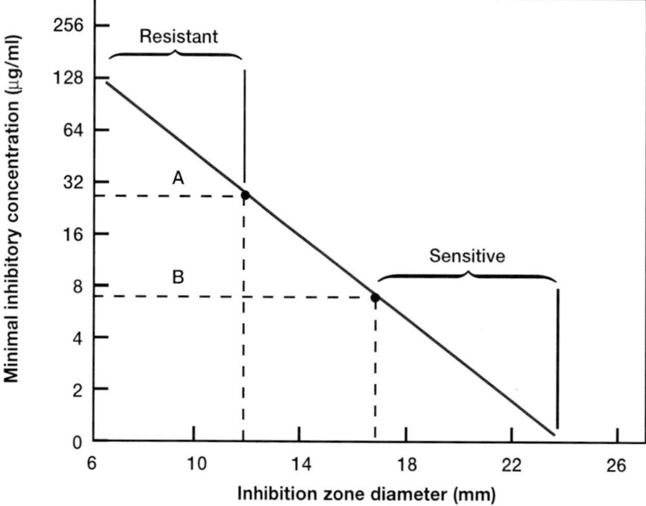

Figure 16.2 Interpretation of Kirby-Bauer Test Results. The relationship between the minimal inhibitory concentrations with a hypothetical drug and the size of the zone around a disk in which microbial growth is inhibited. As the sensitivity of microorganisms to the drug increases, the MIC value decreases and the inhibition zone grows larger. Suppose that this drug varies from 7–28 μg/ml in the body during treatment. Dashed line *A* shows that any pathogen with a zone of inhibition less than 12 mm in diameter will have an MIC value greater than 28 μg/ml and will be resistant to drug treatment. A pathogen with a zone diameter greater than 17 mm will have an MIC less than 7 μg/ml and will be sensitive to the drug (see line *B*). Zone diameters between 12 and 17 mm indicate intermediate sensitivity and usually signify resistance.

Mechanisms of Action of Antimicrobial Agents

The mechanisms of action of specific chemotherapeutic agents are taken up in more detail when individual drugs and groups of drugs are discussed later in this chapter. A few general observations are offered at this point. It is important to know something about the mechanisms of drug action because such knowledge helps explain the nature and degree of selective toxicity of individual drugs and sometimes aids in the design of new chemotherapeutic agents.

Antimicrobial drugs can damage pathogens in several ways, as can be seen in table 16.4, which summarizes the mechanisms of the antibacterial drugs listed in table 16.1. The most selective antibiotics are those that interfere with the synthesis of bacterial cell walls (e.g., penicillins, cephalosporins, and bacitracin). These drugs have a high therapeutic index because bacterial cell walls have a unique structure (*see chapter 3*) not found in eucaryotic cells.

Streptomycin, gentamicin, spectinomycin, clindamycin, chloramphenicol, tetracyclines, erythromycin, and many other antibiotics inhibit protein synthesis by binding with the procaryotic ribosome. Because these drugs discriminate between procaryotic and eucaryotic ribosomes, their therapeutic index is fairly high, but not as favorable as that of cell wall synthesis inhibitors. Some drugs bind to the 30S (small) subunit, while others attach to the 50S (large) ribosomal subunit. Several different steps in the protein synthesis mechanism can be affected: aminoacyl-tRNA binding, peptide bond formation, mRNA reading, and translocation. For example, fusidic acid binds to EF-G and blocks translocation, while mucopirocin inhibits isoleucyl-tRNA synthetase (*see chapter 10*).

The antibacterial drugs that inhibit nucleic acid synthesis or damage cell membranes often are not as selectively toxic as other antibiotics. This is because procaryotes and eucaryotes do not differ as greatly with respect to nucleic acid synthetic mechanisms or cell membrane structure. Good examples of drugs that affect nucleic acid synthesis or membrane structure are quinolones and polymyxins. Quinolones inhibit the DNA gyrase and thus interfere with DNA replication, repair, and transcription. Polymyxins act as detergents or surfactants and disrupt the bacterial plasma membrane.

Several valuable drugs act as **antimetabolites:** they block the functioning of metabolic pathways by competitively inhibiting the use of metabolites by key enzymes. Sulfonamides and several other drugs inhibit folic acid metabolism. Sulfonamides (e.g., sulfanilamide, sulfamethoxazole, and sulfacetamide) have a high therapeutic index because humans cannot synthesize folic acid and must obtain it in their diet. Most bacterial pathogens synthesize their own folic acid and are there-

TABLE 16.4 Mechanisms of Antibacterial Drug Action

Drug	Mechanism of Action
Cell Wall Synthesis Inhibition	
Penicillin Ampicillin Carbenicillin Methicillin Cephalosporins	Inhibit transpeptidation enzymes involved in the cross-linking of the polysaccharide chains of the bacterial cell wall peptidoglycan. Activate cell wall lytic enzymes.
Bacitracin	Inhibits cell wall synthesis by interfering with action of the lipid carrier that transports wall precursors across the cell membrane.
Protein Synthesis Inhibition	
Streptomycin Gentamicin	Binds with the 30S subunit of the bacterial ribosome to inhibit protein synthesis and causes misreading of mRNA.
Chloramphenicol	Binds to the 50S ribosomal subunit and blocks peptide bond formation through inhibition of peptidyl transferase.
Tetracyclines	Bind to the 30S ribosomal subunit and interfere with aminoacyl-tRNA binding.
Erythromycin	Binds to the 50S ribosomal subunit and inhibits peptide chain elongation.
Fusidic acid	Binds to EF-G and blocks translocation.
Nucleic Acid Synthesis Inhibition	
Ciprofloxacin and other quinolones	Inhibit bacterial DNA gyrase and thus interfere with DNA replication, transcription, and other activities involving DNA.
Rifampin	Blocks RNA synthesis by binding to and inhibiting the DNA-dependent RNA polymerase.
Cell Membrane Disruption	
Polymyxin B	Binds to the cell membrane and disrupts its structure and permeability properties.
Metabolic Antagonism	
Sulfonamides	Inhibit folic acid synthesis by competition with p-aminobenzoic acid.
Trimethoprim	Blocks tetrahydrofolate synthesis through inhibition of the enzyme dihydrofolate reductase.
Dapsone	Interferes with folic acid synthesis.
Isoniazid	May disrupt pyridoxal or NAD metabolism and functioning. Inhibits the synthesis of the mycolic acid "cord factor."

fore susceptible to inhibitors of folate metabolism. Antimetabolite drugs also can inhibit other pathways. For example, isoniazid interferes with either pyridoxal or NAD metabolism.

1. Give five ways in which chemotherapeutic agents kill or damage bacterial pathogens.
2. What are antimetabolites?

Factors Influencing the Effectiveness of Antimicrobial Drugs

It is crucial to recognize that drug therapy is not a simple matter. Drugs may be administered in several different ways, and they do not always spread rapidly throughout the body or immediately kill all invading pathogens. A complex array of factors influence the effectiveness of drugs.

First, the drug must actually be able to reach the site of infection. The mode of administration plays an important role. A drug such as penicillin G is not suitable for oral administration because it is relatively unstable in stomach acid. Some antibiotics—for example, gentamicin and other aminoglycosides—are not well absorbed from the intestinal tract and must be injected intramuscularly or given intravenously. Other antibiotics (neomycin, bacitracin) are applied topically to skin lesions. Nonoral routes of administration are often called **parenteral routes.** Even when an agent is administered properly, it may be excluded from the site of infection. For example, blood clots or necrotic tissue can protect bacteria from a drug, either because body fluids containing the agent may not easily reach the pathogens or because the agent is absorbed by materials surrounding it.

Second, the pathogen must be susceptible to the drug. Bacteria in abscesses may be dormant and therefore resistant to chemotherapy, because penicillins and many other agents affect pathogens only if they are actively growing and dividing. A pathogen, even though growing, may simply not be susceptible to a particular agent. For example, penicillins and cephalosporins, which inhibit cell wall synthesis (table 16.4), do not harm mycoplasmas, which lack cell walls.

Third, the chemotherapeutic agent must exceed the pathogen's MIC value if it is going to be effective. The concentration reached will depend on the amount of drug administered, the route of administration and speed of uptake, and

Box 16.1

The Use of Antibiotics in Microbiological Research

Although the use of antibiotics in the treatment of disease is emphasized in this chapter, it should be noted that antibiotics are extremely important research tools. For example, they aid the cultivation of viruses by preventing bacterial contamination. When eggs are inoculated with a virus sample, antibiotics are often included in the inoculum to maintain sterility. Usually, a mixture of antibiotics (e.g., penicillin, amphotericin, and streptomycin) is also added to tissue cultures used for virus cultivation and other purposes.

Researchers often use antibiotics as instruments to dissect metabolic processes by inhibiting or blocking specific steps and observing the consequences. Although selective toxicity is critical when antibiotics are employed therapeutically, specific toxicity is more important in this context: the antibiotic must act by a specific and precisely understood mechanism. A clinically useful antimicrobial agent such as penicillin sometimes may be employed in research, but often an agent with specific toxicity and excellent research potential is too toxic for therapeutic use. The actinomycins, discovered in 1940 by Selman Waksman, are a case in point. They are so toxic to higher organisms that it was suggested they be used as rat poison. Today, actinomycin D is a standard research tool specifically used to block RNA synthesis. Other examples of antibiotics useful in research, with the process inhibited, are the following: chloramphenicol (bacterial protein synthesis), cycloserine (peptidoglycan synthesis), nalidixic acid and novobiocin (bacterial DNA synthesis), rifampin (bacterial RNA synthesis), cycloheximide (eucaryotic protein synthesis), daunomycin (fungal RNA synthesis), mitomycin C (DNA synthesis), polyoxin D (fungal cell wall chitin synthesis), and cerulenin (fatty acid synthesis).

In practice, the antibiotic is administered and changes in cell function are monitored. If one desired to study the dependence of bacterial flagella synthesis on RNA transcription, the flagella could be removed by high speed mixing in a blender, followed by actinomycin D addition to the incubation mixture. The bacterial culture would then be observed for flagella regeneration in the absence of RNA synthesis. The results of such experiments must be interpreted with caution. Flagella synthesis may have been blocked because actinomycin D inhibited some other process, thus affecting flagella regeneration indirectly rather than simply inhibiting transcription of a gene required for flagella synthesis. Furthermore, not all microorganisms respond in the same way to a particular drug.

---

the rate at which the drug is cleared or eliminated from the body. It makes sense that a drug will remain at high concentrations longer if it is absorbed over a long period and excreted slowly.

Finally, chemotherapy has been rendered a less effective and much more complex matter by the spread of drug resistance plasmids. The nature of drug resistance is dealt with later in this chapter.

1. Briefly discuss the factors that influence the effectiveness of antimicrobial drugs.
2. What is parenteral administration of a drug?

Antibacterial Drugs

With the background provided by the preceding introduction to chemotherapy, a few of the more important chemotherapeutic agents can be briefly described. Table 16.5 lists some drugs used to treat a few of the more common bacterial pathogens. Tables 16.1 and 16.4 summarize the properties and mechanism of action of a variety of drugs.

Sulfonamides or Sulfa Drugs

A good way to inhibit or kill pathogens is by use of compounds that are **structural analogues,** molecules structurally similar to metabolic intermediates. These analogues compete with metabolites in metabolic processes because of their similarity, but are just different enough so that they cannot function normally in cellular metabolism. The first antimetabolites to be used successfully as chemotherapeutic agents were the sulfonamides, discovered by G. Domagk. **Sulfonamides** or sulfa drugs are structurally related to sulfanilamide, an analogue of p-aminobenzoic acid (figures 16.3 and 16.4). The latter substance is used in the synthesis of the cofactor folic acid.

When sulfanilamide or another sulfonamide enters a bacterial cell, it competes with p-aminobenzoic acid for the active site of an enzyme involved in folic acid synthesis, and the folate concentration decreases. The decline in folic acid is detrimental to the bacterium because folic acid is essential to the synthesis of purines and pyrimidines, the bases used in the construction of DNA, RNA, and other important cell constituents. The resulting inhibition of purine and pyrimidine synthesis leads to cessation of bacterial growth or death of the pathogen. Sulfonamides are selectively toxic for many patho-

Table 16.5 Chemotherapy of Some Representative Bacterial Pathogens

Pathogen	Representative Diseases	Drugs of Choice[a]
Gram-Positive Bacteria		
Corynebacterium diphtheriae	Diphtheria	Erythromycin, penicillin G
Staphylococcus, penicillinase-positive	Boils, pneumonia, wound infections	A cephalosporin, cloxacillin, dicloxacillin
Staphylococcus, penicillinase-negative	Boils, pneumonia, wound infections	Penicillin G or V, a cephalosporin, vancomycin
Streptococcus, hemolytic	Strep throat, skin infections, sepsis, rheumatic fever	Penicillin[b] G or V, erythromycin, or a cephalosporin
Streptococcus pneumoniae	Pneumonia	Penicillin[b] G or V, erythromycin, or a cephalosporin
Gram-Negative Bacteria		
Bordetella pertussis	Whooping cough or pertussis	Erythromycin, ampicillin
Escherichia coli	Urinary tract infections	A cephalosporin, ampicillin
Haemophilus influenzae type b	Meningitis, pneumoniae	Cefotaxime, ceftriaxone, ampicillin
Klebsiella pneumoniae	Pneumonia, urinary tract infections	Newer cephalosporins[c], gentamicin
Legionella pneumophila	Pneumonia	Erythromycin
Neisseria gonorrhoeae	Gonorrhea	Ceftriaxone, spectinomycin
Pseudomonas aeruginosa	Urinary tract and burn infections, pneumonia	Carbenicillin or ticarcillin
Salmonella typhi	Typhoid fever, septicemia, gastroenteritis	Ceftriaxone, chloramphenicol, ampicillin
Shigella dysenteriae	Dysentery	Trimethoprim-sulfamethoxazole, ciprofloxacin
Vibrio cholerae	Cholera	Tetracycline, trimethoprim-sulfamethoxazole
Acid-Fast Bacteria		
Mycobacterium-avium complex	Pneumonia	Rifampin plus ethambutol
Mycobacterium tuberculosis	Tuberculosis	Isoniazid plus rifampin ± pyrazinamide
Other Bacteria		
Chlamydia trachomatis	Nongonococcal urethritis, trachoma	Tetracycline, erythromycin
Mycoplasma pneumoniae	Pneumonia	Tetracycline, erythromycin
Rickettsia spp.	Rocky Mountain spotted fever, typhus fever	Tetracycline, chloramphenicol
Treponema pallidum subsp. *pallidum*	Syphilis	Penicillin G, tetracycline, ceftriaxone

[a]The drugs are listed in approximate order of preference. A number of others are also used.

[b]Penicillinase-resistant penicillins may have to be employed (e.g., methicillin, nafcillin, oxacillin).

[c]Many newer cephalosporins are available: ceftriaxone, cefoperazone, cefuroxime, among others.

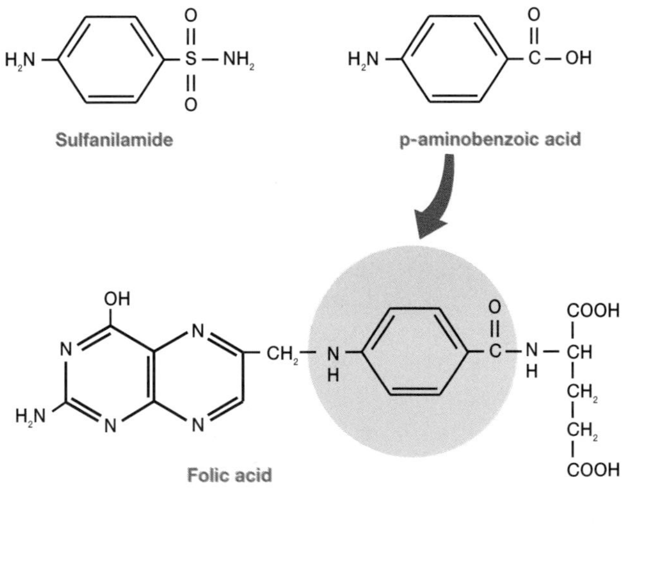

Figure 16.3 Sulfanilamide. Sulfanilamide and its relationship to the structure of folic acid.

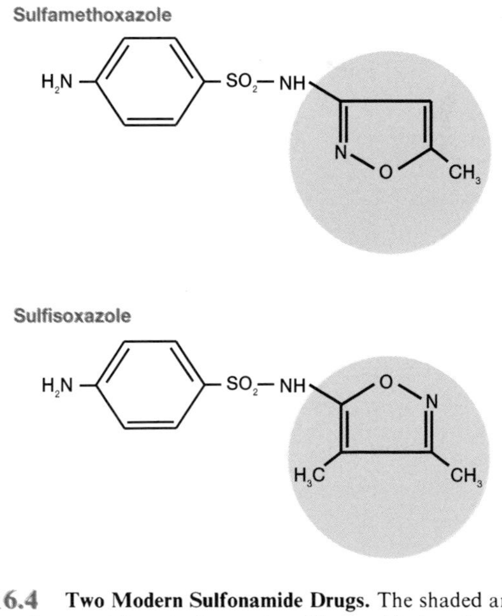

Figure 16.4 Two Modern Sulfonamide Drugs. The shaded areas are side chains substituted for a hydrogen in sulfanilamide (see figure 16.3).

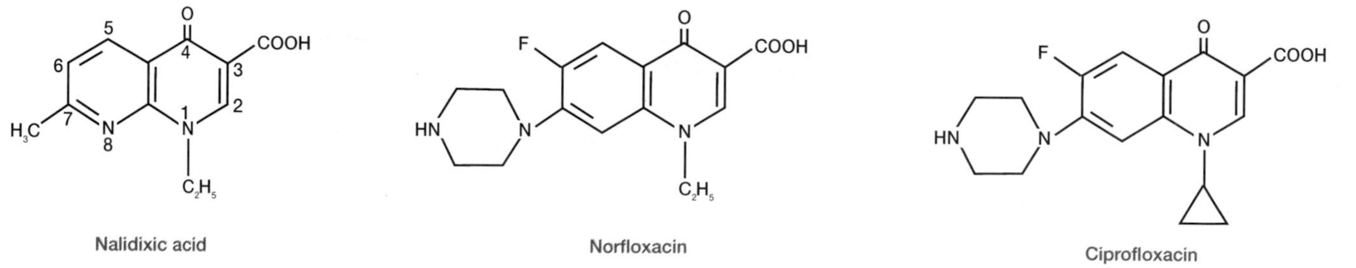

Nalidixic acid

Norfloxacin

Ciprofloxacin

Figure 16.5 **Quinolone Antimicrobial Agents.** Ciprofloxacin and norfloxacin are newer generation fluoroquinolones. The 4-quinolone ring in nalidixic acid has been numbered.

gens because these bacteria manufacture their own folate and cannot effectively take up the cofactor. In contrast, humans cannot synthesize folate and must obtain it in the diet; therefore, sulfonamides will not affect the host.

Sulfonamides and the competitive inhibition of enzymes (pp. 142–43).
Purine and pyrimidine structure and synthesis (pp. 183–84).

The effectiveness of sulfonamides is limited by the increasing sulfonamide resistance of many bacteria. Furthermore, as many as 5% of the patients receiving sulfa drugs experience adverse side effects, chiefly in the form of allergic responses (fever, hives, and rashes).

Quinolones

A second group of synthetic antimicrobial agents are increasingly used to treat a wide variety of infections. The **quinolones** are synthetic drugs that contain the 4-quinolone ring. The first quinolone, nalidixic acid (figure 16.5), was synthesized in 1962. More recently, a family of fluoroquinolones has been produced. Three of these—ciprofloxacin, norfloxacin, and ofloxacin—are currently used in the United States, and more fluoroquinolones are being synthesized and tested. Quinolones are effective when administered orally. They sometimes cause adverse side effects, particularly gastrointestinal upset.

Quinolones act by inhibiting the bacterial DNA gyrase or topoisomerase II, probably by binding to the DNA-gyrase complex. This enzyme introduces negative twists in DNA and helps separate its strands. DNA gyrase inhibition disrupts DNA replication and repair, transcription, bacterial chromosome separation during division, and other cell processes involving DNA. It is not surprising that quinolones are bactericidal.

The mechanism of DNA replication (pp. 198–201).

The quinolones are broad-spectrum drugs. They are highly effective against enteric bacteria such as *E. coli* and *Klebsiella pneumoniae.* They can be used with *Haemophilus, Neisseria, Pseudomonas aeruginosa,* and other gram-negative pathogens. The quinolones are also active against gram-positive bacteria such as *Staphylococcus aureus, Streptococcus pyogenes,* and *Mycobacterium tuberculosis.* Currently they are used in treating urinary tract infections, sexually transmitted diseases caused by *Neisseria* and *Chlamydia,* gastrointestinal infections, respiratory tract infections, skin infections, and osteomyelitis.

Penicillins

Penicillin G or benzylpenicillin, the first antibiotic to be widely used in medicine, has the structural properties characteristic of the penicillin family (figure 16.6). Most **penicillins** are derivatives of 6-aminopenicillanic acid and differ from one another only with respect to the side chain attached to its amino group. The most crucial feature of the molecule is the β-lactam ring, which appears to be essential for activity. **Penicillinase,** the enzyme synthesized by many penicillin-resistant bacteria, destroys penicillin activity by hydrolyzing a bond in this ring (figure 16.6).

The mechanism of action of penicillins is still not completely known. Their structures do resemble that of the terminal D-alanyl-D-alanine found on the peptide side chain of the peptidoglycan subunit. It has been proposed that penicillins inhibit the enzyme catalyzing the transpeptidation reaction because of their structural similarity, which would block the synthesis of a complete, fully cross-linked peptidoglycan and lead to osmotic lysis. The mechanism is consistent with the observation that penicillins act only on growing bacteria that are synthesizing new peptidoglycan. However, more recently it has been discovered that penicillins bind to several penicillin-binding proteins and may destroy bacteria by activating their own autolytic enzymes (autolysins). Apparently, the mechanism of penicillin action is more complex than previously imagined.

Bacterial peptidoglycan structure (pp. 51–53).
Peptidoglycan synthesis (pp. 186–88).

Penicillins differ from each other in several ways. Penicillin G is effective against gonococci, meningococci, and several gram-positive pathogens such as streptococci and staphylococci (table 16.5), but must be administered parenterally because it is destroyed by stomach acid. Penicillin V is similar to penicillin G, but it is more acid resistant and can be given orally. Ampicillin can be administered orally and has a broader spectrum of activity as it is effective against gram-negative bacteria such as *Haemophilus, Salmonella,* and *Shi-*

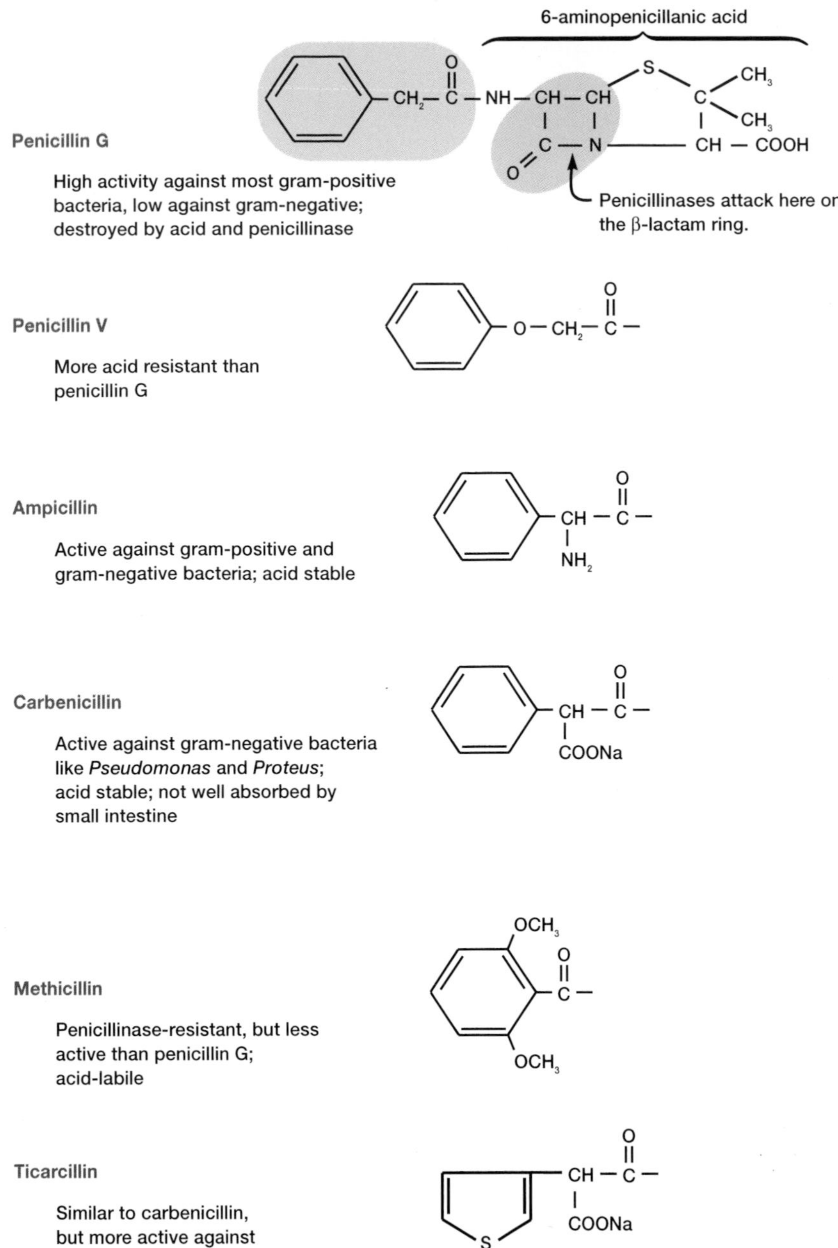

Figure 16.6 **Penicillins.** The structures and characteristics of representative penicillins. All are derivatives of 6-aminopenicillanic acid; in each case, the shaded portion of penicillin G is replaced by the side chain indicated. The β-lactam ring is also shaded, and an arrow points to the bond that is hydrolyzed by penicillinases.

gella. Carbenicillin and ticarcillin are also broad spectrum and particularly potent against *Pseudomonas* and *Proteus.*

An increasing number of bacteria are penicillin resistant. Penicillinase-resistant penicillins such as methicillin (figure 16.6), nafcillin, and oxacillin are frequently employed against these bacterial pathogens.

Although penicillins are the least toxic of the antibiotics, about 1 to 5% of the adults in the United States are allergic to them. Occasionally, a person will die of a violent allergic response; therefore, patients should be questioned about penicillin allergies before treatment is begun.

Cephalosporins

Cephalosporins are a family of antibiotics originally isolated in 1948 from the fungus *Cephalosporium,* and their β-lactam structure is very similar to that of the penicillins (figure 16.7). As might be expected from their structural similarities, cephalosporins resemble penicillins in inhibiting the transpeptidation reaction during peptidoglycan synthesis. They are broad-spectrum drugs frequently given to patients with penicillin allergies.

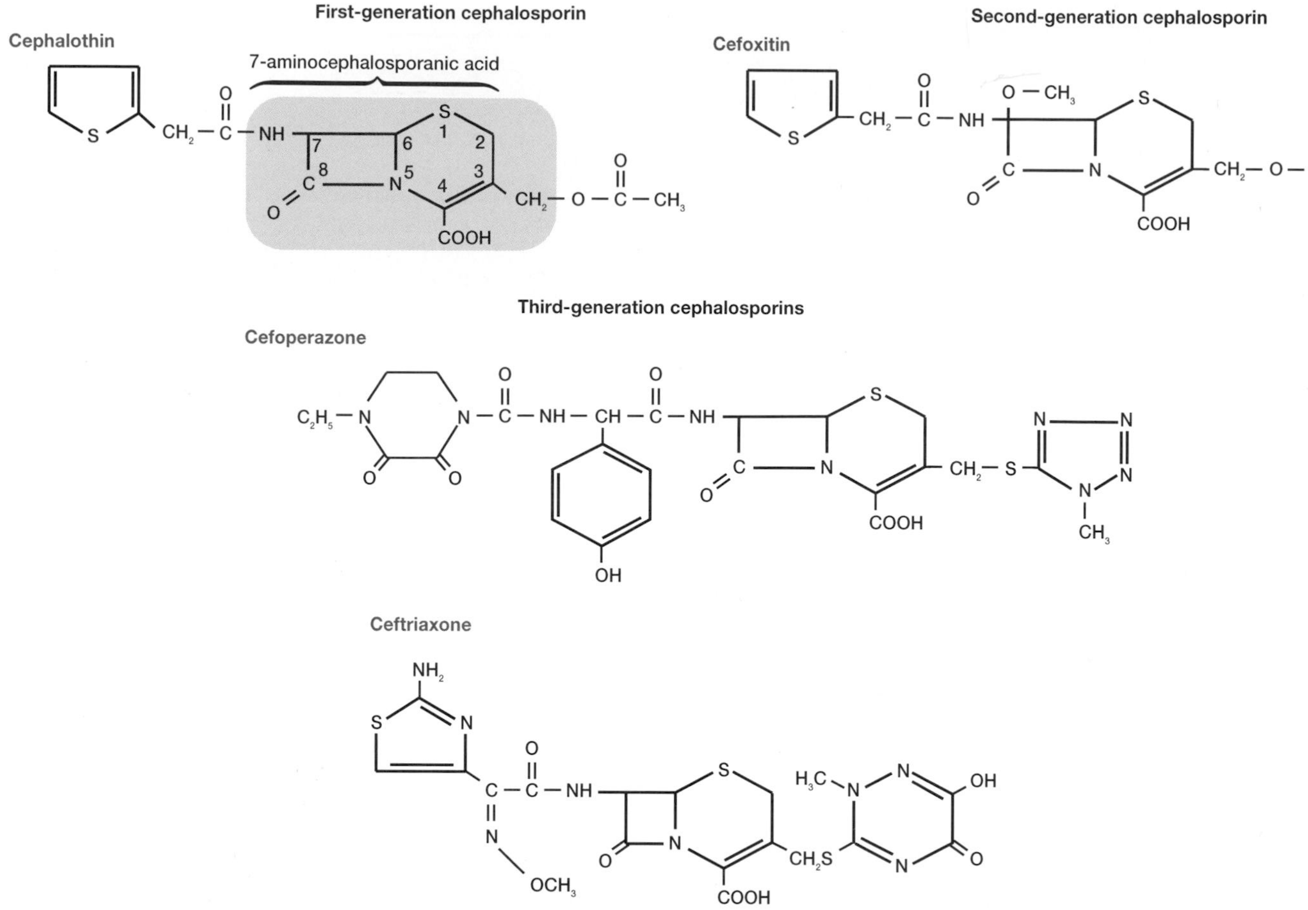

First-generation cephalosporin

Cephalothin

7-aminocephalosporanic acid

Second-generation cephalosporin

Cefoxitin

Third-generation cephalosporins

Cefoperazone

Ceftriaxone

Figure 16.7 Cephalosporin Antibiotics. These drugs are derivatives of 7-aminocephalosporanic acid and contain a β-lactam ring.

Many cephalosporins are in use. There are three groups or generations of these drugs that differ in their spectrum of activity (figure 16.7). First-generation cephalosporins are more effective against gram-positive than gram-negative pathogens. Second-generation drugs act against many gram-negative as well as gram-positive pathogens. Third-generation drugs are particularly effective against gram-negative pathogens, and often also reach the central nervous system.

Most cephalosporins (including cephalothin, cefoxitin, ceftriaxone, and cefoperazone) are administered parenterally. Cefoperazone is resistant to destruction by β-lactamases and effective against many gram-negative bacteria, including *Pseudomonas aeruginosa.* Cephalexine and cefixime are given orally rather than by injection.

The Tetracyclines

The **tetracyclines** are a family of antibiotics with a common four-ring structure to which a variety of side chains are attached (figure 16.8). Oxytetracycline and chlortetracycline are naturally produced by some species of the actinomycete genus *Streptomyces;* others are semisynthetic drugs. These antibiotics inhibit protein synthesis by combining with the small (30S) subunit of the ribosome and inhibiting the binding of aminoacyl-tRNA molecules to the ribosomal A site. Because their action is only bacteriostatic, the effectiveness of treatment depends on active host resistance to the pathogen.

Tetracycline (chlortetracycline, doxycycline)

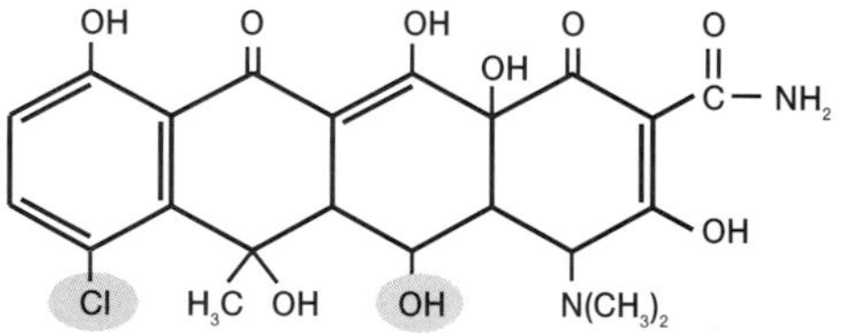

Figure 16.8 Tetracyclines. Three members of the tetracycline family. Tetracycline lacks both of the groups that are shaded. Chlortetracycline (aureomycin) differs from tetracycline in having a chlorine atom (blue); doxycycline consists of tetracycline with an extra hydroxyl (purple).

The mechanism of protein synthesis (pp. 204–13).

Tetracyclines are broad-spectrum antibiotics active against gram-negative bacteria, gram-positive bacteria, rickettsias, chlamydiae, and mycoplasmas (table 16.5). High doses may result in nausea, diarrhea, yellowing of teeth in children, and damage to the liver and kidneys.

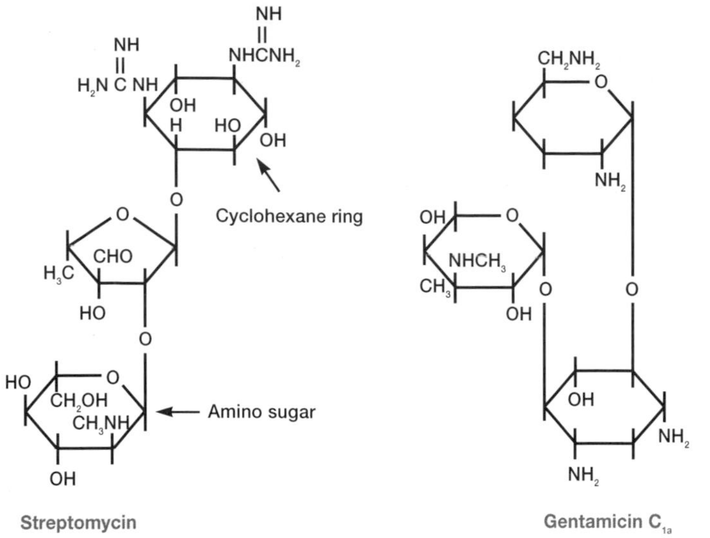

Figure 16.9 **Representative Aminoglycoside Antibiotics.**

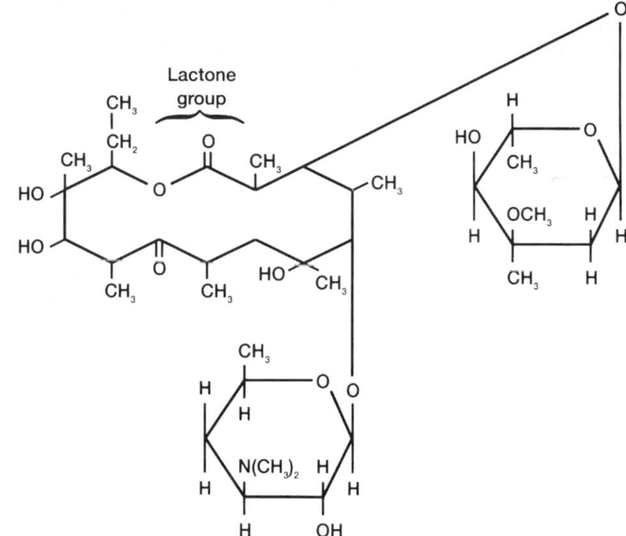

Figure 16.10 **Erythromycin, a Macrolide Antibiotic.** The 14-member lactone ring is connected to two sugars.

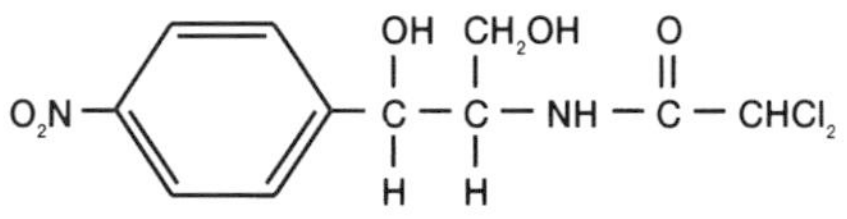

Aminoglycoside Antibiotics

There are several important **aminoglycoside antibiotics. Streptomycin,** kanamycin, neomycin, and tobramycin are synthesized by *Streptomyces,* whereas gentamicin comes from a related bacterium, *Micromonospora purpurea.* Although there is considerable variation in structure among the different aminoglycosides, all contain a cyclohexane ring and amino sugars (figure 16.9). Aminoglycosides bind to the small ribosomal subunit and interfere with protein synthesis in at least two ways. They directly inhibit protein synthesis, and also cause misreading of the genetic message carried by mRNA.

The aminoglycosides are bactericidal and tend to be most active against gram-negative pathogens. Streptomycin's usefulness has decreased greatly due to widespread drug resistance, but it is still effective against tuberculosis and plague. Gentamicin is used to treat *Proteus, Escherichia, Klebsiella,* and *Serratia* infections. Aminoglycosides are quite toxic and can cause deafness, renal damage, loss of balance, nausea, and allergic responses.

Erythromycin

Erythromycin, the most frequently used **macrolide antibiotic,** is synthesized by *Streptomyces erythraeus.* The macrolides contain a 12- to 22-carbon lactone ring linked to one or more sugars (figure 16.10). Erythromycin is usually bacteriostatic and binds with the 23S rRNA of the 50S ribosomal subunit to inhibit peptide chain elongation during protein synthesis.

Erythromycin is a relatively broad-spectrum antibiotic effective against gram-positive bacteria, mycoplasmas, and a few gram-negative bacteria. It is used with patients allergic to penicillins and in the treatment of whooping cough, diphtheria, diarrhea caused by *Campylobacter,* and pneumonia from *Legionella* or *Mycoplasma* infections.

Figure 16.11 **Chloramphenicol.**

Chloramphenicol

Although **chloramphenicol** (figure 16.11) was first produced from cultures of *Streptomyces venezuelae,* it is now made through chemical synthesis. Like erythromycin, chloramphenicol binds to 23S rRNA on the 50S ribosomal subunit. It inhibits the peptidyl transferase and is bacteriostatic.

This antibiotic has a very broad spectrum of activity, but unfortunately is quite toxic. One may see allergic responses or neurotoxic reactions. The most common side effect is a temporary or permanent depression of bone marrow function, leading to aplastic anemia and decreased blood leukocytes. Chloramphenicol is used only in life-threatening situations when no other drug is adequate.

1. For each class of antibacterial drugs presented, give the following information: general chemical composition or structure, mechanism of action, behavioral properties (cidal or static, broad spectrum or narrow spectrum), route of administration and therapeutic uses, significant problems with side effects, and drug resistance.
2. Define structural analogue, 6-aminopenicillanic acid, penicillinase, aminoglycoside, and macrolide.

Box 16.2

Antibiotic Misuse and Drug Resistance

The sale of antimicrobial drugs is big business. In the United States, 18 to 19 million pounds of antibiotics valued at about 270 million dollars are produced annually. Approximately 40 to 50% of these antibiotics are added to livestock feed.

Because of the massive quantities of antibiotics being prepared and used, an increasing number of diseases are resisting treatment due to the spread of drug resistance. A good example is *Neisseria gonorrhoeae,* the causative agent of gonorrhea. Gonorrhea was first treated successfully with sulfonamides in 1936, but by 1942 most strains were resistant and physicians turned to penicillin. Within 16 years, a penicillin-resistant strain had emerged in the Far East. A penicillinase-producing gonococcus reached the United States in 1976 and is still spreading in this country.

In late 1968, an epidemic of dysentery caused by *Shigella* broke out in Guatemala and affected at least 112,000 persons; 12,500 deaths resulted. The strains responsible for this devastation carried an R plasmid giving them resistance to chloramphenicol, tetracycline, streptomycin, and sulfonamide. In 1972, a typhoid epidemic swept through Mexico producing 100,000 infections and 14,000 deaths. It was due to a *Salmonella typhi* strain with the same multiple drug-resistance pattern seen in the previous *Shigella* outbreak.

Haemophilus influenzae type b is responsible for many cases of childhood pneumonia and middle ear infections, as well as respiratory infections and meningitis. It is now becoming increasingly resistant to tetracyclines, ampicillin, and chloramphenicol.

In 1946, almost all strains of *Staphylococcus* were penicillin sensitive. Today, most hospital strains are resistant to penicillin G, and some are now also resistant to methicillin and/or gentamicin.

It is clear from these and many other examples that drug resistance is becoming an extremely serious public health problem. Much of the difficulty arises from drug misuse. Drugs frequently have been overused in the past. It has been estimated that over 50% of the antibiotic prescriptions in hospitals are given without clear evidence of infection or adequate medical indication. Many physicians have administered antibacterial drugs to patients with colds, influenza, viral pneumonia, and other viral diseases. Frequently, antibiotics are prescribed without the pathogen having been cultured or identified or without bacterial sensitivity to the drug having been determined. Toxic, broad-spectrum antibiotics are sometimes given in place of narrow-spectrum drugs as a substitute for culture and sensitivity testing, with the consequent risk of dangerous side effects, superinfections, and the selection of drug-resistant mutants. The situation is made worse by patients not completing their course of medication. When antibiotic treatment is ended too early, drug-resistant mutants may survive. Drugs are frequently available to the public in the third world; people may practice self-administration of antibiotics and further increase the prevalence of drug-resistant strains.

The use of antibiotics in animal feeds is undoubtedly another contributing factor to increasing drug resistance. The addition of low levels of antibiotics to livestock feeds does raise the efficiency and rate of weight gain in cattle, pigs, and chickens (partially because of infection control in overcrowded animal populations). However, this also increases the number of drug-resistant bacteria in animal intestinal tracts. There is evidence for the spread of bacteria such as *Salmonella* from animals to human populations. In 1983, 18 people in four midwestern states were infected with a multiply drug-resistant strain of *Salmonella newport.* Eleven were hospitalized for salmonellosis and one died. All 18 patients had recently been infected by eating hamburger from beef cattle fed subtherapeutic doses of chlortetracycline for growth promotion. Elimination of antibiotic food supplements might well aid in slowing the spread of drug resistance.

Drug Resistance

The spread of drug-resistant pathogens is one of the most serious threats to the successful treatment of microbial disease (Box 16.2). This section describes the ways in which bacteria acquire drug resistance and how resistance spreads within a bacterial population.

Mechanisms of Drug Resistance

Bacteria become drug resistant in several different ways. It should be noted at the beginning that a particular type of resistance mechanism is not confined to a single class of drugs. Two bacteria may use different resistance mechanisms to withstand the same chemotherapeutic agent. Furthermore, resistant mutants arise spontaneously and are then selected. Mutants are not created directly by exposure to a drug.

Pathogens often become resistant simply by preventing entrance of the drug. Many gram-negative bacteria are unaffected by penicillin G because it cannot penetrate the envelope's outer membrane. Changes in penicillin binding proteins also render a cell resistant. A decrease in permeability can lead to sulfonamide resistance.

Many bacterial pathogens resist attack by inactivating drugs through chemical modification. The best-known example is the hydrolysis of the β-lactam ring of many penicillins by the enzyme penicillinase. Drugs are also inactivated by the addition of groups. Resistant organisms may phosphorylate or acetylate aminoglycosides and acetylate chloramphenicol.

Because each chemotherapeutic agent acts on a specific target, resistance arises when the target enzyme or organelle is modified so that it is no longer susceptible to the drug. For example, the affinity of ribosomes for erythromycin and chloramphenicol can be decreased by a change in the 23S rRNA to which they bind. The effects of antimetabolites may be resisted through alteration of susceptible enzymes. In sulfonamide-resistant bacteria, the enzyme that uses p-aminobenzoic acid during folic acid synthesis (the tetrahydropteroic acid synthetase) often has a much lower affinity for sulfonamides.

Resistant bacteria may either use an alternate pathway to bypass the sequence inhibited by the agent or increase the production of the target metabolite. For example, some bacteria are resistant to sulfonamides simply because they use preformed folic acid from their surroundings rather than synthesize it themselves. Other strains increase their rate of folic acid production, and thus counteract sulfonamide inhibition.

The Origin and Transmission of Drug Resistance

The genes for drug resistance are present on both the bacterial chromosome and **plasmids,** small circular DNA molecules that can exist separate from the chromosome or be integrated in it.

Plasmids (pp. 261–64).

Spontaneous mutations in the bacterial chromosome, although they do not occur very often, will make bacteria drug resistant. Usually, such mutations result in a change in the drug receptor; therefore, the antibiotic cannot bind and inhibit (e.g., the streptomycin receptor protein on bacterial ribosomes). Many mutants are probably destroyed by natural host resistance mechanisms. However when a patient is being treated extensively with antibiotics, some resistant mutants may survive and flourish because of their competitive advantage over nonresistant strains.

Frequently, a bacterial pathogen is drug resistant because it has a plasmid bearing one or more resistance genes; such plasmids are called **R plasmids** (resistance plasmids). Plasmid resistance genes often code for enzymes that destroy or modify drugs; for example, the hydrolysis of penicillin or the acetylation of chloramphenicol and aminoglycoside drugs. Plasmid-associated genes have been implicated in resistance to the aminoglycosides, choramphenicol, penicillins and cephalosporins, erythromycin, tetracyclines, sulfonamides, and others. Once a bacterial cell possesses an R plasmid, the plasmid may be transferred to other cells quite rapidly through normal gene exchange processes such as conjugation, transduction, and transformation. Because a single plasmid may carry genes for resistance to several drugs, a pathogen population can become resistant to several antibiotics simultaneously, even though the infected patient is being treated with only one drug.

Conjugation (pp. 268–71).
Transduction (pp. 274–78).
Transformation (pp. 271–74).

Extensive drug treatment favors the development and spread of antibiotic-resistant strains because the antibiotic destroys normal, susceptible bacteria that would usually compete with drug-resistant strains. The result may be the emergence of drug-resistant pathogens leading to a **superinfection.** Superinfections are a significant problem because of the existence of multiple-drug-resistant bacteria that often produce drug-resistant respiratory and urinary tract infections. A classic example of a superinfection resulting from antibiotic administration is the disease pseudomembranous enterocolitis caused by *Clostridium difficile.* When a patient is treated with clindamycin, ampicillin, or cephalosporin, many intestinal bacteria are killed, but *C. difficile* is not. This intestinal inhabitant, which is normally a minor constituent of the population, flourishes in the absence of competition and produces a toxin that stimulates the secretion of a pseudomembrane by intestinal cells. If the superinfection is not treated early with vancomycin, the pseudomembrane must be surgically removed or the patient will die. Fungi, such as the yeast *Candida albicans,* also produce superinfections when bacterial competition is eliminated by antibiotics.

Several strategies can be employed to discourage the emergence of drug resistance. The drug can be given in a high enough concentration to destroy susceptible bacteria and most spontaneous mutants that might arise during treatment. Sometimes two different drugs can be administered simultaneously with the hope that each drug will prevent the emergence of resistance to the other. Finally, chemotherapeutic drugs, particularly broad-spectrum drugs, should be used only when definitely necessary. When possible, the pathogen should be identified, drug sensitivity tests run, and the proper narrow-spectrum drug employed.

1. Briefly describe the four major ways in which bacteria become resistant to drugs and give an example of each.
2. Define plasmid, R plasmid, and superinfection. How are R plasmids involved in the spread of drug resistance?

Antifungal Drugs

Treatment of fungal infections generally has been less successful than that of bacterial infections largely because eucaryotic fungal cells are much more similar to human cells than are bacteria. Many drugs that inhibit or kill fungi are therefore quite toxic for humans. In addition, most fungi have a detoxification system that modifies many antibiotics, probably by hydroxylation. As a result, the added antibiotics are fungistatic only as long as repeated application maintains high levels of unmodified antibiotic. Despite their relatively low therapeutic index, a few drugs are useful in treating many major fungal diseases. Effective antifungal agents frequently either extract membrane sterols or prevent their synthesis. Similarly, because animal cells do not have cell walls, the enzyme chitin synthase is the target for fungal-active antibiotics such as polyoxin D and nikkomycin.

Fungal infections are often subdivided into infections of superficial tissues or superficial mycoses and systemic mycoses. Treatment for these two types of disease is very different. Several drugs are used to treat superficial mycoses.

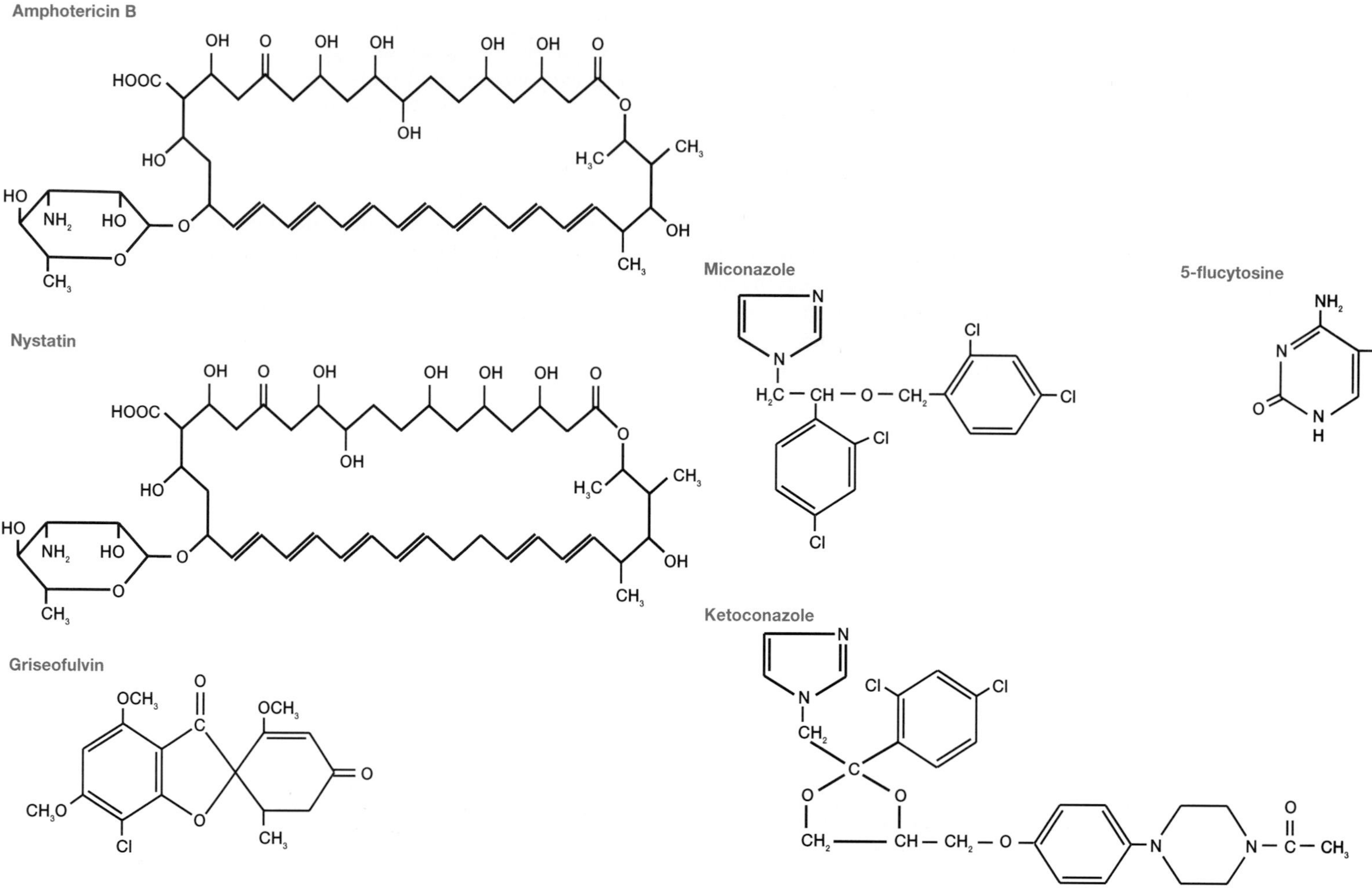

Figure 16.12 Antifungal Drugs. Six commonly used drugs are shown here.

Three drugs containing imidazole—miconazole, ketoconazole (figure 16.12), and clotrimazole—are broad-spectrum agents available as creams and solutions for the treatment of dermatophyte infections such as athlete's foot, and oral and vaginal candidiasis (*see chapter 39*). They are thought to disrupt fungal membrane permeability and inhibit sterol synthesis. Tolnaftate is used topically for the treatment of cutaneous infections, but is not as effective against infections of the skin and hair. **Nystatin** (figure 16.12), a polyene antibiotic from *Streptomyces,* is used to control *Candida* infections of the skin, vagina, or alimentary tract. **Griseofulvin** (figure 16.12), an antibiotic formed by *Penicillium,* is given orally to treat chronic dermatophyte infections. It is thought to disrupt the mitotic spindle and inhibit cell division; it may also inhibit protein and nucleic acid synthesis. Side effects of griseofulvin include headaches, gastrointestinal upset, and allergic reactions.

Superficial and systemic mycoses (pp. 782, 785–87).

Systemic infections are very difficult to control and can be fatal. Two drugs most commonly used against systemic mycoses are **amphotericin B** and 5-flucytosine (figure 16.12). Amphotericin B from *Streptomyces* sp. binds to the sterols in fungal membranes, disrupting membrane permeability and causing leakage of cell constituents. It is quite toxic and used only for serious, life-threatening infections. The synthetic oral antimycotic agent 5-flucytosine (5-fluorocytosine) is effective against most systemic fungi, although drug resistance often develops rapidly. The drug is converted to 5-fluorouracil by the fungi, incorporated into RNA in place of uracil, and disrupts RNA function. Its side effects include skin rashes, diarrhea, nausea, aplastic anemia, and liver damage.

1. Summarize the mechanism of action and the therapeutic use of the following antifungal drugs: miconazole, nystatin, griseofulvin, amphotericin B, and 5-flucytosine.

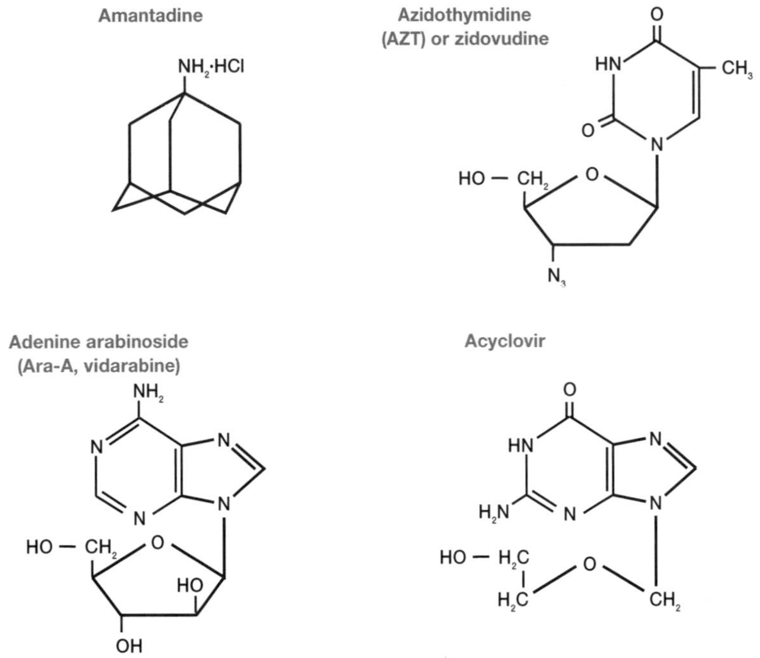

Amantadine

Azidothymidine (AZT) or zidovudine

Adenine arabinoside (Ara-A, vidarabine)

Acyclovir

Figure 16.13 **Representative Antiviral Drugs.**

TABLE 16.6 Chemotherapy of Some Viral Pathogens

Pathogen	Drugs of Choice
Cytomegalovirus	Ganciclovir, foscarnet
Hepatitis B or C virus	α_{2a}- or α_{2b}-interferons
Herpes simplex virus	Acyclovir, vidarabine, foscarnet, trifluridine, idoxuridine
Human immunodeficiency virus	Azidothymidine (zidovudine), didanosine
Influenza A virus	Amantadine
Respiratory syncytial virus	Ribavirin
Varicella-zoster virus	Acyclovir

Antiviral Drugs

For many years, the possibility of treating viral infections with drugs appeared remote because viruses enter host cells and make use of host cell enzymes and constituents to a large extent. A drug that would block virus reproduction also was thought to be toxic for the host. Inhibitors of virus-specific enzymes and life cycle processes have now been discovered, and several drugs are used therapeutically (table 16.6). Four are shown in figure 16.13.

Animal virus life cycles (pp. 384–91).

Most antiviral drugs disrupt either critical stages in the virus life cycle or the synthesis of virus-specific nucleic acids. **Amantadine** can be used to prevent influenza A infections. When given in time, it will reduce the incidence of influenza by 50 to 70% in an exposed population. Amantadine blocks the penetration and uncoating of influenza virus particles (*see chapter 19*). **Adenine arabinoside** or **vidarabine** disrupts the activity of DNA polymerase and several other enzymes involved in DNA and RNA synthesis and function. It is given intravenously or applied as an ointment to treat herpes infections. A third drug, **acyclovir,** is also used in the treatment of herpes infections. Upon phosphorylation, acyclovir resembles deoxy-GTP and inhibits the virus DNA polymerase. Unfortunately, acyclovir-resistant strains of herpes are already developing. Although these three drugs act only against herpes or influenza viruses, drugs active against other viruses are being developed. A notable example is **azidothymidine** (**AZT**) or **zidovudine,** which interferes with the reverse transcriptase enzyme, and therefore with the reproduction of retroviruses such as the human immunodeficiency virus. AZT is one of the few effective AIDS treatments (*see pp. 722–29*). Several newer drugs are now commercially available: didanosine, foscarnet, idoxuridine, and trifluridine.

Probably the most publicized antiviral agents are **interferons.** These small proteins, produced by the host, inhibit virus replication and may be clinically useful in the treatment of influenza, hepatitis, herpes, and colds. *The mechanism of interferon action is discussed in more detail in chapter 30.*

1. Describe two different ways in which antiviral drugs disrupt virus reproduction and give an example of each.

Summary

1. Chemotherapeutic agents are compounds that destroy pathogenic microorganisms or inhibit their growth and are used in the treatment of disease. Most are antibiotics: microbial products or their derivatives that can kill susceptible microorganisms or inhibit their growth.
2. The modern era of chemotherapy began with Paul Ehrlich's work on drugs against African sleeping sickness and syphilis. Other early pioneers were Gerhard Domagk, Alexander Fleming, Howard Florey, Ernst Chain, Norman Heatley, and Selman Waksman.
3. An effective chemotherapeutic agent must have selective toxicity. A drug with great selective toxicity has a high therapeutic index and usually disrupts a structure or process unique to the pathogen. It has fewer side effects.
4. Antibiotics can be classified in terms of the range of target microorganisms (narrow spectrum versus broad spectrum); their source (natural, semisynthetic, or synthetic); and their general effect (static versus cidal).
5. Antibiotic effectiveness can be estimated through the determination of the minimal inhibitory concentration and the minimal lethal concentration with dilution susceptibility tests. Disk diffusion tests like the Kirby-Bauer test are often used to estimate a pathogen's susceptibility to drugs quickly.
6. Chemotherapeutic agents can damage bacterial pathogens in several ways: inhibition of cell wall synthesis, protein synthesis, or nucleic acid synthesis; disruption of membrane structure; and inhibition of key enzymes.
7. A variety of factors can greatly influence the effectiveness of antimicrobial drugs during actual use.
8. Sulfonamides or sulfa drugs resemble p-aminobenzoic acid and competitively inhibit folic acid synthesis.
9. Quinolones are a family of bactericidal synthetic drugs that inhibit DNA gyrase and thus inhibit such processes as DNA replication.
10. Members of the penicillin family contain a β-lactam ring and disrupt bacterial cell wall synthesis, resulting in cell lysis. Some, like penicillin G, are usually administered by injection and are most effective against gram-positive bacteria. Others can be given orally (penicillin V), are broad spectrum (ampicillin, carbenicillin), or are penicillinase resistant (methicillin).
11. Cephalosporins are similar to penicillins, and are given to patients with penicillin allergies.
12. Tetracyclines are broad-spectrum antibiotics having a four-ring nucleus with attached groups. They bind to the small ribosomal subunit and inhibit protein synthesis.
13. Aminoglycoside antibiotics like streptomycin and gentamicin bind to the small ribosomal subunit, inhibit protein synthesis, and are bactericidal.
14. Erythromycin is a bacteriostatic macrolide antibiotic that binds to the large ribosomal subunit and inhibits protein synthesis.
15. Chloramphenicol is a broad-spectrum, bacteriostatic antibiotic that inhibits protein synthesis. It is quite toxic and used only for very serious infections.
16. Bacteria can become resistant to a drug by excluding it from the cell, enzymatically altering it, modifying the target enzyme or organelle to make it less drug sensitive, and so forth. The genes for drug resistance may be found either on the bacterial chromosome or on a plasmid called an R plasmid.
17. Chemotherapeutic agent misuse fosters the increase and spread of drug resistance, and may lead to superinfections.
18. Because fungi are more similar to human cells than bacteria, antifungal drugs generally have a lower therapeutic index than antibacterial agents and produce more side effects.
19. Superficial mycoses can be treated with miconazole, ketoconazole, clotrimazole, tolnaftate, nystatin, and griseofulvin. Amphotericin B and 5-flucytosine are used for systemic mycoses.
20. Antiviral drugs interfere with critical stages in the virus life cycle (amantadine) or inhibit the synthesis of virus-specific nucleic acids (zidovudine, adenine arabinoside, acyclovir). Interferon proteins inhibit virus replication and may be therapeutically useful in the future.

Key Terms

acyclovir *341*
adenine arabinoside or vidarabine *341*
amantadine *341*
aminoglycoside antibiotic *337*
amphotericin B *340*
antibiotic *326*
antimetabolites *330*
azidothymidine (AZT) *341*
broad-spectrum drugs *327*
cephalosporin *335*
chemotherapeutic agent *326*
chloramphenicol *337*
cidal *328*
dilution susceptibility tests *328*
erythromycin *337*
griseofulvin *340*
interferon *341*
Kirby-Bauer method *329*
macrolide antibiotic *337*
minimal inhibitory concentration (MIC) *328*
minimal lethal concentration (MLC) *328*
narrow-spectrum drugs *327*
nystatin *340*
parenteral route *331*
penicillinase *334*
penicillins *334*
plasmids *339*
quinolones *334*
R plasmids *339*
selective toxicity *327*
side effects *327*
static *328*
streptomycin *337*
structural analogues *332*
sulfonamide *332*
superinfection *339*
tetracycline *336*
therapeutic index *327*
zidovudine *341*

Questions for Thought and Review

1. What advantage might soil bacteria and fungi gain from the synthesis of antibiotics?
2. Why do penicillins and cephalosporins have a much higher therapeutic index than most other antibiotics? What are antimetabolites?
3. Would there be any advantage to administering a bacteriostatic agent along with penicillin? Any disadvantage?
4. Why do antifungal drugs have a much lower therapeutic index than do most antibacterial drugs? Why does one often have to apply antifungal medications frequently?
5. What advantages would there be from administering two chemotherapeutic agents simultaneously?
6. Why might it be desirable to prepare a variety of semisynthetic antibiotics?
7. Give several ways in which the development of antibiotic-resistant pathogens can be slowed or prevented.
8. Why is it so difficult to find or synthesize effective antiviral drugs?

Additional Reading

Abraham, E. P. 1981. The beta-lactam antibiotics. *Sci. Am.* 244(6):76–86.

Abramowicz, M., ed. 1990. The choice of antimicrobial drugs. *Medical Letter on Drugs and Therapeutics* 32(817):41–48.

Aharonowitz, Y., and Cohen, G. 1981. The microbiological production of pharmaceuticals. *Sci. Am.* 245(3):141–52.

Balows, A.; Hausler, W. J., Jr.; Herrmann, K. L.; Isenberg, H. D.; and Shadomy, H. J. 1991. *Manual of clinical microbiology,* 5th ed. Washington, D.C.: American Society for Microbiology.

Bean, B. 1992. Antiviral therapy: Current concepts and practices. *Clin. Microbiol. Rev.* 5(2): 146–82.

Bottcher, H. M. 1964. *Wonder drugs: A history of antibiotics.* Philadelphia: J. B. Lippincott.

Brooks, G. F.; Jawetz, E.; Melnick, J. L.; and Adelberg, E. A. 1991. *Medical microbiology,* 19th ed. Norwalk, Conn.: Appleton & Lange.

Clowes, R. C. 1973. The molecule of infectious drug resistance. *Sci. Am.* 228(4):19–27.

Dolin, R. 1985. Antiviral chemotherapy and chemoprophylaxis. *Science* 227:1296–1303.

Eliopoulos, G. M., and Eliopoulos, C. T. 1988. Antibiotic combinations: Should they be tested? *Clin. Microbiol. Rev.* 1(2):139–56.

Franklin, T. J., and Snow, G. A. 1981. *Biochemistry of antimicrobial action,* 3d ed. New York: John Wiley and Sons.

Fromtling, R. A. 1988. Overview of medically important antifungal azole derivatives. *Clin. Microbiol. Rev.* 1(2):187–217.

Gootz, T. D. 1990. Discovery and development of new antimicrobial agents. *Clin. Microbiol. Rev.* 3(1): 13–31.

Hardy, K. 1986. *Bacterial plasmids,* 2d ed. Washington, D.C.: American Society for Microbiology.

Hare, R. 1970. *The birth of penicillin.* Allen and Unwin.

Hirsch, M. S., and Kaplan, J. C. 1987. Antiviral therapy. *Sci. Am.* 256(4):76–85.

Hooper, D. C., and Wolfson, J. S. 1991. Fluoroquinolone antimicrobial agents. *N. Engl. J. Med.* 324(6): 384–92.

Jacoby, G. A., and Archer, G. L. 1991. New mechanisms of bacterial resistance to antimicrobial agents. *N. Engl. J. Med.* 324(9):601–9.

Klugman, K. P. 1990. Pneumococcal resistance to antibiotics. *Clin. Microbiol. Rev.* 3(2):171–96.

Lappe, M. 1982. *Germs that won't die: The medical consequences of the misuse of antibiotics.* Garden City, N.Y.: Doubleday.

Laughlin, C. A.; Black, R. J.; Feinberg, J.; Freeman, D. J.; Ramsey, J.; Ussery, M. A.; and Whitley, R. J. 1991. Resistance to antiviral drugs. *ASM News* 57(10):514–17.

Levy, S. B. 1988. Tetracycline resistance determinants are widespread. *ASM News* 54(8):418–21.

Nayler, J. H. C. 1991. Semi-synthetic approaches to novel penicillins. *Trends Biochem. Sci.* 16:234–37.

Parkman, P. D., and Hopps, H. E. 1988. Viral vaccines and antivirals: Current use and future prospects. *Ann. Rev. Public Health* 9:203–21.

Physician's desk reference. Oradell, N.J.: Medical Economics Books (published annually).

Sanders, C. C. 1991. A problem with antimicrobial susceptibility tests. *ASM News* 57(4):187–90.

Tomasz, A. 1979. The mechanism of the irreversible antimicrobial effects of penicillins: How the beta-lactam antibiotics kill and lyse bacteria. *Ann. Rev. Microbiol.* 33:113–37.

Walsh, T. J. 1988. Recent advances in the treatment of systemic fungal infections: A brief review. *ASM News* 54(5):240–43.

Wolfson, J. S., and Hooper, D. C. 1989. Fluoroquinolone antimicrobial agents. *Clin. Microbiol. Rev.* 2(4):378–424.

Part Five

The Viruses

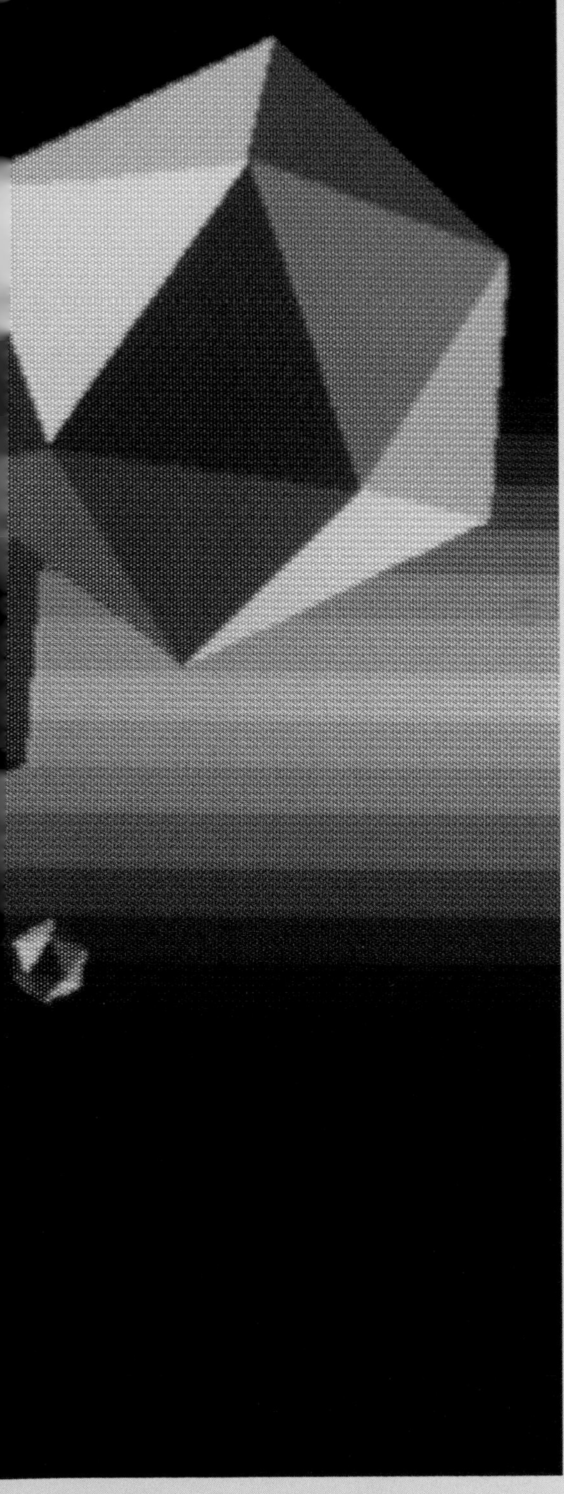

Chapter 17

The Viruses: Introduction and General Characteristics

Chapter 18

The Viruses: Bacteriophages

Chapter 19

The Viruses: Viruses of Eucaryotes

Polioviruses. The poliovirus is a picornavirus. Its icosahedral capsid contains 32 capsomers and lacks an envelope; the genetic material is single-stranded RNA. It is the causative agent of polio. The morphology and reproduction of this virus has been intensively studied.

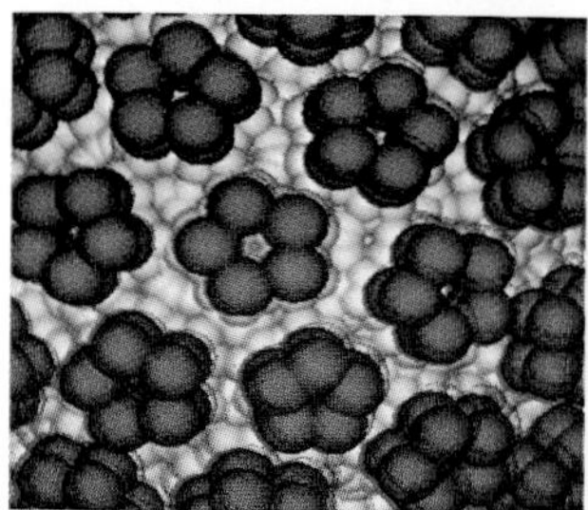

CHAPTER 17
The Viruses: Introduction and General Characteristics

Great fleas have little fleas
upon their backs to bite 'em
And little fleas have lesser
fleas, and so on ad
infinitum

—Augustus De Morgan

Outline

Concepts

1. Viruses are simple, acellular entities consisting of one or more molecules of either DNA or RNA enclosed in a coat of protein (and sometimes, in addition, substances such as lipids and carbohydrates). They can reproduce only within living cells and are obligate intracellular parasites.
2. Viruses are cultured by inoculating living hosts or cell cultures with a virion preparation. Purification depends mainly on their large size relative to cell components, high protein content, and great stability. Virus concentration may be determined from the virion count or from the number of infectious units.
3. All viruses have a nucleocapsid composed of a nucleic acid surrounded by a protein capsid that may be icosahedral, helical, or complex in structure. Capsids are constructed of protomers that self-assemble through noncovalent bonds. A membranous envelope often lies outside the nucleocapsid.
4. More variety is found in the genomes of viruses than in those of procaryotes and eucaryotes; they may be either single-stranded or double-stranded DNA or RNA. The nucleic acid strands can be linear, closed circle, or able to assume either shape.
5. Viruses are classified on the basis of their nucleic acid's characteristics, capsid symmetry, the presence or absence of an envelope, their host, the diseases caused by animal and plant viruses, and other properties.

The next three chapters focus on the viruses. These are infectious agents with fairly simple, acellular organization. They possess only one type of nucleic acid, either DNA or RNA, and cannot reproduce independently of living cells. Clearly, viruses are quite different from procaryotic and eucaryotic microorganisms, and are studied by **virologists.**.

Despite their simplicity in comparison with cellular organisms, viruses are extremely important and deserving of close attention. The study of viruses has contributed significantly to the discipline of molecular biology. Many human viral diseases are already known and more are discovered or arise every year, as demonstrated by the recent appearance of AIDS. The whole field of genetic engineering is based in large part upon discoveries in virology. Thus, it is easy to understand why **virology** (the study of viruses) is such a significant part of microbiology.

This chapter focuses on the broader aspects of virology: its development as a scientific discipline, the general properties and structure of viruses, the ways in which viruses are cultured and studied, and viral taxonomy. Chapter 18 is concerned with the bacteriophages, and chapter 19 is devoted to the viruses of eucaryotes.

Viruses have had enormous impact on humans and other organisms, yet very little was known about their nature until fairly recently. A brief history of their discovery and recognition as uniquely different infectious agents can help clarify their nature.

Early Development of Virology

Although the ancients did not understand the nature of their illnesses, they were acquainted with diseases, such as rabies, that are now known to be viral in origin. In fact, there is some evidence that the great epidemics of A.D. 165 to 180 and A.D. 251 to 266, which severely weakened the Roman Empire and aided its decline, may have been caused by measles and smallpox viruses. Smallpox had an equally profound impact on the New World. Hernán Cortés's conquest of the Aztec Empire in Mexico was made possible by an epidemic that ravaged Mexico City. The virus was probably brought to Mexico in 1520 by the relief expedition sent to join Cortés. Before the smallpox epidemic subsided, it had killed the Aztec King Cuitlahuac (the nephew and son-in-law of the slain emperor, Montezuma II) and possibly a third of the population. Since the Spaniards were not similarly afflicted, it appeared that God's wrath was reserved for the Native Americans, and this disaster was viewed as divine support for the Spanish conquest (see Box 17.1).

The first progress in preventing viral diseases came years before the discovery of viruses. Early in the eighteenth century, Lady Wortley Montagu, wife of the English ambassador to Turkey, observed that Turkish women inoculated their children against smallpox. The children came down with a mild case and subsequently were immune. Lady Montagu tried to educate the English public about the procedure, but without great success. Later in the century, an English country doctor, Edward Jenner, stimulated by a girl's claim that she could not catch smallpox because she had had cowpox, began inoculating humans with material from cowpox lesions. He published the results of 23 successful vaccinations in 1798. Although Jenner did not understand the nature of smallpox, he did manage to successfully protect his patients from the dread disease through exposure to the cowpox virus.

Until well into the nineteenth century, harmful agents were often grouped together and sometimes called viruses (Latin *virus*, poison or venom). Even Louis Pasteur used the term virus for any living infectious disease agent. The development in 1884 of the porcelain bacterial filter by Charles Chamberland, one of Pasteur's collaborators and inventor of the autoclave, made possible the discovery of what are now called viruses. Tobacco mosaic disease was the first to be studied with Chamberland's filter. In 1892, Dimitri Ivanowski published studies showing that leaf extracts from infected plants would induce tobacco mosaic disease even after filtration to remove bacteria. He attributed this to the presence of a toxin. Martinus W. Beijerinck, working independently of Ivanowski, published the results of extensive studies on tobacco mosaic disease in 1898 and 1900. Because the filtered sap of diseased plants was still infectious, he proposed that the disease was caused by an entity different from bacteria, a filterable virus. He observed that the virus would multiply only in living plant cells, but could survive for long periods in a dried state. At the same time, Friedrich Loeffler and Paul Frosch in Germany found that the hoof-and-mouth disease of cattle was also caused by a filterable virus rather than by a toxin. In 1900, Walter Reed began his study of the yellow fever disease whose incidence had been increasing in Cuba. Reed showed that this human disease was due to a filterable virus that was transmitted by mosquitoes. Mosquito control shortly reduced the severity of the yellow fever problem. Thus, by the beginning of this century, it had been established that filterable viruses were different from bacteria and could cause diseases in plants, livestock, and humans.

Shortly after the turn of the century, V. Ellermann and O. Bang in Copenhagen reported that leukemia could be transmitted between chickens by cell-free filtrates and was probably caused by a virus. Three years later in 1911, Peyton Rous from the Rockefeller Institute in New York City reported that a virus was responsible for a malignant muscle tumor in chickens. These studies established that at least some malignancies were caused by viruses.

It was soon discovered that bacteria themselves also could be attacked by viruses. The first published observation suggesting that this might be the case was made in 1915 by Frederick W. Twort. Twort isolated bacterial viruses that could attack and destroy micrococci and intestinal bacilli. Although he speculated that his preparations might contain viruses,

Box 17.1

Disease and the Early Colonization of America

Although the case is somewhat speculative, there is considerable evidence that disease, and particularly smallpox, played a major role in reducing Indian resistance to the European colonization of North America. It has been estimated that Indian populations in Mexico declined about 90% within 100 years of initial contact with the Spanish. Smallpox and other diseases were a major factor in this decline, and there is no reason to suppose that North America was any different. As many as 10 to 12 million Indians may have lived north of the Rio Grande before contact with Europeans. In New England alone, there may have been over 72,000 in A.D. 1600; yet only around 8,600 remained in New England by A.D. 1674, and the decline continued in subsequent years.

Such an incredible catastrophe can be accounted for by consideration of the situation at the time of European contact with the Native Americans. The Europeans, having already suffered major epidemics in the preceding centuries, were relatively immune to the diseases they carried. On the other hand, the Native Americans had never been exposed to diseases like smallpox and were decimated by epidemics. In the sixteenth century, before any permanent English colonies had been established, many contacts were made by missionaries and explorers who undoubtedly brought disease with them and infected the natives. Indeed, the English noted at the end of the century that Indian populations had declined greatly, but attributed it to armed conflict rather than to disease.

Establishment of colonies simply provided further opportunities for infection and outbreak of epidemics. For example, the Huron Indians decreased from a minimum of 32,000 people to 10,000 in 10 years. Between the time of initial English colonization and 1674, the Narraganset Indians declined from around 5,000 warriors to 1,000, and the Massachusetts Indians, from 3,000 to 300. Similar stories can be seen in other parts of the colonies. Some colonists interpreted these plagues as a sign of God's punishment of Indian resistance: the "Lord put an end to this quarrel by smiting them with smallpox. . . . Thus did the Lord allay their quarrelsome spirit and make room for the following part of his army."

It seems clear that epidemics of European diseases like smallpox laid waste Native American populations and prepared the way for colonization of the North American continent. Many American cities—for example, Boston, Philadelphia, and Plymouth—grew upon sites of previous Indian villages.

Twort did not follow up on these observations. It remained for Felix d'Herelle to establish decisively the existence of bacterial viruses. D'Herelle independently isolated bacterial viruses from patients with dysentery, probably caused by *Shigella dysenteriae*. He noted that when a virus suspension was spread on a layer of bacteria growing on agar, clear circular areas containing viruses and lysed cells developed. A count of these clear zones allowed d'Herelle to estimate the number of viruses present (plaque assay, p. 349). D'Herelle demonstrated that these viruses could reproduce only in live bacteria; therefore, he named them bacteriophages because they could eat holes in bacterial "lawns."

The chemical nature of viruses was established when Wendell M. Stanley announced in 1935 that he had crystallized the tobacco mosaic virus (TMV) and found it to be largely or completely protein. A short time later, Frederick C. Bawden and Norman W. Pirie managed to separate the TMV virus particles into protein and nucleic acid. Thus, by the late 1930s it was becoming clear that viruses were complexes of nucleic acids and proteins able to reproduce only in living cells.

1. Describe the major technical advances and discoveries important in the early development of virology.
2. Give the contribution to virology made by each scientist mentioned in this section.

General Properties of Viruses

Viruses are a unique group of infectious agents whose distinctiveness resides in their simple, acellular organization and pattern of reproduction. A complete virus particle or **virion** consists of one or more molecules of DNA or RNA enclosed in a coat of protein, and sometimes also in other layers. These additional layers may be very complex and contain carbohydrates, lipids, and additional proteins. Viruses can exist in two phases: extracellular and intracellular. Virions, the extracellular phase, possess few if any enzymes and cannot reproduce independently of living cells. In the intracellular phase, viruses exist primarily as replicating nucleic acids that induce host metabolism to synthesize virion components; eventually complete virus particles or virions are released.

In summary, viruses differ from living cells in at least three ways: (1) their simple, acellular organization, (2) the absence of both DNA and RNA in the same virion, and (3) their inability to reproduce independently of cells and carry out cell division as procaryotes and eucaryotes do. Although bacteria such as chlamydia and rickettsia (*see chapter 21*) are obligate intracellular parasites like viruses, they do not meet the first two criteria.

The Cultivation of Viruses

Because they are unable to reproduce independently of living cells, viruses cannot be cultured in the same way as bacteria

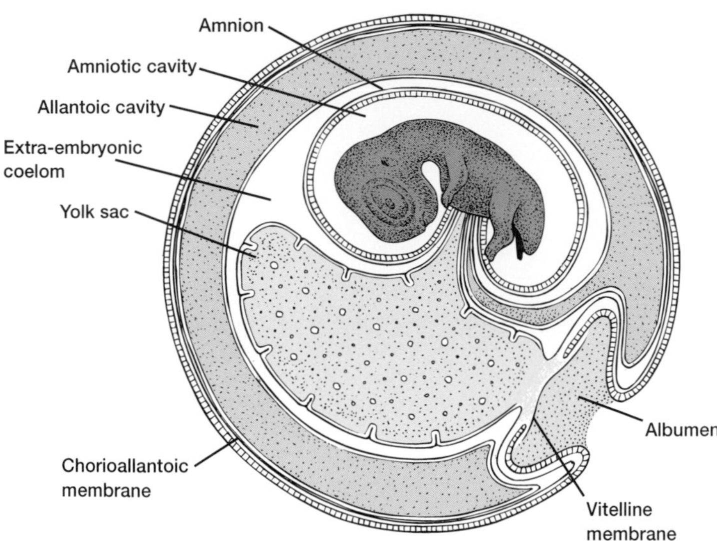

Figure 17.1 **Virus Cultivation Using an Embryonated Egg.** Two sites that are often used to grow animal viruses are the chorioallantoic membrane and the allantoic cavity. The diagram shows a 9-day chicken embryo.

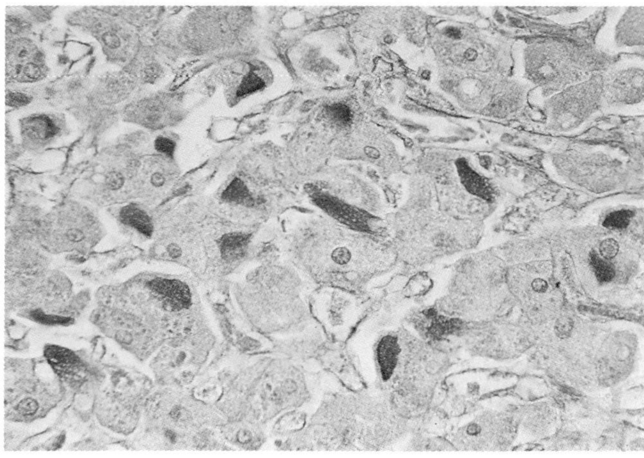

(a)

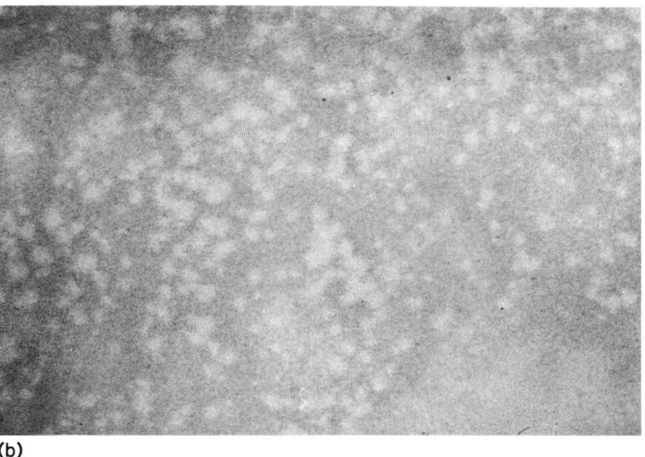

(b)

Figure 17.2 **Virus Plaques.** (*a*) Plaques of viral hepatitis in human liver cells. (*b*) Poliovirus plaques in a monkey kidney cell culture.

and eucaryotic microorganisms. For many years, researchers have cultivated animal viruses by inoculating suitable host animals or embryonated eggs—fertilized chicken eggs incubated about six to eight days after laying (figure 17.1). To prepare the egg for virus cultivation, the shell surface is first disinfected with iodine and penetrated with a small sterile drill. After inoculation, the drill hole is sealed with gelatin and the egg incubated. Viruses may be able to reproduce only in certain parts of the embryo; consequently, they must be injected into the proper region. For example, the myxoma virus grows well on the chorioallantoic membrane, whereas the mumps virus prefers the allantoic cavity. The infection may produce a local tissue lesion known as a pock, whose appearance is often characteristic of the virus.

More recently, animal viruses have been grown in tissue (cell) culture on monolayers of animal cells. This technique is made possible by the development of growth media for animal cells and by the advent of antibiotics that can prevent bacterial and fungal contamination. A layer of animal cells in a specially prepared petri dish is covered with a virus inoculum, and the viruses are allowed time to settle and attach to the cells. The cells are then covered with a thin layer of agar to limit virion spread so that only adjacent cells are infected by newly produced virions. As a result, localized areas of cellular destruction and lysis called **plaques** are often formed (figure 17.2) and may be detected if stained with dyes, such as neutral red or trypan blue, that can distinguish living from dead cells. Viral growth does not always result in the lysis of cells to form a plaque. Animal viruses, in particular, can cause microscopic or macroscopic degenerative changes or abnormalities in host cells and in tissues called **cytopathic effects** (figure 17.3). Cytopathic effects may be lethal, but plaque formation from cell lysis does not always occur.

Bacterial viruses or **bacteriophages** (**phages** for short) are cultivated in either broth or agar cultures of young, actively growing bacterial cells. So many host cells are destroyed that turbid bacterial cultures may clear rapidly because of cell lysis. Agar cultures are prepared by mixing the bacteriophage sample with cool, liquid agar and a suitable bacterial culture. The mixture is quickly poured into a petri dish containing a bottom layer of sterile agar. After hardening, bacteria in the layer of top agar grow and reproduce, forming a continuous, opaque layer or "lawn." Wherever a virion comes to rest in the top agar, the virus infects an adjacent cell and reproduces. Eventually, bacterial lysis generates a plaque or clearing in the lawn (figure 17.4). As can be seen in figure 17.4, plaque appearance is often characteristic of the phage being cultivated.

Plant viruses are cultivated in a variety of ways. Plant tissue cultures, cultures of separated cells, or cultures of protoplasts (*see chapter 3*) may be used. Viruses also can be grown in whole plants. Leaves are mechanically inoculated when rubbed with a mixture of viruses and an abrasive such as carborundum. When the cell walls are broken by the abrasive, the viruses directly contact the plasma membrane and infect

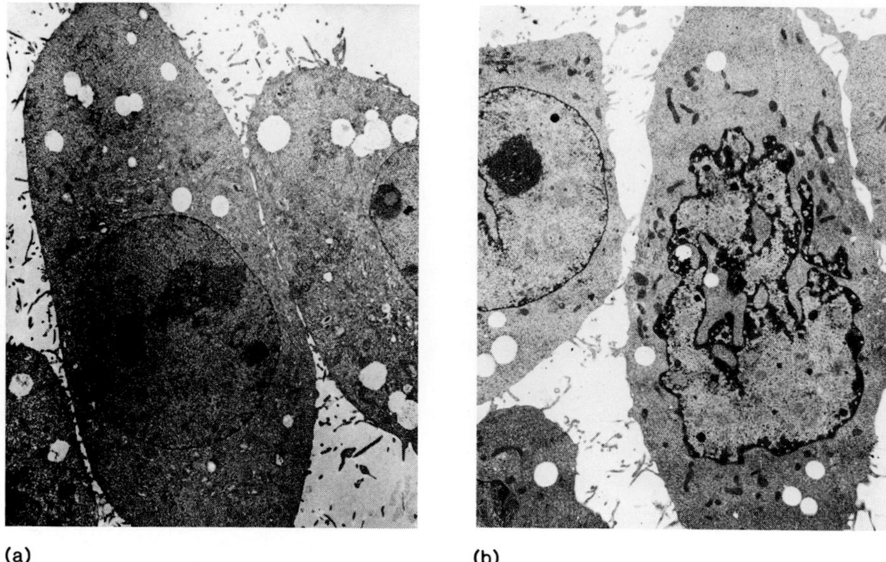

Figure 17.3 **The Cytopathic Effects of Viruses.** (*a*) Normal mammalian cells in tissue culture. (*b*) Appearance of tissue culture cells 18 hours after infection with adenovirus. TEM photomicrographs (×11,000).

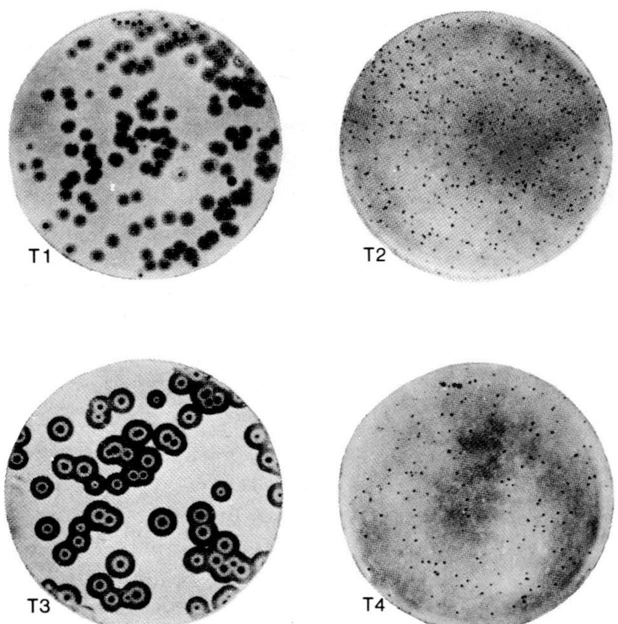

Figure 17.4 **Phage Plaques.** Plaques produced on a lawn of *E. coli* by some of the T coliphages. Note the large differences in plaque appearance. The photographs are about 1/3 full size.

the exposed host cells. (The role of the abrasive is frequently filled by insects that suck or crush plant leaves and thus transmit viruses.) A localized **necrotic lesion** often develops due to the rapid death of cells in the infected area (figure 17.5). Even when lesions do not arise, the infected plant may show symptoms such as changes in pigmentation or leaf shape. Some plant viruses can be transmitted only if a diseased part is grafted onto a healthy plant.

1. What is a virus particle or virion, and how is it different from living organisms?
2. Discuss the ways in which viruses may be cultivated. Define the terms pock, plaque, cytopathic effect, bacteriophage, and necrotic lesion.

Virus Purification and Assays

Virologists must be able to purify viruses and accurately determine their concentrations in order to study virus structure, reproduction, and other aspects of their biology. These methods are so important that the growth of virology as a modern discipline depended on their development.

Virus Purification

Purification makes use of several virus properties. Virions are very large relative to proteins, are often more stable than normal cell components, and have surface proteins. Because of these characteristics, many techniques useful for the isolation of proteins and organelles can be employed in virus isolation. Four of the most widely used approaches are (1) differential and density gradient centrifugation, (2) precipitation of viruses, (3) denaturation of contaminants, and (4) enzymatic digestion of cell constituents.

1. Host cells in later stages of infection that contain mature virions are used as the source of material. Infected cells are first disrupted in a buffer to produce an aqueous suspension or homogenate consisting of cell components and viruses. Viruses can then be isolated by **differential centrifugation,** the centrifugation of a suspension at various speeds to separate particles of

(a)

(b)

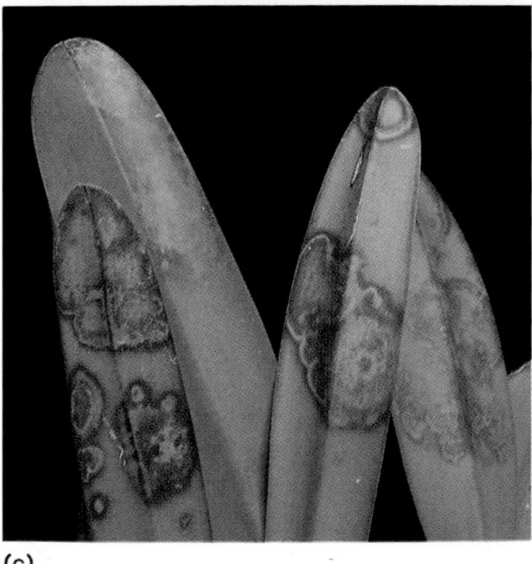
(c)

Figure 17.5 Necrotic Lesions on Plant Leaves. (*a*) Tobacco mosaic virus on *Nicotiana glutinosa*. (*b*) Bean common mosaic disease. Infected leaves show yellow mottling and distortion. (*c*) Tobacco mosaic virus infection of an orchid showing leaf color changes.

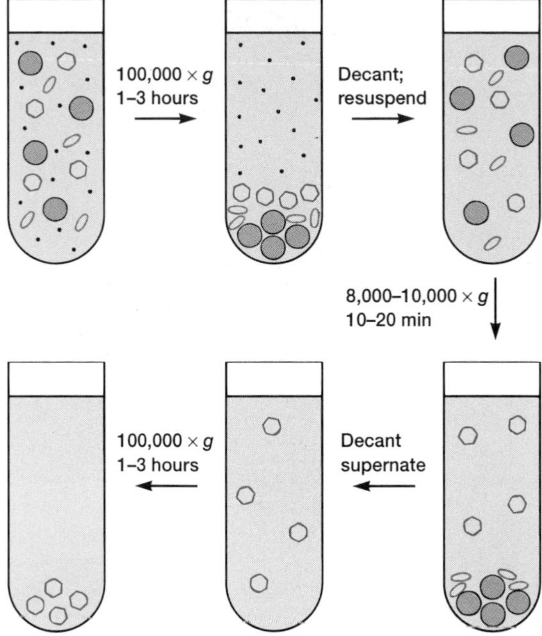

Figure 17.6 The Use of Differential Centrifugation to Purify a Virus. At the beginning, the centrifuge tube contains homogenate and icosahedral viruses (in red). First, the viruses and heavier cell organelles are removed from smaller molecules. After resuspension, the mixture is centrifuged just fast enough to sediment cell organelles while leaving the smaller virus particles in suspension; the purified viruses are then collected. This process can be repeated several times to further purify the virions.

different sizes (figure 17.6). Usually, the homogenate is first centrifuged at high speed to sediment viruses and other large cellular particles, and the supernatant, which contains the homogenate's soluble molecules, is discarded. The pellet is next resuspended and centrifuged at a low speed to remove substances heavier than viruses. Higher speed centrifugation then sediments the viruses. This process may be repeated to purify the virus particles further.

Viruses also can be purified based on their size and density by use of **gradient centrifugation** (figure 17.7). A sucrose solution is poured into a centrifuge tube so that its concentration smoothly and linearly increases between the top and the bottom of the tube. The virus preparation, often after purification by differential centrifugation, is layered on top of the gradient and centrifuged. As shown in figure 17.7*a*, the particles settle under centrifugal force until they come to rest at the level where the gradient's density equals theirs (isopycnic gradient centrifugation). Viruses can be separated from other particles only slightly different in density. Gradients also can separate viruses based on differences in their sedimentation rate (rate zonal gradient centrifugation). When this is done, particles are separated on the basis of both size and density; usually, the largest virus will move most rapidly down the gradient. Figure 17.7*b* shows that viruses differ

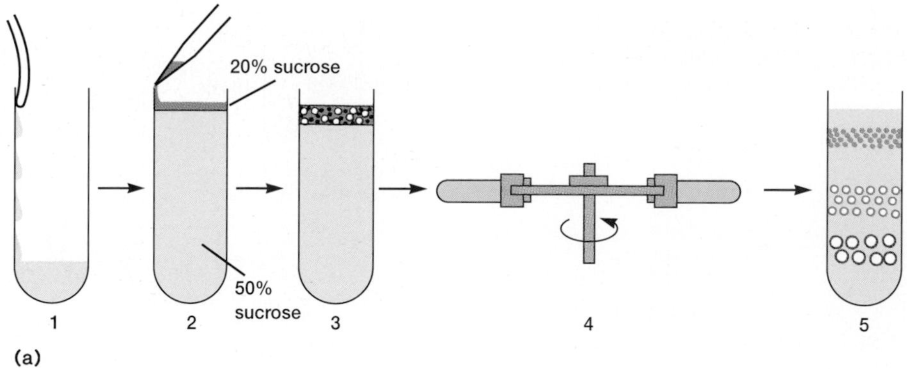

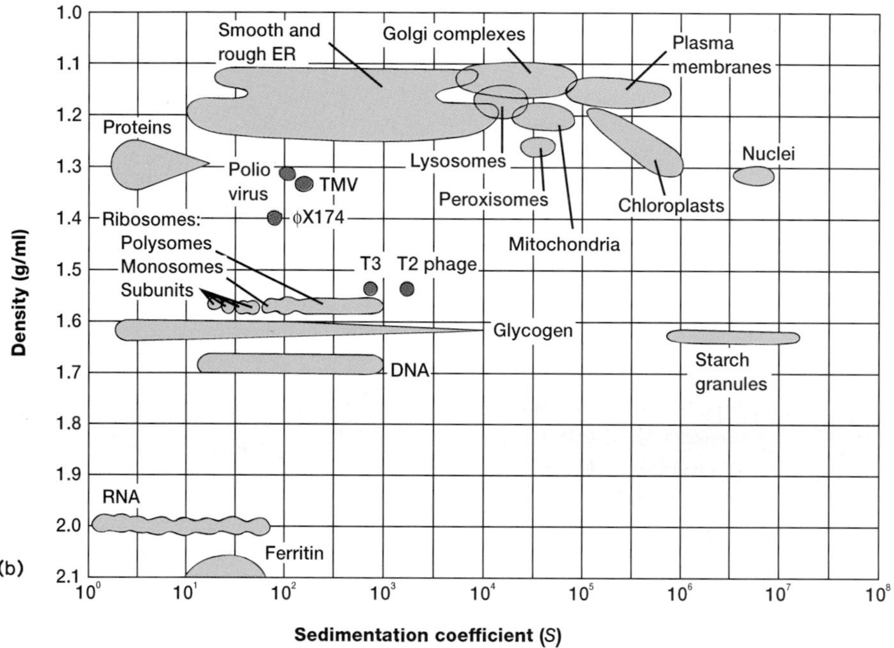

Figure 17.7 Gradient Centrifugation. (*a*) A linear sucrose gradient is prepared, *1*, and the particle mixture is layered on top, *2* and *3*. Centrifugation, *4*, separates the particles on the basis of their density and sedimentation coefficient, 5. In isopycnic gradient centrifugation, the bottom of the gradient is denser than any particle, and each particle comes to rest at a point in the gradient equal to its density. Rate zonal centrifugation separates particles based on their sedimentation coefficient, a function of both size and density, because the bottom of the gradient is less dense than the densest particles and centrifugation is carried out for a shorter time so that particles do not come to rest. The largest, most dense particles travel fastest. (*b*) The densities and sedimentation coefficients of representative viruses (shown in color) and other biological substances. *From P. Sheeler and D. E. Bianchi,* Cell Biology: Structure, Biochemistry, and Function, *2d ed. Copyright © 1980 John Wiley & Sons, Inc., New York, NY. Reprinted by permission of John Wiley & Sons, Inc.*

from one another and cell components with respect to either density (grams per milliliter) or sedimentation coefficient(s). Thus, these two types of gradient centrifugation are very effective in virus purification.

2. Viruses, like many proteins, can be purified through precipitation with concentrated ammonium sulfate. Initially, sufficient ammonium sulfate is added to raise its concentration to a level just below that which will precipitate the virus. After any precipitated contaminants are removed, more ammonium sulfate is added and the precipitated viruses are collected by centrifugation. Viruses sensitive to ammonium sulfate are often purified by precipitation with polyethylene glycol.
3. Viruses frequently are less easily denatured than many normal cell constituents. Contaminants may be denatured and precipitated with heat or a change in pH to purify viruses. Because some viruses also tolerate treatment with organic solvents like butanol and chloroform, solvent treatment can be used to both denature protein contaminants and extract any lipids in the preparation. The solvent is thoroughly mixed with

the virus preparation, then allowed to stand and separate into organic and aqueous layers. The unaltered virus remains suspended in the aqueous phase while lipids dissolve in the organic phase. Substances denatured by organic solvents collect at the interface between the aqueous and organic phases.

Denaturation of enzymes (p. 142).

4. Cellular proteins and nucleic acids can be removed from many virus preparations through enzymatic degradation because viruses are usually more resistant to attack by nucleases and proteases than are free nucleic acids and proteins. For example, ribonuclease and trypsin often degrade cellular ribonucleic acids and proteins while leaving virions unaltered.

Virus Assays

The quantity of viruses in a sample can be determined either by a count of particle number or by measurement of the infectious unit concentration. Although most normal virions are probably potentially infective, many will not infect host cells because they do not contact the proper surface site. Thus, the total particle count may be from two to one million times the infectious unit number depending on the nature of the virion and the experimental conditions. Despite this, both approaches are of value.

Virus particles can be counted directly with the electron microscope. In one procedure, the virus sample is mixed with a known concentration of small latex beads and sprayed on a coated specimen grid. The beads and virions are counted; the virus concentration is calculated from these counts and from the bead concentration (figure 17.8). This technique often works well with concentrated preparations of viruses of known morphology. Viruses can be concentrated by centrifugation before counting if the preparation is too dilute. However, if the beads and viruses are not evenly distributed (as sometimes happens), the final count will be inaccurate.

The most popular indirect method of counting virus particles is the **hemagglutination assay.** Many viruses can bind to the surface of red blood cells (*see figure 33.10*). If the ratio of viruses to cells is large enough, virus particles will join the red blood cells together, forming a network that settles out of suspension or agglutinates. In practice, red blood cells are mixed with a series of virus preparation dilutions and each mixture is examined. The hemagglutination titer is the highest dilution of virus (or the reciprocal of the dilution) that still causes hemagglutination. This assay is an accurate, rapid method for determining the relative quantity of viruses such as the influenza virus. If the actual number of viruses needed to cause hemagglutination is determined by another technique, the assay can be used to ascertain the number of virus particles present in a sample.

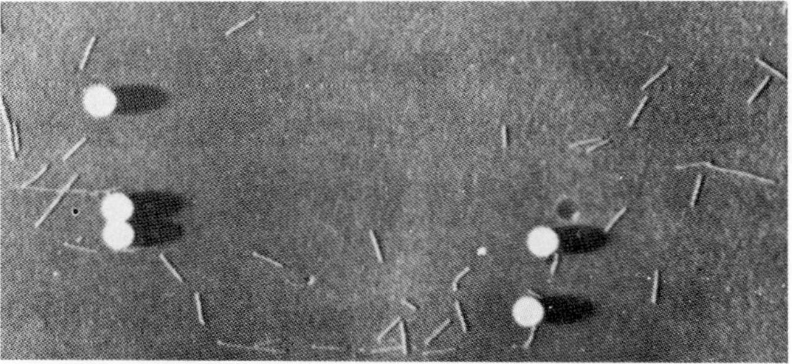

Figure 17.8 **Tobacco Mosaic Virus.** A tobacco mosaic virus preparation viewed in the transmission electron microscope. Latex beads 264 μm in diameter (white spheres) have been added.

A variety of assays analyze virus numbers in terms of infectivity, and many of these are based upon the same techniques used for virus cultivation. For example, in the **plaque assay** several dilutions of bacterial or animal viruses are plated out with appropriate host cells. When the number of viruses plated out are much fewer than the number of host cells available for infection and when the viruses are distributed evenly, each plaque in a layer of bacterial or animal cells is assumed to have arisen from the reproduction of a single virus particle. Therefore, a count of the plaques produced at a particular dilution will give the number of infectious virions or **plaque-forming units** (**PFU**), and the concentration of infectious units in the original sample can be easily calculated. Suppose that 0.10 ml of a 10^{-6} dilution of the virus preparation yields 75 plaques. The original concentration of plaque-forming units is

$$\text{PFU/ml} = (75\ \text{PFU}/0.10\ \text{ml})(10^{6}) = 7.5 \times 10^{8}.$$

Viruses producing different plaque morphology types on the same plate may be counted separately. Although the number of PFU does not equal the number of virus particles, their ratios are proportional: a preparation with twice as many viruses will have twice the plaque-forming units.

The same approach employed in the plaque assay may be used with embryos and plants. Chicken embryos can be inoculated with a diluted preparation or plant leaves rubbed with a mixture of diluted virus and abrasive. The number of pocks on embryonic membranes or necrotic lesions on leaves is multiplied by the dilution factor and divided by the inoculum volume to obtain the concentration of infectious units.

When biological effects are not readily quantified in these ways, the amount of virus required to cause disease or death can be determined by the endpoint method. Organisms or cell cultures are inoculated with serial dilutions of a virus suspension. The results are used to find the endpoint dilution at which 50% of the host cells or organisms are damaged or destroyed (figure 17.9). The **lethal dose** (**LD_{50}**) or **infectious dose** (**ID_{50}**) is the dilution that contains a lethal or infectious dose large enough to destroy or damage 50% of the host cells or organisms.

1. Give the four major approaches by which viruses may be purified, and describe how each works. Distinguish between differential and density gradient centrifugation in terms of how they are carried out.
2. How can one find the virus concentration, both directly and indirectly, by particle counts and measurement of infectious unit concentration? Define plaque-forming units, lethal dose, and infectious dose.

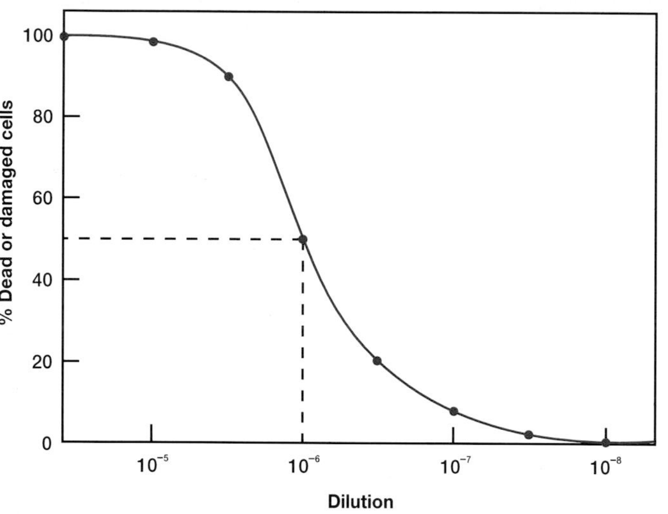

Figure 17.9 **A Hypothetical Dose-response Curve.** The LD_{50} is indicated by the dashed line.

The Structure of Viruses

Virus morphology has been intensely studied over the past decades because of the importance of viruses and the realization that virus structure was simple enough to be understood. Progress has come from the use of several different techniques: electron microscopy, X-ray diffraction, biochemical analysis, and immunology. Although our knowledge is incomplete due to the large number of different viruses, the general nature of virus structure is becoming clear.

Virion Size

Virions range in size from about 10 to 300 or 400 nm in diameter (figure 17.10). The smallest viruses are little larger than ribosomes, whereas the poxviruses, like vaccinia, are about the same size as the smallest bacteria and can be seen in the light microscope. Most viruses, however, are too small to be visible in the light microscope and must be viewed with the scanning and transmission electron microscopes (*see chapter 2*).

General Structural Properties

All virions, even if they possess other constituents, are constructed around a **nucleocapsid** core (indeed, some viruses consist only of a nucleocapsid). The nucleocapsid is composed of a nucleic acid, either DNA or RNA, held within a protein coat called the **capsid,** which protects viral genetic material and aids in its transfer between host cells.

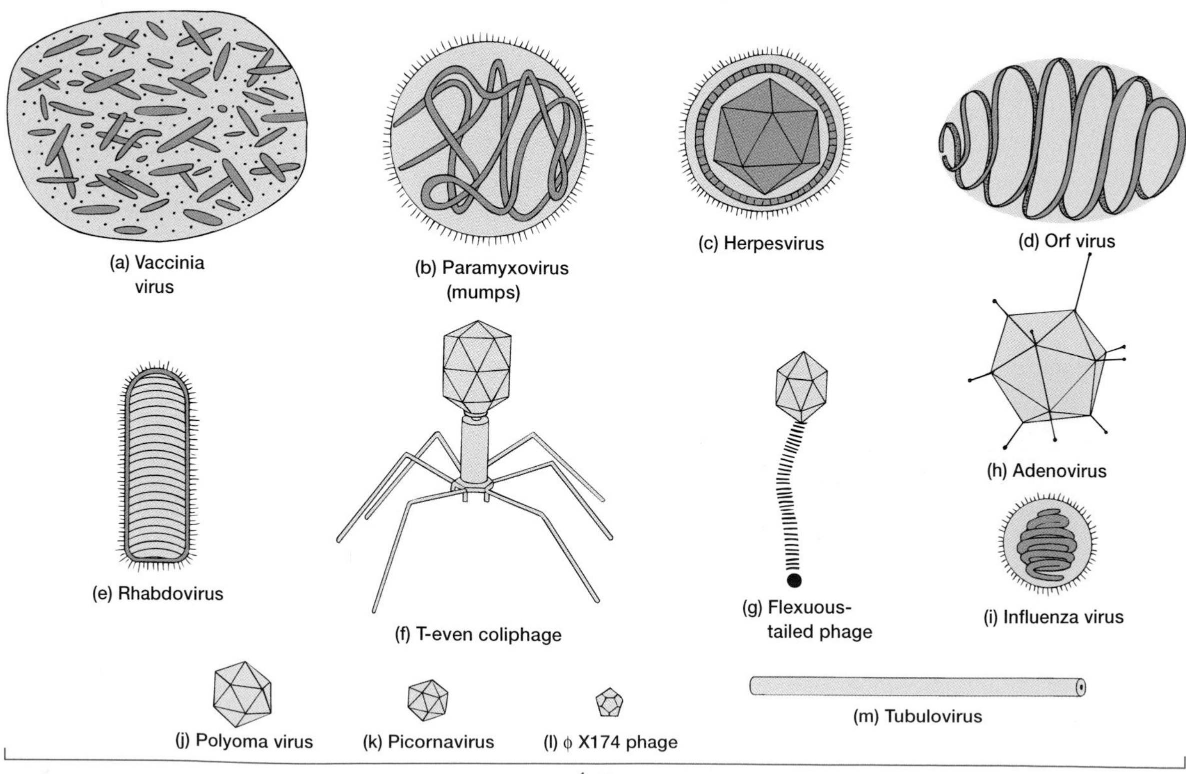

Figure 17.10 **The Size and Morphology of Selected Viruses.** The viruses are drawn to scale. A 1 μm line is provided at the bottom of the figure.

There are four general morphological types of capsids and virion structure.

1. Some capsids are **icosahedral** in shape. An icosahedron is a regular polyhedron with 20 equilateral triangular faces and 12 vertices (figure 17.10*h, j–l*). These capsids appear spherical when viewed at low power in the electron microscope.
2. Other capsids are **helical** and shaped like hollow protein cylinders, which may be either rigid or flexible (figure 17.10*m*).
3. Many viruses have an **envelope,** an outer membranous layer surrounding the nucleocapsid. Enveloped viruses have a roughly spherical but somewhat variable shape even though their nucleocapsid can be either icosahedral or helical (figure 17.10*b, c, i*).
4. **Complex viruses** have capsid symmetry that is neither purely icosahedral nor helical (figure 17.10*a, d, f, g*). They may possess tails and other structures (e.g., many bacteriophages) or have complex, multilayered walls surrounding the nucleic acid (e.g., poxviruses such as vaccinia).

Both helical and icosahedral capsids are large macromolecular structures constructed from many copies of one or a few types of protein subunits or **protomers.** Probably the most important advantage of this design strategy is that the information stored in viral genetic material is used with maximum efficiency. For example, the tobacco mosaic virus (TMV) capsid contains a single type of small subunit possessing 158 amino acids. Only about 474 nucleotides out of 6,000 in the virus RNA are required to code for coat protein amino acids. Unless the same protein is used many times in capsid construction, a large nucleic acid, such as the TMV RNA, cannot be enclosed in a protein coat without using much or all of the available genetic material to code for capsid proteins. If the TMV capsid were composed of six different protomers of the same size as the TMV subunit, about 2,900 of the 6,000 nucleotides would be required for its construction, and much less genetic material would be available for other purposes.

The genetic code and translation (pp. 239–40).

Once formed and exposed to the proper conditions, protomers usually interact specifically with each other and spontaneously associate to form the capsid. Because the capsid is constructed without any outside aid, the process is called self-assembly (p. 58). Some more complex viruses possess genes for special factors that are not incorporated into the virion but are required for its assembly.

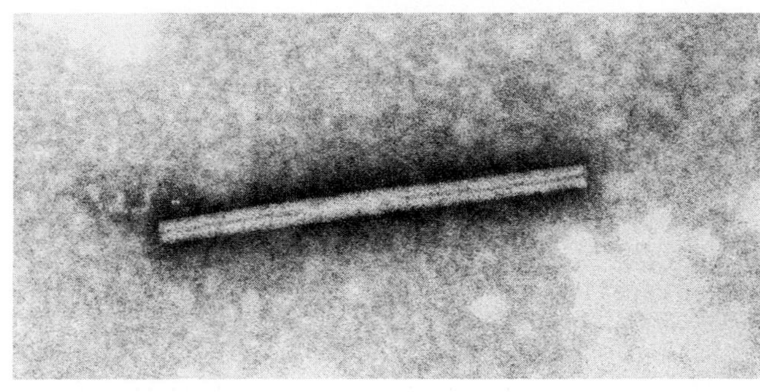

(a)

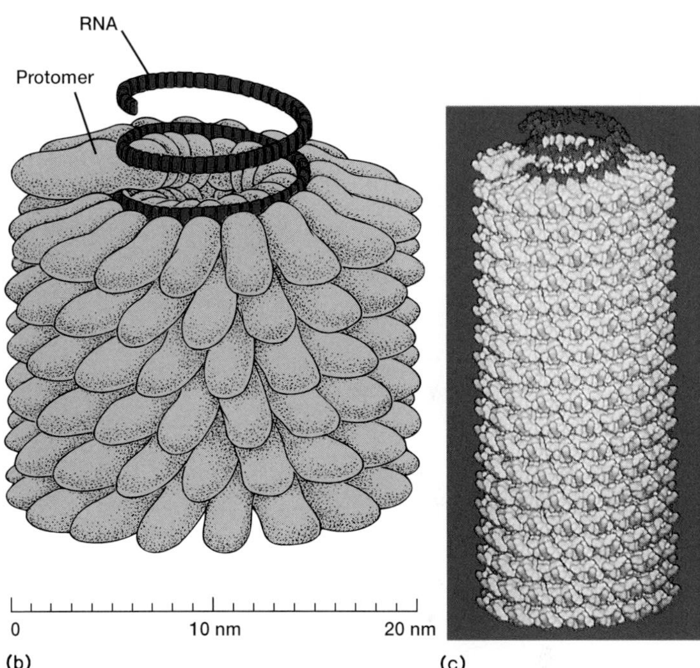

Figure 17.11 **Tobacco Mosaic Virus Structure.** (*a*) An electron migrograph of the negatively stained helical capsid (×400,000). (*b*) Illustration of TMV structure. Note that the nucleocapsid is composed of a helical array of protomers with the RNA spiraling on the inside. (*c*) A model of TMV.

Helical Capsids

Helical capsids are shaped much like hollow tubes with protein walls. The tobacco mosaic virus provides a well-studied example of helical capsid structure (figure 17.11). A single type of protomer associates together in a helical or spiral arrangement to produce a long, rigid tube, 15 to 18 nm in diameter by 300 nm long. The RNA genetic material is wound in a spiral and positioned toward the inside of the capsid where it lies within a groove formed by the protein subunits. Not all helical capsids are as rigid as the TMV capsid. Influenza virus RNAs are enclosed in thin, flexible helical capsids folded within an envelope (figures 17.10*i* and 17.12*a, b*).

The size of a helical capsid is influenced by both its protomers and the nucleic acid enclosed within the capsid. The diameter of the capsid is a function of the size, shape, and interactions of the protomers. The nucleic acid determines helical capsid length because the capsid does not seem to extend much beyond the end of the DNA or RNA.

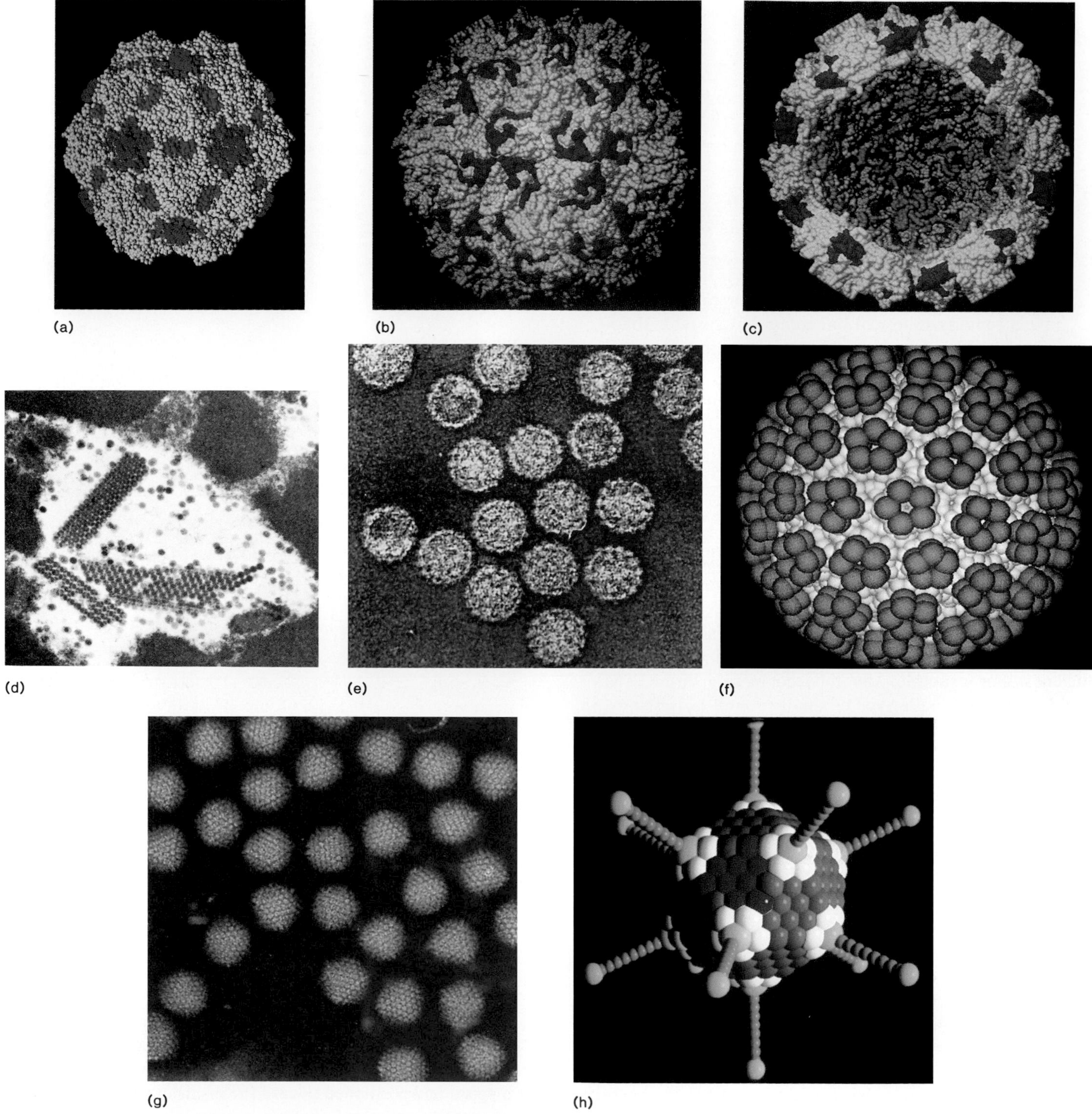

Figure 17.12 Examples of Icosahedral Capsids. (*a*) Canine parvovirus model, 12 capsomers, with the four parts of each capsid polypeptide given different colors. (*b*) and (*c*) Poliovirus model, 32 capsomers, with the four capsid proteins in different colors. The capsid surface is depicted in (*b*) and a cross section, in (*c*). (*d*) Clusters of the human papilloma virus, 72 capsomers (×80,000). (*e*) Simian virus 40 (SV40), 72 capsomers (×340,000). (*f*) Computer-simulated image of the polyomavirus, 72 capsomers, that causes a rare demyelinating disease of the central nervous system. (*g*) Adenovirus, 252 capsomer (×171,000). (*h*) Computer-simulated model of adenovirus.

Icosahedral Capsids

When icosahedral viruses are negatively stained and viewed in the transmission electron microscope, a complex capsid structure is revealed (figure 17.12). The capsids are constructed from ring- or knob-shaped units called **capsomers,** each usually made of five or six protomers. **Pentamers (pentons)** have five subunits; **hexamers (hexons)** possess six. Pentamers are at the vertices of the icosahedron, whereas hexamers form its edges and triangular faces (figure 17.13). The icosahedron in figure 17.13 is constructed of 42 capsomers; larger icosahedra are made if more hexamers are used to form the edges and faces (adenoviruses have a capsid with 252 capsomers as shown in figure 17.12*g, h*). In many plant and bacterial RNA viruses, both the pentamers and hexamers of a capsid are constructed with only one type of subunit, whereas adenovirus pentamers are composed of different proteins than are adenovirus hexamers.

Transmission electron microscopy and negative staining (pp. 32–34).

The study of virus capsid structure can have practical relevance. For example, the complete capsid structure of the rhinovirus, one of the most important cold viruses (*see chapter 36*), has recently been elucidated with the use of X-ray diffraction techniques. The results help explain rhinovirus resistance to human immune defenses. Human antibodies against the virus combine with four types of surface proteins. Unfortunately, there are at least 89 variants of these four antigens, making the production of a cold vaccine from them unlikely. The structural studies have shown that the site that recognizes and binds to cell surface molecules during infection lies at the bottom of a surface cleft about 250 nm deep and 1.2 to 3 nm wide. Thus, the binding site is well protected from the immune system while it carries out its duties, and the preparation of a vaccine directed against the binding site is not feasible. Possibly, drugs that could fit in the cleft and interfere with virus attachment can be designed.

Protomers join to form capsomers through noncovalent bonding. The bonds between proteins within pentamers and hexamers are stronger than those between separate capsomers. Empty capsids can even dissociate into separate capsomers.

Recently, it has been discovered that there is more than one way to build an icosahedral capsid. Although most icosahedral capsids appear to contain both pentamers and hexamers, simian virus 40 (SV40), a small double-stranded DNA polyomavirus, has only pentamers (figure 17.14*a*). The virus is constructed of 72 cylindrical pentamers with hollow centers. Five flexible arms extend from the edge of each pentamer (figure 17.14*b*). Twelve pentamers occupy the icosahedron's vertices and associate with five neighbors, just as they do when hexamers are also present. Each of the 60 nonvertex pentamers associates with its six adjacent neighbors in the way

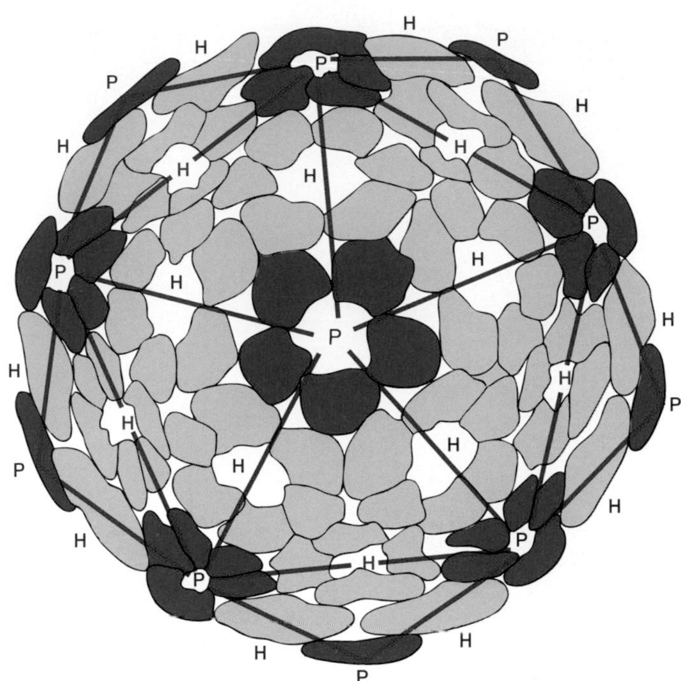

Figure 17.13 The Structure of an Icosahedral Capsid. Pentons are located at the 12 vertices. Hexons form the edges and faces of the icosahedron. This capsid contains 42 capsomers; all protomers are identical.

shown in figure 17.14*c* and *d*. An arm extends toward the adjacent vertex pentamer (pentamer 1) and twists around one of its arms. Three more arms interact in the same way with arms of other nonvertex pentamers (pentamers 3 to 5). The fifth arm binds directly to an adjacent nonvertex pentamer (pentamer 6), but does not attach to one of its arms. An arm does not extend from the central pentamer to pentamer 2; other arms hold pentamer 2 in place. Thus, an icosahedral capsid is assembled without hexamers by using flexible arms as ropes to tie the pentamers together.

Nucleic Acids

Viruses are exceptionally flexible with respect to the nature of their genetic material and employ all four possible nucleic acid types: single-stranded DNA, double-stranded DNA, single-stranded RNA, and double-stranded RNA. All four types are found in animal viruses. Plant viruses most often have single-stranded RNA genomes. Although phages may have single-stranded DNA or single-stranded RNA, bacterial viruses usually contain double-stranded DNA. Table 17.1 summarizes many variations seen in viral nucleic acids. The size of viral genetic material also varies greatly. The smallest genomes (those of the MS2 and Qβ viruses) are around 1×10^6 daltons, just large enough to code for three to four proteins. At the other extreme, T-even bacteriophages, herpesvirus, and vaccinia virus have genomes of 1.0 to 1.6×10^8 daltons and

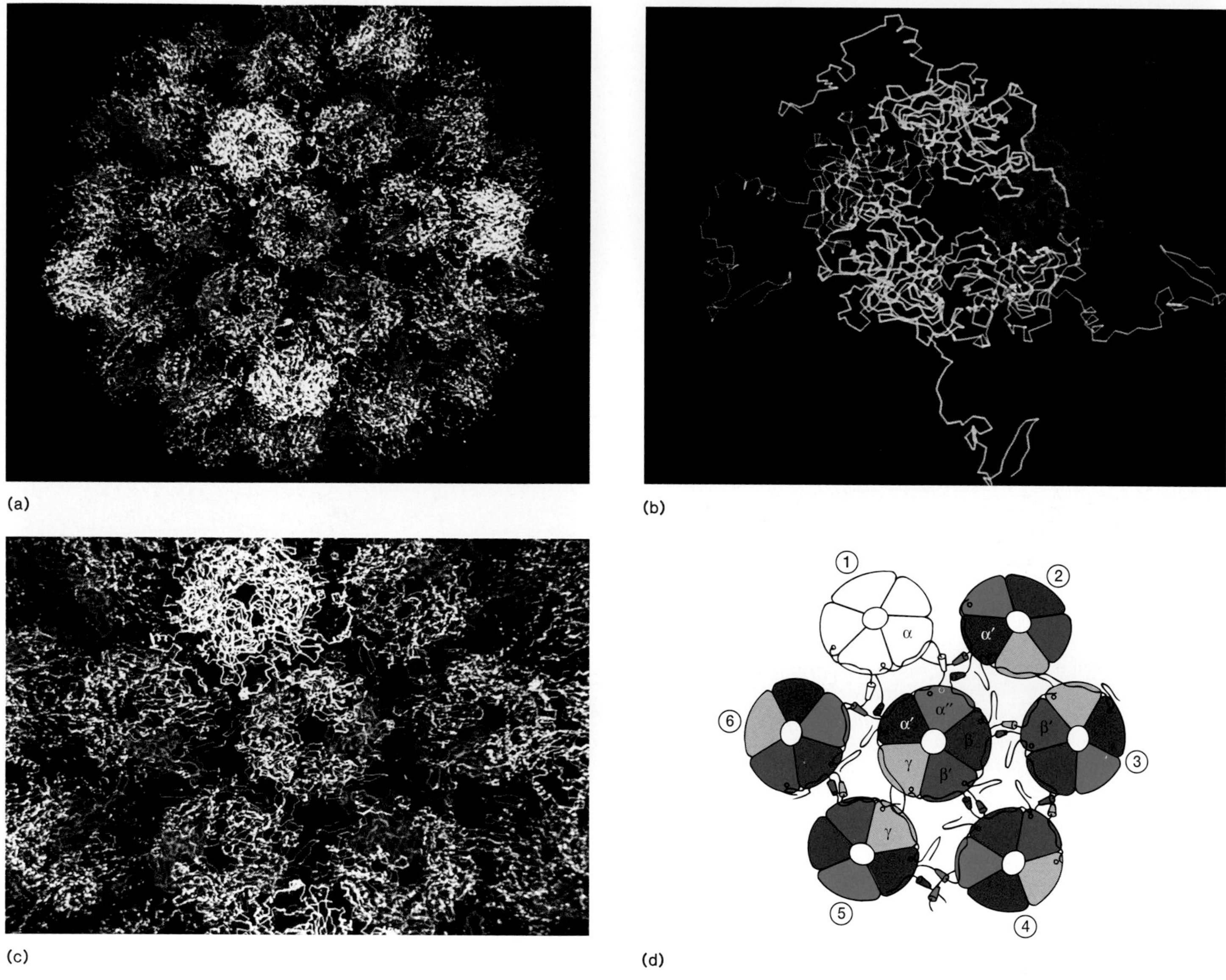

Figure 17.14 An Icosahedral Capsid Constructed of Pentamers. (*a*) The simian virus 40 capsid. The 12 pentamers at the icosahedron vertices are in white. The nonvertex pentamers are shown with each polypeptide chain in a different color. (*b*) A pentamer with extended arms. (*c*) A close-up view of the virion surface showing a central pentamer associated with six neighbors. (*d*) A schematic diagram of the surface structure depicted in part *c*. The body of each pentamer is represented by a five-petaled flower design. Each arm is shown as a line or a line and cylinder (an α-helix) with the same color as the rest of its protomer. The outer protomers are numbered clockwise beginning with the one at the vertex.

may be able to direct the synthesis of over 100 proteins. In the following paragraphs, the nature of each nucleic acid type is briefly summarized.

Nucleic acid structure (pp. 192–95).

Tiny DNA viruses like ϕX174 and M13 bacteriophages or the parvoviruses possess single-stranded DNA (ssDNA) genomes (table 17.1). Some of these viruses have linear pieces of DNA, whereas others use a single, closed circle of DNA for their genome (figure 17.15).

Most DNA viruses use double-stranded DNA (dsDNA) as their genetic material. Linear dsDNA, variously modified, is found in many viruses; others have circular dsDNA. The lambda phage has linear dsDNA with cohesive ends—single-stranded complementary segments 12 nucleotides long—that enable it to cyclize when they base pair with each other (figure 17.16).

Besides the normal nucleotides found in DNA, many virus DNAs contain unusual bases. For example, the T-even phages of *E. coli* (*see chapter 18*) have 5-hydroxymethylcytosine (*see figure 18.7*) instead of cytosine. Glucose is usually attached to the hydroxymethyl group.

Most RNA viruses employ single-stranded RNA (ssRNA) as their genetic material. The RNA base sequence may be

Table 17.1 Types of Viral Nucleic Acids

Nucleic Acid Type	Nucleic Acid Structure	Virus Examples
Single-Stranded DNA		
	Linear single strand	Parvoviruses
	Circular single strand	ϕX174, M13, fd phages
Double-Stranded DNA		
	Linear double strand	Herpesviruses (herpes simplex viruses, cytomegalovirus, Epstein-Barr virus), adenoviruses, T coliphages, lambda phage, and other bacteriophages
	Linear double strand with single chain breaks	T5 coliphage
	Double strand with cross-linked ends	Vaccinia, smallpox
	Closed circular double strand	Papovaviruses (polyoma, human papilloma viruses, SV40), PM2 phage, cauliflower mosaic
Single-Stranded RNA		
	Linear, single stranded, positive strand	Picornaviruses (polio, rhinoviruses), togaviruses, RNA bacteriophages, TMV, and most plant viruses
	Linear, single stranded, negative strand	Rhabdoviruses (rabies), paramyxoviruses (mumps, measles)
	Linear, single stranded, segmented, positive strand	Brome mosaic virus (individual segments in separate virions)
	Linear, single stranded, segmented, diploid (two identical single strands), positive strand	Retroviruses (Rous sarcoma virus, human immunodeficiency virus)
	Linear, single stranded, segmented, negative strand	Paramyxoviruses, orthomyxoviruses (influenza)
Double-Stranded RNA		
	Linear, double stranded, segmented	Reoviruses, wound-tumor virus of plants, cytoplasmic polyhedrosis virus of insects, phage ϕ6, many mycoviruses

Modified from S. E. Luria, et al., *General Virology*, 3d ed. Copyright © 1983 John Wiley & Sons, Inc., New York, NY. Reprinted by permission.

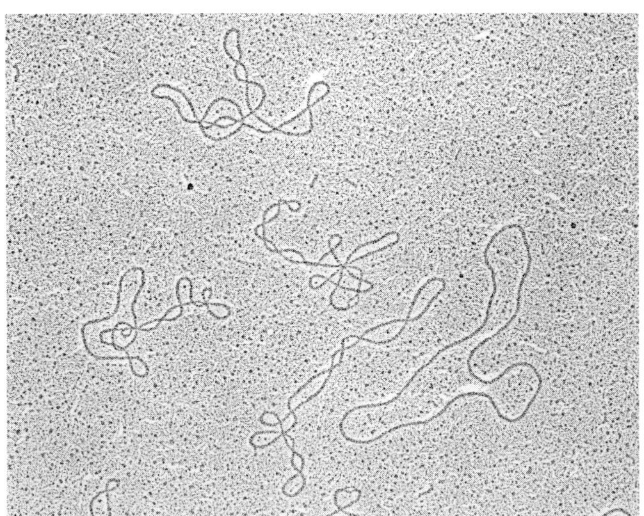

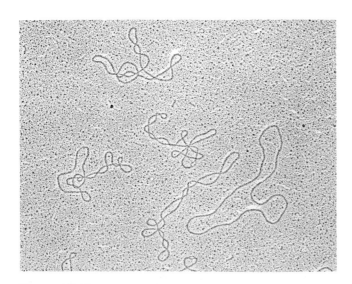

Figure 17.15 Circular Phage DNA. The closed circular DNA of the phage PM2 (×93,000). Note both the relaxed and highly twisted or supercoiled forms.

identical with that of viral mRNA, in which case the RNA strand is called the **plus strand** or **positive strand** (viral mRNA is defined as plus or positive). However, the viral RNA genome may instead be complementary to viral mRNA, and then it is called a **minus** or **negative strand.** Polio, tobacco mosaic, brome mosaic, and Rous sarcoma viruses are all positive strand RNA viruses; rabies, mumps, measles, and influenza viruses are examples of negative strand RNA viruses. Many of these RNA genomes are **segmented genomes;** that is, they are divided into separate parts. It is believed that each fragment or segment codes for one protein. Usually, all segments are probably enclosed in the same capsid even though some virus genomes may be composed of as many as 10 to 12 segments. However, it is not necessary that all segments be located in the same virion for successful reproduction. The brome mosaic virus genome is composed of four segments distributed among three different virus particles. All three of the largest segments are required for infectivity. Despite this complex and seemingly inefficient arrangement, the different brome mosaic virions manage to successfully infect the same host.

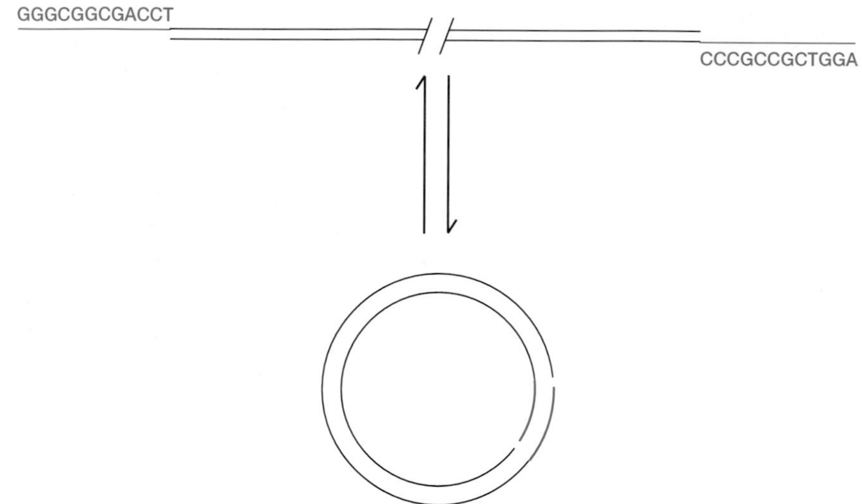

Figure 17.16 **Circularization of Lambda DNA.** The linear DNA of the lambda phage can be reversibly circularized. This is made possible by cohesive ends (in color) that have complementary sequences and can base pair with each other.

Plus strand viral RNA often resembles mRNA in more than the equivalence of its nucleotide sequence. Just as eucaryotic mRNA usually has a 5′ cap of 7-methylguanosine, many plant and animal viral RNA genomes are capped. In addition, most or all plus strand RNA animal viruses also have a poly-A stretch at the 3′ end of their genome, and thus closely resemble eucaryotic mRNA with respect to the structure of both ends. In fact, plus strand RNAs can direct protein synthesis immediately after entering the cell. Strangely enough, a number of single-stranded plant viral RNAs have 3′ ends that resemble eucaryotic transfer RNA, and the genomes of tobacco mosaic virus will actually accept amino acids. Capping is not seen in the RNA bacteriophages.

Eucaryotic mRNA structure and function (pp. 203–4).

A few viruses have double-stranded RNA (dsRNA) genomes. All appear to be segmented; some, such as the reoviruses, have 10 to 12 segments. These dsRNA viruses are known to infect animals, plants, fungi, and even one bacterial species.

Viral Envelopes and Enzymes

Many animal viruses, some plant viruses, and at least one bacterial virus are bounded by an outer membranous layer called an envelope (figure 17.17). Animal virus envelopes usually arise from host cell nuclear or plasma membranes; their lipids and carbohydrates are normal host constituents. In contrast, envelope proteins are coded for by virus genes and may even project from the envelope surface as **spikes** or **peplomers** (figure 17.17*a, b, f*). It is thought that these spikes may be involved in virus attachment to the host cell surface. Since they differ among viruses, they also can be used to identify some viruses. Because the envelope is a flexible, membranous structure, enveloped viruses frequently have a somewhat variable shape and are called pleomorphic. However, the envelopes of viruses like the bullet-shaped rabies virus are firmly attached to the underlying nucleocapsid and endow the virion with a constant, characteristic shape (figure 17.17*c*). In some viruses, the envelope is disrupted by solvents like ether to such an extent that lipid-mediated activities are blocked or envelope proteins are denatured and rendered inactive. The virus is then said to be "ether sensitive."

Influenza virus (figure 17.17*a, b*) is a well-studied example of an enveloped virus. Spikes project about 10 nm from the surface at 7 to 8 nm intervals. Some spikes possess the enzyme neuraminidase, which probably aids the virus in penetrating mucous layers of the respiratory epithelium to reach host cells. Other spikes have hemagglutinin proteins, so named because they can bind the virions to red blood cell membranes and cause hemagglutination (*see figure 33.10*). Hemagglutinins probably participate in virion attachment to host cells. Proteins, like the spike proteins that are exposed on the outer envelope surface, are generally glycoproteins; that is, the proteins have carbohydrate attached to them. A nonglycosylated protein, the M or matrix protein, is found on the inner surface of the envelope and helps stabilize it.

Although it was originally thought that virions had only structural capsid proteins and lacked enzymes, some capsid-associated enzymes have been discovered, particularly in enveloped animal viruses. Many of these are involved in nucleic acid replication. For example, the influenza virus uses RNA as its genetic material and carries an RNA-dependent RNA polymerase that acts as an RNA transcriptase and synthesizes mRNA under the direction of its RNA genome. Although viruses lack true metabolism and cannot reproduce independently of living cells, they may carry one or more enzymes essential to the completion of their life cycles.

Nucleic acid replication and transcription (chapter 10).
Animal virus reproduction (pp. 384–91).

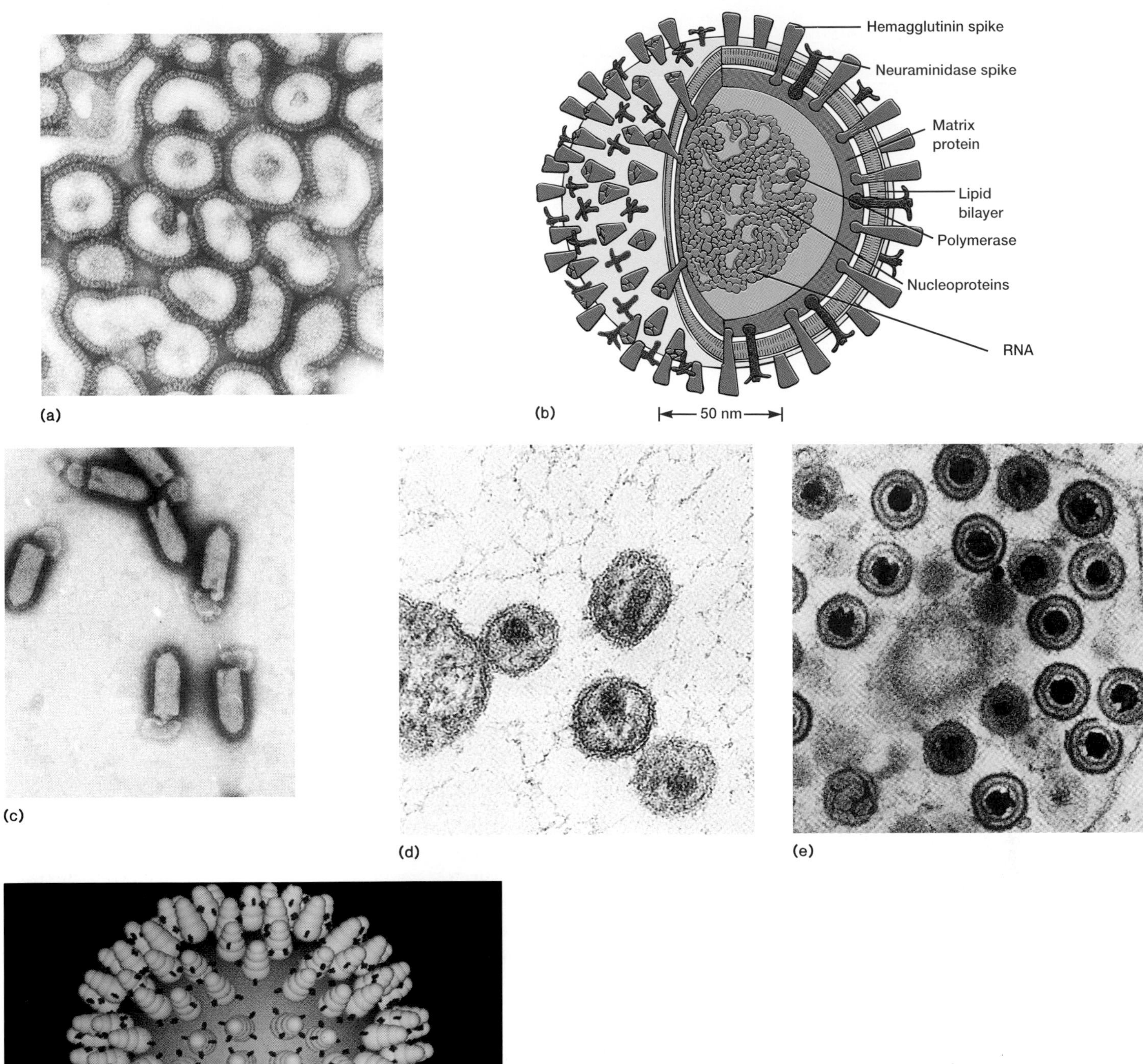

Figure 17.17 Examples of Enveloped Viruses. (*a*) Human influenza virus. Note the flexibility of the envelope and the spikes projecting from its surface (×282,000). (*b*) Diagram of the influenza virion. (*c*) Rhabdovirus particles (×250,000). This is the vesicular stomatitis virus, a relative of the rabies virus, which is similar in appearance. (*d*) Human immunodeficiency viruses (×33,000). (*e*) Herpes viruses (×100,000). (*f*) Computer image of the Semliki Forest virus, a virus that occasionally causes encephalitis in humans. *(b) From Donald Voet and Judith G. Voet,* Biochemistry. *Copyright © 1990 John Wiley & Sons, Inc., New York, NY. Reprinted by permission. Adapted from "The Epidemiology of Influenza" by Martin M. Kaplan and Robert G. Webster. Copyright © 1977 by Scientific American, Inc. All rights reserved.*

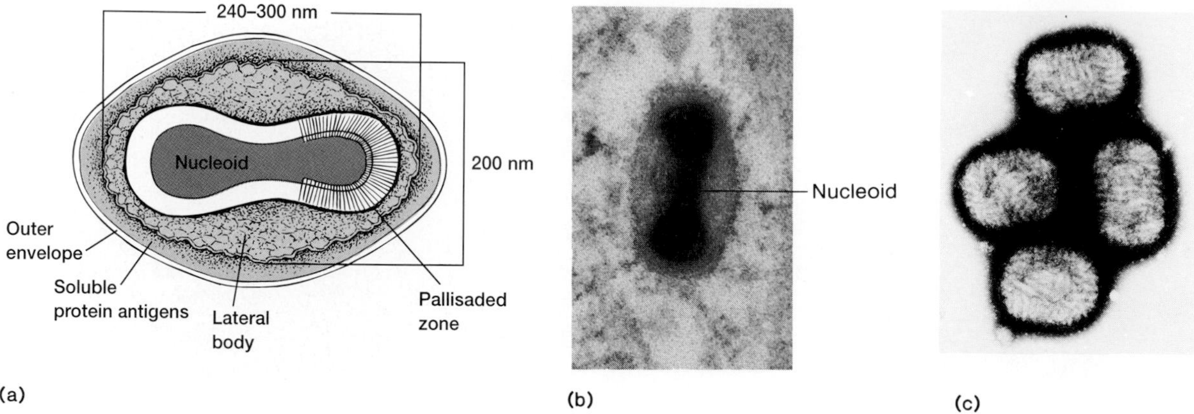

Figure 17.18 **Vaccinia Virus Morphology.** (*a*) Diagram of vaccinia structure. (*b*) Micrograph of the virion clearly showing the nucleoid (×200,000). (*c*) Vaccinia surface structure. An electron micrograph of four virions showing the thick array of surface fibers (×150,000).

Viruses with Capsids of Complex Symmetry

Although most viruses have either icosahedral or helical capsids, there are many viruses that do not fit into either category. The poxviruses and large bacteriophages are two important examples.

The poxviruses are the largest of the animal viruses (about 400 × 240 × 200 nm in size) and can even be seen with a phase-contrast microscope or in stained preparations. They possess an exceptionally complex internal structure with an ovoid- to brick-shaped exterior. The double-stranded DNA is associated with proteins and contained in the nucleoid, a central structure shaped like a biconcave disk and surrounded by a membrane (figure 17.18). Two elliptical or lateral bodies lie between the nucleoid and its outer envelope, a membrane and a thick layer covered by an array of tubules or fibers.

Some large bacteriophages are even more elaborate than the poxviruses. The T2, T4, and T6 phages that infect *E. coli* have been intensely studied. Their head resembles an icosahedron elongated by one or two rows of hexamers in the middle (figure 17.19) and contains the DNA genome. The tail is composed of a collar joining it to the head, a central hollow tube, a sheath surrounding the tube, and a complex base plate. In T-even phages, the base plate is hexagonal and has a pin and a jointed tail fiber at each corner. The tail fibers are responsible for virus attachment to the proper site on the bacterial surface (*see chapter 18*).

There is considerable variation in structure among the large bacteriophages, even those infecting a single host. In contrast with the T-even phages, many coliphages have true icosahedral heads. T1, T5, and lambda phages have sheathless tails that lack a base plate and terminate in rudimentary tail fibers. Coliphages T3 and T7 have short, noncontractile tails without tail fibers. Clearly, these viruses can complete their reproductive cycles using a variety of tail structures.

Complex bacterial viruses with both heads and tails are said to have **binal symmetry** because they possess a combination of icosahedral (the head) and helical (the tail) symmetry.

---

1. Define the following terms: nucleocapsid, capsid, icosahedral capsid, helical capsid, complex virus, protomer, self-assembly, capsomer, pentamer or penton, and hexamer or hexon. How do pentamers and hexamers associate to form a complete icosahedron; what determines helical capsid length and diameter?
2. All four nucleic acid forms can serve as virus genomes. Describe each, the types of virion possessing it, and any distinctive physical characteristics the nucleic acid can have. What are the following: plus strand, minus strand, and segmented genome?
3. What is an envelope? What are spikes or peplomers? Why are some enveloped viruses pleomorphic? Give two functions spikes might serve in the virus life cycle, and the proteins that the influenza virus uses in these processes.
4. What is a complex virus? Binal symmetry? Briefly describe the structure of poxviruses and T-even bacteriophages.

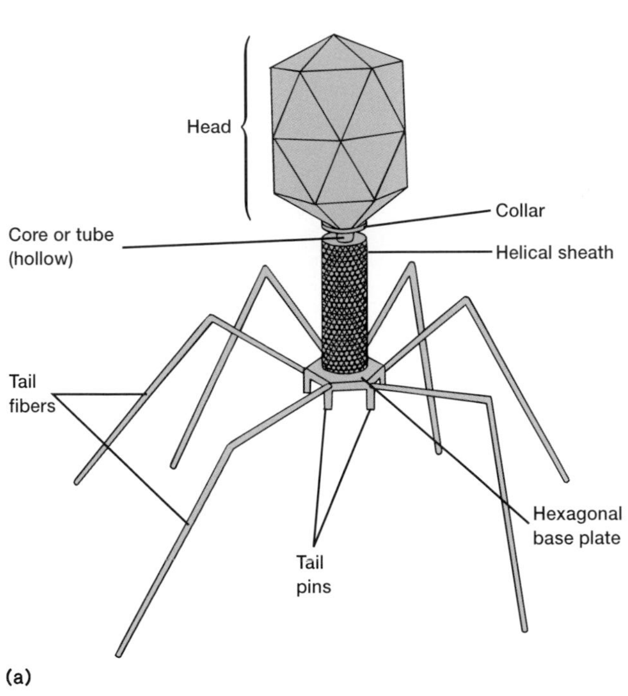

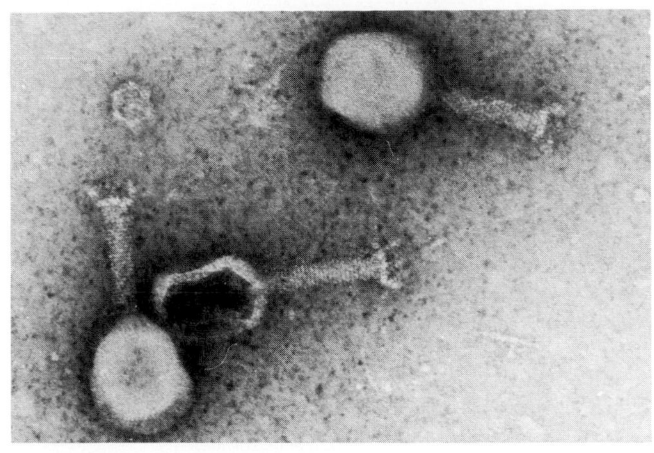

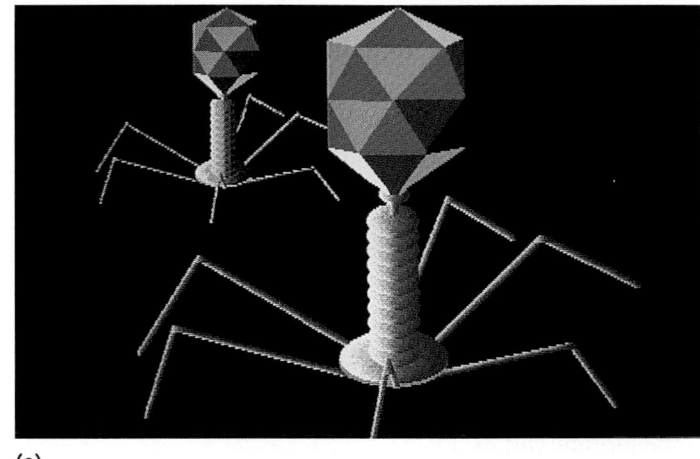

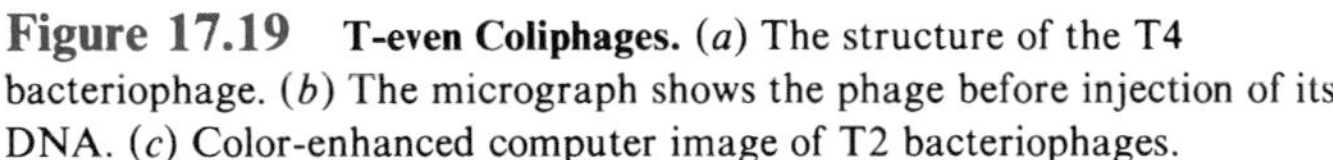

Figure 17.19 **T-even Coliphages.** (*a*) The structure of the T4 bacteriophage. (*b*) The micrograph shows the phage before injection of its DNA. (*c*) Color-enhanced computer image of T2 bacteriophages.

Principles of Virus Taxonomy

The classification of viruses is in a much less satisfactory state than that of either bacteria or eucaryotic microorganisms. In part, this is due to a lack of knowledge of their origin and evolutionary history (see Box 17.2). Usually, viruses are separated into several large groups based on their host preferences: animal viruses, plant viruses, bacterial viruses, bacteriophages, and so forth. Unfortunately, virologists working with these groups have been unable to agree on a uniform system of classification and nomenclature. Recently, the International Committee for Taxonomy of Viruses has developed a uniform classification system and divided viruses into 50 families. The committee places greatest weight on three properties to define families: (1) nucleic acid type, (2) nucleic acid strandedness, and (3) presence or absence of an envelope. Virus family names end in *viridae;* subfamily names, in *virinae;* and genus (and species) names, in *virus.* For example, the poxviruses are in the family *Poxviridae;* the subfamily *Chorodopoxvirinae* contains poxviruses of vertebrates. Within the subfamily are several genera that are distinguished on the basis of immunologic characteristics and host specificity. The genus *Orthopoxvirus* contains several species, among them variola major (the cause of smallpox), vaccinia, and cowpox.

Box 17.2

The Origin of Viruses

The origin and subsequent evolution of viruses are shrouded in mystery, in part because of the lack of a fossil record. However, recent advances in the understanding of virus structure and reproduction have made possible more informed speculation on virus origins. At present, there are two major hypotheses entertained by virologists. It has been proposed that at least some of the more complex enveloped viruses, such as the poxviruses and herpesviruses, arose from small cells, probably procaryotic, that parasitized larger, more complex cells. These parasitic cells would become ever simpler and more dependent on their hosts, much like multicellular parasites have done, in a process known as retrograde evolution. There are several problems with this hypothesis. Viruses are radically different from procaryotes, and it is difficult to envision the mechanisms by which such a transformation might have occurred or the selective pressures leading to it. In addition, one would expect to find some forms intermediate between procaryotes and at least the more complex enveloped viruses, but such forms have not been detected.

The second hypothesis is that viruses represent cellular nucleic acids that have achieved independence and managed to escape the cell. Possibly, a few mutations could convert nucleic acids, which are only synthesized at specific times, into infectious nucleic acids whose replication could not be controlled. This conjecture is supported by the observation that the nucleic acids of retroviruses (*see chapter 19*) and a number of other virions do contain sequences quite similar to those of normal cells, plasmids, and transposons (*see chapter 13*). The small, infectious RNAs called viroids (*see chapter 19*) have base sequences complementary to transposons, the regions around the boundary of mRNA introns (*see chapter 10*), and portions of host DNA. This has led to speculation that they have arisen from introns or transposons.

It is possible that viruses have arisen by way of both mechanisms. Because viruses differ so greatly from one another, it seems likely that they have originated independently many times during the course of evolution. Probably many viruses have evolved from other viruses just as cellular organisms have arisen from specific predecessors. The question of virus origins is complex and quite speculative; future progress in understanding virus structure and reproduction may clarify this question.

Viruses are divided into different taxonomic groups based on a few characteristics, some of the more important of which are:

1. Nature of the host—animal, plant, bacterial, insect, fungal
2. Nucleic acid characteristics—DNA or RNA, single or double stranded, molecular weight
3. Capsid symmetry—icosahedral, helical, binal
4. Presence of an envelope and ether sensitivity
5. Diameter of the virion or nucleocapsid
6. Number of capsomers in icosahedral viruses
7. Immunologic properties
8. Intracellular location of viral replication
9. Disease caused and/or special clinical features, method of transmission

Table 17.2 illustrates the use of some of these properties to describe a few common virus groups. Virus classification is further discussed when bacterial, animal, and plant viruses are considered more specifically, and a fairly complete summary of virus taxonomy is presented in appendix IV.

---

1. List some characteristics used in classifying viruses. Which seem to be the most important?
2. What are the endings for virus families, subfamilies, and genera or species?

TABLE 17.2 Some Common Virus Groups and Their Characteristics

Nucleic Acid	Strandedness	Capsid Symmetry[a]	Presence of Envelope	Size of Capsid (nm)[b]	Number of Capsomers	Virus Group	Host Range[c]
RNA	Single	I	−	22–30	32	*Picornaviridae*	A
		I	+	40–75(e)	32	*Togaviridae*	A
		I?	+	100(e)		*Retroviridae*	A
		H	+	9(h), 80–120(e)		*Orthomyxoviridae*	A
		H	+	18(h), 125–250(e)		*Paramyxoviridae*	A
		H	+	14–16(h), 80–160(e)		*Coronaviridae*	A
		H	+	18(h), 70–80 × 130–240 (bullet shaped)		*Rhabdoviridae*	A
		I	−	25		Bromovirus	P
		I	−	30		Cucumovirus	P
		H	−	18 × 300		Tobamovirus	P
		Bacilliform	−	18 × 18–58		Alfalfa mosaic virus	P
		I	−	26–27	32	*Leviviridae* [Qβ]	B
RNA	Double	I	−	70–80	92	*Reoviridae*	A
		I	−	70		Wound tumor virus	P
		I	−	25–50		Mycoviruses	P
		I	+	100(e)		*Cystoviridae*	B
DNA	Single	I	−	20–25	12	*Parvoviridae*	A
		I	−	18 × 30 (paired particles)		Geminivirus	P
		I	−	25–35		*Microviridae*	B
		H	−	6 × 900–1900		*Inoviridae*	B
DNA	Double	I	−	45–55	72	*Papovaviridae*	A
		I	−	60–90	252	*Adenoviridae*	A
		I	−	130–180		*Iridoviridae*	A
		I	+	100, 180–200(e)	162	*Herpesviridae*	A
		C	+	200–260 × 250–290(e)		*Poxviridae*	A
		H	+	40 × 300(e)		*Baculoviridae*	A
		C	+	28 (core), 42(e)	42	*Hepadnaviridae*	A
		I	−	50		Caulimovirus	P
		I	−	60		*Corticoviridae*	B
		Bi	−	60, 15[d]		*Pedoviridae*	B
		Bi	−	54–65, 140–170[d]		*Styloviridae*	B
		Bi	−	80 × 110, 110[d]		*Myoviridae*	B

[a]Types of symmetry: I, icosahedral; H, helical; Bacilliform, resembles a rodlike bacterium in shape; C, complex; Bi, binal.

[b]Diameter of helical capsid (h); diameter of enveloped virion (e).

[c]Host range: A, animal; P, plant; B, bacterium.

[d]The first number is the head diameter; the second number, the tail length.

Summary

1. People were first protected from a viral disease when Edward Jenner developed a smallpox vaccine in 1798.
2. Chamberland's invention of a porcelain filter that could remove bacteria from virus samples enabled microbiologists to show that viruses were different from bacteria.
3. In the late 1930s, Stanley, Bawden, and Pirie crystallized the tobacco mosaic virus and demonstrated that it was composed only of protein and nucleic acid.
4. A virion is composed of either DNA or RNA enclosed in a coat of protein (and sometimes other substances as well). It cannot reproduce independently of living cells.
5. Viruses are cultivated using tissue cultures, embryonated eggs, bacterial cultures, and other living hosts.
6. Sites of animal viral infection may be characterized by cytopathic effects such as pocks and plaques. Phages produce plaques in bacterial lawns. Plant viruses can cause localized necrotic lesions in plant tissues.
7. Viruses can be purified by techniques such as differential and gradient centrifugation, precipitation, and denaturation or digestion of contaminants.
8. Virus particles can be counted directly with the transmission electron microscope or indirectly by the hemagglutination assay.
9. Infectivity assays can be used to estimate virus numbers in terms of plaque-forming units, lethal dose (LD_{50}), or infectious dose (ID_{50}).

10. All virions have a nucleocapsid composed of a nucleic acid, either DNA or RNA, held within a protein capsid made of one or more types of protein subunits called protomers.
11. There are four types of viral morphology: naked icosahedral, naked helical, enveloped icosahedral and helical, and complex.
12. Helical capsids resemble long hollow protein tubes and may be either rigid or quite flexible. The nucleic acid is coiled in a spiral on the inside of the cylinder.
13. Icosahedral capsids are usually constructed from two types of capsomers: pentamers (pentons) at the vertices and hexamers (hexons) on the edges and faces of the icosahedron.
14. Viral nucleic acids can be either single stranded or double stranded, DNA or RNA. Most DNA viruses have double-stranded DNA genomes that may be linear or closed circles.
15. RNA viruses usually have ssRNA that may be either plus (positive) or minus (negative) when compared with mRNA (positive). Many RNA genomes are segmented.
16. Viruses can have a membranous envelope surrounding their nucleocapsid. The envelope lipids usually come from the host cell; in contrast, many envelope proteins are viral and may project from the envelope surface as spikes or peplomers.
17. Although viruses lack true metabolism, some contain a few enzymes necessary for their reproduction.
18. Complex viruses (e.g., poxviruses and large phages) have complicated morphology not characterized by icosahedral and helical symmetry. Large phages often have binal symmetry: their heads are icosahedral and their tails, helical.
19. Currently, viruses are classified with a taxonomic system placing primary emphasis on the type and strandedness of viral nucleic acids and on the presence or absence of an envelope.

Key Terms

bacteriophage *349*
binal symmetry *362*
capsid *354*
capsomers *357*
complex viruses *355*
cytopathic effects *349*
differential centrifugation *350*
envelope *355*
gradient centrifugation *351*
helical *355*
hemagglutination assay *353*
hexamers (hexons) *357*
icosahedral *355*
infectious dose (ID_{50}) *353*
lethal dose (LD_{50}) *353*
minus or negative strand *359*
necrotic lesion *350*
nucleocapsid *354*
pentamers (pentons) *357*
phage *349*
plaque *349*
plaque assay *353*
plaque-forming units (PFU) *353*
plus or positive strand *359*
protomers *355*
segmented genome *359*
spike or peplomer *360*
virion *348*
virologist *347*
virology *347*
virus *348*

Questions for Thought and Review

1. In what ways do viruses resemble living organisms?
2. Why might virology have developed much more slowly without the use of Chamberland's filter?
3. What advantage would an RNA virus gain by having its genome resemble eucaryotic mRNA?
4. A number of characteristics useful in virus taxonomy are listed on page 364. Can you think of any other properties that might be of considerable importance in future studies on virus taxonomy?

Additional Reading

Bradley, D. E. 1971. A comparative study of the structure and biological properties of bacteriophages. In *Comparative virology*, eds. K. Maramorosch and E. Kurstak, 207–53. New York: Academic Press.

Brown, F., and Wilson, G., eds. 1984. *Topley and Wilson's principles of bacteriology, virology and immunity*, 7th ed., vol. 4, *Virology*. Baltimore: Williams & Wilkins.

Casjens, S. 1985. *Virus structure and assembly*. Boston: Jones and Bartlett.

Davis, B. D.; Dulbecco, R.; Eisen, H. N.; and Ginsberg H. S. 1990. *Microbiology*, 4th ed. Philadelphia: J. B. Lippincott.

Dimmock, N. J., and Primrose, S. B. 1987. *Introduction to modern virology*, 3d ed. London: Blackwell Scientific Publications.

Dulbecco, R., and Ginsberg, H. S. 1988. *Virology*, 2d ed. Philadelphia: J. B. Lippincott.

Fields, B. N., Knipe, D. M.; Chanock, R. M., Hirsch, M. S.; Melnick, J. L.; Monath, T. P.; and Roizman, B., eds. 1990. *Fields virology*, 2d ed. New York: Raven Press.

Fraenkel-Conrat, H.; Kimball, P. C.; and Levy, J. A. 1988. *Virology*, 2d ed. Englewood Cliffs, N.J.: Prentice-Hall.

Harrison, S. C. 1984. Structure of viruses. In *The microbe 1984: Part I, viruses*. 36th Symposium Society for General Microbiology. Cambridge: Cambridge University Press.

Henshaw, N. G. 1988. Identification of viruses by methods other than electron microscopy. *ASM News* 54(9):482–85.

Jennings, F. 1975. *The invasion of America: Indians, colonialism, and the cant of conquest*. Chapel Hill: University of North Carolina Press.

Lechevalier, H. A., and Solotorovsky, M. 1965. *Three centuries of microbiology*. New York: McGraw-Hill.

Lwoff, A., and Tournier, P. 1971. Remarks on the classification of viruses. In *Comparative virology*, eds. K. Maramorosch and E. Kurstak, 1–42. New York: Academic Press.

Luria, S. E.; Darnell, J. E., Jr.; Baltimore, D.; and Campbell, A. 1978. *General virology*, 3d ed. New York: John Wiley and Sons.

Matthews, R. E. F. 1979. Classification and nomenclature of viruses. *Intervirol.* 12(3–5): 132–281.

Matthews, R. E. F. 1985. Viral taxonomy for the nonvirologist. *Ann. Rev. Microbiol.* 39:451–74.

McNeill, W. H. 1976. *Plagues and peoples*. Garden City, N.Y.: Anchor.

Miller, S. E. 1988. Diagnostic virology by electron microscopy. *ASM News* 54(9):475–81.

Morse, S. S. 1989. Emerging viruses. *ASM News* 55(7):358–60.

Nahmias, A. J. 1977. The evolution of viruses. *Ann. Rev. Ecol. Syst.* 8:29–49.

Sanders, F. K. 1981. Viruses. *Carolina Biology Reader*, no. 64, 2d ed. Burlington, N.C.: Carolina Biological Supply Co.

Scott, A. 1985. *Pirates of the cell: The story of viruses from molecule to microbe*. New York: Basil Blackwell.

Stearn, E. W., and Stearn, A. E. 1945. *The effect of smallpox on the destiny of the Amerindian*. Boston: Bruce Humphries.

Strauss, J. H., and Strauss, E. G. 1988. Evolution of RNA viruses. *Ann. Rev. Microbiol.* 42:657–83.

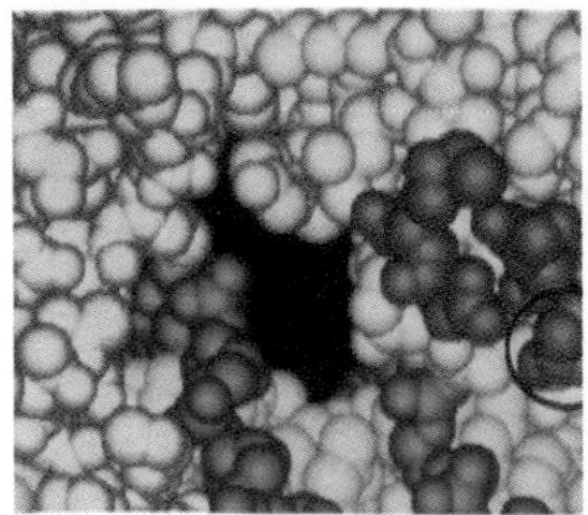

CHAPTER 18
The Viruses: Bacteriophages

You might wonder how such naive outsiders get to know about the existence of bacterial viruses. Quite by accident, I assure you. Let me illustrate by reference to an imaginary theoretical physicist, who knew little about biology in general, and nothing about bacterial viruses in particular. . . . Suppose now that our imaginary physicist, the student of Niels Bohr, is shown an experiment in which a virus particle enters a bacterial cell and 20 minutes later the bacterial cell is lysed and 100 virus particles are liberated. He will say: "How come, one particle has become 100 particles of the same kind in 20 minutes? That is very interesting. Let us find out how it happens! . . . Is this multiplying a trick of organic chemistry which the organic chemists have not yet discovered? Let us find out."

—Max Delbrück

Outline

Concepts

1. Since a bacteriophage cannot independently reproduce itself, the phage takes over its host cell and forces the host to reproduce it.
2. The lytic bacteriophage life cycle is composed of four phases: adsorption of the phage to the host and penetration of virus genetic material, synthesis of virus nucleic acid and capsid proteins, assembly of complete virions, and the release of phage particles from the host.
3. Temperate virus genetic material is able to remain within host cells and reproduce in synchrony with the host for long periods in a relationship known as lysogeny. Usually, the virus genome is found integrated into the host genetic material as a prophage. A repressor protein keeps the prophage dormant and prevents virus reproduction.

Chapter 17 introduces many of the facts and concepts underlying the field of virology, including information about the nature of viruses, their structure and taxonomy, and how they are cultivated and studied. Clearly, the viruses are a complex, diverse, and fascinating group, the study of which has done much to advance disciplines such as genetics and molecular biology.

Chapters 18 and 19 focus on virus diversity. This chapter is concerned with bacterial viruses or **bacteriophages;** the next surveys animal, plant, and insect viruses. The taxonomy, morphology, and reproduction of each group are covered. Where appropriate, the biological and practical importance of viruses is emphasized (Box 18.1), even though viral diseases are examined in chapter 36.

Since the bacteriophages (or simply phages) have been the most intensely studied viruses and are best understood in a molecular sense, this chapter is devoted to them.

Classification of Bacteriophages

Although properties such as host range and immunologic relationships are used in classifying phages, the most important are phage morphology and nucleic acid properties (figure 18.1). The genetic material may be either DNA or RNA; most bacteriophages have DNA (usually double stranded). Most can be placed in one of a few morphological groups: tailless icosahedral phages, viruses with contractile tails, viruses with noncontractile tails, and filamentous phages. There are even a few phages with envelopes. The most complex forms are the phages with contractile tails, for example, the T-even phages of *E. coli*.

1. Briefly describe in general terms the morphology and nucleic acids of the major phage types.

Reproduction of DNA Phages: The Lytic Cycle

After DNA bacteriophages have reproduced within the host cell, many of them are released when the cell is destroyed by lysis. A phage life cycle that culminates with the host cell bursting and releasing virions is called a **lytic cycle.** The events taking place during the lytic cycle will be reviewed in this section, with the primary focus on the T-even phages of *E. coli*. These are double-stranded DNA bacteriophages with complex contractile tails.

The structure of T-even coliphages (p. 362).

The One-Step Growth Experiment

The development of the one-step growth experiment in 1939 by Max Delbrück and Emory Ellis marks the beginning of modern bacteriophage research. In a **one-step growth experiment,** the reproduction of a large phage population is synchronized so that the molecular events occurring during

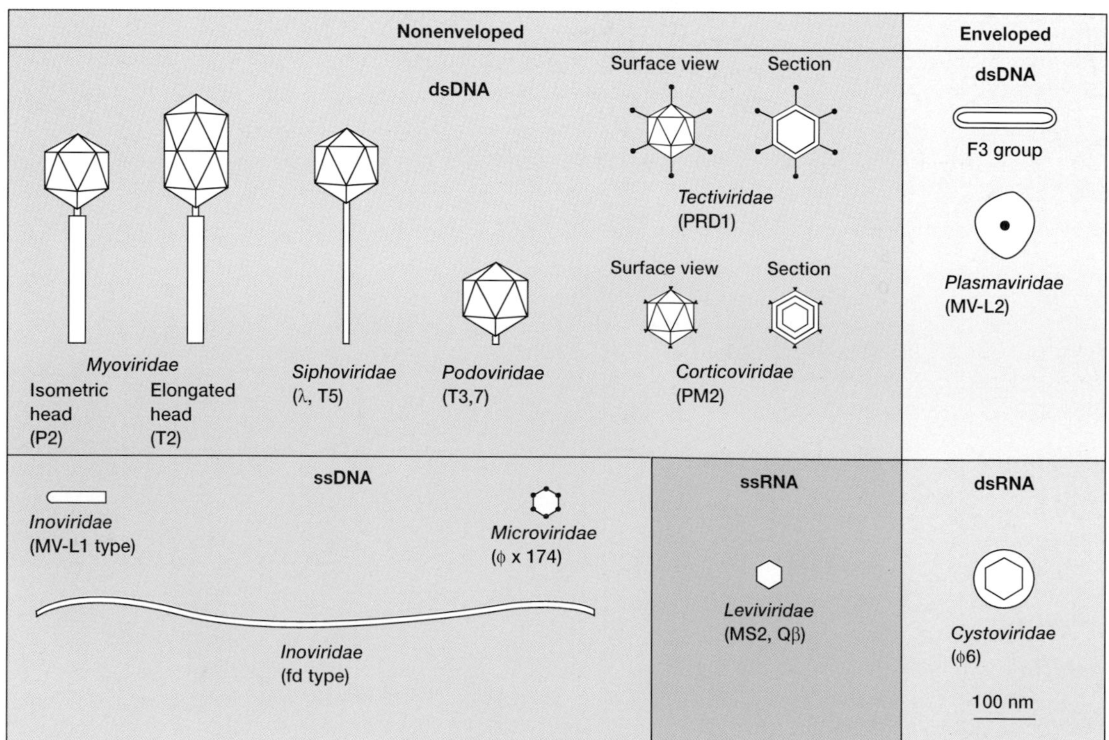

Figure 18.1 Major Bacteriophage Families. The *Myoviridae* are the only family with contractile tails. *Plasmaviridae* are pleomorphic. *Tectiviridae* have distinctive double capsids, while the *Corticoviridae* have complex capsids containing lipid.

Box 18.1

An Ocean of Viruses

Microbiologists have previously searched without success for viruses in marine habitats. Thus, it has been assumed the oceans probably did not contain many viruses. Recent discoveries have changed this view radically. Several groups have either centrifuged seawater at high speeds or passed it through an ultrafilter and then examined the sediment or suspension in an electron microscope (*see figure 41.7*). They have found that marine viruses are about 10 times more plentiful than marine bacteria. Between 10^6 and 10^9 virus particles per milliliter are present at the ocean's surface. It has been estimated that the top one millimeter of the world's oceans could contain a total of over 3×10^{30} virus particles!

Although little detailed work has been done on marine viruses, it appears that many contain double-stranded DNA. Most are probably bacteriophages, and can infect both marine heterotrophs and cyanobacteria. Up to 70% of marine procaryotes may be infected by phages. Viruses that infect diatoms and other major algal components of the marine phytoplankton have also been detected.

Marine viruses may be very important ecologically. Viruses may control marine algal blooms such as red tides (p. 544), and bacteriophages could account for a third or more of the total aquatic bacterial mortality or turnover. If true, this is of major ecological significance because the reproduction of marine bacteria far exceeds marine protozoan grazing capacity. Virus lysis of procaryotic and algal cells may well contribute greatly to carbon and nitrogen cycling in marine food webs. It could reduce the level of marine primary productivity in some situations.

Bacteriophages also may greatly accelerate the flow of genes between marine bacteria. Virus-induced bacterial lysis could generate most of the free DNA present in seawater. Gene transfer between aquatic bacteria by transformation (*see pp. 271–74*) does occur, and bacterial lysis by phages would increase its probability. Furthermore, such high phage concentrations can stimulate gene exchange by transduction (*see pp. 274–78*). These genetic exchanges could have both positive and negative consequences. Genes that enable marine bacteria to degrade toxic pollutants such as those in oil spills could spread through the native population. On the other hand, antibiotic resistance genes in bacteria from raw sewage released into the ocean also might be dispersed.

reproduction can be followed. A culture of susceptible bacteria such as *E. coli* is mixed with bacteriophage particles, and the phages are allowed a short interval to attach to their host cells. The culture is then greatly diluted so that any virus particles released upon host cell lysis will not immediately infect new cells. This strategy works because phages lack a means of seeking out host cells and must contact them during random movement through the solution. Thus, phages are less likely to contact host cells in a dilute mixture. The number of infective phage particles released from bacteria is subsequently determined at various intervals by a plaque count (*see chapter 17*).

A plot of the bacteriophages released from host cells versus time shows several distinct phases (figure 18.2). During the **latent period,** which immediately follows phage addition, there is no release of virions. This is followed by the **rise period** or **burst,** when the host cells rapidly lyse and release infective phages. Finally, a plateau is reached and no more viruses are liberated. The total number of phages released can be used to calculate the **burst size,** the number of viruses produced per infected cell.

The latent period is the shortest time required for virus reproduction and release. During the first part of this phase, host bacteria do not contain any complete, infective virions. This can be shown by lysing them with chloroform. This initial segment of the latent period is called the **eclipse period** because the virions were detectable before infection, but are now concealed or eclipsed. The number of completed, infective

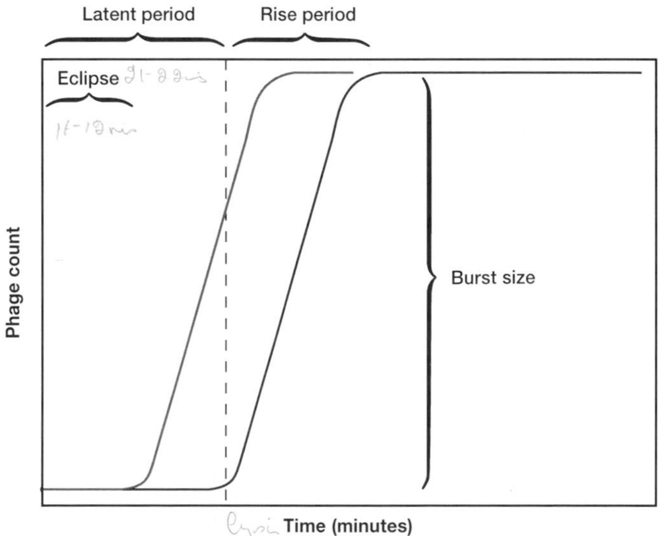

Figure 18.2 The One-step Growth Curve. In the initial part of the latent period, the eclipse period, the host cells do not contain any complete, infective virions. During the remainder of the latent period, an increasing number of infective virions are present, but none are released. The latent period ends with host cell lysis and rapid release of virions during the rise period or burst. In this figure, the (blue) line represents the total number of complete virions. The (red) line is the number of free viruses (the unadsorbed virions plus those released from host cells). When *E. coli* is infected with T2 phage at 37° C, the growth plateau is reached in about 30 minutes and the burst size is approximately 100 or more virions per cell. The eclipse period is 11–12 minutes, and the latent period is around 21–22 minutes in length.

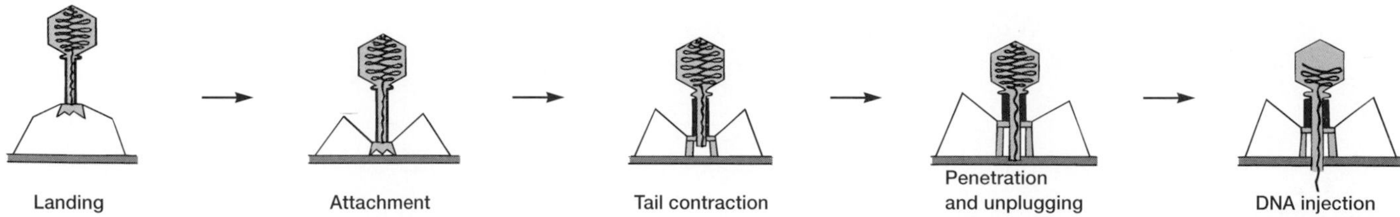

Figure 18.3 **T4 Phage Adsorption and DNA Injection.**

phages within the host increases after the end of the eclipse period, and the host cell is prepared for lysis.

The one-step growth experiment with *E. coli* and phage T2 provides a well-studied example of this process. When the experiment is carried out with actively growing cells in rich medium at 37°C, the growth curve plateau is reached in approximately 30 minutes. Bacteriophage reproduction is an exceptionally rapid process, much faster than animal virus reproduction, which may take hours.

Adsorption to the Host Cell and Penetration

Bacteriophages do not randomly attach to the surface of a host cell; rather, they fasten to specific **receptor sites.** The nature of these receptors varies with the phage; cell wall lipopolysaccharides and proteins, teichoic acids, flagella, and pili can serve as receptors. The T-even phages of *E. coli* use cell wall lipopolysaccharides or proteins as receptors. Variation in receptor properties is at least partly responsible for phage host preferences.

The structure of cell walls, flagella, and pili (pp. 50–55, 57–60).

T-even phage adsorption involves several tail structures (*see figure 17.19*). Phage attachment begins when a tail fiber contacts the appropriate receptor site. As more tail fibers make contact, the baseplate settles down on the surface (figures 18.3 and 18.4). After the baseplate is seated firmly on the cell surface, conformational changes occur in the baseplate and sheath, and the tail sheath contracts. Then the central tube or core is pushed through the bacterial wall. The sheath contains ATP, and this may power the contraction. Finally, the DNA is extruded from the head, through the tail tube, and into the host cell by an unknown mechanism. It is not certain that the tail tube actually penetrates the plasma membrane; the DNA may simply be deposited on the surface of the membrane and then pass through. The penetration mechanisms of other bacteriophages often appear to differ from that of the T-even phages but have not been studied in much detail.

Synthesis of Phage Nucleic Acids and Proteins

Since the T4 phage of *E. coli* has been intensely studied, its reproduction will be used as our example (figure 18.5). Soon after phage DNA injection, the synthesis of host DNA, RNA, and protein is halted, and the cell is forced to make viral constituents. *E. coli* RNA polymerase (*see chapter 10*) starts synthesizing phage mRNA within two minutes. This mRNA and all other **early mRNA** (mRNA transcribed before phage DNA is made) direct the synthesis of the protein factors and enzymes required to take over the host cell and manufacture viral nucleic acids. Some early virus-specific enzymes degrade host DNA to nucleotides, thereby simultaneously halting host gene expression and providing raw material for virus DNA synthesis. Other early proteins and a few phage head proteins that are injected along with phage DNA modify the bacterial RNA polymerase. The polymerase then recognizes promoters on viral DNA rather than bacterial promoter sequences so that virus genes are transcribed in preference to host genes. Within five minutes, the process has progressed to the point that virus DNA synthesis commences.

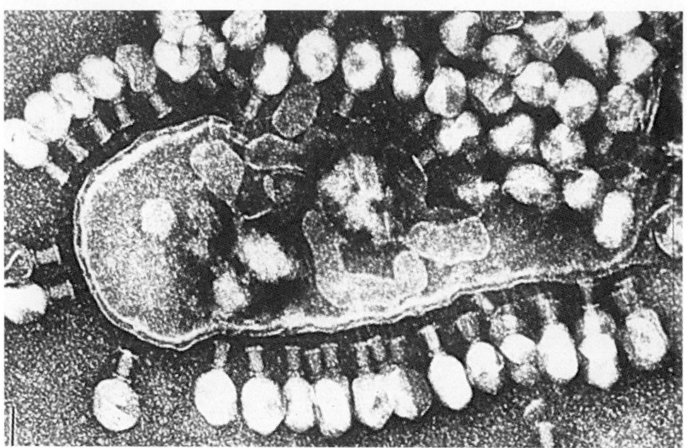

Figure 18.4 **Electron Micrograph of *E. coli* Infected with Phage T4.** Baseplates, contracted sheaths, and tail tubes can be seen (×36,500).

Promoters and transcription (pp. 201–2).

It is clear from the sophisticated control of RNA polymerase and the precise order in which events occur in the reproductive cycle that the expression of T4 genes is tightly regulated. Even the organization of the genome appears suited for efficient control of the life cycle. As can be seen in figure 18.6, genes with related functions—such as the genes for phage head or tail fiber construction—are usually clustered together. Early and late genes are also clustered separately on the genome; they are even transcribed in different directions—early genes in the counterclockwise direction and late genes, clockwise. Since transcription always proceeds in the 5′ to 3′ direction, the early and late genes are located on different DNA strands (*see chapters 10 and 12*).

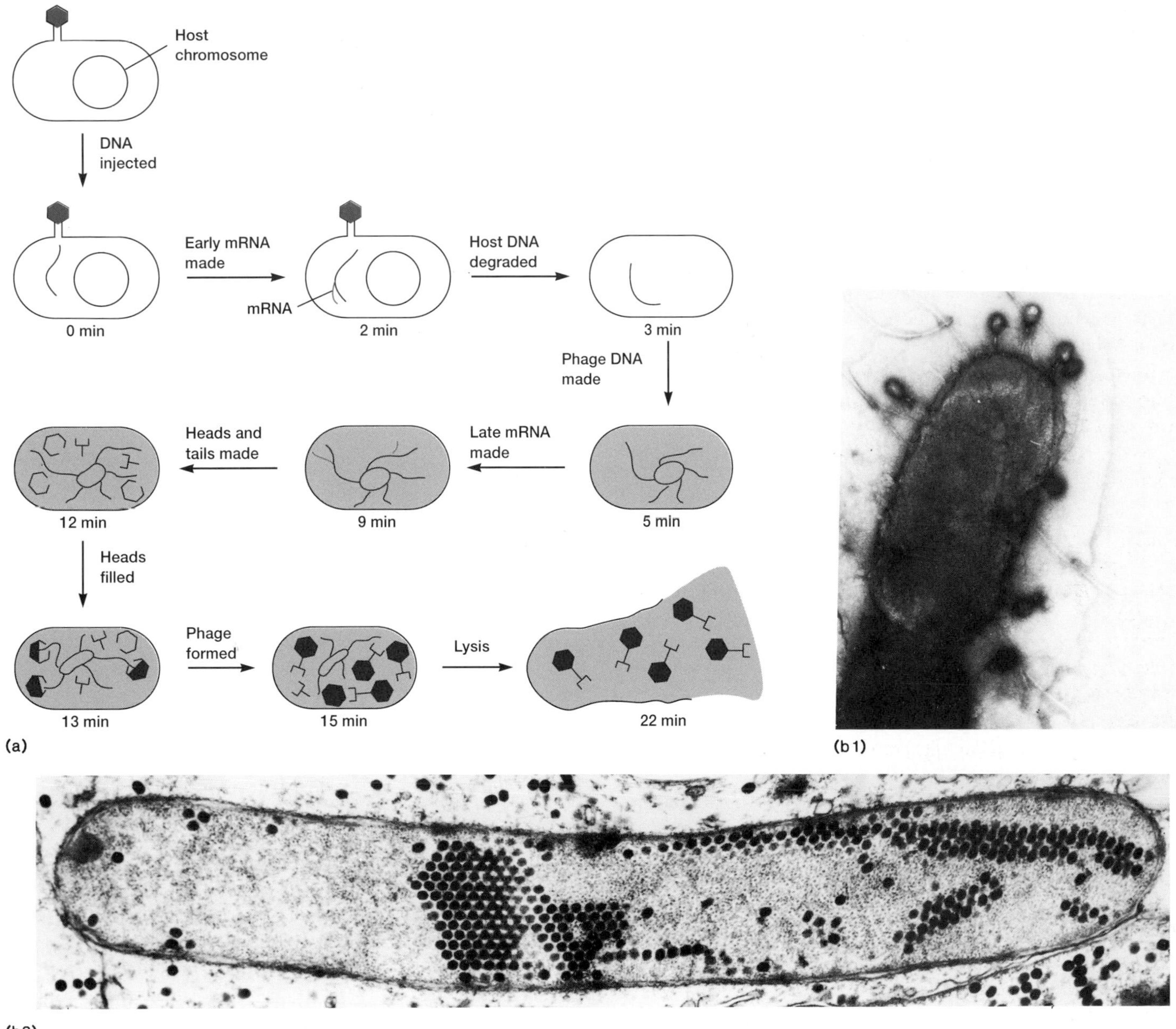

Figure 18.5 **The Life Cycle of Bacteriophage T4.** (*a*) A schematic diagram depicting the life cycle with the minutes after DNA injection given beneath each stage. mRNA is drawn in only at the stage during which its synthesis begins. (*b*) Electron micrographs showing the development of T2 bacteriophages in *E. coli.* (*b1*) Several phages are near the bacterium, and some are attached and probably injecting their DNA. (*b2*) By about 30 minutes after infection, the bacterium contains numerous completed phages.

Considerable preparation is required for synthesis of T4 DNA because it contains **hydroxymethylcytosine (HMC)** instead of cytosine (figure 18.7). HMC must be synthesized by two phage-encoded enzymes before DNA replication can begin. After T4 DNA has been synthesized, it is glucosylated by the addition of glucose to the HMC residues. Glucosylated HMC residues protect T4 DNA from attack by *E. coli* endonucleases called **restriction enzymes,** which would otherwise cleave the viral DNA at specific points and destroy it. This bacterial defense mechanism is called **restriction.** Other groups also can be used to modify phage DNA and protect it against restriction enzymes. For example, methyl groups are added to the amino groups of adenine and cytosine in lambda phage DNA for the same reason. The replication of T4 DNA is an extremely complex process requiring at least seven phage proteins. Its mechanism resembles that described in chapter 10.

Restriction enzymes and genetic engineering (pp. 286–87).

T4 DNA shows what is called terminal redundancy; that is, a base sequence is repeated at both ends of the molecule

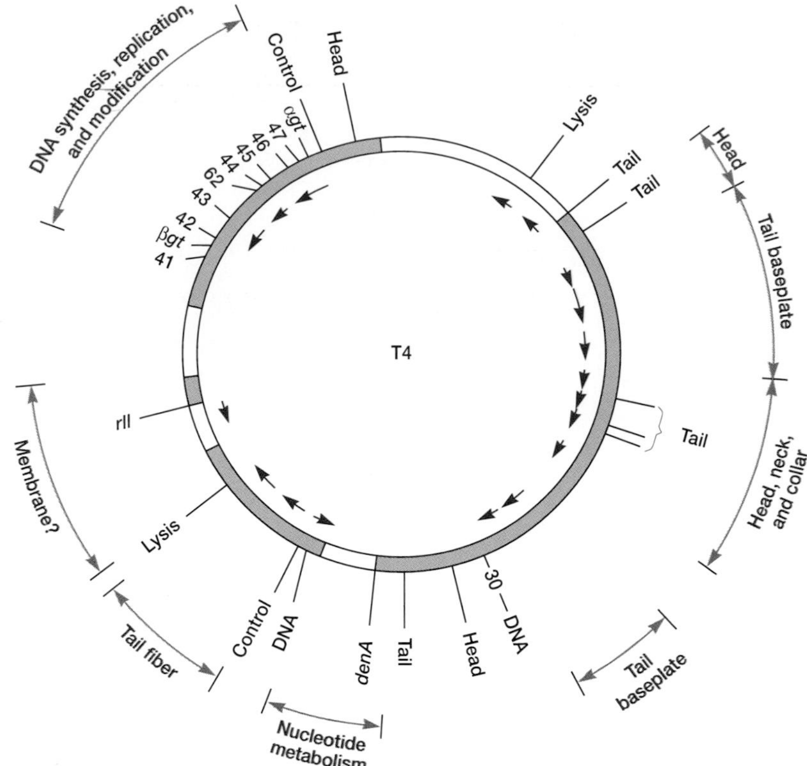

Figure 18.6 **A Map of the T4 Genome.** Some of its genes and their functions are shown. The solid regions of the map are essential to the phage life cycle. The inner colored arrows indicate the origin and direction of gene transcription. Genes with related functions tend to be clustered together.

Figure 18.7 **5-Hydroxymethylcytosine (HMC).** In T4 DNA, the HMC often has glucose attached to its hydroxyl.

(figure 18.8). When many DNA copies have been made, about 6 to 10 copies are joined by their terminally redundant ends with the aid of several enzymes (figure 18.8). These very long DNA strands composed of several units linked together are called **concatemers.** During assembly, concatemers are cleaved in such a way that the genome is slightly longer than the T4 gene set. The genetic map is therefore drawn circular (figure 18.6) because T4 DNA is circularly permuted (figure 18.9). The sequence of genes in each T4 virus of a population is the same but starts with a different gene at the 5′ end. If all the linear pieces of DNA were coiled into circles, the DNA circles would have identical gene sequences.

Thus far, only double-stranded DNA phage reproduction has been discussed, with the lytic phage T4 as an example. Before the process of phage assembly is described, the reproduction of single-stranded DNA phages will be briefly reviewed. The phage ϕX174 is a small ssDNA phage using *E. coli* as its host. Its DNA base sequence is the same as that of the viral mRNA (except that thymine is substituted for uracil) and is therefore positive. The phage DNA must be converted to a double-stranded form before either replication or transcription can occur. When ϕX174 DNA enters the host, it is immediately copied by the bacterial DNA polymerase to form a double-stranded DNA, the **replicative form** or **RF** (figure 18.10). The replicative form then directs the synthesis of more RF copies, mRNA, and copies of the +DNA genome.

The filamentous ssDNA bacteriophages behave quite differently in many respects from ϕX174 and other ssDNA phages. The fd phage is one of the best studied and is shaped like a long fiber about 6 nm in diameter by 900 to 1,900 nm in length (figure 18.1). The circular ssDNA lies in the center of the filament and is surrounded by a tube made of a small coat protein organized in a helical arrangement. The virus infects male *E. coli* cells by attaching to the tip of the pilus; the DNA enters the host along or possibly through the pilus with the aid of a special adsorption protein. A replicative form is first synthesized and then transcribed. A phage-coded protein then aids in replication of the phage DNA by use of the rolling-circle method (*see chapter 10*).

The Assembly of Phage Particles

The assembly of the T4 phage is an exceptionally complex process. **Late mRNA,** or that produced after DNA replication, directs the synthesis of three kinds of proteins: (1) phage structural proteins, (2) proteins that help with phage as-

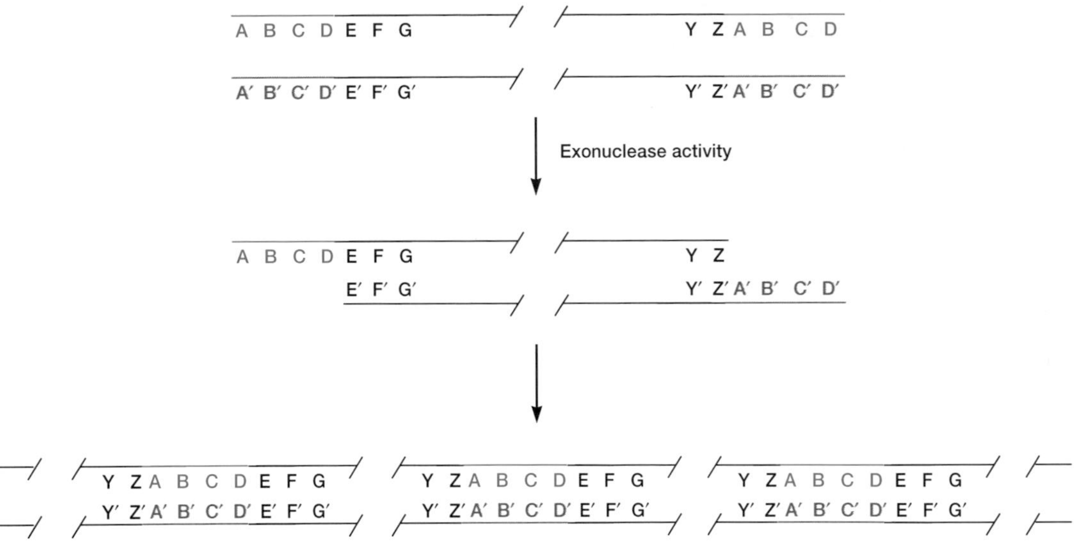

Figure 18.8 An Example of Terminal Redundancy. The gene sequences in color are terminally redundant; they are repeated at each end of the DNA molecule. This makes it possible to join several units together by their redundant ends forming a concatemer. For example, if the 3′ ends of each unit were partially digested by an exonuclease, the complementary 5′ ends would be exposed and could base pair to generate a long chain of repeating units. The breaks between terminal sequences indicate that the DNA molecules are longer than shown here.

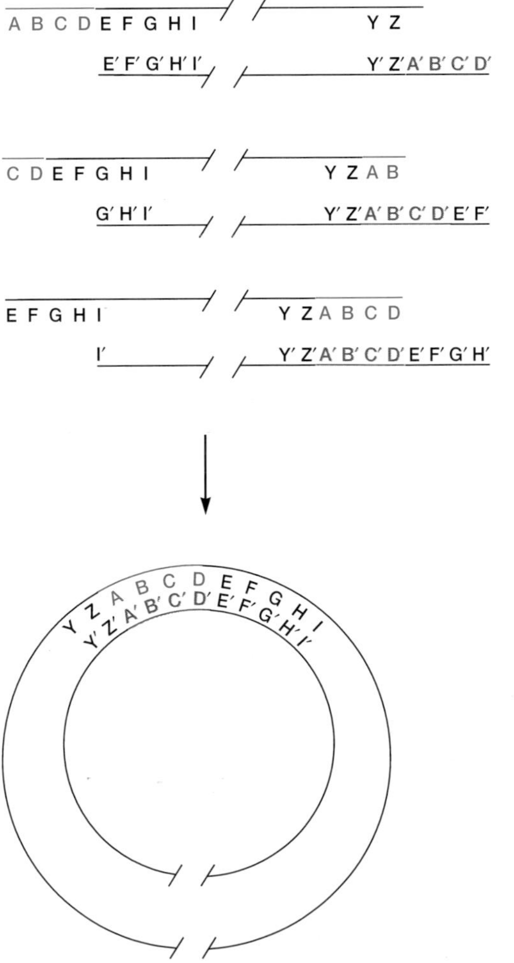

Figure 18.9 Circularly Permuted Genomes Cut from a Concatemer. The concatemer formed in figure 18.8 can be cut at any point into pieces of equal length that contain a complete complement of genes, even though different genes are found at their ends. If each piece has single-stranded cohesive ends as in figure 18.8, it will coil into a circle with the same gene order as the circles produced by other pieces.

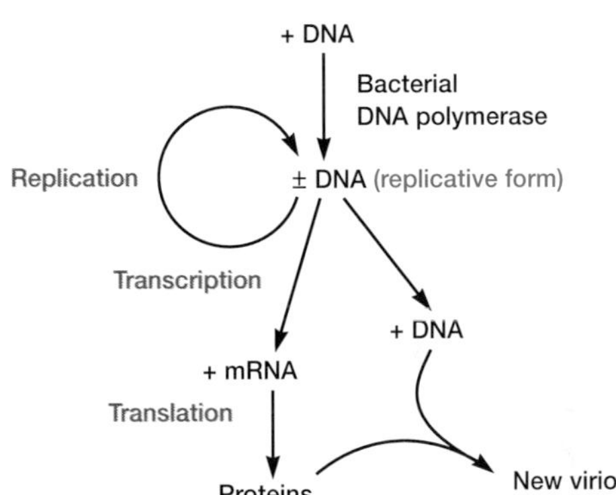

Figure 18.10 The Reproduction of ϕX174, a + Strand DNA Phage.

sembly without becoming part of the virion structure, and (3) proteins involved in cell lysis and phage release. Late mRNA transcription begins about nine minutes after T4 DNA injection into *E. coli*. All the proteins required for phage assembly are synthesized simultaneously and then used in four fairly independent subassembly lines (figure 18.11). The baseplate is constructed of 15 gene products. After the baseplate is finished, the tail tube is built on it and the sheath is assembled around the tube. The phage head shell is constructed separately of more than 10 proteins and then spontaneously combines with the tail assembly. Tail fibers attach to the baseplate after the head and tail have come together. Although many of these steps occur spontaneously, some require special virus enzymes or host cell factors.

DNA packaging within the T4 head is still a somewhat mysterious process. In some way, the DNA is drawn into the completed shell so efficiently that about 500 μm of DNA are packed into a cavity less than 0.1 μm across! It is thought that

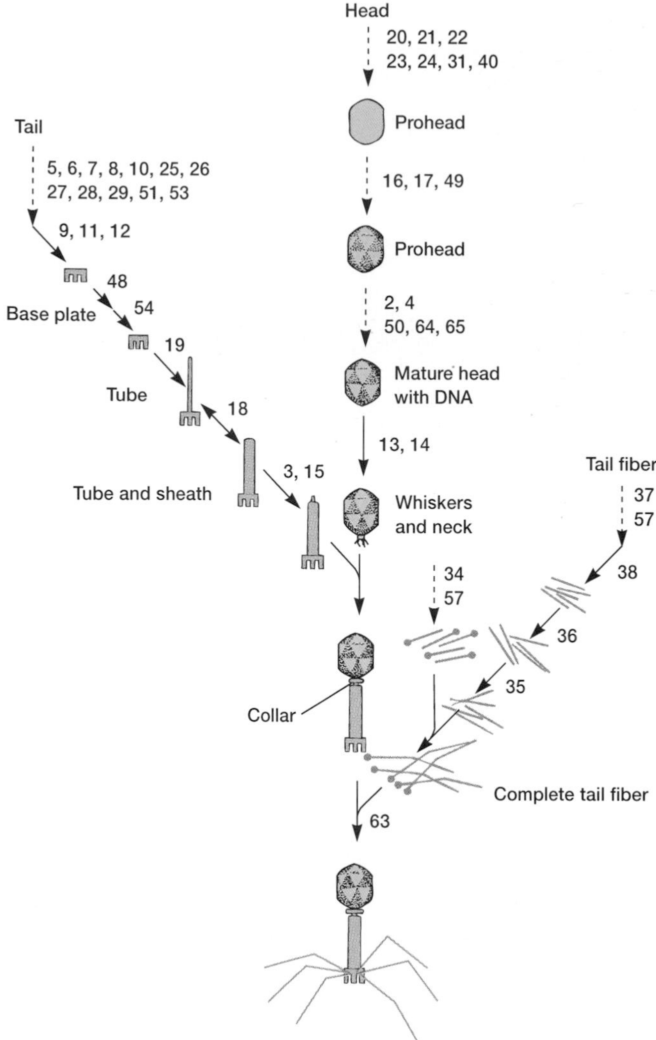

Figure 18.11 **The Assembly of T4 Bacteriophage.** Note the subassembly lines for the baseplate, tail tube and sheath, tail fibers, and head. The numbers are the genes participating in each step of the process.

a long DNA concatemer enters the head shell until it is packed full and contains about 2% more DNA than is needed for the full T4 genome. The concatemer is then cut, and T4 assembly is finished. The first complete T4 particles appear in *E. coli* at 37°C about 15 minutes after infection.

Release of Phage Particles

Many phages lyse their host cells at the end of the intracellular phase. The lysis of *E. coli* takes place after about 22 minutes at 37°C, and approximately 300 T4 particles are released. Several T4 genes are involved in this process. One directs the synthesis of a lysozyme that attacks the cell wall peptidoglycan. Another phage protein both damages the bacterial plasma membrane, allowing lysozyme to reach the cell wall beyond it, and inhibits energy metabolism reactions located in the membrane. The ssDNA virus ϕX174 is also lytic.

It should be noted that some phages are released without lysis of their host cell. For example, the filamentous fd phages do not kill their host cell but establish a symbiotic relationship in which new virions are continually released by a secretory process. Filamentous phage coat proteins are first inserted into the membrane. The coat then assembles around the viral DNA as it is secreted through the host plasma membrane (figure 18.12). The host bacteria grow and divide at a slightly reduced rate.

1. How is a one-step growth experiment carried out? Summarize what occurs in each phase of the resulting growth curve. Define latent period, eclipse period, rise period or burst, and burst size.
2. Be able to describe in some detail what is occurring in each phase of the lytic phage life cycle: adsorption and penetration, nucleic acid and protein synthesis, phage assembly, and phage release. Define the following terms: lytic cycle, receptor site, early mRNA, hydroxymethylcytosine, restriction, restriction enzymes, concatemers, replicative form, and late mRNA.
3. How does the reproduction of the ssDNA phages ϕX174 and fd differ from each other and from the dsDNA T4 phage?

Reproduction of RNA Phages

Many bacteriophages carry their genetic information as single-stranded RNA that can act as a messenger RNA and direct the synthesis of phage proteins. One of the first enzymes synthesized is **RNA replicase,** an RNA-dependent RNA polymerase (figure 18.13). The replicase then copies the original RNA (a plus strand) to produce a double-stranded intermediate (±RNA), which is called the replicative form and is analogous to the ±DNA seen in the reproduction of ssDNA phages. The replicase next uses this replicative form to synthesize thousands of copies of +RNA. Some of these plus strands are used to make more ±RNA in order to accelerate +RNA synthesis. Other +RNA acts as mRNA and directs the synthesis of phage proteins. Finally, +RNA strands are incorporated into maturing virus particles. The genome of these RNA phages serves as both a template for its own replication and an mRNA.

MS2 and Qβ are small, tailless, icosahedral ssRNA phages of *E. coli,* which have been intensely studied (figure 18.1). They attach to the F-pili of their host and enter by an unknown mechanism. These phages have only three or four genes and are genetically the simplest phages known. In MS2, one protein is involved in phage absorption to the host cell (and possibly also in virion construction or maturation). The other three genes code for a coat protein, an RNA replicase, and a protein needed for cell lysis.

Only one dsRNA phage has been discovered, the bacteriophage ϕ6 of *Pseudomonas phaseolicola* (figure 18.1). It is also unusual in possessing a membranous envelope. The icosahedral capsid within its envelope contains an RNA poly-

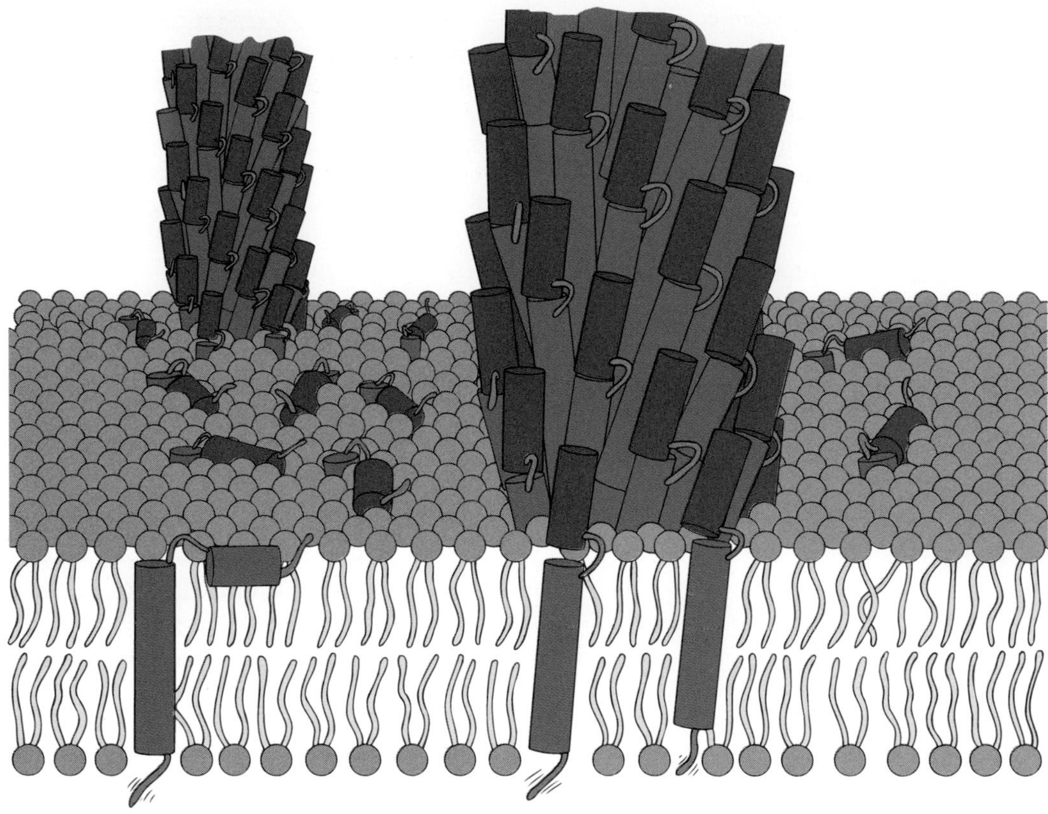

Figure 18.12 **Release of the Pf1 Phage.** The Pf1 phage is a filamentous bacteriophage that is released from *Pseudomonas aeruginosa* without lysis. In this illustration, the blue cylinders are hydrophobic α-helices that span the plasma membrane, and the red cylinders are amphipathic helices that lie on the membrane surface before virus assembly. In each protomer, the two helices are connected by a short, flexible peptide loop (yellow). It is thought that the blue helix binds with viral DNA as it is extruded through the membrane. The red helix simultaneously attaches to the growing virus coat that projects from the membrane surface. Eventually, the blue helix leaves the membrane and also becomes part of the capsid.

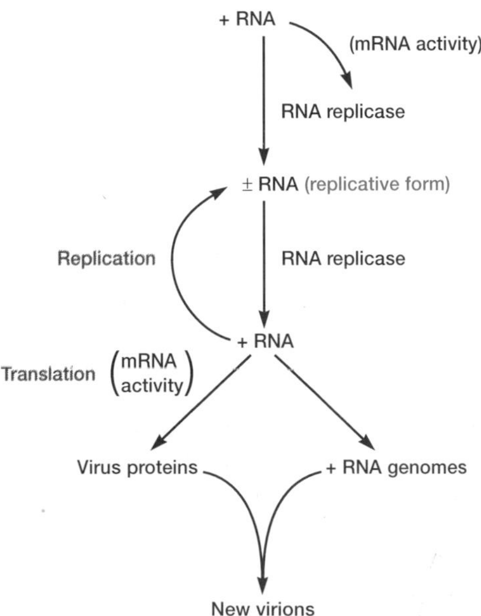

Figure 18.13 **The Reproduction of Single-stranded RNA Bacteriophages.**

merase and three dsRNA segments, each of which directs the synthesis of an mRNA. It is not yet known how the dsRNAs are replicated.

1. How are ssRNA phages reproduced, and what role does RNA replicase play in the process?
2. What is peculiar about the structure of phage $\phi 6$?

Temperate Bacteriophages and Lysogeny

Up to this point, the primary emphasis has been on **virulent bacteriophages;** these are phages that lyse their host cells during the reproductive cycle. Many DNA phages also can establish a different relationship with their host. After adsorption and penetration, the viral genome does not take control of its host and destroy it while producing new phages. Instead, the viral genome remains within the host cell and replicates with the bacterial genome to generate a clone of infected cells that may grow and divide for long periods while appearing perfectly normal. Each of these infected bacteria can produce phages and lyse under appropriate environmental conditions. They cannot, for reasons that will become clear later, be reinfected by the same virus; that is, they have immunity to superinfection. This relationship between the phage and its host is called **lysogeny.** Bacteria having the potential to produce phage particles under some conditions are said to be **lysogens** or **lysogenic,** and phages able to enter into this relationship are **temperate phages.** The latent form of the virus genome that remains within the host but does not destroy it is called the **prophage.** The prophage is usually integrated into the bacterial genome but sometimes exists independently. **Induction** is the process by which phage reproduction is initiated in a lysogenized culture. It leads to the destruction of infected cells and the release of new phages, that is, induction of the lytic cycle. Lysogeny was briefly described earlier in the context of transduction and genetic recombination, but will be discussed in more detail here.

Temperate viruses and transduction (pp. 274–75).

Most bacteriophages that have been studied are temperate, and it appears that there are advantages in being able to lysogenize bacteria. Consider a phage-infected culture that is becoming dormant due to nutrient deprivation. Before bacteria enter dormancy, they degrade their own mRNA and protein. Thus, the phage is in trouble for two reasons: it can only reproduce in actively metabolizing bacteria, and phage reproduction is usually permanently interrupted by the mRNA and protein degradation. This predicament can be avoided if the phage becomes dormant (lysogenic) at the same time as its host; in fact, nutrient deprivation does favor lysogeny. Temperate phages also have an advantage in situations where many viruses per cell initiate an infection, that is, where there is a high multiplicity of infection (MOI). When every cell is infected, the last round of replication will destroy all host cells. Thus there is a risk that the phages may be left without a host and directly exposed to environmental hazards for months or years. This prospect is avoided if lysogeny is favored by a high MOI; some bacteria will survive, carry the virus genome, and synthesize new copies as they reproduce. Not surprisingly, a high MOI does stimulate lysogeny.

A temperate phage may induce a change in the phenotype of its host cell that is not directly related to completion of its life cycle. Such a change is called a **lysogenic conversion** or a conversion and often involves alterations in bacterial surface characteristics or pathogenic properties. For example, when *Salmonella* is infected by an epsilon phage, the structure of its outer lipopolysaccharide layer (*see chapter 3*) may be modified. The phage changes the activities of several enzymes involved in construction of the lipopolysaccharide carbohydrate component and thus alters the antigenic properties of the host. These epsilon-induced changes appear to eliminate surface phage receptors and prevent infection of the lysogen by another epsilon phage. Another example is the temperate phage β of *Corynebacterium diphtheriae,* the cause of diphtheria. Only *C. diphtheriae* that is lysogenized with phage β will produce diphtheria toxin (*see chapters 30 and 37*) because the phage, not the bacterium, carries the toxin gene.

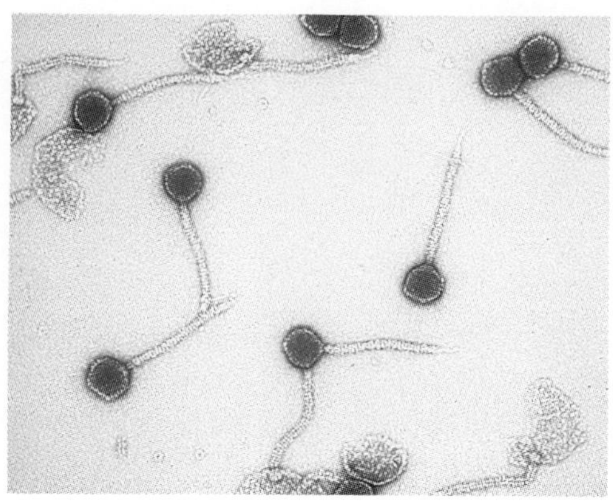
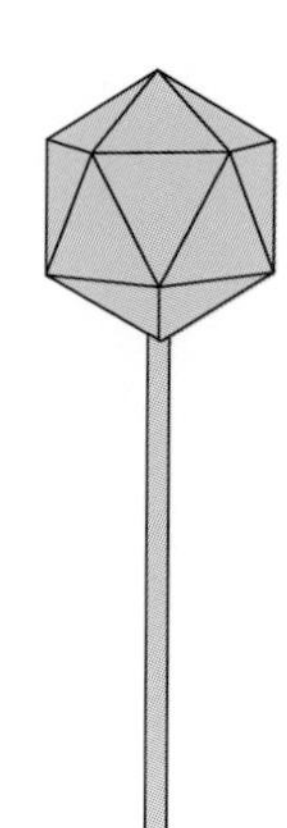

Figure 18.14 **Bacteriophage Lambda.**

Cultures of lactic acid bacteria, called starter cultures, are added to milk during the preparation of many dairy products. For example, *Streptococcus lactis* and *S. cremoris* are used in the production of cheese. One of the greatest problems for the dairy industry is the presence of bacteriophages that destroy these starter cultures. Lactic acid production by a heavily phage-infected starter culture can come to a halt within 30 minutes. The industry tries to overcome this problem by practicing aseptic techniques in order to reduce phage contamination, and by selecting for phage-resistant bacterial cultures. Starter cultures are changed regularly to reduce phage populations by removing susceptible bacterial strains. Up to 80% of the starter culture strains are lysogenized by temperate viruses, and it may be that starter culture destruction is the result of both temperate and lytic phage infections.

The lambda phage that uses the K12 strain of *E. coli* as its host is the best-understood temperate phage and will serve as our example of lysogeny. Lambda is a double-stranded DNA phage possessing an icosahedral head 55 nm in diameter and a noncontractile tail with a thin tail fiber at its end (figure 18.14). The DNA is a linear molecule with cohesive ends—single-stranded stretches, 12 nucleotides long, that have complementary base sequences and can base pair with each other. Because of these cohesive ends, the linear genome cyclizes immediately upon infection (figure 18.15). *E. coli* DNA ligase then seals the breaks, forming a closed circle. The lambda genome has been carefully mapped, and over 40 genes have been located (figure 18.16). Most genes are clustered according to their function, with separate groups involved in head synthesis, tail synthesis, lysogeny and its regulation, DNA replication, and cell lysis.

DNA ligase (p. 199).

Lambda phage can reproduce using a normal lytic cycle. Immediately after lambda DNA enters *E. coli,* it is converted to a covalent circle, and transcription by the host RNA polymerase is initiated. As shown in figure 18.16, the polymerase

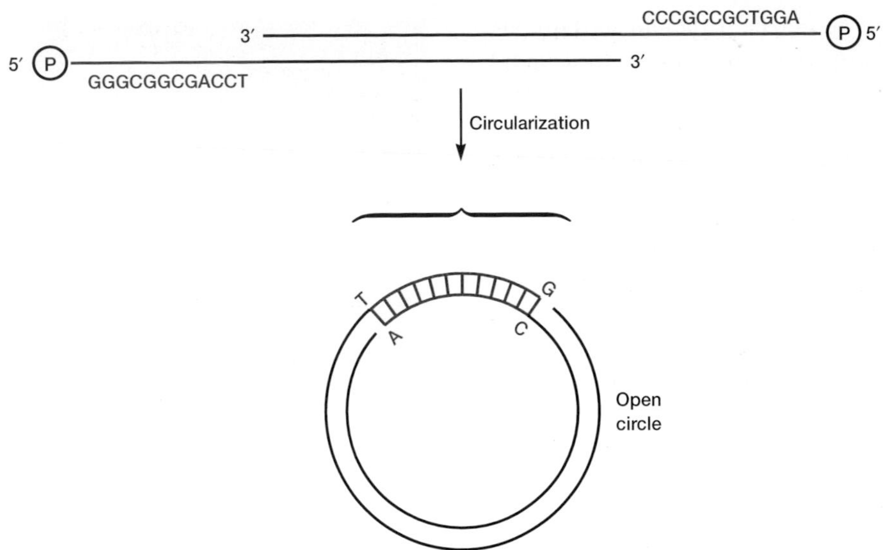

Figure 18.15 Lambda Phage DNA. A diagram of lambda phage DNA showing its 12 base, single-stranded cohesive ends (printed in color) and the circularization their complementary base sequences make possible.

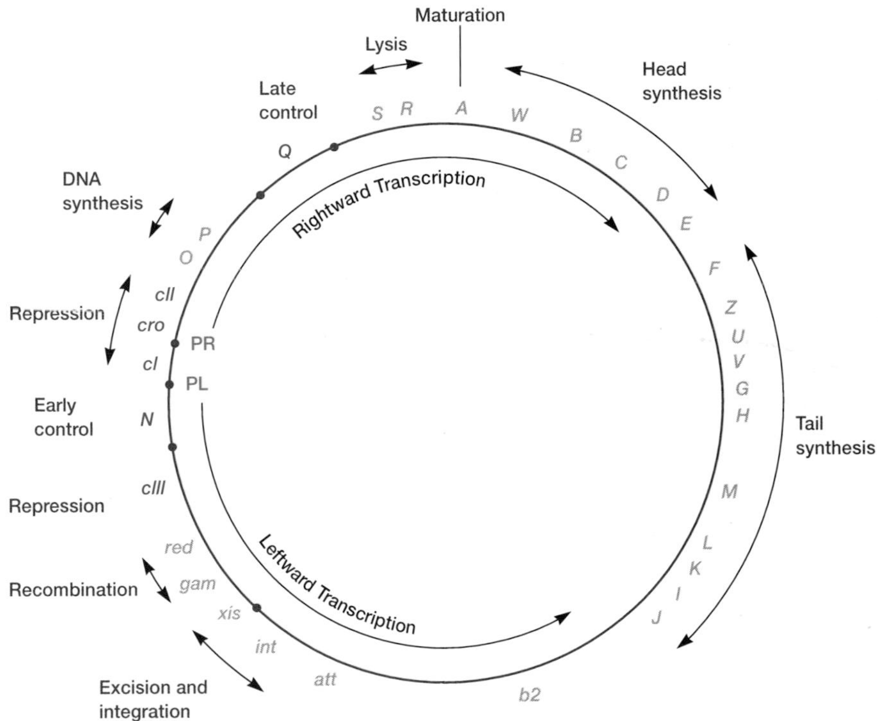

Figure 18.16 The Lambda Phage Genome. The direction of transcription and location of leftward and rightward promoters (*PL* and *PR*) are indicated on the inside of the map. The positions of major regulatory sites are shown by dots on the map and regulatory genes are in color. Lambda DNA is double stranded, and transcription proceeds in opposite directions on opposite strands.

binds to both a rightward promoter (PR) and a leftward promoter (PL) and begins to transcribe in both directions, copying different DNA strands. The first genes that are transcribed code for regulatory proteins that control the lytic cycle: leftward gene *N* and rightward genes *cro* and *cII* (figure 18.16). These and other regulatory genes ensure that virus proteins will be synthesized in an orderly time sequence and will be manufactured only when needed during the life cycle.

Regulation of transcription (pp. 222–28).

Lambda DNA replication and virion assembly are similar to the same processes already described for the T4 phage. One significant difference should be noted. Although initially bidirectional DNA replication is used and theta-shaped intermediates are seen (*see chapter 10*), lambda DNA is primarily synthesized by way of the rolling-circle mechanism to form long concatemers that are finally cleaved to give complete genomes (*see figure 10.9*).

The establishment of lysogeny and the earlier-mentioned immunity of lysogens to superinfection can be accounted for by the presence of the **lambda repressor** coded for by the *cI* gene. The repressor protein chain is 236 amino acids long and folds into a dumbbell shape with globular domains at each end (figure 18.17). One domain is concerned with binding to DNA, while the other binds with another repressor molecule to generate a dimer (the most active form of the lambda repressor). In a lysogen, the repressor is synthesized continuously and binds to the operators O_L and O_R, thereby blocking RNA polymerase activity (figure 18.18*c*). If another lambda phage tries to infect the cell, its mRNA synthesis also will be inhibited. It should be noted that immunity always involves repressor activity. A potential host cell might remain uninfected due to a mutation that alters its phage receptor site. In such an instance, it is said to be resistant, not immune, to the phage.

The sequence of events leading to the initial synthesis of repressor and the establishment of lysogeny is well known (figure 18.18). Immediately after lambda DNA has been circularized and transcription has commenced, the cII and cIII proteins accumulate. The cII protein binds next to the promoter for the *cII* gene (P_{RE}, RE stands for repressor establishment), and stimulates RNA polymerase binding. The cIII protein protects cII from degradation by a host enzyme. Lambda repressor (gpcI) is rapidly synthesized and binds to O_R and O_L, thus turning off mRNA synthesis and the production of the cII and cIII proteins. The *cI* gene continues to be transcribed at a low rate because of the activity of a second promoter (P_{RM}, RM stands for repressor maintenance) that is activated by the repressor itself. This control circuit in which lambda repressor stimulates its own synthesis ensures that lysogeny will normally be stable when once established. Indeed, if this were the whole story, lysogeny would be established every time. However, during this period the **cro protein** (gpcro) has also been accumulating. The cro protein binds to O_R and O_L, turns off the transcription of the repressor gene (as well as inhibiting the expression of other early genes), and represses P_{RM} function (figures 18.18*d* and 18.19). Since the lambda repressor can block *cro* transcription, there is a race between the production of lambda repressor and that of the cro protein. Although cro protein synthesis begins before that of the lambda repressor, gpcro binds to O_R more weakly and must rise to a higher level than the repressor before repressor synthesis is blocked and the lytic cycle started (figure 18.18). The details of this competition are not yet completely clear, but it has been shown that a number of environmental factors influence the outcome of the race and the choice between the lytic and lysogenic pathways.

If the lambda repressor wins the race, the circular lambda DNA is inserted into the *E. coli* genome as first proposed by

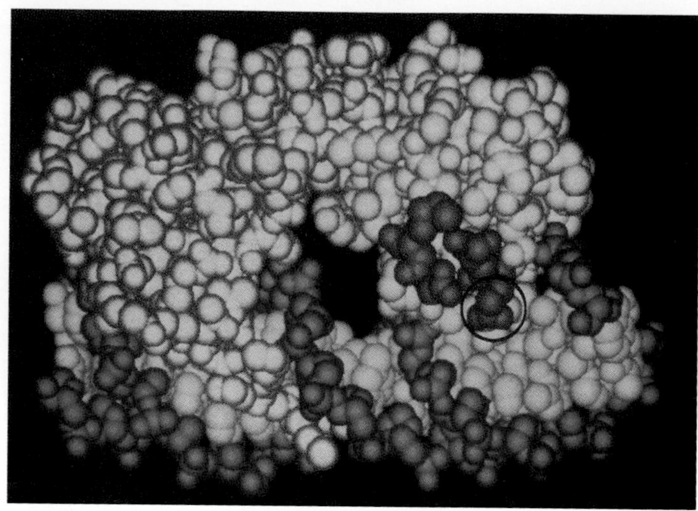

(a)

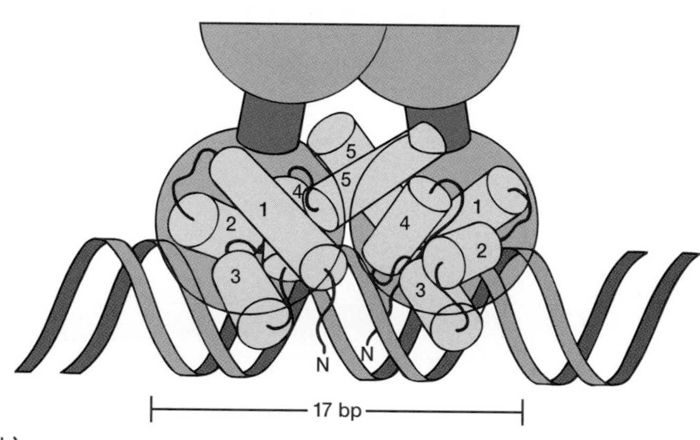

(b)

Figure 18.17 **Lambda Repressor Binding.** (*a*) A computer model of lambda repressor binding to the lambda operator. The lambda repressor dimer (brown and green) is bound to DNA (blue and white). The arms of the dimer wrap around the major grooves of the double helix. (*b*) A diagram of the lambda repressor–DNA complex. The repressor binds to a 17 bp stretch of the operator. The α3-helices make closest contact with the major grooves of the operator (the helices are labeled in order, beginning at the N terminal of the chain). *(b) From Donald Voet and Judith G. Voet,* Biochemistry. *Copyright © 1990 John Wiley & Sons, Inc., New York, NY. Reprinted by permission of John Wiley & Sons, Inc. Reprinted by permission. Originally from M. Ptashne, A Genetic Switch, p. 38, 1986. Reprinted by permission.*

Alan Campbell. **Integration** or insertion is possible because the cII protein stimulates transcription of the *int* gene at the same time as that of the *cI* gene. The *int* gene codes for the synthesis of an **integrase** enzyme, and this protein becomes plentiful before lambda repressor turns off transcription. Lambda DNA has a phage attachment site (the att site) that can base pair with a bacterial attachment site located between the galactose or gal operon and the biotin operon on the *E. coli* chromosome. After these two sites match up, the integrase enzyme, with the aid of a special host protein, catalyzes the physical exchange of viral and bacterial DNA strands (figure 18.20). The circular lambda DNA is integrated into the *E. coli* DNA as a linear region next to the gal operon and is called a prophage.

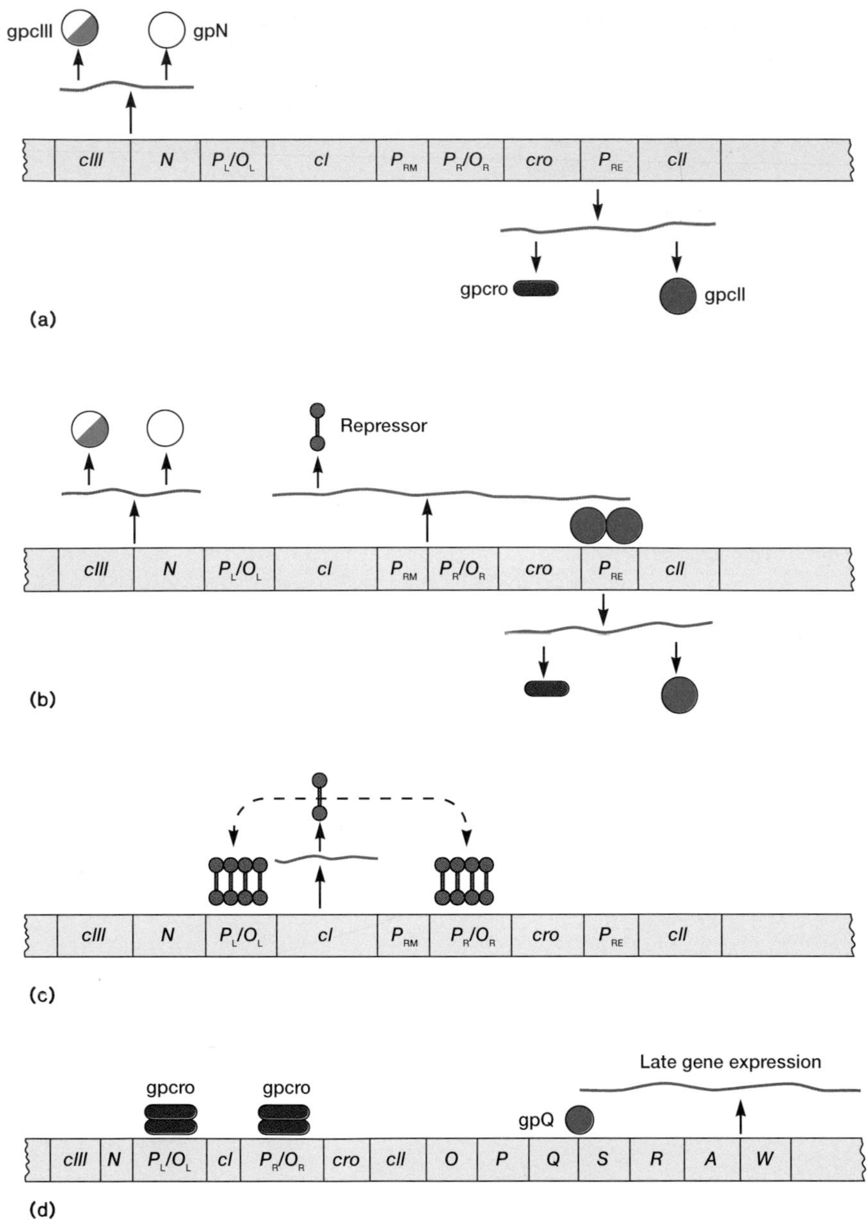

Figure 18.18 **Choice Between Lysogeny and Lysis.** Events involved in the choice between the establishment of lysogeny and continuation of the lytic cycle. The action of *gpN* is ignored for the sake of simplicity, and the scale in part (*d*) differs from that in parts (*a*)–(*c*). The abbreviation gp stands for gene product (gpcro is the product of the *cro* gene). (*a*) and (*b*) illustrate the initial steps leading to lambda repressor synthesis. (*c*) represents the situation when repressor production overcomes cro synthesis and establishes lysogeny. In part (*d*), the cro protein has accumulated more rapidly than lambda repressor and cro protein dimers (the active form) have bound to O_L and O_R. This blocks both *cI* and *cro* gene function, but not late gene expression since gpQ has already accumulated and promoted late mRNA synthesis. See text for further details.

As can be seen in figure 18.20, the linear order of phage genes has been changed or permuted during integration.

The lambda prophage will leave the *E. coli* genome and begin the production of new phages when the host is unable to survive. The process is known as induction and is triggered by a drop in lambda repressor levels. Occasionally, the repressor will spontaneously decline and the lytic cycle commence. However, induction is usually in response to environmental factors such as UV light or chemical mutagens that damage host DNA. This damage causes the recA protein, which usually plays a role in genetic recombination in *E. coli* (*see chapter 12*), to act as a protease and cleave the repressor chain between the two domains. The separated domains cannot assemble to form the normal active repressor dimer, and the lytic cycle genes become active again. There is some recent evidence that activated recA protein may not directly cleave the repressor. RecA may instead bind to the lambda repressor and stimulate it to proteolytically cleave itself. An early gene located next to the *int* gene, the *xis* gene, codes for the synthesis of an **excisionase** protein that binds to the integrase and

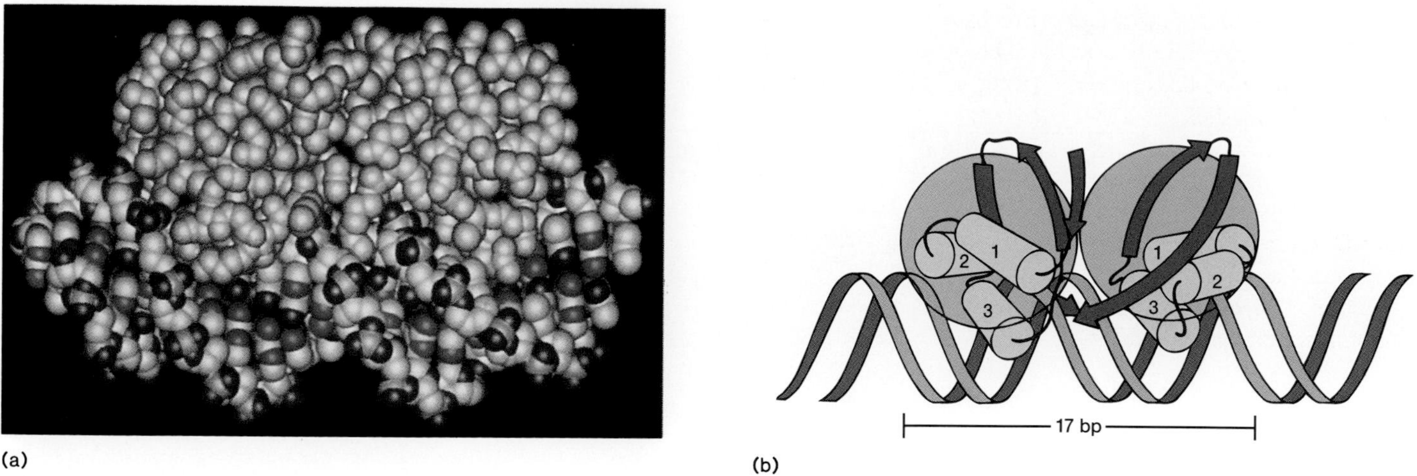

Figure 18.19 Cro Protein Binding. (*a*) A space-filling model of the cro protein–DNA complex. The cro protein is in light blue. (*b*) A diagram of the cro protein dimer–DNA complex. Like the lambda repressor protein, the cro protein binds to two adjacent DNA major grooves. *From Donald Voet and Judith G. Voet,* Biochemistry. *Copyright © 1990 John Wiley & Sons, Inc., New York, NY. Reprinted by permission of John Wiley & Sons, Inc.*

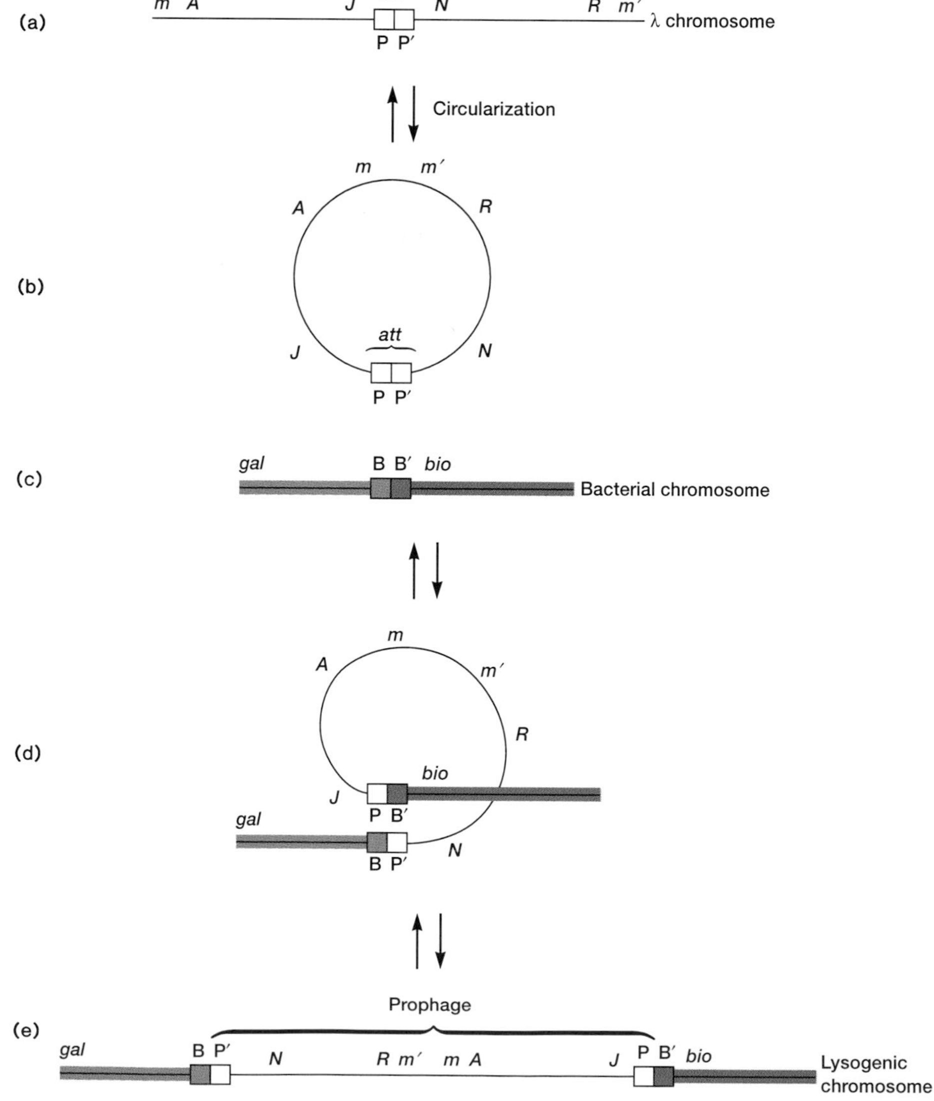

Figure 18.20 Reversible Insertion and Excision of Lambda Phage. After circularization, (*a*) and (*b*), the att site *P,P′* lines up with a corresponding bacterial sequence *B,B′* (*c*) and is integrated between the *gal* and *bio* operons to form the prophage, (*d*) and (*e*). If the process is reversed, the circular lambda chromosome will be restored and can then reproduce.

enables it to reverse the integration process and free the prophage (figure 18.20). The lytic cycle then proceeds normally.

Most temperate phages exist as integrated prophages in the lysogen. Nevertheless, integration is not an absolute requirement for lysogeny. The *E. coli* phage P1 is similar to lambda in that it circularizes after infection and begins to manufacture repressor. However, it remains as an independent circular DNA molecule in the lysogen and is replicated at the same time as the host chromosome. When *E. coli* divides, P1 DNA is apportioned between the daughter cells so that all lysogens contain one or two copies of the phage genome.

1. Define virulent phage, lysogeny, temperate phage, lysogen, prophage, immunity, and induction.
2. What advantages might a phage gain by being capable of lysogeny?
3. Describe lysogenic conversion and its significance.
4. Precisely how, in molecular terms, is a bacterial cell made lysogenic by a temperate phage like lambda?
5. How is a prophage induced to become active again?
6. Be able to describe the roles of the lambda repressor, cro protein, the recA protein, integrase, and excisionase in lysogeny and induction.
7. How does the temperate phage P1 differ from lambda phage?

Summary

1. There are four major morphological groups of phages: tailless icosahedral phages, phages with contractile tails, those with noncontractile tails, and filamentous phages.
2. The lytic cycle of virulent bacteriophages is a life cycle that ends with host cell lysis and virion release.
3. The phage life cycle can be studied with a one-step growth experiment that is divided into an initial eclipse period within the latent period, and a rise period or burst.
4. The life cycle of the T4 phage of *E. coli* is composed of several phases. In the adsorption phase, the phage attaches to a specific receptor site on the bacterial surface. This is followed by penetration of the cell wall and insertion of the viral nucleic acid into the cell.
5. Transcription of T4 DNA first produces early mRNA, which directs the synthesis of the protein factors and enzymes required to take control of the host and manufacture phage nucleic acids.
6. T4 DNA contains hydroxymethylcytosine (HMC) in place of cytosine, and glucose is often added to the HMC in order to protect the phage DNA from attack by host restriction enzymes.
7. T4 DNA replication produces concatemers, long strands of several genome copies linked together.
8. The replication of ssDNA phages proceeds through the formation of a double-stranded replicative form (RF).
9. Late mRNA is produced after DNA replication and directs the synthesis of capsid proteins, proteins involved in phage assembly, and those required for cell lysis and phage release.
10. Complete virions are assembled immediately after the separate components have been constructed.
11. T4, ϕX174, and many other phages are released upon lysis of the host cell. In contrast, the filamentous ssDNA phages are continually released without host cell lysis.
12. When ssRNA bacteriophage RNA enters a bacterial cell, it acts as a messenger and directs the synthesis of RNA replicase, which then produces double-stranded replicative forms and, subsequently, many +RNA copies.
13. The ϕ6 phage is the only dsRNA phage known. It is also unusual in having a membranous envelope.
14. Temperate phages, unlike virulent phages, often reproduce in synchrony with the host genome to yield a clone of virus-infected cells. This relationship is lysogeny, and the infected cell is called a lysogen. The latent form of the phage genome within the lysogen is the prophage.
15. Lysogeny is reversible, and the prophage can be induced to become active again and lyse its host.
16. A temperate phage may induce a change in the phenotype of its host cell that is not directly related to the completion of its life cycle. Such a change is called a conversion.
17. Two of the first proteins to appear after infection with lambda are the lambda repressor and the cro protein. The lambda repressor blocks the transcription of both cro protein and those proteins required for the lytic cycle, while the cro protein inhibits transcription of the lambda repressor gene.
18. There is a race between synthesis of lambda repressor and that of the cro protein. If the cro protein level rises high enough in time, lambda repressor synthesis is blocked and the lytic cycle initiated; otherwise, all genes other than the lambda repressor gene are repressed and the cell becomes a lysogen.
19. The final step in prophage formation is the insertion or integration of the lambda genome into the *E. coli* chromosome catalyzed by a special integrase enzyme.
20. Several environmental factors can lower repressor levels and trigger induction. The prophage becomes active and makes an excisionase protein that causes the integrase to reverse integration, free the prophage, and initiate a lytic cycle.
21. Although most temperate phages exist as integrated prophages, the phage P1 can lysogenize *E. coli* even though its DNA remains separate from the bacterial chromosome.

Key Terms

bacteriophages *368*
burst size *369*
concatemer *372*
cro protein *378*
early mRNA *370*
eclipse period *369*
excisionase *379*
hydroxymethylcytosine (HMC) *371*
induction *376*
integrase *378*
integration *378*
lambda repressor *378*
late mRNA *372*
latent period *369*
lysogen *376*
lysogenic *376*
lysogenic conversion *376*
lysogeny *376*
lytic cycle *368*
one-step growth experiment *368*
prophage *376*
receptor sites *370*
replicative form (RF) *372*
restriction *371*
restriction enzyme *371*
rise period or burst *369*
RNA replicase *374*
temperate phage *376*
virulent bacteriophage *375*

Questions for Thought and Review

1. Explain why the T4 phage genome is circularly permuted.
2. Can you think of a way to simplify further the genomes of the ssRNA phages MS2 and Qβ? Would it be possible to eliminate one of their genes? If so, which one?
3. No temperate RNA phages have yet been discovered. How might this absence be explained?
4. How might a bacterial cell resist phage infections? Give those mechanisms mentioned in the chapter and speculate on other possible strategies.

Additional Reading

Ackermann, H-W. 1987. Bacteriophage taxonomy in 1987. *Microbiol. Sci.* 4(7):214–18.

Bazinet, C., and King, J. 1985. The DNA translocating vertex of dsDNA bacteriophage. *Ann. Rev. Microbiol.* 39:109–29.

Black, L. W. 1989. DNA packaging in dsDNA bacteriophages. *Ann. Rev. Microbiol.* 43:267–92.

Bradley, D. E. 1971. A comparative study of the structure and biological properties of bacteriophages. In *Comparative virology*, ed. K. Maramorosch and E. Kurstak, 207–53. New York: Academic Press.

Campbell, A. M. 1976. How viruses insert their DNA into the DNA of the host cell. *Sci. Am.* 235(6): 103–13.

Davis, B. D.; Dulbecco, R.; Eisen, H. N.; and Ginsberg, H. S. 1990. *Microbiology*, 4th ed. Philadelphia: J. B. Lippincott.

Fiddes, J. C. 1977. The nucleotide sequence of a viral DNA. *Sci. Am.* 237(6):54–67.

Fraenkel-Conrat, H.; Kimball, P. C; and Levy J. A. 1988. *Virology, 2d ed.* Englewood Cliffs, N.J.: Prentice-Hall.

Freifelder, D. 1987. *Molecular biology: A comprehensive introduction to prokaryotes and eukaryotes*, 2d ed. New York: Van Nostrand Reinhold.

Harrison, S. C., and Aggarwal, A. K. 1990. DNA recognition by proteins with the helix-turn-helix motif. *Ann. Rev. Biochem.* 59:933–69.

Kellenberger, E. 1980. Control mechanisms governing protein interactions in assemblies. *Endeavour* 4(1):2–14.

Koerner, J. F., and Snustad, D. P. 1979. Shutoff of host macromolecular synthesis after T-even bacteriophage infection. *Microbiol. Rev.* 43(2):199–223.

Kruger, D. H., and Bickel, T. A. 1983. Bacteriophage survival: Multiple mechanisms for avoiding the deoxyribonucleic acid restriction systems of their hosts. *Microbiol. Rev.* 47(3):345–60.

Lewin, B. 1990. *Genes*, 4th ed. New York: Oxford University Press.

Luria, S. E.; Darnell, J. E., Jr.; Baltimore, D.; and Campbell, A. 1978. *General virology*, 3d ed. New York: John Wiley and Sons.

Ptashne, M. 1992. *A genetic switch*, 2d ed. Cambridge, Mass.: Blackwell Scientific Publications.

Ptashne, M.; Johnson, A. D.; and Pabo, C. O. 1982. A genetic switch in a bacterial virus. *Sci. Am.* 247(5):128–40.

Rabussay, D. 1982. Changes in *Escherichia coli* RNA polymerase after bacteriophage T4 infection. *ASM News* 48(9):398–403.

Stent, G. S., and Calendar, R. 1978. *Molecular genetics: An introductory narrative*, 2d ed. San Francisco: W. H. Freeman.

Watson, J. D.; Hopkins, N. H.; Roberts, J. W.; Steitz, J. A.; and Weiner, A. M. 1987. *Molecular biology of the gene*, 4th ed. Menlo Park, Calif.: Benjamin/Cummings.

Wood, W. B., and Edgar, R. S. 1967. Building a bacterial virus. *Sci. Am.* 217(1):60–74.

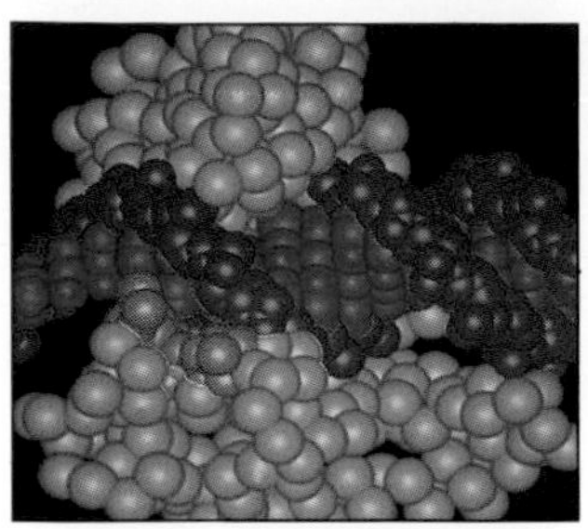

CHAPTER 19
The Viruses: Viruses of Eucaryotes

The Virus

Observe this virus: think

how small

Its arsenal, and yet how

loud its call;

It took my cell, now takes

your cell,

And when it leaves will

take our genes as well.

Genes that are master keys

to growth

That turn it on, or turn it

off, or both;

Should it return to me or

you

It will own the skeleton

keys to do

A number on our

tumblers; stage a coup.

—Michael Newman

Outline

Concepts

1. Although the details differ, animal virus reproduction is similar to that of the bacteriophages in having the same series of phases: adsorption, penetration and uncoating, replication of virus nucleic acids, synthesis and assembly of capsids, and virus release.
2. Viruses may harm their host cells in a variety of ways, ranging from direct inhibition of DNA, RNA, and protein synthesis to the alteration of plasma membranes and formation of inclusion bodies.
3. Not all animal virus infections have a rapid onset and last for a relatively short time. Some viruses establish long-term chronic infections; others are dormant for a while and then become active again. Slow virus infections may take years to develop.
4. Cancer can be caused by a number of factors, including viruses. Viruses may bring oncogenes into a cell, carry promoters that stimulate a cellular oncogene, or in other ways transform cells into tumor cells.
5. Plant viruses are responsible for many important diseases but have not been intensely studied due to technical difficulties. Most are RNA viruses. Insects are the most important transmission agents, and some plant viruses can even multiply in insect tissues before being inoculated into another plant.
6. Members of at least seven virus families infect insects; the most important belong to the *Baculoviridae, Reoviridae,* or *Iridoviridae.* Many insect infections are accompanied by the formation of characteristic inclusion bodies. A number of these viruses show promise as biological control agents for insect pests.
7. Infectious agents simpler than viruses also exist. Viroids are short strands of infectious RNA responsible for several plant diseases. Prions or virinos are somewhat mysterious proteinaceous particles associated with certain degenerative neurological diseases in humans and livestock.

In chapter 18, the bacteriophages are introduced in some detail because they are very important to the fields of molecular biology and genetics, as well as to virology. The present chapter focuses on viruses that use eucaryotic organisms as hosts. Although plant and insect viruses are discussed, particular emphasis is placed on animal viruses since they are very well studied and the causative agents of so many important human diseases. The chapter closes with a brief summary of what is known about infectious agents that are even simpler in construction than viruses: the viroids and prions.

The chapter begins with a discussion of animal viruses. These viruses are not only of great practical importance but are the best studied of the virus groups to be described in this chapter.

Classification of Animal Viruses

When microbiologists first began to classify animal viruses, they naturally thought in terms of features such as the host preferences of each virus. Unfortunately, not all criteria are equally useful. For example, many viruses will infect a variety of animals, and a particular animal can be invaded by several dissimilar viruses. Thus, virus host preferences lack the specificity to distinguish precisely among different viruses. Modern classifications are primarily based upon virus morphology, the physical and chemical nature of virion constituents, and genetic relatedness.

Morphology is probably the most important characteristic in virus classification. Animal viruses can be studied with the transmission electron microscope while still in the host cell or after release. As mentioned in chapter 17, the nature of virus nucleic acids is also extremely important. Nucleic acid properties such as the general type (DNA or RNA), strandedness, size, and segmentation are all useful. Genetic relatedness can be estimated by techniques such as nucleic acid hybridization, nucleic acid and protein sequencing, and by determining the ability to undergo recombination.

A brief diagrammatic description of DNA and RNA animal virus classification is presented in figures 19.1 and 19.2. Figure 19.3 summarizes pictorially much of the same material.

1. List the most important characteristics used in identifying animal viruses (figures 19.1, 19.2, and 19.3). What other properties can be used to establish the relatedness of different viruses?

Reproduction of Animal Viruses

The reproduction of animal viruses is very similar in many ways to that of phages. Animal virus reproduction may be divided into several stages: adsorption, penetration and uncoating, replication of virus nucleic acids, synthesis and assembly of virus capsids, and release of mature viruses. Each of these stages will be briefly described.

Adsorption of Virions

The first step in the animal virus cycle is adsorption to the host cell surface. This occurs through a random collision of the virion with a plasma membrane receptor site protein, frequently a glycoprotein (a protein with carbohydrate covalently attached). Because the capacity of a virus to infect a cell depends greatly upon its ability to bind to the cell, the distribution of these receptor proteins plays a crucial role in the tissue and host specificity of animal viruses. For example, poliovirus receptors are found only in the human nasopharynx, gut, and spinal cord anterior horn cells; in contrast, measles virus receptors are present in most tissues. The dissimilarity in the distribution of receptors for these two viruses helps explain the difference in the nature of polio and measles.

The specific host cell receptor proteins to which viruses attach vary greatly, but they are always surface proteins necessary to the cell. As will be discussed shortly, viruses often enter cells by endocytosis. They trick the host cell by attaching to surface molecules that are normally taken up by endocytosis. Thus, they are passively carried into the cell. These host cell surface proteins usuallly are receptors that bind hormones and other important molecules essential to the cell's function and role in the body (table 19.1).

Table 19.1 Examples of Host Cell Surface Proteins that Serve as Virus Receptors

Virus	Cell Surface Protein
Epstein-Barr virus	Receptor for the C3d complement protein on human B lymphocytes
Hepatitis A virus	Alpha 2-macroglobulin
Herpes simplex virus, type 1	Fibroblast growth factor receptor
Human immunodeficiency virus	CD4 protein on T-helper cells, macrophages, and monocytes
Poliovirus	Neuronal cellular adhesion molecule (NCAM)
Rabies virus	Acetylcholine receptor on neurons
Rhinovirus	Intercellular adhesion molecules (ICAMs) on the surface of respiratory epithelial cells
Reovirus, type 3	β-adrenergic receptor
Vaccinia virus	Epidermal growth factor receptor

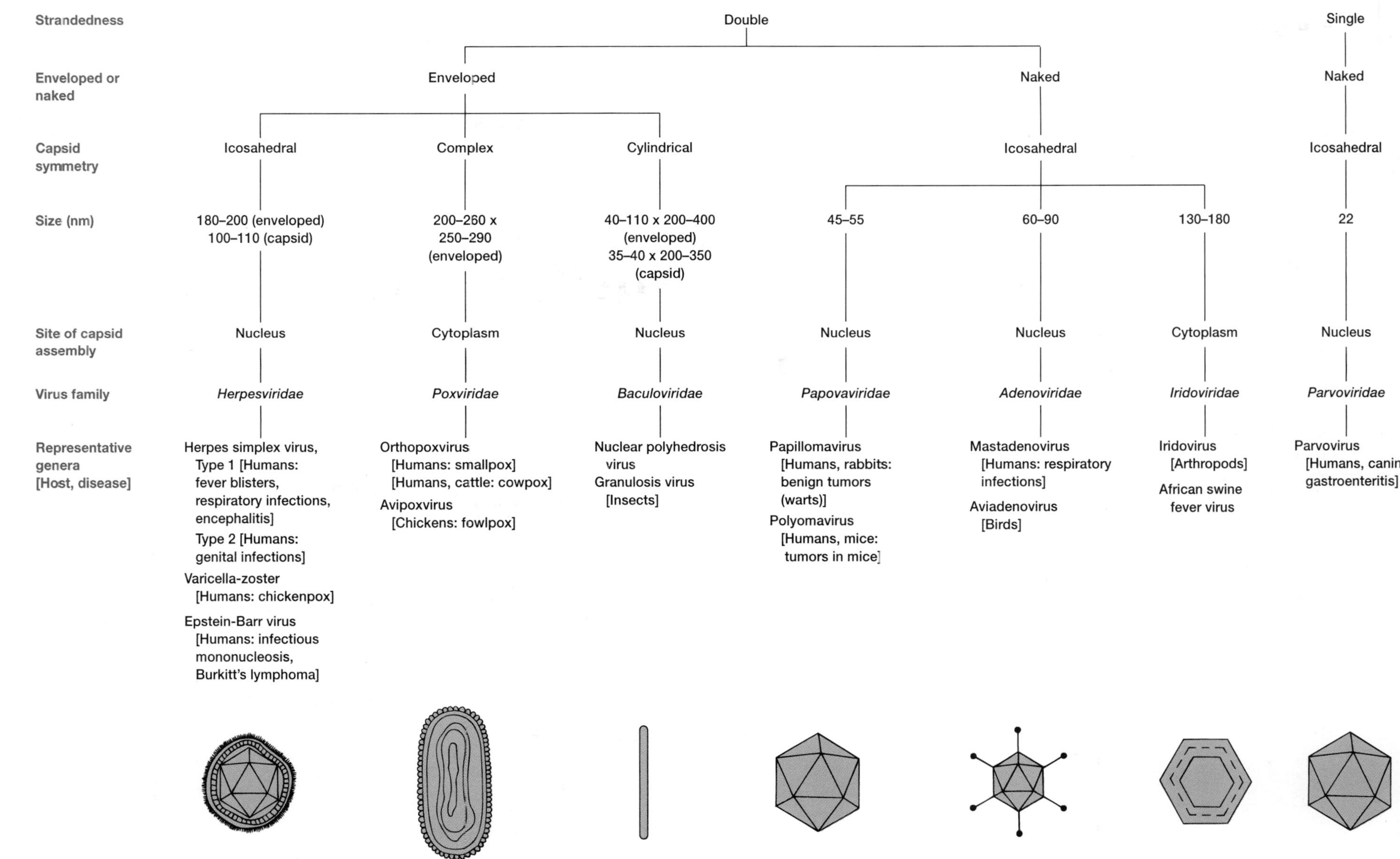
Strandedness
Double
Single
Enveloped or naked
Enveloped
Naked
Naked
Capsid symmetry
Icosahedral
Complex
Cylindrical
Icosahedral
Icosahedral
Size (nm)
180–200 (enveloped) 100–110 (capsid)
200–260 x 250–290 (enveloped)
40–110 x 200–400 (enveloped) 35–40 x 200–350 (capsid)
45–55
60–90
130–180
22
Site of capsid assembly
Nucleus
Cytoplasm
Nucleus
Nucleus
Nucleus
Cytoplasm
Nucleus
Virus family
Herpesviridae
Poxviridae
Baculoviridae
Papovaviridae
Adenoviridae
Iridoviridae
Parvoviridae
Representative genera [Host, disease]
Herpes simplex virus, Type 1 [Humans: fever blisters, respiratory infections, encephalitis]
Type 2 [Humans: genital infections]
Varicella-zoster [Humans: chickenpox]
Epstein-Barr virus [Humans: infectious mononucleosis, Burkitt's lymphoma]
Orthopoxvirus [Humans: smallpox] [Humans, cattle: cowpox]
Avipoxvirus [Chickens: fowlpox]
Nuclear polyhedrosis virus
Granulosis virus [Insects]
Papillomavirus [Humans, rabbits: benign tumors (warts)]
Polyomavirus [Humans, mice: tumors in mice]
Mastadenovirus [Humans: respiratory infections]
Aviadenovirus [Birds]
Iridovirus [Arthropods]
African swine fever virus
Parvovirus [Humans, canines: gastroenteritis]

Figure 19.1 The Taxonomy of DNA Animal Viruses.

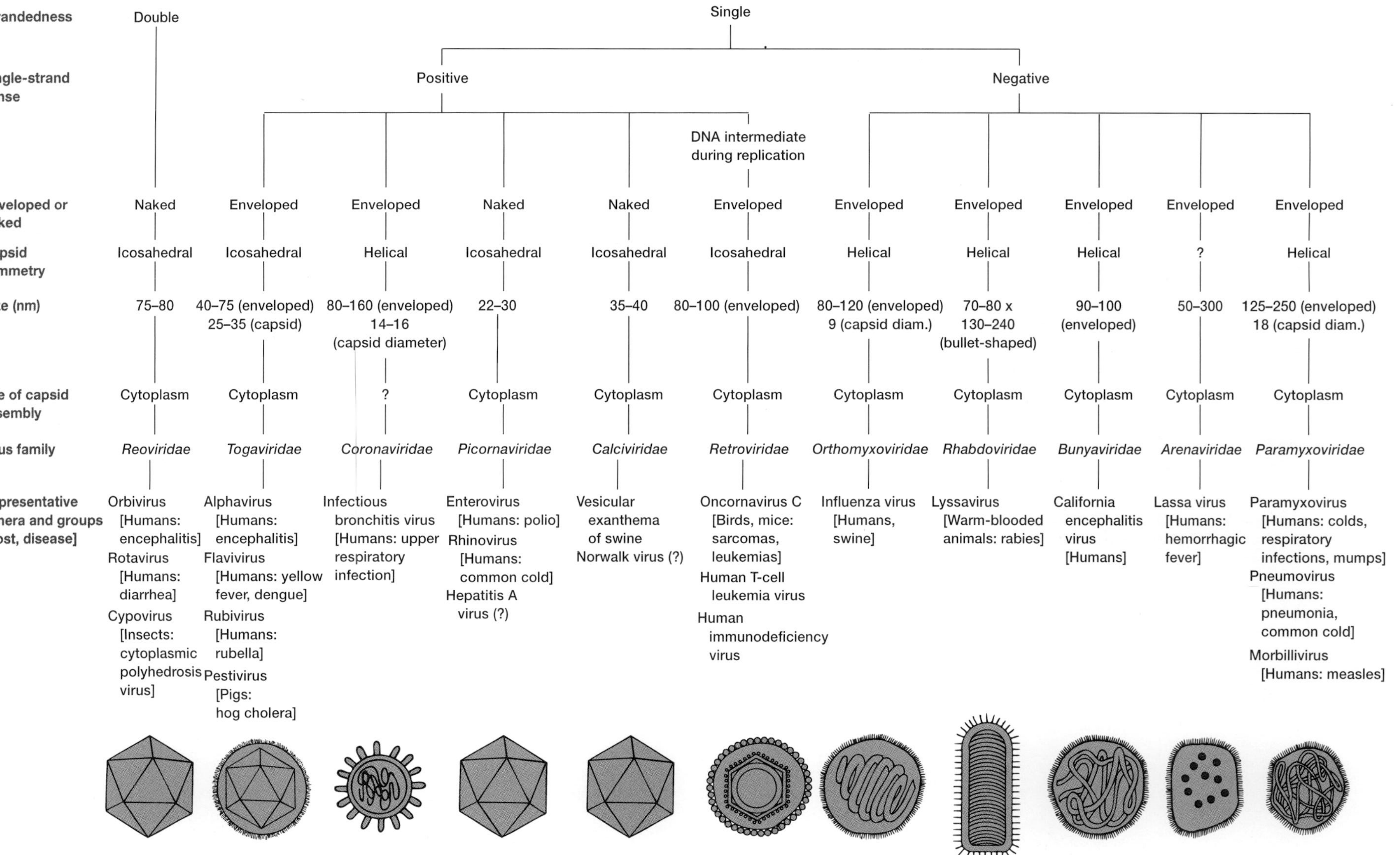
Strandedness
Double
Single
Single-strand sense
Positive
Negative
DNA intermediate during replication
Enveloped or naked
Capsid symmetry
Size (nm)
Site of capsid assembly
Virus family
Representative genera and groups [Host, disease]
Naked
Icosahedral
75–80
Cytoplasm
Reoviridae
Orbivirus [Humans: encephalitis]
Rotavirus [Humans: diarrhea]
Cypovirus [Insects: cytoplasmic polyhedrosis virus]
Enveloped
Icosahedral
40–75 (enveloped) 25–35 (capsid)
Cytoplasm
Togaviridae
Alphavirus [Humans: encephalitis]
Flavivirus [Humans: yellow fever, dengue]
Rubivirus [Humans: rubella]
Pestivirus [Pigs: hog cholera]
Enveloped
Helical
80–160 (enveloped) 14–16 (capsid diameter)
?
Coronaviridae
Infectious bronchitis virus [Humans: upper respiratory infection]
Naked
Icosahedral
22–30
Cytoplasm
Picornaviridae
Enterovirus [Humans: polio]
Rhinovirus [Humans: common cold]
Hepatitis A virus (?)
Naked
Icosahedral
35–40
Cytoplasm
Calciviridae
Vesicular exanthema of swine
Norwalk virus (?)
Enveloped
Icosahedral
80–100 (enveloped)
Cytoplasm
Retroviridae
Oncornavirus C [Birds, mice: sarcomas, leukemias]
Human T-cell leukemia virus
Human immunodeficiency virus
Enveloped
Helical
80–120 (enveloped) 9 (capsid diam.)
Cytoplasm
Orthomyxoviridae
Influenza virus [Humans, swine]
Enveloped
Helical
70–80 x 130–240 (bullet-shaped)
Cytoplasm
Rhabdoviridae
Lyssavirus [Warm-blooded animals: rabies]
Enveloped
Helical
90–100 (enveloped)
Cytoplasm
Bunyaviridae
California encephalitis virus [Humans]
Enveloped
?
50–300
Cytoplasm
Arenaviridae
Lassa virus [Humans: hemorrhagic fever]
Enveloped
Helical
125–250 (enveloped) 18 (capsid diam.)
Cytoplasm
Paramyxoviridae
Paramyxovirus [Humans: colds, respiratory infections, mumps]
Pneumovirus [Humans: pneumonia, common cold]
Morbillivirus [Humans: measles]

Figure 19.2 The Taxonomy of RNA Animal Viruses.

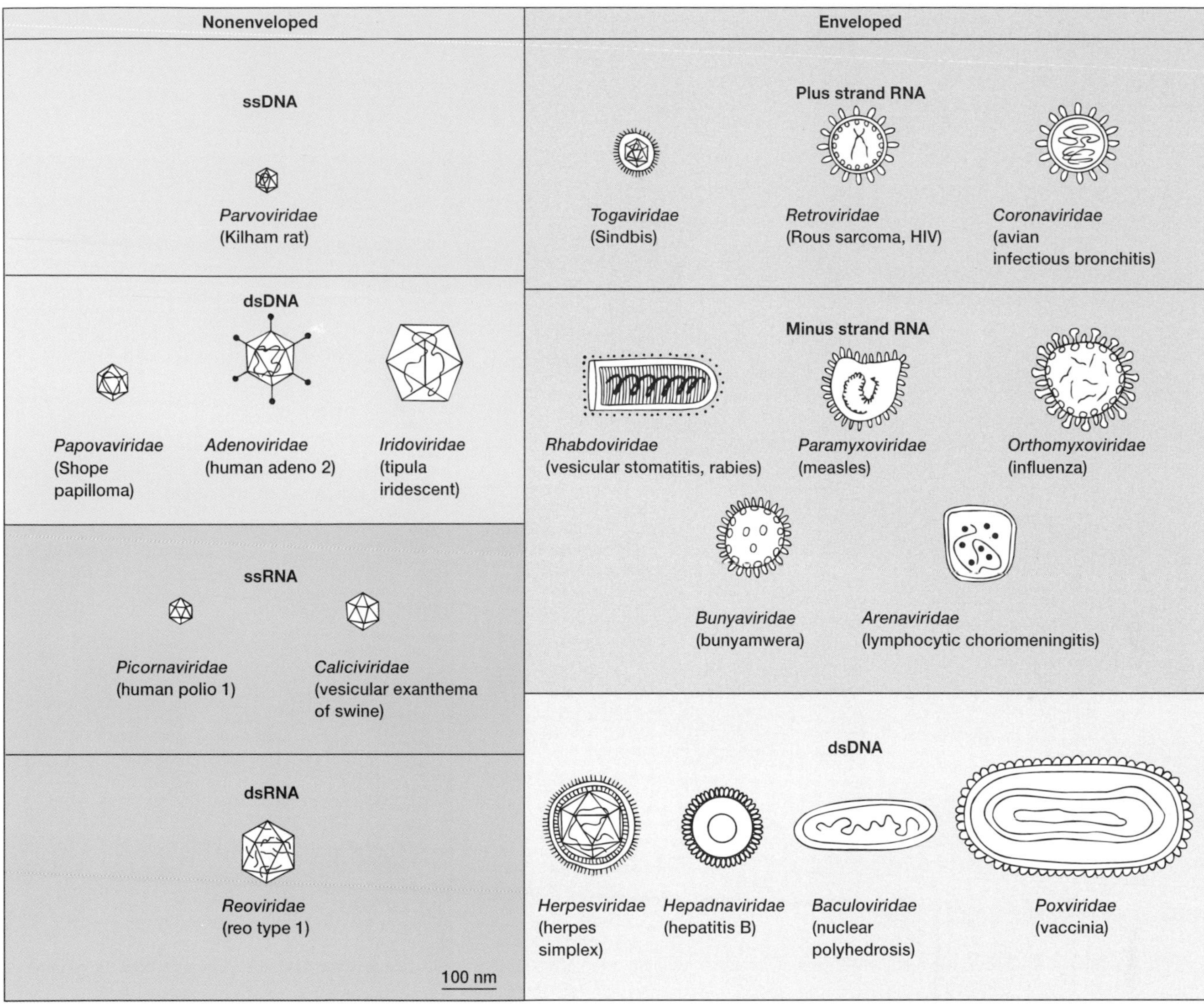

Figure 19.3 **A Diagrammatic Description of the Major Animal Virus Families.**

The surface site on the virus can consist simply of a capsid structural protein or an array of such proteins. In some viruses, for example the poliovirus and rhinoviruses, the binding site is at the bottom of a surface depression or valley. The site can bind to host cell surface projections, but cannot be reached by host antibodies. In other cases, the virus attaches to the host cell through special projections such as the fibers extending from the corners of adenovirus icosahedrons (*see figure 17.12*h) or the spikes of enveloped viruses. As mentioned in chapter 17, the influenza virus has two kinds of spikes: hemagglutinin and neuraminidase (*see figure 17.17*a,b). The hemagglutinin spikes appear to be involved in attachment to the host cell receptor site. Influenza neuraminidase may aid the virus in penetrating nasal and respiratory tract secretions by degrading mucopolysaccharides; it also may directly aid in adsorption.

Human virus diseases (chapter 36).

Penetration and Uncoating

Viruses penetrate the plasma membrane and enter a host cell shortly after adsorption. Virus uncoating, the removal of the capsid and release of viral nucleic acid, occurs during or shortly after penetration. Both events will be reviewed together. The mechanisms of penetration and uncoating must vary with the type of virus because viruses differ so greatly in structure and mode of reproduction. For example, enveloped viruses may enter cells in a different way than naked or unenveloped virions. Furthermore, some viruses only inject their nucleic acid, while others must ensure that a virus-associated RNA or DNA polymerase also enters the host cell along with the virus genome. The entire process from adsorption to final uncoating may take from minutes to several hours.

The detailed mechanisms of penetration and uncoating are still unclear; it is possible that three different modes of entry may be employed (figure 19.4). At least some naked viruses

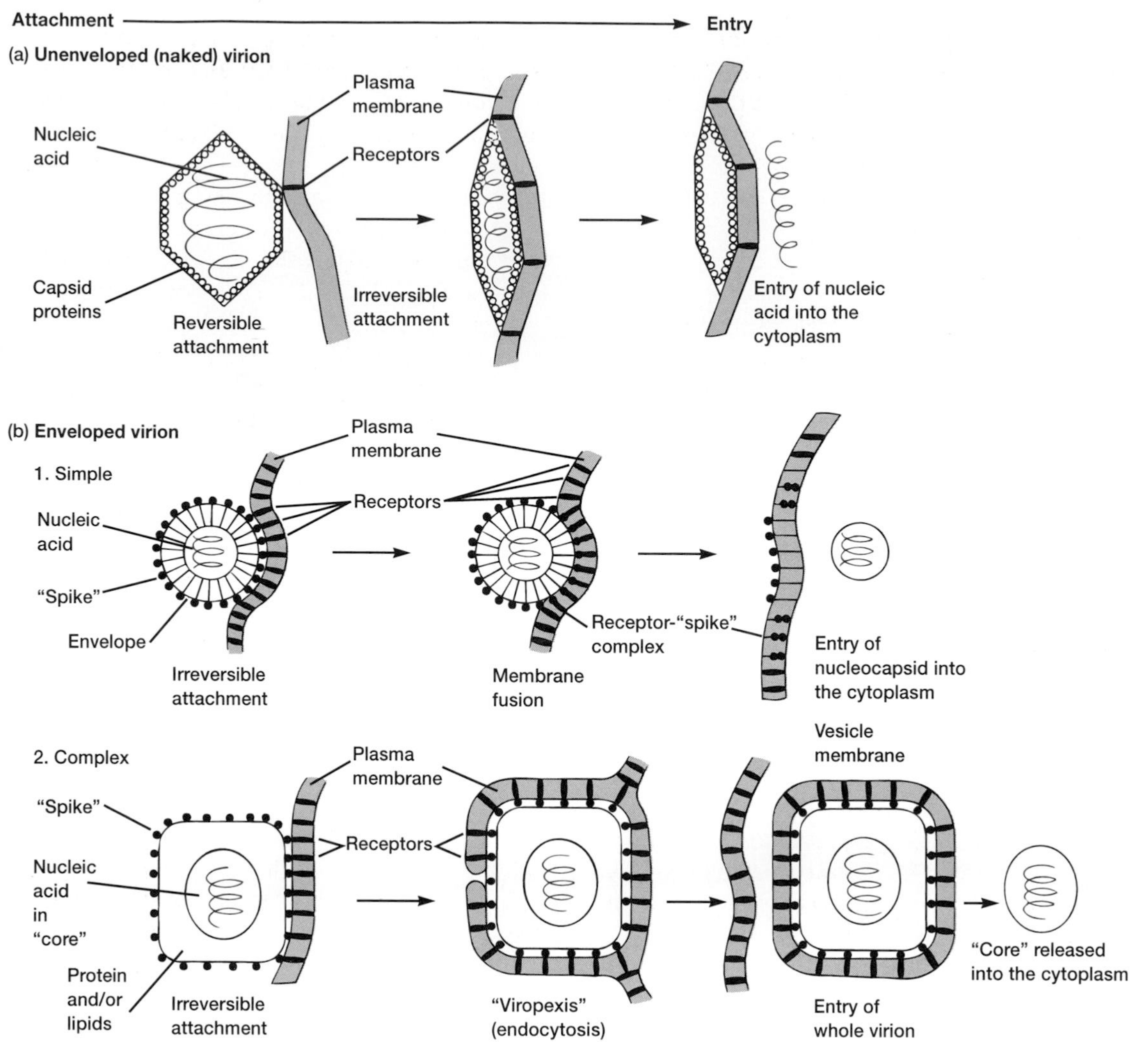

Figure 19.4 **Animal Virus Entry.** Mechanisms of animal virus attachment and entry into host cells. See text for description of the three entry modes.

such as the poliovirus may undergo a major change in capsid structure upon adsorption to the plasma membrane, so that only their nucleic acids are released in the cytoplasm. The envelope of paramyxoviruses, and possibly some other enveloped viruses, seems to fuse directly with the host cell plasma membrane, and the nucleocapsid is deposited in the cytoplasmic matrix. A virus polymerase, associated with the nucleocapsid, then begins transcribing the virus RNA while it is still within the capsid. Most enveloped viruses may enter cells by a third route: through engulfment by receptor-mediated endocytosis to form **coated vesicles.** The virions attach to coated pits, specialized membrane regions coated on the cytoplasmic side with the protein clathrin. The pits then pinch off to form coated vesicles filled with viruses, and these fuse with lysosomes after the clathrin has been removed. Lysosomal enzymes may aid in virus uncoating. In at least some instances, the virus envelope fuses with the lysosomal membrane, and the capsid (which may have been partially removed by lysosomal enzymes) is released into the cytoplasmic matrix. Once in the cytoplasm, viral nucleic acid may be released from the capsid or may function while still attached to capsid components.

Replication and Transcription in DNA Viruses

The early part of the synthetic phase, governed by the **early genes,** is devoted to taking over the host cell and to the synthesis of viral DNA and RNA. Some virulent animal viruses inhibit host cell DNA, RNA, and protein synthesis, though cellular DNA is not usually degraded. In contrast, nonvirulent viruses may actually stimulate the synthesis of host macromolecules. DNA replication usually occurs in the host cell nucleus; poxviruses are exceptions since their genomes are replicated in the cytoplasm. Messenger RNA—at least early mRNA—is transcribed from DNA by host enzymes, except for poxvirus early mRNA, which is synthesized by a viral polymerase. Some examples of DNA virus reproduction will help illustrate these generalizations.

Parvoviruses, with a genome composed of one small, single-stranded DNA molecule about 4,800 bases in size, are the simplest of the DNA viruses (*see figure 17.12*a). The genome is so small that it directs the synthesis of only three polypeptides, all capsid components. Even so, the virus must resort to the use of overlapping genes to fit three genes into such a small molecule. That is, the base sequences that code for the three polypeptide chains overlap each other and are read with different reading frames (*see chapter 12*). Since the genome does not code for any enzymes, the virus must use host cell enzymes for all biosynthetic processes. Thus, viral DNA can only be replicated in the nucleus during the S period of the cell cycle, when the cell replicates its own DNA.

The genetic code and gene structure (pp. 239–44).

Herpesviruses are a large group of icosahedral, enveloped, double-stranded DNA viruses responsible for many important human and animal diseases (*see figure 17.17*e). The genome is a linear piece of DNA about 160,000 base pairs in size and contains at least 50 to 100 genes. Immediately upon uncoating, the DNA is transcribed by host RNA polymerase to form messengers directing the synthesis of several early proteins, principally regulatory proteins and the enzymes required for virus DNA replication. DNA replication with a virus-specific DNA polymerase begins in the cell nucleus within four hours after infection. Host DNA synthesis gradually slows during a lethal virus infection (not all herpes infections result in immediate cell death).

Herpesviruses and disease (pp. 729–31).

Poxviruses such as the vaccinia virus are the largest viruses known and are morphologically complex (*see figure 17.18*). Their double-stranded DNA possesses over 200 genes. The virus enters through receptor-mediated endocytosis in coated vesicles; the central core escapes from the lysosome and enters the cytoplasmic matrix. The core contains both DNA and a DNA-dependent RNA polymerase that synthesizes early mRNAs, one of which directs the production of an enzyme that completes virus uncoating. DNA polymerase and other enzymes needed for DNA replication are also synthesized early in the reproductive cycle, and replication begins about 1.5 hours after infection. About the time DNA replication commences, late mRNA transcription is initiated. Many late proteins are structural proteins used in capsid construction. The complete reproductive cycle in poxviruses takes about 24 hours.

Poxvirus diseases (p. 720).

Replication and Transcription in RNA Viruses

The RNA viruses are much more diverse in their reproductive strategies than are the DNA viruses. Most RNA viruses can be placed in one of four general groups based on their modes of replication and transcription, and their relationship to the host cell genome. Figure 19.5 summarizes the reproductive cycles characteristic of these groups. Transcription patterns are discussed first, and then mechanisms of RNA replication.

Transcription in RNA viruses other than the retroviruses (retroviruses are considered shortly) varies with the nature of the virus genome. The picornaviruses such as poliovirus are the best studied positive strand ssRNA viruses. They use their RNA genome as a giant mRNA, and host ribosomes synthesize an enormous peptide that is then cleaved or processed by both host and viral encoded enzymes to form the proper polypeptides (figure 19.5*a*). In contrast, because their genome is complementary to the mRNA base sequence, negative strand ssRNA viruses (e.g., the orthomyxoviruses and paramyxoviruses) must employ a virus-associated RNA-dependent RNA polymerase or **transcriptase** to synthesize mRNA (figure 19.5*c*). The dsRNA reoviruses carry a virus-associated transcriptase that copies the negative strand of their genome to generate mRNA. Later a new, virus-encoded polymerase continues transcription.

The nature of RNA replication also varies logically with the type of virus genetic material. Viral RNA is replicated in the host cell cytoplasmic matrix. Single-stranded RNA viruses, except retroviruses, use a viral **replicase** that converts the ssRNA into a double-stranded RNA called the **replicative form** (figure 19.5*a,c*). The appropriate strand of this intermediate then directs the synthesis of new viral RNA genomes. Reoviruses differ significantly from this pattern (figure 19.5*b*). The virion contains 10 to 13 different dsRNAs, each coding for an mRNA. Late in the reproductive cycle, a copy of each mRNA associates with the other mRNAs and special proteins to form a large complex. The RNAs in this complex are then copied by the viral replicase to form a double-stranded genome that is incorporated into a new virion.

Retroviruses such as the human immunodeficiency virus (*see figure 17.17*d) possess ssRNA genomes but differ from other RNA viruses in that they synthesize mRNA and replicate their genome by means of DNA intermediates. The virus has an RNA-dependent DNA polymerase or **reverse transcriptase** that copies the +RNA genome to form a −DNA copy (figures 19.5*d* and 19.6*a*). Interestingly, transfer RNA is carried by the virus and serves as the primer required for nucleic acid synthesis (*see chapter 10*). The transformation of RNA into DNA takes place in two steps. First, reverse transcriptase copies the +RNA to form a RNA-DNA hybrid. Then the **ribonuclease H** component of reverse transcriptase degrades the +RNA strand to leave −DNA (figure 19.6*b, c*). After synthesizing −DNA, the reverse transcriptase copies this strand to produce a double-stranded DNA called **proviral DNA,** which can direct the synthesis of mRNA and new +RNA virion genome copies. Notice that during this process genetic information is transferred from RNA to DNA rather than in the normal direction.

The reproduction of retroviruses is remarkable in other ways as well. After proviral DNA has been manufactured, it is converted to a circular form and incorporated or integrated into the host cell chromosome. Virus products are only formed after integration. Sometimes, these integrated viruses can change host cells into tumor cells.

Biology of the HIV virus and AIDS (pp. 722–29).

Synthesis and Assembly of Virus Capsids

Some **late genes** direct the synthesis of capsid proteins, and these spontaneously self-assemble to form the capsid just as in bacteriophage morphogenesis. Recently, the self-assembly

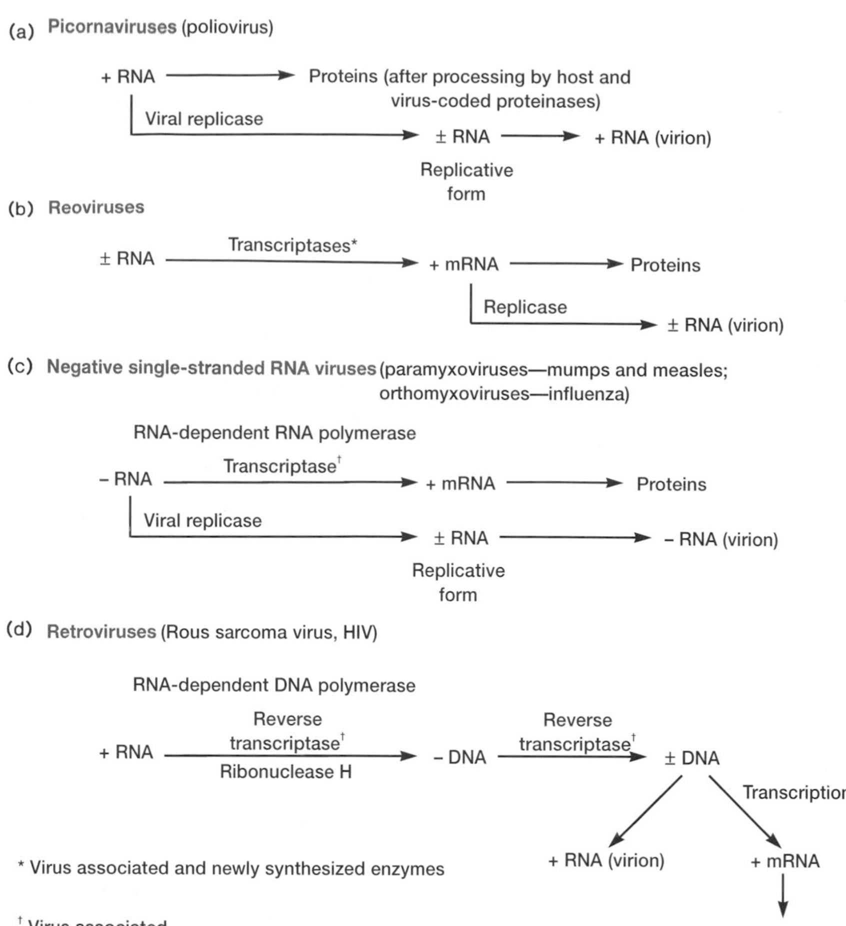

Figure 19.5 **RNA Animal Virus Reproductive Strategies.**

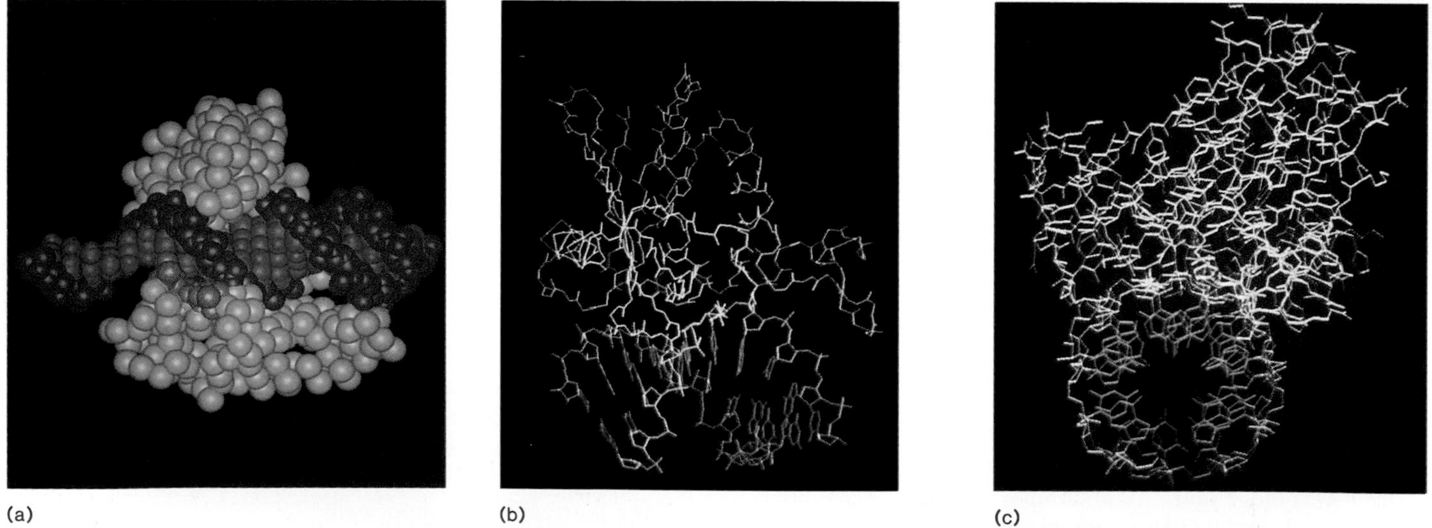

Figure 19.6 **Reverse Transcriptase and Ribonuclease H.** (*a*) A computer generated model of reverse transcriptase bound to a nucleic acid complex. Parts *b* and *c* show a model of *E. coli* ribonuclease H complexed with the RNA · DNA hybrid. (*b*) A side view showing the protein (yellow) complexed with the nucleic acid (DNA backbone in lavender, RNA in pink, bases in blue). (*c*) Top view of complex.

Table 19.2 Intracellular Sites of Animal Virus Reproduction

Virus	Nucleic Acid Replication	Capsid Assembly
DNA Viruses		
Parvoviruses	Nucleus	Nucleus
Papovaviruses	Nucleus	Nucleus
Herpesviruses	Nucleus	At nuclear membrane
Adenoviruses	Nucleus	Nucleus
Poxviruses	Cytoplasm	Cytoplasm
RNA Viruses		
Reoviruses	Cytoplasm	Cytoplasm
Togaviruses	Cytoplasm	At membranes
Picornaviruses	Cytoplasm	Cytoplasm
Paramyxoviruses	Cytoplasm	At membranes
Orthomyxoviruses	Nucleus	At membranes
Rhabdoviruses	Cytoplasm	At membranes
Retroviruses	Cytoplasm and nucleus	At plasma membrane

process has been dramatically demonstrated. The addition of poliovirus RNA to an extract prepared from uninfected human cells (HeLa cells) results in the formation of new, infectious poliovirus virions. It appears that during icosahedral virus assembly empty **procapsids** are first formed; then the nucleic acid is inserted in some unknown way. The site of morphogenesis varies with the virus (table 19.2). Large paracrystalline clusters of complete virions or procapsids are often seen at the site of virus maturation (figure 19.7). The assembly of enveloped virus capsids is generally similar to that of naked virions, except for poxviruses. These are assembled in the cytoplasm by a lengthy, complex process that begins with the enclosure of a portion of the cytoplasmic matrix through construction of a new membrane. Then newly synthesized DNA condenses, passes through the membrane, and moves to the center of the immature virus. Nucleoid and elliptical body construction takes place within the membrane.

Virion Release

Mechanisms of virion release differ between naked and enveloped viruses. Naked virions appear to be released most often by host cell lysis. In contrast, the formation of envelopes and the release of enveloped viruses are usually concurrent processes, and the host cell may continue virion release for some time. First, virus-encoded proteins are incorporated into the plasma membrane. Then the nucleocapsid is simultaneously released and the envelope formed by membrane budding (figures 19.8 and 19.9). Although most envelopes arise from an altered plasma membrane, in herpesviruses, budding and envelope formation usually involves the nuclear envelope.

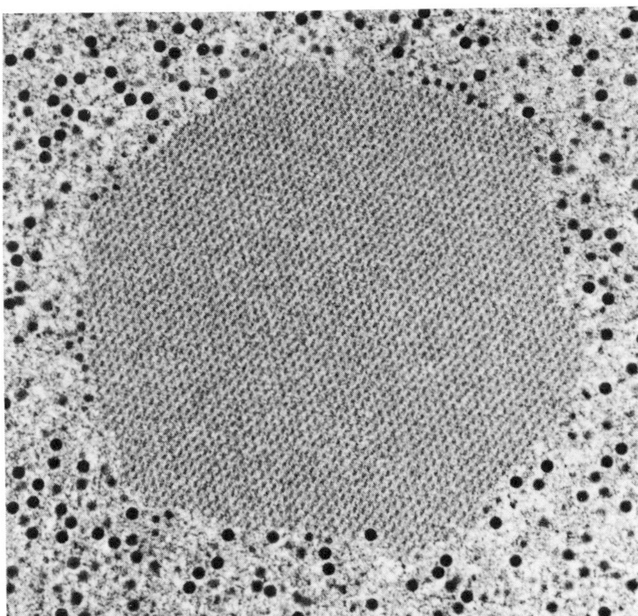

Figure 19.7 **Paracrystalline Clusters.** A crystalline array of adenovirus within the cell nucleus (×35,000).

1. Describe in detail each stage in animal virus reproduction and contrast it with the same stage in the bacteriophage life cycle (*see chapter 18*).
2. What probably plays the most important role in determining the tissue and host specificity of viruses?
3. Discuss how animal viruses can enter host cells.
4. Describe in general terms the replication and transcription processes that take place during animal virus reproduction.
5. Summarize the ways in which picornaviruses, −ssRNA viruses, and reoviruses carry out transcription and replication.
6. Outline in detail the life cycle of the retroviruses. What is proviral DNA?
7. In what two ways are animal viruses released from their host cells?
8. Define the following terms: uncoating, coated vesicles, overlapping genes, replicative form, reverse transcriptase, and late genes.

Cytocidal Infections and Cell Damage

An infection that results in cell death is a **cytocidal infection.** Animal viruses can harm their host cells in many ways; often this leads to cell death. As mentioned earlier (*see chapter 17*), microscopic or macroscopic degenerative changes or abnormalities in host cells and tissues are referred to as cytopathic effects (CPEs), and these often result from virus infections.

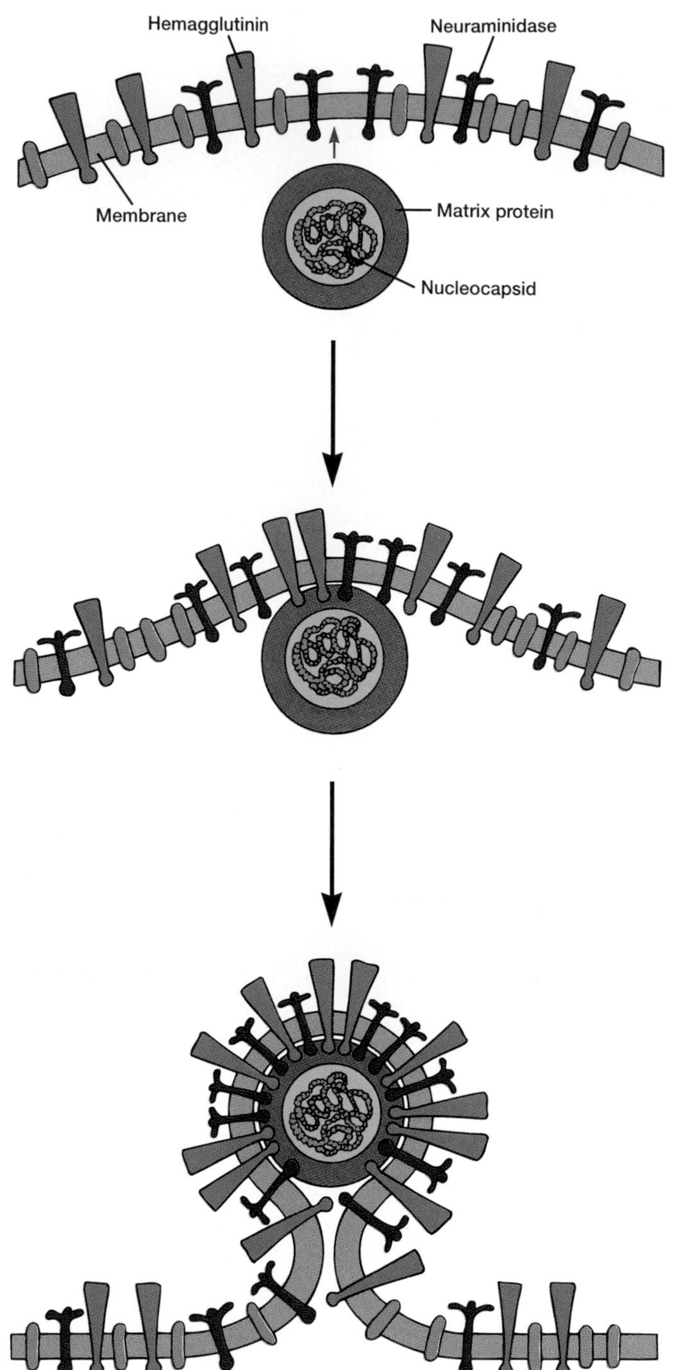

Figure 19.8 Release of Influenza Virus by Plasma Membrane Budding. First, viral envelope proteins (hemagglutinin and neuraminidase) are inserted into the host plasma membrane. Then the nucleocapsid approaches the inner surface of the membrane and binds to it. At the same time, viral proteins collect at the site and host membrane proteins are excluded. Finally, the membrane buds to simultaneously form the viral envelope and release the mature virion. *From Donald Voet and Judith G. Voet,* Biochemistry. *Copyright © 1990 John Wiley & Sons, Inc., New York, NY. Reprinted by permission of John Wiley & Sons, Inc.*

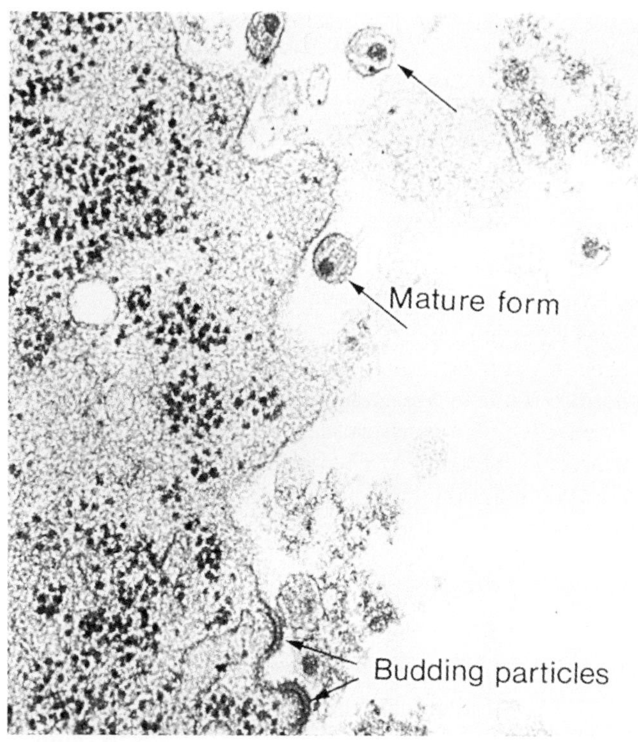

(a)

(b)

Figure 19.9 Human Immunodeficiency Virus (HIV) Release by Plasma Membrane Budding. (*a*) A transmission electron micrograph of HIV particles beginning to bud, as well as some mature particles. (*b*) A scanning electron micrograph view of HIV particles budding from a lymphocyte.

Seven possible mechanisms of host cell damage are briefly described here.

1. Many viruses can inhibit host DNA, RNA, and protein synthesis. Cytocidal viruses (e.g., picornaviruses, herpesviruses, and adenoviruses) are particularly active in this regard. The mechanisms of inhibition are not yet clear.
2. Cell lysosomes may be damaged, resulting in the release of hydrolytic enzymes and cell destruction.
3. Virus infection can drastically alter plasma membranes through the insertion of virus-specific proteins so that the infected cells are attacked by the immune system. When infected by viruses such as herpesviruses and measles virus, as many as 50 to 100 cells may fuse into one abnormal, giant, multinucleate cell called a polykaryocyte. The HIV virus, which causes AIDS (*see chapter 36*), appears to destroy $CD4^+$ T helper cells through its effects on the plasma membrane.
4. High concentrations of proteins from several viruses (e.g., mumps virus and influenza virus) can have a direct toxic effect on cells and organisms.
5. **Inclusion bodies** are formed during many virus infections. These may result from the clustering of subunits or virions within the nucleus or cytoplasm (e.g., the Negri bodies in rabies infections); they may also contain cell components such as ribosomes (arenavirus infections) or chromatin (herpesviruses). Regardless of their composition, these inclusion bodies can directly disrupt cell structure.
6. Chromosomal disruptions result from infections by herpesviruses and others.
7. Finally, the host cell may not be directly destroyed but transformed into a malignant cell. This will be discussed later in the chapter.

It should be emphasized that more than one of these mechanisms may be involved in a cytopathic effect.

1. Outline the ways in which viruses can damage host cells during cytocidal infections.

Persistent, Latent, and Slow Virus Infections

Many virus infections (e.g., influenza) are **acute infections;** that is, they have a fairly rapid onset and last for a relatively short time. However, some viruses can establish **persistent** or **chronic infections** lasting many years. The viruses may reproduce at a slow rate without causing disease symptoms; antibodies to the virus will be found in the blood. Examples of this phenomenon are hepatitis B virus (serum hepatitis), the Epstein-Barr virus, and measles virus. In **latent virus infections,** the virus stops reproducing and remains dormant for a period before becoming active again. During latency, no symptoms, antibodies, or viruses are detectable. Herpes simplex type 1 virus often infects children and then becomes dormant within the nervous system ganglia; years later it can be activated to cause cold sores. The varicella-zoster virus causes chickenpox in children and then, after years of inactivity, may produce the skin disease shingles. These and other examples of such infections are discussed in more detail in chapter 36.

The causes of persistence and latency are probably multiple, although the precise mechanisms are still unclear. The virus genome may be integrated into the host genome. Viruses may become less antigenic and thus less susceptible to attack by the immune system. They may mutate to less virulent and slower reproducing forms. Sometimes a deletion mutation (*see chapter 12*) produces a **defective interfering (DI) particle** that cannot reproduce but slows normal virus reproduction, thereby reducing host damage and establishing a chronic infection.

A small group of viruses cause extremely slowly developing infections, often called **slow virus diseases,** in which symptoms may take years to emerge. Measles virus occasionally produces a slow infection. A child may have a normal case of measles, then 5 to 12 years later develop a degenerative brain disease called subacute sclerosing panencephalitis (SSPE). Many slow viruses may not be normal viruses at all. Several neurological diseases of animals and humans develop slowly. The best studied of these diseases are scrapie, a disease of sheep and goats, and the human diseases kuru, fatal familial insomnia, and the Creutzfeldt-Jakob disease. All are thought to be caused by simple, nonviral agents called prions, which are discussed later in the text.

1. Define the following: acute infection, persistent or chronic infection, latent virus infection, and slow virus disease.
2. Why might an infection be chronic or latent?

Viruses and Cancer

Cancer (Latin *cancer,* crab) is one of our most serious medical problems and the focus of an immense amount of interest and research. A **tumor** (Latin *tumere,* to swell) is a growth or lump of tissue resulting from **neoplasia** or abnormal new cell growth and reproduction due to a loss of regulation. Tumor cells have aberrant shapes and altered plasma membranes that contain distinctive tumor antigens. They may invade surrounding tissues to form unorganized cell masses. They often lose the specialized metabolic activities characteristic of differentiated tissue cells and rely greatly upon anaerobic metabolism. This reversion to a more primitive or less differentiated state is called **anaplasia.**

There are two major types of tumors with respect to overall form or growth pattern. If the tumor cells remain in place to form a compact mass, the tumor is benign. In contrast, malignant or cancerous tumor cells can actively spread throughout the body in a process known as **metastasis,** often by floating in the blood and establishing secondary tumors. Some cancers are not solid, but cell suspensions. For example, leukemias are composed of malignant white blood cells that circulate throughout the body. Indeed, dozens of kinds of cancers arise from a variety of cell types and afflict all kinds of organisms.

As one might expect from the wide diversity of cancers, there are many causes of cancer, only some of which are directly related to viruses. Possibly as many as 30 to 60% of cancers may be related to diet. Many chemicals in our surroundings are carcinogenic and may cause cancer by inducing gene mutations or interfering with normal cell differentiation.

The Ames test for carcinogens (pp. 252–53).

Carcinogenesis is a complex, multistep process. It can be initiated by a chemical, usually a mutagen, but a cancer does not appear to develop until at least one more triggering event (possibly exposure to another chemical carcinogen or a virus) takes place. Often, as is discussed later, cancer-causing genes, or **oncogenes,** are directly involved and may come from the cell itself or be contributed by a virus. Many of these oncogenes seem to be involved in the regulation of cell growth and differentiation; for example, some code for all or part of growth factors that regulate cell growth. It may be that various cancers arise through different combinations of causes. Not surprisingly, the chances of developing cancer rise with age because an older person will have been exposed to carcinogens and other causative factors for a longer time. Cancer surveillance by the immune system (*see chapter 32*) also may be less effective in older people.

Although viruses are known to cause many animal cancers, it is very difficult to prove that this is the case with human cancers since indirect methods of study must be used. One tries to find virus particles and components within tumor cells, using techniques such as electron microscopy, immunologic tests, and enzyme assays. Attempts are also made to isolate suspected cancer viruses by cultivation in tissue culture or other animals. With luck, a good correlation between the presence of a virus and cancer can be detected.

At present, viruses have been implicated in the genesis of at least six human cancers. (1) The Epstein-Barr virus (EBV) is one of the best-studied human cancer viruses. EBV is a herpesvirus and the cause of two cancers. Burkitt's lymphoma is a malignant tumor of the jaw and abdomen found in children of central and western Africa. EBV also causes nasopharyngeal carcinoma. Both the virus particles and the EBV genome have been found within tumor cells; Burkitt's lymphoma patients also have high blood levels of antibodies to EBV. Interestingly, there is reason to believe that a person also must have had malaria to develop Burkitt's lymphoma. Environmental factors must play a role, because EBV does not cause much cancer in the United States despite its prevalence. Possibly this is due to a low incidence of malaria in the United States. (2) Hepatitis B virus appears to be associated with one form of liver cancer (hepatocellular carcinoma) and can be integrated into the human genome. (3) Some strains of human papilloma viruses have been linked to cervical cancer. (4) At least two retroviruses, the human T-cell lymphotropic virus I (HTLV-1) and HTLV-2, seem able to cause cancer, adult T-cell leukemia and hairy-cell leukemia, respectively (*see figure 36.16*). Other retrovirus-associated cancers may well be discovered in the future.

There is considerable interest in the possibility that viruses may be involved in the genesis of many common human diseases not previously associated with viral causation. For example, there is some evidence for virus involvement in the following diseases: juvenile-onset diabetes (the Coxsackie virus), adult diabetes (the lymphocytic choriomeningitis virus), atherosclerosis (herpesviruses), and arthritis (human parvovirus). Just as in studies of human cancer, it is very difficult to implicate a virus conclusively in these diseases since one must depend upon animal studies and examination of diseased tissues. Koch's postulates cannot be satisfied because this would require human experimentation.

It appears that viruses can cause cancer in several ways. They may bring oncogenes into a cell and insert them into its genome. Rous sarcoma virus (a retrovirus) carries an *src* gene that codes for tyrosine kinase. This enzyme is located mainly in the plasma membrane and phosphorylates the amino acid tyrosine in several cellular proteins. It is thought that this alters cell growth and behavior. Since the activity of many proteins is regulated by phosphorylation and several other oncogenes also code for protein kinases, many cancers may result at least partly from altered cell regulation due to changes in kinase activity. The human T-cell lymphotropic viruses, HTLV-1 and HTLV-2, seem to transform T cells (*see chapter 32*) by producing a regulatory protein that sometimes activates genes involved in cell division as well as stimulating virus reproduction. Some oncogenic viruses carry one or more very effective promoters or enhancers (*see chapter 10*). If these viruses integrate themselves next to a cellular oncogene, the promoter or enhancer will stimulate its transcription, leading to cancer. In this case, the oncogene might be necessary for normal cell growth (possibly it codes for a regulatory protein) and only causes cancer when it functions too rapidly or at the wrong time. For example, some chicken retroviruses induce lymphomas when they are integrated next to the *c-myc* cellular oncogene, which codes for a protein that appears to be involved in the induction of either DNA or RNA synthesis. With the possible exception of HTLV-1, it is not yet known how the viruses associated with human cancers actually aid in cancer development.

---

1. What are the major characteristics of cancer?
2. How might viruses cause cancer? Are there other ways in which a malignancy might develop?
3. Define the following terms: tumor, neoplasia, anaplasia, metastasis, and oncogene.

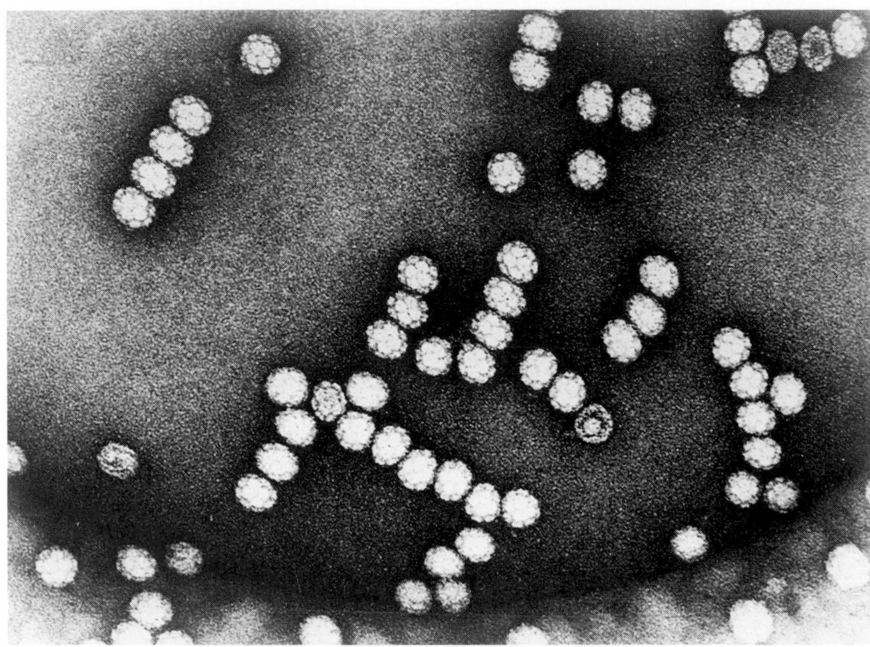

Figure 19.10 **Turnip Yellow Mosaic Virus (TYMV).** An RNA plant virus with icosahedral symmetry.

Plant Viruses

Although it has long been recognized that viruses can infect plants and cause a variety of diseases (*see figure 17.5*), plant viruses generally have not been as well studied as bacteriophages and animal viruses. This is mainly because they are often difficult to cultivate and purify. Some viruses, such as tobacco mosaic virus (TMV), can be grown in isolated protoplasts of plant cells just as phages and some animal viruses are cultivated in cell suspensions. However, many cannot grow in protoplast cultures and must be inoculated into whole plants or tissue preparations. Many plant viruses require insect vectors for transmission; some of these can be grown in monolayers of cell cultures derived from aphids, leafhoppers, or other insects.

Virion Morphology

The essentials of capsid morphology are outlined in chapter 17 and apply to plant viruses since they do not differ significantly in construction from their animal virus and phage relatives. Many have either rigid or flexible helical capsids (tobacco mosaic virus, *see figure 17.11*). Others are icosahedral or have modified the icosahedral pattern with the addition of extra capsomers (turnip yellow mosaic virus, figure 19.10). Most capsids seem composed of one type of protein; no specialized attachment proteins have been detected. Almost all plant viruses are RNA viruses, either single stranded or double stranded (*see tables 17.1 and 17.2*). Caulimoviruses and geminiviruses with their DNA genomes are exceptions to this rule.

Plant Virus Taxonomy

The shape, size, and nucleic acid content of many plant virus groups are summarized in figure 19.11. Like other types of viruses, they are classified according to properties such as nucleic acid type and strandedness, capsid symmetry and size, and envelope presence (*see table 17.2*).

Plant Virus Reproduction

Since tobacco mosaic virus (TMV) has been studied the most extensively, its reproduction will be briefly described. The replication of virus RNA is an essential part of the reproduction process. Most plants contain RNA-dependent RNA polymerases, and it is possible that these normal constituents replicate the virus RNA. However, some plant virus genomes (e.g., turnip yellow virus and cowpea mosaic virus) appear to be copied by a virus-specific RNA replicase. Possibly TMV RNA is also replicated by a viral RNA polymerase, but the evidence is not clear on this matter. Four TMV-specific proteins, one of them the coat protein, are known to be made. Although TMV RNA is plus single-stranded RNA and could directly serve as mRNA, the production of messenger is complex. The coat protein mRNA has the same sequence as the 3′ end of the TMV genome and arises from it by some sort of intracellular processing.

After the coat protein and RNA genome have been synthesized, they spontaneously assemble into complete TMV virions in a highly organized process (figure 19.12). The protomers (*see chapter 17*) come together to form disks composed of two layers of protomers arranged in a helical spiral.

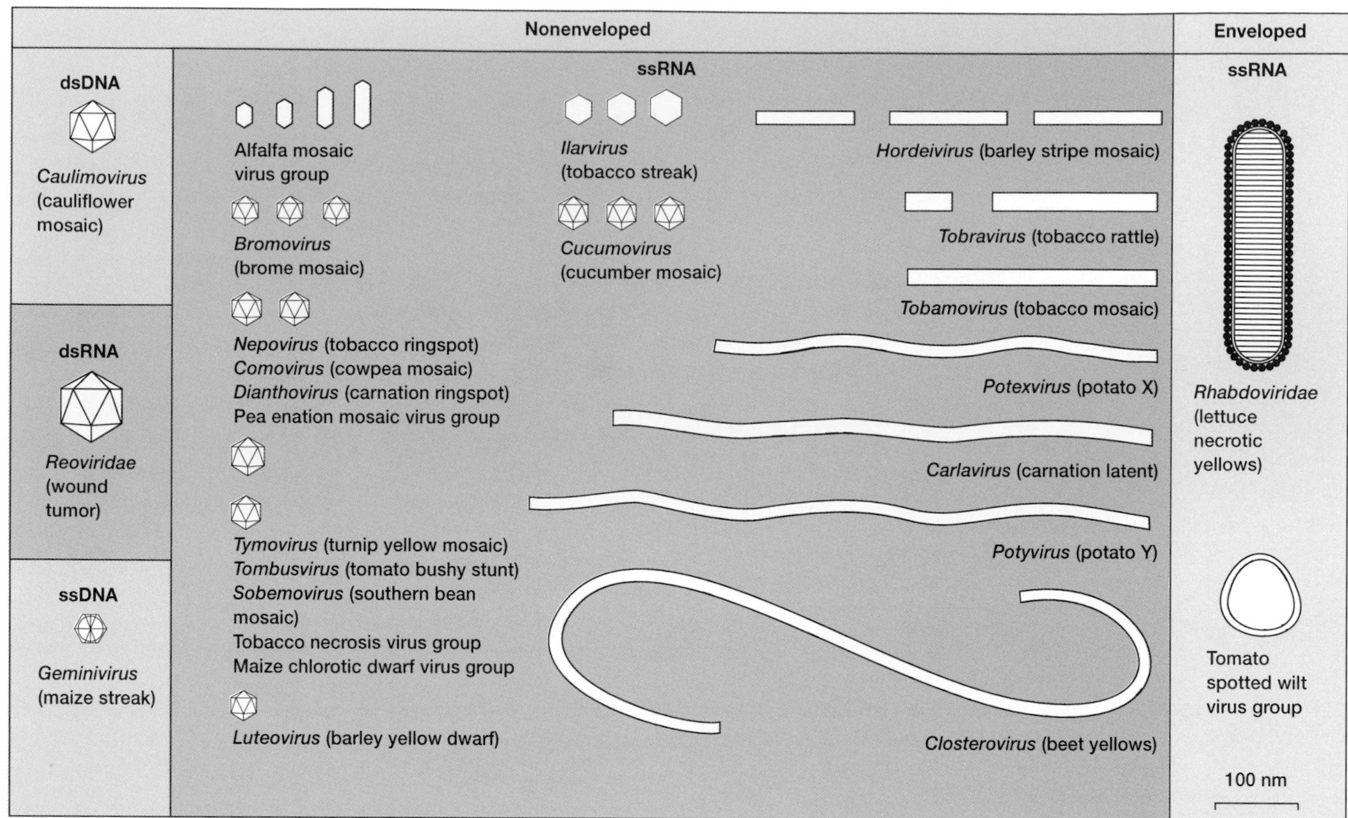

Figure 19.11 **A Diagrammatic Description of the Major Plant Virus Families.** *Reproduced, with permission, from the* Annual Review of Microbiology, *Volume 39,* *© 1985 by Annual Reviews Inc.*

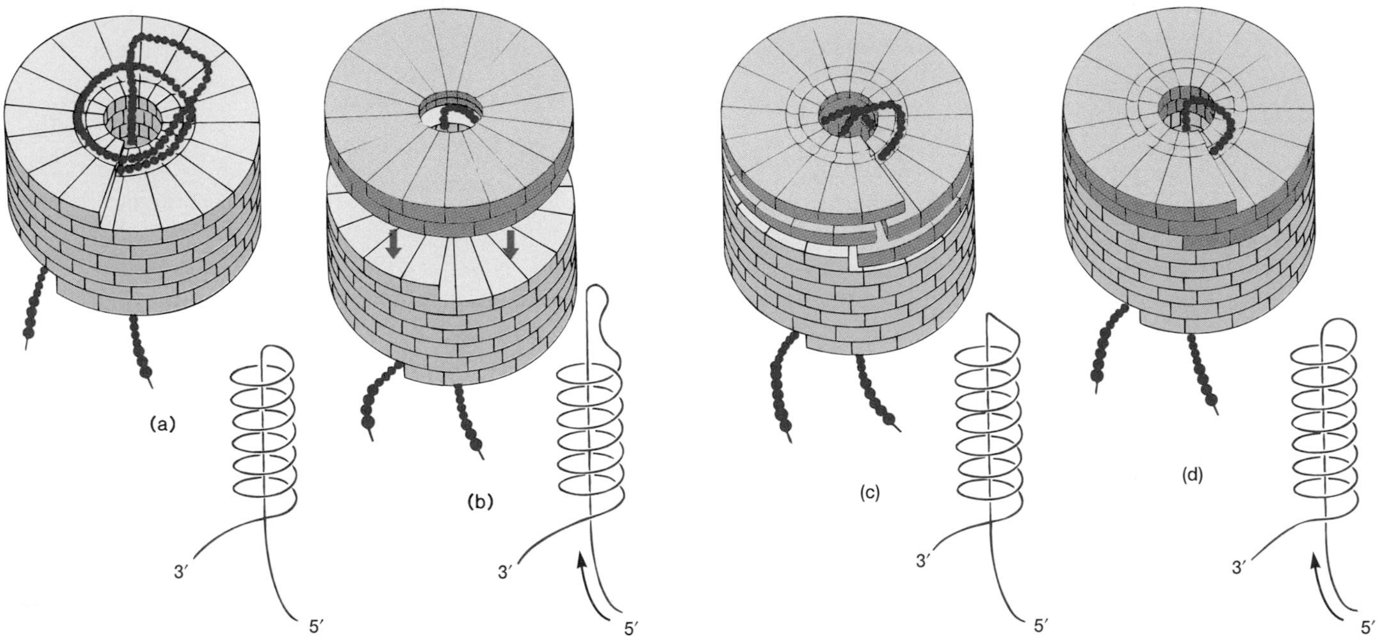

Figure 19.12 **TMV Assembly.** The elongation phase of tobacco mosaic virus nucleocapsid construction. The lengthening of the helical capsid through the addition of a protein disk to its end is shown in a sequence of four illustrations; line drawings depicting RNA behavior are included. The RNA genome inserts itself through the hole of an approaching disk and then binds to the groove in the disk as it locks into place at the end of the cylinder. *From "The Assembly of a Virus" by P. J. Butler and A. Klug. Copyright © 1978 by Scientific American, Inc. All rights reserved.*

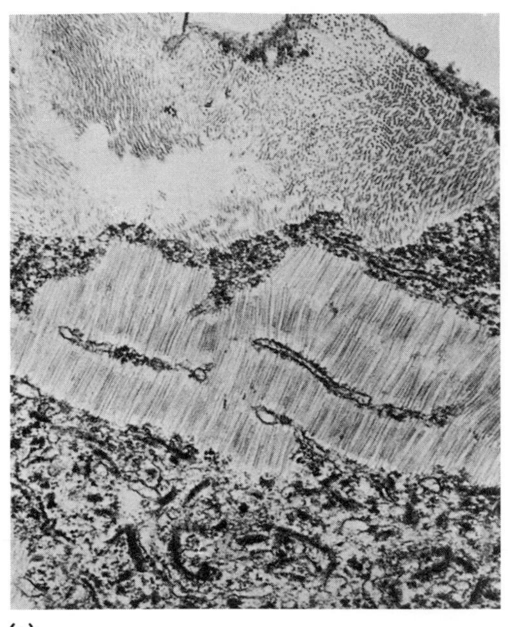

(a)

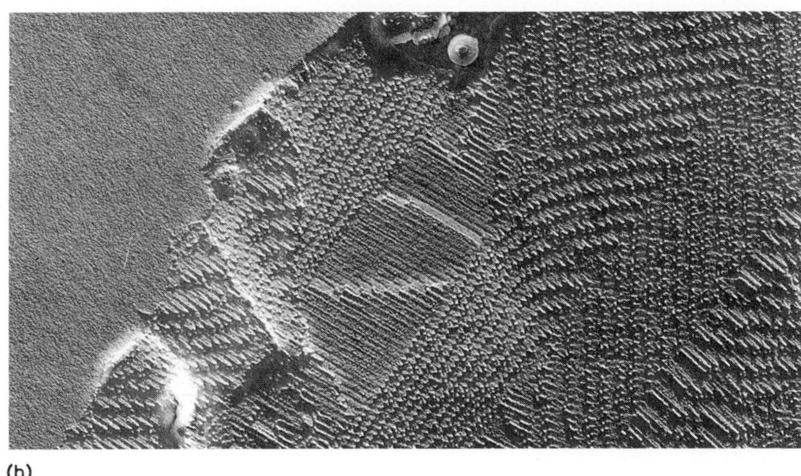

(b)

Figure 19.13 **Intracellular TMV.** (*a*) A crystalline mass of tobacco mosaic virions from a 10-day-old lesion in a *Chenopodium amaranticolor* leaf. (*b*) Freeze-fracture view of a crystalline mass of tobacco mosaic virions in an infected leaf cell. In both views, the particles can be seen longitudinally and in cross section.

Association of coat protein with TMV RNA begins at a special assembly initiation site close to the 3′ end of the genome. The helical capsid grows by the addition of protomers, probably as disks, to the end of the rod. As the rod lengthens, the RNA passes through a channel in its center and forms a loop at the growing end. In this way, the RNA can easily fit as a spiral into the interior of the helical capsid.

The spread of plant viruses in nonvascular tissue is hindered by the presence of tough cell walls. Nevertheless, a virus such as TMV does spread slowly, about 1 mm/day or less. It moves from cell to cell through the plasmodesmata. These are slender cytoplasmic strands extending through holes in adjacent cell walls that join plant cells by narrow bridges. Special viral "movement proteins" are required for movement between cells. The TMV movement protein accumulates in the plasmodesmata, but the way in which it promotes virus movement is not understood.

Several cytological changes can take place in TMV-infected cells. Plant virus infections often produce microscopically visible intracellular inclusions, usually composed of virion aggregates, and hexagonal crystals of almost pure TMV virions sometimes do develop in TMV-infected cells (figure 19.13). The host cell chloroplasts become abnormal and often degenerate, while new chloroplast synthesis is inhibited.

Transmission of Plant Viruses

Since plant cells are protected by cell walls, plant viruses have a considerable obstacle to overcome when trying to establish themselves in a host. TMV and a few other viruses may be carried by the wind or animals and then enter when leaves are mechanically damaged. Some plant viruses are transmitted through contaminated seeds, tubers, or pollen. Soil nematodes can transmit viruses (e.g., the tobacco ringspot virus) while feeding on roots. Tobacco necrosis virus is transmitted by parasitic fungi. However, the most important agents of transmission are insects that feed on plants, particularly sucking insects such as aphids and leafhoppers.

Insects transmit viruses in several ways. They may simply pick up viruses on their mouth parts while feeding on an infected plant, then directly transfer the viruses to the next plant they visit. Viruses may be stored in an aphid's foregut; the aphid will infect plants when regurgitating while it is feeding. Several plant viruses—for example, the wound tumor virus—can multiply in leafhopper tissues before reaching the salivary glands and being inoculated into plants (that is, it uses both insects and plants as hosts).

1. Why have plant viruses not been as well studied as animal and bacterial viruses?
2. Describe in molecular terms the way in which TMV is reproduced.
3. How are plant viruses transmitted between hosts?

Viruses of Fungi and Algae

Most of the fungal viruses or mycoviruses isolated from higher fungi such as *Penicillium* and *Aspergillus* (*see chapter 26*) contain double-stranded RNA and have isometric capsids (that is, their capsids are roughly spherical polyhedra), which are approximately 25 to 50 nm in diameter. Many appear to be latent viruses. Some mycoviruses do induce disease symptoms in hosts such as the mushroom *Agaricus bisporus,* but cytopathic effects and toxic virus products have not yet been observed.

Much less is known about the viruses of lower fungi, and only a few have been examined in any detail. Both dsRNA and dsDNA genomes have been found; capsids are usually isometric or hexagonal and vary in size from 40 to over 200 nm. Unlike the situation in higher fungi, virus reproduction is accompanied by host cell destruction and lysis.

Viruses have been detected during ultrastructural studies of eucaryotic algae, but few have been isolated. Those that have been studied have dsDNA genomes and polyhedral capsids. One virus of the green alga *Uronema gigas* resembles many bacteriophages in having a tail.

1. Describe the major characteristics of the viruses that infect higher fungi, lower fungi, and algae. In what ways do they seem to differ from one another?

Insect Viruses

Members of at least seven virus families (*Baculoviridae, Iridoviridae, Poxviridae, Reoviridae, Parvoviridae, Picornaviridae,* and *Rhabdoviridae*) are known to infect insects and reproduce, or even use them as the primary host (*see table 17.2*). Of these, probably the three most important are the *Baculoviridae, Reoviridae,* and *Iridoviridae.*

The *Iridoviridae* are icosahedral viruses with lipid in their capsids and a linear double-stranded DNA genome. They are responsible for the iridescent virus diseases of the crane fly and some beetles. The group's name comes from the observation that larvae of infected insects can have an iridescent coloration due to the presence of crystallized virions in their fat bodies.

Many insect virus infections are accompanied by the formation of inclusion bodies within the infected cells. Granulosis viruses form granular protein inclusions, usually in the cytoplasm. Nuclear polyhedrosis and cytoplasmic polyhedrosis virus infections produce polyhedral inclusion bodies in the nucleus or the cytoplasm of affected cells. Although all three types of viruses generate inclusion bodies, they belong to two distinctly different families. The cytoplasmic polyhedrosis viruses are reoviruses; they are icosahedral with double shells and have double-stranded RNA genomes. Nuclear polyhedrosis viruses and granulosis viruses are baculoviruses—rod-shaped, enveloped viruses of helical symmetry and with double-stranded DNA.

The inclusion bodies, both polyhedral and granular, are protein in nature and enclose one or more virions (figure 19.14). Insect larvae are infected when they feed on leaves contaminated with inclusion bodies. Polyhedral bodies protect the virions against heat, low pH, and many chemicals; the viruses can remain viable in the soil for years. However, when exposed

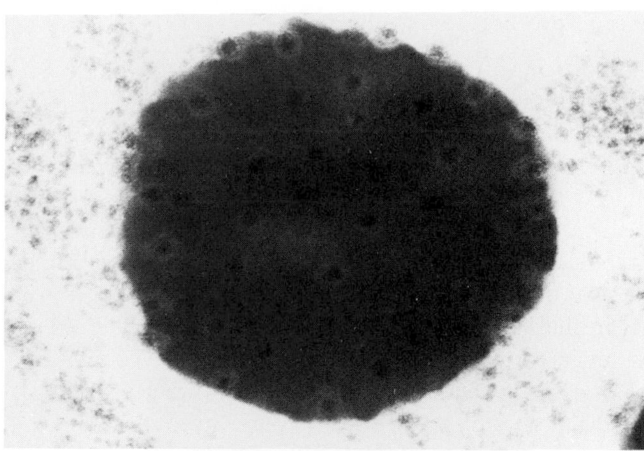

Figure 19.14 **Inclusion Bodies.** A section of a cytoplasmic polyhedron from a zebra caterpillar (*Malanchra picta*). The occluded virus particles with dense cores and external capsids are clearly visible ($\times$50,000).

to alkaline insect gut contents, the inclusion bodies dissolve to liberate the virions, which then infect midgut cells. Some viruses remain in the midgut while others spread throughout the insect. Just as with bacterial and vertebrate viruses, insect viruses can persist in a latent state within the host for generations while producing no disease symptoms. A reappearance of the disease may be induced by chemicals, thermal shock, or even a change in the insect's diet.

Much of the current interest in insect viruses arises from their promise as biological control agents for insect pests (*see chapter 44*). Many people hope that some of these viruses may partially replace the use of toxic chemical pesticides. Baculoviruses have received the most attention for at least three reasons. First, they attack only invertebrates and have considerable host specificity; this means that they should be fairly safe for nontarget organisms. Second, because they are encased in protective inclusion bodies, these viruses have a good shelf life and better viability when dispersed in the environment. Finally, they are well suited for commercial production since they often reach extremely high concentrations in larval tissue (as high as 10^{10} viruses per larva). The use of nuclear polyhedrosis viruses for the control of the cotton bollworm, Douglas fir tussock moth, gypsy moth, alfalfa looper, and European pine sawfly has either been approved by the Environmental Protection Agency or is being considered. The granulosis virus of the codling moth is also useful. Usually, inclusion bodies are sprayed on foliage consumed by the target insects. More sensitive viruses are administered by releasing infected insects to spread the disease.

1. Summarize the nature of granulosis, nuclear polyhedrosis, and cytoplasmic polyhedrosis viruses and the way in which they are transmitted by inclusion bodies.
2. What are baculoviruses and why are they so promising as biological control agents for insect pests?

Viroids and Prions

Although some viruses are exceedingly small and simple, even simpler infectious agents exist. Over 16 different plant diseases—for example, potato spindle-tuber disease, exocortosis of citrus trees, and chrysanthemum stunt disease—are caused by a group of infectious agents called **viroids.** These are circular, single-stranded RNAs, about 250 to 370 nucleotides long (figure 19.15), that can be transmitted between plants through mechanical means or by way of pollen and ovules and are replicated in their hosts. Viroids are found principally in the nucleolus of infected cells; between 200 and 10,000 copies may be present. They do not act as mRNAs to direct protein synthesis, and it is not yet known how they cause disease symptoms. A plant may be infected without showing symptoms; that is, it may have a latent infection. The same viroid, when in another species, might well cause a severe disease. Although viroids could be replicated by an RNA-dependent RNA polymerase, they appear to be synthesized from RNA templates by a host RNA polymerase that mistakes them for a piece of DNA.

The potato spindle-tuber disease agent (PSTV) has been most intensely studied. Its RNA is about 130,000 daltons or 359 nucleotides in size, much smaller than any virus genome. The single-stranded RNA normally exists as a closed circle collapsed into a rodlike shape by intrastrand base pairing, as seems to be the case for all viroids. Several PSTV strains have been isolated, ranging in virulence from ones that cause only mild symptoms to lethal varieties. All variations in pathogenicity are due to a few nucleotide changes in two short regions on the viroid. It is believed that these sequence changes alter the shape of the rod and thus its ability to cause disease.

There is increasing evidence that an infectious agent different from both viruses and viroids can cause disease in livestock and humans. The agent has been called a **prion** (for proteinaceous infectious particle). The best studied of these prions causes a degenerative disorder of the central nervous system in sheep and goats; this disorder is named scrapie. Afflicted animals lose coordination of their movements, tend to scrape or rub their skin, and eventually cannot walk. The agent has been at least partly purified, and no nucleic acid has yet been detected in it. It seems to be a 33 to 35 kDa hydrophobic protein, often called PrP (for **pr**ion **p**rotein). The *PrP* gene is present in many normal vertebrates and invertebrates. Presumably an altered PrP is at least partly responsible for the disease.

Despite the isolation of PrP, the mechanism of the disease continues to elude researchers. Some believe that the disease is transmitted by the PrP alone. They are convinced that the infective pathogen is an abnormal PrP, one that has been either chemically modified or has changed conformation in some way. When abnormal PrP enters a normal brain, it might bind to normal PrP and change its shape, or it could activate enzymes that modify PrP structure. Other researchers feel that this hypothesis is inadequate. They are concerned about the existence of prion strains, presumably genetic, and about the problem of how genetic information can be transmitted between hosts

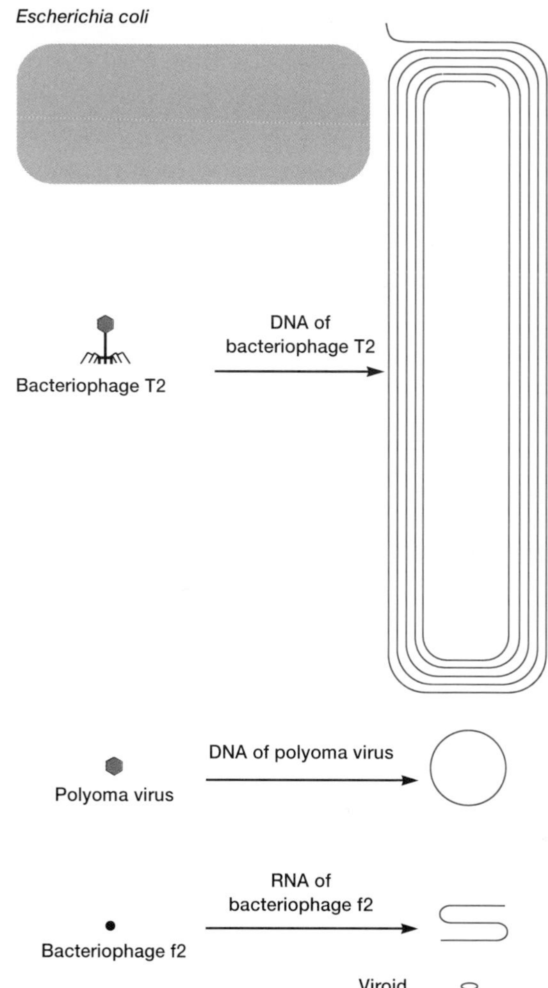

Figure 19.15 **Viroids, Viruses, and Bacteria.** A comparison of *Escherichia coli,* several viruses, and the potato spindle-tuber viroid with respect to size and the amount of nucleic acid possessed. (All dimensions are enlarged approx. ×40,000.)

by a protein. They note that proteins have never been known to carry genetic information. Possibly, the agent is a **virino,** a tiny scrapie-specific nucleic acid that is coated with PrP. The nucleic acid may not be translated but still interact with host cells in some way to cause disease. This hypothesis is consistent with the finding that many strains of the scrapie agent have been isolated. There is also some evidence that a strain can change or mutate. Presumably, further research will clarify the situation.

As mentioned earlier, some slow virus diseases may be due to prions or virinos. This is particularly true of certain neurological diseases of humans and animals. Bovine spongiform encephalopathy ("mad cow disease"), Kuru, fatal familial insomnia, the Creutzfeldt-Jakob disease (CJD), and Gerstmann-Sträussler-Scheinker syndrome (GSS) appear to be prion diseases. They result in progressive degeneration of the brain and eventual death. Mad cow disease has reached epidemic proportions in Great Britain and is spread because cattle are fed bone meal made from cattle. CJD and GSS are rare and cosmopolitan in distribution among middle-aged people, while kuru has been found only in the Fore, an eastern New Guinea tribe. This tribe had a custom of consuming dead kinsmen, and women were given the honor of preparing the brain and practicing this ritual cannibalism. They and their children were infected by handling the diseased brain tissue. (Now that cannibalism has been eliminated, the incidence of kuru has decreased and is found only among the older adults.) Some researchers believe that Alzheimer's disease and multiple sclerosis may be prion diseases.

1. What are viroids and why are they of great interest?
2. How does a viroid differ from a virus?
3. What is a prion or virino? In what way does a prion appear to differ fundamentally from viruses and viroids?

Summary

1. Animal viruses are classified according to many properties; the most important are their morphology and nucleic acids.
2. The first step in the animal virus life cycle is adsorption of the virus to a target cell receptor site; often special capsid structures are involved in this process.
3. Virus penetration of the host cell plasma membrane may be accompanied by capsid removal from the nucleic acid or uncoating. Most often penetration occurs through engulfment to form coated vesicles, but mechanisms such as direct fusion of the envelope with the plasma membrane are also employed.
4. Early mRNA and proteins are involved in taking over the host cell and the synthesis of viral DNA and RNA. DNA replication usually takes place in the host nucleus and mRNA is initially manufactured by host enzymes.
5. Poxviruses differ from other DNA animal viruses in that DNA replication takes place in host cytoplasm and they carry an RNA polymerase. The parvoviruses are so small that they must conserve genome space by using overlapping genes and other similar mechanisms.
6. The genome of positive ssRNA viruses can act as an mRNA, while negative ssRNA virus genomes direct the synthesis of mRNA by a virus-associated transcriptase. Double-stranded RNA reoviruses use both virus-associated and newly synthesized transcriptases to make mRNA.
7. RNA virus genomes are replicated in the host cell cytoplasm. Most ssRNA viruses use a viral replicase to synthesize a dsRNA replicative form that then directs the formation of new genomes.
8. Retroviruses use reverse transcriptase to synthesize a DNA copy of their RNA genome. After the double-stranded proviral DNA has been synthesized, it is integrated into the host genome and directs the formation of virus RNA and protein.
9. Late genes code for proteins needed in (1) capsid construction by a self-assembly process and (2) virus release.
10. Usually, naked virions are released upon cell lysis. In enveloped virus reproduction, virus release and envelope formation normally occur simultaneously after modification of the host plasma membrane, and the cell is not lysed.
11. Viruses can destroy host cells in many ways during cytocidal infections. These include such mechanisms as inhibition of host DNA, RNA, and protein synthesis; lysosomal damage; alteration of host cell membranes; and the formation of inclusion bodies.
12. Although most virus infections are acute or have a rapid onset and last for a relatively short time, some viruses can establish persistent or chronic infections lasting for years. Viruses also can become dormant for a while and then resume activity in what is called a latent infection. Slow viruses may act so slowly that a disease develops over years.
13. Cancer is characterized by the formation of a malignant tumor that metastasizes or invades other tissues and can spread through the body. Carcinogenesis is a complex, multistep process involving many factors.
14. Viruses cause cancer in several ways. For example, they may bring a cancer-causing gene, or oncogene, into a cell, or the virion may insert a promoter next to a cellular oncogene and stimulate it to greater activity.
15. Most plant viruses have an RNA genome and may be either helical or icosahedral. Depending upon the virus, the RNA genome may be replicated by either a host RNA-dependent RNA polymerase or a virus-specific RNA replicase.
16. The TMV nucleocapsid forms spontaneously by self-assembly when disks of coat protein protomers complex with the RNA.
17. Plant viruses are transmitted in a variety of ways. Some enter through lesions in plant tissues, while others are transmitted by contaminated seeds, tubers, or pollen. Most are probably carried and inoculated by plant-feeding insects.
18. Mycoviruses from higher fungi have isometric capsids and dsRNA, while the viruses of lower fungi may have either dsRNA or dsDNA genomes.
19. Members of several virus families—most importantly the *Baculoviridae, Reoviridae,* and *Iridoviridae*—infect insects, and many of these viruses produce inclusion bodies that aid in their transmission.
20. Baculoviruses and other viruses are already finding use as biological control agents for insect pests.
21. Infectious agents simpler than viruses exist. For example, several plant diseases are caused by short strands of infectious RNA called viroids.
22. Prions or virinos are small agents associated with at least six degenerative nervous system disorders: scrapie, bovine spongiform encephalopathy, kuru, fatal familial insomnia, the Gerstmann-Sträussler-Scheinker syndrome, and Creutzfeldt-Jakob disease. The precise nature of prions is not yet clear.

Key Terms

acute infection *393*
anaplasia *393*
cancer *393*
coated vesicles *388*
cytocidal infection *391*
defective interfering (DI) particle *393*
early genes *388*
inclusion bodies *393*
late genes *389*
latent virus infection *393*
metastasis *393*
neoplasia *393*
oncogene *394*
persistent or chronic infections *393*
prion *399*
procapsids *391*
proviral DNA *389*
replicase *389*
replicative form *389*
retrovirus *389*
reverse transcriptase *389*
ribonuclease H *389*
slow virus diseases *393*
transcriptase *389*
tumor *393*
virino *399*
viroid *399*

Questions for Thought and Review

1. Make a list of all the ways you can think of in which viruses differ from their eucaryotic host cells.
2. Compare animal and plant viruses in terms of entry into host cells. Why do they differ so greatly in this regard?
3. Compare lysogeny with the reproductive strategy of retroviruses. What advantage might there be in having one's genome incorporated into that of the host cell?
4. Inclusion bodies are mentioned more than once in the chapter. Where are they found, and what are their functions, if any?
5. Would it be advantageous for a virus to damage host cells? If not, why isn't damage to the host avoided? Is it possible that a virus might become less pathogenic when it has been associated with the host population for a longer time?
6. How does one prove that a virus is causing cancer? Try to think of approaches other than those discussed in the chapter. Give a major reason why it is so difficult to prove that a specific virus causes human cancer.
7. From what you know about cancer, is it likely that a single type of treatment can be used to cure it? What approaches might be effective in preventing cancer?
8. Propose some experiments that might be useful in determining what prions are and how they cause disease.

Additional Reading

Aiken, J. M., and Marsh. R. F. 1990. The search for scrapie agent nucleic acid. *Microbiol. Rev.* 54(3):242–46.

Berns, K. I. 1990. Parvovirus replication. *Microbiol. Rev.* 54(3):316–29.

Bishop, J. M. 1982. Oncogenes. *Sci. Am.* 246(3):81–92.

Braun, M. M.; Heyward, W. L.; and Curran, J. W. 1990. The global epidemiology of HIV infection and AIDS. *Ann. Rev. Microbiol.* 44:555–77.

Buller, R. M. L., and Palumbo, G. J. 1991. Poxvirus pathogenesis. *Microbiol. Rev.* 55(1):80–122.

Butler, P. J. G., and Klug, A. 1978. The assembly of a virus. *Sci. Am.* 239(5):62–69.

Davis, B. D.; Dulbecco, R.; Eisen, H. N.; and Ginsberg, H. S. 1990. *Microbiology,* 4th ed. Philadelphia: J.B. Lippincott.

Diener, T. O. 1987. *The viroids.* New York: Plenum Press.

Eron, C. 1981. *The virus that ate cannibals.* New York: Macmillan.

Fields, B. N.; Knipe, D. M.; Chanock, R. M.; Hirsch, M. S.; Melnick, J. L.; Monath, T. P.; and Roizman, B., eds. 1990. *Fields virology,* 2d ed. New York: Raven Press.

Fraenkel-Conrat, H.; Kimball, P. C.; and Levy, J. A. 1988. *Virology,* 2d ed. Englewood Cliffs, N.J.: Prentice-Hall.

Gallo, R. C. 1986. The first human retrovirus. *Sci. Am.* 255(6):88–98.

Gallo, R. C. 1987. The AIDS virus. *Sci. Am.* 256(1):47–56.

Gallo, R. C., and Montagnier, L. 1988. AIDS in 1988. *Sci. Am.* 259(4):40–48.

Goodenough, U. W. 1991. Deception by pathogens. *Am. Sci.* 79:344–55.

Hazeltine, W. A. 1991. Molecular biology of the human immunodeficiency virus type 1. *FASEB J.* 5(10):2349–60.

Henle, W.; Henle, G.; and Lennette, E. T. 1979. The Epstein-Barr virus. *Sci. Am.* 241(1):48–59.

Hunter, T. 1984. The proteins of oncogenes. *Sci. Am.* 251(2):70–79.

Joklik, W. K.; Willett, H. P.; Amos, D. B.; and Wilfert, C. M. 1992. *Zinsser microbiology*, 20th ed.: Norwalk, CT: Appleton & Lange.

Kaariainen, L.; and Ranki, M. 1984. Inhibition of cell functions by RNA-virus infections. *Ann. Rev. Microbiol.* 38:91–109.

Lai, M. M. C. 1992. RNA recombination in animal and plant viruses. *Microbiol. Rev.* 56(1):61–79.

Luria, S. E.; Darnell, J. E., Jr.; Baltimore, D.; and Campbell, A. 1978. *General virology*, 3d ed. New York: John Wiley and Sons.

Marsh, M., and Helenius, A. 1989. Virus entry into animal cells. *Adv. Virus Res.* 36:107–51.

Matthews, R. E. F. 1991. *Plant virology,* 3d ed. New York: Academic Press.

McGeoch, D. J. 1989. The genomes of the human herpesviruses: Contents, relationships, and evolution. *Ann. Rev. Microbiol.* 43:235–65.

Medori, R., et al. 1992. Fatal familial insomnia, a prion disease with a mutation at codon 178 of the prion protein gene. New. Eng. J. Med. 326(7):444–49.

Miller, L. K.; Lingg, A. J.; and Bulla, L. A., Jr. 1983. Bacterial, viral, and fungal insecticides. *Science* 219:715 21.

Moss, B. 1991. Vaccina virus: A tool for research and vaccine development. *Science* 252:1662–67.

Oldstone, M. B. A. 1989 Viral alteration of cell function. *Sci Am.* 261(2):42–48.

Pardee, A. B., and veer Reddy, G. P. 1982. Cancer: Fundamental ideas. *Carolina Biology Reader,* no. 128. Burlington, N. C.: Carolina Biological Supply Co.

Prusiner, S. B. 1984. Prions. *Sci. Am.* 251(4):50–59.

Prusiner, S. B. 1991. Molecular biology of prion diseases. *Science* 252:1515–22.

Riesner, D., and Gross, H. J. 1985. Viroids. *Ann Rev. Biochem.* 54:531–64.

Sherker, A. H., and Marion, P. L. 1991. Hepadnaviruses and hepatocellular carcinoma. *Ann. Rev. Microbiol.* 45:475–508.

Simons, K.; Garoff, H.; and Helenius, A. 1982. How an animal virus gets into and out of its host cell. *Sci. Am.* 246(2): 58–66.

Steffy, K., and Wong-Staal, F. 1991. Genetic regulation of human immunodeficiency virus. *Microbiol. Rev.* 55(2):193–205.

Stephens, E. B., and Compans, R. W. 1988. Assembly of animal viruses at cellular membranes. *Ann Rev. Microbiol.* 42:489–516.

Strauss, J. H., and Strauss, E. G. 1988. Evolution of RNA viruses. *Ann Rev. Microbiol.* 42:657–83.

Temin, H. M. 1972. RNA-directed DNA synthesis. *Sci. Am.* 226(1):25–33.

Varmus, H. 1987. Reverse transcription. *Sci. Am.* 257(3):56–64.

Varmus, H. 1988. Retroviruses. *Science* 240:1427–35.

Wang, A. L., and Wang, C. C. 1991. Viruses of the protozoa. *Ann. Rev. Microbiol.* 45:251–63.

Webster, R. G.; Bean, W. J.; Gorman, O. T.; Chambers, T. M.; and Kawaoka, Y. 1992. Evolution and ecology of influenza A viruses. *Microbiol. Rev.* 56(1):152–79.

Weinberg, R. A. 1983. A molecular basis of cancer. *Sci. Am.* 249(5):126–42.

Wood, H. A., and Granados, R. R. 1991. Genetically engineered baculoviruses as agents for pest control. *Ann. Rev. Microbiol.* 45:69–87.

PART SIX

The Diversity of the Microbial World

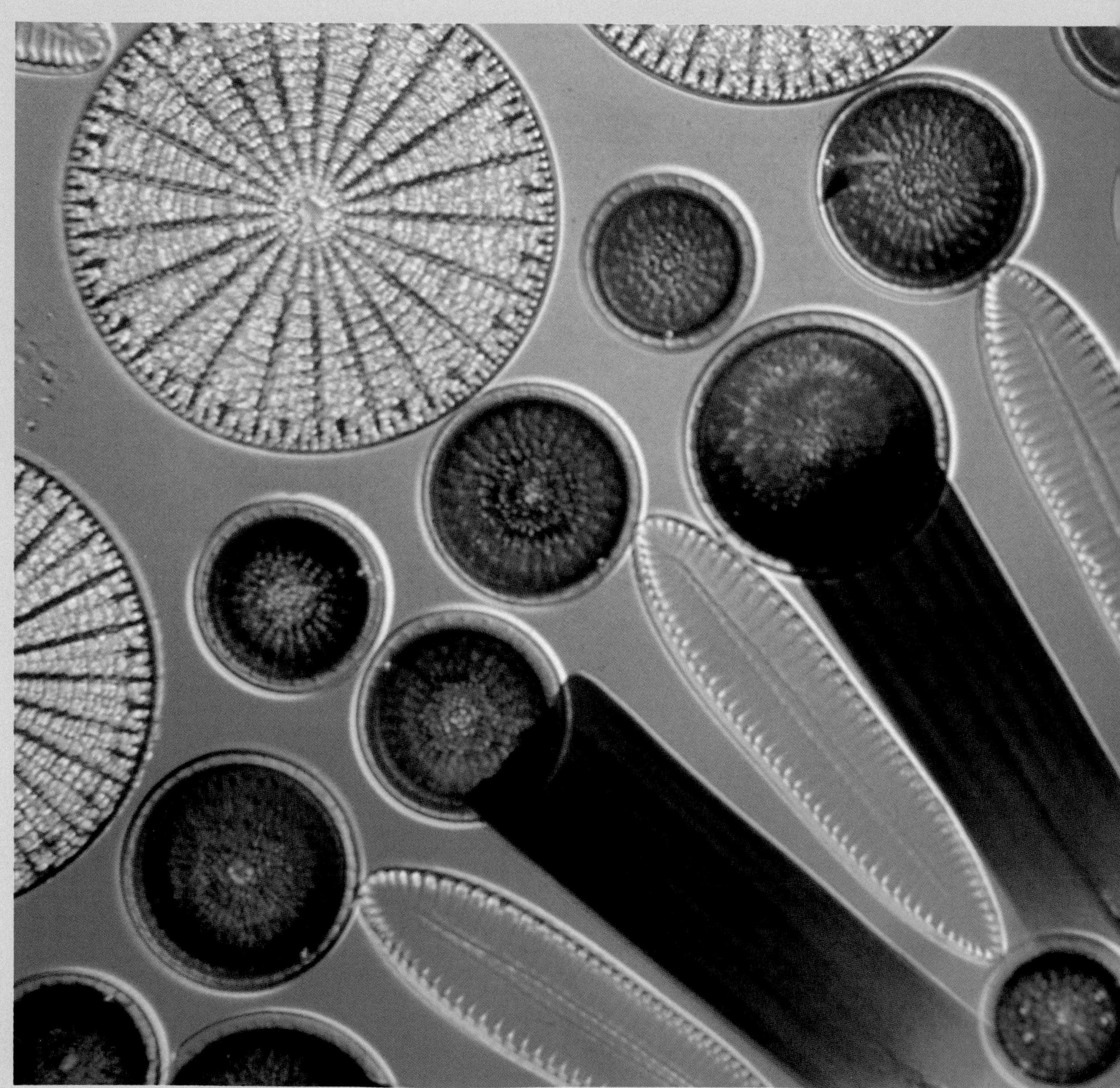

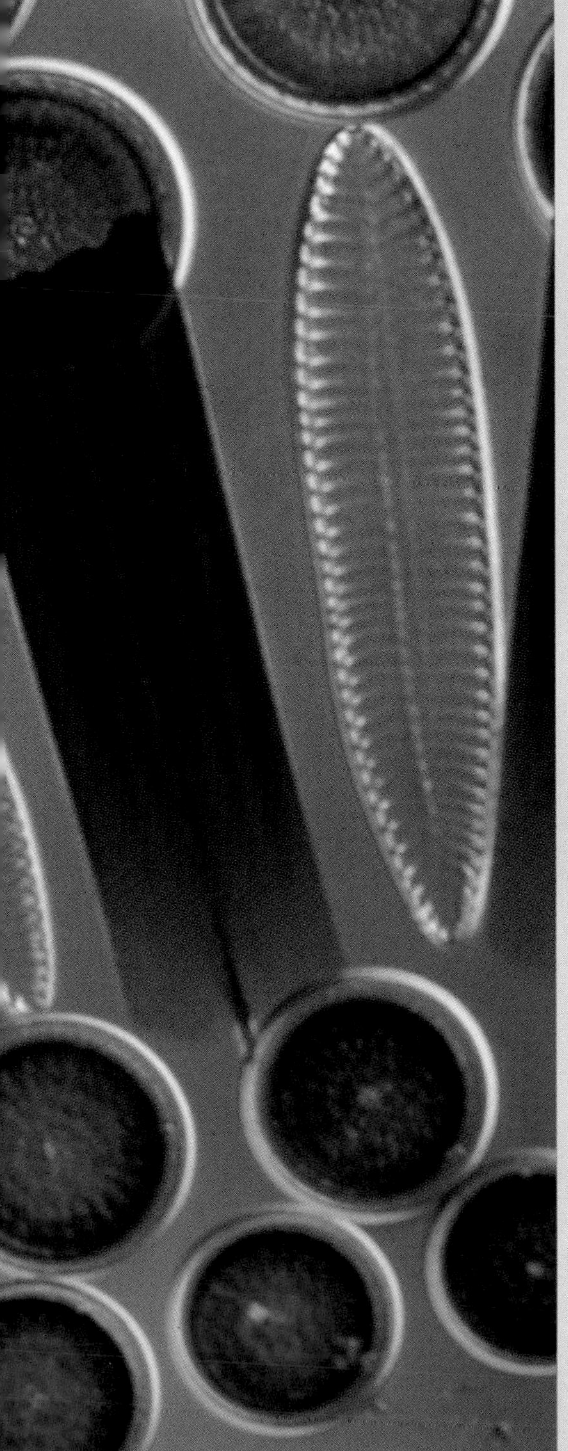

These diatom frustules have been arranged in a geometric pattern and photographed with an interference phase microscope (×450). Diatoms are so numerous in aquatic habitats that they have been called the "grasses of the sea" and are responsible for about 20 to 25% of the world's primary productivity (*see chapter 40*). Diatom frustules contain large quantities of silica and are decorated with an extraordinary array of pores, slits, ridges, spines, and other ornaments. They are among the most beautiful objects in the microbial world.

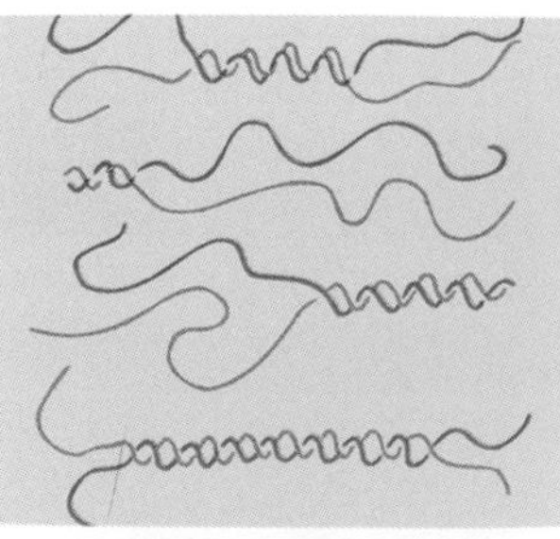

CHAPTER 20

Microbial Taxonomy

What's in a name? that which we call a rose By any other name would smell as sweet. . . .
—W. Shakespeare

It were not best that we should all think alike; it is difference of opinion that makes horse-races.
—Mark Twain

Outline

Concepts

1. In order to make sense of the diversity of organisms, it is necessary to group similar organisms together and organize these groups in a nonoverlapping hierarchical arrangement. Taxonomy is the science of biological classification.
2. The basic taxonomic group is the species, which is defined in terms of either sexual reproduction or general similarity.
3. Classifications are based on an analysis of possible evolutionary relationships (phylogenetic or phyletic classification) or on overall similarity (phenetic classification).
4. Morphological, physiological, metabolic, ecological, genetic, and molecular characteristics are all useful in taxonomy because they reflect the organization and activity of the genome. Nucleic acid structure is probably the best indicator of relatedness because nucleic acids are either the genetic material itself or the products of gene transcription.
5. Organisms can be placed in one of five kingdoms based on cell type, level of organization, and type of nutrition. Alternative systems with only three domains or primary groupings above the kingdom level have been developed.
6. Bacterial taxonomy is rapidly changing due to the acquisition of new data, particularly the use of molecular techniques such as the comparison of ribosomal RNA structure.
7. *Bergey's Manual of Systematic Bacteriology* divides the bacteria between four divisions and seven classes within the kingdom *Procaryotae.*

One of the most fascinating and attractive aspects of the microbial world is its extraordinary diversity. It seems that almost every possible experiment in shape, size, physiology, and life-style has been tried. The sixth part of the text focuses on this diversity. Chapter 20 introduces the general principles of microbial taxonomy. This is followed by a five-chapter (21–25) survey of the most important bacterial groups. The section ends with an extensive introduction to the major types of eucaryotic microorganisms: fungi, algae, and protozoa.

Because of the bewildering diversity of living organisms, it is desirable to classify or arrange them into groups based on their mutual similarities. **Taxonomy** (Greek *taxis,* arrangement or order, and *nomos,* law, or *nemein,* to distribute or govern) is defined as the science of biological classification. In a broader sense, it consists of three separate but interrelated parts: classification, nomenclature, and identification. **Classification** is the arrangement of organisms into groups or **taxa** (singular, **taxon**) based on mutual similarity or evolutionary relatedness. **Nomenclature** is the branch of taxonomy concerned with the assignment of names to taxonomic groups in agreement with published rules. **Identification** is the practical side of taxonomy, the process of determining that a particular isolate belongs to a recognized taxon.

The term **systematics** is often used for taxonomy. However, many taxonomists define it in more general terms as "the scientific study of organisms with the ultimate object of characterizing and arranging them in an orderly manner." Any study of the nature of organisms, when the knowledge gained is used in taxonomy, is a part of systematics. Thus, it encompasses disciplines such as morphology, ecology, epidemiology, biochemistry, molecular biology, and physiology.

Microbial taxonomy is too broad a subject for adequate coverage in a single chapter. Therefore, this chapter emphasizes general principles and uses examples primarily from bacterial taxonomy. The taxonomy of each major eucaryotic microbial group is reviewed where the group is introduced in subsequent chapters.

Taxonomic Ranks

In preparing a classification scheme, one places the microorganism within a small, homogeneous group that is itself a member of larger groups in a nonoverlapping hierarchical arrangement. A category in any rank unites groups in the level below it based on shared properties (figure 20.1). The most commonly used levels or ranks (in ascending order) are species, genera, families, orders, classes, phyla or divisions, and kingdoms. Microbial groups at each level or rank have names with endings or suffixes characteristic of that level (table 20.1). Informal names are often used in place of formal hierarchical ones. Typical examples of such names are purple bacteria, spirochetes, methane-oxidizing bacteria, sulfate-reducing bacteria, and lactic acid bacteria.

The basic taxonomic group in microbial taxonomy is the **species.** Taxonomists working with higher organisms define the term species differently than microbiologists do. Species of higher organisms are groups of interbreeding or potentially interbreeding natural populations that are reproductively isolated from other groups. This is a satisfactory definition for

Table 20.1 An Example of Taxonomic Ranks and Names

Rank	Example
Kingdom	*Procaryotae*
Division (Phylum)	*Tenericutes*
Class	*Mollicutes*
Order	*Mycoplasmatales*
Family	*Mycoplasmataceae*
Genus	*Mycoplasma*
Species	*M. pneumoniae*

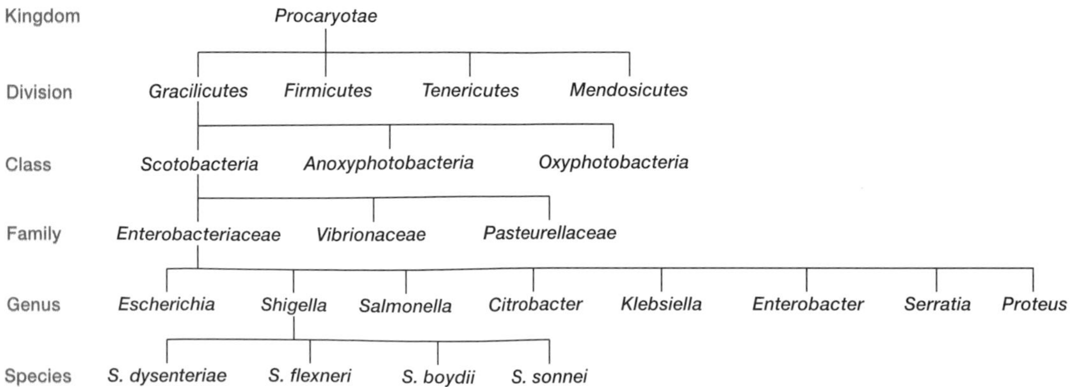

Figure 20.1 **Hierarchical Arrangement in Taxonomy.** In this example, members of the genus *Shigella* are placed within higher taxonomic ranks according to the system presented in *Bergey's Manual of Systematic Bacteriology.* Not all classification possibilities are given for each rank in order to simplify the diagram.

organisms capable of sexual reproduction, but fails with many microorganisms because they do not reproduce sexually. Bacterial species are characterized by phenotypic and genotypic (*see chapter 12*) differences. A bacterial species is a collection of strains that share many stable properties in common and differ significantly from other groups of strains. A **strain** is a population of organisms that descends from a single organism or pure culture isolate. Strains within a species may differ slightly from one another in many ways. **Biovars** are variant bacterial strains characterized by biochemical or physiological differences, **morphovars** differ morphologically, and **serovars** have distinctive antigenic properties. One strain of a species is designated as the **type strain.** It is usually one of the first strains studied and is often more fully characterized than other strains; however, it does not have to be the most representative member. Only those strains very similar to the type strain are included in a species.

Microbiologists name microorganisms by using the **binomial system** of the Swedish botanist Carl von Linné, or Carolus Linneaus as he is often called. The Latinized, italicized name consists of two parts. The first part, which is capitalized, is the generic name, and the second is the uncapitalized specific epithet or name (e.g., *Escherichia coli*). The specific name is stable; the oldest epithet for a particular organism takes precedence and must be used. In contrast, a generic name can change if the organism is assigned to another genus because of new information. For example, *Streptococcus pneumoniae* was originally called *Diplococcus pneumoniae.* Often the name will be shortened by abbreviating the genus name with a single capital letter, for example *E. coli.* A list of approved bacterial names was published in 1980 in the *International Journal of Systematic Bacteriology,* and new valid names are published periodically. *Bergey's Manual of Systematic Bacteriology* has recently been published in four volumes and contains the currently accepted system of bacterial taxonomy.

1. Be able to define the following terms: taxonomy, classification, taxon, nomenclature, identification, systematics, species, strain, type strain, and binomial system.
2. How does the definition of species differ between organisms that reproduce sexually and those not able to do so?

Classification Systems

The most desirable classification system, called a **natural classification,** arranges organisms into groups whose members share many characteristics and reflects as much as possible the biological nature of organisms. Linneaus developed the first natural classification, based largely on anatomical characteristics, in the middle of the eighteenth century. It was a great improvement over previously employed artificial systems because knowledge of an organism's position in the scheme provided information about many of its properties. For example, classification of humans as mammals denotes that they have hair, self-regulating body temperature, and milk-producing mammary glands in the female.

Following the publication in 1859 of Darwin's *On the Origin of Species,* biologists began trying to develop **phylogenetic** or **phyletic classification systems.** These are systems based on evolutionary relationships rather than general resemblance (the term **phylogeny** [Greek *phylon,* tribe or race, and *genesis,* generation or origin] refers to the evolutionary development of a species). This has proven difficult for bacteria and other microorganisms, primarily because of the lack of a good fossil record. The direct comparison of genetic material and gene products such as RNA and proteins overcomes some of these problems.

Many taxonomists maintain that the most natural classification is the one with the greatest information content or predictive value. A good classification should bring order to biological diversity and may even clarify the function of a morphological structure. For example, if motility and flagella are always associated in particular microorganisms, it is reasonable to suppose that flagella are involved in at least some types of motility. When viewed in this way, the best natural classification system may be a **phenetic system,** one that groups organisms together based on the mutual similarity of their phenotypic characteristics. Although phenetic studies can reveal possible evolutionary relationships, they are not dependent upon phylogenetic analysis. They compare many traits without assuming that any features are more phylogenetically important than others; that is, unweighted traits are employed in estimating general similarity. Obviously, the best phenetic classification is one constructed by comparing as many attributes as possible. Organisms sharing many characteristics make up a single group or taxon.

1. What is a natural classification?
2. What are phylogenetic (phyletic) and phenetic classification systems? How do the two systems differ?

Numerical Taxonomy

The development of computers has made possible the quantitative approach known as **numerical taxonomy.** Peter H. A. Sneath and Robert Sokal have defined numerical taxonomy as "the grouping by numerical methods of taxonomic units into taxa on the basis of their character states." Information about the properties of organisms is converted into a form suitable for numerical analysis and then compared by means of a computer. The resulting classification is based on general similarity as judged by comparison of many characteristics, each given equal weight. This approach was not feasible before the advent of computers because of the large number of calculations involved.

The process begins with a determination of the presence or absence of selected characters in the group of organisms under study. A character is usually defined as an attribute about which a single statement can be made. Many characters, at least 50 and preferably several hundred, should be compared for an accurate and reliable classification. It is best to include many different kinds of data: morphological, biochemical, and physiological.

After character analysis, an association coefficient, a function that measures the agreement between characters possessed by two organisms, is calculated for each pair of organisms in the group. The **simple matching coefficient (S_{SM})**, the most commonly used coefficient in bacteriology, is the proportion of characters that match regardless of whether the attribute is present or absent (table 20.2). Sometimes the **Jaccard coefficient (S_J)** is calculated by ignoring any characters that both organisms lack (table 20.2). Both coefficients increase linearly in value from 0.0 (no matches) to 1.0 (100% matches).

The simple matching coefficients, or other association coefficients, are then arranged to form a **similarity matrix.** This is a matrix in which the rows and columns represent organisms and each value is an association coefficient measuring the similarity of two different organisms so that each organism is compared to every other one in the table (figure 20.2*a*). Organisms with great similarity are grouped together and separated from dissimilar organisms (figure 20.2*b*); such groups of organisms are called **phenons** (sometimes called phenoms).

The results of numerical taxonomic analysis are often summarized with a treelike diagram called a **dendrogram** (figures 20.2*c* and 20.3). The diagram is usually placed on its side with the x-axis or abscissa graduated in units of similarity (figure 20.3). Each branch point is at the similarity value relating the two branches. The organisms in the two branches share so many characteristics that the two groups are only seen

TABLE 20.2 The Calculation of Association Coefficients for Two Organisms

In this example, organisms A and B are compared in terms of the characters they do and do not share. The terms in the association coefficient equations are defined as follows:

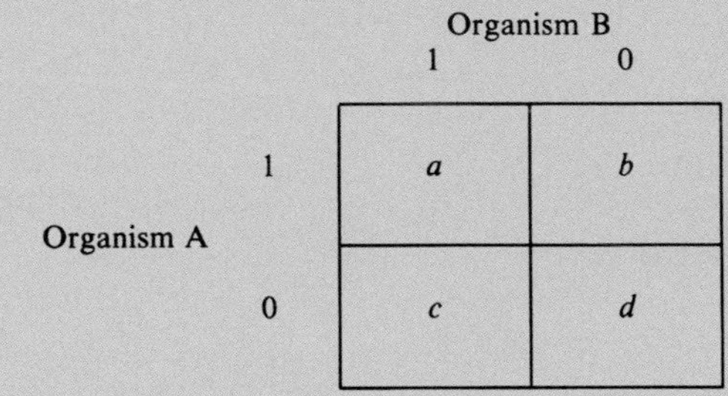

a = number of characters coded as present (1) for both organisms

b and c = numbers of characters differing (1,0 or 0,1) between the two organisms

d = number of characters absent (0) in both organisms

Total number of characters compared = $a + b + c + d$

$$\text{The simple matching coefficient } (S_{SM}) = \frac{a + d}{a + b + c + d}$$

$$\text{The Jaccard coefficient } (S_J) = \frac{a}{a + b + c}$$

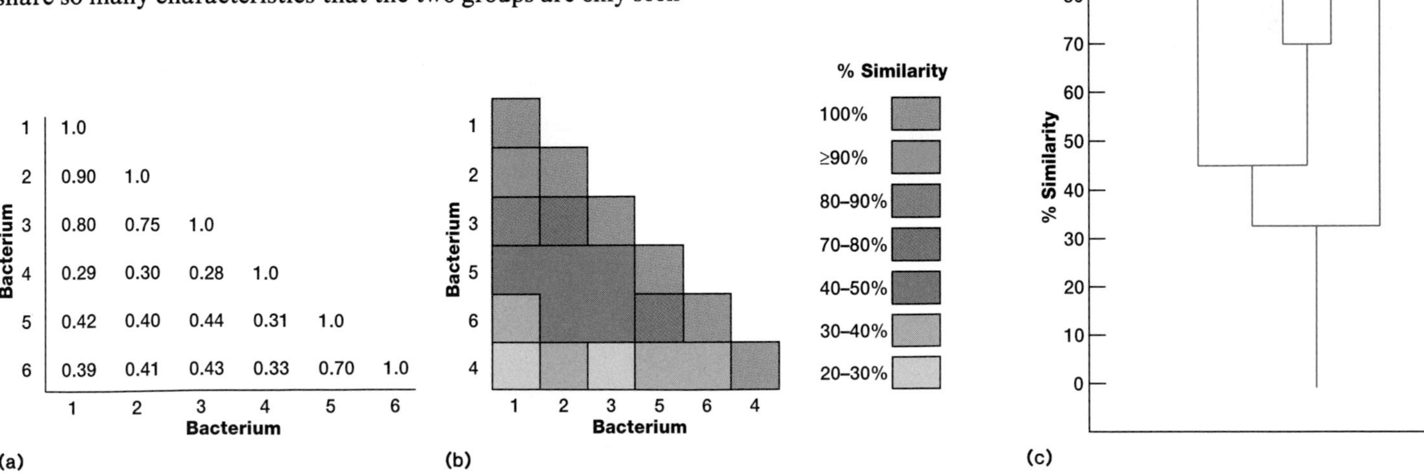

Figure 20.2 Clustering and Dendrograms in Numerical Taxonomy. (*a*) A small similarity matrix that compares six strains of bacteria. The degree of similarity ranges from none (0.0) to complete similarity (1.0). (*b*) The bacteria have been rearranged and joined to form clusters of similar strains. For example, strains 1 and 2 are the most similar. The cluster of 1 plus 2 is fairly similar to strain 3, but not at all to strain 4. (*c*) A dendrogram showing the results of the analysis in part *b*. Strains 1 and 2 are members of a 90-phenon, and strains 1–3 form an 80-phenon (see text). While strains 1–3 may be members of a single species, it is quite unlikely that strains 4–6 belong to the same species as 1–3. *From P. H. A. Sneath,* In Essays in Microbiology, *J. R. Norris and M. H. Richmond (eds.) Copyright © 1978 John Wiley & Sons, Inc., New York, NY.*

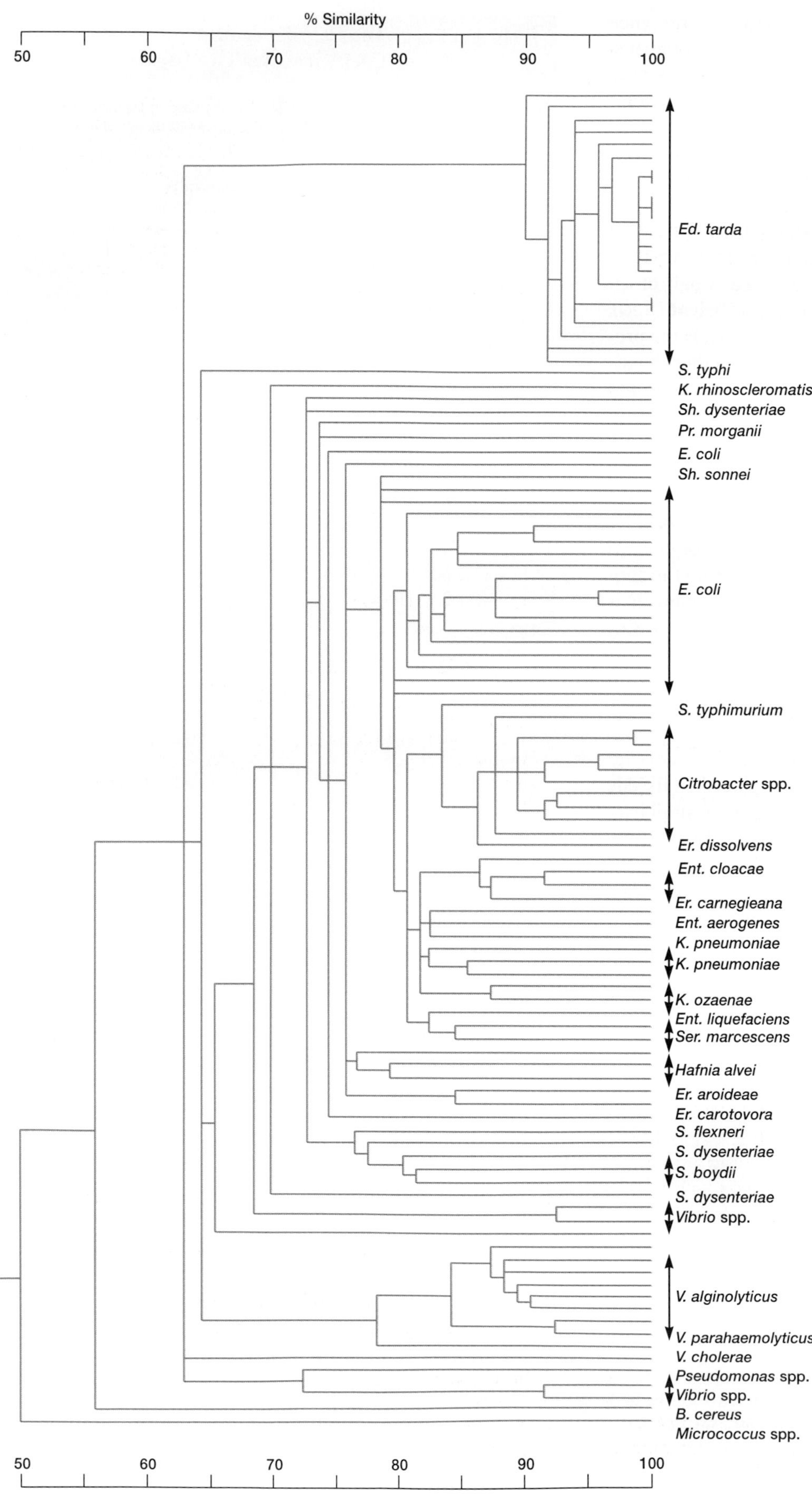

Figure 20.3 The Use of Numerical Taxonomy in Bacteriology. The dendrogram indicates the arrangement of 105 strains of gram-negative fermentative bacteria from the family *Enterobacteriaceae*. The strains were compared using 238 characteristics.

to be separate after examination of association coefficients greater than the magnitude of the branch point value. Below the branch point value, the two groups appear to be one. The ordinate in such a dendrogram has no special significance, and the clusters may be arranged in any convenient order.

The significance of these clusters or phenons in traditional taxonomic terms is not always evident, and the similarity levels at which clusters are labeled species, genera, and so on, are a matter of judgment. Sometimes groups are simply called phenons and preceded by a number showing the similarity level above which they appear (e.g., a 70-phenon is a phenon with 70% or greater similarity among its constituents). Phenons formed at about 80% similarity are often equivalent to bacterial species.

Numerical taxonomy has already proved to be a powerful tool in microbial taxonomy. Although it has often simply reconfirmed already-existing classification schemes, sometimes accepted classifications are found wanting. Its usefulness will continue to increase with the addition of new tests and the accumulation of more data.

1. What is numerical taxonomy and why are computers so important to this approach?
2. Define the following terms: association coefficient, simple matching coefficient, Jaccard coefficient, similarity matrix, phenon, and dendrogram.
3. Which pair of species has more mutual similarity, a pair with an association coefficient of 0.9 or one with a coefficient of 0.6? Why?

Major Characteristics Used in Taxonomy

Many characteristics are used in classifying and identifying microorganisms. This section briefly reviews some of the most taxonomically important properties. Methods often employed in routine laboratory identification of bacteria are covered in the chapter on clinical microbiology (*see chapter 34*).

Morphological Characteristics

Morphological features are important in microbial taxonomy for many reasons. Morphology is easy to study and analyze, particularly in eucaryotic microorganisms and the more complex procaryotes. In addition, morphological comparisons are valuable because structural features depend on the expression of many genes, are usually genetically stable, and normally (at least in eucaryotes) do not vary greatly with environmental changes. Thus, morphological similarity is a good indication of phylogenetic relatedness.

Many different morphological features are employed in the classification and identification of microorganisms (table 20.3). Although the light microscope has always been a very important tool, its resolution limit of about 0.2 μm (*see chapter 2*) reduces its usefulness in viewing smaller microorganisms and structures. The transmission and scanning electron microscopes, with their greater resolution, have immensely aided the study of all microbial groups.

TABLE 20.3 Some Morphological Features Used in Classification and Identification

Feature	Microbial Groups
Cell shape	All major groups[a]
Cell size	All major groups
Colonial morphology	All major groups
Ultrastructural characteristics	All major groups
Staining behavior	Bacteria, some fungi
Cilia and flagella	All major groups
Mechanism of motility	Gliding bacteria, spirochetes
Endospore shape and location	Endospore-forming bacteria
Spore morphology and location	Bacteria, algae, fungi
Cellular inclusions	All major groups
Color	All major groups

[a]Used in classifying and identifying at least some bacteria, algae, fungi, and protozoa.

TABLE 20.4 Some Physiological and Metabolic Characteristics Used in Classification and Identification

Carbon and nitrogen sources
Cell wall constituents
Energy sources
Fermentation products
General nutritional type
Growth temperature optimum and range
Luminescence
Mechanisms of energy conversion
Motility
Osmotic tolerance
Oxygen relationships
pH optimum and growth range
Photosynthetic pigments
Salt requirements and tolerance
Secondary metabolites formed
Sensitivity to metabolic inhibitors and antibiotics
Storage inclusions

Physiological and Metabolic Characteristics

Physiological and metabolic characteristics are very useful because they are directly related to the nature and activity of microbial enzymes and transport proteins. Since proteins are gene products, analysis of these characteristics provides an indirect comparison of microbial genomes. Table 20.4 lists some of the most important of these properties.

Ecological Characteristics

Many properties are ecological in nature since they affect the relation of microorganisms to their environment. Often, these are taxonomically valuable because even very closely related microorganisms can differ considerably with respect to ecological characteristics. Microorganisms living in various parts of the human body markedly differ from one another and from those growing in freshwater, terrestrial, and marine environments. Some examples of taxonomically important ecological properties are life cycle patterns; the nature of symbiotic relationships; the ability to cause disease in a particular host; and habitat preferences such as requirements for temperature, pH, oxygen, and osmotic concentration. Many growth requirements are also considered physiological characteristics (table 20.4).

Genetic Analysis

Because most eucaryotes are able to reproduce sexually, genetic analysis has been of considerable usefulness in the classification of these organisms. As mentioned earlier, the species is defined in terms of sexual reproduction where possible. Although procaryotes do not reproduce sexually, the study of chromosomal gene exchange through transformation and conjugation is sometimes useful in their classification.

Transformation can occur between different bacterial species but only rarely between genera. The demonstration of transformation between two strains provides evidence of a close relationship since bacterial transformation cannot occur unless the genomes are fairly similar. Transformation studies have been carried out with several genera: *Bacillus, Micrococcus, Haemophilus, Rhizobium,* and others. Despite transformation's usefulness, its results are sometimes hard to interpret because an absence of transformation may result from factors other than major differences in DNA sequence.

Transformation (pp. 238, 271–74).
Conjugation (pp. 268–71).

Conjugation studies also yield taxonomically useful data, particularly with the enteric bacteria (*see chapter 21*). For example, *Escherichia* can undergo conjugation with the genera *Salmonella* and *Shigella,* but not with *Proteus* and *Enterobacter.* These observations fit with other data showing that the first three of these genera are more closely related to one another than to *Proteus* and *Enterobacter.*

Plasmids (*see chapter 13*) are undoubtedly important in taxonomy because they are present in most bacterial genera, and many carry genes coding for phenotypic traits. Because plasmids could have a significant effect on classification if they carried the gene for a trait of major importance in the classification scheme, it is best to base a classification on many characters. When the identification of a group is based on a few characteristics and some of these are coded for by plasmid genes, errors may result. For example, hydrogen sulfide production and lactose fermentation are very important in the taxonomy of the enteric bacteria, yet genes for both traits can be borne on plasmids as well as bacterial chromosomes. One must take care to avoid errors as a result of plasmid-borne traits.

Molecular Characteristics

Some of the most powerful approaches to taxonomy are through the study of proteins and nucleic acids. Because these are either direct gene products or the genes themselves, comparisons of proteins and nucleic acids yield considerable information about true relatedness.

Comparison of Proteins

The amino acid sequences of proteins are direct reflections of mRNA sequences and therefore closely related to the structures of the genes coding for their synthesis. For this reason, comparisons of proteins from different microorganisms are very useful taxonomically. There are several ways to compare proteins. The most direct approach is to determine the amino acid sequence of proteins with the same function. The sequences of proteins with dissimilar functions often change at different rates; some sequences change quite rapidly, whereas others are very stable. Nevertheless, if the sequences of proteins with the same function are similar, the organisms possessing them are probably closely related. The sequences of cytochromes and other electron transport proteins, histones, and a variety of enzymes have been used in taxonomic studies. Because protein sequencing is slow and expensive, more indirect methods of comparing proteins have frequently been employed. The electrophoretic mobility of proteins (*see chapter 14*) is useful in studying relationships at the species and subspecies level. Antibodies can discriminate between very similar proteins, and immunologic techniques are used to compare proteins from different microorganisms.

Antibody-antigen reactions in vitro (pp. 656–66).

The physical, kinetic, and regulatory properties of enzymes have been employed in taxonomic studies. Because enzyme behavior reflects amino acid sequence, this approach is useful in studying some microbial groups, and group-specific patterns of regulation have been found.

Nucleic Acid Base Composition

Microbial genomes can be directly compared, and taxonomic similarity can be estimated in many ways. The first, and possibly the simplest, technique to be employed is the determination of DNA base composition. DNA contains four purine and pyrimidine bases: adenine (A), guanine (G), cytosine (C), and thymine (T). In double-stranded DNA, A pairs with T, and G pairs with C. Thus, the **(G + C)/(A + T) ratio** or **G + C content,** the percent of G + C in DNA, reflects the base sequence and varies with sequence changes as follows:

$$\text{Mol\% G + C} = \frac{G + C}{G + C + A + T} \times 100$$

DNA and RNA structure (pp. 192–95).

The base composition of DNA can be determined in several ways. Although the G + C content can be chemically ascertained after hydrolysis of DNA and separation of its bases, physical methods are easier and more often used. The G + C content is often determined from the **melting temperature** (**T_m**) of DNA. In double-stranded DNA, three hydrogen bonds join GC base pairs, and two bonds connect AT base pairs (*see chapter 10*). As a result, DNA with a greater G + C content will have more hydrogen bonds, and its strands will separate only at higher temperatures; that is, it will have a higher melting point. DNA melting can be easily followed spectrophotometrically because the absorbance of 260 nm UV light by DNA increases during strand separation. When a DNA sample is slowly heated, the absorbance increases as hydrogen bonds are broken and reaches a plateau when all the DNA has become single stranded (figure 20.4). The midpoint of the rising curve gives the melting temperature, a direct measure of the G + C content. Since the density of DNA also increases linearly with G + C content, the percent G + C can be obtained by centrifuging DNA in a CsCl density gradient (*see chapter 17*).

The G + C content of many microorganisms has been determined (table 20.5). The G + C content of DNA from animals and higher plants averages around 40% and ranges between 30 and 50%. In contrast, the DNA of both eucaryotic and procaryotic microorganisms varies greatly in G + C content; procaryotic G + C content is the most variable, ranging from around 25 to almost 80%. Despite such a wide range of variation, the G + C content of strains within a particular species is constant. If two organisms differ in their G + C content by more than about 10%, their genomes have quite

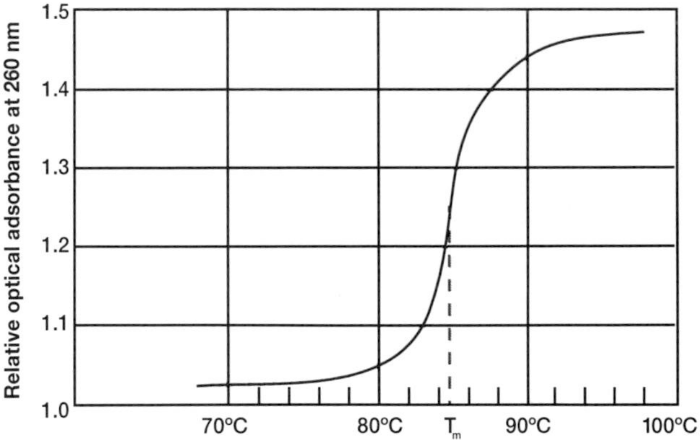

Figure 20.4 **A DNA Melting Curve.** The T_m is indicated.

TABLE 20.5 Representative G + C Contents of Microorganisms

Organism	Percent G + C	Organism	Percent G + C	Organism	Percent G + C
Bacteria		*Streptomyces*	69–73		
Actinomyces	59–73	*Sulfolobus*	31–37	**Slime Molds**	
Anabaena	38–44	*Thermoplasma*	46	*Dictyostelium* spp.	22–25
Bacillus	32–62	*Thiobacillus*	52–68	*Lycogala*	42
Bacteroides	28–61	*Treponema*	25–54	*Physarum polycephalum*	38–42
Bdellovibrio	33–52	**Algae**		**Fungi**	
Caulobacter	63–67	*Acetabularia mediterranea*	37–53	*Agaricus bisporus*	44
Chlamydia	41–44	*Chlamydomonas* spp.	60–68	*Amanita muscaria*	57
Chlorobium	49–58	*Chlorella* spp.	43–79	*Aspergillus niger*	52
Chromatium	48–70	*Cyclotella cryptica*	41	*Blastocladiella emersonii*	66
Clostridium	21–54	*Euglena gracilis*	46–55	*Candida albicans*	33–35
Cytophaga	33–42	*Nitella*	49	*Claviceps purpurea*	53
Deinococcus	62–70	*Nitzschia angularis*	47	*Coprinus lagopus*	52–53
Escherichia	48–52	*Ochromonas danica*	48	*Fomes fraxineus*	56
Halobacterium	66–68	*Peridinium triquetrum*	53	*Mucor rouxii*	38
Hyphomicrobium	59–67	*Scenedesmus*	52–64	*Neurospora crassa*	52–54
Methanobacterium	32–50	*Spirogyra*	39	*Penicillium notatum*	52
Micrococcus	64–75	*Volvox carteri*	50	*Polyporus palustris*	56
Mycobacterium	62–70	**Protozoa**		*Rhizopus nigricans*	47
Mycoplasma	23–40	*Acanthamoeba castellanii*	56–58	*Saccharomyces cerevisiae*	36–42
Myxococcus	68–71	*Amoeba proteus*	66	*Saprolegnia parasitica*	61
Neisseria	47–54	*Paramecium* spp.	29–39		
Nitrobacter	60–62	*Plasmodium berghei*	41		
Oscillatoria	40–50	*Stentor polymorphus*	45		
Prochloron	41	*Tetrahymena* spp.	19–33		
Proteus	38–41	*Trichomonas* spp.	29–34		
Pseudomonas	58–70	*Trypanosoma* spp.	45–59		
Rhodospirillum	62–66				
Rickettsia	29–33				
Salmonella	50–53				
Spirillum	38				
Spirochaeta	51–65				
Staphylococcus	30–38				
Streptococcus	33–44				

different base sequences. On the other hand, it is not safe to assume that organisms with very similar G + C contents also have similar DNA base sequences because two very different base sequences can be constructed from the same proportions of AT and GC base pairs. Only if two microorganisms are also alike phenotypically does their similar G + C content suggest close relatedness.

G + C content data are taxonomically valuable in at least two ways. First, they can confirm a taxonomic scheme developed using other data. If organisms in the same taxon are too dissimilar in G + C content, the taxon probably should be divided. Second, G + C content appears to be useful in characterizing bacterial genera since the variation within a genus is usually less than 10% even though the content may vary greatly between genera. For example, *Staphylococcus* has a G + C content of 30 to 38% while *Micrococcus* DNA has 64 to 75% G + C; yet these two genera of gram-positive cocci have many other features in common.

Nucleic Acid Hybridization

The similarity between genomes can be compared more directly by use of **nucleic acid hybridization** studies. If a mixture of single-stranded DNA formed by heating dsDNA is cooled and held at a temperature about 25°C below the T_m, strands with complementary base sequences will reassociate to form stable dsDNA, while noncomplementary strands will remain single (figure 20.5). Because strands with similar, but not identical, sequences associate to form less temperature stable dsDNA hybrids, incubation of the mixture at 30 to 50°C below the T_m will allow hybrids of more diverse ssDNAs to form. Incubation at 10 to 15°C below the T_m permits hybrid formation only with almost identical strands.

In one of the more widely used hybridization techniques, nitrocellulose filters with bound nonradioactive DNA strands are incubated at the appropriate temperature with single-stranded DNA fragments made radioactive with ^{32}P, ^{3}H, or ^{14}C. After radioactive fragments are allowed to hybridize with the membrane-bound ssDNA, the membrane is washed to remove any nonhybridized ssDNA and its radioactivity is measured. The quantity of radioactivity bound to the filter reflects the amount of hybridization and thus the similarity of the DNA sequences. The degree of similarity or homology is expressed as the percent of experimental DNA radioactivity retained on the filter compared with the percent of homologous DNA radioactivity bound under the same conditions (see table 20.6 for an example).

If DNA molecules are very different in sequence, they will not form a stable, detectable hybrid. Therefore, DNA-DNA hybridization is only used to study closely related microorganisms. More distantly related organisms are compared by carrying out DNA-RNA hybridization experiments using radioactive ribosomal or transfer RNA. Distant relationships can be detected because rRNA and tRNA genes represent only a small portion of the total DNA genome and have not evolved as rapidly as most other microbial genes. The technique is similar to that employed for DNA-DNA hybridization:

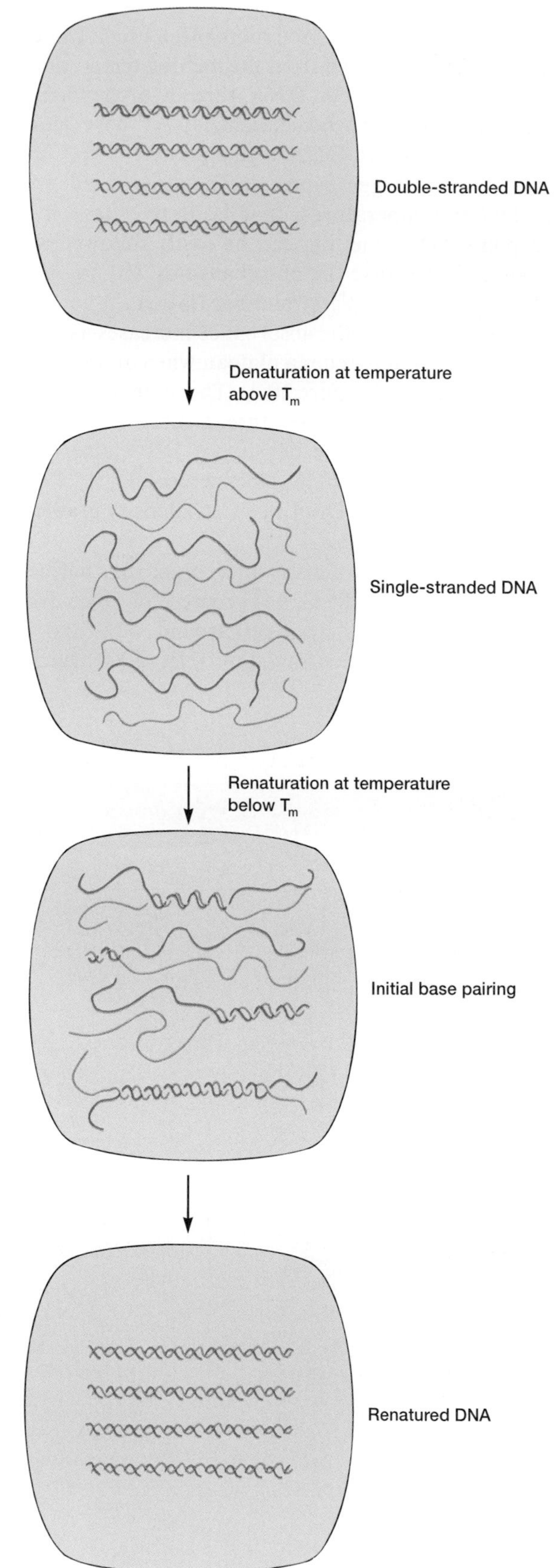

Figure 20.5 Nucleic Acid Melting and Hybridization. Complementary strands are shown in different colors.

Table 20.6 Comparison of *Neisseria* Species by DNA Hybridization Experiments

Membrane-Attached DNA[a]	Percent Homology[b]
Neisseria meningitidis	100
N. gonorrhoeae	78
N. sicca	45
N. flava	35

Source: Data from T. E. Staley and R. R. Colwell, "Application of Molecular Genetics and Numerical Taxonomy to the Classification of Bacteria" in *Annual Review of Ecology and Systematics* 8: 282, 1973

[a]The experimental membrane-attached nonradioactive DNA from each species was incubated with radioactive *N. meningitidis* DNA, and the amount of radioactivity bound to the membrane was measured. The more radioactivity bound, the greater the homology between DNA sequences.

[b] $\frac{\textit{N. meningitidis}\text{ DNA bound to experimental DNA}}{\text{Amount bound to membrane-attached }\textit{N. meningitidis}\text{ DNA}} \times 100$

membrane-bound DNA is incubated with radioactive rRNA, washed, and counted. An even more accurate measurement of homology is obtained by finding the temperature required to dissociate and remove half the radioactive rRNA from the membrane; the higher this temperature, the stronger the rRNA-DNA complex and the more similar the sequences.

Ribosomes and ribosomal RNA (pp. 208–9).
Transfer RNA (pp. 206–7).

Nucleic Acid Sequencing

Despite the usefulness of G + C content determination and nucleic acid hybridization studies, genome structures can only be directly compared by sequencing DNA and RNA. Techniques for rapidly sequencing both DNA and RNA are now available; thus far, RNA sequencing has been used more extensively in microbial taxonomy.

Most attention has been given to sequences of the 5S and 16S rRNAs isolated from the 50S and 30S subunits, respectively, of bacterial ribosomes (*see chapters 3 and 10*). The rRNAs are almost ideal for studies of microbial evolution and relatedness since they are essential to a critical organelle found in all microorganisms. Their functional role is the same in all ribosomes. Furthermore, their structure changes very slowly with time, presumably because of their constant and critical role. Ribosomal RNAs can be characterized in terms of partial sequences by the oligonucleotide cataloging method as follows. Purified, radioactive 16S rRNA is treated with the enzyme T_1 ribonuclease, which cleaves it into fragments. The fragments are separated, and all fragments composed of at least six nucleotides are sequenced. The sequences of corresponding 16S rRNA fragments from different bacteria are then aligned and compared using a computer, and association coefficients (S_{ab} values) are calculated. Complete rRNAs can now be sequenced using the enzyme reverse transcriptase and DNA sequencing techniques. Since rRNA contains variable and stable sequences, both closely related and very distantly related microorganisms can be compared. This is an important advantage because distantly related organisms can only be studied using sequences that change little with time. The results of such studies will be discussed shortly.

1. Summarize the advantages of using each major group of characteristics (morphological, physiological/metabolic, ecological, genetic, and molecular) in classification and identification. How is each group related to the nature and expression of the genome? Give examples of each type of characteristic.
2. What two modes of genetic exchange in bacteria have proved taxonomically useful? Why are plasmids of such importance in bacterial taxonomy?
3. Briefly describe some ways in which proteins from different organisms can be compared.
4. What is the G + C content of DNA and how can it be determined through melting temperature studies and density gradient centrifugation?
5. Discuss the use of G + C content in taxonomy. Why is it not safe to assume that two microorganisms with the same G + C content belong to the same species? In what two ways are G + C content data taxonomically valuable?
6. Describe how nucleic acid hybridization studies are carried out using membrane-bound DNA. Why might one wish to vary the incubation temperature during hybridization? What is the advantage of conducting DNA-RNA hybridization studies?
7. How are rRNA sequencing studies conducted and why is rRNA so suitable for determining relatedness?

The Kingdoms of Organisms

Since the beginning of biology, organisms have been classified as either plants or animals. However, discoveries in microbiology over the past century have shown that the two-kingdom system is oversimplified, and many biologists favor dividing organisms among five kingdoms as first suggested by Robert H. Whittaker. A modified version of Whittaker's five-kingdom system is presented in table 20.7 and figure 20.6. Organisms are placed into five kingdoms based on at least three major criteria: (1) cell type—procaryotic or eucaryotic, (2) level of organization—solitary and colonial unicellular organization or multicellular, and (3) nutritional type (see Box 20.1 for a discussion of the origin of the eucaryotic cell). In this system, the kingdom *Animalia* contains multicellular animals with wall-less eucaryotic cells and primarily ingestive nutrition, whereas the kingdom *Plantae* is composed of multicellular plants with walled eucaryotic cells and primarily photoautotrophic nutrition. Microbiologists study members of the other three kingdoms. The kingdom ***Monera,*** the kingdom ***Procaryotae*** in *Bergey's Manual,* contains all procaryotic organisms. The kingdom ***Fungi*** contains eucaryotic and predominately multinucleate organisms, with nuclei dispersed in a walled and often septate mycelium (*see chapter 26*); their nutrition is absorptive. The fifth kingdom, the ***Protista,*** is the least homogeneous and hardest to define. **Protists** are eucaryotes with

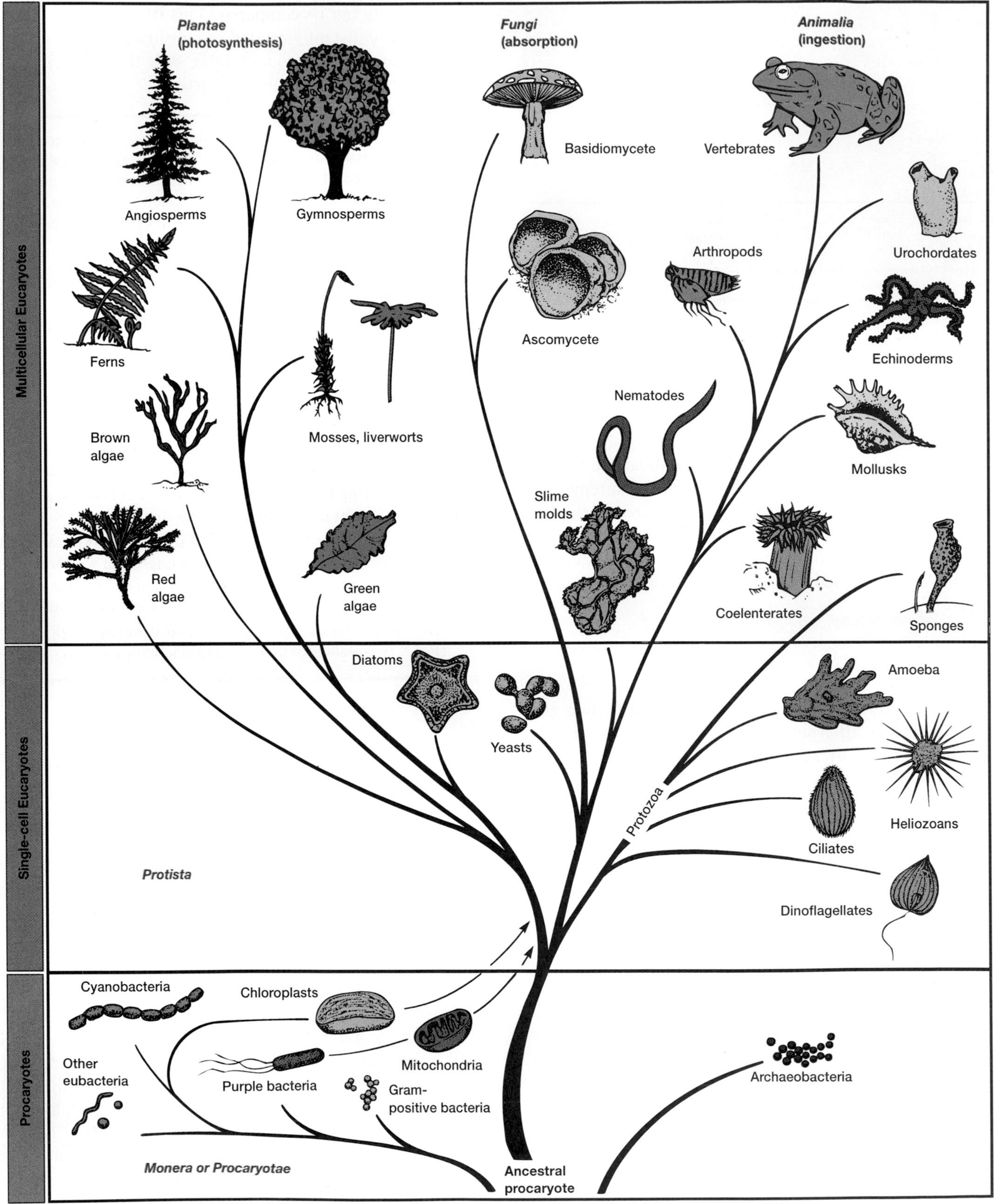

Figure 20.6 Adapted from **The Five-Kingdom System.** A simplified diagram showing the five kingdoms, levels of organization, cell types, and some lines of descent. See text for details *Adapted from Donald Voet and Judith G. Voet,* Biochemistry. *Copyright © 1990 John Wiley & Sons, Inc., New York, NY.*

TABLE 20.7 The Five-Kingdom System

Kingdom *Monera* or *Procaryotae*
Procaryotic cells. Nutrition absorptive, chemosynthetic, photoheterotrophic, or photoautotrophic, with anaerobic, facultative, microaeraphilic, or aerobic metabolism. Reproduction asexual; genetic recombination sometimes occurs. Nonmotile or motile by bacterial flagella composed of flagellin proteins or by gliding. Solitary, unicellular, filamentous, colonial, or mycelial.

Kingdom *Protista*
Eucaryotic cells with solitary and colonial unicellular organization. Usually aerobic. Nutrition ingestive, absorptive, or, if photoautotrophic, in photosynthetic plastids. All forms reproduce asexually; many have true sexual reproduction with meiosis and karyogamy. Nonmotile or motile by cilia, flagella, or other means (e.g., pseudopodia). Lack embryos and complex cell junctions.

Kingdom *Fungi*
Primarily multinucleate organisms with eucaryotic nuclei dispersed in a walled and often septate mycelial syncytium (some forms secondarily unicellular). Usually aerobic. Nutrition heterotrophic and absorptive. Flagella lacking, no motility except protoplasmic streaming. Mycelia are haploid or dikaryotic. Meiosis in zygote; haploid spores produced. No pinocytosis or phagocytosis.

Kingdom *Animalia*
Multicellular animals with wall-less eucaryotic cells. Aerobic. Nutrition primarily ingestive with digestion in an internal cavity, but some forms are absorptive and some lack a digestive cavity; phagocytosis and pinocytosis present. Reproduction mainly sexual with meiosis forming gametes; haploid stages other than gametes almost lacking above lowest phyla. Motility based on contractile fibrils. With one exception, zygote develops into a blastula. Extensive cellular and tissue differentiation with complex cell junctions.

Kingdom *Plantae*
Primarily autotrophic plants, multicellular with walled and frequently vacuolate eucaryotic cells and photosynthetic plastids. Aerobic. Simple multicellular to advanced tissue organization; development from solid embryos. Reproduction primarily sexual, with haploid and diploid stages ("alternation of generations"); haploid stage reduced in higher members of the kingdom. Usually nonmotile.

Adapted from R. H. Whittaker and L. Margulis, *BioSystems* 10:3–18. Copyright © 1978 Elsevier Scientific Publishers-Ireland. Reprinted by permission.

unicellular organization, either in the form of solitary cells or colonies of cells lacking true tissues. They may have ingestive, absorptive, or photoautotrophic nutrition, and they include most of the microorganisms known as algae, protozoa, and many of the simpler fungi. The taxonomy of the major protist and fungal phyla is discussed in more detail in chapters 26–28.

The five-kingdom system is not accepted by all microbiologists. Alternative views will be described in the following section after a discussion of recent phylogenetic studies on bacteria.

1. With what three major criteria did Whittaker divide organisms into five kingdoms?
2. Give the names and main distinguishing characteristics of the five kingdoms.

Bacterial Evolution and Taxonomy

Bacterial taxonomy is now changing rapidly. This is caused by ever-increasing knowledge of the biology of bacteria and remarkable advances in computers and the use of molecular characteristics to determine true phylogenetic relationships between bacterial groups. In this section, phylogenetic and phenetic approaches to bacterial taxonomy are described and compared.

Phylogenetic Studies

Although a variety of molecular techniques are used in estimating the phylogenetic relatedness of bacteria, the comparison of 16S rRNAs isolated from over 500 bacterial species is of particular importance (figure 20.7). The S_{ab} values obtained from rRNA studies are assumed to be a true measure of relatedness; the higher the S_{ab} values obtained from comparing two organisms, the more closely the organisms are related to each other. If all the T_1 ribonuclease fragments from the 16S rRNAs of two organisms are identical in sequence, the S_{ab} value is 1.0. S_{ab} values are also a measure of evolutionary time. A group of bacteria that branched off from other bacteria a long time ago will exhibit a large range of S_{ab} values because it has had more time to diversify than a group that developed more recently. That is, the narrower the range of S_{ab} values in a group of bacteria, the more modern it is. After S_{ab} values have been determined, a computer calculates the relatedness of the organisms and summarizes their relationships in a dendrogram (figure 20.2*c*). As mentioned earlier, it is now possible to sequence complete 16S rRNA molecules rather than fragments. This has given rRNA studies even greater resolving power. Although rRNA comparisons are useful above the species level, DNA similarity studies sometimes are more effective in categorizing individual species.

Ribosomal RNA sequence studies have uncovered a feature of great practical importance. The 16S rRNA of most major phylogenetic groups has one or more characteristic nucleotide sequences called oligonucleotide signatures. **Oligonucleotide signature sequences** are specific oligonucleotide sequences that occur in most or all members of a particular phylogenetic group. They are rarely or never present in other groups, even closely related ones. Thus, signature sequences can be used to place microorganisms in the proper group. Signature sequences have been identified for eubacteria, archaeobacteria, eucaryotes, and many major bacterial groups.

Some tentative generalizations arise from the results of the phylogenetic studies summarized in figures 20.8 through 20.11. First, it appears that there are two quite different groups of bacteria, the eubacteria and the archaeobacteria. The **eubacteria** (Greek *eu,* good, and bacteria) or true bacteria comprise the vast majority of bacteria. Among other properties, eubacteria have cell wall peptidoglycan containing muramic acid (or are related to bacteria with such cell walls) and membrane lipids that resemble eucaryotic membrane lipids in having ester-linked, straight-chained fatty acids (table 20.8).

Box 20.1

The Origin of the Eucaryotic Cell

It appears likely that eucaryotic cells arose from procaryotes about 1.4 billion years ago. There has been considerable speculation about how eucaryotes might have developed from procaryotic ancestors, and two hypotheses have been proposed. According to the first, nuclei, mitochondria, and chloroplasts arose by invagination of the plasma membrane to form double-membrane structures containing genetic material and capable of further development and specialization. The similarities between chloroplasts, mitochondria, and modern bacteria are due to conservation of primitive procaryotic features by the slowly changing organelles.

The more popular endosymbiotic hypothesis proposes that a free-living fermenting bacterium established a permanent symbiotic relationship with photosynthetic bacteria, which then evolved into chloroplasts. Cyanobacteria have been considered the most likely ancestors of chloroplasts. More recently, *Prochloron* has become the favorite candidate. *Prochloron* (p. 476) lives within marine invertebrates and resembles chloroplasts in containing both chlorophyll *a* and *b,* but not phycobilins. The existence of this bacterium suggests that chloroplasts arose from a common ancestor of prochlorophytes and cyanobacteria. Mitochondria arose from an endosymbiotic relationship between the free-living bacterium and bacteria with aerobic respiration (possibly an ancestor of three modern groups: *Agrobacterium, Rhizobium,* and the rickettsias). The symbiotic origin of nuclei has also been proposed, but the evidence is not yet compelling.

The **endosymbiotic hypothesis** has received support from the discovery of an endosymbiotic cyanobacterium that inhabits the biflagellate protist *Cyanophora paradoxa* and acts as its chloroplast. This endosymbiont, called a cyanelle, resembles the cyanobacteria in its photosynthetic pigment system and fine structure and is surrounded by a peptidoglycan layer. It differs from cyanobacteria in lacking the lipopolysaccharide outer membrane characteristic of gram-negative bacteria. The cyanelle may be a recently established endosymbiont that is evolving into a chloroplast.

At present, both hypotheses have supporters. It is possible that new data may help resolve the issue to everyone's satisfaction. However, these hypotheses concern processes that occurred in the distant past and cannot be directly observed. Thus, a complete consensus on the matter may never be reached.

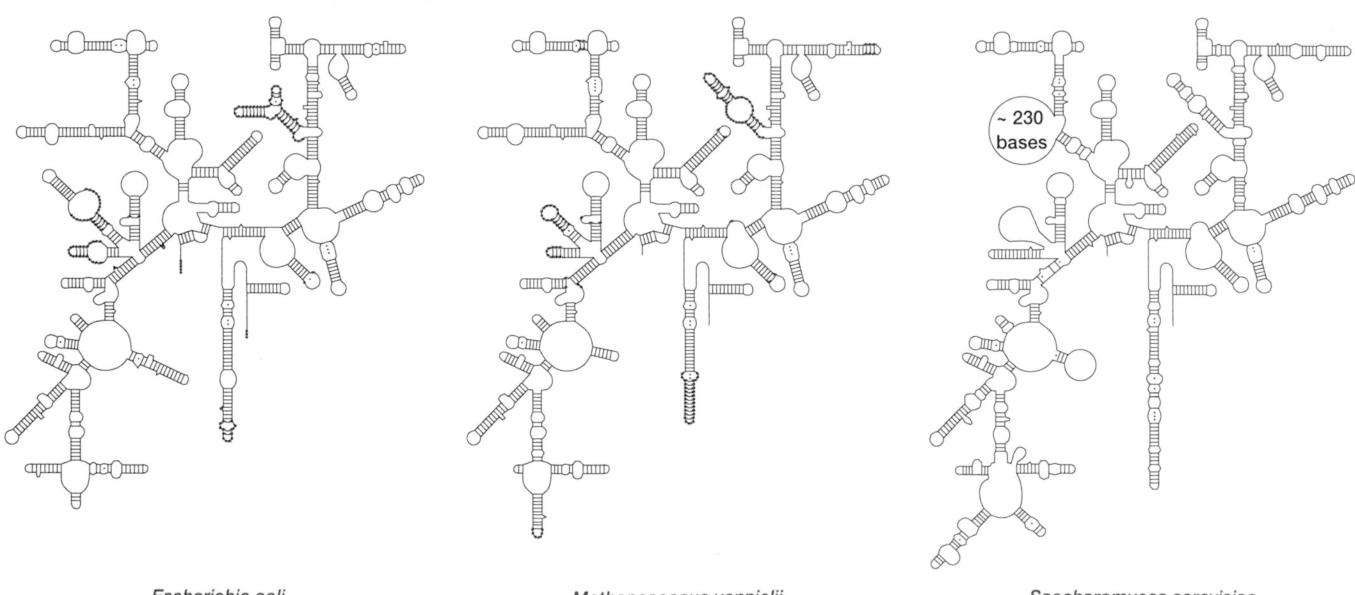

Figure 20.7 Small Ribosomal Subunit RNA. Representative examples of rRNA secondary structures from the three primary domains or kingdoms: eubacteria (*Escherichia coli*), archaeobacteria (*Methanococcus vannielii*), and eucaryotes (*Saccharomyces cerevisiae*). The dots mark positions where eubacteria and archaeobacteria normally differ.

TABLE 20.8 Comparison of Eubacteria, Archaeobacteria, and Eucaryotes

Property	Eubacteria	Archaeobacteria	Eucaryotes
Membrane-Enclosed Nucleus	Absent	Absent	Present
Cell Wall	Peptidoglycan containing muramic acid	Variety of types, no muramic acid	No muramic acid
Membrane Lipid	Have ester-linked, straight-chained fatty acids	Have ether-linked, branched aliphatic chains	Have ester-linked, straight-chained fatty acids
Transfer RNA	Thymine present in most tRNAs	No thymine in T or TψC arm of tRNA	Thymine present
	N-formylmethionine carried by initiator tRNA	Methionine carried by initiator tRNA	Methionine carried by initiator tRNA
mRNA Splicing	Absent	Absent	Present
Ribosomes			
Size	70S	70S	80S (cytoplasmic ribosomes)
Elongation factor 2	Does not react with diphtheria toxin	Reacts	Reacts
Sensitivity to chloramphenicol and kanamycin	Sensitive	Insensitive	Insensitive
Sensitivity to anisomycin	Insensitive	Sensitive	Sensitive
DNA-Dependent RNA Polymerase			
Number of enzymes	One	Several	Three
Structure	Simple subunit pattern (4 subunits)	Complex subunit pattern similar to eucaryotic enzymes (8–12 subunits)	Complex subunit pattern (12–14 subunits)
Rifampicin sensitivity	Sensitive	Insensitive	Insensitive
Methanogenesis	Absent	Present in some species	Absent

The second group, the **archaeobacteria** (Greek *archaios,* ancient, and *bakterion,* a small rod), differ from eubacteria in many respects and resemble eucaryotes in some ways (table 20.8). It should be noted that the spelling archaeobacteria, which will be used in this text, is equivalent to the more popular spelling, archaebacteria. The letter *o* is inserted because this word is formed by combining two Greek words (a point emphasized in *Bergey's Manual*). Although archaeobacteria are described in more detail at a later point, it should be noted that they differ from eubacteria in lacking muramic acid in their cell walls and in possessing (1) membrane lipids with ether-linked branched aliphatic chains, (2) transfer RNAs without thymidine in the T or TψC arm (*see chapter 10*), (3) distinctive RNA polymerase enzymes, and (4) ribosomes of different composition and shape. Thus, although archaeobacteria resemble eubacteria in their procaryotic cell structure, they vary considerably on the molecular level (see Box 20.2).

The archaeobacteria (chapter 24).

The 16S rRNA studies suggest that the eubacteria are composed of approximately 11 major groups or divisions, although more may be discovered with further research. Figures 20.8 and 20.9 summarize eubacterial phylogenetic relationships. Because most of these bacteria are described in chapters 21–25, the eubacterial groups will be only briefly discussed here and representative members mentioned.

1. Gram-positive eubacteria including the mycoplasmas. These bacteria are divided using DNA base composition into two major subdivisions (figure 20.9). The high G + C or actinomycete group contains genera such as *Actinomyces, Arthrobacter, Bifidobacterium, Corynebacterium, Micrococcus, Mycobacterium, Nocardia, Propionibacterium,* and *Streptomyces.* The low G + C group contains *Bacillus, Clostridium, Enterococcus, Lactobacillus, Leuconostoc, Listeria,* mycoplasmas, *Spiroplasma, Staphylococcus,* and *Streptococcus.* The low G + C group is older than the actinomycete group and probably gave rise to it. Mycoplasmas appear to have arisen from the clostridia.

The gram-positive bacteria (chapters 22 and 25).

2. The proteobacteria or purple photosynthetic bacteria and nonphotosynthetic relatives. This division contains most species traditionally thought of as gram-negative

Box 20.2

A New Kingdom of Organisms?

James Lake and his colleagues have found that the ribosomes of sulfur-dependent archaeobacteria differ significantly in shape from those of eubacteria, eucaryotes, and other archaeobacteria. Lake has also analyzed rRNA sequences using a new mathematical procedure and obtained a different phylogentic tree than that shown in figure 20.11 (Box Figure 20.2). He proposes that these sulfur-dependent, thermophilic bacteria constitute a new kingdom of organisms, which he has named eocytes (dawn cells). In this new phylogenetic scheme, there are three major groups of procaryotes. The halobacteria are more closely related to the eubacteria than to the methanogens.

The sulfur-dependent, extremely thermophilic archaeobacteria are renamed eocytes, and the archaeobacteria contain only methanogenic bacteria. In grouping the eocytes with eucaryotes, Lake directly challenges the traditional procaryotic/eucaryotic distinction. He proposes that the early ancestor of the eucaryotes lacked a nucleus, metabolized sulfur, and lived at high temperatures.

It is difficult to forecast the future success of this new proposal. Many microbiologists feel that halobacteria and sulfur-dependent, thermophilic bacteria should be kept among the archaeobacteria because differences in ribosome shape and rRNA sequences are not sufficient to justify such a major taxonomic change. This controversy is a good example of the current excitement and rapid progress in the field of bacterial taxonomy.

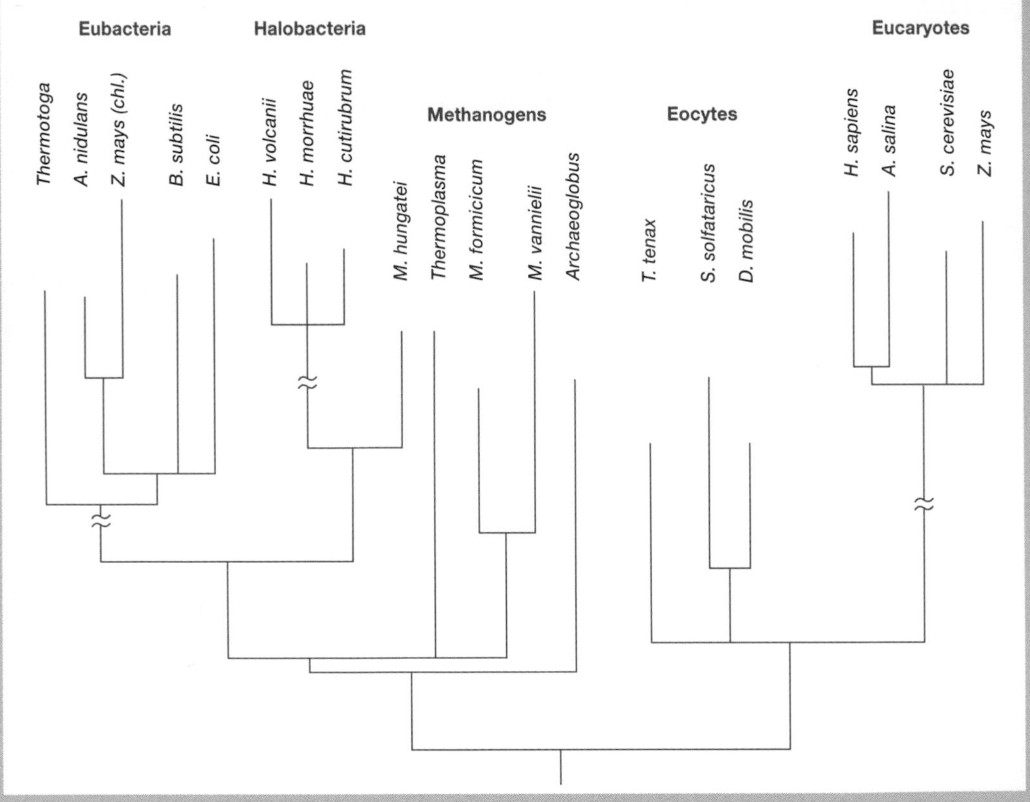

Box Figure 20.2 **The Eocyte Tree.** The length of the most rapidly evolving branches (eubacteria, eucaryotes, and halobacteria) have been shortened to fit the figure. The tree has two deep branches. One branch, which Lake calls the parkaryote branch, contains the eubacteria, halobacteria, and methanogens (plus *Thermoplasma* and *Archaeoglobus*). The other branch, the karyotes, consists of the eocytes and eucaryotes.

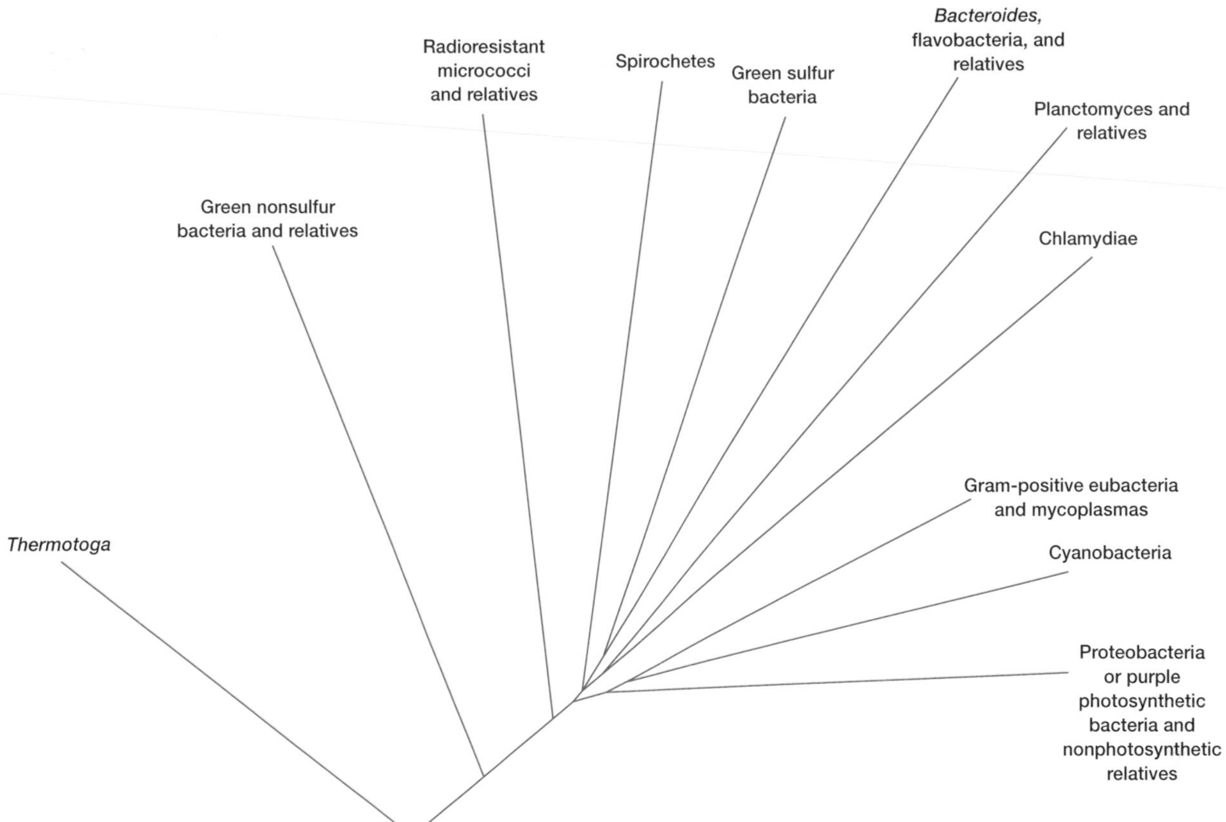

Figure 20.8 **Eubacterial Phylogenetic Tree.** The tree is based on 16S rRNA sequence comparisons. The branch lengths in the tree are proportional to the evolutionary distances between groups.

eubacteria. Sequence data indicate that the ancestral bacterium was a photosynthetic purple bacterium, and photosynthetic genera are scattered throughout the division. Presumably many strains lost photosynthesis when adapting metabolically to new ecological niches. The proteobacteria are exceptionally diverse and are divided into at least five subdivisions as follows:

a. Alpha subdivision: agrobacteria, *Caulobacter, Hyphomicrobium, Methylococcus, Nitrobacter,* purple nonsulfur bacteria, rhizobacteria, and rickettsias.
b. Beta subdivision: *Alcaligenes, Bordetella, Neisseria, Rhodocyclus, Spirillum,* and *Thiobacillus.*
c. Gamma subdivision: *Azotobacter, Beggiatoa, Chromatium,* the family *Enterobacteriaceae, Haemophilus, Legionella, Leucothrix, Pasteurella,* pseudomonads, purple sulfur bacteria, and *Vibrio.*
d. Delta subdivision: *Bdellovibrio, Desulfovibrio* and other sulfur and sulfate reducers, myxobacteria.
e. Epsilon subdivision: *Campylobacter, Wolinella.*

The gram-negative bacteria (chapters 21 and 23).

3. Cyanobacteria. Oxygenic, photosynthetic bacteria. Some representative genera are *Anabaena, Calothrix, Fischerella, Nostoc, Oscillatoria, Scytonema,* and *Synechococcus. Prochloron* is also placed here.

The cyanobacteria (pp. 472–76).

4. Spirochetes and relatives. Gram-negative bacteria with a spiral shape and distinctive axial filament. *Borrelia, Leptospira, Spirochaeta,* and *Treponema.*

The spirochetes (pp. 427–29).

5. The Planctomyces group. These walled, budding bacteria resemble other eubacteria, but lack peptidoglycan. *Planctomyces* and *Isosphaera.*
6. *Bacteroides,* flavobacteria, and relatives. Includes *Bacteroides, Cytophaga, Flavobacterium, Flexibacter,* and *Fusobacterium.*

Bacteroides (p. 441), Cytophaga and Flexibacter (pp. 484–85).

7. Chlamydiae. Intracellular parasites with a unique life cycle and no peptidoglycan. *Chlamydia.*

The order Chlamydiales (pp. 443–44).

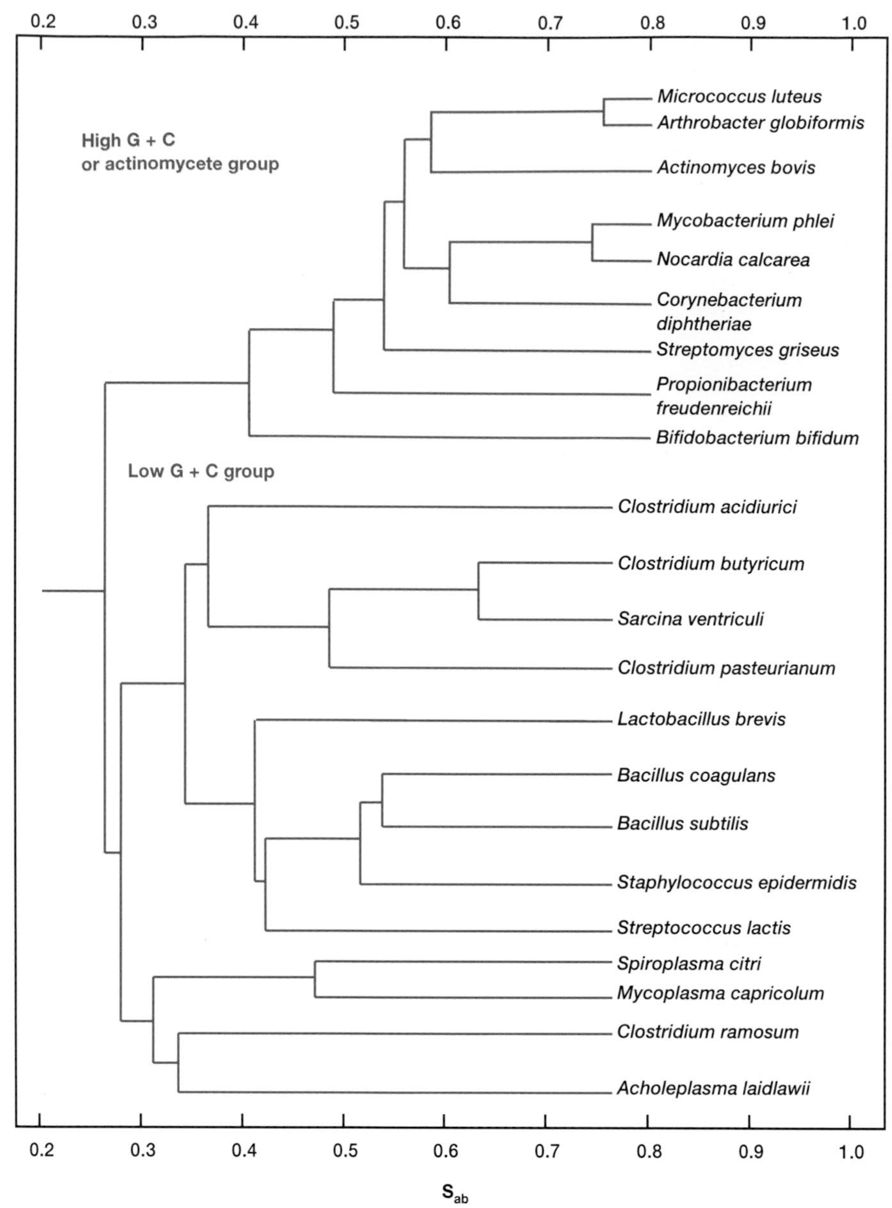

Figure 20.9 **The Phylogeny of Gram-positive Eubacteria.** Data are derived from 16S rRNA studies. See text for further details.

8. Green sulfur bacteria (family *Chlorobiaceae*). Anaerobic, photosynthetic bacteria that can oxidize sulfide and sulfur and have chlorosomes. Includes *Chlorobium* and *Pelodictyon.*

The green sulfur bacteria (p. 471).

9. Green nonsulfur bacteria and relatives. *Chloroflexus, Herpetosiphon,* and *Thermomicrobium.*
10. Radioresistant micrococci and relatives. Members of the family *Deinococcaceae* are gram-positive cocci with great radiation resistance. The thermophilic, gram-negative rods of the genus *Thermus* appear quite different, although the rRNA sequences indicate that they are related to *Deinococcus.*

The Deinococcaceae (p. 455).

11. *Thermotoga* and *Thermosipho.* These thermophilic eubacteria from geothermally heated marine sediments differ from all other eubacteria (figure 20.8), but have not been thoroughly studied.

Archaeobacterial phylogeny is depicted in figure 20.10. The archaeobacteria consist of three main groups: the extreme halophiles, the methanogens (bacteria that reduce carbon dioxide to methane), and sulfur-dependent extreme thermophiles. The tree has two main branches: a cluster of extreme thermophiles such as *Sulfolobus* and the methanogen branch. The genus *Thermoplasma* is thermophilic, but more closely related to the methanogens than to the other thermophiles. The methanogens are placed in three orders, each of which may contain several genera: *Methanococcales* (*Methanococcus*), *Methanobacteriales* (*Methanobacterium, Methano-*

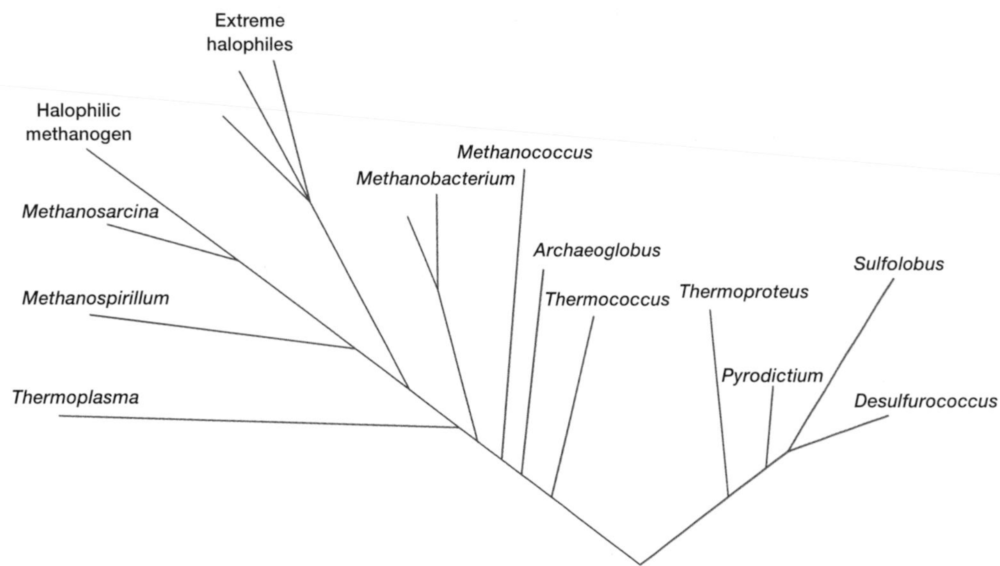

Figure 20.10 **Archaeobacterial Phylogenetic Tree.** The tree is based on 16S rRNA comparisons and has the same meaning as figure 20.8.

brevibacter), and *Methanomicrobiales* (*Methanosarcina, Methanospirillum*). There are at least six genera of extreme halophiles in the order *Halobacteriales.* Archaeobacterial classification will undoubtedly change with further discoveries.

Extensive comparisons of ribosomal RNA sequences by Carl Woese and others have led them to conclude that the five-kingdom system is incorrect. According to Woese, the kingdom *Procaryotae* is composed of two groups, which differ from one another as much as from the eucaryotes. Figure 20.11 depicts a universal phylogenetic tree that reflects these views. The tree is divided into three major branches representing the three primary groups: eubacteria, archaeobacteria, and eucaryotes. The archaeobacteria and eucaryotes are distant relatives. These three primary groups are called domains and placed above the kingdom level. The traditional kingdoms are distributed among these three domains.

Carl Woese, Otto Kandler, and Mark Wheelis have coined new names for the domains. Eucaryotic organisms with primarily glycerol fatty acyl diester membrane lipids and eucaryotic rRNA belong to the eucarya. The domain bacteria contains procaryotic cells with membrane lipids that are primarily diacyl glycerol diesters and with eubacterial rRNA. Procaryotes having isoprenoid glycerol diether or diglycerol tetraether lipids (*see pp. 493–94*) in their membranes and archaeobacterial rRNA compose the third domain, archaea. These new names are intended to replace the more commonly used terms eucaryotes, eubacteria, and archaeobacteria.

At present, there is little agreement about domains and kingdoms. Many microbiologists find the new proposal attractive, while others promote different phylogenetic systems (Box 20.2). A number of biologists support the five-kingdom system. Further research will resolve the issue.

Phenetic Classification and *Bergey's Manual*

Because it has not been possible to classify bacteria satisfactorily based on phylogenetic relationships, the system given in *Bergey's Manual of Systematic Bacteriology* is primarily phenetic. Each of the 33 sections in the four volumes contains bacteria that share a few easily determined characteristics and bears a title that either describes these properties or provides the vernacular names of the bacteria included. The characteristics used to define sections are normally features such as general shape and morphology, Gram-staining properties, oxygen dependence, motility, the presence of endospores, the mode of energy production, and so forth. Bacterial groups are divided among the four volumes in the following manner: (1) gram-negative bacteria of general, medical, or industrial importance; (2) gram-positive bacteria other than actinomycetes; (3) gram-negative bacteria with distinctive properties, cyanobacteria, and archaeobacteria; and (4) actinomycetes (gram-positive filamentous bacteria).

Bacterial classification according to Bergey's Manual of Systematic Bacteriology *(appendix III).*

Although the new edition of *Bergey's Manual* is organized along phenetic and utilitarian lines, R. G. E. Murray, in a theoretical introductory article to the manual, suggests that the kingdom *Procaryotae* be divided into four divisions and seven classes. The four divisions are characterized primarily on (1) the presence or absence of cell walls and (2) the chemical and structural nature of the cell wall. The divisions largely correspond to traditional groups. The ***Gracilicutes*** (thin skin) have gram-negative walls, and members of the ***Firmicutes*** (strong or durable skin) have a gram-positive type cell wall. The ***Tenericutes*** (soft or tender skin) are the mycoplasmas;

Table 20.9 Characteristics of *Gracilicutes, Firmicutes,* and *Tenericutes*

Property	*Gracilicutes*	*Firmicutes*	*Tenericutes*
Cell wall	Gram-negative type wall with inner 2–10 nm peptidoglycan layer and outer membrane (8–10 nm thick) of lipid, protein, and lipopolysaccharide. (There may be a third outermost layer of protein.)	Gram-positive type wall with a homogeneous, thick cell wall (20–80 nm) composed mainly of peptidoglycan. Other polysaccharides and teichoic acids may be present.	Lack a cell wall and peptidoglycan precursors; enclosed by a plasma membrane
Cell shape	Spheres, ovals, straight or curved rods, helices or filaments; some have sheaths or capsules.	Spheres, rods, or filaments; may show true branching	Pleomorphic in shape; may be filamentous, can form branches
Reproduction	Binary fission, sometimes budding	Binary fission	Budding, fragmentation, and/or binary fission
Metabolism	Phototrophic, chemolithoautotropic, or chemoorganoheterotrophic	Usually chemoorganoheterotrophic	Chemoorganoheterotrophic; most require cholesterol and long chain fatty acids for growth.
Motility	Motile or nonmotile. Flagellation can be varied—polar, lophotrichous, peritrichous. Motility may also result from the use of axial filaments (spirochetes) or gliding motility.	Most often nonmotile; have peritrichous flagellation when motile	Usually nonmotile
Appendages	Can produce several types of appendages—pili and fimbriae, prosthecae, stalks	Usually lack appendages (may have spores on hyphae)	Lack appendages
Endospores	Cannot form endospores	Some groups can form endospores.	Cannot form endospores

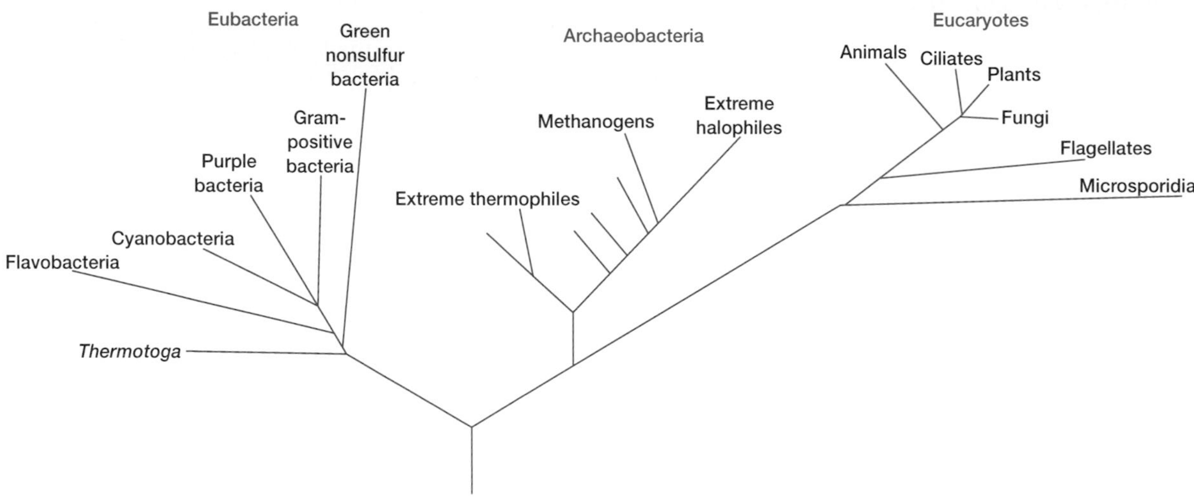

Figure 20.11 **Universal Phylogenetic Tree.** These relationships were determined from rRNA sequence comparisons.

the archaeobacteria are placed in the ***Mendosicutes*** (skin with faults). The divisions differ in many other ways. Some of the most important characteristics of the first three divisions are compared in table 20.9; major features of the archaeobacteria have already been summarized in table 20.8.

Comparison of the phylogenetic scheme outlined previously and pictured in figures 20.8–20.11 with the phenetically based classification of *Bergey's Manual* shows considerable disagreement. Many phenetically defined taxa are not phylogenetically homogeneous and have members distributed among two or more different phylogenetic groups (at least as judged by 16S rRNA studies). Often, characteristics given great weight or importance in *Bergey's Manual* do not appear to be phylogenetically significant. For example, photosynthetic bacteria are found in several different phylogenetic groups together with very closely related nonphotosynthetic bacteria. Thus it may not be appropriate to separate all photosynthetic bacteria from the nonphotosynthetic forms as has

Box 20.3

"Official" Nomenclature Lists: A Letter from *Bergey's**

On a number of occasions lately, the impression has been given that the status of a bacterial taxon in *Bergey's Manual Systematic Bacteriology* or *Bergey's Manual of Determinative Bacteriology* is in some sense official. Similar impressions are frequently given about the status of names in the *Approved List of Bacterial Names* and in the Validation Lists of newly proposed names that appear regularly in the *International Journal of Systematic Bacteriology*. It is therefore important to clarify these matters.

There is no such thing as an official classification. *Bergey's Manual* is not "official"—it is merely the best consensus at the time, and although great care has always been taken to obtain a sound and balanced view, there are also always regions in which data are lacking or confusing, resulting in differing opinions and taxonomic instability. When *Bergey's Manual* disavows that it is an official classification, many bacteriologists may feel that the solid earth is trembling. But many areas are in fact reasonably well established. Yet taxonomy is partly a matter of judgment and opinion, as is all science, and until new information is available, different bacteriologists may legitimately hold different views. They cannot be forced to agree to any "official classification." It must be remembered that, as yet, we know only a small percentage of the bacterial species in nature. Advances in technique also reveal new lights on bacterial relationships. Thus, we must expect that existing boundaries of groups will have to be redrawn in the future, and it is expected that molecular biology, in particular, will imply a good deal of change over the next few decades.

The position with the Approved Lists and the Validation Lists is rather similar. When bacteriologists agreed to make a new start in bacteriological nomenclature, they were faced with tens of thousands of names in the literature of the past. The great majority were useless, because, except for about 2,500 names, it was impossible to tell exactly what bacteria they referred to. These 2,500 were therefore retained in the Approved Lists. The names are only approved in the sense that they were approved for retention in the new bacteriological nomenclature. The remainder lost standing in the nomenclature, which means they do not have to be considered whan proposing new bacterial names (although names can be individually revived for good cause under special provisions).

The new International Code of Nomenclature of Bacteria requires all new names to be validly published to gain standing in the nomenclature, either by being published in papers in the *International Journal of Systematic Bacteriology* or, if published elsewhere, by being announced in the Validation Lists. The names in the Validation Lists are therefore valid only in the sense of being validly published (and therefore they must be taken account of in bacterial nomenclature). The names do not have to be adopted in all circumstances; if users believe the scientific case for the new taxa and validly published names is not strong enough, they need not adopt the names. For example, *Helicobacter pylori* was immediately accepted as a replacement for *Campylobacter pylori* by the scientific community, whereas *Tatlockia micdadei* had not generally been accepted as a replacement for *Legionella micdadei*. Taxonomy remains a matter of scientific judgment and general agreement.

*Sneath, P. H. A., and Brenner, D. J. 1992. "Official" nomenclature lists. *ASM News* 58(4):175.

been done in *Bergey's Manual*. The mycoplasmas are placed in the division *Tenericutes* in *Bergey's Manual,* but rRNA studies show that they are closely related to gram-positive bacteria (division *Firmicutes*) although they lack cell walls. Because rods, cocci, spirals, and other shapes are found scattered among many phylogenetic groups, these morphological variations do not appear to be useful indicators of relatedness.

In the future, phenetic classifications will undoubtedly change substantially to reflect more accurately phylogenetic relationships. It will also be necessary to find more phenotypic traits that are true indicators of phylogenetic relatedness and that are easily studied. Some examples of such traits are peptidoglycan type, fatty acid composition, the presence of teichoic acids, glucose fermentation end products, the relationship to oxygen, and the ability to form endospores.

Despite the uncertainties and problems with the classification in *Bergey's Manual,* it is the most widely accepted and used system for the identification of bacteria (see Box 20.3), and it often does provide phylogenetically meaningful information. Because it is the standard in bacterial taxonomy, the following five chapters survey the bacteria in essentially the same order as that followed in *Bergey's Manual of Systematic Bacteriology*.

1. Discuss the use of S_{ab} values in determining the relatedness of bacteria and evolutionary age of taxonomic groups. What are oligonucleotide signature sequences?
2. What are eubacteria and archaeobacteria, and in what ways do they differ from each other? Very briefly describe the major eubacterial and archaeobacterial groups.
3. How does Woese divide organisms into domains in his universal phylogenetic tree?
4. Name the four divisions of bacteria given in *Bergey's Manual* and state their distinguishing features.

Summary

1. Taxonomy, the science of biological classification, is composed of three parts: classification, nomenclature, and identification.
2. The nature of the species is different for sexually and asexually reproducing organisms. A bacterial species is a collection of strains that have many stable properties in common and differ significantly from other groups of strains.
3. Microorganisms are named according to the binomial system.
4. *Bergey's Manual of Systematic Bacteriology* gives the accepted system of bacterial taxonomy.
5. The two major types of natural classification are phylogenetic systems and phenetic systems.
6. Classifications are often constructed by means of numerical taxonomy, in which the general similarity of organisms is determined by using a computer to calculate and analyze association coefficients.
7. Morphological, physiological, metabolic, and ecological characteristics are widely used in microbial taxonomy.
8. The study of transformation and conjugation in bacteria is sometimes taxonomically useful. Plasmid-borne traits can cause errors in bacterial taxonomy if care is not taken.
9. Proteins are direct reflections of mRNA sequences and may be used to compare genomes from different organisms.
10. The G + C content of DNA is easily determined and taxonomically valuable because it is an indirect reflection of the base sequence.
11. Nucleic acid hybridization studies are used to compare DNA or RNA sequences and thus determine genetic relatedness.
12. Nucleic acid sequencing is the most powerful and direct method for comparing genomes. The sequences of 16S and 5S rRNA are most often used in phylogenetic studies.
13. Organisms may be divided into five kingdoms based on their cell type, level of organization, and nutrition. These are *Animalia, Plantae, Monera* or *Procaryotae, Fungi,* and *Protista.*
14. Studies on the sequence of 16S rRNA and other molecular properties suggest that the procaryotes are divided into two very different groups: eubacteria and archaeobacteria.
15. Eubacteria or true bacteria comprise the vast majority of bacteria and can be separated into at least 11 major subgroups based on 16S rRNA similarity.
16. Archaeobacteria differ from eubacteria in cell wall composition, membrane lipids, tRNA structure, ribosomes, and other properties.
17. *Bergey's Manual of Systematic Bacteriology* provides a primarily phenetic classification, and many taxa are not phylogenetically homogeneous. However, it has been proposed that the kingdom *Procaryotae* be divided between four divisions (*Gracilicutes, Firmicutes, Tenericutes,* and *Mendosicutes*) and seven classes.

Key Terms

archaeobacteria *417*
binomial system *406*
biovar *406*
classification *405*
dendrogram *407*
endosymbiotic hypothesis *416*
eubacteria *415*
Firmicutes *421*
Fungi *413*
G + C content *410*
Gracilicutes *421*
identification *405*
Jaccard coefficient (S_J) *407*
melting temperature (T_m) *411*
Mendosicutes *422*
Monera *413*
morphovar *406*
natural classification *406*
nomenclature *405*
nucleic acid hybridization *412*
numerical taxonomy *406*
oligonucleotide signature sequences *415*
phenetic system *406*
phenons *407*
phylogenetic or phyletic classification systems *406*
phylogeny *406*
Procaryotae *413*
Protista *413*
protists *413*
serovar *406*
similarity matrix *407*
simple matching coefficient (S_{SM}) *407*
species *405*
strain *406*
systematics *405*
taxon *405*
taxonomy *405*
Tenericutes *421*
type strain *406*

Questions for Thought and Review

1. Why are size and shape often less useful in characterizing bacterial species than eucaryotic microbial species?
2. Why might a phylogenetic approach to microbial classification be preferable to a phenetic approach? Give arguments for the use of a phenetic classification. Which do you think is preferable and why?
3. Would a numerical taxonomist favor the phylogenetic or phenetic approach to obtaining a natural classification? Why might it be best to use unweighted characteristics in classification? Are there reasons to give some properties more weight or importance than others?
4. In what way does the simple matching coefficient (S_{SM}) differ from the Jaccard coefficient (S_J)? Give a reason why the former coefficient might be preferable to the latter.
5. What genetic feature of bacteria makes it advisable to use as many characteristics as convenient in classifying and identifying an organism?
6. What properties of the molecules should be considered when selecting RNAs or proteins to sequence for the purpose of determining the relatedness of microorganisms that are only distantly related?
7. Why is the current bacterial classification system likely to change considerably in the future? How would one select the best features to use in identification of unknown microorganisms and determination of relatedness?

Additional Reading

Balows, A.; Truper, H. G.; Dworkin, M.; Harder, W.; and Schleifer, K-H. 1992. *The prokaryotes,* 2d ed. New York: Springer-Verlag.

Cavalier-Smith, T. 1975. The origin of nuclei and of eukaryotic cells. *Nature* 256:463–68.

Fox, G. E., et al. 1980. The phylogeny of prokaryotes. *Science* 209:457–63.

Gibbons, N. E., and Murray, R. G. E. 1978. Proposals concerning the higher taxa of bacteria. *Int. J. Syst. Bacteriol.* 28(1):1–6.

Goodfellow, M.; Jones, D.; and Priest, F. G., eds. 1985. *Computer-assisted bacterial systematics.* New York: Academic Press.

Gray, M. W. 1983. The bacterial ancestry of plastids and mitochondria. *BioScience* 33(11):693–99.

Holt, J. G., ed. 1984–1989. *Bergey's manual of systematic bacteriology.* 4 vols. Baltimore, Md.: Williams & Wilkins.

Johnson, J. L. 1984. Nucleic acids in bacterial classification. In *Bergey's manual of systematic bacteriology,* ed. J. G. Holt, vol. 1, ed. N. R. Krieg, 8–11. Baltimore, Md.: Williams & Wilkins.

Jones, D., and Krieg, N. R. 1984. Serology and chemotaxonomy. In *Bergey's manual of systematic bacteriology,* ed. J. G. Holt, vol. 1, ed. N. R. Krieg, 15–18. Baltimore, Md.: Williams & Wilkins.

Kabnick, K. S., and Peattie, D. A. 1991. *Giardia:* A missing link between prokaryotes and eukaryotes. *American Scientist* 79:34–43.

Knoll, A. H. 1991. End of the proterozoic eon. *Sci. Am.* 265(4):64–73.

Lake, J. A. 1991. Tracing origins with molecular sequences: Metazoan and eukaryotic beginnings. *Trends Biochem. Sci.* 16(2):46–50.

Last, G. A. 1988. Musings on bacterial systematics: How many kingdoms of life? *ASM News* 54(1):22–27.

Margulis, L. 1971. Symbiosis and evolution. *Sci. Am.* 225(2):49–57.

Margulis, L. 1981. How many kingdoms? Current views of biological classification. *Am. Biol. Teach.* 43(9):482–89.

Murray, R. G. E. 1984. The higher taxa, or, a place for everything . . .? In *Bergey's manual of systematic bacteriology,* ed. J. G. Holt, vol. 1, ed. N. R. Krieg, 31–34. Baltimore, Md.: Williams & Wilkins.

Ragan, M. A., and Chapman, D. J. 1978. *A biochemical phylogeny of the protists.* New York: Academic Press.

Schleifer, K. H., and Stackebrandt, E. 1983. Molecular systematics of prokaryotes. *Ann. Rev. Microbiol.* 37:143–87.

Sneath, P. H. A. 1978. Classification of microorganisms. In *Essays in microbiology,* ed. J. R. Norris and M. H. Richmond, 9/1–9/31. New York: John Wiley and Sons.

Sneath, P. H. A. 1984. Numerical taxonomy. In *Bergey's manual of systematic bacteriology,* ed. J. G. Holt, vol. 1, ed. N. R. Krieg, 5–7. Baltimore, Md.: Williams & Wilkins.

Sneath, P. H. A. 1989. Analysis and interpretation of sequence data for bacterial systematics: The view of a numerical taxonomist. *Syst. Appl. Microbiol.* 12:15–31.

Sneath, P. H. A., and Sokal, R. R. 1973. *Numerical taxonomy: The principles and practice of numerical classification.* San Francisco: W. H. Freeman.

Staley, J. T., and Krieg, N. R. 1984. Classification of procaryotic organisms: An overview. In *Bergey's manual of systematic bacteriology,* ed. J. G. Holt, vol. 1, ed. N. R. Krieg, 1–4. Baltimore, Md.: Williams & Wilkins.

Staley, T. E., and Colwell, R. R. 1973. Application of molecular genetics and numerical taxonomy to the classification of bacteria. *Ann. Rev. Ecol. Syst.* 8:273–300.

Vidal, G. 1984. The oldest eukaryotic cells. *Sci. Am.* 250(2):48–57.

Whittaker, R. H., and Margulis, L. 1978. Protist classification and the kingdoms of organisms. *BioSystems* 10:3–18.

Woese, C. R. 1981. Archaebacteria. *Sci. Am.* 244(6):98–122.

Woese, C. R. 1987. Bacterial evolution. *Microbiol. Rev.* 51(2):221–71.

Woese, C. R., Kandler, O., and Wheelis, M. L. 1990. Towards a natural system of organisms: Proposal for the domains archaea, bacteria, and eucarya. *Proc. Natl. Acad. Sci.* 87:4576–79.

Zillig, W.; Palm, P.; Reiter, W.-D.; Gropp, F.; Pühler, G.; and Klenk, H-P. 1988. Comparative evaluation of gene expression in archaebacteria. *Eur. J. Biochem.* 173:473–82.

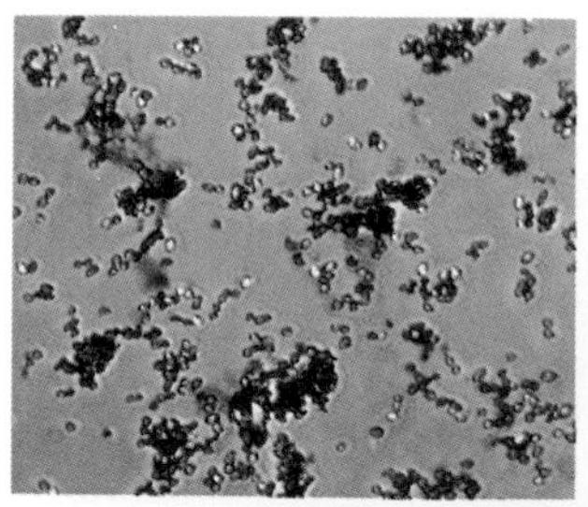

Chapter 21
The Bacteria: Gram-Negative Bacteria of General, Medical, or Industrial Importance

Microbes is a vigitable, an' ivry man is like a conservatory full iv millyons iv these potted plants.

—Finley Peter Dunne

Outline

Concepts

1. Volume 1 of *Bergey's Manual* describes 10 groups of gram-negative bacteria having general, medical, or industrial importance. Bacterial endosymbionts are also included in volume 1.
2. The 10 sections are distinguished on the basis of a few major properties: general shape, the presence or absence of motility, mechanism of motility (spirochetes), oxygen relationships, the absence of a cell wall (mycoplasmas), and the requirement for an intracellular existence (rickettsias and chlamydiae).
3. With the exception of the spirochetes (and possibly the mycoplasmas plus a few other groups), the gram-negative bacteria in volume 1 do not vary drastically in general shape or appearance. In contrast, they are very diverse in their metabolism and life-styles, which range from obligate intracellular parasitism to a free-living existence in soil and aquatic habitats.
4. Many bacteria that specialize in predatory or parasitic modes of existence, such as *Bdellovibrio,* the rickettsias, and the chlamydiae, have relinquished some of their metabolic independence through the loss of metabolic pathways. They use the prey's or host's energy supply and/or cell constituents.
5. Many gram-negative bacteria in volume 1 of *Bergey's Manual* are of considerable importance, either as disease agents or in terms of their effects on the habitat. Others, such as *E. coli,* are major experimental organisms studied in many laboratories.

Chapter 20 offers a review of the general principles of microbial taxonomy and provides an introduction to bacterial taxonomy and *Bergey's Manual of Systematic Bacteriology* (hereafter also called *Bergey's Manual*). Chapters 21 through 25 briefly survey the major characteristics of the most important bacterial groups. Chapter 21 covers the bacteria found in the first volume of *Bergey's Manual,* and the following four chapters are devoted to the bacteria described in the second through fourth volumes. Where possible, the order of *Bergey's Manual* is followed. The archaeobacteria are an exception and will be discussed in chapter 24 regardless of their position in *Bergey's.* For example, coverage of the family *Halobacteriaceae* (volume 1, section 4) is delayed until chapter 24.

Although the treatment of each bacterial group varies somewhat from that of the others, usually an attempt is made to provide a brief review of the group's biology. Such aspects as distinguishing characteristics, morphology, reproduction, physiology, metabolism, and ecology are included. The taxonomy of each major group is summarized, and representative species are discussed. This approach should help one appreciate bacteria as living organisms rather than simply as agents of disease with little interest or importance in other contexts.

Bergey's Manual devotes two volumes to the properties and taxonomy of gram-negative bacteria. Volume 1 focuses on gram-negative bacteria of general, medical, or industrial importance and covers 10 groups plus bacterial endosymbionts. This chapter surveys the bacteria placed in volume 1, except for bacterial endosymbionts (*see chapter 29*). The remaining gram-negative bacteria are described in volume 3 of *Bergey's Manual* and are discussed in chapters 23 and 24.

The Spirochetes

The spirochetes are a group of gram-negative, chemoheterotrophic bacteria distinguished by their structure and mechanism of motility. They are slender, long bacteria (0.1 to 3.0 μm by 5 to 250 μm) with a flexible, helical shape (figure 21.1). Many species are so slim that they are only clearly visible in a light microscope by means of phase-contrast or dark-field optics (*see chapter 2*). Spirochetes differ greatly from other bacteria with respect to motility and can move through very viscous solutions though they lack external rotating flagella. When in contact with a solid surface, they exhibit creeping or crawling movements. Their unique pattern of motility is due to an unusual morphological structure called the axial filament.

The distinctive features of spirochete morphology are evident in electron micrographs (figure 21.2). The central protoplasmic cylinder contains cytoplasm and the nucleoid is bounded by a plasma membrane and gram-negative type cell wall. It corresponds to the body of other gram-negative bacteria. Two to more than a hundred procaryotic flagella, called

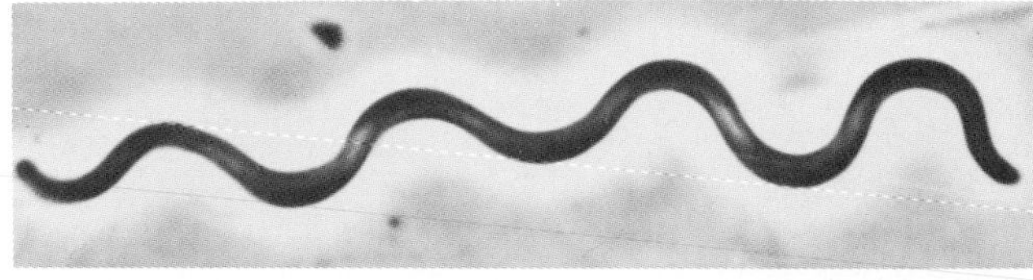

(a)

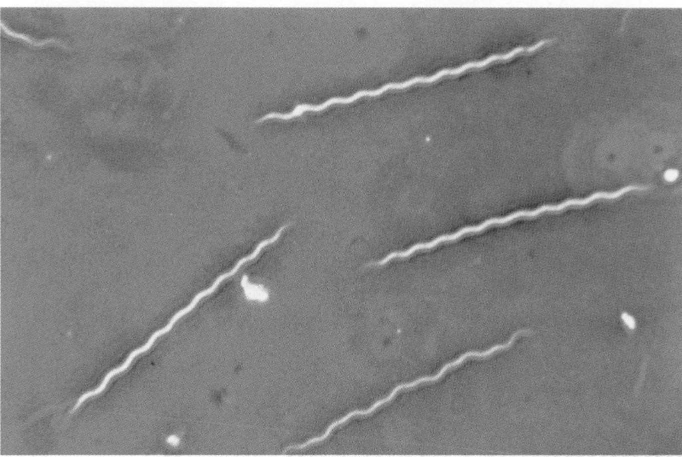

(b)

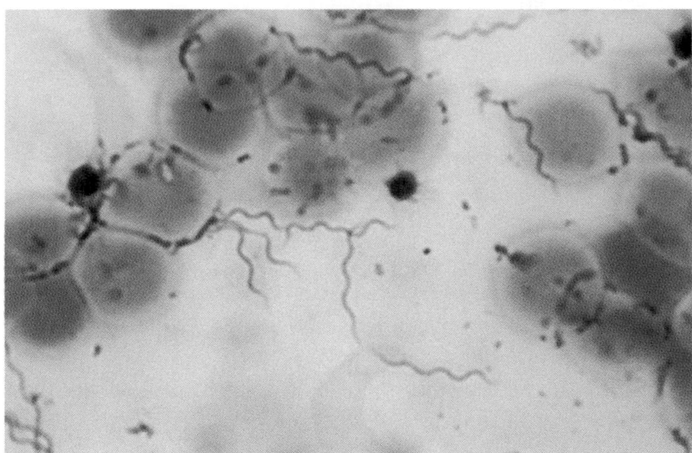

(c)

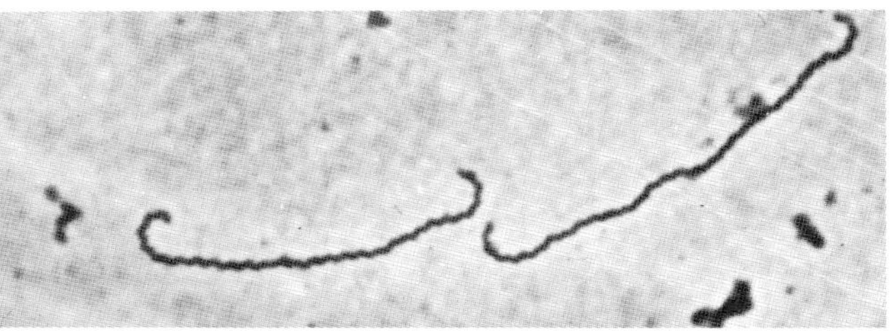

(d)

Figure 21.1 **The Spirochetes.** Representative examples. (*a*) *Cristispira* sp. from the crystalline style of a clam, phase contrast (×2,200). (*b*) *Treponema pallidum* (×1,000). (*c*) *Borrelia duttonii* from human blood (×500). (*d*) *Leptospira interrogans* (×2,200).

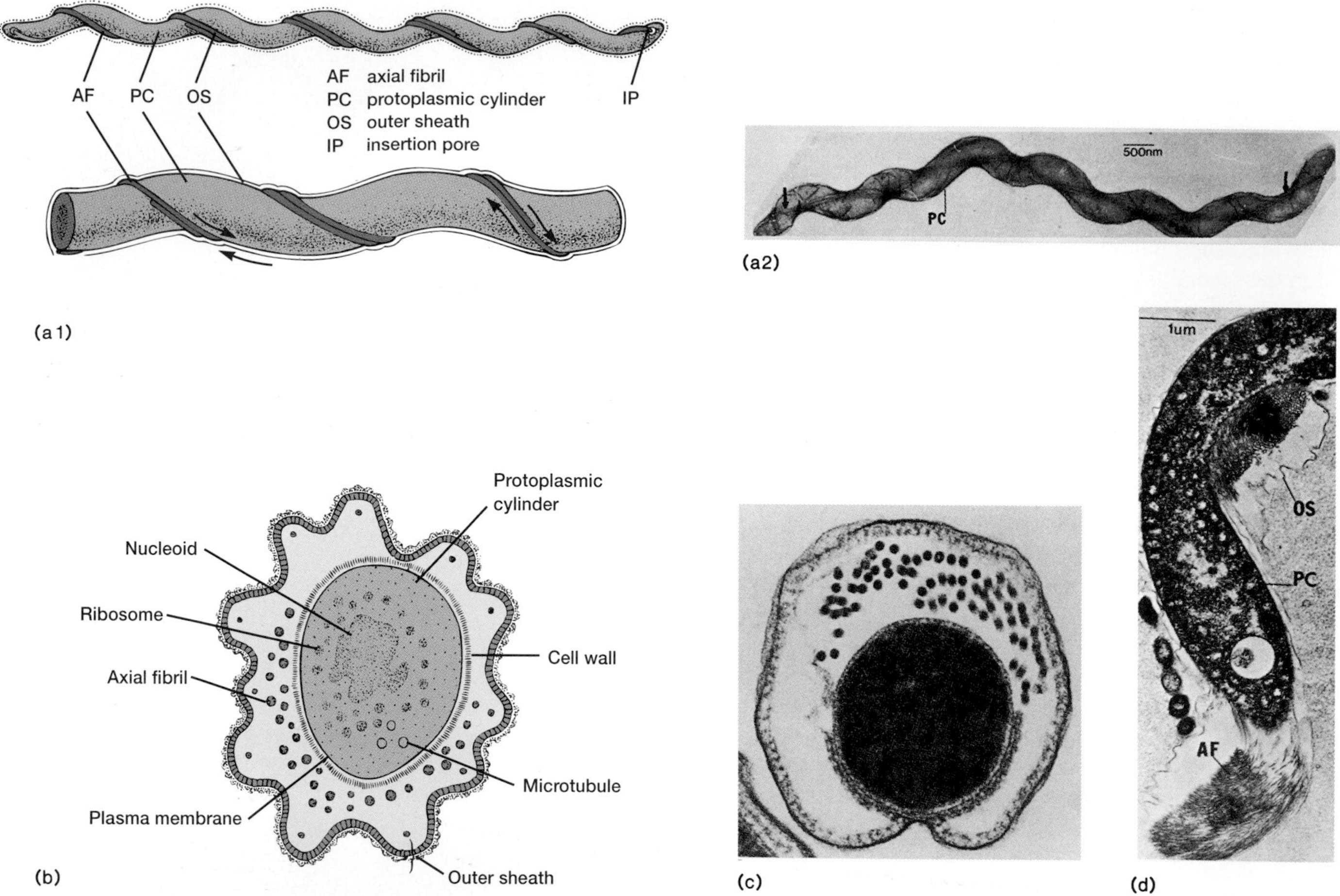

Figure 21.2 **Spirochete Morphology.** (*a1*) A surface view of spirochete structure as interpreted from electron micrographs. (*a2*) A longitudinal view of *T. zuelzerae* with axial fibrils extending most of the cell length. (*b*) A cross section of a typical spirochete showing morphological details. (*c*) Electron micrograph of a cross section of *Clevelandina* from the termite *Reticulitermes flavipes* showing the outer sheath, protoplasmic cylinder, and axial fibrils (×70,000). (*d*) Longitudinal section of *Cristispira* showing the outer sheath (*OS*), the protoplasmic cylinder (*PC*), and the axial fibrils (*AF*).

axial fibrils or **periplasmic flagella,** extend from both ends of the cylinder and often overlap one another in the center third of the cell. The whole complex of periplasmic flagella, the **axial filament,** lies inside a flexible outer sheath or outer membrane. The outer sheath contains lipid, protein, and carbohydrate and varies in structure between different genera. Its precise function is unknown, but it is definitely important because spirochetes will die if it is damaged or removed.

Although the way in which periplasmic flagella propel the cell has not been established, they are responsible for motility because mutants with straight rather than curved flagella are nonmotile. Presumably, the periplasmic flagella rotate like the external flagella of other bacteria. This could cause the corkscrew-shaped outer sheath to rotate and move the cell through the surrounding liquid (figure 21.3). Flagellar rotation could also flex or bend the cell and account for the crawling movement seen on solid surfaces.

Spirochetes can be anaerobic, facultatively anaerobic, or aerobic. Carbohydrates, amino acids, long-chain fatty acids, and long-chain fatty alcohols may serve as carbon and energy sources.

The group is exceptionally diverse ecologically and grows in habitats ranging from mud to the human mouth. Members of the genus *Spirochaeta* are free-living and often grow in anaerobic and sulfide-rich freshwater and marine environments. Some species of the genus *Leptospira* grow in aerobic water and moist soil. In contrast, many spirochetes form symbiotic associations with other organisms and are found in a variety of locations: the hindguts of termites and wood-eating roaches, the digestive tracts of mollusks (*Cristispira*) and mammals, and the oral cavities of animals (*Treponema denticola, T. oralis*). Spirochetes coat the surfaces of many protozoa from termite and wood-eating roach hindguts (figure 21.4). For example, the flagellate *Myxotricha paradoxa* is covered with slender spirochetes (0.15 by 10 μm in length) that are firmly attached and help move the protozoan. Some members of the genera *Treponema, Borrelia,* and *Leptospira* are important pathogens; for example, *Treponema pallidum* (*see figure 2.8*a) causes syphilis (*see chapter 37*), and *Borrelia burgdorferi* (*see figure 37.8*) is responsible for Lyme disease.

The order *Spirochaetales* is divided into two families, the *Spirochaetaceae* with four genera (*Spirochaeta, Cristispira,*

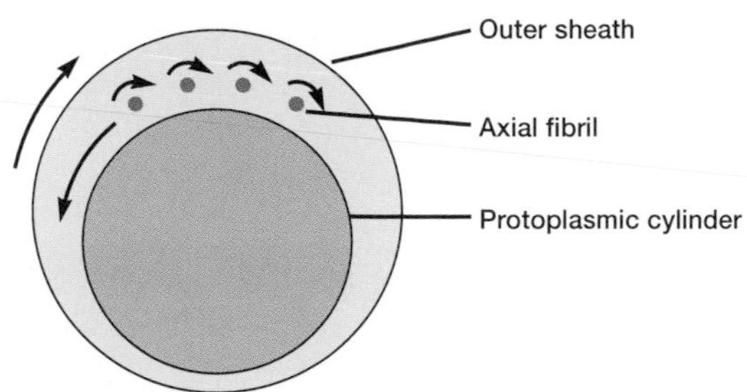

Figure 21.3 Spirochete Motility. A hypothetical mechanism for spirochete motility. See text for details.

Treponema, and *Borrelia*) and the *Leptospiraceae* with only the genus *Leptospira.* Table 21.1 summarizes some more distinctive properties of each genus.

---

1. Give the most important characteristics of the spirochetes.
2. Define the following terms: protoplasmic cylinder, axial fibrils or periplasmic flagella, axial filament, outer sheath or outer membrane. Draw and label a diagram of spirochete morphology, locating these structures.
3. How might spirochetes use their axial filament to move?

Aerobic/Microaerophilic, Motile, Helical/Vibrioid Gram-Negative Bacteria

Section 2 of *Bergey's Manual* contains a loose collection of genera that exhibit several similarities but have not yet been placed in a single family. They all share the following properties:

1. The cells are rigid and helical or vibrioid (curved with less than one complete helical turn).
2. They are motile and have polar flagella.
3. All have respiratory metabolism with oxygen as the normal electron acceptor and are aerobic or microaerophilic.
4. Usually they do not produce indole and are oxidase positive (the oxidase test is positive when bacteria contain cytochrome *c* and enzymes that can oxidize it).
5. Most cannot catabolize carbohydrates; they use amino acids or organic acids as carbon and energy sources.

The group is quite diverse ecologically and is found in soil, freshwater, and marine habitats. Some species are associated with plant roots or grow in the intestinal tract, oral cavity, and reproductive organs of humans and other animals.

Table 21.2 summarizes the major distinguishing properties of the genera in this group. Figure 21.5 presents some examples of the morphology of typical species (*see also figures 2.8c and 2.18b*).

The genera *Spirillum, Aquaspirillum, Oceanospirillum,* and *Azospirillum* are widely dispersed and readily isolated from many environments (table 21.2). Generally, they are a minor part of the bacterial population. *Azospirillum* is associated with the roots of forage grasses, crops such as corn and sorghum, and legumes. It fixes nitrogen at low oxygen tensions and may be important to agricultural productivity under some conditions.

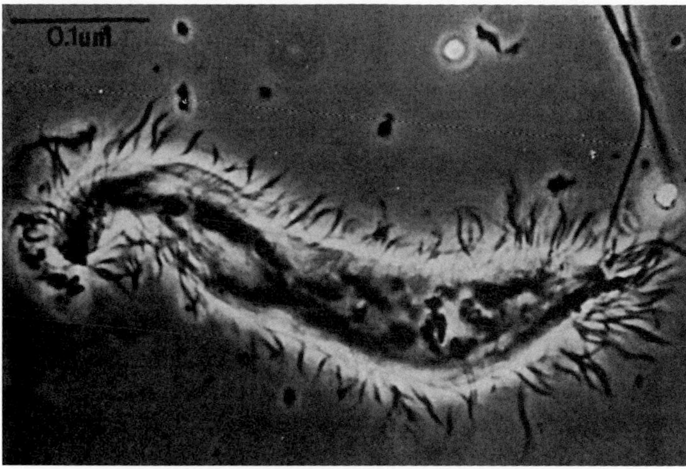

(a)

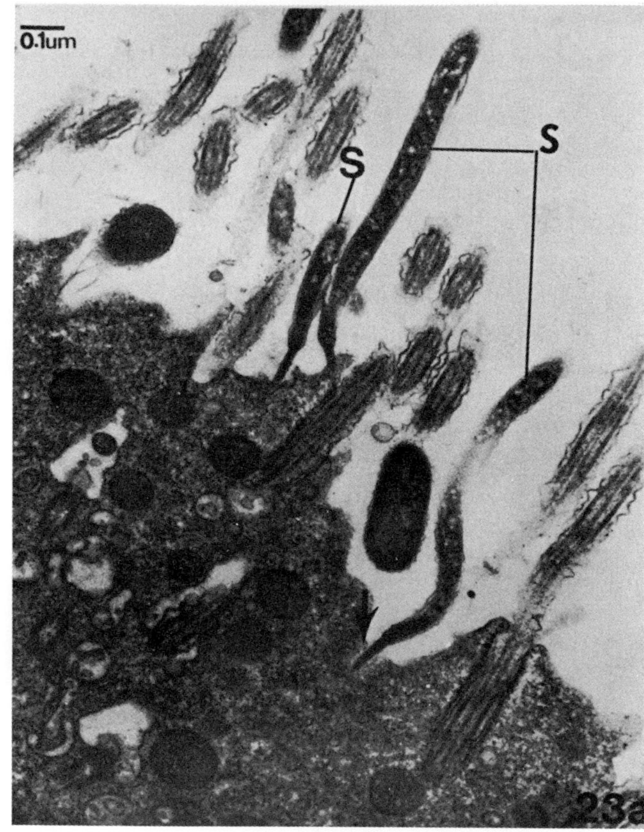

(b)

Figure 21.4 Spirochete-protozoan Associations. The surface spirochetes serve as organs of motility for protozoa (see text). (*a*) The spirochete-*Personympha* association with the spirochetes projecting from the surface. (*b*) Electron micrograph of small spirochetes (*S*) attached to the membrane of the flagellate protozoan *Barbulanympha.*

TABLE 21.1 Characteristics of Spirochete Genera

Genus	Dimensions (μm) and Morphology	G + C Content (mol %)	Oxygen Requirement	Carbon + Energy Source	Habitats
Spirochaeta	0.2–0.75 by 5–250; 2–40 periplasmic flagella (almost always 2)	51–65	Facultatively anaerobic or anaerobic	Carbohydrates	Aquatic and free-living
Cristispira	0.5–3.0 by 30–180; ≥ 100 periplasmic flagella	N. A.[a]	Facultatively anaerobic?	N. A.[a]	Mollusk digestive tract
Treponema	0.1–0.4 by 5–20; 2–16 periplasmic flagella	25–53	Anaerobic or microaerophilic	Carbohydrates or amino acids	Mouth, intestinal tract, and genital areas of animals; some are pathogenic (syphilis, yaws)
Borrelia	0.2–0.5 by 8–30; 30–40 periplasmic flagella	27–32	Anaerobic or microaerophilic	Carbohydrates	Mammals and arthropods; pathogens (relapsing fever, Lyme disease)
Leptospira	0.1 by 6–>12; 2 periplasmic flagella	35–49 (one strain is 53)	Aerobic	Fatty acids and alcohols	Free-living or pathogens of mammals, usually located in the kidney (leptospirosis)

[a]N. A., information not available

TABLE 21.2 Characteristics of Aerobic/Microaerophilic, Motile, Helical/Vibrioid Gram-Negative Bacteria

Genus	Dimensions (μm) and Morphology	G + C Content (mol%)	Oxygen Requirement	Habitats
Spirillum	1.4–1.7 by up to 60; large bipolar bundles of flagella	36–38	Microaerophilic	Stagnant fresh water
Azospirillum	1.0 by 2.1–3.8; single polar flagellum in liquid medium	69–71	Aerobic or microaerophilic	Soil and associated with tree roots; can fix nitrogen
Aquaspirillum	0.2–1.4 μm in diameter; usually helical with bipolar tufts of flagella	49–66	Aerobic (a few may be microaerophilic)	Fresh water
Oceanospirillum	0.3–1.4 μm in diameter; helical; usually with bipolar tufts of flagella	42–51	Aerobic	Marine (coastal waters)
Campylobacter	0.2–0.5 by 0.5–5; vibrioid or helical with a single flagellum at one or both poles	30–38	Microaerophilic	Intestinal tract, reproductive organs, and oral cavity of humans and other animals (*Campylobacter* enteritis)
Bdellovibrio	0.2–0.5 by 0.5–1.4; vibrioid with a single sheathed polar flagellum	33–51.5	Aerobic	Soil, sewage, fresh water and marine; predaceous on other gram-negative bacteria

The genus *Campylobacter* contains both nonpathogens and species pathogenic for humans and other animals. *C. fetus* causes reproductive disease and abortions in cattle and sheep. It is associated with a variety of conditions in humans ranging from **septicemia** (pathogens or their toxins in the blood) to **enteritis** (inflammation of the intestinal tract). *C. jejuni* causes abortion in sheep and enteritis diarrhea in humans.

The genus *Bdellovibrio* (Greek *bdella,* leech) contains gram-negative, curved rods with polar flagella (figure 21.6). The flagellum is unusually thick due to the presence of a sheath that is continuous with the cell wall. *Bdellovibrio* has a distinctive life-style: it preys on other gram-negative bacteria and alternates between a nongrowing predatory phase and an intracellular reproductive phase.

The life cycle of *Bdellovibrio* is complex although it requires only one to three hours for completion (figure 21.7). The free bacterium swims along very rapidly (about 100 cell lengths per second) until it collides violently with its prey. It attaches to the bacterial surface, begins to rotate at rates as high as 100 revolutions per second, and bores a hole through

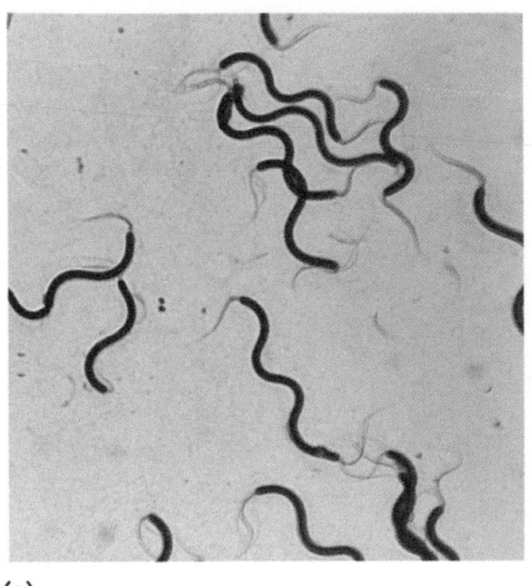
(a)

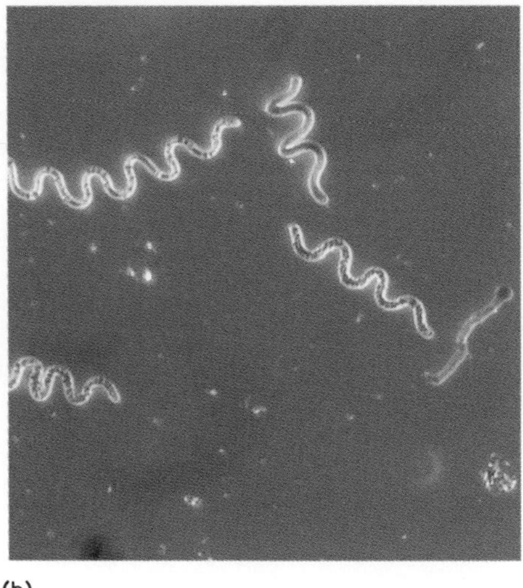
(b)

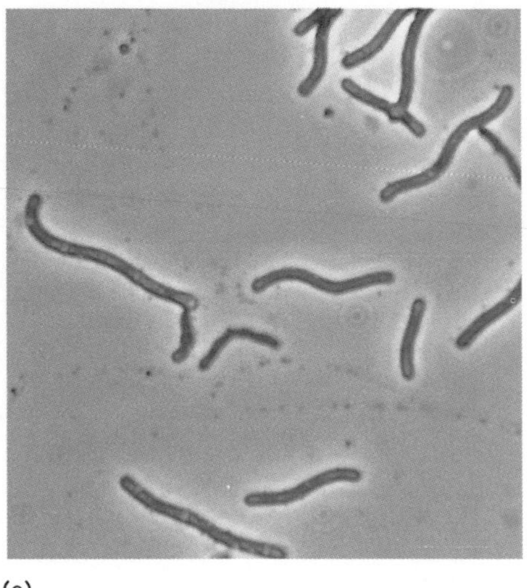
(c)

Figure 21.5 **The Vibrioid Gram-negative Bacteria.** Representative examples. (*a*) *Spirillum volutans* with bipolar flagella visible (×450). (*b*) *Spirillum volutans,* phase contrast (×200). (*c*) *Aquaspirillum sinuosum,* phase contrast (×800).

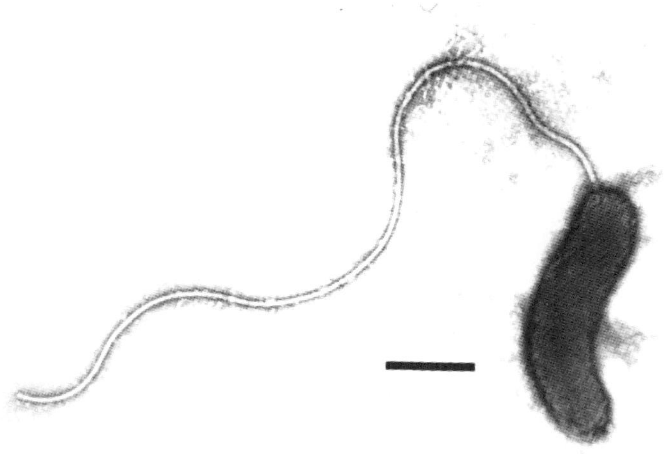

Figure 21.6 ***Bdellovibrio*** **Morphology.** Negatively stained *Bdellovibrio bacteriovorus* with its sheathed polar flagellum. Bar = 0.2 μm.

the host cell wall in 5 to 20 minutes with the aid of several hydrolytic enzymes that it releases. Its flagellum is lost during penetration of the cell.

After entry, *Bdellovibrio* takes control of the host cell and grows in the space between the cell wall and plasma membrane while the host cell loses its shape and rounds up. The predator inhibits host DNA, RNA, and protein synthesis within minutes and disrupts the host's plasma membrane so that cytoplasmic constituents can leak out of the cell. The growing bacterium uses host amino acids as its carbon, nitrogen, and energy source. It employs fatty acids and nucleotides directly in biosynthesis, thus saving carbon and energy. The bacterium rapidly grows into a long filament under the cell wall and then divides into many smaller, flagellated progeny, which escape upon host cell lysis. Such multiple fission is rare in procaryotes.

The *Bdellovibrio* life cycle resembles that of bacteriophages in many ways. Not surprisingly, if a *Bdellovibrio* culture is plated out on agar with host bacteria, plaques will form in the bacterial lawn. This technique is used to isolate pure strains and count the number of viable organisms just as it is with phages.

1. What properties do the bacteria of section 2 in *Bergey's Manual* have in common?
2. Briefly describe the ecological and public health significance of *Azospirillum* and *Campylobacter.* Define septicemia and enteritis.
3. Outline the life cycle of *Bdellovibrio* in detail.

Nonmotile (or Rarely Motile), Gram-Negative Curved Bacteria

Section 3 of *Bergey's Manual* contains several small groups of nonmotile, gram-negative bacteria that are curved to varying degrees. They are widespread in soil, freshwater, and marine habitats; all known species are free-living. At present, many of these bacteria are placed in the family *Spirosomaceae,* with three genera: *Spirosoma, Runella,* and *Flectobacillus.* The remainder are assigned to four other genera. *Microcyclus,* one of the best-studied genera, often grows in a ring shape (figure 21.8) and sometimes has gas vacuoles, presumably for flotation purposes.

1. What is unusual about *Microcyclus*?

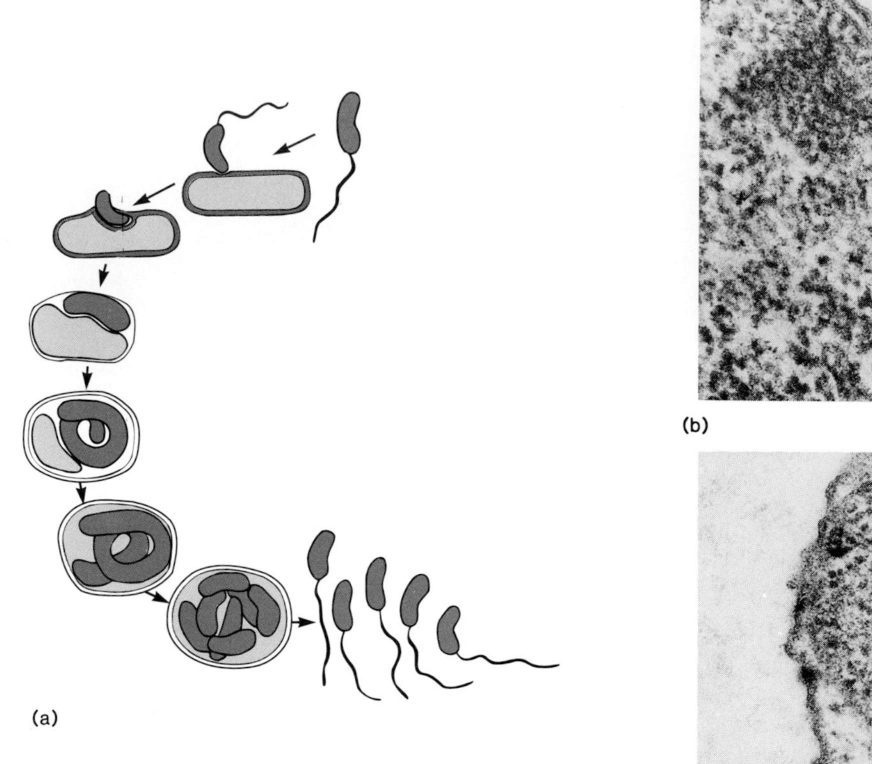

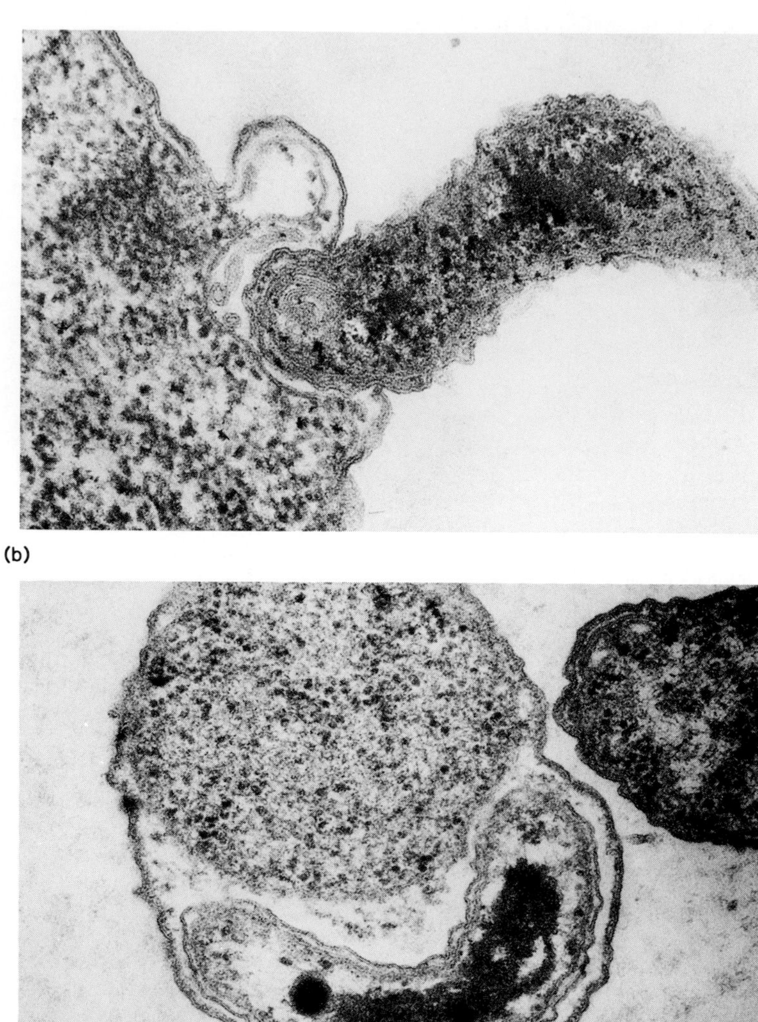

Figure 21.7 The Life Cycle of *Bdellovibrio*. (*a*) A general diagram showing the complete life cycle (see text for details). (*b*) *Bdellovibrio bacteriovorus* penetrating the cell wall of *E. coli* (×55,000). (*c*) A *Bdellovibrio* encapsulated between the cell wall and plasma membrane of *E. coli* (×60,800).

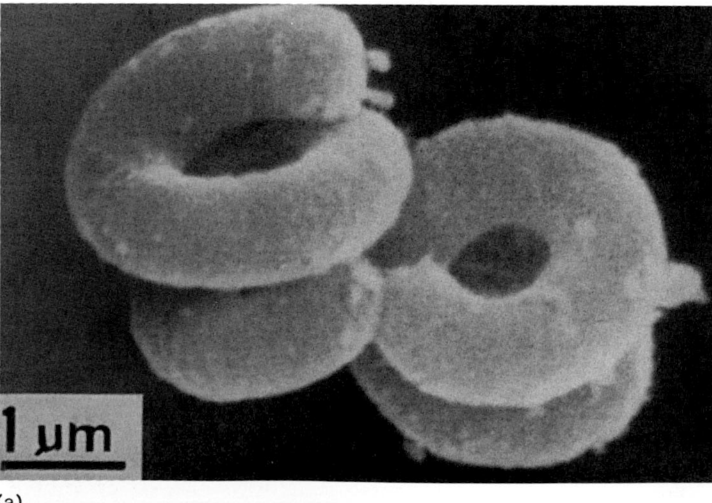

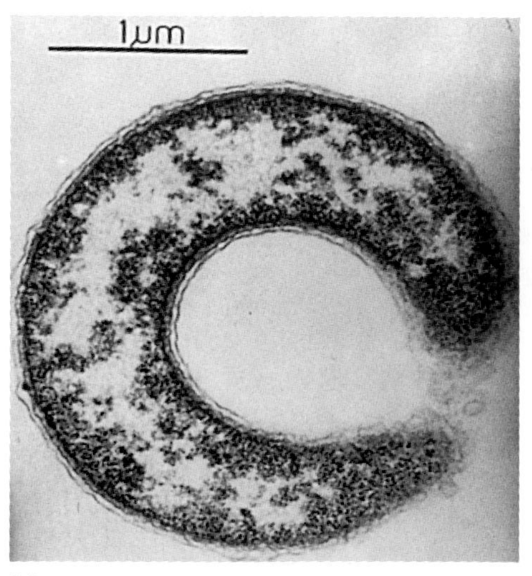

Figure 21.8 The Gram-negative Curved Bacteria. *Microcyclus flavus*. (*a*) A scanning electron micrograph showing ringlike cells. (*b*) A transmission electron micrograph of a longitudinal section.

TABLE 21.3 Characteristics of Gram-Negative Aerobic Rods and Cocci

Family[a]	Morphology	G + C Content (mol %)	Other Distinctive Characteristics	Representative Genera
Pseudomonadaceae	Straight or curved rods, polar flagella; no sheaths or prosthecae	58–71	Metabolism usually respiratory; often can grow with a large variety of carbon sources	*Pseudomonas* *Xanthomonas* *Zoogloea*
Azotobacteraceae	Blunt rods to oval cells; peritrichous or polar flagella (or nonmotile)	52–67.5	Cysts formed in one species; can fix nitrogen	*Azotobacter* *Azomonas*
Rhizobiaceae	Rod-shaped; flagella polar or subpolar, or peritrichous	57–65	Nodules formed on legumes by rhizobia; gall hypertrophies produced by agrobacteria	*Rhizobium* *Agrobacterium*
Methylococcaceae	Rods, vibrios, and cocci; polar flagella or nonmotile	50–64	Can use methane as sole carbon and energy source under aerobic or microaerophilic conditions	*Methylococcus* *Methylomonas*
Acetobacteraceae	Ellipsoidal to rod-shaped; peritrichous, polar, or nonmotile	51–65	Oxidizes ethanol to acetic acid in neutral and acid media	*Acetobacter* *Gluconobacter*
Legionellaceae	Rods 0.3–0.9 μm by 2–20 μm or more; polar or lateral flagella, occasionally nonmotile	39–43	Branched-chain fatty acids in wall; cysteine and iron required for growth	*Legionella*
Neisseriaceae	Cocci (single or in pairs or masses) or rods; nonmotile	38–55	Except for *Acinetobacter*, often parasites of warm-blooded animals	*Neisseria* *Moraxella* *Acinetobacter*

[a]Genera not associated with a family: *Beijerinckia, Derxia, Xanthobacter, Thermus, Thermomicrobium, Halomonas, Alteromonas, Flavobacterium, Alcaligenes, Serpens, Janthinobacterium, Brucella, Bordetella, Franciscella, Paracoccus, Lampropedia.*

Gram-Negative Aerobic Rods and Cocci

Section 4 contains a tremendous variety of gram-negative aerobic rods and cocci, most of which are distributed among 21 genera in eight families (table 21.3). An additional 16 genera are not placed in any family. Although *Brucella* (brucellosis), *Bordetella* (pertussis), and others of these "extra" 16 genera are of considerable medical and practical importance, the focus here is on a few of the more interesting families and some of their genera.

Pseudomonas

The genus *Pseudomonas* contains straight or slightly curved rods, 0.5 to 1.0 μm by 1.5 to 5.0 μm in length, that are motile by one or several polar flagella and lack prosthecae or sheaths (*see figure 3.30*a). These chemoheterotrophs are aerobic and carry out respiratory metabolism with O_2 (and sometimes nitrate) as the electron acceptor. All pseudomonads have a functional tricarboxylic acid cycle and can oxidize substrates to CO_2. Most hexoses are degraded by the Entner-Doudoroff pathway rather than glycolytically.

The Entner-Doudoroff pathway and tricarboxylic acid cycle (pp. 150–51, appendix II).

The genus is an exceptionally heterogeneous taxon composed of 30 species distributed among five rRNA homology groups and over 60 less well examined species. The three best-characterized groups, RNA groups I–III, are subdivided according to properties such as the presence of poly-β-hydroxybutyrate (PHB), the production of a fluorescent pigment, pathogenicity, the presence of arginine dihydrolase, and glucose utilization. For example, the fluorescent subgroup does not accumulate PHB and produces a diffusible, water-soluble, yellow-green pigment that fluoresces under UV radiation (figure 21.9). *Pseudomonas aeruginosa, P. fluorescens, P. putida,* and *P. syringae* are members of this group.

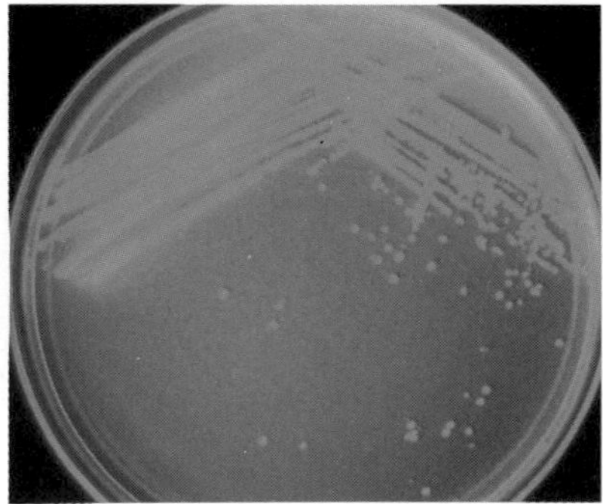

Figure 21.9 ***Pseudomonas* Fluorescence.** *Pseudomonas aeruginosa* colonies fluorescing under ultraviolet light.

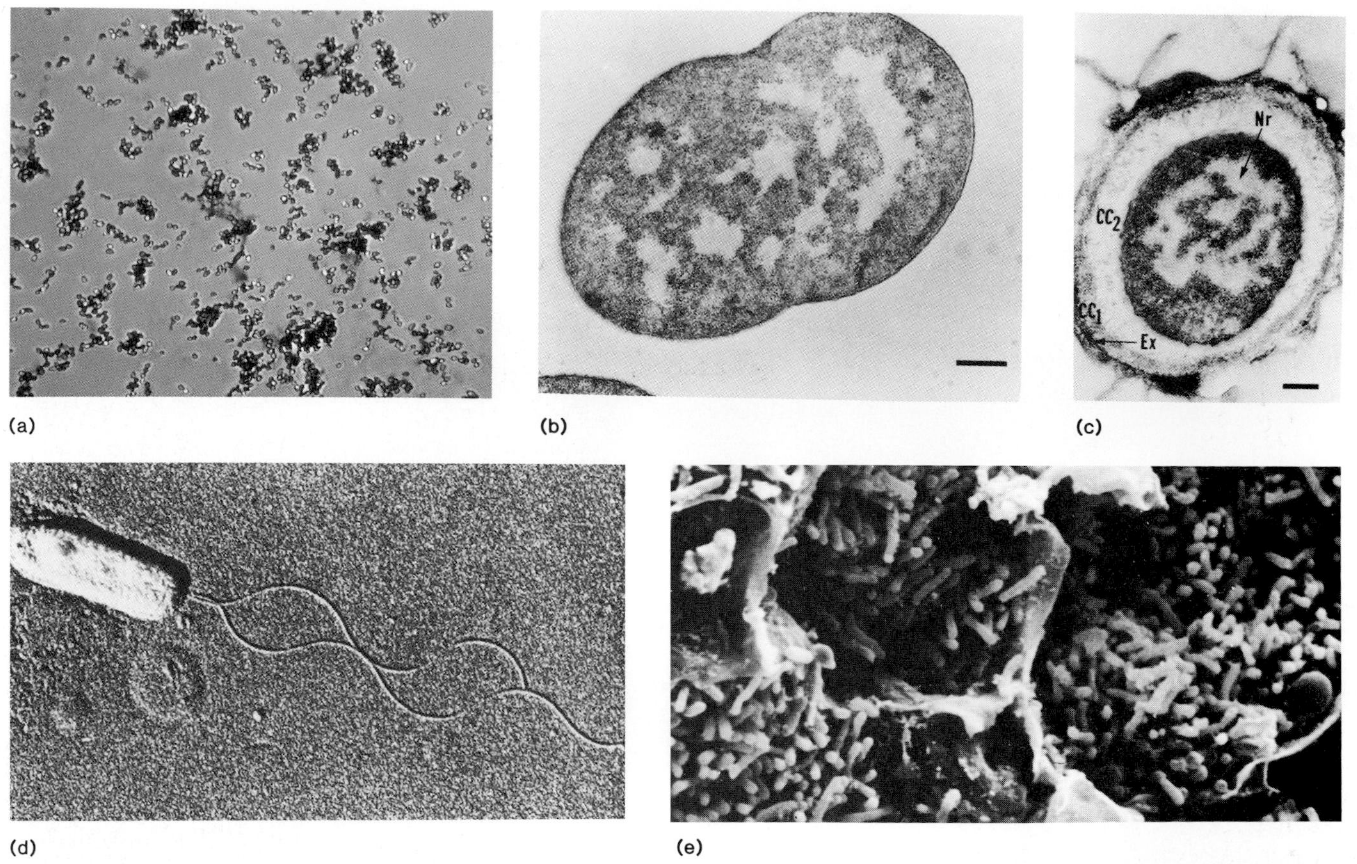

Figure 21.10 **The Gram-negative Aerobic Rods and Cocci.** *Azotobacter* and *Rhizobium.* (*a*) *A. chroococcum* (×120). (*b*) Electron micrograph of *A. chroococcum.* Bar = 0.2 μm. (*c*) *Azotobacter* cyst structure. Bar = 0.2 μm. The nuclear region (*Nr*), exine layers (CC_1 and CC_2), and exosporium (*Ex*) are visible. (*d*) *Rhizobium leguminosarum* with two polar flagella (×14,000). (*e*) Scanning electron micrograph of bacteroids in alfalfa root nodule cells (× 640).

The pseudomonads have a great practical impact in several ways, including the following:

1. Many can degrade an exceptionally wide variety of organic molecules. Thus, they are very important in the **mineralization process** (the microbial breakdown of organic materials to inorganic substances) in nature and in sewage treatment. The fluorescent pseudomonads can use approximately 80 different substances as their carbon and energy sources; *P. cepacia* will degrade over 100 different organic molecules.

Microbial decomposition of organic materials (pp. 815–16; 859–60).

2. Several species (e.g., *P. aeruginosa*) are important experimental subjects. Many advances in microbial physiology and biochemistry have come from their study.
3. Some pseudomonads are major animal and plant pathogens. *P. aeruginosa* infects people with low resistance and invades burn areas or causes urinary tract infections. *P. solanacearum* causes wilts in many plants by producing pectinases, cellulases, and the plant hormones indoleacetic acid and ethylene. *P. syringae* and *P. cepacia* are also important plant pathogens.
4. Pseudomonads such as *P. fluorescens* are involved in the spoilage of refrigerated milk, meat, eggs, and seafood because they grow at 4°C and degrade lipids and proteins.

Azotobacter and *Rhizobium*

Although both *Azotobacter* and *Rhizobium* can carry out nitrogen fixation, they differ in morphology and life-style. The genus *Azotobacter* contains large, usually motile, ovoid bacteria, 1.5 to 2.0 μm in diameter, that are often pleomorphic and form dormant cysts as the culture ages (figure 21.10). Members of the genus *Rhizobium* are 0.5 to 0.9 by 1.2 to 3.0 μm motile rods, often containing poly-β-hydroxybutyrate granules, that become pleomorphic under adverse conditions (figure 21.10). They grow symbiotically within root nodule cells of legumes as nitrogen-fixing bacteroids (figure 21.10*e*). In contrast, *Azotobacter* is a free-living soil genus and fixes atmospheric nitrogen nonsymbiotically.

The biochemistry of nitrogen fixation (pp. 178–80).
The Rhizobium-legume symbiosis (pp. 851–55).

Agrobacterium

The genus *Agrobacterium* is placed in the family *Rhizobiaceae* but differs from *Rhizobium* in not stimulating root nodule formation or fixing nitrogen. Instead, agrobacteria invade the crown, roots, and stems of many plants and transform plant cells into autonomously proliferating tumor cells. The best-studied species is *A. tumefaciens,* which enters many broad-leaved plants through wounds and causes the crown gall disease (figure 21.11). The ability to produce tumors is dependent upon the presence of a large Ti (for tumor-inducing) plasmid. Tumor production by *Agrobacterium* is discussed in greater detail in Box 14.2 and chapter 42.

Plasmids (pp. 261–64).

Figure 21.11 **Agrobacterium.** Crown gall tumor of a tomato plant caused by *Agrobacterium tumefaciens.*

The *Methylococcaceae*

The family *Methylococcaceae* contains rods, vibrios, and cocci that use methane and methanol as their sole carbon and energy sources under aerobic or microaerobic (low oxygen) conditions. That is, they are **methylotrophs.** The family contains two genera: *Methylococcus* (spherical, nonmotile cells) and *Methylomonas* (straight, curved, or branched rods with a single, polar flagellum). When oxidizing methane, the bacteria contain complex arrays of intracellular membranes. Almost all can form some type of resting stage, often a cyst somewhat like that of the azotobacteria. Methylotrophic growth depends on the presence of methane and related compounds. Methanogenesis from substrates such as H_2 and CO_2 is widespread in anaerobic soil and water, and methylotrophic bacteria grow above anaerobic habitats all over the world.

Methane-oxidizing bacteria use methane as a source of both energy and carbon. Methane is first oxidized to methanol by the enzyme methane monooxygenase. The methanol is then oxidized to formaldehyde by methanol dehydrogenase, and the electrons from this oxidation are donated to an electron transport chain for ATP synthesis. Formaldehyde can be assimilated into cell material by the activity of either of two pathways, one involving the formation of the amino acid serine and the other proceeding through the synthesis of sugars such as fructose 6-phosphate and ribulose 5-phosphate.

1. What characteristics do the bacteria of section 4 of *Bergey's Manual* share?
2. List the major distinctive properties of the genera *Pseudomonas, Azotobacter, Rhizobium,* and *Agrobacterium* and the family *Methylococcaceae.*
3. Why are the pseudomonads such important microorganisms? What is mineralization?
4. How do *Azotobacter* and *Rhizobium* differ in life-style?
5. What effect does *Agrobacterium* have on plant hosts?
6. In what habitats would one expect to see the *Methylococcaceae* growing, and why?
7. What is a methylotroph? How do methane-oxidizing bacteria use methane as both an energy source and a carbon source?

Facultatively Anaerobic Gram-Negative Rods

Section 5 of *Bergey's Manual* contains 27 genera of facultatively anaerobic gram-negative rods, 20 of which are distributed among three families. Table 21.4 summarizes the distinguishing properties of the families *Enterobacteriaceae, Vibrionaceae,* and *Pasteurellaceae.* This section briefly describes the three families, with emphasis on the first because of its medical, industrial, and biological importance.

The *Enterobacteriaceae*

The family *Enterobacteriaceae* is the largest of the three families in section 5 of *Bergey's Manual.* It contains gram-negative, peritrichously flagellated or nonmotile, facultatively anaerobic, straight rods with simple nutritional requirements (*see figures 2.14*c, *2.18*a, *3.26*a, *3.29, and 3.30*c). Figure 21.12 depicts the relatedness of several common species of the family.

The metabolic properties of the *Enterobacteriaceae* are very useful in characterizing its constituent genera. Members of the family, often called **enterobacteria** or **enteric bacteria** (Greek *enterikos,* pertaining to the intestine), all degrade sugars by means of the Embden-Meyerhof pathway and cleave

Table 21.4 Characteristics of Families of Facultatively Anaerobic Gram-Negative Rods

Characteristics	*Enterobacteriaceae*	*Vibrionaceae*	*Pasteurellaceae*
Cell dimensions	0.3–1.0 by 1.0–6.0 μm	0.3–1.3 by 1.0–3.5 μm	0.2–0.3 by 0.3–2.0 μm
Morphology	Straight rods; peritrichous flagella or nonmotile	Straight or curved rods; polar flagella	Coccoid to rod-shaped cells, sometimes pleomorphic; nonmotile
Physiology	Oxidase negative	Oxidase positive; all can use D-glucose as sole or principal carbon source	Oxidase positive; heme and/or NAD often required for growth; organic nitrogen source required
G + C content	38–60%	38–63%	38–47%
Symbiotic relationships	Some parasitic on mammals and birds; some species plant pathogens	Most not pathogens (with a few exceptions)	Parasites of mammals and birds
Representative genera	*Escherichia, Shigella, Salmonella, Citrobacter, Klebsiella, Enterobacter, Erwinia, Serratia, Proteus, Yersinia*	*Vibrio, Photobacterium, Aeromonas*	*Pasteurella, Haemophilus*

Modified from *Bergey's Manual of Systematic Bacteriology*, Vol. 1, p. 408, J. G. Holt and N. R. Krieg (eds.). Copyright © 1984 Williams and Wilkins Co., Baltimore, MD. Reprinted by permission.

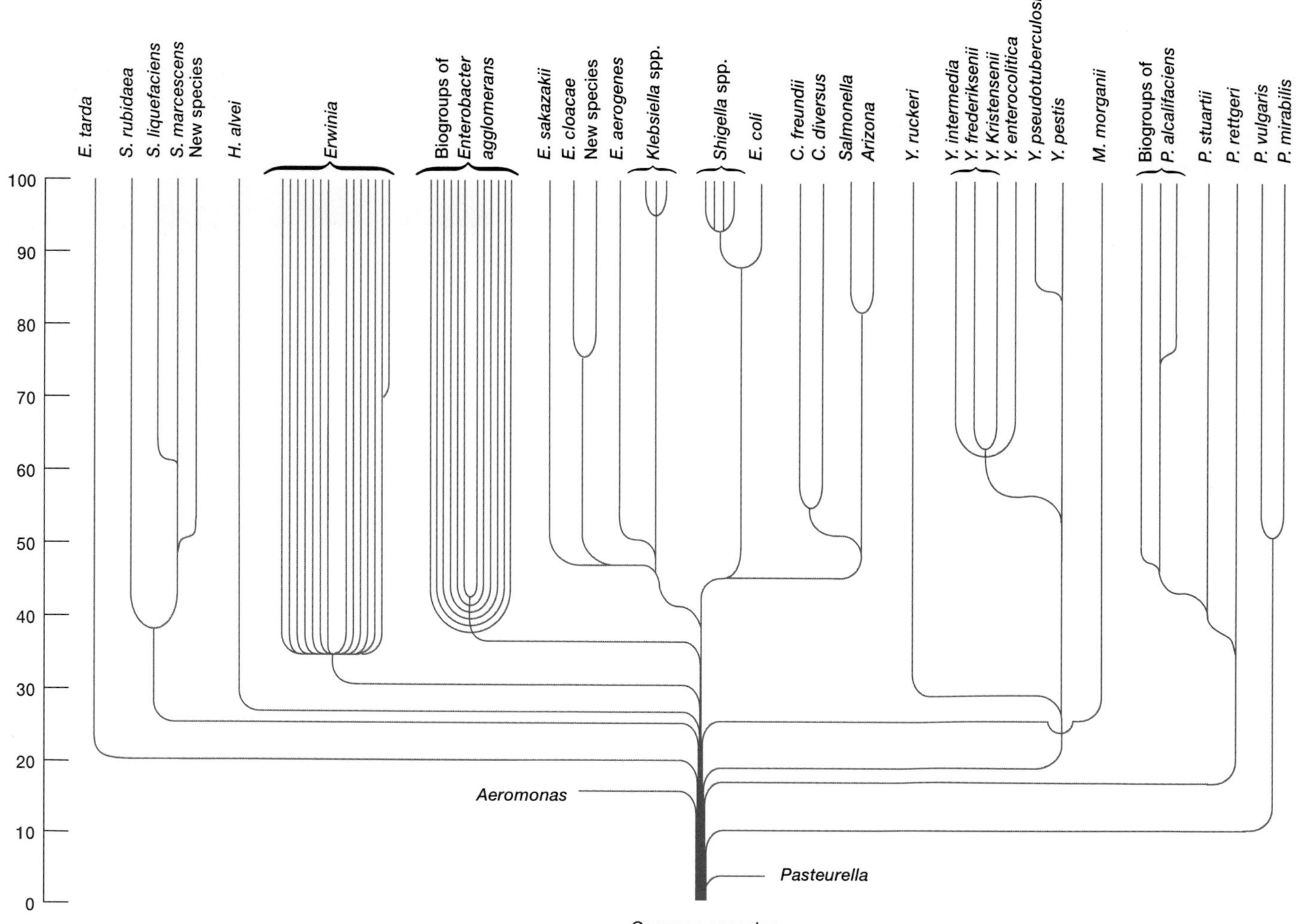

Figure 21.12 Divergence of the family *Enterobacteriaceae*. The DNA relatedness of genera in the family *Enterobacteriaceae*. The ordinate is the percentage of relatedness, and the horizontal branches depict the degree of relatedness of the branching group to all organisms that have not yet branched off. A common ancestor for the family is assumed.

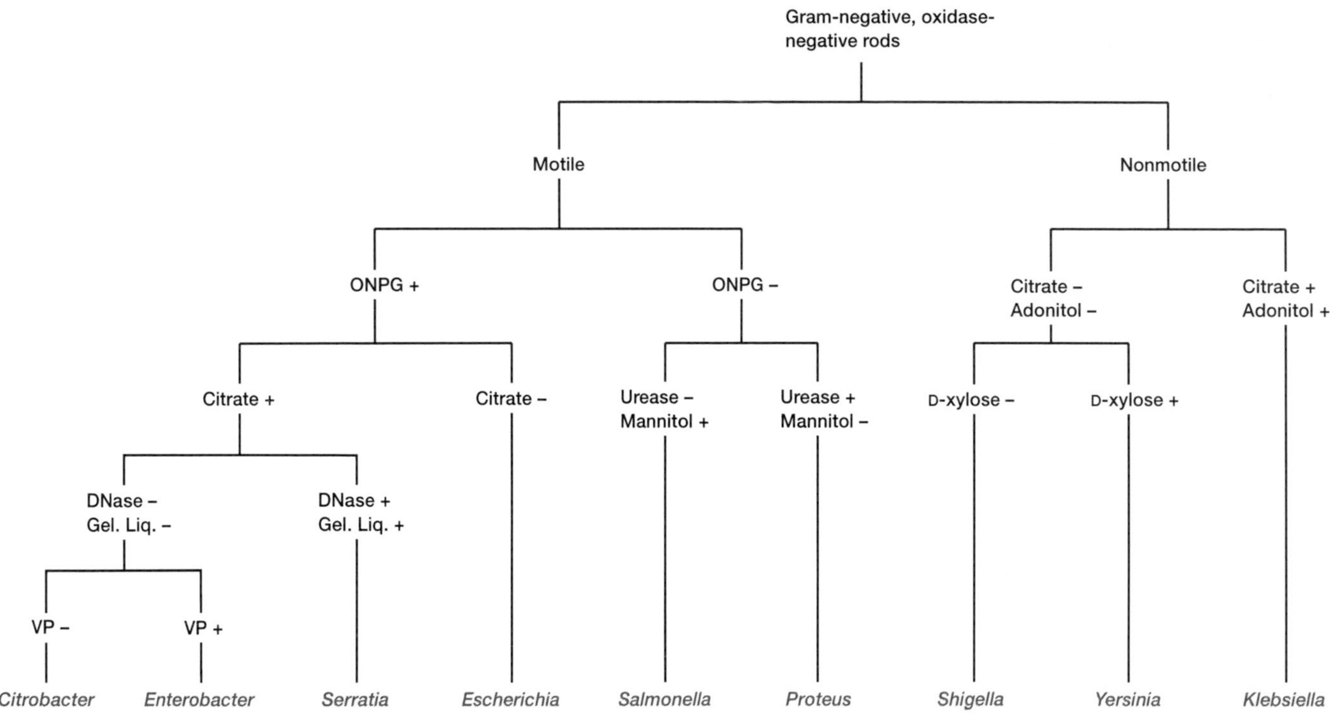

Figure 21.13 **Identification of Enterobacterial Genera.** A dichotomous key to selected genera of enteric bacteria based on motility and biochemical characteristics. The following abbreviations are used: ONPG, o-nitrophenyl-β-D-galactopyranoside (a test for β-galactosidase); DNase, deoxyribonuclease; Gel. Liq., gelatin liquefaction; and VP, Voges-Proskauer (a test for acetoin).

pyruvic acid to yield formic acid in formic acid fermentations. Those enteric bacteria that produce large amounts of gas during sugar fermentation, such as *Escherichia* spp., have the formic hydrogenlyase complex that degrades formic acid to H_2 and CO_2. The family can be divided into two groups based on their fermentation products. The majority (e.g., the genera *Escherichia, Proteus, Salmonella,* and *Shigella*) carry out mixed acid fermentation and produce mainly lactate, acetate, succinate, formate (or H_2 and CO_2), and ethanol. In butanediol fermentation, the major products are butanediol, ethanol, and carbon dioxide. *Enterobacter, Serratia, Erwinia,* and *Klebsiella* are butanediol fermenters. As mentioned previously (*see chapter 8*), the two types of formic acid fermentation are distinguished by the methyl red and Voges-Proskauer tests.

Formic acid fermentation and the family Enterobacteriaceae *(pp. 156–58).*

Because the enteric bacteria are so similar in appearance, biochemical tests are normally used to identify them after a preliminary examination of their morphology, motility, and growth responses (see figure 21.13 for a simple example). Some more commonly used tests are those for the type of formic acid fermentation, lactose and citrate utilization, indole production from tryptophan, urea hydrolysis, and hydrogen sulfide production. For example, lactose fermentation occurs in *Escherichia* and *Enterobacter* but not in *Shigella, Salmonella,* or *Proteus.* Table 21.5 summarizes a few of the biochemical properties useful in distinguishing between genera of enteric bacteria. The mixed acid fermenters are located on the left in this table and the butanediol fermenters on the right. The usefulness of biochemical tests in identifying enteric bacteria is shown by the popularity of commercial identification systems, such as the Enterotube and API 20-E systems, that are based on these tests.

Commercial rapid identification systems (pp. 683–88).

TABLE 21.5 Some Characteristics of Selected Genera in the *Enterobacteriaceae*

Characteristics	*Escherichia*	*Shigella*	*Salmonella*	*Citrobacter*	*Proteus*
Methyl red	+	+	+	+	+
Voges-Proskauer	−	−	−	−	d
Indole production	(+)	d	−	d	d
Citrate use	−	−	(+)	+	d
H_2S production	−	−	(+)	d	(+)
Urease	−	−	−	(+)	+
β-galactosidase	(+)	d	d	+	−
Gas from glucose	+	−	(+)	+	+
Acid from lactose	+	−	(−)	d	−
Phenylalanine deaminase	−	−	−	−	+
Lysine decarboxylase	(+)	−	(+)	−	−
Ornithine decarboxylase	(+)	d	(+)	(+)	d
Motility	d	−	(+)	+	+
Gelatin liquifaction (22°C)	−	−	−	−	+
% G + C	48–52	49–53	50–53	50–52	38–41
Other characteristics	1.1–1.5 by 2.0–6.0 μm; peritrichous when motile	No gas from sugars	Peritrichous flagella	1.0 by 2.0–6.0 μm; peritrichous	0.4–0.8 by 1.0–3.0 μm; peritrichous

[a](+) usually present
[b](−) usually absent
[c]d, strains or species vary in possession of characteristic

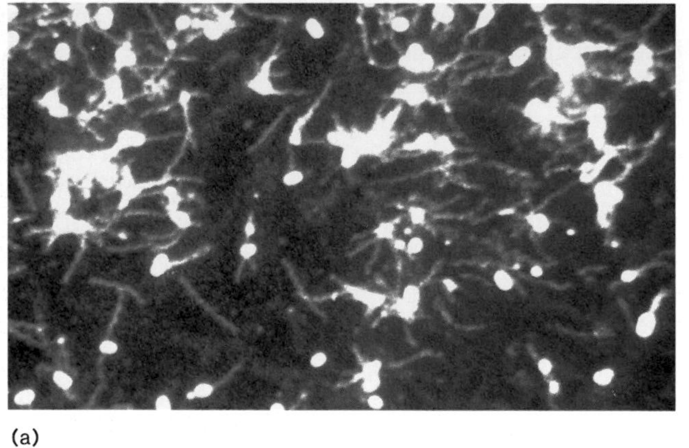
(a)

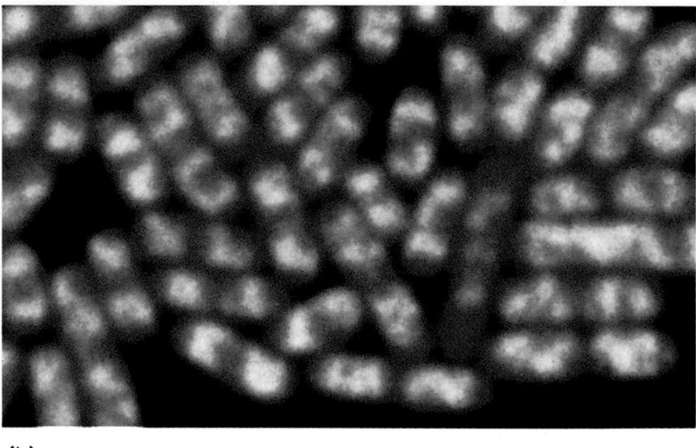
(b)

Figure 21.14 **The *Enterobacteriaceae*.** *Salmonella* treated with fluorescent stains. (*a*) *Salmonella enteritidis* with peritrichous flagella (×500). *S. enteritidis* is associated with gastroenteritis. (*b*) *Salmonella typhi* with acridine orange stain (×2,000). *S. typhi* causes typhoid fever.

Members of the *Enterobacteriaceae* are so common, widespread, and important that they are probably more often seen in most laboratories than any other bacteria. *Escherichia coli* is undoubtedly the best-studied bacterium and the experimental organism of choice for many microbiologists. It is a major inhabitant of the colon of humans and other warm-blooded animals, and it is quite useful in the analysis of water for fecal contamination (*see chapter 41*). Some strains cause gastroenteritis or urinary tract infections. Several enteric genera contain very important human pathogens responsible for a variety of diseases: *Salmonella* (figure 21.14), typhoid fever and gastroenteritis; *Shigella,* bacillary dysentery; *Klebsiella,* pneumonia; *Yersinia,* plague. Members of the genus *Erwinia* are major pathogens of crop plants and cause blights, wilts, and several other plant diseases. These and other members of the family are discussed in more detail at later points in the text (*see chapter 37*).

Yersinia	*Klebsiella*	*Enterobacter*	*Erwinia*	*Serratia*
+	(+)[a]	(−)[b]	+	d[c]
−(37°C)	(+)	+	(+)	+
d	d	−	(−)	(−)
(−)	(+)	+	(+)	+
−	−	−	(+)	−
d	(+)	(−)	−	−
+	(+)	+		+
(−)	(+)	(+)	(−)	d
(−)	(+)	(+)	d	d
−	−	(−)	(−)	−
(−)	(+)	d	−	d
d	−	(+)	−	d
−(37°C)	−	+	+	+
(−)	−	d	d	(+)
46–50	53–58	52–60	50–58	53–59
0.5–0.8 by 1.0–3.0 μm; peritrichous when motile	0.3–1.0 by 0.6–6.0 μm; capsulated	0.6–1.0 by 1.2–3.0 μm; peritrichous	0.5–1.0 by 1.0–3.0 μm; peritrichous; plant pathogens and saprophytes	0.5–0.8 by 0.9–2.0 μm; peritrichous; colonies often pigmented

The *Vibrionaceae*

Members of the family *Vibrionaceae* are gram-negative, straight or curved rods with polar flagella (figure 21.15). Most are oxidase positive, and all use D-glucose as their sole or primary carbon and energy source (see table 21.4). The majority are aquatic microorganisms, widespread in fresh water and the sea. There are four genera in the family: *Vibrio, Photobacterium, Aeromonas,* and *Plesiomonas.* Studies of rRNA homology suggest that *Vibrio* and *Photobacterium* are closely related, whereas the latter two genera differ more widely.

Several vibrios are important pathogens. *V. cholerae* (*see figure 3.1e*) is the causative agent of cholera (*see chapter 37*), and *V. parahaemolyticus* sometimes causes gastroenteritis in humans following consumption of contaminated seafood. *V. anguillarum* and others are responsible for fish diseases.

Several members of the family are unusual in being bioluminescent. *Vibrio fischeri* and at least two species of *Photobacterium* are among the few marine bacteria capable of **bioluminescence** and emit a blue-green light because of the activity of the enzyme luciferase (see Box 21.1). The peak emission of light is usually between 472 and 505 nm, but one strain of *V. fischeri* emits yellow light with a major peak at 545 nm. Although many of these bacteria are free-living, *P. phosphoreum* and *P. leiognathi* live symbiotically in the luminous organs of fish (figure 21.16).

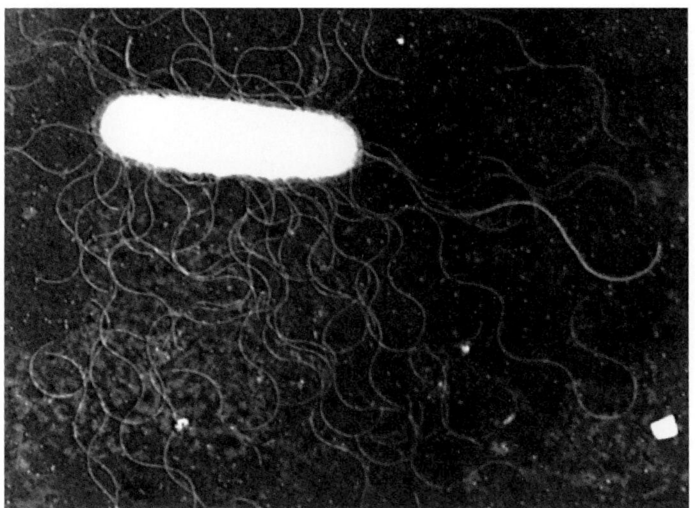

Figure 21.15 **The *Vibrionaceae*.** Electron micrograph of *Vibrio alginolyticus* grown on agar, showing a sheathed polar flagellum and unsheathed lateral flagella (×18,000).

Box 21.1

Bacterial Bioluminescence

Several species in the genera *Vibrio* and *Photobacterium* can emit light of a blue-green color. The enzyme luciferase catalyzes the reaction and uses reduced flavin mononucleotide, molecular oxygen, and a long-chain aldehyde as substrates.

$$FMNH_2 + O_2 + RCHO \xrightarrow{\text{luciferase}} FMN + H_2O + RCOOH + \text{light}$$

The evidence suggests that an enzyme-bound, excited flavin intermediate is the direct source of luminescence. Because the electrons used in light generation are probably diverted from the electron transport chain and ATP synthesis, the bacteria are expending considerable energy on luminescence. Luminescence activity is regulated and can be turned off or on under the proper conditions.

There is much speculation about the role of bacterial luminescence and its value to bacteria, particularly because it is such an energetically expensive process. Luminescent bacteria occupying the special luminous organs of fish do not emit light when they grow as free-living organisms in the seawater. Free-living luminescent bacteria can reproduce and infect young fish. Once settled in a fish's luminous organ, they begin emitting light, which the fish uses for its own purposes (*see chapter 29*). Other luminescent bacteria growing on potential food items such as small crustacea may use light to attract fish to the food source. After ingestion, they could establish a symbiotic relationship in the host's gut.

(a)

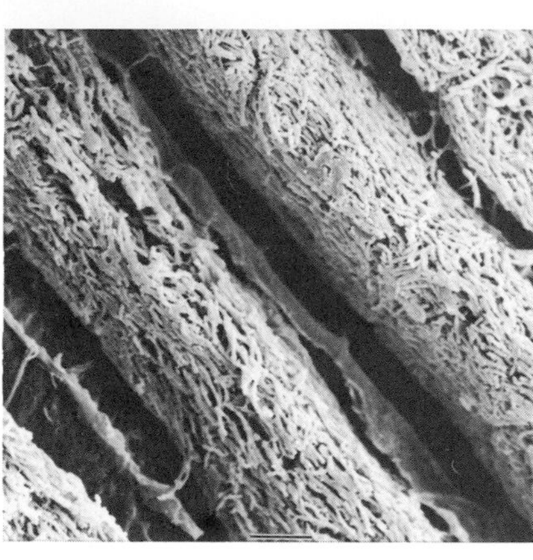

(b)

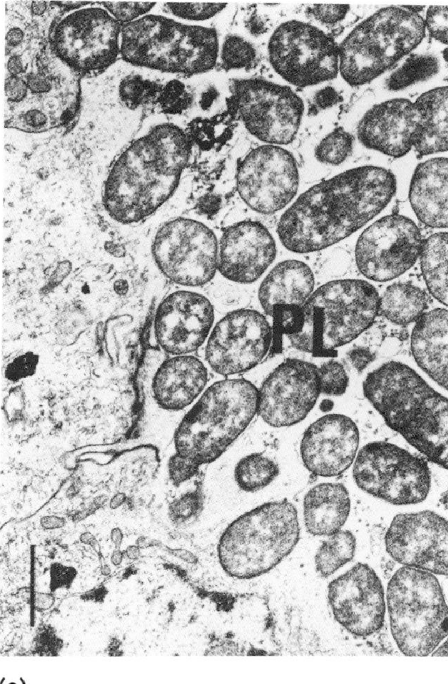

(c)

Figure 21.16 **Bioluminescence.** (*a*) A photograph of the Atlantic flashlight fish *Kryptophanaron alfredi*. The light area under the eye is the fish's luminous organ, which can be covered by a lid of tissue. (*b*) The masses of photobacteria in the SEM view are separated by thin epithelial cells. Line = 6.0 μm. (*c*) Ultrathin section of the luminous organ of a fish, *Equulities novaehollandiae,* with the bioluminescent bacterium *Photobacterium leiognathi, PL.* Bar = 2 μm.

The *Pasteurellaceae*

The *Pasteurellaceae* differ from the other two families in several ways (table 21.4). Most notably, they are small (0.2 to 0.3 μm in diameter) and nonmotile, normally oxidase positive, have complex nutritional requirements of various kinds, and are parasitic in vertebrates. The family contains three genera: *Pasteurella, Haemophilus,* and *Actinobacillus.*

As might be expected, members of this family are best known for the diseases they cause in humans and many animals. *P. multilocida* and *P. haemolytica* are important animal pathogens. *P. multilocida* is responsible for fowl cholera, which destroys many chickens, turkeys, ducks, and geese each year. *P. haemolytica* is at least partly responsible for pneumonia in cattle, sheep, and goats (e.g., "shipping fever" in cattle). *H. influenzae* type b is a major human pathogen that causes a variety of diseases, including meningitis in children (*see chapter 37*).

---

1. List the major distinguishing traits of the families *Enterobacteriaceae, Vibrionaceae,* and *Pasteurellaceae.*
2. Into what two groups can the enteric bacteria be placed based on their fermentation patterns?
3. Give two reasons why the enterobacteria are so important.
4. What major human disease is associated with the *Vibrionaceae,* and what species causes it?
5. Briefly describe bioluminescence and the way it is produced.

Anaerobic Gram-Negative Straight, Curved, and Helical Rods

Section 6 of *Bergey's Manual* contains one family, the *Bacteroidaceae,* and thirteen genera. All species are obligately anaerobic, nonsporing, motile or nonmotile rods of various shapes. They are chemoheterotrophic and usually produce a mixture of organic acids as fermentation end products, but they do not reduce sulfate or other sulfur compounds. The genera are identified according to properties such as general shape, motility and flagellation pattern, and fermentation end products. These bacteria grow in habitats such as the oral cavity and intestinal tract of humans and other animals and the rumen (*see chapter 29*) of ruminants.

Although the difficulty of culturing these anaerobes has hindered the determination of their significance, they are clearly widespread and important. Often they benefit their host. *Bacteroides succinogenes* and *B. ruminicola* are major components of the rumen flora; the former actively degrades cellulose, and the latter ferments starch, pectin, and other carbohydrates. About 30% of the bacteria isolated from human feces are members of the genus *Bacteroides,* and these organisms may provide extra nutrition by degrading cellulose, pectins, and other complex carbohydrates. The family is also involved in human disease. Members of the genus *Bacteroides* are associated with diseases of major organ systems, ranging from the central nervous system to the skeletal system. *B. fragilis* is a particularly common anaerobic pathogen found in abdominal, pelvic, pulmonary, and blood infections.

---

1. Give the major properties of the family *Bacteroidaceae.*
2. How do these bacteria benefit and harm their hosts?

Dissimilatory Sulfate- or Sulfur-Reducing Bacteria

Section 7 of *Bergey's Manual* contains a morphologically diverse group of seven gram-negative genera. These bacteria are united by their strictly anaerobic nature and the ability to use elemental sulfur or sulfate and other oxidized sulfur compounds as electron acceptors during anaerobic respiration (figure 21.17). An electron transport chain generates ATP and reduces sulfur and sulfate to hydrogen sulfide. The best-studied sulfate-reducing genus is *Desulfovibrio; Desulfuromonas* uses only elemental sulfur as an acceptor.

Anaerobic respiration (p. 158).

These bacteria are very important in the cycling of sulfur within the ecosystem. Because significant amounts of sulfate are present in almost all aquatic and terrestrial habitats, sulfate-reducing bacteria are widespread and active in locations made anaerobic by microbial digestion of organic materials. *Desulfovibrio* and other sulfate-reducing bacteria thrive in habitats such as muds and sediments of polluted lakes and streams, sewage lagoons and digesters, and waterlogged soils. *Desulfuromonas* is most prevalent in anaerobic marine and estuarine sediments. It also can be isolated from methane digesters and anaerobic hydrogen-sulfide rich muds of freshwater habitats. It uses elemental sulfur, but not sulfate as its electron acceptor. Often sulfate and sulfur reduction is apparent from the smell of hydrogen sulfide and the blackening of water and sediment by iron sulfide. Hydrogen sulfide production in waterlogged soils can kill animals, plants, and microorganisms. Sulfate-reducing bacteria negatively impact

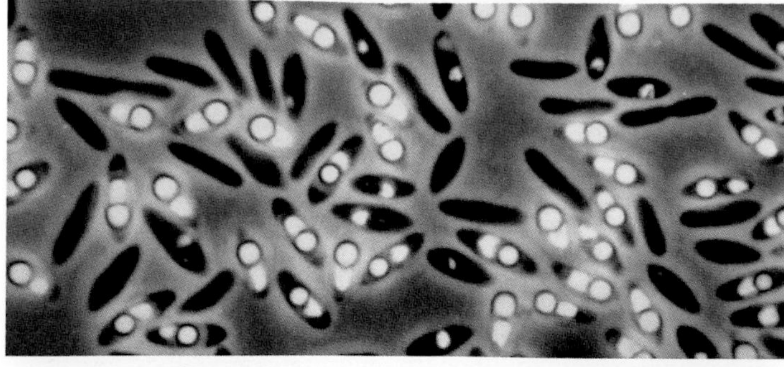

(a)

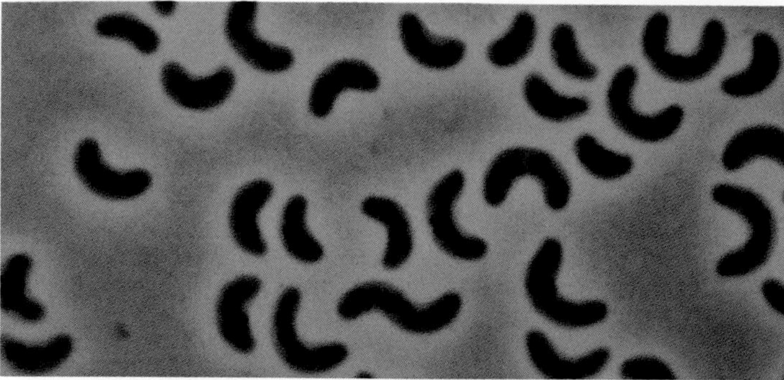

(b)

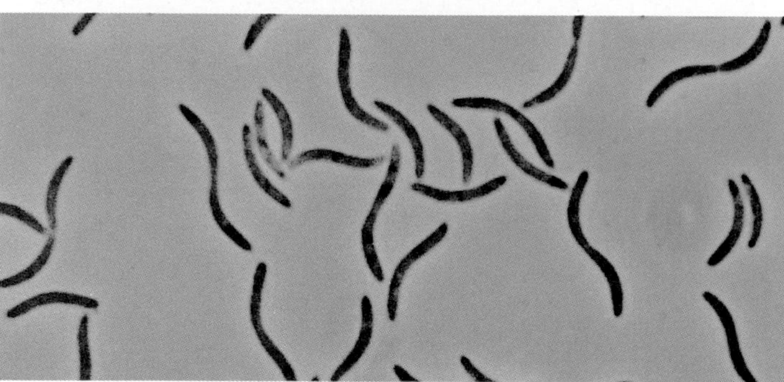

(c)

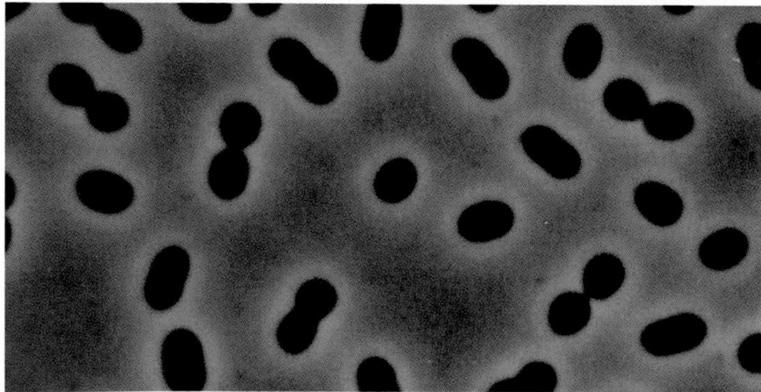

(d)

Figure 21.17 The Dissimilatory Sulfate- or Sulfur-reducing Bacteria. Representative examples. (*a*) Negatively stained *Desulfuromonas acetoxidans* with spores and gas vesicles, phase contrast (×2,000). (*b*) Phase-contrast micrograph of *Desulfovibrio saprovorans* with PHB inclusions (×2,000). (*c*) *Desulfovibrio gigas*, phase contrast (×2,000). (*d*) *Desulfobacter postgatei*, phase contrast (×2,000).

industry because of their primary role in the anaerobic corrosion of iron in pipelines, heating systems, and other structures (*see chapter 44*).

The sulfur cycle (p. 810).

1. Describe the metabolic specialization that characterizes the bacteria in section 7 of *Bergey's Manual*.
2. Explain the distribution of these bacteria in terms of their metabolic properties. Why are they so important?

Anaerobic Gram-Negative Cocci

The family *Veillonellaceae* contains anaerobic, chemoheterotrophic cocci ranging in diameter from about 0.3 to 2.5 μm. Usually, they are diplococci (often with their adjacent sides flattened), but they may exist as single cells, clusters, or chains. All have complex nutritional requirements and ferment substances such as carbohydrates, lactate and other organic acids, and amino acids to produce gas (CO_2 and often H_2) plus a mixture of volatile fatty acids. They are parasites of homeothermic (warm-blooded) animals.

Like many groups of anaerobic bacteria, members of this family have not been thoroughly studied. They are part of the normal flora of the mouth, the gastrointestinal tract, and the urogenital tract of humans and other animals. For example, *Veillonella* is plentiful on the tongue surface and dental plaque of humans, and it can be isolated about 10 to 20% of the time from the vagina. Gram-negative, anaerobic cocci are found in infections of the head, lungs, and the female genital tract, but their precise role in such infections is unclear.

1. What properties distinguish the family *Veillonellaceae*?

The Rickettsias and Chlamydiae

Section 9 of *Bergey's Manual* contains two orders—the *Rickettsiales* and *Chlamydiales*—three families, and ten genera. Despite this diversity, almost all these gram-negative bacteria are obligate intracellular parasites: they only grow and reproduce within host cells. These bacteria are about the same size as the poxviruses (*p. 362*) and resemble viruses in their intracellular existence. They differ from viruses in having both DNA and RNA, a plasma membrane, functioning ribosomes, enzymes participating in metabolic pathways, reproduction by binary fission, and other features.

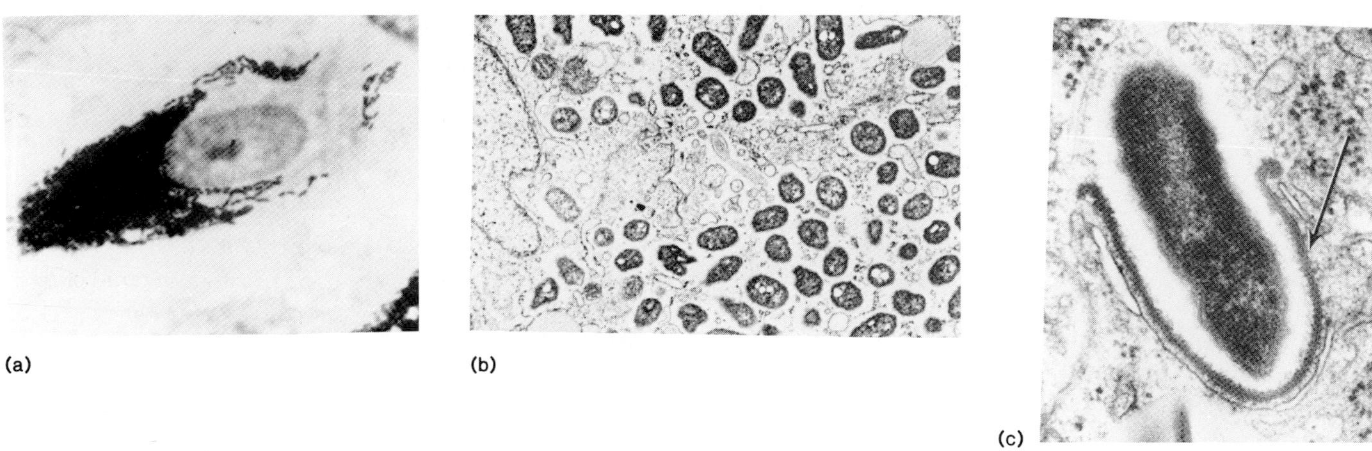

Figure 21.18 **The *Rickettsiales*.** Rickettsial morphology and reproduction. (*a*) A human fibroblast filled with *Rickettsia prowazekii* (×1,200). (*b*) A chicken embryo fibroblast late in infection with free cytoplasmic *R. prowazekii* (×13,600). (*c*) *R. prowazekii* leaving a disrupted phagosome (*arrow*) and entering the cytoplasmic matrix (×46,000).

Order *Rickettsiales*

Almost all members of the order *Rickettsiales* are rod-shaped, coccoid, or pleomorphic bacteria with typical gram-negative walls and no flagella. Although their size varies, these bacteria tend to be very small. For example, *Rickettsia* is 0.3 to 0.5 μm in diameter and 0.8 to 2.0 μm long; *Coxiella* is 0.2 to 0.4 μm by 0.4 to 1.0 μm. All are parasitic or mutualistic (*see chapter 29*). The parasitic forms grow in vertebrate erythrocytes, reticuloendothelial cells (*see figure 30.10*), and vascular endothelial cells. Often, they also live in blood-sucking arthropods such as fleas, ticks, mites, or lice, which serve as vectors (*see chapter 35*) or primary hosts. The families *Rickettsiaceae* and *Bartonellaceae* have regular cell walls, and at least some species in each family can be cultured axenically in special media. The *Rickettsiaceae* grow only in nucleated cells, whereas the *Bartonellaceae* also can live in erythrocytes. The third family, the *Anaplasmataceae,* lacks a cell wall, cannot be cultured axenically, and grows only in erythrocytes.

Because of the importance of the *Rickettsiaceae* as human pathogens, their reproduction and metabolism have been intensively studied. Rickettsias enter the host cell by inducing phagocytosis. Members of the genus *Rickettsia* immediately escape the phagosome and reproduce by binary fission in the cytoplasm (figure 21.18). In contrast, *Coxiella* remains within the phagosome after it has fused with a lysosome and actually reproduces within the phagolysosome. Eventually the host cell bursts, releasing new organisms. Besides incurring damage from cell lysis, the host is harmed by the toxic effects of rickettsial cell walls (wall toxicity appears related to the mechanism of penetration into host cells).

Rickettsias are very different from most other bacteria in physiology and metabolism. They lack the glycolytic pathway and do not use glucose as an energy source, but rather oxidize glutamate and tricarboxylic acid cycle intermediates such as succinate. The rickettsial plasma membrane has carrier-mediated transport systems, and host cell nutrients and coenzymes are absorbed and directly used. For example, rickettsias take up both NAD and uridine diphosphate glucose. The membrane also has an adenylate exchange carrier that exchanges ADP for external ATP. Thus, host ATP may provide much of the energy needed for growth. This metabolic dependence explains why many of these organisms must be cultivated in the yolk sacs of chick embryos or in tissue culture cells.

This order contains many important pathogens. *Rickettsia prowazekii* and *R. typhi* are associated with typhus fever, and *R. rickettsii,* with Rocky Mountain spotted fever. *Coxiella burnetii* causes Q fever in humans. These diseases are discussed later in some detail (*see chapter 38*). It should be noted that rickettsias are also important pathogens of domestic animals such as dogs, horses, sheep, and cattle.

Order *Chlamydiales*

The order *Chlamydiales* has only one family, *Chlamydiaceae,* with one genus, *Chlamydia.* Chlamydiae are nonmotile, coccoid, gram-negative bacteria, ranging in size from 0.2 to

1.5 μm. They can reproduce only within cytoplasmic vesicles of host cells by a unique developmental cycle involving the formation of elementary bodies and reticulate bodies. Although their envelope resembles that of other gram-negative bacteria, the wall differs in lacking muramic acid and a peptidoglycan layer. They achieve osmotic stability by cross-linking their outer membrane proteins with disulfide bonds. Chlamydiae are extremely limited metabolically and are obligate intracellular parasites of mammals and birds. (However, chlamydia-like bacteria have recently been isolated from spiders, clams, and freshwater invertebrates.) The size of their genome is 4 to 6 × 10^8 daltons, one of the smallest of all procaryotes, and the G + C content is 41 to 44%.

Chlamydial reproduction begins with the attachment of an **elementary body (EB)** to the cell surface (figure 21.19). Elementary bodies are 0.2 to 0.4 μm in diameter, contain electron-dense nuclear material and a rigid cell wall, and are infectious (figure 21.20). The host cell phagocytoses the EB, which then prevents the fusion of lysosomes with the phagosome and begins to reorganize itself to form a **reticulate body (RB)** or **initial body.** The RB is specialized for reproduction rather than infection. Reticulate bodies are 0.6 to 1.5 μm in diameter and have less dense nuclear material and more ribosomes than EBs; their walls are also more flexible. About 8 to 10 hours after infection, the reticulate body begins to divide and RB reproduction continues until the host cell dies. A chlamydia-filled vacuole or inclusion can become large enough to be seen in a light microscope and even fill the host cytoplasm. After 20 to 25 hours, RBs begin changing back into infectious EBs and continue this process until the host cell lyses and releases the chlamydiae 48 to 72 hours after infection.

Chlamydial metabolism is very different from that of other gram-negative bacteria. Chlamydiae cannot catabolize carbohydrates or other substances and synthesize ATP. *Chlamydia psittaci,* one of the best-studied species, lacks both flavoprotein and cytochrome electron transport chain carriers, but does have a membrane translocase that acquires host ATP in exchange for ADP. Thus, chlamydiae seem to be energy parasites that are completely dependent upon their hosts for ATP. When supplied with precursors from the host, RBs can synthesize DNA, RNA, and protein. They also can synthesize at least some amino acids and coenzymes. The EBs lack much metabolic activity and cannot take in ATP or synthesize proteins. They seem to be dormant forms concerned exclusively with transmission and infection.

Figure 21.19 **The Chlamydial Elementary Body.** The elementary bodies of *Chlamydia trachomatis* infecting tissue culture cells, fluorescent stain (×1,000).

The two traditionally recognized chlamydial species are important pathogens of humans and other warm-blooded animals. *C. trachomatis* infects humans and mice. In humans, it causes trachoma, nongonococcal urethritis, and other diseases (*see chapter 38*). *C. psittaci* causes psittacosis in humans. However, unlike *C. trachomatis,* it also infects many other animals (e.g., parrots, turkeys, sheep, cattle, and cats) and invades the intestinal, respiratory, and genital tracts; the placenta and fetus; the eye; and the synovial fluid of joints. Recently, a new species has been discovered and named *Chlamydia pneumoniae.* It appears to be a major cause of human pneumonia.

1. Compare rickettsias and chlamydiae with viruses.
2. Give the major characteristics of the orders *Rickettsiales* and *Chlamydiales.*
3. Briefly describe the life cycle of rickettsias.
4. In what way does the physiology and metabolism of the rickettsias differ from that of other bacteria?
5. Name some important rickettsial diseases.
6. What are elementary and reticulate bodies? Briefly describe the steps in the chlamydial life cycle.
7. How does chlamydial metabolism differ from that of other bacteria, including the rickettsias?
8. List two or three chlamydial diseases of humans.

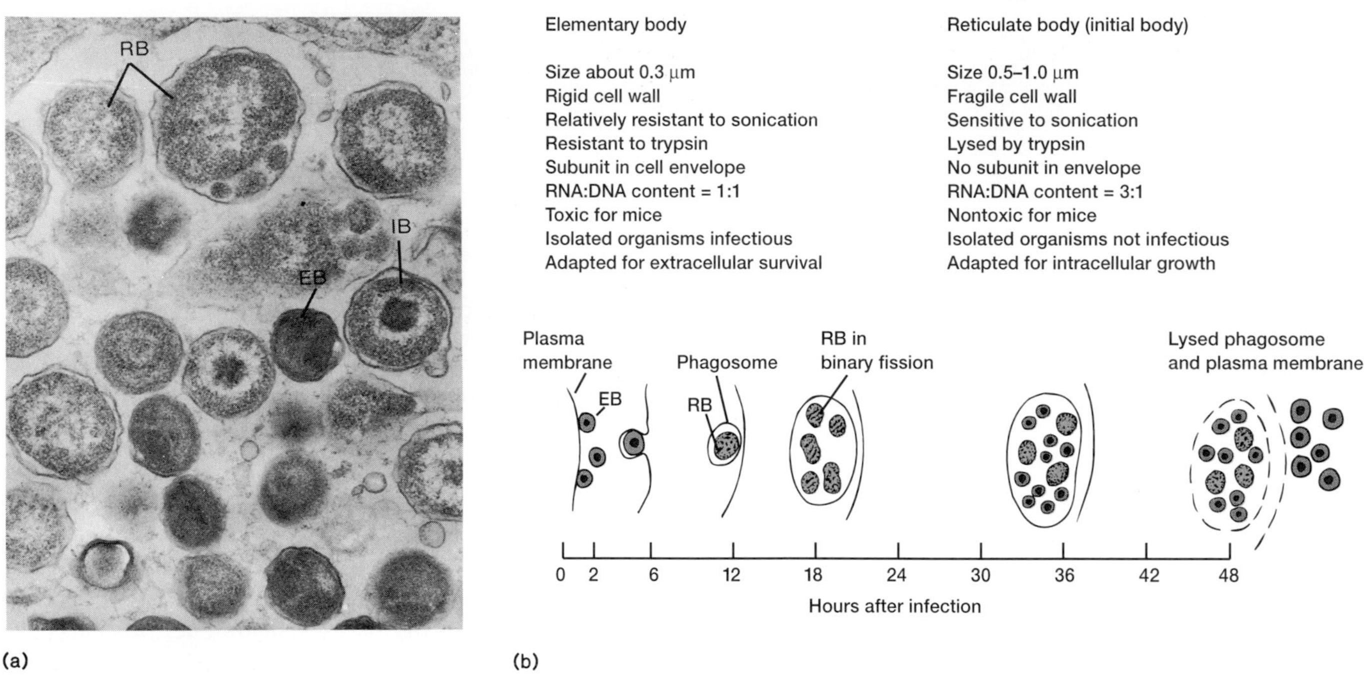

Figure 21.20 The Chlamydial Life Cycle. (*a*) An electron micrograph of a microcolony of *Chlamydia trachomatis* in the cytoplasm of a host cell (×160,000). Three developmental stages are visible: the elementary body, *EB;* reticulate body, *RB;* and intermediate body, *IB,* a chlamydial cell intermediate in morphology between the first two forms. (*b*) A schematic representation of the infectious cycle of chlamydiae.

The Mycoplasmas

Section 10 of *Bergey's Manual* describes the division *Tenericutes,* which contains only one class, *Mollicutes,* and one order, *Mycoplasmatales,* with three families and six genera. **Mycoplasmas** lack cell walls and cannot synthesize peptidoglycan precursors. Thus, they are penicillin resistant but susceptible to lysis by osmotic shock and detergent treatment. Because they are bounded only by a plasma membrane, these procaryotes are pleomorphic and vary in shape from spherical or pear-shaped organisms, about 0.3 to 0.8 μm in diameter, to branched or helical filaments (figure 21.21). They are the smallest bacteria capable of self-reproduction. Although most are nonmotile, some can glide along liquid-covered surfaces. Most species differ from the vast majority of bacteria in requiring sterols for growth. They usually are facultative anaerobes, but a few are obligate anaerobes. When growing on agar, most species will form colonies with a "fried egg" appearance because they grow into the agar surface at the center while spreading outward on the surface at the colony edges (figure 21.22). Their genome is one of the smallest found in procaryotes, about 5 to 10 $\times$ 10^8 daltons; the G + C content ranges from 23 to 41%. Mycoplasmas can be saprophytes, commensals (*see chapter 29*), or parasites, and many are pathogens of plants, animals, or insects.

Table 21.6 summarizes the taxonomy of the class *Mollicutes* as given in *Bergey's Manual.* Families and genera are distinguished primarily by means of the following properties: shape, the requirement for cholesterol, NADH oxidase location, the urease reaction, habitat, and the effect of oxygen and temperature. In addition, the family *Acholeplasmataceae* can synthesize fatty acids from acetate, but the family *Spiroplasmataceae* cannot.

It is currently believed that the mollicutes arose from gram-positive bacteria, probably from the clostridia. *Bergey's Manual* discusses the genus *Thermoplasma* in the context of both the mycoplasmas (volume 1) and the archaeobacteria (volume 3). Recent evidence shows that it is a member of the archaeobacteria, and not a mycoplasma.

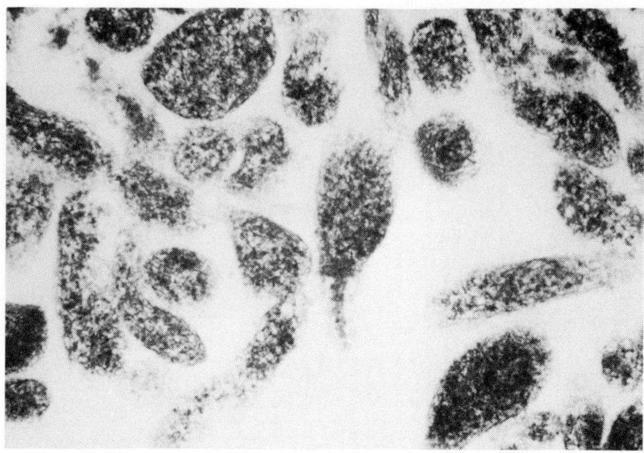

(a)

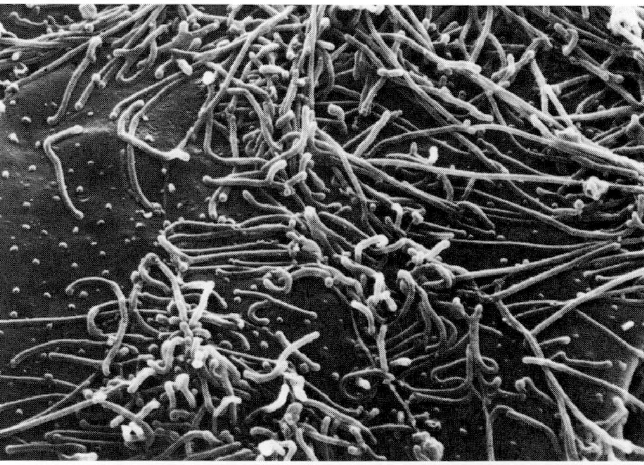

(b)

Figure 21.21 **The Mycoplasmas.** Electron micrographs of *Mycoplasma pneumoniae* showing its pleomorphic nature. (*a*) A transmission electron micrograph of several cells (×47,880). (*b*) A scanning electron micrograph (×26,000).

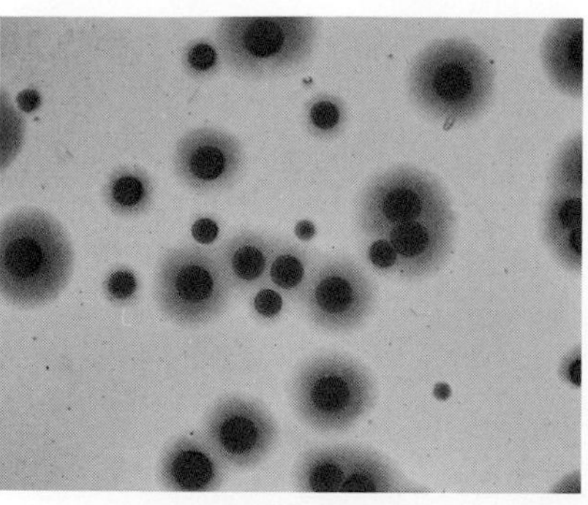

Figure 21.22 ***Mycoplasma* Colonies.** Note the "fried egg" appearance, stained (×100).

The metabolism of mycoplasmas is not particularly unusual, although they are deficient in several biosynthetic sequences and often require sterols, fatty acids, vitamins, amino acids, purines, and pyrimidines. Those genera needing sterols incorporate them into the plasma membrane. Mycoplasmas are usually more osmotically stable than eubacterial protoplasts, and their membrane sterols may be a stabilizing factor. Some produce ATP by electron transport, and others produce it by the Embden-Meyerhof pathway and lactic acid fermentation.

Mycoplasmas appear to be very important, even though their roles in nature and disease are not yet completely clear. They are remarkably widespread and can be isolated from animals, plants, the soil, and even compost piles. Indeed, about 10% of the cell cultures in use are probably contaminated with mycoplasmas, which seriously interfere with tissue culture experiments and are difficult to detect and eliminate. In animals, mycoplasmas colonize mucous membranes and joints and are often associated with diseases of the respiratory and urogenital tracts. Mycoplasmas cause several major diseases in livestock, for example, contagious bovine pleuropneumonia in cattle (*M. mycoides*), chronic respiratory disease in chickens (*M. gallisepticum*), and pneumonia in swine (*M. hyopneu-

TABLE 21.6 Taxonomy and Properties of Organisms Included in the Class *Mollicutes*

Classification	Current Number of Recognized Species	Genome Size (10^8 daltons)	Mol% G + C of DNA	Cholesterol Requirement	Location of NADH Oxidase: Cyt[a]	Location of NADH Oxidase: Mem	Distinctive Properties	Habitat
Order I. *Mycoplasmatales*								
Family I. *Mycoplasmataceae*								
Genus I. *Mycoplasma*	87	4–8	23–41	+	+	−		Animals
Genus II. *Ureaplasma*	5	5–7	27–30	+			Urease positive	Animals
Family II. *Acholeplasmataceae*								
Genus I. *Acholeplasma*	11	10	27–36	−	−	+		Animals and plants
Family III. *Spiroplasmataceae*								
Genus I. *Spiroplasma*	11	10	25–31	+	+	−	Helical filaments	Arthropods and plants
Genera of Uncertain Taxonomic Position								
Genus *Anaeroplasma*	4	10	29–33	+			Anaerobic; some digest bacteria	Rumens of cattle and sheep
Genus *Asteroleplasma*	1	10	40	−			Oxygen sensitive	Rumens of cattle and sheep

Modified from *Bergey's Manual of Systematic Bacteriology,* Vol. 1, p. 741, J. G. Holt and N. R. Krieg (eds.).

[a]Abbreviations: *Cyt,* cytoplasm; *Mem,* plasma membrane; *NADH,* reduced nicotinamide adenine dinucleotide.

moniae). *M. pneumoniae* causes primary atypical pneumonia in humans, and there is increasing evidence that *M. hominis* and *Ureaplasma urealyticum* are also human pathogens. Spiroplasmas have been isolated from insects, ticks, and a variety of plants. They cause disease in citrus plants, cabbage, broccoli, corn, honey bees, and other hosts. Arthropods probably often carry the spiroplasmas between plants. Presumably many more pathogenic mollicutes will be discovered as techniques for their isolation and study improve.

1. What morphological feature distinguishes the mycoplasmas? In what division and class are they found?
2. Give other distinguishing properties of the mycoplasmas.
3. What might mycoplasmas use sterols for?
4. Where are mycoplasmas found in animals? List several animal and human diseases caused by them. What kinds of organisms do spiroplasmas usually infect?

Summary

1. The spirochetes are slender, long, helical, gram-negative bacteria that are motile because of the axial filament underlying an outer sheath or outer membrane.
2. Section 2 of *Bergey's Manual* contains genera that are helical or vibrioid, motile, respiratory, and aerobic or microaerophilic (e.g., *Campylobacter* and *Bdellovibrio*).
3. Section 3 has one family and several other genera of nonmotile (or rarely motile), curved, gram-negative bacteria.
4. Section 4 of *Bergey's Manual* contains a large and diverse collection of gram-negative aerobic rods and cocci.
5. The genus *Pseudomonas* contains straight or slightly curved, gram-negative, aerobic rods that are motile by one or several polar flagella and do not have prosthecae or sheaths.
6. The pseudomonads participate in the mineralization process in nature, are major experimental subjects, cause many diseases, and often spoil refrigerated food.
7. *Azotobacter* and *Rhizobium* carry out nitrogen fixation, while *Agrobacterium* causes the development of plant tumors.
8. The *Methylococcaceae* are methylotrophs; they use methane and methanol as their sole carbon and energy sources.
9. Section 5 of *Bergey's Manual* contains facultatively anaerobic, gram-negative rods divided into three families and seven other genera.
10. The *Enterobacteriaceae,* often called enterobacteria or enteric bacteria, are gram-negative, peritrichously flagellated or nonmotile, facultatively anaerobic, straight rods with simple nutritional requirements.
11. The enteric bacteria are usually identified with a variety of physiological tests and are very important experimental organisms and pathogens of plants and animals.
12. The two other families in section 5 are the oxidase-positive *Vibrionaceae* and the parasitic *Pasteurellaceae.*
13. Obligately anaerobic, gram-negative rods are placed in the family *Bacteroidaceae* when they do not reduce sulfate or other sulfur compounds. Some of these bacteria benefit their hosts (e.g., rumen inhabitants), whereas others are pathogenic.
14. Section 7 of *Bergey's Manual* contains gram-negative bacteria that are anaerobic and can use elemental sulfur and oxidized sulfur compounds as electron acceptors in anaerobic respiration. They are very important in sulfur cycling in the ecosystem.
15. The family *Veillonellaceae* contains anaerobic, gram-negative cocci that are parasites of vertebrates.
16. Section 9 of *Bergey's Manual* describes the orders *Rickettsiales* and *Chlamydiales,* all obligately intracellular parasites responsible for many diseases.
17. Rickettsias have many plasma membrane carriers and make extensive use of host cell nutrients, coenzymes, and ATP.
18. Chlamydiae are nonmotile, coccoid, gram-negative bacteria that reproduce within the cytoplasmic vacuoles of host cells by a life cycle involving elementary bodies (EBs) and reticulate bodies (RBs). They are energy parasites and cannot make ATP.
19. Mycoplasmas are gram-negative bacteria that lack cell walls and cannot synthesize peptidoglycan precursors. Many species require sterols for growth. They are the smallest bacteria capable of self-reproduction and usually grow on agar to give colonies a "fried-egg" appearance.

Key Terms

axial fibrils or periplasmic flagella *428*
axial filament *428*
bioluminescence *439*
elementary body (EB) *444*
enteric bacteria (enterobacteria) *435*
enteritis *430*
initial body *444*
methylotroph *435*
mineralization process *434*
mycoplasmas *445*
reticulate body (RB) *444*
septicemia *430*

Questions for Thought and Review

1. Suppose that you were a microbiologist working on the Mars probe project. How would you determine whether microorganisms are living on the planet, and why?
2. How does *Bdellovibrio* take advantage of its host's metabolism to grow rapidly and efficiently? In what ways does its reproduction resemble that of bacteriophages?
3. Why might the ability to form dormant cysts be of great advantage to *Agrobacterium* but not as much to *Rhizobium*?
4. Some bacterial groups contain a great diversity of bacteria, whereas other groups have fewer species that resemble each other very closely. What might account for this?
5. What advantage might a species gain by specializing in its nutrient requirements or habitat as methylotrophs and halobacteria do? Are there disadvantages to this strategy?
6. Why are biochemical tests of such importance in identifying enteric bacteria?
7. A pond with black water and a smell like rotten eggs might contain an excess of what kind of bacteria?
8. How might one try to culture outside the host a newly discovered rickettsia or chlamydia?

Additional Reading

Balows,A.; Trüper, H. G.; Dworkin, M.; Harder, W.; and Schleifer, K.-H. 1992. *The prokaryotes,* 2d ed. New York: Springer-Verlag.

Canale-Parola, E. 1978. Motility and chemotaxis of spirochetes. *Ann. Rev. Microbiol.* 32:69–99.

Chilton, M-D. 1983. A vector for introducing new genes into plants. *Sci. Am.* 248(6):51–59.

Diedrich, D. L. 1988. Bdellovibrios: Recycling, remodelling and relocalizing components from their prey. *Microbiol. Sci.* 5(4):100–103.

Haber, C. L.; Allen, L. N.; Zhao, S.; and Hanson, R. S. 1983. Methylotrophic bacteria: Biochemical diversity and genetics. *Science* 221:1147–53.

Harwood, C. S., and Canale-Parola, E. 1984. Ecology of spirochetes. *Ann. Rev. Microbiol.* 38:161–92.

Hase, T. 1985. Developmental sequence and surface membrane assembly of rickettsiae. *Ann. Rev. Microbiol.* 39:69–88.

Holt, J. G., editor-in-chief. 1984. *Bergey's manual of systematic bacteriology,* vol. 1, ed. N. R. Krieg. Baltimore, Md.: Williams & Wilkins.

Holt, S. C. 1978. Anatomy and chemistry of spirochetes. *Microbiol. Rev.* 42(1):114–60.

MacDonell, M. T.; Swartz, D. G.; Ortiz-Conde, B. A.; Last, G. A.; and Colwell, R. R. 1986. Ribosomal RNA phylogenies for the vibrio-enteric group of eubacteria. *Microbiol. Sci.* 3:172–78.

Maniloff, J. 1983. Evolution of wall-less procaryotes. *Ann. Rev. Microbiol.* 37:477–99.

Meighen, E. A. 1991. Molecular biology of bacterial bioluminescence. *Microbiol. Rev.* 55(1):123–42.

Moulder, J. W. 1984. Looking at chlamydiae without looking at their hosts. *ASM News* 50(8):353–62.

Moulder, J. W. 1991. Interaction of chlamydiae and host cells in vitro. *Microbiol. Rev.* 55(1):143–90.

Salyers, A. A. 1984. *Bacteroides* of the human lower intestinal tract. *Ann. Rev. Microbiol.* 38:293–313.

Schachter, J., and Caldwell, H. D. 1980. Chlamydiae. *Ann. Rev. Microbiol.* 34:285–309.

Stanbridge, E. J. 1976. A reevaluation of the role of mycoplasmas in human disease. *Ann. Rev. Microbiol.* 30:169–87.

Starr, M. P., and Chatterjee, A. K. 1972. The genus *Erwinia*: Enterobacteria pathogenic to plants and animals. *Ann. Rev. Microbiol.* 26:389–426.

Starr, M. P., and Seidler, R. J. 1971. The bdellovibrios. *Ann. Rev. Microbiol.* 25:649–78.

Walker, R. I., et al. 1986. Pathophysiology of *Campylobacter* enteritis. *Microbiol. Rev.* 50(1):81–94.

Whitcomb, R. F. 1980. The genus *Spiroplasma. Ann. Rev. Microbiol.* 34:677–709.

Winkler, H. H. 1990. Rickettsia species (as organisms). *Ann. Rev. Microbiol.* 44:131–53.

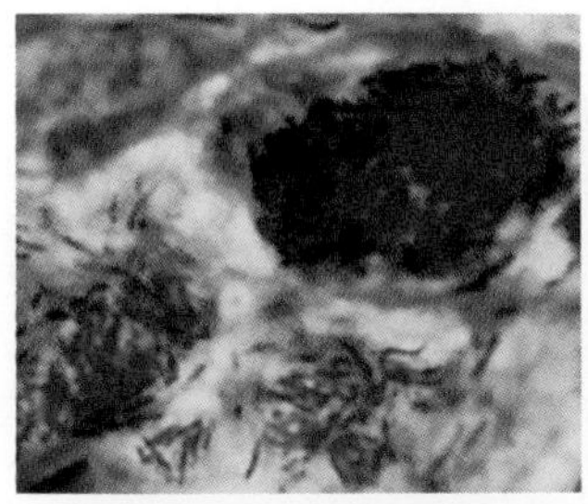

CHAPTER 22
The Bacteria: Gram-Positive Bacteria Other Than Actinomycetes

We noted, after having grown the bacterium through a series of such cultures, each fresh culture being inoculated with a droplet from the previous culture, that the last culture of the series was able to multiply and act in the body of animals in such a way that the animals developed anthrax with all the symptoms typical of this affection.

Such is the proof, which we consider flawless, that anthrax is caused by this bacterium.

—Louis Pasteur

Outline

Concepts

1. Volume 2 of *Bergey's Manual* contains six sections covering all gram-positive bacteria except the actinomycetes. Bacteria are distributed among these sections on the basis of their shape, the ability to form endospores, acid fastness, oxygen relationships, the ability to temporarily form mycelia, and other properties.
2. Peptidoglycan structure varies among different groups in ways that are often useful in identifying specific groups.
3. The six endospore-forming genera are probably not closely related and are grouped together for convenience.
4. Mycobacteria can form filaments, but unlike the actinomycetes, their filaments are easily fragmented into rods or coccoid elements. Actinomycete hyphae are more permanent.
5. Although most gram-positive bacteria are harmless free-living saprophytes, species from all six sections of volume 2 of *Bergey's Manual* are major pathogens of humans, other animals, and plants. Other gram-positive bacteria are very important in the food and dairy industries.

Chapter 22 surveys the bacteria found in volume 2 of *Bergey's Manual of Systematic Bacteriology.* The chapter describes all major groups of gram-positive bacteria except the actinomycetes, which are covered in chapter 25. As in chapter 21, much of this chapter discusses the most significant biological features of each group and describes representatives selected because of their practical importance or their extensive use in microbiological research.

Volume 2 of *Bergey's Manual* contains six sections that describe all gram-positive bacteria except the actinomycetes. Most of these bacteria are distributed among the first four sections on the basis of their general shape (whether they are rods, cocci, or irregular) and their ability to form endospores. The rod-shaped, acid-fast bacteria known as mycobacteria are placed in section 16 (the sections in *Bergey's Manual* are numbered consecutively, beginning with the first section in volume 1). The last section describes the nocardioforms. These are gram-positive filamentous bacteria that form a branching network called a mycelium, which can break into rod-shaped or coccoid forms. The nocardioforms are discussed in more detail in volume 4, section 26, of *Bergey's Manual.* Therefore they are described in chapter 25, which covers the actinomycetes.

The gram-positive bacteria are placed in the division *Firmicutes* as previously noted (*see chapter 20*). Analysis of the phylogenetic relationships within the gram-positive bacteria by comparison of 16S rRNA sequences (*see figure 20.9*) shows that they are divided into a low G + C group and high G + C or actinomycete group. The distribution of genera within and between the groups shown in figure 20.9 differs markedly from the classification in volume 2 of *Bergey's Manual.* Thus the classification system given in volume 2 of the manual is a practical, convenient one that probably does not always reflect true phylogenetic relatedness. At points, the treatment in this chapter will differ from that given in *Bergey's Manual.*

Bacterial evolution and taxonomy (pp. 415–23).
The structure of the gram-positive cell wall (pp. 51–53).

Peptidoglycan structure varies considerably among different gram-positive groups. Most gram-negative bacteria have a peptidoglycan structure in which *meso*-diaminopimelic acid in position 3 is directly linked through its free amino group with the free carboxyl of the terminal D-alanine of an adjacent peptide chain (figure 22.1*a*; *see also figure 3.18*). This same peptidoglycan structure is present in many gram-positive genera, for example, *Bacillus, Clostridium, Lactobacillus, Corynebacterium, Mycobacterium,* and *Nocardia.* In other gram-positive bacteria, lysine is substituted for diaminopimelic acid in position 3, and the peptide subunits of the glycan chains are cross-linked by interpeptide bridges containing monocarboxylic L-amino acids or glycine, or both (figure 22.1*b*). Many genera including *Staphylococcus, Streptococcus, Micrococcus, Lactobacillus,* and *Leuconostoc* have peptidoglycan of this structural type. The genus *Streptomyces* and several other actinomycete genera have replaced *meso*-diaminopimelic acid with L, L-diaminopimelic acid in position 3 and have one glycine residue as the interpeptide bridge. The plant pathogenic corynebacteria provide a final example of peptidoglycan variation. In some of these bacteria, the interpeptide bridge connects positions 2 and 4 of the peptide subunits rather than 3 and 4 (figure 22.1*c*). Because the interpeptide bridge connects the carboxyl groups of glutamic acid and alanine, a diamino acid such as ornithine is used in the bridge. Many other variations in peptidoglycan structure are found, including other interbridge structures and large differences in the frequency of cross-linking between glycan chains. Bacilli and most gram-negative bacteria have fewer cross-links between chains than do gram-positive bacteria such as *Staphylococcus aureus* in which almost every muramic acid is cross-linked to another. These structural variants are often characteristic of particular groups and are therefore taxonomically useful.

1. Describe, in a diagram, the chemical composition and structure of the peptidoglycan found in gram-negative bacteria and many gram-positive genera.
2. Briefly discuss the ways in which three other peptidoglycan types differ from the gram-negative peptidoglycan.
3. How do bacilli and most gram-negative bacteria differ from gram-positive bacteria such as *S. aureus* with respect to cross-linking frequency?

Gram-Positive Cocci

Although section 12 of *Bergey's Manual* contains only chemoheterotrophic, mesophilic, nonsporing gram-positive cocci, the two families and 15 genera are phylogenetically very diverse. Genera are distinguished according to properties such as oxygen relationships, cell arrangement, the presence of catalase and cytochromes, peptidoglycan structure, and G + C content (table 22.1). A few of the most important groups are now described.

The *Micrococcaceae*

The family *Micrococcaceae* contains gram-positive cocci, 0.5 to 2.5 μm in diameter, that divide in more than one plane to form regular or irregular clusters of cells. All are aerobic or facultatively anaerobic. The peptidoglycan diamino acid is L-lysine. The two most important genera are *Micrococcus* and *Staphylococcus.* These genera differ in such aspects as cell arrangement, oxygen relationships, ability to ferment glucose, the presence of oxidase and teichoic acid, and G + C content (tables 22.1 and 22.2).

The genus *Micrococcus* contains aerobic, catalase-positive cocci that occur mainly in pairs, tetrads, or irregular clusters and are usually nonmotile. Micrococci are often yellow,

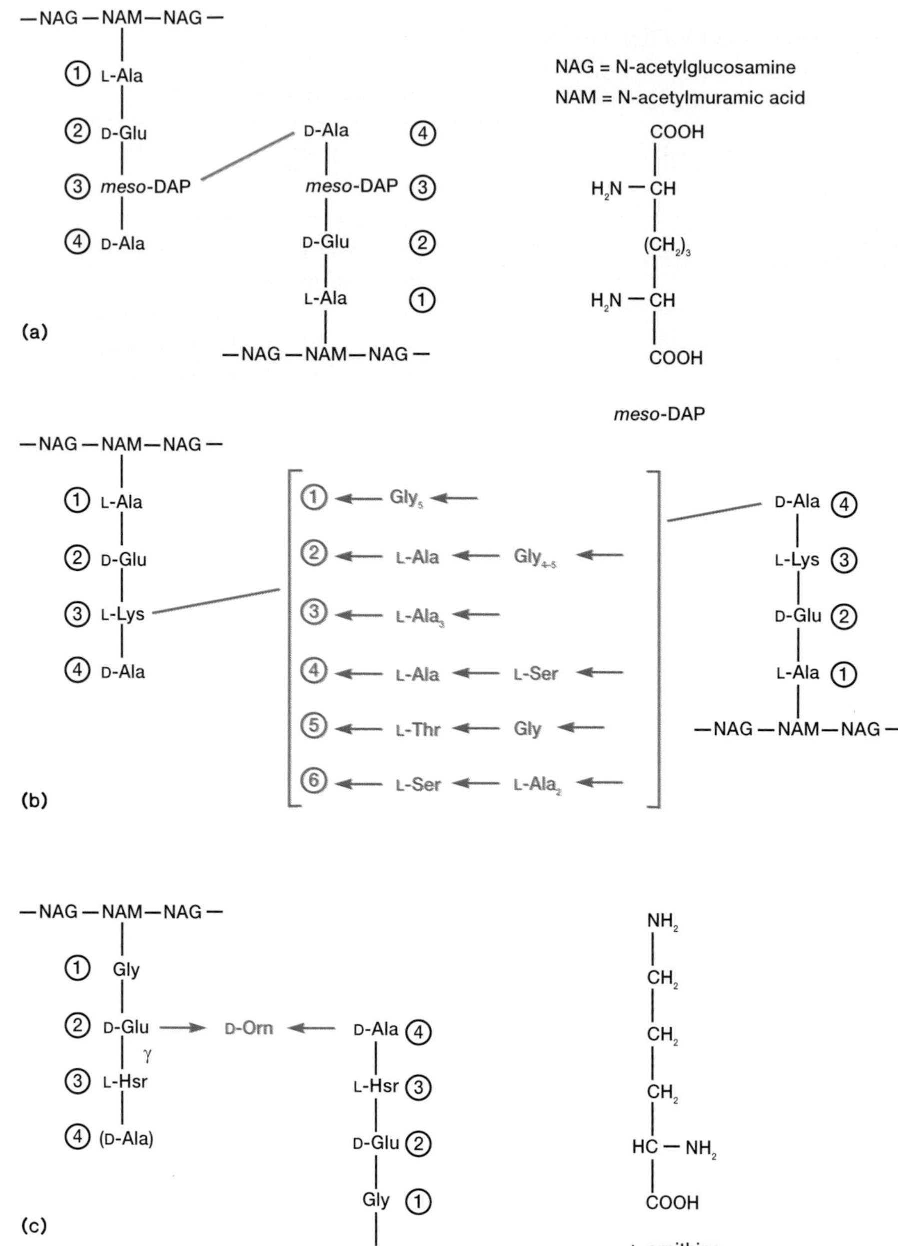

Figure 22.1 **Representative Examples of Peptidoglycan Structure.** (*a*) The peptidoglycan, with a direct cross-linkage between positions 3 and 4 of the peptide subunits, that is present in most gram-negative and many gram-positive bacteria. (*b*) Peptidoglycan with lysine in position 3 and an interpeptide bridge. The bracket contains six typical bridges: (*1*) *Staphylococcus aureus,* (*2*) *S. epidermidis,* (*3*) *Micrococcus roseus* and *Streptococcus thermophilus,* (*4*) *Lactobacillus viridescens,* (*5*) *Streptococcus salvarius,* and (*6*) *Leuconostoc cremoris.* The arrows indicate the polarity of peptide bonds running in the C to N direction. (*c*) An example of the cross-bridge extending between positions 2 and 4 from *Corynebacterium poinsettiae.* The interbridge contains an L-diamino acid like ornithine, and L-homoserine (L-Hsr) is in position 3. The abbreviations and structures of amino acids in the figure are found in appendix I. See text for more detail.

TABLE 22.1 Differential Properties of Representative Genera of Gram-Positive Cocci

	Aerobic Genera		Facultatively Anaerobic Genera			Anaerobic
Characteristics	*Micrococcus*	*Deinococcus*	*Staphylococcus*	*Streptococcus, Enterococcus, Lactococcus*	*Leuconostoc*	*Sarcina*
Predominant arrangement of cells, commonest listed first[a]	Clusters, tetrads	Pairs, tetrads	Clusters, pairs	Chains, pairs	Pairs, chains	Cuboidal packets
Strict aerobes	+[b]	+	−	−	−	−
Facultative anaerobes or microaerophiles	−	−	+	+	+	−
Strict anaerobes	−	−	−	−	−	+
Catalase reaction	+	+	+	−	−	−
Cytochromes present	+	+	+	−	−	ND
Major anaerobic carbohydrate fermentation products	NA	NA	Lactate	Lactate	Lactate, ethanol	Acetate, ethanol, H_2, CO_2, or butyrate
Mol% G + C of DNA	64–75	62–70	30–39	34–46	38–44	28–31

Modified from K. H. Schleifer, *Bergey's Manual of Systematic Bacteriology,* Vol. 2, edited by P. H. A. Sneath, N. S. Mair, and M. E. Sharpe. Copyright © 1986 Williams and Wilkins Co., Baltimore, MD. Reprinted by permission.

[a]Only arrangements other than single cells are listed.

[b]Symbols: +, 90% or more of strains positive; −, 10% or less of strains positive; NA, not applicable; ND, not determined.

TABLE 22.2 Properties of the Genera *Micrococcus* and *Staphylococcus*

Characteristics	*Micrococcus*	*Staphylococcus*
Irregular clusters	+[a]	+
Tetrads	+	−
Capsule	−	−
Motility	−	−
Anaerobic glucose fermentation	−	+
Oxidase and benzidine tests	+	−
Resistance to lysostaphin	R	S
Glycine present in peptidoglycan	−	+
Teichoic acid present in cell wall	−	+
Mol% G + C of DNA	64–75	30–39

Modified from K. H. Schleifer, *Bergey's Manual of Systematic Bacteriology,* Vol. 2, edited by P. H. A. Sneath, N. S. Mair, and M. E. Sharpe. Copyright © 1986 Williams and Wilkins Co., Baltimore, MD. Reprinted by permission.

[a]Symbols: +, 90% or more of strains positive; −, 10% or less of strains positive; R, resistant; and S, sensitive.

orange, or red in color. They are widespread in soil, water, and on mammalian skin; the last habitat may be their normal one. Despite their frequent presence on the skin, micrococci do not seem particularly pathogenic. Their G + C content, cell wall composition, and 16S rRNA sequence indicate that they may be more closely related to the genus *Arthrobacter* than to other gram-positive cocci.

Members of the genus *Staphylococcus* are facultatively anaerobic, nonmotile, gram-positive cocci that usually form irregular clusters (figure 22.2*a,b*; *see also figure 3.1*a). Like the micrococci, they are catalase positive but differ from the former in several ways: they are oxidase negative, ferment glucose anaerobically, and have teichoic acid in their cell walls and DNA with a much lower G + C content (30 to 39%). Staphylococci are normally associated with the skin, skin glands, and mucous membranes of warm-blooded animals. *S. epidermidis* is a common skin resident that is sometimes responsible for endocarditis and infections of patients with lowered resistance (e.g., wound infections, surgical infections, urinary tract infections). *S. aureus* is the most important human staphylococcal pathogen and causes boils, abscesses,

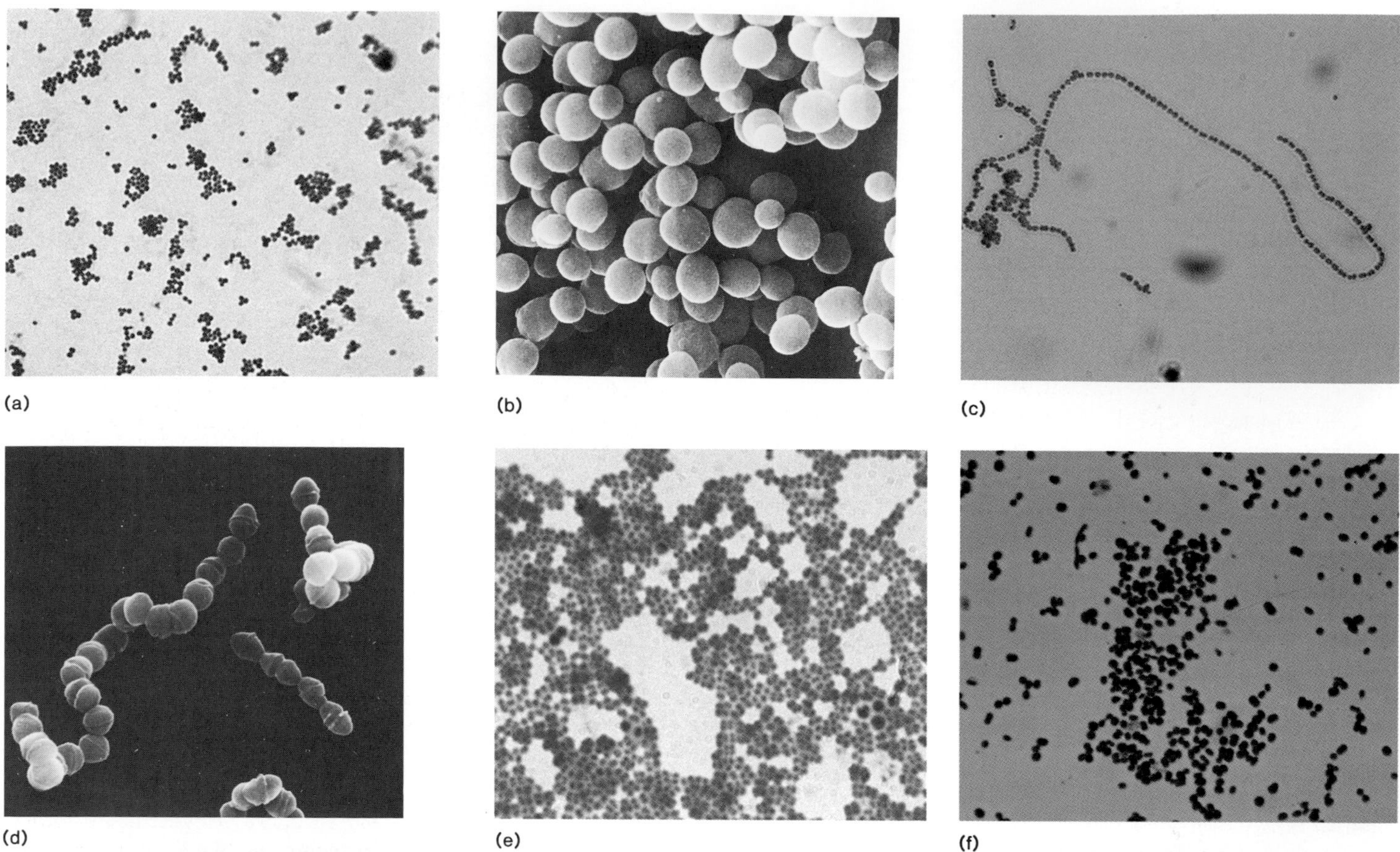

Figure 22.2 **Representative Gram-positive Cocci.** (*a*) *Staphylococcus aureus,* Gram-stained smear (×1,500). (*b*) Staphylococci arranged like a cluster of grapes, scanning electron micrograph (×34,000). (*c*) *Streptococcus pyogenes,* (×400). (*d*) Scanning electron micrograph of *Streptococcus* (×33,000). (*e*) *Micrococcus luteus.* Methylene blue stain (×1,000). (*f*) *Streptococcus pneumoniae* (×400).

wound infections, pneumonia, toxic shock syndrome, food poisoning, and other diseases. Unlike other common staphylococci, it produces the enzyme **coagulase,** which causes blood plasma to clot. Growth patterns on blood agar are also useful in identifying these staphylococci (figure 22.4). *S. aureus* usually grows on the nasal membranes and skin; it is also found in the gastrointestinal and urinary tracts of warm-blooded animals.

Staphylococcal diseases (pp. 756–59, 766–67).

In March 1986, an outbreak of acute gastrointestinal illness occurred following a buffet supper served to 855 people at a New Mexico country club. At least 67 people were ill with diarrhea, nausea, or vomiting, and 24 required emergency medical treatment or hospitalization. The problem was *S. aureus* growing in the turkey and dressing served at the buffet. One or more of the food handlers carried *S. aureus* and had contaminated the food. Because the turkey was cooled for three hours after cooking, there was sufficient time for the bacterium to grow and produce toxins (*see chapters 37 and 43*). This case is not uncommon. Turkey accounts for 10 to 21% of all bacterial food poisoning cases in which the source of poisoning is known, and it must be cooked and handled carefully.

The *Deinococcaceae*

The family *Deinococcaceae* contains only one genus and is quite different from other cocci in 16S rRNA sequence, cell wall structure, and other properties. The deinococci are gram-positive cocci associated in pairs or tetrads (figure 22.3) that are aerobic, catalase positive, and usually able to produce acid from few sugars. Unlike most gram-positive bacteria, their plasma membrane contains a large amount of palmitoleic acid and lacks phosphatidylglycerol phospholipids. The cell wall contains L-ornithine and has several distinct layers, including an outer membrane. Almost all strains of *Deinococcus* are unusually resistant to both desiccation and radiation.

Much remains to be discovered about the biology of these bacteria. Deinococci can be isolated from ground meat, feces, air, fresh water, and other sources, but their natural habitat is not yet known. The cause of their great radiation resistance is also not understood; it is possible to obtain radiation-sensitive mutants, and this should prove useful in clarifying the mechanism of resistance.

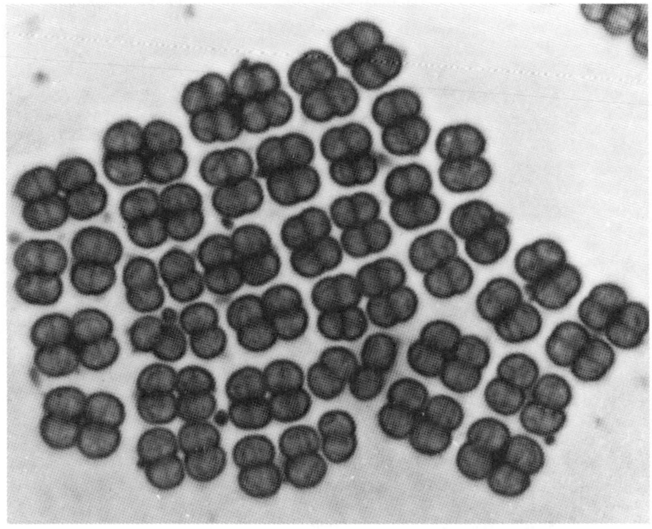

Figure 22.3 **The *Deinococcaceae*.** A *Deinococcus radiodurans* microcolony showing cocci arranged in tetrads (average cell diameter 2.5 μm).

Streptococcus

Section 12 *of Bergey's Manual* contains six genera of facultatively anaerobic or microaerophilic, catalase-negative, gram-positive cocci. The genus *Streptococcus* is an important member of this group. The streptococci occur in pairs or chains when grown in liquid media (figure 22.2*c,d; see also figure 3.1*b), do not form endospores, and are usually nonmotile. They are all chemoheterotrophs that ferment sugars with lactic acid, but no gas, as the major product; that is, they carry out homolactic fermentation (*see chapter 8*). A few species are anaerobic rather than facultative.

The genus is large and diverse, and it is difficult to satisfactorily classify these bacteria. Table 22.3 has been adapted from volume 2 of *Bergey's Manual* and represents the more traditional approach to their classification. The 29 species of genus *Streptococcus* are subdivided into five major groups: (1) pyogenic hemolytic streptococci, (2) oral streptococci, (3) enterococci, (4) lactic acid streptococci, and (5) anaerobic streptococci. Recently, the application of nucleic acid hybridization and sequencing techniques to these bacteria has produced a new classification system. The original single genus is now divided into three separate genera: *Streptococcus, Enterococcus,* and *Lactococcus*. Table 22.4 briefly summarizes the new system. The genus *Streptococcus* still contains most of the important human pathogens and has both pyogenic and oral subdivisions. *Enterococcus* and *Lactococcus* roughly correspond to the enterococci and lactic acid streptococci groups of the traditional classification (table 22.3).

Many characteristics are used to identify these cocci. One of their most important taxonomic characteristics is the ability to lyse erythrocytes when growing on blood agar, an agar medium containing 5% sheep or horse blood (figure 22.4). In **α-hemolysis,** a 1 to 3 mm greenish zone of incomplete hemolysis forms around the streptococcal colony; **β-hemolysis** is characterized by a zone of clearing or complete lysis without a marked color change. In addition, other hemolytic patterns are sometimes seen. Serological studies (*see chapters 33 and 34*) are also very important in identification because streptococci often have distinctive cell wall antigens. Polysaccharide and teichoic acid antigens found in the wall or between the wall and the plasma membrane are used to identify these cocci, particularly pathogenic β-hemolytic streptococci, by the **Lancefield grouping system** (*see Box 33.3*). Biochemical and physiological tests are essential in identification (e.g., growth temperature preferences, carbohydrate fermentation patterns, acetoin production, reduction of litmus milk, sodium chloride and bile salt tolerance, and the ability to hydrolyze arginine, esculin, hippurate, and starch). Sensitivity to bacitracin, sulfa drugs, and optochin (ethylhydrocuprein) are also used to identify particular species.

Table 22.3 Properties of Selected Streptococci and Relatives

Characteristics	Pyogenic Streptococci		Oral Streptococci		Enterococci	Lactic Acid Streptococci
	S. pyogenes	*S. pneumoniae*	*S. sanguis*	*S. mutans*	*S. faecalis*	*S. lactis*
Growth at 10°C	–[a]	–	–	–	+	+
Growth at 45°C	–	–	d	d	+	–
Growth at 6.5% NaCl	–	–	–	–	+	–
Growth at pH 9.6	–	–	–	–	+	–
Growth with 40% bile	–	–	d	d	+	+
α-hemolysis	–	+	+	–	–	d
β-hemolysis	+	–	–	–	+	–
Arginine hydrolysis	+	+	+	–	+	d
Hippurate hydrolysis	–	–	–	–	+	d
Mol% G + C of DNA	35–39	30–39	40–46	36–38	34–38	39

Modified from K. H. Schleifer, *Bergey's Manual of Systematic Bacteriology*, Vol. 2, edited by P. H. A. Sneath, N. S. Mair, and M. E. Sharpe. Copyright © 1986 Williams and Wilkins Co., Baltimore, MD. Reprinted by permission.

[a]Symbols: +, 90% or more of strains positive; –, 10% or less of strains positive; d, 11–89% of strains are positive.

Table 22.4 Classification of the Streptococci, Enterococci, and Lactococci

Characteristics	Streptococcus	Enterococcus	Lactococcus
Growth at 45°C	Variable	+	–
Growth at 10°C	–	Usually +	+
Growth in 6.5% NaCl broth	–	+	Variable
Growth at pH 9.6	–	+	–
Hemolysis	Usually β (pyogenic) or α (oral)	Usually –	Usually –
Serological group (Lancefield)	Variable (A–O)	Usually D	Usually N
Mol% G + C (normal range)	35–46	38–41	34–38
Representative species	Pyogenic streptococci *S. pyogenes* *S. equi* *S. dysgalactiae* Oral streptococei *S. gordonii* *S. salvarius* *S. sanguis* *S. oralis* *S. pneumoniae* *S. mitis* *S. mutans* Other streptococci *S. bovis* *S. saccharolyticus* *S. thermophilus* *S. suis*	*E. faecalis* *E. faecium* *E. avium* *E. durans* *E. gallinarum*	*L. lactis* *L. raffinolactis* *L. plantarum*

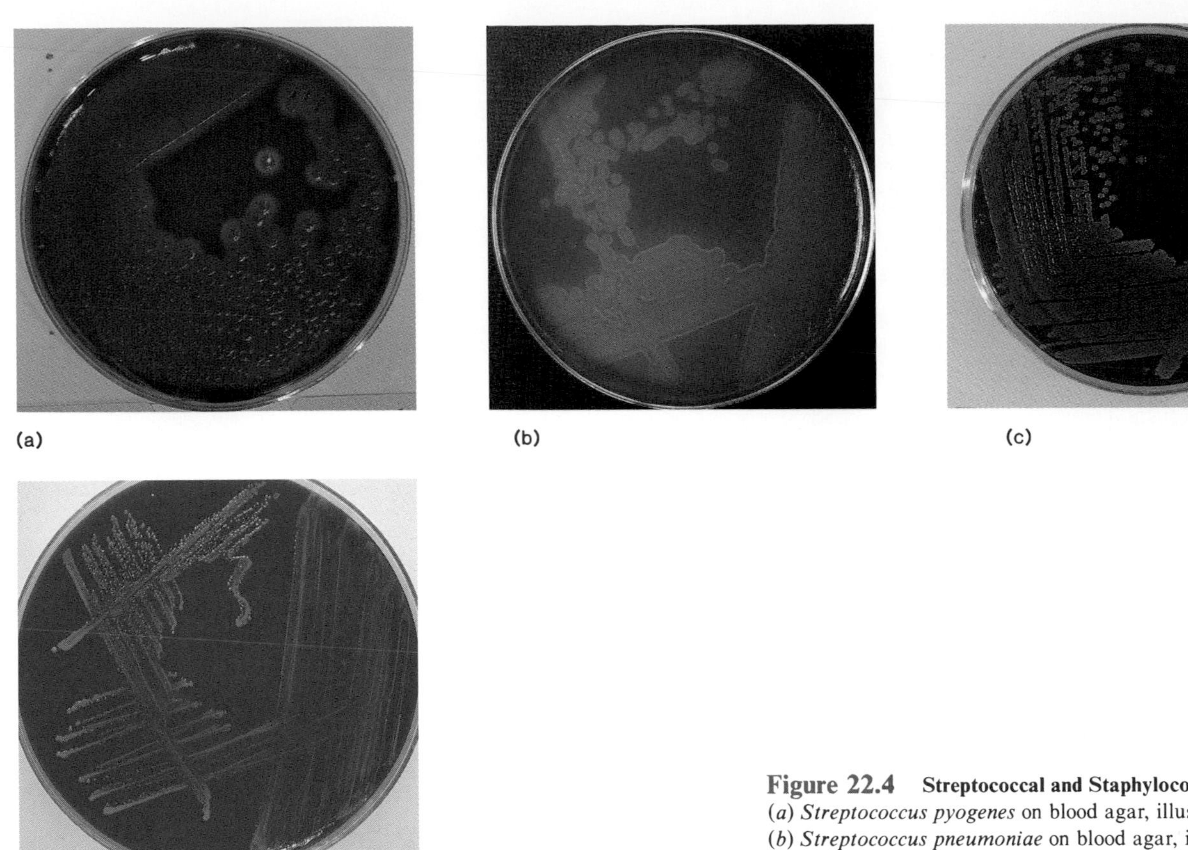

Figure 22.4 **Streptococcal and Staphylococcal Hemolytic Patterns.** (*a*) *Streptococcus pyogenes* on blood agar, illustrating β-hemolysis. (*b*) *Streptococcus pneumoniae* on blood agar, illustrating α-hemolysis. (*c*) *Staphylococcus aureus* on blood agar, illustrating β-hemolysis. (*d*) *Staphylococcus epidermidis* on blood agar with no hemolysis.

Members of the three genera have considerable practical importance. Pyogenic streptococci are usually pathogens and associated with pus formation (pyogenic means pus producing). Most species produce β-hemolysis on blood agar and form chains of cells. The major human pathogen in this group is *S. pyogenes* (streptococcal sore throat, acute glomerulonephritis, rheumatic fever). The normal habitat of oral streptococci is the oral cavity and upper respiratory tract of humans and other animals. In other respects, oral streptococci are not necessarily similar. *S. pneumoniae* is α-hemolytic and grows as pairs of cocci (figures 22.2*f* and 22.4*b*). It is associated with lobar pneumonia. *S. mutans* is associated with the formation of dental caries (*see chapter 38*). The enterococci such as *E. faecalis* are normal residents of the intestinal tracts of humans and most other animals. *E. faecalis* is an opportunistic pathogen that can cause urinary tract infections and endocarditis. Unlike the streptococci, the enterococci will grow in 6.5% sodium chloride. The lactococci ferment sugars to lactic acid and can grow at 10°C but not at 45°C. *L. lactis* is widely used in the production of buttermilk and cheese (*see chapter 43*) because it can curdle milk and add flavor through the synthesis of diacetyl and other products.

Streptococcal diseases (pp. 745–47).

Leuconostoc

The genus *Leuconostoc* contains facultative gram-positive cocci, which may be elongated or elliptical and arranged in pairs or chains (figure 22.5). Leuconostocs lack catalase and cytochromes and carry out **heterolactic fermentation** (*see chapter 8*) by converting glucose to D-lactate and ethanol or

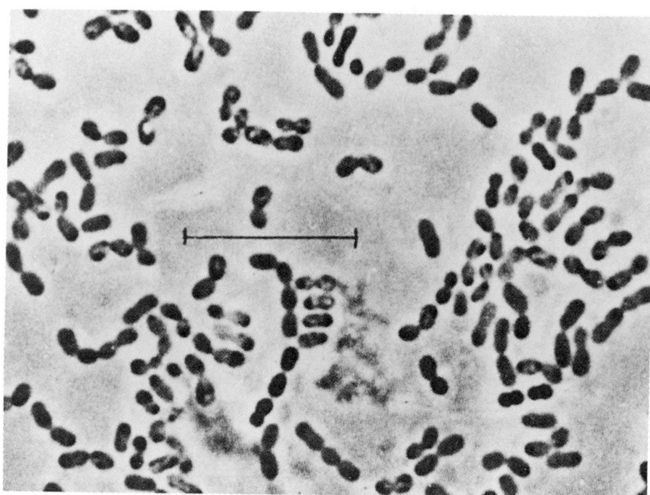

Figure 22.5 ***Leuconostoc.*** *Leuconostoc mesenteroides,* phase-contrast micrograph. Bar = 10 μm.

acetic acid by means of the phosphoketolase pathway (figure 22.6). They can be isolated from plants, silage, and milk. The genus is used in wine production, in the fermentation of vegetables such as cabbage (sauerkraut; *see figure 43.19*) and cucumbers (pickles), and in the manufacture of buttermilk, butter, and cheese. *L. mesenteroides* synthesizes dextrans from sucrose and is important in industrial dextran production. *Leuconostoc* species are involved in food spoilage and tolerate high sugar concentrations so well that they grow in syrup and are a major problem in sugar refineries.

A variety of gram-positive bacteria produce lactic acid as their major or sole fermentation product and are sometimes collectively called **lactic acid bacteria.** *Streptococcus, Enterococcus, Lactococcus,* and *Leuconostoc* are all members of this group. The genus *Lactobacillus* is also included (see p. 462). Lactic acid bacteria are nonsporing and usually nonmotile. They lack cytochromes and obtain energy by substrate-level phosphorylation rather than electron transport and oxidative phosphorylation. They normally depend on sugar fermentation for energy. Nutritionally, they are fastidious and many vitamins, amino acids, purines, and pyrimidines must be supplied because of their limited biosynthetic capabilities. Lactic acid bacteria are usually categorized as facultative anaerobes, but some classify them as aerotolerant anaerobes.

Oxygen relationships (pp. 126–28).

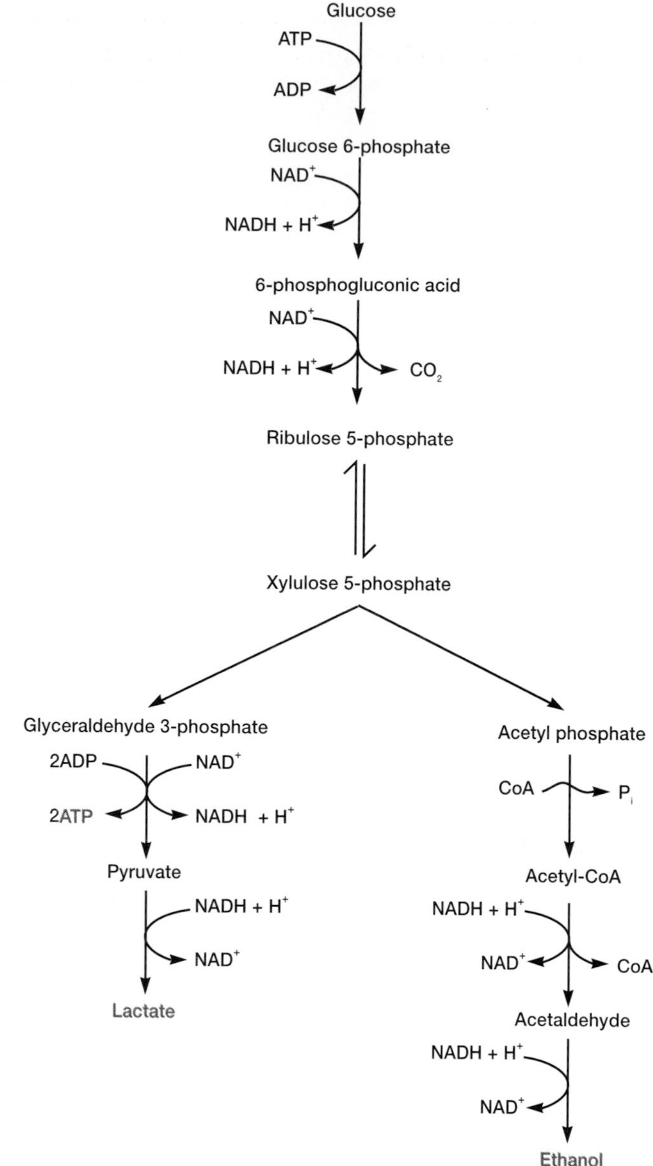

Figure 22.6 Heterolactic Fermentation and the Phosphoketolase Pathway. The phosphoketolase pathway converts glucose to lactate, ethanol, and CO_2.

1. Describe the major distinguishing characteristics of the following taxa: *Micrococcaceae, Micrococcus, Staphylococcus, Deinococcus, Streptococcus, Enterococcus, Lactococcus,* and *Leuconostoc.*
2. How does the pathogen *S. aureus* differ from the common skin resident *S. epidermidis,* and where is it normally found?
3. What are α-hemolysis, β-hemolysis, and the Lancefield grouping system?
4. Give a representative species and its importance for *Streptococcus, Enterococcus,* and *Lactococcus.* Distinguish between pyogenic and oral streptococci.
5. Of what practical importance is *Leuconostoc?* What are lactic acid bacteria?

Endospore-Forming Gram-Positive Rods and Cocci

Section 13 of *Bergey's Manual* contains almost all the gram-positive, endospore-forming bacteria. (Because of its resemblance to the actinomycetes, the endospore-forming genus *Thermoactinomyces* is placed in volume 4 of the manual.) Bacterial endospores are round or oval intracellular objects that have a complex structure with a spore coat, cortex, and inner spore membrane surrounding the protoplast (figure 22.7). They contain dipicolinic acid, are very heat resistant, and can remain dormant and viable for very long periods (Box 22.1). Usually, endospores are observed in the light microscope after spore staining (*see chapter 2*). They can also be detected by heating a culture at 70 to 80°C for 10 minutes followed by incubation in the proper growth medium. Because only endospores and some thermophiles would survive such heating, bacterial growth tentatively confirms their presence.

Bacterial endospore structure (pp. 62–65).

Although endospore-forming bacteria are distributed widely, they are primarily soil inhabitants. Soil conditions are often extremely variable, and endospores are an obvious advantage in surviving periods of dryness or nutrient deprivation.

Section 13 contains six genera of endospore-forming bacteria, five of which are described in table 22.5. The two best-studied and most important genera are *Bacillus* and *Clostridium* (figure 22.8; *see also figures 2.8*d, *2.14*a, *3.1*c, *3.10*, *37.10*a, and *37.19*a). Both contain gram-positive, endospore-forming, chemoheterotrophic rods that are either peritrichously flagellated or nonmotile. The genus *Clostridium* is obligately anaerobic and lacks both catalase and a complete electron transport chain. The genus *Bacillus* is aerobic, or sometimes facultative, and catalase positive.

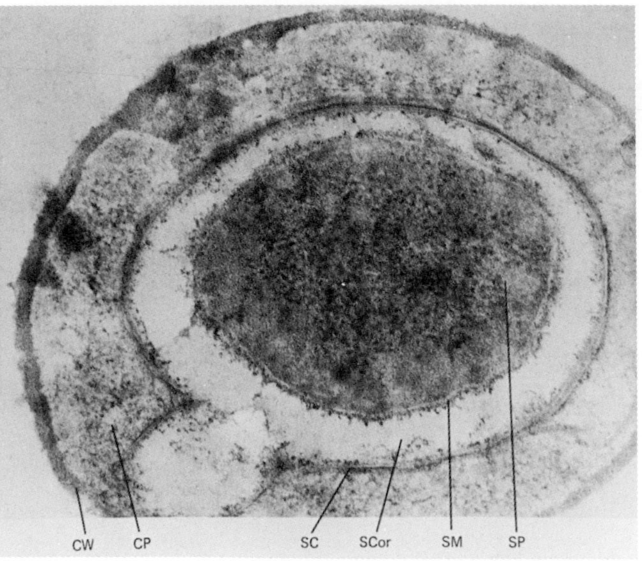

Figure 22.7 **Bacterial Endospores.** A cross section of *Bacillus megaterium* and its endospore within the vegetative cell wall, *CW;* cell protoplast, *CP;* spore coat, *SC;* spore cortex, *SCor;* spore membrane, *SM;* and spore protoplast, *SP* (×120,000).

The taxonomy of these bacteria is still somewhat uncertain. The rRNA data indicate that *Bacillus* and *Clostridium* belong to a procaryotic branch that contains nonsporing bacteria such as *Streptococcus, Staphylococcus, Lactobacillus,* and possibly the mycoplasmas (*see figure 20.9*). Many nonsporing bacteria are similar to spore formers except for their lack of endospores. Furthermore, *Bacillus* and *Clostridium* are very diverse phenotypically and genotypically; eventually, they may be subdivided to form new genera.

These two genera are of considerable practical importance. Members of the genus *Bacillus* produce the antibiotics bacitracin, gramicidin, and polymyxin. *B. cereus* causes some forms of food poisoning and can infect humans. *B. anthracis* is the causative agent of the disease anthrax, which can affect both farm animals and humans (*see chapter 37*). Several species are used as insecticides (*see chapter 44*). For example, *B. thuringiensis* and *B. sphaericus* form a solid protein crystal, the **parasporal body,** next to their spores during endospore

Box 22.1

Spores in Space

During the nineteenth-century argument over the question of the evolution of life, the panspermia hypothesis became popular. According to this hypothesis, life did not evolve from inorganic matter on earth but arrived as viable bacterial spores that had escaped from another planet. More recently, the British astronomer Fred Hoyle has revived the hypothesis based on his study of the absorption of radiation by interstellar dust. Hoyle maintains that dust grains were initially viable bacterial cells and that the beginning of life on earth was due to the arrival of bacterial endospores that had survived their trip through space.

Even more recently, Peter Weber and J. Mayo Greenberg from the University of Leiden in the Netherlands have studied the effect of very high vacuum, low temperature, and UV radiation on the survival of *Bacillus subtilis* endospores. Their data suggest that endospores within an interstellar molecular cloud might be able to survive around 4.5 to 45 million years. Molecular clouds move through space at speeds sufficient to transport spores between solar systems in this length of time. Although these results do not prove the panspermia hypothesis, they are consistent with the possibility that bacteria might be able to travel between planets capable of supporting life.

TABLE 22.5 Differential Characteristics of Selected Endospore-Forming Genera

Characteristics	*Bacillus*	*Clostridium*	*Sporolactobacillus*	*Desulfotomaculum*	*Sporosarcina*
Rod-shaped	+[a]	+	+	+	−
Filaments	−	D	−	−	−
Rods or filaments curved	−	D	−	D	NA
Cocci in tetrads or packets	−	−	−	−	+
Motile	+	+	+	+	+
Stain gram-positive, at least in young cultures	+	+	+	−	+
Strict aerobes	D	−	−	−	+
Facultative anaerobes or microaerophils	D	−	+	−	−
Strict anaerobes	−	+	−	+	−
Homolactic fermentation	D	−	+	−	−
Dissimilatory sulfate reduction to sulfide	−	−	−	+	−
Catalase	+	−	−	−	+
Oxidase	D	−	ND	ND	+
Marked acidity from glucose	+	D	+	−	−
Nitrate reduced to nitrite	D	D	−	ND	D
Mol% G + C of DNA	32–69	24–54	38–40	37–50	40–42

Modified from *Bergey's Manual of Systematic Bacteriology*, Vol. 2, edited by P. H. A. Sneath, N. S. Mair, and M. E. Sharpe. Copyright © 1986 Williams and Wilkins Co., Baltimore, MD. Reprinted by permission.

[a]Symbols: +, 90% or more of strains positive; −, 10% or less of strains positive; D, substantial proportion of species differ; NA, not applicable; ND, not determined.

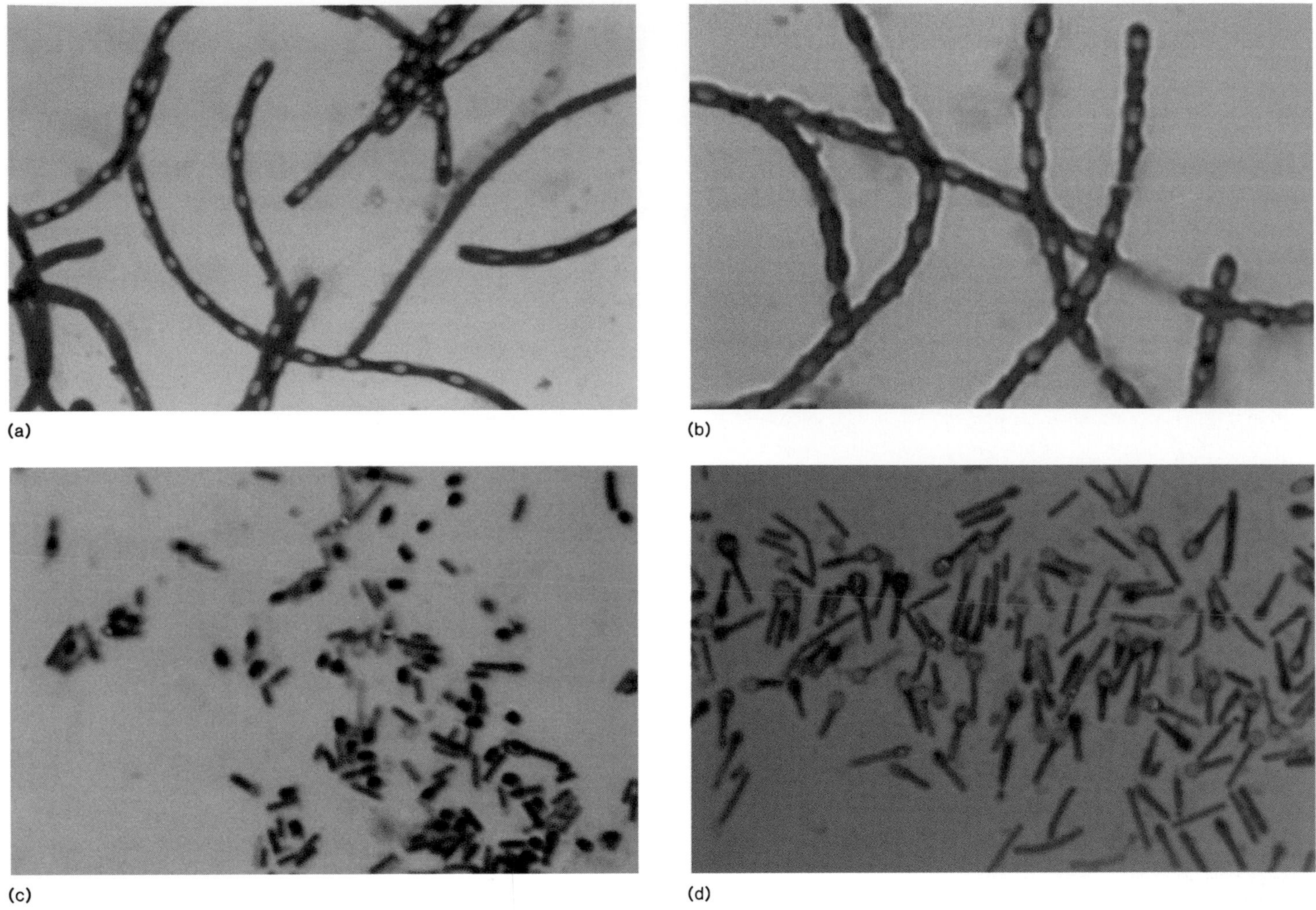

Figure 22.8 **Endospore-forming Bacteria.** Representative members of the genera *Bacillus* and *Clostridium*. (*a*) *B. anthracis*, spores elliptical and central (×700). (*b*) *B. subtilis*, spores elliptical and central. (*c*) *C. botulinum*, spores elliptical and subterminal, cells slightly swollen (×500). (*d*) *C. tetani*, spores round and terminal (×500).

formation. The *B. thuringiensis* parasporal body contains protein toxins that will kill over 100 species of moths by dissolving in the alkaline gut contents of caterpillars and destroying their gut epithelium. The solubilized toxin proteins are cleaved by midgut proteases to smaller toxic polypeptides that attack the epithelial cells. The alkaline gut contents escape into the blood, causing paralysis and death. One of these toxins has been isolated and shown to form pores in the plasma membrane. These channels allow monovalent cations such as potassium to pass. Most *B. thuringiensis* toxin genes are carried on large plasmids. The *B. sphaericus* parasporal body contains proteins toxic for mosquito larvae. Members of the genus *Clostridium* also have great practical impact. Because they are anaerobic and form heat-resistant endospores, they are responsible for many cases of food spoilage, even in canned foods. *C. botulinum* is the causative agent of botulism (*see chapter 37*). Clostridia often can ferment amino acids to produce ATP by oxidizing one amino acid and using another as an electron

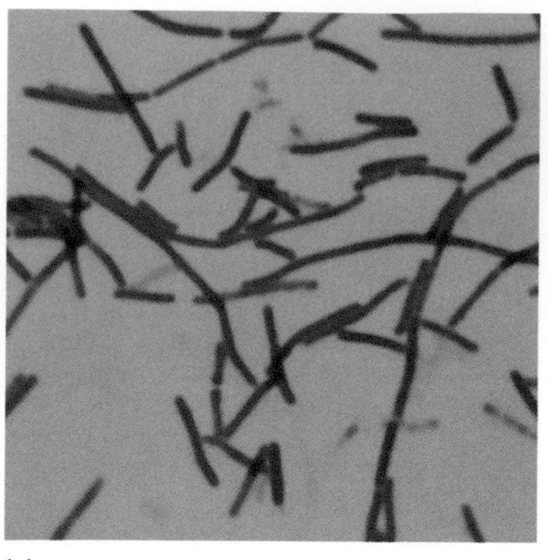
(a)

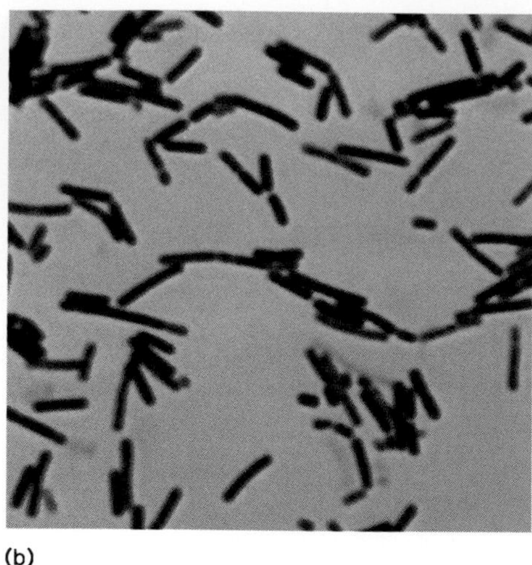
(b)

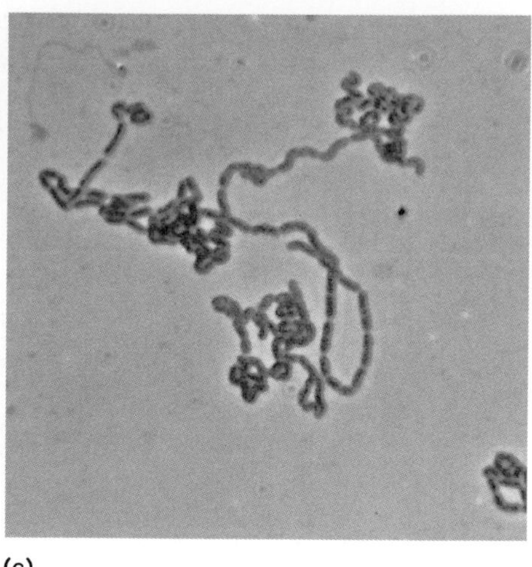
(c)

Figure 22.9 **Regular, Nonsporing Gram-positive Rods.** Representative lactobacilli. (*a*) *L. acidophilus* (×1,000). (*b*) *L. lactis,* Gram stain (×500). (*c*) *L. bulgaricus,* phase contrast (×600).

acceptor in a process called the Stickland reaction. This generates ammonia, hydrogen sulfide, fatty acids, and amines during the anaerobic decomposition of proteins. These products are responsible for many unpleasant odors arising during putrefaction. Several clostridia produce toxins and are major disease agents. *C. tetani* is the causative agent of tetanus, and *C. perfringens,* of gas gangrene and food poisoning. Clostridia are also industrially valuable; for example, *C. acetobutylicum* is used to manufacture butanol.

Microbiology of food (chapter 43).

1. What is a bacterial endospore? Give its most important properties and two ways to demonstrate its presence.
2. List the major properties of the genera *Bacillus* and *Clostridium.*
3. What practical impacts do the two genera have on society? Define parasporal body and Stickland reaction.

Regular, Nonsporing, Gram-Positive Rods

Section 14 of *Bergey's Manual* contains seven genera of regular, nonsporing, and usually nonmotile, gram-positive rods that are mesophilic and chemoheterotrophic and grow only in complex media. The largest genus, *Lactobacillus,* contains rods and sometimes coccobacilli that lack catalase and cytochromes, are usually facultative or microaerophilic, produce lactic acid as their main or sole fermentation product, and have complex nutritional requirements (figure 22.9). Lactobacilli carry out either a homolactic fermentation using the Embden-Meyerhof pathway or a heterolactic fermentation with the pentose phosphate pathway (*see chapter 8*). They grow optimally under slightly acid conditions, when the pH is around 4.5 to 6.4. The genus is found on plant surfaces and in dairy products, meat, water, sewage, beer, fruits, and many other materials. Lactobacilli also are part of the normal flora of the human body in the mouth, intestinal tract, and vagina. They are not usually pathogenic.

Lactobacillus is indispensable to the food and dairy industries (*see chapter 43*). Lactobacilli are used in the production of fermented vegetable foods (sauerkraut, pickles, silage), beverages (beer, wine, juices), sour dough, Swiss cheese and other hard cheeses, yogurt, and sausage. They also are sometimes responsible for spoilage of beer, milk, and meat.

Yogurt is probably the most popular fermented milk product in the United States and is produced commercially and by individuals using yogurt-making kits. In commercial production, nonfat or low-fat milk is pasteurized, cooled to 43°C or lower, inoculated with *Streptococcus thermophilus* and *Lactobacillus bulgaricus. S. thermophilus* grows more rapidly at first and renders the milk anaerobic and weakly acidic. *L. bulgaricus* then acidifies the milk even more. Acting together, the two species ferment almost all the lactose to lactic acid and flavor the yogurt with diacetyl (*S. thermophilus*) and acetaldehyde (*L. bulgaricus*). Fruits or fruit flavors to be added are pasteurized separately and then combined with the yogurt.

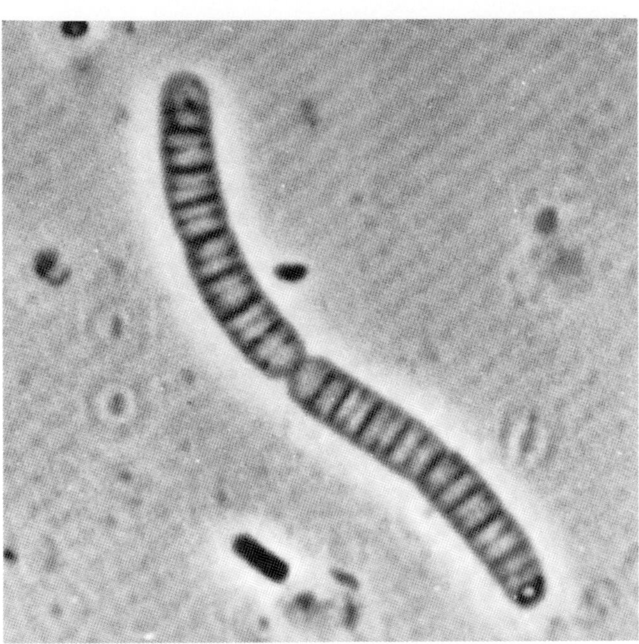

Figure 22.10 ***Caryophanon*** **Morphology.** *Caryophanon latum* in a trichome chain. Note the disk-shaped cells stacked side by side, phase contrast (×3,450).

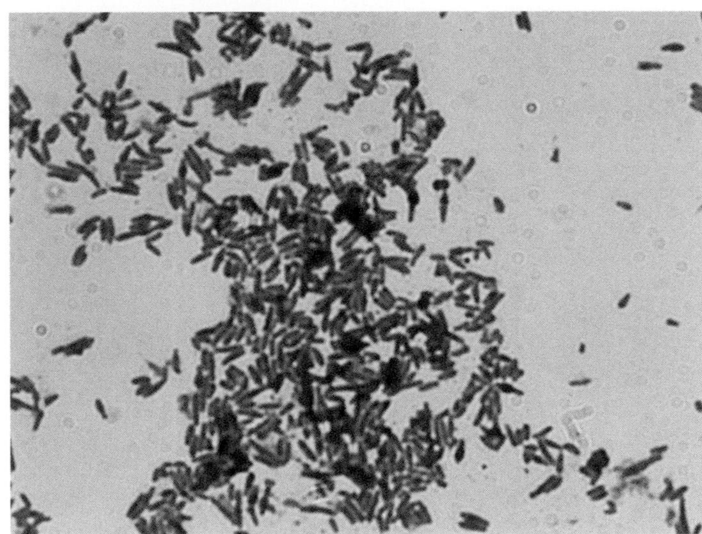

Figure 22.11 **Irregular, Nonsporing, Gram-positive Rods.** *Corynebacterium diphtheriae* (×1,000). Note the irregular shapes of individual cells, the angular associations of pairs of cells, and palisade arrangements.

Several other genera in this section are of interest. The genus *Listeria* has short rods that are aerobic or facultative, catalase positive, and motile by peritrichous flagella. They are widely distributed in nature, particularly in decaying matter. *Listeria monocytogenes* is a pathogen of humans and other animals and causes listeriosis, an important food infection (*p. 871*). *Caryophanon* is strictly aerobic, catalase positive, and motile by peritrichous flagella. Its normal habitat is cow dung. *Caryophanon* morphology is distinctive. Individual cells are disk-shaped (1.5 to 2.0 μm wide by 0.5 to 1.0 μm long) and joined to form rods about 10 to 20 μm long (figure 22.10).

1. List the major properties of the genus *Lactobacillus*. Why is it important in the food and dairy industries?
2. What is distinctive about the morphology of *Caryophanon*?

Irregular, Nonsporing, Gram-Positive Rods

Section 15 of *Bergey's Manual* contains an exceptionally diverse group of 21 genera. Most are irregularly shaped, nonsporing, gram-positive rods with aerobic or facultative metabolism. (However, six genera are anaerobic.) The rods may be straight or slightly curved and usually have swellings, club shapes, or other deviations from normal rod morphology. Three representative genera are briefly described: *Corynebacterium, Arthrobacter,* and *Actinomyces*.

The genus *Corynebacterium* comprises aerobic and facultative, catalase-positive, straight to slightly curved rods, often with tapered ends. Club-shaped forms are also seen. The bacteria often remain partially attached after a distinctive type of binary fission called snapping division, resulting in angular arrangements of the cells, somewhat like Chinese letters, or a palisade arrangement in which rows of cells are lined up side by side (figure 22.11). Corynebacteria form metachromatic granules, and their walls have *meso*-diaminopimelic acid. Although some species are harmless soil and water saprophytes, many corynebacteria are plant or animal pathogens. For example, *C. diphtheriae* is the causative agent of diphtheria in humans (*see chapter 37*).

The genus *Arthrobacter* contains aerobic, catalase-positive rods with respiratory metabolism and lysine in its peptidoglycan. Its most distinctive feature is a rod-coccus growth cycle (figure 22.12). When *Arthrobacter* is growing in exponential phase, the bacteria are irregular, branched rods and may reproduce by snapping division. As they enter stationary phase, the cells change to a coccoid form. Upon transfer to fresh medium, the coccoid cells produce outgrowths and again form actively growing rods. Although arthrobacters are often

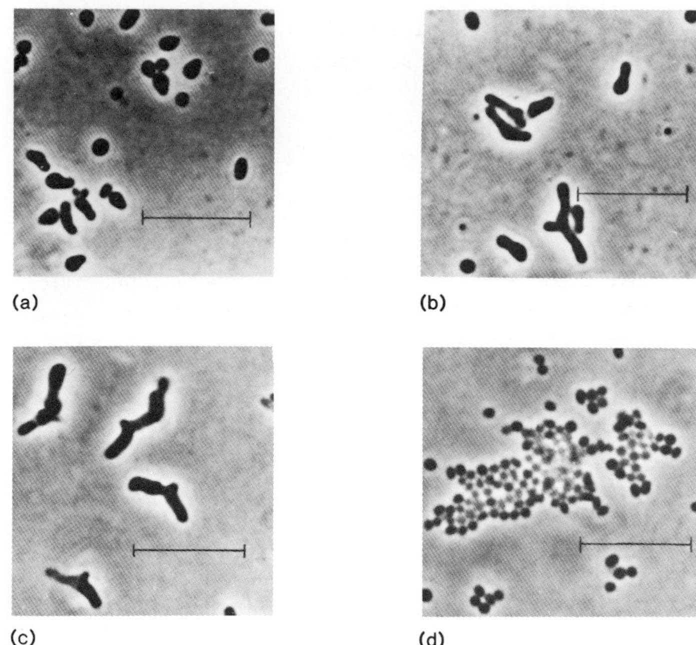

Figure 22.12 **The Rod-coccus Growth Cycle.** The rod-coccus cycle of *Arthrobacter globiformis* when grown at 25° C. (*a*) Rods are outgrowing from cocci 6 hours after inoculation. (*b*) Rods after 12 hours of incubation. (*c*) Bacteria after 24 hours. (*d*) Cells after reaching stationary phase (3 days incubation). The cells used for inoculation resembled these stationary-phase cocci. Bars = 10 μm.

isolated from fish, sewage, and plant surfaces, their most important habitat is the soil, where they constitute a significant component of the microbial flora. They are well adapted to this niche because they are very resistant to desiccation and nutrient deprivation. This genus is unusually flexible nutritionally and can even degrade some herbicides and pesticides; it is probably important in the mineralization of complex organic molecules.

The mechanism of **snapping division** has been studied in *Arthrobacter.* These bacteria have a two-layered cell wall, and only the inner layer grows inward to generate a transverse wall dividing the new cells. The completed transverse wall or septum next thickens and puts tension on the outer wall layer, which still holds the two cells together. Eventually, increasing tension ruptures the outer layer at its weakest point, and a snapping movement tears the outer layer apart around most of its circumference. The new cells now rest at an angle to each other and are held together by the remaining portion of the outer layer that acts as a hinge.

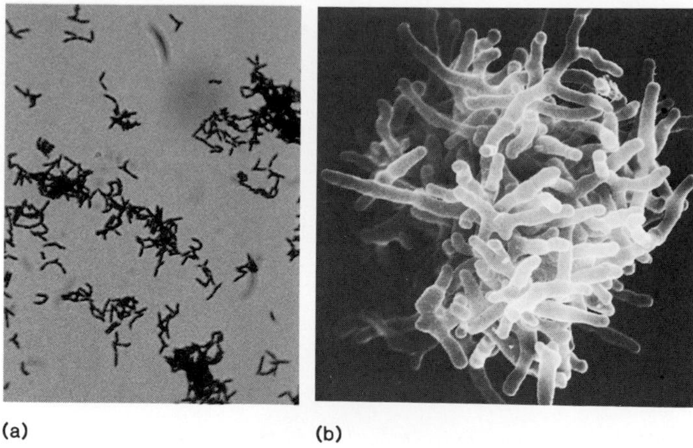

Figure 22.13 **Representatives of the Genus *Actinomyces.*** (*a*) *A. naeslundii,* Gram stain (×1,000). (*b*) *Actinomyces,* scanning electron micrograph (×18,000). Note filamentous nature of the colony.

Members of the genus *Actinomyces* are straight or slightly curved rods that vary considerably in shape and slender filaments with true branching (figure 22.13). The rods and filaments may have swollen, clubbed, or clavate ends. They are either facultative anaerobes or anaerobes, and require CO_2 for best growth. The cell walls contain lysine but not diaminopimelic acid or glycine. *Actinomyces* species are normal inhabitants of mucosal surfaces of humans and other warm-blooded animals; the oral cavity is their preferred habitat. *A. bovis* causes lumpy jaw in cattle. *Actinomyces* is responsible for actinomycoses, ocular infections, and periodontal disease in humans. The most important human pathogen is *A. israelii.*

1. What properties are common to all genera in section 15? Describe the major features of the genera *Corynebacterium, Arthrobacter,* and *Actinomyces,* including their normal habitat and the importance of each.
2. What is snapping division? The rod-coccus growth cycle?

The Mycobacteria

The family *Mycobacteriaceae* contains one genus, *Mycobacterium,* composed of slightly curved or straight rods that

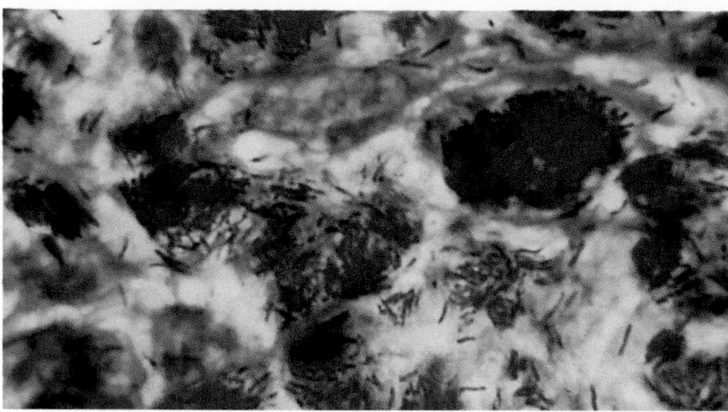

Figure 22.14 **The Mycobacteria.** *Mycobacterium leprae.* Acid-fast stain (×400). Note the masses of red mycobacteria within blue-green host cells.

sometimes branch or form filaments (figure 22.14). Mycobacterial filaments differ from those of actinomycetes in readily fragmenting into rods and coccoid bodies when disturbed. They are aerobic and catalase positive. Mycobacteria grow very slowly and must be incubated for 2 to 40 days after inoculation of a solidified complex medium to form a visible colony. Their cell walls have a very high lipid content and contain waxes with 60 to 90 carbon **mycolic acids.** These are complex fatty acids with a hydroxyl group on the β-carbon and an aliphatic chain attached to the α-carbon. The presence of mycolic acids and other lipids outside the peptidoglycan layer makes mycobacteria **acid-fast** (basic fuchsin dye cannot be removed from the cell by acid alcohol treatment). Extraction of wall lipid with alkaline ethanol destroys acid fastness.

Actinomycete morphology (p. 507).
Acid-fast staining (p. 29).

Although some mycobacteria are free-living saprophytes, they are best known as animal pathogens. *M. bovis* causes tuberculosis in cattle, other ruminants, and primates. Because this bacterium can produce tuberculosis in humans, dairy cattle are tested for the disease yearly; milk pasteurization kills the pathogen and affords further protection against disease transmission. Thus, *M. tuberculosis* is the chief source of tuberculosis in humans. The other major mycobacterial human disease is leprosy, caused by *M. leprae.*

Tuberculosis (pp. 747–50), leprosy (pp. 755–56).

1. Give the distinctive properties of the genus *Mycobacterium,* and indicate how mycobacteria differ from actinomycetes.
2. Define mycolic acid and acid-fast.
3. List two human mycobacterial diseases and their causative agents. Which pathogen causes tuberculosis in cattle?

Summary

1. Bacteria are placed in sections of volume 2 of *Bergey's Manual* according to characteristics such as general shape, the possession of endospores, and their response to acid-fast staining.
2. Based on 16S rRNA analysis, gram-positive bacteria are divided into low G + C and high G + C groups; this system is not always directly related to that in *Bergey's Manual.*
3. Peptidoglycan structure often differs between groups in taxonomically useful ways. Most variations are in amino acid 3 of the peptide subunit or in the interpeptide bridge.
4. Mesophilic, nonsporing, gram-positive cocci are placed in section 12 of *Bergey's Manual.*
5. Members of the genus *Staphylococcus* are facultatively anaerobic, nonmotile, gram-positive cocci that form irregular clusters. They grow on the skin and mucous membranes of warm-blooded animals, and some are important human pathogens.
6. Members of the family *Deinococcaceae* are aerobic, gram-positive cocci that are distinctive in their unusually great resistance to desiccation and radiation.
7. The genera *Streptococcus, Enterococcus,* and *Lactococcus* contain gram-positive cocci arranged in pairs and chains that are usually facultative and carry out homolactic fermentation. Some important species are the pyogenic coccus *S. pyogenes,* the oral streptococci *S. pneumoniae* and *S. mutans,* the enterococcus *E. faecalis,* and the lactococcus *L. lactis.*
8. *Leuconostoc* carries out heterolactic fermentation using the phosphoketolase pathway and is involved in the production of fermented vegetable products, buttermilk, butter, and cheese.
9. Section 13 has endospore-forming gram-positive rods and cocci. *Bacillus* synthesizes antibiotics and insecticides, and causes food poisoning and anthrax. *Clostridium* is responsible for botulism, tetanus, food spoilage, and putrefaction.
10. Section 14 of *Bergey's Manual* contains regular, nonsporing, gram-positive rods such as *Lactobacillus, Listeria,* and *Caryophanon. Lactobacillus* carries out lactic acid fermentation and is extensively used in the food and dairy industries.
11. Most bacteria in section 15 (e.g., *Corynebacterium, Arthrobacter,* and *Actinomyces*) are irregularly shaped, nonsporing, gram-positive rods with aerobic or facultative metabolism.
12. Corynebacteria use snapping division, and many cause disease (e.g., diphtheria) in plants and animals. The genus *Arthrobacter* has a rod-coccus life cycle, and also shows snapping division. Several species of *Actinomyces* are pathogenic for humans and other animals.
13. Mycobacteria form either rods or filaments that readily fragment. Their cell walls have a high lipid content and mycolic acids; the presence of these lipids makes them acid-fast. Mycobacteria cause tuberculosis and leprosy in humans.

Key Terms

acid-fast *465*
coagulase *454*
α-hemolysis *455*
β-hemolysis *455*
heterolactic fermentation *457*
lactic acid bacteria *458*
Lancefield grouping system *455*
mycolic acid *465*
parasporal body *459*
snapping division *464*

Questions for Thought and Review

1. On the basis of the treatment of gram-positive bacteria given in volume 2 of *Bergey's Manual,* is it best to consider the volume as an identification guide, a description of phylogenetic relationships, or some of both? Explain your reasoning and conclusion.
2. Draw a diagram illustrating how gram-positive cocci might divide to produce the various clustering patterns observed (chains, tetrads, cubical packets, grapelike clusters).
3. Describe the characteristics most important in distinguishing between members of the following groups of genera: *Micrococcus, Staphylococcus,* and *Streptococcus; Bacillus* and *Clostridium; Corynebacterium, Arthrobacter,* and *Actinomyces; Actinomyces,* and *Mycobacterium.*
4. Which bacteria are noted for each of the following: the rod-coccus growth cycle, radiation resistance, formation of dental caries, metachromatic granules and snapping division, coagulase, 60 to 90 carbon mycolic acids and acid fastness, pathogenicity for insects, production of fermented foods and cheeses, use of the phosphoketolase pathway, and the generation of odors during putrefaction?
5. Account for the ease with which anaerobic clostridia can be isolated from soil and other generally aerobic niches.
6. Propose an explanation for the observation that mycobacteria are normally only weakly gram positive.

Additional Reading

Aronson, A. I.; Beckman, W.; and Dunn, P. 1986. *Bacillus thuringiensis* and related insect pathogens. *Microbiol. Rev.* 50(1):1–24.

Balows, A.; Truper, H. G.; Dworkin, M.; Harder W.; and Schleifer, K.-H. 1992. *The prokaryotes,* 2d ed. New York: Springer-Verlag.

Braun, V., and Hantke, K. 1974. Biochemistry of bacterial cell envelopes. *Ann. Rev. Biochem.* 43:89–121.

Goren, M. B. 1972. Mycobacterial lipids: Selected topics. *Bacteriol. Rev.* 36(1):1–32.

Holt, J. G., editor-in-chief. 1986. *Bergey's manual of systematic bacteriology,* vol. 2, ed. P. H. A. Sneath, N. S. Mair, and M. E. Sharpe. Baltimore, Md.: Williams & Wilkins.

Hoyle, F., and Wickramasinghe, C. 1981. Where microbes boldly went. *New Scientist* (13 Aug.): 412–15.

Krulwich, T. A., and Pelliccione, N. J. 1979. Catabolic pathways of coryneforms, nocardias, and mycobacteria. *Ann. Rev. Microbiol.* 33: 95–111.

Lambert, B., and Peferoen, M. 1992. Insecticidal promise of *Bacillus thuringiensis*: Facts and mysteries about a successful biopesticide. *BioScience* 42(2):112–22.

Loesche, W. J. 1986. Role of *Streptococcus mutans* in human dental decay. *Microbiol. Rev.* 50(4):353–80.

Murray, B. E. 1990. The life and times of the enterococcus. *Clin. Microbiol. Rev.* 3(1):46–65.

Ross, P. W. 1985. Streptococcal infections in man. *Microbiol. Sci.* 2(6):174–78.

Schleifer, K. H., and Kandler, O. 1972. Peptidoglycan types of bacterial cell walls and their taxonomic implications. *Bacteriol. Rev.* 36(4):407–77.

Schleifer, K. H., and Kilpper-Balz, R. 1987. Molecular and chemotaxonomic approaches to the classification of streptococci, enterococci, and lactococci: A review. *Syst. Appl. Microbiol.* 10:1–19.

Ward, J. B. 1981. Teichoic and teichuronic acids: Biosynthesis, assembly, and location. *Microbiol. Rev.* 45(2):211–43.

Weber, P., and Greenberg, J. M. 1985. Can spores survive in interstellar space? *Nature* 316: 403–07.

Whiteley, H. R., and Schnepf, H. E. 1986. The molecular biology of parasporal crystal body formation in *Bacillus thuringiensis. Ann. Rev. Microbiol.* 40:549–76.

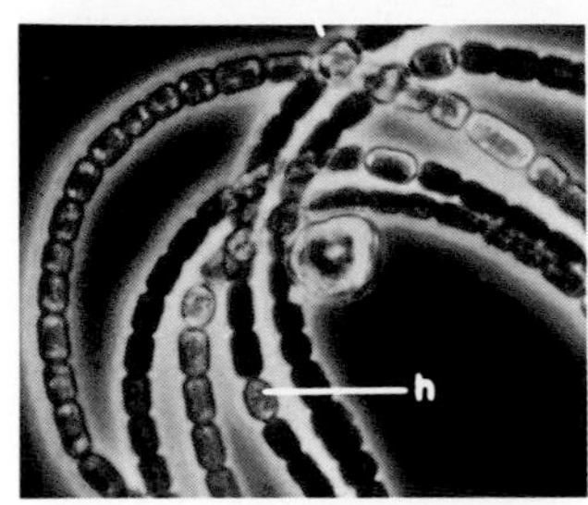

CHAPTER 23
The Bacteria: Remaining Gram-Negative Bacteria and Cyanobacteria

There are wide areas of the bacteriological landscape in which we have so far detected only some of the highest peaks, while the rest of the beautiful mountain range is still hidden in the clouds and the morning fogs of ignorance. The gold is still lying on the ground, but we have to bend down to grasp it.

—Preface to *The Prokaryotes*

Outline

Concepts

1. Volume 3 of *Bergey's Manual* contains eight sections, each describing a different group of gram-negative bacteria. Although few of these bacteria are of major medical or industrial significance, they are interesting biologically and often of considerable ecological importance.
2. There are three major groups of photosynthetic procaryotes: the purple bacteria, the green bacteria, and the cyanobacteria. The cyanobacteria are placed in a separate section because they resemble eucaryotic phototrophs in that they possess photosystem II and carry out oxygenic photosynthesis. The purple and green bacteria use electron donors other than water and carry out anoxygenic photosynthesis.
3. Chemolithotrophic bacteria obtain energy and electrons by oxidizing inorganic compounds rather than the organic nutrients employed by most bacteria. They often have substantial ecological impact because of their ability to oxidize many forms of inorganic nitrogen and sulfur.
4. Bacteria do not always have simple, unsophisticated morphology but may produce prosthecae, stalks, buds, sheaths, or complex fruiting bodies.
5. Gliding motility is widely distributed among bacteria and is very useful to organisms that digest insoluble nutrients or move over the surfaces of solid substrata.
6. A group of bacteria is often placed in a particular section of volume 3 on the basis of a few carefully selected properties such as nutritional type even though it may share many characteristics (e.g., morphological traits) with organisms in a different section.

Chapter 23 introduces the bacteria described in volume 3 of *Bergey's Manual of Systematic Bacteriology.* This volume contains bacteria that differ in some distinctive way from the more common gram-negative and gram-positive bacteria of volumes 1 and 2, respectively. The differences include motility (gliding bacteria), metabolism (phototrophic and chemolithotrophic bacteria), and reproduction or morphology (budding, appendaged, and sheathed bacteria). Although volume 3 also covers the archaeobacteria, this group is so distinctive that it will be described separately in chapter 24. This chapter describes the major biological features of each group and a few selected representative forms of particular interest.

Volume 3 of *Bergey's Manual* contains eight sections, each covering a different group of bacteria. This chapter briefly describes the major features of the organisms in the first seven sections.

Anoxygenic Photosynthetic Bacteria

There are three groups of photosynthetic procaryotes: the purple bacteria, the green bacteria, and the cyanobacteria. The cyanobacteria differ most fundamentally from the green and purple photosynthetic bacteria in being able to carry out **oxygenic photosynthesis.** They use water as an electron donor and generate oxygen during photosynthesis. In contrast, purple and green bacteria use **anoxygenic photosynthesis.** Because they are unable to use water as an electron source, they employ reduced molecules such as hydrogen sulfide, sulfur, hydrogen, and organic matter as their electron source for the generation of NADH and NADPH. Consequently, purple and green bacteria do not produce oxygen but often form sulfur granules. Purple sulfur bacteria accumulate granules within their cells, whereas green sulfur bacteria deposit the sulfur granules outside their cells. The purple nonsulfur bacteria use organic molecules as an electron source. There are also differences in photosynthetic pigments, the organization of photosynthetic membranes, nutritional requirements, and oxygen relationships. These are summarized in table 23.1 and will be dis-

TABLE 23.1 Characteristics of the Major Groups of Photosynthetic Procaryotes

	Anoxygenic Photosynthetic Bacteria				Oxygenic Photosynthetic Bacteria
Characteristic	Green Sulfur	Green Nonsulfur[a]	Purple Sulfur	Purple Nonsulfur	Cyanobacteria
Major photosynthetic pigments	Bacteriochlorophylls *a* plus *c*, *d*, or *e* (the major pigment)	Bacteriochlorophylls *a* and *c*	Bacteriochlorophyll *a* or *b*	Bacteriochlorophyll *a* or *b*	Chlorophyll *a* plus phycobiliproteins
Morphology of photosynthetic membranes	Photosynthetic system partly in chlorosomes that are independent of the plasma membrane	Chlorosomes present when grown anaerobically	Photosynthetic system contained in spherical or lamellar membrane complexes that are continuous with the plasma membrane	Photosynthetic system contained in spherical or lamellar membrane complexes that are continuous with the plasma membrane	Membranes lined with phycobilisomes
Photosynthetic electron donors	H_2, H_2S, S	Photoheterotrophic donors—a variety of sugars, amino acids, and organic acids; photoautotrophic donors—H_2S, H_2	H_2, H_2S, S	Usually organic molecules; sometimes reduced sulfur compounds or H_2	H_2O
Sulfur deposition	Outside of the cell		Inside the cell[b]	Outside of the cell	
Nature of photosynthesis	Anoxygenic	Anoxygenic	Anoxygenic	Anoxygenic	Oxygenic (sometimes facultatively anoxygenic)
General metabolic type	Obligately anaerobic photolithoautotrophs	Usually photoheterotrophic; sometimes photoautotrophic or chemoheterotrophic (when aerobic and in the dark)	Obligately anaerobic photolithoautotrophs	Usually anaerobic photoorganoheterotrophs; some facultative photolithoautotrophs (in dark, chemoorganoheterotrophs)	Aerobic photolithoautotrophs
Motility	Nonmotile; some have gas vesicles	Gliding	Motile with polar flagella; some are peritrichously flagellated	Motile with polar flagella or nonmotile; some have gas vesicles	Nonmotile or with gliding motility; some have gas vesicles
Percent G + C	48–58	53–55	45–70	61–72	35–71

[a]Characteristics of *Chloroflexus.*
[b]With the exception of *Ectothiorhodospira.*

cussed further in the context of individual groups. Because cyanobacterial photosynthesis differs so greatly from that of purple and green bacteria (*see chapter 8*), cyanobacteria are placed in section 19 of *Bergey's Manual* and are described later in this chapter. Section 18 of the manual contains the purple and green anoxygenic photosynthetic bacteria.

The mechanism of bacterial photosynthesis (pp. 162–68).

Purple Bacteria

There are three groups of purple bacteria: the families *Chromatiaceae* (the purple sulfur bacteria) and *Ectothiorhodospiraceae,* and the purple nonsulfur bacteria. The family *Ectothiorhodospiraceae* has only one genus, *Ectothiorhodospira,* with red, spiral-shaped, polarly flagellated cells that deposit sulfur globules externally (figure 23.1; *see also figure 3.9*b). The other two groups are much larger. All purple bacteria use anoxygenic photosynthesis, possess bacteriochlorophylls *a* or *b,* and have their photosynthetic apparatus in membrane systems that are continuous with the plasma membrane. Most are motile by polar flagella.

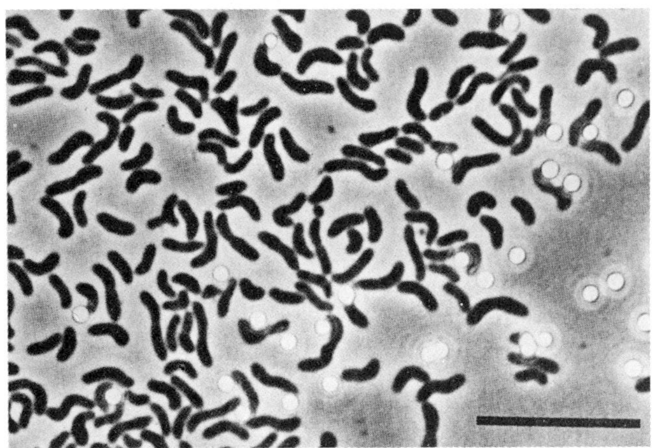

Figure 23.1 **Purple Bacteria.** *Ectothiorhodospira mobilis.* Light micrograph. Bar = 10 μm.

The **purple sulfur bacteria** are strict anaerobes and usually photolithoautotrophs. They oxidize hydrogen sulfide to sulfur and deposit it internally as sulfur granules (usually within invaginated pockets of the plasma membrane); often they eventually oxidize the sulfur to sulfate. Hydrogen may also serve as an electron donor. *Thiospirillum, Thiocapsa,* and *Chromatium* are typical purple sulfur bacteria (figure 23.2). They are found in anaerobic, sulfide-rich zones of lakes (*see chapter 41*).

Because purple and green photosynthetic bacteria grow best in deeper anaerobic zones of aquatic habitats, they cannot effectively use parts of the visible spectrum normally employed by photosynthetic organisms. There is often a dense surface layer of cyanobacteria and algae in lakes and ponds that absorbs a large amount of blue and red light. The bacteriochlorophyll pigments of purple and green bacteria absorb longer wavelength, far-red light (table 23.2) not used by other photosynthesizers (figure 23.3). In addition, the bacteriochlorophyll absorption peaks at about 350 to 550 nm enable them to grow at greater depths because shorter wavelength light can penetrate water farther. As a result, when the water is sufficiently clear, a layer of green and purple bacteria develops in the anaerobic, hydrogen sulfide-rich zone (figure 23.4).

The microbial ecology of lakes (pp. 830–31).

The **purple nonsulfur bacteria** are exceptionally flexible in their choice of an energy source. Normally, they grow anaerobically as photoorganoheterotrophs; they trap light energy and employ organic molecules as both electron and carbon sources. Although they are called nonsulfur bacteria, some species can oxidize very low, nontoxic levels of sulfide to sulfate, but they do not oxidize elemental sulfur to sulfate. In the absence of light, most purple nonsulfur bacteria can grow aerobically as

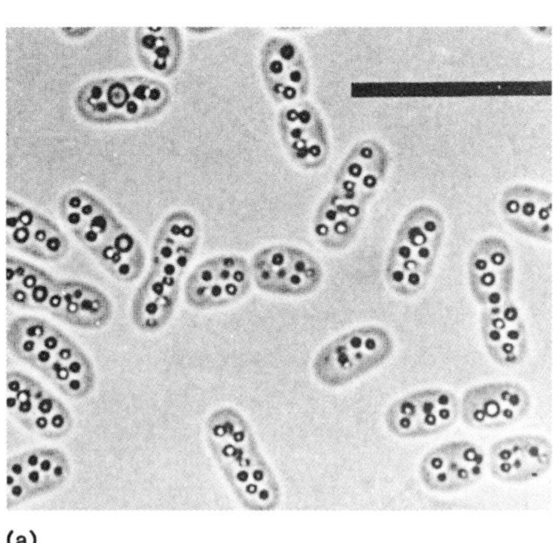

(a)

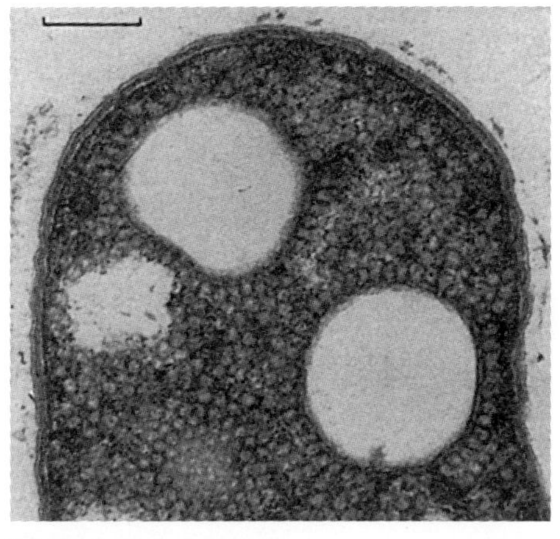

(b)

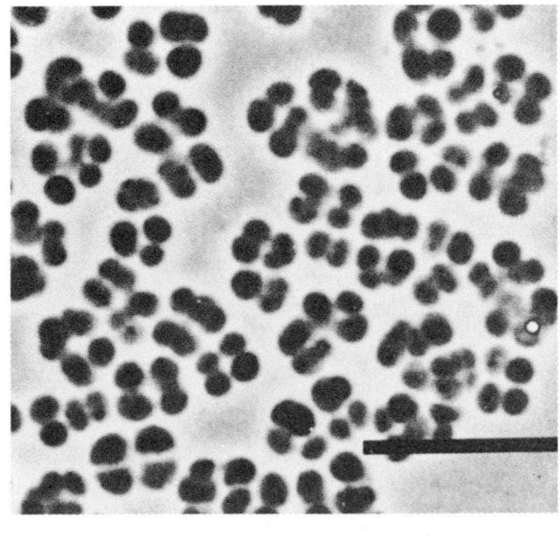

(c)

Figure 23.2 **Typical Purple Sulfur Bacteria.** (*a*) *Chromatium vinosum* with intracellular sulfur granules. Bar = 10 μm. (*b*) Electron micrograph of *C. vinosum.* Note the intracytoplasmic vesicular membrane system. The large white areas are the former sites of sulfur globules. Bar = 0.3μm. (*c*) *Thiocapsa roseopersicina.* Bar = 10 μm.

Table 23.2 Procaryotic Bacteriochlorophyll and Chlorophyll Absorption Maxima

Pigment	Long Wavelength Maxima (nm): *In Ether or Acetone*	*Approximate Range of Values in Cells*
Chlorophyll *a*	665	680–685
Bacteriochlorophyll *a*	775	850–910 (purple bacteria)[a]
Bacteriochlorophyll *b*	790	1,020–1,035
Bacteriochlorophyll *c*	660	745–760
Bacteriochlorophyll *d*	650	725–745
Bacteriochlorophyll *e*	647	715–725

[a]The spectrum of bacteriochlorophyll *a* in green bacteria has a different maximum, 805–810 nm.

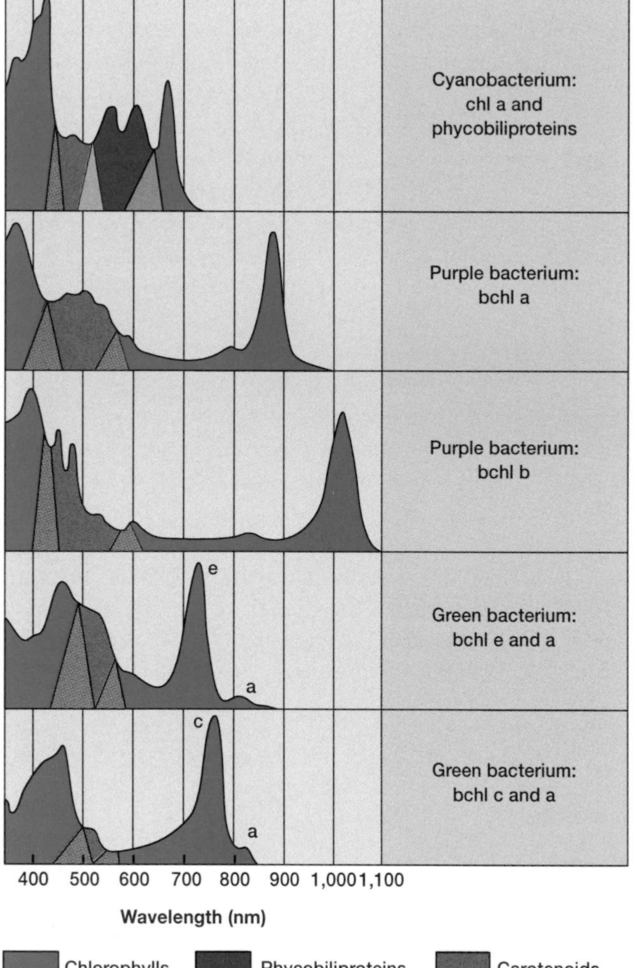

Figure 23.3 Photosynthetic Pigments. Absorption spectra of five photosynthetic bacteria showing the differences in absorption maxima and the contributions of various accessory pigments.

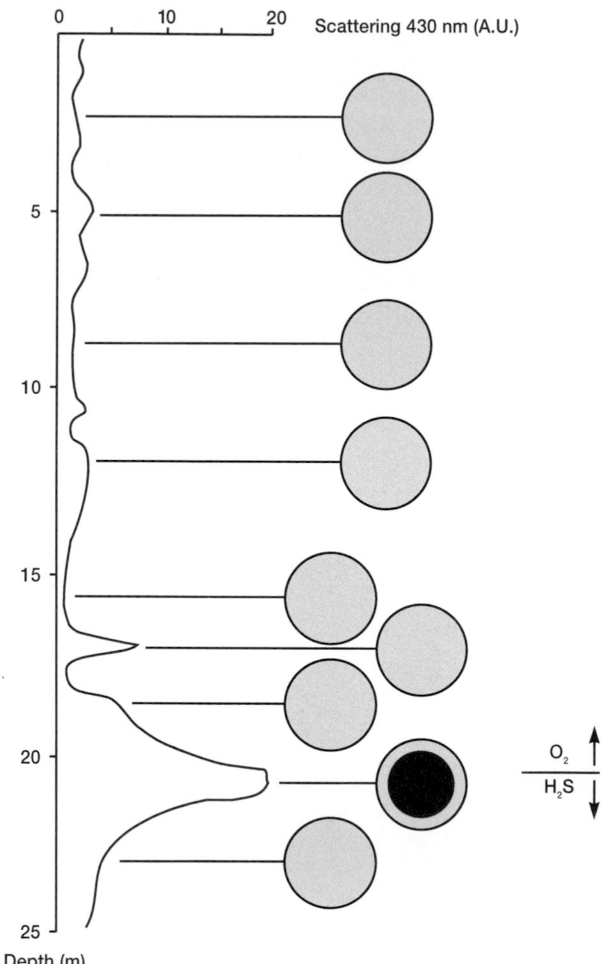

Figure 23.4 The Distribution of Photosynthetic Microorganisms in a Norwegian Fjord. Fifty-milliliter samples taken at various depths have been filtered and the filters mounted on a graph showing the light scattering (an indication of turbidity due to bacterial growth) as a function of depth. The peak at 20.8 m is caused by the purple bacteria *Chromatium* spp. growing at the interface of the aerobic zone and the anaerobic, sulfide-rich layer. The smaller peak at 17.1 m is caused by dinoflagellate growth.

chemoorganoheterotrophs, but some species carry out fermentations and grow anaerobically. Oxygen inhibits bacteriochlorophyll and carotenoid synthesis so that cultures growing aerobically in the dark are colorless.

Purple nonsulfur bacteria vary considerably in morphology (figure 23.5). They may be spirals (*Rhodospirillum*), rods (*Rhodopseudomonas*), half circles or circles (*Rhodocyclus*), or they may even form prosthecae and buds (*Rhodomicrobium*). Because of their metabolism, they are most prevalent in the mud and water of lakes and ponds with abundant organic matter and low sulfide levels. There are also marine species.

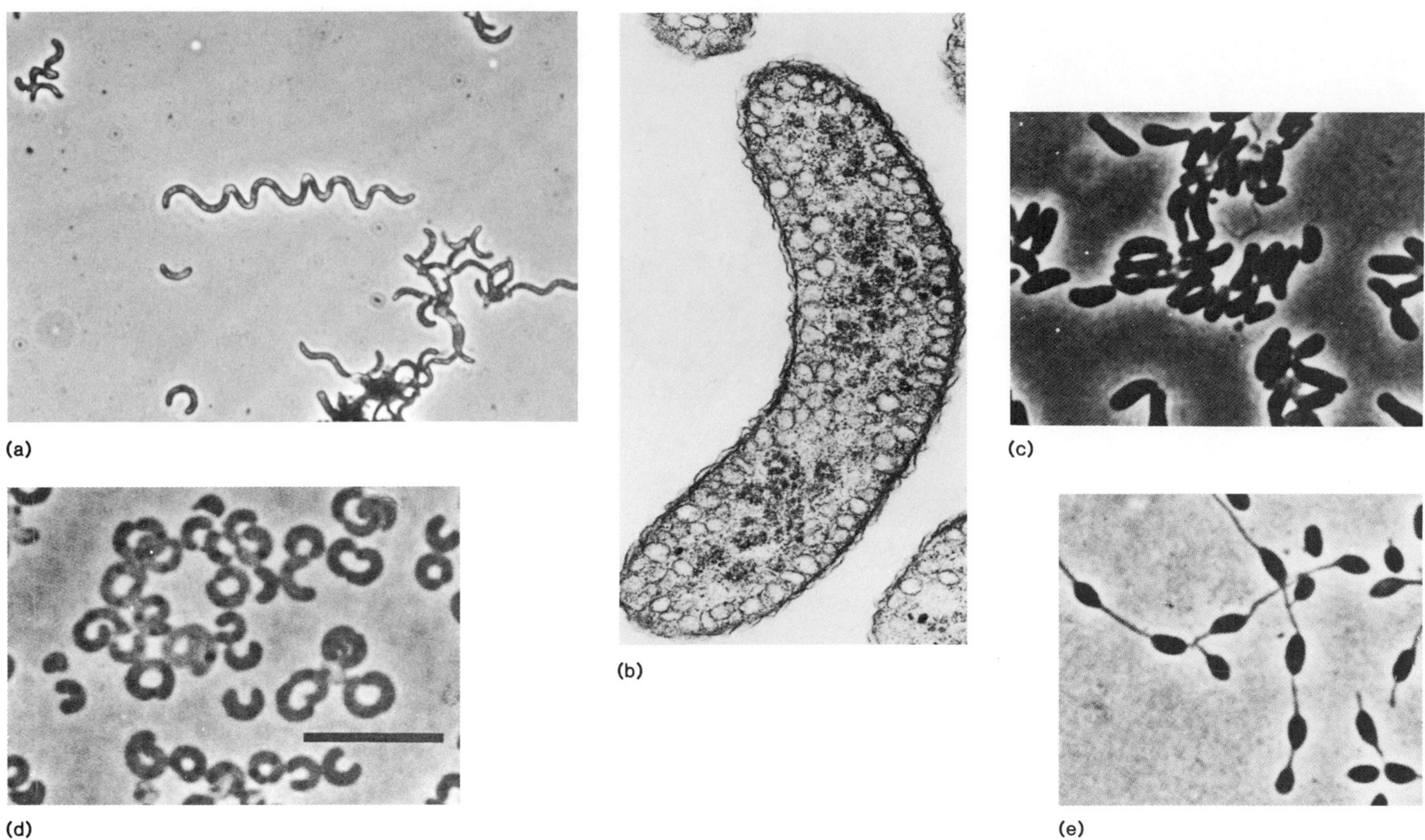

Figure 23.5 Typical Purple Nonsulfur Bacteria. (*a*) *Rhodospirillum rubrum,* phase contrast (×410). (*b*) *R. rubrum* grown anaerobically in the light. Note vesicular invaginations of the cytoplasmic membranes, transmission electron micrograph (×51,000). (*c*) *Rhodopseudomonas acidophila,* phase contrast. (*d*) *Rhodocyclus purpureus,* phase contrast. Bar = 10 μm. (*e*) *Rhodomicrobium vannielli* with vegetative cells and buds, phase contrast.

Green Bacteria

As with the purple bacteria, there are two groups of green bacteria: the **green sulfur bacteria** and the **green nonsulfur bacteria.** The green sulfur bacteria are a small group of obligately anaerobic photolithoautotrophs that use hydrogen sulfide, elemental sulfur, and hydrogen as electron sources. The elemental sulfur produced by sulfide oxidation is deposited outside the cell. Their photosynthetic pigments are located in ellipsoidal vesicles called **chlorosomes** or chlorobium vesicles, which are attached to the plasma membrane but are not continuous with it. The chlorosome membrane is not a normal lipid bilayer or unit membrane. Chlorosomes contain accessory bacteriochlorophyll pigments, but the reaction center bacteriochlorophyll is located in the plasma membrane. These bacteria flourish in the anaerobic, sulfide-rich zones of lakes. Although they lack flagella and are nonmotile, some species have gas vesicles (figure 23.6*c*) to adjust their depth for adequate light and hydrogen sulfide. Those forms without vesicles are found in sulfide-rich muds at the bottom of lakes and ponds.

Like the purple bacteria, the green sulfur bacteria are very diverse morphologically. They may be rods, cocci, or vibrios; some grow singly, and others form chains and clusters (figure 23.6*a,b*). They are either grass-green or chocolate-brown in color. Representative genera are *Chlorobium, Prosthecochloris,* and *Pelodictyon.*

Chloroflexus is the major representative of the green nonsulfur bacteria (*Bergey's Manual* calls the group multicellular filamentous green bacteria). It is a filamentous, gliding, thermophilic bacterium that is often isolated from neutral to alkaline hot springs where it grows in the form of orange-reddish mats, usually in association with cyanobacteria. Although it resembles the green bacteria in ultrastructure and photosyn-

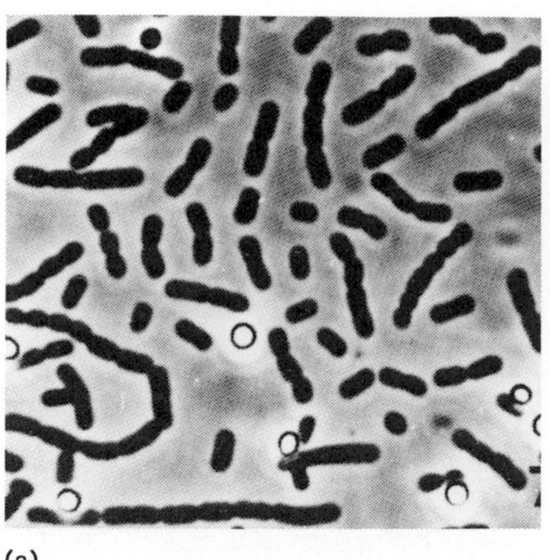
(a)

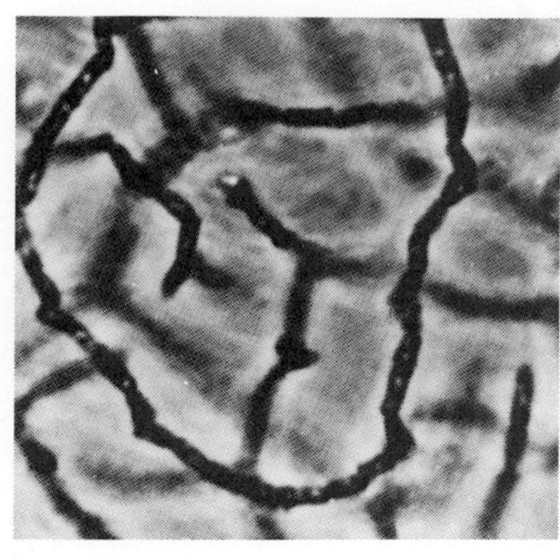
(b)

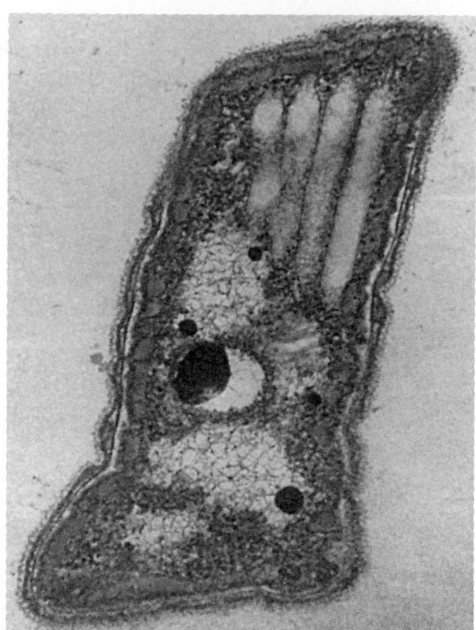
(c)

Figure 23.6 Typical Green Sulfur Bacteria. (*a*) *Chlorobium limicola* with extracellular sulfur granules. (*b*) *Pelodictyon clathratiforme.* (*c*) An electron micrograph of *P. clathratiforme* (×105,000). Note the chlorosomes (dark gray areas) and gas vesicles (light gray areas with pointed ends).

thetic pigments, its metabolism is more similar to that of the purple nonsulfur bacteria. *Chloroflexus* can carry out anoxygenic photosynthesis with organic compounds as carbon sources or grow aerobically as a chemoheterotroph. It doesn't appear closely related to any eubacterial group based on 16S rRNA studies, and its taxonomic position is uncertain.

1. How do oxygenic and anoxygenic photosynthesis differ from each other and why?
2. Give the major characteristics of the following groups: purple sulfur bacteria, purple nonsulfur bacteria, and green sulfur bacteria. How do purple and green bacteria differ? Compare *Chloroflexus* with the green sulfur bacteria.
3. Why are the bacteriochlorophylls of purple and green bacteria so useful beneath the surface of lakes?
4. What are chlorosomes or chlorobium vesicles?

Oxygenic Photosynthetic Bacteria

Two groups of oxygenic photosynthesizers are found in section 19 of *Bergey's Manual,* the cyanobacteria and the order *Prochlorales.* The cyanobacteria are discussed first, followed by a brief survey of the prochlorophytes.

The **cyanobacteria** are the largest and most diverse group of photosynthetic bacteria. There is little agreement about the number of cyanobacterial species. Older classifications had as many as 2,000 or more species. In one recent system, this has been reduced to 62 species and 24 genera. The G + C content of the group ranges from 35 to 71%. Although cyanobacteria are true procaryotes, their photosynthetic system closely resembles that of the eucaryotes because they have chlorophyll *a* and photosystem II, and carry out oxygenic photosynthesis. Like the red algae, cyanobacteria use phycobiliproteins as accessory pigments. Photosynthetic pigments and electron transport chain components are located in thylakoid membranes lined with particles called **phycobilisomes** (figure 23.7). These contain phycobilin pigments, particularly phycocyanin, and transfer energy to photosystem II. Carbon dioxide is assimi-

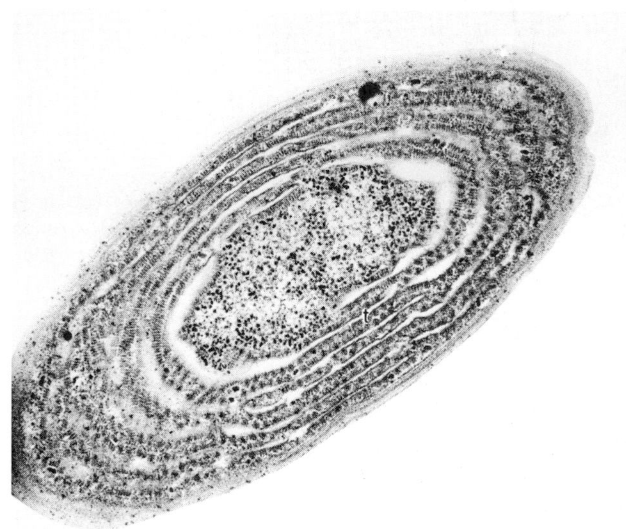

Figure 23.7 Cyanobacterial Thylakoids and Phycobilisomes. *Synechococcus lividus* with an extensive thylakoid system. The phycobilisomes lining these thylakoids are clearly visible as granules at location *t* (×60,000).

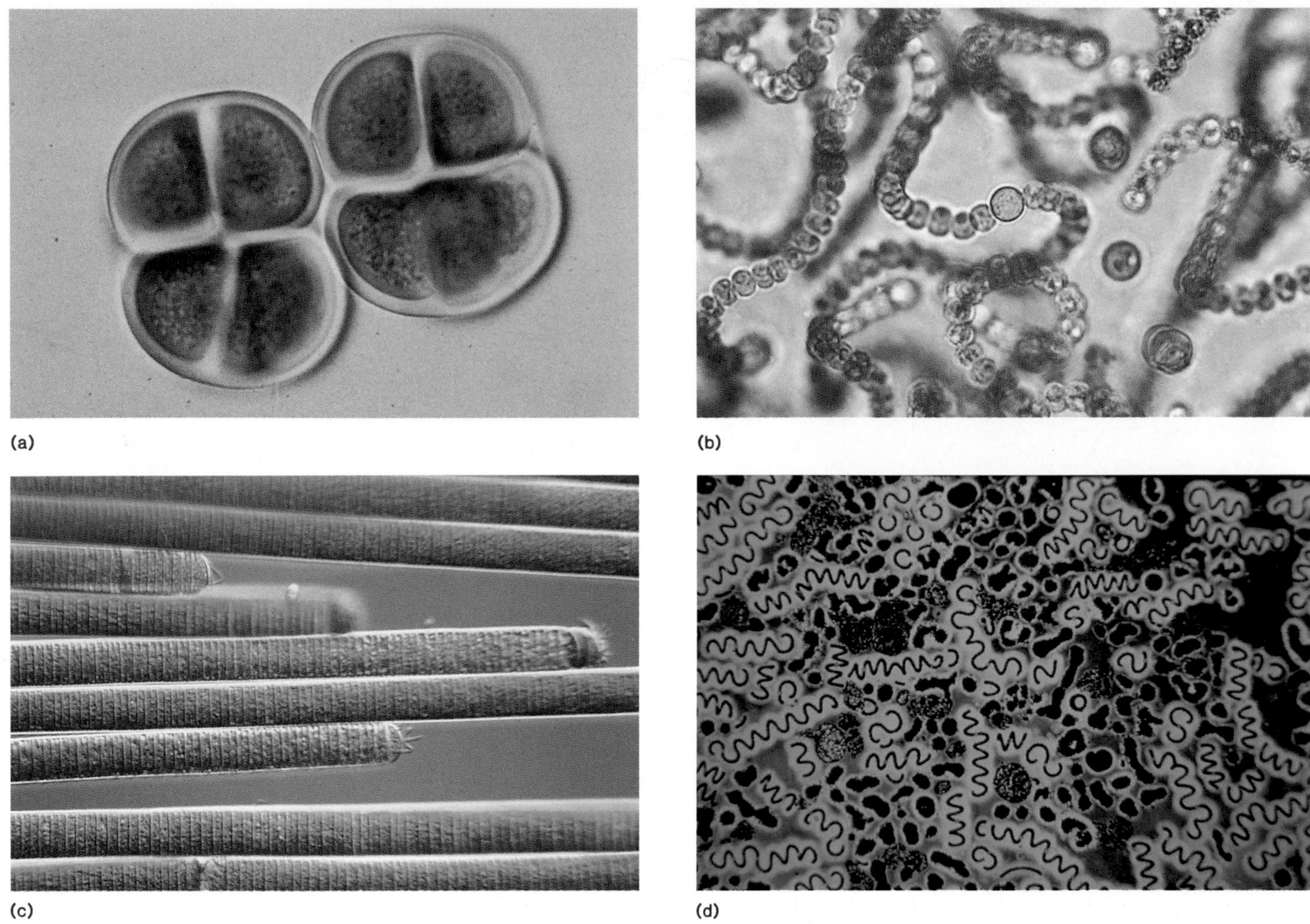

Figure 23.8 **Oxygenic Photosynthetic Bacteria.** Representative cyanobacteria. (*a*) *Chroococcus furgidus,* two colonies of four cells each (×600). *b*) *Nostoc* with heterocysts (×550). (*c*) *Oscillatoria* trichomes seen with Nomarski interference-contrast optics (×250). (*d*) The cyanobacteria *Anabaena spiroides* and *Microcystis aeruginosa.* The spiral *A. spiroides* is covered with a thick gelatinous sheath (×1,000).

lated through the Calvin cycle, and the reserve carbohydrate is glycogen. Sometimes they will store extra nitrogen as polymers of arginine or aspartic acid in cyanophycin granules. Since cyanobacteria lack the enzyme α-ketoglutarate dehydrogenase, they do not have a fully functional citric acid cycle. The pentose phosphate pathway plays a central role in their carbohydrate metabolism. Although many cyanobacteria are obligate photolithoautotrophs, some can grow slowly in the dark as chemoheterotrophs by oxidizing glucose and a few other sugars. Under anaerobic conditions, *Oscillatoria limnetica* oxidizes hydrogen sulfide instead of water and carries out anoxygenic photosynthesis much like the green photosynthetic bacteria. Cyanobacteria are capable of considerable metabolic flexibility.

Cyanobacteria vary greatly in shape and appearance. They range in diameter from about 1 to 10 μm and may be unicellular, exist as colonies of many shapes, or form filaments called trichomes (figure 23.8; *see also figures 3.12–3.14*). A **trichome** is a row of bacterial cells that are in close contact with one another over a large area. In contrast, adjacent cells in a simple chain (such as those commonly found in the genus *Bacillus*) associate by only a small area of contact. Although most appear blue-green because of phycocyanin, a few are red or brown in color because of the red pigment phycoerythrin. Despite this variety, cyanobacteria have typical procaryotic cell structures and a normal gram-negative type cell wall (figure 23.9). They often use gas vesicles to move vertically in the water, and many filamentous species have gliding motility (Box 23.1; *see also chapter 3*). Although cyanobacteria lack flagella, several strains of the marine genus *Synechococcus* are able to move at rates of up to 25 μm/second by means of an unknown mechanism.

Cyanobacteria reproduce by binary fission, budding, fragmentation, and multiple fission. In the last process, a cell en-

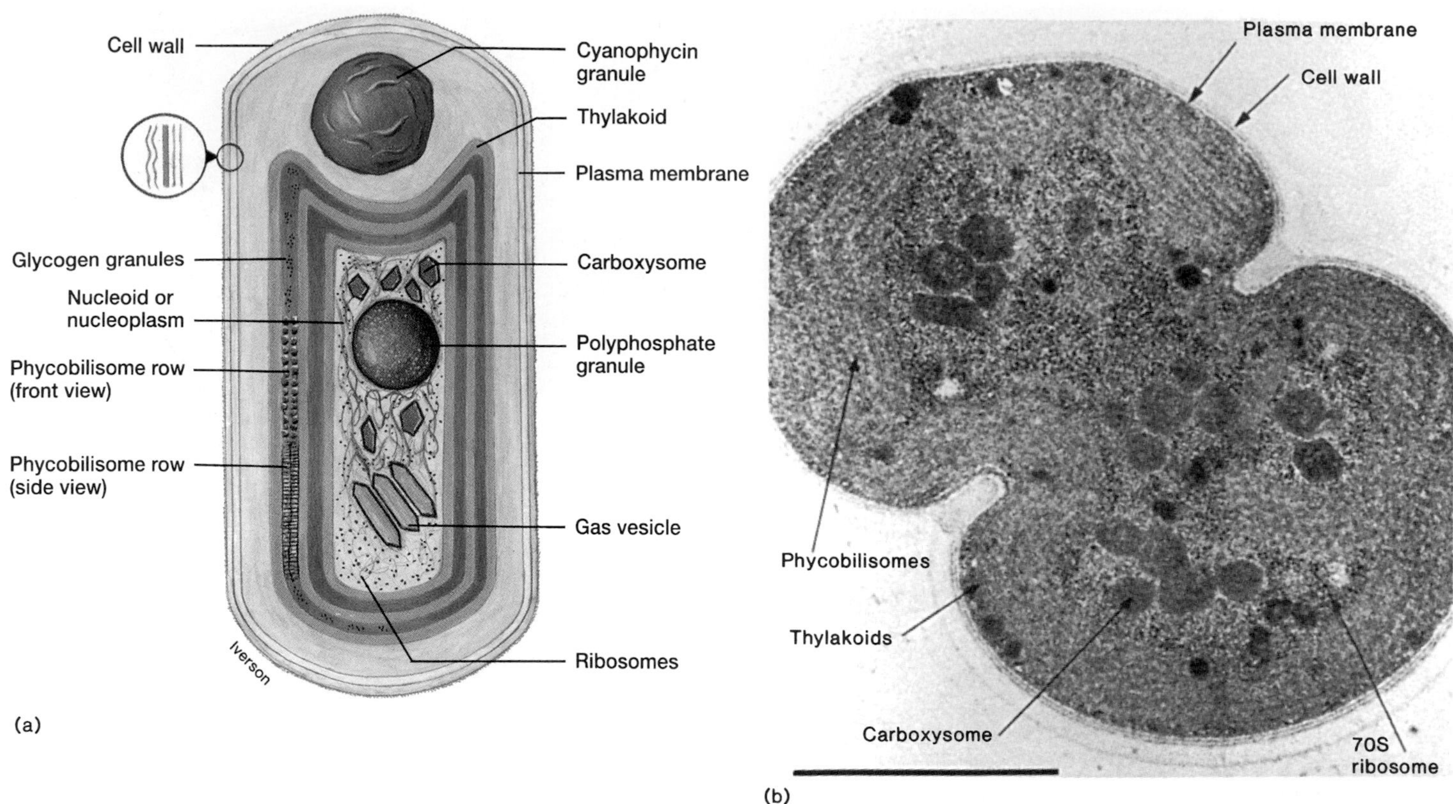

Figure 23.9 **Cyanobacterial Cell Structure.** (*a*) Schematic diagram of a vegetative cell. The insert shows an enlarged view of the envelope with its outer membrane and peptidoglycan. (*b*) Thin section of *Synechocystis* during division (bar = 1 μm). Many structures are visible.

larges and then divides several times to produce many smaller progeny, which are released upon the rupture of the parental cell. Fragmentation of filamentous cyanobacteria can generate small, motile filaments called **hormogonia.** Some species develop **akinetes,** specialized, dormant, thick-walled resting cells that are resistant to desiccation. Often these germinate to form new filaments.

Many trichome-forming or filamentous cyanobacteria fix atmospheric nitrogen by means of special cells called **heterocysts** (figure 23.10). Around 5 to 10% of the cells in a trichome develop into heterocysts when cyanobacteria are deprived of both nitrate and ammonia, their preferred nitrogen sources. When transforming themselves into heterocysts, cyanobacterial cells synthesize a very thick new wall, reorganize their photosynthetic membranes, discard their phycobiliproteins and photosystem II, and synthesize the nitrogen-fixing enzyme nitrogenase. Photosystem I is still functional and produces ATP, but no oxygen arises from noncyclic photophosphorylation because photosystem II is absent. This inability to generate O_2 is critical because the nitrogenase is extremely oxygen sensitive. The heterocyst wall slows or prevents O_2 diffusion into the cell, and any O_2 present is consumed during respiration. The structure and physiology of the heterocyst ensures that it will remain anaerobic; it is totally dedicated to nitrogen fixation. It obtains nutrients from adjacent vegetative cells and contributes fixed nitrogen in the form of the amino acid glutamine.

The biochemistry of nitrogen fixation (pp. 178–80).

The classification of cyanobacteria is still in an unsettled state, partly due to the lack of pure cultures. At present, all taxonomic schemes must be considered tentative. Group I of section 19 in *Bergey's Manual* contains the cyanobacteria and is divided into five orders with 27 genera. Table 23.3 briefly summarizes the major characteristics of the five orders. These are distinguished using colony or trichome morphology and reproductive patterns. Some other properties important in cyanobacterial characterization are cell morphology, ultrastructure, genetic characteristics, physiology and biochemistry, and habitat/ecology (preferred habitat and growth habit). The authors consider genera as tentative and often do not provide species names.

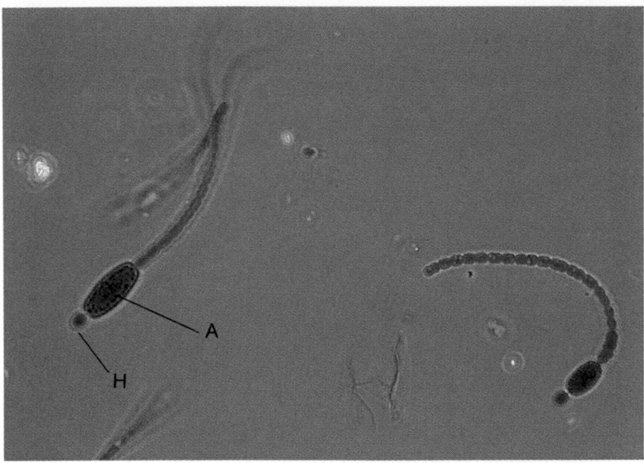

(a)

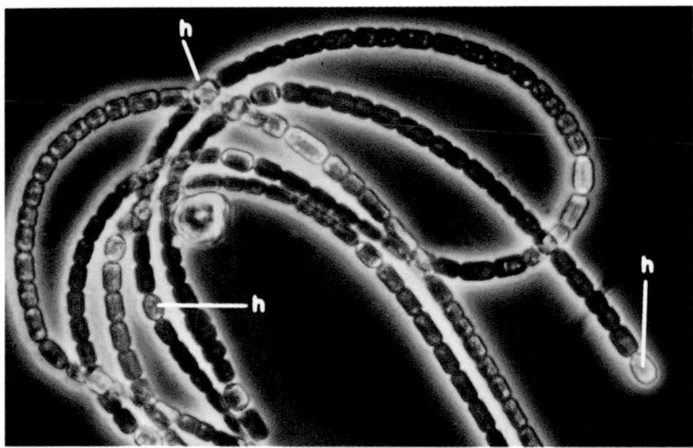

(b)

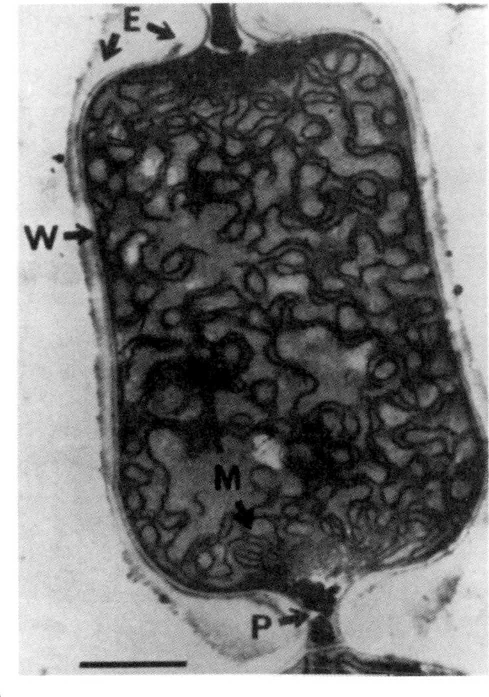

(c)

Figure 23.10 Examples of Heterocysts and Akinetes.
(*a*) *Cylindrospermum* with terminal heterocysts (H) and subterminal akinetes (*A*) (×500). (*b*) *Anabaena,* with heterocysts. (*c*) An electron micrograph of an *Anabaena* heterocyst (bar = 1 μm). Note the cell wall (*W*), additional outer walls (*E*), membrane system (*M*), and a pore channel to the adjacent cell (*P*).

TABLE 23.3 Characteristics of the Cyanobacterial Orders

Order	General Shape	Reproduction and Growth	Heterocysts	% G + C	Other Properties	Representative Genera
Chroococcales	Unicellular rods or cocci; nonfilamentous aggregates	Binary fission, budding	−	40–49	Almost always nonmotile	*Chamaesiphon* *Gloeobacter* *Gloeothece* *Gleocapsa* *Synechococcus*
Pleurocapsales	Unicellular rods or cocci; may be held together in aggregates	Multiple fission to form baeocytes	−	40–46	Only some baeocytes are motile	*Pleurocapsa* *Dermocarpa*
Oscillatoriales	Filamentous, unbranched trichome with only vegetative cells	Binary fission in a single plane, fragmentation	−	40–67	Usually motile	*Lyngbya* *Oscillatoria* *Spirulina* *Pseudanabaena*
Nostocales	Filamentous, unbranched trichome may contain specialized cells	Binary fission in a single plane, fragmentation to form hormogonia	+	38–47	Often motile, may produce akinetes	*Anabaena* *Cylindrospermum* *Nostoc* *Scytonema* *Calothrix*
Stigonematales	Filamentous trichomes either with branches or composed of more than one row of cells	Binary fission in more than one plane, hormogonia formed	+	42–46	May produce akinetes, greatest morphological complexity and differentiation in cyanobacteria	*Fischerella* *Stigonema* *Geitleria*

The *Chroococcales* are unicellular rods or cocci that are almost always nonmotile and reproduce by binary fission or budding. The *Pleurocapsales* are also unicellular, though several individual cells may be held together in an aggregate by an outer wall. Members of this order reproduce by multiple fission to form spherical, very small, reproductive cells, often called **baeocytes,** which escape when the outer wall ruptures. Some baeocytes disperse through gliding motility. The other three orders contain filamentous cyanobacteria. Usually the trichome is unbranched and is often surrounded by a sheath or slime layer. The *Oscillatoriales* form unbranched trichomes composed only of vegetative cells, whereas the *Nostocales* and *Stigonematales* produce heterocysts in the absence of an adequate nitrogen source (they also may form akinetes). Heterocystous cyanobacteria are subdivided into those that form linear filaments (*Nostocales*) and cyanobacteria that divide in a second plane to produce branches or aggregates (*Stigonematales*).

Cyanobacteria are very tolerant of environmental extremes and are present in almost all aquatic and terrestrial environments. Thermophilic species may grow at temperatures of up to 75°C in neutral to alkaline hot springs. Some unicellular forms even grow in the fissures of desert rocks. In nutrient-rich or eutrophic warm ponds and lakes, surface cyanobacteria such as *Anacystis* and *Anabaena* can reproduce rapidly to form blooms. The release of large amounts of organic matter upon the death of the bloom microorganisms stimulates the growth of chemoheterotrophic bacteria that subsequently deplete the available oxygen. This kills fish and other organisms (*see chapter 41*). Other cyanobacteria, for example, *Oscillatoria,* are so pollution resistant and characteristic of fresh water with high organic matter content that they act as water pollution indicators.

Cyanobacteria are particularly successful in establishing symbiotic relationships with other organisms. They are the photosynthetic partner in most lichen associations. Cyanobacteria are symbionts with protozoa, and nitrogen-fixing species form associations with a variety of plants (liverworts, mosses, gymnosperms, and angiosperms).

Types of symbiotic relationships (chapter 29).
Lichens (pp. 566–67).

Group II in section 19 contains the order *Prochlorales* with two genera, *Prochloron* and *Prochlorothrix.* These are oxygenic phototrophic procaryotes that have both chlorophyll *a* and *b,* but lack phycobilins. Thus while they resemble cyanobacteria with respect to chlorophyll *a,* they differ in also possessing chlorophyll *b,* the only procaryotes to do so. Because prochlorophytes lack phycobilin pigments and phycobilisomes, they are grass-green in color. They resemble chloroplasts in their pigments and thylakoid structure, but their 5S and 16S rRNAs show affinities with the cyanobacteria. Possibly a common ancestor gave rise to prochlorophytes, cyanobacteria, and plant chloroplasts.

The two recognized prochlorophyte genera are quite different from one another. *Prochloron* was first discovered as an extracellular symbiont growing either on the surface or within the cloacal cavity of marine colonial ascidian invertebrates (figure 23.11). These bacteria are single-celled, spherical, and from 8 to 30 μm in diameter. Their mol% of G + C is 31 to 41. *Prochlorothrix* is free living, has cylindrical cells that form filaments, and has been found in Dutch lakes. Its DNA has a higher G + C content (53 mol%). Unlike *Prochloron,* it has been cultured in the laboratory.

Another small prochlorophyte, less than 1 μm in diameter, has recently been discovered flourishing about 100 meters below the ocean surface. It differs from other prochlorophytes in having divinyl chlorophyll *a* and α-carotene instead of chlorophyll *a* and β-carotene. During the summer, it reaches concentrations of 5×10^5 cells per milliliter. It is one of the most numerous of the marine plankton, and a significant component of the marine microbial food web.

---

1. Summarize the major characteristics of the cyanobacteria that distinguish them from other photosynthetic organisms.
2. Define or describe the following: phycobilisomes, hormogonia, akinetes, heterocysts, and baeocytes.
3. What is a trichome and how does it differ from a simple chain of cells?
4. Briefly discuss the ways in which cyanobacteria reproduce.
5. How are heterocysts modified to carry out nitrogen fixation? When do cyanobacteria develop heterocysts?
6. Give the features of the five major cyanobacterial groups.
7. List some important positive and negative impacts cyanobacteria have on humans and the environment.
8. Compare the prochlorophytes with cyanobacteria and chloroplasts. Where does one find them?

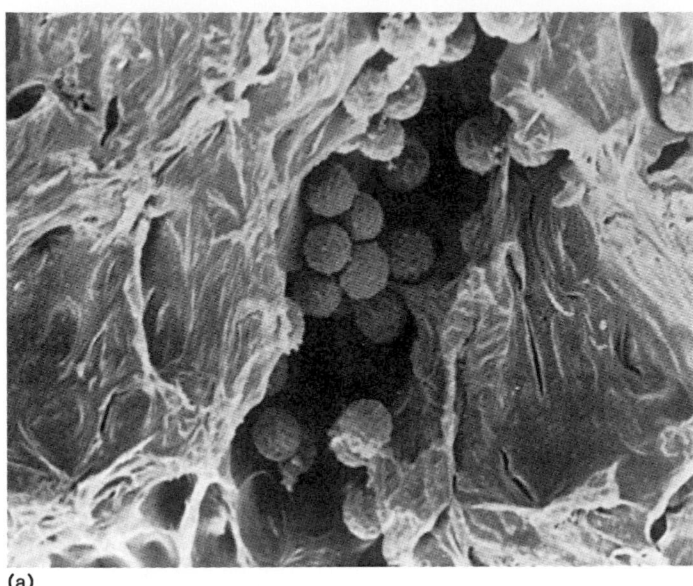
(a)

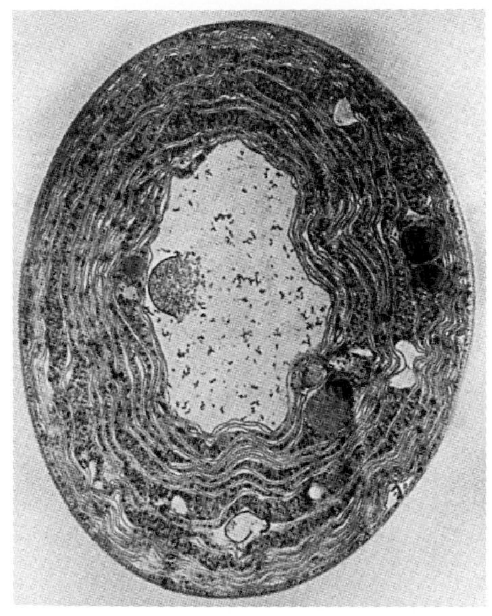
(b)

Figure 23.11 **Prochloron.** (*a*) A scanning electron micrograph of *Prochloron* cells on the surface of a *Didemnum candidum* colony. (*b*) *Prochloron didemni* section (transmission electron micrograph, ×23,500).

Aerobic Chemolithotrophic Bacteria and Associated Organisms

Section 20 of *Bergey's Manual* is devoted to chemolithotrophic bacteria, those bacteria that derive energy and electrons from reduced inorganic compounds. Normally, they employ CO_2 as their carbon source or are chemolithoautotrophs, but some can function as chemolithoheterotrophs and use reduced organic carbon sources. These bacteria are divided into four groups based on the inorganic compounds they prefer to oxidize: nitrifiers, colorless sulfur bacteria (sulfur oxidizers), obligate hydrogen oxidizers, and metal oxidizers. Magnetotactic bacteria (*see Box 3.2 and chapter 40*) are also included in section 20 of *Bergey's Manual.* The metabolism of chemolithotrophs is described in chapter 8 of this text; the focus here is on the biology of the first two groups. Their ecological impact is discussed further in the context of microbial ecology.

The biochemistry of chemolithotrophy (pp. 160–62).

Nitrifying Bacteria

Currently, nine genera of **nitrifying bacteria** are placed in the family *Nitrobacteraceae.* Although all are aerobic, gram-negative organisms without endospores and able to oxidize either ammonia or nitrite, they differ considerably in other properties (table 23.4). Nitrifiers may be rod-shaped, ellipsoidal, spherical, spirillar or lobate, and they may possess either polar or peritrichous flagella (figure 23.12). Often, they have extensive membrane complexes lying in their cytoplasm. Identification is based on properties such as their preference for nitrite or ammonia, their general shape, and the nature of any cytomembranes present.

Nitrifying bacteria are very important ecologically and can be isolated from soil, sewage disposal systems, and freshwater and marine habitats. The genera *Nitrobacter* and *Nitrococcus* oxidize nitrite to nitrate; *Nitrosomonas, Nitrosospira, Nitrosococcus,* and *Nitrosolobus* oxidize ammonia to nitrite. When two genera such as *Nitrobacter* and *Nitrosomonas* grow together in a niche, ammonia is converted to nitrate, a process

TABLE 23.4 Selected Characteristics of Representative Nitrifying Bacteria

Species	Cell Morphology and Size	Reproduction	Motility	Cytomembranes	G + C (Mol %)	Habitat
Ammonia-Oxidizing Bacteria						
Nitrosomonas europaea	Rod; 0.8–1.0 × 1.0–2.0 μm	Binary fission	+ or −; 1 or 2 subpolar flagella	Peripheral, lamellar	47.4–51.0	Soil, sewage, fresh water, marine
Nitrosococcus oceanus	Coccoid; 1.8–2.2 μm in diameter	Binary fission	+; 1 or more subpolar flagella	Centrally located parallel bundle, lamellar	50.5–51.0	Obligately marine
Nitrosospira briensis	Spiral; 0.3–0.4 μm in diameter	Binary fission	+ or −; 1 to 6 peritrichous flagella	Lacking	54.1 (1 strain)	Soil
Nitrosolobus multiformis	Lobular; 1.0–1.5 μm in diameter	Binary fission	+; 1 to 20 peritrichous flagella	Internal, compartmentalizing the cell	53.6–55.1	Soil
Nitrite-Oxidizing Bacteria						
Nitrobacter winogradskyi	Rod; 0.6–0.8 × 1.0–2.0 μm	Budding	+ or −; 1 polar flagellum	Polar cap of flattened vesicles in peripheral region of the cell	60.7–61.7	Soil, fresh water, marine
Nitrococcus mobilis	Coccoid; 1.5–1.8 μm in diameter	Binary fission	+; 1 or 2 subpolar flagella	Tubular cytomembranes randomly arranged throughout cytoplasm	61.2 (1 strain)	Marine

called **nitrification** (*see chapter 8*). Nitrification occurs rapidly in soils treated with fertilizers containing ammonium salts. Nitrate nitrogen is readily used by plants, but it is also rapidly lost through leaching of the water soluble nitrate and by denitrification to nitrogen gas. Thus, nitrification is a mixed blessing.

The nitrogen cycle and microorganisms (pp. 810–12).

Colorless Sulfur Bacteria

Like the nitrifiers, colorless sulfur bacteria are a very diverse group. A number of genera capable of oxidizing hydrogen sulfide are placed in section 23 because they contain filamentous gliding bacteria (*Beggiatoa, Thioploca, Thiothrix*). Unicellular rod-shaped or spiral sulfur-oxidizing bacteria that are nonmotile or motile by flagella are located here (table 23.5). Only some of these bacteria have been isolated and studied in pure culture. Most is known about the genera *Thiobacillus* and *Thiomicrospira*. *Thiobacillus* is a gram-negative rod, and *Thiomicrospira* is a long spiral cell (figure 23.13); both are polarly flagellated. They differ from many of the nitrifying bacteria in that they lack extensive internal membrane systems.

The metabolism of *Thiobacillus* has been intensely studied (*see p. 162*). It grows aerobically by oxidizing a variety of inorganic sulfur compounds (elemental sulfur, hydrogen sulfide, thiosulfate) to sulfate. ATP is produced with a combination of oxidative phosphorylation and substrate-level phosphorylation by means of adenosine 5′-phosphosulfate (*see figure 8.21*). Although *Thiobacillus* normally uses CO_2 as its major carbon source, *T. novellus* and a few other strains can grow heterotrophically. Some species are very flexible metabolically. For example, *Thiobacillus ferrooxidans* also uses ferrous iron as an electron donor and produces ferric iron as well as sulfuric acid. *T. denitrificans* even grows anaerobically by reducing nitrate to nitrogen gas. It should be noted that sulfur-oxidizing bacteria such as *Thiobacterium* and *Macromonas* probably do not derive energy from sulfur oxidation. They may use the process to detoxify metabolically produced hydrogen peroxide.

Sulfur-oxidizing bacteria have a wide distribution and great practical importance. *Thiobacillus* grows in soil and aquatic habitats, both freshwater and marine. In marine habitats, *Thiomicrospira* is more important than *Thiobacillus*. Because of their great acid tolerance (*T. thiooxidans* grows at pH 0.5 and cannot grow above pH 6), these bacteria prosper in habitats they have acidified by sulfuric acid production, even

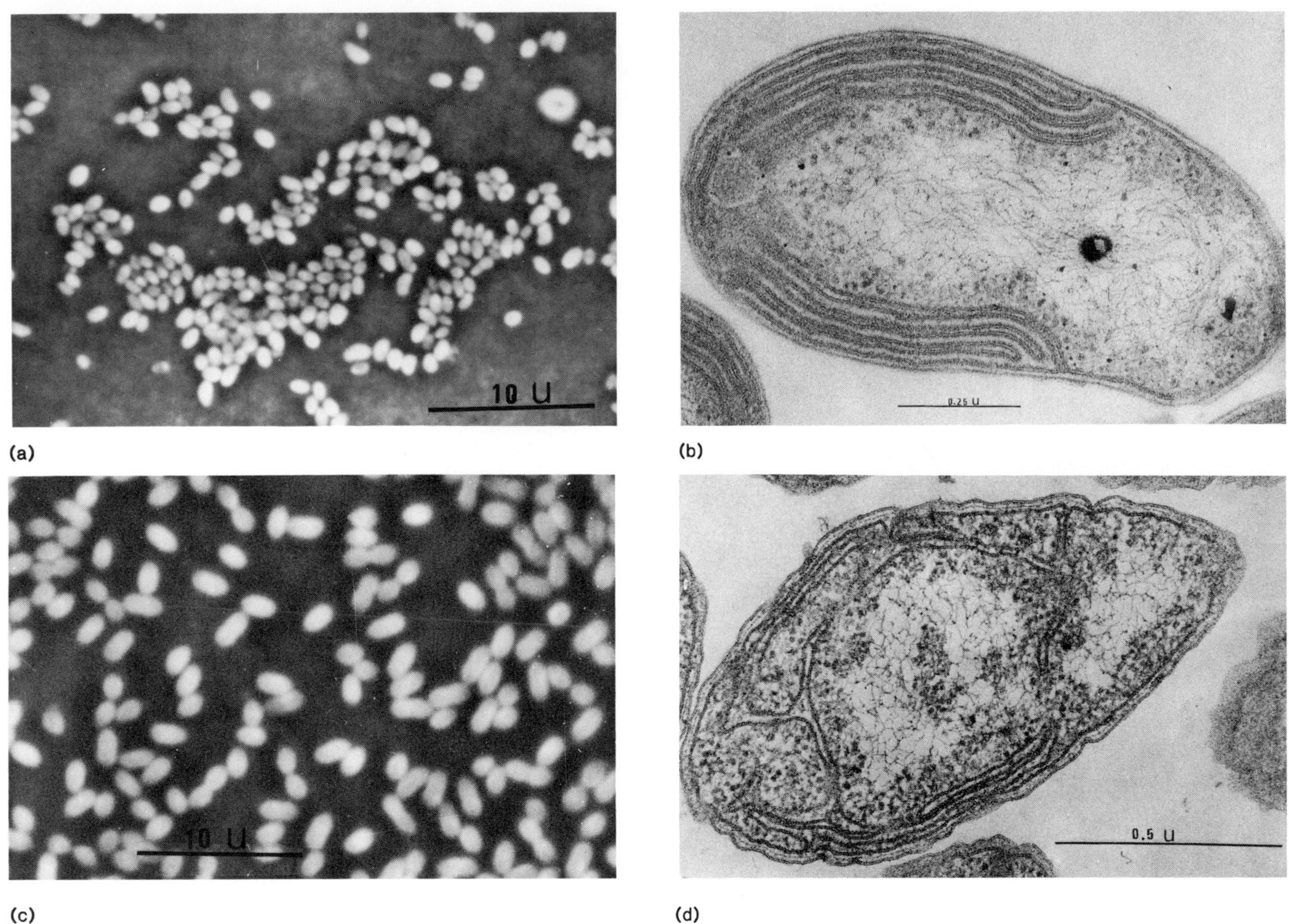

Figure 23.12 **Representative Nitrifying Bacteria.** (*a*) *Nitrobacter winogradskyi,* phase contrast (×2,500). (*b*) *N. winogradskyi.* Note the polar cap of cytomembranes (×213,000). (*c*) *Nitrosomonas europaea,* phase contrast (×2,500). (*d*) *N. europaea* with extensive cytoplasmic membranes (×81,700).

TABLE 23.5 Sulfur-Oxidizing Genera in Section 20 of *Bergey's Manual*

Genus	Cell Shape	Motility, Flagella	% G + C	Location of Sulfur Deposit[a]	Nutritional Type
Thiobacillus	Rods	+; polar	52–68	Extracellular	Obligate or facultative chemolithotroph
Thiomicrospira	Spirals	+; polar	36–44	Extracellular	Obligate chemolithotroph
Thiobacterium	Rods embedded in gelatinous masses	−		Intracellular	Probably chemoorganoheterotroph
Thiospira	Spiral rods, usually with pointed ends	+; polar (single or in tufts)		Intracellular	Unknown
Macromonas	Rods, cylindrical or bean shaped	+; polar	67	Intracellular	Probably chemoorganoheterotroph

[a]When hydrogen sulfide is oxidized to elemental sulfur.

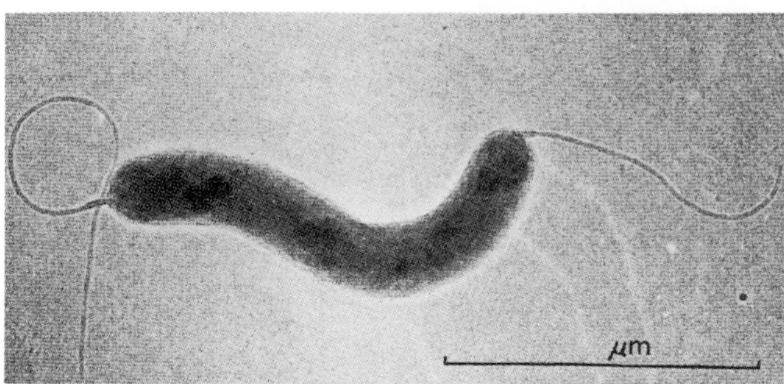

Figure 23.13 **Colorless Sulfur Bacteria.** *Thiomicrospira pelophila* with polar flagella. Bar = 1 μm.

though most other organisms are dying. The production of large quantities of sulfuric acid and ferric iron by *T. ferrooxidans* corrodes concrete and pipe structures. Thiobacilli often cause extensive acid and metal pollution when they release metals from mine wastes. However, sulfur-oxidizing bacteria are also beneficial. They may increase soil fertility when they release unavailable sulfur by oxidizing it to sulfate. Thiobacilli are used in processing low-grade metal ores because of their ability to leach metals from ore.

The sulfur cycle and microorganisms (p. 810).

1. What are chemolithotrophic bacteria? On what basis are they subdivided into four groups?
2. Give the major characteristics of the nitrifying bacteria and discuss their ecological importance. How does the metabolism of *Nitrosomonas* differ from that of *Nitrobacter*?
3. Characterize the colorless sulfur bacteria in section 20 of *Bergey's Manual.*
4. How do colorless sulfur bacteria obtain energy by oxidizing sulfur compounds? What is adenosine 5′-phosphosulfate?
5. List several positive and negative impacts sulfur-oxidizing bacteria have on the environment and human activities.

Budding and/or Appendaged Bacteria

Section 21 of *Bergey's Manual* contains a heterogeneous group of bacteria with at least one of three features: a prostheca, a stalk, or reproduction by budding. A **prostheca** (pl., prosthecae) is an extension of the cell, including the plasma membrane and cell wall, that is narrower than the mature cell. A **stalk** is a nonliving appendage produced by the cell and extending from it. **Budding** is distinctly different from the **binary fission** normally used by bacteria (*see chapter 11*). The bud first appears as a small protrusion at a single point and enlarges to form a mature cell. Most or all of the bud's cell envelope is newly synthesized. In contrast, portions of the parental cell envelope are shared with the progeny cells during binary fission. Finally, the parental cell retains its identity during budding, and the new cell is often smaller than its parent. In binary fission, the parental cell disappears as it forms the progeny.

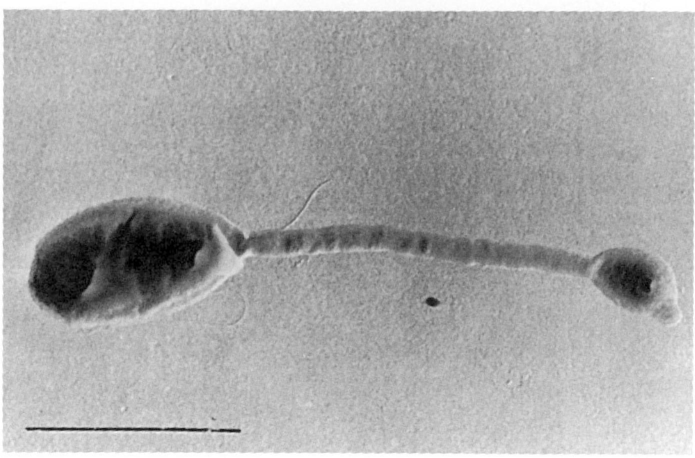

Figure 23.14 **Prosthecate, Budding Bacteria.** *Hyphomicrobium* morphology. *Hyphomicrobium facilis* with hypha and young bud. Bar = 1μm.

The majority of genera in this section are placed in one of two subsections: prosthecate bacteria and nonprosthecate bacteria. Each subsection is further divided into budding and nonbudding bacteria. Chemoheterotrophic genera are described in section 21; members of different nutritional categories are usually placed in other sections with bacteria of the same nutritional type. Two examples of prosthecate bacteria are described here: *Hyphomicrobium* and *Caulobacter.*

The genus *Hyphomicrobium* contains chemoheterotrophic, aerobic, budding bacteria that frequently attach to solid objects in freshwater, marine, and terrestrial environments. (They even grow in laboratory water baths.) The vegetative cell measures about 0.5 to 1.0 by 1 to 3 μm (figure 23.14). At the beginning of the reproductive cycle, the mature cell produces a hypha or prostheca, 0.2 to 0.3 μm in diameter, that grows to several μm in length (figure 23.15). The nucleoid

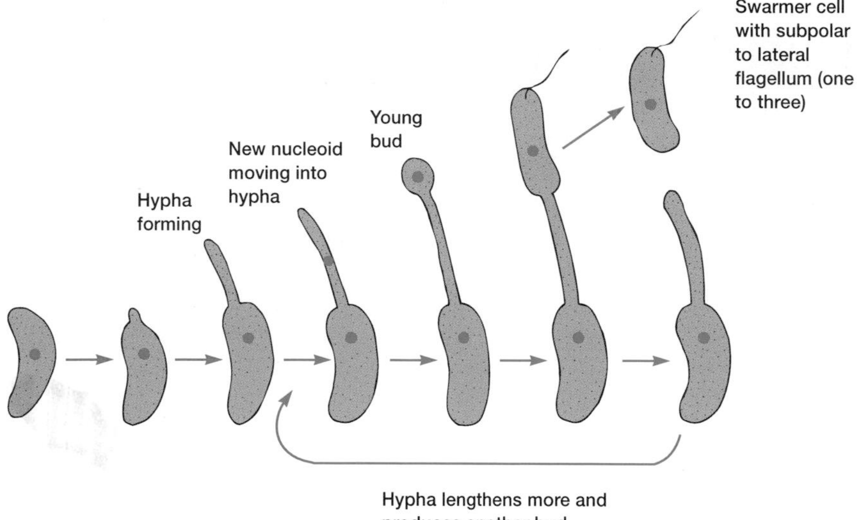

Figure 23.15 **The Life Cycle of *Hyphomicrobium*.**

divides, and a copy moves into the hypha while a bud forms at its end. As the bud matures, it produces one to three flagella, and a septum divides the bud from the hypha. The bud is finally released as an oval- to pear-shaped swarmer cell, which swims off, then settles down and begins budding. The mother cell may bud several times at the tip of its hypha.

Hyphomicrobium is also distinctive in its physiology and nutrition. Sugars and most amino acids do not support abundant growth; instead, *Hyphomicrobium* grows on ethanol and acetate and flourishes with one-carbon compounds such as methanol, formate, and formaldehyde. That is, it is a facultative **methylotroph** and can derive both energy and carbon from reduced one-carbon compounds. It is so efficient at acquiring one-carbon molecules that it grows in medium without an added carbon source (presumably, the medium absorbs sufficient atmospheric carbon compounds). *Hyphomicrobium* may comprise up to 25% of the total bacterial population in oligotrophic or nutrient-poor freshwater habitats.

Bacteria in the genus *Caulobacter* may be polarly flagellated rods or possess a prostheca and **holdfast,** by which they attach to solid substrata (figure 23.16). Caulobacters are usually isolated from freshwater and marine habitats with low nutrient levels, but they are also present in the soil. They often adhere to bacteria, algae, and other microorganisms and may absorb nutrients released by their hosts. The prostheca differs from that of *Hyphomicrobium* in that it lacks cytoplasmic components and is composed almost totally of the plasma membrane and cell wall. It grows longer in nutrient-poor media and can reach more than 10 times the length of the cell body. The prostheca may improve the efficiency of nutrient uptake from dilute habitats by increasing surface area; it also gives the cell extra buoyancy.

The life cycle of *Caulobacter* is unusual (figure 23.17). When ready to reproduce, the cell elongates and a single polar flagellum forms at the end opposite the prostheca. The cell then undergoes asymmetric transverse binary fission to produce a flagellated swarmer cell that swims off. The swarmer eventually comes to rest and forms a new prostheca on the flagellated end while its flagellum disappears. The cycle takes only about two hours for completion.

1. Define the following terms: prostheca, stalk, budding, swarmer cell, methylotroph, and holdfast.
2. Briefly describe the morphology and life cycles of *Hyphomicrobium* and *Caulobacter*.
3. What is unusual about the physiology of *Hyphomicrobium*? How does this influence its ecological distribution?

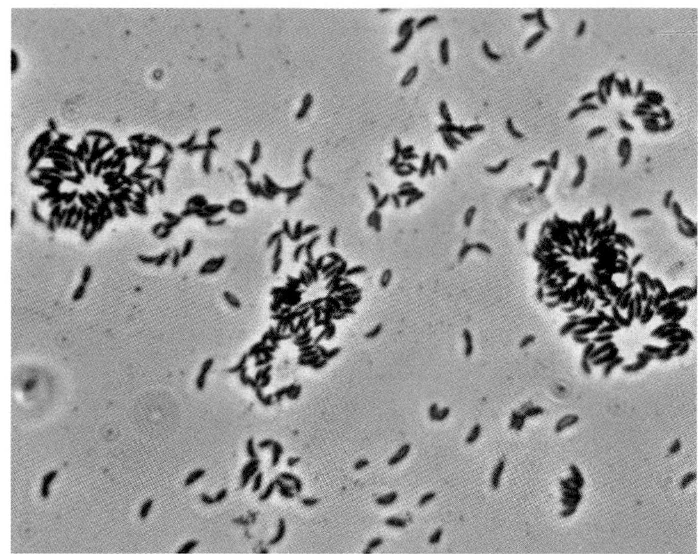

(a)

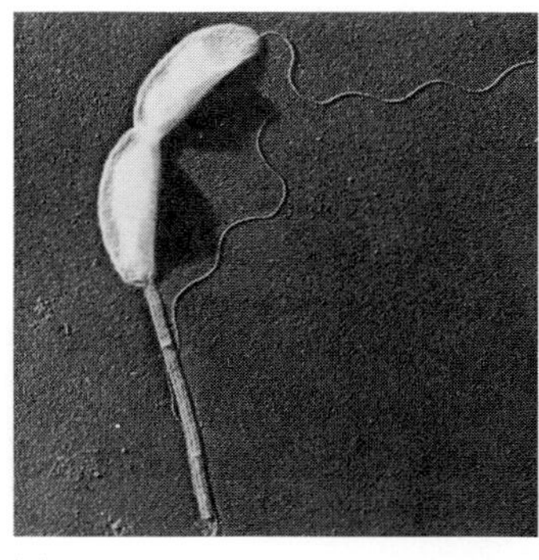

(b)

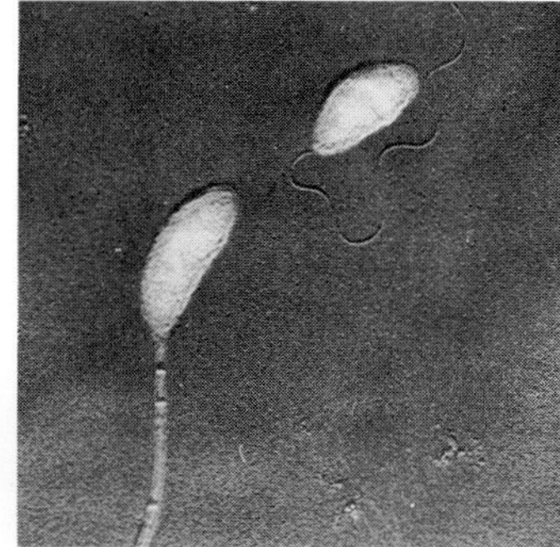

(c)

(d)

Figure 23.16 ***Caulobacter* Morphology and Reproduction.** (*a*) Rosettes of cells adhering to each other by their prosthecae, phase contrast (×600). (*b*) A cell dividing to produce a swarmer (×6,030). Note prostheca and flagellum. (*c*) A cell with a flagellated swarmer (×6,030). (*d*) A rosette as seen in the electron microscope (bar = 1μm).

Sheathed Bacteria

Some bacteria have a **sheath,** a hollow tubelike structure surrounding a chain of cells. Sheaths are often close fitting, but they are never in intimate contact with the cells they enclose and may contain ferric or manganic oxides. They have at least two functions. Sheaths help bacteria attach to solid surfaces and acquire nutrients from slowly running water as it flows past, even if it is nutrient-poor. Sheaths also protect against predators such as protozoa and *Bdellovibrio* (*see chapter 21*). Section 22 of *Bergey's Manual* contains seven genera of gram-negative, aerobic, heterotrophic, sheathed bacteria.

Two of the best-studied genera are *Sphaerotilus* and *Leptothrix* (figures 23.18 and 23.19). *Sphaerotilus* consists of a long sheathed chain of rods, 0.7 to 2.4 by 3 to 10 μm, attached to submerged plants, rocks, and other solid objects, often by a holdfast (figure 23.18). The sheaths are not usually encrusted by metal oxides. Single swarmer cells with a bundle of subpolar flagella escape the filament and form a new chain after attaching to a solid object at another site. *Sphaerotilus* prefers slowly running fresh water polluted with sewage or industrial waste. It grows so well in activated sewage sludge that it sometimes forms tangled masses of filaments and interferes with the proper settling of sludge (*see chapter 41*).

1. What is a sheath and of what advantage is it?
2. How does *Sphaerotilus* maintain its position in running water? How does it reproduce and disperse its progeny?

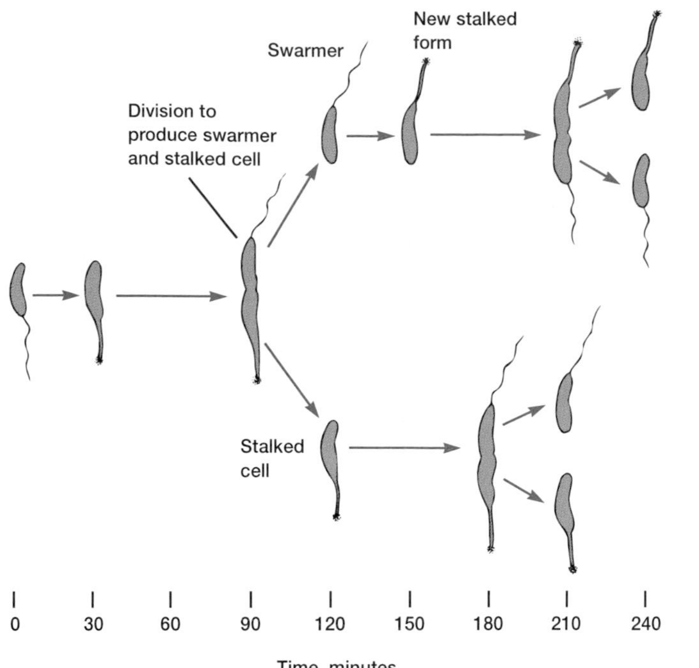

Figure 23.17 ***Caulobacter*** **Life Cycle.** See text for details.

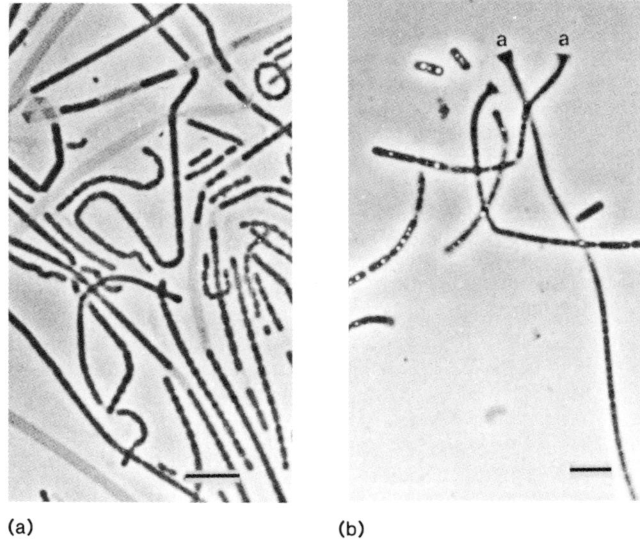

Figure 23.18 **Sheathed Bacteria, *Sphaerotilus natans*.** (*a*) Sheathed chains of cells and empty sheaths. (*b*) Chains with holdfasts (*a*) and individual cells containing poly-β-hydroxybutyrate granules. Bars = 10 μm.

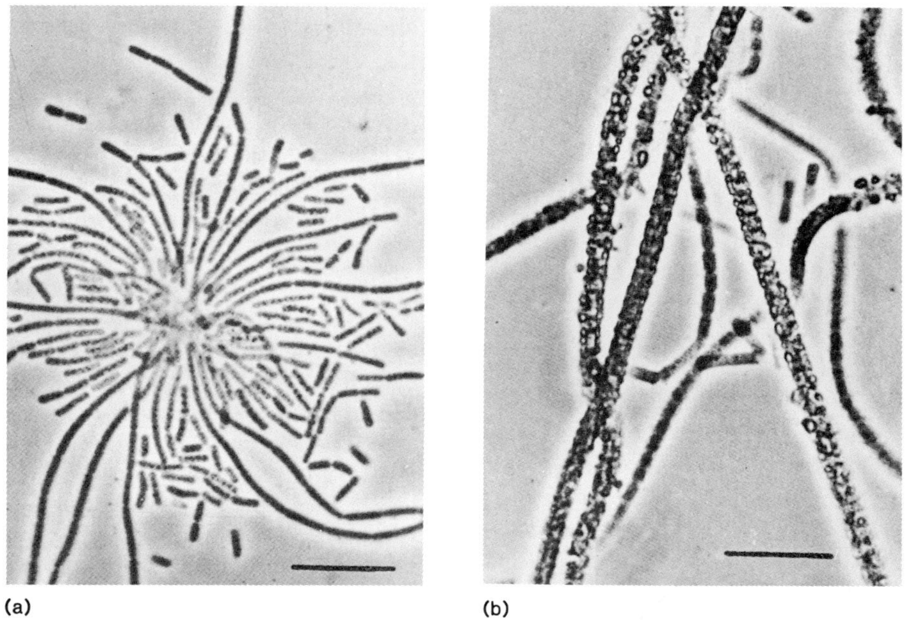

Figure 23.19 **Sheathed Bacteria, *Leptothrix*** Morphology. (*a*) *L. lopholea* trichomes radiating from a collection of holdfasts. (*b*) *L. cholodnii* sheaths encrusted with MnO_2. Bars = 10 μm.

Box 23.1

The Mechanism of Gliding Motility

Gliding motility varies greatly in rate (from about 2 μm per minute to over 600 μm per minute) and in the nature of the motion. Bacteria such as *Myxococcus* and *Flexibacter* glide along in a direction parallel to the longitudinal axis of their cells. Others (*Saprospira*) travel with a screwlike motion or even move in a direction perpendicular to the long axis of the cells in their trichome (*Simonsiella*). *Beggiatoa,* cyanobacteria, and some other bacteria rotate around their longitudinal axis while gliding, but this is not always seen. Many will flex or twitch as well as glide. Such diversity in gliding movement may indicate that more than one mechanism for motility exists. This conclusion is supported by the observation that some gliders (e.g., *Cytophaga* and *Flexibacter*) move attached latex beads over their surface, whereas others such as *Myxococcus* do not. (That is, not all gliding bacteria have moving cell-surface components.) Although slime is required for gliding, it does not appear to propel bacteria directly; rather, it probably attaches them to the substratum and lubricates the surface for more efficient movement.

A variety of mechanisms for gliding motility have been proposed. Cytoplasmic fibrils or filaments are associated with the envelope of many gliding bacteria. In *Oscillatoria,* they seem to be contractile and may produce waves in the outer membrane, resulting in movement. Ringlike protein complexes or rotary assemblies resembling flagellar basal bodies are present in some envelopes. These assemblies may spin and move the bacterium along. There is some evidence that differences in surface tension can propel *Myxococcus xanthus. Myxococcus* may secrete a surfactant at its posterior end (the end opposite the direction of movement) that lowers the surface tension at the rear end of the rod. The cell would be pulled forward by the greater surface tension exerted on its anterior end.

Nonphotosynthetic, Nonfruiting, Gliding Bacteria

Gliding motility is present in a wide diversity of taxa: fruiting and nonfruiting aerobic chemoheterotrophs, cyanobacteria, green nonsulfur bacteria, and at least two gram-positive genera (*Heliobacterium* and *Desulfonema*). Gliding bacteria lack flagella and are stationary while suspended in liquid medium. When in contact with a surface, they glide along, leaving a slime trail; the gliding mechanism is unknown (Box 23.1). Movement can be very rapid; some cytophagas travel 150 μm in a minute, whereas filamentous gliding bacteria may reach speeds of more than 600 μm/minute. Young organisms are the most motile, and motility is often lost with age. Low nutrient levels usually stimulate gliding.

Gliding motility gives a bacterium many advantages. Many aerobic chemoheterotrophic gliding bacteria actively digest insoluble macromolecular substrates such as cellulose and chitin, and gliding motility is ideal for searching these out. Gliding movement is well adapted to drier habitats and to movement within solid masses such as soil, sediments, and rotting wood that are permeated by small channels. Finally, gliding bacteria, like flagellated bacteria, can position themselves at optimal levels of light intensity, oxygen, hydrogen sulfide, temperature, and other factors that influence growth.

Seven genera containing rods or individual filaments are placed in the order *Cytophagales* and family *Cytophagaceae.* Bacteria of the genus *Cytophaga* are slender rods, often with pointed ends (figure 23.20*b*). They differ from the gliding, fruiting myxobacteria in lacking fruiting bodies and having a low G + C ratio. *Sporocytophaga* is similar to *Cytophaga* but forms spherical resting cells called microcysts (figure 23.20*c,d*). Another related genus, *Flexibacter,* produces long, flexible threadlike cells when young (figure 23.20*a*) and is unable to use complex polysaccharides.

Members of the genera *Cytophaga* and *Sporocytophaga* are aerobes that actively degrade complex polysaccharides. Soil cytophagas digest cellulose; both soil and marine forms attack chitin, pectin, and keratin. Some marine species even degrade agar. Cytophagas play a major role in the mineralization of organic matter and can cause great damage to exposed fishing gear and wooden structures. They are also a major component of the bacterial population in sewage treatment plants and presumably contribute significantly to this waste treatment process.

Although most cytophagas are free-living, some can be isolated from vertebrate hosts and are pathogenic. *Cytophaga columnaris* and others cause diseases such as columnaris disease, cold water disease, and fin rot in freshwater and marine

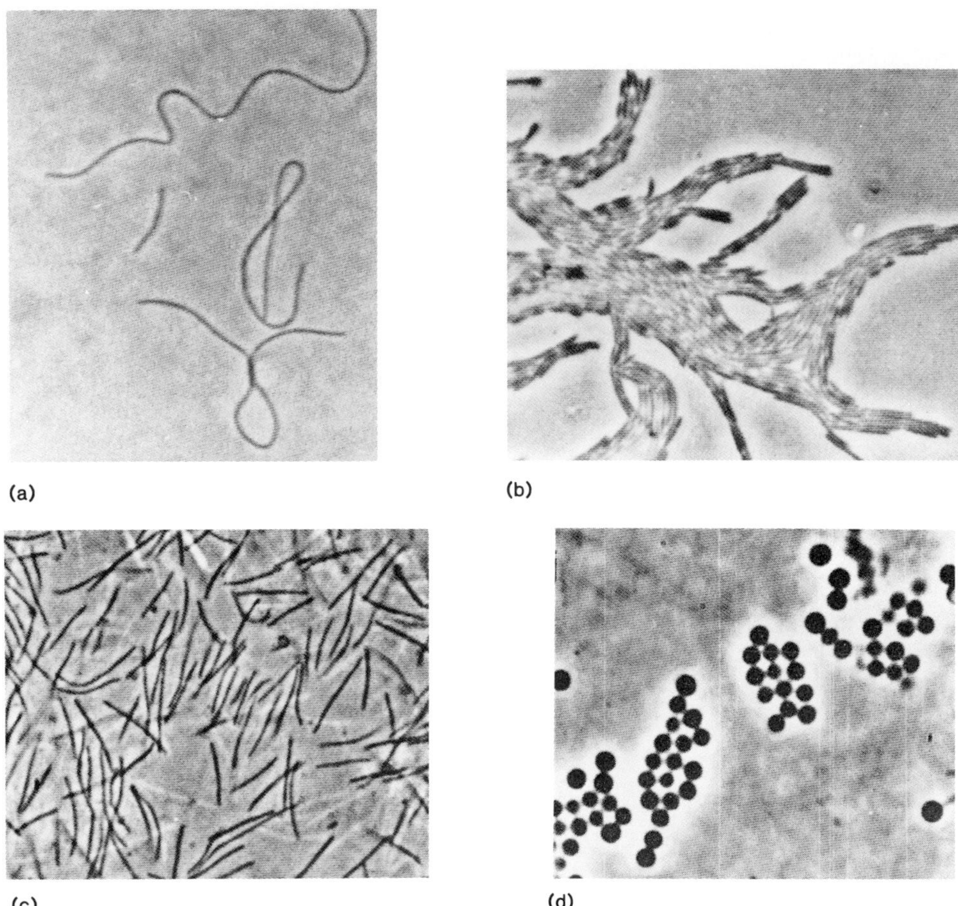

Figure 23.20 Nonphotosynthetic, Nonfruiting, Gliding Bacteria. Representative members of the order *Cytophagales*. (*a*) Long thread cells of *Flexibacter elegans* (×1,100). (*b*) *Cytophaga* sp. (×1,150). (*c*) *Sporocytophaga myxococcoides*, vegetative cells on agar (×1,170). (*d*) *Sporocytophaga myxococcoides*, mature microcysts (×1,750).

fish. *Capnocytophaga* flourishes in the gingival crevice (the space between the tooth surface and the gum) of humans and is unusual among the cytophagas in being anaerobic. It may be involved in the development of periodontal disease.

The order *Beggiatoales* contains gliding bacteria that form trichomes (figure 23.21). Two of the best-studied genera in this order are *Beggiatoa* and *Leucothrix* (figures 23.22 and 23.23). *Beggiatoa* is microaerophilic and grows in sulfide-rich habitats such as sulfur springs, fresh water with decaying plant material, rice paddies, salt marshes, and marine sediments. Its filaments contain short, disklike cells and lack a sheath. *Beggiatoa* is very versatile metabolically. It oxidizes hydrogen sulfide to form large sulfur grains located in pockets formed by invaginations of the plasma membrane. *Beggiatoa* can subsequently oxidize the sulfur to sulfate. The sulfur electrons are used by the electron transport chain in energy production.

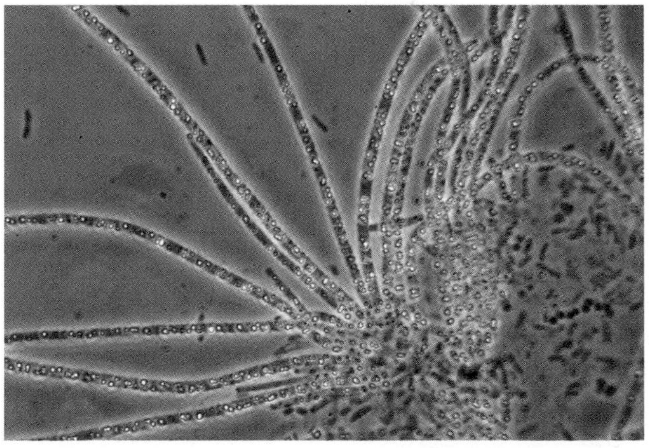

Figure 23.21 ***Thiothrix.*** A *Thiothrix* colony viewed with phase-contrast microscopy (×1,000).

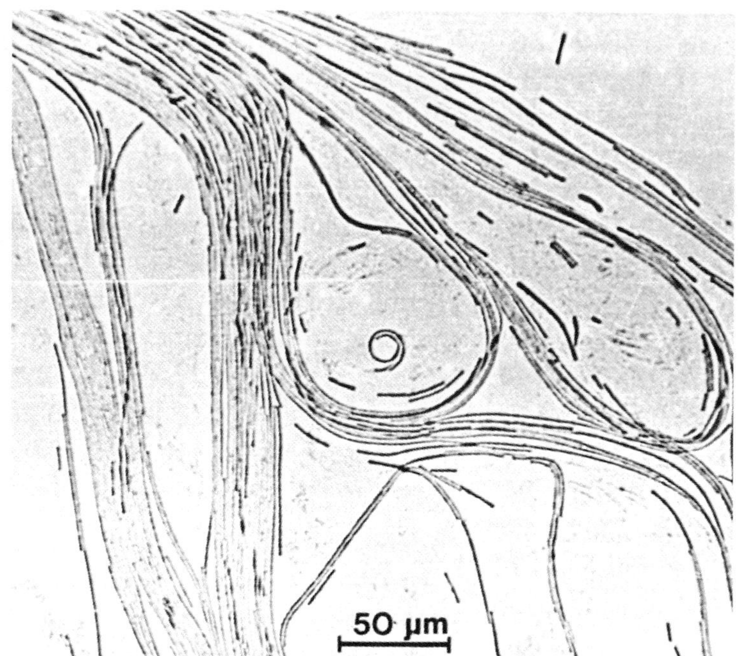

Figure 23.22 ***Beggiatoa.*** *Beggiatoa* sp. colony growing on agar.

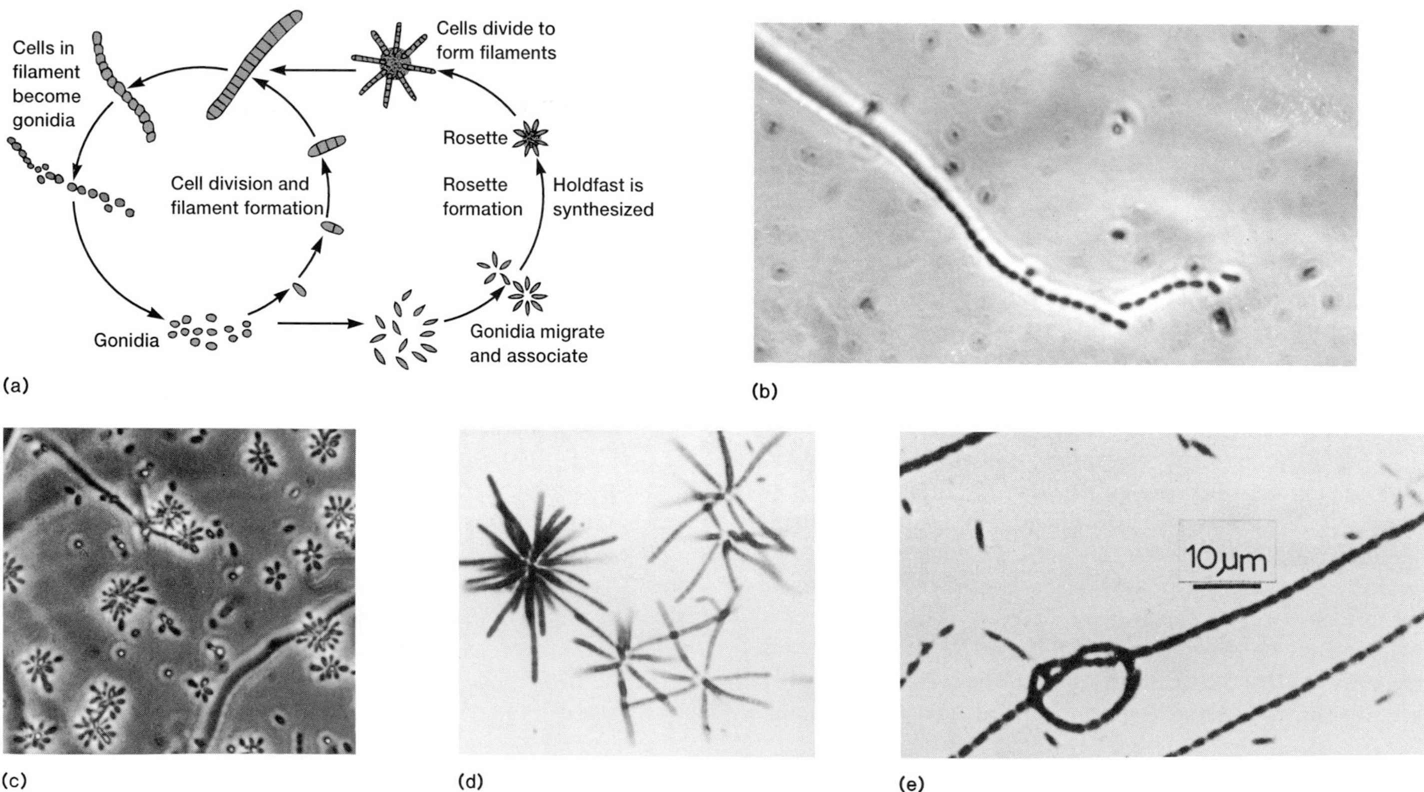

Figure 23.23 **Morphology and Reproduction of *Leucothrix mucor.*** (*a*) Life cycle of *L. mucor.* (*b*) Separation of gonidia from the tip of a mature filament, phase contrast (×1,400). (*c*) Gonidia aggregating to form rosettes, phase contrast (×950). (*d*) Young developing rosettes (×1,500). (*e*) A knot formed by a *Leucothrix* filament.

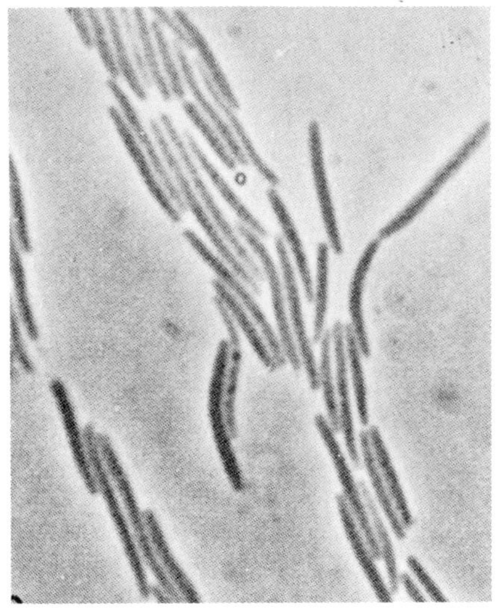
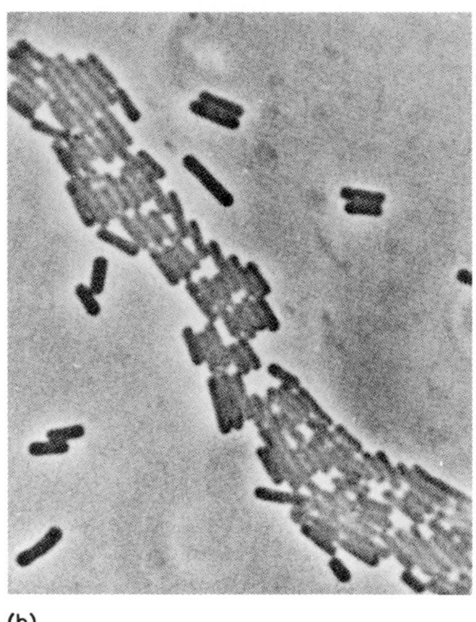
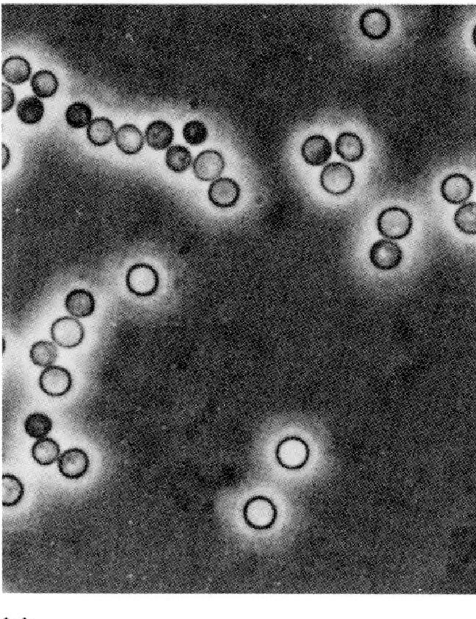

(a) (b) (c)

Figure 23.24 **Gliding, Fruiting Bacteria (Myxobacteria).** Myxobacterial cells and myxospores. (*a*) *Stigmatella aurantiaca* (×1,200). (*b*) *Chondromyces crocatus* (×950). (*c*) Myxospores of *Myxococcus xanthus* (×1,100). All photographs taken with a phase-contrast microscope.

Many strains also can grow heterotrophically with acetate as a carbon source, and some may incorporate CO_2 autotrophically.

1. Give three advantages of gliding motility.
2. Briefly describe the following genera: *Cytophaga, Sporocytophaga,* and *Beggiatoa.*
3. Why are the cytophagas ecologically important?

Gliding, Fruiting Bacteria

The **myxobacteria** are gram-negative, aerobic soil bacteria characterized by gliding motility, a complex life cycle with the production of fruiting bodies, and the formation of dormant myxospores. In addition, their G + C content is around 67 to 71%, significantly higher than that of most gliding bacteria. Myxobacterial cells are rods, about 0.6 to 0.9 by 3 to 8 μm long, and may be either slender with tapered ends or stout with rounded, blunt ends (figure 23.24). The approximately 30 species of myxobacteria in the order *Myxobacteriales* are divided into four families based on the shape of vegetative cells, myxospores, and sporangia.

Most myxobacteria are micropredators or scavengers. They secrete an array of digestive enzymes that lyse bacteria and yeasts. Many myxobacteria also secrete antibiotics, which may kill their prey. The digestion products, primarily small peptides, are absorbed. Most myxobacteria use amino acids as their major source of carbon, nitrogen, and energy. All are chemoheterotrophs with respiratory metabolism.

The myxobacterial life cycle is quite distinctive and in many ways resembles that of the cellular slime molds (figure 23.25). In the presence of a food supply, myxobacteria migrate along a solid surface, feeding and leaving slime trails. During this stage, the cells often form a swarm and move in a coordinated fashion. Some species congregate to produce a sheet of cells that moves rhythmically to generate waves or ripples. When their nutrient supply is exhausted, the myxobacteria aggregate and differentiate into a **fruiting body.** Such bodies range in height from 50 to 500 μm and are often attractively colored red, yellow, or brown by carotenoid pigments. They vary in complexity from simple globular objects (*Myxococcus*) to the elaborate, branching, treelike structures formed by *Stigmatella* and *Chondromyces* (figure 23.26). Some cells develop into dormant **myxospores** that are often enclosed in walled structures called sporangioles or sporangia. Each species forms a characteristic fruiting body.

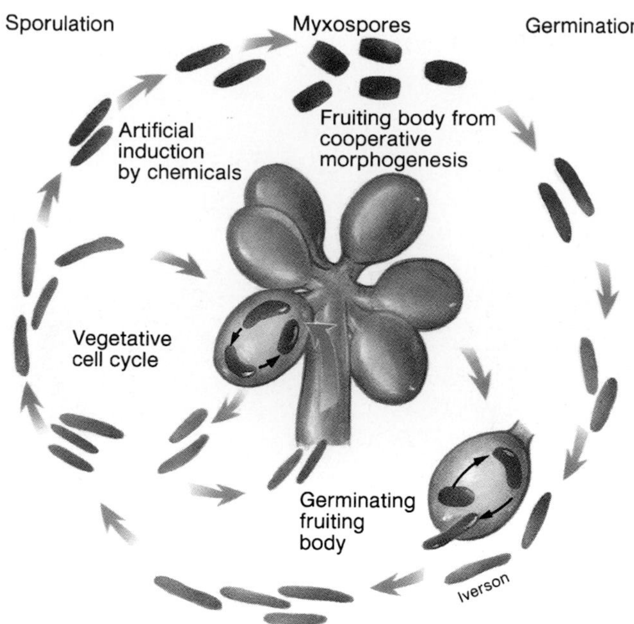

Figure 23.25 Myxobacterial Life Cycle. The outermost sequence depicts the chemical induction of myxospore formation followed by germination. Regular fruiting body production and myxospore germination are also shown.

Myxospores are not only dormant but desiccation-resistant, and they may survive up to 10 years under adverse conditions. They enable myxobacteria to survive long periods of dryness and nutrient deprivation. The use of fruiting bodies provides further protection for the myxospores and assists in their dispersal. (The myxospores are often suspended above the soil surface.) Because myxospores are kept together within the fruiting body, a colony of myxobacteria automatically develops when the myxospores are released and germinate. This communal organization may be advantageous because myxobacteria obtain nutrients by secreting hydrolytic enzymes and absorbing soluble digestive products. A mass of myxobacteria can produce enzyme concentrations sufficient to digest their prey more easily than can an individual cell. Extracellular enzymes diffuse away from their source, and an individual cell will have more difficulty overcoming diffusional losses than a swarm of cells.

Myxobacteria are found in soils worldwide. They are most commonly isolated from neutral soils or decaying plant material such as leaves and tree bark, and from animal dung. Although they grow in habitats as diverse as tropical rain forests and the arctic tundra, they are most abundant in warm areas.

1. Give the major distinguishing characteristics of the myxobacteria. How do they obtain most of their nutrients?
2. Briefly describe the myxobacterial life cycle. What are fruiting bodies, myxospores, and sporangioles?

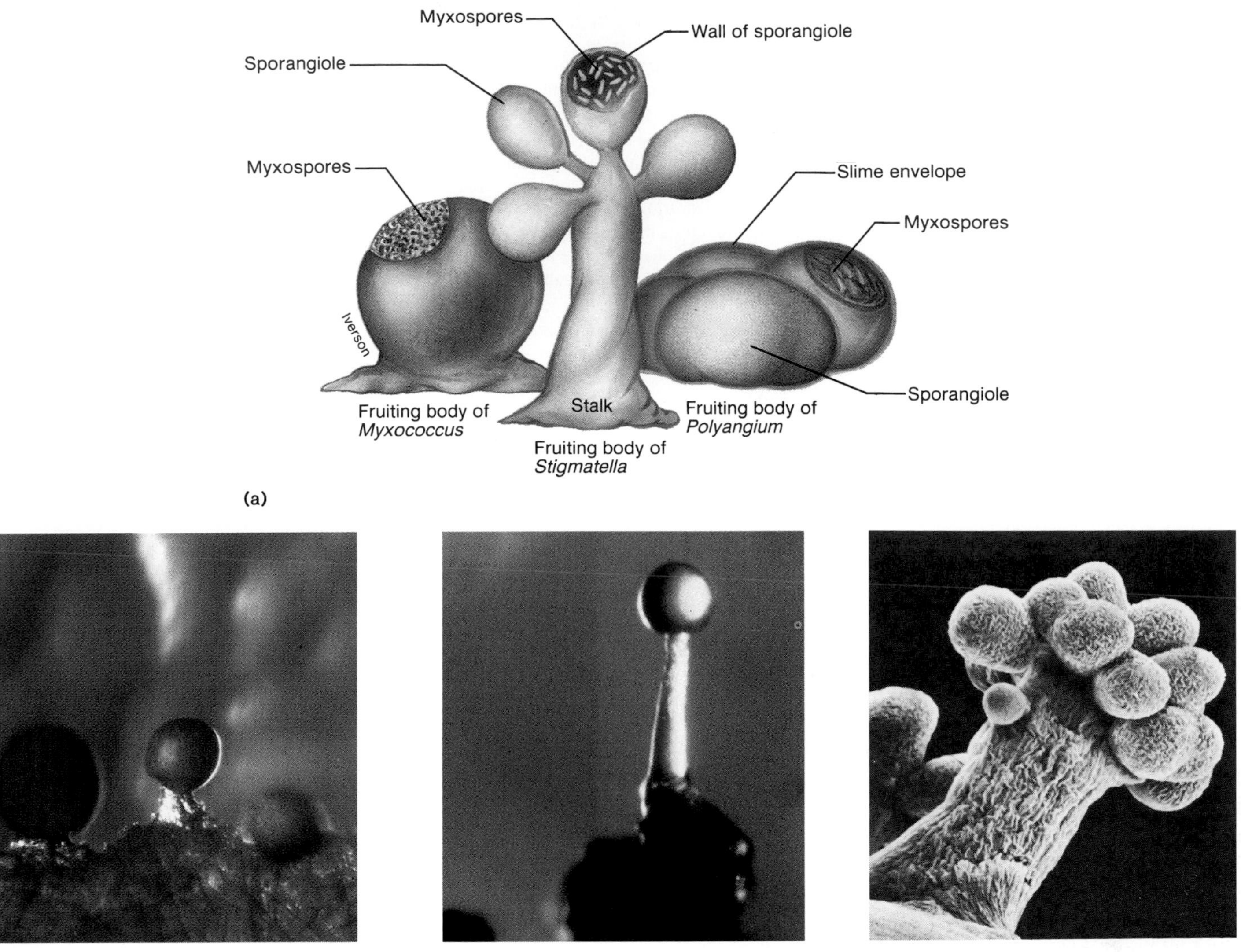

Figure 23.26 **Myxobacterial Fruiting Bodies.** (*a*) An illustration of typical fruiting body structure. (*b*) *Myxococcus fulvus.* Fruiting bodies are about 150–400 μm high. (*c*) *Myxococcus stipitatus.* Stalk is as tall as 200 μm. (*d*) *Chondromyces crocatus* viewed with the SEM. The stalk may reach 700 μm or more in height.

Summary

1. Cyanobacteria carry out oxygenic photosynthesis, and purple and green bacteria use anoxygenic photosynthesis.
2. The four most important groups of purple and green photosynthetic bacteria are the purple sulfur bacteria, the purple nonsulfur bacteria, the green sulfur bacteria, and the green nonsulfur bacteria.
3. The bacteriochlorophyll pigments of purple and green bacteria enable them to live in deeper, anaerobic zones of aquatic habitats.
4. Cyanobacteria carry out oxygenic photosynthesis by means of a photosynthetic apparatus similar to that of the eucaryotes. Like the red algae, they have phycobilisomes.
5. Cyanobacteria reproduce by binary fission, budding, multiple fission, and fragmentation to form hormogonia. Some will produce a dormant akinete.
6. Nitrogen-fixing cyanobacteria usually form heterocysts, specialized cells in which nitrogen fixation occurs.
7. Chemolithotrophic bacteria derive energy and electrons from reduced inorganic compounds. Nitrifying bacteria are aerobic bacteria that oxidize either ammonia or nitrite to nitrate and are responsible for nitrification. Colorless sulfur bacteria oxidize elemental sulfur, hydrogen sulfide, and thiosulfate to sulfate.
8. Section 21 of *Bergey's Manual* contains bacteria with at least one of three features: a prostheca, a stalk, or reproduction by budding.
9. Two examples of the budding and/or appendaged bacteria are *Hyphomicrobium* (budding bacteria that produce swarmer cells), and *Caulobacter* (bacteria with prosthecae and holdfasts).
10. Section 22 contains bacteria such as *Sphaerotilus* with sheaths, hollow tubelike structures that surround chains of cells without being in intimate contact.
11. Gliding motility is present in a diversity of bacteria.
12. Cytophagas degrade proteins and complex polysaccharides and are active in the mineralization of organic matter.
13. Many bacteria exist as trichomes.
14. Myxobacteria are gram-negative, aerobic soil bacteria with gliding motility and a complex life cycle that leads to the production of dormant myxospores held within fruiting bodies.

Key Terms

akinete *474*
anoxygenic photosynthesis *468*
baeocyte *476*
binary fission *480*
budding *480*
chlorosomes *471*
cyanobacteria *472*
fruiting body *487*
gliding motility *484*
green nonsulfur bacteria *471*
green sulfur bacteria *471*
heterocyst *474*
holdfast *481*
hormogonia *474*
methylotroph *481*
myxobacteria *487*
myxospore *487*
nitrification *478*
nitrifying bacteria *477*
oxygenic photosynthesis *468*
phycobilisome *472*
prostheca *480*
purple nonsulfur bacteria *469*
purple sulfur bacteria *469*
sheath *482*
stalk *480*
trichome *473*

Questions for Thought and Review

1. How do cyanobacteria differ from the purple and green photosynthetic bacteria?
2. Relate the physiology of each major group of photosynthetic bacteria to its preferred habitat.
3. Why is nitrogen fixation an oxygen-sensitive process? How are cyanobacteria able to fix nitrogen when they also carry out oxygenic photosynthesis?
4. A number of bacterial groups are found in particular habitats to which they are well adapted. Give the habitat for each of the following groups and discuss the reasons for its preference: *Beggiatoa,* purple and green sulfur bacteria, *Hyphomicrobium,* and *Thiobacillus.*
5. Why are gliding and budding and/or appendaged bacteria distributed among so many different sections in *Bergey's Manual*?
6. Discuss the ways in which the distinctive myxobacterial life cycle may be of great advantage to these organisms.

Additional Reading

Allen, M. M. 1984. Cyanobacterial cell inclusions. *Ann. Rev. Microbiol.* 38:1–25.

Balows, A.; Trüper, H. G.; Dworkin, M.; Harder, W.; and Schleifer, K.-H. 1992. *The prokaryotes,* 2d ed. New York: Springer-Verlag.

Burchart, R. P. 1981. Gliding motility of prokaryotes: Ultrastructure, physiology, and genetics. *Ann. Rev. Microbiol.* 35:497–529.

Dworkin, M., and Kaiser, D. 1985. Cell interactions in myxobacterial growth and development. *Science* 230:18–24.

Ghiorse, W. C. 1984. Biology of iron- and manganese-depositing bacteria. *Ann. Rev. Microbiol.* 38:515–50.

Glazer, A. N. 1983. Comparative biochemistry of photosynthetic light-harvesting systems. *Ann. Rev. Biochem.* 52:125–57.

Hirsch, P. 1974. Budding bacteria. *Ann. Rev. Microbiol.* 28:391–444.

Holt, J. G., editor-in-chief. 1989. *Bergey's manual of systematic bacteriology,* vol. 3, eds. J. T. Staley, M. P. Bryant, and N. Pfennig. Baltimore, Md.: Williams & Wilkins.

Larkin, J. M., and Strohl, W. R. 1983. *Beggiatoa, Thiothrix,* and *Thioploca. Ann. Rev. Microbiol.* 37:341–67.

Mayer, F. 1986. *Cytology and morphogenesis of bacteria.* Berlin: Gebrüder Borntraeger.

Moore, R. L. 1981. The biology of *Hyphomicrobium* and other prosthecate, budding bacteria. *Ann. Rev. Microbiol.* 35:567–94.

Peters, G. A. 1978. Blue-green algae and algal associations. *BioScience* 28(9):580–85.

Poindexter, J. S. 1981. The Caulobacters: Ubiquitous unusual bacteria. *Microbiol. Rev.* 45:123–79.

Reichenbach, H. 1981. Taxonomy of the gliding bacteria. *Ann. Rev. Microbiol.* 35:339–64.

Rogers, L. J., and Gallon, J. R., eds. 1988. *Biochemistry of the algae and cyanobacteria.* New York: Oxford University Press.

Schlegel, H. G., and Bowien, B. 1989. *Autotrophic bacteria.* Madison, Wis.: Science Tech Publishers.

Shapiro, J. A. 1988. Bacteria as multicellular organisms. *Sci. Am.* 258(6):82–89.

Shapiro, L. 1976. Differentiation in the *Caulobacter* cell cycle. *Ann. Rev. Microbiol.* 30:377–407.

Shimkets, L. J. 1990. Social and developmental biology of the myxobacteria. *Microbiol. Rev.* 54(4):473–501.

Stanier, R. Y., and Cohen-Bazire, G. 1977. Phototrophic prokaryotes: The cyanobacteria. *Ann. Rev. Microbiol.* 31:225–74.

Stewart, W. D. P. 1980. Some aspects of structure and function in N_2-fixing cyanobacteria. *Ann. Rev. Microbiol.* 34:497–536.

Wolk, C. P. 1973. Physiology and cytological chemistry of blue-green algae. *Bacteriol. Rev.* 37(1):32–101.

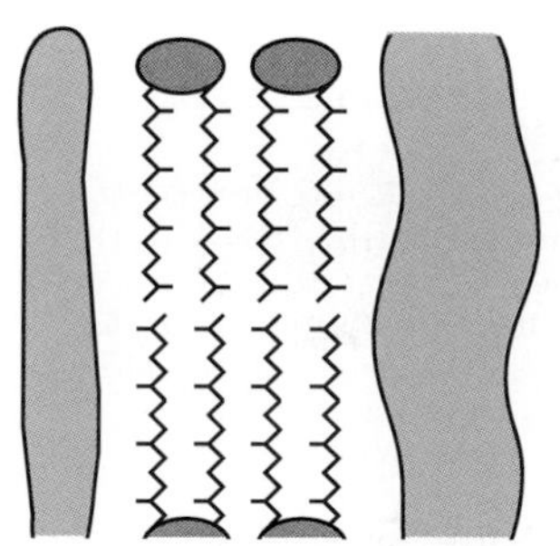

CHAPTER 24
The Bacteria: The Archaeobacteria

As is often the case, epoch-making ideas carry with them implicit, unanalyzed assumptions that ultimately impede scientific progress until they are recognized for what they are. So it is with the prokaryote-eukaryote distinction. Our failure to understand its true nature set the stage for the sudden shattering of the concept when a "third form of life" was discovered in the late 1970s, a discovery that actually left many biologists incredulous. Archaebacteria, as this third form has come to be known, have revolutionized our notion of the prokaryote, have altered and refined the way in which we think about the relationship between prokaryotes and eukaryotes . . . and will influence strongly the view we develop of the ancestor that gave rise to all extant life.

—C. R. Woese and R. S. Wolfe

Outline

Concepts

1. Archaeobacteria differ in many ways from both eubacteria and eucaryotes. These include differences in cell wall structure and chemistry, membrane lipid structure, molecular biology, and metabolism.
2. Archaeobacteria grow in a few restricted or specialized habitats: anaerobic, hypersaline, and high temperature.
3. *Bergey's Manual* divides the archaeobacteria into five major groups: methanogenic archaeobacteria, sulfate reducers, extreme halophiles, cell wall-less archaeobacteria, and extremely thermophilic S^0-metabolizers.
4. Methanogenic and sulfate-reducing archaeobacteria have unique cofactors that participate in methanogenesis.
5. Archaeobacteria have special structural, chemical, and metabolic adaptations that enable them to grow in extreme environments.

Chapters 21 and 22 introduce the bacteria described in volumes 1 and 2 of *Bergey's Manual.* Chapter 23 covers the bacteria found in sections 18 to 24 of volume 3. Although the archaeobacteria are described in section 25 of volume 3, they are placed in a separate chapter here because they are so different from other procaryotes.

This chapter begins with a general introduction to the archaeobacteria. Then it briefly discusses the biology of each major archaeobacterial group.

Comparison of the sequences of rRNA from a great variety of organisms shows that organisms may be divided into three major groups: the eubacteria, archaeobacteria, and eucaryotes (*see figure 20.11*). Some of the most important differences are summarized in table 20.8 (*p. 417*). Because archaeobacteria are different from both eubacteria and eucaryotes, their most distinctive properties will first be described in more detail than earlier and compared with those of the latter two groups.

rRNA sequences and the archaeobacteria (pp. 415–21).

Introduction to the Archaeobacteria

As a group, the **archaeobacteria** (Greek *archaios,* ancient, and *bakterion,* a small rod) are quite diverse, both in morphology and physiology. They can stain either gram positive or gram negative, and may be spherical, rod-shaped, spiral, lobed, or plate-shaped. Some are single cells, while others form filaments or aggregates. They range in diameter from 0.1 to over 15 μm, and some filaments can grow up to 200 μm in length. Multiplication may be by binary fission, budding, fragmentation, or other mechanisms. Archaeobacteria are just as diverse physiologically. They can be aerobic, facultatively anaerobic, or strictly anaerobic. Nutritionally, they range from chemolithoautotrophs to organotrophs. Some are mesophiles, while others are extreme thermophiles and grow above 100°C.

Archaeobacteria usually prefer restricted or extreme aquatic and terrestrial habitats. They are often present in anaerobic, hypersaline, or high-temperature environments. A few are symbionts in animal digestive systems.

Archaeobacterial Cell Walls

Although archaeobacteria can stain either gram positive or gram negative, their wall structure and chemistry differ from that of the eubacteria. There is considerable variety in archaeobacterial wall structure. Many gram-positive archaeobacteria have a wall with a single thick homogeneous layer like gram-positive eubacteria (figure 24.1*a*). Gram-negative archaeobacteria lack the outer membrane and complex peptidoglycan network or sacculus of gram-negative eubacteria. Instead, they usually have a surface layer of protein or glycoprotein subunits (figure 24.1*b*).

Peptidoglycan structure and chemistry (pp. 51–53).

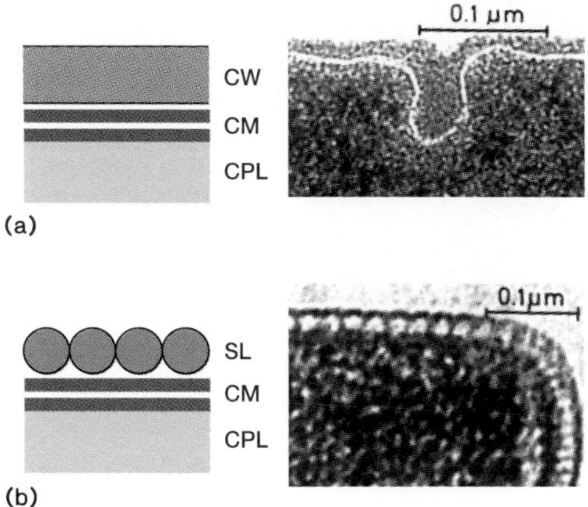

Figure 24.1 **Cell Envelopes of Archaeobacteria.** Schematic representations and electron micrographs of (*a*) *Methanobacterium formicicum,* a typical gram-positive organism, and (*b*) *Thermoproteus tenax,* a gram-negative archaeobacterium. *CW,* cell wall; *SL,* surface layer; *CM,* cell membrane or plasma membrane; *CPL,* cytoplasm.

The chemistry of archaeobacterial cell walls is also quite different from that of the eubacteria. None have the muramic acid and D-amino acids characteristic of eubacterial peptidoglycan. Not surprisingly, all archaeobacteria resist attack by lysozyme and β-lactam antibiotics such as penicillin. Gram-positive archaeobacteria can have a variety of complex polymers in their walls. *Methanobacterium* and some other methanogens have walls containing **pseudomurein,** a peptidoglycanlike polymer that has L-amino acids in its cross-links, N-acetyltalosaminuronic acid instead of N-acetylmuramic acid, and $\beta(1 \rightarrow 3)$ glycosidic bonds instead of $\beta(1 \rightarrow 4)$ glycosidic bonds (figure 24.2). *Methanosarcina* and *Halococcus* contain complex polysaccharides similar to the chondroitin sulfate of animal connective tissue. Other heteropolysaccharides are also found in gram-positive walls.

Gram-negative archaeobacteria have a layer of protein or glycoprotein outside their plasma membrane. The layer may be as thick as 20 to 40 nm. Sometimes there are two layers, a sheath surrounding an electron-dense layer. The chemical content of these walls varies considerably. Some methanogens (*Methanolobus*), *Halobacterium,* and several extreme thermophiles (*Sulfolobus, Thermoproteus,* and *Pyrodictium*) have glycoproteins in their walls. In contrast, other methanogens (*Methanococcus, Methanomicrobium,* and *Methanogenium*) and the extreme thermophile *Desulfurococcus* have protein walls.

Archaeobacterial Lipids and Membranes

As emphasized in table 20.8, a most distinctive feature of the archaeobacteria is the nature of their membrane lipids. They differ from both eubacteria and eucaryotes in having branched chain hydrocarbons attached to glycerol by ether links rather than fatty acids connected by ester links (figure 24.3). Sometimes two glycerol groups are linked to form an extremely long tetraether. Usually the diether side chains are 20 carbons in

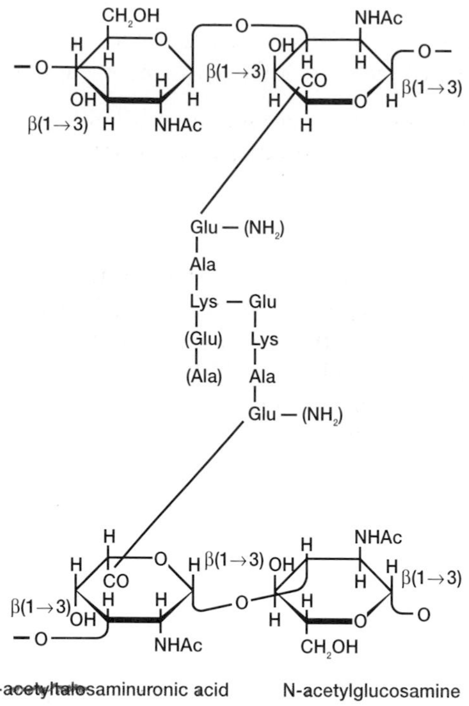

Figure 24.2 The Structure of Pseudomurein. The components in parentheses are not always present. *Ac* represents the acetyl group.

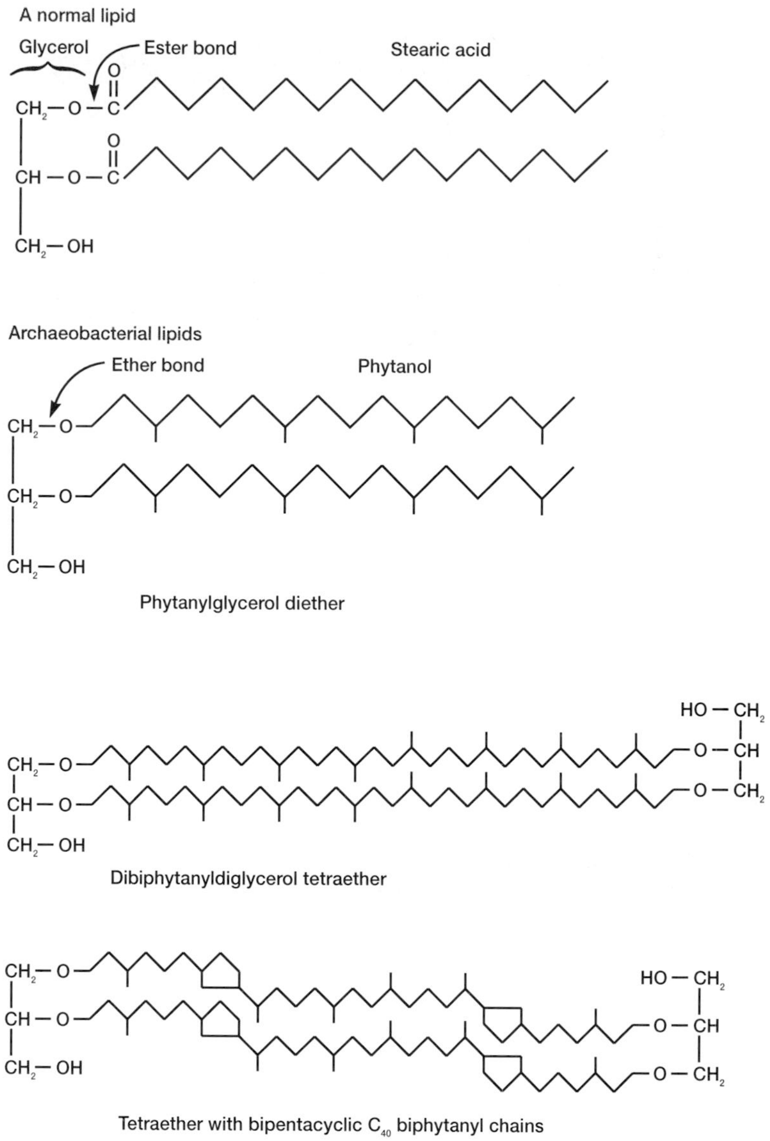

size, and the tetraether chains are 40 carbons. The overall length of the tetraethers can be adjusted by cyclizing the chains to form pentacyclic rings (figure 24.3), and biphytanyl chains may contain from one to four cyclopentyl rings. Polar lipids are also present in archaeobacterial membranes: phospholipids, sulfolipids, and glycolipids. From 7 to 30% of the membrane lipids are nonpolar lipids, which are usually derivatives of squalene (figure 24.4). These lipids can be combined in various ways to yield membranes of different rigidity and thickness. For example, the C_{20} diethers can be used to make a regular bilayer membrane (figure 24.5*a*). A much more rigid monolayer membrane may be constructed of C_{40} tetraether lipids (figure 24.5*b*). Of course, archaeobacterial membranes may contain a mix of diethers, tetraethers, and other lipids. As might be expected from their need for stability, the membranes of extreme thermophiles such as *Thermoplasma* and *Sulfolobus* are almost completely tetraether monolayers.

Genetics and Molecular Biology

Some features of archaeobacterial genetics are similar to those in eubacteria. Their chromosome is a single closed DNA circle. However the genomes of some archaeobacteria are significantly smaller than the normal eubacterium. *E. coli* DNA has a size of about 2.5 × 10^9 daltons, whereas *Thermoplasma acidophilum* DNA is about 0.8 × 10^9 daltons and *Methanobacterium thermoautotrophicum* DNA is 1.1 × 10^9 daltons. The

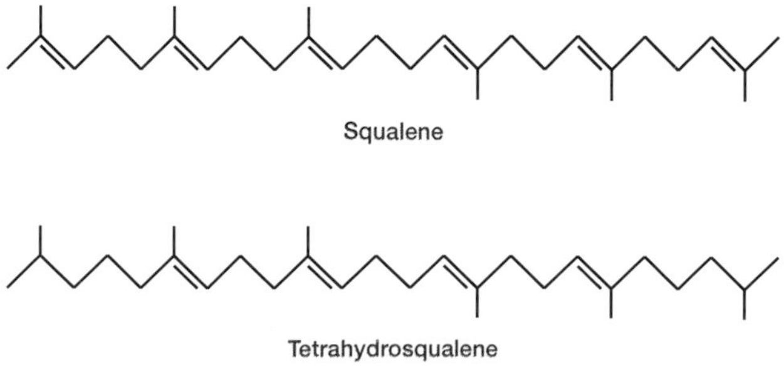

Figure 24.3 Archaeobacterial Membrane Lipids. An illustration of the difference between archaeobacterial lipids and those of other bacteria. Archaeobacterial lipids are derivatives of isopranyl glycerol ethers rather than the usual glycerol fatty acid esters. Three examples of common archaeobacterial glycerolipids are given.

Figure 24.4 Nonpolar Lipids of Archaeobacteria. Two examples of the most predominant nonpolar lipids are the C_{30} isoprenoid squalene and one of its hydroisoprenoid derivatives, tetrahydrosqualene.

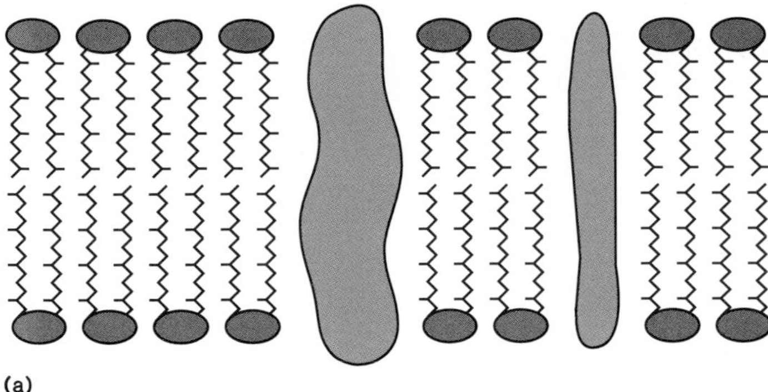

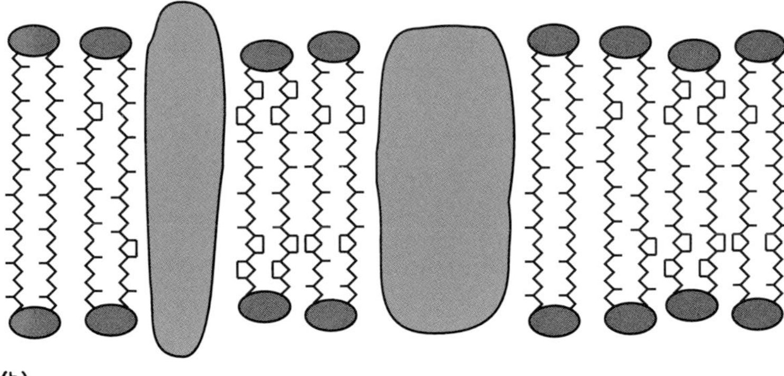

Figure 24.5 Examples of Archaeobacterial Membranes. (*a*) A membrane composed of integral proteins and a bilayer of C_{20} diethers. (*b*) A rigid monolayer composed of integral proteins and C_{40} tetraethers.

variation in G + C content is great, from about 21 to 68 mol%, and is another sign of archaeobacterial diversity. Archaeobacteria have few plasmids.

Archaeobacterial mRNA appears similar to that of eubacteria rather than to eucaryotic mRNA. Polygenic mRNA has been discovered, and there is no evidence for mRNA splicing. Archaeobacterial promoters are similar to those in eubacteria.

Despite the above and other similarities, there are also many differences between archaeobacteria and other organisms. Unlike both eubacteria and eucaryotes, the T ψ C arm (*p. 206*) of archaeobacterial tRNA lacks thymine and contains pseudouridine or 1-methylpseudouridine. The archaeobacterial initiator tRNA carries methionine like the eucaryotic initiator tRNA. Although archaeobacterial ribosomes are 70S like eubacterial ribosomes, electron microscopic studies show that their shape is quite variable and sometimes differs from that of both eubacterial and eucaryotic ribosomes. They do resemble eucaryotic ribosomes in their sensitivity to anisomycin and insensitivity to chloramphenicol and kanamycin. Furthermore, their elongation factor 2 reacts with diphtheria toxin like the eucaryotic EF–2. Finally, archaeobacterial DNA-dependent RNA polymerases resemble the eucaryotic enzymes, not the eubacterial RNA polymerase. They are large, complex enzymes and are insensitive to the drugs rifampin and streptolydigin. These and other differences distinguish the archaeobacteria from both eubacteria and eucaryotes.

Ribosomes and the mechanism of protein synthesis (pp. 208–13).
DNA transcription (pp. 202–6).

Metabolism

Not surprisingly in view of the variety of their life-styles, archaeobacterial metabolism varies greatly between the members of different groups. Some archaeobacteria are organotrophs while others are autotrophic. A few even carry out an unusual form of photosynthesis.

Archaeobacterial carbohydrate metabolism is best understood. The enzyme 6-phosphofructokinase has not been found in archaeobacteria, and they do not appear to degrade glucose by way of the Embden-Meyerhof pathway. Extreme halophiles and thermophiles catabolize glucose using a modified form of the Entner-Doudoroff pathway (*p. 150 and appendix II*) in which the initial intermediates are not phosphorylated. The halophiles have slightly different modifications of the pathway than do the extreme thermophiles, but still produce pyruvate and NADH or NADPH. Methanogens are autotrophs and do not catabolize glucose to any significant extent. In contrast with glucose degradation, gluconeogenesis proceeds by a reversal of the Embden-Meyerhof pathway in halophiles and methanogens. All archaeobacteria that have been studied can oxidize pyruvate to acetyl-CoA. They lack the pyruvate dehydrogenase complex present in eucaryotes and respiratory eubacteria, and use the enzyme pyruvate oxidoreductase for this purpose. Halophiles and the extreme thermophile *Thermoplasma* do seem to have a functional tricarboxylic acid cycle. No methanogen has yet been found with a complete tricarboxylic acid cycle. Evidence for functional cytochrome chains has been obtained in halophiles and thermophiles.

The Embden-Meyerhof pathway and tricarboxylic acid cycle
(pp. 147–48, 150–51, and appendix II).

Very little is known in detail about biosynthetic pathways in the archaeobacteria. Preliminary data suggest that the synthetic pathways for amino acids, purines, and pyrimidines are similar to those in other organisms. Some methanogens can fix atmospheric dinitrogen. Not only do many archaeobacteria use a reversal of the Embden-Meyerhof pathway to synthesize glucose, but at least some methanogens and extreme thermophiles employ glycogen as their major reserve material.

Autotrophy is widespread among the methanogens and extreme thermophiles, and CO_2 fixation occurs in more than one way. *Thermoproteus* and possibly *Sulfolobus* incorporate CO_2 by the reductive tricarboxylic acid cycle (figure 24.6*a*). This pathway is also present in the green sulfur bacteria (*see p. 471*). Methanogenic bacteria and probably most extreme thermophiles incorporate CO_2 by the reductive acetyl-CoA pathway (figure 24.6*b*). A similar pathway is also present in acetogenic bacteria and autotrophic sulfate-reducing bacteria.

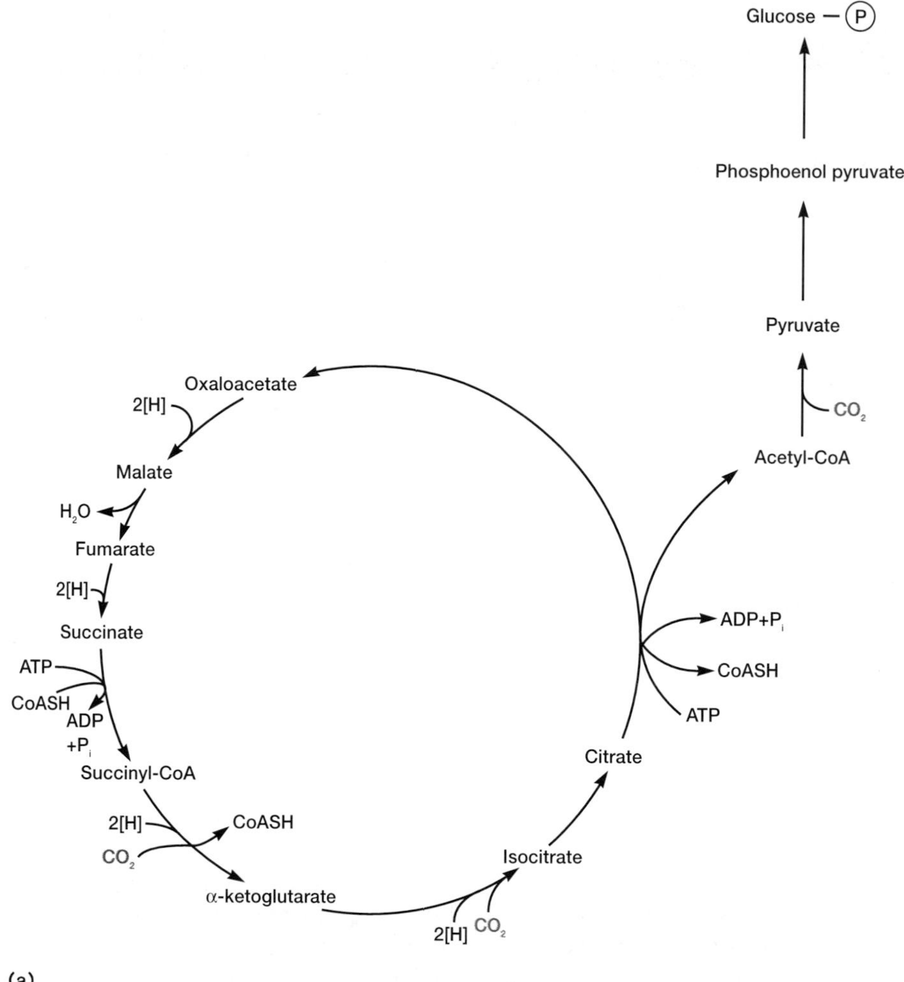

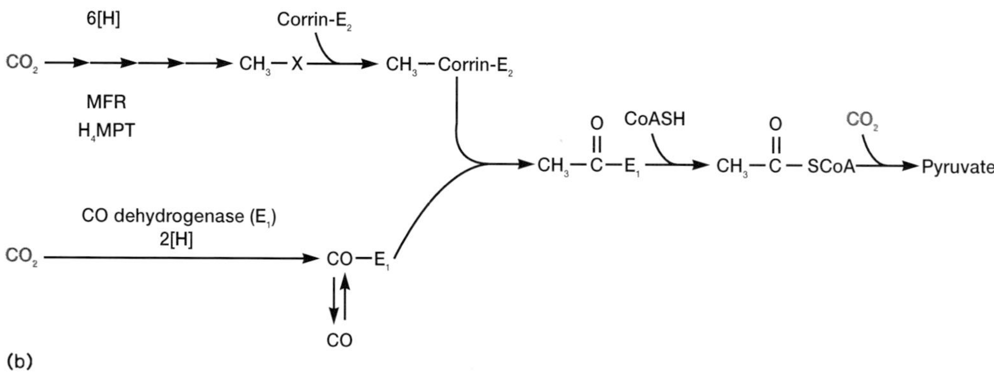

Figure 24.6 Mechanisms of Autotrophic CO_2 Fixation. (*a*) The reductive tricarboxylic acid cycle. The cycle is reversed with ATP and reducing equivalents [H] to form acetyl-CoA from CO_2. The acetyl-CoA may be carboxylated to yield pyruvate, which can then be converted to glucose and other compounds. This sequence appears to function in *Thermoproteus neutrophilus.* (*b*) The synthesis of acetyl-CoA and pyruvate from CO_2 in *Methanobacterium thermoautotrophicum.* One carbon comes from the reduction of CO_2 to a methyl group, and the second is produced by reducing CO_2 to carbon monoxide through the action of the enzyme CO dehydrogenase (E_1). The two carbons are then combined to form an acetyl group. Corrin-E_2 represents the cobamide-containing enzyme involved in methyl transfers. Special methanogen enzymes are described in figures 24.8 and 24.9. *(a) Source: Data in part from Georg Fuchs, "Alternative Pathways of Autotrophic CO_2 Fixation" in Autotrophic Bacteria, 1989.*

TABLE 24.1 Characteristics of the Major Archaeobacterial Groups

Group	General Characteristics	Representative Genera
Methanogenic archaeobacteria	Methane is the major metabolic end product. S^0 may be reduced to H_2S without energy production. Cells possess coenzyme M, factors 420 and 430, and methanopterin.	*Methanobacterium* *Methanococcus* *Methanomicrobium* *Methanosarcina*
Archaeobacterial sulfate reducers	H_2S formed from sulfate by dissimilatory sulfate reduction. Traces of methane also formed. Extremely thermophilic and strictly anaerobic. Possess factor 420 and methanopterin, but not coenzyme M or factor 430.	*Archaeoglobus*
Extremely halophilic archaeobacteria	Rods and regular to very irregular cells. Gram-negative or gram-positive, aerobic or facultatively anaerobic chemoorganotrophs. Require high sodium chloride concentrations for growth ($\geq$ 1.5 M). Neutrophilic or alkalophilic. Mesophilic or slightly thermophilic. Some species contain bacteriorhodopsin and use light for ATP synthesis.	*Halobacterium* *Halococcus* *Natronobacterium*
Cell wall-less archaeobacteria	Coccoid cells lacking a cell envelope. Thermoacidophilic. Aerobic. Plasma membrane contains a mannose-rich glycoprotein and a lipoglycan.	*Thermoplasma*
Extremely thermophilic S^0-metabolizers	Gram-negative rods, filaments, or cocci. Obligately thermophilic (optimum growth temperature between 70–105°C). Aerobic, facultatively aerobic, or strictly anaerobic. Acidophilic or neutrophilic. Autotrophic or heterotrophic. Most are sulfur metabolizers.	*Desulfurococcus* *Methanopyrus* *Pyrodictium* *Sulfolobus* *Thermococcus* *Thermoproteus*

Archaeobacterial Taxonomy

As shown in figure 20.11, the archaeobacteria are quite distinct from other living organisms. Within the group, however, there is great diversity (*see figure 20.10*). Section 25 of *Bergey's Manual* divides the archaeobacteria into five major groups: methanogenic archaeobacteria, archaeobacterial sulfate reducers, extremely halophilic archaeobacteria, cell wall-less archaebacteria, and extremely thermophilic S^0-metabolizers. Table 24.1 summarizes some major characteristics of these five groups and gives representatives of each. The last part of the chapter briefly describes the biology of these groups.

1. What are the archaeobacteria? Briefly describe the major ways in which they differ from eubacteria and eucaryotes.
2. How do archaeobacterial cell walls differ from those of the eubacteria? What is pseudomurein?
3. In what ways do archaeobacterial membrane lipids differ from those of eubacteria and eucaryotes? How might these differences lead to stronger membranes?
4. List the differences between archaeobacteria and other organisms with respect to tRNA, ribosome structure and behavior, EF–2 sensitivity, and RNA polymerases.
5. Briefly describe the way in which archaeobacteria degrade and synthesize glucose. In what two unusual ways do they incorporate CO_2?
6. Characterize the five different groups of archaeobacteria and distinguish between them.

Methanogenic Archaeobacteria

Methanogens are strict anaerobes that obtain energy by converting CO_2, H_2, formate, methanol, acetate, and other compounds to either methane or methane and CO_2. They are autotrophic when growing on H_2 and CO_2. This is the largest group of archaeobacteria. There are at least three orders and 13 genera, which differ greatly in overall shape, 16S rRNA sequence, cell wall chemistry and structure, membrane lipids, and other features. For example, methanogens construct three different types of cell walls. Several genera have walls with pseudomurein as mentioned earlier (figure 24.2). Other walls contain either proteins or heteropolysaccharides. The morphology of typical methanogens is shown in figure 24.7, and selected properties of representative genera are presented in table 24.2.

As might be inferred from the methanogens' ability to produce methane anaerobically, their metabolism is unusual. These bacteria contain several unique cofactors: tetrahydromethanopterin (H_4MPT), methanofuran (MFR), coenzyme M (2-mercaptoethanesulfonic acid), coenzyme F_{420}, and coenzyme F_{430} (figure 24.8). The first three cofactors bear the C_1 unit when CO_2 is reduced to CH_4. F_{420} carries electrons and hydrogens, and F_{430} is a nickel tetrapyrrole serving as a cofactor for the enzyme methyl-CoM methylreductase. The pathway for methane synthesis is thought to function as shown in figure 24.9. It appears that ATP synthesis is linked with methanogenesis by electron transport, proton pumping, and a chemiosmotic mechanism (*see pp. 152–54*). Some methano-

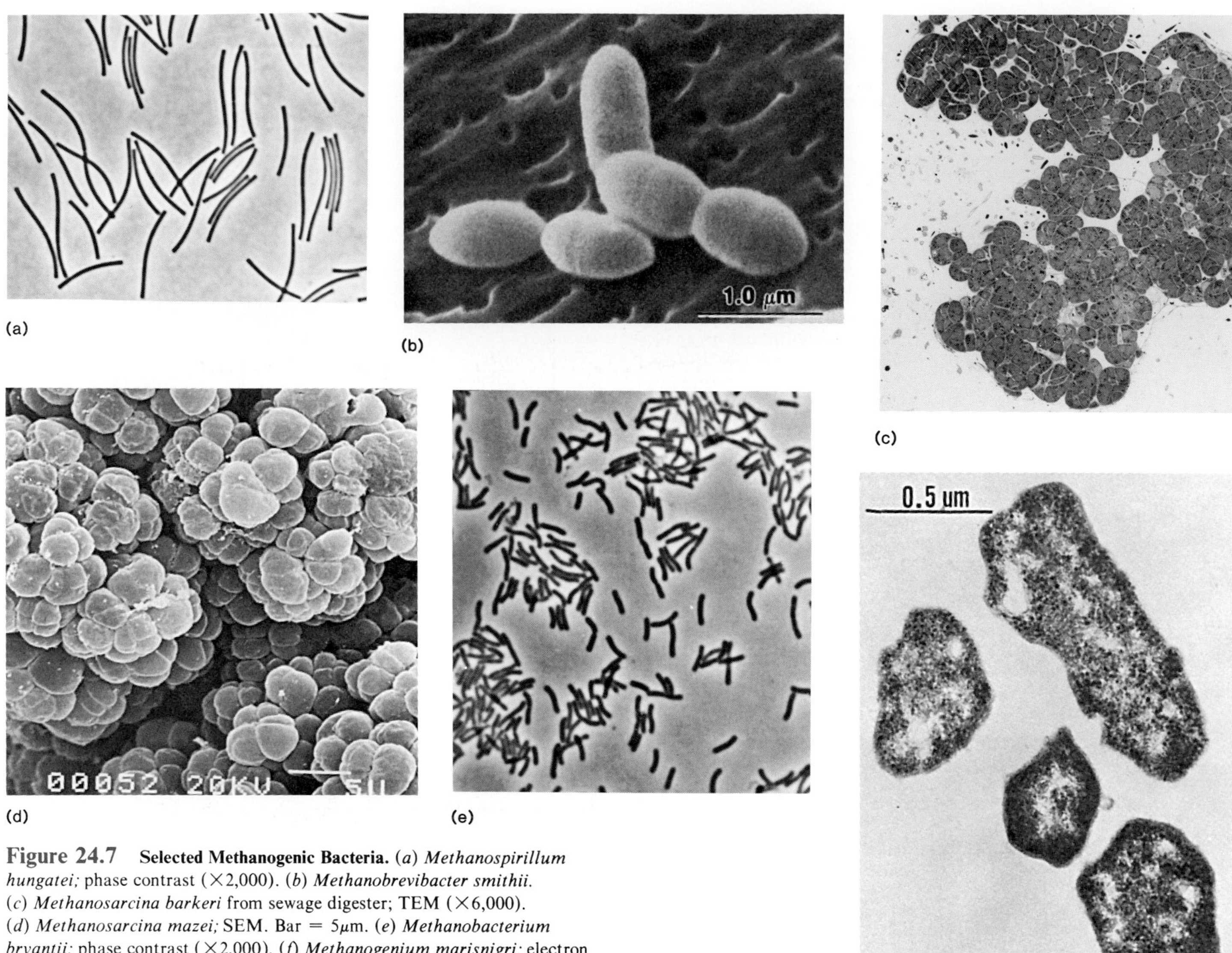

Figure 24.7 **Selected Methanogenic Bacteria.** (*a*) *Methanospirillum hungatei;* phase contrast (×2,000). (*b*) *Methanobrevibacter smithii.* (*c*) *Methanosarcina barkeri* from sewage digester; TEM (×6,000). (*d*) *Methanosarcina mazei;* SEM. Bar = 5μm. (*e*) *Methanobacterium bryantii;* phase contrast (×2,000). (*f*) *Methanogenium marisnigri;* electron micrograph (×45,000).

TABLE 24.2 Selected Characteristics of Representative Genera of Methanogens

Genus	Morphology	% G + C	Wall Composition	Gram Reaction	Motility	Methanogenic Substrates Used
Order I. *Methanobacteriales*						
Methanobacterium	Long rods or filaments	32–61	Pseudomurein	+ to variable	−	$H_2 + CO_2$, formate
Methanothermus	Straight to slightly curved rods	33	Pseudomurein with an outer protein S-layer	+	+	$H_2 + CO_2$
Order II. *Methanococcales*						
Methanococcus	Cocci	29–34	Protein	−	+	$H_2 + CO_2$, formate
Order III. *Methanomicrobiales*						
Methanomicrobium	Short rods	45–49	Protein	−	+	$H_2 + CO_2$, formate
Methanogenium	Cocci	52–61	Protein or glycoprotein	−	+	$H_2 + CO_2$, formate
Methanospirillum	Curved rods or spirilla	45–50	Protein	−	+	$H_2 + CO_2$, formate
Methanosarcina	Cocci, packets	36–43	Heteropolysaccharide or protein	+ to variable	−	$H_2 + CO_2$, methanol, methylamines, acetate

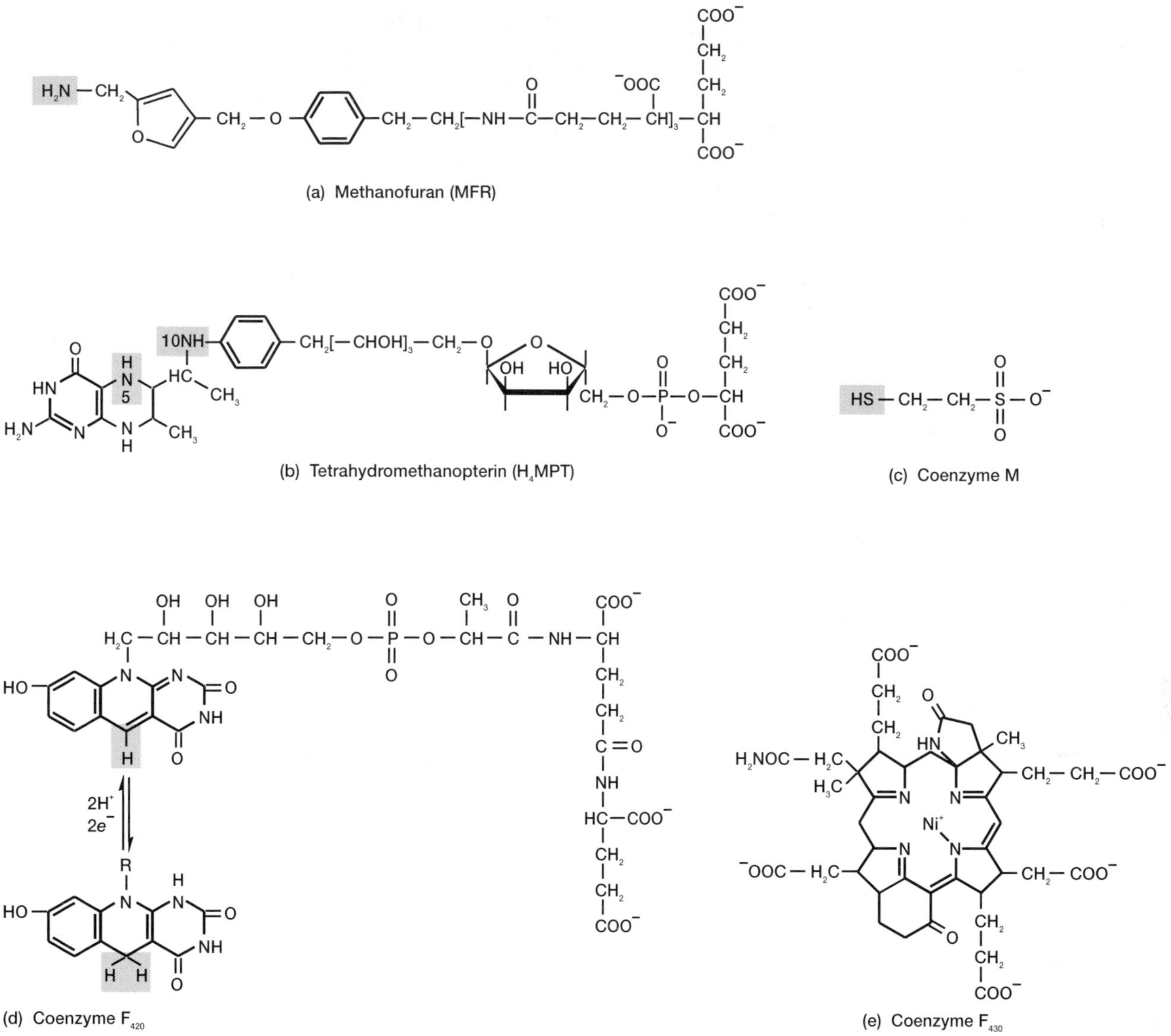

(a) Methanofuran (MFR)

(b) Tetrahydromethanopterin (H_4MPT)

(c) Coenzyme M

(d) Coenzyme F_{420}

(e) Coenzyme F_{430}

Figure 24.8 Methanogen Coenzymes. The portion of F_{420} (part *d*) that is reversibly oxidized and reduced is shown in color. MFR (*a*), H_4MPT (*b*), and coenzyme M (*c*) carry one-carbon units during methanogenesis (MFR and MPT also participate in the synthesis of acetyl-CoA). The places where the carbon units are attached are in color. H_4MPT carries carbon units on nitrogens 5 and 10 in the same way as the coenzyme tetrahydrofolate.

gens can live autotrophically by forming acetyl-CoA from two molecules of CO_2 and then converting the acetyl-CoA to pyruvate and other products (figure 24.6).

Methanogens thrive in anaerobic environments rich in organic matter: the rumen and intestinal system of animals, freshwater and marine sediments, swamps and marshes, hot springs, anaerobic sludge digesters, and even within anaerobic protozoa. Recently, an extremely thermophilic rod-shaped methanogen has been isolated from a marine hydrothermal vent. *Methanopyrus kandleri* has a temperature minimum at 84°C and an optimum of 98°C; it will grow up to 110°C (above the boiling point of water). Methanogens are often of ecological significance. The rate of methane production can be so great that bubbles of methane will sometimes rise to the surface of a lake or pond. A cow can belch 200 to 400 liters of methane a day.

Methanogenic bacteria are potentially of great practical importance since methane is a clean-burning fuel and an excellent energy source. For many years, sewage treatment plants have been using the methane they produce as a source of energy for heat and electricity (*see figure 41.15*). An anaerobic digester will degrade particulate wastes like sewage sludge to H_2, CO_2, and acetate. CO_2-reducing methanogens form CH_4 from CO_2 and H_2, while aceticlastic methanogens cleave acetate to CO_2 and CH_4 (about ⅔ of the methane produced by an anaerobic digester comes from acetate). A kilogram of organic matter can yield up to 600 liters of methane. It is quite likely that future research will greatly increase the efficiency of methane production and make methanogenesis an important source of pollution-free energy.

Methanogenesis can also be an ecological problem (*see chapter 40*). Methane absorbs infrared radiation and thus is a greenhouse gas. There is evidence that atmospheric methane concentrations have been rising over the last 200 years. Methane production may significantly promote future global warming (*see Box 42.1*). Recently, it has been discovered that

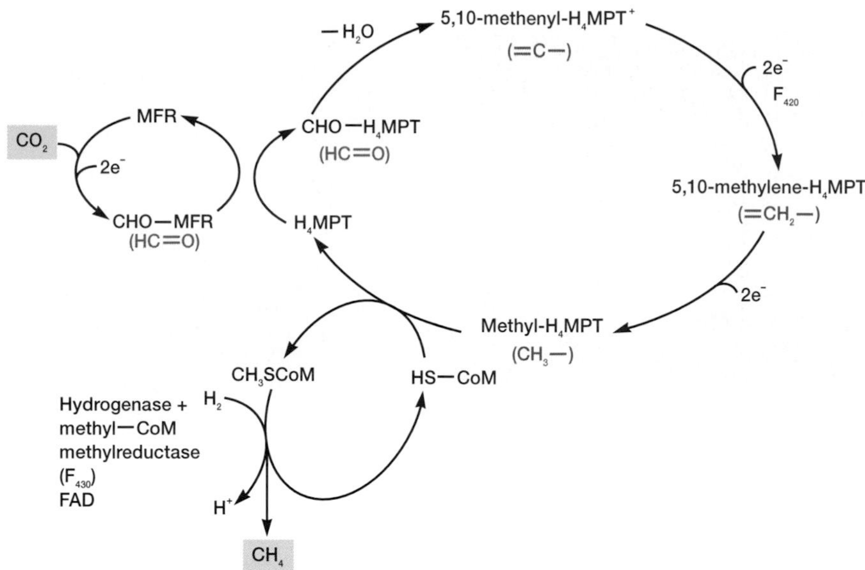

Figure 24.9 **Methane Synthesis.** Pathway for CH_4 synthesis from CO_2 in *M. thermoautotrophicum.* Cofactor abbreviations: methanopterin (*MPT*), methanofuran (*MFR*), and 2-mercaptoethanesulfonic acid or coenzyme M (*CoM*). The nature of the carbon-containing intermediates leading from CO_2 to CH_4 are indicated in parentheses. See text for further details.

methanogens can oxidize Fe^0 and use it to produce methane and energy. This means that methanogens growing around buried or submerged iron pipes and other objects may contribute significantly to their corrosion.

Archaeobacterial Sulfate Reducers

This group contains gram-negative cocci with walls consisting of glycoprotein subunits. At present, only the genus *Archaeoglobus* is known. It can extract electrons from a variety of electron donors (e.g., H_2, lactate, glucose) and reduce sulfate, sulfite, or thiosulfate to sulfide. Elemental sulfur is not used as an acceptor. *Archaeoglobus* is extremely thermophilic (the optimum is about 83°C) and can be isolated from marine hydrothermal vents. The organism is not only unusual in being able to reduce sulfate, unlike other archaeobacteria, but it also possesses the methanogen coenzymes F_{420} and methanopterin. It seems intermediate between the methanogens and the extremely thermophilic S^0-metabolizers.

1. Generally characterize methanogenic archaeobacteria and distinguish them from other groups.
2. Briefly describe how they produce methane and the roles of their unique cofactors in this process.
3. Where does one find methanogens? Discuss their ecological and practical importance.
4. Characterize *Archaeoglobus.* In what way is it similar to the methanogens and how does it differ from other extreme thermophiles?

Extremely Halophilic Archaeobacteria

The **extreme halophiles** are a third major group of archaeobacteria, currently with six genera in one family, the *Halobacteriaceae* (figure 24.10). They are aerobic chemoheterotrophs with respiratory metabolism and require complex nutrients, usually proteins and amino acids, for growth. They are either nonmotile or motile by lophotrichous flagella.

The most obvious distinguishing trait of this family is its absolute dependence on a high concentration of NaCl. These bacteria require at least 1.5 M NaCl (about 8%, wt/vol), and usually have a growth optimum at about 3 to 4 M NaCl (17 to 23%). They will grow at salt concentrations approaching saturation (about 36%). *Halobacterium's* cell wall is so dependent on the presence of NaCl that it disintegrates when the NaCl concentration drops to about 1.5 M. Thus, halobacteria only grow in high-salinity habitats such as marine salterns (*see figure 40.17*a) and salt lakes like the Dead Sea between Israel and Jordan, and the Great Salt Lake in Utah. They also can grow in food products such as salted fish and spoil them. Halobacteria often have red-to-yellow pigmentation from carotenoids that are probably used as protection against strong sunlight. They can reach such high population levels that salt lakes, salterns, and salted fish actually turn red.

The effect of solutes on halophile growth (pp. 122–23).

Probably the best-studied member of the family is *Halobacterium salinarium* (*H. halobium*). This procaryote is unusual because it can trap light energy photosynthetically without the presence of chlorophyll. When exposed to low oxygen levels, some strains of *Halobacterium* synthesize a modified cell membrane called the **purple membrane,** which

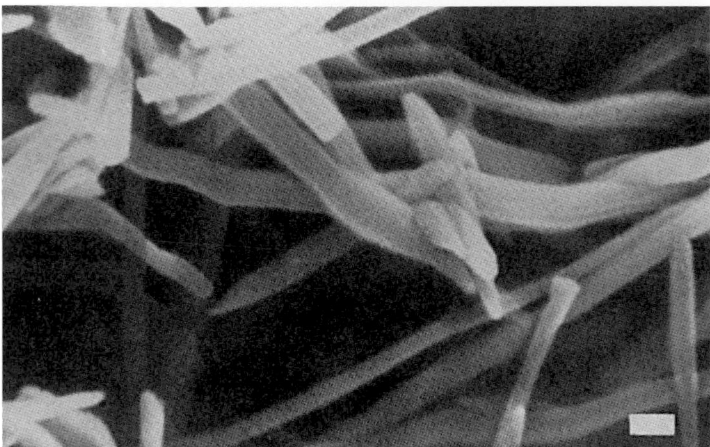

(a)

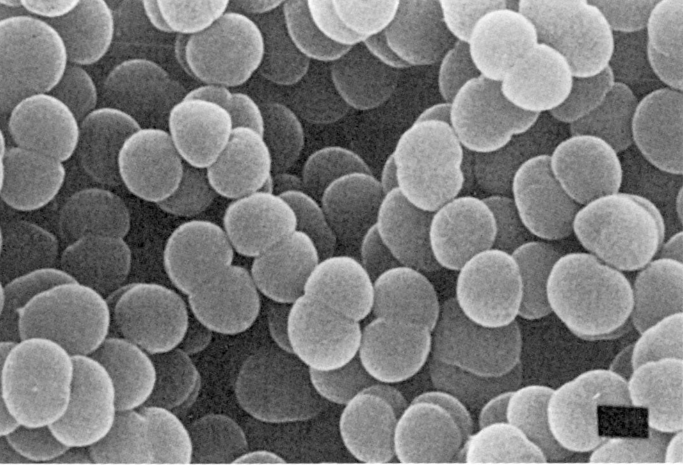

(b)

Figure 24.10 **Examples of Halobacteria.** (*a*) *Halobacterium salinarium.* A young culture that has formed long rods; SEM. Bar = 1 μm. (*b*) *Halococcus morrhuae;* SEM. Bar = 1 μm.

contains the protein **bacteriorhodopsin.** ATP is produced by a unique type of photosynthesis without the participation of bacteriochlorophyll or chlorophyll (Box 24.1).

Cell Wall-less Archaeobacteria

Thermoplasma grows in refuse piles of coal mines. These piles contain large amounts of iron pyrite (FeS), which is oxidized to sulfuric acid by chemolithotrophic bacteria. As a result, the piles become very hot and acidic. This is an ideal habitat for *Thermoplasma* since it grows best at 55 to 59°C and pH 1 to 2. Although it lacks a cell wall, its plasma membrane is strengthened by large quantities of diglycerol tetraethers, lipopolysaccharides, and glycoproteins. The organism's DNA is stabilized by association with a special histonelike protein that condenses the DNA into particles resembling eucaryotic nucleosomes. At 59°C, *Thermoplasma* takes the form of an irregular filament, while at lower temperatures it is spherical (figure 24.11).

Box 24.1

Photosynthesis in *Halobacterium salinarium*

The halophilic bacterium *Halobacterium salinarium* (formerly *H. halobium*) normally depends on respiration for the production of energy. However, under conditions of low oxygen and high light intensity, the bacteria synthesize a deep-purple pigment called bacteriorhodopsin, which closely resembles the sensory pigment rhodopsin from the rods and cones of vertebrate eyes. Bacteriorhodopsin's chromophore is the carotenoid derivative retinal (the aldehyde of vitamin A) that is covalently attached to the pigment protein by a Schiff base with the amino group of a lysine. The protein extends through the plasma membrane, and its retinal rests in the center of the membrane. Individual bacteriorhodopsins aggregate in the membrane to form crystalline patches called the **purple membrane.**

Bacteriorhodopsin functions as a light-driven proton pump. When retinal absorbs light, the double bond between carbons 13 and 14 changes from a *trans* to a *cis* configuration and the Schiff-base loses a proton. Protons are moved across the plasma membrane to the periplasmic space (*see chapter 3*) during these alterations, and it is thought that the Schiff base changes are directly involved in this movement, although the precise mechanism is unclear (Box Figure 24.1). The light-driven proton pumping generates a pH gradient that can be used to power the synthesis of ATP by a chemiosmotic mechanism (*see chapter 8*).

This photosynthetic capacity is particularly useful to *Halobacterium* because oxygen is not very soluble in concentrated salt solutions and may decrease to an extremely low level in its habitat. When the surroundings become temporarily anaerobic, the bacterium uses light energy to synthesize sufficient ATP to remain alive until the oxygen level rises again. *Halobacterium* cannot grow anaerobically because it requires oxygen for continued retinal synthesis, but it can survive the stress of temporary oxygen limitation by means of photosynthesis.

Extremely Thermophilic S^0-Metabolizers

The last group of archaeobacteria contains extremely thermophilic bacteria, many of which are acidophiles and sulfur dependent. The sulfur may be used either as an electron acceptor in anaerobic respiration or as an electron source by lithotrophs. Almost all are strict anaerobes. They grow in

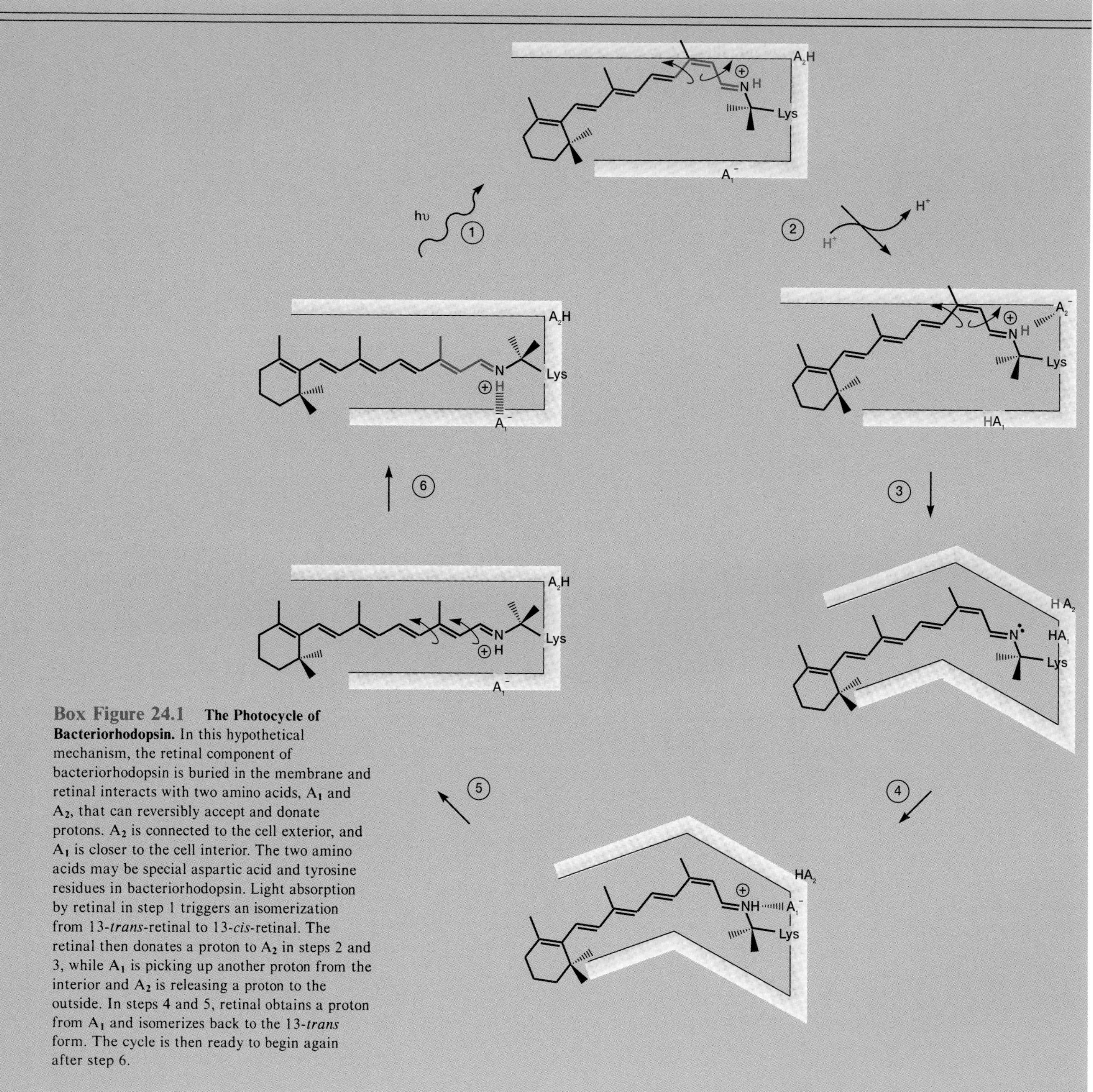

Box Figure 24.1 The Photocycle of Bacteriorhodopsin. In this hypothetical mechanism, the retinal component of bacteriorhodopsin is buried in the membrane and retinal interacts with two amino acids, A_1 and A_2, that can reversibly accept and donate protons. A_2 is connected to the cell exterior, and A_1 is closer to the cell interior. The two amino acids may be special aspartic acid and tyrosine residues in bacteriorhodopsin. Light absorption by retinal in step 1 triggers an isomerization from 13-*trans*-retinal to 13-*cis*-retinal. The retinal then donates a proton to A_2 in steps 2 and 3, while A_1 is picking up another proton from the interior and A_2 is releasing a proton to the outside. In steps 4 and 5, retinal obtains a proton from A_1 and isomerizes back to the 13-*trans* form. The cycle is then ready to begin again after step 6.

geothermally heated water or soils that contain elemental sulfur. These environments are scattered all over the world. Examples are the sulfur-rich hot springs in Yellowstone National Park, Wyoming, and the waters surrounding areas of submarine volcanic activity (*see figures 40.3*a *and 40.17*b). Such habitats are sometimes called solfatara. These archaeobacteria can be very thermophilic. The most extreme example is *Pyrodictium,* a bacterium isolated from geothermally heated sea floors. *Pyrodictium* has a temperature minimum of 82°C, a growth optimum at 105°C, and a maximum at 110°C. Both organotrophic and lithotrophic growth occur in this group. Sulfur and H_2 are the most common electron sources for lith-

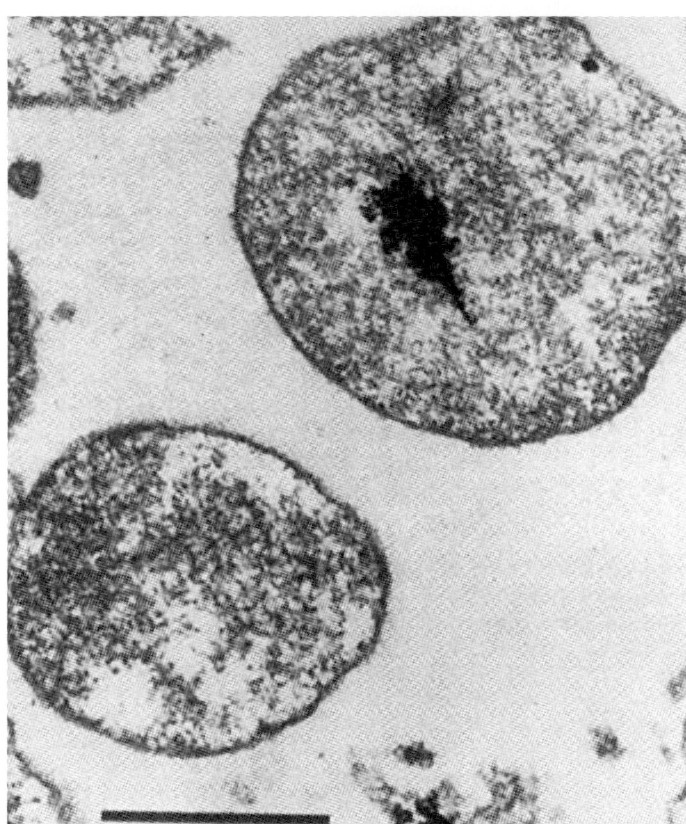

Figure 24.11 ***Thermoplasma.*** Transmission electron micrograph. Bar = 0.5 μm.

otrophs. There are three orders (*Thermococcales, Thermoproteales,* and *Sulfolobales*) and at least nine genera. Two of the better-studied genera are *Thermoproteus* and *Sulfolobus.*

Members of the genus *Sulfolobus* are gram negative, aerobic, irregularly lobed spherical bacteria with a temperature optimum around 70 to 80°C and a pH optimum of 2 to 3 (figure 24.12*a, b*). For this reason, they are classed as **thermoacidophiles,** so called because they grow best at acid pH values and high temperatures. Their cell wall contains lipoprotein and carbohydrate but lacks peptidoglycan. They grow lithotrophically on sulfur granules in hot acid springs and soils while oxidizing the sulfur to sulfuric acid (figure 24.12*b*). Oxygen is the normal electron acceptor, but ferric iron also may be used. Sugars and amino acids such as glutamate also serve as carbon and energy sources.

Thermoproteus is a long thin rod that can be bent or branched (figure 24.12*c*). The cell wall is composed of glycoprotein. *Thermoproteus* is a strict anaerobe and grows at temperatures from 78 to 96°C and pH values between 1.7 and 6.5. It is found in hot springs and other hot aquatic habitats rich in sulfur. It can grow organotrophically and oxidize glucose, amino acids, alcohols, and organic acids with elemental sulfur as the electron acceptor. That is, *Thermoproteus* can carry out anaerobic respiration. It will also grow chemolithotrophically using H_2 and S^0. Carbon monoxide or CO_2 can serve as the sole carbon source.

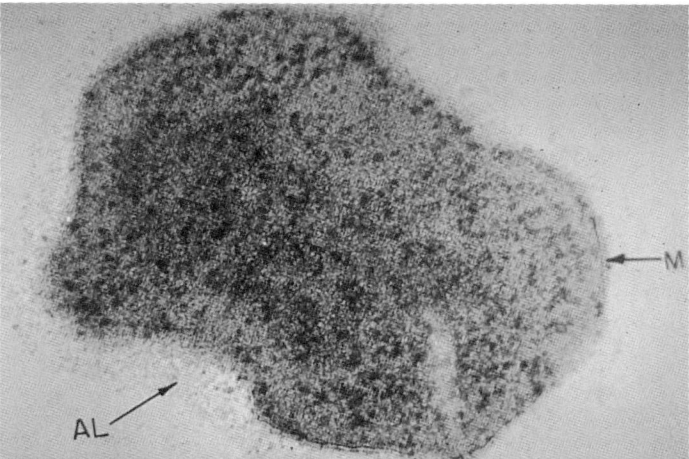

(a)

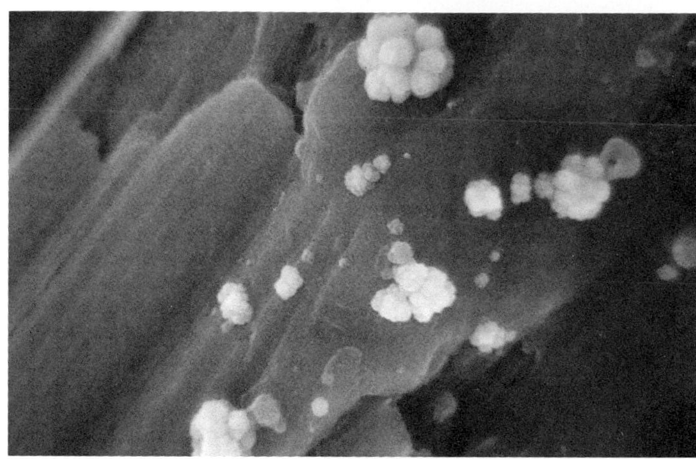

(b)

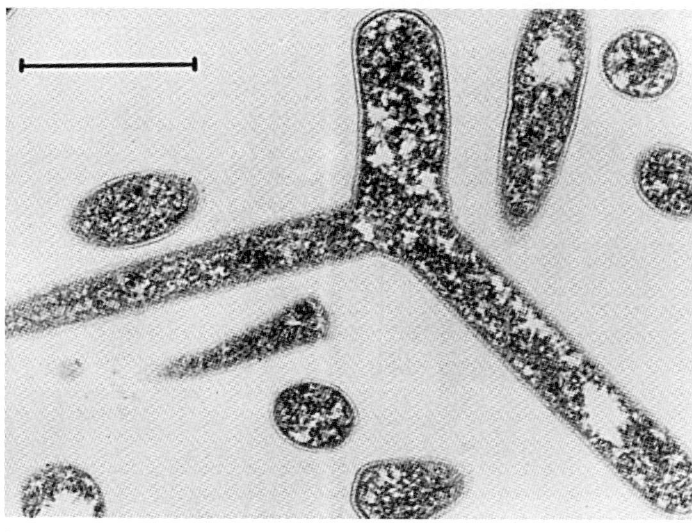

(c)

Figure 24.12 ***Sulfolobus* and *Thermoproteus.*** (*a*) A thin section of *Sulfolobus brierleyi.* The bacterium, about 1 μm in diameter, is surrounded by an amorphous layer (*AL*) instead of a well-defined cell wall; the plasma membrane (*M*) is clearly visible. (*b*) A scanning electron micrograph of a colony of *Sulfolobus* growing on the mineral molybdenite (MoS_2) at 60° C. At pH 1.5–3, the bacterium oxidizes the sulfide component of the mineral to sulfate and solubilizes molybdenum. (*c*) Electron micrograph of *Thermoproteus tenax.* Bar = 1 μm.

1. Where are the extreme halophiles found and what is unusual about their cell walls and growth requirements?
2. Briefly describe how *Halobacterium* carries out photosynthesis. What is the purple membrane and what pigment does it contain?
3. How is *Thermoplasma* able to live in acidic, very hot coal refuse piles when it lacks a cell wall? How is its DNA stabilized?
4. What are thermoacidophiles and where do they grow? In what ways do they use sulfur in their metabolism? Briefly describe *Sulfolobus* and *Thermoproteus*.

Summary

1. Organisms may be divided into three major groups: archaeobacteria, eubacteria, and eucaryotes.
2. Archaeobacteria are highly diverse with respect to morphology, reproduction, physiology, and ecology. They grow in anaerobic, hypersaline, and high-temperature habitats.
3. Archaeobacterial cell walls do not contain peptidoglycan and differ from eubacterial walls in structure. They may be composed of pseudomurein, polysaccharides, glycoproteins, or unaltered proteins.
4. The membrane lipids differ from those of other organisms in having branched chain hydrocarbons connected to glycerol by ether links. Eubacterial and eucaryotic lipids have glycerol connected to fatty acids by ester bonds.
5. Their tRNA, ribosomes, elongation factors, RNA polymerases, and other components distinguish archaeobacteria from eubacteria and eucaryotes.
6. Although much of archaeobacterial metabolism appears similar to that of other organisms, they do differ with respect to glucose catabolism, pathways for CO_2 fixation, and the ability of some to synthesize methane.
7. Archaeobacteria may be divided into five groups: methanogenic bacteria, sulfate reducers, extreme halophiles, cell wall-less archaeobacteria, and extreme thermophiles.
8. Methanogenic bacteria are strict anaerobes that can obtain energy through the synthesis of methane. They have several unique cofactors that are involved in methanogenesis.
9. The extreme thermophile *Archaeoglobus* differs from other archaeobacteria in using a variety of electron donors to reduce sulfate. It also contains the methanogen cofactors F_{420} and methanopterin.
10. Extreme halophiles are aerobic chemoheterotrophs that require at least 1.5 M NaCl for growth. They are found in habitats such as salterns, salt lakes, and salted fish.
11. *Halobacterium salinarium* can carry out photosynthesis without chlorophyll or bacteriochlorophyll by using bacteriorhodopsin, which employs retinal to pump protons.
12. The thermophilic archaeobacterium *Thermoplasma* grows in hot, acid coal refuse piles and survives despite the lack of a cell wall.
13. The extremely thermophilic S^0-metabolizers depend on sulfur for growth and are frequently acidophiles. The sulfur may be used as an electron acceptor in anaerobic respiration or as an electron donor by lithotrophs. They are almost always strict anaerobes and grow in geothermally heated soil and water that is rich in sulfur.

Key Terms

archaeobacteria *493*
bacteriorhodopsin *501*
extreme halophiles *500*
methanogens *497*
pseudomurein *493*
purple membrane *500*
thermoacidophiles *503*

Questions for Thought and Review

1. Discuss how the unusual properties of extreme halophiles and thermophiles reflect the habitats in which they grow.
2. Why do we not classify *Thermoplasma* as a mycoplasma rather than as an archaeobacterium?
3. Discuss what adaptations extremely thermophilic archaeobacteria require to grow at 100°C.
4. Do you think archaeobacteria should be separated from the eubacteria although both groups are procaryotic? Give your reasoning and evidence.
5. Supposing you wished to isolate bacteria from a hot spring in Yellowstone Park, how would you go about it?

Additional Reading

Balch, W. E.; Fox, G. E.; Magrum, L. J.; Woese, C. R.; and Wolfe, R. S. 1979. Methanogens: Reevaluation of a unique biological group. *Microbiol. Rev.* 43:260–96.

Balows, A.; Truper, H. G.; Dworkin, M.; Harder, W.; and Schleifer, K. -H. 1992. *The prokaryotes,* 2d ed. New York: Springer-Verlag.

Birge, R. R. 1990. Nature of the primary photochemical events in rhodopsin and bacteriorhodopsin. *Biochem. Biophys. Acta* 1016:293–327.

Cramer, W. A., and Knaff, D. B. 1991. *Energy transduction in biological membranes.* New York: Springer-Verlag.

Danson, M. J. 1988. Archaeobacteria: The comparative enzymology of their central metabolic pathways. In *Advances in microbial physiology,* eds. A. H. Rose and D. W. Tempest, 165–231. New York: Academic Press.

DiMarco, A. A.; Bobik, T. A.; and Wolfe, R. S. 1990. Unusual coenzymes of methanogenesis. *Ann. Rev. Biochem.* 59:355–94.

Fewson, C. A. 1986. Archaeobacteria. *Biochem. Ed.* 14(3):103–15.

Holt, J. G., editor-in-chief. 1989. *Bergey's manual of systematic bacteriology,* vol. 3, eds. J. T. Staley, M. P. Bryant, and N. Pfennig. Baltimore, Md.: Williams & Wilkins.

Jones, W. J.; Nagle, D. P., Jr.; and Whitman, W. B. 1987. Methanogens and the diversity of archaeobacteria. *Microbiol. Rev.* 51:135–77.

Kandler, O., and Zillig, W., eds. 1986. *Archaeobacteria '85.* New York: Gustav Fischer Verlag.

Oesterhelt, D. 1985. Light-driven proton pumping in halobacteria. *BioScience* 35(1):18–21.

Oesterhelt, D., and Tittor, J. 1989. Two pumps, one principle: Light-driven ion transport in halobacteria. *Trends Biochem. Sci.* 14:57–61.

Pley, U.; Schipka, J.; Gambacorta, A.; Jannasch, H. W.; Fricke, H.; Rachel, R.; and Stetter, K. O. 1991. *Pyrodictium abyssi* sp. nov. represents a novel heterotrophic marine archaeal hyperthermophile growing at 110° C. System. Appl. Microbiol. 14:245–53.

Schlegel, H. G., and Bowien, B. 1989. *Autotrophic bacteria.* Madison, Wis.: Science Tech Publishers.

Stoeckenius, W. 1976. The purple membrane of salt-loving bacteria. *Sci. Am.* 234(6):38–46.

Woese, C. R. 1981. Archaeobacteria. *Sci. Am.* 244(6):98–122.

Woese, C. R., and Wolfe, R. S., eds. 1985. *Archaeobacteria.* Volume VIII of *The bacteria: A treatise on structure and function.* New York: Academic Press.

Wood, H. G.; Radsdale, S. W.; and Pezacka, E. 1986. The acetyl-CoA pathway: A newly discovered pathway of autotrophic growth. *Trends Biochem. Sci.* 11(1):14–17.

Zillig, W.; Palm, P.; Reiter, W.-D.; Gropp, F.; Pühler, G.; and Klenk, H. -P. 1988. Comparative evaluation of gene expression in archaeobacteria. *Eur. J. Biochem.* 173:473–82.

Zinder, S. H. 1984. Microbiology of anaerobic conversion of organic wastes to methane: Recent developments. *ASM News* 50(7):294–98.

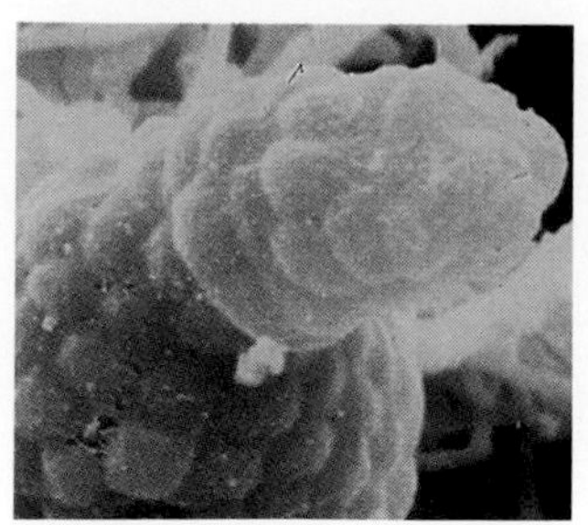

CHAPTER 25
The Bacteria: The Actinomycetes

Actinomycetes are very important from a medical point of view. . . . They may be a nuisance, as when they decompose rubber products, grow in aviation fuel, produce odorous substances that pollute water supplies, or grow in sewage-treatment plants where they form thick clogging foams. . . . In contrast, actinomycetes are the producers of most of the antibiotics.

—H. A. Lechevalier and M. P. Lechevalier

Outline

Concepts

1. Volume 4 of *Bergey's Manual* contains aerobic, gram-positive bacteria—the actinomycetes—that form branching hyphae and asexual spores.
2. The morphology and arrangement of spores, cell wall chemistry, and the types of sugars present in cell extracts are particularly important in actinomycete taxonomy and are used to divide these bacteria into different groups.
3. Actinomycetes have considerable practical impact because they play a major role in the mineralization of organic matter in the soil and are the primary source of most naturally synthesized antibiotics.
4. Actinomycete taxonomy is still developing. The classification of these organisms will probably change considerably in the future.

Chapter 25, the last of the survey chapters on bacteria, describes the bacteria placed in volume 4 of *Bergey's Manual of Systematic Bacteriology.* Actinomycetes are gram positive like the bacteria in volume 2, but are distinctive because they have filamentous hyphae that do not normally undergo fragmentation and produce asexual spores. They closely resemble fungi in overall morphology. Presumably, this resemblance results partly from adaptation to the same habitats. First, the general characteristics of the actinomycetes are summarized. Then representatives are described, with emphasis on morphology, taxonomy, reproduction, and general importance.

Volume 4 of *Bergey's Manual* contains the **actinomycetes,** (s., actinomycete) aerobic, gram-positive bacteria that form branching filaments or hyphae and asexual spores. Although they are a diverse group, the actinomycetes do share many properties.

General Properties of the Actinomycetes

When growing on a solid substratum such as agar, the branching network of hyphae developed by actinomycetes grows both on the surface of the substratum and into it to form a substrate mycelium. Septa usually divide the hyphae into long cells (20 μm and longer) containing several nucleoids. Sometimes a tissuelike mass results and may be called a **thallus.** Many actinomycetes also have an aerial mycelium that extends above the substratum and forms conidia (figure 25.1). Asexual, thin-walled spores called **conidia** (s., conidium) or **conidiospores** are held on the ends of filaments; if the spores are located in a sporangium, they are called **sporangiospores.** The spores can vary greatly in shape (figure 25.2). Actinomycete spores develop by septal formation at filament tips, usually in response to nutrient deprivation. Most are not particularly heat resistant but do withstand desiccation well and thus have considerable adaptive value.

Most actinomycetes are not motile. When motility is present, it is confined to flagellated spores.

Actinomycete cell wall composition varies greatly among different groups and is of considerable taxonomic importance. Four major cell wall types can be distinguished according to three features of peptidoglycan composition and structure: the amino acid in tetrapeptide side chain position 3, the presence of glycine in interpeptide bridges, and peptidoglycan sugar content (table 25.1). Cell extracts of actinomycetes with wall types II, III, and IV also contain characteristic sugars that are useful in identification (table 25.2). Some other taxonomically valuable properties are the morphology and color of mycelia and sporangia, the surface features and arrangement of conidiospores, the percent G + C in DNA, 16S rRNA sequences, the phospholipid composition of cell membranes, and spore heat resistance.

Gram-positive peptidoglycan structure and chemistry (pp. 451–52).

Actinomycetes have considerable practical significance. They are primarily soil inhabitants and are very widely distributed. They can degrade an enormous number and variety of organic compounds and are extremely important in the

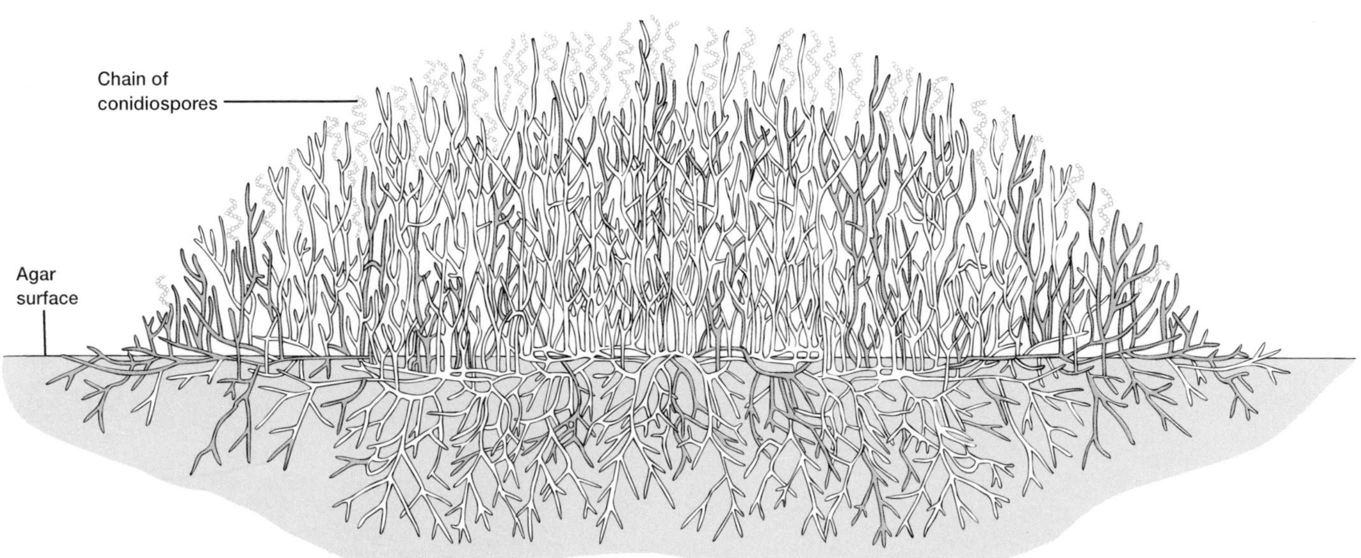

Figure 25.1 **An Actinomycete Colony.** The cross section of an actinomycete colony with living (colored) and dead (white) hyphae. The substrate mycelium and aerial mycelium with chains of conidiospores are shown.

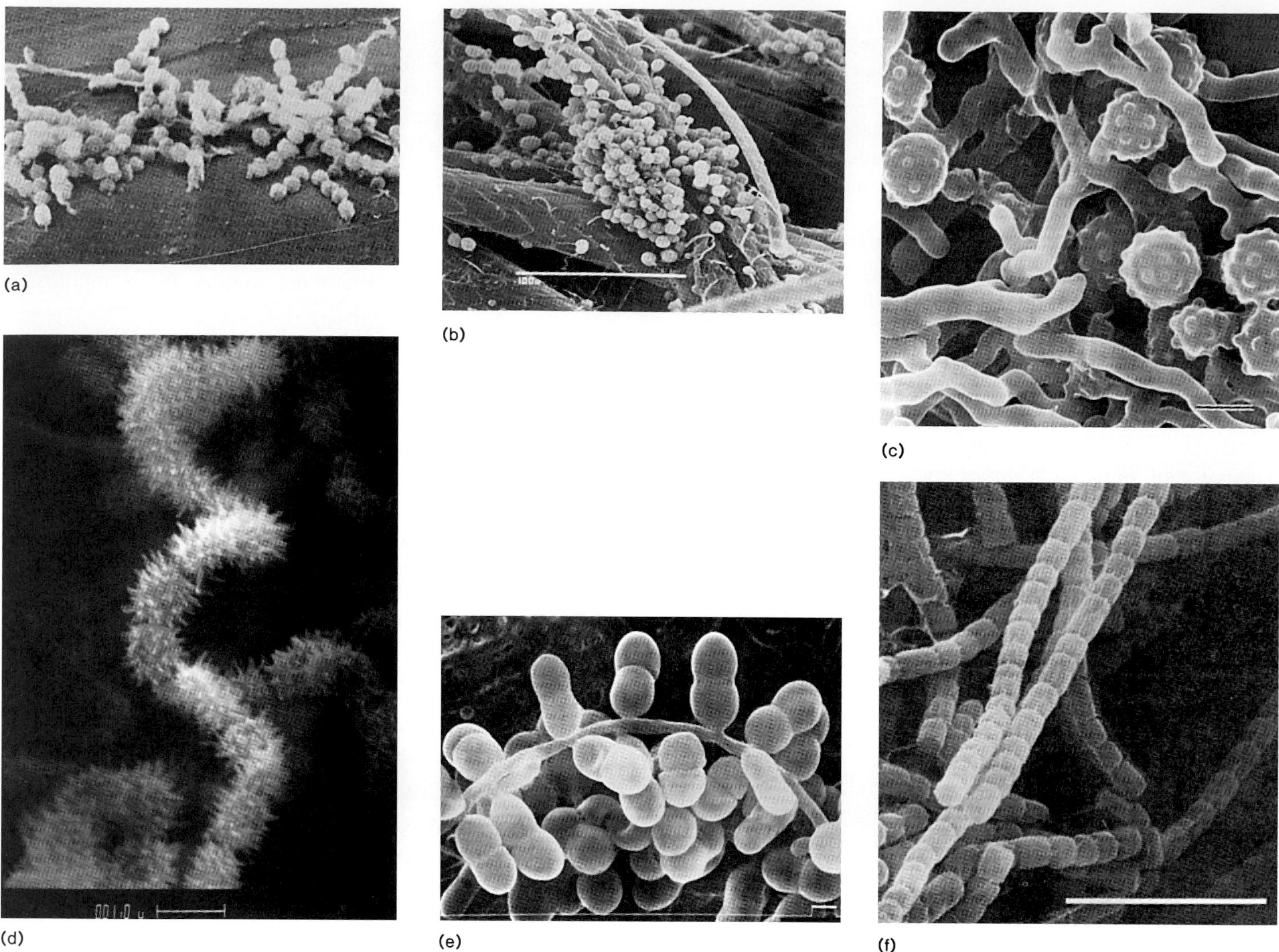

Figure 25.2 Examples of Actinomycete Spores as Seen in the Scanning Electron Microscope. (*a*) Sporulating *Faenia* hyphae (×3,000). (*b*) Sporangia of *Pilimelia columellifera* on mouse hair (×520). (*c*) *Micromonospora echinospora*. Bar = 0.5 μm. (*d*) A chain of hairy streptomycete spores. Bar = 1.0 μm. (*e*) *Microbispora rosea*, paired spores on hyphae. Bar = 10 μm. (*f*) Aerial spores of *Kitasatosporia setae*. Bar = 5 μm.

TABLE 25.1 Actinomycete Cell Wall Types

Cell Wall Type	Diaminopimelic Acid Isomer	Glycine in Interpeptide Bridge	Characteristic Sugars	Representative Genera
I	L, L	+		*Nocardioides, Streptomyces, Streptoverticillium* [HPg]
II	*Meso*	+		*Micromonospora, Pilimelia, Actinoplanes*
III	*Meso*	–		*Thermoactinomyces, Actinomadura, Frankia*
IV	*Meso*	–	Arabinose, galactose	*Saccharomonospora, Faenia, Nocardia*

mineralization of organic matter. Actinomycetes produce most of the medically useful natural antibiotics. Although most actinomycetes are free-living microorganisms, a few are pathogens of humans, other animals, and some plants.

Bergey's Manual divides the actinomycetes into seven sections, primarily based on properties such as cell wall type, conidia arrangement, and the presence or absence of a sporangium (table 25.3). A few genera are not easily placed in sections 26 through 32 and are collected in section 33 under the title "Other Genera."

The sections in volume 4 are not homogeneous, and their composition will undoubtedly change in the future. They do not always fit with 16S rRNA sequence results (figure 25.3). For example, the genus *Actinomadura* is clearly heterogeneous. *Dermatophilus, Promicromonospora,* and *Oerskovia* seem more closely related to bacteria placed in volume 2 than they do to those in volume 4. It has been proposed that the irregular, nonsporing gram-positive rods in volume 2 be clustered with these three genera to form a group called the actinobacteria (figure 25.3).

TABLE 25.2 Actinomycete Whole Cell Sugar Patterns

Sugar Pattern Types[a]	Characteristic Sugars	Representative Genera
A	Arabinose, galactose	*Nocardia, Rhodococcus, Saccharomonospora*
B	Madurose[b]	*Actinomadura, Streptosporangium, Dermatophilus*
C	None	*Thermomonospora, Actinosynnema, Thermoactinomyces, Geodermatophilus*
D	Arabinose, xylose	*Micromonospora, Actinoplanes*

[a]Characteristic sugar patterns are present only in wall types II–IV, those actinomycetes with *meso*-diaminopimelic acid.

[b]Madurose is 3-O-methyl-D-galactose.

1. Define actinomycete, thallus, substrate mycelium, aerial mycelium, conidium, conidiospore, and sporangiospore.
2. Describe how cell wall structure and sugar content are used to classify the actinomycetes. Include a brief description of the four major wall types.
3. Why are the actinomycetes of such practical interest?

Nocardioform Actinomycetes

Most of the eleven genera in section 26 of *Bergey's Manual* are also described in section 17 of volume 2. Because they all resemble members of the genus *Nocardia* (named after Edmond Nocard [1850–1903], French bacteriologist and veterinary pathologist), they are collectively called **nocardi-**

TABLE 25.3 Some Characteristics of Major Actinomycete Groups

Group	Wall Type	Sugar Pattern	Mol % G + C	Spore Arrangement	Presence of Sporangia	Selected Genera
Nocardioform actinomycetes[a]	I, IV, VI[b]	A	59–79	Varies	–	*Nocardia, Rhodococcus, Nocardiodes, Faenia, Oerskovia, Saccharomonospora*
Actinomycetes with multilocular sporangia	III	B, C, D	57–75	Clusters of spores	+(–)[c]	*Geodermatophilus, Dermatophilus, Frankia*
Actinoplanetes	II	D	71–73	Varies	Usually +	*Actinoplanes, Pilimelia, Dactylosporangium, Micromonospora*
Streptomyces and related genera	I	Not of taxonomic value	69–78	Chains of 5 to more than 50 spores	–	*Streptomyces, Streptoverticillium, Sporichthya*
Maduromycetes	III	B, C	64–74	Varies	+ or –	*Actinomadura, Microbispora, Planomonospora, Streptosporangium*
Thermomonospora and related genera	III	C (sometimes B)	64–73	Varies	–	*Thermomonospora, Actinosynnema, Nocardiopsis*
Thermoactinomycetes	III	C	52–55	Single, heat-resistant endospores	–	*Thermoactinomyces*

[a]Several genera have mycolic acid. Filaments readily fragment into rods and coccoid elements.

[b]Type VI cell wall has lysine instead of diaminopimelic acid.

[c]Members have clusters of spores that may not always be surrounded by a sporangial wall.

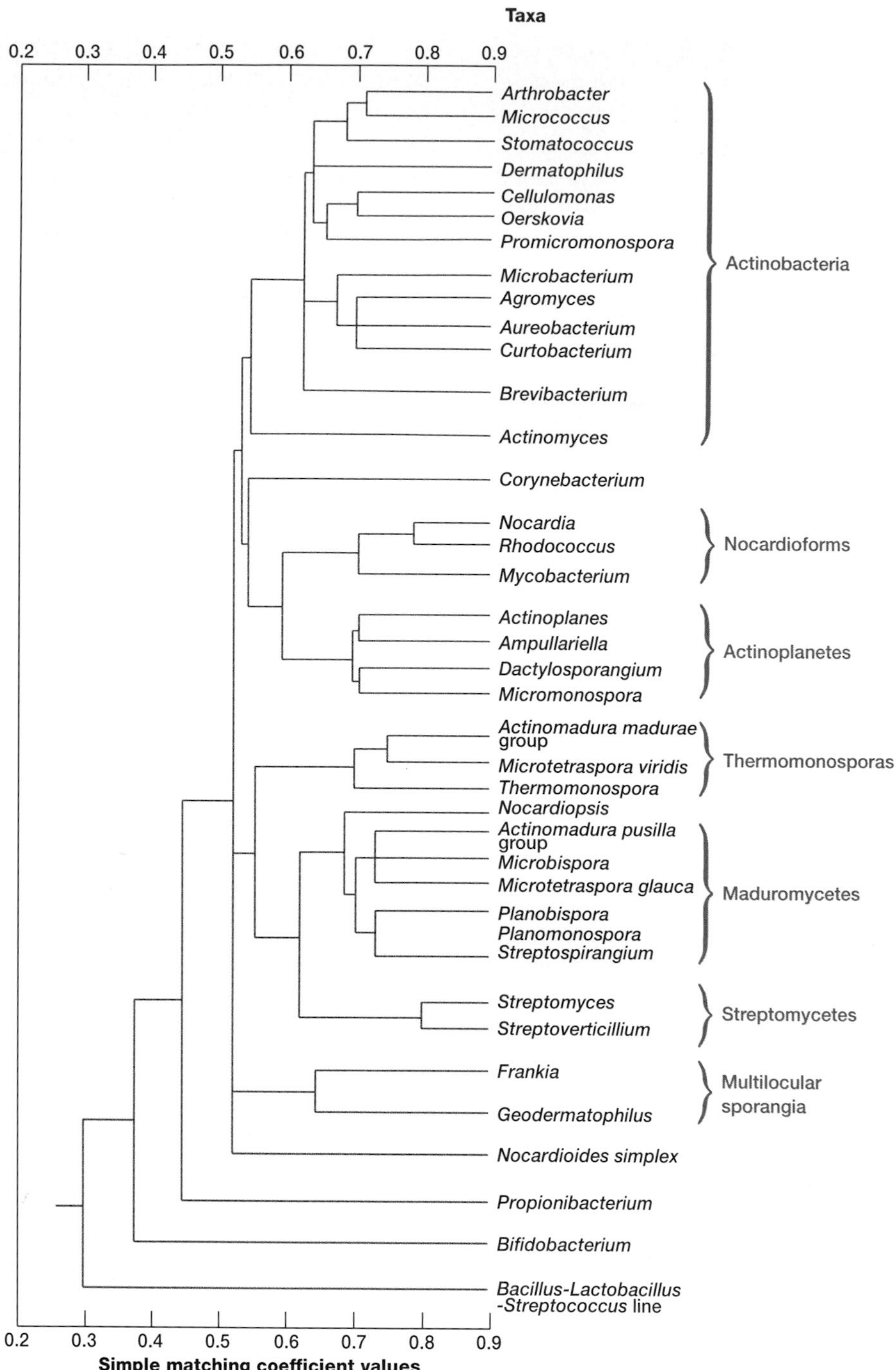

Figure 25.3 **Actinomycete Relationships.** This classification is based on 16S rRNA sequence comparisons.

oforms. Nocardioforms develop a substrate mycelium that readily breaks into rods and coccoid elements (figure 25.4). Several genera also form an aerial mycelium that rises above the substratum and may produce conidia. All genera have a high G + C content like other actinomycetes, and almost all are strict aerobes. Most species have peptidoglycan with *meso*-diaminopimelic acid and no peptide interbridge. The wall usually contains a carbohydrate composed of arabinose and galactose; mycolic acids are present in the genera *Nocardia* and *Rhodococcus*.

The genus *Nocardia* is distributed worldwide in soil and is also found in aquatic habitats. Nocardiae are involved in the degradation of hydrocarbons and waxes, and can contribute to the biodeterioration of rubber joints in water and sewage pipes. Although most are free-living saprophytes, some species, particularly *N. asteroides*, are opportunistic pathogens that cause nocardiosis in humans and other animals. People with low resistance due to other health problems are most at risk. The lungs are most often infected, but the central nervous system and other organs may be invaded.

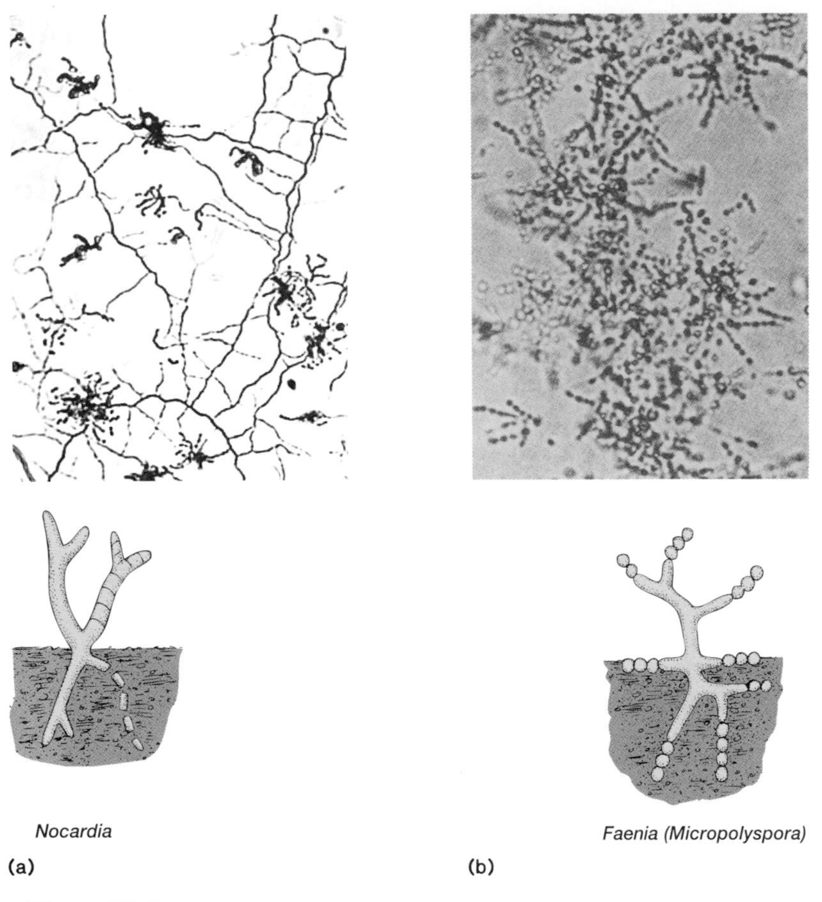

Figure 25.4 **Nocardioform Actinomycetes.** (*a*) *Nocardia asteroides,* substrate mycelium and aerial mycelia with conidia illustration and light micrograph (×1,250). (*b*) A photomicrograph (×650) and illustration of the substrate mycelium of *Faenia rectivirgula* near the growing margin.

1. What is a nocardioform, and how can the group be distinguished from other actinomycetes?
2. Where is *Nocardia* found, and what problems may it cause? Consider both environmental and public health concerns.

Actinomycetes with Multilocular Sporangia

The actinomycetes in section 27 of *Bergey's Manual* form clusters of spores when a hypha divides both transversely and longitudinally. (Multilocular means having many cells or compartments.) All three genera in this section have type III cell walls (table 25.1), although the cell extract sugar patterns differ. The G + C content varies from 57 to 75 mol%. *Geodermatophilus* (type IIIC) has motile spores and is an aerobic soil organism. *Dermatophilus* (type IIIB) also forms packets of motile spores with tufts of flagella, but it is a facultative anaerobe and a parasite of mammals responsible for the skin infection streptothrichosis. *Frankia* (type IIID) forms nonmotile sporangiospores in a sporogenous body (figure 25.5). It grows in symbiotic association with the roots of at least eight families of higher nonleguminous plants (e.g., alder trees) and

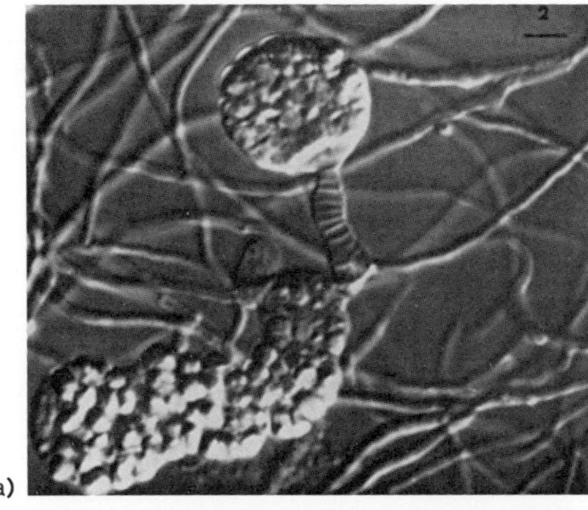

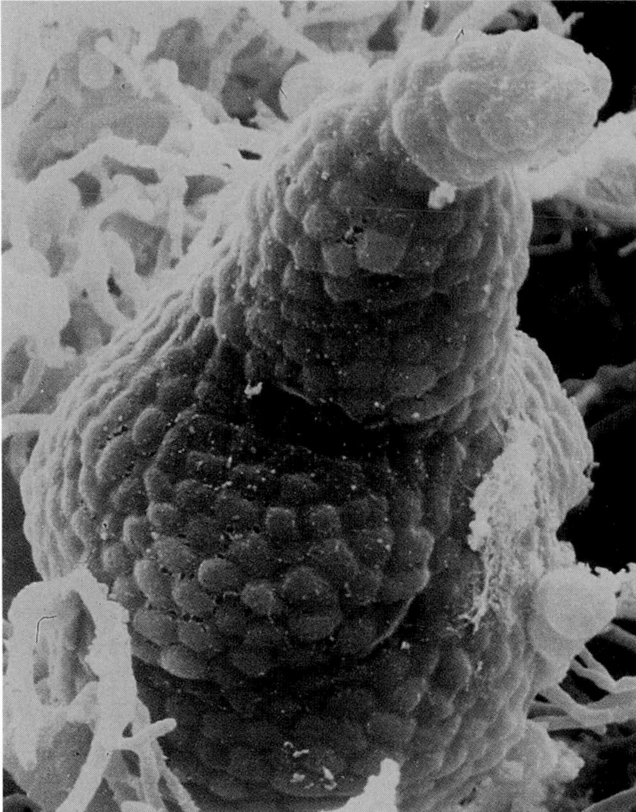

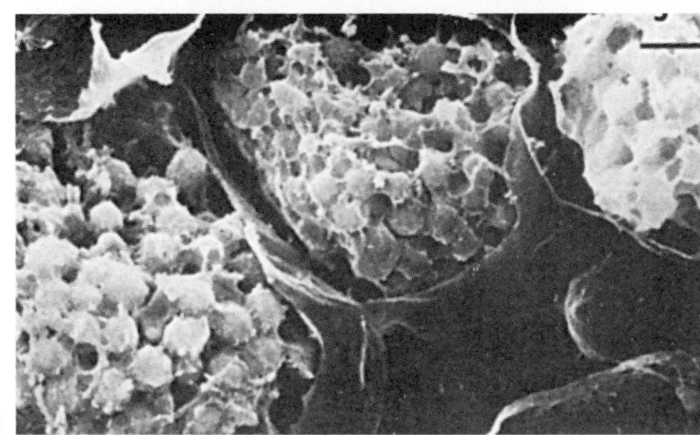

Figure 25.5 **Actinomycetes with Multilocular Sporangia.** *Frankia.* (*a*) An interference contrast micrograph showing hyphae, sporangia, and spores. Bar = 5 μm. (*b*) A scanning electron micrograph of a sporangium surrounded by hyphae. (*c*) A nodule of the alder *Alnus rubra* showing cells filled with vesicles of *Frankia.* Scanning electron microscopy. Bar = 5 μm.

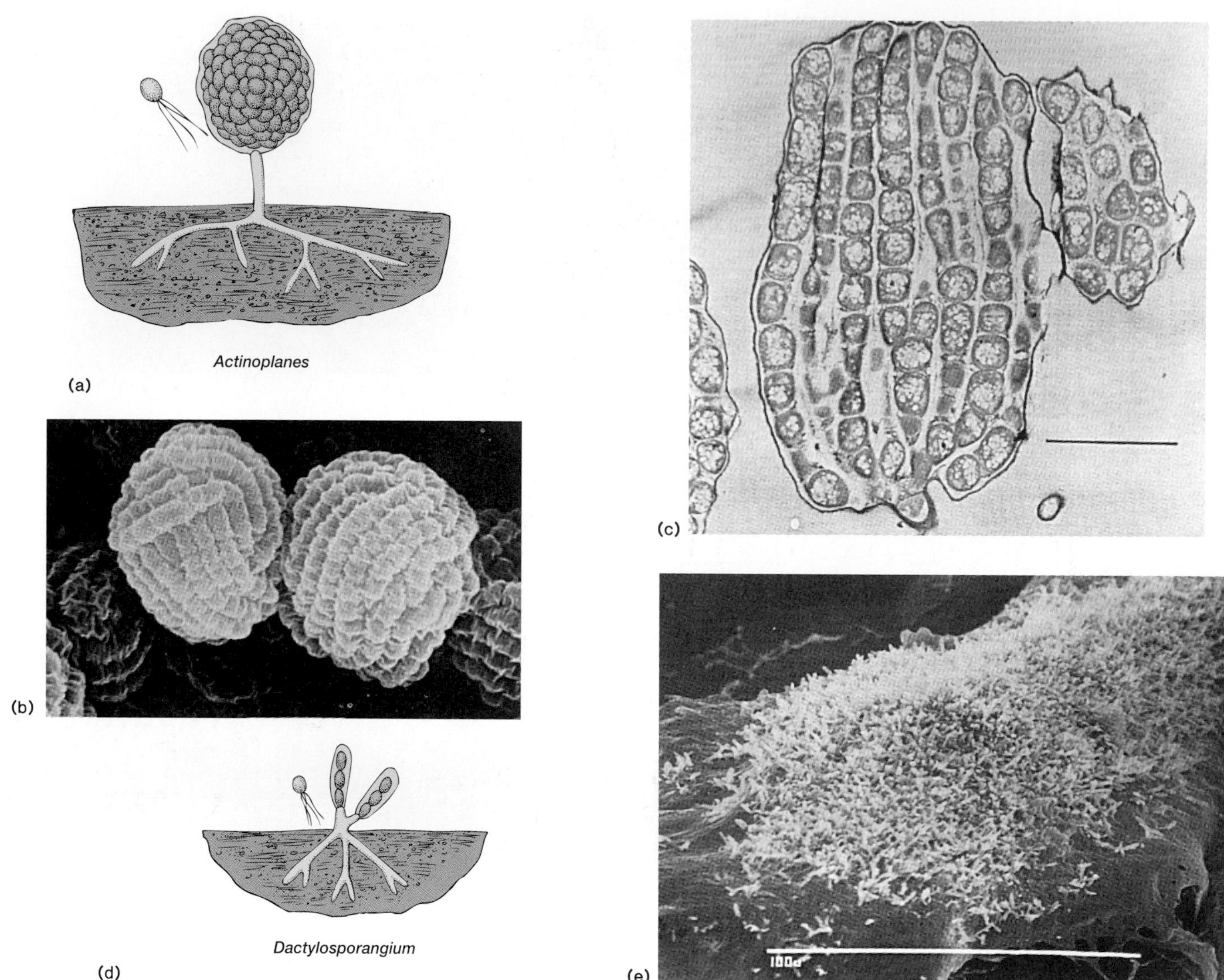

Figure 25.6 **Actinoplanetes.** Actinoplanete morphology. (*a*) *Actinoplanes* structure. (*b*) A scanning electron micrograph of mature *Actinoplanes* sporangia. Bar = 5 μm. (*c*) A section of an *Actinoplanes rectilineatus* sporangium. Bar = 3 μm. (*d*) *Dactylosporangium* structure. (*e*) A *Dactylosporangium* colony covered with sporangia (×700).

is a microaerophile able to fix atmospheric nitrogen (*see chapter 42*).

The roots of infected plants develop nodules that fix nitrogen so efficiently that a plant such as an alder can grow in the absence of combined nitrogen when nodulated. Within the nodule cells, *Frankia* forms branching hyphae with globular vesicles at their ends (figure 25.5*c*). These vesicles may be the sites of nitrogen fixation. The nitrogen-fixation process resembles that of *Rhizobium* in that it is oxygen sensitive and requires molybdenum and cobalt.

1. Give the major distinguishing characteristics of the actinomycetes in section 27 of *Bergey's Manual.*
2. How does *Frankia* differ from *Geodermatophilus* and *Dermatophilus*?

Actinoplanetes

Actinoplanetes (Greek *actinos,* a ray or beam, and *planes,* a wanderer), section 28, have an extensive substrate mycelium and are wall type IID. Normally, an aerial mycelium is absent or rudimentary. Conidiospores are usually formed within a sporangium raised above the surface of the substratum at the end of a special hypha called a sporangiophore. The spores can be either motile or nonmotile. These bacteria vary in the arrangement and development of their spores. Some genera (*Actinoplanes, Ampullariella, Pilimelia*) have spherical, cylindrical, or irregular sporangia with a few to several thousand spores per sporangium (figures 25.2*b* and 25.6). The sporangium develops above the substratum at the tip of a sporangiophore; the spores are arranged in coiled or parallel chains (figure 25.7). *Dactylosporangium* forms club-shaped, fingerlike, or pyriform sporangia with one to six spores (figure

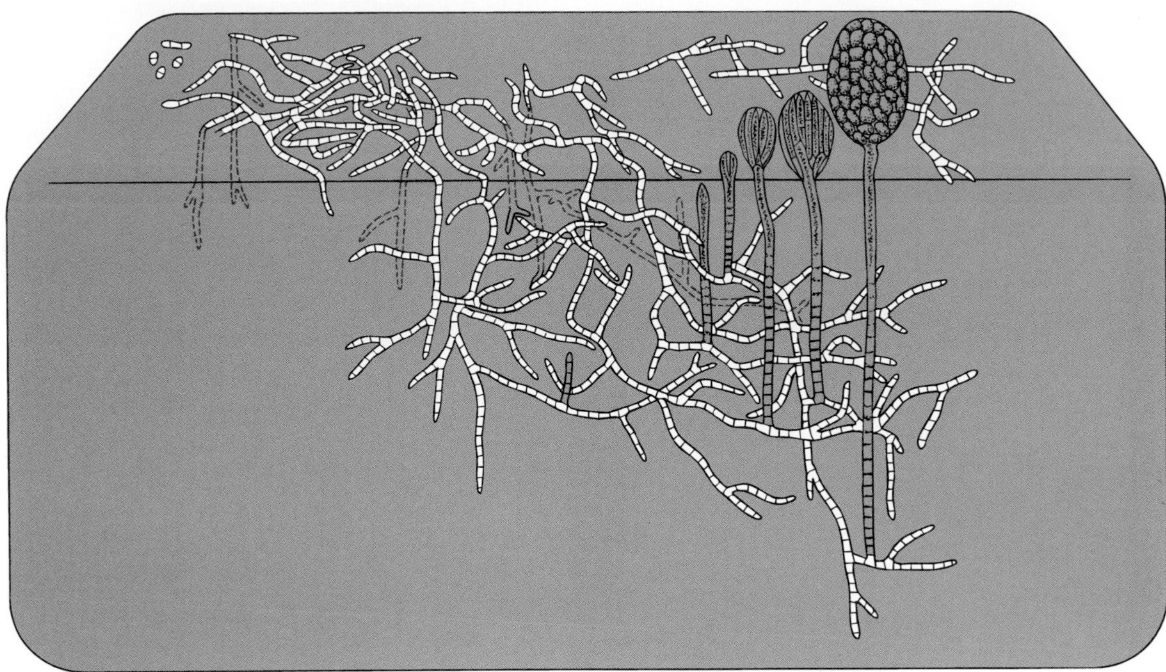

Figure 25.7 **Sporangium Development in an Actinoplanete.** The developing sporangium is shown in color.

25.6*d,e*). *Micromonospora* bears single spores, which often occur in branched clusters of sporophores (figure 25.2*c*). The section presently contains five genera.

Actinoplanetes (s., actinoplanete) grow in almost all soil habitats, ranging from forest litter to beach sand. They also flourish in fresh water, particularly in streams and rivers (probably because of abundant oxygen and plant debris). Some have been isolated from the ocean. The soil-dwelling species may have an important role in the decomposition of plant and animal material. *Pilimelia* grows in association with keratin. *Micromonospora* actively degrades chitin and cellulose, and it can produce antibiotics such as gentamicin.

1. Give the distinguishing properties of the actinoplanetes.
2. Briefly describe the variations in sporangia and sporophore organization of the actinoplanetes.

Streptomyces and Related Genera

Section 29 of *Bergey's Manual* contains four genera whose aerial hyphae divide in a single plane to form chains of 5 to 50 or more nonmotile conidiospores with surface texture ranging from smooth to spiny and warty (figures 25.2*d*, 25.8, and 25.9). All have a type I cell wall and a G + C content of around 69 to 78%. The substrate mycelium, when present, does not undergo fragmentation. Members of the section as a whole are often called **streptomycetes** (Greek *streptos,* bent or twisted, and *myces,* fungus).

Streptomyces is an enormous genus; 378 species are given in the Approved Lists of Bacterial Names. *Bergey's Manual* describes 142 species in detail. Undoubtedly, the number of accepted species will decrease with further progress in streptomycete taxonomy. Members of the genus are strict aerobes, are wall type I, and form chains of nonmotile spores within a thin, fibrous sheath (figure 25.9). The three to many conidia in each chain are often pigmented and can be smooth, hairy, or spiny in texture. *Streptomyces* species are determined by means of a mixture of morphological and physiological characteristics, including the following: the color of the aerial and substrate mycelia, spore arrangement, surface features of individual spores, carbohydrate utilization, melanin production, nitrate reduction, and the hydrolysis of urea and hippuric acid.

Streptomycetes are very important, both ecologically and medically. The natural habitat of most streptomycetes is the soil, where they may constitute from 1 to 20% of the culturable population. In fact, the odor of moist earth is largely the result of streptomycete production of volatile substances such as **geosmin.** Streptomycetes play a major role in mineralization. They are very flexible nutritionally and can aerobically degrade resistant substances such as pectin, lignin, chitin, keratin, latex, and aromatic compounds. Streptomycetes are best known for their synthesis of a vast array of antibiotics, some of which are useful in medicine and biological research. Examples include amphotericin B, chloramphenicol, erythromycin, neomycin, nystatin, streptomycin, tetracycline, and so forth (figure 25.10*a*). Although most streptomycetes are nonpathogenic saprophytes, a few are associated with plant and animal diseases. *Streptomyces scabies* causes scab disease in potatoes and beets (figure 25.10*b*). *S. somaliensis* is the only streptomycete known to be pathogenic for humans. It is associated with **actinomycetoma,** an infection of subcutaneous tissues that produces lesions and leads to swelling, abscesses,

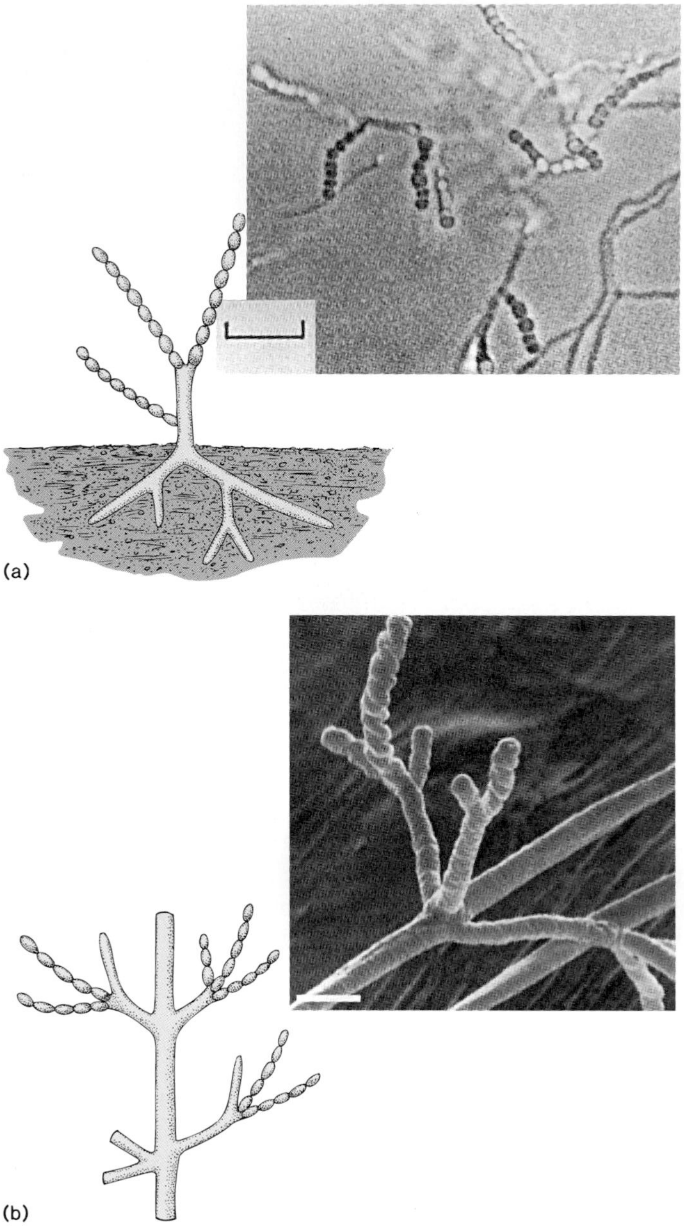

(a)

(b)

Figure 25.8 ***Streptomyces* and Related Genera.** Streptomycete conidia arrangement. (*a*) An illustration of typical *Streptomyces* morphology; a light micrograph of *S. carpinesis* spore chains. Bar = 5 μm. (*b*) *Streptoverticillium* morphology; a scanning electron micrograph of *Sv. salmonis* with developing spore chains. Bar = 2 μm.

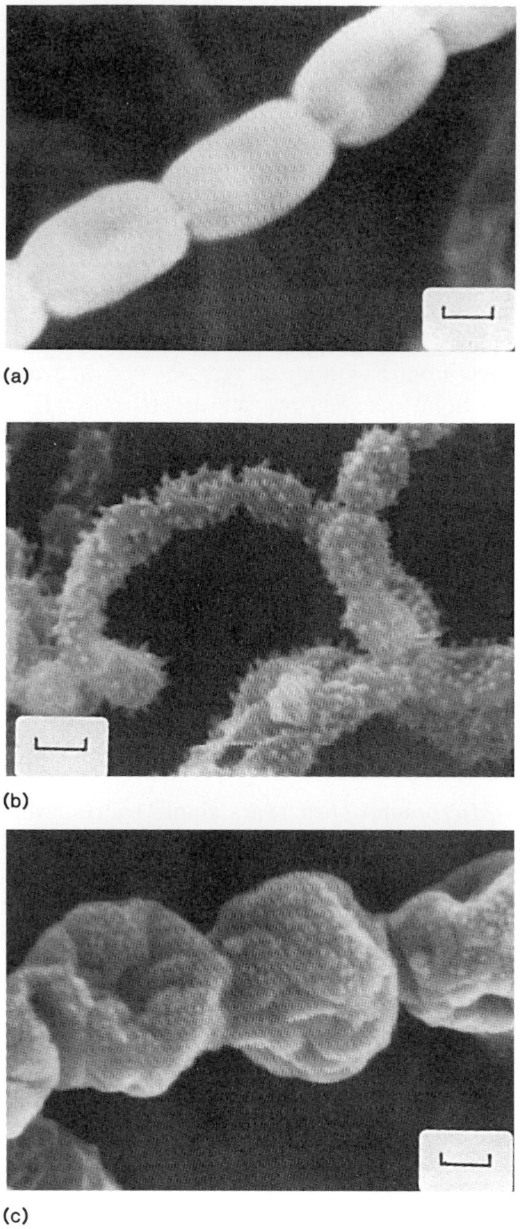

(a)

(b)

(c)

Figure 25.9 **Streptomycete Spores.** (*a*) Smooth spores of *S. niveus;* scanning electron micrograph. Bar = 0.25 μm. (*b*) Spiney spores of *S. viridochromogenes.* Bar = 0.5 μm. (*c*) Warty spores of *S. pulcher.* Bar = 0.25 μm.

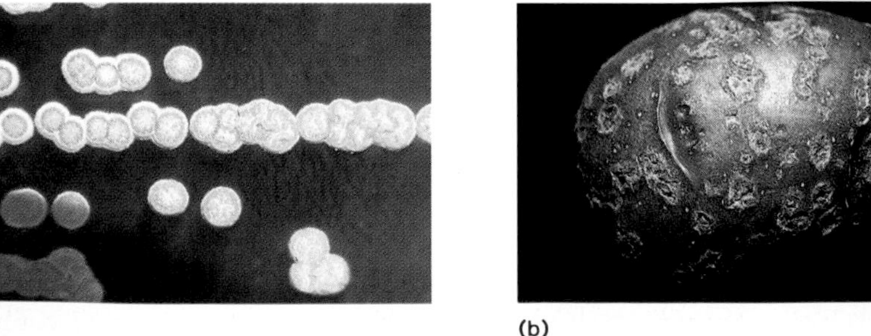

(a) (b)

Figure 25.10 **Streptomycetes with Practical Importance.** (*a*) *Streptomyces griseus.* Colonies of the actinomycete that produces streptomycin. (*b*) *Streptomyces scabies,* an actinomycete growing on a potato.

and even bone destruction if untreated. *S. albus* and other species have been isolated from patients with various ailments and may be pathogenic.

Antibiotics and their properties (chapter 16).

Other genera in this section are easily distinguished from *Streptomyces. Streptoverticillium* has an aerial mycelium with a whorl of three to six short branches that are produced at fairly regular intervals. These branches have secondary branches bearing chains of spores (figure 25.8*b*). *Sporichthya* is one of the strangest of the actinomycetes. It lacks a substrate mycelium. The hyphae remain attached to the substratum by holdfasts and grow upward to form aerial mycelia that release motile, flagellate conidia in the presence of water.

1. What characteristics do the actinomycetes of section 29 have in common? What is a streptomycete?
2. Describe the major properties of the genus *Streptomyces.*
3. Give three ways in which *Streptomyces* is of ecological and medical importance.
4. How do *Streptoverticillium* and *Sporichthya* differ from *Streptomyces*?

Maduromycetes

A collection of seven genera called the maduromycetes are placed in section 30. All genera have type III cell walls and the sugar derivative **madurose** (3-O-methyl-D-galactose) in whole cell homogenates. Their G + C content is 64 to 74 mol%. Aerial mycelia bear pairs or short chains of spores, and the substrate mycelium is branched. Some genera form sporangia. Typical genera are *Actinomadura, Microbispora, Planomonospora,* and *Streptosporangium* (figures 25.2*e* and 25.11). *Actinomadura* is another actinomycete associated with the disease actinomycetoma.

Thermomonospora and Related Genera

The four genera in section 31 all have type III cell walls and usually a type C sugar pattern (*Thermomonospora* can be IIIB). Their spores are not heat resistant. There is considerable variation in morphology and life-style. *Thermomonospora* produces single spores on the aerial mycelium or on both the aerial and substrate mycelia. It has been isolated from high-temperature habitats such as compost piles and hay, and can grow at 40 to 48° C. *Actinosynnema* produces a compact cluster of erect hyphae that resembles a stem; chains of motile conidia are attached to the top of the cluster, much like the branches of a bush or small tree. *Nocardiopsis* has a substrate mycelium that fragments like the nocardioforms, but it lacks mycolic acid and has a different cell wall chemistry. The aerial hyphae assume a zigzag shape and then form long chains of spores.

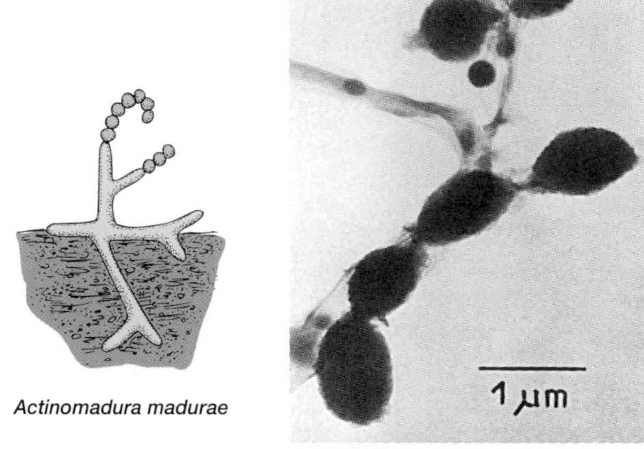

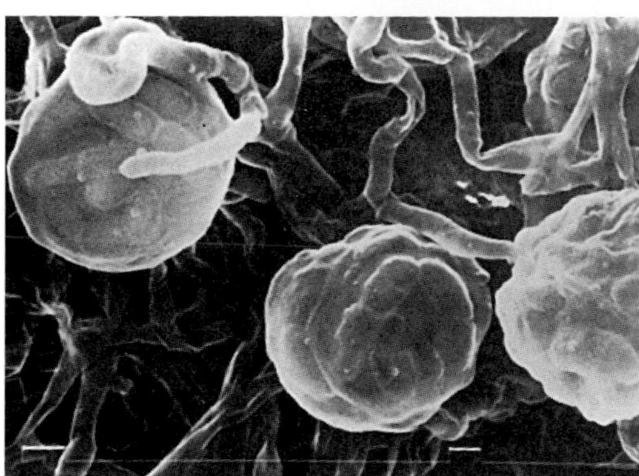

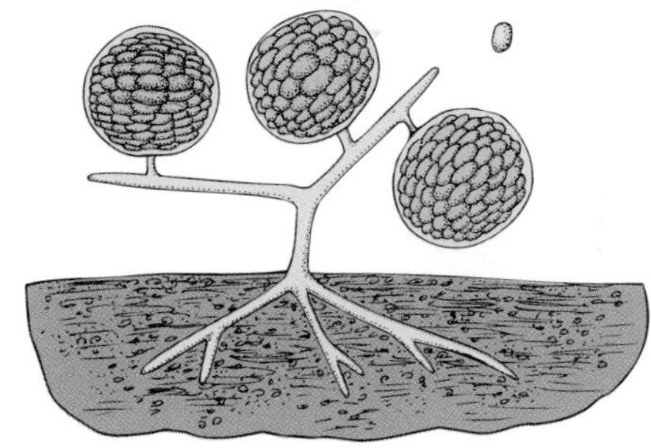

Figure 25.11 **Maduromycetes.** (*a*) *Actinomadura madurae* morphology; illustration and electron micrograph of a spore chain (×16,500). (*b*) *Streptosporangium* morphology; illustration and micrograph of *S. album* on oatmeal agar with sporangia and hyphae; SEM. Bar = 10 μm.

Thermoactinomycetes

Section 32 of *Bergey's Manual* contains one genus, *Thermoactinomyces.* This genus is thermophilic and wall type IIIC, grows between 45 and 60°C, and forms single spores on both its aerial and substrate mycelia (figure 25.12). Its G + C content is lower than that of other actinomycetes (52 to 55 mol%),

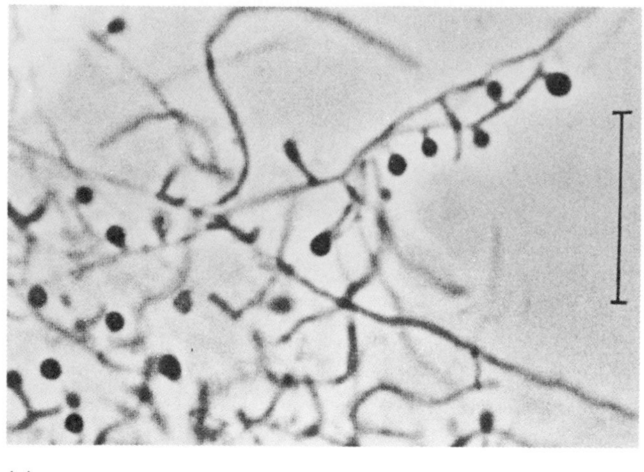

(a)

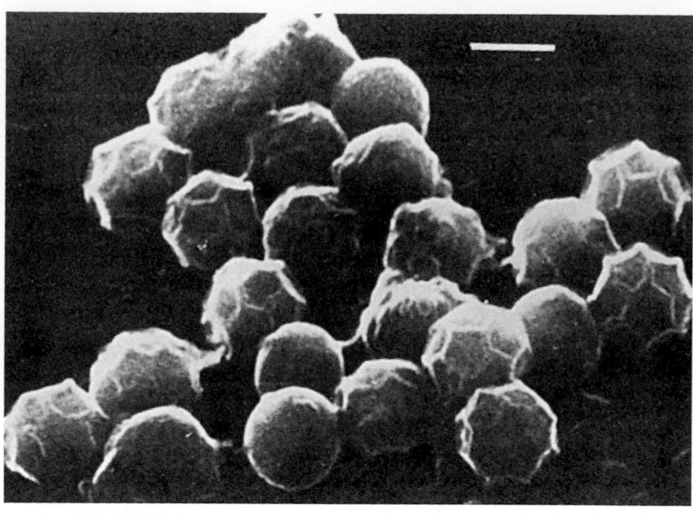

(b)

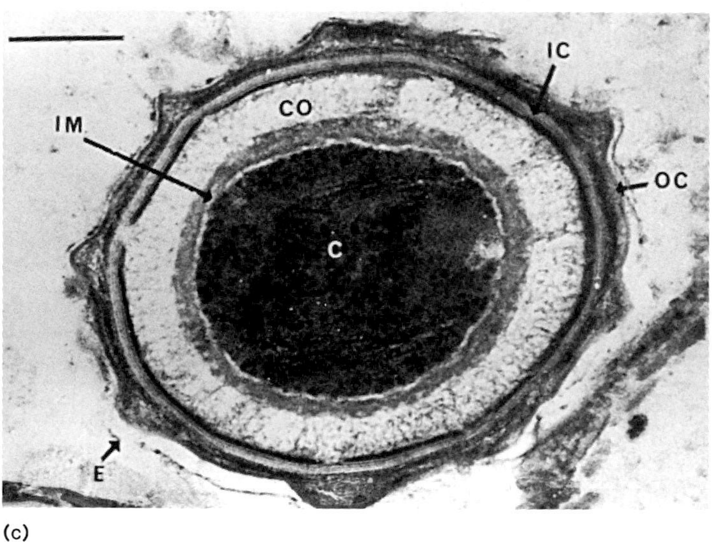

(c)

Figure 25.12 **Thermoactinomycetes.** Thermoactinomycete structure. (*a*) *Thermoactinomyces vulgaris* aerial mycelium with developing endospores at tips of hyphae (Bar = 10 μm). (*b*) Scanning electron micrograph of *T. sacchari* spores. Bar = 1 μm. (*c*) Thin section of *T. sacchari* endospore. Bar = 0.1 μm. *E*, exosporium; *OC*, outer spore coat; *IC*, inner spore coat; *CO*, cortex; *IM*, inner forespore membrane; *C*, core.

and its 16S rRNA sequence suggests a relationship with the genus *Bacillus*. It is commonly found in damp haystacks, compost piles, and other high-temperature habitats. Unlike other actinomycete spores, *Thermoactinomyces* spores (figure 25.12*b,c*) are true endospores and very heat resistant; they can survive at 100°C for more than 20 minutes. They are formed within hyphae and appear to have typical endospore structure, including the presence of calcium and dipicolinic acid. *Thermoactinomyces vulgaris* (figure 25.12*a*) from haystacks, grain storage silos, and compost piles is a causative agent of farmer's lung, an allergic disease of the respiratory system in agricultural workers. Recently, spores from *Thermoactinomyces vulgaris* were recovered from the mud of a Minnesota lake and found to be viable after about 7,500 years of dormancy.

The bacterial endospore (pp. 62–65).

1. List the major properties of bacteria in sections 30 and 31 of *Bergey's Manual.* What is madurose?
2. Briefly describe the genus *Thermoactinomyces*, with particular emphasis on its unique features.

Summary

1. Actinomycetes are aerobic, gram-positive bacteria that form branching, usually nonfragmenting, hyphae and asexual spores.
2. The asexual spores borne on aerial mycelia are called conidiospores or conidia if they are at the tip of hyphae and sporangiospores when they are within sporangia.
3. Actinomycetes have several distinctively different types of cell walls and often also vary in terms of the sugars present in cell extracts. Properties such as color and morphology are also taxonomically useful.
4. *Bergey's Manual* divides the actinomycetes into eight groups based on such properties as conidia arrangement, the presence of a sporangium, cell wall type, and cell extract sugars.
5. Nocardioform actinomycetes, in section 26 of *Bergey's Manual,* have hyphae that readily fragment into rods and coccoid elements, and often form aerial mycelia with spores.
6. Bacteria in section 27 produce clusters of spores at hyphal tips and have type III cell walls. *Frankia,* one of the three genera in this section, grows in symbiotic association with higher nonleguminous plants and fixes nitrogen.
7. Section 28 contains the actinoplanetes. All are type IID and have an extensive substrate mycelium. They are present in the soil, fresh water, and the ocean. The soil forms are probably important in mineralization.
8. *Streptomyces* and related genera are placed in section 29. Their aerial hyphae bear chains of 5 to 50 or more nonmotile conidiospores, and all have a type I cell wall.
9. *Streptomyces,* the largest genus of actinomycetes, has chains of nonmotile conidia within a thin sheath.
10. Streptomycetes are important in the degradation of more resistant organic material in the soil and produce many useful antibiotics. A few cause diseases in plants and animals.
11. Section 30 contains the maduromycetes, actinomycetes with type III cell walls and the sugar derivative madurose.
12. *Thermomonospora* and related genera are normally type IIIC (a few are IIIB), and they are placed in section 31.
13. Section 32 contains *Thermoactinomyces,* a thermophilic actinomycete that forms true endospores.

Key Terms

actinomycete *507*
actinomycetoma *513*
conidia *507*
conidiospores *507*
geosmin *513*
madurose *515*
nocardioforms *509*
sporangiospores *507*
streptomycetes *513*
thallus *507*

Questions for Thought and Review

1. Suppose that you discovered a nodulated plant that could fix atmospheric nitrogen. How might you show that a bacterial symbiont was involved and that *Frankia* rather than *Rhizobium* was responsible?
2. How could one decide whether a newly isolated spore-forming bacterium was producing conidiospores or true endospores?
3. List the properties that are most useful in actinomycete taxonomy, and give some indication of their relative importance.

Additional Reading

Balows, A.; Trüper, H. G.; Dworkin, M.; Harder, W.; and Schleifer, K.-H. 1992. *The prokaryotes,* 2d ed. New York: Springer-Verlag.

Benson, D. R. 1988. The genus *Frankia:* Actinomycete symbionts of plants. *Microbiol. Sci.* 5(1):9–12.

Ensign, J. C. 1978. Formation, properties, and germination of actinomycete spores. *Ann. Rev. Microbiol.* 32:185–219.

Goodfellow, M.; Modarski, M.; and Williams, S. T. eds. 1984. *The biology of the actinomycetes.* New York: Academic Press.

Holt, J. G., editor-in-chief. 1989. *Bergey's manual of systematic bacteriology,* vol. 4, eds. S. T. Williams and M. E. Sharpe. Baltimore, Md.: Williams & Wilkins.

Kalakoutskii, L. V., and Agre, N. S. 1976. Comparative aspects of development and differentiation in actinomycetes. *Bacteriol. Rev.* 40(2):469–524.

Land, G.; McGinnis, M. R.; Staneck, J.; and Gatson, A. 1991. Aerobic pathogenic *Actinomycetales.* In *Manual of clinical microbiology,* 5th ed., eds. A. Balows, W. J. Hausler, Jr., K. L. Herrmann, H. D. Isenberg, and H. J. Shadomy, 340–59. Washington, D.C.: American Society for Microbiology.

Lechevalier, M. P., and Lechevalier, H. A. 1980. The chemotaxonomy of actinomycetes. In *Actinomycete taxonomy,* eds. A. Dietz and D. W. Thayer, *Actinomycete taxonomy,* special publication 6, 227–91. Arlington, Va.: Society for Industrial Microbiology.

Parenti, F., and Coronelli, C. 1979. Members of the genus *Actinoplanes* and their antibiotics. *Ann. Rev. Microbiol.* 33:389–411.

Schwintzer, C. R., and Tjepkema, J. D., eds. 1990. *The biology of Frankia and actinorhizal plants.* San Diego, Calif.: Academic Press.

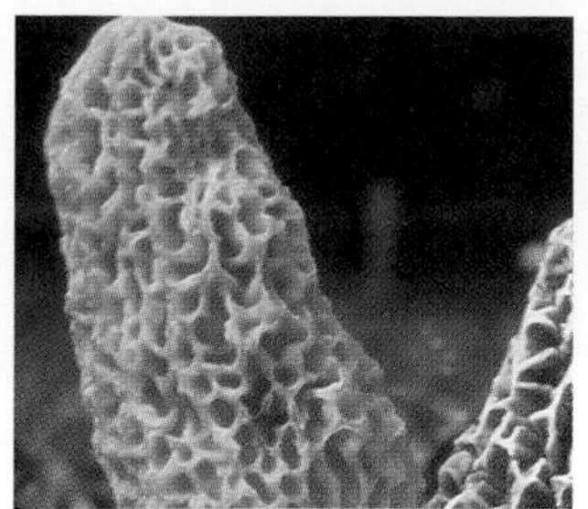

CHAPTER 26
The Fungi, Slime Molds, and Water Molds

Yeasts, molds, mushrooms, mildews, and the other fungi pervade our world. They work great good and terrible evil. Upon them, indeed, hangs the balance of life; for without their presence in the cycle of decay and regeneration, neither man nor any other living thing could survive.

—Lucy Kavaler,

Outline

Concepts

1. Fungi are widely distributed and are found wherever moisture is present. They are of great importance to humans in both beneficial and harmful ways.
2. Fungi exist primarily as filamentous hyphae. A mass of hyphae is called a mycelium.
3. Like some bacteria, fungi digest insoluble organic matter by secreting exoenzymes, then absorbing the soluble nutrients.
4. Two reproductive structures occur in the fungi: (1) sporangia form asexual spores, and (2) gametangia form sexual gametes.
5. The zygomycetes are characterized by resting structures called zygospores—cells in which zygotes are formed.
6. The ascomycetes form zygotes within a characteristic club-shaped structure, the ascus.
7. Yeasts are unicellular fungi—mainly ascomycetes.
8. Basidiomycetes possess dikaryotic hyphae with two nuclei, one of each mating type. The hyphae divide uniquely, forming basidiocarps within which basidia can be found.
9. The deuteromycetes (Fungi Imperfecti) have either lost the capacity for sexual reproduction, or it has never been observed.
10. The slime and water molds resemble the fungi only in appearance and life-style. The three divisions (*Myxomycota, Acrasiomycota,* and *Oomycota*) are now widely regarded as protists.

This chapter is an introduction to the fungi. It describes the members of the kingdom *Fungi,* surveys their diversity, discusses their ecological and commercial impact, and presents some of their typical life cycles. A brief overview is also given of the protists that resemble the fungi—the slime molds and water molds.

According to the five-kingdom system of Whittaker (*see chapter 20*), fungi are placed in the kingdom *Fungi.* Microbiologists use the term **fungus** (pl., fungi; Latin *fungus,* mushroom) to include eucaryotic, spore-bearing organisms with absorptive nutrition and no chlorophyll that reproduce sexually and asexually. Scientists who study fungi are known as **mycologists** (Greek *mykes,* mushroom, and *logos,* discourse), and the scientific discipline dealing with fungi is called **mycology.** The study of fungal toxins and their effects on various organisms is called **mycotoxicology,** and the diseases caused in animals are known as **mycoses** (s., mycosis).

Distribution

Fungi are primarily terrestrial organisms, although a few are freshwater or marine. Many are pathogenic and infect plants and animals. Fungi also form beneficial relationships with other organisms. For example, about three-fourths of all vascular plants form associations (called mycorrhizae) between their roots and fungi. Lichens are associations of fungi and either algae or cyanobacteria.

Mycorrhizae (pp. 855–56).
Lichens (pp. 566–67).

Importance

Fungi are important to humans in both beneficial and harmful ways. With bacteria and a few other groups of heterotrophic organisms, fungi act as decomposers, a role of enormous significance. They degrade complex organic materials in the environment to simple organic compounds and inorganic molecules. In this way, carbon, nitrogen, phosphorus, and other critical constituents of bodies are released and made available for other organisms (*see chapter 40*).

Fungi are the major cause of plant diseases (figure 26.1*a*). Over 5,000 species attack economically valuable crops and garden plants, and also many wild plants. In like manner, many diseases of animals (figure 26.1*b* and table 26.1) and humans (*see chapter 39*) are caused by fungi.

Fungi, especially the yeasts, are essential to many industrial processes involving fermentation (*see chapter 43*). Examples include the making of bread, wine, and beer. Fungi also play a major role in the preparation of some cheeses, soy sauce, and sufu; in the commercial production of many organic acids (citric, gallic) and certain drugs (ergometrine, cortisone); and in the manufacture of many antibiotics (penicillin, griseofulvin) and the immunosuppressive drug cyclosporine.

In addition, fungi are important research tools in the study of fundamental biological processes. Cytologists, geneticists, biochemists, biophysicists, and microbiologists regularly use fungi in their research.

(a)

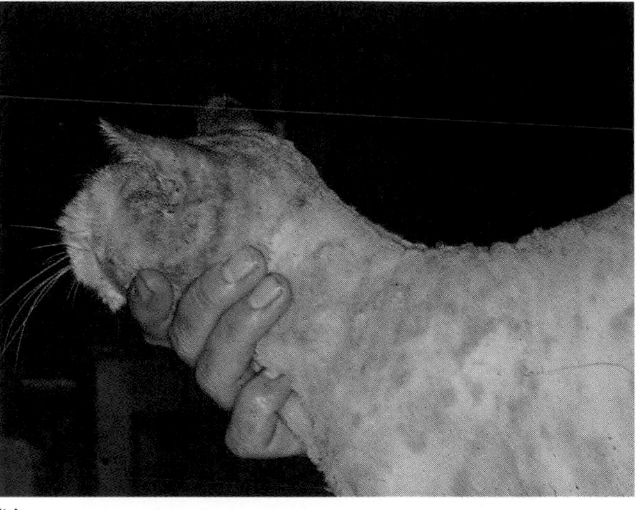

(b)

Figure 26.1 **Fungal Diseases.** (*a*) Apple scab. Midsummer scab lesions on fruit and leaf, which developed from spring infections. These lesions produce spores capable of causing further spread of the disease. (*b*) Dermatomycosis. Ringworm in a kitten due to *Microsporum canis.*

Structure

The body or vegetative structure of a fungus is called a **thallus** (pl., thalli). It varies in complexity and size, ranging from the single-cell microscopic yeasts to multicellular molds, macroscopic puffballs, and mushrooms (figure 26.2). The fungal cell is usually encased in a cell wall of **chitin.** Chitin is a strong

Table 26.1 Some Mycotoxicoses Produced by Fungal Mycotoxins in Domestic Animals

Disease	Fungus	Mycotoxin	Contaminated Foodstuff	Animals Affected
Aflatoxicosis	*Aspergillus flavus*	Aflatoxins	Rice, corn, sorghum, cereals, peanuts, soybeans	Poultry, swine, cattle, sheep, dogs
Ergotism	*Claviceps purpurea*	Ergot alkaloids	Seedheads of many grasses, grains	Cattle, horses, swine, poultry
Mushroom poisoning	*Amanita verna*	Amanitins	Eaten from pastures	Cattle
Poultry hemorrhagic syndrome	*Aspergillus flavus* and others	Aflatoxins	Toxic grain and meal	Chickens
Slobbers	*Rhizoctonia*	Alkaloid slaframine	Red clover	Sheep, cattle
Tall fescue toxicosis	*Acremonium coenophialum* (an endophytic fungus)	Unknown	Endophyte-infected tall fescue plants	Cattle

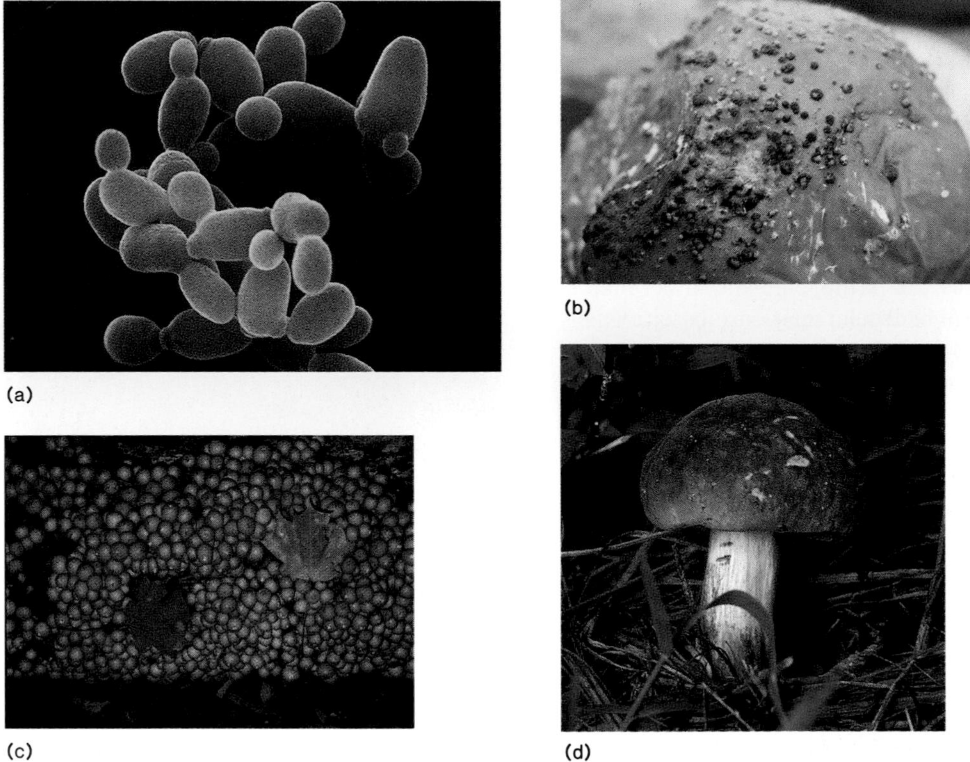

Figure 26.2 **Fungal Thalli.** (*a*) The microscopic, unicellular yeast, *Saccharomyces cerevisiae,* the yeast that makes bread dough rise; SEM($\times$21,000). Notice that some of the yeasts are reproducing by budding. (*b*) The multicellular common mold, *Penicillium,* growing on an apple. (*c*) A large group of puffballs, *Lycoperdon,* growing on a log. (*d*) A mushroom is made up of densely packed hyphae that form the mycelium or visible structure (thallus).

but flexible nitrogen-containing polysaccharide consisting of N-acetylglucosamine residues.

A **yeast** is a unicellular fungus that has a single nucleus and reproduces either asexually by budding and transverse division (figure 26.2*a*) or sexually through spore formation. Each bud that separates can grow into a new yeast, and some group together to form colonies. Generally, yeast cells are larger than bacteria, vary considerably in size, and are commonly spherical to egg shaped. They have no flagella but do possess most of the other eucaryotic organelles (figure 26.3).

A **mold** consists of long, branched, threadlike filaments of cells called **hyphae** (s., hypha; Greek *hyphe,* web) that form a **mycelium** (pl., mycelia), a tangled mass or tissuelike aggregation (figure 26.4). In some fungi, protoplasm streams through hyphae, uninterrupted by cross walls. These hyphae are called **coenocytic** (figure 26.5*a*). The hyphae of other fungi

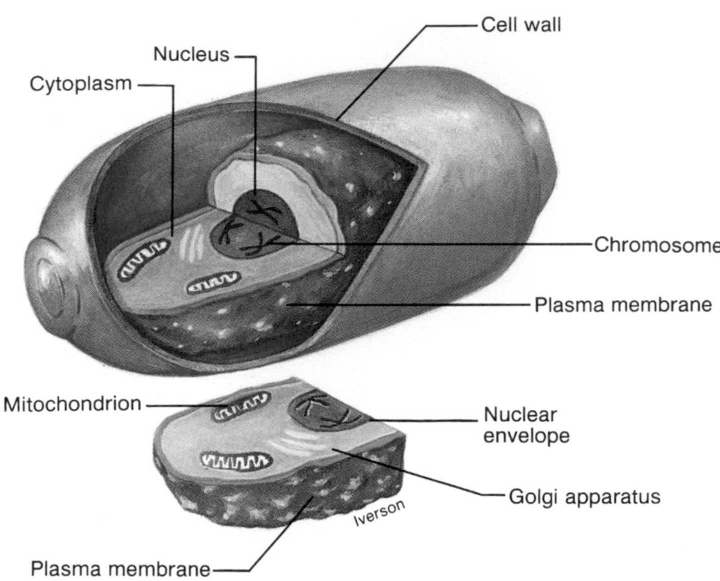

Figure 26.3 **A Yeast.** Diagrammatic drawing of a yeast cell showing typical morphology. For clarity, the plasma membrane has been drawn separated from the cell wall. In a living cell, the plasma membrane adheres tightly to the cell wall.

(figure 26.5*b*) have cross walls called **septa** (s., septum) with either a single pore (figure 26.5*c*) or multiple pores (figure 26.5*d*) that permit protoplasmic streaming. These hyphae are termed **septate.**

Hyphae are composed of an outer cell wall and a hollow inner lumen, which contains the protoplasm and organelles (figure 26.6). A plasma membrane surrounds the protoplasm and lies next to the cell wall.

Many fungi, especially those that cause diseases in humans and animals, are dimorphic (table 26.2); that is, they have two forms. Dimorphic fungi can change from (1) the yeast (Y) form in the animal to (2) the mold or mycelial form (M) in the external environment in response to changes in various environmental factors (nutrients, CO_2 tension, oxidation-reduction potentials, temperature). This shift is called the **YM shift.** In plant-associated fungi, the opposite type of dimorphism exists: the mycelial form occurs in the plant and the yeast form in the external environment.

1. How can a fungus be defined?
2. Where are fungi found?
3. Why are fungi important as decomposers?
4. What are some forms represented by different fungal thalli?
5. What organelles would you expect to find in the cytoplasm of a typical fungus?
6. Describe a typical yeast; a typical mold.

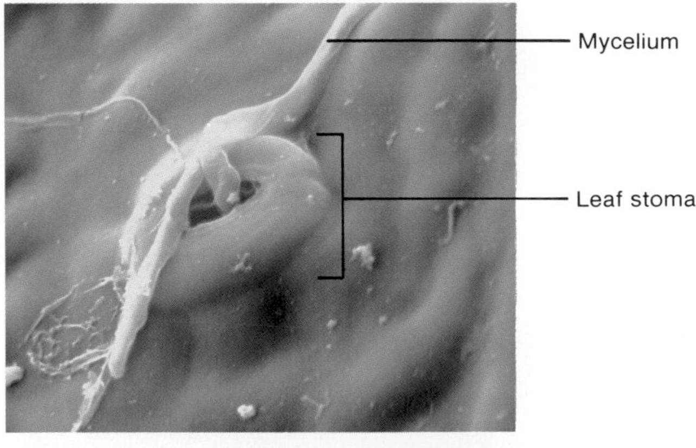

(a)

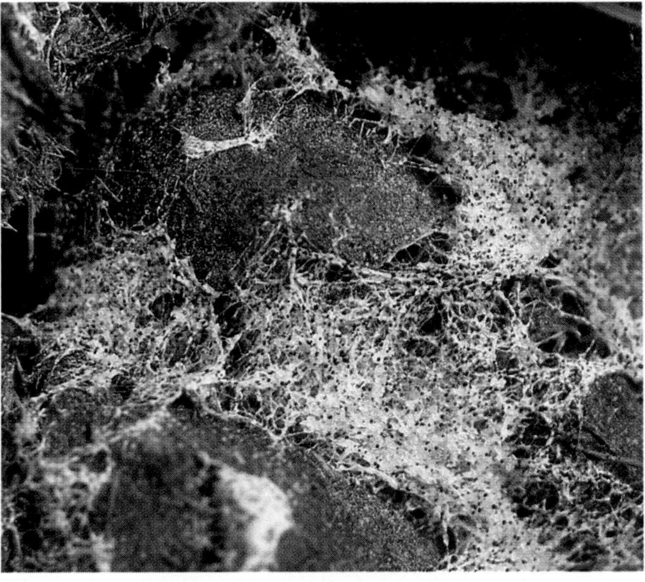

(b)

Figure 26.4 **Mold Mycelia.** (*a*) Scanning electron micrograph of a young mycelial aggregate forming over a leaf stoma (×1,000). (*b*) A very large macroscopic mycelium of a basidiomycete growing on the soil.

Nutrition and Metabolism

Fungi grow best in dark, moist habitats, but they are found wherever organic material is available. Most fungi are **saprophytes,** securing their nutrients from dead organic material. Like many bacteria, fungi can secrete hydrolytic enzymes that digest external substrates. They then absorb the soluble products. They are chemoorganoheterotrophs and use organic material as a source of carbon, electrons, and energy.

Glycogen is the primary storage polysaccharide in fungi. Most fungi use carbohydrates (preferably glucose or maltose) and nitrogenous compounds to synthesize their own amino acids and proteins.

Fungi are usually aerobic. Some yeasts, however, are facultatively anaerobic and can obtain energy by fermentation, such as in the production of ethyl alcohol from glucose. Obligately anaerobic fungi are found in the rumen of cattle.

(a)

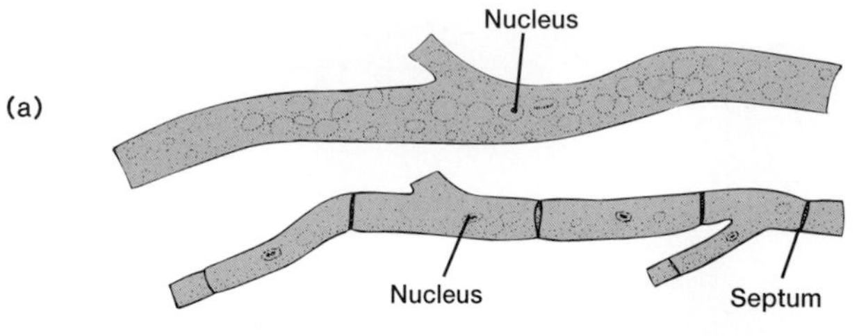

(b)

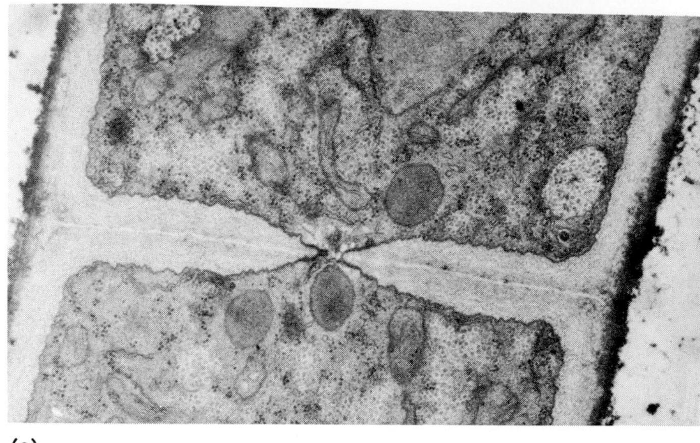

(c)

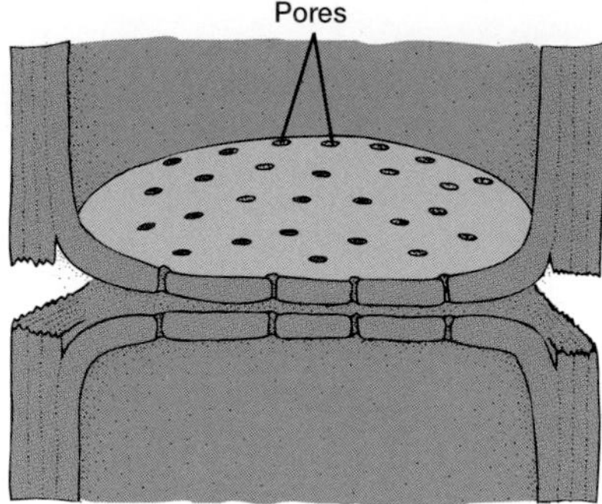

(d)

Figure 26.5 **Hyphae.** Drawings of (*a*) coenocytic hyphae and (*b*) hyphae divided into cells by septa. (*c*) Electron micrograph (×40,000) of a section of *Drechslera sorokiniana* showing wall differentiation and a single pore. (*d*) Drawing of a multiperforate septal wall structure.

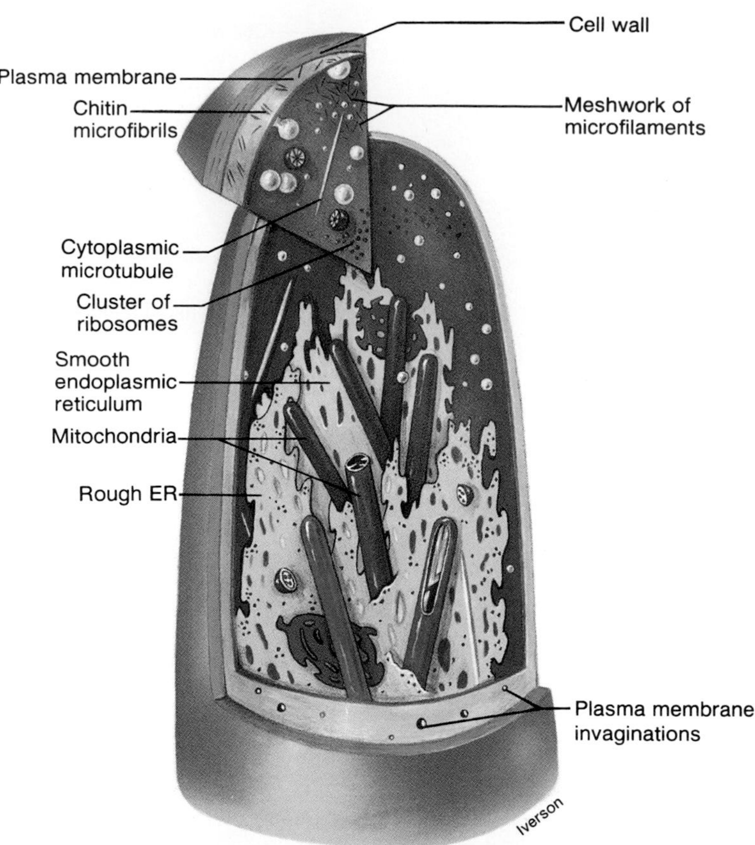

Figure 26.6 **Hyphal Morphology.** Diagrammatic representation of hyphal tip showing typical organelles and other structures.

TABLE 26.2 Some Medically Important Dimorphic Fungi

Fungus	Disease[a]
Blastomyces dermatitidis	Blastomycosis
Candida albicans	Candidiasis
Coccidioides capsulatum	Coccidioidomycosis
Histoplasma capsulatum	Histoplasmosis
Sporothrix schenckii	Sporotrichosis
Paracoccidioides brasiliensis	Paracoccidiodomycosis

[a]See chapter 39 for a discussion of each of these diseases.

Reproduction

Reproduction in fungi can be either asexual or sexual. Asexual reproduction is accomplished in several ways:

1. A parent cell can divide into two daughter cells by central constriction and formation of a new cell wall (figure 26.7*a*).
2. Somatic vegetative cells (figure 26.2*a*) may bud to produce new organisms. This is very common in the yeasts.
3. The most common method of asexual reproduction is the production of spores. Asexual spore formation occurs in an individual fungus through mitosis and subsequent cell division. There are several types of asexual spores:

   a. A hypha can fragment (by the separation of hyphae through splitting of the cell wall or septum) to form cells that behave as spores. These cells are called **arthroconidia** or **arthrospores** (figure 26.7*b*).

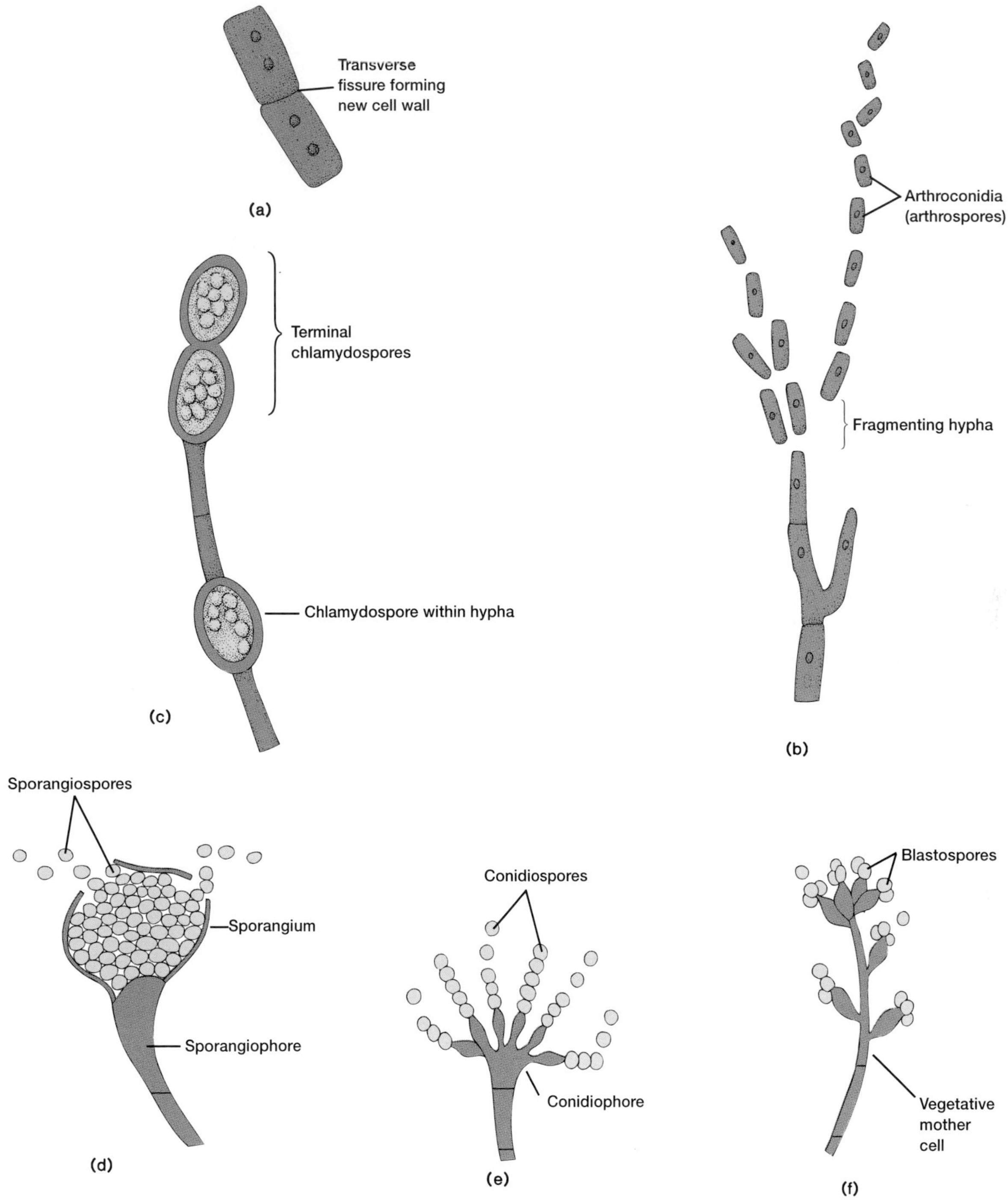

Figure 26.7 Diagrammatic Representation of Asexual Reproduction in the Fungi and Some Representative Spores. (*a*) Transverse fission. (*b*) Hyphal fragmentation resulting in arthroconidia (arthrospores) and (*c*) chlamydospores, (*d*) Sporangiospores in a sporangium. (*e*) Conidiospores arranged in chains at the end of a conidiophore. (*f*) Blastospores are formed from buds off of the parent cell.

b. If the cells are surrounded by a thick wall before separation, they are called **chlamydospores** (figure 26.7*c*).
c. If the spores develop within a sac (**sporangium;** pl., sporangia) at a hyphal tip, they are called **sporangiospores** (figure 26.7*d*).
d. If the spores are not enclosed in a sac but produced at the tips or sides of the hypha, they are termed **conidiospores** (figures 26.7*e* and 26.11).
e. Spores produced from a vegetative mother cell by budding (figure 26.7*f*) are called **blastospores.**

Sexual reproduction in fungi involves the union of compatible nuclei. Some fungal species are self-fertilizing and produce sexually compatible gametes on the same mycelium. Other species require outcrossing between different but sexually compatible mycelia. Depending on the species, sexual fusion may occur between haploid gametes, gamete-producing bodies called **gametangia,** or hyphae. Sometimes, both the cytoplasm and haploid nuclei fuse immediately to produce the diploid zygote. Usually, however, there is a delay between cytoplasmic and nuclear fusion. This produces a **dikaryotic stage** in which cells contain two separate haploid nuclei, one from each parent (figure 26.8). Sexual reproduction can also yield spores. For example, the zygote sometimes develops into a **zygospore** (figure 26.9), an **ascospore** (figure 26.12), or a **basidiospore** (figure 26.14).

Fungal spores are important for several reasons. The size, shape, color, and number of spores are useful in the identification of fungal species. The spores are often small and light and can remain suspended in air for long periods. Thus, they frequently aid in fungal dissemination, a significant factor that explains the wide distribution of many fungi. Fungal spores often spread by adhering to the bodies of insects and other animals. The bright colors and fluffy textures of many molds are often due to their aerial hyphae and spores.

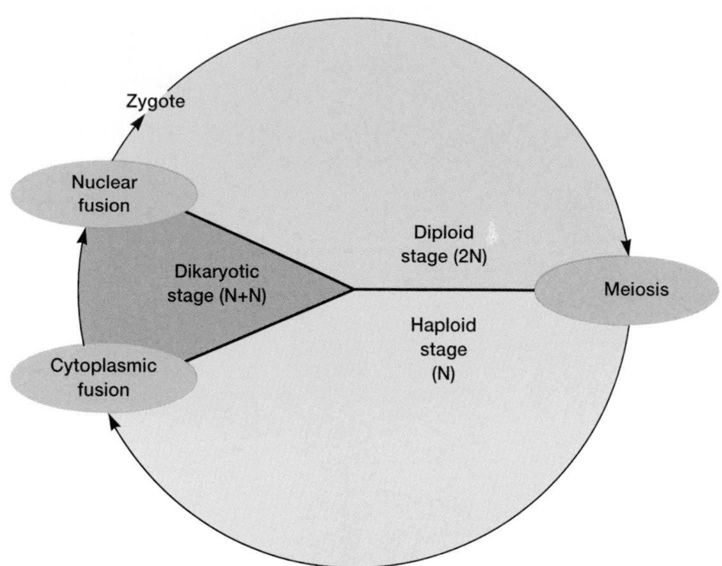

Figure 26.8 **Reproduction in Fungi.** A drawing of the generalized life cycle for fungi showing the alternation of haploid and diploid stages. Some fungal species do not pass through the dikaryotic stage indicated in this drawing. The asexual (haploid) stage is used to produce spores that aid in the dissemination of the species. The sexual (diploid) stage involves spores that are used for surviving adverse environmental conditions (e.g., cold, dryness, heat).

Characteristics of the Fungal Divisions

The taxonomic scheme used in this chapter classifies the fungi into four divisions (table 26.3), based primarily on variations in sexual reproduction. In mycology, a division is equivalent to a phylum in animal classification schemes.

Table 26.3 Divisions of Fungi

Division	Common Name	Approximate Number of Species
Zygomycota	Zygomycetes	600
Ascomycota	Sac fungi	35,000
Basidiomycota	Club fungi	30,000
Deuteromycota	Fungi Imperfecti	30,000

Division *Zygomycota*

The division *Zygomycota* contains the fungi called **zygomycetes.** Most live on decaying plant and animal matter in the soil; a few are parasites of plants, insects, animals, and humans. The hyphae of zygomycetes are coenocytic, with many haploid nuclei. Asexual spores, usually wind dispersed, develop in sporangia at the tips of aerial hyphae. Sexual reproduction produces tough, thick-walled zygotes called zygospores that can remain dormant when the environment is too harsh for growth of the fungus.

The bread mold, *Rhizopus stolonifer,* is a very common member of this division. This fungus grows on the surface of moist, carbohydrate-rich foods, such as breads, fruits, and vegetables. On breads, for example, *Rhizopus*'s hyphae rapidly cover the surface. Special hyphae called rhizoids extend into the bread, and absorb nutrients (figure 26.9). Other hyphae (stolons) become erect, then arch back into the substratum forming new rhizoids. Still others remain erect and produce at their tips asexual sporangia filled with the black spores, giving the mold its characteristic color. Each spore, when liberated, can start a new mycelium.

Rhizopus usually reproduces asexually, but if food becomes scarce or environmental conditions unfavorable, it begins sexual reproduction. Sexual reproduction requires compatible strains of opposite mating types (figure 26.9). These have traditionally been labeled + and − strains because they are not morphologically distinguishable as male and female. When the two mating strains are close, hormones are produced that cause their hyphae to form projections called **progametangia,** and then mature gametangia. After fusion of the gametangia, the nuclei of the two gametes fuse, forming a zygote. The zygote develops a thick, rough, black coat and becomes a dormant zygospore. Meiosis often occurs at the time of germination; the zygospore then cracks open and produces an asexual sporangium and the cycle begins anew.

The zygomycetes also contribute to human welfare. For example, *Rhizopus* is used in Indonesia to produce a food called tempeh from boiled, skinless soybeans. Another zygomycete is used with soybeans in the Orient to make a cheese called sufu (*see chapter 43*). Others are employed in the commercial

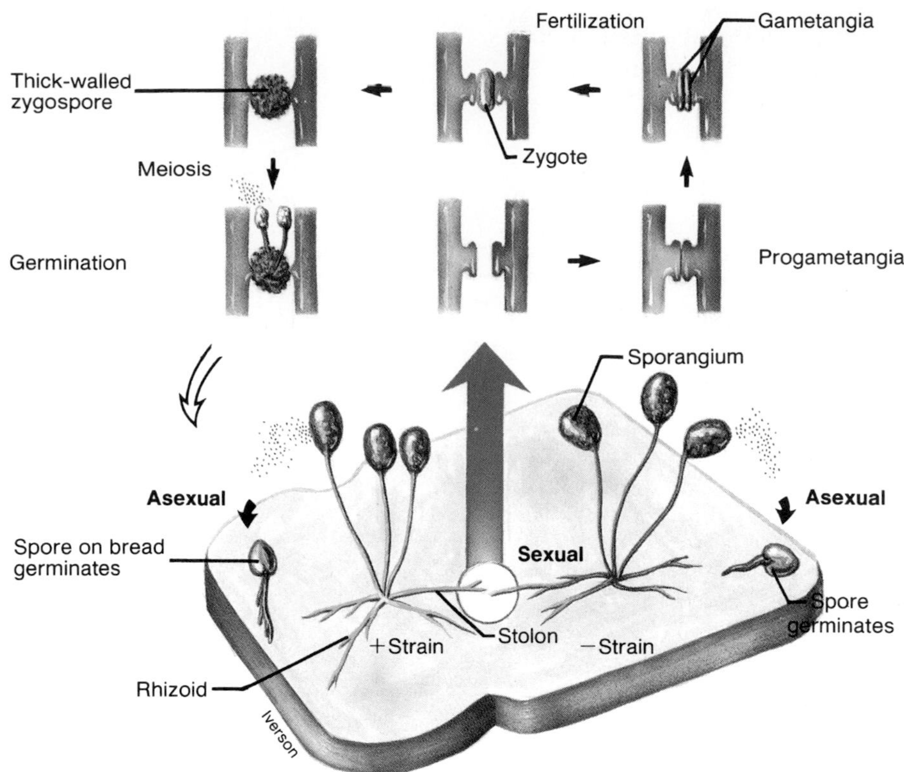

Figure 26.9 **Division *Zygomycota*.** Diagrammatic representation of the life cycle of *Rhizopus stolonifer*. Both the sexual and asexual phases are illustrated.

preparation of some anesthetics, birth control agents, industrial alcohols, meat tenderizers, and the yellow coloring agent in margarine and butter substitutes.

1. What generally governs the ecological distribution of fungi?
2. How is asexual reproduction accomplished in the fungi? Sexual reproduction?
3. Describe each of the following types of asexual fungal spores: sporangiospore, conidiospore, and blastospore.
4. Describe how a typical zygomycete reproduces.
5. Describe some beneficial uses for zygomycetes.

Division *Ascomycota*

The division *Ascomycota* contains the fungi called **ascomycetes,** commonly known as the sac fungi. Many species are quite familiar and economically important (figure 26.10). For example, most of the red, brown, and blue-green molds that cause food spoilage are ascomycetes. The powdery mildews that attack plant leaves and the fungi that cause chestnut blight and Dutch elm disease are ascomycetes. Many yeasts as well as edible morels and truffles are ascomycetes. The pink bread mold *Neurospora crassa,* also an ascomycete, has been a most important research tool in genetics and biochemistry.

Many ascomycetes are parasites on higher plants. *Claviceps purpurea* parasitizes rye and other grasses, causing the disease **ergot. Ergotism,** the toxic condition in humans and animals who eat grain infected with the fungus, is often accompanied by gangrene, psychotic delusions, nervous spasms, abortion, and convulsions. During the Middle Ages, ergotism, then known as St. Anthony's fire, killed thousands of people. For example, over 40,000 deaths from ergot poisoning were recorded in France in the year 943. It has been suggested that the widespread accusations of witchcraft in Salem Village (now Danvers) and other Massachusetts communities in the late 1690s may have resulted from outbreaks of ergotism. The pharmacological activities of ergot are due to its active ingredient, lysergic acid diethylamide (LSD). In controlled dosages, ergot can be used to induce labor, lower blood pressure, and ease migraine headaches.

The ascomycetes are named for their characteristic reproductive structure, the club- or sac-shaped **ascus** (pl., asci; Greek *askos,* sac). The mycelium of the ascomycetes is composed of septate hyphae. Asexual reproduction is common in the ascomycetes and takes place by way of conidiospores (figure 26.11).

Sexual reproduction in the ascomycetes always involves the formation of an ascus containing haploid **ascospores** (figure

(a) (b) (c)

Figure 26.10 **Division *Ascomycota.*** (*a*) The common morel, *Morchella esculenta,* is one of the choicest edible fungi. It fruits in the spring. (*b*) Scarlet cups, *Sarcoscypha coccinea,* with open ascocarps (apothecia). (*c*) The highly prized black truffle, *Tuber brumale.* Technically, truffles are mycorrhizal associations on oak trees.

Figure 26.11 **Asexual Reproduction in Ascomycetes.** Characteristic conidiospores of *Aspergillus* as viewed with the electron microscope (×1,200).

26.12*a*). In the more complex ascomycetes, ascus formation is preceded by the development of special **ascogenous hyphae** into which pairs of nuclei migrate (figure 26.12*b*). One nucleus of each pair originates from a "male" mycelium (**antheridium**) or cell, and the other from a "female" organ or cell (**ascogonium**) that has fused with it. As the ascogenous hyphae grow, the paired nuclei divide so that there is one pair of nuclei in each cell. After the ascogenous hyphae have matured, nuclear fusion occurs at the hyphal tips in the ascus mother cells. The diploid zygote nucleus then undergoes meiosis, and the resulting four haploid nuclei divide mitotically again to produce a row of eight nuclei in each developing ascus. These nuclei are walled off from one another. Thousands of asci may be packed together in a cup- or flask-shaped **ascocarp** (figure 26.10*b*). When the ascospores mature, they are often released from the asci with great force. If, perchance, the mature ascocarp is jarred, it may appear to belch puffs of "smoke" consisting of thousands of ascospores. Upon reaching a suitable environment, the ascospores germinate and start the cycle anew.

Although the term yeast is used in a general sense to refer to all unicellular fungi that reproduce asexually by either budding or binary fission (figure 26.13*a,b*), many yeast genera are classified specifically within the ascomycetes because of their sexual reproduction (figure 26.13*c,d*). Yeasts are present in both terrestrial and aquatic habitats in which a suitable carbon source is available.

Division *Basidiomycota*

The division *Basidiomycota* contains the **basidiomycetes,** commonly known as the club fungi. Examples include the smuts, jelly fungi, rusts, shelf fungi, stinkhorns, puffballs, toadstools, mushrooms, and bird's nest fungi.

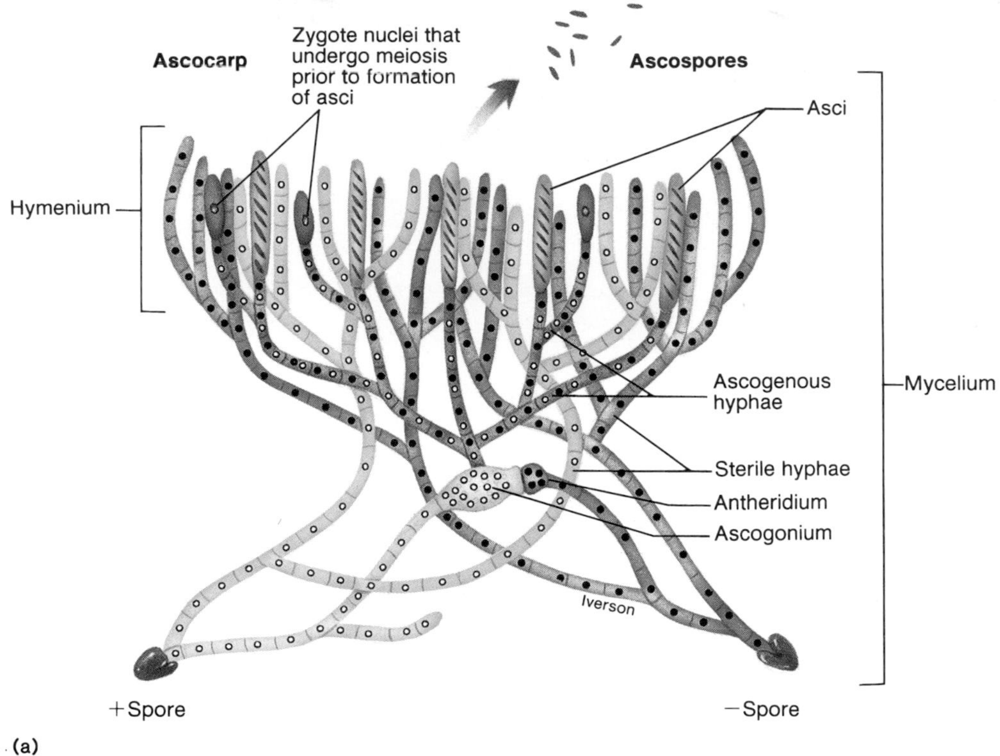

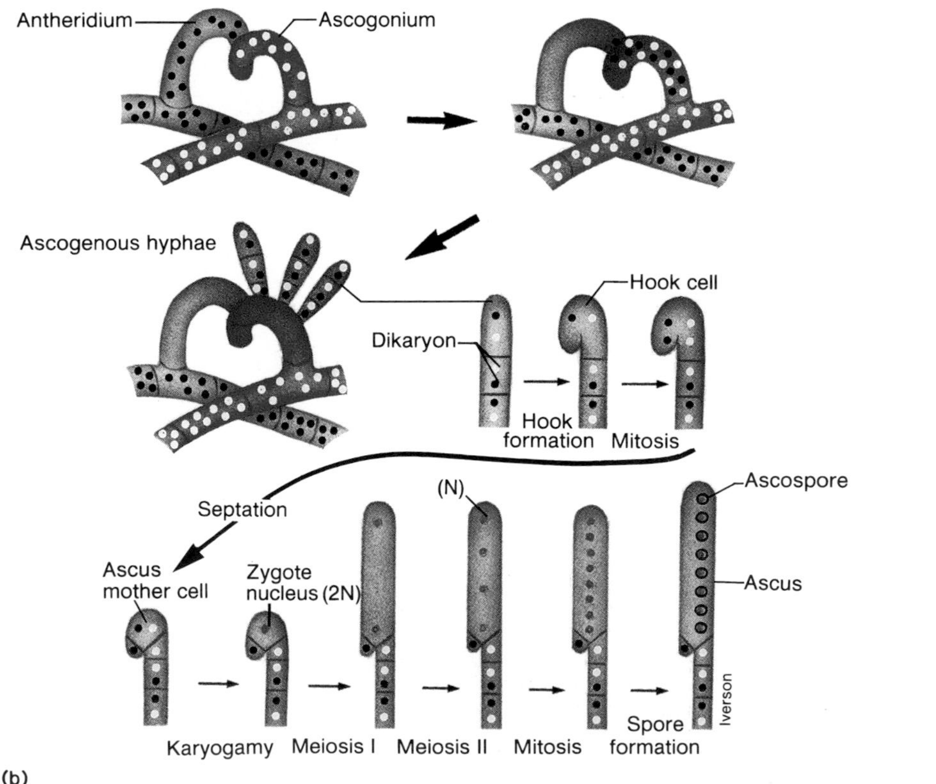

Figure 26.12 The Life Cycle of Ascomycetes. Sexual reproduction involves the formation of asci and ascospores. Within the ascus, karyogamy is followed by meiosis to produce the ascospores. (*a*) Sexual reproduction and ascocarp morphology of a cup fungus. (*b*) The details of sexual reproduction in ascogenous hyphae. The nuclei of the two mating types are represented by unfilled and filled circles. See the text for details.

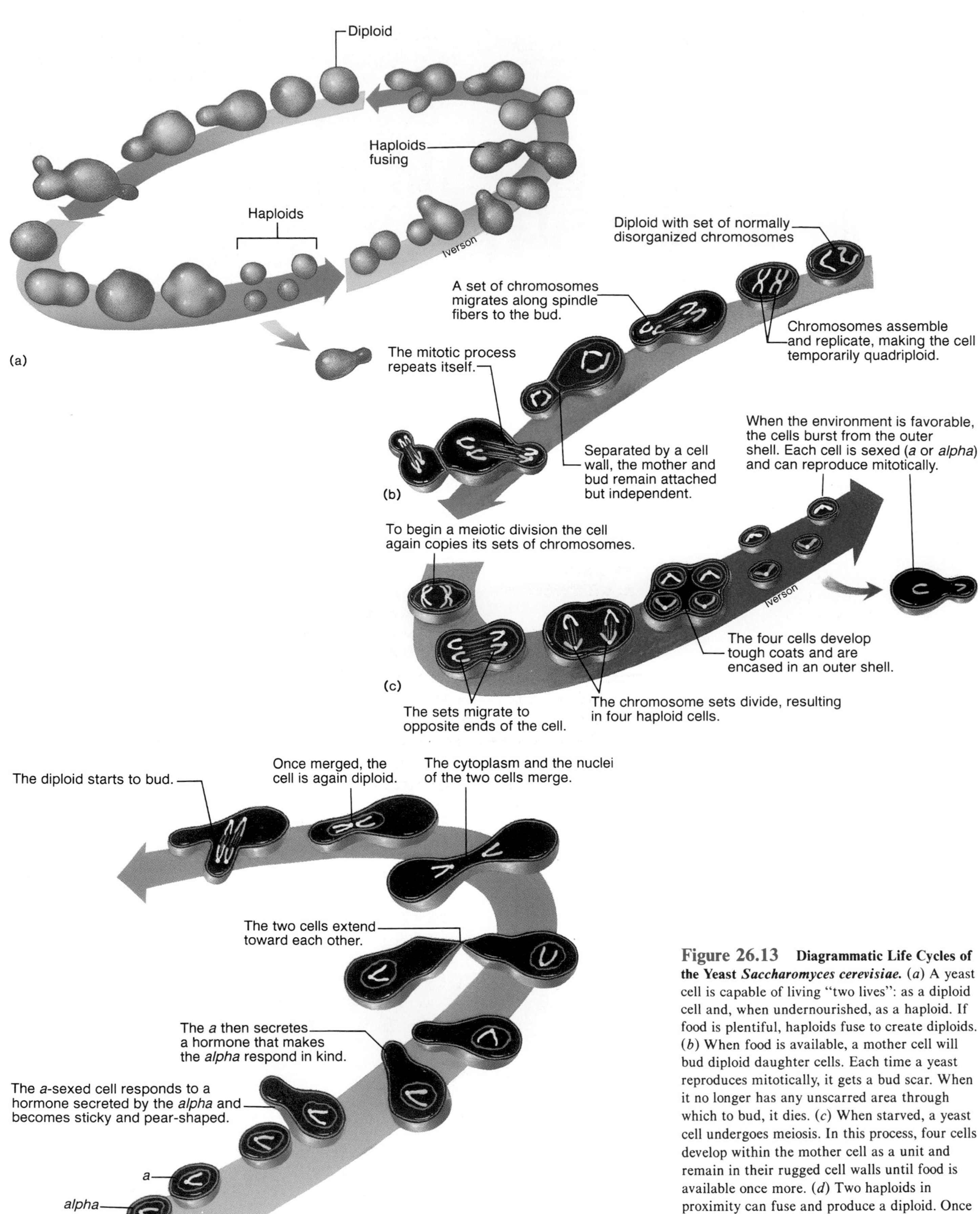
Diploid
Haploids fusing
Haploids
(a)
Iverson
Diploid with set of normally disorganized chromosomes
A set of chromosomes migrates along spindle fibers to the bud.
Chromosomes assemble and replicate, making the cell temporarily quadriploid.
The mitotic process repeats itself.
Separated by a cell wall, the mother and bud remain attached but independent.
(b)
When the environment is favorable, the cells burst from the outer shell. Each cell is sexed (*a* or *alpha*) and can reproduce mitotically.
To begin a meiotic division the cell again copies its sets of chromosomes.
Iverson
The four cells develop tough coats and are encased in an outer shell.
(c)
The sets migrate to opposite ends of the cell.
The chromosome sets divide, resulting in four haploid cells.
The diploid starts to bud.
Once merged, the cell is again diploid.
The cytoplasm and the nuclei of the two cells merge.
The two cells extend toward each other.
The *a* then secretes a hormone that makes the *alpha* respond in kind.
The *a*-sexed cell responds to a hormone secreted by the *alpha* and becomes sticky and pear-shaped.
a
alpha
(d)
Iverson

Figure 26.13 Diagrammatic Life Cycles of the Yeast *Saccharomyces cerevisiae.* (*a*) A yeast cell is capable of living "two lives": as a diploid cell and, when undernourished, as a haploid. If food is plentiful, haploids fuse to create diploids. (*b*) When food is available, a mother cell will bud diploid daughter cells. Each time a yeast reproduces mitotically, it gets a bud scar. When it no longer has any unscarred area through which to bud, it dies. (*c*) When starved, a yeast cell undergoes meiosis. In this process, four cells develop within the mother cell as a unit and remain in their rugged cell walls until food is available once more. (*d*) Two haploids in proximity can fuse and produce a diploid. Once formed, it buds even before it attains its final shape. (*See chapter 4 and figures 4.20 and 4.23 for a complete discussion of mitosis and meiosis.*)

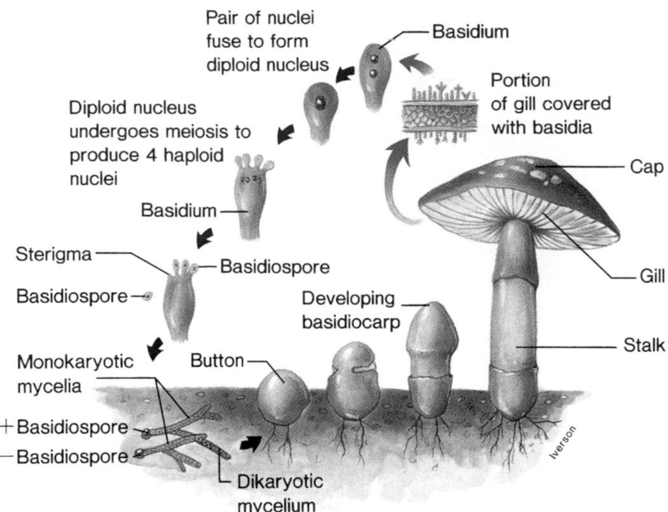

Figure 26.14 **Division *Basidiomycota.*** The life cycle of a typical soil basidiomycete starts with a basidiospore germinating to produce a monokaryotic mycelium (one with a single nucleus in each septate cell). The mycelium quickly grows and spreads throughout the soil. When this primary mycelium meets another monokaryotic mycelium of a different mating type, the two fuse to initiate a new dikaryotic secondary mycelium. The secondary mycelium is divided by septa into cells, each of which contains two nuclei, one of each mating type. This dikaryotic mycelium is eventually stimulated to produce basidiocarps. A solid mass of hyphae forms a button that pushes through the soil, elongates, and develops a cap. The cap contains many platelike gills, each of which is coated with basidia. The two nuclei in the tip of each basidium fuse to form a diploid zygote nucleus, which immediately undergoes meiosis to form four haploid nuclei. These nuclei push their way into the developing basidiospores, which are then released at maturity.

Basidiomycetes are named for their characteristic structure or cell, the **basidium,** that is involved in sexual reproduction (figure 26.14). A basidium (Greek *basidion,* small base) is produced at the tip of hyphae and is normally club shaped. Basidiospores are produced by the basidium, and basidia may be held within fruiting bodies called **basidiocarps.**

The basidiomycetes affect humans in many ways. Most are saprophytes and decompose plant debris, especially cellulose and lignin. Many mushrooms are used as food throughout the world. The cultivation of *Agaricus campestris* is a multimillion-dollar business (*see figure 43.20*). Unfortunately, some mushrooms, such as *Amanita,* are extremely poisonous (*see figure 4.1*g).

Many mushrooms produce specific alkaloids that act as either poisons or hallucinogens. One such example is the "destroying angel" mushroom, *Amanita phalloides.* Two toxins isolated from this species are phalloidin and α-amanitin. Phalloidin primarily attacks liver cells where it binds to cell membranes, causing them to rupture and leak their contents. Alpha-amanitin attacks the cells lining the stomach and small intestine and is responsible for the severe gastrointestinal symptoms associated with mushroom poisoning.

The basidiomycete *Cryptococcus neoformans* is an important human pathogen. It produces the disease called **cryptococcosis,** a systemic infection primarily involving the lungs and central nervous system. Other basidiomycetes, the smuts and rusts, are virulent plant pathogens that cause extensive damage to cereal crops; millions of dollars worth of crops are destroyed annually. In these fungi, large basidiocarps are not formed. Instead, the small basidia arise from hyphae at the surface of the host plant. The mycelia grow either intra- or extracellularly in plant tissue.

Fungal diseases (pp. 782–90).

Division *Deuteromycota*

To a large degree, fungal taxonomy is based on specific patterns of sexual reproduction. When a fungus lacks the sexual phase (perfect stage), or if this phase has not been observed, it is placed within the division *Deuteromycota,* commonly called the Fungi Imperfecti. Once a perfect stage is observed, the fungus is transferred to its proper division.

Most Fungi Imperfecti are terrestrial, with only a few being reported from freshwater and marine habitats. The majority are either saprophytes or parasites of plants. A few are parasitic on other fungi.

Many imperfect fungi directly affect human welfare. Several are human pathogens, causing such diseases as athlete's foot, ringworm, and histoplasmosis (*see chapter 39*). The chemical activities of many Fungi Imperfecti are industrially important. For example, some species of *Penicillium* (*see figure 4.1*d,e) synthesize the well-known antibiotics penicillin and griseofulvin. Other species give characteristic aromas to cheeses such as Gorgonzola, Camembert, and Roquefort. Different species of *Aspergillus* are used to ferment soy sauce and to manufacture citric, gluconic, and gallic acids. *Aspergillus flavus* and *A. parasiticus* produce secondary metabolites, called aflatoxins, that are highly toxic and carcinogenic to animals and humans (*see chapter 43*). Another group of fungal toxins, the trichothecenes, are strong inhibitors of protein synthesis in eucaryotic cells.

1. Describe the ascomycete life cycle. How are the ascomycetes important to humans?
2. How do yeasts reproduce sexually?
3. Describe the life cycle of a typical basidiomycete.
4. How do some Fungi Imperfecti affect humans?

Slime Molds and Water Molds

The **slime molds** and **water molds** resemble fungi in appearance and life-style. In their cellular organization, reproduction, and life cycles, they are most closely related to the protists (*see figure 20.6*).

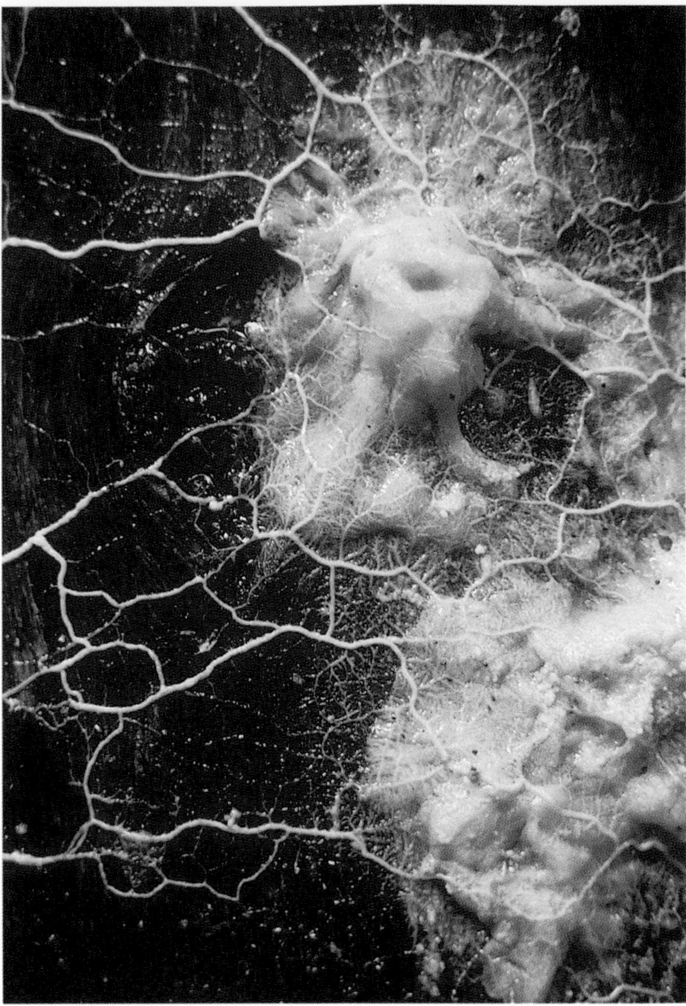

Figure 26.15 **Slime Molds.** Plasmodium of the slime mold *Physarum;* light micrograph (×175).

Division *Myxomycota*

Under appropriate conditions, **plasmodial slime molds** exist as thin, streaming masses of colorful protoplasm that creep along in an amoeboid fashion over moist, rotting logs, leaves, and other organic matter. Feeding is by phagocytosis. Because this streaming mass lacks cell walls, it is called a **plasmodium** (figures 26.15 and 26.16*a*). The plasmodium contains many nuclei, and as the organism grows, the diploid nuclei divide repeatedly (*see figure 4.21*).

When the plasmodium matures or when food and/or moisture are scarce, it moves into a lighted area and develops delicate ornate fruiting bodies (figure 26.16*b–e*). As the fruiting bodies mature, they form spores with cellulose walls that are resistant to environmental extremes. The spores germinate in the presence of adequate moisture to release either nonflagellated amoeboid **myxamoebae** or flagellated **swarm cells.** Initially, the myxamoebae or swarm cells feed and are haploid (figure 26.16*a*); eventually, they fuse to form a diploid zygote. The zygote feeds, grows, and multiplies its nuclei through synchronous mitotic division to form the multinucleate plasmodium.

Division *Acrasiomycota*

The vegetative stage of **cellular slime molds** consists of individual irregular amoeboid cells termed myxamoebae (figure 26.17*a*). The myxamoebae feed phagocytically on bacteria and yeasts. When food is plentiful, they divide repeatedly by mitosis and cytokinesis, producing new daughter myxamoebae. As their food supply is exhausted, the myxamoebae begin to secrete cyclic adenosine monophosphate (cAMP). This attracts other myxamoebae that move toward the cAMP chemotactic source and, in turn, secrete more cAMP. When the individual myxamoebae aggregate (figure 26.17*b*), they secrete a slimy sheath around themselves and form a sluglike **pseudoplasmodium** (figure 26.17*c*). The pseudoplasmodium may move around as a unit for a while, leaving a slime trail, but it eventually becomes sedentary. In the culmination of the asexual phase, pseudoplasmodial cells begin to differentiate into prestalk cells and prespore cells (figure 26.17*b*). A structure called a **sorocarp** forms (figure 26.17*d*) and matures into a sporangium that produces spores (figure 26.17*e*). The spores are eventually released, and when conditions become favorable, they germinate to release haploid amoebae and start the cycle anew.

Division *Oomycota*

Members of the division *Oomycota* are collectively known as **oomycetes** or water molds. Oomycetes resemble fungi in appearance, consisting of finely branched filaments called hyphae. However, oomycetes have cell walls of cellulose, whereas the walls of most fungi are made of chitin.

Oomycota means "egg fungi," a reference to the mode of sexual reproduction in water molds. A relatively large egg cell is fertilized by either a sperm cell or a smaller antheridium. Most oomycetes also produce asexual zoospores that bear two flagella.

Water molds such as *Saprolegnia* and *Achlya* are saprophytes that grow as cottony masses on dead algae and small animals, mainly in freshwater environments. They are important decomposers in aquatic ecosystems. Some water molds are parasitic on the gills of fish. The water mold *Peronospora hyoscyami* is currently responsible for the troublesome "blue mold" of tobacco plants throughout the world. In the United States alone, blue mold produces millions of dollars of damage yearly to tobacco crops. Other oomycetes cause late blight of potatoes (*Phytophthora infestans*) and grape downy mildew (*Plasmopara viticola*).

1. What is a plasmodium?
2. How do the plasmodial slime molds reproduce? Describe the cellular slime mold life cycle.
3. Where would you look for an oomycete?

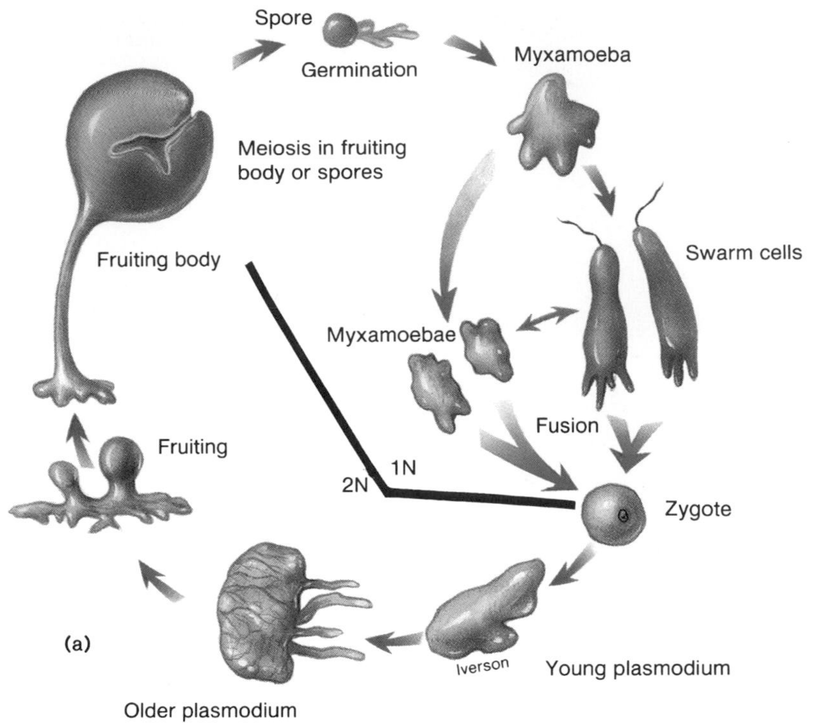

Figure 26.16 **Reproduction in the *Myxomycota.*** (*a*) The life cycle of a plasmodial slime mold. (Different parts of the life cycle are drawn at different magnifications.) Plasmodial slime mold fruiting bodies: (*b*) *Hemitrichia* (×100), (*c*) *Stemonitis* (×100), (*d*) *Physarum polycephalum*, and (*e*) *Arcyria denudata*.

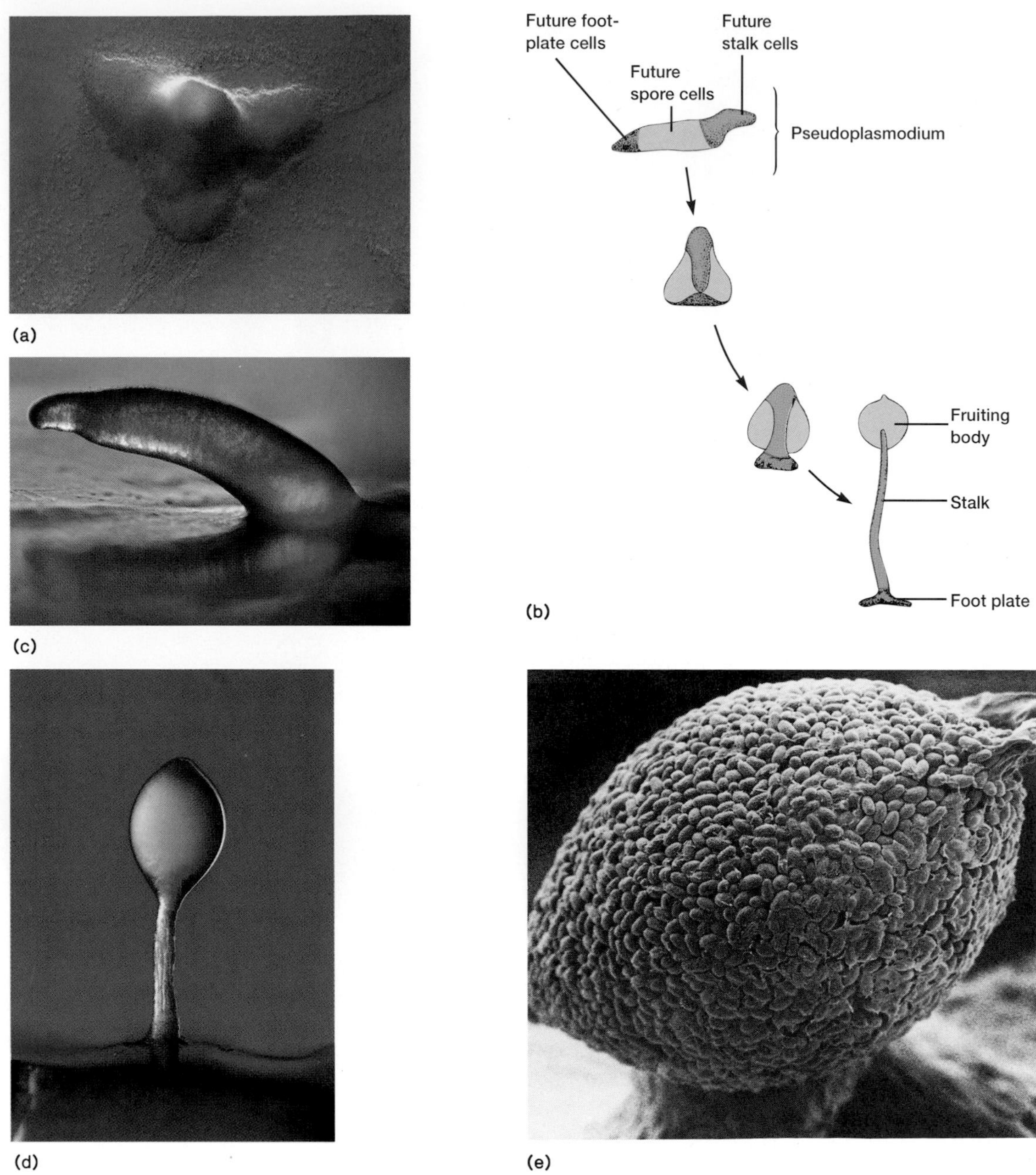

Figure 26.17 **Division *Acrasiomycota*.** *Dictyostelium discoideum,* a cellular slime mold. In the free-living stage, a myxamoeba resembles an irregularly-shaped amoeba. (*a*) Aggregating myxamoebae become polar and begin to move in an oriented direction due to the influence of cAMP. (*b*) Diagrammatic drawing of cell migration involved in the formation of the sorocarp from (*c*) an initial pseudoplasmodium. (*d*) Light micrograph of a mature fruiting body (sorocarp). (*e*) Electron micrograph of a sporangium showing spores (×1,800).

Summary

1. A fungus is a eucaryotic, spore-bearing organism that has absorptive nutrition and lacks chlorophyll; that reproduces asexually, sexually, or by both methods; and that normally has filamentous hyphae surrounded by cell walls, which usually contain chitin.
2. Fungi are omnipresent in the environment, being found wherever water and suitable organic nutrients occur. They secrete enzymes outside their body structure and absorb the digested food.
3. Fungi are important decomposers that break down organic matter; live as parasites on animals, humans, and plants; are the bases of many industrial processes; and are used as research tools in the study of fundamental biological processes.
4. The body or vegetative structure of a fungus is called a thallus.
5. A mold consists of long, branched, threadlike filaments of cells, the hyphae, that form a tangled mass of tissue called a mycelium. Hyphae may be either septate or coenocytic (nonseptate). The mycelium can produce reproductive structures.
6. Yeasts are unicellular fungi that have a single nucleus and reproduce either asexually by budding and transverse division or sexually through spore formation.
7. Many fungi are dimorphic—they alternate between a yeast and a mold form.
8. Asexual reproduction often occurs in the fungi by the production of specific types of spores. Spores are easily dispersed and are resistant to adverse environmental conditions.
9. Sexual reproduction is initiated in the fungi by the fusion of hyphae of different mating strains. In some fungi, the nuclei in the fused hyphae immediately combine to form a zygote. In others the two genetically distinct nuclei remain separate, forming pairs that divide synchronously. Eventually, some nuclei fuse.
10. The zygomycetes are coenocytic. Most are saprophytic. One example is the common bread mold, *Rhizopus stolonifer*. Sexual reproduction occurs through a form of conjugation involving + and − strains.
11. The ascomycetes are known as the sac fungi because they form a sac-shaped reproductive structure called an ascus. Sexual reproduction involves + and − strains. In asexual reproduction, conidiophores produce conidia.
12. The basidiomycetes are the club fungi. They are named after their basidium that produces basidiospores.
13. The deuteromycetes (Fungi Imperfecti) are fungi with no known sexual (perfect) phase.
14. The plasmodial slime molds move about as a plasmodium containing many nuclei that are not separated from one another by membranes. When food or moisture is scarce, these slime molds form sporangia within which spores are produced.
15. The cellular slime molds consist of a vegetative stage called a myxamoeba. The myxamoebae feed until their food supply dwindles, then the cells come together to form a plantlike multicellular structure called a sorocarp. The sorocarp produces haploid spores that germinate when conditions are favorable to form new myxamoebae.
16. The *Oomycota* (water molds), consisting of branched filaments called hyphae, resemble fungi only in appearance.

Key Terms

antheridium *526*
arthroconidia *522*
arthrospore *522*
ascocarp *526*
ascogonium *526*
ascogenous hypha *526*
ascomycetes *525*
ascospore *524*
ascus *525*
basidiocarp *529*
basidiomycetes *526*
basidiospore *524*
basidium *529*
blastospore *523*
cellular slime molds *530*
chitin *519*
chlamydospore *523*
coenocytic *520*
conidiospore *523*
cryptococcosis *529*
dikaryotic stage *524*
ergot *525*
ergotism *525*
fungus *519*
gametangium *524*
hypha *520*
mold *520*
mycelium *520*
mycologist *519*
mycology *519*
mycosis *519*
mycotoxicology *519*
myxamoeba *530*
oomycetes *530*
plasmodial slime mold *530*
plasmodium *530*
progametangium *524*
pseudoplasmodium *530*
saprophyte *521*
septa *521*
septate *521*
slime mold *529*
sorocarp *530*
sporangiospore *523*
sporangium *523*
swarm cell *530*
thallus *519*
water mold *529*
yeast *520*
YM shift *521*
zygomycetes *524*
zygospore *524*

Questions for Thought and Review

1. Some fungi can reproduce both sexually and asexually. What are the advantages and disadvantages of each?
2. Why is nutrition in the fungi a property of membrane function?
3. Why are most fungi confined to a specific ecological niche?
4. Some authorities believe that the slime molds and water molds should be placed in the kingdom *Fungi,* whereas others believe that they should be placed in the kingdom *Protista.* What are the characteristics that these organisms share with both the fungi and protists that have led to this ambiguity?
5. At the present time, those fungi that have no sexual reproduction cannot be classified with their sexually reproducing relatives. Why? Do you think that this is likely to change in the future?
6. Because spores are such a rapid way of reproducing for some fungi, what adaptive "use" is there for an additional sexual phase?
7. Some fungi can be viewed as coenocytic organisms that exhibit little differentiation. When it does occur, such as in the formation of reproductive structures, differentiation is preceded by the septum formation. Why does this occur?
8. The term mushrooming is a proverbial description for expanding rapidly. Why is this an accurate metaphor?
9. Both bacteria and fungi are major environmental decomposers. Obviously, competition exists in a given environment, but fungi usually have an advantage. What is this advantage?
10. Very few antibiotics can control fungi in medicine and agriculture, though there are many for bacterial control. Why is this so?

Additional Reading

Alexopoulos, C. J., and Mims, C. W. 1979. *Introductory mycology,* 3d ed. New York: John Wiley and Sons.

Batra, S. W. T., and Batra, R. L. 1967. The fungus gardens of insects. *Sci. Am.* 217:112–20.

Bonner, J. T. 1969. Hormones in social amoebae and mammals. *Sci. Am.* 220:78–91.

Chang, S. T., and Miles, P. G. 1984. A new look at cultivated mushrooms. *BioScience* 34:358–62.

Cole, G. T. 1986. Models of cell differentiation in conidial fungi. *Micro. Rev.* 50:95–132.

Deacon, J. W. 1980. *Introduction to modern mycology.* New York: John Wiley and Sons.

Ellis, M. B., and Ellis, J. P. 1985. *Microfungi of land plants.* New York: Macmillan.

Herskowitz, I. 1988. Life cycle of budding yeast *Saccharomyces cerevisiae. Microbiol. Rev.* 52(4):536–53.

Kessin, R. H. 1988. Genetics of early *Dictyostelium discoideum* development. *Microbiol. Rev.* 52(1):29–49.

Klionsky, D. J.; Herman, P. K.; and Emr, S. D. 1990. The fungal vacuole: Composition, function, and biogenesis. *Microbiol. Rev.* 54(3):266–92.

Lipke, P., and Kurjan, J. 1992. Sexual agglutination in budding yeast: structure, function, and regulation of adhesion glycoproteins. Microbiol. Rev. 56(1):180–94.

Litten, W. 1975. The most poisonous mushrooms. *Sci. Am.* 232:14–22.

Maresca, B., and Kobayashi, G. S. 1989. Dimorphism in *Histoplasma capsulatum:* A model for the study of cell differentiation in pathogenic fungi. *Microbiol. Rev.* 53(2):186–209.

Martin, G. W., and Alexopoulos, C. J. 1969. *The myxomycetes.* Iowa City: Univ. of Iowa Press.

Matossian, M. K. 1982. Ergot and the Salem witchcraft affair. *Am. Scientist* 70:355–71.

Monmaney, T. 1985. Yeast at work. *Science 85* 6(6):30–36.

Moore-Landecker, E. 1991. *Fundamentals of fungi,* 3rd ed. Englewood Cliffs, N.J.: Prentice-Hall.

Newhouse, J. R. 1990. Chestnut blight. *Sci. Am.* 263:106–11.

Newlon, C. S. 1988. Yeast chromosome replication and segregation. *Microbiol. Rev.* 52(4):568–606.

Orlowski, M. 1991. *Mucor* dimorphism. *Microbial. Rev.* 55(2):234–58.

Phaff, H. J. 1986. My life with yeasts. *Ann. Rev. Microbiol.* 40:1–28.

Phaff, H. J.; Miller, M. W.; and Mrak, E. M. 1978. *The life of yeasts,* 2d ed. Cambridge, Mass.: Harvard University Press.

Ross, I. K. 1979. *Biology of the fungi: Their development, regulation, and associations.* New York: McGraw-Hill.

Smith, A. H.; Smith, H.; and Weber, N. S. 1979. *Gilled mushrooms.* Dubuque, Ia.: Wm. C. Brown.

Smith, A. H.; Smith, H.; and Weber, N. S. 1981. *Non-gilled mushrooms,* 2d ed. Dubuque, Ia.: Wm. C. Brown.

Stevens, R. B., ed. 1974. *Mycology guidebook.* Seattle: Univ. of Washington Press.

Strobel, G. A., and Lanier, G. N. 1981. Dutch elm disease. *Sci. Am.* 245:56–66.

Webster, J. 1980. *Introduction to fungi,* 2d ed. New York: Cambridge Univ. Press.

CHAPTER 27
The Algae

The term algae means different things to different people, and even the professional botanist and biologist find algae embarrassingly elusive of definition. Thus, laymen have given such names as "pond scums," "frog spittle," "water mosses," and "seaweeds," while some professionals shrink from defining them.

—Harold C. Bold and Michael J. Wynne

Outline

Concepts

1. Most algae are found in freshwater and marine environments; a few grow in terrestrial habitats.
2. The algae are not a single, closely related taxonomic group but, instead, are a diverse assemblage of unicellular, colonial, and multicellular eucaryotic organisms.
3. Although algae can be autotrophic or heterotrophic, most are photoautotrophs. They store carbon in a variety of forms, including starch, oils, and various sugars.
4. The body of an alga is called the thallus. Algal thalli range from small solitary cells to large, complex multicellular structures.
5. Algae reproduce asexually and sexually.
6. The following divisions of the algae are discussed: *Chlorophyta* (green algae), *Charophyta* (stoneworts/brittleworts), *Euglenophyta* (euglenoids), *Chrysophyta* (golden-brown and yellow-green algae; diatoms), *Phaeophyta* (brown algae), *Rhodophyta* (red algae), and *Pyrrhophyta* (dinoflagellates).

Chapter 27 presents some general features of algae. The algae are a very diverse group of organisms, ranging in size from microscopic single cells to kelps over 75 m in length. Algae vary considerably in the structure of their cells, the arrangement of cells in colonial and multicellular forms, and in their pigmentation and color. Algae grow in a wide variety of habitats. Within the oceans and freshwater environments, they are the major producers of O_2 and organic material. A few algae live on land in moist soils and other environments.

Phycology or **algology** is the study of algae. The word phycology is derived from the Greek *phykos,* meaning seaweed. The term **algae** (s., alga) was originally used to define simple "aquatic plants." It no longer has any formal significance in classification schemes. Instead, the algae can be described as eucaryotic organisms that lack roots, stems, and leaves but have chlorophyll and other pigments for carrying out oxygen-producing photosynthesis. Many can be placed in the kingdom *Protista.*

Distribution of Algae

Algae most commonly occur in water (fresh, marine, or brackish) in which they may be suspended (**planktonic**) or attached and living on the bottom (**benthic**). A few algae live at the water-atmosphere interface and are termed **neustonic. Plankton** (Greek *plankos,* wandering) consists of free-floating, mostly microscopic aquatic organisms. **Phytoplankton** is made up of algae and small plants, whereas **zooplankton** consists of animals and nonphotosynthetic protists. Some algae grow on moist rocks, wood, trees, and on the surface of moist soil. Algae also live as endosymbionts in various protozoa, mollusks, worms, and corals. Several algae grow as endosymbionts within plants, some are attached to the surface of various structures, and a few lead a parasitic existence. Algae also associate with fungi to form lichens.

Zooxanthella symbiosis (p. 567).
Lichen symbiosis (pp. 566–67).

Classification of Algae

According to the five-kingdom system of Whittaker, the algae belong to seven divisions distributed between two different kingdoms (table 27.1). The primary classification of algae is based on cellular, not organismal, properties. Some more important properties include: (1) cell wall (if present) chemistry and morphology; (2) form in which food or assimilatory products of photosynthesis are stored; (3) chlorophyll molecules and accessory pigments that contribute to photosynthesis; (4) flagella number and the location of their insertion in motile cells; (5) morphology of the cells and/or body (thallus); (6) habitat; (7) reproductive structures; and (8) life history patterns. Based on these properties, the algae are arranged by divisions in table 27.2, which summarizes their more significant characteristics.

Ultrastructure of the Algal Cell

The eucaryotic algal cell (figure 27.1) is surrounded by a thin, rigid cell wall. Some algae have an outer matrix lying outside the cell wall. This is usually flexible and gelatinous, similar to

TABLE 27.1 Classification of Algae

Division (Common Name)	Kingdom
Chrysophyta (yellow-green and golden-brown algae; diatoms)	*Protista* (single cell; eucaryotic)
Euglenophyta (photosynthetic euglenoid flagellates)	*Protista*
Pyrrhophyta (dinoflagellates)	*Protista*
Charophyta (stoneworts)	*Protista*
Chlorophyta (green algae)	*Protista*
Phaeophyta (brown algae)	*Plantae* (multicellular; eucaryotic)
Rhodophyta (red algae)	*Plantae*

TABLE 27.2 Comparative Summary of Some Algal Characteristics

Division	Approximate Number of Species	Common Name	Pigments: Chlorophylls
Chlorophyta	7,500	Green algae	*a, b*
Charophyta	250	Stoneworts (brittleworts)	*a, b*
Euglenophyta	700	Euglenoids	*a, b*
Chrysophyta	6,000	Golden-brown, yellow-green algae; diatoms	*a*, c_1/c_2, rarely *d*
Phaeophyta	1,500	Brown algae	*a, c*
Rhodophyta	3,900	Red algae	*a*, rarely *d*
Pyrrhophyta	1,100	Dinoflagellates	*a*, c_1, c_2

[a] Refers specifically to the vegetative cells. Spores, akinetes, zygotes contain waxes, nonsaponifiable polymers, and phenolic substances.

[b] The following abbreviations are used: fresh water (fw), brackish water (bw), salt water (sw), terrestrial (t).

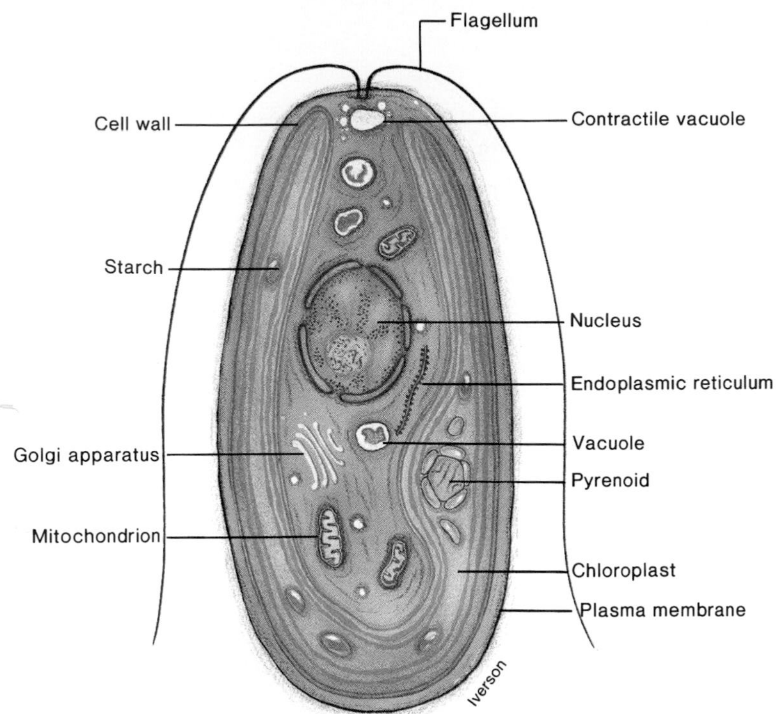

Figure 27.1 **Algal Morphology.** Schematic drawing of a typical eucaryotic algal cell showing some of its organelles and other structures.

Phycobilins (Phycobiliproteins)	Carotenoids	Thylakoids per Stack in Chloroplast	Storage Products	Flagella	Cell Wall[a]	Habitat[b]
—	β-carotene, ± α-carotene, xanthophylls	3–6	Sugars, starch, fructosan	1, 2–8; equal, apical or subapical	Cellulose, mannan, protein, $CaCO_3$	fw, bw, sw, t
—	α-, β-, τ- carotene; xanthophylls	Many	Starch	2; subapical	Cellulose, $CaCO_3$	fw, bw
—	β-carotene, xanthophylls, ± τ-carotene	3	Paramylon, oils, sugars	1–3; slightly apical	Absent	fw, bw, sw, t
—	α-, β-, ϵ- carotene; fucoxanthin; xanthophylls	3	Chrysolaminarin, oils	1–2; equal or unequal, apical; or none	Cellulose, silica, $CaCO_3$, chitin, or absent	fw, bw, sw, t
—	β-carotene, fucoxanthin, xanthophylls	3	Laminarin, mannitol, oils	2; unequal, lateral	Cellulose, alginic acid, fucoidan	bw, sw
C-phycocyanin, allophycocyanin, phycoerythrin	Xanthophylls (β-carotene, zeaxanthine, ± α-carotene)	1	Glycogenlike starch (floridean glycoside)	Absent	Cellulose, xylans, galactans, $CaCO_3$	fw, bw, sw
—	β-carotene, fucoxanthin, peridinin, dinoxanothin	3	Starch, glucan, oils	2; 1 trailing, 1 girdling	Cellulose, or absent	fw, bw, sw

bacterial capsules. When present, the flagella are the locomotor organelles. The nucleus has a typical nuclear envelope with pores; within the nucleus are a nucleolus, chromatin, and karyolymph. The chloroplasts have membrane-bound sacs called thylakoids that carry out the light reactions of photosynthesis. These organelles are embedded in the stroma where the dark reactions of carbon fixation take place. A dense proteinaceous area, the **pyrenoid** that is associated with synthesis and storage of starch may be present in the chloroplasts.

1. Define the word algae.
2. To what two kingdoms do the algae belong?
3. What are some general characteristics of an algal cell?

Algal Nutrition

Algae can be either autotrophic or heterotrophic. Most are photoautotrophic; they require only light and CO_2 as their principal source of energy and carbon. Chemoheterotrophic algae require external organic compounds as carbon and energy sources.

Types of microbial nutrition (pp. 97–99).

Structure of the Algal Thallus (Vegetative Form)

The vegetative body of algae is called the **thallus** (pl., thalli). It varies from the relative simplicity of a single cell to the more striking complexity of multicellular forms, such as the giant kelps. Unicellular algae may be as small as bacteria, whereas kelp can attain a size over 75 m in length.

Algae are unicellular (figure 27.2*a,b*), colonial (figure 27.2 *c,d*), filamentous (figure 27.2*i*), membranous (figure 27.2*e*), or tubular (figure 27.2*f*). Some types become highly differentiated and bladelike (figure 27.2*g*). A few develop primitive rootlike, stemlike, and bladelike structures (figure 27.2*h*) though vascular tissue is absent.

Algal Reproduction

Some unicellular algae reproduce asexually. In this kind of reproduction, gametes do not fuse to form a zygote. There are three basic types of asexual reproduction: fragmentation, spores, and binary fission. In **fragmentation,** the thallus breaks up and each fragmented part grows to form a new thallus. **Spores** can be formed in ordinary vegetative cells or in specialized structures termed sporangia (s., sporangium; Greek *spora,* seed, and *angeion,* vessel). Flagellated motile spores are called **zoospores.** Nonmotile spores produced by sporangia are termed **aplanospores.** In some unicellular algae, **binary fission** occurs (nuclear division followed by division of the cytoplasm).

Other algae reproduce sexually. Eggs are formed within relatively unmodified vegetative cells called **oogonia** (s., oogonium) that function as female structures. Sperm are produced in special male reproductive structures called **antheridia** (s., antheridium). In sexual reproduction, there is fusion of these gametes to produce a diploid **zygote.**

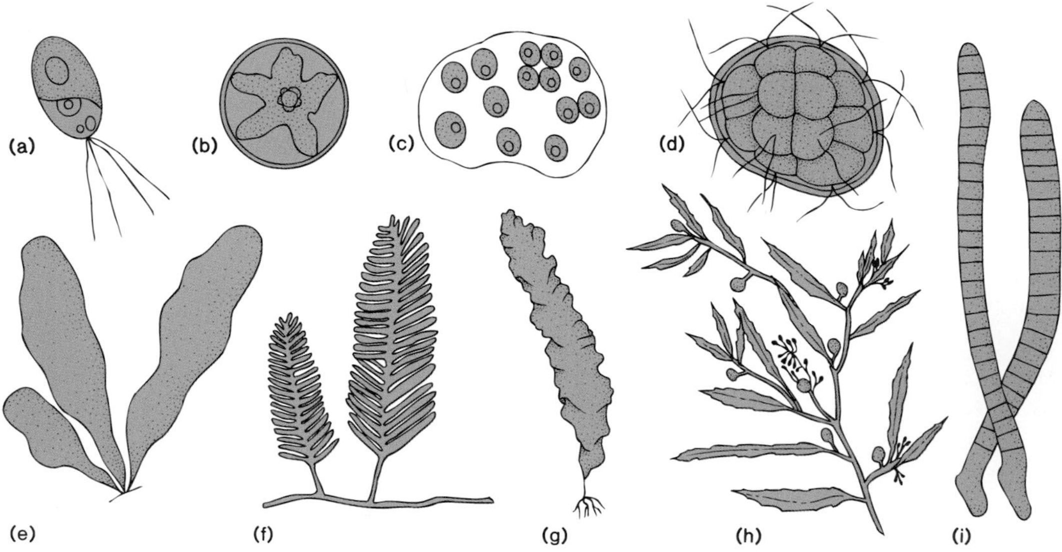

Figure 27.2 Diagrammatic Algal Bodies: (*a*) unicellular, motile; (*b*) unicellular, nonmotile; (*c*) colonial; (*d*) colonial, coenobic; (*e*) membranous; (*f*) tubular; (*g*) bladelike, kelp; (*h*) leafy axis; (*i*) filamentous.

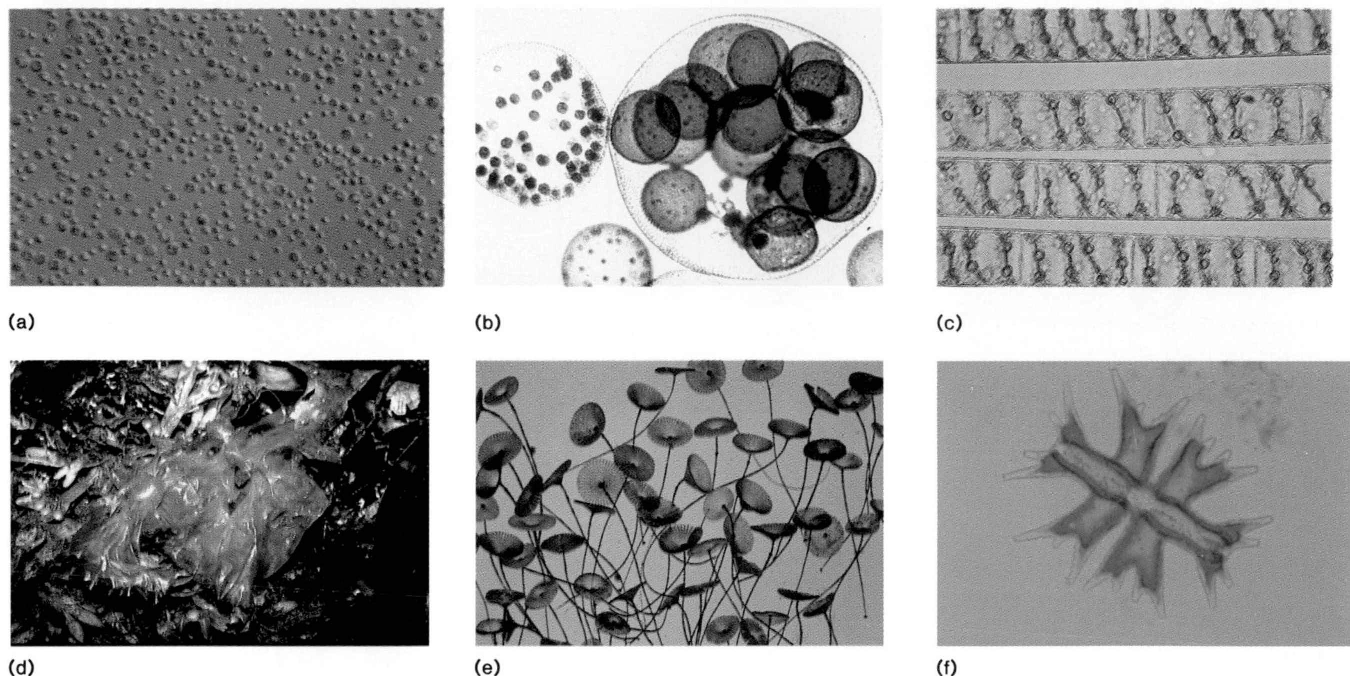

(a) (b) (c) (d) (e) (f)

Figure 27.3 ***Chlorophyta*** **(Green Algae); Light Micrographs.** (*a*) *Chlorella,* a unicellular nonmotile green alga (×160). (*b*) *Volvox,* a typical green algal colony (×450). (*c*) *Spirogyra,* a filamentous green alga (×100). Four filaments are shown. Note the ribbonlike, spiral chloroplasts within each filament. (*d*) *Ulva,* commonly called sea lettuce, has a leafy appearance. (*e*) *Acetabularia,* the mermaid's wine goblet. (*f*) *Micrasterias,* a large desmid (×150).

1. How can algae be classified based on their mode of nutrition?
2. What are the different types of algal thalli?
3. How do algae reproduce asexually?
4. How do algae reproduce sexually?

Characteristics of the Algal Divisions

Chlorophyta (Green Algae)

The *Chlorophyta* or green algae are an extremely varied division. They grow in fresh and salt water, in soil, on other organisms, and within other organisms. The *Chlorophyta* have chlorophylls *a* and *b* along with specific carotenoids, and they store carbohydrates as starch. Many have cell walls of cellulose. They exhibit a wide diversity of body forms, ranging from unicellular to colonial, filamentous, membranous or sheetlike, and tubular types (figure 27.3). Some species have a holdfast structure that anchors them to the substratum. Both asexual and sexual reproduction occur in green algae.

Chlamydomonas is a representative unicellular green alga (figure 27.4). Individuals have two flagella of equal length at the anterior end by which they move rapidly in water. Each cell has a single haploid nucleus, a large chloroplast, a conspicuous pyrenoid, and a **stigma (eyespot)** that aids the cell in phototactic responses. Two small contractile vacuoles at the base of the flagella function as osmotic organelles that continuously remove water. *Chlamydomonas* reproduces asexually by producing zoospores through cell division. The alga also reproduces sexually when some products of cell division act as gametes and fuse to form a four-flagellated diploid zygote that ultimately loses its flagella and enters a resting phase. Meiosis occurs at the end of this resting phase and produces four haploid cells that give rise to adults.

From organisms like *Chlamydomonas,* several distinct lines of evolutionary specialization have evolved in the green algae. The first line contains nonmotile unicellular green algae, such as *Chlorella. Chlorella* (figure 27.3*a*) is widespread both in fresh and salt water and in soil. It only reproduces asexually and lacks flagella, eyespots, and contractile vacuoles; the nucleus is very small.

Motile, colonial organisms such as *Volvox* represent a second major line of evolutionary specialization. A *Volvox* colony (figure 27.3*b*; *see also figure 2.8*b) is a hollow sphere made up of a single layer of 500 to 60,000 individual cells, each containing two flagella and resembling a *Chlamydo-*

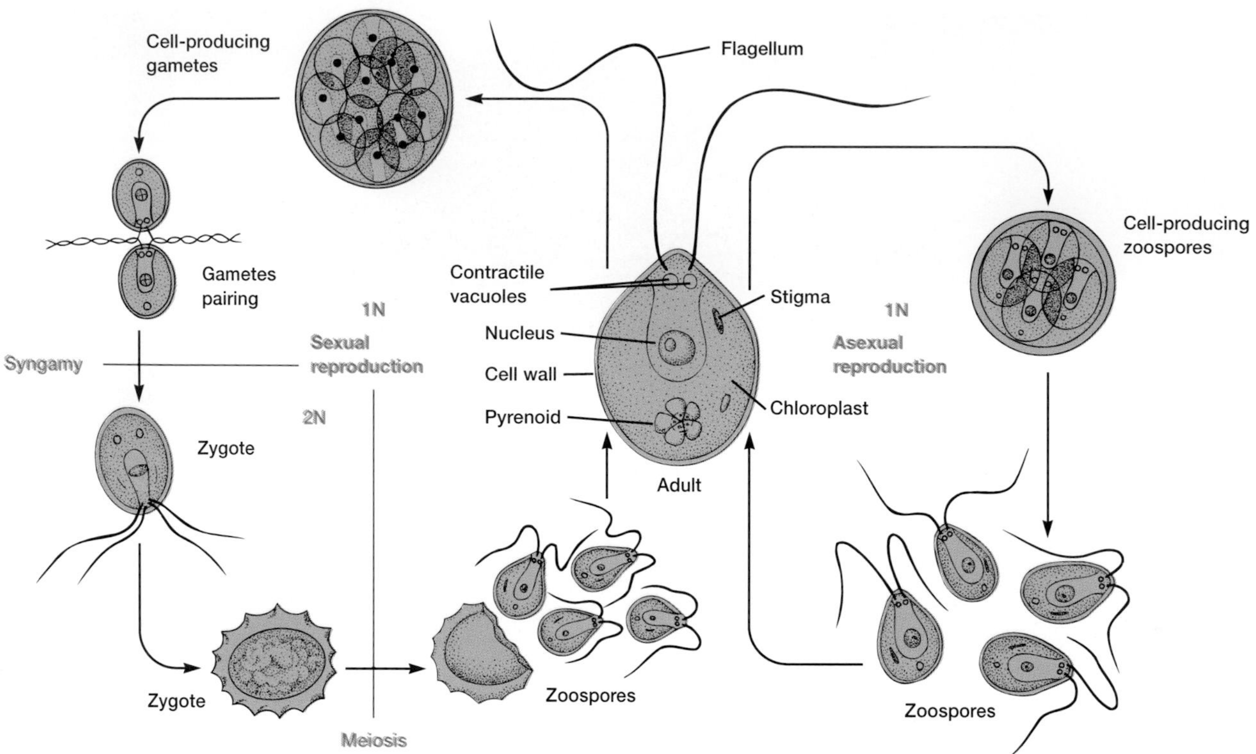

Figure 27.4 ***Chlamydomonas:* The Structure and Life Cycle of This Motile Green Alga.** During asexual reproduction, all structures are haploid; during sexual reproduction, only the zygote is diploid.

monas cell. The flagella of all the cells beat in a coordinated way to rotate the colony in a clockwise direction as it moves through the water. Only a few cells are reproductive, and these are located at the posterior end of the colony. Some divide asexually and produce new colonies. Others produce gametes. After fertilization, the zygote divides to form a daughter colony. In both cases, the daughter colonies stay within the parental colony until it ruptures.

One green alga, *Prototheca moriformis,* causes the disease **protothecosis** in humans and animals. *Prototheca* cells are fairly common in the soil, and it is from this site that most infections occur. Severe systemic infections, such as massive invasion of the bloodstream, have been reported in animal cases. More common in humans is the subcutaneous type of infection. It starts as a small lesion and spreads slowly through the lymph glands, covering large areas of the body.

1. What body forms do the green algae exhibit?
2. How do green algae reproduce?
3. Describe the structure of *Chlamydomonas, Chlorella,* and *Volvox.*
4. Why is the green alga *Prototheca* unique?

Charophyta (Stoneworts/Brittleworts)

The **stoneworts** are abundant in fresh to brackish waters and have a worldwide distribution. Often they appear as a dense covering on the bottom of shallow ponds. Some species precipitate calcium and magnesium carbonate from the water to form a limestone covering, thus giving the *Charophyta* their common names of stoneworts or brittleworts.

Euglenophyta (Euglenoids)

The **euglenoids** share with the *Chlorophyta* and *Charophyta* the presence of chlorophylls *a* and *b* in their chloroplasts. The primary storage product is paramylon, which is unique to euglenoids. They occur in fresh, brackish, and marine waters and on moist soils; they often form water blooms in ponds and cattle water tanks.

The representative genus is *Euglena.* A typical *Euglena* cell (figure 27.5) is elongated and bounded by a plasma membrane. Inside the plasma membrane is a structure called the **pellicle,** which is composed of articulated proteinaceous strips lying side by side. The pellicle is elastic enough to enable turning and flexing of the cell, yet rigid enough to prevent excessive alterations in shape. The several chloroplasts contain chlorophylls *a* and *b* together with carotenoids. The large nucleus contains a prominent nucleolus. The stigma is located near an anterior reservoir. A large contractile vacuole near the reservoir continuously collects water from the cell and empties

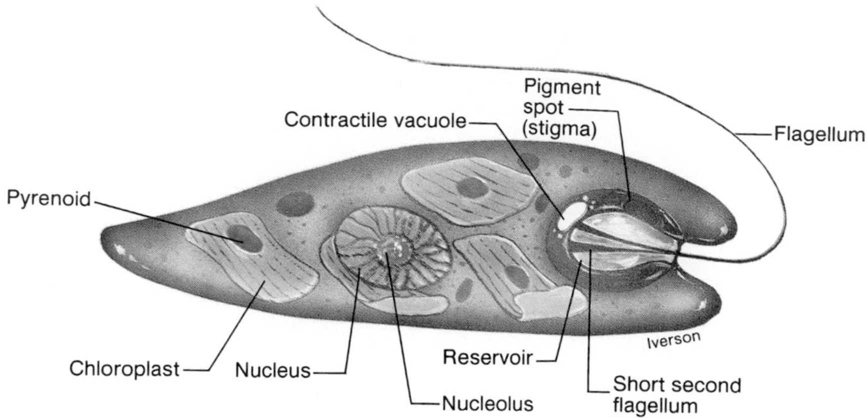

Figure 27.5 ***Euglena.* A Diagram Illustrating the Principal Structures Found in This Euglenoid.** Notice that a short second flagellum does not emerge from the anterior invagination. In some euglenoids, both flagella are emergent.

it into the reservoir, thus regulating the osmotic pressure within the organism. Two flagella arise from the base of the reservoir, although only one emerges from the canal and actively beats to move the cell. Reproduction in euglenoids is by longitudinal mitotic cell division.

Chrysophyta (Golden-Brown and Yellow-Green Algae; Diatoms)

The division *Chrysophyta* is quite diversified with respect to pigment composition, cell wall, and type of flagellated cells. The division is divided into three major classes: golden-brown algae, yellow-green algae, and diatoms. The major photosynthetic pigments are usually chlorophylls *a* and c_1/c_2, and the carotenoid fucoxanthin. When fucoxanthin is the dominant pigment, the cells have a golden-brown color. The major carbohydrate reserve in the *Chrysophyta* is **chrysolaminarin** or **laminarin.**

Some *Chrysophyta* lack cell walls; others have intricately patterned coverings external to the plasma membrane, such as **scales** (figure 27.6*a*), walls, and plates. Diatoms have a distinctive two-piece wall of silica, called a **frustule.** Two anteriorly attached flagella of unequal length are common among *Chrysophyta* (figure 27.6*b*), but some species have no flagella, and others have either one flagellum or two that are of equal length.

Most *Chrysophyta* are unicellular or colonial. Reproduction is usually asexual, but occasionally sexual. Although some marine forms are known, most of the yellow-green and golden-brown algae live in fresh water. Blooms of some species produce unpleasant odors and tastes in drinking water.

The **diatoms** (figure 27.6*c,d; see also figure 4.1*b) are photosynthetic, circular or oblong chrysophyte cells with frustules composed of two halves or thecae that overlap like a petri dish (therefore their name is from the Greek *diatomsos,* cut in two). The larger half is the **epitheca,** and the smaller half is the **hypotheca.** Diatoms grow in fresh water, salt water, and moist soil and comprise a large part of phytoplankton (Box 27.1). The chloroplasts of these chrysophytes contain chlorophylls *a* and *c* as well as carotenoids. Some diatoms are facultative heterotrophs and can absorb carbon-containing molecules through the holes in their walls. The vegetative cells of diatoms are diploid; exist as unicellular, colonial, or filamentous shapes; lack flagella; and have a single large nucleus and smaller plastids. Reproduction consists of the organism dividing asexually, with each half then constructing a new theca within the old one. Because of this mode of reproduction, diatoms get smaller with each reproductive cycle. However, when they diminish to about 30% of their original size, sexual reproduction usually occurs. The diploid vegetative cells undergo meiosis to form gametes, which then fuse to produce a zygote. The zygote develops into an auxospore, which increases in size again and forms a new wall. The mature auxospore eventually divides mitotically to produce vegetative cells with normal frustules.

Diatom frustules are composed of crystallized silica [$Si(OH)_4$] with very fine markings (figure 27.6*c,d*). They have distinctive, and often exceptionally beautiful, patterns that are different for each species. Frustule morphology is very useful in diatom identification. This group of algae is so unique that many phycologists place it in a separate division, the *Bacillariophyta.*

1. Why are the *Charophyta* called stoneworts?
2. How do euglenoids reproduce?
3. How do diatoms reproduce?
4. How do the cells of diatoms differ from those of other organisms?

Box 27.1

Practical Importance of Diatoms

Diatoms have both direct and indirect economic significance for humans. Because diatoms make up most of the phytoplankton of the cooler parts of the ocean, they are the most important ultimate source of food for fish and other marine animals in these regions. It is not unusual for 1 liter of seawater to contain almost a million diatoms.

When diatoms die, their frustules sink to the bottom. Because the siliceous part of the frustule is not affected by the death of the cell, diatom frustules tend to accumulate at the bottom of aquatic environments. These form deposits of material called diatomaceous earth. This material is used as an active ingredient in many commercial preparations, including detergents, fine abrasive polishes, paint removers, decolorizing and deodorizing oils, and fertilizers. Diatomaceous earth is also used extensively as a filtering agent, as a component in insulating (firebrick) and soundproofing products, and as an additive to paint to increase the night visibility of signs and license plates.

The use of diatoms as indicators of water quality and of pollution tolerance is becoming increasingly important. Specific tolerances for given species to various environmental parameters (concentrations of salts, pH, nutrients, nitrogen, temperature) have been compiled.

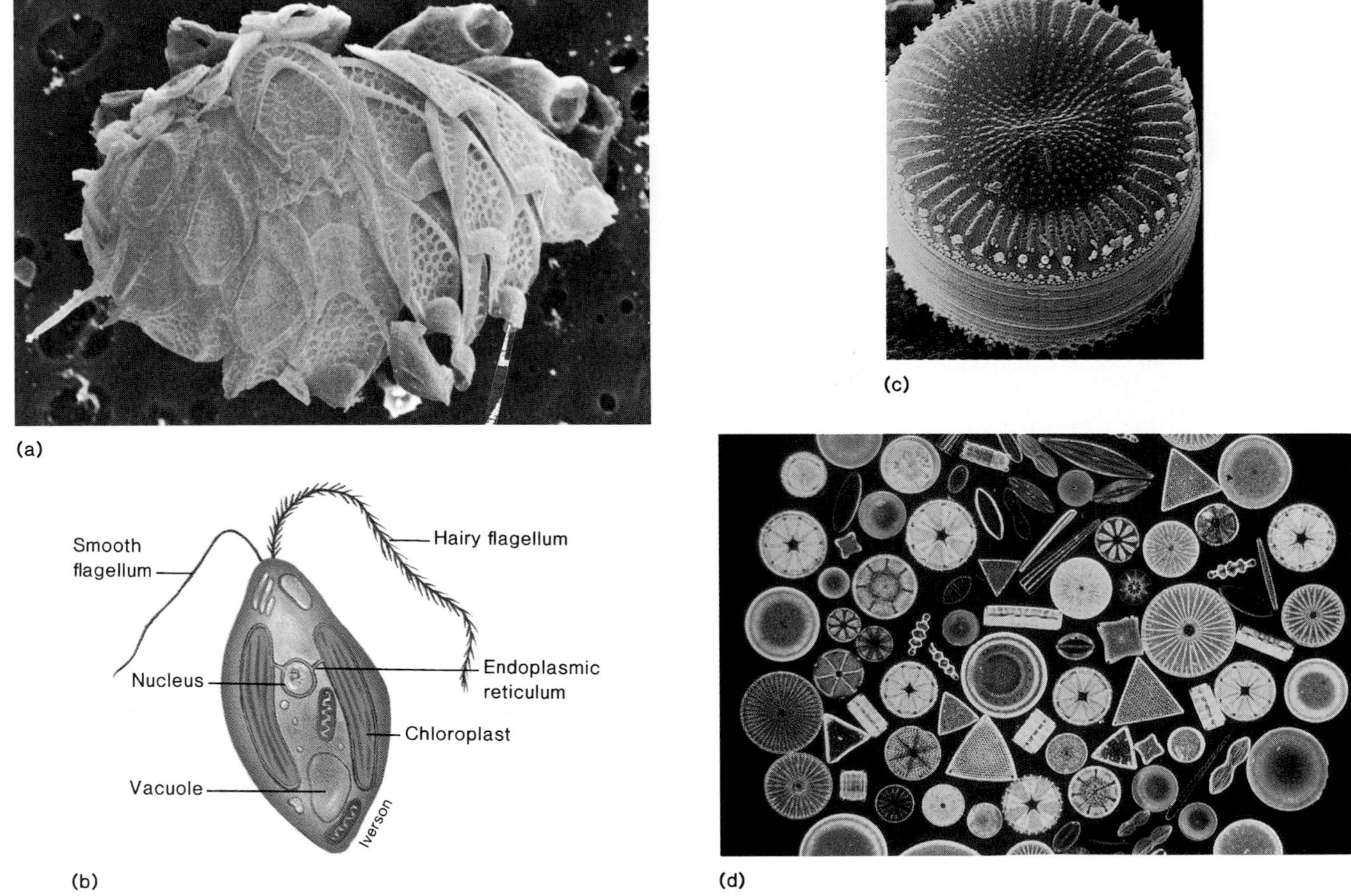

Figure 27.6 ***Chrysophyta*** **(Yellow-green and Golden-brown Algae; Diatoms).** (*a*) Scanning electron micrograph of *Mallomonas*, a chrysophyte, showing its silica scales. The scales are embedded in the pectin wall but synthesized within the Golgi apparatus and transported to the cell surface vesicles (×9,000). (*b*) *Ochromonas*, unicellular chrysophyte. Diagram showing typical cell structure. (*c*) Scanning electron micrograph of a diatom, *Cyclotella meneghiniana* (×750). (*d*) Assorted diatoms as arranged by a light microscopist (×900).

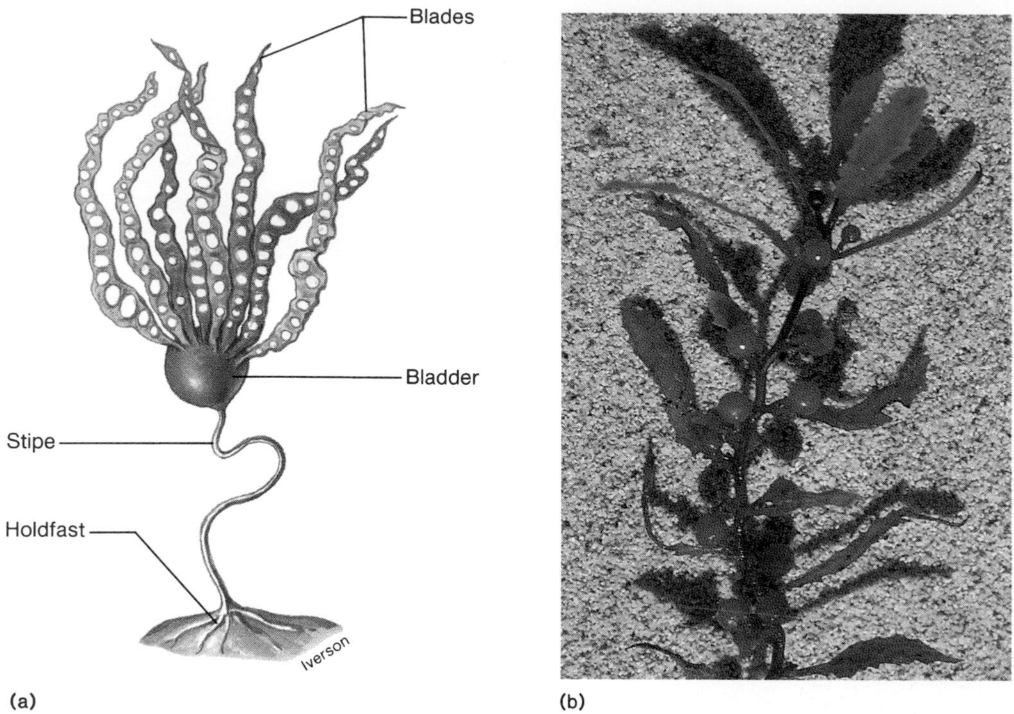

Figure 27.7 ***Phaeophyta* (Brown Algae).** (*a*) Diagram of the parts of the brown alga, *Nereocystis*. Due to the holdfast organ, the heaviest tidal action and surf seldom dislodge brown algae from their substratum. The stipe is a stalk that varies in length; the bladder is a gas-filled float. (*b*) Giant kelp, *Sargassum*, from which the Sargasso Sea got its name.

Phaeophyta (Brown Algae)

The *Phaeophyta* or brown algae (Greek *phaeo*, brown) consist of multicellular organisms that occur almost exclusively in the sea. Most of the conspicuous seaweeds that are brown to olive green in color are assigned to this division. The simplest brown algae consist of small openly branched filaments; the larger, more advanced species have a complex arrangement. Some large **kelps** are conspicuously differentiated into flattened blades, stalks, and holdfast organs that anchor them to rocks (figure 27.7*a*). Some, such as *Sargassum*, form huge floating masses that dominate the Sargasso Sea (figure 27.7*b*). The color of these algae reflects the presence of the brown pigment fucoxanthin, in addition to chlorophylls *a* and *c*, β-carotene, and violaxanthin. The main storage product is laminarin.

Rhodophyta (Red Algae)

The division *Rhodophyta*, the red algae, includes most of the seaweeds (figure 27.8). A few reds are unicellular but most are filamentous and multicellular. Some red algae are up to 1 m long. The stored food is the carbohydrate called floridean starch.

The red algae contain the red pigment phycoerythrin, one of the two types of phycobilins that they possess. The other accessory pigment is the blue pigment phycocyanin. The presence of these pigments explains how the red algae can live at depths of 100 m or more. The wavelengths of light (green, violet, and blue) that penetrate these depths are not absorbed by chlorophyll *a* but, instead, by these phycobilins. Not surprisingly, the concentrations of these pigments often increases with depth as light intensity decreases. The phycobilins, after absorbing the light energy, pass it on to chlorophyll *a*. The algae appear decidedly red when phycoerythrin predominates over the other pigments. When phycoerythrin undergoes photodestruction in bright light, other pigments predominate and the algae take on shades of blue, brown, and dark green.

The cell walls of most red algae include a rigid inner part composed of microfibrils and a mucilaginous matrix. The matrix is composed of sulfated polymers of galactose called agar, funori, porphysan, and carrageenan. It is these four polymers that give the red algae their flexible, slippery texture. Agar is used extensively in the laboratory as a culture medium component (*see chapter 5*). Many red algae also deposit calcium carbonate in their cell walls and play an important role in building coral reefs.

Box 27.2

Dinoflagellate Toxins

The poisonous and destructive **red tides** that occur frequently in coastal areas are often associated with population explosions, or "blooms," of dinoflagellates. *Gymnodinium* and *Gonyaulax* species are the dinoflagellates most often involved. The pigments in the dinoflagellate cells are responsible for the red color of the water. Under these bloom conditions, the dinoflagellates produce a powerful neurotoxin called saxitoxin. The toxin paralyzes the striated respiratory muscles in many vertebrates by inhibiting sodium transport, which is essential to the function of their nerve cells. The toxin does not harm the shellfish that feed on the dinoflagellates. However, the shellfish do accumulate the toxin and are themselves highly poisonous to organisms, such as humans, who consume the shellfish, resulting in a condition known as paralytic shellfish poisoning or neurotoxic shellfish poisoning. Paralytic shellfish poisoning is characterized by numbness of the mouth, lips, face, and extremities. Duration of the illness ranges from a few hours to a few days and usually is not fatal.

Another type of poisoning in humans is called **ciguatera.** It results from eating marine fishes (red bass, moray eels, and gray and Spanish mackerel) that have consumed the dinoflagellate *Gambierdiscus toxicus.* The alga's toxin, called ciguatoxin, accumulates in the flesh of fish. This is one of the most powerful toxins known and remains in the flesh even after it has been cooked. Unfortunately, it cannot be detected in the fishes and they are not visibly affected. In humans, the toxin may cause gastrointestinal disturbances, profuse diarrhea, central nervous system involvement, and respiratory failure.

There are no treatments for the above types of poisonings. Supportive measures are the only therapy.

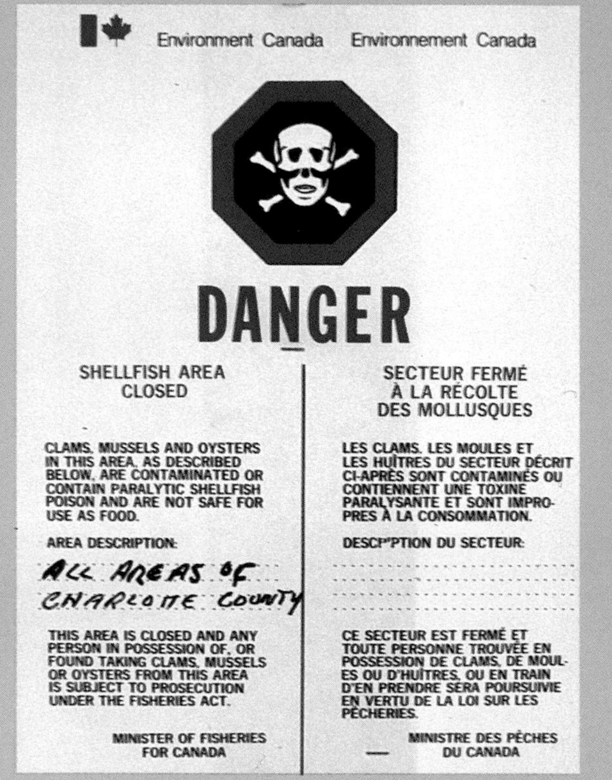

Box Figure 27.2 Sign posted to indicate that shellfishing is prohibited because of a bloom.

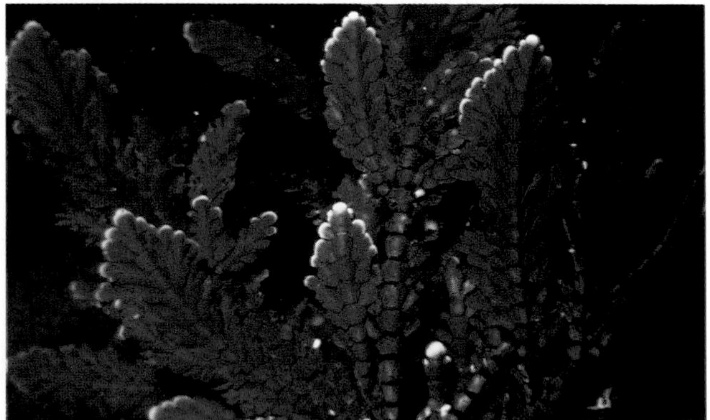

Figure 27.8 ***Rhodophyta*** **(Red Algae).** These algae (e.g., *Corallina gracilis*) are much smaller and more delicate than the brown algae. Most red algae have a filamentous, branched morphology as seen here.

Pyrrhophyta (Dinoflagellates)

The *Pyrrhophyta* or **dinoflagellates** are unicellular, photosynthetic protistan algae. Most dinoflagellates are marine, but some live in fresh water. Along with the chrysophytes and diatoms, the dinoflagellates make up a large part of freshwater and marine plankton and are at the base of many food chains. Species of *Noctiluca, Pyrodinium, Gonyaulax,* and other genera can produce light and are responsible for much of the luminescence (phosphorescence) seen in ocean waters at night. Sometimes dinoflagellate populations reach such high levels that poisonous red tides result (Box 27.2).

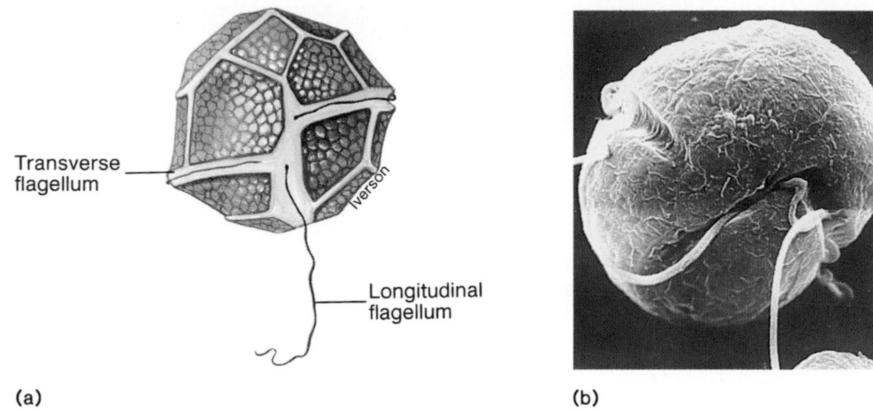

Figure 27.9 **Dinoflagellates.** (*a*) *Ceratium.* (*b*) Scanning electron micrograph of *Gymnodinium* (×4,000). Notice the plates of cellulose and the two flagella: one in the transverse groove and the other projecting outward.

The flagella, protective coats (plates), and biochemistry of the dinoflagellates are distinctive. Many dinoflagellates are armored or clad in stiff, patterned, cellulose plates or thecae, which may become encrusted with silica (figure 27.9). Most have two flagella. In armored dinoflagellates, these flagella beat in two grooves that girdle the cell—one transverse (the cingulum), the other longitudinal (the sulcus). The longitudinal flagellum extends posteriorly as a rudder; the flattened, ribbonlike transverse flagellum propels the cell forward while causing it to spin. Due to this spinning, the dinoflagellates received their name from the Greek *dinein,* "to whirl."

Most dinoflagellates have chlorophylls *a* and *c,* in addition to carotenoids and xanthophylls. As a result, they usually have a yellowish-green to brown color.

Some dinoflagellates can ingest other cells; others are colorless and heterotrophic (osmotrophic). A few even occur as symbionts in many groups of jellyfish, sea anemones, mollusks, and corals. When dinoflagellates form symbiotic relationships, they lose their cellulose plates and flagella, become spherical golden-brown globules in the host cells, and are then termed **zooxanthellae** (*see figure 29.3*a).

1. What is a seaweed?
2. Describe a kelp.
3. Why do red algae appear red?
4. What is unique about the flagellar arrangement in the dinoflagellates?

Summary

1. Phycology (or algology) is the study of algae. Algae are eucaryotic plants or protists that lack roots, stems, and leaves; have chlorophyll and other pigments for carrying out oxygen-producing photosynthesis (or are closely related to photosynthetic species); and lack a sterile covering around the reproductive cells.
2. Algae are found in almost every environmental niche—in fresh, marine, and brackish water; in certain terrestrial environments; and on moist inanimate objects. They may be endosymbionts, parasites, or components of lichens, and make up a large part of the phytoplankton on the earth.
3. Algal classification is based primarily on cellular properties. Examples include cell wall structure and chemistry, form in which nutrients are stored, types of chlorophyll molecules and accessory pigments, flagella, morphology of the thallus, habitat, reproductive structures, and life history patterns.
4. The vegetative structure of the algae varies from the relative simplicity of a single cell to the more striking complexity of organisms such as the giant kelps. Algae can be unicellular, colonial, filamentous, membranous, or tubular.
5. Both sexual and asexual reproduction are present in the algae. Asexual reproduction includes fragmentation, the production of spores, and binary fission. In sexual reproduction, there is fusion of gametes to form a zygote.
6. Green algae (*Chlorophyta*) are a highly diverse group of organisms abundant in the sea, fresh water, and damp terrestrial habitats.
7. Stoneworts/brittleworts (*Charophyta*) have more complex structures than the green algae. They contain whorls of short branches that arise regularly at their nodes. Their gametangia are complex and multicellular. Stoneworts are abundant in fresh to brackish water and are common as fossils.
8. Euglenoids (*Euglenophyta*) have chloroplasts that are biochemically similar to those of the green algae. They have a flexible proteinaceous pellicle inside the plasma membrane.
9. The *Chrysophyta* contain golden-brown algae, yellow-green algae, and diatoms, and vary greatly in pigment composition, cell wall structure, and type of flagellated cells.
10. Brown algae (*Phaeophyta*) are multicellular marine algae, some of which reach lengths of 75 m. The kelps—the largest of the brown algae—contribute greatly to the productivity of the sea as well as to many human needs.
11. Red algae (*Rhodophyta*) have chlorophyll *a* and phycobilins and usually grow at great depths. Some produce agar.
12. Dinoflagellates (*Pyrrhophyta*) are unicellular motile algae responsible for the red tides that are poisonous to many forms of life.

Key Terms

algae *536*
algology *536*
antheridium *538*
aplanospore *538*
benthic *536*
binary fission *538*
chrysolaminarin *541*
ciguatera *544*
diatom *541*
dinoflagellate *544*
epitheca *541*
euglenoid *540*
fragmentation *538*
frustule *541*
hypotheca *541*
kelp *543*
laminarin *541*
neustonic *536*
oogonium *538*
pellicle *540*
phycology *536*
phytoplankton *536*
plankton *536*
planktonic *536*
protothecosis *540*
pyrenoid *538*
red tides *544*
scale *541*
spore *538*
stigma (eyespot) *539*
stonewort *540*
thallus *538*
zooplankton *536*
zoospore *538*
zooxanthellae *545*
zygote *538*

Questions for Thought and Review

1. How can algae be distinguished from the photosynthetic bacteria?
2. What characteristics of algae are used as a basis for algal classification?
3. Although multicellularity must have originated many times in the different multicellular groups, how do present-day green algae provide an especially clear example of the probable origin of multicellularity?
4. What characteristics do euglenoids share with higher plants? With animals? Why are they considered protists?
5. Why do marine algae vary so much more in shape and size than those found in fresh water?
6. The phytoplankton of the open oceans (e.g., the Sargasso Sea) are predominantely algae, and these cells form the base of oceanic food chains. Which morphological characteristics of algae are adaptations to a floating (planktonic) habitat?
7. Freshwater algae are distributed worldwide. They rapidly colonize artificial lakes and water impoundments. How do algae accomplish such widespread dispersal?
8. What problem do diatoms have with continued asexual reproduction? How is this problem solved?
9. How are the red algae similar to the cyanobacteria?
10. What are some important characteristics of the green algae? The red algae? The brown algae? The dinoflagellates?

Additional Reading

Abbott, I. A., and Dawson, E. Y. 1978. *How to know the seaweeds,* 2d ed. Dubuque, Ia.: Wm. C. Brown Publishers.

Arad, S. (Malis), and Yaron, A. 1992. Natural pigments from red microalgae for use in foods and cosmetics. *Trends Food Sci. Technol.* 3:92–97.

Bold, H. C., and Wynne, M. J. 1985. *Introduction to the algae,* 2d ed. Englewood Cliffs, N.J.: Prentice-Hall.

Collins, M. 1978. Algal toxins. *Microbiol. Rev.* 42(3):725–46.

Darley, W. M. 1982. *Algal biology: A physiological approach.* Boston: Blackwell Scientific Publications.

Dodge, J. D. 1973. *The fine structure of algal cells.* New York: Academic Press.

Gibor, A. 1966. *Acetabularia:* A useful giant cell. *Sci. Am.* 215:118–24.

Larkum, A. W. D., and Barrett, J. 1983. Light-harvesting processes in algae. *Adv. Bot. Res.* 10:1–219.

Lee, R. E. 1989. *Phycology.* 2d ed. New York: Cambridge University Press.

Lembi, C. A., and Waaland, J. R., eds. 1988. *Algae and human affairs.* New York: Cambridge University Press.

Lobban, C. S., and Wynne, M. J., eds. 1981. *The biology of seaweeds.* Oxford: Blackwell Scientific Publications.

Noble, R. C. 1990. Death on the half-shell: the health hazards of eating shellfish. Perspect. Biol. Med. *33:*313–22.

Perasso, R. 1989. Origin of the algae. *Nature* 339:142–44.

Pickett-Heaps, J. D. 1975. *Green algae: Structure, reproduction and evolution in selected genera.* Sunderland, Mass.: Sinauer Associates.

Prescott, G. W. 1978. *How to know the freshwater algae,* 3d ed. Dubuque, Ia.: Wm. C. Brown Publishers.

Round, F. E. 1984. *The ecology of algae.* New York: Cambridge University Press.

Saffo, M. B. 1987. New light on seaweeds. *BioScience* 37:170–180.

Steidinger, K. A., and Haddad, K. 1981. Biologic and hydrographic aspects of red tides. *BioScience* 31:814–19.

Stewart, W. D. P., ed. 1975. *Algal physiology and biochemistry.* Oxford: Blackwell Scientific Publications.

Sze, P. 1993. *A biology of the algae,* 2d ed. Dubuque, Ia.: Wm. C. Brown Publishers.

Trainor, F. R. 1978. *Introductory phycology.* New York: John Wiley and Sons.

Van der Meer, J. P. 1983. The domestication of seaweeds. *BioScience* 33:172–76.

Van Etten, J.; Lane, L.; and Meints, R. 1991. Viruses and viruslike particles of eukaryotic algae. *Microbiol. Rev.* 55(4):586–620.

Wells, M. L.; Mayer, L. M.; and Guillard, R. R. L. 1991. Evaluation of iron as a triggering factor for red tide blooms. *Mar. Ecol. Prog. Ser.* 69:93–102.

CHAPTER 28
The Protozoa

And a pleasant sight they are indeed. Their shapes range from teardrops to bells, barrels, cups, cornucopias, stars, snowflakes, and radiating suns, to the common amoebas, which have no real shape at all. Some live in baskets that look as if they were fashioned of exquisitely carved ivory filigree. Others use colored bits of silica to make themselves bright mosaic domes. Some even form graceful transparent containers shaped like vases or wine glasses of fine crystal in which they make their homes.

—Helena Curtis

Outline

Concepts

1. Protozoa are protists exhibiting heterotrophic nutrition and various types of locomotion. They occupy a vast array of habitats and niches and have organelles similar to those found in other eucaryotic cells, and also specialized organelles.
2. The protozoa are divided into seven phyla: *Sarcomastigophora, Labyrinthomorpha, Apicomplexa, Microspora, Ascetospora, Myxozoa,* and *Ciliophora.* These phyla represent four major groups: flagellates, amoebae, ciliates, and sporozoa.
3. Protozoa usually reproduce asexually by binary fission. Some have sexual cycles, involving meiosis and the fusion of gametes or gametic nuclei resulting in a diploid zygote. The zygote is often a thick-walled, resistant, and resting cell called a cyst. Some protozoa undergo conjugation in which nuclei are exchanged between cells.
4. All protozoa have one or more nuclei; some have a macro- and micronucleus.
5. Various protozoa feed by holophytic, holozoic, or saprozoic means; some are predatory or parasitic.

Chapter 28 presents the major biological features of the protists, known as protozoa. The most important groups are the flagellates, amoebae, sporozoa, and ciliates. These protists demonstrate the great adaptive potential of the basic single eucaryotic cell.

The microorganisms called **protozoa** (s., protozoan; Greek *protos,* first, and *zoon,* animal) are studied in the discipline called **protozoology.** A protozoan can be defined as a usually motile eucaryotic unicellular protist. Protozoa are directly related only on the basis of a single negative characteristic—they are not multicellular. All, however, demonstrate the basic body plan of a single protistan eucaryotic cell.

Distribution

Protozoa grow in a wide variety of moist habitats. Moisture is absolutely necessary for the existence of protozoa because they are susceptible to desiccation. Most protozoa are free living and inhabit freshwater or marine environments. Many terrestrial protozoa can be found in decaying organic matter, in soil, and even in beach sand; some are parasitic in plants or animals.

Importance

Protozoa play a significant role in the economy of nature. For example, they make up a large part of **plankton**—small, free-floating aquatic organisms that are an important link in the many aquatic food chains and food webs of aquatic environments. A **food chain** is a series of organisms, each feeding on the preceding one. A **food web** is a complex interlocking series of food chains. Protozoa are also useful in biochemical and molecular biological studies. Many biochemical pathways used by protozoa are present in all eucaryotic cells. Finally, some of the most important diseases of humans (*see chapter 39*) and animals (table 28.1) are caused by protozoa.

Microbial populations in natural environments (pp. 805–7).

Morphology

Because protozoa are eucaryotic cells, in many respects their morphology and physiology are the same as the cells of multicellular animals (*see figures 4.2 and 4.3*). However, because all of life's various functions must be performed within the individual protozoan, some morphological and physiological features are unique to protozoan cells. For example, the cytoplasm of the protozoan vegetative form, the **trophozoite,** is surrounded by a plasma membrane. In some species, the cy-

TABLE 28.1 Pathogenic Protozoa That Cause Major Diseases of Domestic Animals

Protozoan Group	Genus	Host	Preferred Site of Infection	Disease
Amoebae	*Entamoeba*	Mammals	Intestine	Amebiasis
	Iodamoeba	Swine	Intestine	Enteritis
Sporozoa	*Babesia*	Cattle	Blood cells	Babesiosis
	Theileria	Cattle, sheep, goats	Blood cells	Theilariasis
	Sarcocystis	Mammals, birds	Muscles	—
	Toxoplasma	Cats	Intestine	Toxoplasmosis
	Isospora	Dogs	Intestine	Coccidiosis
	Eimeria	Cattle, cats, chickens, swine	Intestine	Coccidiosis
	Plasmodium	Many animals	Bloodstream, liver	Malaria
	Leucocytozoon	Birds	Spleen, lungs, blood	—
	Cryptosporidium	Mammals	Intestine	Cryptosporidiosis
Ciliates	*Balantidium*	Swine	Large intestine	Balantidiasis
Flagellates	*Leishmania*	Dogs, cats, horses, sheep, cattle	Spleen, bone marrow, mucous membranes	Leishmaniasis
	Trypanosoma	Most animals	Blood	Trypanosomiasis
	Trichomonas	Horses, cattle	Genital tract	Trichomoniasis (abortion)
	Histomonas	Birds	Intestine	Blackhead disease
	Giardia	Mammals	Intestine	Giardiasis

toplasm immediately under the plasma membrane is semisolid or gelatinous, giving some rigidity to the cell body. It is termed the **ectoplasm.** The bases of the flagella or cilia and their associated fibrillar structures are embedded in the ectoplasm. The plasma membrane and structures immediately beneath it are called the **pellicle.** Inside the ectoplasm is the area referred to as the **endoplasm,** which is more fluid and granular in composition and contains most of the organelles. Some protozoa have one nucleus, others have two or more identical nuclei. Still other protozoa have two distinct types of nuclei—a macronucleus and one or more micronuclei. The **macronucleus,** when present, is typically larger and associated with trophic activities and regeneration processes. The **micronucleus** is diploid and involved in both genetic recombination during reproduction and the regeneration of the macronucleus.

One or more vacuoles are usually present in the cytoplasm of protozoa. These are differentiated into contractile, secretory, and food vacuoles. **Contractile vacuoles** function as osmoregulatory organelles in those protozoa that live in a hypotonic environment, such as a freshwater lake. Osmotic balance is maintained by continuous water expulsion. Most marine protozoa and parasitic species are isotonic to their environment and lack such vacuoles. **Phagocytic vacuoles** are conspicuous in holozoic and parasitic species and are the sites of food digestion. (*see figure 4.10*). **Secretory vacuoles** usually contain specific enzymes that perform various functions (such as excystation).

Most anaerobic protozoa (such as *Trichonympha,* which lives in the gut of termites; *see figure 29.1*b) have no mitochondria, no cytochromes, and no functional tricarboxylic acid cycle. However, some do have small, membrane-delimited organelles termed **hydrogenosomes.** These structures contain a unique electron transfer pathway in which hydrogenase transfers electrons to protons (which act as the terminal electron acceptors), and molecular hydrogen is formed.

---

1. Describe a typical protozoan.
2. What roles do protozoa play in the trophic structure of their communities and in the organisms with which they associate?
3. What is unique about the nuclei of some protozoa?
4. Where can protozoa be found?
5. What are the functions of contractile, phagocytic, and secretory vacuoles?

Nutrition

Most protozoa are chemoheterotrophic (*see figure 7.1*). There are two types of heterotrophic nutrition found in the protozoa: holozoic and saprozoic. In **holozoic nutrition,** nutrients such as bacteria are acquired by phagocytosis and the subsequent formation of phagocytic vacuoles. Some ciliates have a specialized structure for phagocytosis called the **cytostome** (cell mouth). In **saprozoic nutrition,** nutrients such as amino acids and sugars cross the plasma membrane by pinocytosis, diffusion, or carrier-mediated transport (facilitated diffusion or active transport).

Encystment and Excystment

Many protozoa are capable of **encystation.** They develop into a resting stage called a **cyst,** which is a dormant form marked by the presence of a wall and by the reduction of metabolic activity to a very low level. Cyst formation is particularly common among aquatic, free-living protozoa and parasitic forms. Cysts serve three major functions: (1) they protect against adverse changes in the environment, such as nutrient deficiency, desiccation, adverse pH, and low partial pressure of O_2; (2) they are sites for nuclear reorganization and cell division (reproductive cysts); and (3) they serve as a means of transfer between hosts in parasitic species.

Although the exact stimulus for **excystation** (escape from the cysts) is unknown, excystation generally is triggered by a return to favorable environmental conditions. For example, cysts of parasitic species excyst after ingestion by the host.

Locomotory Organelles

A few protozoa are nonmotile. Most, however, can move by one of three major types of locomotory organelles: pseudopodia, flagella, or cilia. **Pseudopodia** (s., pseudopodium; false feet) are cytoplasmic extensions found in the amoebae that are responsible for the movement and food capture. There are many types of pseudopodia. Flagellates and ciliates move by flagella and cilia. Electron microscopy has shown that protozoan flagella and cilia are structurally the same and identical in function to those of other eucaryotic cells (*see figures 4.24–4.28*).

Reproduction

Most protozoa reproduce asexually, and some also carry out sexual reproduction. The most common method of asexual reproduction is **binary fission.** During this process, the nucleus first undergoes mitosis; then the cytoplasm divides by cytokinesis to form two identical individuals (figure 28.1).

The most common method of sexual reproduction is **conjugation.** In this process, there is an exchange of gametes between paired protozoa of complementary mating types (**conjugants;** *see figure 2.12*b). Conjugation is most prevalent among ciliate protozoa. A well-studied example is *Paramecium caudatum* (figure 28.2). At the beginning of conjugation, two ciliates unite, fusing their pellicles at the contact point. The macronucleus in each is degraded. The individual micronuclei divide twice by meiosis to form four haploid pronuclei, three of which disintegrate. The remaining pronucleus divides again mitotically to form two gametic nuclei, a stationary one and a migratory one. The migratory nuclei pass into the respective conjugates. Then the ciliates separate, the gametic nuclei fuse, and the resulting diploid zygote nucleus undergoes three rounds of mitosis. The eight resulting nuclei have different fates: one nucleus is retained as a micronucleus; three others are destroyed; and the four remaining nuclei develop into macronuclei. Each separated conjugant now undergoes cell division. Eventually, progeny with one macronucleus and one micronucleus are formed.

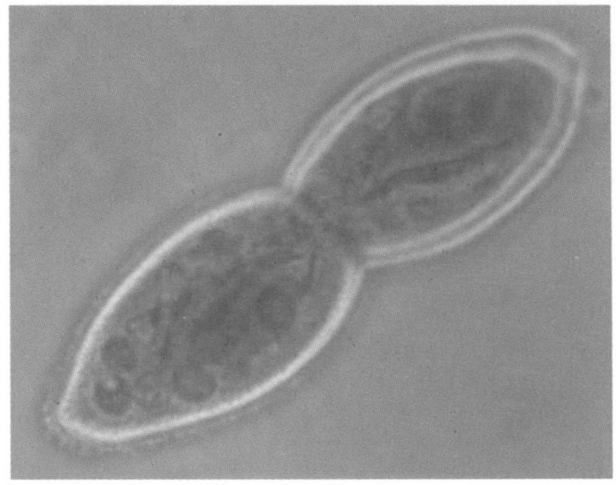

Figure 28.1 Protozoan Reproduction. Binary fission in *Paramecium caudatum* (×100).

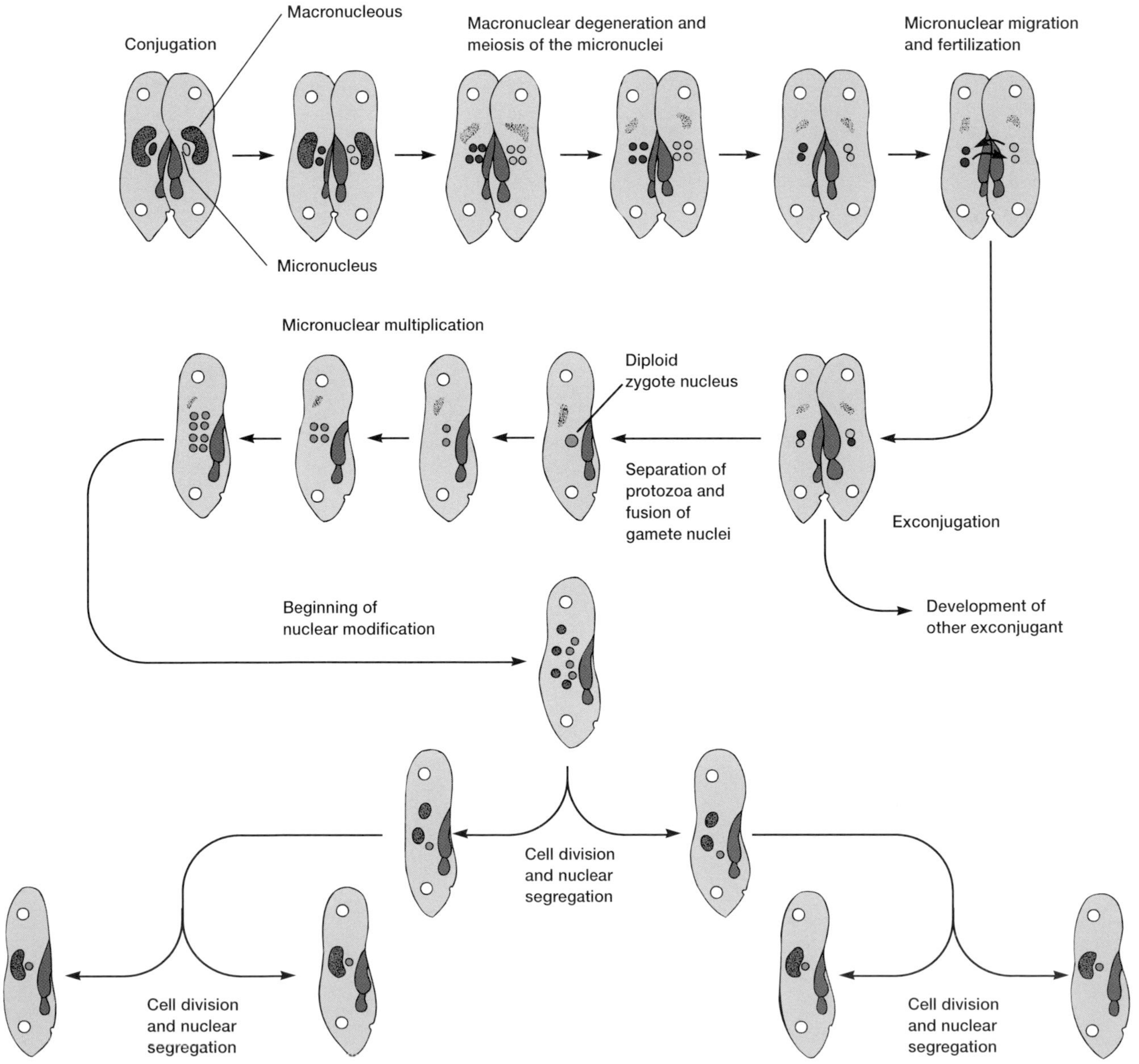

Conjugation
Macronucleous
Micronucleus
Macronuclear degeneration and meiosis of the micronuclei
Micronuclear migration and fertilization
Exconjugation
Development of other exconjugant
Diploid zygote nucleus
Separation of protozoa and fusion of gamete nuclei
Micronuclear multiplication
Beginning of nuclear modification
Cell division and nuclear segregation
Cell division and nuclear segregation
Cell division and nuclear segregation

Figure 28.2 Conjugation in *Paramecium caudatum,* Schematic Drawing. Follow the arrows. After the conjugants separate, only one of the exconjugants is followed; however, a total of eight new protozoa result from the conjugation.

1. What specific nutritional types exist among protozoa?
2. What functions do cysts serve for a typical protozoan? What causes excystation to occur?
3. What is a pseudopodium?
4. How do protozoa reproduce asexually?
5. How do protozoa reproduce sexually? Describe the process of ciliate conjugation.

Classification

Protozoologists now regard the *Protozoa* as a subkingdom, which contains seven of the 14 phyla found within the kingdom *Protista* (table 28.2). The phylum *Sarcomastigophora,* consists of flagellates and amoebae with a single type of nucleus. The phyla *Labyrinthomorpha, Apicomplexa, Microspora, Ascetospora,* and *Myxozoa* have either saprozoic or parasitic species. The phylum *Ciliophora* has ciliated protozoa with two

TABLE 28.2 Current Abbreviated Classification of the Subkingdom *Protozoa*

Taxonomic Group	Characteristics	Examples
Phylum: *Sarcomastigophora*	Locomotion by flagella, pseudopodia, or both; when present, sexual reproduction is essentially syngamy (union of gametes external to the parents); single type of nucleus	
Subphylum: *Mastigophora*	One or more flagella; division by longitudinal binary fission; sexual reproduction in some groups	
Class: *Zoomastigophorea*	Chromatophores absent; one to many flagella; amoeboid forms, with or without flagella; sexuality known in some groups; mainly parasitic	*Trypanosoma* *Giardia* *Trichomonas* *Leishmania* *Trichonympha*
Subphylum: *Sarcodina*	Locomotion primarily by pseudopodia; shells (tests) often present; flagella restricted to reproductive stages when present; asexual reproduction by fission; mostly free living	
Superclass: *Rhizopoda*	Locomotion by pseudopodia or by protoplasmic flow with discrete pseudopodia; some contain tests	*Amoeba* *Elphidium* *Coccodiscus*
Phylum: *Labyrinthomorpha*	Spindle-shaped cells capable of producing mucous tracks; trophic stage as ectoplasmic network; nonamoeboid cells; saprozoic and parasitic on algae and seagrass	*Labyrinthula*
Phylum: *Apicomplexa*	All members have a spore-forming stage in their life cycle; contain an apical complex; sexuality by syngamy; all species parasitic; cysts often present; cilia absent; often called the Sporozoa	*Plasmodium* *Toxoplasma* *Eimeria* *Cryptosporidium* *Pneumocystis*
Phylum: *Microspora*	Unicellular spores with spiroplasm containing polar filaments; obligatory intracellular parasites	*Nosema*
Phylum: *Ascetospora*	Spore with one or more spiroplasms; no polar capsules or polar filaments; all parasitic in invertebrates	*Haplosporidium*
Phylum: *Myxozoa*	Spores of multicellular origin; one or more polar capsules; all parasitic, especially in fish	*Myxosoma*
Phylum: *Ciliophora*	Simple cilia or compound ciliary organelles in at least one stage in the life cycle; two types of nuclei; contractile vacuole present; binary fission transverse; sexuality involving conjugation; most species free living, but many commensal, some parasitic	*Didinium* *Stentor* *Vorticella* *Tetrahymena* *Paramecium* *Tokophrya* *Entodinium* *Nyctotherus* *Balantidium* *Ichthyophthirius*

types of nuclei. The classification of this subkingdom into phyla is based primarily on types of nuclei, mode of reproduction, and mechanism of locomotion.

Representative Types

This section describes some representatives of each group of protozoan protists to present an overview of protozoan diversity and to provide a basis for comparing different groups.

Phylum *Sarcomastigophora*

Protists that have a single type of nucleus and possess flagella (subphylum *Mastigophora*) or pseudopodia (subphylum *Sarcodina*) are placed in the phylum *Sarcomastigophora.* Both sexual and asexual reproduction are seen in this phylum.

The subphylum *Mastigophora* contains both phytoflagellates, chloroplast-bearing flagellates and close relatives, and **zooflagellates.** Zooflagellates do not have chlorophyll and are either holozoic, saprozoic, or symbiotic. Asexual reproduction occurs by longitudinal binary fission along the major body axis. Sexual reproduction is known for a few species, and encystment is common. Zooflagellates are characterized by the presence of one or more flagella. Most members are uninucleate. One major group, the kinetoplastids, has its mitochondrial DNA in a special region called the **kinetoplast** (figure 28.3*a; see also figures 2.12*c *and 4.13*).

Some zooflagellates are free living. The choanoflagellates are a distinctive example in that they have one flagellum, are solitary or colonial, and are on stalks. Other zooflagellates form symbiotic relationships. For example, *Trichonympha* species (*see figure 29.1*b) are found in the intestine of termites and produce enzymes that the termite needs to digest the wood particles on which it feeds.

The protozoan-termite relationship (pp. 565–66).

Many zooflagellates are important human parasites. *Giardia lamblia* (*see figure 39.18*) can be found in the human intestine where it may cause severe diarrhea. It is transmitted through water that has been contaminated with feces (*see chapters 39 and 41*). Trichomonads, such as *Trichomonas vaginalis,* live in the vagina and urethra of women and in the prostate, seminal vesicles, and urethra of men. They are transmitted primarily by sexual intercourse.

Giardiasis (pp. 792–93).
Trichomoniasis (p. 798).

The zooflagellates called **trypanosomes** are important blood pathogens of humans and animals in certain parts of the world. Because they live in the blood, they are also called hemoflagellates. These parasites (figure 28.3*a; see also figure 4.1*c) have a typical zooflagellate structure. A major human trypanosomal disease is African sleeping sickness caused by *Trypanosoma brucei rhodesiense* or *T. brucei gambiense* (*see chapter 39*).

Trypanosomiasis (p. 797).

The subphylum *Sarcodina* contains the amoeboid protists. They are found throughout the world in both fresh and salt water and are abundant in the soil. Several species are parasites of mammals. Simple amoebae (s., amoeba) move almost continually using their pseudopodia (**amoeboid movement**). Many have no definite shape, and their internal structures (figure 28.3*b*) occupy no particular position. The single nucleus, contractile and phagocytic vacuoles, and ecto- and endoplasm shift as the amoebae move. Amoebae engulf a variety of materials (small algae, bacteria, other protozoa) through phagocytosis. Some material moves into and out of the plasma membrane by pinocytosis. Reproduction in the amoebae is by simple asexual binary fission. Some amoebae can form cysts.

Many free-living forms are more complex than simple amoebae. *Arcella* manufactures a loose-fitting shell or **test** for protection (figure 28.4*a*). These amoebae extend their pseudopodia from the test aperture to either feed or creep along. The foraminiferans and radiolarians primarily are marine amoebae, with a few occurring in fresh and brackish water. Most foraminiferans live on the seafloor, whereas radiolarians are usually found in the open sea (Box 28.1). Foraminiferan tests and radiolarian skeletons have many unique and beautiful shapes (figure 28.4*b,c*). They range in diameter from about 20 μm to several cm.

Finally, there are many symbiotic amoebae, most of which live in other animals. Two common genera are *Endamoeba* and *Entamoeba. Endamoeba blattae* is common in the intestine of cockroaches, and related species are present in termites. *Entamoeba histolytica (see figure 39.17)* is an important parasite of humans, in whom it often produces severe amebic dysentery, which may be fatal. Free-living amoebae from two genera, *Naegleria* and *Acanthamoeba,* can cause disease in humans and other mammals (*see chapter 39*).

Amebiasis (pp. 790–92).

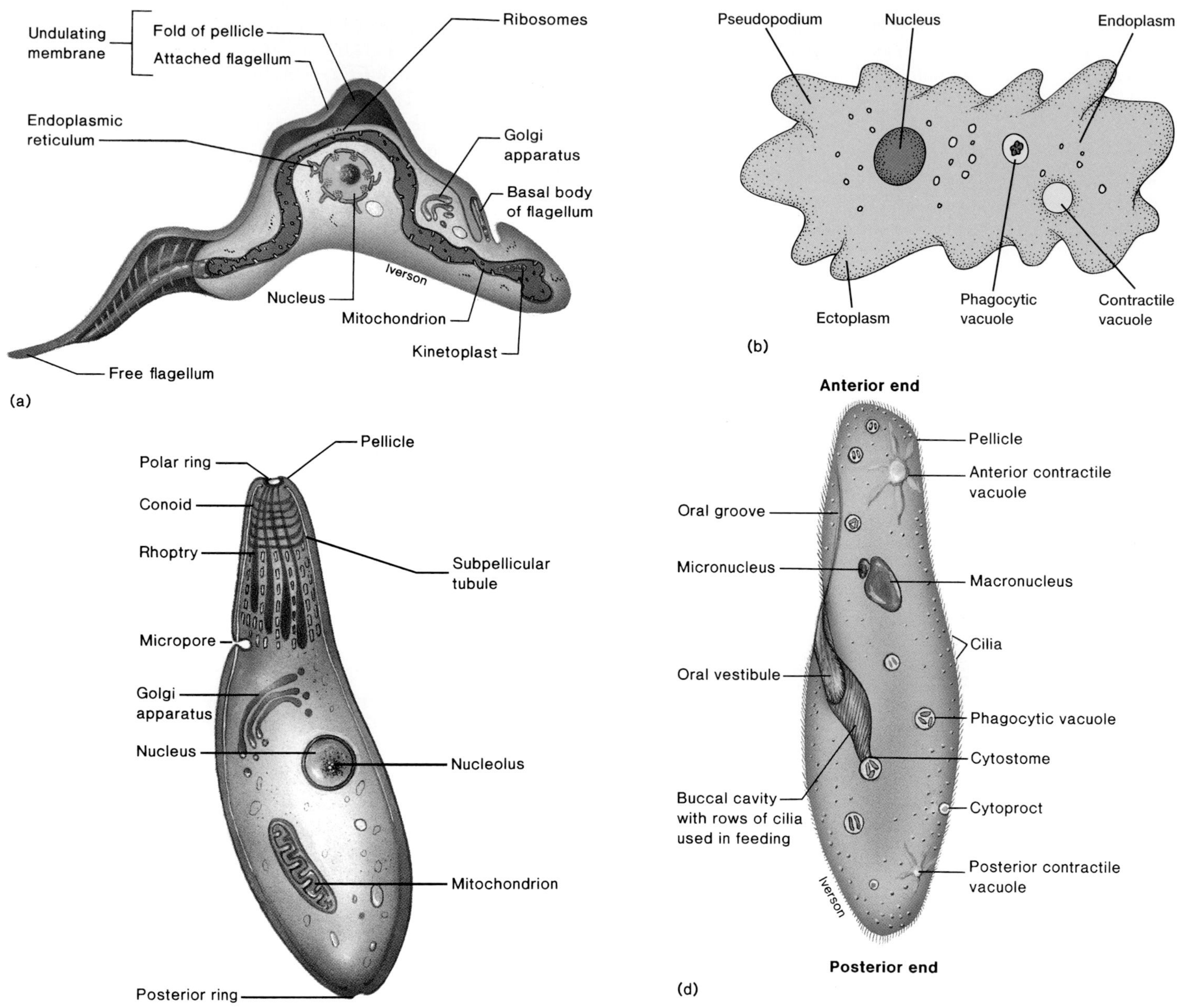

Figure 28.3 **Drawings of Some Representative Protozoa.** (*a*) Structure of the flagellate, *Trypanosoma brucei rhodesiense.* (*b*) The structure of the amoeboid protist, *Amoeba proteus.* (*c*) Structure of an apicomplexan sporozoite. (*d*) Structure of the ciliate *Paramecium caudatum.*

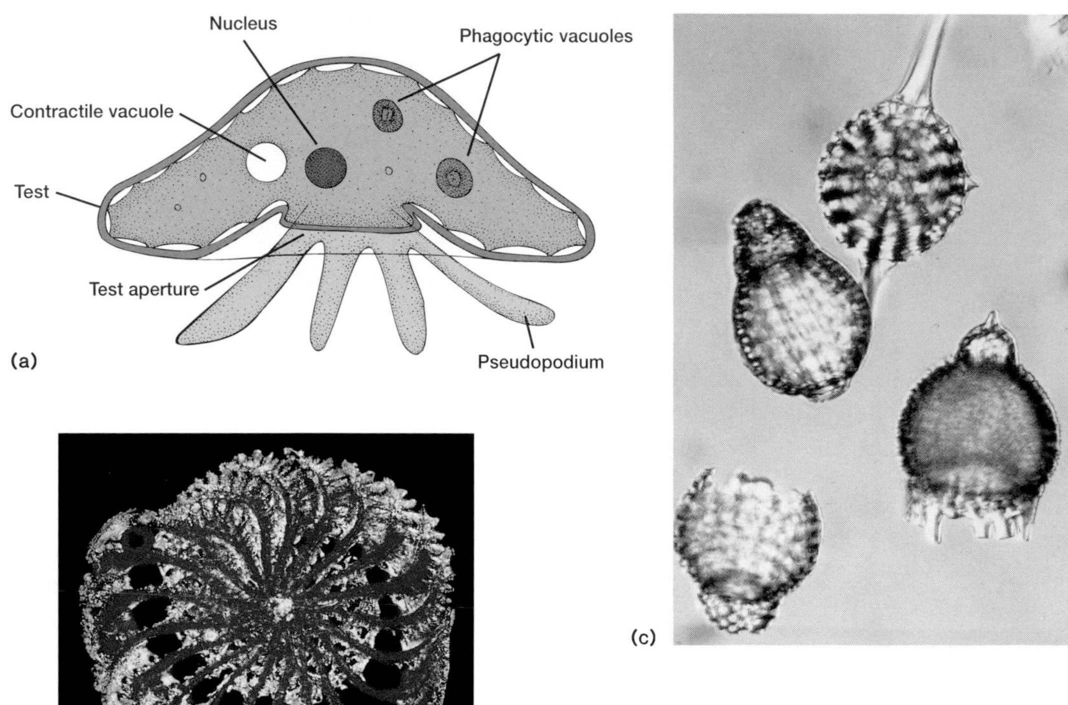

Figure 28.4 **Some Free-living Sarcodines.** (*a*) An illustration of *Arcella*, showing the test or shell that is made of chitinlike material secreted by the protist. (*b*) The test of the foraminiferan, *Elphidium cristum* (×100). (*c*) A group of siliceous radiolarian shells, light micrograph (×63).

Box 28.1

The Importance of Foraminiferans

Of over 40,000 described species of foraminiferans, about 90% are fossil. During the Tertiary period (about 230 million years ago), the foraminiferans contributed massive shell accumulations to geologic formations. They were so abundant that they formed thick deposits, which became uplifted over time and exposed as great beds of limestone in Europe, Asia, and Africa. The White Cliffs of Dover, the famous landmark of southern England, are made up almost entirely of foraminiferan shells. The Egyptian pyramids of Gizeh, near Cairo, are built of foraminiferan limestone. Currently, foraminiferans are important aids to geologists in identifying and correlating rock layers as they search for oil-bearing strata. The calcareous shells of abundant planktonic foraminiferans are today settling and accumulating over much of the ocean floor as thick deposits called "globigerina ooze"—limestone of the distant future.

1. What characteristics would be exhibited by a protozoan that belongs to the phylum *Sarcomastigophora*? What two subphyla does it contain?
2. How would you characterize a zooflagellate? An amoeba?
3. What two human diseases are caused by zooflagellates?
4. Where can two different symbiotic amoebae be found?

Phylum *Labyrinthomorpha*

The very small phylum *Labyrinthomorpha* consists of protists that have spindle-shaped or spherical nonamoeboid vegetative cells. In some genera, amoeboid cells move within a network of mucous tracks using a typical gliding motion. Most members are marine and either saprozoic or parasitic on algae. Several years ago, *Labyrinthula* killed most of the "eel grass" on the Atlantic coast, depriving ducks of their food and starving many of them.

Phylum *Apicomplexa*

The **apicomplexans,** often collectively called the sporozoans, have a spore-forming stage in their life cycle and lack special locomotory organelles (except in the male gametes, and the zygote or ookinete). They are either intra- or intercellular parasites of animals and are distinguished by a unique arrangement of fibrils, microtubules, vacuoles, and other organelles, collectively called the apical complex, which is located at one end of the cell.

The **apical complex** contains several components (figure 28.3*c*). One or two electron-dense polar rings are at the apical end. The **conoid** consists of a cone of spirally arranged fibers lying next to the polar rings. Subpellicular microtubules radiate from the polar rings and probably serve as support elements. Two or more **rhoptries** extend to the plasma membrane and secrete their contents at the cell surface. These secretions aid in the penetration of the host cell. One or more micropores are thought to function in the intake of nutrients.

Apicomplexans have complex life cycles in which certain stages occur in one host (the mammal) and other stages in a different host (often a mosquito). The life cycle has both asexual and sexual phases and is characterized by an alternation of haploid and diploid generations. At some point, an asexual reproduction process called schizogony occurs. **Schizogony** is a rapid series of mitotic events producing many small infective organisms through the formation of uninuclear buds. Sexual reproduction involves the fertilization of a large female macrogamate by a small, flagellated male gamete. The resulting zygote becomes a thick-walled cyst called an **oocyst.** Within the oocyst, meiotic divisions produce infective haploid spores.

The four most important sporozoan parasites are *Plasmodium* (the causative agent of malaria), *Toxoplasma* (the causative agent of toxoplasmosis), *Eimeria* (the causative agent of coccidiosis), and *Pneumocystis* (the causative agent of pneumocystis pneumonia—primarily in AIDS patients).

Malaria (pp. 793–95).

Phylum *Microspora*

The small microsporans (3 to 6 μm) are obligatory intracellular parasites. Included in these protozoa are several species of some economic importance because they parasitize beneficial insects. *Nosema bombicus* parasitizes silkworms (figure 28.5) causing the disease **pebrine,** and *Nosema apis* causes serious dysentery (foul brood) in honeybees. There has been an increased interest in these parasites because of their possible role as biological control agents for certain insects. For example, *Nosema locustae* has been approved and registered by the United States Environmental Protection Agency for use in residual control of rangeland grasshoppers. Recently, four microsporidian genera (*Nosema, Encephalitozoon, Pleistophora, Enterocytozoon*) have been implicated in human diseases with immunosuppressed and AIDS patients.

Phylum *Ascetospora*

Ascetospora is a relatively small phylum that consists exclusively of parasitic protists characterized by spores lacking polar caps or polar filaments. Ascetosporans such as *Haplosporidium* are parasitic primarily in the cells, tissues, and body cavities of mollusks.

Phylum *Myxozoa*

The myxozoans are all parasitic, most on freshwater and marine fish. They have a resistant spore with one to six coiled polar filaments. The most economically important myxozoan is *Myxosoma cerebralis,* which infects the nervous system and

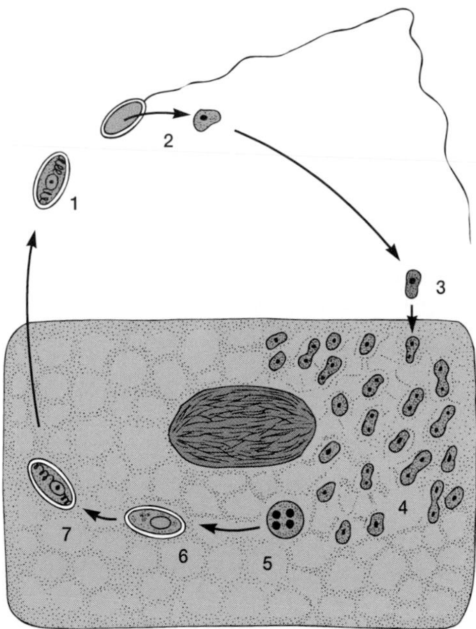

Figure 28.5 The Microsporean *Nosema bombicus,* which is Fatal To Silkworms. (*1*) A typical spore with one coiled filament. (*2*) When ingested, it extrudes the filament. (*3*) The parasite enters an epithelial cell in the intestine of the silkworm and (*4*) divides many times to form small amoebae that eventually fill the cell and kill it. During this phase, some of the amoebae with four nuclei become spores (*5,6,7*). Silkworms are infected by eating leaves contaminated by the feces of infected worms.

auditory organ of trout and salmon (salmonids). Infected fish lose their sense of balance and tumble erratically—thus the name whirling or tumbling disease. **Proliferative kidney disease,** caused by an unclassified myxozoan, has become one of the most important diseases of cultured salmon throughout the world.

1. Describe a typical apicomplexan (sporozoan) protist, including its apical complex.
2. Summarize the sporozoan life cycle. What is schizogony?
3. What are the four most important sporozoan parasites and the diseases they cause?
4. Give one economically important disease caused by a microsporan. What group of animals do myxosporans usually parasitize?

Phylum *Ciliophora*

The phylum *Ciliophora* is the largest of the seven protozoan phyla. There are about 8,000 species of these unicellular, heterotrophic protists that range from about 10 to 3,000 μm long. As their name implies, ciliates employ many cilia as locomotory organelles. The cilia are generally arranged either in longitudinal rows (figure 28.3*d*; *see also figure 4.26*) or in spirals around the body of the organism. They beat with an oblique stroke; therefore, the protist revolves as it swims. Coordination of ciliary beating is so precise that the protist can go either forward or backward.

There is great variation in ciliate shape, and most do not look like the slipper-shaped *Paramecium (see figures 2.8*e *and 4.1*a). In some species (*Vorticella*), there is a long stalk by which the protozoan attaches itself to the substrate. *Stentor* attaches to a substrate and stretches out in a trumpet shape to feed (*see figure 4.1*f). A few species have tentacles for the capture of prey. Some can discharge toxic threadlike darts called toxicysts, which are used in capturing prey.

A most striking feature of ciliates is their ability to capture many particles in a short time by the action of the cilia around the buccal cavity. Food first enters the cytostome and passes into phagocytic vacuoles that fuse with lysosomes after detachment from the cytostome. A vacuole's contents are digested when the vacuole is acidified and lysosomes release digestive enzymes into it. After the digested material has been absorbed into the cytoplasm, the vacuole fuses with a special region of the pellicle called the **cytoproct** and empties its waste material to the outside. Contractile vacuoles are used for osmoregulation and are present chiefly in freshwater species.

Solutes and water activity, chapter 6.

Most ciliates have two types of nuclei: a large macronucleus and a smaller micronucleus. The micronucleus is diploid and contains the normal somatic chromosomes. It divides by mitosis and transmits genetic information through meiosis and sexual reproduction. Macronuclei are derived from micronuclei by a complex series of steps. Within the macronucleus are many chromatin bodies, each containing many copies of only one or two genes. Macronuclei are thus polyploid and divide by elongating, and then by constricting. They produce mRNA to direct protein synthesis, maintain routine cellular functions, and control normal cell metabolism.

Some ciliates reproduce asexually by transverse binary fission, forming two equal daughter protozoa. Many ciliates also reproduce by conjugation as previously described.

Although most ciliates are free living, symbiotic forms do exist. Some ciliated protozoa live as harmless commensals. For example, *Entodinium* is found in the rumen of cattle, and *Nyctotherus* occurs in the colon of frogs. Other ciliates are strict parasites. For example, *Balantidium coli* lives in the intestine of mammals, including humans, where it can produce dysentery. *Ichthyophthirius* lives in fresh water where it can attack many species of fish, producing a disease known as "ick."

1. Describe the morphology of a typical ciliated protozoan.
2. Describe the food-gathering structures found in the ciliated protozoa.
3. In ciliates, what is the function of the macronucleus? The micronucleus?
4. How do ciliates reproduce?
5. Where can the following ciliates be found: *Entodinium, Nyctotherus, Balantidium, Ichthyophthirius?*

Summary

1. Protozoa are protists that can be defined as usually motile eucaryotic unicellular microorganisms.
2. Protozoa are found wherever other organisms exist. They are important components of food chains and food webs. Many are parasitic in humans and animals, and some have become very useful in the study of molecular biology.
3. Because protozoa are eucaryotic cells, in many respects their morphology and physiology resemble those of multicellular animals. However, because all their functions must be performed within the individual protist, many morphological and physiological features are unique to protozoan cells.
4. Some protozoa can secrete a resistant covering and go into a resting stage (encystation) called a cyst. Cysts protect the organism against adverse environments, function as a site for nuclear reorganization, and serve as a means of transmission in parasitic species.
5. Protozoa move by one of three major types of locomotory organelles: pseudopodia, flagella, or cilia. Some have no means of locomotion.
6. Most protozoa reproduce asexually, some use sexual reproduction, and some employ both methods.
7. There are seven protozoan phyla. The phylum *Sarcomastigophora* is characterized by protists that have a single type of nucleus, possess flagella (subphylum *Mastigophora*), pseudopodia (subphylum *Sarcodina*), or both types of locomotory organelles.
8. The subphylum *Sarcodina* consists of the amoeboid protists. They are found throughout the world in both fresh and salt water and in the soil. Some species are parasitic.
9. The phylum *Labyrinthomorpha* contains protists that have spindle-shaped or spherical nonamoeboid vegetative cells. Most members are marine and either saprozoic or parasitic on algae.
10. The phylum *Apicomplexa* consists of sporozoan protists that possess an apical complex, a unique arrangement of fibrils, microtubules, vacuoles, and other organelles at one end of the cell. Representative members include the *Plasmodium* parasites, which cause malaria; *Toxoplasma,* which causes toxoplasmosis; and *Eimiria,* the agent of coccidiosis.
11. The phylum *Microspora* consists of very small protists that are intracellular parasites of every major animal group. They are transmitted from one host to the next as a spore, the form from which the group obtains its name.
12. The phylum *Ascetospora* contains protists that produce spores lacking polar capsules. These protists are primarily parasitic in mollusks.
13. The phylum *Myxozoa* consists entirely of parasitic species, usually found in fish. The spore is characterized by one to six polar filaments.
14. The phylum *Ciliophora* comprises a group of protists that have cilia and two types of nuclei. Conjugation in the *Ciliophora* is a form of sexual reproduction that involves exchange of micronuclear material.

Key Terms

amoeboid movement *553*
apical complex *556*
apicomplexan *556*
binary fission *550*
conjugant *550*
conjugation *550*
conoid *556*
contractile vacuole *550*
cyst *550*
cytoproct *557*
cytostome *550*
ectoplasm *550*
encystation *550*
endoplasm *550*
excystation *550*
food chain *549*
food web *549*
holozoic nutrition *550*
hydrogenosomes *550*
kinetoplast *553*
macronucleus *550*
micronucleus *550*
oocyst *556*
pebrine *556*
pellicle *550*
phagocytic vacuoles *550*
plankton *549*
proliferative kidney disease *557*
protozoa *549*
protozoology *549*
pseudopodia *550*
rhoptry *556*
saprozoic nutrition *550*
schizogony *556*
secretory vacuole *550*
test *553*
trophozoite *549*
trypanosome *553*
zooflagellate *553*

Questions for Thought and Review

1. What is the economic impact or human relevance of the protozoa?
2. What criteria are now used in the classification of protozoa?
3. Seven protozoan phyla have been discussed. What are their distinguishing characteristics?
4. What are some typical organelles found in the protozoa?
5. What advantage is there to those protozoa that are capable of forming a cyst?
6. How do protozoa move? Reproduce?
7. If the diversity within the seven protozoan phyla is considered, which one shows the greatest evolutionary advancement? Defend your answer.
8. The protozoa are said to be grouped together for a negative reason. What does this mean?
9. Describe how DNA is distributed to daughter cells when the ciliate *Paramecium* divides. Include a discussion of both conjugation and binary fission.
10. How does a protozoan cyst differ from a bacterial endospore?

Additional Reading

Adam, R. 1991. The biology of *Giardia* spp. *Microbiol. Rev.* 55(4):706–32.

Bamforth, S. S. 1980. Terrestrial protozoa. *J. Protozool.* 27:33–36.

Band, R. D., ed. 1983. Symposium—the biology of small amoebae. *J. Protozool.* 30:192–214.

Clew, H. R.; Saha, A. K.; Siddhartha, D.; and Remaley, A. T. 1988. Biochemistry of *Leishmania* species. *Microbiol. Rev.* 52(4):412–32.

Corliss, J. O. 1979. *The ciliated protozoa: Characterization, classification and guide to the literature,* 2d ed. New York: Pergamon Press.

Corliss, J. O. 1981. What are the taxonomic and evolutionary relationships of the protozoa to the Protista? *BioSystems* 14:445–59.

Curtis, H. 1968. *The marvelous animals: An introduction to the protozoa.* Garden City, N.Y.: Natural History Press.

Dunelson, J. E., and Turner, M. J. 1985. How the trypanosome changes its coat. *Sci. Am.* 252(2): 44–51.

Edds, K. T. 1981. Cytoplasmic streaming in a heliozoan. *BioSystems* 14:371–76.

Fenchel, T. 1987. *Ecology of protozoa.* New York: Springer-Verlag.

Grell, K. G. 1973. *Protozoology.* Heidelberg: Springer-Verlag.

Gutteridge, W. E., and Coombs, G. H. 1977. *Biochemistry of parasitic protozoa.* Baltimore, Md.: University Park Press.

Jahn, T. L.; Bovee, E. C.; and Jahn, F. F. 1979. *How to know the protozoa.* Dubuque, Ia.: Wm. C. Brown.

Krier, J. P. 1978. *Parasitic protozoa.* New York: Academic Press.

Kudo, R. R. 1966. *Protozoology,* 5th ed. Springfield, Ill.: Charles C. Thomas.

Laybourn-Parry, J. 1984. *A functional biology of free-living protozoa.* Berkeley: University of California Press.

Lee, J. J.; Hunter, S. H.; and Bovee, E. C. 1985. *An illustrated guide to the protozoa.* Lawrence, Kan.: Allen Press, Society of Protozoologists.

Levandowsky, M., and Hunter, S. E., eds. 1979. *Biochemistry and physiology of protozoa.* New York: Academic Press.

Levine, N. D., ed. 1980. A newly revised classification of the protozoa. *J. Protozool.* 27:37–58.

Margulis, L.; Corliss, J. O.; and Melkonian, M. 1990. *Handbook of Protoctista.* Boston: Jones and Bartlett.

Muller, M. 1975. Biochemistry of protozoan microbodies: Peroxisomes, α-glycerophosphate oxidase bodies, hydrogenosomes. *Ann. Rev. Microbiol.* 29:467–83.

Rudzinska, M. A. 1973. Do suctoria really feed by suction? *BioScience* 23(2):87–94.

Ryley, J. F. 1980. Recent developments in coccidian biology: Where do we go from here? *Parasitology* 80:189–209.

Sleigh, M. 1978. *The biology of protozoa.* Amsterdam and New York: Elsevier Biomedical Press.

Wichterman, R. 1986. *The biology of Paramecium,* 2d ed. New York: Plenum Press.

Wolfe, M. 1992. Giardiasis. *Clin. Microbiol. Rev.* 5(1):93–100.

Part Seven
The Nature of Symbiotic Associations

Lichens are one of the best known and most successful symbiotic associations. These colorful and attractive British soldier lichens (*Cladonia cristatella*) are an excellent example of a fruticose lichen.

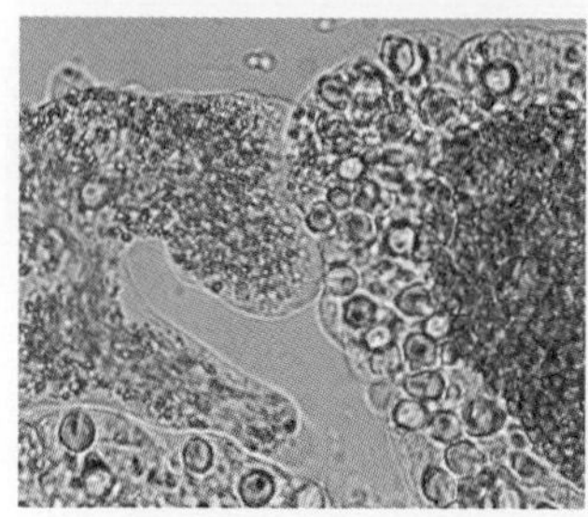

Chapter 29
Symbiotic Associations: Commensalism, Mutualism, and Normal Microbiota of the Human Body

Symbiosis, particularly parasitism, is frequently regarded in distasteful terms by the hygiene-conscious citizen. We have a tendency to think of it as a peculiar and abnormal association of some lower organism with a higher one. There is an element of snobbishness in such a view, which must be quickly abandoned when a discerning look is taken of the living world. In nature there is probably no such thing as a symbiote-free organism. The phenomenon of symbiosis is quite as common as life itself.

—Clark P. Reed

Outline

Concepts

1. Symbiosis is the living together in close association of two or more dissimilar organisms.
2. Commensalism is a relationship in which one organism benefits while the other is neither harmed nor helped.
3. Mutualism is the living together of two or more organisms in an association in which both members benefit.
4. Germfree animals or those that live in association with a few known microorganisms are gnotobiotic.
5. Most microorganisms associated with the human body are bacteria; they normally prefer to colonize specific sites.

Symbiosis is the living together in close association of two or more dissimilar organisms. This chapter focuses on these associations, emphasizing two of the three types of symbiosis: commensalism and mutualism. Parasitism, the third type of symbiosis, is covered in chapter 30. Within the present chapter, examples are given that illustrate each specific symbiotic relationship. However, it is also possible experimentally to have no microorganisms on or in an animal, a condition termed germfree or gnotobiotic life. This special case is discussed. The chapter closes with an overview of the normal microorganisms associated with the human body.

Many microorganisms live much of their lives in a special ecological relationship: an important part of their environment is a member of another species. Any microorganism that spends a portion or all of its life associated with another organism of a different species is called a **symbiont** (or **symbiote**), and the relationship is designated as **symbiosis** (Greek *sym,* together, and *bios,* life).

Types of Symbiosis, Functions, and Examples

There are three types of symbiotic relationships: commensalism, mutualism, and parasitism. Within each category, the association may be either ectosymbiotic or endosymbiotic. In **ectosymbiosis** one organism remains outside the other. In **endosymbiosis** one organism is present within the other.

Commensalism

Commensalism (Latin *com,* together, and *mensa,* table) is a relationship in which one organism, the **commensal,** benefits while the other, the host, is neither harmed nor helped. Often, both the host and the commensal "eat at the same table." The spatial proximity of the two partners permits the commensal to feed on substances captured or ingested by the host. Typically, the commensal also obtains shelter by living either on or in the host. The commensal is not directly dependent on the host metabolically and causes it no particular harm. When the commensal is separated from its host experimentally, it can survive without being provided some factor or factors of host origin. For example, the bacterium *Escherichia coli* lives in the human colon and benefits from the nutrients, warmth, and shelter found there, but usually causes no disease or discomfort. Later in this chapter, the normal bacteria of the human body are considered, and examples of how they may benefit humans in various commensal relationships are presented.

Mutualism

Mutualism (Latin *mutuus,* borrowed or reciprocal) defines the relationship in which some reciprocal benefit accrues to both partners. In this relationship, the **mutualist** and the host are metabolically dependent on each other. Several examples of mutualism are presented next.

The Protozoan-Termite Relationship

A classic example of mutualism is the flagellated protozoa that live in the gut of termites and wood roaches. These flagellates exist entirely on a diet of carbohydrates, acquired as the cellulose wood chips ingested by their host (figure 29.1). The protozoa engulf wood particles, digest the cellulose, and metabolize it to acetate and other products. Termites absorb and oxidize the acetate released by their flagellates. Because the host is incapable of synthesizing cellulases, it is dependent on the mutualistic protozoa for its existence.

(a)

(b)

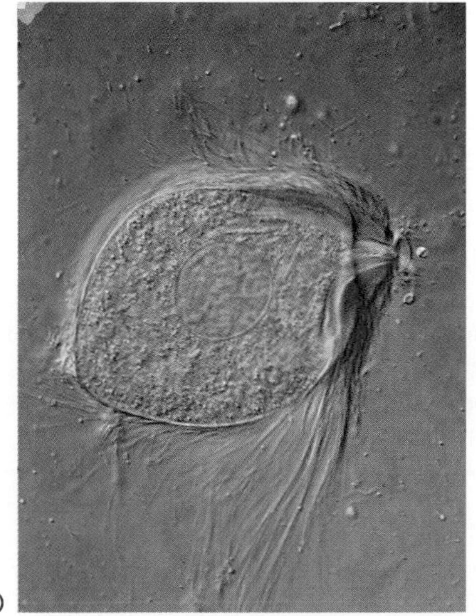

Figure 29.1 **Mutualism.** Light micrographs of (*a*) worker termite of the genus *Reticulitermes* eating wood (×10), and (*b*) *Trichonympha,* a flagellated protozoan from the termite's gut (×135). Notice the many flagella over most of its length. The ability of *Trichonympha* to break down cellulose enables termites to use wood as a food source.

Figure 29.2 **Lichens.** (*a*) Crustose (encrusting) lichens growing on a granite post. (*b*) Foliose (leafy) lichens growing on a rock wall. (*c*) Fruticose (shrubby) lichens commonly called reindeer moss. (*d*) A variety of crustose and fruticose rock lichens, several brightly colored. (*e*) British soldier lichens (fruticose).

This mutualistic relationship can be readily tested in the laboratory if wood roaches are placed in a bell jar containing wood and a high concentration of O_2. Because O_2 is toxic to the flagellates, they die. The wood roaches are unaffected by the high O_2 concentration and continue to ingest wood, but they soon die of starvation due to a lack of cellulases.

1. Define commensalism and mutualism.
2. Give one example of commensalism and mutualism.
3. What is the specific biochemical relationship between termites and their protozoan partners?
4. What would happen to a termite if its gut protozoa were destroyed by a specific antibiotic?

Lichens

Another excellent example of mutualism is provided by lichens. **Lichens** are the association between ascomycetes (the fungus) and certain genera of either green algae or cyanobacteria. In a lichen, the fungal partner is termed the **mycobiont** and the algal or cyanobacterial partner, the **phycobiont.** The remarkable aspect of this mutualistic association is that its morphology and metabolic relationships are so constant that lichens are assigned generic and species names. The characteristic morphology of a given lichen is a property of the mutualistic association and is not exhibited by either symbiont individually. There are three morphological types of lichens: **crustose** lichens are compact and appressed to a substratum such as a rock; **foliose** lichens have a leaflike appearance; and **fruticose** lichens have a shrubby shape. Examples of various lichen types are shown in figure 29.2.

Ascomycetes (pp. 525–26).

Because the phycobiont functions as a photosynthetic organism—dependent only on light, air, and certain mineral nutrients—the fungus can get its organic carbon directly from the alga or cyanobacterium. The fungus often obtains nutrients from its partner by haustoria (projections of fungal hyphae) that penetrate the phycobiont cell wall. It also uses the O_2 produced during phycobiont photosynthesis in carrying out respiration. In turn, the fungus protects the phycobiont

Box 29.1

Lichens and Air Pollution

Lichens are bioindicators of air quality because they are extremely sensitive to two common atmospheric pollutants—ozone (O_3) and sulfur dioxide (SO_2). This sensitivity arises from the lichen's great ability to absorb and concentrate substances dissolved in rain and dew. Because they have no means of excreting absorbed elements, lichens are particularly sensitive to toxic compounds. Lichen distribution thus is inversely related to the amount of SO_2, O_3, and toxic metals in the atmosphere. These pollutants are easily absorbed by lichens and destroy both chlorophylls *a* and *b*. This decreases photosynthesis and disrupts the metabolic balance between the fungus and the alga or cyanobacterium. Eventually this leads to the destruction of the lichen.

Lichens vary in their response to pollutants. The best bioindicators are the moderately sensitive species, such as *Hypogymnia enteromorpha.* This species also can be used as a means to assess pollution by radioactive substances that may occur following the crash of satellites, uranium mining operations, and nuclear fallout from weapon testing.

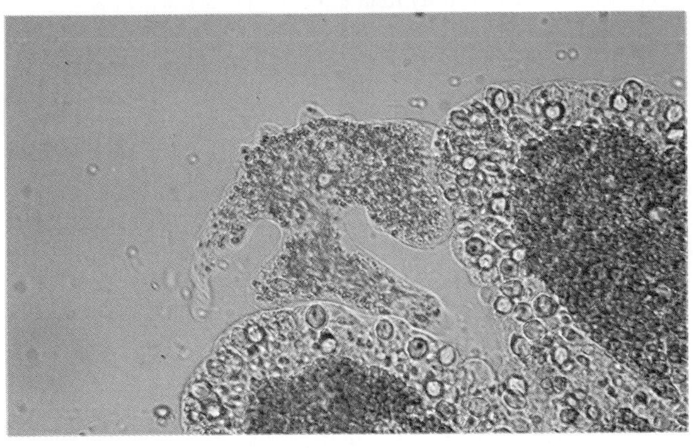
(a)

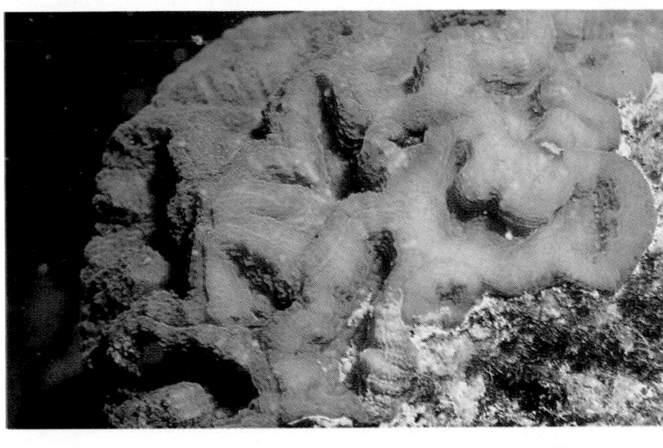
(b)

Figure 29.3 **Zooxanthellae.** (*a*) Zooxanthellae (green) within the tip of a hydra tentacle (×150). (*b*) The green color of this rose coral (*Manilina*) is due to the abundant zooxanthellae within its tissues.

from excess light intensities, supplies water and minerals to it, and provides a firm substratum within which the phycobiont can grow protected from environmental stress (Box 29.1).

Zooxanthellae

Many marine invertebrates (sponges, jellyfish, sea anemones, corals, ciliates) harbor endosymbiotic, spherical algal cells called **zooxanthellae** within their tissue (figure 29.3*a*). These algae may be from the *Chrysophyta, Euglenophyta, Pyrrhophyta, Chlorophyta,* and *Rhodophyta (see tables 27.1 and 27.2).* Because the degree of host dependency on the mutualistic alga is somewhat variable, only one well-known example is presented.

The hermatypic (reef-building) corals (figure 29.3*b*) obtain the bulk of their energy requirements from their zooxanthellae and are unable to use zooplankton from the water. Pigments produced by the coral protect the algae from the harmful effects of ultraviolet radiation. Clearly the zooxanthellae also benefit the coral because the calcification rate is at least 10 times greater in the light than in the dark. Hermatypic corals lacking zooxanthellae have a very low rate of calcification. Based on the stable carbon isotopic composition, it has been determined that most of the organic carbon in the tissues of the hermatypic corals has come from the zooxanthellae. Because of this coral-algal mutualistic relationship—capturing, conserving, and cycling nutrients and energy—coral reefs are among the most productive and successful of known ecosystems.

The Elysia-Codium *Relationship*

Some marine mollusks, notably *Elysia,* engage in what can be termed chloroplast symbiosis. *Elysia* eats a seaweed (*Codium*) and the mollusk's gut cells phagocytose intact chloroplasts from the partially digested algae. In some unknown way, the chloroplasts avoid digestion. Eventually, the chloroplasts reside in the host's cells, photosynthesizing at rates comparable to those attained in the intact seaweed. Photosynthate moves from the chloroplasts to the mollusk and is used, for instance, in the

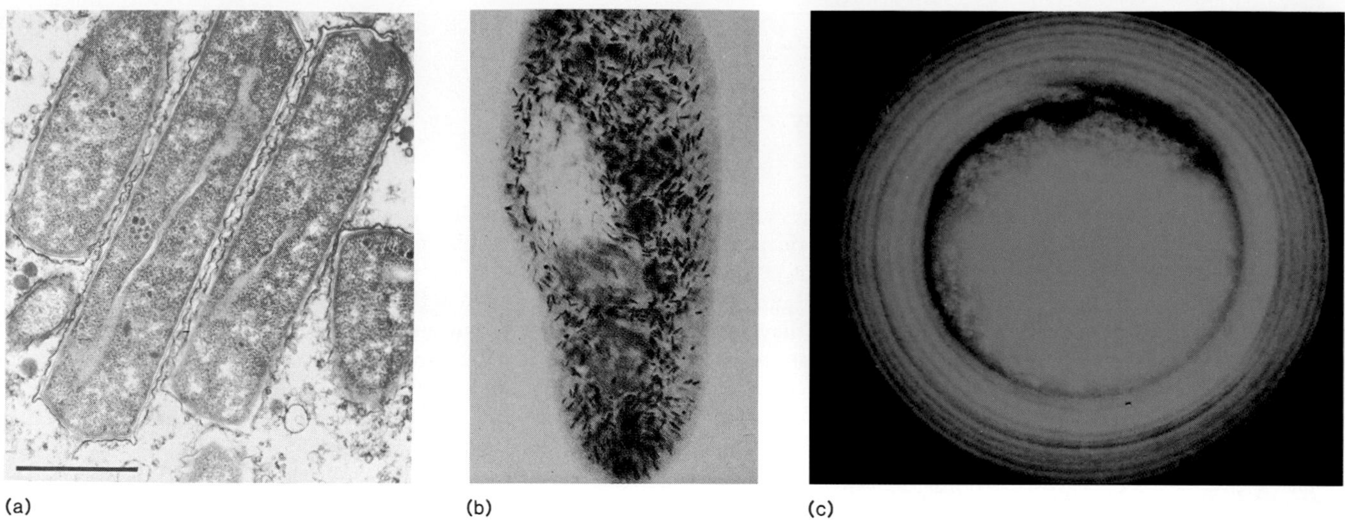

Figure 29.4 **Protozoan Endosymbionts.** (*a*) Several large bacteria found within *Pelomyxa palustris*. Longitudinal section. Bar = 1 μm. (*b*) *Paramecium tetraurelia* bearing kappa endosymbionts shown here as numerous black rods within the cytoplasm. (*c*) Intact R body from *Caedibacter varicaedens* of *Paramecium biaurelia*.

production of mucus. The plant organelles can photosynthesize for three months in this symbiotic relationship, but can neither divide nor synthesize many important plastid components like chlorophyll, glycolipids, membrane proteins, and nucleic acids.

Endosymbionts of Protozoa

Intracellular mutualistic symbioses between bacteria and protozoa are very common. Because these bacteria live within the host they are called **endosymbionts.** Endosymbionts have been described for all groups of protozoa. Two well-known examples are discussed in this section.

Two types of endosymbionts have been found in the giant amoeba *Pelomyxa palustris:* a small bacterium found throughout the cytoplasm, and a large, coccoid or rod-shaped bacterium (figure 29.4*a*). It has been suggested that these endosymbionts function as mitochondria because this organelle is absent in *P. palustris*.

One of the most fascinating protozoan endosymbionts is the **kappa** endosymbiont within *Paramecium aurelia* and other species (figure 29.4*b*). Starting in the 1940s, Tracy M. Sonneborn and coworkers began their classic genetic experiments on the *P. aurelia* complex. They found that some strains of paramecia contained an invisible cytoplasmic heredity unit that they called kappa. In 1974, John Preer and his associates discovered that kappa was an obligate bacterial endosymbiont. It was also found that most strains of *P. aurelia* fall into one of two classes: killers and sensitives.

The killers contain kappa endosymbionts that give them the ability to produce and release toxins lethal for susceptible protozoan strains called sensitives. If the toxin acts only during protozoan conjugation, the protozoa are called mate killers. Furthermore, the endosymbionts also confer upon the host paramecia resistance to the toxins. Obviously, the killer has a selective advantage over the sensitives.

It has been shown that kappa (1) has chemical and morphological properties of a small bacterium, (2) is susceptible to certain antibiotics, (3) contains DNA, and (4) requires the presence of a specific K-gene for reproduction within its host. Some kappa cells contain **R bodies** (figure 29.4*c*) that are refractile under phase-contrast microscopy and are called brights; those that lack R bodies are called nonbrights.

R bodies are unusual bacterial inclusion bodies. They are highly insoluble protein ribbons, typically seen coiled into cylindrical structures within cells. It has been noted that R bodies unwind under certain conditions and are associated with toxicity. Until recently they have been known to occur only in certain species of bacteria that are obligate endosymbionts of paramecia. However, recently, several free-living bacterial species that have the ability to produce R bodies have also been reported.

1. What are three morphological forms of a lichen?
2. Describe the mutualistic relationship between zooxanthellae and corals.
3. Give an example of chloroplast symbiosis.
4. What is the difference between killer and sensitive strains of *P. aurelia?*
5. What is an R body?

The Rumen Ectosymbiosis

Ruminants are a group of herbivorous animals that have a stomach divided into four compartments and chew a cud consisting of regurgitated, partially digested food. Examples include cattle, deer, camels, sheep, goats, and giraffes. This feeding method has evolved in animals that need to eat large amounts of food quickly, chewing being done later at a more comfortable or safer location. More importantly, by using mi-

Box 29.2

Modifying the Rumen Microbiota to Benefit Humans

Cattle raisers can often modify the diet of their animals for specific purposes. For example, if beef cattle are given the antibiotic monensin, most of the normal rumen protozoa that produce hydrogen and CO_2 are killed. Thus, the methane-producing bacteria (e.g., *Methanobrevibacter ruminantium*) do not have hydrogen and CO_2 from which to produce methane. This lack, in turn, increases the production of propionate by the starch decomposers (*Succinimonas amylolytica*) and decreases methane production. Because more propionate is produced and not wasted as CO_2 and hydrogen, more nutrients are available to the cattle; therefore, they eat less food than the nonmonensin-treated animals but gain weight at about the same rate. Increased propionate production stimulates gluconeogenesis, which leads to an increase in animal protein.

croorganisms to degrade the thick cellulose walls of grass and other vegetation, ruminants can digest vast amounts of otherwise unavailable forage. Because ruminants cannot synthesize cellulases, they have evolved a symbiotic relationship with microorganisms that produce these enzymes. Cellulases hydrolyze the β (1→4) linkages between successive D-glucose residues of cellulose and release glucose, which is then fermented to organic acids. These organic acids (acetate, butyrate, propionate) are the true energy source for the ruminant (Box 29.2; *see also figure 8.14*).

The upper portion of a ruminant's stomach expands to form a large pouch called the **rumen** (figure 29.5), and a smaller honeycomblike reticulum. The bottom portion of the stomach consists of an antechamber called the omasum, with the "true" stomach (abomasum) behind it.

The insoluble polysaccharides and cellulose eaten by the ruminant are mixed with saliva and enter the rumen. Within the rumen, food is churned in a constant rotary motion and eventually reduced to a pulpy mass. The rumen contains a large microbial population (10^{12} cells per milliliter) that partially digests and ferments the food it is processing. Later, the food moves into the reticulum. Mouthfuls are then regurgitated as a "cud," which is thoroughly chewed for the first time. The food is mixed with saliva, reswallowed, and reenters the rumen while another cud is passed up to the mouth. As this process continues, the partially digested plant material becomes more liquid in nature. The liquid then begins to flow out of the reticulum and into the lower parts of the stomach: first the omasum and then the abomasum (the true stomach). It is in the abomasum that the food encounters the host's normal digestive enzymes and the digestive process continues in the regular mammalian way.

Recently, anaerobic filamentous fungi of the genus *Neocallimastix* have been discovered in the normal microbial community of the rumen. These fungi appear to participate in the microbial attack on the lignin-cellulose complexes that make up a major part of the organic material in many adult plants.

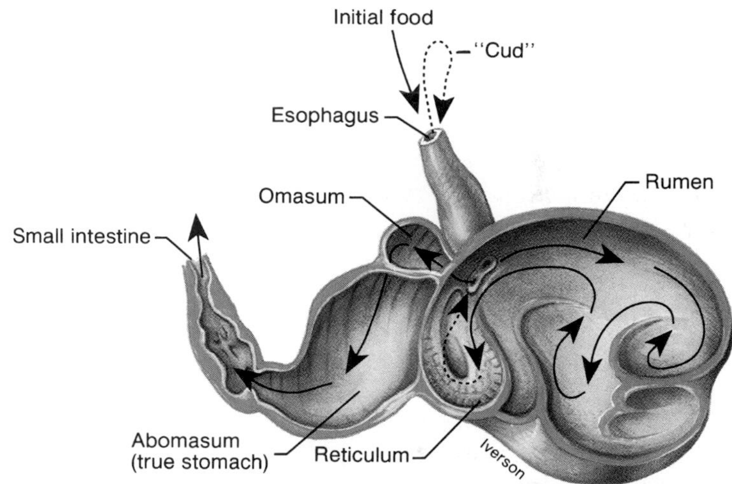

Figure 29.5 **Ruminant Stomach.** Stomach compartments of a cow. Arrows indicate direction of food movement.

Food entering the rumen is quickly attacked by the cellulolytic anaerobic bacteria, fungi, and protozoa. Microorganisms break down the plant material, as illustrated in figure 29.6. Because the oxidation-reduction potential in the rumen is −30mV, all indigenous microorganisms engage in anaerobic metabolism. Specifically, the eubacteria ferment carbohydrates to fatty acids, carbon dioxide, and hydrogen. The archaeobacteria (methanogens) produce methane (CH_4) from CO_2 and H_2.

The dietary carbohydrates degraded in the rumen include soluble sugars, starch, pectin, hemicellulose, and cellulose. The largest percentage of each carbohydrate is fermented to volatile fatty acids (acetic, propionic, butyric, formic, and valeric), CO_2, H_2, and methane. The fatty acids produced by the rumen organisms are absorbed into the bloodstream and are oxidized by the animal as its main source of energy. The CO_2 and methane, produced at a rate of 2 liters/minute, are released by **eructation** (Latin *eructare,* to belch), a continuous, scarcely audible reflex process similar to belching. Energy trapped as ATP during fermentation is used to maintain the

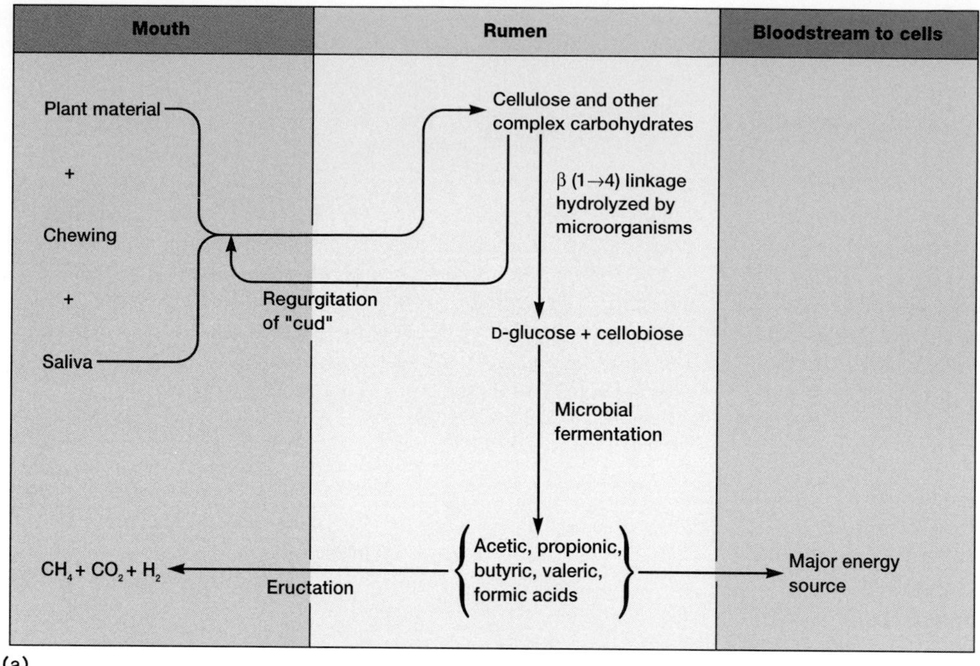

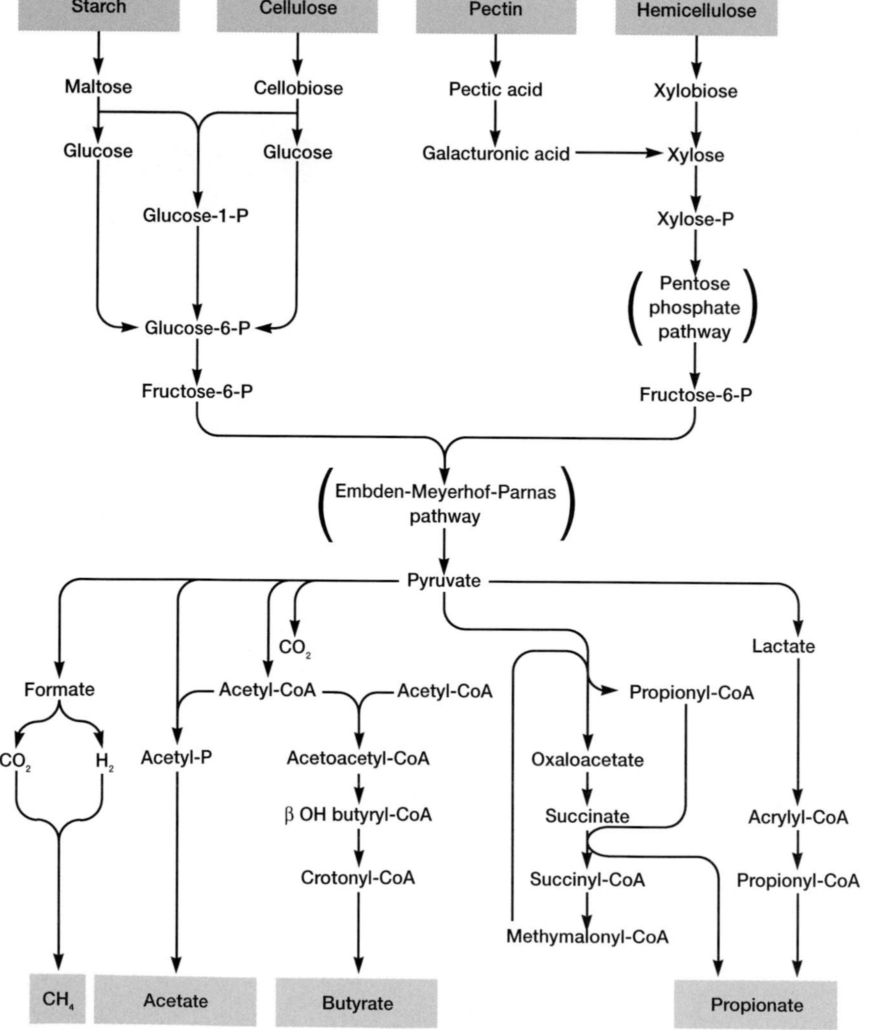

Figure 29.6 Rumen Biochemistry. (*a*) An overview of the biochemical-physiological processes occurring in various parts of a cow's digestive system. (*b*) More specific biochemical pathways involved in rumen fermentation of the major plant carbohydrates. The top boxes represent substrates and the bottom boxes some of the end products.

rumen microorganisms and to support their growth. These microorganisms, in turn, produce most of the vitamins needed by the ruminant. In the remaining two stomachs, the microorganisms, having performed their symbiotic task, are digested by stomach enzymes to yield amino acids, sugars, and other products. These also are absorbed and used by the ruminant.

1. What structural features of the rumen make it suitable for a herbivorous type of diet?
2. Why does a cow chew its cud?
3. What biochemical role do the rumen microorganisms play in this type of symbiosis?

Germfree (Gnotobiotic) Animals

The microorganisms normally associated with a particular tissue or structure are called the **microbiota,** the **indigenous microbial population,** the **microflora,** or the **microbial flora.** For consistency, the term microbiota is used in the following discussion.

The human fetus in utero (as is the case in most mammals) is free from bacteria and other microorganisms. Within hours after birth, it begins to acquire a normal microbiota, which stabilizes during the first week or two of life. From then on, enormous numbers of microorganisms are associated with the human body. Under normal conditions, the human lives in symbiosis with billions of metabolizing microorganisms.

In 1897, Louis Pasteur suggested that animals could not live in the absence of microorganisms. Attempts between 1899 and 1908 to grow germfree chickens had limited success because the birds died within a month. Thus, it was believed that intestinal bacteria were essential for the adequate nutrition and health of the chickens. It was not until 1912 that germfree chickens were shown to be as healthy as normal birds when they were fed an adequate diet.

Since then, germfree animals have become commonplace in research laboratories. Germfree animals or those living in association with one or more known microorganisms are called **gnotobiotic** (Greek *gnotos,* known, and *biota,* the fauna and flora of a region). A gnotobiotic system is one in which the composition of the microbiota is known (figure 29.7).

Establishing a germfree colony of rats, mice, hamsters, rabbits, guinea pigs, or monkeys begins with cesarean sections on pregnant females. The operation is performed under aseptic conditions in a germfree isolator. The newborn animals are then transferred to other germfree isolators in which all entering air, water, and food is sterile. Once the gnotobiotic animals become established under these germfree conditions, normal mating among themselves maintains the germfree colony.

Establishment of a germfree colony is much easier with chickens and other birds than with mammals. Fertile eggs from the birds are first sterilized with a germicide and then placed in sterile, germfree isolators. When the chick hatches, it is sterile and able to feed by itself. To make sure that a colony

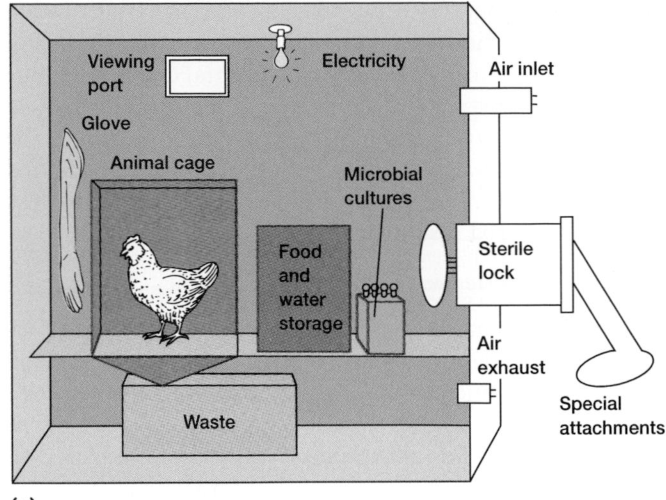

(a)

(b)

Figure 29.7 Raising Germfree Animals. (*a*) Schematic of a germfree isolator. (*b*) Germfree isolators for rearing colonies of small mammals.

is germfree, periodic bacteriologic examination of all exhaust air, waste material, and cages must be done. Microorganisms should not be found.

Germfree animals are not anatomically and physiologically normal. For example, they possess poorly developed lymphoid tissue, a thin intestinal wall, an enlarged cecum, and a low antibody titer (*see chapter 31*). They require high amounts of vitamin K and the B complexes. In normal animals, vitamin K is usually synthesized by *E. coli.* Germfree animals also have reduced cardiac output and lower metabolic rates.

Germfree animals are usually more susceptible to pathogens. With the normal protective microbiota absent, foreign and pathogenic microorganisms establish themselves very easily. The number of microorganisms necessary to infect a germfree animal and produce a diseased state is also much smaller. Conversely, germfree animals are almost completely resistant to the intestinal protozoan *Entamoeba histolytica* that causes amebic dysentery. This resistance results from the absence of the bacteria that *E. histolytica* uses as a food source. Germfree animals also do not show any dental caries or plaque formation (*see chapter 38*). However, if they are inoculated with cariogenic (caries or cavity causing) streptococci of the *Streptococcus mutans-Streptococcus gordonii* group and fed a high-sucrose diet, they will develop caries. (*S. gordonii* was formerly considered a subpopulation of *S. sanguis*.)

Entamoeba histolytica (pp. 790–92)

Germfree animals and gnotobiotic techniques provide good experimental systems to investigate the interactions of animals and specific microorganisms. The comparison of animals containing normal microbiota with germfree animals helps one better understand the many complex symbiotic associations that exist between the host and its microorganisms.

1. Define microbiota and gnotobiotic.
2. How would you establish a germfree colony of mice? Of chickens?
3. Compare a germfree mouse to a normal one with respect to overall health. What benefits does an animal gain from its microbiota?

Distribution of the Normal Microbiota of the Human Body

In the balance of this chapter, the distribution and occurrence of the normal microbiota of the human body are considered (figure 29.8). An overview of the microorganisms native to different regions of the body and an introduction to the types one can expect to find on culture reports is presented (*see figure 34.3*). Because bacteria make up most of the normal microbiota, they are emphasized, whereas the fungi (mainly yeasts) and protozoa are treated cursorily.

There are four main reasons to acquire knowledge of normal human microbiota:

1. An understanding of the different microorganisms at specific locations provides greater insight to the possible infections that might result from injury to these body sites.
2. A knowledge of the native organisms in the infected part of the body gives the physician-investigator perspective on the possible source and significance of microorganisms isolated from the infection site.
3. A knowledge of the indigenous microbiota helps the physician-investigator understand the causes and consequences of overgrowth by microorganisms normally absent at a specific body site.
4. An increased awareness of the role that these indigenous microbiota play in stimulating the host immune response can be gained. This awareness is important because the immune system provides protection against potential pathogens.

Skin

Commensal microorganisms living on or in the skin can be either resident (normal) or transient microbiota. Resident organisms normally grow on or in the skin. Their presence becomes fixed in well-defined distribution patterns. Those microorganisms that are temporarily present are transients. Transients usually do not become firmly entrenched; they are unable to multiply and normally die in a few hours.

The anatomy and physiology of the skin vary from one part of the body to another, and resident microbiota reflect these variations. The skin surface or epidermis is not a favorable environment for colonization by microorganisms. Several factors are responsible for this hostile microenvironment. First, the skin is subject to periodic drying. Lack of moisture drives many resident microbiota into a dormant state. However, in certain parts of the body (scalp, ears, axillary areas, genitourinary and anal regions, perineum, palms), moisture is sufficiently high to support a resident microbiota. Second, the skin has a slightly acidic pH due to the organic acids produced by normal staphylococci and secretions from skin oil and sweat glands. The acidic pH (4 to 6) discourages colonization by many microorganisms. Third, sweat contains a high concentration of sodium chloride. This makes the skin surface hyperosmotic and osmotically stresses most microorganisms. Finally, certain inhibitory substances (bactericidal and/or bacteriostatic) on the skin help control colonization, overgrowth, and infection from resident microorganisms. For example, the sweat glands release lysozyme, which lyses *Staphylococcus epidermidis* and other gram-positive bacteria. The oil glands secrete complex lipids that may be partially degraded by the enzymes from certain gram-positive bacteria (*Propionibacterium acnes*). These bacteria can change the secreted lipids to unsaturated fatty acids such as oleic acid that have strong antimicrobial activity against gram-negative bacteria and some fungi. Some of these fatty acids are volatile and may be associated with a strong odor. Therefore, many body deodorants contain antibacterial substances that act selectively against gram-positive bacteria to reduce the production of aromatic unsaturated fatty acids and body odor. However, deodorants can shift the microbiota to predominately gram-negative bacteria and precipitate subsequent infections.

Most skin bacteria are found on the superficial squamous epithelium, colonizing dead cells, or closely associated with the oil and sweat glands. Excretions from these glands provide the water, amino acids, urea, electrolytes, and specific fatty acids that serve as nutrients primarily for *Staphylococcus epidermidis* and aerobic corynebacteria. Gram-negative bacteria generally are found in the moister regions. The yeasts *Pityrosporum ovale* and *P. orbiculare* normally occur on the scalp. Some dermatophytic fungi may colonize the skin and produce athlete's foot and ringworm.

Athlete's foot disease (p. 784).

The most prevalent bacterium in skin glands is the gram-positive, anaerobic, lipophilic rod *Propionibacterium acnes*.

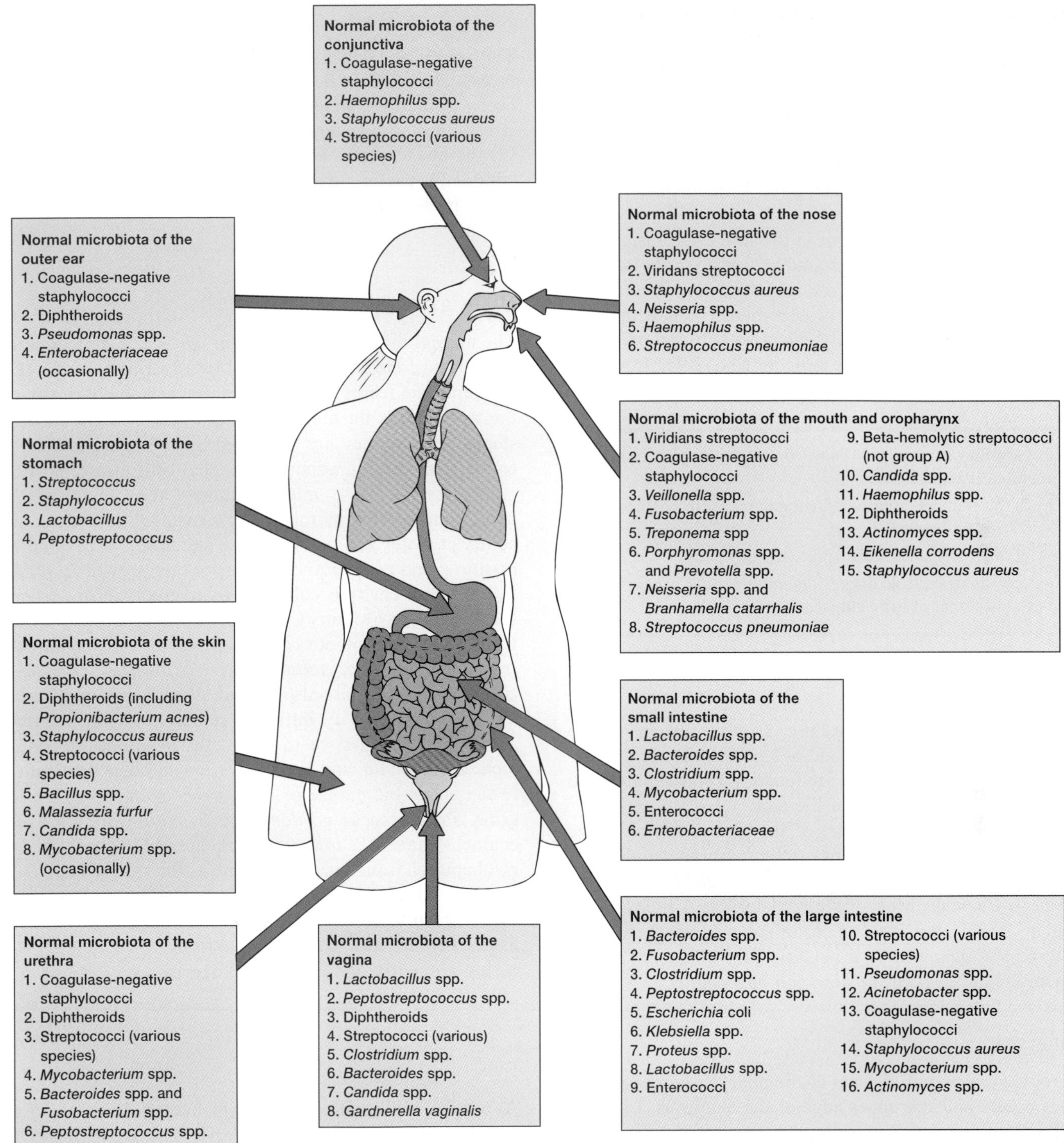

Figure 29.8 **Normal Microbiota of a Human.** A compilation of microorganisms which constitute normal microbiota encountered in various body sites.

This bacterium is usually harmless; however, it has been associated with the skin disease **acne vulgaris.** Acne commonly occurs during adolescence when the endocrine system is very active. Although the male hormone testosterone appears to be the most potent stimulator of oil glands, adrenal and ovarian hormones also can stimulate sebaceous secretions. Hormonal activity stimulates an overproduction of sebum, fluid secreted by the oil glands. A large volume of sebum accumulates within the glands and provides an ideal microenvironment for *P. acnes.* In some individuals, this accumulation triggers an inflammatory response that causes redness and swelling of the gland's duct and produces a **comedo** (pl. comedones), a plug of sebum and keratin in the duct. Inflammatory lesions (papules, pustules, nodules) commonly called "blackheads" or "pimples" can result. *P. acnes* has been identified as the organism that produces lipases which break the sebum triglycerides into free fatty acids. Free fatty acids are especially irritating because they can enter the dermis and promote inflammation. Because *P. acnes* is extremely sensitive to tetracycline, this antibiotic may aid acne sufferers. Accutane, a synthetic form of vitamin A, is also used.

Some pathogens found on or in the skin are transient residents that colonize the area around orifices. *Staphylococcus aureus* is the best example. It resides in the nostrils and perianal region but survives poorly elsewhere. In like manner, *Clostridium perfringens* usually colonizes only the perineum and thighs, especially in those patients who suffer from diabetes.

1. What are four reasons why knowledge of the normal human microbiota is important?
2. Why is the skin not usually a favorable microenvironment for colonization by bacteria?
3. How do microorganisms contribute to body odor?
4. What physiological role does *Propionibacterium acnes* play in the establishment of acne vulgaris?

Nose and Nasopharynx

The normal microbiota of the nose is found just inside the nares. *Staphylococcus aureus* and *S. epidermidis* are the predominant bacteria present, and are found in approximately the same numbers as on the skin of the face.

The nasopharynx, that part of the pharynx lying above the level of the soft palate, may contain small numbers of *Streptococcus pneumoniae, Neisseria meningitidis,* and *Haemophilus influenzae.* Most of these bacteria lack the capsules present in strains causing clinical infection. Diphtheroids, a large group of gram-positive bacteria (*see chapter 22*), are commonly found in both the nose and nasopharynx.

Oropharynx

The oropharynx is that division of the pharynx lying between the soft palate and the upper edge of the epiglottis. Like the nose, large numbers of *Staphylococcus aureus* and *S. epidermidis* inhabit this region. The most important bacteria found in the oropharynx are the various alpha-hemolytic streptococci (*S. oralis, S. milleri, S. gordonii, S. salivarius*); large numbers of diphtheroids; *Branhamella catarrhalis;* and small gram-negative cocci related to *Neisseria meningitidis.* It should be noted that the palatine and pharyngeal tonsils harbor a similar microbiota, except that within the tonsillar crypts, there is an increase in *Micrococcus* and the anaerobes *Porphyromonas, Prevotella,* and *Fusobacterium.* (*Porphyromonas* spp. and *Prevotella* spp. were formerly classified as *Bacteroides.*)

Respiratory Tract

The upper and lower respiratory tracts (trachea, bronchi, bronchioles, alveoli) do not have a normal microbiota. This is because microorganisms are removed by (1) the continuous stream of mucus generated by the ciliated epithelial cells and (2) the phagocytotic action of the alveolar macrophages. In addition, a bactericidal effect is exerted by the enzyme lysozyme, present in nasal mucus.

Mucociliary blanket (p. 592).

Oral Cavity (Mouth)

The normal microbiota of the oral cavity contains organisms able to resist mechanical removal by adhering to surfaces like the gums and teeth. Those that cannot attach are removed by the mechanical flushing of the oral cavity contents to the stomach where they are destroyed by hydrochloric acid. The continuous desquamation of epithelial cells also removes microorganisms. Those microorganisms able to colonize the mouth find a very comfortable environment due to the availability of water and nutrients, the suitability of pH and temperature, and the presence of many other growth factors.

The oral cavity is colonized by microorganisms from the surrounding environment within hours after a human is born. Initially, the microbiota consists mostly of the genera *Streptococcus, Neisseria, Actinomyces, Veillonella,* and *Lactobacillus.* Some yeasts are also present. Most microorganisms that invade the oral cavity initially are aerobes and obligate anaerobes. As the first teeth erupt, the anaerobes (*Porphyromonas, Prevotella,* and *Fusobacterium*) become dominant due to the anaerobic nature of the gingival groove. As the teeth grow, *Streptococcus gordonii* and *S. mutans* attach to their enamel surfaces; *S. salivarius* attaches to the buccal and gingival epithelial surfaces and colonizes the saliva. These streptococci produce a glycocalyx and various adherence factors that enable them to attach to oral surfaces. The presence of these bacteria contributes to the eventual formation of dental plaque, caries, gingivitis, and periodontal disease.

Periodontal disease (pp. 776–78).

Eye

At birth and throughout human life, bacterial commensals are found on the conjunctiva of the eye. The predominant bacterium is *Staphylococcus epidermidis* followed by *S. aureus,* aerobic corynebacteria (diphtheroids), and *Streptococcus pneumoniae.* Cultures from the eyelids or conjunctiva also yield *Branhamella catarrhalis, Escherichia, Klebsiella, Proteus, Enterobacter, Neisseria,* and *Bacillus* species. Few anaerobic organisms are present.

External Ear

The basic microbiota of the external ear resemble that of the skin, with coagulase-negative staphylococci and *Corynebac-*

terium predominating. Less frequently found are *Bacillus, Micrococcus,* and *Neisseria* species. Gram-negative rods such as *Proteus, Escherichia,* and *Pseudomonas* are occasionally seen. Mycological studies show the following fungi to be normal microbiota: *Aspergillus, Alternaria, Penicillium, Candida,* and *Saccharomyces.*

Stomach

As noted earlier, many microorganisms are washed from the oral cavity into the stomach. Owing to the very acidic pH (2 to 3) of the gastric contents, most microorganisms are killed (Box 29.3). As a result, the stomach usually contains less than 10 viable bacteria per milliliter of gastric fluid. These are mainly *Streptococcus, Staphylococcus, Lactobacillus, Peptostreptococcus,* and yeasts such as *Candida* spp. Microorganisms may survive if they pass rapidly through the stomach or if the organisms ingested with food are particularly resistant to gastric pH (mycobacteria). Normally, the number of microorganisms increases after a meal but quickly falls as the acidic pH takes its toll. Changes in the gastric microbiota also occur if there is an increase in gastric pH following intestinal obstruction, which permits a reflux of alkaline duodenal secretions into the stomach. If the gastric pH increases, the microbiota of the stomach are likely to reflect that of the oropharynx and, in addition, contain both gram-negative aerobic and anaerobic bacteria.

Small Intestine

The small intestine is divided into three anatomical areas: the duodenum, jejunum, and ileum. The duodenum (the first 25 cm of the small intestine) contains few microorganisms because of the combined influence of the stomach's acidic juices and the inhibitory action of bile and pancreatic secretions. Of the bacteria present, gram-positive cocci and rods compose most of the microbiota. *Enterococcus faecalis,* lactobacilli, diphtheroids, and the yeast *Candida albicans* are occasionally found in the jejunum. In the distal portion of the small intestine (ileum), the microbiota begin to take on the characteristics of the colon microbiota. It is within the ileum that the pH becomes more alkaline. As a result, anaerobic gram-negative bacteria and members of the family *Enterobacteriaceae* become established.

Large Intestine (Colon)

The colon has the largest microbial population in the body. Microscopic counts of feces approach 10^{12} organisms per gram wet weight. Over 300 different species have been isolated from human feces. The colon can be viewed as a large fermentation vessel, and the microbiota consist primarily of anaerobic, gram-negative, nonsporing bacteria and gram-positive, spore-forming, and nonsporing rods. Not only are the vast majority of microorganisms anaerobic, but many different species are present in large numbers. Several studies have shown that the ratio of anaerobic to facultative anaerobic bacteria is approximately 300 to 1. Even the most abundant of the latter, *Escherichia coli,* is only about 0.1% of the total population.

Besides the many bacteria in the large intestine, the yeast *Candida albicans* and certain protozoa may occur as harmless commensals. *Trichomonas hominis, Entamoeba hartmanni, Endolimax nana,* and *Iodamoeba butschlii* are common inhabitants.

Protozoan diseases (pp. 790–98).

Various physiological processes move the microbiota through the colon so that an adult excretes about 3×10^{13} microorganisms daily. These processes include peristalsis and segmentation, desquamation of the surface epithelial cells to which microorganisms are attached, and continuous flow of mucus that carries adhering microorganisms with it. To maintain homeostasis of the microbiota, the body must continually replace those lost microorganisms. The bacterial population in the human colon usually doubles once or twice a day. Under normal conditions, the resident microbial community is self-regulating. Competition and mutualism between different microorganisms and between the microorganisms and their host serve to maintain a status quo. However, if the intestinal environment is disturbed, the normal flora may change greatly. Disruptive factors include stress, altitude changes, starvation, parasitic organisms, diarrhea, and use of antibiotics (Box 29.3). Finally, it should be emphasized that the actual proportions of the individual bacterial populations within the indigenous microbiota depend largely on a person's diet.

The initial residents of the colon of breast-fed infants are members of the gram-positive *Bifidobacterium* genus, because human milk contains a disaccharide amino sugar that *Bifidobacterium* species require as a growth factor. In formula-fed infants, *Lactobacillus* species (also gram positive) predominate because formula lacks the required growth factor. With the ingestion of solid food, these initial colonizers of the colon are eventually displaced by a typical gram-negative microbiota. Ultimately the composition of the adult's microbiota is established.

Genitourinary Tract

The upper genitourinary tract (kidneys, ureters, and urinary bladder) is usually free of microorganisms. In both the male and female, a few bacteria (*Staphylococcus epidermidis, Enterococcus faecalis,* and *Corynebacterium* spp.) are usually present in the distal portion of the urethra. *Neisseria* and some members of the *Enterobacteriaceae* are occasionally found.

In contrast, the adult female genital tract, because of its large surface area and mucous secretions, has a complex microbiota that constantly changes with the female's menstrual cycle. The major microorganisms are the acid-tolerant *Lactobacillus* spp., called **Döderlein's bacilli,** which ferment the glycogen produced by the vaginal epithelium, forming lactic acid. As a result, the pH of the vagina and cervical os is maintained between 4.4 and 4.6.

Box 29.3

Selective Antimicrobial Modulation

Since the early 1970s, the aggressive chemotherapy used in the treatment of patients suffering from leukemia (more recently, bone marrow transplants) has occasionally resulted in the development of a deficiency of granular white blood cells termed **granulocytopenia** (*granulocyte* and Greek *penia,* poverty). If this condition is prolonged, the normal defenses against bacteria provided by these cells is reduced or lost, and infections usually result. The most important contributors to such infections are *Pseudomonas aeruginosa,* the *Enterobacteriaceae, Staphylococcus aureus,* and *Candida albicans.* No anaerobic bacteria are involved. Most of these microorganisms are exogenous and acquired from the hospital environment. By using trimethoprim/sulfamethoxazole combined with nystatin, or gentamicin combined with nystatin, the aerobic microbiota of the colon can be reduced and the anaerobic microbiota increased. This shift to anaerobic microbiota inhibits the growth of aerobes resulting in colonization resistance. The approach to infection control by this method is called selective antimicrobial modulation (SAM) and has been shown in several studies to reduce significantly the incidence of infections in the compromised granulocytopenic patient.

1. What are the most common microorganisms found in the nose? The oropharynx? The nasopharynx? The tonsillar crypts? The lower respiratory tract? The oral cavity? The eye? The external ear? The stomach? The small intestine? The colon? The genitourinary tract?
2. Why is the colon considered a large fermentation vessel?
3. What physiological processes move the microbiota through the gastrointestinal tract?
4. How do the initial residents of breast-fed infants differ from bottle-fed infants?
5. Describe the microbiota of the upper and lower female genitourinary tract.

Summary

1. Symbiosis is a class of relationships in which organisms of different species closely interact with one another for much of their lives.
2. Commensalism occurs when two species live in a relationship in which one benefits while the other is neither harmed nor helped.
3. When two organisms interact so that both benefit, the relationship is termed mutualism. In the protozoan-termite relationship, the termites provide food and shelter for the protozoa. The latter digest cellulose. Lichens are the mutualistic association between ascomycetes (the fungus) and either green algae or cyanobacteria. The fungal partner is the mycobiont and the algal partner, the phycobiont. The fungus provides protection and some nutrients; the phycobiont provides photosynthetic products. Zooxanthellae are specific algal cells found within various marine invertebrates. The coral-algal mutualistic relationship is a good example. Chloroplast mutualism exists in the marine slug-seaweed (*Elysia-Codium*) relationship. Endosymbionts form mutualistic associations with some protozoa.
4. Ruminants are herbivorous animals that possess a special organ, the rumen, within which the digestion of cellulose and other plant polysaccharides occurs through special activity of anaerobic symbiotic bacteria, protozoa, and fungi.
5. Animals that are germfree or those living with one or more known microorganisms are termed gnotobiotic. Methods are available for rearing germfree colonies. Germfree animals and techniques provide good experimental systems with which the interactions of animals and specific species of microorganisms can be investigated.
6. Commensal microorganisms living on or in the skin can be characterized as either transients or residents. The most prominent bacterium found on the skin is *Staphylococcus epidermidis. Propionibacterium acnes* is associated with the skin glands, and *Staphylococcus aureus* is usually found around orifices.
7. *Staphylococcus aureus* and *S. epidermidis* are the predominant bacteria of the nose, whereas *Streptococcus pneumoniae, Neisseria meningitidis,* and *Haemophilus influenzae* predominate in the nasopharynx. *Staphylococcus aureus, S. epidermidis,* and various alpha-hemolytic streptococci are most common in the oropharynx.
8. The normal microbiota of the oral cavity is composed of those organisms able to resist mechanical removal.
9. The predominant bacterium found in the conjunctiva of the eye is *S. epidermidis.*
10. The microbiota of the external ear resembles that of the skin.
11. The stomach contains very few microorganisms due to its acidic pH.
12. The distal portion of the small intestine and the entire large intestine have the largest microbial populations in the body. Over 300 species have been identified, the vast majority of them anaerobic.
13. The upper genitourinary tract is usually free of microorganisms. In contrast, the adult female genital tract has a complex microbiota.

Key Terms

acne vulgaris *573*
comedo *573*
commensal *565*
commensalism *565*
crustose *566*
Döderlein's bacillus *575*
ectosymbiosis *565*
endosymbiont *568*
endosymbiosis *565*
eructation *569*
foliose *566*
fruticose *566*
gnotobiotic *571*
granulocytopenia *576*
kappa *568*
lichen *566*
microbiota (indigenous microbial population, microflora, or microbial flora) *571*
mutualism *565*
mutualist *565*
mycobiont *566*
phycobiont *566*
R body *568*
rumen *569*
ruminant *568*
symbiont or symbiote *565*
symbiosis *565*
zooxanthellae *567*

Questions for Thought and Review

1. Describe and give an example of mutualism and commensalism.
2. From a biochemical point of view, how do termites benefit the protozoa they harbor in their intestines? How do the protozoa benefit the termites?
3. Describe the symbiotic association between algae and fungi. What does the mycobiont contribute to this relationship?
4. Why is the coral-algal mutualistic relationship such a beneficial one?
5. What is unique about the *Elysia-Codium* symbiosis?
6. What is the difference between killer and sensitive strains of *Paramecium aurelia?*
7. List and describe some benefits of the rumen symbiosis with respect to both partners.
8. Suggest some experiments that might be useful in a germfree colony kept under gnotobiotic conditions. Give details.
9. What are some specific obstacles that must be overcome before a microorganism can colonize a body site?
10. Describe why the microenvironment of the skin is both favorable to some microorganisms and unfavorable to others. Explain your answer.

Additional Reading

Ahmadjian, V. 1963. The fungi of lichens. *Sci. Am.* 208:122–27.

Ahmadjian, V. 1986. *Symbiosis.* Hanover, N.H.: University Press of New England.

Baldwin, R. L. 1984. Digestion and metabolism of ruminants. *BioScience* 34(4):244–49.

Barnett, H. L., and Binder, F. L. 1973. The fungal-host parasite relationship. *Ann. Rev. Phytopathol.* 11:273.

Breznak, J. A. 1975. Intestinal microbiota of termites and other xylophagus insects. *Ann. Rev. Microbiol.* 36:323.

Childress, J. J.; Felbeck, H.; and Somero, G. N. 1987. Symbiosis in the deep sea. *Sci. Am.* 256(2):114–20.

Drasar, B. S., and Barrow, P. A. 1985. *Intestinal microbiology.* Washington, D.C.: American Society for Microbiology.

Gordon, H. A., and Pesti, L. 1971. The gnotobiotic animal as a tool in the study of host microbial relationship. *Bacteriol. Rev.* 35:390–429.

Hungate, R. E. 1975. The rumen microbial ecosystem. *Ann. Rev. Microbiol.* 29:39–47.

Lee, J. J., and Corliss, J. O. 1985. Symposium on "symbiosis" in protozoa. *J. Protozool.* 32(3):371–423.

Mackowiak, P. A. 1982. The normal microbial flora. *N. Engl. J. Med.* 307:83.

Margulis, L. 1981. *Symbiosis in cell evolution: Life and its environment on the early earth.* San Francisco: Freeman.

Margulis, L., and Fester, R., eds. 1991. *Symbiosis as a source of evolutionary innovation: Speciation and morphogenesis.* Cambridge, Mass.: MIT Press.

Marples, M. J. 1969. Life on the human skin. *Sci. Am.* 220:108–29.

Marsh, P., and Martin, M. 1984. *Oral microbiology,* 2d ed. Washington, D.C.: American Society for Microbiology.

Philips, A. W., and Smith, J. E. 1959. Germ-free animal techniques and their application. *Adv. Appl. Microbiol.* 1:141–74.

Pond, F. R.; Gibson, I.; Lalucat, J.; and Quackenbush, R. L. 1989. R-body producing bacteria. *Microbiol. Rev.* 53(1):25–67.

Savage, D. C. 1977. Microbial ecology of the gastrointestinal tract. *Ann. Rev. Microbiol.* 31:107–33.

Sonneborn, T. M. 1975. The *Paramecium aurelia* complex of fourteen sibling species. *Trans. Am. Micros. Soc.* 94:155–78.

Symbiosis: 29th symposium of the society for experimental biology 1975. New York: Cambridge University Press.

Taylor, D. L. 1973. Algal symbionts of invertebrates. *Ann. Rev. Microbiol.* 27:171.

Williams, A. G. 1986. Rumen holotrich ciliate protozoa. *Microbiol. Rev.* 50(1):25–49.

Wolin, M. J. 1979. The rumen fermentation: A model for microbial interactions in anaerobic ecosystems. In *Advance in microbial ecology,* ed. M. Alexander, vol. 3, 49–77. New York: Plenum.

Wolin, M. J. 1981. Fermentation in the rumen and large intestine. *Science* 213:1463–68.

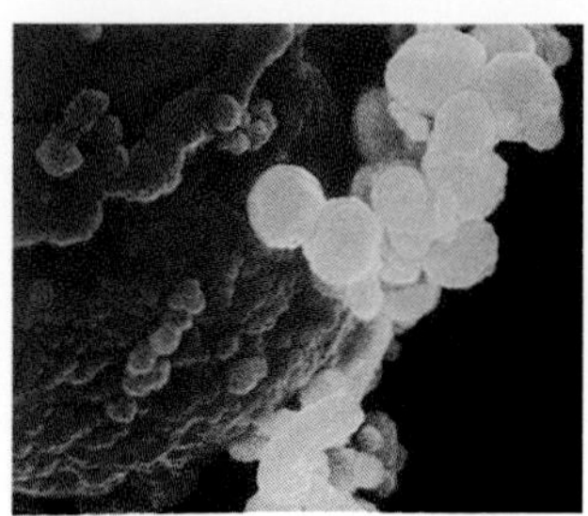

CHAPTER 30

Symbiotic Associations: Parasitism, Pathogenicity, and Resistance

In real life, however, even in our worst circumstances we have always been a relatively minor interest of the vast microbial world. Pathogenicity is not the rule. Indeed, it occurs so infrequently and involves such a relatively small number of species, considering the huge population of bacteria on earth, that it has a freakish aspect. Disease usually results from inconclusive negotiations for symbiosis, an overstepping of the line by one side or the other, a biological misinterpretation of borders.

—Lewis Thomas

Outline

Concepts

1. If a symbiont either harms or lives at the expense of another organism, it is called a parasitic organism and the relationship is termed parasitism. In this relationship, the body of the animal is referred to as the host.
2. Those parasitic organisms capable of causing disease are called pathogens. Disease is any change in the host from a healthy to an unhealthy, abnormal state in which part or all of the host's body is not properly adjusted or capable of carrying on its normal functions.
3. For a parasitic organism to cause disease, it must be transmitted to a suitable host, attach to and/or colonize the host, grow and multiply within or on the host, and interfere with or impair the normal physiological activities of the host. When a parasitic organism is growing and multiplying within or on a host, the host is said to have an infection.
4. The host's ability to resist infection depends on a continuous defense against parasitic invasion.
5. Resistance arises from both innate and acquired body defense mechanisms. The innate, general, or nonspecific immune mechanisms are those with which a host is genetically endowed and include general, physical, chemical, and biological barriers. Acquired resistance mechanisms are those that the host acquires upon contact with either the parasitic organism or its products.

Chapter 29 introduces the concept of symbiosis and deals with two of its subordinate categories: commensalism and mutualism. In chapter 30, the third category, parasitism, is presented along with one of its possible consequences—pathogenicity. Fortunately, most higher animals possess defense mechanisms that offer resistance to the continuous onslaught of parasitic organisms. Some innate, nonspecific forms of resistance include general, physical, chemical, and biological barriers.

The parasitic way of life is so successful that it has evolved independently in nearly all groups of microorganisms. In recent years, concerted efforts to understand microorganisms and their relationships with their hosts have developed within the disciplines of virology, rickettsiology, chlamydiology, bacteriology, mycology, parasitology (protozoology and helminthology), entomology, and zoology. This chapter examines the parasitic way of life in terms of health and disease in the animal body.

Host-Parasite Relationships (Parasitism)

If a symbiont either harms or lives at the expense of another organism (the **host**), it is a **parasitic organism,** and the relationship is called **parasitism.** In this relationship, the body of the host can be viewed as a microenvironment that shelters and supports the growth and multiplication of the parasitic organism. The parasitic organism is usually the smaller of the two partners and is metabolically dependent on the host. There are many parasitic agents or organisms among the viruses, procaryotes, fungi, plants, and animals (table 30.1). By convention, when the word **parasite** is used without qualification, it refers specifically to a protozoan or helminthic (nematode, trematode, cestode) organism (*see appendix V*).

Several types of parasitism are recognized. If an organism lives on the surface of its host, it is an **ectoparasite;** if it lives internally, it is an **endoparasite.** The host on or in which the parasitic organism either attains sexual maturity or reproduces is the **final host.** A host that serves as a temporary but essential environment for development is an **intermediate host.** In contrast, a **transfer host** is not necessary for the completion of the organism's life cycle, but is used as a vehicle for reaching a final host. A host infected with a parasitic organism that also can infect humans is called a **reservoir host.**

Because, by definition, parasitic organisms are dependent on their hosts, the symbiotic relationship between the host and parasite is a dynamic one (figure 30.1). When a parasite is growing and multiplying within or on a host, the host is said to have an **infection.** The nature of an infection can vary widely with respect to severity, location, and number of organisms involved (table 30.2). An infection may or may not result in overt disease. An **infectious disease** is any change from a state of health in which part of all of the host body is not properly adjusted or capable of carrying on its normal functions due to the presence of a parasitic organism or its products. Any parasitic organism or agent that produces such a disease is a **pathogen.** Its ability to cause disease is **pathogenicity.**

At times an infectious organism can enter a latent state in which there is no shedding of the organism and no symptoms present within the host. This latency can be either intermittent or quiescent. Intermittent latency is exemplified by the herpesvirus that causes cold sores or fever blisters. After an initial infection, the symptoms subside. However, the virus remains in local nerve tissue and can be activated weeks or

TABLE 30.1 Categorization of Parasitic Organisms (Agents) by Size

Discipline		Parasitic Group		Approximate Size
Virology		Prions	Agents	350,000 Da
		Viroids		130,000 Da
		Animal viruses		25–300 nm
Bacteriology		Chlamydiae	Microorganisms (microbiota)	0.2–1.5 μm
		Mycoplasmas		0.3–0.8 μm
		Rickettsias		0.5–2 μm
		Bacteria		1–10 μm
Mycology		Fungi		5–10 μm diameter
Protozoology	Parasitology	Protozoa		1–150 μm
Helminthology		Nematodes	Parasites	3 mm–30 cm
		Platyhelminthes (cestodes, trematodes)		1 mm–10 m
Entomology		Ticks and mites	Ectoparasites	0.1–15 mm
Zoology		Horsehair worms		10–20 cm
		Mesozoa		Up to 100 cm
		Leeches		1–5 cm

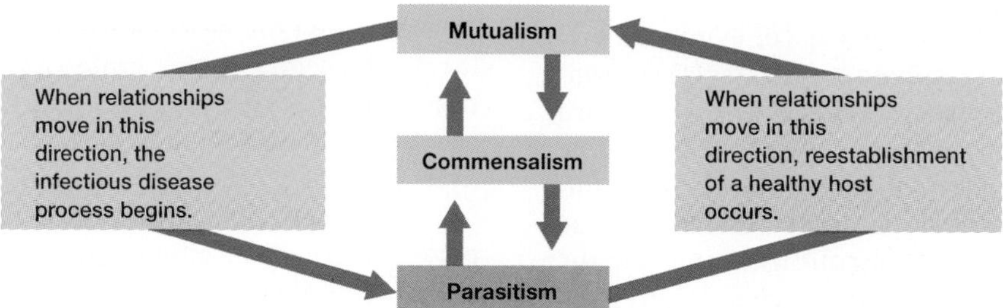

Figure 30.1 **Symbiosis.** All symbiotic relationships are dynamic, and shifts among them can occur as indicated by the arrows. The most beneficial relationship is mutualism; the most destructive is parasitism. Host susceptibility, virulence of the parasitic organism, and number of parasites are factors that influence these relationships. Disease can result from a shift from either mutualism or commensalism to parasitism. Health may be regained by the reestablishment of mutualism or commensalism.

months later by factors such as stress or sunlight. In a quiescent latency, the organism persists but remains inactive for long periods of time, usually for years. For example, the varicella-zoster virus causes chickenpox in children and remains after the disease has subsided. In adulthood, under certain conditions, the same virus may erupt into a disease called shingles.

Cold sores (p. 729)
Chickenpox (varicella) and shingles (herpes zoster) (p. 716).

The outcome of most host-parasite relationships is dependent on three main factors: (1) the number of organisms present in or on the host, (2) the virulence of the organism, and (3) the host's defenses or degree of resistance. Usually, the greater the number of parasitic organisms within a given host, the greater the likelihood of disease. However, a few organisms can cause disease if they are extremely virulent or if the host's resistance is low. A host's resistance can drop so much that its own microbiota may cause disease. Such a disease is sometimes called an endogenous disease because the agent originally comes from within the host's own body. Endogenous diseases can be a serious problem among hospitalized patients with very low resistance (*see chapter 38*).

The term **virulence** (Latin *virulentia,* from *virus,* poison) refers to the degree or intensity of pathogenicity. It is determined by three characteristics of the pathogen: invasiveness, infectivity, and pathogenic potential. **Invasiveness** is the ability of the organism to spread to adjacent or other tissues. **Infectivity** is the ability of the organism to establish a focal point of infection. **Pathogenic potential** refers to the degree that the pathogen causes morbid symptoms. A major aspect of pathogenic potential is toxigenicity. **Toxigenicity** is the pathogen's ability to produce **toxins,** chemical substances that will damage the host and produce disease. Virulence is often measured experimentally by determining the **lethal dose 50 (LD_{50})** or the **infective dose 50 (ID_{50}).** These values refer to the dose or number of pathogens that will either kill or infect, respectively, 50% of an experimental group of hosts within a specified period (figure 30.2).

1. Define parasitic organism, parasitism, infection, infectious disease, pathogenicity, virulence, invasiveness, infectivity, pathogenic potential, and toxigenicity.
2. What types of parasitism are recognized?
3. What factors determine the outcome of most host-parasite relationships?

Determinants of Infectious Disease

To induce an infectious disease, a pathogen must be able to

1. Initially be transported to the host
2. Adhere to, colonize, or invade the host
3. Multiply (grow) or complete its life cycle on or in the host
4. Initially evade host defense mechanisms
5. Possess the mechanical, chemical, or molecular ability to damage the host

The first four factors influence the degree of invasiveness. Toxigenicity plays a major role in the fifth. Each determinant is now discussed in more detail.

Transmissibility of the Pathogen

An essential feature in the development of an infectious disease is the initial transport of the pathogen to the host. The most obvious means is direct contact—from host to host (coughing, sneezing, body contact). Pathogens are also transmitted indirectly in a variety of ways. Infected hosts shed pathogens into their surroundings. Once in the environment, pathogens can be deposited on various surfaces, from which

TABLE 30.2 Various Types of Infections Associated with Parasitic Organisms

Type	Definition
Abscess	A localized infection with a collection of pus surrounded by an inflamed area
Acute	Short but severe course
Bacteremia	Viable bacteria in the bloodstream
Chronic	Persists over a long time
Covert	Subclinical, no symptoms
Cross	Transmitted between hosts infected with different organisms
Focal	Exists in circumscribed areas
Fulminating	Infectious agent multiplies with great intensity
Generalized	Affects many parts of the body
Iatrogenic	Caused as a result of health care
Latent	Persists in tissues for long periods, during most of which there are no symptoms
Localized	Restricted to a limited region or to one or more anatomical areas
Mass	Infectious agent occurs in large numbers in systemic circulation
Mixed	More than one organism present simultaneously
Nosocomial	Develops during a hospital stay
Opportunistic	Due to an agent that does not harm a healthy host but takes advantage of an unhealthy one
Overt	Symptomatic
Phytogenic	Caused by plant pathogens
Primary	First infection that often allows other organisms to appear on the scene
Pyogenic	Results in pus formation
Secondary	Caused by an organism following an initial or primary infection
Septic	Produced by or due to decomposition by microorganisms; relating to or caused by sepsis
Septicemia	Blood poisoning associated with persistence of pathogenic organisms or their toxins in the bloodstream
Sporadic	Occurs only occasionally
Subclinical (inapparent or covert)	No detectable symptoms or manifestations
Systemic	Spread throughout the body
Terminal	Occurs near the end of a disease and frequently causes death
Toxemia	Condition arising from toxins in the blood
Zoonosis	Caused by a parasitic organism that is normally found in animals other than humans

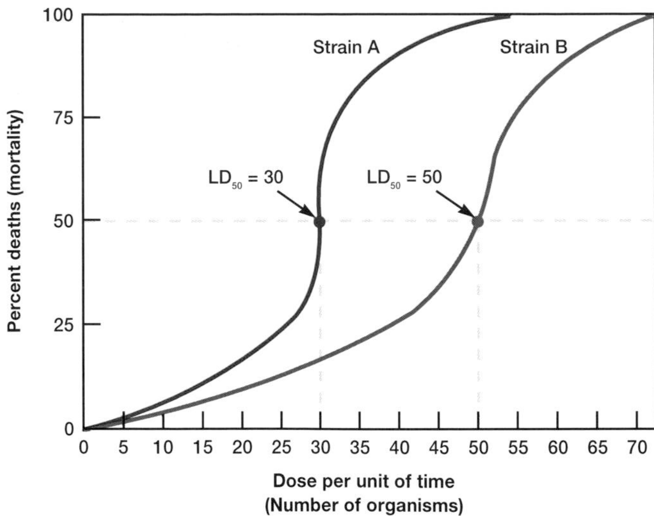

Figure 30.2 Determination of the LD_{50} of a Pathogenic Microorganism. Various doses of a specific pathogen are injected into experimental host animals. Deaths are recorded and a graph constructed. In this example, the graph represents the susceptibility of host animals to two different strains of a pathogen—strain A and strain B. For strain A, the LD_{50} is 30, and for strain B it is 50. Hence, strain A is more virulent than strain B.

they can be either resuspended into the air or directly transmitted to a host later. Soil, water, and food are indirect vehicles that harbor and transmit pathogens to hosts. **Vectors** (organisms that transmit pathogens from one host to another) and **fomites** (inanimate objects that harbor and transmit pathogens) are also involved in the spread of many pathogens.

Attachment and Colonization by the Pathogen

After being transmitted to an appropriate host, the pathogen must be able to adhere to and colonize host cells and tissues. Colonization depends on the ability of the pathogen to compete successfully with the host's normal microbiota for essential nutrients. Specialized structures that allow the pathogen to compete for surface attachment sites are also necessary for colonization.

Pathogens and many nonpathogens adhere with a high degree of specificity to particular tissues. Adherence factors (table 30.3) are one reason for this specificity. These are specialized structures or molecules on the pathogen's cell surface that bind to complementary receptor sites on the host cell surface (figure 30.3).

Entry of the Pathogen

Entry into host cells and tissues is a specialized strategy used by many pathogens for survival and multiplication . Pathogens

TABLE 30.3 Adherence Factors That Play a Role in Infectious Diseases

Adherence Factor	Description
Filamentous hemagglutinin	Causes adherence to erythrocytes
Fimbriae	Filamentous structures that help attach bacteria to solid surfaces
Glycocalyx or capsule	A layer of exopolysaccharide fibers with a distinct outer margin that surrounds many cells; it inhibits phagocytosis and aids in adherence
Lectin	Any carbohydrate-binding protein or glycoprotein of nonimmune origin
Ligand	A low molecular weight molecule that exhibits specific binding to complementary binding site on a high molecular weight molecule, such as a protein
Mucous gel	The glycoprotein layer or mucopolysaccharide layer of glycosaminoglycans covering animal cell mucosal surfaces
Pili	Filamentous structures that bind procaryotes together for the transfer of genetic material
Receptors	Complementary binding sites that bind specific ligands or adhesins
S layer	The outermost regularly structured layer of cell envelopes of archaeobacteria and eubacteria that may promote adherence to surfaces
Slime	A tenacious bacterial film that is less compact than a capsule
Teichoic and lipoteichoic acids	Cell wall components in gram-positive bacteria that aid in adhesion

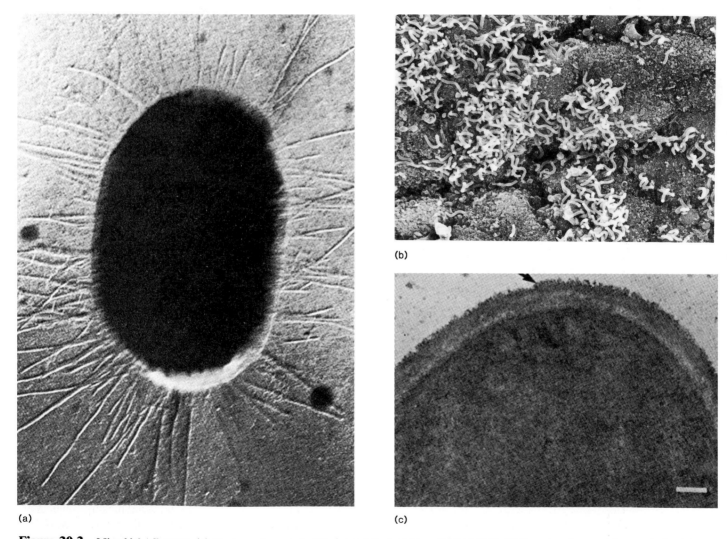

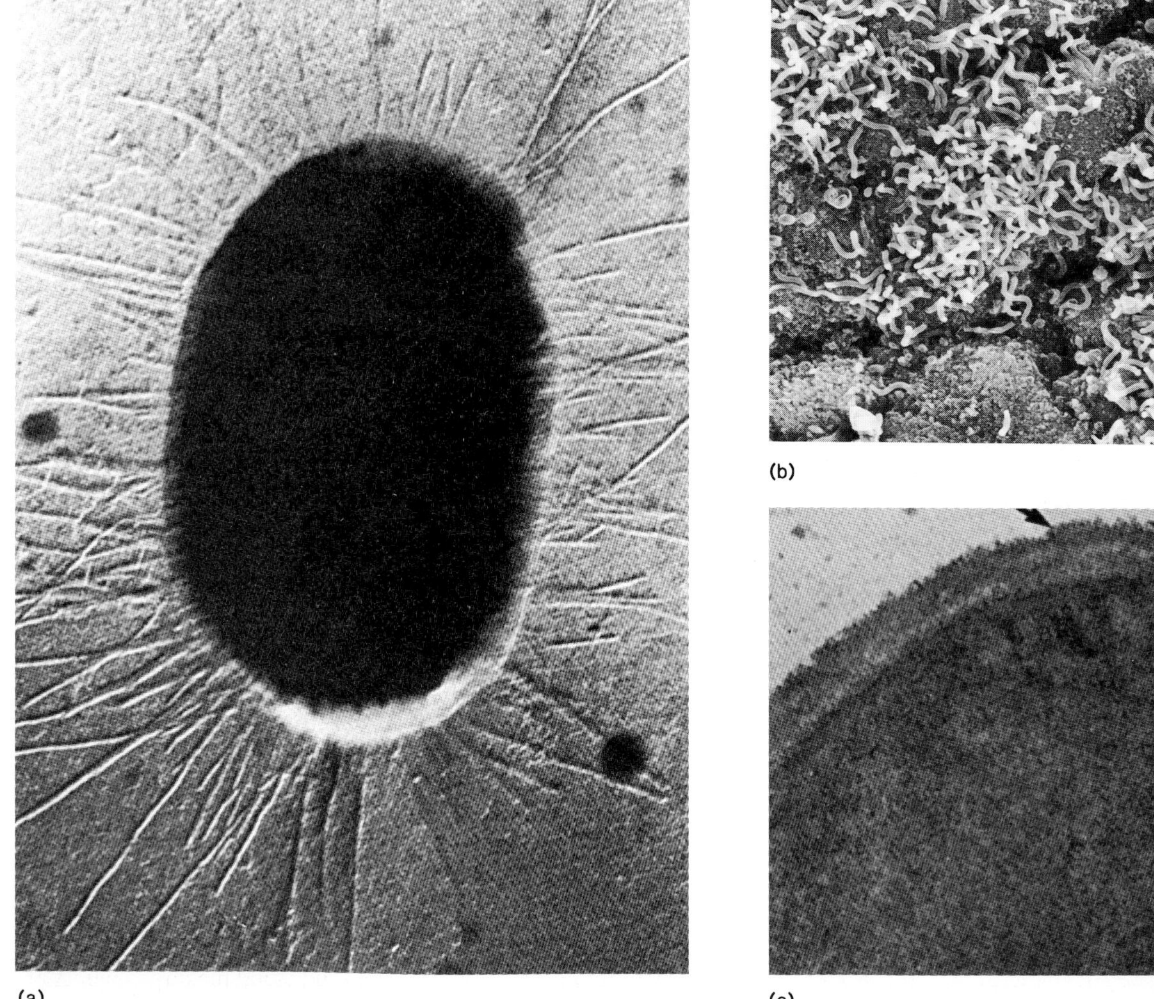

Figure 30.3 **Microbial Adherence.** (*a*) Electron micrograph of fimbriated *Escherichia coli* (×16,625). (*b*) Scanning electron micrograph of epithelial cells with adhering *Vibrio* (×1,200). (*c*) *Candida albicans* fimbriae (arrow) are used to attach the fungus to vaginal epithelial cells. Bar = 0.20 μm.

TABLE 30.4 Products Involved in Pathogen Dissemination Throughout the Body of a Mammalian Host

Product	Organism Involved	Physiology
Coagulase	*Staphylococcus aureus*	Coagulates (clots) the fibrinogen in plasma. The clot protects the pathogen from phagocytosis and isolates it from other host defenses.
Collagenase	*Clostridium* spp.	Breaks down collagen that forms the framework of connective tissues; allows the pathogen to spread.
Deoxyribonuclease (along with calcium and magnesium)	Group A streptococci, staphylococci, *Clostridium perfringens*	Lowers viscosity of exudates, giving the pathogen more mobility.
Elastase and alkaline protease	*Pseudomonas aeruginosa*	Cleave laminin associated with basement membranes.
Hemolysins	Staphylococci, streptococci, *Escherichia coli*, *Clostridium perfringens*	Lyse erythrocytes, causing anemia and weakened host defenses; make iron available for microbial growth.
Hyaluronidase	Groups A, B, C, and G streptococci, staphylococci, clostridia	Hydrolyzes hyaluronic acid, a constituent of the intercellular ground substance that cements cells together and renders the intercellular spaces amenable to passage by the pathogen.
Hydrogen peroxide (H_2O_2) and ammonia (NH_3)	*Mycoplasma* spp., *Ureaplasma* spp.	Are produced as metabolic wastes. These are toxic and damage epithelia in respiratory and urogenital systems.
Immunoglobulin A protease	*Streptococcus pneumoniae*	Cleaves immunoglobulin A into Fab and Fc fragments.
Lecithinase	*Clostridium* spp.	Destroys the lecithin (phosphatidycholine) component of plasma membranes, allowing pathogen to spread.
Leukocidins	Staphylococci, pneumococci, streptococci	Cause degranulation of lysosomes within leukocytes, which decreases host resistance; also kill leukocytes.
Porins	*Salmonella typhimurium*	Inhibit leukocyte phagocytosis by activating the adenylate cyclase system.
Protein A	*Staphylococcus aureus*	Located on cell wall. Immunoglobulin G (IgG) binds to protein A by its Fc end, thereby preventing complement from interacting with bound IgG.
Streptokinase (fibrinolysin, staphylokinase)	Group A, C, and G streptococci, staphylococci	Acts as an enzyme in plasma to convert plasminogen to plasmin, thus digesting fibrin clots; this allows the pathogen to move from the clotted area.

often actively penetrate the host's epithelium after attachment to the epithelial surface. This may be accomplished through production of lytic substances that alter the host tissue by (1) attacking the ground substance and basement membranes of integuments and intestinal linings, (2) degrading carbohydrate-protein complexes between cells or on the cell surface (the glycocalyx), or (3) disrupting the cell surface.

At times, a pathogen can penetrate the epithelial surface by passive mechanisms not related to the pathogen itself. Examples include (1) small breaks, lesions, or ulcers in a mucous membrane that permit initial entry; (2) wounds, abrasions, or burns on the skin's surface; (3) arthropod vectors that create small wounds while feeding; (4) tissue damage caused by other organisms; and (5) existing eucaryotic internalization pathways (e.g., endocytosis).

Once inside the epithelia, the pathogen may penetrate to deeper tissues and continue disseminating throughout the body of the host. One way the pathogen accomplishes this is by producing specific products and/or enzymes that promote spreading (table 30.4). The pathogen also may enter the small terminal lymphatic capillaries that surround epithelial cells. These capillaries merge into large lymphatic vessels that eventually drain into the circulatory system. Once the circulatory system is reached, the pathogen has access to all organs and systems of the host.

Growth and Multiplication of the Pathogen

For a pathogen to be successful in growth and reproduction, it must find an appropriate environment (nutrients, pH, temperature, redox potential) within the host. Those areas of the host's body that provide the most favorable conditions for existence will harbor the pathogen and allow it to grow and multiply to produce an infection. Some pathogens invade specific cells in which they grow and multiply. Many of these intracellular pathogens have evolved such elaborate nutrient-gathering mechanisms that they have become totally dependent on the host's cells. Finally, some pathogens can actively grow and multiply specifically in the blood plasma. Their metabolic waste products are often toxic and produce a condition known as **septicemia** (Greek *septikos,* produced by putrefaction, and *haima,* blood).

1. What are some ways in which pathogens are transmitted to their hosts? Define vector and fomite.
2. Describe several specific adherence factors by which pathogens attach to host cells.
3. How do pathogens actively enter host tissues? How do they passively enter?
4. Once inside the epithelial surface, what are some mechanisms that pathogens possess to promote their dissemination throughout the body of a host?

Toxigenicity

Two distinct categories of disease can be recognized based on the pathogen's role in the disease-causing process: infections and intoxications. An infectious disease results partly from the pathogen's growth and reproduction (or invasiveness) that often produce tissue alterations.

Intoxications are diseases that result from the entrance of a specific toxin into the body of a host. Toxins can even induce disease in the absence of the organism that produced them. A **toxin** (Latin *toxicum,* poison) is a specific substance, often a metabolic product of the organism, that damages the host. The term **toxemia** refers to the condition caused by toxins that have entered the blood of the host. Toxins produced by organisms can be divided into two main categories: exotoxins and endotoxins.

Exotoxins

Exotoxins are soluble, heat-labile, protein toxins that are usually released into the surroundings as the pathogen grows. Often, exotoxins may travel from the site of infection to other body tissues or target cells in which they exert their effects. Exotoxins usually are

1. Synthesized by specific pathogens that often have plasmids or prophages bearing the exotoxin genes
2. Heat-labile proteins inactivated at 60 to 80°C
3. Among the most lethal substances known (toxic in very small doses [microgram per kilogram amounts])
4. Associated with specific diseases
5. Highly immunogenic and stimulate the production of neutralizing antibodies **(antitoxins)**
6. Easily inactivated by formaldehyde, iodine, and other chemicals to form immunogenic **toxoids**
7. Unable to produce a fever in the host directly
8. Often given the name of the disease they produce (e.g., diphtheria toxin, botulinum toxin)
9. Usually categorized as neurotoxins, cytotoxins, or enterotoxins according to their mechanism of action

Although toxins occur in many forms, there is a general structural model to which they frequently conform—the **AB model.** In this model, each toxin is composed of an enzymatic subunit or fragment (A) that is responsible for the toxic effect once inside the host cell and a binding subunit or fragment (B). Isolated A subunits are enzymatically active but lack binding and cell entry capability, whereas isolated B subunits bind to target cells but are nontoxic and biologically inactive. The B subunit interacts with specific receptors on the target cell or tissue such as the gangliosides GM_1 for cholera toxin, GT_1 and/or GD_1 for tetanus toxin, and GD_1 for botulinum toxin.

Several mechanisms for the entry of A subunits or fragments into target cells have been proposed. In one mechanism, the B subunit inserts into the plasma membrane and creates a pore through which the A subunit enters (figure 30.4*a*). In another mechanism, entry is by receptor-mediated endocytosis (figure 30.4*b*).

Exotoxins exert their effects in a variety of ways—for instance, by (1) inhibition of protein synthesis, (2) inhibition of nerve synapse function, (3) disruption of membrane transport, or (4) damage to plasma membranes. The mechanism of action can be quite complex, as shown by the example of diphtheria toxin (figure 30.4*b*). The diphtheria toxin is a protein of about 62,000 mol wt. It binds to cell surface receptors by the B fragment portion and is taken into the cell through the formation of a clathrin-coated vesicle (*see p. 388*). The toxin then enters the vesicle membrane and is cleaved into two parts, one of which, the A fragment, escapes into the cytoplasm. The A fragment is an enzyme and catalyzes the addition of an ADP-ribose group to the eucaryotic elongation factor EF2 that aids in translocation during protein synthesis (*see p. 211*). The substrate for this reaction is the coenzyme NAD^+.

$$NAD^+ + EF2 \rightarrow \text{ADP-ribosyl-EF2} + \text{nicotinamide}$$

The modified EF2 protein cannot participate in the elongation cycle of protein synthesis, and the cell dies because it can no longer synthesize proteins.

Exotoxins vary widely in their relative contribution to the disease process with which they are associated. The general properties of some exotoxins are presented in table 30.5.

Exotoxins may be divided into three categories on the basis of the site affected: **neurotoxins** (nerve tissue), **enterotoxins** (intestinal mucosa), and **cytotoxins** (general tissues). Some of the enteric pathogens that produce these exotoxins are presented in table 30.6.

Neurotoxins are usually ingested as preformed toxins that affect the nervous system and indirectly cause enteric (pertaining to the small intestine) symptoms. Examples include staphylococcal enterotoxin B, *Bacillus cereus* emetic toxin (Greek *emetos,* vomiting), botulinum toxin, ciguatoxin and saxitoxin from dinoflagellates, and tetrodotoxin from puffer fish.

True enterotoxins (Greek *enter,* intestine) have a direct affect on the intestinal mucosa and elicit profuse fluid secretion. The classic enterotoxin, cholera toxin (choleragen), has been studied extensively. It is an AB toxin. The B subunit is made of five parts arranged as a donut-shaped ring. The B subunit ring anchors itself to the epithelial cell's plasma membrane and then inserts the smaller A subunit into the cell. The A subunit activates tissue adenylate cyclase to increase intes-

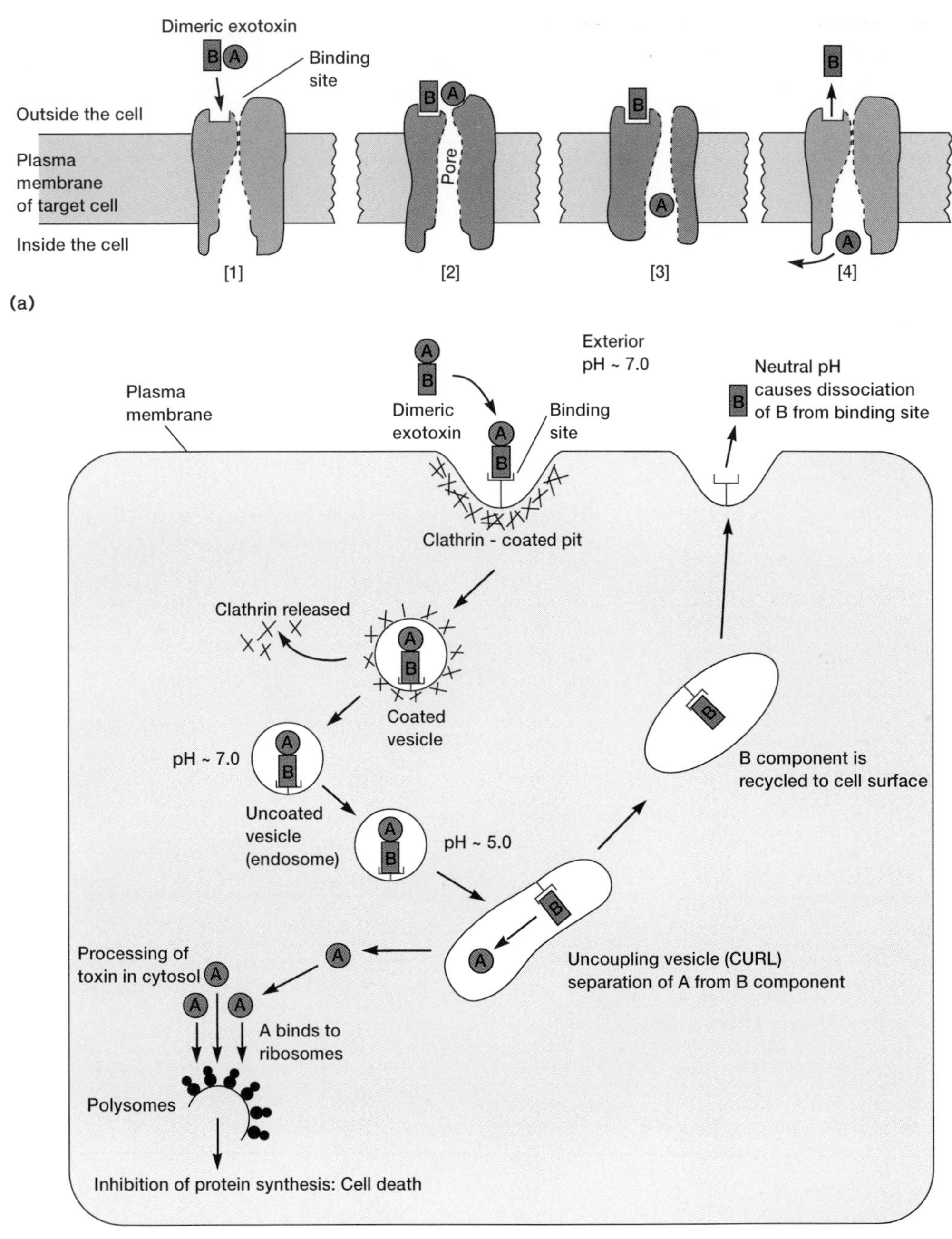

Figure 30.4 Diagrammatic Representation of Two Exotoxin Transport Mechanisms. (*a*) Domain B of the dimeric exotoxin (*AB*) binds to a specific membrane receptor of a target cell [*1*]. A conformational change [*2*] generates a pore [*3*] through which the A domain crosses the membrane and enters the cytosol, followed by recreation [*4*] of the binding site. (*b*) Receptor-mediated endocytosis of the diphtheria toxin involves the dimeric exotoxin binding to a receptor-ligand complex that is internalized in a clathrin-coated pit that pinches off to become a coated vesicle. The clathrin coat depolymerizes resulting in an uncoated endosome vesicle. The pH in the endosome decreases due to the H^+-ATPase activity. The low pH causes A and B components to separate. An endosome in which this separation occurs is sometimes called a CURL (compartment of uncoupling of receptor and ligand). The B domain is then recycled to the cell surface. The A domain moves through the cytosol, binds to ribosomes, and inhibits protein synthesis, leading to the death of the cell.

TABLE 30.5 Properties of Some AB Model Bacterial Exotoxins

Toxin	Organism	Genetic Control	Subunit Structure	Target Cell Receptor	Enzymatic Activity	Biologic Effects
Anthrax toxins	*B. anthracis*	Plasmid	Three separate proteins (EF, LF, PA)[a]	Unknown, probably glycoprotein	EF is a calmodulin-dependent adenylate cyclase; LF enzyme activity is unknown	EF + PA: increase in target cell cAMP level, localized edema; LF + PA: death of target cells and experimental animals
Bordetella adenylate cyclase toxin	*Bordetella* spp.	Chromosomal	A-B[b]	Unknown, probably glycolipid	Calmodulin-activated cyclase	Increase in target cell cAMP level; modified cell function or cell death
Botulinum toxin	*C. botulinum*	Phage	A-B[c]	Possibly ganglioside (GD_{1b})	None known	Decrease in peripheral, presynaptic acetylcholine release; flaccid paralysis
Cholera toxin	*V. cholera*	Chromosomal	A-5B[d]	Ganglioside (GM_1)	ADP ribosylation of adenylate cyclase regulatory protein, G_S	Activation of adenylate cyclase, increase in cAMP level; secretory diarrhea
Diphtheria toxin	*C. diphtheriae*	Phage	A-B[e]	Probably glycoprotein	ADP ribosylation of elongation factor 2	Inhibition of protein synthesis; cell death
Heat-labile enterotoxins[f]	*E. coli*	Plasmid	Similar or identical to cholera toxin			
Pertussis toxin	*B. pertussis*	Chromosomal	A-5B[g]	Unknown, probably glycoprotein	ADP ribosylation of signal-transducing G proteins	Block of signal transduction mediated by target G proteins
Pseudomonas exotoxin A	*P. aeruginosa*	Chromosomal	A-B	Unknown, but different from diphtheria toxin	Similar or identical to diphtheria toxin	
Shiga toxin	*S. dysenteriae*	Chromosomal	A-5B[h]	Glycoprotein or glycolipid	RNA *N*-glycosidase	Inhibition of protein synthesis, cell death
Shiga-like toxins	*Shigella* spp., *E. coli*	Phage	Similar or identical to shiga toxin			
Tetanus toxin	*C. tetani*	Plasmid	A-B[c]	Ganglioside (GT_1 and/or GD_{1b})	None known	Decrease in neurotransmitter release from inhibitory neurons; spastic paralysis

From G. L. Mandell, R. G. Douglas, and J. E. Bennett, *Principles and Practice of Infectious Diseases* 3d ed. Copyright © 1990 Churchill-Livingstone, Inc. Medical Publishers, New York, NY.

[a]The binding component (known as protective antigen [PA]) catalyzes/facilitates the entry of either edema factor (EF) or lethal factor (LF).

[b]Apparently synthesized as a single polypeptide with binding and catalytic (adenylate cyclase) domains.

[c]Holotoxin is apparently synthesized as a single polypeptide and cleaved proteolytically as diphtheria toxin; subunits are referred to as L: light chain, A equivalent; H: heavy chain, B equivalent.

[d]The A subunit is proteolytically cleaved into A_1 and A_2, with A_1 possessing the ADP-ribosyl transferase activity; the binding component is made up of five identical B units.

[e]Holotoxin is synthesized as a single polypeptide and cleaved proteolytically into A and B components held together by disulfide bonds.

[f]The heat-labile enterotoxins of *E. coli* are now recognized to be a family of related molecules with identical mechanisms of action.

[g]The binding portion is made up of two dissimilar heterodimers labeled S2-S3 and S2-S4 that are held together by a bridging peptide, SS.

[h]Subunit composition and structure similar to cholera toxin.

tinal cyclic AMP (cAMP) concentrations. High concentrations of cAMP provoke the movement of massive quantities of water and electrolytes across the intestinal cells into the lumen of the gut. The genes for this enterotoxigenicity reside on the *Vibrio cholera* chromosome.

Cytotoxic products of several enteric pathogens are responsible for the mucosal destruction that often results in inflammatory colitis. For example, *Staphylococcus aureus* produces a delta cytotoxin that impairs water absorption and causes cytotoxic disruption of the intestinal mucosa. *Clostridium difficile* produces a potent cytotoxin that causes hemorrhagic fluid secretion.

The intramuscular injection of botulinum toxin type A can be used to treat many hyperactive muscle disorders. The FDA has licensed botulinum toxin for the treatment of strabismus, blepharospasm, and hemifacial spasm. It is the first microbial toxin to be used by physicians.

1. What is the difference between an infectious disease and an intoxication? Define toxemia.
2. Describe some general characteristics of exotoxins.
3. How do exotoxins get into host cells?
4. Describe the biological effects of several bacterial exotoxins.
5. Discuss the mechanisms by which exotoxins can damage cells.
6. What are the three categories of exotoxins?

Box 30.1

Detection and Removal of Endotoxins

Bacterial endotoxins have plagued the pharmaceutical industry and medical device producers for years. For example, administration of drugs contaminated with endotoxins can result in complications—even death—to patients. Recently, endotoxins have become a problem for individuals and firms working with cell cultures and genetic engineering. The result has been the development of sensitive tests and methods to identify and remove these endotoxins. The procedures must be very sensitive to trace amounts of endotoxins. Most firms have set a limit of 0.25 **endotoxin units (E.U.),** 0.025 ng/ml, or less as a release standard for their drugs, media, or products.

One of the most accurate tests for endotoxins is the in vitro *Limulus* amoebocyte lysate (LAL) assay. The assay is based on the observation that when an endotoxin contacts the clot protein from circulating amoebocytes of *Limulus,* a gel-clot forms. The assay kits available today contain calcium, proclotting enzyme, and procoagulogen. The proclotting enzyme is activated by bacterial endotoxin and calcium to form active clotting enzyme (see Box Figure 30.1). Active clotting enzyme then catalyzes the cleavage of procoagulogen into polypeptide subunits (coagulogen). The subunits join by disulfide bonds to form a gel-clot. Spectrophotometry is then used to measure the protein precipitated by the lysate. The LAL test is sensitive at the nanogram level, but must be standardized against Food and Drug Administration Bureau of Biologics endotoxin reference standards. Results are reported in endotoxin units per milliliter and reference made to the particular reference standards used.

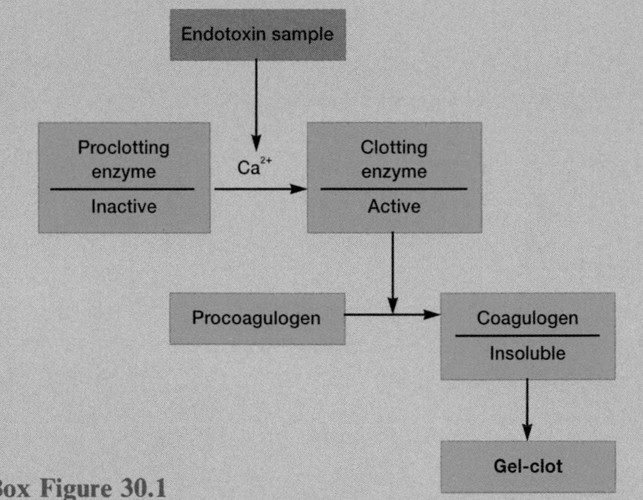

Box Figure 30.1

Removal of endotoxins presents more of a problem than their detection. Those present on glassware or medical devices can be inactivated if the equipment is heated at 250°C for 30 minutes. Those found in solution range in size from molecules with molecular weights of 20,000 to large aggregates with diameters up to 0.1 μm. Thus, they cannot be removed by conventional filtration systems. Manufacturers are currently developing special filtration systems and filtration cartridges that retain these endotoxins and help alleviate contamination problems.

TABLE 30.6 Examples of Some Bacteria That Produce Exotoxins

Category	Bacterial Examples
Neurotoxin	*Clostridium botulinum* *C. tetani* *Staphylococcus aureus* *Bacillus cereus*
Enterotoxin	*Vibrio cholerae* *Escherichia coli* *Salmonella* spp. *Klebsiella* spp. *Clostridium perfringens*
Cytotoxin	*Shigella* spp. *Vibrio parahaemolyticus* *S. aureus* *Clostridium difficile*

Endotoxins

Most gram-negative bacteria have a lipopolysaccharide (LPS) in the outer membrane layer of their cell wall that, under certain circumstances, is toxic to specific hosts. This LPS (*see figures 3.23 and 3.24*) is called an **endotoxin** because it is bound to the bacterium and is released when the microorganism lyses (Box 30.1). Some is also released during bacterial multiplication. The toxin component of the LPS is the lipid portion, called lipid A. Lipid A is not a single macromolecular structure but appears to be a complex array of lipid residues. The lipid A component exhibits all the properties (see characteristic 5 following) associated with endotoxicity and gram-negative bacteremia.

Gram-negative cell wall (pp. 54–55).

Besides the preceding characteristics, endotoxins are

1. Heat stable
2. Toxic only at high doses (milligram per kilogram amounts)

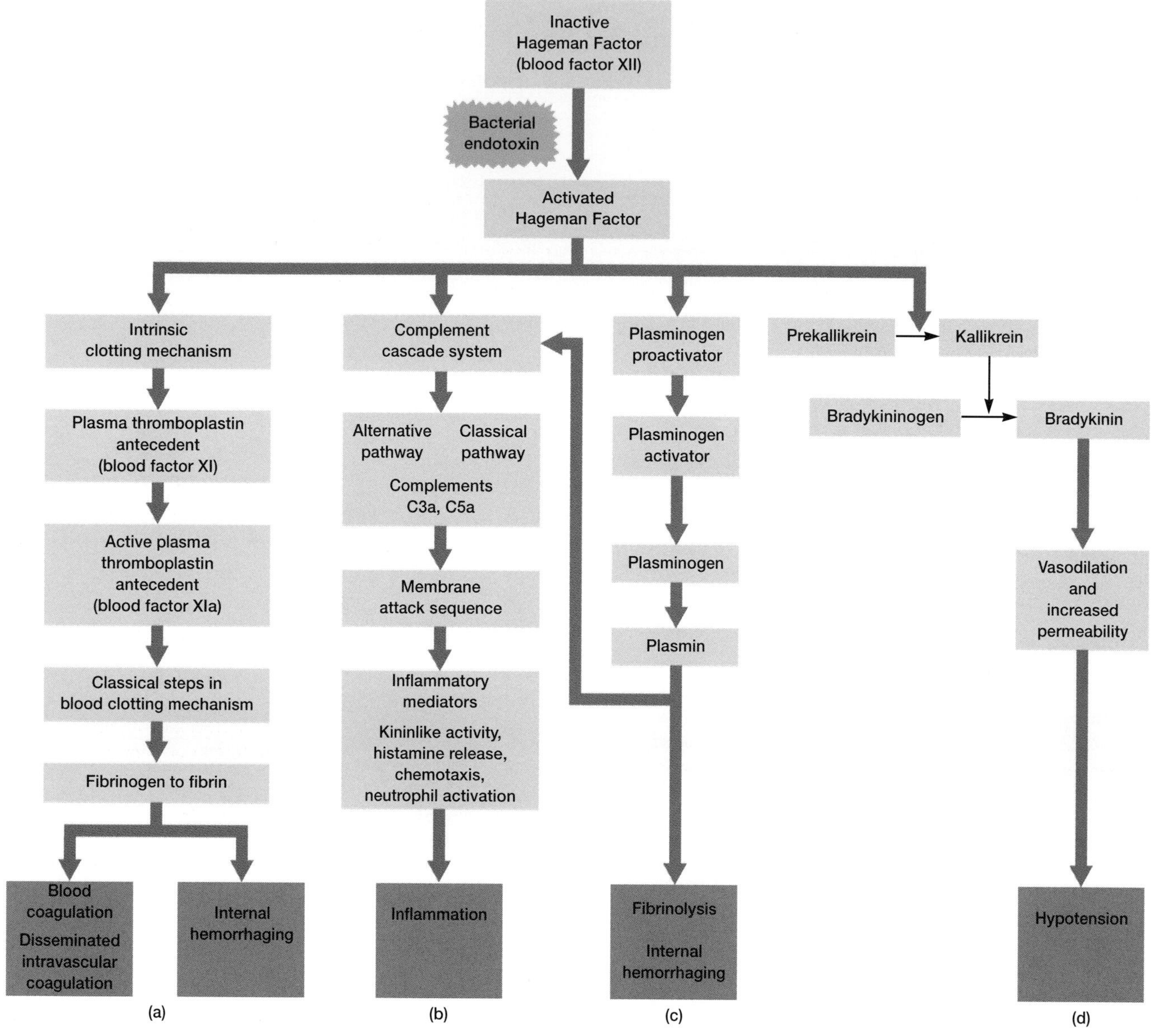

Figure 30.5 **Endotoxins.** The pathophysiological effects of gram-negative bacterial endotoxins (Lipid A) on a mammalian host. Both in vivo and in vitro endotoxins exert four systemic effects. (*a*) They can start the blood clotting cascade that leads to blood coagulation, thrombosis, and acute disseminated intravascular coagulation, which in turn depletes platelets and various clotting factors, which results in clinical bleeding; (*b*) they can activate the complement system, which leads to inflammation; (*c*) they can activate fibrinolysis; and (*d*) they can trigger a series of enzymatic reactions that lead to the release of bradykinins and other vasoactive peptides, which causes hypotension.

3. Weakly immunogenic
4. Generally similar, despite source
5. Usually capable of producing fever (are pyrogenic), shock, blood coagulation, weakness, diarrhea, inflammation, intestinal hemorrhage, and fibrinolysis (enzymatic breakdown of fibrin, the major protein component of blood clots)

Endotoxins initially activate Hageman Factor (blood clotting factor XII), which in turn activates up to four humoral systems: coagulation, complement, fibrinolytic, and kininogen systems.

Activated Hageman Factor can initiate **coagulation** (the intrinsic blood clotting cascade) through the conversion of plasma thromboplastin antecedent to active thromboplastin antecedent (figure 30.5). The rest of the cascade is classic, leading to cross-linked fibrin formation and a blood clot. If this system goes unchecked, thrombosis (plugs in blood vessels) produces disseminated intravascular coagulation. This, in turn, causes depletion of blood platelets as well as clotting factors II, V, and VII. In disseminated intravascular coagulation, the use of platelets and coagulation factors exceeds hemostatic production rates, which leads to internal hemorrhaging and organ failure (lungs, kidneys, liver).

Box 30.2

Escherichia coli Hemolysin

The hemolysin of *Escherichia coli* has attracted much attention recently for three main reasons: (1) it is a significant pathogenic factor; (2) it is the only known protein to be secreted by *E. coli;* and (3) its mode of action appears to be a model for other hemolysins.

The hemolysin determinant consists of four genes. The *hlyC* gene encodes a 20,000 MW polypeptide that promotes posttranslational modification of the protein encoded by the *hlyA* structural gene, rendering this protein hemolytically active. The *hlyA* product is a 70 kDa polypeptide. Two other genes, *hlyB* and *hlyD,* are involved in the secretion of the hemolysin from the bacterial cell.

When the hemolysin contacts an erythrocyte membrane, it inserts into the target membrane lipid bilayer to create hydrophilic transmembrane pores that allow the rapid efflux of hemoglobin from the erythrocyte.

Serum contains a group of proteins called complement that plays an important role in the removal of pathogens. Because the complement system normally acts in concert with specific immune responses, a detailed discussion of it is presented in chapter 33. As a general overview, the activation of complement (figure 30.5*b*) causes the release of inflammatory mediators to induce (1) smooth muscle contraction in postcapillary venules, (2) movement of neutrophils to the infection site, (3) neutrophils to produce leukotrienes, (4) increased vascular permeability, and (5) mast cell degranulation. Inflammation is the result.

The complement system (pp. 648–51)

While Hageman Factor is activating the coagulation and complement pathways, it also activates a countervening fibrinolytic system (figure 30.5*c*). Activated Hageman Factor causes the conversion of plasminogen proactivator to plasminogen activator. The latter stimulates the conversion of plasminogen to plasmin. Plasmin is a very potent fibrinolytic protein that mediates blood clot lysis, which contributes further to internal hemorrhaging. The plasmin formed also can initiate the complement cascade and exacerbate the inflammatory reaction initially triggered by the gram-negative endotoxin.

Finally, the endotoxin-activated Hageman Factor can activate the kininogen system (figure 30.5*d*), leading to hypotension (lowered blood pressure). In this pathway, Hageman Factor causes the conversion of prekallikrein to activated kallikrein, which in turn catalyzes the conversion of bradykininogen to bradykinin. Bradykinin is a potent vasoactive peptide that causes pain, vasodilation, and increased vascular permeability. This vasoactive effect leads to leakage of the fluid portion of the blood into interstitial spaces; therefore, blood volume decreases and hypovolemic hypotension results. Circulatory shock and death can follow.

Gram-negative endotoxins also indirectly induce a fever in the host by causing macrophages to release **endogenous pyrogens** that reset the hypothalamic thermostat. Recent evidence indicates that one important endogenous pyrogen is the lymphokine interleukin-1. Other cytokines released by macrophages, such as the tumor necrosis factor and IL-6, also produce fever and other endotoxin symptoms.

1. Describe the chemical structure of the LPS endotoxin.
2. List some general characteristics of endotoxins.
3. Describe how each of the following operate after being activated by a gram-negative bacterial endotoxin: intrinsic clotting cascade, complement activation, fibrinolytic system, and kininogen system.
4. How do gram-negative endotoxins induce fever in a mammalian host?

Leukocidins and Hemolysins

Some pathogens produce extracellular toxins that may kill phagocytic leukocytes and are termed **leukocidins** (*leuko*cyte and Latin *caedere,* to kill). Most leukocidins are produced by pneumococci, streptococci, and staphylococci. Once liberated, the toxin attaches to leukocyte membranes and triggers a series of changes leading to degranulation and release of lysosomal enzymes into the leukocyte's cytosol (these events may result from changes in plasma membrane permeability due to pore formation). This destroys the leukocyte and, in turn, decreases host resistance. A similar substance, called **leukostatin,** interferes with the ability of leukocytes to engulf the microorganisms releasing it.

Other toxins called **hemolysins** (*haima,* blood, and Greek *lysis,* dissolution) also can be secreted by pathogenic bacteria (Box 30.2). Many hemolysins probably form pores in the plasma membrane of erythrocytes through which hemoglobin and/or ions are released (the erythrocytes lyse or, more specifically, hemolyze). **Streptolysin-O (SLO)** is a hemolysin, produced by *Streptococcus pyogenes,* that is inactivated by O_2 (hence the "O" in its name). SLO causes beta hemolysis of erythrocytes on agar plates incubated anaerobically. A complete zone of clearing around the bacterial colony growing on blood agar is called **beta hemolysis,** and a partial clearing of the blood is called **alpha hemolysis. Streptolysin-S (SLS)** is also produced by *S. pyogenes* but is insoluble and bound to the bacterial cell. It is O_2 stable (hence the "S" in its name) and causes beta hemolysis on aerobically incubated blood-agar plates. SLS also can act as a leukocidin by killing leukocytes that phagocytose the bacterial cell to which it is bound. It should be noted that hemolysins attack the membranes of many cells, not just erythrocytes and leukocytes.

1. What is the mode of action of a leukocidin? Of a hemolysin?
2. Name two specific hemolysins.

General or Nonspecific Host Immune Defense Mechanisms

With few exceptions, a potential pathogen invading a human host immediately confronts a vast array of general or nonspecific host immune (Latin *immunis,* safe) defense mechanisms (figure 30.6). Although the effectiveness of some mechanisms is not great, collectively, their defense against infection is formidable. These general or nonspecific defense mechanisms can be grouped into four major categories: general, physical, chemical, and biological barriers.

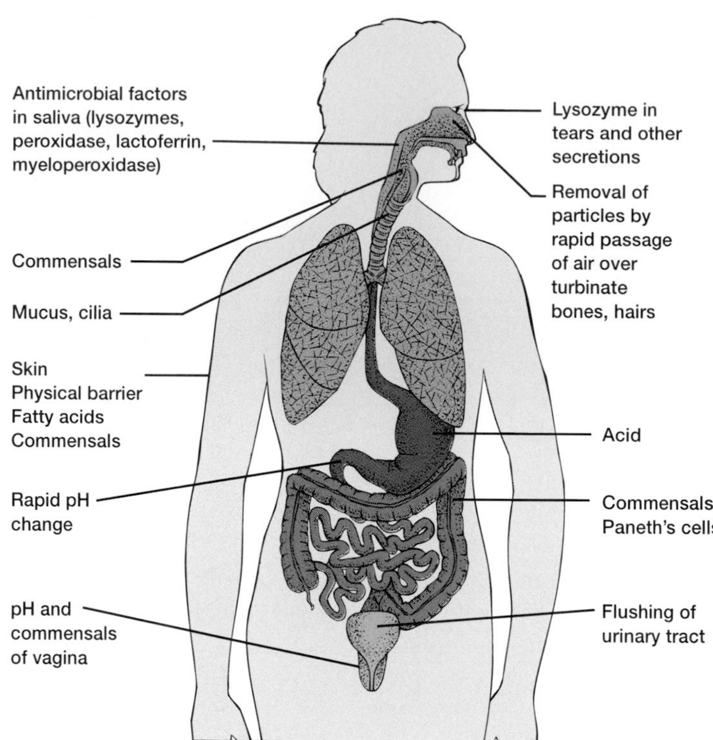

Figure 30.6 **Host Defenses.** Some nonspecific host defense mechanisms that help prevent entry of microorganisms into the host's tissues.

General Barriers

Many direct factors (nutrition, physiology, fever, age, genetics, race) and equally as many indirect factors (personal hygiene, socioeconomic status, living conditions) influence all host-parasite relationships. At times, they favor the establishment of the parasitic organism; at other times, they provide some measure of defense to the host. Because these factors are so ill-defined, they can be viewed as either general or nonspecific barriers to the establishment of an organism. A few of the better known direct factors are discussed.

Nutrition

In general, the more malnourished the host, the greater will be its susceptibility to, and the severity of, infections. This is especially true for very young hosts.

Acute-Phase Reactants

The term **acute-phase reactants** refers to the qualitative and quantitative changes that occur in the host's blood plasma during an acute infection. These changes can decrease the virulence of the pathogen and increase the overall general defense of the host. For example, the virulence of many organisms is enhanced with increased iron availability. Evidence exists that some hosts are able to redistribute iron in an attempt to withhold it **(hypoferremia)** from the microorganism. Conversely, **hyperferremia** can lead to infections with certain normally harmless organisms or provide for their dissemination. Gonococci, for example, spread most often during menstruation, a time in which there is an increased concentration of free iron available to these bacteria.

Gonorrhea (pp. 754–55).

At times, mammals obtain toxic levels of iron. Siderophores (*see chapter 5*) produced by bacteria and yeasts can lessen the toxic effects of iron-overload poisoning. For example, the hydroxamate-type siderophore, Desferal, an actinomycete metabolite, is the drug of choice for clearing the mammalian system of excess iron.

Fever

From a physiological point of view, fever results from disturbances in hypothalamic thermoregulatory activity, leading to an increase of the thermal "set point." In adult humans, fever is defined as an oral temperature above 98.6°F (37°C) or a rectal temperature above 99.5°F (37.5°C). In almost every instance, there is a specific constituent, the exogenous pyrogen (Greek *pyr,* fire, and *gennan,* to produce) of the infecting organism, that directly triggers fever production. Examples include the gram-negative bacterial endotoxins; the N-acetylglucosamine-N-acetylmuramic acid polymer and/or a derived portion of the peptidoglycan cell wall material of both gram-positive and gram-negative bacteria; the soluble enterotoxin secreted by staphylococci; the erythrogenic toxin of group A streptococci; and many yet undefined soluble carbohydrate and protein moieties produced by viruses, bacteria, fungi, protozoa, and helminths.

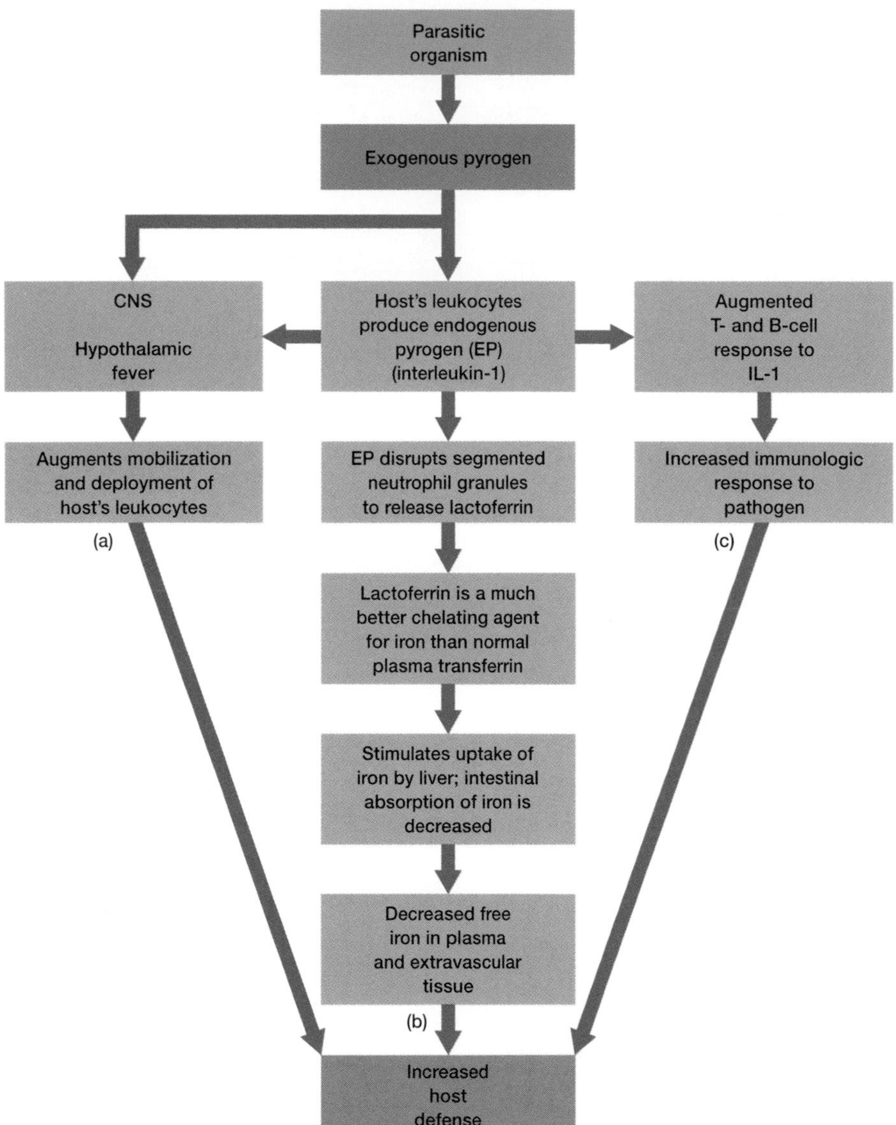

Figure 30.7 **Effects of Fever in A Mammalian Host.** (*a*) One of the first effects of a fever is to accelerate and augment the mobilization and deployment of the host's leukocytes, which, in turn, isolate and help destroy the pathogen. (*b*) A fever leads to a reduction of free plasma iron, which limits the growth of those organisms that require a narrow, crucial concentration of iron for their replication and synthesis of toxins. (*c*) One of the immunologic consequences of fever is the production of endogenous pyrogen (IL-1). IL-1 causes the proliferation, maturation, and activation of lymphocytes (T and B cells), which, in turn, augment the immunologic response of the host to the pathogen.

The fever induced by a parasitic organism augments the host's defenses by three complementary pathways (figure 30.7): (*a*) it stimulates leukocytes so that they can destroy the organism; (*b*) it enhances microbiostasis (growth inhibition) by decreasing iron available to the organism; and (*c*) it enhances the specific activity of the immune system.

Age

Generally, when the host is either very young or very old, susceptibility to infection increases. Babies are at a particular risk after their maternal immunity (*see chapter 31*) has disappeared and before their own immune system has matured. In very old persons, there is a decline in the immune system and in the homeostatic functioning of many organs that reduces host defenses.

Genetic Factors

The importance of genetic factors—both the pathogen's and the host's—to susceptibility has been recognized for many years. Often, one species is resistant to the infectious diseases that affect another species. For example, humans are not susceptible to canine distemper or hog cholera. Species resistance may depend upon genetically determined factors such as (1) temperature of the host (endothermic versus heterothermic animals); (2) genetically controlled metabolic, physiological, and anatomic differences that affect the ability of an organism to cause infection; and (3) food procurement mechanisms (herbivore, carnivore, omnivore). Even within species, genetically associated racial resistance factors affect the host's defense at times. For example, blacks, when compared to whites, are more resistant to falciparum malaria due to the sickle cell trait that they often possess.

Malaria (pp. 793–95).

1. Describe how each of the following general barriers contribute to the general defense of the host: nutrition, acute-phase reactants, fever, age, and genetics.
2. What is a siderophore?

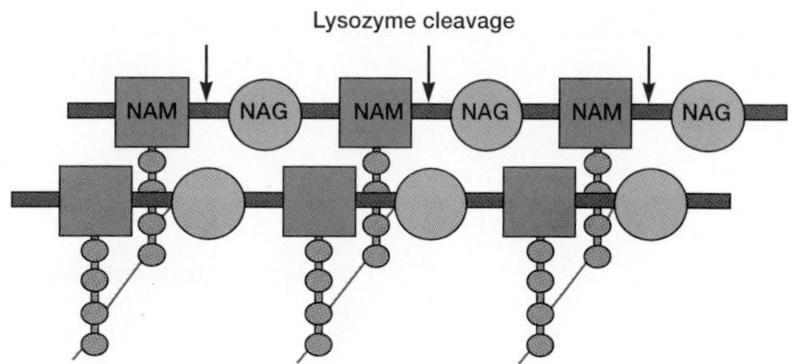

Figure 30.8 Action of Lysozyme on the Cell Wall of Gram-Positive Bacteria. In the structure of the cell wall peptidoglycan backbone, the β $(1\rightarrow4)$ bonds connect alternating N-acetyglucosamine (*NAG*) and N-acetylmuramic acid (*NAM*) residues. The chains are cross-linked through the tetrapeptide side chains. Lysozyme splits the molecule at the places indicated by the arrows.

Physical Barriers

Physical or mechanical barriers, along with the host's secretions (flushing mechanisms), are the first line of defense against parasitic organisms. Protection of the most important body surfaces by this mechanism is discussed next.

Skin and Mucous Membranes

The intact skin forms a very effective mechanical barrier to parasitic invasion. There are several reasons for this:

1. Few organisms have the innate ability to penetrate the skin because its outer layer consists of thick, closely packed keratinized cells (keratins are scleroproteins comprising the main components of hair, nails, and outer skin cells) that organisms cannot enzymatically attack.
2. Continuous shedding of the outer squamous epithelial cells (desquamation) removes those organisms that do manage to adhere.
3. Relative dryness of the skin slows microbial growth.
4. Mild acidity (pH 5 to 6, due to the breakdown of lipids into fatty acids by the normal skin microbiota) inhibits the growth of many organisms.
5. The normal skin microbiota acts antagonistically against many pathogens; it also occupies attachment sites and competes for nutrients.
6. Sebum liberated from the oil (sebaceous) glands forms a protective film over the surface of the skin.
7. Normal washing (by humans) continually removes organisms.

The mucous membranes of the respiratory, digestive, and urogenital systems withstand parasitic organisms because the intact stratified squamous epithelium and mucous secretions form a protective covering that resists penetration and traps many microorganisms. Furthermore, many mucosal surfaces are bathed in specific antiparasitic secretions. For example, cervical mucus, prostatic fluid, and tears are toxic to many bacteria. One antibacterial substance is lysozyme (muramidase), an enzyme that lyses bacteria by hydrolyzing the $\beta(1\rightarrow4)$ bond connecting N-acetylmuramic acid and N-acetylglucosamine in the bacterial cell wall peptidoglycan—especially in gram-positive organisms (figure 30.8). These mucous secretions also contain specific immune proteins that help prevent the attachment of organisms and significant amounts of iron-binding proteins (lactoferrin) that sequester iron away from these organisms.

Respiratory System

The mammalian respiratory system has formidable defense mechanisms. Once inhaled, an organism must first survive and penetrate the air filtration system of the upper and lower respiratory tracts. Because the airflow in these tracts is very turbulent, organisms are deposited on the moist, sticky mucosal surfaces. The cilia in the nasal cavity beat toward the pharynx, so that mucus with its trapped microorganisms is moved toward the mouth and expelled. Humidification of the air by the nasal turbinates causes many hygroscopic organisms to swell and aids the phagocytic process.

The **mucociliary blanket** of the respiratory epithelium traps organisms less than 10 μm in diameter that are deposited on the mucosal surface and transports them by ciliary action away from the lungs. Organisms larger than 10 μm are usually trapped by hairs and cilia lining the nasal cavity. Coughing and sneezing reflexes clear the respiratory system of organisms by expelling air forcefully from the lungs through the mouth and nose, respectively. Salivation also washes organisms from the mouth and nasopharyngeal areas into the stomach.

Intestinal Tract

Once parasitic organisms reach the stomach, many are killed by its gastric juice (a mixture of hydrochloric acid, enzymes, and mucus). The very high acidity of gastric juice (pH 2 to 3) is usually sufficient to destroy most organisms and their toxins, although exceptions exist (protozoan cysts, *Clostridium* and *Staphylococcus* toxins). However, many organisms are protected by food particles and reach the small intestine.

Once in the small intestine, pathogens often are damaged by various pancreatic enzymes, bile, enzymes in intestinal secretions, and secretory IgA antibody (*see chapter 31*). **Peristalsis** (Greek *peri,* around, and *stalsis,* contraction) and the normal loss of columnar epithelial cells act in concert to purge intestinal microorganisms. In addition, the normal microbiota of the large intestine (*see figure 29.8*) is extremely important in preventing the establishment of pathogenic organisms. For example, many normal commensals in the intestinal tract produce metabolic products, such as fatty acids, that prevent "unwanted" organisms from colonizing. Other normal microbiota take up attachment sites and compete for nutrients.

Secretory IgA (p. 615)

Genitourinary Tract

Under normal circumstances, the kidneys, ureters, and urinary bladder of mammals are sterile. Urine within the urinary bladder is also sterile. However, in both the male and female, a few bacteria are usually present in the distal portion of the urethra (*see figure 29.8*).

The factors responsible for this sterility are complex. For example, urine kills some bacteria due to its low pH and the presence of urea and other metabolic end products (uric acid, hippuric acid, indican, fatty acids, mucin, enzymes). The kidney medulla is so hypertonic that few organisms can survive. The lower urinary tract is flushed with urine and some mucus 4 to 10 times each day, eliminating potential pathogens. In males, the anatomical length of the urethra (20 cm) provides a distance barrier that excludes microorganisms from the urinary bladder. Conversely, the short urethra (5 cm) in females is more readily traversed by microorganisms; this explains why general urinary tract infections are 14 times more common in females than in males.

The vagina has another unique defense. Under the influence of estrogens, the vaginal epithelium produces increased amounts of glycogen that acid-tolerant *Lactobacillus* species called Döderlein's bacilli (*see chapter 29*) degrade to form lactic acid. Normal vaginal secretions contain up to 10^8 Döderlein's bacilli per ml. Thus, an acidic environment (pH 3 to 5) unfavorable to most organisms is established. Cervical mucus also has some antibacterial activity.

The Eye

The conjunctiva is a specialized mucus-secreting epithelial membrane that lines the interior surface of each eyelid and the exposed surface of the eyeball. It is kept moist by the continuous flushing action of tears (lacrimal fluid) from the lacrimal glands. Tears contain large amounts of lysozyme and other antimicrobial substances.

---

1. Why is the skin such a good first line of defense against parasitic invasion?
2. How do intact mucous membranes resist parasitic invasion of the host?
3. Describe the different antimicrobial defense mechanisms that operate within the respiratory system of mammals.
4. What factors operate within the gastrointestinal system that help prevent the establishment of pathogenic organisms?
5. Except for the anterior portion of the urethra, why is the genitourinary tract a sterile environment?

Chemical Barriers

Mammalian hosts have a chemical arsenal with which to combat the continuous onslaught of parasitic organisms. Some of these chemicals (gastric juices, salivary glycoproteins, lysozyme, oleic acid on the skin, urea) have already been discussed with respect to the specific body site(s) they protect. In addition, tissue extracts, blood, lymph, and other body fluids contain a potpourri of defensive chemicals such as antibodies and complement (*see chapter 33*), fibronectin, hormones, beta-lysin and other polypeptides, interferons, and bacteriocins.

Fibronectin

Fibronectin is a high molecular weight glycoprotein that can interact with certain bacteria. For example, it binds to surface components of *S. aureus* and groups A, C, and G streptococci. This aids in the nonspecific clearance of the bacteria from the body. Fibronectin also covers the receptors of certain epithelial cells to block the attachment of many bacteria.

Hormones

The effects of various mammalian hormones on host defense mechanisms are just beginning to be unraveled. The depressive effects of the corticosteroids on the inflammatory response and immune system are well known. Estrogen's effect on the microbiota of the vagina varies during the menstrual cycle; nonspecific resistance increases with rises in estrogen concentration. The activity of testosterone and adrenal hormones in acne vulgaris caused by *Propionibacterium acnes* has been previously discussed (*see chapter 29*).

Beta-Lysin and Other Polypeptides

Beta-lysin is a cationic polypeptide released from blood platelets; it can kill some gram-positive bacteria by disrupting their plasma membranes. Other cationic polypeptides include leukins, plakins, and phagocytin. A zinc-containing polypeptide, known as the prostatic antibacterial factor, is an important antimicrobial substance secreted by the prostate gland in males.

Interferons

Interferons are a family of related low molecular weight, regulatory glycoproteins produced by many eucaryotic cells in response to numerous inducers: a virus infection, double-stranded RNA, endotoxins, antigenic stimuli, mitogenic (stimulating mitosis) agents, and many parasitic organisms capable of intracellular growth (*Listeria monocytogenes,* chlamydiae, rickettsias, protozoa). Interferons are usually species specific but virus nonspecific. Currently, human interferons are classified as IFN-α, IFN-β, and IFN-γ. Some of the ways in which

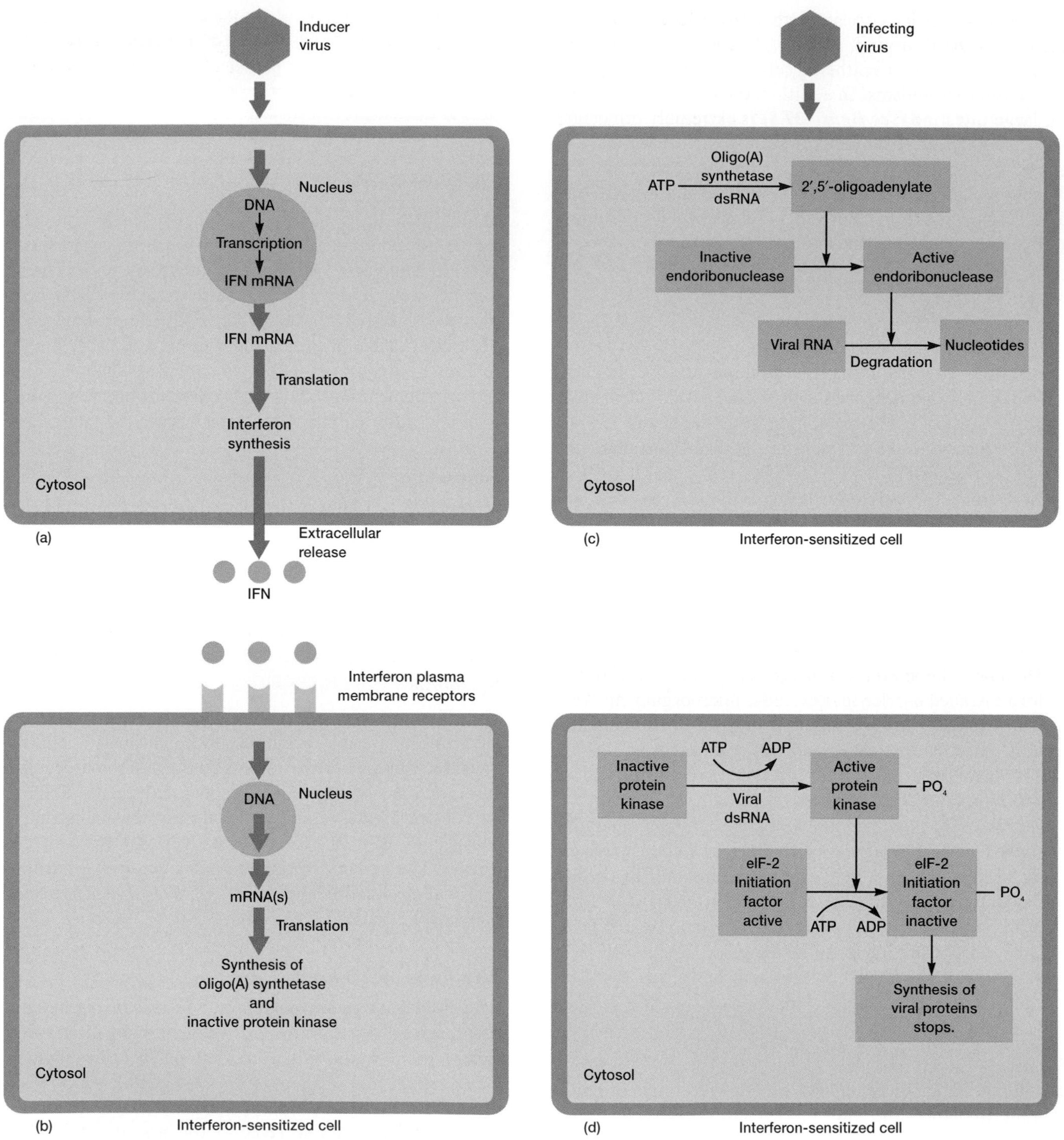

Figure 30.9 The Antiviral Action of Interferon. (*a*) Interferon (IFN) synthesis and release is often induced by a virus infection or double-stranded RNA (dsRNA). (*b*) Interferon binds to a ganglioside receptor on the plasma membrane of a second cell and triggers the production of enzymes that render the cell resistant to virus infection. The two most important such enzymes are oligo(A) synthetase and a special protein kinase. (*c*) When an interferon-stimulated cell is infected, viral protein synthesis is inhibited by an active endoribonuclease that degrades viral RNA. (*d*) An active protein kinase phosphorylates and inactivates the initiation factor eIF-2 required for viral protein synthesis.

interferon renders cells resistant to virus infections are described in figure 30.9.

Tumor Necrosis Factor Alpha

Tumor necrosis factor alpha (TNF-α) is released from monocyte or macrophage lineage cells in response to lipopolysaccharides or bacteria such as *Mycobacterium tuberculosis.* In addition to tumoricidal activity, TNF-α has many biological activities and has been recognized as an important inflammatory mediator. It affects a variety of cell types, including polymorphonuclear cells, endothelial cells, fibroblasts, and macrophages. Recently, it has been found that TNF-α can activate macrophages, which inhibits pathogen multiplication in vitro.

Bacteriocins

As previously noted, the first line of defense against parasitic organisms is the host's anatomical barrier, consisting of the skin and mucous membranes. These surfaces are colonized by normal microbiota, which by themselves provide a biological barrier against uncontrolled proliferation of foreign microorganisms. Many of these normal bacteria synthesize and release plasmid-encoded substances called **bacteriocins** that are lethal to related species. Bacteriocins may give their producers an adaptive advantage against other bacteria. Sometimes, they may increase bacterial virulence by damaging host cells such as mononuclear phagocytes.

Most bacteriocins that have been identified are proteins and are produced by gram-negative bacteria. For example, *E. coli* synthesizes bacteriocins called **colicins,** which are coded for by several different plasmids (ColB, ColE1, ColE2, ColI, and ColV). Some colicins bind to specific receptors on the cell envelope of sensitive target bacteria and cause cell lysis or attack specific intracellular sites such as ribosomes.

1. How does fibronectin prevent the establishment of a parasitic organism?
2. Give a specific example of how a hormone can affect the establishment of a parasitic organism.
3. How does beta-lysin function against gram-positive bacteria?
4. Describe how interferon confers a protective role to a mammalian cell against viral attack.
5. How do bacteriocins function?

Biological Barriers

Once an organism has breached the first line of defense, it encounters the host's second line of defense: cells that can be mobilized against invading organisms to form a living barrier. In addition to the normal indigenous microbiota, this second line of defense involves cells initially derived from bone marrow cells. Most of these are phagocytes of the **mononuclear phagocyte system (MPS)** or **reticuloendothelial system (RES).** This is the collection of macrophages, monocytes, and associated cells that are located in the liver, spleen, lymph nodes, and bone marrow (figures 30.10 and 30.11*a,b*). When pathogens in the blood or lymph pass by these cells, they are usually phagocytosed and destroyed.

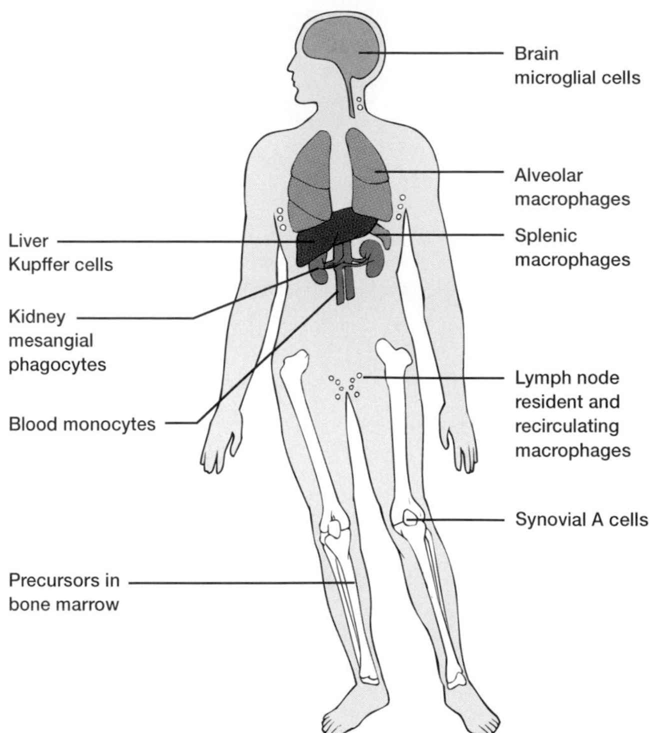

Figure 30.10 **The Reticuloendothelial System.** This system consists of tissue (such as found within the liver, spleen, and lymph nodes) containing "fixed" or immobile phagocytes that have specific names depending on their location.

Normal Indigenous Microbiota

As already mentioned briefly, another important factor in the general defense of the host against infections is the biological barrier established by the normal microbiota. In many parts of the body, conditions favor the growth of normal residents (*see figure 29.8*) and thus suppress colonization by pathogens. Some ways by which normal microbiota may inhibit pathogen colonization include

1. Producing bacteriocins toxic to other bacteria
2. Competing with potential pathogens for space and nutrients
3. Inhibiting infection by preventing pathogens from attaching to host surfaces
4. Influencing specific clearing mechanisms to rid the body or a particular area of pathogens

Inflammation

Inflammation is an important nonspecific defense reaction to tissue injury, such as that caused by a pathogen or a wound. The major events that occur during an inflammatory reaction are summarized in figure 30.12.

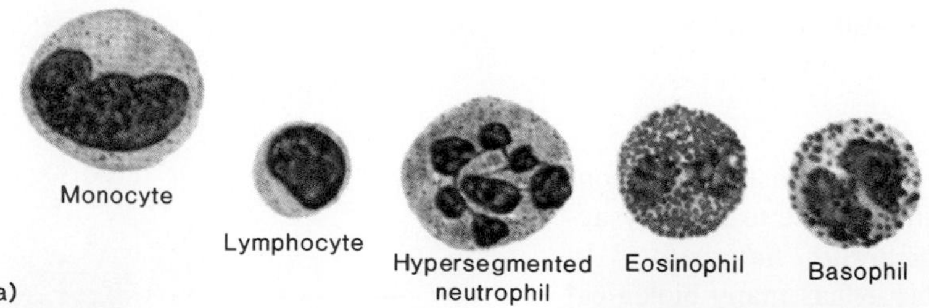

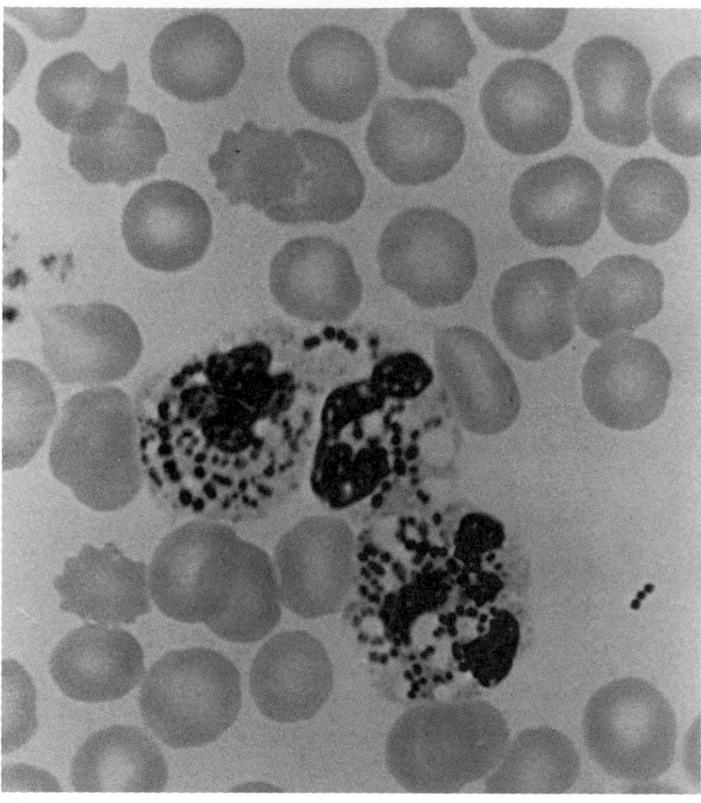
(b)

(c)

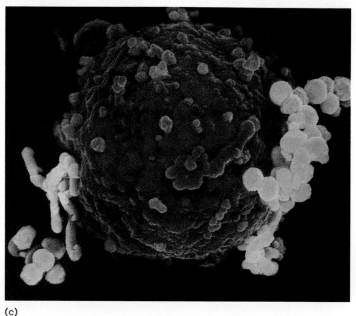

Figure 30.11 **Cells of the Immune System.** (*a*) An illustration of the five common types of leukocytes. (*b*) A light micrograph of neutrophils phagocytosing streptococci (×1,000). (*c*) An artificially colored scanning electron micrograph of a B lymphocyte (red) covered with bacteria (green) (×8,500). (a) *Illustration by Carolina Biological Supply Company.*

The inflammatory response is triggered by a complex set of events induced by pathogen invasion or tissue injury. Cells ruptured in the infected area release their cytoplasmic contents, which, in turn, raises the acidity in the surrounding extracellular fluid. This decrease in pH activates the extracellular enzyme kallikrein, which splits bradykinin from its long precursor chain. Bradykinin binds to receptors on the capillary wall, opening the junctions between cells and allowing fluid and infection-fighting leukocytes to leave the capillary and enter the infected tissue. Simultaneously, bradykinin (figure 30.13) binds to mast cells in the connective tissue associated with most small blood vessels. This activates the mast cells by causing an influx of calcium ions, which leads to degranulation and release of preformed mediators such as histamine. If nerves in the infected area are damaged, they release substance P, which also binds to mast cells, boosting preformed-mediator release. Histamine, in turn, makes the intercellular junctions in the capillary wall wider so that more fluid, leukocytes, kallikrein, and bradykinin precursors move out, causing edema. Bradykinin then binds to nearby capillary cells and stimulates the production of prostaglandins (PGE_2 and $PGF_{2\alpha}$) to promote tissue swelling in the infected area. Prostaglandins also bind to free nerve endings, making them fire and start a pain impulse.

The change in mast cell plasma membrane permeability associated with activation allows phospholipase A_2 to release arachidonic acid. Arachidonic acid is then metabolized by the cyclooxygenase or lipoxygenase pathways, depending on mast cell type. The newly synthesized mediators include prostaglandins E_2 and $F_{2\alpha}$, thromboxane A_2, slow reacting substance (SRS), and leukotrienes ($LTC_4 + LTD_4$). All play various roles in the inflammatory response.

These physiological changes lead to five classic symptoms of inflammation: redness, heat, pain, swelling, and altered function. The net effect is defensive in nature. In summary, the offending pathogen is neutralized and eliminated by a series of events, the most important of which are the following:

1. The increase in blood flow and capillary dilation bring into the area more antimicrobial factors and leukocytes. Dead cells also release antimicrobial factors.

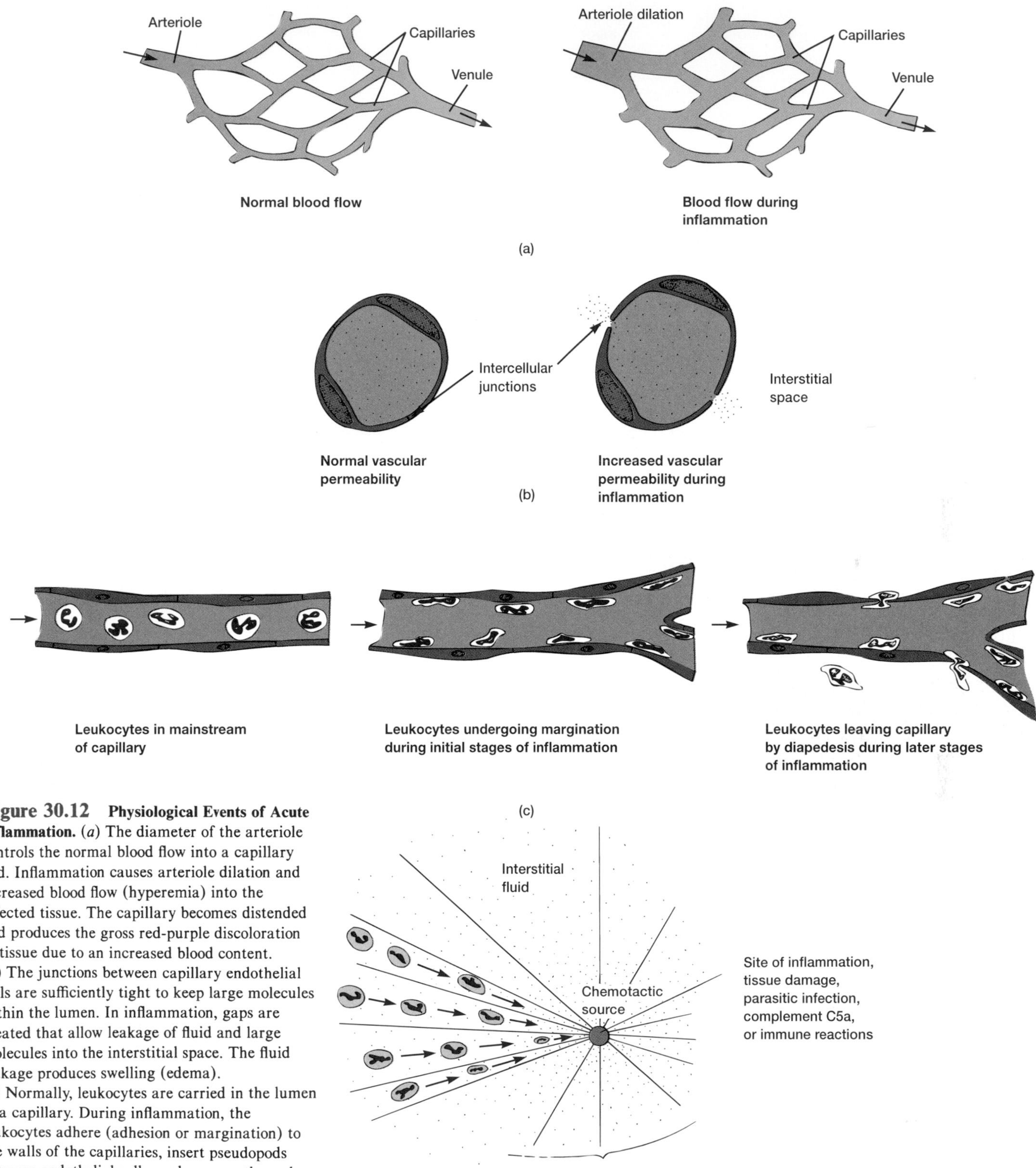

Figure 30.12 Physiological Events of Acute Inflammation. (*a*) The diameter of the arteriole controls the normal blood flow into a capillary bed. Inflammation causes arteriole dilation and increased blood flow (hyperemia) into the affected tissue. The capillary becomes distended and produces the gross red-purple discoloration of tissue due to an increased blood content. (*b*) The junctions between capillary endothelial cells are sufficiently tight to keep large molecules within the lumen. In inflammation, gaps are created that allow leakage of fluid and large molecules into the interstitial space. The fluid leakage produces swelling (edema). (*c*) Normally, leukocytes are carried in the lumen of a capillary. During inflammation, the leukocytes adhere (adhesion or margination) to the walls of the capillaries, insert pseudopods between endothelial cells, and squeeze through by a process called diapedesis. (*d*) After leaving the capillary, the leukocytes are attracted (chemotaxis) to the source of the inflammation by various chemotactic or chemokinetic substances. Once at the infection site, leukocytes phagocytose the microorganisms or dead tissue.

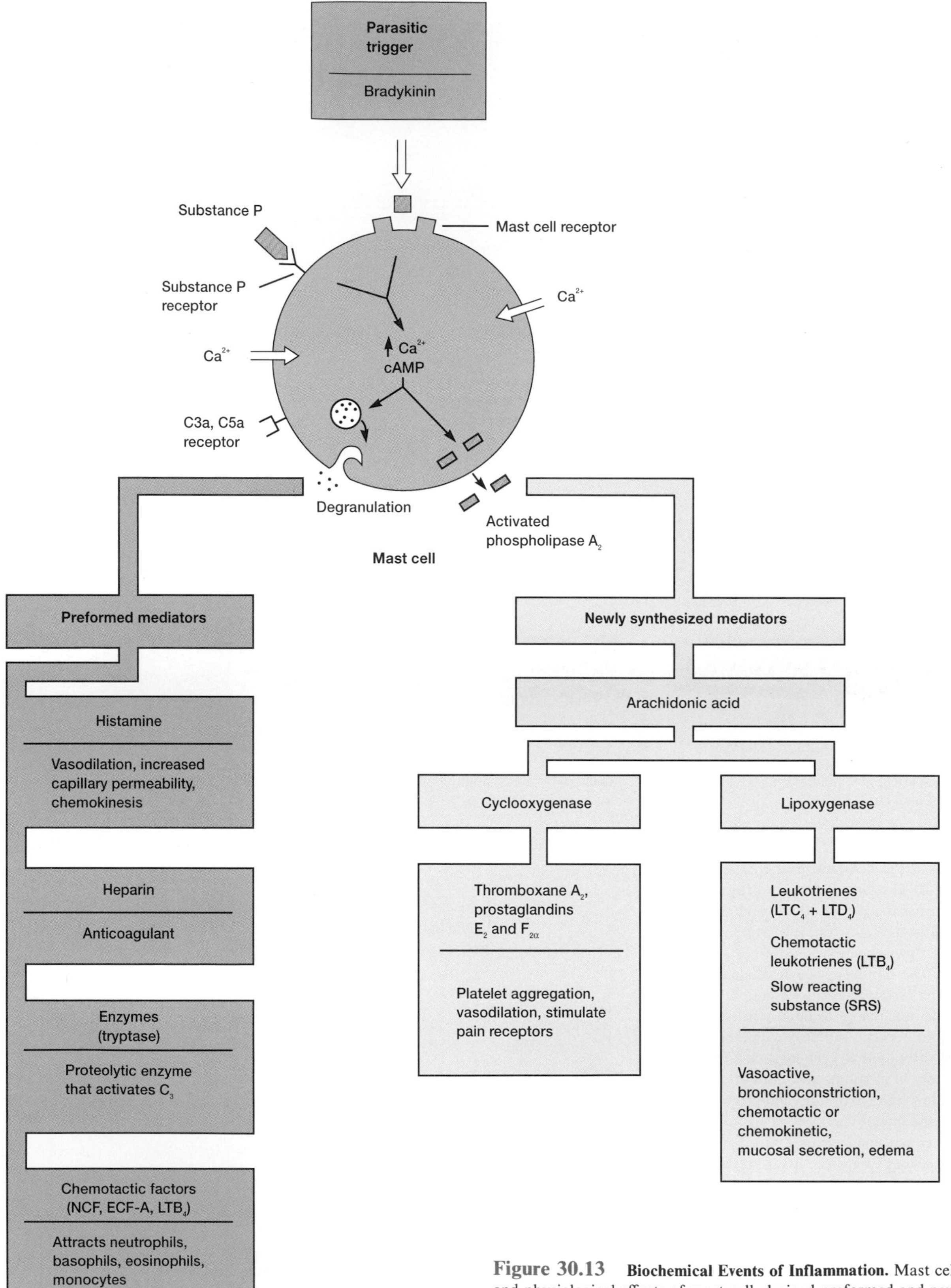

Figure 30.13 **Biochemical Events of Inflammation.** Mast cell activation and physiological effects of mast cell–derived preformed and newly synthesized mediators that lead to the inflammatory response.

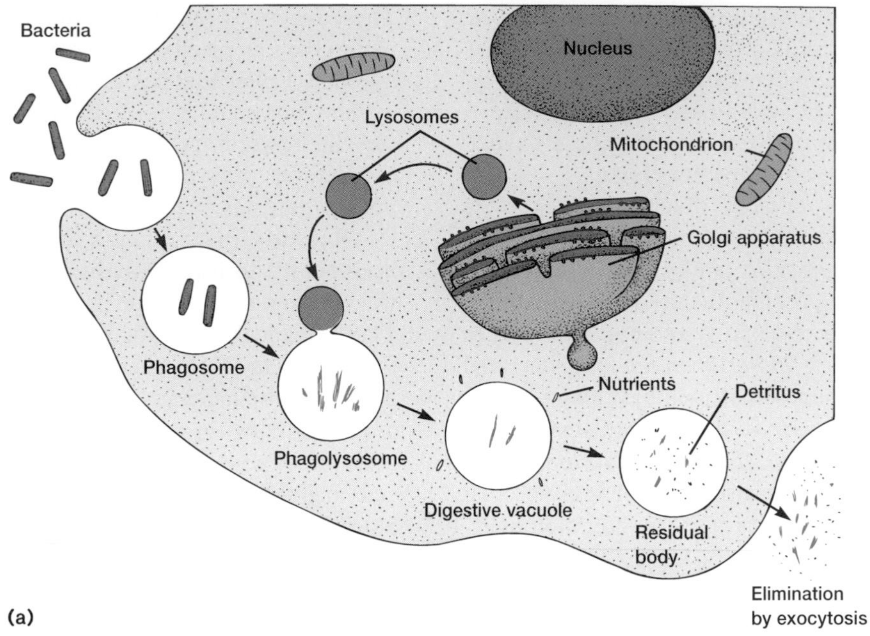

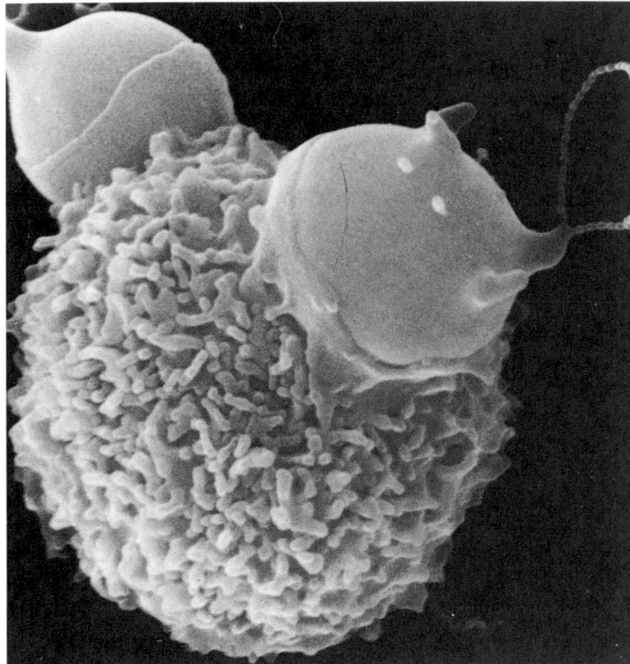

Figure 30.14 Phagocytosis. (*a*) Drawing of phagocytosis, showing ingestion, intracellular digestion, and exocytosis. (*b*) A liver macrophage phagocytoses two old, misshapen erthrocytes. The contents of the phagocytic vacuoles will be digested by lysosomal enzymes. Macrophages can also phagocytose and destroy pathogenic bacteria; scanning electron micrograph (×900).

2. The rise in temperature stimulates the inflammatory response and may inhibit microbial growth.
3. A fibrin clot often forms and may limit the spread of the invaders so that they remain localized.
4. Phagocytes collect in the inflamed area and phagocytose the pathogen. Neutrophils arrive first, followed later by macrophages.

---

1. What are some ways by which the normal indigenous microbiota of a host prevent colonization by pathogens?
2. What major events occur during an inflammatory reaction, and how do they contribute to pathogen destruction?
3. What causes degranulation of mast cells?
4. How do the cyclooxygenase and lipoxygenase pathways work?

Phagocytosis

Phagocytic cells (monocytes, macrophages, tissue macrophages, and neutrophils) must be able to recognize the invading microorganism (figure 30.11*a,b*). They do this with receptors on their surface that allow them to attach nonspecifically to a variety of organisms.

Once ingested by phagocytosis, the membrane-enveloped microorganism is delivered to lysosomes by fusion of the phagocytic vacuole (**phagosome**) with the lysosome membrane, forming a new vacuole called a **phagolysosome** (figure 30.14).

Because lysosomes seem to disappear from the cytoplasm during this fusion, the process is often called degranulation. Lysosomes contribute to the phagolysosome a variety of hydrolases such as lysozyme, phospholipase A_2, ribonuclease, deoxyribonuclease, and proteases. Collectively, these participate in the destruction of the entrapped organism.

Intracellular pathogens employ diverse strategies to survive within macrophages. For example, *Mycobacterium tuberculosis, M. leprae, Legionella pneumophila,* and *Toxoplasma gondii* inhibit phagosome fusion with lysosomes, thereby preventing exposure to toxic lysosomal contents. In contrast, *Trypanosoma cruzi, Listeria monocytogenes,* and *Shigella flexneri* lyse the phagosomal membrane and escape into the cytoplasm. A third group of microorganisms (*Leishmania* spp., *Mycobacterium lepraemurium, Salmonella typhimurium*) is found within the macrophage phagolysosomal compartment, where they apparently resist inactivation by lysosomal factors.

Lysosomes and endocytosis (pp. 76–77).

Besides the oxygen-independent lysosomal hydrolases, macrophage lysosomes contain oxygen-dependent enzymes that can produce reactive oxygen intermediates (ROIs) such as the superoxide radical ($O_2^- \cdot$), hydrogen peroxide (H_2O_2), singlet oxygen ($^1O_2$), hydroxyl radical ($OH \cdot$), and in the case of neutrophils, hypohalite ions (chloride, iodine, and bromide). Some reactions forming these toxic products are shown here.

Superoxide formation

$$NADPH + 2O_2 \xrightarrow{\text{NADPH oxidase}} 2O_2^- \cdot + H^+ + NADP^+$$

Hydrogen peroxide formation

$$2O_2^- \cdot + 2H^+ \xrightarrow{\text{superoxide dismutase}} H_2O_2 + O_2$$

Hypochlorous acid formation

$$H_2O_2 + Cl^- \xrightarrow{\text{myeloperoxidase}} HOCl + OH^-$$

Singlet oxygen formation

$$ClO^- + H_2O_2 \longrightarrow {}^1O_2 + Cl^- + H_2O$$

Hydroxyl radical formation

$$O_2^- \cdot + H_2O_2 \longrightarrow OH \cdot + OH^- + O_2$$

These reactions result from the **respiratory burst** that accompanies the increased oxygen consumption and ATP generation needed for phagocytosis. Because these reactions occur as soon as the phagosome is formed, lysosome fusion is not necessary for the respiratory burst. The toxic oxygen products and hypohalites thus produced are effective in killing invading microorganisms.

Recently, macrophages, neutrophils, and mast cells have been shown to form reactive nitrogen intermediates (RNIs) when stimulated by interferons or the tumor necrosis factor. These molecules include nitric oxide (NO) and its oxidized forms, nitrite (NO_2^-) and nitrate (NO_3^-). The RNIs are very potent cytotoxic agents, and may be either released from cells or generated within cell vacuoles. Nitric oxide is probably the most effective RNI. Macrophages produce it from the amino acid arginine when stimulated by cytokines. Nitric oxide can block cellular respiration by complexing with the iron in electron transport proteins. Macrophage killing of the herpes simplex virus, the protozoa *Toxoplasma gondii* and *Leishmania major,* the opportunistic fungus *Cryptococcus neoformans,* the metazoan pathogen *Schistosoma mansoni,* and tumor cells involves RNIs.

Neutrophils also produce a newly defined family of broad-spectrum antimicrobial peptides called **defensins.** There are four human defensins, called human neutrophil proteins (HNPs): HNP-1, 2, 3, and 4. These defensins are synthesized by myeloid precursor cells during their sojourn in the bone marrow, and are then stored in the cytoplasmic granules of mature cells. This compartmentation strategically locates defensins for extracellular secretion or delivery to phagocytic vacuoles. Susceptible microbial targets include a variety of gram-positive and gram-negative bacteria, yeasts and molds, and some viruses. Defensins act against bacteria and fungi by permeabilizing cell membranes. They form voltage-dependent membrane channels that allow ionic efflux. Antiviral activity involves direct neutralization of enveloped viruses; nonenveloped viruses are not affected by defensins.

1. Once a phagolysosome forms, how is the entrapped organism destroyed?
2. What is the purpose of the respiratory burst that occurs within macrophages?
3. What are defensins? How do they function?

Summary

1. Parasitism is a type of symbiosis between two species in which the smaller organism is physiologically dependent on the larger one, termed the host. The parasitic organism usually harms its host in some way.
2. An infection is the colonization of the host by a parasitic organism. An infectious disease is the result of the interaction between the parasitic organism and its host, causing the host to change from a state of health to a diseased state. Any organism that produces such a disease is a pathogen.
3. Pathogenicity refers to the quality or ability of an organism to produce pathological changes or disease. Virulence refers to the degree or intensity of pathogenicity of an organism and is measured experimentally by the LD_{50} or ID_{50}.
4. Pathogens or their products can be transmitted to a host by either direct or indirect means. Transmissibility is the initial requisite in the establishment of an infectious disease.
5. Special adherence factors allow pathogens to bind to specific receptor sites on host cells and colonize the host.
6. Pathogens can enter host cells by both active and passive mechanisms. Once inside, they can produce specific products and/or enzymes that promote dissemination throughout the body of the host.
7. The pathogen generally seeks out the area of the host's body that provides the most favorable conditions for its growth and multiplication.
8. Intoxications are diseases that result from the entrance of a specific toxin into a host. The toxin can induce the disease in the absence of the toxin-producing organism. Toxins produced by pathogens can be divided into two main categories: exotoxins and endotoxins.
9. Exotoxins are soluble, heat-labile, toxic proteins produced by the pathogen as a result of its normal metabolism. They can be categorized as neurotoxins, cytotoxins, or enterotoxins. Most exotoxins conform to the AB model in which the A subunit or fragment is enzymatic and the B subunit or fragment, the binding portion. Several mechanisms exist by which the A component enters target cells.
10. Endotoxins are heat-stable, toxic substances that are part of the cell wall lipopolysaccharide of some gram-negative bacteria. Most endotoxins function by initially activating Hageman Factor, which, in turn, activates one to four humoral systems. These include the intrinsic blood clotting cascade, complement activation, fibrinolytic system, and kininogen system. Endotoxins also stimulate macrophages to release cytokines such as IL-1, IL-6, and TNF-α.
11. Many direct factors (nutrition, physiology, age, fever, genetics, race) and equally as many indirect factors (personal hygiene, socioeconomic status, environmental living conditions) contribute in some degree to all host-parasite relationships. At times, they favor the establishment of the parasitic organism; at other times they provide some measure of defense to the host.
12. Physical (mechanical) barriers along with host secretions (flushing mechanisms) are the host's first line of defense against pathogens. Examples include the skin and mucous membranes; and the epithelia of the respiratory, gastrointestinal, and genitourinary systems.
13. Mammalian hosts have specific chemical barriers that help combat the continuous onslaught of parasitic organisms. Examples include fibronectin, hormones, interferons, beta-lysin and other cationic polypeptides.
14. Many normal bacteria found on the skin and mucous membranes produce plasmid-encoded proteins called bacteriocins. These are released into the host's environment and are lethal to related bacterial species.
15. The second line of defense that mammalian hosts possess against parasitic invasion consists predominately of phagocytotic cells of the reticuloendothelial system and are widely dispersed throughout the body of the host.
16. Inflammation is one of the host's nonspecific defensive reactions to a tissue injury that may be caused by a pathogen. Some major events that occur are increased blood supply, increased capillary permeability, leukocyte migration, and phagocytosis.
17. Phagocytosis involves the recognition, ingestion, and destruction of parasitic organisms by lysosomal enzymes, superoxide radicals, hydrogen peroxide, defensins, RNIs, and hypohalite ions.

Key Terms

AB model *584*
acute-phase reactant *590*
alpha hemolysis *590*
antitoxin *584*
bacteriocin *595*
beta hemolysis *590*
beta-lysin *593*
coagulation *588*
colicin *595*
cytotoxins *584*
defensins *600*
ectoparasite *579*
endogenous pyrogen *589*
endoparasite *579*
endotoxin *587*
endotoxin units (E.U.) *587*
enterotoxins *584*
exotoxin *584*
fibronectin *593*
final host *579*
fomites *581*
hemolysin *589*
host *579*
hyperferremia *590*
hypoferremia *590*
infection *579*
infectious disease *579*
infective dose 50 (ID_{50}) *580*
infectivity *580*
inflammation *595*
interferon *593*
intermediate host *579*
intoxication *584*
invasiveness *580*
lethal dose 50 (LD_{50}) *580*
leukocidin *589*
leukostatin *589*
mononuclear phagocyte system (MPS) *595*
mucociliary blanket *592*
neurotoxins *584*
parasite *579*
parasitic organism *579*
parasitism *579*
pathogen *579*
pathogenic potential *580*
pathogenicity *579*
peristalsis *593*
phagolysosome *599*
phagosome *599*
reservoir host *579*
respiratory burst *600*
reticuloendothelial system (RES) *595*
septicemia *583*
streptolysin-O (SLO) *589*
streptolysin-S (SLS) *590*
toxemia *584*
toxigenicity *580*
toxin *580*
toxoid *584*
transfer host *579*
vectors *581*
virulence *580*

Questions for Thought and Review

1. Why does a parasitic organism not have to be a parasite?
2. In general, infectious diseases that are commonly fatal are newly evolved relationships between the parasitic organism and the host. Why is this so?
3. What does an organism require to be parasitic?
4. What determinants provide the parasitic organism with the ability to colonize and invade the host?
5. How do the degree and lethality of the pathogenic alterations induced by a parasitic organism influence its establishment?
6. How do most exotoxins enter a host cell?
7. What is the mode of action of several well-known exotoxins?
8. What is the difference between the general properties of endotoxins and exotoxins?
9. How can general nonspecific barriers help to prevent the establishment of a parasitic organism within a host? How could they aid in the establishment of a parasitic organism?
10. What constitutes a mammalian host's first line of defense against invasion by a pathogen?
11. Describe how the physiology of each of the following systems contributes to the defense of the host: respiratory, gastrointestinal, and genitourinary.
12. How do specific chemicals produced by a mammalian host contribute to its general defense against infectious organisms?
13. How can inflammation be beneficial to a host and detrimental to a parasitic organism?
14. What is the relationship between the respiratory burst and phagocytosis?
15. What role do defensins play in host defense?

Additional Reading

Aggeler, J., and Webb, Z. 1982. Initial events during phagocytosis by macrophages viewed from outside and inside the cell. *J. Cell. Biol.* 94:613–23.

Atkins, E. 1984. Fever: The old and the new. *J. Inf. Dis.* 149:339–47.

Ayoub, E. M., ed. 1990. *Microbial determinants of virulence and host response.* Washington, D.C.: American Society for Microbiology.

Balkwill, F. R. 1986. Interferons: From molecular biology to man. Part 2. Interferons and cell function. *Microbiol. Sci.* 3(8):229–33.

Beaman, L., and Beaman, B. L. 1984. The role of oxygen and its derivatives in microbial pathogenesis and host defense. *Ann. Rev. Microbiol.* 38:27–48.

Bhakdi, S., and Tranum-Jensen, J. 1991. Alpha-toxin of *Staphylococcus aureus. Microbiol. Rev.* 55(4): 733–51.

Bizzini, B. 1979. Tetanus toxin. *Microbiol. Rev.* 43: 224–40.

Brubaker, R. R. 1985. Mechanisms of bacterial virulence. *Ann. Rev. Microbiol.* 39:21–50.

Burke, D. C. 1977. The status of interferon. *Sci. Am.* 236:42–62.

Collier, H. J. O. 1962. Kinins. *Sci. Am.* 207:111–18.

Costerton, J. W.: Geesey, G. G.; and Cheng, K.-J. 1978. How bacteria stick. *Sci. Am.* 238:86–91.

Edelson, R. L., and Fink, J. M. 1985. The immunologic function of the skin. *Sci. Am.* 256:46–53.

Eidels, L. R. L., and Hart, D. A. 1983. Membrane receptors for bacterial toxins. *Microbiol. Rev.* 47:596–614.

Gill, D. M. 1982. Bacterial toxins: A table of lethal amounts. *Microbiol. Rev.* 46:86–88.

Gordon, J., and Minks, M. A. 1981. The interferon renaissance: Molecular aspects of induction and action. *Microbiol. Rev.* 45:244–51.

Horowitz, M. A. 1982. Phagocytosis of microorganisms. *Rev. Inf. Dis.* 4:104–8.

Isenberg, H. D. 1988. Pathogenicity and virulence: Another view. *Clin. Microbiol. Rev.* 1(1):40–53.

Johnson, J. R. 1991. Virulence factors in *Escherichia coli* urinary tract infection. *Clin. Microbiol. Rev.* 4(1):80–128.

Lancaster, Jr., J. R. 1992. Nitric oxide in cells. *Am. Scientist* 80(3):248–59.

Lehrer, R. I.; Ganz, T.; and Selsted, M. E. 1990. Defensins: Natural peptide antibiotics from neutrophils. *ASM News* 56(6):315–18.

McNabb, D. C., and Tomasi, T. B. 1981. Host defense mechanisms at mucosal surfaces. *Ann. Rev. Microbiol.* 33:477–96.

Middlebrook, J. L., and Dorland, R. B. 1984. Bacterial toxins: Cellular mechanisms of action. *Microbiol. Rev.* 48:199–221.

Miller, J. F.; Mekalanos, J. J.; and Falkow, S. 1989. Coordinate regulation and sensory transduction in the control of bacterial virulence. *Science* 243: 916–22.

Moody, M. D. 1986. Microorganisms and iron limitation. *BioScience* 36:618–23.

Moss, J., and Vaughan, M. 1990. *ADP-ribosylating toxins and G proteins.* Washington, D.C.: American Society for Microbiology.

Patole, M. S., and Ramasarma, T. 1988. Role of H_2O_2 in phagocytosis and parasitism. *Biochem. Ed.* 16(2):58–62.

Petri, W. A. 1991. Invasive amebiasis and the galactose-specific lectin of *Entamoeba histolytica. ASM News* 57(6):299–306.

Rietschel, E. T., and Brade, H. 1992. Bacterial endotoxins. *Sci. Am.* 267(2):54–61.

Roberts, N. J. 1979. Temperature and host defense. *Microbiol. Rev.* 43:241–45.

Rumyantsev, S. N. 1992. Observations on constitutional resistance to infection. *Immunol. Today* 13(5): 184–87.

Schantz, E. J., and Johnson, E. A. 1992. Properties and use of botulinum toxin and other microbial neurotoxins in medicine. *Microbiol. Rev.* 56(1): 80–99.

Schlesinger, R. B. 1982. Defense mechanisms of the respiratory system. *BioScience* 32(1):45–50.

Slifkin, M., and Doyle, R. J. 1990. Lectins and their application to clinical microbiology. *Clin. Microbiol. Rev.* 3(5):197–218.

Smith, H. 1977. Microbial surfaces in relation to pathogenicity. *Bacteriol. Rev.* 41:475–500.

Stephen, J., and Pietrowski, R. A. 1986. *Bacterial toxins,* 2d ed. Washington, D.C.: American Society for Microbiology.

Sugiyama, H. 1980. *Clostridium botulinum* neurotoxin. *Microbiol. Rev.* 44:419–48.

Taylor, P. W. 1983. Bactericidal and bacteriolytic activity against gram-negative bacteria. *Microbiol. Rev.* 47:46–65.

Waters, V. L., and Crosa, J. H. 1991. Colicin V virulence plasmids. *Microbiol. Rev.* 55(3):437–50.

Part Eight
The Fundamentals of Immunology

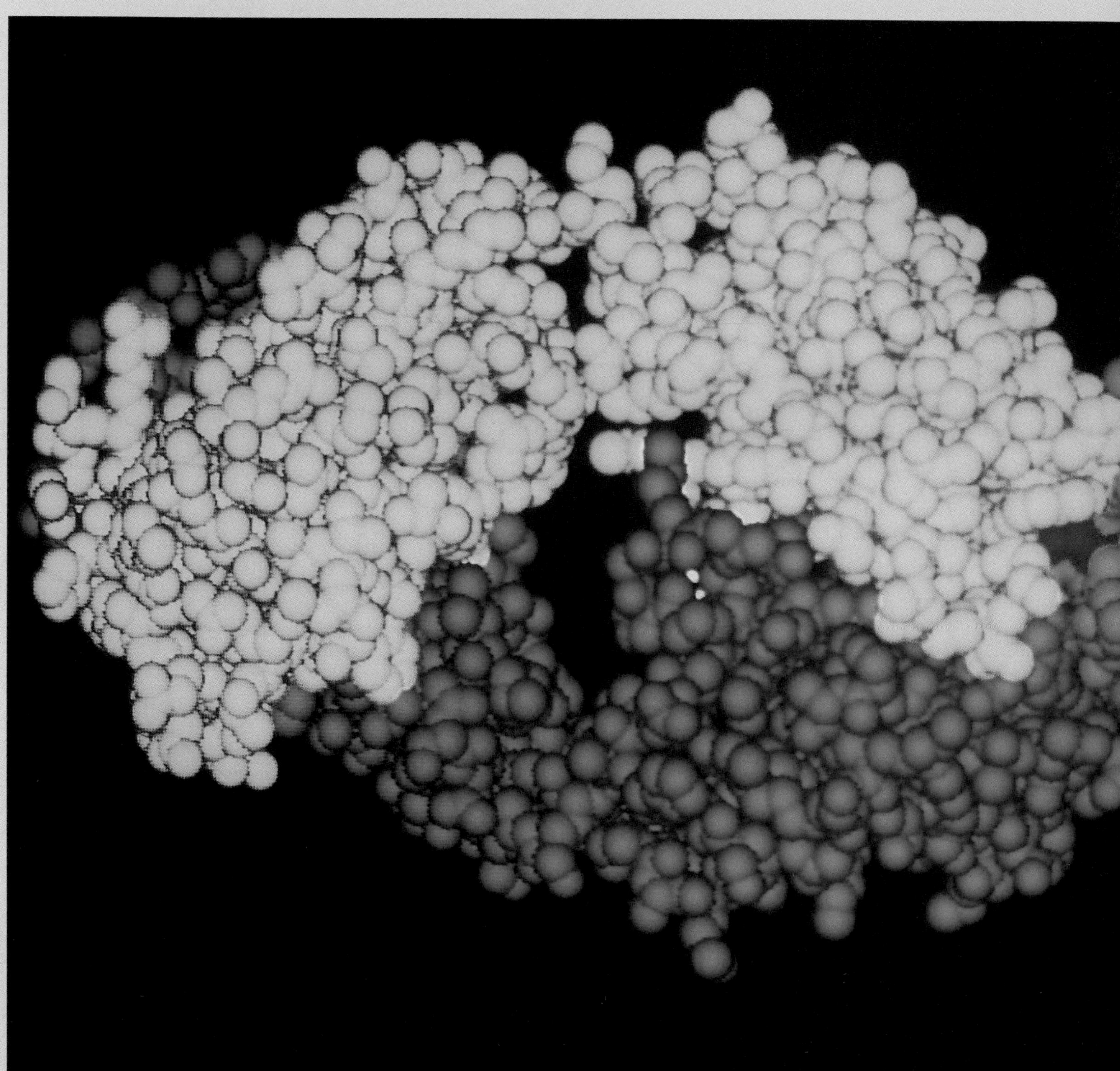

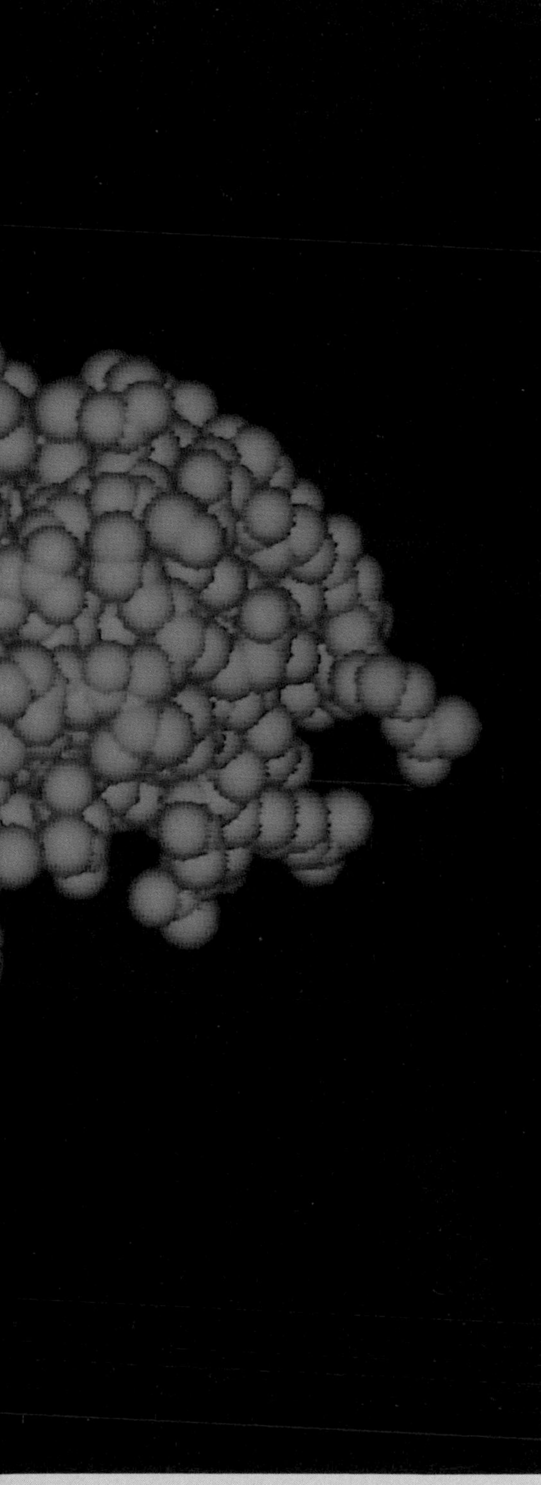

Chapter 31

The Immune Response: Antigens and Antibodies

Chapter 32

The Immune Response: Chemical Mediators, B- and T-Cell Biology, and Immune Disorders

Chapter 33

The Immune Response: Antigen-Antibody Reactions

A model of the Fab part of an antibody molecule and the antigen with which it combines. The heavy chain domains V_H and C_H1 are in blue and the light chain domains V_L and C_L are in yellow. The antigen is shown in green.

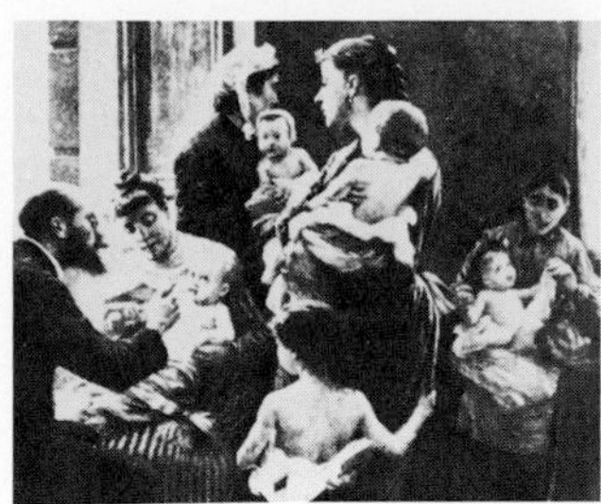

CHAPTER 31

The Immune Response: Antigens and Antibodies

The remarkable capacity of the immune system to respond to many thousands of different substances with exquisite specificity saves us all from certain death by infection.

—Martin C. Raff

Outline

Concepts

1. The major function of the immune response in higher animals is to provide specific protection (immunity) against harmful microorganisms, cancer cells, and certain macromolecules that are collectively termed antigens (immunogens).
2. Terminology has been developed to describe the different types of immunity: nonspecific versus specific, natural versus artificial, and active versus passive.
3. Two major classes of lymphocytes, B cells and T cells, specifically recognize and respond to antigens. B cells form immune products called immunoglobulins (antibodies) whereas T cells become activated or sensitized to perform several functions.
4. Immunoglobulins specifically interact with free antigens or cells bearing antigens and mark them for subsequent removal.
5. The basic structure of the immunoglobulin molecule consists of four polypeptide chains: two heavy chains and two light chains.
6. There are five immunoglobulin (Ig) classes based on physicochemical and biological properties. These are IgG, IgM, IgA, IgD, and IgE.
7. Immunoglobulins can be produced naturally in an animal's body in response to immunizations. They, along with catalytic antibodies, also can be produced in vitro through the use of hybridomas.
8. Two distinguishing characteristics of immunoglobulins are their diversity and specificity.

The nonspecific host defenses that are integral components of an animal's resistance to infectious agents or their products are discussed in the previous chapter. In this chapter, another aspect of defense called specific immunity is examined. Specific immunity, in contrast to nonspecific defenses or resistance, involves the production of a highly specific defensive response when the host is exposed to an infectious agent or microorganism, its products, some tumor cells, or certain macromolecules. The body recognizes these substances as not belonging to itself and develops a specific immune response, leading to destruction or neutralization of the substances. The necessity of this system to human survival and the problems that it can cause under certain circumstances are presented in this chapter and the next.

As discussed in chapter 30, vertebrates (including humans) are continuously exposed to microorganisms, their metabolic products, or other foreign macromolecules that can cause disease. Fortunately, these animals are equipped with an immune system that protects them against this exposure. **Immunity** (Latin *immunis,* free of burden) refers to the general ability of a host to resist a particular disease. The **immune response** that results is a specific and complex series of defensive actions widely distributed throughout the animal's body (figure 31.1). Each particular immune response is a unique local sequence of events, shaped by the nature of the challenge. **Immunology** is the science that deals with these immune responses and the many phenomena responsible for this type of defense.

Nonspecific Resistance

Nonspecific resistance refers to those general mechanisms inherited as part of the innate structure and function of each animal. These act in concert as a first line of defense against infectious microorganisms, their products (toxins), or foreign macromolecules before they cause disease. The various nonspecific defense mechanisms include biological barriers (inflammation, phagocytosis), chemical barriers (enzymatic action, interferons, beta-lysin, fibronectin, complement), general barriers (fever), and physical barriers (skin, mucous membranes). These are discussed in detail in chapter 30.

Specific Immunity

If the nonspecific defenses are breached, specific immunity (the immune response) is called upon to protect the host. This system consists of several immunologic mechanisms in which lymphocytes recognize the presence of particular foreign agents or substances termed antigens (immunogens) and act to eliminate them. Elimination can occur by direct lymphocytic destruction of the antigens or by formation of specialized proteins

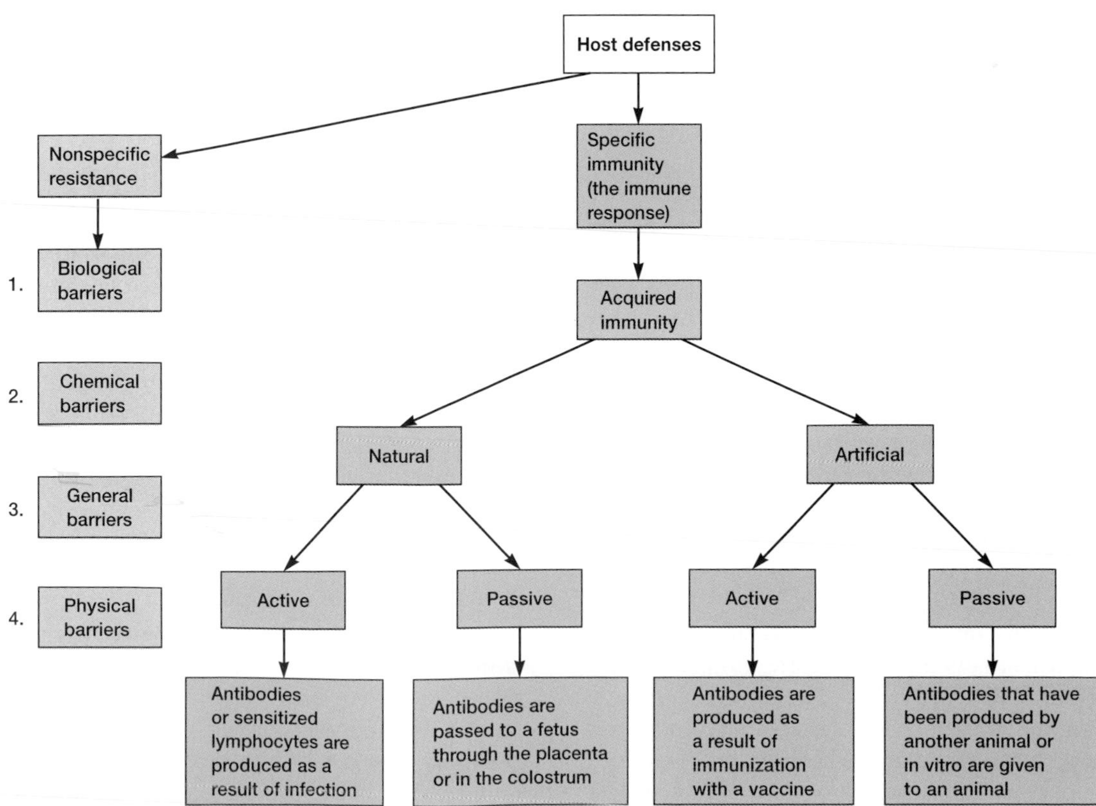

Figure 31.1 Different Host Defenses. This illustration can be used as a guide to the text description of the immune response. The details of the human immune system are astonishingly complex; many events have been omitted so that the overall organization of the immune response can be clearly seen.

TABLE 31.1 Examples of Vaccines Used by the United States to Prevent Viral and Bacterial Diseases in Humans

Disease	Vaccine	Booster	Recommendation
Viral Diseases			
Hepatitis B	HB viral antigen (Engerix-B, Recombivax HB)	None	High-risk medical personnel
Influenza	Inactivated virus or viral components	Yearly	Chronically ill individuals and those over 65
Measles Mumps Rubella	Attenuated viruses (combination MMR vaccine)	None	Children 15–19 months old
Poliomyelitis	Attenuated (oral poliomyelitis vaccine, OPV) or inactivated vaccine	Adults as needed	Children 2–3 years old
Rabies	Inactivated virus	None	For individuals in contact with wildlife, animal control personnel, veterinarians
Yellow fever	Attenuated virus	10 years	Military personnel and individuals traveling to endemic areas
Bacterial Diseases			
Cholera	Fraction of *Vibrio cholerae*	6 months	Individuals in endemic areas
Diptheria Pertussis Tetanus	Diptheria toxoid, killed *Bordetella pertussis,* tetanus toxoid (DPT vaccine)	10 years	Children 2–3 months old
Haemophilus influenzae type b	Polysaccharide-protein conjugate (HbCV) or bacterial polysaccharide (HbPV)	None	Children under 5 years of age
Meningococcal infections	Bacterial polysaccharides of serotypes A/C/Y/W-135	None	Military; high-risk individuals
Plague	Fraction of *Yersinia pestis*	Yearly	Individuals in contact with rodents in endemic areas
Pneumococcal pneumonia	Purified *S. pneumoniae* polysaccharide of 23 pneumococcal types	None	Adults over 50 with chronic disease
Tuberculosis	Attenuated *Mycobacterium bovis* (BCG vaccine)	3–4 years	Individuals exposed to TB for prolonged periods of time
Typhoid fever	Killed *Salmonella typhi*	3–4 years	Individuals in endemic areas
Typhus fever	Killed *Rickettsia prowazekii*	Yearly	Scientists and medical personnel in areas where typhus is endemic

called antibodies that either disrupt antigen function or target the antigen for destruction by other cells.

Acquired Immunity

Acquired immunity refers to the type of specific immunity a host develops after exposure to a suitable antigen, or after transfer of antibodies or lymphocytes from an immune donor. Acquired immunity can be obtained actively or passively by natural or artificial means.

Naturally Acquired Immunity

Naturally acquired active immunity occurs when an individual's immune system contacts an appropriate antigenic stimulus during normal daily activities. The immune system responds by producing antibodies and sensitized lymphocytes that inactivate or destroy the antigen. The immunity produced can be either lifelong, as with measles or chickenpox, or last for only a few years, as with tetanus (*see chapter 37*).

Naturally acquired passive immunity involves the transfer of antibodies from one host to another. For example, some of a pregnant woman's antibodies pass across the placenta to her fetus. If the female is immune to diseases such as polio or diphtheria, this placental transfer also gives the newborn immunity to these diseases. Certain other antibodies can pass from the female to her offspring in the first secretions (called **colostrum**) from the mammary glands. Unfortunately, naturally acquired passive immunity generally lasts only a short time (weeks or months at the most).

Artificially Acquired Immunity

Artificially acquired active immunity results when an animal is immunized with a vaccine. A **vaccine** consists of a preparation of killed microorganisms; living, weakened (attenuated) microorganisms; or inactivated bacterial toxins (toxoids) that are administered to an animal to induce immunity artificially. Table 31.1 summarizes the principal vaccines used to prevent viral and bacterial diseases in humans.

Artificially acquired passive immunity results when antibodies that have been produced either in an animal or by specific methods in vitro are introduced into a host. Although this type of immunity is immediate, it is short lived (about three weeks). An example would be botulinum antitoxin produced in a horse and given to a human suffering from botulism food poisoning (*see chapter 37*).

1. Give several examples of nonspecific resistance.
2. Contrast active and passive immunity.
3. How does naturally acquired immunity occur?
4. What is a vaccine?
5. Why is artificially acquired passive immunity at times lifesaving?

Origin of Lymphocytes

Because lymphocytes (table 31.2) are the pivotal cells of the specific immunologic response, their origin, movement, processing, and general function are summarized here. In chapter 32, their biology is described more thoroughly. Undifferentiated lymphocytes are derived from bone marrow stem cells. They are produced in the bone marrow at a very high rate (10^9 cells per day). Some lymphocytes migrate through the blood or lymphatic circulatory systems to the secondary lymphoid tissue (thymus, spleen, aggregated lymph nodules in the intestines, and lymph nodes) where they produce lymphocyte colonies.

The undifferentiated lymphocytes that migrate to the thymus undergo special processing and become **T cells** or **T lymphocytes.** (The letter T represents the thymus.) Some T cells are transported away from the thymus and enter the bloodstream where they comprise 70 to 80% of the circulating lymphocytes. Other T cells tend to reside in various organs of the lymphatic system, such as the lymph nodes and spleen. This thymus-dependent differentiation of T cells (or **thymocytes**) occurs during early childhood, and by adolescence the secondary lymphoid organs of the body generally contain a full complement of T cells.

B cells or **B lymphocytes** differentiate in the fetal liver and adult bone marrow (*see figure 30.11c*). (The letter B was originally derived from the **b**ursa of Fabricius, a specialized appendage of the cloaca of chickens where these lymphocytes differentiate.) B cells are distributed by the blood and make up 20 to 30% of the circulating lymphocytes (figure 31.2). They also settle in the various lymphoid organs along with the T cells. Figure 31.2 summarizes the development of the T and B cells.

There is a population of lymphoid cells that do not have characteristics of either T or B cells. These are called **null cells** because they lack the specific surface markers of B or T cells, and can be distinguished from them by the presence of cytoplasmic granules. It is currently believed that this population of cells contains most **natural killer (NK) cells** and antibody-dependent cytotoxic T cells (*see chapter 32*). Although these cells are probably of bone marrow origin, their exact lineage is uncertain.

TABLE 31.2 Classes of Lymphocytes

Lymphocyte	Role
T Cells	
T_H (helper) cells	Provide "assistance," or potentiate expression of immune function by other lymphocytes
T_S (suppressor) cells	Suppress or impair expression of immune function by other lymphocytes
T_C (cytotoxic) cells	Bring about cytolysis and cell death of "targets"
T_D cells T_{DTH} (delayed-type hypersensitivity) cells	Recruit and regulate a variety of nonspecific blood cells and macrophages in expression of delayed (Type IV) hypersensitivity reactions
T_R (regulator) cells	Control balance between enhancement and suppression of response to antigen
B Cells	
B lymphocytes	Proliferate/mature into antibody-producing cells
Plasma cells	Are mature, active antibody-producing cells
Null Cells	
Natural killer cells	Bring about cytolysis and death of target cells

Function of Lymphocytes

Plasma cells are fully differentiated antibody-synthesizing cells that are derived from B lymphocytes. They respond to antigens by secreting antibodies into the blood and lymph. Antibodies are glycoproteins produced by plasma cells after the B cells in their lineage have been exposed to antigens. Antibodies are specifically directed against the antigen that caused their formation. Because antibodies are soluble in blood and lymph fluids, they provide **humoral** (Latin *humor,* a liquid) **immunity** or **antibody-mediated immunity.** The humoral immune response defends mostly against bacteria, bacterial toxins, and viruses that enter the body's various fluid systems.

T cells do not secrete antibodies. Instead, they attack (1) host cells that have been parasitized by viruses or microorganisms, (2) tissue cells that have been transplanted from one host to another, and (3) cancer cells. They also produce **lymphokines,** chemical mediators that play specific augmenting and regulatory roles in the immune system. Since T

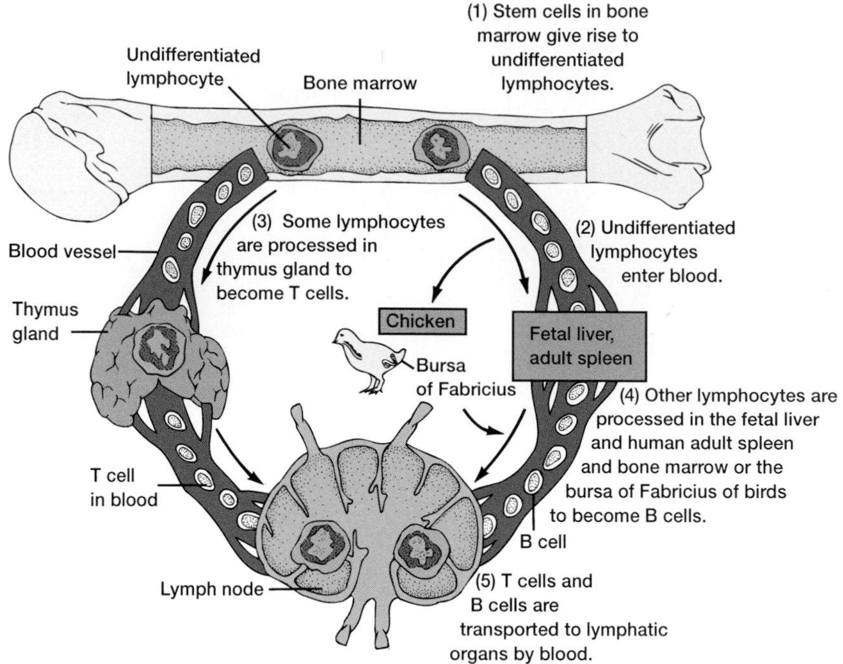

Figure 31.2 Schematic Representation of Lymphocyte Development. Bone marrow releases undifferentiated lymphocytes, which, after processing, become T and B cells.

cells must physically contact foreign cells or infected cells in order to destroy them, they are said to provide **cell-mediated immunity.**

Null cells (and particularly natural killer cells) destroy tumor cells and virus- and other parasite-infected cells. They also help regulate the immune response. Null cells often exhibit antibody-dependent cellular cytotoxicity. The biology of these three major lymphocyte categories is discussed in more detail in chapter 32 (*pp. 628–35*).

Antigens

The immune system distinguishes between "self" and "nonself" through an elaborate recognition process. Prior to birth, the body somehow makes an inventory of the proteins and various other large molecules present (self) and removes most T cells specific for self-determinants. Subsequently, self-substances can be distinguished from nonself substances, and lymphocytes can produce specific immunologic reactions against the latter, leading to their removal.

Nonself foreign substances, such as proteins, nucleoproteins, polysaccharides, and some glycolipids, to which lymphocytes respond are called **antigens** (**anti**body **gen**erator) or **immunogens.** Most antigens are large, complex molecules with a molecular weight generally greater than about 10,000. The ability of a molecule to function as an antigen depends on its size and the complexity of its structure.

Each antigen can have several **antigenic determinant sites** or **epitopes** (figure 31.3). These areas of the molecule stimulate production of and combine with specific antibodies. Antibodies are formed most readily in response to determinants

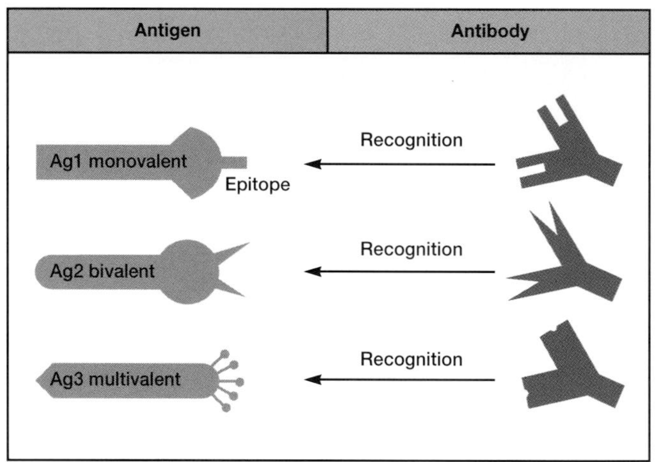

Figure 31.3 Antigens. Antigen (Ag) molecules each have a set of antigenic determinants or epitopes. The epitope on monovalent antigen (Ag1) is different from those on the bivalent antigen (Ag2). Some antigens (Ag3) have repeating epitopes; however, each antibody only recognizes one epitope rather than the whole multivalent antigen.

that project from the foreign molecule or to terminal residues of a specific polymer chain.

The number of antigenic determinant sites on the surface of an antigen is its **valence.** If one determinant site is present, the antigen is monovalent. Most antigens, however, have more than one determinant site and are termed multivalent. Multivalent antigens generally elicit a stronger immune response than do monovalent antigens. Sometimes a multivalent antigen, termed a **heterophile** or **heterologous antigen,** can react with antibodies produced in response to a different antigen.

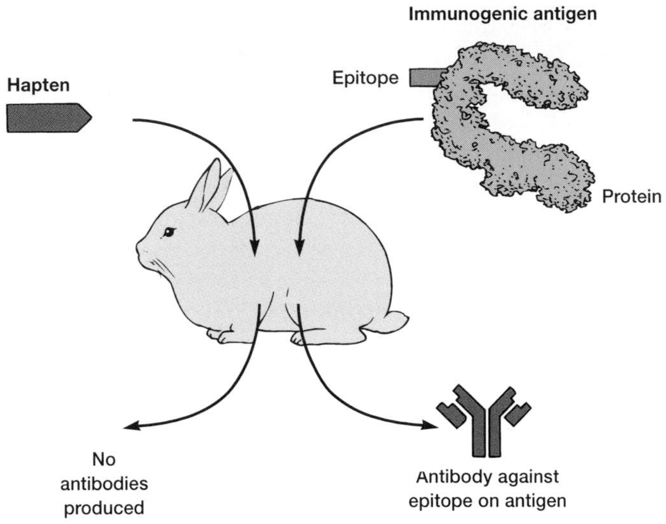

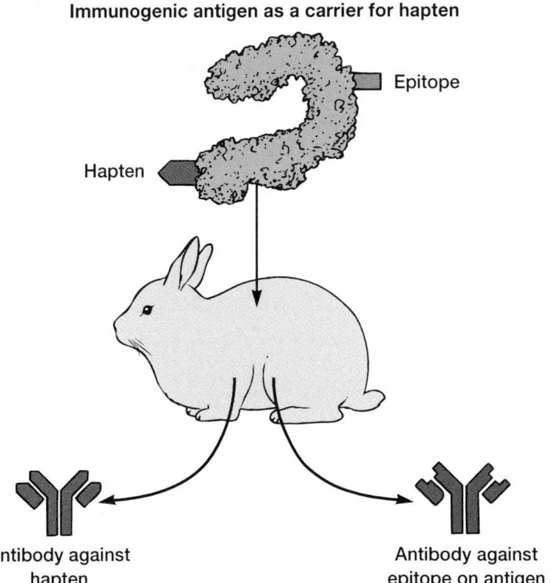

Figure 31.4 Effect of Carrier On Immunogenicity of Hapten. Haptens, as small molecules, are incapable of inducing an immune response when presented alone. This is markedly different from what occurs when a more complex, immunogenic antigen is presented. In this instance, antibody directed against the specific eptiope(s) of that antigen can be produced. However, if hapten is linked to a large molecule (which is used as a carrier), the hapten may then be immunogenic. Under these conditions, antibody is produced, not only against the epitope of the immunogenic carrier molecule but against the epitope of the hapten, as well.

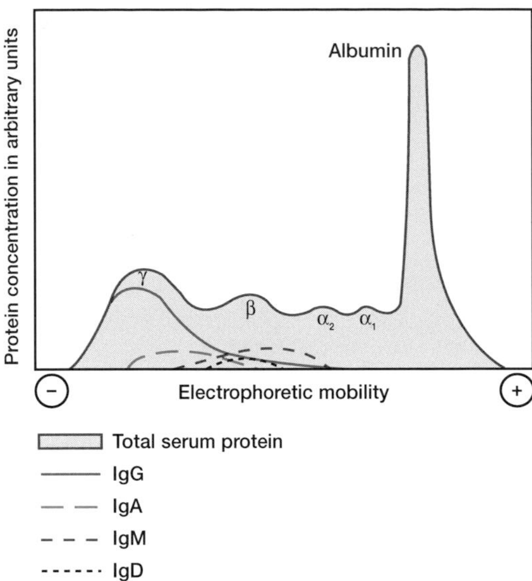

Figure 31.5 Electrophoresis of Human Serum. Schematic representation of electrophoretic results illustrating the distribution of serum proteins and the four major classes of immunoglobulins.

Haptens

Many small organic molecules are not antigenic by themselves but can become antigenic if they bond to a larger carrier molecule such as a protein (figure 31.4). They cannot stimulate antibody formation by themselves, but can react with antibodies once formed. Such small molecules are termed **haptens** (Latin *haptein,* to grasp). When lymphocytes are stimulated by the combined molecule, they can react to either the hapten or the larger carrier molecule. One example of a hapten is penicillin. By itself, penicillin is not antigenic. However, when it combines with certain serum proteins of sensitive individuals, the resulting molecule does initiate a severe and sometimes fatal allergic immune reaction. In these instances, the hapten is acting as an antigenic determinant on the carrier molecule.

1. What is the function of the thymus gland? The bursa of Fabricius? Null cells?
2. Contrast the humoral and cell-mediated immune responses.
3. Distinguish between "self" and "nonself" substances.
4. Define and give several examples of an antigen.
5. What is an antigenic-determinant site or epitope?
6. Describe a hapten.

Antibodies

Antibodies or **immunoglobulins** (**Ig**) are a group of glycoproteins present in the blood serum and tissue fluids of mammals. Serum glycoproteins can be separated according to their charge in an electric field (*see chapter 19, electrophoresis*) and classified as albumin, alpha-1 globulin, alpha-2 globulin, beta globulin, and gamma globulin (figure 31.5). Notice that the gamma globulin band is wide because it contains a heterogenous class of immunoglobulins; namely, IgG, IgA, IgM, and IgD. These differ from each other in molecular size, structure, charge, amino acid composition, and carbohydrate content. Electrophoretically, the immunoglobulins show a broad range of heterogeneity, which extends from the gamma to the alpha-2 fraction of serum. Most antibodies in serum are in the IgG class. A fifth immunoglobulin class, the IgE class, has a similar mobility to IgD but cannot be represented quantitatively because of its low concentration in serum.

Immunoglobulin Structure

Antibodies have more than one antigen-combining site. For example, most human antibodies have two combining sites or are bivalent (figure 31.3). Some bivalent antibody molecules can combine to form multimeric antibodies that have up to ten combining sites.

All immunoglobulin molecules have a basic structure composed of four polypeptide chains (figure 31.6*a*) connected to each other by disulfide bonds. Each light chain usually consists of about 220 amino acids and has a molecular weight of approximately 25,000. Each heavy chain consists of about 440 amino acids and has a molecular weight of 50,000 to 77,000. The heavy chains are structurally distinct for each immunoglobulin class or subclass. Both light and heavy chains contain two different regions. **Constant regions** (C_L and C_H) have amino acid sequences that do not vary significantly between antibodies of the same subclass. The **variable regions** (V_L and V_H) from different antibodies do have different sequences (figure 31.6*b*).

The four chains are arranged in the form of a flexible Y with a hinge region (figure 31.6*b*). This hinge region allows the antibody molecule to assume a T shape. The stalk of the Y is termed the **crystallizable fragment (Fc)** and contains the site at which the antibody molecule can bind to a cell. The top of the Y consists of two **antigen-binding fragments (Fab)** that bind with compatible antigenic determinant sites. The Fc fragments are composed only of constant regions, whereas the Fab fragments have both constant and variable regions. Intrachain disulfide bonds create loops within each antibody chain (figure 31.6*c*). A loop along with approximately 25 amino acids on each side is a **domain.** Interchain disulfide bonds link heavy and light chains together.

More specifically, the light chain exists in two distinct forms called kappa (κ) and lambda (λ). These can be distinguished by the amino acid sequence of the carboxyl portion of the chain (figure 31.7*a*). In human immunoglobulins, the carboxyl-terminal portion of all κ chains is identical; thus, this region is termed the constant (C_L) domain. With respect to the lambda chains, there are four very similar sequences that define the subtypes λ_1, λ_2, λ_3, and λ_4 with their corresponding constant regions $C_\lambda 1$, $C_\lambda 2$, $C_\lambda 3$, and $C_\lambda 4$. Within the light chain variable domain are hypervariable regions that differ in amino acid sequence more frequently than the rest of the variable domain.

In the heavy chain, the NH_2-terminal domain has a pattern of variability similar to that of the V_κ and V_λ domains and is termed the V_H domain. However, notice in figure 31.7*b* that this domain contains four hypervariable regions. The other domains of the heavy chains are termed constant domains and are numbered C_H1, C_H2, C_H3, and sometimes C_H4, starting with the domain next to the variable domain (figure 31.6*c*). The constant domains of the heavy chain form the constant (C_H) region. It is the amino acid sequence of this region that determines the classes of heavy chains. In humans, there are five classes of heavy chains called gamma (γ), alpha (α), mu (μ), delta (δ), and epsilon (ϵ). The properties of these heavy chains determine, respectively, the five immunoglobulin classes—IgG, IgA, IgM, IgD, and IgE. Each immunoglobulin class differs in its general properties, half-life, distribution in the body, and interaction with other components of the host's defensive systems.

Within two of the major immunoglobulin classes, there are variants or subclasses. For example, IgG can be grouped into four subclasses (IgG1, IgG2, IgG3, and IgG4) and IgA into two subclasses (IgA1 and IgA2). Each subclass has differences in the amino acid composition of the heavy chain. These variations can be classified as (1) **isotypes,** referring to the variations in the heavy chain constant regions associated with the different classes and subclasses that are normally present in all individuals (figure 31.8*a*); (2) **allotypes,** the genetically controlled allelic forms of immunoglobulin molecules that are not present in all individuals (figure 31.8*b*); and (3) **idiotypes,** referring to individual specific immunoglobulin molecules that differ in the hypervariable region of the Fab portion (figure 31.8*c*). The many variations of immunoglobulin structure reflect the diversity of antibodies generated by the immune response.

---

1. Name the two isotypes of antibody light chains.
2. What is the function of the Fc region of an antibody? The Fab region?
3. What is the hypervariable region of an antibody? The constant region?
4. What determines the class of heavy chain within an antibody?
5. Name the five immunoglobulin classes.
6. What determines the isotype of IgG?

Immunoglobulin Function

All immunoglobulin molecules are bifunctional. The Fab region is concerned with binding to antigen, whereas the Fc region mediates binding to host tissue, various cells of the immune system, some phagocytic cells, or the first component of the complement system. The binding of an antibody with an antigen usually does not cause destruction of the antigen or of the microorganism, cell, or agent to which it is attached. Rather, the antibody serves to mark and identify the target for immunologic attack and to activate nonspecific immune responses that can destroy the target. For example, bacteria that are covered with antibodies are better targets for phagocytosis by neutrophils and macrophages. This ability of an antibody to stimulate phagocytosis is termed **opsonization**. Immune destruction is also promoted by antibody-induced activation of the complement system. Both opsonization and the complement system are discussed in detail in chapter 33.

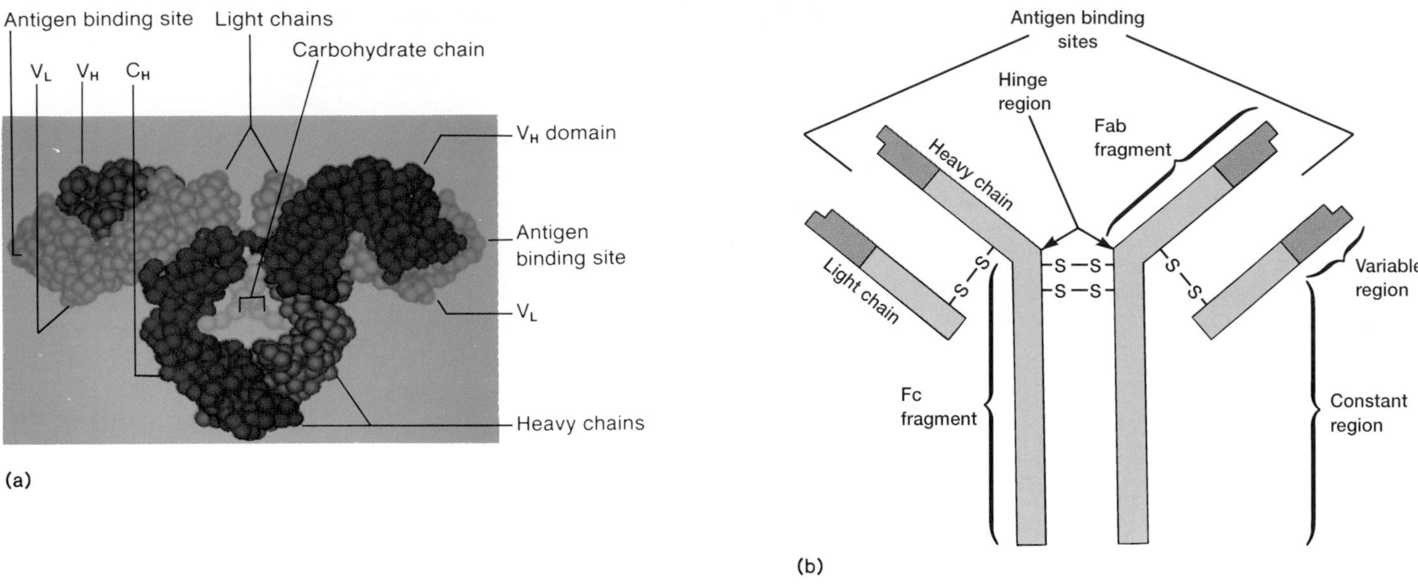

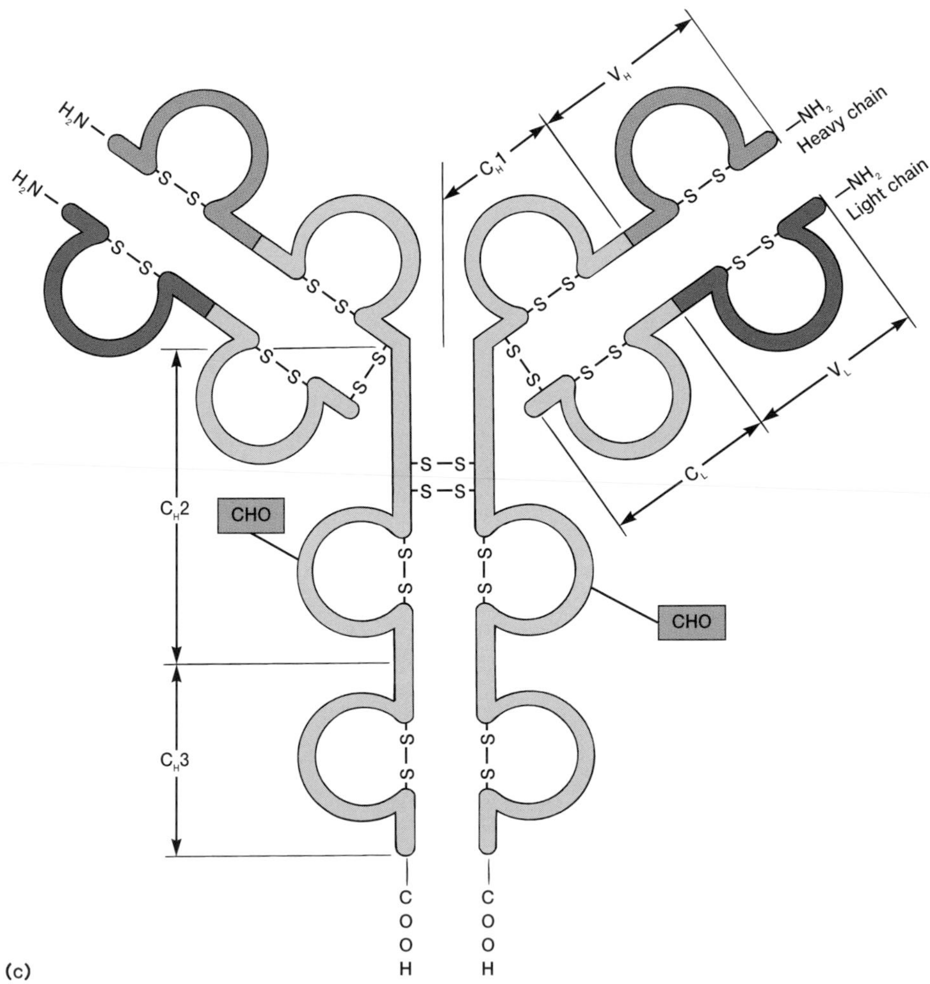

Figure 31.6 **Immunoglobulin (Antibody) Structure.** (*a*) A computer-generated model of antibody structure showing the arrangement of the four polypeptide chains. (*b*) An immunoglobulin molecule. The molecule consists of two identical light chains and two identical heavy chains held together by disulfide bonds. (*c*) Within the immunoglobulin unit structure, intrachain disulfide bonds create loops that form domains. All light chains contain a single variable domain (V_L) and a single constant domain (C_L). Heavy chains contain a variable domain (V_H) and either three or four constant domains (C_H1, C_H2, C_H3, and C_H4).

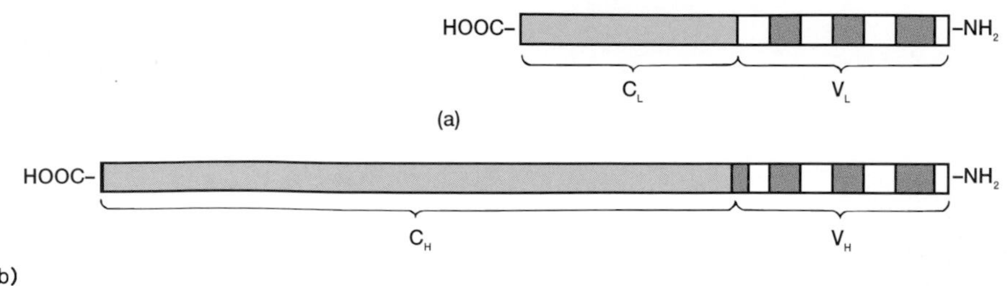

Figure 31.7 Constant and Variable Domains. Location of constant (C) and variable (V) domains within (*a*) light chains and (*b*) heavy chains. The dark yellow bands represent hypervariable regions within the variable domains.

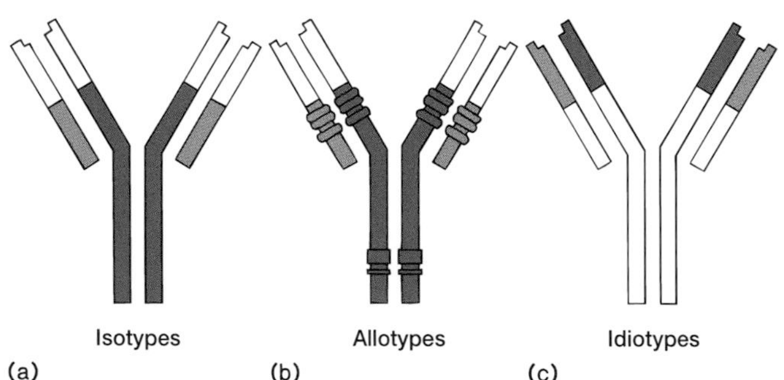

Figure 31.8 Variants or Subclasses of Immunoglobulins. (*a*) Isotypes represent all variants present in serum of a normal individual. (*b*) Allotypes represent alternative forms: genetically controlled so not present in all individuals. (*c*) Idiotypes are individually specific to each immunoglobulin molecule.

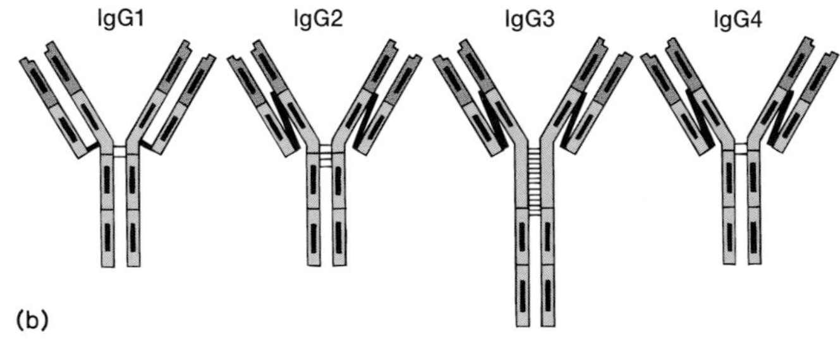

Figure 31.9 Immunoglobulin G. (*a*) The basic structure of human IgG. (*b*) The structure of the four human IgG subclasses. Note the arrangement and numbers of disulfide bonds (shown as thin black lines).

Immunoglobulin Classes

IgG is the major immunoglobulin in human serum, accounting for 70 to 75% of the immunoglobulin pool (figure 31.9*a*). IgG is present in blood plasma and tissue fluids. The IgG class acts against bacteria and viruses by opsonizing the invaders and neutralizing toxins. It is also one of the two immunoglobulin classes that activate complement by the classical pathway (*see chapter 33*). IgG is the only immunoglobulin molecule able to cross the placenta and provide naturally acquired passive immunity for the newborn.

There are about four IgG subclasses (IgG1, IgG2, IgG3, and IgG4) that vary chemically in their chain composition and the number and arrangement of interchain disulfide bonds (figure 31.9*b*). About 65% of the total serum IgG is IgG1, and 23% is IgG2. Differences in biological function have been noted in these subclasses. For example, IgG2 antibodies are opsonic and develop in response to antitoxins. Anti-Rh antibodies (*see chapter 32*) are of the IgG1 or IgG3 subclass. IgG1 and IgG3 also bind best to monocytes and macrophages, and activate complement most effectively. The IgG4 antibodies function as skin-sensitizing immunoglobulins (*see chapter 32*).

IgM accounts for about 10% of the immunoglobulin pool. It is a polymer (pentamer) of five monomeric units, each composed of two heavy chains and two light chains (figure 31.10). The monomers are arranged in a pinwheel array with the Fc ends in the center, held together by a special **J** (joining) **chain.** IgM is the first immunoglobulin made during B-cell maturation and the first secreted into serum during a primary antibody response (figure 31.18*b*). Since IgM is so large, it does not leave the bloodstream or cross the placenta. IgM agglutinates bacteria, activates complement by the classical pathway, and enhances the ingestion of pathogens by phagocytic cells. This class also contains special antibodies such as red blood cell agglutinins and heterophile antibodies.

IgA accounts for about 15% of the immunoglobulin pool. Some IgA is present in the serum as a monomer of two heavy and two light chains. Most IgA, however, occurs in the serum as a polymerized dimer held together by a J chain (figure 31.11). IgA has special features that are associated with secretory mucosal surfaces (Box 31.1). During the transport of IgA from the mucosa-associated lymphoid tissue to mucosal

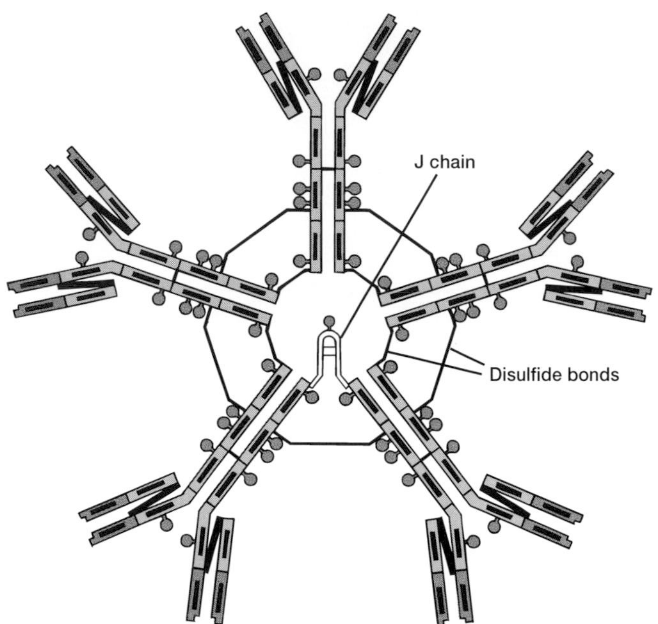

Figure 31.10 **Immunoglobulin M.** The pentameric structure of human IgM. The disulfide bonds linking peptide chains are shown in black; carbohydrate side chains are in green.

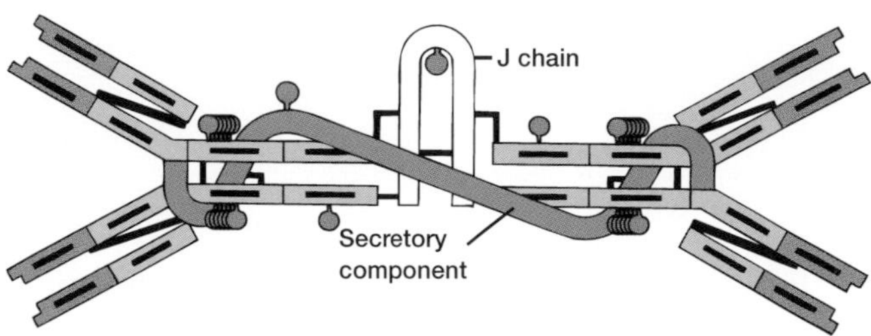

Figure 31.11 **Immunoglobulin A.** The dimeric structure of human secretory IgA. Notice the secretory component (blue) wound around the IgA dimer and attached to the constant domain of each IgA monomer.

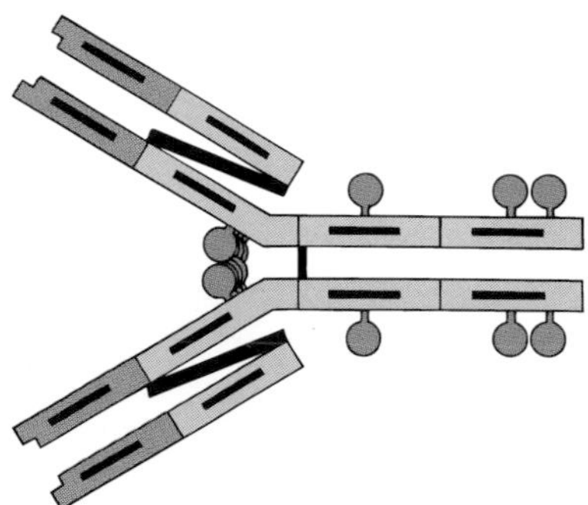

Figure 31.12 **Immunoglobulin D.** The structure of human IgD. The disulfide bonds linking protein chains are shown in black; carbohydrate side chains are in green.

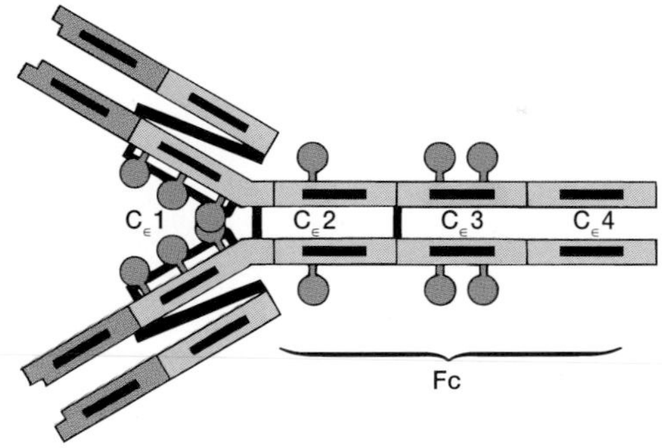

Figure 31.13 **Immunoglobulin E.** The structure of human IgE. $C_\epsilon 1$ — $C_\epsilon 4$ represent the constant domain regions.

surfaces, it acquires a protein termed the secretory component. **Secretory IgA (sIgA),** as the modified molecule is now called, is the primary immunoglobulin of the secretory immune system. This system is found in the gastrointestinal tract, upper and lower respiratory tracts, and genitourinary system. Secretory IgA is also found in saliva, tears, and breast milk. In these fluids and related body areas, sIgA plays a major role in protecting surface tissues against infectious microorganisms. For example, in breast milk sIgA helps protect nursing newborns. In the intestine, sIgA attaches to viruses, bacteria, and protozoan parasites such as *Entamoeba histolytica.* This prevents pathogen adherence to mucosal surfaces and invasion of host tissues, a phenomenon known as immune exclusion. Secretory IgA also plays a role in the alternate complement pathway (*see chapter 33*).

IgD is an immunoglobulin found in trace amounts in the blood serum. It has a monomer structure (figure 31.12) similar to IgG. IgD antibodies do not fix complement and cannot cross the placenta, but they are abundant on the surface of B cells and bind antigens, thus signaling the B cell to start antibody production.

IgE (figure 31.13) makes up only 0.00005% of the total immunoglobulin pool. The classic skin-sensitizing and anaphylactic antibodies belong to this class. IgE molecules have four constant region domains ($C_\epsilon 1$, $C_\epsilon 2$, $C_\epsilon 3$, and $C_\epsilon 4$) and two light chains. The Fc portion of the $C_\epsilon 4$ chain can bind to special Fc_ϵ receptors on mast cells and basophils. When two IgE molecules on the surface of these cells are cross-linked by binding to the same antigen, the cells degranulate. This degranulation releases histamine and other pharmacological mediators of anaphylaxis. It also stimulates eosinophilia and gut hypermotility (increased rate of movement of the intestinal contents) that aid in the elimination of helminthic parasites (*see appendix V*). Thus, though IgE is present in small amounts, this class of antibodies has very potent biological capabilities.

Anaphylaxis (pp. 635–37).

Table 31.3 summarizes some of the more important physicochemical properties of the human immunoglobulin classes.

Box 31.1

The GALT System

The major host defense against food-borne pathogens (e.g., *E. coli, S. aureus, Bacillus cereus, Cryptosporidium,* rotavirus, Norwalk virus) and products such as the *C. difficile* cytotoxin is the secretory immune system, whose major immunoglobulin is sIgA. **G**ut-**a**ssociated **l**ymphoid **t**issue or **GALT,** and particularly **Peyer's patches,** contains a full repertoire of lymphoreticular cells, T and B cells, and macrophages. Commitment to antibody synthesis occurs in the Peyer's patches, but antibody production does not (Box Figure 31.1).

In Peyer's patches, microfold cells sample foreign antigens from the gut lumen, and macrophages process the antigens for presentation to $CD4^+$ (T4) lymphocytes. The T4 lymphocytes produce the cytokines interleukin-2, interleukin-4, and IFN-γ that stimulate follicle B lymphocytes to become IgA-producing cells. The stimulated B lymphocytes migrate through lymph nodes, the thoracic duct, and the bloodstream before they "home" to the lamina propria of the intestine (a connective tissue layer lying beneath the intestinal epithelium). Here they undergo final differentiation because of the influence of interleukin-6 produced by the predominant T4 lamina propria lymphocytes. When these plasma cells produce IgA dimers, the dimers bind to epithelial cell surface receptors and are transported across the cells in vesicles. They are released into the intestinal lumen as sIgAs (figure 31.11) after being connected to secretory components (SCs) by the epithelial cell. The GALT system is exceptionally complex and is composed of several interactive sites. Some immunologists reserve the term "GALT" for the IgA inductive sites (Peyer's patches, appendix, and solitary lymph nodes) and do not consider the IgA effector sites (lamina propria and intraepithelial lymphocytes) to be a part of GALT.

Enteric pathogens may disrupt the homeostasis of the gut microenvironment and suppress the immune response through toxin production. For example, *E. coli* produces two toxins: (1) a cholera-like, heat-labile toxin (LT) that is an adenylate cyclase activator, and (2) a heat-stable toxin (ST) that activates the intestinal membrane-bound guanylate cyclase. Both LT and ST profoundly effect the immune response by increasing the levels of these cyclic nucleotides, which in turn decreases the ability of Peyer's patch cells to generate GALT sIgA. Thus, enterotoxigenic *E. coli* can establish itself and cause disease.

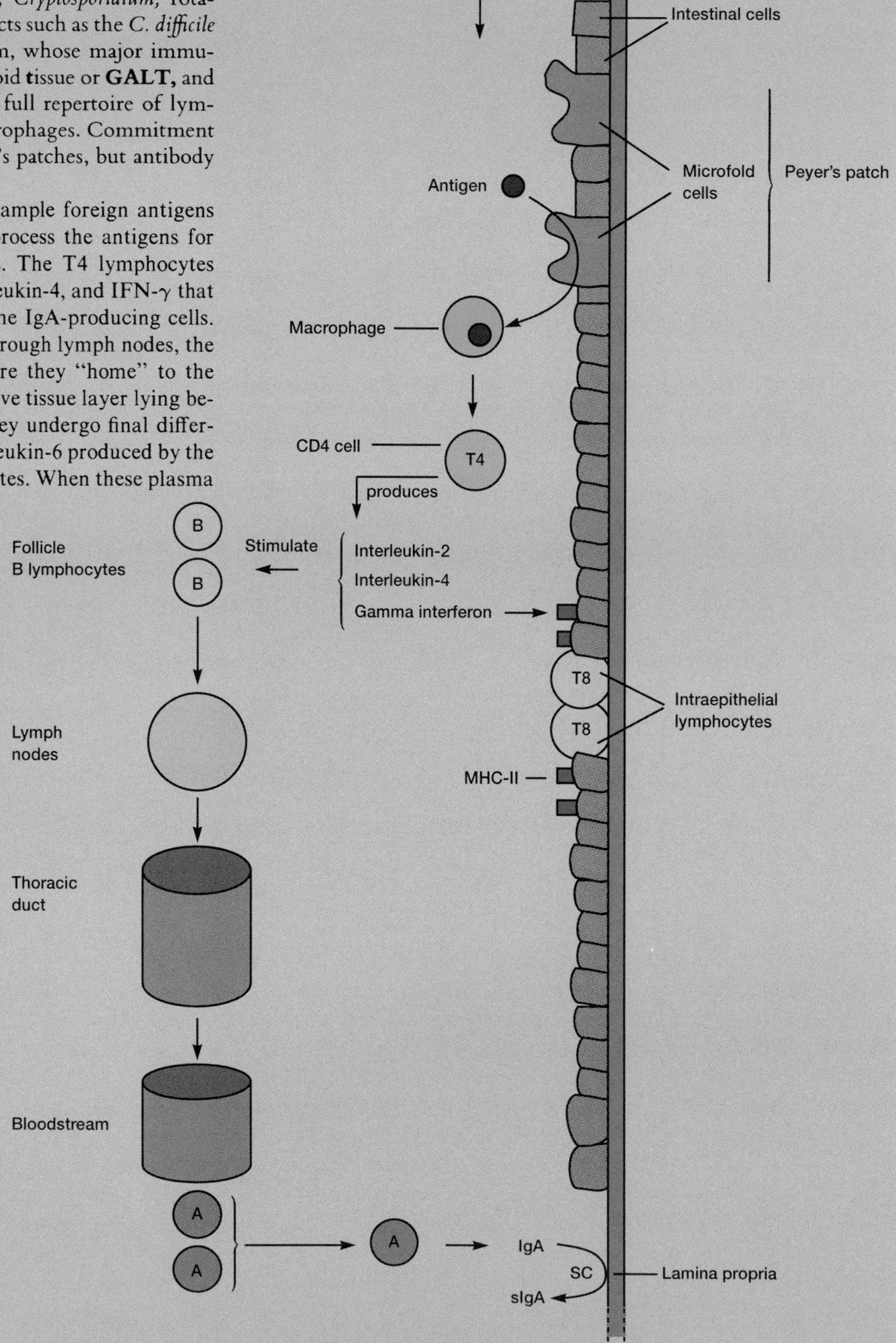

Box Figure 31.1

TABLE 31.3 Physicochemical Properties of Human Immunoglobulin Classes

Property	Immunoglobulin Classes				
	IgG[a]	IgM	IgA[b]	IgD	IgE
Heavy chain	γ_1	μ	α_1	δ	ϵ
Mean serum concentration (mg/ml)	9	1.5	3.0	0.03	0.00005
Valency	2	5(10)	(2–4)	2	2
Molecular weight of heavy chain (10^3)	51	65	56	70	72
Molecular weight of entire molecule (10^3)	146	970	160[d]	184	188
Placental transfer	+	0	0	0	0
Half-life in blood (days)[c]	21	10	6	3	2
Complement activation			0	0	0
Classical pathway	++	+++			
Alternate pathway	0	0	+	0	0
Major characteristics	Most abundant Ig in body fluids; neutralizes toxins, opsonizes bacteria, activates complement, maternal Ab	First to appear after antigen stimulation; very effective agglutinator	Secretory antibody; protects external surfaces	Present on B cell surface B cell recognition of Ag(?)	Anaphylactic-mediating antibody: resistance to helminths
% carbohydrate	3	7–10	7	12	11

[a]Properties of IgG subclass 1.
[b]Properties of IgA subclass 1.
[c]Time required for half of the antibodies to disappear.
[d]sIgA ≅ 360–400 kDa

1. What does antigen binding accomplish in a host?
2. What is opsonization?
3. Give the major functions of each immunoglobulin class.
4. Why is the structure of IgG considered the model for all five immunoglobulin classes?
5. Which immunoglobulin can cross the placenta?
6. Which immunoglobulin is most prevalent in the immunoglobulin pool? The least prevalent?

Diversity of Antibodies

One unique property of antibodies is their remarkable diversity. According to current estimates, each human or mouse can synthesize more than 10 million different kinds of antibodies. How is this diversity generated? The answer is threefold: rearrangement of antibody gene segments, somatic mutations, and generation of different codons during antibody gene splicing.

Susumu Tonegawa of the Massachusetts Institute of Technology won the 1987 Nobel Prize for Physiology or Medicine for his contribution toward understanding how the human body makes the multitude of antibodies that it needs to fight off disease.

Immunoglobulin genes are split or interrupted genes with many exons (*see pp. 203–4*). Embryonic B cells contain a small number of exons, close together on the same chromosome, that determine the constant (C) region of the light chains (figure 31.14). Separated from them, but on the same chromosome, is a larger cluster of exons that determines the variable (V) region of the light chains. During B-cell differentiation, one exon for the constant region is spliced (cut and joined) onto one exon for the variable region. It is this splicing that produces a complete light-chain antibody gene. A similar splicing mechanism also occurs to join the constant and variable exons of the heavy chains.

Because the light-chain genes actually consist of three parts, and the heavy-chain genes consist of four, the formation of a finished antibody molecule is slightly more complicated than previously outlined (figure 31.15). The germ line DNA for the light-chain gene contains multiple coding sequences called V and J (joining) regions. During the differentiation of a B cell, a deletion (which is variable in length) occurs that joins one V exon with one J exon. This DNA joining process is termed combinatorial joining since it can create many combinations of the V and J regions. When the light-chain gene is transcribed, transcription continues through the DNA region that encodes for the constant portion of the gene. RNA splicing subsequently joins the VJ and C regions creating mRNA.

Combinatorial joining in the formation of a heavy-chain gene occurs by means of DNA splicing of the heavy-chain counterparts of V and J along with a third set of D (diversity) sequences (figure 31.16*a*). Initially, all heavy chains have the μ type of constant region. This corresponds to antibody class IgM (figure 31.16*b*). Another DNA splice joins the VDJ region with a different constant region that can subsequently change the class of antibody produced by the B cell (figure 31.16*c*).

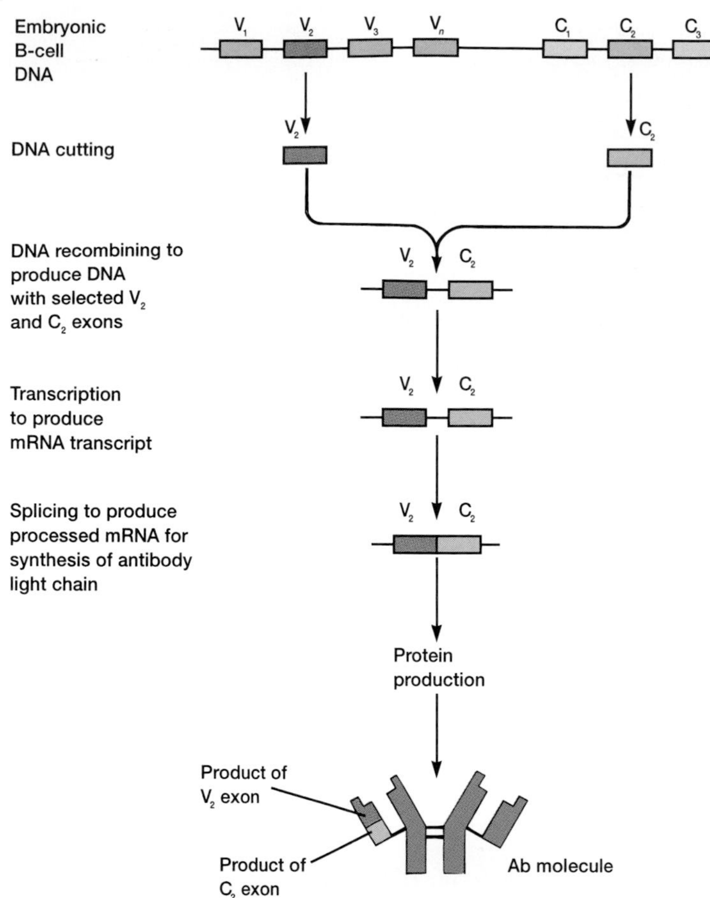

Figure 31.14 Gene Shuffling and Antibody Diversity. Antibody diversity is partly the result of the shuffling of gene sequences that code for both heavy and light chains. This drawing shows the shuffling, cutting, and splicing process to produce an assembled light chain of the antibody molecule.

The amount of antibody diversity in the mouse that can be generated by combinatorial joining is shown in table 31.4. In this animal, the κ light chains are formed from combinations of about 250 V_κ and 4 J_κ regions giving approximately $250 \times 4 = 1{,}000$ different κ chains. The λ chains have their own V_λ and J_λ regions but smaller in number than their κ counterparts ($3 \times 3 = 9$ different λ chains). The heavy chains have approximately 250 V_H, 10 D, and 4 J_H regions giving $250 \times 10 \times 4 = 10{,}000$ different combinations. Because any light chain can combine with any heavy chain, there will be at least $1{,}000 \times 10{,}000 = 10^7$ possible antibody types.

The value of 10^7 different antibodies is actually an underestimate since antibody diversity is further augmented by two processes:

1. The V regions of germ line DNA are susceptible to a high rate of somatic mutation during B-cell development in the bone marrow. These mutations allow B-cell clones to produce different polypeptide sequences.
2. The junction for either VJ or VDJ splicing in combinatorial joining can occur between different nucleotides and thus generate different codons in the spliced gene. For example, one VJ splicing event can join the V sequence CCTCCC with the J sequence TGGTGG in two ways:

   CCTCCC + TGGTGG = CCGTGG,

   which codes for the amino acids proline and tryptophan; and

   CCTCCC + TGGTGG = CCTCGG,

   which codes for proline and arginine. Thus, the same VJ joining could produce polypeptides differing in a single amino acid.

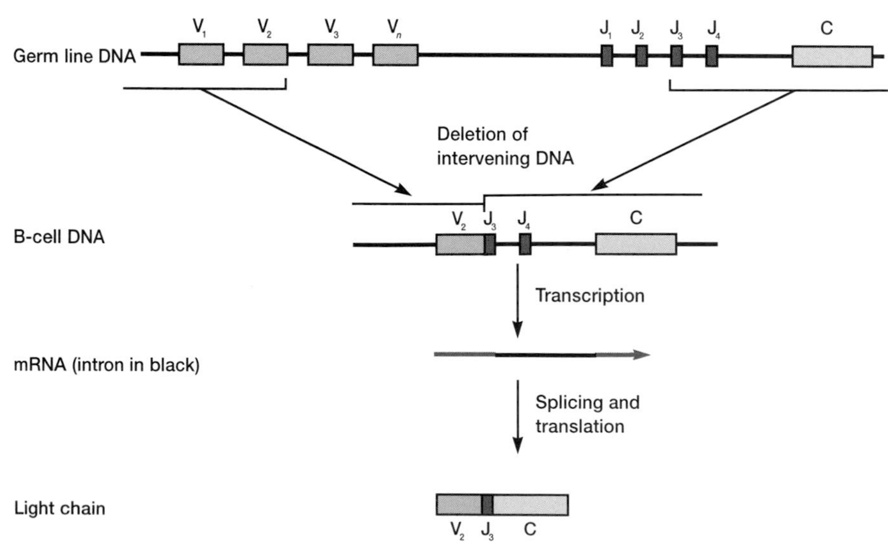

Figure 31.15 Light Chain Production in A Mouse. One V exon is randomly joined with one J-C region by deletion of the intervening DNA. The remaining J exons are eliminated from the RNA transcript during RNA processing. An intron is a segment of DNA occurring between expressed regions of genes.

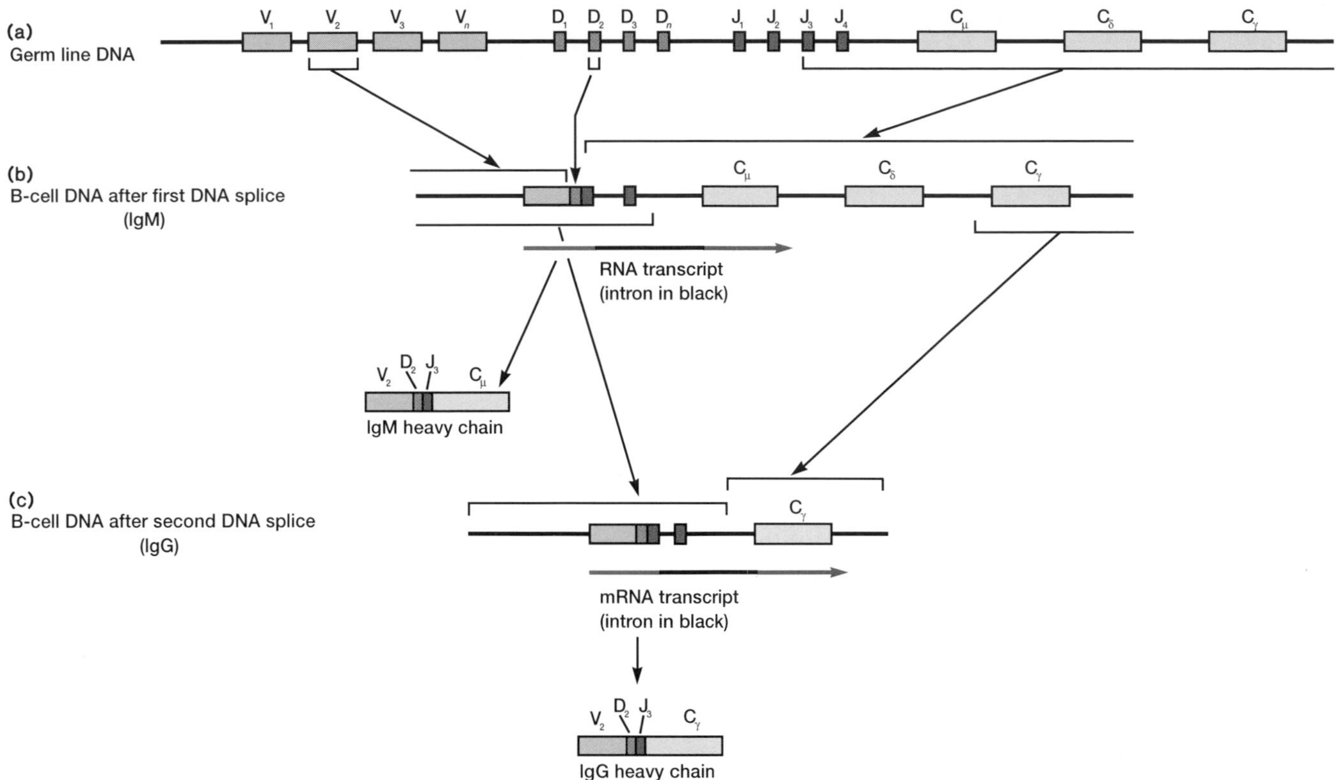

Figure 31.16 **The Formation of A Gene For the Heavy Chain of an Antibody Molecule.** See text for further details.

TABLE 31.4 Number of Antibodies Possible through the Combinatorial Joining of Mouse Germ Line Genes

λ light chains	V regions = 3[a] J regions = 3 Combinations = 3 × 3 = 9
κ light chains	V_κ regions = 250 J_κ regions = 4 Combinations = 250 × 4 = 1,000
Heavy chains	V_H = 250 D = 10 J_H = 4 Combinations = 250 × 10 × 4 = 10,000
Diversity of antibodies	κ-containing: 10,000 × 1,000 = 10^7 λ-containing: 9 × 10,000 = 90,000

[a]*Approximate values.*

1. How many chromosomes encode for antibody production in humans?
2. What is the name of each exon set that encodes for the different regions of antibody chains?
3. Describe what is meant by combinatorial joining of VJ and C regions of a chromosome.
4. In addition to combinatorial joining, what other two processes play a role in antibody diversity?

Specificity of Antibodies

As noted previously, combinatorial joinings, somatic mutations, and variations in the splicing process generate the great variety of antibodies produced by B cells. From a large, diverse B-cell pool, specific cells are stimulated by antigens to reproduce and form a B-cell clone that contains the same genetic information. This is known as the **clonal selection theory,** first set forth in 1957 by Sir MacFarlane Burnet of Australia, David W. Talmage of the United States, and Niels Kaj Jerne of the World Health Organization as a hypothesis to explain immunologic specificity and memory.

The existence of a small B-cell **clone** (a population of cells derived asexually from a single parent) that can respond to one or a few antigens by producing the correct antibody is the first tenet of this theory. The lymphoid system is thus considered to be composed of many B-cell clones, each clone able to recognize a specific antigen. The antigen selects the appropriate clone of B cells (hence the phrase "clonal selection"), and the cells from the other clones are unaffected.

The second tenet proposes that each B-cell clone is genetically programmed to respond to its own distinctive antigen before the antigen is introduced. The particular antibody for which an individual B cell is genetically competent is integrated into the plasma membrane of that B cell and acts as a specific surface receptor for the corresponding antigen molecule. The reaction of the antibody and antigen initiates the differentiation and multiplication of the B cell to form two different cell populations: plasma cells and **memory B cells** (figure 31.17).

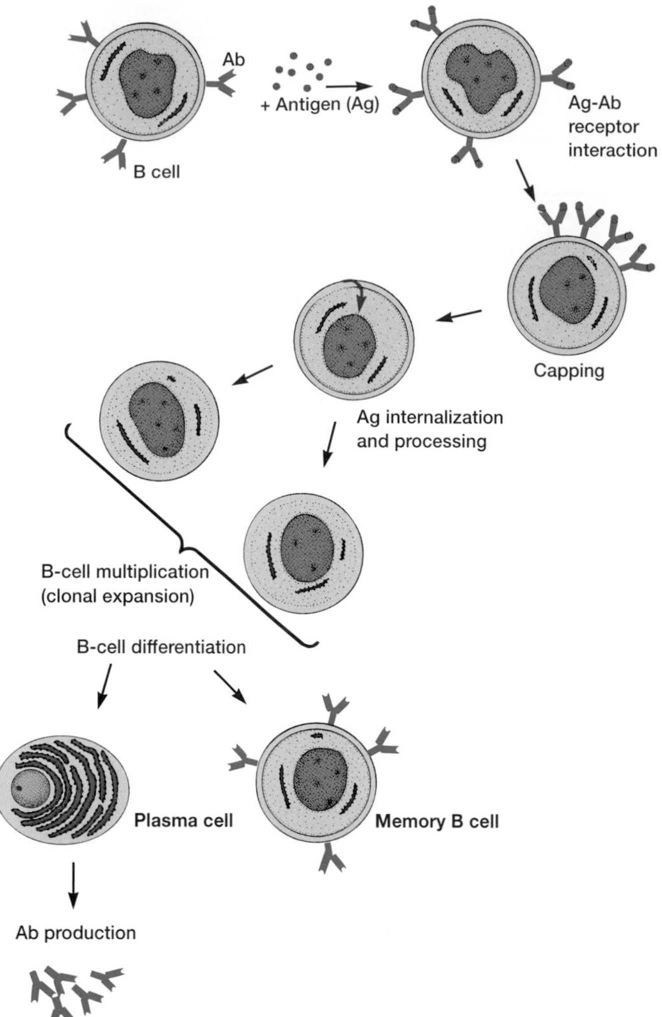

Figure 31.17 Clonal Selection. According to the clonal selection theory, exposure of a B cell to an antigen stimulates the production of B-cell clones and the maturation of some members of the B-cell clones into plasma cells and memory B cells. One step in this sequence is capping, the regional aggregation of antibodies on the surface of the cell following Ag-Ab interaction.

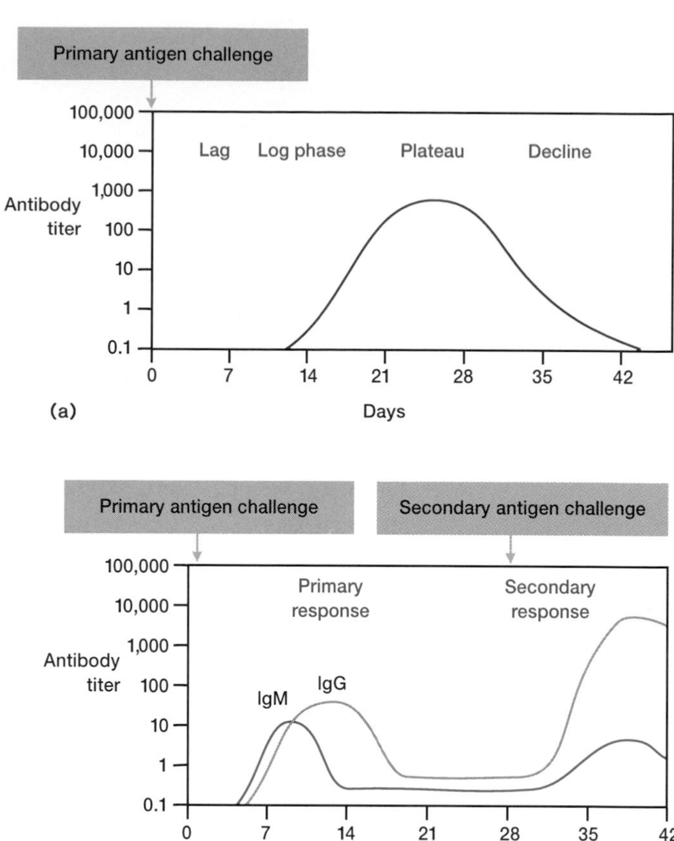

Figure 31.18 Primary and Secondary Antibody Responses. (*a*) The four phases of a primary antibody response. (*b*) Comparison of primary and secondary antibody responses. Note that the secondary response to the antigen is both faster and greater than the primary response. T-cell–mediated responses exhibit similar immunologic memory.

Plasma cells are literally protein factories that produce about 2,000 antibodies per second in their brief five- to seven-day life span. Memory B cells can initiate the antibody-mediated immune response upon detecting the particular antigen molecule for which they are genetically programmed (i.e., they have specificity). These memory cells circulate more actively from blood to lymph and live much longer (years or even decades) than plasma cells. Memory cells are responsible for the immune system's rapid secondary antibody response (figure 31.18*b*) to the same antigen.

Sources of Antibodies

The need for pure homogeneous antibodies has increased dramatically in recent years. Currently, antibodies are produced either naturally by immunization or artificially through hybridoma formation.

Immunization

Specific antibodies can be produced naturally by the immunization of domestic animals or human volunteers (Box 31.2). Whatever the source, purified antigen is injected into the host. The host's immune system recognizes and responds to the antigen, and its B cells proliferate and differentiate to produce specific antibodies. To promote the efficiency of antigen stimulation of antibody production, the antigen is mixed with an **adjuvant** (Latin *adjuvans,* aiding), which enhances the rate and quantity of antibody produced. Following repeated antigen injections at regular intervals, blood is withdrawn from the host and allowed to clot. The fluid that remains after the blood clots is the **serum.** Because this serum has been obtained from an immunized host and contains the desired antibodies, it is called **antiserum.**

Although antiserum is a major and convenient source of antibodies, its usefulness is limited in three ways:

1. Antibodies obtained by this method are polyclonal; they are produced by several B-cell clones and have different specificities. This decreases their sensitivity to particular antigens and results in some degree of cross-reaction with closely related antigen molecules.

Box 31.2

The First Immunizations

Since the time of the ancient Greeks, it has been recognized that people who have recovered from plague, smallpox, yellow fever, and various other infectious diseases rarely contract the diseases again. The first scientific attempts at artificial immunizations were made in the late eighteenth century by the English physician Edward Jenner (1749–1823). Jenner investigated the basis for the widespread belief of the English peasants that anyone who had vaccinia (cowpox) never contracted smallpox. Smallpox was often fatal—10 to 40% of the victims died—and those who recovered had disfiguring pockmarks. Yet most English milkmaids, who were readily infected with cowpox, had clear skin because cowpox was a relatively mild infection that left no scars.

Jenner's first immunization recipient was a healthy, eight-year-old boy known never to have had either cowpox or smallpox. As Jenner expected, immunization with the cowpox virus caused only mild symptoms in the boy. When he subsequently inoculated the boy with smallpox virus, the boy showed no symptoms of the disease.

Jenner then inoculated large numbers of his patients with cowpox pus, as did other physicians in England and on the European continent (Box Figure 31.2). By 1800, the practice known as vaccination (Latin *vacca,* cow) had begun in America, and by 1805, Napoleon Bonaparte had ordered all French soldiers to be vaccinated.

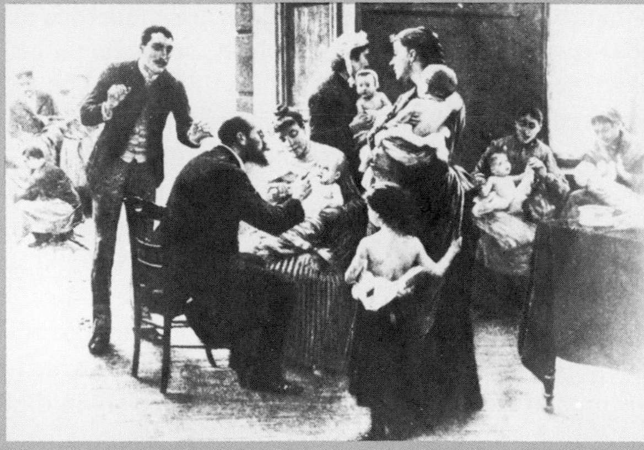

Box Figure 31.2 Nineteenth-century physicians performing vaccinations on children.

Further work on immunization was carried out by Louis Pasteur (1822–1895). Pasteur discovered that neglected, old cultures of chicken cholera bacteria, which had not been regularly transferred to fresh culture medium, produced only a mild attack of cholera when inoculated into chickens. Somehow the old cultures had become less pathogenic (attenuated) for the chickens. He then found that fresh cultures of the bacteria failed to produce cholera in chickens that had been previously inoculated with old, attenuated cultures. To honor Jenner, Pasteur gave the name vaccine to any preparation of a weakened pathogen that was used (as was Jenner's "vaccine virus") to immunize against infectious disease.

2. Second or repeated injections of antiserum from one species to another can cause serious allergic or hypersensitivity reactions.
3. Antiserum contains a mixture of antibodies (figure 31.5), not all of which are of interest in a given immunization.

The Primary Antibody Response

During immunization procedures (and also in naturally acquired immunity), there is an initial lag phase of several days following a primary challenge with an antigen. During the lag phase, no antibody can be detected (figure 31.18*a*). The antibody **titer,** which is the reciprocal of the highest dilution of an antiserum that gives a positive reaction in the test being used, rises logarithmically to a plateau during the second or log phase. In the plateau phase, the antibody titer stabilizes. This is followed by a decline phase, during which antibodies are naturally metabolized or bound to the antigen and cleared from the circulation. During the primary antibody response, IgM appears first, then IgG (figure 31.18*b*). The affinity of the antibodies for the antigen's determinants is low to moderate during this primary antibody response.

The Secondary Antibody Response

The primary antibody response primes the immune system so that it possesses specific immunologic memory through its clones of memory B cells. Upon secondary antigen challenge (figure 31.18*b*), the B cells mount a heightened or **anamnestic** (Greek *anamnesis,* remembrance) **response** to the same antigen. Compared to the primary antibody response, the secondary antibody response has a shorter lag phase, a more rapid log phase, persists for a longer plateau period, attains a higher IgG titer, and produces antibodies with a higher affinity for the antigen (affinity maturation).

Hybridomas

Recently, the limitations of antiserum as a source of antibodies have been overcome with the development of techniques to manipulate and culture various mammalian cells that synthesize antibodies in vitro. Each cell and its progeny normally produce a **monoclonal antibody** of a single specificity. The hybridoma technique for monoclonal antibody production was first developed by Georges Köhler and Cesar Milstein in 1975 in Cambridge, England. For this discovery, they were awarded a Nobel Prize for Physiology or Medicine in 1984.

The methodology of this technique is illustrated in figure 31.19. Animals (usually mice or rats) are immunized with antigens as discussed previously. Once the animals are producing a large quantity of antibodies, their spleens are removed. The spleen cells are separated from each other and fused with **myeloma cells** by the addition of polyethylene glycol, which promotes membrane fusion. Myeloma cells are cancerous plasma cells that can readily be cultivated. Mutant myeloma cells incapable of producing immunoglobulins are used. These fused cells, derived from spleen cells and myeloma cells, are called **hybridomas** (they are hybrids of the two cells).

The fusion mixture is then transferred to a culture medium containing a combination of **h**ypoxanthine, **a**minopterin, and **t**hymidine (HAT). Aminopterin is a poison that blocks a specific metabolic pathway in cells. Myeloma cells lack an enzyme that allows their growth in the presence of aminopterin. However, the pathway is by-passed in spleen cells provided with the intermediate metabolites hypoxanthine and thymidine. As a result, the hybridomas grow in the HAT medium but the myeloma cells die because they have a metabolic defect and cannot employ the bypass pathway.

When the culture is initially established using the HAT medium, it contains spleen cells, myeloma cells, and hybridomas. The unfused spleen cells die naturally in culture within a week or two, and the myeloma cells die in the HAT as just described. In contrast, the fused cells survive because they have the immortality of the myeloma and the metabolic bypass of the spleen cells. Some hybridomas that have the antibody-producing capacity of the original spleen cells are randomly placed in culture wells. The wells are individually tested for production of the desired antibody, and, if positive, the cells within the well are cloned. The clone is immortal and produces monoclonal antibody.

Monoclonal antibodies currently have many applications. For example, they are routinely used in the typing of tissue, in the identification and epidemiological study of infectious microorganisms, in the identification of tumor and other surface antigens, in the classification of leukemias, and in the identification of functional populations of different types of T cells. Anticipated future uses include (1) passive immunizations against infectious agents and toxic drugs, (2) tissue and organ graft protection, (3) stimulation of tumor rejection and elimination, (4) manipulation of the immune response, (5) preparation of more specific and sensitive diagnostic procedures, and (6) delivery of antitumor agents (immunotoxins) to tumor cells (Box 31.3).

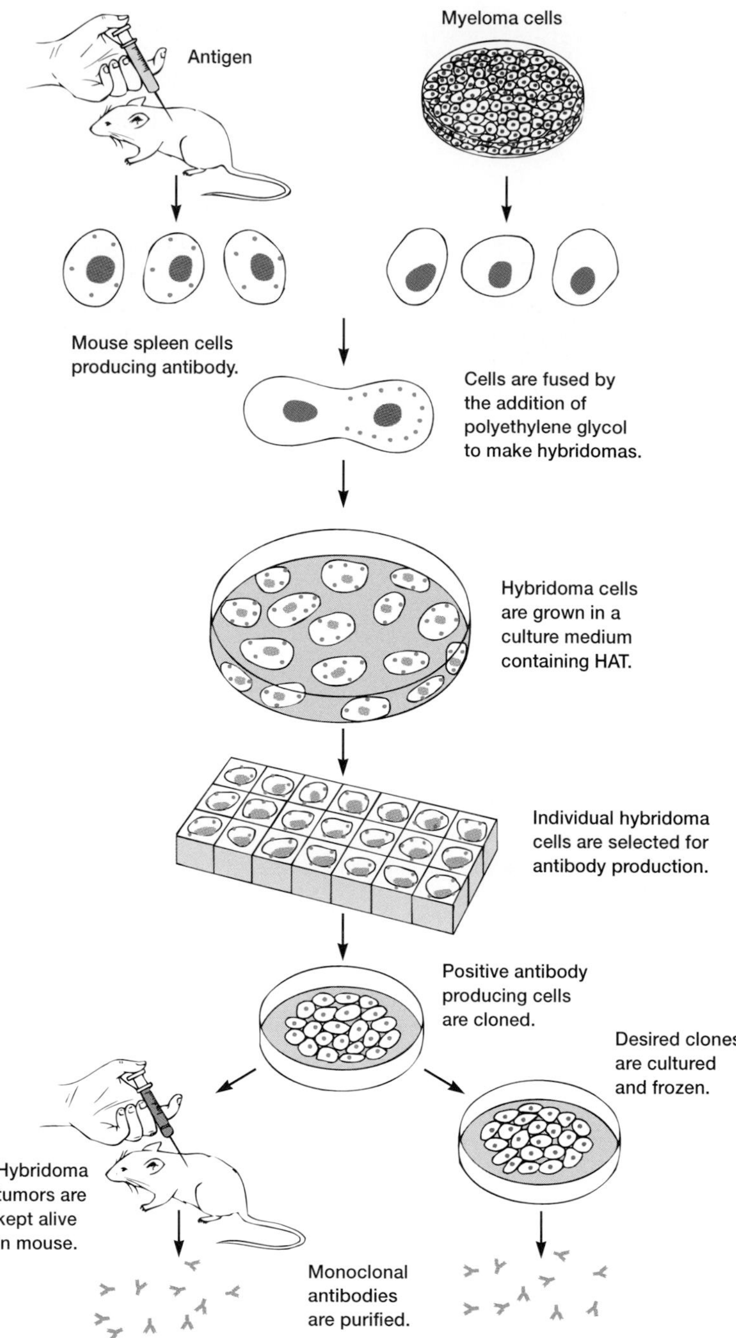

Figure 31.19 Technique for the Production of Monoclonal Antibodies. Antigen-stimulated spleen cells are fused with special mutant myeloma cells, yielding hybridomas. Each of them secretes a single, "monoclonal" antibody. Once the hybridoma secreting the desired antigen is identified, it is cloned to generate many antibody-secreting cells that yield the huge quantity of a single antibody needed in medicine or science. Some hybridoma cells may be stored frozen and later cloned for antibody production or kept alive in laboratory animals. *From Barrett, et al.,* Biology, *© 1986, pp. 338, 832. Reprinted by permission of Prentice-Hall, Inc., Englewood Cliffs, NJ.*

Catalytic Antibodies

In the past several years, immunologists have applied the principles of enzymology to create a new class of antibody molecules—**catalytic antibodies.** Catalytic antibodies catalyze specific chemical reactions by lowering the free energy of transition states (*see figure 7.13*). This is accomplished by the antibody binding reactants in a cleft or crevice on its surface and inducing structural changes in the substrate molecules. Because antibodies to a huge array of biopolymers, natural prod-

Box 31.3

Immunotoxins

One result of hybridoma research is the production of **immunotoxins** (Box Figure 31.3). Immunotoxins are monoclonal antibodies that have been attached to a specific toxin or toxic agent (antibody + toxin = immunotoxin). Immunotoxins kill target cells and no others, because the antibody binds specifically to plasma membrane surface antigens found only on the target cells. This approach is being used to treat certain types of cancer.

In this procedure, cancer cells from a person are injected into mice or rats to stimulate the production of specific antibodies against their plasma membrane antigens. Monoclonal antibodies are produced using hybridomas, purified, and attached to an agent toxic to the cancer cells.

When the immunotoxin is given to a cancer patient, it circulates through the body and binds only to the cancer cells that have the appropriate surface antigens. After binding to the surface, the immunotoxin is taken into cancer cells by receptor-mediated endocytosis (*see chapter 19*), and released inside. The immunotoxin then interferes with the metabolism of the target cells and kills them. Although this procedure is still experimental, it holds great promise in the treatment of certain types of cancer.

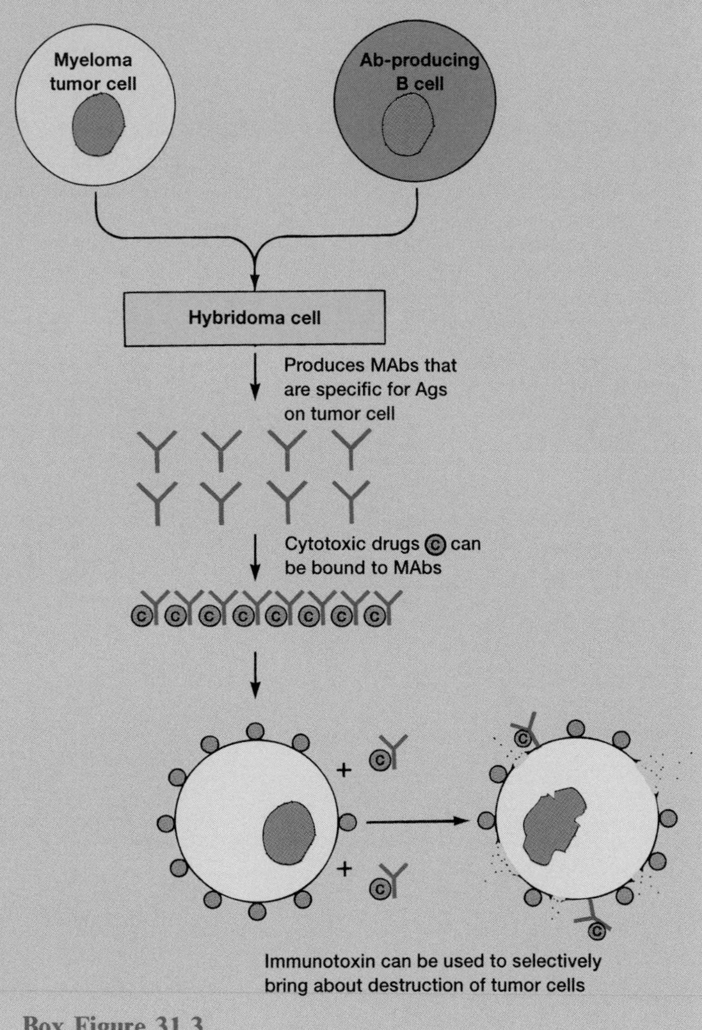

Box Figure 31.3

ucts, and synthetic molecules can be formed, catalytic antibodies offer a unique approach for generating tailor-made, enzymelike catalysts.

Catalytic antibodies are made by coupling a transition-state analogue to a carrier protein and injecting the combination into an experimental animal. Antibody-secreting spleen cells are taken from the animal and fused with myeloma cells (figure 31.19). The hybrid antibody-secreting cells divide indefinitely and generate clones of cells, each hybrid clone secreting a monoclonal catalytic antibody with a unique antigen binding pocket. A clone that makes catalytic antibody specific for the analogue is then selected.

Currently available catalytic antibodies transform relatively simple compounds. Much of the potential of catalytic antibodies for biotechnology and molecular biology depends on the development of catalytic antibodies able to act on proteins or nucleic acids. If this can be accomplished, catalytic antibodies could extend the immune system's innate capacity to defend the body. For example, one might stimulate the immune system of a patient with heart disease to produce antibodies that would break up the proteins in blood clots, forestalling heart attacks.

1. What are the two tenets of the clonal selection theory?
2. What two cell populations are produced by a B cell after it initially reacts with an antigen?
3. What is the function of an adjuvant?
4. What is the difference between serum and antiserum?
5. What is a hybridoma? A catalytic antibody? How is each made?

Summary

1. Nonspecific resistance refers to those general mechanisms inherited as part of the innate structure and function of each animal. Examples include biological, chemical, general, and physical barriers. The specific immune response system consists of certain cells (lymphocytes) that can recognize foreign molecules (antigens) and respond to them by forming specialized proteins (antibodies) that target the molecule or agent for interaction with other components of the immune system.
2. Acquired immunity refers to the type of specific immunity that a host develops after exposure to a suitable antigen. It can be obtained actively or passively by natural or artificial means.
3. Undifferentiated lymphocytes are produced in adult bone marrow by stem cells. Some of these lymphocytes migrate to the thymus and undergo special processing to become T cells (T lymphocytes). Other lymphocytes differentiate in other places in the body to form B cells (B lymphocytes). A third population of lymphoid cells that do not have characteristics of either T or B cells are called null cells. This population contains natural killer and cytotoxic T cells.
4. B cells provide defense against antigens by differentiating into plasma cells that secrete antibodies into the blood and lymph, providing humoral or antibody-mediated immunity. T cells attack host cells that have been parasitized by microorganisms, tissue cells transplanted from one host to another, and cancer cells. Because they must physically contact these foreign cells to destroy them, they are said to provide cell-mediated immunity. T cells also play a major role in regulating the immune response. Null cells nonspecifically kill tumor cells, virus- and other parasite-infected cells.
5. An antigen or immunogen is a foreign substance to which lymphocytes respond. Each antigen can have several antigenic determinant sites or epitopes that stimulate production of and combine with specific antibodies. Haptens are small organic molecules that are not antigenic by themselves but can become antigenic if bonded to a larger carrier molecule.
6. Antibodies or immunoglobulins are a group of glycoproteins present in the blood and tissue fluids of all mammals. All immunoglobulins have a basic structure composed of four chains of polypeptides (two light and two heavy) connected to each other by disulfide bonds. In humans, five immunoglobulin classes exist: IgG, IgA, IgM, IgD, and IgE.
7. Antibody diversity results from the rearrangement of the individual exons on the antibody coding chromosomes, somatic mutations, the generation of different codons during splicing, and the independent assortment of light- and heavy-chain genes.
8. The clonal selection theory is a hypothesis that explains immunologic specificity and memory.
9. Specific antibodies can be produced naturally from immunized animals. The primary antibody response in a host occurs following initial exposure to the antigen. This response has lag, log, plateau, and decline phases. Upon secondary antigen challenge, the B cells mount a heightened and accelerated anamnestic response.
10. Hybridomas result from the fusion of spleen cells with myeloma cells. Some of these cells produce a single monoclonal antibody. Monoclonal antibodies have many uses.
11. Catalytic antibodies catalyze specific chemical reactions by lowering the free energy of transition states. These antibodies can be produced by hybridoma techniques.

Key Terms

acquired immunity *608*
adjuvant *620*
allotype *612*
anamnestic response *621*
antibody *611*
antibody-mediated immunity *609*
antigen *610*
antigen-binding fragment (Fab) *612*
antigenic determinant site *610*
antiserum *620*
artificially acquired active immunity *608*
artificially acquired passive immunity *609*
B cell *609*
B lymphocyte *609*
catalytic antibodies *622*
cell-mediated immunity *610*
clonal selection theory *619*
clone *619*
colostrum *608*
constant region *612*
crystallizable fragment (Fc) *612*
domain *612*
epitope *610*
GALT *616*
hapten *611*
heterologous antigen *610*
heterophile *610*
humoral immunity *609*
hybridoma *622*
idiotype *612*
IgA *614*
IgD *615*
IgE *615*
IgG *614*
IgM *614*
immune response *607*
immunity *607*
immunogen *610*
immunoglobulin (Ig) *611*
immunology *607*
immunotoxin *623*
isotype *612*
J chain *614*
lymphokines *609*
memory B cells *619*
monoclonal antibody *621*
myeloma cell *622*
natural killer (NK) cell *609*
naturally acquired active immunity *608*
naturally acquired passive immunity *608*
nonspecific resistance *607*
null cell *609*
opsonization *612*
Peyer's patches *616*
plasma cells *609*
secretory IgA (sIgA) *615*
serum *620*
T cell *609*
thymocyte *609*
titer *621*
T lymphocyte *609*
vaccine *608*
valence *610*
variable region *612*

Questions for Thought and Review

1. What is the difference between a nonspecific host defense mechanism and a specific defense mechanism? Give examples.
2. What are the major differences between natural and artificial immunity and between active and passive immunity?
3. What is the difference between antibody-mediated and cell-mediated immunity?
4. How does the clonal selection theory explain the ability of the immune response to distinguish between self and nonself antigens?
5. All antibodies are proteins. Is there some property of proteins that has favored the evolution of antibodies within this group of biomolecules rather than within others?
6. How did the clonal selection theory inspire the development of monoclonal antibody techniques?
7. What is the difference in the kinetics of antibody formation in response to a first and second exposure to the same antigen?
8. What is the basis of immunologic memory?
9. What is the derivation of B and T cells?
10. Why are antibodies termed immunoglobulins?
11. Assuming that a person's immunologic memory is intact, how can you explain this person getting a common cold year after year?
12. Why is passive immunity conferred from mother to child only temporary?
13. Draw and label a diagram of IgG1 showing its major structural features. Describe how the other four Ig classes differ in structure from IgG.
14. How do catalytic antibodies function?

Additional Reading

Abruzzo, L. V., and Rowley, D. A. 1983. Homeostasis of the antibody response: Immunoregulation by NK cells. *Science* 222:581–83.

Ada, G. L., and Nossal, G. 1987. The clonal-selection theory. *Sci. Am.* 257(5):62–69.

Alt, F. W.; Blackwell, T. K.; and Yancopoulos, G. D. 1987. Development of the primary antibody response. *Science* 238:1079–88.

Bellanti, J. A. 1985. *Immunology III,* 3d ed. Philadelphia: W. B. Saunders.

Coleman, R. M.; Lombard, M. F.; and Sicard, R. E; 1992. *Fundamental immunology,* 2d ed. Dubuque, Iowa: Wm. C. Brown.

Collier, R. J., and Koplan, D. A. 1984. Immunotoxins. *Sci. Am.* 251(2):56–64.

Cooper, M. D., and Lawton, A. R. 1974. The development of the immune system. *Sci. Am.* 231(5):124–30.

Desowitz, R. S. 1987. The thorn in the starfish: How the human immune system works. New York: Norton.

Edelman, G. M. 1970. The structure and function of antibodies. *Sci. Am.* 223(1):34–41.

Edelson, R. L., and Fink, J. M. 1985. The immunologic function of the skin. *Sci. Am.* 252(2):46–53.

Kennedy, R. C.; Melnick, J. L.; and Dreesman, G. R. 1986. Anti-idiotypes and immunity. *Sci. Am.* 255(5):48–56.

Kuby, J. 1992. *Immunology.* New York: W. H. Freeman.

Lascombe, M. B., and Poljak, R. J. 1988. Three-dimensional structure of antibodies. *Ann. Rev. Immunol.* 6:555–80.

Lerner, R. A., and Tramontano, A. 1988. Catalytic antibodies. *Sci. Am.* 258(6):58–70.

Manser, T.; Huang, S. Y.; and Gefter, M. L. 1984. Influence of clonal selection on the expression of immunoglobulin variable gene regions. *Science* 266:1238–88.

Marrack, P., and Kappler, J. 1987. The T cell receptor. *Science* 238:1073–78.

Milstein, C. 1980. Monoclonal antibodies. *Sci. Am.* 243(4):66–74.

Paul, W. E. 1990. *Fundamental immunology,* 2d ed. New York: Raven Press.

Payne. W. J., Jr.; Marshall, D. L.; Shockley, R. K.; and Martin, W. J. 1988. Clinical laboratory applications of monoclonal antibodies. *Clin. Microbiol. Rev.* 1(3):313–29.

Rajewsky, K.; Forster, I.; and Cumano, A. 1987. Evolution and somatic selection of the antibody repertoire in the mouse. *Science* 238:1088–93.

Roitt, I. M. 1991. *Essential immunology,* 7th ed. Boston: Blackwell Scientific Publications.

Roitt, I. M.; Brostoff, J.; and Male, D. K. 1989. *Immunology,* 2d ed. St. Louis: C. V. Mosby.

Schultz, P. G.; Lerner, R. A.; and Benkovic, S. J. 1990. Catalytic antibodies. *Chem. Eng. News* 68(22):26–40.

Tizard, I. R. 1988. *Immunology: An introduction,* 2d ed. Philadelphia: W. B. Saunders.

Tonegawa, S. 1983. Somatic generation of antibody diversity. *Nature* 302:575–77.

Tonegawa, S. 1985. The molecules of the immune system. *Sci. Am.* 252(3):41–57.

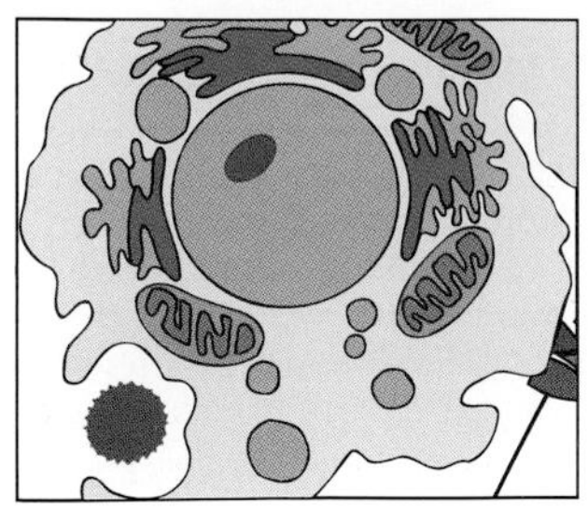

CHAPTER 32
The Immune Response: Chemical Mediators, B- and T-Cell Biology, and Immune Disorders

Autoimmunity—in which the immune system recognizes and attacks the self's own tissues—is not as simple as it seemed. Self-recognition appears to be at the heart of health as well as of certain diseases.
—Irun R. Cohen

Outline

Concepts

1. B cells have receptor immunoglobulins on their plasma membrane surface that are specific for given antigenic determinants. Contact with the antigenic determinant causes the B cell to divide and differentiate into plasma cells and memory B cells. Plasma cells secrete antibodies that specifically interact with the antigens eliciting their production.
2. Antigen binding on the surface of a T cell causes that cell to proliferate and form sensitized lymphocytes that then interact with the antigen, leading to its removal. T cells also have roles as helper and suppressor cells.
3. Lymphocytes often participate in the immune response by secreting nonspecific proteins called lymphokines. In addition, lymphocytes are affected by proteins such as interleukins and interferons.
4. Some individuals experience harmful overreactions of the immune system known as hypersensitivities or allergies. There are four types: type I or anaphylactic hypersensitivities are characterized by the release of physiological mediators from IgE-bound mast cells and basophils; type II or cytotoxic hypersensitivities result from complement-dependent lysis of cells; type III hypersensitivities involve the formation of immune complexes that are deposited on basement membranes; and type IV hypersensitivities arise from the reaction of T_{DTH} cells, lymphokines, and macrophages.
5. At times the body loses tolerance for its own antigens and attacks them. This attack produces autoimmune diseases. At other times, the immune system can become defective, leading to immunodeficiencies.
6. Both type I and type IV hypersensitivities are involved in tissue transplantation rejection. Rejection reactions also play an important role in eliminating cancer cells, in blood transfusion accidents, and in Rh incompatibility.

Chapter 31 presents a basic discussion of antigens and antibodies. Chapter 32 continues this discussion by presenting other chemical mediators that enter into the immune response, B- and T-cell biology, hypersensitivities (allergies), and several types of immune phenomena that can cause problems in the animal body under certain circumstances. For example, there are genetic, developmental, or induced defects in the immune system that may cause over- or underreaction and lead to various immune disorders. In addition, graft rejection, cancer, and blood transfusion reactions are discussed, because these problems are immune response mediated.

Two major classes of lymphocytes, B cells and T cells, specifically recognize immunogens and form immune proteins called immunoglobulins and sensitized lymphocytes, respectively. These are the immune products that interact with antigens. This defense is further aided by chemical mediators, such as lymphokines, that transmit messages between immune system cells. As with all systems in an animal's body, problems can occur within the immune system leading to defects (immunodeficiencies) or over-reactions (hypersensitivities). These aspects of immune system function are presented in this chapter in the continuing discussion of the immune response.

Chemical Mediators

Although antigen-activated T cells do not synthesize and secrete antibodies, they do release a variety of soluble mediator proteins or polypeptides termed **lymphokines** (Latin *lympho,* pure water, and Greek *kinesis,* movement). Lymphokines participate in many functions of T cells and transmit growth, differentiation, and behavioral signals between immune system cells. Lymphokines are a subset of **cytokines,** which includes products of other types of cells (e.g., monokines from monocytes). Lymphokines act at a short distance, are unrelated to immunoglobulins, and are not immunologically specific in their actions.

One of the best-known lymphokines is the **migration inhibition factor (MIF).** It is produced by activated delayed-type hypersensitivity lymphocytes and inhibits the migration of tissue macrophages away from the site of infection. It is easy to see that the activation of only a few specific T cells at the site of a delayed hypersensitivity reaction (e.g., a positive tuberculin skin test) would greatly amplify the reaction by releasing MIF. The released MIF would cause the accumulation of the many nonspecific macrophages characteristic of this type of reaction. Because activated macrophages are far more potent than resting macrophages in phagocytosis and intracellular killing, such a recruitment mechanism would greatly amplify the nonspecific defense against microorganisms and other nonself factors.

The term **interleukin** has been given to those protein or polypeptide mediators that act on or signal between cells of the immune system; T cells are normally affected. Interleukins are immunologically nonspecific and act at a short range. An interleukin, however, need not be a lymphocytic product (that is, some interleukins are not lymphokines). Several interleukins are briefly discussed.

Interleukin-1 (IL-1), is a polypeptide produced mainly by macrophages. It stimulates lymphocyte differentiation and T-cell function. IL-1 is weakly mitogenic by itself, but is a very potent T-cell co-mitogen when present with phytohemagglutinin in vitro. A **mitogen** (causing mitosis) can nonspecifically trigger B- and T-cell division, immunologic activation, and production of small amounts of IgM by B cells. **Phytohemagglutinin** is a proteinaceous hemagglutinin (causing red blood cells to agglutinate or clump together) of plant origin used to induce cell division of lymphocytes. It appears that phytohemagglutinin induces IL-1 production from macrophages. Secreted IL-1, together with the mitogen, greatly enhances polyclonal T-cell activation and multiplication. IL-1 is also identical with the endogenous pyrogen derived from PMNs. It stimulates the development of fever, and is a very effective second signal for T cells.

Interleukin-2 (IL-2) or **T-cell growth factor** is a polypeptide produced mainly by $CD4^+$ T cells and is a true T-cell lymphokine. It promotes the growth and function of T cells, hence the alternate name T-cell growth factor. It is released by T-helper cells under the influence of IL-1. T cells stimulated by mitogen and IL-1 synthesize and secrete IL-2, which is a potent stimulator of activated T cells and causes the production of more IL-2 receptors on T cells. The more receptors on T cells, the more responsive they are to IL-2. For example, cells with many IL-2 receptors can be kept growing in vitro if IL-2 is supplied.

Interleukin-3 (IL-3) is a product of a subset of T-helper cells. It regulates the proliferation of stem cells, and influences the differentiation of bone-marrow-derived mast cells and other leukocytes (see below).

Colony-stimulating factors (CSFs) are another subset of lymphokines produced by $CD4^+$ T cells. CSFs are glycoproteins that regulate the production, differentiation, and activation of phagocytic cells. Four major CSFs have been identified:

1. Interleukin-3 (multi-CSF) induces the formation of granulocytes, macrophages, and eosinophils.
2. Granulocyte-macrophage CSF induces the formation of granulocytes and macrophages.
3. Macrophage CSF induces the formation of macrophages.
4. Granulocyte CSF induces the formation of granulocytes.

The CSFs stimulate host defenses against infectious pathogens. The capacity of CSFs to increase the number of available phagocytes is one mechanism for killing invading microorganisms. In addition to increasing phagocyte numbers, CSFs also modulate secretory capacity, chemotaxis, and phagocyte killing.

Interferons (IFNs) are also cytokines and have been previously discussed with respect to their antiviral properties (*see*

TABLE 32.1 Summary of Some Lymphocyte Mediators and Their Functions

Category	Mediator	Function
Regulation of immune system cells (nonantigen specific)	T-cell activating protein (TAP)	Activates T cells
	Allogenic effect factor (AEF)	Effects lymphocyte production
	T-cell replacing factor (TRF)	Replaces T cells
	Interleukin-1 (IL-1)	Promotes multiplication and activation of T cells; activates B cell proliferation and regulates B cell differentiation
	Interleukin-2 (IL-2)	Causes proliferation of T cells and activation of T_H cells and natural killer cells; activates antibody-producing B cells
	Interleukin-3 (IL-3)	Regulates proliferation of stem cells and differentiation of mast cells
	Interleukin-4 (IL-4)	Effects proliferation and differentiation of T and B cells, macrophages, and mast cells; promotes synthesis of IgG1 and IgE antibodies
	Interleukin-5 (IL-5)	Induces B cell differentiation and IgM and IgA antibody synthesis; promotes cytotoxic T cell production
	Interleukin-6 (IL-6)	Stimulates B cell growth and differentiation
	Interferons	Inhibit viral replication, slow cell growth, enhance natural killer cell activity, reduce antibody formation
	B-cell differentiation factors (BCDFs)	Cause B cells to differentiate into plasma cells
	B-cell growth factors (BCGFs)	Cause B-cell proliferation
	Colony-stimulating factors (CSFs)	Stimulate proliferation of granulocytes and macrophages
Regulation of immune system cells (antigen specific)	T-helper factor (T_HF)	T-cell stimulation
	T-suppressor factor (T_SF)	T-cell suppression
	Skin-reactive factor (SRF)	Skin reactions—induction of inflammation and mononuclear cell recruitment
	Migration inhibitory factor (MIF)	Decreases macrophage migration and stimulates phagocytosis
	Macrophage-activating factor (MAF)	Stimulates macrophage activity
	Leukocyte inhibitory factor (LIF)	Inhibits leukocyte migration
	Macrophage-chemotactic factor (MCF)	Recruits macrophages to tumor sites and sites of inflammation
	Macrophage fusion factor (MFF)	Causes macrophages to fuse
Destruction of nonleukocyte target cells	Lytic tumor factor (LT) or lymphotoxin	Lyses tumors and causes inflammation

pp. 593–95). In addition, interferons have been shown to affect the immune system. The type of IFN generated in the immune response is IFN-γ, produced by T cells after appropriate inductive signals. IFN-γ modulates T- and B-cell functions such as augmentation and diminution of the immune responses; regulation of antibody production and various T-cell functions; expression of cell surface antigens; and regulation of natural killer cell activity and certain macrophage functions. Table 32.1 summarizes some of the better-known lymphocyte mediators and their functions.

1. Describe several characteristics of lymphokines.
2. Describe the function of MIF.
3. What are interleukins? Colony-stimulating factors?
4. Describe the function of interleukin-1 and interleukin-2.
5. What role do interferons play in the immune response?

B-Cell Biology

Stem cells in the bone marrow produce uncommitted B cells that migrate to lymphoid tissue where some mature into cells that circulate freely in the bloodstream and lymph (*see figure 31.2*). These circulating B cells contain surface immunoglobulin M (IgM) and IgD. It is the IgM, however, that is the major receptor for its appropriate antigen. B cells also have receptors for the Fc part of some immunoglobulin classes and may have receptors for complement (C3b). Each B cell may have as many as 10,000 of these specific antigen-receptor sites. The total mature B-cell population of each individual thus carries receptors specific for many antigens; however, each mature B cell possesses receptors specific only for a particular antigenic determinant.

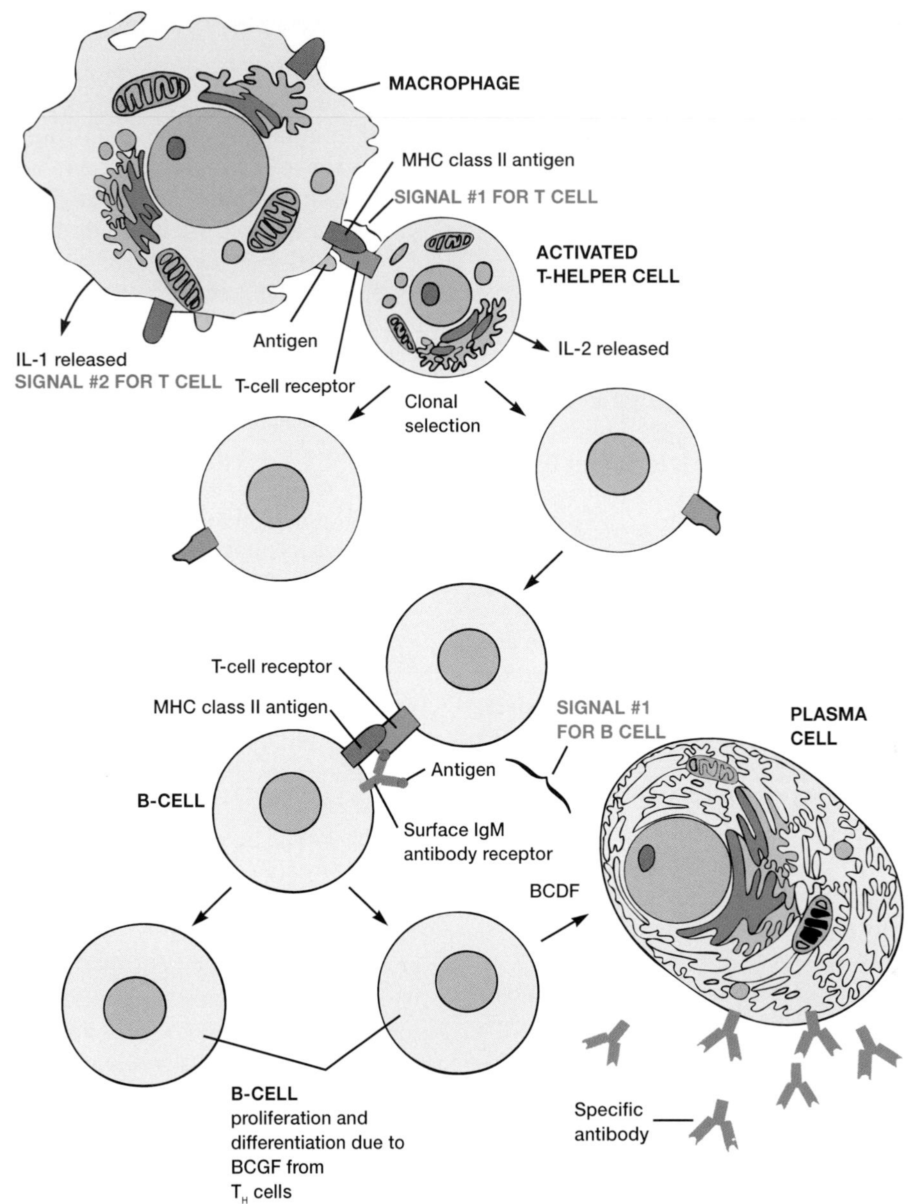

Figure 32.1 **T-Dependent Antigen Triggering of a B Cell.** Schematic diagram of the events occurring in the interactions of macrophages, T-helper cells, and B cells that produce cell-mediated immunity.

B-Cell Activation

The mature B cell divides and/or secretes antibody when triggered by the appropriate signals. Triggering can be antigen specific, activating a particular clone of cells, or nonspecific and polyclonal via B-cell mitogens.

T-Dependent Antigen Triggering

Most antigens have more than one type of antigenic determinant site (epitope) on each molecule (*see figure 31.3*). B cells specific for a given epitope on the antigen (e.g., epitope X) often cannot develop into plasma cells secreting antibody (anti-X) without the collaboration of helper T cells. A **T-helper (T_H)cell** is a T lymphocyte that helps initiate an effector (B cell or macrophage) function. In other words, binding of the epitope X to the B cell may be necessary, but it is not sufficient for B-cell activation. Antigens that are processed with the aid of T-helper cells are called **T-dependent antigens.** Examples include bacteria, foreign red blood cells, certain proteins, and hapten-carrier combinations (*see figure 31.4*).

The basic mechanism for T-dependent antigen triggering of a B cell is illustrated in figure 32.1 and involves three cells: (1) an antigen-presenting macrophage to process and present the antigen; (2) a T-helper cell able to recognize the antigen and to respond to it; and (3) a B cell specific for the antigen. When all cells and antigen are present, the following sequence takes place. The macrophage presents part of the antigen and its own major histocompatibility complex to the T-helper cell (signal #1 to the T cell), and also makes interleukin-1 (signal

#2 to the T cell). Interleukin-1 stimulates the T-cell clone to divide and make interleukin-2. Interleukin-1 also stimulates the hypothalamus to raise the body temperature (causing a fever), which enhances the activity of T cells. Interleukin-2 stimulates T-helper cells to multiply. These helper cells secrete a lymphokine called B-cell growth factor (BCGF), which causes B cells to multiply. As the number of B cells increases, T-helper cells produce another lymphokine, the B-cell differentiation factor (BCDF). This lymphokine instructs some B cells to stop replicating, to differentiate into plasma cells, and to start producing antibodies. At this step, the B cell recognizes the antigen via its surface IgM receptors for the antigen (signal #1 for the B cell) and is subsequently triggered to proliferate and differentiate into plasma cells secreting antibody. All the immune responses that produce IgG, IgA, and IgE involve this T-dependent antigen triggering.

Fever (pp. 590–91).

T-Independent Antigen Triggering

Not all antibody responses require T-cell help. There are antigens that trigger B cells into immunoglobulin production without T-cell cooperation. These are called **T-independent antigens.** Examples include certain tumor-promoting agents, anti-immunoglobulin (anti-Ig), antibodies to certain B-cell differentiation antigens, and bacterial lipopolysaccharides. The T-independent antigens are polymeric, that is, composed of repeating polysaccharide or protein subunits. They elicit almost exclusively IgM antibody formation, and there is little switching to IgG; the resulting antibody generally has a low affinity for antigen.

The mechanism for activation by T-independent antigens probably depends on their polymeric structure. These large molecules present a large array of identical epitopes to a B cell specific for that determinant. The multivalent nature of the cell receptor (surface IgM) is such that cell activation occurs and IgM is secreted. Because there is no T-cell help, the B cell cannot switch to high affinity IgG production.

---

1. Name three types of B-cell receptors.
2. What is a mitogen?
3. Name the three cells involved in T-dependent antigen triggering. How does it occur?
4. How does T-independent antigen triggering occur?

T-Cell Biology

T cells are the elements of the cell-mediated immune response. They are immunologically specific, can carry a vast repertoire of immunologic memory, and can function in a variety of regulatory and effector ways.

T-Cell Receptor Proteins

T cells have specific receptors for antigens on their plasma membrane surface. These receptors are molecules closely related to immunoglobulins, but they cannot bind to free antigens. The receptor site is composed of two parts, an alpha protein chain and a beta protein chain (figure 32.2*a*). Each chain is stabilized by disulfide bridges. The protein receptor is anchored into the plasma membrane and extends into the cytoplasm. The recognition sites of the T-cell receptor extend out from the membrane and have a terminal variable section complementary to antigens. T cells respond to antigens exposed on the surfaces of antigen-presenting cells, most of which are **macrophages.** Macrophages present the antigen, with other macrophage surface markers termed histocompatibility antigens, to the T cells (figure 32.2*b*). Some knowledge of these histocompatibility antigens is required before T-cell-macrophage interactions and T-cell biology can be understood.

Histocompatibility Antigens

Almost all human tissue cells contain **histocompatibility antigens** or **HLAs** (**h**uman **l**eukocyte **a**ntigens) on their surface. These antigens are plasma membrane proteins. The structure of a class I HLA was worked out in 1987 using X-ray crystallography. The antigen (figure 32.3) consists of a complex of two protein chains, one with a molecular weight of 45,000 (the heavy chain) and the other with a molecular weight of 12,000 (the light chain). The two chains contain four regions. The outer segment of the heavy chain can be divided into three functional domains, designated α_1, α_2, and α_3. The β_2-microglobulin (β_2m) protein and α_3 segment of the heavy chain are noncovalently associated with one another and are close to the plasma membrane. A small segment of the heavy chain is attached to the membrane by a short amino acid sequence that extends into the cell interior, but the rest of the protein protrudes to the outside. The α_1 and α_2 domains lie to the outside and form the antigen-binding pocket.

Histocompatibility antigens are coded by a group of genes termed the **major histocompatibility complex (MHC)** genes after the prefix "histo," meaning tissue (figure 32.4). These genes are labeled A, B, C, and D and are codominant. Because each individual has two sets of these genes, one from each parent, a person can potentially have any combination of eight different MHC or HLA glycoprotein molecules. Only one gene of each pair can code for an antigen in a given individual, and these antigens are different from one individual to another. The closer two individuals are related, the more similar are their histocompatibility antigens. It should be noted here that the designations HLA and MHC are similar, but not identical. MHC is the more general term and refers to any histocompatibility antigen complex. HLA applies specifically to human histocompatibility antigens.

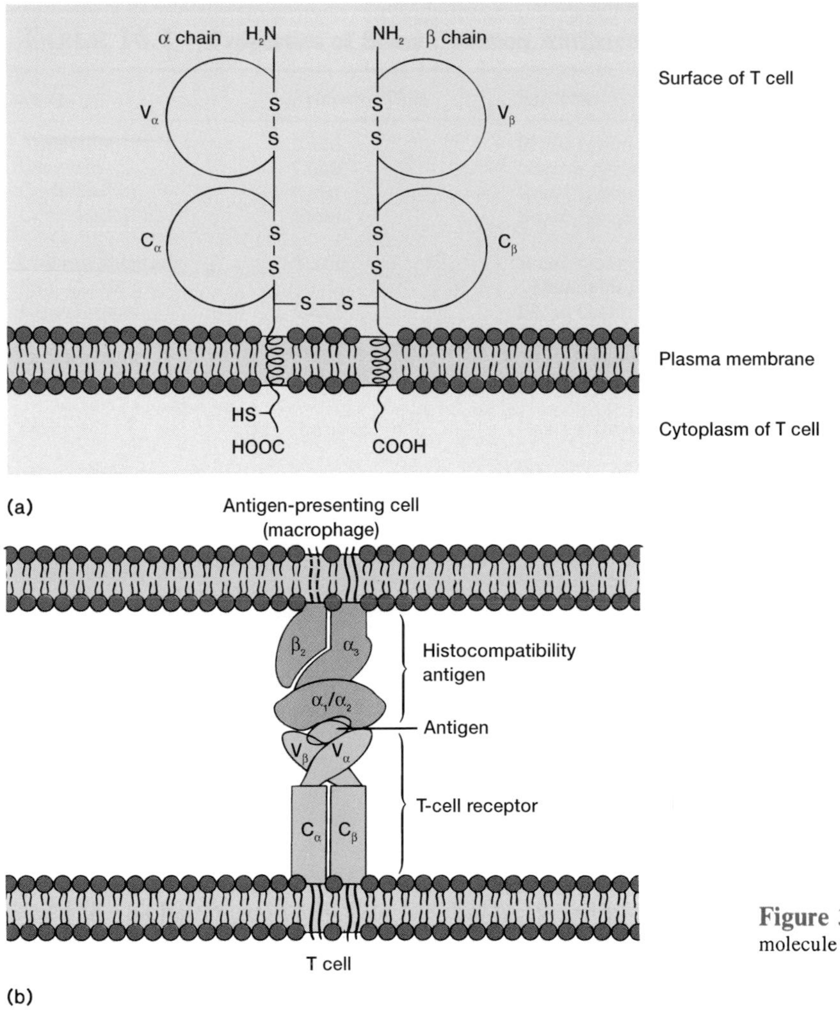

Figure 32.2 T-cell Receptor Proteins and Activation. (*a*) A schematic illustration of the proposed overall structure of the antigen receptor site on a T-cell plasma membrane. (*b*) An antigen-presenting cell begins the activation process by displaying an antigen fragment on its surface as part of a complex with the histocompatibility antigens. A T cell is triggered when the variable region of its receptor (designated V_α and V_β) reacts with the antigen on the presenting cell surface.

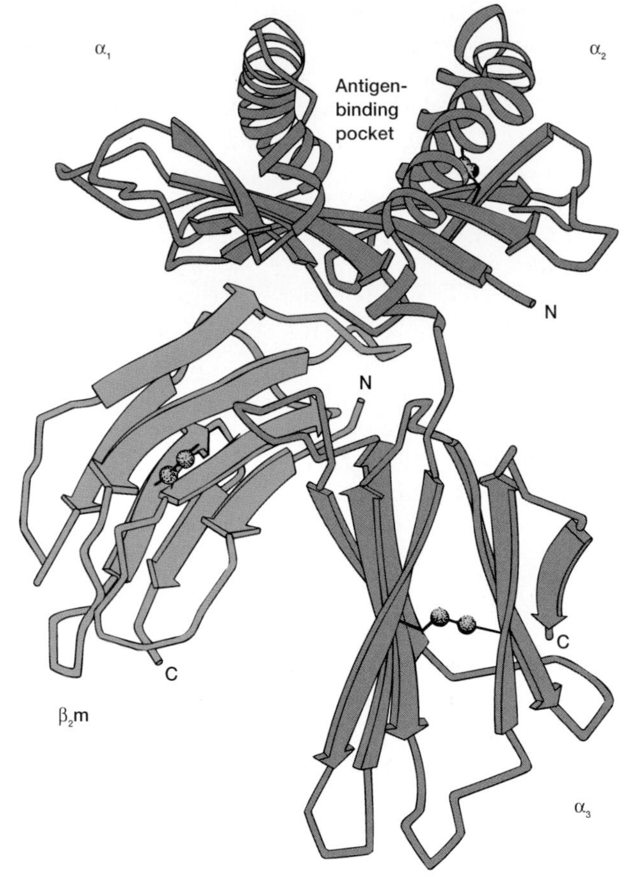

Figure 32.3 Histocompatibility Antigen Structure. This side view of the molecule shows the arrangement of its four regions (α_1, α_2, α_3, and β_2).

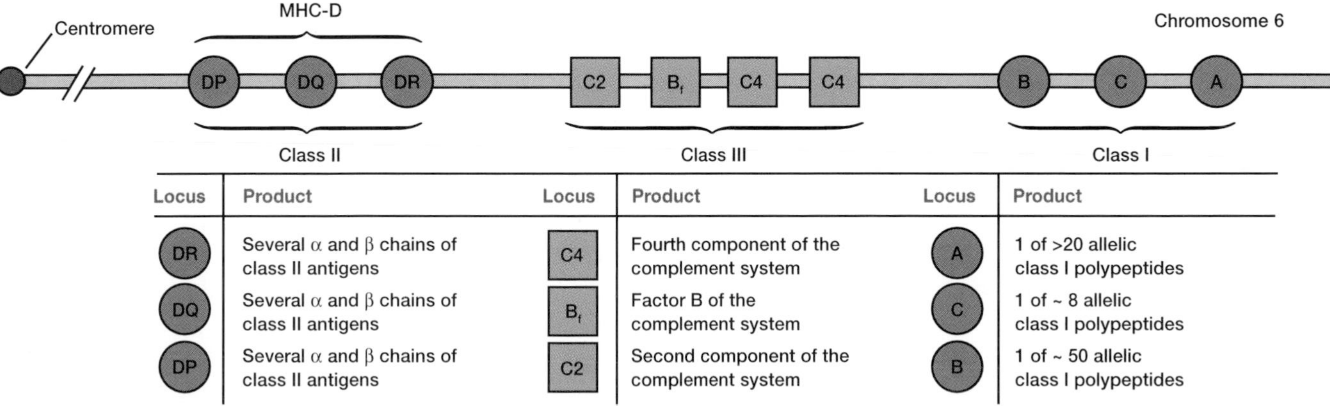

Figure 32.4 Major Histocompatibility Complex. The MHC region of human chromosome 6 and the gene products associated with each locus. DP, DQ, and DR are not single loci but clusters of several loci encoding α and β chains of class II antigens.

Box 32.1

Donor Selection for Tissue or Organ Transplants

The likelihood of tissue or organ transplant acceptance can be increased by using persons who have a great degree of genetic similarity. The greater the similarity, the more likely the persons will have similar MHC complexes. The reason for this is that 77 histocompatibility alleles are determined by just four human histocompatibility genes—HLA-A, HLA-B, HLA-C, and HLA-D (which includes DQ, DR, and DP). These genes are transmitted from parents to offspring according to Mendelian genetics. For example, two siblings (brother and sister) may have approximately a 25% chance of being HLA-B identical, a 50% chance of sharing one HLA-B gene, and only a 25% chance of having completely different HLA-B genes. Therefore, when tissue or organ transplants are performed, a deliberate attempt is made to use tissues from siblings or other genetically related, histocompatible people.

Once a tissue donor has been selected, various in vitro tests are performed to find how closely the HLAs match. The greater the degree of similarity, the greater the chances are that the transplant will be successful. One such test is the lymphocytotoxic cross-match test. In this test, the recipient's serum is mixed with the donor's lymphocytes to determine whether antibodies are present that will cause tissue rejection. The cell-mediated lymphocytosis test, in which recipient leukocytes are mixed with donor serum, is used to test for acute rejection. After a prolonged incubation period, the presence of activated killer cells is an indication that rejection will occur.

The MHC of humans is a cluster of genes located on chromosome 6. The MHC directs the production of three classes of proteins termed class I, class II, and class III.

Class I antigens are made by all cells of the body except red blood cells and comprise HLA types A, B, and C. Class I antigens serve to identify almost all cells of the body as "self." They also stimulate antibody production when introduced into a host with different class I antigens. This is the basis for HLA typing when a patient is being prepared for an organ transplant (Box 32.1).

Class II molecules comprise the D group of HLA and are produced only by activated macrophages, mature B cells, some T cells, and certain cells of other tissues. Class II antigens do not stimulate antibody production but are required for T-cell communication with macrophages and B cells. As discussed in more detail later, part of the T-cell receptor must recognize a class II antigen on the adjacent cell before the T cell can secrete lymphokines necessary for the immune response.

The class III genes encode the second component of complement activation (factor B), which participates in the alternative pathway, and two forms of C4, the fourth component of the complement system (*see figure 33.2*).

Classes I and II of HLA are involved in not only the immunologic recognition of microorganisms but also the individual's susceptibility to particular noninfectious diseases. For example, there is evidence of HLA-linked determinants in tuberculoid leprosy, paralytic poliomyelitis, multiple sclerosis, and acute glomerulonephritis (*see chapter 37*). The effect of HLA antigen presence on the frequency of these diseases is just beginning to be understood.

Regulator T Cells

Regulator T cells control the development of effector T cells. Two subsets exist: T-helper cells and T-suppressor cells. T-helper (T_H) cells are needed for T-cell-dependent antigens to be effectively presented to B cells. For example, when a virus infects a host, it is phagocytosed by a macrophage and partially digested (figure 32.5*a*; *see also figure 30.14*b). The viral antigens are then moved to the surface of the macrophage where they are presented in close association with class II MHC. This combination of viral antigens and MHC is needed for recognition by the receptor protein on the surface of the T-helper cell. T-helper cells that are activated in this way secrete interleukin-2, which activates B cells and cytotoxic T cells. T_H cells also secrete gamma interferon (IFN-γ), which is capable of activating macrophages with enhanced microbicidal activity. Cytotoxic T cells are directly stimulated by antigen presenting cells that process viral antigens and display them with class I HLA (figure 32.5*b*). Bacterial proteins and other extracellular proteins are presented to T cells by macrophages with class II HLA.

T-suppressor (T_S) cells can suppress B-cell and T-cell responses. Subpopulations of T_S cells specific to a given antigen can be stimulated to proliferate by the interleukin-2 from activated T-helper cells. This proliferation occurs at a slow rate to provide negative feedback control for that part of the immune response known as acquired immune tolerance.

Acquired Immune Tolerance

Acquired immune tolerance refers to the body's ability to produce antibodies against nonself antigens while "tolerating" (not producing antibodies against) self-antigens. This tolerance arises early in embryonic life when immunologic competence is being established. Two major mechanisms have been proposed to account for acquired immune tolerance: clonal deletion and functional inactivation. It is very likely that both are partly responsible for immune tolerance.

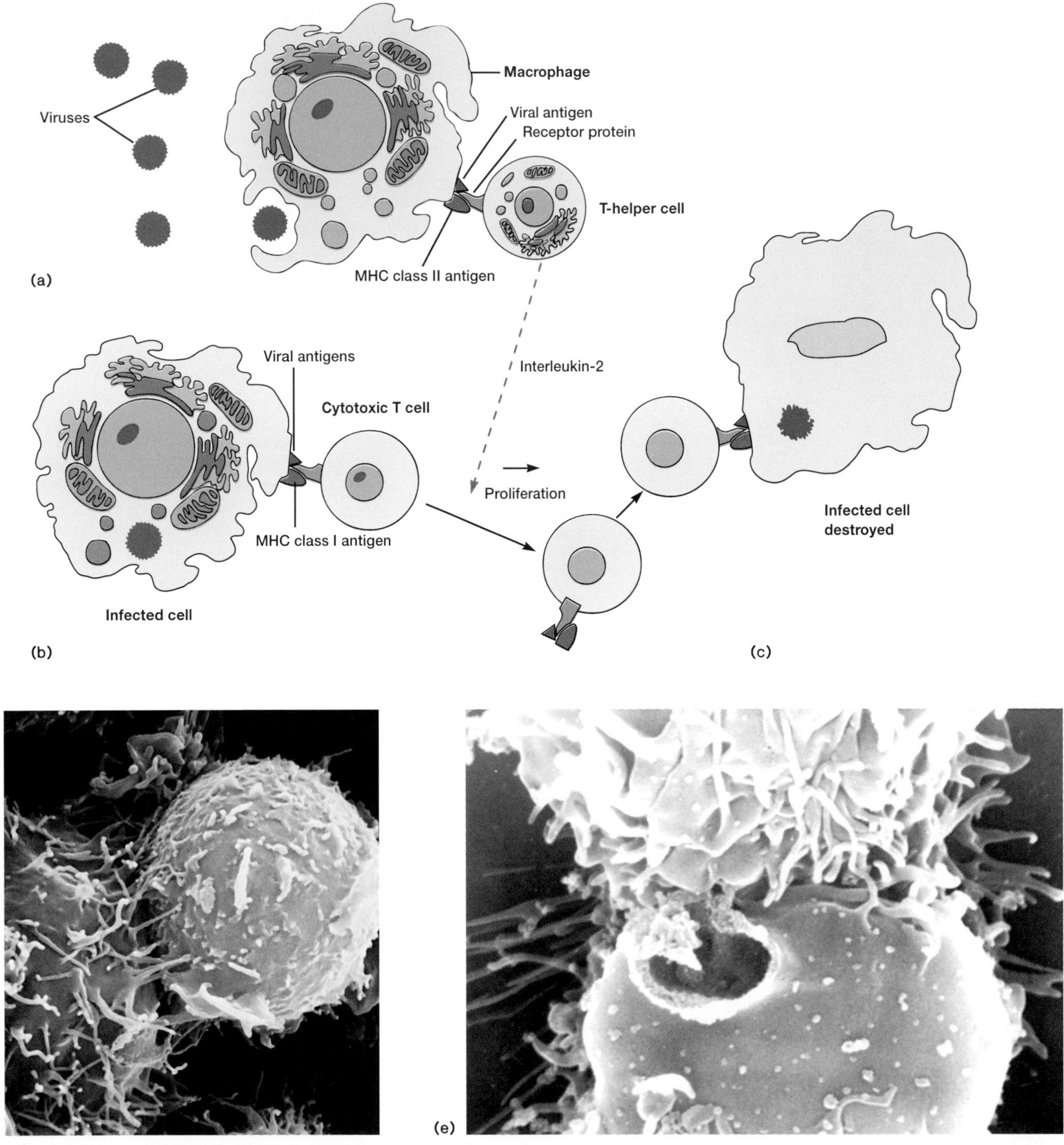

Figure 32.5 Regulator and Effector T Cells. (*a*) A virus is phagocytosed by a macrophage and its antigens presented to T-helper cells in association with class II MHC. (*b*) Once activated, the T_H cell secretes IL-2. It is the IL-2 that regulates the proliferation of cytotoxic T cells. (*c*) Once the T_C cells proliferate, they cause lysis and death of the virus-infected cell. (*d*) A cytotoxic T cell (left) contacts a target cell (right) (×5,700). (*e*) The T cell secretes a protein that forms a hole in the target cell's membrane and causes its contents to leak out (cytolysis) (×45,000). *(a)–(c) Adapted from "The Immune System in AIDS" by Jeffrey Lawrence.*

TABLE 32.2 Comparison of Lymphocytes Involved in the Immune Response

Property	T Cells	B Cells
Origin	Bone marrow in adults	Bone marrow in adults
Maturation and differentiation	Thymus	Lymphoid tissue or bone marrow; bursa of Fabricius in birds
Longevity	Long (months or years)	Short (days to weeks)
Mobility	Great	Very little
Complement receptors	Absent	Present
Surface immunoglobulins	Absent	Present (IgM, IgD)
Proliferation	Upon antigenic stimulation	Upon antigenic stimulation, differentiate into plasma and memory cells
Immunity type	Cell mediated and humoral	Humoral
Distribution	High in blood, lymph, and lymphoid tissue	High in spleen, lymph nodes, bone marrow, and other lymphoid tissue; low in blood
Secretory product	Lymphokines	Antibodies
Subsets and functions	T-helper (T_H) cell: necessary for B-cell activation by T-dependent antigens and T-effector cells	Memory cell: a long-lived cell responsible for the anamnestic response
	T-suppressor (T_S) cell: blocks induction and/or activation of T_H cells and B cells; helps maintain tolerance	Plasma cell: a cell arising from a B cell that manufactures specific antibodies
	T-regulator (T_R) cell: develops into T_H or T_S cells and controls balance between enhancement and suppression of response to antigen	
	Delayed-type hypersensitivity T (T_{DTH}) cell: provides protection against infectious agents, mediates inflammation, and activates macrophages in delayed-type hypersensitivity	
	Cytotoxic T (T_C) cell: lyses cells recognized as nonself and parasite-infected cells	

According to the clonal deletion theory, tolerance to self-antigens is achieved by destruction of lymphocytes that can interact specifically with self-antigens. Recent evidence shows that deletion of those immature T cells able to recognize self-antigens does take place in the thymus. It is not known how this occurs.

According to the functional inactivation theory, some lymphocytes that interact with self-antigens are present throughout life. However, they are normally inhibited from attacking self-antigen in one of two ways. T-suppressor cells could keep T-helper cells in check. The lymphocytes could also enter a functionally unresponsive state in which they do not react to antigens. This condition is sometimes called **clonal anergy.** It appears that clonal anergy arises from the inability of T cells to produce their own growth hormone, interleukin-2, when restimulated by antigens. From this viewpoint, autoimmunity may be due to a defect in T-cell function. Table 32.2 summarizes and compares the lymphocytes involved in the immune response.

Effector T Cells

Effector T cells directly attack specific target cells. Several subsets exist: namely, cytotoxic T cells, delayed-type hypersensitivity T cells, and natural killer cells.

Cytotoxic T (T_C) cells are the most-studied and best-understood subset of T cells. They react with virus-infected cells that display both viral antigens and class I MHC proteins on their surface (figure 32.5*b*). Once recognition has occurred, T_C cells are stimulated to divide and proliferate by the interleukin-2 secreted by T-helper cells (superantigens and interleukin-2; Box 32.2). Thus, activation of T-helper cells by macrophages is required for the full immunologic defense provided by the cytotoxic T cells. Unfortunately, the price paid for this defense is death of the virus-infected cell (figure 32.5*c–e*). For example, much of the hepatic necrosis seen in acute hepatitis B infection (*see chapter 36*) arises from the destruction of liver cells bearing viral antigens—destruction mediated by T_C cells.

Delayed-type hypersensitivity T (T_{DTH}) cells are responsible for initiating type IV delayed-type hypersensitivity reactions through secretion of lymphokines. This type of sensitivity and the function of T_{DTH} cells are covered in detail later in the chapter. T_{DTH} cells also activate macrophages and inhibit their migration away from a site of inflammation.

Natural killer (NK) cells are a small population of large nonphagocytic granular lymphocytes with insignificant amounts of surface immunoglobulin and class I HLA. Although they are discussed in the context of effector T cells, NK cells may not be true T cells. Thus, they are often referred

Box 32.2

Bacterial Superantigens

Bacterial exoproteins (e.g., exotoxins or exoproducts) can exert their immunomodulatory effects on T cells and contribute significantly to bacterial pathogenesis. Several bacterial exoproteins function as superantigens. Superantigens are proteins that stimulate the immune system much more extensively than do normal antigens. Examples include staphylococcal enterotoxin A (SEA), staphylococcal toxic shock syndrome toxin-1 (TSST-1), staphylococcal exfoliative toxin, and streptococcal pyrogenic exotoxin A. Superantigens stimulate the proliferation of many different T-cell clones, as many as one in five T cells. Some consider them T-cell mitogens (substances that stimulate lymphocyte proliferation) rather than antigens. They are unusual in that they act at concentrations several magnitudes lower than do the conventional T-cell mitogens, phytohemagglutinin and concanavalin A. Yet, like normal antigens, these bacterial exoproteins require antigen-presenting cells bearing major histocompatibility complex (MHC) class II molecules to stimulate T cells (figure 32.1). They also may need cytokines produced by accessory cells.

The most prevalent antigen receptor on T lymphocytes is the T-cell receptor (figure 32.2). Superantigens stimulate T cells bearing a particular V_β element of the TCR, regardless of the nature of the V_α receptor element. They bind directly to the outer surface of the MHC molecule without being processed by the antigen-presenting cell. Thus, a superantigen can activate many more T-helper cells than a normal antigen does.

The immune modulation of T cells by microbial superantigens is one way in which bacteria influence the host immune response. T-cell stimulation results in the production of large amounts of lymphokines, including interleukin-2. This may favor pathogen growth because of immune depression resulting from T-cell death after superantigen activation. Some populations of superantigen-activated T cells may attack host tissue and produce autoimmune disease.

to as null cells. NK cells are found in animals never exposed to relevant antigens. Asialo GM_1, a glycosphingolipid, is a characteristic antigenic surface marker present in relatively high density on the plasma membranes of NK cells. They also have Fc receptors for IgG and have been shown to function in vivo in antibody-dependent cell-mediated cytotoxicity against IgG-coated pathogens or tumor cells.

NK cells help in the host's defense by recognizing and destroying tumor cells, virus-infected cells, fungi, bacteria, protozoa, and helminth parasites. This process is called **immune surveillance.** These cells or organisms are killed by cell-mediated lysis (cytolysis) as follows.

NK cells:

1. Are activated by interferons produced by virus-infected cells and/or interleukin-2 from T cells
2. Recognize a class I MHC on target cells and bind to it (figure 32.5*d*)
3. Then undergo a Ca^{2+}-dependent sequence consisting of
   a. Microtubule assembly
   b. Movement of cytoplasmic granules toward the part of the plasma membrane in contact with the target cell
   c. Reorientation of the Golgi apparatus to the target cell, and
   d. Movement of the microtubule-organizing center and other cytoskeletal components into the region of the cytoplasm adjacent to the bound target cell
4. Next insert a pore-forming protein, perforin 1, into the target cell's plasma membrane (figure 32.5*e*)
5. Also produce lysosomal secretions.

As a result of these NK cell activities, the attacked cell lyses. Finally, the NK cell recycles its cytoplasmic components in preparation for another attack on a target cell.

1. What is the function of an antigen-presenting cell?
2. What is a histocompatibility antigen? What are MHCs and HLAs? Describe the roles of the three HLA classes.
3. Outline the functions of a T-helper cell.
4. What is acquired immune tolerance and how does it arise?
5. Name three types of effector T cells and describe their roles and the way in which they operate. What is immune surveillance?
6. Briefly compare and contrast B cells and T cells with respect to their formation, structure, and roles in the immune response (see table 32.2).

Hypersensitivities (Allergies)

Hypersensitivity or **allergy** (the terms are used interchangeably) is an exaggerated immune response that results in tissue damage and is manifested in the individual on second or subsequent contact with an antigen. Hypersensitivity reactions can be classified as either immediate or delayed. Obviously, immediate reactions appear faster than delayed ones, but the main difference between them is in the nature of the immune response to the antigen. Realizing this fact, Peter Gell and Robert Coombs developed a classification system for reactions responsible for hypersensitivities in 1963. Their system correlates clinical symptoms with information about immunologic events that occur during hypersensitivity reactions. The Gell-Coombs classification system divides hypersensitivity into four types: I, II, III, and IV.

Type I (Anaphylaxis) Hypersensitivity

Type I (anaphylaxis) hypersensitivity is characterized by an allergic reaction occurring immediately following an individual's second contact with the responsible antigen (the **allergen**). Upon initial exposure to an allergen, B cells are stimulated to differentiate into plasma cells and produce spe-

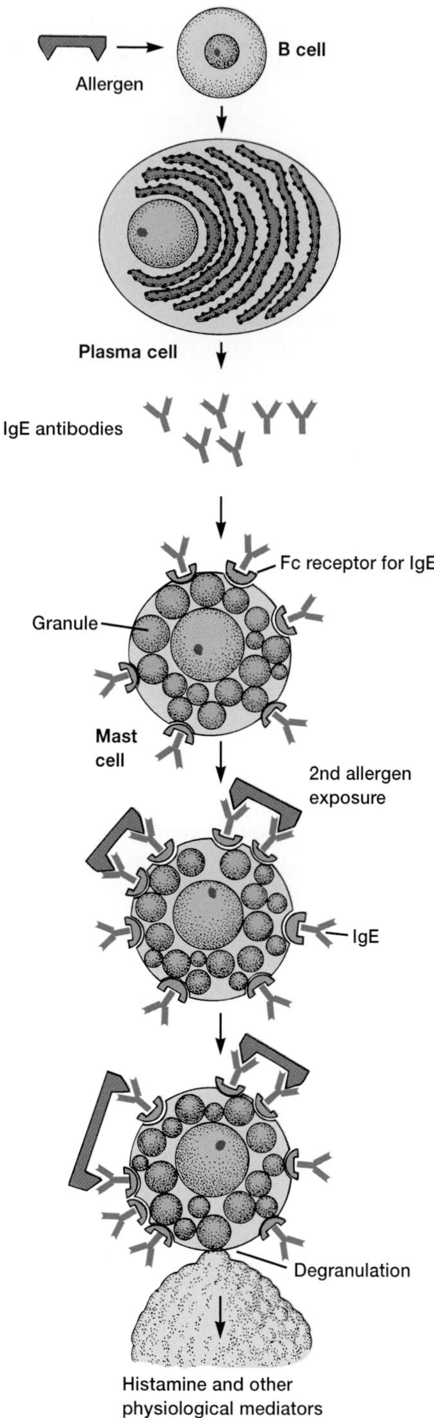

Figure 32.6 **Type I (Anaphylaxis) Hypersensitivity.** This type of hypersensitivity occurs when IgE antibodies attach to mast cells. The combination of these antibodies with allergens stimulates the mast cell (or basophil) to degranulate and produce the physiological mediators that cause the anaphylactic reaction, asthma, or hay fever.

cific IgE with the help of T cells (figure 32.6). This IgE is sometimes called a **reagin,** and the individual has a hereditary predisposition for its production. Once synthesized, IgE binds to the Fc receptors of mast cells or basophils and sensitizes these cells, making the individual allergic to the allergen. When a second exposure to the allergen occurs, the allergen attaches to the surface-bound IgE on the sensitized mast cells causing degranulation. Degranulation releases physiological mediators (*see figure 30.13*) such as histamine, SRS-A (slow reacting substance of anaphylaxis; also called leukotriene), heparin, prostaglandins, PAF (platelet-activation factor), ECF-A (eosinophil chemotactic factor of anaphylaxis), and proteolytic enzymes. These mediators trigger smooth muscle contractions, vasodilation, increased vascular permeability, and mucous secretion. The inclusive term for these responses is **anaphylaxis** (Greek *ana,* up, back, again; and *phylaxis,* protection). Anaphylaxis can be divided into systemic and localized reactions.

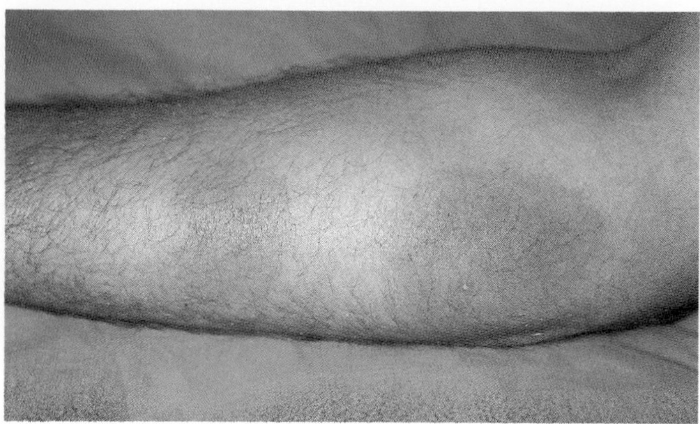

Figure 32.7 **The Anaphylactic Response to Bee Venom.** This person has been stung on the arm by a bee, leading to type I (anaphylaxis) hypersensitivity in a generalized area.

Systemic anaphylaxis is a generalized response that occurs when an individual sensitized to an allergen receives a subsequent exposure to it. The reaction is immediate due to the large amount of mast cell mediators released over a short period. Usually there is respiratory impairment caused by smooth muscle constriction in the bronchioles. The arterioles dilate, which greatly reduces arterial blood pressure and increases capillary permeability with rapid loss of fluid into the tissue spaces (*see figure 30.12*a,b). Because of these reactions, the individual can die within a few minutes from reduced venous return, asphyxiation, reduced blood pressure, and circulatory shock. Common examples of allergens that can produce systemic anaphylaxis include drugs (penicillin), passively administered antisera, and insect venom from the stings or bites of wasps, hornets, or bees (figure 32.7).

Localized anaphylaxis is called an atopic ("out of place") allergy. The symptoms that develop depend primarily on the route by which the allergen enters the body. **Hay fever** (allergic rhinitis) is a good example of an atopic allergy involving the upper respiratory tract. Initial exposure involves airborne allergens—such as plant pollen, fungal spores, animal dander, and house dust mites—that sensitize mast cells located within the mucous membranes. Reexposure to the allergen causes the typical localized anaphylactic response: itchy and tearing eyes, congested nasal passages, coughing, and sneezing. Antihistamine drugs are used to help alleviate these symptoms.

Bronchial asthma (asthma means panting) is an example of an atopic allergy involving the lower respiratory tract. Common allergens are the same as for hay fever. In bronchial asthma, however, the air sacs (alveoli) become overdistended and fill with fluid and mucus; the smooth muscle contracts and narrows the walls of the bronchi. Bronchial constriction produces a wheezing or whistling sound during exhalation. Symptomatic relief is obtained from bronchodilators that help relax the bronchial muscles, and from expectorants and liquefacients that dissolve and expel mucous plugs that accumulate.

Allergens that enter the body via the digestive system may cause food allergies. **Hives** (eruptions of the skin) are a good diagnostic sign of a true food allergy. Once established, type I food allergies are usually permanent but can be partially controlled with antihistamines or by avoidance of the allergen.

Skin testing can be used to identify the allergen responsible for allergies. These tests involve inoculating small amounts of suspect allergen(s) into the skin. Sensitivity to the antigen is shown by a rapid inflammatory reaction characterized by redness, swelling, and itching at the site of inoculation (*see figure 33.5*a). The affected area in which the allergen-mast cell reaction takes place is called a wheal and flare reaction site.

Once the responsible allergen has been identified, the individual should avoid contact with it. At times this is not possible, and **desensitization** is warranted. This procedure consists of a series of allergen doses injected beneath the skin to stimulate the production of IgG antibodies rather than IgE antibodies. The circulating IgG antibodies can then act as blocking antibodies to intercept and neutralize allergens before they have time to react with mast cell-bound IgE. Recent evidence suggests that suppressor T-cell activity also may cause a decrease in IgE synthesis. Allergy injections are about 65 to 75% effective in individuals whose allergies are caused by inhaled allergens.

Type II (Cytotoxic) Hypersensitivity

Type II (cytotoxic) hypersensitivity is generally called a cytolytic or cytotoxic reaction because it results in the destruction of host cells, either by lysis or toxic mediators. In type II hypersensitivity, IgG or IgM antibodies are directed against cell surface or tissue antigens. They usually stimulate the complement pathway and a variety of effector cells (figure 32.8). The antibodies interact with complement (C1q) and the effector cells through their Fc regions. The damage mechanisms are a reflection of the normal physiological processes involved in interaction of the immune system with pathogens. A classic example of type II hypersensitivity is that resulting when a person receives a transfusion with blood from a donor with a different blood group.

Type III (Immune Complex) Hypersensitivity

Type III (immune complex) hypersensitivity involves the formation of immune complexes (figure 32.9*a*; *see also figure 33.7*). Normally these complexes are removed effectively by monocytes of the reticuloendothelial system (*see figure 30.10*). In the presence of excess amounts of some antigens, the antigen-antibody complexes may not be efficiently removed. Their accumulation can lead to a hypersensitivity reaction from complement that triggers a variety of inflammatory processes. This inflammation causes damage, especially of blood vessels (vasculitis; figure 32.9*b*), kidney glomerular basement membranes (glomerulonephritis), joints (arthritis), and skin.

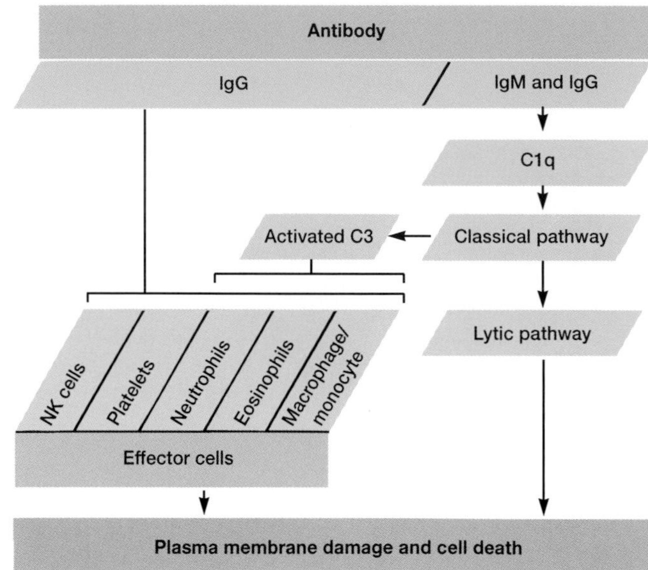

Figure 32.8 **Type II (Cytotoxic) Hypersensitivity.** The action of antibody occurs through effector cells or the membrane attack complex, which damages target cell plasma membranes, causing cell destruction.

Diseases resulting from type III reactions can be placed into three groups. First, a persistent viral, bacterial, or protozoan infection, together with a weak antibody response, leads to chronic immune complex formation and eventual deposition of the complex in host tissues. Second, the continued production of autoantibody to self-antigen during an autoimmune disease can lead to prolonged immune complex formation. This overloads the reticuloendothelial system, and tissue deposition of the complexes occurs (e.g., in the disease **systemic lupus erythematosus**). Third, immune complexes can form at body surfaces (such as the lungs), following repeated inhalation of allergens from molds, plants, or animals. For example, in Farmer's lung disease, an individual has circulating antibodies to fungi after being exposed repeatedly to moldy hay. These antibodies are primarily IgG. When the allergens (fungal spores) enter the alveoli of the lungs, local immune complexes form, leading to inflammation.

Some group A streptococcal infections can produce an immunologically mediated acute glomerulonephritis (*see chapter 37*). Although the mechanism is not completely understood, it is believed that complexes of antibody and streptococcal antigen are deposited within the kidney glomeruli and generate a type III hypersensitivity reaction.

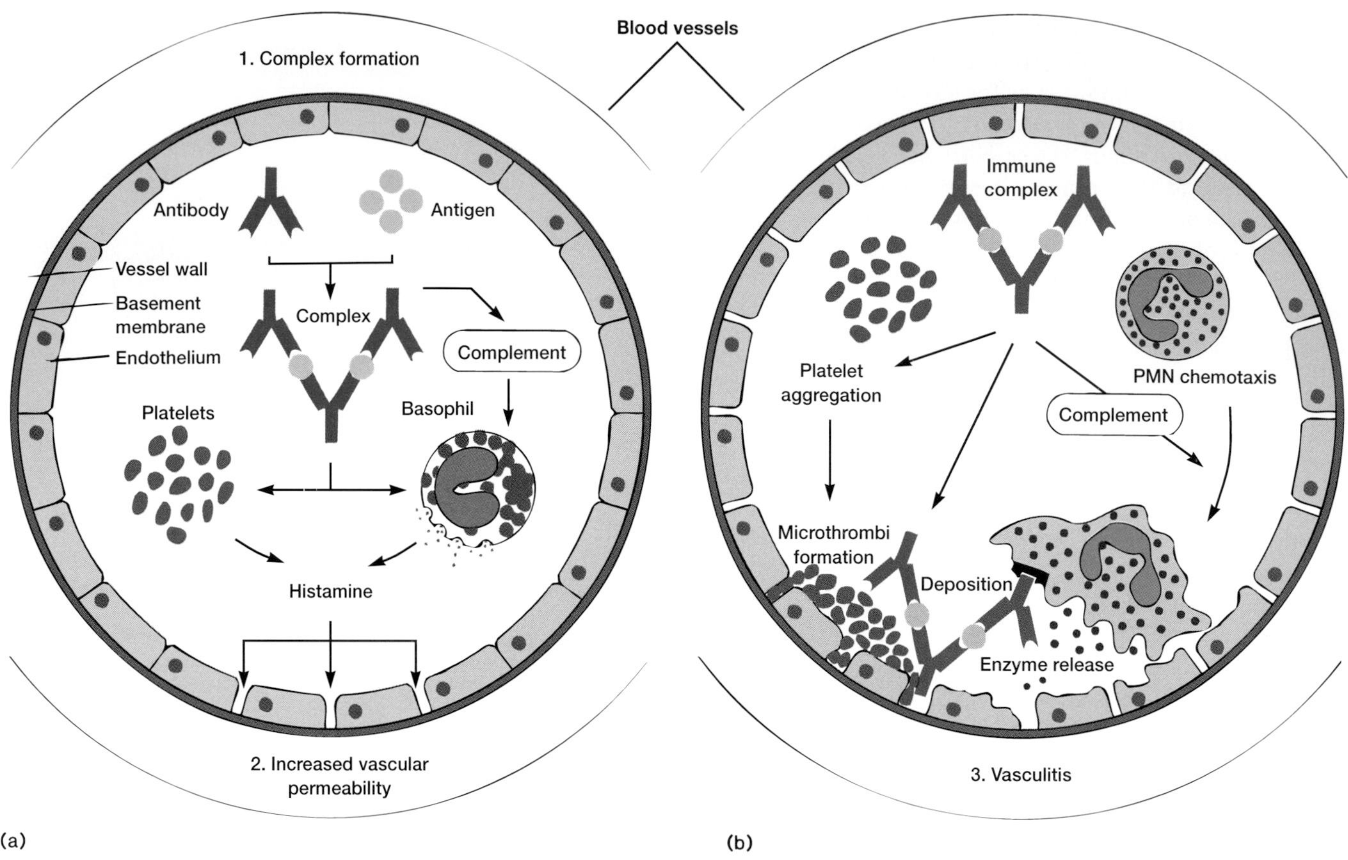

Figure 32.9 **Type III (Immune Complex) Hypersensitivity.** Deposition of immune complexes in blood vessel walls. (*a*) Antibody and antigen combine to form immune complexes. These activate complement, which causes basophils and platelets to degranulate and release histamine and other mediators. These mediators increase vascular permeability. (*b*) The increased permeability allows the immune complexes to be deposited in the blood vessel wall. This induces platelet aggregation to form microthrombi (blood clots) on the vessel wall. PMNs, stimulated by complement, degranulate causing enzymatic damage to the blood vessel wall.

Type IV (Cell-Mediated) Hypersensitivity

Type IV (cell-mediated) hypersensitivity involves delayed T-cell-mediated immune reactions. A major factor in the type IV reaction is the time required for a special subset of T cells called delayed-type hypersensitivity T (T_{DTH}) cells to migrate to and accumulate near the antigens. This usually takes a day or more.

Type IV hypersensitivities can be transferred passively from one individual to another by either intact sensitized T_{DTH} cells or an extract of T_{DTH} cells. The extract is called transfer factor and has been clinically used in humans against mucocutaneous candidiasis (*see chapter 39*).

Type IV reactions occur when antigens, especially those binding to tissue cells, are phagocytosed by macrophages and then presented to receptors on the T_{DTH} cell surface. Contact between the antigen and T_{DTH} cell causes the cell to proliferate and release lymphokines. Lymphokines attract lymphocytes, macrophages, and basophils to the affected tissue. Extensive tissue damage may result. Examples of type IV hypersensitivities include tuberculin hypersensitivity (the **TB skin test**), allergic contact dermatitis, some autoimmune diseases, transplantation rejection, and killing of cancer cells.

In tuberculin hypersensitivity, a partially purified protein called tuberculin, which is obtained from the bacillus that causes tuberculosis (*Mycobacterium tuberculosis; see chapter 37*), is injected into the skin of the forearm (figure 32.10*a*). The response in a tuberculin-positive individual begins in about 8 hours, and a reddened area surrounding the injection site becomes indurated (firm and hard) within 12 to 24 hours. The reaction reaches its peak in 48 hours and then subsides. The size of the induration is directly related to the amount of antigen that was introduced and to the degree of hypersensitivity of the tested individual. Other microbial products used in type IV skin testing are histoplasmin for histoplasmosis, coccidioidin for coccidioidomycosis, lepromin for leprosy, and brucellergen for brucellosis.

Allergic contact dermatitis is caused by haptens (*see figure 31.4*) that combine with proteins in the skin to form the allergen that elicits the immune response. The haptens are the antigenic determinants, and the skin proteins are the carrier

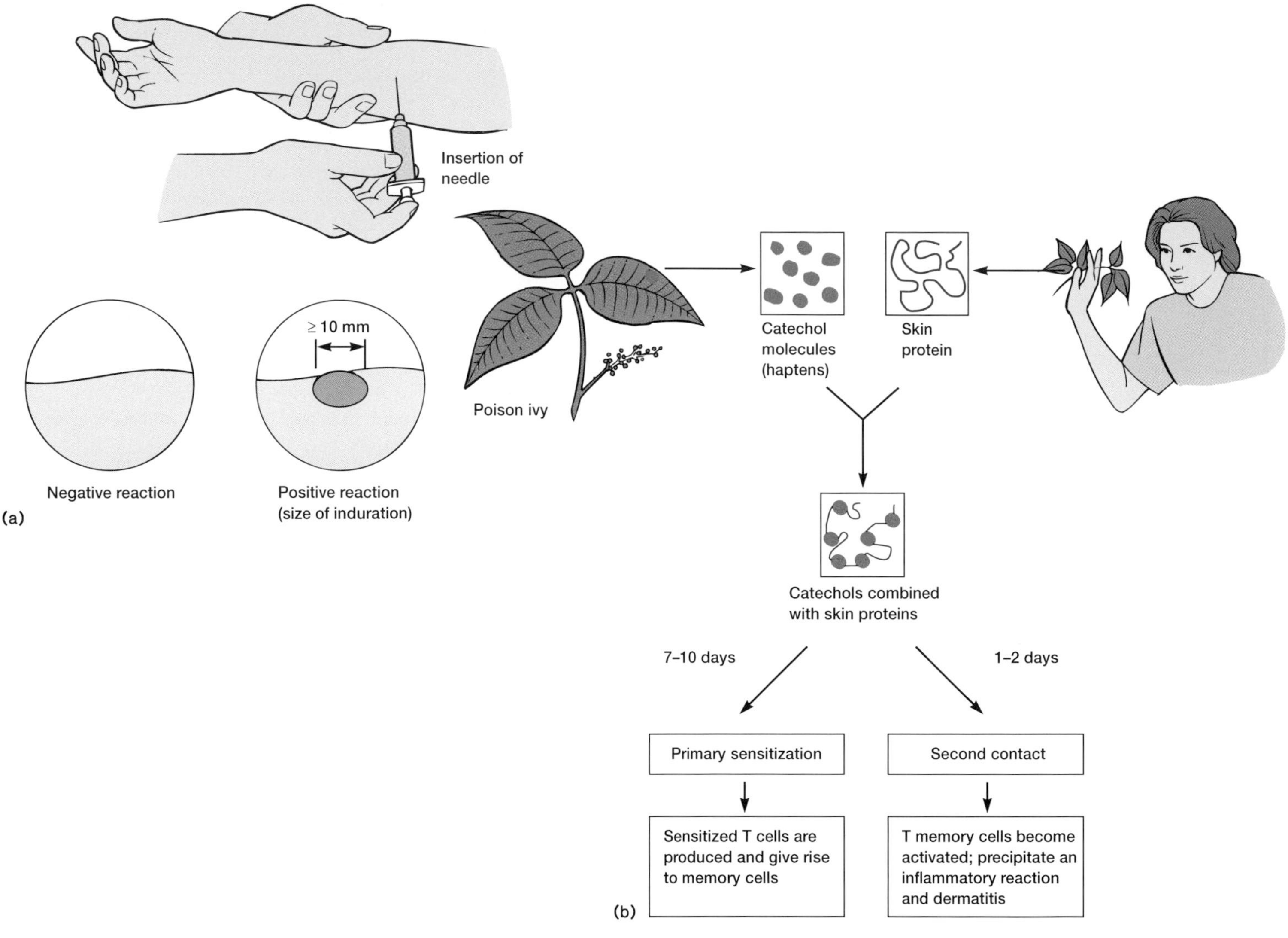

Figure 32.10 Mechanism of Type IV (Cell-mediated) Hypersensitivity. (*a*) The tuberculin skin test. A positive reaction is one in which the diameter of induration is 10 mm or more. (*b*) In this example of contact dermatitis to poison ivy, a person initially becomes exposed to the catechol molecules from the poison ivy plant. The catechols combine with high molecular weight skin proteins and act as haptens. After 7–10 days, sensitized T cells are produced and give rise to memory T cells. Upon second contact, the catechols bind to the same skin proteins, and the T memory cells become activated in only 1–2 days, leading to the inflammatory reaction (contact dermatitis).

molecules for the haptens. Examples of these haptens include cosmetics, plant materials (catechol molecules from poison ivy and poison oak; figure 32.10*b*), topical chemotherapeutic agents, metals, and jewelry (especially that containing nickel).

Several important chronic diseases involve cell and tissue destruction by type IV hypersensitivity reactions. These diseases are caused by viruses, mycobacteria, protozoa, and fungi that produce chronic infections in which the macrophages and T cells are continually stimulated. Examples are leprosy, tuberculosis, leishmaniasis, candidiasis, and herpes simplex lesions.

1. What is the main difference between immediate and delayed hypersensitivity reactions?
2. Discuss the mechanism of type I hypersensitivity reactions and how these can lead to systemic and localized anaphylaxis.
3. What causes a wheal and flare reaction site?
4. Why are type II hypersensitivity reactions called cytolytic or cytotoxic?
5. What characterizes a type III hypersensitivity reaction? Give an example.
6. Characterize a type IV hypersensitivity reaction.
7. What is the TB skin test used for?

TABLE 32.3 Some Autoimmune Diseases in Humans

Disorder	Organ or Structure Affected	Evidence or Mechanism
Highly probable[a]		
Autoimmune thrombocytopenic purpura	Platelets	Phagocytosis of antibody-sensitized platelets
Graves' disease	Thyroid	Thyroid-stimulating hormone receptor antibody
Hashimoto's thyroiditis	Thyroid	Cell-mediated and humoral thyroid cytotoxicity
Insulin resistance	Pancreas	Insulin receptor antibody
Multiple sclerosis	Nerves	Myelin sheath destruction
Myasthenia gravis	Muscles	Acetylcholine receptor antibody
Systemic lupus erythematosus (SLE)	DNA	Circulating immune complexes
Probable		
Adrenergic drug disorder	Lungs	β-adrenergic receptor antibody
Diabetes mellitus (type I)	Pancreas	Cell-mediated and humoral islet cell antibodies
Glomerulonephritis	Kidney	Glomerular basement membrane antibody or immune complexes
Infertility (some cases)	Testes	Antispermatozoal antibodies
Rheumatoid arthritis	Cartilage	Immune complexes in joints
Possible		
Chronic active hepatitis	Liver	Smooth muscle antibody
Vasculitis	Blood vessels	Immunoglobulin and complement in vessel walls, low serum complement
Vitiligo	Skin	Melanocyte antibody

[a]Boldfaced headings indicate the likelihood that the disorder is an autoimmune disease.

Autoimmune Diseases

As discussed earlier, the body is normally able to distinguish its own self-antigens from foreign nonself antigens and does not mount an immunologic attack against the former. This phenomenon is called immune tolerance. At times the body loses tolerance and mounts an abnormal immune attack, either with antibodies or T cells, against a person's own self-tissue antigens. This type of misguided attack produces a group of human disorders termed **autoimmune diseases** or **autoallergies** (table 32.3).

Although their causal mechanism is not well known, these diseases are more common in older people and may involve viral or bacterial infections. Some investigators believe that the release of abnormally large quantities of antigens may occur when the infectious agent causes tissue damage. The same agents also may cause body proteins to change into forms that stimulate antibody production or T-cell activation. Simultaneously, the activity of T-suppressor cells, which normally limits this type of reaction, seems to be repressed. Many autoimmune diseases have a genetic component. For example, there is a well-known association between an individual's susceptibility to Graves' disease (George Bush, the 41st president of the United States, suffers from this disease) or multiple sclerosis and a specific determinant on the major histocompatibility complex.

Transplantation (Tissue) Rejection

Tissue transplant rejection is the third area (after hypersensitivity and autoimmunity) in which the immune system can act detrimentally. It is occasionally desirable to replace a nonfunctional or damaged body part by transplanting a tissue or organ from one person to another. Some transplants do not stimulate an immune response. For example, a transplanted cornea is rarely rejected since lymphocytes do not circulate into the anterior chamber of the eye. This site is considered an immunologically privileged site. It is also possible to transplant privileged tissue and not stimulate an immune response. An example is a heart valve transplanted from a pig to a human. Such a graft between different species is termed a **xenograft** (Greek *xenos,* strayed).

When one's own tissue is transplanted from one part of the body to another, the graft is not rejected. This type of graft is termed an **autograft** (Greek *autos,* self). An example would be skin from the thigh grafted over a burned area on the arm.

Because identical twins have the same genetic constituency, tissues or organs can be transplanted between them without causing an immune response. This type of graft is termed an **isograft** (Greek *iso,* equal).

Usually, however, transplants are done between genetically different individuals within a species. These are termed **allografts** (Greek *allos,* others). With allografts, there is the possibility that the recipient's cells will recognize the donor's tissues as foreign. This triggers the recipient's immune mechanisms, which may destroy the donor tissue. Such a response is termed a tissue rejection reaction. A tissue rejection reaction can occur by two different mechanisms. First, foreign major histocompatibility complex (MHC) class II antigens on the graft stimulate host T-helper cells to aid cytotoxic T cells in graft destruction (figure 32.11*a*). Cytotoxic T cells recognize the graft through the foreign MHC class I antigens. A

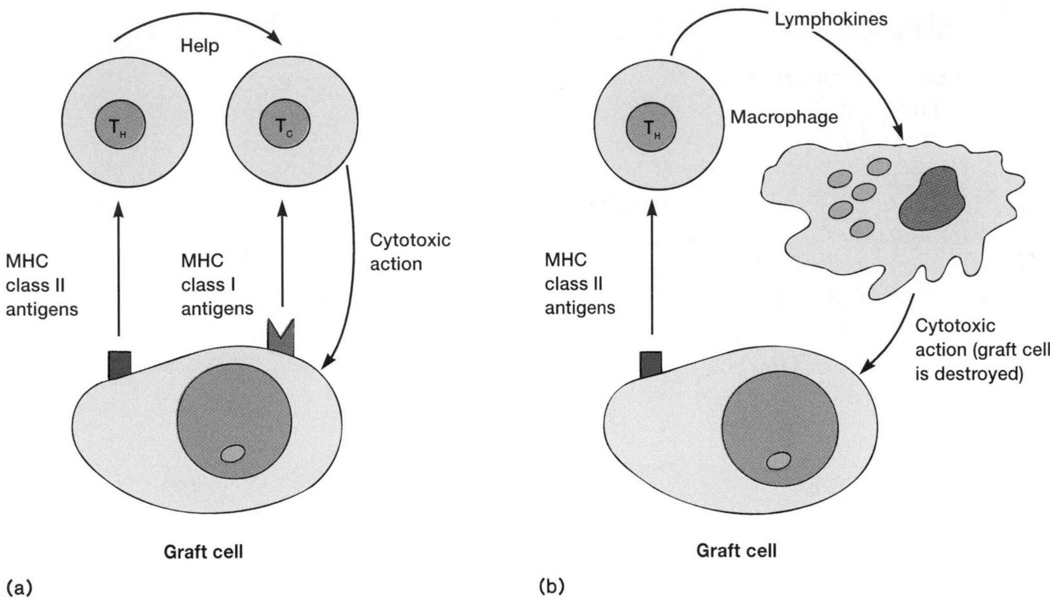

Figure 32.11 **Graft Cell Destruction.** (*a*) Foreign MHC class II antigens on the graft cell stimulate host T-helper cells to help cytotoxic T cells destroy the target graft cell. Cytotoxic T cells recognize the graft cell by its foreign MHC class I antigens. (*b*) T-helper cells reacting to the graft cell release lymphokines that stimulate macrophages to enter the graft and destroy it via cytotoxic action.

second mechanism involves the T-helper cells reacting to the graft and releasing lymphokines (figure 32.11*b*). The lymphokines stimulate macrophages to enter the graft and destroy it.

As presented in figure 32.11*a, b,* the major histocompatibility complex antigens play a dominant role in tissue rejection reactions because of their unique association with the recognition system of T cells. Unlike antibodies, T cells cannot recognize or react directly with non-MHC antigens (viruses, allergens). They recognize these antigens only in association with, or complexed to, an MHC antigen. There are two classes of MHC antigens. Class I MHC antigens are present on every cell in the body and, consequently, are important targets of the rejection reaction. Class II MHC antigens are involved in T-helper cell activation as previously discussed. The greater the antigenic difference between class I antigens of the recipient and donor tissues, the more rapid and severe the rejection reaction is likely to be. However, the reaction can sometimes be minimized if recipient and donor tissues are matched as closely as possible (Box 32.1). This means locating a donor whose tissues are antigenically similar to those of the prospective recipient.

Immunosuppressive drugs (azathioprine, glucocorticoid steroids, cyclosporin-A, cyclophosphamide), antilymphocyte globulin, and irradiation also can be used to reduce the rejection of transplanted tissue. These measures interfere with the recipient's immune mechanism in various ways. For example, they may suppress the formation of antibodies by B cells or the production of T cells, thereby reducing the humoral and cellular responses, respectively. Unfortunately, immunosuppressive measures leave the recipient relatively unprotected against infections and cancer.

Immunodeficiencies

Defects in one or more components of the immune system can result in its failing to recognize and respond properly to antigens. Such **immunodeficiencies** can make a person more prone to infection than those people capable of a complete and active immune response. Despite the increase in knowledge of functional derangements and cellular abnormalities in the various immunodeficiency disorders, the fundamental biological errors responsible for them remain largely unknown. To date, most genetic errors associated with these immunodeficiencies are located on the X chromosome and produce primary or congenital immunodeficiencies (table 32.4). Other immunodeficiencies can be acquired because of infections by immunosuppressive microorganisms (chronic mucocutaneous candidiasis) or by some viruses (HIV).

Tumor Immunity

Oncology (the study of tumors) has revealed that tumor biology is similar to and interrelated with the functions of the immune response. Tumor cells are believed to develop frequently in everyone. Most tumors appear to be clones of single cells that have become transformed. This is similar to the development of lymphocyte clones (*see figure 31.17*) in response to antigens. Lymphocyte clones, however, are under inhibitory control systems—such as control exerted by T-suppressor cells and negative feedback by antibodies. In contrast, the division of tumor cells is not effectively controlled by normal inhibitory mechanisms.

Tumor cells are relatively unspecialized; that is, they dedifferentiate, or become similar to the less specific cells of the

Table 32.4 Some Congenital Immune Deficiencies in Humans

Condition	Symptoms	Cause
Chronic granulomatous disease	Defective monocytes and neutrophils leading to catalase-positive bacterial infections	Failure to produce reactive oxygen intermediates
Congenital hypogammaglobulinemia	B-cell deficiency and inability to produce adequate specific humoral antibodies	Problem with development of the "bursa-equivalent"
DiGeorge syndrome	T-cell deficiency and very poor cell-mediated immunity	Lack of thymus or a poorly developed thymus
Stem cell (Swiss-type) deficiency	Both antibody production and cell-mediated immunity impaired	Apparent absence of stem cells (ancestors of both B cells and T cells)

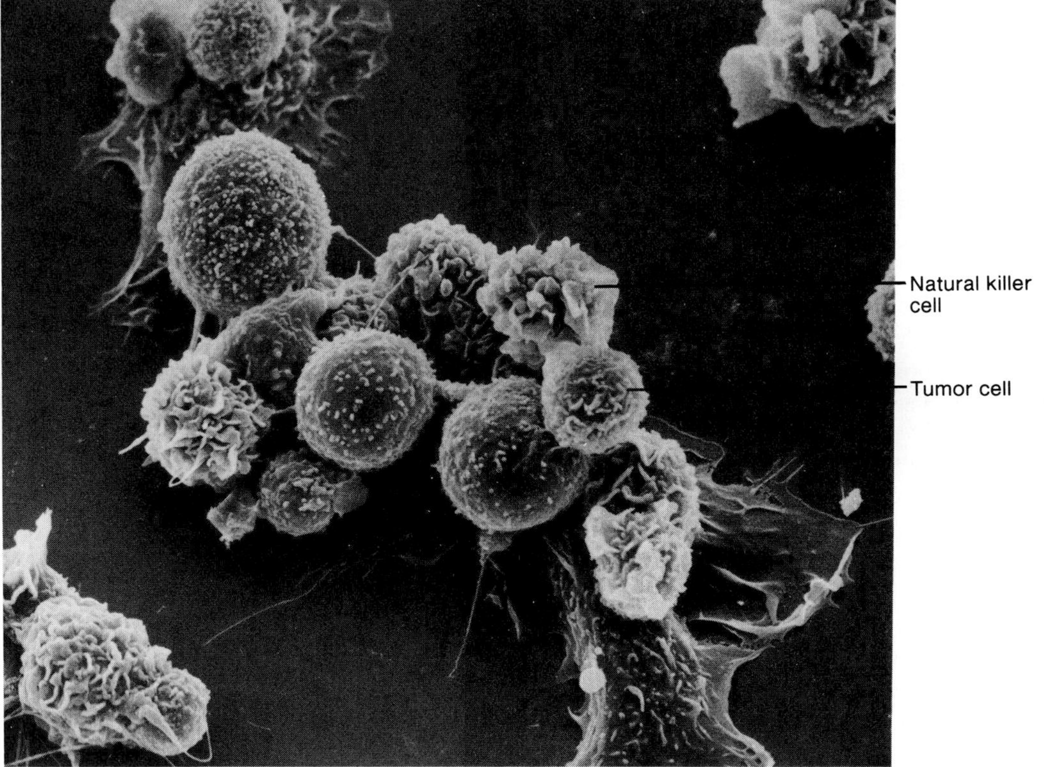

Figure 32.12 **Immune Surveillance.** Scanning electron micrograph of natural killer cells binding to tumor cells. The cancer cells will eventually be destroyed by cell-mediated cytolysis (×4,000).

embryo. As tumor cells dedifferentiate, they reveal surface antigens that can stimulate the immune response and lead to their eventual destruction. Some of these antigens are proteins normally produced only in embryonic or fetal life. Because they are absent at the time immunologic competence is established, they are treated as foreign and subjected to immunologic attack. The release of two such antigens into the blood has provided the basis for laboratory diagnosis of some cancers. For example, carcinoembryonic antigen tests can be used to detect colon cancer and alpha-fetal protein (normally produced only by the fetal liver) to diagnose liver cancer.

Cancer biology and viruses (pp. 393–94).

Once formed, tumors are attacked by the cell-mediated immune system; however, humoral immunity (antibodies) may have a supportive role through complement-dependent cytotoxic reactions. Natural killer cells recognize tumor-specific antigens on the cancer cell surfaces and destroy such cells before a recognizable tumor develops (figure 32.12). This function is termed immune surveillance against cancer.

Once established, some tumors seem to avoid immune surveillance and immune rejection, a phenomenon known as immunologic escape. Several mechanisms have been postulated to explain this. Tumor cells may shed their specific antigens, antigen modulation, and thus evade recognition by the

Table 32.5 Antigens and Antibodies in Human ABO Blood Groups

Blood Type	Antigen Present on Erythrocyte Membranes	Antibody in Plasma	Can Receive Blood from	Can Give Blood to
A	A	Anti-B	A and O	A and AB
B	B	Anti-A	B and O	B and AB
AB	A and B	Neither anti-A nor anti-B	AB; A; B; O	AB only
O	Neither	Anti-A and anti-B	O	O; A; B; AB

immune system. Some tumors also may release factors that eventually suppress the entire cell-mediated immune response.

Human Blood Types

The surface of human red blood cells (erythrocytes) contains genetically determined sets of antigen molecules. People can be divided into **blood type** groups based on the surface antigens that are present. Red blood cell A and B antigens are primarily integral plasma membrane glycoproteins that are attached during red blood cell formation by specific A and B transfer enzymes. These surface antigens are genetically determined, and a person's blood type is inherited. Two major classifications of human erythrocytes are ABO grouping and the Rh system.

ABO blood grouping is based on differences in the type of glycoproteins present on erythrocyte surfaces. Type A people have type A glycoproteins on their erythrocyte surfaces; type B people have type B glycoproteins; type AB people have both glycoproteins; type O people have neither of them. The A and B glycoproteins function as antigens and are recognized by specific antibody molecules, usually of the IgM class. Anti-A antibody and anti-B antibody occur in human plasma and combine specifically with A and B antigens, respectively. As a result, the erythrocytes clump together. This cell clumping is termed hemagglutination and is discussed in more detail in chapter 33.

Humans normally do not produce antibodies to a particular antigen until they are exposed to it; however, the ABO system is an exception. People routinely produce antibodies to those blood group antigens not present on their own erythrocytes without ever being directly exposed to them. It is thought that these antibodies are formed in response to antigens of the intestinal microbiota that are similar to the blood group substances. The antibodies could then cross-react with antigens on the appropriate erythrocyte. Type A people have the anti-B antibody in their plasma; type B people have anti-A antibody; type AB people have neither of these antibodies; and type O people have both antibodies (table 32.5). It is for this reason that blood is typed before being transfused from one person to another. Donor and recipient must be compatible, or a severe hemagglutination reaction may occur. For example, if type A erythrocytes are given to a type O person, the anti-A antibody in the recipient's blood combines with the A antigen on the surface of the donor erythrocytes and causes them to agglutinate. These agglutinates can block blood vessels, causing serious circulatory problems.

The second major classification of human blood types is the **Rh system.** (Rh stands for rhesus; this factor was first identified in the blood of rhesus monkeys.) Rh classification is based upon the presence or absence of the Rh antigen (or D antigen) on erythrocytes. Rh-positive (Rh+) people have the Rh antigen while Rh-negative (Rh−) people do not. About 85% of all people in the United States are Rh+.

In contrast to the situation with the ABO system, Rh− people do not normally have anti-Rh antibodies in their plasma. They must be exposed to the Rh antigen to develop the antibodies. For example, anti-Rh antibodies develop in an Rh− person over a period of several months following an accidental transfusion of Rh+ blood. Once this has happened, a second transfusion can result in a severe hemagglutination reaction.

The Rh factor is also important in pregnancies. A serious problem may occur when an Rh− mother and an Rh+ father have an Rh+ child. The child produces red blood cells with the Rh antigen, which gains access to the mother's circulation when the placenta ruptures during childbirth, or in some cases during an abortion or miscarriage. Responding to the foreign Rh antigens, the mother produces IgG anti-Rh antibodies that are able to cross the placenta. During any subsequent pregnancy, the Rh− mother may produce enough anti-Rh antibodies to agglutinate the red blood cells of the Rh+ child, producing **hemolytic disease of the newborn.** This disease is characterized by anemia (reduction of red blood cells) and jaundice (yellowing of the skin from the breakdown products of the red blood cells).

This disease can be prevented by injecting the Rh− mother with anti-Rh antibody (RhoGAM) within 72 hours after delivery of the child. These passively administered antibodies prevent the mother from developing antibodies against the Rh antigen of the Rh+ red blood cells that have entered her circulation.

1. What is an autoimmune disease and how might it develop?
2. What is an immunologically privileged site and how is it related to transplantation success?
3. How does a tissue rejection reaction occur?
4. Describe an immunodeficiency.
5. Tell what is meant by immune surveillance.
6. Describe the two major classifications of human red blood cells and explain their importance.

Summary

1. T cells release lymphokines. Lymphokines are small proteins or polypeptides that transmit growth, differentiation, and behavioral signals between immune system cells. Other chemical immune mediators include interleukins and interferons.
2. B cells can be stimulated to divide and/or secrete antibody when triggered by the appropriate signals.
3. T cells are pivotal elements of the cell-mediated immune response. T cells have antigen-specific receptor proteins (as well as histocompatibility antigens) for antigens on their plasma membrane surface. These histocompatibility antigens are proteins coded by a group of genes termed the major histocompatibility complex.
4. Regulator T cells control the development of effector cells. Two subsets exist: T-helper cells and T-suppressor cells. T-helper cells are needed for T-cell-dependent antigens to be effectively presented to B cells. T-suppressor cells are involved in slowing responses.
5. Acquired immune tolerance is the ability of a host to produce antibodies against nonself antigens while tolerating (not producing antibodies against) self-antigens. Two theories have been proposed to account for this tolerance: clonal deletion and functional inactivation.
6. Effector T cells directly attack specific target cells. Several subsets exist, namely, cytotoxic T cells, delayed-type hypersensitivity T cells, and natural killer cells.
7. When the immune response occurs in an exaggerated form and results in tissue damage to the individual, the term hypersensitivity or allergy is applied. There are four types of hypersensitivity reactions, designated as types I through IV.
8. The immune system can act detrimentally and reject tissue transplants. Four types of transplants exist: xenografts involve transplants of privileged tissue between different species; autografts involve transplants from one part of the body to another of the same individual; isografts are transplants between identical twins; and allografts are transplants between genetically different individuals of the same species.
9. Immunodeficiency diseases are a diverse group of conditions in which an individual's susceptibility to various infections is increased; several severe diseases can arise because of one or more defects in the specific or nonspecific immune response.
10. Tumor biology is similar to and interrelated with the functions of the immune response. Generally, tumor cells are relatively unspecialized and dedifferentiated.
11. Two major classifications of human red blood cells (erythrocytes) are ABO grouping and the Rh system. ABO grouping is based on the types of glycoproteins present on erythrocyte surfaces. Rh classification is based on the presence or absence of the Rh antigen on erythrocytes.

Key Terms

Questions for Thought and Review

1. Immunology has contributed a new depth of understanding to our knowledge of plasma membrane function. How has the study of lymphocytes deepened our understanding of plasma membrane biology?
2. A person with AIDS has a low T-helper/T-suppressor cell ratio. What problems does this create?
3. How do B and T cells cooperate in the immune response? What is the role of macrophages?
4. What is the difference between an immunodeficiency and an allergy?
5. Describe how IgE can have both beneficial and detrimental effects in the same host. Why have its detrimental effects persisted? Explain your reasoning.
6. In desensitization procedures, the allergist injects more of the same allergen to which the person is allergic. How can this be beneficial?
7. How does the existence of tumor-specific antigens provide insight into immunotherapy for cancer? How might hybridomas fit into this treatment?
8. What is the immunologic basis behind blood transfusions?
9. How does hemolytic disease of the newborn develop?
10. What is agammaglobulinemia? How does it occur?

Additional Reading

References provided at the end of chapter 31 may also be consulted for further information.

Atkinson, M. A., and MacClaren, N. K. 1990. What causes diabetes. *Sci. Am.* 263(1):62–71.

Bochner, B. S., and Lichtenstein, L. M. 1991. Anaphylaxis. *N. Engl. J. Med.* 324(25):1785–90.

Brodt, P. 1983. Cancer immunobiology—Three decades in review. *Ann. Rev. Microbiol.* 37:447–76.

Buisseret, P. D. 1982. Allergy. *Sci. Am.* 247:86–95.

Cohen, I. R. 1988. The self, the world, and autoimmunity. *Sci. Am.* 258(4):52–60.

Crabtree, G. R. 1989. Contingent genetic regulatory events in T lymphocyte activation. *Science* 243:355–61.

Grey, H.; Alessandro, S.; and Buss, S. 1988. How T cells see antigen. *Sci. Am.* 261(5):56–67.

Honjo, T. 1983. Immunoglobulin genes. *Ann. Rev. Immunol.* 1:499–511.

Jerne, N. K. 1973. The immune system. *Sci. Am.* 229(2):52–60.

Johnson, H. M.; Russell, J. K.; and Pontzer, C. H. 1992. Superantigens in human disease. *Sci. Am.* 266(4):92–101.

Kishimoto, T., and Hirano, T. 1988. Molecular regulation of B lymphocyte response. *Ann. Rev. Immunol.* 6:485–512.

Knight, J. G. 1982. Autoimmune diseases: Defects in immune specificity rather than a loss of suppressor cells. *Immunol. Today* 3:326–41.

Koffler, D. 1980. Systemic lupus erythematosus. *Sci. Am.* 243(5):52–61.

Lerner, F. A. 1983. The genetics of antibody diversity. *Sci. Am.* 246(2):102–13.

Marrack, P., and Kappler, J. 1986. The T cell and its receptor. *Sci. Am.* 254(3):36–45.

Marx, J. L. 1983. Chemical signals in the immune system. *Science* 221:1362.

Metzger, H., ed. 1990. *Fc receptors and the action of antibodies.* Washington, D.C.: American Society for Microbiology.

Milstein, C. 1980. Monoclonal antibodies. *Sci. Am.* 243(4):66–74.

Nossal, G. L. V. 1983. Cellular mechanisms of immunologic tolerance. *Ann. Rev. Immunol.* 1:33–68.

Notkins, A. L., and Koprowski, H. 1973. How the immune response to a virus can cause cancer. *Sci. Am.* 228(1):22–31.

Old, L. J. 1977. Cancer immunology. *Sci. Am.* 236(1):62–79.

Old, L. J. 1988. Tumor necrosis factor. *Sci. Am.* 258(3):59–75.

Raff, M. C. 1976. Cell surface immunology. *Sci. Am.* 234(6):30–38.

Ramsdell, F., and Fowlkes, B. 1990. Clonal deletion versus clonal anergy. *Science* 248:342–48.

Rennie, J. 1990. The body against itself. *Sci. Am.* 263(6):106–15.

Rose, N. R. 1981. Autoimmune diseases. *Sci. Am.* 244(2):80–103.

Rosenberg, S. A. 1990. Adoptive immunotherapy for cancer. *Sci. Am.* 262(5):62–69.

Samuelsson, B. 1983. Leukotrienes: Mediators of immediate hypersensitivity and inflammation. *Science* 220:568–660.

Smith, R. A. 1990. Interleukin-2. *Sci. Am.* 262(3): 50–57.

Smith, R. H., and Steimberg, A. D. 1983. Autoimmunity—A perspective. *Ann. Rev. Immunol.* 1:175–210.

von Boehmer, H. 1988. The development biology of T lymphocytes. *Ann Rev. Immunol.* 6:309–26.

von Boehmer, H., and Kisielow, P. 1991. How the immune system learns about self. *Sci. Am.* 265(4):74–81.

Weill, J. C., and Reynaud, C. A. 1987. The chicken B cell compartment. *Science* 238:1094–98.

Young, J. D.-E., and Cohn, Z. A. 1988. How killer cells kill. *Sci. Am.* 258(1):38–44.

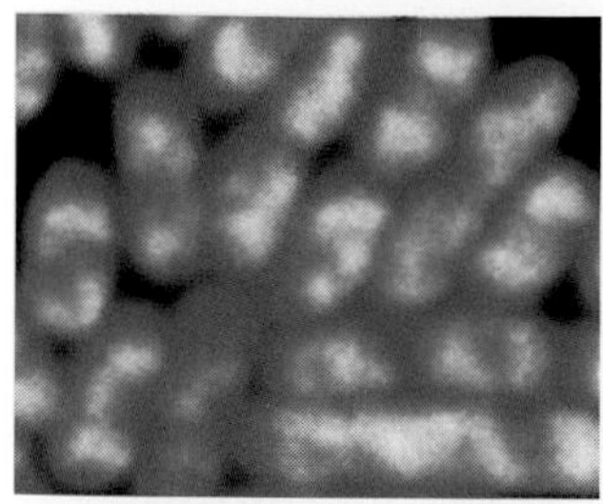

CHAPTER 33

The Immune Response: Antigen-Antibody Reactions

A foreign cell in the body is identified by antibody, but the cell is destroyed by other agents. Among them is "complement," an intricate set of enzymes [proteins].

—Manfred M. Mayer

Outline

Concepts

1. Various types of antigen-antibody reactions occur in higher animals (in vivo) that lead to immune product formation. This union of antigen and antibody initiates the participation of other body elements that determine the ultimate fate of the antigen. For example, the complement system can be activated, leading to cell lysis, phagocytosis, chemotaxis, or stimulation of the inflammatory response. Other defensive reactions include toxin neutralization, antibody-dependent cell-mediated cytotoxicity, opsonization, and immune complex formation.
2. The union of antigen and antibody in vitro produces either a visible reaction or one that can be made visible in a variety of ways. These techniques can be used to identify viruses, microorganisms, and their products; to quantitate and identify antigens and antibodies; to follow the course of a disease; to determine the serotype of a microorganism; and to determine the amount of protection from disease an animal possesses.
3. The older classic tests are named according to what happens to the antigen: agglutination, complement fixation, precipitation, neutralization, and capsular swelling. More recent tests are named according to the technique used: enzyme-linked immunosorbent assay, immunodiffusion, immunoelectrophoresis, immunofluorescence, immunoprecipitation, and radioimmunoassay.

The two previous chapters present a basic discussion of antigens, antibodies, and chemical mediators in both health and disease. This chapter completes the discussion of the immune response by describing the various antigen-antibody reactions that occur in higher animals (in vivo) to protect them against the continuous onslaught of infectious agents, microorganisms and their products, certain macromolecules, and tumor cells. Outside the animal body, tests based on antigen-antibody reactions have become exceptionally useful in the diagnosis of infectious diseases, in the identification of specific viruses and microorganisms, and for monitoring immunologic problems. These in vitro techniques are also discussed.

As presented in chapter 31, the basic Y-shaped antibody molecule is bifunctional. The variable domain (within the Fab regions) binds target antigen while the Fc region interacts with various cells of the immune system, some phagocytic cells, or the first component of the complement system. The principal function of antigen-antibody binding is to identify the target antigen as foreign so that other components of the immune response can react with it and destroy the foreign body.

Antigen-Antibody Binding

An antigen binds to an antibody at the antigen-binding site within the Fab regions of the antibody. More specifically, a pocket is formed by the folding of both the V_H and V_L regions (*see figure 31.6*c). It is at this site that specific amino acids contact the antigen's epitope determinant or haptenic groups and form multiple noncovalent bonds between the antigen and amino acids of the binding site (figure 33.1).

Because binding is due to weak, noncovalent bonds such as hydrogen bonds and electrostatic attractions, the antigen's shape must exactly match that of the antigen-binding site. If the shapes of the epitope (*see figure 31.3*) and binding site are not truly complementary, the antibody will not effectively bind the antigen. Although a lock-and-key mechanism normally may operate, in at least one case the antigen-binding site does change shape when it complexes with the antigen (an induced fit mechanism). Regardless of the precise mechanism, antibody specificity results from the nature of antibody-antigen binding.

(a)

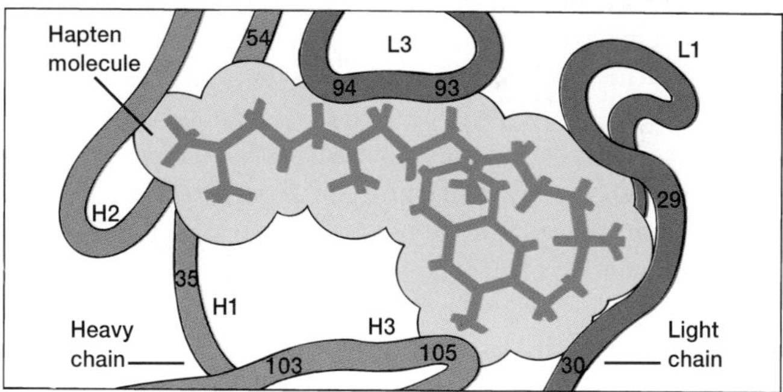

(b)

Figure 33.1 Antigen-antibody Binding. (*a*) An analogy for antigen-antibody binding is represented by the apple (antigen) being held by the fingers (antibody) forming a pocket in which the antigen sits. (*b*) Based on X-ray crystallography, the hapten molecule nestles in a pocket formed by the antibody combining site. In the illustration, the hapten makes contact with only 10–12 amino acids in the hypervariable regions of the light and heavy chains. The numbers represent contact amino acids.

Antigen-Antibody Reactions in the Animal Body (In Vivo)

Higher animals (vertebrates) possess a highly sophisticated, defensive immunologic response mechanism for developing resistance to specific viruses, microorganisms, macromolecules, foreign agents, and cancer cells. This immunologic response is triggered by specific antigen-antibody reactions. Those reactions occurring in the animal's body are discussed first, and then reactions outside the body.

The Complement System

The **complement system** is composed of a group of serum proteins that play a major role in the animal's defensive immune response. For example, complement proteins can lyse antibody-coated eucaryotic cells and bacteria (cytolysis). Complement can mediate inflammation and attract and activate phagocytic cells. Generally, complement proteins amplify the effects of antibodies (e.g., lysis of cells).

The complement cascade is made up of at least 17 complement proteins designated C1 (which has three protein subcomponents) through C9 in addition to Factor B, Factor D, Factor H, Factor I, C4b binding protein, C1 INH complex, S protein, and properdin (table 33.1). The complement system acts in a cascade fashion, the activation of one component resulting in the activation of the next. Collectively, the complement proteins make up much of the globulin fraction of serum (*see figure 31.5*). Within plasma and other body fluids, complement proteins are in an inactive state. They usually are activated after the binding of antibodies to antigens and are specifically directed against the target molecules identified by the antibodies. Because complement activation involves binding of the components to antibody-antigen complexes and to each other with their consequent removal from serum, this event is called **complement fixation.**

There are two pathways of complement activation: the **classical** and **alternative pathways** (figure 33.2). Although they employ similar mechanisms, specific proteins are unique to the first part of each pathway.

Activation of the classical pathway requires initiation by the interaction of antibodies with an antigen that is usually cell bound. The order of effectiveness in activating complement is as follows: IgM > IgG3 > IgG1 >IgG2. However, some microbial products (lipid A of endotoxin and staphylococcal protein A) or plasmin (a proteolytic enzyme that dissolves the fibrin of blood clots) may activate C1 directly without antibody participation. Following binding of antigen to antibody, the C1 complement component, which is composed of three proteins (q, r, and s), attaches to the Fc portion of the antibody molecule through its C1q subcomponent. In the presence of calcium ions, a trimolecular complex (C1qrs · Ag · Ab) that has esterase activity is rapidly formed. The activated C1s subcomponent attacks and cleaves its natural substrates in serum (C2 and C4). This leads to binding of a portion of each molecule (C2b and C4b) to the antigen-antibody-complement complex with the release of small C4a and C2a fragments. (The released C2 fragment traditionally has been called C2a. For consistency, this text will label all larger complement fragments that are bound to the target cell as b fragments.) With the binding of C2b to C4b, an enzyme with trypsinlike proteolytic activity is generated. The natural substrate for this enzyme is C3; thus, it is termed a C3 convertase. Through the activity of $\overline{C4b2b}$ (the bar indicates an active enzyme complex), C3 is cleaved into a bound subcomponent C3b and a C3a soluble component. C3b then absorbs to bound C4b2b, forming the complex C4b2b3b which cleaves C5 into fragments C5a and C5b. C6 and C7 rapidly bind to C5b, forming a C5b67complex that possesses an unstable membrane-binding site; once bound to a membrane, this complex is stable. C8 and C9 then bind, forming the **membrane attack complex** (C5b6789) that creates a pore in the plasma membrane of the target cell (figure 33.3*a,b*). It is believed that the actual pore

TABLE 33.1 Proteins of the Complement Cascade in Serum

Protein	Fragment	Function
Recognition Unit		
C1	q	Binds to the Fc portion of antigen-antibody complexes
	r	Subunit of C1; activates C1s
	s	Cleaves C4 and C2 due to its enzymatic activity
Activation Unit		
C2		Causes viral neutralization
C3	a	Anaphylatoxin, immunoregulatory
	b	Key component of the alternative pathway and major opsonin in serum
	e	Induces leukocytosis
C4	a	Anaphylatoxin
	b	Causes viral neutralization
Membrane Attack Unit		
C5	a	Anaphylatoxin; principal chemotactic factor in serum; induces neutrophil attachment to blood vessel walls
	b	Initates membrane attack
C6, C7, C8, C9		Participate with C5b in formation of the membrane attack complex that lyses targeted cells
Alternative Pathway		
Factor B		Causes macrophage spreading on surfaces; precursor of C3 convertase
Factor $\overline{D}$		Cleaves Factor B to form active C3Bb in alternative pathway
Properdin		Stabilizes alternative pathway convertase
Regulatory Proteins		
Factor H		Promotes C3b breakdown and regulates alternative pathway
Factor I		Degrades C3b and regulates alternative pathway
C4b binding protein		Inhibits assembly and accelerates decay of C4bC2a
C1 INH complex		Binds to and dissociates C1r and C1s from C1
S protein		Binds fluid-phase C5b67; prevents membrane attachment

Classical pathway

Alternative pathway

Initiation

Antigen + Antibody
(Lipid A, protein A, plasmin) (IgM IgG3 IgG1 IgG2)

AgAb + C1q + C1r + C1s + C4 → C4b & C4a; C2 → C2b, C2a

Molecules with repeating chemical structures
• Endotoxins
• Polysaccharides
• Lipopolysaccharides
• Teichoic acid
Or IgA, IgE

Amplification

Common pathway
C3 convertases
C4b2b
C3bBb

$\bar{D}$ + C3b + B + P; B → Bb, Ba

C3 → C3a; C3b

C5 Convertases
C4b2b3b
C3Bb3b

C5 → C5a; C5b

Inflammation
Kinin activity
Histamine release
Chemotaxis
Opsonization
Neutrophil activation
Phagocytosis
Clearing of immune complexes

Membrane attack complex

$\overline{\text{C5b6789}}$

Insertion, cell lysis, death (membrane attack complex)

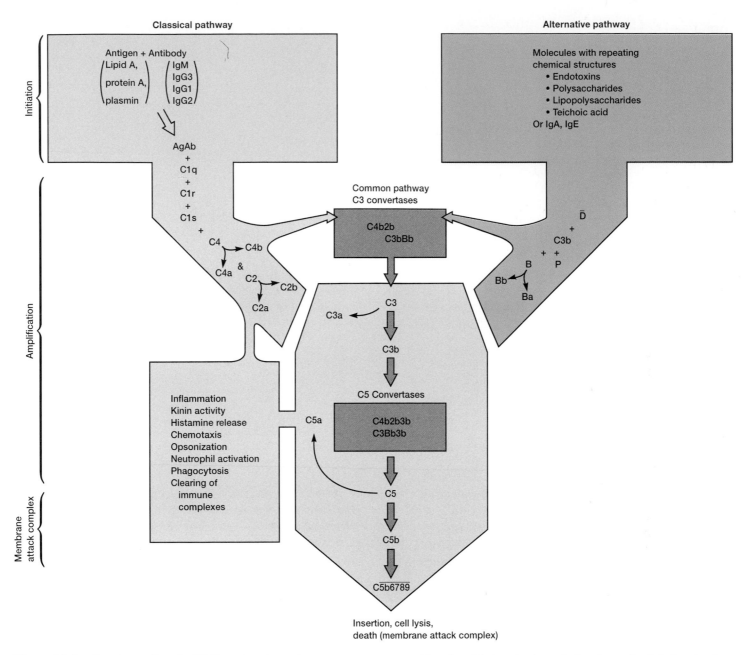

Figure 33.2 Complement Cascade. Within each pathway the components are arranged in order of their activation and aligned opposite their functional and structural analogs in the opposite pathway.

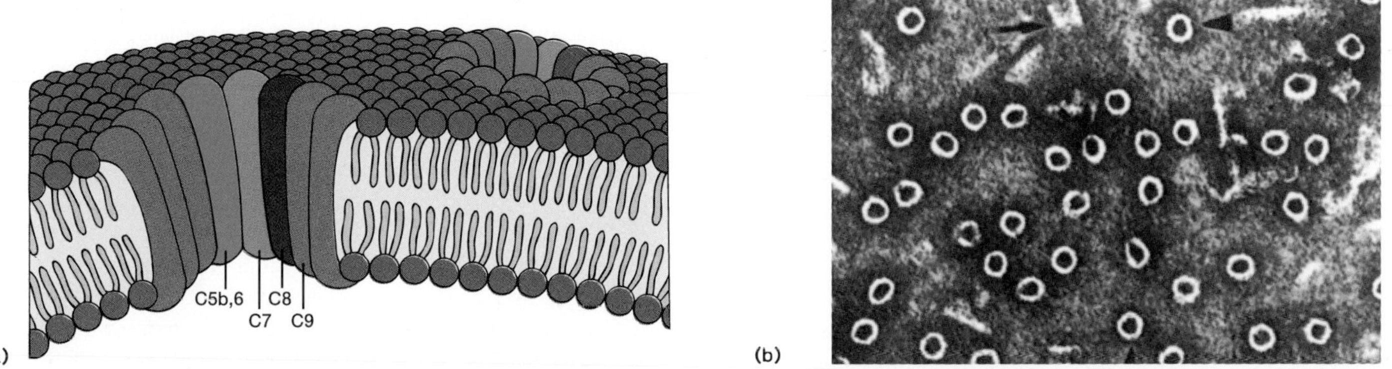

Figure 33.3 The Membrane Attack Complex. (*a*) The membrane attack complex is a tubular structure that forms a transmembrane pore in the target cell's plasma membrane. (*b*) Electron micrograph of this complex seen in side view (arrows) and top view (arrowheads). *(a) From Donald Voet and Judith G. Voet,* Biochemistry. *Copyright © 1990 John Wiley & Sons, Inc., New York, NY.*

Box 33.1

Complement-Resistant Cells

Resistance to complement attack is an attribute of some procaryotic and eucaryotic cells. Complement resistance may be due to factors affecting either the formation of lytic channels or how well the channels lyse target cells. Recent experimental evidence supports both mechanisms. For example, cells may activate complement inefficiently in both the presence and the absence of antibody. Changes in the concentration of membrane antigens may alter the capacity of bound IgG to activate C1. Even after C1 activation, membrane factors may inhibit completion of the activation sequence at the step of C3 and C5 convertase formation or at the steps of C8 and C9 binding. Finally, factors affecting membrane lipid composition and hormonal regulation of lipid synthesis can alter the capacity of the membrane attack complex to insert itself effectively into the plasma membrane and create a lytic channel. Once the latter step has occurred, eucaryotic cells have devised several mechanisms to avoid the lethal effect of the complement channel. The membrane attack complex may be endocytosed by a calcium-dependent process. The complement channel can be actively extruded from neutrophils by exocytosis. Finally, platelets bearing the membrane attack complex demonstrate increased activity of the sodium-potassium pump, a process that repolarizes the membrane during sublethal complement attack.

Recently, investigators have shown that specific membrane-bound proteins also help protect normal host cells from the detrimental effects of complement activation. These proteins provide the basis for discrimination between self and nonself. In addition, nucleated eucaryotic cells are highly resistant to complement-mediated cytolysis. Resistance is correlated with a high replacement rate of cell membrane lipids and the ability to shed the membrane attack complex from the cell surface. Thus, the complement cascade participates in the development of the inflammatory response, elimination of pathogens, and removal of immune complexes without destroying host cells at the same time.

is a doughnut-shaped polymer of C9. (The perforin pores generated by cytotoxic T cells (*see figure 32.5*e) have similar structures since perforin and C9 are homologous but lack an analog of the C5b678 complex.) If the cell is eucaryotic, Na^+ and H_2O enter through the pore leading to osmotic lysis of the cell. If the cell is a gram-negative bacterium, lysozyme from the blood enters through the pore and digests the peptidoglycan cell wall causing the bacterium to lyse osmotically. In contrast, gram-positive bacteria are resistant to the cytolytic action of the membrane attack complex because they lack an outer membrane.

The alternative pathway does not require the binding of antibodies to antigens for its activation. Instead, it plays an important role in the innate, nonspecific immune defense against intravascular invasion by bacteria and some fungi that occurs before the development of specific antibodies. The alternative in pathway begins with cleavage of C3 into fragments C3a and C3b by a blood enzyme. These fragments are produced at a slow rate and do not carry out the next step, the cleavage of C5, because free C3b is rapidly cleaved into inactive fragments. However, when C3b binds to lipopolysaccharide (LPS) of bacterial cell outer membranes, to aggregates of IgA or IgE, or to some endotoxins, it becomes stable (figure 33.2). A protein in blood termed Factor B adsorbs to bound C3b and is cleaved into two fragments by Factor$\overline{D}$, leading to the formation of active enzyme $\overline{C3bBb}$ (this complex is sometimes called the C3 convertase because it cleaves more C3 to C3a and C3b). $\overline{C3bBb}$ is further stabilized by a second blood protein, properdin, and is changed into the C5 convertase ($\overline{C3bBb3b}$). The convertase then cleaves C5 to C5a and C5b. The steps that follow in the alternative pathway are identical with the last steps of the classical pathway.

This overview of complement activation provides a basis for consideration of the function of complement as an integrated system during an animal's defensive effort. Gram-negative bacteria arriving at a local tissue site will interact with components of the alternative pathway, resulting in the generation of biologically active fragments, opsonization of the bacteria, and initiation of the lytic sequence (however, see Box 33.1). If the bacteria persist or if they invade the animal a second time, antibody responses also will activate the classical pathway. The classical pathway is much more rapid and efficient in mediating opsonization and complement fragment generation.

The generation of complement fragments C3a and C5a leads to several important inflammatory effects. Mast cells release their contents, and the blood supply to the area increases markedly (hyperemia) as blood vessels dilate due to released histamine (*see figure 30.13*). These fragments also cause the release of neutrophils from the bone marrow into the circulation. Neutrophils then make their way to the site of hyperemia where, in the presence of C5a, they attach to the endothelium and leave the blood vessels (*see figure 30.12*). C5a induces a directed, chemotactic migration of neutrophils to the site of complement activation. Macrophages in the area can synthesize even more complement components to interact with the bacteria. All these defensive events promote the ingestion and ultimate destruction of the bacteria by the neutrophils and macrophages (*see figure 30.14*).

1. How does antigen-antibody binding occur? What is the basis for antibody specificity?
2. List the major functions of complement.
3. What is complement fixation?
4. How is the classical complement pathway activated? The alternative pathway?
5. What is the membrane attack complex, and how does its formation lead to cell lysis?
6. What role do complement fragments C3a and C5a play in an animal's defense against gram-negative bacteria?

Toxin Neutralization

Immunity to a disease like diphtheria (*see p. 742*) depends on the production of specific antibodies that inactivate the toxins produced by the bacteria. This process is termed **toxin neutralization** (figure 33.4*a,b*). Once neutralized, the toxin-antibody complex is either unable to attach to receptor sites on host target cells or unable to enter the cell. For example, diphtheria toxin inhibits protein synthesis after binding to the cell surface by the B fragment and subsequent passage of the active A fragment into the cytosol of the target cell (*see figure 30.5*a,b). Thus, the antibody blocks the toxic effect by inhibiting the entry of the A fragment or the binding of the B fragment. Antiserum containing neutralizing antibody against a toxin is called **antitoxin.**

Exotoxins (pp. 584–86).

Viral Neutralization

IgG, IgM, and IgA antibodies can bind to some viruses during their extracellular phase and inactivate them. This antibody-mediated viral inactivation is called **viral neutralization.** Fixation of classical pathway complement component C4b to the virus aids the neutralization process. Viral neutralization prevents a viral infection due to the inability of the virus to bind to its target cell (figure 33.4*c*).

Adherence Inhibiting Antibodies

The capacity of bacteria to colonize the mucosal surfaces of mammalian hosts is dependent in part on their ability to adhere to mucosal epithelial cells. Recent studies have demonstrated that secretory IgA (sIgA; *see p. 615*) antibodies inhibit certain bacterial adherence promoting factors. Thus, sIgA has a unique role in protecting the host against infection from certain pathogenic bacteria and perhaps from other microorganisms on mucosal surfaces.

Antibody-Dependent Cell-Mediated Cytotoxicity

When cells are infected with viruses, changes often occur in the plasma membrane proteins that lead to the development of an antibody response against the cells. Cytolysis of the infected cell may be achieved by the membrane attack complex—that is, by the combined action of antibody and complement—or by the binding of natural killer cells to the altered cell after attachment to the Fc region of the antibody (figure 33.5). Antibodies provide the bridging mechanism for an intimate association between the effector and target cell. After attachment, the natural killer cell destroys the target cell by releasing cytotoxic mediators or directly causing cytolysis. The process is called **antibody-dependent cell-mediated cytotoxicity.**

IgE and Parasitic Infections

Immune reactions against protozoan and helminthic parasites are only partially understood. Parasites that have a tissue invasive phase in their life cycle are often associated with both eosinophilia (an excessive number of eosinophils in the blood) and elevated IgE levels. Recent evidence shows that, in the presence of elevated IgE, eosinophils can bind to the parasites and discharge their lysosomal granules. Degranulation releases lytic and inflammatory mediators that destroy the parasites (*see figure 32.6*).

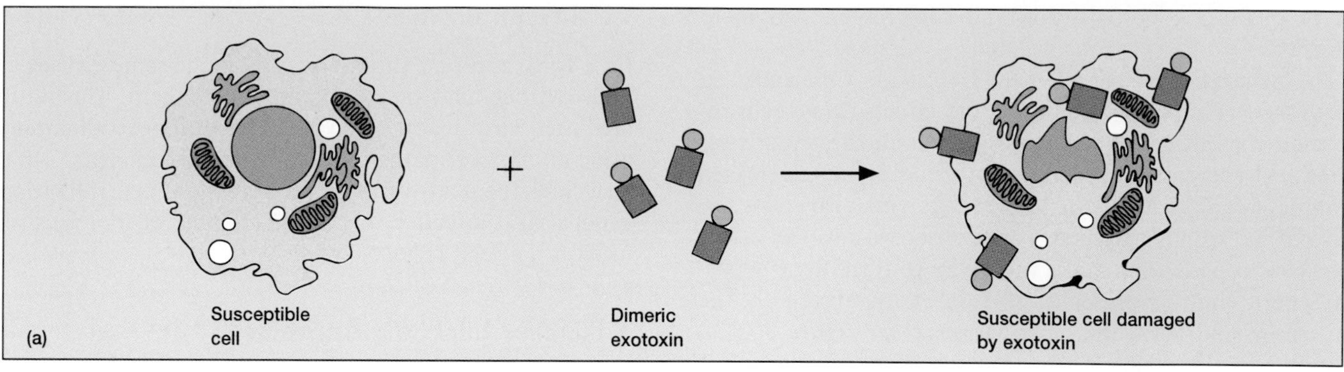

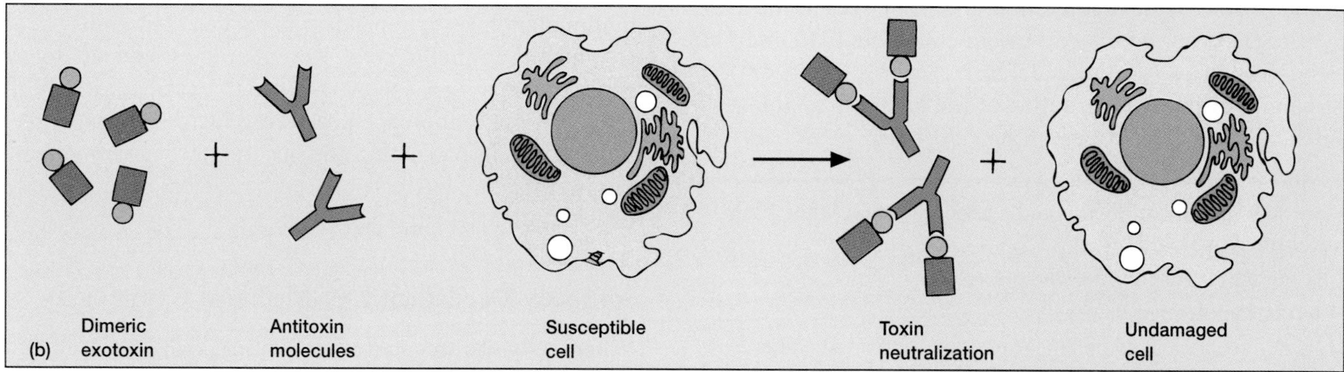

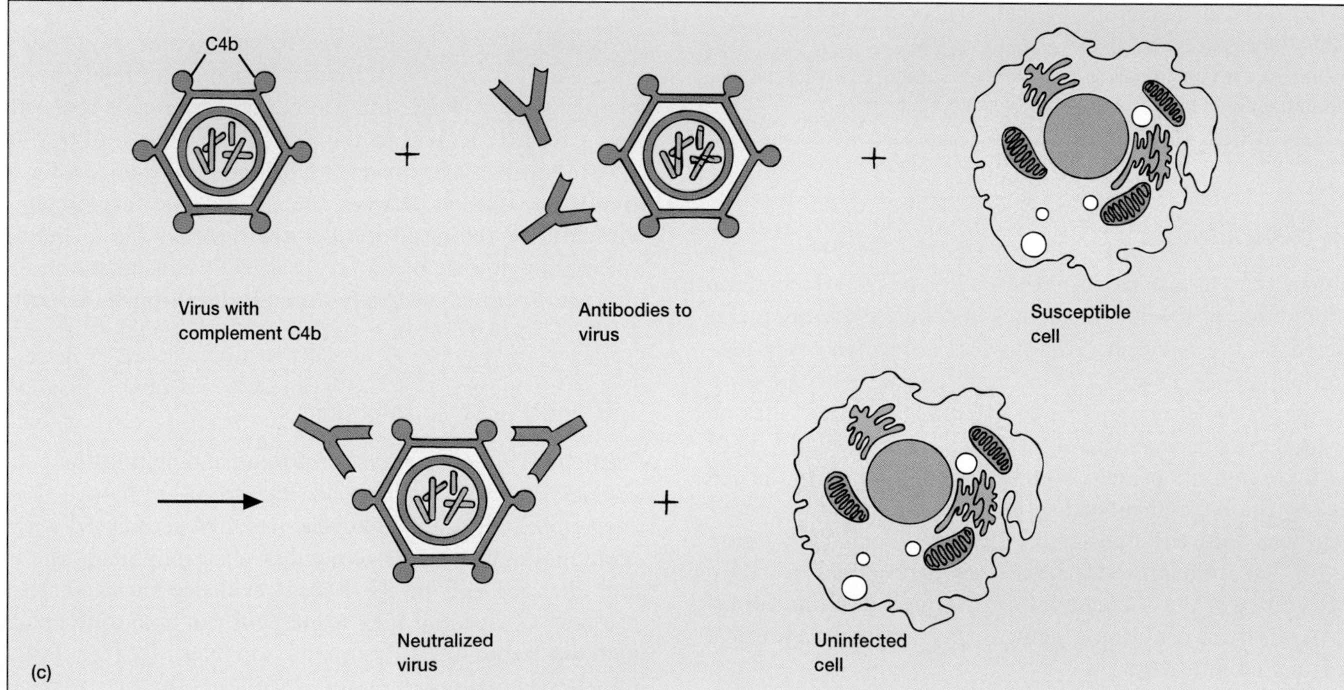

Figure 33.4 Neutralization Reactions. (*a*) Effects of a dimeric exotoxin on a susceptible cell. (*b*) Neutralization of the toxin by antitoxin. (*c*) In viral neutralization the specific antibodies neutralize the virus and prevent it from attaching to the susceptible cell.

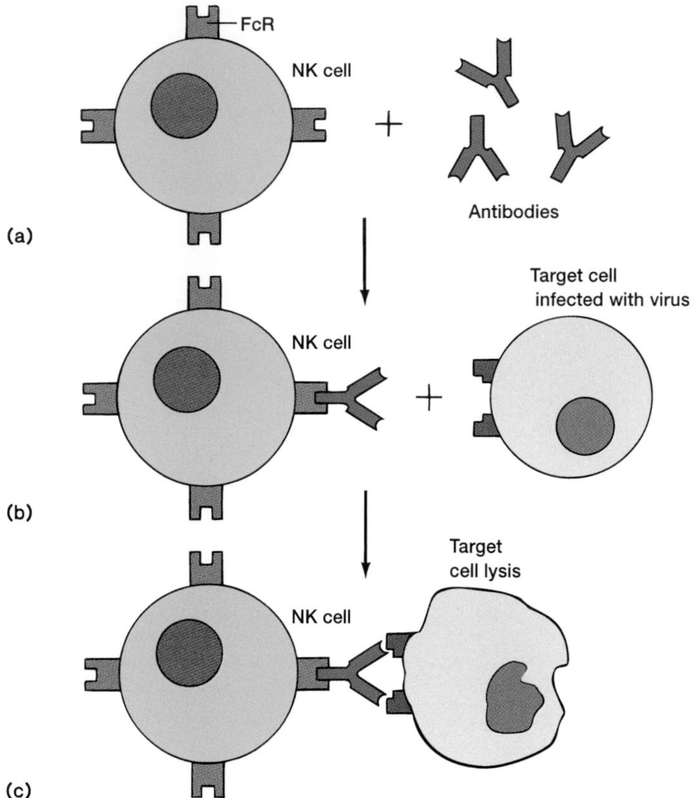

Figure 33.5 Antibody-dependent Cell-mediated Cytotoxicity. (*a*) In this mechanism, the FcR antibody binds to the Fc receptor (FcR) on the Fc region of natural killer (NK) cells. (*b*) These antibodies then can bind to target cell antigens against which they are specific. (*c*) Subsequently, lysis of the target cells occurs.

Opsonization

As noted in chapter 30, phagocytes have an intrinsic ability to bind directly to microorganisms by nonspecific cell surface receptors, form phagosomes, and digest the microorganisms (figure 33.6*a*). This phagocytic process can be greatly enhanced by opsonization. **Opsonization** (Greek *opson,* prepare victims for) is the process in which microorganisms or other particles are coated by antibody and/or complement, thereby being prepared for "recognition" and ingestion by phagocytic cells. Opsonizing antibodies belong to the IgG1, IgG3, or IgM isotypes and have a site in the C_H3 domain (*see figure 31.6*c) that binds to an Fc receptor on the surface of macrophages. This binding forms a bridge between the phagocyte and the antigen (figure 33.6*b*).

Phagocytosis (pp. 599–600).

IgG or IgM antibodies also can activate the complement system, most often through the classical pathway, but sometimes by the alternative pathway. The net result is to assemble the C3b convertase on the surface of the microorganism together with opsonic fragment C3b. Both neutrophils and macrophages possess surface receptors for C3b. A bridge that aids phagocytosis eventually forms between the microorganism, complement, and phagocytic cell (figure 33.6*c*). If both antibody and C3b opsonize, binding is greatly enhanced (figure 33.6*d*).

Opsonizing antibodies can be directed against surface components of bacteria (e.g., capsular polysaccharides, M protein of group A streptococci, and peptidoglycans of staphylococci). Components that can trigger deposition of C3b on the surface of microorganisms by the alternative complement pathway include lipopolysaccharides of gram-negative bacteria, lipoteichoic acids of gram-positive cocci, and fungal polysaccharides. Finally, some bacterial components, staphylococcal protein A or lipid A of endotoxin, can activate the classical complement pathway without the participation of antibody.

Inflammation

The inflammatory response can be triggered by nonimmune and immune factors. In the latter case, the binding of an antibody to an antigen triggers inflammation by two different routes. The first is mediated by IgE's attachment to the surface of mast cells and basophils, which results in the release of histamine (*see figure 32.6*). Histamine causes most of the characteristic events of the inflammatory response.

Inflammation (pp. 595–99).

The second route involves the complement pathway in which the C3a and C5a fragments (*see figure 33.2*) bind to mast cells and platelets, triggering the release of histamine. C5a also contributes to inflammation by being a very potent chemotactic factor. It attracts macrophages, neutrophils, and basophils to the site of complement fixation.

1. How does toxin neutralization occur? Viral neutralization?
2. How does antibody-dependent cell-mediated cytotoxicity occur?
3. Describe the role of IgE in resistance to parasitic infections.
4. What is opsonization and how can it take place?
5. What are the two routes by which antibodies stimulate inflammation?

Phagocytic cell	Degree of binding	Opsonin
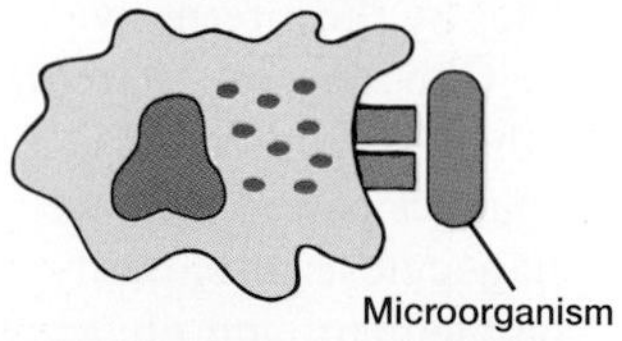 Microorganism (a) Attachment by nonspecific receptors	±	-
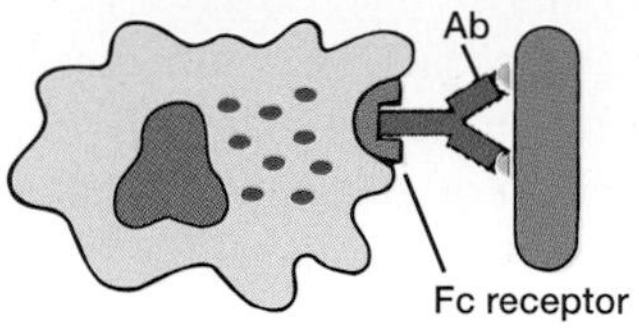 Ab Fc receptor (b)	+	Antibody
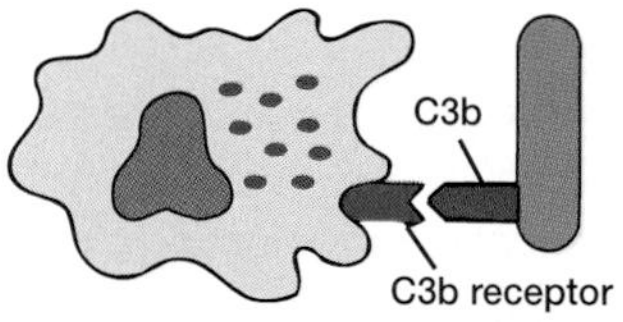 C3b C3b receptor (c)	+ +	Complement C3b
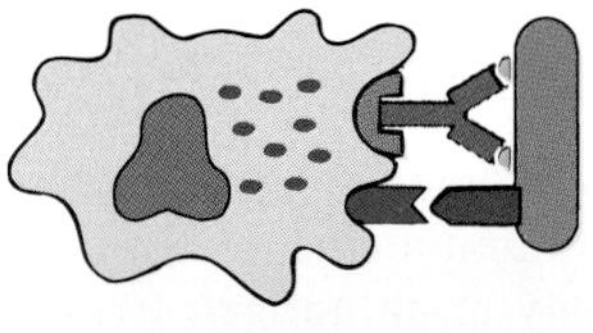 (d)	+ + + +	Antibody and complement C3b

Figure 33.6 Opsonization. (*a*) A phagocytic cell has some intrinsic ability to bind directly to a microorganism through nonspecific receptors. (*b*) This binding ability is enhanced if the microorganism elicits the formation of antibodies (Ab) that act as a bridge to attach the microorganism to the Fc receptor on the phagocytic cell. (*c*) If the microorganism has activated complement (C3b), the degree of binding is further enhanced by the C3b receptor. (*d*) If both antibody and C3b opsonize, binding is greatly enhanced.

Immune Complex Formation

Because antibodies have at least two antigen-binding sites and most antigens have at least two antigenic determinants, cross-linking can occur, producing large aggregates termed **immune complexes** (figure 33.7; *see also figure 32.9*). If the immune complex becomes large enough to settle out of solution, a **precipitation** (Latin *praecipitare,* to cast down) or **precipitin reaction** occurs, with the **precipitin** antibody being responsible for the reaction. When the immune complex involves the cross-linking of cells or particles, an **agglutination reaction** occurs and the responsible antibody is an **agglutinin.** Agglutination specifically involving red blood cells is a **hemagglutination** reaction and is caused by a **hemagglutinin.** These immune complexes are more readily phagocytosed in vivo than are free antigens.

The extent of immune complex formation, whether within an animal or in vitro, depends upon the relative concentrations of antibody and antigen. If there is a large excess of antibody, separate antibody molecules usually bind to each antigenic determinant and a less insoluble network or lattice forms (figure 33.17*a*). When antigen is present in excess, two different antigen molecules bind to each antibody and network development also is inhibited. In the equivalence zone, the ratio of antibody and antigen concentrations is optimal for the formation of a large network of interconnected antibody and antigen molecules (figures 33.7 and 33.17). All antibody and antigen molecules precipitate or agglutinate as an insoluble complex. Precipitin reactions can occur in both solutions and agar gel media. In either case, antibody-antigen equivalence is required for optimal results (e.g., in the immunodiffusion and immunoprecipitation techniques to be discussed later in the chapter).

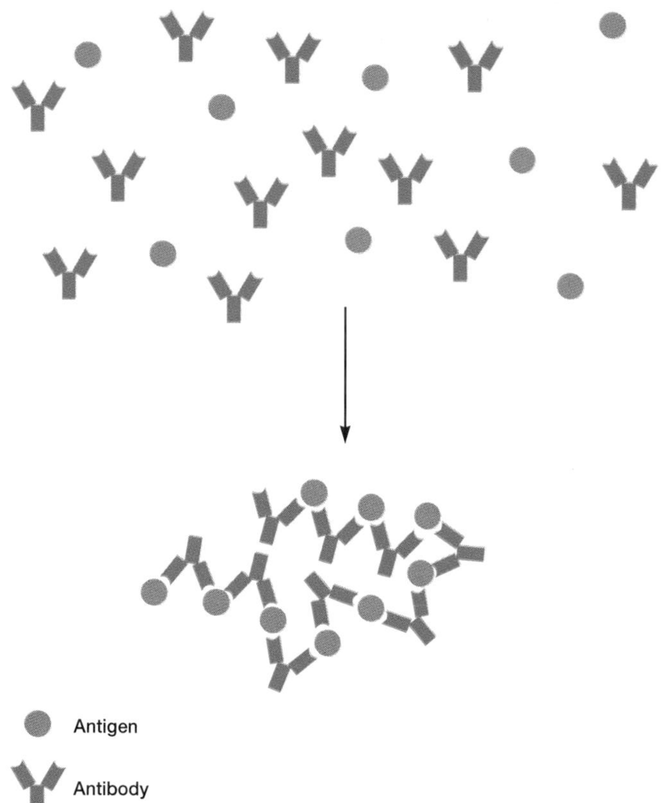

Figure 33.7 Immune Complex Formation. Antibodies cross-link antigens forming an aggregate of antibody and antigen termed an immune complex.

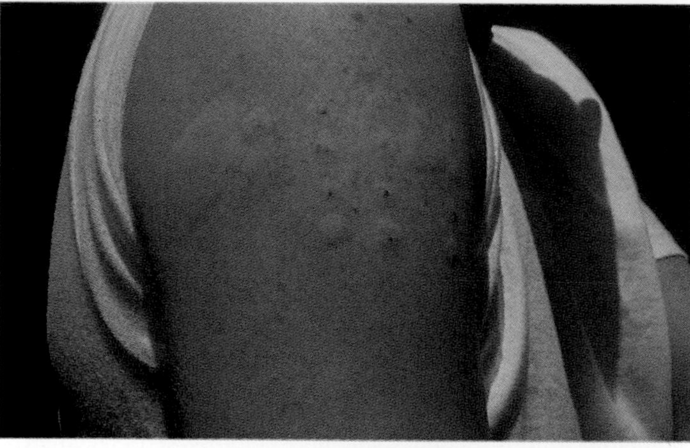

(a)

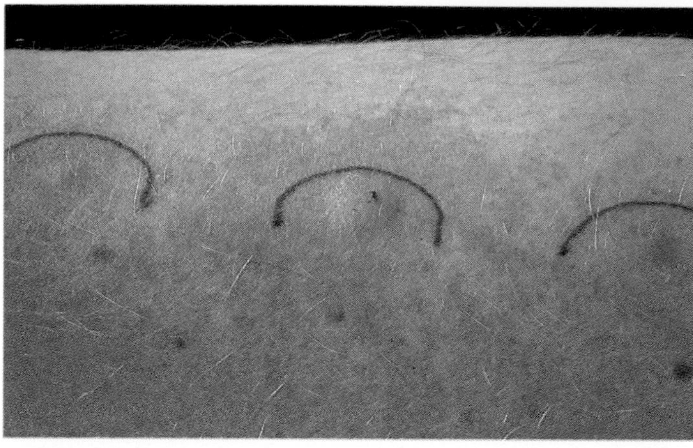

(b)

Figure 33.8 In Vivo Skin Testing. (*a*) Skin prick tests with grass pollen in a person with summer hay fever. Notice the various reactions with increasing dosages (from top to bottom). (*b*) Skin patch test. The surface of the skin (left) is abraded and the suspect allergic extract placed on the skin. After 48 hours (right) it is eczematous and positive for the suspect antigen.

In Vivo Testing

Immunologic reactions in vivo can be of diagnostic value and are the basis for some skin tests. Skin testing is the oldest and most valuable tool for the clinical allergist/immunologist. Two basic tests are used: immediate and delayed.

In immediate skin testing, the antigens to be tested are introduced into the skin of the patient's forearm or back by pricking or scratching the skin with test antigens. Positive tests appear within 20 minutes as raised erythematous (red) lesions (figure 33.8*a*). The degree of sensitivity is determined from measurement of lesion size. Immediate skin testing is done primarily to determine respiratory allergies.

Delayed or cell-mediated skin tests are used to test for the presence of T-cell reactivity to a specific antigen. They are done by either patch testing or intradermal injection of antigen. Patch testing is used to diagnose contact dermatitis or food allergies. The test allergen is applied to an abraded area of the skin and covered with a patch. The patch is removed 48 hours later, and the skin site is evaluated (figure 33.8*b*). If positive, a typical contact eczema or inflammation of the skin characterized by redness, itching, and oozing vesicular lesions develops at the site. Some commercially available antigens for evaluating delayed reactions include those for *Candida albicans, Trichophyton,* mumps, diphtheria-tetanus fluid toxoid, streptokinase-streptodornase, PPD (tuberculin), and histoplasmin.

Antigen-Antibody Reactions In Vitro

The first part of this chapter deals with those antigen-antibody reactions that occur within an animal (in vivo). Many of these same reactions can take place outside the animal (in vitro) under controlled laboratory conditions and are extensively used in diagnostic testing. The branch of immunology concerned with reactions in vitro is **serology** (serum and -ology).

During the last two decades, there has been a marked increase in the number, sensitivity, and specificity of serological tests. The increase results from a better understanding of the cell surface of various lymphocytes, the production of monoclonal antibodies (*see figure 31.19*), the development of radioactive and enzyme-linked assays, and the use of fluorescence technology. In this section, some more common serological tests employed in the diagnosis of microbial and immunologic diseases are presented.

Agglutination

As noted in figure 33.7, when an immune complex is formed by cross-linking cells or particles with specific antibodies, it is called an agglutination reaction. Agglutination reactions usually form visible aggregates or clumps (**agglutinates**) that can be seen with the unaided eye. Direct agglutination reactions are very useful in the diagnosis of certain diseases. For example, the **Widal test** is a reaction involving the agglutination of typhoid bacilli when they are mixed with serum containing typhoid antibodies from an individual who has typhoid fever.

Recently, techniques have been developed that employ microscopic synthetic latex spheres coated with antigens. These coated microspheres are extremely useful in diagnostic agglutination reactions. For example, the modern pregnancy test detects the elevated levels of human chorionic gonadotropin (HCG) hormone that occurs in female urine and blood early in pregnancy (figure 33.9*a,b*). Latex agglutination tests are also used to detect antibodies that develop during certain mycotic, helminthic, and bacterial infections, and in drug testing (Box 33.2).

Hemagglutination usually results from antibodies cross-linking red blood cells through attachment to surface antigens and is routinely used in blood typing (*see table 32.5*). In addition, certain viruses can accomplish **viral hemagglutination.** For example, if a person has a certain viral disease, such as measles, antibodies will be present in the serum to react with the measles viruses and neutralize them. In a positive test, hemagglutination occurs when measles viruses and red blood cells are mixed, but is not seen when the person's serum is added to the mixture; this shows that the serum antibodies have neutralized the measles viruses (figure 33.10*a,b*). This hemagglutination inhibition test is widely used to diagnose influenza, measles, mumps, mononucleosis, and other viral infections (*see chapter 36*).

Agglutination tests are also used to measure antibody titer (*see figure 31.18*). In the tube or well agglutination test, a specific amount of antigen is added to a series of tubes (figure 33.11*a*) or shallow wells in a microtiter plate (figure 33.11*b*). Serial dilutions of serum (1/20, 1/40, 1/80, 1/160, etc.) containing the antibody are then added to each tube or well. The greatest dilution of serum showing an agglutination reaction is determined, and the reciprocal of this dilution is the serum antibody titer.

1. What is an immune complex? A precipitin? An agglutinin?
2. Of what value is in vivo skin testing?
3. What is serology?
4. When would you use the Widal test? Describe how latex agglutination tests work.
5. Why does hemogglutination occur and how can it be used in the clinical laboratory?

Complement Fixation

When complement binds to an antigen-antibody complex, it becomes "fixed" and "used up." Complement fixation tests are very sensitive and can be used to detect extremely small amounts of an antibody for a suspect microorganism in an individual's serum. A known antigen is mixed with test serum lacking complement (figure 33.12*a*). When immune complexes have had time to form, complement is added (figure 33.12*b*) to the mixture. If immune complexes are present, they will fix and consume complement. Afterward, sensitized indicator cells, usually sheep red blood cells previously coated with complement-fixing antibodies, are added to the mixture. Lysis of the indicator cells (figure 33.12*c*) results if immune complexes do not form in part *a* of the test because the antibodies are not present in the test serum. In the absence of antibodies, complement remains and lyses the indicator cells. On the other hand, if the specific antibodies are present in the test serum and complement is consumed by the immune complexes, insufficient amounts of complement will be available to lyse the indicator cells. Absence of lysis shows that specific antibodies are present in the test serum.

Complement fixation was once used in the diagnosis of syphilis (the Wassermann test) and is currently used in the diagnosis of certain viral, fungal, rickettsial, chlamydial, and protozoan diseases.

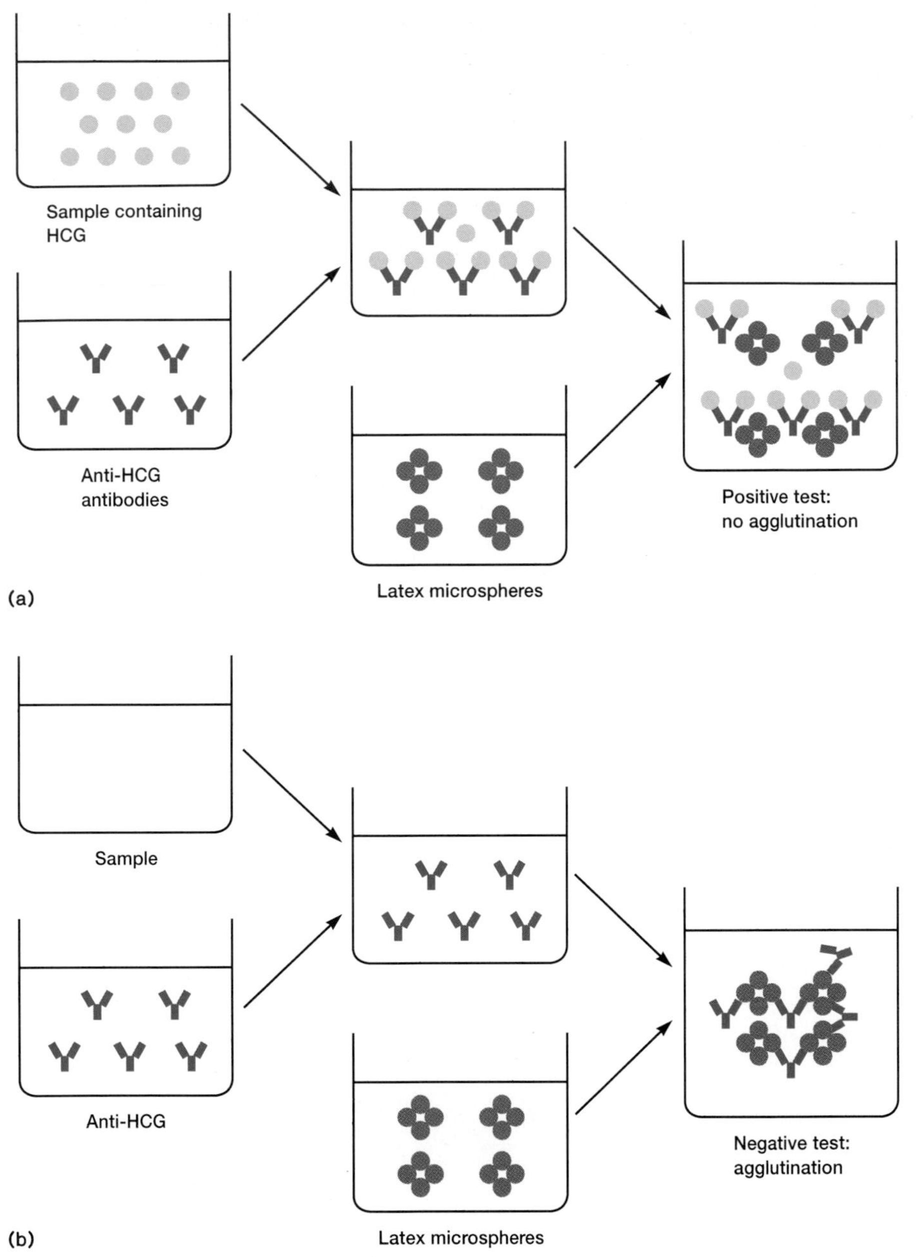

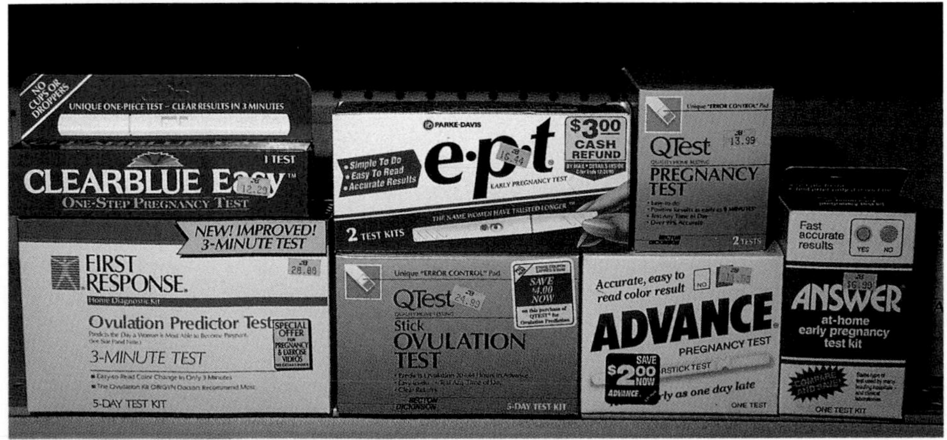

Figure 33.9 Latex Agglutination Test for Pregnancy. (*a*) In a positive test, urine from the female containing HCG is mixed with a solution of antibody specific for HCG. In the second step, latex microspheres coated with HCG are added. If HCG is present, it binds to HCG-specific antibodies, thereby preventing them from agglutinating the microspheres. (*b*) In a negative test, the microspheres coated with antigen are agglutinated by HCG-specific antibody. (*c*) Many rapid pregnancy test kits are now available in drugstores.

Box 33.2

The Rapid Detection of Drugs in Urine

Historically, drug-screening assays were done by classical chemical methods such as thin-layer chromatography (TLC) or liquid chromatography (LC) which, though accurate, are laborious procedures. Assays currently employed in limited drug testing are based on antigen-antibody reactions. Some examples are the radioimmunoassay, enzyme immunoassay, and fluorescence immunoassay methods, all of which depend on sophisticated instrumentation. With the requirement for massive drug testing in sports, federal civilian employment, the military, and other fields, rapid procedures are needed.

One such rapid procedure uses a latex agglutination immunoassay for the detection of cocaine, morphine, barbiturates, THC (marijuana), methadone, pencyclidine, and amphetamines. This method provides the accuracy of the immunoassay approach without the need for expensive equipment, and gives accurate on-site "yes" and "no" results within 3 minutes.

The latex agglutination-inhibition test relies on competition for the antibody between a latex-drug conjugate and any drug that may be present in the urine. A urine sample is placed in the mixing well of a slide containing antibody reagent, buffer, and latex reagent. If the drug is absent, the latex-drug conjugate binds to the antibody and forms large particles that agglutinate. Therefore, agglutination is evidence for the absence of drugs in the urine specimen (Box Figure 33.2*a,c*). If a drug is present in the urine sample, it competes with the latex conjugate for the small amount of available antibody. A sufficient quantity of the drug will prevent the formation of particles and agglutination (Box Figure 33.2*b*), and a positive urine sample does not change the smooth milky appearance of the test mixture (Box Figure 33.2*c*).

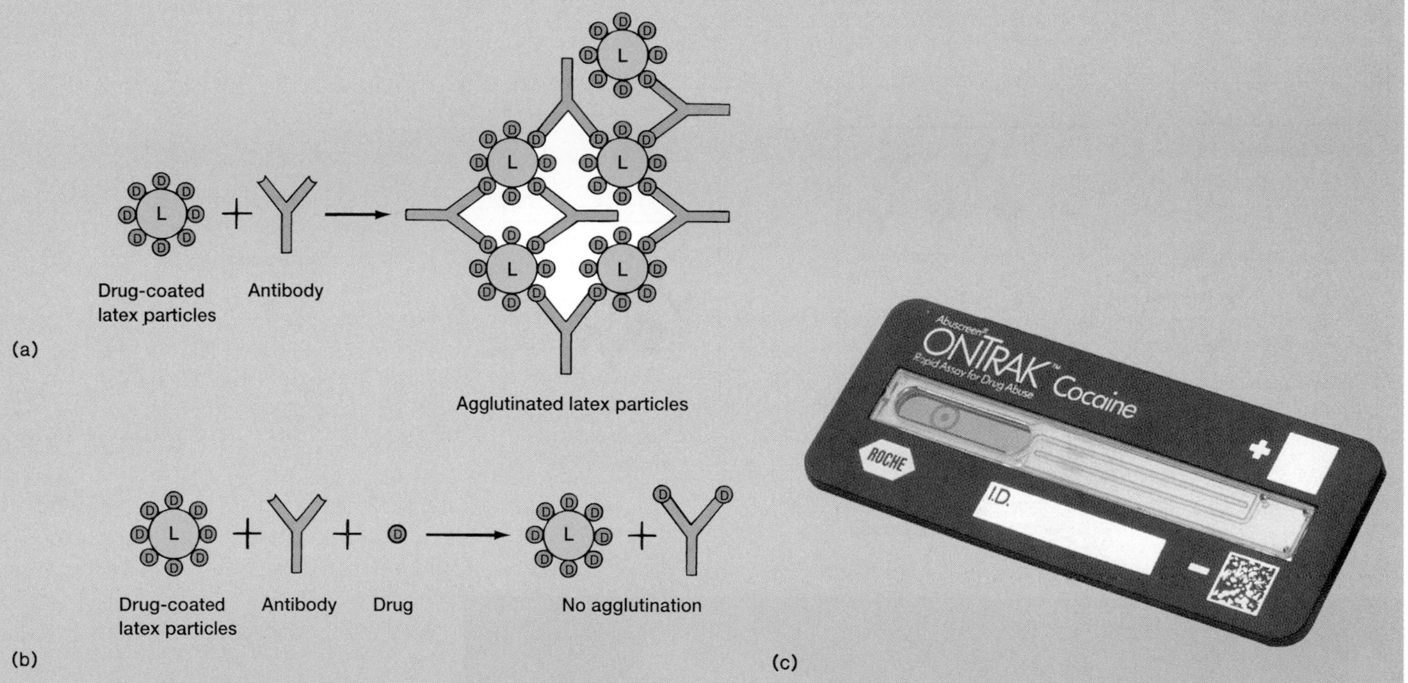

Box Figure 33.2 **Rapid Urine Testing for Drugs.** (*a*) An illustration of a reaction with a negative urine sample. (*b*) A positive urine sample. (*c*) One rapid test for cocaine is called Abuscreen and is manufactured by Roche Diagnostic Systems.

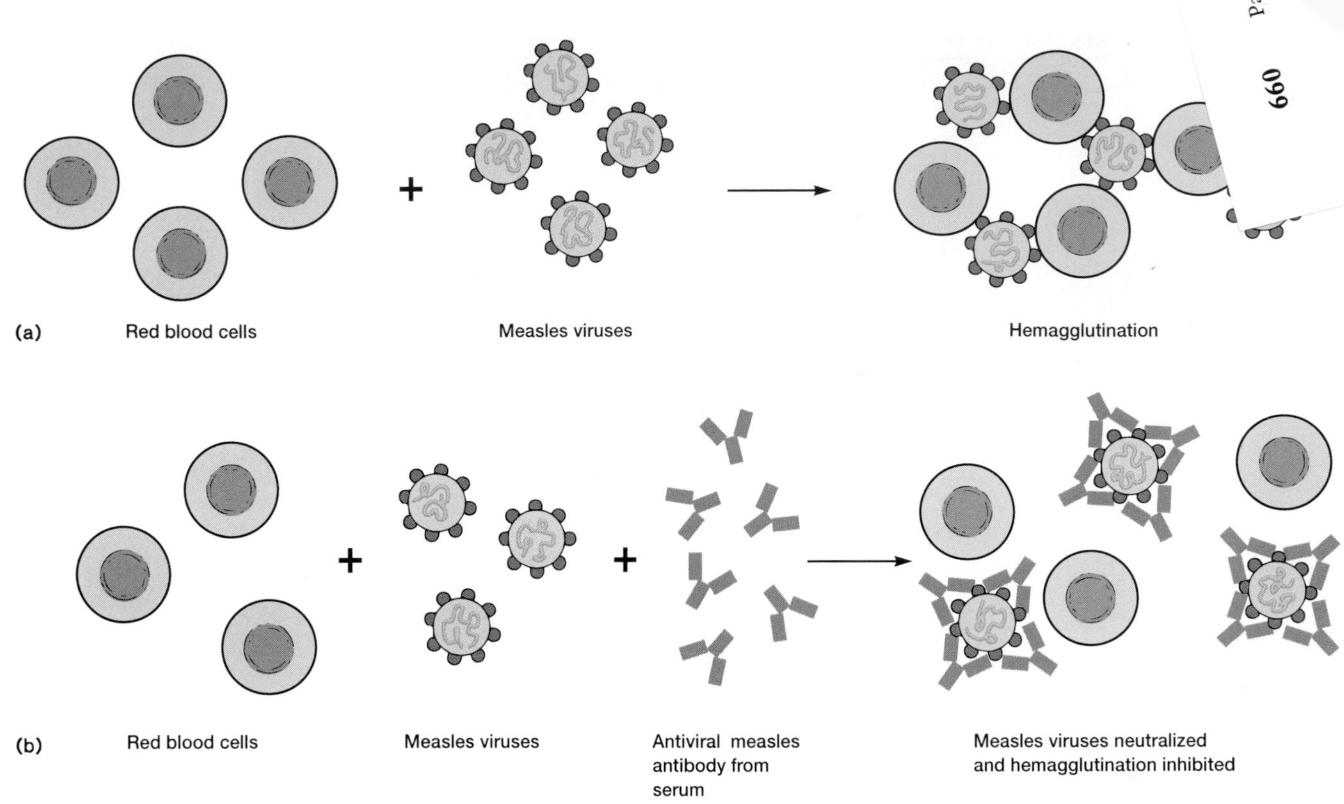

Figure 33.10 **Viral Hemagglutination.** (*a*) Certain viruses can bind to red blood cells causing hemagglutination. (*b*) If serum containing specific antibodies to the virus is mixed with the red blood cells, the antibodies will neutralize the virus and inhibit hemagglutination.

Enzyme-Linked Immunosorbent Assay

The **enzyme-linked immunosorbent assay (ELISA)** or enzyme immunoassay (EIA) has become one of the most widely used serological tests for antibody or antigen detection. This test involves the linking of various "label" enzymes to either antigens or antibodies. Two basic methods are used: the double antibody sandwich assay and the indirect immunosorbent assay.

The double antibody sandwich assay is used for the detection of antigens (figure 33.13*a*). In this assay, specific antibody is placed in wells of a microtiter plate (or it may be attached to a membrane). The antibody is absorbed onto the walls, sensitizing the plate. A test antigen is then added to each well. If the antigen reacts with the antibody, the antigen is retained when the well is washed to remove unbound antigen. An antibody-enzyme conjugate specific for the antigen is then added to each well. The final complex is formed of an outer antibody-enzyme, middle antigen, and inner antibody; that is, it is a layered (Ab-Ag-Ab) sandwich. A substrate that the enzyme will convert to a colored product is then added, and any resulting product is quantitatively measured by optical density scanning of the plate (figure 33.13*c*). If the antigen has reacted with the absorbed antibodies in the first step, the ELISA test is positive. If the antigen is not recognized by the absorbed antibody, the ELISA test is negative because the unattached antigen has been washed away, and no antibody-enzyme is bound. This assay is currently being used for the detection of *Helicobacter pylori* infections and the causative agents of syphilis, brucellosis, salmonellosis, and cholera. Many other antigens also can be detected by the sandwich method. For example, there are ELISA kits on the market that can test for over 90 different food allergens.

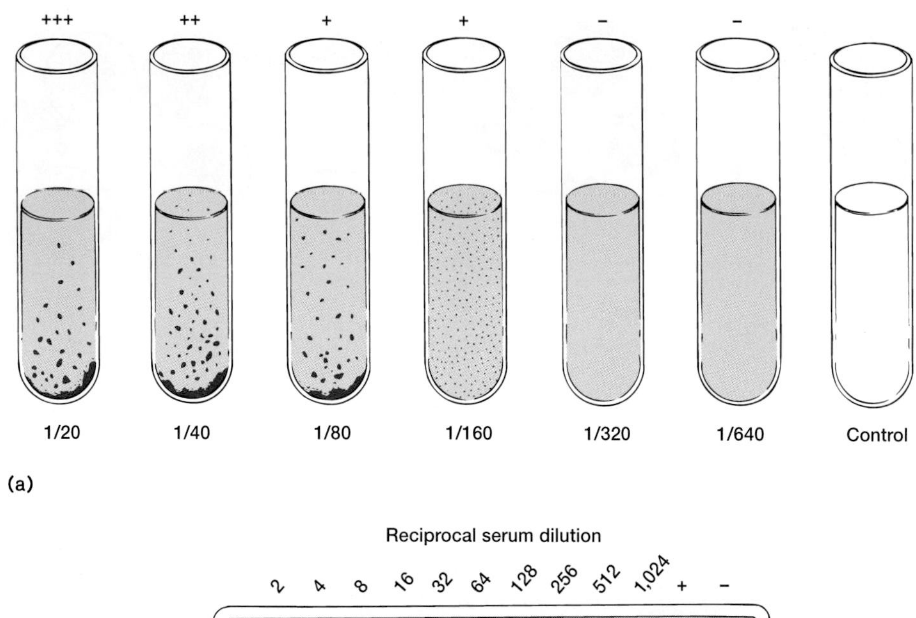

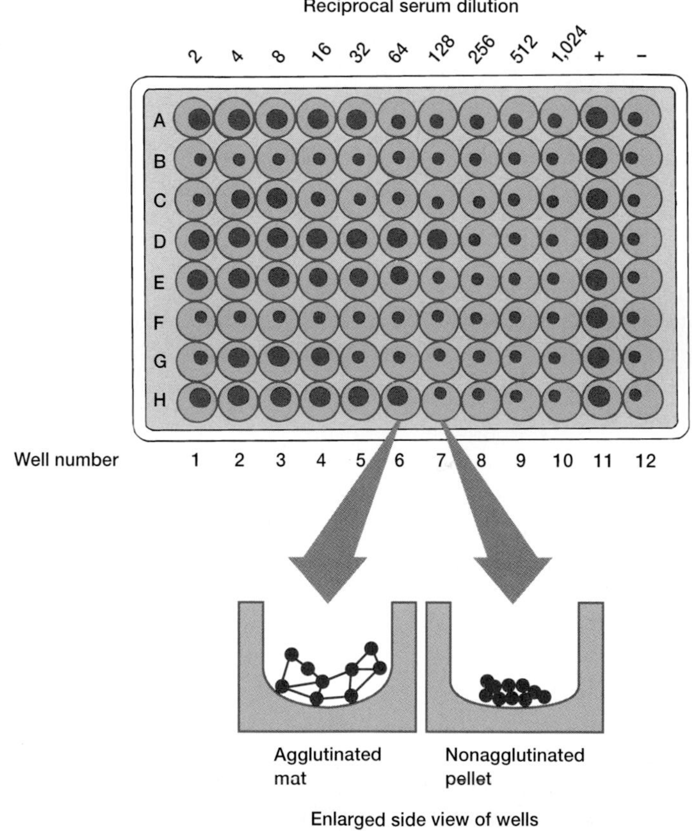

Figure 33.11 **Agglutination Tests.** (*a*) Tube agglutination test for determining antibody titer. The titer in this example is 160 since there is no agglutination in the next tube in the dilution series (1/320). The color in the dilution tubes indicates the presence of the patient's serum. (*b*) A microtiter plate illustrating hemagglutination. The antibody is placed in the wells (1–10). Positive controls (row 11) and negative controls (row 12) are included. Red blood cells are added to each well. If sufficient antibody is present to agglutinate the cells, they sink as a mat to the bottom of the well. If insufficient antibody is present, they form a pellet at the bottom. Can you read the different titers in rows A–H?

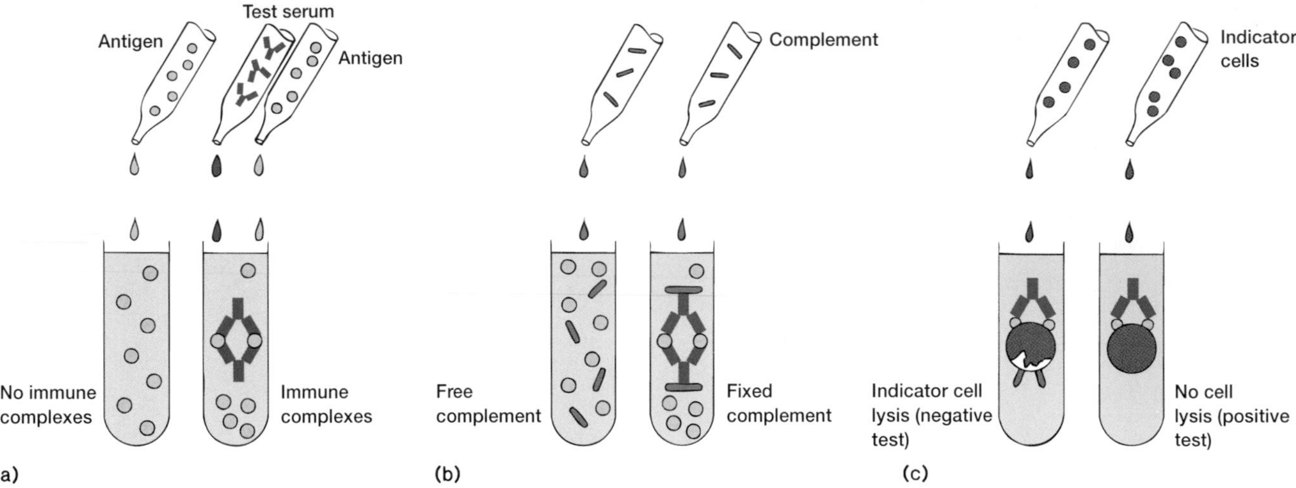

Figure 33.12 **Complement Fixation** (*a*) Test serum (antiserum) is added to one test tube. A fixed amount of antigen is then added to both tubes. If antibody is present in the test serum, immune complexes form. (*b*) When complement is added, if complexes are present, they fix complement and consume it. (*c*) Indicator cells and a small amount of anti-erythrocyte antibody are added to the two tubes. If there is complement present, the indicator cells will lyse (a negative test); if the complement is consumed, no lysis will occur (a positive test).

The indirect immunosorbent assay detects antibodies rather than antigens. In this assay, antigen in appropriate sensitizing buffer is incubated in the wells of a microtiter plate (figure 33.13*b*) and is absorbed onto the walls of the wells. Free antigen is washed away. Test antiserum is added, and if specific antibody is present, it binds to the antigen. Unbound antibody is washed away. Alternatively, the test sample can be incubated with a suspension of latex beads that have the desired antigen attached to their surface. After allowing time for antibody-antigen complex formation, the beads are trapped on a filter and unbound antibody is washed away. An anti-antibody that has been covalently coupled to an enzyme, such as horseradish peroxidase, is added next. The antibody-enzyme complex (the conjugate) binds to the test antibody, and after unbound conjugate is washed away, the attached ligand is visualized by the addition of a chromogen. A **chromogen** is a colorless substrate acted on by the enzyme portion of the ligand to produce a colored product. The amount of test antibody is quantitated in the same way as an antigen is in the double antibody sandwich method. The indirect immunosorbent assay is currently being used to test for antibodies to human immunodeficiency virus (the causative agent of AIDS) and rubella virus (German measles), and to detect certain drugs in serum. For example, antigen-coated latex beads are used in the SUDS HIV-1 test to detect HIV serum antibodies in about 10 minutes (*see table 34.5*).

Immunodiffusion

Immunodiffusion refers to a precipitation reaction that occurs between an antibody and antigen in an agar gel medium. Two techniques are routinely used: single radial immunodiffusion and double diffusion in agar.

The **single radial immunodiffusion (RID) assay** or Mancini technique quantitates antigens. Monospecific antibody is added to agar; then the mixture is poured onto slides and allowed to set. Wells are cut in the agar and known amounts of standard antigen added. The unknown test antigen is added to a separate well (figure 33.13*a*). The slide is left for 24 hours or until equilibrium has been reached, during which time the antigen diffuses out of the wells to form insoluble complexes. The size of the resulting precipitation ring surrounding various dilutions of antigen selected is proportional to the amount of antigen in the well (the wider the ring, the greater the antigen concentration). This is because the antigen's concentration drops as it diffuses farther out into the agar. The antigen forms a precipitin ring in the agar when its level has decreased sufficiently to reach equivalence and combine with the antibody to produce a large, insoluble network. This method is commonly used to quantitate serum immunoglobulins, complement proteins, and other substances.

The **double diffusion agar assay (Ouchterlony technique)** is based on the principle that diffusion of both antibody and

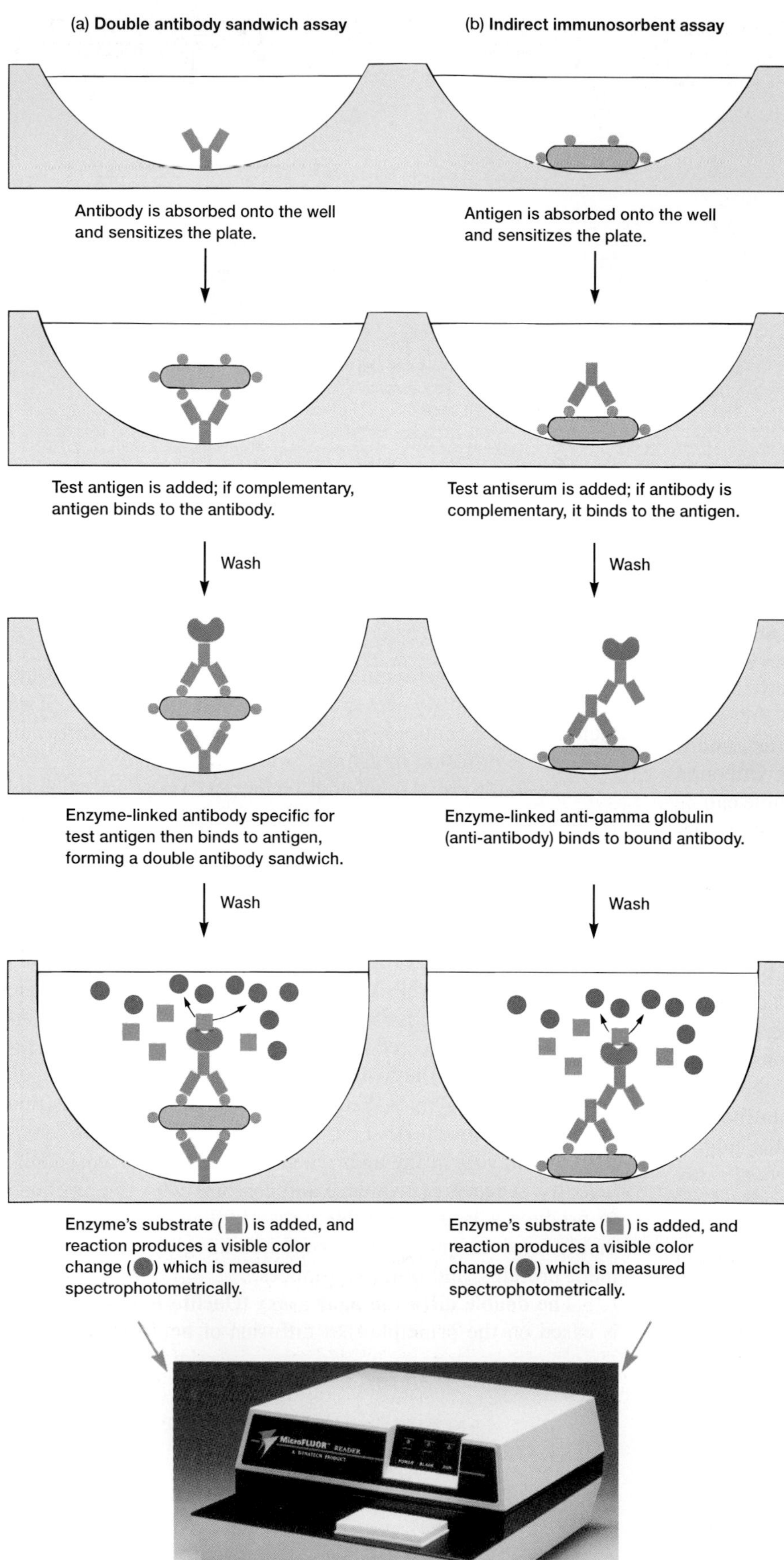

(a) Double antibody sandwich assay
(b) Indirect immunosorbent assay
Antibody is absorbed onto the well and sensitizes the plate.
Antigen is absorbed onto the well and sensitizes the plate.
Test antigen is added; if complementary, antigen binds to the antibody.
Test antiserum is added; if antibody is complementary, it binds to the antigen.
Wash
Wash
Enzyme-linked antibody specific for test antigen then binds to antigen, forming a double antibody sandwich.
Enzyme-linked anti-gamma globulin (anti-antibody) binds to bound antibody.
Wash
Wash
Enzyme's substrate (■) is added, and reaction produces a visible color change (●) which is measured spectrophotometrically.
Enzyme's substrate (■) is added, and reaction produces a visible color change (●) which is measured spectrophotometrically.
MicroFLUOR READER

Figure 33.13 The ELISA or EIA Test. (*a*) The double antibody sandwich method for the detection of antigens. (*b*) The indirect immunosorbent assay for detecting antibodies. See text for details. (*c*) An automated ELISA reader used to spectrophotometrically detect color changes in the microtiter plate wells.

antigen (hence, double diffusion) through agar can form stable and easily observable immune complexes. Test solutions of antigen and antibody are added to the separate wells punched in agar. The solutions diffuse outward, and when antigen and the appropriate antibody meet, they combine and precipitate at the equivalence zone, producing an indicator line (or lines) (figure 33.14*b*). The visible line of precipitation permits a comparison of antigens for identity (same antigenic determinants), partial identity (cross-reactivity), or nonidentity against a given selected antibody. For example, if a V-shaped line of precipitation forms, this demonstrates that the antibodies bind to the same antigenic determinants in each antigen sample and are identical. If one well is filled with a different antigen that shares some but not all determinants with the first antigen, a Y-shaped line of precipitation forms, demonstrating partial identity. In this reaction, the stem of the Y, called a spur, is formed if those antigen or antigenic determinants absent in the first well but present in the second one (antigen A in figure 33.14*b*) react with the diffusing antibodies. If two completely unrelated antigens are added to the wells, either a single straight line of precipitation forms between the two wells, or two separate lines of precipitation form, creating an X-shaped pattern, a reaction of nonidentity.

Immunoelectrophoresis

Some antigen mixtures are too complex to be resolved by simple diffusion and precipitation. Greater resolution is obtained by the technique of **immunoelectrophoresis** in which antigens are first separated based on their electrical charge, then visualized by the precipitation reaction. In this procedure, antigens are separated by electrophoresis in an agar gel. Positively charged proteins move to the negative electrode, and negatively charged proteins move to the positive electrode (figure 33.15*a*). A trough is then cut next to the wells (figure 33.15*b*) and filled with antibody. If the plate is incubated, the antibodies and antigens will diffuse and eventually form precipitation bands (figure 33.15*c*) that can be better visualized by staining (figure 33.15*d*). This assay is used to separate the major blood proteins in serum for certain diagnostic tests.

Electrophoresis (p. 293)

Immunofluorescence

Immunofluorescence is a process in which dyes called fluorochromes are exposed to UV, violet, or blue light to make them fluoresce or emit visible light. Dyes such as rhodamine B or fluorescein isothiocyanate can be coupled to antibody molecules without changing the antibody's capacity to bind to a specific antigen. Fluorochromes also can be attached to antigens. There are two main kinds of fluorescent antibody assays: direct and indirect.

Direct immunofluorescence involves fixing the specimen (cell or microorganism) containing the antigen of interest onto a slide (figure 33.16*a*). Fluorescein-labeled antibodies are then

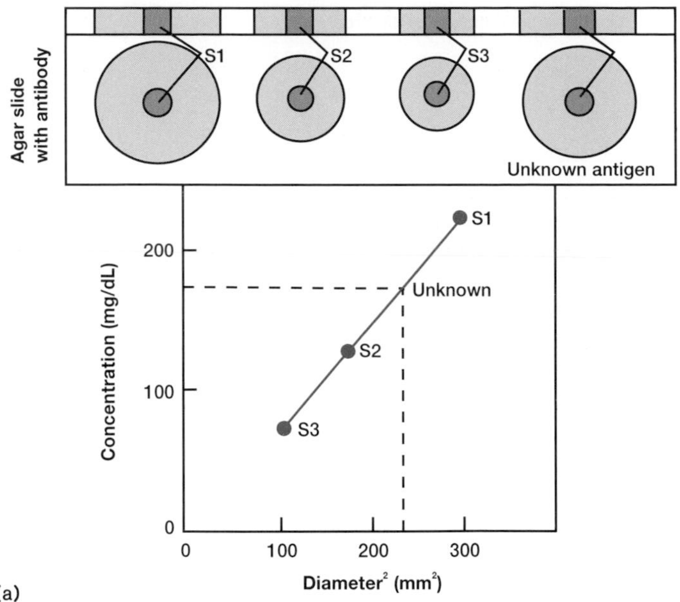

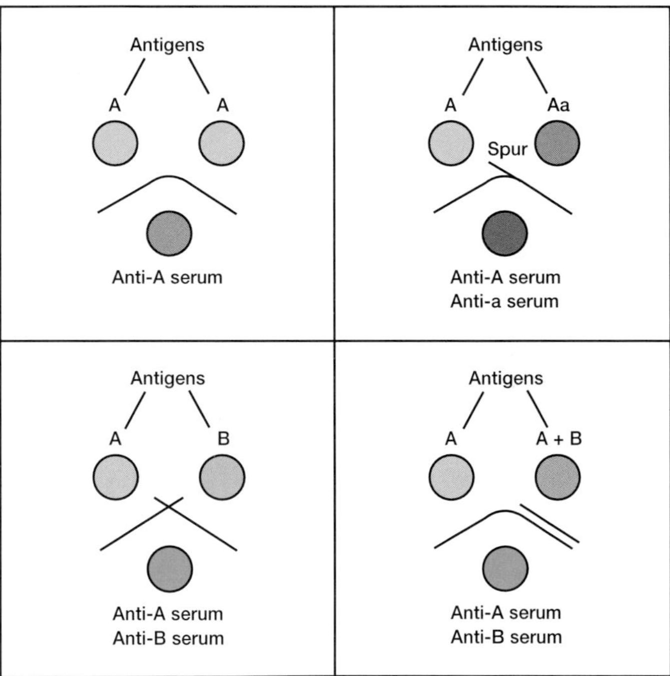

Figure 33.14 **Immunodiffusion.** (*a*) Single radial immunodiffusion assay. Three standard solutions of different antigen concentrations (*S*) and an unknown are placed on agar. After equilibriation, the ring diameters are measured. Usually, the square of the diameter of the standard rings is plotted on the x-axis and the antigen concentration on the y-axis. From this standard curve the concentration of an unknown can be determined. (*b*) Double diffusion agar assay showing characteristics of identity (top left), reaction of nonidentity (bottom left), partial identity (top right), and a complex pattern (bottom right).

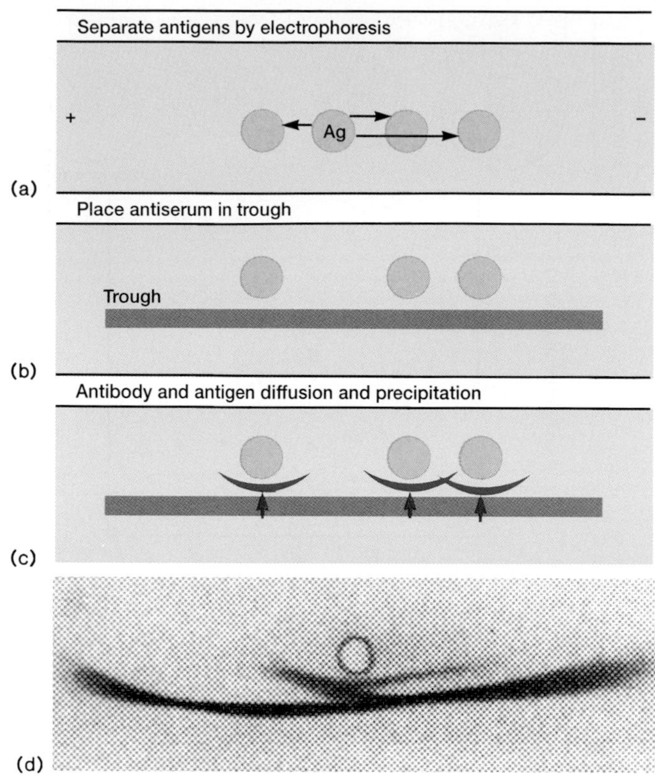

Figure 33.15 Immunoelectrophoresis. (*a*) Antigens are separated in an agar gel by an electrical charge. (*b*) Antibody (antiserum) is then placed in a trough cut parallel to the direction of antigen migration. (*c*) The antigens and antibodies diffuse through the agar and form precipitin arcs. (*d*) After staining, better visualization is possible.

added to the slide and incubated. The slide is washed to remove any unbound antibody and examined with the fluorescence microscope (*see figure 2.11*) for a yellow-green fluorescence. The pattern of fluorescence reveals the antigen's location. Direct immunofluorescence is used to identify antigens such as those found on the surface of group A streptococci and to diagnose enteropathogenic *Escherichia coli, Neisseria meningitidis, Salmonella typhi* (figure 33.16*c*), *Shigella sonnei, Listeria monocytogenes, Haemophilius influenzae* b, and the rabies virus.

The fluorescence microscope (pp. 26–27).

Indirect immunofluorescence (figure 33.16*b*) is used to detect the presence of antibodies in serum following an individual's exposure to microorganisms. In this technique, a known antigen is fixed onto a slide. The test antiserum is then added, and if the specific antibody is present, it reacts with antigen to form a complex. When fluorescein-labeled anti-immunoglobulin is added, it reacts with the fixed antibody.

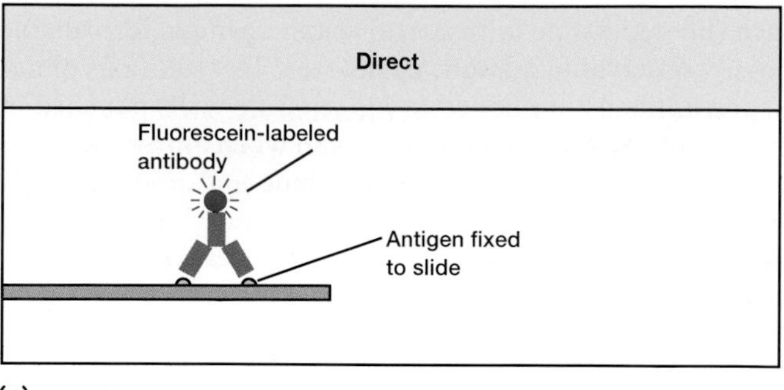

(a)

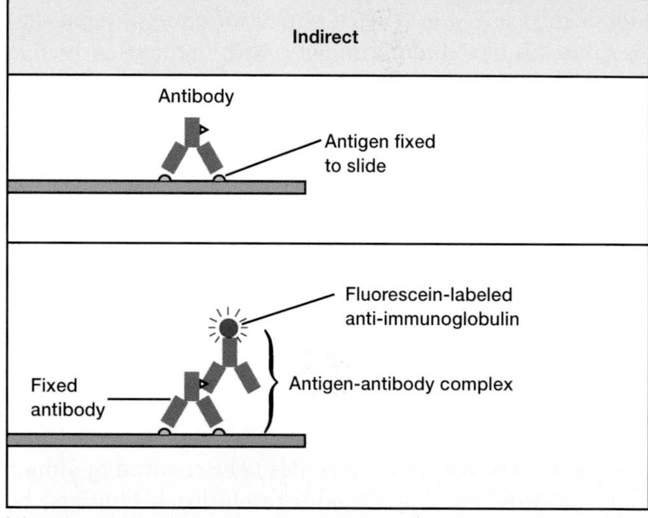

(b)

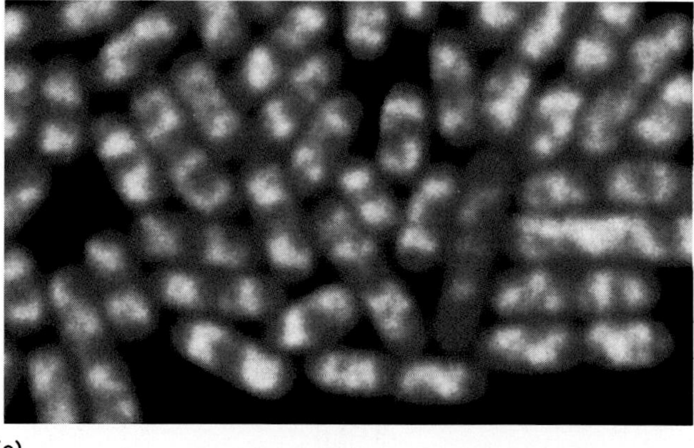

(c)

Figure 33.16 Direct and Indirect Immunofluorescence. (*a*) In the direct fluorescent-antibody (FA) technique, the specimen containing antigen is fixed to a slide. Fluorescenated antibodies that recognize the antigen are then added, and the specimen is examined under a UV microscope for yellow-green fluorescence. (*b*) Indirect fluorescent-antibody technique (IFA). The antigen on a slide reacts with an antibody directed against it. The antigen-antibody complex is located with a fluorescent antibody that recognizes immunoglobulins. (*c*) *Salmonella typhi* stained with acridine orange (×2,000).

After incubation and washing, the slide is examined with the fluorescence microscope. The occurrence of fluorescence shows that antibody specific to the test antigen is present in the serum. Indirect immunofluorescence is used to identify the presence of *Treponema pallidum* antibodies in the diagnosis of syphilis (treponemal antibody absorption, FTA-ABS; *see figure 37.18*) as well as antibodies produced in response to other microorganisms.

Immunoprecipitation

The **immunoprecipitation** technique detects soluble antigens that react with antibodies called precipitins. The precipitin reaction occurs when bivalent or multivalent antibodies and antigens are mixed in the proper proportions. The antibodies link the antigen to form a large antibody-antigen network or lattice that settles out of solution when it becomes sufficiently large (figure 33.17*a*). Immunoprecipitation reactions occur only at the equivalence zone when there is an optimal ratio of antigen to antibody so that a lattice forms. If the precipitin reaction takes place in a test tube (figure 33.17*b*), a precipitation ring forms in the area in which the optimal ratio or equivalence zone develops.

1. What does a negative complement fixation test show? A positive test?
2. What are the two types of ELISA methods and how do they work? What is a chromogen?
3. Name two types of immunodiffusion tests and describe how they operate.
4. Describe the immunoelectrophoresis technique.
5. What is a fluorochrome?
6. Name and describe the two kinds of fluorescent antibody assays.
7. Specifically, when do immunoprecipitation reactions occur?

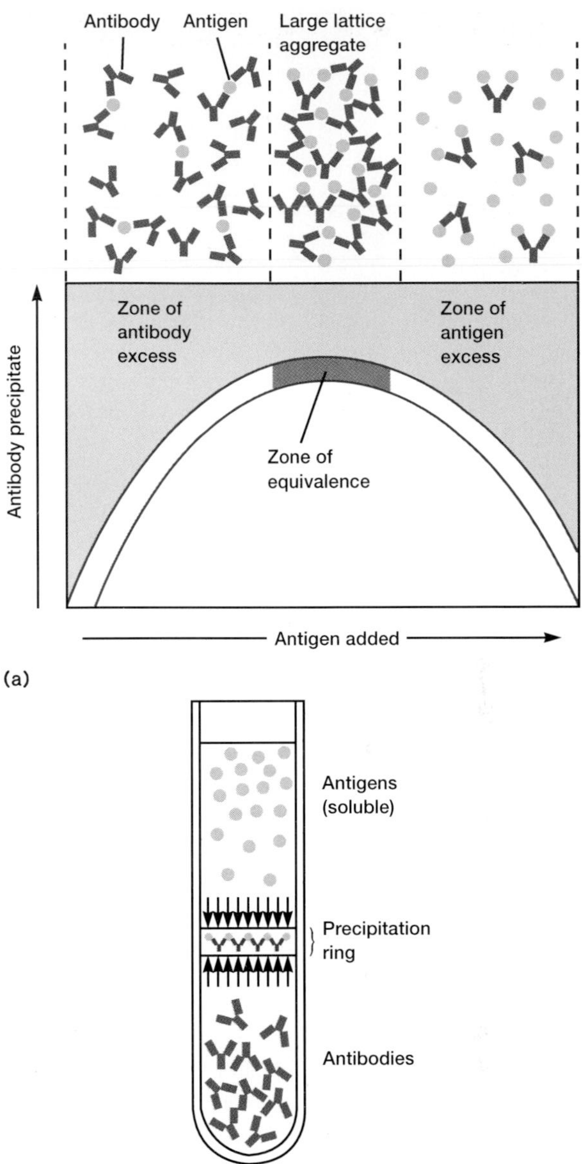

Figure 33.17 Immunoprecipitation. (*a*) Graph showing that a precipitation curve is based on the ratio of antigen to antibody. The zone of equivalence represents the optimal ratio for precipitation. (*b*) A precipitation ring test. Antibodies and antigens diffuse toward each other in a test tube. A precipitation ring is formed at the zone of equivalence.

Neutralization

Neutralization tests are antigen-antibody reactions that determine whether the activity of a toxin or virus has been neutralized by antibody. Laboratory animals or tissue culture cells are used as "indicator systems" in these tests. The toxin or virus to be assayed has known effects on the indicator system. The effect in the animal might be death, paralysis, or skin lesions. For example, when the exotoxin of *Clostridium botulinum* is suspected of causing food poisoning in a person (*see chapter 37*), a sample of either the suspect food or the serum, stools, or vomitus of the ill patient is collected. Two groups of indicator mice are used. The control group receives the botulinum antitoxin and the experimental group does not. Filtrates from the source samples are injected into both groups of mice. If toxin is present, all mice except those receiving the antitoxin will die, and the test is positive for food botulism.

Viral neutralization assays are frequently used to detect viral infections. Suspected blood serum containing viral anti-

Box 33.3

History and Importance of Serotyping

In the early 1930s, Rebecca Lancefield (1895–1981) recognized the importance of serological tests. She developed a classification system for the streptococci based on the antigenic nature of cell wall carbohydrates. Her system is now known as the **Lancefield system** in which each different serotype is identified by a letter (A through O). This scheme is based on specific antibody agglutination reactions with cell wall carbohydrate antigens (C polysaccharides) extracted from the streptococci. Lancefield also showed that further subdividing of the group A streptococci into specific serological types was possible, based on the presence of type-specific M (protein) antigens.

More recently, *Escherichia coli, Salmonella,* and other bacteria have been serotyped with specific antigen-antibody reactions involving flagella (H) antigens, capsular (K) antigens, and somatic (cell wall or O) antigens. Within *E. coli,* there are over 167 different O antigens.

The current value of serotyping may be seen in the fact that *E. coli* O55, O111, and O127 serotypes are the ones most frequently associated with infantile diarrhea. Thus, the serotype of *E. coli* from stool samples of infants with infantile diarrhea is of diagnostic value and aids in identifying the source of the infection.

bodies can be introduced into tissue culture cells or embryonated eggs (*see figure 17.1 and table 34.1*). If antibodies are present against the virus, viral neutralization will occur and prevent the virus from infecting the culture cells. No cytopathic effects will be seen.

Radioimmunoassay

The **radioimmunoassay (RIA)** technique has become an extremely important tool in biomedical research and clinical practice (e.g., in cardiology, blood banking, diagnosis of allergies, and endocrinology). Indeed, Rosalyn Yalow won the 1977 Nobel Prize in Physiology or Medicine for its development. RIA uses a purified antigen that is radioisotope-labeled and competes for antibody with unlabeled standard or antigen in experimental samples. The radioactivity associated with the antibody is then detected by means of radioisotope analyzers and autoradiography (photographic emulsions that show areas of radioactivity). If there is much antigen in an experimental sample, it will compete with the radioisotope-labeled antigen for antigen binding sites on the antibody, and little radioactivity will be bound. A large amount of bound radioactivity indicates that there is little antigen present in the experimental sample.

Serotyping

Serotyping refers to serological procedures used to differentiate strains (serovars or serotypes) of microorganisms that differ in the antigenic composition of a structure or product (Box 33.3). The serological identification of a strain of a pathogen has diagnostic value. Often, the symptoms of infections depend on the nature of cell products released by a pathogen. Genes for virulence often occur in the same clone with genes for antigenic cell wall material. Therefore, it is possible to identify a pathogen serologically by testing for cell wall antigens. For example, there are 84 strains of *Streptococcus pneumoniae,* each differing in the nature of its capsular material. These differences can be detected by capsular swelling (termed the **Quellung reaction**) if antisera specific for the capsular types are used (figure 33.18).

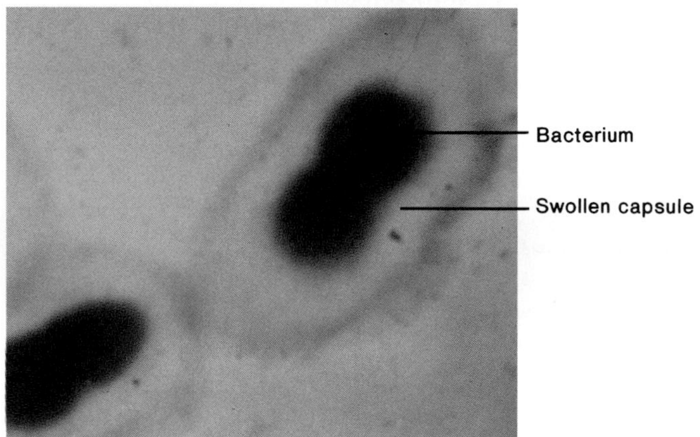

Figure 33.18 **Serotyping.** *Streptococcus pneumoniae* have reacted with a specific pneumonococcal antiserum leading to swelling of their capsules (the Quellung reaction). The capsules seen around the pairs indicate potential virulence.

---

1. What is a neutralization test?
2. Describe the RIA technique. What is serotyping?

Summary

1. The major function of an antigen-antibody reaction is to mark and target antigens for destruction by other components of the immune system.
2. Antigen-antibody complexes activate a system of proteins called the complement system. This results in the formation of a membrane attack complex that can lyse antibody-coated cells. Other components of this pathway mediate inflammation, act as attractants for and activators of phagocytic cells, and can act as opsonins.
3. In toxin neutralization, antibodies react with toxins, preventing either the B fragment of the toxin from binding to the target cell or the A fragment from entering the cytosol of the cell.
4. IgG, IgM, and IgA antibodies can bind to some extracellular viruses and inactivate them by what is termed viral neutralization.
5. Secretory IgA inhibits certain bacterial adherence promoting factors to prevent bacteria from colonizing mucosal surfaces and intiating the disease process.
6. Antibody-dependent cell-mediated cytotoxicity involves an antibody response to certain virus-infected cells. Cytolysis of the infected cells can then be achieved by the action of complement and formation of the membrane attack complex.
7. Elevated IgE levels cause eosinophils to bind to tissue parasites causing degranulation and release of lytic and inflammatory mediators that destroy the parasites.
8. Opsonization is the process by which microorganisms or other particles are coated with antibody and/or complement and thus prepared for ingestion by phagocytic cells.
9. An antigen can trigger inflammation by binding to IgE on the surface of either mast cells or basophils and stimulating histamine release. Antibodies also can trigger inflammation through the production of complement fragments C3a and C5a that bind to mast cells and platelets and trigger the release of histamine.
10. The cross-linking of large numbers of antigens and antibodies produces immune complexes. The responsible antibodies are precipitins, agglutinins, or hemagglutinins, depending on the cell or molecule involved.
11. Two types of immunologic reactions are used for in vitro skin testing: immediate and delayed. Immediate tests detect antibodies and are used to test for respiratory allergies, whereas delayed tests detect cell-mediated immunity and are used to detect contact dermatitis and food allergies.
12. Agglutination reactions in vitro usually form aggregates or clumps (agglutinates) that are visible with the naked eye. Tests have been developed, such as the Widal test, latex microsphere agglutination reaction, hemagglutination, and viral hemagglutination, to detect antigen as well as to determine antibody titer.
13. The complement fixation test can be used to detect a specific antibody for a suspect microorganism in an individual's serum.
14. The enzyme-linked immunosorbent assay (ELISA) involves the linking of various enzymes to either antigens or antibodies. Two basic methods are involved: the double antibody sandwich method and the indirect immunosorbent assay. The first method detects antigens and the latter, antibodies.
15. Immunodiffusion refers to a precipitation reaction that occurs between antibody and antigen in an agar gel medium. Two techniques are routinely used: double diffusion in agar and single radial diffusion.
16. In immunoelectrophoresis antigens are separated based on their electrical charge, then visualized by precipitation and staining.
17. Immunofluorescence is a process in which certain dyes called fluorochromes are irradiated with UV, violet, or blue light to make them fluoresce. These dyes can be coupled to antibody. There are two main kinds of fluorescent antibody assays: direct and indirect.
18. Immunoprecipitation reactions occur only when there is an optimal ratio of antigen and antibody to produce a lattice at the zone of equivalence, which is evidenced by a visible precipitate.
19. Neutralization tests are antigen-antibody reactions that determine whether the activity of a toxin or virus has been neutralized by antibody. Laboratory animals or tissue culture cells are used as indicator systems in these tests.
20. Radioimmunoassays use a purified antigen that is radioisotope-labeled and used to compete with unlabeled standard or antigens in experimental samples for a specific antibody.
21. Serotyping refers to serological procedures used to differentiate strains (serovars or serotypes) of microorganisms that have differences in the antigenic composition of a structure or product.

Key Terms

agglutinates *656*
agglutination reaction *654*
agglutinin *654*
alternative pathway *648*
antibody-dependent cell-mediated cytotoxicity *651*
antitoxin *651*
chromogen *661*
classical pathway *648*
complement fixation *648*
complement system *648*
double diffusion agar assay (Ouchterlony technique) *661*
enzyme-linked immunosorbent assay (ELISA) *659*
hemagglutination *654*
hemagglutinin *654*
immune complex *654*
immunodiffusion *661*
immunoelectrophoresis *663*
immunofluorescence *663*
immunoprecipitation *665*
Lancefield system *666*
membrane attack complex *648*
opsonization *653*
precipitation or precipitin reaction *654*
precipitin *654*
Quellung reaction *666*
radioimmunoassay (RIA) *666*
serology *656*
serotyping *666*
single radial immunodiffusion (RID) assay *661*
toxin neutralization *651*
viral hemagglutination *656*
viral neutralization *651*
Widal test *656*

Questions for Thought and Review

1. If excess complement is added in a complement fixation test, how will this affect the results?
2. Compare and contrast the characteristics of four types of in vivo immunologic reactions. Give an example of each type.
3. Specific chemical markers are conjugated to antibodies or antigens in some immunologic tests. What are the advantages of this technique over such tests as agglutination or complement fixation?
4. Even though the complement system is not an integral part of the immune system, it is usually described in association with it. Why is this so?
5. What is the value in determining an antibody titer?
6. Of what value are neutralization tests performed in animals or in tissue culture cells?
7. What is the importance of serotyping?
8. What are some features of an in vitro antigen-antibody reaction that make the reaction so useful for identification and monitoring tests?
9. How are chromogens used as biological markers?
10. What is the benefit of opsonization?
11. Even though *Streptococcus pneumoniae* has over 80 different serotypes, serological reactions are sensitive enough to differentiate one serotype from another. How is this possible?
12. How can electrophoresis be used in immunologic testing procedures?

Additional Reading

References provided at the end of chapter 31 may also be consulted for further information.

Figueroa, J. E., and Densen, P. 1991. Infectious diseases associated with complement deficiencies. *Clin. Microbiol. Rev.* 4(3):359–95.

Herrmann, J. E. 1986. Enzyme-linked immunoassays for the detection of microbial antigens and their antibodies. *Adv. Appl. Microbiol.* 31:271–89.

Herzenberg, L. A.; Sweet, R. G.; and Herzenberg, L. A. 1976. Fluorescence activated cell sorting. *Sci. Am.* 234(6):108–14.

Jackson, J. B., and Balfour, H. H., Jr. 1988. Practical diagnostic testing for human immunodeficiency virus. *Clin. Microbiol. Rev.* 1(1):124–38.

Joiner, K. A. 1988. Complement evasion by bacteria and parasites. *Ann. Rev. Microbiol.* 42:201–30.

Joiner, K. A.; Brown, E. J.; and Frank, M. M. 1984. Complement and bacteria; Chemistry and biology in host defense. *Ann. Rev. Immunol.* 2:461–88.

Lennette, E. H.; Balows, A.; Hausler, W. J., Jr.; Herrmann, K. L.; and Shadomy, H. J., eds. 1991. *Manual of clinical microbiology,* 5th ed. Washington, D.C.: American Society for Microbiology.

Mayer, M. 1973. The complement system. *Sci. Am.* 229(5):54–66.

Müller-Eberhard, H. J. 1988. Molecular organization and function of the complement system. *Ann. Rev. Biochem.* 57:321–47.

Peterson, E. M. 1981. ELISA: A tool for the clinical microbiologist. *Am. J. Med. Tech.* 47:905–10.

Reid, K. B. M., and Porter, R. R. 1981. The proteolytic activation system of complement. *Ann. Rev. Biochem.* 50:433–49.

Rose, N. R.; Macario, E.; Fahey, J.; Friedman, H.; and Penn, G., eds. 1992. *Manual of clinical laboratory immunology,* 4th ed. Washington, D.C. American Society for Microbiology.

Ross, G. D., ed. 1986. *Immunobiology of the complement system.* New York: Academic Press.

Part Nine
Microbial Diseases

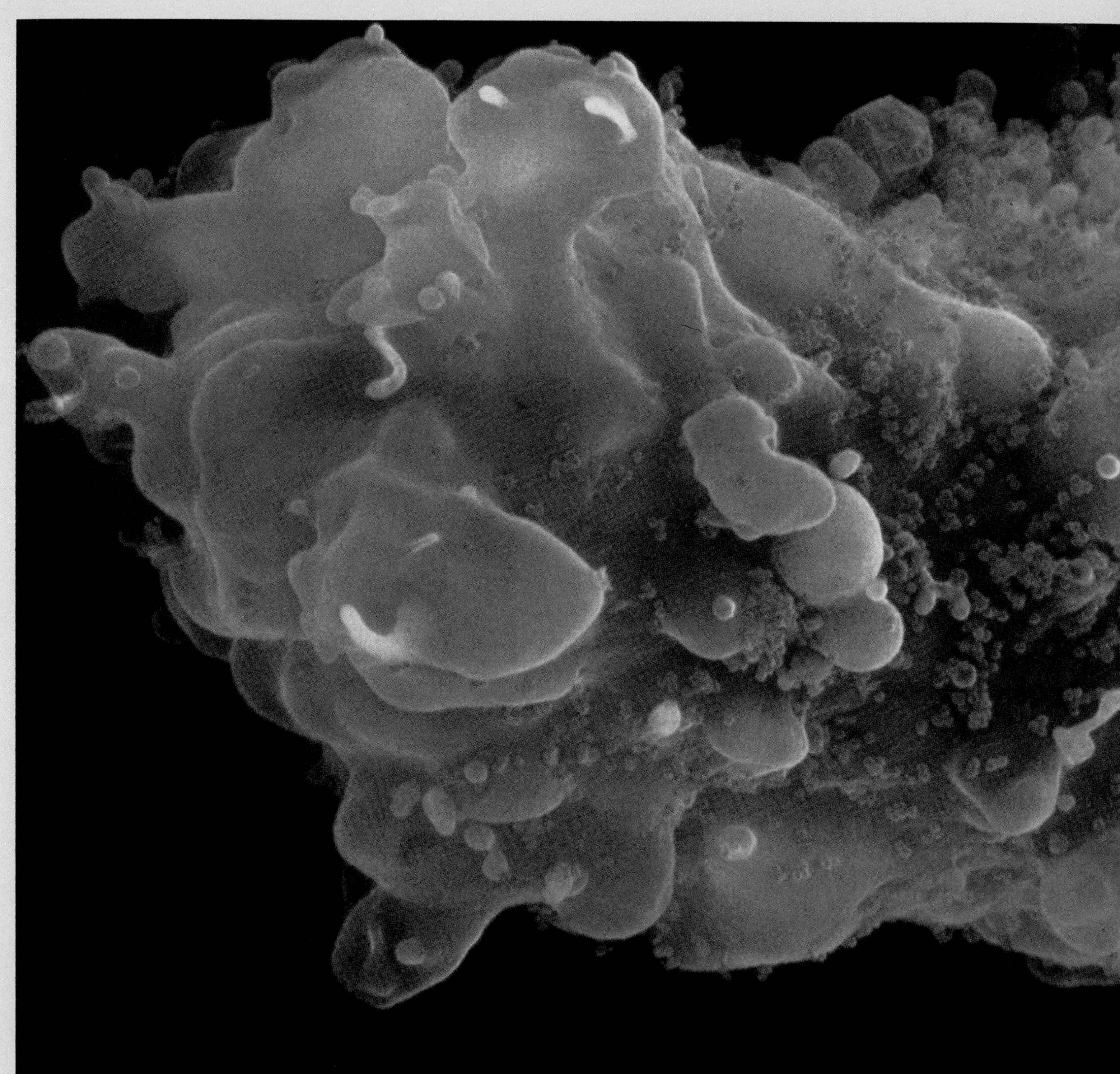

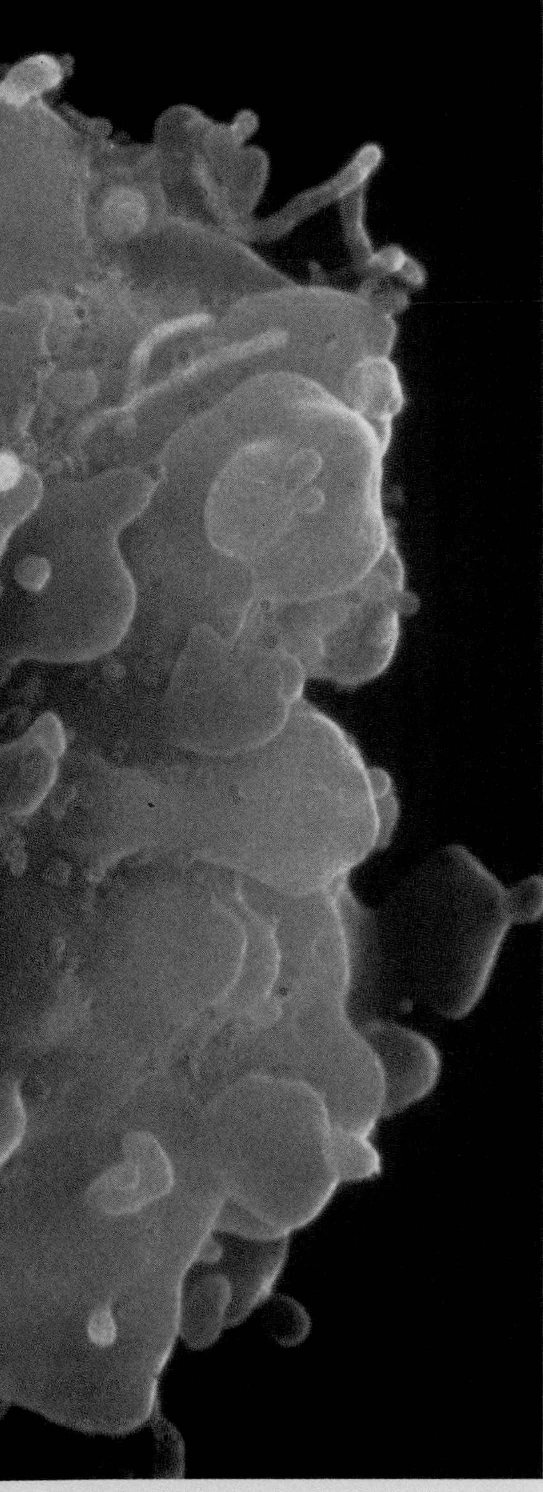

A T cell being attacked by the AIDS virus (human immunodeficiency virus). The virus particles are colored blue.

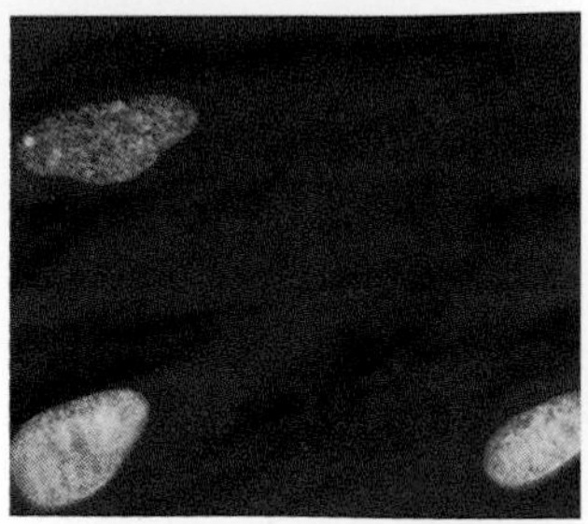

CHAPTER 34
Clinical Microbiology

The specimen is the beginning. All diagnostic information from the laboratory depends upon the knowledge by which specimens are chosen and the care with which they are collected and transported.

—Cynthia A. Needham

Outline

Concepts

1. Clinical microbiologists and clinical microbiology laboratories perform many services, all related to the identification of microorganisms.
2. Success in clinical microbiology depends on (1) using the proper aseptic technique; (2) correctly obtaining the clinical specimen from the infected patient via swabs, needle aspiration, intubation, or catheters; (3) correctly handling the specimen; and (4) quickly transporting the specimen to the laboratory.
3. Once the clinical specimen reaches the laboratory, it is cultured and identified. Identification measures include microscopy; growth on enrichment, selective, differential, or characteristic media; specific biochemical tests; rapid test methods; immunologic techniques; bacteriophage typing; and molecular methods such as DNA probes, gas-liquid chromatography, and plasmid fingerprinting.
4. After the microorganism has been isolated, cultured, and/or identified, samples are used in susceptibility tests to find which method of control will be most effective. The results are provided to the physician as quickly as possible.
5. Computer systems in clinical microbiology are designed to speed identification of the pathogen and communication of results back to the physician.

Pathogens, particularly bacteria, coexist with harmless microorganisms on or in the human host. These pathogens must be properly identified as actual etiologic agents of infectious diseases. This is the purpose of clinical microbiology. The clinical microbiologist identifies agents and organisms (hereafter referred to as microorganisms) based on morphological, biochemical, and immunologic procedures. Time is a significant factor in the identification process, especially in life-threatening situations. Computers and commercially available methods of rapid identification have greatly aided the clinical microbiologist. Molecular methods allow identification of microorganisms based on highly specific biochemical properties. Once isolated and identified, the microorganism can then be subjected to antimicrobial sensitivity tests. In the final analysis, the patient's well-being and health benefit most from clinical microbiology—the subject of this chapter.

The major concern of the **clinical microbiologist** is to isolate and identify microorganisms from clinical specimens rapidly. The purpose of the clinical microbiology laboratory is to provide the physician with information concerning the presence or absence of microorganisms that may be involved in the infectious disease process (figure 34.1). These individuals and facilities also determine the susceptibility of microorganisms to antimicrobial agents. Clinical microbiology makes use of information obtained from research on such diverse topics as microbial biochemistry and physiology, immunology, and the host-parasite relationships involved in the infectious disease process. Computers and rapid tests have assumed an ever-increasing role in clinical microbiology.

Specimens

In clinical microbiology, a clinical specimen (hereafter, specimen) represents a portion or quantity of human material that is tested, examined, or studied to determine the presence or absence of particular microorganisms. Several guidelines on selection, collection, and handling of the specimen need emphasis (Box 34.1):

1. The specimen selected should adequately represent the infectious disease.
2. A quantity of specimen adequate for complete examination should be obtained.

Box 34.1

Universal Precautions for Microbiology Laboratories

Blood and other body fluids from all patients should be considered infective.

1. All specimens of blood and body fluids should be put in a well-constructed container with a secure lid to prevent leaking during transport. Care should be taken when collecting each specimen to avoid contaminating the outside of the container and of the laboratory form accompanying the specimen.
2. All persons processing blood and body-fluid specimens should wear gloves. Masks and protective eyewear should be worn if mucous membrane contact with blood or body fluids is anticipated. Gloves should be changed and hands washed after completion of specimen processing.
3. For routine procedures, such as histologic and pathological studies or microbiologic culturing, a biological safety cabinet is not necessary. However, biological safety cabinets should be used whenever procedures are conducted that have a high potential for generating droplets. These include activities such as blending, sonicating, and vigorous mixing.
4. Mechanical pipetting devices should be used for manipulating all liquids in the laboratory. Mouth pipetting must not be done.
5. Use of needles and syringes should be limited to situations in which there is no alternative, and the recommendations for preventing injuries with needles outlined under universal precautions (*see inside back cover*) should be followed.
6. Laboratory work surfaces should be decontaminated with an appropriate chemical germicide after a spill of blood or other body fluids and when work activities are completed.
7. Contaminated materials used in laboratory tests should be decontaminated before reprocessing or be placed in bags and disposed of in accordance with institutional policies for disposal of infective waste.
8. Scientific equipment that has been contaminated with blood or other body fluids should be decontaminated and cleaned before being repaired in the laboratory or transported to the manufacturer.
9. All persons should wash their hands after completing laboratory activities and should remove protective clothing before leaving the laboratory.
10. There should be no eating, drinking, or smoking in the work area.

Adapted from the Centers for Disease Control Guidelines. 1987. *Morbid. Mortal. Weekly Report.* 36 (Suppl. 2S) 5S–10S.

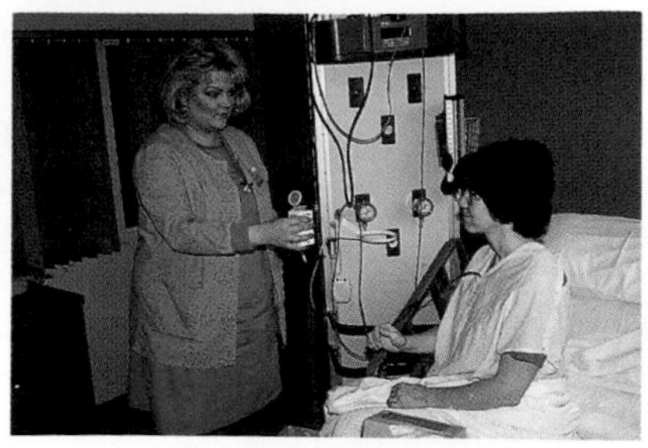

(a) The identification of the organism begins at the patient's bedside. The nurse is giving instructions to the patient on how to obtain a sputum specimen.

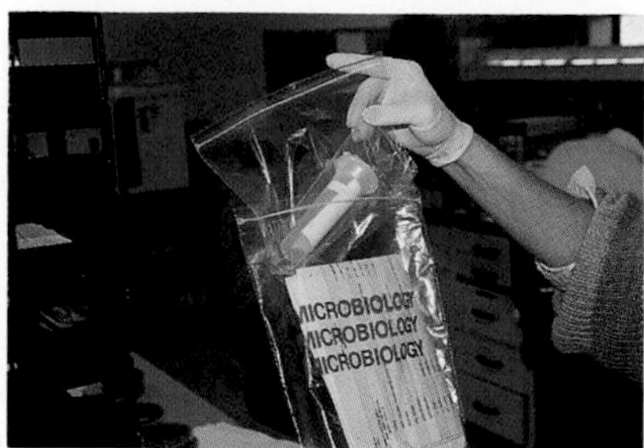

(b) The specimen is sent to the laboratory to be processed. Notice that the specimen and worksheet are in different Ziplock bags.

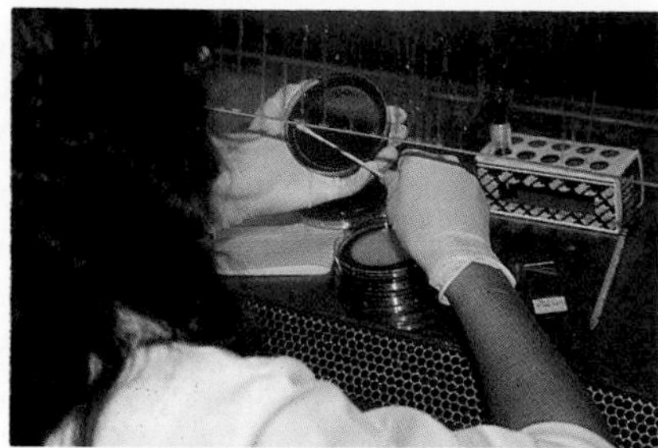

(c) Specimens such as sputum are plated on various types of media under a laminar airflow hood. This is to prevent specimen aerosols from coming in contact with the microbiologist.

(d) Sputum and other specimens are usually Gram stained to determine whether or not bacteria are present and to obtain preliminary results on the nature of any bacteria found.

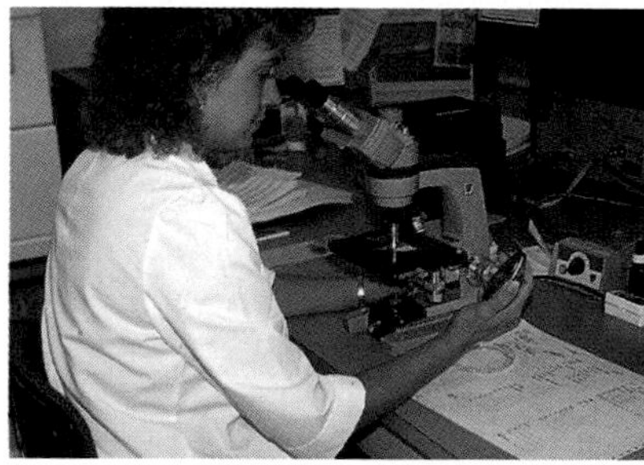

(e) After incubation, the plates are examined for significant isolates. The Gram stain may be reexamined for correlation.

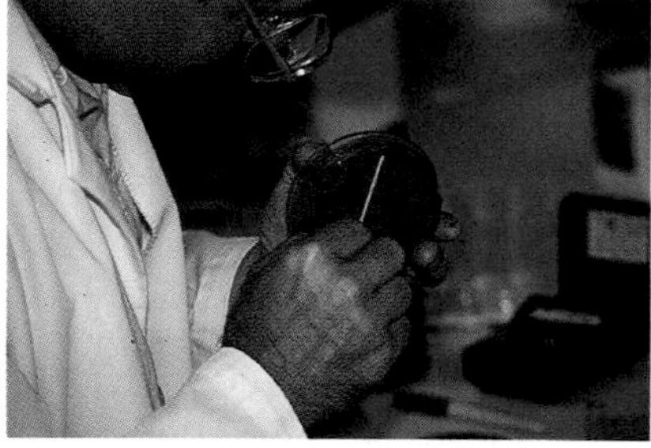

(f) Suspect colonies are picked for biochemical and/or serological testing.

(g) Colonies are prepared for identification by rapid test systems.

(h) In a short period of time, sometimes four hours, computer-generated information is obtained that will consist of biochemical identification and antibiotic susceptibility results.

(i) All information about the specimen is now entered into a computer and the data are transmitted directly to the hospital ward.

Figure 34.1 **Isolation and Identification of Microorganisms in a Clinical Laboratory.**

3. Attention must be given to specimen collection in order to avoid contamination from the many varieties of microorganisms indigenous to the skin and mucous membranes (*see figure 29.8*).
4. The specimen should be forwarded promptly to the clinical laboratory.
5. If possible, the specimen should be obtained before antimicrobial agents have been administered to the patient.

Collection

Specimens may be collected by several methods using aseptic technique. Aseptic technique refers to specific procedures used to prevent unwanted microorganisms from contaminating the clinical specimen. Each method is designed to ensure that only the proper material will be sent to the clinical laboratory.

The most common method used to collect specimens from the skin and mucous membranes (eye, ear, nose, throat, open wounds) is the sterile **swab.** A sterile swab is a rayon- or dacron-tipped polystyrene applicator. Manufacturers of swabs have their own unique container design and instructions for proper use. For example, many commercially manufactured swabs contain a transport medium designed to preserve a variety of microorganisms and to prevent multiplication of rapidly growing members of the population (figure 34.2*a*).

Specimens from blood and cerebrospinal fluid may be collected by **needle aspiration**. After the proper aseptic technique has been followed, a sample of fluid is drawn into a sterile tube that has been treated with an anticoagulant such as heparin or potassium oxalate (figure 34.2*b*). The anticoagulant prevents the microorganisms from being entrapped in a clot, which would make isolation difficult.

Intubation (Latin *in,* into and *tuba,* tube) is the inserting of a tube into a body canal or hollow organ. For example, intubation can be used to collect specimens from the stomach. In this procedure, a long sterile tube is attached to a syringe (figure 34.2*c*), and the tube is either swallowed by the patient or passed through a nostril into the person's stomach. Specimens are then withdrawn periodically into the sterile syringe. The most common intubation tube is the Levin tube.

A **catheter** is a tubular instrument used for withdrawing or introducing fluids from or into the body. For example, urine specimens may be collected with catheters. Three types are commonly used. The hard catheter is used when the urethra is very narrow or has strictures. The French catheter is a soft tube used to obtain a single specimen sample. If multiple samples are required over a prolonged period, a Foley catheter is used (figure 34.2*d*).

Urine also may be obtained by the clean-catch method. After the patient has cleansed the urethral meatus (opening), a small container is used to collect the urine. In the clean-catch midstream method, the first urine voided is not collected because it will be contaminated with those microorganisms normally occurring in the lower portion of the urethra. Only the midstream portion is collected since it most likely will contain those microorganisms found in the urinary bladder.

Most lower respiratory specimens are obtained from sputum. Specifically, **sputum** is the mucous secretion expectorated from the lungs, bronchi, and trachea through the mouth, in contrast to saliva, which is the secretion of the salivary glands. Sputum is collected in specially designed sputum cups (figures 34.1*a* and 34.2*e*).

Handling

Immediately after collection, the specimen must be properly labeled and handled. The person collecting the specimen is responsible for ensuring that the name, registration number, and location of the patient are correctly and legibly written or imprinted on the culture request form (figure 34.3*a*). This information must correspond to that written or imprinted on a label affixed to the specimen container. The type or source of the sample and the choice of tests to be performed also must be specified on the request form (figure 34.3*b*).

Transport

Speed in transporting the specimen to the clinical laboratory after it has been obtained from the patient is of prime importance. Some laboratories refuse to accept specimens if they have been in transit too long.

Microbiological specimens may be transported to the laboratory by various means (figure 34.1*b*). For example, certain specimens should be transported in a medium that preserves the microorganisms and helps maintain the ratio of one organism to another. This is especially important for specimens in which normal microorganisms may be mixed with microorganisms foreign to the body location.

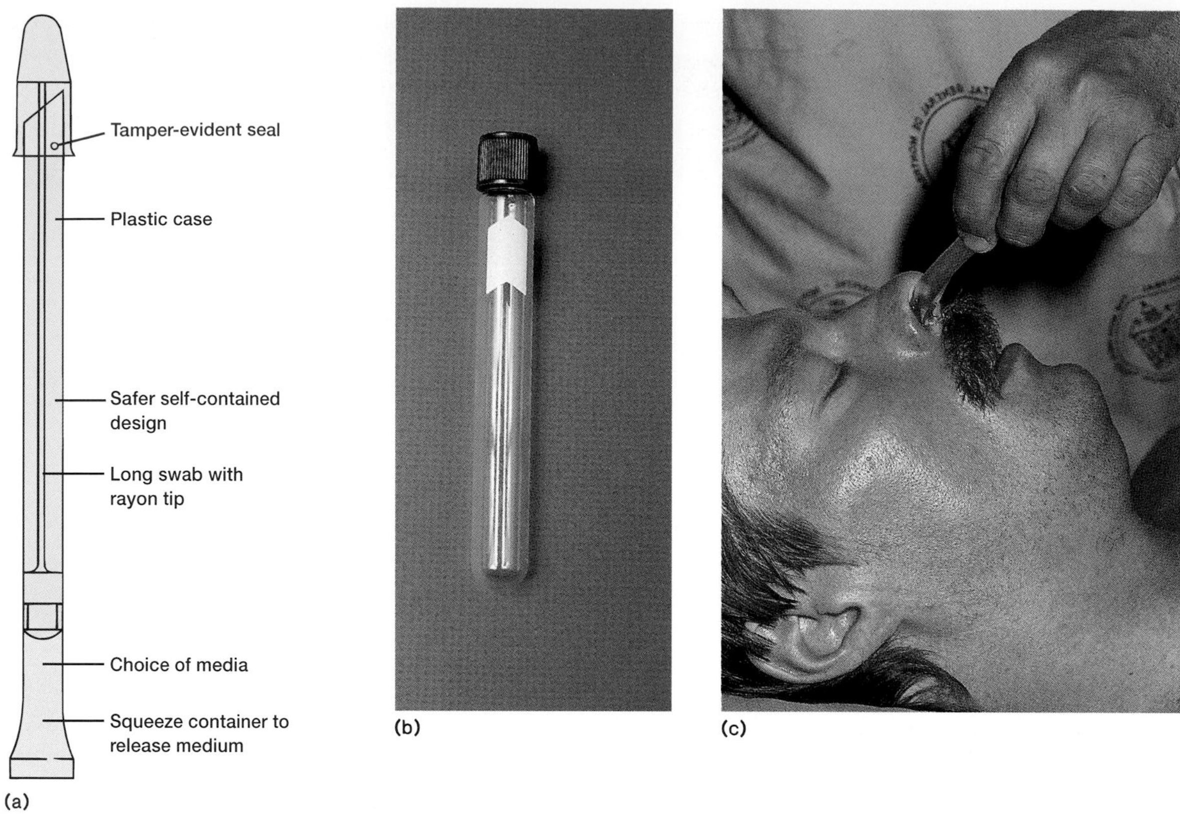

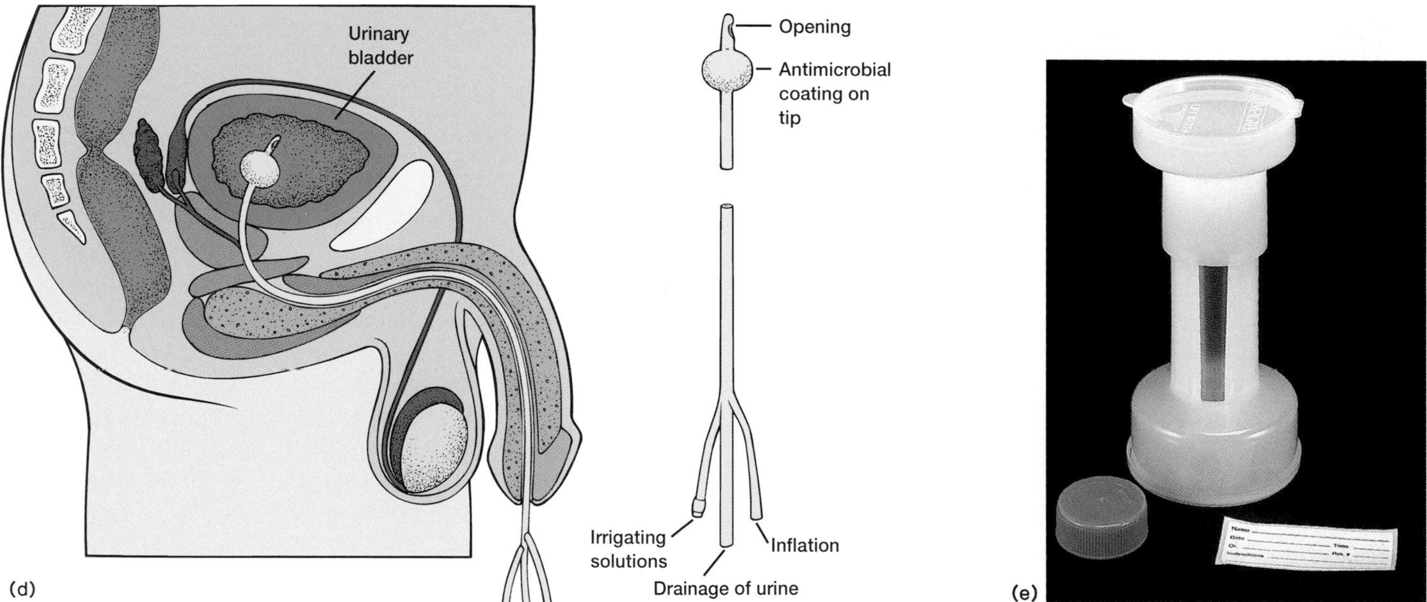

Figure 34.2 Collection of Clinical Specimens. (*a*) A drawing of a sterile swab with a specific transport medium. (*b*) Sterile collection tubes for cerebrospinal fluid. (*c*) Nasotracheal intubation. (*d*) A drawing of a Foley catheter. Notice that three separate lumens are incorporated within the round shaft of the catheter for drainage of urine, inflation, and introducing irrigating solutions into the urinary bladder. After the Foley catheter has been introduced into the urinary bladder, the tip is inflated to prevent it from being expelled. (*e*) This specially designed sputum cup allows the patient to expectorate a clinical specimen directly into the cup. In the laboratory, the cup can be opened from the bottom to reduce the chance of contamination from extraneous pathogens.

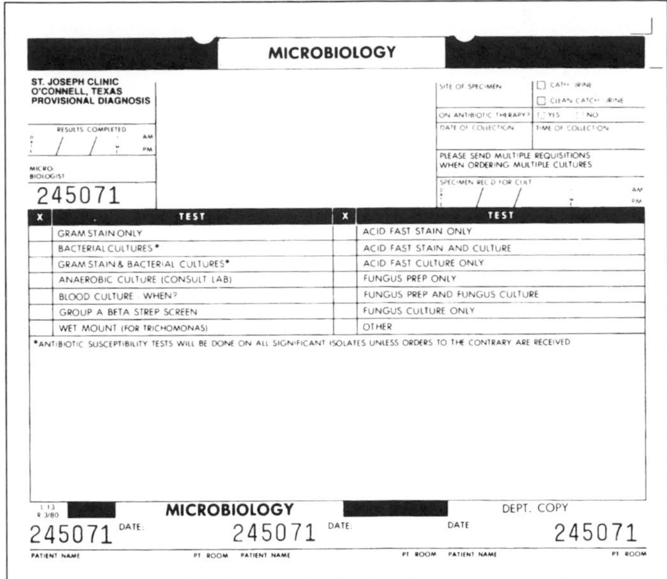

MICROBIOLOGY

ST. JOSEPH CLINIC
O'CONNELL, TEXAS
PROVISIONAL DIAGNOSIS

RESULTS COMPLETED / / AM PM

MICROBIOLOGIST

245071

SITE OF SPECIMEN | ☐ CATH. URINE ☐ CLEAN CATCH URINE
ON ANTIBIOTIC THERAPY? ☐ YES ☐ NO
DATE OF COLLECTION | TIME OF COLLECTION
PLEASE SEND MULTIPLE REQUISITIONS WHEN ORDERING MULTIPLE CULTURES
SPECIMEN REC'D FOR CULT / / AM PM

X	TEST	X	TEST
	GRAM STAIN ONLY		ACID FAST STAIN ONLY
	BACTERIAL CULTURES*		ACID FAST STAIN AND CULTURE
	GRAM STAIN & BACTERIAL CULTURES*		ACID FAST CULTURE ONLY
	ANAEROBIC CULTURE (CONSULT LAB)		FUNGUS PREP ONLY
	BLOOD CULTURE - WHEN?		FUNGUS PREP AND FUNGUS CULTURE
	GROUP A BETA STREP SCREEN		FUNGUS CULTURE ONLY
	WET MOUNT (FOR TRICHOMONAS)		OTHER

*ANTIBIOTIC SUSCEPTIBILITY TESTS WILL BE DONE ON ALL SIGNIFICANT ISOLATES UNLESS ORDERS TO THE CONTRARY ARE RECEIVED

MICROBIOLOGY DEPT. COPY

245071 DATE 245071 DATE DATE 245071

PATIENT NAME PT. ROOM PATIENT NAME PT. ROOM PATIENT NAME PT. ROOM

(a)

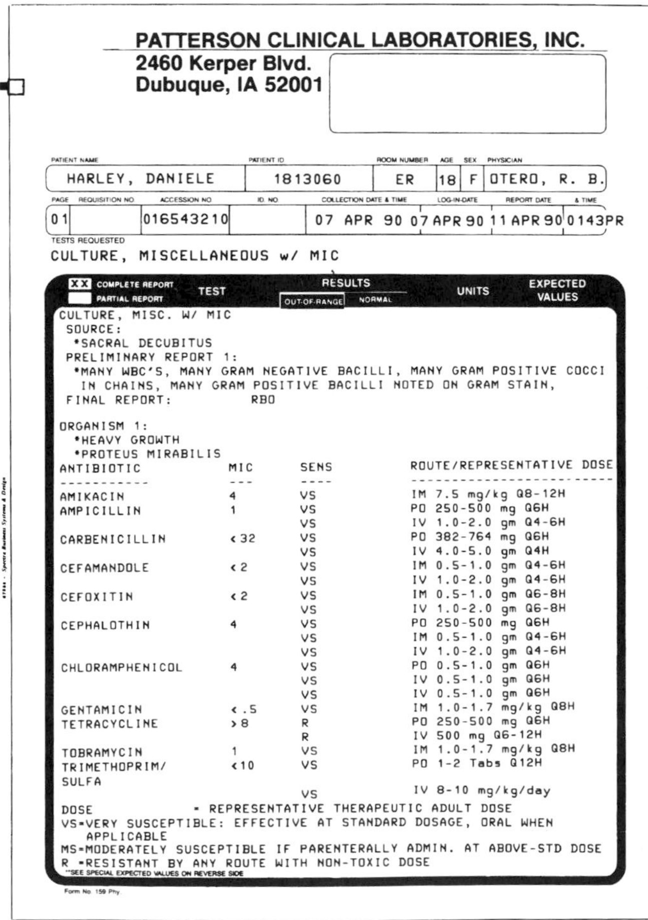

PATTERSON CLINICAL LABORATORIES, INC.
2460 Kerper Blvd.
Dubuque, IA 52001

PATIENT NAME	PATIENT ID	ROOM NUMBER	AGE	SEX	PHYSICIAN
HARLEY, DANIELE	1813060	ER	18	F	OTERO, R. B.

PAGE	REQUISITION NO.	ACCESSION NO.	ID. NO.	COLLECTION DATE & TIME	LOG-IN-DATE	REPORT DATE	& TIME
01		016543210		07 APR 90	07 APR 90	11 APR 90	0143PR

TESTS REQUESTED
CULTURE, MISCELLANEOUS w/ MIC

XX COMPLETE REPORT / PARTIAL REPORT — TEST — RESULTS (OUT-OF-RANGE, NORMAL) — UNITS — EXPECTED VALUES

CULTURE, MISC. W/ MIC
SOURCE:
*SACRAL DECUBITUS
PRELIMINARY REPORT 1:
*MANY WBC'S, MANY GRAM NEGATIVE BACILLI, MANY GRAM POSITIVE COCCI IN CHAINS, MANY GRAM POSITIVE BACILLI NOTED ON GRAM STAIN,
FINAL REPORT: RBO

ORGANISM 1:
*HEAVY GROWTH
*PROTEUS MIRABILIS

ANTIBIOTIC	MIC	SENS	ROUTE/REPRESENTATIVE DOSE
AMIKACIN	4	VS	IM 7.5 mg/kg Q8-12H
AMPICILLIN	1	VS	PO 250-500 mg Q6H
		VS	IV 1.0-2.0 gm Q4-6H
CARBENICILLIN	< 32	VS	PO 382-764 mg Q6H
		VS	IV 4.0-5.0 gm Q4H
CEFAMANDOLE	< 2	VS	IM 0.5-1.0 gm Q4-6H
		VS	IV 1.0-2.0 gm Q4-6H
CEFOXITIN	< 2	VS	IM 0.5-1.0 gm Q6-8H
		VS	IV 1.0-2.0 gm Q6-8H
CEPHALOTHIN	4	VS	PO 250-500 mg Q6H
		VS	IM 0.5-1.0 gm Q4-6H
		VS	IV 1.0-2.0 gm Q4-6H
CHLORAMPHENICOL	4	VS	PO 0.5-1.0 gm Q6H
		VS	IV 0.5-1.0 gm Q6H
		VS	IV 0.5-1.0 gm Q6H
GENTAMICIN	< .5	VS	IM 1.0-1.7 mg/kg Q8H
TETRACYCLINE	> 8	R	PO 250-500 mg Q6H
		R	IV 500 mg Q6-12H
TOBRAMYCIN	1	VS	IM 1.0-1.7 mg/kg Q8H
TRIMETHOPRIM/ SULFA	< 10	VS	PO 1-2 Tabs Q12H
		VS	IV 8-10 mg/kg/day

DOSE = REPRESENTATIVE THERAPEUTIC ADULT DOSE
VS=VERY SUSCEPTIBLE: EFFECTIVE AT STANDARD DOSAGE, ORAL WHEN APPLICABLE
MS=MODERATELY SUSCEPTIBLE IF PARENTERALLY ADMIN. AT ABOVE-STD DOSE
R =RESISTANT BY ANY ROUTE WITH NON-TOXIC DOSE
**SEE SPECIAL EXPECTED VALUES ON REVERSE SIDE

Form No. 159 Phy

(b)

Figure 34.3 Culture Request Forms. (*a*) A typical handwritten culture request form that accompanies a clinical specimen to the laboratory. (*b*) A typical computerized culture request form from a clinical laboratory showing various information such as tests ordered (culture, [white blood cell count], MIC [minimal inhibitory concentration]), source of clinical specimen (sacral decubitus ulcer), preliminary microorganisms, final microorganisms, and antibiotic choice with route and dose.

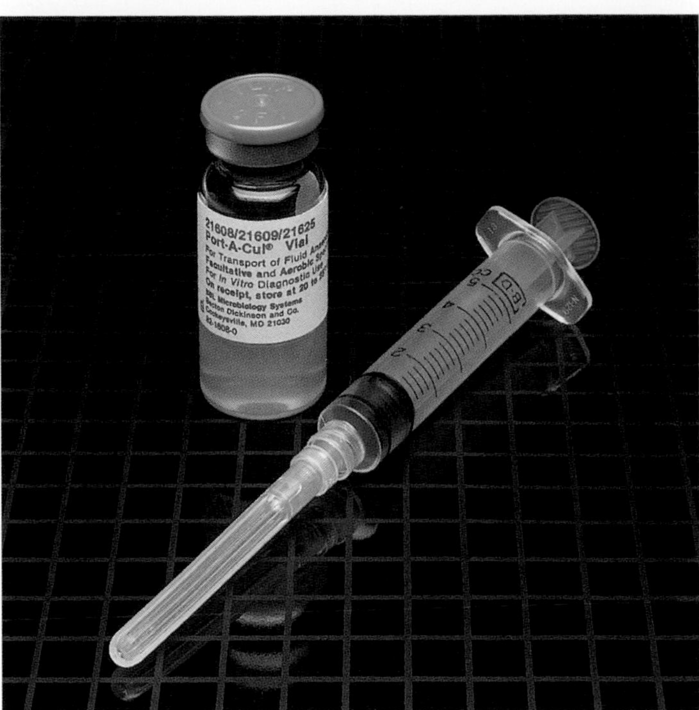

Figure 34.4 Some Anaerobic Transport Systems. A vial and syringe. These systems may contain a non-nutritive transport medium that retards diffusion of oxygen after specimen addition and helps maintain microorganism viability up to 72 hours. A built-in color indicator system is clear and turns lavender in the presence of oxygen.

Special treatment is required for specimens when the microorganism is thought to be anaerobic. The material is aspirated with a needle and syringe. Most of the time it is practical to remove the needle, cap the syringe with its original seal, and bring the specimen directly to the clinical laboratory. Transport of these specimens should take no more than 10 minutes; otherwise, the specimen must be injected immediately into an anaerobic transport vial (figure 34.4). Vials should contain a transport medium with an indicator, such as resazurin, to show that the interior of the vial is anaerobic at the time the specimen is introduced. Swabs for anaerobic culture are usually less satisfactory than aspirates or tissues, even if they are transported in an anaerobic vial.

Many clinical laboratories insist that stool specimens (the fecal discharge from the bowels) for culture be transported in various buffered preservatives. Preparation of these transport media is described in various manuals (see Additional Reading).

Transport of urine specimens to the clinical laboratory must be done as soon as possible. No more than one hour should elapse between the time the specimen is obtained and the time it is examined. If this time schedule cannot be followed, the urine sample must be refrigerated immediately.

---

1. What is the function of the clinical microbiologist? The clinical microbiology laboratory?
2. What general guidelines should be followed in collecting and handling clinical specimens?
3. Define the following terms: specimen, swab, catheter, and sputum.
4. What are some transport problems associated with stool specimens? Anaerobic cultures? Urine specimens?
5. What must be done to a clinical specimen before it goes to the clinical laboratory?

Identification of Microorganisms from Specimens

The clinical microbiology laboratory can provide preliminary or definitive identification of microorganisms based on (1) microscopic examination of specimens, (2) study of the growth and biochemical characteristics of isolated microorganisms, (3) immunologic tests that detect antibodies or microbial antigens, (4) bacteriophage typing, and (5) molecular methods.

Microscopy

Wet-mount, heat-fixed, or chemical-fixed specimens can be examined with an ordinary bright-field microscope. These preparations can be enhanced with either phase-contrast or dark-field microscopy. The latter is the procedure of choice for the detection of spirochetes in skin lesions of people with early syphilis or in cerebrospinal fluid and urine of people with early leptospirosis. The fluorescence microscope can be used to identify certain microorganisms (*Mycobacterium tuberculosis*) after they are stained with fluorochromes (*see chapter 2*). (Some morphological features used in classification and identification of microorganisms are presented in chapter 20 and in table 20.3.)

The light microscope (pp. 21–27).

Many stains that can be used to examine specimens for specific microorganisms have been described. Two of the more widely used are the Gram stain and acid-fast stain. Because these stains are based on the chemical composition of cell walls, they are not useful in identifying bacteria without walls. Refer to standard references, such as the *Manual of Clinical Microbiology* published by the American Society for Microbiology, for details about other reagents and staining procedures.

Growth or Biochemical Characteristics

Typically, microorganisms have been identified by their particular growth or biochemical characteristics. These characteristics vary depending on whether the clinical microbiologist is dealing with viruses, rickettsias, chlamydiae, mycoplasmas, gram-positive or gram-negative bacteria, fungi, or parasites.

Viruses

Viruses are identified by isolation in living cells (table 34.1), by serological tests, or by nucleic acid technology (techniques such as restriction endonuclease analysis, nucleic acid hybridization, and the polymerase chain reaction [*see chapter 14*]). Several types of living cells are available: cell cultures, embryonated hen's eggs, and experimental animals.

Cell cultures are divided into three general classes:

1. Primary cultures consist of cells derived directly from tissues like the kidney and lung.
2. Semicontinuous cell cultures are obtained from subcultures of a primary culture and usually consist of diploid fibroblasts that undergo a finite number of divisions.
3. Continuous cell cultures are derived from transformed cells that are generally epithelial in origin. These cultures grow rapidly, are heteroploid (having a chromosome number that is not a simple multiple of the haploid number), and can be subcultured indefinitely.

Each type of cell culture favors the growth of a different array of viruses, just as bacterial culture media have differing selective and restrictive properties for growth of bacteria (table 34.3).

Viral replication in cell cultures is detected in three ways: (1) by observing cytopathic effects, (2) by hemadsorption, and (3) by using interference.

A **cytopathic effect** is an observable change that occurs in cells because of viral replication (*see chapter 17*). Examples include ballooning, binding together, clustering, or even death of the culture cells (*see figure 17.3*). During the incubation period of a cell culture, red blood cells can be added. Several viruses alter the plasma membrane of infected culture cells so that red blood cells adhere firmly to them. This phenomenon is called **hemadsorption** (*see figure 33.10*). Some viruses produce no cytopathic effects and no hemadsorption. Cells infected with these viruses, although they appear normal, are resistant to infection with certain other viruses, a phenomenon called **interference.**

TABLE 34.1 Some Different Cell Types Used to Isolate and Identify Human Viruses

Virus Family	Virus Type	Suckling Mouse	Embryonated Egg	Primary Monkey Kidney	Primary Human Embryonic Kidney	Semi-continuous Human Diploid	Continuous Human Heteroploid
Picornaviridae	Rhinovirus	0[a]	0	+	±	++	0 to ++
	Polioviruses 1–3	0	0	++	++	++	++
	Coxsackievirus A1–24	++	0	0 to ++[b]	0 to ++[b]	0 to ++[b]	0 to ++[b]
	Coxsackievirus B1–6	++	0	++	++	±	++
	Echovirus 1–32	0 to ++	0	++	+	++	0
Orthomyxoviridae	Influenza A, B, C	0	++	0 to ++	±	±	0
Paramyxoviridae	Parainfluenza 1, 2, 3, 4	0	0	++	+	+	±
	Respiratory syncytial	0	0	+	+	++	++
	Measles	0	0	+	++	0	±
	Mumps	0	++	++	++	+	0
Adenoviridae	Adenovirus types 1–34	0	0	± to +	++	+	+
Herpesviridae	Herpes simplex	0	++	0 to ++	++	++	0 to ++
	Varicella-zoster	0	0	0	+	++	0
	Cytomegalovirus	0	0	0	0	++	0
Togaviridae	Rubella virus	0	0	+	+	0	0
	Eastern equine encephalitis	++	0	±		0	±
	Western equine encephalitis	++	0	+		0	+
	St. Louis encephalitis	++	0	0		0	+
Bunyaviridae	California encephalitis virus	++	0	0		0	+
Reoviridae	Rotavirus	0	0	0	0	0	0
	Colorado tick fever	+	0	0	0	0	0
Poxviridae	Vaccinia	+	++	++	++	++	++
Arenaviridae	Lymphocytic choriomeningitis	++	++	++	++	++	++

From G. L. Mandell, R. G. Douglas, and J. E. Bennett, *Principles and Practice of Infectious Diseases,* 3d ed. Copyright © 1990 Churchill Livingstone Inc. Medical Publishers, New York, N.Y.

[a]0: not suitable for isolation; ±: some strains may be recovered; +: many strains may be recovered; ++: most strains will be recovered.

[b]Coxsackievirus A-7, A-9, and A-16 strains grow well in cell culture.

Embryonated hen's eggs can be used for virus isolation. There are three main routes of egg inoculation for virus isolation: (1) the allantoic cavity, (2) the amniotic cavity, and (3) the chorioallantoic membrane (*see figure 17.1*). Virus replication is recognized by the development of pocks on the chorioallantoic membrane, by the development of hemagglutinins (*see figure 33.10*) in the allantoic and amniotic fluid, and by death of the embryo.

Laboratory animals, especially suckling mice, can be used for virus isolation. Inoculated animals are observed for specific signs of disease or death.

Several new serological tests for viral identification make use of monoclonal antibody-based immunofluorescence (*see figure 33.16*). These tests (figure 34.5) detect viruses such as the cytomegalovirus and herpes simplex virus in tissue-vial cultures.

1. Name two specimens for which microscopy would be used in the initial diagnosis of an infectious disease.
2. Name three general classes of cell cultures.
3. Give the three ways by which the presence of viral replication is detected in cell culture.
4. What are the three main routes of egg inoculation for virus isolation?

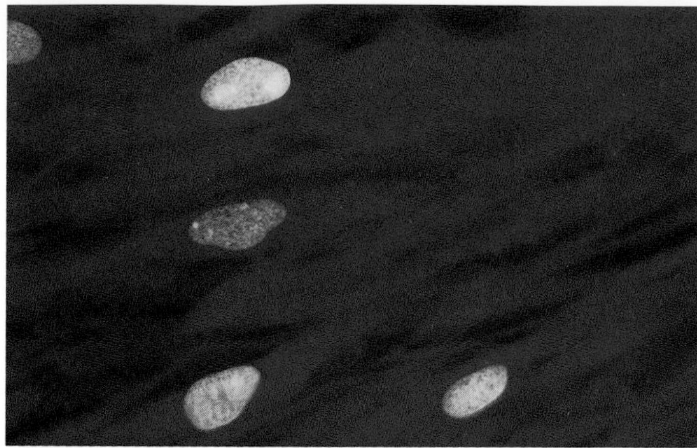

(a)

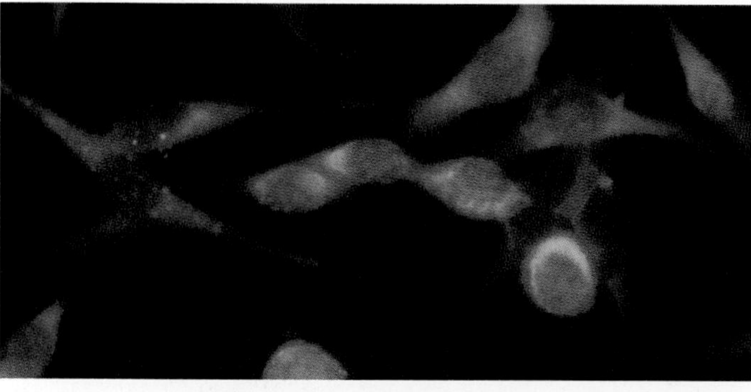

(b)

Figure 34.5 Viral Identification Test Using Immunofluorescence in Tissue Culture. (*a*) Two infected nuclei in a cytomegalovirus (CMV) positive tissue culture. (*b*) Several infected cells in a herpes simplex virus positive tissue culture.

Fungi

The diagnosis of fungi can often be made if a portion of the specimen is mixed with a drop of 10% potassium hydroxide on a glass slide. A coverslip is applied, the slide is flamed gently, and the material is examined microscopically. Culture procedures also can be used, but these require many days for growth and collection of results. Antibody testing on blood plasma is also routinely done, as is the direct examination of specimens by immunofluorescence when specific antisera are available.

Parasites

Wet mounts of stool specimens or urine can be examined microscopically for the presence of eggs, cysts, larvae, or vegetative cells of parasites. Blood smears for malaria and trypanosome parasites are stained with Giemsa stain. Some serological tests are also available. For example, the protozoan *Pneumocystis carinii,* one of the major pathogens involved in AIDS, can now be quickly detected with a specific monoclonal antibody-based immunofluorescence test.

Rickettsias

Although rickettsias, chlamydiae, and mycoplasmas are bacteria, they differ from other bacterial pathogens in a variety of ways. Therefore, the identification of these three groups is discussed separately. Rickettsias can be diagnosed serologically or by isolation of the microorganism. Because isolation is both hazardous to the clinical microbiologist and expensive, serological methods are preferred. Isolation of rickettsias is generally confined to reference and specialized research laboratories.

Chlamydiae

Chlamydiae can be demonstrated in tissues and cell scrapings with Giemsa staining, which detects the characteristic intracellular inclusion bodies (*see figure 21.20*). Immunofluorescent staining of tissues and shed cells with monoclonal antibody reagents is a more sensitive and specific means of diagnosis (Box 34.2; *see also figure 31.19*). The most sensitive method for demonstrating chlamydiae in clinical specimens is growth in cell culture (McCoy cells).

Immunofluorescence (pp. 663–65).

Mycoplasmas

The most routinely used techniques for diagnosis of the mycoplasmas are serological (hemagglutinin) or complement-fixing antigen-antibody reactions. These microorganisms are slow growing; therefore, positive results from isolation procedures are rarely available before 30 days—a long delay with an approach that offers little advantage over standard serological techniques. Recently, nucleic acid hybridization has been applied to the detection of *Mycoplasma pneumoniae* in clinical specimens.

Agglutination (p. 656).

Box 34.2

Monoclonal Antibodies in Clinical Microbiology

An important application of monoclonal antibody technology (*see hybridomas, chapter 31*) is the identification of microorganisms. Monoclonal antibodies have been prepared for a wide range of viruses, bacteria, fungi, and parasites in many research laboratories. If specific monoclonal antibodies are selected, immunologic assays can be created for different types of analyses. For example, monoclonal antibodies of cross-species or cross-genus reactivity have applications in the taxonomy of microorganisms. Those monoclonal antibodies that define species-specific antigens are extremely valuable in diagnostic reagents. Monoclonal antibodies that exhibit more restrictive specificity can be used to identify strains or biotypes within a species, to aid in studies of antigenic drift, and in epidemiological studies involving the matching of microbial strains. In addition, individual antigenic determinants on protein molecules can be mapped.

In the clinical microbiology laboratory, monoclonal antibodies to viral or bacterial antigens are replacing polyclonal antibodies for use in culture confirmation when very accurate, rapid identification is required. With the use of sensitive techniques such as fluorescent antibody assays (*see chapter 33*), it is possible to perform culture identifications with improved accuracy, speed, and fewer microorganisms. The formulation of direct assays with monoclonal antibody reagents, which contain no contaminating antibodies and produce a minimum of artifacts, is now reality. The highly defined and reproducible properties of monoclonal antibodies invite their incorporation into immunoassays being developed for the next generation of instruments that will detect microbial antigens and serum antibodies for the clinical microbiologist.

Bacteria

The presence of bacterial growth can usually be recognized by the development of colonies on solid media or turbidity in liquid media. The time for visible growth to occur is an important variable in the clinical laboratory. For example, most pathogenic bacteria require only a few hours to produce visible growth, whereas it may take weeks for colonies of mycobacteria to become evident. The clinical microbiologist as well as the clinician should be aware of reasonable reporting times for various cultures (table 34.2).

The initial identity of a bacterial organism may be suggested by (1) the source of the culture specimen; (2) its microscopic appearance; (3) its pattern of growth on selective, differential, enrichment, or characteristic media (table 34.3; *see also figure 5.10*); and (4) its hemolytic, metabolic, and fermentative properties on the various media (table 34.3; *see also table 20.4*). After the microscopic and growth characteristics of a bacterium are examined, specific biochemical tests can be performed. These tests represent the most common method for identifying bacteria (table 34.4 and figure 34.6).

Microbial nutrition and types of media (chapter 5).

Table 34.2 Some Average Reporting Times for Microscopic and Culture Procedures Performed in a Clinical Laboratory

Procedure	Time
Microscopic	
Acid-fast stain	1 hour
Gram's stain	0.5 hour
India ink for *Cryptococcus*	0.5 hour
Stains for parasites	0.5 hour
Toluidine blue for *Pneumocystis*	1 hour
Culture	
Actinomyces	10 days
Anaerobic bacteria	2–14 days
Brucella	21 days
Other commonly encountered bacteria	2–7 days
Leptospires	30 days
Mycobacteria	8 weeks
Chlamydiae	2–3 days
Mycoplasmas	30 days
Fungi	4–6 weeks
Viruses	2–14 days

TABLE 34.3 Isolation of Pure Bacterial Cultures from Specimens

Selective Media

A selective medium is prepared by the addition of specific substances to a culture medium that will permit growth of one group of bacteria while inhibiting growth of some other groups. The following are examples:

Salmonella-Shigella agar (*SS*) is used to isolate *Salmonella* and *Shigella* species. Its bile salt mixture inhibits many groups of coliforms. Both *Salmonella* and *Shigella* species produce colorless colonies because they are unable to ferment lactose. Lactose-fermenting bacteria will produce pink colonies.

Mannitol salt agar (*MS*) is used for the isolation of staphylococci. The selectivity is obtained by the high (7.5%) salt concentration that inhibits growth of many groups of bacteria. The mannitol in this medium helps in differentiating the pathogenic from the nonpathogenic staphylococci, in that the former ferment mannitol to form acid while the latter do not.

Bismuth sulfite agar (*BS*) is used for the isolation of *Salmonella typhi*, especially from stool and food specimens. *S. typhi* reduces the sulfite to sulfide, resulting in black colonies with a metallic sheen.

Differential Media

The incorporation of certain chemicals into a medium may result in diagnostically useful growth or visible change in the medium after incubation. The following are examples:

Eosin methylene blue agar (*EMB*) differentiates between lactose fermenters and nonlactose fermenters. EMB contains lactose, salts, and two dyes—eosin and methylene blue. *E. coli*, which is a lactose fermenter, will produce a dark colony or one that has a metallic sheen. *S. typhi*, a nonlactose fermenter, will appear colorless.

MacConkey agar is used for the selection and recovery of *Enterobacteriaceae* and related gram-negative rods. The bile salts and crystal violet this medium contains inhibit the growth of gram-positive bacteria and some fastidious gram-negative bacteria. Because lactose is the sole carbohydrate, lactose-fermenting bacteria produce colonies that are various shades of red, whereas nonlactose fermenters produce colorless colonies.

Hektoen enteric agar is used to increase the yield of *Salmonella* and *Shigella* species relative to other microbiota. The high bile salt concentration inhibits the growth of gram-positive bacteria and retards the growth of many coliform strains.

Enrichment Media

The addition of blood, serum, or extracts to tryptic soy agar or broth will support the growth of many fastidious bacteria. These media are used primarily to isolate bacteria from cerebrospinal fluid, pleural fluid, sputum, and wound abscesses. The following are examples:

Blood agar (can also be a differential medium): addition of citrated blood to tryptic soy agar makes possible variable hemolysis, which permits differentiation of some species of bacteria. Three hemolytic patterns can be observed on blood agar.

1. α-hemolysis—greenish to brownish halo around the colony (e.g., *Streptococcus gordonii, Streptococcus pneumoniae*).
2. β-hemolysis—complete lysis of blood cells resulting in a clearing effect around growth of the colony (e.g., *Staphylococcus aureus* and *Streptococcus pyogenes*).
3. γ-hemolysis—no change in medium (e.g., *Staphylococcus epidermidis* and *Staphylococcus saprophyticus*).

Chocolate agar is made from heated blood, which provides necessary growth factors to support bacteria such as *Haemophilus influenzae* and *Neisseria gonorrhoeae.*

Characteristic Media

Characteristic media are used to test bacteria for particular metabolic activities, products, or requirements. The following are examples:

Urea broth is used to detect the enzyme urease. Some enteric bacteria are able to break down urea, in the presence of urease, into ammonia and CO_2.

Triple sugar iron (*TSI*) *agar* contains lactose, sucrose, and glucose plus ferrous ammonium sulfate and sodium thiosulfate. TSI is used for the identification of enteric organisms by their ability to attack glucose, lactose, or sucrose and to liberate sulfides from ammonium sulfate or sodium thiosulfate.

Citrate agar contains sodium citrate, which serves as the sole source of carbon, and ammonium phosphate, the sole source of nitrogen. Citrate agar is used to differentiate enteric bacteria on the basis of citrate utilization.

Lysine iron agar (*LIA*) is used to differentiate bacteria that can either deaminate or decarboxylate the amino acid lysine. LIA contains lysine, which permits enzyme detection, and ferric ammonium citrate for the detection of H_2S production.

Sulfide, indole, motility (*SIM*) medium can perform three different tests. One can observe the production of sulfides, formation of indole (a metabolic product from tryptophan utilization), and motility. This medium is generally used for the differentiation of enteric organisms.

Table 34.4 Some Biochemical Tests Used to Identify an Unknown Bacterium Taken from the Patient's Specimen

Biochemical Test	Description
Casein hydrolysis	Detects the presence of caseinase, an enzyme able to hydrolyze milk protein casein.
Catalase	Detects the presence of catalase, which converts hydrogen peroxide to water and O_2.
Coagulase	Detects the presence of coagulase. Coagulase causes plasma to clot. This is an important test for *Staphylococcus aureus*.
Lipid hydrolysis	Detects the presence of lipase. Lipase breaks down lipids into simple fatty acids and glycerol.
Gelatin liquefaction	Detects whether or not a bacterium can produce an enzyme that hydrolyzes the gelatin protein.
Hydrogen sulfide production	Detects the formation of hydrogen sulfide from the amino acid cysteine by the presence of cysteine desulfurase; important in *Salmonella* identification.
IMViC (indole; methyl red; Voges-Proskauer; citrate)	The indole test detects the production of indole from the amino acid tryptophan. Methyl red is a pH indicator to determine whether the bacterium has produced acid. Vi (Voges-Proskauer) detects the production of acetoin. The citrate test determines whether or not the bacterium can utilize sodium citrate as a sole source of carbon.
Litmus milk	Determines five bacterial characteristics: (1) a pink color indicates an acid reaction and the fermentation of lactose; (2) a blue color arises from an alkaline reaction and no proteolysis (hydrolysis of protein); (3) a white color indicates litmus has acted as an electron acceptor; (4) curd formation indicates acid produced from lactose or rennin production; and (5) peptonization shows curd (coagulated protein) digestion by casein hydrolysis.
Nitrate reduction	Detects whether a bacterium can use nitrate as an electron acceptor.
Oxidase	Detects the presence of cytochrome *c* oxidase that is able to reduce O_2; especially important in detecting *Neisseria* spp. and pseudomonads.
Starch hydrolysis	Detects the presence of the enzyme amylase, which is able to hydrolyze starch.

Classic dichotomous keys are coupled with the biochemical tests for the identification of bacteria from specimens. Generally, less than 20 tests are required to identify clinical bacterial isolates to the species level (figure 34.7).

1. How can fungi and parasites be detected in a clinical specimen? Rickettsias? Chlamydiae? Mycoplasmas?
2. Why must the clinical microbiologist know what are reasonable reporting times for various microbial specimens?
3. How can a clinical microbiologist determine the initial identity of a bacterium?
4. Describe a dichotomous key that would be used to identify a bacterium.

Rapid Methods of Identification

Clinical microbiology has benefited greatly from technological advances in equipment, computer programs and data bases, molecular biology, and immunochemistry. With respect to the detection of microorganisms in specimens, there has been a shift from the multistep methods previously discussed to unitary procedures and systems that incorporate standardization, speed, reproducibility, miniaturization, mechanization, and automation. These rapid identification methods can be divided into three categories: (1) manual biochemical systems, (2) mechanized/automated systems, and (3) immunologic systems.

(a) **Methyl Red Test** This is a qualitative test of the acidity produced by bacteria grown in MRVP broth. After the addition of several drops of methyl red solution, a bright red color (left tube) is a positive test; a yellow or orange color (right tube) is a negative test.

(b) **Voges-Proskauer Reaction** VP-positive bacteria produce acetylmethylcarbinol or acetoin, which reacts with the reagents to produce a red color (left tube); a VP-negative tube is shown on the right.

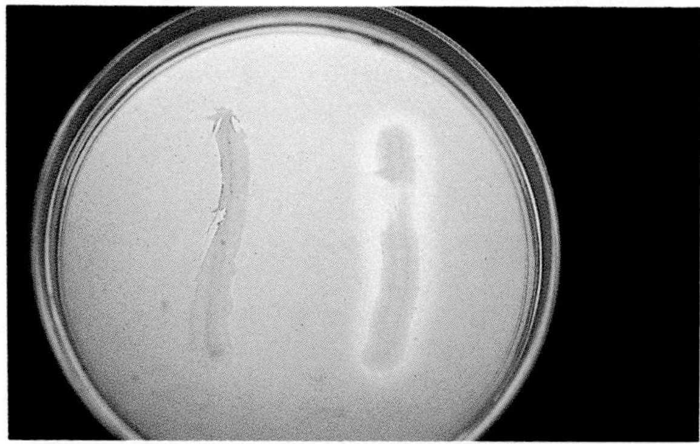

(c) **Starch Hydrolysis** After incubation on starch agar, plates are flooded with iodine solution. A positive test is indicated by the colorless area around the growth (right); a negative test is shown on the left.

(d) **Tube Catalase Test** After incubation of slant cultures, 1 ml of 3% hydrogen peroxide is trickled down the slants. Catalase converts hydrogen peroxide to water and oxygen bubbles (left tube); a negative catalase test is shown in the right tube.

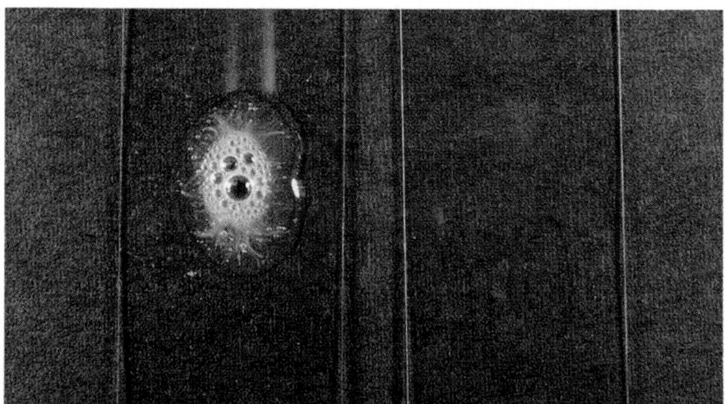

(e) **Slide Catalase Test** A wooden applicator stick (or nichrome wire loop) is used to pick up a colony from a culture plate and place it in a drop of hydrogen peroxide on a glass slide. A positive catalase reaction (left slide) shows gas bubbles; a negative catalase reaction reveals an absence of gas bubbles (right slide).

(f) **Nitrate Reduction** After 24–48 hours of incubation, nitrate reagents are added to culture tubes. The tube on the left illustrates gas formation (a positive reaction for nitrate reduction); the tube in the middle is a positive reaction for nitrate reduction to nitrite as indicated by the red color; the tube on the right is a negative broth control.

Figure 34.6 **Some Common Diagnostic Tests Used in Microbiology.**

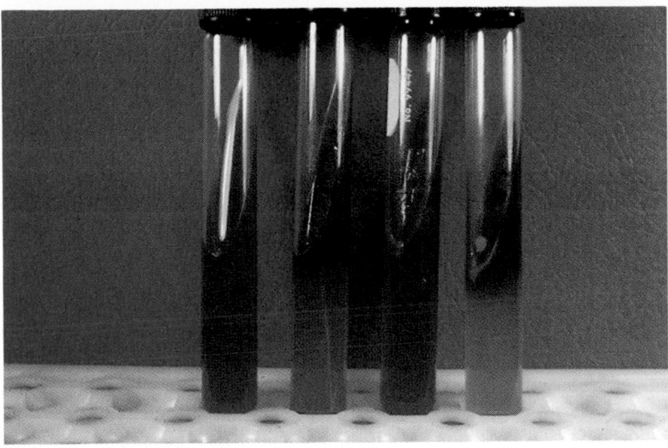

(g) **Test for Amino Acid Decarboxylase** The tube on the left is an uninoculated control; the second tube from the left is lysine decarboxylase negative; the third tube is lysine decarboxylase positive; and the tube on the right is lysine deaminase positive.

(h) **Test for Indole** Tryptophan can be broken down to indole by some bacteria. The presence of indole is detected by adding Kovacs' reagent. A red color on the surface is a positive test for indole (left tube) and an orange-yellow color is a negative test for indole (right tube).

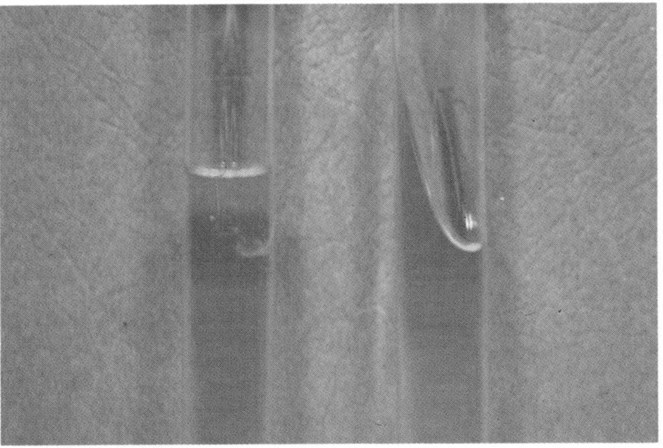

(i) **Gelatin Hydrolysis** If gelatin is hydrolyzed by the enzyme gelatinase, it does not gel when cooled but remains a liquid. Thus, it flows when the culture is tilted backward (right tube). A negative control is on the left. Note that the solid gelatin does not flow when the tube is tilted.

(j) **Phenylalanine Deamination Test** When 10% ferric chloride is added to a phenylalanine deaminase agar slant culture, a dark green color (tube on the right) is a positive test for the enzyme. The tube on the left is an uninoculated control, and the tube in the middle is a negative test.

Figure 34.6 *continued*

(k) **Bile Solubility for Pneumococcus** The tube on the left and the one in the middle contain cultures of pneumococcus (*Streptococcus pneumoniae*). The tube on the right contains α-streptococcus. One-tenth ml of deoxycholate was added to the middle and right tubes, and 0.1 ml of distilled water was added to the left tube. The pneumococci in the center are bile soluble, as indicated by the clear suspension. The bacteria in the right tube are not bile soluble, as indicated by the turbidity.

(l) **Stormy Fermentation of Litmus Milk** The tube on the left shows fermentation; the tube on the right is negative for stormy fermentation.

(m) **Fermentation Reactions** The tube on the left shows acid (yellow color) and gas in the Durham tube. The center tube shows no carbohydrate utilization to produce acid or gas. The tube on the right has less acid formation than that on the left.

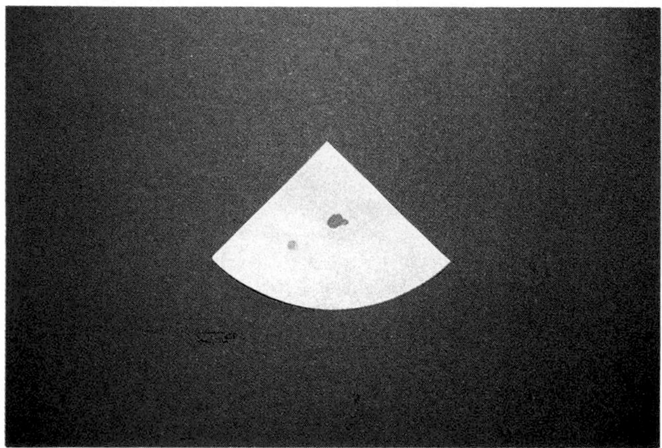

(n) **Oxidase Test** Filter paper is moistened with a few drops of 1% tetramethyl-p-phenylenediamine dihydrochloride. With a wooden applicator, growth from an agar medium is smeared on the paper. A positive test is the development of a purple color within 10 seconds.

(o) **Bacitracin Sensitivity Test** The bacteria on the left are presumptively identified as group A streptococci because of inhibition by the antibiotic bacitracin. The bacteria on the right are bacitracin resistant.

Figure 34.6 *continued*

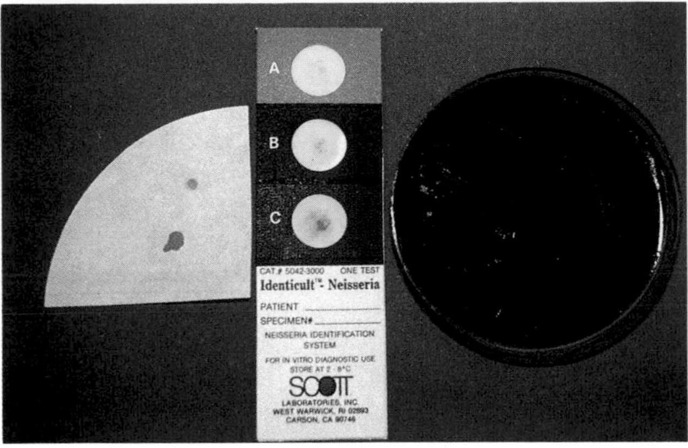

(p) **Identification of *Neisseria gonorrhoeae*** Oxidase-positive test on filter paper is shown on the left. Identicult reaction for *N. gonorrhoeae* (center). The plate on the right shows characteristic growth on modified Thayer-Martin medium.

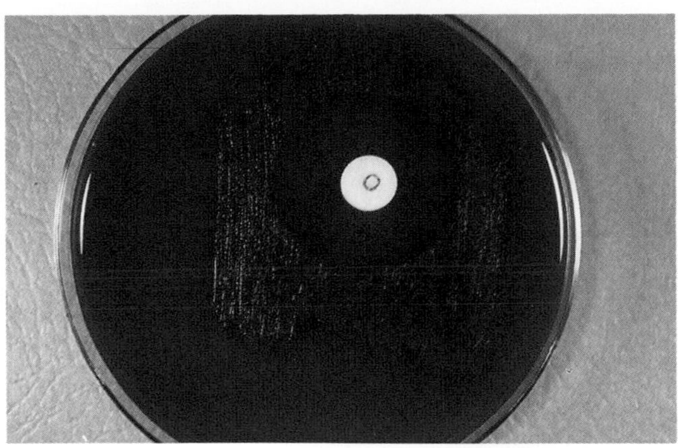

(q) **Optochin Sensitivity** An optochin disk on blood agar specifically inhibits pneumococci. The optochin test for *Streptococcus pneumoniae* is based on the zone of inhibition.

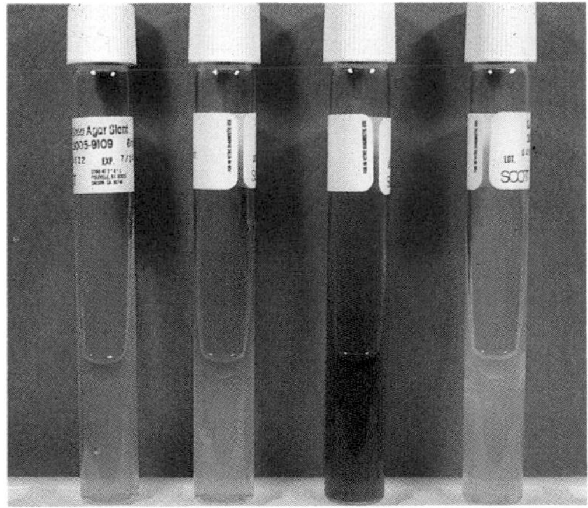

(r) **Urease Production** Christensen urea agar slants. Urease-positive bacteria hydrolyze urea to ammonia, which turns the phenol red indicator red-violet. From left to right: uninoculated control, delayed positive (> 24 hrs), rapidly positive (< 4 hrs), negative reaction.

(s) **Triple Sugar Iron Agar Reactions** Left to right: left tube is an uninoculated control, second tube is K/K (nonfermenter), third tube is A/A with gas indicating lactose or sucrose fermentation, and the right tube is K/A plus H_2S production. A stands for acid production, and K indicates that the medium becomes alkaline.

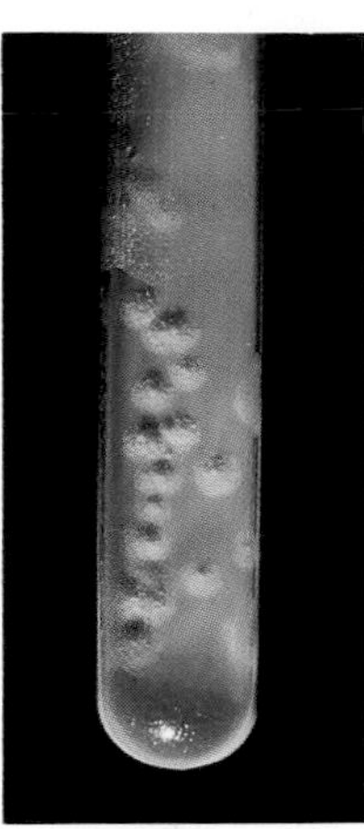

(t) **Lowenstein-Jensen Medium** Growth of *Mycobacterium tuberculosis* showing nodular and nonpigmented growth.

Figure 34.6

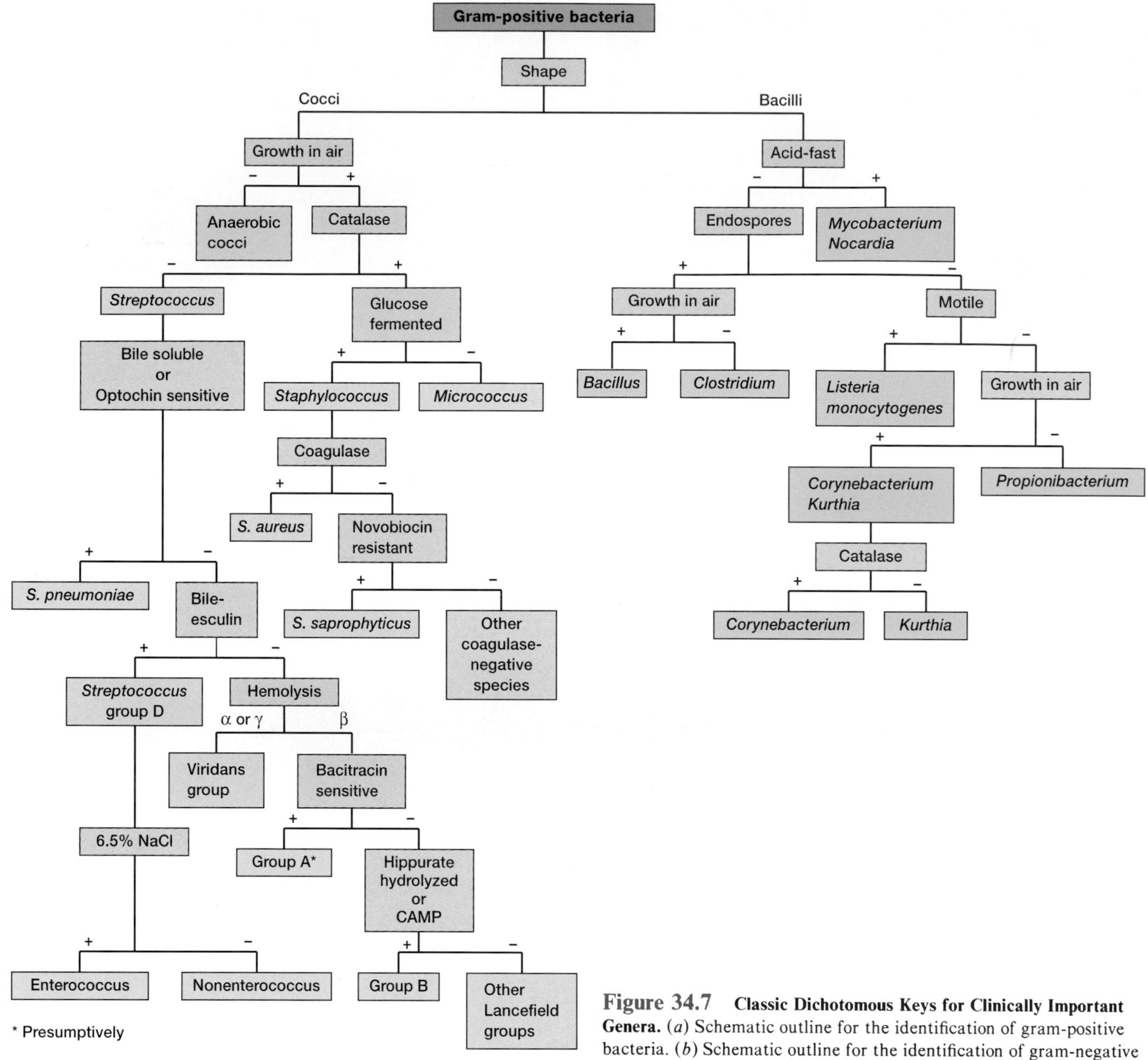

Figure 34.7 Classic Dichotomous Keys for Clinically Important Genera. (*a*) Schematic outline for the identification of gram-positive bacteria. (*b*) Schematic outline for the identification of gram-negative bacteria.

One of the most popular manual biochemical systems for the rapid identification of members of the family *Enterobacteriaceae* and other gram-negative bacteria is the API 20E system. It consists of a plastic strip (figure 34.8) with 20 microtubes containing dehydrated substrates that can detect certain biochemical characteristics. The test substrates in the 20 microtubes are inoculated with bacteria in sterile physiological saline. After 5 hours or overnight incubation, the 20 test results are converted to a seven- or nine-digit profile (figure 34.9). This profile number can be used with a computer or a book called the *API Profile Index* to find the name of the bacterium.

A popular mechanized/automated system is the Autobac IDX (figure 34.10). This system enables the clinical microbiologist to identify *Enterobacteriaceae* and glucose-nonfermenting gram-negative bacilli, screen urine specimens for the presence of bacteria, and determine antimicrobial susceptibilities. The approach of the Autobac IDX is unique, basing identification of gram-negative bacteria on differential growth inhibition by antimicrobial compounds. The effects of these inhibitors, along with six manually obtained parameters, are used to identify microorganisms. Results are printed automatically. Another system, the Biolog automated bacterial identification system identifies bacteria based on their ability to catabolize 95 different carbon sources. The system can identify almost 800 gram-negative and gram-positive species.

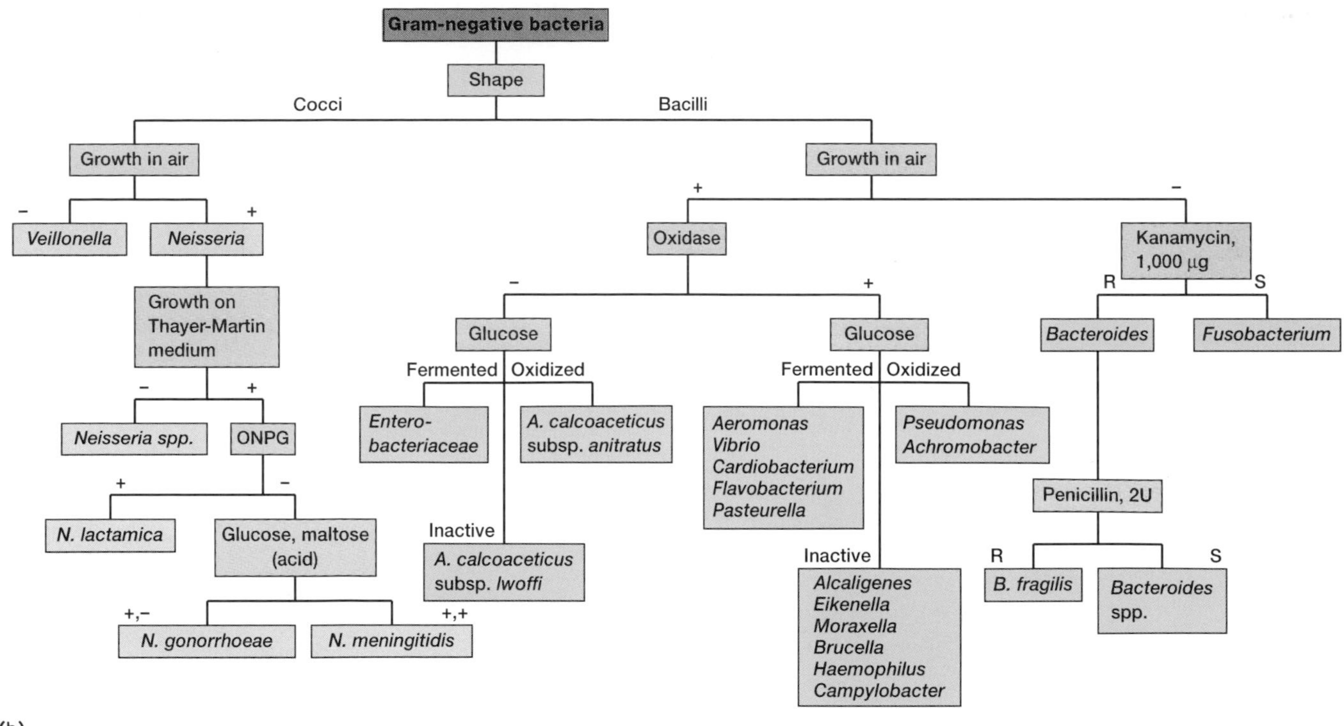

Figure 34.7 *continued*

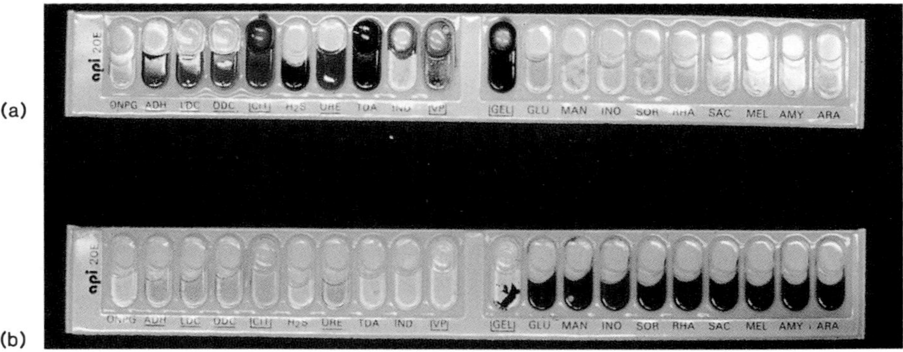

Figure 34.8 **Rapid Tests.** The API 20E manual biochemical system for microbial identification. (*a*) Positive and (*b*) negative results.

Immunologic Techniques

The culturing of certain viruses, bacteria, fungi, and parasites from clinical specimens may not be possible because the methodology remains undeveloped (*Treponema pallidum*; hepatitis A, B, C; and Epstein-Barr virus), is unsafe (rickettsias), or is impractical for all but a few clinical microbiology laboratories (mycobacteria). Cultures also may be negative because of prior antimicrobial therapy or because of the chronic state of the infectious disease in the patient. Under these circumstances, detection of antibodies or antigens may be of considerable diagnostic use.

Immunologic systems for the detection and identification of pathogens from clinical specimens are easy to use, give relatively rapid reaction endpoints, and are sensitive and specific. Some of the more popular immunologic rapid test kits for viruses and bacteria are presented in table 34.5.

Each individual's immunologic response to a microorganism is quite variable. As a result, the interpretation of serological tests is sometimes difficult. For example, a single elevated antibody titer usually does not distinguish between active and past infections. Furthermore, the lack of a measurable antibody titer may reflect either a microorganism's lack

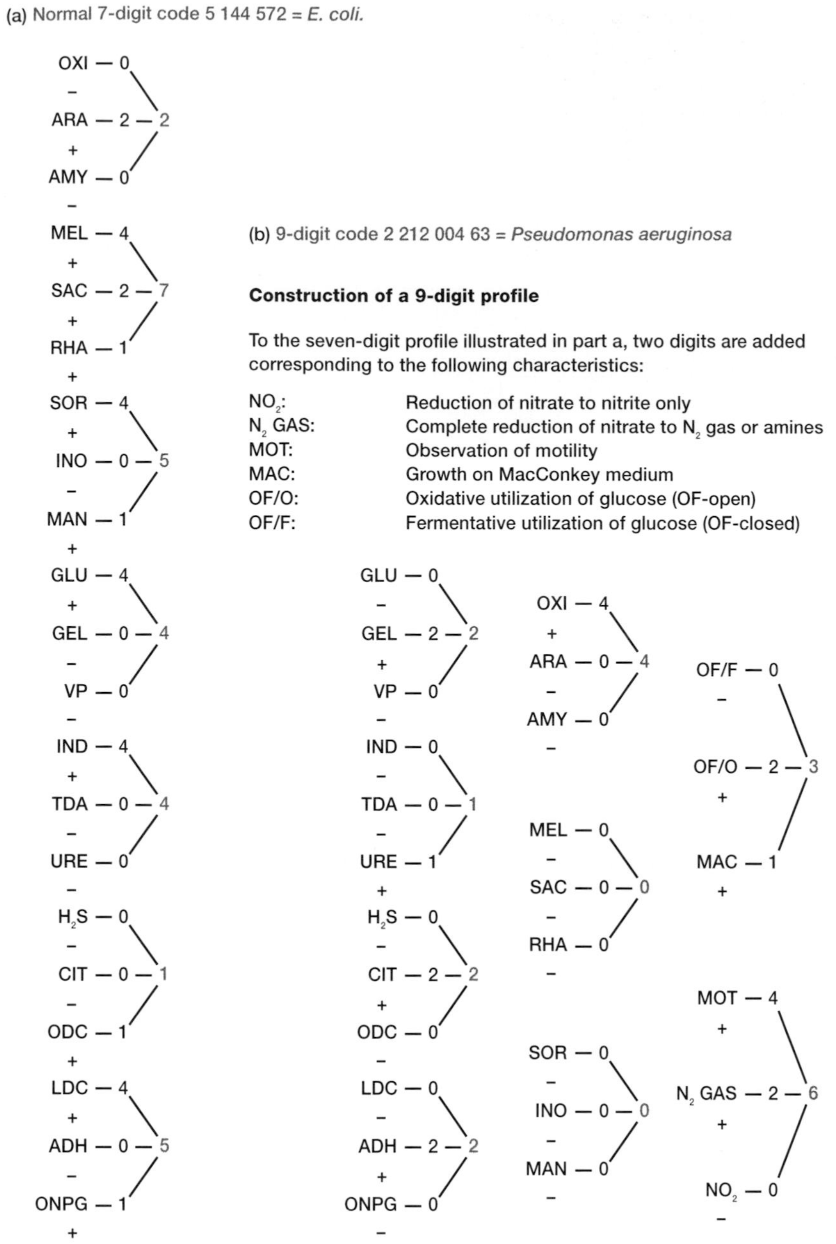

Figure 34.9 The API 20E Profile Number. The conversion of API 20E test results to the codes used in identification of unknown bacteria. The test results read top to bottom (and right to left in part b) correspond to the 7– and 9–digit codes when read in the right-to-left order. The tests required for obtaining a 7–digit code take an 18–24 hour incubation and will identify most members of the *Enterobacteriaceae.* The longer procedure that yields a 9–digit code is required to identify many gram-negative nonfermenting bacteria. The following tests are common to both procedures: ONPG (β-galactosidase); ADH (arginine dihydrolase); LDC (lysine decarboxylase); ODC (ornithine decarboxylase); CIT (citrate utilization); H_2S (hydrogen sulfide production); URE (urease); TDA (tryptophane deaminase); IND (indole production); VP (Voges-Proskauer test for acetoin); GEL (gelatin liquefaction); the fermentation of glucose (GLU), mannitol (MAN), inositol (INO), sorbitol (SOR), rhamnose (RHA), sucrose (SAC), melibiose (MEL), amygdalin (AMY), and arabinose (ARA); and OXI (oxidase test).

Figure 34.10 **An Autobac IDX System for the Rapid Screening of Clinical Specimens for Bacteria.** The system performs identifications of enterics and nonfermenters, simultaneous qualitative and quantitative susceptibilities, direct urine screens, and direct blood culture susceptibilities.

TABLE 34.5 Common Rapid Immunologic Test Kits for the Detection of Bacteria and Viruses in Clinical Specimens

Bactigen (Wampole Laboratories, Cranburg, N.J.)
The Bactigen kit is used for the detection of *Streptococcus pneumoniae, Haemophilus influenzae* type b, and *Neisseria meningitidis* groups A, B, C, and Y from cerebrospinal fluid, serum, and urine.

Culturette Group A Strep ID Kit (Marion Scientific, Kansas City, Mo.)
The Culturette kit is used for the detection of group A streptococci from throat swabs.

Directigen (Hynson, Wescott, and Dunning, Baltimore, Md.)
The Directigen Meningitis Test kit is used to detect *H. influenzae* type b, *S. pneumoniae,* and *N. meningitidis* groups A and C.
The Directigen Group A Strep Test kit is used for the direct detection of group A streptococci from throat swabs.

Fluortec-F and -M (General Diagnostics, Morris Plains, N.J.)
These two kits are used for the rapid detection of the *Bacteroides fragilis* and *Bacteroides melaninogenicus* groups in clinical specimens.

Gono Gen (Micro-Media Systems, San Jose, Calif.)
The Gono Gen kit detects *Neisseria gonorrhoeae.*

QuickVue *H. pylori* Test (Quidel, San Diego, CA)
A seven minute test for detection of IgG antibodies against *Helicobacter pylori* in human serum or plasma.

Staphaurex (Wellcome Diagnostics, Research Triangle Park, N.C.)
Staphaurex screens and confirms *Staphylococcus aureus* in 30 seconds.

Directigen RSV (Becton Dickinson Microbiology Systems, Cockeysville, Md.)
By using a nasopharyngeal swab, the respiratory syncytial virus can be detected in 15 minutes.

Surecell Herpes (HSV) Test (Kodak, Rochester, N.Y.)
Detects the herpes (HSV) 1 and 2 viruses in minutes.

SUDS HIV-1 Test (Murex Corporation, Norcross, GA)
Detects antibodies to HIV-1 antigens in about 10 minutes.

of immunogenicity or an insufficient time for an antibody response to develop following the onset of the infectious disease. For these reasons, test selection and timing of specimen collection are essential to the proper interpretation of immunologic tests.

Antibody titer (pp. 621 and 656).

The most widely used immunologic techniques available to detect microorganisms in clinical specimens are covered in detail in chapter 32. Examples include direct immunofluorescence, complement fixation, neutralization, agglutination, immunoelectrophoresis, and immunoassay; monoclonal antibodies are especially useful in some of these techniques (Box 34.2). No single technique is universally applicable for measuring an individual's immunologic response to all microorganisms. Techniques are therefore chosen based on their selectivity, specificity, ease, speed of performance, and cost-effectiveness.

One of the newest immunologic techniques being used in the clinical microbiology laboratory is **immunoblotting.** Immunoblotting involves polyacrylamide gel electrophoresis of a protein specimen followed by transfer of the separated proteins to nitrocellulose sheets (*see figure 14.5*). Protein bands are then visualized by treating the nitrocellulose sheets with solutions of dye-tagged antibodies. This procedure demonstrates the presence of common and specific proteins among different strains of microorganisms (figure 34.11). Immunoblotting also can be used to show strain-specific immune responses to microorganisms, to serve as an important diagnostic indicator of a recent infection with a particular strain of microorganism, and to allow for prognostic implications with severe infectious diseases.

Another new immunologic technique uses liposomes. A **liposome** (Greek *lipos,* fat, and *soma,* body) is an artificially created microscopic spherical vesicle formed by a lipid bilayer enclosing an aqueous compartment. The aqueous compartment contains a colored dye. The liposome is then sensitized by coupling a specific antibody (or antigen) to the liposome surface. Antibodies specific for the desired pathogen are attached to a membrane in a particular shape, such as a triangle. After a patient's sample is added to the test well (figure 34.12*a*), the analyte (the substance undergoing analysis) instantly binds to the capture antibodies (or antigen). When the dye-filled sensitized liposomes are added (figure 34.12*b*) and link to the immobilized analyte (34.12*c*), a triangle (or other shape) indicates a positive reaction (figure 34.12*d*). If the sample is negative and contains no analyte, no binding occurs in the test area when the liposomes are added and either a

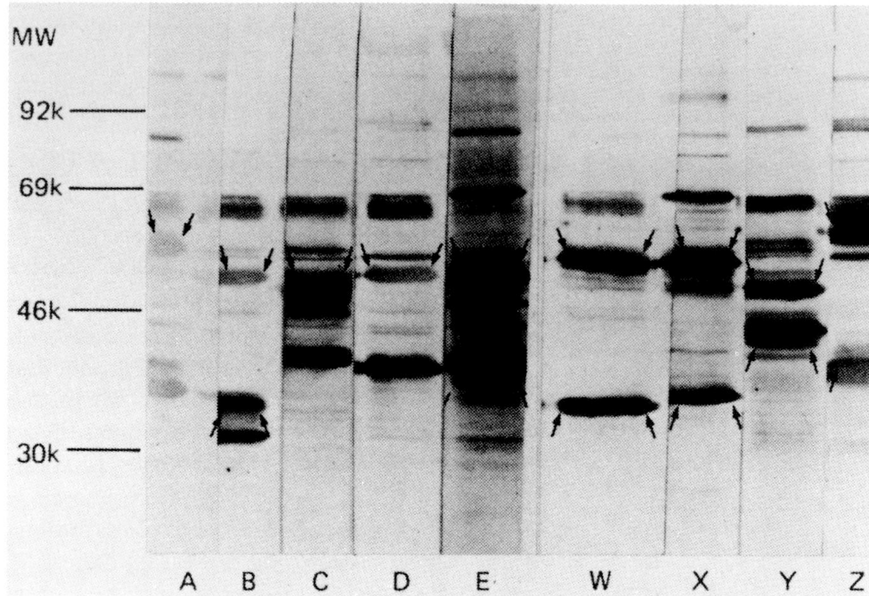

Figure 34.11 **Immunoblotting.** Immunoblot of the standard strains of *Clostridium difficile.* Arrows indicate strain-specific bands.

blank or some other shape appears. Liposome tests are currently available for group A streptococci and the respiratory syncytial virus.

---

1. Describe in general how biochemical tests are used in the API system to identify bacteria.
2. Name the two basic immunologic procedures used in rapid test kits to identify microorganisms.
3. Why might cultures for some microorganisms be unavailable?
4. Why are test selection and timing of specimen collection essential to the proper interpretation of immunologic tests?
5. What are some uses of immunoblotting in the clinical microbiology laboratory? The use of liposomes?

Bacteriophage Typing

Bacteriophages (phages) are viruses that attack members of a particular bacterial species, or strains within a species (*see chapter 18*). **Bacteriophage (phage) typing** is based on the specificity of phage surface receptors for cell surface receptors. Only those bacteriophages that can attach to these surface receptors can infect bacteria and cause lysis. On a petri dish culture, lytic bacteriophages cause plaques on lawns of sensitive bacteria. These plaques represent infection by the virus (*see figures 17.2 and 17.4*).

In bacteriophage typing, the clinical microbiologist inoculates the bacterium to be tested onto a petri plate. The plate is heavily and uniformly inoculated with a cotton swab so that the bacteria will grow to form a solid sheet or lawn of cells. No uninoculated areas should be left. The plate is then marked off into squares (15 to 20 mm per side), and each square is inoculated with a drop of suspension from the different phages available for typing. After the plate is incubated for 24 hours, it is observed for plaques. The phage type is reported as a specific genus and species followed by the types that can infect the bacterium. For example, the series 10/16/24 indicates that this bacterium is sensitive to phages 10, 16, and 24, and belongs to a collection of strains, called a **phagovar,** that have this particular phage sensitivity.

Molecular Methods

With the application of new molecular technology, it is now possible to analyze the molecular characteristics of microorganisms in the clinical laboratory. Some of the most accurate approaches to microbial identification are through the analysis of proteins and nucleic acids. Examples previously discussed (*see chapter 20*) include comparison of proteins; physical, kinetic, and regulatory properties of microbial enzymes; nucleic acid base composition (*see table 20.5*); nucleic acid hybridization; and nucleic acid sequencing. Three other molecular methods being widely used are DNA probes, gas-liquid chromatography, and plasmid fingerprinting.

DNA Probes

A recent development in clinical microbiology is the application of DNA (RNA also can be used) as a diagnostic reagent for the detection and identification of microorganisms. DNA probe technology (*see chapter 14*) identifies a microorganism by probing its genetic composition. The use of cloned

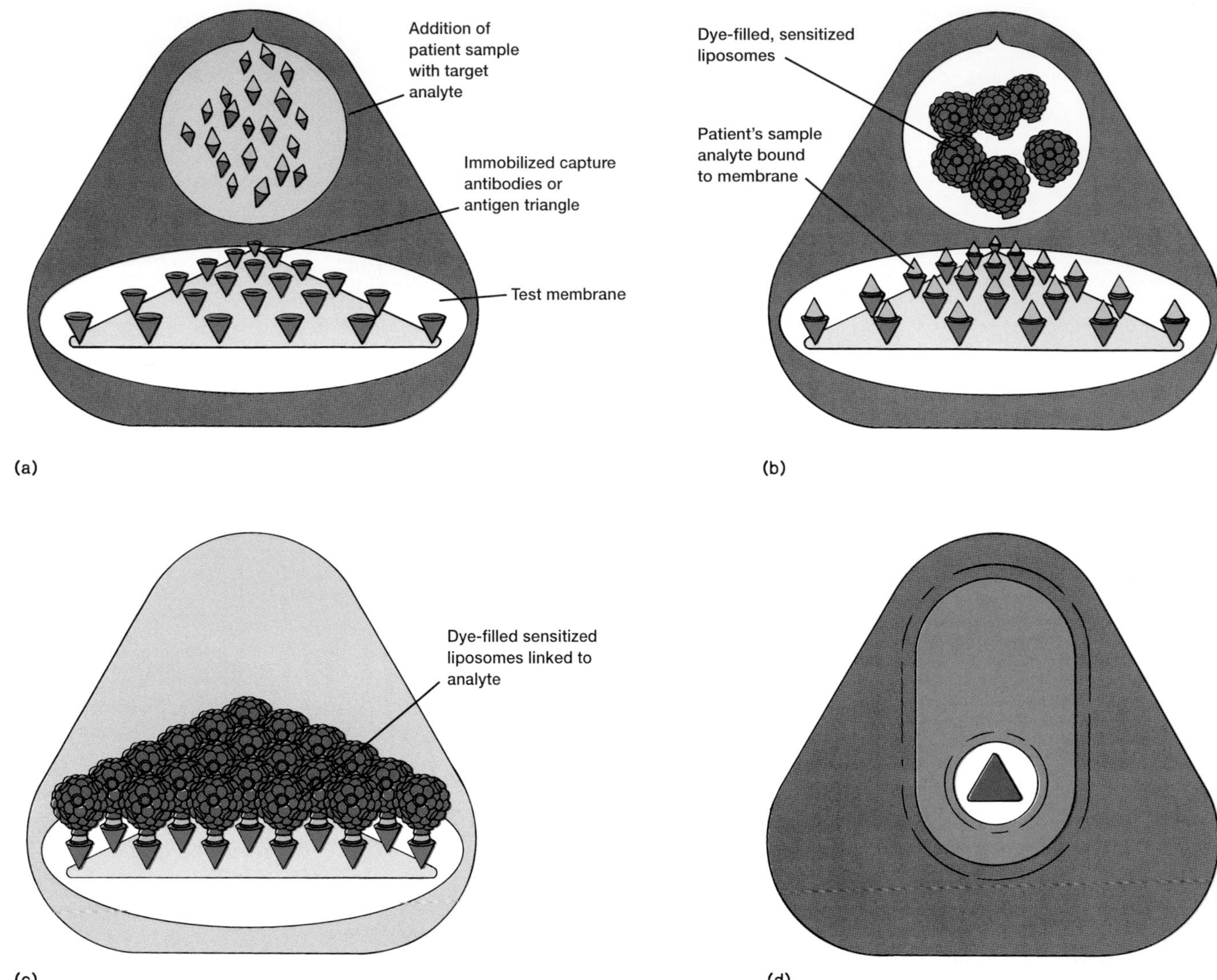

Figure 34.12 The Use of Liposomes in Diagnostic Testing. (*a*) The addition of a patient's sample containing the target analyte. Immobilized capture antibodies or antigen on a test membrane form a geometric design such as a triangle. (*b*) When dye-filled liposomes are added, they bind to the patient's bound analyte (*c*) giving a triangle. (*d*) The triangle indicates a positive reaction.

DNA as a probe is based upon the capacity of single-stranded DNA to bind (hybridize) with a complementary nucleic acid sequence present in test specimens to form a double-stranded DNA hybrid (figure 34.13). Thus, a single-stranded sequence derived from one microorganism (the probe) is used to search for others containing the same sequence. This hybridization reaction may be applied to purified DNA preparations, to bacterial colonies, or to clinical specimens such as tissue, serum, sputum, and pus. Recently, DNA probes have been developed that bind to complementary strands of ribosomal RNA. These DNA:rRNA hybrids are more sensitive than conventional DNA probes, give results in two hours or less, and require the presence of fewer microorganisms. DNA probe sensitivity can be increased by over one million if the target DNA is first amplified using the polymerase chain reaction (*see pp. 291–92*). DNA:rRNA probes have been developed for the rapid detection of *Plasmodium falciparum,* rotaviruses, herpes simplex viruses, cytomegalovirus, enterotoxigenic *E. coli,* mycoplasmas, *Mycobacterium* species, and *Chlamydia trachomatis.*

Gas-Liquid Chromatography

During chromatography, a chemical mixture carried by a liquid or gas is separated into its individual components because of processes such as adsorption, ion-exchange, and partitioning between different solvent phases. In gas-liquid chromatography (GLC), specific microbial metabolites, cellular fatty acids, and products from the pyrolysis (a chemical change

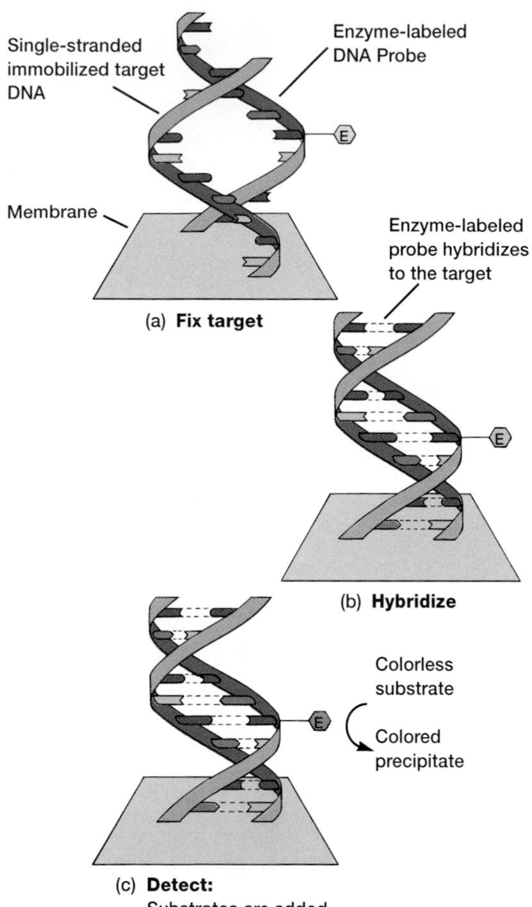

Figure 34.13 **DNA Probes.** (*a*) Single-stranded target nucleic acid is bound to a membrane. A DNA probe with attached enzyme (E) is also employed. (*b*) The probe is added to the membrane. If the probe hybridizes to the target DNA, a double-stranded DNA hybrid is formed. (*c*) A colorless substrate is added. The enzyme attached to the probe converts the substrate to a colored precipitate. This detection system is semiquantitative, in that color intensity is proportional to the quantity of hybridized target nucleic acid present.

caused by heat) of whole bacterial cells are analyzed and identified. These compounds are easily removed from growth media by extraction with an organic solvent such as ether. The ether extract is then injected into the GLC system. Both volatile and nonvolatile acids can be identified. Based on the pattern of fatty acid production, common anaerobes isolated from clinical specimens can be identified.

The reliability, precision, and accuracy of GLC have been improved significantly with continued advances in instrumentation; the introduction of instruments for high-performance liquid chromatography; and the use of mass spectrometry, nuclear magnetic resonance spectroscopy, and associated analytical techniques for the identification of components separated by the chromatographic process. These combined techniques have recently been used to discover specific chemical markers of various infectious disease agents by direct analysis of body fluids.

Plasmid Fingerprinting

As presented in chapter 13, a plasmid is an autonomously replicating extrachromosomal molecule of DNA in bacteria. **Plasmid fingerprinting** identifies microbial isolates of the same or similar strains; related strains often contain the same number of plasmids with the same molecular weights and similar phenotypes. In contrast, microbial isolates that are phenotypically distinct have different plasmid fingerprints. Plasmid fingerprinting of many *E. coli, Salmonella, Campylobacter,* and *Pseudomonas* strains and species has demonstrated that this method is often more accurate than other phenotyping methods such as biotyping, antibiotic resistance patterns, phage typing, and serotyping.

Most plasmids found in clinical isolates are termed cryptic because their gene products have not yet been determined. The distribution of cryptic plasmids in nearly all genera of clinically important bacteria provides a means of marking different strains of the same species. Plasmid fingerprinting exploits the ability of many bacterial species to carry "excess baggage" and has become a powerful tool in the investigation of nosocomial infections and epidemics.

The technique of plasmid fingerprinting involves five steps:

1. The bacterial strains are grown in broth or on agar plates.
2. The cells are harvested and lysed with a detergent.
3. The plasmid DNA is separated from the chromosomal DNA.
4. The plasmid DNA is applied to agarose gels and electrophoretically separated.
5. The gel is stained with methylene blue, which binds to DNA, causing it to fluoresce under UV light. The plasmid DNA bands are then located.

Because the migration rate of plasmid DNA in agarose is inversely proportional to the molecular weight, plasmids of a different size appear as distinct bands in the stained gel. The molecular weight of each plasmid species can then be determined from a plot of the log of the distance that each species has migrated versus the log of the molecular weights of plasmids of known size (table 34.6) that have been electrophoresed simultaneously in the same gel (figure 34.14).

TABLE 34.6 Some Plasmids Useful as Molecular Weight Standards and Their Biological Sources

Bacterial Source	Molecular Weight of Plasmid (× 10^6)
Pseudomonas aeruginosa PA02 (pMg1)	312
P. aeruginosa PA02 (pMG5)	280
Escherichia coli DY78 (TP116)	143
E. coli DT41 (R27)	112
Salmonella typhimurium LT2	60
E. coli K-12 J5–3 (Sa)	26
E. coli JC411 (ColE1)	4

Susceptibility Testing

Many clinical microbiologists believe that determining the susceptibility of a microorganism to specific chemotherapeutic agents (drugs) is one of the most important tests performed in the clinical microbiology laboratory. Results (figure 34.6*o*) can show the agents to which a microorganism is most susceptible and the proper therapeutic dose needed to treat the infectious disease. (Dilution susceptibility tests, disk diffusion tests [Kirby-Bauer method], and drug concentration measurements in the blood are discussed in detail in chapter 16.)

1. What is the basis for bacteriophage typing?
2. How can DNA probes be used by the clinical microbiologist? Gas-liquid chromatography?
3. How can a suspect bacterium be plasmid fingerprinted? Why is susceptibility testing so important in clinical microbiology?

Computers in Clinical Microbiology

Computer systems in the clinical microbiology laboratory are designed primarily to replace the handwritten mode of information acquisition and transmission. Computers improve the efficiency of the laboratory operation and increase the speed and clarity with which results can be reported to physicians. From a work-flow standpoint, the major functions involving the computer are test ordering, result entry, analysis of results, and report preparation (figure 34.1*h,i*).

Test orders may be entered into the computer from the hospital unit or laboratory. Ordering routines should include specific requests (e.g., rule out *Nocardia* and diphtheria), all pertinent patient data, and an accession number. Once the test order has been placed, the system should allow the usual work flow to proceed with the labeled clinical specimen, date, test order, and computer accession number.

After clinical results are obtained in the laboratory, they are entered into a written log and then into the computer. To meet the many needs of microbiological entry, the computer system must be rapid and flexible in its entry modes.

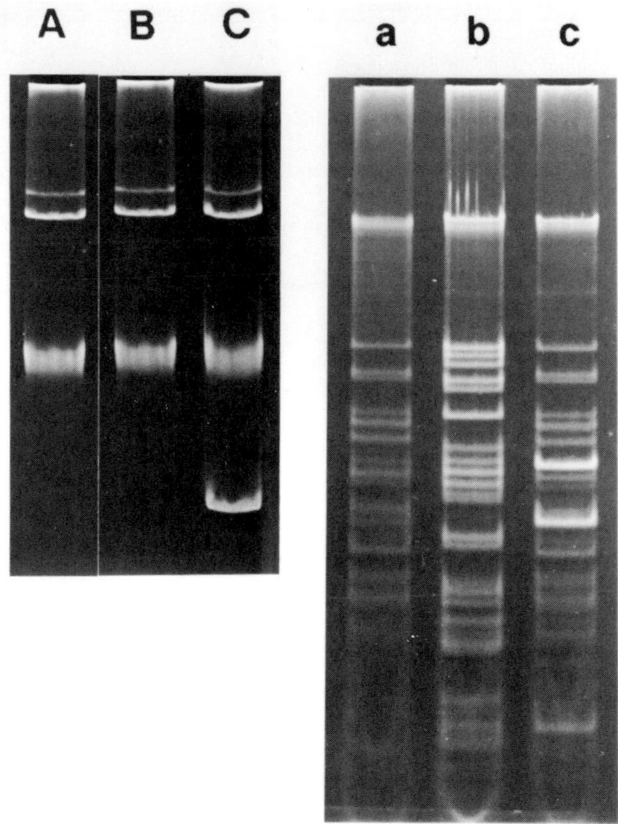

Figure 34.14 **Plasmid Fingerprinting.** Agarose gel electrophoresis of plasmid DNA.

Printed reports of the patient's laboratory findings (figure 34.3*b*) are the product of the computer system. Print programs also should allow flexible formatting of the reports so that additional data can be generated. For example, the computer system should be able to generate cumulative reports that summarize days to weeks of an inpatient stay.

Besides reporting routine laboratory tests, computers can manage specimen logs, reports of overdue tests, quality control statistics, antimicrobial susceptibility probabilities, hospital epidemiological data, and many other items. The computer can be interfaced with different automated instruments in the laboratory for rapid and accurate calculation and transfer of clinical data.

1. What are some different ways in which computers can be used in the clinical microbiology laboratory?
2. From the standpoint of work flow, how can computers be specifically used in a clinical microbiology laboratory?

Summary

1. The major concern of the clinical microbiologist is to isolate and identify microorganisms from clinical specimens rapidly. A clinical specimen represents a portion or quantity of biological material that is tested, examined, or studied to determine the presence or absence of specific microorganisms.
2. Specimens may be collected by various methods that include swabs, needle aspiration, intubation, catheters, and clean-catch techniques. Each method is designed to ensure that only the proper material will be sent to the clinical laboratory.
3. Immediately after collection, the specimen must be properly handled and labeled. Speed in transporting the specimen to the clinical laboratory after it has been collected is of prime importance.
4. The clinical microbiology laboratory can provide preliminary or definitive identification of microorganisms based on (1) microscopic examination of specimens; (2) growth or biochemical characteristics of microorganisms isolated from cultures; and (3) immunologic techniques that detect antibodies or microbial antigens.
5. Viruses are identified by isolation in living cells or serological tests. Several types of living cells are available: cell culture, embryonated hen's eggs, and experimental animals. Rickettsial disease can be diagnosed serologically or by isolation of the organism. Chlamydiae can be demonstrated in tissue and cell scrapings with Giemsa stain, which detects the characteristic intracellular inclusion bodies. The most routinely used techniques for diagnosis of the mycoplasmas are serological. Diagnosis of fungi can often be made if a portion of the specimen is mixed with a drop of 10% potassium hydroxide. Wet mounts of stool specimens or urine can be examined microscopically for the presence of parasites.
6. The initial identity of a bacterial organism may be suggested by (1) the source of the culture specimen; (2) its microscopic appearance; (3) its pattern of growth on selective, differential, enrichment, or characteristic media; and (4) its hemolytic, metabolic, and fermentative properties.
7. Rapid methods for microbial identification can be divided into three categories: (1) manual biochemical systems, (2) mechanized/automated systems, and (3) immunologic systems.
8. Bacteriophage typing for bacterial identification is based on the fact that phage surface receptors bind to specific cell surface receptors. On a petri plate culture, bacteriophages cause plaques on lawns of bacteria with the proper receptors.
9. Various molecular methods also can be used to identify microorganisms. Examples include DNA probes, gas-liquid chromatography, and plasmid fingerprinting.
10. Computer systems in clinical microbiology are designed to replace handwritten information exchange and to speed data evaluation and report preparation.

Key Terms

bacteriophage (phage) typing *692*
catheter *675*
clinical microbiologist *673*
cytopathic effect *678*
hemadsorption *678*
immunoblotting *691*
interference *678*
intubation *675*
liposome *691*
needle aspiration *675*
phagovar *692*
plasmid fingerprinting *694*
sputum *675*
swab *675*

Questions for Thought and Review

1. How can clinical specimens be taken from a patient with various infectious diseases? Give specific examples of procedures used.
2. As more new ways for identifying the characteristics of microorganisms emerge, the number of distinguishable microbial strains also seems to increase. Why do you think this occurs?
3. What precaution must be observed when a culture is obtained from the respiratory system?
4. Why are miniaturized identification systems used in clinical microbiology? Describe one such system and its advantage over classic dichotomous keys.
5. Why is gas-liquid chromatography a useful approach to the identification of anaerobes?
6. How does a clinical microbiologist convert an API 20E test result to a numerical code for bacterial identification?
7. How is a dichotomous key used in bacterial identification?
8. What are some different ways in which biochemical reactions can be used to identify microorganisms?
9. What are some advantages of automation in the clinical microbiology laboratory?
10. When should laboratory animals be used in the identification of microorganisms?
11. Why is plasmid fingerprinting such an accurate method for the identification of microorganisms?

Additional Reading

Aldridge, K. E. 1984. Comparison of rapid identification assays for *Staphylococcus aureus*. *J. Clin. Microbiol.* 19:703–4.

Aldridge, K. E., and Hodges, R. L. 1981. Correlation studies of Entero-Set-20, API 20E, and conventional media systems for *Enterobacteriaceae*. *J. Clin. Microbiol.* 7:507–13.

Balows, A.; Hausler, W. J., Jr.; Herrmann, K. L.; Isenberg, H. D.; and Shadomy, H. J. 1991. *Manual of clinical microbiology,* 5th ed. Washington, D.C.: American Society for Microbiology.

Baron, E. J., and Finegold, S. M. 1990. *Bailey and Scott's Diagnostic Microbiology,* 8th ed. St. Louis: C.V. Mosby.

Chonmaitree, T.; Baldwin, C.; and Lucia, H. 1989. Role of the virology laboratory in diagnosis and management of patients with central nervous system disease. *Clin. Microbiol. Rev.* 2(1):1–14.

Coonrod, J. D.; Kunz, L. J.; and Ferraro, M. J. 1983. *The direct detection of microorganisms in clinical specimens.* San Diego, Calif.: Academic Press.

Finegold, S. M., and Smith, W. J. 1982. *Diagnostic microbiology,* 6th ed. St. Louis: C. V. Mosby.

Forney, J. E., ed. 1987. *Collection, handling, and shipment of microbiological specimens* (PHHS Publication No. 976). Washington, D.C.: U.S. Public Health Service.

Gerhardt, P.; Murray, R. G. E.; Costilow, R. N.; Nester, E. W.; Wood, W. A.; Krieg, N. R.; and Phillips, G. B., eds. 1981. *Manual of methods for general bacteriology.* Washington, D.C.: American Society for Microbiology.

Gray, L. D., and Fedorko, D. P. 1992. Laboratory diagnosis of bacterial meningitis. *Clin. Microbiol. Rev.* 5(2):130–45.

Grimont, P. A. D. 1985. DNA probe specific for *Legionella pneumophilia*. *J. Clin. Microbiol.* 21(3):431–37.

Isenberg, H. D., ed. 1992. *Clinical Microbiology Procedures Handbook.* Washington, D.C.: American Society for Microbiology.

Johnson, F. B. 1990. Transport of viral specimens. *Clin. Microbiol. Rev.* 3(2):120–31.

Kaplan, D. S., and Picciolu, G. L. 1989. Characterization of instrumentation and calibrators for quantitative microfluorometry for immunofluorescence tests. *J. Clin. Microbiol.* 27(3):442–47.

Koneman, E. W.; Allen, S. D.; Dowell, V. R., Jr.; Janda, W. M.; Sommers, H. M.; and Winn, W. C., Jr. 1988. *Color atlas and textbook of diagnostic microbiology,* 3d ed. Philadelphia: J. B. Lippincott.

Manafi, M.; Kneifel, W.; and Bascomb, S. 1991. Fluorogenic and chromogenic substrates used in bacterial diagnostics. *Microbiol. Rev.* 55(3):335–48.

Mayer, L. W. 1988. Use of plasmid profiles in epidemiologic surveillance of disease outbreaks and in tracing the transmission of antibiotic resistance. *Clin. Microbiol. Rev.* 1(2):228–43.

Pezzlo, M. 1988. Detection of urinary tract infections by rapid methods. *Clin. Microbiol. Rev.* 1(3):268–80.

Rotbart, H. A. 1991. Nucleic acid detection systems for enteroviruses. *Clin. Microbiol. Rev.* 4(2):156–68.

Ryon, K. J., and Peebles, J. E. 1982. On-line computer entry of routine and AutoMicrobic System bacteriology results. In *Rapid methods and automation in microbiology,* ed. R. C. Tilton, 23–27. Washington, D.C.: American Society for Microbiology.

Tenover, F. C. 1988. Diagnostic deoxyribonucleic acid probes for infectious diseases. *Clin. Microbiol. Rev.* 1(1):82–101.

Turgeon, M. L. 1990. *Immunology and serology in laboratory medicine.* St. Louis: C. V. Mosby Co.

Washington, J. A., II. 1981. *Laboratory procedures in clinical microbiology.* New York: Springer-Verlag.

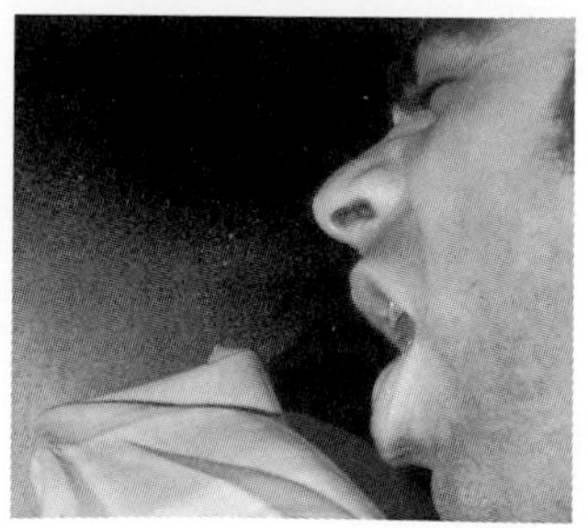

CHAPTER 35
The Epidemiology of Infectious Disease

Epidemics of infectious disease are often compared with forest fires. Once fire has spread through an area, it does not return until new trees have grown up. Epidemics in humans develop when a large population of susceptible individuals is present. If most individuals are immune, then an epidemic will not occur.

—Andrew Cliff and Peter Haggett

Outline

Concepts

1. The science of epidemiology deals with the occurrence and distribution of disease within a given population. Infectious disease epidemiology is concerned with organisms or agents responsible for the spread of infectious diseases in human and other animal populations.
2. Because numbers and time are major epidemiological parameters, statistics is an important working tool in this discipline. Statistics is used to determine morbidity, frequency, and mortality rates.
3. To trace the origin and manner of spread of an infectious disease outbreak, it is necessary to learn what pathogen is responsible.
4. Epidemiologists investigate five links in the infectious disease cycle: (1) characteristics of the pathogen, (2) source and/or reservoir of the pathogen, (3) mode of transmission, (4) susceptibility of the host, and (5) exit mechanisms.
5. The control of nosocomial (hospital) infections has received increasing attention in recent years because of the number of individuals involved, increasing costs, and the length of hospital stays.

This chapter describes the epidemiological parameters that are studied in the infectious disease cycle. The practical goal of epidemiology is to establish effective control, prevention, and eradication measures within a given population. Because nosocomial (hospital) acquired infections have increased in recent years, a brief synopsis of their epidemiology is also presented.

The science of epidemiology originated and evolved in response to the great epidemic diseases such as cholera, typhoid fever, smallpox, and yellow fever (Box 35.1). Today, its scope embraces all diseases: infectious diseases and those resulting from anatomical deformities, genetic abnormalities, metabolic dysfunction, malnutrition, neoplasms, psychiatric disorders, and aging. This chapter emphasizes only infectious disease epidemiology.

By definition, **epidemiology** (Greek *epi,* upon, and *demos,* people or population, and *logy,* study) is the science that evaluates the occurrence, determinants, distribution, and control of health and disease in a defined human population. **Health** is the condition in which the organism (and all of its parts) performs its vital functions normally or properly. It is a state of physical and mental well-being, and not merely the absence of disease. A **disease** (French *des,* from, and *aise,* ease) is an impairment of the normal state of an organism or any of its components that hinders the performance of vital functions. It is a response to environmental factors (e.g., malnutrition, industrial hazards, climate), specific infective agents (e.g., viruses, bacteria, fungi, protozoa, helminths), inherent defects of the body (e.g., various genetic or immunologic anomalies), or combinations of these.

Any individual who practices epidemiology is an **epidemiologist.** Epidemiologists are, in effect, disease detectives. Their major concerns are the discovery of the factors essential to disease occurrence and the development of methods for disease prevention.

Epidemiological Terminology

When a disease occurs occasionally, and at irregular intervals in a human population, it is a **sporadic disease** (e.g., typhoid fever). When it maintains a steady, low-level frequency at a moderately regular interval, it is an **endemic** (Greek *endemos,* dwelling in the same people) **disease** (e.g., the common cold). **Hyperendemic diseases** gradually increase in occurrence frequency beyond the endemic level but not to the epidemic level (e.g., the common cold during winter months). An **epidemic** (Greek *epidemios,* upon the people) is a sudden increase in the occurrence of a disease above the expected level (figure 35.1). The AIDS epidemic is an excellent example. The first case in an epidemic is called the **index case.** An **outbreak,** on the other hand, is the sudden, unexpected occurrence of a disease, usually focally or in a limited segment of a population (e.g., Legionnaires' disease). Although the epidemiology of an outbreak may be no different from that of an epidemic, the community regards the outbreak as less serious. A **pandemic** (Greek *pan,* all) is an increase in disease occurrence within a large population over a very wide region (usually the world). Usually, pandemic diseases spread among continents. The influenza outbreak of the 1960s is a good example.

The discipline that deals with the factors that influence the frequency of a disease in an animal population is known as **epizootiology.** Moderate prevalence of a disease in animals is termed **enzootic,** a sudden outbreak is **epizootic,** and wide dissemination is **panzootic.** Animal diseases that can be transmitted to humans are termed **zoonoses** (Greek *zoon,* animal, and *nosos,* disease).

Measuring Frequency: The Tools of Epidemiologists

Besides epidemiology's historical ties to epidemic diseases, it has also been closely allied with statistics. **Statistics** is the branch of mathematics dealing with the collection, organization, and interpretation of numerical data. As a science particularly concerned with rates and the comparison of rates, epidemiology was the first medical field in which statistical methods were extensively used.

Measures of frequency are usually expressed as fractions. The numerator is the number of individuals experiencing the event—infection or other problem—and the denominator is the number of individuals in whom the event could have occurred, that is, the population at risk. The fraction is a proportion or ratio but is commonly called a rate because a time period is always specified. (A rate also can be expressed as a percentage if the fraction is multiplied by 100.) In population statistics, rates are usually stated per 1,000 individuals, although other powers of 10 may be used for particular diseases (e.g., per 100 for very common diseases and per 10,000 or 100,000 for uncommon diseases).

Box 35.1

John Snow: The First Epidemiologist

Much of what we know today about the epidemiology of cholera is based on the classic studies conducted by the British physician John Snow between 1849 and 1854. During this period, a series of cholera outbreaks occurred in London, England, and Snow set out to find the source of the disease. Some years earlier when he was still a medical apprentice, Snow had been sent to help during an outbreak of cholera among coal miners. His observations convinced him that the disease was usually spread by unwashed hands and shared food, not by "bad" air or casual direct contact.

Thus, when the outbreak of 1849 occurred, Snow believed that cholera was spread among the poor in the same way as among the coal miners. He suspected that water, and not unwashed hands and shared food, was the source of the cholera infection among the wealthier residents. Snow examined official death records and discovered that most of the victims in the Broad Street area had lived close to the Broad Street pump or had been in the habit of drinking from it. He concluded that cholera was spread by drinking water from the Broad Street pump, which was contaminated with raw sewage containing the disease agent. When the pump handle was removed, the number of cholera cases dropped dramatically.

In 1854, another cholera outbreak struck London. Part of the city's water supply came from two different suppliers: the Southwark and Vauxhall Company and the Lambeth Company. Snow interviewed cholera patients and found that most of them purchased their drinking water from the Southwark and Vauxhall Company. He also discovered that this company obtained its water from the Thames River below locations where Londoners had discharged their sewage. In contrast, the Lambeth Company took its water from the Thames before the river reached the city. The death rate from cholera was over eightfold lower in households supplied with Lambeth Company water. Water contaminated by sewage was transmitting the disease. Finally, Snow concluded that the cause of the disease must be able to multiply in water. Thus, he nearly recognized that cholera was caused by a microorganism, though Robert Koch didn't discover the causative bacterium (*Vibrio cholerae*) until 1883.

To commemorate these achievements, the John Snow Pub now stands at the site of the old Broad Street pump. Those who complete the Epidemiologic Intelligence Program at the Centers for Disease Control receive an emblem bearing a replica of a barrel of Whatney's Ale—the brew dispensed at the John Snow Pub.

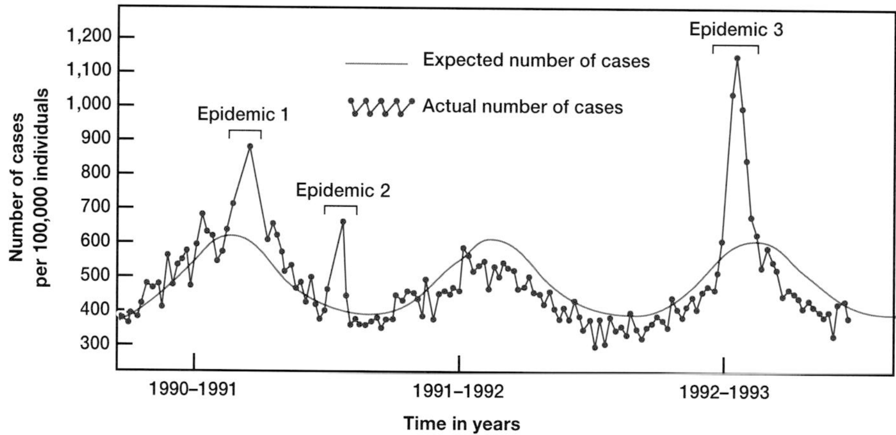

Figure 35.1 A Graph Illustrating Three Epidemics. The solid line indicates the expected number of endemic cases. The connected dots indicate the actual number of cases. Epidemics (marked by brackets) are sharp increases in the number of cases of a disease above that which is normally expected (solid line).

Figure 35.2 **Prevalence of AIDS.** This figure portrays U.S. military recruit data, showing marked geographic variations in HIV–1 prevalence, with the highest rates in the northeast and mid-Atlantic regions. Data are for the period October 1985 to September 1989. *Source: J. F. Brundage, et al., "Tracking the Spread of the HIV Epidemic Among Young Adults in the US" in* Journal of AIDS, *3:1168–1180, 1990.*

A **morbidity rate** measures the number of individuals that become ill because of a specific disease within a susceptible population during a specific period. It is an incidence rate and reflects the number of new cases in a period. The rate is commonly determined when the number of new cases of illness in the general population is known from clinical reports. It is calculated as follows:

$$\text{Morbidity rate} = \frac{\text{number of new cases of a disease during a specific period}}{\text{number of individuals in the population}}$$

For example, if there were 700 new cases of influenza per 100,000 individuals, then the morbidity rate would be expressed as 700 per 100,000 or 0.7%.

The **prevalence rate** refers to the total number of individuals infected in a population at any one time no matter when the disease began. The prevalence rate depends on both the incidence rate and the duration of the illness. For example, figure 35.2 shows the geographic distribution by place of residence of persons with AIDS among U.S. military recruits. From these data, the overall crude **seroprevalence rate** for recruits is 1.42 per 1,000 for men (0.14%) and 0.66 per 1,000 for women (0.06%).

The **mortality rate** is the relationship of the number of deaths from a given disease to the total number of cases of the disease. The mortality rate is a simple statement of the proportion of all deaths that are assigned to a single cause. It is calculated as follows:

$$\text{Mortality rate} = \frac{\text{number of deaths due to a given disease}}{\text{size of the total population with the same disease}}$$

For example, if there were 15,000 deaths due to AIDS in a year, and the total number of people infected was 30,000, the mortality rate would be 15,000 per 30,000 or 1 per 2 or 50%.

The determination of morbidity, prevalence, and mortality rates aids public health personnel in directing health care efforts to control the spread of infectious diseases. For example, a sudden increase in the morbidity rate of a particular disease may forecast a need for action to prevent a future rise in mortality.

---

1. What is epidemiology? Seroprevalence?
2. What terms are used to describe the occurrence of a disease in a human population? In an animal population?
3. Define morbidity rate, prevalence rate, and mortality rate.
4. How would you define a disease? Health?

Box 35.2

Rift Valley Fever: Long-Distance Surveillance

Satellite data can be used to predict outbreaks of a viral disease that affects sheep and cattle and sometimes humans in Africa. The disease, Rift Valley fever, is spread by mosquitoes, which thrive in flood conditions. Although there have been no recent outbreaks in humans, a 1977 epidemic resulted in 18,000 illnesses and 598 deaths.

The epidemiological surveillance of this disease is based on a chain of connections. Weather satellites can measure the extent of green vegetation, which in turn represents the amount of rainfall, which is related to the flood conditions that produce favorable environments for a mosquito population explosion. As noted, mosquitoes carry the Rift Valley fever virus and spread the virus to animals and humans. When conditions are favorable for the virus-carrying mosquitoes, local governments can take steps to control the mosquitoes before they become a serious problem.

Infectious Disease Epidemiology

An infectious disease is due to agents such as viruses, bacteria, fungi, protozoa, and helminths that can be transmitted from one host to another. It can vary from mild to severe to deadly for the host. An epidemiologist studying an infectious disease is concerned with the causative agent, the source and/or reservoir of the disease agent, how it was transmitted, what host and environmental factors could have aided development of the disease within a defined population, and how best to control or eliminate the disease. These factors describe the natural history or cycle of an infectious disease.

Recognition of an Infectious Disease in a Population

Epidemiologists can recognize an infectious disease in a population by using various surveillance methods. Surveillance is a dynamic activity that includes gathering information on the development and occurrence of a disease, collating and analyzing the data, summarizing the findings, and using the information to select control methods (Box 35.2). Some combination of the following surveillance methods is used most often:

1. Generation of morbidity data from case reports
2. Collection of mortality data from death certificates
3. Investigation of actual cases
4. Collection of data from reported epidemics
5. Field investigation of epidemics
6. Review of laboratory results: surveys of a population for antibodies against the agent and specific microbial serotypes, skin tests, cultures, stool analyses, etc.
7. Population surveys using valid statistical sampling to determine who has the disease
8. Use of animal and vector disease data
9. Collection of information on the usage of specific biologics—antibiotics, antitoxins, vaccines, and other prophylactic measures
10. Use of demographic data on population characteristics such as human movements during a specific time of the year

Correlation with a Single Causative Agent

After an infectious disease has been recognized in a population, epidemiologists correlate the disease outbreak with a specific organism—its exact cause must be discovered (Box 35.3). At this point, the clinical or diagnostic microbiology laboratory enters the investigation. Its purpose is to isolate and identify the organism that caused the disease and to determine the pathogen's susceptibility to control agents or methods that may assist in its eradication.

Role of clinical microbiology in disease diagnosis (chapter 34).

Recognition of an Epidemic

As previously noted, an infectious disease epidemic is usually a short-term increase in the occurrence of the disease in a particular population (figure 35.1). Two major types of epidemic are recognized: common source and propagated.

A **common-source epidemic** is characterized by a sharp rise to a peak and then a rapid, but not as pronounced, decline in the number of individuals infected (figure 35.3*a*). This type of epidemic usually results from a single common contami-

Box 35.3

"Typhoid Mary"

In the early 1900s, there were thousands of typhoid fever cases, and many died of the disease. Most of these cases arose when people drank water contaminated with sewage or ate food handled by or prepared by individuals who were shedding the typhoid fever bacterium (*Salmonella typhi*). The most famous carrier of the typhoid bacterium was Mary Mallon.

Between 1896 and 1906, Mary Mallon worked as a cook in seven homes in New York City. Twenty-eight cases of typhoid fever occurred in these homes while she worked in them. As a result, the New York City Health Department had Mary arrested and admitted to a city hospital for patients with infectious diseases. Examination of Mary's stools showed that she was shedding large numbers of typhoid bacteria though she exhibited no external symptoms of the disease. An article published in 1908 in the *Journal of the American Medical Association* referred to her as "Typhoid Mary," an epithet by which she is still known today. After being released when she pledged not to cook for others or serve food to them, Mary changed her name and began to work as a cook again. For five years she managed to avoid capture while continuing to spread typhoid fever. Eventually, the authorities tracked her down. She was held in custody for 23 years until she died in 1938. As a lifetime carrier, Mary Mallon was positively linked with 10 outbreaks of typhoid fever, 53 cases, and 3 deaths.

nated source such as food (food poisoning) or water (Legionnaires' disease).

A **propagated epidemic** is characterized by a relatively slow and prolonged rise and then a gradual decline in the number of individuals infected (figure 35.3*b*). This type of epidemic usually results from the introduction of a single infected individual into a susceptible population. The initial infection is then propagated to others in a gradual fashion until many individuals within the population are infected. An example is the increase in mumps or chickenpox cases that coincides with new populations of sensitive children who arrive in classrooms each fall. Only one infected child is necessary to propagate the epidemic.

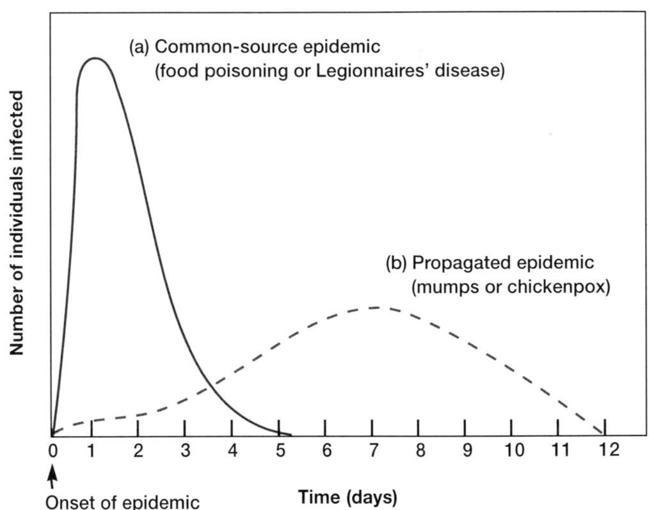

Figure 35.3 Epidemic Curves. (*a*) In a common-source epidemic, there is a rapid increase up to a peak in the number of individuals infected and then a rapid but more gradual decline. Cases are usually reported for a period that equals approximately one incubation period of the disease. (*b*) In a propagated epidemic, the curve has a gradual rise and then a gradual decline. Cases are usually reported over a time interval equivalent to several incubation periods of the disease. *From Brock/Smith/Madigan, Biology of Microorganisms,* 4/e. © 1984, p. 571. Adapted by permission of Prentice-Hall, Inc., Englewood Cliffs, NJ.

To understand how epidemics are propagated, consider figure 35.4. At time 0, all individuals in this population are susceptible to a hypothetical pathogen. The introduction of an infected individual initiates the epidemic outbreak (lower curve), which spreads and reaches a peak (day 15). As individuals recover from the disease, they become immune and no longer transmit the pathogen (upper curve). The number of susceptible individuals therefore decreases. The decline in the number of susceptibles to the threshold density (the minimum number of individuals necessary to continue propagating the disease) coincides with the peak of the epidemic wave, and the incidence of new cases declines because the pathogen cannot propagate itself.

Herd immunity is the resistance of a population to infection and pathogen spread because of the immunity of a large percentage of the population (figure 35.5). The larger the proportion of those immune, the smaller the probability of effective contact between infective and susceptible individuals; that is, many contacts will be with immunes, and thus the population will exhibit a group resistance. A susceptible member of such an immune population enjoys an immunity that is not of his or her own making (not self-made) but, instead, arises because of membership in the group.

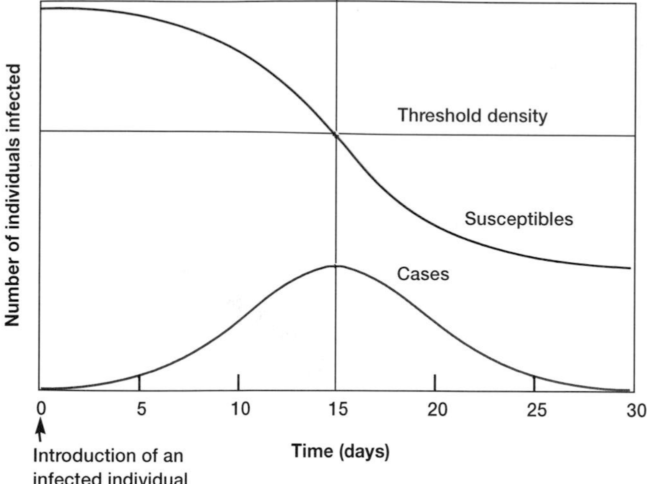

Figure 35.4 Diagrammatic Representation of the Spread of An Imaginary Propagated Epidemic. The lower curve represents the number of cases and the upper curve the number of susceptible individuals. Notice the coincidence of the peak of the epidemic wave with the threshold density of susceptibles.

At times, public health officials immunize large portions of the susceptible population in an attempt to maintain a high level of herd immunity. Any increase in the number of susceptibles may result in an endemic disease becoming epidemic. The proportion of immune individuals to susceptibles must be constantly monitored because new susceptible individuals continually enter a population through migration and birth. In addition, pathogens can change so much through processes such as antigenic shift (see next paragraph) that immune individuals become susceptible again.

Many pathogens do not ordinarily change in nature. They cause endemic diseases because infected humans perpetually transfer them to others (e.g., sexually transmitted diseases) or because they continually reenter the human population from animal reservoirs (e.g., rabies). Other pathogens continue to evolve and may produce epidemics (e.g., AIDS, influenza virus [A strain], and *Legionella* bacteria). One way in which a pathogen changes is by **antigenic shift**, a major genetically determined change in the antigenic character of a pathogen. An antigenic shift can be so extensive that the pathogen is no longer recognized by the host's immune system. For example, influenza viruses frequently change by mutation from one antigenic type to another. Antigenic shift also occurs through the hybridization of different influenza virus serovars; two serovars of a virus intermingle to form a new antigenic type. Hybridization may occur between an animal strain and a human strain of the virus. When resistance in the human population becomes so high that the virus can no longer spread (herd immunity), it is transmitted to animals, where the hybridization takes place. Smaller antigenic changes also can take place in pathogen strains and help the pathogen avoid host immune responses. These smaller changes are called **antigenic drift.**

Most great flu pandemics have originated in China where contact between humans and swine or ducks is frequent and close. The pandemic of 1918 was probably the result of hybridization because the survivors of that pandemic were found to have antibodies to the animal form of swine flu.

Whenever antigenic shift or drift occurs, the population of susceptibles increases because the immune system has not been exposed to the new mutant strain. If the percentage of susceptibles is above the threshold density (figure 35.4), the level of protection provided by herd immunity will decrease and the morbidity rate will increase. For example, the morbidity rates of diphtheria and measles among school children may reach epidemic levels if the number of susceptibles rises above 30% for the whole population. As a result, the goal of public health agencies is to make sure that at least 70% of the population is immunized against these diseases to provide the herd immunity necessary for protection of those who are not immunized.

1. How can epidemiologists recognize an infectious disease in a population?
2. Differentiate between common-source and propagated epidemics.
3. Explain herd immunity.
4. What is the significance of antigenic shift and drift in epidemiology?

The Infectious Disease Cycle: Story of a Disease

To continue to exist, a pathogen must reproduce and be disseminated among its hosts. Thus, an important aspect of infectious disease epidemiology is a consideration of how reproduction and dissemination occur. The **infectious disease cycle** or **chain** represents these events in the form of an intriguing epidemiological mystery story (figure 35.6).

What Pathogen Caused the Disease?

The first link in the infectious disease cycle is the pathogen. After an infectious disease has been recognized in a population, epidemiologists must correlate the disease outbreak with a specific pathogen. The disease's exact cause must be discovered. This is where Koch's postulates (*see chapter 1*), and modifications of them, are used to determine the etiology or cause of an infectious disease. At this point, the clinical or diagnostic microbiology laboratory enters the investigation (*see chapter 34*). Its purpose is to isolate and identify the pathogen that caused the disease and to determine the pathogen's susceptibility to control agents or methods that may assist in its eradication.

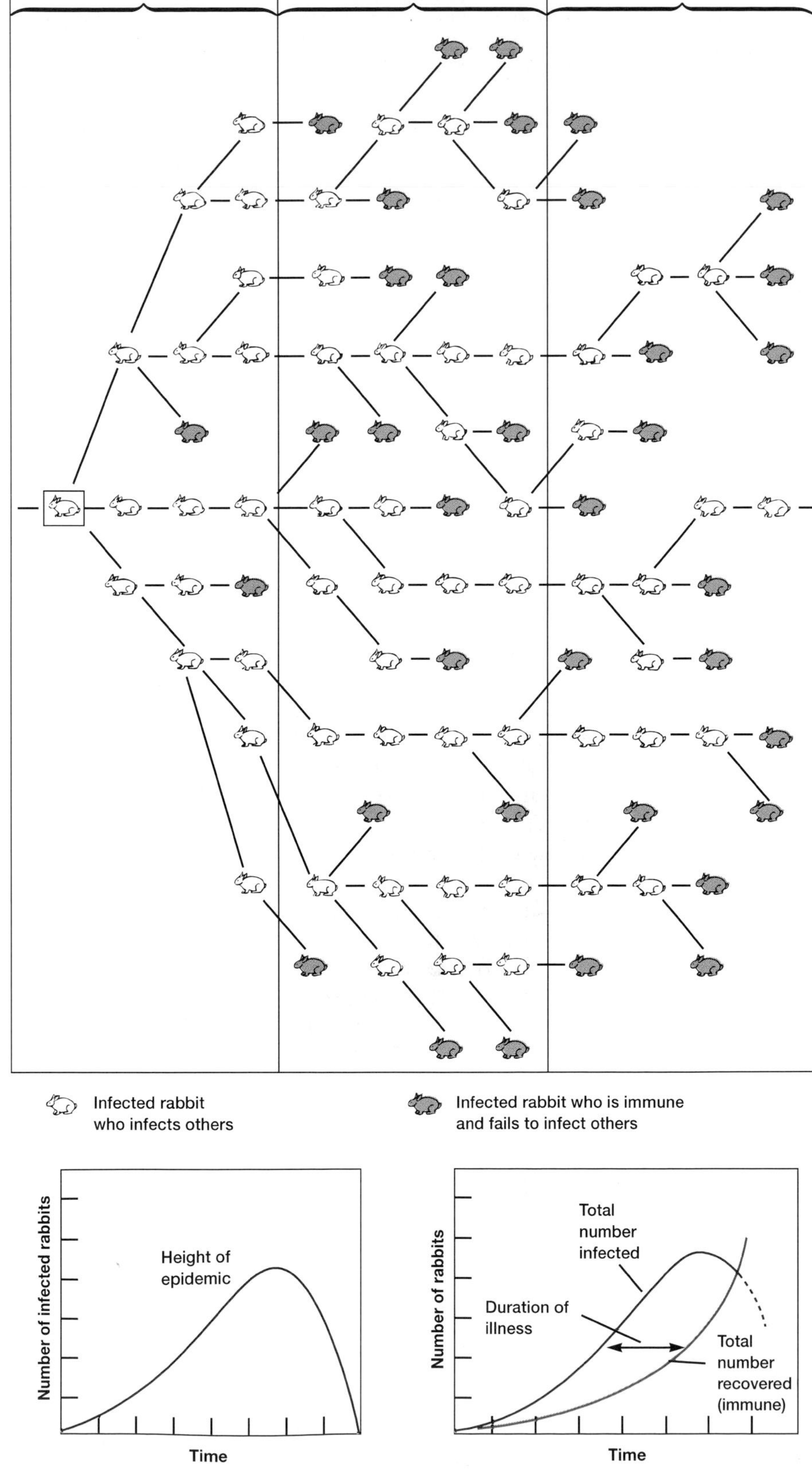

Figure 35.5 **Herd Immunity.** The kinetics of the spread of an infectious disease and the effect of increasing the number of immune individuals in the population in limiting the disease. On day 1, a single infected individual enters the population. The incubation period is 1 day, and recovery occurs in 2 days. The number of susceptibles is the total population on day 1. The number of infected and recovered are illustrated in the two graphs.

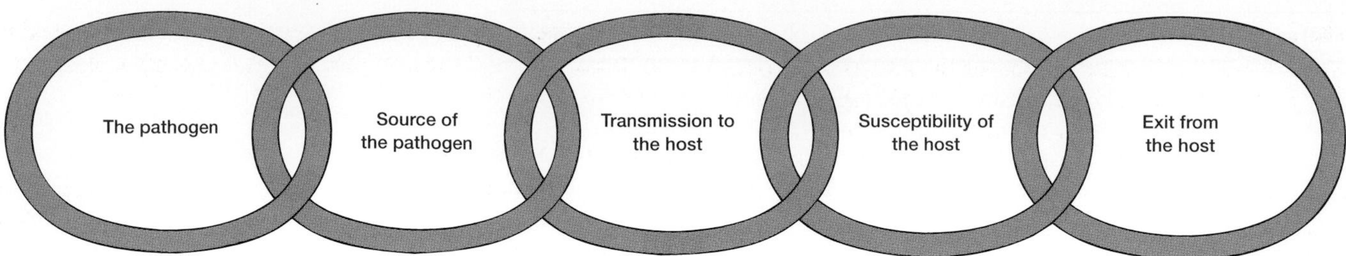

Figure 35.6 **Infectious Disease Cycle.** See text for further details.

Many pathogens can cause infectious diseases in humans and will be presented in detail in chapters 36 to 39. Often these pathogens are transmissible from one human to another. A **communicable disease** is one caused by a pathogen that can be transmitted from one host to another. Pathogens have the potential to produce disease (pathogenicity); however, this potential is a function of such factors as the number of pathogens, their virulence, and the nature and magnitude of host defenses (*pp. 579–80*).

What Was the Source and/or Reservoir of the Pathogen?

The source and/or reservoir of a pathogen is the second link in the infectious disease cycle. Identifying the source and/or reservoir is an important aspect of epidemiology. If the source or reservoir of the infection can be eliminated or controlled, the infectious disease cycle itself will be interrupted and transmission of the pathogen will be prevented (see Boxes 35.1 and 35.3).

A **source** is the location from which the pathogen is immediately transmitted to the host, either directly through the environment or indirectly through an intermediate agent. The source can be either animate (e.g., humans or animals) or inanimate (e.g., water, soil, or food). The **period of infectivity** is the time during which the source is infectious or is disseminating the pathogen.

The **reservoir** is the site or natural environmental location in which the pathogen is normally found living and from which infection of the host can occur. Reservoirs also can be animate or inanimate.

Much of the time, human hosts are the most important animate sources of the pathogen and are called carriers. A **carrier** is an infected individual who is a potential source of infection for others. Carriers play an important role in the epidemiology of disease. Four types of carriers are recognized:

1. An **active carrier** is an individual who has an overt clinical case of the disease.
2. A **convalescent carrier** is an individual who has recovered from the infectious disease but continues to harbor large numbers of the pathogen.
3. A **healthy carrier** is an individual who harbors the pathogen but is not ill.
4. An **incubatory carrier** is an individual who is incubating the pathogen in large numbers but is not yet ill.

Convalescent, healthy, and incubatory carriers may harbor the pathogen for only a brief period (hours, days, or weeks) and then are called **casual, acute,** or **transient carriers.** If they harbor the pathogen for long periods (months, years, or life), they are called **chronic carriers.**

As noted earlier, infectious diseases called zoonoses occur in animals and are occasionally transmitted to humans; thus, these animals also can serve as reservoirs. Humans contract the pathogen by several mechanisms: coming into direct contact with diseased animal flesh (tularemia); drinking contaminated cow's milk (tuberculosis and brucellosis); inhaling dust particles contaminated by animal excreta or products (Q fever, anthrax); or eating insufficiently cooked infected flesh (anthrax, trichinosis). In addition, being bitten by insect **vectors** (organisms that spread disease from one host to another) such as mosquitoes, ticks, fleas, mites, or biting flies (equine encephalomyelitis and malaria, Lyme disease, Rocky Mountain spotted fever, plague, scrub typhus, and tularemia); or being bitten by a diseased animal (rabies) can lead to infection.

Table 35.1 lists some more common zoonoses found in the Western Hemisphere. This table is noninclusive in scope; it merely abbreviates the enormous spectrum of zoonotic diseases that are relevant to human epidemiology. Domestic animals are the most common source of zoonoses because they live in greater proximity to humans than do wild animals. Diseases of wild animals that are transmitted to humans tend to occur sporadically because close contact is infrequent. Other major reservoirs of pathogens are water, soil, and food. These reservoirs are discussed in detail in chapters 41–43.

How Was the Pathogen Transmitted?

To maintain an infectious disease in a human population, the pathogen must be transmitted from one host or source to another. Transmission is the third link in the infectious disease cycle and occurs by four main routes: airborne, contact, vehicle, and vector-borne.

Airborne Transmission

Because air is not a suitable medium for the growth of a pathogen, any pathogen that is airborne must have originated from a source such as humans, other animals, plants, soil, food, or water. In **airborne transmission,** the pathogen is truly suspended in the air and travels over a meter or more from the source to the host. The pathogen can be contained within droplet nuclei or dust. **Droplet nuclei** are small particles, 1 to 4 μm in diameter, that represent what is left from the evaporation of larger particles (10 μm or more in diameter) called droplets. Droplet nuclei can remain airborne for hours or days and travel long distances.

When animals or humans are the source of the airborne pathogen, it is usually propelled from the respiratory tract into the air by an individual's coughing, sneezing, or vocalization. For example, enormous numbers of moisture droplets are aerosoled during a typical sneeze (figure 35.7). Each droplet is about 10 μm in diameter and initially moves about 100 m/second or more than 200 mi/hour!

Dust is also an important route of airborne transmission. At times, a pathogen adheres to dust particles and contributes to the number of airborne pathogens when the dust is resuspended by some disturbance. A pathogen that can survive for relatively long periods in or on dust creates an epidemiological problem, particularly in hospitals, where dust can be the source of hospital-acquired infections. Table 35.2 summarizes some human airborne pathogens and the diseases they cause.

Contact Transmission

Contact transmission implies the coming together or touching of the source or reservoir of the pathogen and the host (Box 35.4). Contact can be direct, indirect, or by droplet spread. Direct contact implies an actual physical interaction with the infectious source. This route is frequently called person-to-person contact. Person-to-person transmission occurs primarily by touching, kissing, or sexual contact (sexually transmitted diseases); by contact with oral secretions or body lesions (herpes and boils); by nursing mothers (staphylococcal infections); and through the placenta (AIDS, syphilis). Some infectious pathogens also can be transmitted by direct contact with animals or animal products (*Salmonella* and *Campylobacter*).

Indirect contact refers to the transmission of the pathogen from the source to the host via an intermediary—most often an inanimate object. The intermediary is usually contaminated by an animate source. Common examples of intermediary inanimate objects include thermometers, eating utensils, drinking cups, and bedding. *Pseudomonas* bacteria are easily transmitted by this route. This mode of transmission is often also considered a form of vehicle transmission (see next section).

In droplet spread, the pathogen is carried on particles larger than 5 μm. The route is through the air but only for a very short distance—usually less than a meter. Because these particles are large, they quickly settle out of the air. As a result, droplet transmission of a pathogen depends on the proximity of the source and the host. Measles is an example of a droplet-spread disease.

Vehicle Transmission

Inanimate materials or objects involved in pathogen transmission are called **vehicles.** In **common vehicle transmission,** a single inanimate vehicle or source serves to spread the pathogen to multiple hosts but does not support its reproduction. Examples include surgical instruments, bedding, and eating utensils. In epidemiology, these common vehicles are called **fomites** (s., fomes or fomite). A single source containing pathogens (blood, drugs, IV fluids) can contaminate a common vehicle that causes multiple infections. Food and water are important common vehicles for many human diseases (*see tables 37.3 and 43.5*).

Vector-Borne Transmission

Living transmitters of a pathogen are called vectors. Most vectors are arthropods (insects, ticks, mites, fleas) or vertebrates (dogs, cats, skunks, bats). **Vector-borne transmission** can be either external or internal. In external (mechanical) transmission, the pathogen is carried on the body surface of a vector. Carriage is passive, with no growth of the pathogen during transmission. An example would be flies carrying *Shigella* organisms on their feet from a fecal source to a plate of food that a person is eating.

In internal transmission, the pathogen is carried within the vector. Here, it can go into either a harborage or biologic transmission phase. In **harborage transmission,** the pathogen does not undergo morphological or physiological changes within the vector. An example would be the transmission of *Yersinia pestis* (the etiologic agent of plague) by the rat flea from rat to human. **Biologic transmission** implies that the pathogen does go through a morphological or physiological change within the vector. An example would be the developmental sequence of the malarial parasite inside its mosquito vector.

Malaria (pp. 793–95).

Why Was the Host Susceptible to the Pathogen?

The fourth link in the infectious disease cycle is the host. The susceptibility of the host to a pathogen depends on both the pathogenicity of the organism and the nonspecific and specific defense mechanisms of the host. These susceptibility factors are the basis for chapters 30 through 33 and are outside the realm of epidemiology.

How Did the Pathogen Leave the Host?

The fifth and last link in the infectious disease cycle is release or exit of the pathogen from the host. It is equally important that the pathogen escapes from its host as it is that the pathogen originally contacts and enters the host. Unless a suc-

TABLE 35.1 Infectious Organisms in Nonhuman Reservoirs That May Be Transmitted to Humans

Disease	Etiologic Agent	Usual or Suspected Nonhuman Host	Usual Method of Human Infection
Anthrax	*Bacillus anthracis*	Cattle, horses, sheep, swine, goats, dogs, cats, wild animals, birds	Inhalation or ingestion of spores; direct contact
Babesiosis	*Babesia bovis, Babesia divergens, Babesia microti*	*Ixodes* ticks of various species	Bite of infected tick
Brucellosis (undulant fever)	*Brucella melitensis, B. abortus, B. suis*	Cattle, goats, swine, sheep, horses, mules, dogs, cats, fowl, deer, rabbits	Milk; direct or indirect contact
Campylobacteriosis	*Campylobacter fetus, C. jejuni*	Cattle, sheep, poultry, swine, pets, other animals	Contaminated water and food
Cat-scratch fever	Unknown	Cats, dogs	Cat or dog scratch
Colorado tick fever	Arbovirus	Squirrels, chipmunks, mice, porcupines	Tick bite
Cowpox	Cowpox virus	Cattle, horses	Skin abrasions
Cryptosporidiosis	*Cryptosporidium* spp.	Calves	Contact with infected calves
Encephalitis (California)	Arbovirus	Rats, squirrels, horses, deer, hares, cows	Mosquito
Encephalitis (St. Louis)	Arbovirus	Birds	Mosquito
Encephalomyelitis (Eastern equine)	Arbovirus	Birds, ducks, fowl, horses	Mosquito
Encephalomyelitis (Venezuelan equine)	Arbovirus	Rodents, horses	Mosquito
Encephalomyelitis (Western equine)	Arbovirus	Birds, snakes, squirrels, horses	Mosquito
Giardiasis	*Giardia lamblia*	Rodents, deer, cattle, dogs, cats	Contaminated water
Glanders	*Pseudomonas mallei*	Horses	Skin contact; inhalation
Herpes B viral encephalitis	*Herpesvirus simiae*	Monkeys	Monkey bite; contact with material from monkeys
Leptospirosis	*Leptospira interrogans*	Dogs, rodents, wild animals	Direct contact with urine, infected tissue, and contaminated water
Listeriosis	*Listeria monocytogenes*	Sheep, cattle, goats, guinea pigs, chickens, horses, rodents, birds, crustaceans	Unknown
Lyme disease	*Borrelia burgdorferi*	Ticks (*Ixodes dammini*) or related ticks	Bite of infected tick
Lymphocytic choriomeningitis	Arbovirus	Mice, rats, dogs, monkeys, guinea pigs	Inhalation of contaminated dust; ingestion of contaminated food
Mediterranean fever (boutonneuse fever, African tick typhus)	*Rickettsia conorii*	Dogs	Tick bite

Modified from Guy Youmans, Philip Paterson, and Hubert Sommers, *The Biologic and Clinical Basis of Infectious Diseases*. Copyright © 1985 W.B. Saunders, Philadelphia, PA. Reprinted by permission.

Disease	Etiologic Agent	Usual or Suspected Nonhuman Host	Usual Method of Human Infection
Melioidosis	*Pseudomonas pseudomallei*	Rats, mice, rabbits, dogs, cats	Arthopod vectors, water, food
Orf (contagious ecthyma)	Virus	Sheep, goats	Through skin abrasions
Pasteurellosis	*Pasteurella multocida*	Fowl, cattle, sheep, swine, goats, mice, rats, rabbits	Animal bite
Plague (bubonic)	*Yersinia pestis*	Domestic rats, many wild rodents	Flea bite
Psittacosis	*Chlamydia psittaci*	Birds	Direct contact, respiratory aerosols
Q fever	*Coxiella burnetii*	Cattle, sheep, goats	Inhalation of infected soil and dust
Rabies	Rabies virus (Rhabdovirus group)	Dogs, bats, opposums, skunks, foxes, cats, cattle	Bite of rabid animal
Rat bite fever	*Spirillum minus*	Rats, mice, cats	Rat bite
	Streptobacillus moniliformis	Rats, mice, squirrels, weasels, turkeys, contaminated food	Rat bite
Relapsing fever (borreliosis)	*Borrelia* spp.	Rodents, porcupines, opposums, armadillos, ticks, lice	Tick or louse bite
Rickettsialpox	*Rickettsia akari*	Mice	Mite bite
Rocky Mountain spotted fever	*Rickettsia rickettsii*	Rabbits, squirrels, rats, mice, groundhogs	Tick bite
Salmonellosis	*Salmonella* spp. (except *S. typhosa*)	Fowl, swine, sheep, cattle, horses, dogs, cats, rodents, reptiles, birds, turtles	Direct contact; food
Scrub typhus	*Rickettsia tsutsugamushi*	Wild rodents, rats	Mite bite
Tuberculosis	*Mycobacterium bovis*	Cattle, horses, cats, dogs	Milk; direct contact
Tularemia	*Francisella tularensis*	Wild rabbits, most other wild and domestic animals	Direct contact with infected carcass, usually rabbit; tick bite, biting flies
Typhus fever (endemic)	*Rickettsia mooseri*	Rats	Flea bite
Vesicular stomatitis	Virus (Rhabdovirus group)	Cattle, swine, horses	Direct contact
Weil's disease (leptospirosis)	*Leptospira interrogans*	Rats, mice, skunks, opposums, wildcats, foxes, raccoons, shrews, bandicoots, dogs, cattle, swine	Through skin, drinking water, eating food
Yellow fever (jungle)	Yellow fever virus	Monkeys, marmosets, lemurs, mosquitoes	Mosquito

Figure 35.7 **A Sneeze.** High-speed photograph of an aerosol generated by an unstifled sneeze. The particles seen are comprised of saliva and mucus laden with microorganisms. These airborne particles may be infectious when inhaled by a susceptible host. Even a surgical mask will not prevent the spread of all particles.

Box 35.4

The First Indications of Person-to-Person Spread of an Infectious Disease

In 1773, Charles White, an English surgeon and obstetrician, published his "Treatise on the Management of Pregnant and Lying-In Women." In it, he appealed for surgical cleanliness to combat childbed or puerperal fever. (**Puerperal fever** is an acute febrile condition that can follow childbirth and is caused by streptococcal infection of the uterus and/or adjacent regions.) In 1795, Alexander Gordon, a Scottish obstetrician, published his "Treatise on the Epidemic Puerperal Fever of Aberdeen," which demonstrated for the first time the contagiousness of the disease. In 1843, Oliver Wendell Holmes, a noted physician and anatomist in the United States, published a paper entitled "On the Contagiousness of Puerperal Fever" and also appealed for surgical cleanliness to combat this disease.

However, the first person to realize that a pathogen could be transmitted from one person to another was the Hungarian physician Ignaz Phillip Semmelweis. Between 1847 and 1849, Semmelweis observed that women who had their babies at the hospital with the help of medical students and physicians were four times as likely to contract puerperal fever as those who gave birth with the help of midwives. He concluded that the physicians and students were infecting women with material remaining on their hands after autopsies and other activities. Semmelweis thus began washing his hands with a calcium chloride solution before examining patients or delivering babies. This simple procedure led to a dramatic decrease in the number of cases of puerperal fever and saved the lives of many women. As a result, Semmelweis is credited with being the pioneer of antisepsis in obstetrics. Unfortunately, in his own time, most of the medical establishment refused to acknowledge his contribution and adopt his procedures. After years of rejection, Semmelweis had a nervous breakdown in 1865. He died a short time later of a wound infection. It is very probable that it was a streptococcal infection, arising from the same pathogen he had struggled against his whole professional life.

TABLE 35.2 Some Airborne Pathogens and the Diseases They Cause in Humans

Microorganism	Disease	Microorganism	Disease
Viruses		**Bacteria**	
Varicella	Chickenpox	*Actinomyces* spp.	Lung infections
Influenza	Flu	*Bordetella pertussis*	Whooping cough
Rubeola	Measles	*Chlamydia* spp.	Psittacosis
Rubella	German measles	*Cornyebacterium diphtheriae*	Diphtheria
Mumps	Mumps	*Mycoplasma pneumoniae*	Pneumonia
Poliomyelitis	Polio	*Mycobacterium tuberculosis*	Tuberculosis
Acute respiratory viruses	Viral pneumonia	*Neisseria* spp.	Meningitis
		Streptococcus spp.	Pneumonia, sore throat
		Fungi	
		Blastomyces spp.	Lung infections
		Candida spp.	Disseminated infections
		Coccidioides spp.	Coccidioidomycosis
		Histoplasma capsulatum	Histoplasmosis

cessful escape occurs, the disease cycle will be interrupted and the pathogenic species will not be perpetuated. Escape can be active and passive, although often a combination of the two occurs. Active escape takes place when a pathogen actively moves to a portal of exit and leaves the host. Examples include the many parasitic helminths that migrate through the body of their host, eventually reaching the surface and exiting. Passive escape occurs when a pathogen or its progeny leaves the host in feces, urine, droplets, saliva, or desquamated cells. Microorganisms usually employ passive escape mechanisms.

1. What are some epidemiologically important characteristics of a pathogen? What is a communicable disease?
2. Define source, reservoir, period of infectivity, and carrier.
3. What types of infectious disease carriers does epidemiology recognize?
4. Describe the four main types of infectious disease transmission and give examples of each.
5. Define droplet nuclei, vehicle, fomite, and vector.

Control of Epidemics

The development of an infectious disease is a complex process involving many factors, as is the design of specific epidemiological control measures. Epidemiologists must consider available resources and time constraints, adverse effects of potential control measures, and human activities that might influence the spread of the infection. Many times control activities reflect compromises among alternatives. To proceed intelligently, one must identify components of the infectious disease cycle that are primarily responsible for a particular epidemic. Control measures should be directed toward that part of the cycle that is most susceptible to control—the weakest link in the chain.

There are three kinds of control measures. The first kind is directed toward reducing or eliminating the source or reservoir of infection:

1. Quarantine and isolation of cases and carriers
2. Destruction of an animal reservoir of infection
3. Treatment of sewage to reduce water contamination
4. Therapy that reduces or eliminates infectivity of the individual

The second kind of control measure is designed to break the connection between the source of the infection and susceptible individuals. Examples include general sanitation measures:

1. Chlorination of water supplies
2. Pasteurization of milk
3. Supervision and inspection of food and people who handle food
4. Destruction of vectors by spraying with insecticides

The third type of control measure reduces the number of susceptible individuals and raises the general level of herd immunity by immunization. Examples include the following:

1. Passive immunization to give a temporary immunity following exposure to a pathogen or when a disease threatens to take an epidemic form
2. Active immunization to protect the individual from the pathogen and the host population from the epidemic

The Role of the Public Health System: Epidemiological Guardian

The control of an infectious disease relies heavily on a well-defined network of nurses, physicians, epidemiologists, and infection control personnel who supply epidemiological information to a network of local, state, national, and international organizations. These individuals and organizations comprise the public health system. For example, each state has a public health laboratory that is involved in infection surveillance and control. The communicable disease section of a state laboratory includes specialized laboratory services for the examination of specimens or cultures submitted by physicians, the local health department, hospital laboratories, sanitarians, epidemiologists, and others. These groups share their findings with other health agencies in the state, the Centers for Disease Control (CDC), and the World Health Organization (WHO).

The Centers for Disease Control of the United States Public Health Service (USPHS), located in Atlanta, Georgia, has a central role in the U.S. public health system. The CDC (1) supports all United States laboratories, the World Health Organization, and other countries in their efforts against infectious diseases; (2) publishes the accumulated state reports in the *Morbidity and Mortality Weekly Reports (MMWR);* (3) establishes criteria for the safe and accurate performance of microbiological procedures; (4) trains public health personnel; and (5) conducts seminars on all aspects of epidemiology.

Nosocomial Infections

Nosocomial infections (Greek *nosos,* disease, and *komeion,* to take care of) are produced by infectious pathogens that develop within a hospital or other type of clinical care facility and are acquired by patients while they are in the facility. Besides harming patients, nosocomial infections can affect nurses, physicians, aides, visitors, salespeople, delivery personnel, custodians, and anyone who has contact with the hospital. Most nosocomial infections become clinically apparent while patients are still hospitalized; however, disease onset can occur after patients have been discharged. Infections that are incubating when patients are admitted to a hospital are not nosocomial; they are community acquired. However, because such infections can serve as a ready source or reservoir of pathogens for other patients or personnel, they are also considered in the total epidemiology of nosocomial infections.

Source

Pathogens that cause nosocomial diseases come from either endogenous or exogenous sources. Endogenous sources are the patient's own microbiota; exogenous sources are microbiota other than the patient's. Endogenous pathogens are either brought into the hospital by the patient or are acquired when the patient becomes colonized after admission. In either case, the pathogen colonizing the patient may subsequently cause a nosocomial disease (e.g., when the pathogen is transported to another part of the body or when the host's resistance drops). If it cannot be determined that the specific pathogen responsible for a nosocomial disease is exogenous or endogenous, then the term autogenous is used. An **autogenous infection** is one by an agent derived from the microbiota of the patient, despite whether it became part of the patient's microbiota following his or her admission to the hospital.

There are many potential exogenous sources in a hospital. Animate sources are the hospital staff, other patients, and visitors. Some examples of inanimate exogenous sources are food, urinary catheters, intravenous and respiratory therapy equipment, and water systems (softeners, dialysis units, and hydrotherapy equipment).

Control, Prevention, and Surveillance

In the United States, nosocomial infections occur in approximately 5% of the patients admitted to acute-care hospitals, prolong hospital stays by 4 to 13 days, result in over two billion dollars a year in direct hospital charges, and lead to over 20,000 direct and 60,000 indirect deaths annually. The enormity of this problem has led most hospitals to allocate substantial resources to the development of methods and programs for the surveillance, prevention, and control of nosocomial infections.

All personnel involved in the care of patients should be familiar with basic infection control measures such as isolation policies of the hospital; aseptic techniques; proper handling of equipment, supplies, food, and excreta; and surgical wound care and dressings. To adequately protect their patients, hospital personnel must practice proper aseptic technique and hand-washing procedures, and must wear gloves when contacting mucous membranes, secretions, and "moist body substances." Patients also should be monitored with respect to the frequency, distribution, symptomatology, and other characteristics common to nosocomial infections. This constant control and surveillance can be invaluable in preventing many nosocomial infections, patient discomfort, extra stays, and further expense. Nosocomial infections are discussed further in chapter 38 (*pp. 778–79*).

The Hospital Epidemiologist

Because of nosocomial infections, all hospitals desiring accreditation by the Joint Commission on Accreditation of Hospitals (JCAH) must have a designated individual directly responsible for developing and implementing policies governing control of infections and communicable diseases. This individual is usually a registered nurse and is known as a hospital epidemiologist, nurse epidemiologist, infection control nurse, or infection control practitioner. He or she must report to an infection control committee composed of various professionals who have expertise in the different aspects of infection control. The infection control committee periodically evaluates laboratory reports, patients' charts, and surveys done by the hospital epidemiologist to determine whether there has been any increase in the frequency of particular infectious diseases or potential pathogens.

Overall, the services provided by the infection control practitioner should include at least the following:

1. Research in infection control
2. Evaluation of disinfectants, rapid test systems, and other products
3. Efforts to encourage appropriate legislation related to infection control, particularly at the state level
4. Efforts to contain hospital operating costs, especially those related to fixed expenses such as the DRGs (diagnostic related groups)
5. Surveillance and comparison of endemic and epidemic infection frequencies
6. Direct participation in a variety of hospital activities relating to infection control
7. Establishment and maintenance of a system for identifying, reporting, investigating, and controlling infections and communicable diseases of patients and hospital personnel
8. Maintenance of a log of incidents related to infections and communicable diseases

Recently, computer software packages have been developed to aid the infection control practitioner. Such packages generate routine reports, cause-and-effect tabulations, and graphics for the daily epidemiological monitoring that must be done.

---

1. In what three general ways can epidemics be controlled? Give one or two specific examples of each type of control measure.
2. What role does the public health system play in epidemiology?
3. Describe a nosocomial infection.
4. What two general sources are responsible for nosocomial infections? Give some specific examples of each general source.
5. Why are nosocomial infections important?
6. What does a hospital epidemiologist do to control nosocomial infections?

Summary

1. Epidemiology is the science that evaluates the determinants, occurrence, distribution, and control of health and disease in a defined population.
2. The epidemiology of an infectious disease involves the determination of what pathogen caused a specific disease, the source and/or reservoir of the pathogen, how the disease was transmitted, and what host and environmental factors could have caused the disease to develop within a defined population. Disease control and elimination are also important functions of the epidemiologist.
3. Epidemiological data can be obtained from such factors as morbidity, prevalence, and mortality rates.
4. Surveillance is necessary for recognizing a specific infectious disease within a given population. This consists of gathering data on the occurrence of the disease, collating and analyzing the data, summarizing the findings, and applying the information to control measures.
5. One purpose of the clinical microbiology laboratory is to isolate and identify the pathogen responsible for an infectious disease outbreak.
6. A common-source epidemic is characterized by a sharp rise to a peak and then a rapid, but not as pronounced, decline in the number of individuals infected. A propagated epidemic is characterized by a relatively slow and prolonged rise and then a gradual decline in the number of individuals infected.
7. Herd immunity is the resistance of a population to infection and pathogen spread because of the immunity of a large percentage of the individuals within the population.
8. The infectious disease cycle or chain involves the characteristics of the pathogen, the source and/or reservoir of the pathogen, the transmission of the pathogen, the susceptibility of the host, the exit mechanism of the pathogen from the body of the host, and its spread to a new reservoir or host.
9. There are four major modes of transmission: airborne, contact, vehicle, and vector-borne.
10. The public health system consists of individuals and organizations that function in the control of infectious diseases and epidemics.
11. Epidemiological control measures can be directed toward reducing or eliminating infection sources, breaking the connection between sources and susceptible individuals, or isolating the susceptible individuals and raising the general level of herd immunity by immunization.
12. Nosocomial infections are infections that develop within a hospital and are produced by a pathogen acquired during a patient's stay. These infections come from either endogenous or exogenous sources.
13. Hospitals must designate an individual to be responsible for identifying and controlling nosocomial infections. This person is known as a nurse epidemiologist, hospital epidemiologist, infection control nurse, or infection control practitioner.

Key Terms

active carrier *706*
acute carrier *706*
airborne transmission *707*
antigenic drift *704*
antigenic shift *704*
autogenous infection *711*
biologic transmission *707*
carrier *706*
casual carrier *706*
chronic carrier *706*
common-source epidemic *702*
common vehicle transmission *707*
communicable disease *706*
contact transmission *707*
convalescent carrier *706*
disease *699*
droplet nuclei *707*
endemic disease *699*
enzootic *699*
epidemic *699*
epidemiologist *699*
epidemiology *699*
epizootic *699*
epizootiology *699*
fomite *707*
harborage transmission *707*
health *699*
healthy carrier *706*
herd immunity *703*
hyperendemic disease *699*
incubatory carrier *706*
index case *699*
infectious disease cycle (chain) *704*
morbidity rate *701*
mortality rate *701*
nosocomial infection *711*
outbreak *699*
pandemic *699*
panzootic *699*
period of infectivity *706*
prevalence rate *701*
propagated epidemic *703*
puerperal fever *710*
reservoir *706*
seroprevalence rate *701*
source *706*
sporadic disease *699*
statistics *699*
transient carrier *706*
vector *706*
vector-borne transmission *707*
vehicle *707*
zoonosis *699*

Questions for Thought and Review

1. Why is international cooperation a necessity in the field of epidemiology?
2. Are any risks involved in the attempt to eliminate a pathogen from the world?
3. Why is a knowledge of statistics important to an epidemiologist?
4. What is the role of the Centers for Disease Control in epidemiology?
5. What common sources of infectious disease are found in your community? How can the etiologic agents of such infectious diseases spread from their source or reservoir to members of your community?
6. How could you experimentally prove the cause of an infectious disease?
7. How do epidemiologists recognize an infectious disease within a given population?
8. How could you prove that an epidemic of a given infectious disease was occurring?
9. Why is the infectious disease cycle also referred to as a chain?
10. How can epidemics be controlled?
11. Why are nosocomial infections so important to humans?
12. Why do some epidemiologists believe it is impossible to prevent all nosocomial infections?
13. What is the difference between a fomite and a vector? A reservoir and a source?
14. How can changes in herd immunity contribute to an outbreak of a disease on an island?
15. What is an index case?

Additional Reading

Anderson, R. M., and May, R. M. 1992. Understanding the AIDS pandemic. *Sci. Am.* 266(5):58–67.

Axnick, K. J., and Yarbrough, M. 1984. *Infection control: An integrated approach.* St. Louis: C. V. Mosby.

Ayliffe, G. A., and Taylor, L. J. 1982. *Hospital acquired infections: Principles and prevention.* Littleton, Mass.: John Wright-PSG.

Bennett, J. V., and Brachman, P. S., eds. 1986. *Hospital infections,* 2d ed. Boston: Little, Brown.

Beyt, B. E., Jr.; Troxler, S. H.; and Guidry, J. L. 1984. Computer assisted hospital surveillance and control of nosocomial infections. *Clinical Research* 32:291A.

Centers for Disease Control. *Morbidity and mortality reports.* (A weekly report that discusses infectious diseases.) Atlanta: Centers for Disease Control.

Centers for Disease Control. *Surveillance.* (Annual summaries of specific infectious diseases.) Atlanta: Centers for Disease Control.

Chesney, P. J.; Bergdoll, M. S.; and Davis, J. P. 1984. The disease spectrum, epidemiology, and etiology of toxic-shock syndrome. *Ann. Rev. Microbiol.* 38: 315–38.

Cliff, A., and Haggett, P. 1984. Island epidemics. *Sci. Am.* 250(5):138–47.

Epidemiology issue. 1986. *Science* 234 (November 24): 921.

Fraser, D. W., and McDade, J. E. 1979. Legionellosis. *Sci. Am.* 241(4):82–99.

Haley, R. W.; Quade, D.; Freeman, H. E.; and Bennett, J. V. 1980. Conceptual model of an infection surveillance and control program. *Am. J. Epidemiol.* 111:608–12.

Harris, A. A.; Levin, S.; and Rrenholme, G. 1984. Selected aspects of nosocomial infections in the 1980s. *Am. J. Med.* 77(1B):3–11.

Heyward, W. L., and Curran, J. W. 1988. The epidemiology of AIDS in the U.S. *Sci. Am.* 259(4):72–81.

Kahn, H. A. 1983. *An introduction to epidemiologic methods.* New York: Oxford University Press.

Kaplan, M. M., and Webster, R. G. 1977. The epidemiology of influenza. *Sci. Am.* 237(6):88–92.

Lilienfeld, A. M., and Lilienfeld, D. E. 1980. *Foundations of epidemiology,* 2d ed. New York: Oxford University Press.

Mann, J. M.; Chin, J.; Piot, P.; and Quinn, T. 1988. The international epidemiology of AIDS. *Sci. Am.* 259(4):82–89.

Mausner, J. S., and Kramer, S. 1985. *Epidemiology: An introductory text,* 2nd ed. Philadelphia, PA: W. B. Saunders.

McEvedy, C. 1988. The bubonic plague. *Sci. Am.* 258(2):118–23.

Murray, J. D. 1987. Modeling the spread of rabies. *American Scientist* 75:280–84.

Pike, R. M. 1979. Laboratory-associated infections: Incidence, fatalities, causes, and prevention. *Ann. Rev. Microbiol.* 33:41–66.

Riley, L. W. 1987. The epidemiologic, clinical, and microbiological features of hemorrhagic colitis. *Ann. Rev. Microbiol.* 41:383–407.

Roueche, B. 1984. *The medical detectives.* 2 vols. New York: Truman Talley Books.

Salk, D. 1980. Eradication of poliomyelitis in the United States. *Rev. Infect. Dis.* 2:228–30.

Schultz, M. G. 1983. Emerging zoonoses. *N. Engl. J. Med.* 308:1285.

Stuart-Harris, C. 1981. The epidemiology and prevention of influenza. *American Scientist* (March–April):166–72.

Thomas, C., and Morgan-Witts. 1982. *Anatomy of an epidemic.* New York: Doubleday.

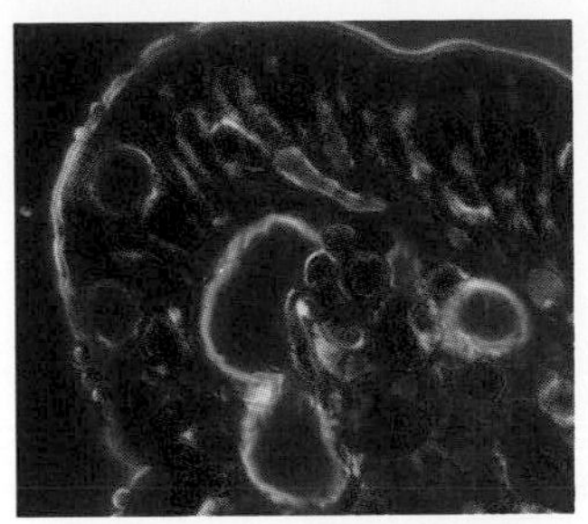

CHAPTER 36
Human Diseases Caused by Viruses

It is a modern plague: the first great pandemic of the second half of the 20th century. The flat, clinical-sounding name given to the disease by epidemiologists—acquired immune deficiency syndrome—has been shortened to the chilling acronym AIDS.

Robert C. Gallo

Outline

Concepts

1. Some viruses can be transmitted through the air and directly or indirectly involve the respiratory system. Most of these viruses are highly communicable and cause diseases such as chickenpox, German measles, influenza, measles, mumps, respiratory syndromes and viral pneumonia, and the now extinct smallpox.
2. The arthropod-borne diseases are transmitted by arthropod vectors from human to human or animal to human. Examples include the various encephalitides, Colorado tick fever, and historically important yellow fever.
3. Some viruses are so sensitive to environmental influences that they are unable to survive for significant periods of time outside their hosts. These viruses are transmitted from host to host by direct contact and cause diseases such as AIDS, cold sores, the common cold, cytomegalovirus inclusion disease, genital herpes, certain leukemias, infectious mononucleosis, rabies, and hepatitis.
4. Viruses that can be transmitted by food and water and usually either grow in or pass through the intestinal system leave the body in the feces and are acquired through the oral route. Examples of such diseases include viral gastroenteritis, infectious hepatitis, and poliomyelitis.
5. The slow virus diseases represent progressive pathological processes caused by viruses or prions that remain clinically silent during a prolonged period of months or years, after which progressive clinical disease becomes apparent, usually ending months later in profound disability or death. Examples include Creutzfeldt-Jakob disease, kuru, progressive multifocal leukoencephalopathy, Gerstmann-Sträussler-Scheinker syndrome, and subacute sclerosing panencephalitis.
6. One other disease associated with viruses but that does not fit into any of the foregoing categories is warts.

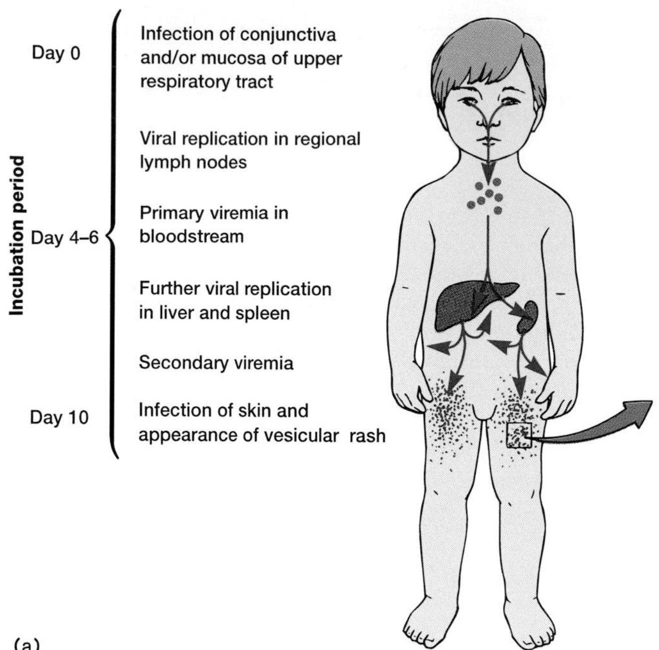

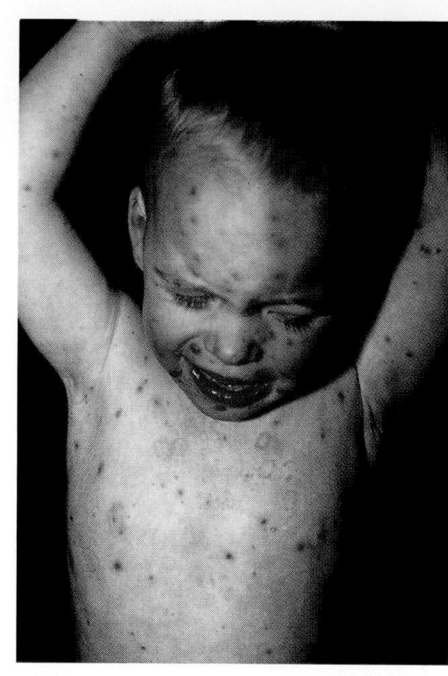

Figure 36.1 **Chickenpox (Varicella).** (*a*) Pathogenesis. (*b*) Typical vesicular skin rash.

Chapters 17, 18, and 19 provide a review of the general biology of viruses and an introduction to basic virology. Chapter 36 continues this coverage by discussing viruses that are pathogenic to humans. Viruses are grouped according to their mode of acquisition and transmission, and viral diseases that occur in the United States are emphasized.

More than 400 different viruses can infect humans. Human diseases caused by viruses are unusually interesting, considering the small amount of genetic information introduced into a host cell. This apparent simplicity belies the severe pathological features, clinical consequences, and death that result from many viral diseases. With few exceptions, only prophylactic or supportive treatment is available. Collectively, these diseases are some of the most common and yet most puzzling of all infectious diseases. The resulting frustration is compounded when year after year familiar diseases of unknown etiology become linked to virus infections.

Airborne Diseases

Because air does not support virus growth, any virus that is airborne must have originated from a source such as another human. When humans are the source of the airborne virus, it is usually propelled from the respiratory tract by an individual's coughing, sneezing, or vocalizing.

Chickenpox (Varicella) and Shingles (Herpes Zoster)

Chickenpox (varicella) is a highly contagious skin disease primarily of children two to seven years of age. About 200,000 cases are reported annually in the United States. The causative agent is the varicella-zoster virus, a member of the family *Herpesviridae,* which is acquired by droplet inhalation into the respiratory system. Following an incubation period of from 10 to 23 days, small vesicles erupt on the face or upper trunk, fill with pus, rupture, and become covered by scabs (figure 36.1). Healing of the vesicles occurs in about 10 days. During this time, intense itching is often present; treatment is with acyclovir. Chickenpox can be prevented or the infection shortened with a live varicella vaccine.

Individuals who recover from chickenpox are subsequently immune to this disease; however, they are not free of the virus. Some viruses reside as viral DNA in a dormant state within the nuclei of sensory neurons in the dorsal root ganglia, producing a latent infection (figure 36.2*a*). During latency, viral DNA is maintained in infected cells but virions cannot be detected. When the infected person becomes immunocompromised (e.g., AIDS) or is under psychological or physiological stress, the viruses may become activated (figure 36.2*b*). They migrate down sensory nerves, initiate viral replication, and produce painful vesicles (figure 36.2*c*) because of sensory nerve damage. This reactivated form of chickenpox is called **shingles (herpes zoster)**. Most cases occur in people over 50 years of age. Heart surgery patients often get shingles because of the stress accompanying such a procedure. Shingles does not require specific therapy; however, in immunocompromised individuals, acyclovir (Zovirax; *see figure 16.13*) is recommended.

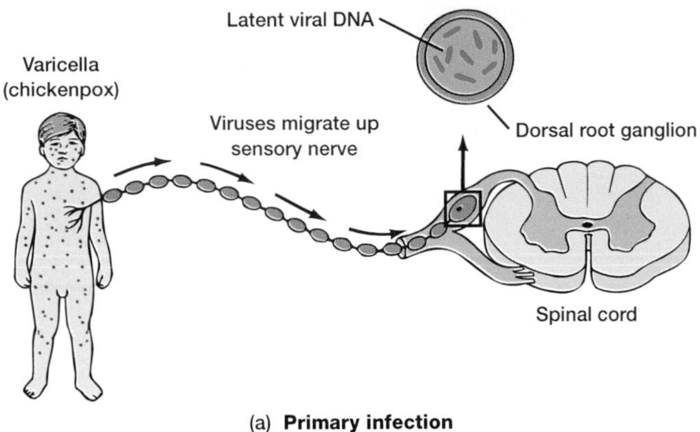

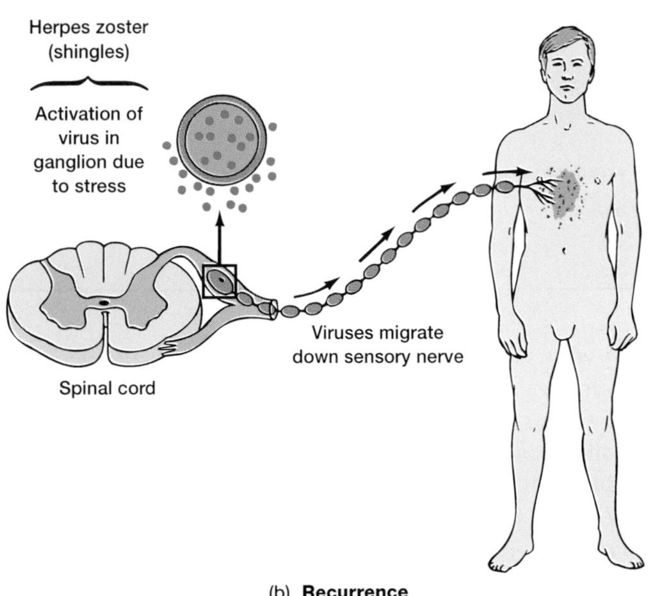

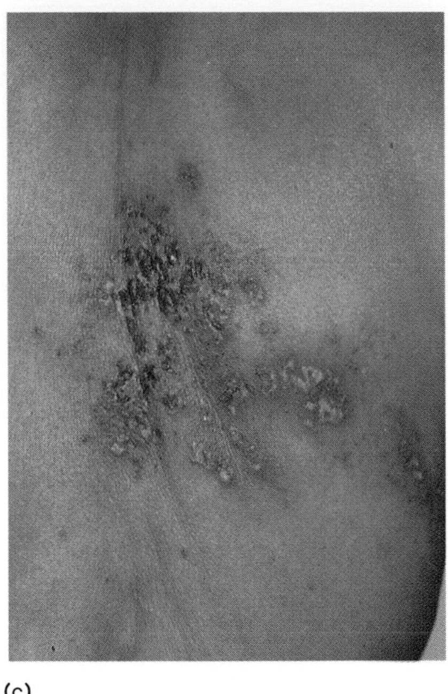

Figure 36.2 Pathogenesis of the Varicella-zoster Virus. (*a*) After an initial infection with varicella (chickenpox), the viruses migrate up sensory peripheral nerves to their ganglia, producing a latent infection. (*b*) When a person becomes immunocompromised or is under psychological or physiological stress, the viruses may be activated. (*c*) They migrate down sensory nerve axons, initiate viral replication, and produce painful vesicles.

German Measles (Rubella)

Rubella (Latin *rubellus,* reddish) was first described in Germany in the 1800s and was subsequently called German measles. It is a moderately contagious skin disease that occurs primarily in children five to nine years of age. It is caused by the rubella virus, a single-stranded RNA virus that is a member of the family *Togaviridae.* Rubella is worldwide in distribution and occurs more frequently during the winter and spring months. This virus is spread in droplets that are shed from the respiratory secretions of infected individuals. Once the virus is inside the body, the incubation period ranges from 12 to 23 days. A rash of small red spots (figure 36.3), usually lasting no more than 3 days, and a light fever are the normal symptoms. The rash appears as immunity develops and the virus disappears from the blood, suggesting that the rash is immunologically mediated and not caused by the virus infecting skin cells.

Rubella can be a disastrous disease (**congenital rubella syndrome**) in the first trimester of pregnancy and can lead to

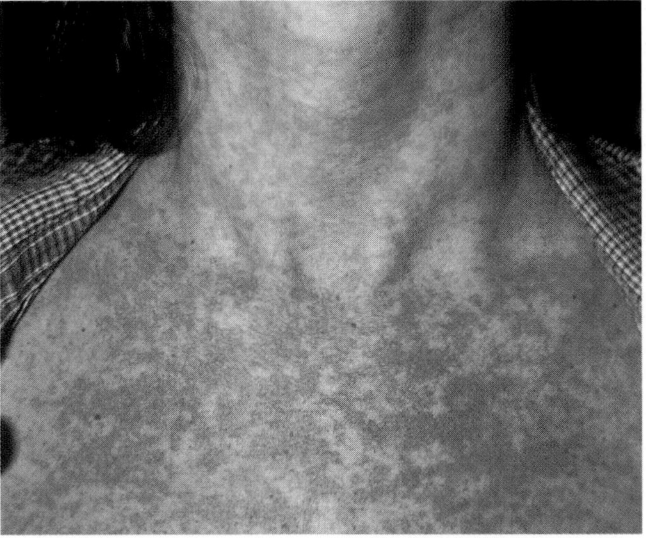

Figure 36.3 German Measles (Rubella). This disease is characterized by a rash of red spots. Notice that the spots are not raised above the surrounding skin as in measles (rubeola; see figure 36.4).

Box 36.1

Reye's and Guillain-Barré Syndromes

An occasional complication of influenza, chickenpox, and a few other viral diseases in children under 14 years of age is **Reye's syndrome.** After the initial infection has disappeared, the child suddenly begins to vomit persistently and experiences convulsions followed by delirium and a coma. Pathologically, the brain swells with injury to the neuronal mitochondria, fatty infiltration into the liver occurs, blood ammonia is elevated, and both serum glutamic oxaloacetic transaminase (SGOT) and serum glutamic pyruvic transaminase (SGPT) are elevated in the blood. Diagnosis is made by the measurement of the levels of these enzymes and ammonia.

The relationship between the initial viral infection and the brain and liver damage is unknown. Treatment is nonspecific and directed toward reducing intracranial pressure and correcting metabolic and electrolyte abnormalities. Some children who recover have residual neurological deficits—impaired mental capacity, seizures, and hemiplegia (paralysis on one side of the body). Mortality ranges from 10 to 40%. It is suspected that the use of aspirin or salicylate-containing products to lower the initial viral fever increases one's chances of acquiring Reye's syndrome.

Another condition that involves the central nervous system and is associated with influenza infections is **Guillain-Barré syndrome** (sometimes called French polio). In this disorder, the individual suffers a delayed reaction (usually within eight weeks) either to the actual virus infection or to vaccines against influenza. The virus or viral antigen in the vaccine damages the Schwann cells that myelinate the peripheral nerves and thus cases demyelination. As a result, a prominent feature of this syndrome is a symmetric weakness of the extremities and sensory loss. Fortunately, recovery is usually complete because Schwann cells eventually proliferate and wrap around the demyelinated nerves.

fetal death, premature delivery, or a wide array of congenital defects that affect the heart, eyes, and ears.

Because rubella is usually such a mild infection, no treatment is indicated. All children and women of childbearing age who have not been previously exposed to rubella should be vaccinated. The live attenuated rubella vaccine (RA 27/3) is recommended. Because routine vaccination began in the United States in 1969, fewer than 1,000 cases of rubella and 10 cases of congenital rubella occur annually.

Influenza (Flu)

Influenza (Italian, to influence), or the **flu,** is a respiratory system disease caused by orthomyxoviruses. Influenza viruses are classified into A, B, and C groups based on the antigens (H and N) of their protein coats. One unique feature of the influenza viruses is the frequency with which changes in antigenicity occur. These changes are called antigenic variation. If the variation is small, it is called antigenic drift; if it is large, it is called antigenic shift. Antigenic variation occurs almost yearly with the influenza A virus, less frequently with the B virus, and has not been demonstrated with the C virus. Alteration of the virus's antigenic structure can lead to infection in humans by variants of the virus to which little or no resistance is present in the population at risk. This explains why influenza continues to be a major epidemic disease and frequently produces worldwide pandemics.

The virus (*see figure 17.17*a) is acquired by inhalation or ingestion of virus-contaminated respiratory secretions. During an incubation period of one to two days, the virus adheres to the epithelium of the respiratory system and hydrolyzes the mucus that covers it using the neuraminidase present in the viral envelope spikes. The virus then attaches to the epithelial cell by its hemagglutinin spike protein and enters the cell. Influenza is characterized by chills, fever, headache, malaise, and general muscular aches and pains. These symptoms arise from the death of respiratory epithelial cells, probably due to attacks by activated T cells. Recovery usually occurs in 3 to 7 days, during which cold-like symptoms appear as the fever subsides. Influenza alone is not usually fatal. However, death may result from pneumonia caused by secondary bacterial invaders such as *Staphylococcus aureus, Streptococcus pneumoniae,* and *Haemophilus influenzae.* Influenza epidemics vary greatly in severity and mortality rate. For example, the pandemics of 1918 and 1957 were severe, while the 1977 pandemic was mild. The most disastrous recorded pandemic occurred in 1918–1919. It has been estimated that 21 million died worldwide, and about 550,000 deaths were recorded in the United States alone.

As with many other viral diseases, only the symptoms of influenza are usually treated. However, the antiviral drug amantadine (*see figure 16.13*) has been shown to reduce the duration and symptoms of type A influenza if administered promptly. Aspirin (salicylic acid) should be avoided in children younger than 14 years to reduce the risk of Reye's syndrome (Box 36.1). The mainstay for prevention of influenza since the late 1940s has been inactivated virus vaccines, especially for the chronically ill, individuals over age 65, and residents of nursing homes.

Antiviral drugs (p. 341).

Influenza A infections usually peak in the winter and involve 10% or more of the population, with rates of 50 to 75% in school-age children. Influenza B usually accounts for only 3% of all flu cases in the United States.

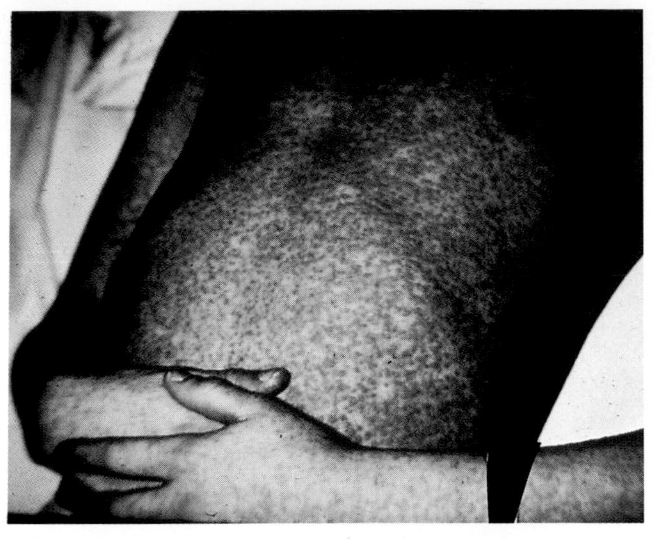

(a)

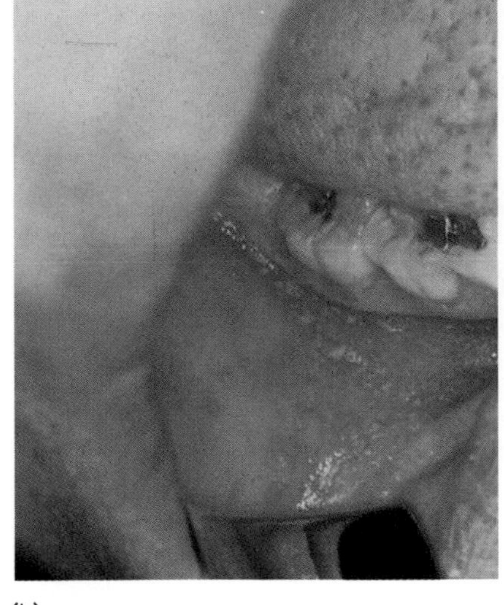

(b)

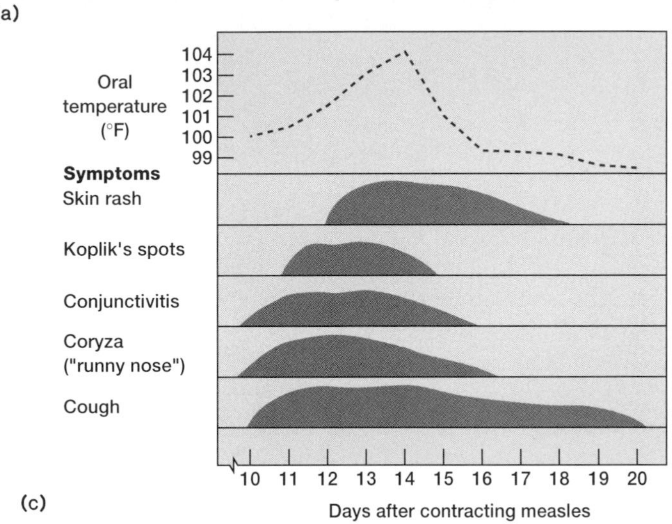

(c)

Figure 36.4 Measles (Rubeola). (*a*) The rash of small, raised spots is typical of measles. (*b*) Koplik's spots in the oral cavity are a characteristic sign of a measles infection. (*c*) Signs and symptoms of a measles infection.

Measles (Rubeola)

Measles (rubeola: Latin *rubeus,* red) is a highly contagious skin disease that is endemic throughout the world. The measles virus, a member of the genus *Morbillivirus* and the family *Paramyxoviridae,* enters the body through the respiratory tract or the conjunctiva of the eyes. The incubation period is usually 10 to 21 days, and the first symptoms begin about the tenth day with a nasal discharge, cough, fever, headache, and conjunctivitis. Within 3 to 5 days, skin eruptions occur as faintly pink macules (figure 36.4) and normally last about 5 to 10 days. Lesions of the oral cavity include the diagnostically useful bright-red **Koplik's spots** with a bluish-white speck in the center of each (figure 36.4*b*). Very infrequently, there occurs a progressive degeneration of the central nervous system called subacute sclerosing panencephalitis (table 36.6). Permanent immunity to measles follows the disease, and no specific treatment is available. The use of attenuated measles vaccine (Attenuvax) or in combination (MMR vaccine; measles, mumps, rubella) is recommended for all children. Since public health immunization programs began in 1966, there has been over a 98% decrease in measles cases. Currently, there are about 2,000 to 3,000 cases a year in the United States, and 90% of these are among unvaccinated individuals. In less well-developed countries, however, the morbidity and mortality in young children from measles infection remain high. It has been estimated that measles kills about 220,000 people a year worldwide.

Mumps

Mumps is an acute generalized disease that occurs primarily in school-age children. It is caused by a paramyxovirus that is transmitted in saliva and respiratory droplets to nonimmune contacts. The portal of entry is the respiratory tract. The most prominent manifestations of mumps are swelling and tenderness of the salivary (parotid) glands 16 to 18 days after infection of the host by the virus (figure 36.5). The swelling usually lasts for 1 to 2 weeks and is accompanied by a low grade fever. Meningitis and inflammation of the epididymis and testes can be important complications associated with this disease—especially in the postpubescent male. Therapy of mumps is limited to symptomatic and supportive measures. A live, attenuated mumps virus vaccine is available. It is usually given as part of the triple MMR vaccine. There are currently

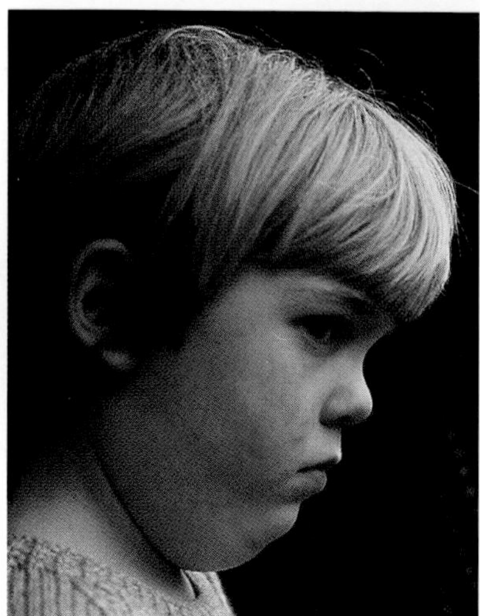

Figure 36.5 **Mumps.** Child with diffuse swelling of the salivary (parotid) glands due to the mumps virus.

about 3,000 cases of mumps in the United States each year. Recently, a few outbreaks have occurred among unvaccinated children in states without a comprehensive school immunization law. Prevention and control involves keeping children with mumps from school or other associated activities for about two weeks.

Respiratory Syndromes and Viral Pneumonia

Acute viral infections of the respiratory system are among the most common causes of human disease. The infectious agents are called the acute respiratory viruses and collectively produce a variety of clinical manifestations, including rhinitis (inflammation of the mucous membrane of the nose), tonsillitis, laryngitis, and bronchitis. The adenoviruses, coxsackievirus A, coxsackievirus B, echovirus, influenza viruses, parainfluenza viruses, poliovirus, respiratory syncytial virus, and reovirus are thought to be responsible. It should be emphasized that for most of these viruses there is a lack of specific correlation between the agent and the clinical manifestation—hence the term syndrome. A **syndrome** is a set of signs and symptoms that occur together and characterize a particular disease. Immunity is not complete, and reinfection is common. The best treatment is care and rest.

Classification of animal viruses (pp. 384–87).

In those cases of pneumonia for which no cause can be identified, viral pneumonia may be assumed if mycoplasmal pneumonia has been ruled out. The clinical picture is nonspecific. Symptoms may be mild, or there may be severe illness and death.

Mycoplasma disease, primary atypical pneumonia (p. 773).

Respiratory syncytial virus (RSV) is often described as the most dangerous cause of respiratory infection in young children. The RSV is a member of the RNA virus family, *Paramyxoviridae.* It is enveloped with two virally specific glycoproteins as part of the structure. One of them, the large glycoprotein or G, is responsible for the binding of the virus to the host cell; the other, the fusion protein or F, permits fusion of the viral envelope with the host cell plasma membrane, leading to entry of the virus. The F protein also induces the fusion of the plasma membranes of infected cells. RSV thus gets its name from the resulting formation of a syncytium or multinucleated mass of fused cells. The source of the RSV is respiratory secretions of humans. Clinical manifestations consist of an acute onset of fever, cough, rhinitis, and nasal congestion. In infants and young children, this often progresses to bronchitis and viral pneumonia. Diagnosis is by a Directigen rapid test kit. The virus is found worldwide and occurs as yearly (winter or early spring) outbreaks lasting several months. Treatment is with aerosolized ribavirin. Prevention and control consists of isolation for RSV-infected individuals, use of gowns when contact with secretions is likely, and strict attention to good handwashing practices.

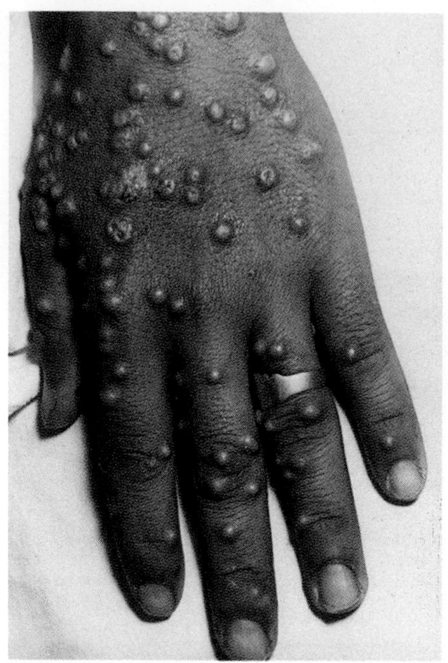

Figure 36.6 **Smallpox.** Back of hand showing single crop of smallpox vesicles.

Smallpox (Variola)

Smallpox (variola) was once one of the most prevalent of all diseases (figure 36.6). Since the advent of immunization with the vaccinia virus (*see figure 17.18*), and because of concerted efforts by the World Health Organization, smallpox has been eradicated throughout the world. This was possible because a disease such as smallpox has obvious clinical features, virtually no asymptomatic carriers, only a human reservoir, and a short period of infectivity (three to four weeks). Because humans are the only hosts, the spread of smallpox virus was prevented until no new cases developed, effectively eradicating the disease. The virus, however, is still maintained in selected laboratories around the world.

Disease (smallpox) and the early colonization of America (p. 348).

Box 36.2

Viral Hemorrhagic Fevers: A Microbial History Lesson

Scientists know of several viruses lurking in the tropics that—with a little help from nature—could wreak far more loss of life than will likely result from the AIDS pandemic. Collectively, these viruses produce what is known as the **hemorrhagic fevers.** The viruses are passed among wild vertebrates which serve as reservoir hosts. Arthropods transmit the viruses among vertebrates, and to humans when they invade the environment of the natural host. These diseases, distributed throughout the world, are known by over 70 names, usually denoting the geographic area where they were first described.

Viral hemorrhagic fevers are most often fatal. Patients suffer headache, muscle pain, flushing of the skin, massive hemorrhaging either locally or throughout the body, circulatory shock, and death.

Despite the lack of public awareness, recent outbreaks may foreshadow much broader outbreaks in the future. For example, in the late 1960s, dozens of scientists in West Germany fell seriously ill, and several died, from a mysterious new disease. Victims suffered from a breakdown of liver function and a bizarre combination of bleeding and blood clots. The World Health Organization traced the outbreak to a batch of fresh monkey cells the scientists had used to grow polio viruses. The cells from the imported Ugandan monkeys were infected with the lethal tropical Marburg virus and the scientists suffered from **Marburg viral hemorrhagic fever.**

In 1977, the *Plebovirus* causing Rift Valley Fever in sheep and cattle moved from these animals into the South African population. The virus, which causes severe weakness, incapacitating headaches, damage to the retina, and hemorrhaging, then made its way to Egypt, where millions of humans became infected and thousands died.

Among the most frightening hemorrhagic outbreak was that of the **Ebola virus hemorrhagic fever** in Zaire and Sudan in 1976. This disease infected more than 1,000 people and left over 500 dead. It became concentrated in hospitals, where it killed many of the Belgian physicians and nurses treating infected patients.

In the United States, in 1989, epidemiologists provided new evidence that rats infected with a potentially deadly hemorrhagic virus are prevalent in Baltimore slums. The virus appears to be taking a previously unrecognized toll on the urban poor by causing **Korean hemorrhagic fever.**

Although to date, these epidemics have not become global, they do provide a humbling vision of humankind's viral vulnerability. History shows that the life-threatening viral hemorrhagic outbreaks have often arisen when humans moved into unexplored terrain or when living conditions deteriorated in ways that generated new viral hosts. In each case, medical and scientific resources have been reactive, not proactive.

1. Why are chickenpox and shingles discussed together? What is their relationship?
2. When is a German measles infection most dangerous and why?
3. Briefly describe the course of an influenza infection and how the virus causes the symptoms associated with the flu. Why has it been difficult to develop a single flu vaccine?
4. What are some common symptoms of measles?
5. What are Koplik's spots?
6. What is one side effect that mumps can cause in a young postpubescent male?
7. Describe some syndromes caused by the acute respiratory viruses. Respiratory syncytial virus.
8. Is viral pneumonia a specific disease? Explain.

Arthropod-Borne Diseases

The arthropod-borne viruses (arboviruses) are transmitted by bloodsucking arthropods from one vertebrate host to another. They multiply in the tissues of the arthropod without producing disease, and the vector acquires a lifelong infection. Diseases produced by the arboviruses can be divided into three clinical syndromes: (1) fevers of an undifferentiated type with or without a rash; (2) encephalitis (inflammation of the brain), often with a high fatality rate; and (3) hemorrhagic fevers, also frequently severe and fatal (Box 36.2). Table 36.1 summarizes the six major human arbovirus diseases that occur in the United States. For all these diseases, immunity is believed to be permanent after a single infection. No vaccines are available for humans, although supportive treatment is beneficial.

TABLE 36.1 Summary of the Six Major Human Arbovirus Diseases That Occur in the United States

Disease	Distribution	Vectors	Mortality Rate
California encephalitis (La Crosse)	North Central, Atlantic, South	Mosquitoes (*Aedes* spp.)	Fatalities rare
Colorado tick fever	Pacific Coast (mountains)	Ticks (*Dermacentor andersoni*)	Fatalities rare
Eastern equine encephalitis (EEE)	Atlantic, Southern Coast	Mosquitoes (*Aedes* spp.)	50–70%
St. Louis encephalitis (SLE)	Widespread	Mosquitoes (*Culex* spp.)	10–30%
Venezuelan equine encephalitis (VEE)	Southern United States	Mosquitoes (*Aedes* spp. and *Culex* spp.)	20–30% (children) <10% (adults)
Western equine encephalitis (WEE)	Mountains of the West	Mosquitoes (*Culex* spp.)	3–7%

Yellow Fever

Yellow fever is not endemic to the United States, but because of its historical importance, it is briefly discussed here. Yellow fever was the first human disease found to be caused by a virus. (Walter Reed discovered this in 1901.) It also provided the first confirmation (by Carlos Juan Finley) that an insect could transmit a virus. Yellow fever is caused by a flavivirus that is endemic in many tropical areas, such as Mexico, South America, and Africa.

The disease received its first name, yellow jack, because jaundice is a prominent sign in severe cases. The jaundice is due to the deposition of bile pigments in the skin and mucous membranes because of damage to the liver. The disease is spread through a population in two epidemiological patterns. In the urban cycle, human-to-human transmission is by *Aedes aegypti* mosquitoes. In the sylvatic cycle, the mosquitoes transmit the virus between monkeys and from monkeys to humans (sylvatic means in the woods or affecting wild animals).

Once inside a person, the virus spreads to local lymph nodes and multiplies; from this site, it spreads to the liver, spleen, kidneys, and heart, where it can persist for days. In the early stages of the disease, the infected person experiences fever, chills, headache, and backache, followed by nausea and vomiting. In severe cases, the virus produces lesions in the infected organs and hemorrhaging occurs.

There is no specific treatment for yellow fever. Diagnosis is by serology. An active immunity to yellow fever results from an initial infection or from vaccines containing the attenuated yellow fever 17D strain or the Dakar strain virus. Prevention and control of this disease involves vaccination and control of the insect vector.

Direct Contact Diseases

Acquired Immune Deficiency Syndrome (AIDS)

It is now recognized that **AIDS** (**a**cquired **i**mmune **d**eficiency **s**yndrome) is the first great pandemic of the second half of the twentieth century. First described in 1981, AIDS is the result of an infection by a new virus, the **human immunodeficiency virus (HIV).** The disease appears to have begun in central Africa as early as the 1950s. Recently, a virus (the simian T-lymphotropic virus type III or STLV-III) related to HIV-1, the strain primarily responsible for AIDS, has been isolated from African green monkeys and is believed to be the ancestor of the AIDS virus. Somehow, this virus entered humans and mutated to the two current African human viruses SBL and LAV-2, which are intermediate viruses; these intermediate viruses may have evolved into the highly virulent HIV-1. Once established, HIV-1 spread to the Caribbean and then to the United States and Europe.

Epidemiologically, AIDS occurs worldwide. It is concentrated in central Africa, Haiti, the United States (*see figure 35.2*), and Europe, but is spreading elsewhere. The World Health Organization estimates that 8 to 10 million people are infected worldwide and that 30 to 40 million people will have been infected by the year 2000. Through June 1992, about 230,000 cases and 152,000 deaths have been reported in the United States. However, according to the Centers for Disease Control, the number of U.S. cases will level off during the next five years, with about 60,000 to 70,000 new cases each year (figure 36.7). The groups most at risk in acquiring AIDS are (in descending order of risk) homosexual/bisexual men; intravenous (IV) drug users; heterosexuals who have intercourse with drug users, prostitutes, and bisexuals; transfusion patients or hemophiliacs who must receive clotting factor preparations made from donated blood; and children born of

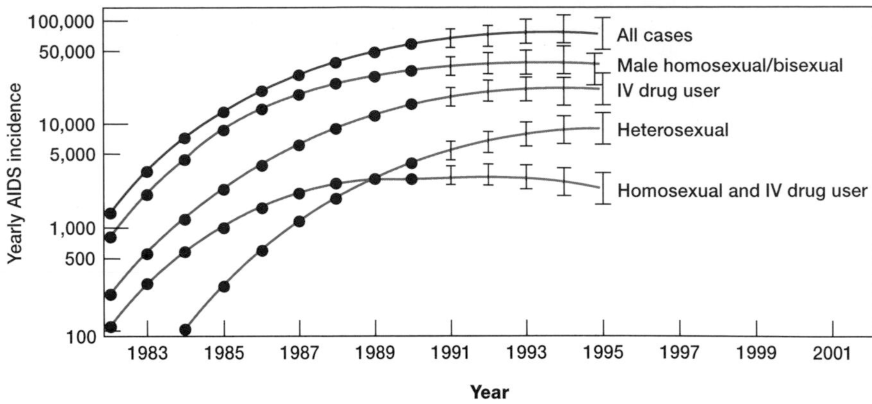

Figure 36.7 **Signs of Slowing in the AIDS Epidemic.** AIDS cases are projected from 1990–1995. Data are corrected for 10% underreporting. Ranges (bars) reflect uncertainty in infection rate and incubation period.

infected mothers. The mortality rate from AIDS is extremely high. About 85 to 90% of those diagnosed as having AIDS before 1986 have already died. By the year 2000, AIDS will become the major killer of children with an estimated 10 million infections worldwide.

AIDS is caused primarily by the HIV-1 virus (some cases result from an HIV-2 infection). This virus is a retrovirus and closely related to HTLV-1, the cause of adult T-cell leukemia, and HTLV-2, which has been isolated from individuals with hairy cell leukemia (*see leukemia, p. 731*). HIV-1 is an enveloped lentivirus (*see chapter 19*) with a cylindrical core inside its capsid (figure 36.8). The core contains two copies of its plus single-stranded RNA genome and several enzymes. Thus far, 10 virus-specific proteins have been discovered. One of them, the gp120 envelope protein, participates in HIV-1 attachment to $CD4^+$ cells (T-helper cells; see below).

One feature that distinguishes the lentiviruses from other retroviruses is the remarkable complexity of their viral genomes. Most retroviruses that are capable of replication contain only three genes—namely, *gag, pol,* and *env*. The *gag* and *env* genes encode the core nucleocapsid polypeptides and surface-coat proteins of the virus, respectively, whereas the *pol* gene codes for the viral reverse transcriptase, protease, integrase, and ribonuclease enzymes. HIV-1 also contains in its 9 kb RNA genome at least six additional genes (*vif, vpu, vpr, tat, rev,* and *nef*) (figure 36.9). It is the distinct and concerted actions of these six genes that probably underlie the profound pathogenicity of HIV-1. From a therapeutic standpoint, this same genome may also be the "Achilles' heel" of the virus. This possibility has spurred a broad-based search for antagonists specific for these HIV-1 gene products.

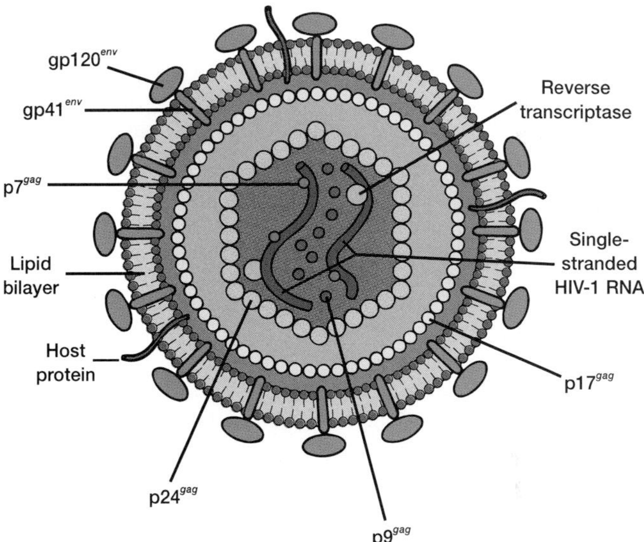

Figure 36.8 **Schematic Diagram of the HIV–1 Virion.** The HIV–1 virion is an icosahedral structure containing 72 external spikes. These spikes are formed by the two major viral-envelope proteins, gp120 and gp41. The HIV–1 lipid bilayer is also studded with various host proteins, including class I and class II histocompatibility antigens, acquired during virion budding. The core of HIV–1 contains four nucleocapsid proteins (p24, p17, p9, p7) each of which is proteolytically cleaved from a 53–kDa (53 kDa)*gag* precursor by the HIV–1 protease. The phosphorylated p24 polypeptide forms the chief component of the inner shelf of the nucleocapsid, whereas the p17 protein is associated with the inner surface of the lipid bilayer and stabilizes the exterior and interior components of the virion. The p7 protein binds directly to the genomic RNA through a zinc-finger structural motif and together with p9 forms the nucleoid core. The retroviral core contains two copies of the single-stranded HIV–1 genomic RNA that is associated with the various preformed viral enzymes, including the reverse transcriptase, integrase, ribonuclease, and protease.

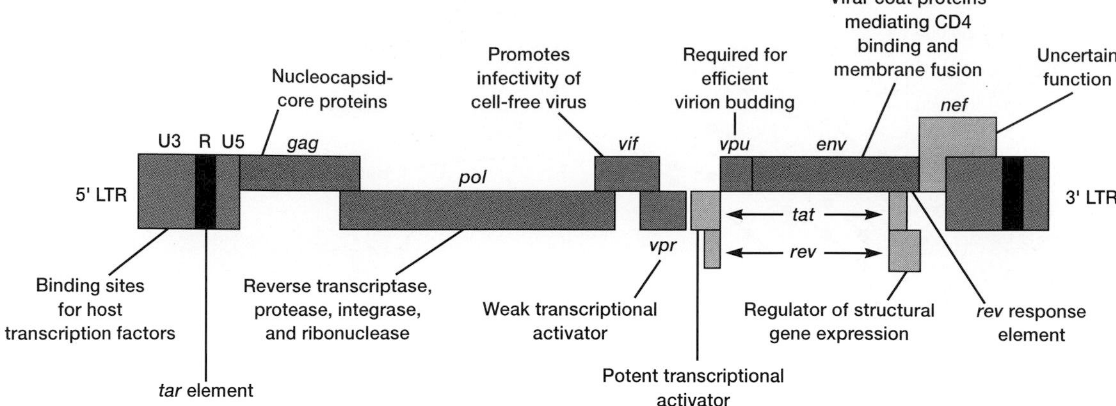

Figure 36.9 **Genomic Structure of HIV–1.** All nine known HIV–1 genes are shown in this schematic drawing and recognized primary functions given. The 5′ and 3′ long terminal repeats (LTRs) containing regulatory sequences recognized by various host transcription factors are also depicted. Notice the positions of the *tat* and *rev* genes, the *tar* (transactivation response) element, and the *rev* response element.

The AIDS virus is acquired by direct exposure of a person's bloodstream to body fluids (blood, semen, vaginal secretions) containing the virus through sexual contact or perinatally from an infected mother to her fetus (table 36.2). It is also possible that a newborn can be infected through breast-feeding. Once inside the body, the virus gp120 envelope protein (figure 36.8) binds to the CD4 glycoprotein plasma membrane receptor on $CD4^+$ T cells, macrophages, and monocytes (figure 36.10). After the envelope has fused with the plasma membrane, the virus releases its core protein and two RNA strands into the cytoplasm.

Inside the infected cell, the core protein remains associated with the RNA as it is copied into a single strand of DNA by the RNA/DNA-dependent DNA polymerase activity of the reverse transcriptase enzyme. The RNA is next degraded by another reverse transcriptase component, ribonuclease H, and the DNA strand is duplicated to form a double-stranded DNA copy of the original RNA genome. A complex of the double-stranded DNA (the provirus) and the integrase enzyme moves into the nucleus. Then the proviral DNA is integrated into the cell's DNA through a complex sequence of reactions catalyzed by the integrase (figure 36.10). The integrated provirus can remain latent, giving no sign of its presence. Alternatively, the provirus can force the cell to synthesize viral mRNA. Some of the RNA is translated to produce viral proteins by the cell's own ribosomes. Viral proteins and the complete HIV-1 RNA genome are then assembled into new virions that bud from the infected host cell (*see figure 19.9*). Eventually, the host cell lyses.

The precise mechanism of AIDS pathogenesis is still not known, and many hypotheses exist. Many believe that AIDS is caused primarily by the direct destruction of T cells, although the exact mechanism is unclear. The cytophathic effect may be due to the disruption of plasma membrane permeability and function by excessive virus budding. Free gp120 proteins may bind to CD4 proteins on uninfected cells, making them targets for attack by immune system cells. Infected cells do fuse with other cells to form large, multinulceate syncytia that eventually die, and this may contribute greatly to cell destruction. Possibly several processes are responsible for cell death.

Table 36.2 HIV-1 Modes of Transmission

Route
Sexual: homosexual, heterosexual
Parenteral: intravenous drug use, transfusion, needle-stick (rare), invasive medical/dental procedure (very rare)
Maternal to child: perinatal, breast-feeding
Cutaneous (very rare): mucous membrane, skin

Predisposing factors
Life-style: large number of sexual partners, sexually promiscuous partner (e.g., prostitute contact), drug abuse with needle sharing
Traumatic sexual practices: "fisting," penile/vaginal abrasion
Sexually transmitted diseases: ulcerative genital lesions, exudative genital infection

Adapted from Blattner, W. A. 1991. HIV epidemiology: past, present, and future. *FASEB Journal* 5(10): 2340-48.

Once a human's $CD4^+$ cells are infected with HIV-1, four types of pathological changes may ensue. First, a mild form of AIDS may develop with symptoms that include fever, malaise, headache, macular rash, weight loss, lymph node enlargement (lymphadenopathy), oral candidiasis (figure 36.11), and the presence of antibodies to HIV-1 (figure 36.12). These symptoms may occur in the first few months after infection, last for one to three weeks, and recur. This is known as **AIDS-related complex (ARC)**. ARC develops into a full-blown case of AIDS in an undetermined proportion of cases.

Second, a true case of AIDS can develop directly upon infection. The mean interval between HIV infection and the onset of AIDS appears to be about 8 to 10 years, although it varies considerably with each individual. At first, a person's

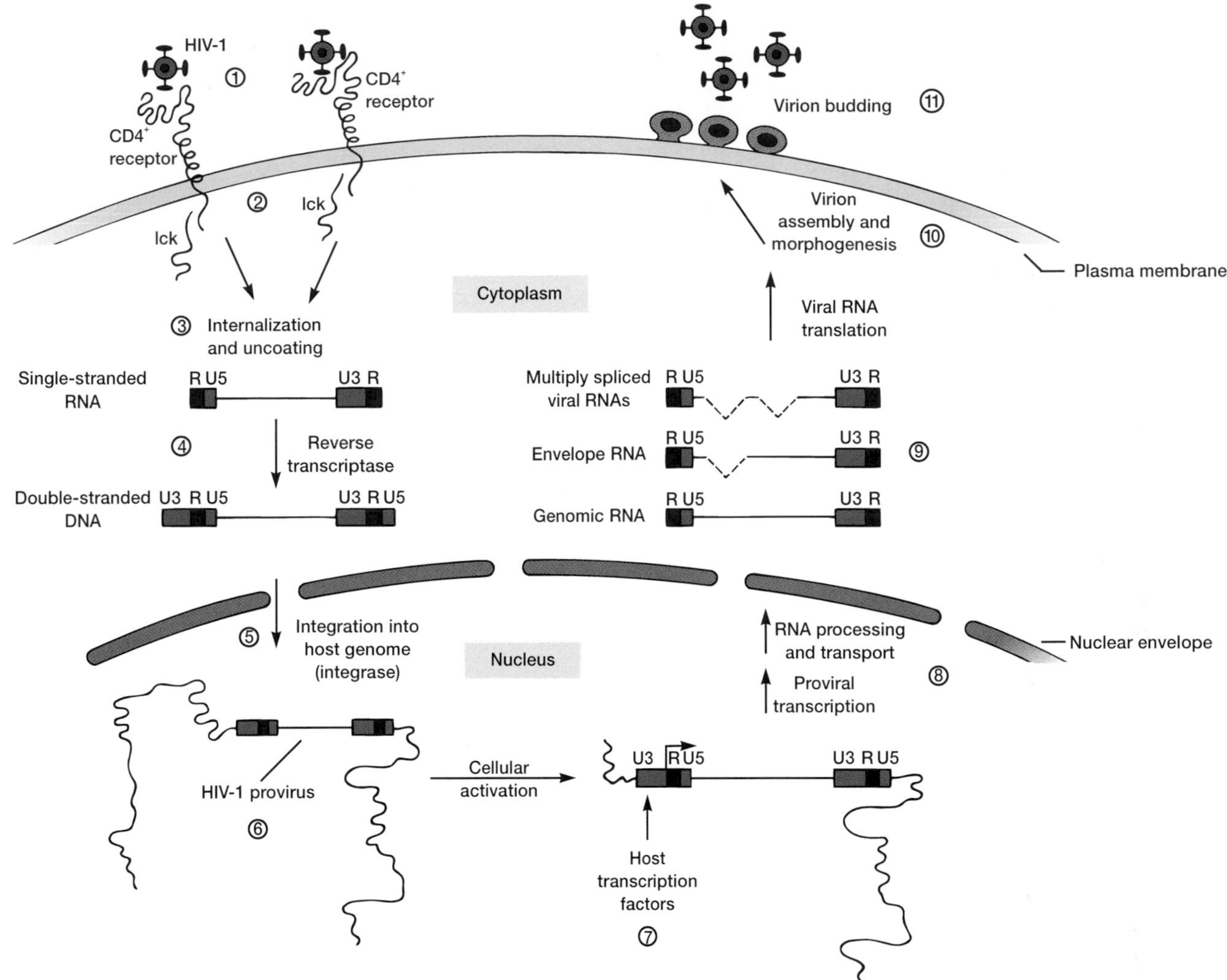

Figure 36.10 **Life Cycle of HIV-1.** (*1*) After interaction of gp120 with the CD4+ cell plasma membrane receptor, gp41–mediated membrane fusion occurs. (*2*) This leads to the entry of HIV–1 into the cell. The *lck* denotes a lymphoid-specific tyrosine kinase that binds to CD4. (*3*) After internalization and uncoating, reverse transcription of viral RNA begins. (*4*) This leads to the production of the double-stranded DNA form of the virus in the presence of appropriate host factors. (*5*) The HIV–1 integrase promotes the insertion of this viral DNA duplex into the CD4 cell's genome after the DNA has entered the nucleus. (*6*) This gives rise to the HIV–1 provirus. (*7*) The expression of the HIV–1 gene is stimulated initially by the action of specific inducible and constitutive host transcription factors with binding sites in the long terminal repeat. Their binding leads to the sequential production of various viral mRNAs. (*8*) The first mRNAs produced correspond to the multiply spliced species of approximately 2.0 kilobases encoding tat, rev, and nef (see figure 36.9) regulatory proteins. (*9*) Subsequently, the viral structural proteins are produced, allowing the (*10*) assembly and morphogenesis of the virions. (*11*) The new HIV–1 virions that are produced by viral budding from the host CD4+ cell can then reinitiate the retroviral life cycle by infecting other CD4+ target cells.

immune system responds to the HIV-1 infection by manufacturing HIV-1 antibodies, but not in sufficient quantities to stop the viral attack. The virus becomes established within primarily $CD4^+$ T-helper cells. Initially, $CD4^+$ T-helper cells proliferate abnormally in the lymph nodes. Thereafter, the lymph nodes' internal structure collapses due to viral replication. This leads to a decline in the number of lymphocytes within the lymph nodes and results in a selective depletion of the $CD4^+$ T-cell subset that is critical to the propagation of the entire T-cell pool. When this $CD4^+$ population declines, interleukin-2 (IL-2) production also decreases. Because IL-2 stimulates the production of T cells in general (*see figure 32.1*), the whole T-cell population may decline. This leaves the infected person open to opportunistic infections: invasion by pathogens that proliferate widely only because the immune system is defective.

It should be noted that factors other than direct T cell destruction also may be involved in AIDS pathogenesis. HIV may reduce the immune response by destroying or disabling dendritic cells, which present foreign antigens to T cells. HIV also mutates exceptionally rapidly and thus could evade and eventually overwhelm the immune system. There is some evidence that HIV can induce T cells to commit suicide. HIV may disrupt the balance between different types of T-helper cells and consequently decrease the killer T cell population. It is possible that several different mechanisms contribute.

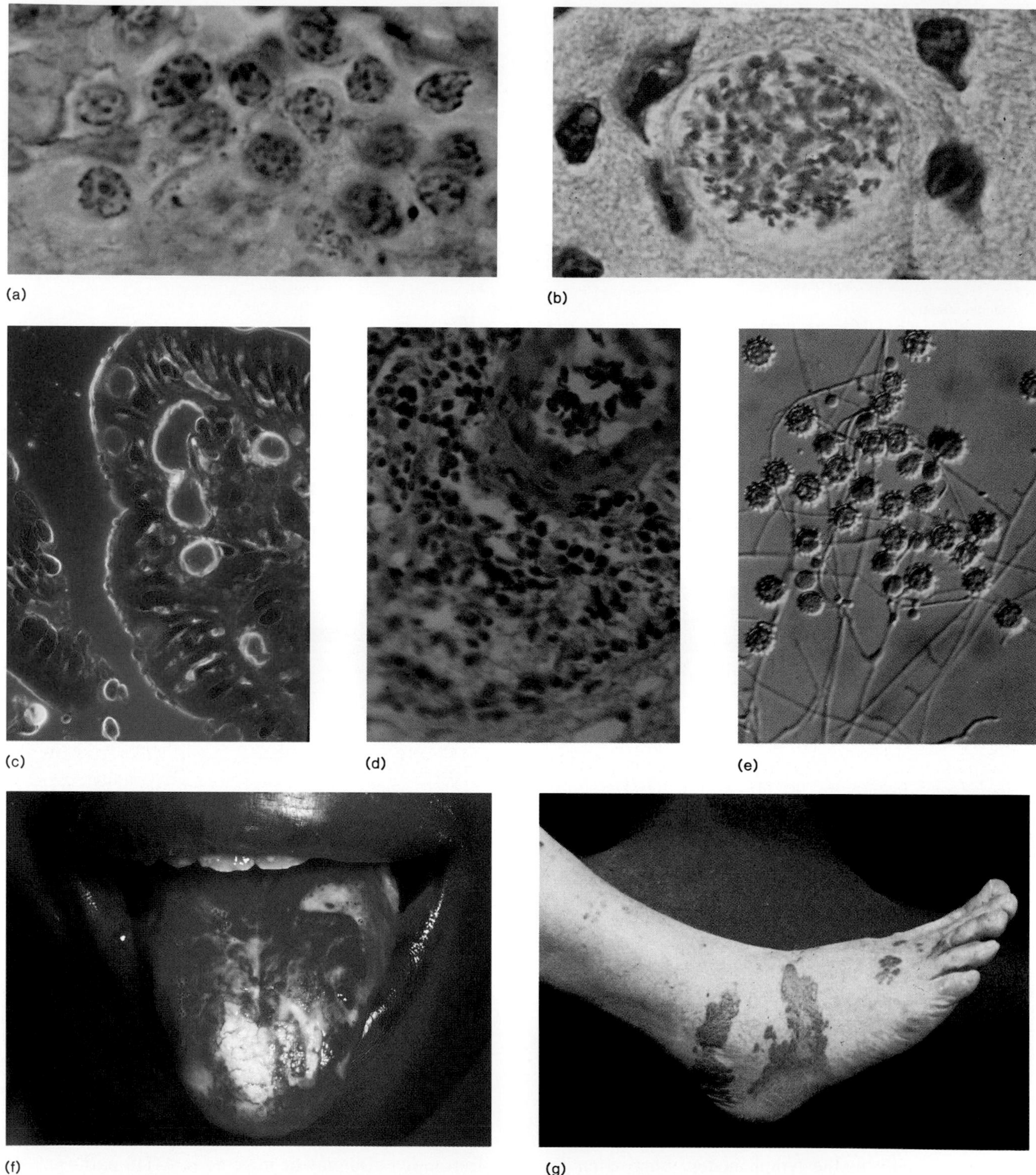

Figure 36.11 Some Pathogens and Diseases Associated with AIDS. (*a*) *Pneumocystis carinii* protozoa in lung tissue; light micrograph (×500). (*b*) *Toxoplasma gondii* protozoa located in a brain cyst; light micrograph (×1,000). (*c*) *Crypotosporidium* attached to the surface of the intestinal epithelium; light micrograph (×1,000). (*d*) *Crypotococcus neoformans* in the central nervous system; light micrograph (×150). (*e*) *Histoplasma capsulatum,* hyphae and sporangia; light micrograph (×125). (*f*) Candidiasis of the oral cavity and tongue (thrush) caused by *Candida albicans.* (*g*) Kaposi's sarcoma on the lower limb of an AIDS patient.

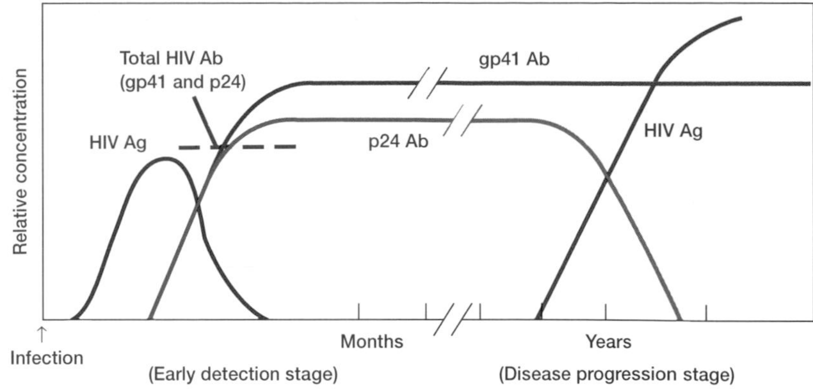

Figure 36.12 **The Typical Serological Pattern in an HIV–1 Infection.** HIV–1 antigen (HIV Ag) is detectable as early as 2 weeks after infection and typically declines at seroconversion. Seroconversion occurs when HIV–1 antibodies have risen to detectable levels. This usually takes place several weeks to months after the HIV–1 infection. The period between HIV–1 infection and seroconversion is often associated with an acute illness. Whether or not the individual has flulike symptoms, the appearance of circulating HIV–1 antigens typically occurs before IgG antibodies against gp4I and gp24 develop. HIV–1 antigen then usually disappears following seroconversion, but reappears in the latter stages of the disease. The reappearance of antigen usually indicates impending clinical deterioration. An asymptomatic HIV–1 antigen-positive individual is six times more likely to develop AIDS within 3 years than a similar individual who is HIV–1 antigen negative. Thus, testing for the presence of the HIV–1 antigen assists clinicians in monitoring the progression of the disease.

The development of a particular opportunistic infection is related to the concentration of $CD4^+$ cells in the blood. Healthy individuals have about 1,000 such cells in every cubic millimeter of blood. In HIV-1 infected individuals, the number declines by an average of about 40 to 80 cells per cubic millimeter a year. When the $CD4^+$ cell count falls to between 400 and 200 per cubic millimeter, the first opportunistic infections (table 36.3) usually appear. Examples of such opportunistic infections and diseases include *Pneumocystis carinii* pneumonia (figure 36.11*a*); *Mycobacterium avium-intracellulare* pneumonia (*see pp. 744–45*); toxoplasmosis (figure 36.11*b*); herpes zoster infection (figure 36.2*b*); chronic diarrhea; cryptococcal meningitis (figure 36.11*d*); *Histoplasma capsulatum* infection (figure 36.11*e*); and tuberculosis (*see figure 37.7*).

The third main type of disease caused by HIV-1 involves the central nervous system since virus-infected macrophages can cross the blood-brain barrier. The classical symptoms of central nervous system disease in AIDS patients are headaches, fevers, subtle cognitive changes, abnormal reflexes, and ataxia (irregularity of muscular action). Dementia and severe sensory and motor changes characterize more advanced stages of the disease. Autoimmune neuropathies, cerebrovascular disease, and brain tumors are also common. Histological changes include inflammation of neurons, nodule formation, and demyelination. All evidence indicates that these neurological changes are correlated with higher levels of HIV-1 antigen (see figure 36.12) and/or the HIV-1 genome in central nervous system tissue. In AIDS dementia, macrophages and glial cells (supporting cells of the nervous system) are primarily infected and bud new viruses. However, it is unlikely that direct infection of these cells by HIV-1 is responsible for the symptoms. More probably the symptoms arise through either the secretion of viral proteins or viral induction of cy-

TABLE 36.3 Infectious Processes Associated with AIDS

Pathogen	Clinical Manifestations
Frequent	
Bacteria	
Mycobacterium avium-intracellulare	Lymphadenopathy, disseminated tuberculosis
Mycobacterium tuberculosis	Tuberculosis
Fungi	
Candida albicans	Esophagitis, oral thrush
Coccidioides immitis	Pneumonia
Cryptococcus neoformans	Meningitis
Histoplasma capsulatum	Pneumonia, fevers, disseminated infections
Protozoa	
Pneumocystis carinii	Pneumonia
Toxoplasma gondii	Central nervous system involvement (toxoplasmic encephalitis), retinochoroiditis, diarrhea
Cryptosporidium spp.	Severe diarrhea
Viruses	
Cytomegalovirus	Dissemination (retina, esophagus, colon)
Epstein-Barr virus	Oral hairy leukoplakia (whitish patches on tongue)
Herpes simplex	Stomatitis, ulcerative mucocutaneous lesions
Herpes-zoster-varicella	Localized skin lesions
Rare	
Bacteria	
Listeria monocytogenes	Meningitis
Nocardia spp.	Lung and brain infections
Fungi	
Aspergillus spp.	None
Mucor or *Rhizopus*	None

Box 36.3

AIDS Testing

Although a cure for AIDS is not yet in sight, effective diagnostic methods and graphic documentation techniques have been developed to identify those who have been infected with the HIV virus. To screen for AIDS, a blood sample is centrifuged to separate the serum from the blood cells. The serum is then subjected to an initial ELISA screening procedure to detect antibodies to the HIV virus. Any blood sample that is reactive on repeat testing is presumed positive for antibodies to the virus, indicating the individual has been infected with the virus. Blood samples testing positive two out of three times are then subjected to confirmation testing by the Western Blot Procedure.

The Western Blot Procedure involves the isolation and testing of individual viral proteins to determine if the antibodies initially discovered in ELISA tests will recognize HIV-specific protein antigen and bind to proteins considered to be definitive signs of infection. In this procedure, component proteins of the HIV virus are separated on polyacrylamide gel by electrophoresis and transblotted onto nitrocellulose strips, which are then incubated with the individual's serum. Any HIV antibodies present bind to viral proteins contained on the test strip. The bound antibodies are then visualized using a conjugate enzyme and chromagen. The intensity of protein band staining depends on the degree of antibody binding to the viral proteins on the strip.

Findings are reported as negative if no virus-specific bands are detected and if the results concur with the initial ELISA screening. Results are reported as positive if the characteristic HIV protein bands are present and the ELISA screening test was also positive. When reactions are observed with other viral bands or when the results of the Western Blot do not correlate with the ELISA test, the findings are reported as indeterminate, and the tests are repeated in six weeks.

After the Western Blot is completed, a permanent visual documentation can be made of each test strip. For example, the figure shows three control strips that are negative, strongly positive, and weakly positive. The remaining 40 test strips show results ranging from negative (laboratory number 13) to strongly positive (number 28).

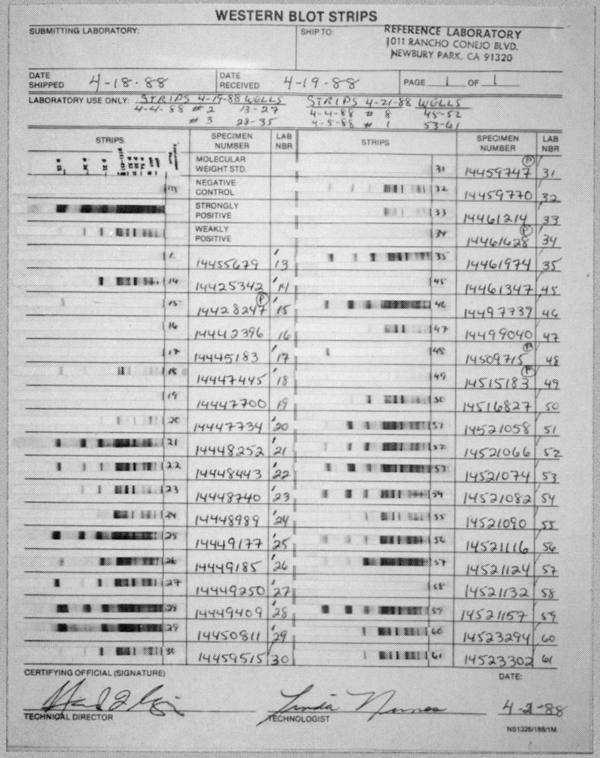

Box Figure 36.3 **Western Blot Strips.** *From Howard I. Kim, Director of Immunology and Infectious Diseases, Damon Clinical Laboratories. Newbury Park. CA 91320. Reprinted by permission.*

tokines that bind to glial cells and neurons. HIV-1 induction of interleukin-1 and tumor necrosis factor-α (TNF-α) may stimulate further viral reproduction and the induction of other cytokines (e.g., interleukin-6, granulocyte-macrophage colony-stimulating factor [GMCSF]). IL-1 and TNF-α in combination with IL-6 and GMCSF could account for the many clinical and histopathological findings in the central nervous system of AIDS individuals.

The fourth result of an HIV infection is cancer. Individuals infected with HIV-1 have an increased risk of three types of tumors: (1) Kaposi's sarcoma (figure 36.11*g*), (2) carcinomas of the mouth and rectum, and (3) B-cell lymphomas or lymphoproliferative disorders. It seems likely that the depression of the initial immune response enables secondary tumor-causing agents to initiate the cancers. Recently, a new DNA-containing herpesvirus (HHV-6) has been isolated from these cancers.

The laboratory diagnosis of AIDS can be by viral isolation and culture or by using assays for viral reverse transcriptase activity or viral antigens (figure 36.12). However, diagnosis is best accomplished through the detection of specific anti-HIV antibodies in the blood. These antibodies may be detected by GENIE or HIVAG-1 (Abbott) rapid tests, the SUDS (Murex) 10 minute assay, enzyme-linked immunoabsorbent assay (*see figure 33.13*), indirect immunofluorescence (*see figure 33.16*), immunoblot (Western Blot; Box 36.3), and radioimmunoprecipitation (*see figure 33.17*) methods.

It has recently been shown that if a person infected with HIV-1 is also infected with HHV-6, the development of AIDS is strikingly enhanced. Both viruses can infect a single $CD4^+$ cell; however, the presence of HHV-6 greatly accelerates the process. This is the first clear evidence of a cofactor being involved in AIDS. (A cofactor is an independent agent or substance that acts synergistically to foster disease.) The scenario is as follows. Ordinarily, the immune system controls the herpes virus. However, if the immune system is suppressed by HIV-1, the herpes virus will start to replicate, unleashing a vicious cycle of destruction. First, it kills T-helper cells; then it infects AIDS virus-infected cells and awakens the virus from its latent state. Thus, immune suppression caused by the HIV activates the HHV-6 virus, the HHV-6 virus in turn activates the HIV-1 virus, and both kill $CD4^+$ T-helper cells. HIV-1 also may enter through the genital skin blisters caused by the herpes infection, and this would aid transmission. Genital herpes is probably a cofactor for HIV-2 as well as HIV-1.

The immunologic defect in AIDS is apparently irreversible and has not yet responded to treatment. Primary treatment is directed at the symptoms, opportunistic infections, and malignancies. Current investigations focus on drugs that interrupt the reverse transcriptase as it synthesizes DNA. One such drug is azidothymidine (AZT) or zidovudine, whose trade name is Retrovir. AZT is a nucleotide base analog that is an effective inhibitor of HIV replication. Since it closely resembles thymidine but lacks the correct attachment point for the next nucleotide in the chain (due to its lack of a hydroxyl group on carbon 3′), it serves as a DNA chain terminator. Thus, replication ceases when it is incorporated into the growing DNA chain. This drug appears to slow the progress of the disease, but does not cure it. It probably does not improve the chances of survival, nor does it have any affect on virus transmission by infected individuals. Unfortunately, some AZT-resistant strains of the virus are now arising.

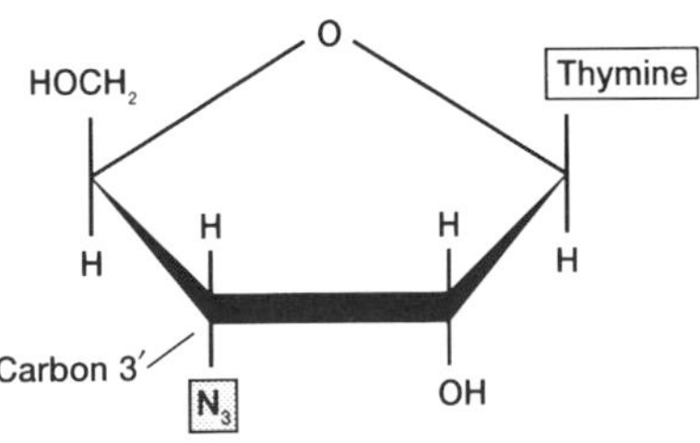

Another avenue of current research is the development of a vaccine that can (1) stimulate the production of neutralizing antibodies which can bind to the virus envelope and prevent it from entering host cells, and (2) promote the destruction of those cells already infected with the virus. The production of an effective vaccine, if possible, is not yet in sight. One difficulty is that the envelope proteins of the virus (figure 36.8) continually change their antigenic properties.

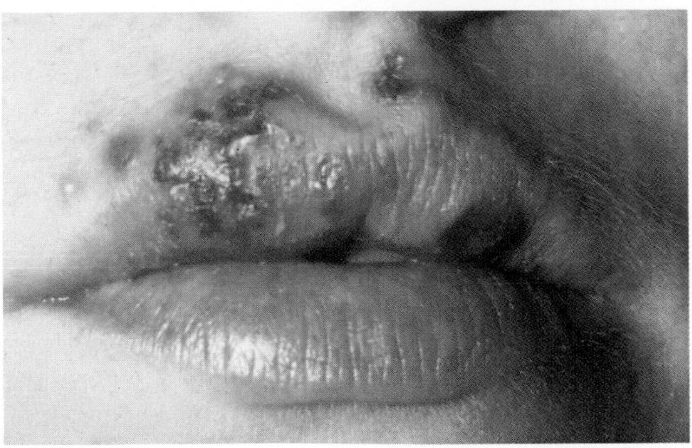

Figure 36.13 **Cold Sores.** Herpes simplex fever blisters on the lip.

Prevention and control of AIDS involves screening of blood and heat treatment of blood products to destroy the virus. Education and protected sexual behavior and practices, including the use of condoms, are keys to comprehensive community-based prevention programs. Education of intravenous drug users concerning the need to avoid sharing needles and syringes is also very important in prevention programs. It must be emphasized that protected sexual behavior and the use of condoms is safer, but not totally safe. The only absolute protection against AIDS and other STDs is abstinence and completely monogamous relationships.

Cold Sores

Cold sores or **fever blisters** (**herpes labialis**) are caused by the herpes simplex type 1 virus (HSV-1). Like all herpesviruses, it is a double-stranded DNA virus with an enveloped, icosahedral capsid. The term herpes is derived from the Greek word meaning "to creep," and clinical descriptions of herpes labialis (lips) go back to the time of Hippocrates (circa 400 B.C.). Transmission is through direct contact of epithelial tissue surfaces with the virus. A blister(s) develops at the inoculation site (figure 36.13) because of host- and viral-mediated tissue destruction. Most blisters involve the epidermis and surface mucous membranes of the lips, mouth, and gums. The blisters generally heal within a week. However, after a primary infection, the virus travels to the trigeminal nerve ganglion, where it remains in a latent state for the lifetime of the infected person. Stressful stimuli such as excessive sunlight, fever, trauma, chilling, emotional stress, and hormonal changes can reactivate the virus. Once reactivated, the virus moves from the trigeminal ganglion down a peripheral nerve to the border of the lip or other parts of the face to produce another fever blister. Primary and recurring infections also may occur in the eyes, causing **herpetic keratitis** (inflammation of the cornea)—currently a major cause of blindness in the United States. The drugs vidarabine and acyclovir are effective against cold sores. By adulthood, 70 to 90% of all people in the United States have been infected and have type 1 herpes antibodies.

Common Cold

The **common cold (coryza:** Greek *koryza,* discharge from the nostrils) is one of the most frequent infections experienced by humans of all ages. About 30% of the cases are caused by rhinoviruses (Greek *rhinos,* nose), single-stranded RNA viruses in the family *Picornaviridae.* There are about 110 distinct serotypes, and each of these antigenic types has a varying capacity to infect the nasal mucosa and to cause a cold. In addition, immunity to many of them is transitory. Several other respiratory viruses are also associated with colds (e.g., coronaviruses and parainfluenza viruses). Thus, colds are common because of the diversity of rhinoviruses, the involvement of other respiratory viruses, and the lack of a durable immunity.

Viral invasion of the upper respiratory tract is the basic mechanism in the pathogenesis of a cold. The clinical manifestations include the familiar nasal stuffiness and/or partial obstruction, sneezing, scratchy throat, and a watery discharge from the nose. The discharge becomes thicker and assumes a yellowish appearance over several days. General malaise is commonly present. The disease usually runs its course in about a week. Diagnosis of the common cold is made from observations of clinical symptoms. There are no procedures for direct examination of clinical specimens or for serological diagnosis.

The source of the cold viruses may be infected individuals excreting viruses in nasal secretions, airborne transmission over short distances by way of moisture droplets, or transmission on contaminated hands or fomites. Epidemiological studies of rhinovirus colds have shown that the familiar explosive, noncontained sneeze (*see figure 35.7*) may not play an important role in virus spread. Rather, hand-to-hand contact between a rhinovirus "donor" and a susceptible "recipient" is more likely. The common cold occurs worldwide with two main seasonal peaks, spring and early autumn. Infection is most common early in life and generally decreases with an increase in age. Nothing is available for treating the common cold except additional rest, extra fluids, and the use of aspirin for alleviating local and systemic discomfort.

Cytomegalovirus Inclusion Disease

Cytomegalovirus inclusion disease is caused by the human cytomegalovirus (HCMV), a member of the herpesvirus family. Most people become infected with this virus at some time during their life; in the United States, as many as 80% of individuals older than 35 years have been exposed to this virus and carry a lifelong infection. Although most HCMV infections are asymptomatic, certain patient groups are at risk to develop serious illness and long-term effects from an HCMV infection. For example, this virus remains the leading cause of congenital virus infection in the United States, a significant cause of transfusion-acquired infections, and a frequent contributor to morbidity and mortality among organ transplant recipients and immunocompromised individuals (especially AIDS patients). Because the virus persists in the body, it is shed for several years in saliva, urine, semen, and cervical secretions.

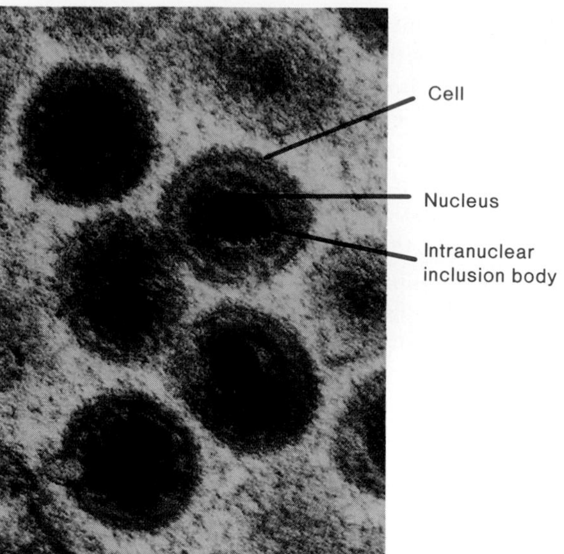

Figure 36.14 **Cytomegalovirus Inclusion Disease.** Electron micrograph of a single animal cell infected with the cytomegalovirus (×66,000). The intranuclear inclusion body has a typical "owl-eyed" appearance.

The HCMV can infect any cell of the body, where it multiplies slowly and causes the host cell to swell in size—hence the prefix cytomegalo, which means "an enlarged cell." Infected cells contain the unique **intranuclear inclusion bodies** and cytoplasmic inclusions (figure 36.14). In fatal cases, cell damage is seen in the gastrointestinal tract, lungs, liver, spleen, and kidneys. Normally, cytomegalovirus inclusion disease symptoms resemble those of infectious mononucleosis.

Laboratory diagnosis is by viral isolation from urine, blood, lung, semen, or tissue. Serological tests (e.g., immunofluorescence, complement fixation, ELISA, and immunohistologic staining) are also available.

Epidemiologically, the virus has a worldwide distribution, especially in developing countries where infection is universal by childhood. The prevalence of this disease increases with a lowering of socioeconomic status and hygienic practices. The only drugs available, ganciclovir and foscarnet, are used only for high-risk patients. Infection can be prevented by avoiding close personal contact (including sexual) with an actively infected individual. Transmission by blood transfusion or organ transplantation can be avoided by using blood or organs from seronegative donors.

Genital Herpes

Genital herpes is caused by the herpes simplex type 2 virus (HSV-2) (figure 36.15*a*). It is most frequently transmitted by sexual contact. The primary infection occurs after an incubation period of about a week. Fever, a burning sensation, and genital soreness are frequently present. Blisters that appear in the infected area (figure 36.15*b*) are the result of cell lysis and

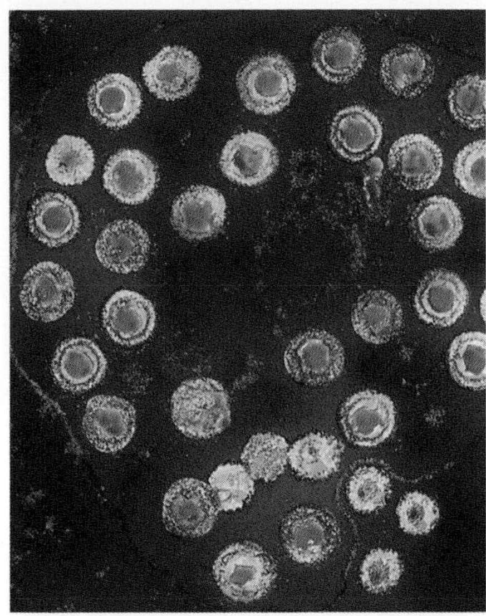
(a)

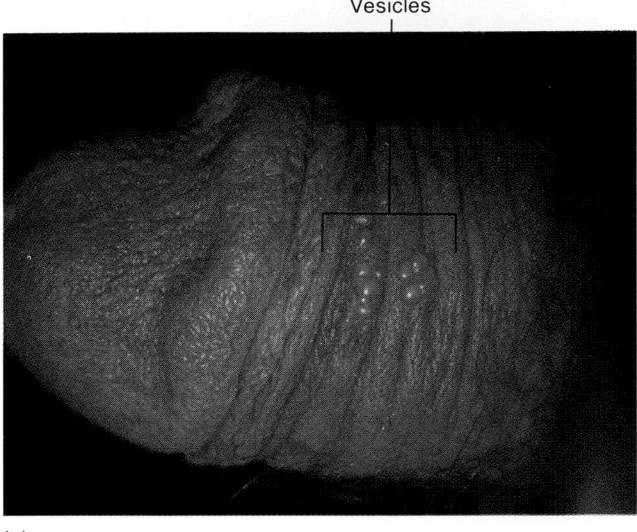

(b)

Figure 36.15 **Genital Herpes.** (*a*) Herpes simplex type 2 virus (yellow and green) inside an infected cell (*b*) Herpes vesicles on the penis. The vesicles contain fluid that is infectious.

the development of a local inflammatory response; they contain fluid and infectious viruses. Although blisters generally heal spontaneously in a few weeks, the viruses retreat to nerve cells in the sacral spinal plexus, where they remain in a latent form. Periodically, the viruses multiply and migrate down nerve fibers to the epithelia, where they produce new blisters. Activation may be due to sunlight, sexual activity, illness accompanied by fever, hormones, or stress.

Besides being transmitted by sexual contact, herpes can be spread to an infant during vaginal delivery, leading to **congenital (neonatal) herpes.** Congenital herpes is one of the most life-threatening of all infections in newborns, affecting approximately 1,500 to 2,200 babies per year in the United States. It can result in neurological involvement as well as blindness. As a result, any female who has had genital herpes should have a caesarean section instead of delivering vaginally. For unknown reasons, genital herpes is also associated with a higher-than-normal rate of cervical cancer and miscarriages.

Although there is no cure for genital herpes, oral use of the antiviral drug acyclovir has proven to be effective in ameliorating the recurring blister outbreaks. Topical acyclovir is also effective in reducing virus shedding, the time until the crusting of blisters occurs, and new lesion formation. Idoxuridine and trifluridine are used to treat herpes infections of the eye.

In the United States, the incidence of genital herpes has increased so much during the past decade that is now a very common sexually transmitted disease. It is estimated that over 25 million Americans are infected with the herpes simplex type 2 virus.

Leukemia

Certain **leukemias** in humans are caused by two retroviruses: human T-cell lymphotropic virus 1 (HTLV-1) and HTLV-2. The viruses are transmitted by transfusions of contaminated blood, among drug addicts sharing needles, by sexual contact, across the placenta, from the mother's milk, or by mosquitoes.

Retroviruses and their replication (p. 389).

Viruses and cancer (pp. 393–94).

HTLV-1 causes **adult T-cell leukemia.** Once within the body, the HTLV-1 virus enters white blood cells and integrates into the cellular genome, where it turns on the growth-promoting genes by acting at a distance from them (trans-acting). The transformed cell proliferates extensively, and death generally results from the explosive proliferation of the leukemia cells or from opportunistic infections. To date, no effective treatment exists.

In 1982 the second human retrovirus (HTLV-2) was shown to be the agent responsible for hairy-cell leukemia. This virus shares the same trans-acting mechanism as HTLV-1. Hairy-cell leukemia gets its name from the many membrane-derived protrusions that give white blood cells the appearance of being "hairy" (figure 36.16). This leukemia is a chronic, progressive lymphoproliferative disease. The malignancy is believed to originate in a stage of B-cell development. The bone marrow, spleen, and liver become infiltrated with malignant cells. This lowers the person's immunity. The primary cause of mortality is bacterial and other opportunistic infections. IFN-α has shown some promise for treatment in certain cases.

Mononucleosis (Infectious)

The Epstein-Barr virus (EBV), a member of the family *Herpesviridae,* is the etiologic agent of **infectious mononucleosis (mono),** a disease whose symptoms closely resemble those of cytomegalovirus-induced mononucleosis. Because the Epstein-Barr virus occurs in oropharyngeal secretions, it can be spread by mouth-to-mouth contact (hence the terminology in-

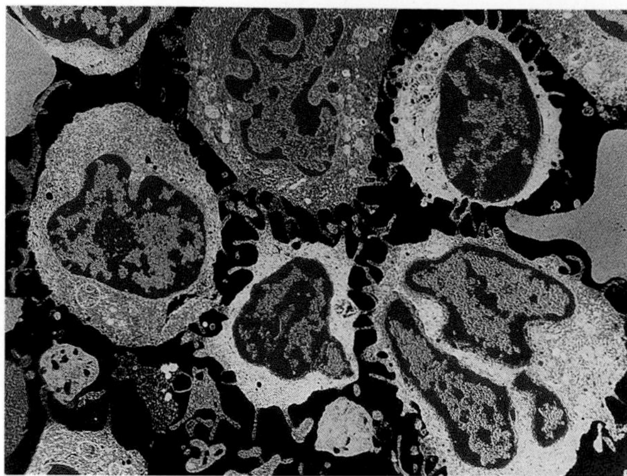

Figure 36.16 Hairy-Cell Leukemia. False-color transmission electron micrograph (×3,100) of abnormal B lymphocytes. Notice that the lymphocytes are covered with characteristic hair-like membrane-derived protrusions.

fectious and kissing disease) or shared drinking bottles and glasses. Once a person has contracted mono, the virus enters lymphatic tissue, multiplies, and infects B cells. Infected B cells rapidly proliferate and take on an atypical appearance (Downey cells) that is useful in diagnosis. The disease is manifested by enlargement of the lymph nodes and spleen, sore throat, headache, nausea, general weakness and tiredness, and a mild fever that usually peaks in the early evening. The disease lasts for one to six weeks and is self-limited.

Treatment of mononucleosis is largely supportive and includes plenty of rest. Specific chemotherapy with phosphonoacetic acid, adenine arabinoside, and acyclovir is nearing clinical application. Diagnosis of mononucleosis is made with a serological test for nonspecific (heterophil) antibodies. Several rapid tests are on the market.

The peak incidence of mononucleosis occurs in people 15 to 25 years of age. Collegiate populations, particularly those in the upper-socioeconomic class, have a high incidence of the disease. About 50% of college students have no immunity, and approximately 15% of these can be expected to contract mononucleosis. People in lower-socioeconomic classes tend to acquire immunity to the disease because of early childhood infection. It has been estimated that over 80% of the United States population are eventually infected.

Recent evidence suggests that EBV is expressed in unusual patterns among individuals with AIDS or who are at risk for developing AIDS. These patterns include elevated titers of antibodies directed against EBV antigens, increased oropharyngeal excretion of EBV, increased number of circulating EBV-infected B cells, and isolation of EBV from the plasma of AIDS patients. It appears that EBV, which exists in a latent state in immunocompetent hosts, is reactivated by AIDS virus-induced immunosuppression. Such EBV infections may be related to the development of hairy-cell leukemia and some of the lymphoproliferative disorders seen in AIDS patients.

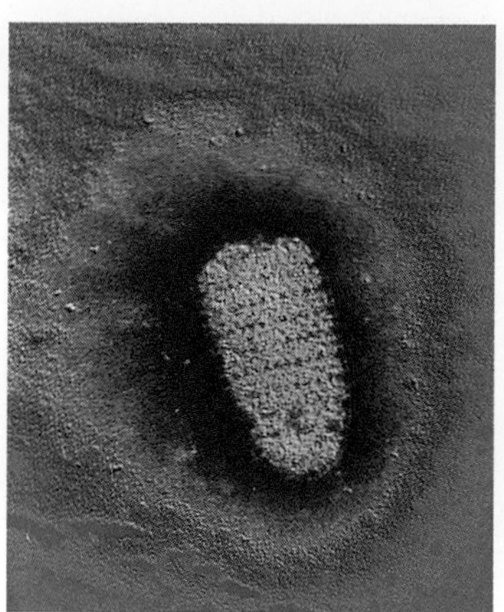

(a)

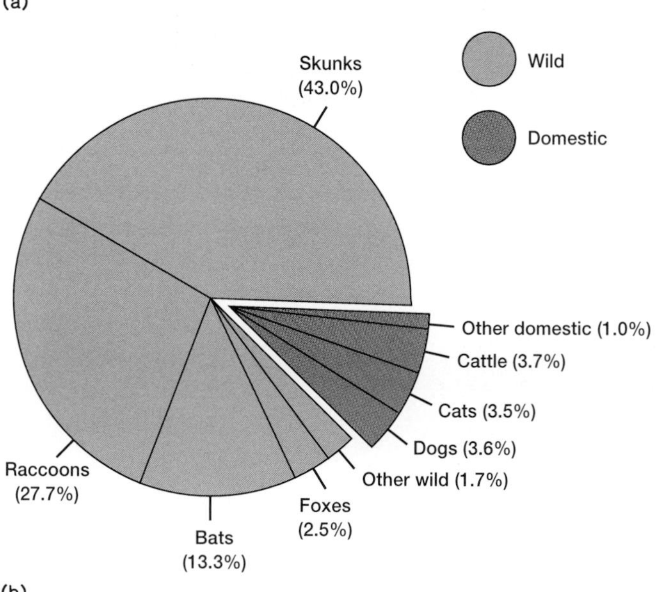

(b)

Figure 36.17 Rabies. (*a*) Electron micrograph (×36,700) of the bullet-shaped rabies virus (yellow). Note their bullet shape. (*b*) Animal rabies in the United States. 1990. *Source for b: The Centers for Disease Control, Atlanta, GA.*

Rabies

Rabies is primarily a disease of carnivorous animals (skunks, foxes, insectivorous bats, raccoons, dogs). It is caused by the rabies virus, a bullet-shaped rhabdovirus (figure 36.17*a; see also figure 17.17*c) that multiplies within the salivary glands of infected animals. It is transmitted to humans by the bite of an infected animal whose saliva contains the virus; aerosols of the virus that can be present in caves where bats roost; or contamination of scratches, abrasions, open wounds, and mucous membranes with saliva from an infected animal.

Once inside the body, the virus multiplies within skeletal muscle and connective tissue. Eventually, it reaches peripheral nerve endings and migrates to the central nervous system. The virus continues to multiply within the brain neurons, producing characteristic **Negri bodies,** masses of viruses or unassembled viral subunits that are visible in the light microscope.

In the past, the diagnosis of rabies consisted of examining nervous tissue for the presence of these bodies. Today, diagnosis is based on fluorescent-antibody tests performed on nervous tissue.

Symptoms of rabies usually begin 2 to 16 weeks after viral exposure and include anxiety, irritability, depression, fatigue, loss of appetite, fever, and a sensitivity to light and sound. The disease quickly progresses to a stage of paralysis. In about 50% of all cases, intense and painful spasms of the throat and chest muscles occur when the victim swallows liquids. The mere sight, thought, or smell of water can set off spasms. Consequently, rabies has been called hydrophobia (fear of water). Death results from destruction of the regions of the brain that regulate breathing.

Safe and effective vaccines (human diploid cell rabies vaccine or duck embryo vaccine) against rabies are available; however, to be effective they must be given soon after the person has been infected. Veterinarians and laboratory personnel, who have a high risk of exposure to rabies, are usually immunized every two years and tested for the presence of a suitable antibody titer. About 30,000 people annually receive this treatment. In the United States, fewer than 10 cases of rabies occur yearly in humans; about 6,000 cases of animal rabies are reported each year (figure 36.17*b*). Prevention and control involves preexposure vaccination of dogs and cats, postexposure vaccination of humans, and preexposure vaccination of humans at special risk.

Serum Hepatitis: Hepatitis B; Hepatitis C; Delta Agent

Any infection that results in inflammation of the liver is called **hepatitis** (Greek *hepaticus,* liver). Three closely related viruses and one agent are responsible (table 36.4). **Hepatitis A (infectious hepatitis)** is transmitted by fecal-oral contamination and is discussed in the section on food-borne and waterborne diseases. The other major types include **hepatitis B (serum hepatitis), hepatitis C** (formerly called **non-A, non-B**), and hepatitis associated with the **Delta agent**.

TABLE 36.4 Characteristics of the Viral Hepatitides[a]

Properties of the Viruses

Agent	*Nucleic Acid Composition*	*Virus Family*
Hepatitis A	Single-stranded linear RNA (positive strand); no envelope	*Picornaviridae* (enterovirus 72)
Hepatitis B	Nicked, circular, mostly double-stranded DNA; envelope present	*Hepadnaviridae*
Hepatitis C	Single-stranded, linear RNA (positive strand); envelope present	*Flaviviridae*
Delta	Single-stranded RNA; envelope present	?

Epidemiology, Transmission, and Vaccines

	A	*B*	*C*	*Delta*
Epidemiological patterns				
Epidemic	Yes	No	Yes	Yes
Sporadic	Yes	Yes	Yes	Yes
Transmission				
Fecal-oral	Yes	Yes	Yes	No
Sexual	Yes	Yes	Yes	?
In utero	No	Yes	?	Yes
Parenteral[b]	Rare	Yes	Yes	Yes
Vaccines	No	Yes[c]	No	No

Clinical Comparison

	A	*B*	*C*	*Delta*
Incubation period (days)	15–40	60–180	28–112	?
Asymptomatic infection	Usual	Common	Common	?
Chronicity	No	Yes (10%)	Yes (50%)	Yes
Long-term pathology	No	Cirrhosis	Cirrhosis	?

[a]Plural of hepatitis.

[b]Not through the alimentary canal but rather by injection through some other route.

[c]Recombivax HB, Engerix-B.

Serum hepatitis is caused by the hepatitis B virus (HBV), a double-stranded DNA virus of complex structure. HBV is classified as a hepadnavirus. Serum from individuals infected with hepatitis B contains three distinct antigenic particles: a spherical 22 nm particle, a 42 nm spherical particle called the **Dane particle,** and tubular or filamentous particles that vary in length (figure 36.18). The small spherical and tubular particles are the unassembled components of the Dane particle—the infective form of the virus. The unassembled particles contain hepatitis B surface antigen (HBsAg) whose presence in the blood is (1) an indicator of hepatitis B infection, (2) the basis for the large-scale screening of blood for the hepatitis B virus, and (3) the basis for the first vaccine for human use developed by recombinant DNA technology.

The hepatitis B virus is normally transmitted through blood transfusions, contaminated equipment, drug users' unsterile needles, or any body secretion (saliva, sweat, semen, breast milk, urine, feces). The virus also can pass from the blood of an infected mother through the placenta to infect the fetus. Each year an estimated 300,000 people in the United States are infected with HBV. About 4,000 persons die yearly from hepatitis-related cirrhosis and about 1,000 die from HBV-related liver cancer.

The clinical signs of hepatitis B vary widely. Most cases are asymptomatic. However, sometimes fever, loss of appetite, abdominal discomfort, nausea, fatigue, and other symptoms gradually appear following an incubation period of one to three months. The virus infects liver hepatic cells and causes liver tissue degeneration and the release of liver-associated enzymes (transaminases) into the bloodstream. This is followed by jaundice, the accumulation of bilirubin (a breakdown product of hemoglobin) in the skin and other tissues with a resulting yellow appearance. Chronic hepatitis B infection also causes the development of primary liver cancer, hepatocellular carcinoma (hepatitis B is second in importance only to tobacco as a known human carcinogen).

General measures for prevention and control involve (1) excluding contact with HBV-infected blood and secretions, and minimizing needle-sticks by scrupulous technique; (2) passive prophylaxis with intramuscular injection of hepatitis B immune globulin within 7 days of exposure; and (3) active prophylaxis with two new recombinant vaccines: Engerix-B, and Recombivax HB. These vaccines are widely used by health professionals who are at increased risk of contacting the hepatitis B virus. Other individuals who should be vaccinated include:

1. Contacts of HBV carriers (e.g., household members, sex partners, institutional inmates)
2. International travelers
3. Sexually active homosexual males
4. Hemodialysis patients
5. Recipients of blood and related products that are possibly HBV contaminated

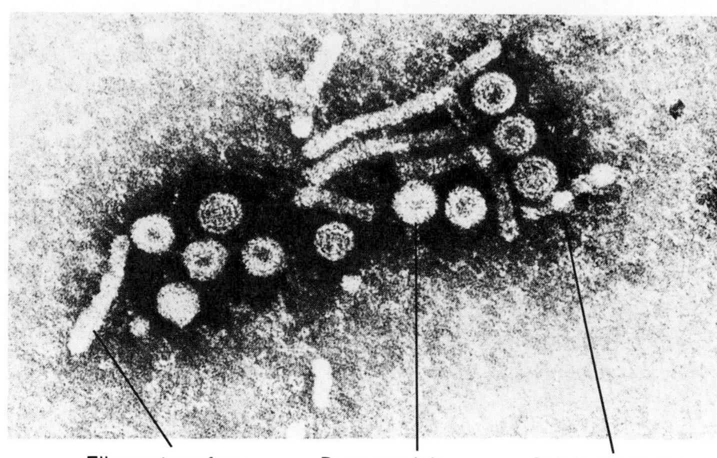

Figure 36.18 **Hepatitis B Virus in Serum.** Electron micrograph (×210,000) showing the three distinct types of hepatitis B antigenic particles. The spherical particles and filamentous forms are small sphere or long filaments without an internal structure, and only two of the three characteristic viral envelope proteins appear on their surface. Dane particles are the complete, infectious virion.

Another form of hepatitis is called hepatitis C (formerly non-A, non-B). The virion has an 80 nm diameter, has a lipid coat, contains a single strand of RNA, and has been designated the hepatitis C virus (HCV) within the family *Flaviviridae.* This virus is transmitted by intimate contact with virus-contaminated blood, by the fecal-oral route, or through organ transplantation. Hepatitis C disease is milder than hepatitis B and similar to hepatitis A, but often without the jaundice. Diagnosis is made by a first-generation enzyme-linked immunosorbent assay (ELISA), which detects serum antibody to a recombinant antigen of HCV. Hepatitis C is found worldwide. In recent years, hepatitis C has accounted for more than 90% of hepatitis cases developing after blood transfusion. Treatment is with recombinant IFN-α three times weekly for 6 months.

In 1977, a cytopathic hepatitis agent termed the Delta agent was discovered. It is transmitted as an infectious agent but cannot cause disease unless the individual is first infected with the hepatitis B virus. The Delta agent replicates only in liver cells in which hepatitis B virus is also actively replicating. The Delta agent consists of a single-stranded RNA core surrounded by HBsAg. It is genetically similar to plant viruses called viroids (*see chapter 19*), and is considered to represent a new class of animal viruses.

The clinical manifestations, pathology, and epidemiology of the Delta agent are the same as for the HBV. However, coinfection of the liver with the Delta agent and HBV may lead to a more serious acute or chronic infection than with HBV alone. In 1990, 25 to 35% of all HBsAg-positive intravenous drug users in the United States were infected with the Delta agent.

1. Describe the AIDS virus and how it cripples the immune system. How is the virus transmitted? What four types of pathological changes can result?
2. Why do people periodically get cold sores? Describe the causative agent.
3. Why do people get the common cold so frequently? How are cold viruses spread?
4. Give two major ways in which herpes simplex type 2 virus is spread. Why do herpes infections become active periodically?
5. What two types of leukemias are caused by viruses?
6. Describe the causative agent and some symptoms of mononucleosis.
7. How does the rabies virus cause death in humans?
8. What are the different causative agents of serum hepatitis and how do they differ from one another? How can one avoid hepatitis?

Food-Borne and Waterborne Diseases

Food and water have been recognized as potential carriers of disease since the beginning of recorded history. Collectively, more infectious diseases occur by these two routes than any other. A few human viral diseases that are food- and waterborne are now discussed.

Water-based diseases (chapter 41).
Diseases and food (chapter 43).

Gastroenteritis (Viral)

Acute viral gastroenteritis (inflammation of the stomach or intestines) is caused by five major categories of viruses: rotaviruses (figure 36.19), enteric adenovirus, Norwalk virus, calicivirus, and astrovirus. The medical importance of these viruses is summarized in table 36.5.

The viruses responsible for gastroenteritis are probably transmitted by the fecal-oral route. Infection is most common during the cooler months in contrast to bacteria-caused diarrheal diseases, which usually occur in the warmer months of the year. Diarrheal diseases are the leading cause of childhood deaths (5 to 10 million deaths per year) in developing countries where malnutrition is common. Current estimates are that viral gastroenteritis produces 30 to 40% of the cases of infectious diarrhea in the United States, far outnumbering documented cases of bacterial and parasitic diarrhea (the cause of approximately 40% of presumed cases of diarrhea remains unknown). In the United States, rotaviruses account for about

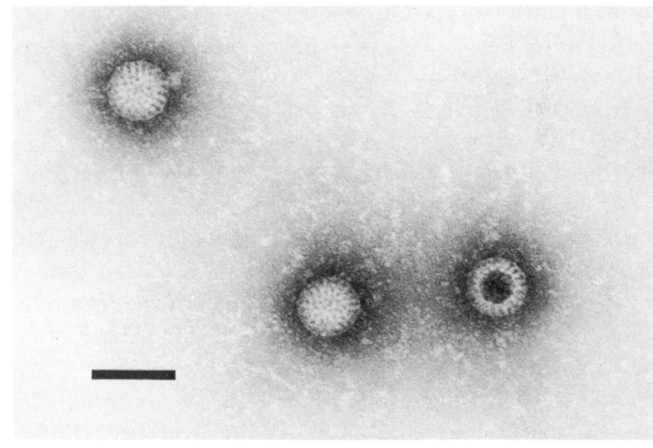

Figure 36.19 **Viral Gastroenteritis.** Electron micrograph of rotaviruses in a human gastroenteritis stool filtrate. Note the spokelike appearance of the capsids. Bar = 100 nm.

TABLE 36.5 Medically Important Gastroenteritis Viruses

Virus	Epidemiological Characteristics	Clinical Characteristics
Rotavirus		
Group A	Endemic diarrhea in infants worldwide	Dehydrating diarrhea for 5–7 days; fever and vomiting common
Group B	Large outbreaks in adults and children in China	Severe watery diarrhea for 3–5 days
Group C	Sporadic cases in children worldwide	Similar to group A
Enteric adenovirus	Endemic diarrhea in infants worldwide	Prolonged diarrhea lasting 5–12 days with fever and vomiting
Norwalk virus	Epidemics of vomiting and diarrhea in older children and adults; occurs in families, communities, and nursing homes; often associated with shellfish, other food, or water	Acute vomiting, fever, myaglia, and headache lasting 1–2 days
Calicivirus	Pediatric diarrhea; associated with shellfish and other foods in adults	Rotaviruslike illness in children; Norwalk-like in adults
Astrovirus	Pediatric diarrhea; reported in nursing homes	Watery diarrhea for 1–3 days

3.5 million cases of illness, resulting in 35% of the hospitalizations for gastroenteritis, and 75 to 150 deaths each year.

Viral gastroenteritis is seen most frequently in infants 1 to 11 months of age, where the virus attacks the upper intestinal epithelial cells of the villus, causing malabsorption, impairment of sodium transport, and diarrhea. The clinical manifestations range from asymptomatic to a relatively mild diarrhea with headache and fever to a severe and occasionally fatal dehydrating disease. Vomiting is almost always present.

Viral gastroenteritis is usually self-limited. Treatment is designed to provide relief through the use of oral fluid replacement with isotonic liquids, analgesics, and antiperistaltic agents.

Infectious Hepatitis

Infectious hepatitis (hepatitis A) is usually transmitted by fecal-oral contamination of food, drink, or shellfish that live in contaminated water and contain the virus in their digestive system. The disease is caused by the hepatitis A virus (HAV) a picornavirus. The hepatitis A virus is an icosahedral, single-stranded RNA virus that lacks an envelope and is quite different from the hepatitis B virus. Once in the digestive system, the viruses multiply within the intestinal epithelium. Usually, only mild intestinal symptoms result. Occasionally, viremia (the presence of viruses in the blood) occurs and the viruses may spread to the liver, kidneys, and spleen. The viruses reproduce in the liver, enter the bile, and are released into the small intestine. This explains why feces are so infectious. Symptoms last from 2 to 20 days and include anorexia, general malaise, nausea, diarrhea, fever, and chills. If the liver becomes infected, jaundice ensues. Laboratory diagnosis is by detection of the hepatitis A antibody. The mortality rate is low (less than 1%), and infections in children are usually asymptomatic. Most cases resolve in four to six weeks and yield a strong immunity. Approximately 40 to 80% of the United States population have serum antibodies though few have been aware of the disease. Control of infection is by simple hygienic measures and the sanitary disposal of excreta.

Poliomyelitis

Poliomyelitis (Greek *polios,* gray, and *myelos,* marrow or spinal cord), **polio,** or **infantile paralysis** is caused by the poliovirus, a member of the family *Picornaviridae* (Box 36.4). The virus is very stable and can remain infectious for relatively long periods in food and water—its main routes of transmission. Once ingested, the virus multiplies in the mucosa of the throat and/or small intestine. From these sites the virus invades the tonsils and lymph nodes of the neck and terminal portion of the small intestine. Generally, there are either no symptoms or a brief illness characterized by fever, headache, sore throat, vomiting, and loss of appetite. The virus sometimes enters the bloodstream and causes a viremia. In most cases (more than 99%), the viremia is transient and clinical disease does not result. In the minority of cases (less than 1%), the viremia persists and the virus enters the central nervous system and causes paralytic polio. The virus has a high affinity for anterior horn motor nerve cells of the spinal cord. Once inside these cells, it multiplies and destroys the cells; this results in motor and muscle paralysis. Since the arrival of the formalin-inactivated Salk vaccine (1954) and the attenuated virus Sabin vaccine (1962), the incidence of polio has decreased markedly. There are fewer than 10 cases per year, and no endogenous reservoir of polioviruses exists in the United States. However, there is a continuing need for vaccination programs in all population groups to limit the spread of poliovirus when it is introduced from other countries. Prevention and control is by vaccination; global eradication of polio is possible by the year 2000.

---

1. What two virus groups are associated with acute viral gastroenteritis? How do they cause the disease's symptoms?
2. Describe some symptoms of infectious hepatitis.
3. Why is hepatitis A called infectious hepatitis?
4. At what specific sites within the body can the poliomyelitis virus multiply? What is the usual outcome of an infection?

Slow Virus Diseases

An introduction to slow diseases caused by viruses and prions appears on pp. 393 and 399–400. A **slow virus disease** can be defined as a progressive pathological process caused by a transmissible agent—virus or prion—that remains clinically silent during a prolonged incubation period of months to years, after which progressive clinical disease becomes apparent. This usually ends months later in disability or death. It is probably inappropriate to say that a slow virus disease or slow disease is caused by a slow virus for two reasons. First, the causative agent of some of these diseases does not fit the conventional definition of a virus. Second, even in those diseases caused by viruses, it is the disease process and not the virus that is slow. Six of these diseases are summarized in table 36.6.

Other Diseases

Several other human diseases (e.g., diabetes mellitus, viral arthritis) have been associated with viruses but do not fit into any of the previous categories. A good example is warts.

Warts

Warts or verrucae (Latin *verruca,* wart) are horny projections on the skin caused by the papillomaviruses (*see figure 17.12*d). At least eight distinct genotypes produce benign epithelial tumors that vary in respect to their location, clinical appearance, and histopathologic features. Warts occur principally in children and young adults and are limited to the skin and mucous membranes. The viruses are spread between people by

Box 36.4

A Brief History of Polio

Like many other infectious diseases, polio is probably of ancient origin. Various Egyptian hieroglyphics dated approximately 2000 B.C. depict individuals with wasting, withered legs and arms (see illustration). In 1840 the German orthopedist Jacob von Heine described the clinical features of poliomyelitis and identified the spinal cord as the problem area. Little further progress was made until 1890, when Oskar Medin, a Swedish pediatrician, portrayed the natural history of the disease as epidemic in form. He also recognized that a systemic phase, characterized by minor symptoms and fever, occurred early and was complicated by paralysis only occasionally. Major progress occurred in 1908, when Karl Landsteiner and William Popper successfully transmitted the disease to monkeys. In the 1930s, much public interest in polio occurred because of the polio experienced by Franklin D. Roosevelt. This led to the founding of the March of Dimes campaign in 1938; the sole purpose of the March of Dimes was to collect money for research on polio. In 1949 John Enders, Thomas Weller, and Frederick Robbins discovered that the poliovirus could be propagated in vitro in cultures of human embryonic tissues of nonneural origin. This was the keystone that later led to the development of vaccines.

In 1952 David Bodian recognized that there were three distinct serotypes of the poliovirus. Jonas Salk successfully immunized humans with formalin-inactivated poliovirus in 1953, and this vaccine (IPV) was licensed in 1955. In 1962 Albert Sabin and others developed the live attenuated poliovirus vaccines (oral polio vaccines, OPV). Both the Salk and Sabin vaccines led to a dramatic decline of paralytic poliomyelitis in most developed countries and, as such, have been rightfully hailed as two of the great accomplishments of medical science.

Box Figure 36.4 **Ancient Egyptian with Polio.** Note the withered leg.

Table 36.6 Slow Virus Diseases of Humans

Disease	Agent	Incubation Period	Nature of Disease
Creutzfeldt-Jakob disease	Prion	Months to years	Spongiform encephalopathy (degenerative changes in the central nervous system)
Kuru	Prion	Months to years	Spongiform encephalopathy
Gerstmann-Sträussler-Scheinker Syndrome (GSS)	Prion	Months to years	Genetic neurodegenerative disease
Fatal familial insomnia	Prion	Months to years	Genetic neurodegenerative disease with progressive, untreatable insomnia
Progressive multifocal leukoencephalopathy	Papovavirus	Years	Central nervous system demyelination
Subacute sclerosing panencephalitis (SSPE)	Measle virus variant	2–20 years	Chronic sclerosing (hard tissue) panencephalitis (involving both white and gray matter of the brain)

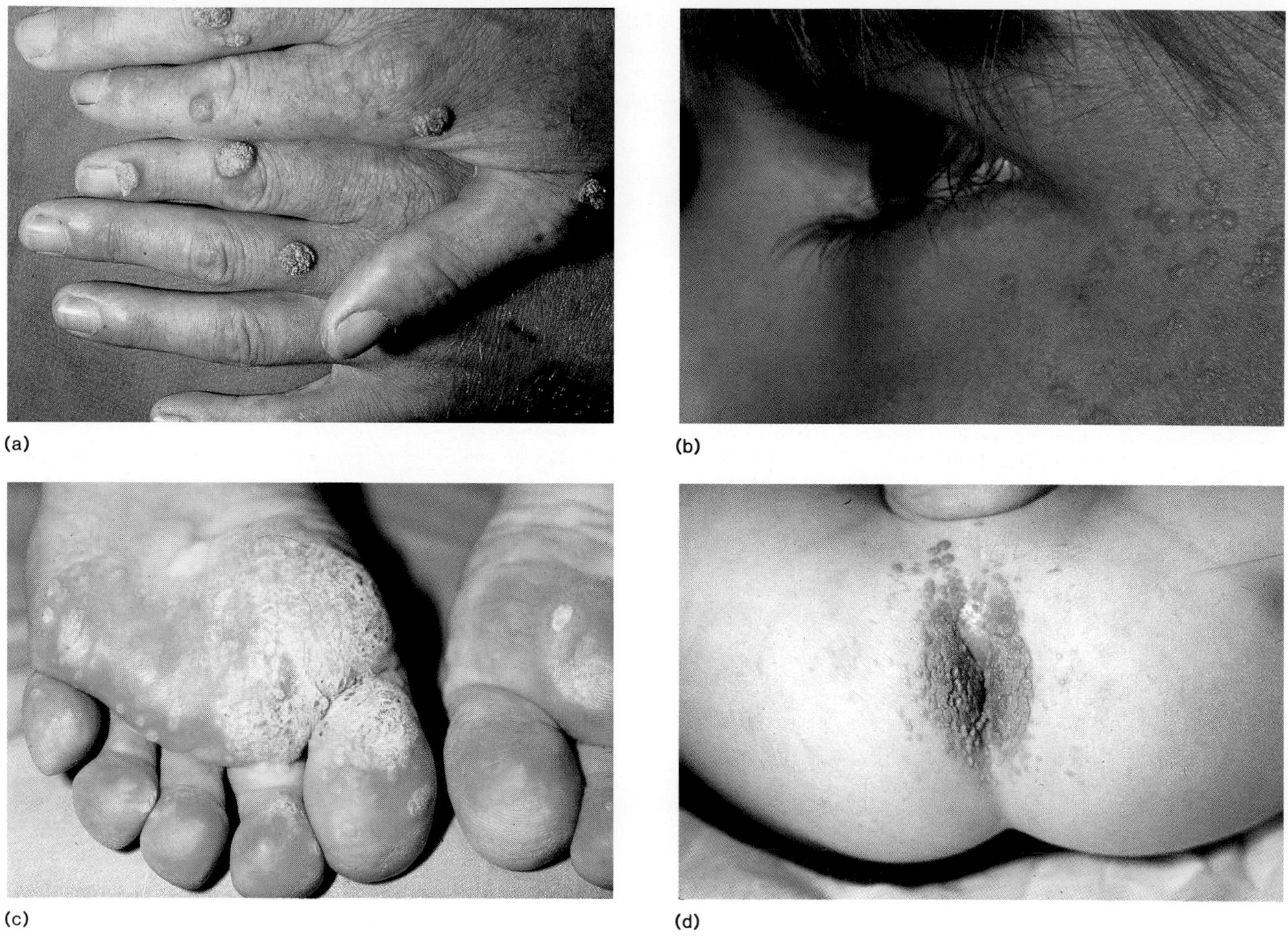

Figure 36.20 **Warts.** (*a*) Common wart on finger. (*b*) Flat warts on the face. (*c*) Plantar warts on the foot. (*d*) Perianal condyloma acuminata.

direct contact; autoinoculation occurs through scratching. Four major kinds of warts are **plantar warts, verrucae vulgaris, flat** or **plane warts,** and **anogenital condylomata (venereal warts)** (figure 36.20). Treatment includes physical destruction of the wart(s) by electrosurgery, cryosurgery with liquid nitrogen or solid CO_2, laser fulguration (drying), direct application of the drug podophyllum to the wart(s), or injection of IFN-α (Intron A).

Anogenital condylomata (venereal warts) are sexually transmitted and caused by the type 6 papillomavirus. Once the virus enters the body, the incubation period is one to six months. The warts (figure 36.20*d*) are soft, pink, cauliflowerlike growths that occur on the external genitalia, in the vagina, on the cervix, or in the rectum. They are often multiple and vary in size. Evidence exists that cervical cancer and perianal cancer occur more frequently in people who have a history of genital warts.

1. Name six slow virus diseases of humans.
2. What kind of viruses cause the formation of warts? Describe venereal warts or anogenital condylomata.

Summary

1. More than 400 different viruses can infect humans. These viruses can be grouped and discussed according to their mode of acquisition/transmission.
2. Most airborne viral diseases involve either directly or indirectly the respiratory system. Examples include chickenpox (varicella), shingles (herpes zoster), German measles (rubella), influenza (flu), measles (rubeola), mumps, the acute respiratory viruses such as the respiratory syncytial virus, the extinct smallpox (variola), and viral pneumonia.
3. The arthropod-borne viral diseases are transmitted by arthropod vectors from human to human or animal to human. Examples include California encephalitis, Colorado tick fever, St. Louis encephalitis, eastern, western, and Venezuelan equine encephalitis, and yellow fever. All these diseases are characterized by fever, headache, nausea, vomiting, and the characteristic encephalitis.
4. Person-to-person contact is another way of acquiring or transmitting a viral disease. Examples of such diseases include AIDS, cold sores, the common cold, cytomegalovirus inclusion disease, genital herpes, certain leukemias, infectious mononucleosis, rabies, and the three types of hepatitis: hepatitis A (infectious hepatitis); hepatitis B (serum hepatitis); and hepatitis C.
5. The viruses that are transmitted in food and water usually grow in the intestinal system and leave the body in the feces. Acquisition is generally by the oral route. Examples of diseases caused by these viruses include acute viral gastroenteritis (rotavirus and others), infectious hepatitis A, and poliomyelitis.
6. A slow virus disease is a pathological process caused by a transmissible agent (a prion or virus) that remains clinically silent for a prolonged period, after which the clinical disease becomes apparent. Examples include Creutzfeldt-Jakob disease, kuru, subacute sclerosing panencephalitis, Gerstmann-Sträussler-Scheinker syndrome, fatal familial insomnia, and progressive multifocal leukoencephalopathy. These diseases are chronic infections of the central nervous system that result in progressive degenerative changes and eventual death.
7. Common skin warts are caused by viruses and can be spread by autoinoculation through scratching or by direct or indirect contact. Anogenital condylomata (venereal warts) are sexually transmitted.

Key Terms

acute viral gastroenteritis *735*
adult T-cell leukemia *731*
AIDS *722*
AIDS-related complex (ARC) *724*
anogenital condylomata (venereal warts) *738*
chickenpox (varicella) *716*
cold sore *729*
common cold *730*
congenital (neonatal) herpes *731*
congenital rubella syndrome *717*
coryza *730*
cytomegalovirus inclusion disease *730*
Dane particle *734*
Delta agent *733*
Ebola virus hemorrhagic fever *721*
fever blister *729*
flat or plane warts *738*
genital herpes *730*
German measles (rubella) *717*
Guillain-Barré syndrome *718*
hemorrhagic fevers *721*
hepatitis *733*
hepatitis A (infectious hepatitis) *733*
hepatitis B (serum hepatitis) *733*
hepatitis C (non-A, non-B) *733*
herpes labialis *729*
herpetic keratitis *729*
human immunodeficiency virus (HIV) *722*
infantile paralysis *736*
infectious mononucleosis (mono) *731*
influenza or flu *718*
intranuclear inclusion body *730*
Koplik's spots *719*
Korean hemorrhagic fever *721*
leukemia *731*
Marburg viral hemorrhagic fever *721*
measles (rubeola) *719*
mumps *719*
Negri bodies *732*
plantar warts *738*
polio *736*
poliomyelitis *736*
rabies *732*
respiratory syncytial virus (RSV) *720*
Reye's syndrome *718*
shingles (herpes zoster) *716*
slow virus disease *736*
smallpox (variola) *720*
syndrome *720*
verrucae vulgaris *738*
wart *736*
yellow fever *722*

Questions for Thought and Review

1. Briefly describe each of the major or most common viral diseases in terms of its causative agent, signs and symptoms, the course of infection, mechanism of pathogenesis, epidemiology, and prevention and/or treatment.
2. Why are slow virus diseases not caused by slow viruses?
3. Which virus is responsible for each of the following: shingles, German measles, influenza, measles, mumps, and smallpox?
4. What are respiratory syndromes?
5. From an epidemiological perspective, why are most arthropod-borne viral diseases hard to control?
6. In terms of molecular genetics, why is the common cold such a prevalent viral infection in humans?
7. What are the differences between the four types of hepatitis?
8. Which viral diseases can be transmitted by sexual contact?
9. Why are the herpesviruses called persistent viruses?
10. What viruses specifically attack the nervous system?
11. Why is rabies such a feared disease?
12. Which viruses can cause encephalitis? Hepatitis?
13. What are two types of mononucleosis?
14. Will it be possible to eradicate many viral diseases in the same way as smallpox? Why or why not?

Additional Reading

Aiken, J. M., and Marsh, R. F. 1990. The search for scrapie agent nucleic acid. *Microbiol. Rev.* 54(3):242–46.

Alcamo, I. E. 1993. *AIDS: The biological basis.* Dubuque, Iowa, Wm. C. Brown Communications, Inc.

Aral, S. O., and Holmes, K. 1991. Sexually transmitted diseases in the AIDS era. *Sci. Am.* 264(2):62–69.

Bean, B. 1992. Antiviral therapy: current concepts and practices. *Clin. Microbiol. Rev.* 5(2):146–82.

Blacklow, N. R., and Greenberg, H. B. 1991. Viral gastroenteritis. *N. Eng. J. Med.* 325(4):252–64.

Blattner, W. A. 1990. *Human retrovirology:* HTLV. New York: Raven Press.

Buller, R. M., and Palumbo, G. 1991. Poxvirus pathogenesis. *Microbiol. Rev.* 55(1):80–122.

Christensen, M. L. 1989. Human viral gastroenteritis. *Clin. Microbiol. Rev.* 2(1):51–89.

Corey, L., and Holmes, K. K. 1983. Genital herpes simplex virus infections: Current concepts in diagnosis, therapy and prevention. *Ann. Intern. Med.* 98:973–81.

Dulbecco, R., and Ginsberg, H. S. 1988. *Virology,* 2d ed. Philadelphia: J. B. Lippincott.

Fields, B. N.; Knipe, D. M.; Chanock, R. M.; Hirsch, M. S.; Melnick, J. L.; Monath, T. P.; and Roizman, B., eds. 1990. *Fields virology,* 2d ed. New York: Raven Press.

Forbes, B. A. 1989. Acquisition of cytomegalovirus infection: an update. *Clin. Microbiol. Rev.* 3(2):204–16.

Friedman-Kien, A. E. 1989. *Color atlas of AIDS.* Philadelphia: W. B. Saunders.

Galasso, G. J. 1990. *Antiviral agents and viral diseases of man.* 3d ed. New York: Raven Press.

Gallo, R. C. 1986. The first human retrovirus. *Sci. Am.* 255(1):88–101.

Gallo, R. C. 1987. The AIDS virus. *Sci. Am.* 256(1):46–73.

Gallo, R. C., and Montagnier, L. 1988. AIDS in 1988. *Sci. Am.* 259(4):40–48.

Greene, W. C. 1991. The molecular biology of human immunodeficiency virus type 1 infection. *N. Eng. J. Med.* 324(5):308–15.

Henchal, E. A., and Putnak, R. J. 1990. The dengue viruses. *Clin. Microbiol. Rev.* 3(4):376–96.

Hazeltine, W. A., and Wang-Staal, F. 1988. The molecular biology of the AIDS virus. *Sci. Am.* 259(4):52–62.

Henderson, D. A. 1976. The eradication of smallpox. *Sci. Am.* 235(1):25–33.

Henle, W.; Henle, G.; and Lennette, E. T. 1979. The Epstein-Barr virus. *Sci. Am.* 241(3):48–59.

Hirsch, M. S., and Kaplan, J. C. 1987. Antiviral therapy. *Sci. Am.* 256(4):76–85.

Hogle, J. M; Chow, M.; and Filman, O. 1987. The structure of the poliovirus. *Sci. Am.* 256(3):42–54.

Holland, J. J. 1974. Slow, inapparent, and recurrent viruses. *Sci. Am.* 230(5):32–40.

Hollinger, F. B. 1990. *Viral hepatitis.* 2d ed. New York: Raven Press.

Huang, A. S. 1977. Viral pathogenesis and molecular biology. *Microbiol. Rev.* 41:811–19.

Jackson, J. B., and Balfour, H. H., Jr. 1988. Practical diagnostic testing for human immunodeficiency virus. *Clin. Microbiol. Rev.* 1(1):124–38.

Joklik, W. K.; Willett, H. P.; Amos, D. B.; and Wilfert, C. M. 1992. *Zinsser microbiology,* 20th ed. E. Norwalk, Conn.: Appleton & Lange.

Kaplan, M. M., and Koprowski, H. 1980. Rabies. *Sci. Am.* 242(2):120–34.

Kaplan, M. M., and Webster, R. G. 1977. The epidemiology of influenza. *Sci. Am.* 237(4):88–106.

Mandell, G. L.: Douglas, R. G., Jr.; and Bennett, J. E. 1990. *Principles and practice of infectious disease,* 3d ed. New York: John Wiley and Sons.

McDougall, J. K., ed. 1990. *Cytomegaloviruses.* New York: Springer-Verlag.

Mills, J., and Masur, H. 1990. AIDS-related infections. *Sci Am.* 263(2):50–59.

Morgan, E. M., and Rapp, F. 1977. Measles virus and its associated diseases. *Microbiol. Rev.* 41:636–41.

Norkin, L. C. 1982. Papoviral persistent infections. *Microbiol. Rev.* 46:384–91.

Okano, M.; Thiele, G. M.; Davis, J. R.; Grierson, H. L.; and Purtilo, D. T. 1988. Epstein-Barr virus and human diseases: Recent advances in diagnosis. *Clin. Microbiol. Rev.* 1(3):300–12.

Parkman, P. D., and Hopps, H. E. 1988. Viral vaccines and antivirals: Current use and future prospects. *Ann. Rev. Public Health* 9:203–21.

Prusiner, S. B. 1984. Prions. *Sci. Am.* 251:50–59.

Redfield, R. R., and Burk, D. S. 1988. HIV infection: The clinical picture. *Sci. Am.* 259(4):90–98.

Roman, A., and Fife, K. 1989. Human papillomviruses: Are we ready to type? *Clin. Microbiol. Rev.* 2:166–90.

Severin, M. 1989. Acquired immunodeficiency syndrome: More than a health related dilemma. *Clin. Microbiol. Rev.* 2(4):425–36.

Shaw, M. W.; Arden, N. H.; and Maassab, H. F. 1992. New aspects of influenza virus. *Clin. Microbiol. Rev.* 5(1):74–92.

Shulman, S. T.; Phair, J. P.; and Sommers, H. M. 1992. *The biologic and clinical basis of infectious disease,* 4th ed. Philadelphia, PA.: W.B. Saunders.

Spector, D. H., and Baltimore, D. 1975. The molecular biology of poliovirus. *Sci. Am.* 232(1):24–31.

Staczek, J. 1990. Animal cytomegaloviruses. *Microbiol. Rev.* 54(3):247–65.

Steffy, K., and Wong-Staal, F. 1991. Genetic regulation of human immunodeficiency virus. *Microbiol. Rev.* 55(2):193–205.

Stevens, J. G. 1989. Human herpesviruses: A consideration of the latent state. *Microbiol. Rev.* 53(3):318–32.

Sweet, C., and Smith, H. 1980. Pathogenicity of influenza viruses. *Microbiol. Rev.* 44:303–9.

The FASEB Journal. 1991. 5(10). AIDS (Thematic Issue).

Tiollais, P., and Buendia, M. A. 1991. Hepatitis B virus. *Sci. Am.* 264(4):116–24.

Welliver, R. C. 1989. Detection, pathogenesis, and therapy of respiratory syncytial virus infections. *Clin. Microbiol. Rev.* 1(1):27–39.

Winkler, W. G., and Vogel, K. 1992. Control of rabies in wildlife. *Sci. Am.* 266(6):86–93.

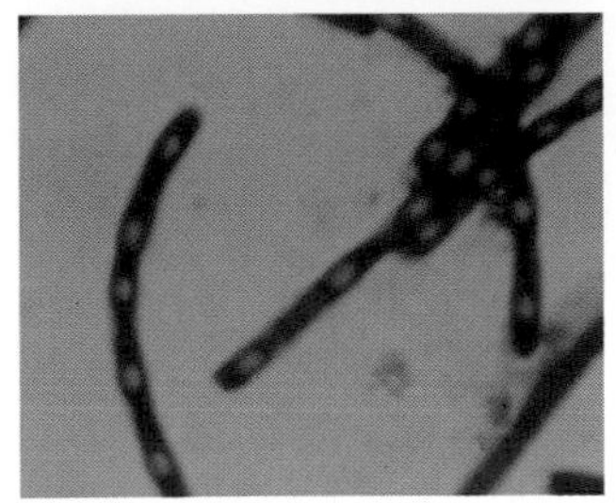

CHAPTER 37

Human Diseases Caused Primarily by Gram-Positive and Gram-Negative Bacteria

Soldiers have rarely won wars. They more often mop up after the barrage of epidemics. And typhus, with its brothers and sisters—plague, cholera, typhoid, dysentery—has decided more campaigns than Caesar, Hannibal, Napoleon, and all the . . . generals of history. The epidemics get the blame for the defeat, the generals the credit for victory. It ought to be the other way around. . . ."

—Hans Zinsser

Outline

Concepts

1. Bacterial diseases of humans can be discussed according to their mode of acquisition/transmission.
2. Most of the airborne diseases caused by bacteria involve the respiratory system. Examples include diphtheria, Legionnaires' disease and Pontiac fever, pertussis, streptococcal diseases, and tuberculosis. Other airborne bacteria can cause skin diseases, including cellulitis, erysipelas, and scarlet fever, or systemic diseases such as meningitis, glomerulonephritis, and rheumatic fever.
3. Although arthropod-borne bacterial diseases are generally rare, they are of interest either historically (plague) or because they have been newly introduced into humans (Lyme disease).
4. Most of the direct contact bacterial diseases involve the skin or underlying tissues. Examples include anthrax, gas gangrene, leprosy, staphylococcal diseases, and syphilis. Others can become disseminated throughout specific regions of the body—for example, gonorrhea, staphylococcal diseases, syphilis, tetanus, and tularemia.
5. The food-borne and waterborne bacterial diseases are contracted when contaminated food or water is ingested. These diseases are essentially of two types: infections and intoxications. An infection occurs when a pathogen enters the gastrointestinal tract and multiplies. Examples include typhoid fever and the gastroenteritis caused by *Vibrio* species. An intoxication occurs because of the ingestion of a toxin produced outside the body. Examples include botulism and staphylococcal food poisoning.

The first two parts of this textbook cover the general biology of bacteria. Chapters 21 through 25 specifically review bacterial morphology and taxonomy. Chapter 37 continues the coverage of bacteria by discussing some of the more important gram-positive and gram-negative bacteria that are pathogenic to humans.

Of all the known bacterial species, only a few are pathogenic to humans. In the following sections, the more important gram-positive and gram-negative disease-causing bacteria are discussed according to their mode of acquisition/transmission.

Airborne Diseases

Most airborne diseases caused by bacteria involve the respiratory system. Other airborne bacteria can cause skin diseases. Some of the better known of these diseases are now discussed.

Diphtheria

Diphtheria (Greek *diphthera,* membrane, and *ia,* condition) is an acute contagious disease caused by the gram-positive *Corynebacterium diphtheriae* (*see figure 22.11*). C. diphtheriae is well adapted to airborne transmission by way of nasopharyngeal secretions and is very resistant to drying. Diphtheria mainly affects poor people living in crowded conditions. Once within the respiratory system, bacteria that carry the prophage β (*see chapter 18*) produce diphtheria toxin, an exotoxin that causes an inflammatory response and the formation of a grayish pseudomembrane on the respiratory mucosa (figure 37.1*a–c*). The exotoxin is also absorbed into the circulatory system and distributed throughout the body, where it may cause destruction of cardiac, kidney, and nervous tissues by inhibiting protein synthesis (*see table 30.5*).

Diphtheria exotoxin (p. 584).

Typical symptoms of diphtheria include a thick mucopurulent (containing both mucus and pus) nasal discharge, fever, and cough. Diagnosis is made by observation of the pseudomembrane in the throat and by bacterial culture. Diphtheria antitoxin is given to neutralize any unabsorbed exotoxin in the affected individual's tissues; penicillin and erythromycin are used to treat the infection. Prevention is by active immunization with the **DPT** (**d**iphtheria-**p**ertussis-**t**etanus) **vaccine.**

C. diphtheriae can also infect the skin, usually at a wound or skin lesion, causing a slow-healing ulceration termed **cutaneous diphtheria** (figure 37.1*d*). Most cases involve people over 30 years of age who have a weakened immunity to the diphtheria toxin and live in tropical areas.

Fewer than 100 diphtheria cases are reported annually in the United States, and most occur in nonimmunized individuals. Because humans are the only reservoirs for *C. diphtheriae,* total eradication theoretically is possible if worldwide immunization becomes a reality, as it did for smallpox.

Legionnaires' Disease and Pontiac Fever

In 1976 the term **Legionnaires' disease,** or **legionellosis,** was coined to describe an outbreak of pneumonia that occurred at the Pennsylvania State American Legion Convention in Philadelphia. The bacterium responsible for the outbreak was described as *Legionella pneumophila,* a nutritionally fastidious aerobic gram-negative rod (figure 37.2). It is now known that this bacterium is part of the natural microbial community of soil and aquatic ecosystems, and it has been found in large numbers in air-conditioning systems and shower stalls.

Infection with *L. pneumophila* results from the airborne spread of bacteria from an environmental reservoir to the human respiratory system. Males over 50 years of age most commonly contract the disease, especially if they are compromised by heavy smoking, alcoholism, or chronic illness. The bacteria reside within the phagosomes of alveolar macrophages, where they multiply and produce localized tissue destruction through export of a cytotoxic exoprotease. Symptoms include a high fever, nonproductive cough (respiratory secretions are not brought up during coughing), headache, neurological manifestations, and severe bronchopneumonia. Diagnosis depends on isolation of the bacterium, documentation of a rise in antibody titer over time, or a rapid test kit using urine to detect antigens. Treatment begins with supportive measures and the administration of erythromycin or rifampin.

Prevention of Legionnaires' disease depends on the identification and elimination of the environmental source of *L. pneumophila* contamination. Chlorination, the heating of water, and the cleaning of water-containing devices can help control the multiplication and spread of *Legionella.* These control measures are effective because the pathogen does not appear to be spread from person to person.

Since the initial outbreak of this disease in 1976, many outbreaks during summer months have been recognized in all parts of the United States. About 800 to 1,000 cases are diagnosed each year, and about 30,000 or more additional mild or subclinical cases are thought to occur. It is estimated that 3 to 6% of all nosocomial pneumonias are due to *L. pneumophila,* especially among immunocompromised patients.

L. pneumophila also causes an illness called **Pontiac fever.** This disease, which resembles an allergic disease more than an infection, is characterized by an abrupt onset of fever, headache, dizziness, and muscle pains. It is indistinguishable clinically from the various respiratory syndromes caused by viruses. Pneumonia does not occur. The disease resolves spontaneously within 2 to 5 days. No deaths from Pontiac fever have been reported.

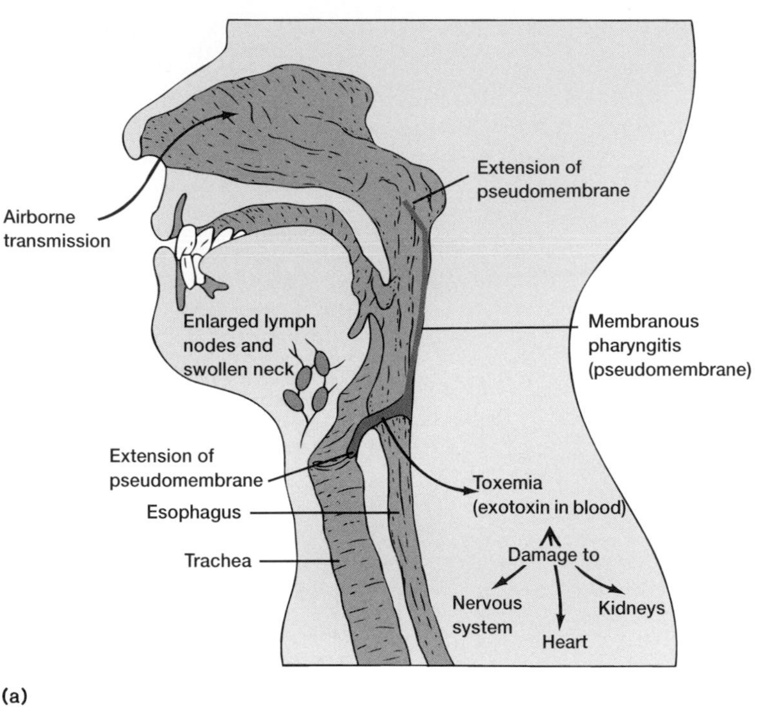

(a)

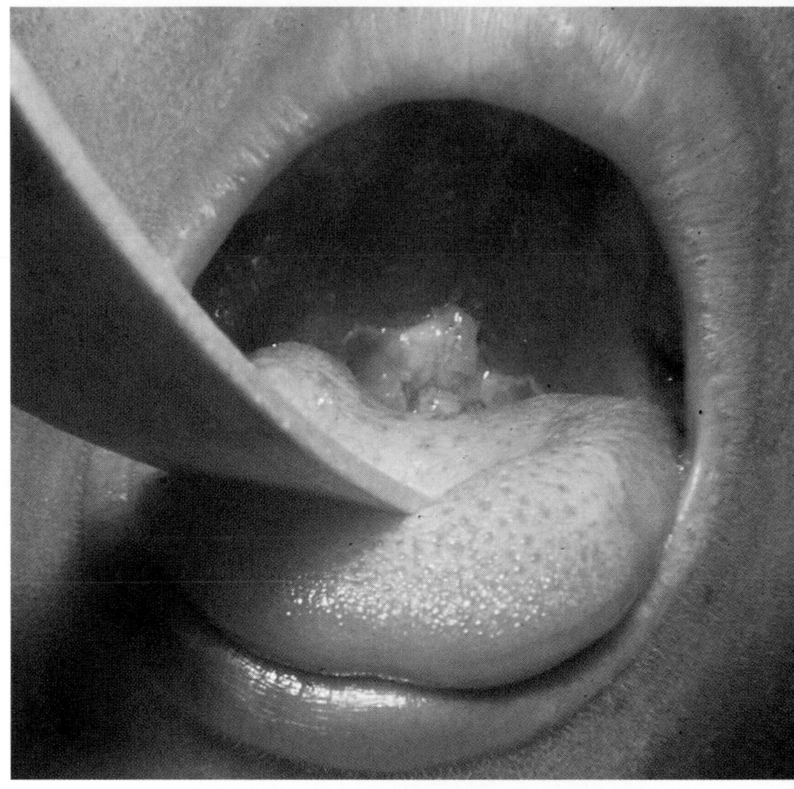

(b)

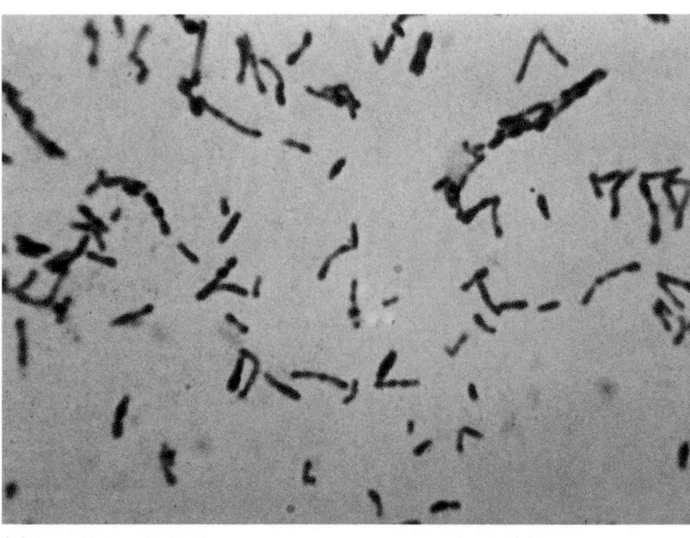

(c)

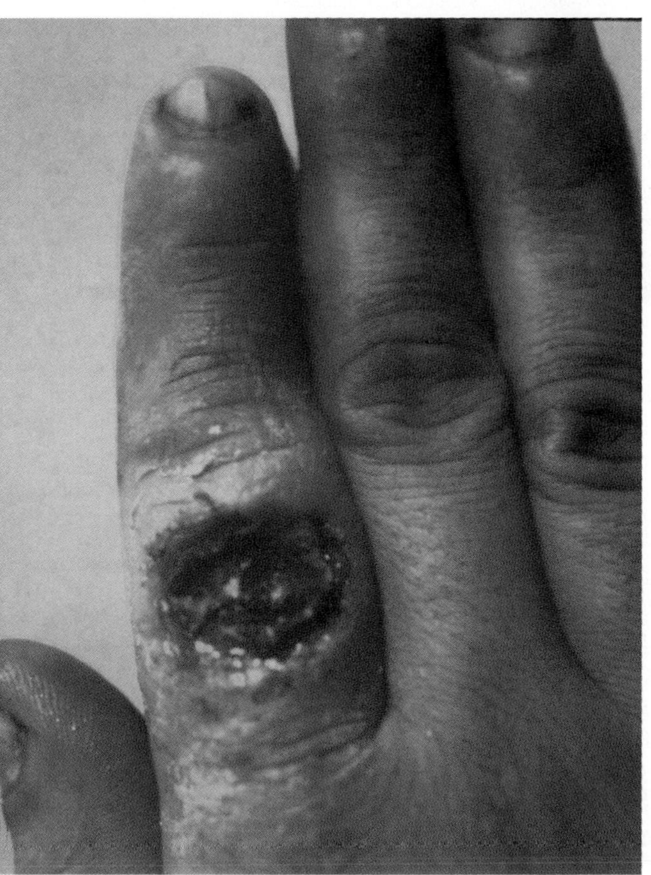

(d)

Figure 37.1 Diphtheria. (*a*) Pathogenesis. (*b*) This tough, grayish pseudomembrane overlying the tonsils may extend to the oropharynx, or even to the lungs. (*c*) *Corynebacterium diphtheriae.* Note the typical clubbed shapes (Greek *coryne,* club) and the V and Y arrangements (× 600). (*d*) Necrotic cutaneous diphtherial lesion about 15 days after onset.

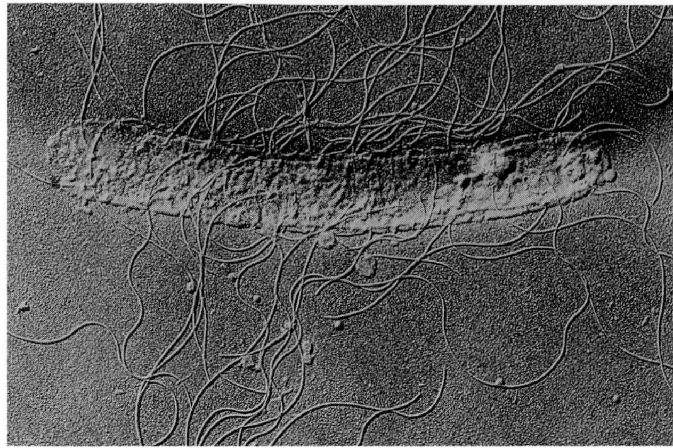

Figure 37.2 **Legionnaires' Disease.** *Legionella pneumophila,* the causative agent of Legionnaires' disease, with many lateral flagella; SEM (× 10,000).

Pontiac fever was first described from an outbreak in a county health department in Pontiac, Michigan. Ninety-five percent of the employees became ill and eventually showed elevated serum titers against *L. pneumophila.* These bacteria were later isolated from the lungs of guinea pigs exposed to the air of the building. The likely source was water from a defective air conditioner.

Meningitis

Meningitis (Greek *meninx,* membrane, and *-itis,* inflammation) is an inflammation of the brain or spinal cord meninges (membranes). Based on the specific cause, it can be divided into **bacterial (septic) meningitis** and the **aseptic meningitis syndrome** (table 37.1). As shown by the table, there are many causes of the aseptic meningitis syndrome, only some of which can be treated with antimicrobials. Thus, accurate identification of the causative agent is essential to proper treatment of the disease. The immediate sources of the bacteria responsible for meningitis are respiratory secretions from carriers or active cases. The bacteria initially colonize the nasopharynx after which they cross the mucosal barrier and enter the bloodstream and cerebrospinal fluid, where they produce inflammation of the meninges.

The usual symptoms of meningitis include an initial respiratory illness or sore throat interrupted by one of the meningeal syndromes: vomiting, headache, lethargy, confusion, and stiffness in the neck and back. Bacterial meningitis can be diagnosed by a Gram stain and culture of the bacteria from cerebrospinal fluid or rapid tests (*see table 34.5*). Once the causative bacterium is identified by culture, specific antibiotics (penicillin, chloramphenicol, cefotaxime, ceftriaxone, ofloxacin) are used for treatment. Vaccines against *S. pneumoniae* and *N. meningitidis* are available for high-risk individuals. Approximately 48% of all bacterial meningitis cases among children in the United States are due to *Haemophilus influenzae* type b (about 15,000 annually), 27% to *Neisseria meningitidis,* (about 2,500) and 11% to *Streptococcus pneumoniae* (about 1,000). *Listeria monocytogenes* is becoming an important cause of meningitis in immunocompromised, immunosuppressed, AIDS, and cancer patients, (*see table 36.3*).

Table 37.1 Causative Agents of Meningitis by Diagnostic Category

Type of Meningitis	Causative Agent
Bacterial Meningitis	*Streptococcus pneumoniae* *Neisseria meningitidis* *Haemophilus influenzae* type b Gram-negative bacilli Group B streptococci *Listeria monocytogenes* *Mycobacterium tuberculosis* *Nocardia asteroides* *Staphylococcus aureus* *Staphylococcus epidermidis*
Aseptic Meningitis Syndrome	
Agents requiring antimicrobials	Fungi Amoebae Syphilis Mycoplasmas
Agents requiring other treatments	Viruses Leptospires Cancers Parasitic cysts Chemicals

A person may have meningitis symptoms, but show no gram-negative bacteria in gram-stained specimens, and have negative cultures. In such a case, the diagnosis often is aseptic meningitis syndrome. Aseptic meningitis is more difficult to treat, and the prognosis is usually poor.

Mycobacterium avium–M. intracellulare Pneumonia

During the past decade, it has been discovered that there is an extremely large group of mycobacteria which are normal inhabitants of soil and water. Two of these have become noteworthy pathogens in the United States. The two, *Mycobacterium avium* and *Mycobacterium intracellulare* are so closely related that they are referred to as the *M. avium–M. intracellulare* (MAI) complex.

These mycobacteria are found worldwide and infect a variety of birds and animals. Both cause a pulmonary infection in humans similar to *M. tuberculosis.* These infections are seen most often in elderly persons with preexisting pulmonary disease.

Shortly after the recognition of AIDS and the associated opportunistic infections (*see table 36.3*), it became apparent that one of the more common infections was caused by MAI. Disseminated infection with MAI occurs in 30 to 50% of persons with AIDS in the United States with $CD4^+$ cell counts of less than 100 per cubic millimeter (*see AIDS, pp. 722–29*). Disseminated infection with MAI produces disabling symptoms, including fever, malaise, weight loss, and diarrhea. Carefully controlled epidemiological studies have shown that MAI shortens survival by 5 to 7 months among persons with AIDS. With more effective antiviral therapy for AIDS and with prolonged survival, the number of cases of disseminated MAI is likely to increase substantially, and its contribution to AIDS mortality will increase concomitantly.

MAI can be isolated from sputum, blood, and aspirates of bone marrow. Acid-fast stains are of value in making a diagnosis. The most sensitive method for detection is the commercially available lysis-centrifugation blood culture system (Wampole Laboratories). Treatment of MAI is similar to that used for tuberculosis, using a combination of drugs such as amikacin, clofazimine, rifampin, and ethambutol.

Pertussis

Pertussis (Latin *per,* intensive, and *tussis,* cough), sometimes called "whooping cough," is caused by the gram-negative bacterium *Bordetella pertussis,* or sometimes *B. parapertussis.* Pertussis is a highly contagious disease that primarily affects children. It has been estimated that over 95% of the world's population has experienced either mild or severe symptoms of the disease. Around 500,000 die from the disease each year. However, there are only about 2,000 to 4,000 cases and less than 10 deaths annually in the United States.

Transmission occurs by inhalation of the bacterium in droplets released from an infectious person. The incubation period is 7 to 14 days. Once inside the upper respiratory tract, the bacteria attach to the ciliated epithelial cells by producing a specific adherence factor called filamentous hemagglutinin (*see table 30.3*), which recognizes a complementary molecule on the cells. After attachment, the bacteria synthesize several toxins (*see table 30.5*) that are responsible for the symptoms. The most important toxin is pertussis toxin, which causes increased tissue susceptibility to histamine and serotonin, and an increased lymphocyte response. *B. pertussis* also produces extracytoplasmic adenylate cyclase and hemolysin, which act as surface-associated toxins, and tracheal cytotoxin, which also destroys epithelial tissue. In addition, the secretion of a thick mucus impedes ciliary action, and often, ciliated epithelial cells die.

Pertussis is divided into three stages. (1) The catarrhal stage, so named because of the mucous membrane inflammation, is insidious and resembles the common cold. (2) Prolonged coughing sieges characterize the paroxysmal stage. During this stage, the infected person tries to cough up the mucous secretions by making 5 to 15 rapidly consecutive coughs followed by the characteristic whoop—a hurried deep inspiration. The catarrhal and paroxysmal stages last about 6 weeks. (3) Final recovery may take several months (the convalescent stage).

Laboratory diagnosis of pertussis is by culture of the bacterium, fluorescent antibody staining of smears from nasopharyngeal swabs, and serological tests. The development of a strong, lasting immunity takes place after an initial infection. Treatment is with erythromycin, tetracycline, or chloramphenicol. Prevention is with the DPT vaccine (*p. 608*); vaccination of children is recommended when they are 2 to 3 months old (*see table 31.1*).

Streptococcal Diseases

Streptococci, commonly called strep, are a heterogenous group of gram-positive bacteria. In this group, *Streptococcus pyogenes* (group A β-hemolytic streptococci; *see pp. 455–57*) is one of the most important bacterial pathogens. The different serotypes produce (1) extracellular enzymes that break down host molecules; (2) streptokinases, enzymes that activate a host-blood factor that dissolves blood clots; (3) the cytolysins streptolysin O and streptolysin S, which kill host leukocytes; and (4) capsules and M protein that help retard phagocytosis.

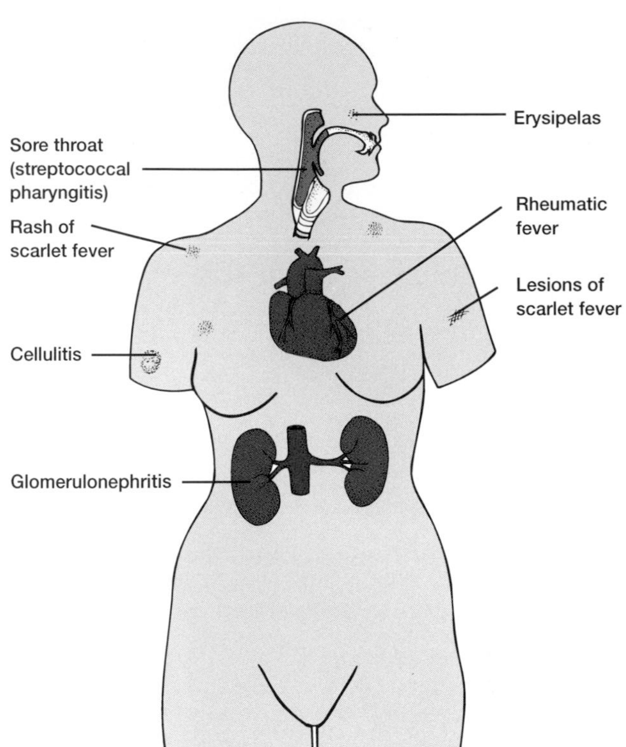

Figure 37.3 **Streptococcal Diseases.** Some of the more prominent diseases associated with group A streptococcal infections, and the body sites affected.

S. pyogenes is widely distributed among humans, but usually people are asymptomatic carriers. Individuals with acute infections may spread the pathogen, and transmission can occur through respiratory droplets, direct, or indirect contact. When highly virulent strains appear in schools, they can cause sharp outbreaks of sore throats and scarlet fever. Due to the cumulative buildup of antibodies to many different *S. pyogenes* serotypes over the years, outbreaks among adults are less frequent.

Diagnosis of a streptococcal infection is based on both clinical and laboratory findings. Several rapid tests are available (*see table 34.5*). Treatment is with penicillin or erythromycin. Vaccines are not available for streptococcal diseases other than streptococcal pneumonia because of the large number of serotypes.

The best control measure is prevention of bacterial transmission. Individuals with a known infection should be isolated and treated. Personnel working with infected patients should follow standard aseptic procedures.

In the following sections, some more important human streptococcal diseases are discussed (figure 37.3).

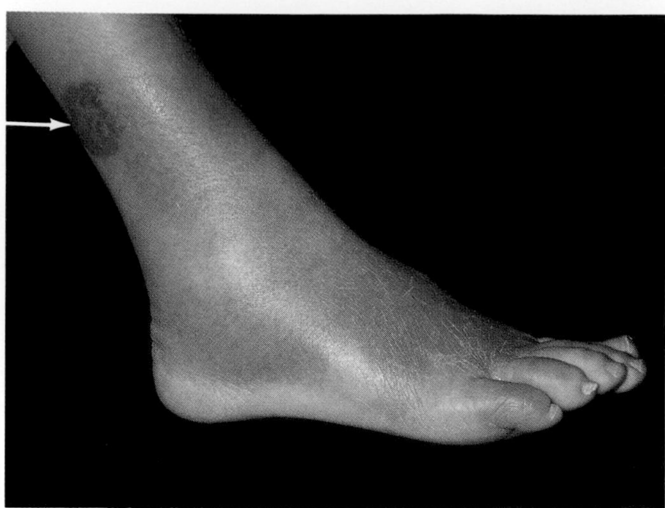

Figure 37.4 **Erysipelas.** Notice the bright, raised, rubbery lesion at the site of initial entry (white arrow) and the spread of the inflammation to the foot. The reddening is caused by toxins produced by the streptococci as they invade new tissue.

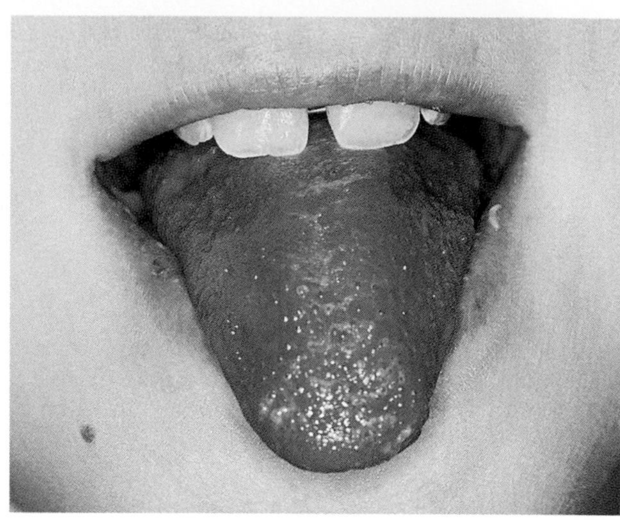

Figure 37.5 **Scarlet Fever.** The "strawberry tongue" of this streptococcal disease.

Cellulitis and Erysipelas

Cellulitis is a diffuse, spreading infection of subcutaneous skin tissue. The resulting inflammation is characterized by a defined area of redness (erythema) and the accumulation of fluid (edema).

Erysipelas (Greek *erythros,* red, and *pella,* skin) is an acute infection and inflammation of the dermal layer of the skin. It occurs primarily in infants and people over 30 years of age with a history of streptococcal sore throat. The skin often develops painful reddish patches that enlarge and thicken with a sharply defined edge (figure 37.4). Recovery usually takes a week or longer if no treatment is given. The drugs of choice for the treatment of erysipelas are erythromycin and penicillin. Erysipelas may recur periodically at the same body site for years.

Poststreptococcal Diseases

The poststreptococcal diseases are glomerulonephritis and rheumatic fever. They occur 1 to 4 weeks after an acute streptococcal infection (hence the term post). Today, these two diseases are the most serious problems associated with streptococcal infections in the United States.

Glomerulonephritis or **Bright's disease** is an inflammatory disease of the renal glomeruli, membranous structures within the kidney where blood is filtered. Damage probably results from the deposition of antigen-antibody complexes, possibly involving the streptococcal M protein, in the glomeruli. Thus, the disease arises from a type III immune complex–mediated hypersensitivity reaction (*see figure 32.9*). The complexes cause destruction of the glomerular membrane allowing proteins and blood to leak into the urine. Clinically, the affected person exhibits edema, fever, hypertension, and hematuria (blood in the urine). The disease occurs primarily among school-age children. Diagnosis is based on the clinical history, physical findings, and confirmatory evidence of prior streptococcal infection. The incidence of glomerulonephritis in the United States is less than 0.5% of streptococcal infections. Penicillin G or erythromycin can be given for any residual streptococci. However, there is no specific therapy once kidney damage has occurred. About 80 to 90% of all cases undergo slow spontaneous healing of the damaged glomeruli, whereas the others develop a chronic form of the disease. The latter may require a kidney transplant or lifelong renal dialysis.

Rheumatic fever is an autoimmune disease characterized by inflammatory lesions involving the heart valves, joints, subcutaneous tissues, and central nervous system. It usually results from a prior streptococcal sore throat infection. The exact mechanism of rheumatic fever development remains unknown. The disease occurs most frequently among children 6 to 15 years of age and manifests itself through a variety of signs and symptoms, making diagnosis difficult. In the United States, rheumatic fever has become very rare (less than 0.05% of streptococcal infections), but it occurs a hundred times more frequently in tropical countries. Therapy is directed at decreasing the inflammation and fever, and controlling cardiac failure. Salicylates and corticosteroids are the mainstays of treatment. Though rheumatic fever is rare, it is still the most common cause of permanent heart valve damage in children.

Scarlet Fever

Scarlet fever (scarlatina) results from a throat infection with a strain of *S. pyogenes* that carries a lysogenic bacteriophage. This codes for the production of an erythrogenic or rash-inducing toxin that causes shedding of the skin. Scarlet fever is a communicable disease spread by inhalation of infective respiratory droplets. After a two-day incubation period, a scarlatinal rash appears on the upper chest and then spreads to the remainder of the body. This rash represents the skin's generalized reaction to the circulating toxin. Along with the rash, the infected individual experiences a sore throat, chills, fever, headache, and a strawberry-colored tongue (figure 37.5). Treatment is with penicillin.

Streptococcal Sore Throat

Streptococcal sore throat is one of the most common bacterial infections of humans and is commonly called strep throat. The β-hemolytic group A streptococci are spread by droplets of saliva or nasal secretions. The incubation period in humans is 2 to 4 days. The incidence of sore throat is greater during the winter and spring months.

The action of the strep bacteria in the throat (**pharyngitis**) or on the tonsils (**tonsillitis**) stimulates an inflammatory response and the lysis of leukocytes and erythrocytes. An inflammatory exudate consisting of cells and fluid is released from the blood vessels and deposited in the surrounding tissue. This is accompanied by a general feeling of discomfort or malaise, fever (usually above 101°F), and headache. Prominent physical manifestations include redness, edema, and lymph node enlargement in the throat. Several common rapid test kits are available for diagnosing strep throat. In the absence of complications, the disease is self-limited and disappears within a week. However, treatment with penicillin G benzathine (or erythromycin for penicillin-allergic people) can shorten the infection and clinical syndromes, and is especially important in children for the prevention of complications such as rheumatic fever and glomerulonephritis. Infections in older children and adults tend to be milder and less frequent due in part to the immunity they have developed against the many serotypes encountered in early childhood. Prevention and control measures include proper disposal or cleaning of objects (e.g., Kleenex, handkerchiefs) contaminated by discharges from the infected individual.

Streptococcal Pneumonia

Streptococcal pneumonia is now considered an **endogenous infection;** that is, it is contracted from one's own normal microbiota (*see figure 29.8*). It is caused by the gram-positive *Streptococcus pneumoniae,* found in the upper respiratory tract (figure 37.6). However, disease usually occurs only in those individuals with predisposing factors such as viral infections of the respiratory tract, physical injury to the tract, alcoholism, or diabetes. About 60 to 80% of all respiratory diseases known as pneumonia are caused by *S. pneumoniae.* An estimated 150,000 to 300,000 people in the United States contract this form of pneumonia annually, and about 13,000 to 66,000 deaths result.

The primary virulence factor of *S. pneumoniae* is its capsular polysaccharide. The capsule prevents binding of antibody to the cell wall and thus inhibits phagocytosis. The pathogenesis is due to the rapid multiplication of the bacteria in alveolar spaces. The alveoli fill with blood cells and fluid and become inflamed. The sputum is often rust colored because of blood coughed up from the lungs. The onset of clinical symptoms is usually abrupt, with chills, hard labored breathing, and chest pain. Diagnosis is by chest X ray, biochemical tests, and culture. Penicillin G, cefotaxime, ofloxacin, and ceftriaxone have contributed to a greatly reduced mortality rate. For individuals who are sensitive to penicillin, erythromycin, or tetracycline can be used. A pneumococcal vaccine (Pneumovax) is available for people who are debilitated (e.g., people in chronic-care facilities). Preventive and control measures include immunization and adequate treatment of infected persons.

Tuberculosis

Over a century ago, Robert Koch (*see figure 1.5*) identified *Mycobacterium tuberculosis* as the causative agent of **tuberculosis (TB).** At the time, TB was rampant, causing 1/7 of all deaths in Europe and 1/3 of deaths among productive young adults. Today, TB remains a global health problem of enormous dimension. It is estimated that there are 1 billion (20% of the world's human population) infected worldwide, with 8 million new cases and over 3 million deaths per year.

In the United States, this disease occurs most commonly among the elderly, malnourished or alcoholic poor males, and Native Americans. More than 20,000 new cases of tuberculosis and over 12,000 deaths are reported annually. Most cases are caused by the acid-fast *Mycobacterium tuberculosis,* acquired from other humans through droplet nuclei and the respiratory route (figure 37.7). Transmission to humans from susceptible animal species and their products (e.g., milk) is also possible. Recently, there has been a steady yearly increase in the number of TB cases as a result of the AIDS epidemic. Available statistics indicate that a close association exists between AIDS and TB. Therefore, further spread of HIV infec-

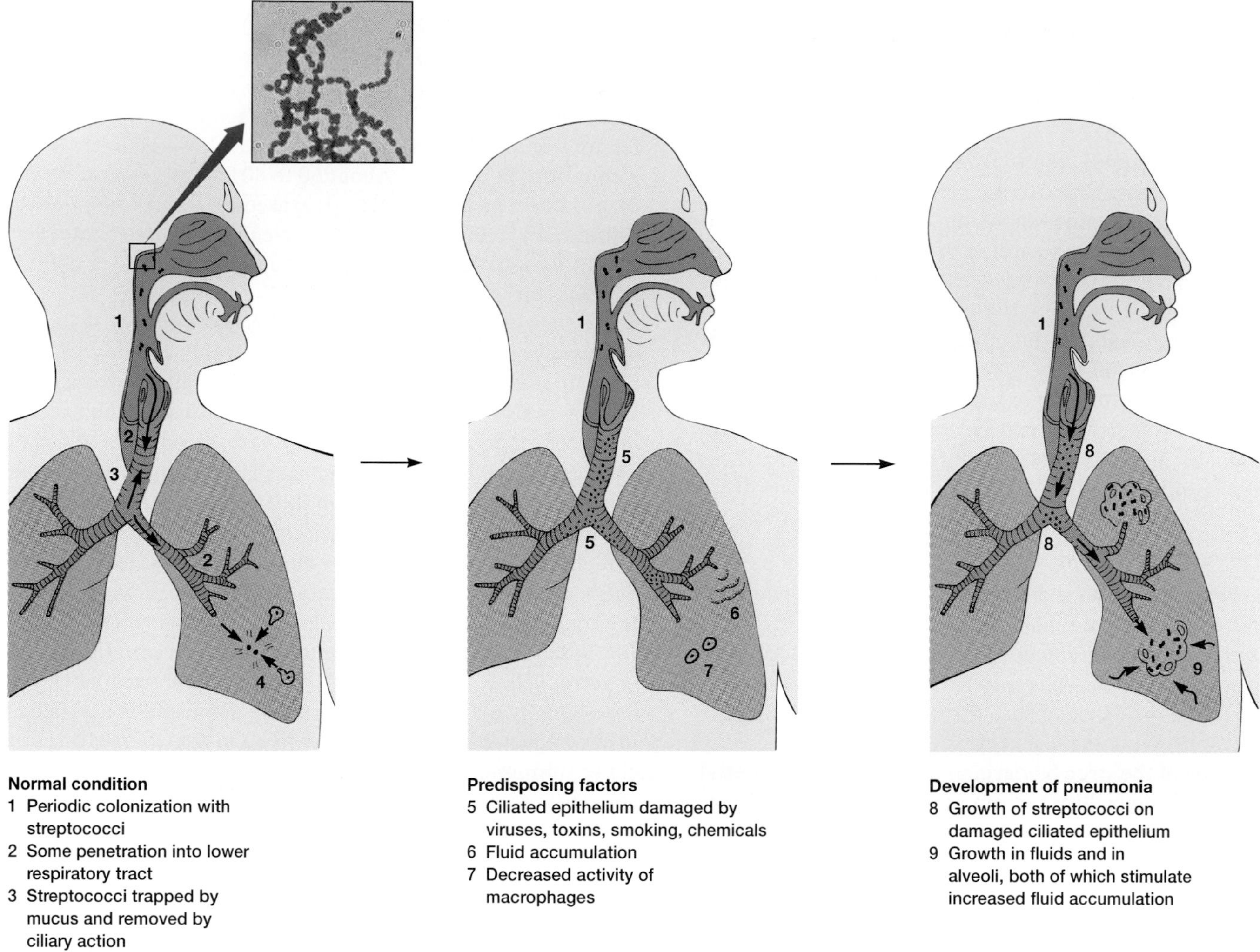

Figure 37.6 **Predisposition to and the Development of Streptococcal Pneumonia.** The insert shows the morphology of *Streptococcus pneumoniae* (×1,000).

tion among the population with a high prevalence of TB infection is resulting in dramatic increases in TB.

AIDS (pp. 722–29).

Once in the lungs, the bacteria grow and eventually are surrounded by lymphocytes, macrophages, and connective tissue in a hypersensitivity response that forms small, hard nodules called **tubercles,** which are characteristic of tuberculosis and give the disease its name. The disease process usually stops at this stage, but the bacteria remain alive within the tubercles. In time, the tubercle may change to a cheeselike consistency and is then called a **caseous lesion.** If such lesions calcify they are termed **Ghon complexes,** which show up prominently in a chest X ray. Sometimes, the tubercle lesions liquefy and form air-filled **tuberculous cavities.** From these cavities the bacteria can spread to new foci of infections throughout the body. This spreading is often called **miliary tuberculosis** due to the many tubercles the size of millet seeds that are formed in the infected tissue. It also may be called **reactivation tuberculosis** because the bacteria have been reactivated in the initial site of infection.

Persons infected with *M. tuberculosis* develop a cell-mediated immunity due to the bacteria being phagocytosed by macrophages. This immunity involves sensitized T cells (figure 37.7*b*) and is the basis for the tuberculin skin test (*see figure 32.10*). In this test a **p**urified **p**rotein **d**erivative (PPD) of *M. tuberculosis* is injected intracutaneously (the mantoux test). If the person has had tuberculosis, sensitized T cells react with these proteins, and a delayed hypersensitivity reaction occurs within 48 hours. This positive skin reaction appears as an induration (hardening) and reddening of the area around the injection site. Multiple puncture tests such as the Tine test are more convenient, but not as accurate.

In a young person, a positive skin test possibly indicates active tuberculosis. In older persons, it may result from previous disease, vaccination, or a false-positive test. In both cases, X rays and bacterial isolation should be completed.

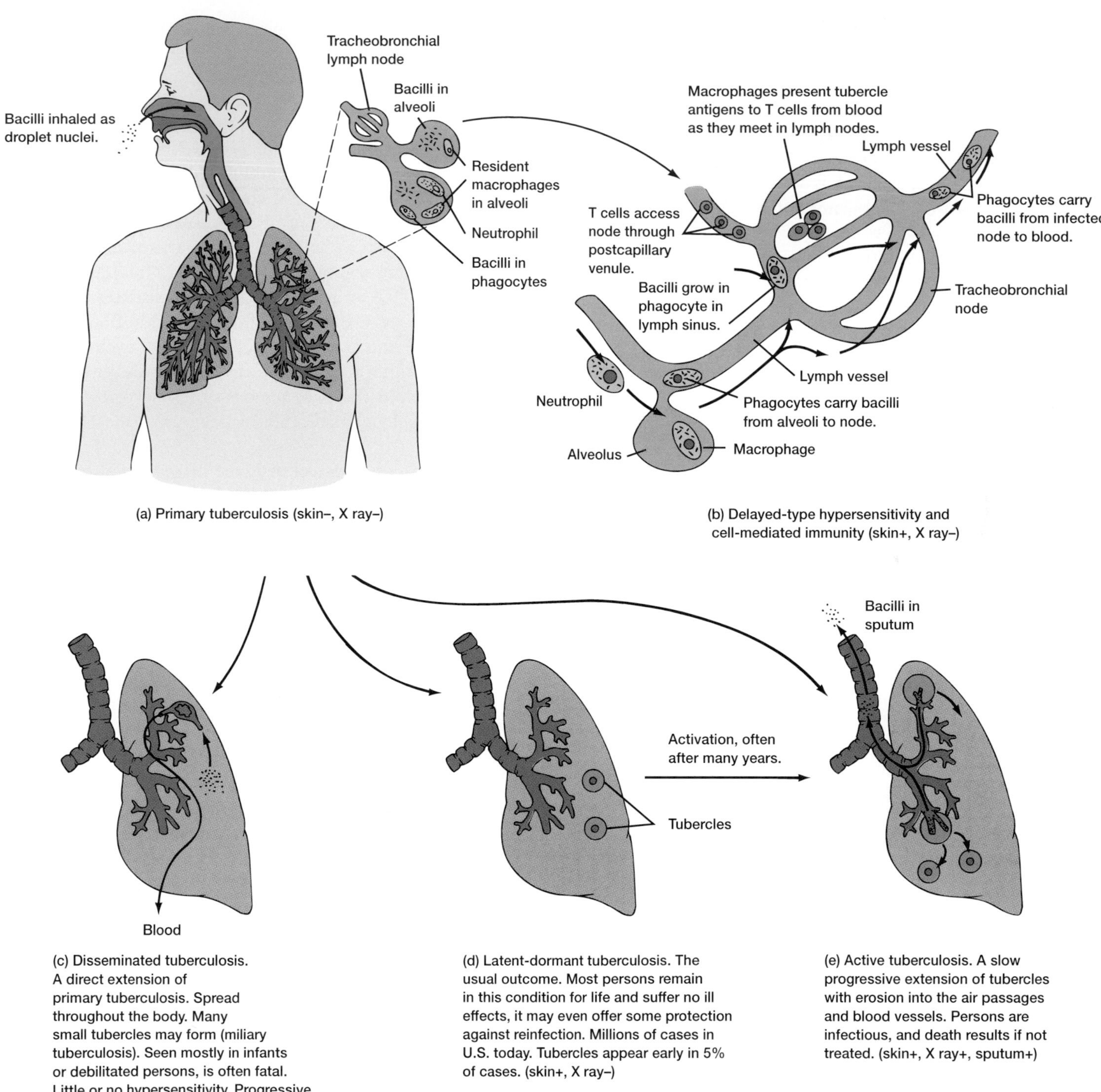

Figure 37.7 **Tuberculosis.** The history of untreated tuberculosis in the human body.

The symptoms of tuberculosis are fever, fatigue, and weight loss. A cough, which is characteristic of pulmonary involvement, may result in expectoration of bloody sputum.

Laboratory diagnosis of tuberculosis is by isolation of the acid-fast bacterium, chest X ray, commercially available DNA probes, and the Mantoux or tuberculin skin test. Both chemotherapy and chemoprophylaxis are carried out by administering isoniazid (INH), streptomycin, ethambutol, or rifampin. Usually two or more drugs are administered simultaneously for 6 to 12 months as a way of decreasing the possibility that the patient develops drug resistance (e.g., INH plus rifampin). Recently, new multi-drug-resistant strains (MDR-TB) have developed and are spreading.

Prevention and control of tuberculosis requires rapid specific therapy to interrupt infectious spread. In many countries, individuals, especially infants, are vaccinated with bacille

Calmette-Guérin (BCG) vaccine to prevent complications such as meningitis. Tuberculosis rates also can be lowered by better public health measures and social conditions, for example, a reduction in homelessness and drug abuse.

1. What causes the typical symptoms of diphtheria and how are individuals protected against this disease?
2. What is the environmental source of the bacterium that causes Legionnaires' disease? Pontiac fever?
3. What are the two major types of meningitis? Why is it so important to determine which type a person has?
4. Name the three stages of pertussis.
5. Name seven human diseases caused by *Streptococcus pyogenes*. How do they differ from one another?
6. How is tuberculosis diagnosed? Describe the various types of lesions and how they are formed.

Arthropod-Borne Diseases

Although arthropod-borne bacterial diseases are generally rare, they are of interest either historically (plague) or because they have been newly introduced into humans (Lyme disease). These two diseases are now discussed.

Lyme Disease

Lyme disease (LD, Lyme borreliosis) was first observed and described in 1975 among people of Old Lyme, Connecticut. It has become the most common tick-borne zoonosis in the United States, with more than 8,000 cases being reported annually. In fact, Lyme disease has reached epidemic proportions, and if it were not for AIDS, Lyme disease would be the most important "new" infectious disease of humans in the United States.

The bacterium responsible for this disease is the spirochete *Borrelia burgdorferi* (figure 37.8*a*), and deer and field mice are the natural hosts. In the Northeast, *B. burgdorferi* is transmitted to humans by the bite of an infected deer tick (*Ixodes scapularis,* formerly *Ixodes dammini;* figure 37.8*b*). On the Pacific Coast, especially in California, the reservoir is a dusky-footed woodrat, and the tick, *I. pacificus.*

Clinically, Lyme disease is a complex illness with three major stages. The initial, acute stage occurs a week to 10 days after an infectious tick bite. The illness usually begins with an expanding, ring-shaped, skin lesion with a red outer border and partial central clearing (figure 37.8*c*). This is often accompanied by flulike symptoms (malaise and fatigue, headache, fever, and chills). Often the tick bite is unnoticed, and the skin lesion may be missed due to skin coloration or its obscure location such as on the scalp. Thus treatment, which is usually effective at this stage, may not be given because the illness is passed off as "just a touch of the flu."

The second stage may appear weeks or months after the initial infection. It consists of several symptoms such as neurological abnormalities, heart inflammation, and bouts of arthritis (usually in the major joints such as the elbows or knees). The inflammation that produces organ damage is initiated and possibly perpetuated by the immune response to one or more spirochetal proteins.

Finally, like syphilis, years later the tertiary stage may appear. Infected individuals may develop demyelination of neurons with symptoms resembling Alzheimer's disease and multiple sclerosis. Behavioral changes also can occur.

Laboratory diagnosis of Lyme's disease is based on (1) the recovery of the spirochete from the patient, (2) use of the polymerase chain reaction (*see chapter 14*) for detection of *B. burgdorferi* DNA in urine, or (3) serological testing (Lyme ELISA or Western Blot) for IgM and IgG antibodies to the pathogen. Treatment with penicillin or tetracycline early in the illness results in prompt recovery and prevents arthritis and other complications. If nervous system involvement is suspected, ceftriaxone is used since it can cross the blood-brain barrier.

Prevention and control of Lyme disease involves environmental modification (clearing and burning tick habitat) and the application of acaricidal compounds (agents that destroy mites and ticks). An individual's risk of acquiring Lyme disease may be greatly reduced by education and personal protection. The following points should be kept in mind whenever a person is active in an area where Lyme disease or other tick-borne zoonoses occur:

1. It takes a minimum of 24 hours of attachment and feeding for transmission to occur; thus, prompt removal of attached ticks will greatly reduce the risk of infection.
2. Because each deer tick life cycle stage is most abundant at a certain time, there are periods when an individual should be most aware of the risk of infection. The most dangerous times are May through July, when the majority of nymphal deer ticks are present and the risk of transmission is greatest.
3. If you must be in the woods, dress accordingly. Wear light-colored pants and good shoes. Tuck the cuffs of your pants into long socks to deny ticks easy entry under your clothes. After coming out of the woods, check all clothes for ticks.
4. Repellants containing high concentrations of DEET (diethyltoluamide) or permanone are available over the counter and are very noxious to ticks.
5. As soon as possible after being in a high-risk area, examine your body for bites or itches. Taking a shower and using lots of soap aids in this examination. Areas such as the scalp, armpits, and groin are difficult to examine effectively, but are preferred sites for tick attachment. Special attention should be given to these parts of the body.

Plague

In the southwestern part of the United States, **plague** (Latin *plaga,* pest) occurs primarily in wild rodents (ground squirrels

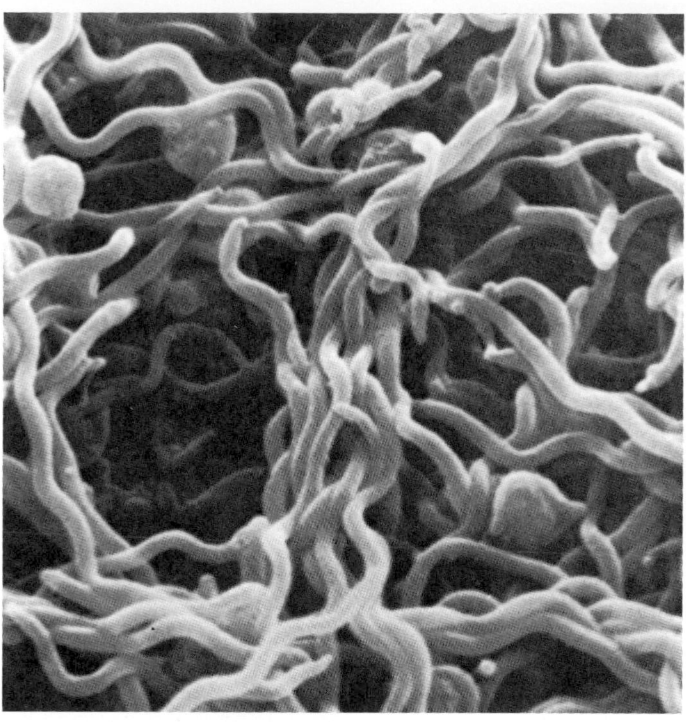

(a)

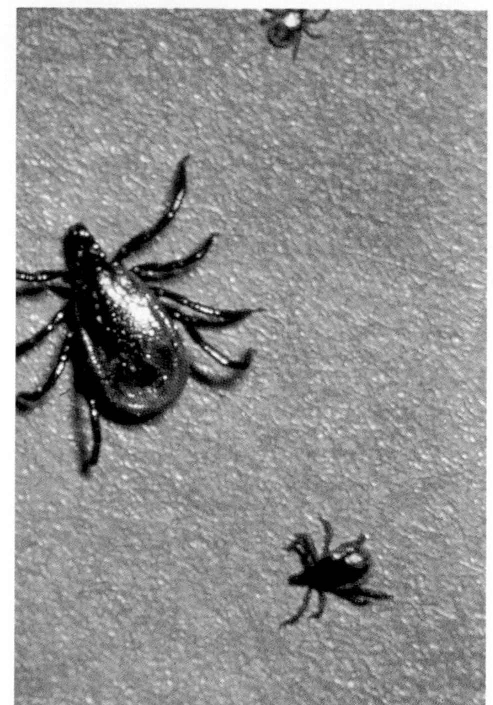

(b)

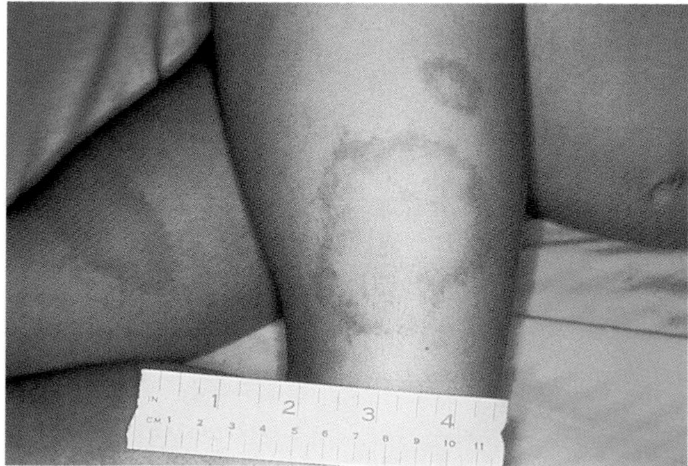

(c)

Figure 37.8 **Lyme Disease.** (*a*) The etiological agent is the spirochete *Borrelia burgdorferi;* SEM. (*b*) The vector in the Northeast is the tick *Ixodes scapularis.* The youngest (nymphal stage, top) is about the size of a poppy seed. An unengorged adult (bottom), and an engorged adult (center) can reach the size of a jelly bean. (*c*) The typical rash showing concentric rings around the initial site of the tick bite.

and prairie dogs). However, massive human epidemics occurred during the Middle Ages, and the disease was known as the Black Death because one of its characteristics is blackish areas on the skin caused by subcutaneous hemorrhages. Infections now occur in humans only sporadically or in limited outbreaks. In the United States, approximately 25 cases are reported annually, and the mortality rate is about 15%

The disease is caused by the gram-negative bacterium *Yersinia pestis.* It is transmitted from rodent to human by the bite of an infected flea, direct contact with infected animals or their products, or inhalation of contaminated airborne droplets (figure 37.9). Once in the human body, the bacteria multiply in the blood and lymph. An important factor in the virulence of *Y. pestis* is its ability to survive and proliferate inside phagocytic cells rather than be killed by them. One of the ways this is accomplished is by the YOPS (yersinal plasmid-encoded outer membrane proteins) that counteract natural defense mechanisms and help the bacteria multiply and disseminate in the host.

Symptoms—besides the subcutaneous hemorrhages—include fever and the appearance of enlarged lymph nodes called **buboes** (hence the old name, bubonic plague). In 50 to 70% of the untreated cases, death follows in 3 to 5 days from toxic conditions caused by the large number of bacilli in the blood.

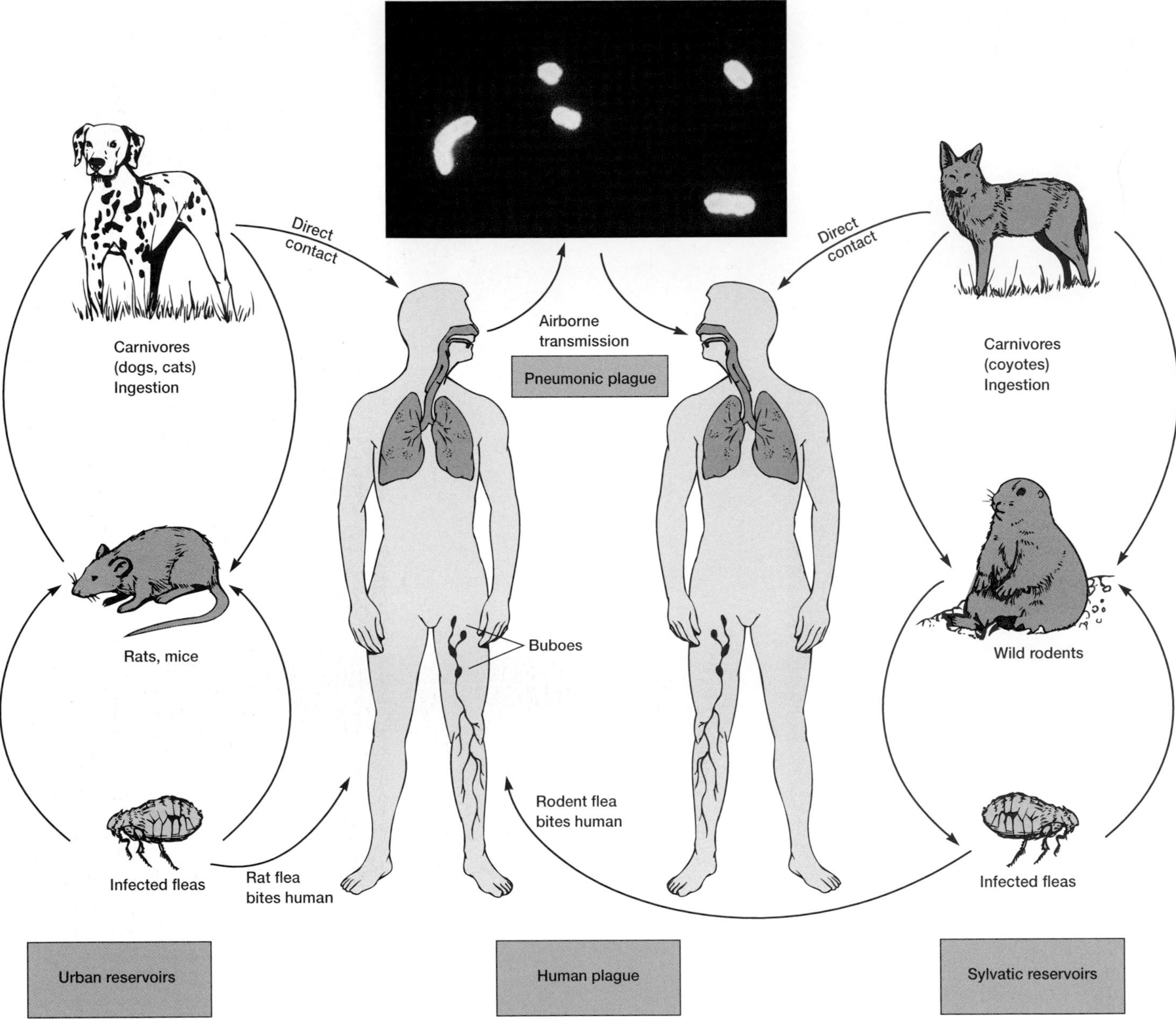

Figure 37.9 Plague. Plague is spread to humans through (1) the urban cycle and rat fleas, (2) the sylvatic cycle centered on wild rodents and their fleas, or (3) by airborne transmission from an infected person leading to pneumonic plague. Dogs, cats, and coyotes can also acquire the bacterium by ingestion of infected animals. The insert shows *Yersinia pestis* stained with fluorescent antibodies.

Laboratory diagnosis of plague is by direct microscopic examination, culture of the bacterium, serological tests, and phage testing. Treatment is with streptomycin or tetracycline, and recovery from the disease gives a good immunity.

If the patient does not receive antibiotics, the bacteria may invade the lungs and produce pneumonic plague, a type of pneumonia. The mortality rate for this kind of plague is almost 100% if it is not recognized within 12 to 24 hours. Obviously, great care must be taken to prevent the spread of airborne infections to personnel taking care of pneumonic plague patients.

Prevention and control involves ectoparasite and rodent control, isolation of human patients, prophylaxis or abortive therapy of exposed persons, and vaccination (USP Plague vaccine) of persons at high risk.

1. What is the causative agent of Lyme disease and how is it transmitted to humans? How does the illness begin? Describe the three stages of Lyme disease.
2. Why is plague sometimes called the Black Death? How is it transmitted? Distinguish between bubonic and pneumonic plague.

Direct Contact Diseases

Most of the direct contact bacterial diseases involve the skin or underlying tissues. Others can become disseminated through specific regions of the body. Several examples of each are now discussed.

Box 37.1

A Brief History of Anthrax

Anthrax is not only of current significance as an infection of animals and humans, but it is also of historical interest for it was investigated extensively by the founders of bacteriology. Robert Koch, (*See Robert Koch and Koch's postulates in chapter 1*) in 1876, was the first to isolate the causative microorganism in pure culture and to reproduce the disease with the culture. Louis Pasteur, Francis Rous, and Charles-Edouard Chamberland, in 1881, demonstrated active immunization with attenuated anthrax cultures in their famous experiment at Pouilly-le-Fort. Their dramatic demonstration before a special French commission of the immunizing properties of attenuated cultures of anthrax bacilli is a significant milestone in the history of bacteriology.

Also of interest is the fact that anthrax in humans was originally called woolsorter's disease because it was especially prevalent in those who sorted sheep wool. In the mid-1800s, the British physician John Bell observed a relationship between the symptoms of anthrax and woolsorting. He inoculated healthy sheep with blood from a human who had died from anthrax and noted that all the sheep died. The anthrax bacillus was then recovered from the dead sheep, which confirmed the relationship between the bacterium and the disease symptoms.

These discoveries prompted the development of preventive measures for stopping the spread of anthrax among woolsorters. Examples include the soaking of wool in formaldehyde solutions before handling by the woolsorters and the vaccination of animals with Pasteur's attenuated anthrax bacillus.

Anthrax

Anthrax (Greek *anthrax,* coal) is a highly infectious animal disease that can be transmitted to humans by direct contact with infected animals (cattle, goats, sheep) or their products. The causative bacterium is the gram-positive *Bacillus anthracis* (figure 37.10*a*). Its endospores can remain viable in soil and animal products for decades. Human infection is usually through a cut or abrasion of the skin, resulting in **cutaneous anthrax;** however, inhaling endospores may result in **pulmonary anthrax,** also known as woolsorter's disease (Box 37.1). If endospores reach the intestine, **gastrointestinal anthrax** may result.

In humans, the incubation period for cutaneous anthrax is 1 to 15 days. It begins with a skin papule that ulcerates and is called an **eschar** (figure 37.10*b*). Headache, fever, and nausea are the major symptoms. Pulmonary anthrax resembles influenza. If the bacteria invade the bloodstream, the disease can be fatal. The signs and symptoms of anthrax are due to anthrax toxins, a complex exotoxin system composed of three proteins (*see table 30.5*).

Diagnosis is by direct microscopic examination, culture of the bacterium, and serology; therapy is with penicillin G or penicillin G plus streptomycin. Vaccination of animals, primarily cattle, is an important control measure. However, people with a high occupational risk, such as those who handle infected animals or their products, including hides and wool, should be immunized with the cell-free vaccine obtainable from the Centers for Disease Control. Fewer than 10 cases of anthrax occur annually in the United States.

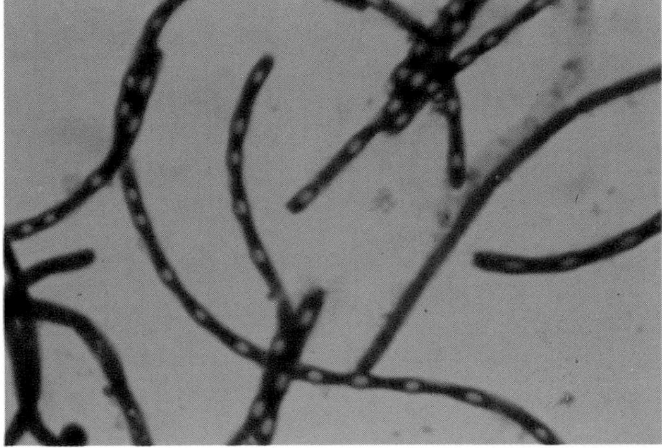

(a)

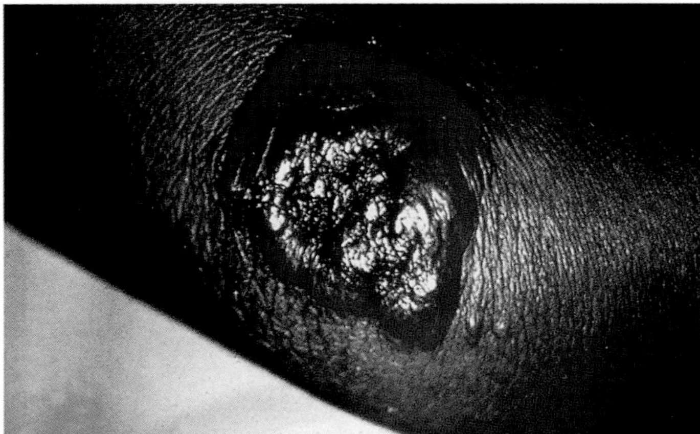

(b)

Figure 37.10 **Anthrax.** (*a*) *Bacillus anthracis,* the causative bacterium, occurs in chains with elliptical endospores; light micrograph (×600). (*b*) The malignant (destructive) eschar pustules of anthrax on the arm of an infected person.

Gas Gangrene or Clostridial Myonecrosis

Clostridium perfringens, C. novyi, and *C. septicum* are gram-positive spore-forming rods termed the histotoxic clostridia.

They can produce a necrotizing infection of skeletal muscle called **gas gangrene** (Greek *gangraina,* an eating sore) or **clostridial myonecrosis** (*myo,* muscles, and *necrosis,* death).

Histotoxic clostridia occur in the soil worldwide, and also are part of the normal endogenous microflora of the human large intestine (*see figure 29.8*). Contamination of injured tissue with spores from soil containing histotoxic clostridia or bowel flora is the usual means of transmission. Infections are commonly associated with wounds resulting from abortions, automobile accidents, military combat, or frostbite.

If the spores germinate in anaerobic tissue, the bacteria grow and secrete α-toxin. Growth often results in the accumulation of gas (mainly hydrogen as a result of carbohydrate fermentation), and of the toxic breakdown products of skeletal muscle tissue.

Clinical manifestations include severe pain, edema, drainage, and muscle necrosis. The pathology arises from progressive skeletal muscle necrosis due to the effects of α-toxin. Other enzymes produced by the bacteria degrade collagen and tissue, facilitating spread of the disease.

Gas gangrene is a medical emergency. Laboratory diagnosis is through recovery of the appropriate species of clostridia accompanied by the characteristic disease symptoms. Treatment is extensive surgical debridement (removal of all dead tissue), the administration of polyvalent antitoxin, and antimicrobial therapy with penicillin and tetracycline. Hyperbaric oxygen therapy (the use of high concentrations of oxygen at elevated pressures) is also considered effective. The oxygen saturates the infected tissue and thereby prevents the growth of the obligately anaerobic clostridia.

Prevention and control includes debridement of contaminated traumatic wounds plus antimicrobial prophylaxis and prompt treatment of all wound infections. Amputation of limbs is often necessary to prevent further spread of the disease.

Gonorrhea

Gonorrhea (Greek *gono,* seed, and *rhein,* to flow) is an acute, infectious, sexually transmitted disease of the mucous membranes of the genitourinary tract, eye, rectum, and throat. It is caused by the gram-negative, oxidase-positive, diplococcus, *Neisseria gonorrhoeae.* These bacteria are also referred to as **gonococci** (pl. of gonococcus; Greek *gono,* seed, and *coccus,* berry) and have a worldwide distribution.

Once inside the body, the gonococci attach to the microvilli of mucosal cells by means of pili and protein II, which function as adhesins. This attachment prevents the bacteria from being washed away by normal vaginal discharges or by the strong flow of urine. They are then phagocytosed by the mucosal cells and may even be transported through the cells to the intercellular spaces and subepithelial tissue. Phagocytes, such as neutrophils, also may contain gonococci (figure 37.11) inside vesicles. Because the gonococci are intracellular at this time, the host's defenses have little effect on the bacteria. Following penetration of the bacteria, the host tissue responds locally by the infiltration of mast cells, more PMNs, and plasma cells. These cells are later replaced by fibrous tissue that may lead to urethral closing, or stricture, in males.

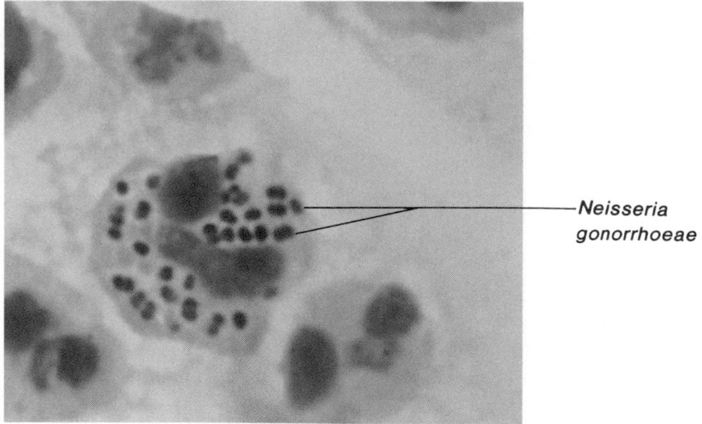

Figure 37.11 Gonorrhea. Gram stain of male urethral exudate showing *Neisseria gonorrhoeae* (diplococci) inside a PMN; light micrograph (×500). Although the presence of gram-negative diplococci in exudates is a probable indication of gonorrhea, the bacterium should be isolated and identified.

In males, the incubation period is 2 to 8 days. The onset consists of a urethral discharge of yellow, creamy pus and frequent, painful urination that is accompanied by a burning sensation. In females, the disease is more insidious in that few individuals are aware of any symptoms. However, some symptoms may begin 7 to 21 days after infection. These are generally mild; some vaginal discharge may occur. The gonococci also can infect the uterine tubes and surrounding tissues, leading to **pelvic inflammatory disease (PID).** This occurs in 10 to 20% of infected females. Gonococcal PID is a major cause of sterility and ectopic pregnancies because of scar formation in the fallopian tubes. Gonococci disseminate most often during menstruation, a time in which there is an increased concentration of free iron available to the bacteria. In both sexes, disseminated gonococcal infection with bacteremia may occur. This can lead to involvement of the joints (gonorrheal arthritis), heart (gonorrheal endocarditis), or pharynx (gonorrheal pharyngitis). Gonorrheal eye infections occur most often in newborns as they pass through an infected birth canal. The resulting disease is called **ophthalmia neonatorum** or **conjunctivitis of the newborn,** which was once a leading cause of blindness in many parts of the world. To prevent this, tetracycline, erythromycin, or silver nitrate in dilute solution is placed in the eyes of newborns. This type of treatment is required by law in the United States.

Laboratory diagnosis of gonorrhea is by Gram stain of the urethral discharge of males and culture of the bacterium from both males and females. Because the gonococci are very sensitive to adverse environmental conditions and survive poorly outside the body, special transport media are necessary. Serological tests and a DNA probe for *N. gonorrhoeae* are currently being developed and used to supplement other diagnostic techniques.

The Centers for Disease Control consider four treatment regimens to be coequal after sensitivity testing has been done:

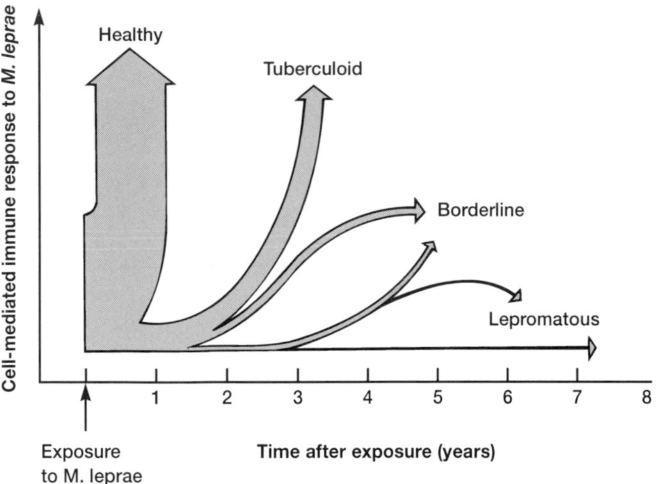

Figure 37.12 **Development of Leprosy.** A schematic representation of the hypothesis of how the development of subclinical infection and various types of leprosy is related to the time of onset of cell-mediated immune response to *M. leprae* antigens after the initial exposure. The thickness of the lines indicates the proportion of individuals from the exposed population that is likely to fall into each category.

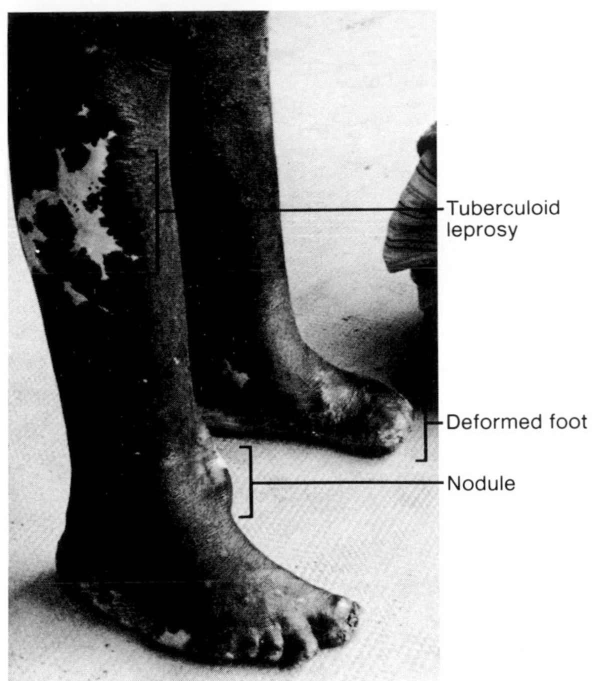

Figure 37.13 **Leprosy.** In tuberculoid leprosy, the skin within the nodule is completely without sensation. The deformed foot is associated with lepromatous leprosy. Note the disfiguring nodule on the ankle.

(1) penicillin G plus probenecid, (2) ampicillin plus probenecid, (3) ceftriaxone or ofloxacin plus doxycycline for 7 days, or (4) spectinomycin.

Penicillin-resistant strains of gonococci have now developed and occur worldwide. Most of these strains carry a plasmid that directs the formation of penicillinase, a β-lactamase enzyme able to inactivate penicillin G and ampicillin. Since 1980, strains of *N. gonorrhoeae* with chromosomally mediated penicillin resistance have developed. Instead of producing a penicillinase, these strains have altered penicillin-binding proteins. Since 1986, tetracycline-resistant *N. gonorrhoeae* have also developed.

The most effective method for control of this sexually transmitted disease is public education, diagnosing and treating the asymptomatic patient, using condoms, and treating infected individuals quickly to prevent further spread of the disease. Slightly under a million cases of gonorrhea are reported in the United States each year, but the actual number of cases is thought to be three to four times greater than that. More than 60% of all cases occur in the 15- to 24-year-old age group. Repeated gonococcal infections are common. Protective immunity to reinfection does not arise because of the antigenic variety of gonococci.

Leprosy

Leprosy (Greek *Lepros,* scaly, scabby, rough) or **Hansen's disease** is a severely disfiguring skin disease caused by *Mycobacterium leprae* (*see figure 22.14*). The only reservoirs of proved significance are humans. The disease most often occurs in tropical countries, where there are more than 10 million cases. An estimated 4,000 cases exist in the United States, with approximately 200 to 300 new cases reported annually.

Transmission of leprosy is most likely to occur when children are exposed for prolonged periods to infected individuals who shed large numbers of *M. leprae.* Nasal secretions are the most likely infectious material for family contacts.

The incubation period is about 3 to 5 years, but may be much longer, and the disease progresses slowly. The bacterium invades peripheral nerve and skin cells and becomes an obligate intracellular parasite. It is most frequently found in the Schwann cells that surround peripheral nerve axons and in mononuclear phagocytes. The earliest symptom of leprosy is usually a slightly pigmented skin eruption several centimeters in diameter. Approximately 75% of all individuals with this early solitary lesion heal spontaneously because of the cell-mediated immune response to *M. leprae.* However, in some individuals this immune response may be so weak that one of two distinct forms of the disease occurs: tuberculoid or lepromatous leprosy (figure 37.12).

Tuberculoid (neural) leprosy is a mild, nonprogressive form of leprosy associated with a delayed-type hypersensitivity reaction (*see chapter 32*) to antigens on the surface of *M. leprae.* It is characterized by damaged nerves and regions of the skin that have lost sensation and are surrounded by a border of nodules (figure 37.13). Afflicted individuals who do not develop hypersensitivity have a relentlessly progressive form of the disease, called **lepromatous (progressive) leprosy,** in which large numbers of *M. leprae* develop in the skin cells. The bacteria kill skin tissue, leading to a progressive loss of facial features, fingers, toes, and other structures. Moreover, disfiguring nodules form all over the body. Nerves are also infected, but are usually less damaged than in tuberculoid leprosy.

Because the leprosy bacillus cannot be cultured in vitro, laboratory diagnosis is supported by the demonstration of the bacterium in biopsy specimens and by acid-fast staining. Serodiagnostic methods, such as the fluorescent leprosy antibody absorption test and ELISA have recently been developed.

Treatment is long-term with the sulfone drug dapson and rifampin, with or without clofazimine. Alternative drugs are ethionamide or protionamide. There is good evidence that the nine-banded armadillo is an animal reservoir for the leprosy bacillus in the United States but plays no role in transmission of leprosy to humans. Identification and treatment of patients with leprosy is the key to control. Children of presumably contagious parents should be given chemoprophylactic drugs until treatment of the parents has made them noninfectious.

Peptic Ulcer Disease

A gram-negative, microaerophilic curved bacillus found in gastric biopsy specimens from patients with histologic gastritis (Greek *gaster,* stomach, and *itis,* inflammation) was successfully cultured in Perth, Australia, in 1982 and was soon named *Campylobacter pylori* (recently changed to *Helicobacter pylori*). It now appears that this bacterium is responsible for most cases of gastritis not associated with another known primary cause (e.g., autoimmune gastritis or eosinophilic gastritis), and it is a leading factor in the pathogenesis of **peptic ulcer disease**. In addition, there are strong positive correlations between gastric cancer rates and *H. pylori* infection rates in certain populations.

It has yet to be determined conclusively whether this is a true pathogen or merely a commensal bacterium on a previously altered mucosa. However, suggestive evidence is mounting. For example, *H. pylori* has been isolated from the gastric mucosa (figure 37.14) of 95% of patients with gastric ulcer disease and virtually 100% of those patients with chronic gastritis, but not in healthy tissue. Of interest is the observation that a healthy volunteer inoculated with *H. pylori* developed gastritis, which cleared after antibiotic therapy. *H. pylori* colonizes only gastric-type epithelium, and surface pili are believed to be the adhesins associated with this process. *H. pylori* is also a strong producer of urease. Urease activity may create an alkaline environment by urea hydrolysis to produce ammonia that protects the bacterium from gastric acid until it becomes established under the layer of mucus in the stomach. The potential virulence factors responsible for epithelial cell damage and inflammation probably include proteases, phospholipases, cytokines, and cytotoxins.

H. pylori is most likely transmitted from person to person, although infection from a common exogenous source cannot be completely ruled out. Support for the person-to-person transmission comes from evidence of clustering within families and from reports of higher than expected prevalences in residents of custodial institutions and nursing homes.

Laboratory diagnosis of *H. pylori* is by culture of gastric biopsy specimens, examination of stained biopsies for the presence of bacteria, or detection of urease activity in the biopsies.

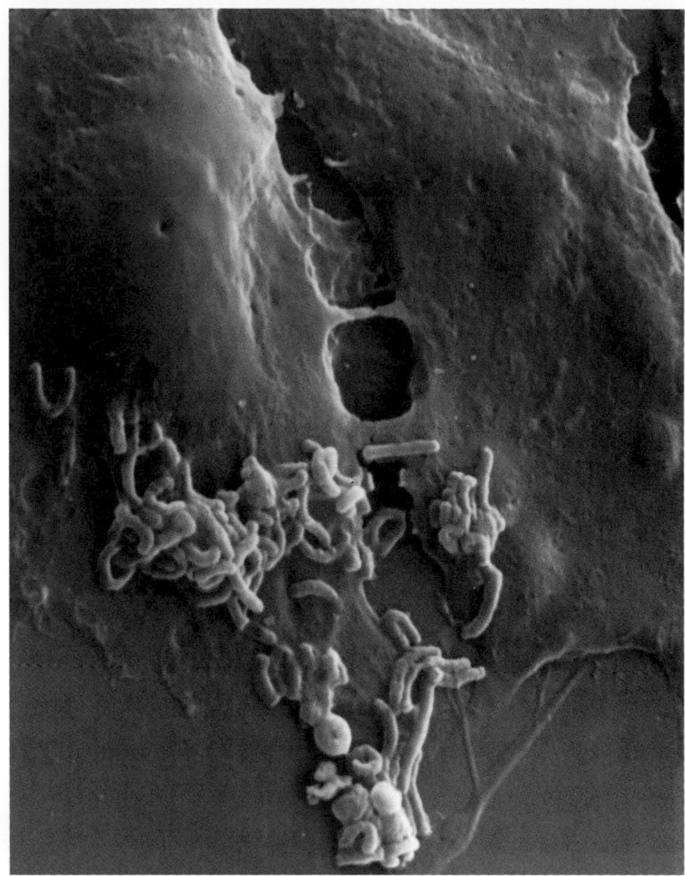

Figure 37.14 **Peptic Ulcer Disease.** Scanning electron micrograph (×3,441) of *Helicobacter pylori* adhering to gastric cells.

Noninvasive methods include detection of serum IgG antibodies to *H. pylori* (a rapid test kit is now available). Treatment is with bismuth subsalicylate (Pepto-Bismol) combined with metronidazole and tetracycline.

Staphylococcal Diseases

Staphylococci are among the most important bacteria that cause disease in humans. They are normal inhabitants of the upper respiratory tract, skin, intestine, and vagina (*see figure 29.8*). Staphylococci, with pneumococci and streptococci, are members of a group of invasive gram-positive bacteria known as the pyogenic (or pus-producing) cocci. These bacteria cause various suppurative, or pus-forming diseases (e.g., boils, carbuncles, folliculitis, impetigo contagiosa, scalded-skin syndrome) in humans.

Staphylococci can be divided into pathogenic and relatively nonpathogenic strains based on the synthesis of the enzyme coagulase. Coagulase-positive strains, classified as *S. aureus* (*see figure 22.2*a,b), often produce a yellow carotenoid pigment—which has led to their being colloquially called golden staph (*see figure 5.8*)—and cause severe chronic infections. Strains that do not produce coagulase, such as *S. epidermidis,* are nonpigmented and are generally less invasive, but they have increasingly been associated, as opportunistic pathogens, with serious nosocomial infections.

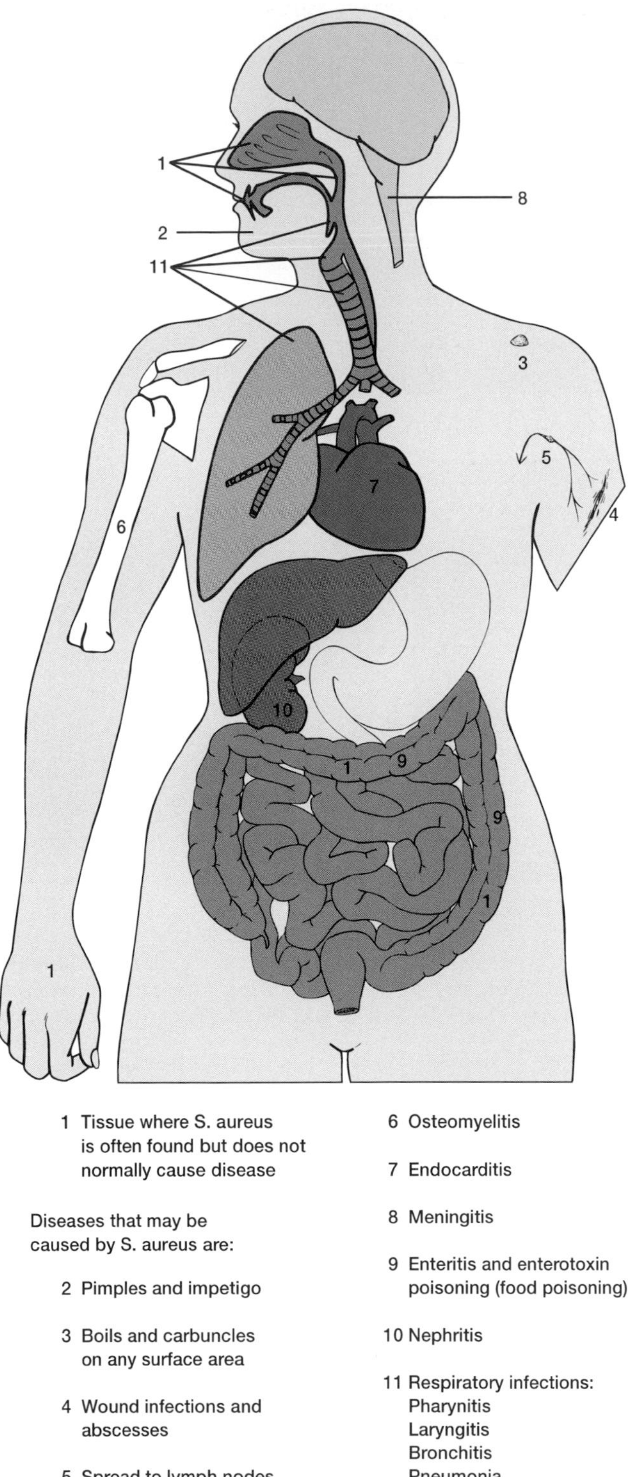

Figure 37.15 **Staphylococcal Diseases.** The sites of the major staphylococcal infections of humans are indicated by the above numbers.

TABLE 37.2 Various Toxins and Enzymes Produced by Staphylococci

Product	Physiological Action
β-lactamase	Breaks down penicillin
Catalase	Converts hydrogen peroxide into water and oxygen and reduces killing by phagocytosis
Coagulase	Reacts with prothrombin to form a complex that can cleave fibrinogen and cause the formation of a fibrin clot; fibrin may also be deposited on the surface of staphylococci, which may protect them from destruction by phagocytic cells; coagulase production is synonymous with invasive pathogenic potential
DNase	Destroys DNA
Enterotoxins	Are divided into heat-stable toxins of six known types (A, B, C1, C2, D, E); responsible for the gastrointestinal upset typical of food poisoning
Exfoliative toxin	Causes loss of the surface layers of the skin in scalded-skin syndrome
Hemolysins	Alpha hemolysin destroys erythrocytes and causes skin destruction Beta hemolysin destroys erythrocytes and sphingomyelin around nerves Gamma hemolysin destroys erythrocytes Delta hemolysin destroys erythrocytes
Hyaluronidase	Also known as spreading factor; breaks down hyaluronic acid located between cells, allowing for penetration and spread of bacteria
Leukocidin	Inhibits phagocytosis by granulocytes and can destroy these cells
Lipases	Break down lipids
Protein A	Is antiphagocytic by competing with neutrophils for the Fc portion of specific opsonins
Proteinases	Break down proteins
Toxic shock syndrome toxin-1	Is associated with the fever, shock, and multisystem involvement of toxic shock syndrome

Staphylococci, harbored by either an asymptomatic carrier or a person with the disease, can be spread by the hands, expulsion from the respiratory tract, and transport in or on animate and inanimate objects. Staphylococci can produce disease in almost every organ and tissue of the body (figure 37.15). However, it should be emphasized that staphylococcal disease, for the most part, occurs in people whose defensive mechanisms have been compromised.

Staphylococci produce disease through their ability to multiply and spread widely in tissues and through their production of many extracellular substances (table 37.2). Some of these substances are exotoxins, while others are enzymes thought to be involved in staphylococcal invasiveness. Many toxin genes are carried on plasmids; in some cases, genes re-

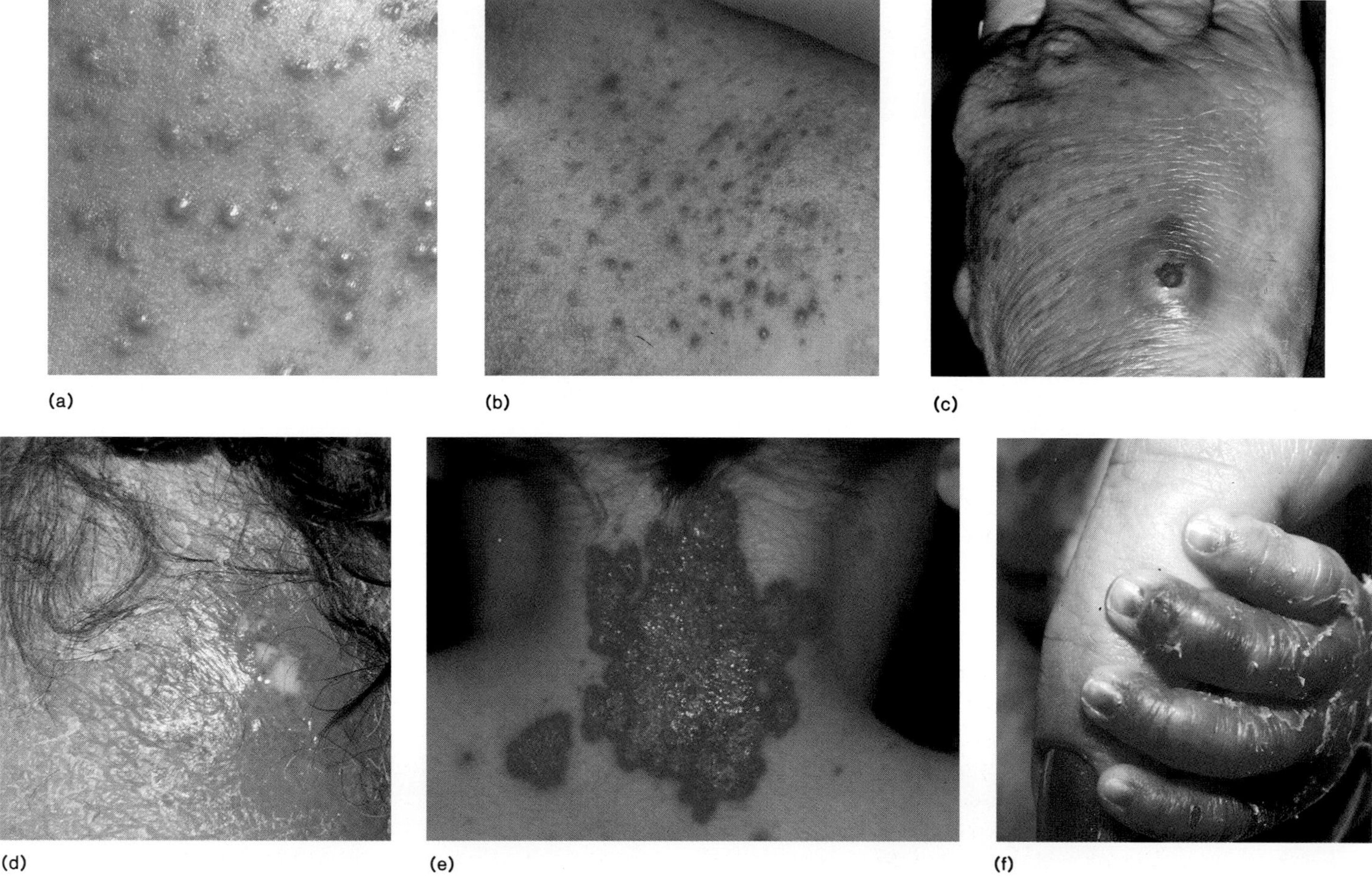

Figure 37.16 **Staphylococcal Skin Infections** (*a*) Superficial folliculitis in which raised, domed pustules form around hair follicles. (*b*) In deep folliculitis, the microorganism invades the deep portion of the follicle and dermis. (*c*) A furuncle arises when a large abscess forms around a hair follicle. (*d*) A carbuncle consists of a multilocular abscess around several hair follicles. (*e*) Impetigo on the neck of a 2-year-old male. (*f*) Scalded skin syndrome in a 1 week old premature male infant. Reddened areas of skin peel off, leaving "scalded"-looking moist areas.

sponsible for pathogenicity reside on both a plasmid and the host chromosome.

The pathogenic capacity of a particular *S. aureus* strain is due to the combined effect of extracellular factors and toxins, together with the invasive properties of the strain. At one end of the disease spectrum is staphylococcal food poisoning, caused solely by the ingestion of preformed enterotoxin. At the other end of the spectrum are staphylococcal bacteremia and disseminated abscesses in most organs of the body.

The classic example of the staphylococcal lesion is the localized abscess (figure 37.16*a–d*).When *S. aureus* becomes established in a hair follicle, tissue necrosis results. Coagulase is produced and forms a fibrin wall around the lesion that limits the spread. Within the center of the lesion, liquefaction of necrotic tissue occurs, and the abscess spreads in the direction of least resistance. The abscess may be either a furuncle or a carbuncle. The central necrotic tissue drains, and healing eventually occurs. However, the bacteria may spread from any focus by the lymphatics and bloodstream to other parts of the body.

Newborn infants and children can develop a superficial skin infection characterized by the presence of encrusted pustules (figure 37.16*e*). This disease, called impetigo contagiosa, is caused by *S. aureus* and group A streptococci. It is contagious and can spread through a nursery or school. It usually occurs in areas where sanitation and personal hygiene are poor.

Toxic shock syndrome (TSS) is a staphylococcal disease with potentially serious consequences. Most cases of this syndrome have occurred in females who use superabsorbent tampons during menstruation. However, the toxin associated with this syndrome is also produced in men and in nonmenstruating women by *S. aureus* present at sites other than the genital area (e.g., in surgical wound infections). Toxic shock syndrome is characterized by low blood pressure, fever, diarrhea, an extensive skin rash, and shedding of the skin. These symptoms are caused by the toxic shock syndrome toxin-1 released by the *S. aureus* (table 37.2), but several other enterotoxins (SEB and SEC_1) also may be involved. Several hundred cases of toxic shock syndrome are reported annually in the United States.

Box 37.2

Resistant Staphylococci

During the late 1950s and early 1960s, *Staphylococcus aureus* caused considerable morbidity and mortality as a nosocomial, or hospital-acquired, pathogen. Since then, penicillinase-resistant, semisynthetic penicillins have proved to be successful antimicrobial agents in the treatment of staphylococcal infections. Unfortunately, methicillin-resistant *S. aureus* (MRSA) strains have recently emerged as a major nosocomial problem. The majority of the strains are resistant to several of the most commonly used antimicrobial agents, including macrolides, aminoglycosides, and the beta-lactam antibiotics, including the latest generation of cephalosporins. Serious infections by methicillin-resistant strains have been most often successfully treated with an older, potentially toxic antibiotic, vancomycin.

Recently, methicillin-resistant *S. epidermidis* strains have also emerged as a nosocomial problem, especially in individuals with prosthetic heart valves or in people who have undergone other forms of cardiac surgery. Resistance to methicillin may also extend to the cephalosporin antibiotics. Difficulties in performing in vitro tests that adequately recognize cephalosporin resistance of these strains continue to exist. Serious infections due to methicillin-resistant *S. epidermidis* have been successfully treated with combination therapy, including vancomycin plus rifampin or an aminoglycoside.

Staphylococcal scalded skin syndrome (SSSS) is a third example of a common staphylococcal disease (figure 37.16*f*). SSSS is caused by strains of *S. aureus* that carry a plasmid-borne gene for the **exfoliative toxin** or **exfoliatin** (sometimes the toxin gene is on the bacterial chromosome instead). In this disease, the epidermis peels off to reveal a red area underneath—thus, the name of the disease. SSSS is seen most commonly in infants and children, and neonatal nurseries occasionally suffer large outbreaks of the disease.

The definitive diagnosis of staphylococcal disease can be made only by isolation and identification of the staphylococcus involved. This requires culture, catalase, and coagulase tests; serology; DNA fingerprinting; and phage typing. Commercial rapid test kits are also available. There is no specific prevention for staphylococcal disease. The mainstay of treatment is the administration of specific antibiotics: penicillin, cloxacillin, methicillin, vancomycin, oxacillin, cefotaxime, ceftriaxone, a cephalosporin, or rifampin and others. Because of the prevalence of drug-resistant strains, all staphylococcal isolates should be tested for antimicrobial susceptibility (Box 37.2). Cleanliness, hygiene, and aseptic management of lesions are the best means of control.

Syphilis

Venereal syphilis (Greek *syn,* together, and *philein,* to love) is a contagious sexually transmitted disease caused by the spirochete *Treponema pallidum* subsp. *pallidum* (*T. pallidum, see figure 21.1*b). **Congenital syphilis** is the disease acquired in utero from the mother.

T. pallidum enters the body through mucous membranes or minor breaks or abrasions of the skin. It migrates to the regional lymph nodes and rapidly spreads throughout the body. The disease is not highly contagious, and there is only about a one in ten chance of acquiring it from a single exposure to an infected sex partner.

Three recognizable stages of syphilis occur in untreated adults. In the primary stage, after an incubation period of about 10 days to 3 weeks or more, the initial symptom is a small, painless, reddened ulcer, or **chancre,** (French *canker,* a destructive sore) with a hard ridge that appears at the infection site (figure 37.17*a*) and contains spirochetes. Contact with the chancre during sexual intercourse may result in disease transmission. In about 1/3 of the cases, the disease does not progress further and the chancre disappears. Serological tests are positive in about 80% of the individuals during this stage (figure 37.18). In the remaining cases, the spirochetes enter the bloodstream and are distributed throughout the body.

Within 2 to 10 weeks after the primary lesion, the disease may enter the secondary stage, which is characterized by a skin rash (figure 37.17*b*). By this time, 100% of the individuals are serologically positive. Other symptoms during this stage include the loss of patches of hair, malaise, and fever. Both the chancre and the rash lesions are infectious.

After several weeks, the disease becomes latent. During the latent period, the disease is not normally infectious, except for possible transmission from mother to fetus (congenital syphilis). After many years, a tertiary stage develops in about 40% of untreated individuals with secondary syphilis. During this stage, degenerative lesions called **gummas** (figure 37.17*c*) form in the skin, bone, and nervous system as the result of hypersensitivity reactions. This stage is also characterized by a great reduction in the number of spirochetes in the body. Involvement of the central nervous system may result in tissue loss that can lead to mental retardation, blindness, a "shuffle" walk (tabes), or insanity. Many of these symptoms have been associated with such well-known people as Al Capone, Francisco Goya, Henry VIII, Adolf Hitler, Scott Joplin, Friedrich Nietzsche, Franz Schubert, Oscar Wilde, and Kaiser Wilhelm (Box 37.3).

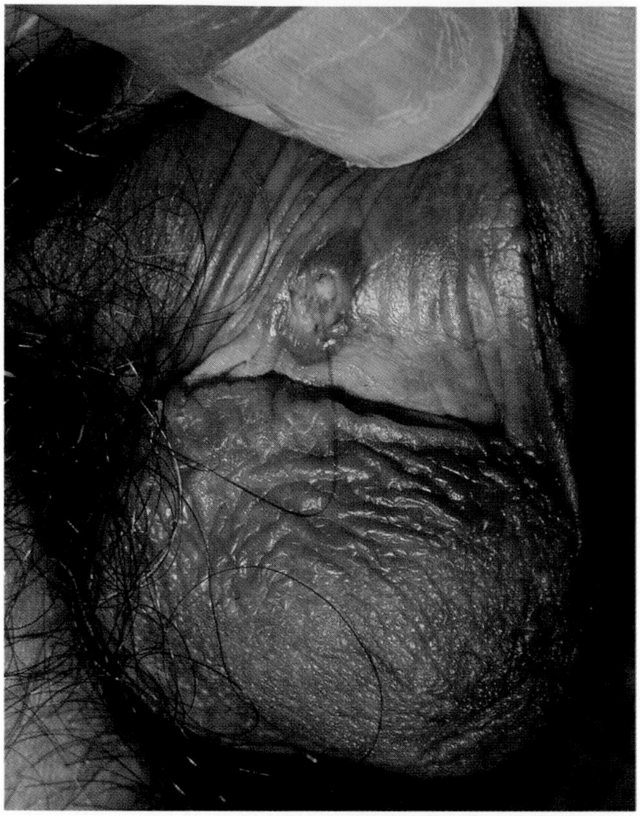

(a)

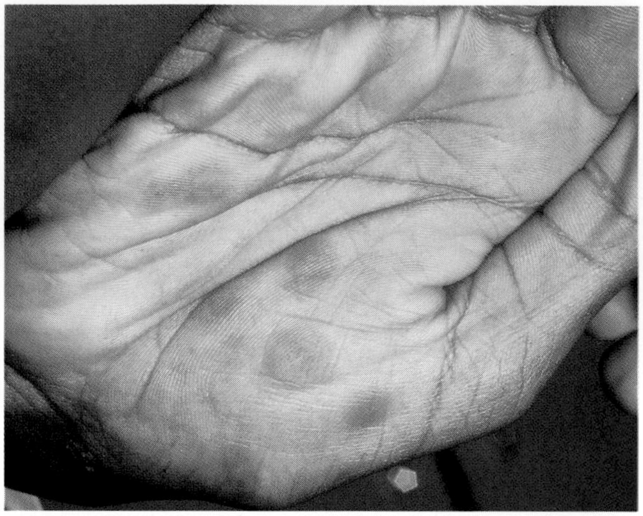

(b)

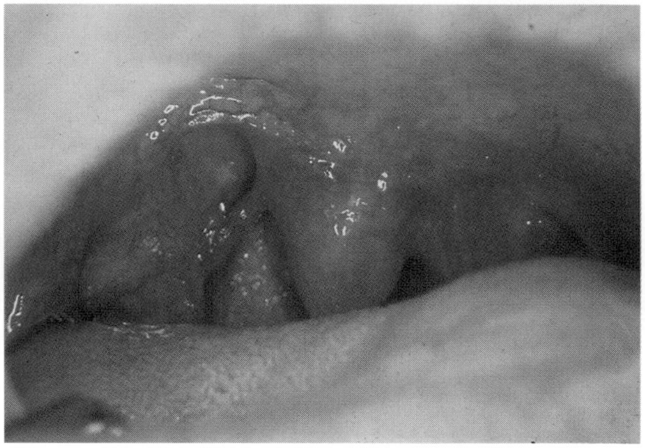

(c)

Figure 37.17 **Syphilis.** (*a*) Primary syphilitic chancre of the penis. (*b*) Palmar lesions of secondary syphilis. (*c*) Ruptured gumma and ulcer of upper hard palate of the mouth.

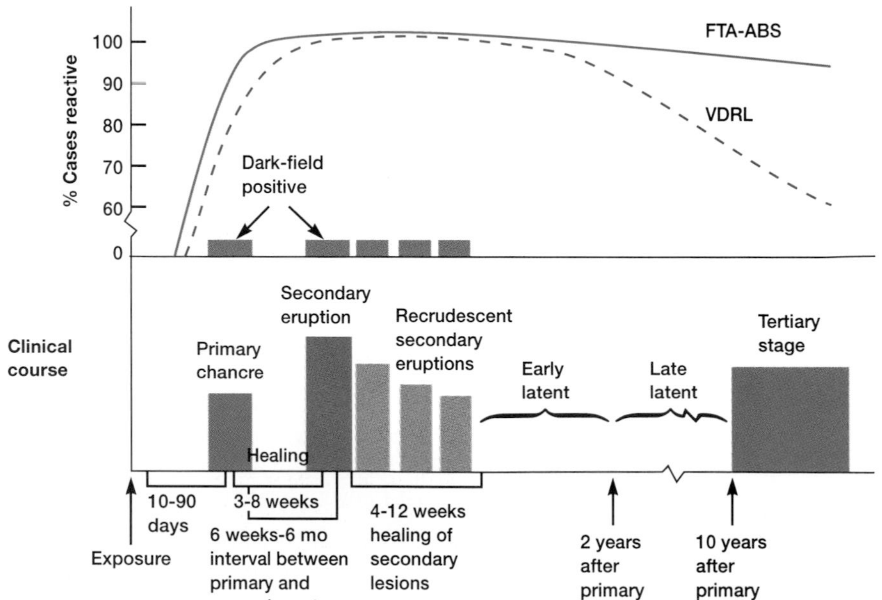

Figure 37.18 **The Course of Untreated Syphilis.** See text for further details.

Box 37.3

A Brief History of Syphilis

Syphilis was first recognized in Europe near the end of the fifteenth century. During this time the disease reached epidemic proportions in the Mediterranean areas. According to one hypothesis, syphilis is of New World origin and Christopher Columbus (1451–1506) and his crew acquired it in the West Indies and introduced it into Spain after returning from their historic voyage. Another hypothesis is that syphilis had been endemic for centuries in Africa and may have been transported to Europe at the same time that vast migrations of the civilian population were occurring (1500).

Syphilis was initially variously called the Italian disease, the French disease, and the great pox as distinguished from smallpox. In 1530 the Italian physician and poet Girolamo Fracastoro wrote *Syphilis sive Morbus Gallicus* (Syphilis or the French Disease). In this poem, a Spanish shepherd named Syphilis is punished for being disrespectful to the gods by being cursed with the disease. Several years later, Fracastoro published a series of papers in which he described the possible mode of transmission of the "seeds" of syphilis through sexual contact.

Its venereal transmission was not definitely shown until the eighteenth century. The term venereal is derived from the name Venus, the Roman goddess of love. Recognition of the different stages of syphilis was demonstrated in 1838 by Philippe Ricord, who reported his observations on more than 2,500 human inoculations. In 1905, Fritz Schaudinn and Erich Hoffmann discovered the causative bacterium, and in 1906, August von Wassermann introduced the diagnostic test that bears his name. In 1909, Paul Ehrlich introduced an arsenic derivative, arsphenamine or salvarsan, as therapy. During this period, an anonymous limerick aptly described the course of this disease:

There was a young man from Black Bay
Who thought syphilis just went away
He believed that a chancre
Was only a canker
That healed in a week and a day.

But now he has "acne vulgaris"—
(Or whatever they call it in Paris);
On his skin it has spread
From his feet to his head,
And his friends want to know where his hair is.

There's more to his terrible plight:
His pupils won't close in the light
His heart is cavorting,
His wife is aborting,
And he squints through his gun-barrel sight.

Arthralgia cuts into his slumber;
His aorta is in need of a plumber;
But now he has tabes,
And saber-shinned babies,
While of gummas he has quite a number.

He's been treated in every known way,
But his spirochetes grow day by day;
He's developed paresis,
Has long talks with Jesus,
And thinks he's the Queen of the May.

Diagnosis of syphilis is through a clinical history, a thorough physical examination, and dark-field and immunofluorescence examination of fluids from the lesions (except oral lesions) for typical motile or fluorescent spirochetes. Because humans respond to *T. pallidum* with the formation of antitreponemal antibody and a complement-fixing reagin, serological tests are very informative. Examples include tests using nontreponemal antigens (VDRL, **V**enereal **D**isease **R**esearch **L**aboratories test; RPR, **R**apid **P**lasma **R**eagin test; complement fixation or the Wassermann test) and treponemal antibody (FTA-ABS, **f**luorescent **t**reponemal **a**ntibody-**abs**orption test; TPI, ***T. pallidum* i**mmobilization; *T. pallidum* complement fixation; TPHA, ***T. pallidum* h**em**a**gglutination).

Treatment in the early stages of the disease is easily accomplished with long-acting benzathine penicillin G or aqueous procaine penicillin. Later stages of syphilis are more difficult to treat with drugs and require much larger doses over a longer period. For example, in neurosyphilis cases, treponemes occasionally survive such drug treatment. Immunity to syphilis is not complete, and subsequent infections can occur once the first infection has spontaneously disappeared or has been eliminated with antibiotics.

Prevention and control of syphilis depends on (1) prompt and adequate treatment of all new cases, (2) follow-up on sources of infection and contact so they can be treated, (3) sexual hygiene, and (4) prophylaxis (condoms) to prevent exposure. At present, the incidence of syphilis, as well as other sexually transmitted diseases, is rising in most parts of the world. In the United States, around 50,000 cases of primary and secondary syphilis in the civilian population and about 1,000 cases of congenital syphilis are reported annually. The highest incidence is among those 20 to 39 years of age.

Tetanus

Tetanus (Greek *tetanos,* to stretch) is caused by *Clostridium tetani,* an anaerobic gram-positive spore former (figure 37.19*a*). This bacterium is commonly found in hospital environments, in soil and dust, and in the feces of many farm animals and humans.

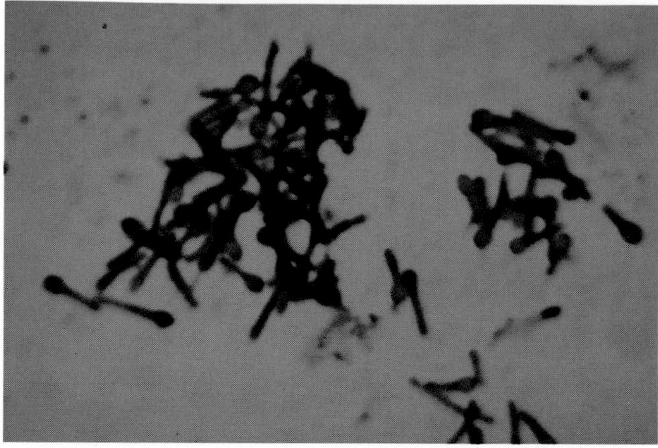

(a)

(b)

Figure 37.19 **Tetanus.** (*a*) The etiologic agent is *Clostridium tetani*, with round, terminal endospores; light micrograph (×900). (*b*) Painting of a soldier dying of tetanus. Note the extreme contraction of the trunk and extremity muscles producing opisthotonos—backward bowing of the back.

Transmission to humans is associated with skin wounds. Any break in the skin can allow *C. tetani* endospores to enter, and if the oxygen tension is low enough, the endospores germinate. When the bacteria die and lyse, the neurotoxin **tetanospasmin** is released. Tetanospasmin blocks inhibitory synapses of the spinal cord motor neurons by preventing the release of the inhibitory mediators gamma-aminobutyric acid and glycine. This allows uncontrolled stimulation of skeletal muscles.

Early in the course of the disease, tetanospasmin causes tension or cramping and twisting in skeletal muscles surrounding the wound and tightness of the jaw muscles. With more advanced disease, there is trismus ("lockjaw"), an inability to open the mouth because of the spasm of the masseter muscles. Facial muscles may go into spasms producing the characteristic expression known as risus sardonicus. Spasms or contractions of the trunk and extremity muscles may be so severe that there is boardlike rigidity, painful tonic convulsions, and opisthotonos (backward bowing of the back so that the heels and back approach each other (figure 37.19*b*). Death usually results from spasms of the diaphragm and intercostal muscles (the respiratory muscles). A second toxin **tetanolysin,** is a hemolysin that aids in tissue destruction.

Testing for tetanus is suggested whenever an individual has a history of wound infection and muscle stiffness. Prevention of tetanus involves the use of the tetanus toxoid. The toxoid, which incorporates an adjuvant (aluminum salts) to increase its immunizing potency, is given routinely with diphtheria toxoid and pertussis vaccine. An initial dose is normally administered a few months after birth, a second dose 4 to 6 months later, and finally a reinforcing dose 6 to 12 months after the second injection. A final booster is given between the ages of 4 to 6 years. For many years booster doses of tetanus toxoid were administered every 3 to 5 years. However, that practice has been discontinued since it has been shown that a single booster dose can provide protection for 10 to 20 years. Serious hypersensitivity reactions have occurred when too many doses of toxoid were administered over a period of years. Booster doses today are generally given only when an individual has sustained a wound infection.

Control measures for tetanus are not possible because of the wide dissemination of the bacterium in the soil and the long survival of its endospores. The case fatality rate in generalized tetanus ranges from 30 to 90% because tetanus treatment is not very effective. Therefore, prevention is all important and depends on (1) active immunization with toxoid, (2) proper care of wounds contaminated with soil, (3) prophylactic use of antitoxin, and (4) administration of penicillin. Around 100 cases of tetanus are reported annually in the United States—the majority of which are intravenous drug users.

Tularemia

The gram-negative bacterium *Francisella tularensis* is widely found in animal reservoirs in the United States and causes the disease **tularemia** (from Tulare, a county in California where the disease was first described). It may be transmitted to humans by biting arthropods (ticks or deer flies), direct contact with infected tissue (rabbits), inhalation of aerosols, or ingestion of contaminated food or water. After an incubation period of 2 to 10 days, a primary ulcerative lesion appears at the infection site, lymph nodes enlarge, and a high fever develops. Diagnosis is by culture of the bacterium and fluorescent antibody and agglutination tests; treatment is with streptomycin, tetracycline, or aminoglycoside antibiotics. Prevention and control involves public education, protective clothing, and vector control. An attenuated live vaccine is available from the United States Army for high-risk laboratory workers. Fewer than 300 cases of tularemia are reported annually in the United States.

1. How can humans acquire anthrax? Gas gangrene? Gonorrhea? Leprosy? Peptic ulcer disease? Describe the major symptoms of each.
2. Define the following terms: cutaneous anthrax, pelvic inflammatory disease, ophthalmia neonatorum, tuberculoid and lepromatous leprosy, gummas, and chancre.
3. Describe several diseases caused by the staphylococci.
4. Name and describe the three stages of syphilis.
5. How is the disease tetanus acquired? What are its symptoms and how do they arise?
6. From what animal can tularemia be contracted?

Food-Borne and Waterborne Diseases

Many bacteria contaminating food and water can cause acute gastroenteritis or inflammation of the stomach and intestinal lining. When food is the source of the pathogen, the condition is often called **food poisoning.** Gastroenteritis can arise in two ways. The bacteria may actually produce a **food-borne infection.** That is, they may first colonize the gastrointestinal tract and grow within it, then either invade host tissues or secrete exotoxins. Alternatively, the pathogen may secrete an exotoxin that contaminates the food and is then ingested by the host. This is sometimes referred to as a **food intoxication** because the toxin is ingested and the presence of living bacteria is not required. Because these toxins disrupt the functioning of the intestinal mucosa they are called **enterotoxins** (*see table 30.6*). Common symptoms of enterotoxin poisoning are nausea, vomiting, and diarrhea.

This section describes several of the more common bacteria associated with gastrointestinal infections, food intoxications, and waterborne diseases. Table 37.3 summarizes many of the bacterial pathogens responsible for food poisoning.

Diseases transmitted by foods (pp. 871–73).
Food spoilage (pp. 873–76).

Worldwide, diarrheal diseases are second only to respiratory diseases as a cause of death; they are the leading cause of childhood death, and in some parts of the world they are responsible for more years of potential life lost than all other causes combined. For example, each year around 5 million children (more than 13,600 a day) die from diarrheal diseases in Asia, Africa, and South America. In the United States, estimates exceed 10,000 deaths per year from diarrhea, and an average of 500 childhood deaths are reported.

Cholera

Throughout history, **cholera** (Greek *chole,* bile) has caused repeated epidemics in various areas of the world, especially in Asia, the Middle East, and Africa. The disease has been rare

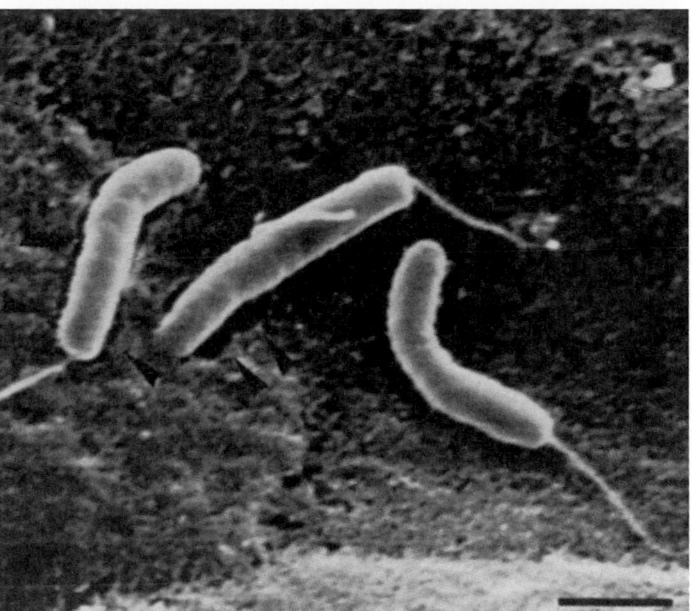

Figure 37.20 **Cholera.** *Vibrio cholerae* adhering to intestinal epithelium; scanning electron micrograph (×12,000). Notice that the bacteria are slightly curved with a single polar flagellum.

in the United States since the 1800s, but an endemic focus is believed to exist on the gulf coast of Louisiana and Texas.

Cholera is caused by the gram-negative *Vibrio cholerae* serogroup O1 bacterium (figure 37.20), which is acquired by ingesting food or water contaminated by fecal material from patients or carriers. (Shellfish and plankton may be the natural reservoirs.) In 1961 the biotype of *V. cholerae,* called *El Tor,* emerged as an important cause of cholera pandemics.

Once the bacteria enter the body, the incubation period is from several hours to 3 or more days. The bacteria adhere to the intestinal mucosa of the small intestine, where they are not invasive but secrete **choleragen,** a cholera enterotoxin. The enterotoxin enters the intestinal epithelial cells and activates the enzyme adenyl cyclase by the addition of an ADP-ribosyl group in a way similar to that employed by diphtheria toxin (*see figure 30.4*). As a result, choleragen stimulates hypersecretion of water and chloride ions while inhibiting absorption of sodium ions. The patient experiences an outpouring of fluid and electrolytes, with associated abdominal muscle cramps, vomiting, fever, and watery diarrhea. The diarrhea can be so profuse that a person can lose 10 to 15 liters of fluid during the infection. Death may result from the elevated concentration of blood proteins, caused by reduced fluid levels, which leads to circulatory shock and collapse.

Laboratory diagnosis is by culture of the bacterium from feces and subsequent identification by agglutination reactions with specific antisera. Treatment is by oral rehydration therapy with NaCl plus sucrose to stimulate water uptake by the intestine; the antibiotics of choice are a tetracycline, trimethoprim-sulfamethoxazole, or ciprofloxacin. The most reliable control methods are based on proper sanitation, especially of

TABLE 37.3 Bacteria That Cause Acute Bacterial Diarrheas and Food Poisonings

Organism	Incubation Period (Hours)	Vomiting	Diarrhea	Fever	Epidemiology	Pathogenesis
Staphylococcus	1–8 (rarely, up to 18)	+++	+	−	Staphylococci grow in meats, dairy and bakery products and produce enterotoxins.	Enterotoxins act on receptors in gut that transmit impulse to medullary centers.
Bacillus cereus	2–16	+++	++	−	Growth in reheated fried rice causes vomiting or diarrhea.	Enterotoxins formed in food or in gut from growth of *B. cereus*.
Clostridium perfringens	8–16	±	+++	−	Clostridia grow in rewarmed meat dishes. Large numbers ingested.	Enterotoxin produced during sporulation in gut, causes hypersecretion.
Clostridium botulinum serotypes A–G	24–96	±	Rare	−	Clostridia grow in anaerobic foods and produce exotoxin.	Exotoxin absorbed from gut blocks acetylcholine release at neuromuscular junction.
Escherichia coli (ETEC)	24–72	±	++	−	Bacteria grow in gut and produce enterotoxin. May also invade superficial epithelium.	Toxin causes hypersecretion of chloride and water in small intestine ("travelers' diarrhea").
Vibrio parahaemolyticus	6–96	+	++	±	Bacteria grow in seafood and in gut and produce enterotoxin, or invade.	Toxin causes hypersecretion; vibrios invade epithelium; stools may be bloody.
Vibrio cholerae (serogroup O1)	4–72	+	+++	−	Bacteria grow in gut and produce enterotoxin.	Enterotoxin causes hypersecretion of chloride and water in small intestine. Infective dose $>10^7$ vibrios.
Shigella spp. (mild cases)	24–72	±	++	+	Bacteria grow in superficial gut epithelium. *S. dysenteriae* produces exotoxin.	Organisms invade epithelial cells, blood, mucus, and PMNs in stools. Infective dose $<10^3$ organisms.
Salmonella typhi.	10–48	±	++	+	Bacteria grow in gut. Do not produce enterotoxin.	Superficial infection of gut, little invasion. Infective dose $>10^5$ organisms.
Clostridium difficile	?	−	+++	+	Antibiotic-associated colitis.	Enterotoxin causes epithelial necrosis in colon; pseudomembranous colitis.
Campylobacter jejuni	1–10 days	−	+++	++	Infection by oral route from foods, animals. Bacteria grow in small intestine.	Invasion of mucous membrane. Enterotoxin production causes inflammation of mucosa.
Yersinia enterocolitica	?	±	++	+	Fecal-oral transmission. Food-borne. Animals infected.	Gastroenteritis or mesenteric adenitis. Occasional bacteremia. Enterotoxin produced occasionally.

Clinical Features
Abrupt onset, intense vomiting for up to 24 hours, recovery in 24–48 hours. Occurs in persons eating the same food. No treatment usually necessary except to restore fluids and electrolytes.
With incubation period of 2–8 hours, mainly vomiting. With incubation period of 8–16 hours, mainly diarrhea.
Abrupt onset of profuse diarrhea; vomiting occasionally. Recovery usual without treatment in 1–4 days. Many clostridia in cultures of food and feces of patients.
Diplopia, dysphagia, dysphonia, respiratory embarrassment. Treatment requires clear airway, ventilation, and intravenous polyvalent antitoxin. Exotoxin present in food and serum. Mortality rate high.
Usually abrupt onset of diarrhea; vomiting rare. A serious infection in newborns. In adults, "traveler's diarrhea" is usually self-limited in 1–3 days. Use diphenoxylate (Lomotil) but no antimicrobials.
Abrupt onset of diarrhea in groups consuming the same food, especially crabs and other seafood. Recovery is usually complete in 1–3 days. Food and stool cultures are positive for *Vibrio.*
Abrupt onset of liquid diarrhea in endemic area. Needs prompt replacement of fluids and electrolytes IV or orally. Tetracycline shortens excretion of vibrios. Stool cultures positive for *Vibrio.*
Abrupt onset of diarrhea, often with blood and pus in stools, cramps, tenesmus, and lethargy. Stool cultures are positive for *Shigella.* Given trimethoprim-sulfamethoxazole or ampicillin or chloramphenicol in severe cases. Do not give opiates. Often mild and self-limited. Restore fluids.
Gradual or abrupt onset of diarrhea and low-grade fever. No antimicrobials unless systemic dissemination is suspected. Stool cultures are positive for *Salmonella.* Prolonged carriage is frequent.
Especially after abdominal surgery, abrupt bloody diarrhea and fever. Exotoxin in stool. Oral vancomycin useful in therapy.
Fever, diarrhea; PMNs and fresh blood in stool, especially in children. Usually self-limited. Special media needed for culture at 43°C. Erythromycin or quinolones in severe cases with invasion. Usual recovery in 5–8 days.
Severe abdominal pain, diarrhea, fever; PMNs and blood in stool; polyarthritis, erythema nodosum, especially in children. If severe, treat with gentamicin. Keep stool specimen at 4°C before culture.

water supplies. The mortality rate without treatment is often over 50%; with treatment and supportive care, it is less than 1%. Fewer than 20 cases of cholera are reported each year in the United States.

Botulism

Food-borne **botulism** (Latin *botulus,* sausage)is a form of food poisoning caused by *Clostridium botulinum* serotypes A–G. *C. botulinum* is an obligate anaerobic endospore-forming, gram-positive rod that is found in soil and aquatic sediments. The most common source of infection is canned food that has not been heated sufficiently to kill contaminating *C. botulinum* endospores. The endospores can germinate, and an exotoxin is produced during vegetative growth. If the food is then eaten without adequate cooking, the exotoxin remains active and the disease results.

The botulinum exotoxin is a neurotoxin that binds to the synapses of motor neurons and prevents the release of the neurotransmitter acetylcholine. As a consequence, muscles do not contract in response to motor neuron activity, and flaccid paralysis results. Symptoms of botulism occur within 2 days of exotoxin ingestion and include blurred vision, difficulty in swallowing and speaking, muscle weakness, nausea, and vomiting. Without adequate treatment, 1/3 of the patients may die within a few days of either respiratory or cardiac failure.

Laboratory diagnosis is by a hemagglutination test or inoculation of mice with the patient's serum, stools, or vomitus to prove toxigenicity. Treatment relies on supportive care and polyvalent antitoxin. Fewer than 100 cases of botulism occur in the United States annually.

Infant botulism is the most common form of botulism in the United States and is confined to infants under a year of age. Approximately 100 cases are reported each year. It appears that ingested endospores, which may be present in honey or other baby foods, germinate in the infant's intestine. *C. botulinum* then multiplies and produces the exotoxin. The infant becomes constipated, listless, generally weak, and eats poorly. Death may result from respiratory failure.

Prevention and control of botulism involves (1) strict adherence to safe food-processing practices by the food industry, (2) educating the public on safe home-preserving (canning) methods for foods, and (3) not feeding honey to infants younger than 1 year of age.

In 1990, type A botulinal exotoxin was approved for treating the eye disorders blepharospasm and cross-eye. In both of these disorders, muscle spasms cause the eyes either to close involuntarily or wander. To control these problems, the toxin is injected directly into specific sites where it blocks muscular response to the nerve impulse responsible for the spasms.

Campylobacter jejuni Gastroenteritis

Campylobacter jejuni is a gram-negative curved rod found in the intestinal tract of animals (figure 37.21). Studies with

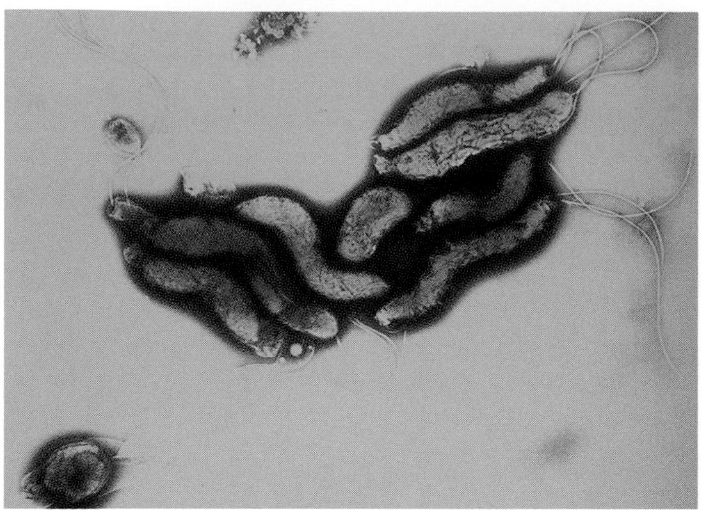

Figure 37.21 ***Campylobacter jejuni;* Gastroenteritis.** *Campylobacter jejuni;* TEM (×5,130). The Campylobacter family comprises curved or spiral Gram-negative bacilli with flagella at one or both poles, giving the bacteria a darting motility.

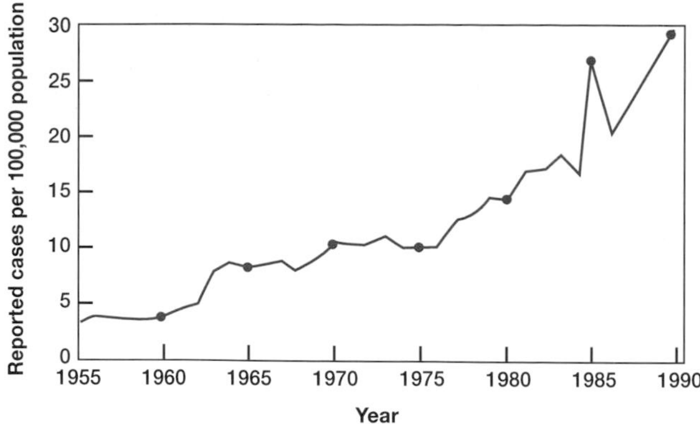

Figure 37.22 **Salmonellosis.** Incidence of cases for the United States. Notice that the trend is upward. *Source: The Centers for Disease Control, Atlanta, GA.*

chickens, turkeys, and cattle have shown that as much as 50 to 100% of a flock or herd of these birds or animals secrete *C. jejuni*. These bacteria also can be isolated in high numbers from surface waters. They are transmitted to humans by contaminated food and water, contact with infected animals, or anal-oral sexual activity. *C. jejuni* causes an estimated 2 million cases of *Campylobacter* **gastroenteritis**—inflammation of the intestine—and subsequent diarrhea in the United States each year.

The incubation period is 1 to 10 days. *C. jejuni* invades the epithelium of the small intestine, causing inflammation, and also secretes an exotoxin that is antigenically similar to the cholera toxin. Symptoms include diarrhea, high fever, severe inflammation of the intestine along with ulceration, and bloody stools.

Laboratory diagnosis is by culture in an atmosphere with reduced O_2 and added CO_2. The disease is self-limited, and treatment is supportive; fluids, electrolyte replacement, and erythromycin or quinolones are used in severe cases. Recovery usually takes from 5 to 8 days. Prevention and control involves good personal hygiene and food handling precautions, including pasteurization of milk and thorough cooking of poultry.

Salmonellosis

Salmonellosis (*Salmonella* gastroenteritis) is caused by over 2,000 *Salmonella* serovars (strains; a subspecies category). The most frequently reported one from humans is *S.* serovar *typhimurium*. This bacterium is a gram-negative, motile, nonspore-forming rod.

The initial source of the bacterium is the intestinal tracts of animals. Humans acquire the bacteria from contaminated foods such as beef products, poultry, eggs, egg products, or water. Around 50,000 cases a year are reported in the United States, but there actually may be as many as 2 to 3 millon cases annually. Overall, the incidence of this food-borne disease has been increasing steadily (figure 37.22).

Once the bacteria are in the body, the incubation time is only 8 to 48 hours. The disease results from a true food-borne infection because the bacteria multiply and invade the intestinal mucosa. Abdominal pain, cramps, diarrhea, and fever are the most prominent symptoms, which usually persist for 2 to 5 days but can last for several weeks. During the acute phase of the disease, as many as 1 billion salmonella can be found per gram of feces. Most adult patients recover, but the loss of fluids can cause problems for children and elderly people.

Laboratory diagnosis is by isolation of the bacterium from food or patients' stools. Treatment is with fluid and electrolyte replacement. Prevention depends on good food-processing practices, proper refrigeration, and adequate cooking.

Staphylococcal Food Poisoning

Staphylococcal food poisoning is the major type of food intoxication in the United States. It is caused by ingestion of improperly stored or cooked food (particularly foods such as ham, processed meats, chicken salad, pastries, ice cream, and hollandaise sauce) in which *Staphylococcus aureus* has grown.

S. aureus (a gram-positive coccus) is very resistant to heat, drying, and radiation; it is found in the nasal passages and on the skin of humans and other mammals worldwide. From these sources it can readily enter food. If the bacteria are allowed to incubate in certain foods, they produce heat-stable enterotoxins that render the food dangerous even though it appears normal. Six different enterotoxins have been identified and are designated A through F.

Typical symptoms include severe abdominal pain, diarrhea, vomiting, and nausea. The onset of symptoms is usually rapid (1 to 6 hours) and of short duration (less than 24 hours). The mortality rate of staphylococcal food poisoning is negligible among healthy individuals.

Diagnosis is based on the symptoms or laboratory diagnosis of the bacteria from foods. Enterotoxins may be detected in foods by animal toxicity tests. Treatment is with fluid and electrolyte replacement. Prevention and control involve avoidance of food contamination, and control of personnel responsible for food preparation and distribution.

Traveler's Diarrhea

Millions of people travel yearly from country to country. Unfortunately, a large percentage of these travelers acquire a rapidly acting, dehydrating condition called **traveler's diarrhea.** This results from an encounter with certain viruses, bacteria, or protozoa normally absent from the traveler's environment. One of the major organisms is enterotoxigenic *Escherichia coli* (ETEC). This bacterium circulates in the local population, usually without symptoms, undoubtedly due to the immunity afforded by previous exposure. Since many of these bacteria are needed to initiate infection, contaminated food and water are the major means by which the disease is spread. This is the basis for the popular warning to international travelers, "Don't drink the local water."

The heat-stable enterotoxins (STs) and heat-labile enterotoxins produced by *E. coli* are among the most important factors responsible for the diarrhea (*see Box 13.1*). Enterotoxigenic *E. coli* releases STs after binding to the brush border through the use of colonization factor antigens such as fimbriae. The STs cause fluid secretion by stimulating guanylate cyclase in intestinal cells.

Diagnosis of traveler's diarrhea is based on past travel history and symptoms. Laboratory diagnosis is by isolation of the bacteria from feces and determination of its virulence factors. Treatment is with fluid and electrolytes plus doxycycline and trimethoprim-sulfamethoxazole. Recovery is usually without complications. Prevention and control involve avoiding contaminated food and water.

Typhoid Fever

Typhoid (Greek *typhodes,* smoke) **fever** is caused by several virulent serovars of *Salmonella typhi* and is acquired by ingestion of food or water contaminated by feces of infected humans or animals. In earlier centuries the disease occurred in great epidemics.

Once in the small intestine the incubation period is 10 to 14 days. The bacteria colonize the small intestine, penetrate the epithelium, and spread to the lymphoid tissue, blood, liver, and gallbladder. Symptoms include fever, headache, abdominal pain, anorexia, and malaise, which last several weeks. After approximately three months, most individuals stop shedding bacteria in their feces. However, a few individuals continue to shed *S. typhi* for extended periods but show no symptoms. In these carriers, the bacteria continue to grow in the gallbladder and reach the intestine through the bile duct (*see Box 35.3 on "Typhoid Mary," one of the most famous typhoid carriers*).

Laboratory diagnosis of typhoid fever is by demonstration of typhoid bacilli in the blood, urine, or stools and serology (the Widal test). Treatment with ceftriaxone, trimethoprim-sulfamethoxazole, or ampicillin has reduced the mortality rate to less than 1%. Recovery from typhoid confers a permanent immunity.

Purification of drinking water, milk pasteurization, prevention of food handling by carriers, and complete isolation of patients are the most successful prophylactic measures. There is a vaccine for high-risk individuals (*see table 31.1*). About 400 to 500 cases of typhoid fever occur annually in the United States.

1. Distinguish between food intoxication, food poisoning, and food-borne infection. What is an enterotoxin?
2. Why is cholera the most severe form of gastroenteritis?
3. How does one acquire botulism? Describe how botulinum toxin causes flaccid paralysis.
4. What is the most common form of gastroenteritis in the United States and how are the symptoms caused?
5. What is the usual source of the bacterium responsible for salmonellosis?
6. Describe the most common type of food poisoning in the United States and how it arises.
7. Describe a typhoid carrier. How does one become a carrier?

Summary

1. Although only a small percentage of all bacteria are responsible for human illness, the suffering and death they cause are significant. Each year, millions of people are infected by pathogenic bacteria using the four major modes of transmission: airborne, arthropod-borne, direct contact, and food-borne and waterborne.
2. As the fields of microbiology, immunology, pathology, pharmacology, and epidemiology have expanded current understanding of the disease process, the incidence of many human illnesses has decreased. Many bacterial infections, once the leading cause of death, have been successfully brought under control in most developed countries. Alternatively, several are increasing in incidence throughout the world.
3. The bacteria emphasized in this chapter and the diseases they cause are as follows:
   a. Airborne diseases
      diphtheria (*Corynebacterium diphtheriae*)
      Legionnaires' disease and Pontiac fever (*Legionella pneumophila*)
      meningitis (*Haemophilus influenzae* b, *Neisseria meningitidis,* and *Streptococcus pneumoniae*)
      M. avium–M. intracellulare pneumonia
      Streptococcal diseases (*Streptococcus pyogenes*)
      tuberculosis (*Mycobacterium tuberculosis*)
      pertussis (*Bordetella pertussis*)
   b. Arthropod-borne diseases
      Lyme disease (*Borrelia burgdorferi*)
      plague (*Yersinia pestis*)
   c. Direct contact diseases
      anthrax (*Bacillus anthracis*)
      gas gangrene or clostridial myonecrosis (*Clostridium perfringens*)
      gonorrhea (*Neisseria gonorrhoeae*)
      leprosy (*Mycobacterium leprae*)
      peptic ulcer disease (*Helicobacter pylori*)
      staphylococcal diseases (*Staphylococcus aureus*)
      syphilis (*Treponema pallidum*)
      tetanus (*Clostridium tetani*)
      tularemia (*Francisella tularensis*)
   d. Food-borne and waterborne diseases
      cholera (*Vibrio cholerae*)
      botulism (*Clostridium botulinum*)
      gastroenteritis (*Campylobacter jejuni* and other bacteria)
      salmonellosis (*Salmonella* serovar *typhimurium*)
      staphylococcal food poisoning (*Staphylococcus aureus*)
      typhoid fever (*Salmonella typhi*)
      traveler's diarrhea (*Escherichia coli* or ETEC)

Key Terms

anthrax *753*
aseptic meningitis syndrome *744*
bacterial (septic) meningitis *744*
botulism *765*
Bright's disease *746*
buboes *751*
caseous lesion *748*
cellulitis *746*
chancre *759*
cholera *763*
choleragen *763*
clostridial myonecrosis *754*
congenital syphilis *759*
cutaneous anthrax *753*
cutaneous diphtheria *742*
diphtheria *742*
DPT vaccine *742*
endogenous infection *747*
enterotoxin *763*
erysipelas *746*
eschar *753*
exfoliative toxin (exfoliatin) *759*
food-borne infection *763*
food intoxication *763*
food poisoning *763*
gastroenteritis *766*
gastrointestinal anthrax *753*
gas gangrene *754*
Ghon complexes *748*
glomerulonephritis *746*
gonococci *754*
gonorrhea *754*
gumma *759*
Legionnaires' disease (legionellosis) *742*
lepromatous (progressive) leprosy *755*
leprosy (Hansen's disease) *755*
Lyme disease (LD, Lyme borreliosis) *750*
meningitis *744*
miliary tuberculosis *748*
ophthalmia neonatorum (conjunctivitis of the newborn) *754*
pelvic inflammatory disease (PID) *754*
peptic ulcer disease *756*
pertussis *745*
pharyngitis *747*
plague *750*
Pontiac fever *742*
pulmonary anthrax *753*
reactivation tuberculosis *748*
rheumatic fever *746*
salmonellosis *766*
scarlet fever (scarlatina) *747*
staphylococcal food poisoning *766*
staphylococcal scalded skin syndrome (SSSS) *759*
streptococcal sore throat *747*
streptococcal pneumonia *747*
tetanolysin *762*
tetanospasmin *762*
tetanus *761*
tonsillitis *747*
toxic shock syndrome (TSS) *758*
traveler's diarrhea *767*
tubercles *748*
tuberculoid (neural) leprosy *755*
tuberculosis (TB) *747*
tuberculous cavities *748*
tularemia *762*
typhoid fever *767*
venereal syphilis *759*

Questions for Thought and Review

1. Briefly describe each of the major or most common bacterial diseases in terms of its causative agent, signs and symptoms, the course of infection, mechanism of pathogenesis, epidemiology, and prevention and/or treatment.
2. Because there are many etiologies of gastroenteritis, how is a definitive diagnosis usually made?
3. Differentiate between the following factors of bacterial intoxication and infection: etiologic agents, onset, duration, symptoms, and treatment.
4. How would you differentiate between salmonellosis and botulism?
5. Why do urinary tract infections frequently occur after catheterization procedures?
6. Why is botulism the most serious form of bacterial food poisoning?
7. Why is tuberculosis still a problem in underdeveloped countries? In the United States?
8. What is the standard treatment for most bacterial diarrheas?
9. Why are most cases of gastroenteritis not treated with antibiotics?
10. How are tetanus, gas gangrene, and botulism related?
11. Give several reasons why treatment of tuberculosis is so difficult.
12. Describe the various bacteria that cause food poisoning, and give each one's most probable source.
13. Explain why Lyme disease is receiving increased attention in the United States.
14. You have been assigned the task of eradicating gonorrhea in your community. Explain how you would accomplish this.
15. You are a park employee. How would you prevent people from getting arthropod-borne diseases?

Additional Reading

Allen, S. D. 1984. Current relevance of anaerobic bacteria. *Clin. Microbiol. Newsl.* 6:147–49.

Aral, S. O., and Holmes, K. K. 1991. Sexually transmitted diseases in the AIDS era. *Sci. Am.* 264(2):62–69.

Archer, D. L., and Young, F. E. 1988. Contemporary issues: Diseases with a food vector. *Clin. Microbiol. Rev.* 1(4):377–98.

Aronon, S. S. 1980. Infant botulism. *Ann. Rev. Med.* 31:541–50.

Barbour, A. G. 1988. Laboratory aspects of Lyme borreliosis. *Clin. Microbiol. Rev.* 1(4):399–414.

Barksdale, L. 1970. *Corynebacterium diphtheriae* and its relatives. *Bacteriol. Rev.* 34:378–422.

Benenson, A. S., ed. 1990. *Control of communicable diaseases in man,* 15th ed. Washington, D.C.: American Public Health Association.

Bisno, A. L. 1991. Group A streptococcal infections and acute rheumatic fever. *N. Eng. J. Med.* 325(11):783–93.

Blake, P. A.; Weaver, R. E.; and Hollis, D. G. 1980. Diseases of humans caused by vibrios. *Ann. Rev. Microbiol.* 34:341–67.

Broome, C. V., and Facklam, R. R. 1981. Epidemiology of clinically significant isolates of *Streptococcus pneumoniae*. *Rev. Infect. Dis.* 3:277–80.

Crissey, J. T., and Denenholz, B. 1984. Syphilis. *Clin. Dermatol.* 4:81–95.

Dennie, C. C. 1962. *A history of syphilis.* Springfield, Ill.: Charles C. Thomas.

Dowling, J. N.; Saha, A. K.; and Glew, R. H. 1992. Virulence factors of the family *Legionellaceae*. *Microbiol. Rev.* 56(1):32–60.

Dowell, V. R. J. 1984. Botulism and tetanus: Selected epidemiologic and microbiologic aspects. *Rev. Infect. Dis.* 6:202–7.

Fischetti, V. A. 1991. Streptococcal M protein. *Sci. Am.* 264(6):58–65.

Fraser, D. W., and McDade, J. E. 1979. Legionellosis. *Sci. Am.* 241(5):82–89.

Friedman, R. L. 1988. Pertussis: The disease and new diagnostic methods. *Clin. Microbiol. Rev.* 1(4):365–76.

Gorbach, S. L. 1975. Traveler's diarrhea and toxigenic *E. coli*. *N. Engl. J. Med.* 292:933–36.

Guerrant, R., and Bobak, D. 1991. Bacterial and protozoal gastroenteritis. *N. Eng. J. Med.* 325(5):327–40.

Habicht, G. S.; Beck, G.; and Benach, J. L. 1987. Lyme disease. *Sci. Am.* 257(3):78–83.

Hastings, R. C.; Gillis, T. P.; Krahenbuhl, J. L.; and Franzblau, S. G. 1988. Leprosy. *Clin. Microbiol. Rev.* 1(3):330–48.

Hatheway, C. L. 1990. Toxigenic clostridia. *Clin. Microbiol. Rev.* 3(1):66–98.

Hirschhorn, N., and Greenough, W. 1991. Progress in oral rehydration therapy. *Sci. Am.* 264(5):50–56.

Hook, E. W., and Marra, C. M. 1992. Acquired Syphilis in adults. *N. Eng. J. Med.* 326(16):1060–69.

Janda, J. M.; Powers, C.; Bryant, R. G.; and Abbott, S. L. 1988. Current perspectives on the epidemiology and pathology of clinically significant *Vibrio* spp. *Clin. Microbiol. Rev.* 1(3):245–67.

Johnson, S. R., and Galask, R. P. 1983. Group B streptococcal disease. *Clin. Obst. Gyn.* 10:105–21.

Joklik, W. K.; Willett, H. P.; Amos, D. B.; and Wilfert, C. M. 1992. *Zinsser microbiology,* 20th ed. E. Norwalk, Conn.: Appleton & Lange.

Lyerly, D. M.; Krivan, H. C.; and Wilkins, T. D. 1988. *Clostridium difficile:* Its diseases and toxins. *Clin. Microbiol. Rev.* 1(1):1–18.

Mandell, G. L.; Douglas, R. G., Jr.; and Bennett, J. E. 1990. *Principles and practices of infectious diseases,* 3d ed. New York: John Wiley and Sons.

McEvedy, C. 1988. The bubonic plague. *Sci. Am.* 258(2):118–23.

Penner, J. L. 1988. The genus *Campylobacter:* A decade of progress. *Clin. Microbiol. Rev.* 1(2):157–72.

Peterson, W. 1991. *Helicobacter pylori* and peptic ulcer disease. *N. Engl. J. Med.* 324(15):1043–48.

Rosebury, T. 1973. *Microbes and morals: The strange story of venereal disease.* New York: Ballantine.

Shulman, S. T.; Phair, J. P.; and Sommers, H. M. 1992. *The biologic and clinical basis of infectious diseases,* 4th ed. Philadelphia, PA.: W. B. Saunders.

Stine, G. J. 1992. *The biology of sexually transmitted diseases.* Dubuque, Iowa: Wm. C. Brown Communications, Inc.

Szczepanski, A., and Benach, J. 1991. Lyme borreliosis: Host response to *Borrelia burgdorferi*. *Microbiol. Rev.* 55(1):21–34.

Todd, J. K. 1988. Toxic shock syndrome. *Clin. Microbiol. Rev.* 1(4):432–46.

Tramont, E. 1989. Gonococcal vaccines. *Clin. Microbiol. Rev.* 2(Suppl.):S74–S77.

Unny, S. K., and Middlebrook, B. L. 1983. Streptococcal rheumatic carditis. *Microbiol. Rev.* 47:97–109.

Wayne, L. G. and Sramek, H. A. 1992. Agents of newly recognized or infrequently encountered mycobacterial diseases. *Clin. Microbiol. Rev.* 5(1):1–25.

Winn, W. C., 1988. Legionnaires disease: Historical perspective. *Clin. Microbiol. Rev.* 1(1):60–81.

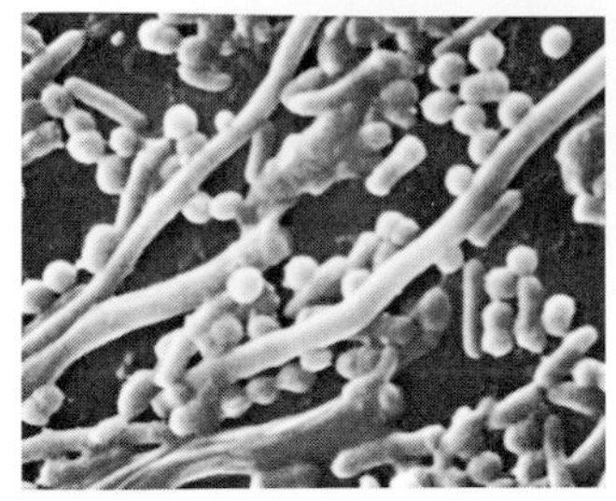

Chapter 38

Human Diseases Caused by Other Bacteria (Chlamydiae, Mycoplasmas, Rickettsias); Dental and Nosocomial Infections

The chlamydiae are a genetically diverse group of bacteria with similarities in morphology, intracellular development, and antigenic properties. Chlamydiae are common causes of sexually transmitted disease (STD) and respiratory infection. Millions of cases of C. trachomatis *infection occur in the United States each year. . . .*

—Robert C. Barnes

Outline

Concepts

1. Two chlamydial species cause human disease: *Chlamydia trachomatis,* which causes inclusion conjunctivitis, lymphogranuloma venereum, nongonococcal urethritis, and trachoma; and *C. psittaci,* which causes psittacosis.
2. Three species of mycoplasmas are human pathogens: *Mycoplasma hominis* and *Ureaplasma urealyticum* cause genitourinary tract disease, whereas *M. pneumoniae* is a major cause of acute respiratory disease and pneumonia.
3. The rickettsias found in the United States can be divided into the typhus group (epidemic typhus caused by *R. prowazekii* and murine typhus caused by *R. typhi*) and the spotted fever group (Rocky Mountain spotted fever caused by *R. rickettsii*), with Q fever (caused by *Coxiella burnetii*) being an exception because it forms endosporelike structures and does not have to use an insect vector as the other rickettsias do.
4. Several bacterial odontopathogens are responsible for the most common bacterial diseases in humans—tooth decay and periodontal disease. Both are the result of plaque formation and the production of lactic and acetic acids by the odontopathogens.
5. Bacteria are the leading cause of hospital-acquired or nosocomial diseases such as bacteremias, burn wound infections, respiratory tract infections, surgical wound infections, and urinary tract infections. These diseases represent a significant proportion of all infectious diseases acquired by humans.

Chapter 21 reviews the general biology of the chlamydiae, mycoplasmas, and rickettsias; chapter 35 presents the epidemiology of nosocomial diseases. Chapter 38 continues this coverage by discussing the important human pathogens noted in these chapters. The microorganisms involved in dental infections are also described.

Several groups of bacterial pathogens are unusual in many respects and do not fit neatly into traditional classification and discussion schemes. Nevertheless, they are important and worthy of discussion. This chapter contains an assemblage of such bacteria and the diseases they cause.

Chlamydial Diseases

The chlamydiae are obligate intracellular parasites of eucaryotic cells (*see figure 21.17*). Three species cause human disease: *Chlamydia trachomatis, C. psittaci,* and *C. pneumoniae.*

The chlamydiae (pp. 443–44).

The clinical spectrum and organ specificity of chlamydial infections are determined by the method of transmission and antigenic properties of the infecting strain. The fifteen serotypes of *C. trachomatis* are divided among three groups: mouse, lymphogranuloma venereum (serotypes L_1–L_3), and trachoma (serotypes A–K). Serotypes A–K infect epithelial cells and are associated with infections of the conjunctiva (membrane covering the eyeball), pharynx, respiratory tract, urethra, cervix of the uterus, and uterine tubes. Serotypes L_1–L_3 cause the systemic disease lymphogranuloma venereum. These human chlamydial diseases are summarized in table 38.1, and several are now discussed in more detail.

Table 38.1 Human Diseases Caused by Chlamydiae

Diseases	Method of Transmission	Serotypes
Chlamydia trachomatis		
Endemic trachoma	Flies, fingers, fomites	A, B, Ba, C
Inclusion conjunctivitis	During birth, sexual contact	D–K
Newborn pneumonia	During birth	D–K
Nongonococcal urethritis	Sexual contact	D–K
Postgonococcal urethritis	Sexual contact	D–K
Epididymitis	Sexual contact	D–K
Proctitis	Sexual contact	D–K
Mucopurulent cervicitis	Sexual contact	D–K
Endometritis	Sexual contact	D–K
Salpingitis	Sexual contact	D–K
Lymphogranuloma venereum	Sexual contact	L_1, L_2, L_3
Chlamydia psittaci		
Psittacosis (ornithosis)	Zoonosis from birds	None
Chlamydia pneumoniae		
Pneumonia	Not known	None

Inclusion Conjunctivitis

Inclusion conjunctivitis is an acute infectious disease caused by *C. trachomatis* serotypes D–K, and it occurs throughout the world. It is characterized by a copious mucous discharge from the eye, an inflamed and swollen conjunctiva, and the presence of large inclusion bodies. In inclusion conjunctivitis of the newborn, the chlamydiae are acquired during passage through an infected birth canal. The disease appears 7 to 12 days after birth. If the chlamydiae colonize an infant's nasopharynx and tracheobronchial tree, pneumonia may result. Adult inclusion conjunctivitis is acquired by contact with infective genital tract discharges.

Without treatment, recovery usually occurs spontaneously over several weeks or months. Therapy involves treatment with tetracycline, erythromycin, or a sulfonamide. The specific diagnosis of *C. trachomatis* can be made by direct immunofluorescence, Giemsa stain, nucleic acid probes, and culture. Genital chlamydial infections and inclusion conjunctivitis are sexually transmitted diseases that are spread by indiscriminate contact with multiple sex partners. Prevention depends upon diagnosis and treatment of all infected individuals.

Lymphogranuloma Venereum

Lymphogranuloma venereum (LGV) is a sexually transmitted disease caused by *C. trachomatis* serotypes L_1–L_3. It has a worldwide distribution but is more common in tropical climates.

LGV proceeds through three phases. (1) In the primary phase, a small ulcer appears several days to several weeks after a person is exposed to the chlamydiae. The ulcer may appear on the penis in males or on the labia or vagina in females. The ulcer heals quickly and leaves no scar. (2) The secondary phase begins two to six weeks after exposure, when the chlamydiae infect lymphoid cells, causing the regional lymph nodes to become enlarged and tender; such nodes are called buboes (figure 38.1). Systemic symptoms such as fever, chills, and anorexia are common. (3) If the disease is not treated, a late phase ensues. This results from fibrotic changes and abnormal lymphatic drainage that produces fistulas (abnormal passages leading from an abscess or a hollow organ to the body surface or from one hollow organ to another) and urethral or rectal strictures (a decrease in size). An untreatable fluid accumulation in the penis, scrotum, or vaginal area may result.

The disease is detected by staining infected cells with iodine to observe inclusions (chlamydia-filled vacuoles), by culture of the chlamydiae from a bubo, or by the detection of a high micro-IF antibody titer to an LGV serotype. Treatment in the early phases consists of aspiration of the buboes and administration of drugs: tetracycline, doxycycline, erythro-

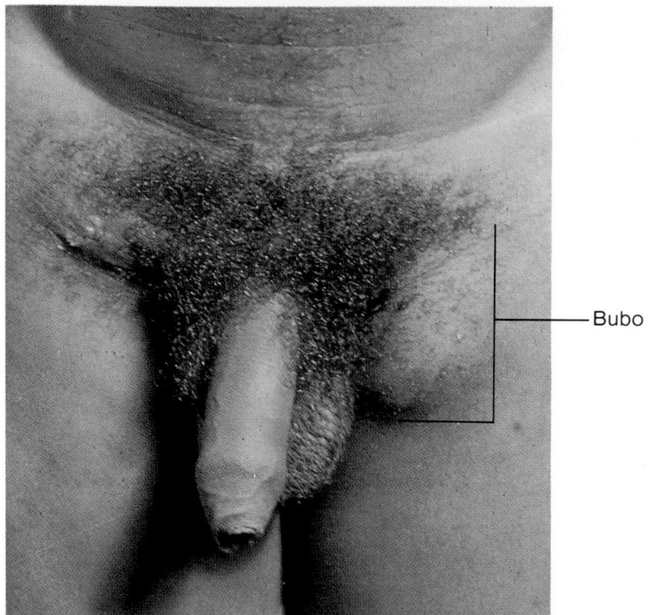

Figure 38.1 **Lymphogranuloma Venereum.** The bubo in the left inguinal area is draining.

mycin, or sulfamethoxazole. The late phase may require surgery. The methods used for the control of LGV are the same as for other sexually transmitted diseases: reduction in promiscuity, use of condoms, and early diagnosis and treatment of infected individuals. About 300 cases of LGV occur annually in the United States.

Nongonococcal Urethritis

Nongonococcal urethritis (NGU) is any inflammation of the urethra not due to the bacterium *Neisseria gonorrhoeae.* This condition is caused both by nonmicrobial factors such as catheters and drugs and by infectious microorganisms. The most important causative agents are *C. trachomatis, Ureaplasma urealyticum, Mycoplasma hominis, Trichomonas vaginalis, Candida albicans,* and herpes simplex viruses. Most infections are acquired sexually, and of these, approximately 50% are *Chlamydia* infections. NGU from chlamydia is probably the most common sexually transmitted disease in the United States, with over 10 million Americans infected. It is endemic elsewhere throughout the world.

Symptoms of NGU vary widely. Males may have few or no manifestations of disease; however, complications can exist. These include a urethral discharge, itching, and inflammation of the male reproductive structures. Females may be asymptomatic or have a severe infection called pelvic inflammatory disease (PID) that often leads to sterility. Chlamydia may account for as many as 200,000 to 400,000 cases of PID annually in the United States. In the pregnant female, a chlamydial infection is especially serious because it is directly related to miscarriage, stillbirth, inclusion conjunctivitis, and infant pneumonia.

Diagnosis of NGU requires the demonstration of a leukocyte exudate and exclusion of urethral gonorrhea by Gram stain and culture. Several rapid tests for detecting *Chlamydia* in urine specimens are also available. Treatment is with tetracycline, doxycycline, erythromycin, or sulfisoxazole.

Psittacosis (Ornithosis)

Psittacosis (ornithosis) is a worldwide infectious disease of birds that is transmissible to humans. It was first described in association with parrots and parakeets, both of which are psittacine birds. The disease is now recognized in many other birds—among them, pigeons, chickens, ducks, and turkeys—and the general term ornithosis (Latin *ornis,* bird) is used.

Ornithosis is caused by *Chlamydia psittaci.* Humans contract this disease either by handling infected birds or by inhaling dried bird excreta that contains viable *C. psittaci.* Ornithosis is recognized as an occupational hazard within the poultry industry, particularly to workers in turkey processing plants.

After entering the respiratory tract, the chlamydiae are transported to the cells of the liver and spleen. They multiply within these cells and then invade the lungs, where they cause inflammation, hemorrhaging, and pneumonia.

Laboratory diagnosis is either by isolation of *C. psittaci* from blood or sputum, or by serological studies. Treatment is with tetracycline. Because of antibiotic therapy, the mortality rate has dropped from 20 to 2 percent. Between 100 and 200 cases of ornithosis are reported annually in the United States. Prevention and control is by chemoprophylaxis (tetracycline) for pet birds and poultry.

Recently, a new strain of *C. psittaci,* the TWAR strain (Taiwan, acute respiratory), has been identified in Taiwan and Iran. This strain has been isolated from human atypical pneumonia and pharyngitis patients, but not from birds or mammals. Age-specific prevalence rates suggest that transmission occurs in childhood and peaks early in life. These TWAR strains appear to be circulating among humans without an avian or mammal reservoir. Based on DNA homology, the TWAR strain has been given the name *C. pneumoniae.*

Trachoma

Trachoma (Greek *trachoma,* roughness) is a contagious disease created by *C. trachomatis* serotypes A–C. It is one of the oldest known infectious diseases of humans and is the greatest single cause of blindness throughout the world. Probably over 500 million people are infected and 20 million blinded each year by this chlamydia. In endemic areas, most children are chronically infected within a few years of birth. Active disease in adults over age 20 is three times as frequent in females as in males because of mother-child contact. Although uncommon in the United States, except among American Indians in the Southwest, trachoma is widespread in Asia, Africa, and South America.

Trachoma is transmitted by contact with inanimate objects such as soap and towels, by hand-to-hand contact that carries *C. trachomatis* from an infected eye to an uninfected eye, or by flies. The disease begins abruptly with an inflamed conjunctiva. This leads to an inflammatory cell exudate and necrotic eyelash follicles beneath the conjunctival surface (figure 38.2). The disease usually heals spontaneously. However, with reinfection, vascularization of the cornea, or **pannus** formation, occurs, leading to scarring of the conjunctiva. If scar tissue accumulates over the cornea, blindness results.

Diagnosis and treatment of trachoma are the same as for inclusion conjunctivitis (previously discussed). However, prevention and control of trachoma lies more in health education and personal hygiene—such as access to clean water for washing—than in treatment.

1. How can one distinguish *Chlamydia trachomatis* from *C. psittaci?* How do chlamydiae differ from most other bacteria?
2. How does an infant acquire inclusion conjunctivitis?
3. Describe the three phases of lymphogranuloma venereum.
4. What is nongonococcal urethritis and what agents can cause it? Describe complications that may develop in the absence of treatment.
5. How do humans contract ornithosis?
6. How does *C. trachomatis*, serotypes A–C, cause trachoma? Describe how it is transmitted, and the way in which blindness may result.

Figure 38.2 **Trachoma.** An active infection showing marked follicular hypertrophy of both eyelids. The inflammatory nodules cover the thickened conjuctiva of the eye.

Mycoplasmal Diseases

The mycoplasmas are pleomorphic, nonmotile microorganisms without cell walls and are among the smallest independently living organisms. They are ubiquitous worldwide as saprophytes and parasites of plants, animals, and humans. Three species are pathogenic to humans: *Ureaplasma urealyticum, Mycoplasma hominis,* and *M. pneumoniae.*

The mycoplasmas (pp. 445–47).

Genitourinary Diseases

Ureaplasma urealyticum and *M. hominis* are common parasitic microorganisms of the genital tract, and their transmission is related to sexual activity. Both mycoplasmas can opportunistically cause inflammation of the reproductive organs of males and females. Because mycoplasmas are not usually cultured by clinicians, management and treatment of these infections depend on a recognition of clinical syndromes and provision for adequate therapy. Tetracyclines are active against most strains; resistant organisms can be treated with erythromycin.

Primary Atypical Pneumonia

Typical pneumonia has a bacterial origin. If a bacterium cannot be isolated, the pneumonia is termed atypical and a virus is usually suspected. If viruses can't be detected, then **mycoplasmal pneumonia** can be considered. This pneumonia is caused by *Mycoplasma pneumoniae* (*see figure 21.21*), a mycoplasma with worldwide distribution. Spread involves close contact and sometimes airborne droplets. The disease is fairly common and mild in infants and small children; serious disease is seen principally in older children and young adults.

Respiratory syndromes and viral pneumonia (p. 720).

M. pneumoniae usually infects the upper respiratory tract, and subsequently moves to the lower respiratory tract, where it attaches to respiratory mucosal cells. It then produces peroxide, which may be a toxic factor, but the exact mechanism of pathogenesis is unknown. An alternation in mucosal cell nucleic acid metabolism has been observed. The manifestations of this disease vary in severity from asymptomatic to a serious pneumonia. The latter is accompanied by death of the surface mucosal cells, lung infiltration, and congestion. Initial symptoms include headache, weakness, a low fever, and a predominant cough. The disease and its symptoms usually persist for weeks. The mortality rate is less than 1%.

Several rapid tests using latex agglutination of *M. pneumoniae* antibodies are available for diagnosis of mycoplasmal pneumonia. When isolated from respiratory secretions, the mycoplasmas form distinct colonies with a "fried egg" appearance (*see figure 21.22*). During the acute stage of the disease, diagnosis must be made by clinical observations. Tetracyclines or erythromycin are effective in treatment. There are no preventive measures.

1. What three species of mycoplasmas cause disease in humans?
2. What two mycoplasmas may cause genitourinary disease?
3. What is meant by primary atypical pneumonia? How is this disease spread?

Rickettsial Diseases

The rickettsias, like the chlamydiae, are fastidious bacteria that are obligate intracellular parasites. They are small, pleo-

Box 38.1

The Hazards of Microbiological Research

The investigation of human pathogens is often a very dangerous matter, and several microbiologists have been killed by the microorganisms they were studying. The study of typhus fever provides a classic example. In 1906, Howard T. Ricketts (1871–1910), an associate professor of pathology at the University of Chicago, became interested in Rocky Mountain spotted fever, a disease that had decimated the Nez Percé and Flathead Indians of Montana. By infecting guinea pigs, he established that a small bacterium was the disease agent and was transmitted by ticks. In late 1909, Ricketts traveled to Mexico to study Mexican typhus.

He discovered that a microorganism similar to the Rocky Mountain spotted fever bacillus could cause the disease in monkeys and be transmitted by lice. Despite his careful technique, he was bitten while transferring lice in his laboratory and died of typhus fever on May 3, 1910. The causative agent of typhus fever was fully described in 1916 by the Brazilian scientist H. da Roche-Lima and named *Rickettsia prowazekii* in honor of Ricketts and Stanislaus von Prowazek, a Czechoslovakian microbiologist who died in 1915 while studying typhus.

Today, modern equipment to control microorganisms, such as laminar airflow hoods (*see figures 15.6 and 34.1c*), have greatly reduced the risks of research on microbial pathogens.

morphic coccobacilli maintained in the environment through a cycle involving mammalian reservoirs and insect vectors. Except for epidemic typhus, humans are only incidental hosts and are not useful in propagating the rickettsias (Box 38.1). Rickettsias can be divided into four groups based on characteristics such as their intracellular location, growth temperature optimum, antigenic properties, DNA composition (mol% G + C), and the diseases they cause. These four are (1) the typhus group—epidemic typhus (*Rickettsia prowazekii*) and murine typhus (*R. typhi*); (2) the spotted fever group—Rocky Mountain spotted fever (*R. rickettsii*) and similar diseases; (3) the scrub typhus group (*R. tsutsugamushi*); and (4) a miscellaneous group that includes Q fever (*Coxiella burnetii*). In the following sections, only rickettsial diseases that occur in the United States are discussed.

The rickettsias (pp. 442–43).

Epidemic (Louse-Borne) Typhus

Epidemic (louse-borne) typhus is caused by *Rickettsia prowazekii,* which is transmitted from person to person by the body louse. In the United States a reservoir of *R. prowazekii* also exists in the southern flying squirrel. When a louse feeds on an infected rickettsemic person, the rickettsias infect the insect's gut and multiply, and large numbers of organisms appear in the feces in about a week. When a louse takes a blood meal, it defecates. The irritation causes the affected individual to scratch the site and contaminate the bite wound with rickettsias. The rickettsias then spread via the bloodstream and infect the endothelial cells of the blood vessels, causing a **vasculitis** (inflammation of the blood vessels). This produces an abrupt headache, fever, and muscle aches. A rash begins on the upper trunk, and spreads. Without treatment, recovery takes about two weeks, though mortality rates are very high (around 50%), especially in the elderly. Recovery from the disease gives a solid immunity and also protects the person from murine typhus.

Diagnosis is by the characteristic rash, symptoms, and the Weil-Felix reaction (Box 38.2). Chloramphenicol and tetracycline are effective against typhus. Control of the human body louse and the conditions that foster its proliferation are mainstays in the prevention of epidemic typhus, although a typhus vaccine is available for high-risk individuals. The importance of louse control and good public hygiene is shown by the prevalence of typhus epidemics during times of war and famine when there is crowding and little attention to the maintenance of proper sanitation. For example, around 30 million cases of typhus fever and 3 million deaths occurred in the Soviet Union and Eastern Europe between 1918 and 1922. The bacteriologist Hans Zinsser believes that Napoleon's retreat from Russia in 1812 may have been partially provoked by typhus and dysentery epidemics that ravaged the French army. Fewer than 25 cases of epidemic typhus are reported in the United States each year.

Endemic (Murine) Typhus

The etiologic agent of **endemic (murine) typhus** is *Rickettsia typhi.* It occurs in isolated areas around the world, including southeastern and Gulf Coast states, especially Texas. The disease occurs sporadically in individuals who come into contact with rats and their fleas. The disease is nonfatal in the rat and is transmitted from rat to rat by fleas. When an infected flea takes a human blood meal, it defecates. Its feces are heavily laden with rickettsias, which infect humans by contaminating the bite wound.

The clinical manifestations of murine (Latin *mus, muris,* mouse or rat) typhus are similar to those of epidemic typhus except that they are milder in degree and the mortality rate is much lower: less than 5%. Diagnosis and treatment are also the same. Rat control and avoidance of rats are preventive measures for the disease. Fewer than 100 cases of endemic typhus are reported in the United States each year.

Box 38.2

The Weil-Felix Reaction

During World War I, Austrian bacteriologists Edmond Weil (1880–1922) and Arthur Felix (1887–1956) were in charge of diagnosing typhus in the Austrian Army. In 1916, they developed the agglutination test for typhus that now bears their names.

The **Weil-Felix reaction** is based on the detection of heterophilic antibodies, that is, cross-reactive antibodies. In this test, antibodies produced in response to a particular rickettsial infection agglutinate bacterial strains of *Proteus,* designated OX–19, OX–2, and OX–K. These cross-reactions occur because rickettsial strains possess cell wall antigens that are similar to polysaccharide O antigens of the *Proteus* strains. The results are considered to give presumptive diagnosis of rickettsial diseases. However, it should be kept in mind that *Proteus* species are indigenous to humans and may cause urinary tract infections. Therefore, the presence of antibodies to *Proteus* OX strains is not of itself an absolute criterion for the presence of rickettsial disease.

Q Fever

Q fever (*Q* for query because the cause of the fever was not known for some time) is an acute zoonotic disease caused by *Coxiella burnetii. C. burnetii* is different from the other rickettsias in its ability to survive outside host cells by forming a resistant endosporelike body. This rickettsia infects both wild animals and livestock. In animals, ticks transmit *C. burnetii*, whereas in humans, transmission is primarily by inhalation of dust contaminated with rickettsias from dried animal feces, urine, or milk. The disease is apt to occur in epidemic form among slaughterhouse workers and sporadically among farmers and veterinarians. Each year, fewer than 100 cases of Q fever are reported in the United States.

In humans, after inhalation of the rickettsias, local proliferation occurs in the lungs. This may result in mild respiratory symptoms similar to those of atypical pneumonia or influenza. Q fever itself is an acute illness characterized by the sudden onset of severe headache, myalgia (muscle pain), and fever, which may remain very high for more than a month if not treated. Unlike the other rickettsial diseases, Q fever is not accompanied by a rash. It is rarely fatal, but endocarditis—inflammation of the heart muscle—occurs in about 10% of the cases. Five to ten years may elapse between the initial infection and the appearance of the endocarditis. During this interval, the rickettsias live in the liver and often cause hepatitis. Diagnosis is most commonly made serologically. Treatment is with chloramphenicol and tetracycline.

Rocky Mountain Spotted Fever

Rocky Mountain spotted fever is caused by *Rickettsia rickettsii.* Although originally detected in the Rocky Mountain area, most cases of this disease now occur east of the Mississippi River. The disease is transmitted by ticks and usually occurs in people who are or have been in tick-infested areas. There are two principal vectors: *Dermacentor andersoni,* the wood tick, is distributed in the Rocky Mountain states and is active during the spring and early summer. *D. variabilis,* the dog tick, has assumed greater importance and is almost exclusively confined to the eastern half of the United States. Unlike the other rickettsias discussed, *R. rickettsii* can pass from generation to generation of ticks through their eggs in a process known as **transovarian passage.** No humans or mammals are needed as reservoirs for the continued propagation of this rickettsia in the environment.

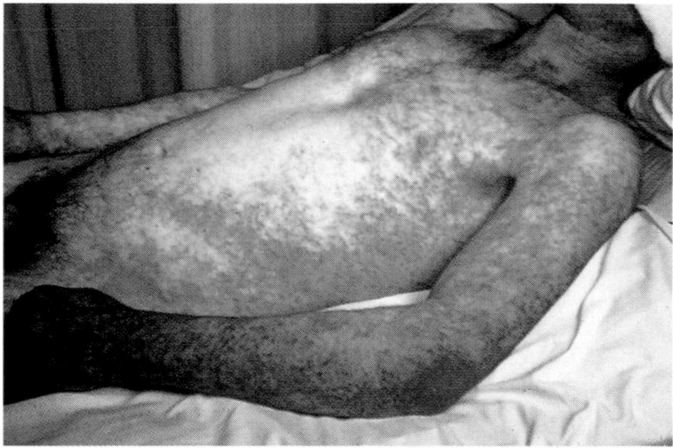

Figure 38.3 **Rocky Mountain Spotted Fever.** Typical rash occurring on the arms and chest consists of generally distributed, sharply defined macules.

When humans contact infected ticks, the rickettsias are either deposited on the skin (if the tick defecates after feeding) and then subsequently rubbed or scratched into the skin, or the rickettsias are deposited into the skin as the tick feeds. Once inside the skin, the rickettsias enter the endothelial cells of small blood vessels, where they multiply and produce a characteristic vasculitis (inflammation of blood vessels).

The disease is characterized by the sudden onset of a headache, high fever, chills, and a skin rash (figure 38.3) that initially appears on the ankles and wrists and then spreads to the trunk of the body. If the disease is not treated, the rickettsias can destroy the blood vessels in the heart, lungs, or kidneys and cause death. Usually, however, severe pathological changes are avoided by antibiotic therapy (chloramphenicol, chlortetracycline), the development of immune resistance, and supportive therapy. Diagnosis is made through observation of symptoms and signs such as the characteristic rash, and by serological tests. The best means of prevention remains the avoidance of tick-infested habitats and animals (*see preventive methods for Lyme disease, p. 750*). There are fewer than 1,000 reported cases of Rocky Mountain spotted fever annually in the United States.

1. What two antibiotics are used against most rickettsial infections?
2. How is epidemic typhus spread? Murine typhus? What are their symptoms?
3. What is unique about *Coxiella burnetii* compared to the other rickettsias?
4. Describe the symptoms of Rocky Mountain spotted fever.
5. How does transovarian passage occur?

Dental Infections

Some microorganisms found in the oral cavity are discussed in chapter 29 and presented in figure 29.8. Of this large number, only a few bacteria can be considered true **odontopathogens,** dental pathogens. These few pathogens are responsible for the most common bacterial diseases in humans: tooth decay and periodontal disease.

Dental Plaque

The human tooth has a naturally occurring defense mechanism against bacterial colonization that complements the protective role of saliva. The hard enamel surface selectively absorbs acidic glycoproteins (mucins) from saliva, forming a membranous layer called the **acquired enamel pellicle.** This pellicle, or organic covering, contains many sulfate (SO_4^{2-}) and carboxylate (—COO^-) groups that confer a net negative charge to the tooth surface. Because most bacteria also have a net negative charge, there is a natural repulsion between the tooth surface and bacteria in the oral cavity. Unfortunately, this natural defense mechanism breaks down when dental plaque formation occurs.

Dental plaque formation begins with the initial colonization of the pellicle by *Streptococcus gordonii* (formerly considered a subpopulation of *S. sanguis*), *S. oralis,* and *S. mitis.* These bacteria selectively adhere to the pellicle by specific ionic, hydrophobic, and lectinlike interactions. Once the tooth surface is colonized, subsequent attachment of other bacteria results from a variety of specific coaggregation reactions. The most important species at this stage are *Actinomyces viscosus, A. naeslundii,* and *S. gordonii.* After these species colonize the pellicle, a microenvironment is created that allows *Streptococcus mutans* and *S. sobrinus* to become established on the tooth surface by attaching to these initial colonizers.

These streptococci produce extracellular enzymes (glucosyltransferases) that polymerize the glucose moiety of sucrose into a heterogeneous group of extracellular water-soluble and water-insoluble glucan polymers and other polysaccharides. The fructose by-product can be used in fermentation. **Glucans** are branched-chain polysaccharides composed of glucose units held together by α-1,6 and α-1,3 linkages (figure 38.4*a*). They act like a cement to bind bacterial cells together, forming a plaque ecosystem (figure 38.4*b,c*). (Dental plaque is one of the most dense collections of bacteria in the body—and perhaps the source of the first human microorganisms to be seen under a microscope by Anton van Leeuwenhoek in the seventeenth century.) Once plaque becomes established, a low oxidation reduction potential is created on the surface of the tooth. This leads to the growth of strict anaerobic bacteria (*Bacteroides melaninogenicus, B. oralis,* and *Veillonella alcalescens*), especially between opposing teeth and the dental-gingival crevices.

After the microbial plaque ecosystem develops, bacteria produce lactic and possibly acetic acids from sucrose and other sugars. Because plaque is not permeable to saliva, the acids are not diluted or neutralized, and they demineralize the enamel to produce a lesion on the tooth. It is this chemical lesion that initiates dental decay.

Dental Decay (Caries)

As described previously, a histologically undetectable chemical lesion caused by the diffusion into the tooth's enamel of undissociated fermentation acids initiates the decay process. Once these acids move below the enamel surface, they dissociate and react with the hydroxyapatite of the enamel to form soluble calcium and phosphate ions. As the ions diffuse outward, some reprecipitate as calcium phosphate salts in the tooth's surface layer to create a histologically sound outer layer overlying a porous subsurface area. Between meals and snacks, the pH returns to neutrality and some calcium phosphate reenters the lesion and crystallizes. The result is a demineralization-remineralization cycle.

When fermentable foods high in sucrose are eaten for prolonged periods, acid production overwhelms the repair process and demineralization is greater than remineralization. This leads to cavitation, otherwise known as dental decay or **caries** (kar'ēz; Latin, rottenness). Once the hard enamel has been breached, bacteria can invade the dentin and pulp of the tooth and cause its death.

No drugs are available to prevent dental caries. The main strategies for prevention include minimal ingestion of sucrose; daily brushing, flossing, and mouthwashes; and professional cleaning at least twice a year to remove plaque. The use of fluorides in toothpaste, drinking water, mouthwashes, or professionally applied to the teeth protects against lactic and acetic acids and reduces tooth decay.

Periodontal Disease

Periodontal disease refers to a diverse group of diseases that affect the periodontium. The **periodontium** (per″e-o-don′she-um) is the supporting structure of a tooth and includes the

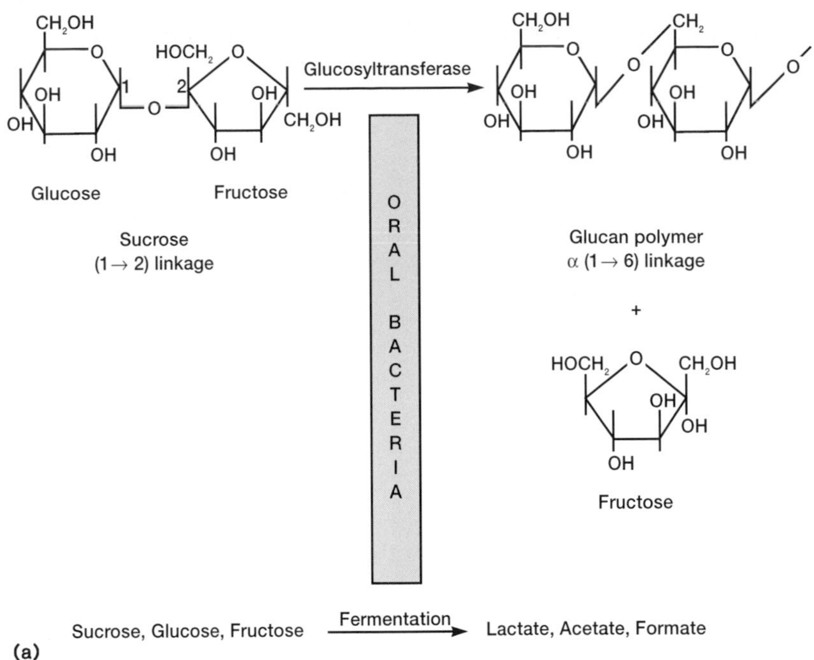

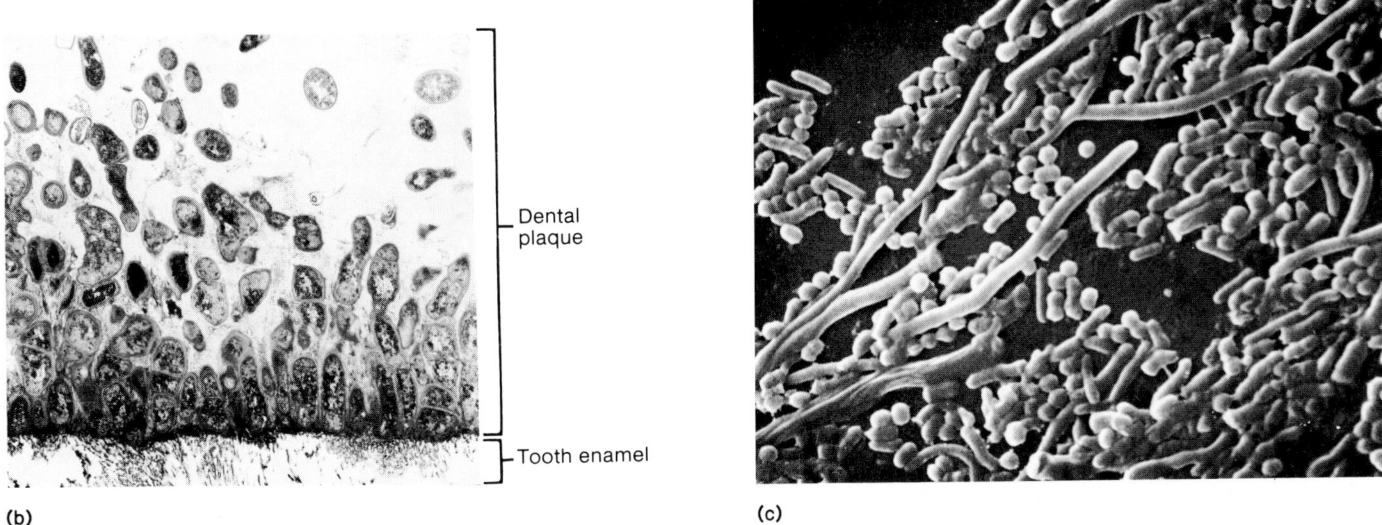

Figure 38.4 The Formation of Dental Plaque. (*a*) The enzyme glucosyltransferase or dextransucrase (produced by oral bacteria) causes the assembly of glucose units from sucrose into glucans, and fructose is released. Sucrose, glucose, and fructose molecules also are metabolized by the oral bacteria to produce lactate and other acids. Lactate is responsible for dental caries. (*b*) Dental plaque consisting of bacteria plus polysaccharides such as glucan is shown attached to the enamel surface of a tooth; transmission electron micrograph (×13,600). (*c*) Scanning electron micrograph of accumulated bacteria on the surface of a tooth contributing to the plaque ecosystem (×12,500).

cementum, the periodontal membrane, the bones of the jaw, and the gingivae (gums). The gingiva (jin-ji'vah) is the tissue, covered by a mucous membrane, that surrounds the necks of the teeth and covers the jaws. Disease is initiated by the formation of **subgingival plaque,** the plaque that forms at the dentogingival margin and extends down into the gingival tissue. Recent evidence shows that *Porphyromonas* (formerly *Bacteroides*) *gingivalis* is the main species in this plaque and is responsible for the breakdown of tissue through the production of a trypsinlike protease. The result is an initial inflammatory reaction known as **periodontitis,** which is caused by the host's immune response to both the plaque bacteria and the tissue destruction. This leads to swelling of the tissue and the formation of periodontal pockets. Bacteria colonize these pockets and cause more inflammation, which leads to the formation of a periodontal abscess, bone destruction or **periodontosis,** inflammation of the gingiva or **gingivitis,** and general tissue necrosis (figure 38.5). If the condition is not treated, the tooth may fall out of its socket.

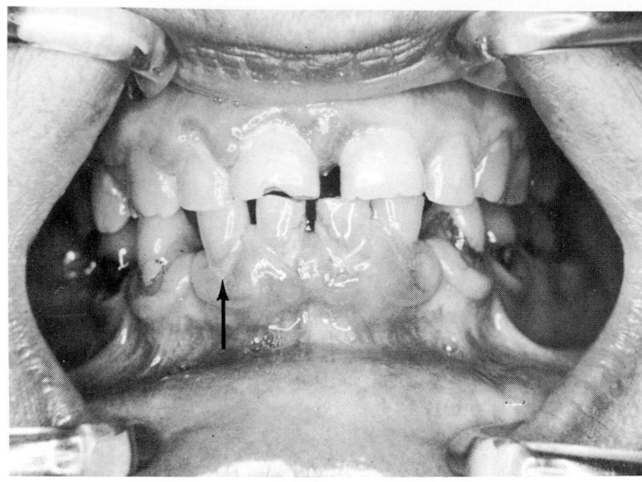

Figure 38.5 **Periodontal Disease.** Notice the plaque on the teeth (arrow), especially at the gum margins, and the inflamed gums.

Periodontal disease can be controlled by frequent plaque removal; by brushing, flossing, and mouthwashes; and at times, oral surgery of the gums and antibiotics.

1. Name some common odontopathogens that are responsible for dental caries, dental plaque, and periodontal disease. Be specific.
2. What is the function of the acquired enamel pellicle?
3. How does plaque formation occur? Dental decay?
4. Describe some pathological manifestations of periodontal disease.
5. How can caries and periodontal diseases be prevented?

Nosocomial Infections

The epidemiology of **nosocomial infections** (nos″o-ko′me-al) is presented in chapter 35 (*pp. 711–12*). What follows is a continuation of that discussion by means of a brief examination of the microorganisms and agents responsible for these infections. The Centers for Disease Control estimate that 5 to 10% of all hospital patients acquire some type of nosocomial infection. Because approximately 40 million people are admitted to hospitals annually, about 2 to 4 million people may develop an infection they did not have upon entering the hospital. Thus, nosocomial infections represent a significant proportion of all infectious diseases acquired by humans.

Nosocomial diseases are usually caused by bacteria, most of which are noninvasive and part of the normal microbiota; viruses, protozoa, and fungi are rarely involved. Figure 38.6 summarizes the most common types of nosocomial infections. The specific bacteria associated with the various body sites and disease processes follow.

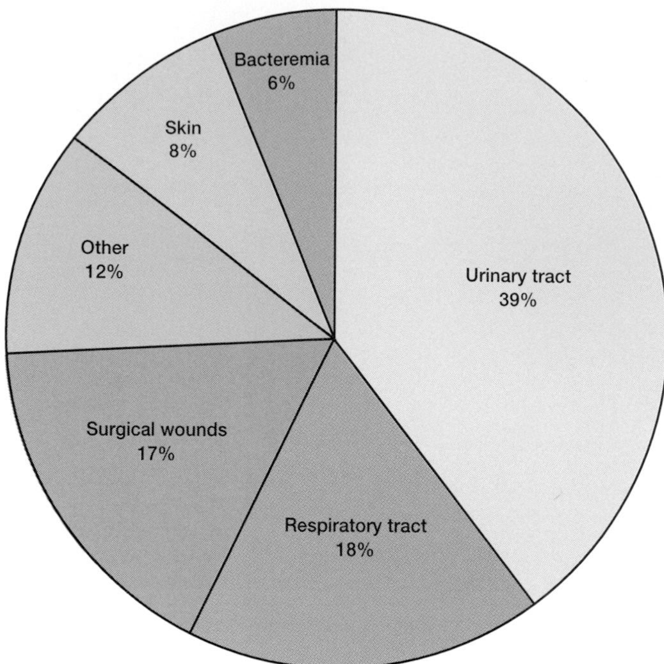

Figure 38.6 **Nosocomial Infections.** Relative frequency by body site.

Bacteremia

Bacteremia (bak″ter-e′me-ah), the transient presence of bacteria in the blood, accounts for about 6% of all nosocomial infections. Two types exist: primary and secondary. A primary bacteremia results from the accidental introduction of bacteria directly into the body by way of intravenous infusion, transducers, respiratory devices, prostheses, catheters, endoscopes, hemodialysis units, or parenteral nutrition. A secondary bacteremia results from an infection at another body site such as the urinary tract, respiratory tract, or surgical wounds. The microorganisms most responsible for bacteremias are *Staphylococcus aureus, Escherichia coli,* and *Klebsiella* spp. Bacteremias occur primarily in neonates and the elderly, whose natural defense mechanisms are compromised.

Burn Wounds

A burn wound is tissue damage caused by heat, chemicals, radiation, or electricity. Because the skin is no longer an effective barrier to microorganisms, many are now able to colonize the burned tissue. *Pseudomonas aeruginosa, Staphylococcus aureus,* and many gram-negative bacilli are most often acquired by the burn victim from the hospital environment.

Physical barriers, the skin (p. 592).

Respiratory Tract Infections

The lower respiratory tract ranks second in incidence (about 18%) of nosocomial infections. More patients die of nosocomial pneumonias than of any other nosocomial disease. Most of these pneumonias are related to respiratory devices that aid breathing or administer medications. The microorganisms most responsible are *Klebsiella spp., S. aureus,* and *Pseudomonas aeruginosa.*

Surgical Wound Infections

Surgical wound infections rank third in incidence (about 17%) of nosocomial infections. It is estimated that 5 to 12% of all surgical patients develop postoperative infections. In operations involving the gastrointestinal, respiratory, and genitourinary tracts, the infection rate approaches 30%. The main reason is that these sites harbor large numbers of microorganisms, and once displaced, they can colonize the surgical wound. The bacteria most frequently isolated from surgical wounds are *S. aureus* and *E. coli.*

Urinary Tract Infections

Urinary tract infections are the most common nosocomial infections. They usually account for about 39% of all cases and are typically related to urinary catheterization. The bacteria most frequently associated with these infections are *E. coli,* group D streptococci, and *P. aeruginosa.*

Miscellaneous Infections

In adults, cutaneous infections are among the least common of all nosocomial infections. However, newborns are particularly susceptible to cutaneous and eye infections. As a result, most nurseries have an infection rate three times greater than any other hospital area. Cutaneous infections result primarily from *S. aureus,* coagulase-negative staphylococci, *P. aeruginosa,* and *E. coli.*

Another group of individuals particularly susceptible to nosocomial infections are those who are immunosuppressed, cancer and AIDS patients, and organ transplant patients receiving immunosuppressive drug therapy.

1. Of what significance are nosocomial infections compared to infectious diseases as a whole?
2. What are two types of bacteremias? Give an example of how each occurs.
3. What are the most frequent pathogens of burn wounds?
4. What causes most nosocomial pneumonias?
5. Why are surgical wound infections so common?
6. Why are urinary tract infections so common in hospitals?

Summary

1. The chlamydiae are obligate intracellular parasites of eucaryotic cells. Three species cause human disease: *Chlamydia trachomatis, C. pneumoniae,* and *C. psittaci. C. trachomatis* causes inclusion conjunctivitis, lymphogranuloma venereum (LGV), nongonococcal urethritis (NGU), and trachoma. All of these diseases are spread by person-to-person contact. Psittacosis (ornithosis) is an infectious disease of birds that is transmitted to humans. It is caused by *C. psittaci. C. pneumoniae* causes a variety of lower respiratory tract diseases.
2. The mycoplasmas are procaryotes without cell walls and are the smallest independently living organisms. Two mycoplasmas, *Ureaplasma urealyticum* and *Mycoplasma hominis,* are common pathogens of the genitourinary tract and spread by sexual contact. They cause limited pathology in humans. Mycoplasmal pneumonia, or primary atypical pneumonia, is caused by *Mycoplasma pneumoniae.* Its spread involves airborne droplets and close contact. The lower respiratory tract is usually colonized where mucosal damage is apparent, and this leads to pneumonia.
3. The rickettsias are fastidious bacteria that are obligate intracellular parasites. They are maintained in the environment through a cycle involving mammalian reservoirs and insect vectors. Humans are only occasional, incidental hosts. Epidemic (louse-borne) typhus is caused by *R. prowazekii.* It is spread from person to person by the body louse. Murine (endemic) typhus is caused by *R. typhi.* It is spread from rats to humans by rat fleas. Q fever is an acute zoonotic disease caused by *Coxiella burnetii.* Humans inhale the rickettsias from contaminated animal products. This leads to an atypical pneumonia and other problems. Rocky Mountain spotted fever is caused by *R. rickettsii.* It is transmitted to humans by ticks and causes a characteristic skin rash.
4. Dental plaque formation begins on a tooth with the initial colonization of the acquired enamel pellicle by *Streptococcus gordonii, S. oralis,* and *S. mitis.* Other bacteria then become attached and form a plaque ecosystem. The bacteria produce acids that cause a chemical lesion on the tooth and initiate dental decay or caries. Periodontal disease is a group of diverse clinical entities that affect the periodontium. Disease is initiated by the formation of subgingival plaque, which leads to tissue inflammation known as periodontitis and to periodontal pockets. Bacteria that colonize these pockets can cause an abscess, periodontosis, gingivitis, and general tissue necrosis.
5. Nosocomial infections are a significant proportion of all infectious diseases acquired by humans. Examples include primary and secondary bacteremias, burn wound infections, respiratory tract infections, surgical wound infections, and urinary tract infections. Newborns are particularly prone to nosocomial cutaneous and eye infections. Individuals who are immunosuppressed, cancer and AIDS patients, and organ transplant patients receiving immunosuppressive drug therapy are also particularly susceptible to nosocomial infections.

Key Terms

acquired enamel pellicle *776*
bacteremia *778*
caries *776*
dental plaque *776*
endemic (murine) typhus *774*
epidemic (louse-borne) typhus *774*
gingivitis *777*
glucans *776*
inclusion conjunctivitis *771*
lymphogranuloma venereum (LGV) *771*
mycoplasmal pneumonia *773*
nongonococcal urethritis (NGU) *772*
nosocomial infections *778*
odontopathogens *776*
pannus *773*
periodontal disease *776*
periodontitis *777*
periodontium *776*
periodontosis *777*
psittacosis (ornithosis) *772*
Q fever *775*
Rocky Mountain spotted fever *775*
subgingival plaque *777*
trachoma *772*
transovarian passage *775*
vasculitis *774*
Weil-Felix reaction *775*

Questions for Thought and Review

1. Briefly describe each of the major or most common diseases covered in this chapter in terms of its causative agent, signs and symptoms, the course of infection, mechanism of pathogenesis, epidemiology, and prevention and/or treatment.
2. What specific bacteria are responsible for most nosocomial infections?
3. Why does the use of specific chemotherapeutic agents often give rise to nosocomial infections?
4. What is the major difference between trachoma and inclusion conjunctivitis?
5. Why is Rocky Mountain spotted fever the most important rickettsial disease in the United States?
6. As a group of diseases, what are some of the common characteristic symptoms and pathological results of rickettsial disease?
7. Why is it difficult to definitely isolate the microorganism responsible for nongonococcal urethritis?
8. Why are dental diseases relatively new infectious diseases of humans?
9. What makes Q fever a unique rickettsial disease?
10. What specific chlamydial and mycoplasmal diseases are transmitted sexually?

Additional Reading

Ayliffe, G. A., and Taylor, L. J. 1982. *Hospital-acquired infections. Principles and prevention.* Littleton, Mass.: John Wright-PSG.

Baca, O. G., and Pretsky, D. 1983. Q fever and *Coxiella burnetii:* A model for host-parasite interactions. *Microbiol. Rev.* 47:127–44.

Barnes, R. C. 1989. Laboratory diagnosis of human chlamydial infections. *Clin. Microbiol. Rev.* 2(2):119–36.

Benenson, A. S., ed. 1990. *Control of communicable diseases in man,* 15th ed. Washington, D.C.: American Public Health Association.

Bennett, J., and Brachman, P. S. 1986. *Hospital infections.* Boston: Little, Brown.

Burgdorfer, W., and Anacker, R. L., eds. 1981. *Rickettsiae and rickettsial disease.* New York: Academic Press.

Cassell, C. H., and Cole, B. C. 1981. Mycoplasmas as agents of human disease. *N. Engl. J. Med.* 304:80–85.

Gibbons, R. J., and van Houte, J. 1975. Bacterial adherence in oral microbiol ecology. *Ann. Rev. Microbiol.* 29:19–44.

Gurevich, I. 1989. *Infectious diseases in critical care nursing.* Rockville, Md.: Aspen.

Hamada, S., and Slade, H. D. 1980. Biology, immunology, and cariogenicity of *Streptococcus mutans. Microbiol. Rev.* 44:331–46.

International symposium on control of nosocomial infection. 1981. *Rev. Infect. Dis.* 3:entire issue.

Loesche, W. J. 1986. Role of *Streptococcus mutans* in human dental decay. *Microbiol. Rev.* 50:353–71.

Marsh, P. D., and Martin, M. V. 1984. *Oral microbiology,* 2d ed. Washington, D.C.: American Society for Microbiology.

McDade, J. E. 1980. Evidence of *Rickettsia prowazekii* infections in the United States. *Am. J. Trop. Med. Hyg.* 29:277–79.

Mergenhan, S. E., and Rosan, B. 1985. *Molecular basis of oral microbial adhesion.* Washington, D.C.: American Society for Microbiology.

Moulder, J. W. 1991. Interaction of chlamydiae and host cells in vitro. *Microbiol. Rev.* 55(1):143–90.

Schachter, J., and Caldwell, H. 1980. Chlamydiae. *Ann. Rev. Microbiol.* 34:285–90.

Smith, P. W. 1985. *Infection control in long-term care facilities.* New York: John Wiley and Sons.

Traub, R.; Wisseman, C. L.; and Farhang-Azad, A. 1978. The ecology of murine typhus—A critical review. *Trop. Dis. Bull.* 75:237–91.

Walker, D. H. 1989. Rocky Mountain spotted fever: A disease in need of microbial concern. *Clin. Microbiol. Rev.* 2(3):227–40.

Walsh, T. J., and Pizzo, P. A. 1988. Nosocomial fungal infections: A classification for hospital-acquired fungal infections and mycoses arising from endogenous flora or reactivation. *Ann. Rev. Microbiol.* 42:517–45.

Wayne, L. G., and Sramek, H. A. 1992. Agents of newly recognized or infrequently encountered mycobacterial diseases. *Clin. Microbiol. Rev.* 5(1):1–25.

Weiss, E. 1982. The biology of the rickettsiae. *Ann. Rev. Microbiol.* 32:345–70.

Wenzel, R. P. 1988. *Prevention and control of nosocomial infections.* Baltimore: Williams & Wilkins.

Zinsser, H., ed. 1935. *Rats, lice, and history.* Boston: Little, Brown.

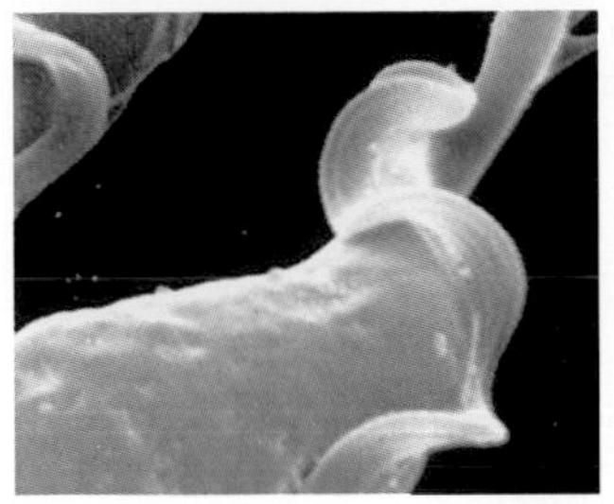

CHAPTER 39
Human Diseases Caused by Fungi and Protozoa

The ability of fungi to cause disease in humans appears to be an accidental phenomenon.

—John W. Rippon

Outline

Concepts

1. Fungal diseases (mycoses) are usually divided into five groups according to the level of infected tissue and mode of entry into the host: (1) superficial, (2) cutaneous, (3) subcutaneous, (4) systemic, and (5) opportunistic infections.
2. The superficial mycoses occur mainly in the tropics and include black piedra, white piedra, and tinea versicolor.
3. The cutaneous mycoses that is, those of the outer layer of the skin, are generally called ringworms, tineas, or dermatophytoses. These diseases occur worldwide and represent the most common fungal diseases in humans.
4. The dermatophytes that cause the subcutaneous—below the skin—mycoses are normal saprophytic inhabitants of the soil. They must be introduced into the body beneath the cutaneous layer. Examples of these diseases include chromomycosis, maduromycosis, and sporotrichosis.
5. The systemic mycoses are the most serious of the fungal infections in the normal host because the responsible fungi can disseminate throughout the body. Examples include blastomycosis, coccidioidomycosis, cryptococcosis, and histoplasmosis.
6. The opportunistic mycoses can create life-threatening situations in the compromised host. Examples of these diseases include aspergillosis and candidiasis.
7. About 20 different protozoa cause human diseases that afflict hundreds of millions of people throughout the world. Examples include amebiasis, giardiasis, malaria, *Pneumocystis carinii* pneumonia, the hemoflagellate diseases, toxoplasmosis, and trichomoniasis.
8. Certain fungal and protozoan diseases are increasing in incidence because of organ transplants, immunosuppressive drugs, and AIDS.

The purpose of this chapter is to describe some of the fungal and protozoan microorganisms that are pathogenic to humans and to discuss the clinical manifestations, diagnosis, epidemiology, pathogenesis, and treatment of the diseases caused by them.

Besides the viruses and bacteria, two other major groups of microorganisms cause infectious diseases in humans: fungi and protozoa. The biology of these organisms is covered in chapters 26 and 28, respectively. This chapter emphasizes the diseases they cause.

Fungal Diseases

Although hundreds of thousands of fungal species are found in the environment, only about 50 can produce disease in humans. **Medical mycology** is the discipline that deals with the fungi that causes human disease. These fungal diseases, known as **mycoses** (s., mycosis; Greek *mykes,* fungus), are divided into five groups according to the type of infected tissue in the host: superficial, cutaneous, subcutaneous, systemic, and opportunistic mycoses.

Superficial Mycoses

The superficial mycoses are extremely rare in the United States, and most occur in the tropics. The fungi responsible are limited to the outer surface of hair and skin, and hence are called superficial. The infections are collectively termed **piedras** (Spanish for stone because they are associated with the hard nodules formed by mycelia [*see figure 26.4*] on the hair shaft). Superficial mycoses are also called **tineas** (Latin for grub, larva, worm), the specific type being designated by a modifying term. For example, **black piedra** (tinea nigra) is caused by *Piedraia hortae* and forms hard black nodules on the hairs of the scalp (figure 39.1). **White piedra** (tinea albigena) is caused by the yeast *Trichosporon beigelii* and forms light-colored nodules on the beard and mustache. **Tinea versicolor** is caused by the yeast *Malassezia furfur* and forms brownish-red scales on the skin of the trunk, neck, face, and arms. Treatment involves removal of the skin scales with a cleansing agent and removal of the infected hairs. Good personal hygiene prevents these infections.

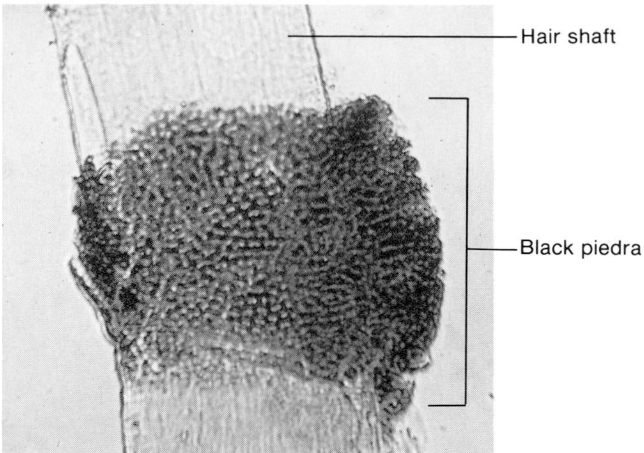

Figure 39.1 **Superficial Mycosis: Black Piedra.** Hair shaft infected with *Piedraia hortae*; light micrograph (×200).

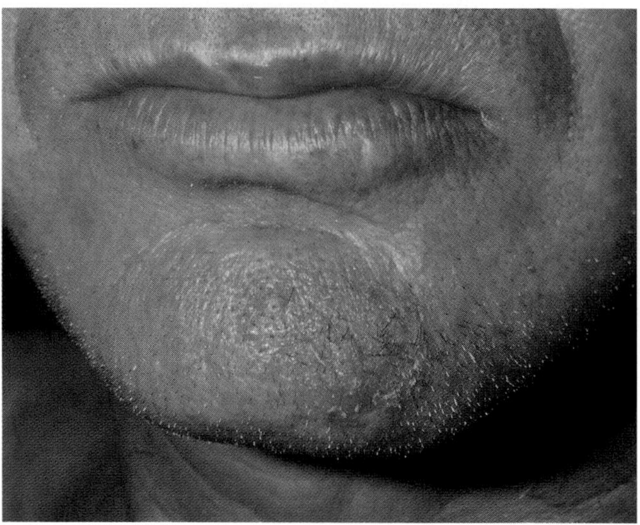

Figure 39.2 **Cutaneous Mycosis: Tinea Barbae.** Ringworm of the beard caused by *Trichophyton mentagrophytes.*

Cutaneous Mycoses

Cutaneous mycoses—also called **dermatophytoses** (Greek *derma,* skin, and *phyto,* plant), **ringworms,** or tineas—occur worldwide and represent the most common fungal diseases in humans. Three genera of cutaneous fungi, or **dermatophytes,** are involved in these mycoses: *Epidermophyton, Microsporum,* and *Trichophyton.* Diagnosis is by microscopic examination of biopsied areas of the skin cleared with 10% potassium hydroxide and by culture on Sabouraud dextrose agar. Treatment is with topical ointments such as miconazole (Monistatderm), tolnaftate (Tinactin), or clotrimazole (Lotrimin) for two to four weeks. Griseofulvin is the only oral fungal agent currently approved by the FDA for treating dermatophytoses.

The mode of action of ketoconazole, miconazole, tolnaftate, and griseofulvin (pp. 339–40).

Tinea barbae (Latin *barba,* the beard) is an infection of the beard hair (figure 39.2) caused by *Trichophyton menta-*

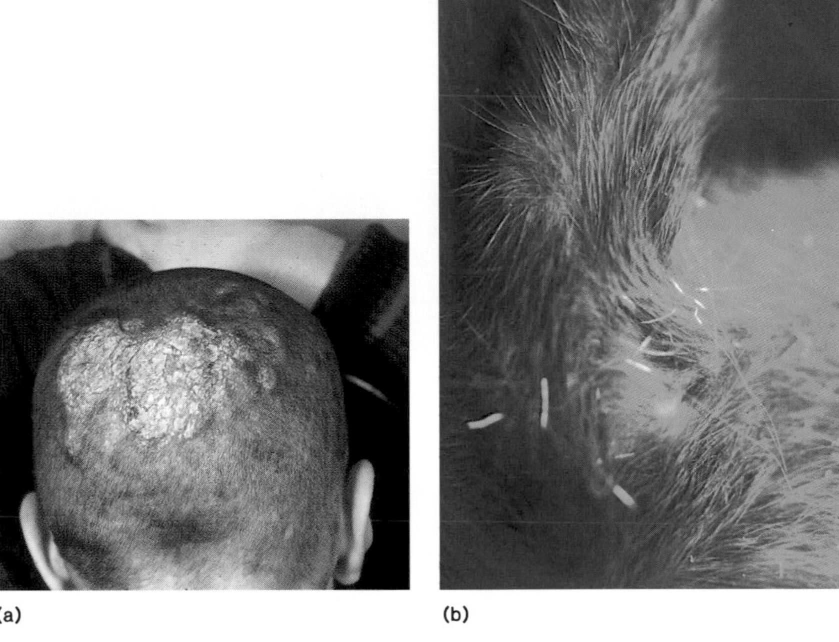

Figure 39.3 **Cutaneous Mycosis: Tinea Capitis.** (*a*) Ringworm of the head caused by *Microsporum audouinii.* (*b*) Close-up utilizing a Wood's light.

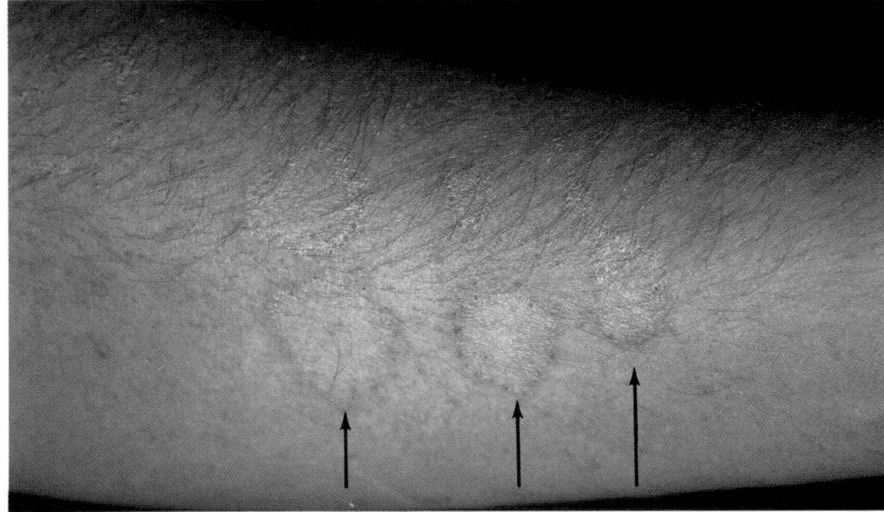

Figure 39.4 **Cutaneous Mycosis: Tinea Corporis.** Ringworm of the body—in this case, the forearm—caused by *Trichophyton mentagrophytes.* Notice the circular patches (arrows).

grophytes or *T. verrucosum.* It is predominantly a disease of men who live in rural areas and acquire the fungus from infected animals.

Tinea capitis (Latin *capita,* the head) is an infection of the scalp hair (figure 39.3*a*). It is characterized by loss of hair, inflammation, and scaling. Tinea capitis is primarily a childhood disease caused by *Trichophyton* or *Microsporum* species. Person-to-person transmission of the fungus occurs frequently when poor hygiene and overcrowded conditions exist. The fungus also occurs in domestic animals, from whom it can be transmitted to humans. A Wood's light (a UV light) can help with the diagnosis of tinea capitis because fungus-infected hair fluoresces when illuminated by UV radiation (figure 39.3*b*).

Tinea corporis (Latin *corpus,* the body) is a dermatophytic infection of the smooth or bare parts of the skin (figure 39.4). The disease is characterized by circular, red, well-demarcated, scaly, vesiculopustular lesions accompanied by itching. Tinea corporis is caused by *Trichophyton rubrum, T. mentagrophytes,* or *Microsporum canis.* Transmission of the disease is by direct contact with infected animals (*see figure 26.1*b) or humans or by indirect contact through fomites.

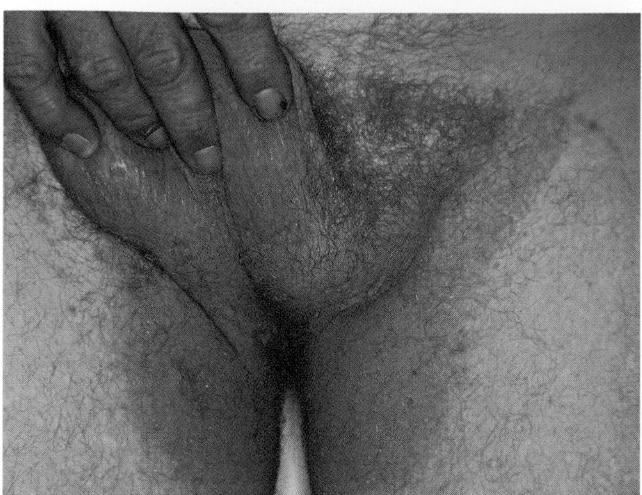

Figure 39.5 **Cutaneous Mycosis: Tinea Cruris.** Ringworm of the groin caused by *Epidermophyton floccosum.*

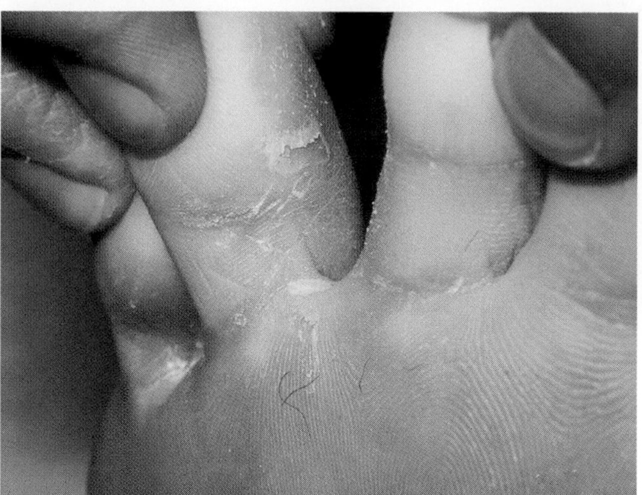

Figure 39.6 **Cutaneous Mycosis: Tinea Pedis.** Ringworm of the foot caused by *Trichophyton rubrum, T. mentagrophytes,* or *Epidermophyton floccosum.*

Tinea cruris (Latin *crura,* the leg) is a dermatophytic infection of the groin (figure 39.5). The pathogenesis and clinical manifestations are similar to those of tinea corporis. The responsible fungi are *Epidermophyton floccosum, T. mentagrophytes,* or *T. rubrum.* Factors predisposing one to recurrent disease are moisture, occlusion, and skin trauma. Wet bathing suits, athletic supporters ("jock itch"), tight-fitting slacks, panty hose, and obesity are frequently contributing factors.

Tinea pedis (Latin *pes,* the foot), also known as "athlete's foot," and **tinea manuum** (Latin *mannus,* the hand) are dermatophytic infections of the feet (figure 39.6) and hands, respectively. Clinical symptoms vary from a fine scale to a vesiculopustular eruption. Itching is frequently present. Warmth, humidity, trauma, and occlusion increase susceptibility to infection. Most infections are caused by *T. rubrum, T. mentagrophytes,* or *E. floccosum.* Tinea pedis and tinea manuum occur throughout the world, are most commonly found in adults, and increase in frequency with age.

Tinea unguium (Latin *unguis,* nail) is a dermatophytic infection of the nail bed (figure 39.7). In this disease, the nail becomes discolored and then thickens. The nail plate rises and separates from the nail bed. *Trichophyton rubrum* or *T. mentagrophytes* are the causative fungi.

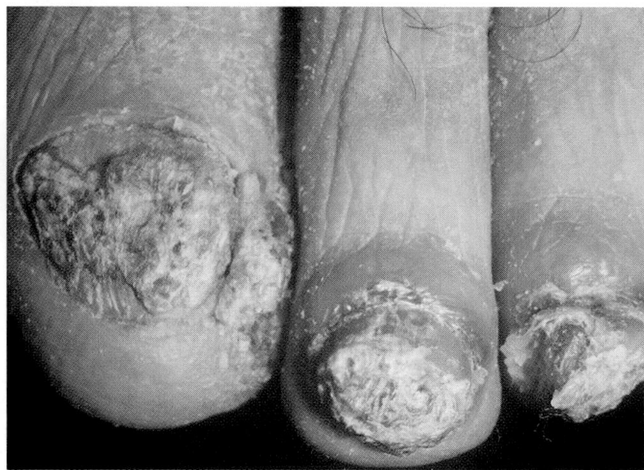

Figure 39.7 **Cutaneous Mycosis: Tinea Unguium.** Ringworm of the nails caused by *Trichophyton rubrum.*

Subcutaneous Mycoses

The dermatophytes that cause subcutaneous mycoses are normal saprophytic inhabitants of soil and decaying vegetation. Because they are unable to penetrate the skin, they must be introduced into the subcutaneous tissue by a puncture wound that has been contaminated with soil containing the fungi. Most infections involve barefooted agricultural workers. Once in the subcutaneous tissue, the disease develops slowly—often over a period of years. During this time, the fungi produce a nodule that eventually ulcerates and the organisms spread along lymphatic channels producing more subcutaneous nodules. At times, such nodules drain to the skin surface. The administration of oral 5-fluorocytosine, iodides, amphotericin B, and surgical excision are the usual treatments. Diagnosis is accomplished by culture of the infected tissue.

The mode of action of amphotericin B (p. 340).

One type of subcutaneous mycosis is **chromomycosis.** The nodules are pigmented a dark brown. This disease is caused by the black molds *Phialophora verrucosa* or *Fonsecaea ped-*

rosoi. These fungi exist worldwide, especially in tropical and subtropical regions. Most infections involve the legs and feet (figure 39.8).

Another subcutaneous mycosis is **maduromycosis,** caused by *Madurella mycetomatis,* which is distributed worldwide and is especially prevalent in the tropics. Because the fungus destroys subcutaneous tissue and produces serious deformities, the resulting infection is often called a **mycetoma** or fungal tumor (figure 39.9). One form of mycetoma, known as Madura foot, occurs through skin abrasions acquired while walking barefoot on contaminated soil.

Sporotrichosis is the subcutaneous mycosis caused by the dimorphic (*see table 26.2*) fungus *Sporothrix schenckii.* The disease occurs throughout the world and is the most common subcutaneous mycotic disease in the United States. The fungus can be found in the soil, on living plants, such as barberry shrubs and roses, or in plant debris, such as sphagnum moss and pine-bark mulch. Infection occurs by a puncture wound from a thorn or splinter contaminated with the fungus. The disease is an occupational hazard to florists, gardeners, and forestry workers. After an incubation period of 1 to 12 weeks, a small red papule arises and begins to ulcerate (figure 39.10). New lesions appear along lymph channels and can remain localized or spread throughout the body, producing **extracutaneous sporotrichosis.**

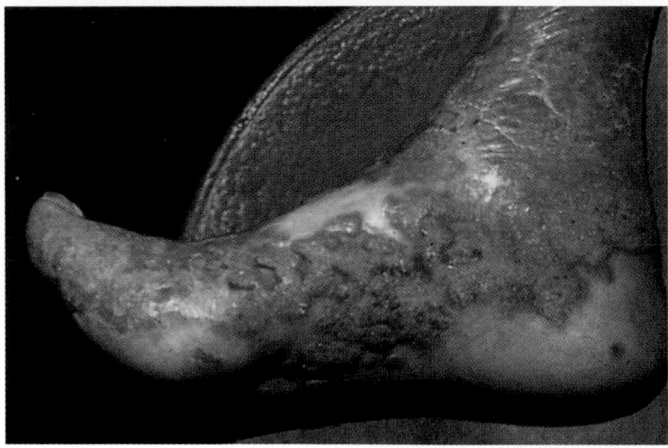

Figure 39.8 **Subcutaneous Mycosis.** Chromomycosis of the foot caused by *Fonsecaea pedrosoi.*

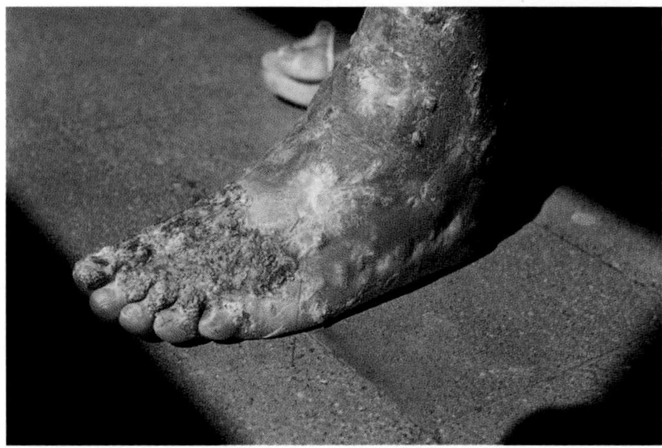

Figure 39.9 **Subcutaneous Mycosis.** Mycotic mycetoma of the foot caused by *Madurella mycetomatis.*

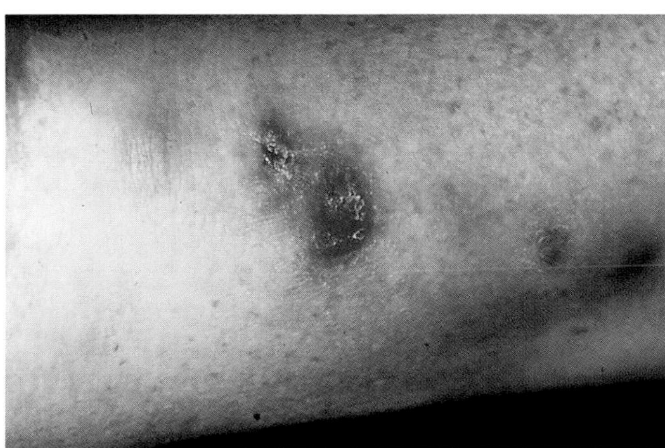

Figure 39.10 **Subcutaneous Mycosis.** Sporotrichosis of the arm caused by *Sporothrix schenckii.*

Systemic Mycoses

Except for *Cryptococcus neoformans,* which has only a yeast form, the fungi that cause the systemic or deep mycoses are dimorphic; that is, they exhibit a parasitic yeastlike phase (Y) and a saprophytic mold or mycelial phase (M). (*See the YM shift, chapter 26.*) Most systemic mycoses are acquired by the inhalation of spores from soil in which free-living fungi reside. If a person inhales enough spores, an infection begins as a lung lesion, becomes chronic, and spreads through the bloodstream to other organs (the target organ varies with the species).

Blastomycosis is the systemic mycosis caused by *Blastomyces dermatitidis,* a fungus that grows as a budding yeast in humans but as a mold on culture media and in the environment. It is found predominately in the soil of the Mississippi and Ohio River basins. The disease occurs in three clinical forms: cutaneous, pulmonary, and disseminated. The initial infection begins when blastospores (*see figure 26.7*f) are inhaled into the lungs. The fungus can then spread rapidly, especially to the skin, where cutaneous ulcers and abscess formation occur (figure 39.11). *B. dermatitidis* can be isolated from pus and biopsy sections. Diagnosis requires the demonstration of thick-walled, yeastlike cells 8 to 15 μm in diameter. Complement-fixation, immunodiffusion, and skin (blastomycin) tests are also useful. Amphotericin B, 2-hydroxystilbamidine, or imidazole derivatives are the drugs of choice for treatment. Surgery may be necessary for the

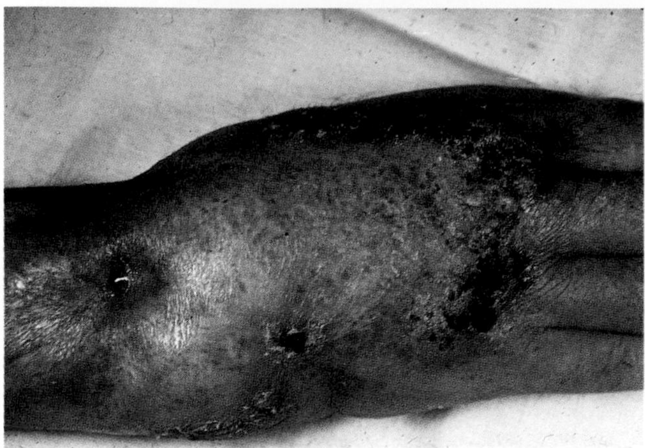

Figure 39.11 **Systemic Mycosis.** Blastomycosis of the forearm caused by *Blastomyces dermatitidis.*

drainage of large abscesses. Approximately 30 to 60 deaths are reported each year in the United States. There are no preventive or control measures.

Coccidioidomycosis, also known as valley fever, San Joaquin fever, or desert rheumatism because of the geographical distribution of the fungus, is caused by *Coccidioides immitis. C. immitis* exists in the dry, highly alkaline soils of North, Central, and South America. It has been estimated that in the United States about 100,000 people are infected annually with 50 to 100 deaths. Endemic areas have been defined by massive skin testing with the antigen coccidioidin (*see chapter 33*). In the soil and on culture media, this fungus grows as a mold that forms arthroconidia (*see figure 26.7*b) at the tips of hyphae. Arthroconidia are so abundant in these endemic areas that by simply moving through such an area, one can acquire the disease by inhaling them. In humans, the fungus grows as a yeast-forming, thick-walled spherule filled with endospores (figure 39.12). Most cases of coccidioidomycosis are asymptomatic or indistinguishable from ordinary upper respiratory infections. Almost all cases resolve themselves in a few weeks, and a lasting immunity results. A few infections result in a progressive chronic pulmonary disease. The fungus also can spread throughout the body, involving almost any organ or site. Diagnosis is accomplished by aspiration and identification of the large spherules (approximately 80 μm in diameter) in pus, sputum, and aspirates. Culturing clinical samples in the presence of penicillin and streptomycin on Sabouraud agar is also diagnostic. Newer methods of rapid confirmation include the testing of supernatants of liquid media cultures for antigens, serology, and skin testing. Miconazole, ketoconazole, and amphotericin B are the drugs of choice for treatment. Prevention involves reducing exposure to dust (soil) in endemic areas.

Cryptococcosis is a systemic mycosis caused by *Cryptococcus neoformans.* This fungus always grows as a large budding yeast. In the environment, *C. neoformans* is a saprophyte

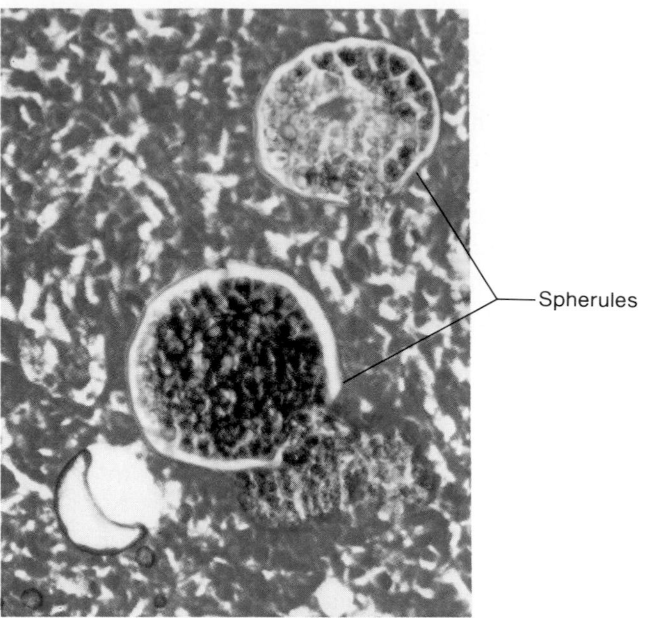

Figure 39.12 **Systemic Mycosis: Coccidioidomycosis.** *Coccidioides immitis* mature spherules filled with endospores within a tissue section; light micrograph (×400).

with a worldwide distribution. Aged, dried pigeon droppings are an apparent source of infection. Cryptococcosis is found in approximately 15% of AIDS patients (*see table 36.3*). The fungus enters the body by the respiratory tract, causing a minor pulmonary infection that is usually transitory. Some pulmonary infections spread to the skin, bones, viscera, and central nervous system. Once the nervous system is involved, meningitis usually results (*see table 37.1 and figure 36.11*d). Diagnosis is accomplished by detection of the thick-walled spherical yeast cells (figure 39.13) in pus, sputum, or exudate smears using India ink to define the organism. The fungus can be easily cultured on Sabouraud dextrose agar. Identification of the fungus in body fluids is made by immunological procedures. Treatment includes a combination of amphotericin B and fluorocytosine. There are no preventive or control measures.

Histoplasmosis is caused by *Histoplasma capsulatum* var. *capsulatum,* a facultative fungus parasite that grows intracellularly. It appears as a small budding yeast in humans and on culture media at 37° C. At 25° C it grows as a mold, producing small microconidia (1 to 5 μm in diameter) that are borne singly at the tips of short conidiophores. Large macroconidia or chlamydospores (8 to 16 μm in diameter) are also formed on conidiophores (figure 39.14*a; see also figure 36.11*e). In humans, the yeastlike form grows within phagocytic cells (figure 39.14*b*). *H. capsulatum* var. *capsulatum* is found as the mycelial form in soils throughout the world and is localized in areas that have been contaminated with bird or bat excrement. The chlamydospores, particularly the micro-

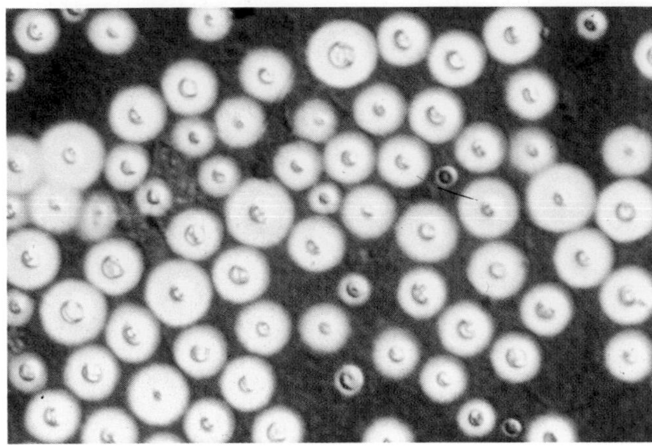

Figure 39.13 **Systemic Mycosis: Cryptococcosis.** India ink preparation showing *Cryptococcus neoformans.* Although these microorganisms are not budding, they can be differentiated from artifacts by their doubly refractile cell walls, distinctly outlined capsules, and refractile inclusions in the cytoplasm; light micrograph (×150).

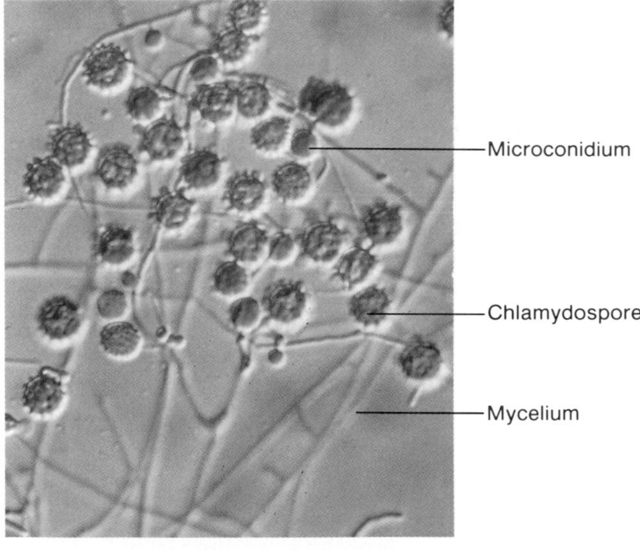

(a)

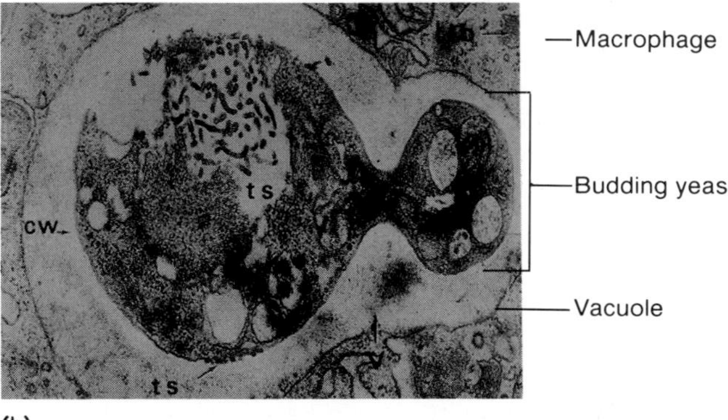

(b)

Figure 39.14 **Morphology of *Histoplasma capsulatum* var. *capsulatum.*** (*a*) Mycelia, microconidia, and chlamydospores as found in the soil. These are the infectious particles; light micrograph (×125). (*b*) Yeastlike cells in a macrophage. Budding *H. capsulatum* within a vacuole. Tubular structures, *ts,* are observed beneath the cell wall, *cw;* electron micrograph (×23,000).

conidia, are easily spread by air currents. Within the United States, histoplasmosis is endemic within the Mississippi, Kentucky, Tennessee, Ohio, and Rio Grande River basins. More than 75% of the people who reside in parts of these areas have antibodies against the fungus. It has been estimated that in endemic areas of the United States, about 500,000 individuals are infected annually; 50,000 to 200,000 become ill; 3,000 require hospitalization; and about 50 die. The total number of infected individuals may be over 40 million in the United States alone. Histoplasmosis is an occupational disease among spelunkers (people who explore caves) and bat guano miners.

Humans acquire histoplasmosis from airborne microconidia that are produced under favorable environmental conditions. Microconidia are most prevalent where bird droppings, or guano—especially from starlings, crows, blackbirds, cowbirds, sea gulls, turkeys, and chickens—have accumulated. It is noteworthy that the birds themselves are not infected because of their high body temperature; their droppings simply provide the nutrients for this fungus. Only bats and humans demonstrate the disease and harbor the fungus.

Histoplasmosis is a disease of the reticuloendothelial system; thus, many organs of the body can be infected (*see figure 30.10*). More than 95% of "histo" cases have either no symptoms or mild symptoms such as coughing, fever, and joint pain. Lesions may appear in the lungs and show calcification; most infections resolve on their own. Only rarely does the disease disseminate.

Laboratory diagnosis is accomplished by complement-fixation tests and isolation of the fungus from tissue specimens. Most individuals with this disease exhibit a hypersensitive state that can be demonstrated by the histoplasmin skin test. Currently, the most effective treatment is with amphotericin B or ketoconazole. Prevention and control involves wearing protective clothing and masks before entering or working in infested habitats. Soil decontamination with 3 to 5% formalin is effective where economically and physically feasible.

Opportunistic Mycoses

An **opportunistic organism** is generally harmless in its normal environment but becomes pathogenic in a compromised host. A **compromised host** is seriously debilitated and has a lowered resistance to infection. There are many causes of this condition, among them the following: malnutrition, alcoholism, cancer, diabetes, leukemia, or another infectious disease; trauma from surgery or injury; an altered microbiota from the prolonged use of antibiotics (e.g., in vaginal candidiasis); and immunosuppression by drugs, viruses (HIV), hormones, or genetic deficiencies. The most important opportunistic mycoses include systemic aspergillosis and candidiasis.

Of all the fungi that cause disease in compromised hosts, none are as widely distributed as the *Aspergillus* species. *Aspergillus* is omnipresent in nature, being found wherever organic debris occurs. *Aspergillus fumigatus* is the usual cause of **aspergillosis.** *A. flavus* is the second most important species, particularly in invasive disease of immunosuppressed patients.

The major portal of entry for *Aspergillus* is the respiratory tract. Inhalation of conidiospores (*see figure 26.7*e) can lead to several types of pulmonary aspergillosis. One type is allergic aspergillosis. Infected individuals may develop an immediate allergic response and suffer typical asthmatic attacks when exposed to fungal antigens on the conidiospores. In bronchopulmonary aspergillosis, the major clinical manifestation of the allergic response is a bronchitis resulting from both type I and type III hypersensitivities (*see figures 32.6 and 32.9*). Although tissue invasion seldom occurs in bronchopulmonary aspergillosis, *Aspergillus* can often be cultured from the sputum. A most common manifestation of pulmonary involvement is the occurrence of colonizing aspergillosis, in which *Aspergillus* forms colonies within the lungs that develop into "fungus balls" called aspergillomas. These consist of a tangled mass of mycelia growing in a circumscribed area. From the pulmonary focus, the fungus may spread, producing disseminated aspergillosis in a variety of tissues and organs (figure 39.15). In patients whose resistance has been severely compromised, invasive aspergillosis may occur and fill the lung with fungal mycelia.

Laboratory diagnosis of aspergillosis depends on identification, either by direct examination of pathological specimens or by isolation and characterization of the fungus. Successful therapy depends on treatment of the underlying disease so that host resistance increases. Unfortunately, *Aspergillus* species are not as susceptible to amphotericin B or imidazole drugs as are most other fungi.

Candidiasis is the mycosis caused by the dimorphic fungus *Candida albicans* (figure 39.16*a*). In contrast to the other pathogenic fungi, *C. albicans* is a member of the normal microbiota within the gastrointestinal tract, respiratory tract, vaginal area, and mouth (*see figures 29.8 and 36.11*f). In healthy individuals, *C. albicans* does not produce disease. Growth is suppressed by other microbiota. However, if anything upsets the normal microbiota, *Candida* may multiply rapidly and produce candidiasis. Recently, *Candida* species have become important nosocomial pathogens. In some hospitals, they may represent almost 10% of nosocomial bloodstream infections.

No other mycotic pathogen produces as diverse a spectrum of disease in humans as does *C. albicans* (Box 39.1). Most infections involve the skin or mucous membranes. This occurs because *C. albicans* is a strict aerobe and finds such surfaces very suitable for growth. Cutaneous involvement usually occurs when the skin becomes overtly moist or damaged.

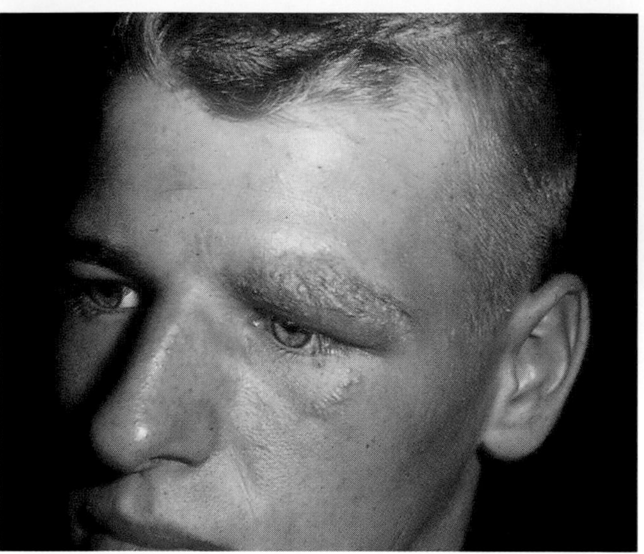

Figure 39.15 **An Opportunistic Mycosis.** Aspergillosis of the eye caused by *Aspergillus fumigatus.*

Oral candidiasis, or **thrush** (figure 39.16*b*), is a fairly common disease in newborns. It is seen as many small white flecks that cover the tongue and mouth. At birth, newborns do not have a normal microbiota in the oropharyngeal area. If the mother's vaginal area is heavily infected with *C. albicans,* the upper respiratory tract of the newborn becomes colonized during passage through the birth canal. Thrush occurs because growth of *C. albicans* cannot be inhibited by the other microbiota. Once the newborn has developed its own normal oropharyngeal microbiota, thrush becomes uncommon. **Paronychia** and **onychomycosis** are associated with *Candida* infections of the subcutaneous tissues of the digits and nails, respectively (figure 39.16*c*). These infections usually result from continued immersion of the appendages in water.

Intertriginous candidiasis involves those areas of the body, usually opposed skin surfaces, that are warm and moist: axillae, groin, skin folds. **Napkin (diaper) candidiasis** is typically found in infants whose diapers are not changed frequently and therefore are not kept dry. **Candidal vaginitis** can result from diabetes, antibiotic therapy, oral contraceptives, pregnancy, or any other factor that compromises the female host. Normally, the omnipresent Lactobacilli (Döderlein's bacilli) can control *Candida* in this area by the low pH the bacilli create. However, if their numbers are decreased by any of the aforementioned factors, *Candida* may proliferate, causing a curdlike, yellow-white discharge from the vaginal area. *Candida* can be transmitted to males during intercourse and lead to **balanitis;** thus, it also can be considered a sexually transmitted disease. Balanitis is a *Candida* infection of the male glans penis and occurs primarily in uncircumcised males. The disease begins as vesicles on the penis that develop into patches and are accompanied by severe itching and burning.

Lactobacilli or Döderlein's bacilli (p. 575).

Box 39.1

The Emergence of Candidiasis

Written descriptions of oral lesions that were probably thrush date back to the early 1800s. In 1839, Bernard Langenbeck in Germany described the organism he found in oral lesions of a patient as a "Typhus-Leichen." By 1841, Emil Berg established the fungal etiology of thrush by infecting healthy babies with what he called "aphthous membrane material." In 1843, Charles Robin gave the organism its first name: *Oidium albicans.* Since then, more than a hundred synonyms have been used for this fungus; of them all, *Candida albicans,* the name Roth Berkhout proposed in 1923, has persisted.

In 1861, Albert Zenker described the first well-documented case of systemic candidiasis. Historically, the most interesting period for candidiasis coincided with the introduction of antibiotics. Since then, there have been documented cases of this fungus involving all tissues and organs of the body, as well as an increase in the overall incidence of candidiasis. Some *Candida*-associated infections and diseases include arthritis, endophthalmitis, meningitis, myocarditis, myositis, and peritonitis. Besides the widespread use of antibiotics, other therapeutic modalities and surgical procedures such as organ transplants and prostheses have been important in the expanding worldwide incidence of candidiasis.

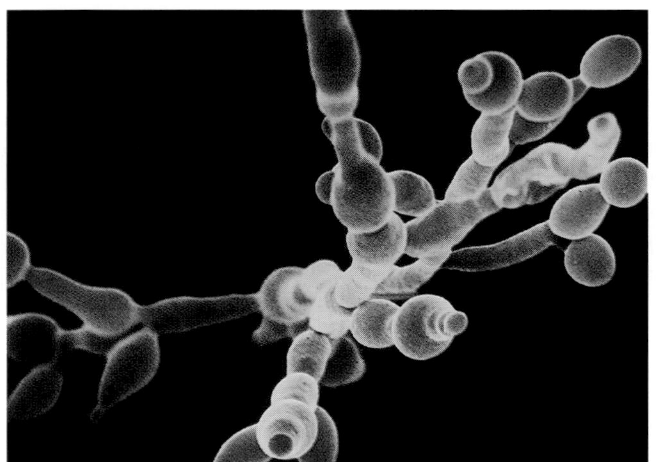

(a)

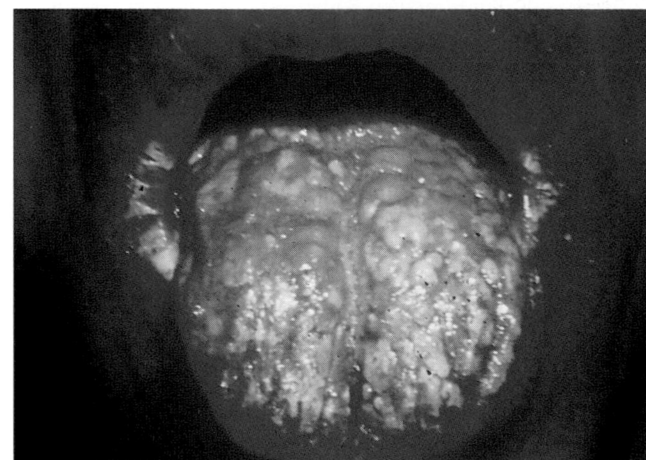

(b)

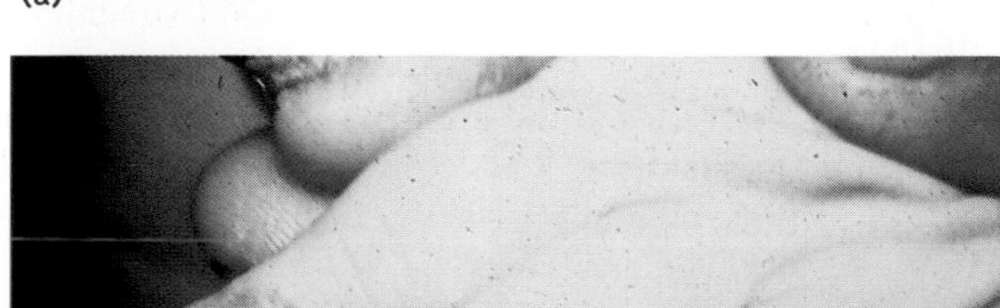

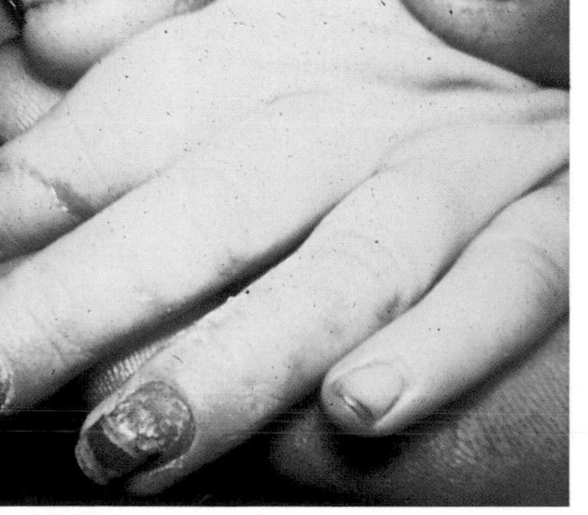

(c)

Figure 39.16 Opportunistic Mycoses Caused by *Candida albicans.* (*a*) Scanning electron micrograph of the yeast form (×10,000). Notice that some of the cells are reproducing by budding. (*b*) Thrush, or oral candidiasis, is characterized by the formation of white patches on the mucous membranes of the tongue and elsewhere in the oropharyngeal area. These patches form a pseudomembrane composed of spherical yeast cells, leukocytes, and cellular debris. (*c*) Paronychia and onychomycosis of the hands.

TABLE 39.1 Examples of Medically Important Protozoa

Phylum	Group	Pathogen	Disease
Sarcomastigophora	Amoebae	*Entamoeba histolytica*	Amebiasis, amebic dysentery
		Acanthamoeba spp., *Naegleria fowleri*	Amebic meningoencephalitis
Apicomplexa	Coccidia	*Cryptosporidium* spp.	Cryptosporidiosis
Ciliophora	Ciliates	*Balantidium coli*	Balantidiasis
Sarcomastigophora	Blood and tissue flagellates	*Leishmania tropica*	Cutaneous leishmaniasis
		L. braziliensis	Mucocutaneous leishmaniasis
		L. donovani	Kala-azar (visceral leishmaniasis)
		Trypanosoma cruzi	American trypanosomiasis
		T. brucei gambiense, T. brucei rhodesiense	African sleeping sickness
Sarcomastigophora	Digestive and genital organ flagellates	*Giardia lamblia*	Giardiasis
		Trichomonas vaginalis	Trichomoniasis
Apicomplexa	Sporozoa	*Plasmodium falciparum, P. malariae, P. ovale, P. vivax*	Malaria
		Pneumocystis carinii	Pneumocystis pneumonia
		Toxoplasma gondii	Toxoplasmosis

Diagnosis of candidiasis is difficult because (1) this fungus is a frequent secondary invader in diseased hosts, (2) a mixed microbiota is most often found in the diseased tissue, and (3) no completely specific immunologic procedures for the identification of *Candida* currently exist. There is no satisfactory treatment for candidiasis. Cutaneous lesions can be treated with topical agents such as sodium caprylate, sodium propionate, gentian violet, nystatin, miconazole, and Trichomycin. Ketoconazole, amphotericin B, fluconazole, and flucytosine also can be used for systemic candidiasis.

Nystatin (p. 340).

1. How are human fungal diseases categorized?
2. What are three types of piedras that infect humans?
3. Briefly describe the six tineas that occur in humans.
4. Describe the three types of subcutaneous mycoses that affect humans.
5. Why is *Histoplasma capsulatum* found in bird feces but not within the birds themselves?
6. Why are some mycotic diseases of humans called opportunistic mycoses?
7. What parts of the human body can be affected by *Candida* infections?

Protozoan Diseases

Protozoa have become adapted to practically every type of habitat on the face of the earth, including the human body. Though fewer than 20 genera of protozoa cause disease in humans (table 39.1), their impact is formidable. For example, there are over 150 million cases of malaria in the world each year. In tropical Africa alone, malaria is responsible each year for the deaths of more than a million children under the age of 14. It is estimated that there are at least 8 million cases of trypanosomiasis, 12 million cases of leishmaniasis, and over 500 million cases of amebiasis yearly. *Pneumocystis carinii* has suddenly emerged as a major cause of death in AIDS patients. The remainder of this chapter discusses some of these protozoan diseases of humans.

Amebiasis

Entamoeba histolytica is the protozoan responsible for **amebiasis (amebic dysentery).** This very common parasite is endemic in warm climates where adequate sanitation and effective personal hygiene is lacking. Within the United States, about 3,000 to 5,000 cases are reported annually. However, it is a major cause of parasitic death worldwide; about 500 million people are infected and as many as 100,000 die of amebiasis each year.

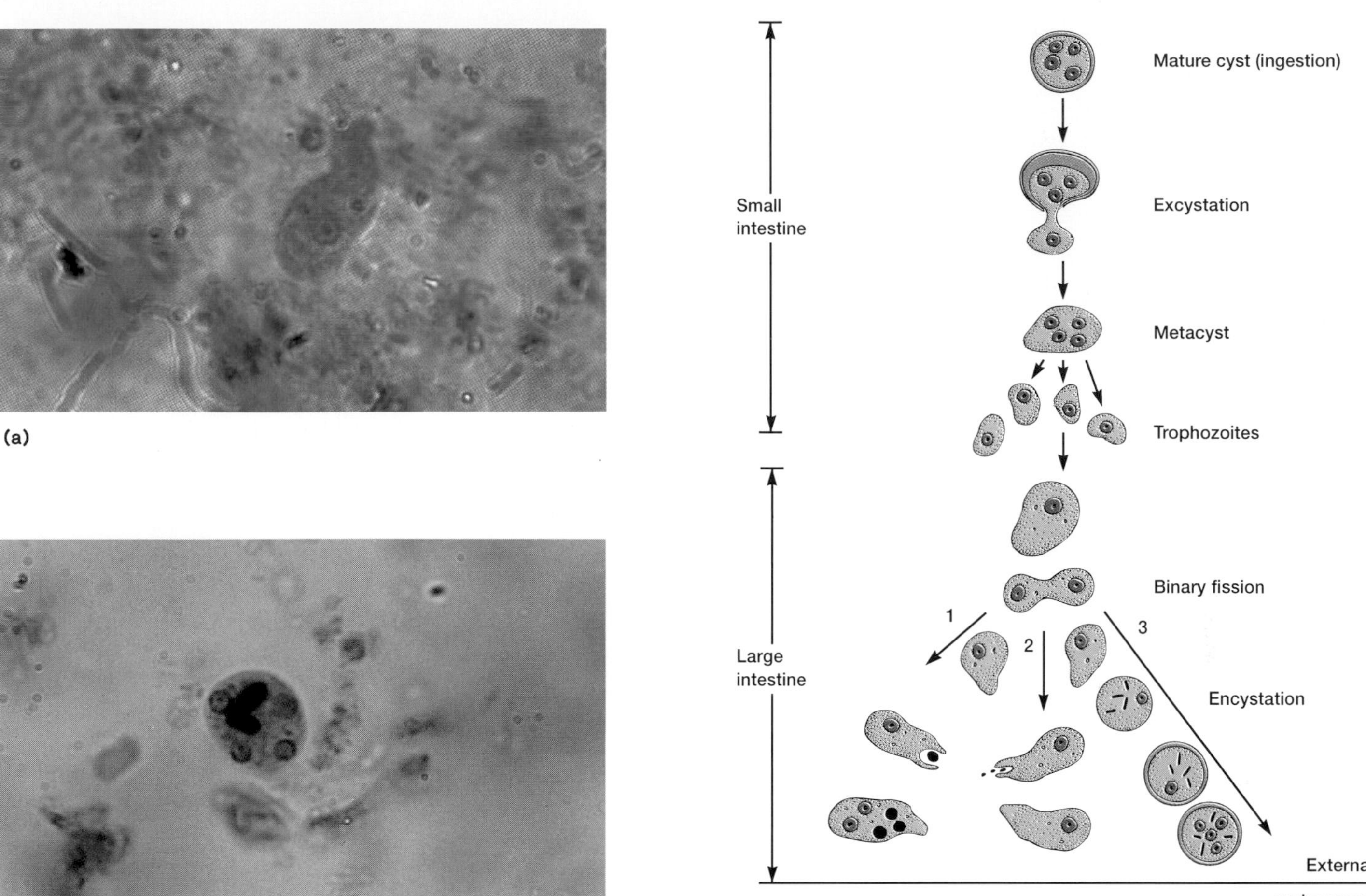

Figure 39.17 Amebiasis Caused by *Entamoeba histolytica.* (*a*) Light micrographs of a trophozoites (×1,000) and (*b*) a cyst (×1,000). (*c*) Life cycle. Infection occurs by the ingestion of a mature cyst of the parasite. Excystation occurs in the lower region of the small intestine and the metacyst rapidly divides to give rise to eight small trophozoites (only four are shown). These enter the large intestine, undergo binary fission, and may (1) invade the host tissues, (2) live in the lumen of the large intestine without invasion, or (3) undergo encystation and pass out of the host in the feces.

Infection occurs by ingestion of mature cysts. After excystation in the lower region of the small intestine, the metacyst divides rapidly to produce eight small trophozoites (figure 39.17). These trophozoites move to the large intestine where they can invade the host tissue, live as commensals in the lumen of the intestine, or undergo encystation.

If the infective trophozoites invade the intestinal tissues, they multiply rapidly and spread laterally, while feeding on erythrocytes, bacteria, and yeasts. The invading trophozoites destroy the epithelial lining of the large intestine by producing proteolytic enzymes. Lesions (ulcers) are characterized by minute points of entry into the mucosa, colonization of the mucosal layer, and extensive enlargement of the lesion after penetration into the submucosa. *E. histolytica* also may invade and produce lesions in extraintestinal foci, especially the liver, to cause hepatic amebiasis. However, all extraintestinal amebic lesions are secondary to the ones established in the large intestine.

The symptoms of amebiasis are highly variable, ranging from an asymptomatic infection to fulminating dysentery, exhaustive diarrhea accompanied by blood and mucus, appendicitis, and abscesses in the liver, lungs, or brain.

Laboratory diagnosis of amebiasis is based upon finding trophozoites in fresh warm stools and cysts in ordinary stools. Serological testing also should be done. The therapy for amebiasis is complex and depends on the location of the infection

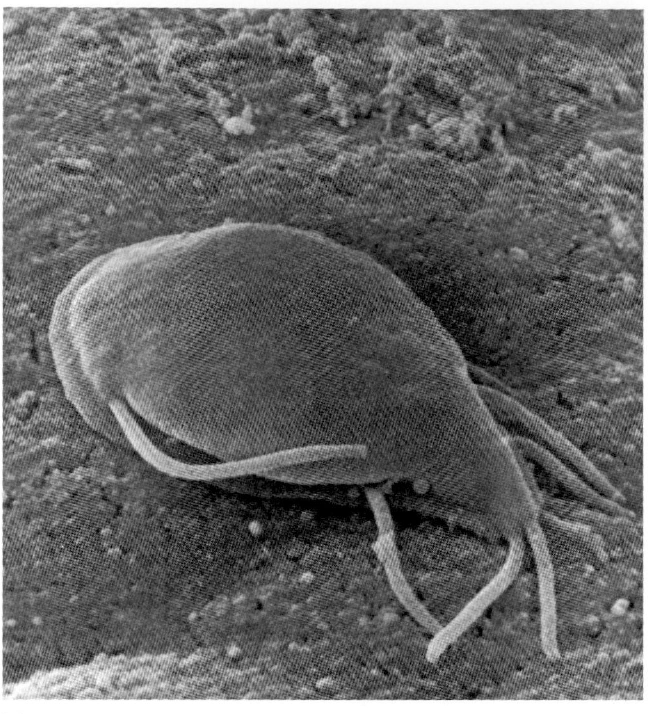
(a)

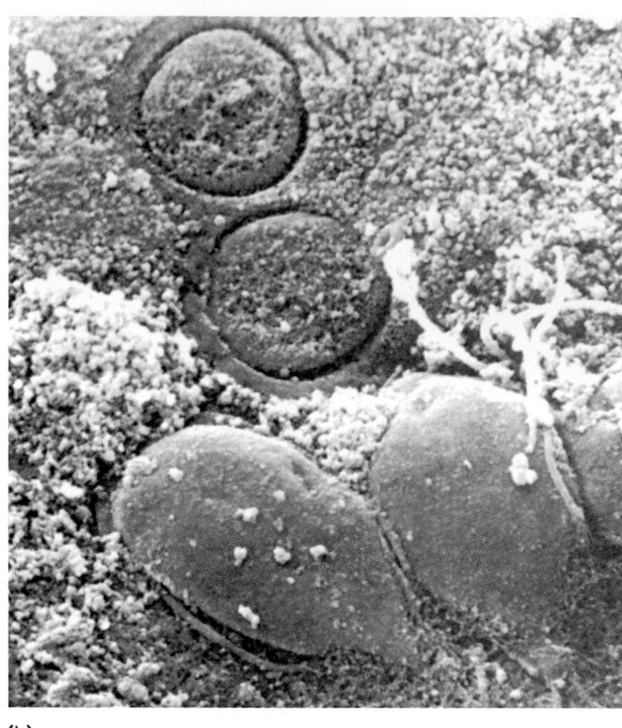
(b)

Figure 39.18 **Giardiasis.** (*a*) *Giardia lamblia* adhering to the epithelium by its sucking disk; scanning electron micrograph. (*b*) Upon detachment from the epithelium, the protozoa often leave clear impressions on the microvillus surface (upper circles); scanning electron micrograph.

within the host and the host's condition. Asymptomatic cyst passers should always be treated because they represent the most important reservoir of the parasite in the population. Amebaquin (diiodohydroxyquin) is the drug of choice for carriers. In symptomatic intestinal amebiasis, metronidazole (Flagyl) or Aralen phosphate are the drugs of choice. Prevention and control of amebiasis is achieved by avoiding water or food that might be contaminated with human feces in endemic areas. Chlorination or freezing of water does not destroy *E. histolytica* cysts.

Freshwater Amoeba Diseases

Free-living amoebae of the genera *Naegleria* and *Acanthamoeba* are facultative parasites responsible for causing **primary amebic meningoencephalitis** (*see table 37.1*) in humans. They are among the most common protozoa found in fresh water and moist soil. In addition, several *Acanthamoeba* spp. are known to infect the eye, causing a chronically progressive ulcerative *Acanthamoeba* **keratitis,** inflammation of the cornea, which may result in blindness. Wearers of soft contact lenses may be predisposed to this infection and should take care to prevent contamination of their lens cleaning and soaking solutions. Diagnosis of these infections is by demonstration of the amoebae in clinical specimens. Most freshwater amoebae are resistant to commonly used antibiotic agents. These amoebae are involved in fewer than 100 human disease cases annually in the United States.

Giardiasis

Giardia lamblia (syn., *G. duodenalis, G. intestinalis*) is a flagellated protozoan (figure 39.18*a*) that causes the very common intestinal disease **giardiasis.** (It was discovered by van Leeuwenhoek when he examined his own stools.) *G. lamblia* is worldwide in distribution, and it affects children more seriously than it does adults. In the United States, this protozoan is the most common cause of epidemic waterborne diarrheal disease (about 30,000 cases yearly). Approximately 7% of the population are healthy carriers and shed cysts in their feces. *G. lamblia* is endemic in child day-care centers in the United States, with estimates of 5 to 15% of diapered children being infected.

Transmission is most frequent with cyst-contaminated water supplies. Epidemic outbreaks have been recorded in wilderness areas, suggesting that humans may be infected from "clean water" with *Giardia* harbored by rodents, deer, cattle, or household pets. This implies that human infections also can be a zoonosis (*see table 35.1*). As many as 200 million humans may be infected worldwide.

Following ingestion, the cysts undergo excystation in the duodenum, forming trophozoites. The trophozoites inhabit the upper portions of the small intestine, where they attach to the intestinal mucosa by means of their sucking disks (figure 39.18*a*). The ability of the trophozoites to adhere to the intestinal epithelium accounts for the fact that they are rarely found in stools. It is thought that the trophozoites feed on

Box 39.2

A Brief History of Malaria

No other single infectious disease has had the impact on humans that malaria has. The first references to its periodic fever and chills can be found in early Chaldean, Chinese, and Hindu writings. In the late fifth century B.C., Hippocrates described certain aspects of malaria. In the fourth century B.C., the Greeks noted an association between individuals exposed to swamp environments and the subsequent development of periodic fever and enlargement of the spleen (splenomegaly). In the seventeenth century, the Italians named the disease *mal' aria* (bad air) because of its association with the ill-smelling vapors from the swamps near Rome. At about the same time, the bark of the quina-quina (cinchoma) tree of South America was used to treat the intermittent fevers, although it was not until the mid-nineteenth century that quinine was identified as the active alkaloid. The major epidemiological breakthrough came in 1880, when French army surgeon Charles Louis Alphonse Laveran observed exflagellated gametocytes in fresh blood. Five years later, the Italian histologist Camillo Golgi observed the multiplication of the asexual blood forms. In the late 1890s, Patrick Manson postulated that malaria was transmitted by mosquitoes. Sir Ronald Ross, a British army surgeon in the Indian Medical Service, subsequently observed developing plasmodia in the intestine of mosquitoes, supporting Manson's theory. Using birds as experimental models, Ross definitively established the major features of the life cycle of *Plasmodium* and received the Nobel Prize in 1902.

Human malaria is known to have contributed to the fall of the ancient Greek and Roman empires. Troops in both the United States Civil War and the Spanish-American War were severely incapacitated by the disease. More than 25% of all hospital admissions during these wars were malaria patients. During World War II, malaria epidemics severely threatened both the Japanese and Allied forces in the Pacific. The same can be said for the military conflicts in Korea and Vietnam.

In the twentieth century, efforts have been directed toward understanding the biochemistry and physiology of malaria, controlling the mosquito vector, and developing antimalarial drugs. In the 1960s, it was demonstrated that resistance to *P. falciparum* among West Africans was associated with the presence of hemoglobin-S in their erythrocytes. Hb-S differs from normal hemoglobin-A with a single amino acid, valine, in each half of the Hb molecule. Consequently, these erythrocytes—responsible for sickle cell disease—have a low binding capacity for oxygen. Because the malarial parasite has a very active aerobic metabolism, it cannot grow and reproduce within these erythrocytes.

In 1955, the World Health Organization began a worldwide malarial eradication program that finally collapsed by 1976. Among the major reasons for failure were the development of resistance to DDT by the mosquito vectors and the development of resistance to chloroquine by strains of *Plasmodium.* Currently, scientists are directing their efforts at new epidemiological approaches, such as the development of vaccines and more potent drugs. For example, in 1984, the gene encoding the sporozoite antigen was cloned, permitting the antigen to be mass-produced by genetic engineering techniques. Overall, no greater achievement for molecular biology could be imagined than the control of malaria—a disease that has caused untold misery throughout the world since antiquity and remains one of the world's most serious infectious diseases.

mucous secretions and reproduce to form such a large population that they interfere with nutrient absorption by the intestinal epithelium.

Giardiasis varies in severity, and asymptomatic carriers are common. The disease can be acute or chronic. Acute giardiasis is characterized by severe diarrhea, epigastric pain, cramps, voluminous flatulence ("passing gas"), and anorexia. Chronic giardiasis is characterized by intermittent diarrhea, with periodic appearance and remission of symptoms. A protein like the cholera toxin has been isolated from *G. lamblia* and is responsible for the diarrhea.

Laboratory diagnosis is based on the identification of trophozoites—only the severest of diarrhea—or cysts in stools. A commercial ELISA test is also available for the detection of *G. lamblia* antigen in stool specimens. Quinacrine hydrochloride (Atabrine) and metronidazole (Flagyl) are the drugs of choice for adults, and furazolidone is used for children because it is available in a pleasant-tasting liquid suspension. Prevention and control involves proper treatment of community water supplies, especially the use of slow sand filtration (*see chapter 41*).

Malaria

The most important human parasite among the sporozoa is *Plasmodium,* the causative agent of **malaria** (Box 39.2). It has been estimated that more than 100 million people are infected, and about one million die annually of malaria in Africa alone. About 1,000 cases are reported each year in the United States.

Human malaria is caused by four species of *Plasmodium: P. falciparum, P. malariae, P. vivax,* and *P. ovale.* The life cycle of *P. vivax* is shown in figure 39.19. The parasite first enters the bloodstream through the bite of an infected female *Anopheles* mosquito. As she feeds, the mosquito injects a small amount of saliva containing an anticoagulant along with small haploid sporozoites (*see figure 28.3*c). The sporozoites in the

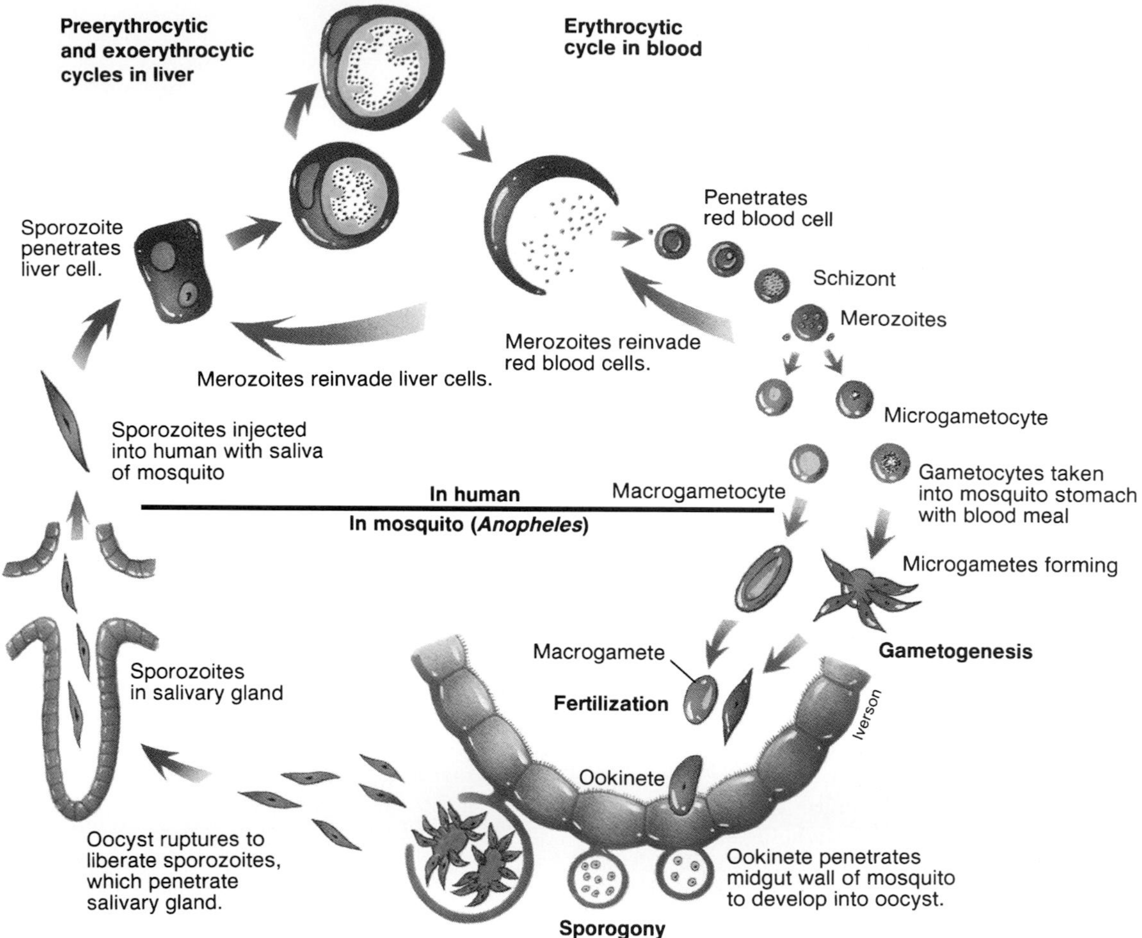

Figure 39.19 **Malaria.** Life cycle of *Plasmodium vivax.*

bloodstream immediately enter hepatic cells of the liver. In the liver, they undergo multiple asexual fission (schizogony) and produce merozoites. After being released from the liver cells, the merozoites either infect other liver cells, thus continuing the preerythrocytic stage, or attach to erythrocytes and penetrate these cells.

Once inside the erythrocyte, the *Plasmodium* begins to enlarge as a uninucleate cell termed a trophozoite. The trophozoite's nucleus then divides asexually to produce a schizont that has 6 to 24 nuclei. The schizont divides and produces mononucleated merozoites. Eventually, the erythrocyte lyses, releasing the merozoites into the bloodstream to infect other erythrocytes. This erythrocytic stage is cyclic and repeats itself approximately every 48 to 72 hours or longer, depending on the species of *Plasmodium* involved. This sudden release of merozoites, toxins, and erythrocyte debris triggers an attack of the chills and fever, so characteristic of malaria. Occasionally, merozoites differentiate into macrogametocytes and microgametocytes, which do not rupture the erythrocyte. When these are ingested by a mosquito, they develop into female and male gametes, respectively. In the mosquito's gut, the infected erythrocytes lyse and gametocytes fuse to form a diploid zygote called the ookinete. The ookinete migrates to the mosquito's gut wall, penetrates, and forms an oocyst on its outer surface. In a process called sporogony, the oocyst undergoes meiosis and forms sporozoites that migrate to the salivary glands of the mosquito. The cycle is now complete, but when the mosquito bites another human host, the cycle begins anew.

The pathological changes caused by malaria involve not only the erythrocytes but also the spleen and other visceral organs. Classic symptoms first develop with the synchronized release of merozoites and erythrocyte debris into the bloodstream, resulting in the malarial paroxysms—shaking chills, then burning fever followed by sweating. Several of these paroxysms constitute an attack. After one attack, there is a remission that lasts from a few weeks to several months. Then there is a relapse. Between paroxysms, the patient feels normal. Anemia can result from the loss of erythrocytes, and the spleen and liver often hypertrophy.

Diagnosis of malaria is made by demonstrating the presence of parasites within Wright- or Giemsa-stained erythrocytes (figure 39.20). When blood smears are negative, serological testing can establish a diagnosis of malaria in individuals. Treatment includes administration of chloroquine,

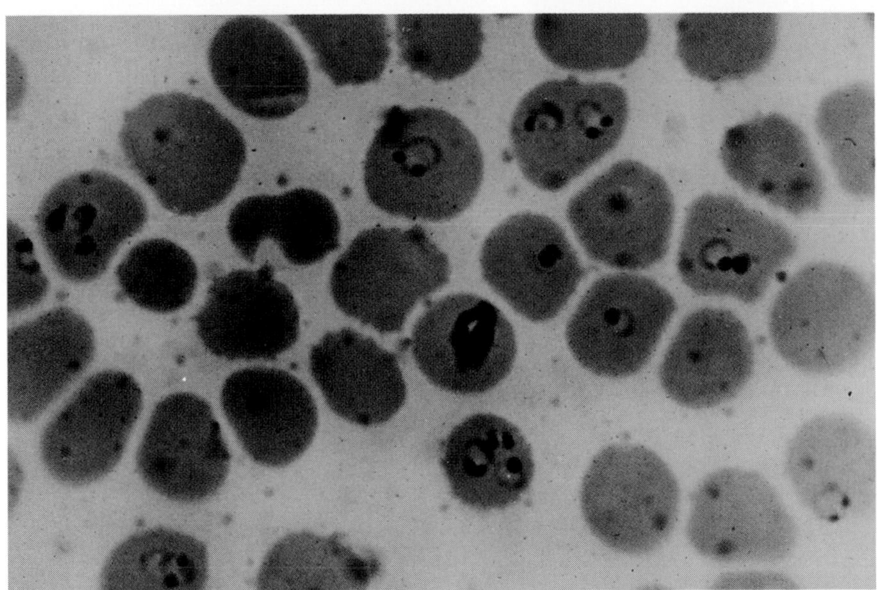

Figure 39.20 Malaria: Erythrocytic Cycle. Trophozoites of *P. falciparum* in circulating erythrocytes; light micrograph (×1,100). The young trophozoites resemble small rings resting in the erythrocyte cytoplasm.

amodiaquine, or mefloquine. These suppressant drugs are effective in eradicating erythrocytic asexual stages. Primaquine has proved satisfactory in eradicating the exoerythrocytic stages. However, because resistance to these drugs is occurring rapidly, more expensive drug combinations are being used. One example is Fansidar, a combination of pyrimethamine and sulfadoxine. It is worth noting that individuals who are traveling to areas where malaria is endemic (figure 39.21) should receive chemoprophylactic treatment with chloroquine. Efforts to develop a vaccine are underway.

Pneumocystis carinii Pneumonia

Pneumocystis carinii is probably a sporozoan parasite (its taxonomy is still under investigation) and is found in the lungs of many mammals, including humans. Although it has a global distribution, its life cycle is not well known. Recent hospital outbreaks suggest that the parasite is transmitted by direct contact between humans through inhalation of infectious material.

The disease that this parasite causes, ***Pneumocystis carinii* pneumonia, (PCP),** occurs almost exclusively in immunocompromised hosts. Extensive use of immunosuppressive drugs and irradiation for the treatment of cancers and following organ transplants accounts for the formidable prevalence rates noted recently. This pneumonia also occurs in more than 80% of AIDS patients. Both the organism and the disease remain localized in the lungs—even in fatal cases. Within the lungs, *Pneumocystis* causes the aveoli to fill with a frothy exudate.

Laboratory diagnosis of pneumocystis pneumonia can be made definitively only by demonstrating the presence of the microorganisms in infected lung material (*see figure 36.11*a). Treatment is by means of oxygen therapy and either a combination of trimethoprim and sulfamethoxazole, or inhalable pentamidine. Prevention and control is through prophylaxis with drugs in susceptible persons.

Hemoflagellate Diseases

The flagellated protozoa that are transmitted by the bites of infected arthropods and infect the blood and tissues of humans are called **hemoflagellates.** There are two major groups of pathogens: the leishmanias and the trypanosomes.

Leishmaniasis

Leishmanias are flagellated protozoa that cause a group of human diseases collectively called **leishmaniasis.** The primary reservoirs of these parasites are canines and rodents. All species of *Leishmania* use sand flies as intermediate hosts. The leishmanias are transmitted from animals to humans or between humans by these sand flies. When an infected sand fly takes a human blood meal, it introduces promastigotes into the skin of the definitive host. Within the skin, the promastigotes are engulfed by macrophages, multiply by binary fission and form small cells called amastigotes. These destroy the host cell, and are engulfed by other macrophages in which they continue to develop and multiply.

Leishmania braziliensis, which has an extensive distribution in the forest regions of tropical America, causes mucocutaneous leishmaniasis (figure 39.22*a*). The disease produces lesions involving the mouth, nose, throat, and skin and results in extensive scarring and disfigurement.

Leishmania donovani is endemic in large areas within northern China, eastern India, the Mediterranean countries, the Sudan, and Latin America. It produces visceral leishmaniasis (kala-azar). The disease involves the reticuloendothelial system and often results in intermittent fever and enlargement

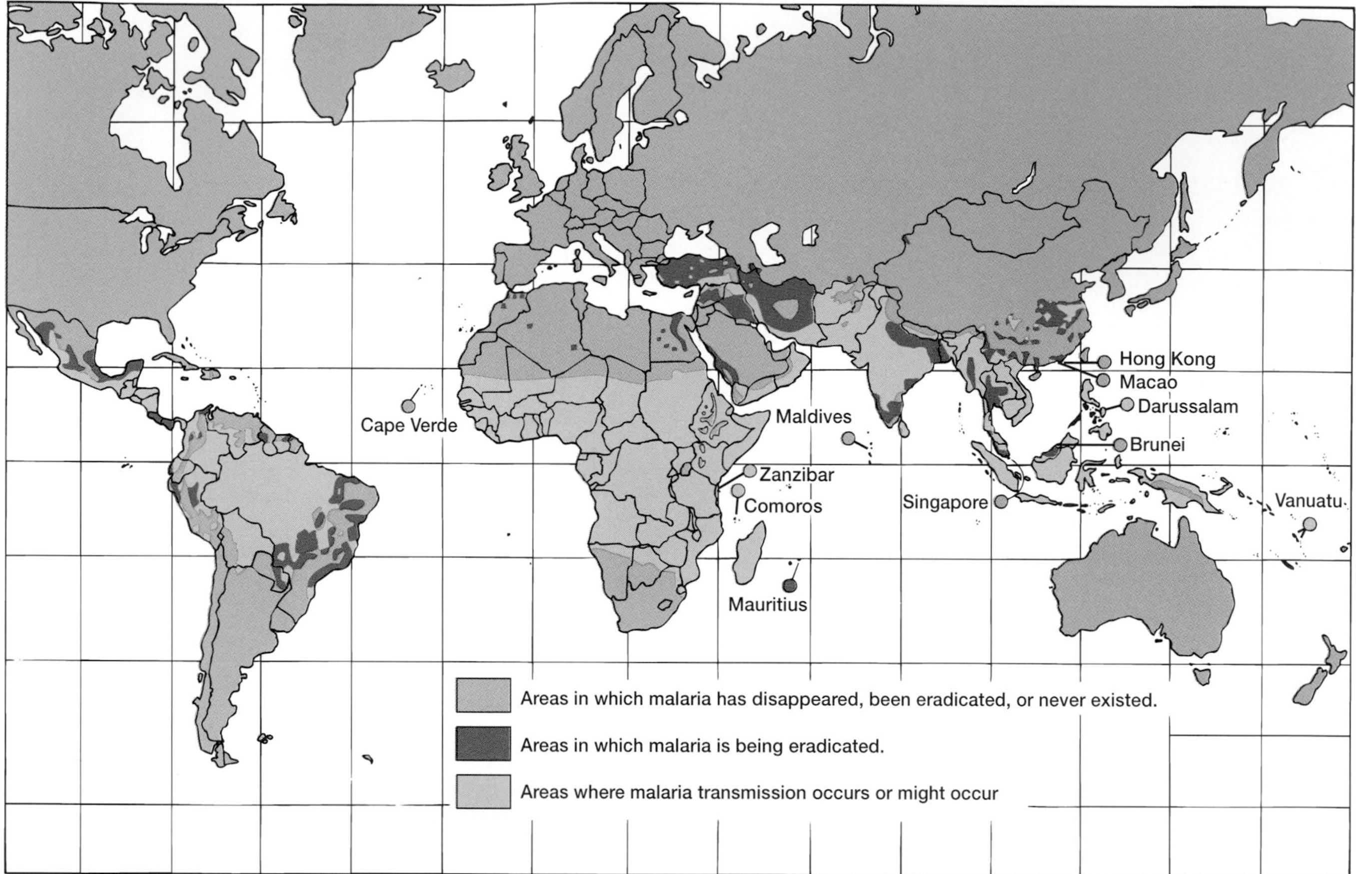

Figure 39.21 Geographic Distribution of Malaria. Notice that malaria is endemic around the equator. From the *World Health Statistics Quarterly* 41:69(1988). Reprinted by permission of the World Health Organization, Switzerland.

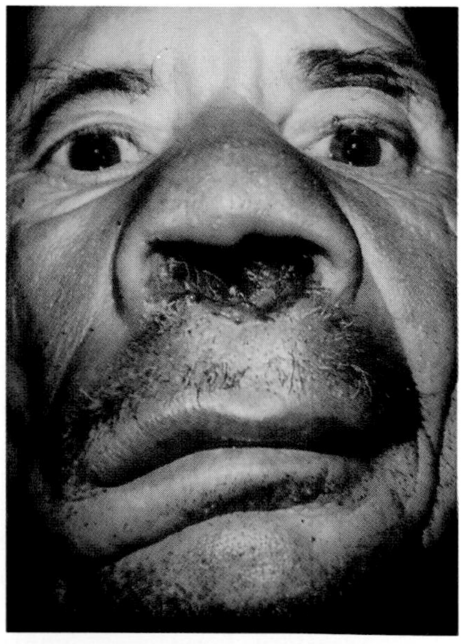
(a)

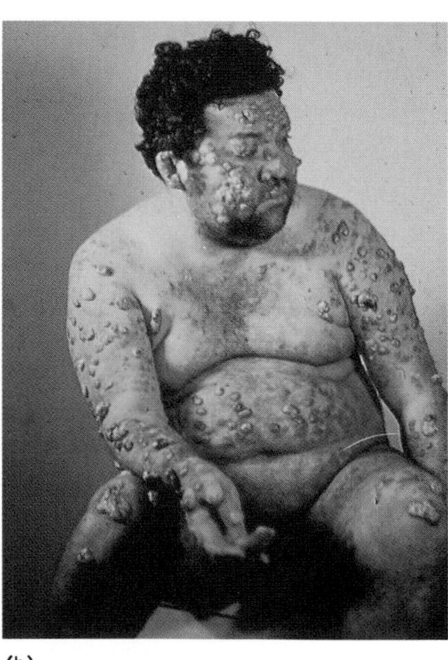
(b)

Figure 39.22 Leishmaniasis. (*a*) A person with mucocutaneous leishmaniasis, which has destroyed the nasal septum and deformed the nose and lips. (*b*) A person with diffuse cutaneous leishmaniasis.

Box 39.3

The Surface Antigens of Parasitic Trypanosomes

Because of the severe debilitation to humans and animals caused by the trypanosomes, a serious effort is now under way by the World Health Organization to produce a vaccine for trypanosome-caused diseases. A major obstacle to such a vaccine is the fact that trypanosomes can change their surface glycoproteins. In the tsetse fly, the trypanosomes are noninfective. Before transfer to a mammalian host, they secrete a thick coat of variant surface glycoprotein (VSG) that protects them from the host's antibodies. Furthermore, they can genetically change this coat into a series of variant antigenic types (VATs) more rapidly than the immune system of the mammalian host can respond.

Each trypanosome possesses genes for 2,000 or more different VSGs in its nucleus. Gene expression occurs in a somewhat predictable order. For example, the expression of one VSG usually occurs after the expression of another specific VSG. As a result, the VSGs in a population of trypanosomes are heterogeneous in any given infection, but there is a single VSG that predominates and against which the host mounts its defensive antibodies. When a new host is infected, the trypanosomes change their repertoire and new VSG variants appear. The mechanism that controls switching from one VAT, in which one VSG is being expressed, to another VAT, with the expression of a different VSG, is being intensely studied. Although the process is not yet completely understood, activation involves gene rearrangement by at least two mechanisms. It is also clear that the switching is independent of the host's immune system. When the tsetse fly takes the trypanosomes up again, the trypanosomes shed their glycoprotein coats and the genetic cycle starts anew.

of the spleen and liver. Individuals who recover develop a permanent immunity.

Leishmania tropica and *L. mexicana* occur in the more arid regions of the Eastern Hemisphere and cause cutaneous leishmaniasis. In this disease, a relatively small red papule forms at the site of each insect bite—the inoculation site. These papules are frequently found on the face and ears. They eventually develop into crustated ulcers (figure 39.22*b*). Healing occurs with scarring and a permanent immunity.

Laboratory diagnosis of leishmaniasis is based on finding the parasite within infected macrophages in stained smears from lesions or infected organs. Treatment includes the pentavalent antimonial compounds (Pentostam, Glucantime).

Trypanosomiasis

Another group of flagellated protozoa called **trypanosomes** (*see figure 28.3*a) cause the aggregate of diseases termed **trypanosomiasis.** *Trypanosoma brucei gambiense,* found in the rain forests of west and central Africa, and *T. brucei rhodesiense,* found in the upland savannas of east Africa, cause African trypanosomiasis. Reservoirs for these trypanosomes are domestic cattle and wild animals, within which the parasites cause severe malnutrition. Both species use tsetse flies as intermediate hosts. The parasites are transmitted through the bite of the fly to humans. Once the protozoa enter the bloodstream, they begin to multiply. They cause interstitial inflammation and necrosis within the lymph nodes and small blood vessels of the brain and heart. In *T. brucei rhodesiense* infection, the disease develops so rapidly that infected individuals often die within a year. In *T. brucei gambiense* infection, the parasites invade the central nervous system, where necrotic damage causes a variety of nervous disorders, including the characteristic sleeping sickness. The name sleeping sickness is derived from the lethargy of the host—characteristically lying prostrate, drooling from the mouth, and insensitive to pain. Usually, the victim dies in 2 to 3 years. Trypanosomiasis is such a problem in parts of Africa that millions of square miles are not fit for human habitation.

T. cruzi causes **American trypanosomiasis (Chagas' disease),** which occurs in the tropics and subtropics of continental America. The parasite uses the triatomid (cone-nosed) bug as a vector. As the triatomid bug takes a blood meal, the parasites are discharged in the insect's feces. Some trypanosomes enter the bloodstream through the wound and invade the liver, spleen, lymph nodes, and central nervous system, where they multiply and destroy the parasitized cells.

Trypanosomiasis is diagnosed by finding motile parasites in fresh blood and by serological testing. Treatment for African trypanosomiasis uses suramin and pentamidine for non-nervous-system involvement and melarsoprol when the nervous system is involved. Currently, there is no drug suitable for Chagas' disease, although nifurtimox (Lampit) has shown some value. Vaccines are not useful because the parasite is able to change its protein coat and evade the immunological response (Box 39.3).

Toxoplasmosis

Toxoplasmosis is a disease caused by the protozoan *Toxoplasma gondii.* This protozoan has been found in cats, dogs, cattle, sheep, and humans. Animals shed cysts in the feces; the cysts enter another host by way of the nose or mouth; and the trophozoites colonize the intestine. Toxoplasmosis also can be transmitted by the ingestion of raw or undercooked meat, congenital transfer, blood transfusion, or a tissue transplant. Originally, toxoplasmosis gained public notice when it was discovered that in pregnant women the protozoan might also infect the fetus, causing serious congenital defects or death.

Most cases of toxoplasmosis are asymptomatic. Adults usually complain of an "infectious mononucleosislike" syndrome. In immunoincompetent or immunosuppressed individuals, it frequently results in fatal disseminated disease with a heavy cerebral involvement.

Acute toxoplasmosis is usually accompanied by lymph node swelling (lymphadenopathy) with reticular cell enlargement or hyperplasia. Pulmonary necrosis, myocarditis, and hepatitis caused by tissue necrosis are common. Retinitis (inflammation of the retina of the eye) is associated with the necrosis due to the proliferation of the parasite within retinal cells. Currently, toxoplasmosis has become a major cause of death in AIDS patients from a unique encephalitis with necrotizing lesions accompanied by inflammatory infiltrates (*see figure 36.11*b).

Laboratory diagnosis of toxoplasmosis is by serological tests. Epidemiologically, toxoplasmosis is ubiquitous in all higher animals. Treatment of toxoplasmosis is with a combination of pyrimethamine (Daraprim) and sulfadiazine. Prevention and control requires minimizing exposure by the following: avoiding eating raw meat, washing hands after working in the soil, cleaning cat litterboxes daily, keeping personal cats indoors if possible, and feeding them commercial food.

Trichomoniasis

Trichomoniasis is a sexually transmitted disease caused by the protozoan flagellate *Trichomonas vaginalis* (figure 39.23). In response to the parasite, the body accumulates leukocytes at the site of the infection. In females, this usually results in a profuse purulent discharge that is yellowish to light cream in color and characterized by a disagreeable odor. The discharge is accompanied by itching. Males are generally asymptomatic because of the trichomonacidal action of prostatic secretions; however, at times a burning sensation occurs during urination. Diagnosis is made in females by microscopic examination of the discharge and identification of the parasite. Infected males will demonstrate the parasite in semen or urine. Treatment is by administration of metronidazole (Flagyl).

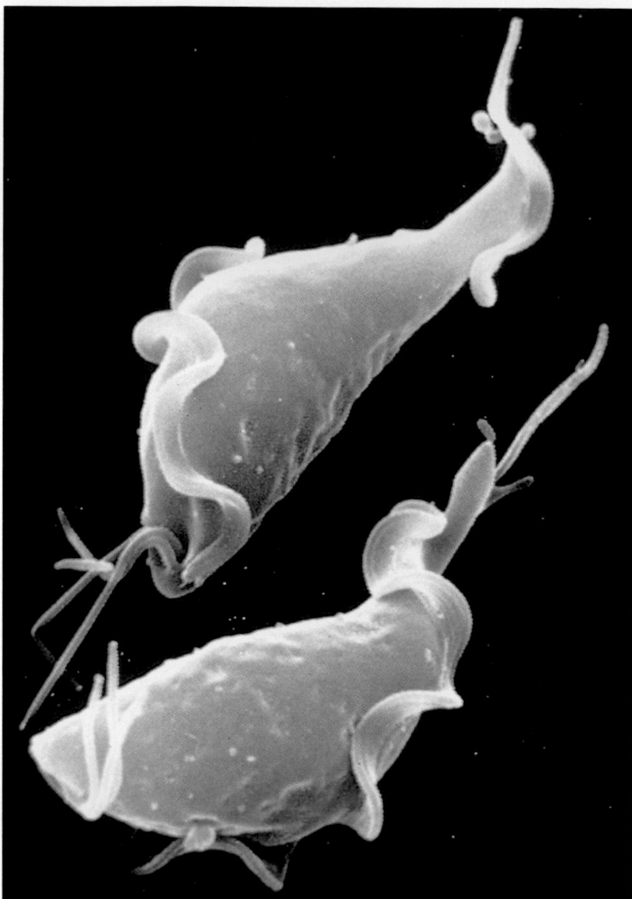

Figure 39.23 **Trichomoniasis.** *Trichomonas vaginalis,* showing the characteristic undulating membranes and flagellae; scanning electron micrograph (×12,000).

1. How do infections caused by *Entamoeba histolytica* occur?
2. What is the most common cause of epidemic waterborne diarrheal disease?
3. Describe in detail the life cycle of the malarial parasite.
4. When is *Pneumocystis carinii* pneumonia likely to occur in humans?
5. What protozoa are represented within the hemoflagellates? What diseases do they cause?
6. In what two ways does *Toxoplasma* affect human health?
7. How would you diagnose trichomoniasis in a female? In a male?

Summary

1. Human fungal diseases, or mycoses, can be divided into five groups according to the level and mode of entry into the host. These are the superficial, cutaneous, subcutaneous, systemic, and opportunistic mycoses.
2. The superficial mycoses are collectively termed piedras. Three types can occur in humans: black piedra (tinea nigra), white piedra (tinea albigena), and tinea versicolor.
3. The cutaneous fungi that parasitize the hair, nails, and outer layer of the skin are called dermatophytes, and their infections are termed dermatophytoses, ringworms, or tineas. Seven types can occur in humans: tinea barbae (ringworm of the beard), tinea capitis (ringworm of the scalp), tinea corporis (ringworm of the body), tinea cruris (ringworm of the groin), tinea pedis (ringworm of the feet), tinea manuum (ringworm of the hands), and tinea unguium (ringworm of the nails).
4. The dermatophytes that cause the subcutaneous mycoses are normal saprophytic inhabitants of soil and decaying vegetation. Three types of subcutaneous mycoses can occur in humans: chromomycosis, maduromycosis, and sporotrichosis.
5. Most systemic mycoses that occur in humans are acquired by inhaling the spores from the soil where the free-living fungi are found. Four types can occur in humans: blastomycosis, coccidioidomycosis, cryptococcosis, and histoplasmosis.
6. An opportunistic organism is one that is generally harmless in its normal environment, but that can become pathogenic in a compromised host. The most important opportunistic mycoses affecting humans include systemic aspergillosis and candidiasis.
7. Protozoa are responsible for some of the most serious human diseases that affect hundreds of millions of people worldwide.
8. *Entamoeba histolytica* is the amoeboid protozoan responsible for amebiasis. This is a very common disease in warm climates throughout the world. It is acquired when one ingests the cysts with contaminated food or water.
9. *Giardia lamblia* is a flagellated protozoan that causes the common intestinal disease giardiasis. This disease is distributed throughout the world, and in the United States it is the most common cause of waterborne diarrheal disease.
10. The most important human parasite among the sporozoa is *Plasmodium*, the causative agent of malaria. Human malaria is caused by four species of *Plasmodium: P. falciparum, P. vivax, P. malariae,* and *P. ovale.*
11. *Pneumocystis carinii* is a sporozoan parasite that can cause pneumonia and death in compromised humans, especially in AIDS patients.
12. The flagellated protozoa that are transmitted by arthropods and infect the blood and tissues of humans are called hemoflagellates. Two major groups occur: the leishmanias, which cause the diseases collectively termed leishmaniasis, and the trypanosomes, which cause trypanosomiasis.
13. Toxoplasmosis is a disease caused by the protozoan *Toxoplasma gondii.* It is one of the major causes of death in AIDS patients.
14. Trichomoniasis is a sexually transmitted disease caused by the protozoan flagellate *Trichomonas vaginalis.*

Key Terms

American trypanosomiasis (Chagas' disease) *797*
amebiasis (amebic dysentery) *790*
aspergillosis *788*
balanitis *788*
black piedra *782*
blastomycosis *785*
candidal vaginitis *788*
candidiasis *788*
chromomycosis *784*
coccidioidomycosis *786*
compromised host *787*
cryptococcosis *786*
dermatophyte *782*
dermatophytosis *782*
extracutaneous sporotrichosis *785*
giardiasis *792*
hemoflagellate *795*
histoplasmosis *786*
intertriginous candidiasis *788*
keratitis *792*
leishmania *795*
leishmaniasis *795*
maduromycosis *785*
malaria *793*
medical mycology *782*
mycetoma *785*
mycosis *782*
napkin (diaper) candidiasis *788*
onychomycosis *788*
opportunistic organism *787*
paronychia *788*
piedra *782*
Pneumocystis carinii pneumonia (PCP) *795*
primary amebic meningoencephalitis *792*
ringworm *782*
sporotrichosis *785*
thrush *788*
tinea *782*
tinea barbae *782*
tinea capitis *783*
tinea corporis *784*
tinea cruris *784*
tinea manuum *784*
tinea pedis *784*
tinea unguium *784*
tinea versicolor *782*
toxoplasmosis *798*
trichomoniasis *798*
trypanosome *797*
trypanosomiasis *797*
white piedra *782*

Questions for Thought and Review

1. Briefly describe each of the major or most common fungal and protozoan diseases in terms of its causative agent, signs and symptoms, the course of infection, mechanism of pathogenesis, epidemiology, and prevention and/or treatment.
2. There are no antibiotics—nothing equivalent to penicillin or streptomycin—to control fungi in the human body. Recalling the mechanisms by which antibiotics affect bacteria, why do you think there has been no similar success in the control of fungal pathogens?
3. Give several reasonable explanations why most fungal diseases in humans are not contagious.
4. What factors determine one's susceptibility to fungal diseases?
5. What is the relationship between AIDS and mycotic and protozoan diseases?
6. Why do most protozoan diseases occur in the tropics?
7. Give several reasons why malaria is still one of the most serious of all infectious diseases that affect humans.
8. What morphological property is the most important in the identification of fungi?
9. Compared to fungal parasites, how are protozoan parasites transmitted?
10. What are dermatomycoses?

Additional Reading

Adam, R. D. 1991. The biology of *Giardia* spp. *Microbiol. Rev.* 55(4):706–32.

Ahearn, D. G. 1978. Medically important yeasts. *Ann. Rev. Microbiol.* 32:59–75.

Ajello, L. 1977. Systemic mycoses in modern medicine. *Contrib. Microbiol. Immunol.* 3:2–12.

Bartlett, M., and Smith, J. 1991. *Pneumocystis carinii,* an opportunist in immunocompromised patients. *Clin. Microbiol. Rev.* 4(2):137–49.

Cabral-Marciano, F. 1988. *Biology* of *Naegleria* spp. *Microbiol. Rev.* 52(1):114–33.

Cheng, T. C. 1986. *General parasitology,* 2d ed. New York: Academic Press.

Cogswell, F. 1992. The hypnozoite and relapse in primate malaria. *Clin. Microbiol. Rev.* 5(1):26–35.

Emmons, C. W.; Chapman, H. B.; Utz, J. P.; and Kwon-Chung, K. J. 1977. *Medical mycology,* 3d ed. Philadelphia: Lea and Febiger.

Fouts, A. C., and Kraus, S. J. 1980. *Trichomonas vaginalis:* Reevaluation of its clinical presentation and laboratory diagnosis. *J. Infect. Dis.* 141:137–41.

Glew, R. H.; Saha, A. K.; Das, S.; and Remaley, A. T. 1988. Biochemistry of the *Leishmania* species. *Microbiol.Rev.* 52(4):412–32.

Goodwin, R. A.; Loyd, J. E.; and Des Prez, R. M. 1981. Histoplasmosis in normal hosts. *Medicine* 60:231–40.

John, D. T. 1982. Primary amoebic meningoencephalitis and the biology of *Naegleria fowleri. Ann. Rev. Microbiol.* 36:101–23.

Joklik, W. K.; Willett, H. P.; Amos, D. B.; and Wilfert, C. M. 1992. *Zinsser microbiology,* 20th ed. Norwalk, Conn.: Appleton & Lange.

Jones, J. M. 1990. Laboratory diagnosis of invasive candidiasis. *Clin. Microbiol. Rev.* 3(1):32–45.

Kretschmer, R. R. 1990. *Amebiasis.* Boca Raton, Fla.: CRC Press.

Markell, E. K. 1986. *Medical parasitology,* 6th ed. Philadelphia: W. B. Saunders.

Mills, J., and Masur, H. 1990. AIDS-related infections. *Sci. Am.* 263(2):50–59.

Mirelman, D. 1987. Ameba-bacterium relationship in amebiasis. *Microbiol. Rev.* 51(2):272–84.

Musial, C. E.; Cockerill, F. R.; and Roberts, G. D. 1988. Fungal infections of the immunocompromised host: Clinical and laboratory aspects. *Clin. Microbiol. Rev.* 1(4):349–64.

Pearson, R. D. 1983. The immunobiology of leishmaniasis. *Rev. Infect. Dis.* 5:907–10.

Petri, W. 1991. Invasive amebiasis and the galactose-specific lectin of *Entamoeba histolytica. ASM News* 57(6):299–306.

Ravidin, J. I., and Guerrant, R. L. 1982. A review of the parasite cellular mechanisms involved in pathogenesis of amebiasis. *Rev. Infect. Dis.* 4:1185–90.

Rippon, J. W. 1988. *Medical mycology,* 3d ed. Philadelphia: W. B. Saunders.

Schmidt, G. D., and Roberts, L. R. 1989. *Foundations of parasitology,* 4th ed. St. Louis: C. V. Mosby.

Seed, J. R.; Hall, J. E.; and Price, C. C. 1983. A physiological mechanism to explain pathogenesis in African trypanosomiasis. *Contrib. Microbiol. Immunol.* 7:83–91.

Smith, J. W., and Wolfe, M. S. 1980. Giardiasis. *Ann. Rev. Med.* 31:373–80.

Stevens, D. D. 1982. Giardiasis: Host-pathogen biology. *Rev. Infect. Dis.* 4:851–60.

Travassos, L. R., and Loyd, K. O. 1980. *Sporothrix schenckii* and related species of ceratocytis. *Microbiol. Rev.* 44:683–90.

Visvsvara, G. S. 1981. *Giardia lamblia:* America's no. 1 intestinal parasite. *Diagn. Med.* 4:24–26.

Wolfe, M. 1992. Giardiasis. *Clin. Microbiol. Rev.* 5(1):93–100.

Wyler, D. J. 1983. Malaria: Resurgence, resistance, and research. *N. Engl. J. Med.* 308:875–934.

Zimmer, B. L. 1990. Serology of coccidioidomycosis. *Clin. Microbiol. Rev.* 3(2):247–68.

Part Ten
Microorganisms and the Environment

We often describe ourselves as the "effluent society," and many of our wastes severely damage the environment. The wastes don't always kill; sometimes they enrich, as in the case of eutrophication. In this process, nutrients such as phosphates and nitrates stimulate the growth of many aquatic microorganisms. The result is often a lake filled with algae such as the one shown. Eutrophic lakes have lost much of their recreational and aesthetic appeal for humans and often suffer from large fish kills.

Chapter 40

Microorganisms as Components of the Environment

Everything is everywhere,

the environment selects.

—M. W. Beijerinck

Outline

Concepts

1. Microorganisms, as populations and communities, are an important part of natural environments. Microorganisms interact with the environment and play important roles in succession.
2. Microbial communities in most environments are complex, with interactions ranging from being competitive to being mutually beneficial.
3. Microbial growth requires nutrients, including carbon, nitrogen, phosphorus, and iron, all of which must be present in usable forms. Substrates containing these nutrients vary widely in their chemical composition, structure, and biodegradability.
4. Most microorganisms normally associated with higher organisms and those grown in the laboratory (including genetically engineered microorganisms) tend to be less able to compete and survive in natural environments.
5. Extreme environments restrict the range of microbial types able to survive and function. This can be the result of physical factors such as temperature, pH, pressure, or salinity. Many microorganisms found in "extreme" environments are especially adapted to survive and function under these particular conditions.

Microorganisms are essential components of every **ecosystem** (an ecosystem is a community of organisms and their physical and chemical environment that functions as an ecological unit). In this chapter, the role of microorganisms as a part of natural environments, the physiological state of microorganisms, nutrient cycling and decomposition processes, and the fundamentals of successional interactions are discussed. Besides these topics, the fate and potential effects of "foreign" and genetically engineered microorganisms in the environment are considered, together with the topic of extreme environments.

As just noted, environments are components of ecosystems. An environment is the total of the external conditions influencing an organism or group of organisms. **Natural environments,** whether they be soils, lakes, rivers, oceans, the deep subsurface, or other habitats, have many common characteristics. Living organisms in these natural environments have two complementary roles: the synthesis of new organic matter from CO_2 and other inorganic compounds during **primary production,** and decomposition of this accumulated organic matter. The latter process is carried out primarily by the microorganisms. A simple natural environment that can form the basis of a self-regulating ecosystem is illustrated in figure 40.1. This consists of an alga and a "halo" of surrounding bacteria that are using the organic matter excreted by the alga. The higher **consumers,** including humans, are chemoheterotrophs. These consumers depend on such basic "life support systems" provided by organisms that accumulate and decompose organic matter. The many contributions of the microorganisms to these natural environments are still not adequately understood. Our challenge is to better appreciate the role of microorganisms in the functioning of varied natural environments.

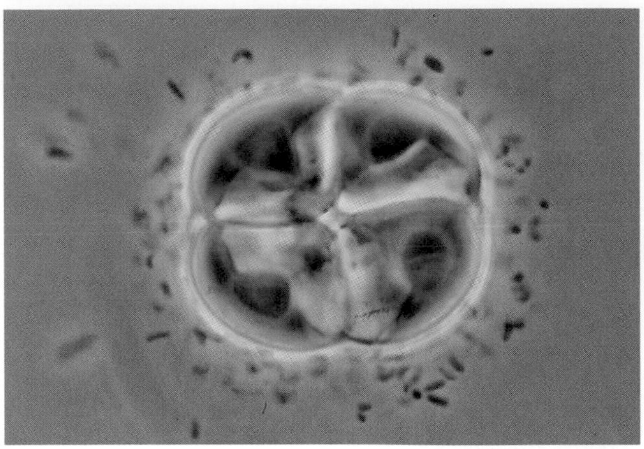

Figure 40.1 **The Beginning of an Ecosystem.** An alga, producing organic matter during photosynthesis, is surrounded by chemoheterotrophs using the excreted carbon (×1,000).

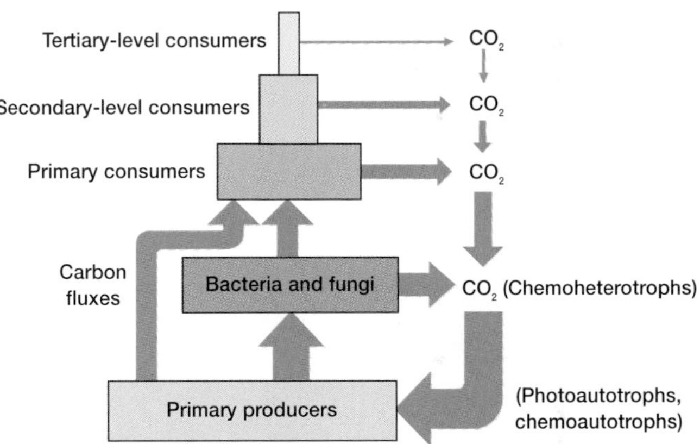

Figure 40.2 **The Ecological Role of Microorganisms.** Microorganisms play a vital role in the functioning of an ecosystem. Carbon is fixed by the primary producers, which use light or chemically bound energy. Chemoheterotrophic bacteria and fungi serve as the main decomposers of organic matter. Some microorganisms (protozoa) also serve as consumers.

Microorganisms and the Structure of Natural Environments

Microorganisms play many important roles in natural environments. The general relationships between the **primary producers** that accumulate organic matter, the heterotrophic **decomposers,** and the consumers are illustrated in figure 40.2. Microorganisms of different types contribute to each of these complementary relationships.

In terrestrial environments, the primary producers are usually vascular plants. In aquatic and marine environments, the cyanobacteria and algae play a similar role. The major energy source driving primary production is light in both habitats.

The biology of algae (chapter 27).
Cyanobacteria (pp. 472–76).

This view of the nature of primary production and the role of microorganisms in these processes has changed since 1977, when deep marine hydrothermal vents releasing hydrogen sulfide were discovered. The vents support large populations of tube-shaped worms and giant mussels (figure 40.3). The source of the organic carbon upon which these consumer organisms depended was found to be chemolithoautotrophic bacteria.

These bacteria are primarily of the genera *Thiobacillus, Thiomicrospira, Thiothrix, and Beggiatoa* and they play the major role in fixation of carbon for use by the consumers. Although chemolithoautotrophic bacteria had been known to contribute minor amounts of organic carbon to ecosystems, this was the first time a major ecosystem was found dependent on the activity of chemolithoautotrophs.

Chemolithotrophic metabolism (pp. 160–62).

Another unique food chain involves methane-fixing microorganisms as the first step in providing organic matter for consumers. In this case, methylotrophs, bacteria capable of using methane, occur as intracellular symbionts of the vent mussels. In these mussels, the thick fleshy gills are filled with bacterialike bodies (figure 40.4). These methane vent-dependent ecosystems have been found in the Gulf of Mexico, off the Oregon coast, and between the Eurasian and Philippine tectonic plates.

(a)

(b)

Figure 40.3 Chemlithoautotrophic Bacteria as Primary Producers. Some microbially based ecosystems can function in the absence of light, using chemically bound energy. Sulfide-emitting "black smokers" in the Galapagos trench (*a*) with a closer view of hydrothermal vent animals (*b*). The tube worm *Riftia pachyptila* and giant mussels use symbiotic bacteria as a nutritional source.

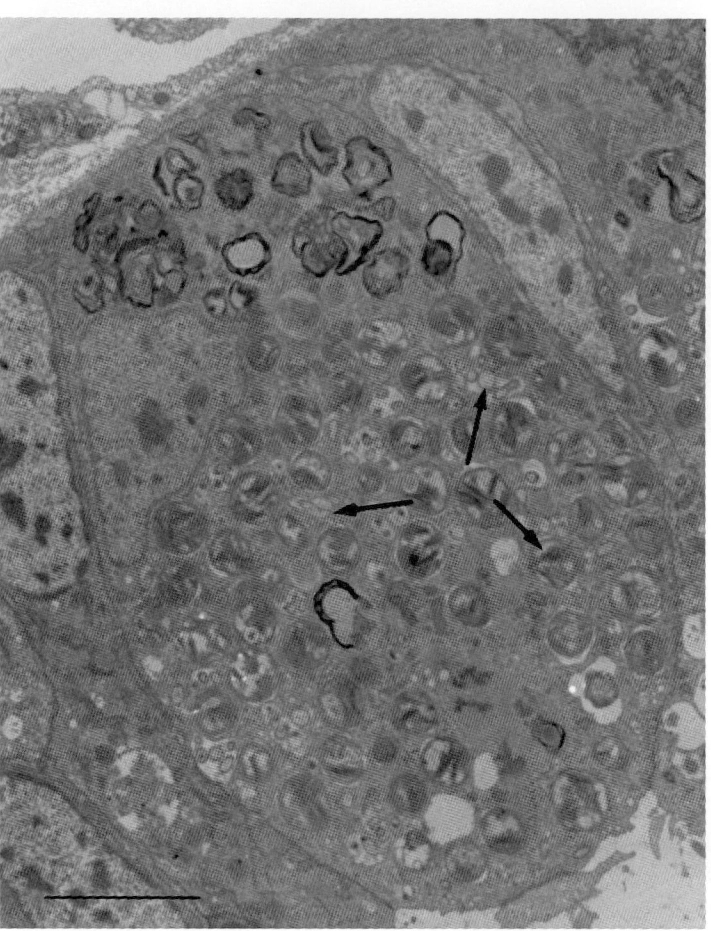

Figure 40.4 Methylotrophs as Nutrient Sources. Methane can serve as a carbon and energy source in recently discovered microbially based ecosystems. Mussels growing above methane seeps in the Florida escarpment (Gulf of Mexico) have bacterial symbionts within their gill epithelial cells. The transmission electron micrographs show two size classes of symbionts (arrows): large coccoid-shaped cells and small coccoid or rod-shaped bacteria. Scale bar = 5 μm.

In 1990, hydrothermal vents were also discovered in a freshwater environment, at the bottom of Lake Baikal, the oldest (25 million years old) and deepest lake in the world. This lake is located in the far east of Russia (figure 40.5*a*), and has the largest volume of any freshwater lake (not the largest area—which is Lake Superior). The bacterial mats, with long white strands, are in the center of the vent field where the highest temperatures are found (figure 40.5*b*). At the edge of the vent field, where the water temperature is lower, the bacterial mat ends, and sponges, gastropods, and other organisms are present (figure 40.5*c*).

Microorganisms, primarily the chemoheterotrophs, also use large amounts of the organic carbon generated by primary producers. The process of organic matter decomposition that releases simpler, inorganic compounds is called **mineralization.**

In addition, microorganisms such as the protozoa act as consumers, just as do insects and animals. Microbial consumers use organic matter produced by higher plants, cyanobacteria, and algae.

Microorganisms play another important role in the functioning of natural environments. Microbial cells, on a dry weight basis, typically consist of approximately 50% carbon, 14% nitrogen, and 3% phosphorus, to name only the major elements, and are nutrient-rich. Microorganisms, after their growth is completed, themselves become a much-desired nutrient source for other organisms in a cyclic process.

Microorganisms thus carry out many important functions in natural environments, including the following:

1. Decomposing (mineralizing) organic substrates.
2. Serving as a nutrient-rich food source for other chemoheterotrophic microorganisms. With the relatively rapid turnover of microbial populations, this results in transformation of organic materials to mineral forms.
3. Serving as a food source for protozoa, nematodes, and soil insects, thus creating a **food web** (a network of interlinked food chains).
4. Modifying substances for use by other organisms.

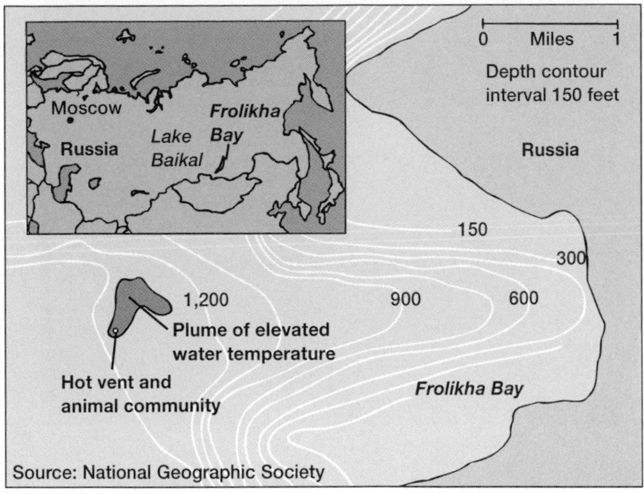

(a)

(b)

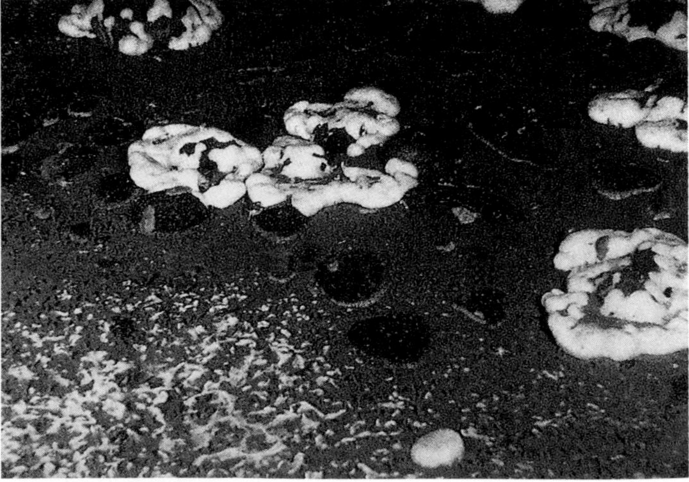

(c)

Figure 40.5 Hydrothermal Vent Ecosystems in Freshwater Environments. Lake Baikal (in Russia) has been found to have low temperature hydrothermal vents. (*a*) Location of Lake Baikal (inset) and Frolikha Bay, site of the hydrothermal vent field. (*b*) Bacterial mat near the center of the vent field. (*c*) Bacterial filaments, sponges, and tubes at the edge of the vent field.

5. Changing the amounts of materials in soluble and gaseous forms. This occurs either directly by metabolic processes or indirectly by modifying the environment.
6. Producing inhibitory compounds that decrease microbial activity or limit the survival and functioning of plants and animals.

Microbial functioning in these processes is conditional; it depends on meeting all of the growth requirements of each particular microorganism.

Nutrient requirements (pp. 97–100).

Microorganisms exist in natural habitats both as **populations** of similar types of organisms, such as a micro-colony growing at a localized site, and as **communities** made up of different types of interactive populations. Microorganisms, with other living organisms, use available energy sources and nutrients, contributing to energy flow and nutrient cycling. In many locations, it is possible to observe massive growth of microorganisms (figure 40.6).

The microbial environment is complex and constantly changing. It is characterized by the presence of overlapping **gradients** of resources, toxic materials, and other limiting factors (figure 40.7). For example, gradients can form when aerobic and anaerobic regions intersect or when a lighted zone changes into a dark region, creating unique **microenvironments.** The study of microorganisms and their relationship to their particular environments is called **microbial ecology.** Where the conditions in a microenvironment are suitable, specialized groups of microorganisms can maintain themselves with minimum competition from other microorganisms that have slightly different functional requirements.

Gradients of essential factors also influence the survival and functioning of microbial populations, an expression of **Liebig's law of the minimum,** which states that the indispensable nutrient in least plentiful supply relative to an organism's requirements will limit its growth. Factors that can limit microorganisms (and other living organisms) include water, energy in the form of light and chemical compounds, temperature, nutrients, pressure, pH, and salinity. Of course, the situation may be more complex than this. Multiple limiting factors can influence particular populations, and the limiting factors can change over time and space. Too much of something (metals, salts, hydrogen ions, heat) also can limit microbial activity, an expression of **Shelford's law of tolerance.**

The Physiological State of Microorganisms in the Environment

Most microorganisms are confronted with deficiencies that limit their activities except in relatively rare instances when excess nutrients allow unlimited growth. Even if adequate amounts of nutrients become available, the rapid growth rate of microorganisms will quickly deplete these nutrients and possibly result in the release of toxic waste products, which will limit further growth.

Microbial growth processes (chapter 6).

(a)

(b)

(c)

(d)

(e)

Figure 40.6 Examples of Massive Microbial Growth in the Environment. (*a*) Purple photosynthetic sulfur bacteria growing in a bog. (*b*) Eutrophic pond with heavy surface growth of algae and cyanobacteria. (*c*) Sewage lagoon with a bloom of purple photosynthetic bacteria. (*d*) Red algae growing on a snowfield. (*e*) A shallow bay in northern Denmark with pinkish-red patches of sulfur bacteria floating amidst green algae.

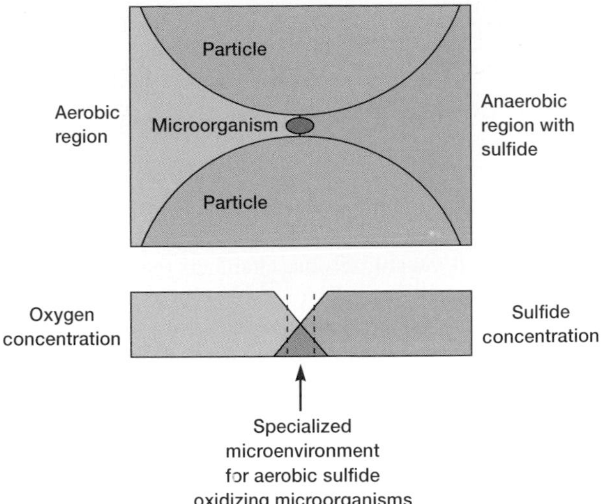

Figure 40.7 The Effect of Chemical Gradients on Microorganisms. Chemical gradients can create microenvironments where microorganisms find specific conditions required for their growth. In this example, oxygen and sulfide mix between two particles, to create a specialized microenvironment for sulfur-oxidizing bacteria.

In response to low nutrient levels and intense competition, many microorganisms become more competitive in nutrient capture and exploitation of available resources. Often the organism's morphology will change in order to increase its surface area and ability to absorb nutrients. This can involve conversion to "mini" and "ultramicro" cells or changes in the morphology of prosthecate (*see pp. 480–81*) bacteria (figure 40.8), for example. Specific and nonspecific attachment to surfaces will increase, allowing microorganisms to use nutrients that are often present at higher localized concentrations on these surfaces. Microorganisms can also sequester critical limiting nutrients, such as iron, making them less available to competing microorganisms. As noted in Box 40.1, many microorganisms found in these low-nutrient environments can survive and grow with extremely low nutrient levels, or under "oligotrophic" conditions.

Natural substances also can directly inhibit microbial growth in low-nutrient environments. These agents include phenolics, tannins, ammonia, ethylene, and volatile sulfur compounds. The presence of such compounds can lead to the general phenomenon of **microbiostasis,** or **bacteriostasis** and **fungistasis,** which affect bacteria and fungi, respectively. This may be a means by which microorganisms avoid expending limited energy reserves until an adequate supply of nutrients becomes available. Natural chemicals are also important in plant pathology and may aid in controlling soil-borne microbial diseases.

1. Define the following terms: primary production, ecosystem, consumer, decomposer, mineralization, food web, and microenvironment.
2. How are Liebig's law of the minimum and Shelford's law of tolerance related?
3. List six functions microorganisms have in natural environments.
4. What limitation usually affects microorganisms in the environment? In what ways do they respond?

Nutrient Cycling Processes

The microbial community plays a major role in **biogeochemical cycling.** As this term suggests, both biological and chemical processes are involved in the cycling and transformations of nutrients important to microorganisms, plants, and animals. This often involves oxidation-reduction reactions (*see chapter 7*) and can change the chemical and physical characteristics of the nutrients.

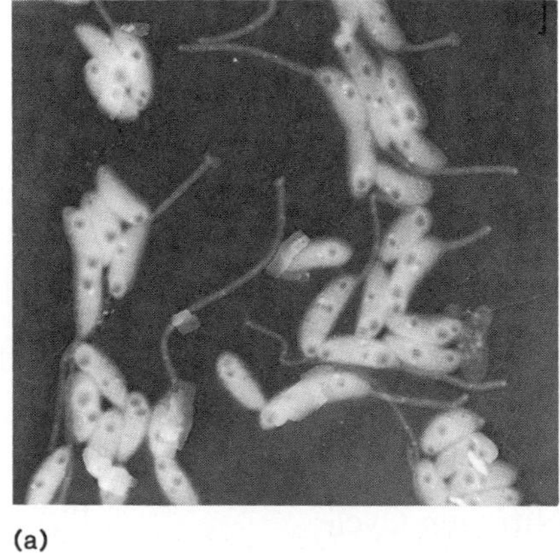

(a)

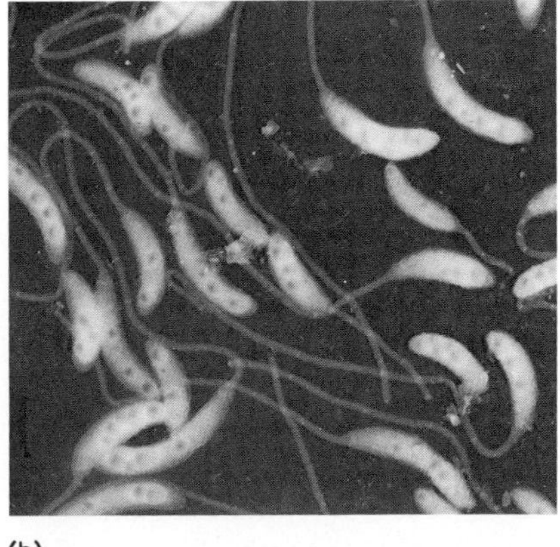

(b)

Figure 40.8 Morphology and Nutrient Absorption. Microorganisms can change their morphology in response to starvation and different limiting factors to improve their ability to survive. (*a*) *Caulobacter* has relatively short stalks when nitrogen is limiting. (*b*) The stalks are extremely long under phosphorus-limited conditions.

Box 40.1

The Challenges of Growing Oligotrophs

The presence in soil and water of large populations of microorganisms that can grow on media containing less than 15 mg/liter of dissolved organic carbon is of continuing interest. In many soil studies, the populations of such oligotrophic or "small nutrition" microorganisms are markedly higher than populations of copiotrophic or "abundant nutrition" microorganisms, commonly grown on media containing 8,000 to 10,000 mg/liter of utilizable nutrients. Microbiologists have observed twofold to tenfold higher populations of oligotrophs than copiotrophs in many soils.

While investigating these oligotrophs, scientists have observed that lower populations of oligotrophs are recovered on media incubated in closed, yet still aerobic, chambers than in petri dishes of the same low-nutrient medium exposed to the laboratory atmosphere. More recently, airborne organic substances have been found to stimulate microbial growth in such dilute media, and this atmospheric enrichment can allow significant populations of microorganisms to develop even in distilled water. The presence of such nutrients can affect experiments in biochemistry and molecular biology, as well as studies of oligotrophs.

Microorganisms are essential in the transformation of carbon, nitrogen, sulfur, and iron. The major reduced and oxidized forms of these elements are noted in table 40.1, together with their valence states. Significant gaseous components occur in the carbon and nitrogen cycles and, to a lesser extent, in the sulfur and phosphorus cycles. Thus, a soil or aquatic microorganism can often fix gaseous forms of carbon and nitrogen compounds. In the "sedimentary" cycles, such as that for iron, there is no gaseous component.

Carbon Cycle

Carbon can be present in reduced forms, such as methane (CH_4) and organic matter, and in more oxidized forms, such as carbon monoxide (CO) and carbon dioxide (CO_2). The major pools present in an integrated carbon cycle are shown in figure 40.9. Reductants (e.g., hydrogen, which is a strong reductant) and oxidants (e.g., O_2) influence the course of biological and chemical reactions involving carbon. Hydrogen can be produced during organic matter degradation, especially under anaerobic conditions when fermentation occurs. If hydrogen and methane are generated, they can move from anaerobic to aerobic areas. This creates an opportunity for aerobic hydrogen and methane oxidizers to function.

Methane levels in the atmosphere have been increasing approximately 1% per year, from 0.7 to 1.6 to 1.7 ppm (volume) in the last 300 years. Major methane sources are ruminants, which can produce 200 to 400 liters of methane per day, together with manure lagoons used to process animal wastes. Other sources are coal mines, sewage treatment plants, landfills, marshes, and rice paddies. Anaerobic microorganisms in the guts of termites also can contribute to methane production.

Physiology of aerobic hydrogen and methane utilizers (pp. 161; 435).

Carbon fixation occurs through the activities of cyanobacteria and green algae, photosynthetic bacteria (e.g., *Chromatium* and *Chlorobium*), and aerobic chemolithoautotrophs.

Sulfur Cycle

Microorganisms contribute greatly to the sulfur cycle, a simplified version of which is shown in figure 40.10. Photosynthetic microorganisms transform sulfur by using sulfide as an electron source. In the absence of light, sulfide can cross into oxidized environments, allowing *Thiobacillus* and similar chemolithoautotrophic genera to function (*see pp. 162 and 478–80*). In contrast, when sulfate diffuses into reduced habitats, it provides an opportunity for different groups of microorganisms to carry out **sulfate reduction.** For example, when a usable organic reductant is present, *Desulfovibrio* can derive energy by using sulfate as an oxidant (*see pp. 158 and 441–42*). This use of sulfate as an external electron acceptor to form sulfide, which accumulates in the environment, is an example of a **dissimilatory reduction** process and anaerobic respiration. In comparison, the reduction of sulfate for use in amino acid and protein biosynthesis is described as an **assimilatory reduction** process (*see chapter 9*). Other organisms have been found to carry out dissimilatory elemental sulfur reduction. These include *Desulfuromonas* (*p. 441*), thermophilic archaeobacteria (*see chapter 24*), and also cyanobacteria in hypersaline sediments.

When pH and oxidation-reduction conditions are favorable, several key transformations in the sulfur cycle also occur as the result of regular chemical reactions. Abiotic oxidation of sulfide to elemental sulfur takes place rapidly at a neutral pH, with a half-life of approximately 10 minutes for sulfide at room temperature.

Nitrogen Cycle

The nitrogen cycle is "simpler" than the carbon or sulfur cycles because there are no distinct and different uses by photosynthetic microorganisms (figure 40.11). Despite this simplification, several important aspects of the nitrogen cycle should be emphasized: the processes of nitrification, denitrification, and nitrogen fixation.

TABLE 40.1 The Major Forms of Carbon, Nitrogen, Sulfur, and Iron Important in Biogeochemical Cycling

Cycle	Significant Gaseous Component Present?	Major Forms and Valences: *Reduced Forms*	*Intermediate Oxidation State Forms*			*Oxidized Forms*
C	Yes	CH_4 (−4)	CO (+2)			CO_2 (+4)
N	Yes	NH_4^+, Organic N (−3)	N_2 (0)	N_2O (+1)	NO_2^- (+3)	NO_3^- (+5)
S	Yes	H_2S, SH groups in organic matter (−2)	S^0 (0)	$S_2O_3^{2-}$ (+2)	SO_3^{2-} (+4)	SO_4^{2-} (+6)
Fe	No	Fe^{2+} (+2)				Fe^{3+} (+3)

Note: The carbon, nitrogen, and sulfur cycles have significant gaseous components, and these are described as gaseous nutrient cycles. The iron cycle does not have a gaseous component, and this is described as a sedimentary nutrient cycle. Major reduced, intermediate oxidation state, and oxidized forms are noted, together with valences.

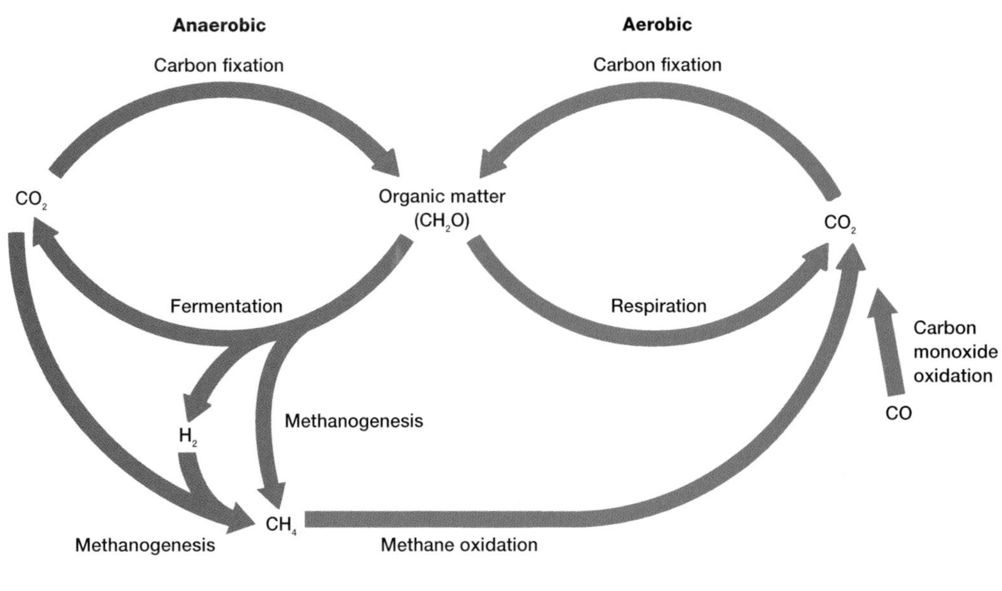

Figure 40.9 The Carbon Cycle in the Environment. Microorganisms play important roles in the environmental carbon cycle. Carbon fixation can occur through photoautotrophic and chemoautotrophic processes. Methane can be produced from inorganic substrates ($CO_2 + H_2$) or from organic matter. Carbon monoxide (CO)—produced by automobiles, industry, etc.—is returned to the carbon cycle by CO-oxidizing bacteria. Aerobic processes are noted with blue arrows, while anaerobic processes are shown with red arrows.

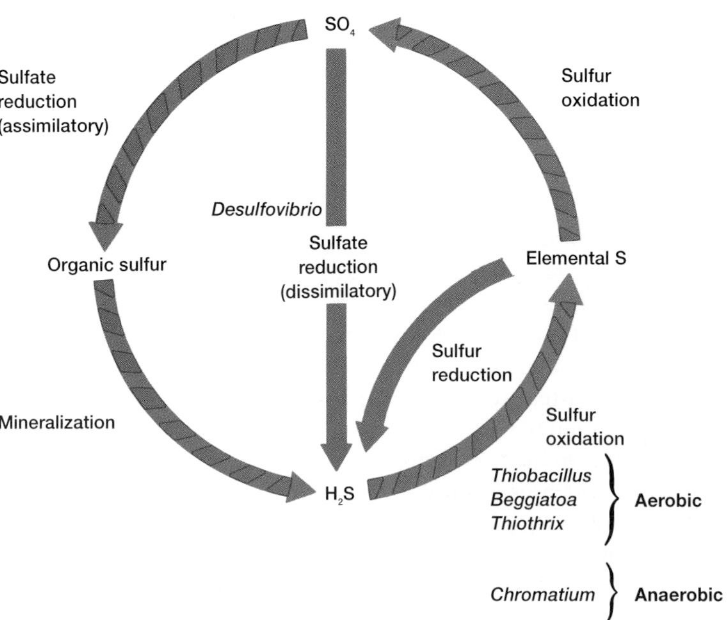

Figure 40.10 The Sulfur Cycle. Photosynthetic and chemosynthetic microorganisms contribute to the environmental sulfur cycle. Anaerobic sulfate reduction by *Desulfovibrio,* a dissimilatory process, is noted with a purple arrow. Sulfate reduction also can occur in assimilatory reactions. Elemental sulfur reduction to sulfide is carried out by desulfuromonas, thermophilic archaeobacteria, or cyanobacteria in hypersaline sediments. Sulfur oxidation can be carried out by a wide range of aerobic chemotrophs and by anaerobic phototrophs.

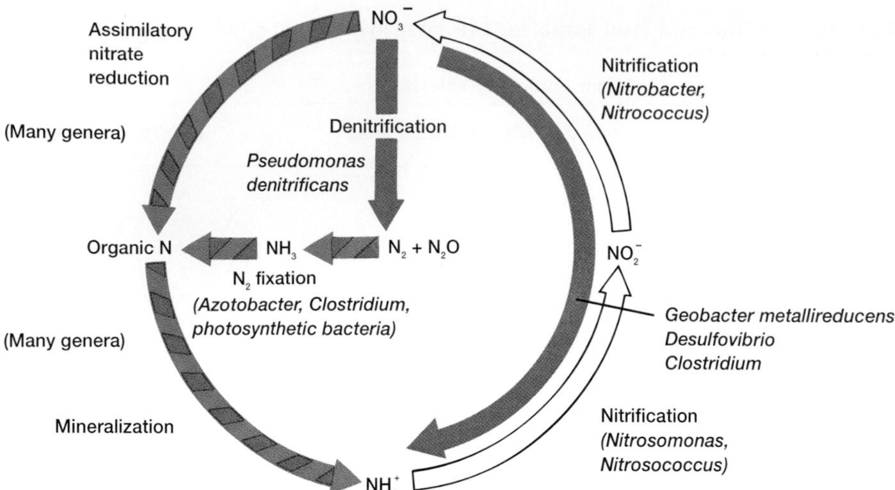

Figure 40.11 The Environmental Nitrogen Cycle. Flows that occur predominantly under aerobic conditions are noted with open arrows. Anaerobic dissimilatory processes are noted with solid bold arrows. Processes occurring under both aerobic and anaerobic conditions are marked with cross-barred arrows. Important genera contributing to the nitrogen cycle are given as examples.

Nitrification is the aerobic process of ammonium ion (NH_4^+) oxidation to nitrite (NO_2^-) and subsequent nitrite oxidation to nitrate (NO_3^-). Bacteria of the genera *Nitrosomonas* and *Nitrosococcus,* for example, play important roles in the first step, and *Nitrobacter* and related chemolithoautotrophic bacteria carry out the second step. In addition, **heterotrophic nitrification** by bacteria and fungi contributes significantly to these processes in more acidic environments, where the chemolithoautotrophic nitrifiers are less able to function.

Nitrification and nitrifiers (p. 161).

The process of **denitrification** requires a different set of environmental conditions. This dissimilatory process, in which nitrate is used as an oxidant in anaerobic respiration, usually involves heterotrophs such as *Pseudomonas denitrificans.* The major products of denitrification include nitrogen gas (N_2) and nitrous oxide (N_2O), although nitrite (NO_2^-) also can accumulate. Nitrite is of environmental concern because it can contribute to the formation of carcinogenic nitrosamines. Finally, nitrate can be transformed to ammonia in dissimilatory reduction by a variety of bacteria, including *Geobacter metallireducens, Desulfovibrio* spp., and *Clostridium.*

Denitrification and anaerobic respiration (p. 158).

Nitrogen assimilation occurs when inorganic nitrogen is used as a nutrient and incorporated into new microbial biomass. Ammonium ion, because it is already reduced, can be directly incorporated without major energy costs. However, when nitrate is assimilated, it must be reduced with a significant energy expenditure. In this process, nitrite may accumulate as a transient intermediate.

The biochemistry of nitrogen assimilation (pp. 177–78).

Nitrogen fixation can be carried out by aerobic or anaerobic bacteria. Under aerobic conditions, a wide range of free-living microbial genera (*Azotobacter, Azospirillum*) contribute to this process. Under anaerobic conditions, the most important free-living nitrogen fixers are members of the genus *Clostridium.* In addition, nitrogen fixation can occur through the activities of bacteria that develop symbiotic associations with plants. These associations include *Rhizobium* and *Bradyrhizobium* with legumes, *Frankia* in association with many woody shrubs, and *Anabaena* with *Azolla,* a water fern important in rice cultivation.

The establishment of the Rhizobium-*legume association (pp. 851–55).*

The nitrogen-fixation process involves a sequence of reduction steps that require major energy expenditures. Reductive processes are extremely sensitive to O_2 and must occur under anaerobic conditions even in aerobic microorganisms. Protection of the nitrogen-fixing enzymes is achieved by means of a variety of mechanisms, including physical barriers, as occurs with heterocysts in some cyanobacteria (*see chapter 23*), O_2 scavenging molecules, and high rates of metabolic activity. Nitrogen fixation is catalyzed by the enzyme **nitrogenase,** using reduced ferredoxin as its immediate source of reducing power. The biochemistry of this process is described in chapter 9.

The biochemistry of nitrogen fixation (pp. 178–80).

Other Cycling Processes

The iron cycle is especially important in terms of microbial function (figure 40.12). The major genera that carry out iron oxidations, transforming ferrous ion (Fe^{2+}) to ferric ion (Fe^{3+}), are *Thiobacillus ferrooxidans* under acidic conditions, *Gallionella* under neutral pH conditions, and *Sulfolobus* under acidic, thermophilic conditions. Much of the earlier literature suggested that additional genera could oxidize iron, including *Sphaerotilus* and *Leptothrix.* Confusion about the role of these genera resulted from the occurrence of the chemical oxidation of ferrous ion to ferric ion (forming insoluble iron precipitates) at neutral pH values, where microorganisms also grow on organic substrates. Many of these microorganisms, formerly described as "iron bacteria," are now classified as chemoheterotrophs.

Iron reduction occurs under anaerobic conditions resulting in the accumulation of ferrous ion. Although many microorganisms can reduce small amounts of iron during their

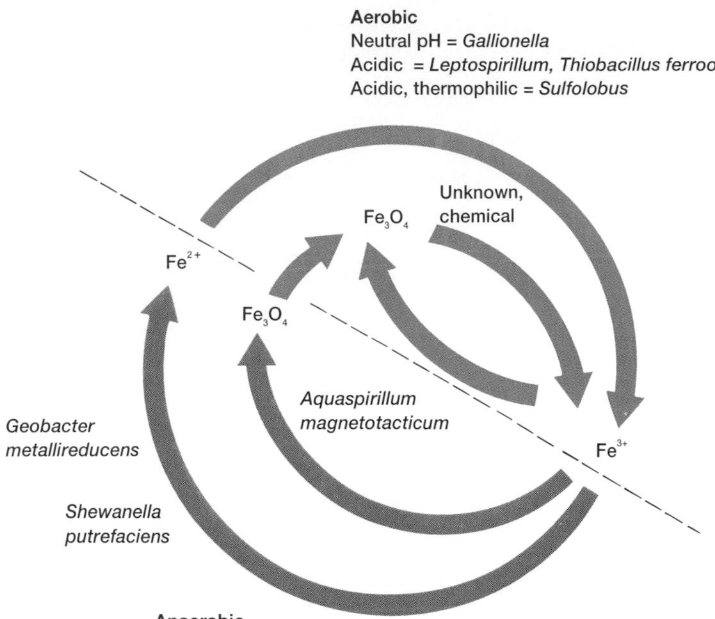

Figure 40.12 **The Iron Cycle.** A simplified iron cycle with examples of microorganisms contributing to these oxidation and reduction processes. In addition to ferrous ion (2+) oxidation and ferric ion (3+) reduction, magnetite (Fe_3O_4), a mixed valence iron compound formed by magnetotactic bacteria, is important in the iron cycle. Different microbial groups carry out the oxidation of ferrous ion, depending on environmental conditions.

metabolism, most iron reduction is carried out by specialized iron-respiring microorganisms such as *Geobacter metallireducens* and *Shewanella putrefaciens,* which can obtain energy for growth from iron reduction.

In addition to these relatively simple reductions to ferrous ion, some magnetotactic bacteria such as *Aquaspirillum magnetotacticum* (*see Box 3.2*) transform extracellular iron to the mixed valence iron oxide mineral magnetite (Fe_3O_4) and construct intracellular magnetic compasses. Furthermore, dissimilatory iron-reducing bacteria accumulate magnetite as an extracellular product.

Magnetite has been detected in sediments, where it is present in particles similar to those found in bacteria, indicating a longer-term contribution of bacteria to iron cycling processes. It has been suggested that Fe^{3+} reduction may have been the first globally significant mechanism for organic matter oxidation to CO_2.

The importance of microorganisms in manganese and phosphorus cycling is becoming much better appreciated. The manganese cycle (figure 40.13) involves the transformation of manganous ion (Mn^{2+}) to MnO_2 (equivalent to manganic ion [Mn^{4+}]), which occurs in hydrothermal vents, bogs, and as an important part of rock varnishes. *Leptothrix, Arthrobacter,* and *Metallogenium* are important in Mn^{2+} oxidation. *Shewanella, Geobacter,* and other chemoorganotrophs can carry out the complementary manganese reduction process. The manganese cycle is very closely linked to the iron cycle.

The microbial transformation of phosphorus involves primarily the transformation of phosphorus (+5 valence) from simple orthophosphate to various more complex forms, including polyphosphates found in metachromatic granules (*see p. 48*). A unique (and assumed microbial) product is phosphine (PH_3) with a −3 valence, which is liberated from swamps, and which ignites when exposed to air. This can then ignite methane produced in the same environment!

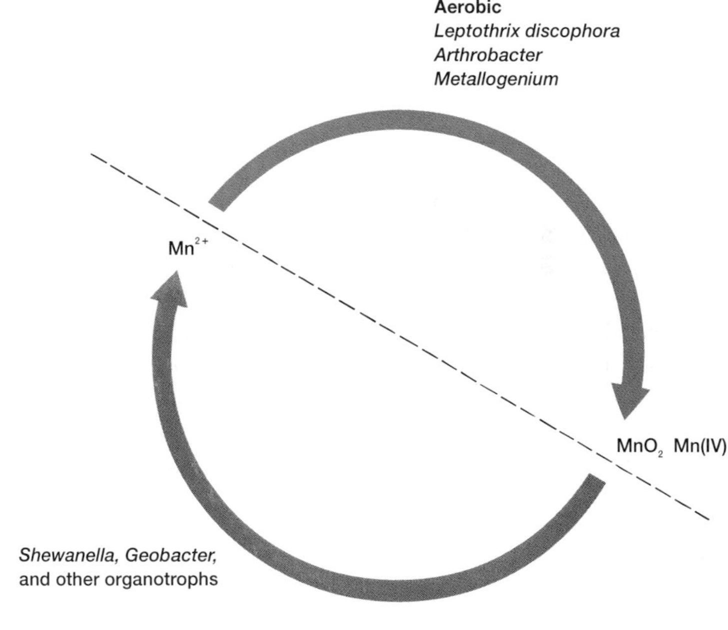

Figure 40.13 **The Manganese Cycle.** Microorganisms make important contributions to the manganese cycle. Manganous ion (2+) is oxidized to manganic oxide (valence equivalent to 4+). Manganous oxide reduction is noted with a solid arrow. Examples of organisms carrying out these processes are given.

---

1. What major oxidized and reduced forms of carbon, nitrogen, and sulfur are important in biogeochemical cycling?
2. By what means can microorganisms contribute to biogeochemical cycling in the environment?
3. Define or describe the following: dissimilatory reduction, assimilatory reduction, nitrification, denitrification, nitrogen fixation, and nitrogenase.

…2 Examples of Microorganism-Metal Interactions and Relations to Effects on Microorganisms and Warm-Blooded Animals

…l Group	Metal		Interactions and Transformations	
			Microorganisms	*Warm-blooded Animals*
Noble metals	Ag Au Pt	Silver Gold Platinum	Microorganisms can reduce ionic forms to the elemental state, which can become associated with cell walls. Low levels of ionized metals released to the environment have antimicrobial activity.	Many of these metals can be reduced to elemental forms and do not tend to cross the blood-brain barrier. Silver reduction can result in argyria and inert deposits in the skin.
Metals that form stable carbon metal bonds	As Hg Se	Arsenic Mercury Selenium	Microorganisms can transform inorganic and organic forms to methylated forms, some of which tend to bioaccumulate in higher tropic levels.	Methylated forms of some metals can cross the blood-brain barrier, resulting in neurological effects or death.
Other metals	Cu Zn Co	Copper Zinc Cobalt	In the ionized form, at higher concentrations, these metals can directly inhibit microorganisms. They are often required at lower concentrations as trace elements.	At higher levels, clearance from higher organisms occurs by reaction with plasma proteins and other mechanisms. Many of these metals serve as trace elements at lower concentrations.

Metals and Microorganisms

Many microorganisms can transform metals and metalloids, the latter being intermediate in properties between metals and nonmetals—for example, silicon, selenium, and arsenic. Although most metals are commonly considered toxic, microorganisms and more complex organisms do not respond similarly to different groups of metals. The general groupings of metals and their different effects on warm-blooded animals and microorganisms are summarized in table 40.2.

The "metals" can be considered in broad categories. The "noble metals" tend not to cross the vertebrate blood-brain barrier but can have distinct effects on microorganisms. Microorganisms also reduce ionic forms of noble metals to their elemental forms.

The second group includes metals or metalloids that microorganisms can methylate to form more mobile products, some of which can cross the blood-brain barrier and affect the central nervous system of vertebrates. The mercury cycle is of particular interest and illustrates many characteristics of this group of metals.

Mercury compounds were widely used as microbial growth inhibitors in pulp and paper mills until the mid-1960s. Mercury, with its ability to form stable, volatile organometallic compounds, is now of major environmental concern. A devastating situation developed in southwestern Japan when large-scale mercury poisoning occurred in the Minamata Bay region because of industrial mercury releases into the marine environment. Inorganic mercury that accumulated in bottom muds of the bay was methylated by anaerobic bacteria of the genus *Desulfovibrio* (figure 40.14). Such methylated mercury forms are volatile and lipid soluble, and the mercury concentrations increased in the food chain (a process known as **biomagnification**). The mercury was ultimately ingested by the human population, the "top consumers," through their primary food source—fish.

The third group of metals occurs in ionic forms directly toxic to microorganisms. The metals in this group also can affect more complex organisms. However, plasma proteins react with the ionic forms of these metals and aid in their excretion unless excessive long-term contact and ingestion occur. Relatively high doses of these metals are required to cause lethal effects. At lower concentrations, many of them serve as required trace elements.

The differing sensitivity of more complex organisms and microorganisms to metals forms the basis of many antiseptic procedures developed over the last 150 years (*see chapter 15*). The noble metals, although microorganisms tend to develop resistance to them, continue to be used in preference to antibiotics in some medical applications. A good example is the treatment of burns with silver-containing antimicrobial compounds.

1. What are different ways in which microorganisms can interact with metals? How can microbial activity render metals more toxic for humans?
2. Why do metals such as mercury have such major effects on higher organisms?

Interactions in Resource Utilization

Microbial community succession can occur when organic and inorganic substrates are used as nutrients by different groups of microorganisms. Under aerobic conditions, oxidized products such as nitrate, sulfate, and carbon dioxide (figure 40.15) will result from microbial activities. In comparison, under anaerobic conditions reduced end products tend to accumulate. Use of waste products of one group of microorganisms by other organisms is a commensalistic relationship (*see chapter 29*) that is often seen. An excellent example of such relationships is the Winogradsky column (*p. 824*).

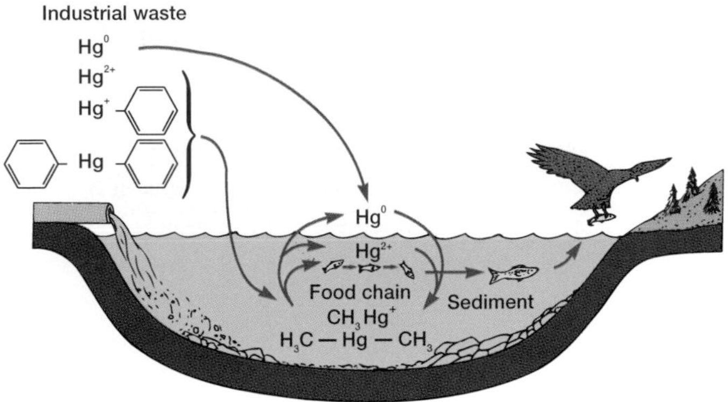

Figure 40.14 The Mercury Cycle. Microorganisms can increase the toxicity of mercury to higher organisms, including humans. Microorganisms in anaerobic sediments, primarily *Desulfovibrio,* can transform mercury to methylated forms that can undergo biomagnification.

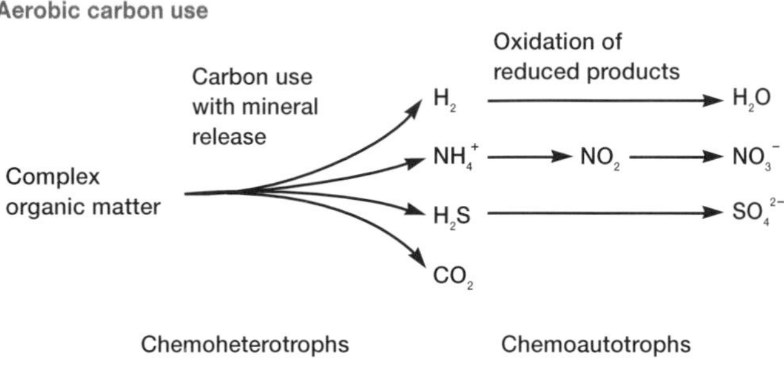

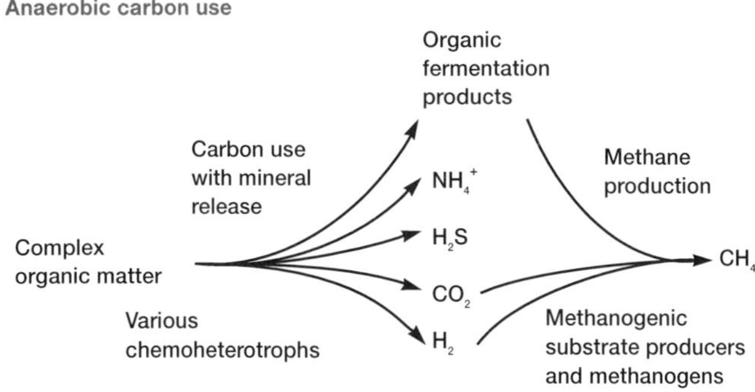

Figure 40.15 The Influence of Oxygen on Carbon Use. Microorganisms form different products when breaking down complex organic matter aerobically than they do under anaerobic conditions. Under aerobic conditions oxidized products accumulate, while reduced products accumulate anaerobically. These reactions also illustrate commensalistic transformations of a substrate.

Microbial interactions and succession also occur when mixtures of electron acceptors—oxidants—are available, as noted in table 40.3. If O_2, nitrate, manganese ion, ferric ion, sulfate, CO_2, or any combination thereof is available in a particular environment, a predictable sequence of oxidant use takes place when an oxidizable substrate is available. Oxygen is employed as an electron acceptor first because it inhibits nitrate use by microorganisms capable of respiration with either O_2 or nitrate. While O_2 is available, sulfate reducers and methanogens are inhibited, if not killed, because these groups are obligate anaerobes.

Electron acceptors and growth energetics (pp. 154–55; 158).

Table 40.3 Order of Oxidant Use by a Mixed Microbial Community under Successive Aerobic and Anaerobic Conditions (pH 7.0)

Oxidant	Oxidant couple reduction potential (V)	Oxygen Present
Oxygen	+0.814	+
Nitrate	+0.741	−
Mn^{4+}	+0.401	−
Fe^{3+}	−0.185	−
SO_4^{2-}	−0.214	−
CO_2	−0.244	−

Source: Data from Smith, K. A., and Arah, J. R. M. 1986. *Anaerobic microenvironments in soil and the occurrence of anaerobic bacteria.* FEMS Symposium 33, p. 258. Barking, Essex: Elsevier.

Once the O_2 and nitrate are exhausted and fermentation products, including hydrogen, have accumulated, competition for use of other oxidants begins. Oxidized forms of manganese and iron will be used first, followed by competition between sulfate reducers and methanogens. This is influenced by the greater energy yield obtained with sulfate as an electron acceptor. Differences in enzymatic affinity for hydrogen, an important substrate used by both groups, also play a critical role. The sulfate reducer *Desulfovibrio* grows rapidly and uses the available hydrogen at a faster rate than *Methanobacterium.* When the sulfate is exhausted, *Desulfovibrio* no longer oxidizes hydrogen, and the hydrogen concentration rises. The methanogens finally dominate the habitat and reduce CO_2.

This order of oxidant use is repeated whenever O_2, nitrate, Mn^{4+}, Fe^{3+}, and/or sulfate again becomes available. On the basis of these interactions and competition for substrates and oxidants, microbial use of available resources can be understood, predicted, and managed.

1. How does commensalism relate to processes of microbial succession?
2. Why is an orderly sequence of oxidant use observed in many natural ecosystems?

Organic Substrate Use by Microorganisms

In the earlier discussion of nutrient cycling, no distinction is made between different types of organic substrates. This is a marked oversimplification because organic substrates differ in their susceptibility to degradation.

Prediction of the potential for a particular organic substrate to be decomposed can be based for the most part on the following criteria:

- Elemental composition
- Structure of basic repeating units
- Linkages between repeating units
- Nutrients present in the environment
- Abiotic conditions (pH, oxidation-reduction potential, O_2, osmotic conditions)
- Microbial community present

TABLE 40.4 Complex Organic Substrate Characteristics That Influence Decomposition and Degradability

Substrate	Basic Subunit	Linkages (if Critical)	Elements Present in Large Quantity					Degradation	
			C	*H*	*O*	*N*	*P*	With O_2	Without O_2
Starch	Glucose	$\alpha(1\rightarrow4)$ $\alpha(1\rightarrow6)$	+	+	+	−	−	+	+
Cellulose	Glucose	$\beta(1\rightarrow4)$	+	+	+	−	−	+	+
Hemicellulose	C6 and C5 sugars	$\beta(1\rightarrow4)$, $\beta(1\rightarrow3)$, $\beta(1\rightarrow6)$	+	+	+	−	−	+	+
Lignin	Phenylpropane	C−C, C−O bonds	+	+	+	−	−	+	−
Chitin	N-acetylglucosamine	$\beta(1\rightarrow4)$	+	+	+	+	−	+	+
Protein	Amino acids	Peptide bonds	+	+	+	+	−	+	+
Hydrocarbon	Aliphatic, cyclic, aromatic		+	+	−	−	−	+	+
Lipids	Glycerol, fatty acids; some contain phosphate and nitrogen	Esters	+	+	+	+	+	+	+
Microbial biomass		Complex	+	+	+	+	+	+	+
Nucleic acids	Purine and pyrimidine bases, sugars, phosphate	Complex	+	+	+	+	+	+	+

The major organic substrate classes used by microorganisms are summarized in table 40.4. Of these potential substrates, only previously grown microbial biomass contains all of the nutrients required for microbial growth. Microbial biomass is an important substrate in sewage sludge, soil, and aquatic environments. It provides a readily available, rich source of nitrogen, phosphorus, and other nutrients for use by chemoheterotrophs.

Of the substrates listed in table 40.4, only chitin, protein, microbial biomass, and nucleic acids contain nitrogen in large amounts. The remaining complex substrates contain primarily or only carbon, hydrogen, and oxygen. If microorganisms are to grow by using these substrates, they must acquire remaining nutrients they need for growth from the environment.

The O_2 relationships for the use of these substrates are also of interest because most of the substrates can be readily degraded with or without oxygen's presence. The major exception is lignin.

Hydrocarbons are unique in that microbial degradation, especially of straight-chained and branched forms, involves the initial addition of molecular O_2. Recently, anaerobic degradation of hydrocarbons with sulfate or nitrate as oxidants has been observed. With sulfate present, organisms of the genus *Desulfovibrio* are active. This occurs only slowly and with microbial communities that have been exposed to these compounds for extended periods. Such degradation may have resulted in the sulfides that are present in "sour gases" associated with petroleum.

Lignin, an important structural component in mature plant materials—composing 1/3 of wood, for example—is a special case in which biodegradability is dependent on O_2 availability. Absence of significant degradation often occurs because most filamentous fungi that degrade lignin can function only with O_2 present. Lignin's lack of biodegradability under anaerobic conditions results in accumulation of lignified materials, including the formation of peat bogs and muck soils. This absence of lignin degradation under anaerobic conditions is also important in construction. Large masonry structures are often built on swampy sites by driving in wood pilings below the water table and placing the building footings on the pilings. As long as the foundations remain water-saturated and anaerobic, the structure is stable. If the water table drops, however, the pilings will begin to rot and the structure will be threatened. In Boston, a dropping water table in the Back Bay area, a landfill, has resulted in subsidence of many older and architecturally important homes, all because of the activity of O_2-dependent filamentous fungi. Moreover, the cleanup of harbors can lead to increased decomposition of costly docks built with wooden pilings.

These patterns of microbial biodegradation are important in many habitats. They contribute to the accumulation of petroleum products, the formation of bogs, and the preservation of valuable historical objects.

1. Which common microbial substrate is not degraded to any great extent under anaerobic conditions?
2. Why is decomposition important for the continued functioning of the environment?

"Foreign" Derived Microorganisms—Survival and Fate

A characteristic of natural habitats, whether soil, water, or other natural systems, is that most bacteria used as indicators of fecal pollution are not able to survive. The natural habitat for many of these organisms is the nutrient-rich intestinal tract of warm-blooded animals, and these microorganisms appear to have lost the ability to survive in the harsh outside world.

These **allochthonous,** or introduced, microorganisms are constantly being added to natural environments as a part of human and animal wastes and in sewage materials. Generally, these vegetative bacteria survive better at lower temperatures (figure 40.16), a phenomenon first observed over 50 years ago.

Box 40.2

Viable but Nonculturable Vegetative Bacteria

For most of microbiology's history, a viable microorganism has been defined as one that is able to grow actively, resulting in the formation of a colony or visible turbidity in a liquid medium. John R. Postgate of the University of Sussex in England was one of the first to note that microorganisms stressed by survival in natural habitats—or in many selective laboratory media—were particularly sensitive to secondary stresses. Such stress can result in the production of viable but nonculturable microorganisms. To determine the growth potential of such microorganisms, Postgate developed what is now called the Postgate Microviability Assay, which allows microorganisms to be cultured in a thin agar film under a coverslip. The potential of a cell to change its morphology, even if it does not grow, indicates that the microorganism does show "life signs."

Since that time many workers, including R. R. Colwell and K. Kogure, have developed additional sensitive microscopic and isotopic procedures to evaluate the presence and significance of these viable but "nonculturable" bacteria. As examples, levels of fluorescent antibody and acridine orange-stained cells are often compared with population counts obtained by the most probable number (MPN) method and plate counts using selective and nonselective media. The release of radioactive-labeled cell materials also is used to monitor stress effects on microorganisms. Despite these advances, the estimation of substrate-responsive viable cells by the microscopic direct viable count (DVC) method, first proposed by Postgate, is still important. These studies emphasize that even when bacteria such as *Escherichia coli, Vibrio cholerae, Klebsiella pneumoniae, Enterobacter aerogenes,* and *Enterococcus faecalis* cannot be cultured using conventional laboratory media and cultural techniques, they still can play a role in infectious disease and the functioning of natural systems.

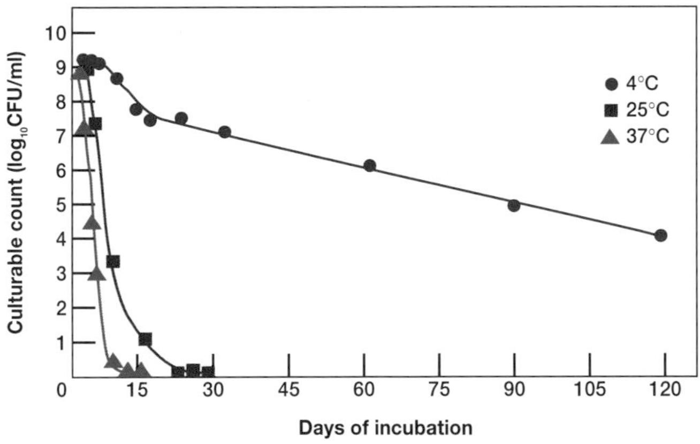

Figure 40.16 Temperature and Microbial Survival. Microorganisms survive better in the environment at lower temperatures. This is illustrated by the survival of *Campylobacter jejuni* in stream water samples held at various temperatures.

At extremely low temperatures, including freezing conditions, the survival time of these "foreign" microorganisms is greatly extended.

The detection of low levels of "foreign" bacteria is a major focus in environmental microbiology. If they are pathogens, they may be present at levels below the limit of test sensitivity. *Campylobacter, Legionella,* and *Vibrio* have been of concern, and such microorganisms confront the environmental microbiologist with new challenges (see Box 40.2).

Many studies have been directed toward learning why "foreign" microorganisms gradually die after being released to the environment. Among the possibilities are predation by protozoa, parasitism by *Bdellovibrio* (*see chapter 21*) and other organisms, lack of space, lack of nutrients, and the presence of toxic substances. After many years of study, it appears that the major reason "foreign" microorganisms die out is that they cannot compete effectively with indigenous microorganisms for the low amounts of nutrients present in the environment.

1. Name several genera of disease-causing microorganisms for which soil or water may be an important route of disease transmission.
2. What is the effect of lower temperatures on the die-out rate of "foreign" microorganisms that are added to the natural environment?
3. Why do "foreign" microorganisms often rapidly die out in a habitat?

Genetically Engineered Microorganisms—Fate and Effects

The procedures by which genetically engineered microorganisms are produced are summarized in chapter 14. These techniques allow the creation of desired microbial variants with specifically altered gene sequences in plasmids or in the genomes of bacteria and fungi. The field of genetic engineering, which has developed since the mid-1970s, will continue to be a driving force in biotechnology. These developments also may have significant effects on the environment.

Before modified microorganisms can affect the environment, a series of steps must first occur. These steps include survival and multiplication of the added microorganisms and transfer of the new genetic sequence to other organisms (*see chapter 13*). Each step will occur at a low-probability level, and the possibility of an effect's eventually becoming observable can be estimated.

Assessing the potential effects of genetically engineered microorganisms on the environment and the microbial community is a difficult process, and the uncertainties are many. Most genetic studies carried out to date have involved transfers between microorganisms of known genera that can be grown in the laboratory. However, most microorganisms present in the environment have not been cultured or characterized. Several factors limit the rates of genetic exchange in natural habitats, including the general lack of nutrients and the corresponding low growth rates of most microorganisms. The physical separation that occurs with microorganism attachment to surfaces and particles and the tendency of free DNA to become adsorbed onto clays and other surfaces will limit interactions. Furthermore, DNA, if released from a genetically engineered microorganism, may be degraded by other microorganisms.

Finally, it should be noted that microorganisms modified by modern genetic engineering techniques may be less fit to compete in low-nutrient-flux natural environments. Additional safeguards can be developed before the microorganisms are released into the environment. Microorganisms can be constructed with special DNA sequences that contain copy blocks, making it less likely that they will replicate the special DNA or exchange it with other microorganisms. Such a modified microorganism also may be less fit to compete and survive because of the additional energetic burden required to maintain the extra DNA. Many genetically modified microorganisms are planned for use in the natural environment. General principles of population genetics, bacterial physiology, epidemiology, and pathogenesis will play important roles in any attempts to assess possible effects of the release of these intentionally altered microorganisms. By the use of molecular techniques, it has been possible to detect transformation and genetic exchange in soils and waters, which increases the importance of studying genetic exchange in the environment.

1. What steps might be required to occur before genetically engineered microorganisms could cause observable effects in the environment?
2. What factors can limit the exchange of genetic information in the environment, especially from genetically engineered microorganisms?

Extreme Environments

Microorganisms vary greatly in their tolerance for pH, oxidized versus reduced environments, temperature, pressure, salinity, water availability, and ionizing radiation (*see chapter 6*). Thus, environmental factors greatly influence the survival of any particular microorganism. An extreme environment is one in which physical or chemical conditions become even more restrictive, resulting in a decrease in the diversity of microbial types that can maintain themselves. Under particularly restrictive conditions (figure 40.17), this process can continue until only a single type of microorganism, essentially a monoculture, exists. From the viewpoint of the successful microorganism, however, there may be less justification for describing

(a)

(b)

(c)

Figure 40.17 Microorganisms Growing in Extreme Environments. Many microorganisms are especially suited to survive in extreme environments. (*a*) Salterns turned red by halophilic algae and halobacteria. (*b*) A hot spring colored green and blue by cyanobacterial growth. (*c*) A source of acid drainage from a mine into a stream. The soil and water have turned red due to the presence of precipitated iron oxides caused by the activity of bacteria such as *Thiobacillus*.

TABLE 40.5 Characteristics of Extreme Environments in which Microorganisms Grow

Stress	Environmental Conditions	Microorganisms Observed
Temperature	110–115° C, deep marine trenches	*Methanopyrus kandleri* *Pyrodictium abyssi*
	85° C, hot springs	*Thermus* *Sulfolobus*
	75° C, sulfur hot springs	*Thermothrix thiopara*
Osmotic stress	13–15% NaCl	*Chlamydomonas*
	25% NaCl	*Halobacterium* *Halococcus*
Acidic pH	pH 3.0 or lower	*Saccharomyces* *Thiobacillus*
Basic pH	pH 10.0 or above	*Bacillus*
Low water availability	a_w = 0.6–0.65	*Torulopsis* *Candida*
Temperature and low pH	85° C, pH 1.0	*Cyanidium* *Sulfolobus acidocaldarum*
Pressure	500–1,035 atm	*Colwellia hadaliensis*

such an environment as extreme. If a particular microorganism can grow and reproduce in a specific habitat, the habitat might be optimum for it and actually required for the microorganism's continued existence. Important extreme environments and major microbial groups that have been observed in them are summarized in table 40.5.

Many microbial genera have specific requirements for survival and functioning in so-called extreme environments. For example, a high sodium ion concentration is required to maintain membrane integrity in many halophilic bacteria, including members of the genus *Halobacterium.* Halobacteria requires a sodium ion concentration of at least 1.5 M, and about 3 to 4 M for optimum growth.

Halophilic bacteria (pp. 123; 500–501).

The bacteria found in deep-sea environments have different pressure requirements, depending on the depth from which they are recovered. These bacteria can be described as **barotolerant bacteria** (growth from approximately 1 to 500 atm), **moderately barophilic bacteria** (growth optimum 5,000 meters, and still able to grow at 1 atm), and **extreme barophilic bacteria,** which require approximately 400 atm or higher for growth (*see chapter 41*).

Intriguing changes in basic physiological processes occur in microorganisms functioning under extreme acidic or alkaline conditions. These acidophilic and alkalophilic microorganisms have markedly different problems in maintaining a more neutral internal pH. Obligately acidophilic microorganisms can grow at a pH of 3.0, and a difference of 4.0 pH units can exist between the interior and exterior of the cell. These acidophiles include members of the genera *Thiobacillus, Bacillus, Sulfolobus,* and *Thermoplasma.* The higher relative internal pH is maintained by a net outward translocation of protons. This may occur as the result of unique membrane lipids, hydrogen ion removal during reduction of oxygen to water, or the pH-dependent characteristics of membrane-bound enzymes.

The extreme alkalophilic microorganisms grow at pH values of 10.0 and higher and must maintain a net inward translocation of protons. These obligate alkalophiles cannot grow below a pH of 8.5 and are often members of the genus *Bacillus; Micrococcus* and *Exiguobacterium* representatives have also been reported. Some photosynthetic cyanobacteria also have similar characteristics. Increased internal proton concentrations may be maintained by means of coordinated hydrogen and sodium ion fluxes.

The need to maintain chemiosmotic processes (*see chapter 8*), despite markedly different relative hydrogen ion concentrations in the environment, has resulted in unique differences in extreme acidophilic and alkalophilic microorganisms. These differences are still only poorly understood, especially in the alkalophiles.

Observations of microbial growth at temperatures approaching 115–120° C in thermal vent areas (Box 40.3, *see also Box 6.1*) reveal that the limits of temperature on microbial growth have not yet been determined. This area will continue to be a fertile field for investigation. For some successful microorganisms, an extreme environment may not be "extreme" but required and even, perhaps, ideal.

Thermophilic microorganisms (pp. 124–26; 501–3).

1. What are the main characteristics of extreme environments and the microbial populations that occur in them?
2. What is the internal pH of extreme acidophiles and alkalophiles?

Methods Used in Environmental Studies

A wide variety of techniques can be used to evaluate the presence, types, and activities of microorganisms in the environment (table 40.6). Many of these techniques were developed during the "golden age of discovery" in the last part of the nineteenth century. These include viable counting, microscopic procedures, and measures of nutrient cycling.

With environmental samples that contain higher levels of nutrients and microorganisms, such as polluted waters and sewage, organic carbon can be measured by the **biochemical oxygen demand (BOD)** and **chemical oxygen demand (COD)** procedures. The organic matter in a sample also can be reacted directly with O_2. The CO_2 is then quickly measured by sensitive infrared techniques or potentiometric procedures to give a measure of the **total organic carbon (TOC).** More specific information on these analyses will be provided in chapter 41.

History of microbiology (pp. 5–15).
Nutrient cycling processes (pp. 809–13).

Box 40.3

The Potential of Microorganisms from Extreme Environments for Use in Modern Biotechnology

There is great interest in the characteristics of bacteria isolated from the outflow mixing regions above deep hydrothermal vents that release water at 250 to 350° C. This is because these bacteria can grow at temperatures around 115° C. The problems in growing these microorganisms, often archaeobacteria, are formidable. For example, to grow some of them, it will be necessary to use culturing chambers of inert metal (gold; see Deming, 1986, in the Additional Reading section), and other specialized equipment to maintain water in the liquid state at these high temperatures.

A major potential application of microorganisms from such extreme environments is based on their exceptionally thermostable enzymes. These enzymes may have important applications in methane production, metal leaching and recovery, and for use in immobilized enzyme systems. In addition, the possibility of selective stereochemical modification of compounds normally not in solution at lower temperatures may provide new routes for directed chemical syntheses. This will be an exciting and expanding area of the modern biological sciences to which environmental microbiologists can make significant contributions.

TABLE 40.6 Methods Used to Study Microorganisms in Different Environments

		Environment			
Characteristic Evaluated	**Technique Employed or Property Measured**	***Marine***	***Freshwater***	***Sewage***	***Soil***
Nutrients	Chemical analysis (C, N, P, etc.)	++	++	++	++
	COD (chemical oxygen demand)	–	+	++	–
	BOD (biochemical oxygen demand)	–	+	++	–
Microbial biomass	Photosynthetic pigments	++	++	–	–
	Filtration and dry weight	++	++	+	–
	Measurement of chemical constituents (ATP, muramic acid, polybetahydroxybutyric [PHB] acid, lipopolysaccharides)	++	++	++	++
	Microscopy and biovolume conversion to biomass using conversion factors	++	++	+	++
	Fumigation incubation/extraction	–	–	+	++
	Glucose-amended respiration	–	–	+	++
Microbial numbers/types	Microscopic procedures—epifluorescence/phase contrast	++	++	++	++
	Immersion/insertion slides	+	++	+	++
	Viable enumeration procedures (cultural, microscopic)	++	++	++	++
	Film contact procedures (plastic film)	+	++	–	–
	Direct microorganism isolation	++	++	++	++
	Thin sections of samples	–	–	–	++
	Scanning electron microscopy	++	++	++	++
	Direct DNA extraction and analysis	++	++	++	++
	Polymerase chain reaction (PCR) for DNA amplification	++	++	++	++
	DNA probe and hybridization techniques	++	++	++	++
Microbial viability and turnover	Nalidixic acid and microscopic observation (inhibitor prevents cell division, resulting in elongated active cells); direct viable counts	++	++	++	++
	Stable and radioactive isotope studies	++	++	+	++
Microbial activity	Microscopy with reducible dyes	++	++	++	++
	Autoradiography	++	++	++	++
	Microcalorimetry	–	–	+	+
	Gas exchange (O_2, CO_2, N_2, CH_4)	++	++	++	++
	Assessment of fungal and bacterial contributions by use of selective antibiotic inhibition.	+	+	–	++
	Substrate utilization rate	++	++	++	++
	Fluorescent substrate hydrolysis	++	++	++	++
Community structure	Microscopic analyses of diversity	++	++	++	++
	Physiological diversity of microbial isolates	++	++	++	++
	16S ribosomal RNA analysis	++	++	++	++

Major uses are noted with two plus signs (++) and minor uses are noted with one plus (+); minimal or no use noted with (–).

Many newer and more sensitive procedures are now available, including the use of radioactive substrates and sophisticated techniques to measure the viability and activity of individual microorganisms. Hybridization techniques can be used to "probe" colonies to determine if they contain specific DNA sequences.

Microbial community diversity can be assessed by several approaches, including molecular phylogeny based on analyses of 16S ribosomal RNA (*see p. 415*). Small amounts of DNA can be recovered from environmental samples and "amplified" by use of the polymerase chain reaction (PCR). With such techniques, it is now possible to study the genetic characteristics of microorganisms without having to grow them in the laboratory.

The polymerase chain reaction (pp. 291–92).

As noted in table 40.6, some of these techniques are limited in terms of the types of samples that can be analyzed. This may be due to low microbial populations (marine and some freshwater samples) or high concentrations of interfering organic matter or particulates in samples. In contrast, the newer molecular procedures, such as direct DNA extraction, 16S RNA-based phylogeny, PCR, and DNA probe and hybridization techniques, are applicable to a variety of samples.

Measurements made by these techniques can span a wide range of time scales and physical dimensions. In marine, freshwater, sewage, and plant root environments, as examples, responses can be measured in seconds and minutes. For deep marine and soil organic matter changes, a time scale of years, decades, or even centuries may be required. The physical scale used in a study may range from a single bacterium and its microenvironment to a lake, ocean, or entire plant-soil system.

Summary

1. Microorganisms exist in populations and communities, and their interactions with the abiotic environment contribute to the development of different ecosystems. The normal physiological state of microorganisms in natural environments is starvation unless resources allowing growth become available.
2. Biogeochemical cycling involves oxidation and reduction processes, and changes in the concentrations of gaseous cycle components, such as carbon, nitrogen, and sulfur can result from microbial activity.
3. Microorganisms serve as primary producers that accumulate organic matter. In addition, many chemoheterotrophs decompose the organic matter that primary producers accumulate and carry out mineralization, the release of inorganic nutrients from organic matter.
4. Major polymers used by microorganisms differ in structure, linkage, elemental composition, and susceptibility to degradation under aerobic and anaerobic conditions. Lignin is only degraded under aerobic conditions, a fact that has important implications in terms of carbon retention in the biosphere.
5. Metals can be considered in three broad groups: the noble metals, metals for which toxic organometallic compounds can be formed, and certain other metals. The second of these groups is of particular concern.
6. Most disease-causing and indicator bacteria from the intestinal tract of humans and other higher organisms do not survive in the environment. At lower temperatures, however, increased survival can occur. Some disease-causing microorganisms may have environmental reservoirs.
7. Genetically engineered microorganisms will be of continuing concern, particularly if foreign DNA is maintained in the environment.
8. Decreased species diversity usually occurs in extreme environments, and many microorganisms that can function in such habitats have specialized growth requirements. For them, extreme environments can be required.
9. Many approaches can be used to study microorganisms in the environment. These include nutrient cycling, biomass, numbers, activity, and community structure analyses. It is now possible to study the genetic characteristics of microorganisms that cannot be grown in the laboratory.

Key Terms

Questions for Thought and Review

1. How might you attempt to grow a microorganism in the laboratory to increase its chances of being a strong competitor when placed back in a natural habitat? Do you think that this is even possible?
2. Considering the possibility of microorganisms functioning at temperatures approaching 120° C, what do you think the limiting factor for microbial growth at higher temperatures will be and why?
3. *Bdellovibrio* was suggested as a predator to eliminate *E. coli* from recreational waters in experiments carried out in the 1970s. What do you think the results of this experiment were, and why?
4. Where in their bodies do most people have noble metals? Why have these been used with more success than nonmetal materials, which have been tested over many decades?
5. Compare the degradation of lignin and cellulose by microorganisms. What different environmental factors are required for these important polymers to be degraded?
6. You have isolated DNA from an environmental sample that contains the genetic information of microorganisms that you cannot grow in the laboratory. What will you do with this DNA and why?

Additional Reading

Alexander, M. 1985. Genetic engineering: Ecological consequences. Reducing the uncertainties. *Issues Sci. Technol.* 1(3):57–68.

Atlas, R. M., and Bartha, R. 1992. *Microbial ecology: Fundamentals and applications.* 3rd ed. Redwood City, CA: Benjamin/Cummings.

Atlas, R. M.; Sayler, G.; Burlage, R. S.; and Bej, A. K. 1992. Molecular approaches for environmental monitoring of microorganisms. *Biotechniques* 12(5):706–17.

Cavanaugh, C. M.; Levering, P. R.; Maki, J. S.; Mitchell, R.; and Lidstrom, M. E. 1987. Symbiosis of methylotropic bacteria and deep-sea mussels. *Nature* 325:346–48.

Childress, J. J.; Felbeck, H.; and Somero, G. N. 1987. Symbiosis in the deep sea. *Sci. Am.* 256(5):115–20.

Cole, J. A., and Ferguson, S. J. 1988. "The Nitrogen and sulfur cycles." In *42d Symposium of the Society for General Microbiology.* New York: Cambridge University Press.

Davis, B. D. 1986. Bacterial domestication: Underlying assumptions. *Science* 235:1329–35.

Deming, J. W. 1986. The biotechnological future for newly described, extremely thermophilic bacteria. *Microbial Ecol.* 12:111–19.

Deming, J. W.; Somers, L. K.; Straube, W. L.; Swartz, D. G.; and Macdonell, M. T. 1988. Isolation of an obligately barophilic bacterium and description of *Colwellia,* new genus. *Syst. Appl. Microbiol.* 10:152–60.

Department of Energy. 1990. *Energy and Climate Change.* Chelsea, Mich.: Lewis Publishers.

Dorn, R. I. 1991. Rock varnish. *American Scientist* 79:452–53.

Edwards, C. 1990. *Microbiology of extreme environments.* New York: McGraw Hill.

Evans, W. C., and Fuchs, G. 1988. Anaerobic degradation of aromatic compounds. *Ann. Rev. Microbiol.* 42:289–317.

Fletcher, M., and Gray. T. R. G., eds. 1987. "Ecology of microbial communities." In *41st Symposium of the Society for General Microbiology.* New York: Cambridge University Press.

Gottschal, J. C., and Prins, R. A. 1991. Thermophiles: A life at elevated temperatures. *Trends. Ecol. & Evol.* 6:157–62.

Jannasch, H., and Taylor, C. D. 1984. Deep-sea microbiology. *Ann. Rev. Microbiol.* 38:487–514.

Jones, M. L., ed. 1985. Hydrothermal vents of the eastern Pacific: An overview. *Bull. Biol. Soc. of Washington,* No. 6.

Kirk, T. K., and Farrell, R. L. 1987. Enzymatic "combustion": The microbial degradation of lignin. *Ann. Rev. Microbiol.* 41:465–505.

Krulwich, T. A., and Ivey, D. M. 1990. Bioenergetics in extreme environments. In *The Bacteria,* vol. XII. eds. I. C. Gunsalus, J. R. Sokatch and L. N. Ornston, 417–35. New York: Academic Press.

Kuenen, J. G.; and Robertson, L. A.; van Gemerden, H. 1985. Microbial interactions among aerobic and anaerobic sulfur-oxidizing bacteria. *Adv. Microbial Ecol.* 8:1–59.

Levin, M. A.; Seidler, R. J.; and Rogul, M. 1991. *Microbial ecology: Principles, methods, and applications.* New York: McGraw-Hill.

Lovley, D. K. 1991. Dissimilatory Fe (III) and Mn (IV) reduction. *Microbiol. Rev.* 55:259–87.

Lynch, J. M., and Hobbie, J. E. 1988. *Micro-organisms in action: Concepts and applications in microbial ecology.* Boston, MA: Blackwell Scientific.

Lynch, J. M., and Poole, N. J., eds. 1979. *Microbial ecology: A conceptual approach.* New York: John Wiley and Sons.

Margulis, L.; Chase, D.; and Guerrero, R. 1986. Microbial communities. *BioScience* 36(3):160–70.

Mitchell, R., ed. 1992. *Environmental microbiology.* New York: John Wiley.

Norris, J. R., and Grigorova, R. 1990. Techniques in microbial ecology. In *Methods in microbiology,* vol. 22. New York: Academic Press.

Poole, R. K., and Gadd, C. M. 1989. *Metal-microbe interactions,* vol. 26. Soc. Gen. Microbiol. Oxford: IRL Press.

Stacey, G.; Burris, R. H.; and Evans, H. J. 1991. *Biological nitrogen fixation.* New York: Chapman and Hall.

Steffan, R. J., and Atlas, R. M. 1991. Polymerase chain reaction: Applications in environmental microbiology. *Ann. Rev. Microbiol.* 45:137–61.

Stotzky, G., and Babich, H. 1986. Survival of, and genetic transfer by, genetically engineered bacteria in natural environments. *Adv. Appl. Microbiol.* 31:93–138.

Tunnicliffe, V. 1992. Hydrothermal-vent communities of the deep sea. *American Scientist* 80:336–49.

Whitman, W. B., and Rogers, J. E. 1991. *Microbial production of greenhouse gases: Methane, nitrogen oxides, and halomethanes.* Washington, D.C.: American Society for Microbiology.

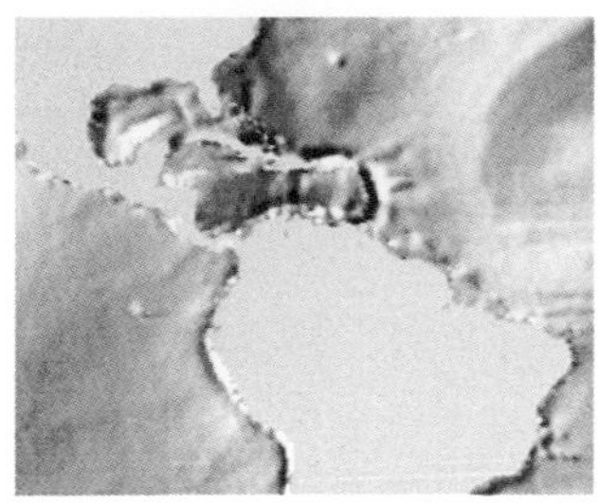

CHAPTER 41
Marine and Freshwater Environments

Water is a very good servant, but it is a cruel master.

—John Bullein

The world turns softly
Not to spill its lakes and rivers
The water is held in its arms
And the sky is held in the water
What is water, That pours silver
And can hold the sky?

—Hilda Conkling

Outline

Concepts

1. Marine environments represent the largest volume of water on the earth. Most of this water is at low temperature (2 to 3° C) and under high pressure.
2. Phytoplankton (photosynthetic microorganisms) form the basis of primary production in most marine and freshwater environments.
3. Deep hydrothermal vents and hydrocarbon seeps provide unique environments for microorganisms to function without light.
4. Microbial mats are found in many marine and freshwater environments. These complex microbial communities also occur as fossils and represent one of the earliest forms of microbial communities, with some being 3.5 billion years old.
5. Dissolved O_2 is present at low concentrations in most waters. If sufficient organic matter is present, microorganisms use O_2 faster than it can be replenished. This can result in cyclic changes in dissolved O_2 or in complete depletion of O_2.
6. The biological utilization of organic wastes follows regular and predictable sequences. Once these sequences are understood, more efficient sewage treatment systems can be created.
7. Indicator organisms, which usually die off at slower rates than many disease-causing microorganisms, can be used to evaluate the microbiological quality of water.
8. Disease-causing viruses and protozoa may be present in water that meets the standards set in terms of indicator organisms.
9. Many important disease-causing microorganisms can be present in water not contaminated by human wastes.
10. Groundwater is an important source of water, especially in rural areas. This resource is of increasing concern to microbiologists.

Of all the water found on earth, 97% is marine. Most of this water is at a temperature of 2 to 3° C and devoid of light; 62% is under high pressure (>100 atm). Microscopic phytoplankton and associated bacteria create a complex food web that can extend over long distances and extreme depths. The marine environment seems so vast that it cannot be affected by pollution; however in coastal areas, human activities are increasingly disrupting microbial processes and damaging water quality.

Fresh waters, although a small part of the waters on earth, are extremely important as a source of drinking water. In many locations the contamination of surface and subsurface waters by domestic and industrial wastes causes environmental problems.

Marine and freshwater environments create unique niches for many specialized microorganisms not found in habitats without a continuous water phase.

The Nature of Marine and Freshwater Environments

The mixing and movement of nutrients, O_2, and waste products that occur in freshwater and marine environments are the dominant factors controlling the microbial community. For example, in deep lakes or oceans, organic matter from the surface can sink to great depths, creating nutrient-rich zones where decomposition takes place. Gases and soluble wastes produced by microorganisms in these deep marine zones can move into upper waters and stimulate the activity of other microbial groups. Similar processes take place on a lesser scale in nutrient-rich lakes, and even in microbial mats (*see p. 826*) where gradients are established on a scale of millimeters.

The **Winogradsky column** illustrates many of these principles (figure 41.1). In this microcosm, a layer of reduced mud is mixed with sodium sulfate, sodium carbonate, and shredded newspaper—a cellulose source—and additional mud and water are placed in the column, which is then incubated in the light. A series of reactions occurs as the column begins to mature, with particular members of the microbial community developing in specific microenvironments in response to chemical gradients.

In the bottom of the column, cellulose is degraded to fermentation products by the genus *Clostridium* (*see chapter 22*). With these fermentation products available as reductants and using sulfate as an oxidant, *Desulfovibrio* produces hydrogen sulfide (*see chapter 21*). The hydrogen sulfide diffuses upward toward the oxygenated zone, creating a stable hydrogen sulfide gradient. In this gradient, the photoautotrophs *Chlorobium* and *Chromatium* develop as visible olive green and purple zones. These microorganisms use hydrogen sulfide as an electron source, and CO_2, from sodium carbonate, as a carbon source. Above this region, the purple nonsulfur bacteria of the genera *Rhodospirillum* and *Rhodopseudomonas* can grow. These photoheterotrophs use organic matter as an electron donor under anaerobic conditions and function in a zone where the sulfide level is lower. Both O_2 and hydrogen sulfide may be present higher in the column, allowing specially adapted microorganisms to function. These include *Beggiatoa* and *Thiothrix*, which use reduced sulfur compounds as a reductant and O_2 as an oxidant. In the upper portion of the column, diatoms and cyanobacteria may be visible.

Green and purple photosynthetic bacteria (pp. 468–72).
Cyanobacteria (pp. 472–76).

These commensalistic microorganisms, which develop sequentially, are dependent on the reductant originally provided as the cellulose or plant materials. When this reductant is exhausted, the column gradually becomes oxidized and the sulfide-dependent photosynthetic microorganisms and other anaerobes can no longer maintain themselves in the microcosm.

Oxygen and carbon dioxide, two important gases in water, also develop concentration gradients. Oxygen is only sparingly soluble in water. Its solubility is influenced by the concentration in the gas phase, the water temperature, the gas pressure, and dissolved salts. Temperature and pressure, especially, control the amount of O_2 available for use by microorganisms (table 41.1). At lower temperatures, the O_2 concentration can be markedly higher. Rapid depletion of dissolved O_2 can occur when nutrients contaminate the water, often leading to fish kills and the transformation of lakes and rivers into open sewers.

The second major gas in water, CO_2, plays many important roles in chemical and biological processes. The carbon-dioxide-bicarbonate-carbonate equilibrium can control the pH in weakly buffered waters, or it can be controlled by the pH of strongly buffered waters. As shown in figure 41.2, the pH of distilled water, which is not buffered, is determined by the dissolved CO_2 in equilibrium with the air, and is approximately 5.0 to 5.5. In comparison, water strongly buffered at pH 8.0 contains CO_2 absorbed from the air, which is present primarily as bicarbonate. When microorganisms such as algae use CO_2, the pH of many waters will be increased.

1. What are the reasons for adding cellulose, sodium sulfate, and sodium carbonate to the Winogradsky column? Discuss this in relation to microbial groups that respond to these materials or their products.
2. What are the two most important gases in aquatic environments? What factors influence their concentrations in water?

The Microbial Community in Marine and Freshwater Environments

Important Microorganisms

Water, because it provides a unique physical environment, favors the existence of many types of microorganisms that are not common in soils (table 41.2). The sulfur and nonsulfur photosynthetic bacteria and the aerobic hydrogen sulfide oxidizers—*Beggiatoa* and *Thiothrix*—occur in waterlogged zones where hydrogen sulfide is present, as noted earlier.

Several unique groups of chemoheterotrophs should also be mentioned. These include the sheathed and sessile bacteria

Important reactions and microorganisms

Component

Water layer — Diatoms and cyanobacteria

O_2-dominated mud (light brown) — Algae and aerobic sulfide oxidizing microorganisms *Beggiatoa* *Thiobacillus* *Thiothrix*

Rust-colored zone — Photoheterotrophs *Rhodospirillum* *Rhodopseudomonas*

Mud

Red zone — *Chromatium*

Green zone — *Chlorobium*

H_2S diffusion

Mud plus sulfate, carbonate, and newspaper (as a cellulose source)

Anaerobic H_2S-dominated zone (black) — Cellulose → fermentation products (*Clostridium*); Fermentation products plus sulfate → sulfide (*Desulfovibrio*)

Figure 41.1 **The Winogradsky Column.** This forms a microcosm where microorganisms and nutrients interact over a vertical gradient. Fermentation products and sulfide migrate up from the reduced lower zone, and oxygen penetrates from the surface. This creates conditions similar to those in a lake with nutrient-rich sediments. Light is provided to simulate the penetration of sunlight into the anaerobic lower region, which allows photosynthetic microorganisms to develop.

TABLE 41.1 The Effects of Water Temperature and Elevation on Dissolved O_2 Levels (mg/l) in Water

Temperature (° C)	Elevation above Sea Level (Meters) 0	1,000	2,000	3,000
0	14.6	12.9	11.4	10.2
5	12.8	11.2	9.9	8.9
10	11.3	9.9	8.8	7.9
15	10.0	8.9	7.9	7.1
25	9.1	8.1	7.1	6.4
30	8.2	7.3	6.4	5.8
35	7.5	6.6	5.9	5.3
40	6.9	6.1	5.4	4.9

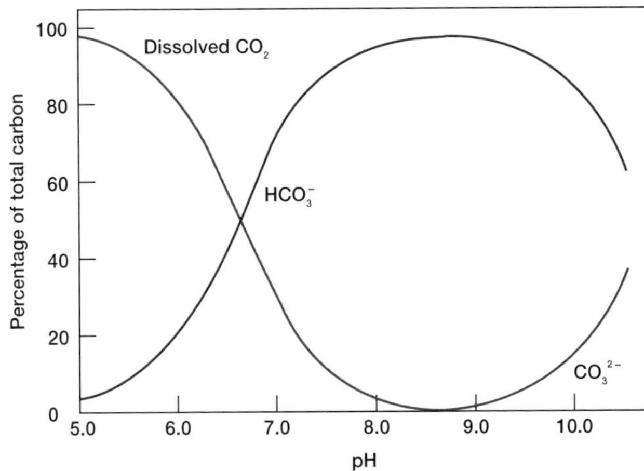

Figure 41.2 **The Relationship of pH to Dissolved CO_2.** Atmospheric gases can affect the physical characteristics of water. The pH of water is influenced by the amount of dissolved carbon dioxide in the water, and the equilibrium of the dissolved carbon dioxide with bicarbonate and carbonate ions.

of the genera *Sphaerotilus* and *Leucothrix;* the prosthecate and budding bacteria of the genera *Caulobacter* and *Hyphomicrobium;* and a wide range of aerobic gliding bacteria, including the genera *Flexithrix* and *Flexibacter.* These organisms are characterized by their exploitation of surfaces and nutrient gradients. They are obligate aerobes, although sometimes they can carry out denitrification, as occurs in the genus *Hyphomicrobium.* In addition, bacteria may be primarily colonizers of submerged surfaces, allowing subsequent development of a surface-fouling biomass.

Sheathed and budding bacteria (pp. 480–83).
Gliding bacteria (pp. 484–87).

Other interesting groups of microorganisms are present besides these morphologically diverse chemoheterotrophic bacteria. These include eucaryotic photosynthetic algae such as diatoms (*see chapter 27*), aquatic fungi, and protozoa such as the foraminiferans and radiolarians (*see chapter 28*). The eucaryotic algae can be a major source of organic carbon used by the chemoorganotrophic microorganisms (*see figure 40.1*).

TABLE 41.2 Important Procaryotic Genera Found Primarily in Marine and Freshwater Environments[a]

Group	Genera
Photoautotrophs	*Chlorobium* *Chloroherpeton* *Chromatium* *Pelodictyon* *Thiodictyon* *Thiopedia*
Photoheterotrophs	*Chloroflexus* *Heliobacterium* *Heliothrix* *Rhodocyclus* *Rhodomicrobium* *Rhodopseudomonas* *Rhodospirillum*
Chemoheterotrophs	*Blastobacter* *Caulobacter* *Flexibacter* *Flexithrix* *Gemmobacter* *Hyphomicrobium* *Leucothrix* *Sphaerotilus*
Chemolithoautotrophs	*Beggiatoa*[b] *Gallionella* *Thioploca*[b] *Thiothrix*[b] *Thiovulum*

[a]Sources: From V. M. Gorlenko, G. A. Dubinina, and S. I. Kuznetov, 1983. "The Ecology of Aquatic Micro-Organisms," *Die Binnengewässer*, Vol. XXVIII, Schweizerbart'sche Verlagsbuchhandlung, Stuttgart, 1983; and B. Rothe et al., 1987. "The Phylogenetic Position of the Budding Bacteria *Blastobacter aggregatus* and *Gemmobacter aquatilis*" in *Archives of Microbiology* 147:92–99, 1987.
[b]Many are mixotrophs.

Figure 41.3 Microbial Mats. Microorganisms, through their metabolic activities, can create environmental gradients resulting in layered ecosystems. A vertical section of a hot spring (55° C) microbial mat, showing the varied layers of microorganisms.

A group of fungi especially adapted to water are the aquatic oomycetes (*see chapter 26*). These water molds have motile swarmer stages which exploit surfaces. Protozoa play a major role in water as well as in soil. By grazing on other groups of microorganisms, protozoa increase nutrient cycling.

In oligotrophic or low-nutrient waters, many microorganisms make significant contributions. These include the sessile, prosthecate, and gliding microorganisms, which attach to and move over surfaces where organic matter is adsorbed. If surfaces are not available, these microorganisms tend to attach to one another, with a resulting increase in the surface area and the possibility of nutrient capture. When microorganisms associate like this, they can form flocs, which are important in sewage treatment.

Microbial mats are found in many freshwater and marine environments. These mats are complex layered microbial communities that can form at the surface of rocks or sediments in hypersaline and freshwater lakes, lagoons, hot springs, and beach areas (figure 41.3). They consist of microbial filaments, including cyanobacteria. A major characteristic of mats is the extreme gradients that are present. Light only penetrates approximately 1 mm into these communities, and below this photosynthetic zone, anaerobic conditions occur (figure 41.4). As is the case at the bottom of a Winogradsky column, sulfate-reducing bacteria play a major role in mats. The sulfide that these organisms produce diffuses to the anaerobic lighted region, allowing sulfur photosynthetic microorganisms to grow. Some believe that microbial mats could have allowed the formation of terrestrial ecosystems prior to the development of vascular plants, and fossil microbial mats, called **stromatolites,** have been dated at over 3.5 billion years old. Molecular techniques and stable isotope measurements (*see table 40.6*) are being used to better understand the development of these unique microbial communities.

1. Waters allow microorganisms to exist that are not common in soils. What are some of these microorganisms and what are their unique characteristics?
2. What are microbial mats? In what types of environments do they occur?
3. What is the estimated age of the oldest stromatolites?

The Marine Environment

In terms of sheer volume, the marine environment represents the largest portion of the biosphere, containing 97% of the earth's water (figure 41.5). Much of this is in the "deep sea" at a depth greater than 1,000 meters, representing 75% of the

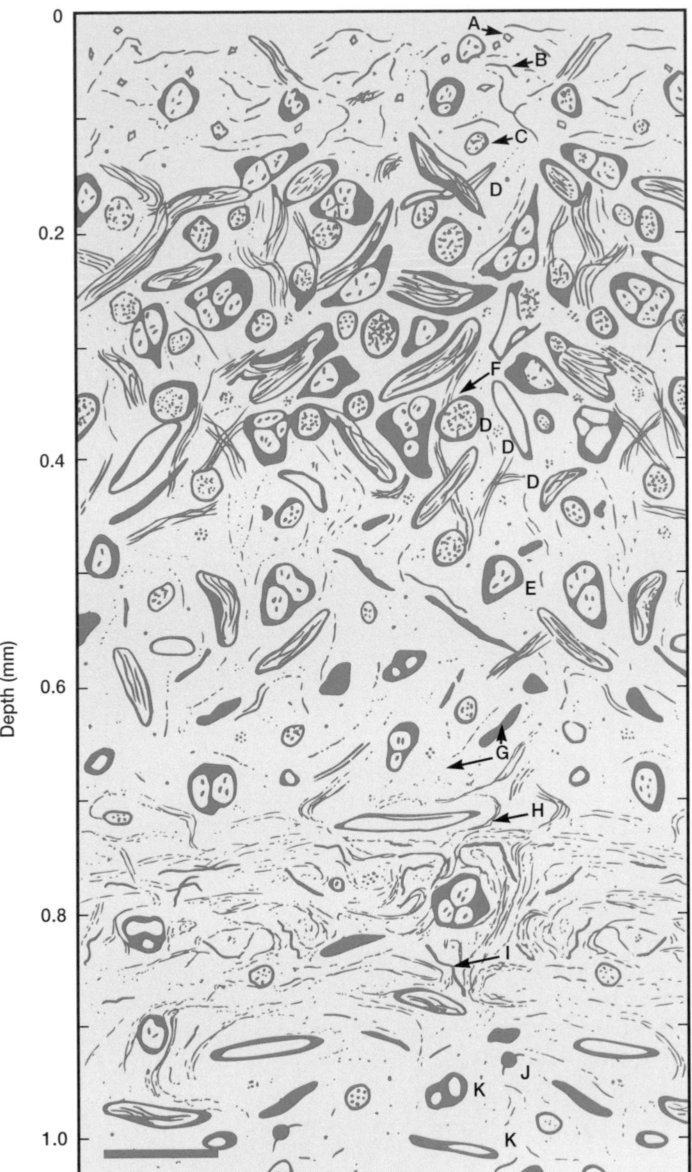

Figure 41.4 A Layered Microbial Ecosystem. In these layered ecosystems, microorganisms can create gradients over microscopic-scale distances. Schematic vertical cross-section of the topmost 1 mm of a marine cyanobacterial mat. The horizontal bar at lower left is 100 μm long. Letters along the right margin refer to the following: *A*, diatoms; *B, Spirulina* sp.; *C, Oscillatoria* spp.; *D, Microcoleus chthonoplastes; E,* nonphotosynthetic bacteria; *F*, unicellular cyanobacteria; *G*, fragments of bacterial mucilage; *H, Chloroflexus* spp. (green bacteria); *I, Beggiatoa* spp. (nonphotosynthetic sulfide-oxidizing bacteria); *J*, unidentified grazer; *K*, abandoned cyanobacterial sheaths.

ocean's volume. The ocean has been called a "high-pressure refrigerator," with most of the volume below 100 meters at a constant 3° C temperature. As noted in figure 41.6, the ocean, at its greatest depth, is slightly more than 11,000 meters deep or equivalent to almost 29 Empire State Buildings (each 1,250 feet or 381 meters in height) stacked on top of one another! The pressure in the marine environment increases approximately 1 atm/10 meters in depth, and pressures are in the vicinity of 1,000 atm at the greatest ocean depths (figure 41.6).

A series of unique pressure relationships is observed among the bacteria growing in this vertically differentiated system. These bacteria fall into three groups. Some bacteria are barotolerant and grow between 0 and 400 atm, but best at atmospheric pressure. Many other bacteria are **barophiles** (Greek, *baro,* weight and *philein,* to love) and prefer higher pressures. Moderate barophiles grow optimally at 400 atm, but still grow at 1 atm; extreme barophiles grow only at higher pressures. It is likely that each group functions best in a particular pressure range because of several factors. Pressure differences influence many biological processes including cell division, flagellar assembly, DNA replication, membrane transport, and protein synthesis. Porins, outer membrane proteins that form channels for diffusion of materials into the periplasm (*see figure 3.23*), also function most effectively at specific pressures.

Pressure and microbial growth (pp. 128–29).

As with most other ecosystems, the major source of organic matter in marine systems is photosynthetic activity, primarily from **phytoplankton** (Greek *phyto,* plant and *planktos,* wandering) in illuminated surface waters. A common planktonic genus is *Synechococcus,* which can reach densities of 10^4 to 10^5 per milliliter at the ocean surface. Picocyanobacteria may represent 20 to 80% of the total phytoplankton biomass upon which grazers depend. In addition, the oceans harbor large populations of viruses, especially in coastal regions (figure 41.7). These viruses may cause major reductions in primary production due to their infection of cyanobacteria (*see Box 18.1*).

Most nutrient cycling in oceans occurs in the top 300 meters where light penetrates. Light allows phytoplankton to grow and fall as a "marine snow" to the seabed. This "trip" can take a month or longer. Most of the organic matter that falls below the 300 meter zone is decomposed, and only 1% of photosynthetically derived material reaches the deep-sea floor unaltered. Because low inputs of organic matter occur in the deep sea, the ability of microorganisms to grow under oligotrophic conditions becomes important (*see Box 40.1*).

The carbon cycle within the ocean environment is only poorly understood (figure 41.8), but clearly microorganisms can profoundly influence the cycle. It is estimated that the large amount of dissolved organic carbon (DOC) in the ocean has a mean age of greater than 1,000 years. Organic matter in deep ocean waters has a similarly long residence time. Besides DOC, massive deposits of methane hydrate occur in ocean sediments. Under the low-temperature, high-pressure conditions found at the ocean floor below 500 meters, methane accumulates in latticelike cages of crystalline water. There may be up to 10,000 billion metric tons of carbon present as methane hydrate worldwide, twice as much as is trapped in known terrestrial coal, oil, and gas reserves.

A major goal of microbiologists is to isolate and culture unique marine microorganisms under natural conditions. New techniques to collect microorganisms without changing temperature or pressure are under development. For example,

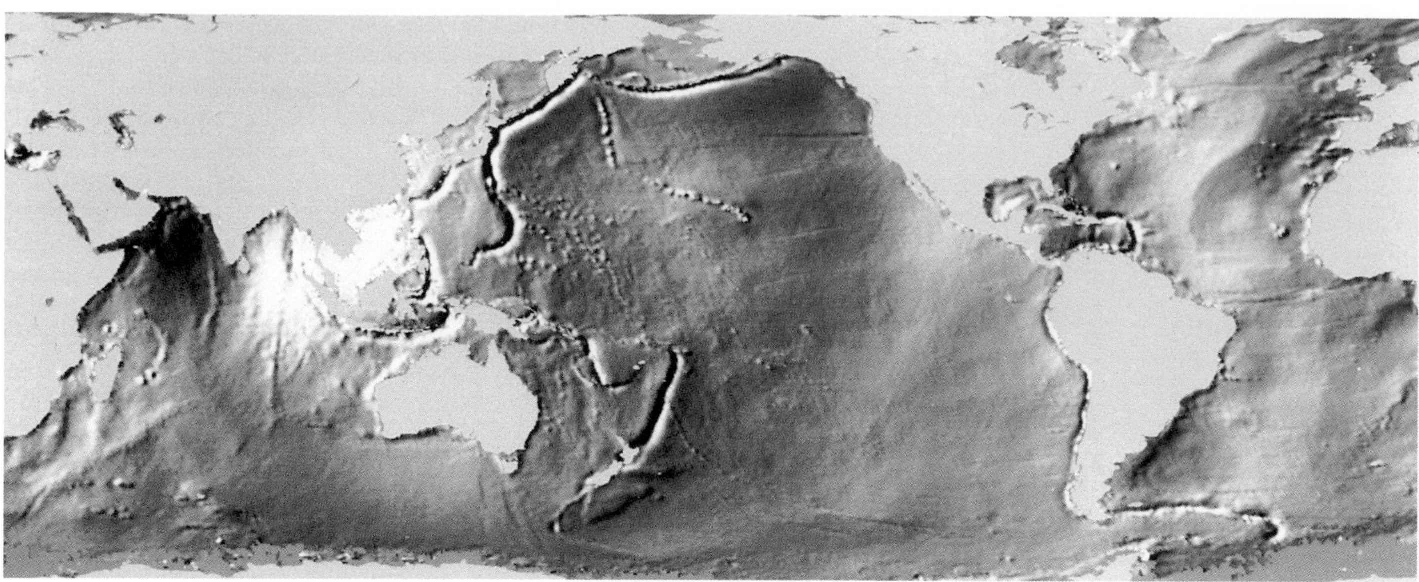

Figure 41.5 The Marine Environment—A Global-scale View. Within this vast volume of water, microorganisms function over a wide range of pressure, temperature, salinity, and nutritional conditions.

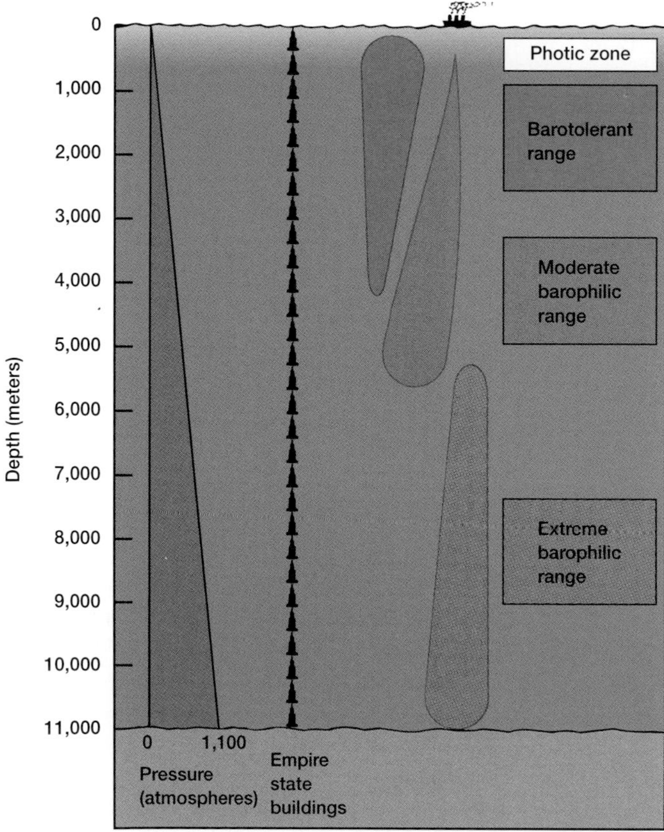

Figure 41.6 The Deep Marine Environment. Microorganisms with specialized pressure relationships are found at various depths. The relative activity and depth relationships of barotolerant, moderate barophilic, and extreme barophilic microorganisms are noted schematically. The pressure can approach 1,100 atmospheres at the deepest ocean depths. The depths are given in meters and "stacked" Empire State Buildings for perspective. Light penetrates only into a relatively shallow surface layer, creating the photic zone.

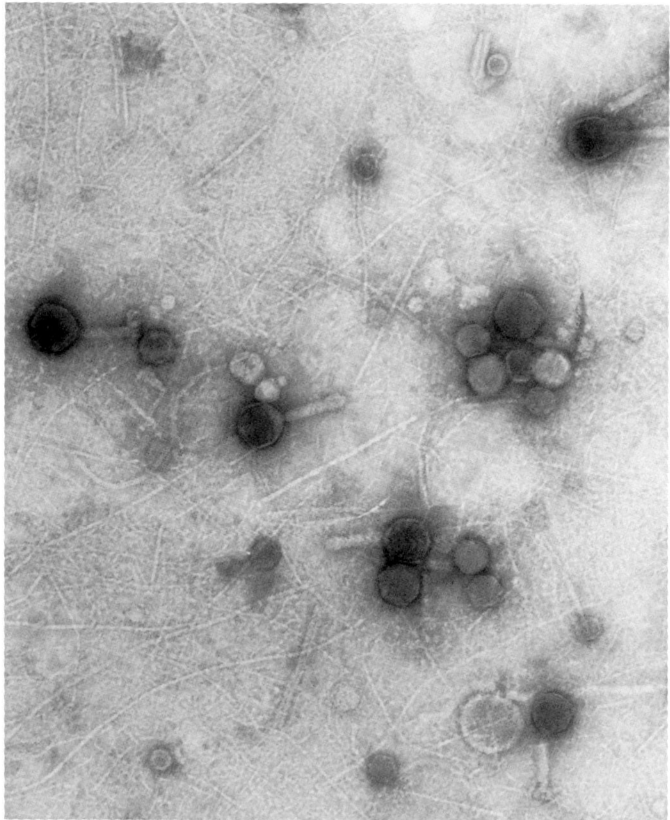

Figure 41.7 Viruses in the Ocean. Viruses have recently been found present in marine environments at surprisingly high levels. A variety of viral shapes are observable, with approximately 10^7 viral particles present per milliliter. Their function and significance in possibly influencing microbial populations and nutrient cycling are still not fully understood.

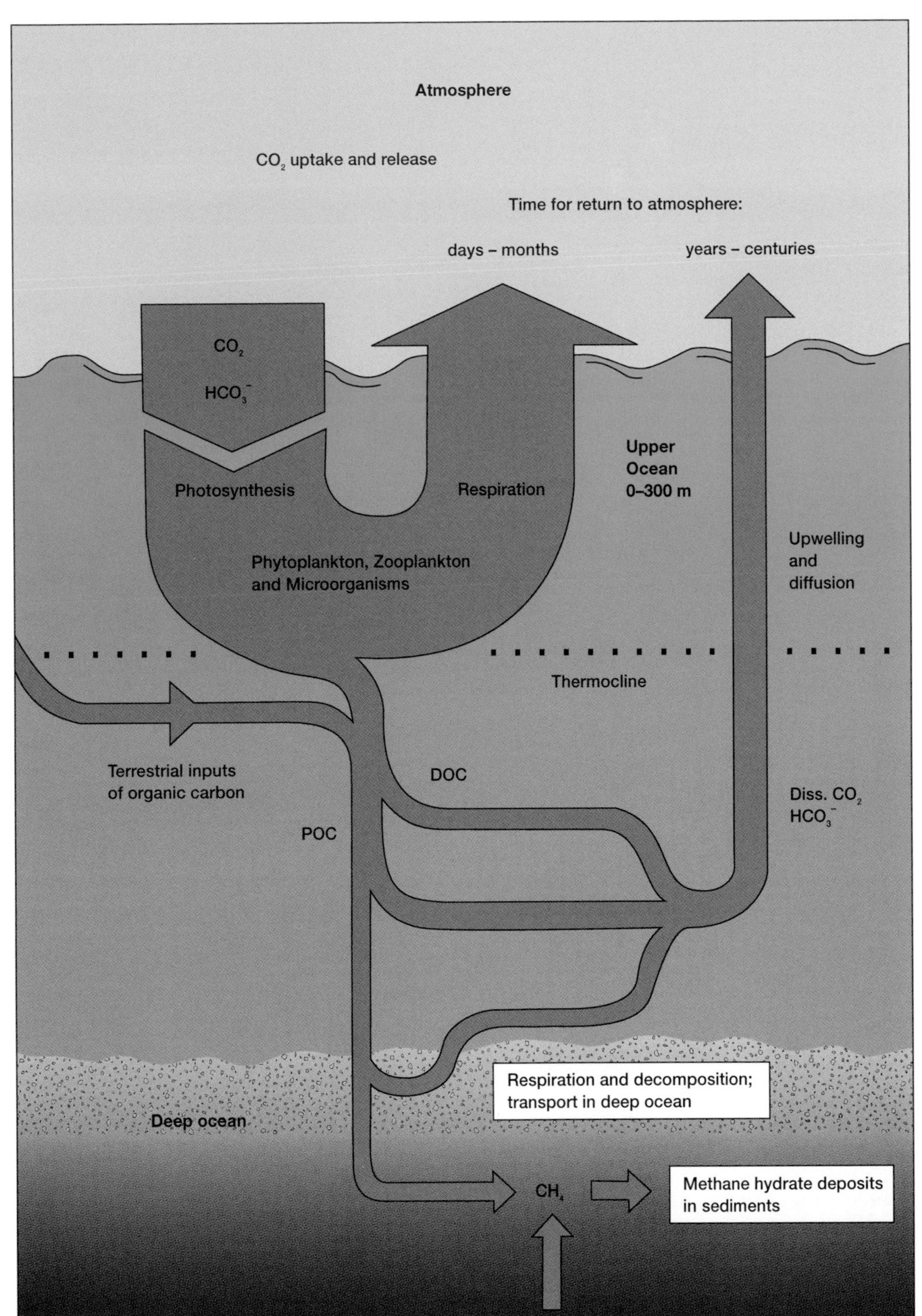

Figure 41.8 Carbon Cycling in the Ocean Environment. Microorganisms in the oceans can influence global carbon cycling and ocean-atmosphere interactions. Most carbon processing occurs in the surface water zone, with particulate organic carbon (POC), dissolved organic carbon (DOC), and methane hydrate (in sediments) being major carbon pools.

Japanese and American scientists are constructing equipment to culture microorganisms that grow at 1,000 atm without decompressing them.

Increased human populations and the urban development occurring in coastal areas around the world is taxing the seemingly inexhaustible ability of oceans to absorb and process pollutants. Coastal areas that have limited mixing with ocean waters (e.g., the Baltic Sea, Long Island Sound, the Mediterranean) are showing signs of nutrient enrichment and microbial pollution. One example is shellfish contamination by runoff waters from urbanized coastal areas. Only a few years ago shellfish could be harvested without delay after major rainfalls; now, one week or longer is needed to allow dieback of polluting microorganisms. This economically impacts individuals who depend on harvesting of shellfish for a livelihood.

Another problem that relates to ocean waters and water mixing in coastal areas is the occurrence of red tides (*see Box 27.2*). This has major economic effects when shellfish cannot be harvested or consumed.

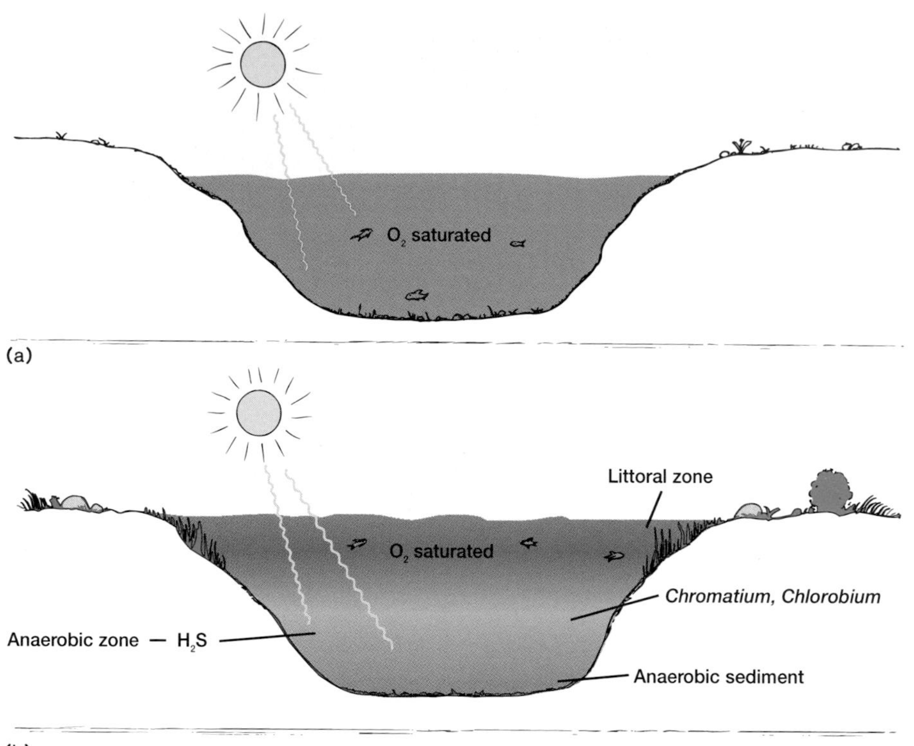

Figure 41.9 Oligotrophic and Eutrophic Lakes. Lakes can have different levels of nutrients, ranging from low nutrient to extremely high nutrient systems. The comparison of (*a*) an oligotrophic (nutrient-poor) lake, which is oxygen saturated and has a low microbial population, with (*b*) a eutrophic (nutrient-rich) lake. The eutrophic lake has a bottom sediment layer and can have an anaerobic hypolimnion. Sulfur photosynthetic microorganisms can develop in this anaerobic region.

1. Why is the marine environment called a "high-pressure" refrigerator?
2. What are the unique pressure relationships of deep-sea microorganisms? Define the terms barotolerant, moderate barophilic, and extreme barophilic.
3. Carbon cycling in the ocean can take thousands of years. What are the possible roles of particulate and dissolved organic matter in this cycle?
4. Microbiological quality of waters in marine coastal areas is important. What are some major concerns?

The Freshwater Environment

Lakes and rivers provide microbial environments that are different from the larger oceanic systems in many important ways. For example, in lakes, mixing and water exchange can be limited. This creates vertical gradients over much shorter distances. Changes in rivers occur over distance and/or time as water flows through river channels.

Lakes vary in nutrient status. Some are oligotrophic or nutrient-poor (figure 41.9*a*), others are **eutrophic** or nutrient-rich (figure 41.9*b*). Nutrient-poor lakes remain aerobic throughout the year, and seasonal temperature shifts do not result in distinct stratification. In contrast, nutrient-rich lakes usually have bottom sediments that contain organic matter. In thermally stratified lakes, the **epilimnion** (warm, upper layer) is aerobic, while the **hypolimnion** (deeper, colder, bottom layer) is often anaerobic (particularly if the lake is nutrient-rich). The epilimnion and hypolimnion are separated by a zone of rapid temperature decrease called the **thermocline,** and there is little mixing of water between the two layers. In the spring and fall, the aerobic surface water and the anaerobic subsurface water will turn over as the result of differences in temperature and specific gravity. After such mixing occurs, motile bacteria and algae migrate within the water column to again find their most suitable environment. When sufficiently large amounts of nutrients are added to water, **eutrophication** (nutrient enrichment) takes place and stimulates the growth of plants, algae, and bacteria. Because nitrogen and phosphorus frequently limit microbial growth in freshwater habitats, the addition of nitrogen and phosphorus compounds has a particularly large impact on aquatic systems. Depending on the body of water and the rate of nutrient addition, eutrophication may either require many centuries or occur very rapidly.

If phosphorus is added to oligotrophic water, cyanobacteria play the major role in nutrient accumulation, even in the absence of extra nitrogen. Several genera, notably *Anabaena, Nostoc,* and *Cylindrospermum,* can fix nitrogen under aerobic conditions (*see chapter 23*). The genus *Oscillatoria,* using hydrogen sulfide as an electron donor for photosynthesis, can fix nitrogen under anaerobic conditions. Even with both nitrogen and phosphorus present, cyanobacteria still can compete with algae. Cyanobacteria function more efficiently with higher pH conditions (8.5 to 9.5) and higher temperatures (30 to 35°C). Eucaryotic algae, in comparison, generally prefer a more neutral pH and have lower optimum temperatures. By using CO_2 at rapid rates, cyanobacteria also increase the pH, making the environment less suitable for eucaryotic algae.

Algae (chapter 27).

Cyanobacteria have additional competitive advantages. Many produce hydroxamates, which bind iron, making this important trace nutrient less available for the eucaryotic algae.

Cyanobacteria also often resist predation because of their production of toxins. In addition, some synthesize odor-producing compounds that affect the quality of drinking water.

Both cyanobacteria and algae can contribute to massive blooms in strongly eutrophied lakes (*see figure 40.6b*). This problem may continue for many years, until the nutrients are eventually lost from the lake by normal water flow through it or by precipitation of nutrients in bottom sediments. Lake management can improve the situation by removing or sealing bottom sediments or adding coagulating agents to speed up sedimentation.

Rivers present a different situation in that there is sufficient horizontal water movement to minimize vertical stratification; in addition, most of the functional microbial biomass is attached to surfaces. Only in the largest rivers will a greater relative portion of the microbial biomass be suspended in the water. Depending on the size of the stream or river, the source of nutrients may vary. The source may be in-stream production (autochthonous) based on photosynthetic microorganisms (figure 41.10*a*). Nutrients also may come from outside the stream (allochthonous), including runoff sediment from riparian areas (the edge of a river), or leaves and other organic matter falling directly into the water (figure 41.10*b*). Chemoorganotrophic microorganisms metabolize the available organic material and provide an energy base for the ecosystem. Under most conditions, the amounts of organic matter added to streams and rivers will not exceed the system's oxidative capacity, and productive, aesthetically pleasing streams and rivers will be maintained.

The capacity of streams and rivers to process such added organic matter is, however, limited. If too much organic matter is added, the water may become anaerobic. This is especially the case with urban and agricultural areas located adjacent to streams and rivers. The release of inadequately treated municipal wastes and other materials from a specific location along a river or stream represents a **point source of pollution.** Such point source additions of organic matter can produce distinct and predictable changes in the microbial community and available oxygen, creating an oxygen sag curve (figure 41.11). Runoff from fields and feedlots, and algal blooms in eutrophic water bodies, are examples of **nonpoint sources of pollution.**

When the amount of organic matter added is not excessive, the algae will grow, using the minerals released from the organic matter. This leads to the production of O_2 during the daylight hours, and respiration will occur at night farther down the river, resulting in **diurnal oxygen shifts.** Eventually the O_2 level approaches saturation, completing the self-purification process. This demand for oxygen is usually expressed in terms of the biochemical oxygen demand (BOD) or the chemical oxygen demand (COD).

The biochemical oxygen demand is an indirect measure of organic matter in aquatic environments. It is the amount of dissolved O_2 needed for microbial oxidation of biodegradable organic matter. When O_2 consumption is measured, the O_2 itself must be present in excess and not limit oxidation of the nutrients (table 41.3). To achieve this, the waste sample is diluted to assure that at least 2 mg/liter of O_2 are used while at least 1 mg/liter of O_2 remains in the test bottle. Ammonia released during organic matter oxidation can also exert an O_2 demand in the BOD test, so nitrification or the **nitrogen oxygen demand (NOD)** is often inhibited by nitrapyrin. In the normal BOD test, which is run for five days at 20° C on untreated samples, nitrification is not a major concern. However, when treated effluents are analyzed, NOD can be a problem.

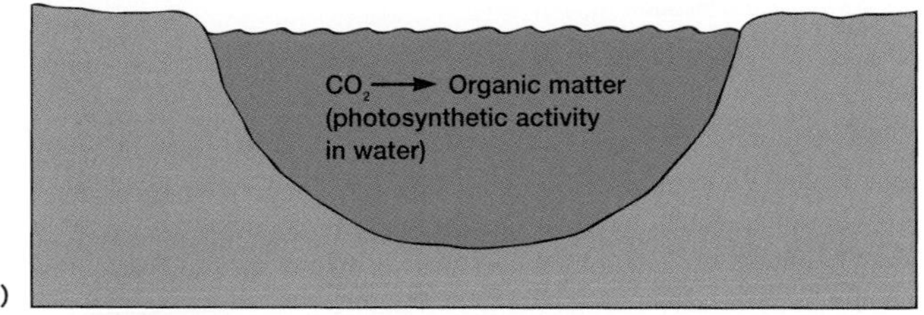

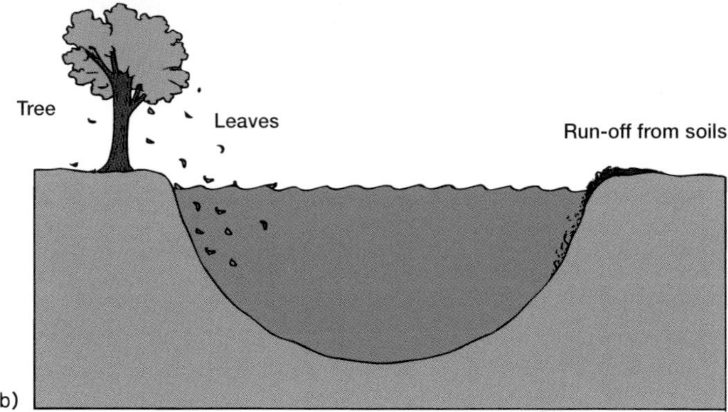

Figure 41.10 Organic Matter Sources for Lakes and Rivers. Organic matter used by microorganisms in lakes and rivers can be synthesized in the water, or can be added to the water from outside. (*a*) Within-stream sources of organic matter (autochthonous sources), primarily photosynthesis, and (*b*) sources of organic matter which enter streams and rivers from outside the water (allochthonous sources). Stream cross sections are shown.

The chemical oxygen demand test involves measuring the amount of an oxidizing agent such as permanganate that is consumed when the organic matter in a water sample is oxidized completely to CO_2. Organic matter in a sample also can be reacted directly with O_2 at a high temperature to produce CO_2. The CO_2 is then quickly measured by sensitive infrared techniques or potentiometric procedures, to give a measure of **total organic carbon** or **TOC.**

Limits on BOD and COD levels in sewage and other waste water have been established. Properly treated low BOD and COD effluents that meet the standards have minimum effects on lakes, rivers, and marine outfall areas.

Microbial biomass measurements, including estimates of the photosynthetic and heterotrophic components in a body of water, are very important in the study of aquatic microorganisms, as noted in table 40.6. The photosynthetic microorganisms in water can be measured by extracting chlorophyll and other photosynthetic pigments. By measuring absorption at a series of wavelengths under neutral and acidic conditions, one can determine the quantity of pheophytin, a chlorophyll molecule with the Mg replaced by two hydrogen atoms. This

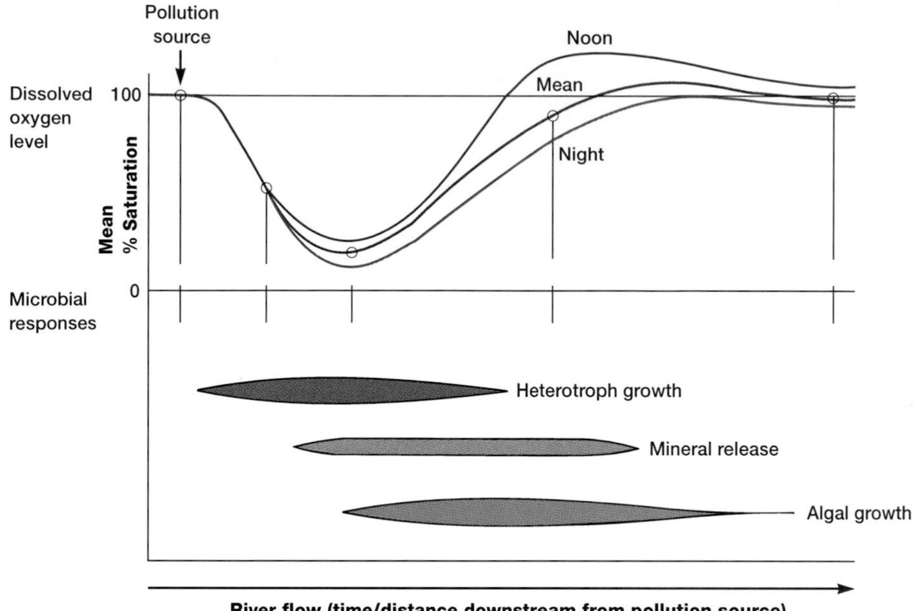

Figure 41.11 **The Dissolved Oxygen Sag Curve.** Microorganisms and their activities can create gradients over distance and time when nutrients are added to rivers. An excellent example is the dissolved oxygen sag curve, caused when organic wastes are added to a clean river system. During the later stages of self-purification, the phototrophic community will again become dominant, resulting in diurnal changes in river oxygen levels.

product is an important index of the physiological stress being placed on the algae.

Microscopic and chemical techniques can also be used to estimate microbial numbers and biomass in water. However, it is often necessary to concentrate the microorganisms by filtration or centrifugation before these measurements can be completed. Bacteria can also be separated from the algae, protozoa, aquatic insects, and animals that may be present by means of centrifugation or filtration.

If too much organic matter (BOD, COD) is added to a river or lake, however, these changes may not be easily reversed and an anaerobic foul-smelling body of water that will not support fish and other oxygen-requiring organisms can be created. At this point, the lake or river is almost "dead" biologically, although important groups of microorganisms are still present and functioning. Once a body of water reaches this point, only major remediation efforts will restore it to its original condition and allow fish and other aquatic animals to survive.

1. Define the following terms: eutrophication, epilimnion, hypolimnion, and thermocline.
2. Define COD and BOD. What are the major methods used to measure microbial biomass in water?
3. Why is filtration often required to concentrate bacteria from water? What often happens when this technique is used with soils?
4. What components should limit the reactions in a BOD test, and what components should not limit reaction rates? Why?

Nutrient Removal from Water

The aerobic self-purification sequence that occurs when organic matter is added to lakes and rivers can be carried out under controlled conditions in large basins. This process of managed nutrient removal from refuse liquids and waste matter, or **sewage treatment** (figure 41.12), involves "cultivating" the microorganisms that would normally develop in a lake or river. Such a process can be effective in minimizing environmental degradation of valuable water resources and in destroying potential human pathogens.

Table 41.3 The Biochemical Oxygen Demand (BOD) Test: A System with Excess and Limiting Components

Components in excess at the end of the incubation period
Nitrogen
Phosphorus
Iron
Trace elements
Microorganisms
Oxygen
Component limiting at the end of the incubation period
Organic matter

Sewage treatment normally involves **primary, secondary,** and **tertiary treatment,** as summarized in table 41.4. Physical, or primary, treatment can remove 20 to 30% of the BOD that is present in particulate form. In this treatment, particulate material is removed by screening, precipitation of small particulates, and settling in basins or tanks. The resulting solid material is usually called **sludge.** Biological, or secondary, treatment is then used for the removal of dissolved organic matter. About 90 to 95% of the BOD and many bacterial pathogens are removed by this process. Tertiary or advanced treatment, which can be physical and/or biological, is used to remove inorganic nitrogen, phosphorus, or other chemicals, including recalcitrant organic compounds. With special procedures, viruses can be removed or inactivated.

Several approaches can be used in secondary treatment to remove dissolved organic matter (figure 41.13). An aerobic

Figure 41.12 **A Modern Sewage Treatment Plant.** Sewage treatment plants allow natural processes of self-purification which occur in rivers and lakes to be carried out under more intense, managed conditions. The figure shows a plant in New Jersey. The trickling filters and activated sludge basins are particularly prominent.

TABLE 41.4 Major Steps in Primary, Secondary, and Tertiary Treatment of Wastes

Treatment Step	Processes
Primary	Removal of insoluble particulate materials by screening, addition of alum and other coagulation agents, and other procedures
Secondary	Biological removal of dissolved organic matter Trickling filters Activated sludge Lagoons Extended aeration systems
Tertiary	Biological removal of inorganic nutrients Chemical removal of inorganic nutrients Virus removal Trace chemical removal

activated sludge system (figure 41.13*a*) involves a horizontal flow of materials with a recycle of sludge—the active biomass that is formed when organic matter is oxidized and degraded by microorganisms. Activated sludge systems can be designed with variations in mixing. In addition, the ratio of organic matter added to the active microbial biomass can be varied. A low-rate system (low nutrient input per unit of microbial biomass) will produce a cleaner effluent. A high-rate system (high nutrient input per unit of microbial biomass) will remove more dissolved organic carbon per unit time, but produce a poorer quality effluent.

Aerobic secondary treatment is also carried out with a trickling filter (figure 41.13*b*). The waste effluent is passed over rocks or other solid materials upon which microbial films have developed, and the microbial community in these films degrades the organic waste. A sewage treatment plant can be operated to produce less sludge by employing the **extended aeration** process (figure 41.13*c*). Microorganisms grow on the dissolved organic matter, and the newly formed microbial biomass is eventually consumed by **endogenous respiration.** This requires extremely large aeration basins and extended aeration times. In addition, with the biological self-utilization of the biomass, minerals originally present in the microorganisms are again released to the water.

When the activated sludge process is run under less than optimum conditions, particularly with lower O_2 levels or with a microbial community that is too young or too old, unsatisfactory floc formation and settling can occur. The result is a **bulking sludge,** caused by the massive development of filamentous bacteria such as *Sphaerotilus* and *Thiothrix,* together with many poorly characterized filamentous organisms (figure 41.14).

All of the aerobic processes produce excess microbial biomass, or sewage sludge, which contains many recalcitrant organics. Often, the sludges from aerobic sewage treatment, together with the materials settled out in primary treatment, are further treated by **anaerobic digestion** (figure 41.15). Anaerobic digesters are large fermentation tanks designed to operate anaerobically with continuous input of untreated sludge and removal of the final, stabilized sludge product. The resulting methane is removed by a vent at the top of the tank. This digestion process involves three steps: (1) the fermentation of the sludge components to form organic acids, including acetate; (2) production of the methanogenic substrates: acetate, CO_2, and hydrogen; and finally, (3) methanogenesis by the methane producers. These methanogenic processes, summarized in table 41.5, involve critical balances between oxidants and reductants. To function most efficiently, the hydrogen concentration must be maintained at a low level. If hydrogen and organic acids accumulate, methane production can be inhibited, resulting in a stuck digester.

Methanogenic bacteria (pp. 497–500).

Anaerobic digestion has many advantages. Most of the microbial biomass produced in aerobic growth is used for methane production. The resulting sludge occupies less volume and can be dried easily. However, heavy metals and other environmental contaminants are often concentrated in the sludge. There may be longer-term environmental and public health effects from disposal of this material on land or in water (Box 41.1).

1. Explain how primary, secondary, and tertiary treatment are accomplished.
2. What is bulking sludge? Name several important microbial groups that contribute to this problem.
3. What are the steps of organic matter processing that occur in anaerobic digestion? Why is acetogenesis such an important step?
4. After anaerobic digestion is completed, why is sludge disposal still of concern?

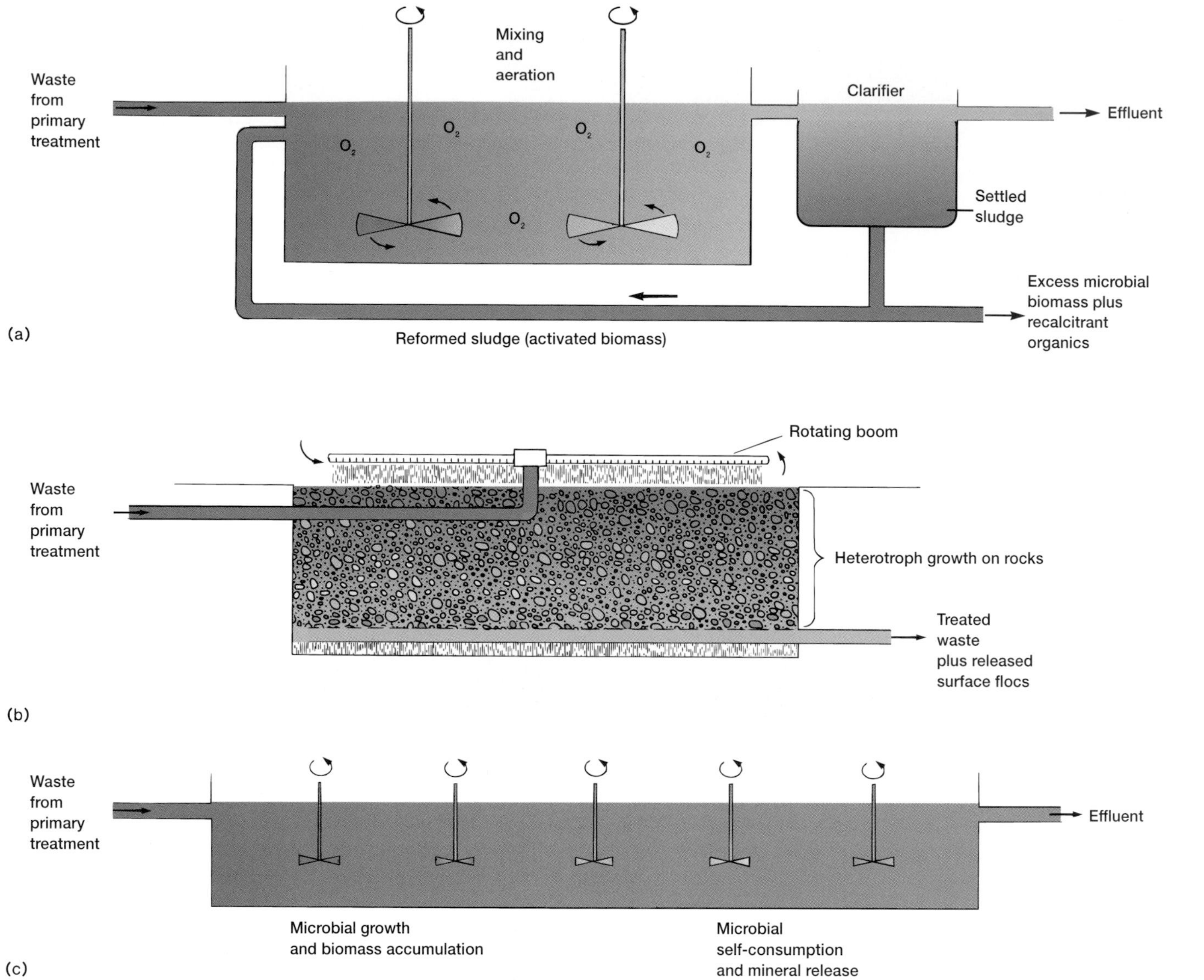

Mixing and aeration
Waste from primary treatment
O_2
Clarifier
Effluent
Settled sludge
Excess microbial biomass plus recalcitrant organics
(a)
Reformed sludge (activated biomass)

Rotating boom
Waste from primary treatment
Heterotroph growth on rocks
Treated waste plus released surface flocs
(b)

Waste from primary treatment
Effluent
Microbial growth and biomass accumulation
Microbial self-consumption and mineral release
(c)

Figure 41.13 Aerobic Secondary Sewage Treatment. Different types of reactors can be used to carry out the microbial utilization of dissolved organic matter in secondary aerobic treatment of sewage. (*a*) Activated sludge with microbial biomass recycling. (*b*) Trickling filter. (*c*) Extended aeration without biomass recycling.

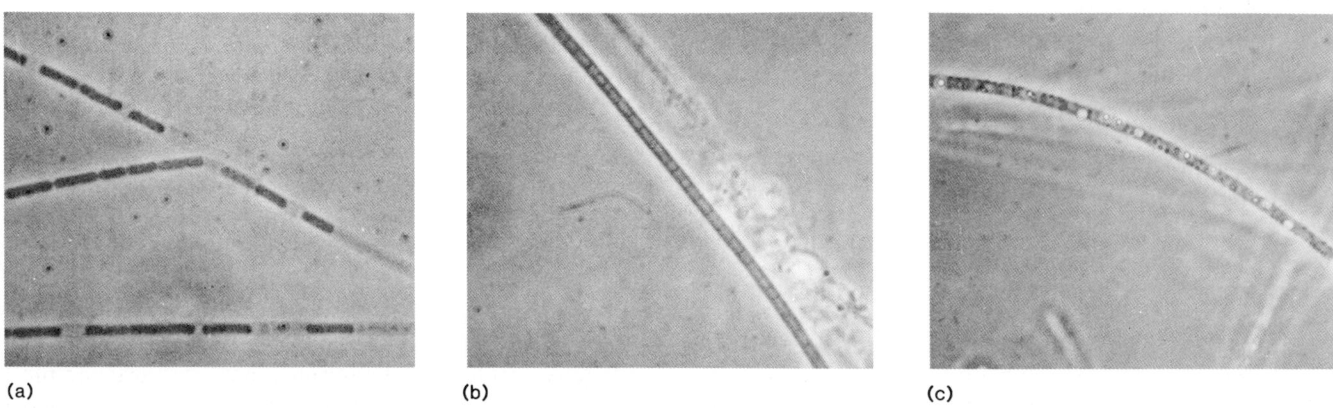

Figure 41.14 Filamentous Microorganisms in Bulking Sludge. If these form major growths during aerobic sludge treatment, they can cause bulking (nonsettling) of sludges. (*a*) *Sphaerotilus.* (*b*) *Thiothrix* without sulfide present. (*c*) *Thiothrix* with sulfide in the environment.

Table 41.5 Sequential Reactions in the Anaerobic Biological Utilization of Organic Wastes

Process Step	Substrates	Products	Major Microorganisms
Fermentation	Organic polymers	Butyrate, propionate, lactate, succinate, ethanol, acetate[a], H_2[a], CO_2[a]	Facultative and obligate anaerobes *Clostridium* *Bacteroides* *Peptostreptococcus* *Peptococcus* *Eubacterium* *Lactobacillus*
Acetogenic reactions	Butyrate, propionate, lactate, succinate, ethanol	Acetate, H_2, CO_2	*Syntrophomonas* *Syntrophobacter* *Acetobacterium*
Methanogenic reactions	Acetate	$CH_4 + CO_2$	*Methanosarcina* *Methanothrix*
	H_2 and HCO_3^-	CH_4	*Methanobrevibacter* *Methanomicrobium* *Methanogenium* *Methanobacterium* *Methanococcus* *Methanospirillum*

[a]Methanogenic substrates produced in the initial fermentation step.

Figure 41.15 High-efficiency Anaerobic Bioreactors Used for Sludge Digestion and Methane Production. These "egg-shaped" units are so well insulated that they maintain their temperature (39° C) without outside heating, and the methane can be burned to provide electricity. These reactors are located at Kiel, Germany.

Water and Disease Transmission

Since the beginning of recorded history water has been recognized as a potential carrier of disease. In order to protect his health, Alexander the Great (356–323 B.C.) had his personal drinking water carried in silver urns. Clearly, the association between noble heavy metals such as silver and the prevention of waterborne diseases was early established through chance observation. The connection between a freshwater supply and the health of an urban population was recognized by the time of the Roman Empire (27 B.C.). However, much of the technology for the protection of the water supply was subsequently lost until the middle of the nineteenth century.

Water Purification

Water purification is a critical link in promoting public health and safety. As shown in figure 41.16, water purification can involve a variety of steps, whose nature depends on the type of impurities in the raw water source. For example, if the raw water contains large amounts of iron and manganese, which will often precipitate when water is exposed to air, it may be necessary to aerate the water and employ other methods to remove these ions early in the purification sequence. Usually, municipal water supplies are purified by a process that consists of at least three or four steps. If the raw water contains a great deal of suspended material, it is often first routed to a **sedimentation basin** and held so that sand and other very large particles can settle out. The partially clarified water is then mixed with chemicals such as alum and lime and moved to a **settling basin** where more material precipitates out. This procedure is called **coagulation** or flocculation and removes microorganisms, organic matter, toxic contaminants, and suspended fine particles. After these steps, the water is further purified by passing it through a filtration unit (figure 41.17). **Rapid sand filters,** which depend on the physical trapping of fine particles and flocs, are usually used for this purpose. This filtration removes up to 99% of the remaining bacteria. After filtration, the water is treated with a disinfectant. This step usually involves chlorination, but ozonation is becoming increasingly popular. When chlorination is employed, the chlorine dose must be large enough to leave residual free chlorine at a concentration of 0.2 to 2.0 mg/liter. A concern is the creation of **halomethanes,** a group of compounds that may be carcinogens, formed when chlorine reacts with organic matter.

The preceding purification process effectively removes or inactivates disease-causing bacteria and indicator organisms (coliforms). Unfortunately, however, the use of coagulants,

Box 41.1

Sewage, Sludge, Long-Term Concerns with Land and Water Disposal: An Environmental Challenge

Sewage treatment plants have allowed large cities to develop near rivers and lakes and still maintain the quality of the water. Large quantities of sludge are produced during sewage treatment and are usually subjected to anaerobic digestion. This process transforms complex organic matter (including microorganisms that grew in the aerobic treatment process) to methane and CO_2. At the same time, heavy metals are concentrated in the residual sludge, and viable cysts of free-living protozoa may be present.

The sludge is disposed of on land or, in large urban areas such as New York, by dumping at designated disposal sites in coastal waters. When such sludges are dumped offshore, free-living protozoa of the genus *Acanthamoeba* may be released in the water, where they may infect bathers. *Acanthamoeba* infections are more widespread than is usually recognized, and the organisms can be common contaminants in water that tests negative for coliforms and fecal coliforms.

The use of water-based waste disposal systems has, of course, allowed major improvements in urban living and the retirement of chamber pots to museums. A product of this advancement in public health is sewage sludge. NIMBY (not in my backyard) has been the usual means of managing this problem.

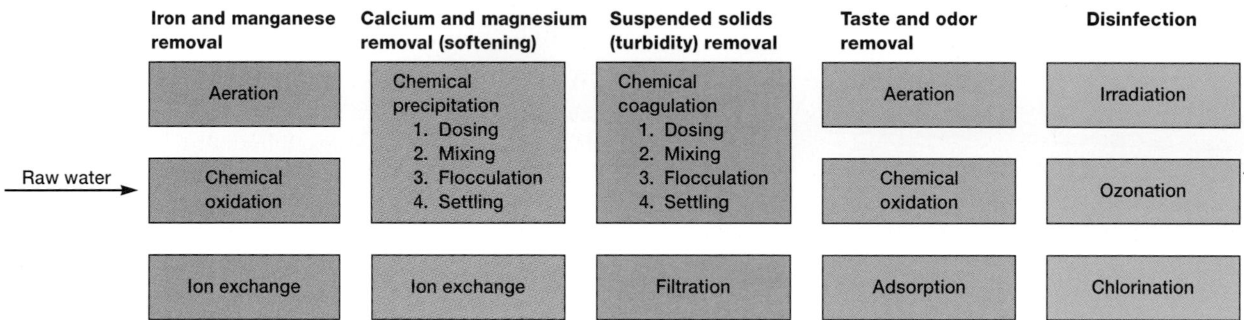

Figure 41.16 Water Purification. Several alternatives can be used for drinking water treatment, depending on the initial water quality. These can include varied physical purification steps and disinfection.

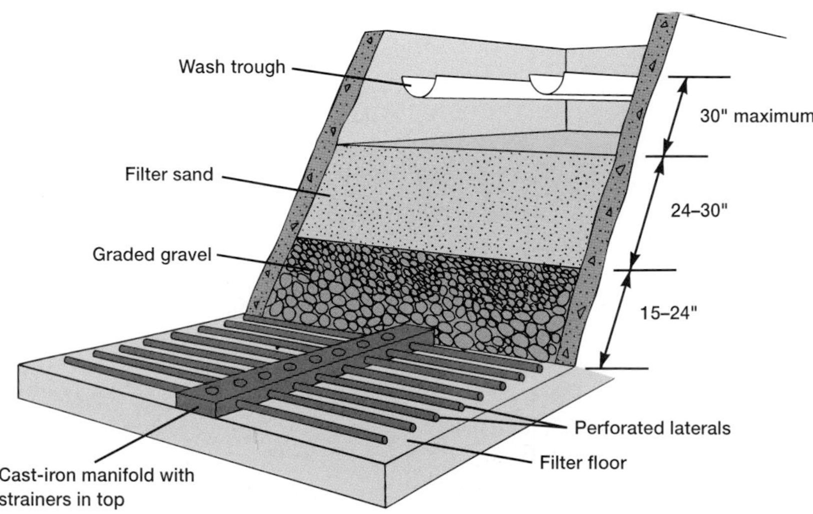

Figure 41.17 Water Filtration. Physical filtration is an important step in drinking water treatment. This is a cross section of a typical sand filter showing layers of sand and graded gravel.

Box 41.2

Waterborne Diseases, Water Supplies, and Slow Sand Filtration: The Return of a Time-Tested Concept in Drinking Water Treatment

Slow sand filtration, in which drinking water is passed through a sand filter that develops a layer of microorganisms on its surface, has had a long and interesting history. After London's severe cholera epidemic of 1849, Parliament, in an act of 1852, required that the entire water supply of London be passed through slow sand filters before use.

The value of this process was shown in 1892, when a major cholera epidemic occurred in Hamburg, Germany, and 10,000 lives were lost. The neighboring town, Altona, which used slow sand filtration, did not have a cholera epidemic. Slow sand filters were installed in many cities in the early 1900s, but the process fell into disfavor with the advent of rapid sand filters, chlorination, and the use of coagulants such as alum. Slow sand filtration, a time-tested process, is coming back into favor because of its filtration effectiveness and lower maintenance costs.

rapid filtration, and chemical disinfection often does not consistently and reliably remove viruses and *Giardia lamblia* cysts. *Giardia* is now recognized as the most common identified waterborne pathogen in the United States. The protozoan, first observed by Leeuwenhoek in 1681, has trophozoite and cyst forms. The disease is often called traveler's disease and is transmitted primarily through untreated stream water or undependable municipal water supplies. More consistent removal of *Giardia* cysts can be achieved with **slow sand filters.** This treatment involves the slow passage of water through a bed of sand in which a microbial layer covers the surface of each sand grain. It is biological rather than physical. Waterborne microorganisms are removed by adhesion to the surface microbial layer (Box 41.2).

Giardia *(pp. 553; 792–93).*

Viruses in drinking water must also be destroyed or removed. Coagulation and filtration do reduce virus levels about 90 to 99%. Further inactivation of viruses by chemical oxidants, high pH, and photooxidation may yield a reduction as great as 99.9%. None of these processes, however, is considered to provide sufficient protection. The World Health Organization (WHO) has recommended that less than one infective unit should be present per 380 liters (100 gallons) of water.

Microbiological Analysis of Water Purity

Monitoring and detection of indicator and disease-causing microorganisms are a major part of sanitary microbiology. By chlorinating drinking water supplies, control of most major disease-causing bacteria can be achieved. The inability to consistently remove viruses and protozoa and the lack of water quality standards for these microorganisms are of continuing concern. Additional problems are caused by recreational bodies of water and areas where shellfish are harvested. In these locations, water treatment barriers against contamination may not exist, and transmission of microbial diseases can occur.

Bacteria from the intestinal tract generally do not survive in the aquatic environment, are under physiological stress, and gradually lose their ability to form colonies on differential and selective media. Their die-out rate depends on the water temperature, the effects of sunlight, the populations of other bacteria present, and the chemical composition of the water. Procedures have been developed to "resuscitate" these stressed coliforms before they are identified using selective and differential media.

A wide range of viral, bacterial, and protozoan diseases result from the contamination of water with human fecal wastes (*see chapters 36, 37, and 39*). Although many of these pathogens can be detected directly, environmental microbiologists have generally used **indicator organisms** as an index of possible water contamination by human pathogens. Researchers are still searching for the "ideal" indicator organism to use in sanitary microbiology. The following are among the suggested criteria for such an indicator:

1. The indicator bacterium should be suitable for the analysis of all types of water: tap, river, ground, impounded, recreational, estuary, sea, and waste.
2. The indicator bacterium should be present whenever enteric pathogens are present.
3. The indicator bacterium should survive longer than the hardiest enteric pathogen.
4. The indicator bacterium should not reproduce in the contaminated water and produce an inflated value.
5. The assay procedure for the indicator should have great specificity; in other words, other bacteria should not give positive results. In addition, the procedure should have high sensitivity and detect low levels of the indicator.
6. The testing method should be easy to perform.
7. The indicator should be harmless to humans.
8. The level of the indicator bacterium in contaminated water should have some direct relationship to the degree of fecal pollution.

Coliforms, including *Escherichia coli,* are members of the family *Enterobacteriaceae*. These bacteria make up approxi-

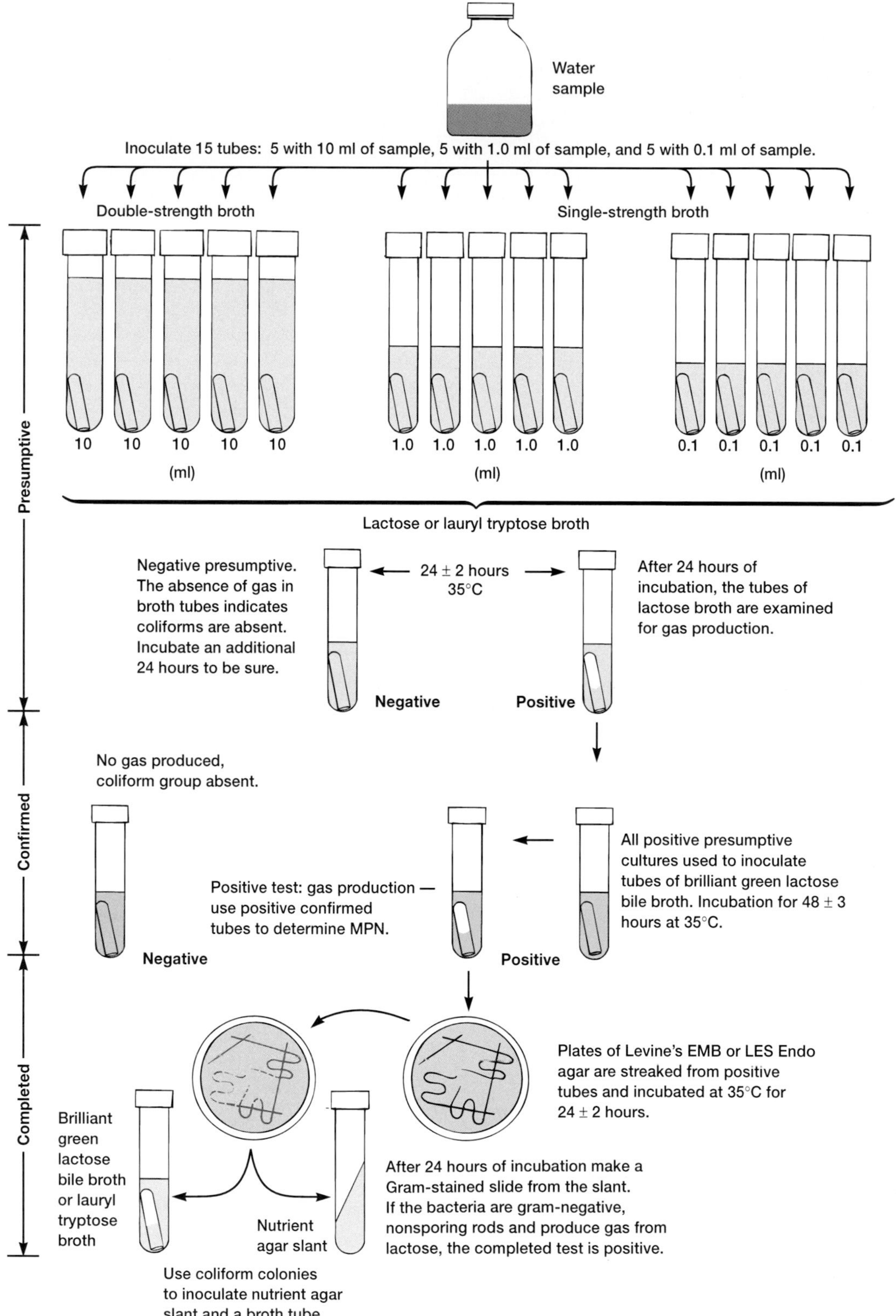

Figure 41.18 The Multiple-Tube Fermentation Test. The multiple-tube fermentation technique has been used for many years for the sanitary analysis of water. Lactose broth tubes are inoculated with different water volumes in the presumptive test. Tubes that are positive for gas production are inoculated into brilliant green lactose bile broth in the confirmed test, and positive tubes are used to calculate the most-probable-number (MPN) value. The completed test is used to establish that coliform bacteria are present.

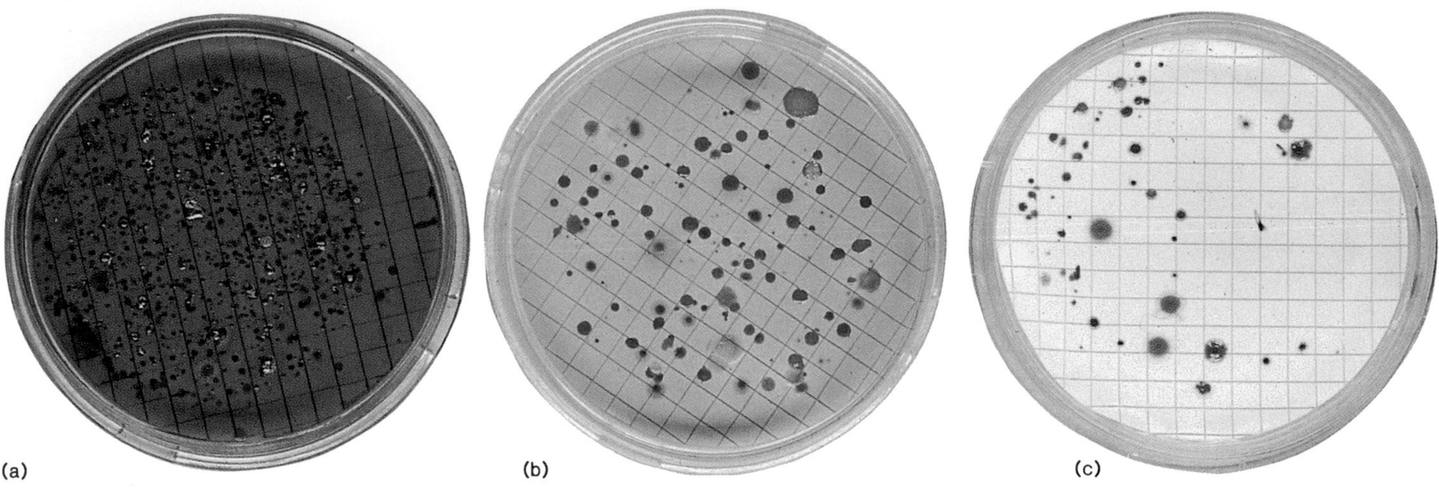

Figure 41.19 **Coliform and Enterococcal Colonies.** Membrane filters made it possible to more rapidly test waters for the presence of coliforms, fecal coliforms, and fecal enterococci by the use of differential media. (*a*) Coliform reactions on an Endo medium. (*b*) Fecal coliform growth on a bile salt medium (m FC agar) containing aniline blue dye. (*c*) Fecal enterococci growing on an azide-containing medium (KF agar) with TTC added to allow better detection of colonies.

mately 10% of the intestinal microorganisms of humans and other animals (*see figure 29.8*) and have found widespread use as indicator organisms. They lose viability in water at slower rates than most of the major intestinal bacterial pathogens. When such "foreign" enteric indicator bacteria are not detectable in a specific volume (100 ml) of water, the water is considered **potable** (Latin *potabilis,* fit to drink), or suitable for human consumption.

The Enterobacteriaceae (pp. 435–38).

The coliform group includes *E. coli, Enterobacter aerogenes,* and *Klebsiella pneumoniae.* Coliforms are defined as facultatively anaerobic, gram-negative, non-sporing, rod-shaped bacteria that ferment lactose with gas formation within 48 hours at 35° C. The original test for coliforms that was used to meet this definition involved the presumptive, confirmed, and completed tests, as shown in figure 41.18. The presumptive step is carried out by means of tubes inoculated with three different sample volumes to give an estimate of the **most probable number (MPN)** of coliforms in the water. The complete process, including the confirmed and completed tests, requires at least four days of incubations and transfers.

Unfortunately, the coliforms include a wide range of bacteria whose primary source may not be the intestinal tract. To deal with this difficulty, tests have been developed that allow waters to be tested for the presence of **fecal coliforms.** These are coliforms derived from the intestine of warm-blooded animals, which can grow at the more restrictive temperature of 44.5° C.

To test for coliforms and fecal coliforms, and more effectively recover stressed coliforms, a variety of simpler and more specific tests have been developed. These include the membrane filtration technique, the **presence-absence (P-A) test** for coliforms, and the related Colilert **defined substrate test** for detecting both coliforms and *E. coli.*

The **membrane filtration technique** (*see figure 6.7*), has become a common and often preferred method of evaluating the microbiological characteristics of water. The water sample is passed through a membrane filter. The filter with its trapped bacteria is transferred to the surface of a solid medium or to an absorptive pad containing the desired liquid medium. Use of the proper medium allows the rapid detection of total coliforms, fecal coliforms, or fecal streptococci by the presence of their characteristic colonies (figure 41.19; *see also figure 6.8*). Samples can be placed on a less selective resuscitative medium, or incubated at a less stressful temperature, prior to growth under the final set of selective conditions. An example of a resuscitation step is the use of a 2-hour incubation on a pad soaked with lauryl sulfate broth, as is carried out in the LES Endo procedure. A resuscitation step is often needed with chlorinated samples, where the microorganisms are especially stressed. The advantages and disadvantages of the membrane filter technique are summarized in table 41.6. Membrane filters have been widely used with water that does not contain high levels of background organisms, sediment, or heavy metals.

More simplified tests for detecting coliforms and fecal coliforms are now available. The presence-absence test (P-A test) can be used for coliforms. This is a modification of the MPN procedure, in which a larger water sample (100 ml) is incubated in a single culture bottle with a triple-strength broth containing lactose broth, lauryl tryptose broth, and bromcresol purple indicator. The P-A test is based on the assumption that no coliforms should be present in 100 ml of drinking water. A positive test results in the production of acid (a yellow color) and constitutes a positive presumptive test requiring confirmation.

To test for both coliforms and *E. coli,* the related Colilert defined substrate test can be used. A water sample of 100 ml is added to a specialized medium containing 0-nitrophenyl-

TABLE 41.6 Advantages and Disadvantages of the Membrane Filter Technique for Evaluation of the Microbial Quality of Water

Advantages

- Good reproducibility
- Single-step results often possible
- Filters can be transferred between different media
- Large volumes can be processed to increase assay sensitivity
- Time savings are considerable
- Ability to complete filtrations on site
- Lower total cost in comparison with MPN procedure

Disadvantages

- High-turbidity waters limit volumes sampled
- High populations of background bacteria cause overgrowth
- Metals and phenols can adsorb to filters and inhibit growth

Source: A. E. Greenberg, R. R. Trussell, and L. S. Clesceri. *Standard Methods for the Examination of Water and Wastewater,* 16th ed. American Public Health Association, Washington, D.C. 1985, p. 886.

β-D-galactopyranoside (**ONPG**) and 4-methylumbelliferyl-β-D-glucuronide (**MUG**) as the only nutrients. If coliforms are present, the medium will turn yellow within 24 hours at 35° C due to the cleavage of the ONPG, as shown in figure 41.20. To check for *E. coli,* the medium is observed under long-wavelength UV light for fluorescence. When *E. coli* is present, the MUG is modified to yield a fluorescent product. If the test is negative for the presence of coliforms, the water is considered acceptable for human consumption. The main change from previous standards is the requirement to have waters free of coliforms and fecal coliforms. If coliforms are present, fecal coliforms or *E. coli* must be tested for.

In the United States, a set of general guidelines for microbiological quality of drinking waters has been developed, including standards for coliforms, viruses, and *Giardia* (table 41.7). If unfiltered surface waters are being used, one coliform test must be run each day when the waters have higher turbidities.

Other indicator microorganisms include **fecal enterococci.** The fecal enterococci are increasingly being used as an indicator of fecal contamination in brackish and marine water. In salt water, these bacteria die back at a slower rate than the fecal coliforms, providing a more reliable indicator of possible recent pollution.

The genus Enterococcus *(pp. 455–57).*

Water-Based Diseases

Many important human pathogens are maintained in association with living organisms other than humans, including many wild animals and birds (*see table 35.1*). Some of these bacterial and protozoan pathogens can survive in water and infect humans. As examples, *Vibrio vulnificus, V. parahaemolyticus,* and *Legionella* are of continuing concern. When waters are used for recreation or are a source of seafood that is consumed uncooked, the possibility for disease transmission certainly exists. A food-borne infection may have serious consequences for immunologically compromised individuals. Some major waterborne diseases are summarized in table 41.8 and discussed in chapter 37.

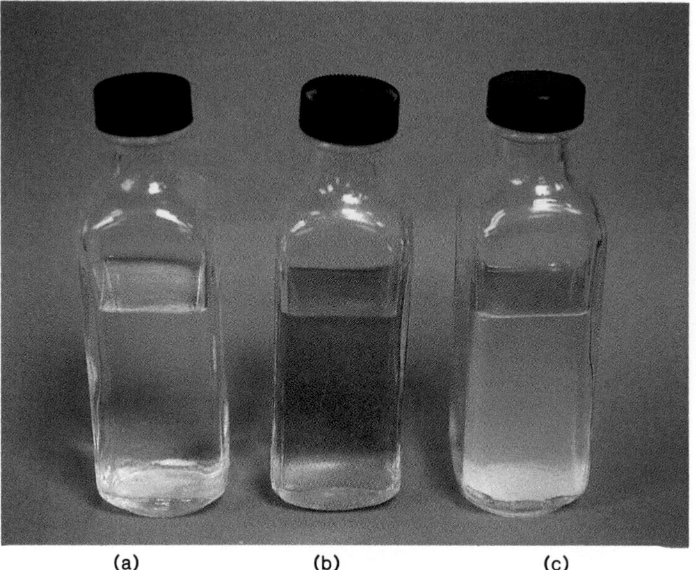

Figure 41.20 The Defined Substrate Test. This much simpler test is now being used to detect coliforms and fecal coliforms in single 100 ml water samples. The medium uses ONPG and MUG (see text) as defined substrates. (*a*) Uninoculated control. (*b*) Yellow color due to the presence of coliforms. (*c*) Fluorescent reaction due to the presence of fecal coliforms.

TABLE 41.7 Current Microbiological Standards for Water in the United States, 1991

Agent	Allowable Limits/100 ml
Coliforms (MCL[a]) per 100 ml	1 positive sample found in <40 samples/month; <5% positive samples found if >40 samples/month
Fecal coliforms	None
Giardia	None
Enterococci	None
Viruses	None

Source: L. S. Clesceri, A. E. Greenberg, and R. R. Trussell, *Standard Methods for the Examination of Water and Wastewater,* 17th ed. American Public Health Association, Washington, DC, 1989.

[a]Maximum concentration limit. Above this limit, one is in violation of the guidelines. The number of analyses to be completed per month depends on the city population.

TABLE 41.8 Water-Based Microbial Pathogens That Can Be Maintained in the Environment Independent of Humans

Organism	Reservoir	Comments
Bacteria		
Aeromonas hydrophila	Free-living	Sometimes associated with gastroenteritis, cellulitis, and other diseases
Chromobacterium violaceum	Soil runoff	Common saprophyte in water—rare in terms of human infection
Mycobacterium	Infected animals and free-living	Complex recovery procedure required
Leptospira	Infected animals	Hemorrhagic effects, jaundice
Legionella pneumophila	Free-living and associated with protozoa	Found in cooling towers, evaporators, condensers, showers, and other water sources.
Vibrio cholera	Free-living	Found in many waters and estuaries
Vibrio parahaemolyticus	Free-living in coastal waters	Causes diarrhea in shellfish consumers
Salmonella enteriditis	Animal intestinal tracts	Common in many waters.
Campylobacter	Bird and animal reservoirs	Major cause of diarrhea; common in processed poultry; a microaerophile
Yersinia enterocolitica	Frequent in animals and in the environment	Waterborne gastroenteritis
Pseudomonas aeruginosa	Free-living	Swimmer's ear and related infections
Protozoa		
Giardia lamblia	Beavers, sheep, dogs, cats	Major cause of early spring diarrhea; important in cold mountain water
Cryptosporidium	Many species of domestic and wild animals	Causes acute enterocolitis; important with immunologically compromised individuals; cysts resistant to chemical disinfection
Acanthamoeba	Sewage sludge disposal areas	Can cause granulomatous amebic encephalitis (GAE); keratitis, corneal ulcers
Naegleria fowleri	Warm water (hot tubs), swimming pools, lakes	Inhalation in nasal passages; central nervous system infection; causes primary amebic meningoencephalitis (PAM)

Waterborne bacterial and viral pathogens are extensively discussed in chapters 36 and 37, and the protozoan *Giardia* is mentioned earlier and in chapter 39. Two other important protozoan pathogens are noted here. The protozoan *Cryptosporidium* occurs in the intestine of many species of domestic and wild animals. This organism produces acute enterocolitis, especially serious for immunologically compromised individuals, such as those with AIDS (*see table 36.3*).

Protozoan diseases (pp. 790–98).
Food-borne and waterborne viral diseases (pp. 735–36).
Food-borne and waterborne bacterial diseases (pp. 763–67).

Another waterborne protozoan disease of increasing concern worldwide is primary amebic meningoencephalitis (PAM), an infection of the central nervous system caused by *Naegleria fowleri.* The disease usually occurs in children or young adults who have been swimming in lakes or pools or have been waterskiing. After nasal infection, the protozoan reaches the brain and initiates an inflammatory response. The result is usually fatal. Warm water and heated industrial effluents promote the growth of this protozoan.

1. Describe the way in which municipal drinking water is usually purified. Why might a slow sand filter be useful in this process?
2. How is a coliform defined? How does this definition relate to presumptive, confirmed, and completed tests? What is an indicator organism, and what properties should it have?
3. How does one differentiate between coliforms and fecal coliforms in the laboratory?
4. In what type of environment is it better to use fecal enterococci rather than fecal coliforms as an indicator organism?
5. What are the advantages and disadvantages of membrane filters for microbiological examinations of water?
6. Why has the defined substrate test with ONPG and MUG been accepted as a test of drinking water quality?
7. Give a few examples of waterborne diseases caused by protozoa and bacteria that are maintained in the environment independent of humans.
8. A *Naegleria fowleri* infection leads to fatal nervous system effects. What is the portal of entry for this organism?

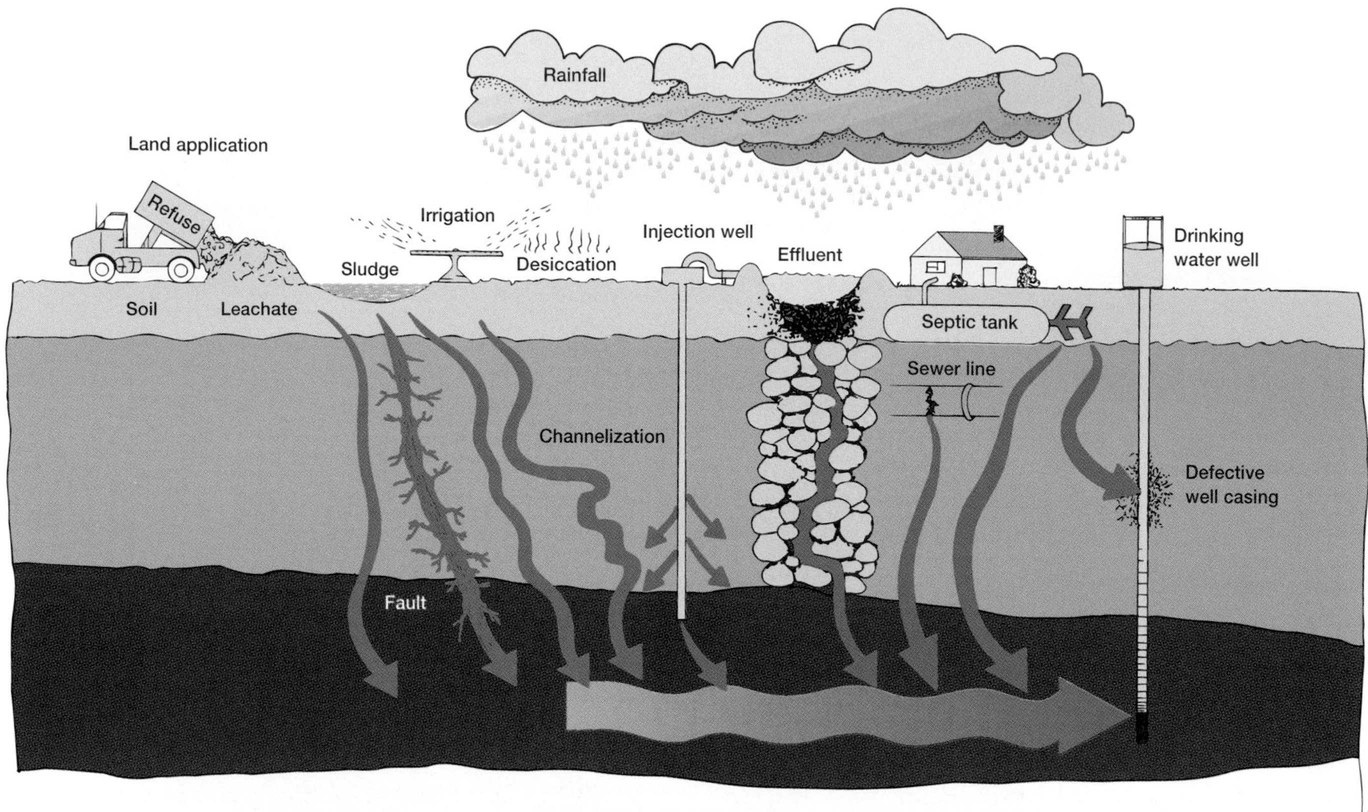

Figure 41.21 **Groundwater Contamination.** Groundwater is used as a drinking water source by many people. This important water source can be affected by many types of chemical pollutants and disease-causing microorganisms. Important soil factors influencing the fate of chemical and microbiological contaminants in water include texture, organics (humic acids), ionic strength, adsorption, structure, pH, and permeability. *From G. Bitton and C. P. Gerba,* Groundwater Pollution Microbiology. *Copyright © John Wiley & Sons, Inc., New York, NY.*

Groundwater Quality and Home Treatment Systems

Groundwater, or water in gravel beds and fractured rocks below the surface soil, is a widely used but often unappreciated water resource. In the United States, groundwater supplies at least 100 million people with drinking water, and in rural and suburban areas beyond municipal water distribution systems, 90 to 95% of all drinking water comes from this source.

The great dependence on this resource has not resulted in a corresponding understanding of microorganisms and microbiological processes that occur in the groundwater environment. Increasing attention is now being given to predicting the fate and effects of groundwater contamination (figure 41.21) on the chemical and microbiological quality of this resource. Pathogenic microorganisms and dissolved organic matter are removed from water during subsurface passage through adsorption and trapping by fine sandy materials, clays, and organic matter. Microorganisms associated with these materials—including predators such as protozoa—can use the trapped pathogens as food. This results in purified water with a lower microbial population.

This combination of adsorption-biological predation is used in home treatment systems (figure 41.22). Conventional **septic tank** systems include an anaerobic liquefaction step that occurs in the septic tank itself. This is followed by organic matter adsorption and entrapment of microorganisms in the aerobic leach-field environment where biological oxidation occurs. A septic tank may not operate correctly for several reasons. It will not function properly if the retention time of the waste in the septic tank is too short. Retention time decreases when the flow is too rapid or when excessive sludge has accumulated in the septic tank. As a result, undigested solids move into the leach field, gradually plugging the system. If the leach field floods and becomes anaerobic, biological oxidation does not occur, and effective treatment ceases.

When a suitable soil is not present and the septic tank outflow drains too rapidly to the deeper subsurface, problems can occur. Fractured rocks and coarse gravel materials provide little effective adsorption or filtration. This may result in the contamination of well water with pathogens and the transmission of disease. In addition, phosphorus from the waste will not be retained effectively and may pollute the groundwater. This often leads to nutrient enrichment of ponds, lakes, and rivers as the subsurface water enters these environmentally sensitive water bodies.

Subsurface zones are also contaminated with pollutants from other sources. Land disposal of sewage sludges, illegal dumping of septic tank pumpage, improper toxic waste disposal, and runoff from agricultural operations all contribute to groundwater contamination with chemicals and microorganisms. Deep-well injection of industrial wastes has raised questions about the longer-term fate and effects of these materials.

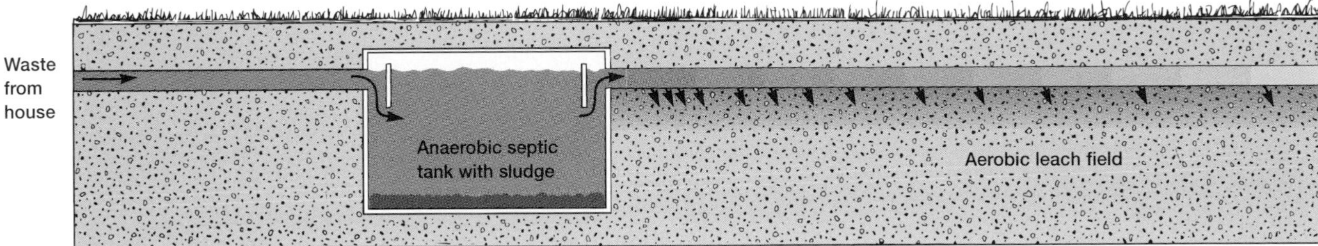

Figure 41.22 **The Septic Tank Home Treatment System.** This system combines an anaerobic waste liquefaction unit (the septic tank) with an aerobic leach field. Biological oxidation of the liquefied waste takes place in the leach field, unless the soil becomes flooded.

The subsurface environment was once thought to be essentially sterile. More recently, viable microorganisms have been discovered down to a depth of 1,000 to 1,200 meters. The subsurface has several distinct limitations for microbial survival and functioning. Nutrients such as nitrogen and phosphorus may be present at very low levels and limit microbial growth. Although most pristine subsurface waters are aerobic, after contamination with organic matter, they quickly become anaerobic. With slow diffusion of O_2 into some subsurface waters, even low levels of organic substrates will render an environment anaerobic. This change cannot be easily corrected. Oxygen, peroxide, nutrients, and microorganisms can be added to enhance degradation in **bioremediation** programs.

Many pollutants that reach the subsurface will persist and may affect the quality of groundwater for extended periods. Much research is being conducted to find ways to treat groundwater, either in place—***in situ*** **treatment**—or above ground. Microorganisms and microbial processes are used in many of these remediation efforts.

1. In rural areas, approximately what percentage of the water used for human consumption is groundwater?
2. What factors can limit microbial activity in subsurface environments? Consider the energetic and nutritional requirements of microorganisms in your answer.
3. How, in principle, are a septic tank system and a leach-field system supposed to work? What factors can reduce the effectiveness of this system?

Summary

1. Marine and freshwater environments are dominated by the water phase, and photosynthetic microorganisms provide much organic matter for use by other organisms.
2. Microbial mats are complex layered communities found in marine and freshwater environments, and also in the fossil record. Some of these fossil mats are more than 3.5 billion years old. Sessile, prosthecate, and gliding microorganisms are also prominent in marine and freshwater environments.
3. Oxygen is present at low levels in water, and if sufficient organic matter is present, heterotrophic microorganisms can use this dissolved O_2 more rapidly than it can be replenished. Respiration or decomposition of algal and cyanobacterial blooms also can cause O_2 depletion.
4. In marine systems, most water is at a low temperature, without light, and under high pressure. Hydrothermal and methane seeps provide unique environments for specialized microorganisms, many of which have developed special symbiotic relations with higher organisms such as clams and tubular worms.
5. The biochemical oxygen demand (BOD) test is crude measure of organic matter that can be oxidized by the aerobic microbial community. In this assay, oxygen should never limit the rate of reaction.
6. With the addition of inorganic nutrients, cyanobacteria and algae play major roles in the accumulation of organic matter in water.
7. Eutrophication can be caused by nutrient releases from rural and urban areas. Sources of nitrogen and phosphorus are particularly important in eutrophication.
8. The addition of organic matter can produce dissolved O_2 sag curves and diurnal O_2 changes in the latter stages of self-purification.
9. Sewage treatment is a controlled intensification of natural self-purification processes, and it can involve primary, secondary, and tertiary treatment.
10. Anaerobic digestion involves the stepwise degradation of organic matter to produce methane.
11. Municipal drinking water is purified through processes such as sedimentation, coagulation with chemicals, sand filtration, and chlorination.
12. Indicator organisms are used to indicate the presence of pathogenic microorganisms. Most probable number (MPN) and membrane filtration procedures are employed to estimate the number of indicator organisms present.
13. Enterococci are better indicator organisms for use in salt water than are coliforms because of their slower relative die-back rates.
14. Presence-absence (P-A) tests for coliforms and defined substrate tests for coliforms and *E. coli* allow 100 ml water volumes to be tested with minimum time and materials.
15. Waterborne bacterial and protozoan diseases are very important in recreational water and water containing shellfish that are eaten raw.
16. Home treatment systems operate on general self-purification principles. The septic tank provides anaerobic liquefaction, whereas the aerobic leach field allows oxidation of the soluble effluent.
17. Groundwater is an important resource that can be affected by pollutants from septic tanks and other sources. This vital water source must be protected and improved.

Key Terms

Questions for Thought and Review

1. You wish to study the characteristics of microorganisms from a 10,000 meter depth in the ocean. What effects might decompression have on these organisms, and what type of equipment would you design to work with them under their natural conditions?
2. Which microorganisms might limit the acceptability of clear high mountain water as drinking water? How might one best treat this water before taking a drink?
3. How might it be possible to rejuvenate an aging eutrophic lake? Consider physical, chemical, and biological approaches.
4. Bulking sludge is a major problem in sewage treatment, especially when activated sludge plants are overloaded. As low O_2 levels are often a cause of this problem, how might existing processes be modified in terms of improving O_2 availability?
5. Why are indicator organisms still being used despite the fact that methods are available for the direct isolation of most pathogens found in water?
6. What alternatives, if any, can one use for protection against microbiological infection when swimming in polluted recreational water? Assume that you are part of a water rescue team.
7. Discuss the threats to the purity of our groundwater and suggest ways in which groundwater quality can be recovered and preserved.

Additional Reading

The references provided at the end of chapter 40 may also be consulted for further information.

Anonymous. 1990. An ocean of viruses may affect global cycles. *ASM News.* 56(12):632–33.

Appenzeller, T. 1991. Fire and ice under the deep-sea floor. *Science* 252:1790–92.

Austin, B. 1988. *Marine microbiology.* New York: Cambridge University Press.

Brown, J.; Colling, A.; Park, A.; Phillips, J.; Rothery, D.; and Wright, J. 1989. *The ocean basins: Their structure and evolution.* New York: Pergamon Press (The Open University).

Childress, J. J.; Felbeck, H.; and Somero, G. N. 1987. Symbiosis in the deep sea. *Sci. Am.* 256:114–20.

Cohen, Y., and Rosenberg, E. 1989. *Microbial mats. Physiological ecology of benthic microbial communities.* Washington, D.C.: American Society for Microbiology.

DesManas, D. J. 1990. Microbial mats and the early evolution of life. *Trends Ecol. & Evol.* 5(5):140–44.

duMoulin, G. C., and Stottmeier, K. D. 1986. Waterborne mycobacteria: An increasing threat to health. *ASM News* 52:525–29.

Gage, J. D., and Tayler, P. A. 1991. *Deep-sea biology: A natural history of organisms at the deep-sea floor.* Cambridge: Cambridge University Press.

Greenberg, A. E.; Clesceri, L. S.; and Eaton, A. D., eds. 1992. *Standard Methods for the examination of water and wastewater,* 18th ed. Washington, D.C.: American Public Health Association.

Heldal, M., and Bratbak, G. 1991. Production and decay of viruses in aquatic environments. *Mar. Ecol. Prog. Ser.* 72:205–12.

Hill, M. 1991. *Nitrates and nitrites in food and water.* Chichester: Ellis Horwood Limited.

Jannasch, H. W., and Taylor, C. D. 1984. Deep-sea microbiology. *Ann. Rev. Microbiol.* 38:487–514.

Kirchman, D. L.; Suzuki, Y.; Garside, C.; and Ducklow, H. W. 1991. High turnover rates of dissolved organic carbon during a spring phytoplankton bloom. *Nature* 352:612–14.

La Riviere, J. W. M. 1977. Microbial ecology of liquid waste treatment. *Adv. Microbial Ecol.* 1:215–59.

McFeters, G. A. 1989. *Drinking water microbiology: Progress and recent developments.* New York: Springer-Verlag.

Marshall, K. 1992. Biofilms; An overview of bacterial adhesion, activity, and control at surfaces. *ASM News* 58(4):202–7.

Montgomery, J. M. 1985. *Water treatment: Principles and design.* New York: John Wiley and Sons.

Myres, F. S., and Anderson, A. 1992. Microbes from 20,000 feet under the sea. *Science* 255:28–29.

Pye, V. I., and Patrick, R. 1983. Groundwater contamination in the United States. *Science* 221:713–18.

Rheinheimer, G. 1991. *Aquatic microbiology* 4th ed. New York: John Wiley and Sons.

Sayles, F. L. 1992. Biogeochemical processes on the sea floor. *Oceanus* 35:68–75.

Sherr, E. B., and Sherr, B. F. 1991. Planktonic microbes: Tiny cells at the base of the ocean's food webs. *Trends Ecol. & Evol.* 6(2):50–54.

Ward, C. H.; Giger, W.; and McCarty, P. L., eds. 1985. *Groundwater quality.* New York: John Wiley and Sons.

Wells, M. L.; Mayer, L. M.; and Guillard, R. R. L. 1991. Evaluation of iron as a triggering factor for red tide blooms. *Mar. Ecol. Prog. Ser.* 69:93–102.

Williamson, P., and Holligan, P. M. 1990. Ocean production and climate change. *Trends Ecol. & Evol.* 5(9):299–303.

CHAPTER 42
The Terrestrial Environment

They [the leaves] that waved so loftily, how contentedly, they return to dust again and are laid low, resigned to lie and decay at the foot of the tree and afford nourishment to new generations of their kind, as well as to flutter on high!

—Henry D. Thoreau

Outline

Concepts

1. Soils are dominated by the solid phase, consisting of organic and inorganic components.
2. Most microorganisms in soils are associated with surfaces, and these surfaces influence microbial use of nutrients and interactions with plants, and with other living organisms.
3. Insects, nematodes, and other soil animals are important parts of the soil. These organisms interact with microorganisms to influence nutrient cycling and other processes.
4. Soils differ in the relationship between organic matter accumulation (primary production) and decomposition.
5. Many plants can develop associations with different types of microorganisms. Such associations make it possible for the plants to compete more effectively for water and nutrients.
6. Composting is a useful technique for maintaining and increasing the fertility of soils.
7. Degradation and degradability are central to an understanding of the fate and effects of pesticides and other chemicals in soils.
8. Soil microorganisms interact with the atmosphere by serving as nucleating agents that can increase precipitation. Soil microorganisms can also degrade gaseous pollutants.

The general characteristics of microorganisms in natural environments are discussed in chapter 40. In this chapter, these general concepts are considered in relation to soils.

In soils, the solid phase (organic and inorganic components) is dominant, and most soils, with some important exceptions, are predominantly aerobic. Vascular plants are the major producers of organic matter. This organic matter accumulates aboveground as leaves and branches, which become litter materials, and belowground, as roots, which grow and die. Algae and cyanobacteria also contribute to the accumulation of organic matter in desert environments.

Soils are not static. They respond to changes in temperature and moisture and to the effects of plowing and other disturbances. Usually, the microorganisms that are a part of soils play important roles in such responses.

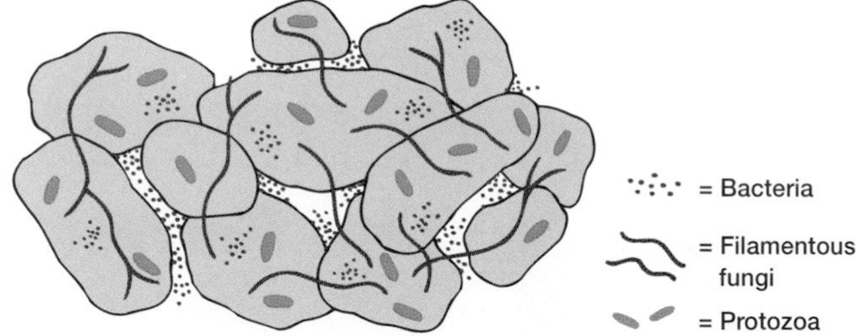

Figure 42.1 The Microenvironment—the World of Microorganisms in Soil. Bacteria tend to be present as isolated microcolonies on surfaces and in pores. Filamentous fungi are able to grow on and between these aggregated particles, or peds. Protozoa move in water films and graze the microorganisms.

Soils are dominated by a solid phase and provide a unique environment for microorganisms. The conditions for microbial life vary widely with differences in soil characteristics and environmental conditions.

Soil is also the habitat for organisms other than bacteria and fungi, including soil-dwelling protozoa, insects, nematodes, and other animals. These organisms and plants also contribute to the formation and maintenance of soils.

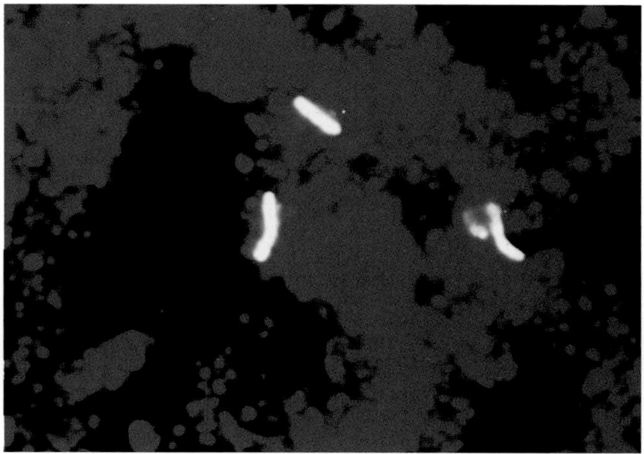

Figure 42.2 Soil Bacteria. Microscopy can be used to observe microorganisms in their natural habitats. Soil bacteria viewed by fluorescence microscopy after treatment with fluorescent stains.

The Environment of Soil Microorganisms

From the viewpoint of the microorganism, the most meaningful way to think of microbial ecology, the soil consists of a variety of surfaces that influence nutrient availability and affect interactions between different microorganisms. Depending on the soil aggregate structure, pores of various sizes are available for exploitation and colonization. A schematic diagram for a typical soil is shown in figure 42.1. As shown in this figure, soils consist of sands, clays, silt, and other particles. The organic matter occurs as freshly added plant, animal, and insect remains, which is gradually transformed into stabilized nutrient-rich humus material. These varied components form heterogeneous aggregates of various sizes called **peds,** which contain a complex network of pores. Bacteria and fungi use different functional strategies to take advantage of this complex physical matrix. Although some soil bacteria (figure 42.2) are motile, most form microcolonies on particle surfaces and require water and nutrients that must be located in their immediate vicinity. Bacteria are found most frequently in the smaller soil pores (2 to 6 μm in diameter). Here, they are probably less liable to be eaten by protozoa, unlike bacteria that are located on the exposed outer surface of a sand grain or organic matter particle.

The filamentous fungi, in contrast, tend to be located on the outside of the aggregates. These organisms, with their filamentous growth, will form bridges between separated regions where moisture is available. The filamentous fungi can move nutrients and water over greater distances in soil. Protozoa, soil insects, nematodes, and other soil animals are also present. Many of these organisms feed on the bacteria and fungi.

The fungi (chapter 26).

Because of limited gas diffusion into and out of these aggregates and the possibility of spaces between aggregates being completely flooded, major changes in the dissolved salts and gases can occur in these smaller pores or microenvironments. Soils have markedly higher concentrations of CO_2, CO, and other gases in comparison with the atmosphere, and a corresponding decrease in O_2 concentration (table 42.1). These changes will be further accentuated in the smaller pores, where many bacteria are found. Oxygen gradients and anaerobic microsites can be formed. When it rains, a soil may quickly change from an aerobic environment with many separated anaerobic microsites to a predominantly anaerobic environment. At lower depths, less O_2 is available, especially in wetter, less permeable soils.

Other physical factors also influence microorganisms that are associated with surfaces. At a neutral pH, most of the solid components of a soil, including the microorganisms, are neg-

TABLE 42.1 Concentrations of Oxygen and Carbon Dioxide in the Atmosphere of a Tropical Soil under Wet and Dry Conditions

	Oxygen Content (%)		Carbon Dioxide Content (%)	
Soil Depth (cm)	Wet	Dry	Wet	Dry
10	13.7	20.7	6.5	0.5
25	12.7	19.8	8.5	1.2
45	12.2	18.8	9.7	2.1
90	7.6	17.3	10.0	3.7
120	7.8	16.4	9.6	5.1

From E. W. Russell, *Soil Conditions and Plant Growth,* 10th ed., table 183, p. 414.

Note: Normal air contains approximately 21% oxygen and 0.035% carbon dioxide.

atively charged. Positively charged ions such as hydrogen and ammonium ions are attracted to negatively charged surfaces. This changes the surface microenvironment. Soil clays and **humus,** which consists of a partially degraded and stabilized organic matter, also attract and bind a variety of organic and inorganic substances. This includes many metal ions and products from the partial decomposition of pesticides. These "foreign" chemical residues associate with the organic matter fraction; however, their longer-term fate and effects are largely unknown.

1. Where are bacteria and fungi found in the soil, and what soil factors determine their distribution?
2. Why can a soil have many anaerobic microsites? Under what conditions will it become completely anaerobic?
3. What is the surface charge of most solid surfaces in soils?
4. Give two examples of substances bound by soil clays and humus.

Soil Microorganisms, Insects, and Other Animals—Contributions to Soils

Soils form under various environmental conditions. When newly exposed geologic materials begin to weather, as after a volcanic event or a simple soil disturbance, microbial colonization occurs. If only subsurface materials are available, phosphorus may be present, but nitrogen and carbon must be imported by biological processes. This can require an extended period, especially under harsh environmental conditions. Under these circumstances, many cyanobacteria, which can fix atmospheric nitrogen and carbon, are active in pioneer-stage nutrient accumulation.

Most soils, once they are formed, are rich sources of nutrients. Nutrients are found in organic matter, microorganisms, soil insects, and other animals. Plants grow, senesce, and die, and at each of these phases, they provide nutrients for soil organisms (figure 42.3). Different plant parts vary in their nutrient content and biomasses (table 42.2). In addition, the turnover times for the various plant parts are quite different.

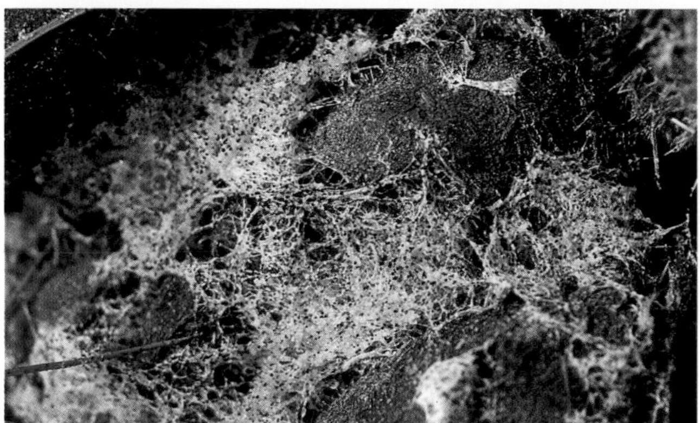

Figure 42.3 Fungi and Organic Matter Decomposition. When environmental conditions are suitable, microorganisms can grow rapidly to decompose organic matter. In this picture, a fungal mycelium has spread over fresh plant litter and horse droppings. Fungi are essential participants in the decomposition of many types of organic matter which are added to soils. The mycelium is producing many sporangia.

The components in the plant-soil system with the lowest carbon-nitrogen ratios (most nutrient-rich) are soil organic matter, microorganisms, soil insects, and other soil animals. The soil organic matter contains the greatest portion of the carbon and nitrogen in a typical soil, but with its slow turnover time (100 to 1,000 years and longer), most of this nutrient resource is not immediately available for plant or microbial use.

Except for the plant component, bacteria and fungi are present in greatest abundance in terms of biomass carbon and nitrogen. These organisms have low carbon-nitrogen ratios and relatively rapid turnover times. Besides their contribution to soil nutrient levels, they actively decompose plant litter. Without soil microorganisms, insects, and other animals, plant materials will accumulate in the environment.

The microbial populations in soils can be very high. In a surface soil, the bacterial population can approach 10^8 to 10^9 per gram dry weight of soil as measured microscopically. Fungi can be present at up to several hundred meters of hyphae per gram of soil (*see figure 26.4*b). It is important to remember, when discussing soils and their microorganisms, that only a minor portion (approximately 10%) of the microscopically observable organisms making up this biomass have been cultured. In terms of biotechnology and basic ecology, the microorganisms that have not been cultured may provide a valuable genetic resource for basic and applied research.

We tend to think that soil fungi are small like the mushrooms sprouting from our lawns. This is natural because most of the fungal thallus lies beneath the soil surface, but such a view sometimes is quite inaccurate. A case in point is the fungus *Armillaria bulbosa,* which lives associated with tree roots in hardwood forests. Recently an individual *Armillaria* clone that covers about thirty acres has been discovered in the Upper Peninsula of Michigan. It is estimated to weigh a minimum of 100 tons (an adult blue whale may weigh 150 tons) and be at least 1,500 years old. Thus some fungal mycelia are among the largest and most ancient living organisms on earth.

TABLE 42.2 Nutrient Resources and Estimated Annual Turnover Times for Major Components in a Grassland Ecosystem

Component	Nutrient Resources[a]		C/N Ratio	Approx. Turnover Time (Years)
	Carbon	Nitrogen		
Above ground green	43.0	1.4	31	1–2
Above ground dead	64.0	2.5	26	2
Plant crowns	160.0	4.6	35	2
Live roots	88.0	2.3	38	3–4
Senescent roots	282.0	14.0	20	3–4
Detrital roots	149.0	9.0	17	4
Soil organic matter	1327.0	127.0	10	100–1,000 and longer
Bacteria	30.4	7.6	4	0.83
Saprophytic fungi	6.3	0.63	10	0.83
Vesicular arbuscular fungi	0.7	0.07	10	0.50
Protozoa	0.2	0.028	7	0.17
Nematodes	0.217	0.022	10	0.50
Insects	0.068	0.004	8	0.40

Modified from R. G. Woodmansee et al., 1978, "Nitrogen Budget of a Shortgrass Prairie," *Oecologia* 34:363–82; and H. W. Hunt et al., 1987, "The Detrital Food Web in a Shortgrass Prairie," *Biol. Fert. Soils* 3:57–68.

[a]In grams per meter square, to a depth of 10 cm.

TABLE 42.3 Easily Cultured Gram-Positive Irregular Branching and Filamentous Bacteria Common in Soils

Bacterial Group	Representative Genera	Comments and Characteristics
Coryneform bacteria	*Arthrobacter*	Rod-coccus cycle
	Cellulomonas	Important in degradation of cellulose
	Corynebacterium	Club-shaped cells
Mycobacteria	*Mycobacterium*	Acid fast
Nocardioforms	*Nocardia*	Rudimentary branching
Actinomycetes	*Streptomyces*	Aerobic filamentous bacteria
	Thermoactinomyces	Higher temperature growth

The gram-positive bacteria, which show varied degrees of branching and mycelial development, are an important and less-studied part of the soil microbial community. They include the coryneform bacteria, the nocardioforms, and the true filamentous bacteria or actinomycetes (table 42.3). These bacteria play a major role in the degradation of hydrocarbons, older plant materials, and soil humus. In addition, some members of these groups actively degrade pesticides. The filamentous actinomycetes, primarily of the genus *Streptomyces,* produce an odor-causing compound called **geosmin,** which gives soils their characteristic earthy odor.

Actinomycetes (chapter 25).

Soil microorganisms can also be categorized on the basis of their preference for either easily available or more resistant substrates. The microorganisms that respond rapidly to the addition of easily utilizable substrates such as sugars and amino acids—for example, members of the genus *Pseudomonas*—are called **zymogenous** microorganisms. Those indigenous forms that tend to use native organic matter to a greater extent are called **autochthonous** microorganisms. Bacteria of the genus *Arthrobacter* and many soil actinomycetes can be considered members of this group. A less understood part of the microbial community in soils and in other natural environments is the **oligotrophs,** microorganisms that can be maintained on media containing less than 15 mg/liter of organic matter (*see Box 40.1*). When soils are inoculated onto such low-nutrient media, oligotrophic microorganisms are often present at higher populations than the easily cultured microorganisms recovered on most normal laboratory media. There is continuing interest in oligotrophs, and the study of these slow-growing organisms is an important area of modern environmental microbiology.

Arthrobacter (pp. 463–64).

Soil insects and other animals such as earthworms also contribute to organic matter transformations in soils. These organisms can serve as **decomposer-reducers.** The term decomposer-reducer indicates that organisms not only carry out decomposition or mineralization but also physically "reduce" the size of organic substrate particles such as plant litter. This increases the surface area and makes organic materials more available for use by bacteria and fungi. The decomposer-reducers mix substrates with their internal gut microflora and enzymes, and this contributes substantially to decomposition. For example, microorganisms in the termite gut provide a source of methane in the atmosphere (Box 42.1; *see also pp. 553, 565–66*).

Protozoa can also influence nutrient cycling by feeding on "palatable" microorganisms. This process of **microbivory,** or use of microorganisms as a food source, results in higher rates of nitrogen and phosphorus mineralization, thus increasing the availability of nutrients for plant growth.

Protozoa (chapter 28).

A significant part of the biological activity of soils arises from enzymes released by plants, insects and other animals,

Box 42.1

Soils, Termites, and Intestinal Microorganisms: A Major Source of Atmospheric Methane

Termites are important components of tropical ecosystems, where their use of cellulosic plant materials allows rapid—sometimes too rapid—recycling of plant materials. Termites harbor significant populations of archaeobacteria that use products of cellulose digestion, including CO_2 and hydrogen, to produce methane.

Termites occur on 2/3 of the earth's land surface, and, based on laboratory studies, 0.77% of the carbon ingested by termites can be released as methane. In tropical wet savannas and cultivated areas, termite populations are increasing rapidly. This increase is being accelerated by the destruction of tropical forests, which results in the accumulation of dead plant materials on the soil surface. This provides an ideal environment for the growth of termites. Termites are estimated to be contributing annually at least 1.5×10^{14} grams of methane, together with hydrogen and CO_2, to the atmosphere. This is believed to be causing measurable increases in the atmospheric methane level. Thus, unseen termites and their associated gut microorganisms may be affecting global warming.

and lysed microorganisms. These free enzymes contribute to many hydrolytic degradation reactions, such as proteolysis; catalase and peroxidase activities also have been detected. Apparently, released enzymes associate with clays and humic materials, which help to protect the enzymes from denaturation and microbial degradation.

1. Describe the contributions of bacteria and fungi to soil nutrient levels.
2. What is the role of protozoa, soil insects, and other animals in the functioning of the microbial community in soils?
3. Where is most of the nitrogen located in a temperate grassland soil? Is it readily available?
4. Describe zymogenous, autochthonous, and oligotrophic microorganisms.

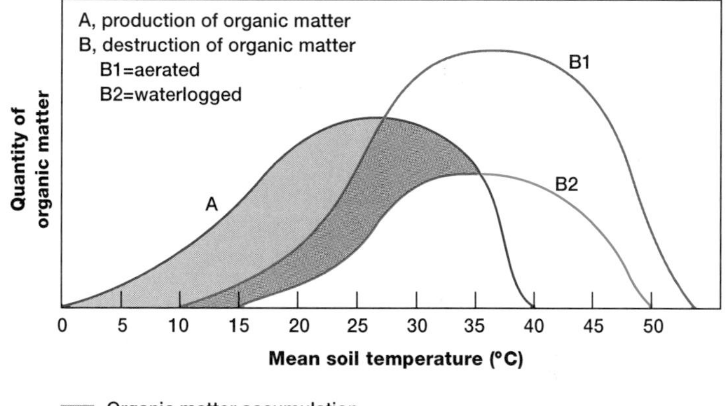

Figure 42.4 Soil Temperature, Moisture, and Organic Matter Decomposition. Temperature and moisture can affect organic matter levels in soils through their control of decomposition. Higher soil temperatures, as occur in the tropics, can lead to more rapid organic matter decomposition and loss of soil fertility. Waterlogging of a soil can protect organic matter from microbial decomposition. See text for discussion. *Source: E. C. J. Mohr, F. A. van Baaren, and J. van Schuylenborgh,* Tropical Soils: A Comprehensive Study of Their Genesis, *3d ed. Mouton-Ichtiar Baru-Van Hoeve, The Hague, 1972.*

Microorganisms and the Formation of Different Soils

In surveying lands across different climatic zones, one can observe significant changes in plant communities and soils. Often, primary production and decomposition are not in equilibrium. The result is excessive accumulation of organic matter or, if decomposition occurs too quickly, a soil poor in organic matter and of low fertility. The production and decomposition of organic matter are influenced by soil temperature and water content, as shown in figure 42.4. Above a mean annual temperature of approximately 25°C in aerobic soils, the rate of decomposition can be higher than that of organic matter accumulation. In terrestrial environments with high soil water contents, however, more organic matter may accumulate because decomposition will be inhibited by lower O_2 availability.

These differences can be observed in soils from tropical, grassland, forest, and bog regions (figure 42.5). In moist tropical soils, with their higher mean temperatures, organic matter is decomposed very quickly, and the mobile inorganic nutrients can be leached out of the surface soil environment, causing a rapid loss of fertility. To limit nutrient loss, many tropical plants have root systems that penetrate the rapidly decomposing litter layer. Thus it is possible to "recycle" nutrients before they are lost with water movement through the soil (figure 42.5*a*). With deforestation, the nutrients are not recycled, leading to their loss from the soil and decreased soil fertility.

In many temperate region soils, in contrast, the decomposition rates are less than that of primary production, leading to litter accumulation. Deep root penetration in temperate grasslands results in the formation of fertile soils, which provide a valuable resource for the growth of crops in intensive agriculture (figure 42.5*b*).

The soils in many cooler coniferous forest environments suffer from an excessive accumulation of organic matter as plant litter (figure 42.5*c*). In winter, when moisture is avail-

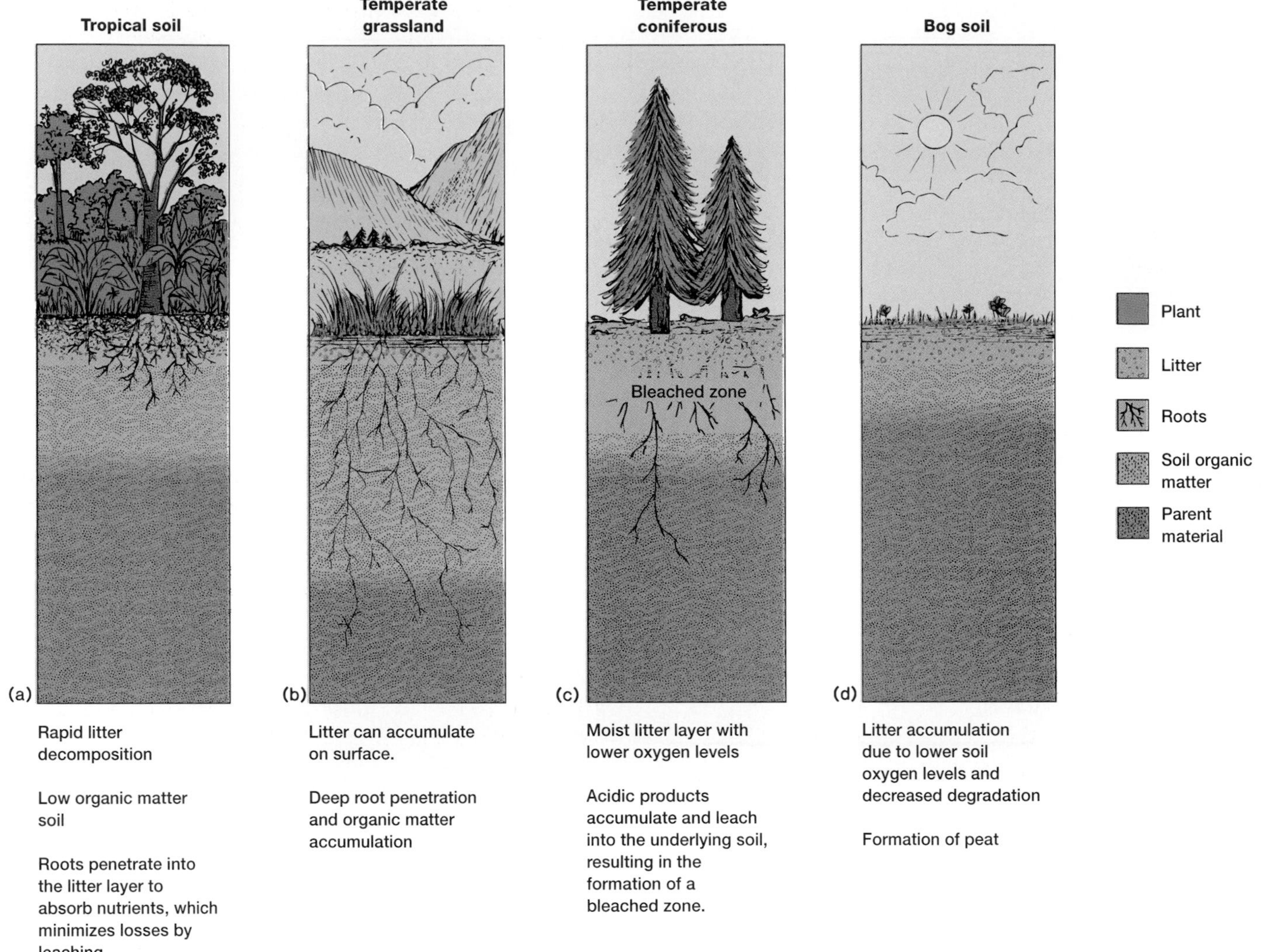

Figure 42.5 Examples of Plant-soil Systems. Climate, parent material, plants, and microorganisms interact over time to form different plant-soil systems. In these figures, the characteristics of tropical, temperate grassland, temperate forest, and bog soils are illustrated.

able, the soils are cool, and this limits decomposition. In summer, when the soils are warm, water is not as available for decomposition. Organic acids are produced in the cool, moist litter layer, and they leach into the underlying soil. These acids solubilize soil components such as aluminum and iron, and a bleached zone may form. Litter continues to accumulate, and fire becomes the major means by which nutrient cycling is maintained. Controlled burns are becoming a more important part of environmental management in this type of plant-soil system.

Bog soils provide a unique set of conditions for microbial growth (figure 42.5*d*). In these soils, the decomposition rate is slowed by the waterlogged, predominantly anoxic conditions, which lead to peat accumulation. When such areas are drained, they become more aerobic and the soil organic matter is degraded, resulting in soil subsidence. Under aerobic conditions, the lignin-cellulose complexes of the accumulated organic matter are more susceptible to decomposition by the filamentous fungi.

This analysis of different soil types shows that the microbial role in soil formation and decomposition is varied but understandable in terms of general environmental factors that influence biological decomposition processes.

1. Describe each major soil type in terms of the balance between primary production and organic matter decomposition.
2. If decomposition is slower than primary production, what are the consequences for plant community functioning?

Soil Microorganism Associations with Plants

Plants, the major source of organic matter upon which soil microorganisms are dependent, are literally covered with microorganisms (figure 42.6). Microorganisms of various kinds are associated with the leaves, stems, flowers, seeds, and roots. The microbial community influences plants in many direct and indirect ways. The presence of microorganisms increases the rate

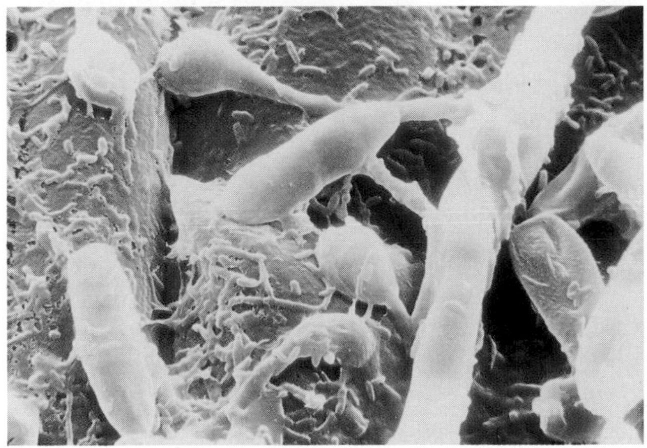

Figure 42.6 Root Surface Microorganisms. Plant roots release nutrients that allow intensive development of bacteria and fungi on and near the plant root surface. A scanning electron micrograph shows bacteria and fungi growing on a root surface.

TABLE 42.4 Compounds Excreted by Axenic Wheat Roots

Volatile Compounds	Low-Molecular-Weight Compounds	High-Molecular-Weight Compounds
CO_2	Sugars	Polysaccharides
Ethanol	Amino acids	Enzymes
Isobutanol	Vitamins	
Isoamyl alcohol	Organic acids	
Acetoin	Nucleotides	
Isobutyric acid		

From J. W. Woldendorp, "The Rhizosphere as Part of the Plant-Soil System" in *Structure and Functioning of Plant Populations* (Amsterdam, Holland: Proceedings, Royal Dutch Academy of Sciences, Natural Sciences Section: 2d Series, 1978) 70:243.

of organic matter release from the roots by **exudation.** With grasses, for example, microorganisms can increase exudation rates by 80 to 100%. Materials released by roots include a wide range of potential microbial substrates, inhibitors, and stimulants (table 42.4). Rates of organic matter release by plants are also increased by stresses, including herbage removal and light changes.

The roles of microorganisms in influencing plant growth through the release of morphology-influencing compounds such as gibberellins and cytokinins are only now beginning to be fully appreciated in terms of their biotechnological potential.

The Rhizosphere

Soil microbial populations respond to the release of organic materials near the plant root, increasing their numbers and changing the characteristics of the microbial community (table 42.5). This region, called the **rhizosphere,** was first described by Lorenz Hiltner in 1904. In more fertile soils, which have higher normal microbial populations, this effect will not be as pronounced as it is with lower fertility soils. The rhizosphere response is especially marked in desert soils that have lower normal populations of microorganisms; a much greater relative stimulation will occur in the nutrient-rich rhizosphere. Rhizosphere microorganisms serve as a labile source of nutrients and also play a critical role in organic matter synthesis and degradation.

The genus *Rhizobium* (*see p. 434*) is a prominent member of the rhizosphere community and can establish a symbiotic association with legumes. *Rhizobium* species only infect and nodulate specific hosts (figure 42.7). This complex process appears to involve host plant recognition molecules called lectins, and the binding of *Rhizobium* to specific sites on root hairs. After bacterial attachment, the root hair curls and the bacteria enter it and induce the plant to form an **infection thread**

TABLE 42.5 Microbial Populations and Their Responses in the Rhizosphere

Organisms	Rhizosphere Soil (R)[a]	Control Soil (S)	Approximate R:S Ratio[b]
Taxonomic Groups			
Unicellular bacteria	$1,200 \times 10^6$	53×10^6	23:1
Actinomycetes	46×10^6	7×10^6	7:1
Fungi	12×10^5	1×10^5	12:1
Protozoa	24×10^2	10×10^2	2:1
Algae	5×10^3	27×10^3	0.2:1
Bacterial Physiological Groups			
Ammonifiers	500×10^6	4×10^6	125:1
Gas-producing anaerobes	39×10^4	3×10^4	13:1
Anaerobes	12×10^6	6×10^6	2:1
Denitrifiers	126×10^6	1×10^5	1,260:1
Aerobic cellulose decomposers	7×10^5	1×10^5	7:1
Anaerobic cellulose decomposers	9×10^3	3×10^3	1:1
Endospore formers	930×10^3	575×10^3	1:1

Modified from T. R. G. Gray and S. T. Williams, *Soil Micro-Organisms*. Copyright © 1971 Longman Group Limited, United Kingdom.

[a]Values are expressed in numbers per gram.

[b]R:S ratio equals ratio of organisms in the rhizophere (R) to those in the control soil (S) not influenced by plant exudates.

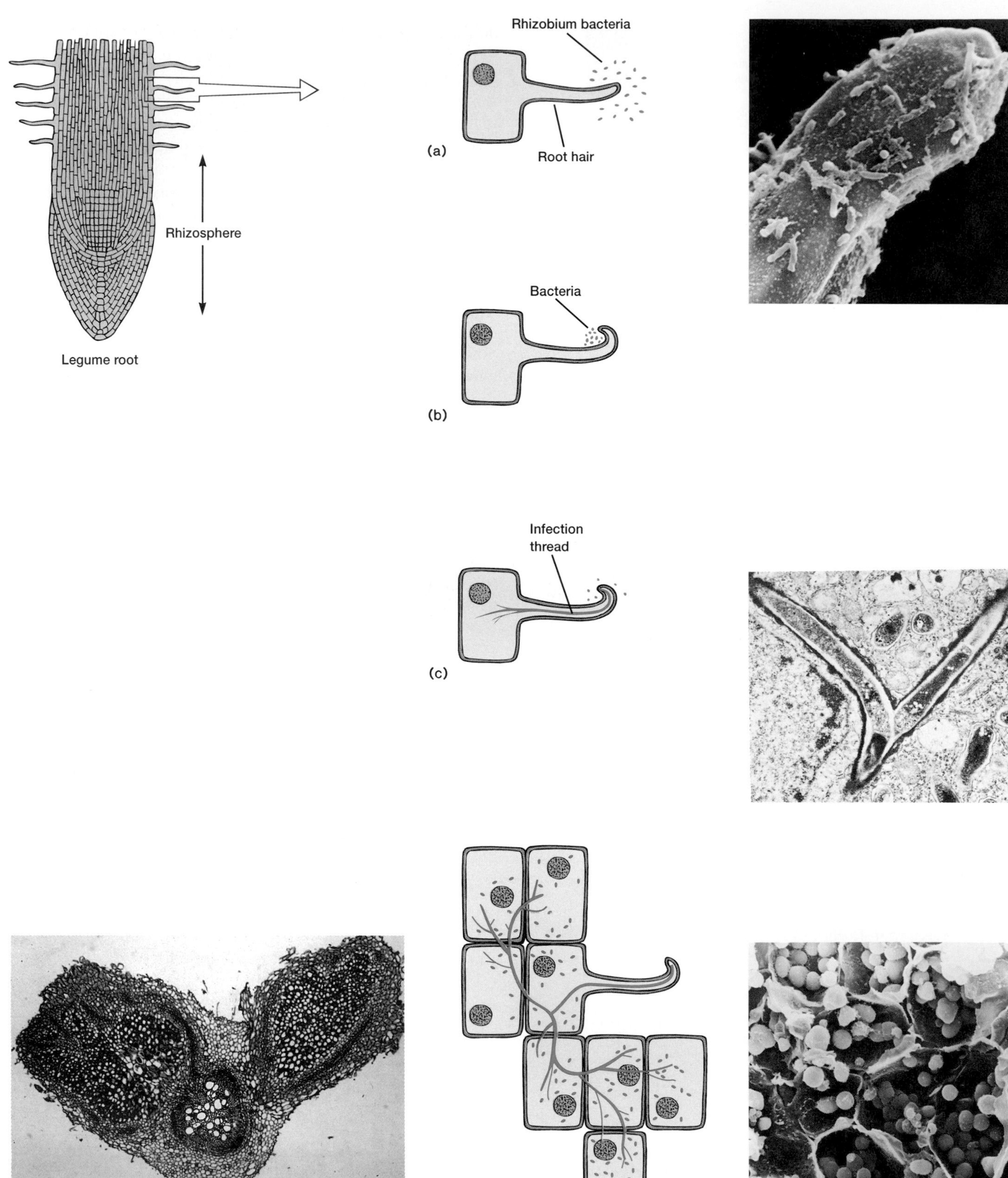

Legume root
Rhizosphere
Rhizobium bacteria
Root hair
(a)
Bacteria
(b)
Infection
thread
(c)
(d1)
(d2)
(d3)

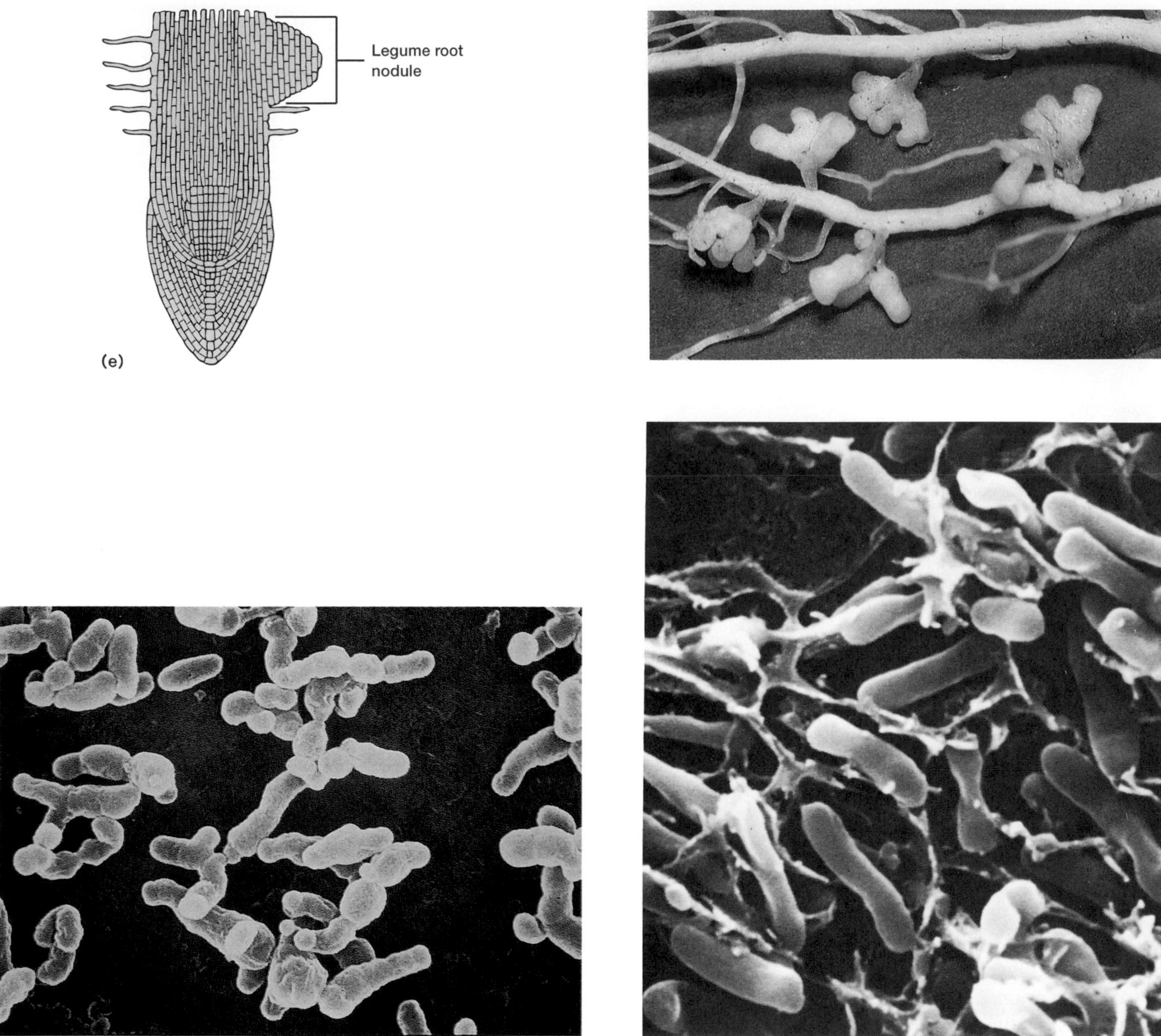

Figure 42.7 Root Nodule Formation. The formation of root nodules on legumes by *Rhizobium* and *Bradyrhizobium* allows atmospheric nitrogen to be fixed for plant use. These drawings and photographs show the events leading to the formation of the nitrogen-fixing symbiosis. (*a*) Establishment of the nitrogen-fixing symbiosis between a legume (soybean, *Glycine max*) and the nitrogen-fixing bacterium *Bradyrhizobium japonicum*. The scanning electron micrograph shows rhizobia attached to a root hair. (*b*) The beginning of infection threads is depicted schematically, and (*c*) shows the growth of an infection thread (containing the rhizobia) into a root cell. The electron micrograph is of a branched infection thread containing bacteroids in soybean. (*d1*) The light micrograph shows a cross section of an infected root and the development of nodules by cell division (×5). (*d2*) The schematic illustrates the continued division and proliferation of bacteroids. (*d3*) The SEM micrograph shows nodule cells of peanuts (*Arachis*) filled with *Bradyrhizobium*. (*e*) A schematic and photograph of *Rhizobium meliloti* nitrogen-fixing nodules on sweet clover (*Trifolium*). (*f*) *Rhizobium leguminosarum* are shown in their free-living form (×2,000). (*g*) *Bradyrhizobium japonicum* bacteroids within a soybean root nodule.

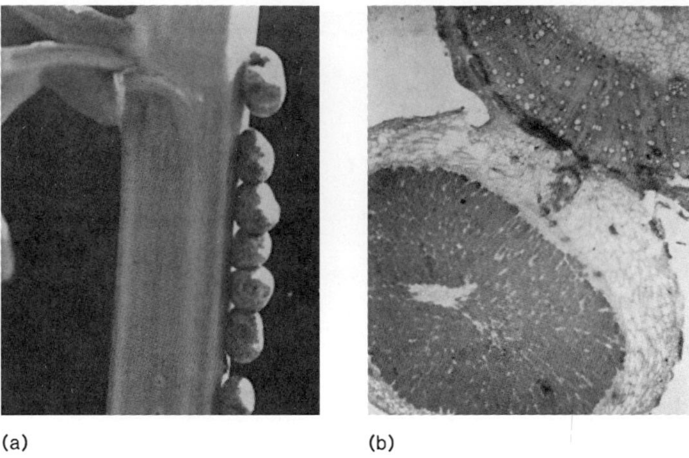

Figure 42.8 **Stem-nodulating Rhizobia.** Nitrogen-fixing microorganisms also can form nodules on stems of some tropical legumes. (*a*) Nodules formed on the stem of a tropical legume by a stem-nodulating *Rhizobium*. (*b*) Cross section of a stem nodule.

that grows down the root hair. *Rhizobium* travels along the infection thread and subsequently infects adjacent root cells. If a cell is tetraploid, it is stimulated to actively divide, and a **root nodule** eventually arises (diploid root cells are usually destroyed). The bacteria multiply within the tetraploid cells and develop into swollen, branched forms called **bacteroids** that can reduce atmospheric nitrogen to ammonia. Host-specific nodule proteins, such as leghemoglobin and nodulins, are also produced to complete this process. The molecular biology of nitrogen fixation by *Rhizobium* is a subject of intense study around the world.

Nitrogen fixation: biochemical aspects (pp. 178–80).
Nitrogen cycle (pp. 810–12).

Mutualistic nitrogen fixation is complemented by **associative nitrogen fixation.** This process is carried out by many microorganisms, including representatives of the genera *Azotobacter* and *Azospirillum,* which use nutrients available in the rhizosphere. These microorganisms contribute to nitrogen accumulation by tropical grasses. Recent evidence suggests that their major contribution may not be in nitrogen fixation but in nitrate reduction, leading to a greater availability of ammonium ions for the plant. This is an area of research that is particularly important in tropical agricultural areas.

Other associations of nitrogen-fixing microorganisms with plants also occur. A particularly interesting association is caused by **stem-nodulating rhizobia,** found primarily in tropical legumes (figure 42.8). These nodules form on the stem just above the soil surface and, because they contain oxygen-producing photosynthetic tissues, have unique mechanisms to protect the oxygen-sensitive nitrogen fixation enzymes.

A major goal of biotechnology is to introduce nitrogen fixation genes into plants that do not normally form such associations. Recently, it has been possible to produce root nodules on nonlegumes such as rice, wheat, and oilseed rape (figure 42.9). It appears that the infection begins with bacterial at-

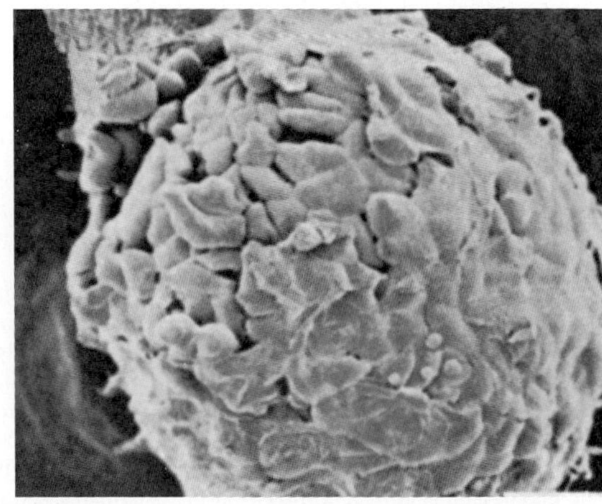

Figure 42.9 **Root Nodules on Nonlegumes.** Recently, *Rhizobium* has been found to form inactive nodules on nonlegumes. (*a*) View of an oilseed rape plant. (*b,c*) Two views of the nodules formed on the root caused by *Rhizobium*.

Box 42.2

Mycorrhizae and the Evolution of Vascular Plants

Fossil evidence shows that endomycorrhizal symbioses were as frequent in vascular plants during the Devonian period, some 387 to 408 million years ago, as they are today. As a result, some botanists have suggested that the evolution of this type of association may have been a critical step in allowing colonization of the land by plants. During this period, soils were poorly developed, and as a result, mycorrhizal fungi were probably significant in aiding the uptake of phosphorus and other nutrients. Even during current times, those plants that start to colonize extremely nutrient-poor soils survive much better if they have endomycorrhizae. Thus, it may have been a symbiotic association of plants and fungi that initially colonized the land and led to our modern vascular plants.

tachment to the root tips. Although these nodules have not yet been found to fix useful amounts of nitrogen, intense work is expected to continue in this area.

Mycorrhizae

Mycorrhizae are fungus-root associations, first discovered by Albert Bernhard Frank in 1885 (Box 42.2). The word *mycorrhizae* comes from the Greek words meaning fungus and roots. These microorganisms contribute to plant functioning in natural environments, agriculture, and reclamation. The roots of about 80% of all kinds of vascular plants are normally involved in symbiotic associations with mycorrhizae.

Two major mycorrhizal associations, **ectomycorrhizal** and **endomycorrhizal,** have been described, and other relationships have also been observed, such as in orchids (figure 42.10). Ectomycorrhizae (figure 42.11) are found primarily in temperate regions associated with certain groups of trees and shrubs. For example, pine trees that grow at the timberline are usually ectomycorrhizal, with the fungal components being basidiomycetes, ascomycetes, or zygomycetes (*see chapter 26*). Ectomycorrhizae grow as an external sheath around the tip of the root, with limited intercellular penetration of the fungus into the cortical regions of the root; they are found predominantly in beech, oak, birch, and coniferous trees.

More than 5,000 species of fungi, predominantly basidiomycetes, are involved in ectomycorrhizal symbiotic relationships. Their extensive mycelia extend far out into the soil and greatly aid the transfer of nutrients to the plant. One of the more important ectomycorrhizal fungi is *Pisolithus tinctorius.* This fungus can be grown in mass culture, with small Styrofoam beads acting as a physical support. The fungal inoculum is then mixed with rooting soil, resulting in improved plant establishment and growth.

Endomycorrhizae (figure 42.12) are of particular interest, as it has not been possible to grow these fungi, usually members of the zygomycetes, without the plant. In this association, the fungal hyphae penetrate the outer cortical cells of the plant root, where they grow intracellularly and form coils, swellings, or minute branches. Endotrophic mycorrhizae are found in wheat, corn, beans, tomatoes, apples, oranges, and many other commercial crops, as well as most pasture and rangeland grasses. Two characteristic intracellular structures are called **vesicles** and **arbuscules;** and thus, endomycorrhizae are often called vesicular-arbuscular mycorrhizae, abbreviated as **VA mycorrhizae.** Recent studies show that plant flavonoids may stimulate spore germination, and this could lead to the development of plant-free cultures of VA mycorrhizae.

Depending on the environment of the plant, mycorrhizae can increase a plant's competitiveness. In wet environments,

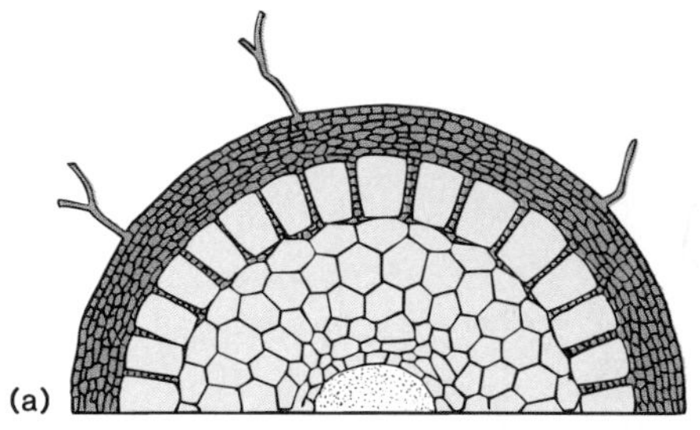

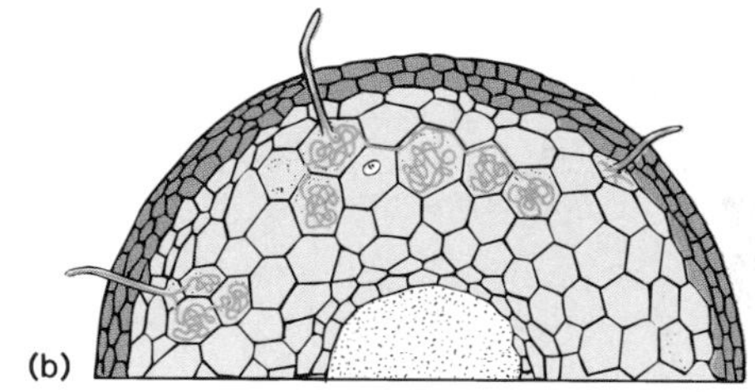

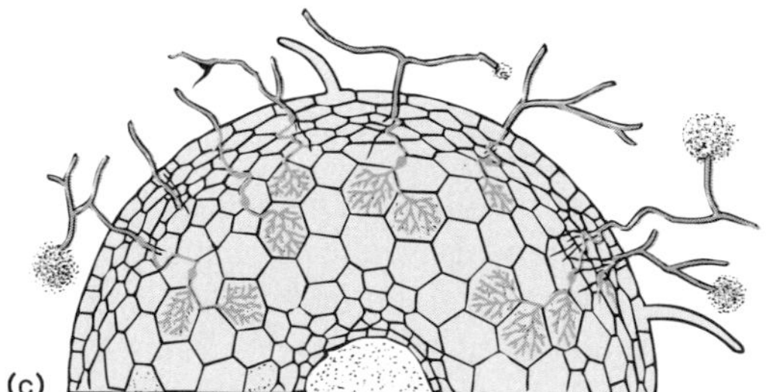

Figure 42.10 **Mycorrhizae.** Fungi also can establish mutually beneficial relationships with plant roots, called mycorrhizae. Root cross sections illustrate different mycorrhizal relationships: (*a*) ectomycorrhizae, as in pines, and two types of endomycorrhizae as found in orchids (*b*) and grasses (*c*).

Figure 42.11 **Ectomycorrhizae.** Ectomycorrhizae form easily observed sheaths around root tips. After washing roots, ectomycorrhizae growing around the roots of *Ceanothus* (Buckbrush) are evident.

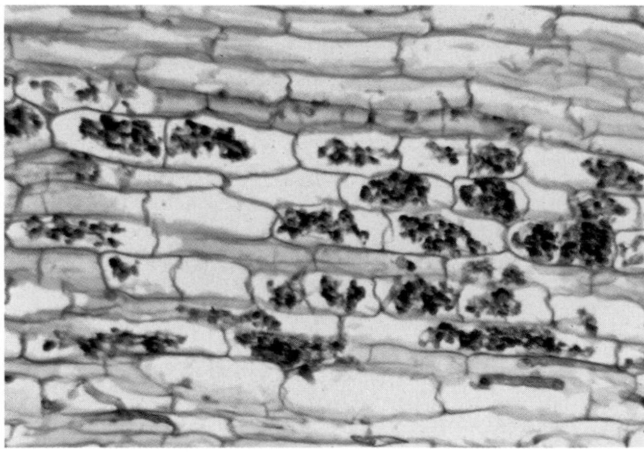

Figure 42.12 **Endomycorrhizae.** Endomycorrhizae, or vesicular-arbuscular mycorrhizae, form characteristic structures within roots. These can be observed with a microscope after the roots are cleared and stained. The arbuscules of *Gigaspora margarita* can be seen inside the root cortex cells of cotton.

they increase the availability of nutrients, especially phosphorus. In arid environments, where nutrients do not limit plant functioning to the same degree, the mycorrhizae aid in water uptake, allowing increased transpiration rates in comparison with nonmycorrhizal plants. These benefits have distinct energy costs for the plant in the form of photosynthate required to support the plant's "mycorrhizal habit." Under certain conditions, the plant is apparently willing to trade photosynthate—produced with the increased water acquisition—for water.

Actinorhizae

Actinomycete associations with plant roots, or **actinorhizal** relationships, also occur. Actinorhizae are formed by the association of *Frankia* strains (*see pp. 511–12*) with eight plant families. They can fix nitrogen and are important in the life of woody, shrublike plants (table 42.6). In comparison, the rhizobia occur in only one family: the *Leguminosae.* These associations are important in areas where Douglas fir forests have been clear-cut, and in bog and heath environments where bayberries and alders are dominant. The nodules of some actinorhizae are marble sized (figure 42.13). Not impressed? Some plants (*Alnus, Ceanothus*) have nodules as large as baseballs. The nodules of *Casuarina* approach soccer ball size!

Members of the genus *Frankia* are slow-growing, and, until 1978, it was impossible to culture the organisms apart from the plant. Since then, this actinomycete has been grown on specialized media supplemented with metabolic intermediates such as pyruvate. Major advances in understanding the physiology, genetics, and molecular biology of these microorganisms are now taking place.

As with the mycorrhizae, the actinorhizal relationship costs the plant energy. However, the plant does benefit and is better able to compete in nature. This association provides a unique opportunity for microbial management to improve plant growth processes.

Tripartite and Tetrapartite Associations

An additional set of interactions occurs when the same plant develops relationships with two or three different types of microorganisms. These more complex interactions are important to a variety of plant types in both temperate and tropical agricultural systems. First described in 1896, these symbiotic associations involve the interaction of the plant-associated microorganisms with each other and the host plant. Several **tripartite associations** are known to occur: the plant plus (1) endomycorrhizae plus rhizobia, including *Rhizobium* and *Bradyrhizobium;* (2) endomycorrhizae and actinorhizae; and (3) ectomycorrhizae and actinorhizae. Nodulated and mycorrhizal plants are better suited for coping with nutrient-deficient environments. **Tetrapartite associations** also occur. These consist of endomycorrhizae, ectomycorrhizae, *Frankia,* and the host plant. These complex associations, in spite of their additional energy costs, provide important benefits for the plant.

Fungal and Bacterial Endophytes of Plants

Specialized fungi and bacteria can live within some plants as **endophytes.** Specialized clavicipitaceous fungi form systemic fungal infections in which the endophyte grows between the plant's cortex cells (figure 42.14). Plants infected with these endophytes may be less susceptible to attack by various chewing insects due to the production of alkaloids, a form of "chemical defense." Not all such relationships are mutualistic (beneficial to both partners). Some are parasitic. Parasitic fungal endophytes can actually reduce the genetic variability of the plant by sterilizing their host (figure 42.15). This decreases the plant's ability to resist the fungal endophyte. The "parasitic castration of plants" by systemic fungi is of major importance in the co-evolution of plants and fungi.

Endophytic bacteria have been discovered in cotton, pears, and potatoes. Some are plant pathogens that can survive for extended periods in a quiescent state. The majority have no

TABLE 42.6 NonLeguminous Nodule-Bearing Plants with *Frankia* Symbioses[a]

Family	Genus	Frankia Isolated?	Isolated Strains Infective?
Casuarinaceae	*Allocasuarina*	+	+
	Casuarina	+	+
	Ceuthostoma	−	−
	Gymnostoma	+	+
Coriariaceae	*Coriaria*	+	−
Datiscaceae	*Datisca*	+	−
Betulaceae	*Alnus*	+	+
Myricaceae	*Comptonia*	+	+
	Myrica	+	+
Elaeagnaceae	*Elaeagnus*	+	+
	Hippophae	+	+
	Shepherdia	+	+
Rhamnaceae	*Ceanothus*	+	−
	Colletia	+	−
	Discaria	−	−
	Kentrothamnus	−	−
	Retanilla	−	−
	Trevoa	−	−
Rosaceae	*Cercocarpus*	+	−
	Chaemabatia	−	−
	Cowania	+	−
	Dryas	−	−
	Purshia	+	−

Source: Dr. D. Baker, Yale University. Personal communication.
[a]*Frankia* isolation from nodules and ability of these isolated strains to initiate nodulation are also noted.

(a)

(b)

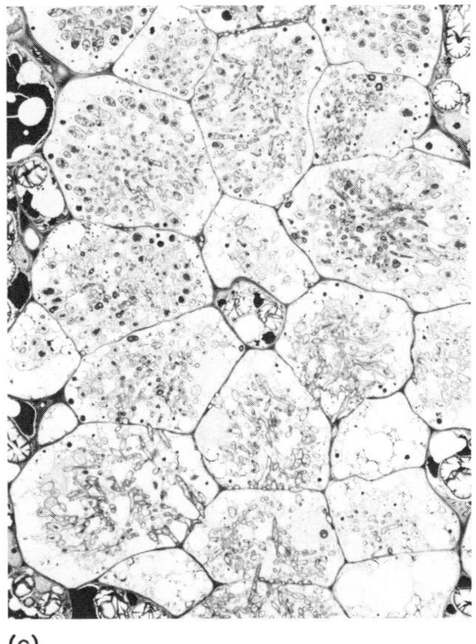
(c)

Figure 42.13 Actinorhizae. Nitrogen-fixing actinomycetes can form associations with woody plants. Actinorhizal nodule development allows symbiotic nitrogen fixation, critical for plant development on infertile sites. (*a*) Light micrograph of actinorhizae around *Comptonia* roots. (*b*) SEM view of two infected cortical cells of *Casuarina*. Note the hyphae from the actinorhizae penetrating the host's cell walls. (*c*) TEM of *Myrica* actinorhizae in nodule cortical cells.

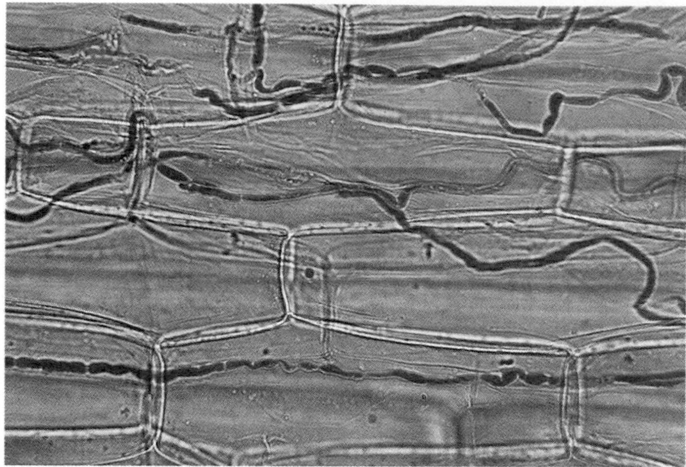

Figure 42.14 **Fungal Endophytes.** Fungi have been found in the upper parts of some plants. A fungal endophyte growing inside the leaf sheath of a grass, tall fescue, is shown.

Figure 42.15 **Parasitic Castration of Plants by Endophytic Fungi.** Stroma of the fungus *Atkinsonella hypoxylon* infecting *Danthonia compressa* and causing abortion of the terminal spikelets.

known positive or deleterious effect on plant growth or development. The use of these bacteria as microbial delivery systems in agriculture is a current topic in agricultural biotechnology.

Agrobacterium, Plant Tumors, and Molecular Biology

Another exciting plant-soil microorganism interaction is the *Agrobacterium* infection that produces tumorlike growths on plants (figure 42.16). Gall formation involves *Agrobacterium* strains that contain the **Ti** (tumor-inducing) **plasmid.** This plasmid may be modified to allow the transfer of genetic characteristics such as herbicide resistance and bioluminescence to plants. With the ability to modify plant DNA using *Agrobacterium*-borne plasmids, rapid advances are being made in plant molecular biology.

Agrobacterium (Box 14.2 and p. 435).

Figure 42.16 ***Agrobacterium.*** This bacterium can be used as a vector to incorporate foreign DNA into susceptible plants. *Agrobacterium*-caused tumor on a *Kalanchoe* sp. plant. *Photo courtesy of Dr. S. Süle, Hungarian Academy of Sciences.*

1. List five characteristics that can be measured in studying the soil microbial community.
2. Define the following terms: rhizosphere, root nodule, bacteroid, associative nitrogen fixation, mycorrhizae, ectomycorrhizae, endomycorrhizae, and actinorhizae.
3. How is a legume root nodule formed by *Rhizobium* activity?
4. What are the major contributions of mycorrhizae and actinorhizae to plant functioning?
5. What are tripartite and tetrapartite associations?
6. What are endophytic microorganisms and why are they important to plants?
7. Discuss the nature and importance of the Ti plasmid.

Soil Organic Matter and Soil Fertility

Soil organic matter helps retain nutrients, maintain soil structure, and hold water for plant use. This important resource is subject to gains and losses, depending on changes in environmental conditions and agricultural management practices. Plowing and other similar disturbances expose the soil organic matter to more oxygen, leading to extensive microbiological degradation of organic matter. Irrigation causes periodic wetting and drying, which can also lead to increased degradation of soil organic matter, especially at higher temperatures.

Several approaches have been developed to overcome management practices that stimulate soil microorganisms and lead to increased degradation of soil organic matter. These approaches include "no till" or "minimum till" agriculture, in which surface soil is minimally disturbed and chemicals are used to control weed growth.

Composting is another, much older but still effective approach to maintaining and augmenting the organic matter content of a soil. In this process, plant materials are allowed to decompose under moist, aerobic conditions, in which the readily utilizable plant fractions are rapidly decomposed. If the compost is too moist or too dry, the desired decomposition process will not occur. The compost pile reaches higher temperatures, allowing thermophilic microorganisms to participate actively in these processes. When the composting is completed, the residual lignin and other more resistant plant materials will have been partially transformed to humus. Such biologically stablized compost, when added to soil, increases the soil organic matter content and does not stimulate the soil microorganisms. Soil organic matter is a dynamic and labile component of the plant-soil system. By direct and indirect management of microbial activities and decomposition processes, it is possible to maintain and augment the fertility of this important natural resource.

Thermophile characteristics (pp. 124–26).

1. What soil management processes can lead to increases in degradation of soil organic matter?
2. Why might composed materials have different effects on soil organic matter than additions of fresh plant materials?

Pesticides and Microorganisms—Fallibility and Recalcitrance

The use of pesticides to control undesired plants, insects, and other organisms has become widespread in the last 40 years. This fact has provoked important questions concerning the nature of pesticide degradation and the degradability of both natural and synthetic chemicals.

Originally, it was assumed, given time and the almost infinite variety of microorganisms, that all organic compounds, including those synthesized in the laboratory, would eventually degrade. This idea of microbial infallibility was suggested in 1952. Observations of natural and synthetic organic compound accumulation in natural environments, however, began to raise questions about the ability of microorganisms to degrade these varied substances and the role of the environment (clays, anaerobic conditions) in protecting some chemicals. With the development of synthetic pesticides, it became distressingly evident that not all organic compounds are immediately biodegradable. This chemical **recalcitrance** (resisting authority or control) resulted from the apparent fallibility of microorganisms, or their inability to degrade some industrially synthesized chemical compounds.

Degradation is subject to several definitions. As shown in figure 42.17, it can be defined as a minor change in a molecule, such as dehalogenation. Degradation can also be thought of as fragmentation, where the original structure of the molecule can still be recognized in the fragments. Finally, biodegradation can mean complete mineralization of a compound to inorganic forms.

Degradation is often promoted by the presence of easily usable energy sources, allowing the modification of an otherwise recalcitrant compound. **Cometabolism** is extremely important in managing the degradation of biologically resistant chemicals such as trichloroethylene (TCE), a widespread contaminant in subsurface environments. By using methane as an energy source, subsurface microorganisms can modify the normally recalcitrant TCE molecule. Dehalogenation of many compounds containing chlorine, bromine, or fluorine occurs faster under anaerobic than under aerobic conditions. Once the dehalogenation steps are completed, degradation of the main structure of many pesticides often proceeds more rapidly

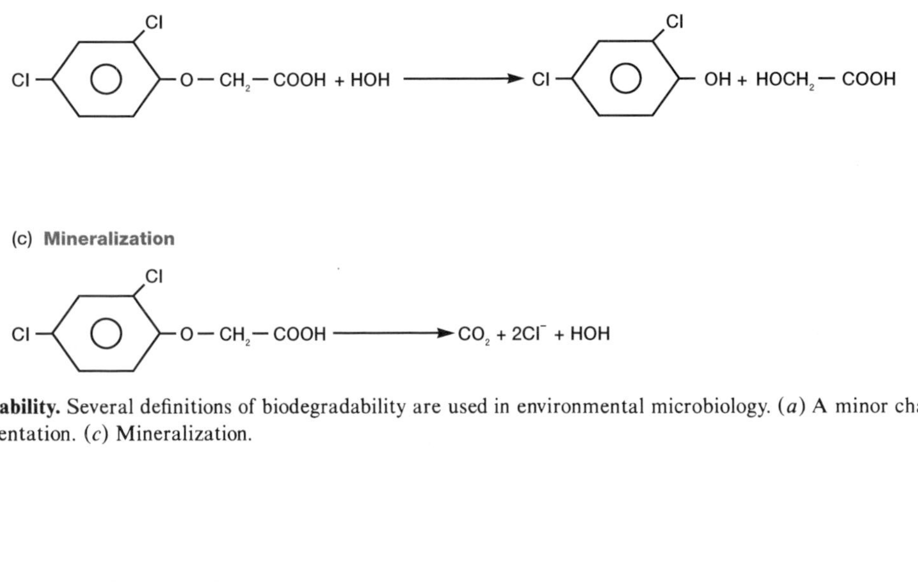

(a) **Minor change (dehalogenation)**

(b) **Fragmentation**

(c) **Mineralization**

Figure 42.17 **Biodegradability.** Several definitions of biodegradability are used in environmental microbiology. (*a*) A minor change in a molecule (dehalogenation). (*b*) Fragmentation. (*c*) Mineralization.

in the presence of O_2. It should be noted that the rate of degradation varies with the location of halogens on the molecule. For example, the presence of substituents in the *meta* position on aromatic compounds can cause a marked inhibition of degradation, as is observed with the herbicide 2,4,5-T, in comparison with 2,4-D (figure 42.18).

The soil microbial community also can change its characteristics upon exposure to complex organic molecules. After the community's repeated exposure to a given chemical, faster rates of degradation may occur (figure 42.19). A microbial community can become so efficient at rapid herbicide decomposition that herbicide effectiveness is diminished. To counteract this process, herbicides can be changed to throw the microbial community off balance, thus preserving the effectiveness of the chemicals. The degradation of many pesticides may also result in the accumulation of organic fragments that bind with organic matter in the soil. The longer-term fate and possible effects of "bound" pesticide residues on the soil system, plants, and higher organisms are largely unknown.

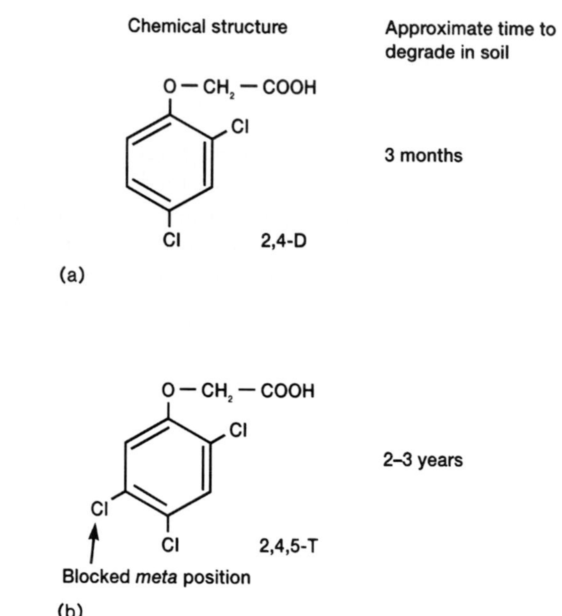

Figure 42.18 The *Meta* Effect and Biodegradation. Minor structural differences can have major effects on the biodegradability of chemicals. The *meta* effect is an important example. (*a*) Readily biodegradable 2,4–D, with an exposed *meta* position. (*b*) Recalcitrant 2,4,5–T, with the *meta* position blocked by a chlorine group.

Soil Microorganism Interactions with the Atmosphere

Soil microorganisms have interesting interactions with the atmosphere. Ice-nucleating bacteria from decomposing litter can increase precipitation. Specific strains of *Pseudomonas syringae* synthesize proteins that can serve as nucleation centers for snow formation in cloud chambers. Such ice-nucleating soil microorganisms may significantly influence weather on a global scale. "Ice-minus" strains of *P. syringae,* developed by recombinant technology, can be sprayed on frost-sensitive crops such as strawberries. In the presence of the "ice-minus" bacteria, water will not freeze on the leaves as readily, making it possible to save frost-sensitive crops from destruction. The release of such genetically modified microorganisms (GEMs) to the environment is of continuing concern (*see pages 304–6*).

Soil microorganisms also influence the atmosphere by degrading airborne pollutants such as hydrogen, CO, benzene, trichloroethylene (TCE), and formaldehyde. They can substantially improve the air in closed buildings (Box 42.3). Although soil microorganisms cannot completely eliminate these pollutants, they can decrease pollutant concentrations to equilibrium levels of approximately 1 to 2 ppm.

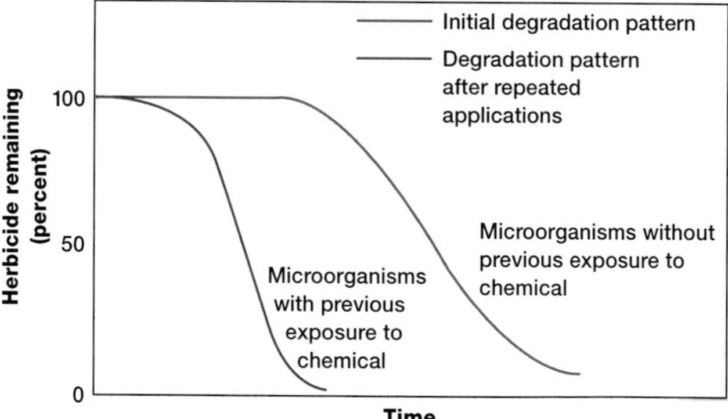

Figure 42.19 Repeated Exposure and Degradation Rate. Addition of an herbicide to a soil can result in changes in the degradative ability of the microbial community. Relative degradation rates for an herbicide after initial addition to a soil, and after repeated exposure to the same chemical.

1. How are recalcitrance, fallibility, and infallibility related?
2. What is the process of cometabolism? Why is it important in microbial modification of many chemicals?
3. What definitions of degradation are used in environmental microbiology?
4. In which ways can soil microorganisms interact with the atmosphere?

Box 42.3

Soil Microorganisms and "Sick Buildings"

A major problem in the development of more energy-efficient homes and office buildings is the potential effect of such closed environments on human health. With many people spending much of their lives in such enclosed environments, the "sick building syndrome" is an increasing concern. While saving energy, these "sick buildings" have higher levels of many volatile compounds, including benzene, trichloroethylene (TCE), formaldehyde, phenolics, and solvents. These are released from rugs, furniture, plastic flooring, paints, and office machines such as photocopiers and printers. An important but still largely unappreciated means of improving the air in such "sick buildings" is through plants and their associated soil microorganisms. Plants not only produce oxygen, but the soil microorganisms degrade many airborne pollutants. It is recommended that one plant be used per 100 square feet of living area. As noted by B. C. Wolverton, "The ultimate solution to the indoor air pollution problem must involve plants, the plant soils and their associated microorganisms." Soil microorganisms, especially in association with plants, can help keep air in closed environments fresher and more healthful (plants are also nice to look at).

Other microorganisms in buildings can cause problems. These include *Legionella* (*see page 742*) which can grow in cooling systems, and allergy-causing molds such as *Aspergillus* (*p. 788*), which can grow in damp areas.

Summary

1. Soils are dominated by the solid phase, consisting of organic and inorganic components.
2. Most microorganisms in soils are associated with surfaces, and these surfaces influence microbial use of nutrients and interactions with plants and other living organisms.
3. Bacteria and fungi in soils have different functional strategies. Fungi tend to be found to a greater extent on the surfaces of aggregates, whereas microcolonies of bacteria are more commonly associated with smaller pores.
4. Insects and other soil animals are also important parts of the soil. These organisms interact with the microorganisms to influence nutrient cycling and other processes.
5. Soil microorganisms, insects, and other animals are nutrient-rich parts of the biomass. With their relatively rapid turnover, they are important sources of nutrients for plants.
6. Soil organic matter has extremely slow turnover times (100 to 1,000 years and longer) and is also a rich nutrient source. In most soils, nutrients from organic matter are not generally available for immediate plant use.
7. Only a small portion of the observable bacteria have been cultured or identified in the laboratory.
8. Insects and other animals serve as decomposer-reducers in the soil environment. These organisms physically reduce the size of materials being decomposed and mix them with the soil.
9. Soils differ in the relationship between organic matter accumulation—primary production—and decomposition.
10. Plants develop associations with many types of microorganisms. These include important associations with rhizobia; actinomycetes, forming actinorhizae; fungi, forming mycorrhizae; and with endophytic fungi and bacteria. *Agrobacterium* forms tumors and is useful in molecular biology.
11. Plants form tripartite and tetrapartite associations with microorganisms when several microbial groups establish associations with the same plant.
12. Fungal and bacterial endophytes can have positive and negative effects upon their host plants.
13. Composting is useful for maintaining and increasing soil fertility.
14. Because microorganisms are fallible, synthetic chemicals are often resistant to biodegradation. Many of these compounds can be modified by microorganisms when an easily utilizable carbon source is present; this process is known as cometabolism.
15. Soil microorganisms can interact with the atmosphere. Some can serve as nucleating agents, while others can degrade airborne pollutants.

Key Terms

actinorhizae *856*
arbuscule *855*
associative nitrogen fixation *854*
autochthonous *848*
bacteroid *854*
cometabolism *859*
decomposer-reducer *848*
ectomycorrhizal *855*
endomycorrhizal *855*
endophyte *856*
exudation *851*
geosmin *848*
humus *847*
infection thread *854*
microbivory *848*
oligotroph *848*
ped *846*
recalcitrance *859*
rhizosphere *851*
root nodule *854*
stem-nodulating rhizobia *854*
tetrapartite association *856*
Ti plasmid *858*
tripartite association *856*
VA mycorrhizae *855*
vesicles *855*
zymogenous *848*

Questions for Thought and Review

1. Why haven't more microbiologists studied the many soil microorganisms that have not been cultured? Why might 16S ribosomal RNA analyses assist with this project?
2. Considering the oxygen gradients, together with nutrient and waste-product gradients present in soils, is there a typical soil microenvironment? Explain.
3. Why might protozoa prefer to use laboratory-grown microorganisms as a nutrient source rather than normal soil microorganisms?
4. Tropical soils throughout the world are under intense pressure in terms of agricultural development. What microbial approaches might be used to better maintain this valuable resource?
5. Microorganisms interact with plants in many ways. How might it be possible to improve these interactions through molecular biology?
6. Plants and their associated soil microorganisms can help in purifying indoor air. Under what conditions could these plants and their associated soils become harmful?

Additional Reading

Alexander, M. 1977. *Introduction to soil microbiology.* New York: John Wiley and Sons.

Allen, M. F. 1991. *The ecology of mycorrhizae.* Cambridge: Cambridge University Press.

Brill, W. J. 1981. Agricultural microbiology. *Sci. Am.* 245(3):198–215.

Burke, M. J., and Lindow, S. E. 1990. Surface properties and size of the ice nucleation site in ice nucleation active bacteria: Theoretical considerations. *Cryobiology* 27:80–84.

Campbell, R. 1985. *Plant microbiology.* London: Edward Arnold.

Chatarpaul, L.; Chakravarty, P.; and Subramaniam, P. 1989. Studies in tetrapartite symbioses. I. Role of ecto- and endomycorrhizal fungi and *Frankia* on the growth performance of *Alnus incana. Plant Soil* 118:145–50.

Cheplick, G. P., and Clay, K. 1988. Acquired chemical defenses in grasses: The role of fungal endophytes. *Oikos* 52:309–18.

Clay, K. 1991. Parasitic castration of plants by fungi. *Trends Ecol. & Evol.* 6(5):162–66.

Daft, M. J.; Clelland, D. M.; and Gardner, I. C. 1985. Symbiosis with endomycorrhizas and nitrogen-fixing organisms. *Proc. Royal Soc. Edinburgh.* 283–98.

Dommergues, Y. R., and Diem, H. G. 1982. Microbiology of tropical soils and plant productivity. In *Developments in plant and soil sciences,* vol. 5. Boston: Martinus Nijhoff/Dr. W. Junk Publishers.

Fenchel, T. 1986. The ecology of heterotrophic microflagellates. *Adv. Microbial Ecol.* 9:57–97.

Frye, R. J.; Welsh, D.; Berry, T. M.; Stevenson, B. A.; and McCallum, T. 1992. Removal of contaminant organic gases from air in closed systems by soil. *Soil Biol. Biochem.* 24(6):607–12.

Halverson, L. J., and Stacey, G. 1986. Signal exchange in plant-microbe interactions. *Microbiol. Rev.* 50:193–225.

Jenny, H. 1980. *The soil resource. Origin and behavior.* New York: Springer-Verlag.

Lathwell, D. J., and Grove, T. L. 1986. Soil-plant relationships in the tropics. *Ann. Rev. Ecol. Syst.* 17:1–16.

McNeil, M. 1964. Lateritic soils. *Sci. Am.* 211:96–102.

Moffat, A. S. 1990. Nitrogen-fixing bacteria find new partners. *Science* 250:910–12.

Mooney, H. A.; Vitousek, P. M.; and Matson, P. A. 1987. Exchange of materials between terrestrial ecosystems and the atmosphere. *Science* 238:926–32.

Nap, J.-P., and T. Bisseling. 1990. Developmental biology of a plant-procaryote symbiosis: The legume root nodule. *Science* 250:948–54.

Oaks, A. 1992. A re-evaluation of nitrogen assimilation in roots. *BioScience* 42(2):103–11.

Paul, E. A., and Clark, F. E. 1989. *Soil microbiology and biochemistry.* New York: Academic Press.

Poole, R. K., and Dow, C. S. 1985. *Microbial gas metabolism. Mechanistic and biotechnological aspects.* Soc. Gen. Microbiol. New York: Academic Press.

Potrykus, I. 1991. Gene transfer to plants: Assessment of published approaches and results. *Annu. Rev. Plant Physiol. Plant Mol. Biol.* 42:205–25.

Reineke, W., and Knackmuss, H. J. 1988. Microbial degradation of haloaromatics. *Ann. Rev. Microbiol.* 42:263–87.

Rochkind-Dubinsky, M. L.; Sayler, G. S.; and Blackburn, J. W. 1987. *Microbiological decomposition of chlorinated aromatic compounds.* New York: Marcel Dekker, Inc.

Schwintzer, C. R., and Tjepkema, J. D. 1990. *The biology of* Frankia *and actinorhizal plants.* New York: Academic Press.

Triplett, E. W.; Roberts, G. P.; Ludden. P. W.; and Handelsman, J. 1989. What's new in nitrogen fixation. *ASM News* 55(1):15–21.

Tsai, S. M., and Phillips, D. A. 1991. Flavonoids released naturally from alfalfa promote development of symbiotic *Glomus* spores in vitro. *Appl. Environ. Microbiol.* 57:1485–88.

Verma, D. P. S. 1991. *Molecular signals in plant-microbe communications.* Boca Raton, Fla.: CRC Press, Inc.

Vitousek, P. M., and Sanford, R. L., Jr. 1986. Nutrient cycling in moist tropical forests. *Ann. Rev. Ecol. Syst.* 17:137–67.

Walden, R., and Schell, J. 1990. Techniques in plant molecular biology—progress and problems. *Eur. J. Biochem.* 192:563–76.

Part Eleven
Food and Industrial Microbiology

Many important industries rely extensively on microbiology and use large fermenters to manufacture their products. In this example, 1,000 gallon fermenters are being used by a California vintner to produce wine.

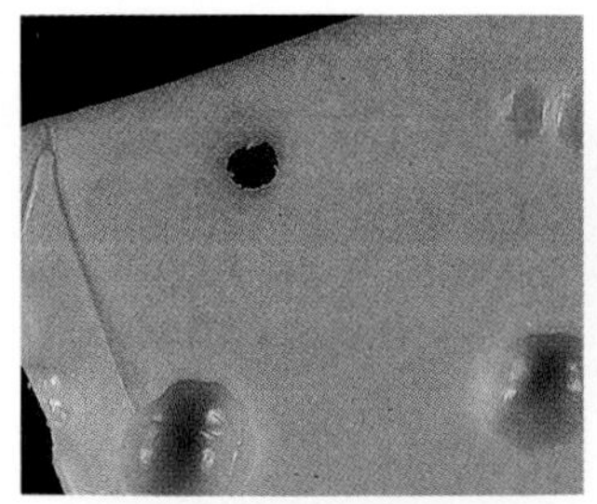

CHAPTER 43
Microbiology of Food

Tell me what you eat, and I will tell you what you are.
—Brillat-Savarin

There is no greater love than love of food. Air and water give us life, but food gives us a way of life.
—George Bernard Shaw

Outline

Concepts

1. Foods often provide an ideal environment for microbial survival and growth.
2. Microbial growth in foods involves successional changes, with intrinsic, or food-related, and extrinsic, or environmental, factors interacting with the microbial community over time.
3. Foods can be preserved by physical, chemical, and biological processes. Refrigeration does not significantly reduce microbial populations but only retards spoilage. Pasteurization results in a pathogen-free product with a longer shelf life.
4. Sterilization can be used to control microorganisms in foods. Heat-resistant spore-forming bacteria are used to test sterilization efficiency.
5. Chemicals can be added to foods to control microbial growth. Such chemicals include sugar, salt (decreasing water availability), and many organic chemicals that affect specific groups of microorganisms.
6. Foods can transmit a wide range of diseases to humans. In food infection, the food serves as a vehicle for the transfer of the pathogen to the consumer, in whom the pathogen grows and causes disease. With food intoxication, the microorganisms grow in the food and produce toxins that can then affect the consumer.
7. Modern molecular techniques are being used to detect disease-causing microorganisms in foods. Rapid and sensitive detection is possible with these procedures.
8. Dairy products, grains, meats, fruits, and vegetables can be fermented. It has been suggested that some fermented dairy products have antimicrobial and anticancer characteristics, especially when particular lactic acid bacteria are used.
9. Wines are produced by the direct fermentation of fruit juices or musts. For fermentation of cereals and grains, starches and proteins contained in these substrates must first be hydrolyzed to provide substrates for the alcoholic fermentation.
10. Microbial cells, grown on a variety of substrates, can be used directly as a nutrient by animals and humans. Popular single-cell protein sources include the cyanobacterium *Spirulina* and yeasts.

Foods are not only of nutritional value to humans but are often ideal culture media for microbial growth. Foods may be preserved by fermentation, or they may spoil, depending on the microorganisms present.

Microorganisms can be used to transform raw foods into gastronomic delights, including wines, cheeses, pickles, sausages, soy sauce, beers, and other alcoholic products. Foods can also serve as a vehicle for disease transmission, and the detection and control of pathogens and food spoilage microorganisms are important parts of food microbiology. During the entire sequence of food handling, from the producer to the final consumer, microorganisms can affect food quality.

Having wholesome, nutrient-rich foods is important for all people. Microbial growth in foods can result in either preservation or spoilage, depending on the microorganisms involved and the food storage conditions.

Undesirable microorganisms can cause illness in two ways when contaminated foods are eaten. A person may be infected by the food-borne pathogens, which then grow in the consumer. In other cases, toxic components may be formed when microorganisms grow in food before it is ingested, resulting in the rapid onset of disease symptoms after consumption. Contamination by disease-causing microorganisms can occur at any point in the food-handling sequence.

Food Spoilage and Preservation Processes

With the beginning of agriculture and a decreasing dependence on hunting and gathering, the need to preserve excess foods from spoilage became essential to survival. The use of salt as a meat preservative and the production of cheeses and curdled milks was introduced in Near Eastern civilization as early as 3000 B.C. The production of wines and the preservation of fish and meat by smoking were also common by this time. Despite a long tradition of efforts to preserve food from spoilage, it was not until the nineteenth century that the microbial spoilage of food was studied rigorously. Louis Pasteur established the modern era of food microbiology in 1857, when he showed that microorganisms cause milk spoilage. Pasteur's work in the 1860s proved that heat could be used to control spoilage organisms in wines and beers (*see chapter 1*).

Control of microorganism growth (chapter 15).

A variety of intrinsic and extrinsic factors determine whether microbial growth will preserve or spoil foods, as shown in figure 43.1. The intrinsic or food-related factors include pH, moisture content, water activity or availability, oxidation-reduction potential, physical structure of the food, available nutrients, and the possible presence of natural antimicrobial agents. Extrinsic or environmental factors include temperature, relative humidity, gases (CO_2, O_2) present, and the types and amounts of microorganisms added to the food.

Intrinsic Factors

The pH of a food is critical because a low pH favors the growth of yeasts and molds (*see chapter 6*). In neutral or alkaline pH foods, such as meats, bacteria are more dominant in the spoilage process. This leads to proteolysis and the anaerobic breakdown of proteins or **putrefaction,** which releases foul-smelling amine compounds. Depending on the major substrate present in a food, different types of spoilage may occur, as summarized in table 43.1.

The presence and availability of water also affect the ability of microorganisms to colonize foods. Simply by drying a food, one can control or eliminate spoilage processes. Water, even if present, can be made less available by adding solutes such as sugar and salt. Water availability is measured in terms of water activity (a_w). This represents the ratio of relative humidity of the air over a test solution compared with that of distilled water. When large quantities of salt or sugar are added to food, most microorganisms are dehydrated by the hypertonic conditions and cannot grow (table 43.2; *see also table 6.3*). Even under these adverse conditions, osmophilic and xerophilic microorganisms may spoil food. **Osmophilic** (Greek *osmus,* impulse, and *philein,* to love) **microorganisms** grow best in or on media with a high osmotic concentration, while **xerophilic** (Greek *xerosis,* dry, and *philein,* to love) **microorganisms** prefer a low a_w environment and may not grow under high a_w conditions.

Water activity and microbial growth (pp. 122–23).

The oxidation-reduction potential of a food also influences spoilage. After they are cooked, meat products, especially broths, often have lower oxidation-reduction potentials. These products with their readily available amino acids, peptides, and growth factors are ideal media for the growth of anaerobes, including *Clostridium* (*see table 37.3*).

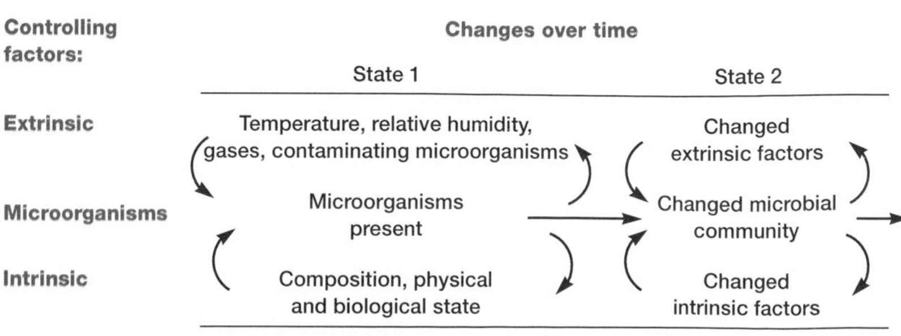

Figure 43.1 **Intrinsic and Extrinsic Factors.** A variety of intrinsic and extrinsic factors can influence microbial growth in foods. Time-related successional changes occur in the microbial community and the food.

TABLE 43.1 Differences in Spoilage Processes in Relation to Food Characteristics

Substrate	Chemical Reactions or Processes[a]	Typical Products and Effects
Pectin	Pectinolysis	Methanol, uronic acids (loss of fruit structure, soft rots)
Proteins	Proteolysis, deamination	Amino acids, peptides, amines, H_2S, ammonia, indole (bitterness, souring, bad odor, sliminess)
Carbohydrates	Hydrolysis, fermentations	Organic acids, CO_2, mixed alcohols (souring, acidification)
Lipids	Hydrolysis, fatty acid degradation	Glycerol and mixed fatty acids (rancidity, bitterness)

[a]Other reactions also occur during the spoilage of these substrates.

TABLE 43.2 Approximate Minimum Water Activity Relationships of Microbial Groups of Importance in Food Spoilage

Organism Group	a_w
Most spoilage bacteria	0.90
Most spoilage yeasts	0.88
Most spoilage molds	0.80
Halophilic bacteria	0.75
Xerophilic molds	0.61
Osmophilic yeasts	0.60

Jay: *MODERN FOOD MICROBIOLOGY*, 3rd ed. Reprinted with permission of the publisher Van Nostrand Reinhold.

The physical structure of a food also can affect the course and extent of spoilage. The grinding and mixing of foods such as sausage and hamburger not only increase the food surface area and alter cellular structure, but also distribute contaminating microorganisms throughout the food. This can result in rapid spoilage if such foods are stored improperly. Vegetables and fruits have outer skins (peels and rinds) that protect them from spoilage. Often, spoilage microorganisms have specialized enzymes that help them weaken and penetrate protective peels and rinds.

Many foods contain natural antimicrobial substances, including complex chemical inhibitors and enzymes. Aldehydic and phenolic compounds are found in cinnamon, mustard, and oregano, and these compounds inhibit microbial growth. Coumarins found in fruits and vegetables exhibit antimicrobial activity. Cow's milk and eggs also contain antimicrobial substances. Eggs are rich in the enzyme lysozyme that can lyse the cell walls of contaminating gram-positive bacteria (*see figure 30.8*).

Extrinsic Factors

Temperature and relative humidity are important extrinsic factors in determining whether a food will spoil. At higher relative humidities, microbial growth is initiated more rapidly, even at lower temperatures (especially when refrigerators are not maintained in a defrosted state). When drier foods are placed in moist environments, moisture absorption can occur on the food surface, eventually allowing microbial growth.

The atmosphere in which the food is stored is also important. This is especially true with shrink-packed foods because many plastic films allow oxygen diffusion, which results in increased growth of surface-associated microbial contaminants. Excess CO_2 can decrease the solution pH, inhibiting microbial growth. Storing meat in a high CO_2 atmosphere inhibits gram-negative bacteria, resulting in a population dominated by the lactobacilli.

1. What are some intrinsic factors that influence food spoilage and how do they exert their effects?
2. Why might sausage and other ground meat products provide a better environment for the growth of food spoilage organisms than raw cuts of meats?
3. List some antimicrobial substances found in foods. What is the mechanism of action of lysozyme?
4. What primary extrinsic factors can determine whether food spoilage will occur?
5. Why might CO_2 concentrations in the food environment modify responses of food spoilage organisms?

Food Preservation Alternatives

Foods can be preserved by a variety of methods (table 43.3). It is vital to eliminate or reduce the populations of spoilage and disease-causing microorganisms and to maintain the microbiological quality of a food with proper storage and packaging. Contamination often occurs after a package or can is opened and just before the food is served. This can provide an ideal opportunity for growth and transmission of pathogens, if care is not taken. Washing utensils with chemical sanitizers such as chlorine or quaternary ammonium compounds and limiting human contact with foods can make a major difference in the microbiological quality of foods.

Sanitizing agent characteristics (pp. 318–21).

Physical Removal of Microorganisms

Microorganisms can be removed from water, wine, beer, juices, soft drinks, and other liquids by filtration. The use of bacteriologic filters can keep bacterial populations low or eliminate them entirely. Prefilters and centrifugation are often used to maximize filter life and effectiveness. Several major brands of beer are filtered rather than pasteurized to better preserve the flavor and aroma of the original product.

Box 43.1

An Army Travels on Its Stomach

The movement and maintenance of large numbers of military personnel have always been limited by food supplies. The need to maintain large numbers of troops under hostile and inclement conditions led the French government in 1795 to offer a prize of 12,000 francs to the individual who could preserve foods for use under field conditions. The prize was awarded to Nicholas Appert, a candy maker, for his development of a heating process in which meats and other products could be preserved under sealed conditions.

Appert's work was based on the assumptions that heating and boiling control "ferments" and that sealing the food in bottles before heating it avoids the effects of air on spoilage. Despite Leeuwenhock's earlier work, Appert did not have the concept of microorganisms to assist him in explaining the effectiveness of his process. His containers were large glass bottles, sealed with laminated corks and fish glue. With extreme care and attention to detail, he was able to heat these bottles in boiling water to provide food that could be stored for several years. Appert's work was an important foundation for the later studies of Louis Pasteur.

Table 43.3 Basic Approaches to Food Preservation

General Technique	Examples of Process
Asepsis, removal of microorganisms	Avoidance of microbial contamination, filtration, centrifugation
Low temperature	Refrigeration, freezing
High temperature	Partial or complete heat inactivation of microorganisms (pasteurization and canning)
Water removal	Lyophilization, use of a spray dryer or heated drum dryer
Water availability decrease	Addition of solutes such as salt or sugar to decrease a_w values
Chemical preservation	Addition of specific inhibitory substances (e.g., organic acids, nitrates, sulfur dioxide)
Radiation	Use of ionizing (UV) and nonionizing (gamma rays) radiation

Figure 43.2 A Canning Operation. Microbial control is important in the processing and preservation of many foods. Worker pouring peas into a large, clean vat during the preparation of vegetable soup. After preparation, the soup is transferred to cans. Each can is heated for a short period, sealed, processed at temperatures around 110–121° C in a canning retort to destroy spoilage microorganisms, and finally cooled.

Temperature Effects

Refrigeration at 5° C retards microbial growth, although with extended storage, psychrophiles and psychrotrophs will eventually grow and produce spoilage. Slow microbial growth at temperatures below −10° C has been described, particularly with fruit juice concentrates, ice cream, and some fruits. Again, extended holding times are required before significant growth occurs. Some microorganisms are very sensitive to cold and their numbers will be reduced by it, but cold does not lead to significant decreases in overall microbial populations.

Temperature effects on microbial growth (pp. 124–26).

Controlling microbial populations in foods by means of high temperatures can significantly limit disease transmission and spoilage. **Pasteurization** uses high temperatures to eliminate disease-causing organisms and reduce microbial populations (*see chapter 15*). Sterilization relies on high temperatures to eliminate all living organisms. Heating processes, first used by Nicholas Appert in 1809 (Box 43.1), provide a safe means of preserving foods, particularly when carried out in commercial **canning** operations (figure 43.2). Canned food is heated in special containers called retorts at about 115° C for intervals ranging from 25 to over 100 minutes. The precise time and temperature depend on the nature of the food. Sometimes canning does not kill all the microorganisms, but only those that will spoil the food (any remaining bacteria are unable to grow). After heat treatment, the cans are cooled as rapidly as possible, usually with cold water. Quality control and processing effectiveness are sometimes compromised, however, in home processing of foods, especially with less acid (pH values greater than 4.6) products such as green beans or meats.

Destruction of microorganisms by heat (pp. 313–16).

Pasteurization involves heating food to a temperature that kills disease-causing microorganisms such as *Mycobacterium tuberculosis* (*see chapters 22 and 37*) and substantially reduces the levels of spoilage organisms. In the processing of milk, beers, and fruit juices by conventional low-temperature holding (LTH) pasteurization, the liquid is maintained at 62.8° C for 30 minutes. Products can also be held at 71° C for 15 seconds, a high-temperature, short-time (HTST) process; milk can be treated at 141° C for 2 seconds for ultra-high-temperature (UHT) processing. Shorter-term processing results in improved flavor and extended product shelf life.

Such heat treatment is based on a statistical probability that the number of remaining viable microorganisms will be below a certain level after a particular heating time at a specific temperature. This process is discussed in detail on pp. 313–16.

Dehydration, used to produce freeze-dried foods, is now a common means of eliminating microbial growth. This modern process is simply an update of older procedures in which grains, meats, fish, and fruits were dried. The combination of free-water loss with an increase in solute concentration in the remaining water makes this type of preservation possible.

Preservation by Chemicals and Radiation

Various chemical agents can be used to preserve foods, and these substances are closely regulated by the U.S. Food and Drug Administration (table 43.4). They include simple organic acids, sulfite, ethylene oxide as a gas sterilant, sodium nitrite, and ethyl formate. The effectiveness of many of these chemical preservatives depends on the food pH. As an example, sodium propionate is most effective at lower pH values, where it is primarily undissociated and able to be taken up by lipids of microorganisms. Breads, with their low pH values, often contain sodium propionate as a preservative. These compounds, "generally recognized as safe" (GRAS), are used with grain, dairy, vegetable, and fruit products. Sodium nitrite is an important chemical used to control germination of *Clostridium* spores and to stabilize the red color of meats. Nitrosamines, formed when nitrites react with secondary amines in the heating of meats, are potential carcinogens and therefore of some concern.

Although prohibited in the United States, nisin and natamycin have been used in other parts of the world to inhibit microbial growth. Nisin, produced naturally by *L. lactis* during cheese production, can be used in cheeses to inhibit the growth of *Clostridium butyricum.*

Radiation, both ionizing and nonionizing, has an interesting history in relation to food preservation. Ultraviolet radiation is used to control populations of microorganisms on the surfaces of laboratory and food-handling equipment, but it does not penetrate food. The major method used for radiation sterilization of food is gamma irradiation from a cobalt-60 source (*see p. 318*). Such electromagnetic radiation has excellent penetrating power and must be used with moist foods because the radiation produces peroxides from water in the microbial cells, resulting in oxidation of sensitive cellular constituents. This process of **radappertization,** named after Nicholas Appert (Box 43.1), can extend the shelf life of seafoods, fruits, and vegetables. To sterilize meat products, commonly 4.5 to 5.6 megarads are used.

Among the more interesting radiation-resistant bacteria that have been studied is *Deinococcus radiodurans* (*see chapter 22*). This bacterium has a complex cell wall structure and tetrad-forming growth patterns (*see figure 22.3*). It also has an extraordinary capacity to withstand high doses of radiation, although the mechanism for its resistance is not understood.

TABLE 43.4 Major Groups of Chemicals Used in Food Preservation

Preservatives	Approximate Maximum Use Range	Organisms Affected	Foods
Propionic acid/propionates	0.32%	Molds	Bread, cakes, some cheeses, inhibitor of ropy bread dough
Sorbic acid/sorbates	0.2%	Molds	Hard cheeses, figs, syrups, salad dressings, jellies, cakes
Benzoic acid/benzoates	0.1%	Yeasts and molds	Margarine, pickle relishes, apple cider, soft drinks, tomato ketchup, salad dressings
Parabens[a]	0.1%	Yeasts and molds	Bakery products, soft drinks, pickles, salad dressings
SO_2/sulfites	200–300 ppm	Insects and microorganisms	Molasses, dried fruits, wine, lemon juice (not to be used in meats or other foods recognized as sources of thiamine)
Ethylene/propylene oxides	700 ppm	Yeasts, molds, vermin	Fumigant for spices, nuts
Sodium diacetate	0.32%	Molds	Bread
Dehydroacetic acid	65 ppm	Insects	Pesticide on strawberries, squash
Sodium nitrite	120 ppm	Clostridia	Meat-curing preparations
Caprylic acid	—	Molds	Cheese wraps
Ethyl formate	15–200 ppm	Yeasts and molds	Dried fruits, nuts

Jay: *MODERN FOOD MICROBIOLOGY*, 3rd. ed. Reprinted with permission of the publisher Van Nostrand Reinhold.

[a]Methyl-propyl-, and heptyl-esters of *p*-hydroxybenzoic acid

1. Describe the major approaches used in food preservation.
2. What types of chemicals can be used to preserve foods?
3. Nitrite is often used to improve the storage characteristics of prepared meats. What toxicological problems may result from the use of this chemical?
4. Under what conditions can ultraviolet light and gamma radiation be used to control microbial populations in foods and in food preparation? What is radappertization?

Diseases and Foods

Diseases Transmitted by Foods

Many diseases transmitted by foods, or food poisonings, are discussed in chapters 36 and 37, and only a few of the more important food-borne bacterial pathogens are mentioned here. There are two primary types of food-related diseases: food-borne infections and food intoxications.

Food-borne and waterborne diseases (pp. 735–36, 763–67.)

A food-borne infection involves the ingestion of the pathogen, followed by growth accompanied by tissue invasion and/or the release of toxins in the host intestine. The major diseases of this type are summarized in table 43.5. Salmonellosis results from ingestion of the organism, and all species and strains of *Salmonella* are pathogenic (*see chapter 37*). Gastroenteritis is the disease of most concern in relation to foods, occurring after an incubation time as short as eight hours. Meats, poultry, and dairy products are the primary sources of the pathogen. *Salmonella* infection can arise from contamination by workers in food-processing plants and restaurants (Box 43.2; *see also Box 35.3*).

Campylobacter jejuni is considered a leading cause of acute bacterial gastroenteritis in humans and can affect persons of all ages. This important pathogen is often transmitted by uncooked or poorly cooked poultry products. For example, transmission often occurs when wooden cutting boards are used for chicken preparation and then for salads. Contamination with as few as 500 viable *Camplyobacter jejuni* can lead to the onset of diarrhea. *Campylobacter jejuni* is also transmitted by raw milk, and the organism has been found on various red meats. Thorough cooking of food prevents this disease transmission problem.

Listeriosis, caused by *Listeria monocytogenes* (*see chapter 22*), is of continuing interest, as shown by the outbreak that occurred in Southern California in 1985. This outbreak was caused by improper pasteurization of milk used in the commercial production of Mexican-style cheeses. At least 86 cases of infection occurred, including 58 cases involving mother-infant pairs. Forty-seven people died. The outbreak was traced to pinhole leaks in the heat exchangers of a pasteurizing unit. The leaks allowed incoming raw milk to contaminate the pasteurized milk before production of the cheese. *Listeria* is difficult to work with because an extended incubation of samples is required for growth and detection.

Escherichia coli is now recognized as an important food-borne disease organism. Enteropathogenic, enteroinvasive, and enterotoxigenic types can cause diarrhea. The enterohemorrhagic *E. coli* 0157:H7, which causes hemorrhagic colitis, is of particular concern.

All these food-borne diseases are associated with poor hygienic practices. Whether by water or food transmission, the fecal-oral route is maintained, with the food providing the vital link between hosts. Fomites, such as sink faucets and drinking cups, also play a role in the maintenance of the fecal-oral route of contamination.

TABLE 43.5 Major Food-Borne Infectious Diseases

Disease	Organism	Incubation Period and Characteristics	Major Foods Involved
Salmonellosis	*S. typhimurium, S. enteritidis*	12–24 hr Enterotoxin and cytotoxins	Meats, poultry, fish, eggs, dairy products
Campylobacteriosis	*Campylobacter jejuni*	Usually 3–5 days Most toxins are heat-labile	Milk, pork, poultry products, water
Listeriosis	*L. monocytogenes*	Varying periods Related to meningitis and abortion; newborns and the elderly especially susceptible	Meat products, especially pork and milk
Escherichia coli diarrhea and colitis	*E. coli*	6–36 hr Enterotoxigenic positive and negative strains; hemorrhagic colitis	Undercooked ground beef, raw milk
Shigellosis	*Shigella sonnei, S. flexneri*	1–7 days	Egg products, puddings
Yersiniosis	*Yersinia enterocolitica*	16–48 hr Some heat-stable toxins	Milk, meat products, tofu
Vibrio parahaemolyticus gastroenteritis	*V. parahaemolyticus*	16–48 hr	Seafood, shellfish

Sources: Data from P. J. VanDemark and B. L. Batzing, *The Microbes*, 1987; *Bergey's Manual*, Vol. 2, 1986; and D. O. Cliver, *Foodborne Diseases*, 1990.

Box 43.2

Typhoid Fever and Canned Meat

Minor errors in canning have led to major typhoid outbreaks. In 1964, canned corned beef produced in South America was cooled, after sterilization, with nonchlorinated water; the vacuum created when the cans were cooled drew *S. typhi* into some of the cans, which were not completely sealed. This contaminated product was later sliced in an Aberdeen, Scotland, food store, and the meat slicer became a continuing contamination source; the result was a major epidemic that involved 400 people. The *S. typhi* was a South American strain, and eventually the contamination was traced to the contaminated water used to cool the cans.

This case emphasizes the importance of careful food processing and handling to control the spread of disease during food production and preparation.

Microbial growth in food products can also result in a **food intoxication,** as summarized in table 37.3 (*pp. 764–65*). Intoxication produces symptoms shortly after the food is consumed because growth of the disease-causing microorganism is not required. Toxins produced in the food can be associated with microbial cells or can be released from the cells.

Most *Staphylococcus aureus* strains cause a staphylococcal enteritis related to the synthesis of extracellular toxins (*see chapter 37*). These are heat-resistant proteins, and heating will not usually render the food safe. The effects of the toxins are quickly felt, with disease symptoms occurring within two to six hours. The main reservoir of *S. aureus* is the human nasal cavity. Frequently, *S. aureus* is transmitted to a person's hands and then is introduced into food during preparation. Growth and enterotoxin production usually occur when contaminated foods are held at room temperature for several hours.

Three gram-positive rods are known to cause food intoxications: *Clostridium botulinum, C. perfringens,* and *Bacillus cereus* (*see table 37.3*). *C. botulinum* poisoning is discussed in chapter 37, and *C. perfringens* intoxication is described here.

Clostridium perfringens food poisoning is one of the more widespread food intoxications. These microorganisms, which produce exotoxins, must grow to levels of approximately 10^6 bacteria per gram or higher in a food to cause disease. At least 10^8 bacteria must be ingested. They are common inhabitants of soil, water, food, spices, and the intestinal tract. Upon ingestion, the cells sporulate in the intestine. The enterotoxin is a spore-specific protein and is produced during the sporulation process. Enterotoxin can be detected in the feces of affected individuals. *Clostridium perfringens* food poisoning is common and occurs after meat products are heated, which results in O_2 depletion. If the foods are cooled slowly, growth of the microorganism can occur. At 45° C, enterotoxin can be detected three hours after growth is initiated.

Bacillus cereus can cause two distinct types of illnesses depending on the type of toxin produced: an emetic illness characterized by nausea and vomiting with an incubation time of 1 to 6 hours, and a diarrheal type, with an incubation of 4 to 16 hours. The emetic type is often associated with boiled or fried rice, while the diarrheal type is associated with a wider range of foods.

Disease Microorganism Detection

A major problem in maintaining food safety is the need to rapidly detect microorganisms in order to curb outbreaks that can affect large populations. This is especially important because of widescale distribution of perishable foods. Standard cultural techniques (*see chapter 5*) may require days to weeks for positive identification of pathogens. Identification is often complicated by the low numbers of pathogens compared with the background microflora. Furthermore, the varied chemical and physical composition of foods can make isolation difficult. Fluorescent antibody, enzyme-linked immunoassays (ELISAs), and radioimmunoassay techniques have proven of value (*see chapter 33*). These can be used to detect small amounts of pathogen-specific antigens. Molecular biology techniques also are increasingly used in identification. The basic methods of analysis and manipulation of DNA and RNA are discussed in chapter 14. These methods are valuable for three purposes: (1) to detect the presence of a single, specific pathogen; (2) to detect viruses that cannot be grown conveniently; and (3) to identify slow-growing or nonculturable pathogens.

Pathogens now can be identified by detecting specific DNA or RNA base sequences with **probes.** These may range from 10 to 10,000 bases in length, but usually probes of 14 to 40 bases are employed. They may be created by generating fragments with restriction endonucleases or through direct chemical synthesis. Probes are labeled by linking them to a variety of enzymatic, isotopic, chromogenic, or luminescent/fluorescent markers. A major advantage of their use is the speed with which specific microorganisms can be detected in a set of cultures, as shown in figure 43.3. In this example, a hydrophobic grid-membrane system has been used. The *Listeria monocytogenes* cultures are radioactive, indicating that they have bound the probe, while other *Listeria* species do not show probe binding.

These techniques also have been extended to allow the detection of a few target cells in large populations of background microorganisms. For example, by using the polymerase chain reaction (*see pp. 291–92*), as few as 10 toxin-producing *E. coli* cells can be detected in a population of 100,000 cells isolated

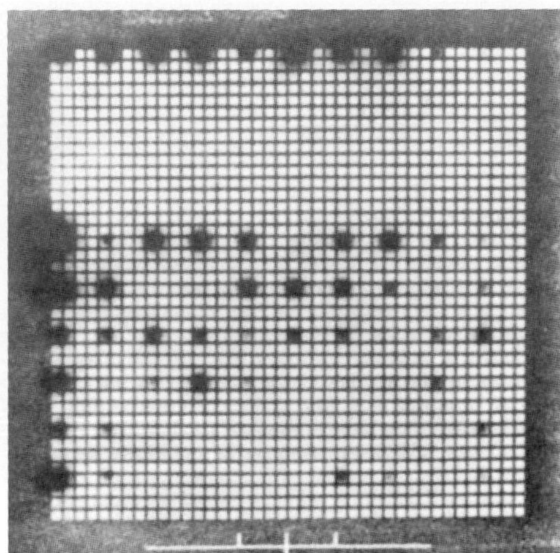

Figure 43.3 Molecular Probes and Food Microbiology. Molecular techniques are finding increasing use in microbial analysis of foods. Autoradiogram of a radioactive-labeled *Listeria monocytogenes* probe against 100 *Listeria* cultures. Only the *Listeria monocytogenes* cultures show sequence homology and binding with the DNA probe, darkening the autoradiogram film. The other *Listeria* spp. do not react with the probe.

Figure 43.4 The Polymerase Chain Reaction and Food Microbiology. The polymerase chain reaction (PCR) can be used to improve detection of microbes and their toxins in foods. PCR of *Escherichia coli* strains isolated from soft cheese. Approximately 100,000 cells of each strain were lysed and subjected to 40 cycles of PCR to detect DNA coding for a heat-stable toxin (STI). Lane 1: 595 base pair (bp) *malB* fragment and 175 bp STI fragment; lane 2: *HinfI* restriction analysis of the 175 bp fragment in lane 1, indicating that this fragment is the STI DNA; lanes 3 and 4: positive controls with LT and ST-producing organisms, respectively; lane 5: molecular weight markers.

from soft cheese samples. A rapid probe of the sample showed that none of the 58 cultures recovered from this cheese produced a heat-labile toxin (LT), whereas one produced a 175 base pair fragment characteristic of the heat-stable toxin (ST). Based on the susceptibility of the *E. coli* DNA to the *Hin*fI restriction enzyme (lane 2), this was identified as DNA related to the ST (lane 4 in figure 43.4).

These brief examples show the exciting potential of applying rapid DNA- and RNA-based techniques to the detection and control of food pathogens.

1. Differentiate between food intoxication and food-borne infection, in terms of disease occurrence.
2. What common food is often related to *Campylobacter*-caused gastroenteritis? What means can be used to control the occurrence of this disease from this source?
3. Why is listeriosis of concern from a medical standpoint?
4. Describe *S. aureus* food intoxication and how it occurs. Will thorough heating prevent the disease?
5. Under what conditions does *C. perfringens* poisoning arise and when is its enterotoxin produced?
6. How can molecular techniques improve the detection of disease-causing microorganisms in foods?

Food Spoilage

The role of microorganisms in the spoilage and deterioration of specific groups of foods has led to interesting turns in history, occasionally being associated with "miracles" and "witchcraft." *Serratia marcescens,* the "miracle organism," for example, can produce a red pigment when it grows on moist food materials. Some historians feel it may have been responsible for reports of "blood" on communion wafers and other bread.

Molds can rapidly grow on grains and corn when these products are held under moist conditions (figure 43.5). Infection of grains by the ascomycete *Claviceps purpura* causes **ergotism,** a toxic condition. Hallucinogenic alkaloids produced by this fungus can lead to altered behavior, abortion, and death if infected grains are eaten. Ergotism is further discussed in chapter 26 (*p. 525*).

One of the more important groups of fungus-derived carcinogens are the **aflatoxins.** These toxins are produced in moist grains and nut products. Aflatoxins were discovered in 1960, when 100,000 turkey poults died from eating fungus-infested peanut meal. *Aspergillus flavus* was found in the infected peanut meal, together with alcohol-extractable toxins termed aflatoxins. These flat-ringed planar compounds intercalate with the cells' nucleic acids and act as frameshift mutagens and carcinogens. This occurs primarily in the liver, where they are converted to unstable derivatives. The major aflatoxin types are shown in figure 43.6. These compounds and their derivatives can be separated by chromatographic procedures and can be recognized under UV light by their characteristic fluorescence (figure 43.7). Besides their importance in grains, they have also been observed in milk, beer, cocoa, raisins, and soybean meal. The U.S. Environmental Protection Agency has established "allowable action levels," and limits of 20 parts per billion (ppb) are used for feeds and nut products. Milk, in comparison, has been assigned an action level of 0.5 ppb.

Mutagens and mutagenic mechanisms (pp. 244–50).

Figure 43.5 **Food Spoilage.** When foods are not stored properly, microorganisms can cause spoilage. Typical examples are fungal spoilage of (*a*) bread and (*b*) corn. Such spoilage of corn is called ear rot. This can result in major economic losses.

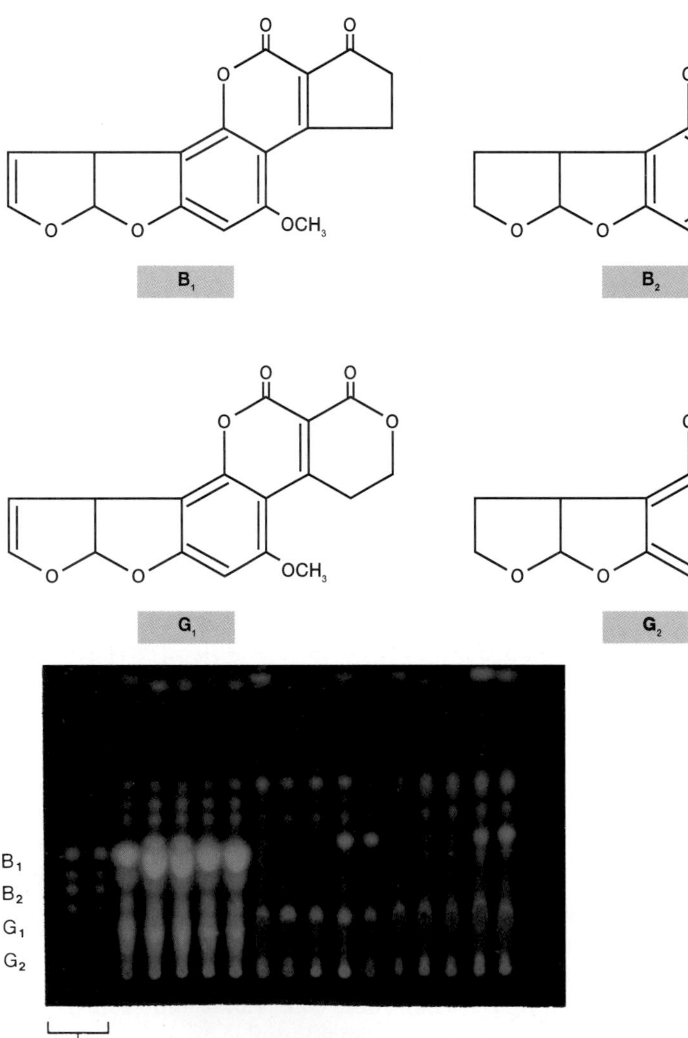

Figure 43.6 **Aflatoxins.** When *Aspergillis flavus* and related fungi grow on foods, carcinogenic aflatoxins can be formed. These have four basic structures. The letter designations refer to the color of the compounds under ultraviolet light after extraction from the grain and separation by chromatography. The B_1 and B_2 compounds fluoresce with a blue color, while the G_1 and G_2 appear green.

Figure 43.7 **Aflatoxin Analysis.** Aflatoxins, if present in foods, can be detected after extraction of the foods with solvents and thin-layer chromatography. A completed chromatographic plate used to test for aflatoxins in foods is shown. The chromatographic plate has been photographed with its bottom down so that the fastest moving components are at the top of the figure. The two left lanes contain the aflatoxin standards B_1, B_2, G_1, and G_2 (B_1 moved the farthest, and G_2 the least). The remaining 15 samples contained varying amounts of the four different aflatoxins, as well as other components.

Meat and dairy products, with their high nutritional value and the presence of easily utilizable carbohydrates, fats, and proteins, are ideal environments for spoilage by microorganisms (figure 43.8). Proteolysis and putrefaction are typical results of microbial spoilage of such high-protein materials. Unpasteurized milk undergoes a predictable four-step succession during spoilage: Acid production by *Lactococcus lactis* subsp. *lactis* (formerly *Streptococcus lactis*) is followed by additional acid production associated with the growth of more acid tolerant organisms such as *Lactobacillus.* At this point, yeasts and molds become dominant and degrade the accumulated lactic acid, and the acidity gradually decreases. Eventually, protein-digesting bacteria become active, resulting in a putrid odor and bitter flavor. The milk, originally opaque, can eventually become clear (figure 43.8*c*).

In comparison, most fruits and vegetables have a much lower protein and fat content and undergo a different kind of spoilage. Readily degradable carbohydrates favor vegetable spoilage by bacteria, especially bacteria that cause soft rots, such as *Erwinia carotovora,* which produces hydrolytic enzymes. The high oxidation-reduction potential and lack of reduced conditions permits aerobes and facultative anaerobes to contribute to the decomposition processes (figure 43.9). Bacteria do not seem important in the initial spoilage of whole fruits; instead, such spoilage is often initiated by molds. These organisms have enzymes that contribute to the weakening and penetration of the protective outer skin.

Food spoilage problems occur with "minimally processed" concentrated frozen citrus products. These are prepared with little or no heat treatment, and major spoilage can be caused by *Lactobacillus* and *Leuconostoc* spp., which produce diacetyl-butter flavors. *Saccharomyces* and *Candida* can also spoil juices. Concentrated juice has a decreased water activity ($a_w = 0.8$ to 0.83), and when kept frozen at about $-9°$ C, the juices can be stored for long periods. However, when concentrated juices are diluted with water that contains spoilage organisms, or if the juice is stored in improperly washed containers, problems can occur. Also, microorganisms in the frozen concentrated juices can begin the spoilage process after addition of water. Ready-to-serve (RTS) juices present other problems as the a_w values are sufficiently high to allow microbial growth. This is especially true with extended storage at refrigeration temperatures. Although pas-

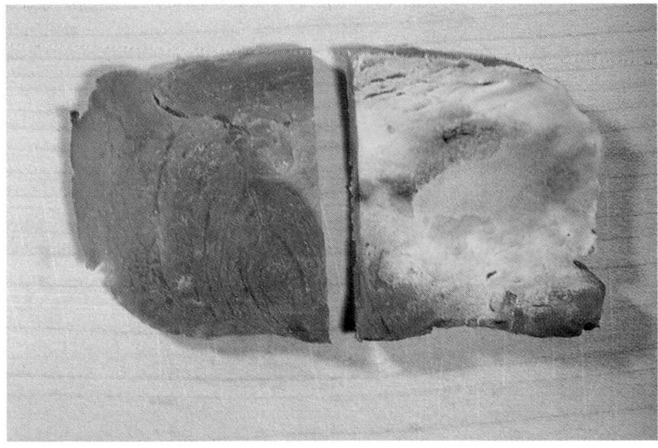

(a)

(b)

(c)

Figure 43.8 Spoilage of Meat and Dairy Products. (*a*) Fresh and spoiled (moldy) meat. (*b*) Moldy cheese. (*c*) Fresh and curdled milk. The curdled milk has undergone a natural sequence of spoilage organism activity, resulting in separated curds and whey.

(a)

(b)

Figure 43.9 Spoilage of Fruits. (*a*) *Penicillium* mold on oranges. (*b*) Mold growing on grapes.

teurization can be used, most consumers are sensitive to the loss of flavor that this process entails.

Spices often possess significant antimicrobial substances. Generally, fungi are more sensitive than most bacteria to spices. Sage and rosemary are two of the most antimicrobial spices. Garlic, which contains allicin, and cloves, which have eugenol, are also important inhibitors of microbial growth. However, spices can also contain pathogenic and spoilage organisms. Coliforms have been detected in most spices, together with *B. cereus, C. perfringens,* and *Salmonella* spp. Microorganisms in spices are eliminated or reduced by ethylene oxide sterilization. This treatment can result in *Salmonella*-free spices and a 90% reduction in the levels of general spoilage organisms.

Despite efforts to eliminate spoilage microorganisms during canning, sometimes canned foods are spoiled (figure 43.10). This may be due to spoilage before canning, underprocessing during canning, and leakage of contaminated water through can seams during cooling. Spoiled food can be altered in such characteristics as color, texture, odor, and taste. Fermentation acids, sulfides, and gases (particularly CO_2 and H_2S) may be produced. In flat sour spoilage, no gas is generated and the can does not swell, but its contents are rendered sour by the presence of fermentation acids. If spoilage microorganisms produce gas, both ends of the can will bulge outward to give a swell. Sometimes the swollen ends can be moved by thumb pressure (soft swells); in other cases, the gas pressure is so great that the ends cannot be dented by hand (hard swells). It should be noted that swelling is not always due to microbial spoilage. Acid in high-acid foods may react with the iron of the can to release hydrogen and generate a hydrogen swell. Hydrogen sulfide production by *Desulfotomaculum* can cause "sulfur stinkers."

Figure 43.10 Food Preservation by Canning. This technique is widely used and very effective. Improper canning can occasionally occur, as shown by this bent and leaking can.

1. What fungal genus produces ergot alkaloids? What conditions are required for the synthesis of these substances?
2. Aflatoxins are produced by which fungal genus? How do they damage animals that eat the contaminated food?
3. Describe in general how food spoilage occurs. What factors influence the nature of the spoilage organisms responsible?
4. Why do concentrated citrus juices present such interesting spoilage problems?
5. How may canned food become spoiled and what changes can occur? Why is swelling not always an indication of microbial spoilage?

Microbiology of Fermented Foods

Over the last several thousand years, fermentation has been a major way of preserving food. Microbial growth, either of natural or inoculated populations, causes chemical and/or textural changes forming a product that can be stored for extended periods. The fermentation process is also used to create new, pleasing food flavors and odors.

Dairy Products

Fermented milk has distinct flavors and aromas, depending on the incubation conditions and the microbial inocula used (table 43.6). All fermented dairy products result from similar manufacturing techniques, in which acid produced through microbial activity causes protein denaturation. To carry out the

TABLE 43.6 Major Fermented Milk Products and Fermenting Microorganisms

Fermented Product	Fermenting Microorganism	Description
Acidophilus milk	*Lactobacillus acidophilus*	Skim milk is sterilized and then inoculated with *L. acidophilus*
Cultured buttermilk	*Lactococcus lactis* subsp. *diacetilactis, Leuconostoc cremoris, Lactococcus cremoris*	Product is made with skimmed or low-fat pasteurized milk
Kefir	*Lactococcus lactis, Lactobacillus bulgaricus, Saccharomyces* spp.	Produced from a mixed lactic acid and alcoholic fermentation
Sour cream	*Lactococcus* spp., *Leuconostoc* spp.	Cream is inoculated and incubated until acidity develops
Yogurt	*Streptococcus thermophilus* and *Lactobacillus bulgaricus*	Product is made from nonfat or low-fat milk to which stabilizers like gelatin are added
Butter	*Lactococcus lactis*	Cream is incubated until the desired acidity is achieved, followed by churning, washing, and salting

process, one usually inoculates milk with the desired culture, incubates it at optimum temperature, and then stops microbial growth by cooling. *Lactobacillus* spp. and *Lactococcus lactis* cultures are used for aroma and acid production. The organism *Lactococcus lactis* subsp. *diacetilactis* converts milk citrate to diacetyl, which gives a special buttery flavor to the finished product. The use of these microorganisms with skim milk produces cultured buttermilk, and when cream is used, sour cream is the result.

The genus Lactobacillus (*p. 462*).
The genus Lactococcus (*pp. 455–57*).

Yogurt is produced by a special starter in which two major bacteria are present in a 1:1 ratio: *S. thermophilus* and *L. bulgaricus*. With these organisms growing in concert, acid is produced by *Streptococcus*, and aroma components are formed by the *Lactobacillus*. Freshly prepared yogurt contains 10^9 bacteria per gram. *Acidophilus milk* is produced by using *Lactobacillus acidophilus*.

Fermented milks may have beneficial effects. *L. acidophilus* may modify the microbial flora in the lower intestine, thus improving general health, and it is often used as a dietary adjunct. Many microorganisms in fermented dairy products stabilize the bowel microflora, and some appear to have antimicrobial properties. The exact nature and extent of health benefits are still unclear, but may involve minimizing lactose intolerance, lowering serum cholesterol, and possibly exhibiting anticancer activity. Several lactobacilli have antitumor compounds in their cell walls. Such findings suggest that diets including lactic acid bacteria, especially *L. acidophilus*, may contribute to the control of colon cancer.

Another interesting group are the bifidobacteria. The genus *Bifidobacterium* contains irregular, nonsporing, gram-positive rods that may be club-shaped or forked at the end (figure 43.11). Bifidobacteria are nonmotile, anaerobic, and ferment lactose and other sugars to acetic and lactic acids. They are typical residents of the human intestinal tract and were discovered in 1906. Many beneficial properties are attributed to them. Bifidobacteria are thought to help maintain the normal intestinal balance, while improving lactose tolerance; to possess antitumorigenic activity; and to reduce serum cholesterol levels. In addition, some believe that they promote calcium absorption and the synthesis of B-complex vitamins. "Bifid"-amended fermented milk products are now available in various parts of the world (figure 43.12).

Cheese is one of the oldest human foods and is thought to have been developed approximately 8,000 years ago. About 2,000 distinct varieties of cheese are produced throughout the world, representing approximately 20 general types (table 43.7 and figure 43.13). Often, cheeses are classified based on texture or hardness as soft cheeses (cottage, cream, Brie), semisoft cheeses (Muenster, Limburger, blue), hard cheeses (cheddar, Colby, Swiss), or very hard cheeses (Parmesan). All cheese results from a lactic acid fermentation of milk. The growth of a **starter culture** produces flavor changes and acid production, which results in coagulation of milk proteins and formation of a curd. Rennin, an enzyme from calf stomachs,

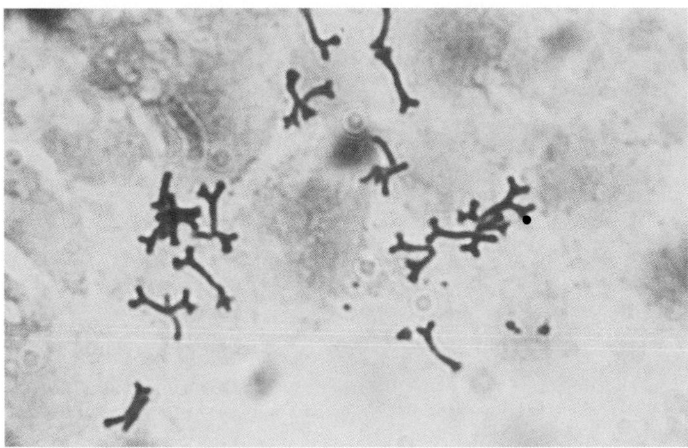

Figure 43.11 ***Bifidobacterium.*** Cultured milks are increasing in popularity. A light micrograph of *Bifidobacterium*, a microorganism suggested to provide many health benefits.

Figure 43.12 **Examples of "Bifid"-amended Dairy Products.** These are produced in many countries.

but now produced by genetically engineered microorganisms, can also be used to promote curd formation. After the curd is formed, it is heated and pressed to remove the watery part of the milk or whey, salted, and then usually ripened (figure 43.14). The cheese curd can be packaged for ripening with or without additional microorganisms. Cheese curd inoculation is used in the manufacture of Roquefort and blue cheese. In this case, *Penicillium roqueforti* spores are added to the curds just before the final cheese processing. Sometimes the surface of an already formed cheese is inoculated at the start of ripening; for example, Camembert cheese is inoculated with spores of *Penicillium camemberti*. The final hardness of the cheese is partially a function of the length of ripening. Soft cheeses are ripened for only about 1 to 5 months, while hard cheeses need 3 to 12 months, and very hard cheeses like Parmesan require 12 to 16 months' ripening.

The ripening process is also critical for Swiss cheese. Gas production by *Propionibacterium* contributes to final flavor development and hole or eye formation in this cheese. Some cheeses are soaked in brine to stimulate the development of specific fungi and bacteria; Limburger is one such cheese.

TABLE 43.7 Major Types of Cheese and Microorganisms Used in Their Production

Cheese (Country of Origin)	Contributing Microorganisms[a]	
	Earlier Stages of Production	Later Stages of Production
Soft, unripened		
Cottage	*Lactococcus lactis*	*Leuconostoc cremoris*
Cream	*L. cremoris,* *L. diacetylactis,* *S. thermophilus,* *Lactobacillus bulgaricus*	
Mozzarella (Italy)	*S. thermophilus,* *Lactobacillus bulgaricus*	
Soft, ripened		
Brie (France)	*Lactococcus lactis,* *L. cremoris*	*Penicillium camemberti,* *P. candidum,* *Brevibacterium linens*
Camembert (France)	*L. lactis,* *L. cremoris*	*Penicillium camemberti,* *Brevibacterium linens*
Semisoft		
Blue (France)	*Lactococcus lactis,* *L. cremoris*	*Penicillium roqueforti*
Brick (United States)	*L. lactis,* *L. cremoris*	*Brevibacterium linens*
Limburger (Belgium)	*L. lactis,* *L. cremoris*	*Brevibacterium linens*
Monterey (United States)	*L. lactis,* *L. cremoris*	
Muenster (United States)	*L. lactis,* *L. cremoris*	*Brevibacterium linens*
Roquefort (France)	*L. lactis,* *L. cremoris*	*Penicillium roqueforti*
Hard, ripened		
Cheddar (Britain)	*Lactococcus lactis,* *L. cremoris,* *E. durans*	*Lactobacillus casei,* *L. plantarum*
Colby (United States)	*L. lactis,* *L. cremoris,* *E. durans*	*L. casei*
Edam (Netherlands)	*L. lactis,* *L. cremoris*	
Gouda (Netherlands)	*L. lactis,* *L. cremoris, L. diacetylactis*	
Swiss (Switzerland)	*L. lactis, L. helveticus,* *S. thermophilus*	*Propionibacterium shermanii,* *P. freudenreichii*
Very hard, ripened		
Parmesan (Italy)	*Lactococcus lactis,* *L. cremoris,* *S. thermophilus*	*Lactobacillus bulgaricus*

[a]*Lactococcus lactis* stands for *L. lactis* subsp. *lactis*. *Lactococcus cremoris* is *L. lactis* subsp. *cremoris*, and *Lactococcus diacetilactis* is *L. lactis* subsp. *diacetilactis*.

Meat and Fish

Besides the fermentation of dairy products, a variety of meat products, especially sausage, can be fermented: country-cured hams, summer sausage, salami, cervelat, Lebanon bologna, fish sauces (processed by halophilic *Bacillus* species), izushi, and katsuobushi. *Pediococcus cerevisiae* and *Lactobacillus plantarum* are most often involved in sausage fermentations. Izushi is based on the fermentation of fresh fish, rice, and vegetables by *Lactobacillus* spp.; katsuobushi results from the fermentation of tuna by *Aspergillus glaucus*. Both meat fermentations originated in Japan.

Wine, Beer, and Other Fermented Alcoholic Beverages

Fermented alcoholic beverages are produced throughout the world from a variety of plant products that contain readily utilizable carbohydrates. To produce such fermentable substrates, called **musts,** it is often only necessary to allow natural fermentation to occur. The must also can be sterilized by pas-

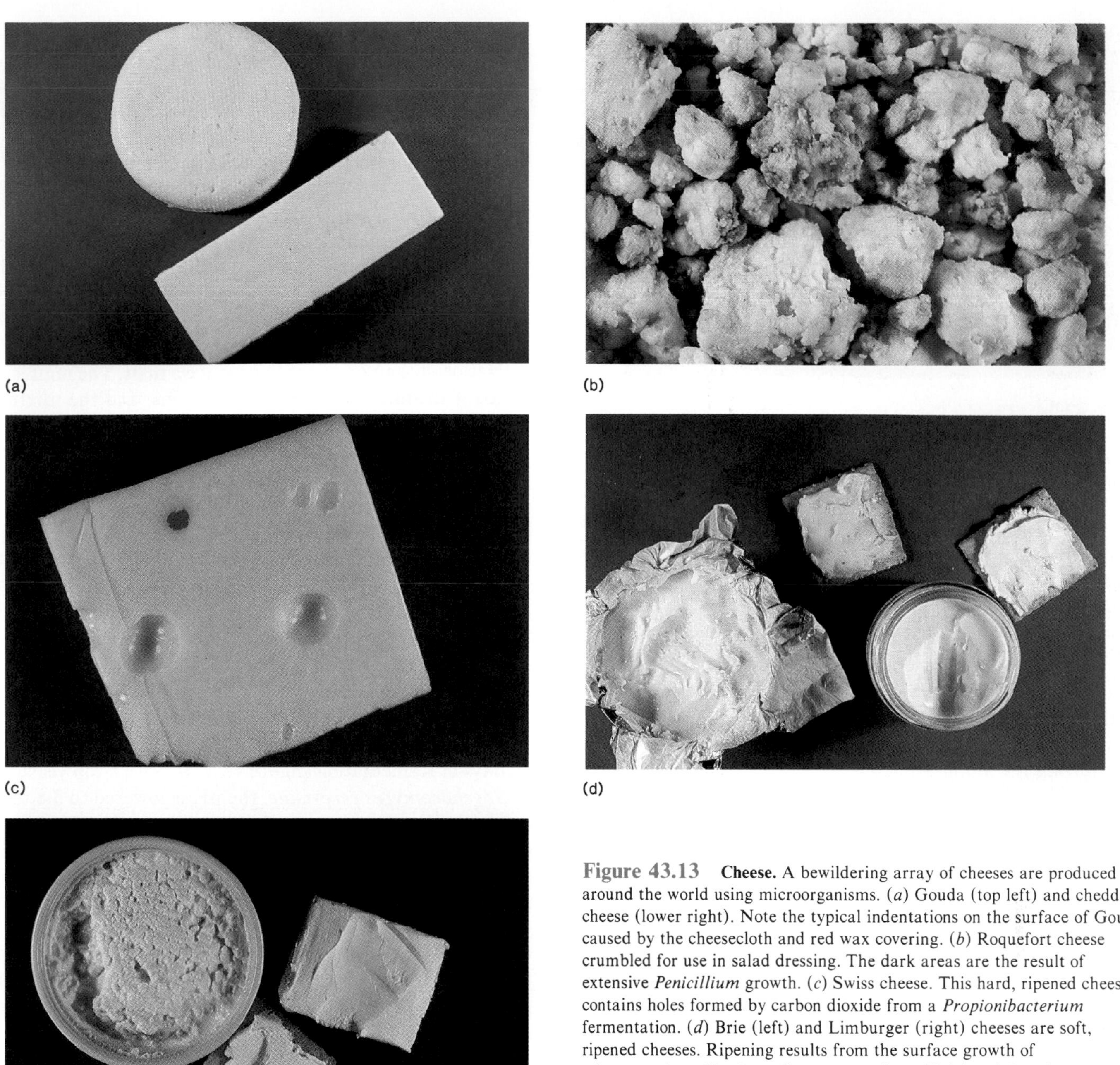

Figure 43.13 **Cheese.** A bewildering array of cheeses are produced around the world using microorganisms. (*a*) Gouda (top left) and cheddar cheese (lower right). Note the typical indentations on the surface of Gouda caused by the cheesecloth and red wax covering. (*b*) Roquefort cheese crumbled for use in salad dressing. The dark areas are the result of extensive *Penicillium* growth. (*c*) Swiss cheese. This hard, ripened cheese contains holes formed by carbon dioxide from a *Propionibacterium* fermentation. (*d*) Brie (left) and Limburger (right) cheeses are soft, ripened cheeses. Ripening results from the surface growth of microorganisms like *Penicillium camemberti* (Brie) and *Brevibacterium linens* (Limburger). (*e*) Cottage cheese and cream cheese (spread on crackers) are soft, unripened cheeses. They are sold immediately after production, and the curd is consumed without further modification by microorganisms.

teurization or the use of sulfur dioxide, and then the desired microbial culture is added.

In contrast, before cereals and other starchy materials can be used as substrates for the production of alcohol, their complex carbohydrates must be degraded. They are mixed with water and incubated in a process called **mashing.** The insoluble material is then removed to yield the **wort,** a clear liquid containing sugars and other simple molecules. Much of the "art" of beer and ale production involves the controlled hydrolysis of protein and carbohydrates to provide the desired body and flavor of the final product.

Wines and Champagnes

Wine production, or the science of enology (Greek *oinos,* wine, and *ology,* the science of), starts with the collection of grapes, continues with their crushing and the separation of the liquid (must) before fermentation, and concludes with a variety of storage and aging steps (figure 43.15). All grapes have white juices. To make a red wine from a red grape, the grape skins are allowed to remain in contact with the must before fermentation to release their skin-coloring components. Wines can be created by using the natural grape skin microorganisms. This natural mixture of bacteria and yeasts gives unpredict-

Figure 43.14 **Cheddar Cheese Production.** Cheddar, a village in England, has given its name to a cheese made in many parts of the world. "Cheddaring" is the process of turning and piling the curd to express whey and develop desired cheese texture.

able fermentation results. To avoid such problems, one can treat the fresh must with a sulfur dioxide fumigant and add a desired strain of *Saccharomyces cerevisiae* or *S. ellipsoideus.* After inoculation, the juice is fermented for three to five days at temperatures varying between 20 and 28° C. Depending on the alcohol tolerance of the yeast strain, the final product may contain 10 to 18% alcohol. Clearing and development of flavor occur during the aging process.

Yeasts (pp. 520; 526–28).

A critical part of wine making involves the choice of whether to produce a dry (no free sugar) or a sweeter (varying amounts of free sugar) wine. This can be controlled by regulating the initial must sugar concentration. With higher levels of sugar, alcohol will accumulate and inhibit the fermentation before the sugar can be completely used, thus producing a sweeter wine. During final fermentation in the aging process, flavoring compounds accumulate and so influence the bouquet of the wine.

Microbial growth during the fermentation process produces sediments, which are removed during **racking.** Racking can be carried out at the time the fermented wine is transferred to bottles or casks for aging or even after the wine is placed in bottles.

Many processing variations can be used during wine production. The wine can be distilled to make a "burned wine" or brandy. *Acetobacter* and *Gluconobacter* can be allowed to oxidize the ethanol to acetic acid and form a **wine vinegar.** In the past, an acetic acid generator was used to recirculate the wine over a bed of wood chips, where the desired microorganisms developed as a surface growth. Today, the process is carried out in large aerobic submerged cultures under much more controlled conditions.

Natural champagnes are produced by continuing the fermentation in bottles to produce a naturally sparkling wine. Sediments that remain are collected in the necks of inverted champagne bottles after the bottles have been carefully turned. The necks of the bottles are then frozen and the corks removed to disgorge the accumulated sediments. The bottles are refilled with clear champagne from another disgorged bottle, and the product is ready for final packaging and labeling.

Beers and Ales

Beer and ale production uses cereal grains such as barley, wheat, and rice. The complex starches and proteins in these grains must be changed to a more readily usable mixture of simpler carbohydrates and amino acids. This process, shown in figure 43.16, involves germination of the barley grains and activation of their enzymes to produce a **malt.** The malt is then mixed with water and the desired grains, and the mixture is transferred to the mash tun or cask in order to hydrolyze the starch to usable carbohydrates. Once this process is completed, the **mash** is heated with hops (dried flowers of the female vine *Humulus lupulis*), which were originally added to the mash to inhibit spoilage microorganisms. The hops also provide flavor and assist in clarification of the wort. In this heating step, the hydrolytic enzymes are inactivated and the wort can be **pitched**—inoculated—with the desired yeast.

Most beers are fermented with **bottom yeasts,** related to *Saccharomyces carlsbergensis,* which settle at the bottom of the fermentation vat. The beer flavor is also influenced by the production of small amounts of glycerol and acetic acid. Bottom yeasts produce beer with a pH of 4.1 to 4.2 and requiring 7 to 12 days of fermentation (figure 43.17). With a top yeast, such as *Saccharomyces cerevisiae,* the pH is lowered to 3.8 to produce ales. Freshly fermented (green) beers are aged or **lagered,** and when they are bottled, CO_2 is usually added. Beer can be pasteurized at 140° F or higher or sterilized by passage through membrane filters to minimize flavor changes.

Distilled Spirits

Distilled spirits are produced by an extension of beer production processes. The fermented liquid is boiled, and the volatile components are condensed to yield a product with a higher alcohol content than beer. Rye and bourbon are examples of whiskeys. Rye whiskey must contain at least 51% rye grain, and bourbon must contain at least 51% corn. Scotch whiskey is made primarily of barley. Usually a **sour mash** is used; the mash is inoculated with a homolactic (lactic acid is the major fermentation product) bacterium such as *Lactobacillus delbrueckii,* which can lower the mash pH to around 3.8 in 6 to 10 hours. This limits the development of undesirable organisms. Vodka and grain alcohols are also produced by distillation. Gin is vodka to which resinous flavoring agents—often juniper berries—have been added to provide a unique aroma and flavor.

Bread and Other Fermented Plant Products

Bread is one of the most ancient of human foods, and the use of yeasts to leaven bread is carefully depicted in paintings from

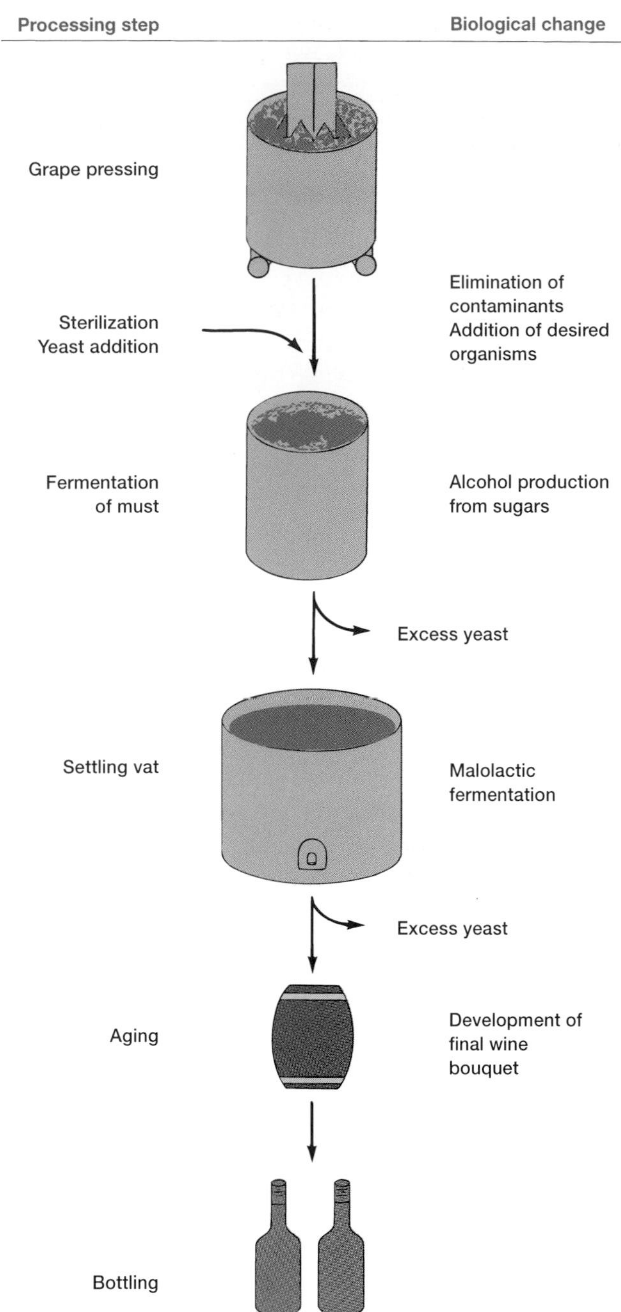

(b)

Figure 43.15 **Wine Making.** Once grapes are pressed, the sugars in the juice (the must) can be immediately fermented to produce wine. (*a*) Must preparation, fermentation, and aging are critical steps. (*b*) A photograph of wine fermentation vats.

ancient Egypt. Samples of bread from 2100 B.C. are on display in the British Museum. In breadmaking, yeast growth is carried out under aerobic conditions. This results in increased CO_2 production and minimum alcohol accumulation. The fermentation of bread involves several steps: alpha- and beta-amylases present in the moistened dough release maltose and sucrose from starch. Then a baker's strain of the yeast *Saccharomyces cerevisiae,* which contains maltase, invertase, and zymase enzymes, is added. The CO_2 produced by the yeast results in the light texture of many breads, and traces of fermentation products contribute to the final flavor. Usually, bakers add sufficient yeast to allow the bread to rise within two hours—the longer the rising time, the more additional growth by contaminating bacteria and fungi and the less desirable the product.

By using more complex assemblages of microorganisms, bakers can produce special breads such as sour doughs. The yeast *Saccharomyces exiguus,* with a *Lactobacillus* species, produces the characteristic acidic flavor and aroma of such breads.

Bread products can be spoiled by *Bacillus* species that produce ropiness. If the dough is baked after these organisms have grown, stringy and ropy bread will result, leading to decreased consumer acceptance.

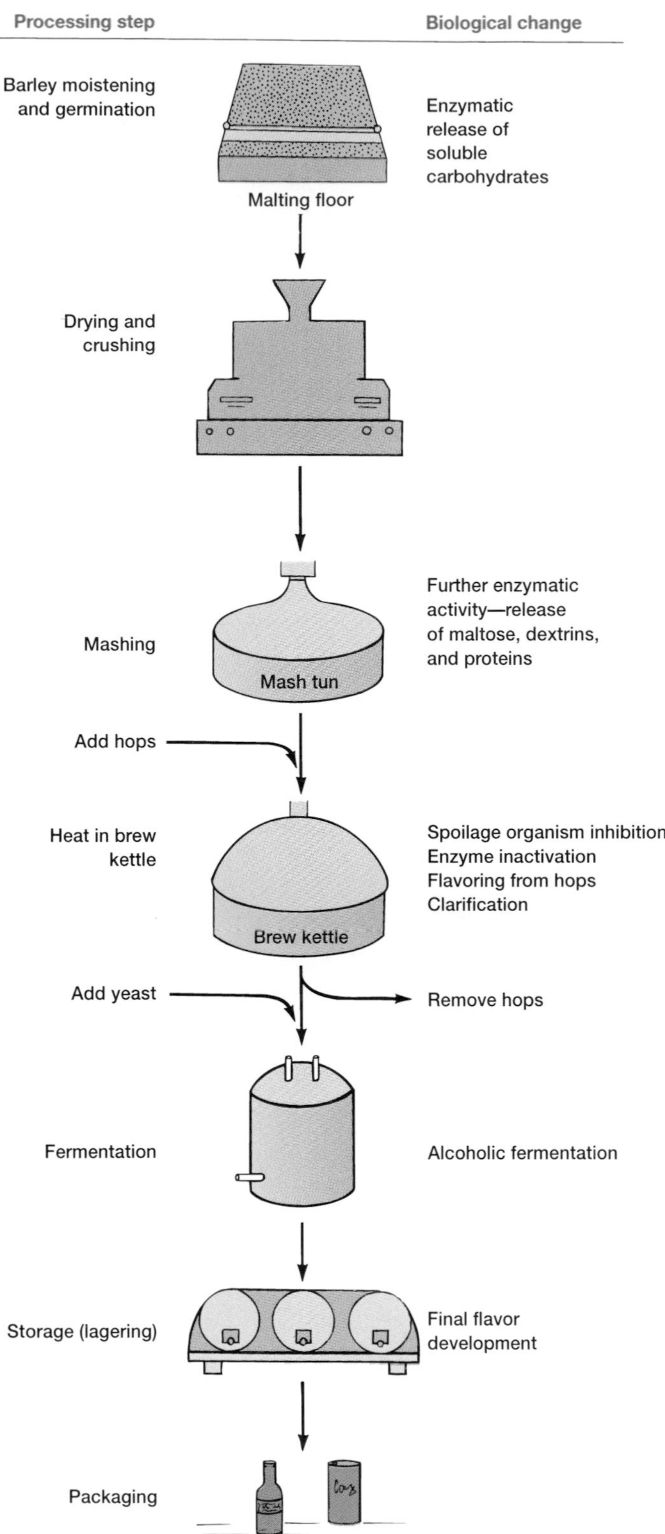

Figure 43.16 Producing Beer. To make beer, the complex carbohydrates in the grain must first be transformed into a fermentable substrate. Beer production thus requires the important steps of malting, and the use of hops and boiling for clarification, flavor development and inactivation of malting enzymes. Only after completion of these steps can the actual fermentation be carried out.

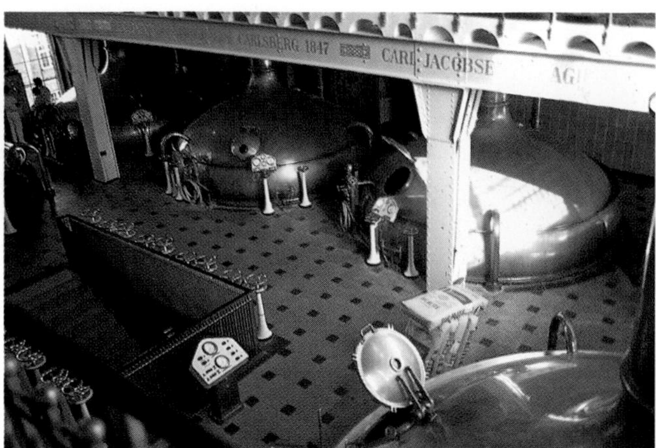

Figure 43.17 Beer Fermentation. Beer making can be carried out on a large scale or in the kitchen. In large-scale processes, copper brewing vats are often used, as shown here in the Carlsberg Brewery, Copenhagen, Denmark.

Many other plant products can be fermented, as summarized in table 43.8. These include sufu, which is produced by the fermentation of tofu, a chemically coagulated soybean milk product. To carry out the fermentation, the tofu curd is cut into small chunks and dipped into a solution of salt and citric acid. After the cubes are heated to pasteurize their surfaces, *Actinimucor elegans* and other *Mucor* species are added. When a white mycelium develops, the cubes, now called pehtze, are aged in salted rice wine. This product has achieved the status of a delicacy in many parts of the Western world. Another popular product is tempeh, a soybean mash fermented by *Rhizopus* (figure 43.18).

Sauerkraut or sour cabbage, is produced from wilted, shredded cabbage, as shown in figure 43.19. Usually, the mixed microbial community of the cabbage is used. A concentration of 2.2 to 2.8% sodium chloride restricts the growth of gram-negative bacteria while favoring the development of the lactic acid bacteria. The primary microorganisms contributing to this product are *Leuconostoc mesenteroides* and *Lactobacillus plantarum*. A predictable microbial succession occurs in sauerkraut's development. The activities of the lactic-acid-producing cocci usually cease when the acid content reaches 0.7 to 1.0%. At this point, *Lactobacillus plantarum* and *Lactobacillus brevis* continue to function. The final acidity is generally 1.6 to 1.8, with lactic acid comprising 1.0 to 1.3% of the total acid in a satisfactory product.

Pickles are produced by placing cucumbers and such components as dill seeds in casks filled with a brine. The sodium chloride concentration begins at 5% and rises to about 16% in six to nine weeks. The salt not only inhibits the growth of undesirable bacteria but also extracts water and water-soluble constituents from the cucumbers. These soluble carbohydrates are converted to lactic acid. The fermentation, which can require 10 to 12 days, involves the development of *Leuconostoc mesenteroides, Enterococcus faecalis, Pediococcus cerevisiae,*

Table 43.8 Fermented Foods Produced from Fruits, Vegetables, Beans, and Related Substrates

Foods	Raw Ingredients	Fermenting Microorganisms	Area
Coffee	Coffee beans	*Erwinia dissolvens, Saccharomyces* spp.	Brazil, Congo, Hawaii, India
Gari	Cassava	*Corynebacterium manihot, Geotrichum* spp.	West Africa
Kenkey	Corn	*Aspergillus* spp., *Penicillium* spp., Lactobacilli, yeasts	Ghana, Nigeria
Kimchi	Cabbage and other vegetables	Lactic acid bacteria	Korea
Miso	Soybeans	*Aspergillus oryzae, Saccharomyces rouxii*	Japan
Ogi	Corn	*Lactobacillus plantarum, Lactococcus lactis, Saccharomyces rouxii*	Nigeria
Olives	Green olives	*Leuconostoc mesenteroides, Lactobacillus plantarum*	Worldwide
Ontjom	Peanut presscake	*Neurospora sitophila*	Indonesia
Pickles	Cucumbers	*Pediococcus cerevisiae, L. plantarum*	Worldwide
Poi	Taro roots	Lactic acid bacteria	Hawaii
Sauerkraut	Cabbage	L. mesenteroides, L. plantarum	Worldwide
Soy sauce	Soybeans	Aspergillus oryzae or *A. soyae, S. rouxii, Lactobacillus delbrueckii*	Japan
Sufu	Soybeans	*Mucor* spp.	China
Tao-si	Soybeans	*A. oryzae*	Philippines
Tempeh	Soybeans	*Rhizopus oligosporus, R. oryzae*	Indonesia, New Guinea, Surinam

Figure 43.18 Fermented Soybean Products. Microorganisms also can be used to process soybean products. Tempeh produced from soybeans and *Rhizopus*. Both the raw cake and the fried food are shown.

Lactobacillus brevis, and *L. plantarum*. *L. plantarum* plays the dominant role in this fermentation process. Sometimes, to achieve more uniform pickle quality, natural microorganisms are first destroyed and the cucumbers are fermented using pure cultures of *P. cerevisiae* and *L. plantarum*.

Grass, chopped corn, and other fresh animal feeds, if stored under moist anaerobic conditions, will undergo a lactic-type mixed fermentation that produces pleasant-smelling **silages.** Trenches or more traditional vertical steel or concrete silos are used to store the silage. The accumulation of organic acids in silage can cause rapid deterioration of these silos. Older wooden stave silos, if not properly maintained, allow the outer portions of the silage to become aerobic, resulting in spoilage of a large portion of the plant material.

1. What gives fermented milk products their flavor?
2. What major steps are used to produce cheese? How is the cheese curd formed in this process? What is whey? How does Swiss cheese get its holes?
3. At what point in the production of Roquefort or blue cheese is a fungus added to the cheese? What fungal genus has been used in the production of these cheeses?
4. Describe and contrast the processes of wine and beer production. How are red wines produced when the juice of all grapes is white? How are champagnes made?
5. Describe how the brewing process is continued to produce distilled spirits like whiskey.
6. How are bread, sauerkraut, and pickles produced? What microorganisms are most important in these fermentations?

Microorganisms as Sources of Food

Besides microorganisms' actions in fermentation as agents of physical and biological change, they themselves can be used as a food source. A variety of bacteria, yeasts, and other fungi have been used as animal and human food sources. Mushrooms (*Agaricus bisporus*) are one of the most important fungi used directly as a food source. Large caves are used to maintain optimal conditions for the production of this delicacy (figure 43.20). Yeasts of the genus *Candida* were grown for human consumption on sulfite waste liquors (lignin fragments and carbohydrates produced during paper manufacture) in

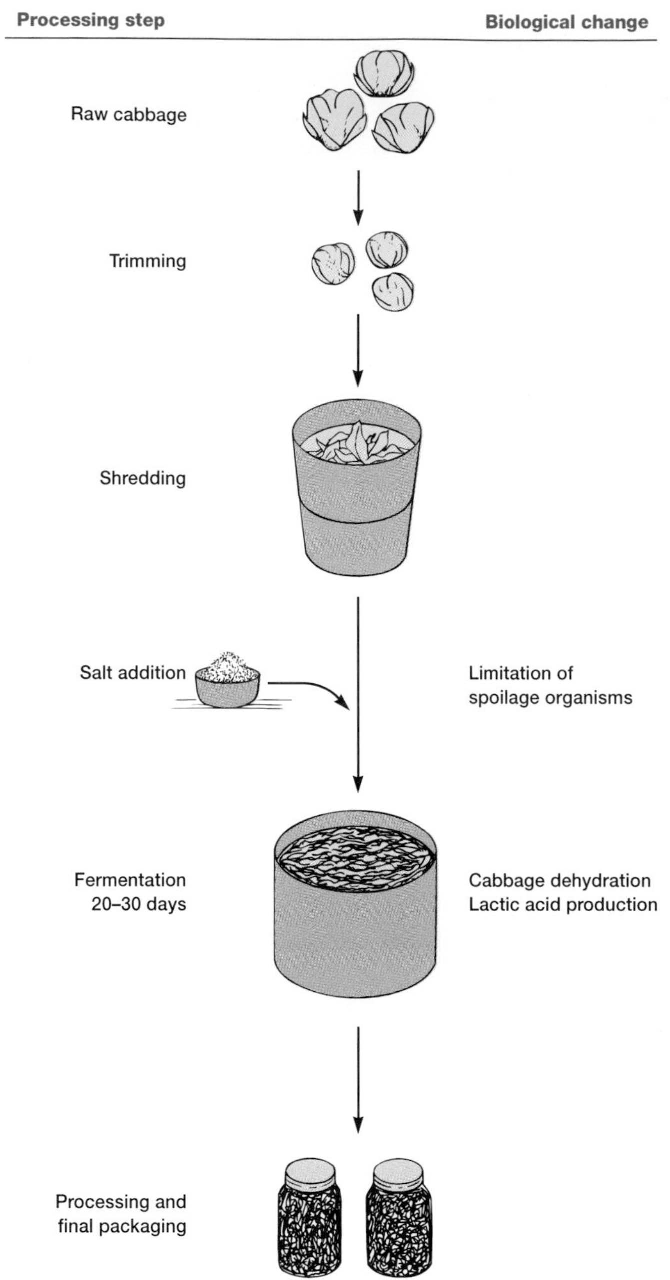

Figure 43.19 **Sauerkraut.** Sauerkraut production employs a lactic acid fermentation. The basic process involves fermentation of shredded cabbage in the presence of 2.25 2.5% by weight of salt to inhibit spoilage organisms.

Figure 43.20 **Mushroom Farming.** Growing mushrooms requires careful preparation of the growth medium and control of environmental conditions. The mushroom bed is a carefully developed compost, which can be steam sterilized to improve mushroom growth.

Germany during World War II. Microorganisms may be used as a protein source and are then called **single-cell protein (SCP).**

Single-cell protein can be used directly as a food source or as a supplement to other foods. A continuing concern is the high level of nucleic acids in microorganisms. Nucleic acid levels may be as high as 16% and result in an accumulation of uric acid, the symptoms of gout, and the formation of kidney stones. Chemical processing can be used to hydrolyze the microbial cell walls, extract the nucleic acids, and decrease the nucleic acid content to less than 2%.

Many substrates and microorganisms can be used for the production of single-cell protein. Substrates for microbial growth include complex organic wastes, methane and other hydrocarbons, and hydrogen plus CO_2. One of the more popular microbial food supplements is the cyanobacterium *Spirulina* (*see chapter 23*). It is used as a food source in Africa and is now being sold in United States health food stores as a dried cake or powdered product.

1. A cyanobacterium is widely used as a food supplement. What is this genus, and in what part of the world was it first used as a significant food source?
2. What is a major problem associated with using microbial cells as a food source?

Summary

1. Most foods, especially when raw, provide an excellent environment for microbial growth. This growth can lead to spoilage or preservation, depending on the microorganisms present and environmental conditions.
2. The course of microbial development in a food is influenced by the characteristics of the food itself—pH, salt content, substrates present, water presence and availability—and the environment, including temperature, relative humidity, and atmospheric composition.
3. Foods can be preserved in a variety of physical and chemical ways, including filtration, alteration of temperature (cooling, pasteurization, sterilization), drying, the addition of chemicals, radiation, and fermentation.
4. Microorganisms in improperly prepared or handled food can produce diseases. These can be either food-borne infections or food intoxications. Usually transmission is by the fecal-oral route with food acting as an intermediary agent.
5. DNA- and RNA-based methods allow rapid detection of disease-causing microorganisms in foods.
6. Fungi, if they can grow in foods, especially cereals and grains, can produce important disease-causing chemicals, including aflatoxins (carcinogens) and ergot alkaloids (mind-altering drugs).
7. Microorganisms can spoil meat, dairy products, fruits, vegetables, and canned goods in several ways. Spices, with their antimicrobial compounds, sometimes protect foods.
8. Dairy products can be fermented to yield cultured liquid products and cheeses. *Lactobacillus acidophilus* and *Bifidobacterium* can modify the intestinal microflora and may contribute to the control of colon cancer.
9. Wines are produced from pressed grapes and can be dry or sweet, depending on the level of free sugar that remains at the end of the alcoholic fermentation. Champagne is produced when the fermentation, resulting in CO_2 formation, is continued in the bottle.
10. Beer and ale are produced from cereals and grains. The starches in these substrates are hydrolyzed, in the processes of malting and mashing, to produce a fermentable wort. *Saccharomyces cerevisiae* is a major yeast used in the production of beer and ale.
11. Many plant products can be fermented with bacteria, yeasts, and fungi. Important products are breads, soy sauce, sufu, and tempeh. Sauerkraut and pickles are produced in a fermentation process in which natural populations of lactobacilli play a major role.
12. Microorganisms themselves can serve as an important food source (single-cell protein). They can be grown on a variety of substrates, including CO_2 (with light), methane, and more complex organic substrates. *Spirulina,* a cyanobacterium, is a popular single-cell protein source.

Key Terms

aflatoxin *873*
bottom yeast *880*
canning *869*
ergotism *873*
food-borne infection *871*
food intoxication *872*
lagered *880*
malt *880*
mash *880*
mashing *879*
must *878*
osmophilic microorganisms *867*
pasteurization *869*
pitched *880*
probe *872*
putrefaction *867*
racking *880*
radappertization *870*
single-cell protein (SCP) *884*
silages *883*
sour mash *880*
starter culture *877*
wine vinegar *880*
wort *879*
xerophilic microorganisms *867*

Questions for Thought and Review

1. You are going through a salad line in a cafeteria at the end of the day. Which type of foods would you tend to avoid, and why?
2. Why does the law only allow foods to be pasteurized a single time? What could be the consequences of repeated pasteurization?
3. Why is it recommended that frozen foods, once thawed, not be frozen again? Suggest some of the reasons for this important principle of food handling.
4. Hamburger is notorious for having relatively high populations of microorganisms. How can these populations be controlled in preparation and storage?
5. Coffee cream is now packaged in small-portion servings that can be held at room temperature for extended periods. How do you think this product is produced and packaged?
6. What factors might limit the application of DNA- and RNA-based molecular techniques to the detection of pathogens in foods? How might these problems be overcome?
7. Why were aflatoxins not discovered before the 1960s? Do you think this was the first time they had grown in a food product to cause disease?
8. Why has gamma irradiation of foods not become a more common means of controlling spoilage of foods?
9. Because of the possible beneficial effects of microorganisms on the intestinal microflora, scientists are becoming increasingly interested in them as dietary supplements. How might modern biotechnology be used to exploit this matter?

Additional Reading

Banwart, G. J. 1989. *Basic food microbiology,* 2d ed. New York: Van Nostrand Reinhold Co.

Board, R. G. 1983. *A modern introduction to food microbiology.* London: Blackwell Scientific Publications.

Brackett, R. E. 1992. Microbiological safety of chilled foods: current issues. *Trends Food Sci. & Technol.* 3(4):81–86.

Brauns, L. A.; Hudson, M. C.; and Oliver, J. D. 1991. Use of the polymerase chain reaction in detection of culturable and nonculturable *Vibrio vulnificus* cells. *Appl. Environ. Microbiol.* 57:2651–55.

Butzler, J. P., and Oosterom, J. 1991. *Campylobacter:* Pathogenicity and significance in foods. *Int. J. Food Microbiol.* 12:1–8.

Chou, C-C.; Ho, F-M.; and Tsai, C-S. 1988. Effects of temperature and relative humidity on the growth and enzyme production by *Actinomucor taiwanensis* during sofu pehtze preparation. *Appl. Environ. Microbiol.* 54:688–92.

Cliver, D. O. 1990. *Food-borne diseases.* San Diego: Academic Press, Inc.

Doyle, M. P. 1989. *Foodborne bacterial pathogens.* New York: Marcel Dekker, Inc.

Doyle, M. P. 1991. *Escherichia coli* 0157:H7 and its significance in foods. *Int. J. Food Microbiol.* 12:289–302.

Elliot, R. P. 1980 (paperback, 1983). *Microbial ecology of foods:* vol. 1, *Factors affecting life and death of microorganisms;* vol. 2. *The food commodities.* The International Commission on Microbial Specifications for Foods, New York: Academic Press.

Farber, J. M., and Peterkin, P. I. 1991. *Listeria monocytogenes,* a food-borne pathogen. *Microbiol. Rev.* 55:476–511.

Fitts, R. A. 1991. Detection of foodborne microorganisms by DNA hybridization. In *Foodborne microorganisms and their toxins: Developing methodology,* ed. M. D. Pierson and N. J. Stern. New York: Marcel Dekker, Inc.

Frazer, W. C., and Westhoff, D. C. 1988. *Food microbiology,* 4th ed. New York: McGraw-Hill.

Hurst, A., and Collins-Thompson, D. L. 1977. Food as a bacterial habitat. *Adv. Microbiol. Ecol.* 3:79–134.

Jay, J. M. 1991. *Modern food microbiology,* 4th ed. New York: Van Nostrand Reinhold Co.

Kapperud, G. 1991. *Yersinia enterocolitica* in food hygiene. *Int. J. Food Microbiol.* 12:53–66.

Kosikowski, F. V. 1985. Cheese. *Sci. Am.* 252(5):88–99.

Marshall, R. T., ed. 1992. *Standard methods for the examination of dairy products.* 16th ed. Washington, D.C.: American Public Health Association.

Moreau, C. 1979. *Moulds, toxins and foods.* New York: John Wiley and Sons.

Nackamkin, I.; Blaser, M. J.; and Tompkins, L. S. 1992. *Campylobacter jejuni: Current status and future trends.* Washington, D.C.: American Society for Microbiology.

National Restaurant Association. 1992. *Applied food service sanitation,* 4th ed. New York: John Wiley and Sons.

Parish, M. E. 1991. Microbiological concerns in citrus juice processing. *Food Tech.* April 1991, pp. 128–32.

Pohland, A. E.; Dowell, V. R., Jr.; and Richard, J. L. 1990. *Microbial toxins in foods and feeds.* New York: Plenum Publishing.

Roberts, T. A.; Hobbs, G.; Christian, J. H. B.; and Skovgaard, N. 1981. *Psychotrophic microorganisms in spoilage and pathogenicity.* New York: Academic Press.

Roberts, T. A., and Skinner, F. A., eds. 1983. *Food microbiology: Advances and prospects.* New York: Academic Press.

Robinson, R. K. 1991. *Therapeutic properties of fermented milk.* New York: Elsevier.

Rose, A. H. 1981. The microbiological production of food and drink. *Sci. Am.* 245(3):126–39.

Sandine, W. E. 1986. Genetic manipulation to improve food fermentation. *Dairy and Food Sanitation* 6(12):548–50.

Shinagawa, K. 1990. Analytical methods for *Bacillus cereus* and other *Bacillus* species. *Int. J. Food Microbiol.* 10:125–42.

Smith, J. E., and Moss, M. O. 1987. *Mycotoxins, formation, analysis, and significance.* New York: John Wiley and Sons.

Steinkraus, H., ed. 1983. *Handbook of indigenous fermented foods,* vol. 9, Microbiology series. New York: Marcel Dekker.

Troller, J. A., and Christian J. H. B. 1978. *Water activity and food.* New York: Academic Press.

Webb, A. D. 1984. The science of making wine. *American Scientist* 72:360–67.

Wolcott, M. J. 1991. DNA-based rapid methods for the detection of foodborne pathogens. *J. Food Prot.* 54: 387–401.

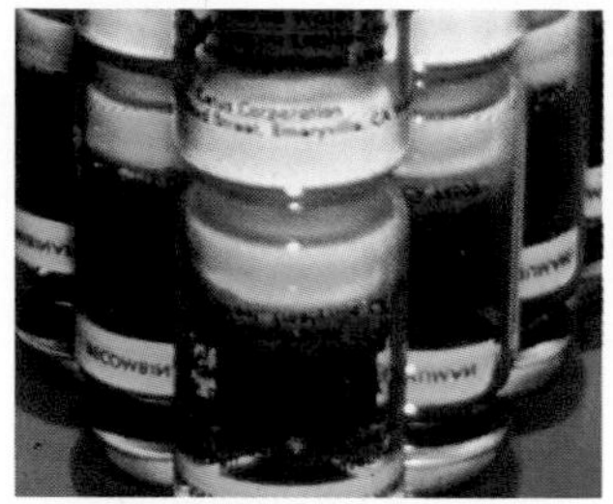

CHAPTER 44
Industrial Microbiology and Biotechnology

Nothing is more agreeable to those devoted to a scientific career than to increase the number of discoveries, but when the results of these observations are demonstrated by practical utility, their joy is complete.

—Louis Pasteur

Biotechnology. The scientific manipulation of living organisms, especially at the molecular genetic level, to produce useful products.

—D. A. Micklos and G. A. Freyer

Outline

Concepts

1. The use of microorganisms in industrial microbiology and as a part of the new biotechnology involves the management of microbial growth processes. Factors limiting microbial growth must also be understood and controlled.
2. In modern biotechnology, microorganisms with specific genetic characteristics can be constructed to meet desired objectives. Earlier, mutation and screening (selection) were the major means of improving cultures from the available gene pool.
3. The media and growth conditions used in industrial processes are often created to provide specific limitations on microbial growth and functioning.
4. Different types of products are produced during and after the completion of microbial growth. Stopping growth at a specific point is often a necessary part of an industrial process.
5. Through recombinant DNA technology, the fields of protein engineering and metabolic engineering have been created.
6. In many situations, it is important to inhibit microbial degradation of materials, or biodegradation. In other situations, it is desirable to facilitate these degradation processes through bioenhancement or bioremediation.
7. Microbial cells and their products can be used in immobilized forms. Such biocatalysts have many uses, including the removal of metals from process streams.
8. Microorganisms are being linked with electronics to produce biosensors. Biosensors have many medical, industrial, and environmental applications.

The characteristics of marine and freshwater environments, soil, and food are discussed in chapters 41, 42, and 43. The major components of each of these environments—water, solid inorganic components, and biologically usable organic matter, respectively—influence microbial communities and their functioning.

When microorganisms are used in industrial microbiology and biotechnology, it is not only possible to choose the particular microorganism or microbial community that will be employed but also to define and control the environment. The environmental conditions can be changed or held constant over time, depending on the goals that have been set for the particular process. Modern gene-cloning techniques also allow a considerable range of possibilities for manipulation of microorganisms and the use of plants and animals as "factories" for the expression of recombinant DNA. Competition between chemical and biological processes in the marketplace, common in the past, is developing into an integrated partnership that continues to provide new opportunities and challenges.

Industrial microbiology has become an important part of the field of microbiology. Industrial microbiology involves the use of microorganisms to produce organic chemicals, antibiotics, other pharmaceuticals, and food supplements (*see Box 8.1*). Microorganisms are also used for the control of insects and other pests, for the recovery of metals, and for the improvement and maintenance of environmental quality. At first, selection and mutation were the major means of improving cultures for use in industrial microbiology. In biotechnology, it is now possible with recombinant DNA techniques to manipulate genetic information and design products such as proteins, or to modify microbial gene expression. In addition, genetic information can be transferred between markedly different groups of organisms. In many situations, the environment in which the microorganism functions and the characteristics of the microorganism itself can be chosen. This chapter describes the accomplishments, present activities, and the future potential of microbial uses in these important areas.

Industrial Microbiology and the New Biotechnology

It is now popular to discuss the area of biotechnology as a new development. The term **biotechnology,** however, is not new. It is a product of the 1930s, a time when efforts were made to use agricultural surpluses to produce plastic and hydrocarbon substrates.

In the earlier development of what became known as industrial microbiology, the primary methods of selecting and improving microorganisms involved what G. Pontecorvo in 1974 described as a "prehistoric" technique: mutation and screening (selection) from the available gene pool. This approach, used with success over many decades, is still important. Recently, a major pharmaceutical company helped establish a 12,000 square mile "genetic preserve" in Central America where a search for unique plants and microorganisms can be undertaken. In modern biotechnology, these classic selection, mutation, and genetic exchange approaches are complemented by the use of recombinant DNA technology. With these procedures, new genetic information can be added intentionally to desired cells. It is now possible to combine DNA sequences from animal and plant cells, as well as from different groups of microorganisms.

Genetic engineering and recombinants (chapter 14).

Table 44.1 Fermentation: A Word with Many Meanings for the Microbiologist

1. Any process involving the mass culture of microorganisms, either aerobic or anaerobic
2. Any biological process that occurs in the absence of O_2
3. Food spoilage
4. The production of alcoholic beverages
5. Use of an organic substrate as the electron donor and acceptor
6. Use of an organic substrate as a reductant, and of the same partially degraded organic substrate as an oxidant (electron acceptor)
7. Growth dependent upon substrate-level phosphorylation

Once a suitable microorganism is available, the microbiologist must still grow it in such a way as to ensure that the desired products are formed. Thus, the use of microorganisms in modern biotechnology is still based on the principles of microbial mass culture developed over decades by industrial microbiologists.

Growth of microorganisms (chapter 6).

The term **fermentation** used in a physiological sense in earlier sections of the book, is employed in a much more general way in relation to industrial microbiology and biotechnology. As noted in table 44.1, the term can have several meanings, including the mass culture of microorganisms (or even plant and animal cells). The development of industrial fermentation processes requires construction of appropriate culture media and the large-scale screening of microorganisms. Often, years are needed to achieve optimum product yields. Many isolates are tested for their ability to produce a new product in the desired quantity. Few are successful.

Fermentation as a physiological process (pp. 155–58).

1. In what different ways can the term fermentation be used?
2. How might industrial microbiology be different from biotechnology?

Microbial Growth Processes

Microbial Culture

Microorganisms for industrial applications are grown in culture tubes, shake flasks, and stirred fermenters or other mass culture systems. Stirred fermenters can range in size from 3 to 4 liters to 100,000 liters or larger, depending on production requirements (figure 44.1). A typical large-scale stirred fermentation unit is illustrated in figure 44.2. This unit requires a large capital investment and skilled operators. All required steps in the growth and harvesting of products must be carried out under aseptic conditions. Not only must the medium be sterilized, but aeration, pH adjustment, sampling, and process monitoring must be carried out under rigorously controlled conditions. When required, foam control agents must be added, especially with high-protein media. Computers are commonly used to monitor outputs from probes that determine microbial biomass, levels of critical metabolic products, pH, input and exhaust gas composition, and other parameters. Such information is needed for precise process and product control.

Besides the traditional stirred aerobic or anaerobic fermenter, other approaches can be used to grow microorganisms. These alternatives, illustrated in figure 44.3, include airlift fermenters (figure 44.3*a*), which eliminate the need for stirrers that can cause problems with filamentous fungi. Also available is solid-state fermentation (figure 44.3*b*), in which the substrate is not diluted in water. In various types of fixed (figure 44.3*c*) and fluidized bed reactors (figure 44.3*d*), the microorganisms are attached to surfaces and medium flows past the fixed or suspended particles.

Dialysis culture units can also be used (figure 44.3*e*). These units allow toxic waste metabolites or end products to diffuse away from the microbial culture and permit new substrates to diffuse through the membrane toward the culture. Continuous culture techniques (figure 44.3*f*) can markedly improve cell outputs and rates of substrate use because microorganisms can be maintained in a continuous logarithmic phase. However, continuous maintenance of an organism in an active growth phase is undesirable in many industrial processes.

Medium Development and Growth Conditions

The medium used to grow a microorganism is critical because it can influence the economic competitiveness of a particular process. Frequently, lower-cost crude materials are used as sources of carbon, nitrogen, and phosphorus (table 44.2). Crude plant hydrolysates are often used as complex sources of carbon, nitrogen, and growth factors. By-products from the brewing industry are often employed because of their lower cost and greater availability. Other useful carbon sources include molasses and whey from the manufacture of cheese.

Microbial growth media (pp. 104–5).

The levels and balance of minerals (especially iron) and growth factors can be critical in medium formulation. For example, biotin and thiamine, by influencing biosynthetic reactions, control product accumulation in many fermentations.

Figure 44.1 Industrial Stirred Fermenters. Modern biotechnology is very capital and labor intensive. With recent advances in molecular biology, close coordination between basic research, process development, manufacturing, quality control, and marketing is required to remain competitive. This picture shows the large fermenters used by a pharmaceutical company in the microbial synthesis of antibiotics.

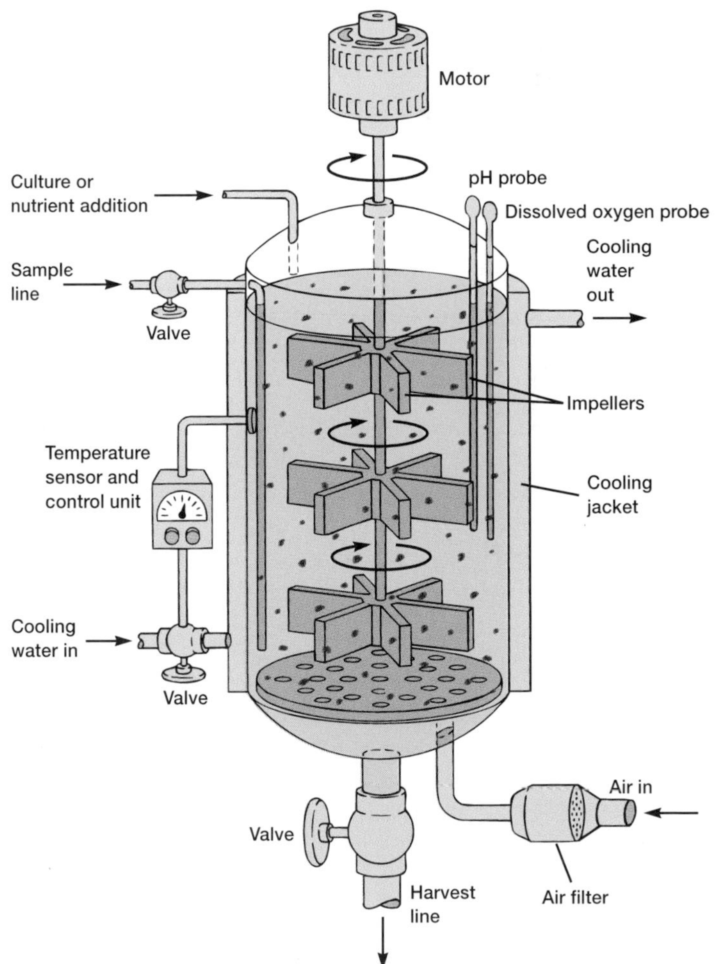

Figure 44.2 A Large-scale Fermentation Unit. For the growth of microorganisms in industrial processes, specially designed stirred fermenters are often used. These can be run under aerobic or anaerobic conditions. Nutrient additions, sampling, and fermentation monitoring can be carried out under aseptic conditions.

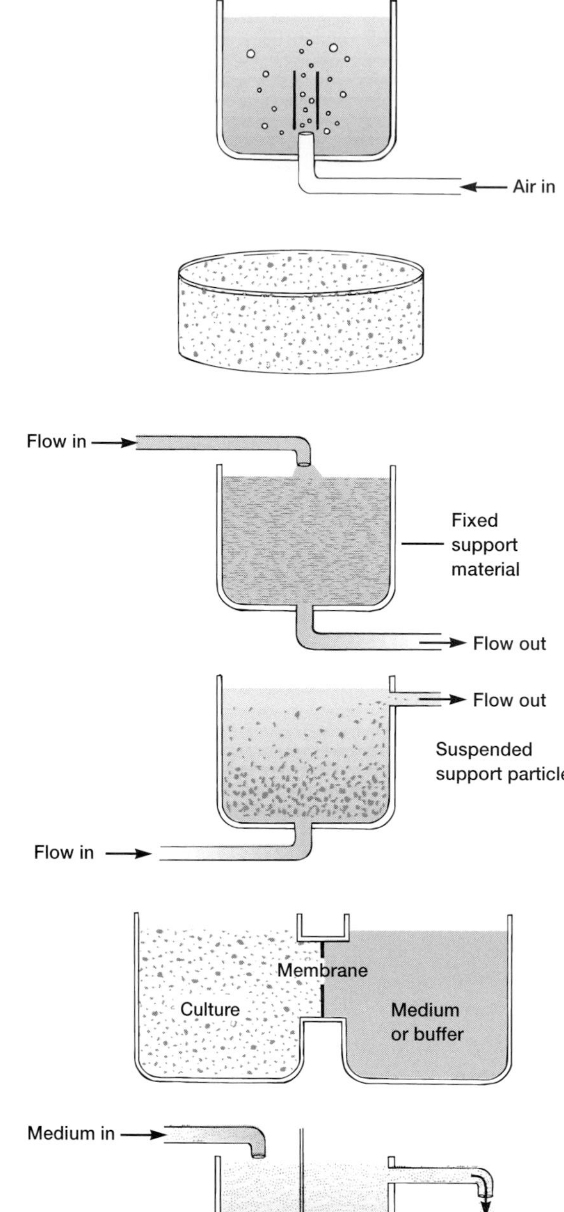

(a) Lift-tube fermenter
Density difference of gas bubbles entrained in medium results in fluid circulation.
Air in
(b) Solid-state fermentation
Growth of culture without presence of added free water
Flow in
(c) Fixed-bed reactor
Microorganisms on surfaces of support material; flow can be up or down.
Fixed support material
Flow out
(d) Fluidized-bed reactor
Microorganisms on surfaces of particles suspended in liquid or gas stream—upward flow
Flow out
Suspended support particles
Flow in
(e) Dialysis culture unit
Waste products diffuse away from the culture. Substrate may diffuse through membrane to the culture.
Membrane
Culture
Medium or buffer
Medium in
(f) Continuous culture unit
Medium in and excess medium to waste with wasted cells
Medium and cells out

Figure 44.3 Alternate Methods for Mass Culture. In addition to stirred fermenters, other methods can be used to culture microorganisms in industrial processes. In many cases, these alternate approaches will have lower operating costs and can provide specialized growth conditions needed for product synthesis.

TABLE 44.2 Major Components of Growth Media Used in Industrial Processes

Source	Raw Material	Source	Raw Material
Carbon and energy	Molasses Whey Grains Agricultural wastes (corncobs)	Vitamins	Crude preparations of plant and animal products
		Iron, trace salts	Crude inorganic chemicals
		Buffers	Chalk or crude carbonates Fertilizer-grade phosphates
Nitrogen	Corn-steep liquor Soybean meal Stick liquor (slaughterhouse products) Ammonia and ammonium salts Nitrates Distiller's solubles	Antifoam agents	Higher alcohols Silicones Natural esters Lard and vegetable oils

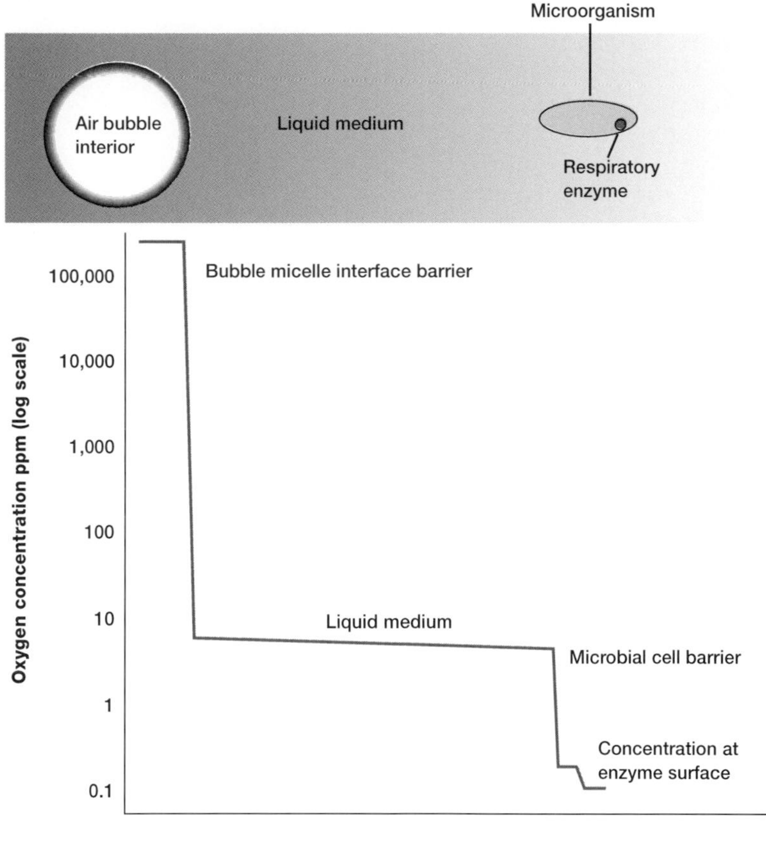

Figure 44.4 Barriers to Oxygen Transfer. Many barriers exist that limit oxygen transfer from air to an enzyme within the microorganism. In spite of having a high oxygen concentration within the air bubble, the barriers to transfer into the liquid medium and finally across the cell membrane result in a low level of oxygen at the enzyme. Only by maintaining a high aeration rate is it possible to assure that oxygen will not limit microbial activity. The oxygen concentration is expressed on a logarithmic scale, and the oxygen present at the site of use can be less than one millionth of that found in the air bubble.

The medium may also be designed so that carbon, nitrogen, phosphorus, iron, or a specific growth factor will become limiting after a given time during the fermentation. In such cases, the limitation often causes a shift from growth to production of desired metabolites.

Frequently, a critical component in the medium, often the carbon source, is added continuously—**continuous feed** or fed-batch mode—so that the microorganism will not have excess substrate available at any given time. An excess of substrate can cause undesirable metabolic waste products to accumulate.

Once a medium is developed, the physical environment for microbial functioning in the mass culture system must be defined. This often involves precise control of agitation, cooling, pH changes, and oxygenation. Phosphate buffers can be used to control pH while also functioning as a source of phosphorus. Oxygen limitations, especially, can be critical in aerobic growth processes. The O_2 concentration and flux rate must be sufficiently high to have O_2 in excess within the cells so that it is not limiting.

Factors limiting O_2 availability can be illustrated in terms of barriers to O_2 movement from a gas bubble to the dissolved O_2 at the site of respiratory enzymes within a microorganism (figure 44.4). Only with a sufficient O_2 concentration and flux rate from the bubble across each barrier will it be possible to avoid O_2 limitation. This is especially true when a dense microbial culture is growing. When filamentous fungi and actinomycetes are cultured, these physical aeration processes can be even further limited by filamentous growth (figure 44.5). Such filamentous growth results in a viscous, plastic medium, known as a **non-Newtonian broth,** which offers even more resistance to stirring and aeration. To avoid this problem, cultures can be grown as pellets or flocs, or bound to artificial particles.

It is essential to assure that these physical factors are not limiting microbial growth. Physical conditions must be considered at the level of the individual microorganism or microenvironment where these important biochemical reactions actually occur. This is most critical during **scaleup,** where a successful procedure developed in a small shake flask is modified for use in a large fermenter. One must understand the microenvironment of the culture and maintain similar conditions near the individual cell despite increases in the culture volume. If a successful transition can be made from a process originally developed in a 250 ml Erlenmeyer flask to a 100,000 liter reactor, then the process of scaleup has been carried out properly.

Strain Selection, Improvement, and Preservation

Throughout most of the development of industrial microbiology, the major sources of microbial cultures were natural materials, such as soil samples, water, and spoiled bread and fruit. Cultures from all areas of the world were examined by

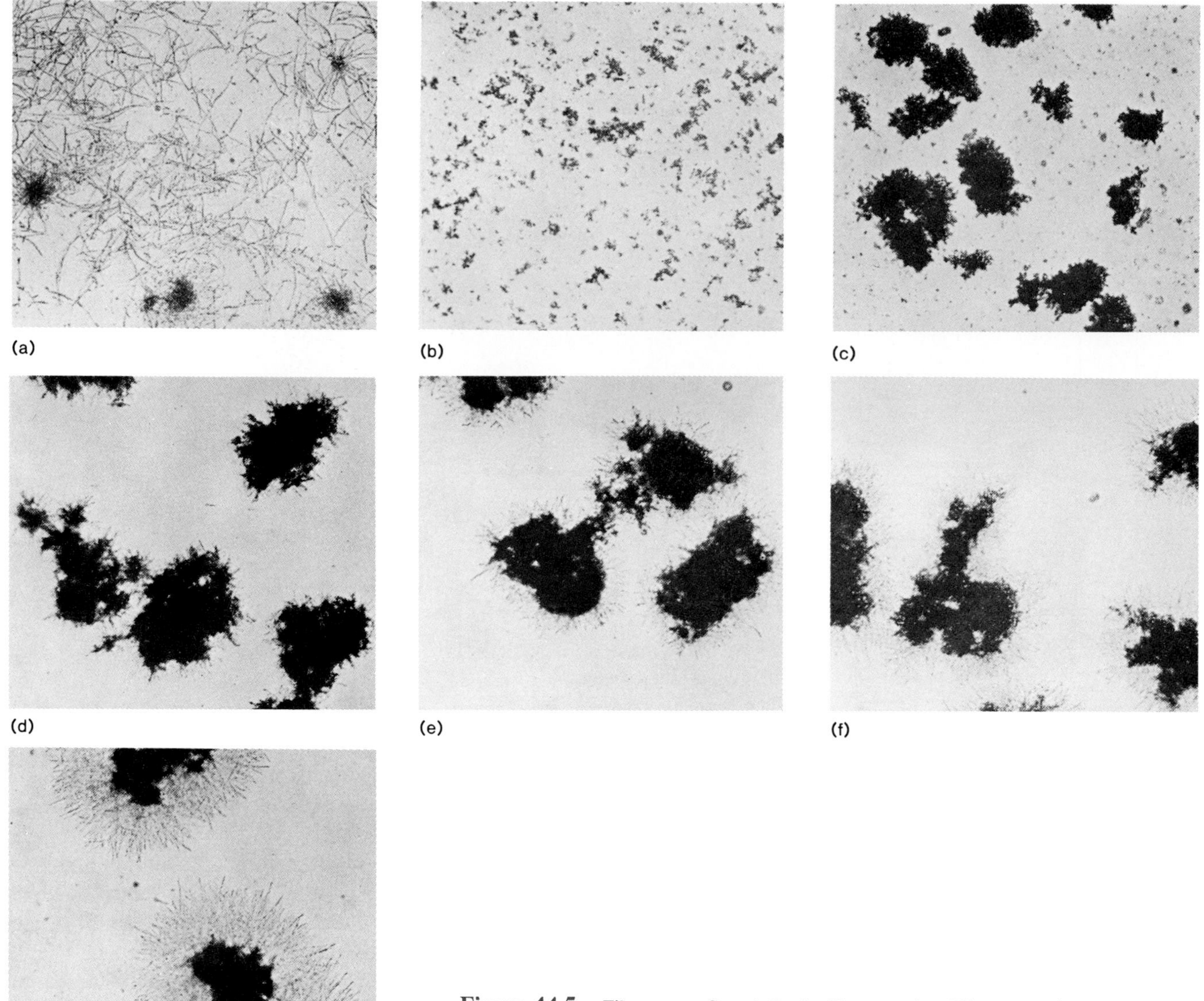

Figure 44.5 **Filamentous Growth During Fermentation.** Filamentous fungi and actinomycetes can change their growth form during the course of a fermentation. The development of pelleted growth by fungi has major effects on oxygen transfer and energy required to agitate the culture. (*a*) Initial culture (time 0). (*b*) 14 hours. (*c*) 16 hours. (*d*) 18 hours. (*e*) 20 hours. (*f*) 22 hours. (*g*) 25 hours.

industrial microbiologists in an attempt to identify isolates with improved characteristics. Once a promising culture was found, a variety of techniques were used for culture improvement, including the use of chemical mutagens and ultraviolet light (*see chapter 12*). As an example, the first cultures of *Penicillium notatum,* which could be grown only under static conditions, yielded low concentrations of penicillin. In 1943, a strain of *Penicillium chrysogenum* was isolated—strain NRRL 1951—which was further improved through screening (selection) and mutation. Today most penicillin is produced with cultures of *Penicillium chrysogenum,* which give 55-fold higher penicillin yields than the original culture when grown in aerobic stirred fermenters.

Genetic manipulations have also been used to identify microorganisms with new and desirable characteristics. The classical methods of microbial genetics (*see chapter 13*) played a vital role in the earlier development of cultures for industrial microbiology. **Protoplast fusion** is now widely used with yeasts and molds. Most of these microorganisms are asexual or of a single mating type, which decreases the chance of random mutations that could lead to strain degeneration. To carry out genetic studies with these microorganisms, protoplasts are prepared by growing the cells in an isotonic solution while inhibiting cell wall growth (figure 44.6). The protoplasts are then mixed with polyethylene glycol (PEG), resulting in partial solubilization of the cell membranes. If fusion occurs to form hybrids, desired recombinants are identified by means of selective plating techniques. After regeneration of the cell wall, the new protoplasm fusion product can be used in further studies.

A major advantage of the protoplast fusion technique is that protoplasts of different microbial species can be fused, even if they are not closely linked taxonomically. For example,

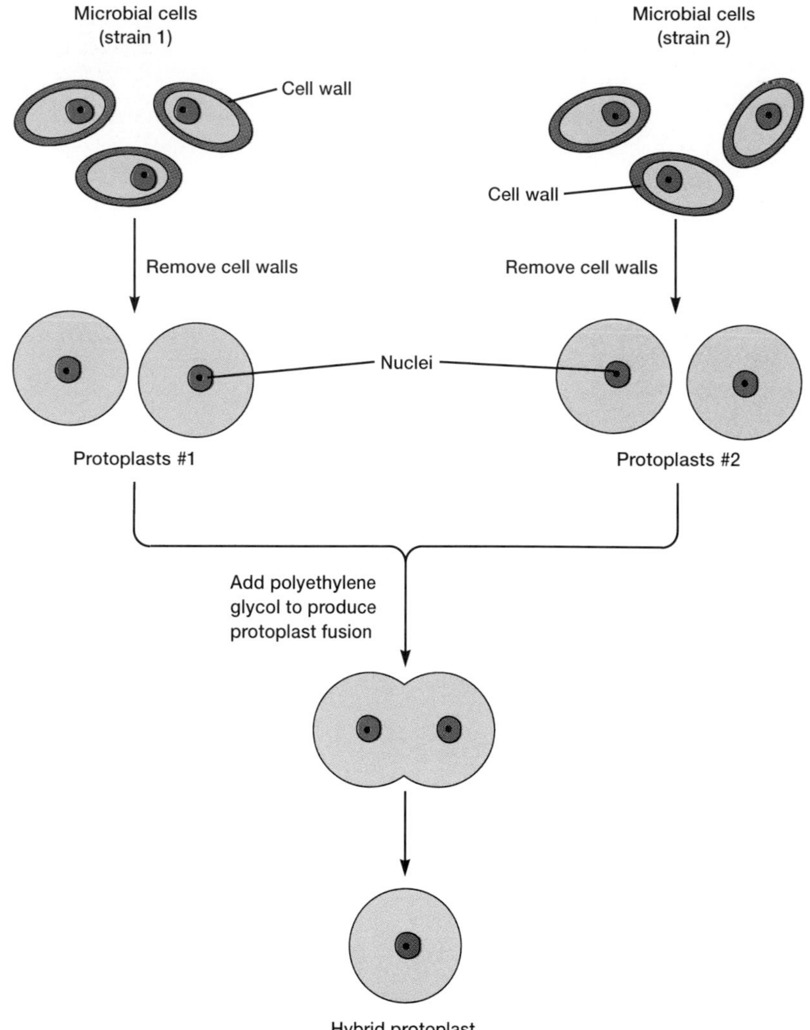

Figure 44.6 Protoplast Fusion and the Genetic Manipulation of Microorganisms. This technique is particularly useful for the genetic improvement of asexual or single-mating-type microorganisms used in many industrial processes.

protoplasts of *Penicillium roquefortii* have been fused with those of *P. chrysogenum.* Even yeast protoplasts and erythrocytes can be fused. Although many of these fusion products are unstable, useful applications continue to be found for them.

Once a desired microorganism has been developed and is ready for use in a particular fermentation, strain stability and preservation are of concern. A variety of culture preservation techniques may be used to maintain desired culture characteristics (table 44.3). **Lyophilization,** or freeze-drying, and storage in liquid nitrogen are frequently employed with microorganisms. Although lyophilization and liquid nitrogen storage are complicated and require expensive equipment, they do allow one to store microbial cultures for years without loss of viability or an accumulation of mutations. Recently, it has been found that many common bacteria can be maintained for extended periods in water agar and common growth media (Box 44.1).

1. What parameters can be monitored in a modern, large-scale industrial fermentation?
2. Besides the aerated, stirred fermenter, what other alternatives are available for the mass culture of microorganisms in industrial processes? What is the principle by which a dialysis culture system functions?
3. What are the major types of materials used as carbon and nitrogen sources in industrial fermentations?
4. Why is it often more difficult to aerate and agitate cultures of actinomycetes and molds in industrial fermentations?
5. Describe the barriers to O_2 transfer from the interior of an air bubble to the active respiratory site within a microorganism.
6. Describe the process of protoplast fusion.
7. What are the major advantages of lyophilization and liquid nitrogen for culture preservation, in comparison with other available techniques?

Box 44.1

Long-Term Survival of Vegetative Bacteria in Dilute Media: Implications for Culture Preservation and Fundamental Physiology

Pseudomonas aeruginosa, Escherichia coli, Salmonella typhimurium, and several other common bacteria have been found to survive in water agar and simple nutrient media for up to 30 years. In comparison with more complex procedures used for culture preservation, such as lyophilization and storage in liquid nitrogen, this preservation approach is only slowly being recognized and accepted by microbiologists. The technique, carried out with washed cultures of bacteria grown in a dilute medium, is the ultimate in simplicity. After preparation, the cultures are sealed with paraffin to minimize moisture loss and stored at room temperature in the dark. If the electricity goes off or the supply of liquid nitrogen is interrupted, which might occur at least once in a period of 15 to 30 years, cultures preserved in dilute media by this simple technique will not be affected. Some microbiologists believe the long survival is due to the use of low-salt media. Reports of successful preservation by these simple procedures have come from China, Hungary, and the United States.

These record survival times for common vegetative bacteria under simple conditions suggest that bacteria may have extraordinary abilities to survive without the continuous addition of nutrients. Such a phenomenon might be considered a form of cryptobiosis, or hidden life. These observations may not only be of help in preserving microorganisms but may provide useful information on basic characteristics of microorganisms as biological entities. The major challenge here is to design experiments that will last 30 years or longer.

Table 44.3 Methods Used to Preserve Cultures of Interest for Industrial Microbiology and Biotechnology

Method	Comments
Periodic transfer	Variables of periodic transfer to new media include transfer frequency, medium used, and holding temperature; this can lead to increased mutation rates and production of variants
Mineral oil slant	A stock culture is grown on a slant and covered with sterilized mineral oil; the slant can be stored at refrigerator temperature
Minimal medium, distilled water, or water agar	Washed cultures are stored under refrigeration; these cultures can be viable for three to five months or longer
Freezing in growth media	Not reliable; can result in damage to microbial structures; with some microorganisms, however, this can be a useful means of culture maintenance
Drying	Cultures are dried on sterile soil (soil stocks), on sterile filter paper disks, or in gelatin drops; these can be stored in a dessicator at refrigeration temperature, or frozen to improve viability
Freeze-drying (lyophilization)	Water is removed by sublimation, in the presence of a cryoprotective agent; sealing in an ampule can lead to long-term viability, with 30 years having been reported
Ultrafreezing	Liquid nitrogen at $-196°$ C is used, and cultures of fastidious microorganisms have been preserved for more than 15 years

TABLE 44.4 Major Microbial Products and Processes of Interest in Industrial Microbiology and Biotechnology

Substances	Microorganisms
Industrial Products	
Ethanol (from glucose)	*Saccharomyces cerevisiae*
Ethanol (from lactose)	*Kluyveromyces fragilis*
Acetone and butanol	*Clostridium acetobutylicum*
2,3-butanediol	*Enterobacter, Serratia*
Enzymes	*Aspergillus, Bacillus, Mucor, Trichoderma*
Agricultural Products	
Gibberellins	*Gibberella fujikuroi*
Food Additives	
Amino acids (lysine, etc.)	*Corynebacterium glutamicum*
Organic acids (citric acid)	*Aspergillus niger*
Nucleotides	*Corynebacterium glutamicum*
Vitamins	*Ashbya, Eremothecium, Blakeslea*
Polysaccharides	*Xanthomonas*
Medical Products	
Antibiotics	*Penicillium, Streptomyces, Bacillus*
Alkaloids	*Claviceps purpurea*
Steroid transformations	*Rhizopus, Arthrobacter*
Insulin, human growth hormone, somatostatin, interferons	*Escherichia coli, Saccharomyces cerevisiae*, and others (recombinant DNA technology)
Biofuels	
Hydrogen	Photosynthetic microorganisms
Methane	*Methanobacterium*
Ethanol	*Zymomonas, Thermoanaerobacter*

Major Products of Industrial Microbiology

Once a suitable microorganism is obtained using these selection and mutation processes, one can produce many compounds and transform a variety of substrates (table 44.4). The discipline responsible for such accomplishments is known as "industrial microbiology." These products include pharmaceutical and medical compounds (antibiotics, hormones, transformed steroids), solvents, organic acids, chemical feedstocks, amino acids, and enzymes. Microorganisms can also be used to produce fuels and energy-related products. The economics of the production of these materials is constantly changing.

Microbial products are often classified as primary and secondary metabolites. As shown in figure 44.7, **primary metabolites** consist of compounds related to the synthesis of microbial cells and are often involved in the growth phase, or **trophophase.** They include amino acids, nucleotides, and fermentation end products such as ethanol and organic acids. In addition, industrially useful enzymes, either associated with the microbial cells or exoenzymes, are often synthesized by microorganisms during growth. These enzymes find many uses in food production and textile finishing.

Secondary metabolites usually accumulate during the period, often called the **idiophase,** that follows the active growth phase. Compounds produced in the idiophase have no direct relationship to the synthesis of cell materials and normal growth. Most antibiotics and the mycotoxins fall into this category.

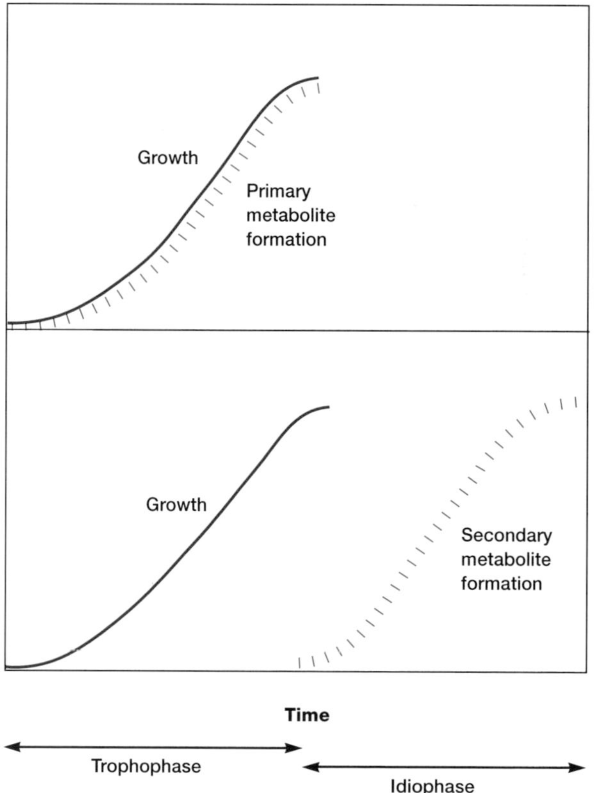

Figure 44.7 Primary and Secondary Metabolites. Depending on the particular organism, the desired product may be formed during or after growth. Primary metabolites are formed during the trophophase (active growth) while secondary metabolites are formed after growth is completed, the idiophase.

Microorganisms, if cultured under ideal conditions without environmental limitations, will attempt to maximize microbial biomass formation. Under conditions of balanced growth, the cells will minimize accumulation of any particular cellular "building blocks" in amounts beyond those required for growth (*see pp. 121–22*). Much of modern industrial microbiology and biochemistry involves "tricking" particular microorganisms into producing large excesses of desired compounds, usually by selecting mutants that have lost the ability to control synthesis of a particular end product. The following sections of this chapter describe examples of changes in genetic characteristics and environmental factors that induce the synthesis of excess amounts of microbial products.

Antibiotics

Many antibiotics are produced by microorganisms, mostly by actinomycetes—predominantly the genus *Streptomyces*—and filamentous fungi (table 44.5). Environmental control is crucial to the production of these important compounds.

Antibiotics in medicine (chapter 16).

Penicillin

Penicillin, produced by *Penicillium chrysogenum,* a fungus that can be grown in stirred fermenters (figure 44.2), is an excellent example of a fermentation for which careful adjustment of the medium composition is used to achieve maximum yields. Rapid production of cells, which can occur when high levels of glucose are used as a carbon source, does not lead to maximum antibiotic yields. Provision of the disaccharide lactose in combination with limited nitrogen availability, stimulates a greater accumulation of penicillin (figure 44.8) because of the slower hydrolysis rate of this sugar. The same result can be achieved by using a slow continuous feed of glucose. If a particular penicillin is needed, the specific precursor is added to the medium. For example, phenylacetic acid is added to

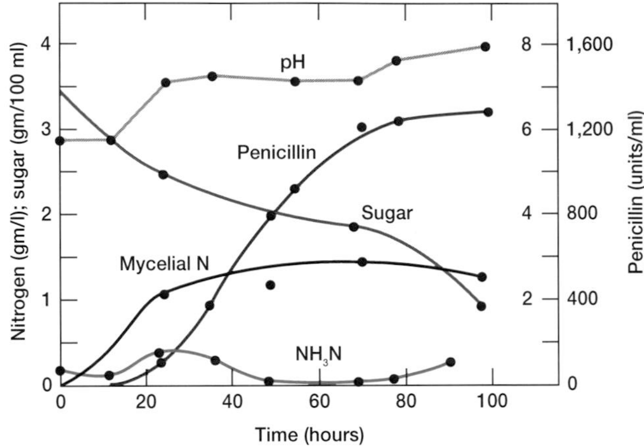

Figure 44.8 The Time Course of the Penicillin Fermentation. This process provides an excellent example of the careful adjustment of limiting factors to maximize yields of a desired antibiotic.

TABLE 44.5 Major Antibiotics Produced by Actinomycetes, Other Bacteria, and Fungi, and the Microorganisms Affected

Microbial/Antibiotic Group	Produced by	Spectrum of Action
Actinomycetes		
Amphotericin-B	*Streptomyces nodosus*	Fungi
Carbomycin	*Streptomyces halstedii*	Gram-positive bacteria
Chlorotetracycline	*Streptomyces aureofaciens*	Broad spectrum
Chloramphenicol	*Streptomyces venezuelae*	Broad spectrum
Cycloheximide	*Streptomyces griseus*	Pathogenic yeasts
Erythromycin	*Streptomyces erythraeus*	Mostly gram-positive bacteria
Kanamycin	*Streptomyces kanamyceticus*	Gram-positive bacteria
Oleandomycin	*Streptomyces antibioticus*	Staphylococci
Oxytetracycline	*Streptomyces rimosus*	Broad spectrum
Neomycin-B	*Streptomyces fradiae*	Broad spectrum
Novobiocin	*Streptomyces niveus*	Gram-positive bacteria
Nystatin	*Streptomyces noursei*	Fungi
Streptomycin	*Streptomyces griseus*	Gram-negative bacteria, *Mycobacterium tuberculosis*
Other Bacteria		
Polymyxin-B	*Bacillus polymyxa*	Gram-negative bacteria
Bacitracin	*Bacillus licheniformis*	Gram-positive bacteria
Fungi		
Cephalosporin	*Cephalosporium acremonium*	Broad spectrum
Fumigillin	*Aspergillus fumigatis*	Amoebae
Griseofulvin	*Penicillium griseofulvum*	Fungi
Penicillin	*Penicillium chrysogenum*	Gram-positive bacteria

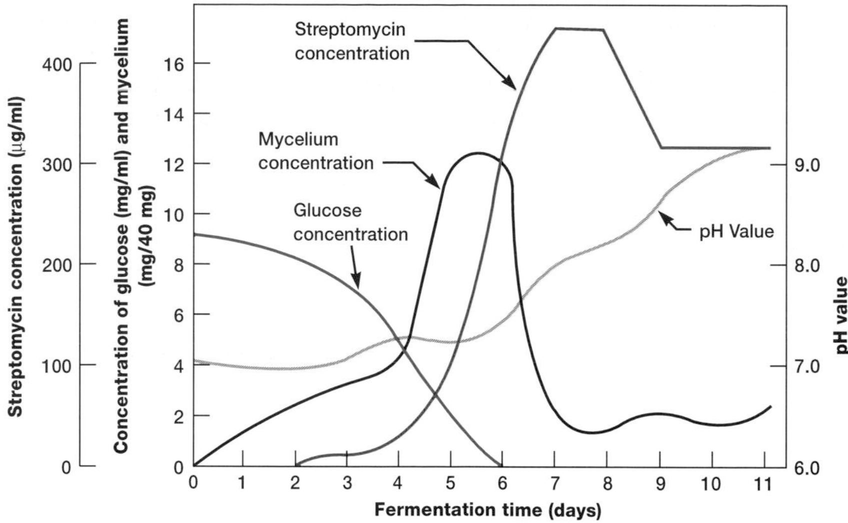

Figure 44.9 **Streptomycin Production by *Streptomyces griseus*.** Sequential use of different carbon sources leads to maximum antibiotic yields.

maximize production of penicillin G, which has a benzyl side chain (*see figure 16.6*). The fermentation pH is maintained around neutrality by the addition of sterile alkali, which assures maximum stability of the newly synthesized penicillin. Once the fermentation is completed, normally in six to seven days, the broth is separated from the fungal mycelium and processed by absorption, precipitation, and crystallization to yield the final product. This basic product can then be modified by chemical procedures to yield a variety of **semisynthetic penicillins.**

Streptomycin

Streptomycin is a secondary metabolite produced by *Streptomyces griseus,* for which changes in environmental conditions and substrate availability also influence final product accumulation. In this fermentation, a soybean-based medium is used with glucose as a carbon source. The nitrogen source is thus in a combined form (soybean meal). After growth, the antibiotic levels in the culture begin to increase (figure 44.9) under conditions of controlled nitrogen limitation.

Amino Acids

Amino acids such as lysine and glutamic acid are used in the food industry as nutritional supplements in bread products and as flavor-enhancing compounds such as monosodium glutamate (MSG).

Amino acid production is typically carried out by means of **regulatory mutants.** The normal microorganism avoids overproduction of biochemical intermediates by the careful regulation of cellular metabolism. Production of glutamic acid and several other amino acids in large quantities is now carried out using mutants of *Corynebacterium glutamicum,* (figure 44.10). A controlled low biotin level and the addition of fatty acid derivatives result in increased membrane permeability and excretion of high concentrations of glutamic acid.

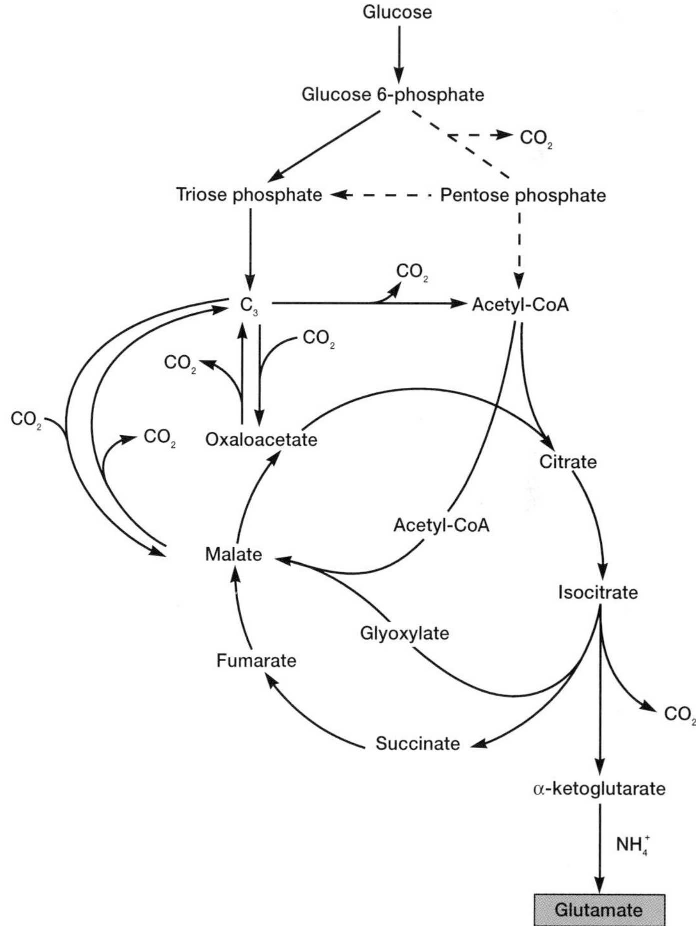

Figure 44.10 **Glutamic Acid Production.** The sequence of biosynthetic reactions leading from glucose to the accumulation of glutamate by *Corynebacterium glutamicum.* By creation of a controlled low biotin level, together with the addition of lipids, membrane permeability is increased. After growth is completed, glutamate becomes the major end product of glucose use by this interesting bacterium. The dashed lines indicate those routes that are of lesser importance in glutamate formation.

The impaired bacteria use the glyoxylate pathway (*see chapter 9*) to meet their needs for essential biochemical intermediates, especially during the growth phase. After growth becomes limited because of changed nutrient availability, an almost complete molar conversion (or 81.7% weight conversion) of isocitrate to glutamate occurs.

Lysine, an essential amino acid used to supplement cereals and breads, was originally produced in a two-step microbial process. This has been replaced by a single-step fermentation in which the bacterium *Corynebacterium glutamicum,* blocked in the synthesis of homoserine, accumulates lysine. Over 44 g/liter can be produced in a three-day fermentation.

Although not used extensively in the United States, microorganisms with related regulatory mutations have been employed to produce a series of 5′ purine nucleotides that serve as flavor enhancers for soups and meat products.

Organic Acids

Organic acid production by microorganisms is important in industrial microbiology and illustrates the effects of trace metal levels and balances on organic acid synthesis and excretion. Citric, acetic, lactic, fumaric, and gluconic acids are major products (table 44.6). Until microbial processes were developed, the major source of citric acid was citrus fruit from Italy. Today, most citric acid is produced by microorganisms; 70% is used in the food and beverage industry, 20% in pharmaceuticals, and the balance in other industrial applications. Citric acid is also used in detergents to replace phosphates, which can contribute to eutrophication.

The essence of citric acid fermentation involves limiting the amounts of trace metals such as manganese and iron to prevent growth of *Aspergillus niger* beyond a certain point. The medium is often treated with ion exchange resins to ensure low and controlled concentrations of available metals. Citric acid fermentation, which earlier was carried out by means of static surface growth, now takes place in aerobic stirred fermenters. Generally, high sugar concentrations (15 to 18%) are used, and copper has been found to counteract the inhibition of citric acid production by iron above 0.2 ppm. The success of this fermentation depends on the regulation and functioning of the glycolytic pathway and the tricarboxylic acid cycle (*see chapter 8*). During the idiophase, when the substrate level is high, citrate synthase activity increases and the activities of aconitase and isocitrate dehydrogenase decrease. This results in citric acid accumulation and excretion by the stressed microorganism.

In comparison, the production of gluconic acid involves a single microbial enzyme, glucose oxidase, found in *Aspergillus niger. A. niger* is grown under optimum conditions in a corn-steep liquor medium. Growth becomes limited by nitrogen, and the resting cells transform the remaining glucose to gluconic acid in a single-step reaction (figure 44.11). Gluconic acid is used as a carrier for calcium and iron and as a component of detergents.

Bioconversion Processes

Various microorganisms can be used to carry out **bioconversions,** also known as **microbial transformations** or **biotransformations.** Such changes are usually specific, minor modifi-

TABLE 44.6 Major Organic Acids Produced by Microbial Processes

Product	Microorganism Used	Representative Uses	Fermentation Conditions
Acetic acid	*Acetobacter* with ethanol solutions	Wide variety of food uses	Single-step oxidation, with 15% solutions produced; 95–99% yields
Citric acid	*Aspergillus niger* in molasses-based medium	Pharmaceuticals, as a food additive	High carbohydrate concentrations and controlled limitation of trace metals; 60–80% yields
Fumaric acid	*Rhizopus nigricans* in sugar-based medium	Resin manufacture, tanning, and sizing	Strongly aerobic fermentation; carbon-nitrogen ratio is critical; zinc should be limited; 60% yields
Gluconic acid	*Aspergillus niger* in glucose-mineral salts medium	A carrier for calcium and sodium	Uses agitation or stirred fermenters; 95% yields
Itaconic acid	*Aspergillus terreus* in molasses-salts medium	Esters can be polymerized to make plastics	Highly aerobic medium, below pH 2.2; 85% yields
Kojic acid	*Aspergillus flavus-oryzae* in carbohydrate-inorganic N medium	The manufacture of fungicides and insecticides when complexed with metals	Iron must be carefully controlled to avoid reaction with kojic acid after fermentation
Lactic acid	Homofermentative *Lactobacillus delbrueckii*	As a carrier for calcium and as an acidifier	Purified medium used to facilitate extraction

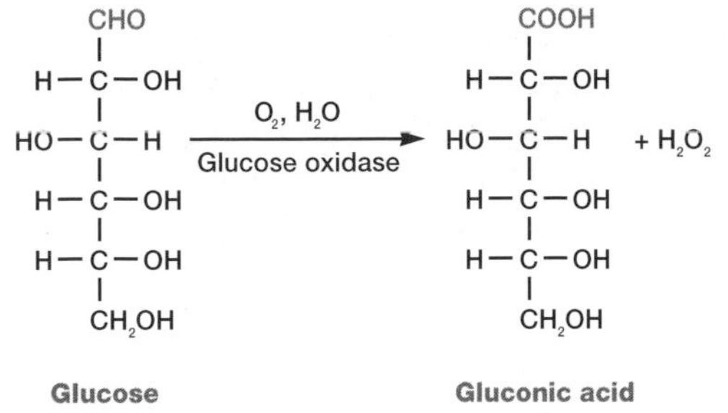

CHO
H—C—OH
HO—C—H
H—C—OH
H—C—OH
CH_2OH

Glucose

O_2, H_2O → Glucose oxidase

COOH
H—C—OH
HO—C—H
H—C—OH
H—C—OH
CH_2OH

Gluconic acid

+ H_2O_2

Figure 44.11 Gluconic Acid Production. This is an example of a transformation carried out by nitrogen-limited nongrowing cells. *Aspergillus niger* is used in this fermentation. The group that is altered is shown in color.

TABLE 44.7 Methods of Producing and Utilizing Biocatalysts in Bioconversion Processes

Biocatalyst Production Procedure
Continuous culture
Batch culture
Biocatalyst Processing Alternatives
Immobilized cells
Dried cells
Permeabilized cells
Resting cells
Cell extracts and enzyme purification
Extracellular enzyme purification

cations of compounds that are not normally used for growth. Bioconversions have major advantages over chemical procedures. Enzymes carry out very specific reactions under mild conditions, and larger water-insoluble molecules can be transformed. Unicellular bacteria, actinomycetes, yeasts, and molds have been used in various bioconversions. The enzymes responsible for these conversions can be intracellular or extracellular.

Biocatalysts can be grown and processed in a variety of ways, as shown in table 44.7. Cells can be produced in batch or continuous culture and then dried for direct use, or they can be prepared in more specific ways to carry out desired bioconversions.

Biotransformations carried out by free enzymes or intact nongrowing cells do have limitations. Reactions that occur in the absence of active metabolism—without reducing power or ATP being available continually—are primarily exergonic reactions (*see chapter 7*). If ATP or reductants are required, an energy source such as glucose must be supplied under carefully controlled nongrowth conditions.

When freely suspended vegetative cells or spores are employed, the microbial biomass is usually used only once. At the end of the process, the cells are discarded. Cells can often be used repeatedly after attaching them to ion exchange resins by ionic interactions or immobilizing them in a polymeric matrix. Major types of **immobilization** techniques are shown in figure 44.12. Ionic, covalent, and physical entrapment approaches can be used to immobilize microbial cells, spores, and enzymes. Microorganisms can also be immobilized on the inner walls of fine tubes. The solution to be modified is then simply passed through the microorganism-lined tubing; this approach is finding applications in many industrial and environmental processes.

Immobilized microorganisms and their associated enzymes are used in bioconversions of steroids, degradation of phenol, and the production of a wide range of antibiotics, enzymes, organic acids, and metabolic intermediates (table 44.8). One application of cells as biocatalysts is the recovery of precious metals from dilute-process streams. As shown in figure 44.13, previously grown microbial cells—in this case an alga—are used to concentrate precious metals, including silver and gold.

1. What are the major differences between primary and secondary metabolites? Define trophophase and idiophase.
2. Why might the use of lactose and glucose in culture media result in different growth rates of a penicillin-producing microorganism?
3. Describe how mutations and the modification of culture conditions are used to stimulate glutamic acid synthesis.
4. What is the principal limitation created to stimulate citric acid accumulation by *Aspergillus niger*?
5. What types of bioconversion processes can be carried out by microbial cells?
6. Describe the major approaches used to immobilize microbial cells or cell components.

Recombinant DNA Techniques in Biotechnology

An array of recombinant DNA techniques are now available for the selection and genetic manipulation of microorganisms (figure 44.14). In these approaches, **biochemical markers,** such as nutrient requirements and antibiotic resistance patterns, are often used to identify mutant or recombinant organisms.

Several techniques can be used to construct desired genotypes, as discussed in chapter 14. These include use of restriction enzymes, preparation of probes and synthetic DNA, site-directed mutagenesis, PCR, and construction of specific cloning vectors. Newer areas of molecular applications in biotechnology include (1) modification of product expression (it is possible to have cells that contain 30% by weight of a specific desired product), (2) custom design of proteins and other products to give them more desirable characteristics, and (3) preparation of vectors to express genes in different organisms. This latter strategy allows the production of specific pro-

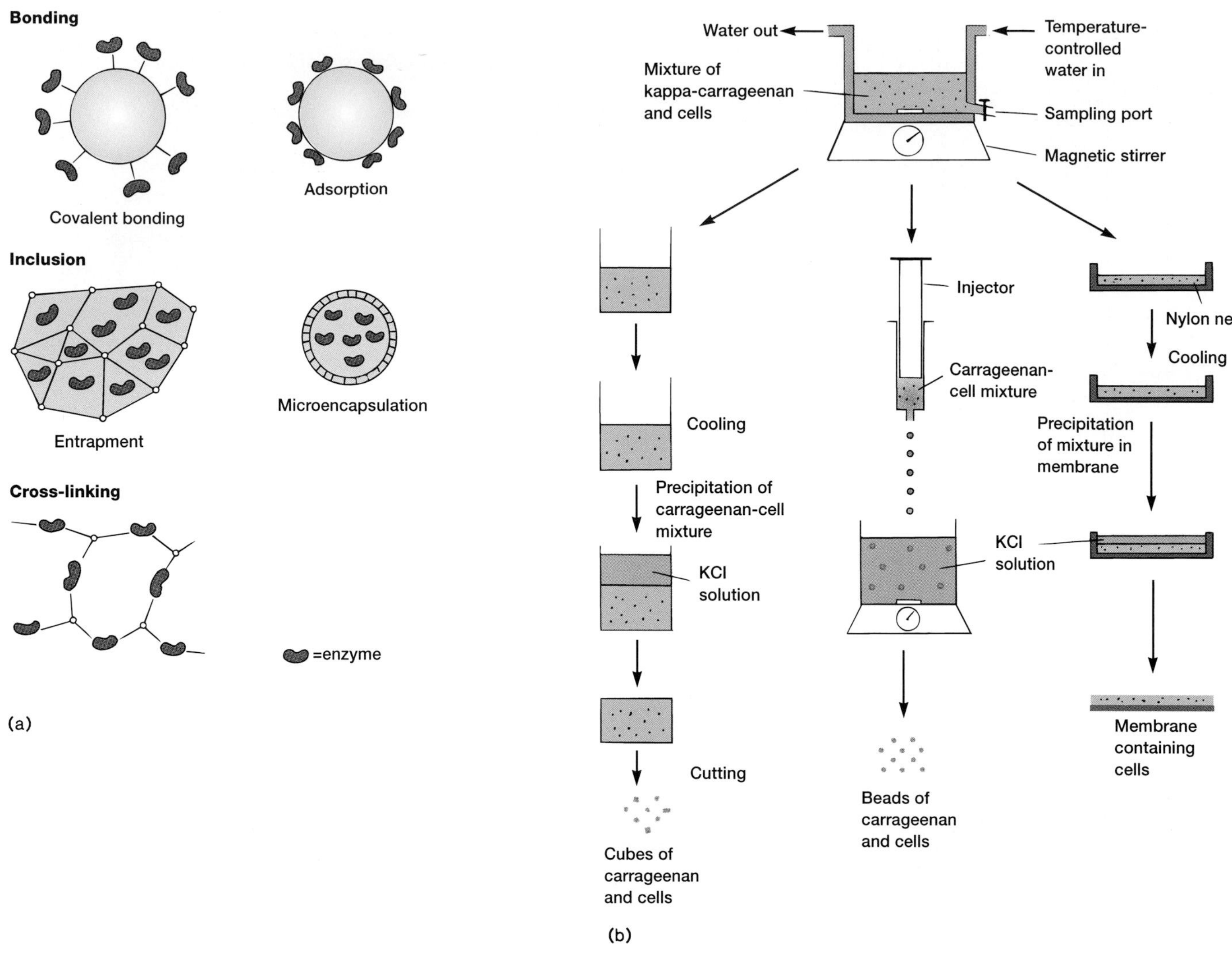

Figure 44.12 Immobilization Techniques. Immobilization of enzymes and whole microorganisms can lead to improved industrial processes. Alternatives for immobilizing (*a*) enzymes and (*b*) whole cells of microorganisms are shown.

TABLE 44.8 Immobilized Microbial Cells: Their Products and Reactions

Process Type	Product or Activity	Microorganisms Used
Antibiotics	Penicillin	*Penicillium chrysogenum*
	Bacitracin	*Bacillus* spp.
	Cephalosporins	*Streptomyces clavuligerus*
Amino acids	L-alanine	*Corynebacterium dismutans*
	L-glutamic acid	*Corynebacterium glutamicum*
	L-tryptophan	*Escherichia coli*
	L-lysine	*Microbacterium ammoniaphila*
Enzymes	Coenzyme A	*Brevibacterium ammoniagenes*
	Protease	*Streptomyces fradiae*
Vitamins	Pantothenic acid	*Escherichia coli*
Steroids	Prednisolone	*Curvularia lunata* and *Corynebacterium simplex*
Fermentation products	Ethanol	*Saccharomyces cerevisiae*
	Lactic acid	*Lactobacillus delbrueckii*
Environmental management	Degradation of paranitrophenol	*Pseudomonas* spp.
	Degradation of phenol	*Candida tropicalis*
	Denitrification	*Micrococcus* spp.
	Heavy-metal sorption (uranium, plutonium)	*Pseudomonas aeruginosa*

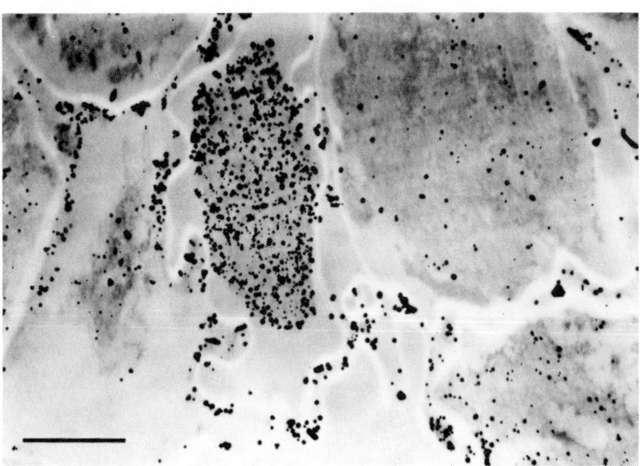

Figure 44.13 The Recovery of Precious Metals. Microorganisms can be used to remove metals from waters, either for water purification or metal recovery. In this electron micrograph, gold has been recovered from a dilute-process stream by cells of *Chlorella vulgaris* (× 26,400). Deposits of gold (black specks) are associated with the green algal cells. Bar = 1 μm.

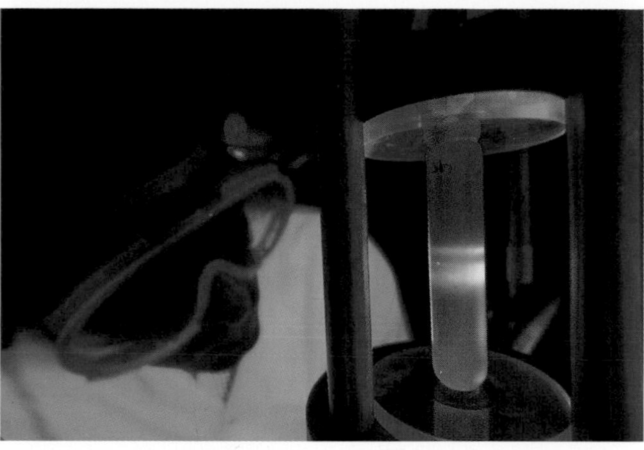

Figure 44.14 Microbial DNA Purification. DNA isolated from microorganisms can be purified by centrifugation in a cesium chloride gradient. The DNA glows under ultraviolet light because it is complexed with the dye ethidium bromide. Recombinant DNA techniques are increasingly used in the pharmaceutical and agricultural industries.

TABLE 44.9 Examples of Recombinant DNA Systems Used to Modify Gene Expression

Product	Microorganism	Change
Actinorhodin	*Streptomyces coelicolor*	Modification of gene transcription
Cellulase	*Clostridium* genes in *Bacillus*	Amplification of secretion through chromosomal DNA amplification
Recombinant protein albumin	*Saccharomyces cerevisiae*	Fusion to a high-production protein
Heterologous protein	*Saccharomyces cerevisiae*	Use of the inducible strong hybrid promoter UAS_{gal}/CYC1

teins and peptides without contamination by similar products that might be synthesized in the original organism. The approach can decrease the time and cost of recovering and purifying a product.

Recombinant DNA Technology (pp. 290–302).

Modification of Gene Expression

With the ability to modify transcription and translation of genetic information, instead of merely selecting random overproducing mutants, it is now possible to make specific changes in control regions as noted in table 44.9. By modifying control mechanisms, enzymes that are normally expressed only in the absence of intermediates or end products can be produced on a constitutive basis. Recombinant plasminogen, for example, can be 20 to 40% of the soluble protein, a tenfold increase in production over the original strain.

Design of Proteins and Peptides

By the use of site-directed mutagenesis and the chemical synthesis of DNA, which can be inserted into the desired organism to produce a modified protein, the field of **protein engineering** has been created. This provides a new and exciting area of molecular biotechnology. Enzymes and bioactive peptides with markedly different characteristics (stability, kinetics, activities) can be created. The molecular basis for the functioning of these modified products also can be better understood.

As summarized in table 44.10, a variety of modified proteins and bioactive peptides has been created. These products can only serve as examples because the field is expanding so rapidly. One of the most interesting areas is the design of enzyme-active sites to promote the modification of "unnatural substrates." This approach may lead to improved transformation of recalcitrant materials, or even the degradation of materials that have previously not been amenable to biological processing.

Progress has reached the point that an entirely new field has emerged: **metabolic engineering.** In this application of recombinant DNA technology, metabolic networks are restructured by the recruitment of proteins from different cells. The result is a change in pathway distribution and rate. Examples of altered small metabolites and protein-related end products are summarized in table 44.11. By combining the metabolic capabilities from two microorganisms, entirely new products and intermediates can be synthesized.

TABLE 44.10 Examples of Modified Proteins Created through Recombinant DNA Technology

Characteristic Modified	Microorganism Used	Result	Mechanism
Glucose isomerase thermostability	*Actinoplanes*	Enhanced stability of soluble and immobilized forms	Amino acid substitution of arginine for lysine at position 253
Protein separation characteristics	*Escherichia coli*	96-fold difference in modified partition coefficient	Fusion with partitioning tetrapeptide
Protein structure and crystallinity	*Escherichia coli*	Unique macromolecular polyamide copolymers	Modification of leader region in message
Biopesticide activity	*Escherichia coli*	Modified moulting of fall army-worm	Insertion of *LacZ* gene and modification of hormone
Enzyme oxidation resistance	*Escherichia coli*	Alpha-antitrypsin more resistant to oxidation	Replacement of a critical oxygen-sensitive methionine with valine
Anticoagulant polypeptide alteration	*Escherichia coli*	Improve the 65 amino acid polypeptide hirudin	Substitution of asparagine residue 47 with lysine or arginine

TABLE 44.11 Heterologous Activities Recruited to Alter Small Metabolite and Protein End Products[a]

Host Organism	Original Metabolite	Enzyme Added (source organism)	New Product
Erwinia herbicola	2,5-DKG[b]	2,5-DKG reductase (*Corynebacterium*)	2-KLG,[c] a vitamin C precursor in a single fermentation
Acremonium chrysogenum	Cephalosporin C	D-amino acid oxidase (*Fusarium solani*), cephalosporin acylase (*Pseudomonas diminuta*)	7ACA,[d] an antibiotic precursor by a single fermentation instead of two steps
Chinese hamster ovary cells	Terminal β-galactosyl residues in cell surface glycoproteins	β-galactoside 2,6-sialyltransferase (rat)	Increased β-galactosyl residues to better resemble human-derived forms

[a]The original metabolite is modified by the combined genetic information from two microorganisms to give a new product.
[b]2,5-DKG = 2,5-diketo-D-gluconic acid.
[c]2-KLG = 2-keto-L-gulonic acid.
[d]7ACA = 7-β-(5-carboxy-5-oxopentanamido)-cephalosporanic acid.

TABLE 44.12 New Therapeutic Synthetic Peptides Expected to Reach the Market in the Next Few Years

Peptide	Therapeutic Use	Comments
LHRH agonists	Prostate cancer	Luteinizing hormone–releasing hormone affects pituitary receptor, blocking release of luteinizing hormone
Octreotide	Diarrhea associated with intestinal tumors	Helps control diarrhea
Immune stimulator	Various cancers	Stimulates immune system to control intracellular recognition of self
RDG peptide	Wound healing	Synthetic tripeptide assists in cell anchoring and healing
HIV-1 blockers	AIDS	Peptides block virus replication inside cell
Pentigetide	Allergy-related problems	Interacts with active peptides to control allergies
Vasoactive intestinal peptide	Sexual dysfunction	Active in men and women
Hirudin	Anticoagulant	Product can compete with warfarin and heparin, two major currently used compounds

Many synthetic medical peptides have been developed through recombinant DNA technology (table 44.12). These include products to promote wound healing and blood coagulation, to treat cancer and AIDS, and to influence sexual dysfunctions. Other products include an intracellular adhesin molecule used to battle the common cold, nervous system growth factors, and compounds useful in hormone disorders. The market for polypeptide hormones is expected to expand especially rapidly.

A major advantage of peptide production with modern biotechnology is that only biologically active stereoisomers are produced. This specificity is required to avoid the harmful side effects of inactive stereoisomers, as occurred in the thalidomide disaster.

Vectors for Product Expression

A biotechnological goal is to express desired genetic information in markedly different microorganisms, often to improve production, product purity, or to final product recovery (figure 44.15). Genetic information from a wide range of biological forms can be inserted into microorganisms. A good example is provided by the foot-and-mouth disease of cattle and other livestock. Genetic information for a foot-and-mouth disease virus antigen can be incorporated into *E. coli,* followed by the expression of this genetic information and synthesis of the gene product for use in vaccine production (figure 44.16).

It is important to note that these systems do not only include the expression of recombinant DNA in microorganisms (table 44.13). Genetic information also can be expressed in baculoviruses (*see p. 398*), which can be replicated in insect larvae to achieve rapid large-scale production. Transgenic plants (*discussed on p. 300*) may be used to manufacture large quantities of a variety of metabolic products. A most imaginative way of incorporating new DNA into a plant is to simply shoot it in using DNA-coated microprojectiles and a gene gun, as discussed in chapter 14.

---

1. Why might one wish to express a gene in a foreign cell?
2. Define the terms protein engineering and metabolic engineering.
3. Why is it often important to produce peptides enzymatically rather than by chemical synthesis?
4. Why might it be as important to modify the genes for control of a product's synthesis as to modify its structural genes?

Other Microbial Applications

Microorganisms continue to find additional applications. For example, they are used as insecticides, biopolymers, and biosensors. The control and stimulation of biodegradation also are important in many areas of environmental management.

Figure 44.15 The Effects of Biotechnology. Products of biotechnology are revolutionizing many areas of medicine. Human interleukin produced by recombinant DNA techniques is shown packaged for shipment. These types of products have led to the formation of many biotechnology companies, whose worth is already in the billions of dollars.

Microbial Insecticides

Many alternatives exist for the use of bacteria, viruses, and fungi as insecticides in **biocontrol** programs (table (44.14). Bacterial insecticides include a variety of *Bacillus* species, primarily *B. thuringiensis* (*see chapter 22*). This bacterium is only weakly toxic to insects as a vegetative cell, but during sporulation, it produces an intracellular protein toxin crystal that acts as a microbial insecticide. Only insects with a sufficiently high pH in their midgut will solubilize the crystal and release the toxin.

B. thuringiensis can be grown in fermenters. When the cells lyse, the spores and crystals are released into the medium. The medium is then centrifuged and made up as a dust or wettable powder. By 1971, commercial preparations of *B. thuringiensis* were registered for use on more than 20 agricultural crops. Currently, attention is focused on the possibility of producing the crystal by means of plasmid-coded genes. There is also great interest in using *B. thuringiensis israelensis* in the control of malaria and river blindness. The toxin from this organism destroys the mosquito vectors of *Plasmodium* and the blackfly vector that carries the larvae of the nematode *Onchocerca volvulus,* the causative agent of river blindness (*see page A38*). A related organism, *Bacillus popilliae,* is used to combat

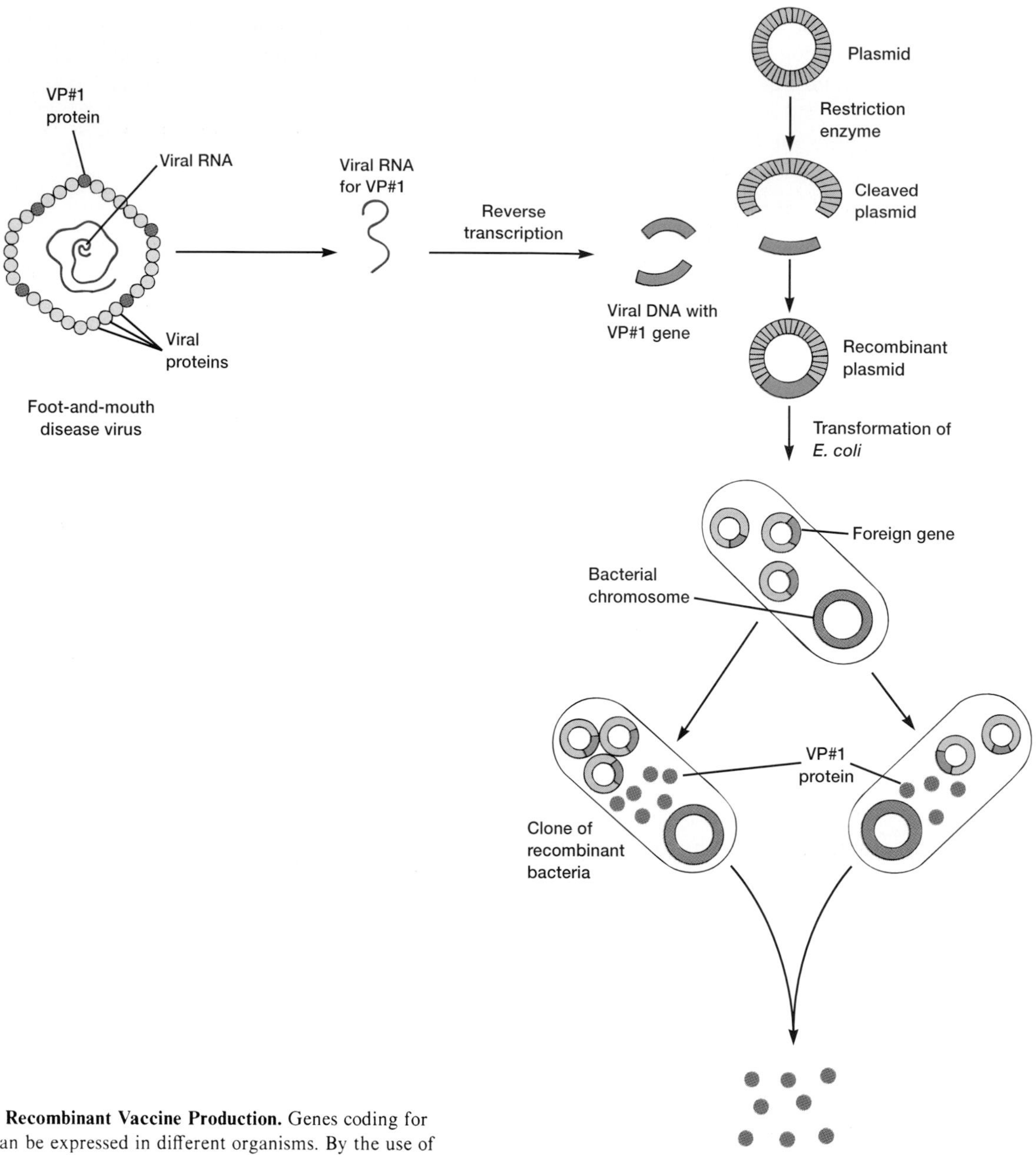

Figure 44.16 **Recombinant Vaccine Production.** Genes coding for desired products can be expressed in different organisms. By the use of recombinant-DNA techniques, a foot-and-mouth disease vaccine is produced through cloning the vaccine genes into *Escherichia coli.*

TABLE 44.13 Examples of Expression of Recombinant DNA in Other Organisms to Improve Process Efficiency

Product	Donor	Recipient	Comment
Hemoglobin	Humans	*Saccharomyces cerevisiae*	Use of a hybrid promoter
Glucose oxidase	*Aspergillus niger*	*Saccharomyces cerevisiae*	Better temperature and pH stability, and no contamination
Interleukin-6	Humans	*Escherichia coli*	N-terminal methionine-free product without purification
Pectinases	*Erwinia*	*Escherichia coli*	Improvement of yield and purity
Monoclonal IgG antibody against *Pseudomonas* lipoprotein I	Mouse light and heavy chain cDNAs	Recombinant baculovirus vector grown in fall army-worms	High level of expression of recombinant monoclonal antibody genes
Alpha and beta portions of hemoglobin	Humans, with tobacco mosaic virus as vector	Tobacco	Large-scale production
Alpha-amylase	*Bacillus licheniformis*	Tobacco	Use of tobacco seeds directly in starch liquifaction, production of an industrial enzyme in plants.

TABLE 44.14 The Use of Bacteria, Viruses, and Fungi As Bioinsecticides: An Older Technology with New Applications

Microbial Group	Major Organisms and Applications
Bacteria	*Bacillus thuringiensis* and *Bacillus popilliae* are the two major organisms of interest. *Bacillus thuringiensis* is used on a wide variety of vegetable and field crops, fruits, shade trees, and ornamentals. *B. popilliae* is used primarily against Japanese beetle larvae. Both bacteria are considered harmless to humans. *Pseudomonas fluorescens,* which contains the toxin-producing gene from *B. thuringiensis,* is used on maize to suppress black cutworms.
Viruses	Three major virus groups that do not appear to replicate in warm-blooded animals are used: nuclear polyhedrosis virus (NPV), granulosis virus (GV), and cytoplasmic polyhedrosis virus (CPV). These occluded viruses are more protected in the environment.
Fungi	Over 500 different fungi are associated with insects. Infection and disease occur primarily through the insect cuticle. Four major genera have been used. *Beauveria bassiana* and *Metarhizium anisopliae* are used for control of the Colorado potato beetle and the froghopper in sugarcane plantations, respectively. *Verticillium lecanii* and *Entomophthora* spp., have been associated with control of aphids in greenhouse and field environments.

the Japanese beetle. This bacterium, however, cannot be grown in fermenters, and inocula must be grown in the living host. The microorganism controls development of larvae, but destruction of the adult beetle requires chemical insecticides.

Viruses that are pathogenic for specific insects include nuclear polyhedrosis viruses (NPV), granulosis viruses (GV), and cytoplasmic polyhedrosis viruses (CPV). Until 1980, only one of these viruses was registered as a control agent, a CPV of the pine caterpillar *Dendrolimus spectabilis.* Currently, 125 types of NPV are known, of which approximately 90% affect the *Lepidoptera*—butterflies and moths. Approximately 50 GVs are known, and they, too, primarily affect butterflies and moths. CPVs are the least host-specific viruses, affecting about 200 different types of insects. The first commercial viral pesticide was marketed under the trade name Elcar for control of the cotton bollworm *Heliothis zea.*

The development of biopesticides is progressing rapidly. One of the most exciting advances involves the use of baculoviruses that have been genetically modified to produce a potent scorpion toxin active against insect larvae. After ingestion by the larvae, viruses are dissolved in the midgut and are released. Because the recombinant baculovirus produces this insect-selective neurotoxin, it acts more rapidly than the parent virus, and the extent of leaf damage can be markedly decreased.

Characteristics of insect viruses (p. 398).

Biopolymers

Biopolymers are produced by many microorganisms and have important uses in the pharmaceutical and food industries because of their ability to modify flow characteristics of liquids and to serve as gelling agents. The advantage of using microbial polysaccharides is that production is independent of climate, political events that can limit raw material supplies, and the depletion of natural resources. Production facilities can be located near sources of inexpensive substrates (e.g., near agricultural areas).

Bacterial exopolysaccharides (p. 56).

At least 75% of all polysaccharides are used as stabilizers, for the dispersion of particulates, as film-forming agents, or to promote water retention in various products (table 44.15).

Polysaccharides help maintain the texture of many frozen foods, such as ice cream, that are subject to drastic temperature changes. These polysaccharides must maintain their properties under the pH conditions in the particular food and be compatible with other polysaccharides. They should not lose their physical characteristics if heated.

Polysaccharides such as scleroglucan are used by the oil industry as drilling mud additives. Xanthan polymers enhance oil recovery by improving water flooding and the displacement of oil. This use of xanthan gum, produced by *Xanthomonas campestris,* represents a large potential market for this microbial product.

Biosensors

A rapidly developing area of biotechnology, arousing intense international scientific interest, is that of **biosensor** production. In this new field of bioelectronics, living microorganisms (or their enzymes or organelles) are linked with electrodes, and biological reactions are converted into electrical currents by these biosensors (figure 44.17). Biosensors are being developed to measure specific components in beer, to monitor pollutants, and to detect flavor compounds in food. It is possible to measure the concentration of substances from many different environments (table 44.16). In the future, biosensors will be able to measure ions and molecules, specific enzymatic activities, and the presence of bacteria. Applications include the detection of glucose, acetic acid, glutamic acid, ethanol, and biochemical oxygen demand. In addition, the application of biosensors to measure cephalosporin, nicotinic acid, and several B vitamins has been described. It appears that biosensors will also be able to detect phenols, methane, and carbon monoxide, which are important in environmental management.

TABLE 44.15 Characteristics and Uses of Commercially Important Microbial Biopolymers

Name	State of Development	Organisms and Applications
Dextran	Production	*Klebsiella, Acetobacter* and *Leuconostoc* are major genera; these alpha-linked glucans are used as blood expanders and absorbants; they can form a hydrophilic layer at burn surfaces
Erwinia exopolysaccharide	Production	Used as a paint component because of its compatibility with basic dyes
Xanthan	Production	Used in foods and in secondary oil recovery programs; produced by *Xanthomonas campestris*
Cellulose microfibrils	Production	Thickener for use in foods; forms films to create textures and barriers; a special *Acetobacter* strain is used.
Pullulan	Development	Produced by *Aureobasidium pullulans*, can serve as biodegradable material for food coating, and may be able to replace starch in selected applications
Scleroglucan	Development	Produced by the fungus *Sclerotium*, and used in drilling muds; forms a viscous gel that exhibits pseudoplasticity
Microbial alginate	Development	*Azotobacter vinlandii* is a major organism; wide range of uses to replace algal alginates; use as a sizing agent and food stabilizer should provide an expanding market
Curdlan	Development	Produced by *Alkaligenes faecalis* var. *myxogenes*; swells in water and produces an irreversible gel; not digested in the intestinal tract
Polyesters	Development	*Pseudomonas oleovorans* produces optically active polyesters under nitrogen-limited conditions. Useful as a chemical feedstock, particularly for specialty plastics.

From I. W. Sutherland and D. C. Ellwood, *Society for General Microbiology Symposium*, 29, p. 128, table 6. Published by Cambridge University Press (1979); data included from R. A. Kent, et al., *Food Technology*, June 1991.

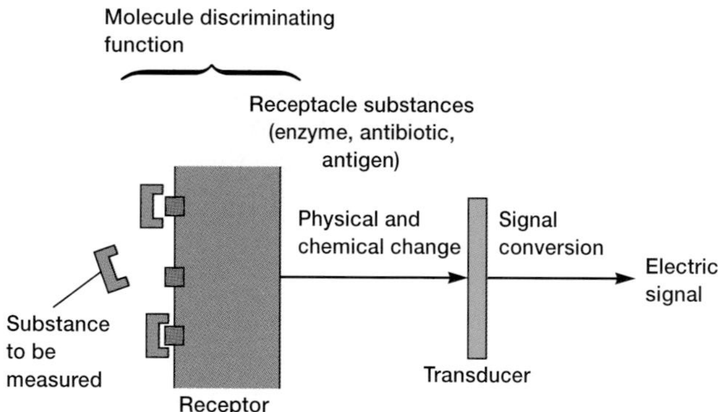

Figure 44.17 Biosensor Design. Biosensors are finding increasing applications in medicine, industrial microbiology, and environmental monitoring. In a biosensor, a biomolecule or whole microorganism carries out a biological reaction, and the reaction products are used to produce an electrical signal.

TABLE 44.16 Biosensors: Potential Biomedical, Industrial, and Environmental Applications

Clinical diagnosis and biomedical monitoring
Agricultural, horticultural, veterinary analysis
Detection of pollution, and microbial contamination of water
Fermentation analysis and control
Monitoring of industrial gases and liquids
Measurement of toxic gas in mining industries
Direct biological measurement of flavors, essences, and pheromones

One of the most interesting applications of this technology is to combine bacterial magnetite (*see page 813 in chapter 40*) with a chromophoric system to measure glucose through a color change (figure 44.18). Rapid advances are being made in biosensor technology to improve the stability of these probes and allow real-time monitoring of glucose and other metabolites, a critical need in modern medicine.

1. What two major bacteria have been used as insecticides?
2. Why might microbial biopolymers be more useful than similar polymers derived from plants and other natural sources? Give some examples of their uses.
3. What are biosensors and how do they detect substances?

Biodeterioration Management

Biodeterioration is a general term referring to the undesired microbial-mediated destruction of paper, paint, metals, textiles, concrete, and other materials. Particularly in humid climates, the effects of microbial growth on stored crops and foods have been estimated to cost over $50 billion per year. Several other examples emphasize the extent of this problem.

Jet Fuels

Some of the earliest difficulties with microbially produced changes in jet fuels resulted from microbial growth in the bottom of storage facilities and in airplane fuel tanks. Trace amounts of water accumulated in the fuel and created a water-hydrocarbon interface where microorganisms grew. Microor-

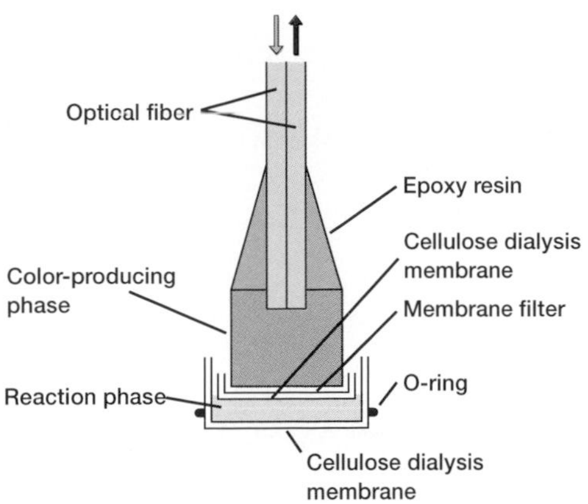

Figure 44.18 Optic Fiber Biosensors. Biosensors can be linked to fiber-optic systems for rapid, real-time detection of metabolites. In this type of system, which can be used to monitor a patient's blood glucose level, glucose oxidase is immobilized onto bacterial magnetite (reaction phase). Levels of glucose (in the range 0.56–22 mM) are linearly proportional to the quantity of H_2O_2 produced, which moves through the inner membrane, and alters the color of a dye, *o*-dianisidine. *Source: Tib Tech, March 1991.*

ganisms in the fuel clogged critical pumps and orifices; the most common resident, *Cladosporium resinae,* has even been called the kerosene fungus. A series of microbial inhibitors have been used to control growth at jet fuel/water interfaces. These inhibitors include organoborons and isothiazolone compounds. A methylchloro/methyl-isothiazolone mixture prevents growth of the fungus at 1 ppm. The problem has also been combated by cleaning storage and aircraft fuel tanks more frequently and using filters in fuel delivery and transfer lines of the aircraft.

Paper

Microbial degradation of stored softwoods used in paper manufacture can cause substantial economic losses. Microbial growth results in a decrease of paper strength, discoloration, and subsequent increased deterioration of the final paper product. Papermaking involves solubilization of lignin and hemicellulose by chemical processes, with the release of microbially utilizable wood sugars. These soluble waste products are removed from the residual cellulose by extensive washing. Microorganisms that grow in the wash solutions produce slimes and surface growths that lower process efficiency and final paper yield and quality. In the past, mercury-containing organic and inorganic compounds were commonly used as **biocides** in pulp and paper manufacturing plants. The long-term effects of these mercury products on lakes and rivers are still being felt in many timber-producing areas of the world (*see chapter 40*). Today, other compounds are used as biocides, including chlorine, phenols, and organosulfide compounds. Extensive washing with hot alkali is often used to maintain papermaking equipment in a slime-free condition.

Computer Chips

Microbial contamination is also a major concern in the computer chip and electronics industries. It can decrease transistor service life by impairing the adhesion of successive coatings on chips during their manufacture. Microbial growth at transistor junctions creates serious problems in critical electronic components. Thus, ultrapure water, free of dissolved organic matter and microorganisms, is a necessity in modern electronics manufacturing. Because of the ability of microorganisms to grow while using low levels of organic compounds leached from plastic tubing or absorbed from the air, even nominally "ultra-pure" water may support microbial growth. Despite careful and repeated membrane filtration, the presence of minicells can lead to microbial contaminants even in water that has been passed through 0.22 μm membrane filters.

Paints

A major challenge in the management of biodeterioration is the control of microbial growth in paint. Microorganisms can grow in the components of paint before paint formulation, in the container, and following application. It is especially necessary to use chemical inhibitors with modern water-soluble latex-based paints. Mercury compounds, primarily phenylmercuric acetate, were added to paint for many years until manufacturers and purchasers became aware of the environmental problems arising from the accumulation of mercury in the environment. The use of such inhibitor-containing paints produces a long-term, low level of volatile mercury in the atmosphere of painted rooms, possibly leading to mercury poisoning of the occupants. Mercury-containing inhibitors have been replaced by other compounds, including quaternary ammonium compounds, barium metaborate, and various chlorinated phenolic products.

Paints used on the hulls of marine vessels can contain high levels of organotin and copper compounds as biocides. These compounds accumulate in harbor muds, changing the metal resistance characteristics of the microorganisms in such environments.

Textiles and Leather

Degradation of cloth and leather materials, especially in tropical or humid environments, is of continuing economic concern. Fungal inhibitors, including a variety of phenolic-based compounds, may be added to natural materials such as cotton and wool. The use of copper compounds in many types of tanning processes leads to a greater resistance of leather products to microbiological degradation.

Metals

The microbial-mediated corrosion of metals is particularly critical where iron pipes are used in waterlogged anaerobic environments or in secondary petroleum recovery processes carried out at older oil fields. In these older fields, water is pumped down a series of wells to force residual petroleum to a central collection point. If the water contains low levels of organic matter and sulfate, anaerobic microbial communities can develop in rust blebs or tubercles (figure 44.19), resulting in punctured iron pipe and loss of critical pumping pressure. Microorganisms that use elemental iron as an electron donor during the reduction of CO_2 in methanogenesis have recently been discovered (Box 44.2). Because of the wide range of in-

BOX 44.2

Methanogens: A New Role for An Ancient Microbial Group

The methanogens, an important group of the archaeobacteria that can produce methane, are considered to be at least 3.5 billion years old. Despite intensive research, new discoveries are still being made concerning these microorganisms. Methanogenic bacteria have now been found to contribute to the anaerobic corrosion of soft iron. Previously, the microbial group usually considered the major "culprit" in the anaerobic corrosion process was the genus *Desulfovibrio,* which can use sulfate as an oxidant and hydrogen produced in the corrosion process as a reductant. Methanogens can use elemental iron as an electron source in their metabolism. It appears that corrosion may occur even without the presence of sulfate, which is required for functioning of *Desulfovibrio.* Rates of iron removal by the methanogens are around 79 mg/1,000 cm^2 of surface area in a 24-hour period. This may not seem a high rate, but in relation to the planned service life of metal structures in muds and subsurface soils—possibly years and decades—such corrosion can become a major problem. Continuous efforts to improve protection of iron structures will be required in view of the diversity of iron-corroding bacteria.

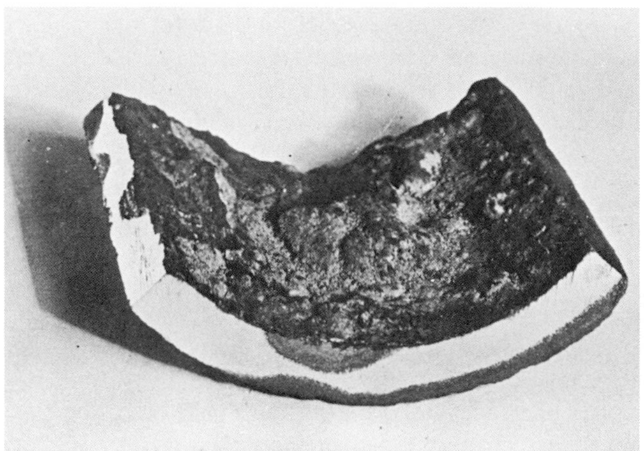
(a)

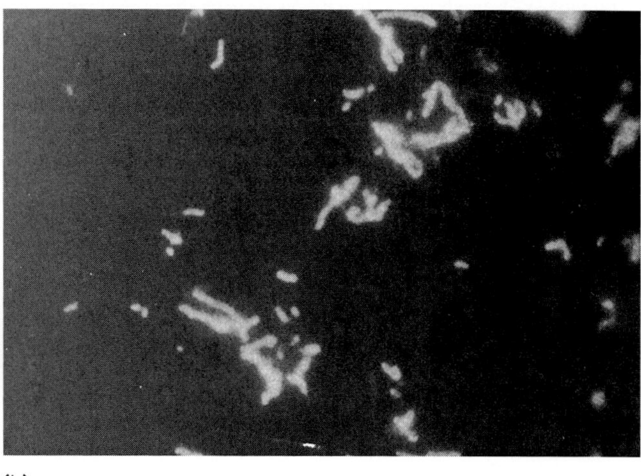
(b)

Figure 44.19 **Microbial-mediated Metal Corrosion.** The microbiological corrosion of iron is a major problem. (*a*) The graphitization of iron under a rust bleb on the pipe surface allows microorganisms, including *Desulfovibrio,* to corrode the inner surface. (*b*) Evidence points to the importance of communities of microorganisms, as opposed to individual species acting alone, as a major factor in microbiologically influenced corrosion. This epifluorescence microscope view (×1,600) is of pipeline steel a few hours after colonization by sulfate-reducing and organic acid-producing bacteria such as species of *Enterobacter* and *Clostridium.*

teractions between microorganisms and metals, the need to develop strategies to deal with corrosion problems is critical.

Concrete

Concrete can be dissolved by thiobacilli (*see chapter 23*). One member of the genus, formerly known as *Thiobacillus concretivorous,* now called *Thiobacillus thiooxidans,* causes major structural damage to highways and concrete sewage lines. In the presence of sufficient moisture and pyritic sulfur compounds of lower oxidation state than sulfate, this microorganism produces sulfuric acid, resulting in the dissolution of concrete. Thiobacilli can be controlled by the selection of cements that do not contain appreciable levels of oxidizable sulfur compounds, the avoidance of contamination of concrete with soil microorganisms, and the use of organic inhibitors.

Biodegradation Enhancement

Because valuable materials such as concrete structures and water and sewage pipelines are subject to biodeterioration, microbial degradation processes must often be inhibited. Under other conditions, microbial activities can be stimulated to restore or maintain environmental quality in **biodegradation enhancement.**

Stimulation of Oil Spill Degradation

To reduce the damage caused by oil spills, the rate of petroleum degradation must be increased. The development of a mixture of suitable microorganisms to degrade the wide variety of aromatic, polycyclic, and aliphatic compounds in most petroleums has been a major goal of researchers.

TABLE 44.17 Requirements for Successful Biostimulation of Hydrocarbon Degradation in a Marine Environment: Interactions of Microbial and Environmental Factors

System Component	Requirement
Microorganism	Ability to degrade hydrocarbons
	Ability to be grown and stored prior to use in a biostimulation process
	Ability to survive and function in a marine environment with its low temperature (5° C or lower) and high salinity
	Ability to assist in bioemulsification
	Possible ability to fix nitrogen
	Ability to grow from low inoculum levels
	No secondary toxic effects
Environment	Accessory nutrients (N, P, K, Fe) available and in same physical location as microorganisms and hydrocarbon substrate
	Absence of toxic chemicals (phenolics, heavy metals, etc.) that can harm hydrocarbon-degrading microorganisms
	Adequate O_2 levels and transfer rates

The first patented bacterial life form—patented by A. M. Chakrabarty in 1974—was a *Pseudomonas* species containing hydrocarbon degradation plasmids. This bacterium is much better able to degrade mixed petroleum residues. The successful functioning of such a microorganism, if it is added to a habitat, depends on several factors (table 44.17). When working with dispersed hydrocarbons in the ocean, contact between the microorganism, the hydrocarbon substrate, and other essential nutrients must be maintained. To achieve this, pellets containing nutrients and an oleophilic (hydrocarbon soluble) preparation have been used. Such a nutritional supplement not only aids inoculated organisms, but also stimulates the natural hydrocarbon-degrading microbial population. This technique has accelerate the degradation of different crude oil slicks by 30 to 40%, in comparison with control oil slicks where the additional nutrients were not available.

A unique challenge for this technology was the *Exxon Valdez* oil spill, which occurred in Alaska in March 1989. Several different approaches were used to increase biodegradation. These included nutrient additions, chemical dispersants, surfactant treatments, and the use of high-pressure steam. High-temperature steam caused a major disruption in the microbial community, which has still not recovered. The use of a microbially produced glycolipid emulsifier has proven helpful, as shown in the laboratory test illustrated in figure 44.20.

Biodegradation can also be enhanced in surface soils and sediments. In one of largest biodegradation enhancement projects carried out, 300,000 yd^3 of hydrocarbon-contaminated soil were treated in California, and a 94% reduction in hydrocarbons was achieved. A mixed population of pigmented microorganisms developed and contributed substantially to this successful project.

Figure 44.20 Microbial Products Can Aid in the Cleanup of Oil Spills. A surfactant glycolipid emulsifier (EM) from *P. aeruginosa* improves removal of crude oil from a solid surface (the interior of a beaker) in a laboratory experiment.

Figure 44.21 PCB Biodegradation. The environment can be managed to promote critical microbial bioremediation reactions. Hudson River (NY) muds are being aerated in cylindrical caissons to stimulate aerobic bacteria to complete the degradation of PCB molecules. Under previously anaerobic conditions, the PCB molecules had been partially dechlorinated.

A unique two-stage degradation process is being used experimentally to degrade PCBs in sediments. The muds in the Hudson River area can carry out a partial dechlorination under anaerobic conditions. Following this first step, the muds are aerated to completely degrade the less-chlorinated anaerobically produced PCP residues (figure 44.21).

Plasmids (pp. 261–64).

Subsurface Biodegradation Enhancement

The same principles can be used to stimulate the degradation of hydrocarbons and other chemical residues in contaminated subsurface environments. The major difference is that geological structures have limited permeability. Although subsurface regions in a pristine state often have O_2 concentrations approaching saturation, the penetration of small amounts of organic matter into these structures can quickly lead to O_2 depletion.

A major gasoline leak occurred in the central northeast coastal region of the United States following a break in a petroleum pipeline in the early 1970s. In this particular subsurface biodegradation enhancement effort, 58 tons of ammonium sulfate and 29 tons of monosodium and disodium phosphates were added to the subsurface and compressed air was pumped into the subsurface zone. An estimated 1,000 barrels of gasoline were degraded in this bioreclamation program. This technology is now being used in a wide range of surface—land farming—and subsurface environments, with and without the use of specific microbial inocula.

An interesting aspect of using microbial inocula in fresh water, marine, and soil environments is the potential attractiveness of laboratory-grown cultures as a food source. The possibility of their more rapid use as a preferred nutrient by protozoa may limit the effectiveness of such inocula in biodegradation enhancement.

Bioleaching of Metals

Another application of biodegradation enhancement is the recovery of metals from ores and mining tailings with metal levels too low for smelting (figure 44.22). Bioleaching, by means of natural populations of *Thiobacillus ferrooxidans* and related thiobacilli, for example, allows recovery of up to 70% of the copper in low-grade ores. Bioleaching, shown in figure 44.23, involves the biological oxidation of copper present in these ores to produce soluble copper sulfate. The copper sulfate can then be recovered by reacting the leaching solution, which contains up to 3.0 g/liter of soluble copper, with iron. The copper sulfate reacts with the elemental iron to form ferrosulfate, and the copper is reduced to the elemental form, which precipitates out in a settling trench. The process is summarized in the following reaction:

$$CuSO_4 + Fe° \rightarrow Cu° + FeSO_4$$

Figure 44.22 Bioleaching of Metals in Mining. Microorganisms can be used to recover metals from low-grade ores and tailings. In this aerial photograph of copper leach dumps at Santa Rita, New Mexico, the presence of iron in the ferrous form (green) and ferric form (orange) creates a mosaic pattern. About 10% of United States copper comes from leaching ore through the activity of bacteria such as *Thiobacillus* and *Leptospirillum*.

Bioleaching may require added phosphorus and nitrogen if these are limiting in the ore materials, and the same process can be used to solubilize uranium.

1. What microbial genus can grow in jet fuels? What problems does this cause and how can these be prevented?
2. What types of inhibitors have been used in the past to control microbial growth in paints?
3. Briefly describe the problems microorganisms pose for paper, computer chips, textiles, leather, metals, and concrete. How can biodegradation be avoided in each case?
4. What factors must one consider when attempting to stimulate the microbial degradation of a massive oil spill in a marine environment?
5. What is the major microbial genus that contributes to bioleaching processes? How, in general terms, does the process work?

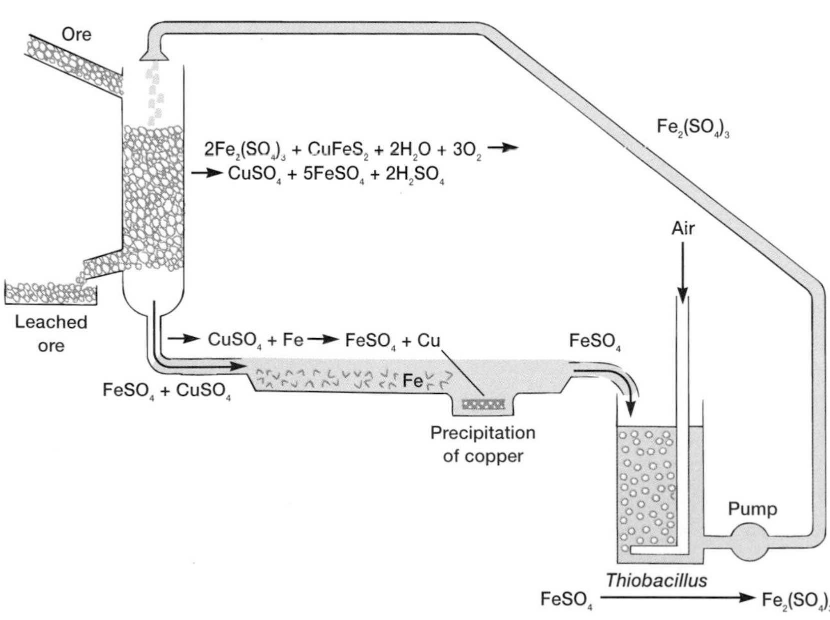

Figure 44.23 Copper Leaching from Low-grade Ores. The chemistry and microbiology of copper ore leaching involve interesting complementary reactions. The microbial contribution is the oxidation of ferrous ion (2+) to ferric ion (3+). *Thiobacillus ferrooxidans* and related thiobacilli are very active in this oxidation. The ferric ion then reacts chemically to solubilize the copper. The soluble copper is recovered by a chemical reaction with elemental iron, which results in an elemental copper precipitate.

Summary

1. Fermentation is a word with different meanings, ranging from a description of a microbial mass culture—aerobic or anaerobic—to rigorous physiological definitions.
2. Microorganisms can be grown in a variety of mass culture systems. Especially with filamentous microorganisms (actinomycetes, molds), it is often necessary to provide high agitation and aeration rates because of the occurrence of viscous plasticlike growth.
3. Fermentation media are usually made from crude materials, such as molasses, corn-steep liquor, and soybean hydrolysates, and are often designed to create specific limitations on microbial growth processes.
4. The mass culture of microorganisms involves transfers of oxygen, heat, and nutrients, requiring the establishment of driving forces or gradients. This is especially critical for O_2.
5. Scaleup is the critical process of carrying out a fermentation in progressively larger reactor vessels.
6. Besides selection, mutation, and screening processes, which have been used for decades in industrial microbiology, modern biotechnology uses molecular techniques and a wide range of vectors to create new DNA constructs for DNA expression.
7. Microbial products can be classified as primary or secondary metabolites, depending on whether or not they are produced during growth. Most antibiotics are secondary metabolites.
8. Most amino acids are produced by carefully selected microbial cultures that have regulatory mutations. This type of impairment, together with careful nutrient stresses, can increase the levels of amino acid excretion.
9. Bioconversions, carried out with nongrowing vegetative microorganisms or free enzymes, can cause minor but important modifications of complex molecules. Cells used in bioconversions can be immobilized.
10. Through recombinant DNA technology, gene expression and genes themselves can be modified. Genes also can be expressed in different microorganisms to improve process efficiency, yield, and purity.
11. Microbial applications of continuing interest include microbial insecticides, biopolymers, and biosensors.
12. Biosensors promise to have expanding applications in industrial microbiology, medicine, and environmental monitoring.
13. Biodeterioration management is a matter of continuing concern. Under special conditions, biocides can be used to control microbial damage to certain materials.
14. Biodegradation enhancement involves the addition of nutrients and microorganisms increase degradation rates. Bioleaching, a related process, can be used to solubilize metals from low-grade ores.

Key Terms

biochemical markers *899*
biocide *907*
biocontrol *903*
bioconversion *898*
biodegradation enhancement *908*
biodeterioration *906*
biopolymer *905*
biosensor *905*
biotechnology *888*
biotransformation *898*
continuous feed *891*
fermentation *888*
idiophase *895*
immobilization *899*
lyophilization *893*
metabolic engineering *901*
microbial transformation *898*
non-Newtonian broth *891*
primary metabolite *895*
protein engineering *901*
protoplast fusion *892*
regulatory mutant *897*
scaleup *891*
secondary metabolite *895*
semisynthetic penicillin *897*
trophophase *895*

Questions for Thought and Review

1. Although continuous culture has many advantages in terms of cell production rates, it is not widely used in industrial processes except for sewage treatment. What are some of the possible reasons for this?
2. Why has scaleup been such a problem in industrial microbiology? Consider mixing, aeration, and cooling processes in a 10 ml culture versus a 100,000 liter fermenter.
3. Most commercial antibiotics are produced by actinomycetes, and only a few are synthesized by fungi and other bacteria. From physiological and environmental viewpoints, how might you attempt to explain this observation?
4. We hear much about the beneficial uses of recombinant DNA technology. What are some of the problems and disadvantages that should be considered when using microorganisms for these applications?
5. As the president of a biotechnology company, you must decide whether a new process for production of a widely used organic acid should be kept secret or whether a patent should be obtained. What factors should you consider in making this decision?
6. Your company has developed a new biocide for use in a paper manufacturing plant. Before this new compound can be released for sale, what types of environmental questions should you ask and what tests should you complete?
7. What are some of the possible advantages of biosensors as opposed to more traditional physical and chemical measurement procedures?

Additional Reading

Aharonowitz, Y., and Cohen, G. 1981. The microbiological production of pharmaceuticals. *Sci. Am.* 245(9):140–52.

American Association for the Advancement of Science. 1991. Frontiers in biotechnology. *Science.* 252:1585–1756.

Andrykovitch, G., and Neihof, R. A. 1987. Fuel-soluble biocides for control of *Cladosporium resinae* in hydrocarbon fuels. *J. Indust. Microbiol.* 2:35–40.

Atkinson, B., and Mavituna, F. 1991. *Biochemical engineering and biotechnology handbook.* New York: Stockton Press.

Biotechnology. 1983. *Science* 219:611–746.

Braun, S., and Vecht-Lifshitz, S. E. 1991. Mycelial morphology and metabolite production. *Trends Biotechnol.* 9:63–68.

Brierley, C. L. 1982. Microbiological mining. *Sci. Am.* 247(2):44–53.

Buckholz, R. G., and Gleeson, M.A.G. 1991. Yeast systems for the commercial production of heterologous proteins. *Bio/Technology.* 9:1067–72.

Collins, C. H., and Beale, A. J. 1992. *Safety in industrial microbiology and biotechnology.* Stoneham, MA: Butterworth-Heinemann.

Crueger, W., and Crueger, A. 1990. *Biotechnology: A textbook of industrial microbiology.* 2d ed., ed. T. D. Brock. Sunderland, Mass.: Sinauer Associates.

Deacon, J. W. 1983. *Microbial control of plant pests and diseases.* Washington, D.C.: American Society for Microbiology.

Dee, N.; McTernan, W. F.; and Kaplan, E. 1987. *Detection, control, and renovation of contaminated groundwater.* New York: American Society of Civil Engineers.

Demain, A. L., and Solomon, N. A. 1986. Industrial microbiology. *Sci. Am.* 245(3):66–75.

Demain, A. L., and Solomon, N. A. 1986. *Manual of industrial microbiology and biotechnology.* Washington, D.C.: American Society for Microbiology.

Eveleigh, D. E. 1981. The microbiological production of industrial chemicals. *Sci. Am.* 245(3):154–78.

Feitelson, J. S.; Payne, J.; and Kim, L. 1992. *Bacillus thuringiensis:* Insects and beyond. *Bio/Technology* 10:271–75.

Ford, T., and Mitchell, R. 1990. The ecology of microbial corrosion. *Adv. Microb. Ecol.* 11:231–62.

Fox, J. L. 1992. Contemplating large-scale use of engineered microbes. *ASM News.* 58(4):191–96.

Harris, T.J.R. 1990. *Protein production by biotechnology.* New York: Elsevier.

Hershberger, C. L.; Queener, S. W.; and Hegeman, G. 1989. *Genetics and molecular biology of industrial microorganisms.* Washington, D.C.: American Society for Microbiology.

Hollenberg, C. P., and Sahm, H., eds. 1988. *Biosensors and environmental biotechnology.* New York: VCH Publishers

Holloway, M. 1991. Soiled Shores. *Sci. Am.* 265(4):102–16.

Hopwood, D. A. 1981. The genetic programming of industrial microorganisms. *Sci. Am.* 245(3):90–102.

Hughes, J., and Qoronfleh, M. W. 1991. Perspectives in plant genetic engineering and biopharmacy. *BioPharm.* 4:18–26.

Hutchins, S. R.; Davidson, M. S.; Brierley, J. A.; and Brierley, C. L. 1986. Microorganisms in reclamation of metals. *Ann. Rev. Microbiol.* 40:311–36.

Ikeda, M., and Kasumata, R. 1992. Metabolic engineering to produce tyrosine or phenylalanine in a tryptophan-producing *Corynebacterium glutamicum* strain. *Appl. Environ. Microbiol.* 58:781–85.

Knight, P. 1991. Baculovirus vectors for making proteins in insect cells. *ASM News.* 57:567–70.

Maramorosch, K. 1991. *Biotechnology for biological control of pests and vectors.* Boca Raton, Fla.: CRC Press.

Matsunaga, T. 1991. Applications of bacterial magnets. *Trends Biotechnol.* 9:91–95.

May, S. W. 1992. Biocatalysis in the 1990s: A perspective. *Enzyme Microb. Technol.* 14:80–84.

Phaff, H. J. 1981. Industrial microorganisms. *Sci. Am.* 245(3):76–89.

Primrose, S. B. 1991. *Molecular Biotechnology,* 2d ed. Oxford: Blackwell Scientific Publishers.

Rose, A. H. 1982. *Microbial biodeterioration. Economic microbiology 6.* New York: Academic Press.

Roy-Sole, M. 1990. Microbes that eat toxins. *Can. Geogr.* June/July:64–69.

Sayler, G. S.; Fox, R.; and Blackburn, J. W. 1991. *Environmental biotechnology for waste treatment.* New York: Plenum Press.

Schonborn, W., ed. 1986. *Microbial degradation. Vol. 8, Biotechnology—A comprehensive treatise.* Deerfield Beach, Fla.: Verlag Chemie International.

Schultz, J. S. 1991. Biosensors. *Sci. Am.* 265(2):64–69.

Smith, J. E. 1985. *Biotechnology principles.* Washington, D.C.: American Society for Microbiology.

Trevan, M. D., et al. 1987. *Biotechnology. The biological principles.* Milton Keynes, UK: Open University Press.

Wagner, C. R., and Benkovic, S. J. 1990. Site directed mutagenesis: A tool for enzyme mechanism dissection. *Trends Biotechnol.* 8:263–70.

Wise, D. N. 1989. *Applied biosensors.* Stoneham, Mass.: Butterworths.

Wood, H. A., and Granados, R. R. 1991. Genetically engineered baculoviruses as agents for pest control. *Ann. Rev. Microbiol.* 45:69–87.

Appendix I
A Review of the Chemistry of Biological Molecules

Appendix I provides a brief summary of the chemistry of organic molecules with particular emphasis on the molecules present in microbial cells. Only basic concepts and terminology are presented; introductory textbooks in biology and chemistry should be consulted for a more extensive treatment of these topics.

Atoms and Molecules

Matter is made of elements that are composed of atoms. An element contains only one kind of atom and cannot be broken down to simpler components by chemical reactions. An atom is the smallest unit characteristic of an element and can exist alone or in combination with other atoms. When atoms combine they form molecules. Molecules are the smallest particles of a substance. They have all the properties of the substance and are composed of two or more atoms.

Although atoms contain many subatomic particles, three directly influence their chemical behavior—protons, neutrons, and electrons. The atom's nucleus is located at its center and contains varying numbers of protons and neutrons (figure AI.1). Protons have a positive charge, and neutrons are uncharged. The mass of these particles and the atoms that they compose is given in terms of the atomic mass unit (AMU), which is 1/12 the mass of the most abundant carbon isotope. Often the term dalton (d) is used to express the mass of molecules. It is also 1/12 the mass of an atom of ^{12}C or 1.661×10^{-24} grams. Both protons and neutrons have a mass of about one dalton. The atomic weight is the actual measured weight of an element and is almost identical to the mass number for the element, the total number of protons and neutrons in its nucleus. The mass number is indicated by a superscripted number preceding the element's symbol (e.g., ^{12}C, ^{16}O, and ^{14}N).

Negatively charged particles called electrons circle the atomic nucleus (figure AI.1). The number of electrons in a neutral atom equals the number of its protons and is given by the atomic number, the number of protons in an atomic nucleus. The atomic number is characteristic of a particular type of atom. For example, carbon has an atomic number of six, hydrogen's number is one, and oxygen's is eight (table AI.1).

The electrons move constantly within a volume of space surrounding the nucleus, even though their precise location in this volume cannot be determined accurately. This volume of space in which an electron is located is called its orbital. Each orbital can contain two electrons. Orbitals are grouped into shells of different energy that surround the nucleus. The first shell is closest to the nucleus and has the lowest energy; it contains only one orbital. The second shell contains four orbitals, one circular and three shaped like dumbbells (figure AI.2*a*). It can contain up to eight electrons. The third shell has even higher energy and holds more than eight electrons. Shells are filled beginning with the innermost and moving outward. For example, carbon has six electrons, two in its first shell and four in the second (figures AI.1 and AI.2*b*). The electrons in the outermost shell are the ones that participate in chemical reactions. The most stable condition is achieved when the outer shell is filled with electrons. Thus the number of bonds an element can form depends upon the number of electrons required to fill the outer shell. Since carbon has four electrons in its outer shell and the shell is filled when it contains eight electrons, it can form four covalent bonds (table AI.1).

Chemical Bonds

Molecules are formed when two or more atoms associate through chemical bonding. Chemical bonds are attractive forces that hold together atoms, ions, or groups of atoms in a molecule or other substance. Many types of chemical bonds are present in organic molecules; three of the most important are covalent bonds, ionic bonds, and hydrogen bonds.

In covalent bonds, atoms are joined together by sharing pairs of electrons (figure AI.3). If the electrons are equally shared between identical atoms (e.g., in a carbon-carbon bond), the covalent bond is strong and nonpolar. When two different atoms such as carbon and oxygen share electrons, the covalent bond formed is polar since the electrons are pulled toward the more electronegative atom, the atom that more strongly attracts electrons. A single pair of electrons is shared in a single bond; a double bond is formed when two pairs of electrons are shared.

Atoms often contain either more or less electrons than the number of protons in their nuclei. When this is the case, they carry a net neg-

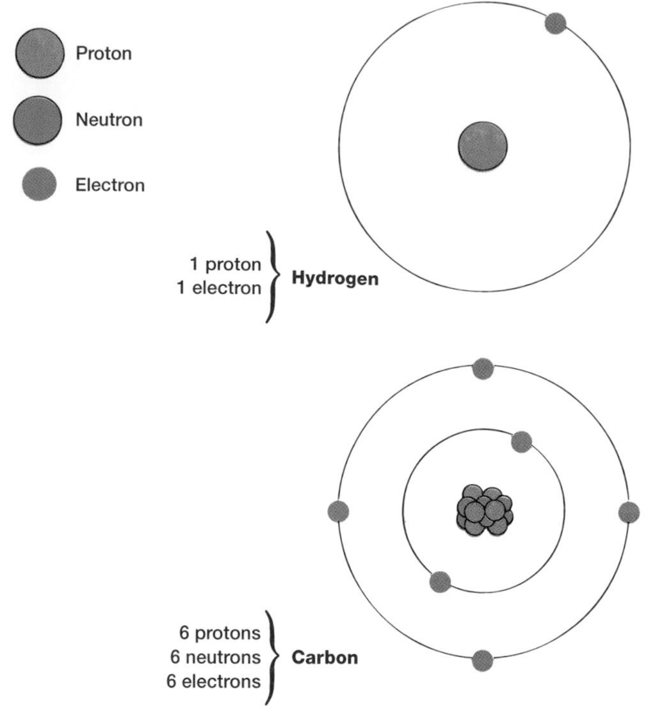

Figure AI.1 **Diagrams of Hydrogen and Carbon Atoms.** The electron orbitals are represented as concentric circles.

TABLE AI.1 Atoms Commonly Present in Organic Molecules

Atom	Symbol	Atomic Number	Atomic Weight	Number of Chemical Bonds
Hydrogen	H	1	1.01	1
Carbon	C	6	12.01	4
Nitrogen	N	7	14.01	3
Oxygen	O	8	16.00	2
Phosphorus	P	15	30.97	5
Sulfur	S	16	32.06	2

ative or positive charge and are called ions. Cations carry positive charges and anions have a net negative charge. When a cation and an anion approach each other, they are attracted by their opposite charges. This ionic attraction that holds two groups together is called an ionic bond. Ionic bonds are much weaker than covalent bonds and are easily disrupted by a polar solvent such as water. For example, the Na^+ cation is strongly attracted to the Cl^- anion in a sodium chloride crystal, but sodium chloride dissociates into separate ions (ionizes) when dissolved in water. Ionic bonds are important in the structure and function of proteins and other biological molecules.

When a hydrogen atom is covalently bonded to a more electronegative atom such as oxygen or nitrogen, the electrons are unequally shared and the hydrogen atom carries a partial positive charge. It will be attracted to an electronegative atom such as oxygen or nitrogen, which carries an unshared pair of electrons; this attraction is called a hydrogen bond (figure AI.4). Although an individual hydrogen bond is weak, there are so many hydrogen bonds in proteins and nucleic acids that they play a major role in determining protein and nucleic acid structure.

Organic Molecules

Most molecules in cells are organic molecules, molecules that contain carbon. Since carbon has four electrons in its outer shell, it tends to form four covalent bonds in order to fill its outer shell with eight electrons. This property makes it possible to form chains and rings of carbon atoms that can also bond with hydrogen and other atoms (figure AI.5). Although adjacent carbons are usually connected by

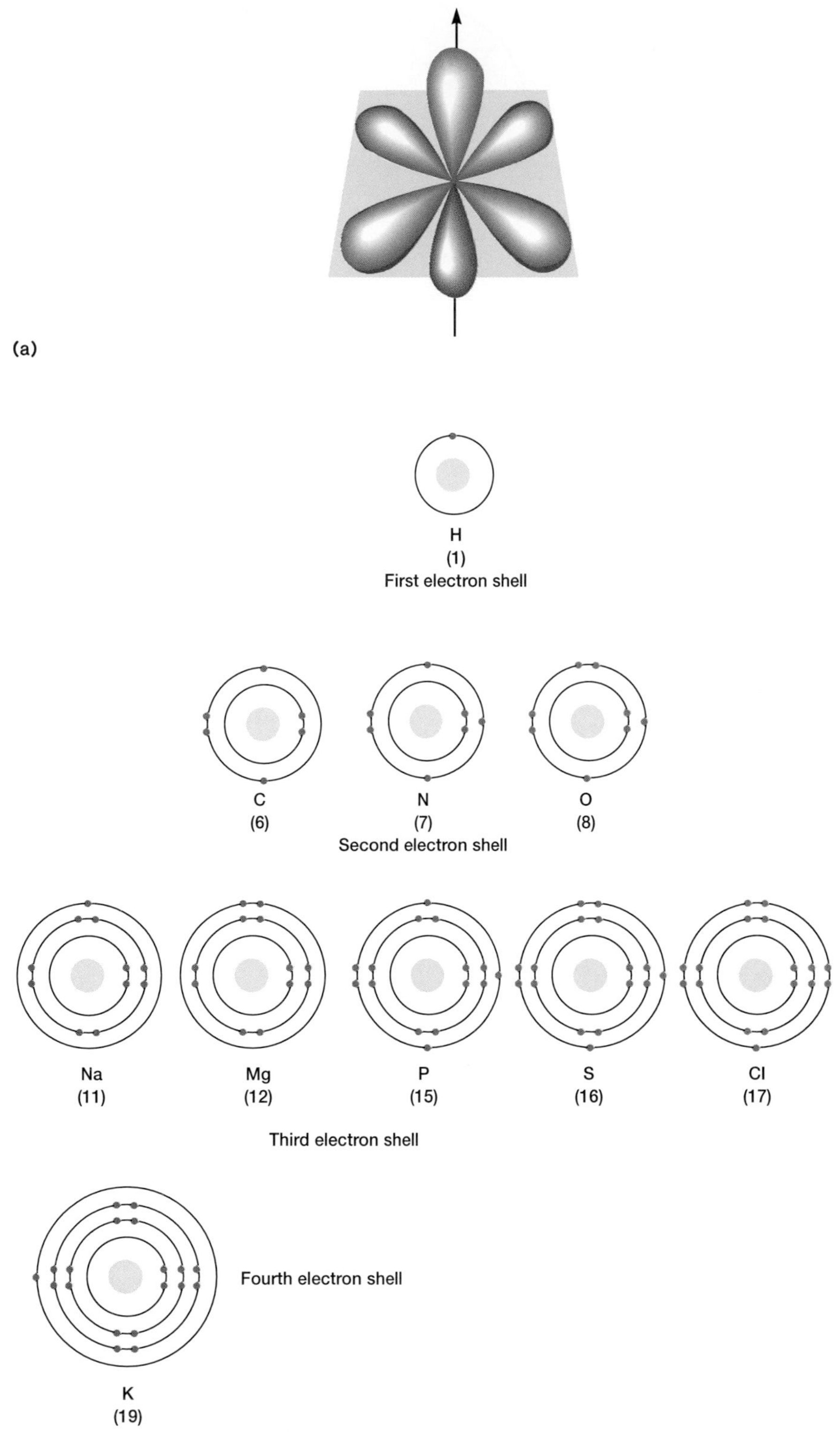

Figure AI.2 **Electron Orbitals.** (*a*) The three dumbbell-shaped orbitals of the second shell. The orbitals lie at right angles to each other. (*b*) The distribution of electrons in some common elements. Atomic numbers are given in parentheses.

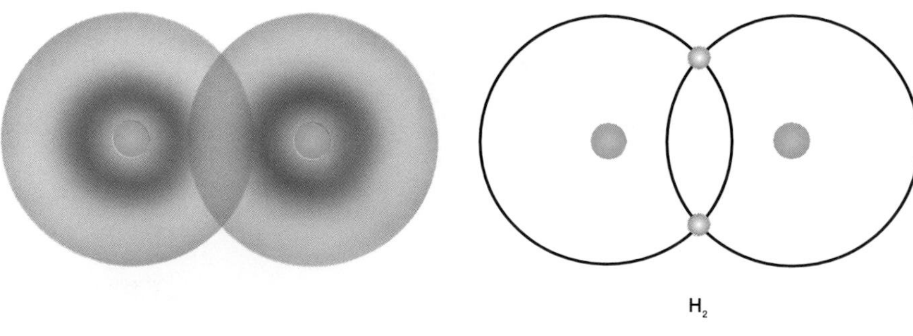

Figure AI.3 The Covalent Bond. A hydrogen molecule is formed when two hydrogen atoms share electrons.

$>C{=}O \cdots H{-}O{-}$

$\geq N \cdots H{-}O{-}C({=}O){-}$

$>C{=}O \cdots H{-}N<$

$-O(H) \cdots H{-}N<$

$\geq N \cdots H{-}N<$

Figure AI.4 Hydrogen Bonds. Representative examples of hydrogen bonds present in biological molecules.

(a) C_6H_{14} (Hexane)

(b) or C_6H_{12} (Cyclohexane)

(c) or C_6H_6 (Benzene)

Figure AI.5 Hydrocarbons. Examples of hydrocarbons that are (*a*) linear, (*b*) cyclic, and (*c*) aromatic.

single bonds, they may be joined by double or triple bonds. Rings that have alternating single and double bonds, like the benzene ring, are called aromatic rings. The hydrocarbon chain or ring provides a chemically inactive skeleton to which more reactive groups of atoms may be attached. These reactive groups with specific properties are known as functional groups. They usually contain atoms of oxygen, nitrogen, phosphorus, or sulfur (figure AI.6) and are largely responsible for most characteristic chemical properties of organic molecules.

Organic molecules are often divided into classes based on the nature of their functional groups. Ketones have a carbonyl group within the carbon chain, while alcohols have a hydroxyl on the chain. Organic acids have a carboxyl group, and amines have an amino group (figure AI.7).

Organic molecules may have the same chemical composition and yet differ in their molecular structure and properties. Such molecules are called isomers. One important class of isomers is the stereoisomers. Stereoisomers have the same atoms arranged in the same nucleus-to-nucleus sequence, but differ in the spatial arrangement of their atoms. For example, an amino acid such as alanine can form stereoisomers (figure AI.8). L-alanine and other L-amino acids are the stereoisomer forms normally present in proteins.

Carbohydrates

Carbohydrates are aldehyde or ketone derivatives of polyhydroxy alcohols. The smallest and least complex carbohydrates are the simple sugars or monosaccharides. The most common sugars have five or six carbons (figure AI.9). A sugar in its ring form has two isomeric structures, the α and β forms, which differ in the orientation of the hydroxyl on the aldehyde or ketone carbon, the anomeric or glycosidic carbon (figure AI.10). Microorganisms have many sugar derivatives in which a hydroxyl is replaced by an amino group or some other functional group (e.g., glucosamine).

Two monosaccharides can be joined by a bond between the anomeric carbon of one sugar and a hydroxyl or the anomeric carbon of the second (figure AI.11). The bond joining sugars is a glycosidic bond and may be either α or β depending upon the orientation of the anomeric carbon. Two sugars linked in this way constitute a disaccharide. Some common disaccharides are maltose (two glucose molecules), lactose (glucose and galactose), and sucrose (glucose and fructose). If ten or more sugars are linked together by glycosidic bonds, a polysaccharide is formed. For example, starch and glycogen are common polymers of glucose that are used as sources of carbon and energy (figure AI.12).

Figure AI.6 Functional Groups. Some common functional groups in organic molecules. The groups are shown in color.

Functional group	Name	Example	
$-OH$	Hydroxyl	$H-CH_2-CH_2-O-H$	Ethanol
$-C(=O)-$	Carbonyl	$H-CH_2-C(=O)-C(=O)-O-H$	Pyruvic acid
$-C(=O)-O-$	Ester	$H_2C-O-C(=O)-CH_2-(CH_2)_{15}-CH_3$ $HC-O-C(=O)-CH_2-(CH_2)_{15}-CH_3$ $H_2C-O-C(=O)-CH_2-(CH_2)_{15}-CH_3$	Tristearyl glycerol (a fat)
$-C(=O)-O-H$	Carboxyl	$H_2N-CH_2-C(=O)-O-H$	Glycine (an amino acid)
$-NH_2$	Amino	$H_2N-CH(CH_3)-C(=O)-O-H$	Alanine (an amino acid)
$-S-H$	Sulfhydryl	$H_2N-CH(CH_2-SH)-C(=O)-O-H$	Cysteine (an amino acid)

Type of molecule	Example
Alcohol	CH_3-CH_2-OH
Aldehyde	$CH_3-C(=O)-H$
Amine	$CH_3-CH_2-NH_2$
Ester	$CH_3-C(=O)-O-CH_2-CH_3$
Ether	$CH_3-CH_2-O-CH_2-CH_3$
Ketone	$CH_3-C(=O)-CH_3$
Organic acid	$CH_3-C(=O)-OH$

Figure AI.7 Types of Organic Molecules. These are classified on the basis of their functional groups.

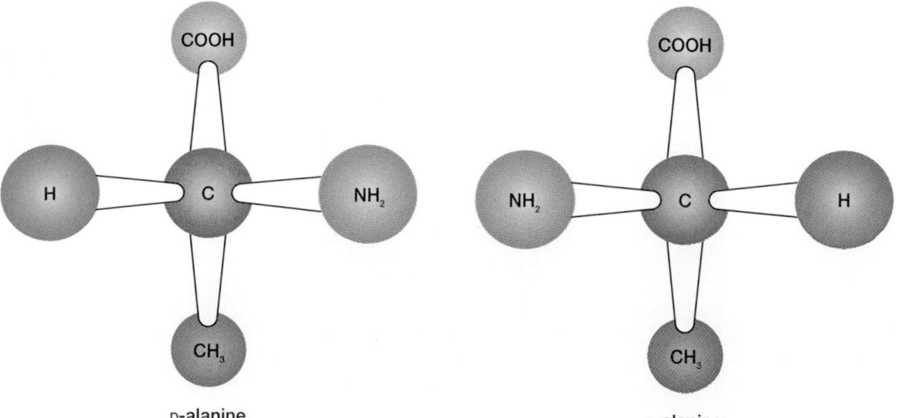

Figure AI.8 **The Stereoisomers of Alanine.** The α-carbon is in color, L-alanine is the form usually present in proteins.

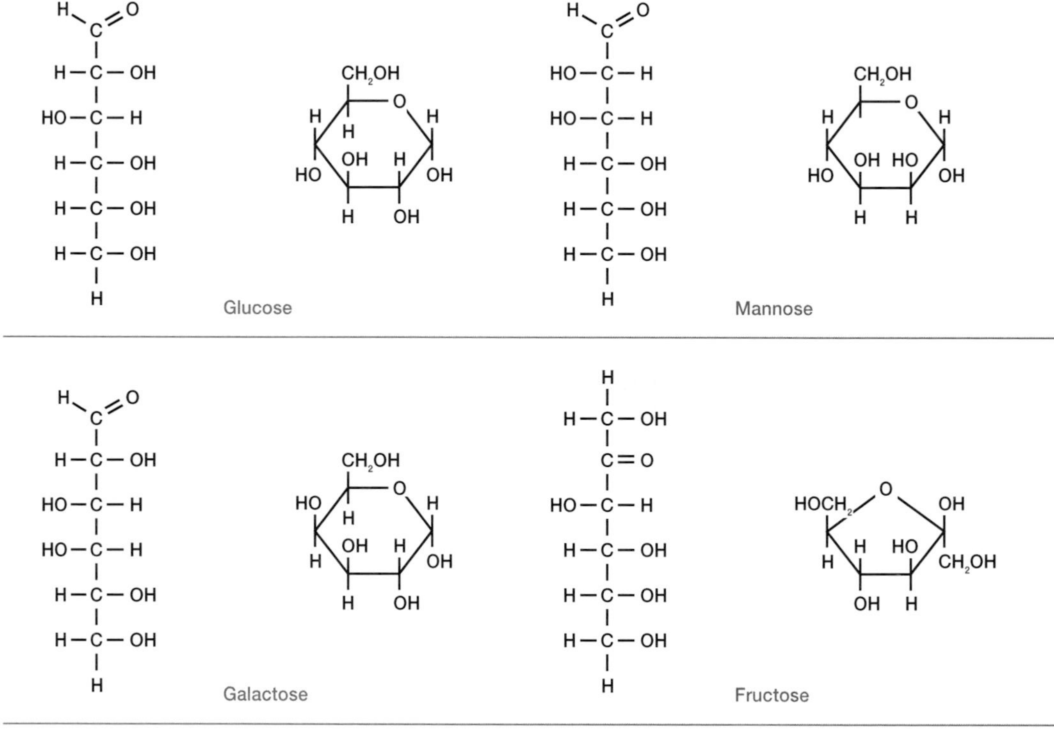

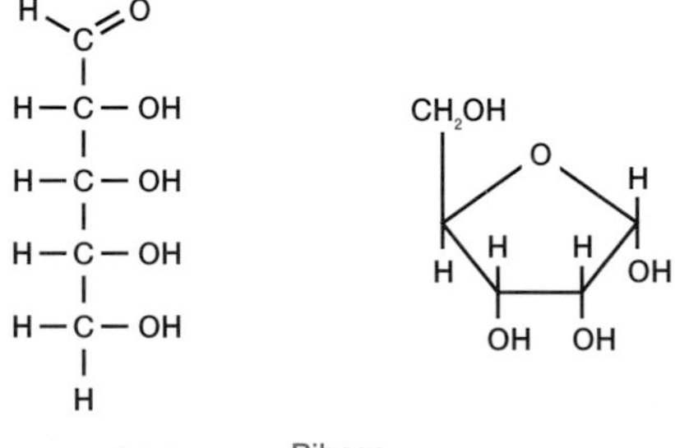

Figure AI.9 **Common Monosaccharides.** Structural formulas for both the open chains and the ring forms are provided.

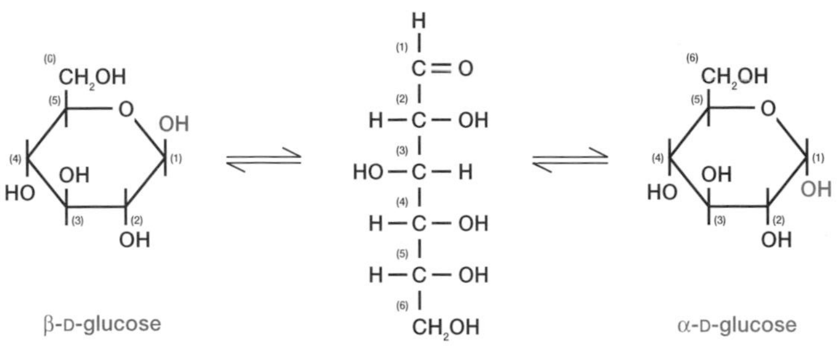

Figure AI.10 **The Interconversion of Monosaccharide Structures.** The open chain form of glucose and other sugars is in equilibrium with closed ring structures (depicted here with Haworth projections). Aldehyde sugars form cyclic hemiacetals, and keto sugars produce cyclic hemiketals. When the hydroxyl on carbon one of cyclic hemiacetals projects above the ring, the form is known as a β form. The α form has a hydroxyl that lies below the plane of the ring. The same convention is used in showing the α and β forms of hemiketals such as those formed by fructose.

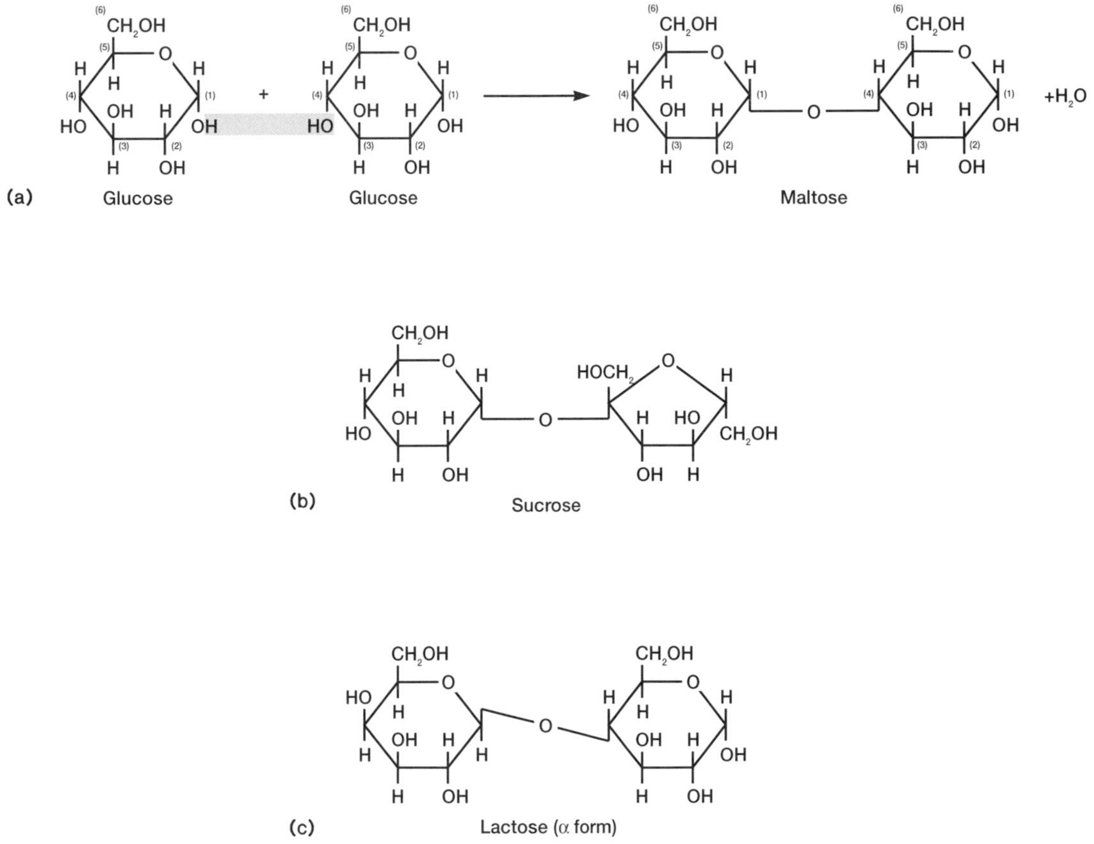

Figure AI.11 **Common Disaccharides.** (*a*) The formation of maltose from two molecules of a α-glucose. The bond connecting the glucose extends between carbons one and four, and involves the α form of the anomeric carbon. Therefore, it is called an α ($1 \rightarrow 4$) glycosidic bond. (*b*) Sucrose is composed of a glucose and a fructose joined to each other through their anomeric carbons, an $\alpha\beta$ ($1 \rightarrow 2$) bond. (*c*) The milk sugar lactose contains galactose and glucose joined by a β ($1 \rightarrow 4$) glycosidic bond.

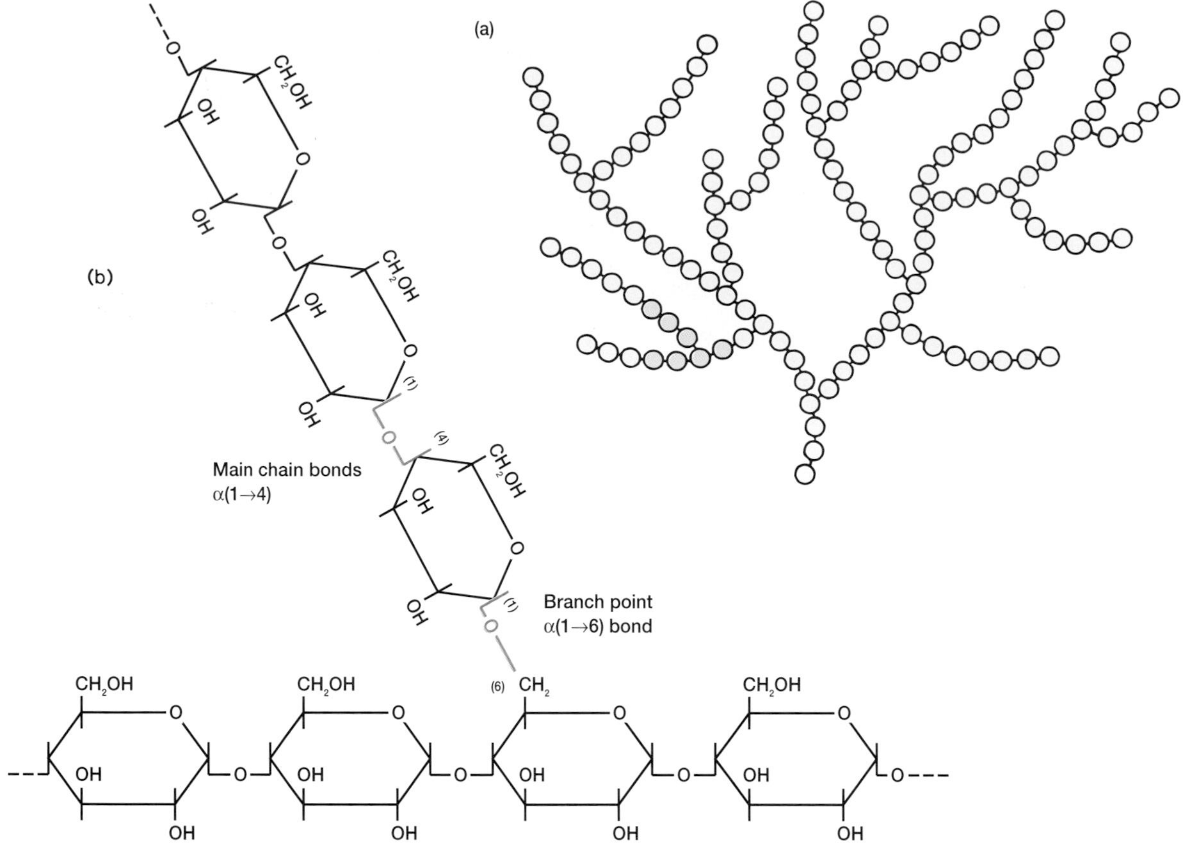

Figure AI.12 Glycogen and Starch Structure. (*a*) An overall view of the highly branched structure characteristic of glycogen and most starch. The circles represent glucose residues. (*b*) A close-up of a small part of the chain (shown in color in part *a*) revealing a branch point with its α (1 → 6) glycosidic bond.

Lipids

All cells contain a heterogeneous mixture of organic molecules that are relatively insoluble in water, but very soluble in nonpolar solvents such as chloroform, ether, and benzene. These molecules are called lipids. Lipids vary greatly in structure and include triacylglycerols, phospholipids, steroids, carotenoids, and many other types. Among other functions, they serve as membrane components, storage forms for carbon and energy, precursors of other cell constituents, and protective barriers against water loss.

Most lipids contain fatty acids, monocarboxylic acids that are often straight chained but may be branched. Saturated fatty acids lack double bonds in their carbon chains, while unsaturated fatty acids have double bonds. The most common fatty acids are 16 or 18 carbons long.

Two good examples of common lipids are triacylglycerols and phospholipids. Triacylglycerols are composed of glycerol esterified to three fatty acids (figure AI.13*a*). They are used to store carbon and energy. Phospholipids are lipids that contain at least one phosphate group and often have a nitrogenous constituent as well. Phosphatidyl ethanolamine is an important phospholipid frequently present in bacterial membranes (figure AI.13*b*). It is composed of two fatty acids esterified to glycerol. The third glycerol hydroxyl is joined with a phosphate group, and ethanolamine is attached to the phosphate. The resulting lipid is very asymmetric with a hydrophobic nonpolar end contributed by the fatty acids and a polar, hydrophilic end. In cell membranes, the hydrophobic end is buried in the interior of the membrane, while the polar-charged end is at the membrane surface and exposed to water.

$$\begin{array}{l} \qquad\qquad\quad O \\ \qquad\qquad\quad \| \\ CH_2-O-C-R \\ |\qquad\qquad\; O \\ |\qquad\qquad\; \| \\ CH-O-C-R \\ |\qquad\qquad\; O \\ |\qquad\qquad\; \| \\ CH_2-O-C-R \end{array}$$

(a)

$$\begin{array}{l} \qquad\qquad\quad O \\ \qquad\qquad\quad \| \\ CH_2-O-C-R \\ |\qquad\qquad\; O \\ |\qquad\qquad\; \| \\ CH_2-O-C-R \\ |\qquad\qquad\; O \\ |\qquad\qquad\; \| \\ CH_2-O-P-O-CH_2-CH_2-\overset{+}{N}H_3 \\ \qquad\qquad\quad | \\ \qquad\qquad\quad O^- \end{array}$$

(b)

Figure AI.13 **Examples of Common Lipids.** (*a*) A triacylglycerol or neutral fat. (*b*) The phospholipid phosphatidyl ethanolamine. The R groups represent fatty acid side chains.

Proteins

The basic building blocks of proteins are amino acids. An amino acid contains a carboxyl group and an amino group on its alpha carbon (figure AI.14). About 20 amino acids are normally found in proteins; they differ from each other with respect to their side chains (figure AI.15). In proteins, amino acids are linked together by peptide bonds between their carboxyls and α-amino groups to form linear polymers called polypeptides (figure AI.16). Each protein is composed of one or more polypeptide chains and has a molecular weight greater than about 6,000 to 7,000.

Proteins have three or four levels of structural organization and complexity. The primary structure of a protein is the sequence of the amino acids in its polypeptide chain or chains. The structure of the polypeptide chain backbone is also considered part of the primary structure. Each different polypeptide has its own amino acid sequence that is a reflection of the nucleotide sequence in the gene that codes for its synthesis. The polypeptide chain can coil along one axis in space into various shapes like the α-helix (figure AI.17). This arrangement of the polypeptide in space around a single axis is called the secondary structure. Secondary structure is formed and stabilized by the interactions of amino acids that are fairly close to one another on the polypeptide chain. The polypeptide with its primary and secondary structure can be coiled or organized in space along three axes to form a more complex, three-dimensional shape (figure AI.18). This level of organization is the tertiary structure (figure AI.19). Amino acids more distant from one another on the polypeptide chain contribute to tertiary structure. Secondary and tertiary structures are

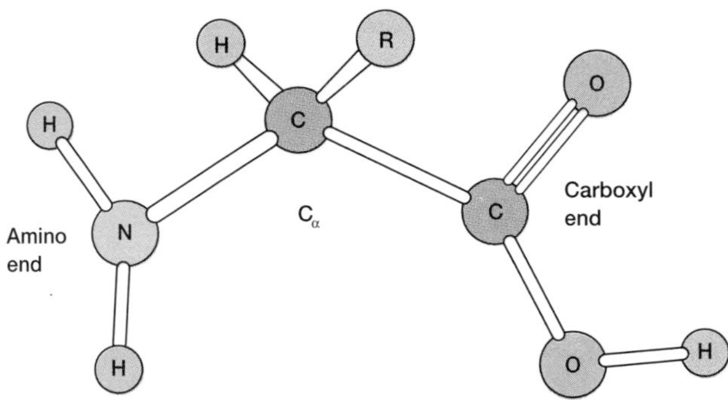

Figure AI.14 **L-Amino Acid Structure.** The uncharged form is shown.

examples of conformation, molecular shape that can be changed by bond rotation and without breaking covalent bonds. When a protein contains more than one polypeptide chain, each chain with its own primary, secondary, and tertiary structure associates with the other chains to form the final molecule. The way in which polypeptides associate with each other in space to form the final protein is called the protein's quaternary structure (figure AI.20).

The final conformation of a protein is ultimately determined by the amino acid sequence of its polypeptide chains. Under proper conditions, a completely unfolded polypeptide will fold into its normal final shape without assistance.

Protein secondary, tertiary, and quaternary structure is largely determined and stabilized by many weak noncovalent forces such as hydrogen bonds and ionic bonds. Because of this, protein shape is often very flexible and easily changed. This flexibility is very important in protein function and in the regulation of enzyme activity. Because of their flexibility, however, proteins readily lose their proper shape and activity when exposed to harsh conditions. The only covalent bond commonly involved in the secondary and tertiary structure of proteins is the disulfide bond. The disulfide bond is formed when two cysteines are linked through their sulfhydryl groups. Disulfide bonds generally strengthen or stabilize protein structure, but are not especially important in directly determining protein conformation.

Nucleic Acids

The nucleic acids, deoxyribonucleic acid (DNA) and ribonucleic acid (RNA), are polymers of deoxyribonucleosides and ribonucleosides joined by phosphate groups. The nucleosides in DNA contain the purines adenine and guanine, and the pyrimidine bases thymine and cytosine. In RNA, the pyrimidine uracil is substituted for thymine. Because of their importance for genetics and molecular biology, the chemistry of nucleic acids is introduced earlier in the text. The structure and synthesis of purines and pyrimidines are discussed in chapter 9 (*pp. 183–84*). The structure of DNA and RNA is described in chapter 10 (*pp. 192–95*).

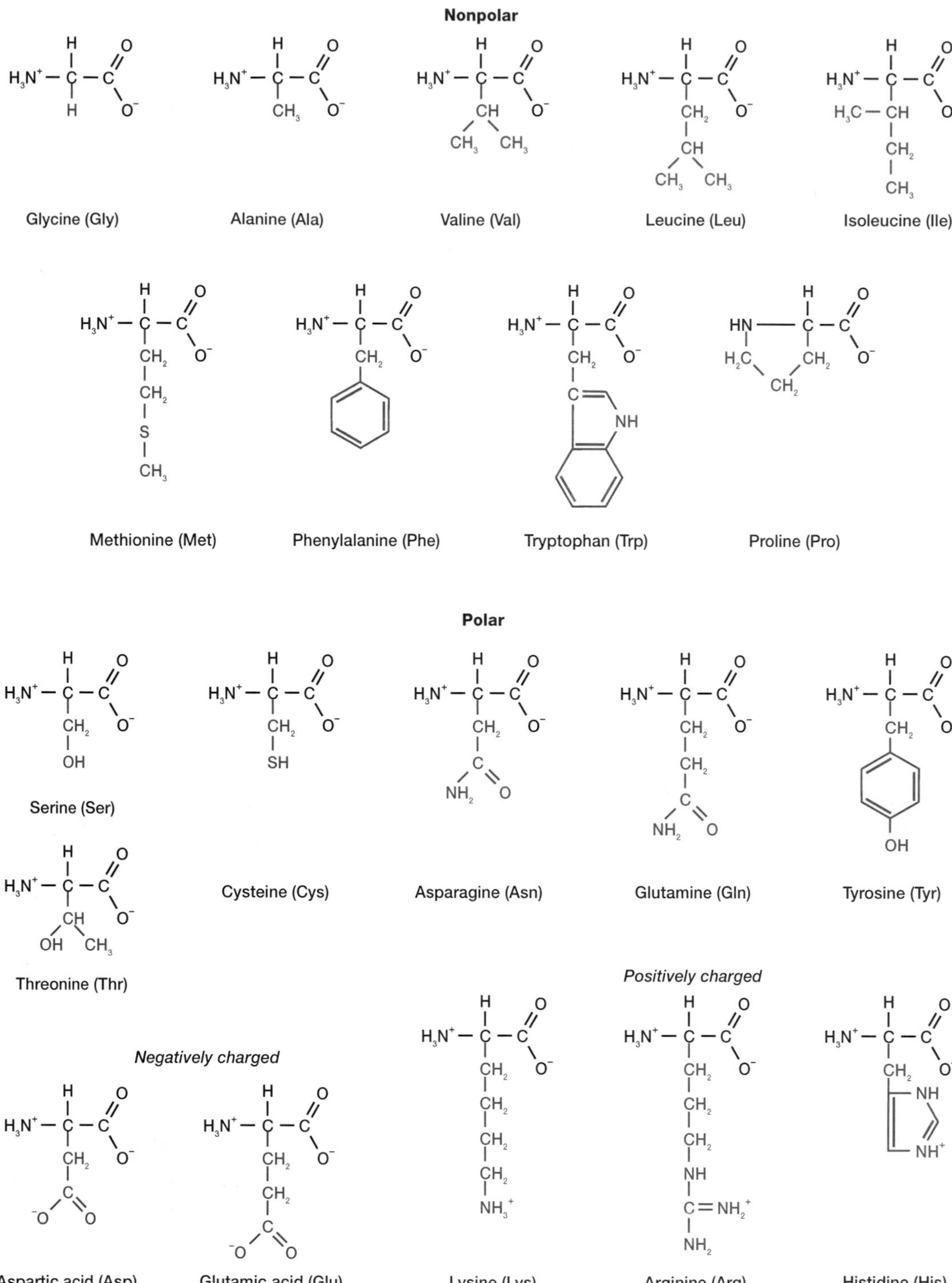

Figure AI.15 The Common Amino Acids. The structures of the α-amino acids normally found in proteins. Their side chains are shown in color, and they are grouped together based on the nature of their side chains—nonpolar, polar, negatively charged (acid), or positively charged (basic). Proline is actually an imino acid rather than an amino acid.

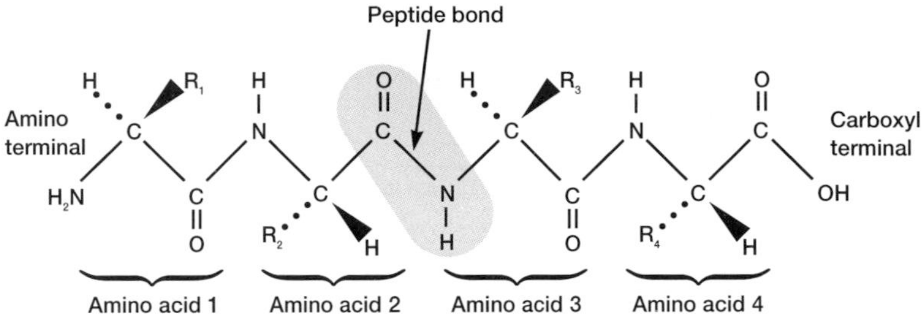

Figure AI.16 **A Tetrapeptide Chain.** The end of the chain with a free α-amino group is the amino or N terminal. The end with the free α-carboxyl is the carboxyl or C terminal. One peptide bond is shown is color.

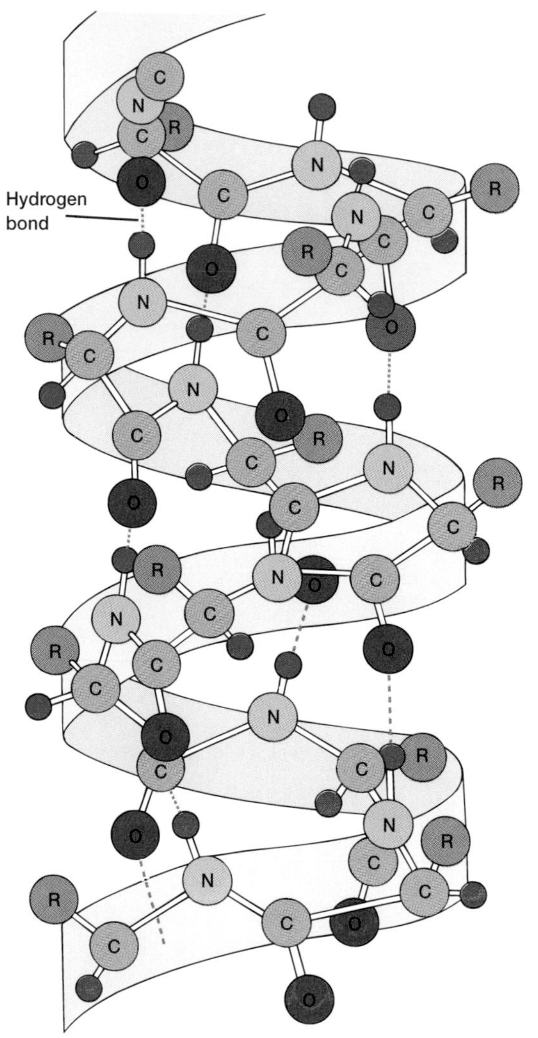

Figure AI.17 **The α-Helix.** A polypeptide twisted into one type of secondary structure, the α-helix. The helix is stabilized by hydrogen bonds joining peptide bonds that are separated by three amino acids.

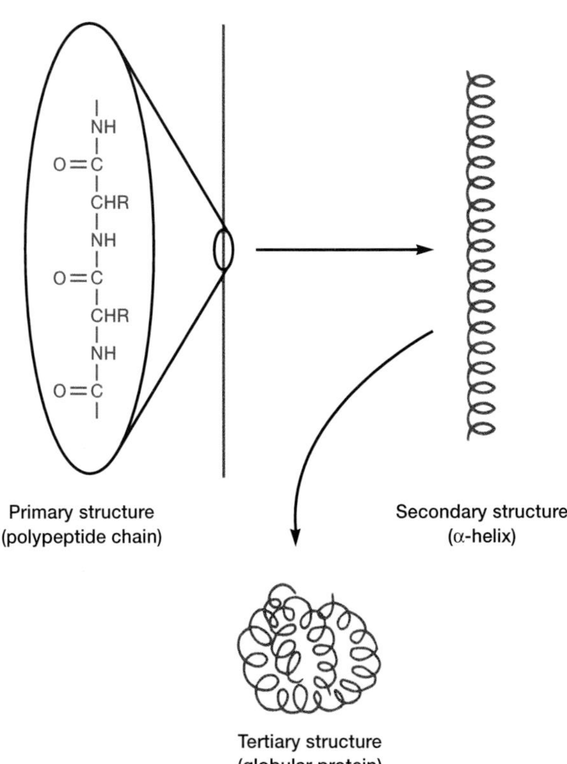

Figure AI.18 **Secondary and Tertiary Protein Structures.** The formation of secondary and tertiary protein structures by folding a polypeptide chain with its primary structure.

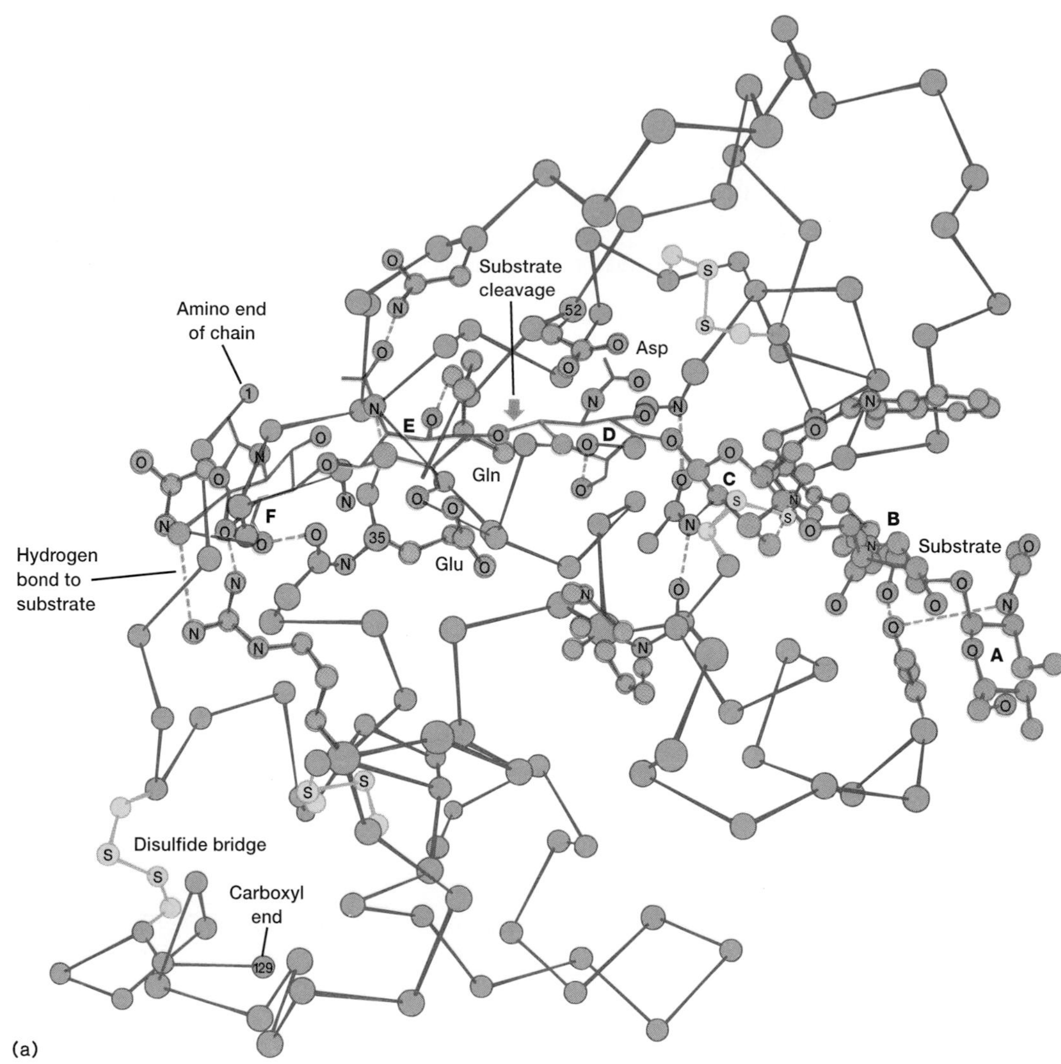

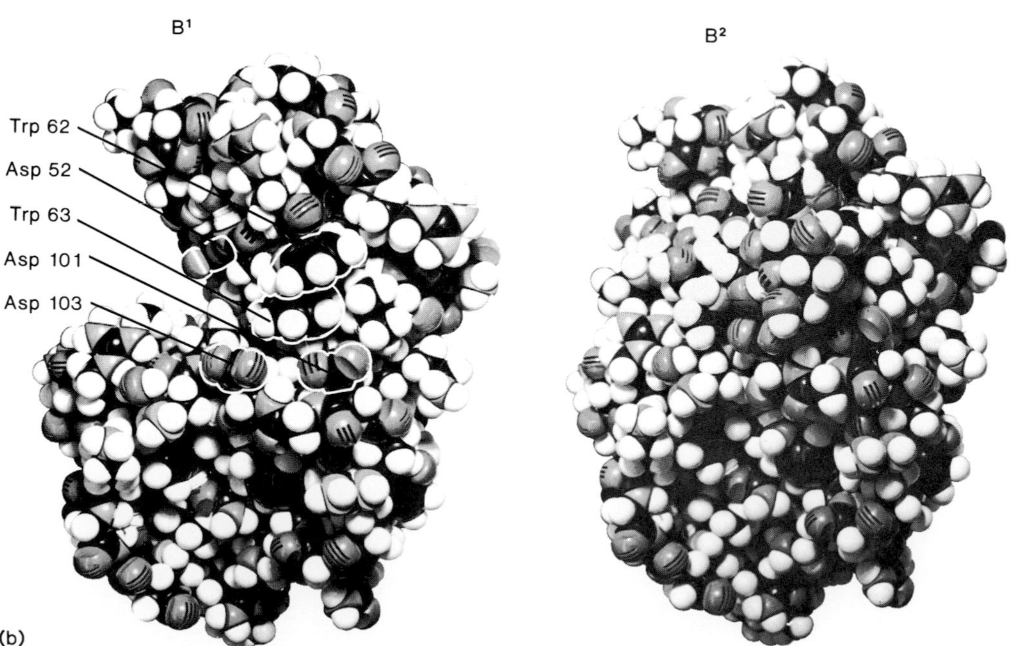

Figure AI.19 Lysozyme. The tertiary structure of the enzyme lysozyme. (*a*) A diagram of the protein's polypeptide backbone with the substrate hexasaccharide shown in color. The point of substrate cleavage is indicated. (*b*) A space-filling model of lysozyme. The left figure shows the empty active site with some of its more important amino acids indicated. On the right, the enzyme has bound its substrate.

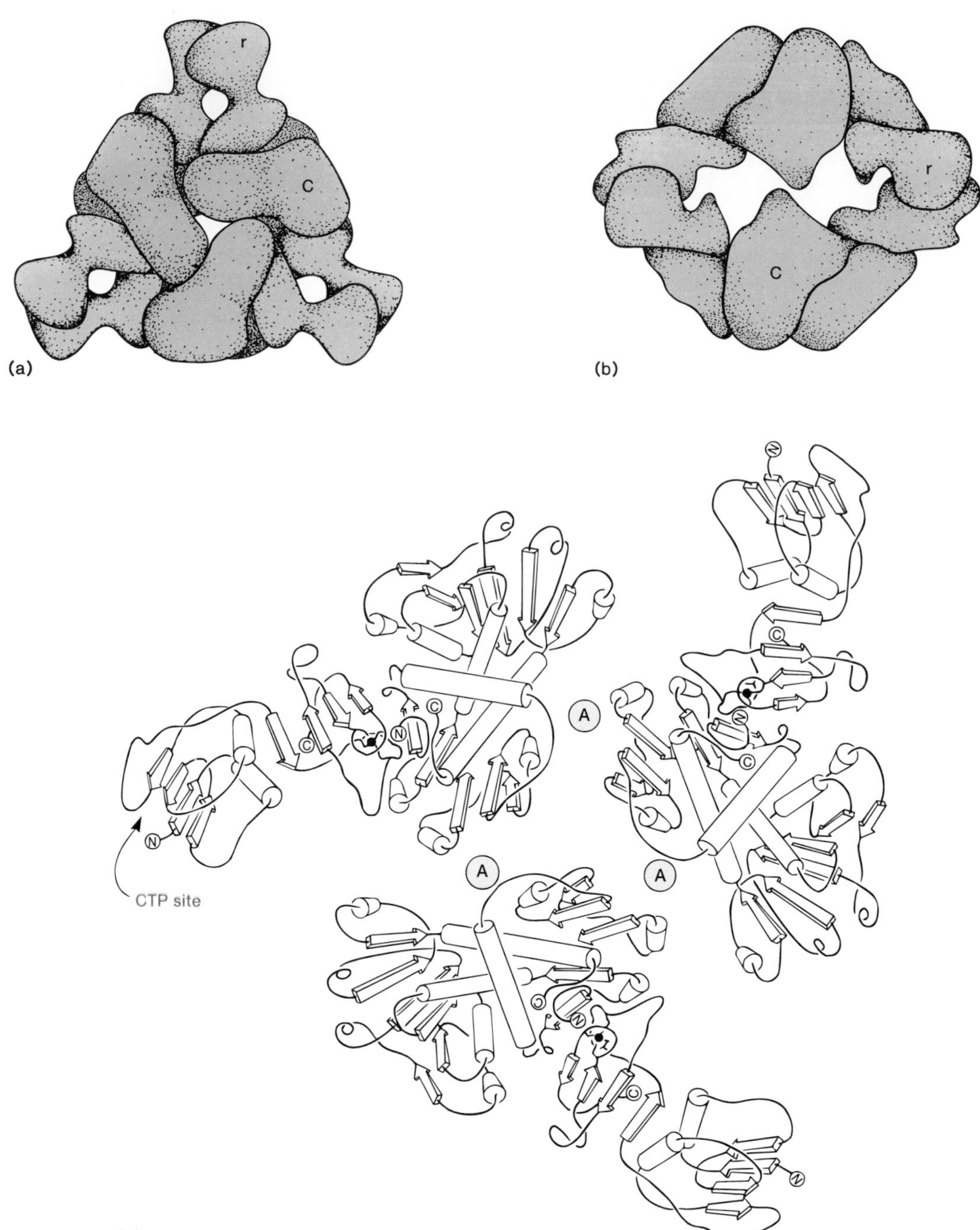

Figure AI.20 **An Example of Quaternary Structure.** The enzyme aspartate carbamoyltransferase from *E. coli* has two types of subunits, catalytic and regulatory. The association between the two types of subunits is shown: (*a*) a top view, and (*b*) a side view of the enzyme. The catalytic (*C*) and regulatory (*r*) subunits are shown in different colors. (*c*) The peptide chains shown when viewed from the top as in (*a*). The active sites of the enzyme are located at the positions indicated by *A*. *(See pp. 218–19 for more details.) From* Biochemistry, *3/E by Lubert Stryer. Copyright © 1975, 1981, 1988. Reprinted with permission of W. H. Freeman and Company.* (a *and* b. *After Krause et al. in* Proceedings of the National Academy of Sciences, *Vol. 82, 1985.* c. *After Kantrowitz et al. in* Trends in Biochemical Science, *Vol. 5, 1980.)*

APPENDIX II
Common Metabolic Pathways

This appendix contains a few of the more important pathways discussed in the text, particularly those involved in carbohydrate catabolism. Enzyme names and final end products are given in color. Consult the text for a description of each pathway and its roles.

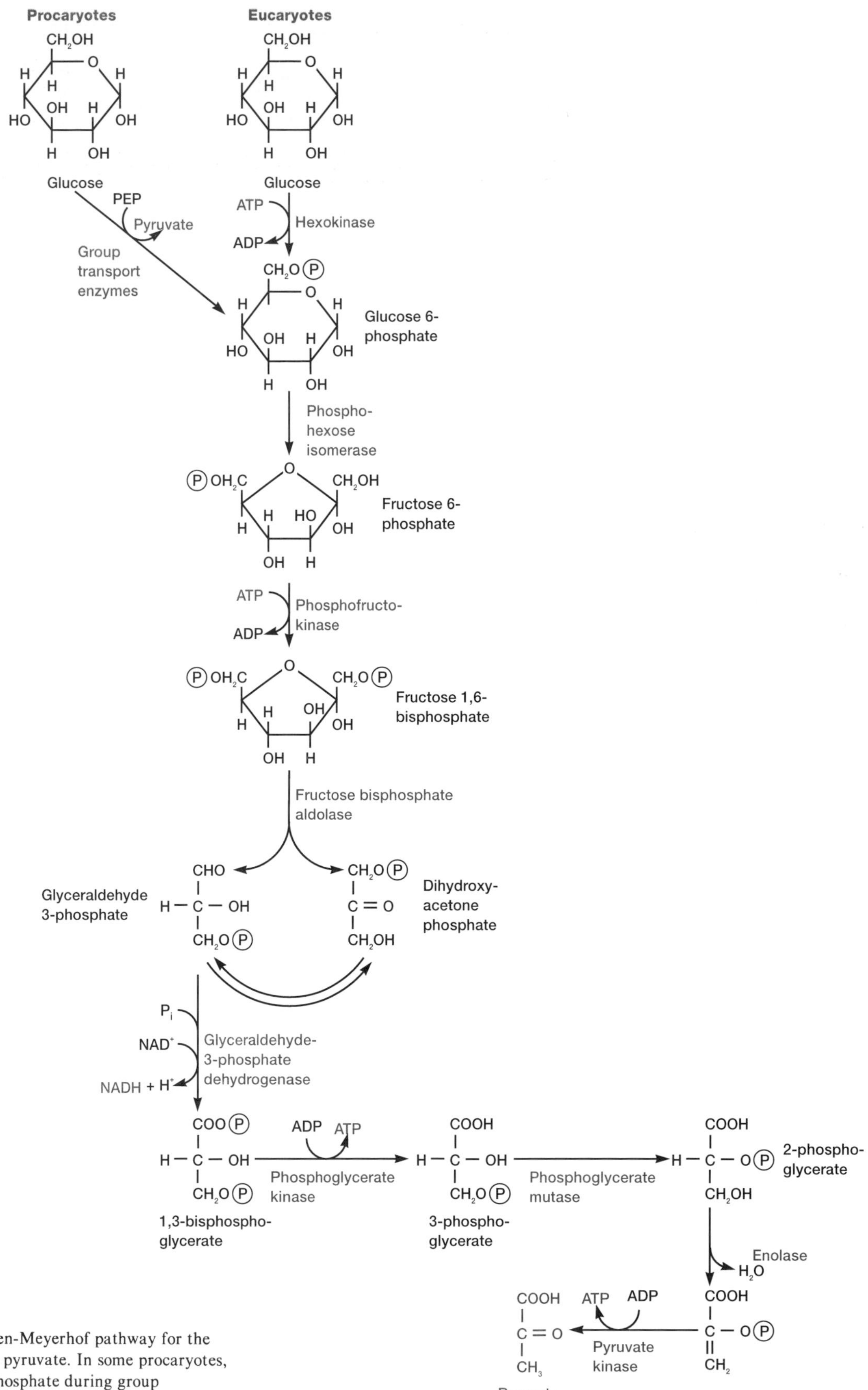

Figure AII.1 **Glycolysis.** The Embden-Meyerhof pathway for the conversion of glucose and other sugars to pyruvate. In some procaryotes, glucose is phosphorylated to glucose 6–phosphate during group translocation transport across the plasma membrane.

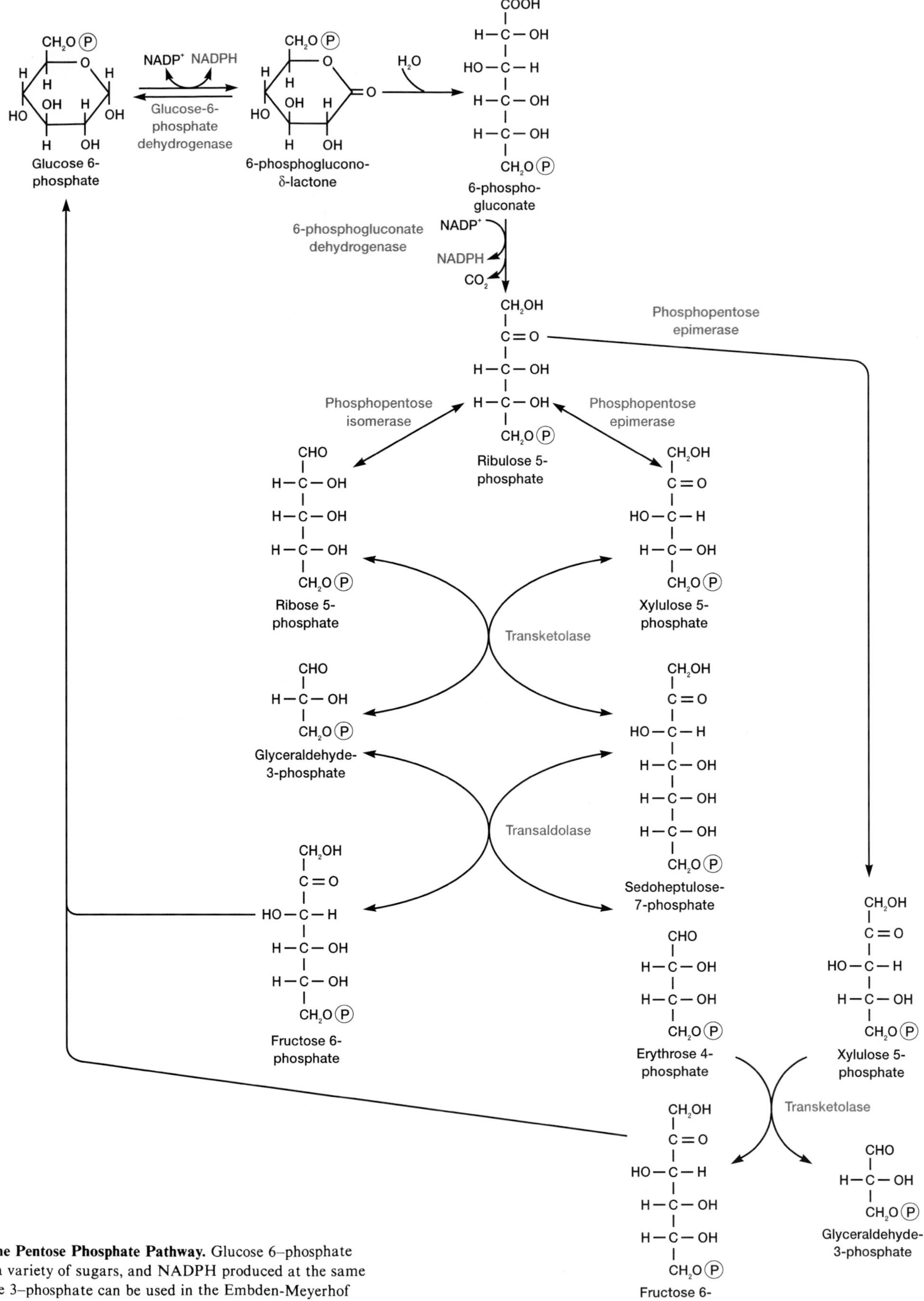

Figure AII.2 **The Pentose Phosphate Pathway.** Glucose 6–phosphate can be converted to a variety of sugars, and NADPH produced at the same time. Glyceraldehyde 3–phosphate can be used in the Embden-Meyerhof pathway or converted to more fructose 6–phosphate.

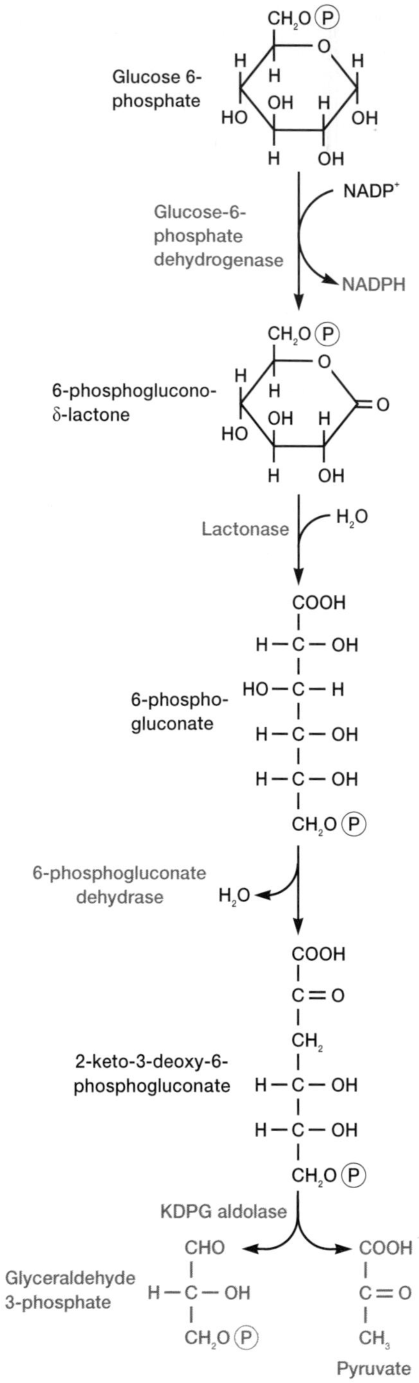

Glucose 6-phosphate

Glucose-6-phosphate dehydrogenase

$NADP^+$ → NADPH

6-phosphoglucono-δ-lactone

Lactonase

H_2O

6-phosphogluconate

COOH
H—C—OH
HO—C—H
H—C—OH
H—C—OH
CH_2Oⓟ

6-phosphogluconate dehydrase

H_2O

2-keto-3-deoxy-6-phosphogluconate

COOH
C=O
CH_2
H—C—OH
H—C—OH
CH_2Oⓟ

KDPG aldolase

Glyceraldehyde 3-phosphate

CHO
H—C—OH
CH_2Oⓟ

COOH
C=O
CH_3

Pyruvate

Figure AII.3 **The Entner-Doudoroff Pathway.**

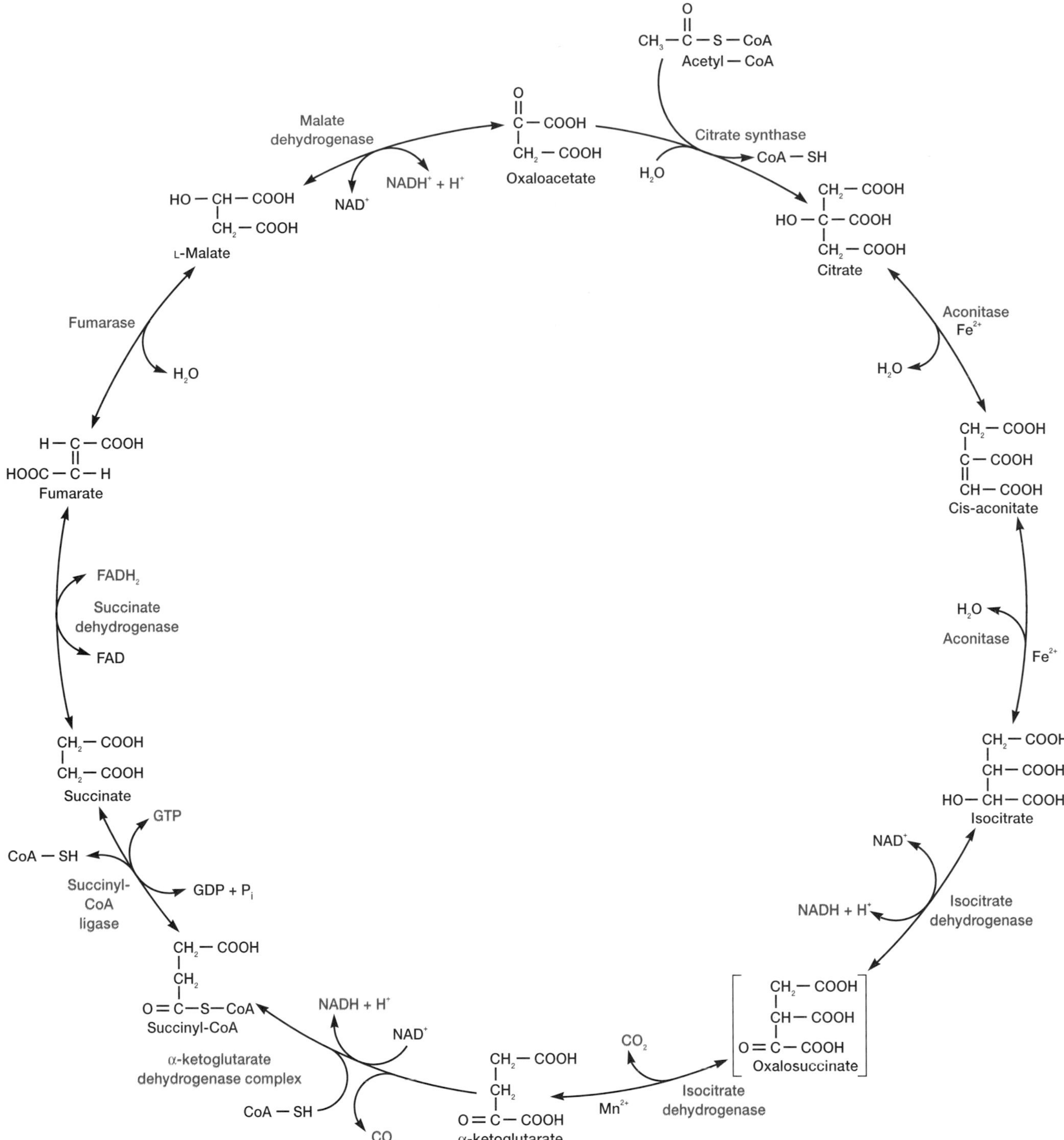

Figure AII.4 **The Tricarboxylic Acid Cycle.** Cis-aconitate and oxalosuccinate remain bound to aconitase and isocitrate dehydrogenase. Oxalosuccinate has been placed in brackets since it is so unstable.

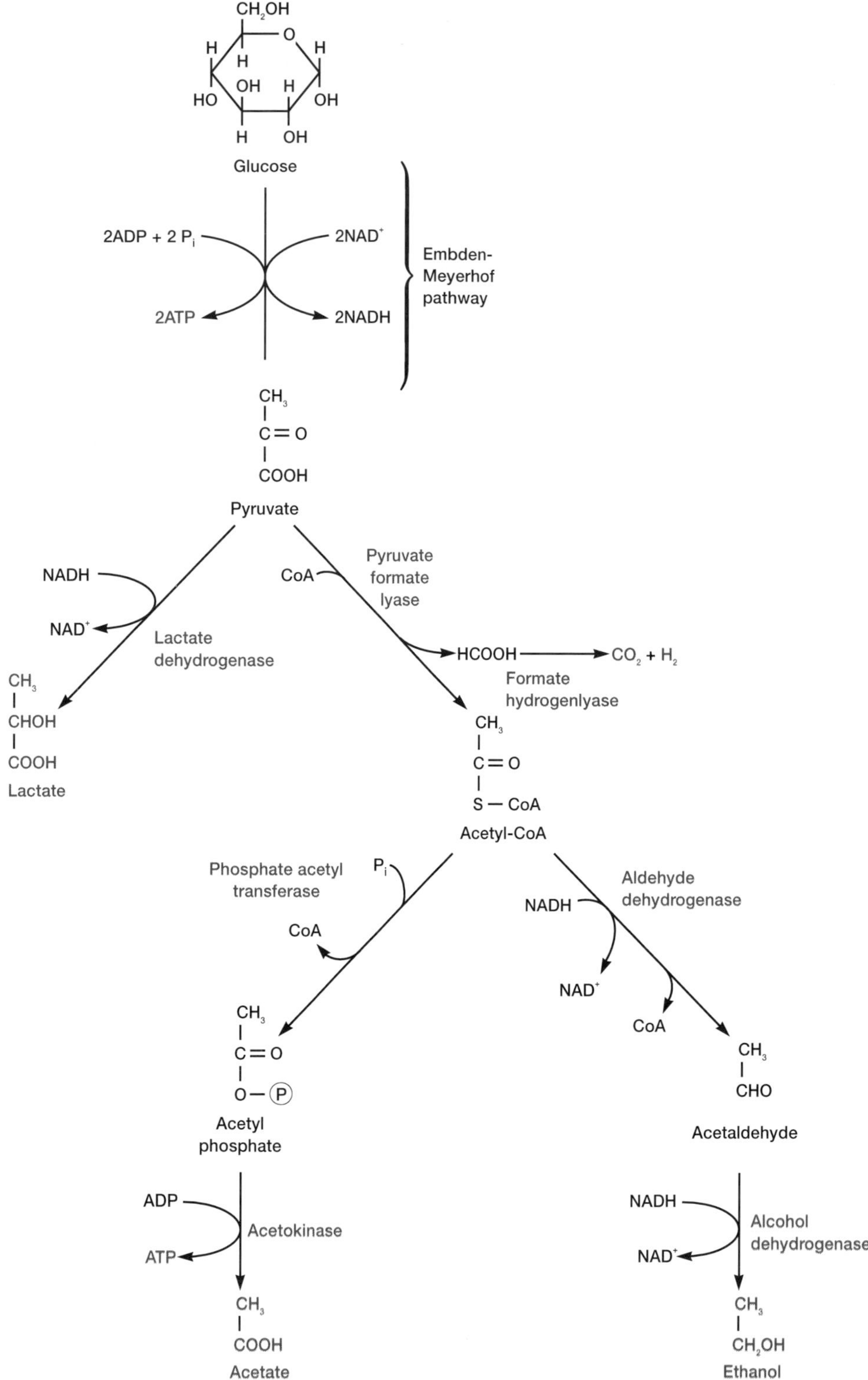

Figure AII.5 **The Mixed Acid Fermentation Pathway.** This pathway is characteristic of many members of the *Enterobacteriaceae* such as *E. coli.*

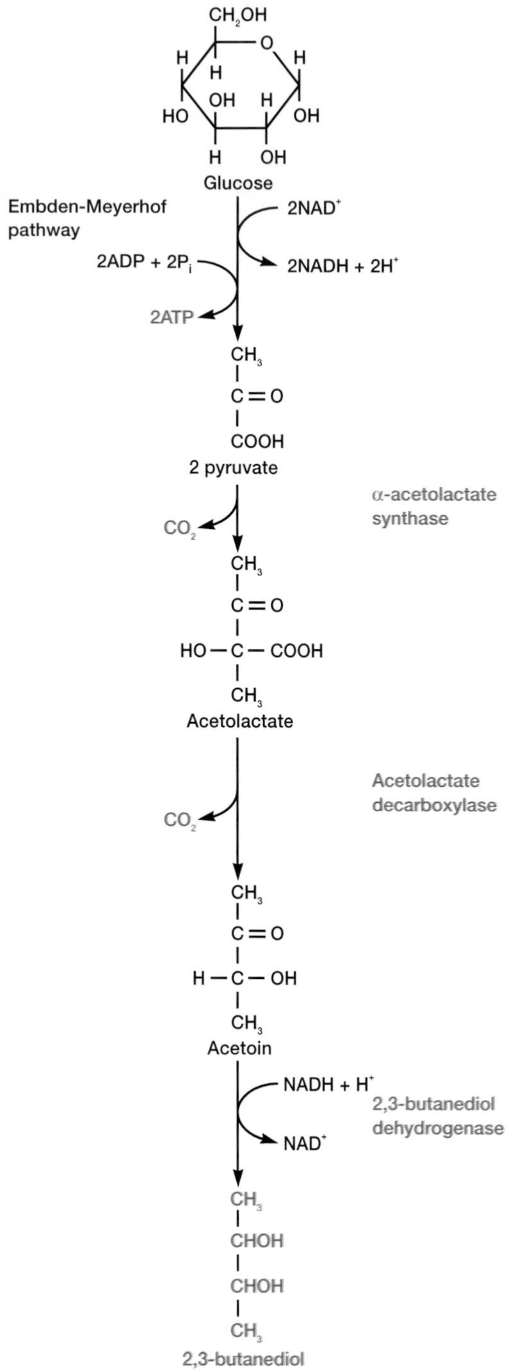

Glucose
Embden-Meyerhof pathway
$2NAD^+$
$2NADH + 2H^+$
$2ADP + 2P_i$
2ATP

CH_3–C=O–COOH
2 pyruvate

α-acetolactate synthase
CO_2

CH_3–C=O–HO–C–COOH–CH_3
Acetolactate

Acetolactate decarboxylase
CO_2

CH_3–C=O–H–C–OH–CH_3
Acetoin

$NADH + H^+$
2,3-butanediol dehydrogenase
NAD^+

CH_3–CHOH–CHOH–CH_3
2,3-butanediol

Figure AII.6 **The Butanediol Fermentation Pathway.** This pathway is characteristic of members of the *Enterobacteriaceae* such as *Enterobacter.* Other products may also be formed during butanediol fermentation.

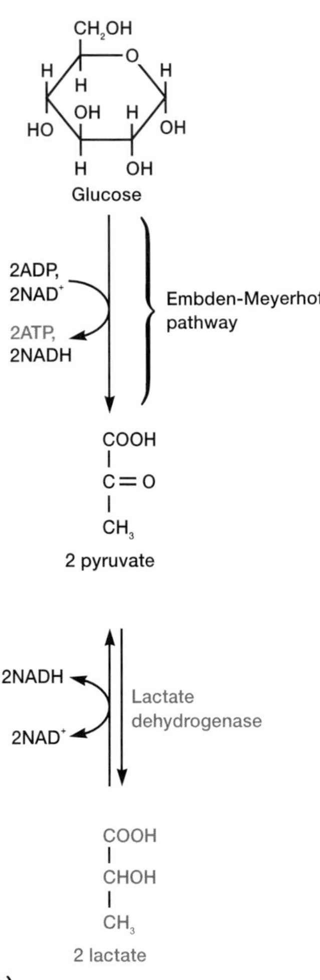

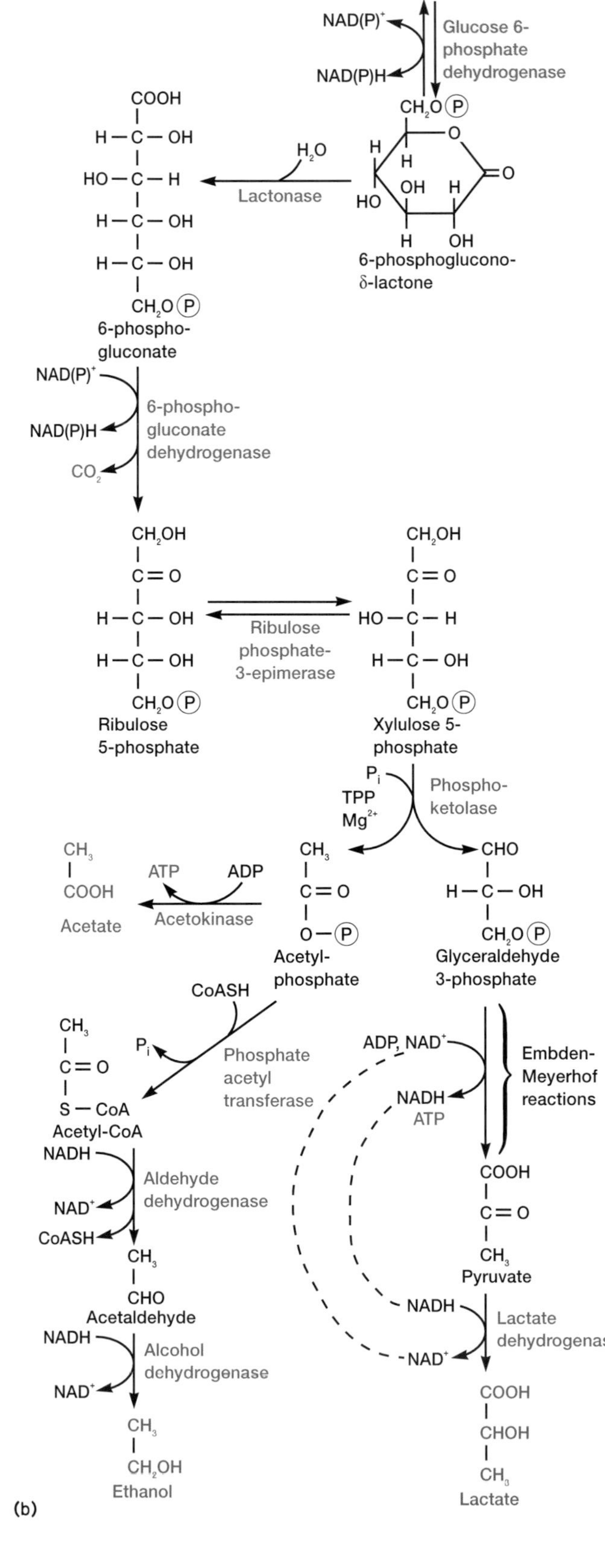

Figure AII.7 Lactic Acid Fermentations. (*a*) Homolactic fermentation pathway. (*b*) Heterolactic fermentation pathway.

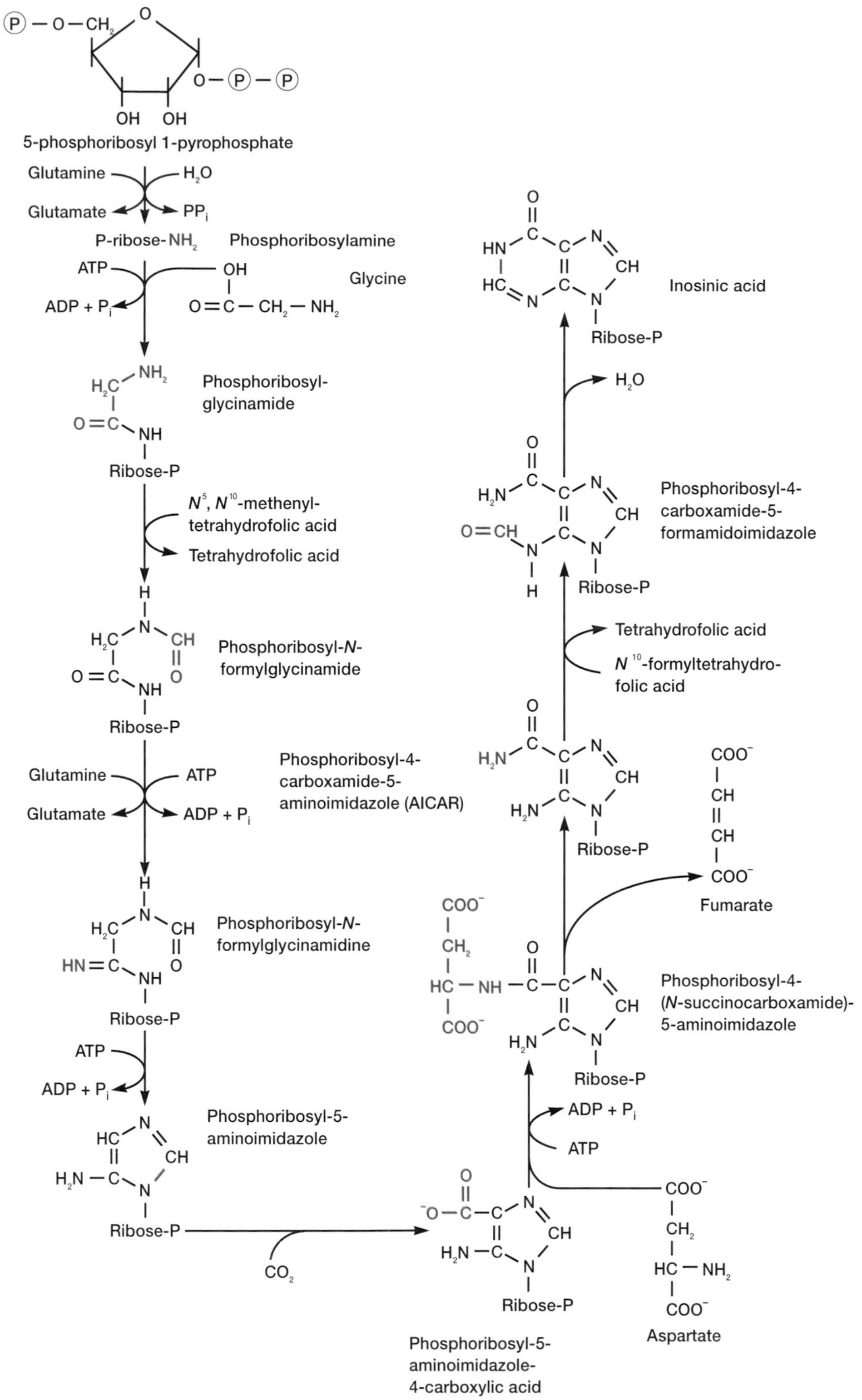

5-phosphoribosyl 1-pyrophosphate

Glutamine, H_2O, Glutamate, PP_i

P-ribose-NH_2 Phosphoribosylamine

ATP, ADP + P_i, Glycine

Phosphoribosyl-glycinamide

N^5, N^{10}-methenyl-tetrahydrofolic acid

Tetrahydrofolic acid

Phosphoribosyl-*N*-formylglycinamide

Glutamine, ATP, Glutamate, ADP + P_i

Phosphoribosyl-*N*-formylglycinamidine

ATP, ADP + P_i

Phosphoribosyl-5-aminoimidazole

CO_2

Phosphoribosyl-5-aminoimidazole-4-carboxylic acid

ATP, ADP + P_i, Aspartate

Phosphoribosyl-4-(*N*-succinocarboxamide)-5-aminoimidazole

Fumarate

Phosphoribosyl-4-carboxamide-5-aminoimidazole (AICAR)

N^{10}-formyltetrahydro-folic acid

Tetrahydrofolic acid

Phosphoribosyl-4-carboxamide-5-formamidoimidazole

H_2O

Inosinic acid

Figure AII.8 The Pathway for Purine Biosynthesis. Inosinic acid is the first purine end product. The purine skeleton is constructed while attached to a ribose phosphate.

APPENDIX III
Classification of Bacteria According to *Bergey's Manual of Systematic Bacteriology*

Volume I*

Section 1

The Spirochetes

Order I *Spirochaetales*
- Family I *Spirochaetaceae*
  - Genus I *Spirochaeta*
  - Genus II *Cristispira*
  - Genus III *Treponema*
  - Genus IV *Borrelia*
- Family II *Leptospiraceae*
  - Genus I *Leptospira*

Other Organisms
- Hindgut Spirochetes of Termites and *Cryptocercus punctulatus*

Section 2

Aerobic/Microaerophilic, Motile, Helical/Vibrioid Gram-Negative Bacteria
- Genus *Aquaspirillum*
- Genus *Spirillum*
- Genus *Azospirillum*
- Genus *Oceanospirillum*
- Genus *Campylobacter*
- Genus *Bdellovibrio*
- Genus *Vampirovibrio*

Section 3

Nonmotile (or Rarely Motile), Gram-Negative Curved Bacteria
- Family I *Spirosomaceae*
  - Genus I *Spirosoma*
  - Genus II *Runella*
  - Genus III *Flectobacillus*
- Other Genera
  - Genus *Microcyclus*
  - Genus *Meniscus*
  - Genus *Brachyarcus*
  - Genus *Pelosigma*

Section 4

Gram-Negative Aerobic Rods and Cocci
- Family I *Pseudomonadaceae*
  - Genus I *Pseudomonas*
  - Genus II *Xanthomonas*
  - Genus III *Frateuria*
  - Genus IV *Zoogloea*
- Family II *Azotobacteraceae*
  - Genus I *Azotobacter*
  - Genus II *Azomonas*
- Family III *Rhizobiaceae*
  - Genus I *Rhizobium*
  - Genus II *Bradyrhizobium*
  - Genus III *Agrobacterium*
  - Genus IV *Phyllobacterium*
- Family IV *Methylococcaceae*
  - Genus I *Methylococcus*
  - Genus II *Methylomonas*
- Family V *Halobacteriaceae*
  - Genus I *Halobacterium*
  - Genus II *Halococcus*
- Family VI *Acetobacteraceae*
  - Genus I *Acetobacter*
  - Genus II *Gluconobacter*
- Family VII *Legionellaceae*
  - Genus I *Legionella*
- Family VIII *Neisseriaceae*
  - Genus I *Neisseria*
  - Genus II *Moraxella*
  - Genus III *Acinetobacter*
  - Genus IV *Kingella*
- Other Genera
  - Genus *Beijerinckia*
  - Genus *Derxia*
  - Genus *Xanthobacter*
  - Genus *Thermus*
  - Genus *Thermomicrobium*
  - Genus *Halomonas*
  - Genus *Alteromonas*
  - Genus *Flavobacterium*
  - Genus *Alcaligenes*
  - Genus *Serpens*
  - Genus *Janthinobacterium*
  - Genus *Brucella*
  - Genus *Bordetella*
  - Genus *Francisella*
  - Genus *Paracoccus*
  - Genus *Lampropedia*

*Material in Appendix III is modified from *Bergey's Manual of Systematic Bacteriology*, J. G. Holt and N. R. Krieg (eds.). Copyright © Williams and Wilkins Co., Baltimore, MD. Reprinted by permission.

Section 5

Facultatively Anaerobic Gram-Negative Rods

Family I *Enterobacteriaceae*
Genus I *Escherichia*
Genus II *Shigella*
Genus III *Salmonella*
Genus IV *Citrobacter*
Genus V *Klebsiella*
Genus VI *Enterobacter*
Genus VII *Erwinia*
Genus VIII *Serratia*
Genus IX *Hafnia*
Genus X *Edwardsiella*
Genus XI *Proteus*
Genus XII *Providencia*
Genus XIII *Morganella*
Genus XIV *Yersina*
Other Genera of the Family *Enterobacteriaceae*
Genus *Obesumbacterium*
Genus *Xenorhabdus*
Genus *Kluyvera*
Genus *Rahnella*
Genus *Cedecea*
Genus *Tatumella*
Family II *Vibrionaceae*
Genus I *Vibrio*
Genus II *Photobacterium*
Genus III *Aeromonas*
Genus IV *Plesiomonas*
Family III *Pasteurellaceae*
Genus I *Pasteurella*
Genus II *Haemophilus*
Genus III *Actinobacillus*
Other Genera
Genus *Zymomonas*
Genus *Chromobacterium*
Genus *Cardiobacterium*
Genus *Calymmatobacterium*
Genus *Gardnerella*
Genus *Eikenella*
Genus *Streptobacillus*

Section 6

Anaerobic Gram-Negative Straight, Curved, and Helical Rods

Family I *Bacteroidaceae*
Genus I *Bacteroides*
Genus II *Fusobacterium*
Genus III *Leptotrichia*
Genus IV *Butyrivibrio*
Genus V *Succinimonas*
Genus VI *Succinivibrio*
Genus VII *Anaerobiospirillum*
Genus VIII *Wolinella*
Genus IX *Selenomonas*
Genus X *Anaerovibrio*
Genus XI *Pectinatus*
Genus XII *Acetivibrio*
Genus XIII *Lachnospira*

Section 7

Dissimilatory Sulfate- or Sulfur-Reducing Bacteria

Genus *Desulfuromonas*
Genus *Desulfovibrio*
Genus *Desulfomonas*
Genus *Desulfococcus*
Genus *Desulfobacter*
Genus *Desulfobulbus*
Genus *Desulfosarcina*

Section 8

Anaerobic Gram-Negative Cocci

Family I *Veillonellaceae*
Genus I *Veillonella*
Genus II *Acidaminococcus*
Genus III *Megasphaera*

Section 9

The Rickettsias and Chlamydias

Order I *Rickettsiales*
Family I *Rickettsiaceae*
Tribe I *Rickettsieae*
Genus I *Rickettsia*
Genus II *Rochalimaea*
Genus III *Coxiella*
Tribe II *Ehrlichieae*
Genus IV *Ehrlichia*
Genus V *Cowdria*
Genus VI *Neorickettsia*
Tribe III *Wolbachieae*
Genus VII *Wolbachia*
Genus VIII *Rickettsiella*
Family II *Bartonellaceae*
Genus I *Bartonella*
Genus II *Grahamella*
Family III *Anaplasmataceae*
Genus I *Anaplasma*
Genus II *Aegyptianella*
Genus III *Haemobartonella*
Genus IV *Eperythrozoon*
Order II *Chlamydiales*
Family I *Chlamydiaceae*
Genus I *Chlamydia*

Section 10

The Mycoplasmas
Division *Tenericutes*
Class I *Mollicutes*

Order I *Mycoplasmatales*
Family I *Mycoplasmataceae*
Genus I *Mycoplasma*
Genus II *Ureaplasma*
Family II *Acholeplasmataceae*
Genus I *Acholeplasma*
Family III *Spiroplasmataceae*
Genus I *Spiroplasma*
Other Genera
Genus *Anaeroplasma*
Genus *Thermoplasma*
Mycoplasma-Like Organisms of Plants and Invertebrates

Section 11

Endosymbionts

A. Endosymbionts of Protozoa
Endosymbionts of ciliates
Endosymbionts of flagellates
Endosymbionts of amoebas
Taxa of endosymbionts
Genus I *Holospora*
Genus II *Caedibacter*
Genus III *Pseudocaedibacter*
Genus IV *Lyticum*
Genus V *Tectibacter*

B. Endosymbionts of Insects
Bloodsucking insects
Plant sap-sucking insects
Cellulose and stored grain feeders
Insects feeding on complex diets
Taxon of endosymbionts:
Genus *Blattabacterium*

C. Endosymbionts of Fungi and Invertebrates Other Than Arthropods
Fungi
Sponges
Coelenterates
Helminthes
Annelids
Marine worms and mollusks

Volume II

Section 12

Gram-Positive Cocci

Family I *Micrococcaceae*
Genus I *Micrococcus*
Genus II *Stomatococcus*
Genus III *Planococcus*
Genus IV *Staphylococcus*
Family II *Deinococcaceae*
Genus I *Deinococcus*
Other Genera
Genus *Streptococcus*
Pyogenic Hemolytic Streptococci
Oral Streptococci
Enterococci
Lactic Acid Streptococci
Anaerobic Streptococci
Other Streptococci
Genus *Leuconostoc*
Genus *Pediococcus*
Genus *Aerococcus*
Genus *Gemella*
Genus *Peptococcus*
Genus *Peptostreptococcus*
Genus *Ruminococcus*
Genus *Coprococcus*
Genus *Sarcina*

Section 13

Endospore-Forming Gram-Positive Rods and Cocci

Genus *Bacillus*
Genus *Sporolactobacillus*
Genus *Clostridium*
Genus *Desulfotomaculum*
Genus *Sporosarcina*
Genus *Oscillospira*

Section 14

Regular, Nonsporing, Gram-Positive Rods

Genus *Lactobacillus*
Genus *Listeria*
Genus *Erysipelothrix*
Genus *Brochothrix*
Genus *Renibacterium*
Genus *Kurthia*
Genus *Caryophanon*

Section 15

Irregular, Nonsporing, Gram-Positive Rods

Genus *Corynebacterium*
Plant Pathogenic Species of *Corynebacterium*
Genus *Gardnerella*
Genus *Arcanobacterium*
Genus *Arthrobacter*
Genus *Brevibacterium*
Genus *Curtobacterium*
Genus *Caseobacter*
Genus *Microbacterium*
Genus *Aureobacterium*
Genus *Cellulomonas*
Genus *Agromyces*
Genus *Arachnia*
Genus *Rothia*
Genus *Propionibacterium*
Genus *Eubacterium*
Genus *Acetobacterium*
Genus *Lachnospira*
Genus *Butyrivibrio*
Genus *Thermoanaerobacter*
Genus *Actinomyces*
Genus *Bifidobacterium*

Section 16

The Mycobacteria

Family *Mycobacteriaceae*
Genus *Mycobacterium*

Section 17

Nocardioforms

Genus *Nocardia*
Genus *Rhodococcus*
Genus *Nocardioides*
Genus *Pseudonocardia*
Genus *Oerskovia*
Genus *Saccharopolyspora*
Genus *Micropolyspora*
Genus *Promicromonospora*
Genus *Intrasporangium*

Volume III

Section 18

Anoxygenic Phototrophic Bacteria

I. Purple bacteria

Family I *Chromatiaceae*
Genus I *Chromatium*
Genus II *Thiocystis*
Genus III *Thiospirillum*
Genus IV *Thiocapsa*
Genus V *Lamprobacter*
Genus VI *Lamprocystis*
Genus VII *Thiodictyon*
Genus VIII *Amoebobacter*
Genus IX *Thiopedia*
Family II *Ectothiorhodospiraceae*
Genus *Ectothiorhodospira*
Purple Nonsulfur Bacteria
Genus *Rhodospirillum*
Genus *Rhodopila*
Genus *Rhodobacter*
Genus *Rhodopseudomonas*
Genus *Rhodomicrobium*
Genus *Rhodocyclus*

II. Green Bacteria

Green Sulfur Bacteria
Genus *Chlorobium*
Genus *Prosthecochloris*
Genus *Pelodictyon*
Genus *Ancalochloris*
Genus *Chloroherpeton*
Multicellular, Filamentous, Green Bacteria
Genus *Chloroflexus*
Genus *Heliothrix*
Genus "*Oscillochloris*"
Genus *Chloronema*

III. Genera Incertae Sedis

Genus *Heliobacterium*
Genus *Erythrobacter*

Section 19

Oxygenic Photosynthetic Bacteria
Group I Cyanobacteria

Subsection I Order *Chroococcales*
1. Genus I *Chamaesiphon*
2. Genus II *Gloeobacter*
3. *Synechococcus*-group
4. Genus III *Gloeothece*
5. *Cyanothece*-group
6. *Gloeocapsa*-group
7. *Synechocystis*-group

Subsection II Order *Pleurocapsales*
1. Genus I *Dermocarpa*
2. Genus II *Xenococcus*
3. Genus III *Dermocarpella*
4. Genus IV *Mxyosarcina*
5. Genus V *Chroococcidiopsis*
6. *Pleurocapsa*-group

Subsection III Order *Oscillatoriales*
Genus I *Spirulina*
Genus II *Arthrospira*
Genus III *Oscillatoria*
Genus IV *Lyngbya*
Genus V *Pseudanabaena*
Genus VI *Starria*
Genus VII *Crinalium*
Genus VIII *Microcoleus*

Subsection IV Order *Nostocales*
Family I *Nostocaceae*
Genus I *Anabaena*
Genus II *Aphanizomenon*
Genus III *Nodularia*
Genus IV *Cylindrospermum*
Genus V *Nostoc*
Family II *Scytonemataceae*
Genus I *Scytonema*
Family III *Rivulariaceae*
Genus I *Calothrix*

Subsection V Order *Stigonematales*
Genus I *Chlorogloeopsis*
Genus II *Fischerella*
Genus III *Stigonema*
Genus IV *Geitleria*

Group II Order *Prochlorales*

Family I *Prochloraceae*
Genus *Prochloron*
Other taxa
Genus "*Prochlorothrix*"

Section 20

Aerobic Chemolithotrophic Bacteria and Associated Organisms

A. Nitrifying Bacteria
Family *Nitrobacteraceae*
Nitrite-oxidizing bacteria
Genus I *Nitrobacter*
Genus II *Nitrospina*
Genus III *Nitrococcus*
Genus IV *Nitrospira*
Ammonia-oxidizing bacteria
Genus V *Nitrosomonas*
Genus VI *Nitrosococcus*
Genus VII *Nitrosospira*
Genus VIII *Nitrosolobus*
Genus IX "*Nitrosovibrio*"
B. Colorless Sulfur Bacteria
Genus *Thiobacterium*
Genus *Macromonas*
Genus *Thiospira*
Genus *Thiovulum*
Genus *Thiobacillus*
Genus *Thiomicrospira*
Genus *Thiosphaera*
Genus *Acidiphilium*
Genus *Thermothrix*
C. Obligately Chemolithotrophic Hydrogen Bacteria
Genus *Hydrogenobacter*
D. Iron- and Manganese-Oxidizing and/or Depositing Bacteria
Family "*Siderocapsaceae*"
Genus I "*Siderocapsa*"
Genus II "*Naumanniella*"
Genus III "*Siderococcus*"
Genus IV "*Ochrobium*"
E. Magnetotactic Bacteria
Genus *Aquaspirillum*
(*A. magnetotacticum*)
Genus "*Bilophococcus*"

Section 21

Budding and/or Appendaged Bacteria

I. Prosthecate Bacteria

A. Budding bacteria
1. Buds produced at tip of prostheca
Genus *Hyphomicrobium*
Genus *Hyphomonas*
Genus *Pedomicrobium*
2. Buds produced on cell surface
Genus *Ancalomicrobium*
Genus *Prosthecomicrobium*
Genus *Labrys*
Genus *Stella*
B. Bacteria that divide by binary transverse fission
Genus *Caulobacter*
Genus *Asticcacaulis*
Genus *Prosthecobacter*

II. Nonprosthecate Bacteria

A. Budding bacteria
1. Lack peptidoglycan
Genus *Planctomyces*
Genus "*Isosphaera*"
2. Contain peptidoglycan
Genus *Ensifer*
Genus *Blastobacter*
Genus *Angulomicrobium*
Genus *Gemmiger*
B. Nonbudding, stalked bacteria
Genus *Gallionella*
Genus *Nevskia*
C. Other bacteria
1. Nonspinate bacteria
Genus *Seliberia*
Genus "*Metallogenium*"
Genus "*Thiodendron*"
2. Spinate bacteria

Section 22

Sheathed Bacteria

Genus *Sphaerotilus*
Genus *Leptothrix*
Genus *Haliscomenobacter*
Genus "*Lieskeella*"
Genus "*Phragmidiothrix*"
Genus *Crenothrix*
Genus "*Clonothrix*"

Section 23

Nonphotosynthetic, Nonfruiting Gliding Bacteria

Order I *Cytophagales*
Family I *Cytophagaceae*
Genus I *Cytophaga*
Genus II *Capnocytophaga*
Genus III *Flexithrix*
Genus IV *Sporocytophaga*
Other genera
Genus *Flexibacter*
Genus *Microscilla*
Genus *Chitinophaga*
Genus *Saprospira*
Order II *Lysobacterales*
Family I *Lysobacteraceae*
Genus I *Lysobacter*
Order III *Beggiatoales*
Family I *Beggiatoaceae*
Genus I *Beggiatoa*
Genus II *Thiothrix*
Genus III *Thioploca*
Genus IV "*Thiospirillopsis*"
Other families and genera
Family *Simonsiellaceae*
Genus I *Simonsiella*
Genus II *Alysiella*
Family "*Pelonemataceae*"
Genus I "*Pelonema*"
Genus II "*Achroonema*"
Genus III "*Peloploca*"
Genus IV "*Desmanthos*"
Other genera
Genus *Toxothrix*
Genus *Leucothrix*
Genus *Vitreoscilla*
Genus *Desulfonema*
Genus *Achromatium*
Genus *Agitococcus*
Genus *Herpetosiphon*

Section 24

Fruiting Gliding Bacteria: The Myxobacteria

Order *Myxococcales*
Family I *Myxococcaceae*
Genus *Myxococcus*
Family II *Archangiaceae*
Genus *Archangium*
Family III *Cystobacteraceae*
Genus I *Cystobacter*
Genus II *Melittangium*
Genus III *Stigmatella*
Family IV *Polyangiaceae*
Genus I *Polyangium*
Genus II *Nannocystis*
Genus III *Chondromyces*

Section 25

Archaeobacteria

Group I Methanogenic Archaeobacteria

Order I *Methanobacteriales*
Family I *Methanobacteriaceae*
Genus I *Methanobacterium*
Genus II *Methanobrevibacter*
Family II *Methanothermaceae*
Genus *Methanothermus*
Order II *Methanococcales*
Family *Methanococcaceae*
Genus *Methanococcus*
Order III *Methanomicrobiales*
Family I *Methanomicrobiaceae*
Genus I *Methanomicrobium*
Genus II *Methanospirillum*
Genus III *Methanogenium*
Family II *Methanosarcinaceae*
Genus I *Methanosarcina*
Genus II *Methanolobus*
Genus III *Methanothrix*
Genus IV *Methanococcoides*

Other Taxa
Family *Methanoplanaceae*
Genus *Methanoplanus*
Other Genus *Methanosphaera*

Group II Archaeobacterial Sulfate Reducers

Order "*Archaeoglobales*"
Family "*Archaeoglobaceae*"
Genus *Archaeoglobus*

Group III Extremely Halophilic Archaeobacteria

Order *Halobacteriales*
Family *Halobacteriaceae*
Genus I *Halobacterium*
Genus II *Haloarcula*
Genus III *Haloferax*
Genus IV *Halococcus*
Genus V *Natronobacterium*
Genus VI *Natronococcus*

Group IV Cell Wall-less Archaeobacteria

Genus *Thermoplasma*

Group V Extremely Thermophilic S^0-Metabolizers

Order I *Thermococcales*
Family *Thermococcaceae*
Genus I *Thermococcus*
Genus II *Pyrococcus*
Order II *Thermoproteales*
Family I *Thermoproteaceae*
Genus I *Thermoproteus*
Genus II *Thermofilum*
Family II *Desulfurococcaceae*
Genus *Desulfurococcus*
Other bacteria
Genus *Staphylothermus*
Genus *Pyrodictium*
Order III *Sulfolobales*
Family *Sulfolobaceae*
Genus I *Sulfolobus*
Genus II *Acidianus*

Volume IV

Section 26

Nocardioform Actinomycetes

Genus *Nocardia*
Genus *Rhodococcus*
Genus *Nocardioides*
Genus *Pseudonocardia*
Genus *Oerskovia*
Genus *Saccharopolyspora*
Genus *Faenia*
Genus *Promicromonospora*
Genus *Intrasporangium*
Genus *Actinopolyspora*
Genus *Saccharomonospora*

Section 27

Actinomycetes with Multilocular Sporangia

Genus *Geodermatophilus*
Genus *Dermatophilus*
Genus *Frankia*

Section 28

Actinoplanetes

Genus *Actinoplanes*
Genus *Ampullariella*
Genus *Pilimelia*
Genus *Dactylosporangium*
Genus *Micromonospora*

Section 29

Streptomycetes and Related Genera

Genus *Streptomyces*
Genus *Streptoverticillium*
Genus *Kineosporia*
Genus *Sporichthya*

Section 30

Maduromycetes

Genus *Actinomadura*
Genus *Microbispora*
Genus *Microtetraspora*
Genus *Planobispora*
Genus *Planomonospora*
Genus *Spirillospora*
Genus *Streptosporangium*

Section 31

Thermomonospora and Related Genera

Genus *Thermomonospora*
Genus *Actinosynnema*
Genus *Nocardiopsis*
Genus *Streptoalloteichus*

Section 32

Thermoactinomycetes

Genus *Thermoactinomyces*

Section 33

Other Genera

Genus *Glycomyces*
Genus *Kibdelosporangium*
Genus *Kitasatospora*
Genus *Saccharothrix*
Genus *Pasteuria*

Appendix IV
Classification of Viruses

In this appendix, selected groups of viruses are briefly described and a sketch of each is included. The illustrations are not to scale but should provide a general idea of the morphology of each group. The viruses are separated into six sections based on their host preferences. The following material has been adapted with permission from chapter 20 of *Introduction to Modern Virology,* 3d ed. by N. J. Dimmock and S. B. Primrose, copyright © 1987 by Blackwell Scientific Publications, Ltd.

Viruses Multiplying in Vertebrates and Other Hosts

Note that some or all of the *Reoviridae, Bunyaviridae, Rhabdoviridae* and *Togaviridae* multiply in *both* vertebrates and other hosts. Other families include genera that multiply solely in vertebrates.

1. Family: *Iridoviridae*

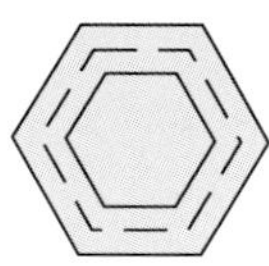

Icosahedral particle (125 to 300 nm) with internal shell of lipid (dashed line) and no glycoproteins. Double-stranded DNA of mol wt 100 to 250 $\times$ 10^6, which is circularly permuted with direct terminal repeats. Contain no enzymes. RNA transcribed by cellular polymerase II. mRNAs have no polyA tails. Cytoplasmic.

Genera:

Iridovirus (small 120 nm blue iridescent viruses of insects)
Chloriridovirus (large 180 nm iridescent viruses of insects)

2. Family: *Poxviridae*

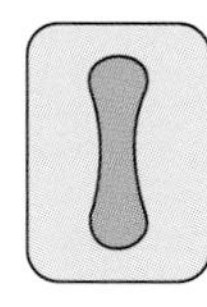

Double-stranded DNA of mol wt 85 to 240 $\times$ 10^6 with inverted terminal repeats. Largest viruses 170 to 260 $\times$ 300 to 450 nm. Complex structure composed of several layers and includes lipid. Core contains all enzymes required for mRNA synthesis. Cytoplasmic multiplication.

Genera:

Orthopoxvirus (vaccinia and related viruses)	Poxviruses of vertebrates
Avipoxvirus (fowlpox and related viruses)	
Parapoxvirus (milker's node and related viruses)	
Entomopoxvirus	Poxviruses of insects

3. Family: *Parvoviridae*

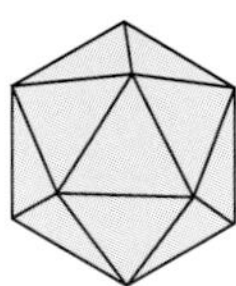

Single-stranded DNA of mol wt 1.5 to 2.2 $\times$ 10^6. Particle is an 18 to 26 nm icosahedron. Multiplication has a nuclear stage.

Genera:

Parvovirus—viruses of vertebrates including humans. Virions mostly −DNA.

Dependovirus—adeno-associated virus. Infects vertebrates. Particles contain either +DNA or −DNA, which forms a double strand upon extraction. Require helper adenovirus or herpesvirus.

Densovirus—viruses of insects. Virions −DNA or +DNA. Helper not required.

4. Family: *Reoviridae*

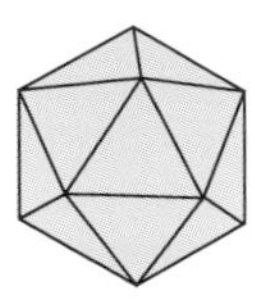

Ten to twelve segments of double-stranded RNA of total mol wt 12 to 20 $\times 10^6$. Particle is a 60 to 80 nm icosahedron. Has an isometric nucleocapsid with transcriptase activity. Cytoplasmic multiplication.

Genera:

Reovirus—of vertebrates
Orbivirus—of vertebrates, but also multiply in insects
Rotavirus—of vertebrates
Cytoplasmic polyhedrosis viruses—of insects
Phytoreovirus—clover wound tumor virus
Fijivirus—Fiji disease of plants

5. Family: *Picornaviridae*

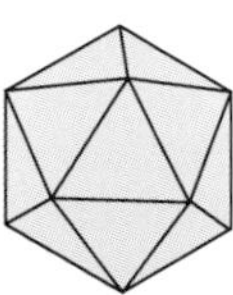

Single-stranded RNA of mol wt 2.5 $\times 10^6$. Particle is 22 to 30 nm with cubic symmetry. Multiplication is cytoplasmic.

Genera:

Enterovirus (acid-resistant, primarily viruses of gastrointestinal tract)
Rhinovirus (acid-labile, mainly viruses of upper respiratory tract)
Aphthovirus (foot-and-mouth disease virus)
Cardiovirus (EMC virus of mice)
Also various viruses of insects

6. Family: *Togaviridae*

Single-stranded RNA of mol wt 4 $\times 10^6$. Enveloped particles 40 to 70 nm diameter contain an icosahedral nucleocapsid. Hemagglutinate. Cytoplasmic, budding from plasma membrane. Have a subgenomic mRNA.

Genera:

Alphavirus (arboviruses, e.g., Semliki Forest virus)
Pestivirus (hog cholera and related viruses) } Not arboviruses
Rubivirus (rubella virus) } Not arboviruses
Arterivirus (equine arteritis virus) } Not arboviruses

7. Family: *Flaviviridae*

Single-stranded RNA of mol wt 3.8 $\times 10^6$. Enveloped particles 40 to 70 nm diameter. Differ from *Alphaviridae* by presence of a matrix protein, the lack of intracellular subgenomic mRNAs and budding from the endoplasmic reticulum. Hemagglutinate. Cytoplasmic.

Genus:

Flavivirus (arboviruses, e.g., yellow fever virus)

8. Family: *Rhabdoviridae*

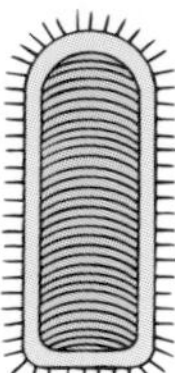

Single-stranded RNA of mol wt 3.5 to 4.6 $\times 10^6$ complementary to mRNA. The bullet-shaped or bacilliform (130 to 380 $\times$ 70 nm) particle is enveloped with 5 to 10 nm spikes. Inside is a helical nucleocapsid with transcriptase activity. Cytoplasmic, budding from plasma membrane.

Genera:

Vesiculovirus (vesicular stomatitis virus group) viruses of vertebrates and insects
Lyssavirus (viruses of vertebrates, e.g., rabies, and of insects, e.g., sigma)
Plant viruses (e.g., lettuce necrotic yellows, potato yellow dwarf, and many others)

9. Family: *Bunyaviridae*

Three segments (large, medium, and small) of single-stranded RNA of total mol wt 6 $\times 10^6$. Enveloped 100 nm particles with spikes and three internal ribonucleoprotein filaments 2 nm wide. Cytoplasmic, budding from the Golgi apparatus. Arthropod-transmitted except *Hantavirus* genus.

Genera:

Bunyavirus (Bunyamwera and 150 or so related viruses)
Hantavirus (Korean hemorrhagic fever or Hantaan virus) *not* arboviruses

Viruses Multiplying Only in Vertebrates

1. Family: *Herpesviridae*

Double-stranded DNA of mol wt 80 to 150 $\times 10^6$. Particle is a 130 nm icosahedron enclosed in a lipid envelope. Buds from nuclear membrane. Latency for the lifetime of the host is common.

Subfamily: *Alphaherpesvirinae*

Genera:

Simplexvirus, e.g., human (alpha) herpesvirus 1 and 2 (formerly herpes simplex virus types 1 and 2)
Poikilovirus, e.g., suid (alpha) herpesvirus 1 (formerly pseudorabies virus)
Also unclassified is human (alpha) herpesvirus 3 (formerly varicella-zoster virus)

Subfamily: *Betaherpesvirinae* (cytomegaloviruses)

Genera:

Human cytomegalovirus group, e.g., human (beta) herpesvirus 5
Mouse cytomegalovirus group

Subfamily: *Gammaherpesvirinae* (lymphoproliferative viruses)

Genera:

Lymphocryptovirus, e.g., human (gamma) herpesvirus 4 (formerly Epstein-Barr virus)
Thetalymphocryptovirus, e.g., gallid herpesvirus 1 (formerly Marek's disease virus)

2. Family: *Adenoviridae*

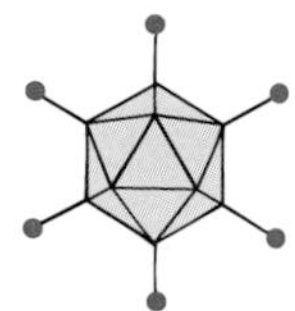

Double-stranded DNA of mol wt 20 to 30 $\times 10^6$. Particle is a 70 to 90 nm icosahedron, which is assembled in the nucleus.

Genera:

Mastodenovirus (adenoviruses of mammals)
Aviadenovirus (adenoviruses of birds)

3. Family: *Papovaviridae*

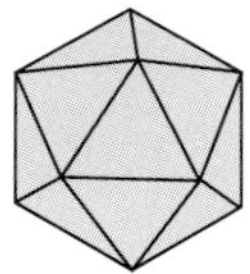

Double-stranded circular DNA. Particles have 72 capsomeres in a skew arrangement and are assembled in the nucleus. Hemagglutinate.

Genera:

Papillomavirus (producing papillomas in several mammalian species) 55 nm particle; DNA 5×10^6 mol wt
Polyomavirus (found in rodents, humans, and other primates) 45 nm particle; DNA 3×10^6 mol wt. Includes SV40 and polyoma virus itself.

4. Family: *Hepadnaviridae*

One complete DNA minus strand of mol wt 1×10^6 with a 5′ terminal protein. DNA is circularized by an incomplete plus strand of variable length (50 to 100%), which overlaps the 3′ and 5′ termini of DNA minus. There is a 42 nm enveloped particle containing a core with DNA polymerase and protein kinase activities. Includes hepatitis B of humans, Pekin duck hepatitis, beechy ground squirrel hepatitis, and woodchuck hepatitis viruses. HBV is strongly associated with liver cancer.

5. Family: *Calicivirus*

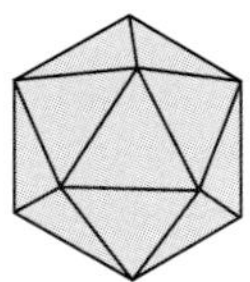

Single-stranded RNA of mol wt 2.7×10^6. Icosahedral 37 nm particle with calix-like (cup-shaped) surface depressions.

Genus:

Calicivirus (vesicular exanthema of swine virus and others causing gastroenteritis in humans, cats, and calves)

6. Family: *Arenaviridae*

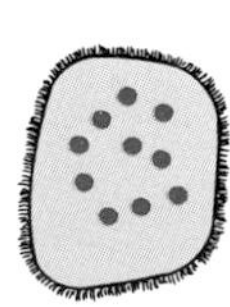

Two (a large and a small) segments of single-stranded RNA of mol wt 3×10^6 and 1.3×10^6. Latter RNA is ambisense. Enveloped 60 to 300 nm particles with spikes. Contain ribosomes which have no known function. Cytoplasmic multiplication; buds from plasma membrane.

Genus:

Arenavirus (lymphocytic choriomeningitis virus and related viruses)

7. Family: *Paramyxoviridae*

Single-stranded RNA of mol wt 5 to 7×10^6. Enveloped 150 nm particles have spikes and contain a nucleocapsid 12 to 17 nm in diameter with transcriptase activity. Cytoplasmic, budding from plasma membrane.

Genera:

Paramyxovirus (Newcastle disease virus group) Only this genus has a neuraminidase activity which is on the same protein (HN) as the hemagglutination activity.
Morbillivirus (measles virus group) Hemagglutinate
Pneumovirus (respiratory syncytial virus group)

8. Family: *Orthomyxoviridae*

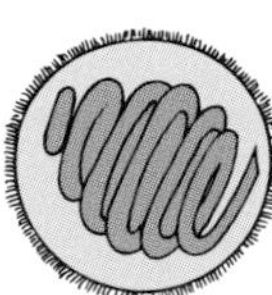

Eight segments of single-stranded RNA of total mol wt 4×10^6. Enveloped 100 nm particles have spikes and contain a helical nucleocapsid 9 nm in diameter with transcriptase activity. Only A and B virions have separate hemagglutinin and neuraminidase proteins. Multiplication requires the nucleus. RNA segments in a mixed infection readily assort to form genetically stable hybrids within a virus. Buds from plasma membrane.

Genus:

Influenzavirus (influenza types A and B virus) Influenza type C virus. Has seven RNA segments and a receptor-destroying activity (a sialic acid—O-acetyl esterase), which is on the hemagglutinin protein.

9. Family: *Filoviridae*

New group containing Marburg and Ebola viruses, which are highly pathogenic for humans. Long filamentous particles 800 to 900 (sometimes 14,000) $\times$ 80 nm with helical nucleocapsid of 50 nm diameter. RNA of mol wt 4.2×10^6. Buds from plasma membrane.

10. Family: *Retroviridae*

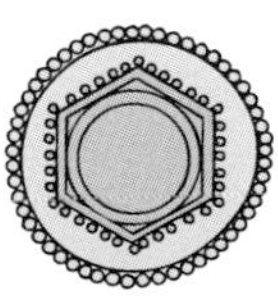

Single-stranded "diploid" RNA with unique sequence having a mol wt 1 to 3×10^6. Enveloped 100 nm particles containing an icosahedral nucleocapsid. Contain RNA-dependent DNA polymerase. The DNA provirus is nuclear.

Subfamily: *Oncovirinae* (RNA tumor virus group) Only these are oncogenic.
Subfamily: *Spumavirinae* (foamy viruses) Cause persistent infections *in vivo* and foamy vacuolated syncytia *in vitro*. Do not transform or cause tumors.
Subfamily: *Lentivirinae* (visna/maedi virus group) Cause slowly progressive disease of the lungs (sheep) and central nervous system (sheep, humans [HIV]).

Viruses Multiplying Only in Invertebrates

Viruses occur not only in insects, crustacea, and molluscs but probably in all groups of invertebrates. The *Poxviridae, Reoviridae, Parvoviridae, Rhabdoviridae,* and *Togaviridae* (see earlier) have representatives that multiply in invertebrates. Some plant viruses are transmitted by, but do not multiply in these vectors.

1. Family: *Baculoviridae*

Double-stranded circular DNA of mol wt 60 to 110 × 10^6. Bacilliform particles 40 to 60 nm × 200 to 400 nm with an outer membrane. May be occluded in a protein inclusion body containing usually one particle (granulosis viruses) or in a polyhedra containing many particles (polyhedrosis viruses).

Genus:
Baculovirus (Bombyx mori nuclear polyhedrosis virus group)

2. Family: *Nudaurelia β virus group*

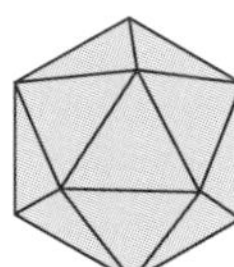

Single-stranded RNA of mol wt 1.8 × 10^6 in a 35 nm particle. T = 4 (whereas in *Picornaviridae* T = 1). All isolated from Lepidoptera.

Viruses Multiplying Only in Plants

Knowledge of virus multiplication in plants is relatively rudimentary since the cell culture systems are less manageable than the animal cell cultures. Work has concentrated on physical properties and disease characteristics. Designation into families or genera has not yet been decided. The *Reoviridae* and *Rhabdoviridae* have members which multiply in both plants and invertebrates. The plant viruses listed below are not known to multiply in their invertebrate vector.

1. Caulimovirus—cauliflower mosaic virus group

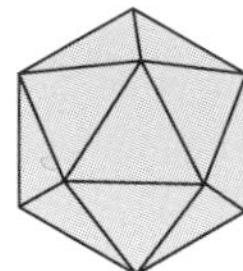

Double-stranded DNA of mol wt 4 to 5 × 10^6. Isometric 50 nm particles. Aphid vectors.

2. Geminivirus

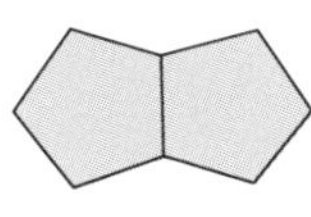

Circular single-stranded DNA of mol wt 0.7 to 1 × 10^6. Quasi-isometric 18 nm particles occurring in pairs and usually found in the nucleus. One molecule of DNA per pair of particles. Persistent in whitefly or leafhopper vectors.

3. Tomato spotted wilt virus group

Four single-stranded RNAs of mol wt 2.6, 1.9, 1.7, and 1.3 × 10^6. Isometric 82 nm particles that contain an inner shell of lipid and a coil of ribonucleoprotein internal to that. Matures by budding through internal cellular membranes of the Golgi apparatus. Thrips vector.

4. Luteovirus—barley yellow dwarf virus group

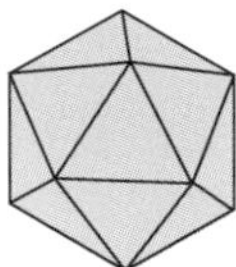

Single-stranded RNA of mol wt 2 × 10^6. Isometric 25 to 30 nm particle. Persistent retention by aphid vectors.

5. Machlovirus—maize chlorotic dwarf virus group

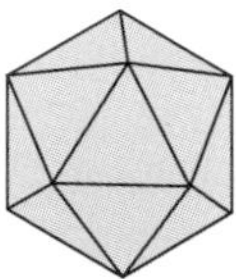

Single-stranded RNA of mol wt 3.2 × 10^6 in 30 nm particle. Only infects grasses and transmitted by leafhoppers.

6. Necrovirus—tobacco necrosis virus group

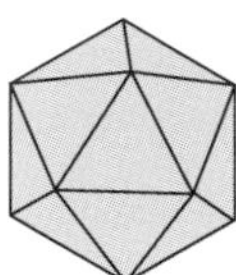

Single-stranded RNA of mol wt 1.5 × 10^6. Isometric 28 nm particle. Cytoplasmic. Fungal vector.

7. Tombusvirus—tomato bushy stunt virus group

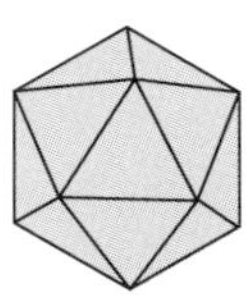

Single-stranded RNA of mol wt 1.5 × 10^6. Particle 30 nm in diameter. Cytoplasmic, nuclear, and sometimes nucleolar location. Transmitted through the soil.

8. Bromovirus—brome mosaic virus group

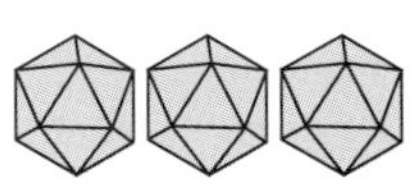

Four single-stranded RNAs of mol wt 1.1, 1.0, 0.8, and 0.3 × 10^6. Particles 26 nm diameter. Infectivity requires the three largest RNAs. Each is in a different particle. Assembly in cytoplasm. Some with a beetle vector.

9. Cucumovirus—cucumber mosaic virus group

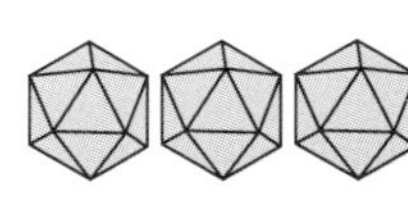

Single-stranded RNAs of mol wt 1.3, 1.1, 0.8, and 0.3 × 10^6. Particles 29 nm in size. Infectivity requires the three largest RNAs, each of which is in a separate particle. Cytoplasmic. Nonpersistent in aphid vector.

10. Alfalfa mosaic virus group

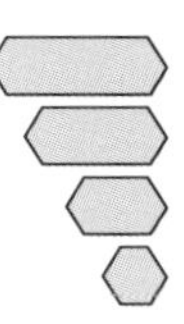

Four bacilliform particles 18 × 58, 18 × 48, 18 × 36, and 18 × 28 nm. Three largest contain single-stranded RNA of mol wt 1.1, 0.8, 0.7, and 0.3 × 10^6 all needed for infectivity. Smallest contains two molecules of the coat protein mRNA. Cytoplasmic. Nonpersistent in aphid vector.

11. Comovirus—cowpea mosaic virus group

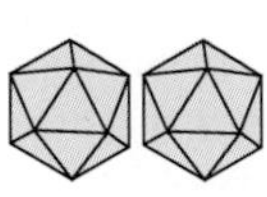

Two 28 nm particles containing single-stranded RNA of mol wt 2.4 or 1.4 × 10^6. Two coat polypeptides encoded by the smaller RNA. Both RNAs needed for infectivity. Cytoplasmic. Beetle vector.

12. Nepovirus—tobacco ringspot virus group

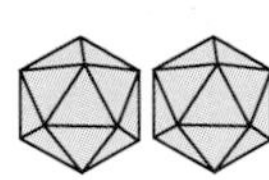

Two 28 nm particles containing single-stranded RNA of mol wt 2.8 and 1.3 to 2.4 × 10^6. Both RNAs needed for infectivity. One coat protein encoded on larger RNA. Some nepoviruses have a third particle containing two molecules of the second RNA. Cytoplasmic. Nematode vectors.

13. Tobamovirus—tobacco mosaic virus group

Single-stranded RNA of mol wt 2 × 10^6. Rigid cylindrical particles 300 × 18 nm. Transmitted mechanically or by seed.

14. Tobravirus—tobacco rattle virus group

Two straight tubular particles. The larger is about 200 nm long with an RNA of 2.4 × 10^6 mol wt. Shorter particles are of variable length and contain RNA of 0.6 to 1.4 × 10^6, which specifies the coat protein. The larger RNA is infectious alone, but both RNAs are needed for synthesis of new particles. Cytoplasmic. Nematode vector.

Viruses Multiplying Only in Fungi

1. Family: *Totiviridae*

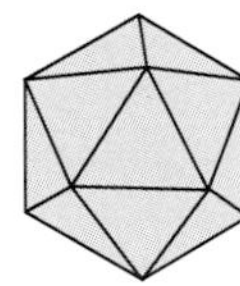

Isometric 40 nm particles with a genome consisting of a single molecule of double-stranded RNA of mol wt 4 × 10^6. One major capsid protein. Single shell. Cytoplasmic.

Genus:
Totivirus (*Saccharomyces cerevisiae* virus LA)

2. Family: *Partitiviridae*

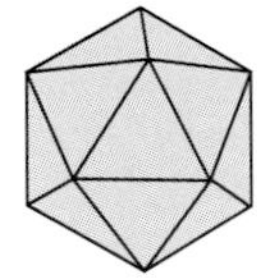

Isometric 32 nm particles with a genome consisting of two molecules of double-stranded RNA each of mol wt 0.9 to 1.6 × 10^6. Each separately encapsidated and both required for infectivity. Cytoplasmic.

Genus:
Partitivirus (*Gaeumannomyces graminis* virus 019/6–A)

Viruses Multiplying Only in Bacteria

Surprisingly little is known of the comparative biology of bacterial viruses as only a few representatives have been studied in detail.

1. Family: *Myoviridae* (T-even phage group)

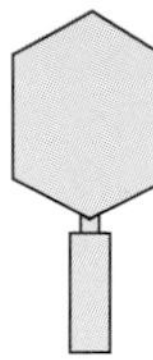

Linear double-stranded DNA of mol wt 120 × 10^6. Head 110 × 80 nm; complex contractile tail 113 nm long. Includes T2, T4, T6, PBS1, SP8, SP50, P2.

2. Family: *Styloviridae* (λ-phage group)

Linear double-stranded DNA of mol wt 33 × 10^6. Head 60 nm diameter; long noncontractile tail up to 539 nm. Includes λ, χ (chi), and ϕ80.

3. Family: *Podoviridae* (T7 and related phage groups)

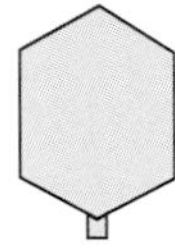

Linear double-stranded DNA of mol wt 25 × 10^6. Head 65 nm diameter. Short (20 nm) tail. Includes P22.

4. Family: *Tectiviridae* (PRD1 phage group)

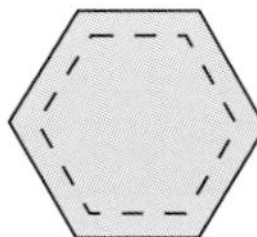

Linear double-stranded DNA of mol wt 9 × 10^6 in 65 nm particles. Contains internal lipid. After injection of DNA a tail structure appears.

5. Family: *Plasmaviridae* (mycoplasma virus type 2 phage group)

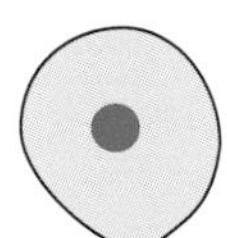

Circular double-stranded DNA of 9 × 10^6 mol wt in lipid-containing envelope. 50 to 90 nm particle with small dense core. Formed by budding. Infects mycoplasmas.

6. Family: *Corticoviridae* (PM2 phage group)

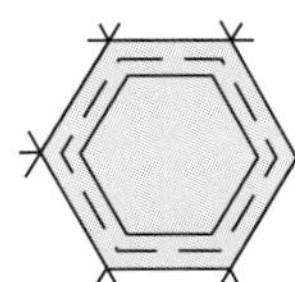

Circular double-stranded DNA of mol wt 6 × 10^6, isometric 60 nm particle, lipid between protein shells, no tail. Spikes at vertices. Infects *Pseudomonas*.

7. Family: *Microviridae* (ϕX174 group)

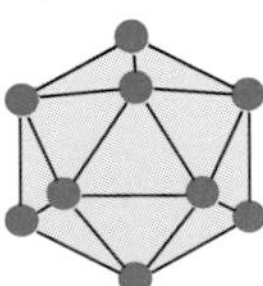

Circular single-stranded DNA of mol wt 1.7 × 10^6. Icosahedron (27 nm) with knobs on 12 vertices. Includes G4.

8. Family: *Inoviridae*

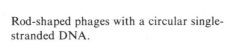

Rod-shaped phages with a circular single-stranded DNA.

Genus:

Inovirus

DNA of 1.9 to 2.7 × 10^6, long flexible filamentous particle up to 1,950 × 6 nm. Host bacteria not lysed. Includes M13 and fd phages.

Genus:

Mycoplasma virus type 1 phages.

DNA 1.5 × 10^6 mol wt. Short rods of 84 × 14 nm.

9. Family: *Cystoviridae* (phage ϕ6 group)

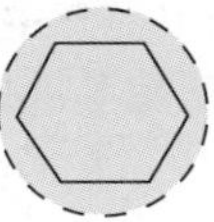

Three segments of linear double-stranded RNA of mol wt 2.3, 3.1, and 5.0 × 10^6. Isometric 75 nm particle with lipid envelope. Infects *Pseudomonas*.

10. Family: *Leviviridae* (single-stranded RNA phage group)

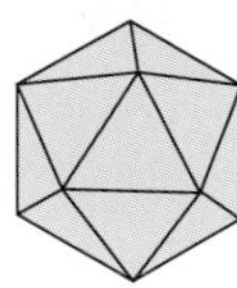

Linear single-stranded RNA of mol wt 1.2 × 10^6. Icosahedral capsid is 23 nm. Includes R17, MS2, and Qβ.

Appendix V
Helminth Diseases

The primary purpose of this appendix is to present a brief overview of the common **helminths** (Greek *helminis,* worm) that parasitize humans. The helminths are divided into two phyla: (1) the Platyhelminthes, which includes the cestodes and the trematodes, and (2) the Nematoda, which includes the nematodes.

Helminths are unique among the infectious organisms of humans because of their size, their prevalence, the complexity of their life cycles and migration within the host, their ability to induce an eosinophilia, and, particularly, because of their inability to undergo direct reproduction in the definitive human host. Whereas viruses, bacteria, fungi, and protozoa must be viewed under light or electron microscopes, all the helminths are visible to the naked eye, ranging in length from 2 to over 10,000 mm.

Helminths may well be the most prevalent of the infectious organisms of humans. For example, it has been estimated that there are almost as many helminthiases as people, considering multiple infections. Two of the helminths probably account for 1 billion infections each (*Ascaris* and *Enterobius*), two for 0.5 billion each (*Trichuris* and hookworms), and two for 0.25 billion each (the schistosomes and filariae). It has been recently estimated that there are currently about 54 million helminth infections in the United States.

Cestodes

Cestodes (tapeworms) are flattened, segmented worms, lacking a gut or mouth, that live as adults attached to the intestinal walls of their hosts. Within the intestine, they may reach 6 to 8 meters in length. The resulting disease state is due to the fact that tapeworms absorb significant quantities of nutrients and vitamins, excrete toxic wastes, and interfere with the normal passage of food through the intestine.

Tapeworms have a series of similar body segments called **proglottids** (figure AV.1*a*). The anterior portion of the worm is called the **scolex** and has hooks and suckers (figure AV.1*b*) that permit anchoring in the host's intestinal wall. In most tapeworms, immature proglottids continue to grow from the neck area and push older, maturing proglottids away from the neck. The mature proglottids contain a full set of male and female reproductive structures, but no digestive structures. Tapeworms obtain their nutrients by direct absorption across their tegument (the surface layer of cells).

Sexual reproduction occurs by either self-fertilization or by cross-fertilization between proglottids. Each proglottid produces thousands of eggs and usually stores them in a uterus. The proglottid expands and becomes a gravid proglottid, which may detach and leave the host's body in the feces (eggs may also be continuously liberated from the uterus in some species).

After leaving the host, the proglottids rupture and eggs are released. In a typical life cycle, if an egg is swallowed by the appropriate vertebrate intermediate host, the larva or oncosphere hatches and is carried by the circulation to skeletal muscle or another organ where it forms an encysted larva termed a **bladder worm (cysticercus).** If a final host eats the skeletal muscle raw or undercooked, the bladder worm excysts in the intestine and develops into a new tapeworm that attaches to the intestinal wall and completes the life cycle.

Some tapeworms found in humans are listed in table AV.1. The most common tapeworm infection is caused by *Vampirolepis nana* (dwarf tapeworm), followed by, in order of prevalence, *Taenia solium* (pork tapeworm), *Diphyllobothrium latum* (the fish tapeworm), and *Taeniarhynchus saginata* (beef tapeworm). In Canada and Alaska, the larval stage of *Echinococcus* species can form **hydatid cysts** in humans. These large cysts occur in visceral organs such as the liver and lungs.

Human tapeworm infections can be prevented by thorough cooking of all meat, because the cysticerci are killed by temperatures above 55°C. Niclosamide is the drug of choice for treatment. Surgery is necessary to remove hydatid cysts.

Nematodes

The **nematodes (roundworms)** are elongated worms that have similar body plans (figure AV.2*a–c*). They are unsegmented, cylindrical, and tapered at both ends. The worm is covered by a thick cuticle and possesses a body cavity (pseudocoelom) that separates the body musculature from the internal organs. Nematodes can infect the intestines, liver, kidneys, eyes, bloodstream, and subcutaneous tissue. Most nematode infections occur in tropical or subtropical parts of the world. In the United States, pinworm infections caused by *Enterobius vermicularis,* are the most prevalent. Worldwide, the most common nematode infection in humans is **ascariasis,** caused by *Ascaris lumbricoides.*

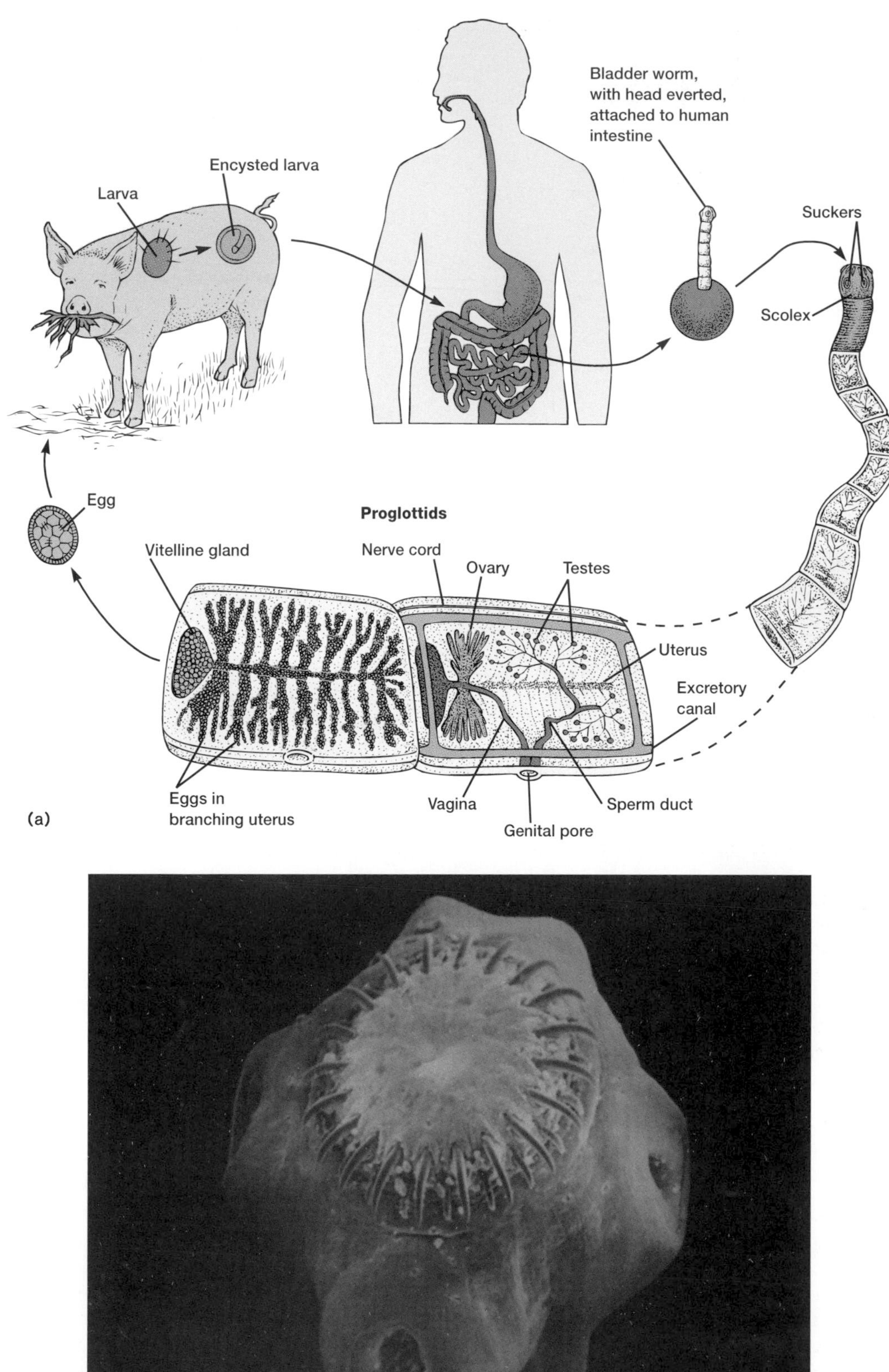

Figure AV.1 The Life Cycle and Morphology of a Common Tapeworm, *Taenia solium*. (*a*) A typical life cycle of the tapeworm showing a mature adult (right), the morphology of mature and ripe proglottids (bottom), and the bladder worm in a pig. (*b*) A scanning electron micrograph of a tapeworm scolex (×100).

Table AV.1 Some Cestode Infections in Humans

Helminth	Disease	Mode of Transmission
Vampirolepis nana (dwarf tapeworm)	Hymenolepiasis	Ingestion of parasite eggs
Taenia solium (pork tapeworm)	Taeniasis	Undercooked pork
Diphyllobothrium latum (fish tapeworm)	Diphyllobothriasis	Undercooked freshwater fish
Taeniarhynchus saginata (beef tapeworm)	Taeniasis	Undercooked beef
Echinococcus spp.	Hydatid	Contact with dogs, foxes; ingestion of eggs

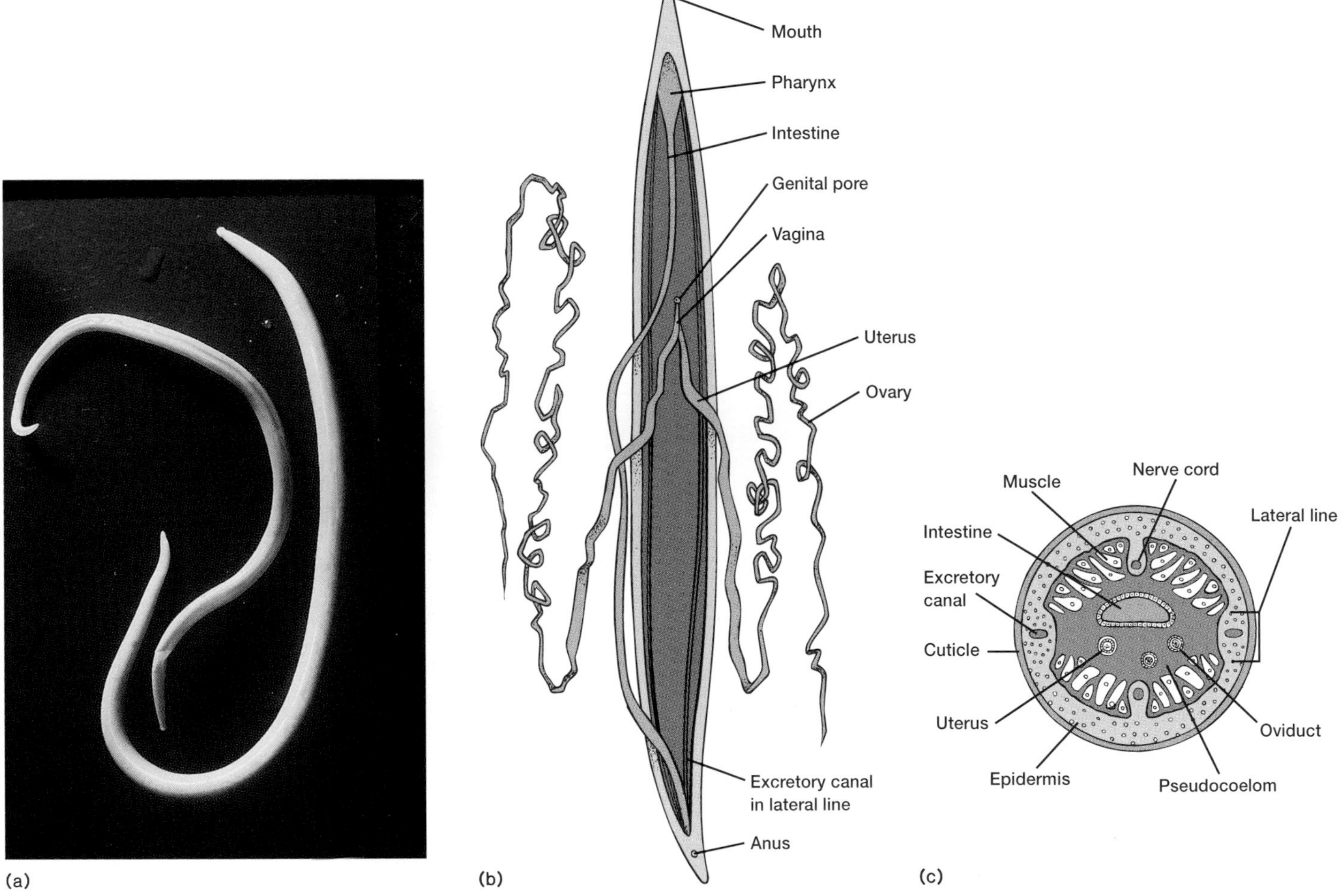

Figure AV.2 Nematode Morphology. (*a*) Male and female *Ascaris* roundworms. The female is larger. (*b*) The internal anatomy of a female *Ascaris*. (*c*) Cross-sectional view of a female *Ascaris*.

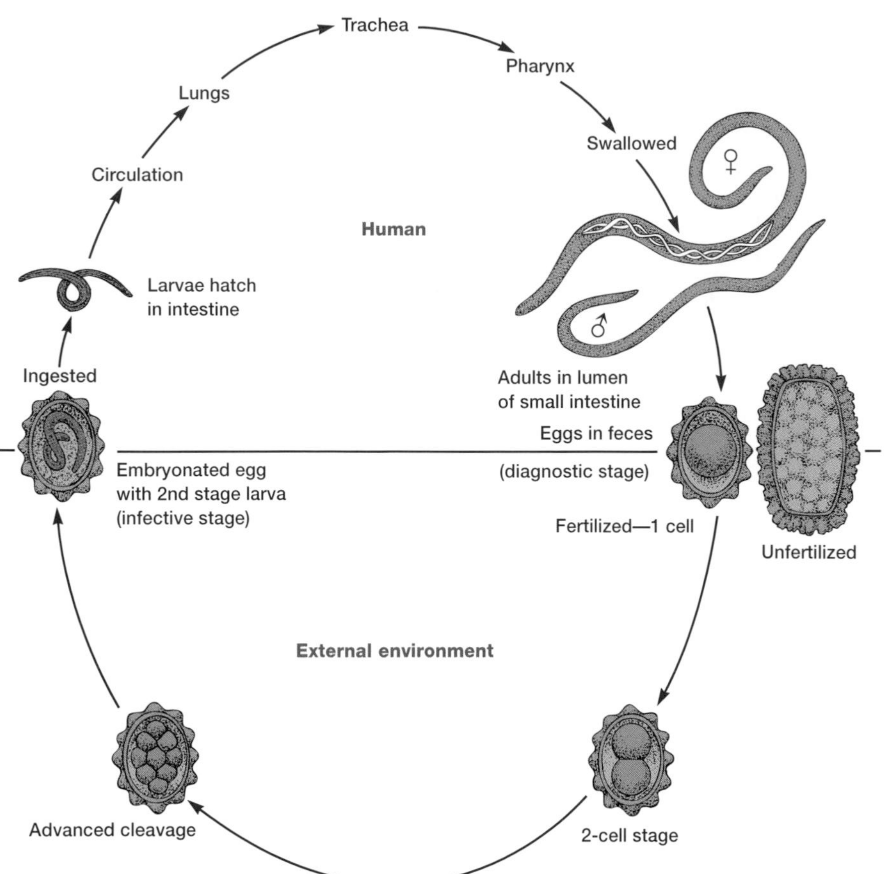

Figure AV.3 **The Life Cycle of the Nematode *Ascaris lumbricoides*.**

An *Ascaris* infection begins when a human ingests soil contaminated with *Ascaris* eggs (figure AV.3). The eggs hatch in the small intestine, penetrate the intestinal wall, enter the bloodstream, and are carried to the lungs. After a brief period in the lungs, the juvenile worms crawl up the trachea and move down the pharynx, eventually reaching the small intestine. Once back in the small intestine, they mature, mate, and begin producing eggs. In the intestine, large numbers of worms may cause intestinal obstruction requiring surgery.

Another human nematode parasite that occurs worldwide is *Enterobius vermicularis* (pinworm) that causes **enterobiasis.** Pinworm infections cause itching in the anal area but little real damage. Other human nematode parasites, however, can cause more serious diseases. For example, **trichinosis** is caused by the roundworm, *Trichinella spiralis.* After ingestion, juveniles escape their cysts, penetrate the intestinal mucosa, mature, and reproduce. The new juveniles are carried throughout the body in the blood. When they reach the skeletal muscle, they invade and form cysts (figure AV.4). The host is damaged by the initial penetration of the worms, the migration of juve-

Figure AV.4 ***Trichinella spiralis* Larvae.** The light micrograph shows larval cysts in skeletal muscle. The larvae could infect a human that ate this meat raw or undercooked (×100).

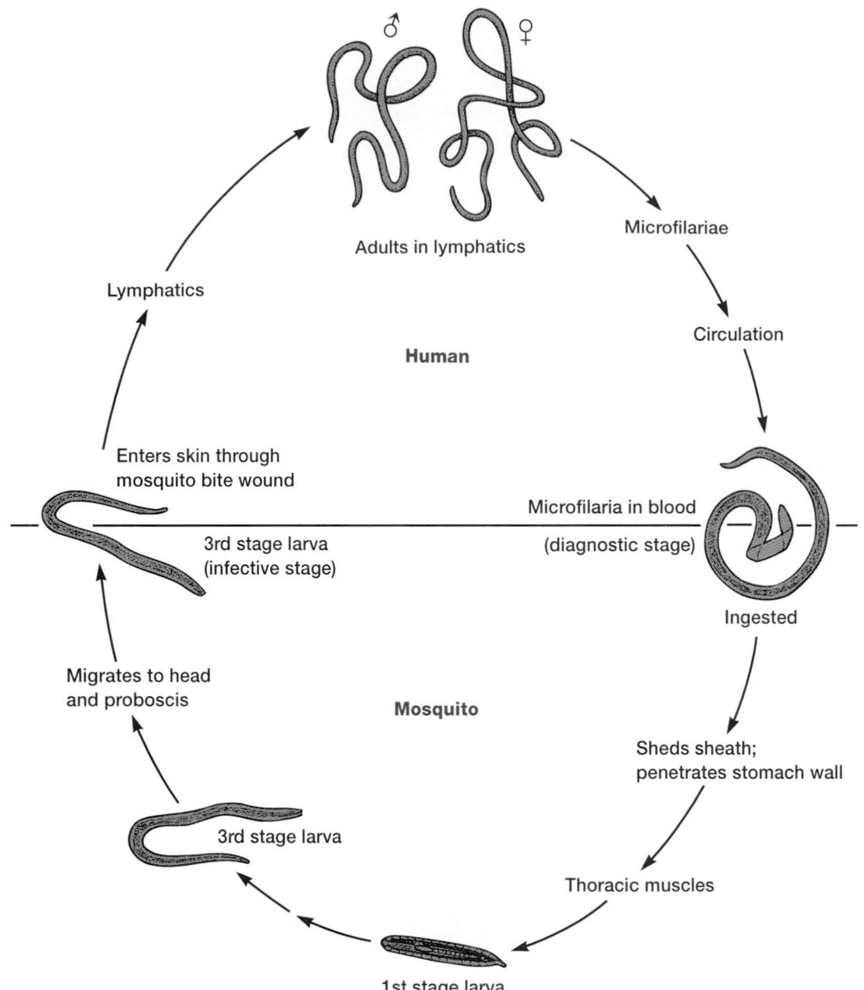

Figure AV.5 The Life Cycle of the Filarial Nematode *Wuchereria bancrofti*. This nematode is the causative agent of elephantiasis. *Source: Melvin,* Common Blood and Tissue Parasites of Man. *Department of Health, Education, and Welfare, Public Health Publication No. 1234, The Centers for Disease Control, Atlanta, GA.*

niles, and cyst formation in skeletal muscle. If the initial dose of juveniles is too large, death results. Americans are usually infected upon eating contaminated, undercooked pork and pork products such as sausage. In the tropics, the **microfilariae** (small larvae) of **filarial worms** (*Loa loa, Onchocerca volvulus, Wuchereria bancrofti*) are transmitted by insect vectors from human to human (figure AV.5). *Wuchereria bancrofti* causes the disease **filariasis.** The adult worms enter the lymphatic system, block lymphatic vessels, and cause fluid accumulation and swelling in various parts of the body. This condition is known as **elephantiasis** (figure AV.6). *Loa loa* causes **loiasis,** which is manifested by swellings on the body; the worms can also migrate to the eye, causing blurred vision. *Onchocerca volvulus* causes blindness (onchocerciasis), dermatitis, and formation of nodules on the body.

The drugs of choice for the intestinal nematodes are mebendazole or thiobendazole; for the filarial worms, diethylcarbamazine or ivermectin. Table AV.2 summarizes some of the more common nematode parasites of humans.

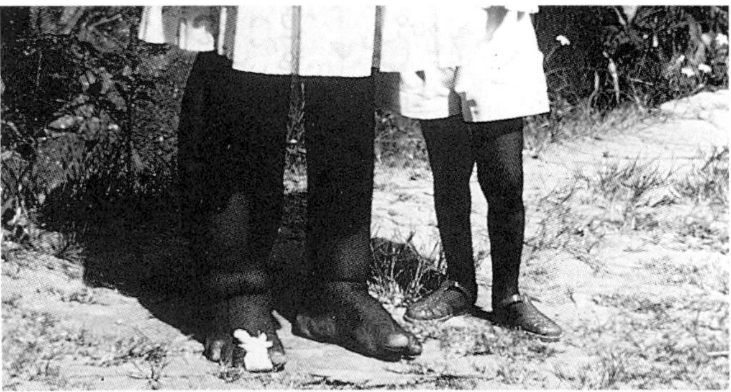

Figure AV.6 Elephantiasis. The person on the left is suffering from elephantiasis. This disease results when adult filarial worms block lymphatic vessels. Fluid accumulates and causes the extreme swelling of certain body regions.

Trematodes

The **trematodes (flukes)** are flatworms that exhibit bilateral symmetry and have one or two suckers. Most flukes are endoparasites (living inside the body of their host), and they use their muscular pharynx to suck nutrients into their digestive system (figure AV.7*a*). Much of the space inside a fluke is occupied by reproductive structures. Except for a group called the schistosomes, most species are **hermaphroditic** (having both male and female reproductive organs) and can produce zygotes either by self-fertilization or cross-fertilization. A fluke can produce large numbers of eggs every day and remain alive for years within its final host. Eggs usually pass into the digestive tract and exit the body with the feces.

Table AV.2 Some Nematode Infections in Humans

Helminth	Disease	Mode of Transmission
Ancylostoma duodenale (hookworm)	Ancylostomiasis	Penetration of skin by infective larvae from contaminated soil
Ascaris lumbricoides (roundworm)	Ascariasis	Ingestion of eggs from feces-contaminated soil
Enterobius vermicularis (pinworm)	Enterobiasis	By hand; passing eggs from anus to mouth
Necator americanus (hookworm)	Hookworm	Penetration of the skin or by ingestion
Strongyloides stercoralis	Strongyloidiasis	Penetration of skin by infective larvae
Toxocara canis (dog worm)	Visceral larva migrans	Ingestion of eggs
Trichuris trichiura (whipworm)	Trichuriasis	Ingestion of eggs
Trichinella spiralis (pork worm)	Trichinosis	Ingestion of undercooked pork
Loa loa (eye worm)	Loiasis	Bite of tabanid fly
Onchocerca volvulus	Onchocerciasis (river blindness)	Bite of black fly
Wuchereria bancrofti	Filariasis	Bite of mosquito

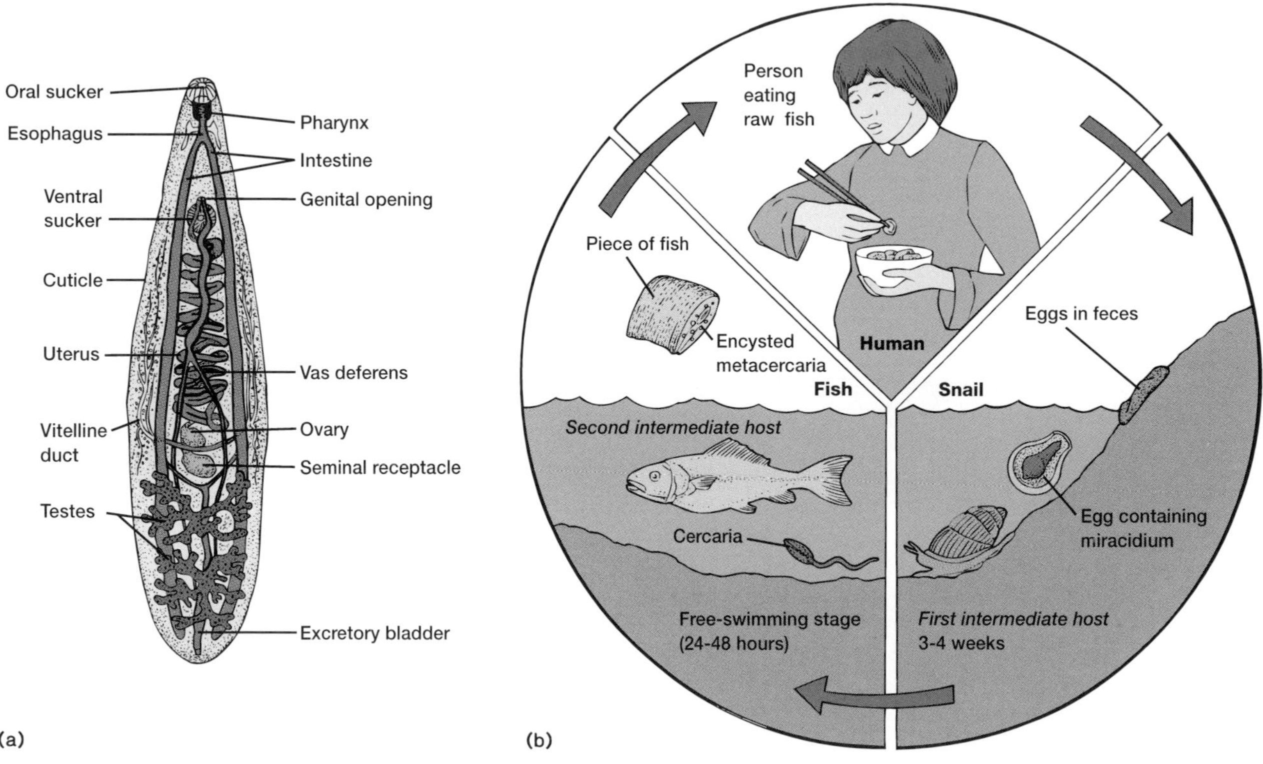

Figure AV.7 **The Human Liver Fluke, *Clonorchis sinensis*.** (*a*) Anatomy of an adult worm. Each adult has both male and female reproductive organs. (*b*) A simplified life cycle of the human liver fluke.

In the life cycle of the human liver fluke (figure AV.7*b*), the eggs leave the host in fecal material and enter the water. The eggs hatch and release a **miracidium** larva after being eaten by a suitable snail host. (In contrast, *Schistosoma* eggs hatch upon contact with water to release a ciliated, free-swimming miracidium larva.) The miracidium penetrates the snail (the first intermediate host), burrows into its liver, and develops into a saclike form called the **sporocyst** in which a series of asexual reproductions take place to generate another intermediate form, the **redia.** Eventually, the rediae produce a large number of larvae termed **cercariae** that leave the snail.

The cercariae swim about actively until they find a suitable second intermediate host such as a fish, plant, or crustacean. Once contact is made, they burrow into the second intermediate host or attach to it and form a cyst termed a **metacercaria.** The metacercaria can remain alive within the second intermediate host for years. If the second intermediate host is eaten by a final host, the metacercaria excysts, the worm makes its way to a specific organ, matures, reproduces, and begins producing eggs. In some life cycles, the cercaria actively penetrates the skin of the final host, such as a human. Once inside the body, the cercaria loses its tail, reaches the intestinal blood vessels, matures, and begins producing eggs.

The adult *S. mansoni* live in the mesenteric venules of the large and small intestines. Schistosomes have separate sexes, and the male and female mate continuously (figure AV.8). Some of the eggs produced mechanically penetrate the gut wall and enter the feces (figure AV.9), and others accumulate in body tissues. The damage caused by these eggs accounts for many of the symptoms of schistosomiasis (dysentery, anemia, general weakness, and greatly reduced resistance to other infections).

The drug of choice for schistosomiasis is niridazole. Table AV.3 summarizes some of the more common trematode parasites of humans.

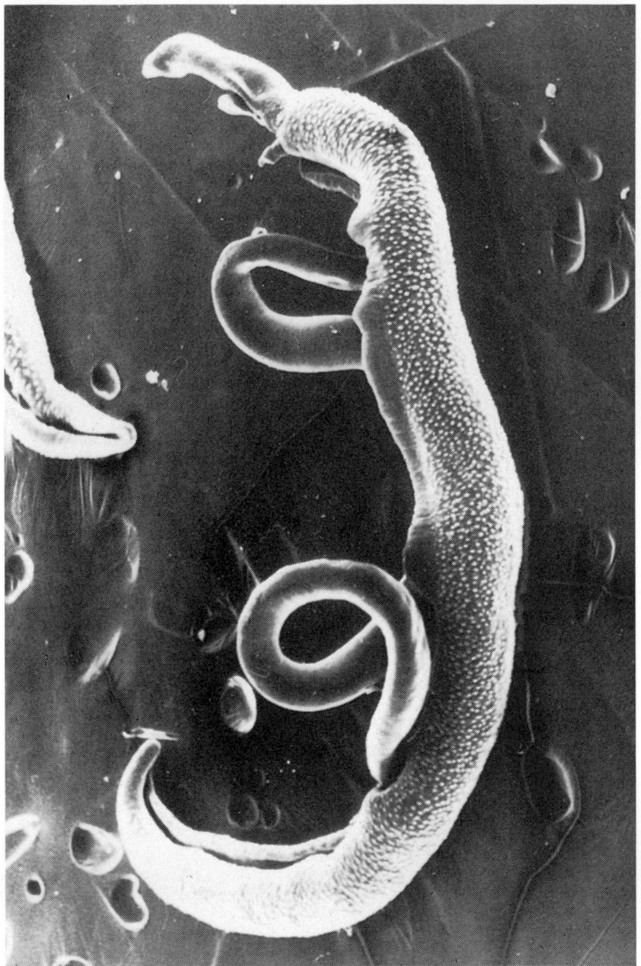

Figure AV.8 **A Pair of *Schistosoma mansoni* Adults.** This scanning electron micrograph shows a large male holding the female in a ventral groove called the gynecophoral canal.

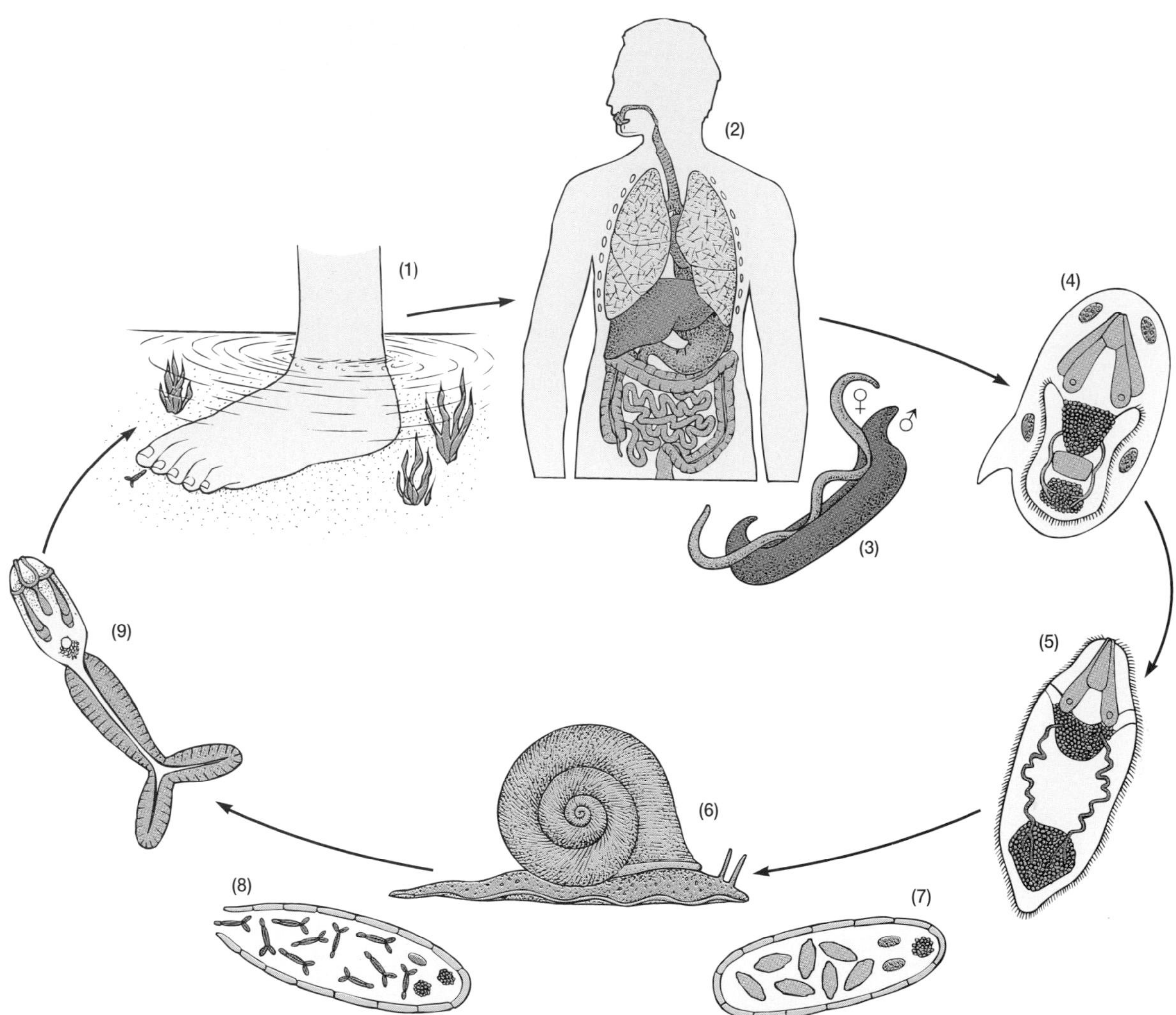

Figure AV.9 The Life Cycle of *Schistosoma mansoni*. (*1*) Cercariae in water penetrate human skin. (*2*) Adult worms are located in the mesenteric venules of the intestines. (*3*) Copulating adult worms. (*4*) An egg containing a developing miracidium larva. (*5*) A ciliated miracidium larva is released when the egg hatches. (*6*) The miracidium penetrates a host snail and develops into (*7*) a sporocyst, which produces (*8*) daughter sporocysts. Each daughter sporocyst forms many cercariae (*9*) that leave the snail, enter the water, and search for a human host. (The illustrations are drawn to different scales.)

TABLE AV.3 Some Trematode Infections of Humans

Helminth	Disease	Mode of Transmission
Clonorchis sinensis (oriental liver fluke)	Clonorchiasis	Eating raw or undercooked fish
Fasciola hepatica (liver fluke)	Fascioliasis	Eating raw watercress
Paragonimus westermani (lung fluke)	Paragonimiasis	Eating raw crabs
Schistosoma haematobium	Schistosomiasis	Cercariae penetrate skin
Schistosoma japonicum	Schistosomiasis	Cercariae penetrate skin
Schistosoma mansoni	Schistosomiasis	Cercariae penetrate skin

Glossary

Pronunciation Guide

Many of the boldface terms in this glossary are followed by a phonetic spelling in parentheses. These pronunciation aids usually come from *Dorland's Illustrated Medical Dictionary.* The following rules are taken from this dictionary and will help in using its phonetic spelling system.

1. An unmarked vowel ending a syllable (an open syllable) is long; thus *ma* represents the pronunciation of *may; ne,* that of *knee; ri,* of *wry; so,* of *sew; too,* of *two;* and *vu,* of *view.*
2. An unmarked vowel in a syllable ending with a consonant (a closed syllable) is short; thus *kat* represents *cat; bed, bed; hit, hit; not, knot; foot, foot;* and *kusp, cusp.*
3. A long vowel in a closed syllable is indicated by a macron; thus *māt* stands for *mate; sēd,* for *seed; bīl,* for *bile; mōl,* for *mole; fūm,* for *fume;* and *fo͞ol,* for *fool.*
4. A short vowel that ends or itself constitutes a syllable is indicated by a breve; thus *ĕ-fekt'* for *effect, ĭ-mūn'* for *immune,* and *ŏ-klo͞od'* for *occlude.*

Primary (') and secondary ('') accents are shown in polysyllabic words. Unstressed syllables are followed by hyphens.

AB model The model that describes the structure and activity of many exotoxins. The B portion of the toxin is responsible for toxin binding to a cell, but does not directly harm it. The A portion enters the cell and disrupts its function. *584*

ABO blood grouping The major human blood typing system that reflects the presence or absence of the A and B glycoprotein antigens in/on erythrocyte plasma membranes. *643*

accessory pigments Photosynthetic pigments such as carotenoids and phycobiliproteins that aid chlorophyll in trapping light energy. *163*

acetyl coenzyme A (acetyl-CoA) A combination of acetic acid and coenzyme A that is energy rich; it is produced by many catabolic pathways and is the substrate for the tricarboxylic acid cycle, fatty acid biosynthesis, and other pathways. *150*

acid dyes Dyes that are anionic or have negatively charged groups such as carboxyls. *28*

acid fast Refers to bacteria like the mycobacteria that cannot be easily decolorized with acid alcohol after being stained with dyes such as basic fuchsin. *465*

acid-fast staining A staining procedure that differentiates between bacteria based on their ability to retain a dye when washed with an acid alcohol solution. *29*

acidophile (as'id-o-fīl'') A microorganism that has its growth optimum between about pH 1.0 and 5.5. *123*

acne vulgaris A chronic inflammatory disease of the sebaceous glands, the lesions occurring most frequently on the face, chest, and back. The inflamed glands may form small pink papules, which sometimes surround comedones so as to have black centers, or form pustules. *573*

acquired enamel pellicle A membranous layer on the tooth enamel surface formed by selectively adsorbing glycoproteins (mucins) from saliva. This pellicle confers a net negative charge to the tooth surface. *776*

acquired immune deficiency syndrome (AIDS) An infectious disease syndrome caused by the human immunodeficiency retrovirus that is characterized by the loss of a normal immune response, followed by increased susceptibility to opportunistic infections and an increased risk of some cancers. *722*

acquired immune tolerance The ability to produce antibodies against nonself antigens while "tolerating" (not producing antibodies against) self antigens. *632*

acquired immunity Refers to the type of specific immunity that develops after exposure to a suitable antigen or is produced after antibodies are transferred from one individual to another. *608*

actinomycete (ak''tĭ-no-mi'sēt) An aerobic, gram-positive bacterium that forms branching filaments (hyphae) and asexual spores. *507*

actinorhizal Referring to an association of an actinomycete and a plant root. *856*

activation energy The energy required to bring reacting molecules together to reach the transition state in a chemical reaction. *141*

active carrier An individual who has an overt clinical case of a disease and who can transmit the infection to others. *706*

active site The part of an enzyme that binds the substrate to form an enzyme-substrate complex and catalyze the reaction. Also called the **catalytic site.** *141*

active transport The transport of solute molecules across a membrane against a concentration and/or electrical gradient; it requires a carrier protein and the input of energy. *102*

acute carrier *See* casual carrier. *706*

acute infections Virus infections with a fairly rapid onset that last for a relatively short time. *393*

acute-phase reactant Refers to the qualitative and quantitative changes that occur in the host's blood plasma during the early stages of an acute infection. These changes are an important part of the host's general or nonspecific defense and do not involve immunoglobulins. *590*

acute viral gastroenteritis An inflammation of the stomach and intestines, normally caused by either the Norwalk viruses or rotaviruses. *735*

acyclovir (a-si'klo-vir) A synthetic purine nucleoside derivative with antiviral activity against the herpes simplex virus. *341*

adenine (ad'e-nēn) A purine derivative, 6-aminopurine, found in nucleosides, nucleotides, coenzymes, and nucleic acids. *183*

adenosine diphosphate (ADP; ah-den'o-sēn) The nucleoside diphosphate usually formed upon the breakdown of ATP when it provides energy for work. *134*

adenosine 5′-triphosphate (ATP) The triphosphate of the nucleoside adenosine, which is a high energy molecule and serves as the cell's major form of energy currency. *134*

adjuvant (aj'ĕ-vant) Material added to an antigen to increase its immunogenicity. Common examples are alum, killed *Bordetella pertussis,* and an oil emulsion of the antigen, either alone (Freund's incomplete adjuvant) or with killed mycobacteria (Freund's complete adjuvant). *620*

adult T-cell leukemia A type of white blood cell cancer caused by the HTLV-1 virus. *731*

aerobe (a'er-ōb) An organism that grows in the presence of atmospheric oxygen. *126*

aerobic respiration (res"pĭ-ra'shun) A metabolic process in which molecules, often organic, are oxidized with oxygen as the final electron acceptor. *134*

aerotolerant anaerobes Microorganisms that grow equally well whether or not oxygen is present. *127*

aflatoxin (af"lah-tok'sin) Fungal metabolite that can cause cancer. *873*

agar (ahg'ar) A complex sulfated polysaccharide, usually extracted from red algae, that is used as a solidifying agent in the preparation of culture media. *105*

agglutinates The visible aggregates formed by an agglutination reaction. *656*

agglutination reaction (ah-gloo"tĭ-na'shun) The formation of an insoluble immune complex by the cross-linking of cells or particles. *654*

agglutinin (ah-gloo"tĭ-nin) The antibody responsible for an agglutination reaction. *654*

AIDS *See* acquired immune deficiency syndrome. *722*

AIDS-related complex (ARC) A collection of symptoms such as lymphadenopathy (swollen lymph glands), fever, malaise, fatigue, loss of appetite, and weight loss. It results from an HIV infection and may progress to frank AIDS. *724*

airborne transmission The type of infectious organism transmission in which the pathogen is truly suspended in the air and travels over a meter or more from the source to the host. *707*

akinetes Specialized, nonmotile, dormant, thick-walled resting cells formed by some cyanobacteria. *474*

alcoholic fermentation A fermentation process that produces ethanol and CO_2 from sugars. *156*

alga (al'gah) A common term for a series of unrelated groups of photosynthetic eucaryotic microorganisms lacking multicellular sex organs (except for the charophytes) and conducting vessels. *536*

algicide (al'jĭ-sīd) An agent that kills algae. *311*

algology (al-gol'o-je) The scientific study of algae. *536*

alkalophile A microorganism that grows best at pHs around 8.5 to 11.5. *123*

allergen (al'er-jen) A substance capable of inducing allergy or specific susceptibility. *635*

allergic contact dermatitis An allergic reaction caused by haptens that combine with proteins in the skin to form the allergen that produces the immune response. *638*

allergy (al'er-je) *See* hypersensitivity. *635*

allochthonous Microorganisms or substances that are not normally present in a particular habitat and that have been introduced from the outside. *816*

allograft (al'o-graft) A graft between individuals of the same species that differ in genotype. *640*

allosteric enzyme (al"o-ster'ik) An enzyme whose activity is altered by the binding of a small effector or modulator molecule at a regulatory site separate from the catalytic site; effector binding causes a conformational change in the enzyme and its catalytic site, which leads to enzyme activation or inhibition. *218*

allotype Allelic variants of antigenic determinant(s) found on antibody chains of some, but not all, members of a species, which are inherited as simple Mendelian traits. *612*

alpha hemolysis A greenish zone of partial clearing around a bacterial colony growing on blood agar. *590*

alternative pathway An antibody-independent pathway of complement activation that includes the C3–C9 components of the classical pathway and several other serum protein factors (e.g., properdin). *648*

amantadine (ah-man'tah-den) An antiviral compound used to prevent type A influenza infections. *341*

amebiasis (am"e-bi'ah-sis) An infection with amoebae, often resulting in dysentery; usually it refers to an infection by *Entamoeba histolytica.* *790*

American trypanosomiasis (Chagas' disease) *See* trypanosomiasis. *797*

Ames test A test that uses a special *Salmonella* strain to test chemicals for mutagenicity and potential carcinogenicity. *252*

amino acid activation The initial stage of protein synthesis in which amino acids are attached to transfer RNA molecules. *206*

aminoacyl or **acceptor site (A site)** The site on the ribosome that contains an aminoacyl-tRNA at the beginning of the elongation cycle during protein synthesis; the growing peptide chain is transferred to the aminoacyl-tRNA and grows by an amino acid. *211*

aminoglycoside antibiotics (am"ĭ-no-gli'ko-sīd) A group of antibiotics synthesized by *Streptomyces* and *Micromonospora,* which contain a cyclohexane ring and amino sugars; all aminoglycoside antibiotics bind to the small ribosomal subunit and inhibit protein synthesis. *337*

amoeboid movement Moving by means of cytoplasmic flow and the formation of pseudopodia (temporary cytoplasmic protrusions of the cytoplasm). *553*

amphibolic pathways (am"fe-bol'ik) Metabolic pathways that function both catabolically and anabolically. *146*

amphitrichous (am-fit'rĕ-kus) A cell with a single flagellum at each end. *57*

amphotericin B (am"fo-ter'i-sin) An antibiotic from a strain of *Streptomyces nodosus* that is used to treat systemic fungal infections; it is also used topically to treat candidiasis. *340*

anabolism (ah-nab'o-lizm") The synthesis of complex molecules from simpler molecules with the input of energy. *146*

anaerobe (an-a'er-ōb) An organism that grows in the absence of free oxygen. *126*

anaerobic digestion (an"a-er-o'bik) The microbiological treatment of sewage wastes under anaerobic conditions to produce methane. *833*

anaerobic respiration (an"a-er-o'bik) An energy-yielding process in which the electron transport chain acceptor is an inorganic molecule other than oxygen. *158*

anamnestic response (an"am-nes'tik) The recall, or the remembering, by the immune system of a prior response to a given antigen. *621*

anaphylaxis (an"ah-fĭ-lak'sis) An immediate (type I) hypersensitivity reaction following exposure of a sensitized individual to the appropriate antigen. Mediated by reagin antibodies, chiefly IgE. *636*

anaplerotic reactions (an"ah-plĕ-rot'ik) Reactions that replenish depleted tricarboxylic acid cycle intermediates. *182*

anogenital condylomata (venereal warts) (kon"dĭ-lo"mah-tah) Warts that are sexually transmitted and caused by the type 6 papilloma virus. Usually occur around the cervix, vulva, perineum, anus, anal canal, urethra, or glans penis. *738*

anoxygenic photosynthesis Photosynthesis that does not oxidize water to produce oxygen; the form of photosynthesis characteristic of purple and green photosynthetic bacteria. *167, 468*

antheridium (an"ther-id'e-um; pl., **antheridia**) A male gamete-producing organ, which may be unicellular or multicellular. *526, 538*

anthrax (an'thraks) An infectious disease of animals caused by ingesting *Bacillus anthracis* spores. Can also occur in humans and is sometimes called **woolsorter's disease.** *753*

antibiotic (an″tĭ-bi-ot′ik) A microbial product or its derivative that kills susceptible microorganisms or inhibits their growth. *326*

antibody (an′tĭ-bod″e) A glycoprotein produced in response to the introduction of an antigen; it has the ability to combine with the antigen that stimulated its production. Also known as an **immunoglobulin** (Ig). *611*

antibody-dependent cell-mediated cytotoxicity An antibody provides the bridging mechanism for the intimate association between the effector and target cell. The effector cell then releases its cytotoxic mediators or directly causes cytolysis—either of which kills the target cell. *651*

antibody-mediated immunity *See* humoral immunity. *609*

anticodon triplet The base triplet on a tRNA that is complementary to the triplet codon on mRNA. *206*

antigen (an′tĭ-jen) A foreign (nonself) substance (such as protein, nucleoprotein, polysaccharide, and some glycolipids) to which lymphocytes respond; also known as an **immunogen** because it induces the immune response. *610*

antigen-binding fragment (Fab) The top of the Y portion of an antibody molecule; contains the sites that bind with compatible antigenic determinants. *612*

antigenic determinant site *See* epitope. *610*

antigenic shift A major change in the antigenic character of an organism that alters it to an antigenic strain unrecognized by host immune mechanisms. A minor antigenic variation in a strain is called **antigenic drift.** *704*

antimetabolite (an″tĭ-mĕ-tab′o-līt) A compound that blocks metabolic pathway function by competitively inhibiting a key enzyme's use of a metabolite because it closely resembles the normal enzyme substrate. *330*

antimicrobial agent An agent that kills microorganisms or inhibits their growth. *312*

antisense RNA A single-stranded RNA with a base sequence complementary to a segment of another RNA molecule that can specifically bind to the target RNA and inhibit its activity. *228*

antiseptic (an″tĭ-sep′tik) Chemical agents applied to tissue to prevent infection by killing or inhibiting pathogens. *311*

antiserum (an″tĭ-se′rum) Serum containing induced antibodies. *620*

antitoxin (an″tĭ-tok′sin) An antibody to a microbial toxin, usually a bacterial exotoxin, that combines specifically with the toxin, in vivo and in vitro, neutralizing the toxin. *584, 651*

apical complex (ap′ĭ-kal) A set of organelles characteristic of members of the phylum *Apicomplexa:* polar rings, subpellicular microtubules, conoid, rhoptries, and micronemes. *556*

apicomplexan (a′pĭ-kom-plek′san) A sporozoan protist that lacks special locomotor organelles but has an apical complex and a spore-forming stage. It is either an intra- or extracellular parasite of animals; a member of the phylum *Apicomplexa.* *556*

aplanospore (a′plan-o-spor) A nonflagellated, nonmotile spore that is involved in asexual reproduction. *538*

apoenzyme (ap″o-en′zīm) The protein part of an enzyme that also has a nonprotein component. *140*

arbuscules Branched, treelike structures formed in cells of plant roots colonized by endotrophic mycorrhizal fungi. *855*

archaeobacteria or **archaebacteria** (ar″keo-bak-tē′re-ah) Bacteria that lack muramic acid in their cell walls, have membrane lipids with ether-linked branched chain fatty acids, and differ in many other ways from eubacteria. *417, 493*

arthroconidium (ar′thro-ko-nid′e-um; pl., **arthroconidia**) A thallic conidium released by the fragmentation or lysis of hypha. It is not notably larger than the parental hypha, and separation occurs at a septum. *522*

arthrospore (ar′thro-spōr) A spore resulting from the fragmentation of a hypha. *522*

artificially acquired active immunity The type of immunity that results from immunizing an animal with a vaccine. *608*

artificially acquired passive immunity The type of immunity that results from introducing into an animal antibodies that have been produced either in another animal or by in vitro methods. *609*

ascariasis (as″kah-ri′ah-sis) Infection by the nematode *Ascaris,* usually by *A. lumbricoides.* *A34*

ascocarp (as′ko-karp) A multicellular structure in ascomycetes lined with specialized cells called asci in which nuclear fusion and meiosis produce ascospores. An ascocarp can be open or closed and may be referred to as a fruiting body. *526*

ascogenous hypha A specialized hypha that gives rise to one or more asci. *526*

ascogonium (as″ko-go′ne-um; pl., **ascogonia**) The receiving (female) organ in ascomycetous fungi which, after fertilization, gives rise to ascogenous hyphae and later to asci and ascospores. *526*

ascomycetes (as″ko-mi-se′tēz) A division of fungi that form ascospores. *525*

ascospore (as′ko-spor) A spore contained or produced in an ascus. *524*

ascus (as′kus) A specialized cell, characteristic of the ascomycetes, in which two haploid nuclei fuse to produce a zygote, which immediately divides by meiosis; at maturity, an ascus will contain ascospores. *525*

aseptic meningitis syndrome *See* meningitis. *744*

aspergillosis (as″per-jil-o′sis) A fungal disease caused by species of *Aspergillus.* *788*

assimilatory reduction The reduction of an inorganic molecule to incorporate it into organic material. No energy is made available during this process. *810*

associative nitrogen fixation Nitrogen fixation by bacteria in the plant root zone (rhizosphere). *854*

athlete's foot *See* tinea pedis. *784*

attenuation (ah-ten″u-a′shun) 1. A mechanism for the regulation of transcription of some bacterial operons by aminoacyl-tRNAs. 2. A procedure that reduces or abolishes the virulence of a pathogen without altering its immunogenicity. *227, 608*

attenuator A rho-independent termination site in the leader sequence that is involved in attenuation. *227*

autoallergy *See* autoimmune disease. *640*

autochthonous (aw-tok′tho-nus) Pertaining to substances or microorganisms indigenous to a particular environment. *848*

autoclave (aw′to-klāv) An apparatus for sterilizing objects by the use of steam under pressure. *314*

autogenous infection (aw-toj′e-nus) An infection that results from a patient's own microbiota, regardless of whether the infecting organism became part of the patient's microbiota subsequent to admission to a clinical care facility. *711*

autograft (aw′to-graft) A graft involving the transplanting of one's own tissue from one part of the body to another part of the same body. *640*

autoimmune disease An immune response to the body's own cells and tissues. Also known as an **autoallergy.** *640*

autolysins (aw-tol′ĭ-sins) Enzymes that partially digest peptidoglycan in growing bacteria so that the peptidoglycan can be enlarged. *188*

autotroph (aw′to-trōf) An organism that uses CO_2 as its sole or principal source of carbon. *97*

auxotroph (awk′so-trōf) A mutated prototroph that lacks the ability to synthesize an essential nutrient and therefore must obtain it or a precursor from its surroundings. *98, 244*

axial filament The organ of motility in spirochetes. It is made of axial fibrils or periplasmic flagella that extend from each end of the protoplasmic cylinder and overlap in the middle of the cell. The outer sheath lies outside the axial filament. *60, 428*

bacillus (bah-sil′lus) A rod-shaped bacterium. *42*

bacteremia (bak″ter-e′me-ah) The presence of bacteria in the blood. *778*

bacterial (septic) meningitis *See* meningitis. *744*

bactericide (bak-tēr′ĭ-sid) An agent that kills bacteria. *311*

bacteriochlorophyll (bak-te″re-o-klo′ro-fil) A modified chlorophyll that serves as the primary light-trapping pigment in purple and green photosynthetic bacteria. *167*

bacteriocin (bak-te′re-o-sin) A protein produced by a bacterial strain that kills other closely related strains. *595, 262*

bacteriophage (bak-te′re-o-fāj″) A virus that uses bacteria as its host; often called a **phage.** *349, 368*

bacteriophage (phage) typing A technique in which strains of bacteria are identified based on their susceptibility to a variety of bacteriophages. *692*

bacteriostasis (bak-te″re-os′tah-sis) *See* bacteriostatic. *809*

bacteriostatic (bak-te″re-o-stat′ik) Inhibiting the growth and reproduction of bacteria. *311*

bacteroid (bak'tĕ-roid) A modified, often pleomorphic, bacterial cell living within the root nodule cells of legumes; it carries out nitrogen fixation. *854*

baeocytes Small, spherical, reproductive cells produced by pleurocapsalean cyanobacteria through multiple fission. *476*

balanced growth Microbial growth in which all cellular constituents are synthesized at constant rates relative to each other. *121*

balanitis (bal''ah-ni'tis) Inflammation of the glans penis usually associated with *Candida* fungi; a sexually transmitted disease. *788*

barophilic (bar''o-fil'ik) or **barophile** Organisms that prefer or require high pressures for growth and reproduction. *129, 827*

barotolerant Organisms that can grow and reproduce at high pressures but do not require them. *129*

basal body The cylindrical structure at the base of procaryotic and eucaryotic flagella that attaches them to the cell. *57, 88*

base analogs Molecules that resemble normal DNA nucleotides and can substitute for them during DNA replication, leading to mutations. *246*

basic dyes Dyes that are cationic, or have positively charged groups, and bind to negatively charged cell structures. Usually sold as chloride salts. *28*

basidiocarp (bah-sid'e-o-karp'') The fruiting body of a basidiomycete that contains the basidia. *529*

basidiomycetes (bah-sid''e-o-mi-se'tēz) A division of fungi in which the spores are born on club-shaped organs called **basidia.** *526*

basidiospore (bah-sid'e-o-spōr) A spore born on the outside of a basidium following karyogamy and meiosis. *524*

basidium (bah-sid'e-um; pl., **basidia**) A structure that bears on its surface a definite number of basidiospores (typically four) that are formed following karyogamy and meiosis. Basidia are found in the basidiomycetes and are usually club-shaped. *529*

batch culture A culture of microorganisms produced by inoculating a closed culture vessel containing a single batch of medium. *113*

B cell, also known as a **B lymphocyte** A type of lymphocyte derived from bone marrow stem cells that matures into an immunologically competent cell under the influence of the bursa of Fabricius in the chicken, and bone marrow in nonavian species. Following interaction with antigen, it becomes a plasma cell, which synthesizes and secretes antibody molecules involved in humoral immunity. *609, 628*

benthic (ben'thic) Pertaining to the bottom of the sea or another body of water. *536*

beta hemolysis A zone of complete clearing around a bacterial colony growing on blood agar. The zone does not change significantly in color. *590*

beta-lysin (ba''tah-li'sin) A polypeptide released from blood platelets that can kill some gram-positive bacteria by disrupting their plasma membranes. *593*

binal symmetry The symmetry of some virus capsids (e.g., those of complex phages) that is a combination of icosahedral and helical symmetry. *362*

binary fission Asexual reproduction in which a cell or an organism separates into two cells. *480, 538, 550*

binomial system The nomenclature system in which an organism is given two names; the first is the capitalized generic name, and the second is the uncapitalized specific epithet. *406*

biochemical markers Biochemical characteristics of microorganisms that can be monitored to detect genetic exchange (e.g., traits such as antibiotic resistance and changes in nutrient requirements). *899*

biochemical oxygen demand (BOD) The amount of oxygen used by organisms in water under certain standard conditions; it provides an index of the amount of microbially oxidizable organic matter present. *819, 831*

biocide A substance that can kill organisms. *907*

biocontrol The use of biological agents such as microorganisms to control undesired plants, insects, and other animals. *903*

bioconversion The use of living organisms to modify substances that are not normally used for growth. Also known as **biotransformation** or **microbial transformation.** *898*

biodegradation enhancement (bi''o-deg''rah-da'shun) The modification of environmental conditions, including inoculation with microorganisms, to enhance degradation of particular compounds. *908*

biodeterioration The biologically mediated deterioration of materials that usually results from microbial activity. *906*

biogeochemical cycling The oxidation and reduction of substances carried out by living organisms and/or abiotic processes that results in the cycling of elements within and between different parts of the ecosystem (the soil, aquatic environment, and atmosphere). *809*

biologic transmission A type of vector-borne transmission in which an infectious organism goes through some morphological or physiological change within the vector. *707*

bioluminescence (bi''o-loo''mi-nes'ens) The production of light by living cells, often through the oxidation of molecules by the enzyme luciferase. *439*

biomagnification The increase in concentration of a substance in higher level consumer organisms. *814*

biopolymer (bi''o-pol'i-mer) A polymer produced by microorganisms or other organisms; used in food and industrial microbiology. *905*

bioremediation The use of biological processes to remedy problems, particularly those caused by pollution. *843*

biosensor The coupling of a biological process with production of an electrical signal or light to detect the presence of particular substances. *905*

biosynthesis *See* anabolism. *146*

biotransformation *See* bioconversion. *898*

black peidra (pe-a'drah) A fungal infection (tinea nigra) caused by *Piedraia hortai* that forms hard black nodules on the hairs of the scalp. *782*

bladder worm (cysticercus) A tapeworm larva found in intermediate hosts; it is a fluid-filled sac containing an invaginated scolex. *A34*

blastomycosis (blas''to-mi-ko'sis) A systemic fungal infection caused by *Blastomyces dermatitidis* and marked by suppurating tumors in the skin or by lesions in the lungs. *785*

blastospore (blas'to-spōr) A spore formed by budding from a hypha. *523*

blood type One of the classes into which blood can be separated based on the presence or absence of specific erythrocyte glycoprotein antigens. *643*

B lymphocyte *See* B cell. *609*

bottom yeast Yeast used in beer production that tends to settle to the bottom of the fermentation vessel. *880*

botulism (boch'oo-lizm) A form of food poisoning caused by a neurotoxin (botulin) produced by *Clostridium botulinum;* sometimes found in improperly canned or preserved food. *765*

bright-field microscope A microscope that illuminates the specimen directly with bright light and forms a dark image on a brighter background. *21*

broad-spectrum drugs Chemotherapeutic agents that are effective against many different kinds of pathogens. *327*

bronchial asthma An example of an atopic allergy involving the lower respiratory tract. *637*

bubo (bu'bo) A tender, inflamed, enlarged lymph node that results from a variety of infections. *751*

bubonic plague *See* plague. *349, 750*

budding A vegetative outgrowth of yeast and some bacteria as a means of asexual reproduction; the daughter cell is smaller than the parent. *480*

bulking sludge Sludges produced in sewage treatment that do not settle properly, usually due to the development of filamentous microorganisms. *833*

burst *See* rise period. *369*

burst size The number of phages released by a host cell during the lytic life cycle. *369*

butanediol fermentation A type of fermentation most often found in the family *Enterobacteriaceae* in which 2,3-butanediol is a major product; acetoin is an intermediate in the pathway and may be detected by the Voges-Proskauer test. *156*

Calvin cycle The main pathway for the fixation (or reduction and incorporation) of CO_2 into organic material during photosynthesis; it is also found in chemolithoautotrophs. *173*

cancer (kan'ser) A malignant tumor that expands locally by invasion of surrounding tissues, and systemically by metastasis. *393*

candidal vaginitis Vaginitis caused by *Candida.* *788*

candidiasis (kan″dĭ-di′ah-sis) An infection caused by *Candida* species of dimorphic fungi, commonly involving the skin. *788*

canning A method of food preservation in which the food is placed in containers, hermetically sealed, and heated to destroy pathogens and spoilage microorganisms. *869*

capsid (kap′sid) The protein coat or shell that surrounds a virion's nucleic acid. *354*

capsomer (kap′so-mer) The ring-shaped morphological unit of which icosahedral capsids are constructed; a capsomer may have five protomers (a **penton** or **pentamer**) or six protomers (a **hexon** or **hexamer**). *357*

capsule A layer of well-organized material, not easily washed off, lying outside the bacterial cell wall. *56*

carboxysomes Polyhedral inclusion bodies that contain the CO_2 fixation enzyme ribulose 1,5-bisphosphate carboxylase; found in cyanobacteria, nitrifying bacteria, and thiobacilli. *48, 173*

caries (ka′re-ēz) Bone or tooth decay. *776*

carotenoids (kah-rot′e-noids) Pigment molecules, usually yellowish in color, that are often used to aid chlorophyll in trapping light energy during photosynthesis. *163*

carrier An infected individual who is a potential source of infection for others and plays an important role in the epidemiology of a disease. *706*

caseous lesion (ka′se-us) A lesion resembling cheese or curd; cheesy. *748*

casual carrier An individual who harbors an infectious organism for only a short period. *706*

catabolism (kah-tab′o-lizm) That part of metabolism in which larger, more complex molecules are broken down into smaller, simpler molecules with the release of energy. *146*

catabolite repression (kah-tab′o-līt) Inhibition of the synthesis of several catabolic enzymes by a metabolite such as glucose. *225*

catalyst (kat′ah-list) A substance that accelerates a reaction without being permanently changed itself. *140*

catalytic antibody An antibody that catalyzes a specific chemical reaction by lowering the free energy of activation. *622*

catalytic site *See* active site. *141*

catheter (kath′ĕ-ter) A tubular surgical instrument for withdrawing fluids from a cavity of the body, especially one for introduction into the bladder through the urethra for the withdrawal of urine. *675*

cell cycle The sequence of events in a cell's growth-division cycle between the end of one division and the end of the next. In eucaryotic cells, it is composed of the G_1 period, the S period in which DNA and histones are synthesized, the G_2 period, and the M period (mitosis). *83, 230*

cell-mediated immunity The type of immunity that results from T cells coming into close contact with foreign cells or infected cells to destroy them; it can be transferred to a nonimmune individual by the transfer of cells. *610*

cellular slime molds Slime molds with a vegetative phase consisting of amoeboid cells that aggregate to form a multicellular pseudoplasmodium; they belong to the division *Acrasiomycota*. *530*

cellulitis (sel″u-li′tis) A diffuse spreading infection of subcutaneous skin tissue caused by streptococci, staphylococci, or other organisms. The tissue is inflamed with edema, redness, pain, and interference with function. *746*

cell wall The strong layer or structure that lies outside the plasma membrane; it supports and protects the membrane and gives the cell shape. *87*

cephalosporin (sef″ah-lo-spōr′in) A group of β-lactam antibiotics derived from the fungus *Cephalosporium*, which share the 7-aminocephalosporanic acid nucleus. *335*

cercaria The final free-swimming larval stage of a trematode. *A40*

cestode (tapeworm) A flattened, segmented worm, lacking a mouth or gut, that lives as an adult in its host's intestine; a member of the class *Cestoidea*, and usually of the subclass *Cestoda*. *A34*

chancre (shang′ker) The primary lesion of syphilis, occurring at the site of entry of the infection. *759*

chemical oxygen demand (COD) The amount of chemical oxidation required to convert organic matter in water and wastewater to CO_2. *819, 831*

chemiosmotic hypothesis (kem″e-o-os-mot′ik) The hypothesis that a proton gradient and an electrochemical gradient are generated by electron transport and then used to drive ATP synthesis by oxidative phosphorylation. *153*

chemoautotroph (ke″mo-aw′to-trōf) *See* chemolithotrophic autotrophs. *98*

chemoheterotroph (ke″mo-het′er-o-trōf″) *See* chemoorganotrophic heterotrophs. *98*

chemolithotroph (ke″mo-lith′o-trōf) *See* chemolithotrophic autotrophs. *98, 160*

chemolithotrophic autotrophs Microorganisms that oxidize reduced inorganic compounds to derive both energy and electrons; CO_2 is their carbon source. *98*

chemoorganotrophic heterotrophs Organisms that use organic compounds as sources of energy, hydrogen, electrons, and carbon for biosynthesis. *98*

chemoreceptors Special protein receptors in the plasma membrane or periplasmic space that bind chemicals and trigger the appropriate chemotaxic response. *62*

chemostat (ke′mo-stat) A continuous culture apparatus that feeds medium into the culture vessel at the same rate as medium containing microorganisms is removed; the medium in a chemostat contains one essential nutrient in limiting quantities. *120*

chemotaxis (ke″mo-tak′sis) The pattern of microbial behavior in which the microorganism moves toward chemical attractants and away from repellents. *60*

chemotherapeutic agents (ke″mo-ther-ah-pu′tik) Compounds used in the treatment of disease that destroy pathogens or inhibit their growth at concentrations low enough to avoid doing undesirable damage to the host. *326*

chemotrophs (ke′mo-trōfs) Organisms that obtain energy from the oxidation of chemical compounds. *98*

chickenpox (varicella; chik′en-poks) A highly contagious skin disease, usually affecting two- to seven-year-old children; it is caused by the varicella-zoster virus, which is acquired by droplet inhalation into the respiratory system. *716*

chimera (ki-me′rah) A recombinant plasmid containing foreign DNA, which is used as a cloning vector in genetic engineering. *298*

chitin (ki′tin) A tough, resistant, nitrogen-containing polysaccharide forming the walls of certain fungi, the exoskeleton of arthropods, and the epidermal cuticle of other surface structures of certain protists and animals. *519*

chlamydospore (klam′ĭ-do-spōr″) An asexually-produced, thick-walled resting spore formed by some fungi. *523*

chloramphenicol (klo″ram-fen′ĭ-kol) A broad spectrum antibiotic that is produced by *Streptomyces venezuelae* or synthetically; it binds to the large ribosomal subunit and inhibits the peptidyl transferase reaction. *337*

chlorophyll (klor′o-fil) The green photosynthetic pigment that consists of a large tetrapyrrole ring with a magnesium atom in the center. *162*

chloroplast (klo′ra-plast) A eucaryotic plastid that contains chlorophyll and is the site of photosynthesis. *79*

cholera (kol′er-ah) An acute infectious enteritis endemic and epidemic in Asia and periodically spreading to the Middle East, Africa, Southern Europe, and South America; caused by *Vibrio cholerae*. *763*

choleragen (kol′er-ah-gen) The cholera enterotoxin; an extremely potent protein molecule elaborated by strains of *Vibrio cholerae* in the small intestine after ingestion of feces-contaminated water or food. It acts on epithelial cells to cause hypersecretion of chloride and bicarbonate and an outpouring of large quantities of fluid from the mucosal surface. *584, 763*

chromatin (kro′mah-tin) The DNA-containing portion of the eucaryotic nucleus; the DNA is almost always complexed with histones. It can be very condensed (heterochromatin) or more loosely organized and genetically active (euchromatin). *82*

chromogen (kro′me-jen) A colorless substrate that is acted on by an enzyme to produce a colored end product. *661*

chromomycosis (kro″mo-mi-ko′sis) A chronic fungal infection of the skin, producing wartlike nodules that may ulcerate. It is caused by the black molds *Phialophora verrucosa* or *Fonsecaea pedrosoi*. *784*

chromophore group (kro′mo-fōr) A chemical group with double bonds that absorbs visible light and gives a dye its color. *28*

chromosomes (kro′mo-somz) The bodies that have most or all of the cell's DNA and contain most of its genetic information (mitochondria and chloroplasts also contain DNA and genes). *471*

chronic carrier An individual who harbors an infectious organism for a long time. *706*

chrysolaminarin The polysaccharide storage product of the chrysophytes and diatoms. *541*

-cide A suffix indicating that the substance kills. *311*

ciguatera (se″gwah-ta′rah) A form of ichthyosarcotoxism, marked by gastrointestinal and neurological symptoms from ingestion of marine fish (e.g., grouper and snapper) that store the toxin in their tissues; it occurs in tropical and subtropical coastal areas. *544*

cilia (sil′e-ah) Threadlike appendages extending from the surface of some protozoa that beat rhythmically to propel them; cilia are membrane-bound cylinders with a complex internal array of microtubules, usually in a 9 + 2 pattern. *87*

citric acid cycle *See* tricarboxylic acid (TCA) cycle. *150*

classical pathway The antibody-dependent pathway of complement activation; it leads to the lysis of pathogens and stimulates phagocytosis and other host defenses. *648*

classification The arrangement of organisms into groups based on mutual similarity or evolutionary relatedness. *405*

clinical microbiologist A person who isolates and identifies microorganisms from clinical specimens. *673*

clonal anergy The functionally unresponsive tolerant state that arises during fetal development whereby the body produces antibodies against nonself antigens while "tolerating" (not producing antibodies against) self-antigens. In clonal anergy, the T cell is incapable of producing its own growth hormone interleukin-2. B cells may also become anergic. *634*

clonal selection theory The theory that clones of effector B and T cells arise from single cells (or very small clones) that are stimulated to reproduce by antigen binding to their receptors. *619*

clone (klōn) A group of genetically identical cells or organisms derived by asexual reproduction from a single parent. *237, 624*

cloning vector A DNA molecule that can replicate (a replicon) and is used to transport a piece of inserted foreign DNA, such as a gene, into a recipient cell. It may be a plasmid, phage, or cosmid. *288, 298*

clostridial myonecrosis (klo-strid′e-al mi″o-ne-kro′sis) Death of individual muscle cells caused by clostridia. Also called **gas gangrene.** *754*

coagulase (ko-ag′u-las) An enzyme that induces blood clotting; it is characteristically produced by pathogenic staphylococci. *454*

coagulation (ko-ag″u-la′shun) 1. The process of blood clot formation. 2. In municipal water purification, it is the process that precipitates microorganisms and organic material out of water by the addition of chemicals. *588, 835*

coccidioidomycosis (kok-sid″e-oi″do-mi-ko′sis) A fungal disease caused by *Coccidioides immitis* that exists in dry, highly alkaline soils. Also known as valley fever, San Joaquin fever, or desert rheumatism. *786*

coccus (kok′us, pl. **cocci,** kok′si) A roughly spherical bacterial cell. *42*

code degeneracy The genetic code is organized in such a way that often there is more than one codon for each amino acid. *239*

codon (ko′don) A sequence of three nucleotides in mRNA that directs the incorporation of an amino acid during protein synthesis or signals the start or stop of translation. *239*

coenocytic (se″no-sit′ik) Refers to a multinucleate cell or hypha formed by repeated nuclear divisions not accompanied by cell divisions. *113, 520*

coenzyme (ko-en′zīm) A loosely bound cofactor that often dissociates from the enzyme active site after product has been formed. *140*

cofactor The nonprotein component of an enzyme; it is required for catalytic activity. *140*

cold sore A lesion caused by the herpes simplex virus; usually occurs on the border of the lips or nares. Also known as a **fever blister** or **herpes labialis.** *729*

colicin (kol′i-sin) A plasmid-encoded protein that is produced by enteric bacteria and binds to specific receptors on the cell envelope of sensitive target bacteria, where it may cause lysis or attack specific intracellular sites such as ribosomes. *595*

coliform (ko′li-form) A gram-negative, nonsporing, facultative rod that ferments lactose with gas formation within 48 hours at 35°C. *837*

colony A cluster or assemblage of microorganisms growing on a solid surface such as the surface of an agar culture medium; the assemblage is often directly visible, but may be seen only microscopically. *106*

colony forming units (CFU) The number of microorganisms that can form colonies in spread plates or pour plates, an indication of the number of viable microorganisms in a sample. *117*

colony-stimulating factor (CSF) A lymphocyte mediator that increases the availability of phagocytes. *627*

colostrum (kō-los′trum) The thin milky fluid secreted by the mammary gland a few days before or after parturition. *608*

comedo (kom′ĕ-do; pl., **comedones**) A plug of dried sebum in an excretory duct of the skin. *573*

cometabolism The modification of a compound not used for growth by a microorganism, which occurs in the presence of another organic material that serves as a carbon and energy source. *859*

commensal (kŏ-men′sal) Living on or within another organism without injuring or benefiting the other organism. *565*

commensalism (kŏ-men′sal-izm″) A type of symbiosis in which one individual gains from the association and the other is neither harmed nor benefited. *565*

common cold An acute, self-limiting, and highly contagious virus infection of the upper respiratory tract that produces inflammation, profuse discharge, and other symptoms. *730*

common-source epidemic An epidemic that is characterized by a sharp rise to a peak and then a rapid, but not pronounced, decline in the number of individuals infected; it usually results from a single contaminated source. *702*

common vehicle transmission The transmission of a pathogen to a host by means of an inanimate medium or vehicle. *707*

communicable disease A disease associated with an agent that can be transmitted from one host to another. *706*

community An assemblage of different types of organisms or a mixture of different microbial populations. *807*

competent A bacterial cell that can take up free DNA fragments and incorporate them into its genome during transformation. *271*

competitive inhibitor A molecule that inhibits enzyme activity by competing with the substrate at the enzyme's active site. *142*

complementary DNA (cDNA) A DNA copy of an RNA molecule (e.g., a DNA copy of an mRNA). *288*

complement fixation The activation of the complement system with the binding of the components to the activator substance and to each other, with their consequent removal from serum. *648*

complement system A group of plasma proteins that plays a major role in an animal's defensive immune response. *648*

complex medium Culture medium that contains some ingredients of unknown chemical composition. *104*

complex viruses Viruses with capsids having a complex symmetry that is neither icosahedral nor helical. *355*

compromised host A host with lowered resistance to infection and disease for any of several reasons. The host may be seriously debilitated (due to malnutrition, cancer, diabetes, leukemia, or another infectious disease), traumatized (from surgery or injury), immunosuppressed, or have an altered microbiota due to prolonged use of antibiotics. *787*

concatemer A long DNA molecule consisting of several genomes linked together in a row. *372*

conditional mutations Mutations that are expressed only under certain environmental conditions. *244*

conformational change hypothesis The hypothesis that the energy from electron transport is used to induce changes in the shape of the enzyme that synthesizes ATP and thereby drives ATP synthesis. *153*

congenital (neonatal) herpes An infection of a newborn caused by transmission of the herpesvirus during vaginal delivery. *731*

congenital rubella syndrome A wide array of congenital defects affecting the heart, eyes, and ears of a fetus during the first trimester of pregnancy and caused by the rubella virus. *717*

congenital syphilis Syphilis that is acquired *in utero* from the mother. *759*

conidiospore (ko-nid′e-o-spōr) An asexual, thin-walled spore borne on hyphae and not contained within a sporangium; it may be produced singly or in chains. *523*

conidium (ko-nid′e-um; pl., **conidia**) *See* conidiospore. *507*

conjugants (kon'joo-gants) Complementary mating types that participate in a form of protozoan sexual reproduction called conjugation. *550*

conjugation (kon''ju-ga'shun) 1. The form of gene transfer and recombination in bacteria that requires direct cell-to-cell contact. 2. A complex form of sexual reproduction commonly employed by protozoa. *268, 550*

conjugative plasmid A plasmid that carries the genes for sex pili and can transfer copies of itself to other bacteria during conjugation. *261*

conoid (ko'noid) A hollow cone of spirally coiled filaments in the anterior tip of certain apicomplexan protozoa. *556*

constant region The part of an antibody molecule that does not vary greatly in amino acid sequence among molecules of the same class, subclass, or type. *612*

constitutive mutant A strain that produces an inducible enzyme continually, regardless of need, because of a mutation in either the operator or regulator gene. *223*

consumer An organism that feeds directly on living or dead animals, by ingestion or by phagocytosis. *805*

contact transmission Transmission of the infectious agent by contact of the source or reservoir of the agent with the host. *707*

continuous culture system A culture system with constant environmental conditions maintained through continual provision of nutrients and removal of wastes. *120*

continuous feed The continuous addition of a substance to a culture during growth. *891*

contractile vacuole (vak'u-ōl) In protists and some animals, a clear fluid-filled cell vacuole that takes up water from within the cell and then contracts, releasing it to the outside through a pore in a cyclical manner. Contractile vacuoles function primarily in osmoregulation and excretion. *550*

convalescent carrier (kon''vah-les'ent) An individual who has recovered from an infectious disease but continues to harbor large numbers of the infectious organism. *706*

corepressor (ko''re-pre'sor) A small molecule that inhibits the synthesis of a repressible enzyme. *222*

cortex (kor'teks) The layer of a bacterial endospore that is thought to be particularly important in conferring heat resistance on the endospore. *63*

coryza (kō-ri'zah) *See* common cold. *730*

cosmid (koz'mid) A plasmid vector with lambda phage cos sites that can be packaged in a phage capsid; it is useful for cloning large DNA fragments. *299*

cristae (kris'te) Infoldings of the inner mitochondrial membrane. *74*

crossing-over A process in which segments of two adjacent DNA strands are exchanged; breaks occur in both strands, and the exposed ends of each strand join to those of the opposite segment on the other strand. *259*

crustose Lichens that are compact and appressed to a substratum. *566*

cryptococcosis (krip''to-kok-o'sis) An infection caused by the basidiomycete, *Cryptococcus neoformans,* which may involve the skin, lungs, brain, or meninges. *529, 786*

crystallizable fragment (Fc) The stem of the Y portion of an antibody molecule. Cells such as macrophages bind to the Fc region, and it is also involved in complement activation. *612*

cutaneous anthrax (ku-ta'ne-us an'thraks) A form of anthrax involving the skin. *753*

cutaneous diphtheria (ku-ta'ne-us dif-the're-ah) A skin disease caused by *Corynebacterium diphtheriae* that infects wound or skin lesions, causing a slow-healing ulceration. *742*

cyanobacteria (si''ah-no-bak-te're-ah) A large group of photosynthetic bacteria with oxygenic photosynthesis and a photosynthetic system like that present in eucaryotic photosynthetic organisms. *472*

cyclic photophosphorylation (fo''to-fos''for-ĭ-la'shun) The formation of ATP when light energy is used to move electrons cyclically through an electron transport chain during photosynthesis; only photosystem I participates. *164*

cyst (sist) A general term used for a specialized microbial cell enclosed in a wall. Cysts are formed by protozoa and a few bacteria. They may be dormant, resistant structures formed in response to adverse conditions or reproductive cysts that are a normal stage in the life cycle. *550*

cytochromes (si'to-krōms) Heme proteins that carry electrons, usually as members of electron transport chains. *138*

cytokine (si'to-kīn) A general term for nonantibody proteins, released by a cell in response to antigens, which are mediators that influence other cells. *627*

cytokinesis (si''to-ki-ne'sis) The division of a cell's cytoplasm into two halves after nuclear division. *85*

cytomegalovirus inclusion disease (si''to-meg''ah-lo-vi'rus) An infection caused by the cytomegalovirus and marked by nuclear inclusion bodies in enlarged infected cells. *730*

cytopathic effect (si''to-path'ik) The observable change that occurs in cells as a result of viral replication. Examples include ballooning, binding together, clustering, or even death of the cultured cells. *349, 678*

cytoplasmic matrix (si''to-plaz'mik) The protoplasm of a cell that lies within the plasma membrane and outside any other organelles. In bacteria, it is the substance between the cell membrane and the nucleoid. *47, 72*

cytoproct (si'to-prokt) Site on a protozoan where undigestible matter is expelled. *557*

cytosine (si'to-sēn) A pyrimidine 2-oxy-4-aminopyrimidine that is found in nucleosides, nucleotides, and nucleic acids. *183*

cytoskeleton (si''to-skel'ĕ-ton) A network of microfilaments, microtubules, intermediate filaments, and other components in the cytoplasm of eucaryotic cells that helps give them shape. *73*

cytostome (si'to-stōm) A permanent site in the ciliate body through which food is ingested. *550*

cytotoxic T (T_C) cell (si''to-tok'sik) A cell that is capable of recognizing virus-infected cells through the major histocompatability antigen and destroying the infected cell. *634*

cytotoxin (si'to-tok'sin) A toxin or antibody that has a specific toxic action upon cells; cytotoxins are named according to the cell for which they are specific (e.g., nephrotoxin). *584*

Dane particle A 42 nm spherical particle that is one of three that make up the antigenic portion of the hepatitis B virus. *734*

dark-field microscopy Microscopy in which the specimen is brightly illuminated while the background is dark. *24*

dark reactivation The excision and replacement of thymine dimers in DNA that occurs in the absence of light. *130*

deamination (de-am''i-na'shun) The removal of amino groups from amino acids. *160*

death phase The phase of microbial growth in a batch culture when the viable microbial population declines. *114*

decimal reduction time (D) The time required to kill 90% of the microorganisms or spores in a sample at a specified temperature. *314*

decomposer An organism that breaks down complex materials into simpler ones, including the release of simple inorganic products. *805*

decomposer-reducer An organism, such as an insect or earthworm, that carries out decomposition and physically reduces the size of substrate particles. *848*

defensin (de-fens'sin) Specific peptides produced by neutrophils that permeabilize the outer and inner membranes of certain microorganisms, thus killing them. *600*

defined medium Culture medium made with components of known composition. *104*

delayed-type hypersensitivity T (T_{DTH}) cell A cell that is responsible for initiating type IV delayed-type hypersensitivity reactions through secretion of lymphokines. *634*

Delta agent A defective RNA virus that is transmitted as an infectious agent, but cannot cause disease unless the individual is also infected with the hepatitis B virus. *733*

denaturation (de-na''chur-a'shun) A change in the shape of an enzyme that destroys its activity; the term is also applied to changes in nucleic-acid shape. *142*

dendrogram A treelike diagram that is used to graphically summarize mutual similarities and relationships between organisms. *407*

denitrification (de-ni''trĭ-fĭ-ka'shun) The reduction of nitrate to nitrogen gas during anaerobic respiration. *158, 812*

dental plaque (plak) A thin film on the surface of teeth consisting of bacteria embedded in a matrix of bacterial polysaccharides, salivary glycoproteins, and other substances. *776*

deoxyribonucleic acid (DNA; de-ok''se-ri''bo-nu-kle'ik) The nucleic acid that constitutes the genetic material of all cellular organisms. It is a polynucleotide composed of deoxyribonucleotides connected by phosphodiester bonds. *50, 192*

dermatophyte (der'mah-to-fīt") A fungus parasitic upon the skin. *782*

dermatophytosis (der"mah-to-fi-to'sis) A fungal infection of the skin; the term is a general term that comprises the various forms of tinea, and it is sometimes used to specifically refer to athlete's foot (tinea pedis). *782*

desensitization (de-sen"si-ti-za'shun) To make a sensitized or hypersensitive individual insensitive or nonreactive to a sensitizing agent. *637*

detergent (de-ter'jent) An organic molecule, other than a soap, that serves as a wetting agent and emulsifier; it is normally used as cleanser, but some may be used as antimicrobial agents. *320*

diatoms (di'ah-toms) Algal protists with siliceous cell walls called frustules. They constitute a substantial subfraction of the phytoplankton. *541*

diauxic growth (di-awk'sik) A biphasic growth pattern or response in which a microorganism, when exposed to two nutrients, initially uses one of them for growth and then alters its metabolism to make use of the second. *225*

dictyosomes (dik'te-o-sōms) Individual stacks of cisternae in a Golgi apparatus. *75*

differential media (dif"er-en'shal) Culture media that distinguish between groups of microorganisms based on differences in their biological characteristics. *105*

differential staining procedures Staining procedures that divide bacteria into separate groups based on staining properties. *29*

diffusion (di-fu'zhun) *See* facilitated diffusion and passive diffusion. *101*

dikaryotic stage (di-kar-e-ot'ik) In fungi, having pairs of nuclei within cells or compartments. Each cell contains two separate haploid nuclei, one from each parent. *524*

dinoflagellate (di"no-flaj'ĕ-lāt) An algal protist characterized by two flagella used in swimming in a spinning pattern. Many are bioluminescent and an important part of marine phytoplankton. *544*

diphtheria (dif-the're-ah) An acute, highly contagious childhood disease that generally affects the membranes of the throat and, less frequently, the nose. It is caused by *Corynebacterium diphtheriae*. *742*

dipicolinic acid A substance present in great quantities in the bacterial endospore. It is thought to contribute to the endospore's heat resistance. *63*

diplococcus (dip"lo-kok'us) A pair of cocci. *42*

diploid (dip'loid) Having double the basic chromosome number. *85*

disease (di-zez) A deviation or interruption of the normal structure or function of any part of the body that is manifested by a characteristic set of symptoms and signs. *699*

disinfectant (dis"in-fek'tant) An agent, usually chemical, that disinfects; normally, it is employed only with inanimate objects. *311*

disinfection (dis"in-fek'shun) The killing, inhibition, or removal of microorganisms that may cause disease. *311*

dissimilatory reduction The use of a substance as an electron acceptor in energy generation. The acceptor (e.g., sulfate or nitrate) is reduced but not incorporated into organic matter during biosynthetic processes. *177, 810*

diurnal oxygen shifts (di-er'nal) The changes in oxygen levels that occur in waters when algae produce and use oxygen on a cyclic basis during day and night. *831*

DNA ligase An enzyme that joins two DNA fragments together through the formation of a new phosphodiester bond. *201*

DNA polymerase (pol-im'er-ās) An enzyme that synthesizes new DNA using a parental DNA strand as a template. *199*

Döderlein's bacillus (ded'er-līnz) Large gram-positive bacteria (*Lactobacillus* spp.) found in the vagina that break down the glycogen produced by the vaginal epithelium and form lactic acid. *575*

domain (do-mān') A loop, along with approximately 25 amino acids on each side of the loop, of an antibody chain. It forms a compact, globular section. *612*

double diffusion agar assay (Ouchterlony technique) An immunodiffusion reaction in which both antibody and antigen diffuse through agar to form stable immune complexes, which can be observed visually. *661*

doubling time *See* generation time. *114*

DPT (diphtheria-pertussis-tetanus) vaccine A vaccine containing three antigens that is used to immunize people against diphtheria, pertussis or whooping cough, and tetanus. *742*

droplet nuclei Small particles (1 to 4 μm in diameter) that represent what is left from the evaporation of larger particles (10 μm or more in diameter) called droplets. *707*

***D* value** *See* decimal reduction time. *314*

early mRNA Messenger RNA produced early in a virus infection that codes for proteins needed to take over the host cell and manufacture viral nucleic acids. *370*

eclipse period (e-klips') The initial part of the latent period in which infected host bacteria do not contain any complete virions. *369*

ecosystem (ek"o-sis'tem) The total community of living organisms and their associated physical and chemical environment. *805*

ectomycorrhizal Referring to a mutualistic association between fungi and plant roots in which the fungus surrounds the root tip with a sheath. *855*

ectoparasite (ek"to-par'ah-sīt) A parasite that lives on the surface of its host. *579*

ectoplasm (ek'to-plazm) The outer stiffer portion or region of the cytoplasm of a cell, which may be differentiated in texture from the inner portion or endoplasm. *550*

ectosymbiosis A type of symbiosis in which one organism remains outside of the other organism. *565*

effector T cell The general term for a group of cells that directly attack specific target cells. *634*

electron transport chain A series of electron carriers that operate together to transfer electrons from donors such as NADH and $FADH_2$ to acceptors such as oxygen. *152*

electrophoresis (e-lek"tro-fo-re'sis) A technique that separates substances through differences in their migration rate in an electrical field due to variations in the number and kinds of charged groups they have. *293*

elementary body A small, dormant body that serves as the agent of transmission between host cells in the chlamydial life cycle. *444*

elephantiasis (el"ĕ-fan-ti'ah-sis) A chronic disease that results from the filarial nematode *Wuchereria bancrofti* living in the lymphatic system; the lymphatics become inflamed and obstructed. *A38*

ELISA *See* enzyme-linked immunosorbent assay. *659*

elongation cycle The cycle in protein synthesis that results in the addition of an amino acid to the growing end of a peptide chain. *211*

Embden-Meyerhof pathway (em'den mi'er-hof) A pathway that degrades glucose to pyruvate; the six-carbon stage converts glucose to fructose 1,6-bisphosphate, and the three-carbon stage produces ATP while changing glyceraldehyde 3-phosphate to pyruvate. *147*

Embola virus hemorrhagic fever An acute infection caused by a virus that produces varying degrees of hemorrhage, shock, and sometimes death. *721*

encystation (en-sis-'ta'shen) The formation of a cyst. *550*

endemic (murine) typhus (mu'rin ti'fus) A form of typhus fever caused by the rickettsia *Rickettsia typhi* that occurs sporadically in individuals who come into contact with rats and their fleas. *774*

endemic disease (en-dem'ik) A disease that is commonly or constantly present in a population, usually at a relatively steady low frequency. *699*

endergonic reaction (end"er-gon'ik) A reaction that does not spontaneously go to completion as written; the standard free energy change is positive, and the equilibrium constant is less than one. *136*

endocytosis (en"do-si-to'sis) The process in which a cell takes up solutes or particles by enclosing them in vesicles pinched off from its plasma membrane. *76*

endogenote (en"do-je'nōt) The recipient bacterial cell's own genetic material into which the donor DNA can integrate. *259*

endogenous infection (en-doj'ĕ-nus in-fek'shun) An infection by a member of an individual's own normal body microbiota. *747*

endogenous pyrogen (en-doj'ĕ-nus pi'ro-jen) A substance such as the lymphokine interleukin-1, which is produced by host cells and induces a fever response in the host. *589*

endogenous respiration (en-doj'ĕ-nus res"pi-ra'shun) The use of nutrients, usually internally stored reserves, to maintain an organism without growth. *833*

endomycorrhizal Referring to a mutualistic association of fungi and plant roots in which the fungus penetrates into the root cells and arbuscules and vesicles are formed. *855*

endoparasite (en″do-par′ah-sīt) A parasite that lives inside the body of its host. *579*

endophyte (en′do-fit) A microorganism living within a plant, but not necessarily parasitic upon it. *856*

endoplasm (en′do-plazm) The central portion of the cytoplasm of a cell. *550*

endoplasmic reticulum (en″do-plas′mik rē-tik′u-lum) A system of membranous tubules and flattened sacs (cisternae) in the cytoplasmic matrix of eucaryotic cells. Rough or granular endoplasmic reticulum (RER or GER) bears ribosomes on its surface; smooth or agranular endoplasmic reticulum (SER or AER) lacks them. *74*

endosome (en′do-sōm) A membranous vesicle formed by endocytosis. *76*

endospore (en′do-spōr) An extremely heat- and chemical-resistant, dormant, thick-walled spore that develops within bacteria. *62*

endosymbiont (en″do-sim′be-ont) An organism that lives within the body of another organism in a symbiotic association. *568*

endosymbiosis (en″do-sim″bi-o′sis) A type of symbiosis in which one organism is found within another organism. *565*

endosymbiotic theory or **hypothesis** The theory that eucaryotic organelles such as mitochondria and chloroplasts arose when bacteria established an endosymbiotic relationship with the eucaryotic ancestor and then evolved into eucaryotic organelles. *81, 416*

endotoxin (en″do-tox′sin) The heat-stable lipopolysaccharide in the outer membrane of the cell wall of gram-negative bacteria that is released when the bacterium lyses, or sometimes during growth, and is toxic to the host. *55, 587*

endotoxin units (E.U.) The units in which the concentration of an endotoxin is measured. *587*

end product inhibition *See* feedback inhibition. *221*

energy The capacity to do work or cause particular changes. *134*

enteric bacteria (enterobacteria; en-ter′ik) Members of the family *Enterobacteriaceae* (gram-negative, peritrichous or nonmotile, facultatively anaerobic, straight rods with simple nutritional requirements); also used for bacteria that live in the intestinal tract. *435*

enterobiasis (en″ter-o-bi′ah-sis) An infection with members of the nematode genus *Enterobius*, usually with *E. vermicularis*. *A37*

enterotoxin (en″ter-o-tok′sin) A toxin specifically affecting the cells of the intestinal mucosa, causing vomiting and diarrhea. *584, 763*

enthalpy (en′thal-pe) The heat content of a system; in a biological system it is essentially equivalent to the total energy of the system. *135*

Entner-Doudoroff pathway A pathway that converts glucose to pyruvate and glyceraldehyde 3-phosphate by producing 6-phosphogluconate and then dehydrating it. *150*

entropy (en′tro-pe) A measure of the randomness or disorder of a system; a measure of that part of the total energy in a system that is unavailable for useful work. *135*

envelope (en′vē-lōp) 1. All the structures outside the plasma membrane in bacterial cells. 2. In virology, it is an outer membranous layer that surrounds the nucleocapsid in some viruses. *50, 355*

enzootic (en″zo-ot′ik) The moderate prevalence of a disease in a given animal population. *699*

enzyme (en′zīm) A protein catalyst with specificity for both the reaction catalyzed and its substrates. *140*

enzyme-linked immunosorbent assay (ELISA) A technique used for detecting and quantifying specific serum antibodies. *659*

epidemic (ep″ĭ-dem′ik) A disease that suddenly increases in occurrence above the normal level in a given population. *699*

epidemic (louse-borne) typhus (ep″i-dem′ik ti′fus) A disease caused by *Rickettsia prowazekii* that is transmitted from person to person by the body louse. *774*

epidemiologist (ep″ĭ-de″me-ol′o-jist) A person who specializes in epidemiology. *699*

epidemiology (epi″-de″me-ol′o-je) The study of the factors determining and influencing the frequency and distribution of disease, injury, and other health-related events and their causes in defined human populations. *699*

epilimnion (ep″e-lim′ne-on″) The warmer aerobic layer at the surface of a thermally stratified lake. It lies above the thermocline. *830*

episome (ep′ĭ-sōm) A plasmid that can exist either independently of the host cell's chromosome or be integrated into it. *261*

epitheca (ep″ĭ-the′kah) The larger of two halves of a diatom frustule (shell). *541*

epitope (ep′i-tōp) An area of the antigen molecule that stimulates the production of, and combines with, specific antibodies; also known as **antigenic determinant site.** *610*

epizootic (ep″ĭ-zo-ot′ik) A sudden outbreak of a disease in an animal population. *699*

epizootiology (ep″i-zo-ot″e-ol′o-je) The field of science that deals with factors determining the frequency and distribution of a disease within an animal population. *699*

equilibrium (e″kwi-lib′re-um) The state of a system in which no net change is occurring and free energy is at a minimum; in a chemical reaction at equilibrium, the rates in the forward and reverse directions exactly balance each other out. *136*

equilibrium constant The constant that is characteristic of a chemical reaction at equilibrium and is based on the relative equilibrium concentrations or activities of products and reactants. *136*

ergot (er′got) The dried sclerotium of *Claviceps purpurea*. Also, an ascomycete that parasitizes rye and other higher plants causing the disease called ergot. *525*

ergotism (er′got-izm) The disease or toxic condition caused by eating grain infected with ergot; it is often accompanied by gangrene, psychotic delusions, nervous spasms, abortion, and convulsions in humans and in animals. *525, 873*

eructation (ĕ-ruk-ta′shun) To belch forth; what ruminants do in releasing CO_2 and other gases. *569*

erysipelas (er″ĭ-sip′ĕ-las) An acute inflammation of the dermal layer of the skin, occurring primarily in infants and persons over 30 years of age with a history of streptococcal sore throat. *746*

erythromycin (ĕ-rith″ro-mi′sin) An intermediate spectrum macrolide antibiotic produced by *Streptomyces erythreus*. *337*

eschar (es′kar) A slough produced on the skin by a thermal burn, gangrene, or the anthrax bacillus. *753*

eubacteria (u″bak-te′re-ah) The large majority of bacteria that have cell wall peptidoglycan containing muramic acid (or are related to such bacteria) and membrane lipids with ester-linked straight chained fatty acids. *415*

eucaryotic cells (u″kar-e-ot′ik) Cells that have a membrane-delimited nucleus and differ in many other ways from procaryotic cells; protists, algae, fungi, plants, and animals are all eucaryotic. *15, 89*

euglenoids (u-gle′noids) A group of algae (the division *Euglenophyta*) or protozoa (order *Euglenida*) that normally have chloroplasts containing chlorophyll *a* and *b*. They usually have a stigma and one or two flagella emerging from an anterior gullet. *540*

eurythermal (u″re-ther′mal) Organisms that grow well in a wide range of temperatures. *125*

eutrophic (u-trof′ik) A nutrient-enriched environment. *830*

eutrophication (u″tro-fĭ-ka′shun) The enrichment of an aquatic environment with nutrients. *830*

excystation (ek″sis-ta′shun) The escape of one or more cells or organisms from a cyst. *550*

exergonic reaction (ek″ser-gon′ik) A reaction that spontaneously goes to completion as written; the standard free energy change is negative, and the equilibrium constant is greater than one. *136*

exfoliatin (eks-fō″le-a′tin) An exotoxin produced by *Staphylococcus aureus* that causes the separation of epidermal layers and produces the symptoms of the scalded skin syndrome. *759*

exfoliative toxin (eks-fo′le-a″tiv) A toxin that causes the loss of the surface layers of the skin. *759*

exoenzymes (ek″so-en′zīms) Enzymes that are secreted by cells. *50*

exogenote (eks″o-je′nōt) The piece of donor DNA that enters a bacterial cell during gene exchange and recombination. *259*

exon (eks′on) The region in a split or interrupted gene that codes for RNA which ends up in the final product (e.g., mRNA). *203*

exotoxin (ek″so-tok′sin) A heat-labile, toxic protein produced by a bacterium as a result of its normal metabolism or because of the acquisition of a plasmid or prophage that redirects its metabolism. It is usually released into the bacterium's surroundings. *584*

exponential phase (eks″po-nen′shul) The phase of the growth curve during which the microbial population is growing at a constant and maximum rate, dividing and doubling at regular intervals. *113*

expression vector A special cloning vector used to express recombinant genes in host cells; the recombinant gene is transcribed and its protein synthesized. *300*

extended aeration The aeration of sewage to allow self-digestion of microorganisms that grew during use of a soluble waste. *833*

extracutaneous sporotrichosis (spo″ro-tri-ko′sis) An infection by the fungus *Sporothrix schenckii* that spreads throughout the body. *785*

exudation (eks″u-da′shun) The release of soluble substances by plant roots and other living organisms. *851*

eyespot *See* stigma. *539*

F factor The fertility factor, a plasmid that carries the genes for bacterial conjugation and makes its *E. coli* host cell the gene donor during conjugation. *261*

F′ plasmid An F plasmid that carries some bacterial genes and transmits them to recipient cells when the F′ cell carries out conjugation; the transfer of bacterial genes in this way is often called sexduction. *271*

F_1 factors Particles on the inner mitochondrial membrane, which are the sites of ATP synthesis by oxidative phosphorylation. *79*

facilitated diffusion Diffusion across the plasma membrane that is aided by a carrier. *101*

facultative anaerobes (fak′ul-ta″tiv an-a′er-ōbs) Microorganisms that do not require oxygen for growth, but do grow better in its presence. *127*

facultative psychrophile (fak′ul-ta″tiv si′kro-fīl) *See* psychrotroph. *126*

FAD *See* flavin adenine dinucleotide. *134*

fatty acid A monocarboxylic acid that may be unbranched or branched, and saturated or unsaturated; the fatty acids most common in lipids are 16 or 18 carbons long. *185*

fatty acid synthetase (sin′thĕ-tās) The multienzyme complex that makes fatty acids; the product is usually palmitic acid. *185*

fecal coliform test (fe′kal ko′li-form) Test for coliforms that are able to grow at 44.5°C, whose normal habitat is the intestinal tract. *839*

fecal enterococci (fe′kal en″ter-o-kok′si) Enterococci found in the intestine of humans and other warm-blooded animals. They are used as indicators of the fecal pollution of water. *840*

feedback inhibition A negative feedback mechanism in which a pathway end product inhibits the activity of an enzyme in the sequence leading to its formation; when the end product accumulates in excess, it inhibits its own synthesis. *221*

fermentation (fer″men-ta′shun) An energy-yielding process in which organic molecules serve as both electron donors and acceptors. *155, 888*

fever blister *See* cold sore. *729*

fibronectin (fi″bro-nek′tin) A high molecular weight adhesive glycoprotein. It can interact with bacteria and mediate their nonspecific clearance from the body. Fibronectin is also involved in cell–cell interactions. *593*

filarial worms (fi-la′re-al) Threadlike parasitic nematodes that reproduce within tissues and body cavities, where the female produces larvae called microfilariae that are ingested by bloodsucking insects. *A38*

filariasis (fil″ah-ri′ah-sis) A disease arising from a filarial nematode infection and the presence of microfilariae. *A38*

fimbria (fim′bre-ah; pl., **fimbriae**) A fine, hairlike protein appendage on some gram-negative bacteria that helps attach them to surfaces. *57*

final host The host on/in which a parasitic organism either attains sexual maturity or reproduces. *579*

Firmicutes (fer-mik′u-tēz) A division in the kingdom *Procaryotae* containing bacteria with gram-positive type cell walls. *421*

first law of thermodynamics Energy can be neither created nor destroyed (even though it can be changed in form or redistributed). *134*

fixation (fik-sa′shun) The process in which the internal and external structures of cells and organisms are preserved and fixed in position. *27*

flagellin (flaj′ĕ-lin) The protein used to construct the filament of a bacterial flagellum. *58*

flagellum (flah-jel′um; pl., **flagella**) A thin, threadlike appendage on many procaryotic and eucaryotic cells that is responsible for their motility. *57, 87*

flat or plane warts Small, smooth, slightly raised warts. *738*

flavin adenine dinucleotide (**FAD**; fla′vin ad′ĕ-nēn) An electron carrying cofactor often involved in energy production (for example, in the tricarboxylic acid cycle and the β-oxidation pathway). *134*

fluid mosaic model The currently accepted model of cell membranes in which the membrane is a lipid bilayer with integral proteins buried in the lipid, and peripheral proteins more loosely attached to the membrane surface. *45*

fluorescent light (floo″o-res′ent) The light emitted by a substance when it is irradiated with light of a different wavelength. *27*

fluorescence microscope A microscope that exposes a specimen to light of a specific wavelength and then forms an image from the fluorescent light produced. Usually the specimen is stained with a fluorescent dye or fluorochrome. *27*

foliose Lichens that have a leaflike appearance. *566*

fomite (fo′mīt; pl., **fomites**) An object that is not in itself harmful, but is able to harbor and transmit pathogenic organisms. Also called **fomes.** *581, 707*

food chain The flow of energy and matter in living organisms through a producer-consumer sequence. *549*

food-borne infection Gastrointestinal illness caused by ingestion of microorganisms, followed by their growth within the host. Symptoms arise from tissue invasion and/or toxin production. *763, 871*

food intoxication Food poisoning caused by microbial toxins produced in a food prior to consumption. The presence of living bacteria is not required. *763, 872*

food poisoning A general term usually referring to a gastrointestinal disease caused by the ingestion of food contaminated by pathogens or their toxins. *763*

food web A network of many interlinked food chains, encompassing primary producers, consumers, decomposers, and detritivores. *549, 806*

fractional steam sterilization *See* tyndallization. *316*

fragmentation (frag″men-ta′shun) A type of asexual reproduction in which a thallus breaks into two or more parts, each of which forms a new thallus. *538*

frameshift mutations Mutations arising from the loss or gain of a base or DNA segment, leading to a change in the codon reading frame and thus a change in the amino acids incorporated into protein. *249*

free energy change The total energy change in a system that is available to do useful work as the system goes from its initial state to its final state. *136*

fruiting body A specialized structure that holds sexually or asexually produced spores; found in fungi and in some bacteria. *487*

frustule (frus′tūl) A silicified cell wall in the diatoms. *541*

fruticose Lichens that have a shrubby shape. *566*

fucoxanthin (fu′ko-zan′thin) A brownish carotenoid found in the *Chrysophyta* (brown algae). *541*

fungicide (fun′ji-sīd) An agent that kills fungi. *311*

fungistasis (fun-ji-sta′sis) *See* fungistatic. *809*

fungistatic (fun″ji-stat′ik) Inhibiting the growth and reproduction of fungi. *311*

fungus (fung′gus; pl., **fungi**) Achlorophyllous, heterotrophic, spore-bearing eucaryotes with absorptive nutrition; usually, they have a walled thallus. *413, 519*

***F* value** The time in minutes at a specific temperature (usually 250°F) needed to kill a population of cells or spores. *314*

GALT (**gut associated lymphoid tissue**) *See* Peyer's patches. *616*

gametangium (gam-ĕ-tan′je-um; pl., **gametangia**) A structure that contains gametes or in which gametes are formed. *524*

gas gangrene (gang′grēn) A type of gangrene that arises from dirty, lacerated wounds infected by anaerobic bacteria, especially species of *Clostridium.* As the bacteria grow, they release toxins and ferment carbohydrates to produce carbon dioxide and hydrogen gas. *754*

gas vacuole A gas-filled vacuole found in cyanobacteria and some other aquatic bacteria that helps its possessor float. It is composed of gas vesicles, which are made of protein. *48*

gastritis (gas-tri'tis) Inflammation of the stomach. *756*

gastroenteritis (gas"tro-en-ter-i'tis) An acute inflammation of the lining of the stomach and intestines, characterized by anorexia, nausea, diarrhea, abdominal pain, and weakness. It has various causes including food poisoning due to such organisms as *E. coli, S. aureus,* and *Salmonella* species; consumption of irritating food or drink; or psychological factors such as anger, stress, and fear. Also called **enterogastritis.** *766*

gene (jēn) A DNA segment or sequence that codes for a polypeptide, rRNA, or tRNA. *202, 240*

generalized transduction The transfer of any part of a bacterial genome when the DNA fragment is packaged within a phage capsid by mistake. *275*

general recombination Recombination involving a reciprocal exchange of a pair of homologous DNA sequences; it can occur any place on the chromosome. *259*

generation time The time required for a microbial population to double in number. *114*

genetic engineering The process of deliberate modification of an organism's genetic information by directly changing its nucleic acid genome. *286*

genital herpes (her'pēz) A sexually transmitted disease caused by the herpes simplex type 2 virus. *730*

genome (je'nōm) The full set of genes present in a cell or virus; all the genetic material in an organism; a haploid set of genes in a cell. *237*

geosmin A group of compounds produced by actinomycetes and cyanobacteria that cause the characteristic odor of soils and waters. *513, 847*

German measles (rubella) A moderately contagious skin disease that occurs primarily in children five to nine years of age that is caused by the rubella virus, which is acquired by droplet inhalation into the respiratory system. *717*

germicide (jer'mi-sīd) An agent that kills pathogens and many nonpathogens, but not necessarily bacterial endospores. *311*

germination (jer"mĭ-na'shun) The stage following bacterial endospore activation in which the endospore breaks its dormant state. Germination is followed by outgrowth. *65*

Ghon complex (gon) The initial focus of parenchymal infection in primary pulmonary tuberculosis. *748*

giardiasis (je"ar-di'ah-sis) A common intestinal disease caused by the parasitic protozoan *Giardia lamblia.* *792*

gingivitis (jin-jĭ-vi'tis) Inflammation of the gingival tissue. *777*

gliding motility A type of motility in which a microbial cell glides along when in contact with a solid surface. *60, 484*

glomerulonephritis (glo-mer"u-lo-nē-fri'tis) An inflammatory disease of the renal glomeruli. *746*

glucans Polysaccharides composed of glucose units held together by glycosidic linkages. Some types of glucans have α-1,3 and α-1,6 linkages and bind bacterial cells together on teeth forming a plaque ecosystem. *776*

gluconeogenesis (gloo"ko-ne"o-jen'e-sis) The synthesis of glucose from noncarbohydrate precursors such as lactate and amino acids. *175*

glycocalyx (gli"ko-kal'iks) A network of polysaccharides extending from the surface of bacteria and other cells. *56*

glycogen (gli'ko-jen) A highly branched polysaccharide containing glucose, which is used to store carbon and energy. *47*

glycolysis (gli-kol'ĭ-sis) The anaerobic conversion of glucose to lactic acid by use of the Embden-Meyerhof pathway. *147*

glycolytic pathway (gli"ko-lit'ik) See Embden-Meyerhof pathway. *147*

glyoxylate cycle (gli-ok'sĭ-lāt) A modified tricarboxylic acid cycle in which the decarboxylation reactions are bypassed by the enzymes isocitrate lyase and malate synthase; it is used to convert acetyl-CoA to succinate and other metabolites. *182*

gnotobiotic (no"to-bi-ot'ik) Animals that are germfree (microorganism free) or live in association with one or more known microorganisms. *571*

Golgi apparatus (gol'je) A membranous eucaryotic organelle composed of stacks of flattened sacs (cisternae), which is involved in packaging and modifying materials for secretion and many other processes. *75*

gonococci (gon'o-kok'si) Bacteria of the species *Neisseria gonorrhoeae* the organisms causing gonorrhea. *754*

gonorrhea (gon"o-re'ah) An acute infectious sexually transmitted disease of the mucous membranes of the genitourinary tract, eye, rectum, and throat. It is caused by *Neisseria gonorrhoeae.* *754*

Gracilicutes (gras"ĭ-lik'u-tez) A division in the kingdom *Procaryotae* containing bacteria with gram-negative type cell walls. *421*

gradient (gra'dē-ent) The change in concentration of a substance over distance. *351, 807*

Gram stain A differential staining procedure that divides bacteria into gram-positive and gram-negative groups based on their ability to retain crystal violet when decolorized with an organic solvent such as ethanol. *29*

grana (gra'nah) A stack of thylakoids in the chloroplast stroma. *82*

granulocytopenia (gran"u-lo-si"to-pe'ne-ah) Deficiency of granulocytes in the blood. *576*

griseofulvin (gris"e-o-ful'vin) An antibiotic from *Penicillium griseofulvum* given orally to treat chronic dermatophyte infections of skin and nails. *340*

groundwater Water within a part of the ground that is totally saturated; it supplies wells and springs. *842*

group translocation A transport process in which a molecule is moved across a membrane by carrier proteins while being chemically altered at the same time. *103*

growth An increase in cellular constituents. *113*

growth factors Organic compounds that must be supplied in the diet for growth because they are essential cell components or precursors of such components and cannot be synthesized by the organism. *99*

growth yield (Y) A quantitative measure of the amount of microbial mass produced from a nutrient. *119*

guanine (gwan'in) A purine derivative, 2-amino-6-oxypurine, found in nucleosides, nucleotides, and nucleic acids. *183*

Guillain-Barré syndrome (ge-yan'bar-ra') A relatively rare disease affecting the peripheral nervous system, especially the spinal nerves, but also the cranial nerves. The cause is unknown, but it most often occurs after an influenza infection or flu vaccination. Also called **French Polio.** *718*

gumma (gum'ah) A soft, gummy tumor occurring in tertiary syphilis. *759*

halomethanes Chlorinated organic compounds formed during the process of water chlorination. *835*

halophile (hal'o-fīl) A microorganism that requires high levels of sodium chloride for growth. *123, 500*

haploid (hap'loid) Having half the number of chromosomes present in a somatic cell or a single set of chromosomes. *85*

hapten (hap'ten) A molecule not immunogenic by itself but that, when coupled to a macromolecular carrier, can elicit antibodies directed against itself. *611*

harborage transmission The mode of transmission in which an infectious organism does not undergo morphological or physiological changes within the vector. *707*

hay fever Allergic rhinitis; a type of atopic allergy involving the upper respiratory tract. *636*

health (helth) A state of optimal physical, mental, and social well-being, and not merely the absence of disease and infirmity. *699*

healthy carrier An individual who harbors an infectious organism, but is not ill. *706*

helical (hel'ĭ-kal) In virology, this refers to a virus with a helical capsid surrounding its nucleic acid. *355*

helicases Enzymes that use ATP energy to unwind DNA ahead of the replication fork. *199*

helminth (hel'minth) A parasitic worm. *A34*

hemadsorption (hem"ad-sorp'shun) The adherence of red blood cells to the surface of something, such as another cell or a virus. *678*

hemagglutination (hem"ah-gloo"tĭ-na'shun) The agglutination of red blood cells by antibodies. *654*

hemagglutinin (hem"ah-gloo'tĭ-nin) The antibody responsible for a hemagglutination reaction. *654*

hemoflagellate (he"mo-flaj'ē-lāt) A flagellated protozoan parasitc that is found in the bloodstream. *553, 795*

hemolysin (he-mol'i-sin) A substance that causes hemolysis (the lysis of red blood cells). At least some hemolysins are enzymes that destroy the phospholipids in erythrocyte plasma membranes. *589*

hemolysis (he-mol'i-sis) The disruption of red blood cells and release of their hemoglobin. There are several types of hemolysis when bacteria such as streptococci and staphylococci grow on blood agar. In α-hemolysis, a narrow greenish zone of incomplete hemolysis forms around the colony. A clear zone of complete hemolysis without any obvious color change is formed during **β-hemolysis.** *455*

hemolytic disease of the newborn A hemolytic anemia of a newborn child that results from the destruction of the infant's red blood cells by antibodies produced by the mother; usually the antibodies are due to Rh blood type incompatibility. Also called **erythroblastosis fetalis.** *643*

hemorrhagic fever A fever usually caused by a specific virus that may lead to hemorrhage, shock, and sometimes death. *721*

hepatitis (hep"ah-ti'tis) Any infection that results in inflammation of the liver. *733*

hepatitis A (infectious hepatitis; hep"ah-ti'tis) A type of hepatitis that is transmitted by fecal-oral contamination; it primarily affects children and young adults, especially in environments where there is poor sanitation and overcrowding. It is caused by the hepatitis A virus, a single-stranded RNA virus. *733, 736*

hepatitis B (serum hepatitis; hep"ah-ti'tis) This form of hepatitis is caused by a double-stranded DNA virus (HBV) formerly called the "Dane particle." The virus is transmitted by body fluids. *733*

hepatitis C About 85% of all cases of viral hepatitis can be traced to either HAV or HBV. The remaining 10 to 15% are believed to be caused by one and possibly several other types of viruses. At least one of these is hepatitis C (formerly non-A, non-B). *733*

herd immunity The resistance of a population to infection and spread of an infectious organism due to the immunity of a high percentage of the population. *703*

hermaphroditic (her-maf'ro-dīt-ik) Refers to an organism possessing both male and female sex organs. *A38*

herpes labialis *See* cold sore. *729*

herpetic keratitis (her-pet'ik ker"ah-ti'tis) An inflammation of the cornea and conjunctiva of the eye resulting from a herpes simplex virus infection. *729*

heterocysts Specialized cells produced by cyanobacteria that are the sites of nitrogen fixation. *474*

heteroduplex DNA A double-stranded stretch of DNA formed by two slightly different strands that are not completely complementary. *259*

heterogeneous nuclear RNA (hnRNA) The RNA transcript of DNA made by RNA polymerase II; it is then processed to form mRNA. *203*

heterolactic fermenters (het"er-o-lak'tik) Microorganisms that ferment sugars to form lactate, and also other products such as ethanol and CO_2. *156, 457*

heterologous antigen *See* heterophile. *610*

heterophile (het"er-o-fīl) An antigen that cross-reacts with antibodies produced in response to a different antigen; also called a heterologous antigen. Usually the term refers to antigens from unrelated species that cross-react because they possess identical or similar epitopes. *610*

heterotroph (het'er-o-trōf") An organism that uses reduced, preformed organic molecules as its principal carbon source. *97*

heterotrophic nitrification Nitrification carried out by chemoheterotrophic microorganisms. *812*

hexose monophosphate pathway (hek'sōs mon"o-fos'fāt) *See* pentose phosphate pathway. *148*

Hfr strain A bacterial strain that donates its genes with high frequency to a recipient cell during conjugation because the F factor is integrated into the bacterial chromosome. *271*

high-energy molecule A molecule whose hydrolysis under standard conditions yields a large amount of free energy (the standard free energy change is more negative than about −7 kcal/mole); a high-energy molecule readily decomposes and transfers groups such as phosphate to acceptors. *136*

histone (his'tōn) A small basic protein with large amounts of lysine and arginine that is associated with eucaryotic DNA in chromatin. *196*

histoplasmosis (his"to-plaz-mo'sis) A systemic fungal infection caused by *Histoplasma capsulatum*. *786*

hives (hivz) An eruption of the skin. *637*

holdfast A structure produced by some bacteria and algae that attaches the cell to a solid object. *481*

holoenzyme A complete enzyme consisting of the apoenzyme plus a cofactor. *140*

holozoic nutrition (hol"o-zo'ik) In this type of nutrition, nutrients (such as bacteria) are acquired by phagocytosis and the subsequent formation of a food vacuole or phagosome. *550*

homolactic fermenters (ho"mo-lak'tik) Organisms that ferment sugars almost completely to lactic acid. *156*

hormogonia Small motile fragments produced by fragmentation of filamentous cyanobacteria; used for asexual reproduction and dispersal. *474*

host (hōst) The body of an organism that harbors another organism. It can be viewed as a microenvironment that shelters and supports the growth and multiplication of a parasitic organism. *579*

host restriction The degradation of foreign genetic material by nucleases after the genetic material enters a host cell. *261*

human immunodeficiency virus (HIV) A retrovirus that is associated with the onset of AIDS. *722*

human leukocyte antigen (HLA) An antigen on the surface of cells of human tissues and organs that is recognized by the immune system cells and therefore is important in graft rejection and regulation of the immune response. *630*

humoral immunity (hu'mor-al) The type of immunity that results from the presence of soluble antibodies being soluble in blood and lymph; also known as **antibody-mediated immunity.** *609*

humus (hu'mus) The organic portion of the soil that remains after microbial degradation of plant and animal matter. *847*

hybridoma (hi"bri-do'mah) A fast-growing cell line produced by fusing a cancer cell (myeloma) to another cell, such as an antibody-producing cell. *622*

hydatid cysts (hy'dah-tid) Larval cysts formed by tapeworms of the genus *Echinococcus* that contain daughter cysts, each holding many scoleces. *A34*

hydrogenosome (hi-dro-jen'osom) A microbodylike organelle that contains a unique electron transfer pathway in which hydrogenase transfers electrons to protons (which act as the terminal electron acceptors) and molecular hydrogen is formed. *550*

hydrophilic (hi"dro-fil'ik) A polar substance that has a strong affinity for water (or is readily soluble in water). *44*

hydrophobic (hi"dro-fo'bik) A nonpolar substance lacking affinity for water (or which is not readily soluble in water). *44*

hyperendemic disease (hi"per-en-dem'ik) A disease that has a gradual increase in occurrence beyond the endemic level, but not at the epidemic level, in a given population; also may refer to a disease that is equally endemic in all age groups. *699*

hyperferremia (hi"per-fer-re'me-ah) An excess of iron in the blood. *590*

hypersensitivity (hi"per-sen'si-tiv"i-te) A condition of increased immune sensitivity in which the body reacts to an antigen with an exaggerated immune response that usually harms the individual. Also termed an **allergy.** *635*

hypha (hi'fah; pl., **hyphae**) The unit of structure of most fungi and some bacteria; a tubular filament. *520*

hypoferremia (hi"po-fĕ-re'me-ah) Deficiency of iron in the blood. *590*

hypolimnion The part of a lake lying below the thermocline (a layer of rapid temperature change) that usually has a relatively uniform, cold temperature and a low oxygen level. *830*

hypotheca (hi-po-theca) The smaller half of a diatom frustule. *541*

hypothesis A tentative assumption or educated guess developed to explain a set of observations. *12*

icosahedral In virology, this term refers to a virus with an icosahedral capsid, which has the shape of a regular polyhedron having 20 equilateral triangular faces and 12 corners. *355*

identification (i-den"ti-fi-ka'shun) The process of determining that a particular isolate or organism belongs to a recognized taxon. *405*

idiophase The period in a batch culture in which secondary metabolites are synthesized in preference to primary metabolites; it generally corresponds to the stationary phase and the end of the log phase. *895*

idiotype (id'e-o-tīp') A set of one or more unique epitopes in the variable region of an immunoglobulin that distinguishes it from immunoglobulins produced by different plasma cells. *612*

IgA Immunoglobulin A; the class of immunoglobulins that is present in dimeric form in many body secretions (e.g., saliva, tears, and bronchial and intestinal secretions) and protects body surfaces. IgA also is present in serum. *614*

IgD Immunoglobulin D; the class of immunoglobulins found on the surface of many B lymphocytes; thought to serve as an antigen receptor in the stimulation of antibody synthesis. *615*

IgE Immunoglobulin E; the immunoglobulin class that binds to mast cells and basophils, and is responsible for type I or anaphylactic hypersensitivity reactions such as hay fever and asthma. IgE is also involved in resistance to helminth parasites. *615*

IgG Immunoglobulin G; the predominant immunoglobulin class in serum. Has functions such as neutralizing toxins, opsonizing bacteria, activating complement, and crossing the placenta to protect the fetus and neonate. *614*

IgM Immunoglobulin M; the class of serum antibody first produced during an infection. It is a large, pentameric molecule that is active in agglutinating pathogens and activating complement. The monomeric form is present on the surface of some B lymphocytes. *614*

immobilization (im-mo"bil-i-za'shun) The incorporation of a simple, soluble substance into the body of an organism, making it unavailable for use by other organisms. *899*

immune (i-mūn') Protected against a particular disease by either specific or nonspecific defenses; a process or substance concerned with the immune response. *590, 607*

immune complex (i-mūn kom'pleks) The product of an antigen-antibody reaction, which may also contain components of the complement system. *654*

immune response (i-mūn re-spons') The response of the body to contact with an antigen that leads to the formation of antibodies and sensitized lymphocytes. The response is designed to render harmless the antigen and the pathogen producing it. *607*

immune surveillance (i-mūn sur-vāl'ans) The still somewhat hypothetical process in which lymphocytes such as natural killer (NK) cells recognize and destroy tumor cells; other cells with abnormal surface antigens (e.g., virus-infected cells) may also be destroyed. *635*

immunity (i-mu'ni-te) Refers to the overall general ability of a host to resist a particular disease; the condition of being immune. *607*

immunization (im"u-ni-za'shun) The induction of protective immunity by administration of either (1) a vaccine or toxoid (active immunization) or (2) preformed antibodies (passive immunization). *620*

immunoblotting The electrophoretic transfer of proteins from polyacrylamide gels to nitrocellulose sheets in order to demonstrate the presence of specific proteins through reaction with labeled antibodies. *691*

immunodeficiency (im"u-no-dē-fish'en-se; pl., **immunodeficiencies**) The inability to produce a normal complement of antibodies or immunologically sensitized T cells in response to specific antigens. *641*

immunodiffusion A technique involving diffusion of antigen and/or antibody through a semisolid gel to produce a precipitin reaction when they meet. *661*

immunoelectrophoresis (i-mu"no-e-lek"tro-fo-re'sis; pl., **immunoelectrophoreses**) The electrophoretic separation of proteins followed by diffusion and precipitation in gels using antibodies against the separated proteins. *663*

immunofluorescence (im"u-no-floo"o-res'ens) A technique used to identify particular antigens microscopically in cells or tissues by the binding of a fluorescent antibody conjugate. *663*

immunogen (im'u-no-jen) *See* antigen. *610*

immunoglobulin (**Ig**; im"u-no-glob'u-lin) *See* antibody. *611*

immunology (im"u-nol'o-je) The branch of science that deals with the immune system and attempts to understand the many phenomena that are responsible for both acquired and innate immunity. It also includes the use of antibody-antigen reactions in other laboratory work (serology and immunochemistry). *607*

immunoprecipitation (im"u-no-pre-sip"i-ta'shun) A reaction involving soluble antigens reacting with antibodies to form a large aggregate that precipitates out of solution. *665*

immunotoxin (im'u-no-tok"sin) A monoclonal antibody that has been attached to a specific toxin or toxic agent (antibody + toxin = immunotoxin) and can kill specific target cells. *623*

inclusion bodies Granules of organic or inorganic material lying in the cytoplasmic matrix of bacteria. *47, 393*

inclusion conjunctivitis (in-klu'zhun kon-junk"tī-vi'tis) An acute infectious disease that occurs throughout the world. It is caused by *Chlamydia trachomatis* that infects the eye and causes inflammation and the occurrence of large inclusion bodies. *771*

incubatory carrier An individual who is incubating an infectious organism in large numbers but is not yet ill. *706*

index case The first disease case in an epidemic within a given population. *699*

indicator organism An organism whose presence indicates the condition of a substance or environment, for example, the potential presence of pathogens. Coliforms are used as indicators of fecal pollution. *837*

inducer (in-dūs'er) A small molecule that stimulates the synthesis of an inducible enzyme. *222*

inducible enzyme An enzyme whose level rises in the presence of a small molecule that stimulates its synthesis. *222*

infantile paralysis (in'fan-til pah-ral'i-sis) *See* poliomyelitis. *736*

infection (in-fek'shun) The invasion of a host by a microorganism with subsequent establishment and multiplication of the agent. An infection may or may not lead to overt disease. *579*

infection thread A tubular structure formed during the infection of a root by nitrogen-fixing bacteria. The bacteria enter the root by way of the infection thread and stimulate the formation of the root nodule. *854*

infectious disease Any change from a state of health in which part or all of the host's body cannot carry on its normal functions because of the presence of an infectious agent or its products. *579*

infectious disease cycle (chain) The chain or cycle of events that describes how an infectious organism grows, reproduces, and is disseminated. *704*

infectious mononucleosis (**mono**; mon"o-nu"kle-o'sis) An acute, self-limited infectious disease of the lymphatic system caused by the Epstein-Barr virus and characterized by fever, sore throat, lymph node and spleen swelling, and the proliferation of monocytes and abnormal lymphocytes. *731*

infective dose 50 (ID_{50}) Refers to the dose or number of organisms that will infect 50% of an experimental group of hosts within a specified time period. *580*

infectivity (in"fek-tiv'i-te) Infectiousness; the state or quality of being infectious or communicable. *580*

inflammation (in"flah-ma'shun) A localized protective response to tissue injury or destruction. Acute inflammation is characterized by pain, heat, swelling, and redness in the injured area. *595*

influenza or **flu** (in"flu-en'zah) An acute viral infection of the respiratory tract, occurring in isolated cases, epidemics, and pandemics. Influenza is caused by three strains of influenza virus, labeled types A, B, and C, based on the antigens of their protein coats. *718*

initial body *See* reticulate body (RB). *444*

insertion sequence (in-ser'shun se'kwens) A simple transposon that contains genes only for those enzymes, such as the transposase, that are required for transposition. *264*

***in situ* treatment** (in si'tu) Chemical and/or biological treatment that is carried out in place in a particular environment. *843*

integration The incorporation of one DNA segment into a second DNA molecule to form a new hybrid DNA. Integration occurs during such processes as genetic recombination, episome incorporation into host DNA, and prophage insertion into the bacterial chromosome. *378*

intercalating agents Molecules that can be inserted between the stacked bases of a DNA double helix, thereby distorting the DNA and inducing insertion and deletion mutations. *246*

interference (in"ter-fēr'ens) In virology, the resistance a cell possesses against other viruses when already infected with a specific virus. *678*

interferons (**IFNs**; in"ter-fēr'ons) A group of glycoproteins that have nonspecific antiviral activity by stimulating cells to produce antiviral proteins, which inhibit the synthesis of viral RNA and proteins. Interferons also regulate the growth, differentiation, and/or function of a variety of immune system cells. Their production may be stimulated by virus infections, intracellular pathogens (chlamydiae and rickettsias), protozoan parasites, endotoxins, and other agents. *341, 593, 627*

intertriginous candidiasis A skin infection caused by *Candida* species. Involves those areas of the body, usually opposed skin surfaces, that are warm and moist (axillae, groin, skin folds). *788*

interleukin (in″ter-loo′kin) A group of proteins produced by macrophages and T cells that regulate growth and differentiation, particularly of lymphocytes. They promote cellular and humoral immune responses. *627*

interleukin-1 (IL-1) A product of antigen-presenting macrophages. It stimulates the production of interleukin-2 by T cells. *627*

interleukin-2 (IL-2) An interleukin that is released by T-helper cells under the influence of interleukin-1. Also known as **T-cell growth factor.** It stimulates the proliferation of T cells and their activity. IL-2 also promotes B-cell growth. *627*

interleukin-3 (IL-3) An interleukin produced by T lymphocytes that has broad effects. It stimulates the proliferation of stem cells that are precursors of erythrocytes, leukocytes, and mast cells. IL-3 also induces the division of mature macrophages and mast cells. *627*

intermediate host (in″ter-me′de-it hōst) The host that serves as a temporary but essential environment for development of a parasite and completion of its life cycle. *579*

intermediate filaments (in″ter-me′de-it fil′ah-ments) Small protein filaments, about 8 to 10 nm in diameter, in the cytoplasmic matrix of eucaryotic cells that are important in cell structure. *73*

intoxication (in-tok″si-ka′shun) A disease that results from the entrance of a specific toxin into the body of a host. The toxin can induce the disease in the absence of the toxin-producing organism. *584*

intranuclear inclusion body (in″trah-nu′kel-ar) A structure found within cells infected with the cytomegalovirus. *730*

intron (in′tron) A noncoding intervening sequence in a split or interrupted gene, which codes for RNA that is missing from the final RNA product. *204*

intubation (in″tu-ba′shun) The introduction of a tube into a hollow organ, such as the trachea or intestine, to keep it open if obstructed. *675*

invasiveness (in-va′siv-nes) The ability of a microorganism to enter a host, grow and reproduce within the host, and spread throughout its body. *580*

ionizing radiation Radiation of very short wavelength or high energy that causes atoms to lose electrons or ionize. *130, 318*

isograft (i′so-graft) A graft involving identical twins, which have the same genetic constituency. *640*

isotype (i′so-tīp) A variant form of an immunoglobulin (e.g., an immunoglobulin class, subclass, or type) that occurs in every normal individual of a particular species. Usually the characteristic antigenic determinant is in the constant region of H and L chains. *612*

J chain A polypeptide present in polymeric IgM and IgA that links the subunits together. *614*

Jaccard coefficient (S_J) An association coefficient used in numerical taxonomy; it is the proportion of characters that match, excluding those that both organisms lack. *407*

kappa (kap′ah) A cytoplasmic bacterial endosymbiont within *Paramecium aurelia* that endows its possessor with the ability to kill sensitive paramecia. *568*

kelp (kelp) A common name for any of the larger members of the order *Laminariales* of the brown algae. *543*

keratitis (ker″ah-ti′tis) Inflammation of the cornea of the eye. *792*

kinetoplast (ki-ne′to-plast) A special structure in the mitochondrion of kinetoplastid protozoa. It contains the mitochondrial DNA. *553*

Kirby-Bauer method A disk diffusion test to determine the susceptibility of a microorganism to chemotherapeutic agents. *329*

Koch's postulates (koks pos′tu-lāts) A set of rules for proving that a microorganism causes a particular disease. *11*

Koplik's spots (kop′liks) Lesions of the oral cavity caused by the measles (rubeola) virus that are characterized by a bluish white speck in the center of each. *719*

Korean hemorrhagic fever An acute infection caused by a virus that produces varying degrees of hemorrhage, shock, and sometimes death. *721*

lactic acid fermentation (lak′tik) A fermentation that produces lactic acid as the sole or primary product. *156*

lagered Pertaining to the process of aging beers to allow flavor development. *880*

lag phase A period following the introduction of microorganisms into fresh culture medium when there is no increase in cell numbers or mass. *113*

laminarin (lam″i-na′rin) One of the principal storage products of the brown algae; a polymer of glucose. *541*

Lancefield system (group; lans′feld) One of the serologically distinguishable groups (as group A, group B) into which streptococci can be divided. *455, 666*

late mRNA Messenger RNA produced later in a virus infection, which codes for proteins needed in capsid construction and virus release. *369*

latent period (la′tent) The initial phase in the one-step growth experiment in which no phages are released. *369*

latent virus infections Virus infections in which the virus quits reproducing and remains dormant for a period before becoming active again. *393*

leader sequence A nontranslated sequence at the 5′ end of mRNA that lies between the operator and the initiation codon; it aids in the initiation and regulation of transcription. *202, 242*

Legionnaires' disease (legionellosis) A pulmonary form of legionellosis, resulting from infection with *Legionella pneumophila.* *742*

leishmanias (lēsh″ma′ne-ās) Zooflagellates, members of the genus *Leishmania,* that cause the disease leishmaniasis. *795*

leishmaniasis (lēsh″mah-ni′ah-sis) The disease caused by the protozoa called leishmanias. *795*

lepromatous (progressive) leprosy (lep-ro′mah-tus lep′ro-se) A relentless, progressive form of leprosy in which large numbers of *Mycobacterium leprae* develop in skin cells, killing the skin cells and resulting in the loss of features. Disfiguring nodules form all over the body. *755*

leprosy (lep′ro-se) A severe disfiguring skin disease caused by *Mycobacterium leprae.* Also known as **Hansen's disease.** *755*

lethal dose 50 (LD_{50}) Refers to the dose or number of organisms that will kill 50% of an experimental group of hosts within a specified time period. *353, 580*

leukemia (loo-ke′me-ah) A progressive, malignant disease of blood-forming organs, marked by distorted proliferation and development of leukocytes and their precursors in the blood and bone marrow. *731*

leukocidin (loo″ko-si′din) A microbial toxin that can damage or kill leukocytes. *589*

L forms Pleomorphic bacterial cells formed by the complete or partial loss of their cell walls. Cell wall loss may be reversible or permanent. *55*

lichen (li′ken) An organism composed of a fungus and an alga in a symbiotic association. *566*

Liebig's law of the minimum (le′bigz) Living organisms and populations will grow until some factor begins to limit further growth. *807*

lipopolysaccharide (lip″o-pol″e-sak′ah-rīd) A molecule containing both lipid and polysaccharide, which is important in the outer membrane of the gram-negative cell wall. *54*

liposome (lip′o-som) A spherical particle formed by a lipid bilayer enclosing an aqueous solution. It may be used to administer chemotherapeutic agents. *691*

lithotroph (lith′o-trōf) An organism that uses reduced inorganic compounds as its electron source. *98, 160*

log phase *See* exponential phase. *113*

loiasis (lo-i′ah-sis) An infection with nematodes of the genus *Loa.* *A38*

lophotrichous (lo-fot′ri-kus) A cell with a cluster of flagella at one or both ends. *57*

Lyme disease (LD, Lyme borreliosis; līm) A tick-borne disease caused by the spirochete *Borrella burgdorferi.* *750*

lymphocyte (lim′fo-sīt) A nonphagocytic, mononuclear leukocyte (white blood cell) that is an immunologically competent cell, or its precursor. Lymphocytes are present in the blood, lymph, and lymphoid tissues. *See* B cell and T cell. *596, 609*

lymphogranuloma venereum (LGV; lim″fo-gran″u-lo′mah) A sexually transmitted disease caused by *Chlamydia trachomatis* serotypes L_1–L_3, which affect the lymph organs in the genital area. *771*

lymphokine (lim′fo-kin) A biologically active protein (e.g., IL-1) secreted by activated lymphocytes, especially sensitized T cells. It acts as an intercellular mediator of the immune response and transmits growth, differentiation, and behavioral signals. *609, 627*

lyophilization (li-of″i-li-za′shun) A method of freeze-drying used to preserve microbial cultures. *893*

lysis (li′sis) The rupture or physical disintegration of a cell. *55*

lysogenic (li-so-jen′ik) *See* lysogens. *275*

lysogens (li′so-jens) Bacteria that are carrying a viral prophage and have the potential of producing bacteriophages under the proper conditions. *275, 376*

lysogeny (li-soj′e-ne) The state in which a phage genome remains within the bacterial host cell after infection and reproduces along with it rather than taking control of the host and destroying it. *275, 376*

lysosome (li′so-sōm) A spherical membranous eucaryotic organelle that contains hydrolytic enzymes and is responsible for the intracellular digestion of substances. *76*

lysozyme (li′so-zīm) An enzyme that degrades peptidoglycan by hydrolyzing the $\beta(1 \rightarrow 4)$ bond that joins N-acetylmuramic acid and N-acetylglucosamine. *55*

lytic cycle (lit′ik) A virus life cycle that results in the lysis of the host cell. *368*

macrolide antibiotic (mak′ro-līd) An antibiotic containing a macrolide ring, a large lactone ring with multiple keto and hydroxyl groups, linked to one or more sugars. *337*

macromolecule (mak″ro-mol′ĕ-kūl) A large molecule that is a polymer of smaller units joined together. *172*

macronucleus (mak″ro-nu′kle-us) The larger of the two nuclei in ciliate protozoa. It is normally polyploid and directs the routine activities of the cell. *550*

macrophage (mak′ro-fāj) The name for a large mononuclear phagocytic cell, present in blood, lymph, and other tissues. Macrophages are derived from monocytes. They phagocytose and destroy pathogens; some macrophages also activate B cells and T cells. *630*

maduromycosis (mah-du′ro-mi-ko′sis) A subcutaneous fungal infection caused by *Madurella mycetoma*; also termed a **mycetoma.** *785*

madurose The sugar derivative 3-O-methyl-D-galactose, which is characteristic of several actinomycete genera that are collectively called maduromycetes. *515*

magnetosomes Magnetite particles in magnetotactic bacteria that are tiny magnets and allow the bacteria to orient themselves in magnetic fields. *61*

magnetotactic bacteria Bacteria that can orient themselves in the earth's magnetic field. *61*

maintenance energy The energy a cell requires simply to maintain itself or remain alive and functioning properly. It does not include the energy needed for either growth or reproduction. *120*

major histocompatibility complex (MHC) A large set of cell surface antigens in each individual, encoded by a family of genes, that serves as a unique biochemical marker of individual identity. It can trigger T-cell responses that may lead to rejection of transplanted tissues and organs. MHC antigens are also involved in the regulation of the immune response and the interactions between immune cells. *630*

malaria (mah-la′re-ah) A serious infectious illness caused by the parasitic protozoan *Plasmodium.* Malaria is characterized by bouts of high chills and fever that occur at regular intervals. *793*

malt (mawlt) Grain soaked in water to soften it, induce germination, and activate its enzymes. The malt is then used in brewing and distilling. *880*

Marburg viral hemorrhagic fever An acute infection caused by a virus that produces varying degrees of hemorrhage, shock, and sometimes death. *721*

mash The soluble materials released from germinated grains and prepared as a microbial growth medium. *880*

mashing The process in which cereals are mixed with water and incubated in order to degrade their complex carbohydrates (e.g., starch) to more readily utilizable forms such as simple sugars. *879*

mast cell A type of connective tissue cell that produces and secretes histamine, heparin, and other biologically active products. It participates in immediate type hypersensitivity reactions and in the inflammatory response. *596, 615, 636*

mean growth rate constant (k) The rate of microbial population growth expressed in terms of the number of generations per unit time. *114*

measles (rubeola; me′zelz) A highly contagious skin disease that is endemic throughout the world. It is caused by a morbilli virus in the family *Paramyxoviridae,* which enters the body through the respiratory tract or through the conjunctiva. *719*

medical mycology (mi-kol′o-je) The discipline that deals with the fungi that cause human disease. *782*

meiosis (mi-o′sis) The process in which a diploid cell divides and forms two haploid cells. *85*

melting temperature (T_m) of DNA The temperature at which double-stranded DNA separates into individual strands; it is dependent on the G + C content of the DNA and is used to compare genetic material in microbial taxonomy. *411*

membrane attack complex (MAC) The complex complement components (C5b–C9) that create a pore in the plasma membrane or a target cell and leads to cell lysis. C9 probably forms the actual pore. *648*

membrane filter A thin filter made of cellulose or synthetic polymers that has a carefully controlled pore size and is used to filter out microorganisms. *117, 316*

membrane filter technique The use of a thin porous filter made from cellulose acetate or some other polymer to collect microorganisms from water, air, and food. *839*

memory B cell A lymphocyte capable of initiating the antibody-mediated immune response upon detection of a specific antigen molecule for which it is genetically programmed. It circulates freely in the blood and lymph and may live for years. *619*

Mendosicutes (men″do-sik′utēz) A division in the kingdom *Procaryotae* composed of bacteria that usually have a wall, but lack peptidoglycan with muramic acid. *422*

meningitis (men″in-ji′tis) A condition that refers to inflammation of the brain or spinal cord meninges (membranes). The disease can be divided into **bacterial (septic) meningitis** (caused by bacteria) and **aseptic meningitis syndrome** (caused by nonbacterial sources). *744*

mesophile (mes′o-fīl) A microorganism with a growth optimum around 20 to 45°C, a minimum of 15 to 20°C, and a maximum about 45°C or lower. *126*

messenger RNA (mRNA) Single-stranded RNA synthesized from a DNA template during transcription that binds to ribosomes and directs the synthesis of protein. *197*

metabolic channeling (mĕt″ah-bol′ik) The localization of metabolites and enzymes in different parts of a cell. *217*

metabolic engineering The rational modification of metabolic pathways and products by the use of molecular biology. *901*

metabolism (me-tab′o-lizm) The total of all chemical reactions in the cell; almost all are enzyme catalyzed. *146*

metacercaria (met″ah-ser-ka′re-ah) The encysted stage of a trematode parasite that is found within the tissues of an intermediate host or on vegetation. It is usually infective for the final or definitive host. *A40*

metachromatic granules (met″ah-kro-mat′ik) Granules of polyphosphate in the cytoplasm of some bacteria that appear a different color when stained with a blue basic dye. They are reserves of phosphate. *48*

metastasis (mĕ-tas′tah-sis) The transfer of a disease like cancer from one organ to another not directly connected with it. *393*

methanogenic bacteria (meth″ah-no-jen′ik) Strictly anaerobic bacteria that derive energy by converting CO_2, H_2, formate, acetate, and other compounds to either methane or methane and CO_2; they are archaeobacteria. *497*

methylotroph A bacterium that uses one-carbon compounds such as methane and methanol as its sole source of carbon and energy. *435, 481*

Michaelis constant (K_m; mi-ka′lis) A kinetic constant for an enzyme reaction that equals the substrate concentration required for the enzyme to operate at half maximal velocity. *142*

microaerophile (mi″kro-a′er-o-fīl) A microorganism that requires low levels of oxygen for growth, around 2 to 10%, but is damaged by normal atmospheric oxygen levels. *127*

microbial ecology The study of microorganisms in their natural environments, with a major emphasis on physical conditions, processes, and interactions that occur on the scale of individual microbial cells. *807, 846*

microbial transformation (mi-kro'be-al) *See* bioconversion. *898*

microbiology (mi"kro-bi-ol'o-je) The study of organisms that are usually too small to be seen with the naked eye. Special techniques are required to isolate and grow them. *5*

microbiostasis The inhibition of microbial growth and reproduction. *809*

microbiota (also **indigenous microbial population, microflora, microbial flora;** mi"kro-bi-o'tah) The microorganisms normally associated with a particular tissue or structure. *571*

microbivory The use of microorganisms as a food source by organisms that can ingest or phagocytose them. *848*

microenvironment (mi"kro-en-vi'ron-ment) The immediate environment surrounding a microbial cell or other structure, such as a root. *807*

microfilaments (mi"kro-fil'ah-ments) Protein filaments, about 4 to 7 nm in diameter, that are present in the cytoplasmic matrix of eucaryotic cells and play a role in cell structure and motion. *72*

microfilariae (mi"kro-fi-la're-e) First-stage juveniles of ovoviviparous filariid nematodes. They are usually located in the blood and tissue fluids of the definitive host. *A38*

micronucleus (mi"kro-nu'kle-us) The smaller of the two nuclei in ciliate protozoa. Micronuclei are diploid and involved only in genetic recombination and the regeneration of macronuclei. *550*

microorganism (mi"kro-or'gan-izm) An organism that is too small to be seen clearly with the naked eye. *5*

microtubules (mi"kro-tu'buls) Small cylinders, about 25 nm in diameter, made of tubulin proteins and present in the cytoplasmic matrix and flagella of eucaryotic cells; they are involved in cell structure and movement. *72*

migration inhibition factor (MIF) A lymphokine produced by T cells that inhibits the movement of macrophages away from the MIF source, thus promoting macrophage accumulation in the area. *627*

miliary tuberculosis (mil'e-a-re) An acute form of tuberculosis in which small tubercles are formed in a number of organs of the body because of dissemination of the bacilli throughout the body by the bloodstream. Also known as **reactivation tuberculosis.** *748*

mineralization process The microbial breakdown of organic materials to inorganic substances. *434, 859*

minimal inhibitory concentration (MIC) The lowest concentration of a drug that will prevent the growth of a particular microorganism. *328*

minimal lethal concentration (MLC) The lowest concentration of a drug that will kill a particular microorganism. *328*

minus, or negative, strand The virus nucleic acid strand that is complementary in base sequence to the viral mRNA. *359*

miracidium (mi-rah-sid'e-um) The ciliated first stage larva of a trematode; it enters an intermediate host, usually a snail, and undergoes further development. *A40*

missense mutation A single base substitution in DNA that changes a codon for one amino acid into a codon for another. *248*

mitochondrion (mi"to-kon'dre-on) The eucaryotic organelle that is the site of electron transport, oxidative phosphorylation, and pathways such as the Krebs cycle; it provides most of a nonphotosynthetic cell's energy under aerobic conditions. It is constructed of an outer membrane and an inner membrane, which contains the electron transport chain. *78*

mitogen (mi'to-jen) A substance that induces mitosis. *627*

mitosis (mi-to'sis) A process that takes place in the nucleus of a eucaryotic cell and results in the formation of two new nuclei, each with the same number of chromosomes as the parent. *83*

mixed acid fermentation A type of fermentation carried out by members of the family *Enterobacteriaceae* in which ethanol and a complex mixture of organic acids are produced. *156*

mixotrophic (mik"so-trof'ik) Refers to microorganisms that combine autotrophic and heterotrophic metabolic processes (they use inorganic electron sources and organic carbon sources). *98*

mold Any of a large group of fungi that cause mold or moldiness and that exist as multicellular filamentous colonies; also the deposit or growth caused by such fungi. Molds do not produce macroscopic fruiting bodies. *520*

Monera (mo-ne'rah) *See Procaryotae. 413*

monoclonal antibody (mon"o-klōn'al) An antibody of a single type that is produced by a population of genetically identical plasma cells (a clone); a monoclonal antibody is typically produced from a cell culture derived from the fusion product of a cancer cell and an antibody-producing cell. *621*

mononuclear phagocyte system (MPS; mon"o-nu'kle-ar) The body's collection of macrophages, monocytes, and their precursors. The system's major function is the removal of foreign material, including pathogens, and cell debris. *595*

monotrichous (mon-ot'ri-kus) Having a single flagellum. *57*

morbidity rate (mor-bid'i-te) Measures the number of individuals who become ill as a result of a particular disease within a susceptible population during a specific time period. *701*

mordant (mor'dant) A substance that helps fix dye on or in a cell. *29*

mortality rate (mor-tal'i-te) The ratio of the number of deaths from a given disease to the total number of cases of the disease. *701*

most probable number (MPN) The statistical estimation of the probable population in a liquid by diluting and determining end points for microbial growth. *839*

mucociliary blanket The layer of cilia and mucus that lines certain portions of the respiratory system; it traps microorganisms up to 10 μm in diameter and then transports them by ciliary action away from the lungs. *602*

mumps An acute generalized disease that occurs primarily in school-age children and is caused by a paramyxovirus that is transmitted in saliva and respiratory droplets. The principal manifestation is swelling of the parotid salivary glands. *719*

murein *See* peptidoglycan. *50*

must The juices of fruits, including grapes, that can be fermented for the production of alcohol. *878*

mutagen (mu'tah-jen) A chemical or physical agent that causes mutations. *244*

mutation (mu-ta'shun) A permanent, heritable change in the genetic material. *244*

mutualism (mu'tu-al-izm") A type of symbiosis in which both partners gain from the association and are unable to survive without it. The mutualist and the host are metabolically dependent on each other. *565*

mutualist (mu'tu-al-ist) An organism associated with another in a relationship that is beneficial to both. *565*

mycelium (mi-se'le-um) A mass of branching hyphae found in fungi and some bacteria. *42, 520*

mycetoma (mi"se-to'mah) *See* maduromycosis. *785*

mycobiont The fungal partner in a lichen. *566*

mycolic acids Complex 60 to 90 carbon fatty acids with a hydroxyl on the β-carbon and an aliphatic chain on the α-carbon; found in the cell walls of mycobacteria. *465*

mycologist (mi-kol'o-jist) A person specializing in mycology; a student of mycology. *519*

mycology (mi-kol'o-je) The science and study of fungi. *519*

mycoplasma (mi"ko-plaz'mah) Bacteria that are members of the class *Mollicutes* and order *Mycoplasmatales*; they lack cell walls and cannot synthesize peptidoglycan precursors; most require sterols for growth; they are the smallest organisms capable of independent reproduction. *445*

mycoplasmal pneumonia (mi"ko-plaz'mal nu-mo'ne-ah) A type of pneumonia caused by *Mycoplasma pneumoniae*. Spread involves airborne droplets and close contact. *773*

mycosis (mi-ko'sis; pl., **mycoses)** Any disease caused by a fungus. *519, 782*

mycotoxicology (mi-ko'tok"si-kol'o-je) The study of fungal toxins and their effects on various organisms. *519*

myeloma cell (mi"e-lo'mah) A tumor cell that is similar to the cell type found in bone marrow. Also, a malignant, neoplastic plasma cell that produces large quantities of antibodies and can be readily cultivated. *622*

myxameba (mik-sah-me'bah; pl., **myxamebae)** A free-living amoeboid cell that can aggregate with other myxameba to form a plasmodium or pseudoplasmodium. Found in cellular slime molds and the myxomycetes. *530*

myxobacteria A group of gram-negative, aerobic soil bacteria characterized by gliding motility, a complex life cycle with the production of fruiting bodies, and the formation of myxospores. *487*

myxospores (mik'so-spōrs) Special dormant spores formed by the myxobacteria. *487*

napkin (diaper) candidiasis Typically found in infants whose diapers are not changed frequently and are therefore not kept dry. Caused by *Candida* species of fungi. *788*

narrow-spectrum drugs Chemotherapeutic agents that are effective only against a limited variety of microorganisms. *327*

natural classification A classification system that arranges organisms into groups whose members share many characteristics and reflect as much as possible the biological nature of organisms. *406*

natural environment An environment that has not been modified by human activity. *805*

natural killer (NK) cell A non-T, non-B lymphocyte present in nonimmunized individuals that exhibits MHC-independent cytolytic activity against tumor cells. *609, 634*

naturally acquired active immunity The type of immunity that develops when an individual's immunologic system comes into contact with an appropriate antigenic stimulus during the course of normal activities; it usually arises as the result of recovering from an infection. *608*

naturally acquired passive immunity The type of immunity that involves the transfer of antibodies from one individual to another. *608*

needle aspiration Using proper aseptic technique, the withdrawal of fluid into a sterile tube that has been treated with an anticoagulant. *675*

negative staining A staining procedure in which a dye is used to make the background dark while the specimen is unstained. *30*

Negri bodies (na'gre) A mass of viruses or unassembled viral subunits found within the brain neurons of rabies-infected animals. *732*

nematode (roundworm; nem'ah-tōd) A class of helminths commonly known as the roundworms; many are parasitic. *A34*

neurotoxin (nu''ro-tok'sin) A toxin that is poisonous to or destroys nerve tissue; especially the toxins secreted by *C. tetani, Corynebacterium diphtheriae,* and *Shigella dysenteriae.* *584*

neustonic (nu'ston''ik) The microorganisms that live at the water-atmosphere interface. *536*

neutralization (nu''tral-i-za'shun) The process in which an antigen-antibody reaction neutralizes a toxin or virus, that is, renders it inactive and unable to harm the host. *651, 665*

neutrophil (nu'tro-fil) A granular, amoeboid leukocyte (granulocyte) with a heavily lobed nucleus that phagocytoses and destroys pathogens. It is the most abundant leukocyte in the blood and is also involved in the inflammatory response. Often called a polymorphonuclear neutrophilic leukocyte (PMN). *123*

neutrophile (nu'tro-fīl'') Microorganisms that grow best at a neutral pH range between pH 5.5 and 8.0. *123*

nicotinamide adenine dinucleotide (NAD^+; nik''o-tin'ah-mīd) An electron-carrying coenzyme; it is particularly important in catabolic processes and usually donates its electrons to the electron transport chain under aerobic conditions. *137*

nicotinamide adenine dinucleotide phosphate ($NADP^+$; nik''o-tin'ah-mīd) An electron-carrying coenzyme that most often participates as an electron carrier in biosynthetic metabolism. *137*

nitrification (ni''trī-fī-ka'shun) The oxidation of ammonia to nitrate. *161, 478, 812*

nitrifying bacteria (ni'trī-fī''ing) Chemolithotrophic, gram-negative bacteria that are members of the family *Nitrobacteriaceae* and convert ammonia to nitrate and nitrite to nitrate. *161, 477*

nitrogen fixation The metabolic process in which atmospheric molecular nitrogen is reduced to ammonia; carried out by cyanobacteria, *Rhizobium,* and other nitrogen-fixing bacteria. *178, 812*

nitrogen oxygen demand (NOD) The demand for oxygen in sewage treatment, caused by nitrifying microorganisms. *831*

nitrogenase (ni'tro-jen-ās) The enzyme that catalyzes biological nitrogen fixation. *178, 812*

nocardioforms Bacteria that resemble members of the genus *Nocardia*; they develop a substrate mycelium that readily breaks up into rods and coccoid elements. *509*

nomenclature (no'men-kla''tūr) The branch of taxonomy concerned with the assignment of names to taxonomic groups in agreement with published rules. *405*

noncompetitive inhibitor Inhibition of enzyme activity that results from the inhibitor binding at a site other than the active site and altering the enzyme's shape to make it less active. *143*

noncyclic photophosphorylation (fo''to-fos''for-i-la'shun) The process in which light energy is used to make ATP when electrons are moved from water to $NADP^+$ during photosynthesis; both photosystem I and photosystem II are involved. *166*

nongonococcal urethritis (NGU) Any inflammation of the urethra not caused by *Neisseria gonorrhoeae.* *772*

non-Newtonian broth A liquid whose viscosity can vary with the rate of agitation; it usually has plasticlike characteristics. *891*

nonsense codon A codon that does not code for an amino acid but is a signal to terminate protein synthesis. *211*

nonsense mutation A mutation that converts a sense codon to a nonsense or stop codon. *249*

nonspecific resistance Refers to those general defense mechanisms that are inherited as part of the innate structure and function of each animal; also known as **nonspecific immunity.** *607*

nosocomial infection (nos''o-ko'me al) An infection that develops within a hospital (or other type of clinical care facility) and is produced by an infectious organism acquired during the stay of the patient. *711, 778*

nuclear envelope (nu'kle-ar) The complex double-membrane structure forming the outer boundary of the eucaryotic nucleus. It is covered by pores through which substances enter and leave the nucleus. *82*

nucleic acid hybridization (nu-kle'ik) The process of forming a hybrid double-stranded DNA molecule using a heated mixture of single-stranded DNAs from two different sources; if the sequences are fairly complementary, stable hybrids will form. *412*

nucleocapsid (nu''kle-o-kap'sid) The nucleic acid and its surrounding protein coat or capsid; the basic unit of virion structure. *354*

nucleoid (nu'kle-oid) An irregularly shaped region in the procaryotic cell that contains its genetic material. *50*

nucleolus (nu-kle'o-lus) The organelle, located within the eucaryotic nucleus and not bounded by a membrane, that is the location of ribosomal RNA synthesis and the assembly of ribosomal subunits. *83*

nucleoside (nu'kle-o-sīd'') A combination of ribose or deoxyribose with a purine or pyrimidine base. *183*

nucleosome (nu'kle-o-sōm'') A complex of histones and DNA found in eucaryotic chromatin; the DNA is wrapped around the surface of the beadlike histone complex. *196*

nucleotide (nu'kle-o-tīd) A combination of ribose or deoxyribose with phosphate and a purine or pyrimidine base; a nucleoside plus one or more phosphates. *183*

nucleus (nu'kle-us) The eucaryotic organelle enclosed by a double-membrane envelope that contains the cell's chromosomes. *82*

null cell A lymphoid cell that does not have characteristics of either T or B cells; also known as **third population cell.** *609*

numerical aperture The property of a microscope lens that determines how much light can enter it and how great a resolution it can provide. *23*

numerical taxonomy The grouping by numerical methods of taxonomic units into taxa based on their character states. *406*

nutrient (nu'tre-ent) A substance that supports growth and reproduction. *97*

nystatin (nis'tah-tin) A polyene antibiotic from *Streptomyces noursei* that is used in the treatment of *Candida* infections of the skin, vagina, and alimentary tract. *340*

O antigen A polysaccharide antigen extending from the outer membrane of some gram-negative bacterial cell walls; it is part of the lipopolysaccharide. *54*

obligate aerobes Organisms that grow only in the presence of air or oxygen. *127*

obligate anaerobes Microorganisms that cannot tolerate the presence of oxygen and die when exposed to it. *127*

odontopathogens Dental pathogens. *776*

Okazaki fragments Short stretches of polynucleotides produced during discontinuous DNA replication. *199*

oligotroph A microorganism that can survive and function in a low-nutrient environment. *848*

oligotrophic (ol″ĭ-go-trof′ik) Bodies of water containing low levels of nutrients, particularly nutrients that support microbial growth. *826*

oncogene (ong′ko-jēn) A gene whose activity is associated with the conversion of normal cells to cancer cells. *394*

oncology (ong-kol′o-je) The study of cancer or tumors. *641*

one-step growth experiment An experiment used to study the reproduction of lytic phages in which one round of phage reproduction occurs and ends with the lysis of the host bacterial population. *368*

onychomycosis (on″ĭ-ko-mi-ko′sis) A fungal infection of the nail plate producing nails that are opaque, white, thickened, friable, and brittle. Also called **ringworm of the nails** and **tinea unguium.** Caused by *Trichophyton* and other fungi. *788*

oocyst (o′o-sist) Cyst formed around a zygote of malaria and related protozoa. *556*

oogonia (o″o-go′ne-a) Mitotically dividing female germ line cells that produce primary oocytes. *538*

oomycetes (o″o-mi-se′tēz) A collective name for members of the division *Oomycota;* also known as the water molds. *530*

operator The segment of DNA to which the repressor protein binds; it controls the expression of the genes adjacent to it. *223*

operon (op′er-on) The sequence of bases in DNA that contains one or more structural genes together with the operator controlling their expression. *224*

ophthalmia neonatorum (of-thal′me-ah ne″o-nat-or-um) A gonorrheal eye infection in a newborn, which may lead to blindness. Also called **conjunctivitis of the newborn.** *754*

opportunistic organism An organism that is generally harmless but becomes pathogenic in a compromised host. *787*

opsonization (op″so-ni-za′shun) The action of opsonins in making bacteria and other cells more readily phagocytosed. Antibodies, complement (especially C3b), and fibronectin are potent opsonins. *612, 653*

organelle (or″gah-nel′) A structure within or on a cell that performs specific functions and is related to the cell in a way similar to that of an organ to the body. *71*

organotrophs Organisms that use reduced organic compounds as their electron source. *98*

osmophilic microorganisms (oz″mo-fil′ik) Microorganisms that grow best in or on media of high solute concentration. *867*

osmosis (oz-mo′sis) The movement of water across a selectively permeable membrane from a dilute solution (higher water concentration) to a more concentrated solution. *55*

osmotolerant Organisms that grow over a fairly wide range of water activity or solute concentration. *122*

outbreak The sudden, unexpected occurrence of a disease in a given population. *699*

outer membrane A special membrane located outside the peptidoglycan layer in the cell walls of gram-negative bacteria. *50*

β-oxidation pathway The major pathway of fatty acid oxidation to produce NADH, $FADH_2$, and acetyl coenzyme A. *159*

oxidation-reduction (redox) reactions Reactions involving electron transfers; the reductant donates electrons to an oxidant. *137*

oxidative phosphorylation (fos″for-ĭ-la′shun) The synthesis of ATP from ADP using energy made available during electron transport. *152*

oxidizing agent or **oxidant** (ok′sĭ-dant) The electron acceptor in an oxidation-reduction reaction. *137*

oxygenic photosynthesis Photosynthesis that oxidizes water to form oxygen; the form of photosynthesis characteristic of eucaryotic algae and cyanobacteria. *167, 468*

pacemaker enzyme The enzyme in a metabolic pathway that catalyzes the slowest or rate-limiting reaction; if its rate changes, the pathway's activity changes. *221*

pandemic (pan-dem′ik) An increase in the occurrence of a disease within a large and geographically widespread population (often refers to a worldwide epidemic). *699*

pannus (pan′us) A superficial vascularization of the cornea with infiltration of granulation tissue. *773*

panzootic (pan″zo-ot′ik) The wide dissemination of a disease in an animal population. *699*

parasite (par′ah-sīt) An organism that lives on or within another organism (the host) and benefits from the association while harming its host. Often the parasite obtains nutrients from the host. *579*

parasitic organism *See* parasite. *579*

parasitism (par′ah-si″tizm) A type of symbiosis in which one organism adversely affects the other (the host), but cannot live without it. *579*

parenteral route (pah-ren′ter-al) A route of drug administration that is nonoral (e.g., by injection). *331*

parfocal (par-fo′kal) A microscope that retains proper focus when the objectives are changed. *22*

paronychia (par″o-nik′e-ah) Inflammation involving the folds of tissue surrounding the nail. *788*

passive diffusion The process in which molecules move from a region of higher concentration to one of lower concentration as a result of random thermal agitation. *101*

passive immunity Temporary immunity acquired by transfer of lymphocytes or preformed antibodies to the recipient. *608*

Pasteur effect (pas-tur′) The decrease in the rate of sugar catabolism and change to aerobic respiration that occurs when microorganisms are switched from anaerobic to aerobic conditions. *154*

pasteurization (pas″ter-ĭ-za′shun) The process of heating milk and other liquids to destroy microorganisms that can cause spoilage or disease. *316, 869*

pathogen (path′o-jen) A disease-producing organism or material. *579*

pathogenic potential The degree that a pathogen causes morbid signs and symptoms. *580*

pathogenicity (path″o-je-nis′ĭ-te) The ability to cause disease. *579*

pébrine (pa-brēn′) An infectious disease of silkworms caused by the protozoan *Nosema bombycis.* *556*

ped A natural soil aggregate, formed partly through bacterial and fungal growth in the soil. *846*

pellicle (pel′ĭ-k′l) A relatively rigid layer of proteinaceous elements just beneath the plasma membrane in many protozoa and algae. The plasma membrane is sometimes considered part of the pellicle. *87, 540, 550*

pelvic inflammatory disease (PID) A severe infection of the female reproductive organs. The disease that results when gonococci and chlamydiae infect the fallopian tubes and surrounding tissue. *754, 772*

penicillins (pen″ĭ-sil-ins) A group of antibiotics containing a β-lactam ring, which are active against gram-positive bacteria. *55, 334*

pentose phosphate pathway (pen′tōs) The pathway that oxidizes glucose 6-phosphate to ribulose 5-phosphate and then converts it to a variety of three to seven carbon sugars; it forms several important products (NADPH for biosynthesis, pentoses, and other sugars) and also can be used to degrade glucose to CO_2. *148*

peplomer or **spike** (pep′lo-mer) A protein or protein complex that extends from the virus envelope and is often important in virion attachment to the host cell surface. *360*

peptic ulcer disease A gastritis caused by *Helicobacter pylori.* *756*

peptide interbridge (pep′tīd) A short peptide chain that connects the tetrapeptide chains in some peptidoglycans. *51*

peptidoglycan (pep″ti-do-gli′kan) A large polymer composed of long chains of alternating N-acetylglucosamine and N-acetylmuramic acid residues. The polysaccharide chains are linked to each other through connections between tetrapeptide chains attached to the N-acetylmuramic acids. It provides much of the strength and rigidity possessed by bacterial cell walls. *50, 451*

peptidyl or **donor site (P site)** The site on the ribosome that contains the peptidyl-tRNA at the beginning of the elongation cycle during protein synthesis. *211*

peptidyl transferase The enzyme that catalyzes the transpeptidation reaction in protein synthesis; in this reaction, an amino acid is added to the growing peptide chain. *211*

peptones (pep′tōns) Water soluble digests or hydrolysates of proteins that are used in the preparation of culture media. *105*

period of infectivity Refers to the time during which the source of an infectious disease is infectious or is disseminating the organism. *706*

periodontal disease (per"e-o-don'tal) A disease located around the teeth or in the periodontium—the tissue investing and supporting the teeth, including the cementum, periodontal ligament, alveolar bone, and gingiva. *776*

periodontitis (per"e-o-don-ti'tis) An inflammation of the periodontium. *777*

periodontium (per"e-o-don'she-um) *See* periodontal disease. *776*

periodontosis (per"e-o-don-to'sis) A degenerative, noninflammatory condition of the periodontium, which is characterized by destruction of tissue. *777*

periplasmic space (per"i-plas'mik) or **periplasm** (per'i-plazm) The space between the plasma membrane and the outer membrane in gram-negative bacteria, and between the plasma membrane and the cell wall in gram-positive bacteria. *50*

peristalsis (per"i-stal'sis) The progressive wavelike or wormlike movement seen in tubes such as the small intestine. This movement propels the contents of the tube down its length. *593*

peritrichous (pĕ-rit'rĭ-kus) A cell with flagella evenly distributed over its surface. *57*

permease (per'me-ās) A membrane carrier protein. *101*

pertussis (per-tus'is) An acute, highly contagious infection of the respiratory tract, most frequently affecting young children, usually caused by *Bordetella pertussis.* Consists of peculiar paroxysms of coughing, ending in a prolonged crowing or whooping respiration; hence the name **whooping cough.** *745*

petri dish (pe'tre) A shallow dish consisting of two round, overlapping halves that is used to grow microorganisms on solid culture medium; the top is larger than the bottom of the dish to prevent contamination of the culture. *108*

Peyer's patches (pi'erz) Lymphatic aggregates in the submucosa of the small intestine, which contain a full repertoire of lymphoreticular cells, T and B cells, and macrophages. Is also a part of the **GALT** system. *616*

phage (fāj) *See* bacteriophage. *349*

phagocytic vacuole (fag"o-sit'ik vak'u-ol) A membrane-delimited vacuole produced by cells carrying out phagocytosis. It is formed by the invagination of the plasma membrane and contains solid material. Also called a **phagosome** (fag'o-som). *550*

phagocytosis (fag"o-si-to'sis) The endocytotic process in which a cell encloses large particles in a membrane-delimited phagocytic vacuole or phagosome and engulfs them. *76, 599*

phagolysosome (fag"o-li'so-sōm) The vacuole that results from the fusion of a phagosome with a lysosome. *599*

phagovar (fag'o-var) A specific phage type. *692*

pharyngitis (far"in-ji'tis) Inflammation of the pharynx. *747*

phase-contrast microscope A microscope that converts slight differences in refractive index and cell density into easily observed differences in light intensity. *24*

phenetic system A classification system that groups organisms together based on their mutual similarity. *406*

phenol coefficient test A test to measure the effectiveness of disinfectants by comparing their activity against test bacteria with that of phenol. *322*

phosphatase (fos'fah-tās") An enzyme that catalyzes the hydrolytic removal of phosphate from molecules. *177*

photoautotroph (fo'to-aw'to-trōf) *See* photolithotrophic autotrophs. *98*

photolithotrophic autotrophs Organisms that use light energy, an inorganic electron source (e.g., H_2O, H_2, H_2S), and CO_2 as a carbon source. *98*

photoorganotrophic heterotrophs Microorganisms that use light energy and organic electron donors, and also employ simple organic molecules rather than CO_2 as their carbon source. *98*

photoreactivation (fo"to-re-ak"tĭ-va'shun) The process in which blue light is used by a photoreactivating enzyme to repair thymine dimers in DNA by splitting them apart. *130, 254*

photosynthesis (fo"to-sin'thĕ-sis) The trapping of light energy and its conversion to chemical energy, which is then used to reduce CO_2 and incorporate it into organic form. *134, 162*

photosynthetic carbon reduction (PCR) cycle *See* Calvin cycle. *173*

photosystem I The photosystem in eucaryotic cells that absorbs longer wavelength light, usually greater than about 680 nm, and transfers the energy to chlorophyll P700 during photosynthesis; it is involved in both cyclic photophosphorylation and noncyclic photophosphorylation. *164*

photosystem II The photosystem in eucaryotic cells that absorbs shorter wavelength light, usually less than 680 nm, and transfers the energy to chlorophyll P680 during photosynthesis; it participates in noncyclic photophosphorylation. *164*

phototrophs Organisms that use light as their energy source. *98*

phycobiliproteins Photosynthetic pigments that are composed of proteins with attached tetrapyrroles; they are often found in cyanobacteria and red algae. *163*

phycobilisomes Special particles on the membranes of cyanobacteria that contain photosynthetic pigments and electron transport chains. *472*

phycobiont (fi"ko-bi'ont) The algal partner in a lichen. *566*

phycocyanin (fi"ko-si'an-in) A blue phycobiliprotein pigment used to trap light energy during photosynthesis. *163*

phycoerythrin (fi"ko-er'i-thrin) A red photosynthetic phycobiliprotein pigment used to trap light energy. *163*

phycology (fi-kol'o-je) The study of algae; **algology.** *536*

phylogenetic or **phyletic classification system** (fi"lo-jĕ-net'ik, fi-let'ik) A classification system based on evolutionary relationships rather than general resemblance. *406*

phytohemagglutinin (fi"to-hem"ah-gloo'tĭ-nin) A lectin of plant origin used to induce cell division of lymphocytes; it also agglutinates erythrocytes (that is, it is a hemagglutinin). *627*

phytoplankton (fi"to-plank'ton) A community of floating photosynthetic organisms, largely composed of algae and cyanobacteria. *536, 827*

piedra (pe-a'drah) A fungal disease of the hair in which white or black nodules of fungi form on the shafts. *782*

pinocytosis (pi"no-si-to'sis) The endocytotic process in which a cell encloses a small amount of the surrounding liquid and its solutes in tiny pinocytotic vesicles or pinosomes. *76*

pitched Pertaining to inoculation of a nutrient medium with yeast, for example, in beer brewing. *880*

plague (plāg) An acute febrile, infectious disease, caused by the bacillus *Yersinia pestis,* which has a high mortality rate; the two major types are **bubonic plague** and **pneumonic plague.** *750*

plankton (plank'ton) Free-floating, mostly microscopic microorganisms that can be found in almost all waters; a collective name. *536, 549*

planktonic (adj.) *See* plankton. *536*

plantar wart A wart that occurs on the sole of the foot. *738*

plaque (plak) 1. A clear area in a lawn of bacteria or a localized area of cell destruction in a layer of animal cells that results from the lysis of the bacteria by bacteriophages or the destruction of the animal cells by animal viruses. 2. The term also refers to dental plaque, a film of food debris, polysaccharides, and dead cells that cover the teeth. It provides a medium for the growth of bacteria (which may be considered a part of the plaque), leading to a microbial plaque ecosystem that can produce dental decay. *349, 750*

plasma cell A mature, differentiated B lymphocyte chiefly occupied with antibody synthesis and secretion; a plasma cell lives for only five to seven days. *609*

plasma membrane The selectively permeable membrane surrounding the cell's cytoplasm; also called the cell membrane, plasmalemma, or cytoplasmic membrane. *44*

plasmid (plaz'mid) A circular, double-stranded DNA molecule that can exist and replicate independently of the chromosome or may be integrated with it. A plasmid is stably inherited, but is not required for the host cell's growth and reproduction. *50, 261, 339*

plasmid fingerprinting A technique used to identify microbial isolates as belonging to the same strain because they contain the same number of plasmids with the identical molecular weights and similar phenotypes. *694*

plasmodial slime mold (plaz-mo'de-al) A member of the division *Myxomycota* that exists as a thin, streaming, multinucleate mass of protoplasm, which creeps along in an amoeboid fashion. *530*

plasmodium (plaz-mo'de-um; pl., **plasmodia**) A stage in the life cycle of myxomycetes (plasmodial slime molds); a multinucleate mass of protoplasm surrounded by a membrane. Also, a parasite of the genus *Plasmodium.* *530*

plasmolysis (plaz-mol'ĭ-sis) The process in which water osmotically leaves a cell, which causes the cytoplasm to shrivel up and pull the plasma membrane away from the cell wall. *55*

plastid (plas'tid) A cytoplasmic organelle of algae and higher plants that contains pigments such as chlorophyll, stores food reserves, and often carries out processes such as photosynthesis. *76*

pleomorphic (ple''o-mor'fik) Refers to bacteria that are variable in shape and lack a single, characteristic form. *43*

plus or positive strand The virus nucleic-acid strand that is equivalent in base sequence to the viral mRNA. *359*

pneumocystis pneumonia, *Pneumocystis carinii* *pneumonia;* (noo''mo-sis-tis) A type of pneumonia caused by the sporozoan parasite *Pneumocystis carinii.* *795*

point mutation A mutation that affects only a single base pair in a specific location. *248*

polar flagellum A flagellum located at one end of an elongated cell. *57*

polio *See* poliomyelitis. *736*

poliomyelitis (po''le-o-mi''ĕ-li'tis) An acute, contagious viral disease that attacks the central nervous system, injuring or destroying the nerve cells that control the muscles and sometimes causing paralysis; also called polio or infantile paralysis. *736*

poly-β-hydroxybutyrate (hi-drok''se-bu'tĭ-rāt) A linear polymer of β-hydroxybutyrate used as a reserve of carbon and energy by many bacteria. *47*

polymerase chain reaction (PCR) An in vitro technique used to synthesize large quantities of specific nucleotide sequences from small amounts of DNA. It employs oligonucleotide primers complementary to specific sequences in the target gene and special heat-stable DNA polymerases. *291*

polyribosome (pol''e-ri'bo-sōm) A complex of several ribosomes with a messenger RNA; each ribosome is translating the same message. Also called a **polysome** (pol'e-sōm). *78, 206*

Pontiac fever A bacterial disease caused by *Legionella pneumophila* that resembles an allergic disease more than an infection. First described from Pontiac, Michigan. *See* Legionnaires' disease. *742*

population An assemblage of organisms of the same type. *807*

porin proteins Proteins that form channels across the outer membrane of gram-negative bacterial cell walls. Small molecules are transported through these channels. *55*

posttranscriptional modification The processing of the initial RNA transcript, heterogeneous nuclear RNA, to form mRNA. *203*

potable (po'tah-b'l) Refers to water suitable for drinking. *839*

pour plate A petri dish of solid culture medium with isolated microbial colonies growing both on its surface and within the medium, which has been prepared by mixing microorganisms with cooled, liquid medium and then allowing the medium to harden. *106*

precipitation (or **precipitin**) **reaction** (pre-sip''ĭ-ta'shun) The reaction of an antibody with a soluble antigen to form an insoluble precipitate. *654*

precipitin (pre-sip'ĭ-tin) The antibody responsible for a precipitation reaction. *654*

presence-absence test (P-A test) A simplified test for the presence of coliforms in water, in which a single 100 ml sample is incubated in a lactose broth. *839*

prevalence rate Refers to the total number of individuals infected at any one time in a given population regardless of when the disease began. *701*

Pribnow box A special base sequence in the promoter that is recognized by the RNA polymerase and is the site of initial polymerase binding. *203, 242*

primary amebic meningoencephalitis An infection of the meninges of the brain by the free-living amoebae *Naegleria* and *Acanthamoeba.* *792*

primary metabolites Microbial metabolites produced during the growth phase of an organism. *895*

primary producer Photoautotrophic and chemoautotrophic organisms that incorporate carbon dioxide into organic carbon and thus provide new biomass for the ecosystem. *805*

primary production The incorporation of carbon dioxide into organic matter by photosynthetic organisms and chemoautotrophic bacteria. *805*

primary treatment The first step of sewage treatment, in which physical settling and screening are used to remove particulate materials. *832*

prion (pri'on) An infectious particle that is the cause of slow diseases like scrapie in sheep and goats; it has a protein component, but no nucleic acid has yet been detected. *399*

probe (prōb) A short, labeled nucleic acid segment complementary in base sequence to part of another nucleic acid, which is used to identify or isolate the particular nucleic acid from a mixture through its ability to bind specifically with the target nucleic acid. *288, 872*

Procaryotae (pro-kar''e-o'te) A kingdom comprising all the procaryotic microorganisms (those lacking true membrane-enclosed nuclei). *413*

procaryotic cells (pro''kar-e-ot'ik) Cells that lack a true, membrane-enclosed nucleus; bacteria are procaryotic and have their genetic material located in a nucleoid. *15, 89*

product The substance formed in an enzyme reaction. *140*

progametangium (pro-gam-ĕ''tan'je-um; pl., **progametangia**) The cell that gives rise to a gametangium and a proximal suspensor during the early stages of sexual reproduction in zygomycetous fungi. *524*

proglottid (pro-glot'id) A segment in the body of a tapeworm that contains both male and female reproductive organs. *A34*

proliferative kidney disease (pro-lif'er-a-tiv) A protozoan disease caused by an unclassified myxozoan in salmonids throughout the world. *557*

promoter The region on DNA at the start of a gene that the RNA polymerase binds to before beginning transcription. *203, 241*

propagated epidemic An epidemic that is characterized by a relatively slow and prolonged rise and then a gradual decline in the number of individuals infected. It usually results from the introduction of an infected individual into a susceptible population. *703*

prophage (pro'fāj) The latent form of a temperate phage that remains within the lysogen, usually integrated into the host chromosome. *275, 376*

prostheca (pros-the'kah) An extension of a bacterial cell, including the plasma membrane and cell wall, that is narrower than the mature cell. *480*

prosthetic group (pros-thet'ik) A tightly bound cofactor that remains at the active site of an enzyme during its catalytic activity. *140*

protease (pro'te-ās) An enzyme that hydrolyzes proteins to their constituent amino acids. Also called a proteinase. *160*

protein engineering (pro'tēn) The rational design of proteins by constructing specific amino acid sequences through molecular techniques, with the objective of modifying protein characteristics. *901*

Protista (pro-tis'tah) A kingdom containing eucaryotes with unicellular organization, either in the form of solitary cells or colonies of cells lacking true tissues; members of the kingdom are called protists. *413*

protomer An individual subunit of a viral capsid; a capsomer is made of protomers. *355*

protonmotive force (PMF) The force arising from a gradient of protons and a membrane potential that is thought to power ATP synthesis and other processes. *153*

protoplast (pro'to-plast) A bacterial or fungal cell with its cell wall completely removed. It is spherical in shape and osmotically sensitive. *47, 55*

protoplast fusion The joining of cells that have had their walls removed. *892*

protothecosis (pro''to-the-ko'sis; pl., **protothecoses**) A disease of humans and animals produced by the green alga *Prototheca moriformis.* *540*

prototroph (pro'to-trōf) A microorganism that requires the same nutrients as the majority of naturally occurring members of its species. *98, 244*

protozoan or **protozoon** (pro''to-zo'an, pl. **protozoa**) A microorganism belonging to the *Protozoa* subkingdom. A unicellular or acellular eucaryotic protist whose organelles have the functional role of organs and tissues in more complex forms. Protozoa vary greatly in size, morphology, nutrition, and life cycle. *549*

protozoology (pro''to-zo-ol'o-je) The study of protozoa. *549*

proviral DNA Viral DNA that has been integrated into host cell DNA. In retroviruses, it is the double-stranded DNA copy of the RNA genome. *389*

pseudomurein A modified peptidoglycan lacking D-amino acids and containing N-acetyltalosaminuronic acid instead of N-acetylmuramic acid; found in methanogenic bacteria. *493*

pseudoplasmodium (soo″do-plaz-mo′de-um; pl., **pseudoplasmodia**) A sausage-shaped amoeboid structure consisting of many myxamebae and behaving as a unit; the result of myxamebal aggregation in the cellular slime molds; also called a slug. *530*

pseudopodium or **pseudopod** (soo″do-po′de-um) A nonpermanent cytoplasmic extension of the cell body by which amoebae and amoeboid organisms move and feed. *550*

psittacosis (**ornithosis**; sit″ah-ko′sis) A disease due to a strain of *Chlamydia psittaci,* first seen in parrots and later found in other birds and domestic fowl (in which it is called ornithosis). It is transmissible to humans. *772*

psychrophile (si′kro-fīl) A microorganism that grows well at 0°C and has an optimum growth temperature of 15°C or lower and a temperature maximum around 20°C. *126*

psychrotroph A microorganism that grows at 0°C, but has a growth optimum between 20 and 30°C, and a maximum of about 35°C. *126*

puerperal fever (pu-er′per-al) An acute, febrile condition following childbirth; it is characterized by infection of the uterus and/or adjacent regions and is caused by streptococci. *710*

pulmonary anthrax (pul′mo-ner″e) A form of anthrax involving the lungs. Also known as **woolsorter's disease.** *753*

pure culture A population of cells that are identical because they arise from a single cell. *106*

purine (pu′rin) A basic, heterocyclic, nitrogen-containing molecule with two joined rings that occurs in nucleic acids and other cell constituents; most purines are oxy or amino derivatives of the purine skeleton. The most important purines are adenine and guanine. *183*

purple membrane An area of the plasma membrane of *Halobacterium* that contains bacteriorhodopsin and is active in photosynthetic light energy trapping. *500*

putrefaction (pu″trē-fak′shun) The microbial decomposition of organic matter, especially the anaerobic breakdown of proteins, with the production of foul-smelling compounds such as hydrogen sulfide and amines. *867*

pyrenoid (pi′rē-noid) The differentiated region of the chloroplast that is a center of starch formation in green algae and stoneworts. *82, 538*

pyrimidine (pi-rim′i-dēn) A basic, heterocyclic, nitrogen-containing molecule with one ring that occurs in nucleic acids and other cell constituents; pyrimidines are oxy or amino derivatives of the pyrimidine skeleton. The most important pyrimidines are cytosine, thymine, and uracil. *183*

pyrogen (pi′ro-jen) A fever-producing substance. *591*

Q fever An acute zoonotic disease caused by the rickettsia *Coxiella burnetii.* *775*

Quellung reaction The increase in visibility or the swelling of the capsule of a microorganism in the presence of antibodies against capsular antigens. *666*

rabies (ra′bēz) An acute infectious disease of the central nervous system, which affects all warm-blooded animals (including humans). It is caused by an RNA virus belonging to the rhabdovirus group. *732*

racking The removal of sediments from wine bottles. *880*

radappertization The use of gamma rays from a cobalt source for control of microorganisms in foods. *870*

radioimmunoassay (**RIA**; ra″de-o-im″u-no-as′a) A very sensitive assay technique that uses a purified radioisotope-labeled antigen or antibody to compete for antibody or antigen with unlabeled standard and samples to determine the concentration of a substance in the samples. *666*

rapid sand filter A bed of sand through which water is passed at a rapid rate to trap physical impurities and purify the water. *835*

R body A refractile structure found within kappa particles that appears to confer killer characteristics on the ciliate protozoan host. *568*

reactivation tuberculosis *See* miliary tuberculosis. *748*

reagin (re′ah-jin) Antibody that mediates immediate hypersensitivity reactions. IgE is the major reagin in humans. *636*

recalcitrance Stubbornness; the resistance of a substance to modification or degradation by microorganisms. *859*

recombinant DNA technology The techniques used in carrying out genetic engineering; they involve the identification and isolation of a specific gene, the insertion of the gene into a vector such as a plasmid to form a recombinant molecule, and the production of large quantities of the gene and its products. *286*

recombination (re″kom-bi-na′shun) The process in which a new recombinant chromosome is formed by combining genetic material from two organisms. *259*

recombination repair A DNA repair process that repairs damaged DNA when there is no remaining template; a piece of DNA from a sister molecule is used. *255*

redia (re′de-ah) A larval form of a digenetic trematode that is produced asexually within a miracidium, sporocyst, or mother redia. It develops within a snail host and can produce rediae or cercariae. *A40*

reducing agent or **reductant** (re-duk′tant) The electron donor in an oxidation-reduction reaction. *137*

reduction potential A measure of the tendency of a reductant to donate electrons to an acceptor in an oxidation-reduction (redox) reaction. The more negative the reduction potential of a compound, the better electron donor it is. *137*

refraction (re-frak′shun) The deflection of a light ray from a straight path as it passes from one medium (e.g., glass) to another (e.g., air). *21*

refractive index (re-frak′tiv) The ratio of the velocity of light in the first of two media to that in the second as it passes from the first to the second. *21*

regularly structured layer (**S layer**) A highly ordered layer of protein or glycoprotein that is present on the surface of many bacteria. *56*

regulator T cell Controls the development of effector T cells. *632*

regulatory mutants Mutant organisms that have lost the ability to limit synthesis of a product, which normally occurs by regulation of activity of an earlier step in the biosynthetic pathway. *897*

replica plating A technique for isolating mutants from a population by plating cells from each colony growing on a nonselective agar medium onto plates with selective media or environmental conditions, such as the lack of a nutrient or the presence of an antibiotic or a phage; the location of mutants on the original plate can be determined from growth patterns on the replica plates. *250*

replication (rep″li-ka′shun) The process in which an exact copy of parental DNA or RNA is made with the parental molecule serving as a template. *197*

replication fork The Y-shaped structure where DNA is replicated. The arms of the Y contain template strand and a newly synthesized DNA copy. *197*

replicative form A double-stranded form of nucleic acid that is formed from a single-stranded virus genome and used to synthesize new copies of the genome. *372, 389*

replicon (rep′li-kon) A unit of the genome that contains an origin for the initiation of replication and in which DNA is replicated. *197, 261*

repressible enzyme An enzyme whose level drops in the presence of a small molecule, usually an end product of its metabolic pathway. *222*

repressor protein (re-pres′or) A protein coded for by a regulator gene that can bind to the operator and inhibit transcription; it may be active by itself or only when the corepressor is bound to it. *223*

reservoir (rez′er-vwar) A site, alternate host, or carrier that normally harbors pathogenic organisms and serves as a source from which other individuals can be infected. *706*

reservoir host An organism other than a human that is infected with a pathogen that can also infect humans. *579*

resolution (rez″o-lu′shun) The ability of a microscope to separate or distinguish between small objects that are close together. *23*

respiration (res″pi-ra′shun) An energy-yielding process in which an electron donor is oxidized using an inorganic electron acceptor. The acceptor may be either oxygen (aerobic respiration) or another inorganic acceptor (anaerobic respiration). *134, 154, 158*

respiratory syncytial virus (RSV; sin-sish′al) A member of the RNA *Paramyxoviridae* family that causes respiratory infections in children. *720*

restricted transduction A transduction process in which only a specific set of bacterial genes are carried to another bacterium by a temperate phage; the bacterial genes are acquired because of a mistake in the excision of a prophage during the lysogenic life cycle. *276*

restriction enzymes Enzymes produced by host cells that cleave virus DNA at specific points and thus protect the cell from virus infection; they are used in carrying out genetic engineering. *287, 371*

reticulate body (RB) The form in the chlamydial life cycle whose role is growth and reproduction within the host cell. *444*

reticuloendothelial system (RES; re-tik″u-lo-en″do-the′le-al) A system of phagocytic cells (including macrophages, monocytes, and specialized endothelial cells) that is located in the liver, spleen, lymph nodes, and bone marrow. It is an important component of the host's general or nonspecific defense against pathogens. *595*

retroviruses (re″tro-vi′rus-es) A group of viruses with RNA genomes that carry the enzyme reverse transcriptase and form a DNA copy of their genome during their reproductive cycle. *389*

reverse transcriptase An RNA-dependent DNA polymerase that uses a viral RNA genome as a template to form a DNA copy; this is a reverse of the normal flow of genetic information, which proceeds from DNA to RNA. *389*

reversible covalent modification A mechanism of enzyme regulation in which the enzyme's activity is either increased or decreased by the reversible covalent addition of a group such as phosphate or AMP to the protein. *219*

Reye's syndrome An acute, potentially fatal disease of childhood that is characterized by severe edema of the brain and increased intracranial pressure, vomiting, hypoglycemia, and liver dysfunction. The cause is unknown but is almost always associated with a previous viral infection (e.g., influenza or varicella-zoster virus infections). *718*

R factors Plasmids bearing one or more drug resistant genes. *261*

rheumatic fever (roo-mat′ik) An autoimmune disease characterized by inflammatory lesions involving the heart valves, joints, subcutaneous tissues, and central nervous system. The disease is associated with hemolytic streptococci in the body. It is called rheumatic fever because two common symptoms are fever and pain in the joints similar to that of rheumatism. *746*

rhizosphere A region around the plant root where materials released from the root increase the microbial population and its activities. *851*

rho factor (ro) The protein that helps RNA polymerase dissociate from the terminator after it has stopped transcription. *203*

rhoptry Saclike, electron dense structure in the anterior portion of a zoite of a member of the phylum *Apicomplexa;* perhaps involved in the penetration of host cells. *556*

Rh system The classification of human blood types based on the presence of the Rh blood group antigens on erythrocytes. The Rh1 antigen is responsible for the **hemolytic disease of the newborn.** *643*

ribonucleic acid (RNA; ri″bo-nu-kle′ik) A polynucleotide composed of ribonucleotides joined by phosphodiester bridges. *192*

ribosomal RNA (rRNA) The RNA present in ribosomes; ribosomes contain several sizes of single-stranded rRNA that contribute to ribosome structure and are also directly involved in the mechanism of protein synthesis. *83, 202*

ribosome (ri′bo-sōm) The organelle where protein synthesis occurs; the message encoded in mRNA is translated here. *49*

ribulose-1,5-bisphosphate carboxylase (ri′bu-lōs) The enzyme that catalyzes the incorporation of CO_2 in the Calvin cycle. *173*

ringworm (ring′werm) The common name for a fungal infection of the skin, even though it is not caused by a worm and is not always ring-shaped in appearance. *782*

rise period or **burst** The period during the one-step growth experiment when host cells lyse and release phage particles. *369*

RNA polymerase The enzyme that catalyzes the synthesis of mRNA under the direction of a DNA template. *202*

Rocky Mountain spotted fever A disease caused by *Rickettsia rickettsii.* *775*

rolling-circle mechanism A mode of DNA replication in which the replication fork moves around a circular DNA molecule, displacing a strand to give a tail that is also copied to produce a new double-stranded DNA. *197*

root nodule Gall-like structures on roots that contain endosymbiotic nitrogen-fixing bacteria (e.g., *Rhizobium* or *Bradyrhizobium* is present in legume nodules). *854*

rumen (roo-men) The expanded upper portion or first compartment of the stomach of ruminants. *569*

ruminant (roo′mi-nant) An herbivorous animal characterized as having a stomach divided into four compartments and chewing a cud consisting of regurgitated, partially digested food. *568*

run The straight line movement of a bacterium. *62*

salmonellosis (sal″mo-nel-o′sis) An infection with certain species of the genus *Salmonella,* usually caused by ingestion of food containing salmonellae or their products. Also known as ***Salmonella* gastroenteritis** or ***Salmonella* food poisoning.** *766*

sanitization (san″i-ti-za′shun) Reduction of the microbial population on an inanimate object to levels judged safe by public health standards; usually, the object is cleaned. *311*

saprophyte (sap′ro-fit) A saprophytic organism. *See* saprophytic. *521*

saprophytic (sap″ro-fit′ik) Of the nature of or pertaining to a saprophyte; taking up non-living, organic nutrients in dissolved form, and usually growing on decomposing organic matter. *521*

saprozoic nutrition (sap″ro-zo′ik) Having the type of nutrition in which organic nutrients are taken up in dissolved form; normally refers to animals or animal-like organisms. *550*

scale (skāl) A platelike organic structure found on the surface of some cells (chrysophytes). *541*

scaleup The process of establishing conditions for growth of larger volumes of microorganisms, based on experience with smaller-scale cultures. *891*

scanning electron microscope An electron microscope that scans a beam of electrons over the surface of a specimen and forms an image of the surface from the electrons that are emitted by it. *35*

scarlatina (skahr″la-te′nah) *See* scarlet fever. *747*

scarlet fever (scarlatina; skar′let) A disease that results from infection with a strain of *Streptococcus pyogenes* that produces an erythrogenic (rash-inducing) toxin which causes shedding of the skin. This is a communicable disease spread by respiratory droplets. *747*

schistosomiasis (shis″to-so-mi′ah-sis) The state of being infected with trematodes of the genus *Schistosoma.* *A40*

schizogony (ski-zog′o-ne) Multiple asexual fission. *556*

scolex (sko′leks) The anterior attachment organ or head of a tapeworm. *A34*

secondary metabolites Products of metabolism that are synthesized after growth has been completed. *895*

secondary treatment The biological utilization of dissolved organic matter in the process of sewage treatment; the organic material is either mineralized or changed to removable solids. *832*

second law of thermodynamics Physical and chemical processes proceed in such a way that the entropy of the universe (the system and its surroundings) increases to the maximum possible. *135*

secretory IgA (sIgA) The primary immunoglobulin of the secretory immune system. *See* IgA. *615*

secretory vacuole In protists and some animals, these organelles usually contain specific enzymes that perform various functions such as excystation. Their contents are released to the cell exterior during exocytosis. *550*

sedimentation basin A large basin often used in the initial stages of municipal water purification. Raw water is held in the basin until large particulate material settles out. *835*

segmented genome A virus genome that is divided into several parts or fragments, each probably coding for the synthesis of a single polypeptide; segmented genomes are very common among the RNA viruses. *359*

selective media Culture media that favor the growth of specific microorganisms; this may be accomplished by inhibiting the growth of undesired microorganisms. *105*

selective toxicity The ability of a chemotherapeutic agent to kill or inhibit a microbial pathogen while damaging the host as little as possible. *327*

self-assembly The spontaneous formation of a complex structure from its component molecules without the aid of special enzymes or factors. *58, 173*

semisynthetic penicillins Penicillins that have been modified after biosynthesis, usually by chemical processes. *897*

sense strand The DNA strand that RNA polymerase copies to produce mRNA, rRNA, or tRNA. *202, 241*

septate (sep'tāt) Divided by a septum or cross wall; also with more or less regular occurring cross walls. *521*

septicemia (sep″ti-se'me-ah) The presence in the blood of bacterial toxins. *583, 430*

septic tank (sep'tik) A tank used to process small quantities of domestic sewage. Solid material settles out and is partially degraded by bacteria as sewage slowly flows through the tank. The outflow is further treated or dispersed in the soil. *842*

septum (sep'tum; pl., **septa**) A wall or partition dividing a body space or cavity. *507, 521*

serology (se-rol'o-je) The branch of immunology that is concerned with in vitro reactions involving one or more serum constituents (e.g., antibodies and complement). *656*

seroprevalence rate The number of individuals that have antibodies in their sera against a specific antigen. *701*

serotyping A technique or serological procedure that is used to differentiate strains (serovars or serotypes) of microorganisms that have differences in the antigenic composition of a structure or product. *666*

serum (se'rum; pl., **serums** or **sera**) The clear, fluid portion of blood lacking both blood cells and fibrinogen. It is the fluid remaining after coagulation of plasma, the noncellular liquid faction of blood. *620*

serum hepatitis *See* hepatitis B.

settling basin A basin used during water purification to chemically precipitate out fine particles, microorganisms, and organic material by **coagulation** or flocculation. *835*

sewage treatment The use of physical and biological processes to remove particulate and dissolved material from sewage. *832*

sex pilus (pi'lus) A thin protein appendage required for bacterial mating or conjugation. The cell with sex pili donates DNA to recipient cells. *57, 269*

sheath (shēth) A hollow tubelike structure surrounding a chain of cells and present in several genera of bacteria. *482*

Shine-Dalgarno sequence A segment in the leader of procaryotic mRNA that binds to a special sequence on the 16S rRNA of the small ribosomal subunit. This helps properly orient the mRNA on the ribosome. *242*

shingles (herpes zoster; shing'g'lz) A reactivated form of chickenpox caused by the herpes zoster virus. *716*

siderophore (sid'er-o-for″) A small molecule that complexes with ferric iron and supplies it to a cell by aiding in its transport across the plasma membrane. *103*

sigma factor A protein that helps the RNA polymerase core enzyme recognize the promoter at the start of a gene. *203*

silent mutation A mutation that does not result in a change in the organism's proteins or phenotype even though the DNA base sequence has been changed. *248*

simple matching coefficient (S_{SM}) An association coefficient used in numerical taxonomy; the proportion of characters that match regardless of whether or not the attribute is present. *407*

single radial immunodiffusion (RID) assay An immunodiffusion technique that quantitates antigens by following their diffusion through a gel containing antibodies directed against the test antigens. *661*

single-cell protein Refers to the microbial cells grown for use primarily as a source of protein in human and animal diets. *884*

site-specific recombination Recombination of nonhomologous genetic material with a chromosome at a specific site. *259*

slime layer A layer of diffuse, unorganized, easily removed material lying outside the bacterial cell wall. *56*

slime mold A common term for members of the divisions *Acrasiomycota* and *Myxomycota*. *530*

slow sand filter A bed of sand through which water slowly flows; the gelatinous microbial layer on the sand grain surface removes waterborne microorganisms, particularly *Giardia*, by adhesion to the gel. This type of filter is used in some water purification plants. *837*

slow virus disease A progressive, pathological process caused by a transmissible agent (virus or prion) that remains clinically silent during a prolonged incubation period of months to years after which progressive clinical disease becomes apparent. *393, 736*

sludge (sluj) A general term for the precipitated solid matter produced during water and sewage treatment; solid particles composed of organic matter and microorganisms that are involved in aerobic sewage treatment (activated sludge). *832*

smallpox (variola; smawl'poks) A highly contagious, often fatal disease caused by a poxvirus. Its most noticeable symptom is the appearance of blisters and pustules on the skin. Vaccination has eradicated smallpox throughout the world. *720*

snapping division A distinctive type of binary fission resulting in an angular or a palisade arrangement of cells, which is characteristic of the genera *Arthrobacter* and *Corynebacterium*. *464*

sorocarp The fructification of the *Acrasiomycetes*. *530*

SOS repair A complex, inducible repair process that is used to repair DNA when extensive damage has occurred. *255*

source The location or object from which a pathogen is immediately transmitted to the host, either directly or through an intermediate agent. *706*

sour mash A mash that has been inoculated with lactic-acid producing microorganisms to control undesirable growth and create a distinct final product taste. *880*

Southern blotting technique The procedure that isolates and identifies DNA fragments from a complex mixture. The isolated, denatured fragments are transferred from an agarose electrophoretic gel to a nitrocellulose filter and identified by hybridization with probes. *288*

specialized transduction *See* restricted transduction. *276*

species (spe'shēz) Species of higher organisms are groups of interbreeding or potentially interbreeding natural populations that are reproductively isolated. Bacterial species are collections of strains that have many stable properties in common and differ significantly from other groups of strains. *405*

spheroplast (sfēr'o-plast) A relatively spherical cell formed by the weakening or partial removal of the rigid cell wall component (e.g., by penicillin treatment of gram-negative bacteria). Spheroplasts are usually osmotically sensitive. *55*

spike *See* peplomer. *360*

spirillum (spi-ril'um) A rigid, spiral-shaped bacterium. *42*

spirochete (spi'ro-kēt) A flexible, spiral-shaped bacterium with periplasmic flagella. *42*

split or **interrupted gene** A structural gene with DNA sequences that code for the final RNA product (expressed sequences or exons) separated by regions coding for RNA absent from the mature RNA (intervening sequences or introns). *203*

spontaneous generation (spon-ta'ne-us) The hypothesis that living organisms can arise from nonliving matter. *8*

sporadic disease (spo-rad'ik) A disease that occurs occasionally and at random intervals in a population. *699*

sporangiophore (spo-ran'je-o-fōr) A fungal hypha that bears a sporangium. *523*

sporangiospore (spo-ran'je-o-spōr) A spore born within a sporangium. *507, 523*

sporangium (spo-ran'je-um; pl., **sporangia**) A saclike structure or cell, the contents of which are converted into an indefinite number of spores. *63, 523*

spore (spōr) A reproductive cell, usually unicellular, capable of developing into an adult without fusion with another cell or of acting as a gamete. Spores may be produced asexually or sexually and are of many types. *538*

sporogenesis (spor'o-jen'ĕ-sis) *See* sporulation. *64*

sporogony (spo-rog'o-ne) Sporulation involving multiple fission of a zygote or sporont to produce sporozoites or sporocysts. Occurs in sporozoan protozoa. *794*

sporotrichosis (spo″ro-tri-ko'sis) A subcutaneous fungal infection caused by the dimorphic fungus *Sporothrix schenchii*. *785*

sporozoite (spo″ro-zo'īt) A motile, infective stage resulting from sporogony in gregarine and coccidian protozoa. In malaria, the sporozoites are liberated from the oocysts in the mosquito, accumulate in the salivary glands, and are transferred to humans when a mosquito takes a blood meal. *794*

sporulation (spor″u-la'shun) The process of spore formation. *64*

spread plate A petri dish of solid culture medium with isolated microbial colonies growing on its surface, which has been prepared by spreading a dilute microbial mixture evenly over the agar surface. *106*

sputum (spu'tum) The mucous secretion from the lungs, bronchi, and trachea that is ejected (expectorated) through the mouth. *675*

stalk (stawk) A nonliving bacterial appendage produced by the cell and extending from it. *480*

standard free-energy change The free-energy change of a reaction at 1 atmosphere pressure when all reactants and products are present in their standard states; usually the temperature is 25°C. *136*

standard reduction potential A measure of the tendency of a reductant to lose electrons in an oxidation-reduction reaction. See reduction potential. *137*

staphylococcal food poisoning (staf''ĭ-lo-kok'al) A type of food poisoning caused by ingestion of improperly stored or cooked food in which *Staphylococcus aureus* has grown. The bacteria produce exotoxins that accumulate in the food. *766*

staphylococcal scalded skin syndrome (SSSS) A disease caused by staphylococci that produce exfoliative toxin. The skin becomes red (erythema) and sheets of epidermis may separate from the underlying tissue. *759*

starter culture An inoculum, consisting of a mixture of carefully selected microorganisms, used to start a commercial fermentation. *877*

-static A suffix indicating that the substance named prevents microbial growth. *311, 328*

stationary phase (sta'shun-er''e) The phase of microbial growth in a batch culture when population growth ceases and the growth curve levels off. *113*

statistics (stah-tis'tiks) The mathematics of the collection, organization, and interpretation of numerical data. *699*

stenothermal (sten''o-ther'mal) Organisms that can grow well only over a narrow temperature range. *125*

sterilization (ster''ĭ-lĭ-za'shun) The process by which all living cells, viable spores, viruses, and viroids are either destroyed or removed from an object or habitat. *311*

stigma (stig'mah) A light-sensitive eyespot, which is found in some algae and photosynthetic protozoa; it is believed to be involved in phototaxis, at least in some cases. *539*

stoneworts A group of approximately 250 species of algae that have a complex growth pattern, with nodal regions from which whorls of branches arise; they are abundant in fresh to brackish waters. *540*

strain A population of organisms that descends from a single organism or pure culture isolate. *237, 406*

streak plate A petri dish of solid culture medium with isolated microbial colonies growing on its surface, which has been prepared by spreading a microbial mixture over the agar surface, using an inoculating loop. *106*

streptococcal sore throat (strep''to-kok'al) One of the most common bacterial infections of humans. It is commonly referred to as "strep throat." The disease is spread by droplets of saliva or nasal secretions and is caused by *Streptococcus* spp. (particularly group A streptococci). *747*

streptococcal pneumonia An endogenous infection of the lungs caused by *Streptococcus pneumoniae* that occurs in predisposed individuals. *747*

streptolysin-O (SLO) (strep-tol'ĭ-sin) A specific hemolysin produced by *Streptococcus pyogenes* that is inactivated by oxygen (hence the "O" in its name). SLO causes beta hemolysis of blood cells on agar plates incubated anaerobically. *589*

streptolysin-S (SLS) A product produced by *Streptococcus pyogenes* that is bound to the bacterial cell but may sometimes be released. SLS causes beta hemolysis on aerobically incubated blood-agar plates and can act as a leukocidin by killing leukocytes that phagocytose the bacterial cell to which it is bound. *590*

streptomycin (strep'to-mi''sin) A bactericidal aminoglycoside antibiotic produced by *Streptomyces griseus*. *337*

strict anaerobes *See* obligate anaerobes. *127*

stroma (stro'mah) The chloroplast matrix that is the location of the photosynthetic carbon dioxide fixation reactions. *79*

stromatolite (stro''mah-to'līt) A fossilized microbial mat community. *826*

structural gene A gene that codes for the synthesis of a polypeptide or polynucleotide with a nonregulatory function. *224*

subgingival plaque (sub-jin'ji-val) The plaque that forms at the dentogingival margin and extends down into the gingival tissue. *777*

substrate (sub'strāt) The substance an enzyme acts upon. *140*

substrate-level phosphorylation The synthesis of ATP from ADP by phosphorylation coupled with the exergonic breakdown of a high-energy organic substrate molecule. *148*

sulfate reduction (sul'fāt) The process of sulfate use as an oxidizing agent, which results in the accumulation of reduced forms of sulfur such as sulfide, or incorporation of sulfur into organic molecules, usually as sulfhydryl groups. *158, 810*

sulfonamide (sul-fon'ah-mīd) A chemotherapeutic agent that has the SO_2-NH_2 group and is a derivative of sulfanilamide. *332*

superinfection (soo''per-in-fek'shun) A new bacterial or fungal infection of a patient that is resistant to the drug(s) being used for treatment. *339*

superoxide dismutase (soo''per-ok'sīd dis-mu'tas) An enzyme that protects many microorganisms by catalyzing the destruction of the toxic superoxide radical. *128*

suppressor mutation A mutation that overcomes the effect of another mutation and produces the normal phenotype. *248*

Svedberg unit (sfed'berg) The unit used in expressing the sedimentation coefficient; the greater a particle's Svedberg value, the faster it travels in a centrifuge. *49*

swab (swahb) A wad of absorbent material usually wound around one end of a small stick and used for applying medication or for removing material from an area; also, a dacron-tipped polystyrene applicator. *675*

swarm cell A flagellated cell; the term is usually applied to the motile cells of the *Myxomycota*. *530*

symbiont (sim'bi-ont) An organism that lives in a state of symbiosis; also known as a **symbiote.** *565*

symbiosis (sim''bi-o'sis) The living together or close association of two dissimilar organisms, each of these organisms being known as a symbiont. *565*

symbiote (sim'bi-ōt) *See* symbiont. *565*

syndrome (sin'drom) A set of signs and symptoms that occur together and characterize a particular disease. *720*

synthetic medium *See* defined medium. *104*

syphilis (sif'ĭ-lis) *See* venereal syphilis.

systematics (sis''te-mat'iks) The scientific study of organisms with the ultimate objective being to characterize and arrange them in an orderly manner; often considered synonymous with taxonomy. *405*

systemic lupus erythematosus (loo'pus er''ĭ-them-ah-to'sus) An autoimmune, inflammatory disease that may affect every tissue of the body. *637*

taxon (tak'son) A group into which related organisms are classified. *405*

taxonomy (tak-son'o-me) The science of biological classification; it consists of three parts: classification, nomenclature, and identification. *405*

TB skin test Tuberculin hypersensitivity test for a previous or current infection with *Mycobacterium tuberculosis*. *638*

T cell A type of lymphocyte derived from bone marrow stem cells that matures into an immunologically competent cell under the influence of the thymus. T cells are involved in a variety of cell-mediated immune reactions; also known as a **T lymphocyte.** *609, 630*

T-cell growth factor *See* interleukin-2. *627*

T-dependent antigen An antigen that effectively stimulates B-cell response only with the aid of T-helper cells that produce interleukin-2 and B-cell growth factor. *629*

teichoic acids (ti-ko'ik) Polymers of glycerol or ribitol joined by phosphates; they are found in the cell walls of gram-positive bacteria. *51*

temperate phages Bacteriophages that can infect bacteria and establish a lysogenic relationship rather than immediately lysing their hosts. *275, 376*

template (tem'plat) A strand of DNA or RNA that specifies the base sequence of a newly synthesized complementary strand of DNA or RNA. *202*

Tenericutes (ten''er-ik'u-tēz) A division in the kingdom *Procaryotae* containing bacteria that lack an outer cell wall and do not synthesize peptidoglycan. *421*

terminator A sequence that marks the end of a gene and stops transcription. *203, 242*

tertiary treatment (ter'she-er-e) The removal from sewage of inorganic nutrients, heavy metals, viruses, etc., by chemical and biological means after microorganisms have degraded dissolved organic material during secondary sewage treatment. *832*

test A loose-fitting shell of an amoeba. *553*

tetanolysin (tet''ah-nol'ĭ-sin) A hemolysin that aids in tissue destruction and is produced by *Clostridium tetani.* *762*

tetanospasmin (tet''ah-no-spaz'min) The neurotoxic component of the tetanus toxin, which causes the muscle spasms of tetanus. Tetanospasmin production is under the control of a plasmid gene. *762*

tetanus (tet'ah-nus) An often fatal disease caused by the anaerobic, spore-forming bacillus *Clostridium tetani,* and characterized by muscle spasms and convulsions. *761*

tetracyclines (tet''rah-si'klēns) A family of antibiotics with a common four-ring structure, which are isolated from the genus *Streptomyces* or produced semisynthetically; all are related to chlortetracycline or oxytetracycline. *336*

tetrapartite associations (tet''rah-par'tīt) A mutualistic association of the same plant with three different types of microorganisms. *856*

thallus (thal'us) A type of body that is devoid of root, stem, or leaf; characteristic of some algae, many fungi, and lichens. *507, 519, 538*

T-helper (T_H) cell A cell that is needed for T-cell–dependent antigens to be effectively presented to B cells. It also promotes cell-mediated immune responses. *629*

theory A set of principles and concepts that have survived rigorous testing and that provide a systematic account of some aspect of nature. *12*

thermal death time (TDT) The shortest period of time needed to kill all the organisms in a microbial population at a specified temperature and under defined conditions. *314*

thermoacidophiles A group of bacteria that grow best at acid pHs and high temperatures; they are members of the *Archaeobacteria.* *503*

thermocline The layer in a thermally stratified lake having the greatest rate of temperature change with depth; it separates the upper warmer epilimnion zone from the lower colder oxygen-poor hypolimnion. *830*

thermodynamics (ther''mo-di-nam'iks) The scientific discipline that deals with heat and energy, and their interconversion. *134*

thermophile (ther'mo-fīl) A microorganism that can grow at temperatures of 55°C or higher; the minimum is usually around 45°C. *126*

thrush (thrush) Infection of the oral mucous membrane by the fungus *Candida albicans;* also known as **oral candidiasis.** *788*

thylakoid (thi'lah-koid) A flattened sac in the chloroplast stroma that contains photosynthetic pigments and the photosynthetic electron transport chain; light energy is trapped and used to form ATP and NAD(P)H in the thylakoid membrane. *79*

thymine (thi'min) The pyrimidine 5-methyluracil that is found in nucleosides, nucleotides, and DNA. *183*

thymocyte (thi'mo-sīt) A lymphocyte in the thymus. *609*

T-independent antigen An antigen that triggers a B cell into immunoglobulin production without T-cell cooperation. *630*

tinea (tin'e-ah) A name applied to many different kinds of fungal infections of the skin, the specific type (depending on characteristic appearance, etiologic agent, and site) usually being designated by a modifying term. *782*

tinea barbae A fungal infection of beard hair caused by *Trichophyton mentagrophytes* or *T. verrucosum.* *782*

tinea capitis A fungal infection of scalp hair caused by species of *Trichophyton* or *Microsporum.* *783*

tinea corporis A fungal infection of the smooth parts of the skin caused by either *Trichophyton rubrum, T. mentagrophytes,* or *Microsporum canis.* *784*

tinea cruris A fungal infection of the groin caused by either *Epidermophyton floccosum, Trichophyton mentagrophytes,* or *T. rubrum;* also known as **jock itch.** *784*

tinea manuum A fungal infection of the hand caused by *Trichophyton rubrum.* *784*

tinea pedis A fungal infection of the foot caused by *Trichophyton rubrum;* also known as **athlete's foot.** *784*

tinea unguium A fungal infection of the nail bed caused by either *Trichophyton rubrum* or *T. mentagrophytes.* Also called **onychomycosis.** *784*

tinea versicolor A fungal infection caused by the yeast, *Malassezia furfur,* that forms brownish-red scales on the skin of the trunk, neck, face, and arms. *782*

Ti plasmid A plasmid obtained from *Agrobacterium tumefaciens* that is used to insert genes into plant cells. *303, 858*

titer (ti'ter) Reciprocal of the highest dilution of an antiserum that gives a positive reaction in the test being used. *621*

T lymphocyte *See* T cell. *609*

tobacco mosaic disease A viral disease that affects over 150 plant genera and is caused by the tobacco mosaic virus (TMV). *347, 355*

tonsillitis (ton'si-li'tis) Inflammation of the tonsils, especially the palatine tonsils. *747*

total organic carbon (TOC) The organic carbon in a biological sample, including both biologically utilizable and nonutilizable materials. *819, 831*

toxemia (tok-se'me-ah) The condition caused by toxins in the blood of the host. *584*

toxic shock syndrome (tok'sik) A staphylococcal disease that most commonly affects females who use certain types of tampons during menstruation. It is associated with the production of toxic shock syndrome toxin by certain strains of *Staphylococcus aureus.* *758*

toxigenicity (tok''si-jĕ-nis'i-tē) The capacity of an organism to produce a toxin. *580*

toxin (tok'sin) A microbial product or component that can injure another cell or organism at low concentrations. Often the term refers to a poisonous protein, but toxins may be lipids and other substances. *580*

toxin neutralization The inactivation of toxins by specific antibodies, called antitoxins, that react with them. *651*

toxoid (tok'soid) A bacterial exotoxin that has been modified so that it is no longer toxic, but will still stimulate antitoxin formation when injected into a person or animal. *584*

toxoplasmosis (tok''so-plaz-mo'sis) A disease of animals and humans caused by the parasitic protozoan, *Toxoplasma gondii.* *798*

trachoma (trah-ko'mah) A chronic infectious disease of the conjunctiva and cornea, producing pain, inflammation and sometimes blindness. It is caused by *Chlamydia trachomatis* serotypes A–C. *772*

transamination (trans''am-i-na'shun) The removal of amino acid's amino group by transferring it to an α-keto acid acceptor. *160*

transcriptase (trans-krip'tās) An enzyme that catalyzes transcription; in viruses with RNA genomes, this enzyme is an RNA-dependent RNA polymerase that is used to make RNA copies of the RNA genomes. *389*

transcription (trans-krip'shun) The process in which single-stranded RNA with a base sequence complementary to the template strand of DNA or RNA is synthesized. *197*

transduction (trans-duk'shun) The transfer of genes between bacteria by bacteriophages. *275*

transfer host (trans'fer) A host that is not necessary for the completion of a parasite's life cycle, but is used as a vehicle for reaching a final host. *579*

transfer RNA (tRNA) A small RNA that binds an amino acid and delivers it to the ribosome for incorporation into a polypeptide chain during protein synthesis. *202*

transformation (trans''for-ma'shun) A mode of gene transfer in bacteria in which a piece of free DNA is taken up by a bacterial cell and integrated into the recipient genome. *238, 271*

transgenic animal or **plant** An animal or plant that has gained new genetic information from the insertion of foreign DNA. It may be produced by such techniques as injecting DNA into animal eggs, **electroporation** of mammalian cells and plant cell protoplasts, or shooting DNA into plant cells with a **gene gun.** *300*

transient carrier *See* casual carrier. *706*

transition mutations (tran-zish'un) Mutations that involve the substitution of a different purine base for the purine present at the site of the mutation or the substitution of a different pyrimidine for the normal pyrimidine. *244*

translation (trans-la'shun) Protein synthesis; the process by which the genetic message carried by mRNA directs the synthesis of polypeptides with the aid of ribosomes and other cell constituents. *197*

transmission electron microscope (trans-mish'un) A microscope that forms an image by passing an electron beam through a specimen and focusing the scattered electrons with magnetic lenses. *32*

transovarian passage (trans''o-va're-an) The passage of a microorganism from generation to generation of hosts through their eggs. *775*

transpeptidation The reaction that forms the peptide cross-links during peptidoglycan synthesis. *188, 211*

transposable elements *See* transposon. *264*

transposition (trans''po-zish'un) The movement of a piece of DNA around the chromosome. *264*

transposon (tranz-po'zon) A DNA segment that carries the genes required for transposition and moves about the chromosome; if it contains genes other than those required for transposition, it may be called a composite transposon. Often, the name is reserved only for transposable elements that also contain genes unrelated to transposition. *264*

transversion mutations (trans-ver'zhun) Mutations that result from the substitution of a purine base for the normal pyrimidine or a pyrimidine for the normal purine. *245*

traveler's diarrhea A type of diarrhea resulting from an encounter with certain viruses, bacteria, or protozoa normally absent from the traveler's environment. One of the major pathogens is enterotoxigenic *Escherichia coli* (ETEC). *767*

trematode (**fluke**; trem'ah-tōd) A member of the class *Trematoda* and the phylum *Platyhelminthes.* All trematodes are flatworm parasites of vertebrates, require a mollusk as their first intermediate host, and have complex life cycles. *A38*

tricarboxylic acid cycle (TCA) The cycle that oxidizes acetyl coenzyme A to CO_2 and generates NADH and $FADH_2$ for oxidation in the electron transport chain; the cycle also supplies carbon skeletons for biosynthesis. *150*

trichinosis (trik''ĭ-no'sis) A disease resulting from infection by *Trichinella spiralis* larvae upon eating undercooked meat. It is characterized by muscular pain, fever, edema, and other symptoms. *A37*

trichome (tri'kōm) A row or filament of bacterial cells that are in close contact with one another over a large area. *473*

trichomoniasis (trik''o-mo-ni'ah-sis) A sexually transmitted disease caused by the parasitic protozoan *Trichomonas vaginalis.* *798*

tripartite associations (tri-par'tīt) A mutualistic association of the same plant with two types of microorganisms. *856*

trophophase The phase of active microbial growth in batch culture; primary metabolism (growth-directed metabolism) is dominant. *895*

trophozoite (trof''o-zo'īt) The active, motile feeding stage of a protozoan organism; in the malarial parasite, the stage of schizogony between the ring stage and the schizont. *549*

trypanosome (tri-pan'o-sōm) A protozoan of the genus *Trypanosoma.* Trypanosomes are parasitic flagellate protozoans that often live in the blood of humans and other vertebrates and are transmitted by insect bites. *553, 797*

trypanosomiasis (tri-pan''o-so-mi'ah-sis) An infection with trypanosomes that live in the blood and lymph of the infected host. *797*

T-suppressor (T_S) cell A cell that is involved in suppressing an immune response. *632*

tubercle (too'ber-k'l) A small, rounded nodular lesion produced by *Mycobacterium tuberculosis.* *748*

tuberculoid (neural) leprosy (too-ber'ku-loid) A mild, nonprogressive form of leprosy that is associated with delayed-type hypersensitivity to antigens on the surface of *Mycobacterium leprae.* It is characterized by early nerve damage and regions of the skin that have lost sensation and are surrounded by a border of nodules. *755*

tuberculosis (too-ber''ku-lo'sis) An infectious disease of humans and other animals resulting from an infection by a species of *Mycobacterium* and characterized by the formation of tubercles and tissue necrosis, primarily as a result of host hypersensitivity and inflammation. Infection is usually by inhalation, and the disease commonly affects the lungs (pulmonary tuberculosis), although it may occur in any part of the body. *747*

tuberculous cavity (too-ber'ku-lus) An air-filled cavity that results from a tubercle lesion. *748*

tularemia (too''lah-re'me-ah) A plaguelike disease of animals caused by the bacterium *Francisella tularensis,* which may be transmitted to humans. *762*

tumble Random turning or tumbling movements made by bacteria when they stop moving in a straight line. *62*

tumor (too'mor) A growth of tissue resulting from abnormal new cell growth and reproduction (neoplasia). *393*

turbidostat A continuous culture system equipped with a photocell that adjusts the flow of medium through the culture vessel so as to maintain a constant cell density or turbidity. *120*

twiddle *See* tumble. *62*

tyndallization (fractional steam sterilization; tyn''dal-i-za'shun) A sterilization process in which a container is heated at 90 to 100°C for 30 minutes on three successive days and incubated at 37°C in between. *316*

type I (anaphylaxis) hypersensitivity A form of immediate hypersensitivity arising from the binding of antigen to IgE attached to mast cells, which then release anaphylaxis mediators such as histamine. Examples: hay fever, asthma, and food allergies. *635*

type II (cytotoxic) hypersensitivity A form of immediate hypersensitivity involving the binding of antibodies to antigens on cell surfaces followed by destruction of the target cells (e.g., through complement attack, phagocytosis, or agglutination). *637*

type III (immune complex) hypersensitivity A form of immediate hypersensitivity resulting from the exposure to excessive amounts of antigens in which antibodies bind to the antigens and produce antibody-antigen complexes. These activate complement and trigger an acute inflammatory response with subsequent tissue damage. Examples: poststreptococcal glomerulonephritis, serum sickness, and farmer's lung disease. *637*

type IV (cell-mediated) hypersensitivity A delayed hypersensitivity response (it appears 24 to 48 hours after antigen exposure). It results from the binding of antigen to activated T lymphocytes, which then release lymphokines and trigger inflammation and macrophage attacks that damage tissue. Type IV hypersensitivity is seen in contact dermititis from poison ivy, leprosy, and tertiary syphilis. *638*

typhoid fever (ti-foid) A bacterial infection transmitted by contaminated food, water, milk, or shellfish. The causative organism is *Salmonella typhi,* which is present in human feces. *767*

ultraviolet radiation (UV; ul''trah-vi'o-let) Radiation of fairly short wavelength, about 10 to 400 nm, and high energy. *130*

uracil (u'rah-sil) The pyrimidine 2,4-dioxypyrimidine, which is found in nucleosides, nucleotides, and RNA. *183*

vaccine (vak'sēn) A preparation of either killed microorganisms; living, weakened (attenuated) microorganisms; or inactivated bacterial toxins (toxoids). It is administered to induce development of the immune response and protect the individual against a pathogen or a toxin. *608*

valence (va'lens) The number of antigenic determinant sites on the surface of an antigen or the number of antigen-binding sites possessed by an antibody molecule. *610*

VA mycorrhizae Endomycorrhizae, characterized by having vesicles and arbuscules present in the plant roots. *855*

variable region The region at the N-terminal end of immunoglobulin heavy and light chains whose amino acid sequence varies between antibodies of different specificity. Variable regions form the antigen binding site. *612*

vasculitis (vas''ku-li'tis) Inflammation of a blood vessel. *774*

vector (vek'tor) 1. In genetic engineering, another name for **cloning vector.** *See* cloning vector. 2. In epidemiology, it is a living organism, usually an arthropod or other animal, that transfers an infective agent from one host to another. *288, 581, 707*

vector-borne transmission The transmission of an infectious pathogen between hosts by means of a **vector**. *707*

vehicle (ve'ĭ-k'l) An inanimate substance or medium involved in the transmission of a pathogen. *707*

venereal syphilis (ve-ne're-al sif'ĭ-lis) A contagious, sexually transmitted disease caused by the spirochete *Treponema pallidum.* *759*

verrucae vulgaris (vĕ-roo'se vul-ga'ris; s. **verruca vulgaris**) The common wart; a raised, epidermal lesion with horny surface caused by an infection with a human papillomavirus. *738*

vesicles (ves'i-k'ls) Small membrane-bound bodies. In endomycorrhizal plants infected with mycorrhizal fungi, swollen areas in the colonizing hyphae. *855*

vibrio (vib're-o) A rod-shaped bacterial cell that is curved to form a comma or an incomplete spiral. *42*

viral hemagglutination (vi'ral hem''ah-gloo''ti-na'shun) The clumping or agglutination of red blood cells caused by some viruses. *656*

viral neutralization An antibody-mediated process in which IgA, IgM, and IgA antibodies bind to some viruses during their extracellular phase and inactivate or neutralize them. *651*

viricide (vir'i-sīd) An agent that inactivates viruses so that they cannot reproduce within host cells. *311*

virino *See* prion. *399*

virion (vi're-on) A complete virus particle that represents the extracellular phase of the virus life cycle; at the simplest, it consists of a protein capsid surrounding a single nucleic acid molecule. *348*

viroid (vi'roid) An infectious agent of plants that is a single-stranded RNA not associated with any protein; the RNA does not code for any proteins and is not translated. *399*

virology (vi-rol'o-je) The branch of microbiology that is concerned with viruses and viral diseases. *347*

virulence (vir'u-lens) The degree or intensity of pathogenicity of an organism as indicated by case fatality rates and/or ability to invade host tissues and cause disease. *580*

virulent bacteriophages (vir'u-lent bak-te're-o-fājs'') Bacteriophages that lyse their host cells during the reproductive cycle. *375*

virus (vi'rus) An infectious agent having a simple acellular organization with a protein coat and a single type of nucleic acid, lacking independent metabolism, and reproducing only within living host cells. *348*

vitamin (vi'tah-min) An organic compound required by organisms in minute quantities for growth and reproduction because it cannot be synthesized by the organism; vitamins often serve as enzyme cofactors or parts of cofactors. *99*

volutin granules (vo-lu'tin) *See* metachromatic granules. *48*

wart (wort) An epidermal tumor of viral origin. *736*

water mold A common term for a member of the division *Oomycota.* *529*

water activity (a_w) A quantitative measure of water availability in the habitat; the water activity of a solution is one-hundredth its relative humidity. *122*

Weil-Felix reaction A test for the diagnosis of typhus and certain other rickettsial diseases. In this test, the blood serum of a patient with suspected rickettsial disease is tested against certain strains of *Proteus vulgaris* (OX-2, OX-19, OX-K). The agglutination reactions, based on antigens common to both organisms, determine the presence and type of rickettsial infection. *775*

white piedra A fungal infection (tinea albigena) caused by the yeast *Trichosporon beigelii* that forms light-colored nodules on the beard and mustache. *782*

Widal test (ve-dahl') A test involving agglutination of typhoid bacilli when they are mixed with serum containing typhoid antibodies from an individual having typhoid fever; used to detect the presence of *Salmonella typhi* and *paratyphi.* *656*

wine vinegar Vinegar produced by the oxidation of alcohol in wine by members of the genus *Acetobacter.* *880*

Winogradsky column A glass column with an anaerobic lower zone and an aerobic upper zone, which allows growth of microorganisms under conditions similar to those found in a nutrient-rich lake. *824*

wort The filtrate of malted grains used as the substrate for the production of beer and ale by fermentation. *879*

xenograft (zen'o-graft) A tissue graft between animals of different species. *640*

xerophilic microorganisms (ze''ro-fil'ik) Microorganisms that grow best under low a_w conditions, and may not be able to grow at high a_w values. *867*

yeast (yēst) A unicellular fungus that has a single nucleus and reproduces either asexually by budding or fission, or sexually through spore formation. *520*

yellow fever An acute infectious disease caused by a flavivirus, which is transmitted to humans by mosquitoes. The liver is affected and the skin turns yellow in this disease. *722*

YM shift The change in shape by dimorphic fungi when they shift from the yeast (Y) form in the animal body to the mold or mycelial form (M) in the environment. *521*

zooflagellates (zo''o-flăj'-e-lāts) Flagellate protozoa that do not have chlorophyll and are either holozoic, saprozoic, or symbiotic. *553*

zoonosis (zo''o-no'sis; pl. **zoonoses**) A disease of animals that can be transmitted to humans. *699*

zooplankton (zo''o-plank'ton) A community of floating, aquatic, minute animals and nonphotosynthetic protists. *536*

zoospore (zo'o-spōr) A motile, flagellated spore. *538*

***z* value** The increase in temperature required to reduce the decimal reduction time to one-tenth of its initial value. *314*

zooxanthella (zo''o-zan-thel'ah) A dinoflagellate found living symbiotically within cnidarians and other invertebrates. *545, 577*

zygomycetes (zi''go-mi-se'tez) A division of fungi that usually has a coenocytic mycelium with chitinous cell walls. Sexual reproduction normally involves the formation of zygospores. The group lacks motile spores. *524*

zygospore (zi'go-spōr) A thick-walled, sexual, resting spore characteristic of the zygomycetous fungi. *524*

zygote (zi'gōt) The diploid (2n) cell resulting from the fusion of male and female gametes. *538*

zymogenous (zi-moj'ĕ-nus) Fermentation producing. In environmental microbiology, the term refers to microorganisms, often transient or alien, that respond rapidly by enzyme production and growth when simple organic substrates become available. *848*

Credits

Photos

Part Openers

1: © CNRI/SPL/Photo Researchers, Inc.; **2:** Wm. C. Brown Communications, Inc., Ray Otero photographer. **3:** © Oxford Molecular Biophysics Lab/SPL/Photo Researchers, Inc.; **4:** © Wm. C. Brown Communications, Inc., Ray Otero photographer; **5:** © Alfred Pasieka/Peter Arnold, Inc.; **6:** © Biophoto Assoc./Photo Researchers, Inc.; **7:** © John Shaw/Tom Stack & Assoc.; **8:** From A. G. Amit and Roberto J. Poljak, "Three-Dimensional Structure of an Antigen-Antibody Complex at 2.8 A Resolution" *Science* 233:747, August 15, 1986. Copyright 1986 by the AAAS.; **9:** Copyright Boehringer-Ingelheim, photo by Lennart Nilsson; **10:** © Doug Sokell/Tom Stack & Assoc.; **11:** © Sylvain Grandadam/Photo Researchers, Inc.

Chapter 1

1.1a: Bettmann Archive; **1.1c:** Historic VU/Visuals Unlimited; **1.2, 1.4, 1.5:** Bettmann Archive; **1.6:** From *ASM News* 47(7):392, 1981. American Society for Microbiology; **1.7:** Bettmann Archive; **1.8, 1.9:** National Library of Medicine.

Chapter 2

2.3: Reichert Scientific Instruments; **2.4:** Nikon Scientific Instruments; **2.8a:** © Arthur M. Siegelman/Visuals Unlimited; **2.8b:** © Robert Calentine/Visuals Unlimited; **2.8c:** © Mike Abbey/Visuals Unlimited; **2.8d:** © George J. Wilder/Visuals Unlimited; **2.8e:** © Mike Abbey/Visuals Unlimited; **2.12a:** © Arthur M. Siegelman/Visuals Unlimited; **2.12b:** Joan H. Sonneborn; **2.12c:** Kallestad Diagnostics; **2.14a:** © Arthur M. Siegelman/Visuals Unlimited; **2.14b:** © Michael A. Gabridge/Visuals Unlimited; **2.14c:** © Arthur M. Siegelman/Visuals Unlimited; **2.14d:** © George J. Wilder/Visuals Unlimited; **2.15, 2.16:** © John Cunningham/Visuals Unlimited; **2.17:** © Manfred Kage/Peter Arnold, Inc.; **2.18:** © John D. Cunningham/Visuals Unlimited; **2.20a:** © George J. Wilder/Visuals Unlimited; **2.20b:** © Biology Media/Photo Researchers, Inc.; **2.21:** © William Ormerod, Jr./Visuals Unlimited; **2.23 a–b:** © Fred Hossler/Visuals Unlimited; **2.25:** E. J. Laishley, University of Calgary; **2.27a:** © David Phillips/Photo Researchers, Inc.; **2.27b:** © Paul W. Johnson/Biological Photo Service; **Box 2.1a:** Courtesy of International Business Machines; **Box 2.1b:** © Driscoll, Youngquist, and Baldeschwieler, Caltech/SPL/ Photo Researchers, Inc.

Chapter 3

3.1a: © LeBeau/Biological Photo Service; **3.1b:** © Arthur M. Siegelman/Visuals Unlimited; **3.1c:** © George J. Wilder/Visuals Unlimited; **3.1d:** © Thomas Tottleben/Tottleben Scientific Company; **3.1e:** © CDC/Biological Photo Service; **3.2 a–c:** © David M. Phillips/Visuals Unlimited; **3.2d:** From J. G. Holt (Ed.) *The Shorter Bergey's Manual of Determinative Bacteriology,* 1977. Williams and Wilkins Co., Baltimore; **3.2e:** From Walther Stoeckenius, *Walsby's Square Bacterium: Fine Structures of an Orthogonal Procaryote*; **3.8:** © S. C. Holt, U. of Texas Health Science Center/Biological Photo Service; **3.9a:** From R. G. E. Murray and S. W. Watson, *Journal of Bacteriology* 89:1597, 1968, American Society for Microbiology; **3.9b:** From J. G. Holt (Ed.), *The Shorter Bergey's Manual of Determinative Bacteriology,* 1977. Williams & Wilkins Co., Baltimore; **3.10:** © Ralph A. Slepecky/Visuals Unlimited; **3.11 a–b:** From A. E. Walsby, *Microbiological Reviews* 36:2, 1972. American Society for Microbiology; **3.12 a–b:** From A. E. Walsby, *Microbiological Reviews* 36:26, 1972. American Society for Microbiology; **3.13:** Daniel Branton, Harvard University; **3.14a:** From J. G. Holt (Ed.), *The Shorter Bergey's Manual of Determinative Bacteriology,* 1977. American Society for Microbiology; **3.14b:** From W. C. Dierksheide & R. M. Pfister, *Canadian Journal of Microbiology* 19(1):151, 1973. National Research Council of Canada; **3.15 a, c:** T. J. Beveridge, Ph.D.; **3.19b:** From H. Formanek et al., *Eur. J. Biochemistry* 46:279–294, 1974, Springer-Verlag; **3.20:** M. R. J. Salton, NYU Medical Center; **3.26 a–b:** © John D. Cunningham/Visuals Unlimited; **3.27:** © George Musil/Visuals Unlimited; **3.28:** R. G. E. Murray, University of Western Ontario; **3.29:** © Fred Hossler/Visuals Unlimited; **3.30 a–b:** © E. C. S. Chan/Visuals Unlimited; **3.30c:** © George J. Wilder/Visuals Unlimited; **3.31 a–b, 3.35, 3.36:** Julius Adler; **Box 3.2a:** D. Balkwill and D. Maratea; **Box 3.2b:** Y. Gorby; **Box 3.2c:** A. Spormann, University of Illinois at Urbana–Champaign; **3.40:** From Moberly, Shafa and Gerhardt, *Journal of Bacteriology* 92:223, 1966. American Society of Microbiology; **3.43 a–f:** From A. N. Barker et al. (Eds.), *Spore Research,* 1971, Academic Press; **3.44:** From J. Hoeniger and C. Headley, *Journal of Bacteriology* 96:1844, 1968. American Society for Microbiology.

Chapter 4

4.1a: © Eric Grave/Photo Researchers, Inc.; **4.1b:** Carolina Biological Supply; **4.1c:** © John D. Cunningham/Visuals Unlimited; **4.1d:** © Arthur M. Siegelman/Visuals Unlimited; **4.1e:** © John D. Cunningham/Visuals Unlimited; **4.1f:** © Tom E. Adams/Visuals Unlimited; **4.1g:** © John D. Cunningham/Visuals Unlimited; **4.2a:** © Richard Rodewald/Biological Photo Service; **4.2b:** © W. L. Dentler, U. of Kansas/Biological Photo Service; **4.5 a–b, 4.6 a–b, 4.7a:** © Manfred Schliwa/Visuals Unlimited; **4.8:** © B. F. King/Biological Photo Service; **4.9a:** © Henry C. Aldrich/Visuals Unlimited; **4.10b:** James Burbach, Univ. of South Dakota; **4.12:** From M. J. Weiss and L. B. Chen, "Rhodamine 123: A Lipophilic, Cationic, Mitochondrial-Specific, Vital Dye," *Kodak Laboratory Chemicals Bulletin,* 55(2), 1984. Reprinted by permission of Eastman Kodak Company; **4.14b:** From Keiichi Tanaka, "Scanning Electron Microscopy of Intracellular Structures," *International Review of Cytology* 68:97–112. Academic Press, 1980.; **4.14c:** © Keith Porter/Photo Researchers, Inc.; **4.15a:** © Michael J. Dykstra/Visuals Unlimited; **4.15b:** © Manfred Schliwa/Visuals Unlimited; **4.16a:** From Fraenkel-Conrat, Kimball, and Levy, *Virology,* 2/E, pp. 202–204. Prentice Hall, Inc. Englewood Cliffs, N.J.; **4.17:** From B. P. Eyden, *Journal of Proto 2001* 22(3): 336–344. © Thomas E. Simpson; **4.18:** Dr. Garry T. Cole, Univ. of Texas at Austin; **4.21:** © Henry C. Aldrich/Visuals Unlimited; **4.25:** From P. R. Desjardins et al., *Canadian Journal of Bot.* 47:1077–1097, 1969. By permission of the National Research Council of Canada; **4.26:** © Karl Aufderheide/Visuals Unlimited; **4.27a:** © Michael C. Webb/Visuals Unlimited; **4.27b:** © K. G. Murti/Visuals Unlimited; **4.29a:** © Ralph A. Slepecky/Visuals Unlimited; **4.29b:** © W. L. Dentler, U. of Kansas/Biological Photo Service.

Chapter 5

5.8: © Larry Jensen/Visuals Unlimited; **5.11a:** © David M. Phillips/Visuals Unlimited; **5.11b:** E. Louise Springer and Ivan L. Roth; **5.11 c–e:** © David M. Phillips/Visuals Unlimited; **5.11f:** © Fred Hossler/Visuals Unlimited.

Chapter 6

6.5a: Courtesy Society for Industrial Microbiology; **6.5b:** Courtesy of Dynatech Labs, Photo by Martin-Schaeffer Advertising; **6.6b:** Millipore Corporation; **6.8a:** Courtesy Nalge Company; **6.8b:** © B. Otero/Visuals Unlimited; **6.8 c–d:** Courtesy Nalge Company; **6.12b:** Courtesy of B. Braun Biotech, Inc.; **Box 6.1:** Science VU-D. Foster, Woods Hole Oceanographic Institution/ Visuals Unlimited.

Chapter 7

7.15 a–b: From Voet-Voet: *Biochemistry,* 1/e, 1990 John Wiley and Sons. Courtesy of Donald Voet.

Chapter 8

8.24 a–b: From J. Deisenhofer and H. Michel, *Les Prix Nobel,* 1989.

Chapter 9

9.14: From A. S. Moffat "Nitrogenase Structure Revealed," *Science* 250:1513, December 14, 1990. Copyright 1990 by the AAAS. Photo by M. M. Georgiadis and D. C. Rees, Caltech.

Chapter 10

10.2 c-1: Copyright Irving Geis; **10.2 c-2:** From Voet-Voet: *Biochemistry,* 1/e, 1990 John Wiley and Sons. Courtesy of Donald Voet.; **10.5 a-1:** Evangelos N. Mourdrianakis; **10.13:** Courtesy of Prof. Thomas A. Steitz, Yale University; **10.22b:** From S. H. Kim, "Three-Dimensional Tertiary Structure of Yeast Phenylelanine Transfer RNA," *Science* 185:436, August 2, 1974. Copyright 1974 by the AAAS; **10.23:** From R. Rould and T. Steitz, "Structure of E. coli Glutaminyl-tRNA Syntetase Complexed with tRNAgln and ATP at 2.8 A Resolution," *Science* 246:1135–1142, Dec. 1, 1989. Copyright 1989 by the AAAS.

Chapter 11

11.5 b-1, c-1: Reprinted by permission W. N. Lipscomb, Harvard University; **11.7 a–b:** Courtesy of David Eisenberg, UCLA; **11.14c:** From S. C. Schultz, G. C. Shields, and T. A. Steitz, "Crystal Structure of a CAP-DNA Complex: The DNA Is Bent by 90 Degrees," *Science* 253:1001–1007, Aug. 30, 1991. Copyright 1991 by the AAAS.

Chapter 13

13.4: Charles C. Brinton, Jr., and Judith Carnahan; **13.22 b–c:** From *Molecular Biology of Bacterial Viruses* by Gunther S. Stent. Copyright © 1963 W. H. Freeman and Company. Reprinted with permission.; **13.23a:** From R. B. Inman, *Journal of Molecular Biology* 51: 61, 1970. Reprinted by permission of Consultants Bureau.

Chapter 14

14.1b: J. M. Rosenberg; **14.9:** Courtesy of The Perkin-Elmer Corporation; **14.10b:** Bio-Rad Laboratories; **14.10c:** Bethesda Research Laboratories, 1985 Catalog, pp. 58 & 59; **14.11 a–b:** Huntington Potter and David Dressler, *Life Magazine,* 1980, Time, Inc.; **Box 14.1:** Courtesy of Keith V. Wood; **Box 14.2b:** From D. W. Ow et al., "Transient and Stable Expression of the Firefly Luciferase Gene in Plant Cells and Transgenic Plants," *Science* 234:856–859, Nov. 14, 1986. Copyright 1986 by the AAAS.

Chapter 15

15.3a: Courtesy of AMSCO Scientific, Apex, NC; **15.4b:** Courtesy Millipore Corporation; **15.5a:** Pall Ultrafine Filter Corporation; **15.5b:** © Fred Hossler/Visuals Unlimited; **15.6a:** Science-VU-Forma/Visuals Unlimited.

Chapter 16

16.1a: Courtesy BBL Microbiology Systems; **16.1b:** © Larry Jensen/Visuals Unlimited.

Chapter 17

17.2a: © M. Abbey/Visuals Unlimited; **17.2b:** © Terry C. Hazen/Visuals Unlimited; **17.3 a–b:** © David M. Phillips/Visuals Unlimited; **17.4 a–d:** From S. E. Luria, *General Virology,* 1978, John Wiley, NY; **17.5a:** © Runk/Schoenberger/Grant Heilman Photography, Inc.; **17.5b:** Douglas Maxwell, University of Wisconsin–Madison; **17.5c:** © Charles Marden Fitch/Taurus Photos; **17.8:** Janey S. Symington; **17.11a:** © Dennis Kunkel/Phototake; **17.11c:** Courtesy of Gerald Stubbs and Keiichi Namba, Vanderbilt University; and Donald Caspar, Brandeis University; **17.12a:** Courtesy of Michael G. Rossman, Purdue University; **17.12 b–c:** From J. M. Hogle et al., "Three-Dimensional Structure of Poliovirus at 2.9 A Resolution," *Science* 229: 1360, Sept. 27, 1985. Copyright 1985 by the AAAS; **17.12d:** George Musil/Visuals Unlimited; **17.12e:** Courtesy of Harold Fisher, University of Rhode Island and Robley Williams, University of California at Berkeley; **17.12f:** © R. Feldman-Dan McCoy/Rainbow; **17.12g:** Courtesy of Harold Fisher, University of Rhode Island and Robley Williams, University of California at Berkeley; **17.12h:** Science VU-NIH, R. Feldman/Visuals Unlimited; **17.14 a–c:** Courtesy of Robert C. Liddington and Stephen C. Harrison, Harvard University.; **17.15, 17.17 a, c:** © K. G. Murti/Visuals Unlimited; **17.17d:** © Dr. Steven Baum/Peter Arnold, Inc.; **17.17e:** © CDC/Science Source/Photo Researchers, Inc.; **17.17f:** © R. Feldman-Dan McCoy/Rainbow; **17.18b:** Center for Disease Control; **17.18c:** © K. G. Murti/Visuals Unlimited; **17.19b:** © Thomas Broker/Phototake; **17.19c:** © Alfred Pasieka/Peter Arnold, Inc.

Chapter 18

18.4: © Lee D. Simon/Photo Researchers, Inc.; **18.5 b-1:** © Fred Hossler/Visuals Unlimited; **18.5 b-2:** George Chapman, Georgetown University; **18.14a:** © M. Wurtz/Photo Researchers, Inc.; **18.17a:** From F. D. Bushman, C. Shang, and M. Ptashne, "A Single Glutamic Acid Residue Plays a Key Role in the Transcriptional Activation Function of Lambda Repressor," *Cell* 58:1163–1171, September 22, 1989. Cell Press; **18.19a:** From A. K. Aggarwal, D. W. Rodgers, M. Drottar, M. Ptashne, and S. C. Harrison, "Recognition of a DNA Operator by the Repressor of Phage 434: A View at High Resolution," *Science* 242:899–907, Nov. 11, 1988. © 1988 by the American Association for the Advancement of Science.

Chapter 19

19.6a: © Will & Deni McIntyre/Photo Researchers, Inc.; **19.6 b–c:** Courtesy of Wayne Hendrickson, Columbia University; **19.7:** © K. G. Murti/Visuals Unlimited; **19.9 a–b:** Center for Disease Control; **19.10:** J. T. Finch-J. M. Kaper, USDA, Agricultural Research Service; **19.13 a–b:** Russell L. Steere, Advanced Biotechnologies, Inc.; **19.14:** J. R. Adams, USDA.

Chapter 21

21.1a: From J. G. Holt (Ed.), *The Shorter Bergey's Manual of Determinative Bacteriology,* 1977. Williams & Wilkins Co., Baltimore; **21.1 b–c:** © A. M. Siegelman/Visuals Unlimited; **21.1d:** From J. G. Holt (Ed.), *The Shorter Bergey's Manual of Determinative Bacteriology,* 1977. Williams & Wilkins Co., Baltimore; **21.2 a-2:** From S. C. Holt, *Microbiological Reviews* 42(1):117, 1978. American Society for Microbiology; **21.2c:** From M. P. Starr et al. (Eds.), *The Prokaryotes,* Springer-Verlag.; **21.2d:** From S. C. Holt, *Microbiological Reviews* 42(1):122, 1978. American Society for Microbiology; **21.4a:** From S. C. Holt, *Microbiological Reviews* 42(1):148, 1978. American Society for Microbiology; **21.4b:** From S. C. Holt, *Microbiological Reviews* 42(1)150, 1978. American Society for Microbiology; **21.5a:** © Runk/Schoenberger/Grant Heilman Photography; **21.5b:** © Tom Tottleben/Tottleben Scientific Company; **21.5c:** © George J. Wilder/Visuals Unlimited; **21.6:** Jeffrey C. Burnham; **21.7 b–c:** From J. C. Burnham and S. F. Conti, "Genus Bdellovibrio" in *Bergey's Manual of Systematic Bacteriology,* Vol. 1, pp. 118–119, edited by N. R. Kreig and J. G. Holt. Copyright © 1984 Williams & Wilkins Co., Baltimore; **21.8 a–b:** Harkisan D. Raj/Visuals Unlimited; **21.9:** © Christine Case/Visuals Unlimited; **21.10a:** © David M. Phillips/Visuals Unlimited; **21.10 b–d:** From N. R. Krieg and J. G. Holt (Eds.), *Bergey's Manual of Systematic Bacteriology,* 1984. Williams and Wilkins Co., Baltimore; **21.10e:** © Carroll P. Vance/Visuals Unlimited; **21.11:** © John D. Cunningham/Visuals Unlimited; **21.14a:** © Arthur M. Siegelman/Visuals Unlimited; **21.14b:** © E. S. Anderson/Photo Researchers, Inc.; **21.15:** From N. R. Krieg and J. G. Holt (Eds.), *Bergey's Manual of Systematic Bacteriology,* 1984. Williams and Wilkins Co., Baltimore; **21.16a:** © Kenneth Lucas, Steinhart Aquarium/Biological Photo Service; **21.16 b–c:** James G. Morin, University of California, Los Angeles; **21.17 a–d:** © F. Widdel/Visuals Unlimited; **21.18 a–c:** From N. R. Kreig and J. G. Holt (Eds.), *Bergey's Manual of Systematic Bacteriology,* 1984. Williams and Wilkins Co., Baltimore; **21.19:** © Martin M. Rotker/Taurus Photos; **21.20a:** © David M. Phillips/Visuals Unlimited; **21.21a:** © Michael G. Gabridge/Visuals Unlimited; **21.21b:** © David M. Phillips/Visuals Unlimited; **21.22:** © Michael Gabridge/Visuals Unlimited.

Chapter 22

22.2a: © Bruce Iverson, Photomicrography; **22.2b:** © David M. Phillips/Visuals Unlimited; **22.2c:** © Tom Tottleben/Tottleben Scientific Company; **22.2d:** © David M. Phillips/Visuals Unlimited; **22.2e:** © Runk/Schoenberger/Grant Heilman Photography; **22.2f:** © M. Abbey/Visuals Unlimited; **22.3:** From J. G. Holt et al. (Eds.), *Bergey's Manual of Systematic Bacteriology,* 1986. Williams and Wilkins Co., Baltimore; **22.4 a–c:** © Fred E. Hossler/Visuals Unlimited; **22.4d:** © Carroll H. Weiss/Camera M.D. Studios; **22.5, 22.7:** From M. P. Starr et al. (Eds.), *The Prokaryotes,* Springer-Verlag; **22.8a:** © A. M. Siegelman/Visuals Unlimited; **22.8b:** © Cabisco/Visuals Unlimited; **22.8 c–d, 22.9 a–b:** © A. M. Siegelman/Visuals Unlimited; **22.9c:** © George J. Wilder/Visuals Unlimited; **22.10:** From J. G. Holt et al. (Eds.), *Bergey's Manual of Systematic Bacteriology,* 1986. Williams and Wilkins Co., Baltimore; **22.11:** © Grant Heilman Photography; **22.12 a–d:** From J. G. Holt et al. (Eds.), *Bergey's Manual of Systematic Bacteriology,* 1986. Williams and Wilkins Co., Baltimore; **22.13a:** © E. C. S. Chan/Visuals Unlimited; **22.13b:** © David M. Phillips/Visuals Unlimited; **22.14:** © John D. Cunningham/Visuals Unlimited.

Chapter 23

23.1, 23.2a: From M. P. Starr et al. (Eds.), *The Prokaryotes,* Springer-Verlag; **23.2b:** From J. T. Staley, M. P. Bryant, N. Pfennig, and J. G. Holt (Eds.), *Bergey's Manual of Systematic Bacteriology,* Vol. 3. Copyright 1989 Williams and Wilkins Co., Baltimore.; **23.2c:** From M. P. Starr et al. (Eds.), *The Prokaryotes,* Springer-Verlag; **23.5a:** George J. Wilder/Visuals Unlimited; **23.5 b–c:** From J. G. Holt (Ed.), *The Shorter Bergey's Manual of Determinative Bacteriology,* 1977. Williams and Wilkins Co., Baltimore; **23.5d:** From M. P. Starr et al. (Eds.), *The Prokaryotes,* Springer-Verlag; **23.5e, 23.6 a–b:** From J. G. Holt (Ed.), *The Shorter Bergey's Manual of Determinative Bacteriology,* 1977. Williams and Wilkins Co., Baltimore; **23.6c:** From J. T. Staley, M. P. Bryant, N. Pfennig, and J. G. Holt (Eds.), *Bergey's Manual of Systematic Bacteriology,* Vol. 3. Copyright 1989 Williams and Wilkins Co., Baltimore. Micrograph G. Cohen-Bazire.; **23.7:** Elizabeth Gentt; **23.8a:** © T. E. Adams/Visuals Unlimited; **23.8b:** © Ron Dengler/Visuals Unlimited; **23.8c:** © M. I. Walker/Photo Researchers, Inc.; **23.8d:** © Tom E. Adams/Visuals Unlimited; **23.9b:** From Carlsberg Research Communications 42:77–98, 1977, © Carlsberg Laboratories.; **23.10a:** © George J. Wilder/Visuals Unlimited; **23.10b:** Michael Richard, Colorado State University; **23.10c:** P. Fay and N. J. Lang, *Proceedings of the Royal Society Series B* 178:185–192, 1971; **23.11a:** From J. T. Staley, M. P. Bryant, N. Pfennig, and J. G. Holt (Eds.), *Bergey's Manual of Systematic Bacteriology,* Vol. 3. Copyright 1989 Williams and Wilkins Co., Baltimore. Micrograph courtesy of Ralph Lewin and L. Cheng.; **23.11b:** Jean Whatley, *New Phytology* 79:309–313, 1977.; **23.12 a–d:** S. W. Watson, Woods Hole Oceanographic Institution; **23.13:** Reprinted by permission of Kluwer Academic Publishers from J. G. Kuenen and H. Veldkamp, Antonie van Leeuwenhoek.; **23.14:** From J. T. Staley, M. P. Bryant, N. Pfennig, and J. G. Holt (Eds.), *Bergey's Manual of Systematic Bacteriology,* Vol. 3. Copyright 1989 Williams and Wilkins Co., Baltimore.; **23.16a:** © George J. Wilder/Visuals Unlimited; **23.16 b–c:** Jeanne S. Poindexter, Long Island University; **23.16d:** From J. T. Staley, M. P. Bryant, N. Pfennig, and J. G. Holt (Eds.), *Bergey's Manual of Systematic Bacteriology,* Vol. 3. Copyright 1989 Williams and Wilkins Co., Baltimore; **23.18 a–b, 23.19 a–b, 23.20 a–d:** From M. P. Starr et al. (Eds.), *The Prokaryotes,* Springer-Verlag; **23.21:** Michael Richard, Colorado State University; **23.22:** From M. P. Starr et al. (Eds.), *The Prokaryotes,* Springer-Verlag; **23.23 b–d:** Ruth L. Harold, National Jewish Center for Immunology; Dr. Harkisan D. Raj; **23.24 a–c:** From M. P. Starr et al (Eds.), *The Prokaryotes,* Springer-Verlag; **23.26 b–c:** © M. Dworkin-H. Reichenbach/Phototake; **23.26d:** © Patricia L. Grilione/Phototake.

Chapter 24

24.1 a-2, b-2: From O. Kandler and H. Konig, "Cell Envelopes of Archaebacteria," *The Bacteria*, Vol. 8, 1985, Academic Press; **24.7a:** From J. G. Holt (Ed.), *The Shorter Bergey's Manual of Determinative Bacteriology*, 1977. Williams and Wilkins Co., Baltimore; **24.7b:** From J. T. Staley, M. P. Bryant, N. Pfennig, and J. G. Holt (Eds.), *Bergey's Manual of Systematic Bacteriology*, Vol. 3. Copyright 1989 Williams and Wilkins Co., Baltimore.; **24.7c:** From J. G. Holt (Ed.), *The Shorter Bergey's Manual of Determinative Bacteriology*, 1977. Williams and Wilkins Co., Baltimore; **24.7d:** From J. T. Staley, M. P. Bryant, N. Pfennig, and J. G. Holt (Eds.), *Bergey's Manual of Systematic Bacteriology*, Vol. 3. Copyright 1989 Williams and Wilkins Co., Baltimore. R. Robinson, Dept. of Microbiology, U. of California, Los Angeles; **24.7e:** From J. G. Holt (Ed.) *The Shorter Bergey's Manual of Determinative Bacteriology*, 1977. Williams and Wilkins Co., Baltimore; **24.7f:** From M. P. Starr et al. (Eds.), *The Prokaryotes*, Springer-Verlag; **24.10a:** From J. T. Staley, M. P. Bryant, N. Pfennig, and J. G. Holt (Eds.), *Bergey's Manual of Systematic Bacteriology*, Vol. 3. Copyright 1989 Williams and Wilkins Co., Baltimore. Prepared by G. Bentzen; photographed by the Laboratory of Clinical Electron Microscopy, U. of Bergen; **24.10b:** From J. T. Staley, M. P. Bryant, N. Pfennig, and J. G. Holt (Eds.), *Bergey's Manual of Systematic Bacteriology*, Vol. 3. Copyright 1989 Williams and Wilkins Co., Baltimore. Prepared by A. L. Ustad, Photography by Dept. of Biophysics, Norwegian Institute of Technology; **24.11:** From J. T. Staley, M. P. Bryant, N. Pfennig, and J. G. Holt, (Eds.), *Bergey's Manual of Systematic Bacteriology*, Vol. 3. Copyright 1989 Williams and Wilkins Co., Baltimore.; **24.12 a–b:** From Corale L. Brierley/Visuals Unlimited; **24.12c:** From J. T. Staley, M. P. Bryant, N. Pfennig, and J. G. Holt (Eds.), *Bergey's Manual of Systematic Bacteriology*, Vol. 3. Copyright 1989 Williams and Wilkins Co., Baltimore. Micrograph courtesy of D. Janekovic and W. Zillig.

Chapter 25

25.2 a–c: From S. T. Williams, M. E. Sharpe, and J. G. Holt (Eds.), *Bergey's Manual of Systematic Bacteriology*, Vol. 4. Copyright 1989 Williams and Wilkins Co., Baltimore.; **25.2d:** Frederick P. Mertz, Lilly Research Lab; **25.2e:** From S. T. Williams, M. E. Sharpe, and J. G. Holt (Eds.), *Bergey's Manual of Systematic Bacteriology*, Vol. 4. Copyright 1989 Williams and Wilkins Co., Baltimore.; **25.2f:** Courtesy Yoko U. Takahashi; **25.4 a-1:** From J. G. Holt et al. (Eds.), *Bergey's Manual of Systematic Bacteriology*, 1986. Williams and Wilkins Co., Baltimore; **25.4 b-1, 25.5a:** From S. T. Williams, M. E. Sharpe, and J. G. Holt (Eds.), *Bergey's Manual of Systematic Bacteriology*, Vol. 4. Copyright 1989 Williams and Wilkins Co., Baltimore; **25.5b:** © R. Howard Berg/Visuals Unlimited; **25.5c:** From S. T. Williams, M. E. Sharpe, and J. G. Holt (Eds.), *Bergey's Manual of Systematic Bacteriology*, Vol. 4. Copyright 1989 Williams and Wilkins Co., Baltimore; **25.6b:** Dr. Akio Seino, *Hakko to Kogyo (Fermentation and Industry)* 41(3):3–4, 1983; **25.6c:** From S. T. Williams, M. E. Sharpe, and J. G. Holt (Eds.), *Bergey's Manual of Systematic Bacteriology*, Vol. 4. Copyright 1989 Williams and Wilkins Co., Baltimore. Courtesy of Dr. H. A. Lechevalier.; **25.6e:** From S. T. Williams, M. E. Sharpe, and J. G. Holt (Eds.), *Bergey's Manual of Systematic Bacteriology*, Vol. 4. Copyright 1989 Williams and Wilkins Co., Baltimore; **25.8 a-2:** From S. T. Williams, M. E. Sharpe, and J. G. Holt (Eds.), *Bergey's Manual of Systematic Bacteriology*, Vol. 4. Copyright 1989 Williams and Wilkins Co., Baltimore. Micrograph from T. Cross, U. of Bradford, Bradford, U.K.; **25.8 b-2:** Courtesy of R. Locci and B. Petrolini Baldan *Rivista di Patologia Vegetale* 7 (Suppl):3–19, 1971; **25.9 a–c:** From S. T. Williams, M. E. Sharpe, and J. G. Holt (Eds.), *Bergey's Manual of Systematic Bacteriology*, Vol. 4. Copyright 1989 Williams and Wilkins Co., Baltimore; **25.10a:** © Christine L. Case/Visuals Unlimited; **25.10b:** © Sherman Thompson/Visuals Unlimited; **25.11 a-2, b-2:** From S. T. Williams, M. E. Sharpe, and J. G. Holt (Eds.), *Bergey's Manual of Systematic Bacteriology*, Vol. 4. Copyright 1989 Williams and Wilkins Co., Baltimore; **25.12a:** From M. P. Starr et al. (Eds.), *The Prokaryotes*, Springer-Verlag; **25.12 b–c:** From S. T. Williams, M. E. Sharpe, and J. G. Holt (Eds.), *Bergey's Manual of Systematic Bacteriology*, Vol. 4. Copyright 1989 Williams and Wilkins Co., Baltimore.

Chapter 26

26.1a: Science VU-USDA/Visuals Unlimited; **26.1b:** © Everett Beneke/Visuals Unlimited; **26.2a:** © David M. Phillips/Visuals Unlimited; **26.2b:** © Sherman Thompson/Visuals Unlimited; **26.2c:** © Richard Thom/Visuals Unlimited; **26.2d:** © William J. Weber/Visuals Unlimited; **26.4a:** © C. Gerald Van Dyke/Visuals Unlimited; **26.4b:** © John D. Cunningham/Visuals Unlimited; **26.5c:** Dr. Garry T. Cole, Univ. of Texas at Austin; **26.10a:** © John D. Cunningham/Visuals Unlimited; **26.10b:** © Robert Calentine/Visuals Unlimited; **26.10c:** © John D. Cunningham/Visuals Unlimited; **26.11:** © David M. Phillips/Visuals Unlimited; **26.15:** © B. Beatty/Visuals Unlimited; **26.16b:** © Victor Duran/Visuals Unlimited; **26.16c:** © Sylvia Sharnoff/Visuals Unlimited; **26.16 d–e:** © Ed Degginger/Bruce Coleman, Inc.; **26.17 a, c, d:** Carolina Biological Supply; **26.17e:** © David Scharf/Peter Arnold Inc.

Chapter 27

27.3a: © M. I. Walker/Photo Researchers, Inc.; **27.3b:** © John D. Cunningham/Visuals Unlimited; **27.3c:** © Manfred Kage/Peter Arnold, Inc.; **27.3 d–e:** © John D. Cunningham/Visuals Unlimited; **27.3f:** © Bruce Iverson/Visuals Unlimited; **27.6 a, c:** © Dr. Anne Smith/SPL/Photo Researchers, Inc.; **27.6d:** © John D. Cunningham/Visuals Unlimited; **27.7b:** © W. H. Hodge/Peter Arnold, Inc.; **27.8:** © John D. Cunningham/Visuals Unlimited; **27.9b:** © David M. Phillips/Visuals Unlimited.

Chapter 28

28.1: © Arthur M. Siegelman/Visuals Unlimited; **28.4b:** Science VU-R. Oldfield-Polaroid/Visuals Unlimited; **28.4c:** © Arthur M. Siegelman/Visuals Unlimited.

Chapter 29

29.1a: © William J. Weber/Visuals Unlimited; **29.1b:** © Mike Abbey/Visuals Unlimited; **29.2 a–e:** © John D. Cunningham/Visuals Unlimited; **29.3a:** © Stan Elems/Visuals Unlimited; **29.3b:** Bob DeGoursey/Visuals Unlimited; **29.4a:** From J. G. Holt (Ed.), *The Shorter Bergey's Manual of Determinative Bacteriology*, 1977. Williams and Wilkins Co., Baltimore; **29.4 b–c:** John R. Preer; **29.7b:** © H. Oscar/Visuals Unlimited.

Chapter 30

30.3a: From Rita M. Gander and Virginia L. Thomas, "Utilization of Anion-Exchange Chromatography and Monoclonal Antibodies to Characterize Multiple Pilus Types on a Uropathogenic Escherichia coli 06 Isolate," *Infection and Immunity*, 51(2):385–393, Feb. 1986. American Society for Microbiology.; **30.3b:** © Veronika Burmeister/Visuals Unlimited; **30.3c:** From M. Persi, J. C. Burnham, and J. L. Duhring, "Effects of Carbon Dioxide and pH on Adhesion of Candida albicans to Vaginal Epithelial Cells," *Infection and Immunity*, 30(1):82–90, Oct. 1985. American Society for Microbiology; **30.11b:** Robert J. Krasner, Ph.D., Dept. of Biology, Providence College; **30.11c:** © Lennart Nilsson; **30.14b:** Jean-Paul Revel.

Chapter 31

31.6a: © R. Feldman-Dan McCoy/Rainbow; **Box 31.2:** Historic VU-NIH/Visuals Unlimited.

Chapter 32

32.5 d–e: From John Ding-E Young and Zanvil A. Cohn, "How Killer Cells Kill," *Scientific American*, 258(1):38–47, Jan. 1988 Scanning electron microscopy by Dr. Gilla Kaplan, The Rockefeller University; **32.7:** © K. Greer/Visuals Unlimited; **32.12:** © W. J. Johnson/Visuals Unlimited.

Chapter 33

33.1a: © Bob Coyle; **33.3b:** Zanvil Cohn, Rockefeller University; **33.8a:** © Stan Elms/Visuals Unlimited; **33.8b:** © Biophoto Associates/Photo Researchers, Inc.; **33.9c:** © Richard Gross, Biological Photography; **Box 33.2c:** Courtesy of Hoffman-La Roche, Inc.; **33.13c:** Courtesy of Dynatech Labs, photo by Martin-Schaeffer Advertising; **33.15d:** From N. R. Rose et al., *Manual of Clinical Laboratory Immunology*, 1992. American Society for Microbiology.; **33.16c:** © E. S. Anderson/Photo Researchers, Inc.; **33.18:** © Raymond B. Otero/Visuals Unlimited.

Chapter 34

34.1 a–i: © Raymond B. Otero/Visuals Unlimited; **34.2b:** © Fred Marsik/Visuals Unlimited; **34.2c:** © Michael English/Custom Medical Stock Photo; **34.2e:** Courtesy of Evergreen Scientific, Los Angeles; **34.4:** Courtesy of Becton Dickinson Microbiology Systems; **34.5 a–b:** Courtesy of Syva Company; **34.6 a–b:** © Raymond B. Otero/Visuals Unlimited; **34.6c:** © Elmer Koneman/Visuals Unlimited; **34.6 d–s:** © Raymond B. Otero/Visuals Unlimited; **34.6t:** © CNRI/Phototake; **34.8:** Analytab Products, A division of Sherwood Medical; **34.10:** Organon Teknika; **34.11:** From Soad Tabaqchali, *Journal of Clinical Microbiology*, 1986, p. 380. American Society for Microbiology; **34.14:** Kristin Birkness, Center for Disease Control.

Chapter 35

35.7: © Runk/Schoenberger/Grant Heilman Photography.

Chapter 36

36.1b: © John D. Cunningham/Visuals Unlimited; **36.2c, 36.3:** © Carroll H. Weiss/Camera M.D. Studios; **36.4a:** Armed Forces Institute of Pathology; **36.4b:** Centers for Disease Control; **36.5:** © Biophoto Associates/Photo Researchers; **36.6:** Armed Forces Institute of Pathology; **36.11a:** © Arthur M. Siegelman/Visuals Unlimited; **36.11b:** © James Webb/Bruce Coleman; **36.11c:** © Cecil H. Fox/Photo Researchers, Inc.; **36.11 d–e:** © Arthur M. Siegelman/Visuals Unlimited; **36.11f:** © Carroll H. Weiss/Camera M.D. Studios; **Box 36.3:** From Howard I. Kim, Director of Immunology and Infectious Diseases, Damon Clinical Laboratories, Newbury Park, CA 91320. Reprinted by permission; **36.13:** © Carroll H. Weiss/Camera M.D. Studios; **36.14:** © Veronika Burmeister/Visuals Unlimited; **36.15a:** © CDC/Science Source/Photo Researchers, Inc.; **36.15b:** © Carroll H. Weiss/Camera M.D. Studios; **36.16:** © Dr. Brian Eyden/SPL/Photo Researchers, Inc.; **36.17a:** © Tektoff-RM/CNRI/SPL/Custom Medical Stock Photo; **36.18:** National Institute of Health; **36.19:** Fred P. Williams, Jr., U.S. Environmental Protection Agency; **Box 36.4:** Bettmann Archive; **36.20a:** © Kenneth E. Greer/Visuals Unlimited; **36.20b:** © Carroll H. Weiss/Camera M.D. Studios; **36.20 c–d:** Kenneth E. Greer/Visuals Unlimited.

Chapter 37

37.1b: Center for Disease Control; **37.1c:** Science VU/Visuals Unlimited; **37.1d:** Armed Forces Institute of Pathology, A-4497–1; **37.2:** © Fred Hossler/Visuals Unlimited; **37.4:** © Carroll H. Weiss/Camera M.D. Studios; **37.5:** Armed Forces Institute of Pathology; **37.6b:** © Grant Heilman Photography; **37.8a:** From *ASM News* 55(2):cover, 1989. American Society for Microbiology; **37.8 b–c:** © CDC/Peter Arnold, Inc.; **37.9b:** Centers for Disease Control; **37.10a:** © Arthur M. Siegelman/Visuals Unlimited; **37.10b:** Science VU/Charles Stratton/Visuals Unlimited; **37.11:** © A. M. Siegelman/Visuals Unlimited; **37.13:** Science VU-WHO/Visuals Unlimited; **37.14:** From V. Neman-Simha and F. Megraud, "In Vitro Model for Campylobacter pylori Adherence Properties," *Infection and Immunity*, 56(12):3329–3333, Dec. 1988. American Society for Microbiology; **37.16 a–e:** © Carroll H. Weiss/Camera M.D. Studios; **37.16f:** Charles Stoer/Camera M.D. Studios; **37.17 a–c:** © Carroll H. Weiss/Camera M.D. Studios; **37.19a:** © John D. Cunningham/Visuals Unlimited; **37.19b:** Courtesy of The Royal

College of Surgeons Museum, Edinburgh, Scotland; **37.20:** From Jacob S. Teppema, "In Vivo Adherence and Colonization of Vibrio cholerae Strains That Differ in Hemagglutinating Activity and Motility, *Journal of Infection and Immunity,* 55(9):2093–2102, Sept. 1987. Reprinted by permission of American Society for Microbiology; **37.21:** © Heather Davies/SPL/Photo Researchers, Inc.

Chapter 38

38.1, 38.2: Armed Forces Institute of Pathology; **38.3:** Department of Health & Human Services, courtesy of Dr. W. Burgdorfer; **38.4b:** © Max Listgarten, University of Pennsylvania/Biological Photo Service; **38.4c:** © Fred Hossler/Visuals Unlimited; **38.5:** © E. C. S. Chan/ Visuals Unlimited.

Chapter 39

39.1: © Everett S. Beneke/Visuals Unlimited; **39.2:** © Carroll H. Weiss/Camera M.D. Studios; **39.3 a–b:** © Everett S. Beneke/Visuals Unlimited; **39.4, 39.5, 39.6, 39.7:** © Carroll H. Weiss/Camera M.D. Studios; **39.8:** © Everett S. Beneke/Visuals Unlimited; **39.9:** Reprinted by permission of Upjohn Co. from E. S. Beneke et al., *Human Mycosis,* 1984; **39.10:** © Everett S. Beneke/Visuals Unlimited; **39.11:** Reprinted by permission of Upjohn Co. from E. S. Beneke et al., *Human Mycosis,*; **39.12:** © E. C. S. Chan/Visuals Unlimited; **39.13, 39.14a:** © Arthur M. Siegelman/Visuals Unlimited; **39.14b:** Armed Forces Institute of Pathology; **39.15:** © Everett S. Beneke/ Visuals Unlimited; **39.16a:** © David M. Phillips/Visuals Unlimited; **39.16 b–c:** © Everett S. Beneke/Visuals Unlimited; **39.17a:** © Larry Jensen/Visuals Unlimited; **39.17b:** © Robert Calentine/Visuals Unlimited; **39.18 a–b:** Dr. Stanley L. Erlansden, Washington University School of Medicine; **39.20:** Centers for Disease Control; **39.22 a–b:** Armed Forces Institute of Pathology; **39.23:** © David M. Phillips/Visuals Unlimited.

Chapter 40

40.1: Michael Richard, Colorado State University; **40.3a:** Science VU-D. Foster, Woods Hole Oceanographic Institution; **40.3b:** © Christine L. Case/ Visuals Unlimited; **40.4:** From C. M. Cavanaugh, P. R. Levering, J. S. Maki, R. Mitchell, and M. E. Lidstrom, "Symbiosis of Melhylotrophic Bacteria and Deep-Sea Mussels," *Nature* 325:347, Jan. 22, 1987. Reprinted by permission of Macmillan Magazines, Ltd.; **40.5 b–c:** Crane, Hecker, and Goluhev, "Heat Flow and Hydrothermal Vents in Lake Baikal, USSR" 72(52):585, Dec. 24, 1991. EOS Trans. American Geophysical Union; **40.6a:** From *ASM News* 53(2):cover, 1987, American Society for Microbiology. Photo by H. Kaltwasser; **40.6b:** © John D. Cunningham/Visuals Unlimited; **40.6c:** Shirley Sparling; **40.6d:** © Pat Armstrong/Visuals Unlimited; **40.6e:** Dr. H. Kaltwasser, Universitas Sara Viensis; **40.8 a–b:** Jean S. Poindexter, Long Island University; **40.17a:** © Pat Armstrong/Visuals Unlimited; **40.17b:** © Dan McCoy/Rainbow; **40.17c:** © John D. Cunningham/Visuals Unlimited.

Chapter 41

41.3: Y. Cohen and E. Rosenberg, *Microbial Mats,* Fig. 1A, p. 4, 1989. American Society for Microbiology; **41.5:** NASA; **41.7:** From Curtis Suttle, "An Ocean of Viruses May Affect Global Cycles," *ASM News,* 56(12):633, 1990, American Society for Microbiology. Photo copyright Curtis Suttle.; **41.12:** © Donald A. Klein; **41.14 a–c, 41.15:** From D. Jenkins et al., *Manual on the Causes and Control of Activated Bulking and Forming,* 1986. U.S. Environmental Protection Agency; **41.19 a–c:** © Raymond B. Otero/Visuals Unlimited; **41.20:** © Donald A. Klein.

Chapter 42

42.2: © C. Gerald Van Dyke/Visuals Unlimited; **42.3:** © John D. Cunningham/Visuals Unlimited; **42.6:** © Sherman Thompson/Visuals Unlimited; **42.7a-2:** Ray Tully, U.S. Department of Agriculture; **42.7c-2:** Reprinted by permission of Ralph W. F. Hardy and National Research Council of Canada; **42.7d-1:** © John D. Cunningham/Visuals Unlimited; **42.7d-3:** Ray Tully, U.S. Department of Agriculture; **42.7 e-2:** © John D. Cunningham/Visuals Unlimited; **42.7f:** © Dr. Jeremy Burgess/SPL/Photo Researchers, Inc.; **42.7g:** Ray Tully, U.S. Department of Agriculture; **42.8a:** B. Dreyfus; **42.8b:** From Y. R. Dommergues and H. G. Diem (eds.), *Microbiology of Tropical Soils and Plant Productivity,* 1982. Martinus Nijhoff/Dr. W. Junk Publishers, The Hague; **42.9a:** Monsanto Corporation; **42.9b:** Michael Davey; **42.9c:** From A. S. Moffat, "Nitrogen Fixing Bacteria Find New Partners," *Science* 250:910, 1990 AAAS; **42.11:** © John Cunningham/ Visuals Unlimited; **42.12:** © R. S. Hussey/Visuals Unlimited; **42.13a:** Dan Richter/Visuals Unlimited; **42.13 b–c:** © R. Howard Berg/Visuals Unlimited; **42.14, 42.15:** Courtesy of Keith Clay, Indiana University–Bloomington.; **42.16:** Dr. Sandor Sule, Plant Protection Institute, Hungary Academy of Sciences.

Chapter 43

43.2: © Mark Seliger, Campbell Soup Company; **43.3:** From Peterkin, Idzigk and Sharpe, "Screening DNA Probes Using the Hydrophonic Probe Grid-Membrane Filter, *Food Microbiology* 6:281–284, 1989. Academic Press, Inc. (London).; **43.4:** Candrian et al., "Detection of E. coli and Identification of Enterotoxigenic Strains" *International Journal of Food Microbiology* 12: 339–351, 1991. Elsevier Science Publishers B. V.; **43.5a:** © Tom E. Adams/Peter Arnold, Inc.; **43.5b:** © Martha Powell/Visuals Unlimited; **43.7:** Edwin J. Bowers, U.S. Department of Agriculture; **43.8a:** © E. R. Degginger/ Bruce Coleman, Inc.; **43.8b:** © David Newman/Visuals Unlimited; **43.8c:** © Donald A. Klein; **43.9a:** © Alec Duncan/Taurus Photos; **43.9b:** © Kevin Schaefer/Peter Arnold, Inc.; **43.10:** U.S. Department of Agriculture; **43.11:** © Elmer Koheman/Visuals Unlimited; **43.12:** From D. B. Hughes and D. G. Hoover, *Food Technology,* April 1991, Fig. 3, p. 79; **43.13 a–e:** © John D. Cunningham/Visuals Unlimited; **43.14:** © Joe Munroe/Photo Researchers, Inc.; **43.15b:** © Christiana Dittmann/Rainbow; **43.17:** © Vance Henry/Taurus Photos; **43.18:** C. W. Hesselting, U.S. Department of Agriculture; **43.20:** © Paul Moylett/Taurus Photos.

Chapter 44

44.1: Society for Industrial Microbiology; **44.5 a–g:** From B. Atkinson and Daoud, *Advanced Biochemical Engineering* 4:83, Springer-Verlag; **44.13:** Glenn W. Bedell, Bio-Recovery Systems, Inc.; **44.14:** © Dan McCoy/Rainbow; **44.15:** Society for Industrial Microbiology; **44.19a:** From J. R. Postgate, *The Sulphate-Reducing Bacteria,* Cambridge University Press; **44.19b:** Daniel H. Pope, Bioindustrial Technologies, Inc.; **44.20:** S. Harry et al., "Enhanced Oil Removal of Exxon Valdez Spilled Oil," *Journal of Biotechnology,* March 1990, p. 228. Elsevier Science Publishers BV.; **44.21:** Courtesy of General Electric Research and Development Center; **44.22:** Corale L. Brierley, Advanced Mineral Technologies, Inc.

Chapter AI

AI.19b: Copyright Irving Geis.

Chapter AV

AV.1b: © Ching Y. Shih/Peter Arnold, Inc.; **AV.2a:** © Larry Jensen/Visuals Unlimited; **AV.4:** © Robert Calentine/Visuals Unlimited; **AV.6:** © Frank Lambrecht/Visuals Unlimited; **AV.8:** National Institute of Health.

Illustrators

Art enhanced and colorized by **Rolin Graphics** for the second edition text.

Carlyn Iverson

Figures 23.9a, 23.25, 23.26a, 26.3, 26.6, 26.12a–b, 26.13a–d, 26.14, 26.16a, 27.7a, 27.9a, BX 27.2, 28.3a, 29.5, 39.19.

Line Art and Text

Chapter 2

2.22: From *Cell Ultrastructure* by William A. Jensen and Roderic B. Park. © 1967 by Wadsworth Publishing Company, Inc. Reprinted by permission of the publisher.

Chapter 3

3.32: From D. G. Smith, "The Bacterial Motility and Chemotaxis" in *Companion to Microbiology.* A. T. Bull and P. M. Meadow (eds.). Copyright © 1978 Longman Group Limited, Essex, England. Reprinted by permission. **3.33:** From F. C. Neidhart et al., *Physiology of the Bacterial Cell: A Molecular Approach.* Copyright © 1990 Sinauer Associates, Inc., Sunderland, MA. Reprinted by permission.

Chapter 4

4.13a: From J. J. Paulin, *Experimental Parasitology,* 41:283–289. Copyright © 1977 Academic Press, Orlando, FL. Reprinted by permission. **4.13b:** Reproduced from the *Journal of Cell Biology,* 66: 404–413, 1975 by copyright permission of the Rockefeller University Press. **4.20:** From W. T. Keeton, *Biological Science,* 3d ed. Copyright © 1980 W. W. Norton & Company, Inc., New York, NY. **4.23:** From W. T. Keeton, *Biological Science,* 3d ed. Copyright © 1980 W. W. Norton & Company, Inc., New York. **4.24:** From "How Cilia Move" copyright © by Scientific American, Inc./George Kelvin, All Rights Reserved. **4.28:** Adapted from "How Cilia Move" copyright © 1974 by Scientific American, Inc./George Kelvin, All Rights Reserved.

Chapter 5

5.5c: From Colin Norman, "How Microorganisms Transport Iron" in *Science,* 225:401–402, 27 July 1984. Copyright 1984 by the AAAS. Reprinted by permission. **5.10:** From H. J. Conn, editor, *Manual of Microbiological Methods.* Copyright © 1957 McGraw-Hill, Inc., New York, NY. Reprinted by permission.

Chapter 6

6.4: Reprinted with the permission of Macmillan Publishing Company from *Microbiology* by George A. Wistreich and Max D. Lechtman. Copyright © 1988 by Macmillan Publishing Company. **6.10:** From A. L. Koch, *Manual of Methods for General Bacteriology.* Copyright © 1981 American Society for Microbiology, Washington, DC. Reprinted by permission of the publisher and the author.

Chapter 7

7.2b: From *Biochemistry,* Third Edition, by Lubert Stryer. Copyright © 1988 by Lubert Stryer. Reprinted by permission of W. H. Freeman and Company.

Chapter 8

8.9: From Y. Anrako and R. B. Gennis, "The Aerobic Respiratory Chain of Escherichia coli" in *Trends in Biochemical Sciences,* 12:262–266, July 1987. Copyright © 1987 Elsevier Science Publishers, Cambridge, England. Reprinted by permission. **8.12:** From Darnell et al., *Molecular Cell Biology,* 2d ed. Copyright © W. H. Freeman and Company, New York, NY. Courtesy of B. Trumpower, **8.27:** From L. A. Staehelin and C. J. Arntzen, *Photosynthesis III Encyclopedia of Plant Physiology,* New Series Vol. 19. Copyright © 1986 Springer-Verlag, Inc., New York, NY. Reprinted by permission. **8.29:** From L. A. Staehelin and C. J. Arntzen, *Photosynthesis III Encyclopedia of Plant Physiology,* New Series Vol. 19. Copyright © 1986 Springer-Verlag, Inc., New York, NY. Reprinted by permission.

Chapter 9

9.4: Bassham/Calvin, *The Path of Carbon Photosynthesis,* © 1957, renewed 1985, p. 57. Reprinted by permission of Prentice-Hall, Inc., Englewood Cliffs, NJ. **9.19:** From Geoffrey Zubay, *Biochemistry,* 2d ed. Copyright © 1988 Macmillan Publishing Company. Reprinted by permission of Wm. C. Brown Communications, Inc., Dubuque, Iowa. All rights reserved. Reprinted by permission.

Chapter 10

10.2a: From: Benjamin Lewin, *Genes,* 4th ed. Copyright © 1990 Cell Press, Inc., Cambridge, MA. Reprinted by permission of the author. **10.2b:** From R. A. Kelln and J. R. Gear, *BioScience,* 30(2):110–111. Copyright © 1980 Ames Institute of Biological Sciences. All rights reserved. Reprinted by permission. **10.5a:** From R. W. Burlingame et al., "Crystallographic Structure of the Octameric Histone Core of the Nucleosome at a Resolution of 3.3A" in *Science,* 228 (4699):551. Copyright 1985 by the AAAS. **10.5b:** From Geoffrey Zubay, *Biochemistry.* Copyright © 1983 Macmillan Publishing Company. Reprinted by permission of Wm. C. Brown Communications, Inc., Dubuque, Iowa. All Rights Reserved. Reprinted by permission. **10.9:** From Benjamin Lewin, *Genes,* 4th ed. Copyright © 1990 Cell Press, Inc., Cambridge, MA. Reprinted by permission of the author. **10.22:** From *Biology of the Cell,* 2/E, by Stephen L. Wolfe © 1981 by Wadsworth, Inc. Reprinted by permission of the publisher. **10.26a–c:** From Robert Weaver and Philip Hedrick, *Genetics.* Copyright © 1989 Wm. C. Brown Communications, Inc., Dubuque, Iowa. All Rights Reserved. Reprinted by permission. **10.26d:** From C. Bernaben and J. A. Lake, 1982, *Proceedings of the National Academy of Sciences,* 79:3111–3115. Reprinted by permission of the author.

Chapter 11

11.5 (line art): Reprinted with permission of William N. Lipscomb on behalf of the authors.

Chapter 12

12.2: From G. D. Elseth and K. D. Baumgardner, *Genetics.* Copyright © 1984 Addison-Wesley Publishing Company, Menlo Park, CA. Reprinted by permission. **12.5a:** From *Growth of the Bacterial Cell* (1983), by J. Ingraham, O. Maaloe, and F. C. Neidhardt. Copyright © 1983 Sinauer Associates, Inc., Sunderland, MA. Reprinted by permission. **12.5b:** From D. Freifelder, *Microbial Genetics,* 1987. Boston: Jones and Bartlett Publishers, p. 130. Used with permission. **12.8:** From *Genetics,* 2/e by Peter J. Russell. Copyright © 1986 by Peter J. Russell. Reprinted by permission of HarperCollins Publishers. **12.19:** From G. D. Elseth and K. D. Baumgardner, *Genetics.* Copyright © 1984 Addison-Wesley Publishing Company, Menlo Park, CA. Reprinted by permission. **12.20:** From G. D. Elseth and K. D. Baumgardner, *Genetics.* Copyright © 1984 Addison-Wesley Publishing Company, Menlo Park, CA. Reprinted by permission.

Chapter 13

13.3: Reproduced from *The Journal of General Physiology,* 49(Suppl.):183, 1966, by copyright permission of the Rockefeller University Press. **13.7:** From Robert F. Weaver and Philip W. Hedrick, *Genetics,* 2d ed. Copyright © 1992 Wm. C. Brown Communications, Inc., Dubuque, Iowa. All Rights Reserved. Reprinted by permission. **13.8:** From Robert F. Weaver and Philip W. Hedrick, *Genetics.* Copyright © 1989 Wm. C. Brown Communications, Inc., Dubuque, Iowa. All Rights Reserved. Reprinted by permission. **13.9a:** From Stanley N. Cohen, *DNA Insertion Elements, Plasmids, and Episomes.* Copyright © 1977 Cold Spring Harbor Laboratory Press, Cold Spring Harbor, NY. Reprinted by permission. **13.9b:** From K. G. Hardy, *Bacterial Plasmids,* 2d ed. Copyright © American Society for Microbiology, Washington, DC. Reprinted by permission. **13.12:** From Leland G. Johnson, *Biology,* 2d ed. Copyright © 1987 Wm. C. Brown Communications, Inc., Dubuque, Iowa. All Rights Reserved. Reprinted by permission. **13.13:** From Leland G. Johnson, *Biology,* 2d ed. Copyright © 1987 Wm. C. Brown Communications, Inc., Dubuque, Iowa. All Rights Reserved. Reprinted by permission. **13.14:** From Leland G. Johnson, *Biology,* 2d ed. Copyright © 1987 Wm. C. Brown Communications, Inc., Dubuque, Iowa. All Rights Reserved. Reprinted by permission. **13.16:** From Robert F. Weaver and Philip W. Hedrick, *Genetics,* 2d ed. Copyright © 1992 Wm. C. Brown Communications, Inc., Dubuque, Iowa. All Rights Reserved. Reprinted by permission. **13.17:** From Leland G. Johnson, *Biology,* 2d ed. Copyright © 1987 Wm. C. Brown Communications, Inc., Dubuque, Iowa. All Rights Reserved. Reprinted by permission. **13.18:** From Leland G. Johnson, *Biology,* 2d ed. Copyright © 1987 Wm. C. Brown Communications, Inc., Dubuque, Iowa. All Rights Reserved. Reprinted by permission. **13.20 (adaptation):** From *Molecular Genetics,* 2/E by Gunther S. Stent and Richard Calendar. Copyright © 1971, 1978 W. H. Freeman and Company. Reprinted with permission. (After Jacob and Wollman, *Sexuality and the Genetics of Bacteria,* 1961.) **13.21:** Reproduced with the permission of Macmillan Publishing Company. Copyright © 1985 by Monroe W. Strickberger.

Chapter 14

14.10a: From D. Freifelder, *Microbial Genetics,* 1987. Boston: Jones and Bartlett Publishers, p. 50. Used with permission. **14.14:** From G. D. Elseth and K. D. Baumgardner, *Genetics.* Copyright © 1984 Addison-Wesley Publishing Company, Menlo Park, CA. Reprinted by permission. **14.16:** From F. Bolivar et al., *Gene,* 2:95. Copyright © 1977 Elsevier Science Publishers BV, Amsterdam. Reprinted by permission. **14.20:** From K. Itakura, R. Hirose, F. Bolivar, and H. W. Boyer, "Expression in ESCHERICHIA COLI of a Chemically Synthesized Gene for the Hormone Somatostatin" in *Science,* 198:1056–1063. Copyright 1977 by the AAAS. Reprinted by permission.

Chapter 15

15.2: Jay: *MODERN FOOD MICROBIOLOGY,* 3rd ed. Reprinted with permission of the publisher Van Nostrand Reinhold. All rights reserved. **15.4a:** Reprinted by permission of Millipore Corporation. **15.6b:** Courtesy of Fisher Scientific.

Chapter 17

17.1: From B. M. Patten, *Foundations of Embryology,* 2d ed. Copyright © 1964 McGraw-Hill, Inc., New York, NY. Reprinted by permission. **17.7a:** From G. Karp, *Cell Biology,* 2d ed. Copyright © 1984 McGraw-Hill, Inc., New York, NY. Reprinted by permission. **17.13:** Figure from *Microbiology* 3/ed. by Bernard D. Davis et al. Copyright © 1980 by Harper & Row, Publisher, Inc. Reprinted by permission of HarperCollins Publishers. **17.14d:** Reprinted by permission from *Nature,* vol. 354(6351):278–284. Copyright © 1991 Macmillan Magazines Limited. **17.18a:** From Westwood et al., *Journal of General Microbiology,* 34:67, 1964. Published with permission from the Society for General Microbiology. **17.19a:** Reproduced with permission from Sanders, F. K., *Viruses 1981.* Carolina Biology Reader Series, Carolina Biological Supply Company, Burlington, NC.

Chapter 18

18.1: From R. E. F. Matthews, "Classification and Nomenclature of Viruses" in *Intervirology,* 12(3–5). Copyright © 1979 S. Karger AG, Basel, Switzerland. Reprinted by permission. **18.3:** From C. K. Matthews, *Comprehensive Virology,* vol. 7, H. L. Fraenkel-Conrat and R. R. Wagner (eds.). Copyright © 1977 Plenum Publishing Corporation, New York, NY. Reprinted by permission. **18.5a:** From D. Freifelder, *Molecular Biology,* 1983. Boston: Jones and Barlett Publishers, pp. 616–617. Used by permission. **18.6:** From D. Freifelder, *Molecular Biology,* 1983. Boston: Jones and Barlett Publishers, p. 614. Used by permission. **18.11:** From D. Freifelder, *Molecular Biology,* 1983. Boston: Jones and Barlett Publishers, p. 627. Used by permission. **18.12:** From R. Nambudripad et al., "Membrane-Mediated Assembly of Filamentous Bacteriophage Pf1 Coat Protein" in *Science,* 252: 1305–1308. Copyright 1991 by the AAAS. **18.15:** From D. Freifelder, *Molecular Biology,* 1983. Boston: Jones and Barlett Publishers, p. 638. Used by permission. **18.16:** From D. Freifelder, *Molecular Biology,* 1983. Boston: Jones and Barlett Publishers, p. 639. Used by permission. **18.20:** From Allan Campbell, "Genetic Structure" in *The Bacterial Phage Lambda,* A. D. Herschy (ed.). Copyright © 1971 Cold Spring Harbor Laboratory Press, Cold Spring Harbor, NY. Reprinted by permission.

Chapter 19

19.3: From R. E. F. Matthews, "Classification and Nomenclature of Viruses" in *Intervirology,* 12(3–5). Copyright © 1979 S. Karger AG, Basel, Switzerland. Reprinted by permission. **19.4:** Fraenkel-Conrat/Kimball, *Virology,* © 1982, p. 288. Reprinted by permission of Prentice-Hall, Inc., Englewood Cliffs, NJ. **19.15:** From Leland G. Johnson, *Biology,* 2d ed. Copyright © 1987 Wm. C. Brown Communications, Inc., Dubuque, Iowa. All Rights Reserved. Reprinted by permission.

Chapter 20

20.3: From R. R. Colwell et al., *International Journal of Systematic Bacteriology,* 24(4):422–433. Copyright © 1974 American Society for Microbiology, Washington, DC. Reprinted by permission. **20.4:** From Benjamin Lewin, *Genes,* 4th ed. Copyright © 1990 Cell Press, Inc., Cambridge, MA. Reprinted by permission of the author. **20.5:** From Benjamin Lewin, *Genes,* 4th ed. Copyright © 1990 Cell Press, Inc., Cambridge, MA. Reprinted by permission of the author. **20.7:** From C. R. Woese, *Microbiological Reviews,* 51(2):221–271. Copyright © 1987 American Society for Microbiology, Washington, DC. Reprinted by permission. **20.8:** From C. R. Woese, *Microbiological Reviews,* 51(2):221–271. Copyright © 1987 American Society for Microbiology, Washington, DC. Reprinted by permission. **20.10:** From C. R. Woese, *Microbiological Reviews,* 51(2):221–271. Copyright © 1987 American Society for Microbiology, Washington, DC. Reprinted by permission. **20.11:** From Carl R. Woese et al., *Proceedings of the National Academy of Sciences,* 87:4576–4579. Copyright © 1990 Carl R. Woese. Reprinted by permission. **Box Figure 20.2:** From J. A. Lake, *Trends in Biochemical Sciences,* 16(2):46–50. Copyright © 1991 Elsevier Science Publishers, Amsterdam, Netherlands. Reprinted by permission. **Box 20.3:** From P. H. A. Sneath and D. J. Brenner, "Official Nomenclature Lists" in *American Society for Microbiology News,* 58(4):175. Copyright © 1992 American Society for Microbiology, Washington, DC. Reprinted by permission.

Chapter 21

21.7a: From J. C. Burnham and S. F. Conti, "Genus Bdellovibrio" in *Bergey's Manual of Systematic Bacteriology,* 9th ed., Vol. 1:118–119, ed. by N. R. Krieg and J. G. Holt. Copyright © 1984 Williams and Wilkins Co., Baltimore, MD. Reprinted by permission. **21.12:** From D. J. Brenner, "Family I: Enterobacteriaceae" in *Bergey's Manual of Systematic Bacteriology,* 9th ed., Vol. 1:410, ed. by N. R. Krieg and J. G. Holt. Copyright © 1984 Williams and Wilkins Co. Baltimore, MD. Reprinted by permission.

Chapter 23

23.3: Stanier/Ingraham/Wheelis/Painter, *The Microbial World,* 5/E, © 1986, page 354. Reprinted by permission of Prentice-Hall, Inc., Englewood Cliffs, NJ. **23.17:** From J. S. Poindexter, *Microbial Reviews,* 28(3):256. Copyright © 1964 American Society for Microbiology, Washington, DC. Reprinted by permission. **23.23a:** From T. D. Brock, "The Genus Leucothrix" in *The Prokaryotes,* M. P. Starr et al. (Eds.). Copyright © 1981 Springer-Verlag, Inc., New York, NY. Reprinted by permission.

Chapter 24

24.1: From O. Kandler and H. Konig, "Cell Envelopes of Archaeobacteria" in *The Bacteria,* Vol. 8:419. Copyright © 1985 Academic Press, Orlando, FL. Reprinted by permission. **24.2:** From O. Kandler and H. Konig, "Cell Envelopes of Archaeobacteria" in *The Bacteria,* Vol. 8:419. Copyright © 1985 Academic Press, Orlando, FL. Reprinted by permission. **24.5:** Reprinted with permission from *Biochemical Education,* 14(3):109, C. A. Fewson, "Archaebacteria." Copyright 1986, Pergamon Press, plc. **24.6b:** From Jones, Nagle, and Whitman, "Methenogens and the Diversity of Archaeobacteria" in *Microbiological Reviews,* 51(1):162. Copyright © 1987 American Society for Microbiology, Washington, DC. Reprinted by permission. **Box Figure 24.1:** From W. A. Cramer and D. B. Knoff, *Energy Transduction in Biological Membranes.* Copyright © 1989 Springer-Verlag, New York, NY. Reprinted by permission.

Chapter 25

25.3: From M. Goodfellow, *Bergey's Manual of Systematic Bacteriology,* Vol. 4. Copyright © 1989 The Williams and Wilkins Co., Baltimore, MD. Reprinted by permission.

Chapter 27

27.2: Bold/Wynne, *Introduction to Algae,* 2/E, © 1985, p. 8. Reprinted by permission of Prentice-Hall, Inc., Englewood Cliffs, NJ.

Chapter 28

28.5: From Theodore L. Jahn, Eugene C. Bovee, and Frances F. Jahn, *How to Know the Protozoa,* 2d ed. Copyright © 1979 Wm. C. Brown Communications, Inc., Dubuque, Iowa. All Rights Reserved. Reprinted by permission.

Chapter 29

29.6: Reprinted with the permission of Macmillan Publishing Company from *Microbiology* by Ronald M. Atlas. Copyright © 1984 by Macmillan Publishing Company. **29.8:** From James H. Jorgensen and Michael Q. Rinaldi, *Dictionary of Bacteria and Fungi.* Copyright © 1986 Eli Lilly and Company, Inc., Indianapolis, IN.

Chapter 31

31.4: From Robert M. Coleman et al., *Fundamental Immunology,* 2d ed. Copyright © 1992 Wm. C. Brown Communications, Inc., Dubuque, Iowa. All Rights Reserved. Reprinted by permission. **31.5:** From Ivan Roitt, Jonathan Brostoff, and David Male, *Immunology.* Copyright © 1985 Gower Medical Publishers, England. Reprinted by permission. **31.8:** From Ivan Roitt, Jonathan Brostoff, and David Male, *Immunology.* Copyright © 1985 Gower Medical Publishers, England. Reprinted by permission. **31.9:** From Ivan Roitt, Jonathan Brostoff, and David Male, *Immunology.* Copyright © 1985 Gower Medical Publishers, England. Reprinted by permission. **31.10:** From Ivan Roitt, Jonathan Brostoff, and David Male, *Immunology.* Copyright © 1985 Gower Medical Publishers, England. Reprinted by permission. **31.11:** From Ivan Roitt, Jonathan Brostoff, and David Male, *Immunology.* Copyright © 1985 Gower Medical Publishers, England. Reprinted by permission. **31.12:** From Ivan Roitt, Jonathan Brostoff, and David Male, *Immunology.* Copyright © 1985 Gower Medical Publishers, England. Reprinted by permission. **31.13:** From Ivan Roitt, Jonathan Brostoff, and David Male, *Immunology.* Copyright © 1985 Gower Medical Publishers, England. Reprinted by permission. **31.15:** From D. Freifelder, *Microbial Genetics,* 1987. Boston: Jones and Bartlett Publishers, p. 484. Used with permission. **31.16:** From D. Freifelder, *Microbial Genetics,* 1987. Boston: Jones and Bartlett Publishers, p. 485. Used with permission. **31.18:** From Ivan Roitt, Jonathan Brostoff, and David Male, *Immunology.* Copyright © 1985 Gower Medical Publishers, England. Reprinted by permission. **Box Figure 31.3:** From Robert M. Coleman et al., *Fundamental Immunology,* 2d ed. Copyright © 1992 Wm. C. Brown Communications, Inc., Dubuque, Iowa. All Rights Reserved. Reprinted by permission.

Chapter 32

32.2a: Reprinted by permission from *Nature,* Vol. 309, p. 754. Copyright © 1984 Macmillan Magazines Ltd. **32.2b:** Reproduced with permission from the *Annual Review of Immunology,* Volume 7, © 1989 by Annual Reviews, Inc. **32.3:** Reprinted by permission from *Nature,* Vol. 329, p. 590. Copyright © 1987 Macmillan Magazines Ltd. **32.8:** From Ivan Roitt, Jonathan Brostoff, and David Male, *Immunology.* Copyright © 1985 Gower Medical Publishers, England. Reprinted by permission. **32.9:** From Ivan Roitt, Jonathan Brostoff, and David Male, *Immunology.* Copyright © 1985 Gower Medical Publishers, England. Reprinted by permission.

Chapter 33

33.1b: From Ivan Roitt, Jonathan Brostoff, and David Male, *Immunology.* Copyright © 1985 Gower Medical Publishers, England. Reprinted by permission. **33.5:** From Robert M. Coleman et al., *Fundamental Immunology,* 2d ed. Copyright © 1992 Wm. C. Brown Communications, Inc., Dubuque, Iowa. All Rights Reserved. Reprinted by permission. **33.14:** From K. J. Bloch and J. E. Salvaggio, "Use and Interpretation of Diagnostic Laboratory Tests" in *Journal of the American Medical Association,* 248:2738. Copyright © 1982 American Medical Association. Reprinted by permission.

Chapter 34

34.7: From John A. Washington (ed.), *Laboratory Procedures in Clinical Microbiology.* Copyright © 1981 Springer-Verlag, New York, NY. Reprinted by permission. **34.12:** Used with permission of Becton Dickinson and Company.

Chapter 35

35.4: From Burrows, *Textbook of Microbiology.* Copyright © 1973 W. B. Saunders, Philadelphia, PA. Reprinted by permission. **Box Figure 35.2:** From Howard I. Kim, Director of Immunology and Infectious Diseases, Damon Clinical Laboratories, Newbury Park, CA. Reprinted by permission.

Chapter 36

36.4c: Reproduced by permission from Krugman, Saul, and Katz, Samuel L.: *Infectious Diseases of Children,* ed. 7, St. Louis, 1981, The C. V. Mosby Co. **36.7:** From Ronald Brookmeyer, "Reconstruction and Future Trends of the AIDS Epidemic in the United States" in *Science,* 253:41. Copyright 1991 by the AAAS. **36.8:** Reprinted with permission from *The New England Journal of Medicine,* 324(5):309, 1991. **36.9:** Reprinted with permission from *The New England Journal of Medicine,* 324(5):309, 1991. **36.10:** Reprinted with permission from *The New England Journal of Medicine,* 324(5):309, 1991. **36.12:** Reproduction of *Abbott Laboratories' illustration of a serological pattern in an HIV-1 infection* has been granted with approval of Abbott Laboratories; All rights reserved by Abbott Laboratories.

Chapter 37

37.12: From Godal et al., *Bulletin of the Institute Pasteur,* 72:273. Copyright © 1974 Institute Pasteur, Paris, France. Reprinted by permission. **37.18:** Reprinted with permission from Joklik, Willett, Amos: *Zinsser Microbiology,* 17th Edition, Appleton-Century-Crofts, New York, 1980.

Chapter 39

39.21: From the *World Health Statistics Quarterly,* 41:69, 1988. Reprinted by permission of the World Health Organization, Switzerland.

Chapter 40

40.5a: Copyright © 1990 by The New York Times Company. Reprinted by permission. **40.16:** From Rollins and Colwell, *Applied and Environmental Microbiology,* 52:531–538. Copyright © 1986 American Society for Microbiology. Reprinted by permission of the publisher and the author.

Chapter 41

41.2: From J. M. Lynch and N. J. Poole, *Microbial Ecology: A Conceptual Approach.* Copyright © 1979 Blackwell Scientific Publications Limited. Reprinted by permission. **41.4:** From D. J. Des Manas, "Microbial Mats and the Early Evolution of Life" in *Trends in Ecology and Evolution,* 5(5):140–144. Copyright © 1990 Elsevier Trends Journals, Cambridge, England. **41.8:** From P. Williamson and P. M. Hollingan, "Ocean Productivity and Climate Change" in *Trends in Ecology and Evolution,* 5(9):300. Copyright © 1991 Elsevier Trends Journals, Cambridge, England. Reprinted by permission. **41.16:** From *Elements of Bioenvironmental Engineering* by Gaudy and Gaudy, and is used by permission of Engineering Press, Inc., the copyright owner. **41.17:** Illustration from *Water Supply and Pollution Control* 3/ed. by John W. Clark et al. Copyright © 1965, 1971, 1977 by Harper & Row, Publishers, Inc. Reprinted by permission of HarperCollins Publishers.

Chapter 42

42.10: From T. R. G. Gray and S. T. Williams, *Soil Microorganisms.* Copyright © 1971 Longman Group Limited, Essex, England. Reprinted by permission.

Chapter 44

44.8: From Caside, *Industrial Microbiology.* Copyright © 1968 Wiley Eastern Limited, Publishers, New Delhi, India. Reprinted by permission. **44.9:** Reprinted with the permission of Macmillan Publishing Company for *Microbiology* by Ronald M. Atlas. Copyright © 1984 by Macmillan Publishing Company. **44.10:** Reed, Prescott, and Dunne: *INDUSTRIAL MICROBIOLOGY,* 4th ed. Reprinted with permission of the publisher Van Nostrand Reinhold. All rights reserved.

Appendixes

AI.6: From Leland G. Johnson, *Biology,* 2d ed. Copyright © 1987 Wm. C. Brown Communications, Inc., Dubuque, Iowa. All Rights Reserved. Reprinted by permission. **AI.10:** From Leland G. Johnson, *Biology,* 2d ed. Copyright © 1987 Wm. C. Brown Communications, Inc., Dubuque, Iowa. All Rights Reserved. Reprinted by permission. **AI.11:** From Leland G. Johnson, *Biology,* 2d ed. Copyright © 1987 Wm. C. Brown Communications, Inc., Dubuque, Iowa. All Rights Reserved. Reprinted by permission. **AI.12:** From Stuart Ira Fox, *Human Physiology,* 3d ed. Copyright © 1990 Wm. C. Brown Communications, Inc., Dubuque, Iowa. All Rights Reserved. Reprinted by permission. **AI.15:** From Leland G. Johnson, *Biology,* 2d ed. Copyright © 1987 Wm. C. Brown Communications, Inc., Dubuque, Iowa. All Rights Reserved. Reprinted by permission. **AI.17:** From Stuart Ira Fox, *Human Physiology,* 3d ed. Copyright © 1990 Wm. C. Brown Communications, Inc., Dubuque, Iowa. All Rights Reserved. Reprinted by permission. **AI.18:** From Leland G. Johnson, *Biology,* 2d ed. Copyright © 1987 Wm. C. Brown Communications, Inc., Dubuque, Iowa. All Rights Reserved. Reprinted by permission. **AI.19a:** From R. E. Dickerson and I. Geis, *The Structure and Action of Proteins,* Benjamin/Cummings Publishing Company, Menlo Park, CA. © 1969 by R. E. Dickerson and I. Geis. Adapted by permission. **AII.4:** Reproduced, with permission, from Harper, H. A.; Rodwell, V. W.; Mayes, P. A.: *Review of Physiological Chemistry,* 17th edition, copyright Lange Medical Publications, 1979. **AV.7:** From Leland G. Johnson, *Biology,* 2d ed. Copyright © 1987 Wm. C. Brown Communications, Inc., Dubuque, Iowa. All Rights Reserved. Reprinted by permission.

INDEX

Page numbers in **boldface** indicate a figure; numbers in *italics* indicate tabular material; numbers preceded by the letter A indicate material in the APPENDIX.

Summary of Universal Precautions and Laboratory Safety Procedures[a]

Universal Precautions

Since medical history and examination cannot reliably identify all patients infected with HIV or other blood-borne pathogens, blood and body-fluid precautions should be consistently used for *all* patients.

1. All health-care workers should routinely use appropriate barrier precautions to prevent skin and mucous-membrane exposure when contact with blood or other body fluids of any patient is anticipated. Gloves should be worn for touching blood and body fluids, mucous membranes, or non-intact skin of all patients, for handling items or surfaces soiled with blood or body fluids, and for performing venipuncture and other vascular access procedures. Gloves should be changed after contact with each patient. Masks and protective eyewear or face shields should be worn during procedures that are likely to generate droplets of blood or other body fluids to prevent exposure of mucous membranes of the mouth, nose, and eyes. Gowns or aprons should be worn during procedures that are likely to generate splashes of blood or other body fluids.
2. Hands and other skin surfaces should be washed immediately and thoroughly if contaminated with blood or other body fluids. Hands should be washed immediately after gloves are removed.
3. All health-care workers should take precautions to prevent injuries caused by needles, scalpels, and other sharp instruments or devices during procedures; when cleaning used instruments; during disposal of used needles; and when handling sharp instruments after procedures. To prevent needlestick injuries, needles should not be recapped, purposely bent or broken by hand, removed from disposable syringes, or otherwise manipulated by hand. After they are used, disposable syringes and needles, scalpel blades, and other sharp items should be placed in puncture-resistant containers for disposal.
4. Although saliva has not been implicated in HIV transmission, to minimize the need for emergency mouth-to-mouth resuscitation, mouthpieces, resuscitation bags, or other ventilation devices should be available for use in areas in which the need for resuscitation is predictable.
5. Health-care workers who have exudative lesions or weeping dermatitis should refrain from all direct patient care and from handling patient-care equipment.
6. The following procedure should be used to clean up spills of blood or blood-containing fluids. (1) Put on gloves and any other necessary barriers. (2) Wipe up excess material with disposable towels and place the towels in a container for sterilization. (3) Disinfect the area with either a commercial EPA-approved germicide or household bleach (sodium hypochlorite). The latter should be diluted from 1:100 (smooth surfaces) to 1:10 (porous or dirty surfaces); the dilution should be no more than 24 hours old. When dealing with large spills or those containing sharp objects such as broken glass, first cover the spill with disposable toweling. Then saturate the toweling with commercial germicide or a 1:10 bleach solution and allow it to stand for at least 10 minutes. Finally clean as described above.

[a]Adapted from the "Centers for Disease Control Guidelines" *Morbidity and Mortality Weekly Report,* 36(Supplement 25): 55–105, 1987.